ALGEBRA

Lines

Slope of the line through $P_1 = (x_1, y_1)$ and $P_2 = (x_2, y_2)$:

$$m = \frac{y_2 - y_1}{x_2 - x_1}$$

Slope-intercept equation of line with slope m and y-intercept b:

$$y = mx + b$$

Point-slope equation of line through $P_1 = (x_1, y_1)$ with slope m:

$$y - y_1 = m(x - x_1)$$

Point-point equation of line through $P_1 = (x_1, y_1)$ and $P_2 = (x_2, y_2)$:

$$y - y_1 = m(x - x_1) \quad \text{where } m = \frac{y_2 - y_1}{x_2 - x_1}$$

Lines of slope m_1 and m_2 are parallel if and only if $m_1 = m_2$.
Lines of slope m_1 and m_2 are perpendicular if and only if $m_1 = -\frac{1}{m_2}$.

Circles

Equation of the circle with center (a, b) and radius r:

$$(x - a)^2 + (y - b)^2 = r^2$$

Distance and Midpoint Formulas

Distance between $P_1 = (x_1, y_1)$ and $P_2 = (x_2, y_2)$:

$$d = \sqrt{(x_2 - x_1)^2 + (y_2 - y_1)^2}$$

Midpoint of $\overline{P_1 P_2}$: $\left(\frac{x_1 + x_2}{2}, \frac{y_1 + y_2}{2}\right)$

Laws of Exponents

$$x^m x^n = x^{m+n} \qquad \frac{x^m}{x^n} = x^{m-n} \qquad (x^m)^n = x^{mn}$$

$$x^{-n} = \frac{1}{x^n} \qquad (xy)^n = x^n y^n \qquad \left(\frac{x}{y}\right)^n = \frac{x^n}{y^n}$$

$$x^{1/n} = \sqrt[n]{x} \qquad \sqrt[n]{xy} = \sqrt[n]{x}\,\sqrt[n]{y} \qquad \sqrt[n]{\frac{x}{y}} = \frac{\sqrt[n]{x}}{\sqrt[n]{y}}$$

$$x^{m/n} = \sqrt[n]{x^m} = \left(\sqrt[n]{x}\right)^m$$

Special Factorizations

$$x^2 - y^2 = (x + y)(x - y)$$
$$x^3 + y^3 = (x + y)(x^2 - xy + y^2)$$
$$x^3 - y^3 = (x - y)(x^2 + xy + y^2)$$

Binomial Theorem

$$(x + y)^2 = x^2 + 2xy + y^2$$
$$(x - y)^2 = x^2 - 2xy + y^2$$
$$(x + y)^3 = x^3 + 3x^2y + 3xy^2 + y^3$$
$$(x - y)^3 = x^3 - 3x^2y + 3xy^2 - y^3$$

$$(x + y)^n = x^n + nx^{n-1}y + \frac{n(n-1)}{2}x^{n-2}y^2 + \cdots + \binom{n}{k}x^{n-k}y^k + \cdots + nxy^{n-1} + y^n$$

where $\binom{n}{k} = \frac{n(n-1)\cdots(n-k+1)}{1 \cdot 2 \cdot 3 \cdot \cdots \cdot k}$

Quadratic Formula

If $ax^2 + bx + c = 0$, then $x = \frac{-b \pm \sqrt{b^2 - 4ac}}{2a}$.

Inequalities and Absolute Value

If $a < b$ and $b < c$, then $a < c$.

If $a < b$, then $a + c < b + c$.

If $a < b$ and $c > 0$, then $ca < cb$.

If $a < b$ and $c < 0$, then $ca > cb$.

$|x| = x \quad$ if $x \geq 0$

$|x| = -x \quad$ if $x \leq 0$

$|x| < a$ means $-a < x < a$.

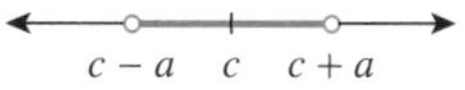

$|x - c| < a$ means $c - a < x < c + a$.

GEOMETRY

Formulas for area A, circumference C, and volume V

Triangle
$A = \frac{1}{2}bh$
$= \frac{1}{2}ab\sin\theta$

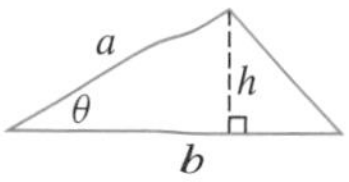

Circle
$A = \pi r^2$
$C = 2\pi r$

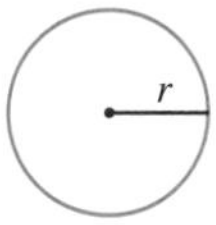

Sector of Circle
$A = \frac{1}{2}r^2\theta$
$s = r\theta$
(θ in radians)

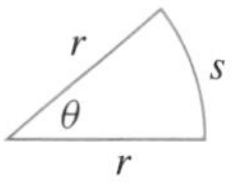

Sphere
$V = \frac{4}{3}\pi r^3$
$A = 4\pi r^2$

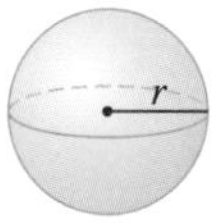

Cylinder
$V = \pi r^2 h$

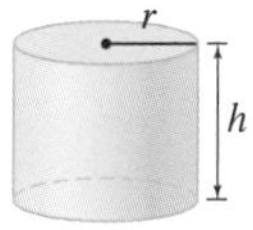

Cone
$V = \frac{1}{3}\pi r^2 h$
$A = \pi r\sqrt{r^2 + h^2}$

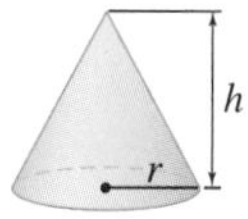

Cone with arbitrary base
$V = \frac{1}{3}Ah$
where A is the area of the base

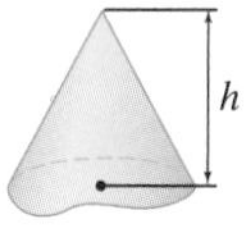

Pythagorean Theorem: For a right triangle with hypotenuse of length c and legs of lengths a and b, $c^2 = a^2 + b^2$.

TRIGONOMETRY

Angle Measurement

π radians $= 180°$

$1° = \dfrac{\pi}{180}$ rad $\quad$ 1 rad $= \dfrac{180°}{\pi}$

$s = r\theta \quad (\theta \text{ in radians})$

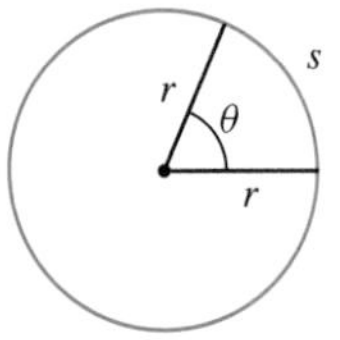

Right Triangle Definitions

$\sin\theta = \dfrac{\text{opp}}{\text{hyp}} \qquad \cos\theta = \dfrac{\text{adj}}{\text{hyp}}$

$\tan\theta = \dfrac{\sin\theta}{\cos\theta} = \dfrac{\text{opp}}{\text{adj}} \qquad \cot\theta = \dfrac{\cos\theta}{\sin\theta} = \dfrac{\text{adj}}{\text{opp}}$

$\sec\theta = \dfrac{1}{\cos\theta} = \dfrac{\text{hyp}}{\text{adj}} \qquad \csc\theta = \dfrac{1}{\sin\theta} = \dfrac{\text{hyp}}{\text{opp}}$

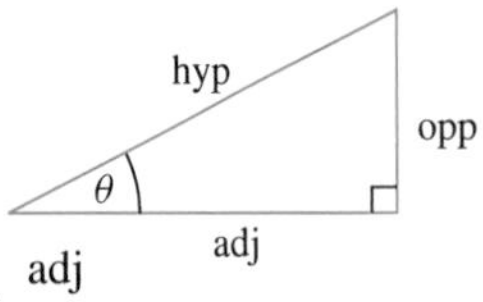

Trigonometric Functions

$\sin\theta = \dfrac{y}{r} \qquad \csc\theta = \dfrac{r}{y}$

$\cos\theta = \dfrac{x}{r} \qquad \sec\theta = \dfrac{r}{x}$

$\tan\theta = \dfrac{y}{x} \qquad \cot\theta = \dfrac{x}{y}$

$\displaystyle\lim_{\theta\to 0}\frac{\sin\theta}{\theta} = 1 \qquad \lim_{\theta\to 0}\frac{1-\cos\theta}{\theta} = 0$

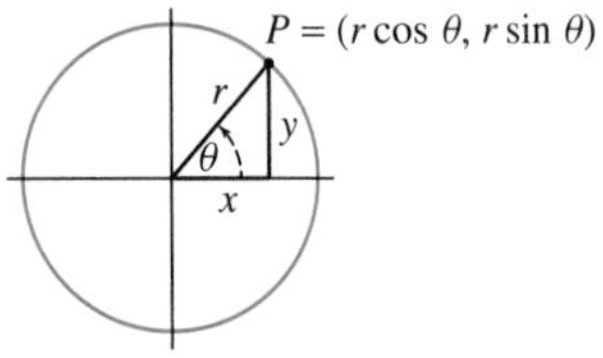

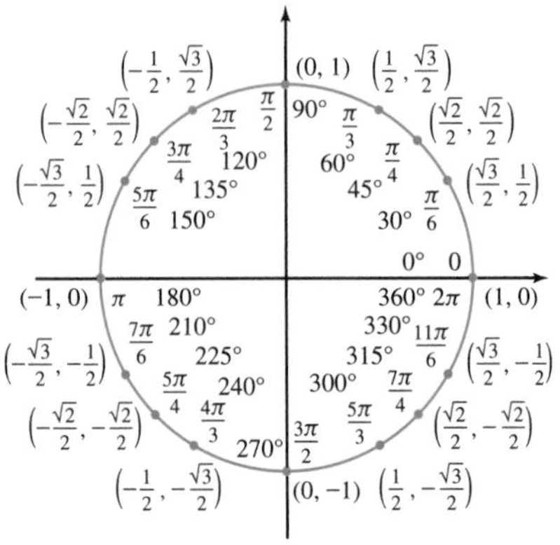

Fundamental Identities

$\sin^2\theta + \cos^2\theta = 1 \qquad \sin(-\theta) = -\sin\theta$

$1 + \tan^2\theta = \sec^2\theta \qquad \cos(-\theta) = \cos\theta$

$1 + \cot^2\theta = \csc^2\theta \qquad \tan(-\theta) = -\tan\theta$

$\sin\left(\dfrac{\pi}{2} - \theta\right) = \cos\theta \qquad \sin(\theta + 2\pi) = \sin\theta$

$\cos\left(\dfrac{\pi}{2} - \theta\right) = \sin\theta \qquad \cos(\theta + 2\pi) = \cos\theta$

$\tan\left(\dfrac{\pi}{2} - \theta\right) = \cot\theta \qquad \tan(\theta + \pi) = \tan\theta$

The Law of Sines

$$\frac{\sin A}{a} = \frac{\sin B}{b} = \frac{\sin C}{c}$$

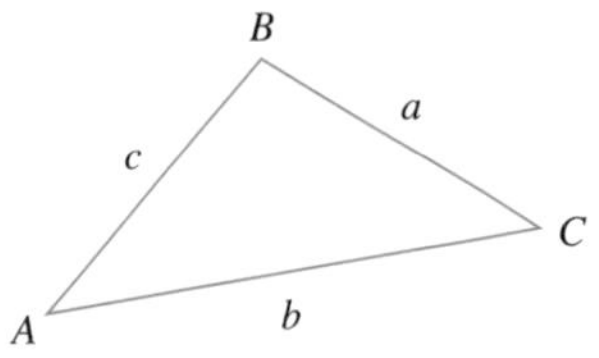

The Law of Cosines

$$a^2 = b^2 + c^2 - 2bc\cos A$$

Addition and Subtraction Formulas

$\sin(x + y) = \sin x\cos y + \cos x\sin y$

$\sin(x - y) = \sin x\cos y - \cos x\sin y$

$\cos(x + y) = \cos x\cos y - \sin x\sin y$

$\cos(x - y) = \cos x\cos y + \sin x\sin y$

$\tan(x + y) = \dfrac{\tan x + \tan y}{1 - \tan x\tan y}$

$\tan(x - y) = \dfrac{\tan x - \tan y}{1 + \tan x\tan y}$

Double-Angle Formulas

$\sin 2x = 2\sin x\cos x$

$\cos 2x = \cos^2 x - \sin^2 x = 2\cos^2 x - 1 = 1 - 2\sin^2 x$

$\tan 2x = \dfrac{2\tan x}{1 - \tan^2 x}$

$\sin^2 x = \dfrac{1 - \cos 2x}{2} \qquad \cos^2 x = \dfrac{1 + \cos 2x}{2}$

Graphs of Trigonometric Functions

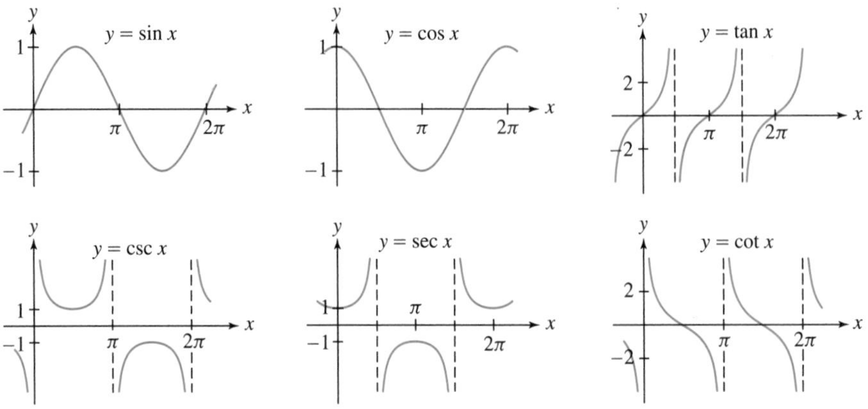

CALCULUS

SECOND EDITION

CALCULUS

SECOND EDITION

JON ROGAWSKI

University of California, Los Angeles

W. H. FREEMAN AND COMPANY
New York

Publisher: Ruth Baruth
Senior Acquisitions Editor: Terri Ward
Development Editor: Tony Palermino
Development Editor: Julie Z. Lindstrom
Associate Editor: Katrina Wilhelm
Editorial Assistant: Tyler Holzer
Market Development: Steven Rigolosi
Executive Marketing Manager: Jennifer Somerville
Media Editor: Laura Capuano
Assistant Editor: Catriona Kaplan
Senior Media Acquisitions Editor: Roland Cheyney
Photo Editor: Ted Szczepanski
Photo Researcher: Julie Tesser
Cover and Text Designer: Blake Logan
Illustrations: Network Graphics and Techsetters, Inc.
Illustration Coordinator: Bill Page
Production Coordinator: Paul W. Rohloff
Composition: Techsetters, Inc.
Printing and Binding: RR Donnelley

Library of Congress Control Number 2011921789
ISBN-13: 978-1-4292-6374-0
ISBN-10: 1-4292-6374-1

Printed in the United States of America
First printing

W. H. Freeman and Company, 41 Madison Avenue, New York, NY 10010
Houndmills, Basingstoke RG21 6XS, England
www.whfreeman.com

To Julie

CONTENTS | CALCULUS

ABOUT JON ROGAWSKI

As a successful teacher for more than 30 years, Jon Rogawski has listened to and learned much from his own students. These valuable lessons have made an impact on his thinking, his writing, and his shaping of a calculus text.

Jon Rogawski received his undergraduate and master's degrees in mathematics simultaneously from Yale University, and he earned his PhD in mathematics from Princeton University, where he studied under Robert Langlands. Before joining the Department of Mathematics at UCLA in 1986, where he is currently a full professor, he held teaching and visiting positions at the Institute for Advanced Study, the University of Bonn, and the University of Paris at Jussieu and at Orsay.

Jon's areas of interest are number theory, automorphic forms, and harmonic analysis on semisimple groups. He has published numerous research articles in leading mathematics journals, including the research monograph *Automorphic Representations of Unitary Groups in Three Variables* (Princeton University Press). He is the recipient of a Sloan Fellowship and an editor of the *Pacific Journal of Mathematics* and the *Transactions of the AMS.*

Jon and his wife, Julie, a physician in family practice, have four children. They run a busy household and, whenever possible, enjoy family vacations in the mountains of California. Jon is a passionate classical music lover and plays the violin and classical guitar.

PREFACE

ABOUT *CALCULUS* by Jon Rogawski

On Teaching Mathematics

As a young instructor, I enjoyed teaching but I didn't appreciate how difficult it is to communicate mathematics effectively. Early in my teaching career, I was confronted with a student rebellion when my efforts to explain epsilon-delta proofs were not greeted with the enthusiasm I anticipated. Experiences of this type taught me two basic principles:

1. We should try to teach students as much as possible, but not more.

2. As math teachers, how we say it is as important as what we say.

The formal language of mathematics is intimidating to the uninitiated. By presenting concepts in everyday language, which is more familiar but not less precise, we open the way for students to understand the underlying ideas and integrate them into their way of thinking. Students are then in a better position to appreciate the need for formal definitions and proofs and to grasp their logic.

On Writing a Calculus Text

I began writing *Calculus* with the goal of creating a text in which exposition, graphics, and layout would work together to enhance all facets of a student's calculus experience: mastery of basic skills, conceptual understanding, and an appreciation of the wide range of applications. I also wanted students to be aware, early in the course, of the beauty of the subject and the important role it will play, both in their further studies and in their understanding of the wider world. I paid special attention to the following aspects of the text:

(a) Clear, accessible exposition that anticipates and addresses student difficulties.

(b) Layout and figures that communicate the flow of ideas.

(c) Highlighted features in the text that emphasize concepts and mathematical reasoning: Conceptual Insight, Graphical Insight, Assumptions Matter, Reminder, and Historical Perspective.

(d) A rich collection of examples and exercises of graduated difficulty that teach basic skills, problem-solving techniques, reinforce conceptual understanding, and motivate calculus through interesting applications. Each section also contains exercises that develop additional insights and challenge students to further develop their skills.

Encouraged by the enthusiastic response to the First Edition, I approached the new edition with the aim of further developing these strengths. Every section of text was carefully revised. During the revision process, I paid particular attention to feedback from adopters, reviewers, and students who have used the book. Their insights and creative suggestions brought numerous improvements to the text.

Calculus has a deservedly central role in higher education. It is not only the key to the full range of quantitative disciplines; it is also a crucial component in a student's intellectual development. I hope this new edition will continue to play a role in opening up for students the multifaceted world of calculus.

My textbook follows a largely traditional organization, with a few exceptions. One such exception is the placement of Taylor polynomials in Chapter 9.

Placement of Taylor Polynomials

Taylor polynomials appear in Chapter 9, before infinite series in Chapter 11. My goal is to present Taylor polynomials as a natural extension of the linear approximation. When I teach infinite series, the primary focus is on convergence, a topic that many students find challenging. After studying the basic convergence tests and convergence of power series, students are ready to tackle the issues involved in representing a function by its Taylor series. They can then rely on their previous work with Taylor polynomials and the Error Bound from Chapter 9. However, the section on Taylor polynomials is designed so that you can cover it together with the material on power series and Taylor series in Chapter 11 if you prefer this order.

CAREFUL, PRECISE DEVELOPMENT

W. H. Freeman is committed to high quality and precise textbooks and supplements. From this project's inception and throughout its development and production, quality and precision have been given significant priority. We have in place unparalleled procedures to ensure the accuracy of all facets of the text:

- Exercises and Examples
- Exposition
- Figures
- Editing
- Composition

Together, these procedures far exceed prior industry standards to safeguard the quality and precision of a calculus textbook.

New to the Second Edition

Enhanced Exercise Sets...with Approximately 25% New and Revised Problems: To refine this strong feature of the text, the exercise sets were extensively reviewed by outside reviewers. Based in part on this feedback, the author carefully revised the exercises to enhance their quality and quantity. The Second Edition features thousands of new and updated problems.

New and Larger Variety of Applications: The Second Edition contains many fresh examples and problems centered on innovative, contemporary applications from engineering, the life sciences, physical sciences, business, economics, medicine, and the social sciences.

Content Changes In Response to Users and Reviewers, including:

- Chapter 2: The topic "Limits at Infinity" has been moved forward from Chapter 4 to Section 2.7.
- Chapter 3: Differentiation–Coverage of differentials has been expanded.
- Chapter 8: Numerical Integration has been moved to the end of the chapter, after the coverage of all integration techniques.
- New Section 8.7: Probability and Integration. This section introduces a basic application of integration which is of importance in the physical sciences as well as in business and the social sciences.
- The multivariable chapters, praised for their strength in the First Edition, have been further revised and polished.
- New Section 16.5: Applications of Multiple Integrals
- Extensive revision and enhancement of graphics throughout the text.

SUPPLEMENTS

For Instructors

- Instructor's Solutions Manual
  Brian Bradie, Christopher Newport University; and Greg Dresden, Washington and Lee University
  Single Variable ISBN: 1-4292-4313-9
  Multivariable ISBN: 1-4292-5501-3
  Contains worked-out solutions to all exercises in the text.

- Test Bank
  Printed, ISBN: 1-4292-4311-2
  CD-ROM, ISBN: 1-4292-4310-4
  Includes multiple-choice and short-answer test items.

- Instructor's Resource Manual
  ISBN: 1-4292-4315-5
  Provides suggested class time, key points, lecture material, discussion topics, class activities, worksheets, and group projects corresponding to each section of the text.

- Instructor's Resource CD-ROM
  ISBN: 1-4292-4314-7
  Search and export all resources by key term or chapter. Includes text images, Instructor's Solutions Manual, Instructor's Resource Manual, and Test Bank.

For Students

- Student Solutions Manual
  Brian Bradie, Christopher Newport University; and Greg Dresden, Washington and Lee University
  Single Variable ISBN: 1-4292-4290-6
  Multivariable ISBN: 1-4292-5508-0
  Offers worked-out solutions to all odd-numbered exercises in the text.

- Software Manuals
  Software manuals covering Maple and Mathematica are offered within CalcPortal. These manuals are available in printed versions through custom publishing. They serve as basic introductions to popular mathematical software options and guides for their use with *Calculus*, Second Edition.

- Companion website at **www.whfreeman.com/rogawski2e**

MEDIA

Online Homework Options

WebAssign Premium

http://www.webassign.net/whfreeman

W. H. Freeman has partnered with WebAssign to provide a powerful, convenient online homework option, making it easy to assign algorithmically generated homework and quizzes for Rogawski's *Calculus*, Second Edition. WebAssign Premium for the new edition of *Calculus* offers thousands of exercises, plus tutorial videos. It will also be available with a full eBook option.

CALCPORTAL

www.yourcalcportal.com

CalcPortal combines a fully customizable eBook with exceptional student and instructor resources, including precalculus diagnostic quizzes, interactive applets, student solutions, review questions, and homework management tools, all in one affordable, easy-to-use, and fully customizable learning space. This new iteration of CalcPortal for *Calculus*, Second Edition, represents a dramatic step forward for online teaching and learning, with innovations that make it both more powerful and easier to use. It will include a turnkey solution with a prebuilt complete course, featuring ready-made assignments for you to use as is or modify.

WeBWorK

http://webwork.maa.org

Developed by the University of Rochester, this open-source homework system is available to students free of charge. For adopters of *Calculus*, Second Edition, W. H. Freeman will increase the current first edition offering to include approximately 2400 algorithmically generated questions with full solutions from the text, plus access to a shared national library test bank with thousands of additional questions, including 1500 problem sets correlated to the table of contents.

ADDITIONAL MEDIA

Freeman SolutionMaster

SolutionMaster

SolutionMaster is an innovative new digital tool to help instructors provide selected, secure solutions to their students. With SolutionMaster, instructors can easily create solutions for any assignment from the textbook

Interactive eBook

The Interactive eBook integrates a complete and customizable online version of the text with its media resources. Students can quickly search the text, and they can personalize the eBook just as they would the print version, with highlighting, bookmarking, and note-taking features. Instructors can add, hide, and reorder content, integrate their own material, and highlight key text.

dynamicbooks

Dynamic Book

Rogawski's *Calculus*, Second Edition, is available as an innovative, customizable, and editable DynamicBook eBook. In DynamicBooks an instructor can easily customize the text presentation by adding, hiding, and modifying content to meet their specific teaching approach to calculus. In addition to highlighting and adding notes, students can link to interactive graphical applets, videos, and other digital assets. Rogawski's DynamicBook can be viewed online, downloaded to a local computer, and downloaded to an iPhone or iPad. Students also have the option to purchase a printed, bound version with the instructor's changes included.

FEATURES

Conceptual Insights encourage students to develop a conceptual understanding of calculus by explaining important ideas clearly but informally.

> **CONCEPTUAL INSIGHT** Leibniz notation is widely used for several reasons. First, it reminds us that the derivative df/dx, although not itself a ratio, is in fact a *limit* of ratios $\Delta f/\Delta x$. Second, the notation specifies the independent variable. This is useful when variables other than x are used. For example, if the independent variable is t, we write df/dt. Third, we often think of d/dx as an "operator" that performs differentiation on functions. In other words, we apply the operator d/dx to f to obtain the derivative df/dx. We will see other advantages of Leibniz notation when we discuss the Chain Rule in Section 3.7.

Ch. 3, p. 111

Graphical Insights enhance students' visual understanding by making the crucial connections between graphical properties and the underlying concepts.

> **GRAPHICAL INSIGHT** Keep the graphical interpretation of limits in mind. In Figure 4(A), $f(x)$ approaches L as $x \to c$ because for any $\epsilon > 0$, we can make the gap less than ϵ by taking δ sufficiently small. By contrast, the function in Figure 4(B) has a jump discontinuity at $x = c$. The gap cannot be made small, no matter how small δ is taken. Therefore, the limit does not exist.

Ch. 2, p. 95

Reminders are margin notes that link the current discussion to important concepts introduced earlier in the text to give students a quick review and make connections with related ideas.

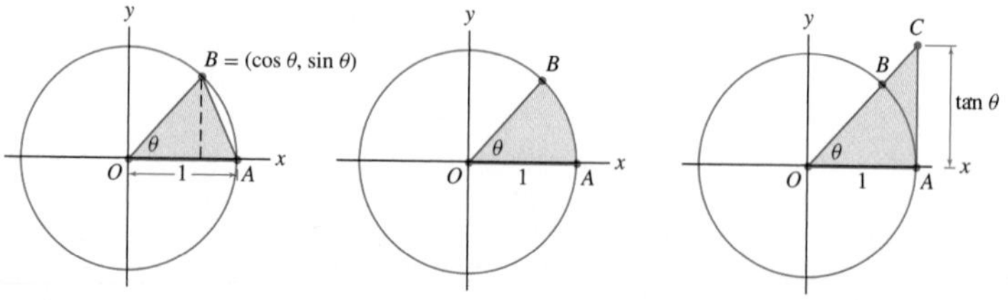

Area of triangle = $\frac{1}{2}\sin\theta$ Area of sector = $\frac{1}{2}\theta$ Area of triangle = $\frac{1}{2}\tan\theta$

FIGURE 5

Proof Assume first that $0 < \theta < \frac{\pi}{2}$. Our proof is based on the following relation between the areas in Figure 5:

$$\text{Area of } \triangle OAB < \text{area of sector } BOA < \text{area of } \triangle OAC \qquad \boxed{2}$$

← REMINDER *Let's recall why a sector of angle θ in a circle of radius r has area $\frac{1}{2}r^2\theta$. A sector of angle θ represents a fraction $\frac{\theta}{2\pi}$ of the entire circle. The circle has area πr^2, so the sector has area $\left(\frac{\theta}{2\pi}\right)\pi r^2 = \frac{1}{2}r^2\theta$. In the unit circle ($r = 1$), the sector has area $\frac{1}{2}\theta$.*

Note: Our proof of Theorem 3 uses the formula $\frac{1}{2}\theta$ for the area of a sector, but this formula is based on the formula πr^2 for the area of a circle, a complete proof of which requires integral calculus.

Let's compute these three areas. First, $\triangle OAB$ has base 1 and height $\sin\theta$, so its area is $\frac{1}{2}\sin\theta$. Next, recall that a sector of angle θ has area $\frac{1}{2}\theta$. Finally, to compute the area of $\triangle OAC$, we observe that

$$\tan\theta = \frac{\text{opposite side}}{\text{adjacent side}} = \frac{AC}{OA} = \frac{AC}{1} = AC$$

Thus, $\triangle OAC$ has base 1, height $\tan\theta$, and area $\frac{1}{2}\tan\theta$. We have shown, therefore, that

$$\underbrace{\frac{1}{2}\sin\theta}_{\text{Area } \triangle OAB} \le \underbrace{\frac{1}{2}\theta}_{\text{Area of sector}} \le \underbrace{\frac{1}{2}\frac{\sin\theta}{\cos\theta}}_{\text{Area } \triangle OAC} \qquad \boxed{3}$$

The first inequality yields $\sin\theta \le \theta$, and because $\theta > 0$, we obtain

$$\frac{\sin\theta}{\theta} \le 1 \qquad \boxed{4}$$

Ch. 2, p. 78

Caution Notes warn students of common pitfalls they may encounter in understanding the material.

*CAUTION The Power Rule applies only to the power functions $y = x^n$. It does not apply to exponential functions such as $y = 2^x$. The derivative of $y = 2^x$ **is not** $x2^{x-1}$. We will study the derivatives of exponential functions later in this section.*

We make a few remarks before proceeding:

- It may be helpful to remember the Power Rule in words: To differentiate x^n, "bring down the exponent and subtract one (from the exponent)."

$$\frac{d}{dx} x^{\text{exponent}} = (\text{exponent})\, x^{\text{exponent}-1}$$

- The Power Rule is valid for all exponents, whether negative, fractional, or irrational:

$$\frac{d}{dx} x^{-3/5} = -\frac{3}{5} x^{-8/5}, \qquad \frac{d}{dx} x^{\sqrt{2}} = \sqrt{2}\, x^{\sqrt{2}-1}$$

Ch. 3, p. 112

Historical Perspectives are brief vignettes that place key discoveries and conceptual advances in their historical context. They give students a glimpse into some of the accomplishments of great mathematicians and an appreciation for their significance.

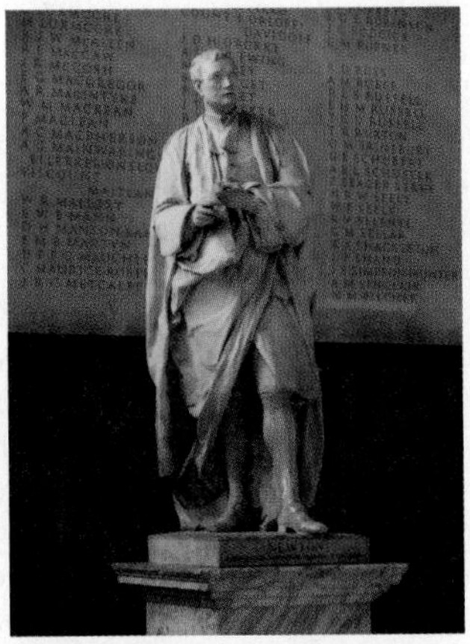

This statue of Isaac Newton in Cambridge University was described in *The Prelude*, a poem by William Wordsworth (1770–1850):

"Newton with his prism and silent face,
The marble index of a mind for ever
Voyaging through strange seas of Thought, alone."

HISTORICAL PERSPECTIVE

Philosophy is written in this grand book—I mean the universe—which stands continually open to our gaze, but it cannot be understood unless one first learns to comprehend the language ... in which it is written. It is written in the language of mathematics ...

—GALILEO GALILEI, 1623

The scientific revolution of the sixteenth and seventeenth centuries reached its high point in the work of Isaac Newton (1643–1727), who was the first scientist to show that the physical world, despite its complexity and diversity, is governed by a small number of universal laws. One of Newton's great insights was that the universal laws are dynamical, describing how the world changes over time in response to forces, rather than how the world actually is at any given moment in time. These laws are expressed best in the language of calculus, which is the mathematics of change.

More than 50 years before the work of Newton, the astronomer Johannes Kepler (1571–1630) discovered his three laws of planetary motion, the most famous of which states that the path of a planet around the sun is an ellipse. Kepler arrived at these laws through a painstaking analysis of astronomical data, but he could not explain why they were true. According to Newton, the motion of any object—planet or pebble—is determined by the forces acting on it. The planets, if left undisturbed, would travel in straight lines. Since their paths are elliptical, some force—in this case, the gravitational force of the sun—must be acting to make them change direction continuously. In his magnum opus *Principia Mathematica*, published in 1687, Newton proved that Kepler's laws follow from Newton's own universal laws of motion and gravity.

For these discoveries, Newton gained widespread fame in his lifetime. His fame continued to increase after his death, assuming a nearly mythic dimension and his ideas had a profound influence, not only in science but also in the arts and literature, as expressed in the epitaph by British poet Alexander Pope: "Nature and Nature's Laws lay hid in Night. God said, *Let Newton be!* and all was Light."

Ch. 2, p. 41

Assumptions Matter uses short explanations and well-chosen counterexamples to help students appreciate why hypotheses are needed in theorems.

Section Summaries summarize a section's key points in a concise and useful way and emphasize for students what is most important in each section.

Section Exercise Sets offer a comprehensive set of exercises closely coordinated with the text. These exercises vary in difficulty from routine, to moderate, to more challenging. Also included are icons indicating problems that require the student to give a written response or require the use of technology T.

Chapter Review Exercises offer a comprehensive set of exercises closely coordinated with the chapter material to provide additional problems for self-study or assignments.

ACKNOWLEDGMENTS

Jon Rogawski and W. H. Freeman and Company are grateful to the many instructors from across the United States and Canada who have offered comments that assisted in the development and refinement of this book. These contributions included class testing, manuscript reviewing, problems reviewing, and participating in surveys about the book and general course needs.

ALABAMA Tammy Potter, *Gadsden State Community College*; David Dempsey, *Jacksonville State University*; Douglas Bailer, *Northeast Alabama Community College*; Michael Hicks, *Shelton State Community College*; Patricia C. Eiland, *Troy University, Montgomery Campus*; James L. Wang, *The University of Alabama*; Stephen Brick, *University of South Alabama*; Joerg Feldvoss, *University of South Alabama* **ALASKA** Mark A. Fitch, *University of Alaska Anchorage*; Kamal Narang, *University of Alaska Anchorage*; Alexei Rybkin, *University of Alaska Fairbanks*; Martin Getz, *University of Alaska Fairbanks* **ARIZONA** Stefania Tracogna, *Arizona State University*; Bruno Welfert, *Arizona State University*; Light Bryant, *Arizona Western College*; Daniel Russow, *Arizona Western College*; Jennifer Jameson, *Coconino College*; George Cole, *Mesa Community College*; David Schultz, *Mesa Community College*; Michael Bezusko, *Pima Community College, Desert Vista Campus*; Garry Carpenter, *Pima Community College, Northwest Campus*; Paul Flasch, *Pima County Community College*; Jessica Knapp, *Pima Community College, Northwest Campus*; Roger Werbylo, *Pima County Community College*; Katie Louchart, *Northern Arizona University*; Janet McShane, *Northern Arizona University*; Donna M. Krawczyk, *The University of Arizona* **ARKANSAS** Deborah Parker, *Arkansas Northeastern College*; J. Michael Hall, *Arkansas State University*; Kevin Cornelius, *Ouachita Baptist University*; Hyungkoo Mark Park, *Southern Arkansas University*; Katherine Pinzon, *University of Arkansas at Fort Smith*; Denise LeGrand, *University of Arkansas at Little Rock*; John Annulis, *University of Arkansas at Monticello*; Erin Haller, *University of Arkansas, Fayetteville*; Daniel J. Arrigo, *University of Central Arkansas* **CALIFORNIA** Harvey Greenwald, *California Polytechnic State University, San Luis Obispo*; Charles Hale, *California Polytechnic State University*; John M. Alongi, *California Polytechnic State University, San Luis Obispo*; John Hagen, *California Polytechnic State University, San Luis Obispo*; Colleen Margarita Kirk, *California Polytechnic State University, San Luis Obispo*; Lawrence Sze, *California Polytechnic State University, San Luis Obispo*; Raymond Terry, *California Polytechnic State University, San Luis Obispo*; James R. McKinney, *California State Polytechnic University, Pomona*; Robin Wilson, *California State Polytechnic University, Pomona*; Charles Lam, *California State University, Bakersfield*; David McKay, *California State University, Long Beach*; Melvin Lax, *California State University, Long Beach*; Wallace A. Etterbeek, *California State University, Sacramento*; Mohamed Allali, *Chapman University*; George Rhys, *College of the Canyons*; Janice Hector, *DeAnza College*; Isabelle Saber, *Glendale Community College*; Peter Stathis, *Glendale Community College*; Douglas B. Lloyd, *Golden West College*; Thomas Scardina, *Golden West College*; Kristin Hartford, *Long Beach City College*; Eduardo Arismendi-Pardi, *Orange Coast College*; Mitchell Alves, *Orange Coast College*; Yenkanh Vu, *Orange Coast College*; Yan Tian, *Palomar College*; Donna E. Nordstrom, *Pasadena City College*; Don L. Hancock, *Pepperdine University*; Kevin Iga, *Pepperdine University*; Adolfo J. Rumbos, *Pomona College*; Carlos de la Lama, *San Diego City College*; Matthias Beck, *San Francisco State University*; Arek Goetz, *San Francisco State University*; Nick Bykov, *San Joaquin Delta College*; Eleanor Lang Kendrick, *San Jose City College*; Elizabeth Hodes, *Santa Barbara City College*; William Konya, *Santa Monica College*; John Kennedy, *Santa Monica College*; Peter Lee, *Santa Monica College*; Richard Salome, *Scotts Valley High School*; Norman Feldman, *Sonoma State University*; Elaine McDonald, *Sonoma State University*; John D. Eggers, *University of California, San Diego*; Bruno Nachtergaele, *University of California, Davis*; Boumediene Hamzi, *University of California, Davis*; Richard Leborne, *University of California, San Diego*; Peter Stevenhagen, *University of California, San Diego*; Jeffrey Stopple, *University of California, Santa Barbara*; Guofang Wei, *University of California, Santa Barbara*; Rick A. Simon, *University of La Verne*; Mohamad A. Alwash, *West Los Angeles College*; Calder Daenzer, *University of California, Berkeley*; Jude Thaddeus Socrates, *Pasadena City College*; Cheuk Ying Lam, *California State University Bakersfield*; Borislava Gutarts, *California State University, Los Angeles*; Daniel Rogalski, *University of California, San Diego*; Don Hartig, *California Polytechnic State University*; Anne Voth, *Palomar College*; Jay Wiestling, *Palomar College*; Lindsey Bramlett-Smith, *Santa Barbara City College*; Dennis Morrow, *College of the Canyons*; Sydney Shanks, *College of the Canyons*; Bob Tolar, *College of the Canyons*; Gene W. Majors, *Fullerton College*; Robert Diaz, *Fullerton College*; Gregory Nguyen, *Fullerton College*; Paul Sjoberg, *Fullerton College*; Deborah Ritchie, *Moorpark College*; Maya Rahnamaie, *Moorpark College*; Kathy Fink, *Moorpark College*; Christine Cole, *Moorpark College*; K. Di Passero, *Moorpark College*; Sid Kolpas, *Glendale Community College*; Miriam Castrconde, *Irvine Valley College*;

Ilkner Erbas-White, *Irvine Valley College*; Corey Manchester, *Grossmont College*; Donald Murray, *Santa Monica College*; Barbara McGee, *Cuesta College*; Marie Larsen, *Cuesta College*; Joe Vasta, *Cuesta College*; Mike Kinter, *Cuesta College*; Mark Turner, *Cuesta College*; G. Lewis, *Cuesta College*; Daniel Kleinfelter, *College of the Desert*; Esmeralda Medrano, *Citrus College*; James Swatzel, *Citrus College*; Mark Littrell, *Rio Hondo College*; Rich Zucker, *Irvine Valley College*; Cindy Torigison, *Palomar College*; Craig Chamberline, *Palomar College*; Lindsey Lang, *Diablo Valley College*; Sam Needham, *Diablo Valley College*; Dan Bach, *Diablo Valley College*; Ted Nirgiotis, *Diablo Valley College*; Monte Collazo, *Diablo Valley College*; Tina Levy, *Diablo Valley College*; Mona Panchal, *East Los Angeles College*; Ron Sandvick, *San Diego Mesa College*; Larry Handa, *West Valley College*; Frederick Utter, *Santa Rose Junior College*; Farshod Mosh, *DeAnza College*; Doli Bambhania, *DeAnza College*; Charles Klein, *DeAnza College*; Tammi Marshall, *Cauyamaca College*; Inwon Leu, *Cauyamaca College*; Michael Moretti, *Bakersfield College*; Janet Tarjan, *Bakersfield College*; Hoat Le, *San Diego City College*; Richard Fielding, *Southwestern College*; Shannon Gracey, *Southwestern College*; Janet Mazzarella, *Southwestern College*; Christina Soderlund, *California Lutheran University*; Rudy Gonzalez, *Citrus College*; Robert Crise, *Crafton Hills College*; Joseph Kazimir, *East Los Angeles College*; Randall Rogers, *Fullerton College*; Peter Bouzar, *Golden West College*; Linda Ternes, *Golden West College*; Hsiao-Ling Liu, *Los Angeles Trade Tech Community College*; Yu-Chung Chang-Hou, *Pasadena City College*; Guillermo Alvarez, *San Diego City College*; Ken Kuniyuki, *San Diego Mesa College*; Laleh Howard, *San Diego Mesa College*; Sharareh Masooman, *Santa Barbara City College*; Jared Hersh, *Santa Barbara City College*; Betty Wong, *Santa Monica College*; Brian Rodas, *Santa Monica College* **COLORADO** Tony Weathers, *Adams State College*; Erica Johnson, *Arapahoe Community College*; Karen Walters, *Arapahoe Community College*; Joshua D. Laison, *Colorado College*; G. Gustave Greivel, *Colorado School of Mines*; Jim Thomas, *Colorado State University*; Eleanor Storey, *Front Range Community College*; Larry Johnson, *Metropolitan State College of Denver*; Carol Kuper, *Morgan Community College*; Larry A. Pontaski, *Pueblo Community College*; Terry Chen Reeves, *Red Rocks Community College*; Debra S. Carney, *University of Denver*; Louis A. Talman, *Metropolitan State College of Denver*; Mary A. Nelson, *University of Colorado at Boulder*; J. Kyle Pula, *University of Denver*; Jon Von Stroh, *University of Denver*; Sharon Butz, *University of Denver*; Daniel Daly, *University of Denver*; Tracy Lawrence, *Arapahoe Community College*; Shawna Mahan, *University of Colorado Denver*; Adam Norris, *University of Colorado at Boulder*; Anca Radulescu, *University of Colorado at Boulder*; Mike Kawai, *University of Colorado Denver*; Janet Barnett, *Colorado State University–Pueblo*; Byron Hurley, *Colorado State University–Pueblo*; Jonathan Portiz, *Colorado State University–Pueblo*; Bill Emerson, *Metropolitan State College of Denver*; Suzanne Caulk, *Regis University*; Anton Dzhamay, *University of Northern Colorado* **CONNECTICUT** Jeffrey McGowan, *Central Connecticut State University*; Ivan Gotchev, *Central Connecticut State University*; Charles Waiveris, *Central Connecticut State University*; Christopher Hammond, *Connecticut College*; Kim Ward, *Eastern Connecticut State University*; Joan W. Weiss, *Fairfield University*; Theresa M. Sandifer, *Southern Connecticut State University*; Cristian Rios, *Trinity College*; Melanie Stein, *Trinity College*; Steven Orszag, *Yale University* **DELAWARE** Patrick F. Mwerinde, *University of Delaware* **DISTRICT OF COLUMBIA** Jeffrey Hakim, *American University*; Joshua M. Lansky, *American University*; James A. Nickerson, *Gallaudet University* **FLORIDA** Abbas Zadegan, *Florida International University*; Gerardo Aladro, *Florida International University*; Gregory Henderson, *Hillsborough Community College*; Pam Crawford, *Jacksonville University*; Penny Morris, *Polk Community College*; George Schultz, *St. Petersburg College*; Jimmy Chang, *St. Petersburg College*; Carolyn Kistner, *St. Petersburg College*; Aida Kadic-Galeb, *The University of Tampa*; Constance Schober, *University of Central Florida*; S. Roy Choudhury, *University of Central Florida*; Kurt Overhiser, *Valencia Community College*; Jiongmin Yong, *University of Central Florida*; Giray Okten, *The Florida State University*; Frederick Hoffman, *Florida Atlantic University*; Thomas Beatty, *Florida Gulf Coast University*; Witny Librun, *Palm Beach Community College North*; Joe Castillo, *Broward County College*; Joann Lewin, *Edison College*; Donald Ransford, *Edison College*; Scott Berthiaume, *Edison College*; Alexander Ambrioso, *Hillsborough Community College*; Jane Golden, *Hillsborough Community College*; Susan Hiatt, *Polk Community College–Lakeland Campus*; Li Zhou, *Polk Community College–Winter Haven Campus*; Heather Edwards, *Seminole Community College*; Benjamin Landon, *Daytona State College*; Tony Malaret, *Seminole Community College*; Lane Vosbury, *Seminole Community College*; William Rickman, *Seminole Community College*; Cheryl Cantwell,

Seminole Community College; Michael Schramm, *Indian River State College*; Janette Campbell, *Palm Beach Community College–Lake Worth* **GEORGIA** Thomas T. Morley, *Georgia Institute of Technology*; Ralph Wildy, *Georgia Military College*; Shahram Nazari, *Georgia Perimeter College*; Alice Eiko Pierce, *Georgia Perimeter College, Clarkson Campus*; Susan Nelson, *Georgia Perimeter College, Clarkson Campus*; Laurene Fausett, *Georgia Southern University*; Scott N. Kersey, *Georgia Southern University*; Jimmy L. Solomon, *Georgia Southern University*; Allen G. Fuller, *Gordon College*; Marwan Zabdawi, *Gordon College*; Carolyn A. Yackel, *Mercer University*; Shahryar Heydari, *Piedmont College*; Dan Kannan, *The University of Georgia*; Abdelkrim Brania, *Morehouse College*; Ying Wang, *Augusta State University*; James M. Benedict, *Augusta State University*; Kouong Law, *Georgia Perimeter College*; Rob Williams, *Georgia Perimeter College*; Alvina Atkinson, *Georgia Gwinnett College*; Amy Erickson, *Georgia Gwinnett College* **HAWAII** Shuguang Li, *University of Hawaii at Hilo*; Raina B. Ivanova, *University of Hawaii at Hilo* **IDAHO** Charles Kerr, *Boise State University*; Otis Kenny, *Boise State University*; Alex Feldman, *Boise State University*; Doug Bullock, *Boise State University*; Ed Korntved, *Northwest Nazarene University* **ILLINOIS** Chris Morin, *Blackburn College*; Alberto L. Delgado, *Bradley University*; John Haverhals, *Bradley University*; Herbert E. Kasube, *Bradley University*; Marvin Doubet, *Lake Forest College*; Marvin A. Gordon, *Lake Forest Graduate School of Management*; Richard J. Maher, *Loyola University Chicago*; Joseph H. Mayne, *Loyola University Chicago*; Marian Gidea, *Northeastern Illinois University*; Miguel Angel Lerma, *Northwestern University*; Mehmet Dik, *Rockford College*; Tammy Voepel, *Southern Illinois University Edwardsville*; Rahim G. Karimpour, *Southern Illinois University*; Thomas Smith, *University of Chicago*; Laura DeMarco, *University of Illinois*; Jennifer McNeilly, *University of Illinois at Urbana-Champaign*; Manouchehr Azad, *Harper College*; Minhua Liu, *Harper College*; Mary Hill, *College of DuPage*; Arthur N. DiVito, *Harold Washington College* **INDIANA** Julie A. Killingbeck, *Ball State University*; John P. Boardman, *Franklin College*; Robert N. Talbert, *Franklin College*; Robin Symonds, *Indiana University Kokomo*; Henry L. Wyzinski, *Indiana University Northwest*; Melvin Royer, *Indiana Wesleyan University*; Gail P. Greene, *Indiana Wesleyan University*; David L. Finn, *Rose-Hulman Institute of Technology* **IOWA** Nasser Dastrange, *Buena Vista University*; Mark A. Mills, *Central College*; Karen Ernst, *Hawkeye Community College*; Richard Mason, *Indian Hills Community College*; Robert S. Keller, *Loras College*; Eric Robert Westlund, *Luther College*; Weimin Han, *The University of Iowa* **KANSAS** Timothy W. Flood, *Pittsburg State University*; Sarah Cook, *Washburn University*; Kevin E. Charlwood, *Washburn University*; Conrad Uwe, *Cowley County Community College* **KENTUCKY** Alex M. McAllister, *Center College*; Sandy Spears, *Jefferson Community & Technical College*; Leanne Faulkner, *Kentucky Wesleyan College*; Donald O. Clayton, *Madisonville Community College*; Thomas Riedel, *University of Louisville*; Manabendra Das, *University of Louisville*; Lee Larson, *University of Louisville*; Jens E. Harlander, *Western Kentucky University*; Philip McCartney, *Northern Kentucky University*; Andy Long, *Northern Kentucky University*; Omer Yayenie, *Murray State University*; Donald Krug, *Northern Kentucky University* **LOUISIANA** William Forrest, *Baton Rouge Community College*; Paul Wayne Britt, *Louisiana State University*; Galen Turner, *Louisiana Tech University*; Randall Wills, *Southeastern Louisiana University*; Kent Neuerburg, *Southeastern Louisiana University*; Guoli Ding, *Louisiana State University*; Julia Ledet, *Louisiana State University* **MAINE** Andrew Knightly, *The University of Maine*; Sergey Lvin, *The University of Maine*; Joel W. Irish, *University of Southern Maine*; Laurie Woodman, *University of Southern Maine*; David M. Bradley, *The University of Maine*; William O. Bray, *The University of Maine* **MARYLAND** Leonid Stern, *Towson University*; Mark E. Williams, *University of Maryland Eastern Shore*; Austin A. Lobo, *Washington College*; Supawan Lertskrai, *Harford Community College*; Fary Sami, *Harford Community College*; Andrew Bulleri, *Howard Community College* **MASSACHUSETTS** Sean McGrath, *Algonquin Regional High School*; Norton Starr, *Amherst College*; Renato Mirollo, *Boston College*; Emma Previato, *Boston University*; Richard H. Stout, *Gordon College*; Matthew P. Leingang, *Harvard University*; Suellen Robinson, *North Shore Community College*; Walter Stone, *North Shore Community College*; Barbara Loud, *Regis College*; Andrew B. Perry, *Springfield College*; Tawanda Gwena, *Tufts University*; Gary Simundza, *Wentworth Institute of Technology*; Mikhail Chkhenkeli, *Western New England College*; David Daniels, *Western New England College*; Alan Gorfin, *Western New England College*; Saeed Ghahramani, *Western New England College*; Julian Fleron, *Westfield State College*; Brigitte Servatius, *Worcester Polytechnic Institute*; John Goulet, *Worcester Polytechnic Institute*; Alexander Martsinkovsky, *Northeastern University*; Marie Clote, *Boston*

College **MICHIGAN** Mark E. Bollman, *Albion College*; Jim Chesla, *Grand Rapids Community College*; Jeanne Wald, *Michigan State University*; Allan A. Struthers, *Michigan Technological University*; Debra Pharo, *Northwestern Michigan College*; Anna Maria Spagnuolo, *Oakland University*; Diana Faoro, *Romeo Senior High School*; Andrew Strowe, *University of Michigan–Dearborn*; Daniel Stephen Drucker, *Wayne State University*; Christopher Cartwright, *Lawrence Technological University*; Jay Treiman, *Western Michigan University* **MINNESOTA** Bruce Bordwell, *Anoka-Ramsey Community College*; Robert Dobrow, *Carleton College*; Jessie K. Lenarz, *Concordia College–Moorhead Minnesota*; Bill Tomhave, *Concordia College*; David L. Frank, *University of Minnesota*; Steven I. Sperber, *University of Minnesota*; Jeffrey T. McLean, *University of St. Thomas*; Chehrzad Shakiban, *University of St. Thomas*; Melissa Loe, *University of St. Thomas*; Nick Christopher Fiala, *St. Cloud State University*; Victor Padron, *Normandale Community College*; Mark Ahrens, *Normandale Community College*; Gerry Naughton, *Century Community College*; Carrie Naughton, *Inver Hills Community College* **MISSISSIPPI** Vivien G. Miller, *Mississippi State University*; Ted Dobson, *Mississippi State University*; Len Miller, *Mississippi State University*; Tristan Denley, *The University of Mississippi* **MISSOURI** Robert Robertson, *Drury University*; Gregory A. Mitchell, *Metropolitan Community College–Penn Valley*; Charles N. Curtis, *Missouri Southern State University*; Vivek Narayanan, *Moberly Area Community College*; Russell Blyth, *Saint Louis University*; Blake Thornton, *Saint Louis University*; Kevin W. Hopkins, *Southwest Baptist University*; Joe Howe, *St. Charles Community College*; Wanda Long, *St. Charles Community College*; Andrew Stephan, *St. Charles Community College* **MONTANA** Kelly Cline, *Carroll College*; Richard C. Swanson, *Montana State University*; Nikolaus Vonessen, *The University of Montana* **NEBRASKA** Edward G. Reinke Jr., *Concordia University*; Judith Downey, *University of Nebraska at Omaha* **NEVADA** Rohan Dalpatadu, *University of Nevada, Las Vegas*; Paul Aizley, *University of Nevada, Las Vegas* **NEW HAMPSHIRE** Richard Jardine, *Keene State College*; Michael Cullinane, *Keene State College*; Roberta Kieronski, *University of New Hampshire at Manchester*; Erik Van Erp, *Dartmouth College* **NEW JERSEY** Paul S. Rossi, *College of Saint Elizabeth*; Mark Galit, *Essex County College*; Katarzyna Potocka, *Ramapo College of New Jersey*; Nora S. Thornber, *Raritan Valley Community College*; Avraham Soffer, *Rutgers, The State University of New Jersey*; Chengwen Wang, *Rutgers, The State University of New Jersey*; Stephen J. Greenfield, *Rutgers, The State University of New Jersey*; John T. Saccoman, *Seton Hall University*; Lawrence E. Levine, *Stevens Institute of Technology*; Barry Burd, *Drew University*; Penny Luczak, *Camden County College*; John Climent, *Cecil Community College*; Kristyanna Erickson, *Cecil Community College*; Eric Compton, *Brookdale Community College*; John Atsu-Swanzy, *Atlantic Cape Community College* **NEW MEXICO** Kevin Leith, *Central New Mexico Community College*; David Blankenbaker, *Central New Mexico Community College*; Joseph Lakey, *New Mexico State University*; Kees Onneweer, *University of New Mexico*; Jurg Bolli, *The University of New Mexico* **NEW YORK** Robert C. Williams, *Alfred University*; Timmy G. Bremer, *Broome Community College State University of New York*; Joaquin O. Carbonara, *Buffalo State College*; Robin Sue Sanders, *Buffalo State College*; Daniel Cunningham, *Buffalo State College*; Rose Marie Castner, *Canisius College*; Sharon L. Sullivan, *Catawba College*; Camil Muscalu, *Cornell University*; Maria S. Terrell, *Cornell University*; Margaret Mulligan, *Dominican College of Blauvelt*; Robert Andersen, *Farmingdale State University of New York*; Leonard Nissim, *Fordham University*; Jennifer Roche, *Hobart and William Smith Colleges*; James E. Carpenter, *Iona College*; Peter Shenkin, *John Jay College of Criminal Justice/CUNY*; Gordon Crandall, *LaGuardia Community College/CUNY*; Gilbert Traub, *Maritime College, State University of New York*; Paul E. Seeburger, *Monroe Community College Brighton Campus*; Abraham S. Mantell, *Nassau Community College*; Daniel D. Birmajer, *Nazareth College*; Sybil G. Shaver, *Pace University*; Margaret Kiehl, *Rensselaer Polytechnic Institute*; Carl V. Lutzer, *Rochester Institute of Technology*; Michael A. Radin, *Rochester Institute of Technology*; Hossein Shahmohamad, *Rochester Institute of Technology*; Thomas Rousseau, *Siena College*; Jason Hofstein, *Siena College*; Leon E. Gerber, *St. Johns University*; Christopher Bishop, *Stony Brook University*; James Fulton, *Suffolk County Community College*; John G. Michaels, *SUNY Brockport*; Howard J. Skogman, *SUNY Brockport*; Cristina Bacuta, *SUNY Cortland*; Jean Harper, *SUNY Fredonia*; Kelly Black, *Union College*; Thomas W. Cusick, *University at Buffalo/The State University of New York*; Gino Biondini, *University at Buffalo/The State University of New York*; Robert Koehler, *University at Buffalo/The State University of New York*; Robert Thompson, *Hunter College*; Ed Grossman, *The City College of New York* **NORTH CAROLINA** Jeffrey Clark, *Elon University*; William L. Burgin, *Gaston College*; Manouchehr H. Misaghian, *Johnson C. Smith*

University; Legunchim L. Emmanwori, *North Carolina A&T State University*; Drew Pasteur, *North Carolina State University*; Demetrio Labate, *North Carolina State University*; Mohammad Kazemi, *The University of North Carolina at Charlotte*; Richard Carmichael, *Wake Forest University*; Gretchen Wilke Whipple, *Warren Wilson College*; John Russell Taylor, *University of North Carolina at Charlotte*; Mark Ellis, *Piedmont Community College* **NORTH DAKOTA** Anthony J. Bevelacqua, *The University of North Dakota*; Richard P. Millspaugh, *The University of North Dakota*; Thomas Gilsdorf, *The University of North Dakota*; Michele Iiams, *The University of North Dakota* **OHIO** Christopher Butler, *Case Western Reserve University*; Pamela Pierce, *The College of Wooster*; Tzu-Yi Alan Yang, *Columbus State Community College*; Greg S. Goodhart, *Columbus State Community College*; Kelly C. Stady, *Cuyahoga Community College*; Brian T. Van Pelt, *Cuyahoga Community College*; David Robert Ericson, *Miami University*; Frederick S. Gass, *Miami University*; Thomas Stacklin, *Ohio Dominican University*; Vitaly Bergelson, *The Ohio State University*; Robert Knight, *Ohio University*; John R. Pather, *Ohio University, Eastern Campus*; Teresa Contenza, *Otterbein College*; Ali Hajjafar, *The University of Akron*; Jianping Zhu, *The University of Akron*; Ian Clough, *University of Cincinnati Clermont College*; Atif Abueida, *University of Dayton*; Judith McCrory, *The University at Findlay*; Thomas Smotzer, *Youngstown State University*; Angela Spalsbury, *Youngstown State University*; James Osterburg, *The University of Cincinnati*; Frederick Thulin, *University of Illinois at Chicago*; Weimin Han, *The Ohio State University*; Critchton Ogle, *The Ohio State University*; Jackie Miller, *The Ohio State University*; Walter Mackey, *Owens Community College*; Jonathan Baker, *Columbus State Community College* **OKLAHOMA** Michael McClendon, *University of Central Oklahoma*; Teri Jo Murphy, *The University of Oklahoma*; Shirley Pomeranz, *University of Tulsa* **OREGON** Lorna TenEyck, *Chemeketa Community College*; Angela Martinek, *Linn-Benton Community College*; Tevian Dray, *Oregon State University*; Mark Ferguson, *Chemekata Community College*; Andrew Flight, *Portland State University* **PENNSYLVANIA** John B. Polhill, *Bloomsburg University of Pennsylvania*; Russell C. Walker, *Carnegie Mellon University*; Jon A. Beal, *Clarion University of Pennsylvania*; Kathleen Kane, *Community College of Allegheny County*; David A. Santos, *Community College of Philadelphia*; David S. Richeson, *Dickinson College*; Christine Marie Cedzo, *Gannon University*; Monica Pierri-Galvao, *Gannon University*; John H. Ellison, *Grove City College*; Gary L. Thompson, *Grove City College*; Dale McIntyre, *Grove City College*; Dennis Benchoff, *Harrisburg Area Community College*; William A. Drumin, *King's College*; Denise Reboli, *King's College*; Chawne Kimber, *Lafeyette College*; David L. Johnson, *Lehigh University*; Zia Uddin, *Lock Haven University of Pennsylvania*; Donna A. Dietz, *Mansfield University of Pennsylvania*; Samuel Wilcock, *Messiah College*; Neena T. Chopra, *The Pennsylvania State University*; Boris A. Datskovsky, *Temple University*; Dennis M. DeTurck, *University of Pennsylvania*; Jacob Burbea, *University of Pittsburgh*; Mohammed Yahdi, *Ursinus College*; Timothy Feeman, *Villanova University*; Douglas Norton, *Villanova University*; Robert Styer, *Villanova University*; Peter Brooksbank, *Bucknell University*; Larry Friesen, *Butler County Community College*; Lisa Angelo, *Bucks County College*; Elaine Fitt, *Bucks County College*; Pauline Chow, *Harrisburg Area Community College*; Diane Benner, *Harrisburg Area Community College*; Emily B. Dryden, *Bucknell University* **RHODE ISLAND** Thomas F. Banchoff, *Brown University*; Yajni Warnapala-Yehiya, *Roger Williams University*; Carol Gibbons, *Salve Regina University*; Joe Allen, *Community College of Rhode Island*; Michael Latina, *Community College of Rhode Island* **SOUTH CAROLINA** Stanley O. Perrine, *Charleston Southern University*; Joan Hoffacker, *Clemson University*; Constance C. Edwards, *Coastal Carolina University*; Thomas L. Fitzkee, *Francis Marion University*; Richard West, *Francis Marion University*; John Harris, *Furman University*; Douglas B. Meade, *University of South Carolina*; George Androulakis, *University of South Carolina*; Art Mark, *University of South Carolina Aiken*; Sherry Biggers, *Clemson University*; Mary Zachary Krohn, *Clemson University*; Andrew Incognito, *Coastal Carolina University*; Deanna Caveny, *College of Charleston* **SOUTH DAKOTA** Dan Kemp, *South Dakota State University* **TENNESSEE** Andrew Miller, *Belmont University*; Arthur A. Yanushka, *Christian Brothers University*; Laurie Plunk Dishman, *Cumberland University*; Beth Long, *Pellissippi State Technical Community College*; Judith Fethe, *Pellissippi State Technical Community College*; Andrzej Gutek, *Tennessee Technological University*; Sabine Le Borne, *Tennessee Technological University*; Richard Le Borne, *Tennessee Technological University*; Jim Conant, *The University of Tennessee*; Pavlos Tzermias, *The University of Tennessee*; Jo Ann W. Staples, *Vanderbilt University*; Dave Vinson, *Pellissippi State Community College*; Jonathan Lamb, *Pellissippi State Community College*

TEXAS Sally Haas, *Angelina College*; Michael Huff, *Austin Community College*; Scott Wilde, *Baylor University and The University of Texas at Arlington*; Rob Eby, *Blinn College*; Tim Sever, *Houston Community College–Central*; Ernest Lowery, *Houston Community College–Northwest*; Shirley Davis, *South Plains College*; Todd M. Steckler, *South Texas College*; Mary E. Wagner-Krankel, *St. Mary's University*; Elise Z. Price, *Tarrant County College, Southeast Campus*; David Price, *Tarrant County College, Southeast Campus*; Michael Stecher, *Texas A&M University*; Philip B. Yasskin, *Texas A&M University*; Brock Williams, *Texas Tech University*; I. Wayne Lewis, *Texas Tech University*; Robert E. Byerly, *Texas Tech University*; Ellina Grigorieva, *Texas Woman's University*; Abraham Haje, *Tomball College*; Scott Chapman, *Trinity University*; Elias Y. Deeba, *University of Houston Downtown*; Jianping Zhu, *The University of Texas at Arlington*; Tuncay Aktosun, *The University of Texas at Arlington*; John E. Gilbert, *The University of Texas at Austin*; Jorge R. Viramontes-Olivias, *The University of Texas at El Paso*; Melanie Ledwig, *The Victoria College*; Gary L. Walls, *West Texas A&M University*; William Heierman, *Wharton County Junior College*; Lisa Rezac, *University of St. Thomas*; Raymond J. Cannon, *Baylor University*; Kathryn Flores, *McMurry University*; Jacqueline A. Jensen, *Sam Houston State University*; James Galloway, *Collin County College*; Raja Khoury, *Collin County College*; Annette Benbow, *Tarrant County College–Northwest*; Greta Harland, *Tarrant County College–Northeast*; Doug Smith, *Tarrant County College–Northeast*; Marcus McGuff, *Austin Community College*; Clarence McGuff, *Austin Community College*; Steve Rodi, *Austin Community College*; Vicki Payne, *Austin Community College*; Anne Pradera, *Austin Community College*; Christy Babu, *Laredo Community College*; Deborah Hewitt, *McLennan Community College*; W. Duncan, *McLennan Community College*; Hugh Griffith, *Mt. San Antonio College* **UTAH** Jason Isaac Preszler, *The University of Utah*; Ruth Trygstad, *Salt Lake City Community College* **VIRGINIA** Verne E. Leininger, *Bridgewater College*; Brian Bradie, *Christopher Newport University*; Hongwei Chen, *Christopher Newport University*; John J. Avioli, *Christopher Newport University*; James H. Martin, *Christopher Newport University*; Mike Shirazi, *Germanna Community College*; Ramon A. Mata-Toledo, *James Madison University*; Adrian Riskin, *Mary Baldwin College*; Josephine Letts, *Ocean Lakes High School*; Przemyslaw Bogacki, *Old Dominion University*; Deborah Denvir, *Randolph-Macon Woman's College*; Linda Powers, *Virginia Tech*; Gregory Dresden, *Washington and Lee University*; Jacob A. Siehler, *Washington and Lee University*; Nicholas Hamblet, *University of Virginia*; Lester Frank Caudill, *University of Richmond* **VERMONT** David Dorman, *Middlebury College*; Rachel Repstad, *Vermont Technical College* **WASHINGTON** Jennifer Laveglia, *Bellevue Community College*; David Whittaker, *Cascadia Community College*; Sharon Saxton, *Cascadia Community College*; Aaron Montgomery, *Central Washington University*; Patrick Averbeck, *Edmonds Community College*; Tana Knudson, *Heritage University*; Kelly Brooks, *Pierce College*; Shana P. Calaway, *Shoreline Community College*; Abel Gage, *Skagit Valley College*; Scott MacDonald, *Tacoma Community College*; Martha A. Gady, *Whitworth College*; Wayne L. Neidhardt, *Edmonds Community College*; Simrat Ghuman, *Bellevue College*; Jeff Eldridge, *Edmonds Community College*; Kris Kissel, *Green River Community College*; Laura Moore-Mueller, *Green River Community College*; David Stacy, *Bellevue College*; Eric Schultz, *Walla Walla Community College*; Julianne Sachs, *Walla Walla Community College* **WEST VIRGINIA** Ralph Oberste-Vorth, *Marshall University*; Suda Kunyosying, *Shepard University*; Nicholas Martin, *Shepherd University*; Rajeev Rajaram, *Shepherd University*; Xiaohong Zhang, *West Virginia State University*; Sam B. Nadler, *West Virginia University* **WYOMING** Claudia Stewart, *Casper College*; Pete Wildman, *Casper College*; Charles Newberg, *Western Wyoming Community College*; Lynne Ipina, *University of Wyoming*; John Spitler, *University of Wyoming* **WISCONSIN** Paul Bankston, *Marquette University*; Jane Nichols, *Milwaukee School of Engineering*; Yvonne Yaz, *Milwaukee School of Engineering*; Terry Nyman, *University of Wisconsin–Fox Valley*; Robert L. Wilson, *University of Wisconsin–Madison*; Dietrich A. Uhlenbrock, *University of Wisconsin–Madison*; Paul Milewski, *University of Wisconsin–Madison*; Donald Solomon, *University of Wisconsin–Milwaukee*; Kandasamy Muthuvel, *University of Wisconsin–Oshkosh*; Sheryl Wills, *University of Wisconsin–Platteville*; Kathy A. Tomlinson, *University of Wisconsin–River Falls*; Joy Becker, *University of Wisconsin-Stout*; Jeganathan Sriskandarajah , *Madison Area Tech College*; Wayne Sigelko, *Madison Area Tech College* **CANADA** Don St. Jean, *George Brown College*; Len Bos, *University of Calgary*; Tony Ware, *University of Calgary*; Peter David Papez, *University of Calgary*; John O'Conner, *Grant MacEwan University*; Michael P. Lamoureux, *University of Calgary*; Yousry Elsabrouty, *University of Calgary*; Douglas Farenick, *University of Regina*

It is a pleasant task to thank the many people whose guidance and support were crucial in bringing this new edition to fruition. I was fortunate that Tony Palermino continued on as my developmental editor. I am happy to thank him again for the wisdom and dedication he brought to the job, and for improvements too numerous to detail.

I wish to thank the many mathematicians who generously shared valuable insights, constructive criticism, and innovative exercises. I am particularly grateful to Professors Elka Block, Brian Bradie, C. K. Cheung, Greg Dresden, Stephen Greenfield, John Kennedy, Frank Purcell, and Jude Socrates, and to Frances Hammock, Don Larson, Nikki Meshkat, and Jane Sherman for invaluable assistance. I would also like to thank Ricardo Chavez and Professors Elena Galaktionova, Istvan Kovacs, and Jiri Lebl for helpful and perceptive comments.

Warmest thanks go to Terri Ward for managing the Second Edition with great skill and grace, and to Julie Lindstrom for overseeing the revision process. I remain indebted to Craig Bleyer for signing this project and standing behind it through the years. I am grateful to Ruth Baruth for bringing her vast knowledge and publishing experience to the project, to Steve Rigolosi for expert market development, and to Katrina Wilhelm for editorial assistance. My thanks are also due to W. H. Freeman's superb production team: Blake Logan, Bill Page, Paul Rohloff, Ted Szczepanski and Vivien Weiss, as well as to John Rogosich and Carol Sawyer at Techsetters, Inc. for their expert composition, and to Ron Weickart at Network Graphics for his skilled and creative execution of the art program.

To my dearest wife, Julie, I owe more than I can say. Thank you for everything. To our wonderful children Rivkah, Dvora, Hannah, and Akiva, thank you for putting up with the calculus book through all these years. And to my mother Elise, and my late father Alexander Rogawski, MD ז"ל, thank you for your love and support from the beginning.

TO THE STUDENT

Although I have taught calculus for more than 30 years, when I enter the classroom on the first day of a new semester, I always have a feeling of excitement, as if a great drama is about to unfold. Does the word *drama* seem out of place in a discussion of mathematics?

Most people would agree that calculus is useful—it is applied across the sciences and engineering to everything from space flight and weather prediction to nanotechnology and financial modeling. But what is dramatic about it?

For me, one part of the drama lies in the conceptual and logical development of calculus. Calculus is based on just a few fundamental concepts (such as limits, tangent lines, and approximations). But as the subject develops, we find that these concepts are adequate to build, step-by-step, a mathematical discipline capable of solving innumerable problems of great practical importance. Along the way, there are high points and moments of suspense—for example, computing a derivative using limits for the first time or learning from the Fundamental Theorem of Calculus that the two branches of calculus (differential and integral) are much more closely related than we might have expected. We also discover that calculus provides the right language for expressing our most fundamental and universal laws of nature, not just Newton's laws of motion, but also the laws of electromagnetism and even the quantum laws of atomic structure.

Another part of the drama is the learning process itself—the personal voyage of discovery. Certainly, one aspect of learning calculus is developing various technical skills. You will learn how to compute derivatives and integrals, solve optimization problems, and so on. These skills are necessary for applying calculus in practical situations, and they provide a foundation for further study of more advanced branches of mathematics. But perhaps more importantly, you will become acquainted with the fundamental ideas on which calculus is based. These ideas are central in the sciences and in all quantitative disciplines, and so they will open up for you a world of new opportunities. The distinguished mathematician I. M. Gelfand put it this way: "The most important thing a student can get from the study of mathematics is the attainment of a higher intellectual level."

This text is designed to develop both skills and conceptual understanding. In fact, the two go hand in hand. As you become proficient in problem solving, you will come to appreciate the underlying ideas. And it is equally true that a solid understanding of the concepts will make you a more effective problem solver. You are likely to devote much of your time to studying the examples in the text and working the exercises. However, the text also contains numerous down-to-earth explanations of the underlying concepts, ideas, and motivations (sometimes under the heading "Conceptual Insight" or "Graphical Insight"). I urge you to take the time to read these explanations and think about them.

Learning calculus will always be a challenge, and it will always require effort. According to legend, Alexander the Great once asked the mathematician Menaechmus to show him an easy way to learn geometry. Menaechmus replied, "There is no royal road to geometry." Even kings must work hard to learn geometry, and the same is true of calculus.

One of the main challenges in writing this textbook was finding a way to present calculus as clearly as possible, in a style that students would find comprehensible and interesting. While writing, I continually asked myself: Can it be made simpler? Have I assumed something the student may not be aware of? Can I explain the deeper significance of an underlying concept without confusing a student who is learning the subject for the first time?

I hope my efforts have resulted in a textbook that is not only student friendly but also encourages you to see the big picture—the beautiful and elegant ideas that hold the entire structure of calculus together. Please let me know if you have any comments or suggestions for improving the text. I look forward to hearing from you.

Best wishes and good luck!

Jon Rogawski

1 PRECALCULUS REVIEW

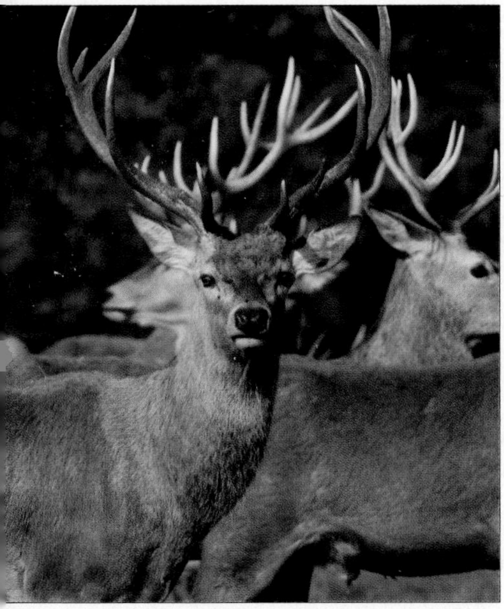

Functions are one of our most important tools for analyzing phenomena. Biologists have studied the antler weight of male red deer as a function of age (see p. 6).

Calculus builds on the foundation of algebra, analytic geometry, and trigonometry. In this chapter, therefore, we review some concepts, facts, and formulas from precalculus that are used throughout the text. In the last section, we discuss ways in which technology can be used to enhance your visual understanding of functions and their properties.

1.1 Real Numbers, Functions, and Graphs

We begin with a short discussion of real numbers. This gives us the opportunity to recall some basic properties and standard notation.

A **real number** is a number represented by a decimal or "decimal expansion." There are three types of decimal expansions: finite, repeating, and infinite but nonrepeating. For example,

$$\frac{3}{8} = 0.375, \qquad \frac{1}{7} = 0.142857142857\ldots = 0.\overline{142857}$$

$$\pi = 3.141592653589793\ldots$$

The number $\frac{3}{8}$ is represented by a finite decimal, whereas $\frac{1}{7}$ is represented by a *repeating* or *periodic* decimal. The bar over 142857 indicates that this sequence repeats indefinitely. The decimal expansion of π is infinite but nonrepeating.

The set of all real numbers is denoted by a boldface **R**. When there is no risk of confusion, we refer to a real number simply as a *number*. We also use the standard symbol $\in$ for the phrase "belongs to." Thus,

$$a \in \mathbf{R} \qquad \text{reads} \qquad \text{“}a \text{ belongs to } \mathbf{R}\text{”}$$

Additional properties of real numbers are discussed in Appendix B.

The set of integers is commonly denoted by the letter **Z** (this choice comes from the German word *Zahl*, meaning "number"). Thus, $\mathbf{Z} = \{\ldots, -2, -1, 0, 1, 2, \ldots\}$. A **whole number** is a nonnegative integer—that is, one of the numbers $0, 1, 2, \ldots$.

A real number is called **rational** if it can be represented by a fraction p/q, where p and q are integers with $q \neq 0$. The set of rational numbers is denoted **Q** (for "quotient"). Numbers that are not rational, such as π and $\sqrt{2}$, are called **irrational**.

We can tell whether a number is rational from its decimal expansion: Rational numbers have finite or repeating decimal expansions, and irrational numbers have infinite, nonrepeating decimal expansions. Furthermore, the decimal expansion of a number is unique, apart from the following exception: Every finite decimal is equal to an infinite decimal in which the digit 9 repeats. For example,

$$1 = 0.999\ldots, \qquad \frac{3}{8} = 0.375 = 0.374999\ldots, \qquad \frac{47}{20} = 2.35 = 2.34999\ldots$$

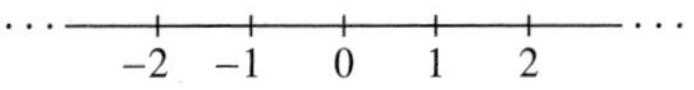

FIGURE 1 The set of real numbers represented as a line.

We visualize real numbers as points on a line (Figure 1). For this reason, real numbers are often referred to as **points**. The point corresponding to 0 is called the **origin**.

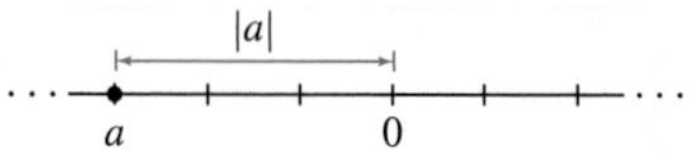

FIGURE 2 $|a|$ is the distance from a to the origin.

The **absolute value** of a real number a, denoted $|a|$, is defined by (Figure 2)

$$|a| = \textit{distance from the origin} = \begin{cases} a & \text{if } a \geq 0 \\ -a & \text{if } a < 0 \end{cases}$$

For example, $|1.2| = 1.2$ and $|-8.35| = 8.35$. The absolute value satisfies

$$|a| = |-a|, \qquad |ab| = |a|\,|b|$$

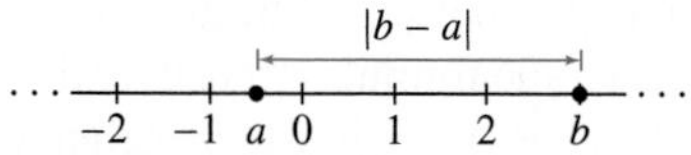

FIGURE 3 The distance from a to b is $|b - a|$.

The **distance** between two real numbers a and b is $|b - a|$, which is the length of the line segment joining a and b (Figure 3).

Two real numbers a and b are close to each other if $|b - a|$ is small, and this is the case if their decimal expansions agree to many places. More precisely, *if the decimal expansions of a and b agree to k places (to the right of the decimal point), then the distance $|b - a|$ is at most* 10^{-k}. Thus, the distance between $a = 3.1415$ and $b = 3.1478$ is at most 10^{-2} because a and b agree to two places. In fact, the distance is exactly $|3.1478 - 3.1415| = 0.0063$.

Beware that $|a + b|$ is not equal to $|a| + |b|$ unless a and b have the same sign or at least one of a and b is zero. If they have opposite signs, cancellation occurs in the sum $a + b$, and $|a + b| < |a| + |b|$. For example, $|2 + 5| = |2| + |5|$ but $|-2 + 5| = 3$, which is less than $|-2| + |5| = 7$. In any case, $|a + b|$ is never larger than $|a| + |b|$ and this gives us the simple but important **triangle inequality**:

$$|a + b| \leq |a| + |b| \qquad \boxed{1}$$

We use standard notation for intervals. Given real numbers $a < b$, there are four intervals with endpoints a and b (Figure 4). They all have length $b - a$ but differ according to which endpoints are included.

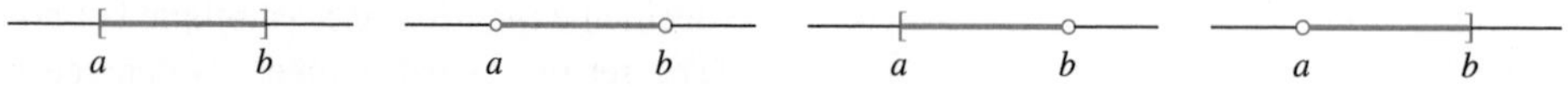

Closed interval $[a, b]$ (endpoints included) — Open interval (a, b) (endpoints excluded) — Half-open interval $[a, b)$ — Half-open interval $(a, b]$

FIGURE 4 The four intervals with endpoints a and b.

The **closed interval** $[a, b]$ is the set of all real numbers x such that $a \leq x \leq b$:

$$[a, b] = \{x \in \mathbf{R} : a \leq x \leq b\}$$

We usually write this more simply as $\{x : a \leq x \leq b\}$, it being understood that x belongs to $\mathbf{R}$. The **open** and **half-open intervals** are the sets

$$\underbrace{(a, b) = \{x : a < x < b\}}_{\text{Open interval (endpoints excluded)}}, \qquad \underbrace{[a, b) = \{x : a \leq x < b\}}_{\text{Half-open interval}}, \qquad \underbrace{(a, b] = \{x : a < x \leq b\}}_{\text{Half-open interval}}$$

The infinite interval $(-\infty, \infty)$ is the entire real line $\mathbf{R}$. A half-infinite interval is closed if it contains its finite endpoint and is open otherwise (Figure 5):

$$[a, \infty) = \{x : a \leq x < \infty\}, \qquad (-\infty, b] = \{x : -\infty < x \leq b\}$$

$[a, \infty)$ — $(-\infty, b]$

FIGURE 5 Closed half-infinite intervals.

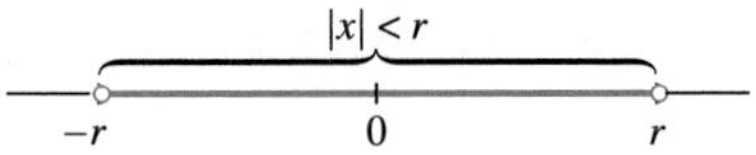

FIGURE 6 The interval $(-r, r) = \{x : |x| < r\}$.

Open and closed intervals may be described by inequalities. For example, the interval $(-r, r)$ is described by the inequality $|x| < r$ (Figure 6):

$$|x| < r \quad \Leftrightarrow \quad -r < x < r \quad \Leftrightarrow \quad x \in (-r, r) \tag{2}$$

More generally, for an interval symmetric about the value c (Figure 7),

$$|x - c| < r \quad \Leftrightarrow \quad c - r < x < c + r \quad \Leftrightarrow \quad x \in (c - r, c + r) \tag{3}$$

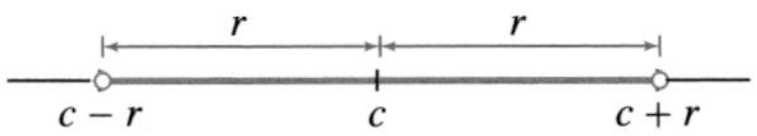

FIGURE 7 $(a, b) = (c - r, c + r)$, where

$$c = \frac{a+b}{2}, \qquad r = \frac{b-a}{2}$$

Closed intervals are similar, with $<$ replaced by $\le$. We refer to r as the **radius** and to c as the **midpoint** or **center**. The intervals (a, b) and $[a, b]$ have midpoint $c = \frac{1}{2}(a + b)$ and radius $r = \frac{1}{2}(b - a)$ (Figure 7).

■ **EXAMPLE 1** Describe [7, 13] using inequalities.

Solution The midpoint of the interval [7, 13] is $c = \frac{1}{2}(7 + 13) = 10$ and its radius is $r = \frac{1}{2}(13 - 7) = 3$ (Figure 8). Therefore,

$$[7, 13] = \{x \in \mathbf{R} : |x - 10| \le 3\}$$ ■

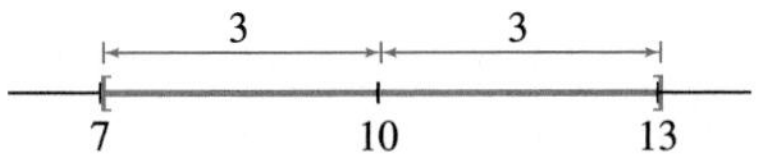

FIGURE 8 The interval [7, 13] is described by $|x - 10| \le 3$.

■ **EXAMPLE 2** Describe the set $S = \{x : |\frac{1}{2}x - 3| > 4\}$ in terms of intervals.

Solution It is easier to consider the opposite inequality $|\frac{1}{2}x - 3| \le 4$ first. By (2),

$$\left|\frac{1}{2}x - 3\right| \le 4 \quad \Leftrightarrow \quad -4 \le \frac{1}{2}x - 3 \le 4$$

$$-1 \le \frac{1}{2}x \le 7 \qquad \text{(add 3)}$$

$$-2 \le x \le 14 \qquad \text{(multiply by 2)}$$

In Example 2 we use the notation ∪ to denote "union": The union $A \cup B$ of sets A and B consists of all elements that belong to either A or B (or to both).

Thus, $|\frac{1}{2}x - 3| \le 4$ is satisfied when x belongs to $[-2, 14]$. The set S is the *complement*, consisting of all numbers x *not in* $[-2, 14]$. We can describe S as the union of two intervals: $S = (-\infty, -2) \cup (14, \infty)$ (Figure 9). ■

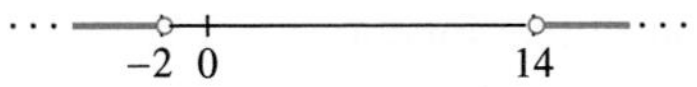

FIGURE 9 The set $S = \{x : |\frac{1}{2}x - 3| > 4\}$.

Graphing

Graphing is a basic tool in calculus, as it is in algebra and trigonometry. Recall that rectangular (or Cartesian) coordinates in the plane are defined by choosing two perpendicular axes, the x-axis and the y-axis. To a pair of numbers (a, b) we associate the point P located at the intersection of the line perpendicular to the x-axis at a and the line perpendicular to the y-axis at b [Figure 10(A)]. The numbers a and b are the x- and y-**coordinates** of P. The x-coordinate is sometimes called the "abscissa" and the y-coordinate the "ordinate." The **origin** is the point with coordinates (0, 0).

The term "Cartesian" refers to the French philosopher and mathematician René Descartes (1596–1650), whose Latin name was Cartesius. He is credited (along with Pierre de Fermat) with the invention of analytic geometry. In his great work La Géométrie, *Descartes used the letters x, y, z for unknowns and a, b, c for constants, a convention that has been followed ever since.*

The axes divide the plane into four quadrants labeled I–IV, determined by the signs of the coordinates [Figure 10(B)]. For example, quadrant III consists of points (x, y) such that $x < 0$ and $y < 0$.

The distance d between two points $P_1 = (x_1, y_1)$ and $P_2 = (x_2, y_2)$ is computed using the Pythagorean Theorem. In Figure 11, we see that $\overline{P_1P_2}$ is the hypotenuse of a right triangle with sides $a = |x_2 - x_1|$ and $b = |y_2 - y_1|$. Therefore,

$$d^2 = a^2 + b^2 = (x_2 - x_1)^2 + (y_2 - y_1)^2$$

We obtain the distance formula by taking square roots.

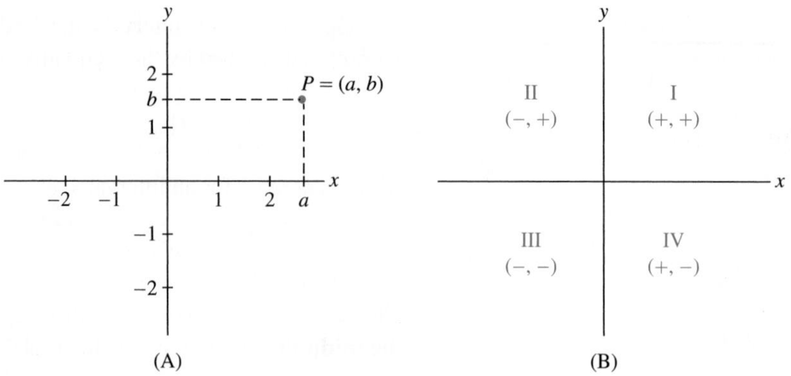

FIGURE 10 Rectangular coordinate system.

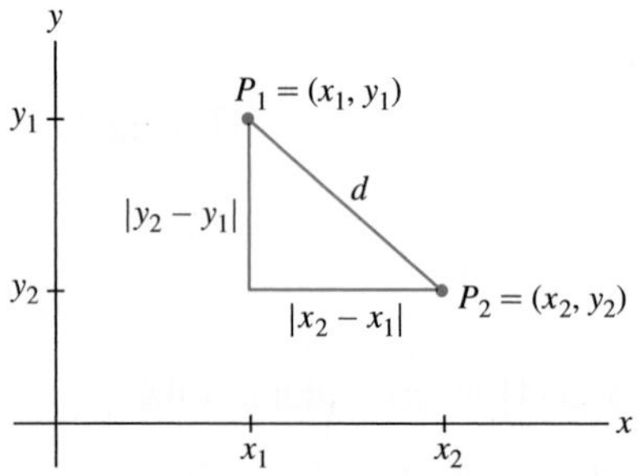

FIGURE 11 Distance d is given by the distance formula.

Distance Formula The distance between $P_1 = (x_1, y_1)$ and $P_2 = (x_2, y_2)$ is equal to

$$d = \sqrt{(x_2 - x_1)^2 + (y_2 - y_1)^2}$$

Once we have the distance formula, we can derive the equation of a circle of radius r and center (a, b) (Figure 12). A point (x, y) lies on this circle if the distance from (x, y) to (a, b) is r:

$$\sqrt{(x - a)^2 + (y - b)^2} = r$$

Squaring both sides, we obtain the standard equation of the circle:

$$(x - a)^2 + (y - b)^2 = r^2$$

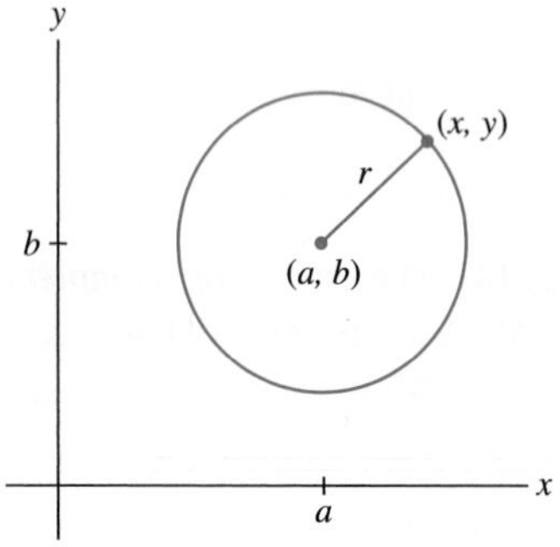

FIGURE 12 Circle with equation $(x - a)^2 + (y - b)^2 = r^2$.

We now review some definitions and notation concerning functions.

DEFINITION A **function** f from a set D to a set Y is a rule that assigns, to each element x in D, a unique element $y = f(x)$ in Y. We write

$$f : D \to Y$$

The set D, called the **domain** of f, is the set of "allowable inputs." For $x \in D$, $f(x)$ is called the **value** of f at x (Figure 13). The **range** R of f is the subset of Y consisting of all values $f(x)$:

$$R = \{y \in Y : f(x) = y \text{ for some } x \in D\}$$

Informally, we think of f as a "machine" that produces an output y for every input x in the domain D (Figure 14).

A function $f : D \to Y$ is also called a "map." The sets D and Y can be arbitrary. For example, we can define a map from the set of living people to the set of whole numbers by mapping each person to his or her year of birth. The range of this map is the set of years in which a living person was born. In multivariable calculus, the domain might be a set of points in three-dimensional space and the range a set of numbers, points, or vectors.

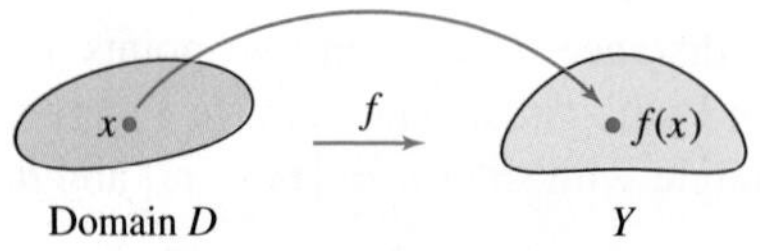

FIGURE 13 A function assigns an element $f(x)$ in Y to each $x \in D$.

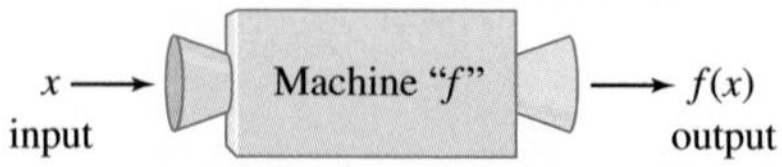

FIGURE 14 Think of f as a "machine" that takes the input x and produces the output $f(x)$.

The first part of this text deals with *numerical* functions f, where both the domain and the range are sets of real numbers. We refer to such a function interchangeably as f or $f(x)$. The letter x is used often to denote the **independent variable** that can take on any value in the domain D. We write $y = f(x)$ and refer to y as the **dependent variable** (because its value depends on the choice of x).

When f is defined by a formula, its natural domain is the set of real numbers x for which the formula is meaningful. For example, the function $f(x) = \sqrt{9-x}$ has domain $D = \{x : x \le 9\}$ because $\sqrt{9-x}$ is defined if $9 - x \ge 0$. Here are some other examples of domains and ranges:

$f(x)$	**Domain** D	**Range** R
x^2	**R**	$\{y : y \ge 0\}$
$\cos x$	**R**	$\{y : -1 \le y \le 1\}$
$\dfrac{1}{x+1}$	$\{x : x \ne -1\}$	$\{y : y \ne 0\}$

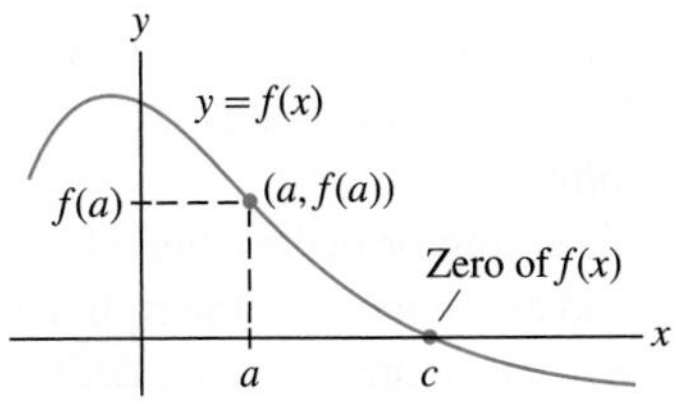

FIGURE 15

The **graph** of a function $y = f(x)$ is obtained by plotting the points $(a, f(a))$ for a in the domain D (Figure 15). If you start at $x = a$ on the x-axis, move up to the graph and then over to the y-axis, you arrive at the value $f(a)$. The absolute value $|f(a)|$ is the distance from the graph to the x-axis.

A **zero** or **root** of a function $f(x)$ is a number c such that $f(c) = 0$. The zeros are the values of x where the graph intersects the x-axis.

In Chapter 4, we will use calculus to sketch and analyze graphs. At this stage, to sketch a graph by hand, we can make a table of function values, plot the corresponding points (including any zeros), and connect them by a smooth curve.

■ **EXAMPLE 3** Find the roots and sketch the graph of $f(x) = x^3 - 2x$.

Solution First, we solve

$$x^3 - 2x = x(x^2 - 2) = 0.$$

The roots of $f(x)$ are $x = 0$ and $x = \pm\sqrt{2}$. To sketch the graph, we plot the roots and a few values listed in Table 1 and join them by a curve (Figure 16). ■

TABLE 1

x	$x^3 - 2x$
−2	−4
−1	1
0	0
1	−1
2	4

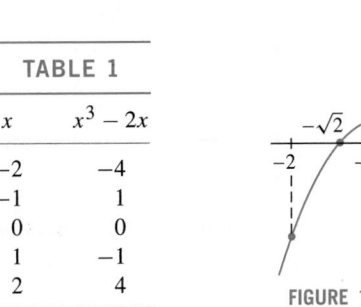

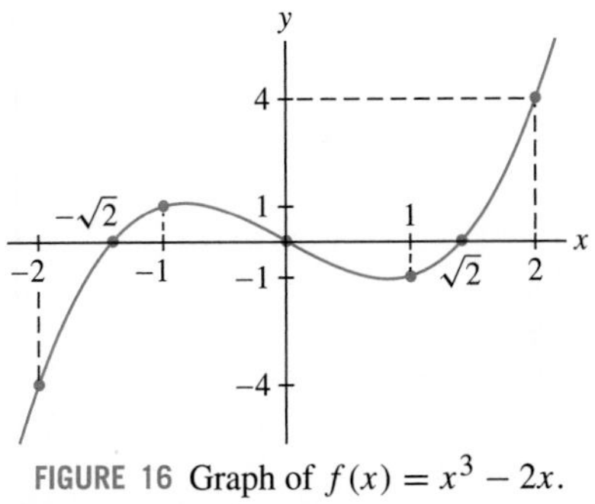

FIGURE 16 Graph of $f(x) = x^3 - 2x$.

Functions arising in applications are not always given by formulas. For example, data collected from observation or experiment define functions for which there may be no exact formula. Such functions can be displayed either graphically or by a table of values. Figure 17 and Table 2 display data collected by biologist Julian Huxley (1887–1975) in a study of the antler weight W of male red deer as a function of age t. We will see that many of the tools from calculus can be applied to functions constructed from data in this way.

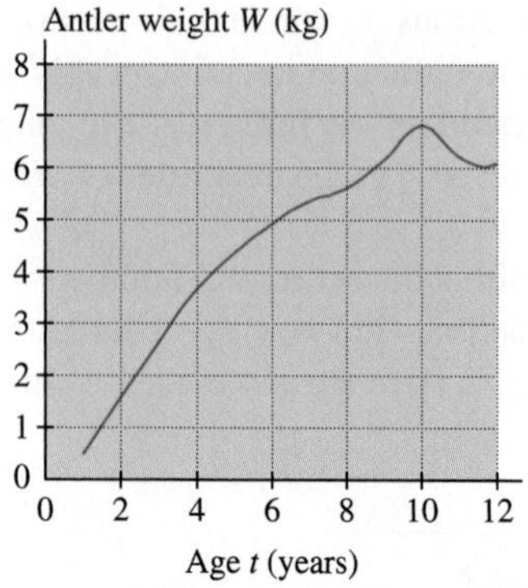

FIGURE 17 Male red deer shed their antlers every winter and regrow them in the spring. This graph shows average antler weight as a function of age.

TABLE 2

t (years)	W (kg)	t (years)	W (kg)
1	0.48	7	5.34
2	1.59	8	5.62
3	2.66	9	6.18
4	3.68	10	6.81
5	4.35	11	6.21
6	4.92	12	6.1

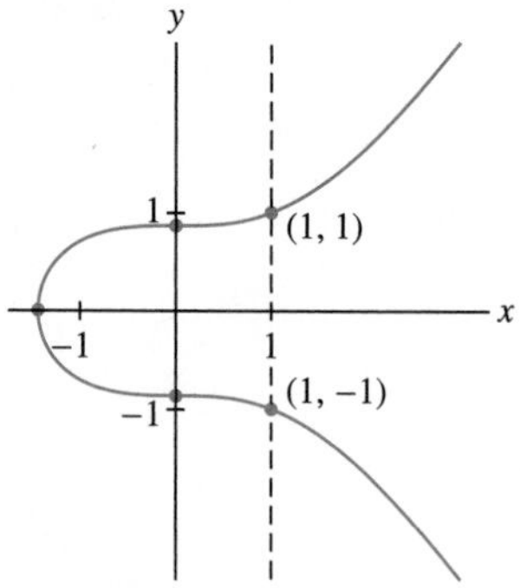

FIGURE 18 Graph of $4y^2 - x^3 = 3$. This graph fails the Vertical Line Test, so it is not the graph of a function.

We can graph not just functions but, more generally, any equation relating y and x. Figure 18 shows the graph of the equation $4y^2 - x^3 = 3$; it consists of all pairs (x, y) satisfying the equation. This curve is not the graph of a function because some x-values are associated with two y-values. For example, $x = 1$ is associated with $y = \pm 1$. A curve is the graph of a function if and only if it passes the **Vertical Line Test**; that is, every vertical line $x = a$ intersects the curve in at most one point.

We are often interested in whether a function is increasing or decreasing. Roughly speaking, a function $f(x)$ is increasing if its graph goes up as we move to the right and is decreasing if its graph goes down [Figures 19(A) and (B)]. More precisely, we define the notion of increase/decrease on an open interval:

- **Increasing** on (a, b) if $f(x_1) < f(x_2)$ for all $x_1, x_2 \in (a, b)$ such that $x_1 < x_2$
- **Decreasing** on (a, b) if $f(x_1) > f(x_2)$ for all $x_1, x_2 \in (a, b)$ such that $x_1 < x_2$

We say that $f(x)$ is **monotonic** if it is either increasing or decreasing. In Figure 19(C), the function is not monotonic because it is neither increasing nor decreasing for all x.

A function $f(x)$ is called **nondecreasing** if $f(x_1) \le f(x_2)$ for $x_1 < x_2$ (defined by $\le$ rather than a strict inequality $<$). **Nonincreasing** functions are defined similarly. Function (D) in Figure 19 is nondecreasing, but it is not increasing on the intervals where the graph is horizontal.

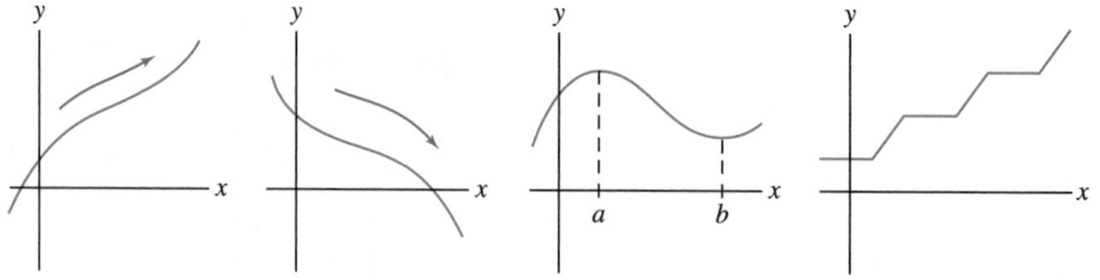

(A) Increasing (B) Decreasing (C) Decreasing on (a, b) but not decreasing everywhere (D) Nondecreasing but not increasing

FIGURE 19

Another important property is **parity**, which refers to whether a function is even or odd:

- $f(x)$ is **even** if $f(-x) = f(x)$
- $f(x)$ is **odd** if $f(-x) = -f(x)$

The graphs of functions with even or odd parity have a special symmetry:

- **Even function:** graph is symmetric about the y-axis. This means that if $P = (a, b)$ lies on the graph, then so does $Q = (-a, b)$ [Figure 20(A)].
- **Odd function:** graph is symmetric with respect to the origin. This means that if $P = (a, b)$ lies on the graph, then so does $Q = (-a, -b)$ [Figure 20(B)].

Many functions are neither even nor odd [Figure 20(C)].

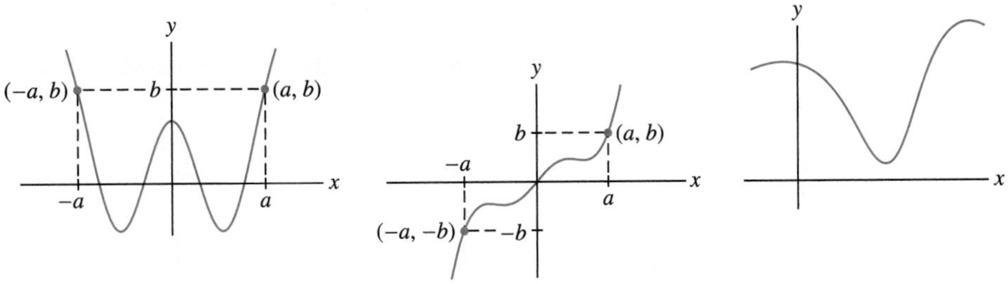

(A) Even function: $f(-x) = f(x)$ Graph is symmetric about the y-axis.

(B) Odd function: $f(-x) = -f(x)$ Graph is symmetric about the origin.

(C) Neither even nor odd

FIGURE 20

EXAMPLE 4 Determine whether the function is even, odd, or neither.

(a) $f(x) = x^4$ **(b)** $g(x) = x^{-1}$ **(c)** $h(x) = x^2 + x$

Solution

(a) $f(-x) = (-x)^4 = x^4$. Thus, $f(x) = f(-x)$ and $f(x)$ is even.
(b) $g(-x) = (-x)^{-1} = -x^{-1}$. Thus, $g(-x) = -g(x)$, and $g(x)$ is odd.
(c) $h(-x) = (-x)^2 + (-x) = x^2 - x$. We see that $h(-x)$ is not equal to $h(x)$ or to $-h(x) = -x^2 - x$. Therefore, $h(x)$ is neither even nor odd. ■

EXAMPLE 5 Using Symmetry Sketch the graph of $f(x) = \dfrac{1}{x^2+1}$.

Solution The function $f(x)$ is positive $[f(x) > 0]$ and even $[f(-x) = f(x)]$. Therefore, the graph lies above the x-axis and is symmetric with respect to the y-axis. Furthermore, $f(x)$ is decreasing for $x \ge 0$ (because a larger value of x makes the denominator larger). We use this information and a short table of values (Table 3) to sketch the graph (Figure 21). Note that the graph approaches the x-axis as we move to the right or left because $f(x)$ gets smaller as $|x|$ increases. ■

TABLE 3

x	$\dfrac{1}{x^2+1}$
0	1
± 1	$\frac{1}{2}$
± 2	$\frac{1}{5}$

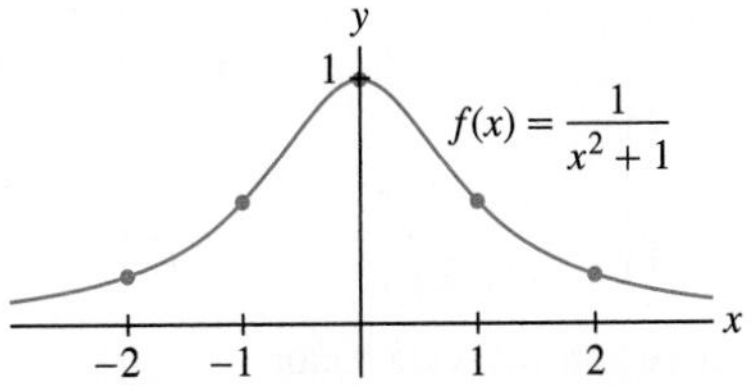

FIGURE 21

Two important ways of modifying a graph are **translation** (or **shifting**) and **scaling**. Translation consists of moving the graph horizontally or vertically:

DEFINITION Translation (Shifting)

- **Vertical translation** $y = f(x) + c$: shifts the graph by $|c|$ units *vertically*, upward if $c > 0$ and c units downward if $c < 0$.
- **Horizontal translation** $y = f(x + c)$: shifts the graph by $|c|$ units *horizontally*, to the right if $c < 0$ and c units to the left if $c > 0$.

Remember that $f(x) + c$ and $f(x + c)$ are different. The graph of $y = f(x) + c$ is a vertical translation and $y = f(x + c)$ a horizontal translation of the graph of $y = f(x)$.

Figure 22 shows the effect of translating the graph of $f(x) = 1/(x^2 + 1)$ vertically and horizontally.

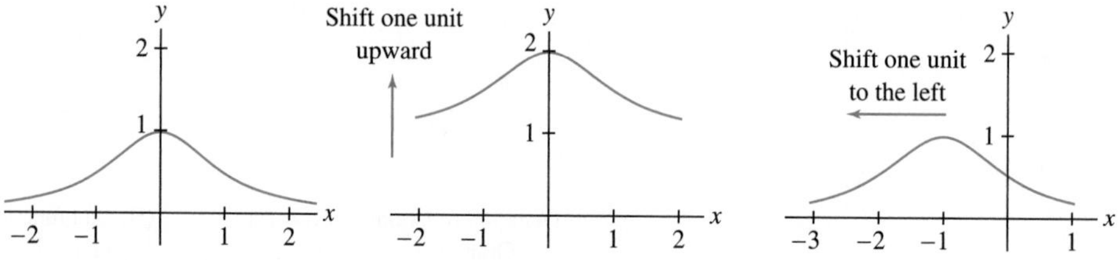

(A) $y = f(x) = \dfrac{1}{x^2 + 1}$ (B) $y = f(x) + 1 = \dfrac{1}{x^2 + 1} + 1$ (C) $y = f(x + 1) = \dfrac{1}{(x + 1)^2 + 1}$

FIGURE 22

■ **EXAMPLE 6** Figure 23(A) is the graph of $f(x) = x^2$, and Figure 23(B) is a horizontal and vertical shift of (A). What is the equation of graph (B)?

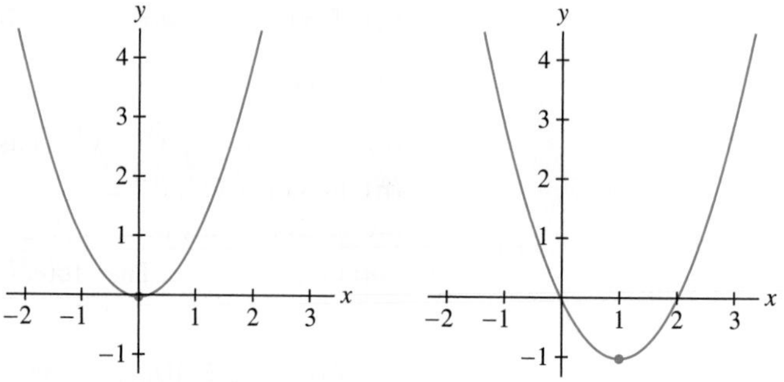

(A) $f(x) = x^2$ (B)

FIGURE 23

Solution Graph (B) is obtained by shifting graph (A) one unit to the right and one unit down. We can see this by observing that the point (0, 0) on the graph of $f(x)$ is shifted to $(1, -1)$. Therefore, (B) is the graph of $g(x) = (x - 1)^2 - 1$. ■

Scaling (also called **dilation**) consists of compressing or expanding the graph in the vertical or horizontal directions:

DEFINITION Scaling

- **Vertical scaling** $y = kf(x)$: If $k > 1$, the graph is expanded vertically by the factor k. If $0 < k < 1$, the graph is compressed vertically. When the scale factor k is negative ($k < 0$), the graph is also reflected across the x-axis (Figure 24).
- **Horizontal scaling** $y = f(kx)$: If $k > 1$, the graph is compressed in the horizontal direction. If $0 < k < 1$, the graph is expanded. If $k < 0$, then the graph is also reflected across the y-axis.

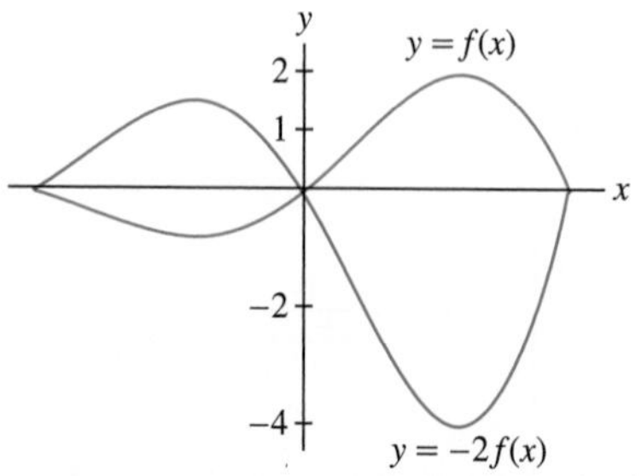

FIGURE 24 Negative vertical scale factor $k = -2$.

We refer to the vertical size of a graph as its **amplitude**. Thus, vertical scaling changes the amplitude by the factor $|k|$.

Remember that $kf(x)$ and $f(kx)$ are different. The graph of $y = kf(x)$ is a vertical scaling, and $y = f(kx)$ a horizontal scaling, of the graph of $y = f(x)$.

EXAMPLE 7 Sketch the graphs of $f(x) = \sin(\pi x)$ and its dilates $f(3x)$ and $3f(x)$.

Solution The graph of $f(x) = \sin(\pi x)$ is a sine curve with period 2. It completes one cycle over every interval of length 2—see Figure 25(A).

- The graph of $f(3x) = \sin(3\pi x)$ is a compressed version of $y = f(x)$, completing three cycles instead of one over intervals of length 2 [Figure 25(B)].
- The graph of $y = 3f(x) = 3\sin(\pi x)$ differs from $y = f(x)$ only in amplitude: It is expanded in the vertical direction by a factor of 3 [Figure 25(C)]. ■

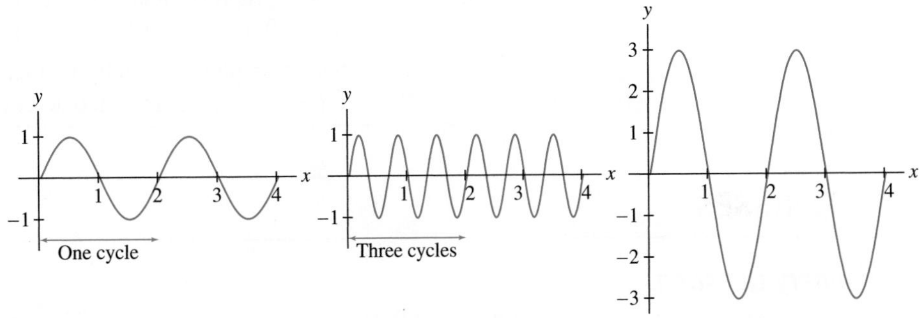

(A) $y = f(x) = \sin(\pi x)$ (B) Horizontal compression: $y = f(3x) = \sin(3\pi x)$ (C) Vertical expansion: $y = 3f(x) = 3\sin(\pi x)$

FIGURE 25 Horizontal and vertical scaling of $f(x) = \sin(\pi x)$.

1.1 SUMMARY

- Absolute value: $|a| = \begin{cases} a & \text{if } a \ge 0 \\ -a & \text{if } a < 0 \end{cases}$
- Triangle inequality: $|a + b| \le |a| + |b|$
- Four intervals with endpoints a and b:

$$(a, b), \quad [a, b], \quad [a, b), \quad (a, b]$$

- Writing open and closed intervals using inequalities:

$$(a, b) = \{x : |x - c| < r\}, \qquad [a, b] = \{x : |x - c| \le r\}$$

where $c = \frac{1}{2}(a + b)$ is the midpoint and $r = \frac{1}{2}(b - a)$ is the radius.

- Distance d between (x_1, y_1) and (x_2, y_2):

$$d = \sqrt{(x_2 - x_1)^2 + (y_2 - y_1)^2}$$

- Equation of circle of radius r with center (a, b):

$$(x - a)^2 + (y - b)^2 = r^2$$

- A *zero* or *root* of a function $f(x)$ is a number c such that $f(c) = 0$.

- Vertical Line Test: A curve in the plane is the graph of a function if and only if each vertical line $x = a$ intersects the curve in at most one point.
-

Increasing:	$f(x_1) < f(x_2)$ if $x_1 < x_2$
Nondecreasing:	$f(x_1) \le f(x_2)$ if $x_1 < x_2$
Decreasing:	$f(x_1) > f(x_2)$ if $x_1 < x_2$
Nonincreasing:	$f(x_1) \ge f(x_2)$ if $x_1 < x_2$

- Even function: $f(-x) = f(x)$ (graph is symmetric about the y-axis).
- Odd function: $f(-x) = -f(x)$ (graph is symmetric about the origin).
- Four ways to transform the graph of $f(x)$:

$f(x) + c$	Shifts graph vertically $\|c\|$ units (upward if $c > 0$, downward if $c < 0$)
$f(x + c)$	Shifts graph horizontally $\|c\|$ units (to the right if $c < 0$, to the left if $c > 0$)
$kf(x)$	Scales graph vertically by factor k; if $k < 0$, graph is reflected across x-axis
$f(kx)$	Scales graph horizontally by factor k (compresses if $k > 1$); if $k < 0$, graph is reflected across y-axis

1.1 EXERCISES

Preliminary Questions

1. Give an example of numbers a and b such that $a < b$ and $|a| > |b|$.

2. Which numbers satisfy $|a| = a$? Which satisfy $|a| = -a$? What about $|-a| = a$?

3. Give an example of numbers a and b such that $|a + b| < |a| + |b|$.

4. What are the coordinates of the point lying at the intersection of the lines $x = 9$ and $y = -4$?

5. In which quadrant do the following points lie?

(a) $(1, 4)$ **(b)** $(-3, 2)$ **(c)** $(4, -3)$ **(d)** $(-4, -1)$

6. What is the radius of the circle with equation $(x - 9)^2 + (y - 9)^2 = 9$?

7. The equation $f(x) = 5$ has a solution if (choose one):

(a) 5 belongs to the domain of f.

(b) 5 belongs to the range of f.

8. What kind of symmetry does the graph have if $f(-x) = -f(x)$?

Exercises

1. Use a calculator to find a rational number r such that $|r - \pi^2| < 10^{-4}$.

2. Which of (a)–(f) are true for $a = -3$ and $b = 2$?

(a) $a < b$ **(b)** $|a| < |b|$ **(c)** $ab > 0$

(d) $3a < 3b$ **(e)** $-4a < -4b$ **(f)** $\frac{1}{a} < \frac{1}{b}$

In Exercises 3–8, express the interval in terms of an inequality involving absolute value.

3. $[-2, 2]$ **4.** $(-4, 4)$ **5.** $(0, 4)$

6. $[-4, 0]$ **7.** $[1, 5]$ **8.** $(-2, 8)$

In Exercises 9–12, write the inequality in the form $a < x < b$.

9. $|x| < 8$ **10.** $|x - 12| < 8$

11. $|2x + 1| < 5$ **12.** $|3x - 4| < 2$

In Exercises 13–18, express the set of numbers x satisfying the given condition as an interval.

13. $|x| < 4$ **14.** $|x| \le 9$

15. $|x - 4| < 2$ **16.** $|x + 7| < 2$

17. $|4x - 1| \le 8$ **18.** $|3x + 5| < 1$

In Exercises 19–22, describe the set as a union of finite or infinite intervals.

19. $\{x : |x - 4| > 2\}$ **20.** $\{x : |2x + 4| > 3\}$

21. $\{x : |x^2 - 1| > 2\}$ **22.** $\{x : |x^2 + 2x| > 2\}$

23. Match (a)–(f) with (i)–(vi).

(a) $a > 3$ **(b)** $|a - 5| < \frac{1}{3}$

(c) $\left|a - \frac{1}{3}\right| < 5$ **(d)** $|a| > 5$

(e) $|a - 4| < 3$ **(f)** $1 \le a \le 5$

(i) a lies to the right of 3.

(ii) a lies between 1 and 7.

(iii) The distance from a to 5 is less than $\frac{1}{3}$.

(iv) The distance from a to 3 is at most 2.

(v) a is less than 5 units from $\frac{1}{3}$.

(vi) a lies either to the left of -5 or to the right of 5.

24. Describe $\left\{x : \dfrac{x}{x+1} < 0\right\}$ as an interval.

25. Describe $\{x : x^2 + 2x < 3\}$ as an interval. *Hint:* Plot $y = x^2 + 2x - 3$.

26. Describe the set of real numbers satisfying $|x - 3| = |x - 2| + 1$ as a half-infinite interval.

27. Show that if $a > b$, then $b^{-1} > a^{-1}$, provided that a and b have the same sign. What happens if $a > 0$ and $b < 0$?

28. Which x satisfy both $|x - 3| < 2$ and $|x - 5| < 1$?

29. Show that if $|a - 5| < \frac{1}{2}$ and $|b - 8| < \frac{1}{2}$, then $|(a + b) - 13| < 1$. *Hint:* Use the triangle inequality.

30. Suppose that $|x - 4| \le 1$.

(a) What is the maximum possible value of $|x + 4|$?

(b) Show that $|x^2 - 16| \le 9$.

31. Suppose that $|a - 6| \le 2$ and $|b| \le 3$.

(a) What is the largest possible value of $|a + b|$?

(b) What is the smallest possible value of $|a + b|$?

32. Prove that $|x| - |y| \le |x - y|$. *Hint:* Apply the triangle inequality to y and $x - y$.

33. Express $r_1 = 0.\overline{27}$ as a fraction. *Hint:* $100r_1 - r_1$ is an integer. Then express $r_2 = 0.2666\ldots$ as a fraction.

34. Represent $1/7$ and $4/27$ as repeating decimals.

35. The text states: *If the decimal expansions of numbers a and b agree to k places, then* $|a - b| \le 10^{-k}$. Show that the converse is false: For all k there are numbers a and b whose decimal expansions *do not agree at all* but $|a - b| \le 10^{-k}$.

36. Plot each pair of points and compute the distance between them:

(a) $(1, 4)$ and $(3, 2)$ **(b)** $(2, 1)$ and $(2, 4)$

(c) $(0, 0)$ and $(-2, 3)$ **(d)** $(-3, -3)$ and $(-2, 3)$

37. Find the equation of the circle with center $(2, 4)$:

(a) with radius $r = 3$.

(b) that passes through $(1, -1)$.

38. Find all points with integer coordinates located at a distance 5 from the origin. Then find all points with integer coordinates located at a distance 5 from $(2, 3)$.

39. Determine the domain and range of the function

$$f : \{r, s, t, u\} \to \{A, B, C, D, E\}$$

defined by $f(r) = A$, $f(s) = B$, $f(t) = B$, $f(u) = E$.

40. Give an example of a function whose domain D has three elements and whose range R has two elements. Does a function exist whose domain D has two elements and whose range R has three elements?

In Exercises 41–48, find the domain and range of the function.

41. $f(x) = -x$ **42.** $g(t) = t^4$

43. $f(x) = x^3$ **44.** $g(t) = \sqrt{2 - t}$

45. $f(x) = |x|$ **46.** $h(s) = \dfrac{1}{s}$

47. $f(x) = \dfrac{1}{x^2}$ **48.** $g(t) = \cos \dfrac{1}{t}$

In Exercises 49–52, determine where $f(x)$ is increasing.

49. $f(x) = |x + 1|$ **50.** $f(x) = x^3$

51. $f(x) = x^4$ **52.** $f(x) = \dfrac{1}{x^4 + x^2 + 1}$

In Exercises 53–58, find the zeros of $f(x)$ and sketch its graph by plotting points. Use symmetry and increase/decrease information where appropriate.

53. $f(x) = x^2 - 4$ **54.** $f(x) = 2x^2 - 4$

55. $f(x) = x^3 - 4x$ **56.** $f(x) = x^3$

57. $f(x) = 2 - x^3$ **58.** $f(x) = \dfrac{1}{(x - 1)^2 + 1}$

59. Which of the curves in Figure 26 is the graph of a function?

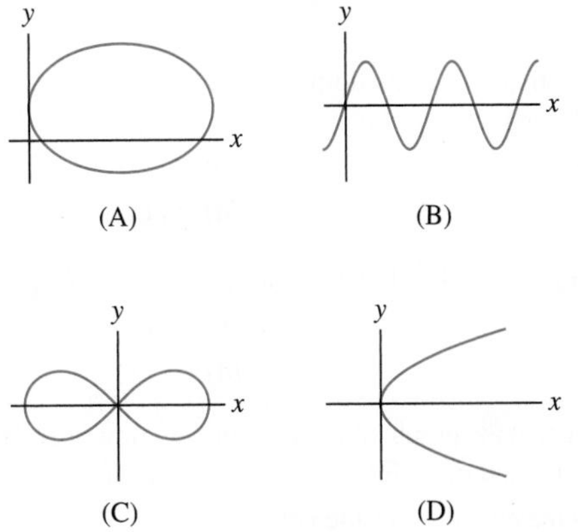

FIGURE 26

60. Determine whether the function is even, odd, or neither.

(a) $f(x) = x^5$ **(b)** $g(t) = t^3 - t^2$

(c) $F(t) = \dfrac{1}{t^4 + t^2}$

61. Determine whether the function is even, odd, or neither.

(a) $f(t) = \dfrac{1}{t^4 + t + 1} - \dfrac{1}{t^4 - t + 1}$ **(b)** $g(t) = 2^t - 2^{-t}$

(c) $G(\theta) = \sin\theta + \cos\theta$ **(d)** $H(\theta) = \sin(\theta^2)$

62. Write $f(x) = 2x^4 - 5x^3 + 12x^2 - 3x + 4$ as the sum of an even and an odd function.

63. Determine the interval on which $f(x) = \dfrac{1}{x-4}$ is increasing or decreasing.

64. State whether the function is increasing, decreasing, or neither.

(a) Surface area of a sphere as a function of its radius

(b) Temperature at a point on the equator as a function of time

(c) Price of an airline ticket as a function of the price of oil

(d) Pressure of the gas in a piston as a function of volume

In Exercises 65–70, let $f(x)$ be the function shown in Figure 27.

65. Find the domain and range of $f(x)$?

66. Sketch the graphs of $f(x+2)$ and $f(x)+2$.

67. Sketch the graphs of $f(2x)$, $f\left(\frac{1}{2}x\right)$, and $2f(x)$.

68. Sketch the graphs of $f(-x)$ and $-f(-x)$.

69. Extend the graph of $f(x)$ to $[-4, 4]$ so that it is an even function.

70. Extend the graph of $f(x)$ to $[-4, 4]$ so that it is an odd function.

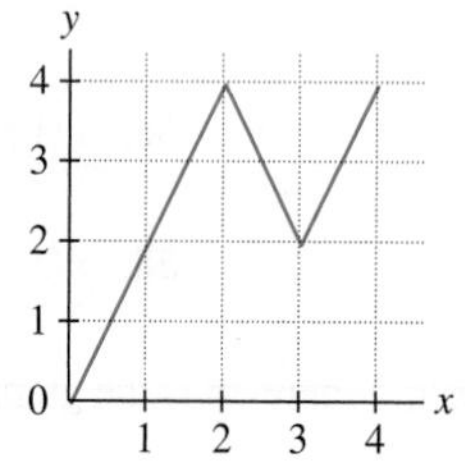

FIGURE 27

71. Suppose that $f(x)$ has domain $[4, 8]$ and range $[2, 6]$. Find the domain and range of:

(a) $f(x) + 3$ **(b)** $f(x+3)$

(c) $f(3x)$ **(d)** $3f(x)$

72. Let $f(x) = x^2$. Sketch the graph over $[-2, 2]$ of:

(a) $f(x+1)$ **(b)** $f(x) + 1$

(c) $f(5x)$ **(d)** $5f(x)$

73. Suppose that the graph of $f(x) = \sin x$ is compressed horizontally by a factor of 2 and then shifted 5 units to the right.

(a) What is the equation for the new graph?

(b) What is the equation if you first shift by 5 and then compress by 2?

(c) GU Verify your answers by plotting your equations.

74. Figure 28 shows the graph of $f(x) = |x| + 1$. Match the functions (a)–(e) with their graphs (i)–(v).

(a) $f(x-1)$ **(b)** $-f(x)$ **(c)** $-f(x)+2$

(d) $f(x-1)-2$ **(e)** $f(x+1)$

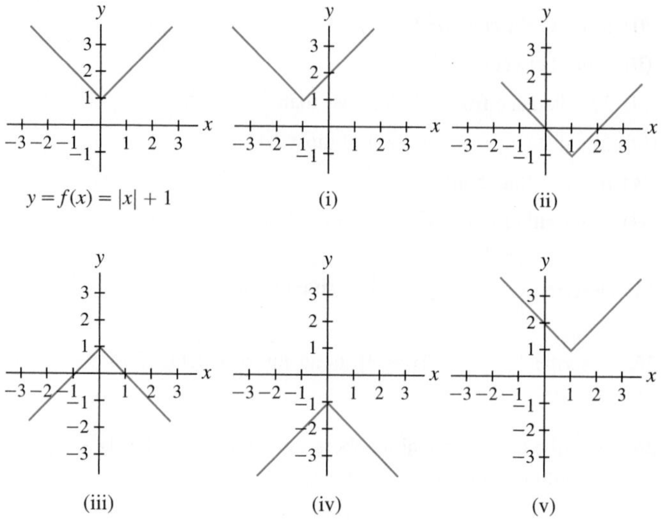

FIGURE 28

75. Sketch the graph of $f(2x)$ and $f\left(\frac{1}{2}x\right)$, where $f(x) = |x| + 1$ (Figure 28).

76. Find the function $f(x)$ whose graph is obtained by shifting the parabola $y = x^2$ three units to the right and four units down, as in Figure 29.

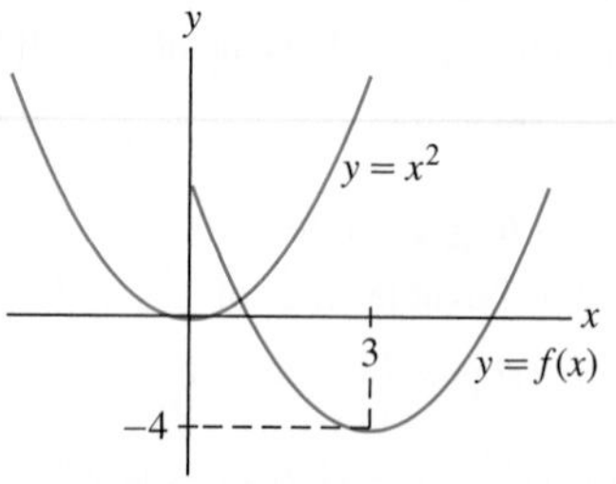

FIGURE 29

77. Define $f(x)$ to be the larger of x and $2 - x$. Sketch the graph of $f(x)$. What are its domain and range? Express $f(x)$ in terms of the absolute value function.

78. For each curve in Figure 30, state whether it is symmetric with respect to the y-axis, the origin, both, or neither.

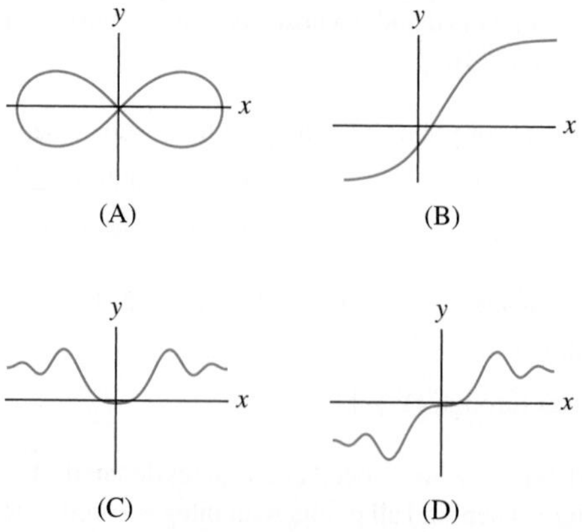

FIGURE 30

79. Show that the sum of two even functions is even and the sum of two odd functions is odd.

80. Suppose that $f(x)$ and $g(x)$ are both odd. Which of the following functions are even? Which are odd?

(a) $f(x)g(x)$ **(b)** $f(x)^3$

(c) $f(x) - g(x)$ **(d)** $\dfrac{f(x)}{g(x)}$

81. Prove that the only function whose graph is symmetric with respect to both the y-axis and the origin is the function $f(x) = 0$.

Further Insights and Challenges

82. Prove the triangle inequality by adding the two inequalities

$$-|a| \le a \le |a|, \qquad -|b| \le b \le |b|$$

83. Show that a fraction $r = a/b$ in lowest terms has a *finite* decimal expansion if and only if

$$b = 2^n 5^m \quad \text{for some } n, m \ge 0.$$

Hint: Observe that r has a finite decimal expansion when $10^N r$ is an integer for some $N \ge 0$ (and hence b divides 10^N).

84. Let $p = p_1 \dots p_s$ be an integer with digits $p_1, \dots, p_s$. Show that

$$\frac{p}{10^s - 1} = 0.\overline{p_1 \dots p_s}$$

Use this to find the decimal expansion of $r = \frac{2}{11}$. Note that

$$r = \frac{2}{11} = \frac{18}{10^2 - 1}$$

85. A function $f(x)$ is symmetric with respect to the vertical line $x = a$ if $f(a - x) = f(a + x)$.

(a) Draw the graph of a function that is symmetric with respect to $x = 2$.

(b) Show that if $f(x)$ is symmetric with respect to $x = a$, then $g(x) = f(x + a)$ is even.

86. Formulate a condition for $f(x)$ to be symmetric with respect to the point $(a, 0)$ on the x-axis.

1.2 Linear and Quadratic Functions

Linear functions are the simplest of all functions, and their graphs (lines) are the simplest of all curves. However, linear functions and lines play an enormously important role in calculus. For this reason, you should be thoroughly familiar with the basic properties of linear functions and the different ways of writing an equation of a line.

Let's recall that a **linear function** is a function of the form

$$f(x) = mx + b \quad (m \text{ and } b \text{ constants})$$

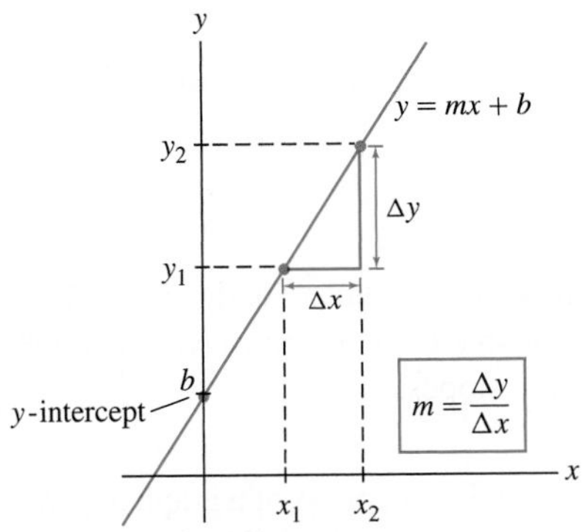

FIGURE 1 The slope m is the ratio "rise over run."

The graph of $f(x)$ is a line of slope m, and since $f(0) = b$, the graph intersects the y-axis at the point $(0, b)$ (Figure 1). The number b is called the y-intercept, and the equation $y = mx + b$ for the line is said to be in **slope-intercept form**.

We use the symbols Δx and Δy to denote the *change* (or *increment*) in x and $y = f(x)$ over an interval $[x_1, x_2]$ (Figure 1):

$$\Delta x = x_2 - x_1, \qquad \Delta y = y_2 - y_1 = f(x_2) - f(x_1)$$

The slope m of a line is equal to the ratio

$$m = \frac{\Delta y}{\Delta x} = \frac{\text{vertical change}}{\text{horizontal change}} = \frac{\text{rise}}{\text{run}}$$

This follows from the formula $y = mx + b$:

$$\frac{\Delta y}{\Delta x} = \frac{y_2 - y_1}{x_2 - x_1} = \frac{(mx_2 + b) - (mx_1 + b)}{x_2 - x_1} = \frac{m(x_2 - x_1)}{x_2 - x_1} = m$$

The slope m measures the *rate of change* of y with respect to x. In fact, by writing

$$\Delta y = m\Delta x$$

we see that a one-unit increase in x (i.e., $\Delta x = 1$) produces an m-unit change Δy in y. For example, if $m = 5$, then y increases by five units per unit increase in x. The rate-of-change interpretation of the slope is fundamental in calculus. We discuss it in greater detail in Section 2.1.

Graphically, the slope m measures the steepness of the line $y = mx + b$. Figure 2(A) shows lines through a point of varying slope m. Note the following properties:

- **Steepness:** The larger the absolute value $|m|$, the steeper the line.
- **Negative slope:** If $m < 0$, the line slants downward from left to right.
- $f(x) = mx + b$ is increasing if $m > 0$ and decreasing if $m < 0$.
- The **horizontal line** $y = b$ has slope $m = 0$ [Figure 2(B)].
- A **vertical line** has equation $x = c$, where c is a constant. The slope of a vertical line is undefined. It is not possible to write the equation of a vertical line in slope-intercept form $y = mx + b$.

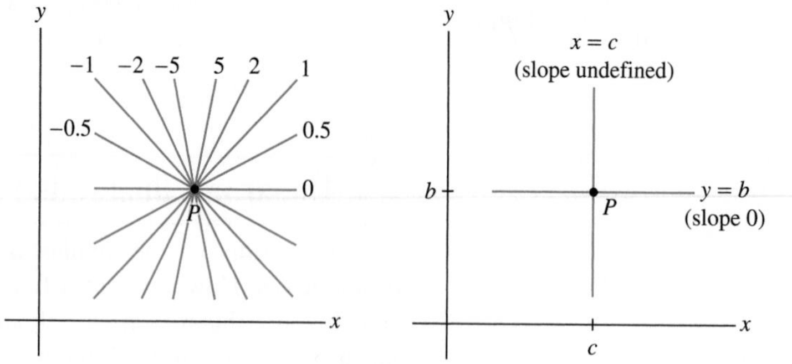

(A) Lines of varying slopes through P (B) Horizontal and vertical lines through P

FIGURE 2

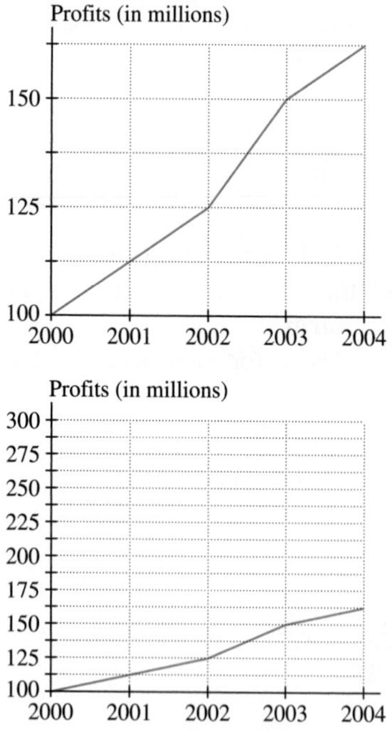

FIGURE 3 Growth of company profits.

CAUTION: Graphs are often plotted using different scales for the x- and y-axes. This is necessary to keep the sizes of graphs within reasonable bounds. However, when the scales are different, lines do not appear with their true slopes.

Scale is especially important in applications because the steepness of a graph depends on the choice of units for the x- and y-axes. We can create very different *subjective* impressions by changing the scale. Figure 3 shows the growth of company profits over a four-year period. The two plots convey the same information, but the upper plot makes the growth look more dramatic.

Next, we recall the relation between the slopes of parallel and perpendicular lines (Figure 4):

- Lines of slopes m_1 and m_2 are **parallel** if and only if $m_1 = m_2$.
- Lines of slopes m_1 and m_2 are **perpendicular** if and only if

$$m_1 = -\frac{1}{m_2} \qquad (\text{or } m_1 m_2 = -1).$$

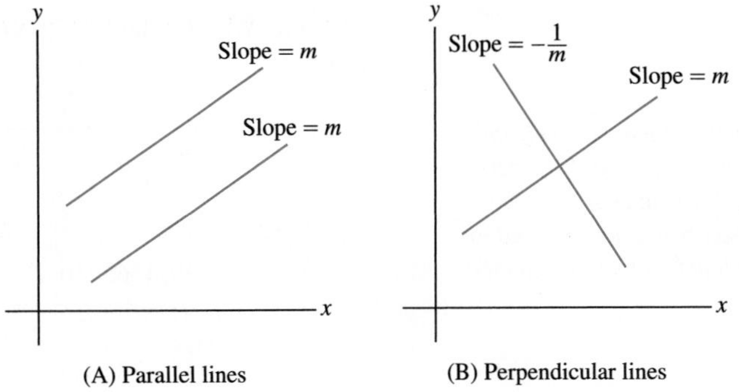

FIGURE 4 Parallel and perpendicular lines.

CONCEPTUAL INSIGHT The increments over an interval $[x_1, x_2]$:

$$\Delta x = x_2 - x_1, \qquad \Delta y = f(x_2) - f(x_1)$$

are defined for any function $f(x)$ (linear or not), but the ratio $\Delta y/\Delta x$ may depend on the interval (Figure 5). The characteristic property of a linear function $f(x) = mx + b$ is that $\Delta y/\Delta x$ has the same value m for every interval. In other words, y has a constant rate of change with respect to x. We can use this property to test if two quantities are related by a linear equation.

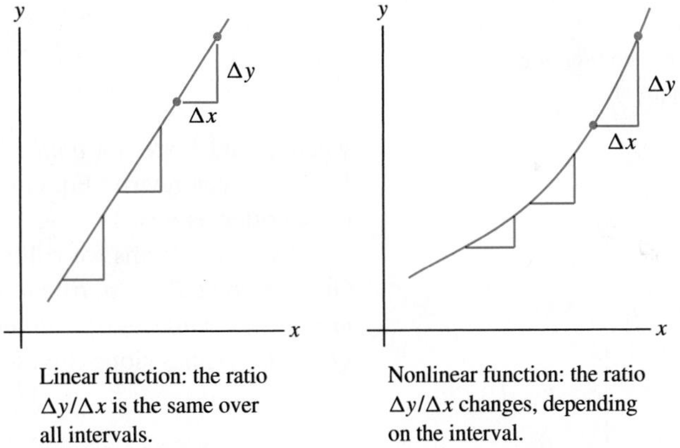

FIGURE 5

■ **EXAMPLE 1 Testing for a Linear Relationship** Do the data in Table 1 suggest a linear relation between the pressure P and temperature T of a gas?

TABLE 1

Temperature (°C)	**Pressure** (kPa)
40	1365.80
45	1385.40
55	1424.60
70	1483.40
80	1522.60

Solution We calculate $\Delta P/\Delta T$ at successive data points and check whether this ratio is constant:

Real experimental data are unlikely to reveal perfect linearity, even if the data points do essentially lie on a line. The method of "linear regression" is used to find the linear function that best fits the data.

(T_1, P_1)	(T_2, P_2)	$\dfrac{\Delta P}{\Delta T}$
(40, 1365.80)	(45, 1385.40)	$\dfrac{1385.40 - 1365.80}{45 - 40} = 3.92$
(45, 1385.40)	(55, 1424.60)	$\dfrac{1424.60 - 1385.40}{55 - 45} = 3.92$
(55, 1424.60)	(70, 1483.40)	$\dfrac{1483.40 - 1424.60}{70 - 55} = 3.92$
(70, 1483.40)	(80, 1522.60)	$\dfrac{1522.60 - 1483.40}{80 - 70} = 3.92$

Because $\Delta P/\Delta T$ has the constant value 3.92, the data points lie on a line with slope $m = 3.92$ (this is confirmed in the plot in Figure 6). ■

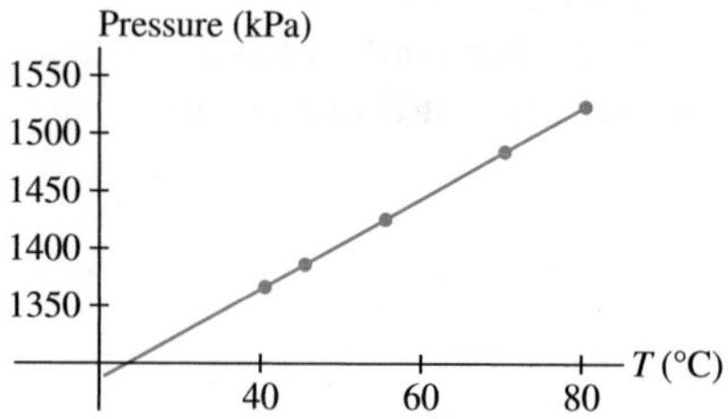

FIGURE 6 Line through pressure-temperature data points.

As mentioned above, it is important to be familiar with the standard ways of writing the equation of a line. The general **linear equation** is

$$ax + by = c \qquad (1)$$

where a and b are not *both* zero. For $b = 0$, we obtain the vertical line $ax = c$. When $b \neq 0$, we can rewrite Eq. (1) in slope-intercept form. For example, $-6x + 2y = 3$ can be rewritten as $y = 3x + \frac{3}{2}$.

Two other forms we will use frequently are the **point-slope** and **point-point** forms. Given a point $P = (a, b)$ and a slope m, the equation of the line through P with slope m is $y - b = m(x - a)$. Similarly, the line through two distinct points $P = (a_1, b_1)$ and $Q = (a_2, b_2)$ has slope (Figure 7)

$$m = \frac{b_2 - b_1}{a_2 - a_1}$$

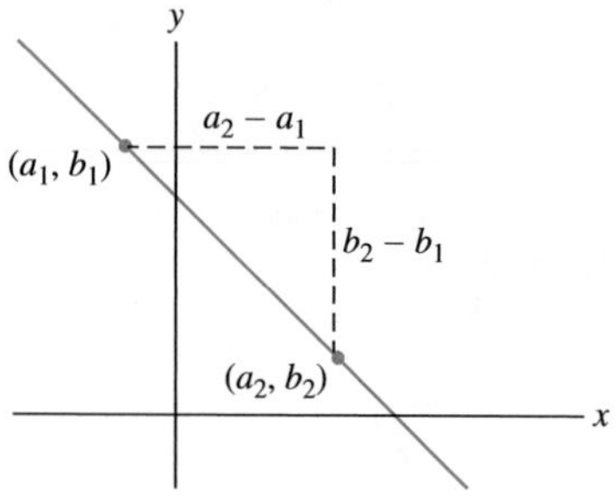

FIGURE 7 Slope of the line between $P = (a_1, b_1)$ and $Q = (a_2, b_2)$ is $m = \dfrac{b_2 - b_1}{a_2 - a_1}$.

Therefore, we can write its equation as $y - b_1 = m(x - a_1)$.

Equations for Lines

1. **Point-slope form** of the line through $P = (a, b)$ with slope m:

$$y - b = m(x - a)$$

2. **Point-point form** of the line through $P = (a_1, b_1)$ and $Q = (a_2, b_2)$:

$$y - b_1 = m(x - a_1) \qquad \text{where } m = \frac{b_2 - b_1}{a_2 - a_1}$$

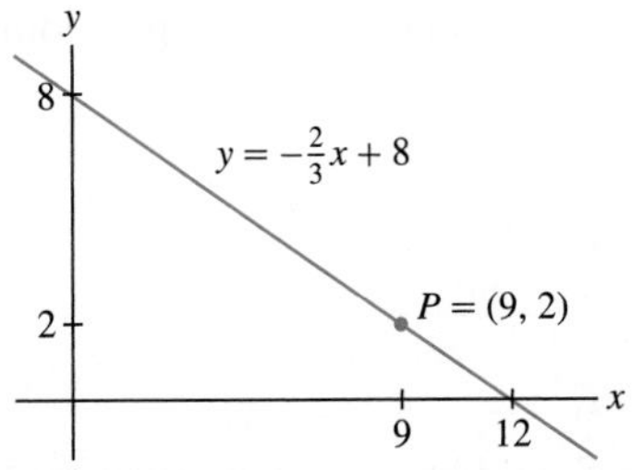

FIGURE 8 Line through $P = (9, 2)$ with slope $m = -\frac{2}{3}$.

EXAMPLE 2 Line of Given Slope Through a Given Point Find the equation of the line through (9, 2) with slope $-\frac{2}{3}$.

Solution In point-slope form:

$$y - 2 = -\frac{2}{3}(x - 9)$$

In slope-intercept form: $y = -\frac{2}{3}(x - 9) + 2$ or $y = -\frac{2}{3}x + 8$. See Figure 8. ■

EXAMPLE 3 Line Through Two Points Find the equation of the line through (2, 1) and (9, 5).

Solution The line has slope

$$m = \frac{5 - 1}{9 - 2} = \frac{4}{7}$$

Because (2, 1) lies on the line, its equation in point-slope form is $y - 1 = \frac{4}{7}(x - 2)$. ■

A **quadratic function** is a function defined by a quadratic polynomial

$$f(x) = ax^2 + bx + c \quad (a, b, c, \text{ constants with } a \neq 0)$$

The graph of $f(x)$ is a **parabola** (Figure 9). The parabola opens upward if the leading coefficient a is positive and downward if a is negative. The **discriminant** of $f(x)$ is the quantity

$$D = b^2 - 4ac$$

The roots of $f(x)$ are given by the **quadratic formula** (see Exercise 56):

$$\text{Roots of } f(x) = \frac{-b \pm \sqrt{b^2 - 4ac}}{2a} = \frac{-b \pm \sqrt{D}}{2a}$$

The sign of D determines whether or not $f(x)$ has real roots (Figure 9). If $D > 0$, then $f(x)$ has two real roots, and if $D = 0$, it has one real root (a "double root"). If $D < 0$, then $\sqrt{D}$ is imaginary and $f(x)$ has no real roots.

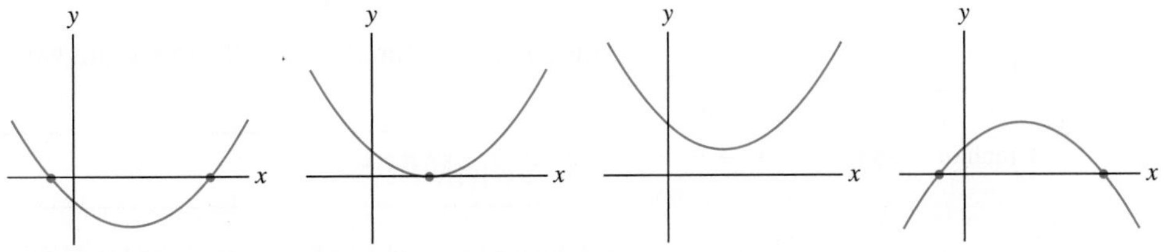

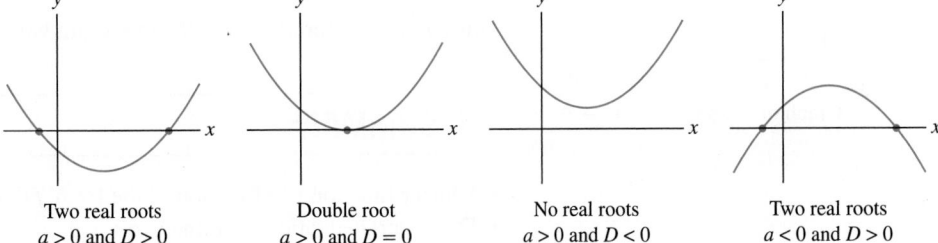

FIGURE 9 Graphs of quadratic functions $f(x) = ax^2 + bx + c$.

When $f(x)$ has two real roots r_1 and r_2, then $f(x)$ factors as

$$f(x) = a(x - r_1)(x - r_2)$$

For example, $f(x) = 2x^2 - 3x + 1$ has discriminant $D = b^2 - 4ac = 9 - 8 = 1 > 0$, and by the quadratic formula, its roots are $(3 \pm 1)/4$ or 1 and $\frac{1}{2}$. Therefore,

$$f(x) = 2x^2 - 3x + 1 = 2(x - 1)\left(x - \frac{1}{2}\right)$$

The technique of **completing the square** consists of writing a quadratic polynomial as a multiple of a square plus a constant:

$$ax^2 + bx + c = a\underbrace{\left(x + \frac{b}{2a}\right)^2}_{\text{Square term}} + \underbrace{\frac{4ac - b^2}{4a}}_{\text{Constant}} \qquad \boxed{2}$$

It is not necessary to memorize this formula, but you should know how to carry out the process of completing the square.

Cuneiform texts written on clay tablets show that the method of completing the square was known to ancient Babylonian mathematicians who lived some 4000 years ago.

■ **EXAMPLE 4 Completing the Square** Complete the square for the quadratic polynomial $4x^2 - 12x + 3$.

Solution First factor out the leading coefficient:

$$4x^2 - 12x + 3 = 4\left(x^2 - 3x + \frac{3}{4}\right)$$

Then complete the square for the term $x^2 - 3x$:

$$x^2 + bx = \left(x + \frac{b}{2}\right)^2 - \frac{b^2}{4}, \qquad x^2 - 3x = \left(x - \frac{3}{2}\right)^2 - \frac{9}{4}$$

Therefore,

$$4x^2 - 12x + 3 = 4\left(\left(x - \frac{3}{2}\right)^2 - \frac{9}{4} + \frac{3}{4}\right) = 4\left(x - \frac{3}{2}\right)^2 - 6 \qquad ■$$

The method of completing the square can be used to find the minimum or maximum value of a quadratic function.

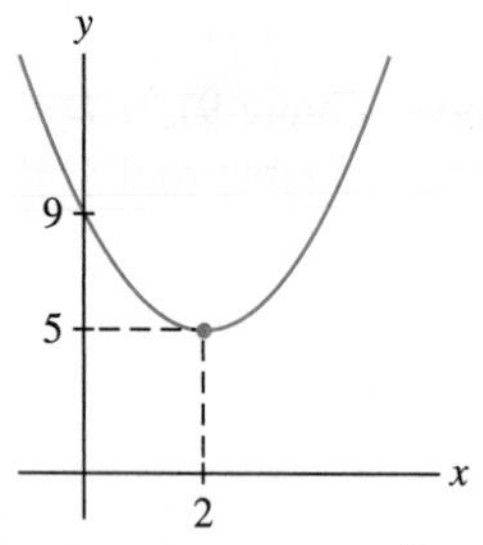

FIGURE 10 Graph of $f(x) = x^2 - 4x + 9$.

■ **EXAMPLE 5 Finding the Minimum of a Quadratic Function** Complete the square and find the minimum value of $f(x) = x^2 - 4x + 9$.

Solution We have

$$f(x) = x^2 - 4x + 9 = (x-2)^2 - 4 + 9 = \overbrace{(x-2)^2}^{\text{This term is } \geq 0} + 5$$

Thus, $f(x) \geq 5$ for all x, and the minimum value of $f(x)$ is $f(2) = 5$ (Figure 10). ■

1.2 SUMMARY

- A linear function is a function of the form $f(x) = mx + b$.
- The general equation of a line is $ax + by = c$. The line $y = c$ is horizontal and $x = c$ is vertical.
- Three convenient ways of writing the equation of a nonvertical line:
  - Slope-intercept form: $y = mx + b$ (slope m and y-intercept b)
  - Point-slope form: $y - b = m(x - a)$ [slope m, passes through (a, b)]
  - Point-point form: The line through two points $P = (a_1, b_1)$ and $Q = (a_2, b_2)$ has slope $m = \dfrac{b_2 - b_1}{a_2 - a_1}$ and equation $y - b_1 = m(x - a_1)$.
- Two lines of slopes m_1 and m_2 are parallel if and only if $m_1 = m_2$, and they are perpendicular if and only if $m_1 = -1/m_2$.

• Quadratic function: $f(x) = ax^2 + bx + c$. The roots are $x = (-b \pm \sqrt{D})/(2a)$, where $D = b^2 - 4ac$ is the discriminant. The roots are real and distinct if $D > 0$, there is a double root if $D = 0$, and there are no real roots if $D < 0$.
• Completing the square consists of writing a quadratic function as a multiple of a square plus a constant.

1.2 EXERCISES

Preliminary Questions

1. What is the slope of the line $y = -4x - 9$?

2. Are the lines $y = 2x + 1$ and $y = -2x - 4$ perpendicular?

3. When is the line $ax + by = c$ parallel to the y-axis? To the x-axis?

4. Suppose $y = 3x + 2$. What is Δy if x increases by 3?

5. What is the minimum of $f(x) = (x + 3)^2 - 4$?

6. What is the result of completing the square for $f(x) = x^2 + 1$?

Exercises

In Exercises 1–4, find the slope, the y-intercept, and the x-intercept of the line with the given equation.

1. $y = 3x + 12$

2. $y = 4 - x$

3. $4x + 9y = 3$

4. $y - 3 = \frac{1}{2}(x - 6)$

In Exercises 5–8, find the slope of the line.

5. $y = 3x + 2$

6. $y = 3(x - 9) + 2$

7. $3x + 4y = 12$

8. $3x + 4y = -8$

In Exercises 9–20, find the equation of the line with the given description.

9. Slope 3, y-intercept 8

10. Slope -2, y-intercept 3

11. Slope 3, passes through $(7, 9)$

12. Slope -5, passes through $(0, 0)$

13. Horizontal, passes through $(0, -2)$

14. Passes through $(-1, 4)$ and $(2, 7)$

15. Parallel to $y = 3x - 4$, passes through $(1, 1)$

16. Passes through $(1, 4)$ and $(12, -3)$

17. Perpendicular to $3x + 5y = 9$, passes through $(2, 3)$

18. Vertical, passes through $(-4, 9)$

19. Horizontal, passes through $(8, 4)$

20. Slope 3, x-intercept 6

21. Find the equation of the perpendicular bisector of the segment joining $(1, 2)$ and $(5, 4)$ (Figure 11). *Hint:* The midpoint Q of the segment joining (a, b) and (c, d) is $\left(\frac{a+c}{2}, \frac{b+d}{2}\right)$.

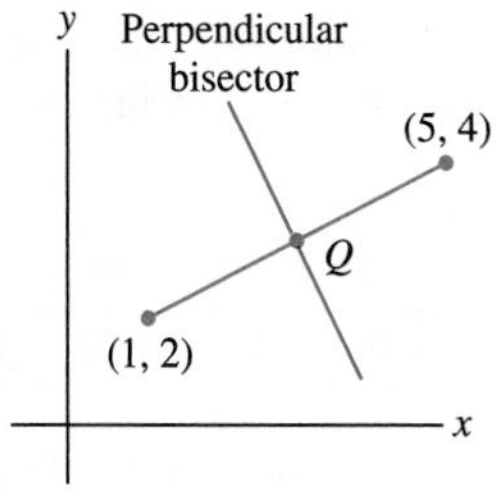

FIGURE 11

22. **Intercept-Intercept Form** Show that if $a, b \neq 0$, then the line with x-intercept $x = a$ and y-intercept $y = b$ has equation (Figure 12)

$$\frac{x}{a} + \frac{y}{b} = 1$$

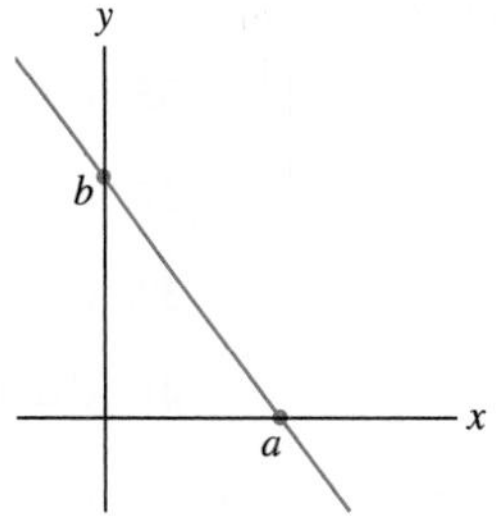

FIGURE 12

23. Find an equation of the line with x-intercept $x = 4$ and y-intercept $y = 3$.

24. Find y such that $(3, y)$ lies on the line of slope $m = 2$ through $(1, 4)$.

25. Determine whether there exists a constant c such that the line $x + cy = 1$:
(a) Has slope 4
(b) Passes through $(3, 1)$
(c) Is horizontal
(d) Is vertical

26. Assume that the number N of concert tickets that can be sold at a price of P dollars per ticket is a linear function $N(P)$ for $10 \le P \le 40$. Determine $N(P)$ (called the demand function) if $N(10) = 500$ and $N(40) = 0$. What is the decrease ΔN in the number of tickets sold if the price is increased by $\Delta P = 5$ dollars?

27. Materials expand when heated. Consider a metal rod of length L_0 at temperature T_0. If the temperature is changed by an amount ΔT, then the rod's length changes by $\Delta L = \alpha L_0 \Delta T$, where α is the thermal expansion coefficient. For steel, $\alpha = 1.24 \times 10^{-5}\,{}^\circ\mathrm{C}^{-1}$.

(a) A steel rod has length $L_0 = 40$ cm at $T_0 = 40°$C. Find its length at $T = 90°$C.

(b) Find its length at $T = 50°$C if its length at $T_0 = 100°$C is 65 cm.

(c) Express length L as a function of T if $L_0 = 65$ cm at $T_0 = 100°$C.

28. Do the points (0.5, 1), (1, 1.2), (2, 2) lie on a line?

29. Find b such that $(2, -1)$, $(3, 2)$, and $(b, 5)$ lie on a line.

30. Find an expression for the velocity v as a linear function of t that matches the following data.

t (s)	0	2	4	6
v (m/s)	39.2	58.6	78	97.4

31. The period T of a pendulum is measured for pendulums of several different lengths L. Based on the following data, does T appear to be a linear function of L?

L (cm)	20	30	40	50
T (s)	0.9	1.1	1.27	1.42

32. Show that $f(x)$ is linear of slope m if and only if

$$f(x+h) - f(x) = mh \quad \text{(for all } x \text{ and } h\text{)}$$

33. Find the roots of the quadratic polynomials:

(a) $4x^2 - 3x - 1$ **(b)** $x^2 - 2x - 1$

In Exercises 34–41, complete the square and find the minimum or maximum value of the quadratic function.

34. $y = x^2 + 2x + 5$

35. $y = x^2 - 6x + 9$

36. $y = -9x^2 + x$

37. $y = x^2 + 6x + 2$

38. $y = 2x^2 - 4x - 7$

39. $y = -4x^2 + 3x + 8$

40. $y = 3x^2 + 12x - 5$

41. $y = 4x - 12x^2$

42. Sketch the graph of $y = x^2 - 6x + 8$ by plotting the roots and the minimum point.

43. Sketch the graph of $y = x^2 + 4x + 6$ by plotting the minimum point, the y-intercept, and one other point.

44. If the alleles A and B of the cystic fibrosis gene occur in a population with frequencies p and $1 - p$ (where p is a fraction between 0 and 1), then the frequency of heterozygous carriers (carriers with both alleles) is $2p(1-p)$. Which value of p gives the largest frequency of heterozygous carriers?

45. For which values of c does $f(x) = x^2 + cx + 1$ have a double root? No real roots?

46. Let $f(x)$ be a quadratic function and c a constant. Which of the following statements is correct? Explain graphically.

(a) There is a unique value of c such that $y = f(x) - c$ has a double root.

(b) There is a unique value of c such that $y = f(x - c)$ has a double root.

47. Prove that $x + \frac{1}{x} \geq 2$ for all $x > 0$. *Hint:* Consider $(x^{1/2} - x^{-1/2})^2$.

48. Let $a, b > 0$. Show that the *geometric mean* $\sqrt{ab}$ is not larger than the *arithmetic mean* $(a+b)/2$. *Hint:* Use a variation of the hint given in Exercise 47.

49. If objects of weights x and w_1 are suspended from the balance in Figure 13(A), the cross-beam is horizontal if $bx = aw_1$. If the lengths a and b are known, we may use this equation to determine an unknown weight x by selecting w_1 such that the cross-beam is horizontal. If a and b are not known precisely, we might proceed as follows. First balance x by w_1 on the left as in (A). Then switch places and balance x by w_2 on the right as in (B). The average $\bar{x} = \frac{1}{2}(w_1 + w_2)$ gives an estimate for x. Show that $\bar{x}$ is greater than or equal to the true weight x.

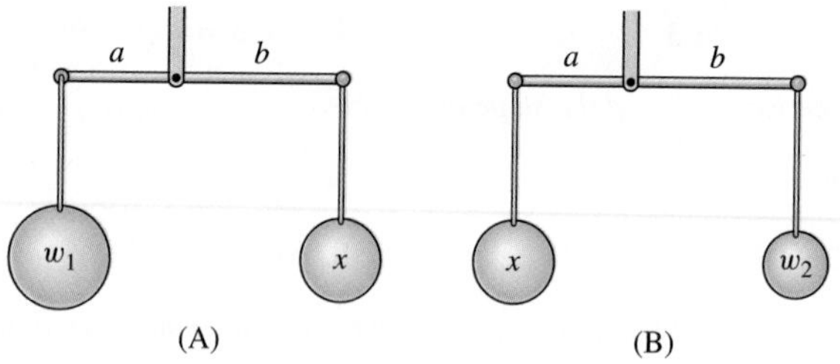

FIGURE 13

50. Find numbers x and y with sum 10 and product 24. *Hint:* Find a quadratic polynomial satisfied by x.

51. Find a pair of numbers whose sum and product are both equal to 8.

52. Show that the parabola $y = x^2$ consists of all points P such that $d_1 = d_2$, where d_1 is the distance from P to $\left(0, \frac{1}{4}\right)$ and d_2 is the distance from P to the line $y = -\frac{1}{4}$ (Figure 14).

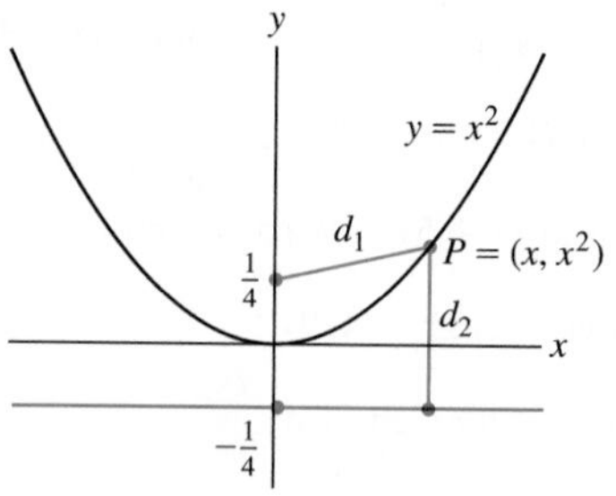

FIGURE 14

Further Insights and Challenges

53. Show that if $f(x)$ and $g(x)$ are linear, then so is $f(x)+g(x)$. Is the same true of $f(x)g(x)$?

54. Show that if $f(x)$ and $g(x)$ are linear functions such that $f(0)=g(0)$ and $f(1)=g(1)$, then $f(x)=g(x)$.

55. Show that $\Delta y/\Delta x$ for the function $f(x)=x^2$ over the interval $[x_1, x_2]$ is not a constant, but depends on the interval. Determine the exact dependence of $\Delta y/\Delta x$ on x_1 and x_2.

56. Use Eq. (2) to derive the quadratic formula for the roots of $ax^2+bx+c=0$.

57. Let $a, c \neq 0$. Show that the roots of

$$ax^2+bx+c=0 \quad \text{and} \quad cx^2+bx+a=0$$

are reciprocals of each other.

58. Show, by completing the square, that the parabola

$$y = ax^2+bx+c$$

is congruent to $y=ax^2$ by a vertical and horizontal translation.

59. Prove **Viète's Formulas**: The quadratic polynomial with α and β as roots is x^2+bx+c, where $b=-\alpha-\beta$ and $c=\alpha\beta$.

1.3 The Basic Classes of Functions

It would be impossible (and useless) to describe all possible functions $f(x)$. Since the values of a function can be assigned arbitrarily, a function chosen at random would likely be so complicated that we could neither graph it nor describe it in any reasonable way. However, calculus makes no attempt to deal with all functions. The techniques of calculus, powerful and general as they are, apply only to functions that are sufficiently "well-behaved" (we will see what well-behaved means when we study the derivative in Chapter 3). Fortunately, such functions are adequate for a vast range of applications.

Most of the functions considered in this text are constructed from the following familiar classes of well-behaved functions:

polynomials rational functions algebraic functions

exponential functions trigonometric functions

logarithmic functions inverse trigonometric functions

We shall refer to these as the **basic functions**.

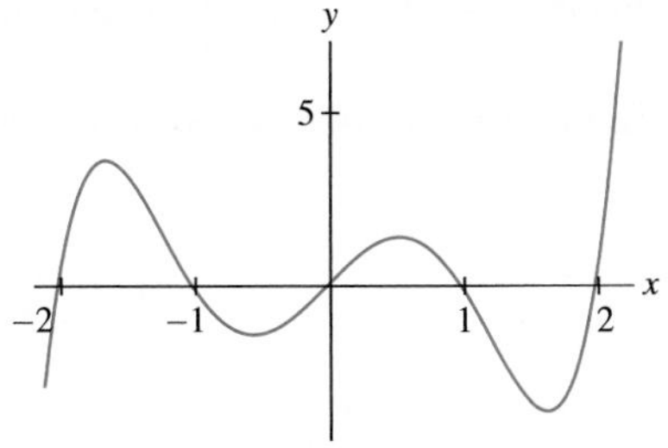

FIGURE 1 The polynomial $y=x^5-5x^3+4x$.

- **Polynomials:** For any real number m, $f(x)=x^m$ is called the **power function** with exponent m. A polynomial is a sum of multiples of power functions with whole-number exponents (Figure 1):

$$f(x)=x^5-5x^3+4x, \qquad g(t)=7t^6+t^3-3t-1$$

Thus, the function $f(x)=x+x^{-1}$ is not a polynomial because it includes a power function x^{-1} with a negative exponent. The general polynomial in the variable x may be written

$$P(x)=a_nx^n+a_{n-1}x^{n-1}+\cdots+a_1x+a_0$$

  - The numbers $a_0, a_1, \ldots, a_n$ are called **coefficients.**
  - The **degree** of $P(x)$ is n (assuming that $a_n \neq 0$).
  - The coefficient a_n is called the **leading coefficient**.
  - The domain of $P(x)$ is **R**.

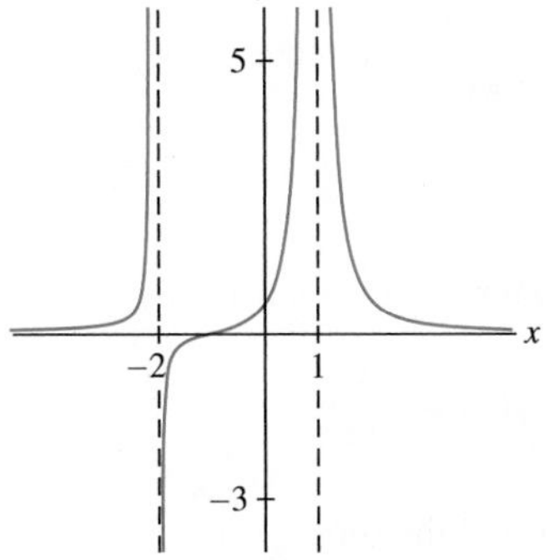

FIGURE 2 The rational function $f(x)=\dfrac{x+1}{x^3-3x+2}$.

- A **rational function** is a *quotient* of two polynomials (Figure 2):

$$f(x)=\frac{P(x)}{Q(x)} \qquad [P(x) \text{ and } Q(x) \text{ polynomials}]$$

The domain of $f(x)$ is the set of numbers x such that $Q(x) \neq 0$. For example,

$$f(x) = \frac{1}{x^2} \qquad \text{domain } \{x : x \neq 0\}$$

$$h(t) = \frac{7t^6 + t^3 - 3t - 1}{t^2 - 1} \qquad \text{domain } \{t : t \neq \pm 1\}$$

Every polynomial is also a rational function [with $Q(x) = 1$].

- An **algebraic function** is produced by taking sums, products, and quotients of *roots* of polynomials and rational functions (Figure 3):

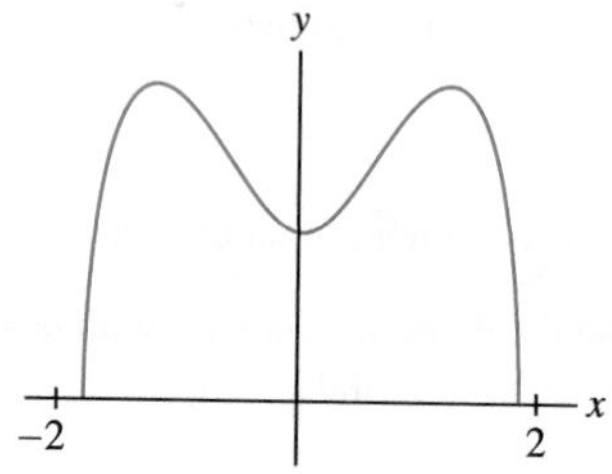

FIGURE 3 The algebraic function $f(x) = \sqrt{1 + 3x^2 - x^4}$.

$$f(x) = \sqrt{1 + 3x^2 - x^4}, \qquad g(t) = (\sqrt{t} - 2)^{-2}, \qquad h(z) = \frac{z + z^{-5/3}}{5z^3 - \sqrt{z}}$$

A number x belongs to the domain of f if each term in the formula is defined and the result does not involve division by zero. For example, $g(t)$ is defined if $t \geq 0$ and $\sqrt{t} \neq 2$, so the domain of $g(t)$ is $D = \{t : t \geq 0 \text{ and } t \neq 4\}$. More generally, algebraic functions are defined by polynomial equations between x and y. In this case, we say that y is **implicitly defined** as a function of x. For example, the equation $y^4 + 2x^2y + x^4 = 1$ defines y implicitly as a function of x.

- **Exponential functions:** The function $f(x) = b^x$, where $b > 0$, is called the exponential function with base b. Some examples are

Any function that is not algebraic is called ***transcendental***. *Exponential and trigonometric functions are examples, as are the Bessel and gamma functions that appear in engineering and statistics. The term "transcendental" goes back to the 1670s, when it was used by Gottfried Wilhelm Leibniz (1646–1716) to describe functions of this type.*

$$f(x) = 2^x, \qquad g(t) = 10^t, \qquad h(x) = \left(\frac{1}{3}\right)^x, \qquad p(t) = (\sqrt{5})^t$$

Exponential functions and their *inverses*, the **logarithmic functions**, are treated in greater detail in Chapter 7.

- **Trigonometric functions** are functions built from $\sin x$ and $\cos x$. These functions are discussed in the next section.

Constructing New Functions

Given functions f and g, we can construct new functions by forming the sum, difference, product, and quotient functions:

$$(f + g)(x) = f(x) + g(x), \qquad (f - g)(x) = f(x) - g(x)$$

$$(fg)(x) = f(x)\,g(x), \qquad \left(\frac{f}{g}\right)(x) = \frac{f(x)}{g(x)} \quad (\text{where } g(x) \neq 0)$$

For example, if $f(x) = x^2$ and $g(x) = \sin x$, then

$$(f + g)(x) = x^2 + \sin x, \qquad (f - g)(x) = x^2 - \sin x$$

$$(fg)(x) = x^2 \sin x, \qquad \left(\frac{f}{g}\right)(x) = \frac{x^2}{\sin x}$$

We can also multiply functions by constants. A function of the form

$$c_1 f(x) + c_2 g(x) \qquad (c_1, c_2 \text{ constants})$$

is called a **linear combination** of $f(x)$ and $g(x)$.

Composition is another important way of constructing new functions. The composition of f and g is the function $f \circ g$ defined by $(f \circ g)(x) = f(g(x))$. The domain of $f \circ g$ is the set of values of x in the domain of g such that $g(x)$ lies in the domain of f.

Example 1 shows that the composition of functions is not *commutative: The functions $f \circ g$ and $g \circ f$ may be (and usually are) different.*

■ **EXAMPLE 1** Compute the composite functions $f \circ g$ and $g \circ f$ and discuss their domains, where

$$f(x) = \sqrt{x}, \qquad g(x) = 1 - x$$

Solution We have

$$(f \circ g)(x) = f(g(x)) = f(1 - x) = \sqrt{1 - x}$$

The square root $\sqrt{1 - x}$ is defined if $1 - x \geq 0$ or $x \leq 1$, so the domain of $f \circ g$ is $\{x : x \leq 1\}$. On the other hand,

$$(g \circ f)(x) = g(f(x)) = g(\sqrt{x}) = 1 - \sqrt{x}$$

The domain of $g \circ f$ is $\{x : x \geq 0\}$. ■

Elementary Functions

Inverse functions are discussed in Section 7.2.

As noted above, we can produce new functions by applying the operations of addition, subtraction, multiplication, division, and composition. It is convenient to refer to a function constructed in this way from the basic functions listed above as an **elementary function**. The following functions are elementary:

$$f(x) = \sqrt{2x + \sin x}, \qquad f(x) = 10^{\sqrt{x}}, \qquad f(x) = \frac{1 + x^{-1}}{1 + \cos x}$$

1.3 SUMMARY

- For m a real number, $f(x) = x^m$ is called the *power function* with exponent m. A polynomial $P(x)$ is a sum of multiples of power functions x^m, where m is a whole number:

$$P(x) = a_n x^n + a_{n-1} x^{n-1} + \cdots + a_1 x + a_0$$

This polynomial has degree n (assuming that $a_n \neq 0$) and a_n is called the leading coefficient.
- A rational function is a quotient $P(x)/Q(x)$ of two polynomials.
- An algebraic function is produced by taking sums, products, and nth roots of polynomials and rational functions.
- Exponential function: $f(x) = b^x$, where $b > 0$ (b is called the base).
- The composite function $f \circ g$ is defined by $(f \circ g)(x) = f(g(x))$. The domain of $f \circ g$ is the set of x in the domain of g such that $g(x)$ belongs to the domain of f.

1.3 EXERCISES

Preliminary Questions

1. Give an example of a rational function.

2. Is $|x|$ a polynomial function? What about $|x^2 + 1|$?

3. What is unusual about the domain of the composite function $f \circ g$ for the functions $f(x) = x^{1/2}$ and $g(x) = -1 - |x|$?

4. Is $f(x) = \left(\frac{1}{2}\right)^x$ increasing or decreasing?

5. Give an example of a transcendental function.

Exercises

In Exercises 1–12, determine the domain of the function.

1. $f(x) = x^{1/4}$

2. $g(t) = t^{2/3}$

3. $f(x) = x^3 + 3x - 4$

4. $h(z) = z^3 + z^{-3}$

5. $g(t) = \dfrac{1}{t+2}$

6. $f(x) = \dfrac{1}{x^2+4}$

7. $G(u) = \dfrac{1}{u^2-4}$

8. $f(x) = \dfrac{\sqrt{x}}{x^2-9}$

9. $f(x) = x^{-4} + (x-1)^{-3}$

10. $F(s) = \sin\left(\dfrac{s}{s+1}\right)$

11. $g(y) = 10^{\sqrt{y}+y^{-1}}$

12. $f(x) = \dfrac{x + x^{-1}}{(x-3)(x+4)}$

In Exercises 13–24, identify each of the following functions as polynomial, rational, algebraic, or transcendental.

13. $f(x) = 4x^3 + 9x^2 - 8$

14. $f(x) = x^{-4}$

15. $f(x) = \sqrt{x}$

16. $f(x) = \sqrt{1-x^2}$

17. $f(x) = \dfrac{x^2}{x + \sin x}$

18. $f(x) = 2^x$

19. $f(x) = \dfrac{2x^3+3x}{9-7x^2}$

20. $f(x) = \dfrac{3x - 9x^{-1/2}}{9-7x^2}$

21. $f(x) = \sin(x^2)$

22. $f(x) = \dfrac{x}{\sqrt{x}+1}$

23. $f(x) = x^2 + 3x^{-1}$

24. $f(x) = \sin(3^x)$

25. Is $f(x) = 2^{x^2}$ a transcendental function?

26. Show that $f(x) = x^2 + 3x^{-1}$ and $g(x) = 3x^3 - 9x + x^{-2}$ are rational functions—that is, quotients of polynomials.

In Exercises 27–34, calculate the composite functions $f \circ g$ and $g \circ f$, and determine their domains.

27. $f(x) = \sqrt{x}, \quad g(x) = x + 1$

28. $f(x) = \dfrac{1}{x}, \quad g(x) = x^{-4}$

29. $f(x) = 2^x, \quad g(x) = x^2$

30. $f(x) = |x|, \quad g(\theta) = \sin\theta$

31. $f(\theta) = \cos\theta, \quad g(x) = x^3 + x^2$

32. $f(x) = \dfrac{1}{x^2+1}, \quad g(x) = x^{-2}$

33. $f(t) = \dfrac{1}{\sqrt{t}}, \quad g(t) = -t^2$

34. $f(t) = \sqrt{t}, \quad g(t) = 1 - t^3$

35. The population (in millions) of a country as a function of time t (years) is $P(t) = 30 \cdot 2^{0.1t}$. Show that the population doubles every 10 years. Show more generally that for any positive constants a and k, the function $g(t) = a2^{kt}$ doubles after $1/k$ years.

36. Find all values of c such that $f(x) = \dfrac{x+1}{x^2+2cx+4}$ has domain **R**.

Further Insights and Challenges

In Exercises 37–43, we define the first difference δf *of a function $f(x)$ by $\delta f(x) = f(x+1) - f(x)$.*

37. Show that if $f(x) = x^2$, then $\delta f(x) = 2x + 1$. Calculate δf for $f(x) = x$ and $f(x) = x^3$.

38. Show that $\delta(10^x) = 9 \cdot 10^x$ and, more generally, that $\delta(b^x) = (b-1)b^x$.

39. Show that for any two functions f and g, $\delta(f+g) = \delta f + \delta g$ and $\delta(cf) = c\delta(f)$, where c is any constant.

40. Suppose we can find a function $P(x)$ such that $\delta P = (x+1)^k$ and $P(0) = 0$. Prove that $P(1) = 1^k$, $P(2) = 1^k + 2^k$, and, more generally, for every whole number n,

$$P(n) = 1^k + 2^k + \cdots + n^k \qquad \boxed{1}$$

41. First show that

$$P(x) = \frac{x(x+1)}{2}$$

satisfies $\delta P = (x+1)$. Then apply Exercise 40 to conclude that

$$1 + 2 + 3 + \cdots + n = \frac{n(n+1)}{2}$$

42. Calculate $\delta(x^3)$, $\delta(x^2)$, and $\delta(x)$. Then find a polynomial $P(x)$ of degree 3 such that $\delta P = (x+1)^2$ and $P(0) = 0$. Conclude that $P(n) = 1^2 + 2^2 + \cdots + n^2$.

43. This exercise combined with Exercise 40 shows that for all whole numbers k, there exists a polynomial $P(x)$ satisfying Eq. (1). The solution requires the Binomial Theorem and proof by induction (see Appendix C).

(a) Show that $\delta(x^{k+1}) = (k+1)x^k + \cdots$, where the dots indicate terms involving smaller powers of x.

(b) Show by induction that there exists a polynomial of degree $k+1$ with leading coefficient $1/(k+1)$:

$$P(x) = \frac{1}{k+1}x^{k+1} + \cdots$$

such that $\delta P = (x+1)^k$ and $P(0) = 0$.

1.4 Trigonometric Functions

We begin our trigonometric review by recalling the two systems of angle measurement: **radians** and **degrees**. They are best described using the relationship between angles and rotation. As is customary, we often use the lowercase Greek letter θ ("theta") to denote angles and rotations.

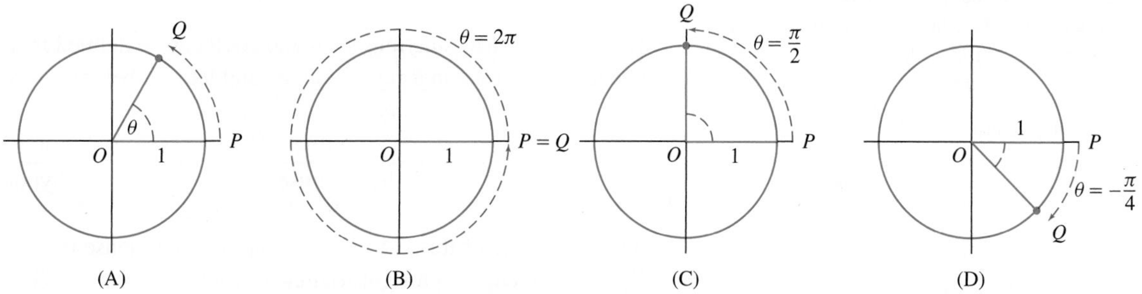

FIGURE 1 The radian measure θ of a counterclockwise rotation is the length along the unit circle of the arc traversed by P as it rotates into Q.

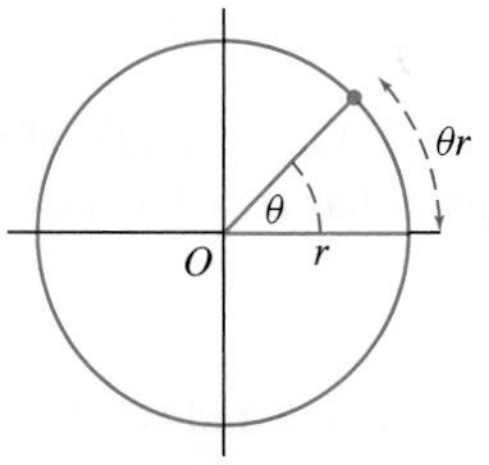

FIGURE 2 On a circle of radius r, the arc traversed by a counterclockwise rotation of θ radians has length θr.

TABLE 1

Rotation through	Radian measure
Two full circles	4π
Full circle	2π
Half circle	π
Quarter circle	$2\pi/4 = \pi/2$
One-sixth circle	$2\pi/6 = \pi/3$

Figure 1(A) shows a unit circle with radius $\overline{OP}$ rotating counterclockwise into radius $\overline{OQ}$. *The radian measure of this rotation is the length θ of the circular arc traversed by P as it rotates into Q.* On a circle of radius r, the arc traversed by a counterclockwise rotation of θ radians has length θr (Figure 2).

The unit circle has circumference 2π. Therefore, a rotation through a full circle has radian measure $\theta = 2\pi$ [Figure 1(B)]. The radian measure of a rotation through one-quarter of a circle is $\theta = 2\pi/4 = \pi/2$ [Figure 1(C)] and, in general, the rotation through one-nth of a circle has radian measure $2\pi/n$ (Table 1). A negative rotation (with $\theta < 0$) is a rotation in the *clockwise* direction [Figure 1(D)]. The unit circle has circumference 2π (by definition of the number π).

The radian measure of an angle such as $\angle POQ$ in Figure 1(A) is defined as the radian measure of a rotation that carries $\overline{OP}$ to $\overline{OQ}$. Notice, however, that the radian measure of an angle is not unique. The rotations through θ and $\theta + 2\pi$ both carry $\overline{OP}$ to $\overline{OQ}$. Therefore, θ and $\theta + 2\pi$ represent the same angle even though the rotation through $\theta + 2\pi$ takes an extra trip around the circle. In general, *two radian measures represent the same angle if the corresponding rotations differ by an integer multiple of 2π.* For example, $\pi/4$, $9\pi/4$, and $-15\pi/4$ all represent the same angle because they differ by multiples of 2π:

$$\frac{\pi}{4} = \frac{9\pi}{4} - 2\pi = -\frac{15\pi}{4} + 4\pi$$

Every angle has a unique radian measure satisfying $0 \le \theta < 2\pi$. With this choice, the angle θ subtends an arc of length θr on a circle of radius r (Figure 2).

Degrees are defined by dividing the circle (not necessarily the unit circle) into 360 equal parts. A degree is $\frac{1}{360}$ of a circle. A rotation through θ degrees (denoted θ°) is a rotation through the fraction $\theta/360$ of the complete circle. For example, a rotation through 90° is a rotation through the fraction $\frac{90}{360}$, or $\frac{1}{4}$, of a circle.

Radians	Degrees
0	0°
$\frac{\pi}{6}$	30°
$\frac{\pi}{4}$	45°
$\frac{\pi}{3}$	60°
$\frac{\pi}{2}$	90°

As with radians, the degree measure of an angle is not unique. Two degree measures represent that same angle if they differ by an integer multiple of 360. For example, the angles -45° and 675° coincide because $675 = -45 + 2(360)$. Every angle has a unique degree measure θ with $0 \le \theta < 360$.

To convert between radians and degrees, remember that 2π rad is equal to 360°. Therefore, 1 rad equals $360/2\pi$ or $180/\pi$ degrees.

- To convert from radians to degrees, multiply by $180/\pi$.
- To convert from degrees to radians, multiply by $\pi/180$.

Radian measurement is usually the better choice for mathematical purposes, but there are good practical reasons for using degrees. The number 360 has many divisors ($360 = 8 \cdot 9 \cdot 5$), and consequently, many fractional parts of the circle can be expressed as an integer number of degrees. For example, one-fifth of the circle is 72°, two-ninths is 80°, three-eighths is 135°, etc.

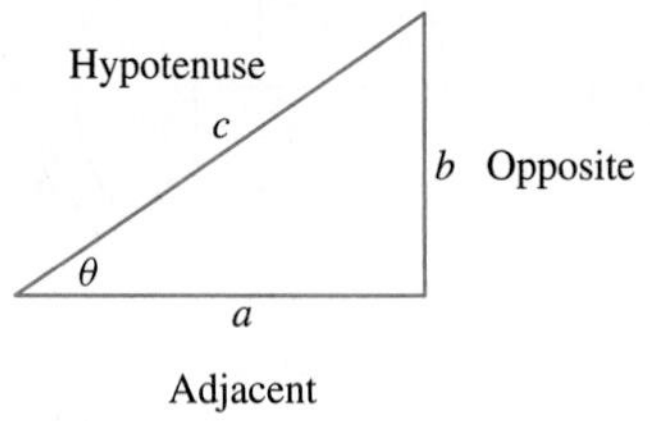

FIGURE 3

■ **EXAMPLE 1** Convert **(a)** 55° to radians and **(b)** 0.5 rad to degrees.

Solution

(a) $55^\circ \times \dfrac{\pi}{180} \approx 0.9599$ rad **(b)** $0.5 \text{ rad} \times \dfrac{180}{\pi} \approx 28.648^\circ$ ■

Convention: *Unless otherwise stated, we always measure angles in radians.*

The trigonometric functions $\sin\theta$ and $\cos\theta$ can be defined in terms of right triangles. Let θ be an acute angle in a right triangle, and let us label the sides as in Figure 3. Then

$$\sin\theta = \frac{b}{c} = \frac{\text{opposite}}{\text{hypotenuse}}, \qquad \cos\theta = \frac{a}{c} = \frac{\text{adjacent}}{\text{hypotenuse}}$$

A disadvantage of this definition is that it makes sense only if θ lies between 0 and $\pi/2$ (because an angle in a right triangle cannot exceed $\pi/2$). However, sine and cosine can be defined for all angles in terms of the unit circle. Let $P = (x, y)$ be the point on the unit circle corresponding to the angle θ as in Figures 4(A) and (B), and define

$$\cos\theta = x\text{-coordinate of } P, \qquad \sin\theta = y\text{-coordinate of } P$$

This agrees with the right-triangle definition when $0 < \theta < \frac{\pi}{2}$. On the circle of radius r (centered at the origin), the point corresponding to the angle θ has coordinates

$$(r\cos\theta, r\sin\theta)$$

Furthermore, we see from Figure 4(C) that $\sin\theta$ is an odd function and $\cos\theta$ is an even function:

$$\sin(-\theta) = -\sin\theta, \qquad \cos(-\theta) = \cos\theta$$

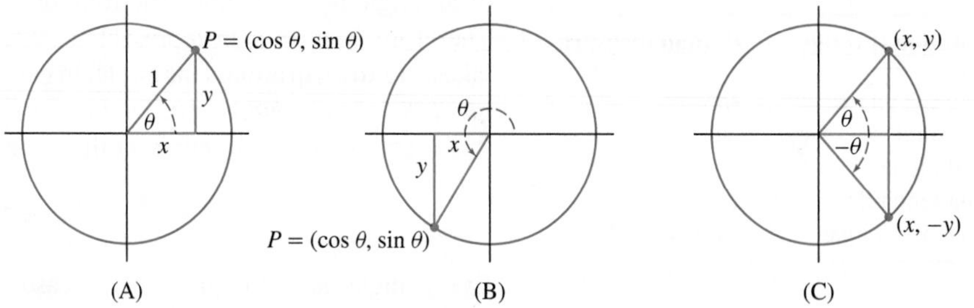

FIGURE 4 The unit circle definition of sine and cosine is valid for all angles θ.

Although we use a calculator to evaluate sine and cosine for general angles, the standard values listed in Figure 5 and Table 2 appear often and should be memorized.

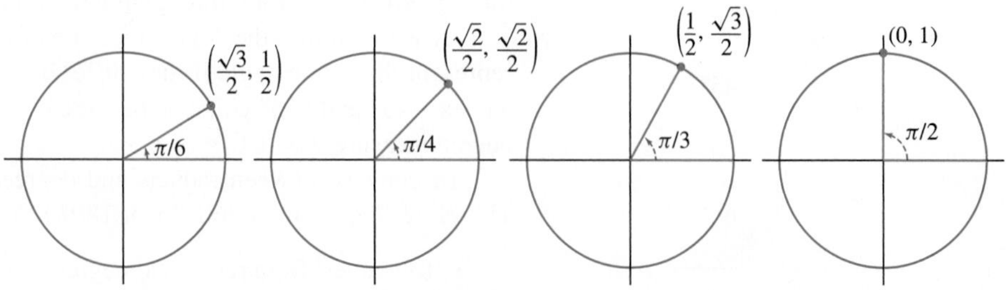

FIGURE 5 Four standard angles: The x- and y-coordinates of the points are $\cos\theta$ and $\sin\theta$.

TABLE 2

θ	0	$\frac{\pi}{6}$	$\frac{\pi}{4}$	$\frac{\pi}{3}$	$\frac{\pi}{2}$	$\frac{2\pi}{3}$	$\frac{3\pi}{4}$	$\frac{5\pi}{6}$	π
$\sin\theta$	0	$\frac{1}{2}$	$\frac{\sqrt{2}}{2}$	$\frac{\sqrt{3}}{2}$	1	$\frac{\sqrt{3}}{2}$	$\frac{\sqrt{2}}{2}$	$\frac{1}{2}$	0
$\cos\theta$	1	$\frac{\sqrt{3}}{2}$	$\frac{\sqrt{2}}{2}$	$\frac{1}{2}$	0	$-\frac{1}{2}$	$-\frac{\sqrt{2}}{2}$	$-\frac{\sqrt{3}}{2}$	-1

The graph of $y = \sin\theta$ is the familiar "sine wave" shown in Figure 6. Observe how the graph is generated by the y-coordinate of the point $P = (\cos\theta, \sin\theta)$ moving around the unit circle.

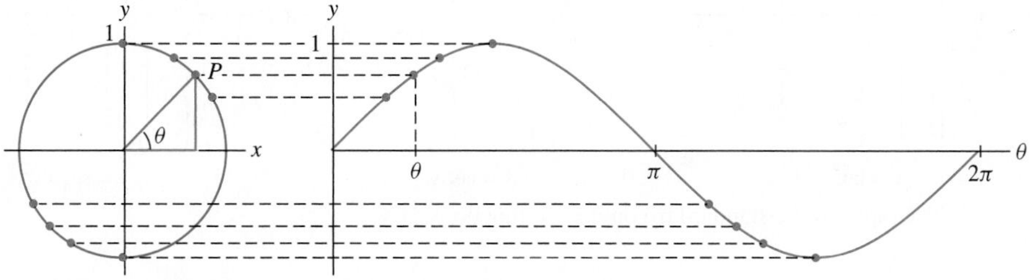

FIGURE 6 The graph of $y = \sin\theta$ is generated as the point $P = (\cos\theta, \sin\theta)$ moves around the unit circle.

The graph of $y = \cos\theta$ has the same shape but is shifted to the left $\pi/2$ units (Figure 7). The signs of $\sin\theta$ and $\cos\theta$ vary as $P = (\cos\theta, \sin\theta)$ changes quadrant.

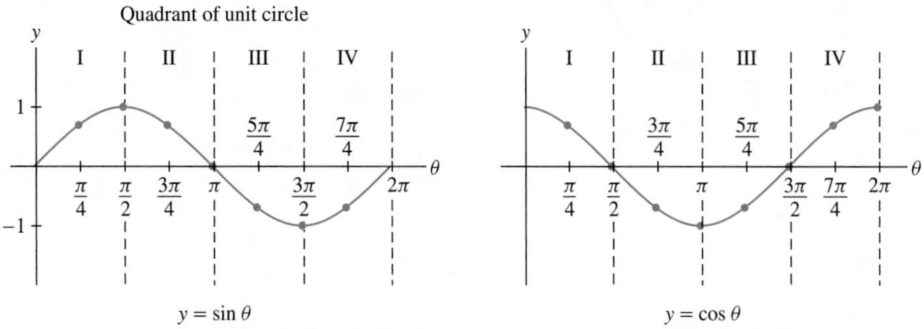

FIGURE 7 Graphs of $y = \sin\theta$ and $y = \cos\theta$ over one *period* of length 2π.

A function $f(x)$ is called **periodic** with period T if $f(x + T) = f(x)$ (for all x) and T is the smallest positive number with this property. The sine and cosine functions are periodic with period $T = 2\pi$ (Figure 8) because the radian measures x and $x + 2\pi k$ correspond to the same point on the unit circle for any integer k:

We often write $\sin x$ and $\cos x$, using x instead of θ. Depending on the application, we may think of x as an angle or simply as a real number.

$$\sin x = \sin(x + 2\pi k), \qquad \cos x = \cos(x + 2\pi k)$$

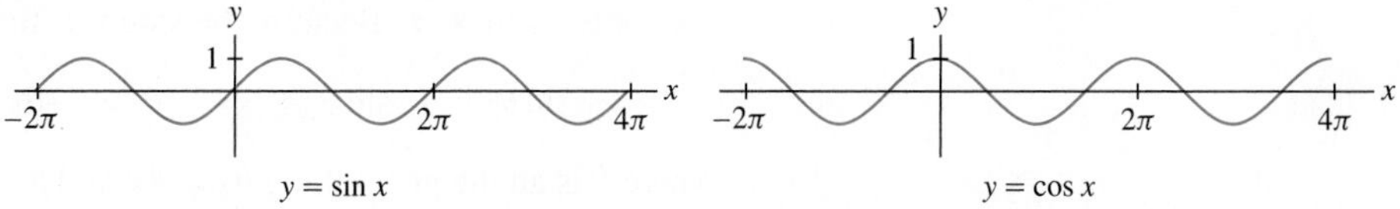

FIGURE 8 Sine and cosine have period 2π.

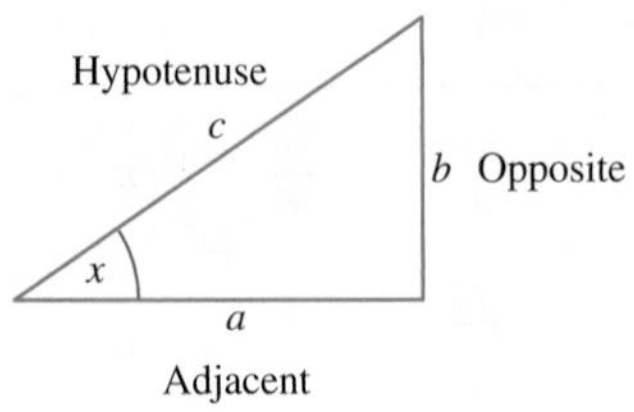

FIGURE 9

There are four other standard trigonometric functions, each defined in terms of $\sin x$ and $\cos x$ or as ratios of sides in a right triangle (Figure 9):

Tangent:	$\tan x = \dfrac{\sin x}{\cos x} = \dfrac{b}{a}$,	Cotangent:	$\cot x = \dfrac{\cos x}{\sin x} = \dfrac{a}{b}$
Secant:	$\sec x = \dfrac{1}{\cos x} = \dfrac{c}{a}$,	Cosecant:	$\csc x = \dfrac{1}{\sin x} = \dfrac{c}{b}$

These functions are periodic (Figure 10): $y = \tan x$ and $y = \cot x$ have period π; $y = \sec x$ and $y = \csc x$ have period 2π (see Exercise 55).

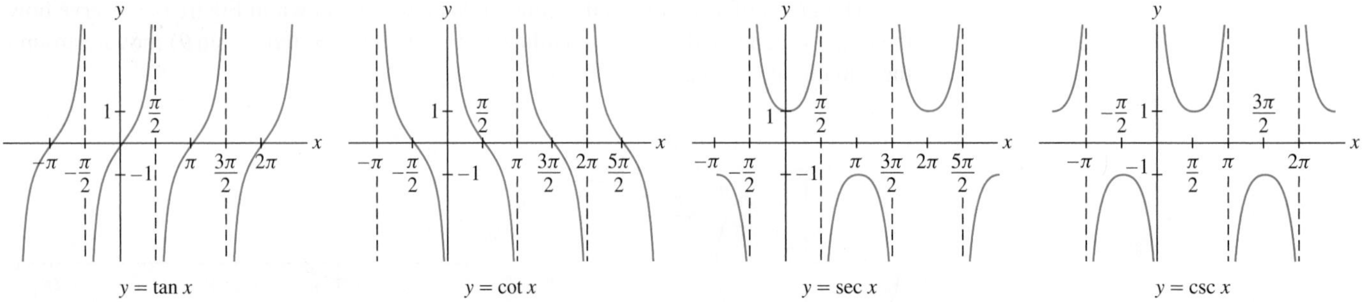

FIGURE 10 Graphs of the standard trigonometric functions.

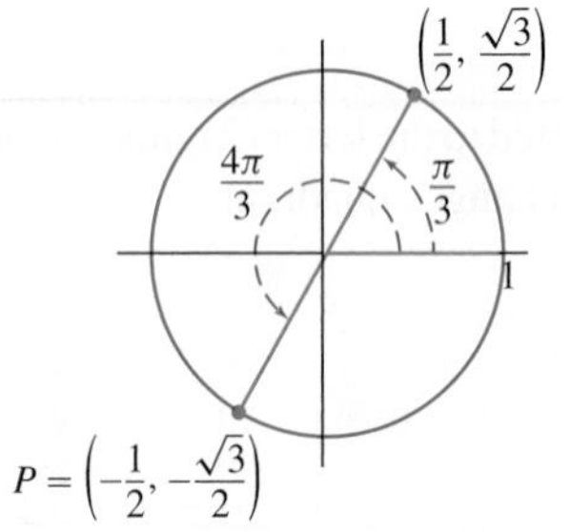

FIGURE 11

■ **EXAMPLE 2 Computing Values of Trigonometric Functions** Find the values of the six trigonometric functions at $x = 4\pi/3$.

Solution The point P on the unit circle corresponding to the angle $x = 4\pi/3$ lies opposite the point with angle $\pi/3$ (Figure 11). Thus, we see that (refer to Table 2)

$$\sin\frac{4\pi}{3} = -\sin\frac{\pi}{3} = -\frac{\sqrt{3}}{2}, \qquad \cos\frac{4\pi}{3} = -\cos\frac{\pi}{3} = -\frac{1}{2}$$

The remaining values are

$$\tan\frac{4\pi}{3} = \frac{\sin 4\pi/3}{\cos 4\pi/3} = \frac{-\sqrt{3}/2}{-1/2} = \sqrt{3}, \quad \cot\frac{4\pi}{3} = \frac{\cos 4\pi/3}{\sin 4\pi/3} = \frac{\sqrt{3}}{3}$$

$$\sec\frac{4\pi}{3} = \frac{1}{\cos 4\pi/3} = \frac{1}{-1/2} = -2, \qquad \csc\frac{4\pi}{3} = \frac{1}{\sin 4\pi/3} = \frac{-2\sqrt{3}}{3}$$

■

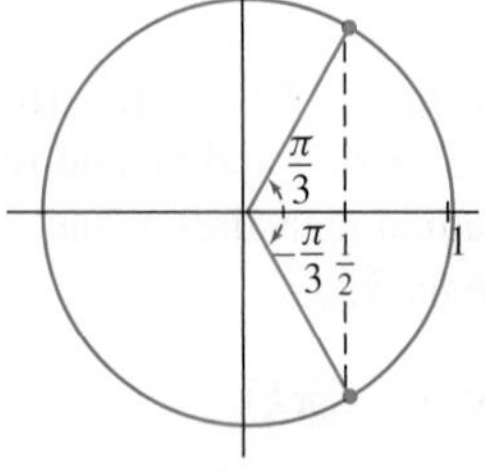

FIGURE 12 $\cos x = \frac{1}{2}$ for $x = \pm\frac{\pi}{3}$

■ **EXAMPLE 3** Find the angles x such that $\sec x = 2$.

Solution Because $\sec x = 1/\cos x$, we must solve $\cos x = \frac{1}{2}$. From Figure 12 we see that $x = \pi/3$ and $x = -\pi/3$ are solutions. We may add any integer multiple of 2π, so the general solution is $x = \pm\pi/3 + 2\pi k$ for any integer k. ■

■ **EXAMPLE 4 Trigonometric Equation** Solve $\sin 4x + \sin 2x = 0$ for $x \in [0, 2\pi)$.

Solution We must find the angles x such that $\sin 4x = -\sin 2x$. First, let's determine when angles θ_1 and θ_2 satisfy $\sin\theta_2 = -\sin\theta_1$. Figure 13 shows that this occurs if $\theta_2 = -\theta_1$ or $\theta_2 = \theta_1 + \pi$. Because the sine function is periodic with period 2π,

$$\sin\theta_2 = -\sin\theta_1 \quad \Leftrightarrow \quad \theta_2 = -\theta_1 + 2\pi k \quad \text{or} \quad \theta_2 = \theta_1 + \pi + 2\pi k$$

where k is an integer. Taking $\theta_2 = 4x$ and $\theta_1 = 2x$, we see that

$$\sin 4x = -\sin 2x \quad \Leftrightarrow \quad 4x = -2x + 2\pi k \quad \text{or} \quad 4x = 2x + \pi + 2\pi k$$

The first equation gives $6x = 2\pi k$ or $x = (\pi/3)k$ and the second equation gives $2x = \pi + 2\pi k$ or $x = \pi/2 + \pi k$. We obtain eight solutions in $[0, 2\pi)$ (Figure 14):

$$x = 0, \quad \frac{\pi}{3}, \quad \frac{2\pi}{3}, \quad \pi, \quad \frac{4\pi}{3}, \quad \frac{5\pi}{3} \quad \text{and} \quad x = \frac{\pi}{2}, \quad \frac{3\pi}{2}$$ ■

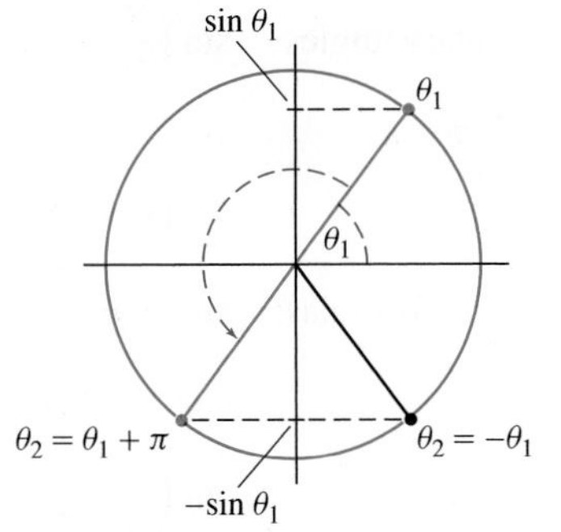

FIGURE 13 $\sin\theta_2 = -\sin\theta_1$ when $\theta_2 = -\theta_1$ or $\theta_2 = \theta_1 + \pi$.

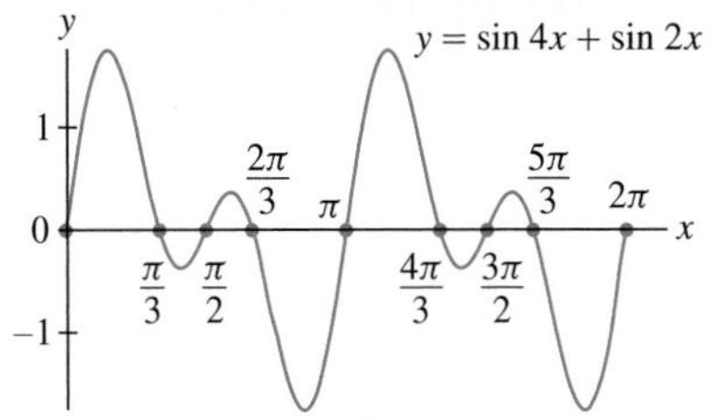

FIGURE 14 Solutions of $\sin 4x + \sin 2x = 0$.

■ **EXAMPLE 5** Sketch the graph of $f(x) = 3\cos\left(2\left(x + \frac{\pi}{2}\right)\right)$ over $[0, 2\pi]$.

CAUTION To shift the graph of $y = \cos 2x$ to the left $\pi/2$ units, we must replace x by $x + \frac{\pi}{2}$ to obtain $\cos\left(2\left(x + \frac{\pi}{2}\right)\right)$. It is incorrect to take $\cos\left(2x + \frac{\pi}{2}\right)$.

Solution The graph is obtained by scaling and shifting the graph of $y = \cos x$ in three steps (Figure 15):

- Compress horizontally by a factor of 2: $y = \cos 2x$
- Shift to the left $\pi/2$ units: $y = \cos\left(2\left(x + \frac{\pi}{2}\right)\right)$
- Expand vertically by a factor of 3: $y = 3\cos\left(2\left(x + \frac{\pi}{2}\right)\right)$ ■

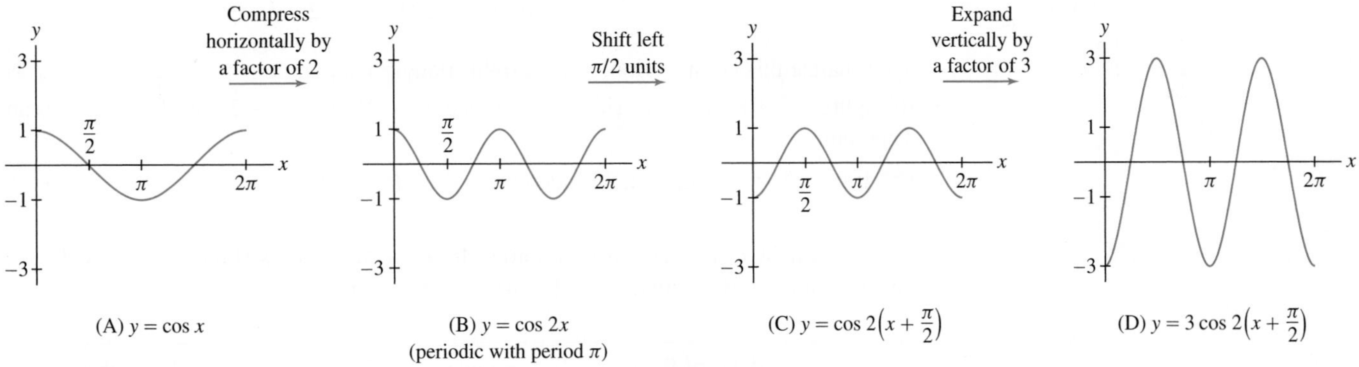

FIGURE 15

Trigonometric Identities

A key feature of trigonometric functions is that they satisfy a large number of identities. First and foremost, sine and cosine satisfy a fundamental identity, which is equivalent to the Pythagorean Theorem:

The expression $(\sin x)^k$ is usually denoted $\sin^k x$. For example, $\sin^2 x$ is the square of $\sin x$. We use similar notation for the other trigonometric functions.

$$\sin^2 x + \cos^2 x = 1 \qquad (1)$$

Equivalent versions are obtained by dividing Eq. (1) by $\cos^2 x$ or $\sin^2 x$:

$$\tan^2 x + 1 = \sec^2 x, \qquad 1 + \cot^2 x = \csc^2 x \qquad (2)$$

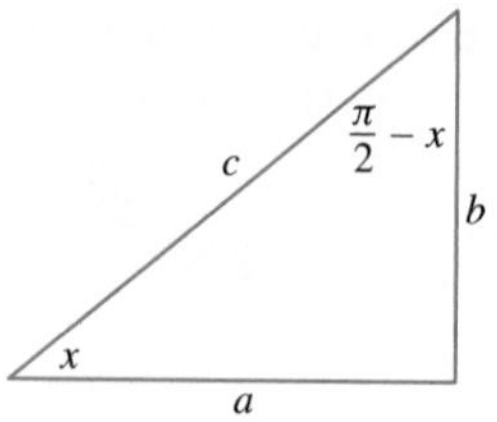

FIGURE 16 For complementary angles, the sine of one is equal to the cosine of the other.

Here is a list of some other commonly used identities. The identities for complementary angles are justified by Figure 16.

Basic Trigonometric Identities

Complementary angles:	$\sin\left(\frac{\pi}{2} - x\right) = \cos x, \quad \cos\left(\frac{\pi}{2} - x\right) = \sin x$
Addition formulas:	$\sin(x + y) = \sin x \cos y + \cos x \sin y$
	$\cos(x + y) = \cos x \cos y - \sin x \sin y$
Double-angle formulas:	$\sin^2 x = \frac{1}{2}(1 - \cos 2x), \quad \cos^2 x = \frac{1}{2}(1 + \cos 2x)$
	$\cos 2x = \cos^2 x - \sin^2 x, \quad \sin 2x = 2 \sin x \cos x$
Shift formulas:	$\sin\left(x + \frac{\pi}{2}\right) = \cos x, \quad \cos\left(x + \frac{\pi}{2}\right) = -\sin x$

■ **EXAMPLE 6** Suppose that $\cos\theta = \frac{2}{5}$. Calculate $\tan\theta$ in the following two cases: **(a)** $0 < \theta < \frac{\pi}{2}$ and **(b)** $\pi < \theta < 2\pi$.

Solution First, using the identity $\cos^2\theta + \sin^2\theta = 1$, we obtain

$$\sin\theta = \pm\sqrt{1 - \cos^2\theta} = \pm\sqrt{1 - \frac{4}{25}} = \pm\frac{\sqrt{21}}{5}$$

(a) If $0 < \theta < \frac{\pi}{2}$, then $\sin\theta$ is positive and we take the positive square root:

$$\tan\theta = \frac{\sin\theta}{\cos\theta} = \frac{\sqrt{21}/5}{2/5} = \frac{\sqrt{21}}{2}$$

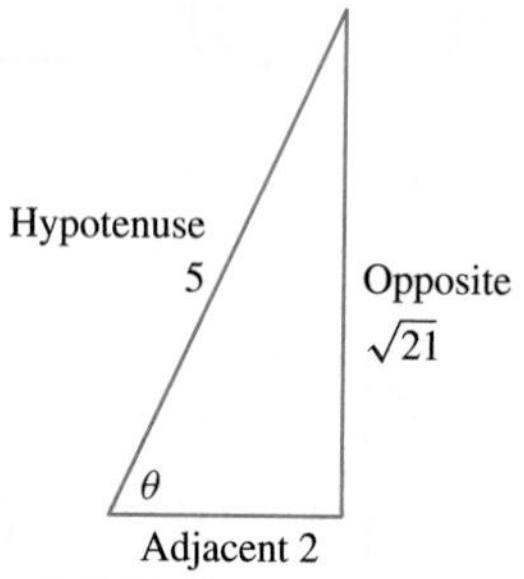

FIGURE 17

To visualize this computation, draw a right triangle with angle θ such that $\cos\theta = \frac{2}{5}$ as in Figure 17. The opposite side then has length $\sqrt{21} = \sqrt{5^2 - 2^2}$ by the Pythagorean Theorem.

(b) If $\pi < \theta < 2\pi$, then $\sin\theta$ is negative and $\tan\theta = -\frac{\sqrt{21}}{2}$. ■

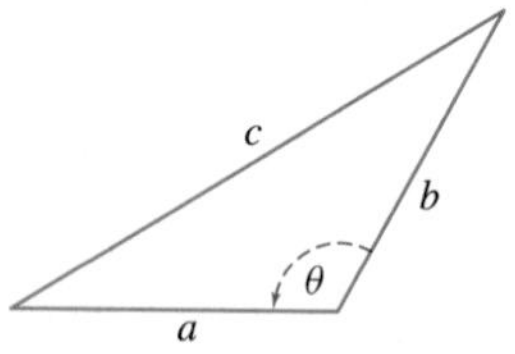

FIGURE 18

We conclude this section by quoting the **Law of Cosines** (Figure 18), which is a generalization of the Pythagorean Theorem (see Exercise 58).

THEOREM 1 Law of Cosines If a triangle has sides a, b, and c, and θ is the angle opposite side c, then

$$c^2 = a^2 + b^2 - 2ab\cos\theta$$

If $\theta = 90°$, then $\cos\theta = 0$ and the Law of Cosines reduces to the Pythagorean Theorem.

1.4 SUMMARY

- An angle of θ radians subtends an arc of length θr on a circle of radius r.
- To convert from radians to degrees, multiply by $180/\pi$.

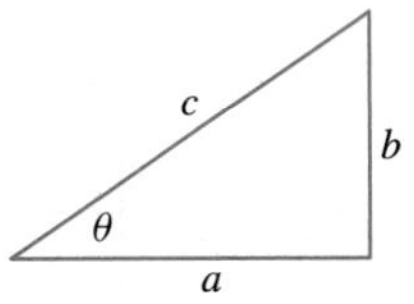

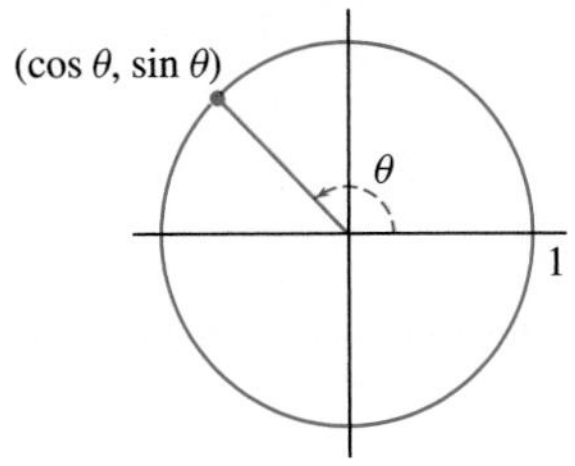

FIGURE 19

- To convert from degrees to radians, multiply by $\pi/180$.
- Unless otherwise stated, all angles in this text are given in radians.
- The functions $\cos\theta$ and $\sin\theta$ are defined in terms of right triangles for acute angles and as coordinates of a point on the unit circle for general angles (Figure 19):

$$\sin\theta = \frac{b}{c} = \frac{\text{opposite}}{\text{hypotenuse}}, \qquad \cos\theta = \frac{a}{c} = \frac{\text{adjacent}}{\text{hypotenuse}}$$

- Basic properties of sine and cosine:
  - Periodicity: $\sin(\theta + 2\pi) = \sin\theta$, $\cos(\theta + 2\pi) = \cos\theta$
  - Parity: $\sin(-\theta) = -\sin\theta$, $\cos(-\theta) = \cos\theta$
  - Basic identity: $\sin^2\theta + \cos^2\theta = 1$
- The four additional trigonometric functions:

$$\tan\theta = \frac{\sin\theta}{\cos\theta}, \qquad \cot\theta = \frac{\cos\theta}{\sin\theta}, \qquad \sec\theta = \frac{1}{\cos\theta}, \qquad \csc\theta = \frac{1}{\sin\theta}$$

1.4 EXERCISES

Preliminary Questions

1. How is it possible for two different rotations to define the same angle?

2. Give two different positive rotations that define the angle $\pi/4$.

3. Give a negative rotation that defines the angle $\pi/3$.

4. The definition of $\cos\theta$ using right triangles applies when (choose the correct answer):

(a) $0 < \theta < \dfrac{\pi}{2}$ **(b)** $0 < \theta < \pi$ **(c)** $0 < \theta < 2\pi$

5. What is the unit circle definition of $\sin\theta$?

6. How does the periodicity of $\sin\theta$ and $\cos\theta$ follow from the unit circle definition?

Exercises

1. Find the angle between 0 and 2π equivalent to $13\pi/4$.

2. Describe $\theta = \pi/6$ by an angle of negative radian measure.

3. Convert from radians to degrees:

(a) 1 **(b)** $\dfrac{\pi}{3}$ **(c)** $\dfrac{5}{12}$ **(d)** $-\dfrac{3\pi}{4}$

4. Convert from degrees to radians:

(a) 1° **(b)** 30° **(c)** 25° **(d)** 120°

5. Find the lengths of the arcs subtended by the angles θ and ϕ radians in Figure 20.

6. Calculate the values of the six standard trigonometric functions for the angle θ in Figure 21.

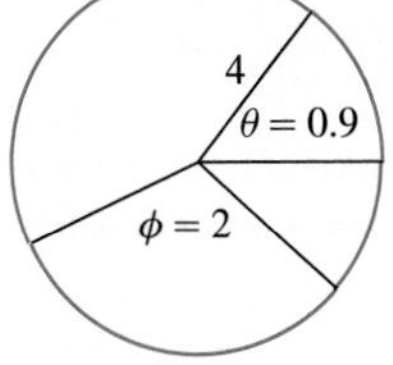

FIGURE 20 Circle of radius 4.

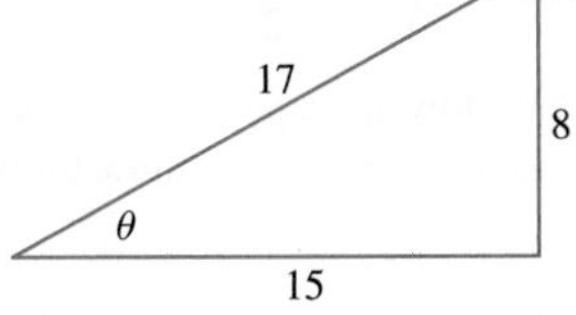

FIGURE 21

7. Fill in the remaining values of $(\cos\theta, \sin\theta)$ for the points in Figure 22.

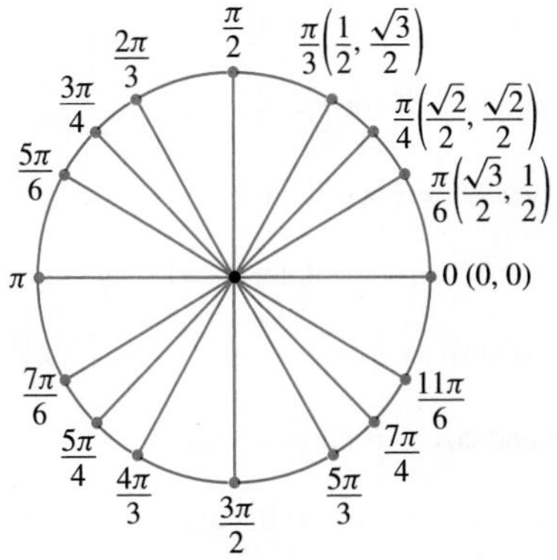

FIGURE 22

8. Find the values of the six standard trigonometric functions at $\theta = 11\pi/6$.

In Exercises 9–14, use Figure 22 to find all angles between 0 and 2π satisfying the given condition.

9. $\cos\theta = \dfrac{1}{2}$ **10.** $\tan\theta = 1$

11. $\tan\theta = -1$

12. $\csc\theta = 2$

13. $\sin x = \dfrac{\sqrt{3}}{2}$

14. $\sec t = 2$

15. Fill in the following table of values:

θ	$\frac{\pi}{6}$	$\frac{\pi}{4}$	$\frac{\pi}{3}$	$\frac{\pi}{2}$	$\frac{2\pi}{3}$	$\frac{3\pi}{4}$	$\frac{5\pi}{6}$
$\tan\theta$							
$\sec\theta$							

16. Complete the following table of signs:

θ	$\sin\theta$	$\cos\theta$	$\tan\theta$	$\cot\theta$	$\sec\theta$	$\csc\theta$
$0 < \theta < \frac{\pi}{2}$	+	+				
$\frac{\pi}{2} < \theta < \pi$						
$\pi < \theta < \frac{3\pi}{2}$						
$\frac{3\pi}{2} < \theta < 2\pi$						

17. Show that if $\tan\theta = c$ and $0 \le \theta < \pi/2$, then $\cos\theta = 1/\sqrt{1+c^2}$. *Hint:* Draw a right triangle whose opposite and adjacent sides have lengths c and 1.

18. Suppose that $\cos\theta = \frac{1}{3}$.

(a) Show that if $0 \le \theta < \pi/2$, then $\sin\theta = 2\sqrt{2}/3$ and $\tan\theta = 2\sqrt{2}$.

(b) Find $\sin\theta$ and $\tan\theta$ if $3\pi/2 \le \theta < 2\pi$.

In Exercises 19–24, assume that $0 \le \theta < \pi/2$.

19. Find $\sin\theta$ and $\tan\theta$ if $\cos\theta = \frac{5}{13}$.

20. Find $\cos\theta$ and $\tan\theta$ if $\sin\theta = \frac{3}{5}$.

21. Find $\sin\theta$, $\sec\theta$, and $\cot\theta$ if $\tan\theta = \frac{2}{7}$.

22. Find $\sin\theta$, $\cos\theta$, and $\sec\theta$ if $\cot\theta = 4$.

23. Find $\cos 2\theta$ if $\sin\theta = \frac{1}{5}$.

24. Find $\sin 2\theta$ and $\cos 2\theta$ if $\tan\theta = \sqrt{2}$.

25. Find $\cos\theta$ and $\tan\theta$ if $\sin\theta = 0.4$ and $\pi/2 \le \theta < \pi$.

26. Find $\cos\theta$ and $\sin\theta$ if $\tan\theta = 4$ and $\pi \le \theta < 3\pi/2$.

27. Find $\cos\theta$ if $\cot\theta = \frac{4}{3}$ and $\sin\theta < 0$.

28. Find $\tan\theta$ if $\sec\theta = \sqrt{5}$ and $\sin\theta < 0$.

29. Find the values of $\sin\theta$, $\cos\theta$, and $\tan\theta$ for the angles corresponding to the eight points in Figure 23(A) and (B).

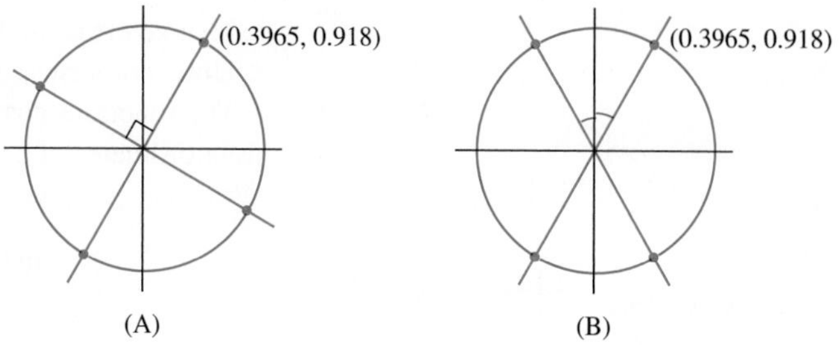

FIGURE 23

30. Refer to Figure 24(A). Express the functions $\sin\theta$, $\tan\theta$, and $\csc\theta$ in terms of c.

31. Refer to Figure 24(B). Compute $\cos\psi$, $\sin\psi$, $\cot\psi$, and $\csc\psi$.

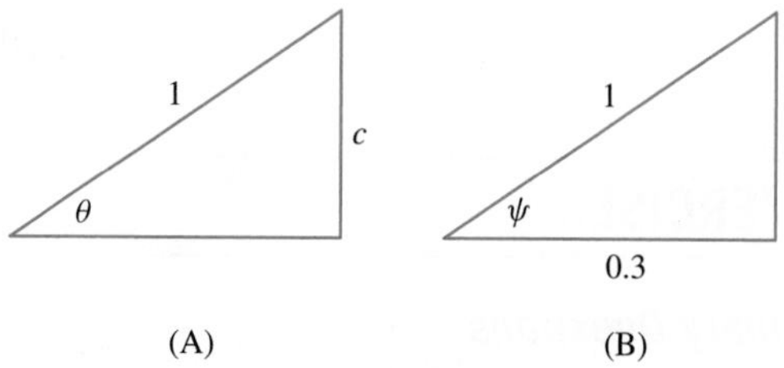

FIGURE 24

32. Express $\cos\left(\theta + \frac{\pi}{2}\right)$ and $\sin\left(\theta + \frac{\pi}{2}\right)$ in terms of $\cos\theta$ and $\sin\theta$. *Hint:* Find the relation between the coordinates (a, b) and (c, d) in Figure 25.

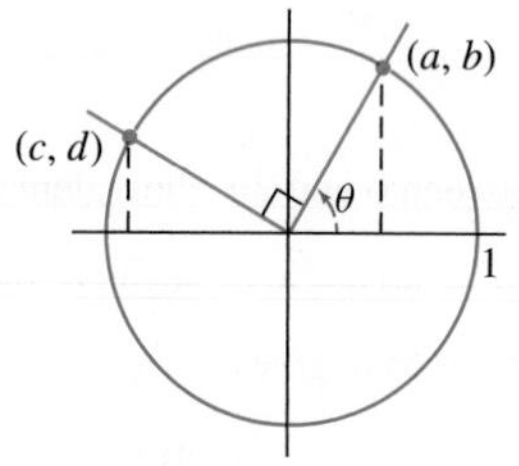

FIGURE 25

33. Use the addition formula to compute $\cos\left(\frac{\pi}{3} + \frac{\pi}{4}\right)$ exactly.

34. Use the addition formula to compute $\sin\left(\frac{\pi}{3} - \frac{\pi}{4}\right)$ exactly.

In Exercises 35–38, sketch the graph over $[0, 2\pi]$.

35. $2\sin 4\theta$

36. $\cos\left(2\left(\theta - \frac{\pi}{2}\right)\right)$

37. $\cos\left(2\theta - \frac{\pi}{2}\right)$

38. $\sin\left(2\left(\theta - \frac{\pi}{2}\right) + \pi\right) + 2$

39. How many points lie on the intersection of the horizontal line $y = c$ and the graph of $y = \sin x$ for $0 \le x < 2\pi$? *Hint:* The answer depends on c.

40. How many points lie on the intersection of the horizontal line $y = c$ and the graph of $y = \tan x$ for $0 \le x < 2\pi$?

In Exercises 41–44, solve for $0 \le \theta < 2\pi$ (see Example 4).

41. $\sin 2\theta + \sin 3\theta = 0$

42. $\sin\theta = \sin 2\theta$

43. $\cos 4\theta + \cos 2\theta = 0$

44. $\sin\theta = \cos 2\theta$

In Exercises 45–54, derive the identity using the identities listed in this section.

45. $\cos 2\theta = 2\cos^2\theta - 1$

46. $\cos^2\dfrac{\theta}{2} = \dfrac{1+\cos\theta}{2}$

47. $\sin\dfrac{\theta}{2} = \sqrt{\dfrac{1-\cos\theta}{2}}$

48. $\sin(\theta+\pi) = -\sin\theta$

49. $\cos(\theta+\pi) = -\cos\theta$

50. $\tan x = \cot\left(\dfrac{\pi}{2} - x\right)$

51. $\tan(\pi - \theta) = -\tan\theta$

52. $\tan 2x = \dfrac{2\tan x}{1-\tan^2 x}$

53. $\tan x = \dfrac{\sin 2x}{1+\cos 2x}$

54. $\sin^2 x\cos^2 x = \dfrac{1-\cos 4x}{8}$

55. Use Exercises 48 and 49 to show that $\tan\theta$ and $\cot\theta$ are periodic with period π.

56. Use trigonometric identities to compute $\cos\frac{\pi}{15}$, noting that $\frac{\pi}{15} = \frac{1}{2}\left(\frac{\pi}{3} - \frac{\pi}{5}\right)$.

57. Use the Law of Cosines to find the distance from P to Q in Figure 26.

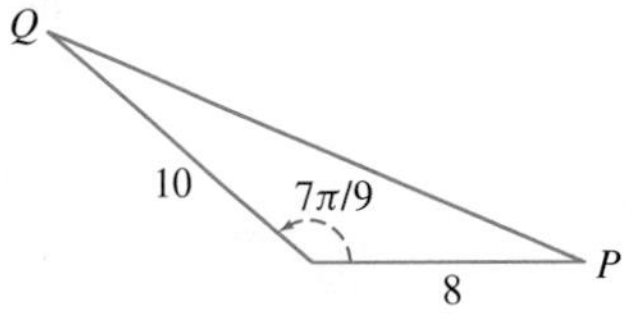

FIGURE 26

Further Insights and Challenges

58. Use Figure 27 to derive the Law of Cosines from the Pythagorean Theorem.

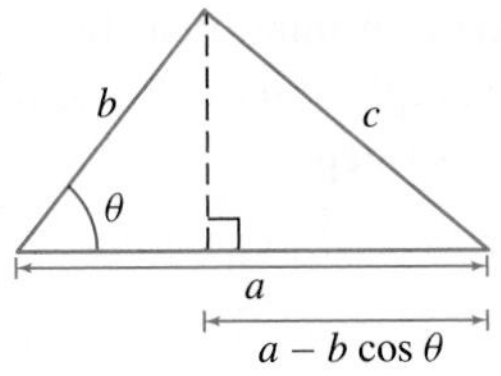

FIGURE 27

59. Use the addition formula to prove

$$\cos 3\theta = 4\cos^3\theta - 3\cos\theta$$

60. Use the addition formulas for sine and cosine to prove

$$\tan(a+b) = \frac{\tan a + \tan b}{1 - \tan a \tan b}$$

$$\cot(a-b) = \frac{\cot a \cot b + 1}{\cot b - \cot a}$$

61. Let θ be the angle between the line $y = mx + b$ and the x-axis [Figure 28(A)]. Prove that $m = \tan\theta$.

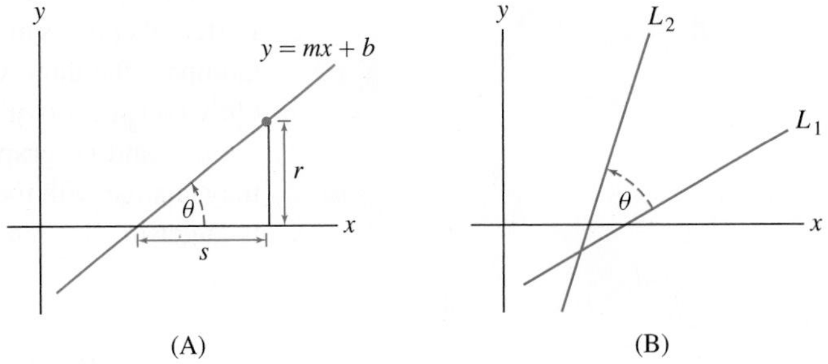

FIGURE 28

62. Let L_1 and L_2 be the lines of slope m_1 and m_2 [Figure 28(B)]. Show that the angle θ between L_1 and L_2 satisfies $\cot\theta = \dfrac{m_2 m_1 + 1}{m_2 - m_1}$.

63. Perpendicular Lines Use Exercise 62 to prove that two lines with nonzero slopes m_1 and m_2 are perpendicular if and only if $m_2 = -1/m_1$.

64. Apply the double-angle formula to prove:

(a) $\cos\dfrac{\pi}{8} = \dfrac{1}{2}\sqrt{2+\sqrt{2}}$

(b) $\cos\dfrac{\pi}{16} = \dfrac{1}{2}\sqrt{2+\sqrt{2+\sqrt{2}}}$

Guess the values of $\cos\dfrac{\pi}{32}$ and of $\cos\dfrac{\pi}{2^n}$ for all n.

1.5 Technology: Calculators and Computers

Computer technology has vastly extended our ability to calculate and visualize mathematical relationships. In applied settings, computers are indispensable for solving complex systems of equations and analyzing data, as in weather prediction and medical imaging. Mathematicians use computers to study complex structures such as the Mandelbrot Set (Figures 1 and 2). We take advantage of this technology to explore the ideas of calculus visually and numerically.

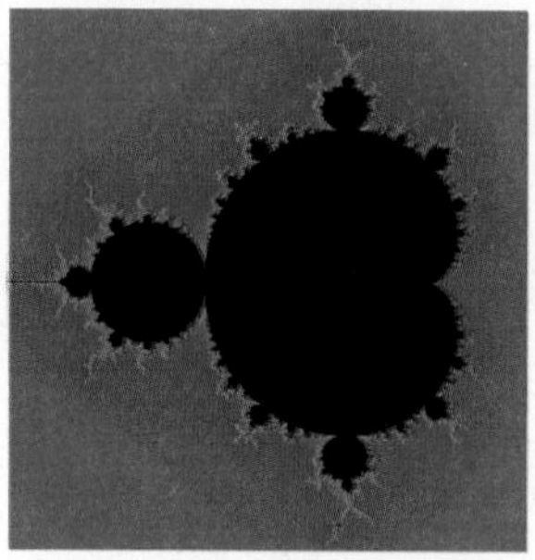

FIGURE 1 Computer-generated image of the Mandelbrot Set, which occurs in the mathematical theory of chaos and fractals.

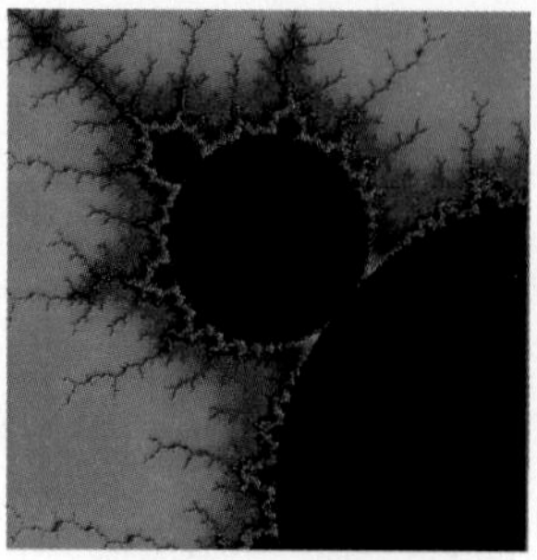

FIGURE 2 Even greater complexity is revealed when we zoom in on a portion of the Mandelbrot Set.

When we plot a function with a graphing calculator or computer algebra system, the graph is contained within a **viewing rectangle**, the region determined by the range of x- and y-values in the plot. We write $[a, b] \times [c, d]$ to denote the rectangle where $a \le x \le b$ and $c \le y \le d$.

The appearance of the graph depends heavily on the choice of viewing rectangle. Different choices may convey very different impressions which are sometimes misleading. Compare the three viewing rectangles for the graph of $f(x) = 12 - x - x^2$ in Figure 3. Only (A) successfully displays the shape of the graph as a parabola. In (B), the graph is cut off, and no graph at all appears in (C). Keep in mind that the scales along the axes may change with the viewing rectangle. For example, the unit increment along the y-axis is larger in (B) than in (A), so the graph in (B) is steeper.

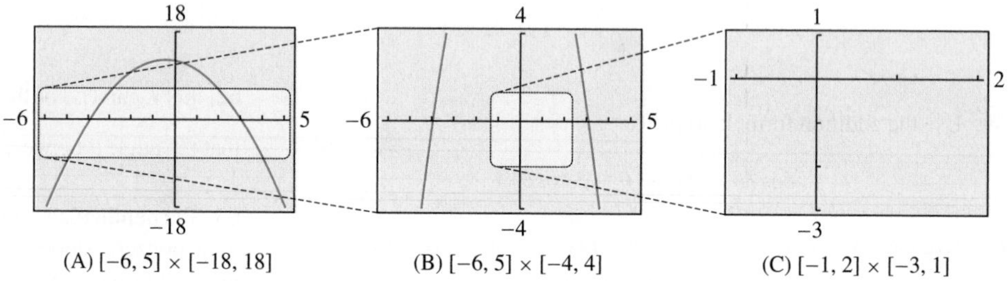

FIGURE 3 Viewing rectangles for the graph of $f(x) = 12 - x - x^2$.

There is no single "correct" viewing rectangle. The goal is to select the viewing rectangle that displays the properties you wish to investigate. This usually requires experimentation.

Technology is indispensable but also has its limitations. When shown the computer-generated results of a complex calculation, the Nobel prize–winning physicist Eugene Wigner (1902–1995) is reported to have said: It is nice to know that the computer understands the problem, but I would like to understand it too.

■ **EXAMPLE 1** **How Many Roots and Where?** How many real roots does the function $f(x) = x^9 - 20x + 1$ have? Find their approximate locations.

Solution We experiment with several viewing rectangles (Figure 4). Our first attempt (A) displays a cut-off graph, so we try a viewing rectangle that includes a larger range of y-values. Plot (B) shows that the roots of $f(x)$ lie somewhere in the interval $[-3, 3]$, but it does not reveal how many real roots there are. Therefore, we try the viewing rectangle in (C). Now we can see clearly that $f(x)$ has three roots. A further zoom in (D) shows that these roots are located near -1.5, 0.1, and 1.5. Further zooming would provide their locations with greater accuracy. ■

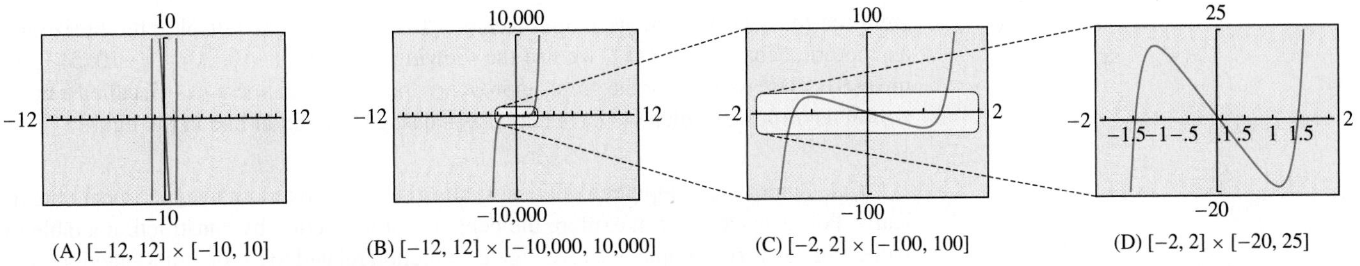

FIGURE 4 Graphs of $f(x) = x^9 - 20x + 1$.

■ **EXAMPLE 2 Does a Solution Exist?** Does $\cos x = \tan x$ have a solution? Describe the set of all solutions.

Solution The solutions of $\cos x = \tan x$ are the x-coordinates of the points where the graphs of $y = \cos x$ and $y = \tan x$ intersect. Figure 5(A) shows that there are two solutions in the interval $[0, 2\pi]$. By zooming in on the graph as in (B), we see that the first positive root lies between 0.6 and 0.7 and the second positive root lies between 2.4 and 2.5. Further zooming shows that the first root is approximately 0.67 [Figure 5(C)]. Continuing this process, we find that the first two roots are $x \approx 0.666$ and $x \approx 2.475$.

Since $\cos x$ and $\tan x$ are periodic, the picture repeats itself with period 2π. All solutions are obtained by adding multiples of 2π to the two solutions in $[0, 2\pi]$:

$$x \approx 0.666 + 2\pi k \qquad \text{and} \qquad x \approx 2.475 + 2\pi k \quad \text{(for any integer } k) \qquad ■$$

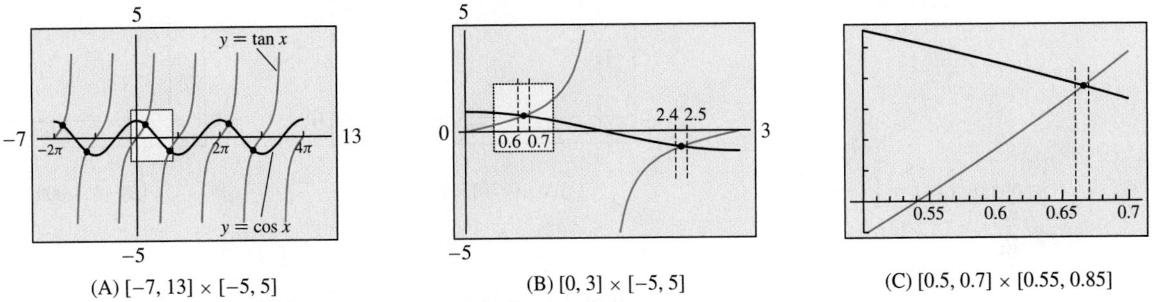

FIGURE 5 Graphs of $y = \cos x$ and $y = \tan x$.

■ **EXAMPLE 3 Functions with Asymptotes** Plot the function $f(x) = \dfrac{1 - 3x}{x - 2}$ and describe its asymptotic behavior.

Solution First, we plot $f(x)$ in the viewing rectangle $[-10, 20] \times [-5, 5]$ as in Figure 6(A). The vertical line $x = 2$ is called a **vertical asymptote**. Many graphing calculators display this line, but it is *not* part of the graph (and it can usually be eliminated by choosing a smaller range of y-values). We see that $f(x)$ tends to ∞ as x approaches 2

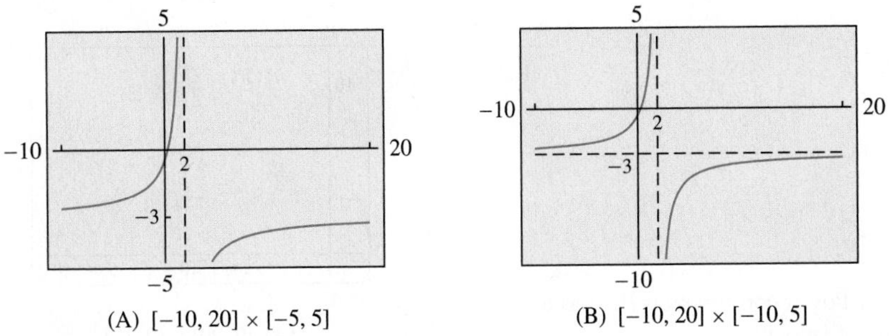

FIGURE 6 Graphs of $f(x) = \dfrac{1 - 3x}{x - 2}$.

from the left, and to $-\infty$ as x approaches 2 from the right. To display the horizontal asymptotic behavior of $f(x)$, we use the viewing rectangle $[-10, 20] \times [-10, 5]$ [Figure 6(B)]. Here we see that the graph approaches the horizontal line $y = -3$, called a **horizontal asymptote** (which we have added as a dashed horizontal line in the figure). ■

Calculators and computer algebra systems give us the freedom to experiment numerically. For instance, we can explore the behavior of a function by constructing a table of values. In the next example, we investigate a function related to exponential functions and compound interest (see Section 7.5).

■ **EXAMPLE 4 Investigating the Behavior of a Function** How does $f(n) = (1 + 1/n)^n$ behave for large whole-number values of n? Does $f(n)$ tend to infinity as n gets larger?

Solution First, we make a table of values of $f(n)$ for larger and larger values of n. Table 1 suggests that $f(n)$ does not tend to infinity. Rather, as n grows larger, $f(n)$ appears to get closer to some value near 2.718. This is an example of limiting behavior that we will discuss in Chapter 2. Next, replace n by the variable x and plot the function $f(x) = (1 + 1/x)^x$. The graphs in Figure 7 confirm that $f(x)$ approaches a limit of approximately 2.7. We will prove that $f(n)$ approaches the number e as n tends to infinity in Section 7.1. ■

TABLE 1

n	$\left(1+\frac{1}{n}\right)^n$
10	2.59374
10^2	2.70481
10^3	2.71692
10^4	2.71815
10^5	2.71827
10^6	2.71828

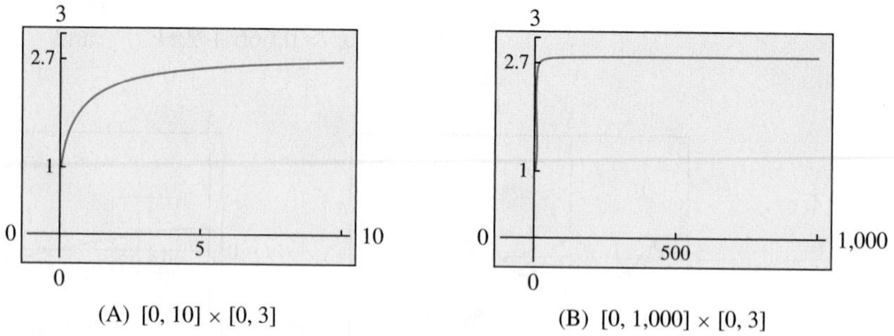

FIGURE 7 Graphs of $f(x) = \left(1+\frac{1}{x}\right)^x$.

■ **EXAMPLE 5 Bird Flight: Finding a Minimum Graphically** According to one model of bird flight, the power consumed by a pigeon flying at velocity v (in meters per second) is $P(v) = 17v^{-1} + 10^{-3}v^3$ (in joules per second). Use a graph of $P(v)$ to find the velocity that minimizes power consumption.

Solution The velocity that minimizes power consumption corresponds to the lowest point on the graph of $P(v)$. We plot $P(v)$ first in a large viewing rectangle (Figure 8). This figure reveals the general shape of the graph and shows that $P(v)$ takes on a minimum value for v somewhere between $v = 8$ and $v = 9$. In the viewing rectangle $[8, 9.2] \times [2.6, 2.65]$, we see that the minimum occurs at approximately $v = 8.65$ m/s. ■

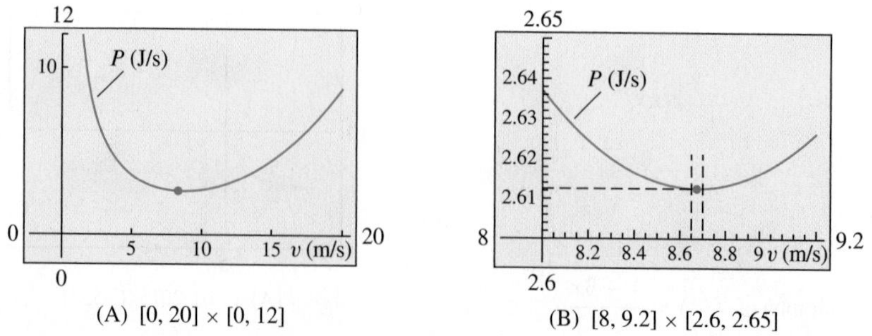

FIGURE 8 Power consumption $P(v)$ as a function of velocity v.

Local linearity is an important concept in calculus that is based on the idea that many functions are *nearly linear* over small intervals. Local linearity can be illustrated effectively with a graphing calculator.

■ **EXAMPLE 6 Illustrating Local Linearity** Illustrate local linearity for the function $f(x) = x^{\sin x}$ at $x = 1$.

Solution First, we plot $f(x) = x^{\sin x}$ in the viewing window of Figure 9(A). The graph moves up and down and appears very wavy. However, as we zoom in, the graph straightens out. Figures (B)–(D) show the result of zooming in on the point $(1, f(1))$. When viewed up close, the graph looks like a straight line. This illustrates the local linearity of $f(x)$ at $x = 1$. ■

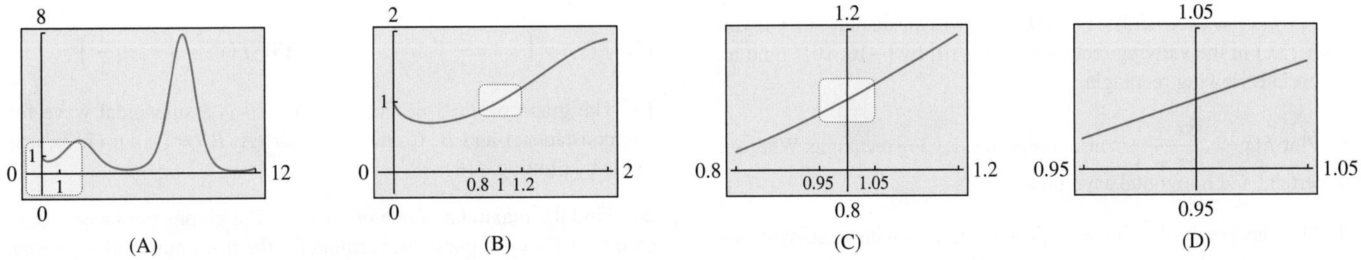

FIGURE 9 Zooming in on the graph of $f(x) = x^{\sin x}$ near $x = 1$.

1.5 SUMMARY

- The appearance of a graph on a graphing calculator depends on the choice of viewing rectangle. Experiment with different viewing rectangles until you find one that displays the information you want. Keep in mind that the scales along the axes may change as you vary the viewing rectangle.
- The following are some ways in which graphing calculators and computer algebra systems can be used in calculus:
  - Visualizing the behavior of a function
  - Finding solutions graphically or numerically
  - Conducting numerical or graphical experiments
  - Illustrating theoretical ideas (such as local linearity)

1.5 EXERCISES

Preliminary Questions

1. Is there a definite way of choosing the optimal viewing rectangle, or is it best to experiment until you find a viewing rectangle appropriate to the problem at hand?

2. Describe the calculator screen produced when the function $y = 3 + x^2$ is plotted with viewing rectangle:

(a) $[-1, 1] \times [0, 2]$ **(b)** $[0, 1] \times [0, 4]$

3. According to the evidence in Example 4, it appears that $f(n) = (1 + 1/n)^n$ never takes on a value greater than 3 for $n > 0$. Does this evidence *prove* that $f(n) \le 3$ for $n > 0$?

4. How can a graphing calculator be used to find the minimum value of a function?

Exercises

The exercises in this section should be done using a graphing calculator or computer algebra system.

1. Plot $f(x) = 2x^4 + 3x^3 - 14x^2 - 9x + 18$ in the appropriate viewing rectangles and determine its roots.

2. How many solutions does $x^3 - 4x + 8 = 0$ have?

3. How many *positive* solutions does $x^3 - 12x + 8 = 0$ have?

4. Does $\cos x + x = 0$ have a solution? A positive solution?

5. Find all the solutions of $\sin x = \sqrt{x}$ for $x > 0$.

6. How many solutions does $\cos x = x^2$ have?

7. Let $f(x) = (x - 100)^2 + 1000$. What will the display show if you graph $f(x)$ in the viewing rectangle $[-10, 10]$ by $[-10, 10]$? Find an appropriate viewing rectangle.

8. Plot $f(x) = \dfrac{8x+1}{8x-4}$ in an appropriate viewing rectangle. What are the vertical and horizontal asymptotes?

9. Plot the graph of $f(x) = x/(4 - x)$ in a viewing rectangle that clearly displays the vertical and horizontal asymptotes.

10. Illustrate local linearity for $f(x) = x^2$ by zooming in on the graph at $x = 0.5$ (see Example 6).

11. Plot $f(x) = \cos(x^2) \sin x$ for $0 \le x \le 2\pi$. Then illustrate local linearity at $x = 3.8$ by choosing appropriate viewing rectangles.

12. If P_0 dollars are deposited in a bank account paying 5% interest compounded monthly, then the account has value $P_0\left(1 + \frac{0.05}{12}\right)^N$ after N months. Find, to the nearest integer N, the number of months after which the account value doubles.

In Exercises 13–18, investigate the behavior of the function as n or x grows large by making a table of function values and plotting a graph (see Example 4). Describe the behavior in words.

13. $f(n) = n^{1/n}$

14. $f(n) = \dfrac{4n+1}{6n-5}$

15. $f(n) = \left(1 + \dfrac{1}{n}\right)^{n^2}$

16. $f(x) = \left(\dfrac{x+6}{x-4}\right)^x$

17. $f(x) = \left(x \tan \dfrac{1}{x}\right)^x$

18. $f(x) = \left(x \tan \dfrac{1}{x}\right)^{x^2}$

19. The graph of $f(\theta) = A\cos\theta + B\sin\theta$ is a sinusoidal wave for any constants A and B. Confirm this for $(A, B) = (1, 1)$, $(1, 2)$, and $(3, 4)$ by plotting $f(\theta)$.

20. Find the maximum value of $f(\theta)$ for the graphs produced in Exercise 19. Can you guess the formula for the maximum value in terms of A and B?

21. Find the intervals on which $f(x) = x(x + 2)(x - 3)$ is positive by plotting a graph.

22. Find the set of solutions to the inequality $(x^2 - 4)(x^2 - 1) < 0$ by plotting a graph.

Further Insights and Challenges

23. CAS Let $f_1(x) = x$ and define a sequence of functions by $f_{n+1}(x) = \frac{1}{2}(f_n(x) + x/f_n(x))$. For example, $f_2(x) = \frac{1}{2}(x + 1)$. Use a computer algebra system to compute $f_n(x)$ for $n = 3, 4, 5$ and plot $f_n(x)$ together with $\sqrt{x}$ for $x \ge 0$. What do you notice?

24. Set $P_0(x) = 1$ and $P_1(x) = x$. The **Chebyshev polynomials** (useful in approximation theory) are defined inductively by the formula $P_{n+1}(x) = 2xP_n(x) - P_{n-1}(x)$.

(a) Show that $P_2(x) = 2x^2 - 1$.

(b) Compute $P_n(x)$ for $3 \le n \le 6$ using a computer algebra system or by hand, and plot $P_n(x)$ over $[-1, 1]$.

(c) Check that your plots confirm two interesting properties: (a) $P_n(x)$ has n real roots in $[-1, 1]$ and (b) for $x \in [-1, 1]$, $P_n(x)$ lies between -1 and 1.

CHAPTER REVIEW EXERCISES

1. Express $(4, 10)$ as a set $\{x : |x - a| < c\}$ for suitable a and c.

2. Express as an interval:

(a) $\{x : |x - 5| < 4\}$

(b) $\{x : |5x + 3| \le 2\}$

3. Express $\{x : 2 \le |x - 1| \le 6\}$ as a union of two intervals.

4. Give an example of numbers x, y such that $|x| + |y| = x - y$.

5. Describe the pairs of numbers x, y such that $|x + y| = x - y$.

6. Sketch the graph of $y = f(x + 2) - 1$, where $f(x) = x^2$ for $-2 \le x \le 2$.

In Exercises 7–10, let $f(x)$ be the function shown in Figure 1.

7. Sketch the graphs of $y = f(x) + 2$ and $y = f(x + 2)$.

8. Sketch the graphs of $y = \frac{1}{2}f(x)$ and $y = f\left(\frac{1}{2}x\right)$.

9. Continue the graph of $f(x)$ to the interval $[-4, 4]$ as an even function.

10. Continue the graph of $f(x)$ to the interval $[-4, 4]$ as an odd function.

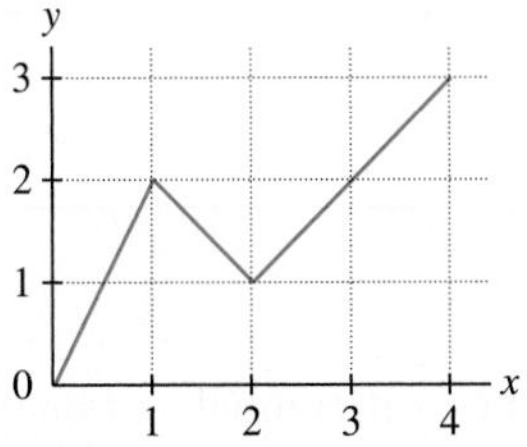

FIGURE 1

In Exercises 11–14, find the domain and range of the function.

11. $f(x) = \sqrt{x+1}$

12. $f(x) = \dfrac{4}{x^4+1}$

13. $f(x) = \dfrac{2}{3-x}$

14. $f(x) = \sqrt{x^2 - x + 5}$

15. Determine whether the function is increasing, decreasing, or neither:

(a) $f(x) = 3^{-x}$

(b) $f(x) = \dfrac{1}{x^2+1}$

(c) $g(t) = t^2 + t$

(d) $g(t) = t^3 + t$

16. Determine whether the function is even, odd, or neither:

(a) $f(x) = x^4 - 3x^2$

(b) $g(x) = \sin(x+1)$

(c) $f(x) = 2^{-x^2}$

In Exercises 17–22, find the equation of the line.

17. Line passing through $(-1, 4)$ and $(2, 6)$

18. Line passing through $(-1, 4)$ and $(-1, 6)$

19. Line of slope 6 through $(9, 1)$

20. Line of slope $-\frac{3}{2}$ through $(4, -12)$

21. Line through $(2, 3)$ parallel to $y = 4 - x$

22. Horizontal line through $(-3, 5)$

23. Does the following table of market data suggest a linear relationship between price and number of homes sold during a one-year period? Explain.

Price (thousands of \$)	180	195	220	240
No. of homes sold	127	118	103	91

24. Does the following table of revenue data for a computer manufacturer suggest a linear relation between revenue and time? Explain.

Year	2001	2005	2007	2010
Revenue (billions of \$)	13	18	15	11

25. Find the roots of $f(x) = x^4 - 4x^2$ and sketch its graph. On which intervals is $f(x)$ decreasing?

26. Let $h(z) = 2z^2 + 12z + 3$. Complete the square and find the minimum value of $h(z)$.

27. Let $f(x)$ be the square of the distance from the point $(2, 1)$ to a point $(x, 3x+2)$ on the line $y = 3x + 2$. Show that $f(x)$ is a quadratic function, and find its minimum value by completing the square.

28. Prove that $x^2 + 3x + 3 \geq 0$ for all x.

In Exercises 29–34, sketch the graph by hand.

29. $y = t^4$

30. $y = t^5$

31. $y = \sin\dfrac{\theta}{2}$

32. $y = 10^{-x}$

33. $y = x^{1/3}$

34. $y = \dfrac{1}{x^2}$

35. Show that the graph of $y = f\left(\frac{1}{3}x - b\right)$ is obtained by shifting the graph of $y = f\left(\frac{1}{3}x\right)$ to the right $3b$ units. Use this observation to sketch the graph of $y = \left|\frac{1}{3}x - 4\right|$.

36. Let $h(x) = \cos x$ and $g(x) = x^{-1}$. Compute the composite functions $h(g(x))$ and $g(h(x))$, and find their domains.

37. Find functions f and g such that the function

$$f(g(t)) = (12t + 9)^4$$

38. Sketch the points on the unit circle corresponding to the following three angles, and find the values of the six standard trigonometric functions at each angle:

(a) $\dfrac{2\pi}{3}$

(b) $\dfrac{7\pi}{4}$

(c) $\dfrac{7\pi}{6}$

39. What is the period of the function $g(\theta) = \sin 2\theta + \sin\frac{\theta}{2}$?

40. Assume that $\sin\theta = \frac{4}{5}$, where $\pi/2 < \theta < \pi$. Find:

(a) $\tan\theta$

(b) $\sin 2\theta$

(c) $\csc\dfrac{\theta}{2}$

41. Give an example of values a, b such that

(a) $\cos(a+b) \neq \cos a + \cos b$

(b) $\cos\dfrac{a}{2} \neq \dfrac{\cos a}{2}$

42. Let $f(x) = \cos x$. Sketch the graph of $y = 2f\left(\frac{1}{3}x - \frac{\pi}{4}\right)$ for $0 \leq x \leq 6\pi$.

43. Solve $\sin 2x + \cos x = 0$ for $0 \leq x < 2\pi$.

44. How does $h(n) = n^2/2^n$ behave for large whole-number values of n? Does $h(n)$ tend to infinity?

45. GU Use a graphing calculator to determine whether the equation $\cos x = 5x^2 - 8x^4$ has any solutions.

46. GU Using a graphing calculator, find the number of real roots and estimate the largest root to two decimal places:

(a) $f(x) = 1.8x^4 - x^5 - x$

(b) $g(x) = 1.7x^4 - x^5 - x$

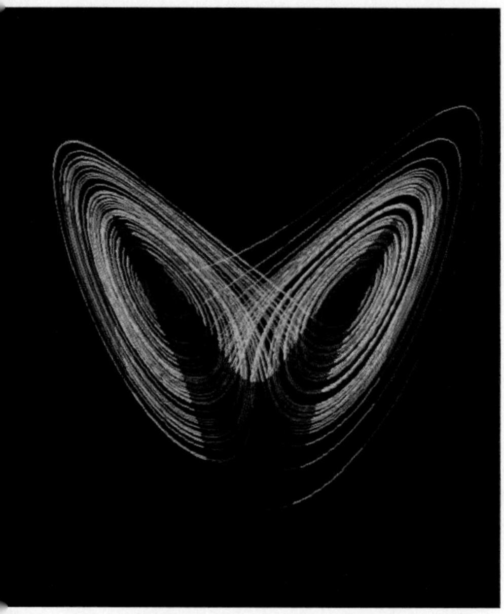

This "strange attractor" represents limit behavior that appeared first in weather models studied by meteorologist E. Lorenz in 1963.

2 LIMITS

Calculus is usually divided into two branches, differential and integral, partly for historical reasons. The subject grew out of efforts in the seventeenth century to solve two important geometric problems: finding tangent lines to curves (differential calculus) and computing areas under curves (integral calculus). However, calculus is a broad subject with no clear boundaries. It includes other topics, such as the theory of infinite series, and it has an extraordinarily wide range of applications. What makes these methods and applications part of calculus is that they all rely on the concept of a limit. We will see throughout the text how limits allow us to make computations and solve problems that cannot be solved using algebra alone.

This chapter introduces the limit concept and sets the stage for our study of the derivative in Chapter 3. The first section, intended as motivation, discusses how limits arise in the study of rates of change and tangent lines.

2.1 Limits, Rates of Change, and Tangent Lines

Rates of change play a role whenever we study the relationship between two changing quantities. Velocity is a familiar example (the rate of change of position with respect to time), but there are many others, such as

- The infection rate of an epidemic (*newly infected individuals per month*)
- Inflation rate (*change in consumer price index per year*)
- Rate of change of atmospheric temperature with respect to altitude

Roughly speaking, if y and x are related quantities, the rate of change should tell us how much y changes in response to a unit change in x. For example, if an automobile travels at a velocity of 80 km/hr, then its position changes by 80 km for each unit change in time (the unit being 1 hour). If the trip lasts only half an hour, its position changes by 40 km, and in general, the change in position is $80t$ km, where t is the change in time (that is, the time elapsed in hours). In other words,

$$\text{Change in position} = \text{velocity} \times \text{change in time}$$

However, this simple formula is not valid or even meaningful if the velocity is not constant. After all, if the automobile is accelerating or decelerating, which velocity would we use in the formula?

The problem of extending this formula to account for changing velocity lies at the heart of calculus. As we will learn, differential calculus uses the limit concept to define *instantaneous velocity*, and integral calculus enables us to compute the change in position in terms of instantaneous velocity. But these ideas are very general. They apply to all rates of change, making calculus an indispensable tool for modeling an amazing range of real-world phenomena.

In this section, we discuss velocity and other rates of change, emphasizing their graphical interpretation in terms of *tangent lines*. Although at this stage, we cannot define precisely what a tangent line is—this will have to wait until Chapter 3—you can think of a tangent line as a line that *skims* a curve at a point, as in Figures 1(A) and (B) but not (C).

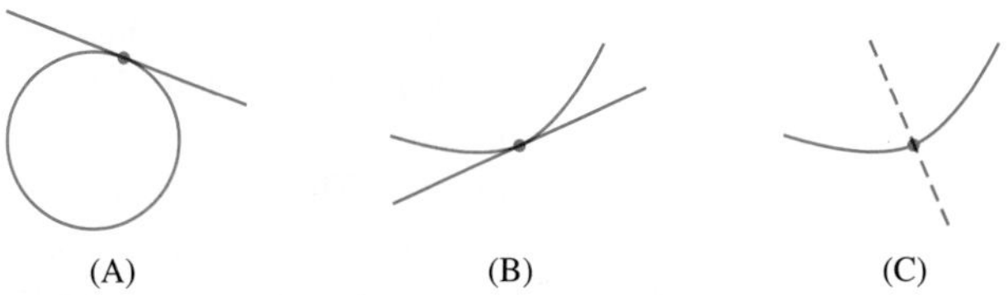

FIGURE 1 The line is tangent in (A) and (B) but not in (C).

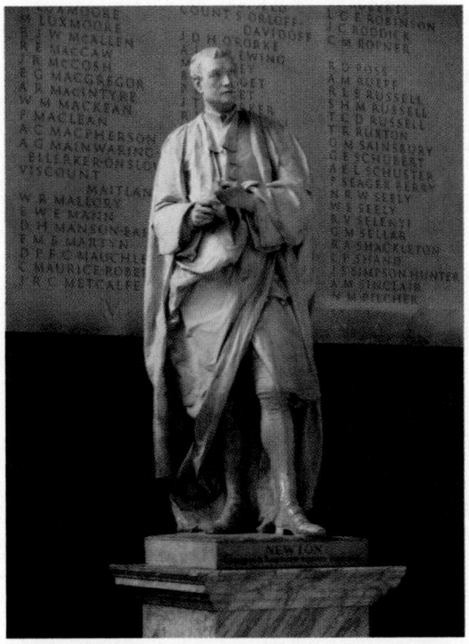

This statue of Isaac Newton in Cambridge University was described in *The Prelude*, a poem by William Wordsworth (1770–1850):

"Newton with his prism and silent face,
The marble index of a mind for ever
Voyaging through strange seas of Thought,
alone."

HISTORICAL PERSPECTIVE

Philosophy is written in this grand book—I mean the universe—which stands continually open to our gaze, but it cannot be understood unless one first learns to comprehend the language ... in which it is written. It is written in the language of mathematics ...

—Galileo Galilei, 1623

The scientific revolution of the sixteenth and seventeenth centuries reached its high point in the work of Isaac Newton (1643–1727), who was the first scientist to show that the physical world, despite its complexity and diversity, is governed by a small number of universal laws. One of Newton's great insights was that the universal laws are dynamical, describing how the world changes over time in response to forces, rather than how the world actually is at any given moment in time. These laws are expressed best in the language of calculus, which is the mathematics of change.

More than 50 years before the work of Newton, the astronomer Johannes Kepler (1571–1630) discovered his three laws of planetary motion, the most famous of which states that the path of a planet around the sun is an ellipse. Kepler arrived at these laws through a painstaking analysis of astronomical data, but he could not explain why they were true. According to Newton, the motion of any object—planet or pebble—is determined by the forces acting on it. The planets, if left undisturbed, would travel in straight lines. Since their paths are elliptical, some force—in this case, the gravitational force of the sun—must be acting to make them change direction continuously. In his magnum opus *Principia Mathematica*, published in 1687, Newton proved that Kepler's laws follow from Newton's own universal laws of motion and gravity.

For these discoveries, Newton gained widespread fame in his lifetime. His fame continued to increase after his death, assuming a nearly mythic dimension and his ideas had a profound influence, not only in science but also in the arts and literature, as expressed in the epitaph by British poet Alexander Pope: "Nature and Nature's Laws lay hid in Night. God said, *Let Newton be!* and all was Light."

Velocity

In linear motion, velocity may be positive or negative (indicating the direction of motion). Speed, by definition, is the absolute value of velocity and is always positive.

When we speak of velocity, we usually mean *instantaneous* velocity, which indicates the speed and direction of an object at a particular moment. The idea of instantaneous velocity makes intuitive sense, but care is required to define it precisely.

Consider an object traveling in a straight line (linear motion). The **average velocity** over a given time interval has a straightforward definition as the ratio

$$\text{Average velocity} = \frac{\text{change in position}}{\text{length of time interval}}$$

For example, if an automobile travels 200 km in 4 hours, then its average velocity during this 4-hour period is $\frac{200}{4} = 50$ km/h. At any given moment the automobile may be going faster or slower than the average.

We cannot define instantaneous velocity as a ratio because we would have to divide by the length of the time interval (which is zero). However, we should be able to estimate instantaneous velocity by computing average velocity over successively smaller time intervals. The guiding principle is: *Average velocity over a very small time interval is very close to instantaneous velocity.* To explore this idea further, we introduce some notation.

The Greek letter Δ (Delta) is commonly used to denote the *change* in a function or variable. If $s(t)$ is the position of an object (distance from the origin) at time t and $[t_0, t_1]$ is a time interval, we set

$$\Delta s = s(t_1) - s(t_0) = \text{change in position}$$

$$\Delta t = \quad t_1 - t_0 \quad = \text{change in time (length of time interval)}$$

The change in position Δs is also called the **displacement**, or **net change** in position. For $t_1 \neq t_0$,

$$\text{Average velocity over } [t_0, t_1] = \frac{\Delta s}{\Delta t} = \frac{s(t_1) - s(t_0)}{t_1 - t_0}$$

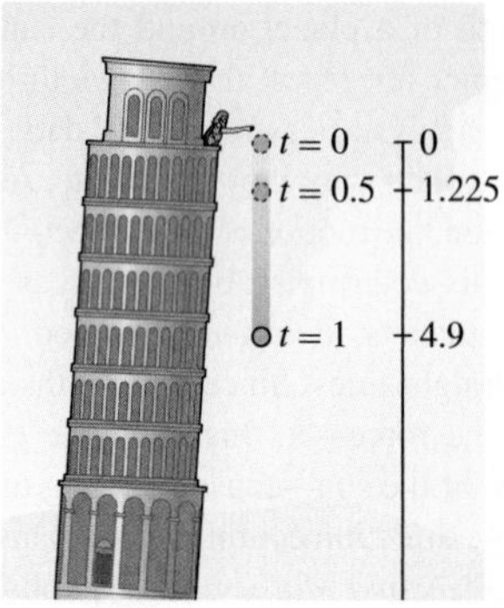

FIGURE 2 Distance traveled by a falling object after t seconds is $s(t) = 4.9t^2$ meters.

One motion we will study is the motion of an object falling to earth under the influence of gravity (assuming no air resistance). Galileo discovered that if the object is released at time $t = 0$ from a state of rest (Figure 2), then the distance traveled after t seconds is given by the formula

$$s(t) = 4.9t^2 \text{ m}$$

EXAMPLE 1 A stone, released from a state of rest, falls to earth. Estimate the instantaneous velocity at $t = 0.8$ s.

Solution We use Galileo's formula $s(t) = 4.9t^2$ to compute the average velocity over the five short time intervals listed in Table 1. Consider the first interval $[t_0, t_1] = [0.8, 0.81]$:

TABLE 1

Time intervals	Average velocity
[0.8, 0.81]	7.889
[0.8, 0.805]	7.8645
[0.8, 0.8001]	7.8405
[0.8, 0.80005]	7.84024
[0.8, 0.800001]	7.840005

$$\Delta s = s(0.81) - s(0.8) = 4.9(0.81)^2 - 4.9(0.8)^2 \approx 3.2149 - 3.1360 = 0.7889 \text{ m}$$

$$\Delta t = 0.81 - 0.8 = 0.01 \text{ s}$$

The average velocity over [0.8, 0.81] is the ratio

$$\frac{\Delta s}{\Delta t} = \frac{s(0.81) - s(0.8)}{0.81 - 0.8} = \frac{0.07889}{0.01} = 7.889 \text{ m/s}$$

There is nothing special about the particular time intervals in Table 1. We are looking for a trend, and we could have chosen any intervals $[0.8, t]$ *for values of* t *approaching* 0.8. *We could also have chosen intervals* $[t, 0.8]$ *for* $t < 0.8$.

Table 1 shows the results of similar calculations for intervals of successively shorter lengths. It looks like these average velocities are getting closer to 7.84 m/s as the length of the time interval shrinks:

7.889, 7.8645, **7.840**5, **7.840**24, **7.840**005

This suggests that 7.84 m/s is a good candidate for the instantaneous velocity at $t = 0.8$. ■

We express our conclusion in the previous example by saying that *average velocity* **converges** *to instantaneous velocity* or that *instantaneous velocity is the* **limit** *of average velocity* as the length of the time interval shrinks to zero.

Graphical Interpretation of Velocity

The idea that average velocity converges to instantaneous velocity as we shorten the time interval has a vivid interpretation in terms of secant lines. The term **secant line** refers to a line through two points on a curve.

Consider the graph of position $s(t)$ for an object traveling in a straight line (Figure 3). The ratio defining average velocity over $[t_0, t_1]$ is nothing more than the slope of the secant line through the points $(t_0, s(t_0))$ and $(t_1, s(t_1))$. For $t_1 \neq t_0$,

$$\text{Average velocity} = \text{slope of secant line} = \frac{\Delta s}{\Delta t} = \frac{s(t_1) - s(t_0)}{t_1 - t_0}$$

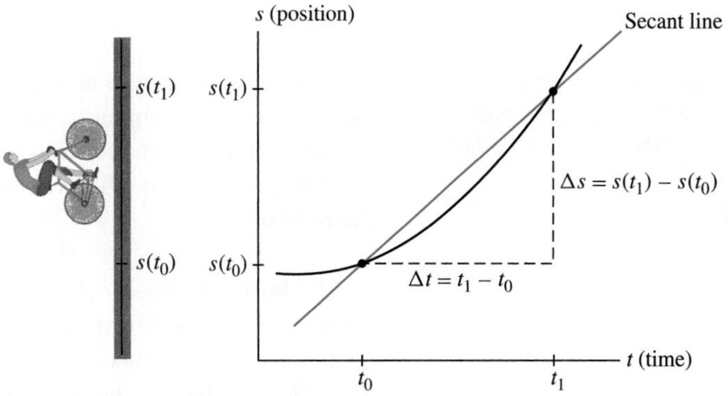

FIGURE 3 The average velocity over $[t_0, t_1]$ is equal to the slope of the secant line.

By interpreting average velocity as a slope, we can visualize what happens as the time interval gets smaller. Figure 4 shows the graph of position for the falling stone of Example 1, where $s(t) = 4.9t^2$. As the time interval shrinks, the secant lines get closer to—*and seem to rotate into*—the tangent line at $t = 0.8$.

Time interval	Average velocity
[0.8, 0.805]	7.8645
[0.8, 0.8001]	7.8405
[0.8, 0.80005]	7.84024

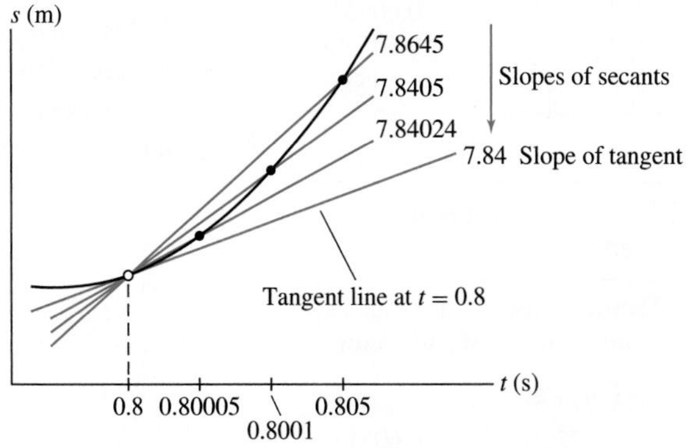

FIGURE 4 The secant lines "rotate into" the tangent line as the time interval shrinks. *Note:* The graph is not drawn to scale.

And since the secant lines approach the tangent line, the slopes of the secant lines get closer and closer to the slope of the tangent line. In other words, the statement

> As the time interval shrinks to zero, the average velocity approaches the instantaneous velocity.

has the graphical interpretation

> As the time interval shrinks to zero, the slope of the secant line approaches the slope of the tangent line.

We conclude that *instantaneous velocity is equal to the slope of the tangent line to the graph of position as a function of time*. This conclusion and its generalization to other rates of change are of fundamental importance in differential calculus.

Other Rates of Change

Velocity is only one of many examples of a rate of change. Our reasoning applies to any quantity y that depends on a variable x—say, $y = f(x)$. For any interval $[x_0, x_1]$, we set

$$\Delta f = f(x_1) - f(x_0), \qquad \Delta x = x_1 - x_0$$

Sometimes, we write Δy and $\Delta y/\Delta x$ instead of Δf and $\Delta f/\Delta x$.

For $x_1 \neq x_0$, the **average rate of change** of y with respect to x over $[x_0, x_1]$ is the ratio

$$\text{Average rate of change} = \frac{\Delta f}{\Delta x} = \underbrace{\frac{f(x_1) - f(x_0)}{x_1 - x_0}}_{\text{Slope of secant line}}$$

The word "instantaneous" is often dropped. When we use the term "rate of change," it is understood that the instantaneous rate is intended.

The **instantaneous rate of change** at $x = x_0$ is the limit of the average rates of change. We estimate it by computing the average rate over smaller and smaller intervals.

In Example 1 above, we considered only right-hand intervals $[x_0, x_1]$. In the next example, we compute the average rate of change for intervals lying to both the left and the right of x_0.

■ **EXAMPLE 2 Speed of Sound in Air** The formula $v = 20\sqrt{T}$ provides a good approximation to the speed of sound v in dry air (in m/s) as a function of air temperature T (in kelvins). Estimate the instantaneous rate of change of v with respect to T when $T = 273$ K. What are the units of this rate?

Solution To estimate the instantaneous rate of change at $T = 273$, we compute the average rate for several intervals lying to the left and right of $T = 273$. For example, the average rate of change over [272.5, 273] is

$$\frac{v(273) - v(272.5)}{273 - 272.5} = \frac{20\sqrt{273} - 20\sqrt{272.5}}{0.5} \approx 0.60550$$

TABLE 2 **Left-Hand Intervals**

Temperature interval	Average rate of change
[272.5, 273]	0.60550
[272.8, 273]	0.60534
[272.9, 273]	0.60528
[272.99, 273]	0.60523

TABLE 3 **Right-Hand Intervals**

Temperature interval	Average rate of change
[273, 273.5]	0.60495
[273, 273.2]	0.60512
[273, 273.1]	0.60517
[273, 273.01]	0.60522

Tables 2 and 3 suggest that the instantaneous rate is approximately 0.605. This is the rate of increase in speed per degree increase in temperature, so it has units of m/s-K, or *meters per second per kelvin*. The secant lines corresponding to the values in the tables are shown in Figures 5 and 6. ■

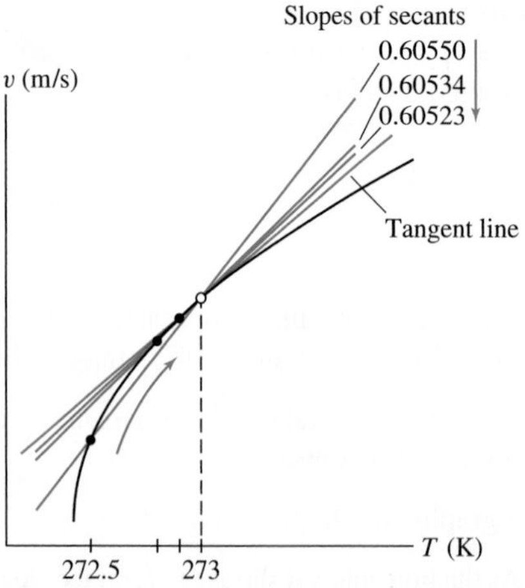

FIGURE 5 Secant lines for intervals lying to the left of $T = 273$.

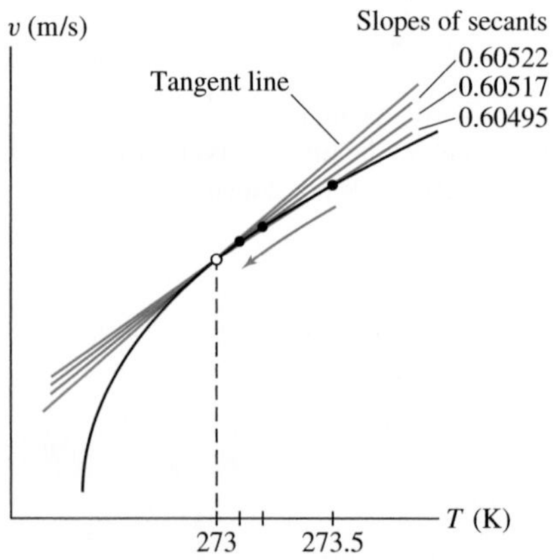

FIGURE 6 Secant lines for intervals lying to the right of $T = 273$.

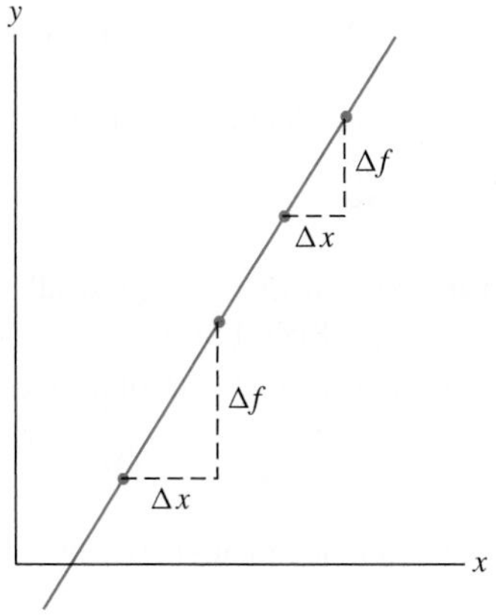

FIGURE 7 For a linear function $f(x) = mx + b$, the ratio $\Delta f/\Delta x$ is equal to the slope m for every interval.

To conclude this section, we recall an important point discussed in Section 1.2: For any linear function $f(x) = mx + b$, *the average rate of change over every interval is equal to the slope m* (Figure 7). We verify as follows:

$$\frac{\Delta f}{\Delta x} = \frac{f(x_1) - f(x_0)}{x_1 - x_0} = \frac{(mx_1 + b) - (mx_0 + b)}{x_1 - x_0} = \frac{m(x_1 - x_0)}{x_1 - x_0} = m$$

The instantaneous rate of change at $x = x_0$, which is the limit of these average rates, is also equal to m. This makes sense graphically because all secant lines and all tangent lines to the graph of $f(x)$ coincide with the graph itself.

2.1 SUMMARY

- The *average rate of change* of $y = f(x)$ over an interval $[x_0, x_1]$:

$$\text{Average rate of change} = \frac{\Delta f}{\Delta x} = \frac{f(x_1) - f(x_0)}{x_1 - x_0} \qquad (x_1 \neq x_0)$$

- The *instantaneous rate of change* is the limit of the average rates of change.
- Graphical interpretation:
  - Average rate of change is the slope of the secant line through the points $(x_0, f(x_0))$ and $(x_1, f(x_1))$ on the graph of $f(x)$.
  - Instantaneous rate of change is the slope of the tangent line at x_0.
- To estimate the instantaneous rate of change at $x = x_0$, compute the average rate of change over several intervals $[x_0, x_1]$ (or $[x_1, x_0]$) for x_1 close to x_0.
- The velocity of an object in linear motion is the rate of change of position $s(t)$.
- Linear function $f(x) = mx + b$: The average rate of change over every interval and the instantaneous rate of change at every point are equal to the slope m.

2.1 EXERCISES

Preliminary Questions

1. Average velocity is equal to the slope of a secant line through two points on a graph. Which graph?

2. Can instantaneous velocity be defined as a ratio? If not, how is instantaneous velocity computed?

3. What is the graphical interpretation of instantaneous velocity at a moment $t = t_0$?

4. What is the graphical interpretation of the following statement? The average rate of change approaches the instantaneous rate of change as the interval $[x_0, x_1]$ shrinks to x_0.

5. The rate of change of atmospheric temperature with respect to altitude is equal to the slope of the tangent line to a graph. Which graph? What are possible units for this rate?

Exercises

1. A ball dropped from a state of rest at time $t = 0$ travels a distance $s(t) = 4.9t^2$ m in t seconds.

(a) How far does the ball travel during the time interval $[2, 2.5]$?

(b) Compute the average velocity over $[2, 2.5]$.

(c) Compute the average velocity for the time intervals in the table and estimate the ball's instantaneous velocity at $t = 2$.

Interval	$[2, 2.01]$	$[2, 2.005]$	$[2, 2.001]$	$[2, 2.00001]$
Average velocity				

2. A wrench released from a state of rest at time $t = 0$ travels a distance $s(t) = 4.9t^2$ m in t seconds. Estimate the instantaneous velocity at $t = 3$.

3. Let $v = 20\sqrt{T}$ as in Example 2. Estimate the instantaneous rate of change of v with respect to T when $T = 300$ K.

4. Compute $\Delta y/\Delta x$ for the interval $[2, 5]$, where $y = 4x - 9$. What is the instantaneous rate of change of y with respect to x at $x = 2$?

In Exercises 5–6, a stone is tossed vertically into the air from ground level with an initial velocity of 15 m/s. *Its height at time t is* $h(t) = 15t - 4.9t^2$ m.

5. Compute the stone's average velocity over the time interval $[0.5, 2.5]$ and indicate the corresponding secant line on a sketch of the graph of $h(t)$.

6. Compute the stone's average velocity over the time intervals $[1, 1.01]$, $[1, 1.001]$, $[1, 1.0001]$ and $[0.99, 1]$, $[0.999, 1]$, $[0.9999, 1]$, and then estimate the instantaneous velocity at $t = 1$.

7. With an initial deposit of \$100, the balance in a bank account after t years is $f(t) = 100(1.08)^t$ dollars.

(a) What are the units of the rate of change of $f(t)$?

(b) Find the average rate of change over $[0, 0.5]$ and $[0, 1]$.

(c) Estimate the instantaneous rate of change at $t = 0.5$ by computing the average rate of change over intervals to the left and right of $t = 0.5$.

8. The position of a particle at time t is $s(t) = t^3 + t$. Compute the average velocity over the time interval $[1, 4]$ and estimate the instantaneous velocity at $t = 1$.

9. Figure 8 shows the estimated number N of Internet users in Chile, based on data from the United Nations Statistics Division.

(a) Estimate the rate of change of N at $t = 2003.5$.

(b) Does the rate of change increase or decrease as t increases? Explain graphically.

(c) Let R be the average rate of change over $[2001, 2005]$. Compute R.

(d) Is the rate of change at $t = 2002$ greater than or less than the average rate R? Explain graphically.

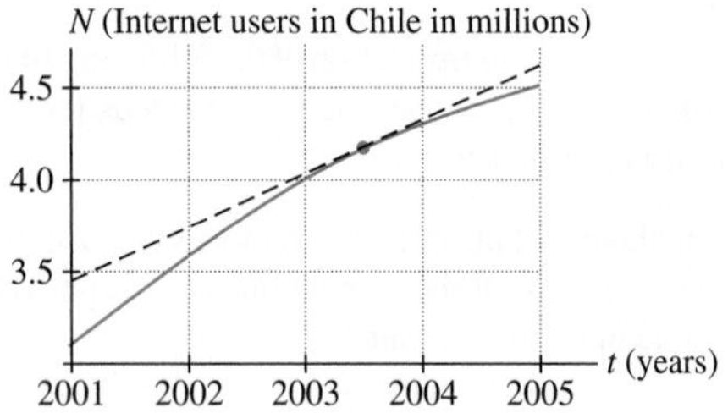

FIGURE 8

10. The **atmospheric temperature** T (in °C) at altitude h meters above a certain point on earth is $T = 15 - 0.0065h$ for $h \leq 12{,}000$ m. What are the average and instantaneous rates of change of T with respect to h? Why are they the same? Sketch the graph of T for $h \leq 12{,}000$.

In Exercises 11–18, estimate the instantaneous rate of change at the point indicated.

11. $P(x) = 3x^2 - 5$; $x = 2$

12. $f(t) = 12t - 7$; $t = -4$

13. $y(x) = \dfrac{1}{x+2}$; $x = 2$

14. $y(t) = \sqrt{3t + 1}$; $t = 1$

15. $f(x) = 3^x$; $x = 0$

16. $f(x) = 3^x$; $x = 3$

17. $f(x) = \sin x$; $x = \dfrac{\pi}{6}$

18. $f(x) = \tan x$; $x = \dfrac{\pi}{4}$

19. The height (in centimeters) at time t (in seconds) of a small mass oscillating at the end of a spring is $h(t) = 8\cos(12\pi t)$.

(a) Calculate the mass's average velocity over the time intervals $[0, 0.1]$ and $[3, 3.5]$.

(b) Estimate its instantaneous velocity at $t = 3$.

20. The number $P(t)$ of *E. coli* cells at time t (hours) in a petri dish is plotted in Figure 9.

(a) Calculate the average rate of change of $P(t)$ over the time interval $[1, 3]$ and draw the corresponding secant line.

(b) Estimate the slope m of the line in Figure 9. What does m represent?

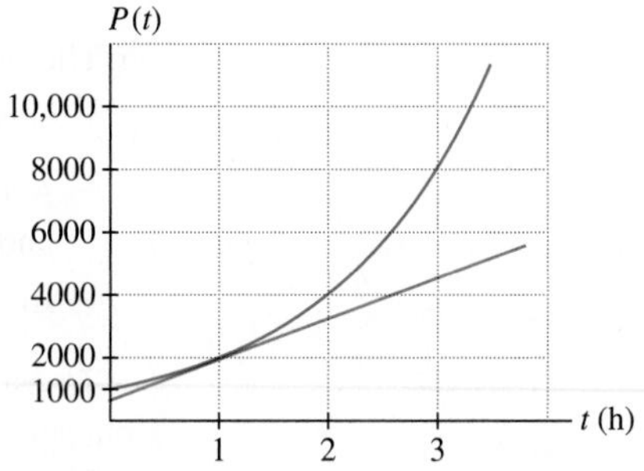

FIGURE 9 Number of *E. coli* cells at time t.

21. Assume that the period T (in seconds) of a pendulum (the time required for a complete back-and-forth cycle) is $T = \frac{3}{2}\sqrt{L}$, where L is the pendulum's length (in meters).

(a) What are the units for the rate of change of T with respect to L? Explain what this rate measures.

(b) Which quantities are represented by the slopes of lines A and B in Figure 10?

(c) Estimate the instantaneous rate of change of T with respect to L when $L = 3$ m.

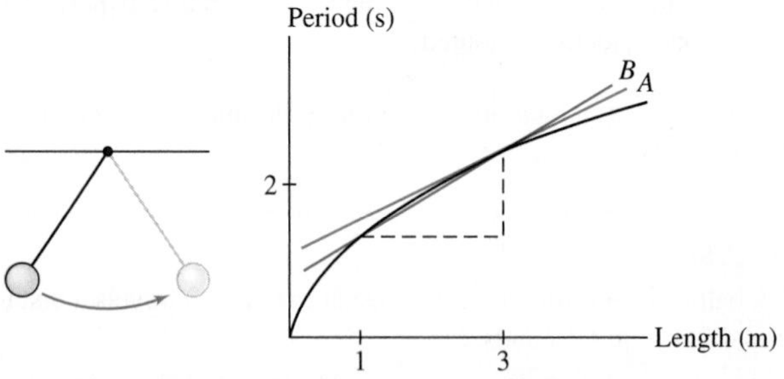

FIGURE 10 The period T is the time required for a pendulum to swing back and forth.

22. The graphs in Figure 11 represent the positions of moving particles as functions of time.

(a) Do the instantaneous velocities at times t_1, t_2, t_3 in (A) form an increasing or a decreasing sequence?

(b) Is the particle speeding up or slowing down in (A)?

(c) Is the particle speeding up or slowing down in (B)?

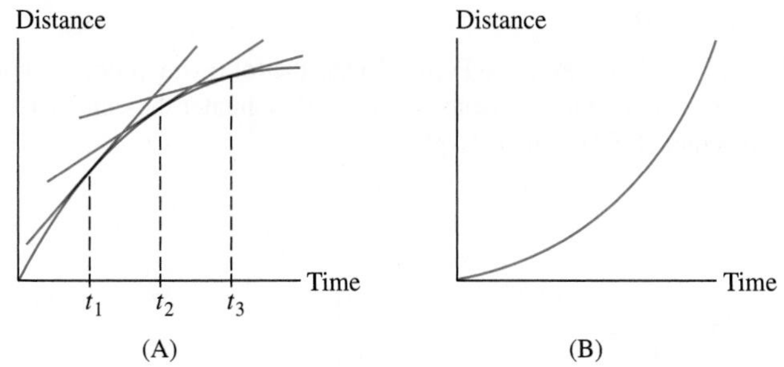

FIGURE 11

23. GU An advertising campaign boosted sales of Crunchy Crust frozen pizza to a peak level of S_0 dollars per month. A marketing study showed that after t months, monthly sales declined to

$$S(t) = S_0 g(t), \quad \text{where } g(t) = \frac{1}{\sqrt{1+t}}.$$

Do sales decline more slowly or more rapidly as time increases? Answer by referring to a sketch the graph of $g(t)$ together with several tangent lines.

24. The fraction of a city's population infected by a flu virus is plotted as a function of time (in weeks) in Figure 12.

(a) Which quantities are represented by the slopes of lines A and B? Estimate these slopes.

(b) Is the flu spreading more rapidly at $t = 1, 2$, or 3?

(c) Is the flu spreading more rapidly at $t = 4, 5$, or 6?

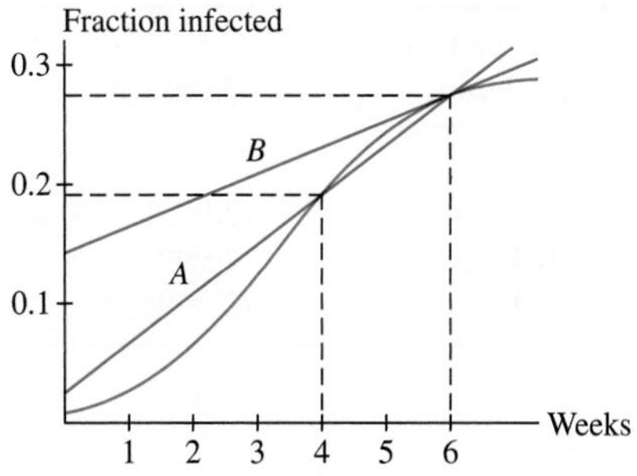

FIGURE 12

25. The graphs in Figure 13 represent the positions s of moving particles as functions of time t. Match each graph with a description:

(a) Speeding up

(b) Speeding up and then slowing down

(c) Slowing down

(d) Slowing down and then speeding up

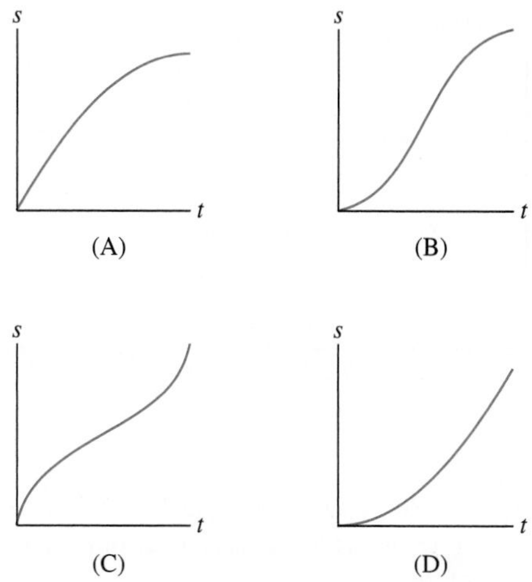

FIGURE 13

26. An epidemiologist finds that the percentage $N(t)$ of susceptible children who were infected on day t during the first three weeks of a measles outbreak is given, to a reasonable approximation, by the formula (Figure 14)

$$N(t) = \frac{100t^2}{t^3 + 5t^2 - 100t + 380}$$

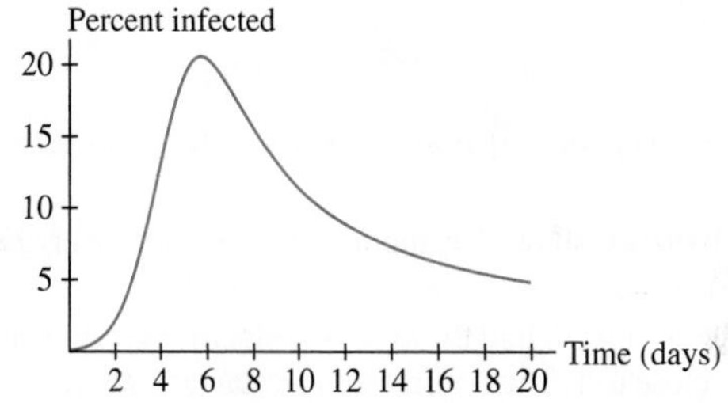

FIGURE 14 Graph of $N(t)$.

(a) Draw the secant line whose slope is the average rate of change in infected children over the intervals [4, 6] and [12, 14]. Then compute these average rates (in units of percent per day).

(b) Is the rate of decline greater at $t = 8$ or $t = 16$?

(c) Estimate the rate of change of $N(t)$ on day 12.

27. The fungus *Fusarium exosporium* infects a field of flax plants through the roots and causes the plants to wilt. Eventually, the entire field is infected. The percentage $f(t)$ of infected plants as a function of time t (in days) since planting is shown in Figure 15.

(a) What are the units of the rate of change of $f(t)$ with respect to t? What does this rate measure?

(b) Use the graph to rank (from smallest to largest) the average infection rates over the intervals [0, 12], [20, 32], and [40, 52].

(c) Use the following table to compute the average rates of infection over the intervals [30, 40], [40, 50], [30, 50].

Days	0	10	20	30	40	50	60
Percent infected	0	18	56	82	91	96	98

(d) Draw the tangent line at $t = 40$ and estimate its slope.

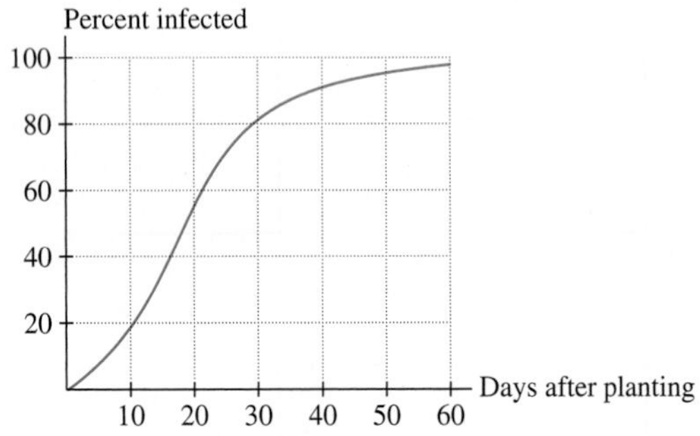

FIGURE 15

28. Let $v = 20\sqrt{T}$ as in Example 2. Is the rate of change of v with respect to T greater at low temperatures or high temperatures? Explain in terms of the graph.

29. If an object in linear motion (but with changing velocity) covers Δs meters in Δt seconds, then its average velocity is $v_0 = \Delta s/\Delta t$ m/s. Show that it would cover the same distance if it traveled at constant velocity v_0 over the same time interval. This justifies our calling $\Delta s/\Delta t$ the *average velocity*.

30. Sketch the graph of $f(x) = x(1-x)$ over $[0, 1]$. Refer to the graph and, without making any computations, find:

(a) The average rate of change over $[0, 1]$

(b) The (instantaneous) rate of change at $x = \frac{1}{2}$

(c) The values of x at which the rate of change is positive

31. Which graph in Figure 16 has the following property: For all x, the average rate of change over $[0, x]$ is greater than the instantaneous rate of change at x. Explain.

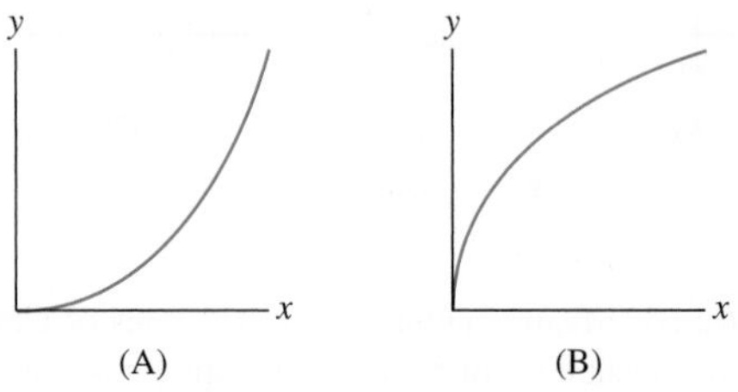

FIGURE 16

Further Insights and Challenges

32. The height of a projectile fired in the air vertically with initial velocity 25 m/s is

$$h(t) = 25t - 4.9t^2 \text{ m.}$$

(a) Compute $h(1)$. Show that $h(t) - h(1)$ can be factored with $(t-1)$ as a factor.

(b) Using part (a), show that the average velocity over the interval $[1, t]$ is $20.1 - 4.9t$.

(c) Use this formula to find the average velocity over several intervals $[1, t]$ with t close to 1. Then estimate the instantaneous velocity at time $t = 1$.

33. Let $Q(t) = t^2$. As in the previous exercise, find a formula for the average rate of change of Q over the interval $[1, t]$ and use it to estimate the instantaneous rate of change at $t = 1$. Repeat for the interval $[2, t]$ and estimate the rate of change at $t = 2$.

34. Show that the average rate of change of $f(x) = x^3$ over $[1, x]$ is equal to

$$x^2 + x + 1.$$

Use this to estimate the instantaneous rate of change of $f(x)$ at $x = 1$.

35. Find a formula for the average rate of change of $f(x) = x^3$ over $[2, x]$ and use it to estimate the instantaneous rate of change at $x = 2$.

36. Let $T = \frac{3}{2}\sqrt{L}$ as in Exercise 21. The numbers in the second column of Table 4 are increasing, and those in the last column are decreasing. Explain why in terms of the graph of T as a function of L. Also, explain graphically why the instantaneous rate of change at $L = 3$ lies between 0.4329 and 0.4331.

TABLE 4 Average Rates of Change of T with Respect to L

Interval	Average rate of change	Interval	Average rate of change
$[3, 3.2]$	0.42603	$[2.8, 3]$	0.44048
$[3, 3.1]$	0.42946	$[2.9, 3]$	0.43668
$[3, 3.001]$	0.43298	$[2.999, 3]$	0.43305
$[3, 3.0005]$	0.43299	$[2.9995, 3]$	0.43303

2.2 Limits: A Numerical and Graphical Approach

The goal in this section is to define limits and study them using numerical and graphical techniques. We begin with the following question: *How do the values of a function $f(x)$ behave when x approaches a number c, whether or not $f(c)$ is defined?*

To explore this question, we'll experiment with the function

$$f(x) = \frac{\sin x}{x} \quad (x \text{ in radians})$$

The undefined expression 0/0 is referred to as an "indeterminate form."

Notice that $f(0)$ is not defined. In fact, when we set $x = 0$ in

$$f(x) = \frac{\sin x}{x}$$

we obtain the undefined expression $0/0$ because $\sin 0 = 0$. Nevertheless, we can compute $f(x)$ for values of x *close* to 0. When we do this, a clear trend emerges.

To describe the trend, we use the phrase "x approaches 0" or "x tends to 0" to indicate that x takes on values (both positive and negative) that get closer and closer to 0. The notation for this is $x \to 0$, and more specifically we write

- $x \to 0+$ if x approaches 0 from the right (through positive values).
- $x \to 0-$ if x approaches 0 from the left (through negative values).

Now consider the values listed in Table 1. The table gives the unmistakable impression that $f(x)$ gets closer and closer to 1 as $x \to 0+$ and as $x \to 0-$.

This conclusion is supported by the graph of $f(x)$ in Figure 1. The point (0, 1) is missing from the graph because $f(x)$ is not defined at $x = 0$, but the graph approaches this missing point as x approaches 0 from the left and right. We say that the *limit* of $f(x)$ as $x \to 0$ is equal to 1, and we write

$$\lim_{x \to 0} f(x) = 1$$

We also say that $f(x)$ *approaches* or *converges to* 1 as $x \to 0$.

TABLE 1

x	$\frac{\sin x}{x}$	x	$\frac{\sin x}{x}$
1	0.841470985	−1	0.841470985
0.5	0.**9**58851077	−0.5	0.**9**58851077
0.1	0.**99**8334166	−0.1	0.**99**8334166
0.05	0.**999**583385	−0.05	0.**999**583385
0.01	0.**9999**83333	−0.01	0.**9999**83333
0.005	0.**99999**5833	−0.005	0.**99999**5833
0.001	0.**999999**833	−0.001	0.**999999**833
$x \to 0+$	$f(x) \to 1$	$x \to 0-$	$f(x) \to 1$

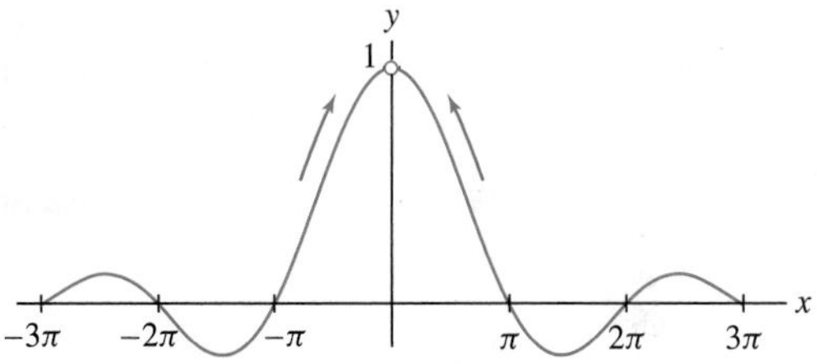

FIGURE 1 Graph of $f(x) = \dfrac{\sin x}{x}$.

CONCEPTUAL INSIGHT The numerical and graphical evidence may convince us that $f(x) = \dfrac{\sin x}{x}$ converges to 1 as $x \to 0$, but since $f(0)$ yields the undefined expression $0/0$, could we not arrive at this conclusion more simply by saying that $0/0$ is equal to 1? The answer is no. *Algebra does not allow us to divide by* 0 *under any circumstances*, and it is not correct to say that $0/0$ equals 1 or any other number.

What we have learned, however, is that a function $f(x)$ may approach a limit as $x \to c$ even if the formula for $f(c)$ produces the undefined expression $0/0$. The limit of $f(x) = \dfrac{\sin x}{x}$ turns out to be 1. We will encounter other examples where $f(x)$ produces $0/0$ but the limit is a number other than 1 (or the limit does not exist).

Definition of a Limit

To define limits, let us recall that the distance between two numbers a and b is the absolute value $|a - b|$, so we can express the idea that $f(x)$ is close to L by saying that $|f(x) - L|$ is small.

The limit concept was not fully clarified until the nineteenth century. The French mathematician Augustin-Louis Cauchy (1789–1857, pronounced Koh-shee) gave the following verbal definition: "When the values successively attributed to the same variable approach a fixed value indefinitely, in such a way as to end up differing from it by as little as one could wish, this last value is called the limit of all the others. So, for example, an irrational number is the limit of the various fractions which provide values that approximate it more and more closely." (Translated by J. Grabiner)

DEFINITION Limit Assume that $f(x)$ is defined for all x in an open interval containing c, but not necessarily at c itself. We say that

the limit of $f(x)$ as x approaches c is equal to L

if $|f(x) - L|$ becomes arbitrarily small when x is any number sufficiently close (but not equal) to c. In this case, we write

$$\lim_{x \to c} f(x) = L$$

We also say that $f(x)$ *approaches* or *converges to* L as $x \to c$ (and we write $f(x) \to L$).

If the values of $f(x)$ do not converge to any limit as $x \to c$, we say that $\lim_{x \to c} f(x)$ *does not exist*. It is important to note that the value $f(c)$ itself, which may or may not be defined, plays no role in the limit. All that matters are the values of $f(x)$ for x close to c. Furthermore, if $f(x)$ approaches a limit as $x \to c$, then the limiting value L is unique.

EXAMPLE 1 Use the definition above to verify the following limits:

(a) $\lim_{x \to 7} 5 = 5$ **(b)** $\lim_{x \to 4} (3x + 1) = 13$

Solution

(a) Let $f(x) = 5$. To show that $\lim_{x \to 7} f(x) = 5$, we must show that $|f(x) - 5|$ becomes arbitrarily small when x is sufficiently close (but not equal) to 7. But observe that $|f(x) - 5| = |5 - 5| = 0$ *for all* x, so what we are required to show is automatic (and it is not necessary to take x close to 7).

(b) Let $f(x) = 3x + 1$. To show that $\lim_{x \to 4} (3x + 1) = 13$, we must show that $|f(x) - 13|$ becomes arbitrarily small when x is sufficiently close (but not equal) to 4. We have

$$|f(x) - 13| = |(3x + 1) - 13| = |3x - 12| = 3|x - 4|$$

Because $|f(x) - 13|$ is a multiple of $|x - 4|$, we can make $|f(x) - 13|$ arbitrarily small by taking x sufficiently close to 4. ■

Reasoning as in Example 1 but with arbitrary constants, we obtain the following simple but important results:

THEOREM 1 For any constants k and c, (a) $\lim_{x \to c} k = k$, (b) $\lim_{x \to c} x = c$.

To deal with more complicated limits and especially, to provide mathematically rigorous proofs, a more precise version of the above limit definition is needed. This more precise version is discussed in Section 2.9, where inequalities are used to pin down the exact meaning of the phrases "arbitrarily small" and "sufficiently close."

Here is one version of the rigorous definition of a limit: $\lim_{x \to c} f(x) = L$ if, for every number n, we can find a value of m such that $|f(x) - L| < 10^{-n}$ for all x such that $0 < |x - c| < 10^{-m}$.

Graphical and Numerical Investigation

Our goal in the rest of this section is to develop a better intuitive understanding of limits by investigating them graphically and numerically.

Graphical Investigation Use a graphing utility to produce a graph of $f(x)$. The graph should give a visual impression of whether or not a limit exists. It can often be used to estimate the value of the limit.

Numerical Investigation We write $x \to c-$ to indicate that x approaches c through values less than c, and we write $x \to c+$ to indicate that x approaches c through values greater than c. To investigate $\lim_{x\to c} f(x)$,

(i) Make a table of values of $f(x)$ for x close to but less than c—that is, as $x \to c-$.

(ii) Make a second table of values of $f(x)$ for x close to but greater than c—that is, as $x \to c+$.

(iii) If both tables indicate convergence to the same number L, we take L to be an estimate for the limit.

Keep in mind that graphical and numerical investigations provide evidence for a limit, but they do not prove that the limit exists or has a given value. This is done using the Limit Laws established in the following sections.

The tables should contain enough values to reveal a clear trend of convergence to a value L. If $f(x)$ approaches a limit, the successive values of $f(x)$ will generally agree to more and more decimal places as x is taken closer to c. If no pattern emerges, then the limit may not exist.

■ **EXAMPLE 2** Investigate $\lim_{x\to 9} \dfrac{x-9}{\sqrt{x}-3}$ graphically and numerically.

Solution The function $f(x) = \dfrac{x-9}{\sqrt{x}-3}$ is undefined at $x = 9$ because the formula for $f(9)$ leads to the undefined expression $0/0$. Therefore, the graph in Figure 9 has a gap at $x = 6$. However, the graph suggests that $f(x)$ approaches 6 as $x \to 9$.

For numerical evidence, we consider a table of values of $f(x)$ for x approaching 9 from both the left and the right. Table 2 confirms our impression that

$$\lim_{x\to 9} \frac{x-9}{\sqrt{x}-3} = 6$$ ■

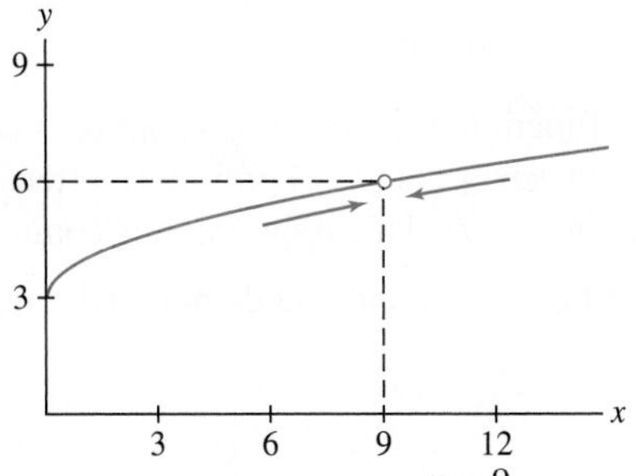

FIGURE 2 Graph of $f(x) = \dfrac{x-9}{\sqrt{x}-3}$.

TABLE 2

$x \to 9-$	$\dfrac{x-9}{\sqrt{x}-3}$	$x \to 9+$	$\dfrac{x-9}{\sqrt{x}-3}$
8.9	**5.98**329	9.1	**6.0**1662
8.99	**5.99**833	9.01	**6.00**1666
8.999	**5.999**83	9.001	**6.000**167
8.9999	**5.9999**833	9.0001	**6.0000**167

■ **EXAMPLE 3 Limit Equals Function Value** Investigate $\lim_{x\to 4} x^2$.

Solution Figure 3 and Table 3 both suggest that $\lim_{x\to 4} x^2 = 16$. But $f(x) = x^2$ is defined at $x = 4$ and $f(4) = 16$, so in this case, *the limit is equal to the function value*. This pleasant conclusion is valid whenever $f(x)$ is a *continuous* function, a concept treated in Section 2.4. ■

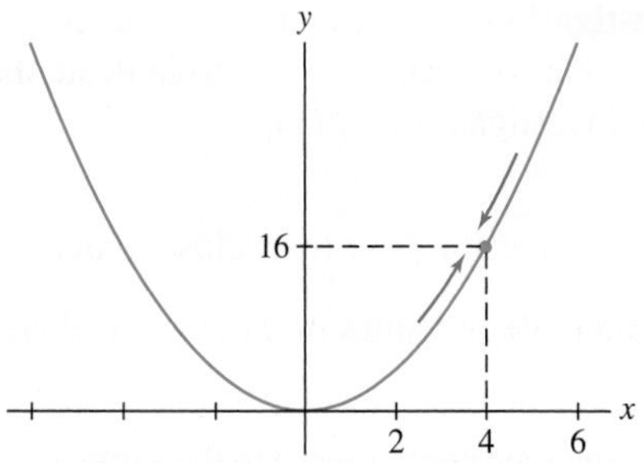

FIGURE 3 Graph of $f(x) = x^2$. The limit is equal to the function value $f(4) = 16$.

TABLE 3

$x \to 4-$	x^2	$x \to 4+$	x^2
3.9	15.21	4.1	16.81
3.99	**15.9**201	4.01	**16.0**801
3.999	**15.992**001	4.001	**16.008**001
3.9999	**15.9992**0001	4.0001	**16.0008**0001

■ **EXAMPLE 4** Investigate $\lim_{h \to 0} \frac{2^h - 1}{h}$.

Solution The function $f(h) = (2^h - 1)/h$ is undefined at $h = 0$, but both Figure 4 and Table 4 suggest that $\lim_{h \to 0} (2^h - 1)/h \approx 0.693$. ■

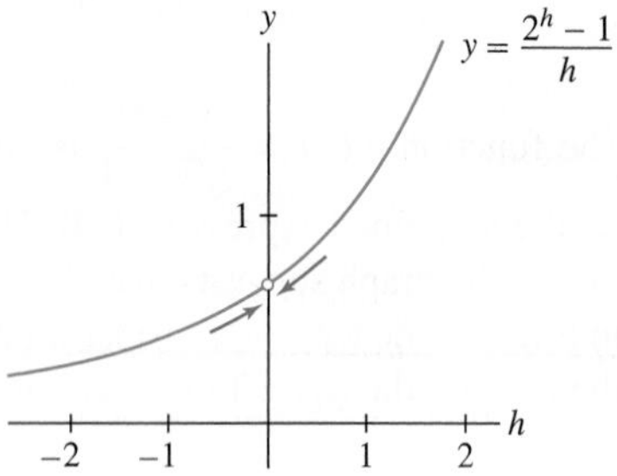

FIGURE 4

TABLE 4

$h \to 0-$	$\frac{2^h - 1}{h}$	$h \to 0+$	$\frac{2^h - 1}{h}$
−0.005	0.69195	0.005	0.69435
−0.001	0.69291	0.001	0.69339
−0.0001	0.69312	0.0001	0.69317
−0.00001	0.69314	0.00001	0.69315

CAUTION Numerical investigations are often suggestive, but may be misleading in some cases. If, in Example 5, we had chosen to evaluate $f(x) = \sin \frac{\pi}{x}$ at the values $x = 0.1, 0.01, 0.001, \ldots$, we might have concluded incorrectly that $f(x)$ approaches the limit 0 as $x \to 0$. The problem is that $f(10^{-n}) = \sin(10^n \pi) = 0$ for every whole number n, but $f(x)$ itself does not approach any limit.

■ **EXAMPLE 5 A Limit That Does Not Exist** Investigate $\lim_{x \to 0} \sin \frac{\pi}{x}$ graphically and numerically.

Solution The function $f(x) = \sin \frac{\pi}{x}$ is not defined at $x = 0$, but Figure 5 suggests that it oscillates between $+1$ and -1 infinitely often as $x \to 0$. It appears, therefore, that $\lim_{x \to 0} \sin \frac{\pi}{x}$ does not exist. This impression is confirmed by Table 5, which shows that the values of $f(x)$ bounce around and do not tend toward any limit L as $x \to 0$. ■

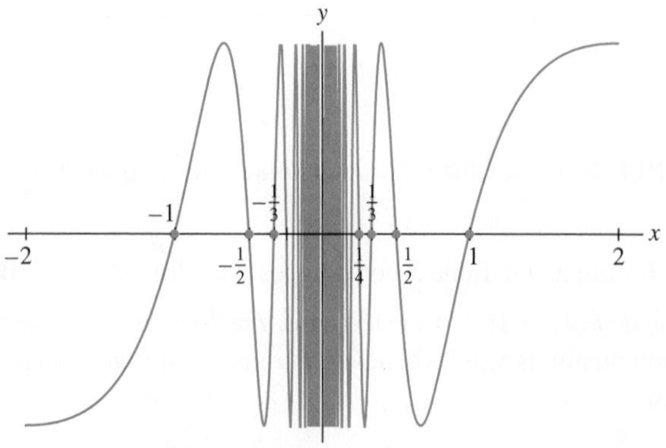

FIGURE 5 Graph of $f(x) = \sin \frac{\pi}{x}$.

TABLE 5 The Function $f(x) = \sin \frac{\pi}{x}$ Does Not Approach a Limit as $x \to 0$

$x \to 0-$	$\sin \frac{\pi}{x}$	$x \to 0+$	$\sin \frac{\pi}{x}$
−0.1	0	0.1	0
−0.03	0.866	0.03	−0.866
−0.007	−0.434	0.007	0.434
−0.0009	0.342	0.0009	−0.342
−0.00065	−0.935	0.00065	0.935

One-Sided Limits

The limits discussed so far are *two-sided*. To show that $\lim_{x\to c} f(x) = L$, it is necessary to check that $f(x)$ converges to L as x approaches c through values both larger and smaller than c. In some instances, $f(x)$ may approach L from one side of c without necessarily approaching it from the other side, or $f(x)$ may be defined on only one side of c. For this reason, we define the one-sided limits

$$\lim_{x\to c-} f(x) \quad \text{(left-hand limit)}, \qquad \lim_{x\to c+} f(x) \quad \text{(right-hand limit)}$$

The limit itself exists if both one-sided limits exist and are equal.

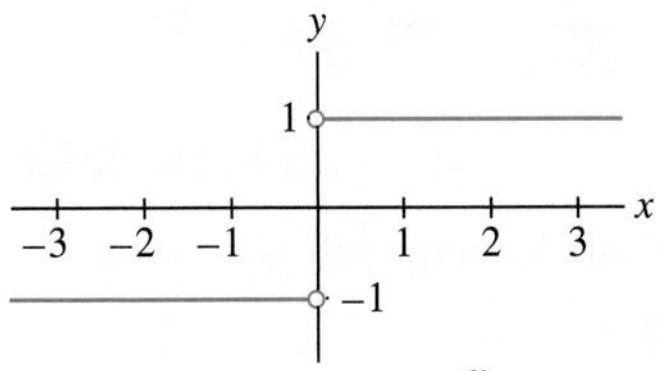

FIGURE 6 Graph of $f(x) = \dfrac{x}{|x|}$.

EXAMPLE 6 Left- and Right-Hand Limits Not Equal Investigate the one-sided limits of $f(x) = \dfrac{x}{|x|}$ as $x \to 0$. Does $\lim_{x\to 0} f(x)$ exist?

Solution Figure 6 shows what is going on. For $x < 0$,

$$f(x) = \frac{x}{|x|} = \frac{x}{-x} = -1$$

Therefore, the left-hand limit is $\lim_{x\to 0-} f(x) = -1$. But for $x > 0$,

$$f(x) = \frac{x}{|x|} = \frac{x}{x} = 1$$

Therefore, $\lim_{x\to 0+} f(x) = 1$. These one-sided limits are not equal, so $\lim_{x\to 0} f(x)$ does not exist. ■

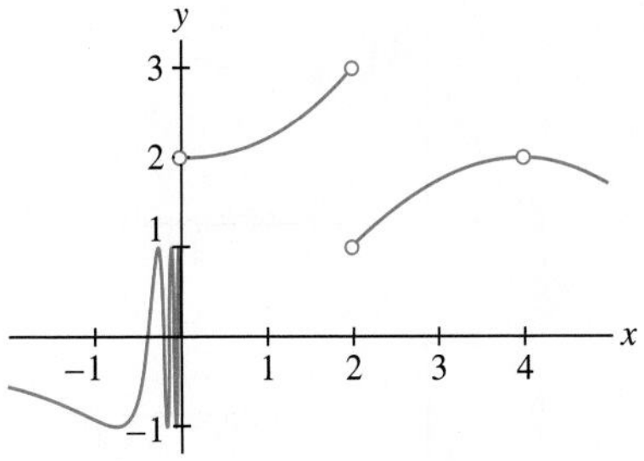

FIGURE 7

EXAMPLE 7 The function $f(x)$ in Figure 7 is not defined at $c = 0, 2, 4$. Investigate the one- and two-sided limits at these points.

Solution

- $c = 0$: The left-hand limit $\lim_{x\to 0-} f(x)$ does not seem to exist because $f(x)$ appears to oscillate infinitely often to the left of $x = 0$. On the other hand, $\lim_{x\to 0+} f(x) = 2$.
- $c = 2$: The one-sided limits exist but are not equal:

$$\lim_{x\to 2-} f(x) = 3 \qquad \text{and} \qquad \lim_{x\to 2+} f(x) = 1$$

  Therefore, $\lim_{x\to 2} f(x)$ does not exist.
- $c = 4$: The one-sided limits exist and both have the value 2. Therefore, the two-sided limit exists and $\lim_{x\to 4} f(x) = 2$. ■

Infinite Limits

Some functions $f(x)$ tend to ∞ or $-\infty$ as x approaches a value c. If so, $\lim_{x\to c} f(x)$ does not exist, but we say that $f(x)$ has an *infinite limit*. More precisely, we write

- $\lim_{x\to c} f(x) = \infty$ if $f(x)$ increases without bound as $x \to c$.
- $\lim_{x\to c} f(x) = -\infty$ if $f(x)$ decreases without bound as $x \to c$.

Here, "decrease without bound" means that $f(x)$ becomes negative and $|f(x)| \to \infty$. One-sided infinite limits are defined similarly. When using this notation, keep in mind that ∞ and $-\infty$ are not numbers.

When $f(x)$ approaches ∞ or $-\infty$ as x approaches c from one or both sides, the line $x = c$ is called a **vertical asymptote**. In Figure 8, the line $x = 2$ is a vertical asymptote in (A), and $x = 0$ is a vertical asymptote in both (B) and (C).

In the next example, the notation $x \to c\pm$ is used to indicate that the left- and right-hand limits are to be considered separately.

■ **EXAMPLE 8** GU Investigate the one-sided limits graphically:

(a) $\displaystyle\lim_{x\to 2\pm} \frac{1}{x-2}$ **(b)** $\displaystyle\lim_{x\to 0\pm} \frac{1}{x^2}$

Solution

(a) Figure 8(A) suggests that

$$\lim_{x\to 2-} \frac{1}{x-2} = -\infty, \qquad \lim_{x\to 2+} \frac{1}{x-2} = \infty$$

The vertical line $x = 2$ is a vertical asymptote. Why are the one-sided limits different? Because $f(x) = \dfrac{1}{x-2}$ is negative for $x < 2$ (so the limit from the left is $-\infty$) and $f(x)$ is positive for $x > 2$ (so the limit from the right is ∞).

(b) Figure 8(B) suggests that $\displaystyle\lim_{x\to 0} \frac{1}{x^2} = \infty$. Indeed, $f(x) = \dfrac{1}{x^2}$ is positive for all $x \neq 0$ and becomes arbitrarily large as $x \to 0$ from either side. The line $x = 0$ is a vertical asymptote. ■

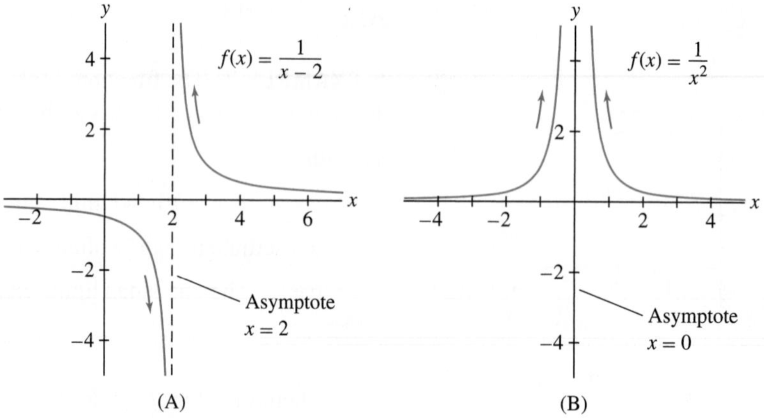

FIGURE 8

CONCEPTUAL INSIGHT You should not think of an infinite limit as a true limit. The notation $\displaystyle\lim_{x\to c} f(x) = \infty$ is merely a shorthand way of saying that $f(x)$ increases beyond all bounds as x approaches c. The limit itself does not exist. We must be careful when using this notation because ∞ and $-\infty$ *are not numbers*, and contradictions can arise if we try to manipulate them as numbers. For example, if ∞ were a number, it would be larger than any finite number, and presumably, $\infty + 1 = \infty$. But then

$$\infty + 1 = \infty$$
$$(\infty + 1) - \infty = \infty - \infty$$
$$1 = 0 \qquad \text{(contradiction!)}$$

To avoid errors, keep in mind the ∞ is not a number but rather a convenient shorthand notation.

2.2 SUMMARY

- By definition, $\lim_{x\to c} f(x) = L$ if $|f(x) - L|$ becomes arbitrarily small when x is any number sufficiently close (but not equal) to c. We say that
  - *The limit of $f(x)$ as x approaches c is L*, or
  - *$f(x)$ approaches (or converges) to L as x approaches c.*
- If $f(x)$ approaches a limit as $x \to c$, then the limit value L is unique.
- If $f(x)$ does not approach a limit as $x \to c$, we say that $\lim_{x\to c} f(x)$ does not exist.
- The limit may exist even if $f(c)$ is not defined.
- *One-sided limits*:
  - $\lim_{x\to c-} f(x) = L$ if $f(x)$ converges to L as x approaches c through values less than c.
  - $\lim_{x\to c+} f(x) = L$ if $f(x)$ converges to L as x approaches c through values greater than c.
- The limit exists if and only if both one-sided limits exist and are equal.
- *Infinite limits*: $\lim_{x\to c} f(x) = \infty$ if $f(x)$ increases beyond bound as x approaches c, and $\lim_{x\to c} f(x) = -\infty$ if $f(x)$ becomes arbitrarily large (in absolute value) but negative as x approaches c.
- In the case of a one- or two-sided infinite limit, the vertical line $x = c$ is called a *vertical asymptote*.

2.2 EXERCISES

Preliminary Questions

1. What is the limit of $f(x) = 1$ as $x \to \pi$?

2. What is the limit of $g(t) = t$ as $t \to \pi$?

3. Is $\lim_{x\to 10} 20$ equal to 10 or 20?

4. Can $f(x)$ approach a limit as $x \to c$ if $f(c)$ is undefined? If so, give an example.

5. What does the following table suggest about $\lim_{x\to 1-} f(x)$ and $\lim_{x\to 1+} f(x)$?

x	0.9	0.99	0.999	1.1	1.01	1.001
$f(x)$	7	25	4317	3.0126	3.0047	3.00011

6. Can you tell whether $\lim_{x\to 5} f(x)$ exists from a plot of $f(x)$ for $x > 5$? Explain.

7. If you know in advance that $\lim_{x\to 5} f(x)$ exists, can you determine its value from a plot of $f(x)$ for all $x > 5$?

Exercises

In Exercises 1–4, fill in the tables and guess the value of the limit.

1. $\lim_{x\to 1} f(x)$, where $f(x) = \dfrac{x^3 - 1}{x^2 - 1}$.

x	$f(x)$	x	$f(x)$
1.002		0.998	
1.001		0.999	
1.0005		0.9995	
1.00001		0.99999	

2. $\lim_{t\to 0} h(t)$, where $h(t) = \dfrac{\cos t - 1}{t^2}$. Note that $h(t)$ is even; that is, $h(t) = h(-t)$.

t	± 0.002	± 0.0001	± 0.00005	± 0.00001
$h(t)$				

3. $\lim_{y\to 2} f(y)$, where $f(y) = \dfrac{y^2 - y - 2}{y^2 + y - 6}$.

y	$f(y)$	y	$f(y)$
2.002		1.998	
2.001		1.999	
2.0001		1.9999	

4. $\lim\limits_{\theta\to 0} f(\theta)$, where $f(\theta) = \dfrac{\sin\theta - \theta}{\theta^3}$.

θ	± 0.002	± 0.0001	± 0.00005	± 0.00001
$f(\theta)$				

5. Determine $\lim\limits_{x\to 0.5} f(x)$ for $f(x)$ as in Figure 9.

6. Determine $\lim\limits_{x\to 0.5} g(x)$ for $g(x)$ as in Figure 10.

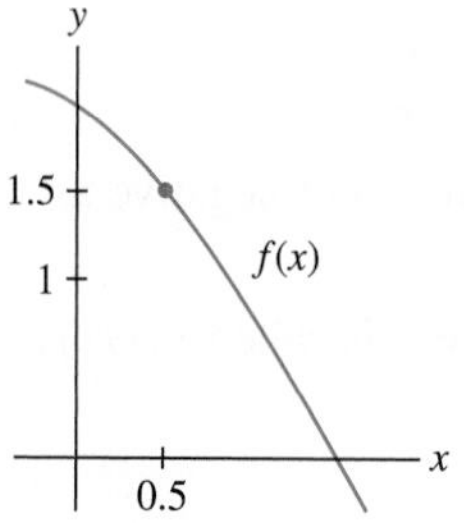

FIGURE 9

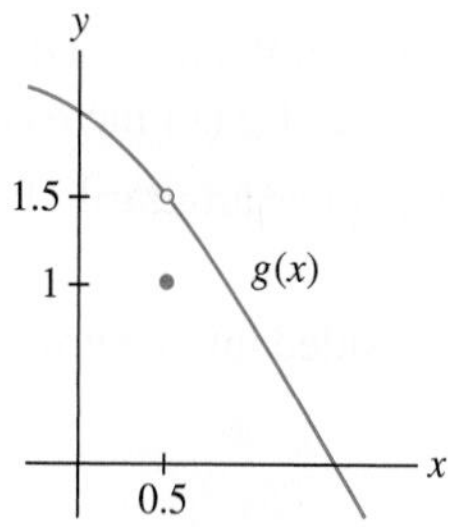

FIGURE 10

In Exercises 7–8, evaluate the limit.

7. $\lim\limits_{x\to 21} x$

8. $\lim\limits_{x\to 4.2} \sqrt{3}$

In Exercises 9–16, verify each limit using the limit definition. For example, in Exercise 9, show that $|3x - 12|$ can be made as small as desired by taking x close to 4.

9. $\lim\limits_{x\to 4} 3x = 12$

10. $\lim\limits_{x\to 5} 3 = 3$

11. $\lim\limits_{x\to 3} (5x + 2) = 17$

12. $\lim\limits_{x\to 2} (7x - 4) = 10$

13. $\lim\limits_{x\to 0} x^2 = 0$

14. $\lim\limits_{x\to 0} (3x^2 - 9) = -9$

15. $\lim\limits_{x\to 0} (4x^2 + 2x + 5) = 5$

16. $\lim\limits_{x\to 0} (x^3 + 12) = 12$

In Exercises 17–36, estimate the limit numerically or state that the limit does not exist. If infinite, state whether the one-sided limits are ∞ or $-\infty$.

17. $\lim\limits_{x\to 1} \dfrac{\sqrt{x} - 1}{x - 1}$

18. $\lim\limits_{x\to -4} \dfrac{2x^2 - 32}{x + 4}$

19. $\lim\limits_{x\to 2} \dfrac{x^2 + x - 6}{x^2 - x - 2}$

20. $\lim\limits_{x\to 3} \dfrac{x^3 - 2x^2 - 9}{x^2 - 2x - 3}$

21. $\lim\limits_{x\to 0} \dfrac{\sin 2x}{x}$

22. $\lim\limits_{x\to 0} \dfrac{\sin 5x}{x}$

23. $\lim\limits_{\theta\to 0} \dfrac{\cos\theta - 1}{\theta}$

24. $\lim\limits_{x\to 0} \dfrac{\sin x}{x^2}$

25. $\lim\limits_{x\to 4} \dfrac{1}{(x - 4)^3}$

26. $\lim\limits_{x\to 1-} \dfrac{3 - x}{x - 1}$

27. $\lim\limits_{x\to 3+} \dfrac{x - 4}{x^2 - 9}$

28. $\lim\limits_{h\to 0} \dfrac{3^h - 1}{h}$

29. $\lim\limits_{h\to 0} \sin h \cos \dfrac{1}{h}$

30. $\lim\limits_{h\to 0} \cos \dfrac{1}{h}$

31. $\lim\limits_{x\to 0} |x|^x$

32. $\lim\limits_{x\to 0} \dfrac{2^x - 3^x}{x}$

33. $\lim\limits_{\theta\to \frac{\pi}{4}} \dfrac{\tan\theta - 2\sin\theta\cos\theta}{\theta - \frac{\pi}{4}}$

34. $\lim\limits_{r\to 0} (1 + r)^{1/r}$

35. $\lim\limits_{\theta\to 0} \dfrac{1 - \cos\theta}{\theta^2}$

36. $\lim\limits_{\theta\to 0} \dfrac{1 - \cos\theta}{\theta^3}$

37. The **greatest integer function** is defined by $[x] = n$, where n is the unique integer such that $n \le x < n + 1$. Sketch the graph of $y = [x]$. Calculate, for c an integer:

(a) $\lim\limits_{x\to c-} [x]$ **(b)** $\lim\limits_{x\to c+} [x]$

38. Determine the one-sided limits at $c = 1$, 2, and 4 of the function $g(x)$ shown in Figure 11, and state whether the limit exists at these points.

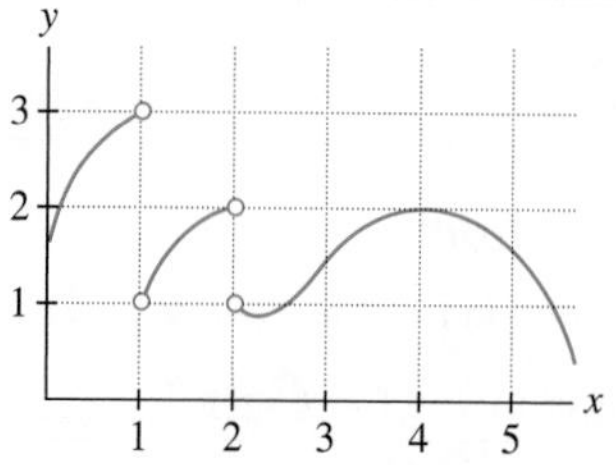

FIGURE 11

In Exercises 39–46, determine the one-sided limits numerically or graphically. If infinite, state whether the one-sided limits are ∞ or $-\infty$, and describe the corresponding vertical asymptote. In Exercise 46, $[x]$ is the greatest integer function defined in Exercise 37.

39. $\lim\limits_{x\to 0\pm} \dfrac{\sin x}{|x|}$

40. $\lim\limits_{x\to 0\pm} |x|^{1/x}$

41. $\lim\limits_{x\to 0\pm} \dfrac{x - \sin|x|}{x^3}$

42. $\lim\limits_{x\to 4\pm} \dfrac{x + 1}{x - 4}$

43. $\lim\limits_{x\to -2\pm} \dfrac{4x^2 + 7}{x^3 + 8}$

44. $\lim\limits_{x\to -3\pm} \dfrac{x^2}{x^2 - 9}$

45. $\lim\limits_{x\to 1\pm} \dfrac{x^5 + x - 2}{x^2 + x - 2}$

46. $\lim\limits_{x\to 2\pm} \cos\left(\dfrac{\pi}{2}(x - [x])\right)$

47. Determine the one-sided limits at $c = 2$, 4 of the function $f(x)$ in Figure 12. What are the vertical asymptotes of $f(x)$?

48. Determine the infinite one- and two-sided limits in Figure 13.

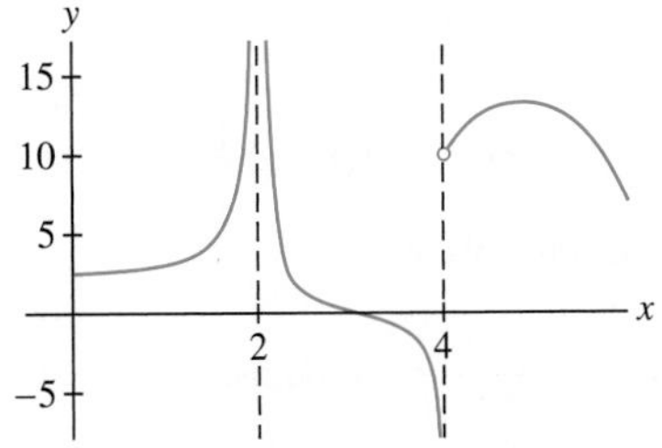

FIGURE 12

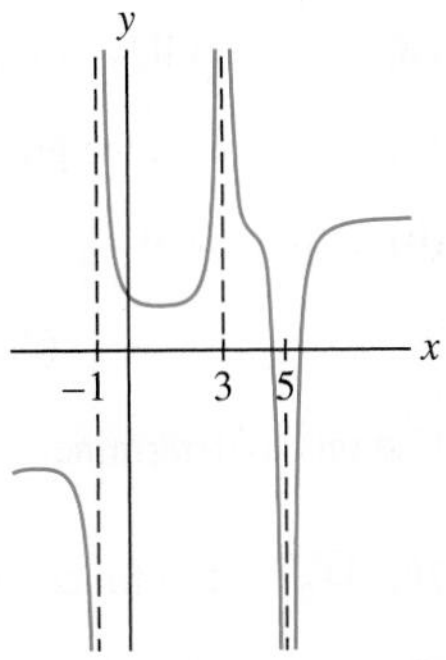

FIGURE 13

In Exercises 49–52, sketch the graph of a function with the given limits.

49. $\lim_{x\to 1} f(x) = 2, \quad \lim_{x\to 3-} f(x) = 0, \quad \lim_{x\to 3+} f(x) = 4$

50. $\lim_{x\to 1} f(x) = \infty, \quad \lim_{x\to 3-} f(x) = 0, \quad \lim_{x\to 3+} f(x) = -\infty$

51. $\lim_{x\to 2+} f(x) = f(2) = 3, \quad \lim_{x\to 2-} f(x) = -1,$
$\lim_{x\to 4} f(x) = 2 \neq f(4)$

52. $\lim_{x\to 1+} f(x) = \infty, \quad \lim_{x\to 1-} f(x) = 3, \quad \lim_{x\to 4} f(x) = -\infty$

53. Determine the one-sided limits of the function $f(x)$ in Figure 14, at the points $c = 1, 3, 5, 6$.

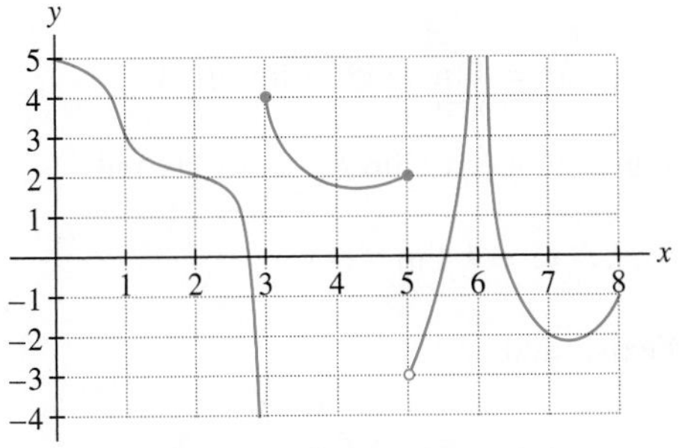

FIGURE 14 Graph of $f(x)$

54. Does either of the two oscillating functions in Figure 15 appear to approach a limit as $x \to 0$?

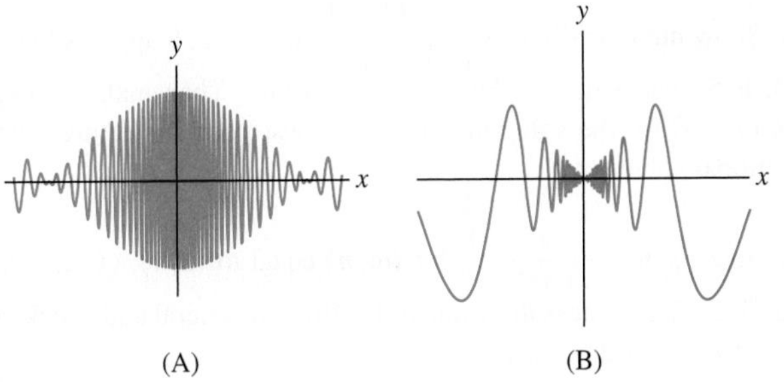

FIGURE 15

GU *In Exercises 55–60, plot the function and use the graph to estimate the value of the limit.*

55. $\lim_{\theta\to 0} \dfrac{\sin 5\theta}{\sin 2\theta}$

56. $\lim_{x\to 0} \dfrac{12^x - 1}{4^x - 1}$

57. $\lim_{x\to 0} \dfrac{2^x - \cos x}{x}$

58. $\lim_{\theta\to 0} \dfrac{\sin^2 4\theta}{\cos\theta - 1}$

59. $\lim_{\theta\to 0} \dfrac{\cos 7\theta - \cos 5\theta}{\theta^2}$

60. $\lim_{\theta\to 0} \dfrac{\sin^2 2\theta - \theta \sin 4\theta}{\theta^4}$

61. Let n be a positive integer. For which n are the two infinite one-sided limits $\lim_{x\to 0\pm} 1/x^n$ equal?

62. Let $L(n) = \lim_{x\to 1} \left(\dfrac{n}{1-x^n} - \dfrac{1}{1-x}\right)$ for n a positive integer. Investigate $L(n)$ numerically for several values of n, and then guess the value of of $L(n)$ in general.

63. GU In some cases, numerical investigations can be misleading. Plot $f(x) = \cos \frac{\pi}{x}$.

(a) Does $\lim_{x\to 0} f(x)$ exist?

(b) Show, by evaluating $f(x)$ at $x = \frac{1}{2}, \frac{1}{4}, \frac{1}{6}, \ldots$, that you might be able to trick your friends into believing that the limit exists and is equal to $L = 1$.

(c) Which sequence of evaluations might trick them into believing that the limit is $L = -1$.

Further Insights and Challenges

64. Light waves of frequency λ passing through a slit of width a produce a **Fraunhofer diffraction pattern** of light and dark fringes (Figure 16). The intensity as a function of the angle θ is

$$I(\theta) = I_m\left(\frac{\sin(R\sin\theta)}{R\sin\theta}\right)^2$$

where $R = \pi a/\lambda$ and I_m is a constant. Show that the intensity function is not defined at $\theta = 0$. Then choose any two values for R and check numerically that $I(\theta)$ approaches I_m as $\theta \to 0$.

65. Investigate $\lim_{\theta\to 0} \dfrac{\sin n\theta}{\theta}$ numerically for several values of n. Then guess the value in general.

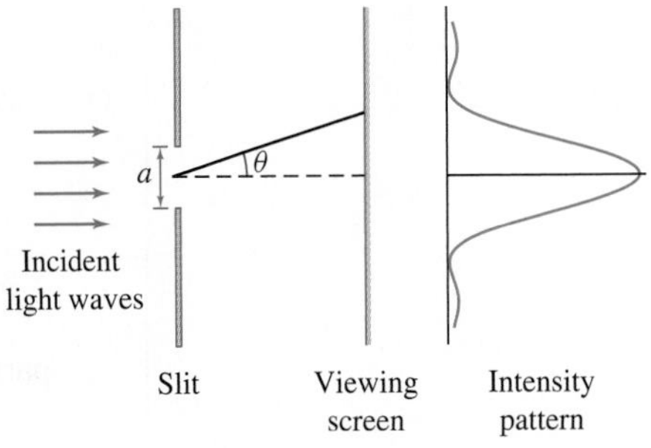

FIGURE 16 Fraunhofer diffraction pattern.

66. Show numerically that $\lim_{x\to 0} \frac{b^x - 1}{x}$ for $b = 3, 5$ appears to equal $\ln 3$, $\ln 5$, where $\ln x$ is the natural logarithm. Then make a conjecture (guess) for the value in general and test your conjecture for two additional values of b.

67. Investigate $\lim_{x\to 1} \frac{x^n - 1}{x^m - 1}$ for (m, n) equal to $(2, 1)$, $(1, 2)$, $(2, 3)$, and $(3, 2)$. Then guess the value of the limit in general and check your guess for two additional pairs.

68. Find by numerical experimentation the positive integers k such that $\lim_{x\to 0} \frac{\sin(\sin^2 x)}{x^k}$ exists.

69. GU Plot the graph of $f(x) = \frac{2^x - 8}{x - 3}$.

(a) Zoom in on the graph to estimate $L = \lim_{x\to 3} f(x)$.

(b) Explain why

$$f(2.99999) \le L \le f(3.00001)$$

Use this to determine L to three decimal places.

70. GU The function $f(x) = \frac{2^{1/x} - 2^{-1/x}}{2^{1/x} + 2^{-1/x}}$ is defined for $x \neq 0$.

(a) Investigate $\lim_{x\to 0+} f(x)$ and $\lim_{x\to 0-} f(x)$ numerically.

(b) Plot the graph of f and describe its behavior near $x = 0$.

2.3 Basic Limit Laws

In Section 2.2 we relied on graphical and numerical approaches to investigate limits and estimate their values. In the next four sections we go beyond this intuitive approach and develop tools for computing limits in a precise way. The next theorem provides our first set of tools.

The proof of Theorem 1 is discussed in Section 2.9 and Appendix D. To illustrate the underlying idea, consider two numbers such as 2.99 and 5.001. Observe that 2.99 is close to 3 and 5.0001 is close to 5, so certainly the sum $2.99 + 5.0001$ is close to $3 + 5$ and the product $(2.99)(5.0001)$ is close to $(3)(5)$. In the same way, if $f(x)$ approaches L and $g(x)$ approaches M as $x \to c$, then $f(x) + g(x)$ approaches the sum $L + M$, and $f(x)g(x)$ approaches the product LM. The other laws are similar.

THEOREM 1 Basic Limit Laws If $\lim_{x\to c} f(x)$ and $\lim_{x\to c} g(x)$ exist, then

(i) Sum Law: $\lim_{x\to c} \big(f(x) + g(x)\big)$ exists and

$$\lim_{x\to c} \big(f(x) + g(x)\big) = \lim_{x\to c} f(x) + \lim_{x\to c} g(x)$$

(ii) Constant Multiple Law: For any number k, $\lim_{x\to c} kf(x)$ exists and

$$\lim_{x\to c} kf(x) = k \lim_{x\to c} f(x)$$

(iii) Product Law: $\lim_{x\to c} f(x)g(x)$ exists and

$$\lim_{x\to c} f(x)g(x) = \Big(\lim_{x\to c} f(x)\Big)\Big(\lim_{x\to c} g(x)\Big)$$

(iv) Quotient Law: If $\lim_{x\to c} g(x) \neq 0$, then $\lim_{x\to c} \frac{f(x)}{g(x)}$ exists and

$$\lim_{x\to c} \frac{f(x)}{g(x)} = \frac{\lim_{x\to c} f(x)}{\lim_{x\to c} g(x)}$$

(v) Powers and Roots: If p, q are integers with $q \neq 0$, then $\lim_{x\to c} [f(x)]^{p/q}$ exists and

$$\lim_{x\to c} [f(x)]^{p/q} = \Big(\lim_{x\to c} f(x)\Big)^{p/q}$$

Assume that $\lim_{x\to c} f(x) \ge 0$ if q is even, and that $\lim_{x\to c} f(x) \neq 0$ if $p/q < 0$. In particular, for n a positive integer,

$$\lim_{x\to c} [f(x)]^n = \Big(\lim_{x\to c} f(x)\Big)^n, \qquad \lim_{x\to c} \sqrt[n]{f(x)} = \sqrt[n]{\lim_{x\to c} f(x)}$$

In the second limit, assume that $\lim_{x\to c} f(x) \ge 0$ if n is even.

Before proceeding to the examples, we make some useful remarks.

- The Sum and Product Laws are valid for any number of functions. For example,

$$\lim_{x\to c}\big(f_1(x)+f_2(x)+f_3(x)\big)=\lim_{x\to c}f_1(x)+\lim_{x\to c}f_2(x)+\lim_{x\to c}f_3(x)$$

- The Sum Law has a counterpart for differences:

$$\lim_{x\to c}\big(f(x)-g(x)\big)=\lim_{x\to c}f(x)-\lim_{x\to c}g(x)$$

This follows from the Sum and Constant Multiple Laws (with $k=-1$):

$$\lim_{x\to c}\big(f(x)-g(x)\big)=\lim_{x\to c}f(x)+\lim_{x\to c}\big(-g(x)\big)=\lim_{x\to c}f(x)-\lim_{x\to c}g(x)$$

- Recall two basic limits from Theorem 1 in Section 2.2:

$$\lim_{x\to c}k=k,\qquad \lim_{x\to c}x=c$$

Applying Law (v) to $f(x)=x$, we obtain

$$\boxed{\lim_{x\to c}x^{p/q}=c^{p/q}}\qquad \boxed{1}$$

for integers p,q. Assume that $c\ge 0$ if q is even and that $c>0$ if $p/q<0$.

■ **EXAMPLE 1** Use the Basic Limit Laws to evaluate:

(a) $\lim_{x\to 2}x^3$ **(b)** $\lim_{x\to 2}(x^3+5x+7)$ **(c)** $\lim_{x\to 2}\sqrt{x^3+5x+7}$

Solution

(a) By Eq. (1), $\lim_{x\to 2}x^3=2^3=8$.

(b)

$$\begin{aligned}\lim_{x\to 2}(x^3+5x+7)&=\lim_{x\to 2}x^3+\lim_{x\to 2}5x+\lim_{x\to 2}7 &&\text{(Sum Law)}\\ &=\lim_{x\to 2}x^3+5\lim_{x\to 2}x+\lim_{x\to 2}7 &&\text{(Constant Multiple Law)}\\ &=8+5(2)+7=25\end{aligned}$$

(c) By Law (v) for roots and (b),

$$\lim_{x\to 2}\sqrt{x^3+5x+7}=\sqrt{\lim_{x\to 2}(x^3+5x+7)}=\sqrt{25}=5$$ ■

You may have noticed that each of the limits in Examples 1 and 2 could have been evaluated by a simple substitution. For example, set $t=-1$ to evaluate

$$\lim_{t\to -1}\frac{t+6}{2t^4}=\frac{-1+6}{2(-1)^4}=\frac{5}{2}$$

Substitution is valid when the function is ***continuous****, a concept we shall study in the next section.*

■ **EXAMPLE 2** Evaluate **(a)** $\lim_{t\to -1}\dfrac{t+6}{2t^4}$ and **(b)** $\lim_{t\to 3}t^{-1/4}(t+5)^{1/3}$.

Solution

(a) Use the Quotient, Sum, and Constant Multiple Laws:

$$\lim_{t\to -1}\frac{t+6}{2t^4}=\frac{\lim_{t\to -1}(t+6)}{\lim_{t\to -1}2t^4}=\frac{\lim_{t\to -1}t+\lim_{t\to -1}6}{2\lim_{t\to -1}t^4}=\frac{-1+6}{2(-1)^4}=\frac{5}{2}$$

(b) Use the Product, Powers, and Sum Laws:

$$\lim_{t\to 3} t^{-1/4}(t+5)^{1/3} = \left(\lim_{t\to 3} t^{-1/4}\right)\left(\lim_{t\to 3} \sqrt[3]{t+5}\right) = \left(3^{-1/4}\right)\left(\sqrt[3]{\lim_{t\to 3} t+5}\right)$$

$$= 3^{-1/4}\sqrt[3]{3+5} = 3^{-1/4}(2) = \frac{2}{3^{1/4}} \qquad ■$$

The next example reminds us that the Basic Limit Laws apply only when the limits of both $f(x)$ and $g(x)$ exist.

■ **EXAMPLE 3 Assumptions Matter** Show that the Product Law cannot be applied to $\lim_{x\to 0} f(x)g(x)$ if $f(x) = x$ and $g(x) = x^{-1}$.

Solution For all $x \neq 0$ we have $f(x)g(x) = x \cdot x^{-1} = 1$, so the limit of the product exists:

$$\lim_{x\to 0} f(x)g(x) = \lim_{x\to 0} 1 = 1$$

However, $\lim_{x\to 0} x^{-1}$ does not exist because $g(x) = x^{-1}$ approaches ∞ as $x \to 0+$ and it approaches $-\infty$ as $x \to 0-$. Therefore, the Product Law cannot be applied and its conclusion does not hold:

$$\left(\lim_{x\to 0} f(x)\right)\left(\lim_{x\to 0} g(x)\right) = \left(\lim_{x\to 0} x\right)\underbrace{\left(\lim_{x\to 0} x^{-1}\right)}_{\text{Does not exist}} \qquad ■$$

2.3 SUMMARY

- The Basic Limit Laws: If $\lim_{x\to c} f(x)$ and $\lim_{x\to c} g(x)$ both exist, then

(i) $\lim_{x\to c} \big(f(x) + g(x)\big) = \lim_{x\to c} f(x) + \lim_{x\to c} g(x)$

(ii) $\lim_{x\to c} kf(x) = k \lim_{x\to c} f(x)$

(iii) $\lim_{x\to c} f(x)\,g(x) = \left(\lim_{x\to c} f(x)\right)\left(\lim_{x\to c} g(x)\right)$

(iv) If $\lim_{x\to c} g(x) \neq 0$, then $\lim_{x\to c} \dfrac{f(x)}{g(x)} = \dfrac{\lim_{x\to c} f(x)}{\lim_{x\to c} g(x)}$

(v) If p, q are integers with $q \neq 0$,

$$\lim_{x\to c} [f(x)]^{p/q} = \left(\lim_{x\to c} f(x)\right)^{p/q}$$

For n a positive integer,

$$\lim_{x\to c} [f(x)]^n = \left(\lim_{x\to c} f(x)\right)^n, \qquad \lim_{x\to c} \sqrt[n]{f(x)} = \sqrt[n]{\lim_{x\to c} f(x)}$$

- If $\lim_{x\to c} f(x)$ or $\lim_{x\to c} g(x)$ does not exist, then the Basic Limit Laws cannot be applied.

2.3 EXERCISES

Preliminary Questions

1. State the Sum Law and Quotient Law.

2. Which of the following is a verbal version of the Product Law (assuming the limits exist)?

(a) The product of two functions has a limit.

(b) The limit of the product is the product of the limits.

(c) The product of a limit is a product of functions.

(d) A limit produces a product of functions.

3. Which statement is correct? The Quotient Law does not hold if:

(a) The limit of the denominator is zero.

(b) The limit of the numerator is zero.

Exercises

In Exercises 1–24, evaluate the limit using the Basic Limit Laws and the limits $\lim_{x\to c} x^{p/q} = c^{p/q}$ *and* $\lim_{x\to c} k = k$.

1. $\lim_{x\to 9} x$

2. $\lim_{x\to -3} 14$

3. $\lim_{x\to \frac{1}{2}} x^4$

4. $\lim_{z\to 27} z^{2/3}$

5. $\lim_{t\to 2} t^{-1}$

6. $\lim_{x\to 5} x^{-2}$

7. $\lim_{x\to 0.2} (3x+4)$

8. $\lim_{x\to \frac{1}{3}} (3x^3+2x^2)$

9. $\lim_{x\to -1} (3x^4-2x^3+4x)$

10. $\lim_{x\to 8} (3x^{2/3}-16x^{-1})$

11. $\lim_{x\to 2} (x+1)(3x^2-9)$

12. $\lim_{x\to \frac{1}{2}} (4x+1)(6x-1)$

13. $\lim_{t\to 4} \dfrac{3t-14}{t+1}$

14. $\lim_{z\to 9} \dfrac{\sqrt{z}}{z-2}$

15. $\lim_{y\to \frac{1}{4}} (16y+1)(2y^{1/2}+1)$

16. $\lim_{x\to 2} x(x+1)(x+2)$

17. $\lim_{y\to 4} \dfrac{1}{\sqrt{6y+1}}$

18. $\lim_{w\to 7} \dfrac{\sqrt{w+2}+1}{\sqrt{w-3}-1}$

19. $\lim_{x\to -1} \dfrac{x}{x^3+4x}$

20. $\lim_{t\to -1} \dfrac{t^2+1}{(t^3+2)(t^4+1)}$

21. $\lim_{t\to 25} \dfrac{3\sqrt{t}-\frac{1}{5}t}{(t-20)^2}$

22. $\lim_{y\to \frac{1}{3}} (18y^2-4)^4$

23. $\lim_{t\to \frac{3}{2}} (4t^2+8t-5)^{3/2}$

24. $\lim_{t\to 7} \dfrac{(t+2)^{1/2}}{(t+1)^{2/3}}$

25. Use the Quotient Law to prove that if $\lim_{x\to c} f(x)$ exists and is nonzero, then

$$\lim_{x\to c} \frac{1}{f(x)} = \frac{1}{\lim_{x\to c} f(x)}$$

26. Assuming that $\lim_{x\to 6} f(x) = 4$, compute:

(a) $\lim_{x\to 6} f(x)^2$ **(b)** $\lim_{x\to 6} \dfrac{1}{f(x)}$ **(c)** $\lim_{x\to 6} x\sqrt{f(x)}$

In Exercises 27–30, evaluate the limit assuming that $\lim_{x\to -4} f(x) = 3$ *and* $\lim_{x\to -4} g(x) = 1$.

27. $\lim_{x\to -4} f(x)g(x)$

28. $\lim_{x\to -4} (2f(x)+3g(x))$

29. $\lim_{x\to -4} \dfrac{g(x)}{x^2}$

30. $\lim_{x\to -4} \dfrac{f(x)+1}{3g(x)-9}$

31. Can the Quotient Law be applied to evaluate $\lim_{x\to 0} \dfrac{\sin x}{x}$? Explain.

32. Show that the Product Law cannot be used to evaluate the limit $\lim_{x\to \pi/2} \left(x-\frac{\pi}{2}\right)\tan x$.

33. Give an example where $\lim_{x\to 0} (f(x)+g(x))$ exists but neither $\lim_{x\to 0} f(x)$ nor $\lim_{x\to 0} g(x)$ exists.

Further Insights and Challenges

34. Show that if both $\lim_{x\to c} f(x)\,g(x)$ and $\lim_{x\to c} g(x)$ exist and $\lim_{x\to c} g(x) \neq 0$, then $\lim_{x\to c} f(x)$ exists. *Hint:* Write $f(x) = \dfrac{f(x)\,g(x)}{g(x)}$.

35. Suppose that $\lim_{t\to 3} tg(t) = 12$. Show that $\lim_{t\to 3} g(t)$ exists and equals 4.

36. Prove that if $\lim_{t\to 3} \frac{h(t)}{t} = 5$, then $\lim_{t\to 3} h(t) = 15$.

37. Assuming that $\lim_{x\to 0} \frac{f(x)}{x} = 1$, which of the following statements is necessarily true? Why?

(a) $f(0) = 0$ **(b)** $\lim_{x\to 0} f(x) = 0$

38. Prove that if $\lim_{x\to c} f(x) = L \neq 0$ and $\lim_{x\to c} g(x) = 0$, then the limit $\lim_{x\to c} \frac{f(x)}{g(x)}$ does not exist.

39. Suppose that $\lim_{h\to 0} g(h) = L$.

(a) Explain why $\lim_{h\to 0} g(ah) = L$ for any constant $a \neq 0$.

(b) If we assume instead that $\lim_{h\to 1} g(h) = L$, is it still necessarily true that $\lim_{h\to 1} g(ah) = L$?

(c) Illustrate (a) and (b) with the function $f(x) = x^2$.

40. Assume that $L(a) = \lim_{x\to 0} \dfrac{a^x - 1}{x}$ exists for all $a > 0$. Assume also that $\lim_{x\to 0} a^x = 1$.

(a) Prove that $L(ab) = L(a) + L(b)$ for $a, b > 0$. *Hint:* $(ab)^x - 1 = a^x(b^x - 1) + (a^x - 1)$. This shows that $L(a)$ "behaves" like a logarithm. We will see that $L(a) = \ln a$ in Section 7.3.

(b) Verify numerically that $L(12) = L(3) + L(4)$.

2.4 Limits and Continuity

In everyday speech, the word "continuous" means having no breaks or interruptions. In calculus, continuity is used to describe functions whose graphs have no breaks. If we imagine the graph of a function f as a wavy metal wire, then f is continuous if its graph consists of a single piece of wire as in Figure 1. A break in the wire as in Figure 2 is called a **discontinuity**.

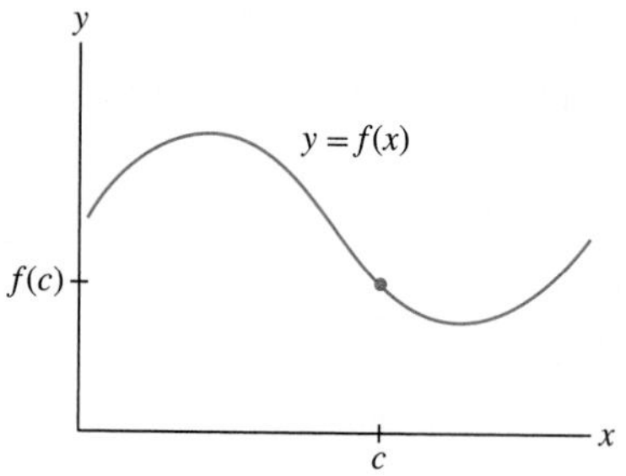

FIGURE 1 $f(x)$ is continuous at $x = c$.

Now observe in Figure 2 that the break in the graph occurs because the left- and right-hand limits as x approaches c are not equal and thus $\lim_{x\to c} g(x)$ does not exist. By contrast, in Figure 1, $\lim_{x\to c} f(x)$ exists and is equal to the function value $f(c)$. This suggests the following definition of continuity in terms of limits.

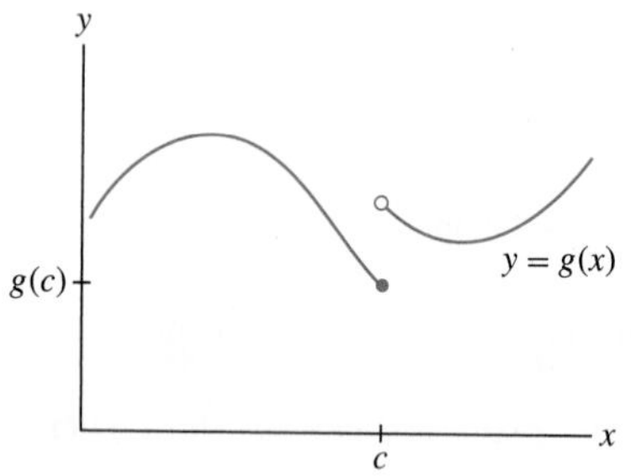

FIGURE 2 Discontinuity at $x = c$: The left- and right-hand limits as $x \to c$ are not equal.

DEFINITION Continuity at a Point Assume that $f(x)$ is defined on an open interval containing $x = c$. Then f is **continuous** at $x = c$ if

$$\lim_{x\to c} f(x) = f(c)$$

If the limit does not exist, or if it exists but is not equal to $f(c)$, we say that f has a **discontinuity** (or is **discontinuous**) at $x = c$.

A function $f(x)$ may be continuous at some points and discontinuous at others. If $f(x)$ is continuous at all points in an interval I, then $f(x)$ is said to be continuous on I. If I is an interval $[a, b]$ or $[a, b)$ that includes a as a left endpoint, we require that $\lim_{x\to a+} f(x) = f(a)$. Similarly, we require that $\lim_{x\to b-} f(x) = f(b)$ if I includes b as a right endpoint. If $f(x)$ is continuous at all points in its domain, then $f(x)$ is simply called continuous.

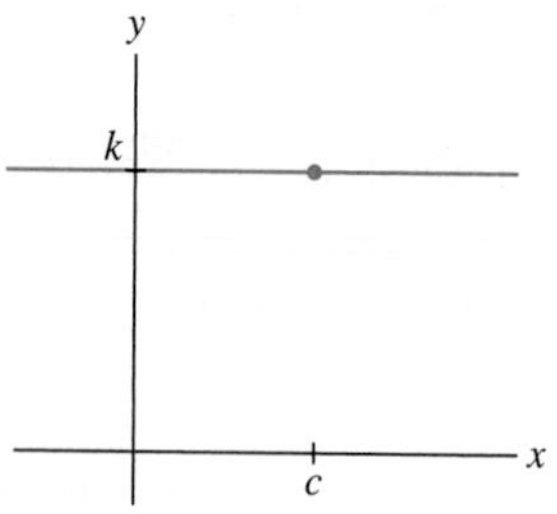

FIGURE 3 The function $f(x) = k$ is continuous.

■ **EXAMPLE 1** Show that the following functions are continuous:

(a) $f(x) = k$ (k any constant) **(b)** $g(x) = x^n$ (n a whole number)

Solution

(a) We have $\lim_{x\to c} f(x) = \lim_{x\to c} k = k$ and $f(c) = k$. The limit exists and is equal to the function value for all c, so $f(x)$ is continuous (Figure 3).

(b) By Eq. (1) in Section 2.3, $\lim_{x\to c} g(x) = \lim_{x\to c} x^n = c^n$ for all c. Also $g(c) = c^n$, so again, the limit exists and is equal to the function value. Therefore, $g(x)$ is continuous. (Figure 4 illustrates the case $n = 1$). ■

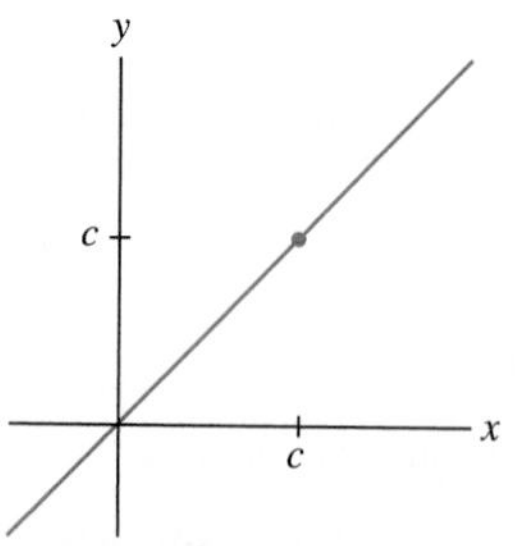

FIGURE 4 The function $g(x) = x$ is continuous.

Examples of Discontinuities

To understand continuity better, let's consider some ways in which a function can fail to be continuous. Keep in mind that continuity at a point $x = c$ requires more than just the existence of a limit. Three conditions must hold:

1. $f(c)$ is defined. **2.** $\lim_{x\to c} f(x)$ exists. **3.** They are equal.

If $\lim_{x\to c} f(x)$ exists but is not equal to $f(c)$, we say that f has a **removable discontinuity** at $x = c$. The function in Figure 5(A) has a removable discontinuity at $c = 2$ because

$$\underbrace{f(2) = 10 \qquad \text{but} \qquad \lim_{x\to 2} f(x) = 5}_{\text{Limit exists but is not equal to function value}}$$

Removable discontinuities are "mild" in the following sense: We can make f continuous at $x = c$ by redefining $f(c)$. In Figure 5(B), $f(2)$ has been redefined as $f(2) = 5$, and this makes f continuous at $x = 2$.

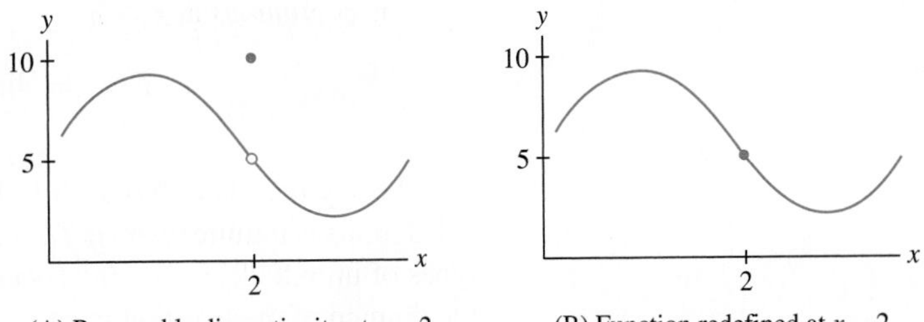

FIGURE 5 Removable discontinuity: The discontinuity can be removed by redefining $f(2)$.

A "worse" type of discontinuity is a **jump discontinuity**, which occurs if the one-sided limits $\lim_{x\to c-} f(x)$ and $\lim_{x\to c+} f(x)$ exist but are not equal. Figure 6 shows two functions with jump discontinuities at $c = 2$. Unlike the removable case, we cannot make $f(x)$ continuous by redefining $f(c)$.

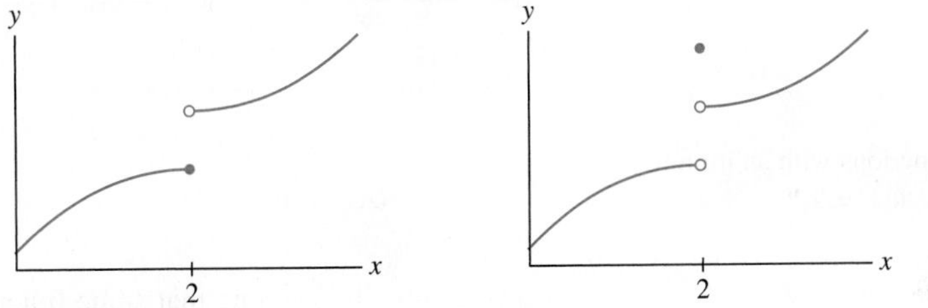

FIGURE 6 Jump discontinuities.

In connection with jump discontinuities, it is convenient to define *one-sided continuity*.

> **DEFINITION One-Sided Continuity** A function $f(x)$ is called:
>
> - **Left-continuous** at $x = c$ if $\lim_{x\to c-} f(x) = f(c)$
> - **Right-continuous** at $x = c$ if $\lim_{x\to c+} f(x) = f(c)$

In Figure 6 above, the function in (A) is left-continuous but the function in (B) is neither left- nor right-continuous. The next example explores one-sided continuity using a piecewise-defined function—that is, a function defined by different formulas on different intervals.

■ **EXAMPLE 2 Piecewise-Defined Function** Discuss the continuity of

$$F(x) = \begin{cases} x & \text{for } x < 1 \\ 3 & \text{for } 1 \le x \le 3 \\ x & \text{for } x > 3 \end{cases}$$

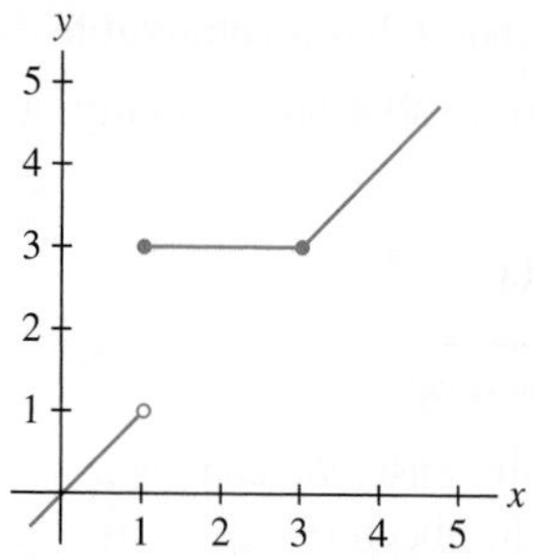

FIGURE 7 Piecewise-defined function $F(x)$ in Example 2.

Solution The functions $f(x) = x$ and $g(x) = 3$ are continuous, so $F(x)$ is also continuous, except possibly at the transition points $x = 1$ and $x = 3$, where the formula for $F(x)$ changes (Figure 7).

- At $x = 1$, the one-sided limits exist but are not equal:

$$\lim_{x\to 1-} F(x) = \lim_{x\to 1-} x = 1, \qquad \lim_{x\to 1+} F(x) = \lim_{x\to 1+} 3 = 3$$

  Thus $F(x)$ has a jump discontinuity at $x = 1$. However, the right-hand limit is equal to the function value $F(1) = 3$, so $F(x)$ is *right-continuous* at $x = 1$.
- At $x = 3$, the left- and right-hand limits exist and both are equal to $F(3)$, so $F(x)$ is *continuous* at $x = 3$:

$$\lim_{x\to 3-} F(x) = \lim_{x\to 3-} 3 = 3, \qquad \lim_{x\to 3+} F(x) = \lim_{x\to 3+} x = 3$$ ■

We say that $f(x)$ has an **infinite discontinuity** at $x = c$ if one or both of the one-sided limits is infinite (even if $f(x)$ itself is not defined at $x = c$). Figure 8 illustrates three types of infinite discontinuities occurring at $x = 2$. Notice that $x = 2$ does not belong to the domain of the function in cases (A) and (B).

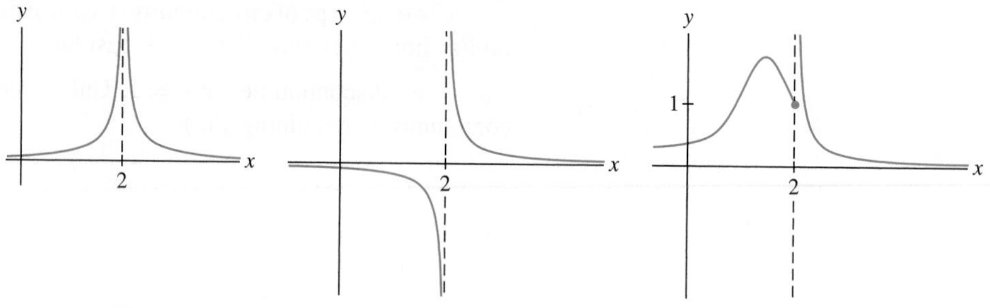

FIGURE 8 Functions with an infinite discontinuity at $x = 2$.

Finally, we note that some functions have more "severe" types of discontinuity than those discussed above. For example, $f(x) = \sin\frac{1}{x}$ oscillates infinitely often between $+1$ and -1 as $x \to 0$ (Figure 9). Neither the left- nor the right-hand limit exists at $x = 0$, so this discontinuity is not a jump discontinuity. See Exercises 88 and 89 for even stranger examples. Although of interest from a theoretical point of view, these discontinuities rarely arise in practice.

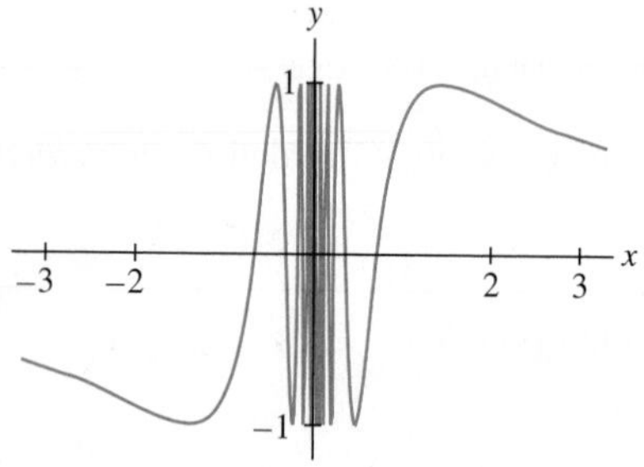

FIGURE 9 Graph of $y = \sin\frac{1}{x}$. The discontinuity at $x = 0$ is not a jump, removable, or infinite discontinuity.

Building Continuous Functions

Having studied some examples of discontinuities, we focus again on continuous functions. How can we show that a function is continuous? One way is to use the **Laws of Continuity**, which state, roughly speaking, that a function is continuous if it is built out of functions that are known to be continuous.

> **THEOREM 1 Basic Laws of Continuity** If $f(x)$ and $g(x)$ are continuous at $x = c$, then the following functions are also continuous at $x = c$:
>
> **(i)** $f(x) + g(x)$ and $f(x) - g(x)$
> **(ii)** $kf(x)$ for any constant k
> **(iii)** $f(x)\,g(x)$
> **(iv)** $f(x)/g(x)$ if $g(c) \neq 0$

Proof These laws follow directly from the corresponding Basic Limit Laws (Theorem 1, Section 2.3). We illustrate by proving the first part of (i) in detail. The remaining laws are proved similarly. By definition, we must show that $\lim_{x\to c}(f(x)+g(x)) = f(c)+g(c)$. Because $f(x)$ and $g(x)$ are both continuous at $x=c$, we have

$$\lim_{x\to c} f(x) = f(c), \qquad \lim_{x\to c} g(x) = g(c)$$

The Sum Law for limits yields the desired result:

$$\lim_{x\to c}(f(x)+g(x)) = \lim_{x\to c} f(x) + \lim_{x\to c} g(x) = f(c)+g(c) \qquad \blacksquare$$

In Section 2.3, we noted that the Basic Limit Laws for Sums and Products are valid for an arbitrary number of functions. The same is true for continuity; that is, if $f_1(x), \ldots, f_n(x)$ are continuous, then so are the functions

$$f_1(x)+f_2(x)+\cdots+f_n(x), \qquad f_1(x)\cdot f_2(x)\cdots f_n(x)$$

When a function $f(x)$ is defined and continuous for all values of x, we say that $f(x)$ is continuous on the real line.

The basic functions are continuous on their domains. Recall (Section 1.3) that the term *basic function* refers to polynomials, rational functions, nth-root and algebraic functions, trigonometric functions and their inverses, and exponential and logarithmic functions.

← REMINDER A rational function is a quotient of two polynomials $P(x)/Q(x)$.

THEOREM 2 Continuity of Polynomial and Rational Functions Let $P(x)$ and $Q(x)$ be polynomials. Then:

- $P(x)$ is continuous on the real line.
- $P(x)/Q(x)$ is continuous on its domain (at all values $x=c$ such that $Q(c)\neq 0$).

Proof The function x^m is continuous for all whole numbers m by Example 1. By Continuity Law (ii), ax^m is continuous for every constant a. A polynomial

$$P(x) = a_nx^n + a_{n-1}x^{n-1} + \cdots + a_1x + a_0$$

is a sum of continuous functions, so it too is continuous. By Continuity Law (iv), a quotient $P(x)/Q(x)$ is continuous at $x=c$, provided that $Q(c)\neq 0$. $\blacksquare$

This result shows, for example, that $f(x) = 3x^4 - 2x^3 + 8x$ is continuous for all x and that

$$g(x) = \frac{x+3}{x^2-1}$$

is continuous for $x\neq \pm 1$. Note that if n is a positive integer, then $f(x)=x^{-n}$ is continuous for $x\neq 0$ because $f(x)=x^{-n}=1/x^n$ is a rational function.

The continuity of the nth-root, trigonometric, and exponential functions should not be surprising because their graphs have no visible breaks (Figure 10). Similarly, logarithmic functions (introduced in Section 7.2) are continuous. However, complete proofs of continuity are somewhat technical and are omitted.

← REMINDER The domain of $y = x^{1/n}$ is the real line if n is odd and the half-line $[0,\infty)$ if n is even.

THEOREM 3 Continuity of Some Basic Functions

- $y = x^{1/n}$ is continuous on its domain for n a natural number.
- $y = \sin x$ and $y = \cos x$ are continuous on the real line.
- $y = b^x$ is continuous on the real line (for $b>0$, $b\neq 1$).
- $y = \log_b x$ is continuous for $x>0$ (for $b>0$, $b\neq 1$).

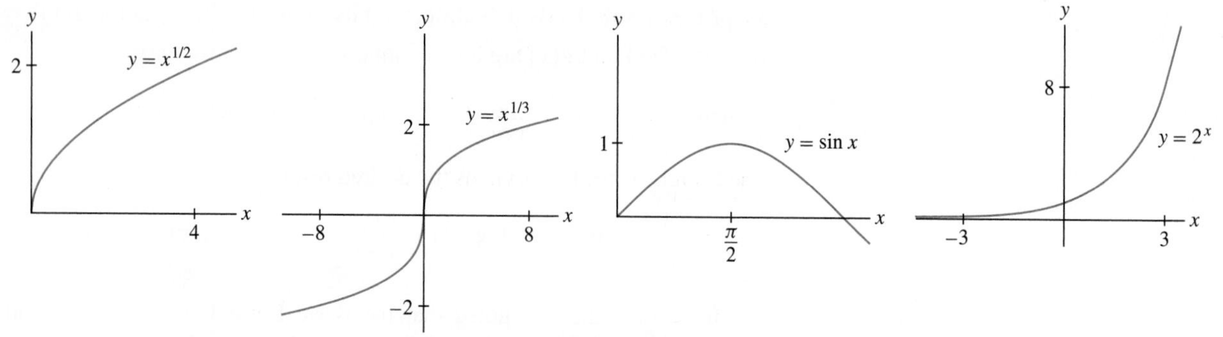

FIGURE 10 As the graphs suggest, these functions are continuous on their domains.

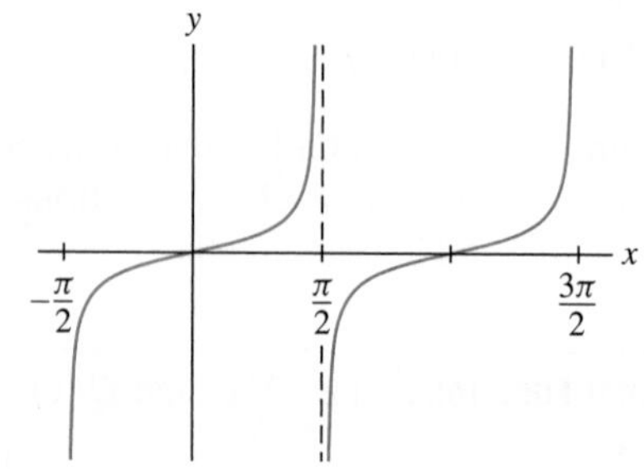

FIGURE 11 Graph of $y = \tan x$.

Because $\sin x$ and $\cos x$ are continuous, Continuity Law (iv) for Quotients implies that the other standard trigonometric functions are continuous on their domains, consisting of the values of x where their denominators are nonzero:

$$\tan x = \frac{\sin x}{\cos x}, \qquad \cot x = \frac{\cos x}{\sin x}, \qquad \sec x = \frac{1}{\cos x}, \qquad \csc x = \frac{1}{\sin x}$$

They have infinite discontinuities at points where their denominators are zero. For example, $\tan x$ has infinite discontinuities at the points (Figure 11)

$$x = \pm\frac{\pi}{2}, \quad \pm\frac{3\pi}{2}, \quad \pm\frac{5\pi}{2}, \ldots$$

Finally, it is important to know that a composition of continuous functions is again continuous. The following theorem is proved in Appendix D.

THEOREM 4 Continuity of Composite Functions If g is continuous at $x = c$, and f is continuous at $x = g(c)$, then the composite function $F(x) = f(g(x))$ is continuous at $x = c$.

For example, $F(x) = (x^2 + 9)^{1/3}$ is continuous because it is the composite of the continuous functions $f(x) = x^{1/3}$ and $g(x) = x^2 + 9$. Similarly, $F(x) = \cos(x^{-1})$ is continuous for all $x \neq 0$, and $F(x) = 2^{\sin x}$ is continuous for all x.

More generally, an **elementary function** is a function that is constructed out of basic functions using the operations of addition, subtraction, multiplication, division, and composition. Since the basic functions are continuous (on their domains), an elementary function is also continuous on its domain by the laws of continuity. An example of an elementary function is

$$F(x) = \sin\left(\frac{x^2 + \cos(2^x + 9)}{x - 8}\right)$$

This function is continuous on its domain $\{x : x \neq 8\}$.

Substitution: Evaluating Limits Using Continuity

It is easy to evaluate a limit when the function in question is known to be continuous. In this case, by definition, the limit is equal to the function value:

$$\lim_{x \to c} f(x) = f(c)$$

We call this the **Substitution Method** because the limit is evaluated by "plugging in" $x = c$.

■ **EXAMPLE 3** Evaluate **(a)** $\lim_{y \to \frac{\pi}{3}} \sin y$ and **(b)** $\lim_{x \to -1} \dfrac{3^x}{\sqrt{x+5}}$.

Solution

(a) We can use substitution because $f(y) = \sin y$ is continuous.

$$\lim_{y \to \frac{\pi}{3}} \sin y = \sin \frac{\pi}{3} = \frac{\sqrt{3}}{2}$$

(b) The function $f(x) = 3^x/\sqrt{x+5}$ is continuous at $x = -1$ because the numerator and denominator are both continuous at $x = -1$ and the denominator $\sqrt{x+5}$ is nonzero at $x = -1$. Therefore, we can use substitution:

$$\lim_{x \to -1} \frac{3^x}{\sqrt{x+5}} = \frac{3^{-1}}{\sqrt{-1+5}} = \frac{1}{6} \qquad ■$$

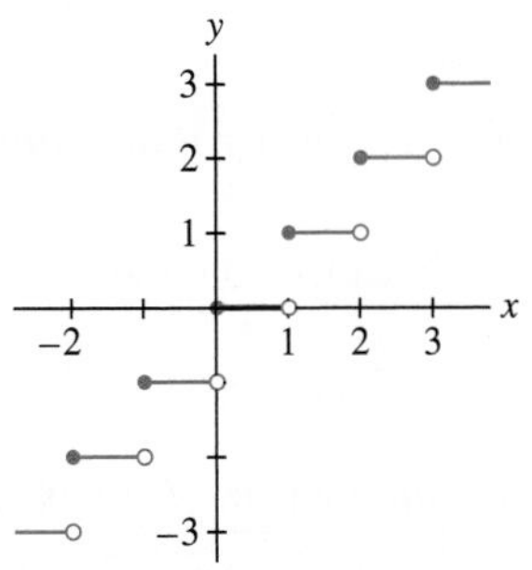

FIGURE 12 Graph of $f(x) = [x]$.

The **greatest integer function** $[x]$ is the function defined by $[x] = n$, where n is the unique integer such that $n \le x < n + 1$ [Figure 12]. For example, $[4.7] = 4$.

■ **EXAMPLE 4 Assumptions Matter** Can we evaluate $\lim_{x \to 2} [x]$ using substitution?

Solution Substitution cannot be applied because $f(x) = [x]$ is not continuous at $x = 2$. Although $f(2) = 2$, $\lim_{x \to 2} [x]$ does not exist because the one-sided limits are not equal:

$$\lim_{x \to 2+} [x] = 2 \quad \text{and} \quad \lim_{x \to 2-} [x] = 1 \qquad ■$$

CONCEPTUAL INSIGHT Real-World Modeling by Continuous Functions Continuous functions are used often to represent physical quantities such as velocity, temperature, and voltage. This reflects our everyday experience that change in the physical world tends to occur continuously rather than through abrupt transitions. However, mathematical models are at best approximations to reality, and it is important to be aware of their limitations.

In Figure 13, atmospheric temperature is represented as a continuous function of altitude. This is justified for large-scale objects such as the earth's atmosphere because the reading on a thermometer appears to vary continuously as altitude changes. However, temperature is a measure of the average kinetic energy of molecules. At the microscopic level, it would not be meaningful to treat temperature as a quantity that varies continuously from point to point.

Similarly, the size $P(t)$ of a population is usually treated as a continuous function of time t. Strictly speaking, $P(t)$ is a whole number that changes by ± 1 when an individual is born or dies, so it is not continuous, but if the population is large, the effect of an individual birth or death is small, and it is both reasonable and convenient to treat $P(t)$ as a continuous function.

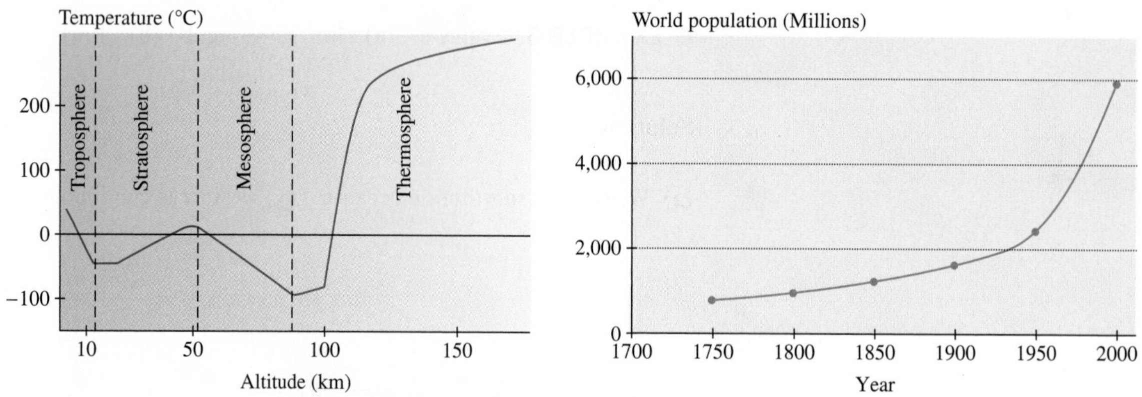

FIGURE 13 Atmospheric temperature and world population are represented by continuous graphs.

2.4 SUMMARY

- Definition: $f(x)$ is *continuous* at $x = c$ if $\lim_{x\to c} f(x) = f(c)$.
- If $\lim_{x\to c} f(x)$ does not exist, or if it exists but does not equal $f(c)$, then f is *discontinuous* at $x = c$.
- If $f(x)$ is continuous at all points in its domain, f is simply called *continuous*.
- *Right-continuous* at $x = c$: $\lim_{x\to c+} f(x) = f(c)$.
- *Left-continuous* at $x = c$: $\lim_{x\to c-} f(x) = f(c)$.
- Three common types of discontinuities: *removable discontinuity* $\left[\lim_{x\to c} f(x)\right.$ exists but does not equal $\left.f(c)\right]$, *jump discontinuity* (the one-sided limits both exist but are not equal), and *infinite discontinuity* (the limit is infinite as x approaches c from one or both sides).
- Laws of Continuity: Sums, products, multiples, inverses, and composites of continuous functions are again continuous. The same holds for a quotient $f(x)/g(x)$ at points where $g(x) \neq 0$.
- Basic functions: Polynomials, rational functions, nth-root and algebraic functions, trigonometric functions and their inverses, exponential and logarithmic functions. Basic functions are continuous on their domains.
- Substitution Method: If $f(x)$ is known to be continuous at $x = c$, then the value of the limit $\lim_{x\to c} f(x)$ is $f(c)$.

2.4 EXERCISES

Preliminary Questions

1. Which property of $f(x) = x^3$ allows us to conclude that $\lim_{x\to 2} x^3 = 8$?

2. What can be said about $f(3)$ if f is continuous and $\lim_{x\to 3} f(x) = \frac{1}{2}$?

3. Suppose that $f(x) < 0$ if x is positive and $f(x) > 1$ if x is negative. Can f be continuous at $x = 0$?

4. Is it possible to determine $f(7)$ if $f(x) = 3$ for all $x < 7$ and f is right-continuous at $x = 7$? What if f is left-continuous?

5. Are the following true or false? If false, state a correct version.

(a) $f(x)$ is continuous at $x = a$ if the left- and right-hand limits of $f(x)$ as $x \to a$ exist and are equal.

(b) $f(x)$ is continuous at $x = a$ if the left- and right-hand limits of $f(x)$ as $x \to a$ exist and equal $f(a)$.

(c) If the left- and right-hand limits of $f(x)$ as $x \to a$ exist, then f has a removable discontinuity at $x = a$.

(d) If $f(x)$ and $g(x)$ are continuous at $x = a$, then $f(x) + g(x)$ is continuous at $x = a$.

(e) If $f(x)$ and $g(x)$ are continuous at $x = a$, then $f(x)/g(x)$ is continuous at $x = a$.

Exercises

1. Referring to Figure 14, state whether $f(x)$ is left- or right-continuous (or neither) at each point of discontinuity. Does $f(x)$ have any removable discontinuities?

Exercises 2–4 refer to the function $g(x)$ in Figure 15.

2. State whether $g(x)$ is left- or right-continuous (or neither) at each of its points of discontinuity.

3. At which point c does $g(x)$ have a removable discontinuity? How should $g(c)$ be redefined to make g continuous at $x = c$?

4. Find the point c_1 at which $g(x)$ has a jump discontinuity but is left-continuous. How should $g(c_1)$ be redefined to make g right-continuous at $x = c_1$?

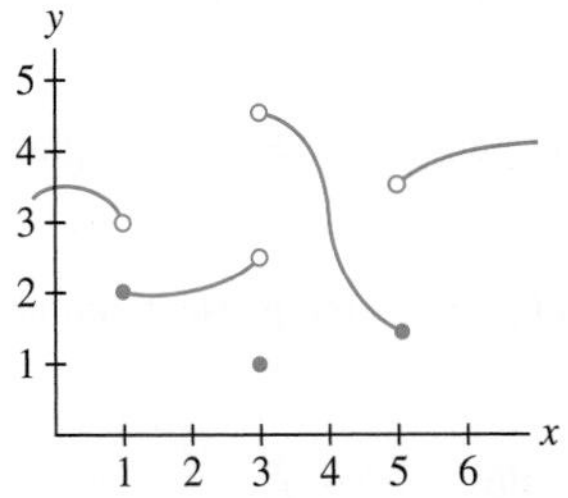

FIGURE 14 Graph of $y = f(x)$

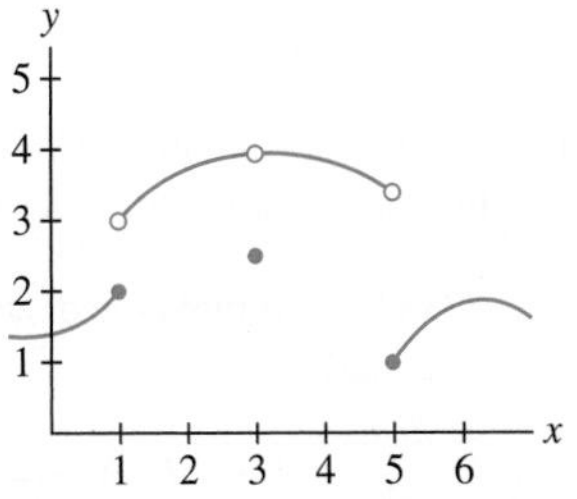

FIGURE 15 Graph of $y = g(x)$

5. In Figure 16, determine the one-sided limits at the points of discontinuity. Which discontinuity is removable and how should f be redefined to make it continuous at this point?

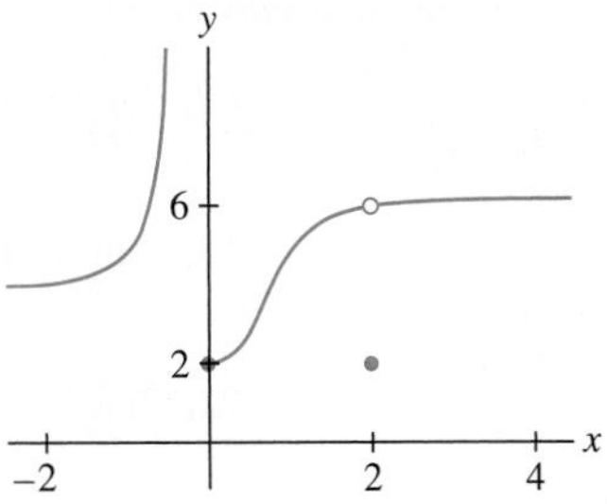

FIGURE 16

6. Suppose that $f(x) = 2$ for $x < 3$ and $f(x) = -4$ for $x > 3$.
(a) What is $f(3)$ if f is left-continuous at $x = 3$?
(b) What is $f(3)$ if f is right-continuous at $x = 3$?

In Exercises 7–16, use the Laws of Continuity and Theorems 2 and 3 to show that the function is continuous.

7. $f(x) = x + \sin x$

8. $f(x) = x \sin x$

9. $f(x) = 3x + 4 \sin x$

10. $f(x) = 3x^3 + 8x^2 - 20x$

11. $f(x) = \dfrac{1}{x^2 + 1}$

12. $f(x) = \dfrac{x^2 - \cos x}{3 + \cos x}$

13. $f(x) = \cos(x^2)$

14. $f(x) = \tan(4^x)$

15. $f(x) = 2^x \cos 3x$

16. $f(x) = \tan\left(\dfrac{1}{x^2 + 1}\right)$

In Exercises 17–34, determine the points of discontinuity. State the type of discontinuity (removable, jump, infinite, or none of these) and whether the function is left- or right-continuous.

17. $f(x) = \dfrac{1}{x}$

18. $f(x) = |x|$

19. $f(x) = \dfrac{x - 2}{|x - 1|}$

20. $f(x) = [x]$

21. $f(x) = \left[\dfrac{1}{2}x\right]$

22. $g(t) = \dfrac{1}{t^2 - 1}$

23. $f(x) = \dfrac{x + 1}{4x - 2}$

24. $h(z) = \dfrac{1 - 2z}{z^2 - z - 6}$

25. $f(x) = 3x^{2/3} - 9x^3$

26. $g(t) = 3t^{-2/3} - 9t^3$

27. $f(x) = \begin{cases} \dfrac{x - 2}{|x - 2|} & x \neq 2 \\ -1 & x = 2 \end{cases}$

28. $f(x) = \begin{cases} \cos \dfrac{1}{x} & x \neq 0 \\ 1 & x = 0 \end{cases}$

29. $g(t) = \tan 2t$

30. $f(x) = \csc(x^2)$

31. $f(x) = \tan(\sin x)$

32. $f(x) = \cos(\pi[x])$

33. $f(x) = \dfrac{1}{2^x - 2^{-x}}$

34. $f(x) = 2[x/2] - 4[x/4]$

In Exercises 35–48, determine the domain of the function and prove that it is continuous on its domain using the Laws of Continuity and the facts quoted in this section.

35. $f(x) = 2 \sin x + 3 \cos x$

36. $f(x) = \sqrt{x^2 + 9}$

37. $f(x) = \sqrt{x} \sin x$

38. $f(x) = \dfrac{x^2}{x + x^{1/4}}$

39. $f(x) = x^{2/3} 2^x$

40. $f(x) = x^{1/3} + x^{3/4}$

41. $f(x) = x^{-4/3}$

42. $f(x) = \cos^3 x$

43. $f(x) = \tan^2 x$

44. $f(x) = \cos(2^x)$

45. $f(x) = (x^4 + 1)^{3/2}$

46. $f(x) = 3^{-x^2}$

47. $f(x) = \dfrac{\cos(x^2)}{x^2 - 1}$

48. $f(x) = 9^{\tan x}$

49. Show that the function

$$f(x) = \begin{cases} x^2 + 3 & \text{for } x < 1 \\ 10 - x & \text{for } 1 \le x \le 2 \\ 6x - x^2 & \text{for } x > 2 \end{cases}$$

is continuous for $x \neq 1, 2$. Then compute the right- and left-hand limits at $x = 1, 2$, and determine whether $f(x)$ is left-continuous, right-continuous, or continuous at these points (Figure 17).

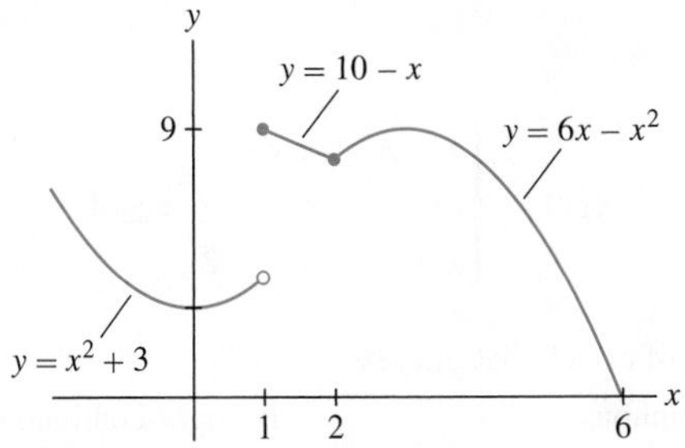

FIGURE 17

50. Sawtooth Function Draw the graph of $f(x) = x - [x]$. At which points is f discontinuous? Is it left- or right-continuous at those points?

In Exercises 51–54, sketch the graph of $f(x)$. At each point of discontinuity, state whether f is left- or right-continuous.

51. $f(x) = \begin{cases} x^2 & \text{for } x \le 1 \\ 2 - x & \text{for } x > 1 \end{cases}$

52. $f(x) = \begin{cases} x + 1 & \text{for } x < 1 \\ \dfrac{1}{x} & \text{for } x \ge 1 \end{cases}$

53. $f(x) = \begin{cases} \dfrac{x^2 - 3x + 2}{|x - 2|} & x \ne 2 \\ 0 & x = 2 \end{cases}$

54. $f(x) = \begin{cases} x^3 + 1 & \text{for } -\infty < x \le 0 \\ -x + 1 & \text{for } 0 < x < 2 \\ -x^2 + 10x - 15 & \text{for } x \ge 2 \end{cases}$

55. Show that the function

$$f(x) = \begin{cases} \dfrac{x^2 - 16}{x - 4} & x \ne 4 \\ 10 & x = 4 \end{cases}$$

has a removable discontinuity at $x = 4$.

56. GU Define $f(x) = x \sin \frac{1}{x} + 2$ for $x \ne 0$. Plot $f(x)$. How should $f(0)$ be defined so that f is continuous at $x = 0$?

In Exercises 57–59, find the value of the constant (a, b, or c) that makes the function continuous.

57. $f(x) = \begin{cases} x^2 - c & \text{for } x < 5 \\ 4x + 2c & \text{for } x \ge 5 \end{cases}$

58. $f(x) = \begin{cases} 2x + 9x^{-1} & \text{for } x \le 3 \\ -4x + c & \text{for } x > 3 \end{cases}$

59. $f(x) = \begin{cases} x^{-1} & \text{for } x < -1 \\ ax + b & \text{for } -1 \le x \le \frac{1}{2} \\ x^{-1} & \text{for } x > \frac{1}{2} \end{cases}$

60. Define

$$g(x) = \begin{cases} x + 3 & \text{for } x < -1 \\ cx & \text{for } -1 \le x \le 2 \\ x + 2 & \text{for } x > 2 \end{cases}$$

Find a value of c such that $g(x)$ is

(a) left-continuous **(b)** right-continuous

In each case, sketch the graph of $g(x)$.

61. Define $g(t) = 2^{1/(t-1)}$ for $t \ne 1$. Answer the following questions, using a plot if necessary.

(a) Can $g(1)$ be defined so that $g(t)$ is continuous at $t = 1$?

(b) How should $g(1)$ be defined so that $g(t)$ is left-continuous at $t = 1$?

62. Each of the following statements is *false*. For each statement, sketch the graph of a function that provides a counterexample.

(a) If $\lim_{x\to a} f(x)$ exists, then $f(x)$ is continuous at $x = a$.

(b) If $f(x)$ has a jump discontinuity at $x = a$, then $f(a)$ is equal to either $\lim_{x\to a-} f(x)$ or $\lim_{x\to a+} f(x)$.

In Exercises 63–66, draw the graph of a function on $[0, 5]$ with the given properties.

63. $f(x)$ is not continuous at $x = 1$, but $\lim_{x\to 1+} f(x)$ and $\lim_{x\to 1-} f(x)$ exist and are equal.

64. $f(x)$ is left-continuous but not continuous at $x = 2$ and right-continuous but not continuous at $x = 3$.

65. $f(x)$ has a removable discontinuity at $x = 1$, a jump discontinuity at $x = 2$, and

$$\lim_{x\to 3-} f(x) = -\infty, \qquad \lim_{x\to 3+} f(x) = 2$$

66. $f(x)$ is right- but not left-continuous at $x = 1$, left- but not right-continuous at $x = 2$, and neither left- nor right-continuous at $x = 3$.

In Exercises 67–80, evaluate using substitution.

67. $\lim_{x\to -1} (2x^3 - 4)$

68. $\lim_{x\to 2} (5x - 12x^{-2})$

69. $\lim_{x\to 3} \dfrac{x + 2}{x^2 + 2x}$

70. $\lim_{x\to \pi} \sin\left(\dfrac{x}{2} - \pi\right)$

71. $\lim_{x\to \frac{\pi}{4}} \tan(3x)$

72. $\lim_{x\to \pi} \dfrac{1}{\cos x}$

73. $\lim_{x\to 4} x^{-5/2}$

74. $\lim_{x\to 2} \sqrt{x^3 + 4x}$

75. $\lim_{x\to -1} (1 - 8x^3)^{3/2}$

76. $\lim_{x\to 2} \left(\dfrac{7x + 2}{4 - x}\right)^{2/3}$

77. $\lim_{x\to 3} 10^{x^2 - 2x}$

78. $\lim_{x\to -\frac{\pi}{2}} 3^{\sin x}$

79. $\lim_{x\to \frac{\pi}{3}} \sin^2(\pi \sin^2 x)$

80. $\lim_{x\to 1} \tan\left(e^{x-1}\right)$

81. Suppose that $f(x)$ and $g(x)$ are discontinuous at $x = c$. Does it follow that $f(x) + g(x)$ is discontinuous at $x = c$? If not, give a counterexample. Does this contradict Theorem 1 (i)?

82. Prove that $f(x) = |x|$ is continuous for all x. *Hint:* To prove continuity at $x = 0$, consider the one-sided limits.

83. Use the result of Exercise 82 to prove that if $g(x)$ is continuous, then $f(x) = |g(x)|$ is also continuous.

84. Which of the following quantities would be represented by continuous functions of time and which would have one or more discontinuities?

(a) Velocity of an airplane during a flight

(b) Temperature in a room under ordinary conditions

(c) Value of a bank account with interest paid yearly

(d) The salary of a teacher

(e) The population of the world

85. In 2009, the federal income tax $T(x)$ on income of x dollars (up to \$82,250) was determined by the formula

$$T(x) = \begin{cases} 0.10x & \text{for } 0 \le x < 8350 \\ 0.15x - 417.50 & \text{for } 8350 \le x < 33{,}950 \\ 0.25x - 3812.50 & \text{for } 33{,}950 \le x < 82{,}250 \end{cases}$$

Sketch the graph of $T(x)$. Does $T(x)$ have any discontinuities? Explain why, if $T(x)$ had a jump discontinuity, it might be advantageous in some situations to earn *less* money.

Further Insights and Challenges

86. If $f(x)$ has a removable discontinuity at $x = c$, then it is possible to redefine $f(c)$ so that $f(x)$ is continuous at $x = c$. Can this be done in more than one way?

87. Give an example of functions $f(x)$ and $g(x)$ such that $f(g(x))$ is continuous but $g(x)$ has at least one discontinuity.

88. Continuous at Only One Point Show that the following function is continuous only at $x = 0$:

$$f(x) = \begin{cases} x & \text{for } x \text{ rational} \\ -x & \text{for } x \text{ irrational} \end{cases}$$

89. Show that $f(x)$ is a discontinuous function for all x where $f(x)$ is defined as follows:

$$f(x) = \begin{cases} 1 & \text{for } x \text{ rational} \\ -1 & \text{for } x \text{ irrational} \end{cases}$$

Show that $f(x)^2$ is continuous for all x.

2.5 Evaluating Limits Algebraically

Substitution can be used to evaluate limits when the function in question is known to be continuous. For example, $f(x) = x^{-2}$ is continuous at $x = 3$, and therefore,

$$\lim_{x\to 3} x^{-2} = 3^{-2} = \frac{1}{9}$$

When we study derivatives in Chapter 3, we will be faced with limits $\lim_{x\to c} f(x)$, where $f(c)$ is not defined. In such cases, substitution cannot be used directly. However, many of these limits can be evaluated if we use algebra to rewrite the formula for $f(x)$.

To illustrate, consider the limit (Figure 1).

$$\lim_{x\to 4} \frac{x^2 - 16}{x - 4}$$

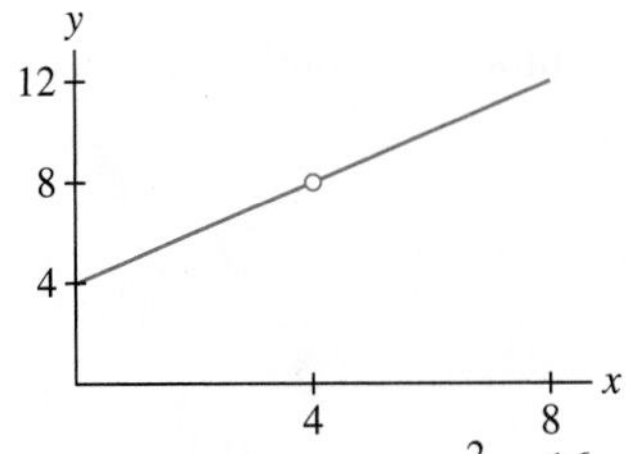

FIGURE 1 Graph of $f(x) = \dfrac{x^2 - 16}{x - 4}$. This function is undefined at $x = 4$, but the limit as $x \to 4$ exists.

The function $f(x) = \dfrac{x^2 - 16}{x - 4}$ is not defined at $x = 4$ because the formula for $f(4)$ produces the undefined expression $0/0$. However, the numerator of $f(x)$ factors:

$$\frac{x^2 - 16}{x - 4} = \frac{(x + 4)(x - 4)}{x - 4} = x + 4 \quad (\text{valid for } x \ne 4)$$

This shows that $f(x)$ coincides with the *continuous* function $x + 4$ for all $x \ne 4$. Since the limit depends only on the values of $f(x)$ for $x \ne 4$, we have

$$\lim_{x\to 4} \frac{x^2 - 16}{x - 4} = \underbrace{\lim_{x\to 4} (x + 4) = 8}_{\text{Evaluate by substitution}}$$

We say that $f(x)$ has an **indeterminate form** (or is **indeterminate**) at $x = c$ if the formula for $f(c)$ yields an undefined expression of the type

Other indeterminate forms are 1^∞, ∞^0, and 0^0. These are treated in Section 7.7.

$$\frac{0}{0}, \quad \frac{\infty}{\infty}, \quad \infty \cdot 0, \quad \infty - \infty$$

Our strategy, when this occurs, is to *transform $f(x)$ algebraically, if possible, into a new expression that is defined and continuous at $x = c$, and then evaluate the limit by substitution ("plugging in")*. As you study the following examples, notice that the critical step is to cancel a common factor from the numerator and denominator at the appropriate moment, thereby removing the indeterminacy.

EXAMPLE 1 Calculate $\lim_{x\to 3} \frac{x^2 - 4x + 3}{x^2 + x - 12}$.

Solution The function has the indeterminate form 0/0 at $x = 3$ because

Numerator at $x = 3$: $\quad x^2 - 4x + 3 = 3^2 - 4(3) + 3 = 0$

Denominator at $x = 3$: $\quad x^2 + x - 12 = 3^2 + 3 - 12 = 0$

***Step 1.* Transform algebraically and cancel.**

$$\frac{x^2 - 4x + 3}{x^2 + x - 12} = \underbrace{\frac{(x-3)(x-1)}{(x-3)(x+4)}}_{\text{Cancel common factor}} = \underbrace{\frac{x-1}{x+4}}_{\text{Continuous at } x=3} \qquad (\text{if } x \neq 3) \qquad \boxed{1}$$

***Step 2.* Substitute (evaluate using continuity).**

Because the expression on the right in Eq. (1) is *continuous* at $x = 3$,

$$\lim_{x\to 3} \frac{x^2 - 4x + 3}{x^2 + x - 12} = \underbrace{\lim_{x\to 3} \frac{x-1}{x+4} = \frac{2}{7}}_{\text{Evaluate by substitution}}$$

■

EXAMPLE 2 The Form $\frac{\infty}{\infty}$ Calculate $\lim_{x\to \frac{\pi}{2}} \frac{\tan x}{\sec x}$.

Solution As we see in Figure 2, both $\tan x$ and $\sec x$ have infinite discontinuities at $x = \frac{\pi}{2}$, so this limit has the indeterminate form ∞/∞ at $x = \frac{\pi}{2}$.

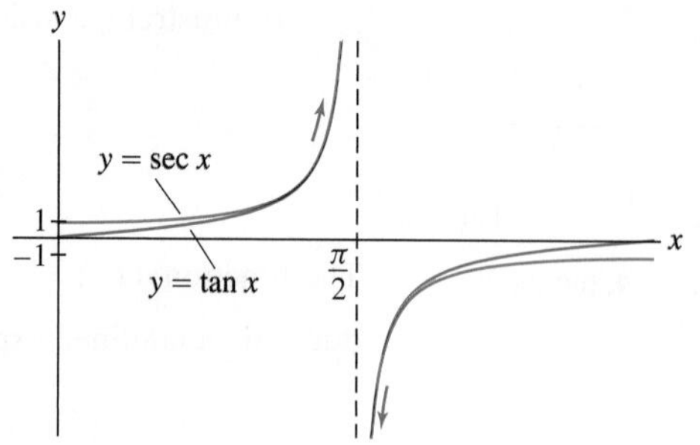

FIGURE 2

***Step 1.* Transform algebraically and cancel.**

$$\frac{\tan x}{\sec x} = \frac{(\sin x)\left(\dfrac{1}{\cos x}\right)}{\dfrac{1}{\cos x}} = \sin x \qquad (\text{if } \cos x \neq 0)$$

Step 2. **Substitute (evaluate using continuity).**
Because $\sin x$ is continuous,

$$\lim_{x\to\frac{\pi}{2}} \frac{\tan x}{\sec x} = \lim_{x\to\frac{\pi}{2}} \sin x = \sin\frac{\pi}{2} = 1$$ ■

The next example illustrates the algebraic technique of "multiplying by the conjugate," which can be used to treat some indeterminate forms involving square roots.

■ **EXAMPLE 3 Multiplying by the Conjugate** Evaluate $\lim_{x\to 4} \frac{\sqrt{x}-2}{x-4}$.

Solution We check that $f(x) = \frac{\sqrt{x}-2}{x-4}$ has the indeterminate form $0/0$ at $x = 4$:

Numerator at $x = 4$: $\quad \sqrt{x}-2 = \sqrt{4}-2 = 0$

Denominator at $x = 4$: $\quad x-4 = 4-4 = 0$

Step 1. **Multiply by the conjugate and cancel.**

Note, in Step 1, that the conjugate of $\sqrt{x}-2$ is $\sqrt{x}+2$, so $(\sqrt{x}-2)(\sqrt{x}+2) = x-4$.

$$\left(\frac{\sqrt{x}-2}{x-4}\right)\left(\frac{\sqrt{x}+2}{\sqrt{x}+2}\right) = \frac{x-4}{(x-4)(\sqrt{x}+2)} = \frac{1}{\sqrt{x}+2} \qquad (\text{if } x \neq 4)$$

Step 2. **Substitute (evaluate using continuity).**
Because $1/(\sqrt{x}+2)$ is continuous at $x = 4$,

$$\lim_{x\to 4} \frac{\sqrt{x}-2}{x-4} = \lim_{x\to 4} \frac{1}{\sqrt{x}+2} = \frac{1}{4}$$

■ **EXAMPLE 4** Evaluate $\lim_{h\to 5} \frac{h-5}{\sqrt{h+4}-3}$.

Solution We note that $f(h) = \frac{h-5}{\sqrt{h+4}-3}$ yields $0/0$ at $h = 5$:

Numerator at $h = 5$: $\quad h-5 = 5-5 = 0$

Denominator at $h = 5$: $\quad \sqrt{h+4}-3 = \sqrt{5+4}-3 = 0$

The conjugate of $\sqrt{h+4}-3$ is $\sqrt{h+4}+3$, and

$$\frac{h-5}{\sqrt{h+4}-3} = \left(\frac{h-5}{\sqrt{h+4}-3}\right)\left(\frac{\sqrt{h+4}+3}{\sqrt{h+4}+3}\right) = \frac{(h-5)\left(\sqrt{h+4}+3\right)}{\left(\sqrt{h+4}-3\right)\left(\sqrt{h+4}+3\right)}$$

The denominator is equal to

$$\left(\sqrt{h+4}-3\right)\left(\sqrt{h+4}+3\right) = \left(\sqrt{h+4}\right)^2 - 9 = h-5$$

Thus, for $h \neq 5$,

$$f(h) = \frac{h-5}{\sqrt{h+4}-3} = \frac{(h-5)\left(\sqrt{h+4}+3\right)}{h-5} = \sqrt{h+4}+3$$

We obtain

$$\lim_{h\to 5} \frac{h-5}{\sqrt{h+4}-3} = \lim_{h\to 5} \left(\sqrt{h+4}+3\right) = \sqrt{9}+3 = 6$$ ■

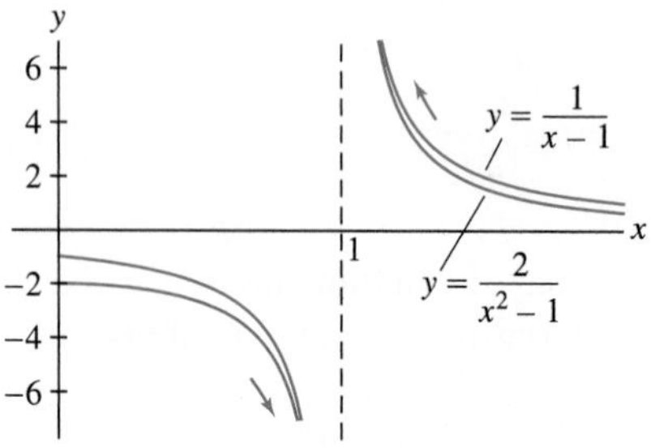

FIGURE 3

■ **EXAMPLE 5** **The Form $\infty - \infty$** Calculate $\lim_{x\to 1}\left(\frac{1}{x-1} - \frac{2}{x^2-1}\right)$.

Solution As we see in Figure 3, $\frac{1}{x-1}$ and $\frac{2}{x^2-1}$ both have infinite discontinuities at $x = 1$, so this limit has the indeterminate form $\infty - \infty$.

Step 1. **Transform algebraically and cancel.**

Combine the fractions and simplify (for $x \neq 1$):

$$\frac{1}{x-1} - \frac{2}{x^2-1} = \frac{x+1}{x^2-1} - \frac{2}{x^2-1} = \frac{x-1}{x^2-1} = \frac{x-1}{(x-1)(x+1)} = \frac{1}{x+1}$$

Step 2. **Substitute (evaluate using continuity).**

$$\lim_{x\to 1}\left(\frac{1}{x-1} - \frac{2}{x^2-1}\right) = \lim_{x\to 1}\frac{1}{x+1} = \frac{1}{1+1} = \frac{1}{2}$$ ■

In the next example, the function has the undefined form $a/0$ with a nonzero. This is *not* an indeterminate form (it is not of the form $0/0$, ∞/∞, etc.).

■ **EXAMPLE 6** **Infinite But Not Indeterminate** Evaluate $\lim_{x\to 2}\frac{x^2-x+5}{x-2}$.

Solution The function $f(x) = \frac{x^2-x+5}{x-2}$ is undefined at $x = 2$ because the formula for $f(2)$ yields $7/0$:

Numerator at $x = 2$: $\quad x^2 - x + 5 = 2^2 - 2 + 5 = 7$

Denominator at $x = 2$: $\quad x - 2 = 2 - 2 = 0$

But $f(x)$ *is not indeterminate* at $x = 2$ because $7/0$ is not an indeterminate form. Figure 4 suggests that the one-sided limits are infinite:

$$\lim_{x\to 2-}\frac{x^2-x+5}{x-2} = -\infty, \qquad \lim_{x\to 2+}\frac{x^2-x+5}{x-2} = \infty$$

The limit itself does not exist. ■

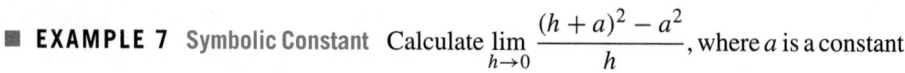

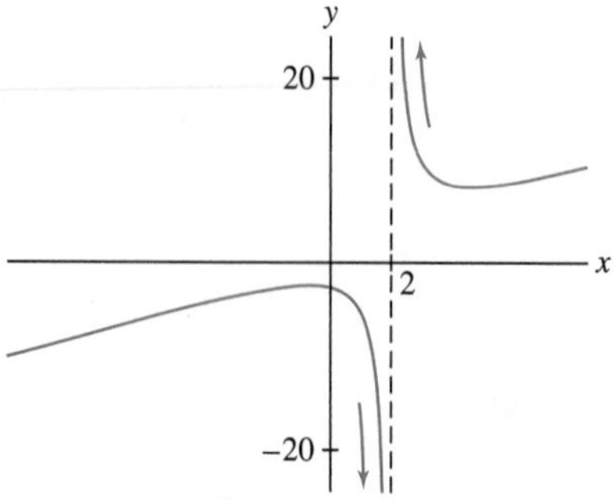

FIGURE 4 Graph of $f(x) = \frac{x^2-x+5}{x-2}$.

As preparation for the derivative in Chapter 3, we evaluate a limit involving a symbolic constant.

■ **EXAMPLE 7** **Symbolic Constant** Calculate $\lim_{h\to 0}\frac{(h+a)^2-a^2}{h}$, where a is a constant.

Solution We have the indeterminate form $0/0$ at $h = 0$ because

Numerator at $h = 0$: $\quad (h+a)^2 - a^2 = (0+a)^2 - a^2 = 0$

Denominator at $h = 0$: $\quad h = 0$

Expand the numerator and simplify (for $h \neq 0$):

$$\frac{(h+a)^2-a^2}{h} = \frac{(h^2+2ah+a^2)-a^2}{h} = \frac{h^2+2ah}{h} = \frac{h(h+2a)}{h} = h + 2a$$

The function $h + 2a$ is continuous (for any constant a), so

$$\lim_{h\to 0}\frac{(h+a)^2-a^2}{h} = \lim_{h\to 0}(h+2a) = 2a$$ ■

2.5 SUMMARY

- When $f(x)$ is known to be continuous at $x = c$, the limit can be evaluated by substitution: $\lim_{x\to c} f(x) = f(c)$.
- We say that $f(x)$ is *indeterminate* (or has an *indeterminate form*) at $x = c$ if the formula for $f(c)$ yields an undefined expression of the type

$$\frac{0}{0}, \quad \frac{\infty}{\infty}, \quad \infty\cdot 0, \quad \infty - \infty$$

- If $f(x)$ is indeterminate at $x = c$, try to transform $f(x)$ algebraically into a new expression that is defined and continuous at $x = c$. Then evaluate by substitution.

2.5 EXERCISES

Preliminary Questions

1. Which of the following is indeterminate at $x = 1$?

$$\frac{x^2+1}{x-1}, \quad \frac{x^2-1}{x+2}, \quad \frac{x^2-1}{\sqrt{x+3}-2}, \quad \frac{x^2+1}{\sqrt{x+3}-2}$$

2. Give counterexamples to show that these statements are false:

(a) If $f(c)$ is indeterminate, then the right- and left-hand limits as $x \to c$ are not equal.

(b) If $\lim_{x\to c} f(x)$ exists, then $f(c)$ is not indeterminate.

(c) If $f(x)$ is undefined at $x = c$, then $f(x)$ has an indeterminate form at $x = c$.

3. The method for evaluating limits discussed in this section is sometimes called "simplify and plug in." Explain how it actually relies on the property of continuity.

Exercises

In Exercises 1–4, show that the limit leads to an indeterminate form. Then carry out the two-step procedure: Transform the function algebraically and evaluate using continuity.

1. $\displaystyle\lim_{x\to 6} \frac{x^2-36}{x-6}$

2. $\displaystyle\lim_{h\to 3} \frac{9-h^2}{h-3}$

3. $\displaystyle\lim_{x\to -1} \frac{x^2+2x+1}{x+1}$

4. $\displaystyle\lim_{t\to 9} \frac{2t-18}{5t-45}$

In Exercises 5–34, evaluate the limit, if it exists. If not, determine whether the one-sided limits exist (finite or infinite).

5. $\displaystyle\lim_{x\to 7} \frac{x-7}{x^2-49}$

6. $\displaystyle\lim_{x\to 8} \frac{x^2-64}{x-9}$

7. $\displaystyle\lim_{x\to -2} \frac{x^2+3x+2}{x+2}$

8. $\displaystyle\lim_{x\to 8} \frac{x^3-64x}{x-8}$

9. $\displaystyle\lim_{x\to 5} \frac{2x^2-9x-5}{x^2-25}$

10. $\displaystyle\lim_{h\to 0} \frac{(1+h)^3-1}{h}$

11. $\displaystyle\lim_{x\to -\frac{1}{2}} \frac{2x+1}{2x^2+3x+1}$

12. $\displaystyle\lim_{x\to 3} \frac{x^2-x}{x^2-9}$

13. $\displaystyle\lim_{x\to 2} \frac{3x^2-4x-4}{2x^2-8}$

14. $\displaystyle\lim_{h\to 0} \frac{(3+h)^3-27}{h}$

15. $\displaystyle\lim_{t\to 0} \frac{4^{2t}-1}{4^t-1}$

16. $\displaystyle\lim_{h\to 4} \frac{(h+2)^2-9h}{h-4}$

17. $\displaystyle\lim_{x\to 16} \frac{\sqrt{x}-4}{x-16}$

18. $\displaystyle\lim_{t\to -2} \frac{2t+4}{12-3t^2}$

19. $\displaystyle\lim_{y\to 3} \frac{y^2+y-12}{y^3-10y+3}$

20. $\displaystyle\lim_{h\to 0} \frac{\frac{1}{(h+2)^2}-\frac{1}{4}}{h}$

21. $\displaystyle\lim_{h\to 0} \frac{\sqrt{2+h}-2}{h}$

22. $\displaystyle\lim_{x\to 8} \frac{\sqrt{x-4}-2}{x-8}$

23. $\displaystyle\lim_{x\to 4} \frac{x-4}{\sqrt{x}-\sqrt{8-x}}$

24. $\displaystyle\lim_{x\to 4} \frac{\sqrt{5-x}-1}{2-\sqrt{x}}$

25. $\displaystyle\lim_{x\to 4} \left(\frac{1}{\sqrt{x}-2}-\frac{4}{x-4}\right)$

26. $\displaystyle\lim_{x\to 0+} \left(\frac{1}{\sqrt{x}}-\frac{1}{\sqrt{x^2+x}}\right)$

27. $\displaystyle\lim_{x\to 0} \frac{\cot x}{\csc x}$

28. $\displaystyle\lim_{\theta\to \frac{\pi}{2}} \frac{\cot\theta}{\csc\theta}$

29. $\displaystyle\lim_{t\to 2} \frac{2^{2t}+2^t-20}{2^t-4}$

30. $\displaystyle\lim_{x\to 1} \left(\frac{1}{1-x}-\frac{2}{1-x^2}\right)$

31. $\displaystyle\lim_{x\to \frac{\pi}{4}} \frac{\sin x-\cos x}{\tan x-1}$

32. $\displaystyle\lim_{\theta\to \frac{\pi}{2}} (\sec\theta-\tan\theta)$

33. $\displaystyle\lim_{\theta\to \frac{\pi}{4}} \left(\frac{1}{\tan\theta-1}-\frac{2}{\tan^2\theta-1}\right)$

34. $\displaystyle\lim_{x\to \frac{\pi}{3}} \frac{2\cos^2 x+3\cos x-2}{2\cos x-1}$

35. GU Use a plot of $f(x) = \dfrac{x-4}{\sqrt{x}-\sqrt{8-x}}$ to estimate $\lim_{x\to 4} f(x)$ to two decimal places. Compare with the answer obtained algebraically in Exercise 23.

36. GU Use a plot of $f(x) = \dfrac{1}{\sqrt{x}-2} - \dfrac{4}{x-4}$ to estimate $\lim_{x\to 4} f(x)$ numerically. Compare with the answer obtained algebraically in Exercise 25.

In Exercises 37–42, evaluate using the identity

$$a^3 - b^3 = (a-b)(a^2+ab+b^2)$$

37. $\lim_{x\to 2} \dfrac{x^3-8}{x-2}$

38. $\lim_{x\to 3} \dfrac{x^3-27}{x^2-9}$

39. $\lim_{x\to 1} \dfrac{x^2-5x+4}{x^3-1}$

40. $\lim_{x\to -2} \dfrac{x^3+8}{x^2+6x+8}$

41. $\lim_{x\to 1} \dfrac{x^4-1}{x^3-1}$

42. $\lim_{x\to 27} \dfrac{x-27}{x^{1/3}-3}$

43. Evaluate $\lim_{h\to 0} \dfrac{\sqrt[4]{1+h}-1}{h}$. *Hint:* Set $x = \sqrt[4]{1+h}$ and rewrite as a limit as $x \to 1$.

44. Evaluate $\lim_{h\to 0} \dfrac{\sqrt[3]{1+h}-1}{\sqrt[2]{1+h}-1}$. *Hint:* Set $x = \sqrt[6]{1+h}$ and rewrite as a limit as $x \to 1$.

In Exercises 45–54, evaluate in terms of the constant a.

45. $\lim_{x\to 0} (2a+x)$

46. $\lim_{h\to -2} (4ah+7a)$

47. $\lim_{t\to -1} (4t-2at+3a)$

48. $\lim_{h\to 0} \dfrac{(3a+h)^2-9a^2}{h}$

49. $\lim_{h\to 0} \dfrac{2(a+h)^2-2a^2}{h}$

50. $\lim_{x\to a} \dfrac{(x+a)^2-4x^2}{x-a}$

51. $\lim_{x\to a} \dfrac{\sqrt{x}-\sqrt{a}}{x-a}$

52. $\lim_{h\to 0} \dfrac{\sqrt{a+2h}-\sqrt{a}}{h}$

53. $\lim_{x\to 0} \dfrac{(x+a)^3-a^3}{x}$

54. $\lim_{h\to a} \dfrac{\frac{1}{h}-\frac{1}{a}}{h-a}$

Further Insights and Challenges

In Exercises 55–58, find all values of c such that the limit exists.

55. $\lim_{x\to c} \dfrac{x^2-5x-6}{x-c}$

56. $\lim_{x\to 1} \dfrac{x^2+3x+c}{x-1}$

57. $\lim_{x\to 1} \left(\dfrac{1}{x-1} - \dfrac{c}{x^3-1}\right)$

58. $\lim_{x\to 0} \dfrac{1+cx^2-\sqrt{1+x^2}}{x^4}$

59. For which sign $\pm$ does the following limit exist?

$$\lim_{x\to 0} \left(\frac{1}{x} \pm \frac{1}{x(x-1)}\right)$$

2.6 Trigonometric Limits

In our study of the derivative, we will need to evaluate certain limits involving transcendental functions such as sine and cosine. The algebraic techniques of the previous section are often ineffective for such functions, and other tools are required. In this section, we discuss one such tool—the Squeeze Theorem—and use it to evaluate the trigonometric limits needed in Section 3.6.

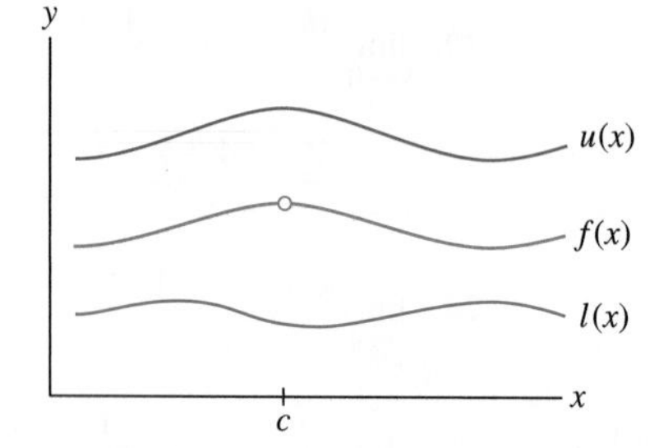

FIGURE 1 $f(x)$ is trapped between $l(x)$ and $u(x)$ (but not squeezed at $x = c$).

The Squeeze Theorem

Consider a function $f(x)$ that is "trapped" between two functions $l(x)$ and $u(x)$ on an interval I. In other words,

$$l(x) \le f(x) \le u(x) \quad \text{for all } x \in I$$

Thus, the graph of $f(x)$ lies between the graphs of $l(x)$ and $u(x)$ (Figure 1).

The Squeeze Theorem applies when $f(x)$ is not just trapped but **squeezed** at a point $x = c$ (Figure 2). By this we mean that for all $x \ne c$ in some open interval containing c,

$$l(x) \le f(x) \le u(x) \quad \text{and} \quad \lim_{x\to c} l(x) = \lim_{x\to c} u(x) = L$$

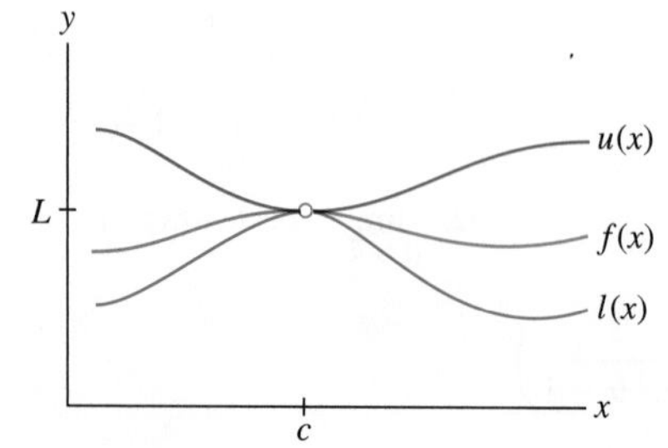

FIGURE 2 $f(x)$ is squeezed by $l(x)$ and $u(x)$ at $x = c$.

We do not require that $f(x)$ be defined at $x = c$, but it is clear graphically that $f(x)$ must approach the limit L, as stated in the next theorem. See Appendix D for a proof.

THEOREM 1 Squeeze Theorem Assume that for $x \neq c$ (in some open interval containing c),

$$l(x) \leq f(x) \leq u(x) \quad \text{and} \quad \lim_{x\to c} l(x) = \lim_{x\to c} u(x) = L$$

Then $\lim_{x\to c} f(x)$ exists and $\lim_{x\to c} f(x) = L$.

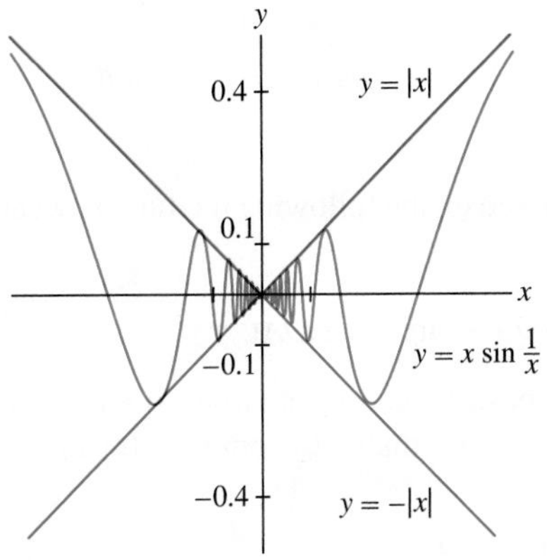

FIGURE 3

EXAMPLE 1 Show that $\lim_{x\to 0} x \sin \frac{1}{x} = 0$.

Solution Although $f(x) = x \sin \frac{1}{x}$ is a product of two functions, we cannot use the Product Law because $\lim_{x\to 0} \sin \frac{1}{x}$ does not exist. However, the sine function takes on values between 1 and −1, and therefore $\left|\sin \frac{1}{x}\right| \leq 1$ for all $x \neq 0$. Multiplying by $|x|$, we obtain $\left|x \sin \frac{1}{x}\right| \leq |x|$ and conclude that (Figure 3)

$$-|x| \leq x \sin \frac{1}{x} \leq |x|$$

Because

$$\lim_{x\to 0} |x| = 0 \quad \text{and} \quad \lim_{x\to 0} (-|x|) = -\lim_{x\to 0} |x| = 0$$

we can apply the Squeeze Theorem to conclude that $\lim_{x\to 0} x \sin \frac{1}{x} = 0$. ■

In Section 2.2, we found numerical and graphical evidence suggesting that the limit $\lim_{\theta\to 0} \frac{\sin\theta}{\theta}$ is equal to 1. The Squeeze Theorem will allow us to prove this fact.

THEOREM 2 Important Trigonometric Limits

$$\lim_{\theta\to 0} \frac{\sin\theta}{\theta} = 1, \qquad \lim_{\theta\to 0} \frac{1-\cos\theta}{\theta} = 0$$

Note that both $\frac{\sin\theta}{\theta}$ and $\frac{\cos\theta-1}{\theta}$ are indeterminate at $\theta = 0$, so Theorem 2 cannot be proved by substitution.

To apply the Squeeze Theorem, we must find functions that squeeze $\frac{\sin\theta}{\theta}$ at $\theta = 0$. These are provided by the next theorem (Figure 4).

THEOREM 3

$$\cos\theta \leq \frac{\sin\theta}{\theta} \leq 1 \quad \text{for} \quad -\frac{\pi}{2} < \theta < \frac{\pi}{2}, \quad \theta \neq 0 \qquad \boxed{1}$$

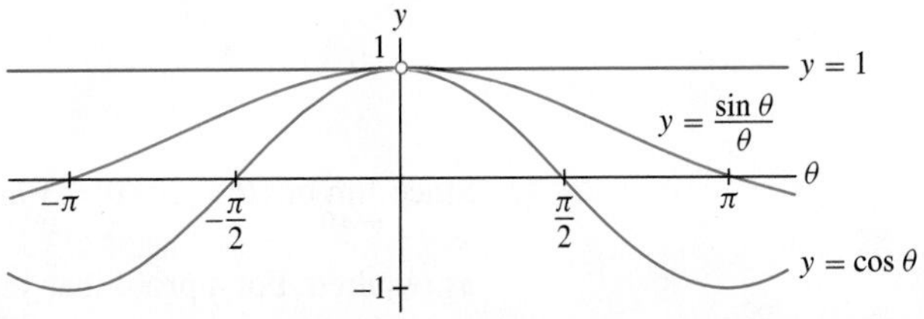

FIGURE 4 Graph illustrating the inequalities of Theorem 3.

Area of triangle $= \frac{1}{2}\sin\theta$ Area of sector $= \frac{1}{2}\theta$ Area of triangle $= \frac{1}{2}\tan\theta$

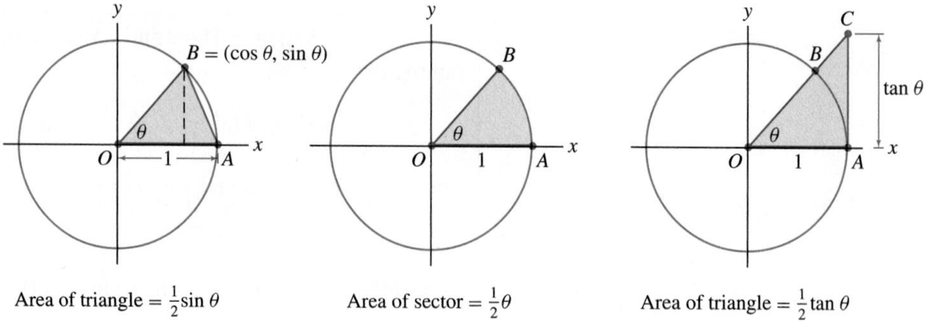

FIGURE 5

Proof Assume first that $0 < \theta < \frac{\pi}{2}$. Our proof is based on the following relation between the areas in Figure 5:

$$\text{Area of } \triangle OAB < \text{area of sector } BOA < \text{area of } \triangle OAC \qquad \boxed{2}$$

Let's compute these three areas. First, $\triangle OAB$ has base 1 and height $\sin\theta$, so its area is $\frac{1}{2}\sin\theta$. Next, recall that a sector of angle θ has area $\frac{1}{2}\theta$. Finally, to compute the area of $\triangle OAC$, we observe that

$$\tan\theta = \frac{\text{opposite side}}{\text{adjacent side}} = \frac{AC}{OA} = \frac{AC}{1} = AC$$

> **REMINDER** *Let's recall why a sector of angle θ in a circle of radius r has area $\frac{1}{2}r^2\theta$. A sector of angle θ represents a fraction $\frac{\theta}{2\pi}$ of the entire circle. The circle has area πr^2, so the sector has area $\left(\frac{\theta}{2\pi}\right)\pi r^2 = \frac{1}{2}r^2\theta$. In the unit circle $(r = 1)$, the sector has area $\frac{1}{2}\theta$.*

Thus, $\triangle OAC$ has base 1, height $\tan\theta$, and area $\frac{1}{2}\tan\theta$. We have shown, therefore, that

$$\underbrace{\frac{1}{2}\sin\theta}_{\text{Area } \triangle OAB} \leq \underbrace{\frac{1}{2}\theta}_{\text{Area of sector}} \leq \underbrace{\frac{1}{2}\frac{\sin\theta}{\cos\theta}}_{\text{Area } \triangle OAC} \qquad \boxed{3}$$

The first inequality yields $\sin\theta \leq \theta$, and because $\theta > 0$, we obtain

$$\frac{\sin\theta}{\theta} \leq 1 \qquad \boxed{4}$$

> *Note: Our proof of Theorem 3 uses the formula $\frac{1}{2}\theta$ for the area of a sector, but this formula is based on the formula πr^2 for the area of a circle, a complete proof of which requires integral calculus.*

Next, multiply the second inequality in (3) by $\dfrac{2\cos\theta}{\theta}$ to obtain

$$\cos\theta \leq \frac{\sin\theta}{\theta} \qquad \boxed{5}$$

The combination of (4) and (5) gives us (1) when $0 < \theta < \frac{\pi}{2}$. However, the functions in (1) do not change when θ is replaced by $-\theta$ because both $\cos\theta$ and $\dfrac{\sin\theta}{\theta}$ are even functions. Indeed, $\cos(-\theta) = \cos\theta$ and

$$\frac{\sin(-\theta)}{-\theta} = \frac{-\sin\theta}{-\theta} = \frac{\sin\theta}{\theta}$$

Therefore, (1) holds for $-\frac{\pi}{2} < \theta < 0$ as well. This completes the proof of Theorem 3. ∎

Proof of Theorem 2 According to Theorem 3,

$$\cos\theta \leq \frac{\sin\theta}{\theta} \leq 1$$

Since $\lim\limits_{\theta\to 0}\cos\theta = \cos 0 = 1$ and $\lim\limits_{\theta\to 0} 1 = 1$, the Squeeze Theorem yields $\lim\limits_{\theta\to 0}\dfrac{\sin\theta}{\theta} = 1$, as required. For a proof that $\lim\limits_{\theta\to 0}\dfrac{1-\cos\theta}{\theta} = 0$, see Exercises 51 and 58. ∎

In the next example, we evaluate another trigonometric limit. The key idea is to rewrite the function of h in terms of the new variable $\theta = 4h$.

■ **EXAMPLE 2 Evaluating a Limit by Changing Variables** Investigate $\lim_{h\to 0} \frac{\sin 4h}{h}$ numerically and then evaluate it exactly.

h	$\frac{\sin 4h}{h}$
±1.0	−0.75680
±0.5	1.81859
±0.2	3.58678
±0.1	**3.8**9 418
±0.05	**3.9**7 339
±0.01	**3.99** 893
±0.005	**3.999** 73

Solution The table of values at the left suggests that the limit is equal to 4. To evaluate the limit exactly, we rewrite it in terms of the limit of $\frac{\sin\theta}{\theta}$ so that Theorem 2 can be applied. Thus, we set $\theta = 4h$ and write

$$\frac{\sin 4h}{h} = 4\left(\frac{\sin 4h}{4h}\right) = 4\frac{\sin\theta}{\theta}$$

The new variable θ tends to zero as $h \to 0$ because θ is a multiple of h. Therefore, we may change the limit as $h \to 0$ into a limit as $\theta \to 0$ to obtain

$$\lim_{h\to 0}\frac{\sin 4h}{h} = \lim_{\theta\to 0} 4\frac{\sin\theta}{\theta} = 4\left(\lim_{\theta\to 0}\frac{\sin\theta}{\theta}\right) = 4(1) = 4 \qquad \boxed{6}$$

■

2.6 SUMMARY

- We say that $f(x)$ is *squeezed* at $x = c$ if there exist functions $l(x)$ and $u(x)$ such that $l(x) \le f(x) \le u(x)$ for all $x \ne c$ in an open interval I containing c, and

$$\lim_{x\to c} l(x) = \lim_{x\to c} u(x) = L$$

The Squeeze Theorem states that in this case, $\lim_{x\to c} f(x) = L$.

- Two important trigonometric limits:

$$\lim_{\theta\to 0}\frac{\sin\theta}{\theta} = 1, \qquad \lim_{\theta\to 0}\frac{1-\cos\theta}{\theta} = 0$$

2.6 EXERCISES

Preliminary Questions

1. Assume that $-x^4 \le f(x) \le x^2$. What is $\lim_{x\to 0} f(x)$? Is there enough information to evaluate $\lim_{x\to \frac{1}{2}} f(x)$? Explain.

2. State the Squeeze Theorem carefully.

3. If you want to evaluate $\lim_{h\to 0}\frac{\sin 5h}{3h}$, it is a good idea to rewrite the limit in terms of the variable (choose one):

(a) $\theta = 5h$ **(b)** $\theta = 3h$ **(c)** $\theta = \frac{5h}{3}$

Exercises

1. State precisely the hypothesis and conclusions of the Squeeze Theorem for the situation in Figure 6.

2. In Figure 7, is $f(x)$ squeezed by $u(x)$ and $l(x)$ at $x = 3$? At $x = 2$?

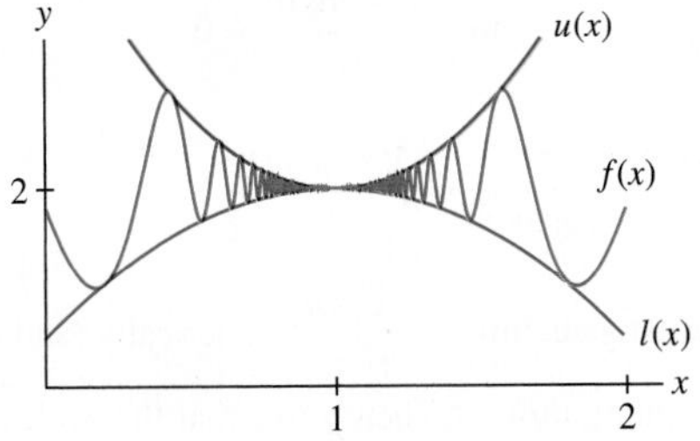

FIGURE 6

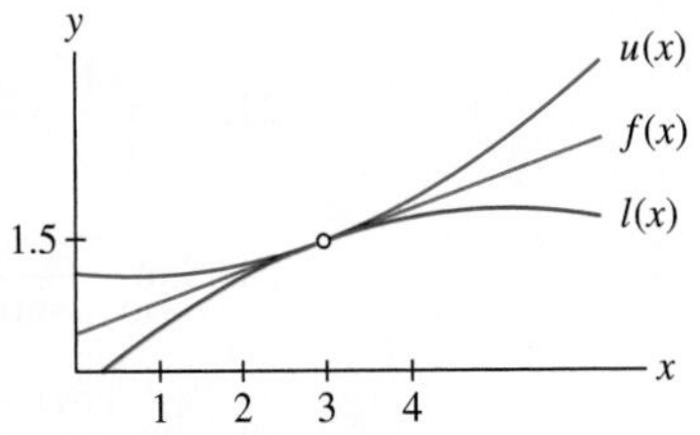

FIGURE 7

3. What does the Squeeze Theorem say about $\lim_{x\to 7} f(x)$ if $\lim_{x\to 7} l(x) = \lim_{x\to 7} u(x) = 6$ and $f(x)$, $u(x)$, and $l(x)$ are related as in Figure 8? The inequality $f(x) \le u(x)$ is not satisfied for all x. Does this affect the validity of your conclusion?

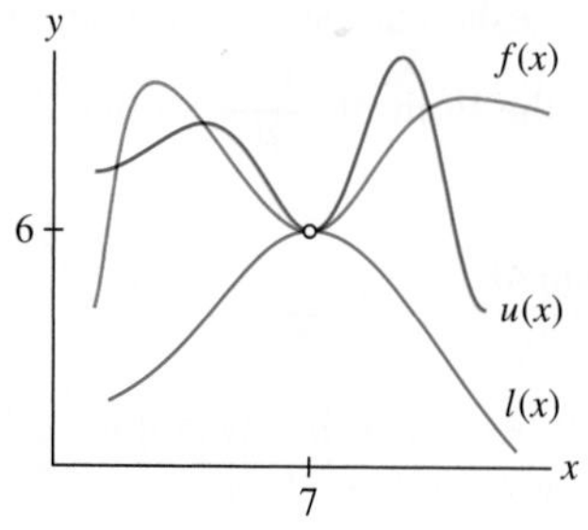

FIGURE 8

4. Determine $\lim_{x\to 0} f(x)$ assuming that $\cos x \le f(x) \le 1$.

5. State whether the inequality provides sufficient information to determine $\lim_{x\to 1} f(x)$, and if so, find the limit.

(a) $4x - 5 \le f(x) \le x^2$

(b) $2x - 1 \le f(x) \le x^2$

(c) $4x - x^2 \le f(x) \le x^2 + 2$

6. GU Plot the graphs of $u(x) = 1 + \left|x - \frac{\pi}{2}\right|$ and $l(x) = \sin x$ on the same set of axes. What can you say about $\lim_{x\to \frac{\pi}{2}} f(x)$ if $f(x)$ is squeezed by $l(x)$ and $u(x)$ at $x = \frac{\pi}{2}$?

In Exercises 7–16, evaluate using the Squeeze Theorem.

7. $\lim_{x\to 0} x^2 \cos \frac{1}{x}$

8. $\lim_{x\to 0} x \sin \frac{1}{x^2}$

9. $\lim_{x\to 1} (x - 1) \sin \frac{\pi}{x - 1}$

10. $\lim_{x\to 3} (x^2 - 9) \frac{x - 3}{|x - 3|}$

11. $\lim_{t\to 0} (2^t - 1) \cos \frac{1}{t}$

12. $\lim_{x\to 0+} \sqrt{x}\, 4^{\cos(\pi/x)}$

13. $\lim_{t\to 2} (t^2 - 4) \cos \frac{1}{t - 2}$

14. $\lim_{x\to 0} \tan x \cos\left(\sin \frac{1}{x}\right)$

15. $\lim_{\theta\to \frac{\pi}{2}} \cos\theta \cos(\tan\theta)$

16. $\lim_{t\to 0-} \sin^2\left(\frac{1}{t}\right) 3^{(1/t)}$

In Exercises 17–26, evaluate using Theorem 2 as necessary.

17. $\lim_{x\to 0} \frac{\tan x}{x}$

18. $\lim_{x\to 0} \frac{\sin x \sec x}{x}$

19. $\lim_{t\to 0} \frac{\sqrt{t^3 + 9} \sin t}{t}$

20. $\lim_{t\to 0} \frac{\sin^2 t}{t}$

21. $\lim_{x\to 0} \frac{x^2}{\sin^2 x}$

22. $\lim_{t\to \frac{\pi}{2}} \frac{1 - \cos t}{t}$

23. $\lim_{\theta\to 0} \frac{\sec\theta - 1}{\theta}$

24. $\lim_{\theta\to 0} \frac{1 - \cos\theta}{\sin\theta}$

25. $\lim_{t\to \frac{\pi}{4}} \frac{\sin t}{t}$

26. $\lim_{t\to 0} \frac{\cos t - \cos^2 t}{t}$

27. Let $L = \lim_{x\to 0} \frac{\sin 14x}{x}$.

(a) Show, by letting $\theta = 14x$, that $L = \lim_{\theta\to 0} 14 \frac{\sin\theta}{\theta}$.

(b) Compute L.

28. Evaluate $\lim_{h\to 0} \frac{\sin 9h}{\sin 7h}$. *Hint:* $\frac{\sin 9h}{\sin 7h} = \left(\frac{9}{7}\right)\left(\frac{\sin 9h}{9h}\right)\left(\frac{7h}{\sin 7h}\right)$.

In Exercises 29–48, evaluate the limit.

29. $\lim_{h\to 0} \frac{\sin 9h}{h}$

30. $\lim_{h\to 0} \frac{\sin 4h}{4h}$

31. $\lim_{h\to 0} \frac{\sin h}{5h}$

32. $\lim_{x\to \frac{\pi}{6}} \frac{x}{\sin 3x}$

33. $\lim_{\theta\to 0} \frac{\sin 7\theta}{\sin 3\theta}$

34. $\lim_{x\to 0} \frac{\tan 4x}{9x}$

35. $\lim_{x\to 0} x \csc 25x$

36. $\lim_{t\to 0} \frac{\tan 4t}{t \sec t}$

37. $\lim_{h\to 0} \frac{\sin 2h \sin 3h}{h^2}$

38. $\lim_{z\to 0} \frac{\sin(z/3)}{\sin z}$

39. $\lim_{\theta\to 0} \frac{\sin(-3\theta)}{\sin(4\theta)}$

40. $\lim_{x\to 0} \frac{\tan 4x}{\tan 9x}$

41. $\lim_{t\to 0} \frac{\csc 8t}{\csc 4t}$

42. $\lim_{x\to 0} \frac{\sin 5x \sin 2x}{\sin 3x \sin 5x}$

43. $\lim_{x\to 0} \frac{\sin 3x \sin 2x}{x \sin 5x}$

44. $\lim_{h\to 0} \frac{1 - \cos 2h}{h}$

45. $\lim_{h\to 0} \frac{\sin(2h)(1 - \cos h)}{h^2}$

46. $\lim_{t\to 0} \frac{1 - \cos 2t}{\sin^2 3t}$

47. $\lim_{\theta\to 0} \frac{\cos 2\theta - \cos\theta}{\theta}$

48. $\lim_{h\to \frac{\pi}{2}} \frac{1 - \cos 3h}{h}$

49. Calculate $\lim_{x\to 0-} \frac{\sin x}{|x|}$.

50. Use the identity $\sin 3\theta = 3\sin\theta - 4\sin^3\theta$ to evaluate the limit $\lim_{\theta\to 0} \frac{\sin 3\theta - 3\sin\theta}{\theta^3}$.

51. Prove the following result stated in Theorem 2:

$$\lim_{\theta\to 0} \frac{1 - \cos\theta}{\theta} = 0 \qquad \boxed{7}$$

Hint: $\frac{1 - \cos\theta}{\theta} = \frac{1}{1 + \cos\theta} \cdot \frac{1 - \cos^2\theta}{\theta}$.

52. GU Investigate $\lim_{h\to 0} \frac{1 - \cos h}{h^2}$ numerically (and graphically if you have a graphing utility). Then prove that the limit is equal to $\frac{1}{2}$. *Hint:* See the hint for Exercise 51.

In Exercises 53–55, evaluate using the result of Exercise 52.

53. $\lim_{h\to 0} \frac{\cos 3h - 1}{h^2}$

54. $\lim_{h\to 0} \frac{\cos 3h - 1}{\cos 2h - 1}$

55. $\lim_{t\to 0} \frac{\sqrt{1-\cos t}}{t}$

56. Use the Squeeze Theorem to prove that if $\lim_{x\to c} |f(x)| = 0$, then $\lim_{x\to c} f(x) = 0$.

Further Insights and Challenges

57. Use the result of Exercise 52 to prove that for $m \neq 0$,

$$\lim_{x\to 0} \frac{\cos mx - 1}{x^2} = -\frac{m^2}{2}$$

58. Using a diagram of the unit circle and the Pythagorean Theorem, show that

$$\sin^2\theta \le (1-\cos\theta)^2 + \sin^2\theta \le \theta^2$$

Conclude that $\sin^2\theta \le 2(1-\cos\theta) \le \theta^2$ and use this to give an alternative proof of Eq. (7) in Exercise 51. Then give an alternative proof of the result in Exercise 52.

59. **(a)** Investigate $\lim_{x\to c} \frac{\sin x - \sin c}{x - c}$ numerically for the five values $c = 0, \frac{\pi}{6}, \frac{\pi}{4}, \frac{\pi}{3}, \frac{\pi}{2}$.

(b) Can you guess the answer for general c?

(c) Check that your answer to (b) works for two other values of c.

2.7 Limits at Infinity

So far we have considered limits as x approaches a number c. It is also important to consider limits where x approaches ∞ or $-\infty$, which we refer to as **limits at infinity**. In applications, limits at infinity arise naturally when we describe the "long-term" behavior of a system as in Figure 1.

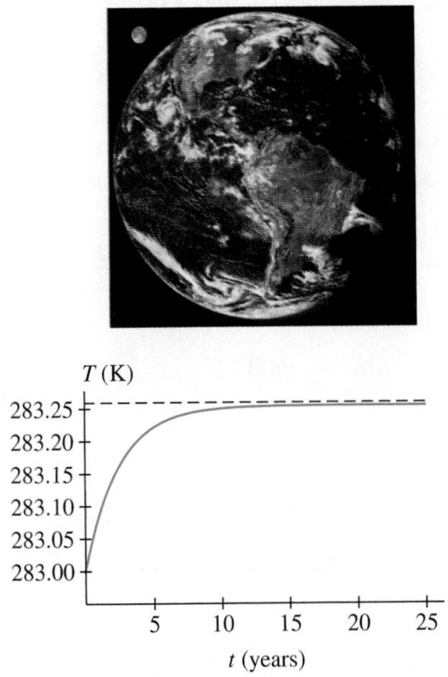

FIGURE 1 The earth's average temperature (according to a simple climate model) in response to an 0.25% increase in solar radiation. According to this model, $\lim_{t\to\infty} T(t) = 283.255$.

The notation $x \to \infty$ indicates that x increases without bound, and $x \to -\infty$ indicates that x decreases (through negative values) without bound. We write

- $\lim_{x\to\infty} f(x) = L$ if $f(x)$ gets closer and closer to L as $x \to \infty$.
- $\lim_{x\to -\infty} f(x) = L$ if $f(x)$ gets closer and closer to L as $x \to -\infty$.

As before, "closer and closer" means that $|f(x) - L|$ becomes arbitrarily small. In either case, the line $y = L$ is called a **horizontal asymptote**. We use the notation $x \to \pm\infty$ to indicate that we are considering both infinite limits, as $x \to \infty$ and as $x \to -\infty$.

Infinite limits describe the **asymptotic behavior** of a function, which is behavior of the graph as we move out to the right or the left.

■ **EXAMPLE 1** Discuss the asymptotic behavior in Figure 2.

Solution The function $g(x)$ approaches $L = 7$ as we move to the right and it approaches $L = 3$ as we move to left, so

$$\lim_{x\to\infty} g(x) = 7, \qquad \lim_{x\to -\infty} g(x) = 3$$

Accordingly, the lines $y = 7$ and $y = 3$ are horizontal asymptotes of $g(x)$. ■

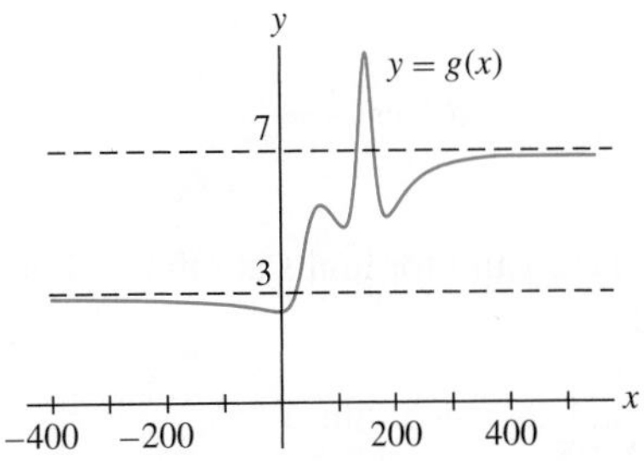

FIGURE 2 The lines $y = 7$ and $y = 3$ are horizontal asymptotes of $g(x)$.

A function may approach an infinite limit as $x \to \pm\infty$. We write

$$\lim_{x\to\infty} f(x) = \infty \qquad \text{or} \qquad \lim_{x\to -\infty} f(x) = \infty$$

if $f(x)$ becomes arbitrarily large as $x \to \infty$ or $-\infty$. Similar notation is used if $f(x)$ approaches $-\infty$ as $x \to \pm\infty$. For example, we see in Figure 3(A) that

$$\lim_{x\to\infty} 2^x = \infty, \qquad \lim_{x\to -\infty} 2^x = 0$$

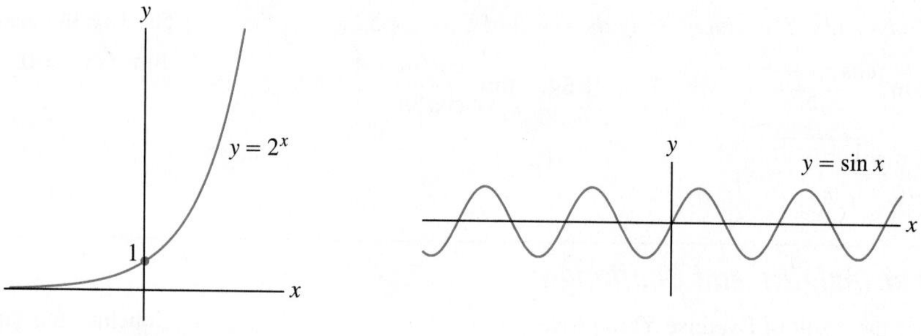

FIGURE 3

However, limits at infinity do not always exist. For example, $f(x) = \sin x$ oscillates indefinitely [Figure 3(B)], so

$$\lim_{x \to \infty} \sin x \qquad \text{and} \qquad \lim_{x \to -\infty} \sin x$$

do not exist.

The limits at infinity of the power functions $f(x) = x^n$ are easily determined. If $n > 0$, then x^n certainly increases without bound as $x \to \infty$, so (Figure 4)

$$\lim_{x \to \infty} x^n = \infty \qquad \text{and} \qquad \lim_{x \to \infty} x^{-n} = \lim_{x \to \infty} \frac{1}{x^n} = 0$$

To describe the limits as $x \to -\infty$, assume that n is a whole number so that x^n is defined for $x < 0$. If n is even, then x^n becomes large and positive as $x \to -\infty$, and if n is odd, it becomes large and negative. In summary,

THEOREM 1 For all $n > 0$,

$$\lim_{x \to \infty} x^n = \infty, \qquad \lim_{x \to \infty} x^{-n} = \lim_{x \to \infty} \frac{1}{x^n} = 0$$

If n is a whole number,

$$\lim_{x \to -\infty} x^n = \begin{cases} \infty & \text{if } n \text{ is even} \\ -\infty & \text{if } n \text{ is odd} \end{cases} \qquad \text{and} \qquad \lim_{x \to -\infty} x^{-n} = \lim_{x \to -\infty} \frac{1}{x^n} = 0$$

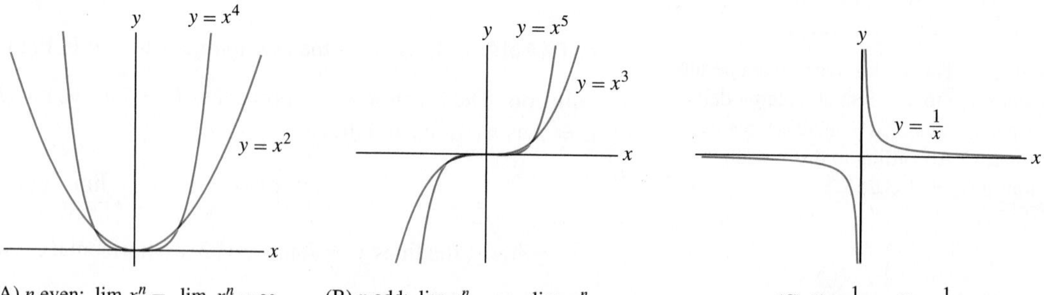

(A) n even: $\lim_{x \to \infty} x^n = \lim_{x \to -\infty} x^n = \infty$ (B) n odd: $\lim_{x \to \infty} x^n = \infty$, $\lim_{x \to -\infty} x^n = -\infty$ (C) $\lim_{x \to \infty} \frac{1}{x} = \lim_{x \to -\infty} \frac{1}{x} = 0$

FIGURE 4

The Basic Limit Laws (Theorem 1 in Section 2.3) are valid for limits at infinity. For example, the Sum and Constant Multiple Laws yield:

$$\lim_{x \to \infty} \left(3 - 4x^{-3} + 5x^{-5}\right) = \lim_{x \to \infty} 3 - 4 \lim_{x \to \infty} x^{-3} + 5 \lim_{x \to \infty} x^{-5}$$
$$= 3 + 0 + 0 = 3$$

EXAMPLE 2 Calculate $\displaystyle\lim_{x\to\pm\infty} \frac{20x^2 - 3x}{3x^5 - 4x^2 + 5}$.

Solution It would be nice if we could apply the Quotient Law directly, but this law is valid only if the denominator has a finite, nonzero limit. Our limit has the indeterminate form ∞/∞ because

$$\lim_{x\to\infty} (20x^2 - 3x) = \infty \qquad \text{and} \qquad \lim_{x\to\infty} (3x^5 - 4x^2 + 5) = \infty$$

The way around this difficulty is to divide the numerator and denominator by x^5 (the highest power of x in the denominator):

$$\frac{20x^2 - 3x}{3x^5 - 4x^2 + 5} = \frac{x^{-5}(20x^2 - 3x)}{x^{-5}(3x^5 - 4x^2 + 5)} = \frac{20x^{-3} - 3x^{-4}}{3 - 4x^{-3} + 5x^{-5}}$$

Now we can use the Quotient Law:

$$\lim_{x\to\pm\infty} \frac{20x^2 - 3x}{3x^5 - 4x^2 + 5} = \frac{\lim_{x\to\pm\infty} \left(20x^{-3} - 3x^{-4}\right)}{\lim_{x\to\pm\infty} \left(3 - 4x^{-3} + 5x^{-5}\right)} = \frac{0}{3} = 0$$

In general, if

$$f(x) = \frac{a_n x^n + a_{n-1}x^{n-1} + \cdots + a_0}{b_m x^m + b_{m-1}x^{m-1} + \cdots + b_0}$$

where $a_n \neq 0$ and $b_m \neq 0$, divide the numerator and denominator by x^m:

$$\begin{aligned} f(x) &= \frac{a_n x^{n-m} + a_{n-1}x^{n-1-m} + \cdots + a_0 x^{-m}}{b_m + b_{m-1}x^{-1} + \cdots + b_0 x^{-m}} \\ &= x^{n-m}\left(\frac{a_n + a_{n-1}x^{-1} + \cdots + a_0 x^{-n}}{b_m + b_{m-1}x^{-1} + \cdots + b_0 x^{-m}}\right) \end{aligned}$$

The quotient in parenthesis approaches the finite limit a_n/b_m because

$$\lim_{x\to\infty} (a_n + a_{n-1}x^{-1} + \cdots + a_0 x^{-n}) = a_n$$

$$\lim_{x\to\infty} (b_m + b_{m-1}x^{-1} + \cdots + b_0 x^{-m}) = b_m$$

Therefore,

$$\lim_{x\to\pm\infty} f(x) = \lim_{x\to\pm\infty} x^{n-m} \lim_{x\to\pm\infty} \frac{a_n + a_{n-1}x^{-1} + \cdots + a_0 x^{-n}}{b_m + b_{m-1}x^{-1} + \cdots + b_0 x^{-m}} = \frac{a_n}{b_m} \lim_{x\to\pm\infty} x^{n-m}$$

THEOREM 2 Limits at Infinity of a Rational Function The asymptotic behavior of a rational function depends only on the leading terms of its numerator and denominator. If $a_n, b_m \neq 0$, then

$$\lim_{x\to\pm\infty} \frac{a_n x^n + a_{n-1}x^{n-1} + \cdots + a_0}{b_m x^m + b_{m-1}x^{m-1} + \cdots + b_0} = \frac{a_n}{b_m} \lim_{x\to\pm\infty} x^{n-m}$$

Here are some examples:

- $n = m$: $\displaystyle\lim_{x\to\infty} \frac{3x^4 - 7x + 9}{7x^4 - 4} = \frac{3}{7} \lim_{x\to\infty} x^0 = \frac{3}{7}$

- $n < m$: $\displaystyle \lim_{x\to\infty} \frac{3x^3 - 7x + 9}{7x^4 - 4} = \frac{3}{7} \lim_{x\to\infty} x^{-1} = 0$
- $n > m$, $n - m$ odd: $\displaystyle \lim_{x\to-\infty} \frac{3x^8 - 7x + 9}{7x^3 - 4} = \frac{3}{7} \lim_{x\to-\infty} x^5 = -\infty$
- $n > m$, $n - m$ even: $\displaystyle \lim_{x\to-\infty} \frac{3x^7 - 7x + 9}{7x^3 - 4} = \frac{3}{7} \lim_{x\to-\infty} x^4 = \infty$

Our method can be adapted to noninteger exponents and algebraic functions.

EXAMPLE 3 Calculate the limits **(a)** $\displaystyle \lim_{x\to\infty} \frac{3x^{7/2} + 7x^{-1/2}}{x^2 - x^{1/2}}$ **(b)** $\displaystyle \lim_{x\to\infty} \frac{x^2}{\sqrt{x^3+1}}$

Solution

The Quotient Law is valid if $\lim_{x\to c} f(x) = \infty$ and $\lim_{x\to c} g(x) = L$, where $L \neq 0$:

$$\lim_{x\to c} \frac{f(x)}{g(x)} = \frac{\lim_{x\to c} f(x)}{\lim_{x\to c} g(x)} = \begin{cases} \infty & \text{if } L > 0 \\ -\infty & \text{if } L < 0 \end{cases}$$

(a) As before, divide the numerator and denominator by x^2, which is the highest power of x occurring in the denominator (this means: multiply by x^{-2}):

$$\frac{3x^{7/2} + 7x^{-1/2}}{x^2 - x^{1/2}} = \left(\frac{x^{-2}}{x^{-2}}\right) \frac{3x^{7/2} + 7x^{-1/2}}{x^2 - x^{1/2}} = \frac{3x^{3/2} + 7x^{-5/2}}{1 - x^{-3/2}}$$

$$\lim_{x\to\infty} \frac{3x^{7/2} + 7x^{-1/2}}{x^2 - x^{1/2}} = \frac{\lim_{x\to\infty} (3x^{3/2} + 7x^{-5/2})}{\lim_{x\to\infty} (1 - x^{-3/2})} = \frac{\infty}{1} = \infty$$

(b) The key is to observe that the denominator of $\dfrac{x^2}{\sqrt{x^3+1}}$ "behaves" like $x^{3/2}$:

$$\sqrt{x^3+1} = \sqrt{x^3(1 + x^{-3})} = x^{3/2}\sqrt{1 + x^{-3}} \qquad (\text{for } x > 0)$$

This suggests that we divide the numerator and denominator by $x^{3/2}$:

$$\frac{x^2}{\sqrt{x^3+1}} = \left(\frac{x^{-3/2}}{x^{-3/2}}\right) \frac{x^2}{x^{3/2}\sqrt{1+x^{-3}}} = \frac{x^{1/2}}{\sqrt{1+x^{-3}}}$$

Then apply Quotient Law:

$$\lim_{x\to\infty} \frac{x^2}{\sqrt{x^3+1}} = \lim_{x\to\infty} \frac{x^{1/2}}{\sqrt{1+x^{-3}}} = \frac{\lim_{x\to\infty} x^{1/2}}{\lim_{x\to\infty} \sqrt{1+x^{-3}}}$$

$$= \frac{\infty}{1} = \infty$$

■

EXAMPLE 4 Calculate the limits at infinity of $f(x) = \dfrac{12x + 25}{\sqrt{16x^2 + 100x + 500}}$.

Solution Divide numerator and denominator by x (multiply by x^{-1}), but notice the difference between x positive and x negative. For $x > 0$,

$$x^{-1}\sqrt{16x^2 + 100x + 500} = \sqrt{x^{-2}}\sqrt{16x^2 + 100x + 500} = \sqrt{16 + \frac{100}{x} + \frac{500}{x^2}}$$

$$\lim_{x\to\infty} \frac{12x + 25}{\sqrt{16x^2 + 100x + 500}} = \frac{\lim_{x\to\infty} \left(12 + \frac{25}{x}\right)}{\lim_{x\to\infty} \sqrt{16 + \frac{100}{x} + \frac{500}{x^2}}} = \frac{12}{\sqrt{16}} = 3$$

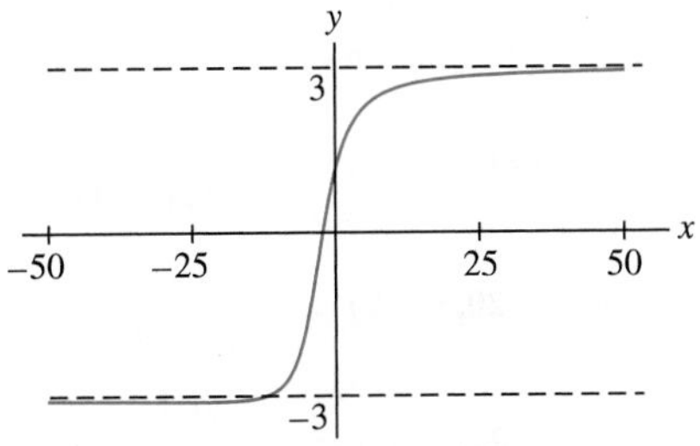

FIGURE 5 Graph of $f(x) = \dfrac{12x+25}{\sqrt{16x^2+100x+500}}$.

However, if $x < 0$, then $x = -\sqrt{x^2}$ and

$$x^{-1}\sqrt{16x^2+100x+500} = -\sqrt{x^{-2}}\sqrt{16x^2+100x+500} = -\sqrt{16 + \frac{100}{x} + \frac{500}{x^2}}$$

So the limit as $x \to -\infty$ is -3 instead of 3 (Figure 5):

$$\lim_{x\to-\infty} \frac{12x+25}{\sqrt{16x^2+100x+500}} = \frac{\lim\limits_{x\to-\infty}\left(12+\frac{25}{x}\right)}{-\lim\limits_{x\to-\infty}\sqrt{16+\frac{100}{x}+\frac{500}{x^2}}} = \frac{12}{-\sqrt{16}} = -3 \quad ■$$

2.7 SUMMARY

- *Limits as infinity*:
  - $\lim\limits_{x\to\infty} f(x) = L$ if $|f(x) - L|$ becomes arbitrarily small as x increases without bound
  - $\lim\limits_{x\to-\infty} f(x) = L$ if $|f(x) - L|$ becomes arbitrarily small as x decreases without bound.
- A horizontal line $y = L$ is a *horizontal asymptote* if

$$\lim_{x\to\infty} f(x) = L \qquad \text{and/or} \qquad \lim_{x\to-\infty} f(x) = L$$

- If $n > 0$, then $\lim\limits_{x\to\infty} x^n = \infty$ and $\lim\limits_{x\to\pm\infty} x^{-n} = 0$. If $n > 0$ is a whole number, then

$$\lim_{x\to-\infty} x^n = \begin{cases} \infty & \text{if } n \text{ is even} \\ -\infty & \text{if } n \text{ is odd} \end{cases} \qquad \text{and} \qquad \lim_{x\to-\infty} x^{-n} = 0$$

- If $f(x) = \dfrac{a_n x^n + a_{n-1}x^{n-1} + \cdots + a_0}{b_m x^m + b_{m-1}x^{m-1} + \cdots + b_0}$ with $a_n, b_m \neq 0$, then

$$\lim_{x\to\pm\infty} f(x) = \frac{a_n}{b_m} \lim_{x\to\pm\infty} x^{n-m}$$

2.7 EXERCISES

Preliminary Questions

1. Assume that

$$\lim_{x\to\infty} f(x) = L \quad \text{and} \quad \lim_{x\to L} g(x) = \infty$$

Which of the following statements are correct?

(a) $x = L$ is a vertical asymptote of $g(x)$.
(b) $y = L$ is a horizontal asymptote of $g(x)$.
(c) $x = L$ is a vertical asymptote of $f(x)$.
(d) $y = L$ is a horizontal asymptote of $f(x)$.

2. What are the following limits?

(a) $\lim\limits_{x\to\infty} x^3$ **(b)** $\lim\limits_{x\to-\infty} x^3$ **(c)** $\lim\limits_{x\to-\infty} x^4$

3. Sketch the graph of a function that approaches a limit as $x \to \infty$ but does not approach a limit (either finite or infinite) as $x \to -\infty$.

4. What is the sign of a if $f(x) = ax^3 + x + 1$ satisfies $\lim\limits_{x\to-\infty} f(x) = \infty$?

5. What is the sign of the leading coefficient a_7 if $f(x)$ is a polynomial of degree 7 such that $\lim\limits_{x\to-\infty} f(x) = \infty$?

6. Explain why $\lim\limits_{x\to\infty} \sin\frac{1}{x}$ exists but $\lim\limits_{x\to 0} \sin\frac{1}{x}$ does not exist. What is $\lim\limits_{x\to\infty} \sin\frac{1}{x}$?

Exercises

1. What are the horizontal asymptotes of the function in Figure 6?

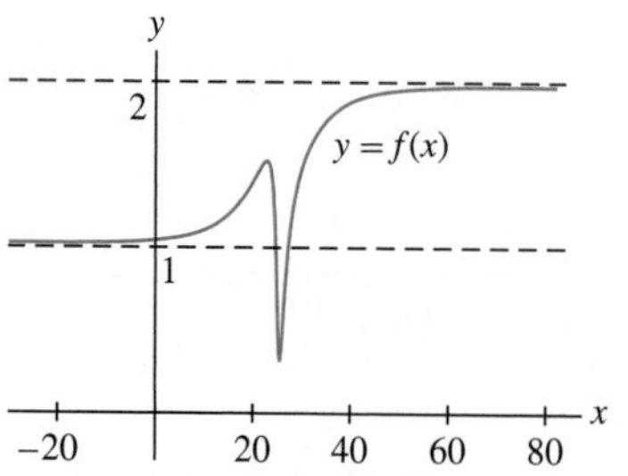

FIGURE 6

2. Sketch the graph of a function $f(x)$ that has both $y = -1$ and $y = 5$ as horizontal asymptotes.

3. Sketch the graph of a function $f(x)$ with a single horizontal asymptote $y = 3$.

4. Sketch the graphs of two functions $f(x)$ and $g(x)$ that have both $y = -2$ and $y = 4$ as horizontal asymptotes but $\lim_{x\to\infty} f(x) \neq \lim_{x\to\infty} g(x)$.

5. GU Investigate the asymptotic behavior of $f(x) = \dfrac{x^3}{x^3 + x}$ numerically and graphically:

(a) Make a table of values of $f(x)$ for $x = \pm 50, \pm 100, \pm 500, \pm 1000$.

(b) Plot the graph of $f(x)$.

(c) What are the horizontal asymptotes of $f(x)$?

6. GU Investigate $\displaystyle\lim_{x\to\pm\infty} \frac{12x+1}{\sqrt{4x^2+9}}$ numerically and graphically:

(a) Make a table of values of $f(x) = \dfrac{12x+1}{\sqrt{4x^2+9}}$ for $x = \pm 100, \pm 500, \pm 1000, \pm 10{,}000$.

(b) Plot the graph of $f(x)$.

(c) What are the horizontal asymptotes of $f(x)$?

In Exercises 7–16, evaluate the limit.

7. $\displaystyle\lim_{x\to\infty} \frac{x}{x+9}$

8. $\displaystyle\lim_{x\to\infty} \frac{3x^2+20x}{4x^2+9}$

9. $\displaystyle\lim_{x\to\infty} \frac{3x^2+20x}{2x^4+3x^3-29}$

10. $\displaystyle\lim_{x\to\infty} \frac{4}{x+5}$

11. $\displaystyle\lim_{x\to\infty} \frac{7x-9}{4x+3}$

12. $\displaystyle\lim_{x\to\infty} \frac{9x^2-2}{6-29x}$

13. $\displaystyle\lim_{x\to-\infty} \frac{7x^2-9}{4x+3}$

14. $\displaystyle\lim_{x\to-\infty} \frac{5x-9}{4x^3+2x+7}$

15. $\displaystyle\lim_{x\to-\infty} \frac{3x^3-10}{x+4}$

16. $\displaystyle\lim_{x\to-\infty} \frac{2x^5+3x^4-31x}{8x^4-31x^2+12}$

In Exercises 17–22, find the horizontal asymptotes.

17. $f(x) = \dfrac{2x^2-3x}{8x^2+8}$

18. $f(x) = \dfrac{8x^3-x^2}{7+11x-4x^4}$

19. $f(x) = \dfrac{\sqrt{36x^2+7}}{9x+4}$

20. $f(x) = \dfrac{\sqrt{36x^4+7}}{9x^2+4}$

21. $f(t) = \dfrac{3^t}{1+3^{-t}}$

22. $f(t) = \dfrac{t^{1/3}}{(64t^2+9)^{1/6}}$

In Exercises 23–30, evaluate the limit.

23. $\displaystyle\lim_{x\to\infty} \frac{\sqrt{9x^4+3x+2}}{4x^3+1}$

24. $\displaystyle\lim_{x\to\infty} \frac{\sqrt{x^3+20x}}{10x-2}$

25. $\displaystyle\lim_{x\to-\infty} \frac{8x^2+7x^{1/3}}{\sqrt{16x^4+6}}$

26. $\displaystyle\lim_{x\to-\infty} \frac{4x-3}{\sqrt{25x^2+4x}}$

27. $\displaystyle\lim_{t\to\infty} \frac{t^{4/3}+t^{1/3}}{(4t^{2/3}+1)^2}$

28. $\displaystyle\lim_{t\to\infty} \frac{t^{4/3}-9t^{1/3}}{(8t^4+2)^{1/3}}$

29. $\displaystyle\lim_{x\to-\infty} \frac{|x|+x}{x+1}$

30. $\displaystyle\lim_{t\to-\infty} \frac{4+6e^{2t}}{5-9e^{3t}}$

31. Determine the limits at infinity of $g(t) = 5^{-1/t^2}$.

32. Show that $\lim_{x\to\infty}(\sqrt{x^2+1} - x) = 0$. *Hint:* Observe that

$$\sqrt{x^2+1} - x = \frac{1}{\sqrt{x^2+1}+x}$$

33. According to the **Michaelis–Menten equation** (Figure 7), when an enzyme is combined with a substrate of concentration s (in millimolars), the reaction rate (in micromolars/min) is

$$R(s) = \frac{As}{K+s} \qquad (A, K \text{ constants})$$

(a) Show, by computing $\lim_{s\to\infty} R(s)$, that A is the limiting reaction rate as the concentration s approaches ∞.

(b) Show that the reaction rate $R(s)$ attains one-half of the limiting value A when $s = K$.

(c) For a certain reaction, $K = 1.25$ mM and $A = 0.1$. For which concentration s is $R(s)$ equal to 75% of its limiting value?

Leonor Michaelis 1875–1949 | Maud Menten 1879–1960

FIGURE 7 Canadian-born biochemist Maud Menten is best known for her fundamental work on enzyme kinetics with German scientist Leonor Michaelis. She was also an accomplished painter, clarinetist, mountain climber, and master of numerous languages.

34. Suppose that the average temperature of the earth is $T(t) = 283 + 3(1 - 10^{-0.13t})$ kelvins, where t is the number of years since 2000.

(a) Calculate the long-term average $L = \lim_{t\to\infty} T(t)$.

(b) At what time is $T(t)$ within one-half a degree of its limiting value?

In Exercises 35–40, calculate the limit.

35. $\lim_{x\to\infty} (\sqrt{4x^4 + 9x} - 2x^2)$

36. $\lim_{x\to\infty} (\sqrt{9x^3 + x} - x^{3/2})$

37. $\lim_{x\to\infty} (2\sqrt{x} - \sqrt{x+2})$

38. $\lim_{x\to\infty} \left(\frac{1}{x} - \frac{1}{x+2}\right)$

39. $\lim_{t\to\infty} \tan\left(\frac{\pi 3^t + 1}{4 - 3^{t+1}}\right)$

40. $\lim_{t\to-\infty} 2\left(\frac{8t}{t+1} + 10^{t+1}\right)$

41. Let $P(n)$ be the perimeter of an n-gon inscribed in a unit circle (Figure 8).

(a) Explain, intuitively, why $P(n)$ approaches 2π as $n \to \infty$.

(b) Show that $P(n) = 2n \sin\left(\frac{\pi}{n}\right)$.

(c) Combine (a) and (b) to conclude that $\lim_{n\to\infty} \frac{n}{\pi} \sin\left(\frac{\pi}{n}\right) = 1$.

(d) Use this to give another argument that $\lim_{\theta\to 0} \frac{\sin\theta}{\theta} = 1$.

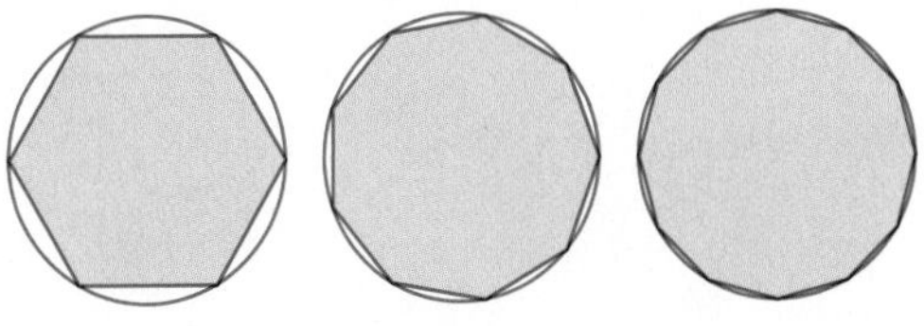

$n = 6$ $n = 9$ $n = 12$

FIGURE 8

42. Physicists have observed that Einstein's theory of **special relativity** reduces to Newtonian mechanics in the limit as $c \to \infty$, where c is the speed of light. This is illustrated by a stone tossed up vertically from ground level so that it returns to earth one second later. Using Newton's Laws, we find that the stone's maximum height is $h = g/8$ meters ($g = 9.8$ m/s^2). According to special relativity, the stone's mass depends on its velocity divided by c, and the maximum height is

$$h(c) = c\sqrt{c^2/g^2 + 1/4} - c^2/g$$

Prove that $\lim_{c\to\infty} h(c) = g/8$.

Further Insights and Challenges

43. Every limit as $x \to \infty$ can be rewritten as a one-sided limit as $t \to 0+$, where $t = x^{-1}$. Setting $g(t) = f(t^{-1})$, we have

$$\lim_{x\to\infty} f(x) = \lim_{t\to 0+} g(t)$$

Show that $\lim_{x\to\infty} \frac{3x^2 - x}{2x^2 + 5} = \lim_{t\to 0+} \frac{3 - t}{2 + 5t^2}$, and evaluate using the Quotient Law.

44. Rewrite the following as one-sided limits as in Exercise 43 and evaluate.

(a) $\lim_{x\to\infty} \frac{3 - 12x^3}{4x^3 + 3x + 1}$

(b) $\lim_{x\to\infty} 2^{1/x}$

(c) $\lim_{x\to\infty} x \sin\frac{1}{x}$

(d) $\lim_{x\to\infty} \ln\left(\frac{x+1}{x-1}\right)$

45. Let $G(b) = \lim_{x\to\infty} (1 + b^x)^{1/x}$ for $b \ge 0$. Investigate $G(b)$ numerically and graphically for $b = 0.2, 0.8, 2, 3, 5$ (and additional values if necessary). Then make a conjecture for the value of $G(b)$ as a function of b. Draw a graph of $y = G(b)$. Does $G(b)$ appear to be continuous? We will evaluate $G(b)$ using L'Hôpital's Rule in Section 7.7 (see Exercise 65 in Section 7.7).

2.8 Intermediate Value Theorem

The **Intermediate Value Theorem (IVT)** says, roughly speaking, that *a continuous function cannot skip values*. Consider a plane that takes off and climbs from 0 to 10,000 meters in 20 minutes. The plane must reach every altitude between 0 and 10,000 meters during this 20-minute interval. Thus, at some moment, the plane's altitude must have been exactly 8371 meters. Of course, this assumes that the plane's motion is continuous, so its altitude cannot jump abruptly from, say, 8000 to 9000 meters.

To state this conclusion formally, let $A(t)$ be the plane's altitude at time t. The IVT asserts that for every altitude M between 0 and 10,000, there is a time t_0 between 0 and 20 such that $A(t_0) = M$. In other words, the graph of $A(t)$ must intersect the horizontal line $y = M$ [Figure 1(A)].

By contrast, a discontinuous function can skip values. The greatest integer function $f(x) = [x]$ in Figure 1(B) satisfies $[1] = 1$ and $[2] = 2$, but it does not take on the value 1.5 (or any other value between 1 and 2).

| *A proof of the IVT is given in Appendix B.*

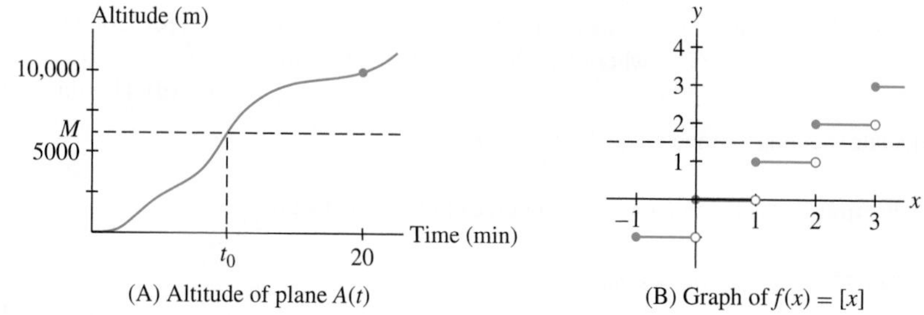

FIGURE 1

> **THEOREM 1 Intermediate Value Theorem** If $f(x)$ is continuous on a closed interval $[a, b]$ and $f(a) \neq f(b)$, then for every value M between $f(a)$ and $f(b)$, there exists at least one value $c \in (a, b)$ such that $f(c) = M$.

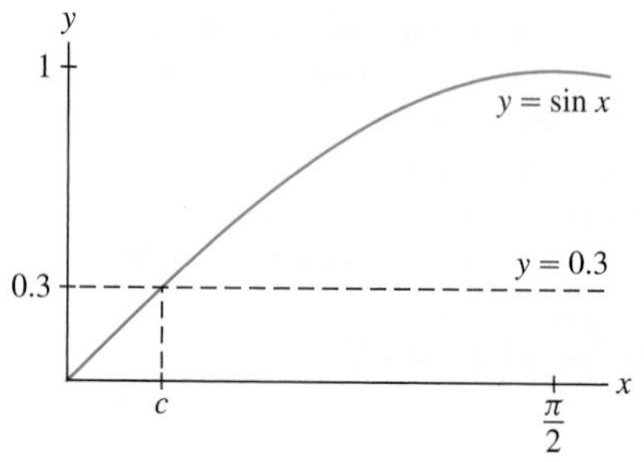

FIGURE 2

A zero or root of a function is a value c such that $f(c) = 0$. Sometimes the word "root" is reserved to refer specifically to the zero of a polynomial.

EXAMPLE 1 Prove that the equation $\sin x = 0.3$ has at least one solution.

Solution We may apply the IVT because $\sin x$ is continuous. We choose an interval where we suspect that a solution exists. The desired value 0.3 lies between the two function values

$$\sin 0 = 0 \qquad \text{and} \qquad \sin \frac{\pi}{2} = 1$$

so the interval $\left[0, \frac{\pi}{2}\right]$ will work (Figure 2). The IVT tells us that $\sin x = 0.3$ has at least one solution in $\left(0, \frac{\pi}{2}\right)$. ■

The IVT can be used to show the existence of zeros of functions. If $f(x)$ is continuous and takes on both positive and negative values—say, $f(a) < 0$ and $f(b) > 0$—then the IVT guarantees that $f(c) = 0$ for some c between a and b.

> **COROLLARY 2 Existence of Zeros** If $f(x)$ is continuous on $[a, b]$ and if $f(a)$ and $f(b)$ are nonzero and have opposite signs, then $f(x)$ has a zero in (a, b).

We can locate zeros of functions to arbitrary accuracy using the **Bisection Method**, as illustrated in the next example.

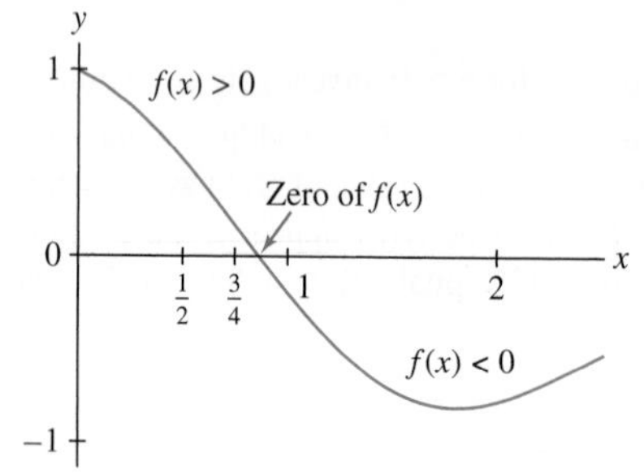

FIGURE 3 Graph of $f(x) = \cos^2 x - 2\sin \frac{x}{4}$.

Computer algebra systems have built-in commands for finding roots of a function or solving an equation numerically. These systems use a variety of methods, including more sophisticated versions of the Bisection Method. Notice that to use the Bisection Method, we must first find an interval containing a root.

EXAMPLE 2 The Bisection Method Show that $f(x) = \cos^2 x - 2\sin \frac{x}{4}$ has a zero in $(0, 2)$. Then locate the zero more accurately using the Bisection Method.

Solution Using a calculator, we find that $f(0)$ and $f(2)$ have opposite signs:

$$f(0) = 1 > 0, \qquad f(2) \approx -0.786 < 0$$

Corollary 2 guarantees that $f(x) = 0$ has a solution in $(0, 2)$ (Figure 3).

To locate a zero more accurately, divide $[0, 2]$ into two intervals $[0, 1]$ and $[1, 2]$. At least one of these intervals must contain a zero of $f(x)$. To determine which, evaluate $f(x)$ at the midpoint $m = 1$. A calculator gives $f(1) \approx -0.203 < 0$, and since $f(0) = 1$, we see that

$$f(x) \text{ takes on opposite signs at the endpoints of } [0, 1]$$

Therefore, $(0, 1)$ must contain a zero. We discard $[1, 2]$ because both $f(1)$ and $f(2)$ are negative.

The Bisection Method consists of continuing this process until we narrow down the location of the zero to any desired accuracy. In the following table, the process is carried out three times:

Interval	Midpoint of interval	Function values	Conclusion
$[0, 1]$	$\frac{1}{2}$	$f(\frac{1}{2}) \approx 0.521$ $f(1) \approx -0.203$	Zero lies in $(\frac{1}{2}, 1)$
$[\frac{1}{2}, 1]$	$\frac{3}{4}$	$f(\frac{3}{4}) \approx 0.163$ $f(1) \approx -0.203$	Zero lies in $(\frac{3}{4}, 1)$
$[\frac{3}{4}, 1]$	$\frac{7}{8}$	$f(\frac{7}{8}) \approx -0.0231$ $f(\frac{3}{4}) \approx 0.163$	Zero lies in $(\frac{3}{4}, \frac{7}{8})$

We conclude that $f(x)$ has a zero c satisfying $0.75 < c < 0.875$. ■

CONCEPTUAL INSIGHT The IVT seems to state the obvious, namely that a continuous function cannot skip values. Yet its proof (given in Appendix B) is subtle because it depends on the *completeness property* of real numbers. To highlight the subtlety, observe that the IVT is *false* for functions defined only on the *rational numbers*. For example, $f(x) = x^2$ is continuous, but it does not have the intermediate value property if we restrict its domain to the rational numbers. Indeed, $f(0) = 0$ and $f(2) = 4$, but $f(c) = 2$ *has no solution* for c rational. The solution $c = \sqrt{2}$ is "missing" from the set of rational numbers because it is irrational. No doubt the IVT was always regarded as obvious, but it was not possible to give a correct proof until the completeness property was clarified in the second half of the nineteenth century.

2.8 SUMMARY

- The Intermediate Value Theorem (IVT) says that a continuous function cannot *skip* values.
- More precisely, if $f(x)$ is continuous on $[a, b]$ with $f(a) \neq f(b)$, and if M is a number between $f(a)$ and $f(b)$, then $f(c) = M$ for some $c \in (a, b)$.
- Existence of zeros: If $f(x)$ is continuous on $[a, b]$ and if $f(a)$ and $f(b)$ take opposite signs (one is positive and the other negative), then $f(c) = 0$ for some $c \in (a, b)$.
- Bisection Method: Assume f is continuous and that $f(a)$ and $f(b)$ have opposite signs, so that f has a zero in (a, b). Then f has a zero in $[a, m]$ or $[m, b]$, where $m = (a + b)/2$ is the midpoint of $[a, b]$. A zero lies in (a, m) if $f(a)$ and $f(m)$ have opposite signs and in (m, b) if $f(m)$ and $f(b)$ have opposite signs. Continuing the process, we can locate a zero with arbitrary accuracy.

2.8 EXERCISES

Preliminary Questions

1. Prove that $f(x) = x^2$ takes on the value 0.5 in the interval $[0, 1]$.

2. The temperature in Vancouver was 8°C at 6 AM and rose to 20°C at noon. Which assumption about temperature allows us to conclude that the temperature was 15°C at some moment of time between 6 AM and noon?

3. What is the graphical interpretation of the IVT?

4. Show that the following statement is false by drawing a graph that provides a counterexample:

If $f(x)$ is continuous and has a root in $[a, b]$, then $f(a)$ and $f(b)$ have opposite signs.

5. Assume that $f(t)$ is continuous on $[1, 5]$ and that $f(1) = 20$, $f(5) = 100$. Determine whether each of the following statements is always true, never true, or sometimes true.

(a) $f(c) = 3$ has a solution with $c \in [1, 5]$.
(b) $f(c) = 75$ has a solution with $c \in [1, 5]$.
(c) $f(c) = 50$ has no solution with $c \in [1, 5]$.
(d) $f(c) = 30$ has exactly one solution with $c \in [1, 5]$.

Exercises

1. Use the IVT to show that $f(x) = x^3 + x$ takes on the value 9 for some x in $[1, 2]$.

2. Show that $g(t) = \dfrac{t}{t+1}$ takes on the value 0.499 for some t in $[0, 1]$.

3. Show that $g(t) = t^2 \tan t$ takes on the value $\frac{1}{2}$ for some t in $\left[0, \frac{\pi}{4}\right]$.

4. Show that $f(x) = \dfrac{x^2}{x^7 + 1}$ takes on the value 0.4.

5. Show that $\cos x = x$ has a solution in the interval $[0, 1]$. *Hint:* Show that $f(x) = x - \cos x$ has a zero in $[0, 1]$.

6. Use the IVT to find an interval of length $\frac{1}{2}$ containing a root of $f(x) = x^3 + 2x + 1$.

In Exercises 7–16, prove using the IVT.

7. $\sqrt{c} + \sqrt{c+2} = 3$ has a solution.

8. For all integers n, $\sin nx = \cos x$ for some $x \in [0, \pi]$.

9. $\sqrt{2}$ exists. *Hint:* Consider $f(x) = x^2$.

10. A positive number c has an nth root for all positive integers n.

11. For all positive integers k, $\cos x = x^k$ has a solution.

12. $2^x = bx$ has a solution if $b > 2$.

13. $2^x + 3^x = 4^x$ has a solution.

14. $\tan x = x$ has infinitely many solutions.

15. $2^x + 1/x = -4$ has a solution.

16. $x = \sin x + \cos x$ has a solution.

17. Carry out three steps of the Bisection Method for $f(x) = 2^x - x^3$ as follows:

(a) Show that $f(x)$ has a zero in $[1, 1.5]$.
(b) Show that $f(x)$ has a zero in $[1.25, 1.5]$.
(c) Determine whether $[1.25, 1.375]$ or $[1.375, 1.5]$ contains a zero.

18. Figure 4 shows that $f(x) = x^3 - 8x - 1$ has a root in the interval $[2.75, 3]$. Apply the Bisection Method twice to find an interval of length $\frac{1}{16}$ containing this root.

19. Find an interval of length $\frac{1}{4}$ in $[1, 2]$ containing a root of the equation $x^7 + 3x - 10 = 0$.

20. Show that $\tan^3\theta - 8\tan^2\theta + 17\tan\theta - 8 = 0$ has a root in $[0.5, 0.6]$. Apply the Bisection Method twice to find an interval of length 0.025 containing this root.

In Exercises 21–24, draw the graph of a function $f(x)$ on $[0, 4]$ with the given property.

21. Jump discontinuity at $x = 2$ and does not satisfy the conclusion of the IVT.

22. Jump discontinuity at $x = 2$ and satisfies the conclusion of the IVT on $[0, 4]$.

23. Infinite one-sided limits at $x = 2$ and does not satisfy the conclusion of the IVT.

24. Infinite one-sided limits at $x = 2$ and satisfies the conclusion of the IVT on $[0, 4]$.

25. Can Corollary 2 be applied to $f(x) = x^{-1}$ on $[-1, 1]$? Does $f(x)$ have any roots?

Further Insights and Challenges

26. Take any map and draw a circle on it anywhere (Figure 5). Prove that at any moment in time there exists a pair of diametrically opposite points A and B on that circle corresponding to locations where the temperatures at that moment are equal. *Hint:* Let θ be an angular coordinate along the circle and let $f(\theta)$ be the difference in temperatures at the locations corresponding to θ and $\theta + \pi$.

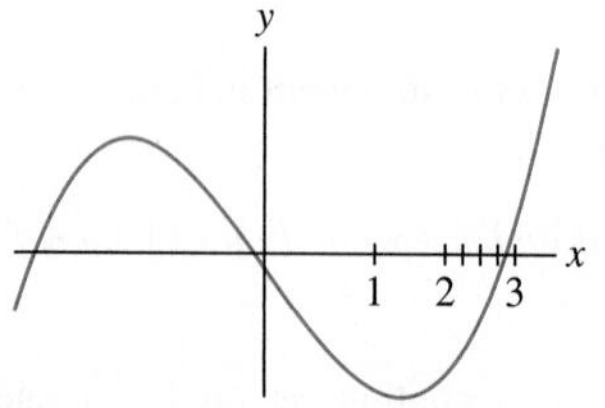

FIGURE 4 Graph of $y = x^3 - 8x - 1$.

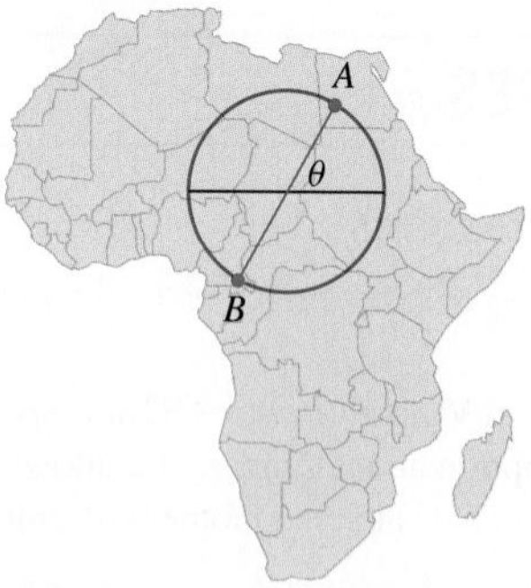

FIGURE 5 $f(\theta)$ is the difference between the temperatures at A and B.

27. Show that if $f(x)$ is continuous and $0 \le f(x) \le 1$ for $0 \le x \le 1$, then $f(c) = c$ for some c in $[0, 1]$ (Figure 6).

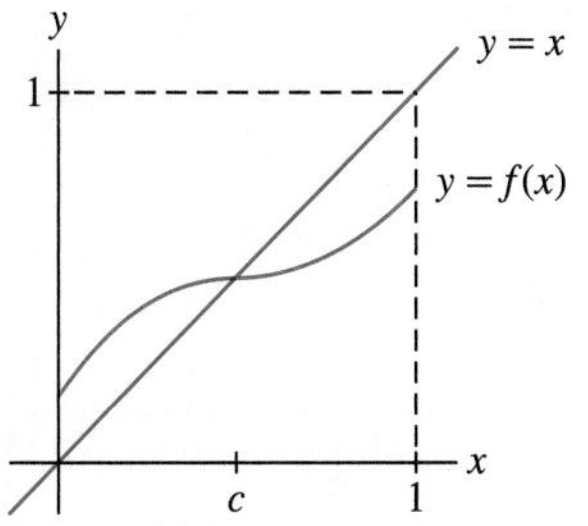

FIGURE 6 A function satisfying $0 \le f(x) \le 1$ for $0 \le x \le 1$.

28. Use the IVT to show that if $f(x)$ is continuous and one-to-one on an interval $[a, b]$, then $f(x)$ is either an increasing or a decreasing function.

29. **Ham Sandwich Theorem** Figure 7(A) shows a slice of ham. Prove that for any angle θ $(0 \le \theta \le \pi)$, it is possible to cut the slice in half with a cut of incline θ. *Hint:* The lines of inclination θ are given by the equations $y = (\tan\theta)x + b$, where b varies from $-\infty$ to ∞. Each such line divides the slice into two pieces (one of which may be empty). Let $A(b)$ be the amount of ham to the left of the line minus the amount to the right, and let A be the total area of the ham. Show that $A(b) = -A$ if b is sufficiently large and $A(b) = A$ if b is sufficiently negative. Then use the IVT. This works if $\theta \ne 0$ or $\frac{\pi}{2}$. If $\theta = 0$, define $A(b)$ as the amount of ham above the line $y = b$ minus the amount below. How can you modify the argument to work when $\theta = \frac{\pi}{2}$ (in which case $\tan\theta = \infty$)?

30. Figure 7(B) shows a slice of ham on a piece of bread. Prove that it is possible to slice this open-faced sandwich so that each part has equal amounts of ham and bread. *Hint:* By Exercise 29, for all $0 \le \theta \le \pi$ there is a line $L(\theta)$ of incline θ (which we assume is unique) that divides the ham into two equal pieces. Let $B(\theta)$ denote the amount of bread to the left of (or above) $L(\theta)$ minus the amount to the right (or below). Notice that $L(\pi)$ and $L(0)$ are the same line, but $B(\pi) = -B(0)$ since left and right get interchanged as the angle moves from 0 to π. Assume that $B(\theta)$ is continuous and apply the IVT. (By a further extension of this argument, one can prove the full "Ham Sandwich Theorem," which states that if you allow the knife to cut at a slant, then it is possible to cut a sandwich consisting of a slice of ham and two slices of bread so that all three layers are divided in half.)

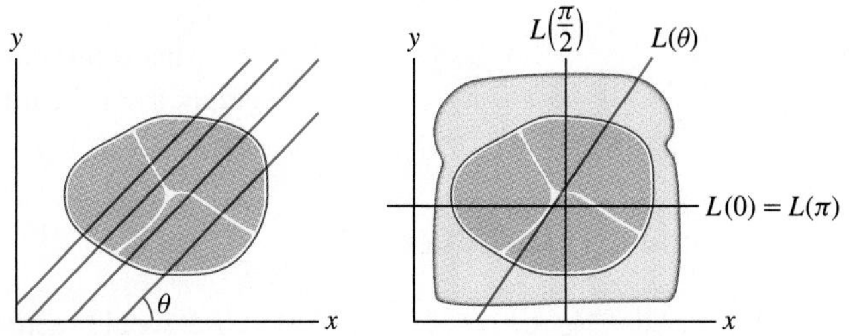

(A) Cutting a slice of ham at an angle θ

(B) A slice of ham on top of a slice of bread

FIGURE 7

2.9 The Formal Definition of a Limit

A "rigorous proof" in mathematics is a proof based on a complete chain of logic without any gaps or ambiguity. The formal limit definition is a key ingredient of rigorous proofs in calculus. A few such proofs are included in Appendix D. More complete developments can be found in textbooks on the branch of mathematics called "analysis."

In this section, we reexamine the definition of a limit in order to state it in a more rigorous and precise fashion. Why is this necessary? In Section 2.2, we defined limits by saying that $\lim_{x\to c} f(x) = L$ if $|f(x) - L|$ becomes arbitrarily small when x is sufficiently close (but not equal) to c. The problem with this definition lies in the phrases "arbitrarily small" and "sufficiently close." We must find a way to specify just how close is sufficiently close.

The Size of the Gap

Recall that the distance from $f(x)$ to L is $|f(x) - L|$. It is convenient to refer to the quantity $|f(x) - L|$ as the *gap* between the value $f(x)$ and the limit L.

Let's reexamine the trigonometric limit

$$\lim_{x\to 0} \frac{\sin x}{x} = 1 \qquad \boxed{1}$$

In this example, $f(x) = \dfrac{\sin x}{x}$ and $L = 1$, so Eq. (1) tells us that the gap $|f(x) - 1|$ gets arbitrarily small when x is sufficiently close, but not equal, to 0 [Figure 1(A)].

Suppose we want the gap $|f(x) - 1|$ to be less than 0.2. How close to 0 must x be? Figure 1(B) shows that $f(x)$ lies within 0.2 of $L = 1$ for all values of x in the interval $[-1, 1]$. In other words, the following statement is true:

$$\left|\frac{\sin x}{x} - 1\right| < 0.2 \qquad \text{if} \qquad 0 < |x| < 1$$

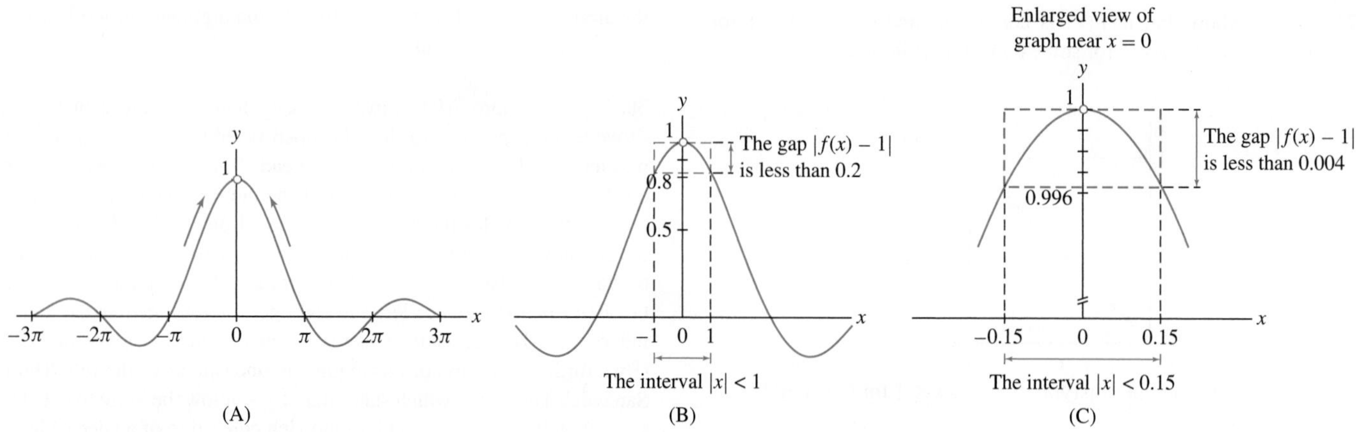

FIGURE 1 Graphs of $y = \dfrac{\sin x}{x}$. To shrink the gap from 0.2 to 0.004, we require that x lie within 0.15 of 0.

If we insist instead that the gap be smaller than 0.004, we can check by zooming in on the graph, as in Figure 1(C), that

$$\left|\frac{\sin x}{x} - 1\right| < 0.004 \qquad \text{if} \qquad 0 < |x| < 0.15$$

It would seem that this process can be continued: By zooming in on the graph, we can find a small interval around $c = 0$ where the gap $|f(x) - 1|$ is smaller than any prescribed positive number.

To express this in a precise fashion, we follow time-honored tradition in using the Greek letters ϵ (epsilon) and δ (delta) to denote small numbers specifying the sizes of the gap and the quantity $|x - c|$, respectively. In our case, $c = 0$ and $|x - c| = |x|$. The precise meaning of Eq. (1) is that for every choice of $\epsilon > 0$, there exists some δ (depending on ϵ) such that

$$\left|\frac{\sin x}{x} - 1\right| < \epsilon \qquad \text{if} \qquad 0 < |x| < \delta$$

The number δ pins down just how close is "sufficiently close" for a given ϵ. With this motivation, we are ready to state the formal definition of the limit.

The formal definition of a limit is often called the ϵ-δ definition. The tradition of using the symbols ϵ and δ originated in the writings of Augustin-Louis Cauchy on calculus and analysis in the 1820s.

FORMAL DEFINITION OF A LIMIT Suppose that $f(x)$ is defined for all x in an open interval containing c (but not necessarily at $x = c$). Then

$$\lim_{x \to c} f(x) = L$$

if for all $\epsilon > 0$, there exists $\delta > 0$ such that

$$|f(x) - L| < \epsilon \qquad \text{if} \qquad 0 < |x - c| < \delta$$

If the symbols ϵ and δ seem to make this definition too abstract, keep in mind that we can take $\epsilon = 10^{-n}$ and $\delta = 10^{-m}$. Thus, $\lim_{x\to c} f(x) = L$ if, for any n, there exist $m > 0$ such that $|f(x) - L| < 10^{-n}$, provided that $0 < |x - c| < 10^{-m}$.

The condition $0 < |x - c| < \delta$ in this definition excludes $x = c$. In other words, the limit depends only on values of $f(x)$ near c but not on $f(c)$ itself. As we have seen in previous sections, the limit may exist even when $f(c)$ is not defined.

■ **EXAMPLE 1** Let $f(x) = 8x + 3$.

(a) Prove that $\lim_{x\to 3} f(x) = 27$ using the formal definition of the limit.

(b) Find values of δ that work for $\epsilon = 0.2$ and 0.001.

Solution

(a) We break the proof into two steps.

***Step 1.* Relate the gap to $|x - c|$.**

We must find a relation between two absolute values: $|f(x) - L|$ for $L = 27$ and $|x - c|$ for $c = 3$. Observe that

$$\underbrace{|f(x) - 27|}_{\text{Size of gap}} = |(8x + 3) - 27| = |8x - 24| = 8|x - 3|$$

Thus, the gap is 8 times as large as $|x - 3|$.

***Step 2.* Choose δ (in terms of ϵ).**

We can now see how to make the gap small: If $|x - 3| < \frac{\epsilon}{8}$, then the gap is less than $8\left(\frac{\epsilon}{8}\right) = \epsilon$. Therefore, for any $\epsilon > 0$, we choose $\delta = \frac{\epsilon}{8}$. With this choice, the following statement holds:

$$|f(x) - 27| < \epsilon \quad \text{if} \quad 0 < |x - 3| < \delta, \quad \text{where } \delta = \frac{\epsilon}{8}$$

Since we have specified δ for all $\epsilon > 0$, we have fulfilled the requirements of the formal definition, thus proving rigorously that $\lim_{x\to 3} (8x + 3) = 27$.

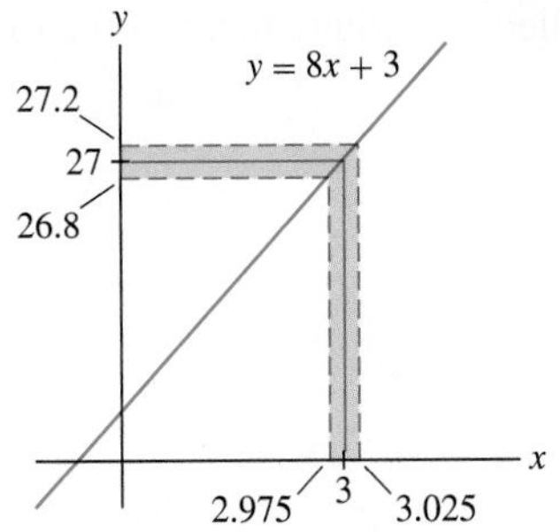

FIGURE 2 To make the gap less than 0.2, we may take $\delta = 0.025$ (not drawn to scale).

(b) For the particular choice $\epsilon = 0.2$, we may take $\delta = \frac{\epsilon}{8} = \frac{0.2}{8} = 0.025$:

$$|f(x) - 27| < 0.2 \quad \text{if} \quad 0 < |x - 3| < 0.025$$

This statement is illustrated in Figure 2. But note that *any positive δ smaller than* 0.025 *will also work.* For example, the following statement is also true, although it places an unnecessary restriction on x:

$$|f(x) - 27| < 0.2 \quad \text{if} \quad 0 < |x - 3| < 0.019$$

Similarly, to make the gap less than $\epsilon = 0.001$, we may take

$$\delta = \frac{\epsilon}{8} = \frac{0.001}{8} = 0.000125$$ ■

The difficulty in applying the limit definition lies in trying to relate $|f(x) - L|$ to $|x - c|$. The next two examples illustrate how this can be done in special cases.

■ **EXAMPLE 2** Prove that $\lim_{x\to 2} x^2 = 4$.

Solution Let $f(x) = x^2$.

***Step 1.* Relate the gap to $|x - c|$.**

In this case, we must relate the gap $|f(x) - 4| = |x^2 - 4|$ to the quantity $|x - 2|$ (Figure 3). This is more difficult than in the previous example because the gap is not a constant multiple of $|x - 2|$. To proceed, consider the factorization

$$|x^2 - 4| = |x + 2|\,|x - 2|$$

Because we are going to require that $|x - 2|$ be small, we may as well assume from the outset that $|x - 2| < 1$, which means that $1 < x < 3$. In this case, $|x + 2|$ is less than 5 and the gap satisfies

$$|x^2 - 4| = |x + 2|\,|x - 2| < 5\,|x - 2| \qquad \text{if} \qquad |x - 2| < 1 \tag{2}$$

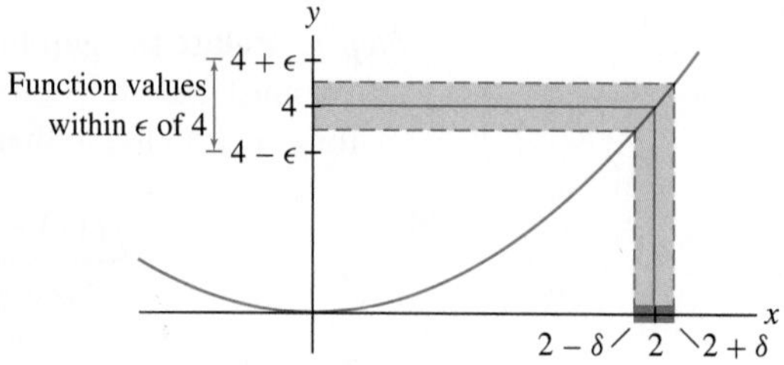

FIGURE 3 Graph of $f(x) = x^2$. We may choose δ so that $f(x)$ lies within ϵ of 4 for all x in $[2 - \delta, 2 + \delta]$.

***Step 2.* Choose δ (in terms of ϵ).**
We see from Eq. (2) that if $|x - 2|$ is smaller than both $\frac{\epsilon}{5}$ and 1, then the gap satisfies

$$|x^2 - 4| < 5|x - 2| < 5\left(\frac{\epsilon}{5}\right) = \epsilon$$

Therefore, the following statement holds for all $\epsilon > 0$:

$$|x^2 - 4| < \epsilon \qquad \text{if} \qquad 0 < |x - 2| < \delta, \quad \text{where } \delta \text{ is the smaller of } \tfrac{\epsilon}{5} \text{ and } 1$$

We have specified δ for all $\epsilon > 0$, so we have fulfilled the requirements of the formal limit definition, thus proving that $\lim_{x\to 2} x^2 = 4$. ■

■ **EXAMPLE 3** Prove that $\lim_{x\to 3} \frac{1}{x} = \frac{1}{3}$.

Solution

***Step 1.* Relate the gap to $|x - c|$.**
The gap is equal to

$$\left|\frac{1}{x} - \frac{1}{3}\right| = \left|\frac{3 - x}{3x}\right| = |x - 3|\,\left|\frac{1}{3x}\right|$$

REMINDER If $a > b > 0$, then $\frac{1}{a} < \frac{1}{b}$. Thus, if $3x > 6$, then $\frac{1}{3x} < \frac{1}{6}$.

Because we are going to require that $|x - 3|$ be small, we may as well assume from the outset that $|x - 3| < 1$—that is, that $2 < x < 4$. Now observe that if $x > 2$, then $3x > 6$ and $\frac{1}{3x} < \frac{1}{6}$, so the following inequality is valid if $|x - 3| < 1$:

$$\left|f(x) - \frac{1}{3}\right| = \left|\frac{3 - x}{3x}\right| = \left|\frac{1}{3x}\right|\,|x - 3| < \frac{1}{6}\,|x - 3| \tag{3}$$

***Step 2.* Choose δ (in terms of ϵ).**
By Eq. (3), if $|x - 3| < 1$ and $|x - 3| < 6\epsilon$, then

$$\left|\frac{1}{x} - \frac{1}{3}\right| < \frac{1}{6}\,|x - 3| < \frac{1}{6}(6\epsilon) = \epsilon$$

Therefore, given any $\epsilon > 0$, we let δ be the smaller of the numbers 6ϵ and 1. Then

$$\left|\frac{1}{x} - \frac{1}{3}\right| < \epsilon \quad \text{if} \quad 0 < |x - 3| < \delta, \quad \text{where } \delta \text{ is the smaller of } 6\epsilon \text{ and } 1$$

Again, we have fulfilled the requirements of the formal limit definition, thus proving rigorously that $\lim_{x \to 3} \frac{1}{x} = \frac{1}{3}$. ■

GRAPHICAL INSIGHT Keep the graphical interpretation of limits in mind. In Figure 4(A), $f(x)$ approaches L as $x \to c$ because for any $\epsilon > 0$, we can make the gap less than ϵ by taking δ sufficiently small. By contrast, the function in Figure 4(B) has a jump discontinuity at $x = c$. The gap cannot be made small, no matter how small δ is taken. Therefore, the limit does not exist.

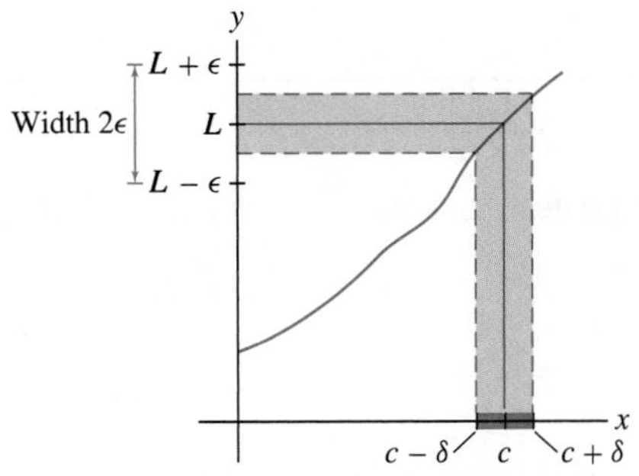

(A) The function is continuous at $x = c$. By taking δ sufficiently small, we can make the gap smaller than ϵ.

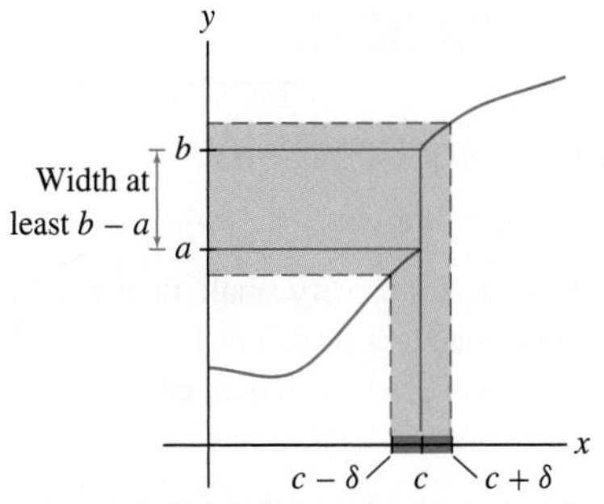

(B) The function is not continuous at $x = c$. The gap is always larger than $(b - a)/2$, no matter how small δ is.

FIGURE 4

Proving Limit Theorems

In practice, the formal limit definition is rarely used to evaluate limits. Most limits are evaluated using the Basic Limit Laws or other techniques such as the Squeeze Theorem. However, the formal definition allows us to prove these laws in a rigorous fashion and thereby ensure that calculus is built on a solid foundation. We illustrate by proving the Sum Law. Other proofs are given in Appendix D.

Proof of the Sum Law Assume that

$$\lim_{x \to c} f(x) = L \quad \text{and} \quad \lim_{x \to c} g(x) = M$$

We must prove that $\lim_{x \to c} (f(x) + g(x)) = L + M$.

Apply the Triangle Inequality (see margin) with $a = f(x) - L$ and $b = g(x) - M$:

← REMINDER The Triangle Inequality [Eq. (1) in Section 1.1] states

$$|a + b| \le |a| + |b|$$

for all a and b.

$$|(f(x) + g(x)) - (L + M)| \le |f(x) - L| + |g(x) - M| \qquad \boxed{4}$$

Each term on the right in (4) can be made small by the limit definition. More precisely, given $\epsilon > 0$, we can choose δ such that $|f(x) - L| < \frac{\epsilon}{2}$ and $|g(x) - M| < \frac{\epsilon}{2}$ if $0 < |x - c| < \delta$ (in principle, we might choose different δ's for f and g, but we may then use the smaller of the two δ's). Thus, Eq. (4) gives

$$|f(x) + g(x) - (L + M)| < \frac{\epsilon}{2} + \frac{\epsilon}{2} = \epsilon \quad \text{if} \quad 0 < |x - c| < \delta \qquad \boxed{5}$$

This proves that

$$\lim_{x\to c}\big(f(x)+g(x)\big) = L + M = \lim_{x\to c} f(x) + \lim_{x\to c} g(x)$$ ∎

2.9 SUMMARY

- Informally speaking, the statement $\lim_{x\to c} f(x) = L$ means that the gap $|f(x) - L|$ tends to 0 as x approaches c.
- The *formal definition* (called the ϵ-δ definition): $\lim_{x\to c} f(x) = L$ if, for all $\epsilon > 0$, there exists a $\delta > 0$ such that

$$|f(x) - L| < \epsilon \quad \text{if} \quad 0 < |x - c| < \delta$$

2.9 EXERCISES

Preliminary Questions

1. Given that $\lim_{x\to 0} \cos x = 1$, which of the following statements is true?

(a) If $|\cos x - 1|$ is very small, then x is close to 0.
(b) There is an $\epsilon > 0$ such that $|x| < 10^{-5}$ if $0 < |\cos x - 1| < \epsilon$.
(c) There is a $\delta > 0$ such that $|\cos x - 1| < 10^{-5}$ if $0 < |x| < \delta$.
(d) There is a $\delta > 0$ such that $|\cos x| < 10^{-5}$ if $0 < |x - 1| < \delta$.

2. Suppose it is known that for a given ϵ and δ, $|f(x) - 2| < \epsilon$ if $0 < |x - 3| < \delta$. Which of the following statements must also be true?

(a) $|f(x) - 2| < \epsilon$ if $0 < |x - 3| < 2\delta$
(b) $|f(x) - 2| < 2\epsilon$ if $0 < |x - 3| < \delta$
(c) $|f(x) - 2| < \frac{\epsilon}{2}$ if $0 < |x - 3| < \frac{\delta}{2}$
(d) $|f(x) - 2| < \epsilon$ if $0 < |x - 3| < \frac{\delta}{2}$

Exercises

1. Based on the information conveyed in Figure 5(A), find values of L, ϵ, and $\delta > 0$ such that the following statement holds: $|f(x) - L| < \epsilon$ if $|x| < \delta$.

2. Based on the information conveyed in Figure 5(B), find values of c, L, ϵ, and $\delta > 0$ such that the following statement holds: $|f(x) - L| < \epsilon$ if $0 < |x - c| < \delta$.

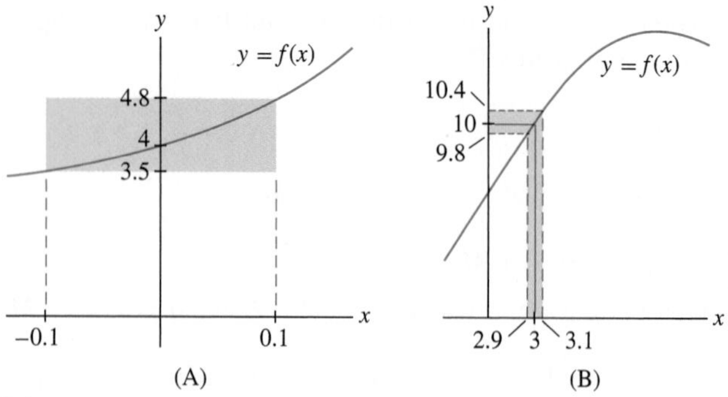

FIGURE 5

3. Consider $\lim_{x\to 4} f(x)$, where $f(x) = 8x + 3$.

(a) Show that $|f(x) - 35| = 8|x - 4|$.
(b) Show that for any $\epsilon > 0$, $|f(x) - 35| < \epsilon$ if $0 < |x - 4| < \delta$, where $\delta = \frac{\epsilon}{8}$. Explain how this proves rigorously that $\lim_{x\to 4} f(x) = 35$.

4. Consider $\lim_{x\to 2} f(x)$, where $f(x) = 4x - 1$.

(a) Show that $|f(x) - 7| < 4\delta$ if $0 < |x - 2| < \delta$.
(b) Find a δ such that

$$|f(x) - 7| < 0.01 \quad \text{if} \quad 0 < |x - 2| < \delta$$

(c) Prove rigorously that $\lim_{x\to 2} f(x) = 7$.

5. Consider $\lim_{x\to 2} x^2 = 4$ (refer to Example 2).

(a) Show that $|x^2 - 4| < 0.05$ if $0 < |x - 2| < 0.01$.
(b) Show that $|x^2 - 4| < 0.0009$ if $0 < |x - 2| < 0.0002$.
(c) Find a value of δ such that $|x^2 - 4|$ is less than 10^{-4} if $0 < |x - 2| < \delta$.

6. With regard to the limit $\lim_{x\to 5} x^2 = 25$,

(a) Show that $|x^2 - 25| < 11|x - 5|$ if $4 < x < 6$. *Hint:* Write $|x^2 - 25| = |x + 5| \cdot |x - 5|$.
(b) Find a δ such that $|x^2 - 25| < 10^{-3}$ if $0 < |x - 5| < \delta$.
(c) Give a rigorous proof of the limit by showing that $|x^2 - 25| < \epsilon$ if $0 < |x - 5| < \delta$, where δ is the smaller of $\frac{\epsilon}{11}$ and 1.

7. Refer to Example 3 to find a value of $\delta > 0$ such that

$$\left|\frac{1}{x} - \frac{1}{3}\right| < 10^{-4} \quad \text{if} \quad 0 < |x - 3| < \delta$$

8. Use Figure 6 to find a value of $\delta > 0$ such that the following statement holds: $\left|1/x^2 - \frac{1}{4}\right| < \epsilon$ if $0 < |x - 2| < \delta$ for $\epsilon = 0.03$. Then find a value of δ that works for $\epsilon = 0.01$.

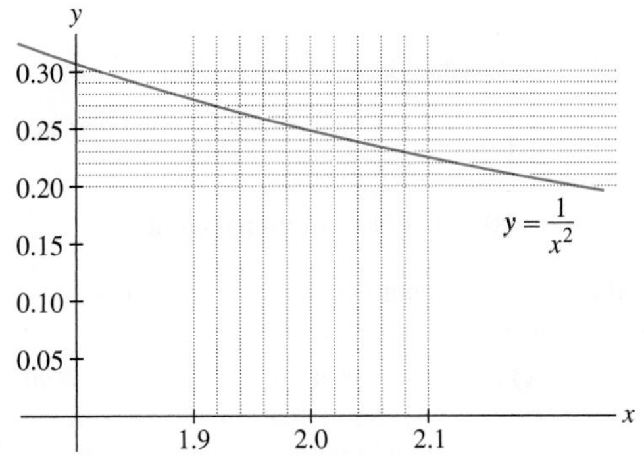

FIGURE 6

9. GU Plot $f(x) = \sqrt{2x - 1}$ together with the horizontal lines $y = 2.9$ and $y = 3.1$. Use this plot to find a value of $\delta > 0$ such that $|\sqrt{2x-1} - 3| < 0.1$ if $0 < |x - 5| < \delta$.

10. GU Plot $f(x) = \tan x$ together with the horizontal lines $y = 0.99$ and $y = 1.01$. Use this plot to find a value of $\delta > 0$ such that $|\tan x - 1| < 0.01$ if $0 < \left|x - \frac{\pi}{4}\right| < \delta$.

11. GU The function $f(x) = 2^{-x^2}$ satisfies $\lim_{x\to 0} f(x) = 1$. Use a plot of f to find a value of $\delta > 0$ such that $|f(x) - 1| < 0.001$ if $0 < |x| < \delta$.

12. GU Let $f(x) = \dfrac{4}{x^2+1}$ and $\epsilon = 0.5$. Using a plot of $f(x)$, find a value of $\delta > 0$ such that $\left|f(x) - \frac{16}{5}\right| < \epsilon$ for $0 < \left|x - \frac{1}{2}\right| < \delta$. Repeat for $\epsilon = 0.2$ and 0.1.

13. Consider $\lim_{x\to 2} \dfrac{1}{x}$.

(a) Show that if $|x - 2| < 1$, then

$$\left|\frac{1}{x} - \frac{1}{2}\right| < \frac{1}{2}|x - 2|$$

(b) Let δ be the smaller of 1 and 2ϵ. Prove:

$$\left|\frac{1}{x} - \frac{1}{2}\right| < \epsilon \quad \text{if} \quad 0 < |x - 2| < \delta$$

(c) Find a $\delta > 0$ such that $\left|\frac{1}{x} - \frac{1}{2}\right| < 0.01$ if $0 < |x - 2| < \delta$.

(d) Prove rigorously that $\lim_{x\to 2} \dfrac{1}{x} = \dfrac{1}{2}$.

14. Consider $\lim_{x\to 1} \sqrt{x+3}$.

(a) Show that $|\sqrt{x+3} - 2| < \frac{1}{2}|x - 1|$ if $|x - 1| < 4$. *Hint:* Multiply the inequality by $|\sqrt{x+3} + 2|$ and observe that $|\sqrt{x+3} + 2| > 2$.

(b) Find $\delta > 0$ such that $|\sqrt{x+3} - 2| < 10^{-4}$ for $0 < |x - 1| < \delta$.

(c) Prove rigorously that the limit is equal to 2.

15. Let $f(x) = \sin x$. Using a calculator, we find:

$$f\left(\frac{\pi}{4} - 0.1\right) \approx 0.633, \quad f\left(\frac{\pi}{4}\right) \approx 0.707, \quad f\left(\frac{\pi}{4} + 0.1\right) \approx 0.774$$

Use these values and the fact that $f(x)$ is increasing on $\left[0, \frac{\pi}{2}\right]$ to justify the statement

$$\left|f(x) - f\left(\frac{\pi}{4}\right)\right| < 0.08 \quad \text{if} \quad 0 < \left|x - \frac{\pi}{4}\right| < 0.1$$

Then draw a figure like Figure 3 to illustrate this statement.

16. Adapt the argument in Example 1 to prove rigorously that $\lim_{x\to c} (ax + b) = ac + b$, where a, b, c are arbitrary.

17. Adapt the argument in Example 2 to prove rigorously that $\lim_{x\to c} x^2 = c^2$ for all c.

18. Adapt the argument in Example 3 to prove rigorously that $\lim_{x\to c} x^{-1} = \frac{1}{c}$ for all $c \neq 0$.

In Exercises 19–24, use the formal definition of the limit to prove the statement rigorously.

19. $\lim_{x\to 4} \sqrt{x} = 2$

20. $\lim_{x\to 1} (3x^2 + x) = 4$

21. $\lim_{x\to 1} x^3 = 1$

22. $\lim_{x\to 0} (x^2 + x^3) = 0$

23. $\lim_{x\to 2} x^{-2} = \dfrac{1}{4}$

24. $\lim_{x\to 0} x \sin \dfrac{1}{x} = 0$

25. Let $f(x) = \dfrac{x}{|x|}$. Prove rigorously that $\lim_{x\to 0} f(x)$ does not exist. *Hint:* Show that for any L, there always exists some x such that $|x| < \delta$ but $|f(x) - L| \geq \frac{1}{2}$, no matter how small δ is taken.

26. Prove rigorously that $\lim_{x\to 0} |x| = 0$.

27. Let $f(x) = \min(x, x^2)$, where $\min(a, b)$ is the minimum of a and b. Prove rigorously that $\lim_{x\to 1} f(x) = 1$.

28. Prove rigorously that $\lim_{x\to 0} \sin \frac{1}{x}$ does not exist.

29. First, use the identity

$$\sin x + \sin y = 2 \sin\left(\frac{x+y}{2}\right) \cos\left(\frac{x-y}{2}\right)$$

to verify the relation

$$\sin(a + h) - \sin a = h \frac{\sin(h/2)}{h/2} \cos\left(a + \frac{h}{2}\right) \qquad \boxed{6}$$

Then use the inequality $\left|\dfrac{\sin x}{x}\right| \leq 1$ for $x \neq 0$ to show that $|\sin(a + h) - \sin a| < |h|$ for all a. Finally, prove rigorously that $\lim_{x\to a} \sin x = \sin a$.

Further Insights and Challenges

30. Uniqueness of the Limit Prove that a function converges to at most one limiting value. In other words, use the limit definition to prove that if $\lim_{x\to c} f(x) = L_1$ and $\lim_{x\to c} f(x) = L_2$, then $L_1 = L_2$.

In Exercises 31–33, prove the statement using the formal limit definition.

31. The Constant Multiple Law [Theorem 1, part (ii) in Section 2.3, p. 58]

32. The Squeeze Theorem. (Theorem 1 in Section 2.6, p. 77)

33. The Product Law [Theorem 1, part (iii) in Section 2.3, p. 58]. *Hint:* Use the identity

$$f(x)g(x) - LM = (f(x) - L)\,g(x) + L(g(x) - M)$$

34. Let $f(x) = 1$ if x is rational and $f(x) = 0$ if x is irrational. Prove that $\lim_{x\to c} f(x)$ does not exist for any c.

35. Here is a function with strange continuity properties:

$$f(x) = \begin{cases} \dfrac{1}{q} & \text{if } x \text{ is the rational number } p/q \text{ in lowest terms} \\ 0 & \text{if } x \text{ is an irrational number} \end{cases}$$

(a) Show that $f(x)$ is discontinuous at c if c is rational. *Hint:* There exist irrational numbers arbitrarily close to c.

(b) Show that $f(x)$ is continuous at c if c is irrational. *Hint:* Let I be the interval $\{x : |x - c| < 1\}$. Show that for any $Q > 0$, I contains at most finitely many fractions p/q with $q < Q$. Conclude that there is a δ such that all fractions in $\{x : |x - c| < \delta\}$ have a denominator larger than Q.

CHAPTER REVIEW EXERCISES

1. The position of a particle at time t (s) is $s(t) = \sqrt{t^2 + 1}$ m. Compute its average velocity over $[2, 5]$ and estimate its instantaneous velocity at $t = 2$.

2. The "wellhead" price p of natural gas in the United States (in dollars per 1000 ft^3) on the first day of each month in 2008 is listed in the table below.

J	F	M	A	M	J
6.99	7.55	8.29	8.94	9.81	10.82

J	A	S	O	N	D
10.62	8.32	7.27	6.36	5.97	5.87

Compute the average rate of change of p (in dollars per 1000 ft^3 per month) over the quarterly periods January–March, April–June, and July–September.

3. For a whole number n, let $P(n)$ be the number of *partitions* of n, that is, the number of ways of writing n as a sum of one or more whole numbers. For example, $P(4) = 5$ since the number 4 can be partitioned in five different ways: 4, $3 + 1$, $2 + 2$, $2 + 1 + 1$, and $1 + 1 + 1 + 1$. Treating $P(n)$ as a continuous function, use Figure 1 to estimate the rate of change of $P(n)$ at $n = 12$.

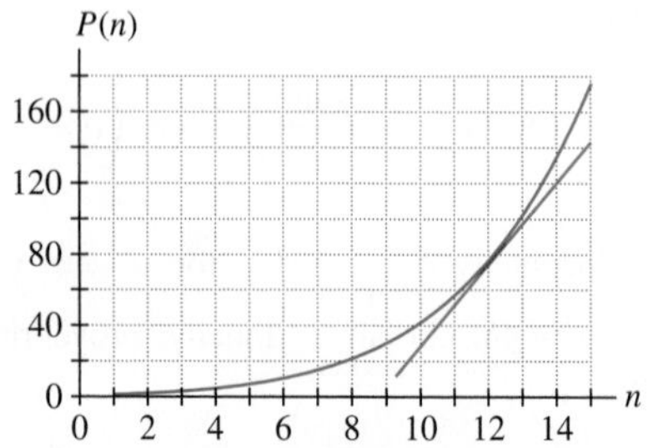

FIGURE 1 Graph of $P(n)$.

4. The average velocity v (m/s) of an oxygen molecule in the air at temperature T (°C) is $v = 25.7\sqrt{273.15 + T}$. What is the average speed at $T = 25°$ (room temperature)? Estimate the rate of change of average velocity with respect to temperature at $T = 25°$. What are the units of this rate?

In Exercises 5–10, estimate the limit numerically to two decimal places or state that the limit does not exist.

5. $\displaystyle\lim_{x\to 0} \frac{1 - \cos^3(x)}{x^2}$

6. $\displaystyle\lim_{x\to 1} x^{1/(x-1)}$

7. $\displaystyle\lim_{x\to 2} \frac{x^x - 4}{x^2 - 4}$

8. $\displaystyle\lim_{x\to 2} \frac{x - 2}{2^x - 4}$

9. $\displaystyle\lim_{x\to 1} \left(\frac{7}{1 - x^7} - \frac{3}{1 - x^3}\right)$

10. $\displaystyle\lim_{x\to 2} \frac{3^x - 9}{5^x - 25}$

In Exercises 11–50, evaluate the limit if it exists. If not, determine whether the one-sided limits exist (finite or infinite).

11. $\displaystyle\lim_{x\to 4} (3 + x^{1/2})$

12. $\displaystyle\lim_{x\to 1} \frac{5 - x^2}{4x + 7}$

13. $\displaystyle\lim_{x\to -2} \frac{4}{x^3}$

14. $\displaystyle\lim_{x\to -1} \frac{3x^2 + 4x + 1}{x + 1}$

15. $\displaystyle\lim_{t\to 9} \frac{\sqrt{t} - 3}{t - 9}$

16. $\displaystyle\lim_{x\to 3} \frac{\sqrt{x + 1} - 2}{x - 3}$

17. $\displaystyle\lim_{x\to 1} \frac{x^3 - x}{x - 1}$

18. $\displaystyle\lim_{h\to 0} \frac{2(a + h)^2 - 2a^2}{h}$

19. $\displaystyle\lim_{t\to 9} \frac{t - 6}{\sqrt{t} - 3}$

20. $\displaystyle\lim_{s\to 0} \frac{1 - \sqrt{s^2 + 1}}{s^2}$

21. $\displaystyle\lim_{x\to -1+} \frac{1}{x + 1}$

22. $\displaystyle\lim_{y\to \frac{1}{3}} \frac{3y^2 + 5y - 2}{6y^2 - 5y + 1}$

23. $\lim_{x\to 1} \dfrac{x^3 - 2x}{x - 1}$

24. $\lim_{a\to b} \dfrac{a^2 - 3ab + 2b^2}{a - b}$

25. $\lim_{x\to 0} \dfrac{4^{3x} - 4^x}{4^x - 1}$

26. $\lim_{\theta\to 0} \dfrac{\sin 5\theta}{\theta}$

27. $\lim_{x\to 1.5} \dfrac{[x]}{x}$

28. $\lim_{\theta\to \frac{\pi}{4}} \sec\theta$

29. $\lim_{z\to -3} \dfrac{z+3}{z^2 + 4z + 3}$

30. $\lim_{x\to 1} \dfrac{x^3 - ax^2 + ax - 1}{x - 1}$

31. $\lim_{x\to b} \dfrac{x^3 - b^3}{x - b}$

32. $\lim_{x\to 0} \dfrac{\sin 4x}{\sin 3x}$

33. $\lim_{x\to 0} \left(\dfrac{1}{3x} - \dfrac{1}{x(x+3)}\right)$

34. $\lim_{\theta\to \frac{1}{4}} 3^{\tan(\pi\theta)}$

35. $\lim_{x\to 0-} \dfrac{[x]}{x}$

36. $\lim_{x\to 0+} \dfrac{[x]}{x}$

37. $\lim_{\theta\to \frac{\pi}{2}} \theta\sec\theta$

38. $\lim_{y\to 3} \left(\sin\dfrac{\pi}{y}\right)^{-1/2}$

39. $\lim_{\theta\to 0} \dfrac{\cos\theta - 2}{\theta}$

40. $\lim_{x\to 4.3} \dfrac{1}{x - [x]}$

41. $\lim_{x\to 2-} \dfrac{x-3}{x-2}$

42. $\lim_{t\to 0} \dfrac{\sin^2 t}{t^3}$

43. $\lim_{x\to 1+} \left(\dfrac{1}{\sqrt{x-1}} - \dfrac{1}{\sqrt{x^2-1}}\right)$

44. $\lim_{t\to \frac{\pi}{2}} \sqrt{2t}\,(\cos t - 1)$

45. $\lim_{x\to \frac{\pi}{2}} \tan x$

46. $\lim_{t\to 0} \cos\dfrac{1}{t}$

47. $\lim_{t\to 0+} \sqrt{t}\cos\dfrac{1}{t}$

48. $\lim_{x\to 5+} \dfrac{x^2 - 24}{x^2 - 25}$

49. $\lim_{x\to 0} \dfrac{\cos x - 1}{\sin x}$

50. $\lim_{\theta\to 0} \dfrac{\tan\theta - \sin\theta}{\sin^3\theta}$

51. Find the left- and right-hand limits of the function $f(x)$ in Figure 2 at $x = 0, 2, 4$. State whether $f(x)$ is left- or right-continuous (or both) at these points.

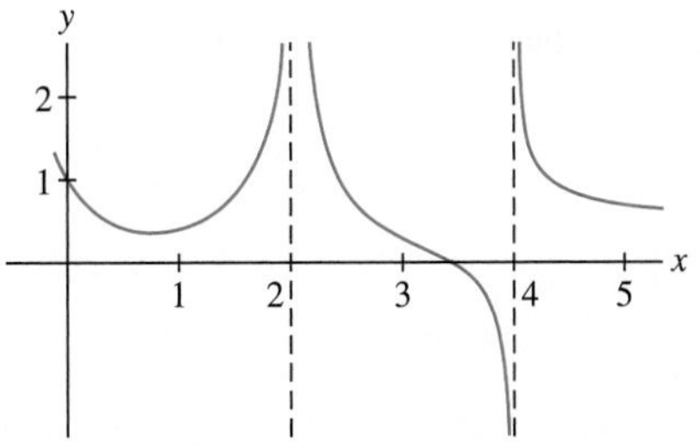

FIGURE 2

52. Sketch the graph of a function $f(x)$ such that

(a) $\lim_{x\to 2-} f(x) = 1, \qquad \lim_{x\to 2+} f(x) = 3$

(b) $\lim_{x\to 4} f(x)$ exists but does not equal $f(4)$.

53. Graph $h(x)$ and describe the type of discontinuity:

$$h(x) = \begin{cases} 2^x & \text{for } x \le 0 \\ x^{-1/2} & \text{for } x > 0 \end{cases}$$

Is $h(x)$ left- or right-continuous?

54. Sketch the graph of a function $g(x)$ such that

$$\lim_{x\to -3-} g(x) = \infty, \qquad \lim_{x\to -3+} g(x) = -\infty, \qquad \lim_{x\to 4} g(x) = \infty$$

55. Find the points of discontinuity of

$$g(x) = \begin{cases} \cos\left(\dfrac{\pi x}{2}\right) & \text{for } |x| < 1 \\ |x - 1| & \text{for } |x| \ge 1 \end{cases}$$

Determine the type of discontinuity and whether $g(x)$ is left- or right-continuous.

56. Show that $f(x) = x2^{\sin x}$ is continuous on its domain.

57. Find a constant b such that $h(x)$ is continuous at $x = 2$, where

$$h(x) = \begin{cases} x + 1 & \text{for } |x| < 2 \\ b - x^2 & \text{for } |x| \ge 2 \end{cases}$$

With this choice of b, find all points of discontinuity.

In Exercises 58–63, find the horizontal asymptotes of the function by computing the limits at infinity.

58. $f(x) = \dfrac{9x^2 - 4}{2x^2 - x}$

59. $f(x) = \dfrac{x^2 - 3x^4}{x - 1}$

60. $f(u) = \dfrac{8u - 3}{\sqrt{16u^2 + 6}}$

61. $f(u) = \dfrac{2u^2 - 1}{\sqrt{6 + u^4}}$

62. $f(x) = \dfrac{3x^{2/3} + 9x^{3/7}}{7x^{4/5} - 4x^{-1/3}}$

63. $f(t) = \dfrac{t^{1/3} - t^{-1/3}}{(t - t^{-1})^{1/3}}$

64. Calculate (a)–(d), assuming that

$$\lim_{x\to 3} f(x) = 6, \qquad \lim_{x\to 3} g(x) = 4$$

(a) $\lim_{x\to 3} (f(x) - 2g(x))$

(b) $\lim_{x\to 3} x^2 f(x)$

(c) $\lim_{x\to 3} \dfrac{f(x)}{g(x) + x}$

(d) $\lim_{x\to 3} (2g(x)^3 - g(x)^{3/2})$

65. Assume that the following limits exist:

$$A = \lim_{x\to a} f(x), \qquad B = \lim_{x\to a} g(x), \qquad L = \lim_{x\to a} \frac{f(x)}{g(x)}$$

Prove that if $L = 1$, then $A = B$. *Hint:* You cannot use the Quotient Law if $B = 0$, so apply the Product Law to L and B instead.

66. GU Define $g(t) = (1 + 2^{1/t})^{-1}$ for $t \ne 0$. How should $g(0)$ be defined to make $g(t)$ left-continuous at $t = 0$?

67. In the notation of Exercise 65, give an example where L exists but neither A nor B exists.

68. True or false?

(a) If $\lim_{x\to 3} f(x)$ exists, then $\lim_{x\to 3} f(x) = f(3)$.

(b) If $\lim_{x\to 0} \dfrac{f(x)}{x} = 1$, then $f(0) = 0$.

(c) If $\lim_{x\to -7} f(x) = 8$, then $\lim_{x\to -7} \dfrac{1}{f(x)} = \dfrac{1}{8}$.

(d) If $\lim_{x\to 5+} f(x) = 4$ and $\lim_{x\to 5-} f(x) = 8$, then $\lim_{x\to 5} f(x) = 6$.

(e) If $\lim_{x\to 0} \dfrac{f(x)}{x} = 1$, then $\lim_{x\to 0} f(x) = 0$.

(f) If $\lim_{x\to 5} f(x) = 2$, then $\lim_{x\to 5} f(x)^3 = 8$.

69. Let $f(x) = x\left[\frac{1}{x}\right]$, where $[x]$ is the greatest integer function. Show that for $x \neq 0$,

$$\frac{1}{x} - 1 < \left[\frac{1}{x}\right] \le \frac{1}{x}$$

Then use the Squeeze Theorem to prove that

$$\lim_{x\to 0} x\left[\frac{1}{x}\right] = 1$$

Hint: Treat the one-sided limits separately.

70. Let r_1 and r_2 be the roots of $f(x) = ax^2 - 2x + 20$. Observe that $f(x)$ "approaches" the linear function $L(x) = -2x + 20$ as $a \to 0$. Because $r = 10$ is the unique root of $L(x)$, we might expect one of the roots of $f(x)$ to approach 10 as $a \to 0$ (Figure 3). Prove that the roots can be labeled so that $\lim_{a\to 0} r_1 = 10$ and $\lim_{a\to 0} r_2 = \infty$.

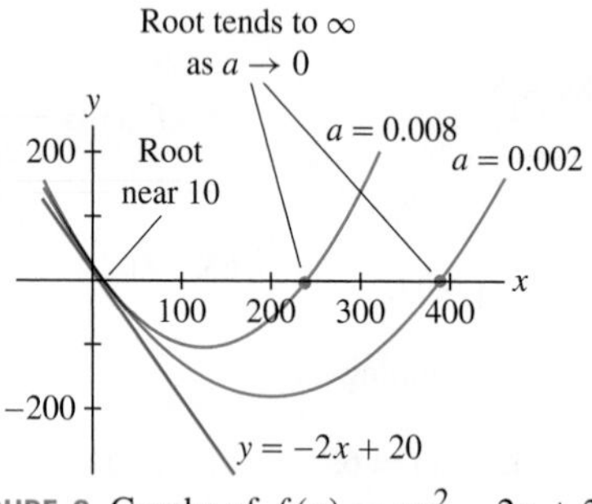

FIGURE 3 Graphs of $f(x) = ax^2 - 2x + 20$.

71. Use the IVT to prove that the curves $y = x^2$ and $y = \cos x$ intersect.

72. Use the IVT to prove that $f(x) = x^3 - \dfrac{x^2+2}{\cos x + 2}$ has a root in the interval $[0, 2]$.

73. Use the IVT to show that $2^{-x^2} = x$ has a solution on $(0, 1)$.

74. Use the Bisection Method to locate a solution of $x^2 - 7 = 0$ to two decimal places.

75. Give an example of a (discontinuous) function that does not satisfy the conclusion of the IVT on $[-1, 1]$. Then show that the function

$$f(x) = \begin{cases} \sin\dfrac{1}{x} & x \neq 0 \\ 0 & x = 0 \end{cases}$$

satisfies the conclusion of the IVT on every interval $[-a, a]$, even though f is discontinuous at $x = 0$.

76. Let $f(x) = \dfrac{1}{x+2}$.

(a) Show that $\left|f(x) - \frac{1}{4}\right| < \dfrac{|x-2|}{12}$ if $|x - 2| < 1$. *Hint:* Observe that $|4(x+2)| > 12$ if $|x - 2| < 1$.

(b) Find $\delta > 0$ such that $\left|f(x) - \frac{1}{4}\right| < 0.01$ for $|x - 2| < \delta$.

(c) Prove rigorously that $\lim_{x\to 2} f(x) = \frac{1}{4}$.

77. GU Plot the function $f(x) = x^{1/3}$. Use the zoom feature to find a $\delta > 0$ such that $|x^{1/3} - 2| < 0.05$ for $|x - 8| < \delta$.

78. Use the fact that $f(x) = 2^x$ is increasing to find a value of δ such that $|2^x - 8| < 0.001$ if $|x - 2| < \delta$. *Hint:* Find c_1 and c_2 such that $7.999 < f(c_1) < f(c_2) < 8.001$.

79. Prove rigorously that $\lim_{x\to -1} (4 + 8x) = -4$.

80. Prove rigorously that $\lim_{x\to 3} (x^2 - x) = 6$.

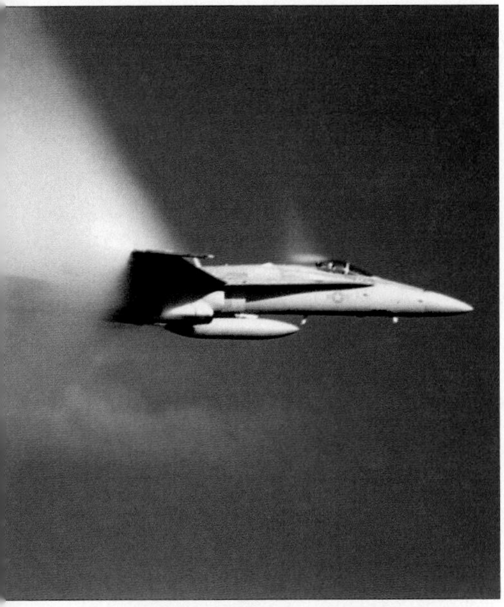

Calculus is the foundation for all of our understanding of motion, including the aerodynamic principles that made supersonic flight possible.

3 DIFFERENTIATION

Differential calculus is the study of the derivative, and differentiation is the process of computing derivatives. What is a derivative? There are three equally important answers: A derivative is a rate of change, it is the slope of a tangent line, and (more formally), it is the limit of a difference quotient, as we will explain shortly. In this chapter, we explore all three facets of the derivative and develop the basic rules of differentiation. When you master these techniques, you will possess one of the most useful and flexible tools that mathematics has to offer.

3.1 Definition of the Derivative

We begin with two questions: What is the precise definition of a tangent line? And how can we compute its slope? To answer these questions, let's return to the relationship between tangent and secant lines first mentioned in Section 2.1.

The secant line through distinct points $P = (a, f(a))$ and $Q = (x, f(x))$ on the graph of a function $f(x)$ has slope [Figure 1(A)]

$$\boxed{\frac{\Delta f}{\Delta x} = \frac{f(x) - f(a)}{x - a}}$$

where

$$\Delta f = f(x) - f(a) \qquad \text{and} \qquad \Delta x = x - a$$

The expression $\dfrac{f(x) - f(a)}{x - a}$ is called the **difference quotient**.

*REMINDER A **secant line** is any line through two points on a curve or graph.*

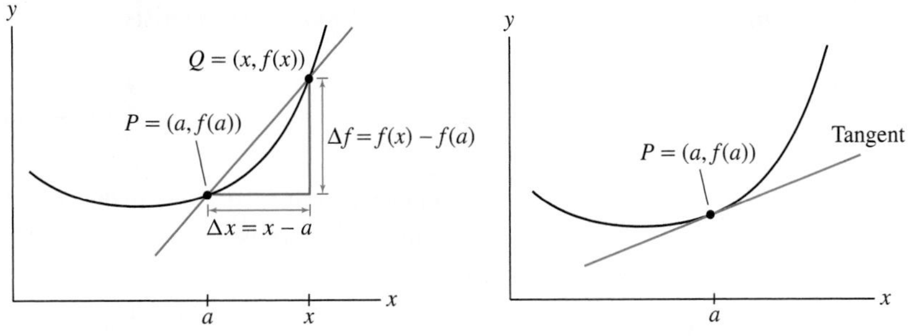

FIGURE 1 The secant line has slope $\Delta f/\Delta x$. Our goal is to compute the slope of the tangent line at $(a, f(a))$.

Now observe what happens as Q approaches P or, equivalently, as x approaches a. Figure 2 suggests that the secant lines get progressively closer to the tangent line. If we imagine Q moving toward P, then the secant line appears to rotate into the tangent line as in (D). Therefore, we may expect the slopes of the secant lines to approach the slope of the tangent line.

Based on this intuition, we define the **derivative** $f'(a)$ (which is read "f prime of a") as the limit

$$f'(a) = \underbrace{\lim_{x \to a} \frac{f(x) - f(a)}{x - a}}_{\text{Limit of slopes of secant lines}}$$

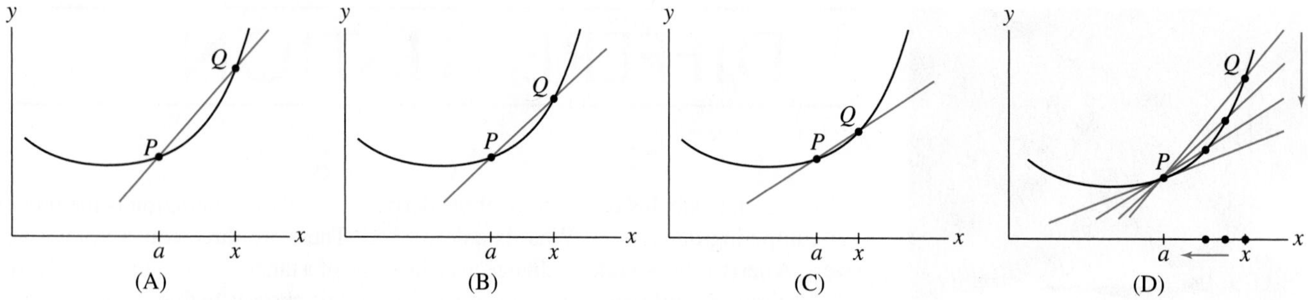

FIGURE 2 The secant lines approach the tangent line as Q approaches P.

There is another way of writing the difference quotient using a new variable h:

$$h = x - a$$

We have $x = a + h$ and, for $x \neq a$ (Figure 3),

$$\frac{f(x) - f(a)}{x - a} = \frac{f(a+h) - f(a)}{h}$$

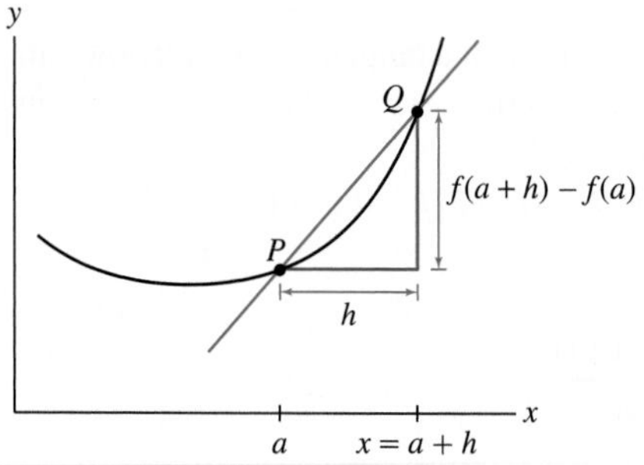

FIGURE 3 The difference quotient can be written in terms of h.

The variable h approaches 0 as $x \to a$, so we can rewrite the derivative as

$$f'(a) = \lim_{h \to 0} \frac{f(a+h) - f(a)}{h}$$

Each way of writing the derivative is useful. The version using h is often more convenient in computations.

DEFINITION The Derivative The derivative of $f(x)$ at $x = a$ is the limit of the difference quotients (if it exists):

$$f'(a) = \lim_{h \to 0} \frac{f(a+h) - f(a)}{h} \qquad \boxed{1}$$

When the limit exists, we say that f is **differentiable** at $x = a$. An equivalent definition of the derivative is

$$f'(a) = \lim_{x \to a} \frac{f(x) - f(a)}{x - a} \qquad \boxed{2}$$

We can now define the tangent line in a precise way, as the line of slope $f'(a)$ through $P = (a, f(a))$.

REMINDER The equation of the line through $P = (a, b)$ of slope m in point-slope form:

$$y - b = m(x - a)$$

DEFINITION Tangent Line Assume that $f(x)$ is differentiable at $x = a$. The tangent line to the graph of $y = f(x)$ at $P = (a, f(a))$ is the line through P of slope $f'(a)$. The equation of the tangent line in point-slope form is

$$y - f(a) = f'(a)(x - a) \qquad \boxed{3}$$

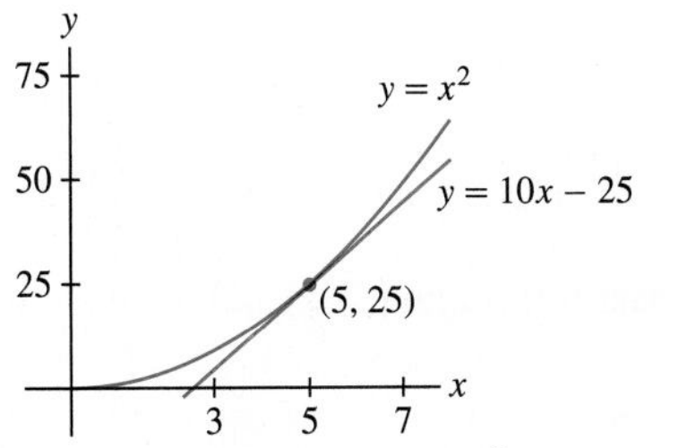

FIGURE 4 Tangent line to $y = x^2$ at $x = 5$.

■ **EXAMPLE 1** **Equation of a Tangent Line** Find an equation of the tangent line to the graph of $f(x) = x^2$ at $x = 5$.

Solution First, we must compute $f'(5)$. We are free to use either Eq. (1) or Eq. (2). Using Eq. (2), we have

$$f'(5) = \lim_{x\to 5} \frac{f(x) - f(5)}{x - 5} = \lim_{x\to 5} \frac{x^2 - 25}{x - 5} = \lim_{x\to 5} \frac{(x-5)(x+5)}{x - 5}$$

$$= \lim_{x\to 5} (x + 5) = 10$$

Next, we apply Eq. (3) with $a = 5$. Because $f(5) = 25$, an equation of the tangent line is $y - 25 = 10(x - 5)$, or, in slope-intercept form: $y = 10x - 25$ (Figure 4). ■

Isaac Newton referred to calculus as the "method of fluxions" (from the Latin word for "flow"), but the term "differential calculus", introduced in its Latin form "calculus differentialis" by Gottfried Wilhelm Leibniz, eventually won out and was adopted universally.

The next two examples illustrate differentiation (the process of computing the derivative) using Eq. (1). For clarity, we break up the computations into three steps.

■ **EXAMPLE 2** Compute $f'(3)$, where $f(x) = x^2 - 8x$.

Solution Using Eq. (1), we write the difference quotient at $a = 3$ as

$$\frac{f(a+h) - f(a)}{h} = \frac{f(3+h) - f(3)}{h} \qquad (h \neq 0)$$

***Step 1.* Write out the numerator of the difference quotient.**

$$f(3+h) - f(3) = \big((3+h)^2 - 8(3+h)\big) - \big(3^2 - 8(3)\big)$$

$$= \big((9 + 6h + h^2) - (24 + 8h)\big) - (9 - 24)$$

$$= h^2 - 2h$$

***Step 2.* Divide by *h* and simplify.**

$$\frac{f(3+h) - f(3)}{h} = \frac{h^2 - 2h}{h} = \underbrace{\frac{h(h-2)}{h} = h - 2}_{\text{Cancel } h}$$

***Step 3.* Compute the limit.**

$$f'(3) = \lim_{h\to 0} \frac{f(3+h) - f(3)}{h} = \lim_{h\to 0} (h - 2) = -2$$ ■

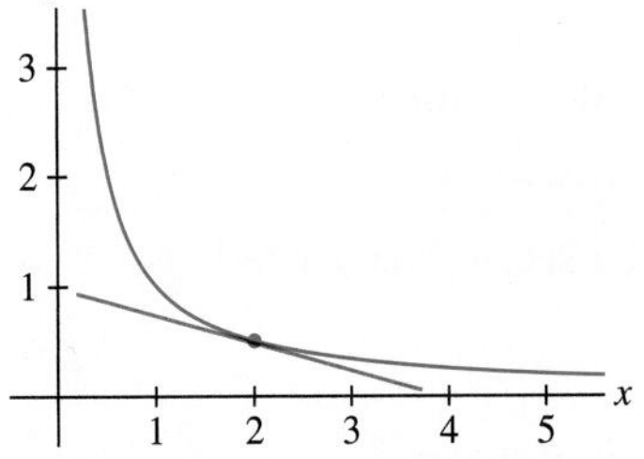

FIGURE 5 Graph of $f(x) = \frac{1}{x}$. The tangent line at $x = 2$ has equation $y = -\frac{1}{4}x + 1$.

■ **EXAMPLE 3** Sketch the graph of $f(x) = \dfrac{1}{x}$ and the tangent line at $x = 2$.

(a) Based on the sketch, do you expect $f'(2)$ to be positive or negative?

(b) Find an equation of the tangent line at $x = 2$.

Solution The graph and tangent line at $x = 2$ are shown in Figure 5.

(a) We see that the tangent line has negative slope, so $f'(2)$ must be negative.

(b) We compute $f'(2)$ in three steps as before.

***Step 1.* Write out the numerator of the difference quotient.**

$$f(2+h) - f(2) = \frac{1}{2+h} - \frac{1}{2} = \frac{2}{2(2+h)} - \frac{2+h}{2(2+h)} = -\frac{h}{2(2+h)}$$

***Step 2.* Divide by *h* and simplify.**

$$\frac{f(2+h) - f(2)}{h} = \frac{1}{h} \cdot \left(-\frac{h}{2(2+h)}\right) = -\frac{1}{2(2+h)}$$

Step 3. **Compute the limit.**

$$f'(2) = \lim_{h\to 0} \frac{f(2+h) - f(2)}{h} = \lim_{h\to 0} \frac{-1}{2(2+h)} = -\frac{1}{4}$$

The function value is $f(2) = \frac{1}{2}$, so the tangent line passes through $\left(2, \frac{1}{2}\right)$ and has equation

$$y - \frac{1}{2} = -\frac{1}{4}(x-2)$$

In slope-intercept form, $y = -\frac{1}{4}x + 1$. ■

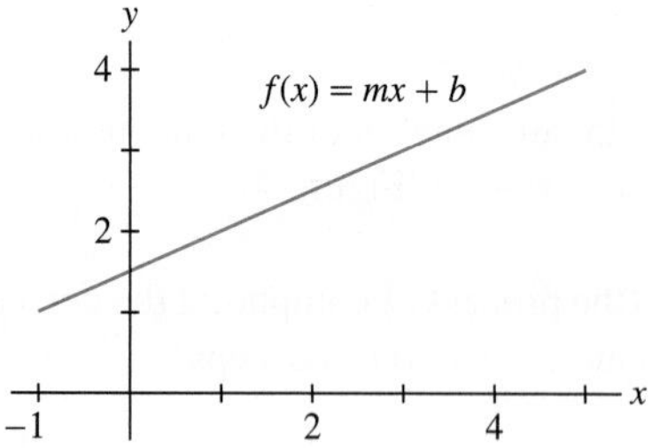

FIGURE 6 The derivative of $f(x) = mx + b$ is $f'(a) = m$ for all a.

The graph of a linear function $f(x) = mx + b$ (where m and b are constants) is a line of slope m. The tangent line at any point coincides with the line itself (Figure 6), so we should expect that $f'(a) = m$ for all a. Let's check this by computing the derivative:

$$f'(a) = \lim_{h\to 0} \frac{f(a+h) - f(a)}{h} = \lim_{h\to 0} \frac{(m(a+h)+b) - (ma+b)}{h}$$
$$= \lim_{h\to 0} \frac{mh}{h} = \lim_{h\to 0} m = m$$

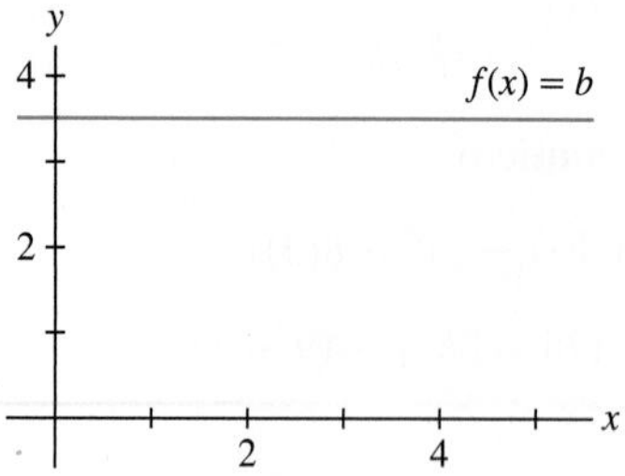

FIGURE 7 The derivative of a constant function $f(x) = b$ is $f'(a) = 0$ for all a.

If $m = 0$, then $f(x) = b$ is constant and $f'(a) = 0$ (Figure 7). In summary,

> **THEOREM 1 Derivative of Linear and Constant Functions**
>
> - If $f(x) = mx + b$ is a linear function, then $f'(a) = m$ for all a.
> - If $f(x) = b$ is a constant function, then $f'(a) = 0$ for all a.

■ **EXAMPLE 4** Find the derivative of $f(x) = 9x - 5$ at $x = 2$ and $x = 5$.

Solution We have $f'(a) = 9$ for all a. Hence, $f'(2) = f'(5) = 9$. ■

Estimating the Derivative

Approximations to the derivative are useful in situations where we cannot evaluate $f'(a)$ exactly. Since the derivative is the limit of difference quotients, the difference quotient should give a good numerical approximation when h is sufficiently small:

$$f'(a) \approx \frac{f(a+h) - f(a)}{h} \qquad \text{if } h \text{ is small}$$

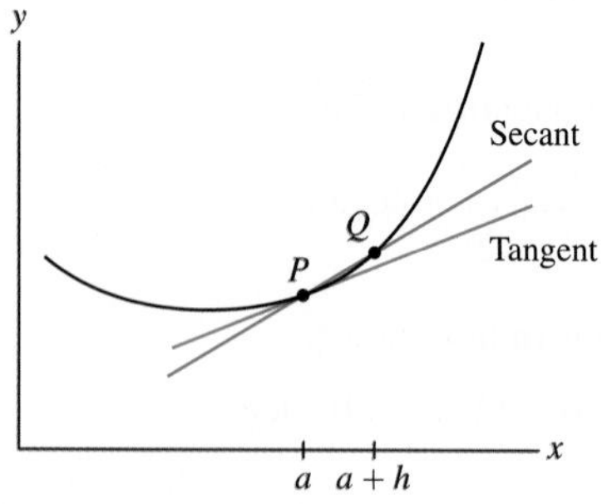

FIGURE 8 When h is small, the secant line has nearly the same slope as the tangent line.

Graphically, this says that for small h, the slope of the secant line is nearly equal to the slope of the tangent line (Figure 8).

■ **EXAMPLE 5** Estimate the derivative of $f(x) = \sin x$ at $x = \frac{\pi}{6}$.

Solution We calculate the difference quotient for several small values of h:

$$\frac{\sin\left(\frac{\pi}{6} + h\right) - \sin\frac{\pi}{6}}{h} = \frac{\sin\left(\frac{\pi}{6} + h\right) - 0.5}{h}$$

Table 1 on the next page suggests that the limit has a decimal expansion beginning 0.866. In other words, $f'\left(\frac{\pi}{6}\right) \approx 0.866$. ■

TABLE 1 Values of the Difference Quotient for Small h

$h > 0$	$\dfrac{\sin\left(\frac{\pi}{6}+h\right)-0.5}{h}$	$h < 0$	$\dfrac{\sin\left(\frac{\pi}{6}+h\right)-0.5}{h}$
0.01	0.863511	−0.01	0.868511
0.001	0.865775	−0.001	0.866275
0.0001	**0.8660** 00	−0.0001	**0.866** 050
0.00001	**0.8660** 229	−0.00001	**0.8660** 279

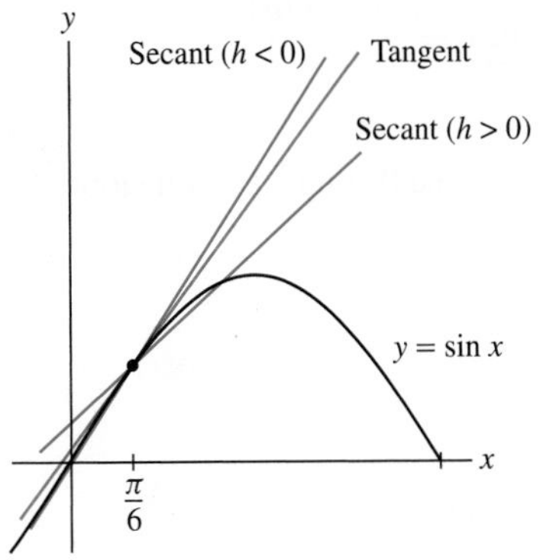

FIGURE 9 The tangent line is squeezed in *between* the secant lines with $h > 0$ and $h < 0$.

This technique of estimating an unknown quantity by showing that it lies between two known values ("squeezing it") is used frequently in calculus.

In the next example, we use graphical reasoning to determine the accuracy of the estimates obtained in Example 5.

■ **EXAMPLE 6** GU **Determining Accuracy Graphically** Let $f(x) = \sin x$. Show that the approximation $f'\left(\frac{\pi}{6}\right) \approx 0.8660$ is accurate to four decimal places.

Solution Observe in Figure 9 that the position of the secant line relative to the tangent line depends on whether h is positive or negative. When $h > 0$, the slope of the secant line is *smaller* than the slope of the tangent line, but it is *larger* when $h < 0$. This tells us that the difference quotients in the second column of Table 1 are smaller than $f'\left(\frac{\pi}{6}\right)$ and those in the fourth column are greater than $f'\left(\frac{\pi}{6}\right)$. From the last line in Table 1 we may conclude that

$$0.866022 \le f'\left(\frac{\pi}{6}\right) \le 0.866028$$

It follows that the estimate $f'\left(\frac{\pi}{6}\right) \approx 0.8660$ is accurate to four decimal places. In Section 3.6, we will see that the exact value is $f'\left(\frac{\pi}{6}\right) = \cos\left(\frac{\pi}{6}\right) = \sqrt{3}/2 \approx 0.8660254$, just about midway between 0.866022 and 0.866028. ■

CONCEPTUAL INSIGHT **Are Limits Really Necessary?** It is natural to ask whether limits are really necessary. The tangent line is easy to visualize. Is there perhaps a better or simpler way to find its equation? History gives one answer: The methods of calculus based on limits have stood the test of time and are used more widely today than ever before.

History aside, we can see directly why limits play such a crucial role. The slope of a line can be computed if the coordinates of *two* points $P = (x_1, y_1)$ and $Q = (x_2, y_2)$ on the line are known:

$$\text{Slope of line} = \frac{y_2 - y_1}{x_2 - x_1}$$

This formula cannot be applied to the tangent line because we know only that it passes through the single point $P = (a, f(a))$. Limits provide an ingenious way around this obstacle. We choose a point $Q = (a + h, f(a + h))$ on the graph near P and form the secant line. The slope of this secant line is just an approximation to the slope of the tangent line:

$$\text{Slope of secant line} = \frac{f(a+h) - f(a)}{h} \approx \text{slope of tangent line}$$

But this approximation improves as $h \to 0$, and by taking the limit, we convert our approximations into the exact slope.

3.1 SUMMARY

- The *difference quotient*:

$$\frac{f(a+h) - f(a)}{h}$$

The difference quotient is the slope of the secant line through the points $P = (a, f(a))$ and $Q = (a+h, f(a+h))$ on the graph of $f(x)$.

- The *derivative* $f'(a)$ is defined by the following equivalent limits:

$$f'(a) = \lim_{h\to 0} \frac{f(a+h) - f(a)}{h} = \lim_{x\to a} \frac{f(x) - f(a)}{x - a}$$

If the limit exists, we say that f is *differentiable* at $x = a$.

- By definition, the tangent line at $P = (a, f(a))$ is the line through P with slope $f'(a)$ [assuming that $f'(a)$ exists].
- Equation of the tangent line in point-slope form:

$$y - f(a) = f'(a)(x - a)$$

- To calculate $f'(a)$ using the limit definition:

Step 1. Write out the numerator of the difference quotient.

Step 2. Divide by h and simplify.

Step 3. Compute the derivative by taking the limit.

- For small values of h, we have the estimate $f'(a) \approx \dfrac{f(a+h) - f(a)}{h}$.

3.1 EXERCISES

Preliminary Questions

1. Which of the lines in Figure 10 are tangent to the curve?

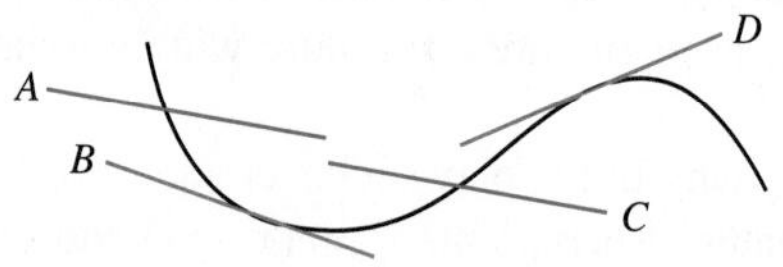

FIGURE 10

2. What are the two ways of writing the difference quotient?

3. Find a and h such that $\dfrac{f(a+h) - f(a)}{h}$ is equal to the slope of the secant line between $(3, f(3))$ and $(5, f(5))$.

4. Which derivative is approximated by $\dfrac{\tan\left(\frac{\pi}{4} + 0.0001\right) - 1}{0.0001}$?

5. What do the following quantities represent in terms of the graph of $f(x) = \sin x$?

(a) $\sin 1.3 - \sin 0.9$ **(b)** $\dfrac{\sin 1.3 - \sin 0.9}{0.4}$ **(c)** $f'(0.9)$

Exercises

1. Let $f(x) = 5x^2$. Show that $f(3+h) = 5h^2 + 30h + 45$. Then show that

$$\frac{f(3+h) - f(3)}{h} = 5h + 30$$

and compute $f'(3)$ by taking the limit as $h \to 0$.

2. Let $f(x) = 2x^2 - 3x - 5$. Show that the secant line through $(2, f(2))$ and $(2+h, f(2+h))$ has slope $2h + 5$. Then use this formula to compute the slope of:

(a) The secant line through $(2, f(2))$ and $(3, f(3))$

(b) The tangent line at $x = 2$ (by taking a limit)

In Exercises 3–6, compute $f'(a)$ in two ways, using Eq. (1) and Eq. (2).

3. $f(x) = x^2 + 9x$, $a = 0$

4. $f(x) = x^2 + 9x$, $a = 2$

5. $f(x) = 3x^2 + 4x + 2$, $a = -1$

6. $f(x) = x^3$, $a = 2$

In Exercises 7–10, refer to Figure 11.

7. Find the slope of the secant line through $(2, f(2))$ and $(2.5, f(2.5))$. Is it larger or smaller than $f'(2)$? Explain.

8. Estimate $\dfrac{f(2+h)-f(2)}{h}$ for $h=-0.5$. What does this quantity represent? Is it larger or smaller than $f'(2)$? Explain.

9. Estimate $f'(1)$ and $f'(2)$.

10. Find a value of h for which $\dfrac{f(2+h)-f(2)}{h}=0$.

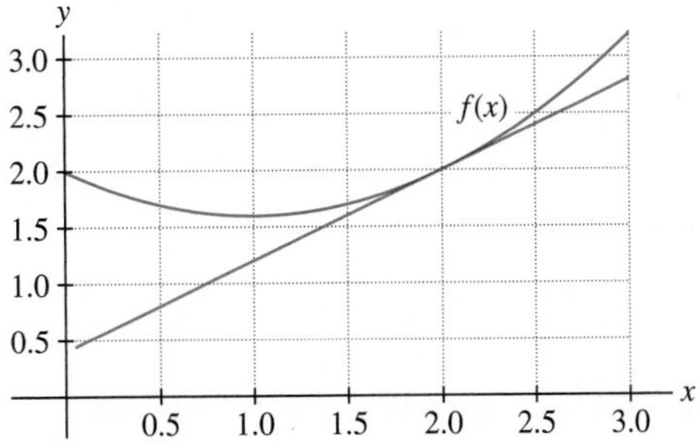

FIGURE 11

In Exercises 11–14, refer to Figure 12.

11. Determine $f'(a)$ for $a = 1, 2, 4, 7$.

12. For which values of x is $f'(x) < 0$?

13. Which is larger, $f'(5.5)$ or $f'(6.5)$?

14. Show that $f'(3)$ does not exist.

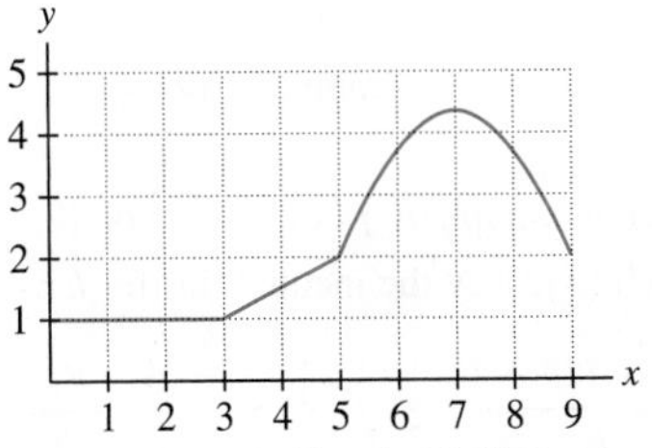

FIGURE 12 Graph of $f(x)$.

In Exercises 15–18, use the limit definition to calculate the derivative of the linear function.

15. $f(x) = 7x - 9$

16. $f(x) = 12$

17. $g(t) = 8 - 3t$

18. $k(z) = 14z + 12$

19. Find an equation of the tangent line at $x = 3$, assuming that $f(3) = 5$ and $f'(3) = 2$?

20. Find $f(3)$ and $f'(3)$, assuming that the tangent line to $y = f(x)$ at $a = 3$ has equation $y = 5x + 2$.

21. Describe the tangent line at an arbitrary point on the "curve" $y = 2x + 8$.

22. Suppose that $f(2+h) - f(2) = 3h^2 + 5h$. Calculate:
(a) The slope of the secant line through $(2, f(2))$ and $(6, f(6))$
(b) $f'(2)$

23. Let $f(x) = \dfrac{1}{x}$. Does $f(-2+h)$ equal $\dfrac{1}{-2+h}$ or $\dfrac{1}{-2}+\dfrac{1}{h}$? Compute the difference quotient at $a = -2$ with $h = 0.5$.

24. Let $f(x) = \sqrt{x}$. Does $f(5+h)$ equal $\sqrt{5+h}$ or $\sqrt{5}+\sqrt{h}$? Compute the difference quotient at $a = 5$ with $h = 1$.

25. Let $f(x) = 1/\sqrt{x}$. Compute $f'(5)$ by showing that

$$\frac{f(5+h)-f(5)}{h} = -\frac{1}{\sqrt{5}\sqrt{5+h}(\sqrt{5+h}+\sqrt{5})}$$

26. Find an equation of the tangent line to the graph of $f(x) = 1/\sqrt{x}$ at $x = 9$.

In Exercises 27–44, use the limit definition to compute $f'(a)$ and find an equation of the tangent line.

27. $f(x) = 2x^2 + 10x, \quad a = 3$

28. $f(x) = 4 - x^2, \quad a = -1$

29. $f(t) = t - 2t^2, \quad a = 3$

30. $f(x) = 8x^3, \quad a = 1$

31. $f(x) = x^3 + x, \quad a = 0$

32. $f(t) = 2t^3 + 4t, \quad a = 4$

33. $f(x) = x^{-1}, \quad a = 8$

34. $f(x) = x + x^{-1}, \quad a = 4$

35. $f(x) = \dfrac{1}{x+3}, \quad a = -2$

36. $f(t) = \dfrac{2}{1-t}, \quad a = -1$

37. $f(x) = \sqrt{x+4}, \quad a = 1$

38. $f(t) = \sqrt{3t+5}, \quad a = -1$

39. $f(x) = \dfrac{1}{\sqrt{x}}, \quad a = 4$

40. $f(x) = \dfrac{1}{\sqrt{2x+1}}, \quad a = 4$

41. $f(t) = \sqrt{t^2+1}, \quad a = 3$

42. $f(x) = x^{-2}, \quad a = -1$

43. $f(x) = \dfrac{1}{x^2+1}, \quad a = 0$

44. $f(t) = t^{-3}, \quad a = 1$

45. Figure 13 displays data collected by the biologist Julian Huxley (1887–1975) on the average antler weight W of male red deer as a function of age t. Estimate the derivative at $t = 4$. For which values of t is the slope of the tangent line equal to zero? For which values is it negative?

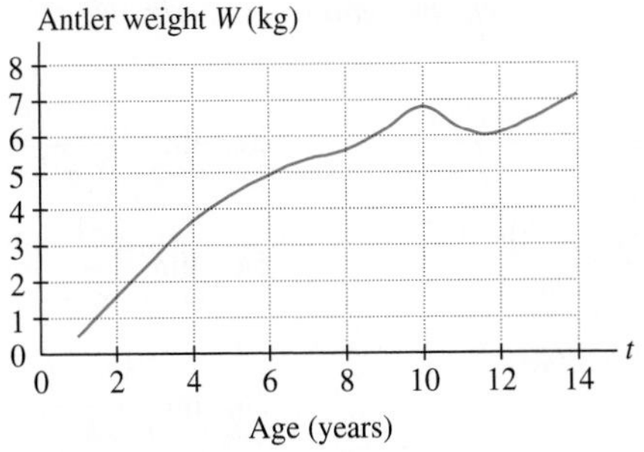

FIGURE 13

46. Figure 14(A) shows the graph of $f(x) = \sqrt{x}$. The close-up in Figure 14(B) shows that the graph is nearly a straight line near $x = 16$. Estimate the slope of this line and take it as an estimate for $f'(16)$. Then compute $f'(16)$ and compare with your estimate.

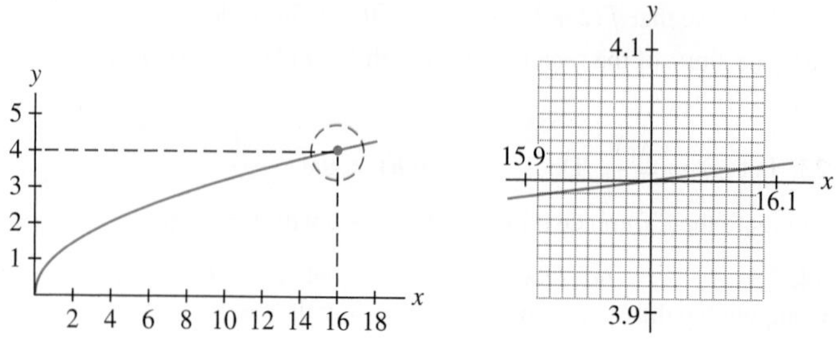

(A) Graph of $y = \sqrt{x}$ (B) Zoom view near (16, 4)

FIGURE 14

47. GU Let $f(x) = \dfrac{4}{1 + 2^x}$.

(a) Plot $f(x)$ over $[-2, 2]$. Then zoom in near $x = 0$ until the graph appears straight, and estimate the slope $f'(0)$.

(b) Use (a) to find an approximate equation to the tangent line at $x = 0$. Plot this line and $f(x)$ on the same set of axes.

48. GU Let $f(x) = \cot x$. Estimate $f'\left(\frac{\pi}{2}\right)$ graphically by zooming in on a plot of $f(x)$ near $x = \frac{\pi}{2}$.

49. Determine the intervals along the x-axis on which the derivative in Figure 15 is positive.

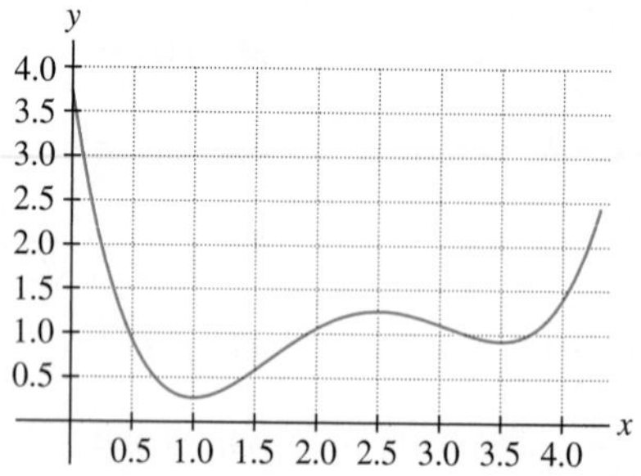

FIGURE 15

50. Sketch the graph of $f(x) = \sin x$ on $[0, \pi]$ and guess the value of $f'\left(\frac{\pi}{2}\right)$. Then calculate the difference quotient at $x = \frac{\pi}{2}$ for two small positive and negative values of h. Are these calculations consistent with your guess?

In Exercises 51–56, each limit represents a derivative $f'(a)$. Find $f(x)$ and a.

51. $\displaystyle\lim_{h\to 0} \frac{(5+h)^3 - 125}{h}$

52. $\displaystyle\lim_{x\to 5} \frac{x^3 - 125}{x - 5}$

53. $\displaystyle\lim_{h\to 0} \frac{\sin\left(\frac{\pi}{6} + h\right) - 0.5}{h}$

54. $\displaystyle\lim_{x\to \frac{1}{4}} \frac{x^{-1} - 4}{x - \frac{1}{4}}$

55. $\displaystyle\lim_{h\to 0} \frac{5^{2+h} - 25}{h}$

56. $\displaystyle\lim_{h\to 0} \frac{5^h - 1}{h}$

57. Apply the method of Example 6 to $f(x) = \sin x$ to determine $f'\left(\frac{\pi}{4}\right)$ accurately to four decimal places.

58. Apply the method of Example 6 to $f(x) = \cos x$ to determine $f'\left(\frac{\pi}{5}\right)$ accurately to four decimal places. Use a graph of $f(x)$ to explain how the method works in this case.

59. For each graph in Figure 16, determine whether $f'(1)$ is larger or smaller than the slope of the secant line between $x = 1$ and $x = 1 + h$ for $h > 0$. Explain.

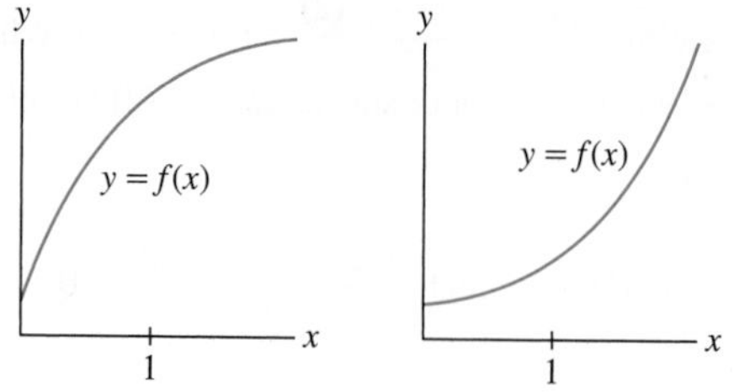

(A) (B)

FIGURE 16

60. Refer to the graph of $f(x) = 2^x$ in Figure 17.

(a) Explain graphically why, for $h > 0$,

$$\frac{f(-h) - f(0)}{-h} \le f'(0) \le \frac{f(h) - f(0)}{h}$$

(b) Use (a) to show that $0.69314 \le f'(0) \le 0.69315$.

(c) Similarly, compute $f'(x)$ to four decimal places for $x = 1, 2, 3, 4$.

(d) Now compute the ratios $f'(x)/f'(0)$ for $x = 1, 2, 3, 4$. Can you guess an approximate formula for $f'(x)$?

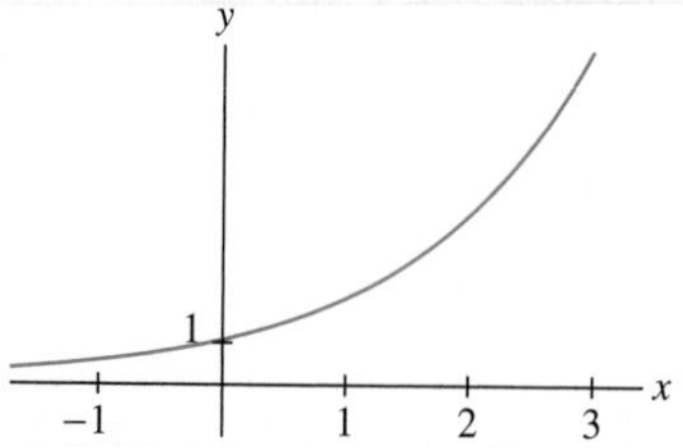

FIGURE 17 Graph of $f(x) = 2^x$.

61. GU Sketch the graph of $f(x) = x^{5/2}$ on $[0, 6]$.

(a) Use the sketch to justify the inequalities for $h > 0$:

$$\frac{f(4) - f(4-h)}{h} \le f'(4) \le \frac{f(4+h) - f(4)}{h}$$

(b) Use (a) to compute $f'(4)$ to four decimal places.

(c) Use a graphing utility to plot $f(x)$ and the tangent line at $x = 4$, using your estimate for $f'(4)$.

62. GU Verify that $P = \left(1, \frac{1}{2}\right)$ lies on the graphs of both $f(x) = 1/(1 + x^2)$ and $L(x) = \frac{1}{2} + m(x - 1)$ for every slope m. Plot $f(x)$ and $L(x)$ on the same axes for several values of m until you find a value of m for which $y = L(x)$ appears tangent to the graph of $f(x)$. What is your estimate for $f'(1)$?

63. GU Use a plot of $f(x) = x^x$ to estimate the value c such that $f'(c) = 0$. Find c to sufficient accuracy so that

$$\left|\frac{f(c+h) - f(c)}{h}\right| \le 0.006 \qquad \text{for} \quad h = \pm 0.001$$

64. GU Plot $f(x) = x^x$ and $y = 2x + a$ on the same set of axes for several values of a until the line becomes tangent to the graph. Then estimate the value c such that $f'(c) = 2$.

In Exercises 65–71, estimate derivatives using the ***symmetric difference quotient*** *(SDQ), defined as the average of the difference quotients at h and −h:*

$$\frac{1}{2}\left(\frac{f(a+h)-f(a)}{h} + \frac{f(a-h)-f(a)}{-h}\right) = \frac{f(a+h)-f(a-h)}{2h} \qquad \boxed{4}$$

The SDQ usually gives a better approximation to the derivative than the difference quotient.

65. The vapor pressure of water at temperature T (in kelvins) is the atmospheric pressure P at which no net evaporation takes place. Use the following table to estimate $P'(T)$ for $T = 303, 313, 323, 333, 343$ by computing the SDQ given by Eq. (4) with $h = 10$.

T (K)	293	303	313	323	333	343	353
P (atm)	0.0278	0.0482	0.0808	0.1311	0.2067	0.3173	0.4754

66. Use the SDQ with $h = 1$ year to estimate $P'(T)$ in the years 2000, 2002, 2004, 2006, where $P(T)$ is the U.S. ethanol production (Figure 18). Express your answer in the correct units.

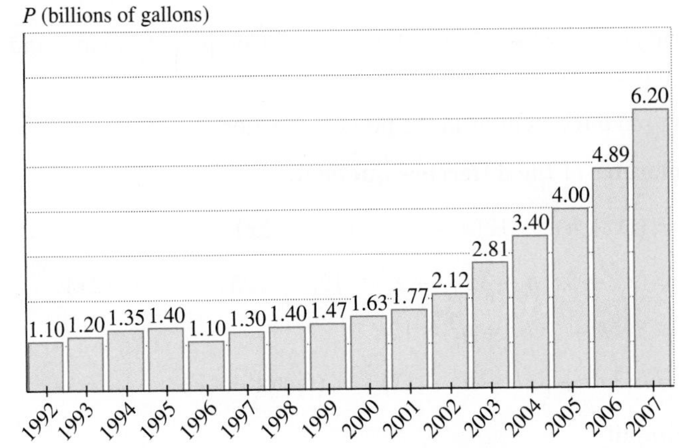

FIGURE 18 U.S. Ethanol Production

In Exercises 67–68, traffic speed S along a certain road (in km/h) varies as a function of traffic density q (number of cars per km of road). Use the following data to answer the questions:

q (density)	60	70	80	90	100
S (speed)	72.5	67.5	63.5	60	56

67. Estimate $S'(80)$.

68. Explain why $V = qS$, called *traffic volume*, is equal to the number of cars passing a point per hour. Use the data to estimate $V'(80)$.

Exercises 69–71: The current (in amperes) at time t (in seconds) flowing in the circuit in Figure 19 is given by Kirchhoff's Law:

$$i(t) = Cv'(t) + R^{-1}v(t)$$

where $v(t)$ is the voltage (in volts), C the capacitance (in farads), and R the resistance (in ohms, Ω).

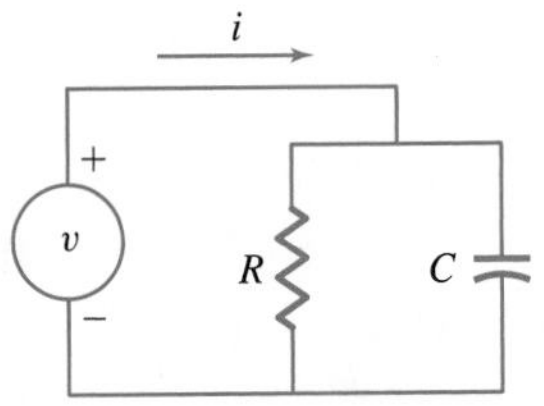

FIGURE 19

69. Calculate the current at $t = 3$ if

$$v(t) = 0.5t + 4 \text{ V}$$

where $C = 0.01$ F and $R = 100\ \Omega$.

70. Use the following data to estimate $v'(10)$ (by an SDQ). Then estimate $i(10)$, assuming $C = 0.03$ and $R = 1{,}000$.

t	9.8	9.9	10	10.1	10.2
$v(t)$	256.52	257.32	258.11	258.9	259.69

71. Assume that $R = 200\ \Omega$ but C is unknown. Use the following data to estimate $v'(4)$ (by an SDQ) and deduce an approximate value for the capacitance C.

t	3.8	3.9	4	4.1	4.2
$v(t)$	388.8	404.2	420	436.2	452.8
$i(t)$	32.34	33.22	34.1	34.98	35.86

Further Insights and Challenges

72. The SDQ usually approximates the derivative much more closely than does the ordinary difference quotient. Let $f(x) = 2^x$ and $a = 0$. Compute the SDQ with $h = 0.001$ and the ordinary difference quotients with $h = \pm 0.001$. Compare with the actual value, which is $f'(0) = \ln 2$.

73. Explain how the symmetric difference quotient defined by Eq. (4) can be interpreted as the slope of a secant line.

74. Which of the two functions in Figure 20 satisfies the inequality

$$\frac{f(a+h)-f(a-h)}{2h} \le \frac{f(a+h)-f(a)}{h}$$

for $h > 0$? Explain in terms of secant lines.

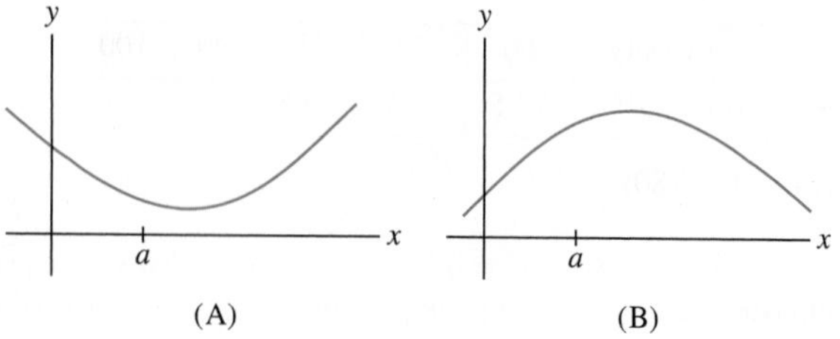

FIGURE 20

75. Show that if $f(x)$ is a quadratic polynomial, then the SDQ at $x = a$ (for any $h \neq 0$) is *equal* to $f'(a)$. Explain the graphical meaning of this result.

76. Let $f(x) = x^{-2}$. Compute $f'(1)$ by taking the limit of the SDQs (with $a = 1$) as $h \to 0$.

3.2 The Derivative as a Function

In the previous section, we computed the derivative $f'(a)$ for specific values of a. It is also useful to view the derivative as a function $f'(x)$ whose value at $x = a$ is $f'(a)$. The function $f'(x)$ is still defined as a limit, but the fixed number a is replaced by the variable x:

$$f'(x) = \lim_{h\to 0} \frac{f(x+h) - f(x)}{h} \qquad \boxed{1}$$

If $y = f(x)$, we also write y' or $y'(x)$ for $f'(x)$.

Often, the domain of $f'(x)$ is clear from the context. If so, we usually do not mention the domain explicitly.

The domain of $f'(x)$ consists of all values of x in the domain of $f(x)$ for which the limit in Eq. (1) exists. We say that $f(x)$ is **differentiable** on (a, b) if $f'(x)$ exists for all x in (a, b). When $f'(x)$ exists for all x in the interval or intervals on which $f(x)$ is defined, we say simply that $f(x)$ is differentiable.

EXAMPLE 1 Prove that $f(x) = x^3 - 12x$ is differentiable. Compute $f'(x)$ and find an equation of the tangent line at $x = -3$.

Solution We compute $f'(x)$ in three steps as in the previous section.

***Step 1.* Write out the numerator of the difference quotient.**

$$\begin{aligned} f(x+h) - f(x) &= \big((x+h)^3 - 12(x+h)\big) - \big(x^3 - 12x\big) \\ &= (x^3 + 3x^2h + 3xh^2 + h^3 - 12x - 12h) - (x^3 - 12x) \\ &= 3x^2h + 3xh^2 + h^3 - 12h \\ &= h(3x^2 + 3xh + h^2 - 12) \qquad \text{(factor out } h\text{)} \end{aligned}$$

***Step 2.* Divide by h and simplify.**

$$\frac{f(x+h) - f(x)}{h} = \frac{h(3x^2 + 3xh + h^2 - 12)}{h} = 3x^2 + 3xh + h^2 - 12 \qquad (h \neq 0)$$

***Step 3.* Compute the limit.**

$$f'(x) = \lim_{h\to 0} \frac{f(x+h) - f(x)}{h} = \lim_{h\to 0} (3x^2 + 3xh + h^2 - 12) = 3x^2 - 12$$

In this limit, x is treated as a constant because it does not change as $h \to 0$. We see that the limit exists for all x, so $f(x)$ is differentiable and $f'(x) = 3x^2 - 12$.

Now evaluate:

$$f(-3) = (-3)^3 - 12(-3) = 9$$

$$f'(-3) = 3(-3)^2 - 12 = 15$$

An equation of the tangent line at $x = -3$ is $y - 9 = 15(x + 3)$ (Figure 1). ■

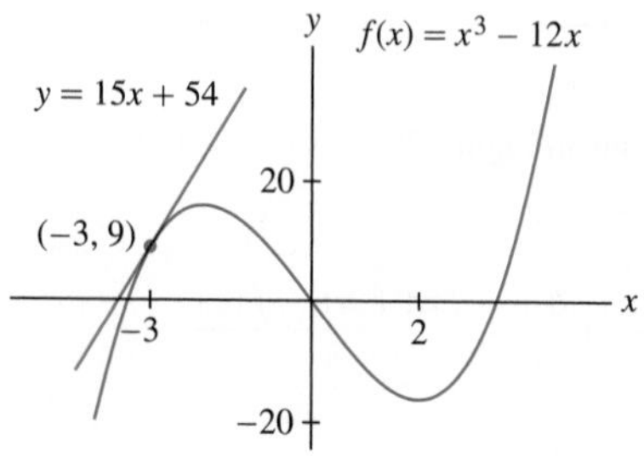

FIGURE 1 Graph of $f(x) = x^3 - 12x$.

■ **EXAMPLE 2** Prove that $y = x^{-2}$ is differentiable and calculate y'.

Solution The domain of $f(x) = x^{-2}$ is $\{x : x \neq 0\}$, so assume that $x \neq 0$. We compute $f'(x)$ directly, without the separate steps of the previous example:

$$y' = \lim_{h\to 0} \frac{f(x+h) - f(x)}{h} = \lim_{h\to 0} \frac{\dfrac{1}{(x+h)^2} - \dfrac{1}{x^2}}{h}$$

$$= \lim_{h\to 0} \frac{\dfrac{x^2 - (x+h)^2}{x^2(x+h)^2}}{h} = \lim_{h\to 0} \frac{1}{h}\left(\frac{x^2 - (x+h)^2}{x^2(x+h)^2}\right)$$

$$= \lim_{h\to 0} \frac{1}{h}\left(\frac{-h(2x+h)}{x^2(x+h)^2}\right) = \lim_{h\to 0} -\frac{2x+h}{x^2(x+h)^2} \qquad \text{(cancel } h\text{)}$$

$$= -\frac{2x+0}{x^2(x+0)^2} = -\frac{2x}{x^4} = -2x^{-3}$$

The limit exists for all $x \neq 0$, so y is differentiable and $y' = -2x^{-3}$. ■

Leibniz Notation

The "prime" notation y' and $f'(x)$ was introduced by the French mathematician Joseph Louis Lagrange (1736–1813). There is another standard notation for the derivative that we owe to Leibniz (Figure 2):

$$\frac{df}{dx} \quad \text{or} \quad \frac{dy}{dx}$$

In Example 2, we showed that the derivative of $y = x^{-2}$ is $y' = -2x^{-3}$. In Leibniz notation, we would write

$$\frac{dy}{dx} = -2x^{-3} \quad \text{or} \quad \frac{d}{dx}x^{-2} = -2x^{-3}$$

To specify the value of the derivative for a fixed value of x, say, $x = 4$, we write

$$\left.\frac{df}{dx}\right|_{x=4} \quad \text{or} \quad \left.\frac{dy}{dx}\right|_{x=4}$$

You should not think of dy/dx as the fraction "dy divided by dx." The expressions dy and dx are called **differentials**. They play a role in some situations (in linear approximation and in more advanced calculus). At this stage, we treat them merely as symbols with no independent meaning.

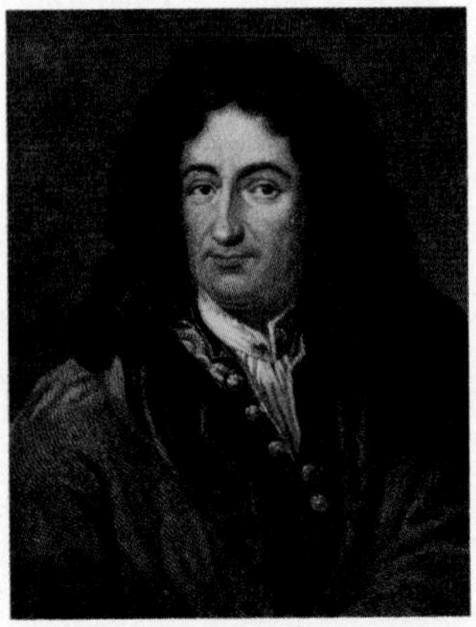

FIGURE 2 Gottfried Wilhelm von Leibniz (1646–1716), German philosopher and scientist. Newton and Leibniz (pronounced "Libe-nitz") are often regarded as the inventors of calculus (working independently). It is more accurate to credit them with developing calculus into a general and fundamental discipline, because many particular results of calculus had been discovered previously by other mathematicians.

CONCEPTUAL INSIGHT Leibniz notation is widely used for several reasons. First, it reminds us that the derivative df/dx, although not itself a ratio, is in fact a *limit* of ratios $\Delta f/\Delta x$. Second, the notation specifies the independent variable. This is useful when variables other than x are used. For example, if the independent variable is t, we write df/dt. Third, we often think of d/dx as an "operator" that performs differentiation on functions. In other words, we apply the operator d/dx to f to obtain the derivative df/dx. We will see other advantages of Leibniz notation when we discuss the Chain Rule in Section 3.7.

A main goal of this chapter is to develop the basic rules of differentiation. These rules enable us to find derivatives without computing limits.

The Power Rule is valid for all exponents. We prove it here for a whole number n (see Exercise 95 for a negative integer n and p. 183 for arbitrary n).

THEOREM 1 The Power Rule For all exponents n,

$$\frac{d}{dx}x^n = nx^{n-1}$$

Proof Assume that n is a whole number and let $f(x) = x^n$. Then

$$f'(a) = \lim_{x\to a} \frac{x^n - a^n}{x - a}$$

To simplify the difference quotient, we need to generalize the following identities:

$$x^2 - a^2 = (x-a)(x+a)$$
$$x^3 - a^3 = (x-a)(x^2 + xa + a^2)$$
$$x^4 - a^4 = (x-a)(x^3 + x^2a + xa^2 + a^3)$$

The generalization is

$$x^n - a^n = (x-a)(x^{n-1} + x^{n-2}a + x^{n-3}a^2 + \cdots + xa^{n-2} + a^{n-1}) \qquad \boxed{2}$$

To verify Eq. (2), observe that the right-hand side is equal to

$$x(x^{n-1} + x^{n-2}a + x^{n-3}a^2 + \cdots + xa^{n-2} + a^{n-1})$$
$$- a(x^{n-1} + x^{n-2}a + x^{n-3}a^2 + \cdots + xa^{n-2} + a^{n-1})$$

When we carry out the multiplications, all terms cancel except the first and the last, so only $x^n - a^n$ remains, as required.

Equation (2) gives us

$$\frac{x^n - a^n}{x - a} = \underbrace{x^{n-1} + x^{n-2}a + x^{n-3}a^2 + \cdots + xa^{n-2} + a^{n-1}}_{n \text{ terms}} \quad (x \neq a) \qquad \boxed{3}$$

Therefore,

$$f'(a) = \lim_{x\to a} (x^{n-1} + x^{n-2}a + x^{n-3}a^2 + \cdots + xa^{n-2} + a^{n-1})$$
$$= a^{n-1} + a^{n-2}a + a^{n-3}a^2 + \cdots + aa^{n-2} + a^{n-1} \quad (n \text{ terms})$$
$$= na^{n-1}$$

This proves that $f'(a) = na^{n-1}$, which we may also write as $f'(x) = nx^{n-1}$. ■

We make a few remarks before proceeding:

*CAUTION The Power Rule applies only to the power functions $y = x^n$. It does not apply to exponential functions such as $y = 2^x$. The derivative of $y = 2^x$ **is not** $x2^{x-1}$. We will study the derivatives of exponential functions later in this section.*

- It may be helpful to remember the Power Rule in words: To differentiate x^n, "bring down the exponent and subtract one (from the exponent)."

$$\frac{d}{dx}x^{\text{exponent}} = (\text{exponent})\, x^{\text{exponent}-1}$$

- The Power Rule is valid for all exponents, whether negative, fractional, or irrational:

$$\frac{d}{dx}x^{-3/5} = -\frac{3}{5}x^{-8/5}, \qquad \frac{d}{dx}x^{\sqrt{2}} = \sqrt{2}\,x^{\sqrt{2}-1}$$

- The Power Rule can be applied with any variable, not just x. For example,

$$\frac{d}{dz}z^2 = 2z, \qquad \frac{d}{dt}t^{20} = 20t^{19}, \qquad \frac{d}{dr}r^{1/2} = \frac{1}{2}r^{-1/2}$$

Next, we state the Linearity Rules for derivatives, which are analogous to the linearity laws for limits.

THEOREM 2 Linearity Rules Assume that f and g are differentiable. Then

Sum and Difference Rules: $f + g$ and $f - g$ are differentiable, and

$$(f+g)' = f' + g', \qquad (f-g)' = f' - g'$$

Constant Multiple Rule: For any constant c, cf is differentiable and

$$(cf)' = cf'$$

Proof To prove the Sum Rule, we use the definition

$$(f+g)'(x) = \lim_{h\to 0} \frac{(f(x+h)+g(x+h)) - (f(x)+g(x))}{h}$$

This difference quotient is equal to a sum ($h \neq 0$):

$$\frac{(f(x+h)+g(x+h)) - (f(x)+g(x))}{h} = \frac{f(x+h)-f(x)}{h} + \frac{g(x+h)-g(x)}{h}$$

Therefore, by the Sum Law for limits,

$$(f+g)'(x) = \lim_{h\to 0} \frac{f(x+h)-f(x)}{h} + \lim_{h\to 0} \frac{g(x+h)-g(x)}{h}$$
$$= f'(x) + g'(x)$$

as claimed. The Difference and Constant Multiple Rules are proved similarly. ■

■ **EXAMPLE 3** Find the points on the graph of $f(t) = t^3 - 12t + 4$ where the tangent line is horizontal (Figure 3).

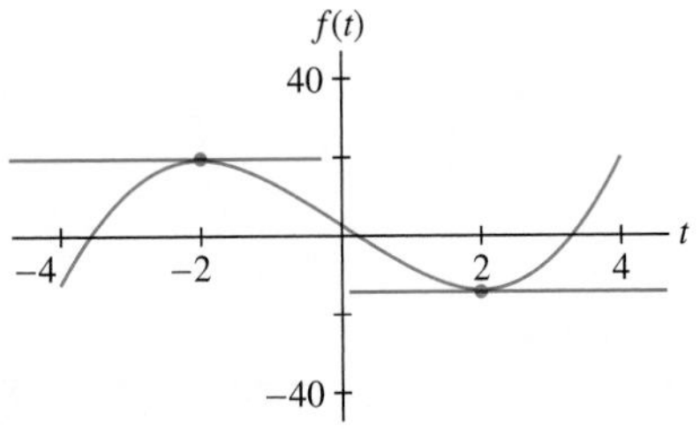

FIGURE 3 Graph of $f(t) = t^3 - 12t + 4$. Tangent lines at $t = \pm 2$ are horizontal.

Solution We calculate the derivative:

$$\begin{aligned}
\frac{df}{dt} &= \frac{d}{dt}(t^3 - 12t + 4) \\
&= \frac{d}{dt}t^3 - \frac{d}{dt}(12t) + \frac{d}{dt}4 && \text{(Sum and Difference Rules)} \\
&= \frac{d}{dt}t^3 - 12\frac{d}{dt}t + 0 && \text{(Constant Multiple Rule)} \\
&= 3t^2 - 12 && \text{(Power Rule)}
\end{aligned}$$

Note in the second line that the derivative of the constant 4 is zero. The tangent line is horizontal at points where the slope $f'(t)$ is zero, so we solve

$$f'(t) = 3t^2 - 12 = 0 \quad \Rightarrow \quad t = \pm 2$$

Now $f(2) = -12$ and $f(-2) = 20$. Hence, the tangent lines are horizontal at $(2, -12)$ and $(-2, 20)$. ■

EXAMPLE 4 Calculate $\left.\dfrac{dg}{dt}\right|_{t=1}$, where $g(t) = t^{-3} + 2\sqrt{t} - t^{-4/5}$.

Solution We differentiate term-by-term using the Power Rule without justifying the intermediate steps. Writing $\sqrt{t}$ as $t^{1/2}$, we have

$$\frac{dg}{dt} = \frac{d}{dt}\left(t^{-3} + 2t^{1/2} - t^{-4/5}\right) = -3t^{-4} + 2\left(\frac{1}{2}\right)t^{-1/2} - \left(-\frac{4}{5}\right)t^{-9/5}$$

$$= -3t^{-4} + t^{-1/2} + \frac{4}{5}t^{-9/5}$$

$$\left.\frac{dg}{dt}\right|_{t=1} = -3 + 1 + \frac{4}{5} = -\frac{6}{5}$$

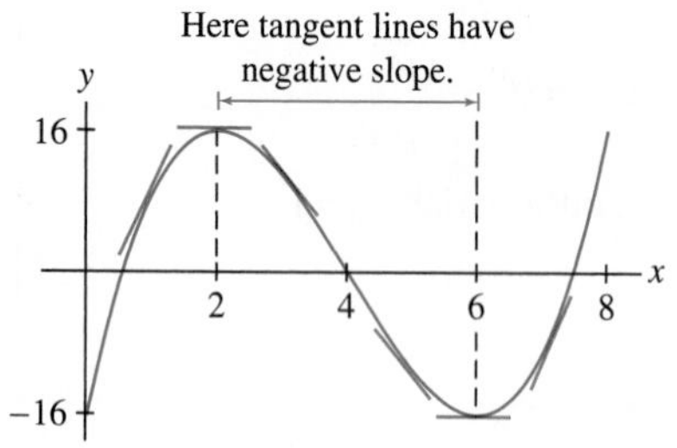

(A) Graph of $f(x) = x^3 - 12x^2 + 36x - 16$

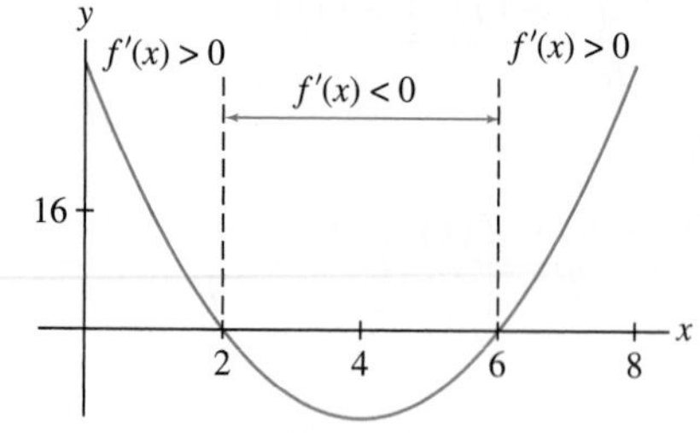

(B) Graph of the derivative $f'(x) = 3x^2 - 24x + 36$

FIGURE 4

The derivative $f'(x)$ gives us important information about the graph of $f(x)$. For example, the sign of $f'(x)$ tells us whether the tangent line has positive or negative slope, and the magnitude of $f'(x)$ reveals how steep the slope is.

EXAMPLE 5 Graphical Insight How is the graph of $f(x) = x^3 - 12x^2 + 36x - 16$ related to the derivative $f'(x) = 3x^2 - 24x + 36$?

Solution The derivative $f'(x) = 3x^2 - 24x + 36 = 3(x-6)(x-2)$ is negative for $2 < x < 6$ and positive elsewhere [Figure 4(B)]. The following table summarizes this sign information [Figure 4(A)]:

Property of $f'(x)$	Property of the Graph of $f(x)$
$f'(x) < 0$ for $2 < x < 6$	Tangent line has negative slope for $2 < x < 6$.
$f'(2) = f'(6) = 0$	Tangent line is horizontal at $x = 2$ and $x = 6$.
$f'(x) > 0$ for $x < 2$ and $x > 6$	Tangent line has positive slope for $x < 2$ and $x > 6$.

Note also that $f'(x) \to \infty$ as $|x|$ becomes large. This corresponds to the fact that the tangent lines to the graph of $f(x)$ get steeper as $|x|$ grows large.

EXAMPLE 6 Identifying the Derivative The graph of $f(x)$ is shown in Figure 5(A). Which graph (B) or (C), is the graph of $f'(x)$?

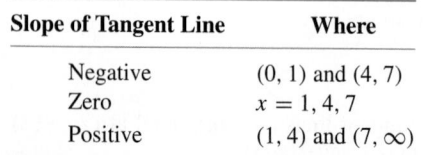

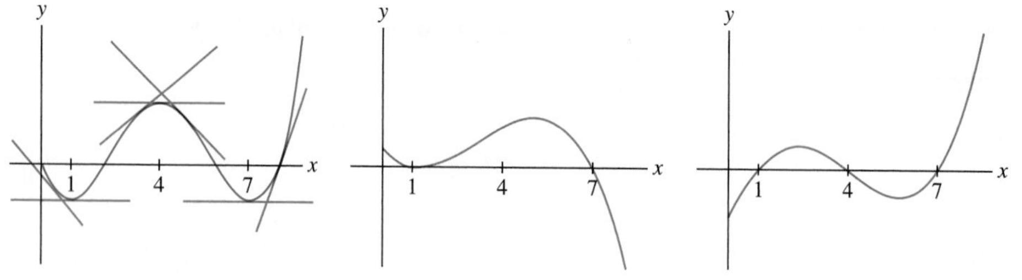

(A) Graph of $f(x)$ (B) (C)

FIGURE 5

Slope of Tangent Line	Where
Negative	(0, 1) and (4, 7)
Zero	$x = 1, 4, 7$
Positive	(1, 4) and $(7, \infty)$

Solution In Figure 5(A) we see that the tangent lines to the graph have negative slope on the intervals (0, 1) and (4, 7). Therefore $f'(x)$ is negative on these intervals. Similarly (see the table in the margin), the tangent lines have positive slope (and $f'(x)$ is positive) on the intervals (1, 4) and $(7, \infty)$. Only (C) has these properties, so (C) is the graph of $f'(x)$.

Differentiability, Continuity, and Local Linearity

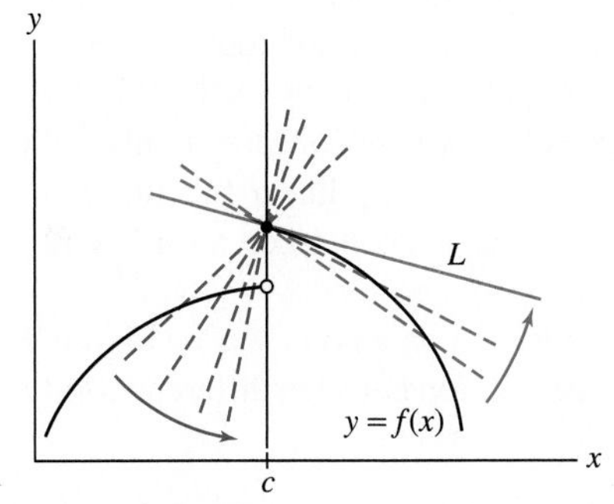

FIGURE 6 Secant lines at a jump discontinuity.

In the rest of this section, we examine the concept of **differentiability** more closely. We begin by proving that a differentiable function is necessarily continuous. In particular, a differentiable function cannot have any jumps. Figure 6 shows why: Although the secant lines from the right approach the line L (which is tangent to the right half of the graph), the secant lines from the left approach the vertical (and their slopes tend to ∞).

> **THEOREM 3 Differentiability Implies Continuity** If f is differentiable at $x = c$, then f is continuous at $x = c$.

Proof By definition, if f is differentiable at $x = c$, then the following limit exists:

$$f'(c) = \lim_{x \to c} \frac{f(x) - f(c)}{x - c}$$

We must prove that $\lim_{x \to c} f(x) = f(c)$, because this is the definition of continuity at $x = c$. To relate the two limits, consider the equation (valid for $x \neq c$)

$$f(x) - f(c) = (x - c)\,\frac{f(x) - f(c)}{x - c}$$

Both factors on the right approach a limit as $x \to c$, so

$$\begin{aligned}\lim_{x \to c}\big(f(x) - f(c)\big) &= \lim_{x \to c}\left((x - c)\,\frac{f(x) - f(c)}{x - c}\right)\\ &= \left(\lim_{x \to c}(x - c)\right)\left(\lim_{x \to c}\frac{f(x) - f(c)}{x - c}\right)\\ &= 0 \cdot f'(c) = 0\end{aligned}$$

by the Product Law for limits. The Sum Law now yields the desired conclusion:

$$\lim_{x \to c} f(x) = \lim_{x \to c}(f(x) - f(c)) + \lim_{x \to c} f(c) = 0 + f(c) = f(c) \qquad \blacksquare$$

Most of the functions encountered in this text are differentiable, but exceptions exist, as the next example shows.

All differentiable functions are continuous by Theorem 3, but Example 7 shows that the converse is false. A continuous function is not necessarily differentiable.

■ **EXAMPLE 7 Continuous But Not Differentiable** Show that $f(x) = |x|$ is continuous but not differentiable at $x = 0$.

Solution The function $f(x)$ is continuous at $x = 0$ because $\lim_{x \to 0} |x| = 0 = f(0)$. On the other hand,

$$f'(0) = \lim_{h \to 0} \frac{f(0 + h) - f(0)}{h} = \lim_{h \to 0} \frac{|0 + h| - |0|}{h} = \lim_{h \to 0} \frac{|h|}{h}$$

This limit does not exist [and hence $f(x)$ is not differentiable at $x = 0$] because

$$\frac{|h|}{h} = \begin{cases} 1 & \text{if } h > 0 \\ -1 & \text{if } h < 0 \end{cases}$$

and thus the one-sided limits are not equal:

$$\lim_{h \to 0+} \frac{|h|}{h} = 1 \quad \text{and} \quad \lim_{h \to 0-} \frac{|h|}{h} = -1 \qquad \blacksquare$$

GRAPHICAL INSIGHT Differentiability has an important graphical interpretation in terms of local linearity. We say that f is **locally linear** at $x = a$ if the graph looks more and more like a straight line as we zoom in on the point $(a, f(a))$. In this context, the adjective *linear* means "resembling a line," and *local* indicates that we are concerned only with the behavior of the graph near $(a, f(a))$. The graph of a locally linear function may be very wavy or *nonlinear*, as in Figure 7. But as soon as we zoom in on a sufficiently small piece of the graph, it begins to appear straight.

Not only does the graph look like a line as we zoom in on a point, but as Figure 7 suggests, the "zoom line" is the tangent line. Thus, the relation between differentiability and local linearity can be expressed as follows:

If $f'(a)$ exists, then f is locally linear at $x = a$: As we zoom in on the point $(a, f(a))$, the graph becomes nearly indistinguishable from its tangent line.

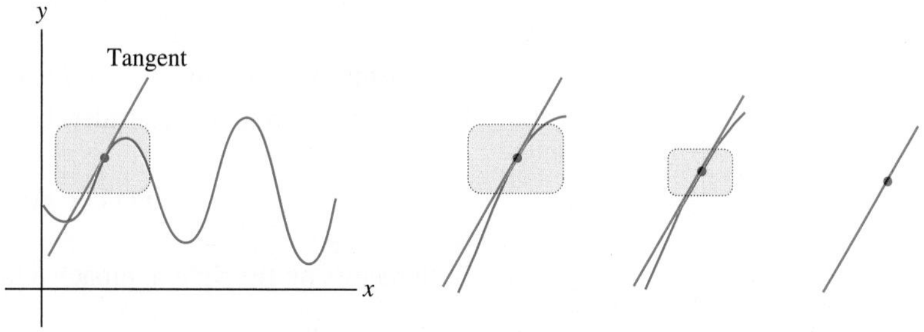

FIGURE 7 Local linearity: The graph looks more and more like the tangent line as we zoom in on a point.

Local linearity gives us a graphical way to understand why $f(x) = |x|$ is not differentiable at $x = 0$ (as shown in Example 7). Figure 8 shows that the graph of $f(x) = |x|$ has a corner at $x = 0$, and this corner *does not disappear*, no matter how closely we zoom in on the origin. Since the graph does not straighten out under zooming, $f(x)$ is not locally linear at $x = 0$, and we cannot expect $f'(0)$ to exist.

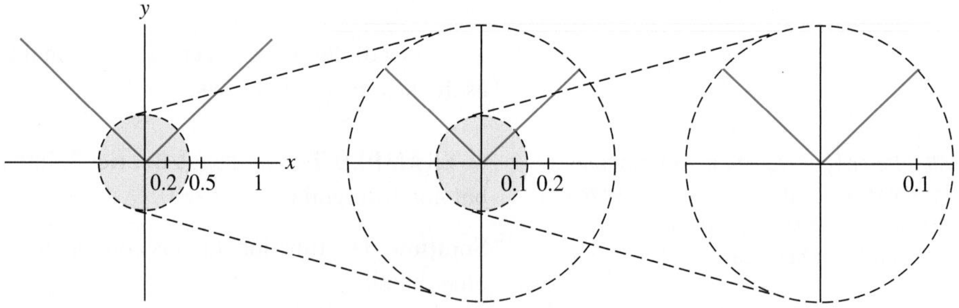

FIGURE 8 The graph of $f(x) = |x|$ is not locally linear at $x = 0$. The corner does not disappear when we zoom in on the origin.

Another way that a continuous function can fail to be differentiable is if the tangent line exists but is vertical (in which case the slope of the tangent line is undefined).

■ **EXAMPLE 8 Vertical Tangents** Show that $f(x) = x^{1/3}$ is not differentiable at $x = 0$.

Solution The limit defining $f'(0)$ is infinite:

$$\lim_{h\to 0} \frac{f(h) - f(0)}{h} = \lim_{h\to 0} \frac{h^{1/3} - 0}{h} = \lim_{h\to 0} \frac{h^{1/3}}{h} = \lim_{h\to 0} \frac{1}{h^{2/3}} = \infty$$

Therefore, $f'(0)$ does not exist (Figure 9). ■

As a final remark, we mention that there are more complicated ways in which a continuous function can fail to be differentiable. Figure 10 shows the graph of $f(x) = x \sin \frac{1}{x}$. If we define $f(0) = 0$, then f is continuous but not differentiable at $x = 0$. The secant lines keep oscillating and never settle down to a limiting position (see Exercise 97).

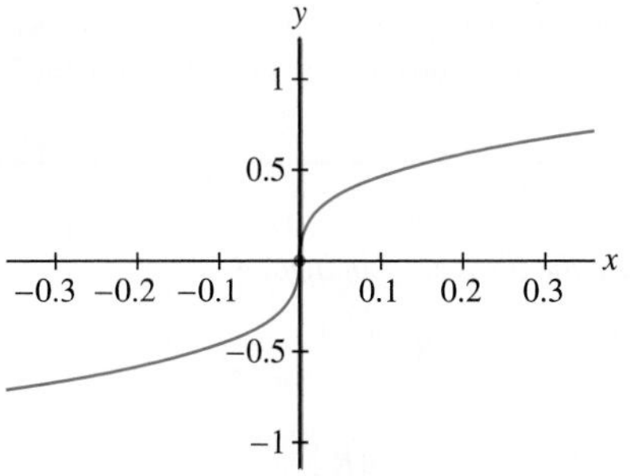

FIGURE 9 The tangent line to the graph of $f(x) = x^{1/3}$ at the origin is the (vertical) y-axis. The derivative $f'(0)$ does not exist.

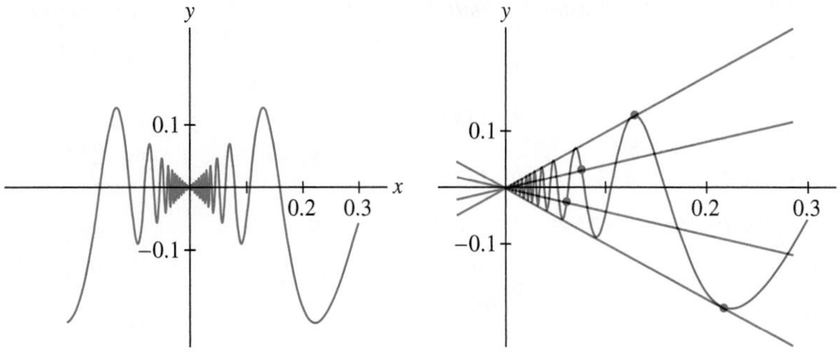

(A) Graph of $f(x) = x \sin \frac{1}{x}$

(B) Secant lines do not settle down to a limiting position.

FIGURE 10

3.2 SUMMARY

- The derivative $f'(x)$ is the function whose value at $x = a$ is the derivative $f'(a)$.
- We have several different notations for the derivative of $y = f(x)$:

$$y', \quad y'(x), \quad f'(x), \quad \frac{dy}{dx}, \quad \frac{df}{dx}$$

The value of the derivative at $x = a$ is written

$$y'(a), \quad f'(a), \quad \left.\frac{dy}{dx}\right|_{x=a}, \quad \left.\frac{df}{dx}\right|_{x=a}$$

- The Power Rule holds for all exponents n:

$$\frac{d}{dx}x^n = nx^{n-1}$$

- The Linearity Rules allow us to differentiate term by term:

Sum Rule: $(f + g)' = f' + g'$, Constant Multiple Rule: $(cf)' = cf'$

- Differentiability implies continuity: If $f(x)$ is differentiable at $x = a$, then $f(x)$ is continuous at $x = a$. However, there exist continuous functions that are not differentiable.
- If $f'(a)$ exists, then f is locally linear in the following sense: As we zoom in on the point $(a, f(a))$, the graph becomes nearly indistinguishable from its tangent line.

3.2 EXERCISES

Preliminary Questions

1. What is the slope of the tangent line through the point $(2, f(2))$ if $f'(x) = x^3$?

2. Evaluate $(f - g)'(1)$ and $(3f + 2g)'(1)$ assuming that $f'(1) = 3$ and $g'(1) = 5$.

3. To which of the following does the Power Rule apply?

(a) $f(x) = x^2$ **(b)** $f(x) = 2^\pi$

(c) $f(x) = x^\pi$ **(d)** $f(x) = \pi^x$

(e) $f(x) = x^x$ **(f)** $f(x) = x^{-4/5}$

4. Choose (a) or (b). The derivative does not exist if the tangent line is: (a) horizontal (b) vertical.

5. If $f(x)$ is differentiable at $x = c$, is $f(x)$ necessarily continuous at $x = c$? Do there exist continuous functions that are not differentiable?

Exercises

In Exercises 1–6, compute $f'(x)$ using the limit definition.

1. $f(x) = 3x - 7$

2. $f(x) = x^2 + 3x$

3. $f(x) = x^3$

4. $f(x) = 1 - x^{-1}$

5. $f(x) = x - \sqrt{x}$

6. $f(x) = x^{-1/2}$

In Exercises 7–14, use the Power Rule to compute the derivative.

7. $\left.\dfrac{d}{dx}x^4\right|_{x=-2}$

8. $\left.\dfrac{d}{dt}t^{-3}\right|_{t=4}$

9. $\left.\dfrac{d}{dt}t^{2/3}\right|_{t=8}$

10. $\left.\dfrac{d}{dt}t^{-2/5}\right|_{t=1}$

11. $\dfrac{d}{dx}x^{0.35}$

12. $\dfrac{d}{dx}x^{14/3}$

13. $\dfrac{d}{dt}t^{\sqrt{17}}$

14. $\dfrac{d}{dt}t^{-\pi^2}$

In Exercises 15–18, compute $f'(x)$ and find an equation of the tangent line to the graph at $x = a$.

15. $f(x) = x^4, \quad a = 2$

16. $f(x) = x^{-2}, \quad a = 5$

17. $f(x) = 5x - 32\sqrt{x}, \quad a = 4$

18. $f(x) = \sqrt[3]{x}, \quad a = 8$

19. Find an equation of the tangent line to $y = 1/\sqrt{x}$ at $x = 9$.

20. Find a point on the graph of $y = \sqrt{x}$ where the tangent line has slope 10.

In Exercises 21–32, calculate the derivative.

21. $f(x) = 2x^3 - 3x^2 + 5$

22. $f(x) = 2x^3 - 3x^2 + 2x$

23. $f(x) = 4x^{5/3} - 3x^{-2} - 12$

24. $f(x) = x^{5/4} + 4x^{-3/2} + 11x$

25. $g(z) = 7z^{-5/14} + z^{-5} + 9$

26. $h(t) = 6\sqrt{t} + \dfrac{1}{\sqrt{t}}$

27. $f(s) = \sqrt[4]{s} + \sqrt[3]{s}$

28. $W(y) = 6y^4 + 7y^{2/3}$

29. $g(x) = \pi^2$

30. $f(x) = x^{\pi}$

31. $h(t) = \sqrt{2}\,t^{\sqrt{2}}$

32. $R(z) = \dfrac{z^{5/3} - 4z^{3/2}}{z}$ *Hint:* Simplify.

In Exercises 33–36, calculate the derivative by expanding or simplifying the function.

33. $P(s) = (4s - 3)^2$

34. $Q(r) = (1 - 2r)(3r + 5)$

35. $g(x) = \dfrac{x^2 + 4x^{1/2}}{x^2}$

36. $s(t) = \dfrac{1 - 2t}{t^{1/2}}$

In Exercises 37–42, calculate the derivative indicated.

37. $\left.\dfrac{dT}{dC}\right|_{C=8}, \quad T = 3C^{2/3}$

38. $\left.\dfrac{dP}{dV}\right|_{V=-2}, \quad P = \dfrac{7}{V}$

39. $\left.\dfrac{ds}{dz}\right|_{z=2}, \quad s = 4z - 16z^2$

40. $\left.\dfrac{dR}{dW}\right|_{W=1}, \quad R = W^{\pi}$

41. $\left.\dfrac{dr}{dt}\right|_{r=4}, \quad r = \dfrac{t^2 + 1}{t^{1/2}}$

42. $\left.\dfrac{dp}{dh}\right|_{h=32}, \quad p = 16h^{0.2} + 8h^{-0.8}$

43. Match the functions in graphs (A)–(D) with their derivatives (I)–(III) in Figure 11. Note that two of the functions have the same derivative. Explain why.

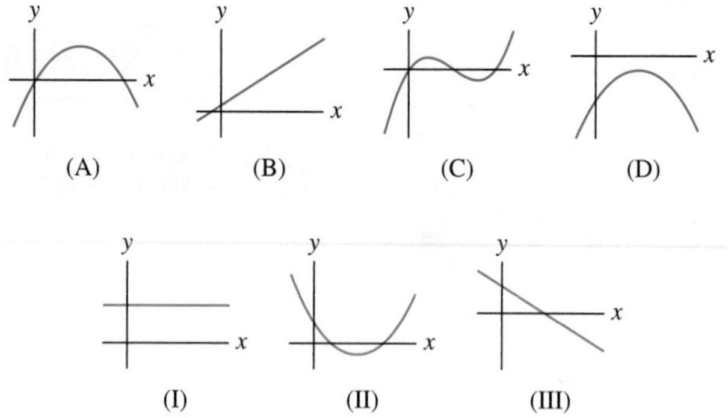

FIGURE 11

44. Of the two functions f and g in Figure 12, which is the derivative of the other? Justify your answer.

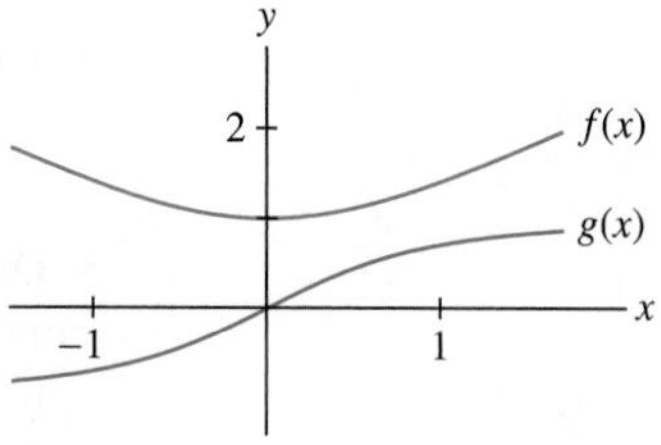

FIGURE 12

45. Assign the labels $f(x)$, $g(x)$, and $h(x)$ to the graphs in Figure 13 in such a way that $f'(x) = g(x)$ and $g'(x) = h(x)$.

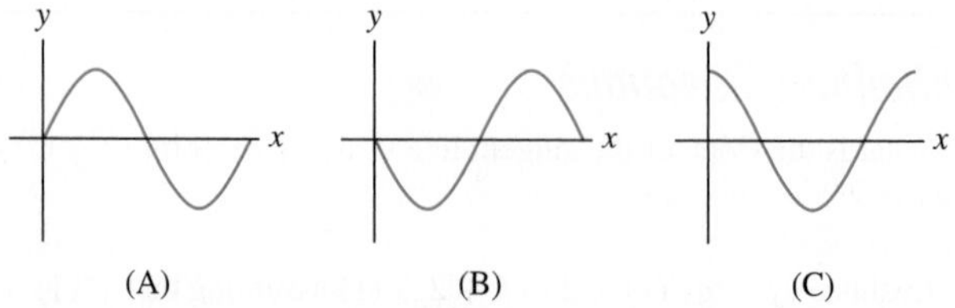

FIGURE 13

46. According to the *peak oil theory*, first proposed in 1956 by geophysicist M. Hubbert, the total amount of crude oil $Q(t)$ produced worldwide up to time t has a graph like that in Figure 14.

(a) Sketch the derivative $Q'(t)$ for $1900 \le t \le 2150$. What does $Q'(t)$ represent?

(b) In which year (approximately) does $Q'(t)$ take on its maximum value?

(c) What is $L = \lim_{t\to\infty} Q(t)$? And what is its interpretation?

(d) What is the value of $\lim_{t\to\infty} Q'(t)$?

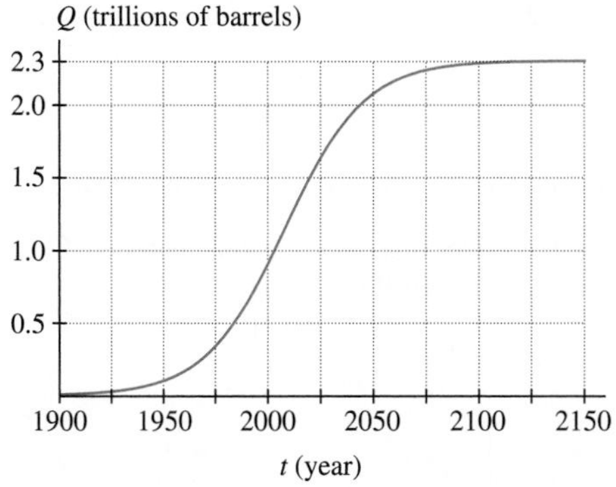

FIGURE 14 Total oil production up to time t

47. Use the table of values of $f(x)$ to determine which of (A) or (B) in Figure 15 is the graph of $f'(x)$. Explain.

x	0	0.5	1	1.5	2	2.5	3	3.5	4
$f(x)$	10	55	98	139	177	210	237	257	268

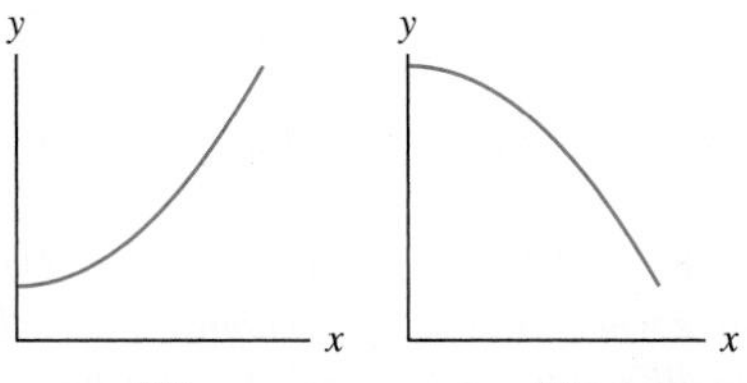

FIGURE 15 Which is the graph of $f'(x)$?

48. Let R be a variable and r a constant. Compute the derivatives:

(a) $\dfrac{d}{dR} R$ **(b)** $\dfrac{d}{dR} r$ **(c)** $\dfrac{d}{dR} r^2 R^3$

49. Compute the derivatives, where c is a constant.

(a) $\dfrac{d}{dt} ct^3$ **(b)** $\dfrac{d}{dz}(5z + 4cz^2)$

(c) $\dfrac{d}{dy}(9c^2y^3 - 24c)$

50. Find the points on the graph of $f(x) = 12x - x^3$ where the tangent line is horizontal.

51. Find the points on the graph of $y = x^2 + 3x - 7$ at which the slope of the tangent line is equal to 4.

52. Find the values of x where $y = x^3$ and $y = x^2 + 5x$ have parallel tangent lines.

53. Determine a and b such that $p(x) = x^2 + ax + b$ satisfies $p(1) = 0$ and $p'(1) = 4$.

54. Find all values of x such that the tangent line to $y = 4x^2 + 11x + 2$ is steeper than the tangent line to $y = x^3$.

55. Let $f(x) = x^3 - 3x + 1$. Show that $f'(x) \ge -3$ for all x and that, for every $m > -3$, there are precisely two points where $f'(x) = m$. Indicate the position of these points and the corresponding tangent lines for one value of m in a sketch of the graph of $f(x)$.

56. Show that the tangent lines to $y = \frac{1}{3}x^3 - x^2$ at $x = a$ and at $x = b$ are parallel if $a = b$ or $a + b = 2$.

57. Compute the derivative of $f(x) = x^{3/2}$ using the limit definition. *Hint:* Show that

$$\frac{f(x+h) - f(x)}{h} = \frac{(x+h)^3 - x^3}{h}\left(\frac{1}{\sqrt{(x+h)^3} + \sqrt{x^3}}\right)$$

58. Sketch the graph of a continuous function on $(0, 5)$ that is differentiable except at $x = 1$ and $x = 4$.

59. Show, using the limit definition of the derivative, that $f(x) = |x^2 - 4|$ is not differentiable at $x = 2$.

60. The average speed (in meters per second) of a gas molecule is

$$v_{avg} = \sqrt{\frac{8RT}{\pi M}}$$

where T is the temperature (in kelvins), M is the molar mass (in kilograms per mole), and $R = 8.31$. Calculate dv_{avg}/dT at $T = 300$ K for oxygen, which has a molar mass of 0.032 kg/mol.

61. Biologists have observed that the pulse rate P (in beats per minute) in animals is related to body mass (in kilograms) by the approximate formula $P = 200m^{-1/4}$. This is one of many *allometric scaling laws* prevalent in biology. Is $|dP/dm|$ an increasing or decreasing function of m? Find an equation of the tangent line at the points on the graph in Figure 16 that represent goat ($m = 33$) and man ($m = 68$).

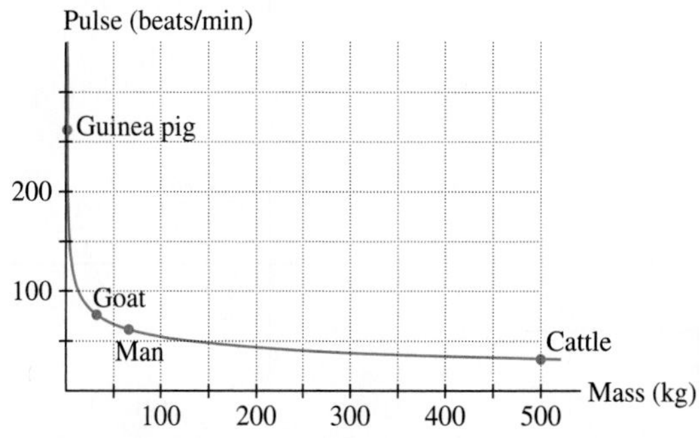

FIGURE 16

62. Some studies suggest that kidney mass K in mammals (in kilograms) is related to body mass m (in kilograms) by the approximate formula $K = 0.007m^{0.85}$. Calculate dK/dm at $m = 68$. Then calculate the derivative with respect to m of the relative kidney-to-mass ratio K/m at $m = 68$.

63. The Clausius–Clapeyron Law relates the *vapor pressure* of water P (in atmospheres) to the temperature T (in kelvins):

$$\frac{dP}{dT} = k\frac{P}{T^2}$$

where k is a constant. Estimate dP/dT for $T = 303, 313, 323, 333, 343$ using the data and the approximation

$$\frac{dP}{dT} \approx \frac{P(T+10) - P(T-10)}{20}$$

T (K)	293	303	313	323	333	343	353
P (atm)	0.0278	0.0482	0.0808	0.1311	0.2067	0.3173	0.4754

Do your estimates seem to confirm the Clausius–Clapeyron Law? What is the approximate value of k?

64. Let L be the tangent line to the hyperbola $xy = 1$ at $x = a$, where $a > 0$. Show that the area of the triangle bounded by L and the coordinate axes does not depend on a.

65. In the setting of Exercise 64, show that the point of tangency is the midpoint of the segment of L lying in the first quadrant.

66. Match functions (A)–(C) with their derivatives (I)–(III) in Figure 17.

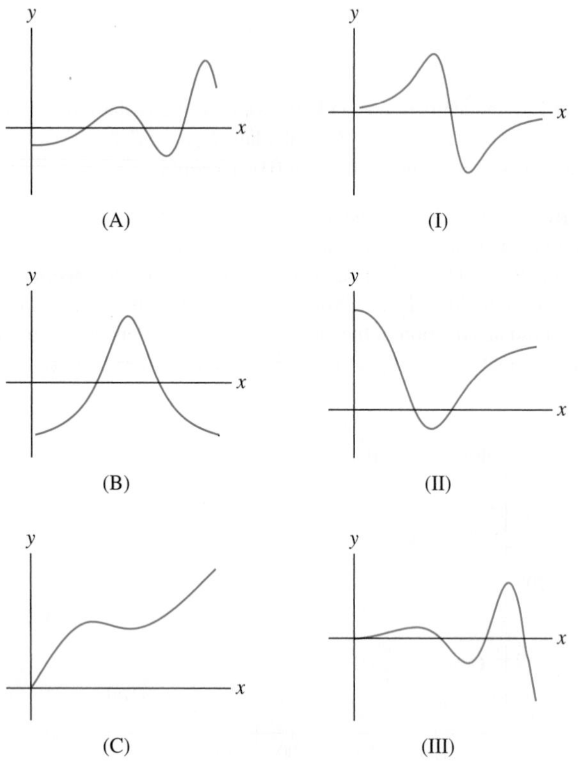

FIGURE 17

67. Make a rough sketch of the graph of the derivative of the function in Figure 18(A).

68. Graph the derivative of the function in Figure 18(B), omitting points where the derivative is not defined.

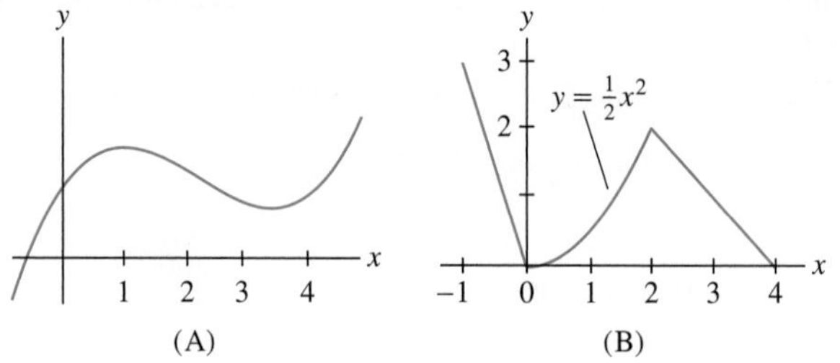

FIGURE 18

69. Sketch the graph of $f(x) = x\,|x|$. Then show that $f'(0)$ exists.

70. Determine the values of x at which the function in Figure 19 is: (a) discontinuous, and (b) nondifferentiable.

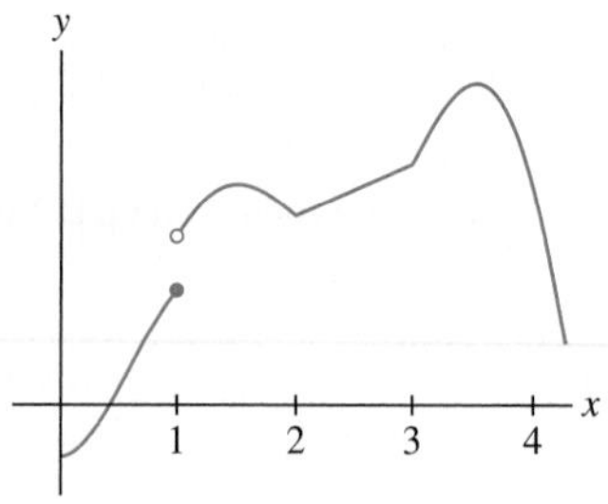

FIGURE 19

In Exercises 71–76, find the points c (if any) such that $f'(c)$ does not exist.

71. $f(x) = |x - 1|$

72. $f(x) = [x]$

73. $f(x) = x^{2/3}$

74. $f(x) = x^{3/2}$

75. $f(x) = |x^2 - 1|$

76. $f(x) = |x - 1|^2$

GU *In Exercises 77–82, zoom in on a plot of $f(x)$ at the point $(a, f(a))$ and state whether or not $f(x)$ appears to be differentiable at $x = a$. If it is nondifferentiable, state whether the tangent line appears to be vertical or does not exist.*

77. $f(x) = (x - 1)|x|, \quad a = 0$

78. $f(x) = (x - 3)^{5/3}, \quad a = 3$

79. $f(x) = (x - 3)^{1/3}, \quad a = 3$

80. $f(x) = \sin(x^{1/3}), \quad a = 0$

81. $f(x) = |\sin x|, \quad a = 0$

82. $f(x) = |x - \sin x|, \quad a = 0$

83. GU Plot the derivative $f'(x)$ of $f(x) = 2x^3 - 10x^{-1}$ for $x > 0$ (set the bounds of the viewing box appropriately) and observe that $f'(x) > 0$. What does the positivity of $f'(x)$ tell us about the graph of $f(x)$ itself? Plot $f(x)$ and confirm this conclusion.

84. Find the coordinates of the point P in Figure 20 at which the tangent line passes through (5, 0).

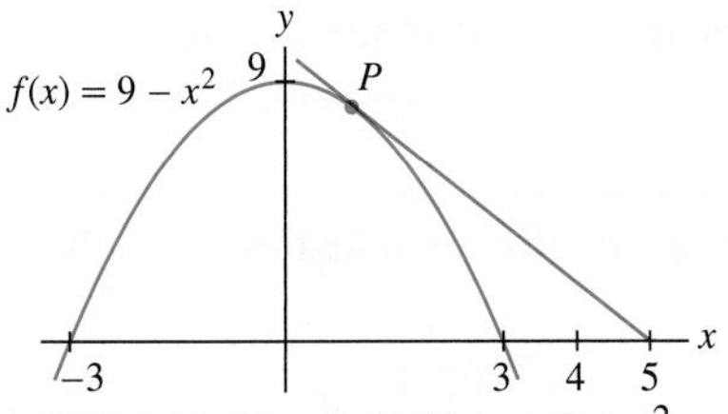

FIGURE 20 Graph of $f(x) = 9 - x^2$.

Exercises 85–88 refer to Figure 21. Length QR is called the subtangent *at P, and length RT is called the* subnormal.

85. Calculate the subtangent of

$$f(x) = x^2 + 3x \quad \text{at } x = 2$$

86. Show that for $n \neq 0$, the subtangent of $f(x) = x^n$ at $x = c$ is equal to c/n.

87. Prove in general that the subnormal at P is $|f'(x)f(x)|$.

88. Show that $\overline{PQ}$ has length $|f(x)|\sqrt{1 + f'(x)^{-2}}$.

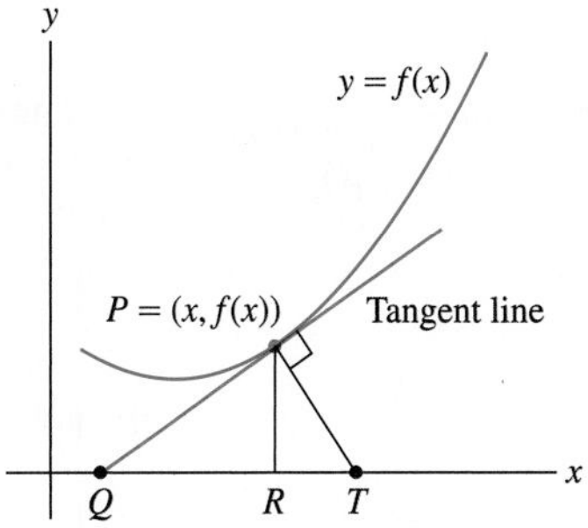

FIGURE 21

89. Prove the following theorem of Apollonius of Perga (the Greek mathematician born in 262 BCE who gave the parabola, ellipse, and hyperbola their names): The subtangent of the parabola $y = x^2$ at $x = a$ is equal to $a/2$.

90. Show that the subtangent to $y = x^3$ at $x = a$ is equal to $\frac{1}{3}a$.

91. Formulate and prove a generalization of Exercise 90 for $y = x^n$.

Further Insights and Challenges

92. Two small arches have the shape of parabolas. The first is given by $f(x) = 1 - x^2$ for $-1 \le x \le 1$ and the second by $g(x) = 4 - (x-4)^2$ for $2 \le x \le 6$. A board is placed on top of these arches so it rests on both (Figure 22). What is the slope of the board? *Hint:* Find the tangent line to $y = f(x)$ that intersects $y = g(x)$ in exactly one point.

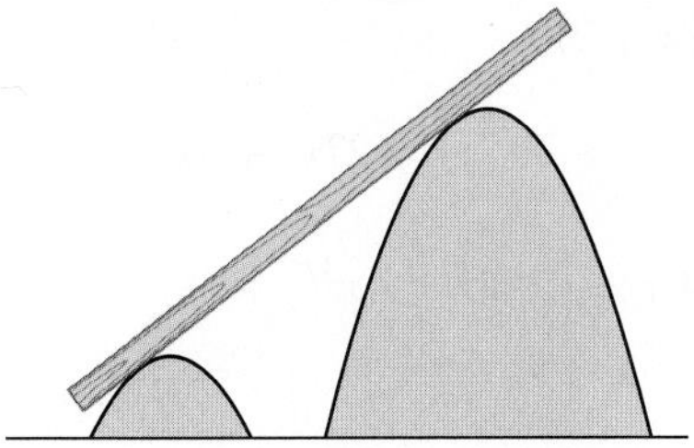

FIGURE 22

93. A vase is formed by rotating $y = x^2$ around the y-axis. If we drop in a marble, it will either touch the bottom point of the vase or be suspended above the bottom by touching the sides (Figure 23). How small must the marble be to touch the bottom?

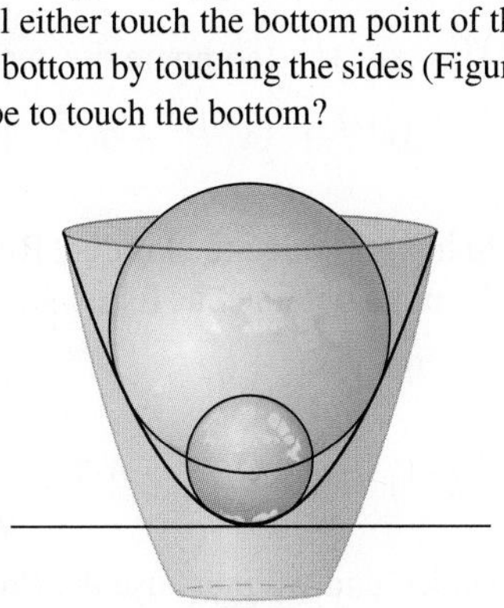

FIGURE 23

94. Let $f(x)$ be a differentiable function, and set $g(x) = f(x + c)$, where c is a constant. Use the limit definition to show that $g'(x) = f'(x + c)$. Explain this result graphically, recalling that the graph of $g(x)$ is obtained by shifting the graph of $f(x)$ c units to the left (if $c > 0$) or right (if $c < 0$).

95. Negative Exponents Let n be a whole number. Use the Power Rule for x^n to calculate the derivative of $f(x) = x^{-n}$ by showing that

$$\frac{f(x+h) - f(x)}{h} = \frac{-1}{x^n(x+h)^n} \, \frac{(x+h)^n - x^n}{h}$$

96. Verify the Power Rule for the exponent $1/n$, where n is a positive integer, using the following trick: Rewrite the difference quotient for $y = x^{1/n}$ at $x = b$ in terms of

$$u = (b+h)^{1/n} \quad \text{and} \quad a = b^{1/n}$$

97. Infinitely Rapid Oscillations Define

$$f(x) = \begin{cases} x \sin \dfrac{1}{x} & x \neq 0 \\ 0 & x = 0 \end{cases}$$

Show that $f(x)$ is continuous at $x = 0$ but $f'(0)$ does not exist (see Figure 10).

98. For which values of c does the equation $x^2 + 4 = cx$ have a unique solution? *Hint:* Draw a graph.

3.3 Product and Quotient Rules

REMINDER *The product function fg is defined by $(fg)(x) = f(x)\,g(x)$.*

This section covers the **Product Rule** and **Quotient Rule** for computing derivatives. These two rules, together with the Chain Rule and implicit differentiation (covered in later sections), make up an extremely effective "differentiation toolkit."

THEOREM 1 Product Rule If f and g are differentiable functions, then fg is differentiable and

$$(fg)'(x) = f(x)\,g'(x) + g(x)\,f'(x)$$

It may be helpful to remember the Product Rule in words: The derivative of a product is equal to *the first function times the derivative of the second function plus the second function times the derivative of the first function*:

$$\text{First} \cdot (\text{Second})' + \text{Second} \cdot (\text{First})'$$

We prove the Product Rule after presenting three examples.

EXAMPLE 1 Find the derivative of $h(x) = x^2(9x + 2)$.

Solution This function is a product:

$$h(x) = \overbrace{x^2}^{\text{First}}\ \overbrace{(9x + 2)}^{\text{Second}}$$

By the Product Rule (in Leibniz notation),

$$h'(x) = \overbrace{x^2}^{\text{First}}\ \overbrace{\frac{d}{dx}(9x + 2)}^{\text{Second}'} + \overbrace{(9x + 2)}^{\text{Second}}\ \overbrace{\frac{d}{dx}(x^2)}^{\text{First}'}$$
$$= (x^2)(9) + (9x + 2)(2x) = 27x^2 + 4x$$

EXAMPLE 2 Find the derivative of $y = (2 + x^{-1})(x^{3/2} + 1)$.

Solution Use the Product Rule:

Note how the prime notation is used in the solution to Example 2. We write $(x^{3/2} + 1)'$ to denote the derivative of $x^{3/2} + 1$, etc.

$$y' = \overbrace{(2 + x^{-1})(x^{3/2} + 1)' + (x^{3/2} + 1)(2 + x^{-1})'}^{\text{First} \cdot (\text{Second})' + \text{Second} \cdot (\text{First})'}$$
$$= (2 + x^{-1})\left(\tfrac{3}{2}x^{1/2}\right) + (x^{3/2} + 1)(-x^{-2}) \quad \text{(compute the derivatives)}$$
$$= 3x^{1/2} + \tfrac{3}{2}x^{-1/2} - x^{-1/2} - x^{-2} = 3x^{1/2} + \tfrac{1}{2}x^{-1/2} - x^{-2} \quad \text{(simplify)}$$

In the previous two examples, we could have avoided the Product Rule by expanding the function. Thus, the result of Example 2 can be obtained as follows:

$$y = (2 + x^{-1})(x^{3/2} + 1) = 2x^{3/2} + 2 + x^{1/2} + x^{-1}$$
$$y' = \frac{d}{dx}(2x^{3/2} + 2 + x^{1/2} + x^{-1}) = 3x^{1/2} + \tfrac{1}{2}x^{-1/2} - x^{-2}$$

In many cases, the function cannot be expanded and we must use the Product Rule. One such function is $f(x) = x\cos x$ whose derivative we find in Section 3.6.

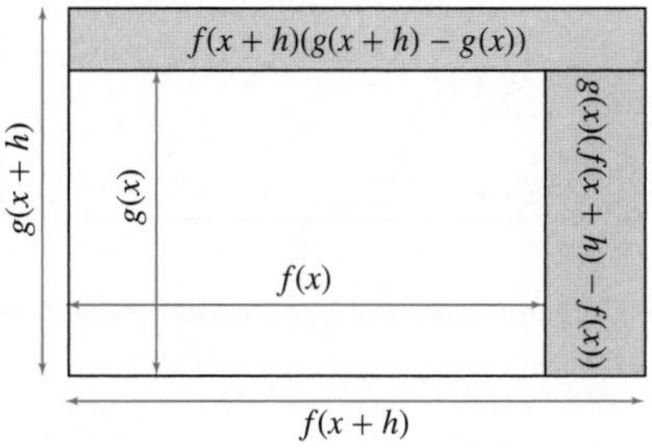

FIGURE 1

Proof of the Product Rule According to the limit definition of the derivative,

$$(fg)'(x) = \lim_{h \to 0} \frac{f(x+h)g(x+h) - f(x)g(x)}{h}$$

We can interpret the numerator as the area of the shaded region in Figure 1: The area of the larger rectangle $f(x+h)g(x+h)$ minus the area of the smaller rectangle $f(x)g(x)$. This shaded region is the union of two rectangular strips, so we obtain the following identity (which you can check directly):

$$f(x+h)g(x+h) - f(x)g(x) = f(x+h)\big(g(x+h) - g(x)\big) + g(x)\big(f(x+h) - f(x)\big)$$

Now use this identity to write $(fg)'(x)$ as a sum of two limits:

$$(fg)'(x) = \underbrace{\lim_{h \to 0} f(x+h)\frac{g(x+h) - g(x)}{h}}_{\text{Show that this equals } f(x)g'(x).} + \underbrace{\lim_{h \to 0} g(x)\frac{f(x+h) - f(x)}{h}}_{\text{Show that this equals } g(x)f'(x).} \qquad \boxed{1}$$

The use of the Sum Law is valid, provided that each limit on the right exists. To check that the first limit exists and to evaluate it, we note that $f(x)$ is continuous (because it is differentiable) and that $g(x)$ is differentiable. Thus

$$\begin{aligned} \lim_{h \to 0} f(x+h)\frac{g(x+h) - g(x)}{h} &= \lim_{h \to 0} f(x+h) \lim_{h \to 0} \frac{g(x+h) - g(x)}{h} \\ &= f(x)\, g'(x) \end{aligned} \qquad \boxed{2}$$

The second limit is similar:

$$\lim_{h \to 0} g(x)\frac{f(x+h) - f(x)}{h} = g(x) \lim_{h \to 0} \frac{f(x+h) - f(x)}{h} = g(x)\, f'(x) \qquad \boxed{3}$$

Using Eq. (2) and Eq. (3) in Eq. (1), we conclude that fg is differentiable and that $(fg)'(x) = f(x)g'(x) + g(x)f'(x)$ as claimed. ■

CONCEPTUAL INSIGHT The Product Rule was first stated by the 29-year-old Leibniz in 1675, the year he developed some of his major ideas on calculus. To document his process of discovery for posterity, he recorded his thoughts and struggles, the moments of inspiration as well as the mistakes. In a manuscript dated November 11, 1675, Leibniz suggests *incorrectly* that $(fg)'$ equals $f'g'$. He then catches his error by taking $f(x) = g(x) = x$ and noticing that

$$(fg)'(x) = \big(x^2\big)' = 2x \qquad \text{is } \textit{not} \text{ equal to} \qquad f'(x)g'(x) = 1 \cdot 1 = 1$$

Ten days later, on November 21, Leibniz writes down the correct Product Rule and comments "*Now this is a really noteworthy theorem.*"

With the benefit of hindsight, we can point out that Leibniz might have avoided his error if he had paid attention to units. Suppose $f(t)$ and $g(t)$ represent distances in meters, where t is time in seconds. Then $(fg)'$ has units of m^2/s. This cannot equal $f'g'$, which has units of $(m/s)(m/s) = m^2/s^2$.

The next theorem states the rule for differentiating quotients. Note, in particular, that $(f/g)'$ is *not* equal to the quotient f'/g'.

← *REMINDER The quotient function f/g is defined by*

$$\left(\frac{f}{g}\right)(x) = \frac{f(x)}{g(x)}$$

THEOREM 2 Quotient Rule If f and g are differentiable functions, then f/g is differentiable for all x such that $g(x) \neq 0$, and

$$\left(\frac{f}{g}\right)'(x) = \frac{g(x)f'(x) - f(x)g'(x)}{g(x)^2}$$

The numerator in the Quotient Rule is equal to *the bottom times the derivative of the top minus the top times the derivative of the bottom*:

$$\frac{\text{Bottom} \cdot (\text{Top})' - \text{Top} \cdot (\text{Bottom})'}{\text{Bottom}^2}$$

The proof is similar to that of the Product Rule (see Exercises 58–60).

■ **EXAMPLE 3** Compute the derivative of $f(x) = \dfrac{x}{1+x^2}$.

Solution Apply the Quotient Rule:

$$f'(x) = \frac{\overbrace{(1+x^2)}^{\text{Bottom}}\ \overbrace{(x)'}^{\text{Top}'} - \overbrace{(x)}^{\text{Top}}\ \overbrace{(1+x^2)'}^{\text{Bottom}'}}{(1+x^2)^2} = \frac{(1+x^2)(1) - (x)(2x)}{(1+x^2)^2}$$

$$= \frac{1+x^2-2x^2}{(1+x^2)^2} = \frac{1-x^2}{(1+x^2)^2}$$ ■

■ **EXAMPLE 4** Find the tangent line to the graph of $f(x) = \dfrac{3x^2+x-2}{4x^3+1}$ at $x = 1$.

Solution

$$f'(x) = \frac{d}{dx}\left(\frac{3x^2+x-2}{4x^3+1}\right) = \frac{\overbrace{(4x^3+1)}^{\text{Bottom}}\ \overbrace{(3x^2+x-2)'}^{\text{Top}'} - \overbrace{(3x^2+x-2)}^{\text{Top}}\ \overbrace{(4x^3+1)'}^{\text{Bottom}'}}{(4x^3+1)^2}$$

$$= \frac{(4x^3+1)(6x+1) - (3x^2+x-2)(12x^2)}{(4x^3+1)^2}$$

$$= \frac{(24x^4+4x^3+6x+1) - (36x^4+12x^3-24x^2)}{(4x^3+1)^2}$$

$$= \frac{-12x^4-8x^3+24x^2+6x+1}{(4x^3+1)^2}$$

At $x = 1$,

$$f(1) = \frac{3+1-2}{4+1} = \frac{2}{5}$$

$$f'(1) = \frac{-12-8+24+6+1}{5^2} = \frac{11}{25}$$

An equation of the tangent line at $\left(1, \frac{2}{5}\right)$ is

$$y - \frac{2}{5} = \frac{11}{25}(x-1) \qquad \text{or} \qquad y = \frac{11}{25}x - \frac{1}{25}$$ ■

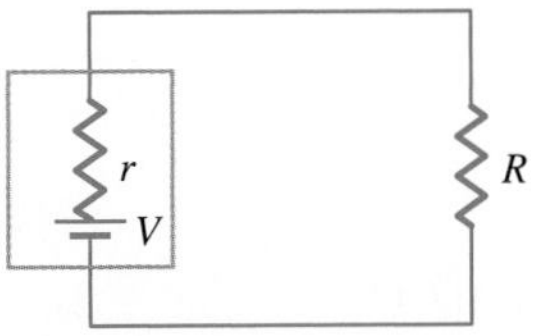

FIGURE 2 Apparatus of resistance R attached to a battery of voltage V.

■ **EXAMPLE 5 Power Delivered by a Battery** The power that a battery supplies to an apparatus such as a laptop depends on the *internal resistance* of the battery. For a battery of voltage V and internal resistance r, the total power delivered to an apparatus of resistance R (Figure 2) is

$$P = \frac{V^2 R}{(R + r)^2}$$

(a) Calculate dP/dR, assuming that V and r are constants.

(b) Where, in the graph of P versus R, is the tangent line horizontal?

Solution

(a) Because V is a constant, we obtain (using the Quotient Rule)

$$\frac{dP}{dR} = V^2 \frac{d}{dR}\left(\frac{R}{(R + r)^2}\right) = V^2 \frac{(R + r)^2 \frac{d}{dR} R - R \frac{d}{dR}(R + r)^2}{(R + r)^4} \qquad \boxed{4}$$

We have $\frac{d}{dR} R = 1$, and $\frac{d}{dR} r = 0$ because r is a constant. Thus,

$$\begin{aligned} \frac{d}{dR}(R + r)^2 &= \frac{d}{dR}(R^2 + 2rR + r^2) \\ &= \frac{d}{dR} R^2 + 2r \frac{d}{dR} R + \frac{d}{dR} r^2 \\ &= 2R + 2r + 0 = 2(R + r) \end{aligned} \qquad \boxed{5}$$

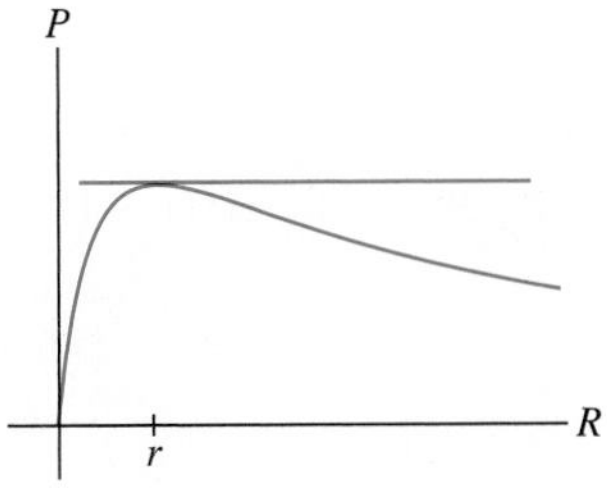

FIGURE 3 Graph of power versus resistance:

$$P = \frac{V^2 R}{(R + r)^2}$$

Using Eq. (5) in Eq. (4), we obtain

$$\frac{dP}{dR} = V^2 \frac{(R + r)^2 - 2R(R + r)}{(R + r)^4} = V^2 \frac{(R + r) - 2R}{(R + r)^3} = V^2 \frac{r - R}{(R + r)^3} \qquad \boxed{6}$$

(b) The tangent line is horizontal when the derivative is zero. We see from Eq. (6) that the derivative is zero when $r - R = 0$—that is, when $R = r$. ■

GRAPHICAL INSIGHT Figure 3 shows that the point where the tangent line is horizontal is the *maximum point* on the graph. This proves an important result in circuit design: Maximum power is delivered when the resistance of the load (apparatus) is equal to the internal resistance of the battery.

3.3 SUMMARY

- Two basic rules of differentiation:

Product Rule: $(fg)' = fg' + gf'$

Quotient Rule: $\left(\frac{f}{g}\right)' = \frac{gf' - fg'}{g^2}$

- Remember: The derivative of fg is *not* equal to $f'g'$. Similarly, the derivative of f/g is *not* equal to f'/g'.

3.3 EXERCISES

Preliminary Questions

1. Are the following statements true or false? If false, state the correct version.

(a) fg denotes the function whose value at x is $f(g(x))$.

(b) f/g denotes the function whose value at x is $f(x)/g(x)$.

(c) The derivative of the product is the product of the derivatives.

(d) $\left.\dfrac{d}{dx}(fg)\right|_{x=4} = f(4)g'(4) - g(4)f'(4)$

(e) $\left.\dfrac{d}{dx}(fg)\right|_{x=0} = f(0)g'(0) + g(0)f'(0)$

2. Find $(f/g)'(1)$ if $f(1) = f'(1) = g(1) = 2$ and $g'(1) = 4$.

3. Find $g(1)$ if $f(1) = 0$, $f'(1) = 2$, and $(fg)'(1) = 10$.

Exercises

In Exercises 1–6, use the Product Rule to calculate the derivative.

1. $f(x) = x^3(2x^2 + 1)$

2. $f(x) = (3x - 5)(2x^2 - 3)$

3. $f(x) = \sqrt{x}(1 - x^3)$

4. $f(x) = (3x^4 + 2x^6)(x - 2)$

5. $\left.\dfrac{dh}{ds}\right|_{s=4}$, $h(s) = (s^{-1/2} + 2s)(7 - s^{-1})$

6. $y = (t - 8t^{-1})(t + t^2)$

In Exercises 7–12, use the Quotient Rule to calculate the derivative.

7. $f(x) = \dfrac{x}{x-2}$

8. $f(x) = \dfrac{x+4}{x^2+x+1}$

9. $\left.\dfrac{dg}{dt}\right|_{t=-2}$, $g(t) = \dfrac{t^2+1}{t^2-1}$

10. $\left.\dfrac{dw}{dz}\right|_{z=9}$, $w = \dfrac{z^2}{\sqrt{z}+z}$

11. $g(x) = \dfrac{1}{1+x^{3/2}}$

12. $h(s) = \dfrac{s^{3/2}}{s^2+1}$

In Exercises 13–16, calculate the derivative in two ways. First use the Product or Quotient Rule; then rewrite the function algebraically and apply the Power Rule directly.

13. $f(t) = (2t + 1)(t^2 - 2)$

14. $f(x) = x^2(3 + x^{-1})$

15. $h(t) = \dfrac{t^2-1}{t-1}$

16. $g(x) = \dfrac{x^3 + 2x^2 + 3x^{-1}}{x}$

In Exercises 17–38, calculate the derivative.

17. $f(x) = (x^3 + 5)(x^3 + x + 1)$

18. $f(x) = \left(\dfrac{1}{x} - x^2\right)(x^3 + 1)$

19. $\left.\dfrac{dy}{dx}\right|_{x=3}$, $y = \dfrac{1}{x+10}$

20. $\left.\dfrac{dz}{dx}\right|_{x=-2}$, $z = \dfrac{x}{3x^2+1}$

21. $f(x) = (\sqrt{x} + 1)(\sqrt{x} - 1)$

22. $f(x) = \dfrac{9x^{5/2} - 2}{x}$

23. $\left.\dfrac{dy}{dx}\right|_{x=2}$, $y = \dfrac{x^4 - 4}{x^2 - 5}$

24. $f(x) = \dfrac{x^4 + x^{-1}}{x+1}$

25. $\left.\dfrac{dz}{dx}\right|_{x=1}$, $z = \dfrac{1}{x^3+1}$

26. $f(x) = \dfrac{3x^3 - x^2 + 2}{\sqrt{x}}$

27. $h(t) = \dfrac{t}{(t+1)(t^2+1)}$

28. $f(x) = x^{3/2}(2x^4 - 3x + x^{-1/2})$

29. $f(t) = 3^{1/2} \cdot 5^{1/2}$

30. $h(x) = \pi^2(x - 1)$

31. $f(x) = (x + 3)(x - 1)(x - 5)$

32. $h(s) = s(s + 4)(s^2 + 1)$

33. $f(x) = \dfrac{x^{3/2}(x^2+1)}{x+1}$

34. $g(z) = \dfrac{(z-2)(z^2+1)}{z}$

35. $g(z) = \left(\dfrac{z^2-4}{z-1}\right)\left(\dfrac{z^2-1}{z+2}\right)$ *Hint:* Simplify first.

36. $\dfrac{d}{dx}\Big((ax + b)(abx^2 + 1)\Big)$ (a, b constants)

37. $\dfrac{d}{dt}\left(\dfrac{xt-4}{t^2-x}\right)$ (x constant)

38. $\dfrac{d}{dx}\left(\dfrac{ax+b}{cx+d}\right)$ (a, b, c, d constants)

In Exercises 39–42, calculate the derivative using the values:

$f(4)$	$f'(4)$	$g(4)$	$g'(4)$
10	−2	5	−1

39. $(fg)'(4)$ and $(f/g)'(4)$.

40. $F'(4)$, where $F(x) = x^2 f(x)$.

41. $G'(4)$, where $G(x) = g(x)^2$.

42. $H'(4)$, where $H(x) = \dfrac{x}{g(x)f(x)}$.

43. Calculate $F'(0)$, where

$$F(x) = \frac{x^9 + x^8 + 4x^5 - 7x}{x^4 - 3x^2 + 2x + 1}$$

Hint: Do not calculate $F'(x)$. Instead, write $F(x) = f(x)/g(x)$ and express $F'(0)$ directly in terms of $f(0)$, $f'(0)$, $g(0)$, $g'(0)$.

44. Proceed as in Exercise 43 to calculate $F'(0)$, where

$$F(x) = \left(1 + x + x^{4/3} + x^{5/3}\right)\frac{3x^5 + 5x^4 + 5x + 1}{8x^9 - 7x^4 + 1}$$

45. Verify the formula $(x^3)' = 3x^2$ by writing $x^3 = x \cdot x \cdot x$ and applying the Product Rule.

46. GU Plot the derivative of $f(x) = x/(x^2 + 1)$ over $[-4, 4]$. Use the graph to determine the intervals on which $f'(x) > 0$ and $f'(x) < 0$. Then plot $f(x)$ and describe how the sign of $f'(x)$ is reflected in the graph of $f(x)$.

47. GU Plot $f(x) = x/(x^2 - 1)$ (in a suitably bounded viewing box). Use the plot to determine whether $f'(x)$ is positive or negative on its domain $\{x : x \neq \pm 1\}$. Then compute $f'(x)$ and confirm your conclusion algebraically.

48. Let $P = V^2R/(R + r)^2$ as in Example 5. Calculate dP/dr, assuming that r is variable and R is constant.

49. Find all values of a such that the tangent line to

$$f(x) = \frac{x - 1}{x + 8} \quad \text{at } x = a$$

passes through the origin (Figure 4).

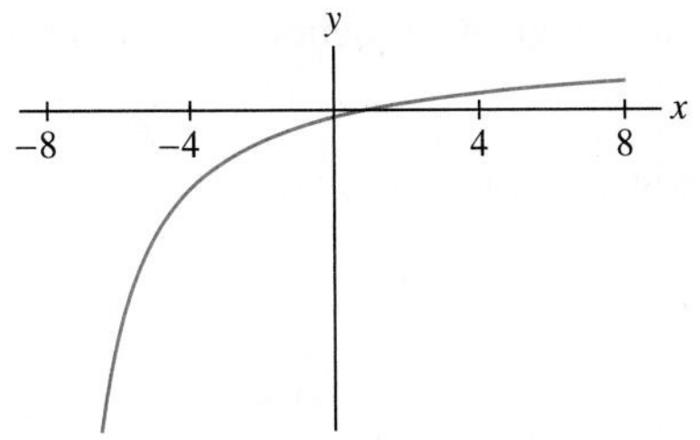

FIGURE 4

50. Current I (amperes), voltage V (volts), and resistance R (ohms) in a circuit are related by Ohm's Law, $I = V/R$.

(a) Calculate $\left.\dfrac{dI}{dR}\right|_{R=6}$ if V is constant with value $V = 24$.

(b) Calculate $\left.\dfrac{dV}{dR}\right|_{R=6}$ if I is constant with value $I = 4$.

51. The revenue per month earned by the Couture clothing chain at time t is $R(t) = N(t)S(t)$, where $N(t)$ is the number of stores and $S(t)$ is average revenue per store per month. Couture embarks on a two-part campaign: (A) to build new stores at a rate of 5 stores per month, and (B) to use advertising to increase average revenue per store at a rate of \$10,000 per month. Assume that $N(0) = 50$ and $S(0) = \$150{,}000$.

(a) Show that total revenue will increase at the rate

$$\frac{dR}{dt} = 5S(t) + 10{,}000N(t)$$

Note that the two terms in the Product Rule correspond to the separate effects of increasing the number of stores on the one hand, and the average revenue per store on the other.

(b) Calculate $\left.\dfrac{dR}{dt}\right|_{t=0}$.

(c) If Couture can implement only one leg (A or B) of its expansion at $t = 0$, which choice will grow revenue most rapidly?

52. The **tip speed ratio** of a turbine (Figure 5) is the ratio $R = T/W$, where T is the speed of the tip of a blade and W is the speed of the wind. (Engineers have found empirically that a turbine with n blades extracts maximum power from the wind when $R = 2\pi/n$.) Calculate dR/dt (t in minutes) if $W = 35$ km/h and W decreases at a rate of 4 km/h per minute, and the tip speed has constant value $T = 150$ km/h.

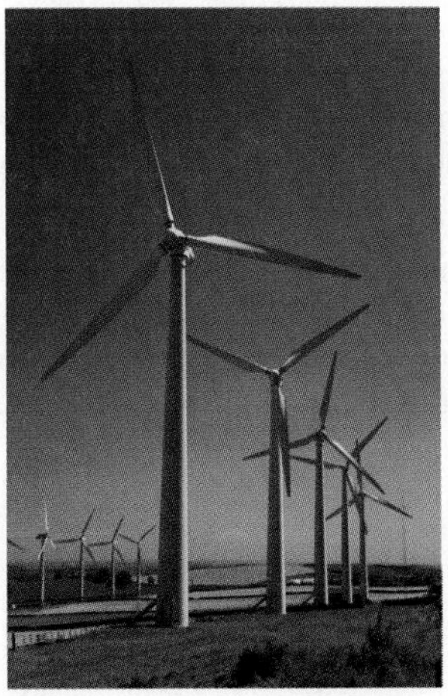

FIGURE 5 Turbines on a wind farm

53. The curve $y = 1/(x^2 + 1)$ is called the *witch of Agnesi* (Figure 6) after the Italian mathematician Maria Agnesi (1718–1799), who wrote one of the first books on calculus. This strange name is the result of a mistranslation of the Italian word *la versiera*, meaning "that which turns." Find equations of the tangent lines at $x = \pm 1$.

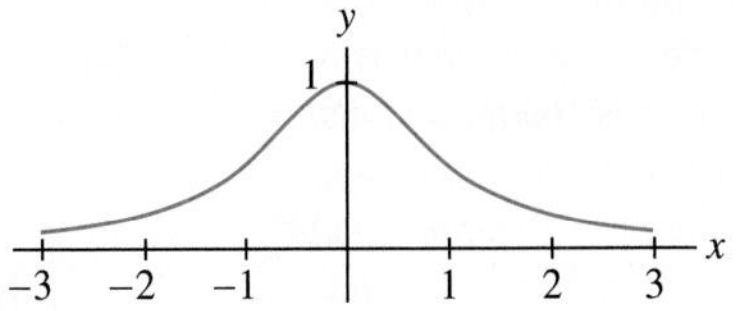

FIGURE 6 The witch of Agnesi.

54. Let $f(x) = g(x) = x$. Show that $(f/g)' \neq f'/g'$.

55. Use the Product Rule to show that $(f^2)' = 2ff'$.

56. Show that $(f^3)' = 3f^2f'$.

Further Insights and Challenges

57. Let f, g, h be differentiable functions. Show that $(fgh)'(x)$ is equal to

$$f(x)g(x)h'(x) + f(x)g'(x)h(x) + f'(x)g(x)h(x)$$

Hint: Write fgh as $f(gh)$.

58. Prove the Quotient Rule using the limit definition of the derivative.

59. Derivative of the Reciprocal Use the limit definition to prove

$$\frac{d}{dx}\left(\frac{1}{f(x)}\right) = -\frac{f'(x)}{f^2(x)} \tag{7}$$

Hint: Show that the difference quotient for $1/f(x)$ is equal to

$$\frac{f(x) - f(x+h)}{hf(x)f(x+h)}$$

60. Prove the Quotient Rule using Eq. (7) and the Product Rule.

61. Use the limit definition of the derivative to prove the following special case of the Product Rule:

$$\frac{d}{dx}(xf(x)) = xf'(x) + f(x)$$

62. Carry out Maria Agnesi's proof of the Quotient Rule from her book on calculus, published in 1748: Assume that f, g, and $h = f/g$ are differentiable. Compute the derivative of $hg = f$ using the Product Rule, and solve for h'.

63. The Power Rule Revisited If you are familiar with *proof by induction*, use induction to prove the Power Rule for all whole numbers n. Show that the Power Rule holds for $n = 1$; then write x^n as $x \cdot x^{n-1}$ and use the Product Rule.

*Exercises 64 and 65: A basic fact of algebra states that c is a root of a polynomial $f(x)$ if and only if $f(x) = (x - c)g(x)$ for some polynomial $g(x)$. We say that c is a **multiple root** if $f(x) = (x - c)^2h(x)$, where $h(x)$ is a polynomial.*

64. Show that c is a multiple root of $f(x)$ if and only if c is a root of both $f(x)$ and $f'(x)$.

65. Use Exercise 64 to determine whether $c = -1$ is a multiple root:

(a) $x^5 + 2x^4 - 4x^3 - 8x^2 - x + 2$

(b) $x^4 + x^3 - 5x^2 - 3x + 2$

66. Figure 7 is the graph of a polynomial with roots at A, B, and C. Which of these is a multiple root? Explain your reasoning using Exercise 64.

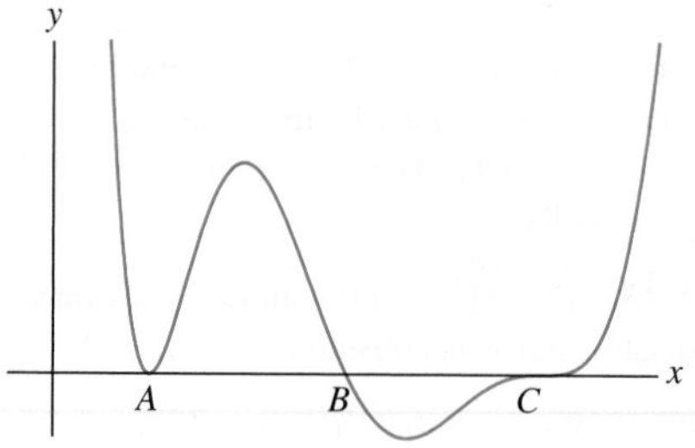

FIGURE 7

3.4 Rates of Change

Recall the notation for the average rate of change of a function $y = f(x)$ over an interval $[x_0, x_1]$:

$$\Delta y = \text{change in } y = f(x_1) - f(x_0)$$

$$\Delta x = \text{change in } x = x_1 - x_0$$

$$\text{Average Rate of Change} = \frac{\Delta y}{\Delta x} = \frac{f(x_1) - f(x_0)}{x_1 - x_0}$$

We usually omit the word "instantaneous" and refer to the derivative simply as the rate of change. This is shorter and also more accurate when applied to general rates, because the term "instantaneous" would seem to refer only to rates with respect to time.

In our prior discussion in Section 2.1, limits and derivatives had not yet been introduced. Now that we have them at our disposal, we can define the **instantaneous** rate of change of y with respect to x at $x = x_0$:

$$\text{Instantaneous Rate of Change} = f'(x_0) = \lim_{\Delta x \to 0} \frac{\Delta y}{\Delta x} = \lim_{x_1 \to x_0} \frac{f(x_1) - f(x_0)}{x_1 - x_0}$$

Keep in mind the geometric interpretations: The average rate of change is the slope of the secant line (Figure 1), and the instantaneous rate of change is the slope of the tangent line (Figure 2).

Leibniz notation dy/dx is particularly convenient because it specifies that we are considering the rate of change of y with respect to the independent variable x. The rate

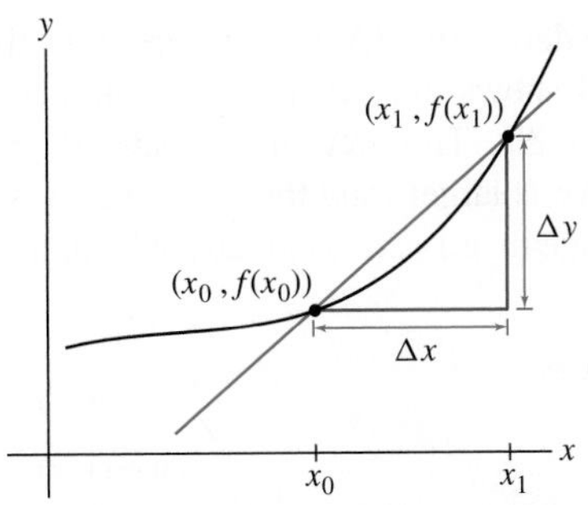

FIGURE 1 The average rate of change over $[x_0, x_1]$ is the slope of the secant line.

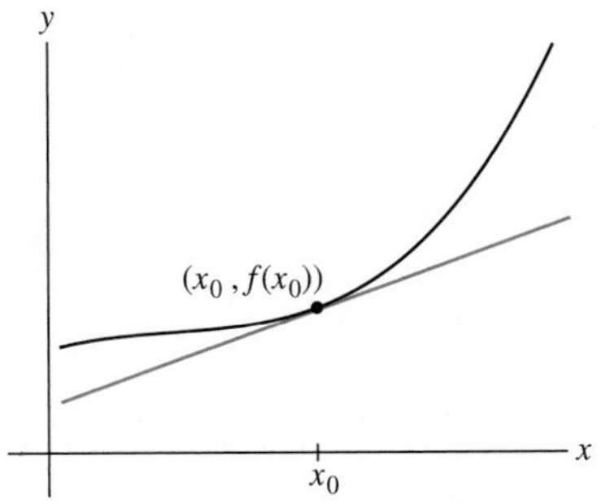

FIGURE 2 The instantaneous rate of change at x_0 is the slope of the tangent line.

dy/dx is measured in units of y per unit of x. For example, the rate of change of temperature with respect to time has units such as degrees per minute, whereas the rate of change of temperature with respect to altitude has units such as degrees per kilometer.

TABLE 1 Data from Mars Pathfinder Mission, July 1997

Time	Temperature (°C)
5:42	−74.7
6:11	−71.6
6:40	−67.2
7:09	−63.7
7:38	−59.5
8:07	−53
8:36	−47.7
9:05	−44.3
9:34	−42

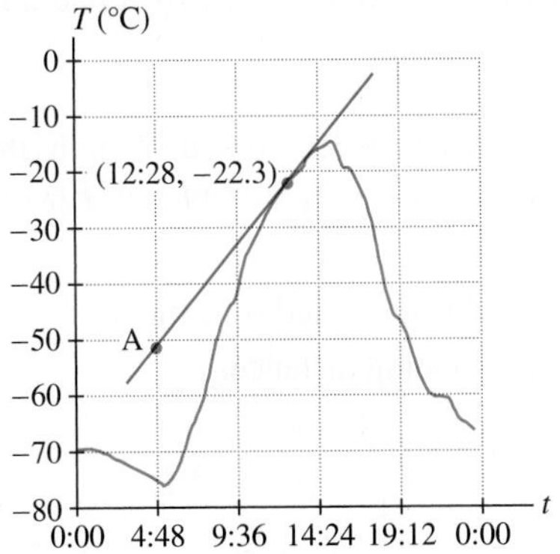

FIGURE 3 Temperature variation on the surface of Mars on July 6, 1997.

■ **EXAMPLE 1** Table 1 contains data on the temperature T on the surface of Mars at Martian time t, collected by the NASA Pathfinder space probe.

(a) Calculate the average rate of change of temperature T from 6:11 AM to 9:05 AM.

(b) Use Figure 3 to estimate the rate of change at $t = $ 12:28 PM.

Solution

(a) The time interval [6:11, 9:05] has length 2 h, 54 min, or $\Delta t = 2.9$ h. According to Table 1, the change in temperature over this time interval is

$$\Delta T = -44.3 - (-71.6) = 27.3°\text{C}$$

The average rate of change is the ratio

$$\frac{\Delta T}{\Delta t} = \frac{27.3}{2.9} \approx 9.4°\text{C/h}$$

(b) The rate of change is the derivative dT/dt, which is equal to the slope of the tangent line through the point (12:28, −22.3) in Figure 3. To estimate the slope, we must choose a second point on the tangent line. Let's use the point labeled A, whose coordinates are approximately (4:48, −51). The time interval from 4:48 AM to 12:28 PM has length 7 h, 40 min, or $\Delta t = 7.67$ h, and

$$\frac{dT}{dt} = \text{slope of tangent line} \approx \frac{-22.3 - (-51)}{7.67} \approx 3.7°\text{C/h}$$ ■

■ **EXAMPLE 2** Let $A = \pi r^2$ be the area of a circle of radius r.

(a) Compute dA/dr at $r = 2$ and $r = 5$.

(b) Why is dA/dr larger at $r = 5$?

Solution The rate of change of area with respect to radius is the derivative

$$\frac{dA}{dr} = \frac{d}{dr}(\pi r^2) = 2\pi r \qquad \boxed{1}$$

By Eq. (1), dA/dr is equal to the circumference $2\pi r$. We can explain this intuitively as follows: Up to a small error, the area ΔA of the band of width Δr in Figure 4 is equal to the circumference $2\pi r$ times the width Δr. Therefore, $\Delta A \approx 2\pi r \Delta r$ and

$$\frac{dA}{dr} = \lim_{\Delta r \to 0} \frac{\Delta A}{\Delta r} = 2\pi r$$

(a) We have

$$\left.\frac{dA}{dr}\right|_{r=2} = 2\pi(2) \approx 12.57 \quad \text{and} \quad \left.\frac{dA}{dr}\right|_{r=5} = 2\pi(5) \approx 31.42$$

(b) The derivative dA/dr measures how the area of the circle changes when r increases. Figure 4 shows that when the radius increases by Δr, the area increases by a band of thickness Δr. The area of the band is greater at $r = 5$ than at $r = 2$. Therefore, the derivative is larger (and the tangent line is steeper) at $r = 5$. In general, for a fixed Δr, the change in area ΔA is greater when r is larger. ■

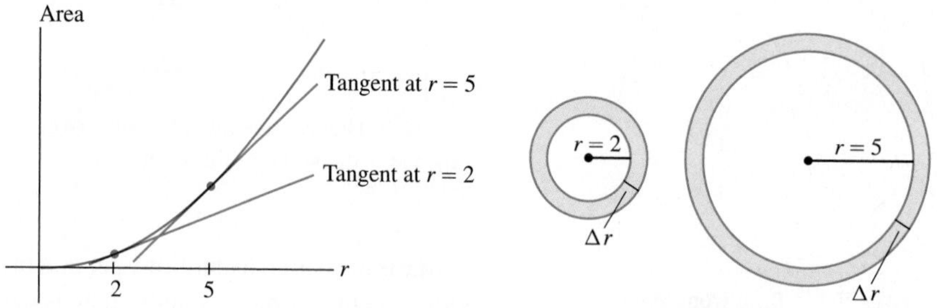

FIGURE 4 The pink bands represent the change in area when r is increased by Δr.

The Effect of a One-Unit Change

For small values of h, the difference quotient is close to the derivative itself:

$$f'(x_0) \approx \frac{f(x_0 + h) - f(x_0)}{h} \qquad \boxed{2}$$

This approximation generally improves as h gets smaller, but in some applications, the approximation is already useful with $h = 1$. Setting $h = 1$ in Eq. (2) gives

$$\boxed{f'(x_0) \approx f(x_0 + 1) - f(x_0)} \qquad \boxed{3}$$

In other words, *$f'(x_0)$ is approximately equal to the change in f caused by a one-unit change in x when $x = x_0$.*

■ **EXAMPLE 3** **Stopping Distance** For speeds s between 30 and 75 mph, the stopping distance of an automobile after the brakes are applied is approximately $F(s) = 1.1s + 0.05s^2$ ft. For $s = 60$ mph:

(a) Estimate the change in stopping distance if the speed is increased by 1 mph.

(b) Compare your estimate with the actual increase in stopping distance.

Solution

(a) We have

$$F'(s) = \frac{d}{ds}(1.1s + 0.05s^2) = 1.1 + 0.1s \text{ ft/mph}$$

$$F'(60) = 1.1 + 6 = 7.1 \text{ ft/mph}$$

Using Eq. (3), we estimate

$$\underbrace{F(61) - F(60)}_{\text{Change in stopping distance}} \approx F'(60) = 7.1 \text{ ft}$$

Thus, when you increase your speed from 60 to 61 mph, your stopping distance increases by roughly 7 ft.

(b) The actual change in stopping distance is $F(61) - F(60) = 253.15 - 246 = 7.15$, so the estimate in (a) is fairly accurate. ■

Marginal Cost in Economics

Although $C(x)$ is meaningful only when x is a whole number, economists often treat $C(x)$ as a differentiable function of x so that the techniques of calculus can be applied.

Let $C(x)$ denote the dollar cost (including labor and parts) of producing x units of a particular product. The number x of units manufactured is called the **production level**. To study the relation between costs and production, economists define the **marginal cost** at production level x_0 as the cost of producing one additional unit:

$$\text{Marginal cost} = C(x_0 + 1) - C(x_0)$$

In this setting, Eq. (3) usually gives a good approximation, so we take $C'(x_0)$ as an estimate of the marginal cost.

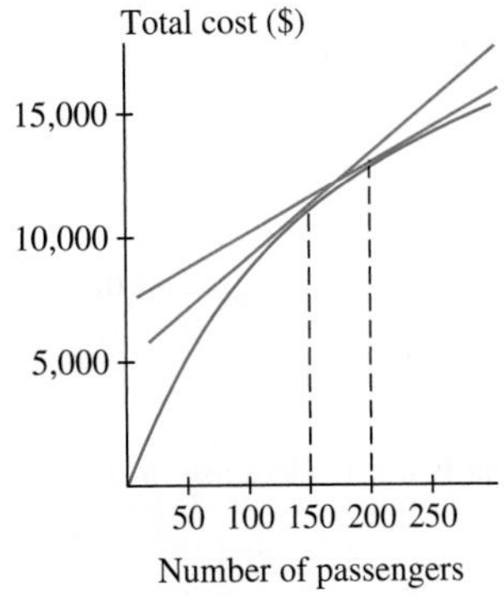

FIGURE 5 Cost of an air flight. The slopes of the tangent lines are decreasing, so marginal cost is decreasing.

■ **EXAMPLE 4 Cost of an Air Flight** Company data suggest that the total dollar cost of a certain flight is approximately $C(x) = 0.0005x^3 - 0.38x^2 + 120x$, where x is the number of passengers (Figure 5).

(a) Estimate the marginal cost of an additional passenger if the flight already has 150 passengers.
(b) Compare your estimate with the actual cost of an additional passenger.
(c) Is it more expensive to add a passenger when $x = 150$ or when $x = 200$?

Solution The derivative is $C'(x) = 0.0015x^2 - 0.76x + 120$.

(a) We estimate the marginal cost at $x = 150$ by the derivative

$$C'(150) = 0.0015(150)^2 - 0.76(150) + 120 = 39.75$$

Thus, it costs approximately $39.75 to add one additional passenger.
(b) The actual cost of adding one additional passenger is

$$C(151) - C(150) \approx 11{,}177.10 - 11{,}137.50 = 39.60$$

Our estimate of $39.75 is close enough for practical purposes.
(c) The marginal cost at $x = 200$ is approximately

$$C'(200) = 0.0015(200)^2 - 0.76(200) + 120 = 28$$

Since $39.75 > 28$, it is more expensive to add a passenger when $x = 150$. ■

Linear Motion

In his famous textbook Lectures on Physics, *Nobel laureate Richard Feynman (1918–1988) uses a dialogue to make a point about instantaneous velocity:*

Policeman: "My friend, you were going 75 miles an hour."

Driver: "That's impossible, sir, I was traveling for only seven minutes."

Recall that *linear motion* is motion along a straight line. This includes horizontal motion along a straight highway and vertical motion of a falling object. Let $s(t)$ denote the position or distance from the origin at time t. Velocity is the rate of change of position with respect to time:

$$v(t) = \text{velocity} = \frac{ds}{dt}$$

The *sign* of $v(t)$ indicates the direction of motion. For example, if $s(t)$ is the height above ground, then $v(t) > 0$ indicates that the object is rising. **Speed** is defined as the absolute value of velocity $|v(t)|$.

Figure 6 shows the position of a car as a function of time. Remember that the height of the graph represents the car's distance from the point of origin. The slope of the tangent line is the velocity. Here are some facts we can glean from the graph:

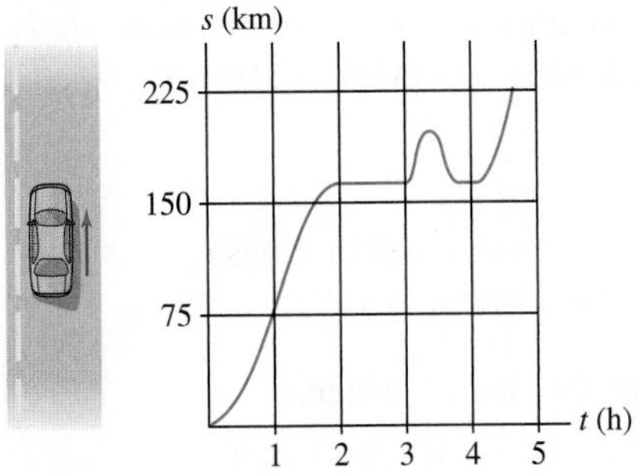

FIGURE 6 Graph of distance versus time.

- **Speeding up or slowing down?** The tangent lines get steeper in the interval [0, 1], so *the car was speeding up during the first hour*. They get flatter in the interval [1, 2], so the car slowed down in the second hour.

- **Standing still** The graph is horizontal over [2, 3] (perhaps the driver stopped at a restaurant for an hour).
- **Returning to the same spot** The graph rises and falls in the interval [3, 4], indicating that the driver returned to the restaurant (perhaps she left her cell phone there).
- **Average velocity** The graph rises more over [0, 2] than over [3, 5], so the average velocity was greater over the first two hours than over the last two hours.

■ **EXAMPLE 5** A truck enters the off-ramp of a highway at $t = 0$. Its position after t seconds is $s(t) = 25t - 0.3t^3$ m for $0 \le t \le 5$.

(a) How fast is the truck going at the moment it enters the off-ramp?

(b) Is the truck speeding up or slowing down?

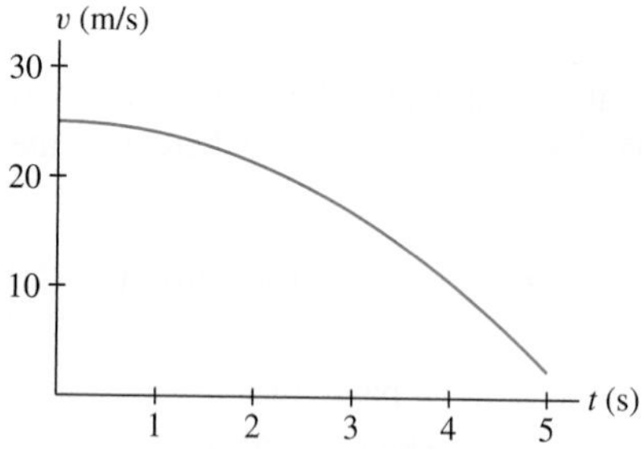

FIGURE 7 Graph of velocity $v(t) = 25 - 0.9t^2$.

Solution The truck's velocity at time t is $v(t) = \frac{d}{dt}(25t - 0.3t^3) = 25 - 0.9t^2$.

(a) The truck enters the off-ramp with velocity $v(0) = 25$ m/s.

(b) Since $v(t) = 25 - 0.9t^2$ is decreasing (Figure 7), the truck is slowing down. ■

Motion Under the Influence of Gravity

Galileo discovered that the height $s(t)$ and velocity $v(t)$ at time t (seconds) of an object tossed vertically in the air near the earth's surface are given by the formulas

Galileo's formulas are valid only when air resistance is negligible. We assume this to be the case in all examples.

$$s(t) = s_0 + v_0 t - \frac{1}{2}gt^2, \qquad v(t) = \frac{ds}{dt} = v_0 - gt \tag{4}$$

The constants s_0 and v_0 are the *initial values*:

- $s_0 = s(0)$, the position at time $t = 0$.
- $v_0 = v(0)$, the velocity at $t = 0$.
- $-g$ is the acceleration due to gravity on the surface of the earth (negative because the up direction is positive), where

$$g \approx 9.8 \text{ m/s}^2 \qquad \text{or} \qquad g \approx 32 \text{ ft/s}^2$$

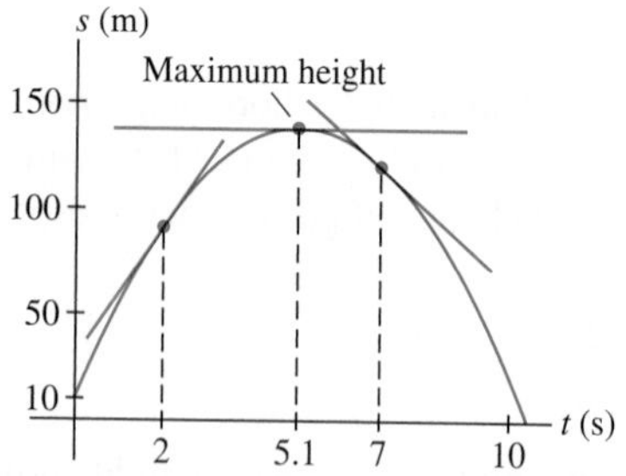

FIGURE 8 Maximum height occurs when $s'(t) = v(t) = 0$, where the tangent line is horizontal.

A simple observation enables us to find the object's maximum height. Since velocity is positive as the object rises and negative as it falls back to earth, the object reaches its maximum height at the moment of transition, when it is no longer rising and has not yet begun to fall. At that moment, its velocity is zero. In other words, *the maximum height is attained when* $v(t) = 0$. At this moment, the tangent line to the graph of $s(t)$ is horizontal (Figure 8).

■ **EXAMPLE 6 Finding the Maximum Height** A stone is shot with a slingshot vertically upward with an initial velocity of 50 m/s from an initial height of 10 m.

(a) Find the velocity at $t = 2$ and at $t = 7$. Explain the change in sign.

(b) What is the stone's maximum height and when does it reach that height?

Galileo's formulas:

$$s(t) = s_0 + v_0 t - \frac{1}{2}gt^2$$

$$v(t) = \frac{ds}{dt} = v_0 - gt$$

Solution Apply Eq. (4) with $s_0 = 10$, $v_0 = 50$, and $g = 9.8$:

$$s(t) = 10 + 50t - 4.9t^2, \qquad v(t) = 50 - 9.8t$$

(a) Therefore,

$$v(2) = 50 - 9.8(2) = 30.4 \text{ m/s}$$
$$v(7) = 50 - 9.8(7) = -18.6 \text{ m/s}$$

At $t = 2$, the stone is rising and its velocity $v(2)$ is positive (Figure 8). At $t = 7$, the stone is already on the way down and its velocity $v(7)$ is negative.

(b) Maximum height is attained when the velocity is zero, so we solve

$$v(t) = 50 - 9.8t = 0 \quad \Rightarrow \quad t = \frac{50}{9.8} \approx 5.1$$

The stone reaches maximum height at $t = 5.1$ s. Its maximum height is

$$s(5.1) = 10 + 50(5.1) - 4.9(5.1)^2 \approx 137.6 \text{ m}$$ ■

In the previous example, we specified the initial values of position and velocity. In the next example, the goal is to determine initial velocity.

How important are units? In September 1999, the $125 million Mars Climate Orbiter *spacecraft burned up in the Martian atmosphere before completing its scientific mission. According to Arthur Stephenson, NASA chairman of the Mars* Climate Orbiter *Mission Failure Investigation Board, 1999, "The 'root cause' of the loss of the spacecraft was the failed translation of English units into metric units in a segment of ground-based, navigation-related mission software."*

■ **EXAMPLE 7 Finding Initial Conditions** What initial velocity v_0 is required for a bullet, fired vertically from ground level, to reach a maximum height of 2 km?

Solution We need a formula for maximum height as a function of initial velocity v_0. The initial height is $s_0 = 0$, so the bullet's height is $s(t) = v_0 t - \frac{1}{2}gt^2$ by Galileo's formula. Maximum height is attained when the velocity is zero:

$$v(t) = v_0 - gt = 0 \quad \Rightarrow \quad t = \frac{v_0}{g}$$

The maximum height is the value of $s(t)$ at $t = v_0/g$:

$$s\left(\frac{v_0}{g}\right) = v_0\left(\frac{v_0}{g}\right) - \frac{1}{2}g\left(\frac{v_0}{g}\right)^2$$
$$= \frac{v_0^2}{g} - \frac{1}{2}\frac{v_0^2}{g} = \frac{v_0^2}{2g}$$

Now we can solve for v_0 using the value $g = 9.8$ m/s^2 (note that 2 km = 2000 m).

In Eq. (5), distance must be in meters because our value of g has units of m/s^2.

$$\text{Maximum height} = \frac{v_0^2}{2g} = \frac{v_0^2}{2(9.8)} = 2000 \text{ m} \qquad \boxed{5}$$

This yields $v_0 = \sqrt{(2)(9.8)2000} \approx 198$ m/s. In reality, the initial velocity would have to be considerably greater to overcome air resistance. ■

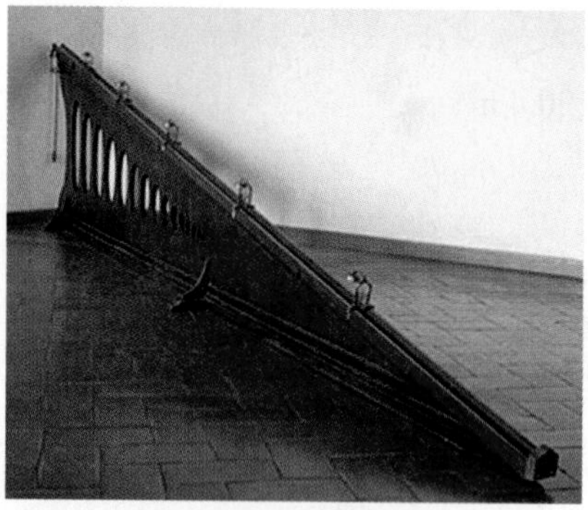

FIGURE 9 Apparatus of the type used by Galileo to study the motion of falling objects.

HISTORICAL PERSPECTIVE

Galileo Galilei (1564–1642) discovered the laws of motion for falling objects on the earth's surface around 1600. This paved the way for Newton's general laws of motion. How did Galileo arrive at his formulas? The motion of a falling object is too rapid to measure directly, without modern photographic or electronic apparatus. To get around this difficulty, Galileo experimented with balls rolling down an incline (Figure 9). For a sufficiently flat incline, he was able to measure the motion with a water clock and found that the velocity of the rolling ball is proportional to time. He then reasoned that motion in free-fall is just a faster version of motion down an incline and deduced the formula $v(t) = -gt$ for falling objects (assuming zero initial velocity).

Prior to Galileo, it had been assumed incorrectly that heavy objects fall more rapidly than lighter ones. Galileo realized that this was not true (as long as air resistance is negligible), and indeed, the formula $v(t) = -gt$ shows that the velocity depends on time but not on the weight of the object. Interestingly, 300 years later, another great physicist, Albert Einstein, was deeply puzzled by Galileo's discovery that all objects fall at the same rate regardless of their weight. He called this the Principle of Equivalence and sought to understand why it was true. In 1916, after a decade of intensive work, Einstein developed the General Theory of Relativity, which finally gave a full explanation of the Principle of Equivalence in terms of the geometry of space and time.

3.4 SUMMARY

- The (instantaneous) rate of change of $y = f(x)$ with respect to x at $x = x_0$ is defined as the derivative

$$f'(x_0) = \lim_{\Delta x \to 0} \frac{\Delta y}{\Delta x} = \lim_{x_1 \to x_0} \frac{f(x_1) - f(x_0)}{x_1 - x_0}$$

- The rate dy/dx is measured in *units of y per unit of x*.
- For linear motion, velocity $v(t)$ is the rate of change of position $s(t)$ with respect to time—that is, $v(t) = s'(t)$.
- In some applications, $f'(x_0)$ provides a good estimate of the change in f due to a one-unit increase in x when $x = x_0$:

$$f'(x_0) \approx f(x_0 + 1) - f(x_0)$$

- Marginal cost is the cost of producing one additional unit. If $C(x)$ is the cost of producing x units, then the marginal cost at production level x_0 is $C(x_0 + 1) - C(x_0)$. The derivative $C'(x_0)$ is often a good estimate for marginal cost.
- Galileo's formulas for an object rising or falling under the influence of gravity near the earth's surface (s_0 = initial position, v_0 = initial velocity):

$$s(t) = s_0 + v_0 t - \frac{1}{2}gt^2, \qquad v(t) = v_0 - gt$$

where $g \approx 9.8$ m/s^2, or $g \approx 32$ ft/s^2. Maximum height is attained when $v(t) = 0$.

3.4 EXERCISES

Preliminary Questions

1. Which units might be used for each rate of change?

(a) Pressure (in atmospheres) in a water tank with respect to depth

(b) The rate of a chemical reaction (change in concentration with respect to time with concentration in moles per liter)

2. Two trains travel from New Orleans to Memphis in 4 hours. The first train travels at a constant velocity of 90 mph, but the velocity of the second train varies. What was the second train's average velocity during the trip?

3. Estimate $f(26)$, assuming that $f(25) = 43$, $f'(25) = 0.75$.

4. The population $P(t)$ of Freedonia in 2009 was $P(2009) = 5$ million.

(a) What is the meaning of $P'(2009)$?

(b) Estimate $P(2010)$ if $P'(2009) = 0.2$.

Exercises

In Exercises 1–8, find the rate of change.

1. Area of a square with respect to its side s when $s = 5$.

2. Volume of a cube with respect to its side s when $s = 5$.

3. Cube root $\sqrt[3]{x}$ with respect to x when $x = 1, 8, 27$.

4. The reciprocal $1/x$ with respect to x when $x = 1, 2, 3$.

5. The diameter of a circle with respect to radius.

6. Surface area A of a sphere with respect to radius r ($A = 4\pi r^2$).

7. Volume V of a cylinder with respect to radius if the height is equal to the radius.

8. Speed of sound v (in m/s) with respect to air temperature T (in kelvins), where $v = 20\sqrt{T}$.

In Exercises 9–11, refer to Figure 10, the graph of distance $s(t)$ from the origin as a function of time for a car trip.

9. Find the average velocity over each interval.

(a) $[0, 0.5]$ **(b)** $[0.5, 1]$ **(c)** $[1, 1.5]$ **(d)** $[1, 2]$

10. At what time is velocity at a maximum?

11. Match the descriptions (i)–(iii) with the intervals (a)–(c).

(i) Velocity increasing

(ii) Velocity decreasing

(iii) Velocity negative

(a) $[0, 0.5]$ **(b)** $[2.5, 3]$ **(c)** $[1.5, 2]$

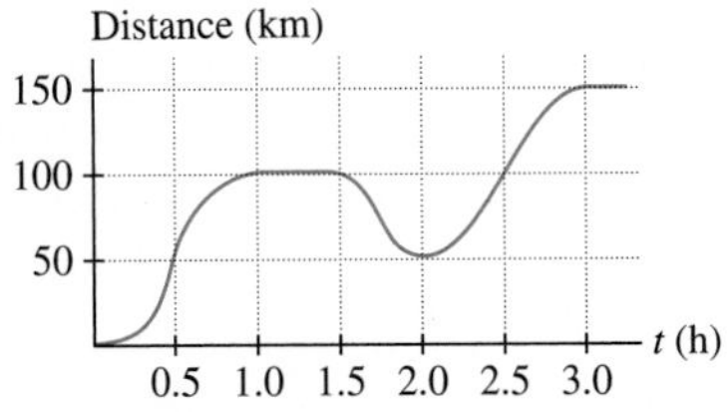

FIGURE 10 Distance from the origin versus time for a car trip.

12. Use the data from Table 1 in Example 1 to calculate the average rate of change of Martian temperature T with respect to time t over the interval from 8:36 AM to 9:34 AM.

13. Use Figure 3 from Example 1 to estimate the instantaneous rate of change of Martian temperature with respect to time (in degrees Celsius per hour) at $t = 4$ AM.

14. The temperature (in °C) of an object at time t (in minutes) is $T(t) = \frac{3}{8}t^2 - 15t + 180$ for $0 \le t \le 20$. At what rate is the object cooling at $t = 10$? (Give correct units.)

15. The velocity (in cm/s) of blood molecules flowing through a capillary of radius 0.008 cm is $v = 6.4 \times 10^{-8} - 0.001r^2$, where r is the distance from the molecule to the center of the capillary. Find the rate of change of velocity with respect to r when $r = 0.004$ cm.

16. Figure 11 displays the voltage V across a capacitor as a function of time while the capacitor is being charged. Estimate the rate of change of voltage at $t = 20$ s. Indicate the values in your calculation and include proper units. Does voltage change more quickly or more slowly as time goes on? Explain in terms of tangent lines.

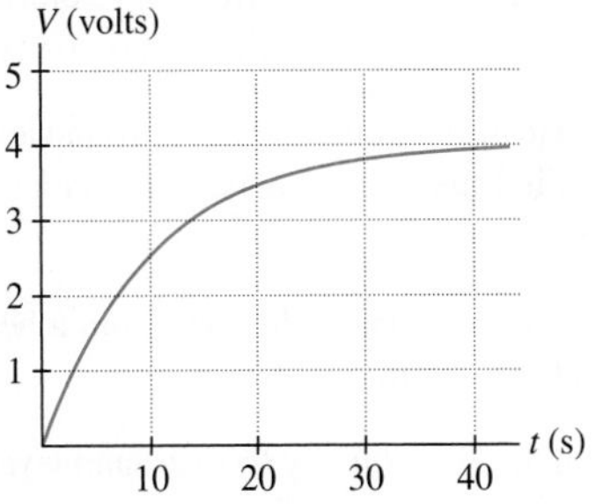

FIGURE 11

17. Use Figure 12 to estimate dT/dh at $h = 30$ and 70, where T is atmospheric temperature (in degrees Celsius) and h is altitude (in kilometers). Where is dT/dh equal to zero?

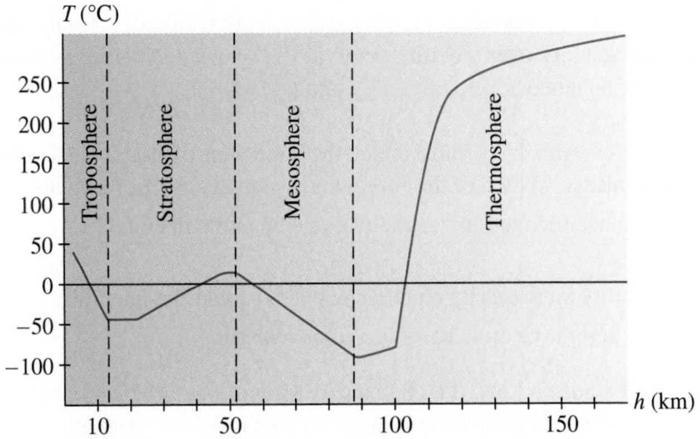

FIGURE 12 Atmospheric temperature versus altitude.

18. The earth exerts a gravitational force of $F(r) = (2.99 \times 10^{16})/r^2$ newtons on an object with a mass of 75 kg located r meters from the center of the earth. Find the rate of change of force with respect to distance r at the surface of the earth.

19. Calculate the rate of change of escape velocity $v_{esc} = (2.82 \times 10^7)r^{-1/2}$ m/s with respect to distance r from the center of the earth.

20. The power delivered by a battery to an apparatus of resistance R (in ohms) is $P = 2.25R/(R + 0.5)^2$ watts. Find the rate of change of power with respect to resistance for $R = 3\ \Omega$ and $R = 5\ \Omega$.

21. The position of a particle moving in a straight line during a 5-s trip is $s(t) = t^2 - t + 10$ cm. Find a time t at which the instantaneous velocity is equal to the average velocity for the entire trip.

22. The height (in meters) of a helicopter at time t (in minutes) is $s(t) = 600t - 3t^3$ for $0 \le t \le 12$.

(a) Plot $s(t)$ and velocity $v(t)$.

(b) Find the velocity at $t = 8$ and $t = 10$.

(c) Find the maximum height of the helicopter.

23. A particle moving along a line has position $s(t) = t^4 - 18t^2$ m at time t seconds. At which times does the particle pass through the origin? At which times is the particle instantaneously motionless (that is, it has zero velocity)?

24. GU Plot the position of the particle in Exercise 23. What is the farthest distance to the left of the origin attained by the particle?

25. A bullet is fired in the air vertically from ground level with an initial velocity 200 m/s. Find the bullet's maximum velocity and maximum height.

26. Find the velocity of an object dropped from a height of 300 m at the moment it hits the ground.

27. A ball tossed in the air vertically from ground level returns to earth 4 s later. Find the initial velocity and maximum height of the ball.

28. Olivia is gazing out a window from the tenth floor of a building when a bucket (dropped by a window washer) passes by. She notes that it hits the ground 1.5 s later. Determine the floor from which the bucket was dropped if each floor is 5 m high and the window is in the middle of the tenth floor. Neglect air friction.

29. Show that for an object falling according to Galileo's formula, the average velocity over any time interval $[t_1, t_2]$ is equal to the average of the instantaneous velocities at t_1 and t_2.

30. An object falls under the influence of gravity near the earth's surface. Which of the following statements is true? Explain.

(a) Distance traveled increases by equal amounts in equal time intervals.

(b) Velocity increases by equal amounts in equal time intervals.

(c) The derivative of velocity increases with time.

31. By Faraday's Law, if a conducting wire of length ℓ meters moves at velocity v m/s perpendicular to a magnetic field of strength B (in teslas), a voltage of size $V = -B\ell v$ is induced in the wire. Assume that $B = 2$ and $\ell = 0.5$.

(a) Calculate dV/dv.

(b) Find the rate of change of V with respect to time t if $v = 4t + 9$.

32. The voltage V, current I, and resistance R in a circuit are related by Ohm's Law: $V = IR$, where the units are volts, amperes, and ohms. Assume that voltage is constant with $V = 12$ volts. Calculate (specifying units):

(a) The average rate of change of I with respect to R for the interval from $R = 8$ to $R = 8.1$

(b) The rate of change of I with respect to R when $R = 8$

(c) The rate of change of R with respect to I when $I = 1.5$

33. Ethan finds that with h hours of tutoring, he is able to answer correctly $S(h)$ percent of the problems on a math exam. Which would you expect to be larger: $S'(3)$ or $S'(30)$? Explain.

34. Suppose $\theta(t)$ measures the angle between a clock's minute and hour hands. What is $\theta'(t)$ at 3 o'clock?

35. To determine drug dosages, doctors estimate a person's body surface area (BSA) (in meters squared) using the formula $\text{BSA} = \sqrt{hm}/60$, where h is the height in centimeters and m the mass in kilograms. Calculate the rate of change of BSA with respect to mass for a person of constant height $h = 180$. What is this rate at $m = 70$ and $m = 80$? Express your result in the correct units. Does BSA increase more rapidly with respect to mass at lower or higher body mass?

36. The atmospheric CO_2 level $A(t)$ at Mauna Loa, Hawaii at time t (in parts per million by volume) is recorded by the Scripps Institution of Oceanography. The values for the months January–December 2007 were

382.45,	383.68,	384.23,	386.26,	386.39,	385.87,
384.39,	381.78,	380.73,	380.81,	382.33,	383.69

(a) Assuming that the measurements were made on the first of each month, estimate $A'(t)$ on the 15th of the months January–November.

(b) In which months did $A'(t)$ take on its largest and smallest values?

(c) In which month was the CO_2 level most nearly constant?

37. The tangent lines to the graph of $f(x) = x^2$ grow steeper as x increases. At what rate do the slopes of the tangent lines increase?

38. Figure 13 shows the height y of a mass oscillating at the end of a spring. through one cycle of the oscillation. Sketch the graph of velocity as a function of time.

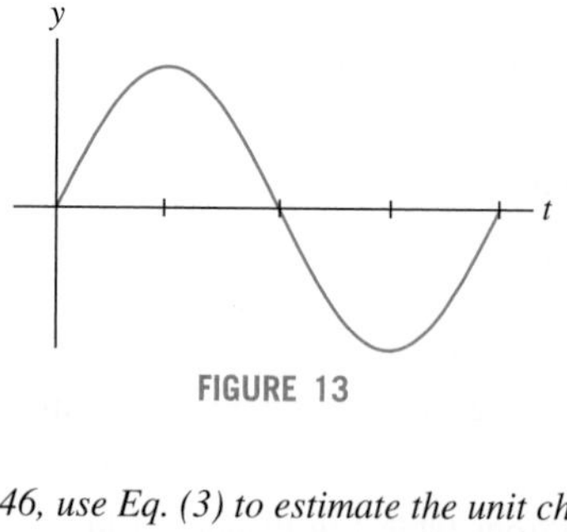

FIGURE 13

In Exercises 39–46, use Eq. (3) to estimate the unit change.

39. Estimate $\sqrt{2} - \sqrt{1}$ and $\sqrt{101} - \sqrt{100}$. Compare your estimates with the actual values.

40. Estimate $f(4) - f(3)$ if $f'(x) = 2^{-x}$. Then estimate $f(4)$, assuming that $f(3) = 12$.

41. Let $F(s) = 1.1s + 0.05s^2$ be the stopping distance as in Example 3. Calculate $F(65)$ and estimate the increase in stopping distance if speed is increased from 65 to 66 mph. Compare your estimate with the actual increase.

42. According to Kleiber's Law, the metabolic rate P (in kilocalories per day) and body mass m (in kilograms) of an animal are related by a *three-quarter-power law* $P = 73.3m^{3/4}$. Estimate the increase in metabolic rate when body mass increases from 60 to 61 kg.

43. The dollar cost of producing x bagels is $C(x) = 300 + 0.25x - 0.5(x/1000)^3$. Determine the cost of producing 2000 bagels and estimate the cost of the 2001st bagel. Compare your estimate with the actual cost of the 2001st bagel.

44. Suppose the dollar cost of producing x video cameras is $C(x) = 500x - 0.003x^2 + 10^{-8}x^3$.

(a) Estimate the marginal cost at production level $x = 5000$ and compare it with the actual cost $C(5001) - C(5000)$.

(b) Compare the marginal cost at $x = 5000$ with the average cost per camera, defined as $C(x)/x$.

45. Demand for a commodity generally decreases as the price is raised. Suppose that the demand for oil (per capita per year) is $D(p) = 900/p$ barrels, where p is the dollar price per barrel. Find the demand when $p = \$40$. Estimate the decrease in demand if p rises to \$41 and the increase if p declines to \$39.

46. The reproduction rate f of the fruit fly *Drosophila melanogaster*, grown in bottles in a laboratory, decreases with the number p of flies in the bottle. A researcher has found the number of offspring per female per day to be approximately $f(p) = (34 - 0.612p)p^{-0.658}$.

(a) Calculate $f(15)$ and $f'(15)$.

(b) Estimate the decrease in daily offspring per female when p is increased from 15 to 16. Is this estimate larger or smaller than the actual value $f(16) - f(15)$?

(c) GU Plot $f(p)$ for $5 \le p \le 25$ and verify that $f(p)$ is a decreasing function of p. Do you expect $f'(p)$ to be positive or negative? Plot $f'(p)$ and confirm your expectation.

47. According to Stevens' Law in psychology, the perceived magnitude of a stimulus is proportional (approximately) to a power of the actual intensity I of the stimulus. Experiments show that the *perceived brightness* B of a light satisfies $B = kI^{2/3}$, where I is the light intensity, whereas the *perceived heaviness* H of a weight W satisfies $H = kW^{3/2}$ (k is a constant that is different in the two cases). Compute dB/dI and dH/dW and state whether they are increasing or decreasing functions. Then explain the following statements:

(a) A one-unit increase in light intensity is felt more strongly when I is small than when I is large.

(b) Adding another pound to a load W is felt more strongly when W is large than when W is small.

48. Let $M(t)$ be the mass (in kilograms) of a plant as a function of time (in years). Recent studies by Niklas and Enquist have suggested that a remarkably wide range of plants (from algae and grass to palm trees) obey a *three-quarter-power growth law*—that is, $dM/dt = CM^{3/4}$ for some constant C.

(a) If a tree has a growth rate of 6 kg/yr when $M = 100$ kg, what is its growth rate when $M = 125$ kg?

(b) If $M = 0.5$ kg, how much more mass must the plant acquire to double its growth rate?

Further Insights and Challenges

Exercises 49–51: The ***Lorenz curve*** $y = F(r)$ *is used by economists to study income distribution in a given country (see Figure 14). By definition,* $F(r)$ *is the fraction of the total income that goes to the bottom* r*th part of the population, where* $0 \le r \le 1$*. For example, if* $F(0.4) = 0.245$*, then the bottom 40% of households receive 24.5% of the total income. Note that* $F(0) = 0$ *and* $F(1) = 1$.

49. Our goal is to find an interpretation for $F'(r)$. The average income for a group of households is the total income going to the group divided by the number of households in the group. The national average income is $A = T/N$, where N is the total number of households and T is the total income earned by the entire population.

(a) Show that the average income among households in the bottom rth part is equal to $(F(r)/r)A$.

(b) Show more generally that the average income of households belonging to an interval $[r, r + \Delta r]$ is equal to

$$\left(\frac{F(r + \Delta r) - F(r)}{\Delta r}\right) A$$

(c) Let $0 \le r \le 1$. A household belongs to the $100r$th percentile if its income is greater than or equal to the income of $100r$ % of all households. Pass to the limit as $\Delta r \to 0$ in (b) to derive the following interpretation: A household in the $100r$th percentile has income $F'(r)A$. In particular, a household in the $100r$th percentile receives more than the national average if $F'(r) > 1$ and less if $F'(r) < 1$.

(d) For the Lorenz curves L_1 and L_2 in Figure 14(B), what percentage of households have above-average income?

50. The following table provides values of $F(r)$ for Sweden in 2004. Assume that the national average income was $A = 30,000$ euros.

r	0	0.2	0.4	0.6	0.8	1
$F(r)$	0	0.01	0.245	0.423	0.642	1

(a) What was the average income in the lowest 40% of households?

(b) Show that the average income of the households belonging to the interval [0.4, 0.6] was 26,700 euros.

(c) Estimate $F'(0.5)$. Estimate the income of households in the 50th percentile? Was it greater or less than the national average?

51. Use Exercise 49 (c) to prove:

(a) $F'(r)$ is an increasing function of r.

(b) Income is distributed equally (all households have the same income) if and only if $F(r) = r$ for $0 \le r \le 1$.

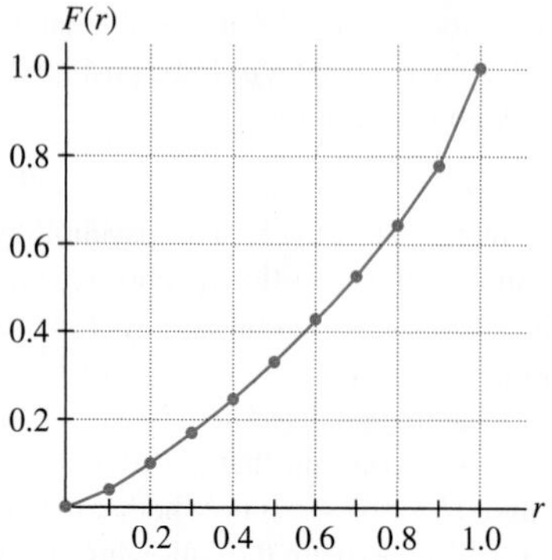

(A) Lorenz curve for Sweden in 2004

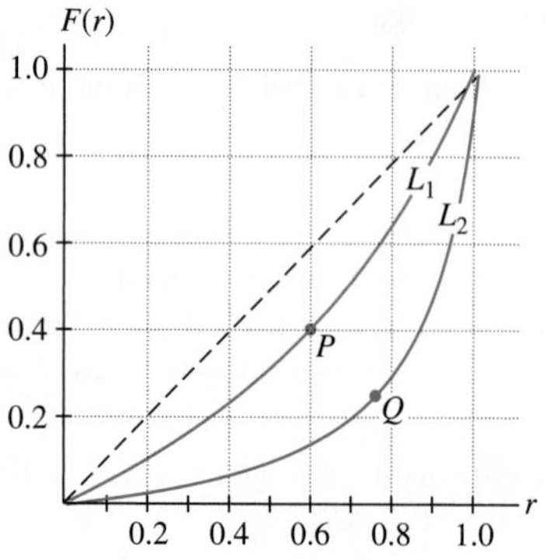

(B) Two Lorenz curves: The tangent lines at P and Q have slope 1.

FIGURE 14

52. CAS Studies of Internet usage show that website popularity is described quite well by Zipf's Law, according to which the nth most popular website receives roughly the fraction $1/n$ of all visits. Suppose that on a particular day, the nth most popular site had approximately $V(n) = 10^6/n$ visitors (for $n \le 15{,}000$).

(a) Verify that the top 50 websites received nearly 45% of the visits. *Hint:* Let $T(N)$ denote the sum of $V(n)$ for $1 \le n \le N$. Use a computer algebra system to compute $T(45)$ and $T(15{,}000)$.

(b) Verify, by numerical experimentation, that when Eq. (3) is used to estimate $V(n+1) - V(n)$, the error in the estimate decreases as n grows larger. Find (again, by experimentation) an N such that the error is at most 10 for $n \ge N$.

(c) Using Eq. (3), show that for $n \ge 100$, the nth website received at most 100 more visitors than the $(n+1)$st website.

In Exercises 53 and 54, the average cost per unit at production level x is defined as $C_{\text{avg}}(x) = C(x)/x$, where $C(x)$ is the cost function. Average cost is a measure of the efficiency of the production process.

53. Show that $C_{\text{avg}}(x)$ is equal to the slope of the line through the origin and the point $(x, C(x))$ on the graph of $C(x)$. Using this interpretation, determine whether average cost or marginal cost is greater at points A, B, C, D in Figure 15.

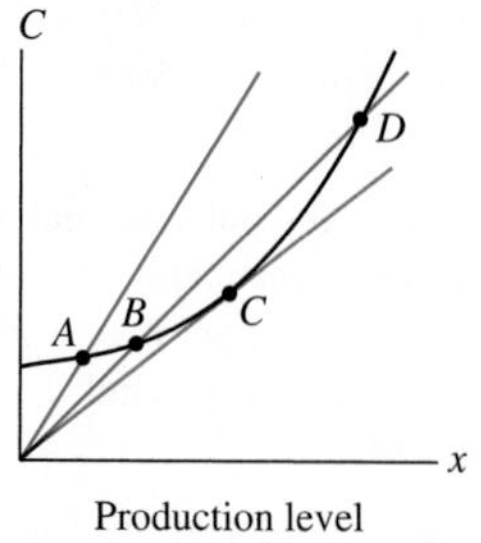

FIGURE 15 Graph of $C(x)$.

54. The cost in dollars of producing alarm clocks is $C(x) = 50x^3 - 750x^2 + 3740x + 3750$ where x is in units of 1000.

(a) Calculate the average cost at $x = 4, 6, 8$, and 10.

(b) Use the graphical interpretation of average cost to find the production level x_0 at which average cost is lowest. What is the relation between average cost and marginal cost at x_0 (see Figure 16)?

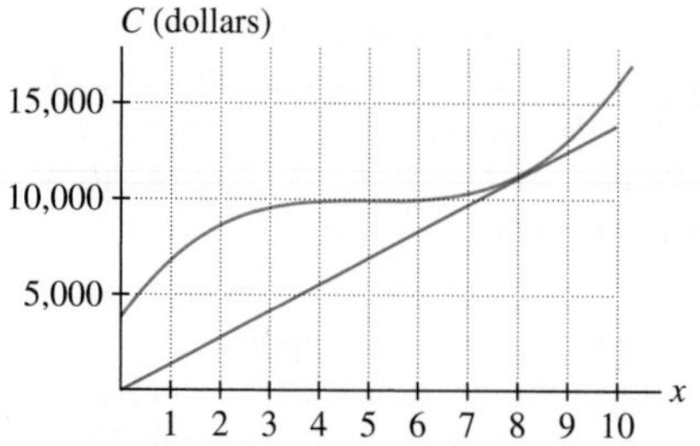

FIGURE 16 Cost function $C(x) = 50x^3 - 750x^2 + 3740x + 3750$.

3.5 Higher Derivatives

Higher derivatives are obtained by repeatedly differentiating a function $y = f(x)$. If f' is differentiable, then the **second derivative**, denoted f'' or y'', is the derivative

$$f''(x) = \frac{d}{dx}\left(f'(x)\right)$$

The second derivative is the rate of change of $f'(x)$. The next example highlights the difference between the first and second derivatives.

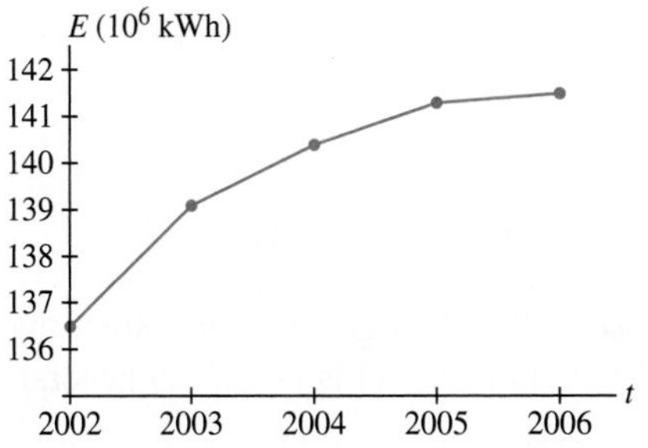

FIGURE 1 Household energy consumption $E(t)$ in Germany in million kilowatt-hours.

EXAMPLE 1 Figure 1 and Table 1 describe the total household energy consumption $E(t)$ in Germany in year t. Discuss $E'(t)$ and $E''(t)$.

TABLE 1 Household Energy Consumption in Germany

Year	2002	2003	2004	2005	2006
Consumption (10^6 kWh)	136.5	139.1	140.4	141.3	141.5
Yearly increase		2.6	1.3	0.9	0.2

Solution We will show that $E'(t)$ is positive but $E''(t)$ is negative. According to Table 1, the consumption each year was greater than the previous year, so the rate of change $E'(t)$ is certainly positive. However, the amount of increase declined from 2.6 million in 2003 to 0.2 in 2006. So although $E'(t)$ is positive, $E'(t)$ decreases from one year to the next, and therefore its rate of change $E''(t)$ is negative. Figure 1 supports this conclusion: The slopes of the segments in the graph are decreasing. ■

- *dy/dx has units of y per unit of x.*
- *d^2y/dx^2 has units of dy/dx per unit of x or units of y per unit of x squared.*

The process of differentiation can be continued, provided that the derivatives exist. The third derivative, denoted $f'''(x)$ or $f^{(3)}(x)$, is the derivative of $f''(x)$. More generally, the nth derivative $f^{(n)}(x)$ is the derivative of the $(n-1)$st derivative. We call $f(x)$ the zeroeth derivative and $f'(x)$ the first derivative. In Leibniz notation, we write

$$\frac{df}{dx}, \quad \frac{d^2 f}{dx^2}, \quad \frac{d^3 f}{dx^3}, \quad \frac{d^4 f}{dx^4}, \ldots$$

EXAMPLE 2 Calculate $f'''(-1)$ for $f(x) = 3x^5 - 2x^2 + 7x^{-2}$.

Solution We must calculate the first three derivatives:

$$f'(x) = \frac{d}{dx}\left(3x^5 - 2x^2 + 7x^{-2}\right) = 15x^4 - 4x - 14x^{-3}$$

$$f''(x) = \frac{d}{dx}\left(15x^4 - 4x - 14x^{-3}\right) = 60x^3 - 4 + 42x^{-4}$$

$$f'''(x) = \frac{d}{dx}\left(60x^3 - 4 + 42x^{-4}\right) = 180x^2 - 168x^{-5}$$

At $x = -1$, $f'''(-1) = 180 + 168 = 348$. ■

Polynomials have a special property: Their higher derivatives are eventually the zero function. More precisely, if $f(x)$ is a polynomial of degree k, then $f^{(n)}(x)$ is zero for $n > k$. Table 2 illustrates this property for $f(x) = x^5$. By contrast, the higher derivatives of a nonpolynomial function are never the zero function (see Exercise 89, Section 4.8).

TABLE 2 Derivatives of x^5

$f(x)$	$f'(x)$	$f''(x)$	$f'''(x)$	$f^{(4)}(x)$	$f^{(5)}(x)$	$f^{(6)}(x)$
x^5	$5x^4$	$20x^3$	$60x^2$	$120x$	120	0

EXAMPLE 3 Calculate the first four derivatives of $y = x^{-1}$. Then find the pattern and determine a general formula for $y^{(n)}$.

Solution By the Power Rule,

$$y'(x) = -x^{-2}, \quad y'' = 2x^{-3}, \quad y''' = -2(3)x^{-4}, \quad y^{(4)} = 2(3)(4)x^{-5}$$

We see that $y^{(n)}(x)$ is equal to $\pm n!\,x^{-n-1}$. Now observe that the sign alternates. Since the odd-order derivatives occur with a minus sign, the sign of $y^{(n)}(x)$ is $(-1)^n$. In general, therefore, $y^{(n)}(x) = (-1)^n n!\,x^{-n-1}$. ■

REMINDER n-factorial is the number

$$n! = n(n-1)(n-2)\cdots(2)(1)$$

Thus

$$1! = 1, \quad 2! = (2)(1) = 2$$
$$3! = (3)(2)(1) = 6$$

By convention, we set $0! = 1$.

EXAMPLE 4 Find an equation of the tangent line to $y = f'(x)$ at $x = 4$ where $f(x) = x^{3/2}$.

Solution The slope of the tangent line to $y = f'(x)$ at $x = 4$ is the derivative $f''(4)$. So we compute the first two derivatives and their values at $x = 4$:

$$f'(x) = \frac{3}{2}x^{1/2}, \qquad f'(4) = \frac{3}{2}(4)^{1/2} = 3$$

$$f''(x) = \frac{3}{4}x^{-1/2}, \qquad f''(4) = \frac{3}{4}(4)^{-1/2} = \frac{3}{8}$$

Therefore, an equation of the tangent line is

$$y - f'(4) = f''(4)(x-4) \quad \Rightarrow \quad y - 3 = \frac{3}{8}(x-4)$$

In slope-intercept form, the equation is $y = \frac{3}{8}x + \frac{3}{2}$. ■

It is not always possible to find a simple formula for the higher derivatives of a function. In most cases, they become increasingly complicated.

One familiar second derivative is acceleration. An object in linear motion with position $s(t)$ at time t has velocity $v(t) = s'(t)$ and acceleration $a(t) = v'(t) = s''(t)$. Thus, acceleration is the rate at which velocity changes and is measured in units of velocity per unit of time or "distance per time squared" such as m/s^2.

EXAMPLE 5 Acceleration Due to Gravity Find the acceleration $a(t)$ of a ball tossed vertically in the air from ground level with an initial velocity of 12 m/s. How does $a(t)$ describe the change in the ball's velocity as it rises and falls?

Solution The ball's height at time t is $s(t) = s_0 + v_0 t - 4.9t^2$ m by Galileo's formula [Figure 2(A)]. In our case, $s_0 = 0$ and $v_0 = 12$, so $s(t) = 12t - 4.9t^2$ m. Therefore, $v(t) = s'(t) = 12 - 9.8t$ m/s and the ball's acceleration is

$$a(t) = s''(t) = \frac{d}{dt}(12 - 9.8t) = -9.8 \text{ m/s}^2$$

As expected, the acceleration is constant with value $-g = -9.8$ m/s^2. As the ball rises and falls, its velocity decreases from 12 to -12 m/s at the constant rate $-g$ [Figure 2(B)]. ■

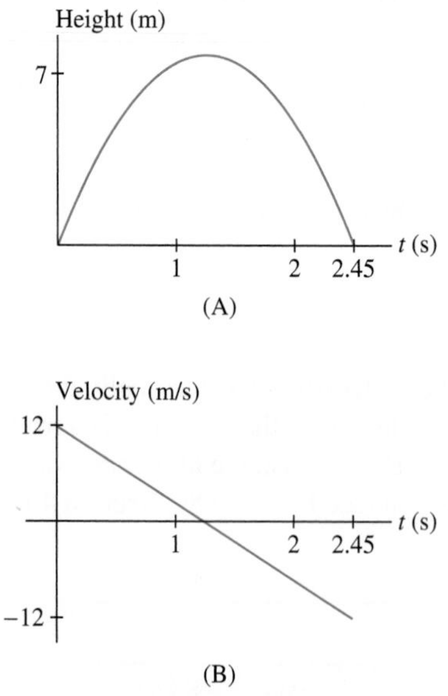

FIGURE 2 Height and velocity of a ball tossed vertically with initial velocity 12 m/s.

GRAPHICAL INSIGHT Can we visualize the rate represented by $f''(x)$? The second derivative is the rate at which $f'(x)$ is changing, so $f''(x)$ is large if the slopes of the tangent lines change rapidly, as in Figure 3(A) on the next page. Similarly, $f''(x)$ is small if the slopes of the tangent lines change slowly—in this case, the curve is relatively flat, as in Figure 3(B). If f is a linear function [Figure 3(C)], then the tangent line does not change at all and $f''(x) = 0$. Thus, $f''(x)$ measures the "bending" or concavity of the graph.

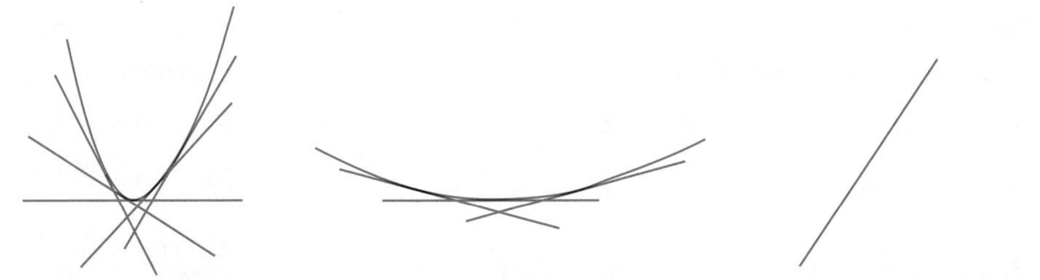

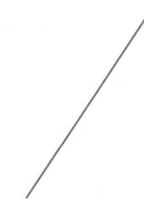

(A) Large second derivative: Tangent lines turn rapidly.

(B) Smaller second derivative: Tangent lines turn slowly.

(C) Second derivative is zero: Tangent line does not change.

FIGURE 3

■ **EXAMPLE 6** Identify curves I and II in Figure 4(B) as the graphs of $f'(x)$ or $f''(x)$ for the function $f(x)$ in Figure 4(A).

Solution The slopes of the tangent lines to the graph of $f(x)$ are *increasing* on the interval $[a, b]$. Therefore $f'(x)$ is an increasing function and its graph must be II. Since $f''(x)$ is the rate of change of $f'(x)$, $f''(x)$ is positive and its graph must be I. ■

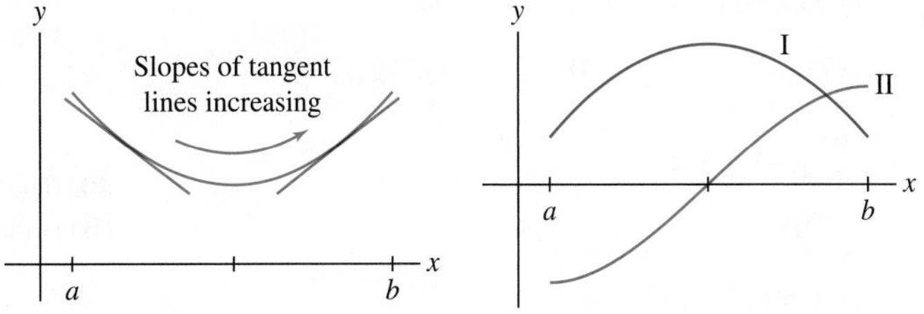

(A) Graph of $f(x)$

(B) Graph of first two derivatives

FIGURE 4

3.5 SUMMARY

- The higher derivatives $f', f'', f''', \ldots$ are defined by successive differentiation:

$$f''(x) = \frac{d}{dx} f'(x), \qquad f'''(x) = \frac{d}{dx} f''(x), \ldots$$

The nth derivative is denoted $f^{(n)}(x)$.

- The second derivative plays an important role: It is the rate at which f' changes. Graphically, f'' measures how fast the tangent lines change direction and thus measures the "bending" of the graph.
- If $s(t)$ is the position of an object at time t, then $s'(t)$ is velocity and $s''(t)$ is acceleration.

3.5 EXERCISES

Preliminary Questions

1. On September 4, 2003, the *Wall Street Journal* printed the headline "Stocks Go Higher, Though the Pace of Their Gains Slows." Rephrase this headline as a statement about the first and second time derivatives of stock prices and sketch a possible graph.

2. True or false? The third derivative of position with respect to time is zero for an object falling to earth under the influence of gravity. Explain.

3. Which type of polynomial satisfies $f'''(x) = 0$ for all x?

4. What is the sixth derivative of $f(x) = x^6$?

Exercises

In Exercises 1–16, calculate y'' and y'''.

1. $y = 14x^2$

2. $y = 7 - 2x$

3. $y = x^4 - 25x^2 + 2x$

4. $y = 4t^3 - 9t^2 + 7$

5. $y = \frac{4}{3}\pi r^3$

6. $y = \sqrt{x}$

7. $y = 20t^{4/5} - 6t^{2/3}$

8. $y = x^{-9/5}$

9. $y = z - \frac{4}{z}$

10. $y = 5t^{-3} + 7t^{-8/3}$

11. $y = \theta^2(2\theta + 7)$

12. $y = (x^2 + x)(x^3 + 1)$

13. $y = \frac{x - 4}{x}$

14. $y = \frac{1}{1 - x}$

15. $y = s^{-1/2}(s + 1)$

16. $y = (r^{1/2} + r)(1 - r)$

In Exercises 17–26, calculate the derivative indicated.

17. $f^{(4)}(1)$, $f(x) = x^4$

18. $g'''(-1)$, $g(t) = -4t^{-5}$

19. $\left.\frac{d^2y}{dt^2}\right|_{t=1}$, $y = 4t^{-3} + 3t^2$

20. $\left.\frac{d^4f}{dt^4}\right|_{t=1}$, $f(t) = 6t^9 - 2t^5$

21. $\left.\frac{d^4x}{dt^4}\right|_{t=16}$, $x = t^{-3/4}$

22. $f'''(4)$, $f(t) = 2t^2 - t$

23. $f'''(-3)$, $f(x) = \frac{12}{x} - x^3$

24. $f''(1)$, $f(t) = \frac{t}{t+1}$

25. $h''(1)$, $h(w) = \frac{1}{\sqrt{w} + 1}$

26. $g''(1)$, $g(s) = \frac{\sqrt{s}}{s+1}$

27. Calculate $y^{(k)}(0)$ for $0 \le k \le 5$, where $y = x^4 + ax^3 + bx^2 + cx + d$ (with a, b, c, d the constants).

28. Which of the following satisfy $f^{(k)}(x) = 0$ for all $k \ge 6$?

(a) $f(x) = 7x^4 + 4 + x^{-1}$

(b) $f(x) = x^3 - 2$

(c) $f(x) = \sqrt{x}$

(d) $f(x) = 1 - x^6$

(e) $f(x) = x^{9/5}$

(f) $f(x) = 2x^2 + 3x^5$

29. Use the result in Example 3 to find $\frac{d^6}{dx^6}x^{-1}$.

30. Calculate the first five derivatives of $f(x) = \sqrt{x}$.

(a) Show that $f^{(n)}(x)$ is a multiple of $x^{-n+1/2}$.

(b) Show that $f^{(n)}(x)$ alternates in sign as $(-1)^{n-1}$ for $n \ge 1$.

(c) Find a formula for $f^{(n)}(x)$ for $n \ge 2$. *Hint:* Verify that the coefficient is $\pm 1 \cdot 3 \cdot 5 \cdots \frac{2n-3}{2^n}$.

In Exercises 31–36, find a general formula for $f^{(n)}(x)$.

31. $f(x) = x^{-2}$

32. $f(x) = (x + 2)^{-1}$

33. $f(x) = x^{-1/2}$

34. $f(x) = x^{-3/2}$

35. $f(x) = \frac{x+1}{x^2}$

36. $f(x) = \frac{x-1}{\sqrt{x}}$

37. **(a)** Find the acceleration at time $t = 5$ min of a helicopter whose height is $s(t) = 300t - 4t^3$ m.

(b) Plot the acceleration $h''(t)$ for $0 \le t \le 6$. How does this graph show that the helicopter is slowing down during this time interval?

38. Find an equation of the tangent to the graph of $y = f'(x)$ at $x = 3$, where $f(x) = x^4$.

39. Figure 5 shows f, f', and f''. Determine which is which.

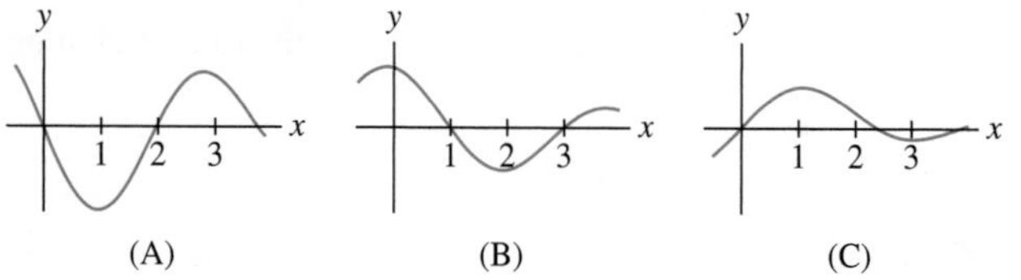

FIGURE 5

40. The second derivative f'' is shown in Figure 6. Which of (A) or (B) is the graph of f and which is f'?

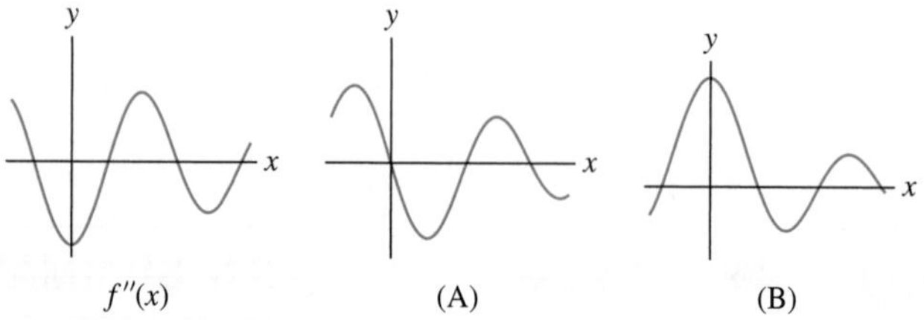

FIGURE 6

41. Figure 7 shows the graph of the position s of an object as a function of time t. Determine the intervals on which the acceleration is positive.

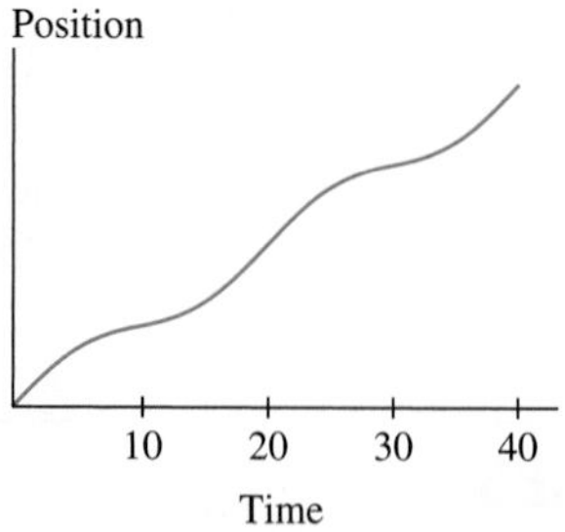

FIGURE 7

42. Find a polynomial $f(x)$ that satisfies the equation $xf''(x) + f(x) = x^2$.

43. Find all values of n such that $y = x^n$ satisfies

$$x^2y'' - 2xy' = 4y$$

44. Which of the following descriptions could *not* apply to Figure 8? Explain.

(a) Graph of acceleration when velocity is constant

(b) Graph of velocity when acceleration is constant

(c) Graph of position when acceleration is zero

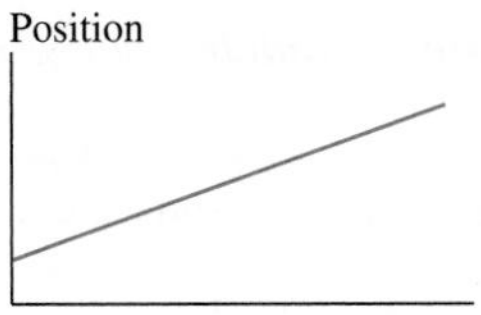

FIGURE 8

45. According to one model that takes into account air resistance, the acceleration $a(t)$ (in m/s^2) of a skydiver of mass m in free fall satisfies

$$a(t) = -9.8 + \frac{k}{m}v(t)^2$$

where $v(t)$ is velocity (negative since the object is falling) and k is a constant. Suppose that $m = 75$ kg and $k = 14$ kg/m.

(a) What is the object's velocity when $a(t) = -4.9$?

(b) What is the object's velocity when $a(t) = 0$? This velocity is the object's terminal velocity.

46. According to one model that attempts to account for air resistance, the distance $s(t)$ (in meters) traveled by a falling raindrop satisfies

$$\frac{d^2s}{dt^2} = g - \frac{0.0005}{D}\left(\frac{ds}{dt}\right)^2$$

where D is the raindrop diameter and $g = 9.8$ m/s^2. Terminal velocity v_{term} is defined as the velocity at which the drop has zero acceleration (one can show that velocity approaches v_{term} as time proceeds).

(a) Show that $v_{\text{term}} = \sqrt{2000gD}$.

(b) Find v_{term} for drops of diameter 10^{-3} m and 10^{-4} m.

(c) In this model, do raindrops accelerate more rapidly at higher or lower velocities?

47. A servomotor controls the vertical movement of a drill bit that will drill a pattern of holes in sheet metal. The maximum vertical speed of the drill bit is 4 in./s, and while drilling the hole, it must move no more than 2.6 in./s to avoid warping the metal. During a cycle, the bit begins and ends at rest, quickly approaches the sheet metal, and quickly returns to its initial position after the hole is drilled. Sketch possible graphs of the drill bit's vertical velocity and acceleration. Label the point where the bit enters the sheet metal.

In Exercises 48 and 49, refer to the following. In a 1997 study, Boardman and Lave related the traffic speed S on a two-lane road to traffic density Q (number of cars per mile of road) by the formula

$$S = 2882Q^{-1} - 0.052Q + 31.73$$

for $60 \le Q \le 400$ *(Figure 9).*

48. Calculate dS/dQ and d^2S/dQ^2.

49. (a) Explain intuitively why we should expect that $dS/dQ < 0$.

(b) Show that $d^2S/dQ^2 > 0$. Then use the fact that $dS/dQ < 0$ and $d^2S/dQ^2 > 0$ to justify the following statement: *A one-unit increase in traffic density slows down traffic more when Q is small than when Q is large.*

(c) GU Plot dS/dQ. Which property of this graph shows that $d^2S/dQ^2 > 0$?

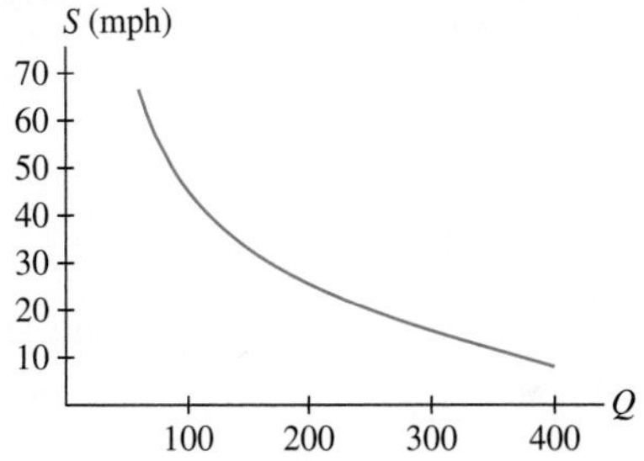

FIGURE 9 Speed as a function of traffic density.

50. CAS Use a computer algebra system to compute $f^{(k)}(x)$ for $k = 1, 2, 3$ for the following functions.

(a) $f(x) = (1 + x^3)^{5/3}$ **(b)** $f(x) = \dfrac{1 - x^4}{1 - 5x - 6x^2}$

51. CAS Let $f(x) = \dfrac{x+2}{x-1}$. Use a computer algebra system to compute the $f^{(k)}(x)$ for $1 \le k \le 4$. Can you find a general formula for $f^{(k)}(x)$?

Further Insights and Challenges

52. Find the 100th derivative of

$$p(x) = (x + x^5 + x^7)^{10}(1 + x^2)^{11}(x^3 + x^5 + x^7)$$

53. What is $p^{(99)}(x)$ for $p(x)$ as in Exercise 52?

54. Use the Product Rule twice to find a formula for $(fg)''$ in terms of f and g and their first and second derivatives.

55. Use the Product Rule to find a formula for $(fg)'''$ and compare your result with the expansion of $(a+b)^3$. Then try to guess the general formula for $(fg)^{(n)}$.

56. Compute

$$\Delta f(x) = \lim_{h \to 0} \frac{f(x+h) + f(x-h) - 2f(x)}{h^2}$$

for the following functions:

(a) $f(x) = x$ **(b)** $f(x) = x^2$ **(c)** $f(x) = x^3$

Based on these examples, what do you think the limit Δf represents?

3.6 Trigonometric Functions

CAUTION In Theorem 1 we are differentiating with respect to x measured in radians. The derivatives of sine and cosine with respect to degrees involves an extra, unwieldy factor of $\pi/180$ (see Example 7 in Section 3.7).

We can use the rules developed so far to differentiate functions involving powers of x, but we cannot yet handle the trigonometric functions. What is missing are the formulas for the derivatives of $\sin x$ and $\cos x$. Fortunately, their derivatives are simple—each is the derivative of the other up to a sign.

Recall our convention: *Angles are measured in radians, unless otherwise specified.*

THEOREM 1 Derivative of Sine and Cosine The functions $y = \sin x$ and $y = \cos x$ are differentiable and

$$\frac{d}{dx}\sin x = \cos x \quad \text{and} \quad \frac{d}{dx}\cos x = -\sin x$$

Proof We must go back to the definition of the derivative:

$$\frac{d}{dx}\sin x = \lim_{h\to 0}\frac{\sin(x+h)-\sin x}{h} \qquad \boxed{1}$$

REMINDER Addition formula for $\sin x$:

$$\sin(x+h) = \sin x\cos h + \cos x \sin h$$

We cannot cancel the h by rewriting the difference quotient, but we can use the addition formula (see marginal note) to write the numerator as a sum of two terms:

$$\begin{aligned}\sin(x+h)-\sin x &= \sin x\cos h + \cos x\sin h - \sin x \qquad \text{(addition formula)}\\ &= (\sin x\cos h - \sin x) + \cos x \sin h\\ &= \sin x(\cos h - 1) + \cos x\sin h\end{aligned}$$

This gives us

$$\frac{\sin(x+h)-\sin x}{h} = \frac{\sin x\,(\cos h-1)}{h} + \frac{\cos x\,\sin h}{h}$$

$$\begin{aligned}\frac{d\sin x}{dx} = \lim_{h\to 0}\frac{\sin(x+h)-\sin x}{h} &= \lim_{h\to 0}\frac{\sin x\,(\cos h-1)}{h} + \lim_{h\to 0}\frac{\cos x\,\sin h}{h}\\ &= (\sin x)\underbrace{\lim_{h\to 0}\frac{\cos h-1}{h}}_{\text{This equals 0.}} + (\cos x)\underbrace{\lim_{h\to 0}\frac{\sin h}{h}}_{\text{This equals 1.}}\end{aligned} \qquad \boxed{2}$$

Here, we can take $\sin x$ and $\cos x$ outside the limits in Eq. (2) because they do not depend on h. The two limits are given by Theorem 2 in Section 2.6,

$$\lim_{h\to 0}\frac{\cos h-1}{h} = 0 \qquad \text{and} \qquad \lim_{h\to 0}\frac{\sin h}{h} = 1$$

Therefore, Eq. (2) reduces to the formula $\frac{d}{dx}\sin x = \cos x$, as desired. The formula $\frac{d}{dx}\cos x = -\sin x$ is proved similarly (see Exercise 53). ■

■ **EXAMPLE 1** Calculate $f''(x)$, where $f(x) = x\cos x$.

Solution By the Product Rule,

$$f'(x) = x\frac{d}{dx}\cos x + \cos x\frac{d}{dx}x = x(-\sin x) + \cos x = \cos x - x\sin x$$

$$f''(x) = (\cos x - x\sin x)' = -\sin x - \big(x(\sin x)' + \sin x\big) = -2\sin x - x\cos x \quad ■$$

GRAPHICAL INSIGHT The formula $(\sin x)' = \cos x$ is made plausible when we compare the graphs in Figure 1. The tangent lines to the graph of $y = \sin x$ have positive slope on the interval $\left(-\frac{\pi}{2}, \frac{\pi}{2}\right)$, and on this interval, the derivative $y' = \cos x$ is positive. Similarly, the tangent lines have negative slope on the interval $\left(\frac{\pi}{2}, \frac{3\pi}{2}\right)$, where $y' = \cos x$ is negative. The tangent lines are horizontal at $x = -\frac{\pi}{2}, \frac{\pi}{2}, \frac{3\pi}{2}$, where $\cos x = 0$.

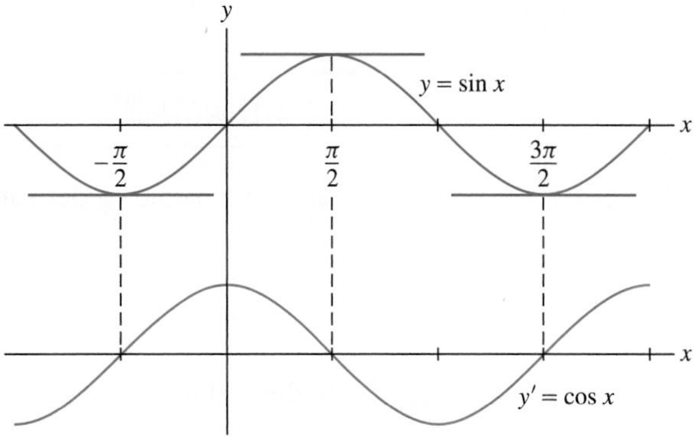

FIGURE 1 Compare the graphs of $y = \sin x$ and its derivative $y' = \cos x$.

REMINDER The standard trigonometric functions are defined in Section 1.4.

The derivatives of the other **standard trigonometric functions** can be computed using the Quotient Rule. We derive the formula for $(\tan x)'$ in Example 2 and leave the remaining formulas for the exercises (Exercises 35–37).

THEOREM 2 Derivatives of Standard Trigonometric Functions

$$\frac{d}{dx}\tan x = \sec^2 x, \qquad \frac{d}{dx}\sec x = \sec x \tan x$$

$$\frac{d}{dx}\cot x = -\csc^2 x, \qquad \frac{d}{dx}\csc x = -\csc x \cot x$$

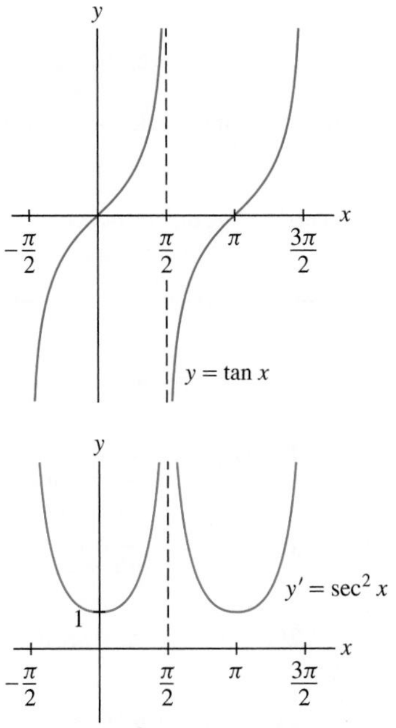

FIGURE 2 Graphs of $y = \tan x$ and its derivative $y' = \sec^2 x$.

■ **EXAMPLE 2** Verify the formula $\frac{d}{dx}\tan x = \sec^2 x$ (Figure 2).

Solution Use the Quotient Rule and the identity $\cos^2 x + \sin^2 x = 1$:

$$\frac{d}{dx}\tan x = \left(\frac{\sin x}{\cos x}\right)' = \frac{\cos x \cdot (\sin x)' - \sin x \cdot (\cos x)'}{\cos^2 x}$$

$$= \frac{\cos x \cos x - \sin x(-\sin x)}{\cos^2 x}$$

$$= \frac{\cos^2 x + \sin^2 x}{\cos^2 x} = \frac{1}{\cos^2 x} = \sec^2 x$$ ■

■ **EXAMPLE 3** Find the tangent line to the graph of $y = \tan\theta \sec\theta$ at $\theta = \frac{\pi}{4}$.

Solution By the Product Rule,

$$y' = \tan\theta\,(\sec\theta)' + \sec\theta\,(\tan\theta)' = \tan\theta\,(\sec\theta\,\tan\theta) + \sec\theta\,\sec^2\theta$$

$$= \tan^2\theta\,\sec\theta + \sec^3\theta$$

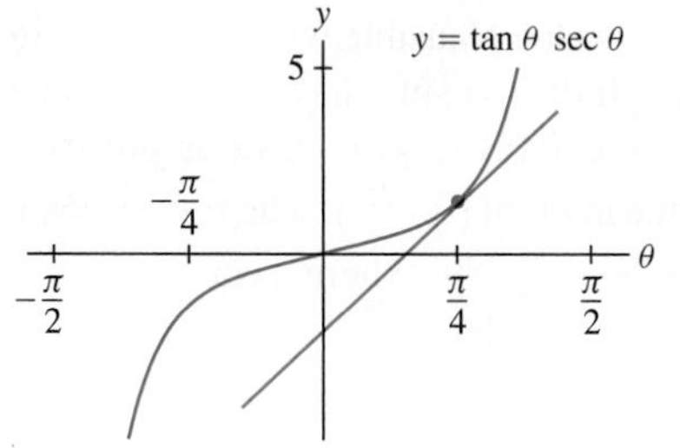

FIGURE 3 Tangent line to $y = \tan\theta\sec\theta$ at $\theta = \frac{\pi}{4}$.

Now use the values $\sec\frac{\pi}{4} = \sqrt{2}$ and $\tan\frac{\pi}{4} = 1$ to compute

$$y\left(\frac{\pi}{4}\right) = \tan\left(\frac{\pi}{4}\right)\sec\left(\frac{\pi}{4}\right) = \sqrt{2}$$

$$y'\left(\frac{\pi}{4}\right) = \tan^2\left(\frac{\pi}{4}\right)\sec\left(\frac{\pi}{4}\right) + \sec^3\left(\frac{\pi}{4}\right) = \sqrt{2} + 2\sqrt{2} = 3\sqrt{2}$$

An equation of the tangent line (Figure 3) is $y - \sqrt{2} = 3\sqrt{2}\left(\theta - \frac{\pi}{4}\right)$. ■

3.6 SUMMARY

- Basic trigonometric derivatives:

$$\frac{d}{dx}\sin x = \cos x, \qquad \frac{d}{dx}\cos x = -\sin x$$

- Additional formulas:

$$\frac{d}{dx}\tan x = \sec^2 x, \qquad \frac{d}{dx}\sec x = \sec x\tan x$$

$$\frac{d}{dx}\cot x = -\csc^2 x, \qquad \frac{d}{dx}\csc x = -\csc x\cot x$$

3.6 EXERCISES

Preliminary Questions

1. Determine the sign (+ or −) that yields the correct formula for the following:

(a) $\frac{d}{dx}(\sin x + \cos x) = \pm\sin x \pm \cos x$

(b) $\frac{d}{dx}\sec x = \pm\sec x\tan x$

(c) $\frac{d}{dx}\cot x = \pm\csc^2 x$

2. Which of the following functions can be differentiated using the rules we have covered so far?

(a) $y = 3\cos x\cot x$ **(b)** $y = \cos(x^2)$ **(c)** $y = 2^x\sin x$

3. Compute $\frac{d}{dx}(\sin^2 x + \cos^2 x)$ without using the derivative formulas for $\sin x$ and $\cos x$.

4. How is the addition formula used in deriving the formula $(\sin x)' = \cos x$?

Exercises

In Exercises 1–4, find an equation of the tangent line at the point indicated.

1. $y = \sin x, \quad x = \frac{\pi}{4}$

2. $y = \cos x, \quad x = \frac{\pi}{3}$

3. $y = \tan x, \quad x = \frac{\pi}{4}$

4. $y = \sec x, \quad x = \frac{\pi}{6}$

In Exercises 5–24, compute the derivative.

5. $f(x) = \sin x\cos x$

6. $f(x) = x^2\cos x$

7. $f(x) = \sin^2 x$

8. $f(x) = 9\sec x + 12\cot x$

9. $H(t) = \sin t\sec^2 t$

10. $h(t) = 9\csc t + t\cot t$

11. $f(\theta) = \tan\theta\sec\theta$

12. $k(\theta) = \theta^2\sin^2\theta$

13. $f(x) = (2x^4 - 4x^{-1})\sec x$

14. $f(z) = z\tan z$

15. $y = \dfrac{\sec\theta}{\theta}$

16. $G(z) = \dfrac{1}{\tan z - \cot z}$

17. $R(y) = \dfrac{3\cos y - 4}{\sin y}$

18. $f(x) = \dfrac{x}{\sin x + 2}$

19. $f(x) = \dfrac{1 + \tan x}{1 - \tan x}$

20. $f(\theta) = \theta\tan\theta\sec\theta$

21. $f(x) = \dfrac{\sin x + 1}{\sin x - 1}$

22. $h(t) = \dfrac{\csc^2 t}{t}$

23. $R(\theta) = \dfrac{\cos\theta}{4 + \cos\theta}$

24. $g(z) = \dfrac{\cot z}{3 - 3\sin z}$

In Exercises 25–34, find an equation of the tangent line at the point specified.

25. $y = x^3 + \cos x, \quad x = 0$

26. $y = \tan\theta, \quad \theta = \frac{\pi}{6}$

27. $y = \sin x + 3\cos x, \quad x = 0$

28. $y = \dfrac{\sin t}{1 + \cos t}, \quad t = \frac{\pi}{3}$

29. $y = 2(\sin\theta + \cos\theta), \quad \theta = \frac{\pi}{3}$

30. $y = \csc x - \cot x, \quad x = \frac{\pi}{4}$

31. $y = (\cot t)(\cos t), \quad t = \dfrac{\pi}{3}$

32. $y = x\cos^2 x, \quad x = \dfrac{\pi}{4}$

33. $y = x^2(1 - \sin x), \quad x = \dfrac{3\pi}{2}$

34. $y = \dfrac{\sin\theta - \cos\theta}{\theta}, \quad \theta = \dfrac{\pi}{4}$

In Exercises 35–37, use Theorem 1 to verify the formula.

35. $\dfrac{d}{dx}\cot x = -\csc^2 x$

36. $\dfrac{d}{dx}\sec x = \sec x \tan x$

37. $\dfrac{d}{dx}\csc x = -\csc x \cot x$

38. Show that both $y = \sin x$ and $y = \cos x$ satisfy $y'' = -y$.

In Exercises 39–42, calculate the higher derivative.

39. $f''(\theta), \quad f(\theta) = \theta\sin\theta$

40. $\dfrac{d^2}{dt^2}\cos^2 t$

41. $y'', \quad y''', \quad y = \tan x$

42. $y'', \quad y''', \quad y = t^2 \sin t$

43. Calculate the first five derivatives of $f(x) = \cos x$. Then determine $f^{(8)}$ and $f^{(37)}$.

44. Find $y^{(157)}$, where $y = \sin x$.

45. Find the values of x between 0 and 2π where the tangent line to the graph of $y = \sin x \cos x$ is horizontal.

46. GU Plot the graph $f(\theta) = \sec\theta + \csc\theta$ over $[0, 2\pi]$ and determine the number of solutions to $f'(\theta) = 0$ in this interval graphically. Then compute $f'(\theta)$ and find the solutions.

47. GU Let $g(t) = t - \sin t$.

(a) Plot the graph of g with a graphing utility for $0 \le t \le 4\pi$.

(b) Show that the slope of the tangent line is nonnegative. Verify this on your graph.

(c) For which values of t in the given range is the tangent line horizontal?

48. CAS Let $f(x) = (\sin x)/x$ for $x \ne 0$ and $f(0) = 1$.

(a) Plot $f(x)$ on $[-3\pi, 3\pi]$.

(b) Show that $f'(c) = 0$ if $c = \tan c$. Use the numerical root finder on a computer algebra system to find a good approximation to the smallest *positive* value c_0 such that $f'(c_0) = 0$.

(c) Verify that the horizontal line $y = f(c_0)$ is tangent to the graph of $y = f(x)$ at $x = c_0$ by plotting them on the same set of axes.

49. Show that no tangent line to the graph of $f(x) = \tan x$ has zero slope. What is the least slope of a tangent line? Justify by sketching the graph of $(\tan x)'$.

50. The height at time t (in seconds) of a mass, oscillating at the end of a spring, is $s(t) = 300 + 40\sin t$ cm. Find the velocity and acceleration at $t = \frac{\pi}{3}$ s.

51. The horizontal range R of a projectile launched from ground level at an angle θ and initial velocity v_0 m/s is $R = (v_0^2/9.8)\sin\theta\cos\theta$. Calculate $dR/d\theta$. If $\theta = 7\pi/24$, will the range increase or decrease if the angle is increased slightly? Base your answer on the sign of the derivative.

52. Show that if $\frac{\pi}{2} < \theta < \pi$, then the distance along the x-axis between θ and the point where the tangent line intersects the x-axis is equal to $|\tan\theta|$ (Figure 4).

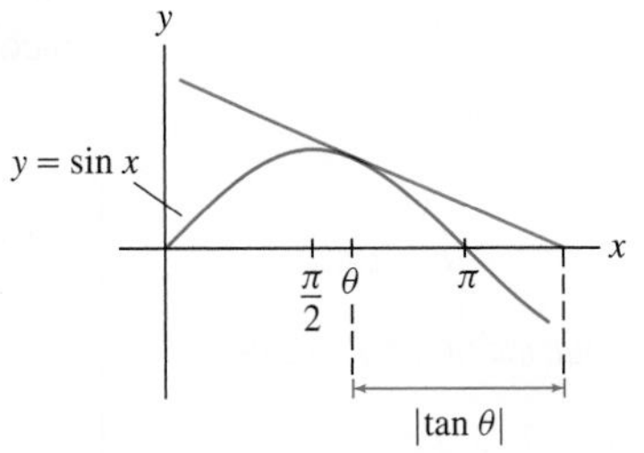

FIGURE 4

Further Insights and Challenges

53. Use the limit definition of the derivative and the addition law for the cosine function to prove that $(\cos x)' = -\sin x$.

54. Use the addition formula for the tangent

$$\tan(x + h) = \frac{\tan x + \tan h}{1 + \tan x \tan h}$$

to compute $(\tan x)'$ directly as a limit of the difference quotients. You will also need to show that $\lim\limits_{h\to 0} \dfrac{\tan h}{h} = 1$.

55. Verify the following identity and use it to give another proof of the formula $(\sin x)' = \cos x$.

$$\sin(x + h) - \sin x = 2\cos\left(x + \tfrac{1}{2}h\right)\sin\left(\tfrac{1}{2}h\right)$$

Hint: Use the addition formula to prove that $\sin(a + b) - \sin(a - b) = 2\cos a \sin b$.

56. Show that a nonzero polynomial function $y = f(x)$ *cannot* satisfy the equation $y'' = -y$. Use this to prove that neither $\sin x$ nor $\cos x$ is a polynomial. Can you think of another way to reach this conclusion by considering limits as $x \to \infty$?

57. Let $f(x) = x\sin x$ and $g(x) = x\cos x$.

(a) Show that $f'(x) = g(x) + \sin x$ and $g'(x) = -f(x) + \cos x$.

(b) Verify that $f''(x) = -f(x) + 2\cos x$ and $g''(x) = -g(x) - 2\sin x$.

(c) By further experimentation, try to find formulas for all higher derivatives of f and g. *Hint:* The kth derivative depends on whether $k = 4n, 4n + 1, 4n + 2$, or $4n + 3$.

58. Figure 5 shows the geometry behind the derivative formula $(\sin\theta)' = \cos\theta$. Segments $\overline{BA}$ and $\overline{BD}$ are parallel to the x- and y-axes. Let $\Delta \sin\theta = \sin(\theta + h) - \sin\theta$. Verify the following statements.

(a) $\Delta \sin\theta = BC$

(b) $\angle BDA = \theta$ *Hint:* $\overline{OA} \perp AD$.

(c) $BD = (\cos\theta)AD$

Now explain the following intuitive argument: If h is small, then $BC \approx BD$ and $AD \approx h$, so $\Delta \sin\theta \approx (\cos\theta)h$ and $(\sin\theta)' = \cos\theta$.

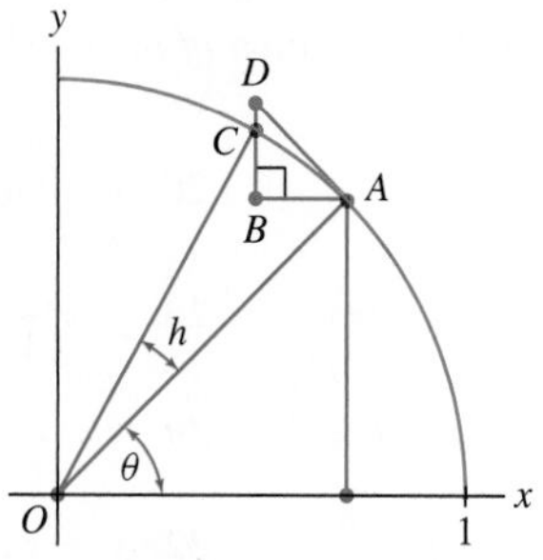

FIGURE 5

3.7 The Chain Rule

The **Chain Rule** is used to differentiate composite functions such as $y = \cos(x^3)$ and $y = \sqrt{x^4 + 1}$.

Recall that a *composite function* is obtained by "plugging" one function into another. The composite of f and g, denoted $f \circ g$, is defined by

$$(f \circ g)(x) = f\big(g(x)\big)$$

For convenience, we call f the *outside* function and g the *inside* function. Often, we write the composite function as $f(u)$, where $u = g(x)$. For example, $y = \cos(x^3)$ is the function $y = \cos u$, where $u = x^3$.

In verbal form, the Chain Rule says

$$\big(f(g(x))\big)' = \textit{outside}'\,(\textit{inside}) \cdot \textit{inside}'$$

A proof of the Chain Rule is given at the end of this section.

THEOREM 1 Chain Rule If f and g are differentiable, then the composite function $(f \circ g)(x) = f(g(x))$ is differentiable and

$$\big(f(g(x))\big)' = f'\big(g(x)\big)\, g'(x)$$

EXAMPLE 1 Calculate the derivative of $y = \cos(x^3)$.

Solution As noted above, $y = \cos(x^3)$ is a composite $f(g(x))$ where

$$f(u) = \cos u, \qquad u = g(x) = x^3$$
$$f'(u) = -\sin u, \qquad g'(x) = 3x^2$$

Note that $f'(g(x)) = -\sin(x^3)$, so by the Chain Rule,

$$\frac{d}{dx}\cos(x^3) = \underbrace{-\sin(x^3)}_{f'(g(x))}\ \underbrace{(3x^2)}_{g'(x)} = -3x^2\sin(x^3)$$ ■

EXAMPLE 2 Calculate the derivative of $y = \sqrt{x^4 + 1}$.

Solution The function $y = \sqrt{x^4 + 1}$ is a composite $f(g(x))$ where

$$f(u) = u^{1/2}, \qquad u = g(x) = x^4 + 1$$

$$f'(u) = \frac{1}{2}u^{-1/2}, \qquad g'(x) = 4x^3$$

Note that $f'(g(x)) = \frac{1}{2}(x^4 + 1)^{-1/2}$, so by the Chain Rule,

$$\frac{d}{dx}\sqrt{x^4 + 1} = \underbrace{\frac{1}{2}(x^4 + 1)^{-1/2}}_{f'(g(x))} \underbrace{(4x^3)}_{g'(x)} = \frac{4x^3}{2\sqrt{x^4 + 1}}$$ ■

EXAMPLE 3 Calculate $\dfrac{dy}{dx}$ for $y = \tan\left(\dfrac{x}{x+1}\right)$.

Solution The outside function is $f(u) = \tan u$. Because $f'(u) = \sec^2 u$, the Chain Rule gives us

$$\frac{d}{dx}\tan\left(\frac{x}{x+1}\right) = \sec^2\left(\frac{x}{x+1}\right) \underbrace{\frac{d}{dx}\left(\frac{x}{x+1}\right)}_{\text{Derivative of inside function}}$$

Now, by the Quotient Rule,

$$\frac{d}{dx}\left(\frac{x}{x+1}\right) = \frac{(x+1)\frac{d}{dx}x - x\frac{d}{dx}(x+1)}{(x+1)^2} = \frac{1}{(x+1)^2}$$

We obtain

$$\frac{d}{dx}\tan\left(\frac{x}{x+1}\right) = \sec^2\left(\frac{x}{x+1}\right)\frac{1}{(x+1)^2} = \frac{\sec^2\left(\frac{x}{x+1}\right)}{(x+1)^2}$$ ■

It is instructive to write the Chain Rule in Leibniz notation. Let

$$y = f(u) = f(g(x))$$

Then, by the Chain Rule,

$$\frac{dy}{dx} = f'(u)\,g'(x) = \frac{df}{du}\frac{du}{dx}$$

or

$$\boxed{\frac{dy}{dx} = \frac{dy}{du}\frac{du}{dx}}$$

Christiaan Huygens (1629–1695), one of the greatest scientists of his age, was Leibniz's teacher in mathematics and physics. He admired Isaac Newton greatly but did not accept Newton's theory of gravitation. He referred to it as the "improbable principle of attraction," because it did not explain how two masses separated by a distance could influence each other.

CONCEPTUAL INSIGHT In Leibniz notation, it appears as if we are multiplying fractions and the Chain Rule is simply a matter of "canceling the du." Since the symbolic expressions dy/du and du/dx are not fractions, this does not make sense literally, but it does suggest that derivatives behave *as if they were fractions* (this is reasonable because a derivative is a *limit* of fractions, namely of the difference quotients). Leibniz's form also emphasizes a key aspect of the Chain Rule: *Rates of change multiply*. To illustrate, suppose that (thanks to your knowledge of calculus) your salary increases twice as fast as your friend's. If your friend's salary increases \$4000 per year, your salary will increase at the rate of 2 × 4000 or \$8000 per year. In terms of derivatives,

$$\frac{d(\text{your salary})}{dt} = \frac{d(\text{your salary})}{d(\text{friend's salary})} \times \frac{d(\text{friend's salary})}{dt}$$

$$\$8000/\text{yr} = \quad 2 \quad \times \quad \$4000/\text{yr}$$

■ **EXAMPLE 4** Imagine a sphere whose radius r increases at a rate of 3 cm/s. At what rate is the volume V of the sphere increasing when $r = 10$ cm?

Solution Because we are asked to determine the rate at which V is increasing, we must find dV/dt. What we are given is the rate dr/dt, namely $dr/dt = 3$ cm/s. The Chain Rule allows us to express dV/dt in terms of dV/dr and dr/dt:

$$\underbrace{\frac{dV}{dt}}_{\substack{\text{Rate of change of volume}\\ \text{with respect to time}}} = \underbrace{\frac{dV}{dr}}_{\substack{\text{Rate of change of volume}\\ \text{with respect to radius}}} \times \underbrace{\frac{dr}{dt}}_{\substack{\text{Rate of change of radius}\\ \text{with respect to time}}}$$

To compute dV/dr, we use the formula for the volume of a sphere, $V = \frac{4}{3}\pi r^3$:

$$\frac{dV}{dr} = \frac{d}{dr}\left(\frac{4}{3}\pi r^3\right) = 4\pi r^2$$

Because $dr/dt = 3$, we obtain

$$\frac{dV}{dt} = \frac{dV}{dr}\frac{dr}{dt} = 4\pi r^2(3) = 12\pi r^2$$

For $r = 10$,

$$\left.\frac{dV}{dt}\right|_{r=10} = (12\pi)10^2 = 1200\pi \approx 3770 \text{ cm}^3/\text{s}$$ ■

We now discuss some important special cases of the Chain Rule.

THEOREM 2 General Power Rule If $g(x)$ is differentiable, then for every number n,

$$\frac{d}{dx}g(x)^n = n(g(x))^{n-1}g'(x)$$

Proof Let $f(u) = u^n$. Then $g(x)^n = f(g(x))$, and the Chain Rule yields

$$\frac{d}{dx} g(x)^n = f'(g(x))g'(x) = n(g(x))^{n-1} g'(x)$$

■ **EXAMPLE 5 General Power and Exponential Rules** Find the derivatives of **(a)** $y = (x^2 + 7x + 2)^{-1/3}$ and **(b)** $y = \sec^4 t$.

Solution Apply $\frac{d}{dx} g(x)^n = n g(x)^{n-1} g'(x)$ in (A) and $\frac{d}{dx} e^{g(x)} = g'(x) e^{g(x)}$ in (B).

(a)
$$\frac{d}{dx}(x^2 + 7x + 2)^{-1/3} = -\frac{1}{3}(x^2 + 7x + 2)^{-4/3} \frac{d}{dx}(x^2 + 7x + 2)$$
$$= -\frac{1}{3}(x^2 + 7x + 2)^{-4/3}(2x + 7)$$

(b)
$$\frac{d}{dt} \sec^4 t = 4 \sec^3 t \frac{d}{dt} \sec t = 4 \sec^3 t (\sec t \tan t) = 4 \sec^4 t \tan t$$

The Chain Rule applied to $f(kx + b)$ yields another important special case:

$$\frac{d}{dx} f(kx + b) = f'(kx + b) \frac{d}{dx}(kx + b) = k f'(kx + b)$$

THEOREM 3 Shifting and Scaling Rule If $f(x)$ is differentiable, then for any constants k and b,

$$\frac{d}{dx} f(kx + b) = k f'(kx + b)$$

For example,

$$\frac{d}{dx} \sin\left(2x + \frac{\pi}{4}\right) = 2 \cos\left(2x + \frac{\pi}{4}\right)$$

$$\frac{d}{dx}(9x - 2)^5 = (9)(5)(9x - 2)^4 = 45(9x - 2)^4$$

$$\frac{d}{dt} \sin(-4t) = -4 \cos(-4t)$$

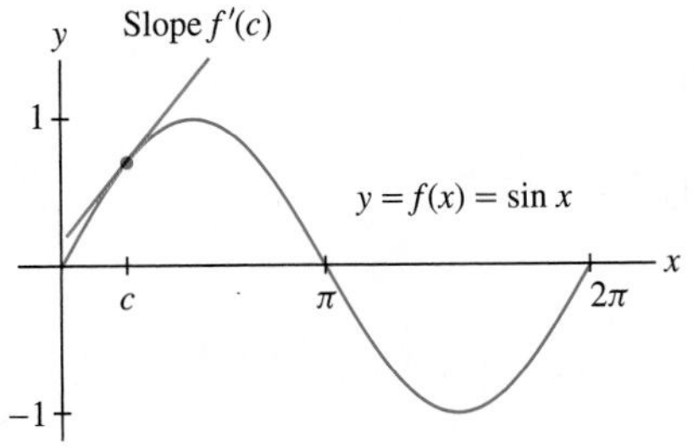

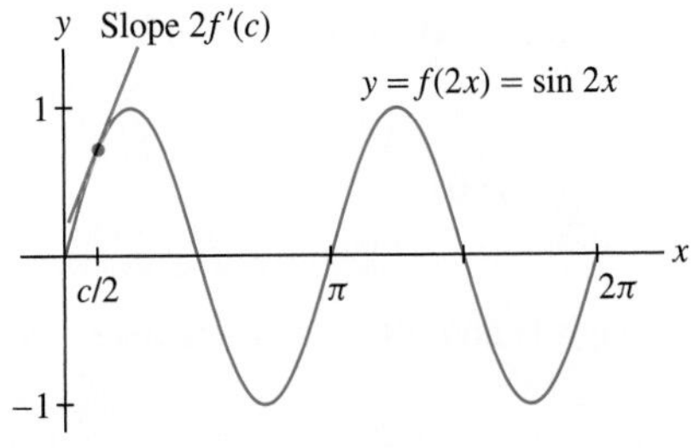

FIGURE 1 The derivative of $f(2x)$ at $x = c/2$ is twice the derivative of $f(x)$ at $x = c$.

GRAPHICAL INSIGHT To understand Theorem 3 graphically, recall that the graphs of $f(kx + b)$ and $f(x)$ are related by shifting and scaling (Section 1.1). For example, if $k > 1$, then the graph of $f(kx + b)$ is a compressed version of the graph of $f(x)$ that is steeper by a factor of k. Figure 1 illustrates a case with $k = 2$.

When the inside function is itself a composite function, we apply the Chain Rule more than once, as in the next example.

■ **EXAMPLE 6** Using the Chain Rule Twice Calculate $\dfrac{d}{dx}\sqrt{1+\sqrt{x^2+1}}$.

Solution First apply the Chain Rule with inside function $u = 1 + \sqrt{x^2+1}$:

$$\frac{d}{dx}\left(1+(x^2+1)^{1/2}\right)^{1/2} = \frac{1}{2}\left(1+(x^2+1)^{1/2}\right)^{-1/2}\frac{d}{dx}\left(1+(x^2+1)^{1/2}\right)$$

Then apply the Chain Rule again to the remaining derivative:

$$\frac{d}{dx}\left(1+(x^2+1)^{1/2}\right)^{1/2} = \frac{1}{2}(1+(x^2+1)^{1/2})^{-1/2}\left(\frac{1}{2}(x^2+1)^{-1/2}(2x)\right)$$

$$= \frac{1}{2}x(x^2+1)^{-1/2}\left(1+(x^2+1)^{1/2}\right)^{-1/2} \qquad ■$$

According to our convention, $\sin x$ denotes the sine of x radians, and with this convention, the formula $(\sin x)' = \cos x$ holds. In the next example, we derive a formula for the derivative of the sine function when x is measured in degrees.

■ **EXAMPLE 7** Trigonometric Derivatives in Degrees Calculate the derivative of the sine function as a function of degrees rather than radians.

Solution To solve this problem, it is convenient to use an underline to indicate a function of degrees rather than radians. For example,

$$\underline{\sin} x = \text{sine of } x \text{ degrees}$$

The functions $\sin x$ and $\underline{\sin} x$ are different, but they are related by

$$\underline{\sin} x = \sin\left(\frac{\pi}{180}x\right)$$

because x degrees corresponds to $\frac{\pi}{180}x$ radians. By Theorem 3,

$$\frac{d}{dx}\underline{\sin} x = \frac{d}{dx}\sin\left(\frac{\pi}{180}x\right) = \left(\frac{\pi}{180}\right)\cos\left(\frac{\pi}{180}x\right) = \left(\frac{\pi}{180}\right)\underline{\cos} x \qquad ■$$

A similar calculation shows that the factor $\frac{\pi}{180}$ appears in the formulas for the derivatives of the other standard trigonometric functions with respect to degrees. For example,

$$\frac{d}{dx}\underline{\tan} x = \left(\frac{\pi}{180}\right)\underline{\sec}^2 x$$

Proof of the Chain Rule The difference quotient for the composite $f \circ g$ is

$$\frac{f(g(x+h)) - f(g(x))}{h} \qquad (h \neq 0)$$

Our goal is to show that $(f \circ g)'$ is the product of $f'(g(x))$ and $g'(x)$, so it makes sense to write the difference quotient as a product:

$$\frac{f(g(x+h)) - f(g(x))}{h} = \frac{f(g(x+h)) - f(g(x))}{g(x+h) - g(x)} \times \frac{g(x+h) - g(x)}{h} \qquad \boxed{1}$$

This is legitimate only if the denominator $g(x+h) - g(x)$ is nonzero. Therefore, to continue our proof, we make the extra assumption that $g(x+h) - g(x) \neq 0$ for all h near but not equal to 0. This assumption is not necessary, but without it, the argument is more technical (see Exercise 105).

Under our assumption, we may use Eq. (1) to write $(f \circ g)'(x)$ as a product of two limits:

$$(f \circ g)'(x) = \underbrace{\lim_{h \to 0} \frac{f(g(x+h)) - f(g(x))}{g(x+h) - g(x)}}_{\text{Show that this equals } f'(g(x)).} \times \underbrace{\lim_{h \to 0} \frac{g(x+h) - g(x)}{h}}_{\text{This is } g'(x).}$$

The second limit on the right is $g'(x)$. The Chain Rule will follow if we show that the first limit equals $f'(g(x))$. To verify this, set

$$k = g(x+h) - g(x)$$

Then $g(x+h) = g(x) + k$ and

$$\frac{f(g(x+h)) - f(g(x))}{g(x+h) - g(x)} = \frac{f(g(x)+k) - f(g(x))}{k}$$

The function $g(x)$ is continuous because it is differentiable. Therefore, $g(x+h)$ tends to $g(x)$ and $k = g(x+h) - g(x)$ tends to zero as $h \to 0$. Thus, we may rewrite the limit in terms of k to obtain the desired result:

$$\lim_{h \to 0} \frac{f(g(x+h)) - f(g(x))}{g(x+h) - g(x)} = \lim_{k \to 0} \frac{f(g(x)+k) - f(g(x))}{k} = f'(g(x)) \qquad \blacksquare$$

3.7 SUMMARY

- The Chain Rule expresses $(f \circ g)'$ in terms of f' and g':

$$(f(g(x)))' = f'(g(x))\, g'(x)$$

- In Leibniz notation: $\dfrac{dy}{dx} = \dfrac{dy}{du}\dfrac{du}{dx}$, where $y = f(u)$ and $u = g(x)$
- General Power Rule: $\dfrac{d}{dx} g(x)^n = n(g(x))^{n-1} g'(x)$
- Shifting and Scaling Rule: $\dfrac{d}{dx} f(kx+b) = kf'(kx+b)$

3.7 EXERCISES

Preliminary Questions

1. Identify the outside and inside functions for each of these composite functions.

(a) $y = \sqrt{4x + 9x^2}$ **(b)** $y = \tan(x^2 + 1)$

(c) $y = \sec^5 x$ **(d)** $y = (1 + x^{12})^4$

2. Which of the following can be differentiated easily *without* using the Chain Rule?

(a) $y = \tan(7x^2 + 2)$ **(b)** $y = \dfrac{x}{x+1}$

(c) $y = \sqrt{x} \cdot \sec x$ **(d)** $y = \sqrt{x \cos x}$

(e) $y = x \sec \sqrt{x}$ **(f)** $y = \tan(4x)$

3. Which is the derivative of $f(5x)$?

(a) $5f'(x)$ **(b)** $5f'(5x)$ **(c)** $f'(5x)$

4. Suppose that $f'(4) = g(4) = g'(4) = 1$. Do we have enough information to compute $F'(4)$, where $F(x) = f(g(x))$? If not, what is missing?

Exercises

In Exercises 1–4, fill in a table of the following type:

$f(g(x))$	$f'(u)$	$f'(g(x))$	$g'(x)$	$(f \circ g)'$

1. $f(u) = u^{3/2}, \quad g(x) = x^4 + 1$

2. $f(u) = u^3, \quad g(x) = 3x + 5$

3. $f(u) = \tan u, \quad g(x) = x^4$

4. $f(u) = u^4 + u, \quad g(x) = \cos x$

In Exercises 5 and 6, write the function as a composite $f(g(x))$ and compute the derivative using the Chain Rule.

5. $y = (x + \sin x)^4$ **6.** $y = \cos(x^3)$

7. Calculate $\dfrac{d}{dx} \cos u$ for the following choices of $u(x)$:

(a) $u = 9 - x^2$ **(b)** $u = x^{-1}$ **(c)** $u = \tan x$

8. Calculate $\dfrac{d}{dx} f(x^2 + 1)$ for the following choices of $f(u)$:

(a) $f(u) = \sin u$ **(b)** $f(u) = 3u^{3/2}$ **(c)** $f(u) = u^2 - u$

9. Compute $\dfrac{df}{dx}$ if $\dfrac{df}{du} = 2$ and $\dfrac{du}{dx} = 6$.

10. Compute $\left.\dfrac{df}{dx}\right|_{x=2}$ if $f(u) = u^2$, $u(2) = -5$, and $u'(2) = -5$.

In Exercises 11–22, use the General Power Rule or the Shifting and Scaling Rule to compute the derivative.

11. $y = (x^4 + 5)^3$ **12.** $y = (8x^4 + 5)^3$

13. $y = \sqrt{7x - 3}$ **14.** $y = (4 - 2x - 3x^2)^5$

15. $y = (x^2 + 9x)^{-2}$ **16.** $y = (x^3 + 3x + 9)^{-4/3}$

17. $y = \cos^4 \theta$ **18.** $y = \cos(9\theta + 41)$

19. $y = (2\cos\theta + 5\sin\theta)^9$ **20.** $y = \sqrt{9 + x + \sin x}$

21. $y = \sin\left(\sqrt{x^2 + 2x + 9}\right)$ **22.** $y = \tan(4 - 3x)\sec(3 - 4x)$

In Exercises 23–26, compute the derivative of $f \circ g$.

23. $f(u) = \sin u, \quad g(x) = 2x + 1$

24. $f(u) = 2u + 1, \quad g(x) = \sin x$

25. $f(u) = u + u^{-1}, \quad g(x) = \tan x$

26. $f(u) = \dfrac{u}{u-1}, \quad g(x) = \csc x$

In Exercises 27 and 28, find the derivatives of $f(g(x))$ and $g(f(x))$.

27. $f(u) = \cos u, \quad u = g(x) = x^2 + 1$

28. $f(u) = u^3, \quad u = g(x) = \dfrac{1}{x+1}$

In Exercises 29–42, use the Chain Rule to find the derivative.

29. $y = \sin(x^2)$ **30.** $y = \sin^2 x$

31. $y = \sqrt{t^2 + 9}$ **32.** $y = (t^2 + 3t + 1)^{-5/2}$

33. $y = (x^4 - x^3 - 1)^{2/3}$ **34.** $y = (\sqrt{x+1} - 1)^{3/2}$

35. $y = \left(\dfrac{x+1}{x-1}\right)^4$ **36.** $y = \cos^3(12\theta)$

37. $y = \sec \dfrac{1}{x}$ **38.** $y = \tan(\theta^2 - 4\theta)$

39. $y = \tan(\theta + \cos\theta)$ **40.** $y = \sqrt{\cot^9 \theta + 1}$

41. $y = \csc(9 - 2\theta^2)$ **42.** $y = \cot(\sqrt{\theta} - 1)$

In Exercises 43–72, find the derivative using the appropriate rule or combination of rules.

43. $y = \tan(x^2 + 4x)$ **44.** $y = \sin(x^2 + 4x)$

45. $y = x\cos(1 - 3x)$ **46.** $y = \sin(x^2)\cos(x^2)$

47. $y = (4t + 9)^{1/2}$ **48.** $y = (z + 1)^4(2z - 1)^3$

49. $y = (x^3 + \cos x)^{-4}$ **50.** $y = \sin(\cos(\sin x))$

51. $y = \sqrt{\sin x \cos x}$ **52.** $y = (9 - (5 - 2x^4)^7)^3$

53. $y = (\cos 6x + \sin x^2)^{1/2}$

54. $y = \dfrac{(x+1)^{1/2}}{x+2}$

55. $y = \tan^3 x + \tan(x^3)$

56. $y = \sqrt{4 - 3\cos x}$

57. $y = \sqrt{\dfrac{z+1}{z-1}}$

58. $y = (\cos^3 x + 3\cos x + 7)^9$

59. $y = \dfrac{\cos(1+x)}{1+\cos x}$

60. $y = \sec(\sqrt{t^2 - 9})$

61. $y = \cot^7(x^5)$

62. $y = \dfrac{\cos(1/x)}{1+x^2}$

63. $y = \left(1 + \cot^5(x^4+1)\right)^9$

64. $y = \sqrt{\cos 2x + \sin 4x}$

65. $y = (1 - \csc^2(1 - x^3))^6$

66. $y = \sin\left(\sqrt{\sin\theta + 1}\right)$

67. $y = \left(x + \dfrac{1}{x}\right)^{-1/2}$

68. $y = \sec\left(1 + (4+x)^{-3/2}\right)$

69. $y = \sqrt{1 + \sqrt{1 + \sqrt{x}}}$

70. $y = \sqrt{\sqrt{x+1} + 1}$

71. $y = (kx+b)^{-1/3}$; k and b any constants

72. $y = \dfrac{1}{\sqrt{kt^4 + b}}$; k, b constants, not both zero

In Exercises 73–76, compute the higher derivative.

73. $\dfrac{d^2}{dx^2}\sin(x^2)$

74. $\dfrac{d^2}{dx^2}(x^2+9)^5$

75. $\dfrac{d^3}{dx^3}(9-x)^8$

76. $\dfrac{d^3}{dx^3}\sin(2x)$

77. The average molecular velocity v of a gas in a certain container is given by $v = 29\sqrt{T}$ m/s, where T is the temperature in kelvins. The temperature is related to the pressure (in atmospheres) by $T = 200P$. Find $\left.\dfrac{dv}{dP}\right|_{P=1.5}$.

78. The power P in a circuit is $P = Ri^2$, where R is the resistance and i is the current. Find dP/dt at $t = \frac{1}{3}$ if $R = 1000\ \Omega$ and i varies according to $i = \sin(4\pi t)$ (time in seconds).

79. An expanding sphere has radius $r = 0.4t$ cm at time t (in seconds). Let V be the sphere's volume. Find dV/dt when (a) $r = 3$ and (b) $t = 3$.

80. From a 2005 study by the Fisheries Research Services in Aberdeen, Scotland, we infer that the average length in centimeters of the species *Clupea harengus* (Atlantic herring) as a function of age t (in years) can be modeled by the function

$$L(t) = 32\left(1 - (1 + 0.37t + 0.068t^2 + 0.0085t^3 + 0.0009t^4)^{-1}\right)$$

for $0 \le t \le 13$. See Figure 2.

(a) How fast is the average length changing at age $t = 6$ years?

(b) GU Use a plot of $g'(t)$ to estimate the age t at which average length is changing at a rate of 5 cm/yr.

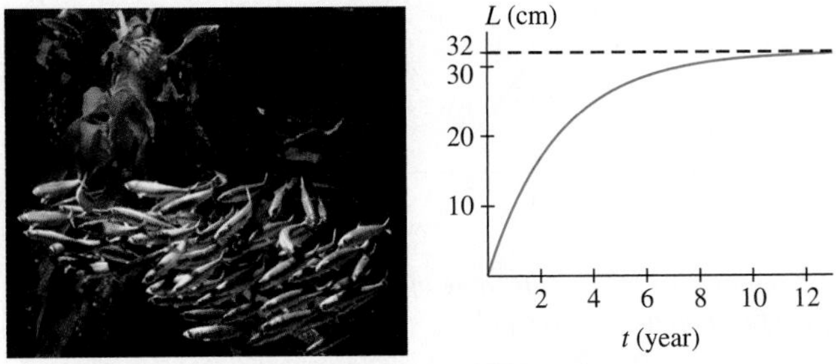

FIGURE 2 Average length of the species *Clupea harengus*

81. According to a 1999 study by Starkey and Scarnecchia, the average weight (in kilograms) at age t (in years) for channel catfish in the Lower Yellowstone River (Figure 3) is approximated by the function

$$W(t) = \left(0.14 + 0.115t - 0.002t^2 + 0.000023t^3\right)^{3.4}$$

Find the rate at which the average weight is changing at $t = 10$ years.

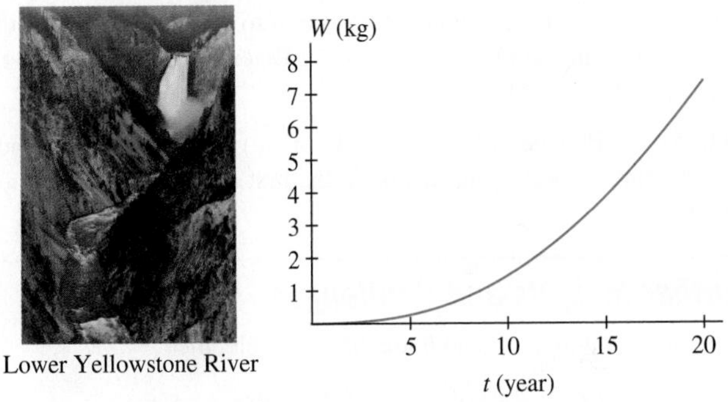

FIGURE 3 Average weight of channel catfish at age t

82. Calculate $M'(0)$ in terms of the constants a, b, k, and m, where

$$M(t) = \left(a + (b-a)\left(1 + kmt + \frac{1}{2}(kmt)^2\right)\right)^{1/m}$$

83. With notation as in Example 7, calculate

(a) $\left.\dfrac{d}{d\theta}\underline{\sin\theta}\right|_{\theta=60^\circ}$

(b) $\left.\dfrac{d}{d\theta}\left(\theta + \underline{\tan\theta}\right)\right|_{\theta=45^\circ}$

84. Assume that

$$f(0) = 2, \quad f'(0) = 3, \quad h(0) = -1, \quad h'(0) = 7$$

Calculate the derivatives of the following functions at $x = 0$:

(a) $(f(x))^3$ **(b)** $f(7x)$ **(c)** $f(4x)h(5x)$

85. Compute the derivative of $h(\sin x)$ at $x = \frac{\pi}{6}$, assuming that $h'(0.5) = 10$.

86. Let $F(x) = f(g(x))$, where the graphs of f and g are shown in Figure 4. Estimate $g'(2)$ and $f'(g(2))$ and compute $F'(2)$.

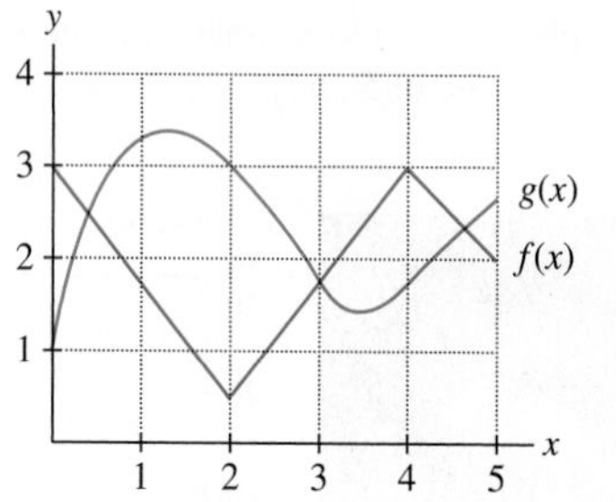

FIGURE 4

In Exercises 87–90, use the table of values to calculate the derivative of the function at the given point.

x	1	4	6
$f(x)$	4	0	6
$f'(x)$	5	7	4
$g(x)$	4	1	6
$g'(x)$	5	$\frac{1}{2}$	3

87. $f(g(x)), \quad x = 6$

88. $e^{f(x)}, \quad x = 4$

89. $g(\sqrt{x}), \quad x = 16$

90. $f(2x + g(x)), \quad x = 1$

91. The price (in dollars) of a computer component is $P = 2C - 18C^{-1}$, where C is the manufacturer's cost to produce it. Assume that cost at time t (in years) is $C = 9 + 3t^{-1}$. Determine the rate of change of price with respect to time at $t = 3$.

92. GU Plot the "astroid" $y = (4 - x^{2/3})^{3/2}$ for $0 \le x \le 8$. Show that the part of every tangent line in the first quadrant has a constant length 8.

93. According to the U.S. standard atmospheric model, developed by the National Oceanic and Atmospheric Administration for use in aircraft and rocket design, atmospheric temperature T (in degrees Celsius), pressure P (kPa = 1,000 pascals), and altitude h (in meters) are related by these formulas (valid in the troposphere $h \le 11{,}000$):

$$T = 15.04 - 0.000649h, \qquad P = 101.29 + \left(\frac{T + 273.1}{288.08}\right)^{5.256}$$

Use the Chain Rule to calculate dP/dh. Then estimate the change in P (in pascals, Pa) per additional meter of altitude when $h = 3{,}000$.

94. Climate scientists use the **Stefan-Boltzmann Law** $R = \sigma T^4$ to estimate the change in the earth's average temperature T (in kelvins) caused by a change in the radiation R (in joules per square meter per second) that the earth receives from the sun. Here $\sigma = 5.67 \times 10^{-8}$ $\text{Js}^{-1}\text{m}^{-2}\text{K}^{-4}$. Calculate dR/dt, assuming that $T = 283$ and $\frac{dT}{dt} = 0.05$ K/yr. What are the units of the derivative?

95. In the setting of Exercise 94, calculate the yearly rate of change of T if $T = 283$ K and R increases at a rate of 0.5 $\text{Js}^{-1}\text{m}^{-2}$ per year.

96. CAS Use a computer algebra system to compute $f^{(k)}(x)$ for $k = 1, 2, 3$ for the following functions:

(a) $f(x) = \cot(x^2)$

(b) $f(x) = \sqrt{x^3 + 1}$

97. Use the Chain Rule to express the second derivative of $f \circ g$ in terms of the first and second derivatives of f and g.

98. Compute the second derivative of $\sin(g(x))$ at $x = 2$, assuming that $g(2) = \frac{\pi}{4}$, $g'(2) = 5$, and $g''(2) = 3$.

Further Insights and Challenges

99. Show that if f, g, and h are differentiable, then

$$[f(g(h(x)))]' = f'(g(h(x)))g'(h(x))h'(x)$$

100. Show that differentiation reverses parity: If f is even, then f' is odd, and if f is odd, then f' is even. *Hint:* Differentiate $f(-x)$.

101. **(a)** Sketch a graph of any even function $f(x)$ and explain graphically why $f'(x)$ is odd.

(b) Suppose that $f'(x)$ is even. Is $f(x)$ necessarily odd? *Hint:* Check whether this is true for linear functions.

102. Power Rule for Fractional Exponents Let $f(u) = u^q$ and $g(x) = x^{p/q}$. Assume that $g(x)$ is differentiable.

(a) Show that $f(g(x)) = x^p$ (recall the laws of exponents).

(b) Apply the Chain Rule and the Power Rule for whole-number exponents to show that $f'(g(x))\,g'(x) = px^{p-1}$.

(c) Then derive the Power Rule for $x^{p/q}$.

103. Prove that for all whole numbers $n \ge 1$,

$$\frac{d^n}{dx^n} \sin x = \sin\left(x + \frac{n\pi}{2}\right)$$

Hint: Use the identity $\cos x = \sin\left(x + \frac{\pi}{2}\right)$.

104. A Discontinuous Derivative Use the limit definition to show that $g'(0)$ exists but $g'(0) \neq \lim_{x\to 0} g'(x)$, where

$$g(x) = \begin{cases} x^2 \sin \dfrac{1}{x} & x \neq 0 \\ 0 & x = 0 \end{cases}$$

105. Chain Rule This exercise proves the Chain Rule without the special assumption made in the text. For any number b, define a new function

$$F(u) = \frac{f(u) - f(b)}{u - b} \quad \text{for all } u \neq b$$

(a) Show that if we define $F(b) = f'(b)$, then $F(u)$ is continuous at $u = b$.

(b) Take $b = g(a)$. Show that if $x \neq a$, then for all u,

$$\frac{f(u) - f(g(a))}{x - a} = F(u)\frac{u - g(a)}{x - a} \qquad \boxed{2}$$

Note that both sides are zero if $u = g(a)$.

(c) Substitute $u = g(x)$ in Eq. (2) to obtain

$$\frac{f(g(x)) - f(g(a))}{x - a} = F(g(x))\frac{g(x) - g(a)}{x - a}$$

Derive the Chain Rule by computing the limit of both sides as $x \to a$.

3.8 Implicit Differentiation

We have developed the basic techniques for calculating a derivative dy/dx when y is given in terms of x by a formula—such as $y = x^3 + 1$. But suppose that y is determined instead by an equation such as

$$y^4 + xy = x^3 - x + 2 \tag{1}$$

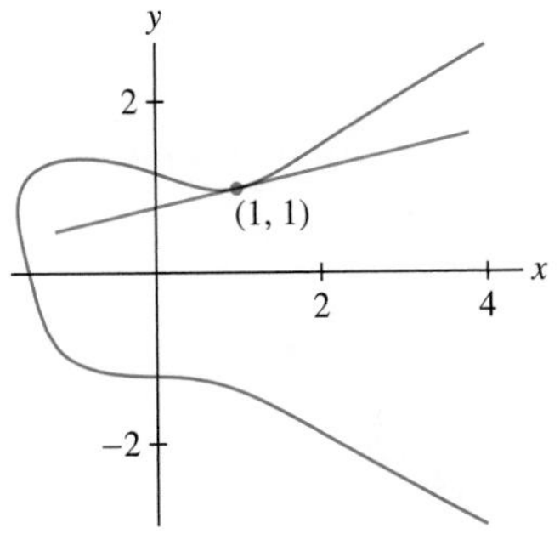

FIGURE 1 Graph of $y^4 + xy = x^3 - x + 2$

In this case, we say that y is defined *implicitly*. How can we find the slope of the tangent line at a point on the graph (Figure 1)? Although it may be difficult or even impossible to solve for y explicitly as a function of x, we can find dy/dx using the method of **implicit differentiation.**

To illustrate, consider the equation of the unit circle (Figure 2):

$$x^2 + y^2 = 1$$

Compute dy/dx by taking the derivative of both sides of the equation:

$$\frac{d}{dx}(x^2 + y^2) = \frac{d}{dx}(1)$$

$$\frac{d}{dx}(x^2) + \frac{d}{dx}(y^2) = 0$$

$$2x + \frac{d}{dx}(y^2) = 0 \tag{2}$$

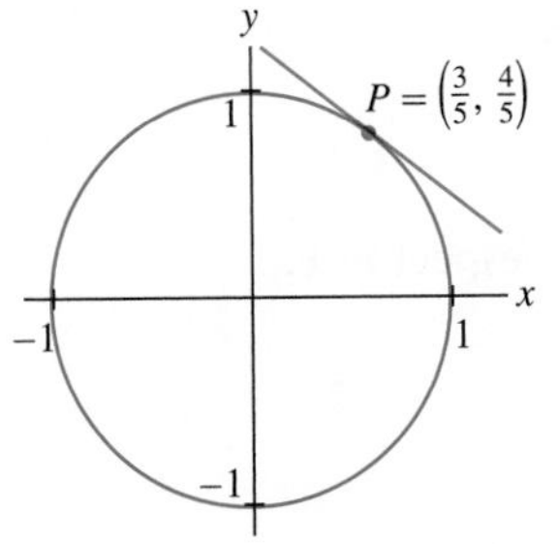

FIGURE 2 The tangent line to the unit circle $x^2 + y^2 = 1$ at P has slope $-\frac{3}{4}$.

How do we handle the term $\frac{d}{dx}(y^2)$? We use the Chain Rule. Think of y as a function $y = f(x)$. Then $y^2 = f(x)^2$ and by the Chain Rule,

$$\frac{d}{dx}y^2 = \frac{d}{dx}f(x)^2 = 2f(x)\frac{df}{dx} = 2y\frac{dy}{dx}$$

Equation (2) becomes $2x + 2y\frac{dy}{dx} = 0$, and we can solve for $\frac{dy}{dx}$ if $y \neq 0$:

$$\boxed{\frac{dy}{dx} = -\frac{x}{y}} \tag{3}$$

■ **EXAMPLE 1** Use Eq. (3) to find the slope of the tangent line at the point $P = \left(\frac{3}{5}, \frac{4}{5}\right)$ on the unit circle.

Solution Set $x = \frac{3}{5}$ and $y = \frac{4}{5}$ in Eq. (3):

$$\left.\frac{dy}{dx}\right|_P = -\frac{x}{y} = -\frac{\frac{3}{5}}{\frac{4}{5}} = -\frac{3}{4}$$ ■

In this particular example, we could have computed dy/dx directly, without implicit differentiation. The upper semicircle is the graph of $y = \sqrt{1 - x^2}$ and

$$\frac{dy}{dx} = \frac{d}{dx}\sqrt{1 - x^2} = \frac{1}{2}\left(1 - x^2\right)^{-1/2}\frac{d}{dx}\left(1 - x^2\right) = -\frac{x}{\sqrt{1 - x^2}}$$

This formula expresses dy/dx in terms of x alone, whereas Eq. (3) expresses dy/dx in terms of both x and y, as is typical when we use implicit differentiation. The two formulas agree because $y = \sqrt{1 - x^2}$.

Notice what happens if we insist on applying the Chain Rule to $\frac{d}{dy}\sin y$. The extra factor appears, but it is equal to 1:

$$\frac{d}{dy}\sin y = (\cos y)\frac{dy}{dy} = \cos y$$

Before presenting additional examples, let's examine again how the factor dy/dx arises when we differentiate an expression involving y with respect to x. It would not appear if we were differentiating with respect to y. Thus,

$$\frac{d}{dy}\sin y = \cos y \quad \text{but} \quad \frac{d}{dx}\sin y = (\cos y)\frac{dy}{dx}$$

$$\frac{d}{dy}y^4 = 4y^3 \quad \text{but} \quad \frac{d}{dx}y^4 = 4y^3\frac{dy}{dx}$$

Similarly, the Product Rule applied to xy yields

$$\frac{d}{dx}(xy) = x\frac{dy}{dx} + y\frac{dx}{dx} = x\frac{dy}{dx} + y$$

The Quotient Rule applied to t^2/y yields

$$\frac{d}{dt}\left(\frac{t^2}{y}\right) = \frac{y\frac{d}{dt}t^2 - t^2\frac{dy}{dt}}{y^2} = \frac{2ty - t^2\frac{dy}{dt}}{y^2}$$

■ **EXAMPLE 2** Find an equation of the tangent line at the point $P = (1, 1)$ on the curve (Figure 1)

$$y^4 + xy = x^3 - x + 2$$

Solution We break up the calculation into two steps.

***Step 1.* Differentiate both sides of the equation with respect to x.**

$$\frac{d}{dx}y^4 + \frac{d}{dx}(xy) = \frac{d}{dx}(x^3 - x + 2)$$

$$4y^3\frac{dy}{dx} + \left(x\frac{dy}{dx} + y\right) = 3x^2 - 1 \qquad \boxed{4}$$

***Step 2.* Solve for $\frac{dy}{dx}$.**

Move the terms involving dy/dx in Eq. (4) to the left and place the remaining terms on the right:

$$4y^3\frac{dy}{dx} + x\frac{dy}{dx} = 3x^2 - 1 - y$$

Then factor out dy/dx and divide:

$$(4y^3 + x)\frac{dy}{dx} = 3x^2 - 1 - y$$

$$\frac{dy}{dx} = \frac{3x^2 - 1 - y}{4y^3 + x} \qquad \boxed{5}$$

To find the derivative at $P = (1, 1)$, apply Eq. (5) with $x = 1$ and $y = 1$:

$$\left.\frac{dy}{dx}\right|_{(1,1)} = \frac{3 \cdot 1^2 - 1 - 1}{4 \cdot 1^3 + 1} = \frac{1}{5}$$

An equation of the tangent line is $y - 1 = \frac{1}{5}(x - 1)$ or $y = \frac{1}{5}x + \frac{4}{5}$. ■

CONCEPTUAL INSIGHT The graph of an equation does not always define a function because there may be more than one y-value for a given value of x. Implicit differentiation works because the graph is generally made up of several pieces called **branches**, each of which does define a function (a proof of this fact relies on the Implicit Function Theorem from advanced calculus). For example, the branches of the unit circle $x^2 + y^2 = 1$ are the graphs of the functions $y = \sqrt{1 - x^2}$ and $y = -\sqrt{1 - x^2}$. Similarly, the graph in Figure 3 has an upper and a lower branch. In most examples, the branches are differentiable except at certain exceptional points where the tangent line may be vertical.

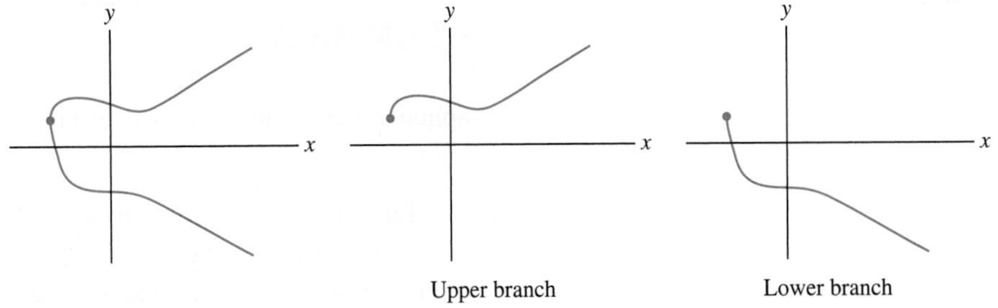

FIGURE 3 Each branch of the graph of $y^4 + xy = x^3 - x + 2$ defines a function of x.

■ **EXAMPLE 3** Calculate dy/dx at the point $\left(\frac{\pi}{4}, \frac{\pi}{4}\right)$ on the curve

$$\sqrt{2}\cos(x + y) = \cos x - \cos y$$

Solution We follow the steps of the previous example, this time writing y' for dy/dx:

$$\frac{d}{dx}\left(\sqrt{2}\cos(x + y)\right) = \frac{d}{dx}\cos x - \frac{d}{dx}\cos y$$

$$-\sqrt{2}\sin(x + y) \cdot (1 + y') = -\sin x + (\sin y)\, y' \qquad \text{(Chain Rule)}$$

$$-\sqrt{2}\sin(x + y) - \sqrt{2}\, y' \sin(x + y) = -\sin x + y' \sin y$$

$$-y'\left(\sin y + \sqrt{2}\sin(x + y)\right) = \sqrt{2}\sin(x + y) - \sin x \text{ (place } y'\text{-terms on the left)}$$

$$y' = \frac{\sin x - \sqrt{2}\sin(x + y)}{\sin y + \sqrt{2}\sin(x + y)}$$

The derivative at the point $\left(-\frac{\pi}{4}, \frac{\pi}{4}\right)$ is

$$\left.\frac{dy}{dx}\right|_{\left(-\frac{\pi}{4}, \frac{\pi}{4}\right)} = \frac{\sin\frac{\pi}{4} - \sqrt{2}\sin\left(\frac{\pi}{4} + \frac{\pi}{4}\right)}{\sin\frac{\pi}{4} + \sqrt{2}\sin\left(\frac{\pi}{4} + \frac{\pi}{4}\right)} = \frac{\sqrt{2}/2 - \sqrt{2}}{\sqrt{2}/2 + \sqrt{2}} = -\frac{1}{3}$$ ■

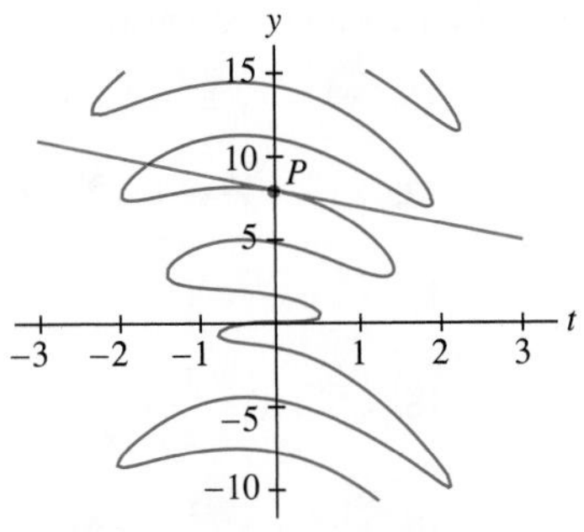

FIGURE 4 Graph of $y\cos(y + t + t^2) = t^3$. The tangent line at $P = \left(0, \frac{5\pi}{2}\right)$ has slope -1.

■ **EXAMPLE 4 Shortcut to Derivative at a Specific Point** Calculate $\left.\frac{dy}{dt}\right|_P$ at the point $P = \left(0, \frac{5\pi}{2}\right)$ on the curve (Figure 4):

$$y\cos(y + t + t^2) = t^3$$

Solution As before, differentiate both sides of the equation (we write y' for dy/dt):

$$\frac{d}{dt}y\cos(y + t + t^2) = \frac{d}{dt}t^3$$

$$y'\cos(y + t + t^2) - y\sin(y + t + t^2)(y' + 1 + 2t) = 3t^2$$

6

We could continue to solve for y', but that is not necessary. Instead, we can substitute $t = 0$, $y = \frac{5\pi}{2}$ directly in Eq. (6) to obtain

$$y' \cos\left(\frac{5\pi}{2} + 0 + 0^2\right) - \left(\frac{5\pi}{2}\right) \sin\left(\frac{5\pi}{2} + 0 + 0^2\right)(y' + 1 + 0) = 0$$

$$0 - \left(\frac{5\pi}{2}\right)(1)(y' + 1) = 0$$

This gives us $y' + 1 = 0$ or $y' = -1$. ■

3.8 SUMMARY

- Implicit differentiation is used to compute dy/dx when x and y are related by an equation.

Step 1. Take the derivative of both sides of the equation with respect to x.
Step 2. Solve for dy/dx by collecting the terms involving dy/dx on one side and the remaining terms on the other side of the equation.

- Remember to include the factor dy/dx when differentiating expressions involving y with respect to x. For instance,

$$\frac{d}{dx} \sin y = (\cos y) \frac{dy}{dx}$$

3.8 EXERCISES

Preliminary Questions

1. Which differentiation rule is used to show $\frac{d}{dx} \sin y = \cos y \frac{dy}{dx}$?

2. One of (a)–(c) is incorrect. Find and correct the mistake.

(a) $\frac{d}{dy} \sin(y^2) = 2y \cos(y^2)$ **(b)** $\frac{d}{dx} \sin(x^2) = 2x \cos(x^2)$

(c) $\frac{d}{dx} \sin(y^2) = 2y \cos(y^2)$

3. On an exam, Jason was asked to differentiate the equation

$$x^2 + 2xy + y^3 = 7$$

Find the errors in Jason's answer: $2x + 2xy' + 3y^2 = 0$

4. Which of (a) or (b) is equal to $\frac{d}{dx}(x \sin t)$?

(a) $(x \cos t)\frac{dt}{dx}$ **(b)** $(x \cos t)\frac{dt}{dx} + \sin t$

Exercises

1. Show that if you differentiate both sides of $x^2 + 2y^3 = 6$, the result is $2x + 6y^2 \frac{dy}{dx} = 0$. Then solve for dy/dx and evaluate it at the point $(2, 1)$.

2. Show that if you differentiate both sides of $xy + 4x + 2y = 1$, the result is $(x + 2)\frac{dy}{dx} + y + 4 = 0$. Then solve for dy/dx and evaluate it at the point $(1, -1)$.

In Exercises 3–8, differentiate the expression with respect to x, assuming that $y = f(x)$.

3. x^2y^3 **4.** $\frac{x^3}{y^2}$ **5.** $(x^2 + y^2)^{3/2}$

6. $\tan(xy)$ **7.** $\frac{y}{y+1}$ **8.** $\sin \frac{y}{x}$

In Exercises 9–26, calculate the derivative with respect to x.

9. $3y^3 + x^2 = 5$ **10.** $y^4 - 2y = 4x^3 + x$

11. $x^2y + 2x^3y = x + y$ **12.** $xy^2 + x^2y^5 - x^3 = 3$

13. $x^3R^5 = 1$ **14.** $x^4 + z^4 = 1$

15. $\frac{y}{x} + \frac{x}{y} = 2y$ **16.** $\sqrt{x+s} = \frac{1}{x} + \frac{1}{s}$

17. $y^{-2/3} + x^{3/2} = 1$ **18.** $x^{1/2} + y^{2/3} = -4y$

19. $y + \frac{1}{y} = x^2 + x$ **20.** $\sin(xt) = t$

21. $\sin(x + y) = x + \cos y$

22. $\tan(x^2 y) = (x + y)^3$

23. $\tan(x + y) = \tan x + \tan y$

24. $x \sin y - y \cos x = 2$

25. $x + \cos(3x - y) = xy$

26. $2x^2 - x - y = \sqrt{x^4 + y^4}$

27. Show that $x + yx^{-1} = 1$ and $y = x - x^2$ define the same curve (except that $(0, 0)$ is not a solution of the first equation) and that implicit differentiation yields $y' = yx^{-1} - x$ and $y' = 1 - 2x$. Explain why these formulas produce the same values for the derivative.

28. Use the method of Example 4 to compute $\left.\frac{dy}{dx}\right|_P$ at $P = (2, 1)$ on the curve $y^2x^3 + y^3x^4 - 10x + y = 5$.

In Exercises 29 and 30, find dy/dx at the given point.

29. $(x + 2)^2 - 6(2y + 3)^2 = 3, \quad (1, -1)$

30. $\sin^2(3y) = x + y, \quad \left(\frac{2 - \pi}{4}, \frac{\pi}{4}\right)$

In Exercises 31–38, find an equation of the tangent line at the given point.

31. $xy + x^2y^2 = 5, \quad (2, 1)$

32. $x^{2/3} + y^{2/3} = 2, \quad (1, 1)$

33. $x^2 + \sin y = xy^2 + 1, \quad (1, 0)$

34. $\sin(x - y) = x \cos\left(y + \frac{\pi}{4}\right), \quad \left(\frac{\pi}{4}, \frac{\pi}{4}\right)$

35. $2x^{1/2} + 4y^{-1/2} = xy, \quad (1, 4)$

36. $\dfrac{x}{x+1} + \dfrac{y}{y+1} = 1, \quad (1, 1)$

37. $\sin(2x - y) = \dfrac{x^2}{y}, \quad (0, \pi)$

38. $x + \sqrt{x} = y^2 + y^4, \quad (1, 1)$

39. Find the points on the graph of $y^2 = x^3 - 3x + 1$ (Figure 5) where the tangent line is horizontal.

(a) First show that $2yy' = 3x^2 - 3$, where $y' = dy/dx$.

(b) Do not solve for y'. Rather, set $y' = 0$ and solve for x. This yields two values of x where the slope may be zero.

(c) Show that the positive value of x does not correspond to a point on the graph.

(d) The negative value corresponds to the two points on the graph where the tangent line is horizontal. Find their coordinates.

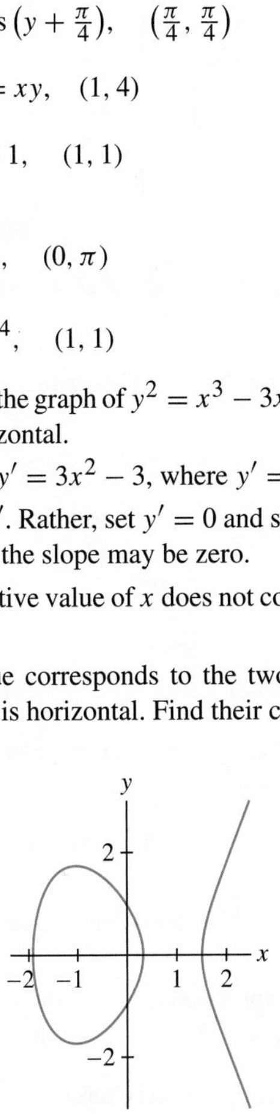

FIGURE 5 Graph of $y^2 = x^3 - 3x + 1$.

40. Show, by differentiating the equation, that if the tangent line at a point (x, y) on the curve $x^2y - 2x + 8y = 2$ is horizontal, then $xy = 1$. Then substitute $y = x^{-1}$ in $x^2y - 2x + 8y = 2$ to show that the tangent line is horizontal at the points $\left(2, \frac{1}{2}\right)$ and $\left(-4, -\frac{1}{4}\right)$.

41. Find all points on the graph of $3x^2 + 4y^2 + 3xy = 24$ where the tangent line is horizontal (Figure 6).

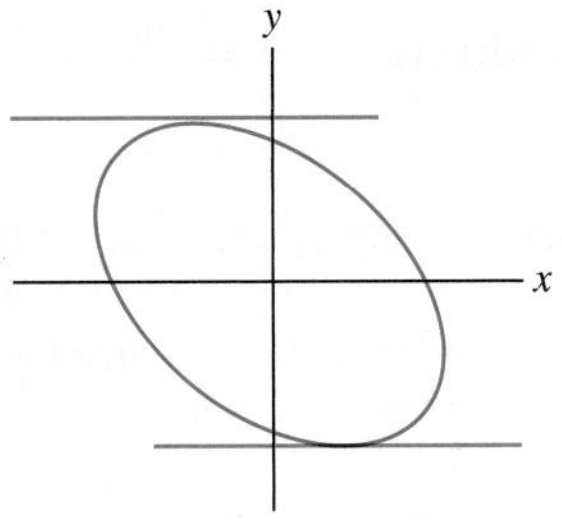

FIGURE 6 Graph of $3x^2 + 4y^2 + 3xy = 24$.

42. Show that no point on the graph of $x^2 - 3xy + y^2 = 1$ has a horizontal tangent line.

43. Figure 1 shows the graph of $y^4 + xy = x^3 - x + 2$. Find dy/dx at the two points on the graph with x-coordinate 0 and find an equation of the tangent line at $(1, 1)$.

44. Folium of Descartes The curve $x^3 + y^3 = 3xy$ (Figure 7) was first discussed in 1638 by the French philosopher-mathematician René Descartes, who called it the folium (meaning "leaf"). Descartes's scientific colleague Gilles de Roberval called it the jasmine flower. Both men believed incorrectly that the leaf shape in the first quadrant was repeated in each quadrant, giving the appearance of petals of a flower. Find an equation of the tangent line at the point $\left(\frac{2}{3}, \frac{4}{3}\right)$.

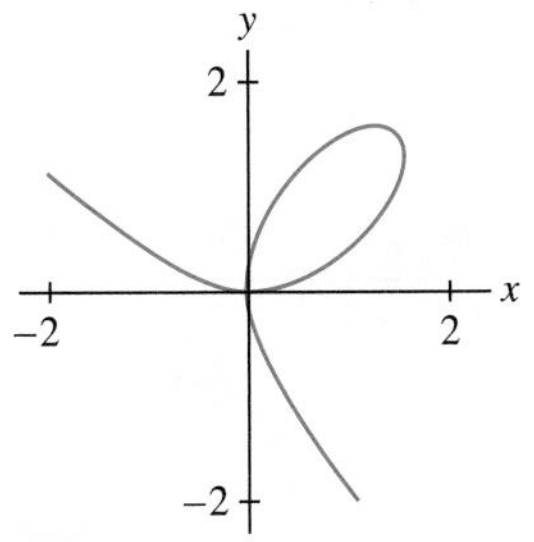

FIGURE 7 Folium of Descartes: $x^3 + y^3 = 3xy$.

45. Find a point on the folium $x^3 + y^3 = 3xy$ other than the origin at which the tangent line is horizontal.

46. GU Plot $x^3 + y^3 = 3xy + b$ for several values of b and describe how the graph changes as $b \to 0$. Then compute dy/dx at the point $(b^{1/3}, 0)$. How does this value change as $b \to \infty$? Do your plots confirm this conclusion?

47. Find the x-coordinates of the points where the tangent line is horizontal on the *trident curve* $xy = x^3 - 5x^2 + 2x - 1$, so named by Isaac Newton in his treatise on curves published in 1710 (Figure 8). *Hint:* $2x^3 - 5x^2 + 1 = (2x - 1)(x^2 - 2x - 1)$.

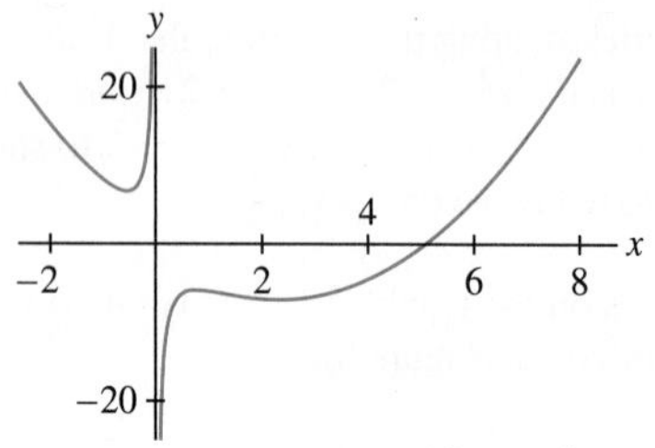

FIGURE 8 Trident curve: $xy = x^3 - 5x^2 + 2x - 1$.

48. Find an equation of the tangent line at each of the four points on the curve $(x^2 + y^2 - 4x)^2 = 2(x^2 + y^2)$ where $x = 1$. This curve (Figure 9) is an example of a *limaçon of Pascal*, named after the father of the French philosopher Blaise Pascal, who first described it in 1650.

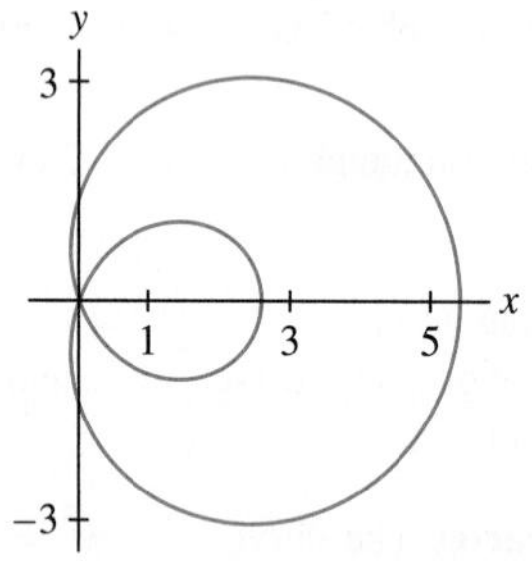

FIGURE 9 Limaçon: $(x^2 + y^2 - 4x)^2 = 2(x^2 + y^2)$.

49. Find the derivative at the points where $x = 1$ on the folium $(x^2 + y^2)^2 = \frac{25}{4}xy^2$. See Figure 10.

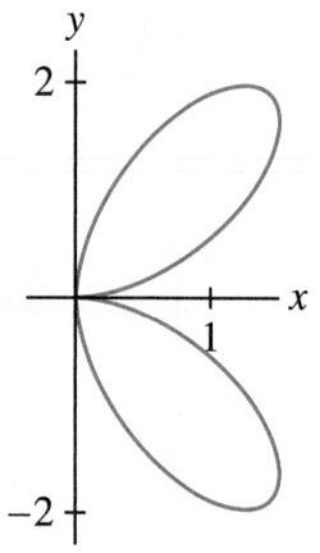

FIGURE 10 Folium curve: $(x^2 + y^2)^2 = \dfrac{25}{4}xy^2$

50. CAS Plot $(x^2 + y^2)^2 = 12(x^2 - y^2) + 2$ for $-4 \le x \le 4$, $4 \le y \le 4$ using a computer algebra system. How many horizontal tangent lines does the curve appear to have? Find the points where these occur.

Exercises 51–53: If the derivative dx/dy (instead of $dy/dx = 0$) exists at a point and $dx/dy = 0$, then the tangent line at that point is vertical.

51. Calculate dx/dy for the equation $y^4 + 1 = y^2 + x^2$ and find the points on the graph where the tangent line is vertical.

52. Show that the tangent lines at $x = 1 \pm \sqrt{2}$ to the *conchoid* with equation $(x - 1)^2(x^2 + y^2) = 2x^2$ are vertical (Figure 11).

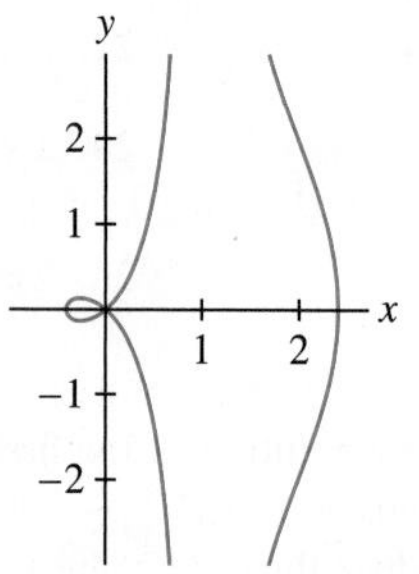

FIGURE 11 Conchoid: $(x - 1)^2(x^2 + y^2) = 2x^2$.

53. CAS Use a computer algebra system to plot

$$y^2 = x^3 - 4x \quad \text{for } -4 \le x \le 4,\ 4 \le y \le 4$$

Show that if $dx/dy = 0$, then $y = 0$. Conclude that the tangent line is vertical at the points where the curve intersects the x-axis. Does your plot confirm this conclusion?

54. Show that for all points P on the graph in Figure 12, the segments $\overline{OP}$ and $\overline{PR}$ have equal length.

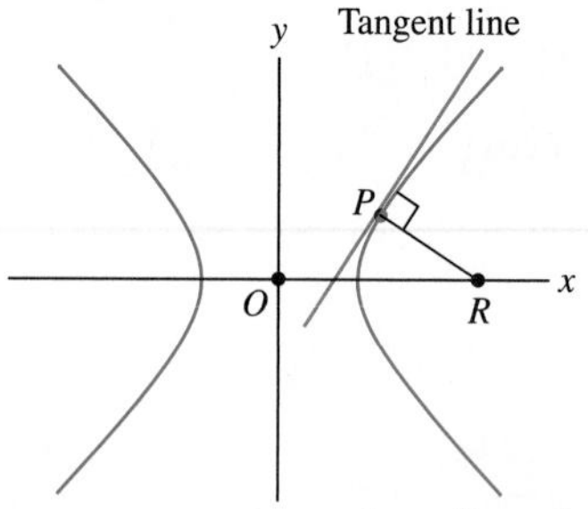

FIGURE 12 Graph of $x^2 - y^2 = a^2$.

In Exercises 55–58, use implicit differentiation to calculate higher derivatives.

55. Consider the equation $y^3 - \frac{3}{2}x^2 = 1$.

(a) Show that $y' = x/y^2$ and differentiate again to show that

$$y'' = \frac{y^2 - 2xyy'}{y^4}$$

(b) Express y'' in terms of x and y using part (a).

56. Use the method of the previous exercise to show that $y'' = -y^{-3}$ on the circle $x^2 + y^2 = 1$.

57. Calculate y'' at the point $(1, 1)$ on the curve $xy^2 + y - 2 = 0$ by the following steps:

(a) Find y' by implicit differentiation and calculate y' at the point $(1, 1)$.

(b) Differentiate the expression for y' found in (a). Then compute y'' at $(1, 1)$ by substituting $x = 1$, $y = 1$, and the value of y' found in (a).

58. Use the method of the previous exercise to compute y'' at the point $(1, 1)$ on the curve $x^3 + y^3 = 3x + y - 2$.

In Exercises 59–61, x and y are functions of a variable t and use implicit differentiation to relate dy/dt and dx/dt.

59. Differentiate $xy = 1$ with respect to t and derive the relation $\frac{dy}{dt} = -\frac{y}{x}\frac{dx}{dt}$.

60. Differentiate $x^3 + 3xy^2 = 1$ with respect to t and express dy/dt in terms of dx/dt, as in Exercise 59.

61. Calculate dy/dt in terms of dx/dt.

(a) $x^3 - y^3 = 1$ **(b)** $y^4 + 2xy + x^2 = 0$

62. The volume V and pressure P of gas in a piston (which vary in time t) satisfy $PV^{3/2} = C$, where C is a constant. Prove that

$$\frac{dP/dt}{dV/dt} = -\frac{3}{2}\frac{P}{V}$$

The ratio of the derivatives is negative. Could you have predicted this from the relation $PV^{3/2} = C$?

Further Insights and Challenges

63. Show that if P lies on the intersection of the two curves $x^2 - y^2 = c$ and $xy = d$ (c, d constants), then the tangents to the curves at P are perpendicular.

64. The *lemniscate curve* $(x^2 + y^2)^2 = 4(x^2 - y^2)$ was discovered by Jacob Bernoulli in 1694, who noted that it is "shaped like a figure 8, or a knot, or the bow of a ribbon." Find the coordinates of the four points at which the tangent line is horizontal (Figure 13).

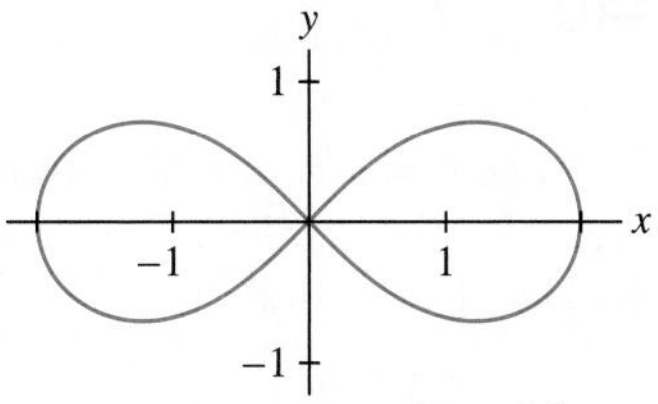

FIGURE 13 Lemniscate curve: $(x^2 + y^2)^2 = 4(x^2 - y^2)$.

65. Divide the curve (Figure 14)

$$y^5 - y = x^2y + x + 1$$

into five branches, each of which is the graph of a function. Sketch the branches.

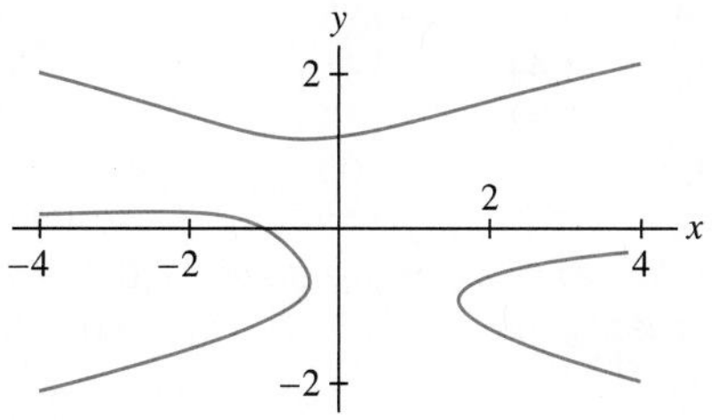

FIGURE 14 Graph of $y^5 - y = x^2y + x + 1$.

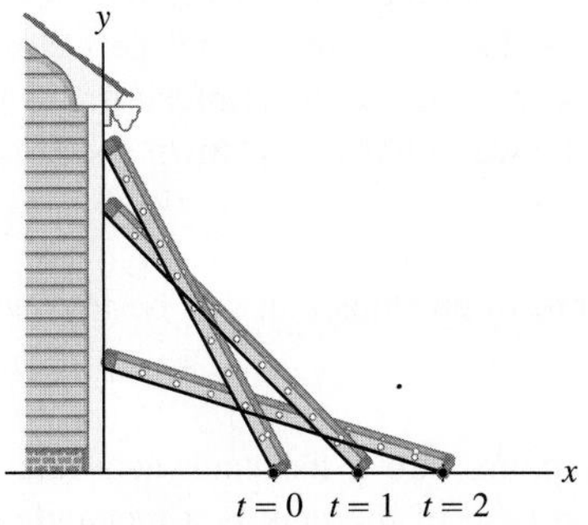

FIGURE 1 Positions of a ladder at times $t = 0, 1, 2$.

3.9 Related Rates

In *related-rate* problems, the goal is to calculate an unknown rate of change in terms of other rates of change that are known. The "sliding ladder problem" is a good example: A ladder leans against a wall as the bottom is pulled away at constant velocity. *How fast does the top of the ladder move*? What is interesting and perhaps surprising is that the top and bottom travel at different speeds. Figure 1 shows this clearly: The bottom travels the same distance over each time interval, but the top travels farther during the second time interval than the first. In other words, the top is speeding up while the bottom moves at a constant speed. In the next example, we use calculus to find the velocity of the ladder's top.

■ **EXAMPLE 1 Sliding Ladder Problem** A 5-meter ladder leans against a wall. The bottom of the ladder is 1.5 meters from the wall at time $t = 0$ and slides away from the wall at a rate of 0.8 m/s. Find the velocity of the top of the ladder at time $t = 1$.

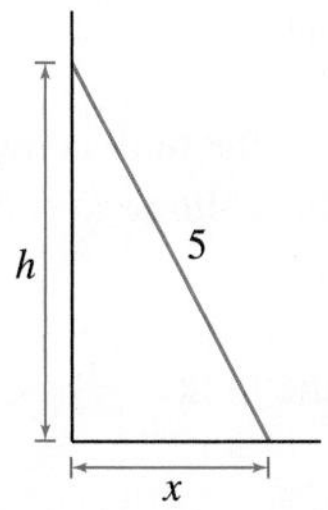

FIGURE 2 The variables x and h.

Solution The first step in any related-rate problem is to choose variables for the relevant quantities. Since we are considering how the top and bottom of the ladder change position, we use variables (Figure 2):

- $x = x(t)$ distance from the bottom of the ladder to the wall
- $h = h(t)$ height of the ladder's top

Both x and h are functions of time. The velocity of the bottom is $dx/dt = 0.8$ m/s. The unknown velocity of the top is dh/dt, and the initial distance from the bottom to the wall is $x(0) = 1.5$, so we can restate the problem as

$$\text{Compute } \frac{dh}{dt} \text{ at } t = 1 \qquad \text{given that} \qquad \frac{dx}{dt} = 0.8 \text{ m/s and } x(0) = 1.5 \text{ m}$$

To solve this problem, we need an equation relating x and h (Figure 2). This is provided by the Pythagorean Theorem:

$$x^2 + h^2 = 5^2$$

To calculate dh/dt, we differentiate both sides of this equation *with respect to t*:

$$\frac{d}{dt}x^2 + \frac{d}{dt}h^2 = \frac{d}{dt}5^2$$

$$2x\frac{dx}{dt} + 2h\frac{dh}{dt} = 0$$

Therefore $\frac{dh}{dt} = -\frac{x}{h}\frac{dx}{dt}$, and because $\frac{dx}{dt} = 0.8$ m/s, the velocity of the top is

$$\frac{dh}{dt} = -0.8\frac{x}{h} \text{ m/s} \tag{1}$$

t	x	h	dh/dt
0	1.5	4.77	−0.25
1	2.3	4.44	−0.41
2	3.1	3.92	−0.63
3	3.9	3.13	−1.00

This table of values confirms that the top of the ladder is speeding up.

To apply this formula, we must find x and h at time $t = 1$. Since the bottom slides away at 0.8 m/s and $x(0) = 1.5$, we have $x(1) = 2.3$ and $h(1) = \sqrt{5^2 - 2.3^2} \approx 4.44$. We obtain (note that the answer is negative because the ladder top is falling):

$$\left.\frac{dh}{dt}\right|_{t=1} = -0.8\frac{x(1)}{h(1)} \approx -0.8\frac{2.3}{4.44} \approx -0.41 \text{ m/s}$$ ■

CONCEPTUAL INSIGHT A puzzling feature of Eq. (1) is that the velocity dh/dt, which is equal to $-0.8x/h$, becomes infinite as $h \to 0$ (as the top of the ladder gets close to the ground). Since this is impossible, our mathematical model must break down as $h \to 0$. In fact, the ladder's top loses contact with the wall on the way down and from that moment on, the formula is no longer valid.

In the next examples, we divide the solution into three steps that can be followed when working the exercises.

■ **EXAMPLE 2 Filling a Rectangular Tank** Water pours into a fish tank at a rate of 0.3 m^3/min. How fast is the water level rising if the base of the tank is a rectangle of dimensions 2×3 meters?

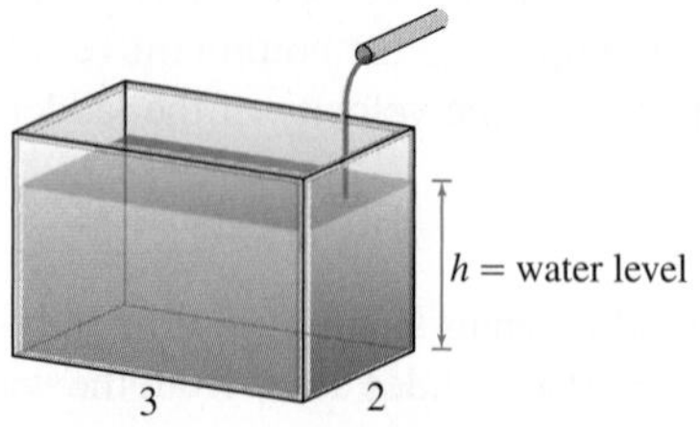

FIGURE 3 V = water volume at time t.

It is helpful to choose variables that are related to or traditionally associated with the quantity represented, such as V for volume, θ for an angle, h or y for height, and r for radius.

Solution To solve a related-rate problem, it is useful to draw a diagram if possible. Figure 3 illustrates our problem.

***Step 1.* Assign variables and restate the problem.**

First, we must recognize that the rate at which water pours into the tank is the derivative of water volume with respect to time. Therefore, let V be the volume and h the height of the water at time t. Then

$$\frac{dV}{dt} = \text{rate at which water is added to the tank}$$

$$\frac{dh}{dt} = \text{rate at which the water level is rising}$$

Now we can restate our problem in terms of derivatives:

$$\text{Compute } \frac{dh}{dt} \qquad \text{given that} \qquad \frac{dV}{dt} = 0.3 \text{ m}^3\text{/min}$$

Step 2. **Find an equation relating the variables and differentiate.**
We need a relation between V and h. We have $V = 6h$ since the tank's base has area 6 m^2. Therefore,

$$\frac{dV}{dt} = 6\frac{dh}{dt} \quad \Rightarrow \quad \frac{dh}{dt} = \frac{1}{6}\frac{dV}{dt}$$

Step 3. **Use the data to find the unknown derivative.**
Because $dV/dt = 0.3$, the water level rises at the rate

$$\frac{dh}{dt} = \frac{1}{6}\frac{dV}{dt} = \frac{1}{6}(0.3) = 0.05 \text{ m/min}$$

Note that dh/dt has units of meters per minute because h and t are in meters and minutes, respectively. ■

The set-up in the next example is similar but more complicated because the water tank has the shape of a circular cone. We use similar triangles to derive a relation between the volume and height of the water. We also need the formula $V = \frac{1}{3}\pi h r^2$ for the volume of a circular cone of height h and radius r.

■ **EXAMPLE 3** **Filling a Conical Tank** Water pours into a conical tank of height 10 m and radius 4 m at a rate of $6 \text{ m}^3\text{/min}$.

(a) At what rate is the water level rising when the level is 5 m high?

(b) As time passes, what happens to the rate at which the water level rises?

Solution

(a) ***Step 1.*** **Assign variables and restate the problem.**
As in the previous example, let V and h be the volume and height of the water in the tank at time t. Our problem, in terms of derivatives, is

$$\text{Compute } \frac{dh}{dt} \text{ at } h = 5 \qquad \text{given that} \qquad \frac{dV}{dt} = 6 \text{ m}^3\text{/min}$$

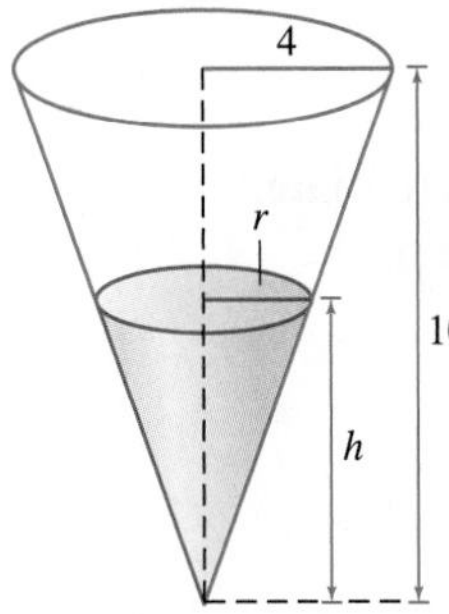

FIGURE 4 By similar triangles,
$$\frac{r}{h} = \frac{4}{10}$$

Step 2. **Find an equation relating the variables and differentiate.**
When the water level is h, the volume of water in the cone is $V = \frac{1}{3}\pi h r^2$, where r is the radius of the cone at height h, but *we cannot use this relation unless we eliminate the variable* r. Using similar triangles in Figure 4, we see that

$$\frac{r}{h} = \frac{4}{10}$$

or

$$r = 0.4h$$

Therefore,

$$V = \frac{1}{3}\pi h(0.4h)^2 = \left(\frac{0.16}{3}\right)\pi h^3$$

$$\frac{dV}{dt} = (0.16)\pi h^2 \frac{dh}{dt} \tag{2}$$

CAUTION A common mistake is substituting the particular value $h = 5$ in Eq. (2). Do not set $h = 5$ until the end of the problem, after the derivatives have been computed. This applies to all related-rate problems.

Step 3. **Use the data to find the unknown derivative.**

We are given that $\frac{dV}{dt} = 6$. Using this in Eq. (2), we obtain

$$(0.16)\pi h^2 \frac{dh}{dt} = 6$$

$$\frac{dh}{dt} = \frac{6}{(0.16)\pi h^2} \approx \frac{12}{h^2} \qquad \boxed{3}$$

When $h = 5$, the level is rising at a rate of $\frac{dh}{dt} \approx 12/5^2 = 0.48$ m/min.

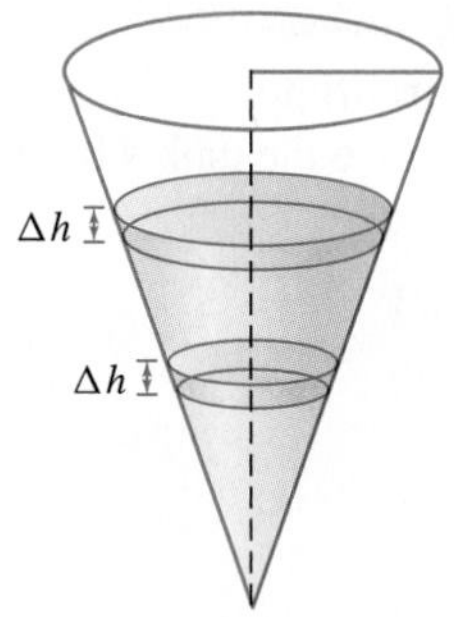

FIGURE 5 When h is larger, it takes more water to raise the level by an amount Δh.

(b) Eq. (3) shows that dh/dt is inversely proportional to h^2. As h increases, the water level rises more slowly. This is reasonable if you consider that a thin slice of the cone of width Δh has more volume when h is large, so more water is needed to raise the level when h is large (Figure 5). ■

■ **EXAMPLE 4 Tracking a Rocket** A spy uses a telescope to track a rocket launched vertically from a launching pad 6 km away, as in Figure 6. At a certain moment, the angle θ between the telescope and the ground is equal to $\frac{\pi}{3}$ and is changing at a rate of 0.9 rad/min. What is the rocket's velocity at that moment?

Solution

Step 1. **Assign variables and restate the problem.**

Let y be the height of the rocket at time t. Our goal is to compute the rocket's velocity dy/dt when $\theta = \frac{\pi}{3}$ so we can restate the problem as follows:

Compute $\left.\frac{dy}{dt}\right\|_{\theta=\frac{\pi}{3}}$ given that $\frac{d\theta}{dt} = 0.9$ rad/min when $\theta = \frac{\pi}{3}$

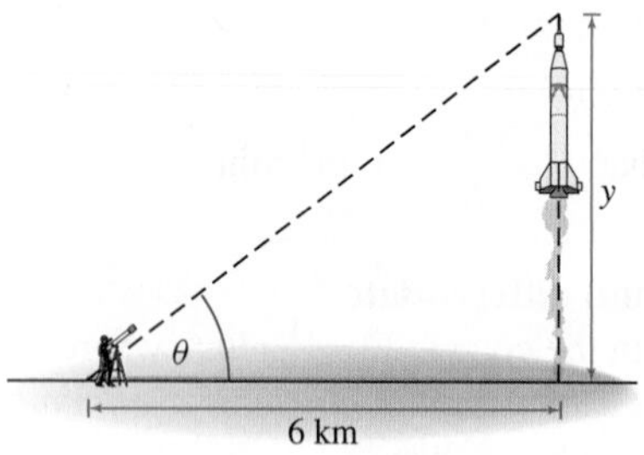

FIGURE 6 Tracking a rocket through a telescope.

Step 2. **Find an equation relating the variables and differentiate.**

We need a relation between θ and y. As we see in Figure 6,

$$\tan\theta = \frac{y}{6}$$

Now differentiate with respect to time:

$$\sec^2\theta \frac{d\theta}{dt} = \frac{1}{6}\frac{dy}{dt}$$

$$\frac{dy}{dt} = \frac{6}{\cos^2\theta}\frac{d\theta}{dt} \qquad \boxed{4}$$

Step 3. **Use the given data to find the unknown derivative.**

At the given moment, $\theta = \frac{\pi}{3}$ and $d\theta/dt = 0.9$, so Eq. (4) yields

$$\frac{dy}{dt} = \frac{6}{\cos^2(\pi/3)}(0.9) = \frac{6}{(0.5)^2}(0.9) = 21.6 \text{ km/min}$$

The rocket's velocity at this moment is 21.6 km/min, or approximately 1296 km/h. ■

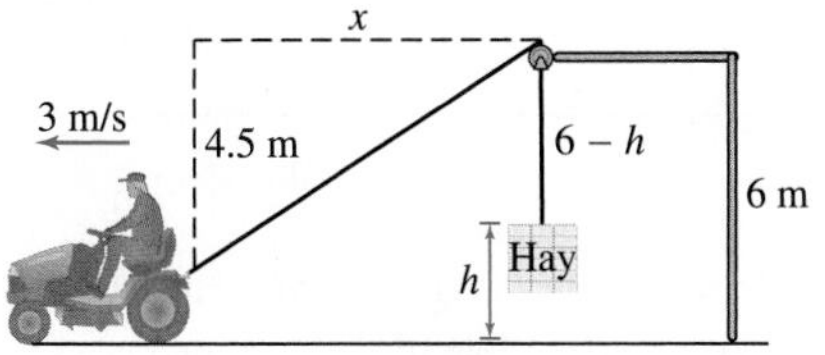

FIGURE 7

■ **EXAMPLE 5** Farmer John's tractor, traveling at 3 m/s, pulls a rope attached to a bale of hay through a pulley. With dimensions as indicated in Figure 7, how fast is the bale rising when the tractor is 5 m from the bale?

Solution

Step 1. **Assign variables and restate the problem.** Let x be the horizontal distance from the tractor to the bale of hay, and let h be the height above ground of the top of the bale. The tractor is 5 m from the bale when $x = 5$, so we can restate the problem as follows:

$$\text{Compute } \left.\frac{dh}{dt}\right|_{x=5} \qquad \text{given that} \qquad \frac{dx}{dt} = 3 \text{ m/s}$$

Step 2. **Find an equation relating the variables and differentiate.**
Let L be the total length of the rope. From Figure 7 (using the Pythagorean Theorem),

$$L = \sqrt{x^2 + 4.5^2} + (6 - h)$$

Although the length L is not given, it is a constant, and therefore $dL/dt = 0$. Thus,

$$\frac{dL}{dt} = \frac{d}{dt}\left(\sqrt{x^2 + 4.5^2} + (6 - h)\right) = \frac{x\frac{dx}{dt}}{\sqrt{x^2 + 4.5^2}} - \frac{dh}{dt} = 0 \qquad \boxed{5}$$

Step 3. **Use the given data to find the unknown derivative.**
Apply Eq. (5) with $x = 5$ and $dx/dt = 3$. The bale is rising at the rate

$$\frac{dh}{dt} = \frac{x\frac{dx}{dt}}{\sqrt{x^2 + 4.5^2}} = \frac{(5)(3)}{\sqrt{5^2 + 4.5^2}} \approx 2.23 \text{ m/s}$$ ■

3.9 SUMMARY

- Related-rate problems present us with situations in which two or more variables are related and we are asked to compute the rate of change of one of the variables in terms of the rates of change of the other variable(s).
- Draw a diagram if possible. It may also be useful to break the solution into three steps:

Step 1. Assign variables and restate the problem.

Step 2. Find an equation that relates the variables and differentiate.

This gives us an equation relating the known and unknown derivatives. Remember not to substitute values for the variables until after you have computed all derivatives.

Step 3. Use the given data to find the unknown derivative.

- The two facts from geometry arise often in related-rate problems: Pythagorean Theorem and the Theorem of Similar Triangles (ratios of corresponding sides are equal).

3.9 EXERCISES

Preliminary Questions

1. Assign variables and restate the following problem in terms of known and unknown derivatives (but do not solve it): How fast is the volume of a cube increasing if its side increases at a rate of 0.5 cm/s?

2. What is the relation between dV/dt and dr/dt if $V = \left(\frac{4}{3}\right)\pi r^3$?

In Questions 3 and 4, water pours into a cylindrical glass of radius 4 cm. Let V and h denote the volume and water level respectively, at time t.

3. Restate this question in terms of dV/dt and dh/dt: How fast is the water level rising if water pours in at a rate of 2 cm^3/min?

4. Restate this question in terms of dV/dt and dh/dt: At what rate is water pouring in if the water level rises at a rate of 1 cm/min?

Exercises

In Exercises 1 and 2, consider a rectangular bathtub whose base is 18 ft^2.

1. How fast is the water level rising if water is filling the tub at a rate of 0.7 ft^3/min?

2. At what rate is water pouring into the tub if the water level rises at a rate of 0.8 ft/min?

3. The radius of a circular oil slick expands at a rate of 2 m/min.

(a) How fast is the area of the oil slick increasing when the radius is 25 m?

(b) If the radius is 0 at time $t = 0$, how fast is the area increasing after 3 min?

4. At what rate is the diagonal of a cube increasing if its edges are increasing at a rate of 2 cm/s?

In Exercises 5–8, assume that the radius r of a sphere is expanding at a rate of 30 cm/min. *The volume of a sphere is* $V = \frac{4}{3}\pi r^3$ *and its surface area is* $4\pi r^2$. *Determine the given rate.*

5. Volume with respect to time when $r = 15$ cm.

6. Volume with respect to time at $t = 2$ min, assuming that $r = 0$ at $t = 0$.

7. Surface area with respect to time when $r = 40$ cm.

8. Surface area with respect to time at $t = 2$ min, assuming that $r = 10$ at $t = 0$.

In Exercises 9–12, refer to a 5-meter ladder sliding down a wall, as in Figures 1 and 2. The variable h is the height of the ladder's top at time t, and x is the distance from the wall to the ladder's bottom.

9. Assume the bottom slides away from the wall at a rate of 0.8 m/s. Find the velocity of the top of the ladder at $t = 2$ s if the bottom is 1.5 m from the wall at $t = 0$ s.

10. Suppose that the top is sliding down the wall at a rate of 1.2 m/s. Calculate dx/dt when $h = 3$ m.

11. Suppose that $h(0) = 4$ and the top slides down the wall at a rate of 1.2 m/s. Calculate x and dx/dt at $t = 2$ s.

12. What is the relation between h and x at the moment when the top and bottom of the ladder move at the same speed?

13. A conical tank has height 3 m and radius 2 m at the top. Water flows in at a rate of 2 m^3/min. How fast is the water level rising when it is 2 m?

14. Follow the same set-up as Exercise 13, but assume that the water level is rising at a rate of 0.3 m/min when it is 2 m. At what rate is water flowing in?

15. The radius r and height h of a circular cone change at a rate of 2 cm/s. How fast is the volume of the cone increasing when $r = 10$ and $h = 20$?

16. A road perpendicular to a highway leads to a farmhouse located 2 km away (Figure 8). An automobile travels past the farmhouse at a speed of 80 km/h. How fast is the distance between the automobile and the farmhouse increasing when the automobile is 6 km past the intersection of the highway and the road?

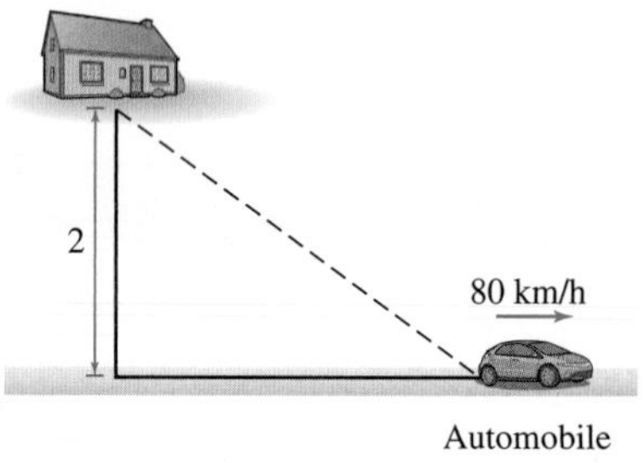

FIGURE 8

17. A man of height 1.8 meters walks away from a 5-meter lamppost at a speed of 1.2 m/s (Figure 9). Find the rate at which his shadow is increasing in length.

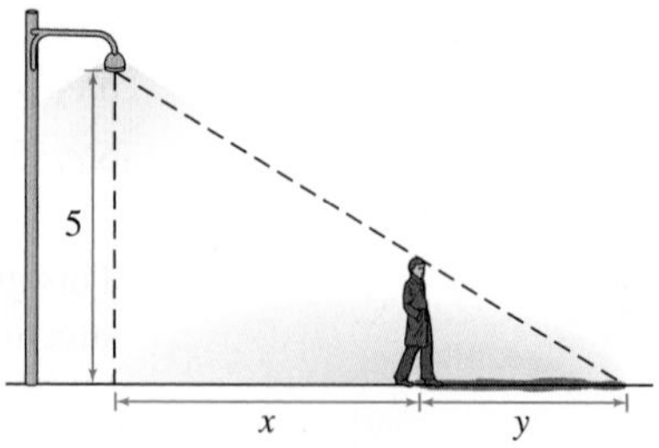

FIGURE 9

18. As Claudia walks away from a 264-cm lamppost, the tip of her shadow moves twice as fast as she does. What is Claudia's height?

19. At a given moment, a plane passes directly above a radar station at an altitude of 6 km.

(a) The plane's speed is 800 km/h. How fast is the distance between the plane and the station changing half an hour later?

(b) How fast is the distance between the plane and the station changing when the plane passes directly above the station?

20. In the setting of Exercise 19, let θ be the angle that the line through the radar station and the plane makes with the horizontal. How fast is θ changing 12 min after the plane passes over the radar station?

21. A hot air balloon rising vertically is tracked by an observer located 4 km from the lift-off point. At a certain moment, the angle between the observer's line of sight and the horizontal is $\frac{\pi}{5}$, and it is changing at a rate of 0.2 rad/min. How fast is the balloon rising at this moment?

22. A laser pointer is placed on a platform that rotates at a rate of 20 revolutions per minute. The beam hits a wall 8 m away, producing a dot of light that moves horizontally along the wall. Let θ be the angle between the beam and the line through the searchlight perpendicular to the wall (Figure 10). How fast is this dot moving when $\theta = \frac{\pi}{6}$?

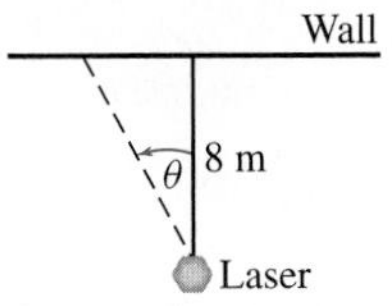

FIGURE 10

23. A rocket travels vertically at a speed of 1,200 km/h. The rocket is tracked through a telescope by an observer located 16 km from the launching pad. Find the rate at which the angle between the telescope and the ground is increasing 3 min after lift-off.

24. Using a telescope, you track a rocket that was launched 4 km away, recording the angle θ between the telescope and the ground at half-second intervals. Estimate the velocity of the rocket if $\theta(10) = 0.205$ and $\theta(10.5) = 0.225$.

25. A police car traveling south toward Sioux Falls at 160 km/h pursues a truck traveling east away from Sioux Falls, Iowa, at 140 km/h (Figure 11). At time $t = 0$, the police car is 20 km north and the truck is 30 km east of Sioux Falls. Calculate the rate at which the distance between the vehicles is changing:

(a) At time $t = 0$ **(b)** 5 minutes later

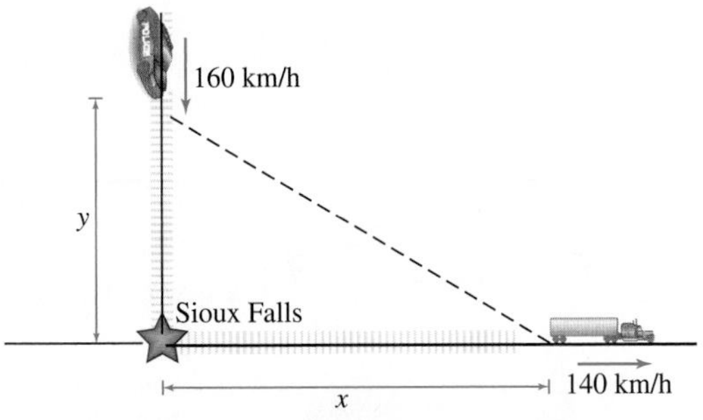

FIGURE 11

26. A car travels down a highway at 25 m/s. An observer stands 150 m from the highway.

(a) How fast is the distance from the observer to the car increasing when the car passes in front of the observer? Explain your answer without making any calculations.

(b) How fast is the distance increasing 20 s later?

27. In the setting of Example 5, at a certain moment, the tractor's speed is 3 m/s and the bale is rising at 2 m/s. How far is the tractor from the bale at this moment?

28. Placido pulls a rope attached to a wagon through a pulley at a rate of q m/s. With dimensions as in Figure 12:

(a) Find a formula for the speed of the wagon in terms of q and the variable x in the figure.

(b) Find the speed of the wagon when $x = 0.6$ if $q = 0.5$ m/s.

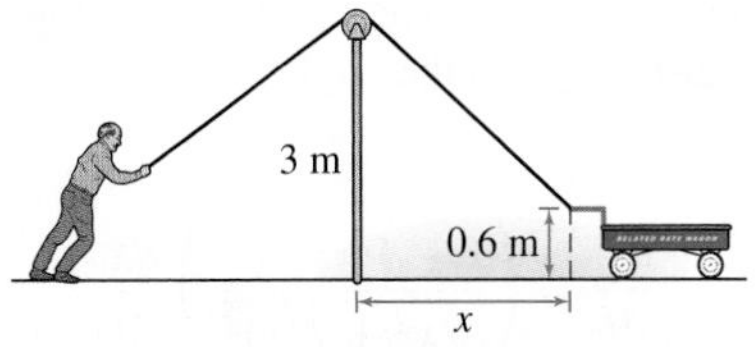

FIGURE 12

29. Julian is jogging around a circular track of radius 50 m. In a coordinate system with origin at the center of the track, Julian's x-coordinate is changing at a rate of -1.25 m/s when his coordinates are (40, 30). Find dy/dt at this moment.

30. A particle moves counterclockwise around the ellipse with equation $9x^2 + 16y^2 = 25$ (Figure 13).

(a) In which of the four quadrants is $dx/dt > 0$? Explain.

(b) Find a relation between dx/dt and dy/dt.

(c) At what rate is the x-coordinate changing when the particle passes the point (1, 1) if its y-coordinate is increasing at a rate of 6 m/s?

(d) Find dy/dt when the particle is at the top and bottom of the ellipse.

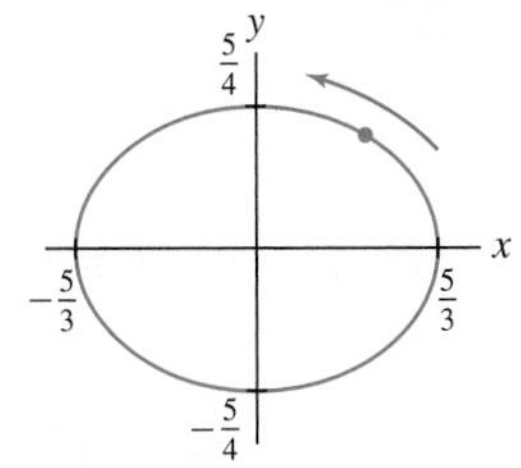

FIGURE 13

In Exercises 31 and 32, assume that the pressure P (in kilopascals) and volume V (in cubic centimeters) of an expanding gas are related by $PV^b = C$, where b and C are constants (this holds in an adiabatic *expansion, without heat gain or loss).*

31. Find dP/dt if $b = 1.2$, $P = 8$ kPa, $V = 100$ cm^2, and $dV/dt = 20$ cm^3/min.

32. Find b if $P = 25$ kPa, $dP/dt = 12$ kPa/min, $V = 100$ cm^2, and $dV/dt = 20$ cm^3/min.

33. The base x of the right triangle in Figure 14 increases at a rate of 5 cm/s, while the height remains constant at $h = 20$. How fast is the angle θ changing when $x = 20$?

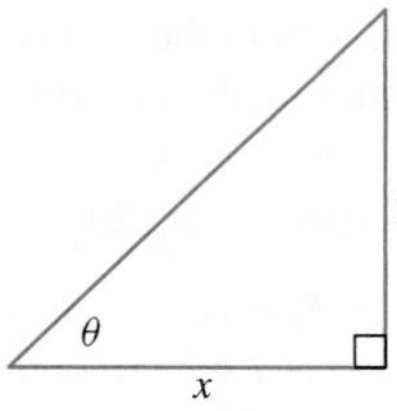

FIGURE 14

34. Two parallel paths 15 m apart run east-west through the woods. Brooke jogs east on one path at 10 km/h, while Jamail walks west on the other path at 6 km/h. If they pass each other at time $t = 0$, how far apart are they 3 s later, and how fast is the distance between them changing at that moment?

35. A particle travels along a curve $y = f(x)$ as in Figure 15. Let $L(t)$ be the particle's distance from the origin.

(a) Show that

$$\frac{dL}{dt} = \left(\frac{x + f(x)f'(x)}{\sqrt{x^2 + f(x)^2}}\right)\frac{dx}{dt}$$

if the particle's location at time t is $P = (x, f(x))$.

(b) Calculate $L'(t)$ when $x = 1$ and $x = 2$ if $f(x) = \sqrt{3x^2 - 8x + 9}$ and $dx/dt = 4$.

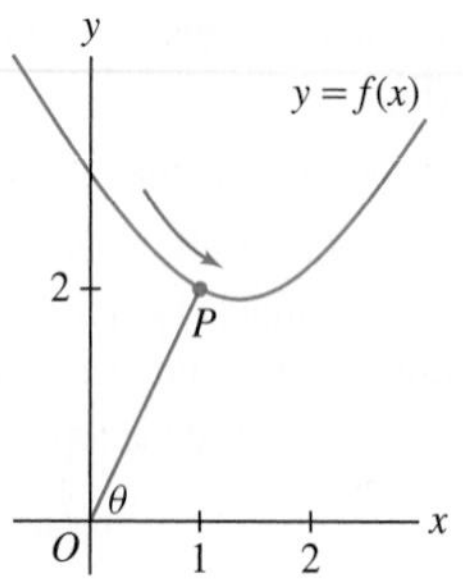

FIGURE 15

36. Let θ be the angle in Figure 15, where $P = (x, f(x))$. In the setting of the previous exercise, show that

$$\frac{d\theta}{dt} = \left(\frac{xf'(x) - f(x)}{x^2 + f(x)^2}\right)\frac{dx}{dt}$$

Hint: Differentiate $\tan\theta = f(x)/x$ and observe that $\cos\theta = x/\sqrt{x^2 + f(x)^2}$.

Exercises 37 and 38 refer to the baseball diamond (a square of side 90 ft) in Figure 16.

37. A baseball player runs from home plate toward first base at 20 ft/s. How fast is the player's distance from second base changing when the player is halfway to first base?

38. Player 1 runs to first base at a speed of 20 ft/s while Player 2 runs from second base to third base at a speed of 15 ft/s. Let s be the distance between the two players. How fast is s changing when Player 1 is 30 ft from home plate and Player 2 is 60 ft from second base?

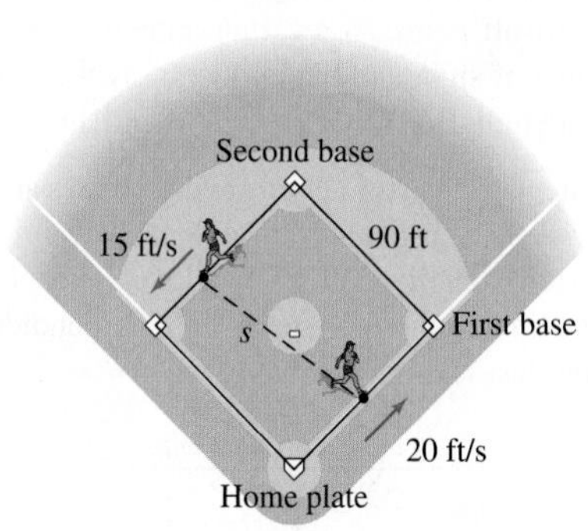

FIGURE 16

39. The conical watering pail in Figure 17 has a grid of holes. Water flows out through the holes at a rate of kA m^3/min, where k is a constant and A is the surface area of the part of the cone in contact with the water. This surface area is $A = \pi r\sqrt{h^2 + r^2}$ and the volume is $V = \frac{1}{3}\pi r^2 h$. Calculate the rate dh/dt at which the water level changes at $h = 0.3$ m, assuming that $k = 0.25$ m.

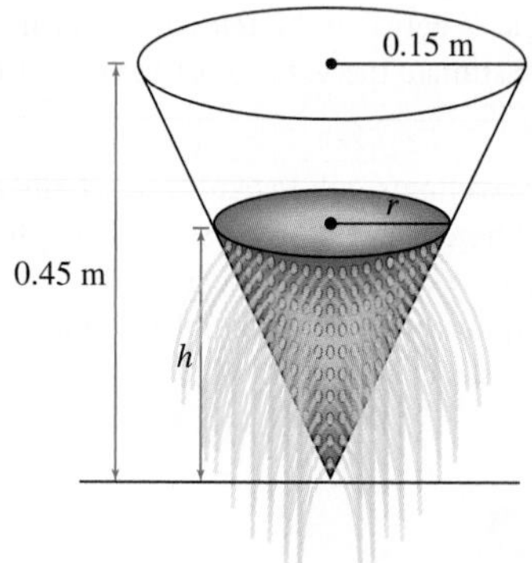

FIGURE 17

Further Insights and Challenges

40. A bowl contains water that evaporates at a rate proportional to the surface area of water exposed to the air (Figure 18). Let $A(h)$ be the cross-sectional area of the bowl at height h.

(a) Explain why $V(h + \Delta h) - V(h) \approx A(h)\Delta h$ if Δh is small.

(b) Use (a) to argue that $\frac{dV}{dh} = A(h)$.

(c) Show that the water level h decreases at a constant rate.

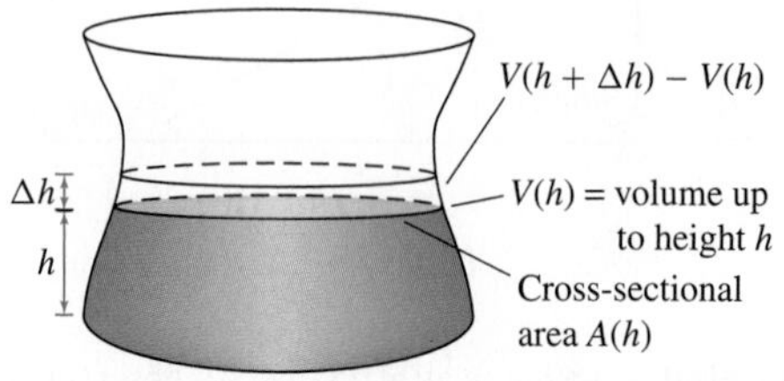

FIGURE 18

41. A roller coaster has the shape of the graph in Figure 19. Show that when the roller coaster passes the point $(x, f(x))$, the vertical velocity of the roller coaster is equal to $f'(x)$ times its horizontal velocity.

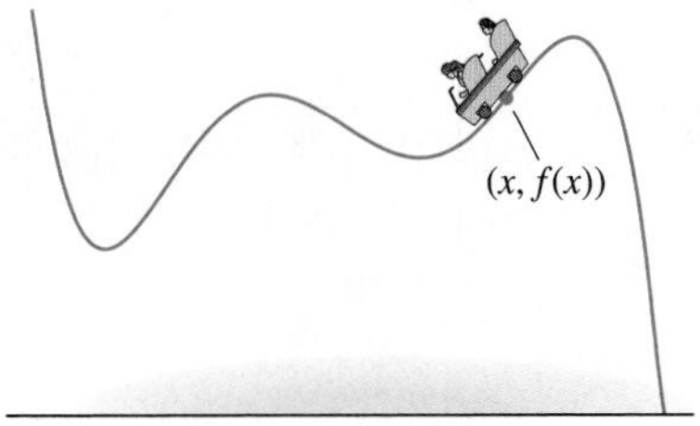

FIGURE 19 Graph of $f(x)$ as a roller coaster track.

42. Two trains leave a station at $t = 0$ and travel with constant velocity v along straight tracks that make an angle θ.

(a) Show that the trains are separating from each other at a rate $v\sqrt{2 - 2\cos\theta}$.

(b) What does this formula give for $\theta = \pi$?

43. As the wheel of radius r cm in Figure 20 rotates, the rod of length L attached at point P drives a piston back and forth in a straight line. Let x be the distance from the origin to point Q at the end of the rod, as shown in the figure.

(a) Use the Pythagorean Theorem to show that

$$L^2 = (x - r\cos\theta)^2 + r^2\sin^2\theta \qquad \boxed{6}$$

(b) Differentiate Eq. (6) with respect to t to prove that

$$2(x - r\cos\theta)\left(\frac{dx}{dt} + r\sin\theta\frac{d\theta}{dt}\right) + 2r^2\sin\theta\cos\theta\frac{d\theta}{dt} = 0$$

(c) Calculate the speed of the piston when $\theta = \frac{\pi}{2}$, assuming that $r = 10$ cm, $L = 30$ cm, and the wheel rotates at 4 revolutions per minute.

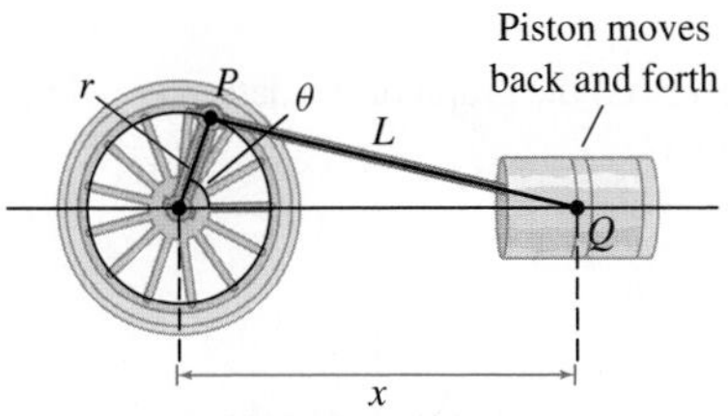

FIGURE 20

44. A spectator seated 300 m away from the center of a circular track of radius 100 m watches an athlete run laps at a speed of 5 m/s. How fast is the distance between the spectator and athlete changing when the runner is approaching the spectator and the distance between them is 250 m? *Hint:* The diagram for this problem is similar to Figure 20, with $r = 100$ and $x = 300$.

CHAPTER REVIEW EXERCISES

In Exercises 1–4, refer to the function $f(x)$ whose graph is shown in Figure 1.

1. Compute the average rate of change of $f(x)$ over [0, 2]. What is the graphical interpretation of this average rate?

2. For which value of h is $\dfrac{f(0.7+h) - f(0.7)}{h}$ equal to the slope of the secant line between the points where $x = 0.7$ and $x = 1.1$?

3. Estimate $\dfrac{f(0.7+h) - f(0.7)}{h}$ for $h = 0.3$. Is this number larger or smaller than $f'(0.7)$?

4. Estimate $f'(0.7)$ and $f'(1.1)$.

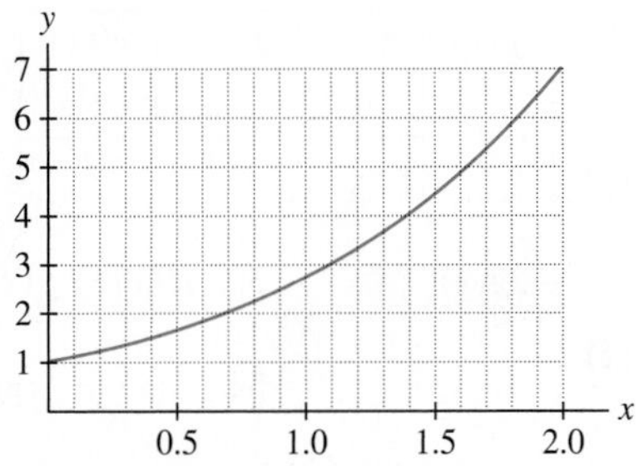

FIGURE 1

In Exercises 5–8, compute $f'(a)$ using the limit definition and find an equation of the tangent line to the graph of $f(x)$ at $x = a$.

5. $f(x) = x^2 - x, \quad a = 1$

6. $f(x) = 5 - 3x, \quad a = 2$

7. $f(x) = x^{-1}, \quad a = 4$

8. $f(x) = x^3, \quad a = -2$

In Exercises 9–12, compute dy/dx using the limit definition.

9. $y = 4 - x^2$

10. $y = \sqrt{2x+1}$

11. $y = \dfrac{1}{2 - x}$

12. $y = \dfrac{1}{(x-1)^2}$

In Exercises 13–16, express the limit as a derivative.

13. $\displaystyle\lim_{h\to 0} \frac{\sqrt{1+h} - 1}{h}$

14. $\displaystyle\lim_{x\to -1} \frac{x^3 + 1}{x + 1}$

15. $\displaystyle\lim_{t\to \pi} \frac{\sin t \cos t}{t - \pi}$

16. $\displaystyle\lim_{\theta\to \pi} \frac{\cos\theta - \sin\theta + 1}{\theta - \pi}$

17. Find $f(4)$ and $f'(4)$ if the tangent line to the graph of $f(x)$ at $x = 4$ has equation $y = 3x - 14$.

18. Each graph in Figure 2 shows the graph of a function $f(x)$ and its derivative $f'(x)$. Determine which is the function and which is the derivative.

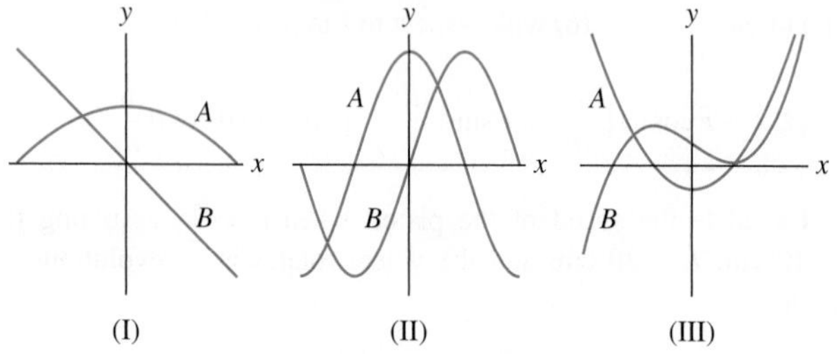

FIGURE 2 Graph of $f(x)$.

19. Is (A), (B), or (C) the graph of the derivative of the function $f(x)$ shown in Figure 3?

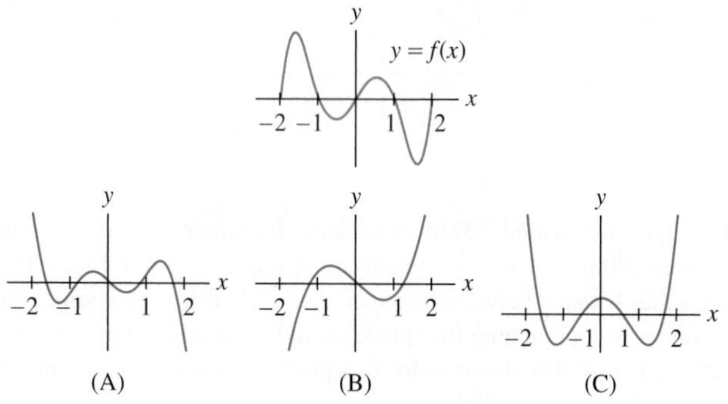

FIGURE 3

20. Let $N(t)$ be the percentage of a state population infected with a flu virus on week t of an epidemic. What percentage is likely to be infected in week 4 if $N(3) = 8$ and $N'(3) = 1.2$?

21. A girl's height $h(t)$ (in centimeters) is measured at time t (in years) for $0 \le t \le 14$:

52, 75.1, 87.5, 96.7, 104.5, 111.8, 118.7, 125.2,
131.5, 137.5, 143.3, 149.2, 155.3, 160.8, 164.7

(a) What is the average growth rate over the 14-year period?

(b) Is the average growth rate larger over the first half or the second half of this period?

(c) Estimate $h'(t)$ (in centimeters per year) for $t = 3, 8$.

22. A planet's period P (number of days to complete one revolution around the sun) is approximately $0.199A^{3/2}$, where A is the average distance (in millions of kilometers) from the planet to the sun.

(a) Calculate P and dP/dA for Earth using the value $A = 150$.

(b) Estimate the increase in P if A is increased to 152.

In Exercises 23 and 24, use the following table of values for the number $A(t)$ of automobiles (in millions) manufactured in the United States in year t.

t	1970	1971	1972	1973	1974	1975	1976
$A(t)$	6.55	8.58	8.83	9.67	7.32	6.72	8.50

23. What is the interpretation of $A'(t)$? Estimate $A'(1971)$. Does $A'(1974)$ appear to be positive or negative?

24. Given the data, which of (A)–(C) in Figure 4 could be the graph of the derivative $A'(t)$? Explain.

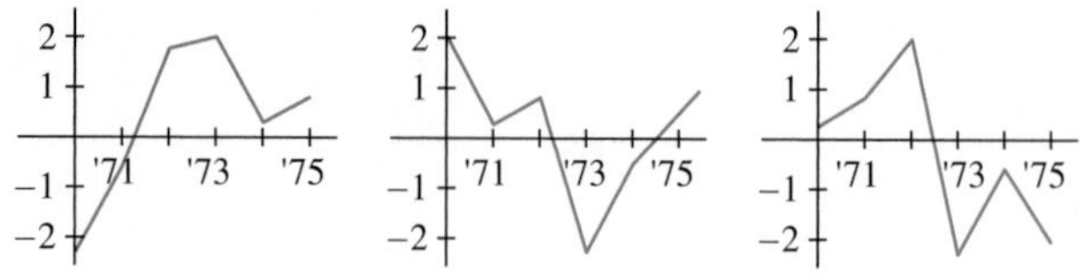

FIGURE 4

In Exercises 25–50, compute the derivative.

25. $y = 3x^5 - 7x^2 + 4$

26. $y = 4x^{-3/2}$

27. $y = t^{-7.3}$

28. $y = 4x^2 - x^{-2}$

29. $y = \dfrac{x+1}{x^2+1}$

30. $y = \dfrac{3t-2}{4t-9}$

31. $y = (x^4 - 9x)^6$

32. $y = (3t^2 + 20t^{-3})^6$

33. $y = (2 + 9x^2)^{3/2}$

34. $y = (x+1)^3(x+4)^4$

35. $y = \dfrac{z}{\sqrt{1-z}}$

36. $y = \left(1 + \dfrac{1}{x}\right)^3$

37. $y = \dfrac{x^4 + \sqrt{x}}{x^2}$

38. $y = \dfrac{1}{(1-x)\sqrt{2-x}}$

39. $y = \sqrt{x + \sqrt{x + \sqrt{x}}}$

40. $h(z) = \left(z + (z+1)^{1/2}\right)^{-3/2}$

41. $y = \tan(t^{-3})$

42. $y = 4\cos(2 - 3x)$

43. $y = \sin(2x)\cos^2 x$

44. $y = \sin\left(\dfrac{4}{\theta}\right)$

45. $y = \dfrac{t}{1 + \sec t}$

46. $y = z\csc(9z + 1)$

47. $y = \dfrac{8}{1 + \cot\theta}$

48. $y = \tan(\cos x)$

49. $y = \tan(\sqrt{1 + \csc\theta})$

50. $y = \cos(\cos(\cos(\theta)))$

In Exercises 51–56, use the following table of values to calculate the derivative of the given function at $x = 2$.

x	$f(x)$	$g(x)$	$f'(x)$	$g'(x)$
2	5	4	−3	9
4	3	2	−2	3

51. $S(x) = 3f(x) - 2g(x)$

52. $H(x) = f(x)g(x)$

53. $R(x) = \dfrac{f(x)}{g(x)}$

54. $G(x) = f(g(x))$

55. $F(x) = f(g(2x))$

56. $K(x) = f(x^2)$

57. Find the points on the graph of $x^3 - y^3 = 3xy - 3$ where the tangent line is horizontal.

58. Find the points on the graph of $x^{2/3} + y^{2/3} = 1$ where the tangent line has slope 1.

59. Find a such that the tangent lines $y = x^3 - 2x^2 + x + 1$ at $x = a$ and $x = a + 1$ are parallel.

In Exercises 60–63, let $f(x) = x^3 - 3x^2 + x + 4$.

60. Find the points on the graph of $f(x)$ where the tangent line has slope 10.

61. For which values of x are the tangent lines to $y = f(x)$ horizontal?

62. Find all values of b such that $y = 25x + b$ is tangent to the graph of $f(x)$.

63. Find all values of k such that $f(x)$ has only one tangent line of slope k.

64. Use the table to compute the average rate of change of Candidate A's percentage of votes over the intervals from day 20 to day 15, day 15 to day 10, and day 10 to day 5. If this trend continues over the last 5 days before the election, will Candidate A win?

Days Before Election	20	15	10	5
Candidate A	44.8%	46.8%	48.3%	49.3%
Candidate B	55.2%	53.2%	51.7%	50.7%

In Exercises 65–70, calculate y''.

65. $y = 12x^3 - 5x^2 + 3x$

66. $y = x^{-2/5}$

67. $y = \sqrt{2x + 3}$

68. $y = \dfrac{4x}{x + 1}$

69. $y = \tan(x^2)$

70. $y = \sin^2(4x + 9)$

In Exercises 71–76, compute $\dfrac{dy}{dx}$.

71. $x^3 - y^3 = 4$

72. $4x^2 - 9y^2 = 36$

73. $y = xy^2 + 2x^2$

74. $\dfrac{y}{x} = x + y$

75. $y = \sin(x + y)$

76. $\tan(x + y) = xy$

77. In Figure 5, label the graphs f, f', and f''.

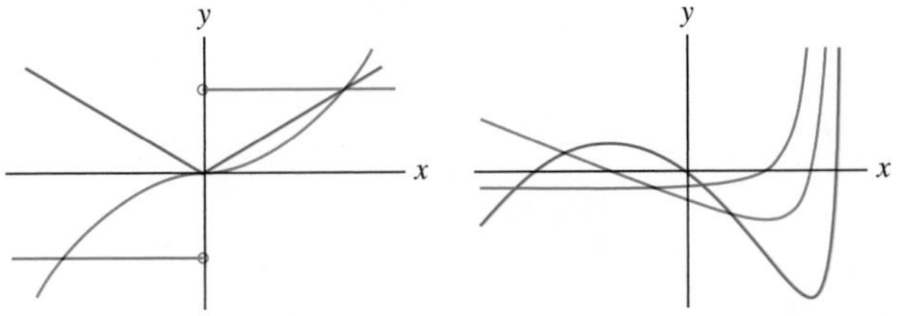

FIGURE 5

78. Let $f(x) = x^2 \sin(x^{-1})$ for $x \neq 0$ and $f(0) = 0$. Show that $f'(x)$ exists for all x (including $x = 0$) but that $f'(x)$ is not continuous at $x = 0$ (Figure 6).

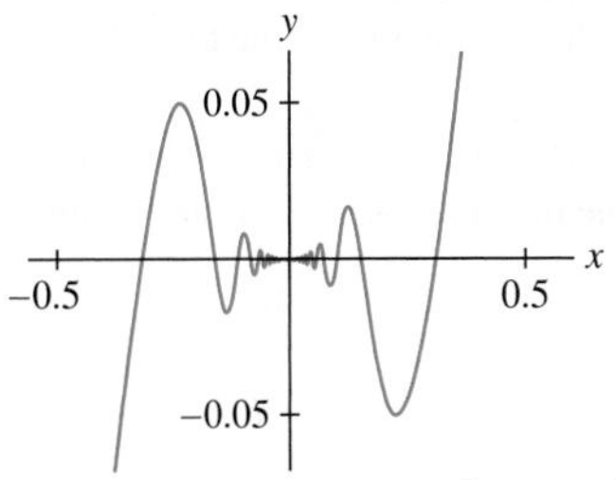

FIGURE 6 Graph of $f(x) = x^2 \sin(x^{-1})$.

Exercises 79–81: Let q be the number of units of a product (cell phones, barrels of oil, etc.) that can be sold at the price p. The ***price elasticity of demand*** *E is defined as the percentage rate of change of q with respect to p. In terms of derivatives,*

$$E = \frac{p}{q}\frac{dq}{dp} = \lim_{\Delta p \to 0} \frac{(100\Delta q)/q}{(100\Delta p)/p}$$

79. Show that the total revenue $R = pq$ satisfies $\dfrac{dR}{dp} = q(1 + E)$.

80. A commercial bakery can sell q chocolate cakes per week at price \$$p$, where $q = 50p(10 - p)$ for $5 < p < 10$.

(a) Show that $E(p) = \dfrac{2p - 10}{p - 10}$.

(b) Show, by computing $E(8)$, that if $p = \$8$, then a 1% increase in price reduces demand by approximately 3%.

81. The monthly demand (in thousands) for flights between Chicago and St. Louis at the price p is $q = 40 - 0.2p$. Calculate the price elasticity of demand when $p = \$150$ and estimate the percentage increase in number of additional passengers if the ticket price is lowered by 1%.

82. How fast does the water level rise in the tank in Figure 7 when the water level is $h = 4$ m and water pours in at 20 m^3/min?

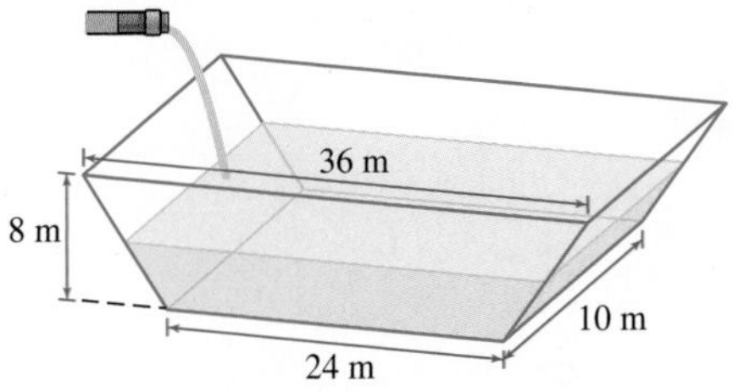

FIGURE 7

83. The minute hand of a clock is 8 cm long, and the hour hand is 5 cm long. How fast is the distance between the tips of the hands changing at 3 o'clock?

84. Chloe and Bao are in motorboats at the center of a lake. At time $t = 0$, Chloe begins traveling south at a speed of 50 km/h. One minute later, Bao takes off, heading east at a speed of 40 km/h. At what rate is the distance between them increasing at $t = 12$ min?

85. A bead slides down the curve $xy = 10$. Find the bead's horizontal velocity at time $t = 2$ s if its height at time t seconds is $y = 400 - 16t^2$ cm.

86. In Figure 8, x is increasing at 2 cm/s, y is increasing at 3 cm/s, and θ is decreasing such that the area of the triangle has the constant value 4 cm^2.

(a) How fast is θ decreasing when $x = 4$, $y = 4$?

(b) How fast is the distance between P and Q changing when $x = 4$, $y = 4$?

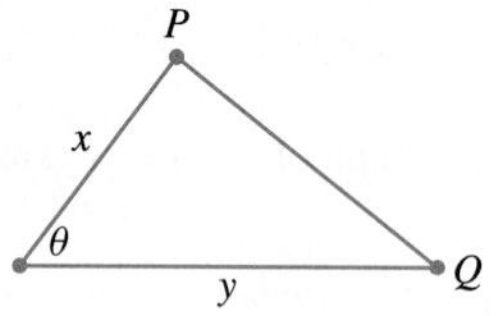

FIGURE 8

87. A light moving at 0.8 m/s approaches a man standing 4 m from a wall (Figure 9). The light is 1 m above the ground. How fast is the tip P of the man's shadow moving when the light is 7 m from the wall?

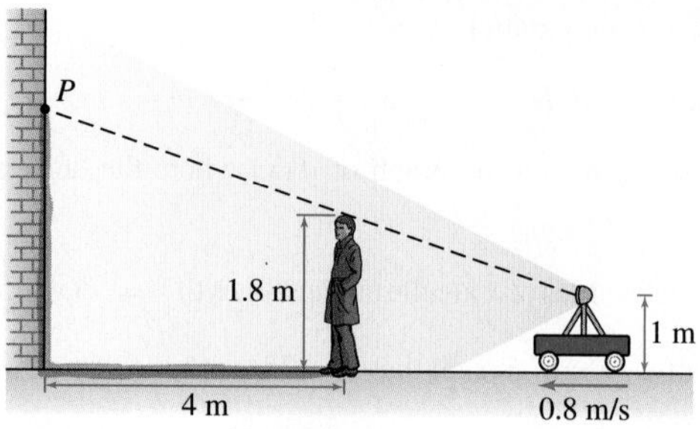

FIGURE 9

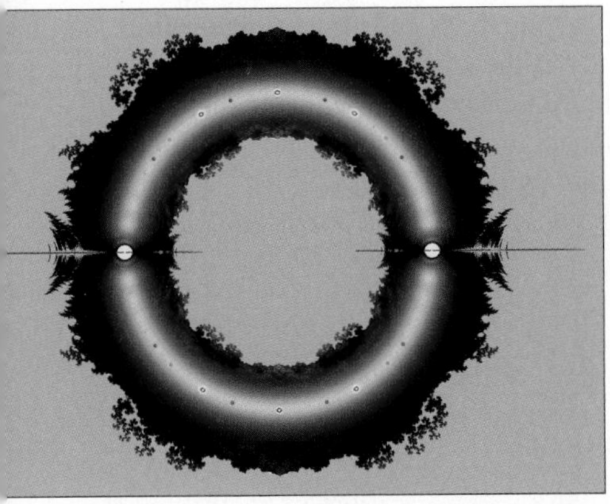

This striking image, created by Sam Derbyshire while an undergraduate student at the University of Warwick in England, is a density plot of the roots (real or complex) of all polynomials of degree 24 whose coefficients are +1 or −1.

4 APPLICATIONS OF THE DERIVATIVE

This chapter puts the derivative to work. The first and second derivatives are used to analyze functions and their graphs and to solve optimization problems (finding minimum and maximum values of a function). Newton's Method in Section 4.7 employs the derivative to approximate solutions of equations. In Section 4.8, we introduce antidifferentiation, the inverse operation to differentiation, to prepare for the study of integration in Chapter 5.

4.1 Linear Approximation and Applications

In some situations we are interested in determining the "effect of a small change." For example:

- How does a small change in angle affect the distance of a basketball shot? (Exercise 39)
- How are revenues at the box office affected by a small change in ticket prices? (Exercise 29)
- The cube root of 27 is 3. How much larger is the cube root of 27.2? (Exercise 7)

In each case, we have a function $f(x)$ and we're interested in the change

$$\Delta f = f(a + \Delta x) - f(a)$$

where Δx is small. The **Linear Approximation** uses the derivative to estimate Δf without computing it exactly. By definition, the derivative is the limit

$$f'(a) = \lim_{\Delta x \to 0} \frac{f(a + \Delta x) - f(a)}{\Delta x} = \lim_{\Delta x \to 0} \frac{\Delta f}{\Delta x}$$

So when Δx is small, we have $\Delta f/\Delta x \approx f'(a)$, and thus,

$$\Delta f \approx f'(a)\Delta x$$

← *REMINDER The notation ≈ means "approximately equal to." The accuracy of the Linear Approximation is discussed at the end of this section.*

Linear Approximation of Δf If f is differentiable at $x = a$ and Δx is small, then

$$\Delta f \approx f'(a)\Delta x \qquad \boxed{1}$$

where $\Delta f = f(a + \Delta x) - f(a)$.

Keep in mind the different roles played by Δf and $f'(a)\Delta x$. The quantity of interest is the *actual change* Δf. We estimate it by $f'(a)\,\Delta x$. The Linear Approximation tells us that up to a small error, Δf is directly proportional to Δx when Δx is small.

GRAPHICAL INSIGHT The Linear Approximation is sometimes called the **tangent line approximation**. Why? Observe in Figure 1 that Δf is the vertical change in the graph from $x = a$ to $x = a + \Delta x$. For a straight line, the vertical change is equal to the slope times the horizontal change Δx, and since the tangent line has slope $f'(a)$, its vertical change is $f'(a)\Delta x$. So the Linear Approximation approximates Δf by the vertical change in the tangent line. When Δx is small, the two quantities are nearly equal.

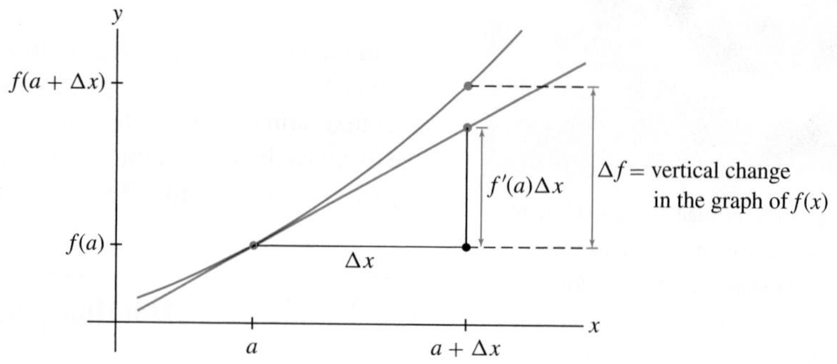

FIGURE 1 Graphical meaning of the Linear Approximation $\Delta f \approx f'(a)\Delta x$.

Linear Approximation:

$$\Delta f \approx f'(a)\Delta x$$

where $\Delta f = f(a + \Delta x) - f(a)$

■ **EXAMPLE 1** Use the Linear Approximation to estimate $\frac{1}{10.2} - \frac{1}{10}$. How accurate is your estimate?

Solution We apply the Linear Approximation to $f(x) = \frac{1}{x}$ with $a = 10$ and $\Delta x = 0.2$:

$$\Delta f = f(10.2) - f(10) = \frac{1}{10.2} - \frac{1}{10}$$

We have $f'(x) = -x^{-2}$ and $f'(10) = -0.01$, so Δf is approximated by

$$f'(10)\Delta x = (-0.01)(0.2) = -0.002$$

In other words,

$$\frac{1}{10.2} - \frac{1}{10} \approx -0.002$$

The error in the Linear Approximation is the quantity

$$\textit{Error} = |\Delta f - f'(a)\Delta x|$$

A calculator gives the value $\frac{1}{10.2} - \frac{1}{10} \approx -0.00196$ and thus our error is less than 10^{-4}:

$$\text{Error} \approx |-0.00196 - (-0.002)| = 0.00004 < 10^{-4}$$ ■

Differential Notation The Linear Approximation to $y = f(x)$ is often written using the "differentials" dx and dy. In this notation, dx is used instead of Δx to represent the change in x, and dy is the corresponding vertical change in the tangent line:

$$\boxed{dy = f'(a)dx} \qquad (2)$$

Let $\Delta y = f(a + dx) - f(a)$. Then the Linear Approximation says

$$\boxed{\Delta y \approx dy} \qquad (3)$$

This is simply another way of writing $\Delta f \approx f'(a)\Delta x$.

■ **EXAMPLE 2 Differential Notation** How much larger is $\sqrt[3]{8.1}$ than $\sqrt[3]{8} = 2$?

Solution We are interested in $\sqrt[3]{8.1} - \sqrt[3]{8}$, so we apply the Linear Approximation to $f(x) = x^{1/3}$ with $a = 8$ and change $\Delta x = dx = 0.1$.

Step 1. **Write out Δy.**

$$\Delta y = f(a + dx) - f(a) = \sqrt[3]{8 + 0.1} - \sqrt[3]{8} = \sqrt[3]{8.1} - 2$$

Step 2. **Compute dy.**

$$f'(x) = \frac{1}{3}x^{-2/3} \quad \text{and} \quad f'(8) = \left(\frac{1}{3}\right)8^{-2/3} = \left(\frac{1}{3}\right)\left(\frac{1}{4}\right) = \frac{1}{12}$$

Therefore, $dy = f'(8)\,dx = \frac{1}{12}(0.1) \approx 0.0083$.

Step 3. **Use the Linear Approximation.**

$$\Delta y \approx dy \quad \Rightarrow \quad \sqrt[3]{8.1} - 2 \approx 0.0083$$

Thus $\sqrt[3]{8.1}$ is larger than $\sqrt[3]{8}$ by the amount 0.0083, and $\sqrt[3]{8.1} \approx 2.0083$. ■

FIGURE 2 Cable position transducer (manufactured by Space Age Control, Inc.). In one application, a transducer was used to compare the changes in throttle position on a Formula 1 race car with the shifting actions of the driver.

When engineers need to monitor the change in position of an object with great accuracy, they may use a cable position transducer (Figure 2). This device detects and records the movement of a metal cable attached to the object. Its accuracy is affected by changes in temperature because heat causes the cable to stretch. The Linear Approximation can be used to estimate these effects.

■ **EXAMPLE 3 Thermal Expansion** A thin metal cable has length $L = 12$ cm when the temperature is $T = 21°$C. Estimate the change in length when T rises to 24°C, assuming that

$$\frac{dL}{dT} = kL \qquad \boxed{4}$$

where $k = 1.7 \times 10^{-5}\,°\mathrm{C}^{-1}$ (k is called the coefficient of thermal expansion).

Solution How does the Linear Approximation apply here? We will use the differential dL to estimate the actual change in length ΔL when T increases from 21° to 24°—that is, when $dT = 3°$. By Eq. (2), the differential dL is

$$dL = \left(\frac{dL}{dT}\right) dT$$

By Eq. (4), since $L = 12$,

$$\left.\frac{dL}{dT}\right|_{L=12} = kL = (1.7 \times 10^{-5})(12) \approx 2 \times 10^{-4}\ \text{cm/°C}$$

The Linear Approximation $\Delta L \approx dL$ tells us that the change in length is approximately

$$\Delta L \approx \underbrace{\left(\frac{dL}{dT}\right) dT}_{dL} \approx (2 \times 10^{-4})(3) = 6 \times 10^{-4}\ \text{cm}$$ ■

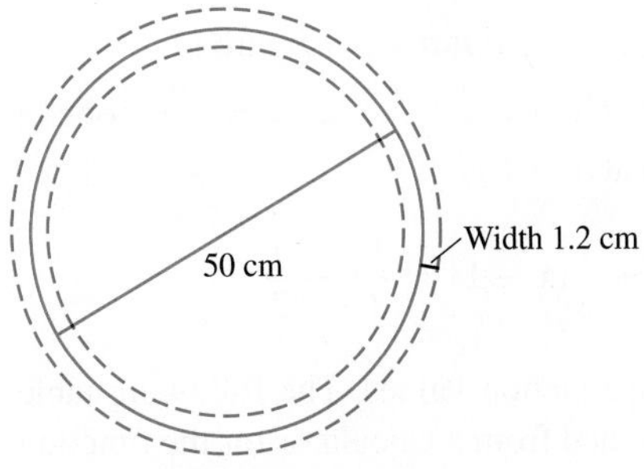

FIGURE 3 The border of the actual pizza lies between the dashed circles.

Suppose that we measure the *diameter* D of a circle and use this result to compute the *area* of the circle. If our measurement of D is inexact, the area computation will also be inexact. What is the effect of the measurement error on the resulting area computation? This can be estimated using the Linear Approximation, as in the next example.

■ **EXAMPLE 4 Effect of an Inexact Measurement** The Bonzo Pizza Company claims that its pizzas are circular with diameter 50 cm (Figure 3).

(a) What is the area of the pizza?

(b) Estimate the quantity of pizza lost or gained if the diameter is off by at most 1.2 cm.

Solution First, we need a formula for the area A of a circle in terms of its diameter D. Since the radius is $r = D/2$, the area is

$$A(D) = \pi r^2 = \pi\left(\frac{D}{2}\right)^2 = \frac{\pi}{4}D^2$$

(a) If $D = 50$ cm, then the pizza has area $A(50) = \left(\frac{\pi}{4}\right)(50)^2 \approx 1963.5\ \text{cm}^2$.

(b) If the actual diameter is equal to $50 + \Delta D$, then the loss or gain in pizza area is $\Delta A = A(50 + \Delta D) - A(50)$. Observe that $A'(D) = \frac{\pi}{2}D$ and $A'(50) = 25\pi \approx 78.5$ cm, so the Linear Approximation yields

In this example, we interpret ΔA as the possible error in the computation of $A(D)$. This should not be confused with the error in the Linear Approximation. This latter error refers to the accuracy in using $A'(D)\,\Delta D$ to approximate ΔA.

$$\Delta A = A(50 + \Delta D) - A(50) \approx A'(D)\Delta D \approx (78.5)\,\Delta D$$

Because ΔD is at most ± 1.2 cm, the loss or gain in pizza is no more than around

$$\Delta A \approx \pm(78.5)(1.2) \approx \pm 94.2\ \text{cm}^2$$

This is a loss or gain of approximately 4.8%. ■

Linearization

To approximate the function $f(x)$ itself rather than the change Δf, we use the linearization $L(x)$ "centered at $x = a$," defined by

$$L(x) = f'(a)(x - a) + f(a)$$

Notice that $y = L(x)$ is the equation of the tangent line at $x = a$ (Figure 4). For values of x close to a, $L(x)$ provides a good approximation to $f(x)$.

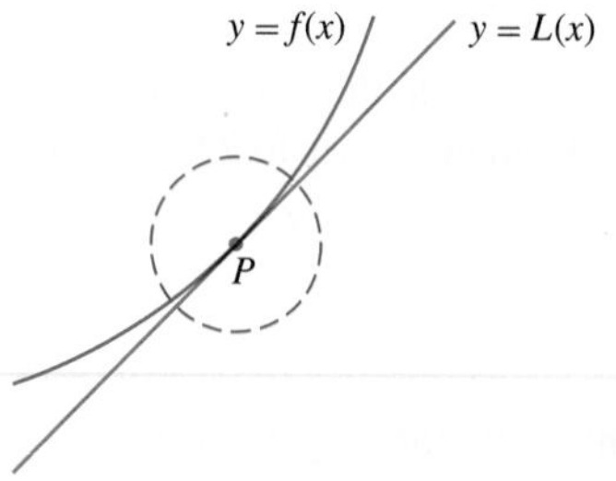

FIGURE 4 The tangent line is a good approximation in a small neighborhood of $P = (a, f(a))$.

Approximating $f(x)$ by Its Linearization If f is differentiable at $x = a$ and x is close to a, then

$$f(x) \approx L(x) = f'(a)(x - a) + f(a)$$

CONCEPTUAL INSIGHT Keep in mind that the linearization and the Linear Approximation are two ways of saying the same thing. Indeed, when we apply the linearization with $x = a + \Delta x$ and re-arrange, we obtain the Linear Approximation:

$$f(x) \approx f(a) + f'(a)(x - a)$$

$$f(a + \Delta x) \approx f(a) + f'(a)\,\Delta x \qquad (\text{since } \Delta x = x - a)$$

$$f(a + \Delta x) - f(a) \approx f'(a)\Delta x$$

■ **EXAMPLE 5** Compute the linearization of $f(x) = \sqrt{x}$ at $a = 1$ (Figure 5).

Solution We evaluate $f(x)$ and its derivative $f'(x) = \frac{1}{2}x^{-1/2}$ at $a = 1$ to obtain $f(1) = \sqrt{1} = 1$ and $f'(1) = \frac{1}{2}$. The linearization at $a = 1$ is

$$L(x) = f(1) + f'(1)(x - 1) = 1 + \frac{1}{2}(x - 1) = \frac{1}{2}x + \frac{1}{2}$$ ■

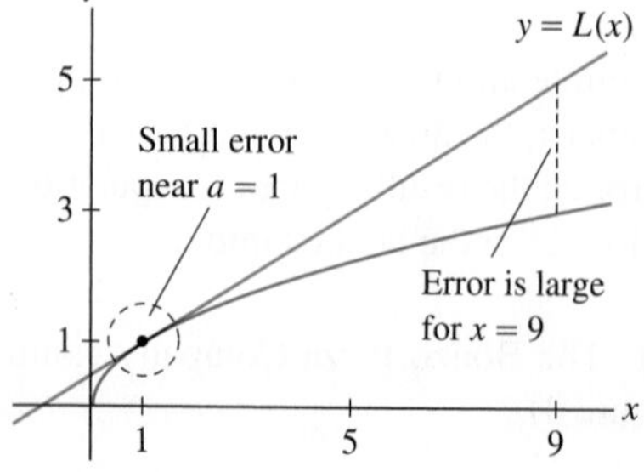

FIGURE 5 Graph of linearization of $f(x) = \sqrt{x}$ at $a = 1$.

The linearization can be used to approximate function values. The following table compares values of the linearization to values obtained from a calculator for the function $f(x) = \sqrt{x}$ of the previous example. Note that the error is large for $x = 9$, as expected, because 9 is not close to the center $a = 1$ (Figure 5).

x	$\sqrt{x}$	**Linearization at $a = 1$**	**Calculator**	**Error**
1.1	$\sqrt{1.1}$	$L(1.1) = \frac{1}{2}(1.1) + \frac{1}{2} = 1.05$	1.0488	0.0012
0.98	$\sqrt{0.98}$	$L(0.98) = \frac{1}{2}(0.98) + \frac{1}{2} = 0.99$	0.98995	$5 \cdot 10^{-5}$
9	$\sqrt{9}$	$L(9) = \frac{1}{2}(9) + \frac{1}{2} = 5$	3	2

In the next example, we compute the **percentage error**, which is often more important than the error itself. By definition,

$$\text{Percentage error} = \left|\frac{\text{error}}{\text{actual value}}\right| \times 100\%$$

EXAMPLE 6 Estimate $\tan\left(\frac{\pi}{4} + 0.02\right)$ and compute the percentage error.

Solution We find the linearization of $f(x) = \tan x$ at $a = \frac{\pi}{4}$:

$$f\left(\frac{\pi}{4}\right) = \tan\left(\frac{\pi}{4}\right) = 1, \qquad f'\left(\frac{\pi}{4}\right) = \sec^2\left(\frac{\pi}{4}\right) = \left(\sqrt{2}\right)^2 = 2$$

$$L(x) = f\left(\frac{\pi}{4}\right) + f'\left(\frac{\pi}{4}\right)\left(x - \frac{\pi}{4}\right) = 1 + 2\left(x - \frac{\pi}{4}\right)$$

At $x = \frac{\pi}{4} + 0.02$, the linearization yields the estimate

$$\tan\left(\frac{\pi}{4} + 0.02\right) \approx L\left(\frac{\pi}{4} + 0.02\right) = 1 + 2(0.02) = 1.04$$

A calculator gives $\tan\left(\frac{\pi}{4} + 0.02\right) \approx 1.0408$, so

$$\text{Percentage error} \approx \left|\frac{1.0408 - 1.04}{1.0408}\right| \times 100 \approx 0.08\%$$

The Size of the Error

The examples in this section may have convinced you that the Linear Approximation yields a good approximation to Δf when Δx is small, but if we want to rely on the Linear Approximation, we need to know more about the size of the error:

$$E = \text{Error} = \left|\Delta f - f'(a)\Delta x\right|$$

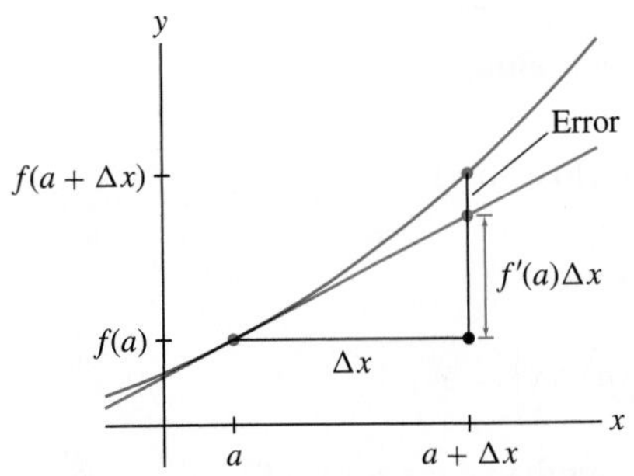

FIGURE 6 Graphical interpretation of the error in the Linear Approximation.

Remember that the error E is simply the vertical gap between the graph and the tangent line (Figure 6). In Section 11.7, we will prove the following **Error Bound**:

$$E \le \frac{1}{2}K\,(\Delta x)^2 \qquad (5)$$

where K is the maximum value of $|f''(x)|$ on the interval from a to $a + \Delta x$.

Error Bound:

$$E \leq \frac{1}{2} K (\Delta x)^2$$

where K is the max of $|f''|$ on the interval $[a, a + \Delta x]$.

The Error Bound tells us two important things. First, it says that the error is small when the second derivative (and hence K) is small. This makes sense, because $f''(x)$ measures how quickly the tangent lines change direction. When $|f''(x)|$ is smaller, the graph is flatter and the Linear Approximation is more accurate over a larger interval around $x = a$ (compare the graphs in Figure 7).

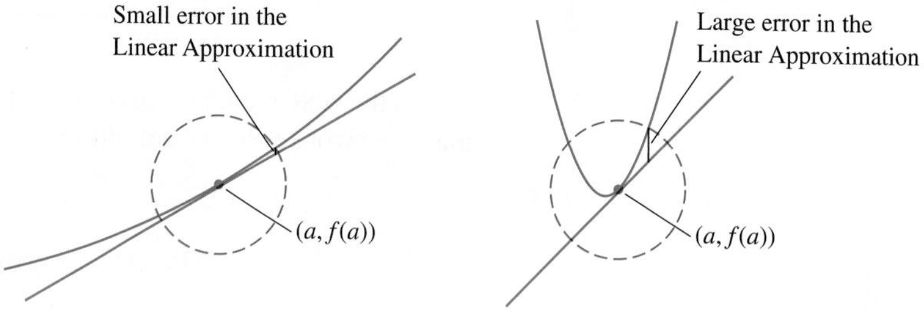

FIGURE 7 The accuracy of the Linear Approximation depends on how much the curve bends.

(A) Graph flat, $f''(x)$ is small. (B) Graph bends a lot, $f''(x)$ is large.

Second, the Error Bound tells us that the error is of *order two* in Δx, meaning that E is no larger than a constant times $(\Delta x)^2$. So if Δx is small, say $\Delta x = 10^{-n}$, then E has substantially smaller order of magnitude $(\Delta x)^2 = 10^{-2n}$. In particular, $E/\Delta x$ tends to zero (because $E/\Delta x < K\Delta x$), so the Error Bound tells us that the graph becomes nearly indistinguishable from its tangent line as we zoom in on the graph around $x = a$. This is a precise version of the "local linearity" property discussed in Section 3.2.

4.1 SUMMARY

- Let $\Delta f = f(a + \Delta x) - f(a)$. The *Linear Approximation* is the estimate

$$\Delta f \approx f'(a)\Delta x \qquad \text{(for } \Delta x \text{ small)}$$

- Differential notation: dx is the change in x, $dy = f'(a)dx$, $\Delta y = f(a + dx) - f(a)$. In this notation, the Linear Approximation reads

$$\Delta y \approx dy \qquad \text{(for } dx \text{ small)}$$

- The *linearization* of $f(x)$ at $x = a$ is the function

$$L(x) = f'(a)(x - a) + f(a)$$

- The Linear Approximation is equivalent to the approximation

$$f(x) \approx L(x) \qquad \text{(for } x \text{ close to } a\text{)}$$

- The error in the Linear Approximation is the quantity

$$\text{Error} = \left|\Delta f - f'(a)\Delta x\right|$$

In many cases, the percentage error is more important than the error itself:

$$\text{Percentage error} = \left|\frac{\text{error}}{\text{actual value}}\right| \times 100\%$$

4.1 EXERCISES

Preliminary Questions

1. True or False? The Linear Approximation says that the vertical change in the graph is approximately equal to the vertical change in the tangent line.

2. Estimate $g(1.2) - g(1)$ if $g'(1) = 4$.

3. Estimate $f(2.1)$ if $f(2) = 1$ and $f'(2) = 3$.

4. Complete the sentence: The Linear Approximation shows that up to a small error, the change in output Δf is directly proportional to ....

Exercises

In Exercises 1–6, use Eq. (1) to estimate $\Delta f = f(3.02) - f(3)$.

1. $f(x) = x^2$

2. $f(x) = x^4$

3. $f(x) = x^{-1}$

4. $f(x) = \dfrac{1}{x+1}$

5. $f(x) = \sqrt{x+6}$

6. $f(x) = \tan \dfrac{\pi x}{3}$

7. The cube root of 27 is 3. How much larger is the cube root of 27.2? Estimate using the Linear Approximation.

8. Estimate $\sin\left(\dfrac{\pi}{2} + 0.1\right) - \sin \dfrac{\pi}{2}$ using differentials.

In Exercises 9–12, use Eq. (1) to estimate Δf. Use a calculator to compute both the error and the percentage error.

9. $f(x) = \sqrt{1+x}, \quad a = 3, \quad \Delta x = 0.2$

10. $f(x) = 2x^2 - x, \quad a = 5, \quad \Delta x = -0.4$

11. $f(x) = \dfrac{1}{1+x^2}, \quad a = 3, \quad \Delta x = 0.5$

12. $f(x) = \tan\left(\dfrac{x}{4} + \dfrac{\pi}{4}\right), \quad a = 0, \quad \Delta x = 0.01$

In Exercises 13–16, estimate Δy using differentials [Eq. (3)].

13. $y = \cos x, \quad a = \frac{\pi}{6}, \quad dx = 0.014$

14. $y = \tan^2 x, \quad a = \frac{\pi}{4}, \quad dx = -0.02$

15. $y = \dfrac{10 - x^2}{2 + x^2}, \quad a = 1, \quad dx = 0.01$

16. $y = \dfrac{3 - \sqrt{x}}{\sqrt{x+3}}, \quad a = 1, \quad dx = -0.1$

In Exercises 17–24, estimate using the Linear Approximation and find the error using a calculator.

17. $\sqrt{26} - \sqrt{25}$

18. $16.5^{1/4} - 16^{1/4}$

19. $\dfrac{1}{\sqrt{101}} - \dfrac{1}{10}$

20. $\dfrac{1}{\sqrt{98}} - \dfrac{1}{10}$

21. $9^{1/3} - 2$

22. $\dfrac{1}{(27.5)^{5/3}} - \dfrac{1}{243}$

23. $\sin(0.023)$

24. $\tan\left(\dfrac{\pi}{4} + 0.01\right) - 1$

25. Estimate $f(4.03)$ for $f(x)$ as in Figure 8.

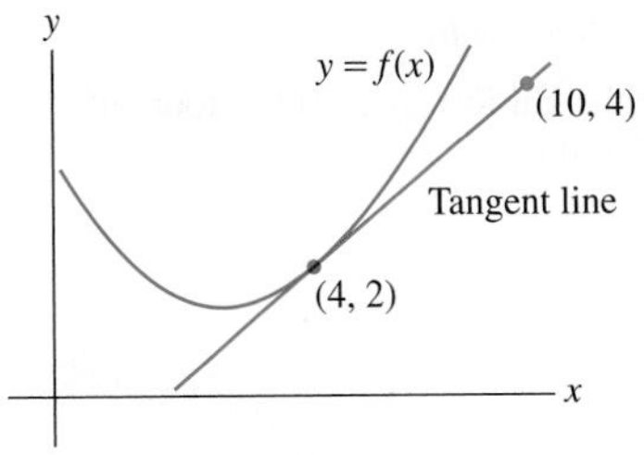

FIGURE 8

26. At a certain moment, an object in linear motion has velocity 100 m/s. Estimate the distance traveled over the next quarter-second, and explain how this is an application of the Linear Approximation.

27. Which is larger: $\sqrt{2.1} - \sqrt{2}$ or $\sqrt{9.1} - \sqrt{9}$? Explain using the Linear Approximation.

28. Estimate $\sin 61° - \sin 60°$ using the Linear Approximation. *Hint:* Express $\Delta\theta$ in radians.

29. Box office revenue at a multiplex cinema in Paris is $R(p) = 3600p - 10p^3$ euros per showing when the ticket price is p euros. Calculate $R(p)$ for $p = 9$ and use the Linear Approximation to estimate ΔR if p is raised or lowered by 0.5 euros.

30. The *stopping distance* for an automobile is $F(s) = 1.1s + 0.054s^2$ ft, where s is the speed in mph. Use the Linear Approximation to estimate the change in stopping distance per additional mph when $s = 35$ and when $s = 55$.

31. A thin silver wire has length $L = 18$ cm when the temperature is $T = 30°$C. Estimate ΔL when T decreases to 25°C if the coefficient of thermal expansion is $k = 1.9 \times 10^{-5}°\text{C}^{-1}$ (see Example 3).

32. At a certain moment, the temperature in a snake cage satisfies $dT/dt = 0.008°$C/s. Estimate the rise in temperature over the next 10 seconds.

33. The atmospheric pressure P at altitude $h = 20$ km is $P = 5.5$ kilopascals. Estimate P at altitude $h = 20.5$ km assuming that

$$\frac{dP}{dh} = -0.87$$

(a) Estimate ΔP at $h = 20$ when $\Delta h = 0.5$.

(b) Compute the actual change, and compute the percentage error in the Linear Approximation.

34. The resistance R of a copper wire at temperature $T = 20°\text{C}$ is $R = 15\ \Omega$. Estimate the resistance at $T = 22°\text{C}$, assuming that $dR/dT|_{T=20} = 0.06\ \Omega/°\text{C}$.

35. Newton's Law of Gravitation shows that if a person weighs w pounds on the surface of the earth, then his or her weight at distance x from the center of the earth is

$$W(x) = \frac{wR^2}{x^2} \qquad (\text{for } x \geq R)$$

where $R = 3{,}960$ miles is the radius of the earth (Figure 9).

(a) Show that the weight lost at altitude h miles above the earth's surface is approximately $\Delta W \approx -(0.0005w)h$. *Hint:* Use the Linear Approximation with $dx = h$.

(b) Estimate the weight lost by a 200-lb football player flying in a jet at an altitude of 7 miles.

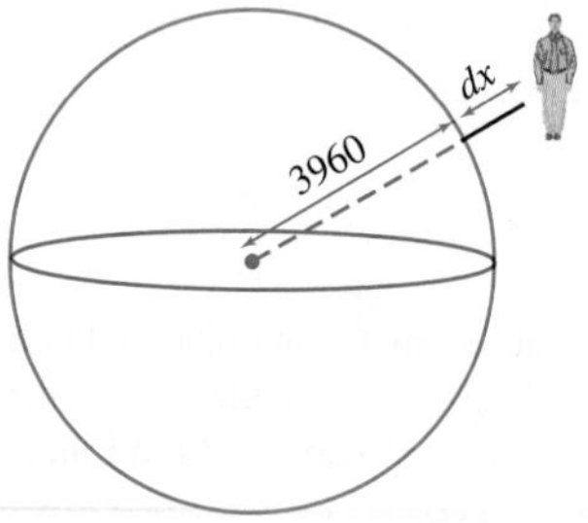

FIGURE 9 The distance to the center of the earth is $3{,}960 + h$ miles.

36. Using Exercise 35(a), estimate the altitude at which a 130-lb pilot would weigh 129.5 lb.

37. A stone tossed vertically into the air with initial velocity v cm/s reaches a maximum height of $h = v^2/1960$ cm.

(a) Estimate Δh if $v = 700$ cm/s and $\Delta v = 1$ cm/s.

(b) Estimate Δh if $v = 1{,}000$ cm/s and $\Delta v = 1$ cm/s.

(c) In general, does a 1 cm/s increase in v lead to a greater change in h at low or high initial velocities? Explain.

38. The side s of a square carpet is measured at 6 m. Estimate the maximum error in the area A of the carpet if s is accurate to within 2 centimeters.

In Exercises 39 and 40, use the following fact derived from Newton's Laws: An object released at an angle θ with initial velocity v ft/s travels a horizontal distance

$$s = \frac{1}{32}v^2 \sin 2\theta \text{ ft (Figure 10)}$$

39. A player located 18.1 ft from the basket launches a successful jump shot from a height of 10 ft (level with the rim of the basket), at an angle $\theta = 34°$ and initial velocity $v = 25$ ft/s.)

(a) Show that $\Delta s \approx 0.255\Delta\theta$ ft for a small change of $\Delta\theta$.

(b) Is it likely that the shot would have been successful if the angle had been off by 2°?

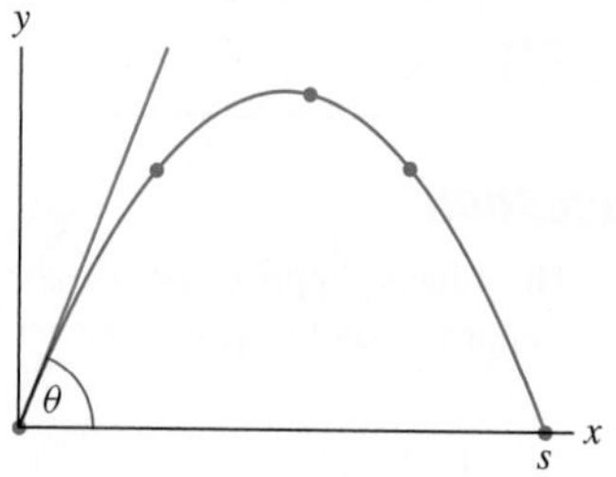

FIGURE 10 Trajectory of an object released at an angle θ.

40. Estimate Δs if $\theta = 34°$, $v = 25$ ft/s, and $\Delta v = 2$.

41. The radius of a spherical ball is measured at $r = 25$ cm. Estimate the maximum error in the volume and surface area if r is accurate to within 0.5 cm.

42. The dosage D of diphenhydramine for a dog of body mass w kg is $D = 4.7w^{2/3}$ mg. Estimate the maximum allowable error in w for a cocker spaniel of mass $w = 10$ kg if the percentage error in D must be less than 3%.

43. The volume (in liters) and pressure P (in atmospheres) of a certain gas satisfy $PV = 24$. A measurement yields $V = 4$ with a possible error of ± 0.3 L. Compute P and estimate the maximum error in this computation.

44. In the notation of Exercise 43, assume that a measurement yields $V = 4$. Estimate the maximum allowable error in V if P must have an error of less than 0.2 atm.

In Exercises 45–54, find the linearization at $x = a$.

45. $f(x) = x^4, \quad a = 1$

46. $f(x) = \dfrac{1}{x}, \quad a = 2$

47. $f(\theta) = \sin^2\theta, \quad a = \frac{\pi}{4}$

48. $g(x) = \dfrac{x^2}{x-3}, \quad a = 4$

49. $y = (1+x)^{-1/2}, \quad a = 0$

50. $y = (1+x)^{-1/2}, \quad a = 3$

51. $y = (1+x^2)^{-1/2}, \quad a = 0$

52. $y = \dfrac{1-x}{1+x}, \quad a = 4$

53. $y = \dfrac{\sin x}{x}, \quad a = \dfrac{\pi}{2}$

54. $y = \dfrac{\sin x}{x}, \quad a = \dfrac{\pi}{4}$

55. What is $f(2)$ if the linearization of $f(x)$ at $a = 2$ is $L(x) = 2x + 4$?

56. Compute the linearization of $f(x) = 3x - 4$ at $a = 0$ and $a = 2$. Prove more generally that a linear function coincides with its linearization at $x = a$ for all a.

57. Estimate $\sqrt{16.2}$ using the linearization $L(x)$ of $f(x) = \sqrt{x}$ at $a = 16$. Plot $f(x)$ and $L(x)$ on the same set of axes and determine whether the estimate is too large or too small.

58. GU Estimate $1/\sqrt{15}$ using a suitable linearization of $f(x) = 1/\sqrt{x}$. Plot $f(x)$ and $L(x)$ on the same set of axes and determine whether the estimate is too large or too small. Use a calculator to compute the percentage error.

In Exercises 59–67, approximate using linearization and use a calculator to compute the percentage error.

59. $\frac{1}{\sqrt{17}}$ **60.** $\frac{1}{101}$ **61.** $\frac{1}{(10.03)^2}$

62. $(17)^{1/4}$ **63.** $(64.1)^{1/3}$ **64.** $(1.2)^{5/3}$

65. $\tan(0.04)$ **66.** $\cos\left(\frac{3.1}{4}\right)$ **67.** $\frac{(3.1)/2}{\sin(3.1/2)}$

68. GU Compute the linearization $L(x)$ of $f(x) = x^2 - x^{3/2}$ at $a = 4$. Then plot $f(x) - L(x)$ and find an interval I around $a = 4$ such that $|f(x) - L(x)| \le 0.1$ for $x \in I$.

69. Show that the Linear Approximation to $f(x) = \sqrt{x}$ at $x = 9$ yields the estimate $\sqrt{9+h} - 3 \approx \frac{1}{6}h$. Set $K = 0.001$ and show that $|f''(x)| \le K$ for $x \ge 9$. Then verify numerically that the error E satisfies Eq. (5) for $h = 10^{-n}$, for $1 \le n \le 4$.

70. GU The Linear Approximation to $f(x) = \tan x$ at $x = \frac{\pi}{4}$ yields the estimate $\tan\left(\frac{\pi}{4} + h\right) - 1 \approx 2h$. Set $K = 6.2$ and show, using a plot, that $|f''(x)| \le K$ for $x \in [\frac{\pi}{4}, \frac{\pi}{4} + 0.1]$. Then verify numerically that the error E satisfies Eq. (5) for $h = 10^{-n}$, for $1 \le n \le 4$.

Further Insights and Challenges

71. Compute dy/dx at the point $P = (2, 1)$ on the curve $y^3 + 3xy = 7$ and show that the linearization at P is $L(x) = -\frac{1}{3}x + \frac{5}{3}$. Use $L(x)$ to estimate the y-coordinate of the point on the curve where $x = 2.1$.

72. Apply the method of Exercise 71 to $P = (0.5, 1)$ on $y^5 + y - 2x = 1$ to estimate the y-coordinate of the point on the curve where $x = 0.55$.

73. Apply the method of Exercise 71 to $P = (-1, 2)$ on $y^4 + 7xy = 2$ to estimate the solution of $y^4 - 7.7y = 2$ near $y = 2$.

74. Show that for any real number k, $(1 + \Delta x)^k \approx 1 + k\Delta x$ for small Δx. Estimate $(1.02)^{0.7}$ and $(1.02)^{-0.3}$.

75. Let $\Delta f = f(5 + h) - f(5)$, where $f(x) = x^2$. Verify directly that $E = |\Delta f - f'(5)h|$ satisfies (5) with $K = 2$.

76. Let $\Delta f = f(1 + h) - f(1)$ where $f(x) = x^{-1}$. Show directly that $E = |\Delta f - f'(1)h|$ is equal to $h^2/(1 + h)$. Then prove that $E \le 2h^2$ if $-\frac{1}{2} \le h \le \frac{1}{2}$. *Hint:* In this case, $\frac{1}{2} \le 1 + h \le \frac{3}{2}$.

4.2 Extreme Values

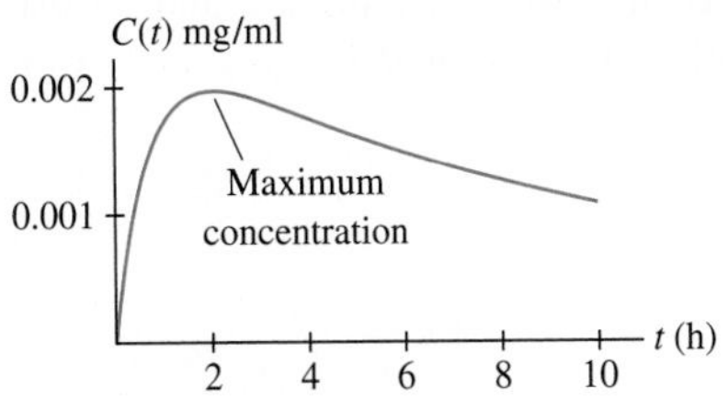

FIGURE 1 Drug Concentration in bloodstream (see Exercise 66).

Often, we drop the word "absolute" and speak simply of the min or max on an interval I. When no interval is mentioned, it is understood that we refer to the extreme values on the entire domain of the function.

In many applications it is important to find the minimum or maximum value of a function $f(x)$. For example, a physician needs to know the maximum drug concentration in a patient's bloodstream when a drug is administered. This amounts to finding the highest point on the graph of $C(t)$, the concentration at time t (Figure 1).

We refer to the maximum and minimum values (max and min for short) as **extreme values** or **extrema** (singular: extremum) and to the process of finding them as **optimization**. Sometimes, we are interested in finding the min or max for x in a particular interval I, rather than on the entire domain of $f(x)$.

DEFINITION Extreme Values on an Interval Let $f(x)$ be a function on an interval I and let $a \in I$. We say that $f(a)$ is the

- **Absolute minimum** of $f(x)$ on I if $f(a) \le f(x)$ for all $x \in I$.
- **Absolute maximum** of $f(x)$ on I if $f(a) \ge f(x)$ for all $x \in I$.

Does every function have a minimum or maximum value? Clearly not, as we see by taking $f(x) = x$. Indeed, $f(x) = x$ increases without bound as $x \to \infty$ and decreases without bound as $x \to -\infty$. In fact, extreme values do not always exist even if we restrict ourselves to an interval I. Figure 2 illustrates what can go wrong if I is open or f has a discontinuity.

- **Discontinuity:** (A) shows a discontinuous function with no maximum value. The values of $f(x)$ get arbitrarily close to 3 from below, but 3 is not the maximum value because $f(x)$ never actually takes on the value 3.

- **Open interval:** In (B), $g(x)$ is defined on the *open* interval (a, b). It has no max because it tends to ∞ on the right, and it has no min because it tends to 10 on the left without ever reaching this value.

Fortunately, our next theorem guarantees that extreme values exist when the function is continuous and I is closed [Figure 2(C)].

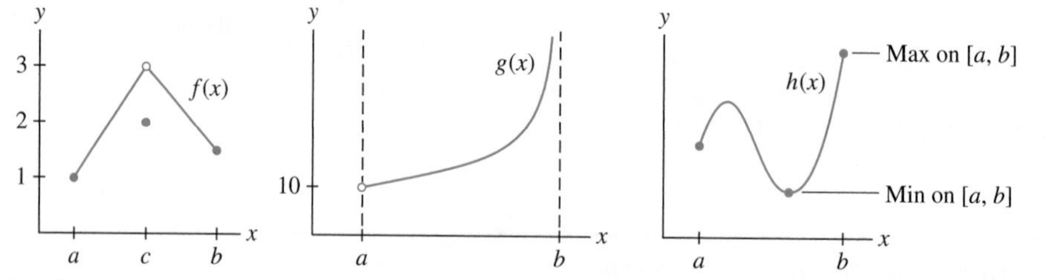

(A) Discontinuous function with no max on $[a, b]$.

(B) Continuous function with no min or max on the open interval (a, b).

(C) Every continuous function on a closed interval $[a, b]$ has both a min and a max on $[a, b]$.

FIGURE 2

REMINDER A closed, bounded interval is an interval $I = [a, b]$ (endpoints included), where a and b are both finite. Often, we drop the word "bounded" and refer to I more simply as a closed interval. An open interval (a, b) (endpoints not included) may have one or two infinite endpoints.

THEOREM 1 Existence of Extrema on a Closed Interval A continuous function $f(x)$ on a closed (bounded) interval $I = [a, b]$ takes on both a minimum and a maximum value on I.

CONCEPTUAL INSIGHT Why does Theorem 1 require a closed interval? Think of the graph of a continuous function as a string. If the interval is closed, the string is pinned down at the two endpoints and cannot fly off to infinity (or approach a min/max without reaching it) as in Figure 2(B). Intuitively, therefore, it must have a highest and lowest point. However, a rigorous proof of Theorem 1 relies on the *completeness property* of the real numbers (see Appendix D).

Local Extrema and Critical Points

We focus now on the problem of finding extreme values. A key concept is that of a local minimum or maximum.

When we get to the top of a hill in an otherwise flat region, our altitude is at a local maximum, but we are still far from the point of absolute maximum altitude, which is located at the peak of Mt. Everest. That's the difference between local and absolute extrema.

Adapted from "Stories About Maxima and Minima," V. M. Tikhomirov, AMS (1990)

DEFINITION Local Extrema We say that $f(x)$ has a

- **Local minimum** at $x = c$ if $f(c)$ is the minimum value of f on some open interval (in the domain of f) containing c.
- **Local maximum** at $x = c$ if $f(c)$ is the maximum value of $f(x)$ on some open interval (in the domain of f) containing c.

A local max occurs at $x = c$ if $(c, f(c))$ is the highest point on the graph within some small box [Figure 3(A)]. Thus, $f(c)$ is greater than or equal to all other *nearby* values, but it does not have to be the absolute maximum value of $f(x)$. Local minima are similar. Figure 3(B) illustrates the difference between local and absolute extrema: $f(a)$ is the absolute max on $[a, b]$ but is not a local max because $f(x)$ takes on larger values to the left of $x = a$.

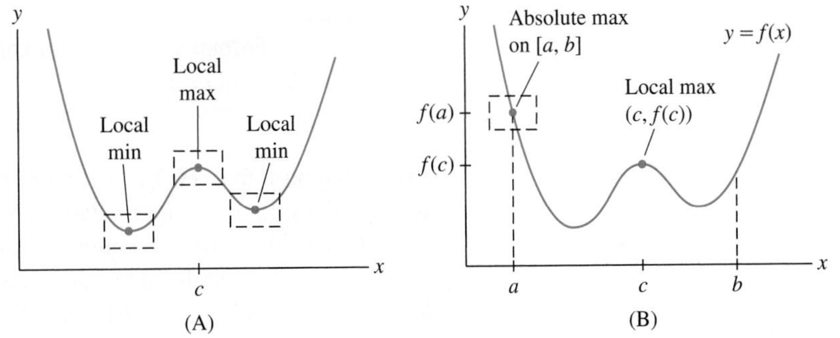

FIGURE 3

How do we find the local extrema? The crucial observation is that *the tangent line at a local min or max is horizontal* [Figure 4(A)]. In other words, if $f(c)$ is a local min or max, then $f'(c) = 0$. However, this assumes that $f(x)$ is differentiable. Otherwise, the tangent line may not exist, as in Figure 4(B). To take both possibilities into account, we define the notion of a critical point.

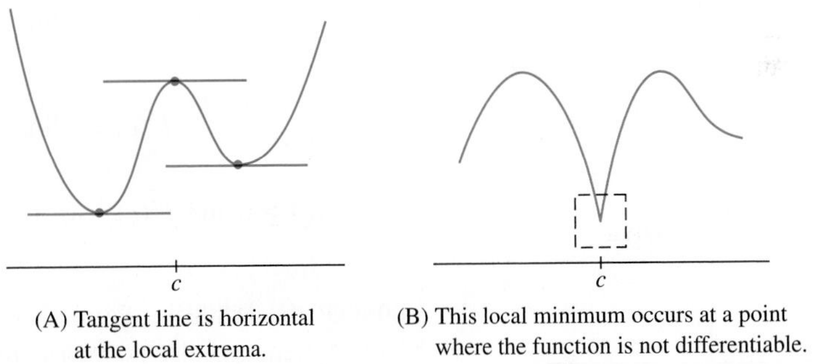
c
(A) Tangent line is horizontal
at the local extrema.
c
(B) This local minimum occurs at a point
where the function is not differentiable.

FIGURE 4

DEFINITION Critical Points A number c in the domain of f is called a **critical point** if *either* $f'(c) = 0$ or $f'(c)$ does not exist.

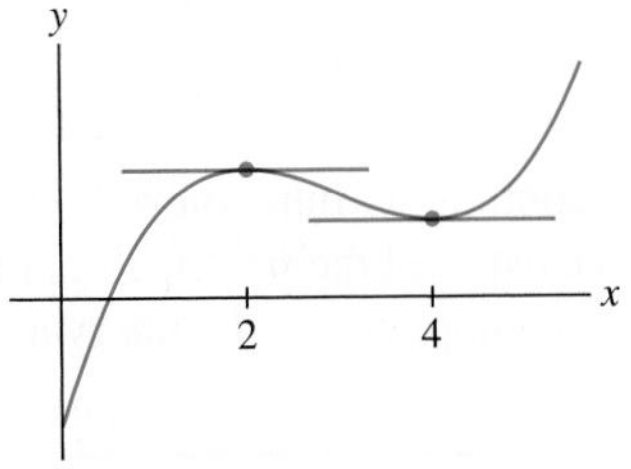

FIGURE 5 Graph of $f(x) = x^3 - 9x^2 + 24x - 10$.

■ **EXAMPLE 1** Find the critical points of $f(x) = x^3 - 9x^2 + 24x - 10$.

Solution The function $f(x)$ is differentiable everywhere (Figure 5), so the critical points are the solutions of $f'(x) = 0$:

$$f'(x) = 3x^2 - 18x + 24 = 3(x^2 - 6x + 8)$$
$$= 3(x-2)(x-4) = 0$$

The critical points are the roots $c = 2$ and $c = 4$. ■

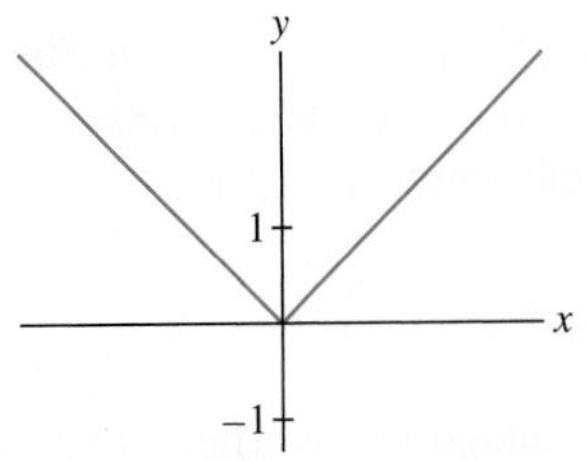

FIGURE 6 Graph of $f(x) = |x|$.

■ **EXAMPLE 2 Nondifferentiable Function** Find the critical points of $f(x) = |x|$.

Solution As we see in Figure 6, $f'(x) = -1$ for $x < 0$ and $f'(x) = 1$ for $x > 0$. Therefore, $f'(x) = 0$ has no solutions with $x \neq 0$. However, $f'(0)$ does not exist. Therefore $c = 0$ is a critical point. ■

The next theorem tells us that we can find local extrema by solving for the critical points. It is one of the most important results in calculus.

THEOREM 2 Fermat's Theorem on Local Extrema If $f(c)$ is a local min or max, then c is a critical point of $f(x)$.

Proof Suppose that $f(c)$ is a local minimum (the case of a local maximum is similar). If $f'(c)$ does not exist, then c is a critical point and there is nothing more to prove. So assume that $f'(c)$ exists. We must then prove that $f'(c) = 0$.

Because $f(c)$ is a local minimum, we have $f(c+h) \geq f(c)$ for all sufficiently small $h \neq 0$. Equivalently, $f(c+h) - f(c) \geq 0$. Now divide this inequality by h:

$$\frac{f(c+h) - f(c)}{h} \geq 0 \quad \text{if } h > 0 \tag{1}$$

$$\frac{f(c+h) - f(c)}{h} \leq 0 \quad \text{if } h < 0 \tag{2}$$

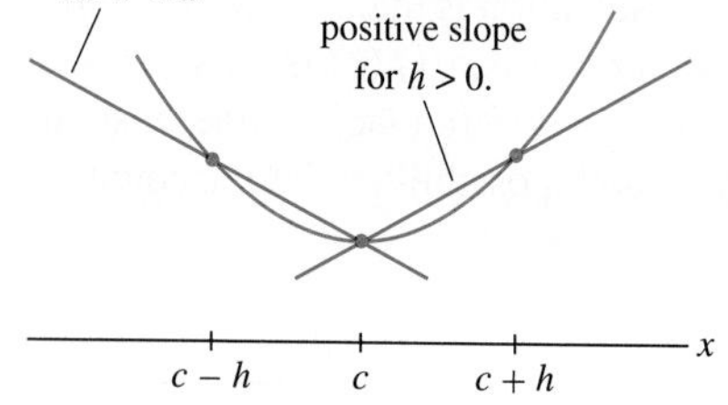

FIGURE 7

Figure 7 shows the graphical interpretation of these inequalities. Taking the one-sided limits of both sides of (1) and (2), we obtain

$$f'(c) = \lim_{h\to 0+} \frac{f(c+h) - f(c)}{h} \geq \lim_{h\to 0+} 0 = 0$$

$$f'(c) = \lim_{h\to 0-} \frac{f(c+h) - f(c)}{h} \leq \lim_{h\to 0-} 0 = 0$$

Thus $f'(c) \geq 0$ and $f'(c) \leq 0$. The only possibility is $f'(c) = 0$ as claimed. ■

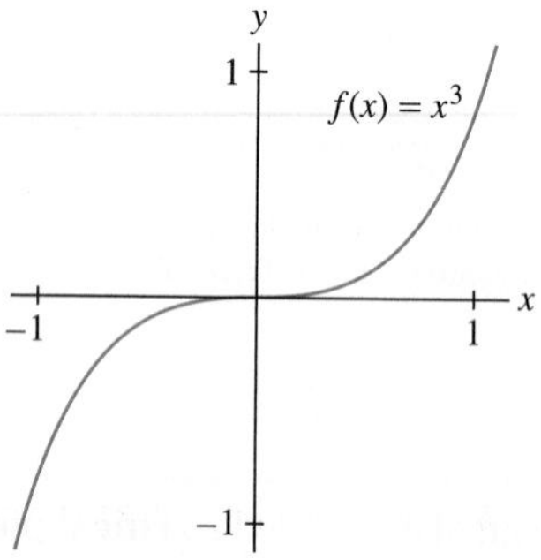

FIGURE 8 The tangent line at (0, 0) is horizontal, but $f(0)$ is not a local min or max.

CONCEPTUAL INSIGHT Fermat's Theorem *does not claim* that all critical points yield local extrema. "False positives" may exist—that is, we might have $f'(c) = 0$ without $f(c)$ being a local min or max. For example, $f(x) = x^3$ has derivative $f'(x) = 3x^2$ and $f'(0) = 0$, but $f(0)$ is neither a local min nor max (Figure 8). The origin is a point of inflection (studied in Section 4.4), where the tangent line crosses the graph.

Optimizing on a Closed Interval

Finally, we have all the tools needed for optimizing a continuous function on a closed interval. Theorem 1 guarantees that the extreme values exist, and the next theorem tells us where to find them, namely among the critical points or endpoints of the interval.

In this section, we restrict our attention to closed intervals because in this case extreme values are guaranteed to exist (Theorem 1). Optimization on open intervals is discussed in Section 4.6.

THEOREM 3 Extreme Values on a Closed Interval Assume that $f(x)$ is continuous on $[a, b]$ and let $f(c)$ be the minimum or maximum value on $[a, b]$. Then c is either a critical point or one of the endpoints a or b.

Proof If c is one of the endpoints a or b, there is nothing to prove. If not, then c belongs to the open interval (a, b). In this case, $f(c)$ is also a local min or max because it is the min or max on (a, b). By Fermat's Theorem, c is a critical point. ■

■ **EXAMPLE 3** Find the extrema of $f(x) = 2x^3 - 15x^2 + 24x + 7$ on $[0, 6]$.

Solution The extreme values occur at critical points or endpoints by Theorem 3, so we can break up the problem neatly into two steps.

Step 1. **Find the critical points.**
The function $f(x)$ is differentiable, so we find the critical points by solving

$$f'(x) = 6x^2 - 30x + 24 = 6(x-1)(x-4) = 0$$

The critical points are $c = 1$ and 4.

Step 2. **Compare values at the critical points and endpoints.**

x-value	Value of f	
1 (critical point)	$f(1) = 18$	
4 (critical point)	$f(4) = -9$	min
0 (endpoint)	$f(0) = 7$	
6 (endpoint)	$f(6) = 43$	max

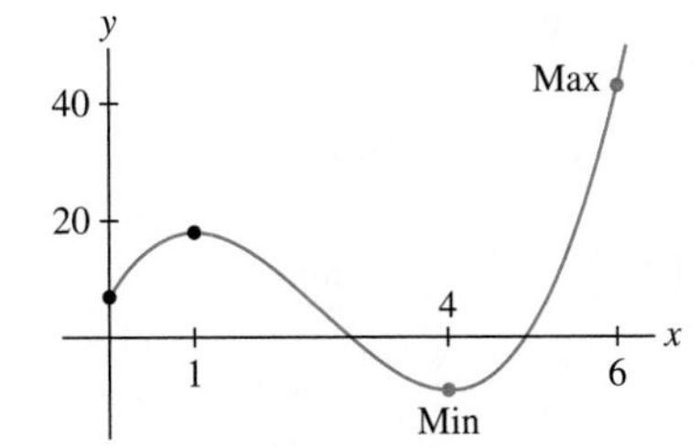

FIGURE 9 Extreme values of $f(x) = 2x^3 - 15x^2 + 24x + 7$ on $[0, 6]$.

The maximum of $f(x)$ on $[0, 6]$ is the largest of the values in this table, namely $f(6) = 43$. Similarly, the minimum is $f(4) = -9$. See Figure 9. ∎

■ **EXAMPLE 4 Function with a Cusp** Find the max of $f(x) = 1 - (x-1)^{2/3}$ on $[-1, 2]$.

Solution First, find the critical points:

$$f'(x) = -\frac{2}{3}(x-1)^{-1/3} = -\frac{2}{3(x-1)^{1/3}}$$

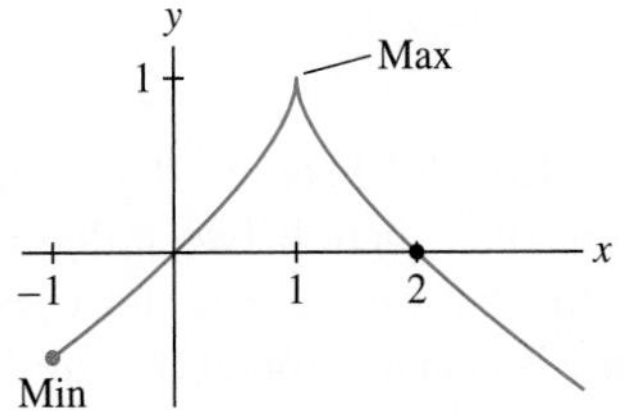

FIGURE 10 Extreme values of $f(x) = 1 - (x-1)^{2/3}$ on $[-1, 2]$.

The equation $f'(x) = 0$ has no solutions because $f'(x)$ is never zero. However, $f'(x)$ does not exist at $x = 1$, so $c = 1$ is a critical point (Figure 10).

Next, compare values at the critical points and endpoints.

x-value	Value of f	
1 (critical point)	$f(1) = 1$	max
−1 (endpoint)	$f(-1) \approx -0.59$	min
2 (endpoint)	$f(2) = 0$	

∎

In the next example, the critical points lie outside the interval $[a, b]$, so they are not relevant to the problem.

■ **EXAMPLE 5 A Critical Point Lying Outside the Interval** Find the extreme values of $f(x) = \dfrac{x}{x^2+5}$ on $[4, 8]$ (Figure 11).

Solution Compute $f'(x)$ using the quotient rule and solve for the critical points:

$$f'(x) = \frac{(x^2+5)(x)' - x(x^2+5)'}{(x^2+5)^2} = \frac{(x^2+5) - x(2x)}{(x^2+5)^2} = \frac{5-x^2}{(x^2+5)^2} = 0$$

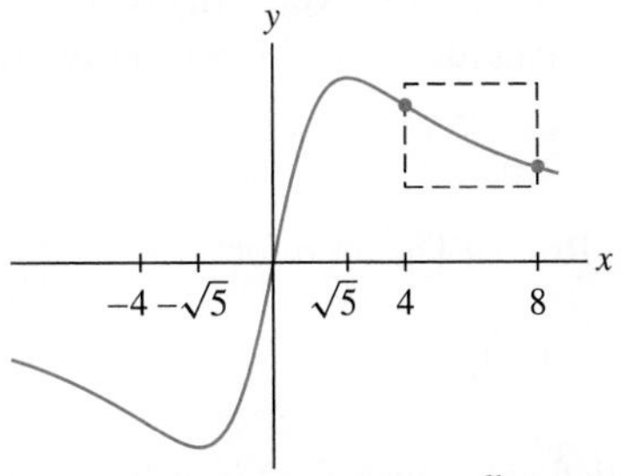

FIGURE 11 Graph of $f(x) = \dfrac{x}{x^2+5}$.

We see that $f'(x) = 0$ if $5 - x^2 = 0$. Hence, the critical points are $c = \pm\sqrt{5} \approx \pm 2.2$, both of which lie *outside* the interval $[4, 8]$. Therefore, the min and max of $f(x)$ on $[4, 8]$ occur at the endpoints. In fact, $f(4) = \frac{4}{21} \approx 0.19$ and $f(8) = \frac{8}{69} \approx 0.116$. Therefore $f(4)$ is the max and $f(8)$ is the min on the interval $[4, 8]$. ∎

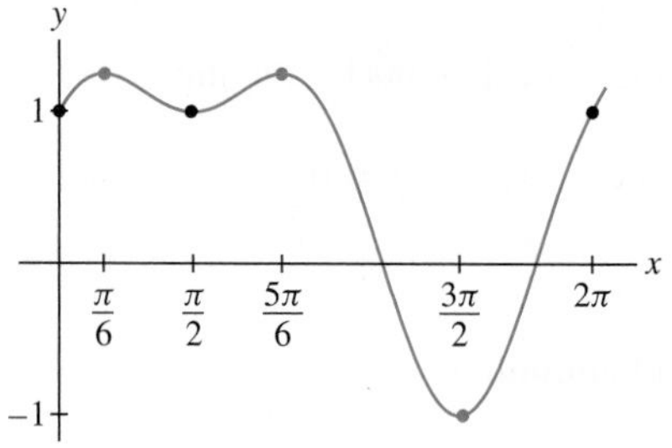

FIGURE 12 $f(x)$ attains a max at $\frac{\pi}{6}$ and $\frac{5\pi}{6}$ and a min at $\frac{3\pi}{2}$.

■ **EXAMPLE 6 Trigonometric Function** Find the min and max of the function $f(x) = \sin x + \cos^2 x$ on $[0, 2\pi]$ (Figure 12).

Solution First, solve for the critical points:

$$f'(x) = \cos x - 2\sin x\cos x = \cos x(1 - 2\sin x) = 0 \quad \Rightarrow \quad \cos x = 0 \text{ or } \sin x = \frac{1}{2}$$

$$\cos x = 0 \quad \Rightarrow \quad x = \frac{\pi}{2}, \frac{3\pi}{2} \qquad \text{and} \qquad \sin x = \frac{1}{2} \quad \Rightarrow \quad x = \frac{\pi}{6}, \frac{5\pi}{6}$$

Then compare the values of $f(x)$ at the critical points and endpoints:

x-value	Value of f	
$\frac{\pi}{2}$ (critical point)	$f(\frac{\pi}{2}) = 1 + 0^2 = 1$	
$\frac{3\pi}{2}$ (critical point)	$f(\frac{3\pi}{2}) = -1 + 0^2 = -1$	min
$\frac{\pi}{6}$ (critical point)	$f(\frac{\pi}{6}) = \frac{1}{2} + \left(\frac{\sqrt{3}}{2}\right)^2 = \frac{5}{4}$	max
$\frac{5\pi}{6}$ (critical point)	$f(\frac{5\pi}{6}) = \frac{1}{2} + \left(-\frac{\sqrt{3}}{2}\right)^2 = \frac{5}{4}$	max
0 and 2π (endpoints)	$f(0) = f(2\pi) = 1$	

■

Rolle's Theorem

As an application of our optimization methods, we prove Rolle's Theorem: If $f(x)$ takes on the same value at two different points a and b, then somewhere between these two points the derivative is zero. Graphically: If the secant line between $x = a$ and $x = b$ is horizontal, then at least one tangent line between a and b is also horizontal (Figure 13).

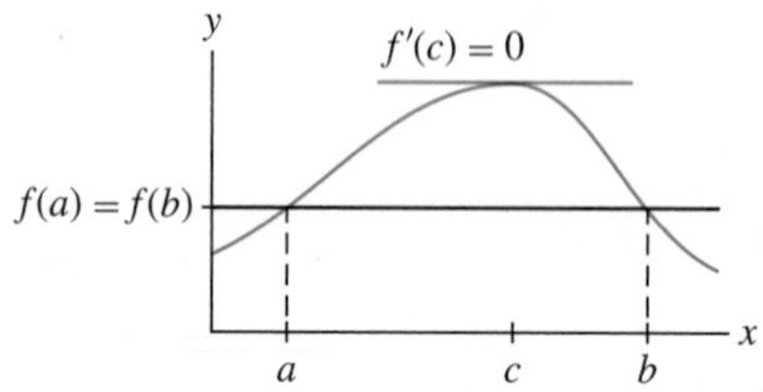

FIGURE 13 Rolle's Theorem: If $f(a) = f(b)$, then $f'(c) = 0$ for some c between a and b.

THEOREM 4 Rolle's Theorem Assume that $f(x)$ is continuous on $[a, b]$ and differentiable on (a, b). If $f(a) = f(b)$, then there exists a number c between a and b such that $f'(c) = 0$.

Proof Since $f(x)$ is continuous and $[a, b]$ is closed, $f(x)$ has a min and a max in $[a, b]$. Where do they occur? If either the min or the max occurs at a point c in the *open* interval (a, b), then $f(c)$ is a local extreme value and $f'(c) = 0$ by Fermat's Theorem (Theorem 2). Otherwise, both the min and the max occur at the endpoints. However, $f(a) = f(b)$, so in this case, the min and max coincide and $f(x)$ is a constant function with zero derivative. Therefore, $f'(c) = 0$ for all $c \in (a, b)$. ■

■ **EXAMPLE 7 Illustrating Rolle's Theorem** Verify Rolle's Theorem for

$$f(x) = x^4 - x^2 \qquad \text{on} \qquad [-2, 2]$$

Solution The hypotheses of Rolle's Theorem are satisfied because $f(x)$ is differentiable (and therefore continuous) everywhere, and $f(2) = f(-2)$:

$$f(2) = 2^4 - 2^2 = 12, \qquad f(-2) = (-2)^4 - (-2)^2 = 12$$

We must verify that $f'(c) = 0$ has a solution in $(-2, 2)$, so we solve $f'(x) = 4x^3 - 2x = 2x(2x^2 - 1) = 0$. The solutions are $c = 0$ and $c = \pm 1/\sqrt{2} \approx \pm 0.707$. They all lie in $(-2, 2)$, so Rolle's Theorem is satisfied with three values of c. ■

■ **EXAMPLE 8** **Using Rolle's Theorem** Show that $f(x) = x^3 + 9x - 4$ has precisely one real root.

Solution First, we note that $f(0) = -4$ is negative and $f(1) = 6$ is positive. By the Intermediate Value Theorem (Section 2.8), $f(x)$ has *at least* one root a in $[0, 1]$. If $f(x)$ had a second root b, then $f(a) = f(b) = 0$ and Rolle's Theorem would imply that $f'(c) = 0$ for some $c \in (a, b)$. This is not possible because $f'(x) = 3x^2 + 9 \geq 9$, so $f'(c) = 0$ has no solutions. We conclude that a is *the only* real root of $f(x)$ (Figure 14). ■

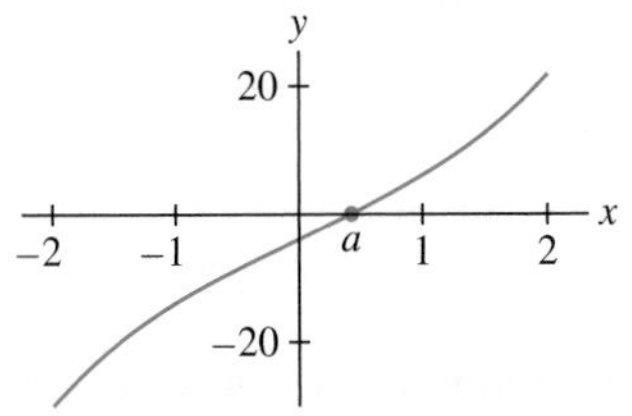

FIGURE 14 Graph of $f(x) = x^3 + 9x - 4$. This function has one real root.

HISTORICAL PERSPECTIVE

Pierre de Fermat (1601–1665)

René Descartes (1596–1650)

Sometime in the 1630's, in the decade before Isaac Newton was born, the French mathematician Pierre de Fermat invented a general method for finding extreme values. Fermat said, in essence, that if you want to find extrema, you must set the derivative equal to zero and solve for the critical points, just as we have done in this section. He also described a general method for finding tangent lines that is not essentially different from our method of derivatives. For this reason, Fermat is often regarded as an inventor of calculus, together with Newton and Leibniz.

We can hardly expect a more general method.... This method never fails and could be extended to a number of beautiful problems; with its aid we have found the centers of gravity of figures bounded by straight lines or curves, as well as those of solids, and a number of other results which we may treat elsewhere if we have the time to do so.

—From Fermat's *On Maxima and Minima and on Tangents*

At around the same time, René Descartes (1596-1650) developed a different but less effective approach to finding tangent lines. Descartes, after whom Cartesian coordinates are named, was a profound thinker—the leading philosopher and scientist of his time in Europe. He is regarded today as the father of modern philosophy and the founder (along with Fermat) of analytic geometry. A dispute developed when Descartes learned through an intermediary that Fermat had criticized his work on optics. Sensitive and stubborn, Descartes retaliated by attacking Fermat's method of finding tangents and only after some third-party refereeing did he admit that Fermat was correct. He wrote:

...Seeing the last method that you use for finding tangents to curved lines, I can reply to it in no other way than to say that it is very good and that, if you had explained it in this manner at the outset, I would have not contradicted it at all.

However, in subsequent private correspondence, Descartes was less generous, referring at one point to some of Fermat's work as "*le galimatias le plus ridicule*"—the most ridiculous gibberish. Today Fermat is recognized as one of the greatest mathematicians of his age who made far-reaching contributions in several areas of mathematics.

4.2 SUMMARY

- The *extreme values* of $f(x)$ on an interval I are the minimum and maximum values of $f(x)$ for $x \in I$ (also called *absolute extrema* on I).
- Basic Theorem: If $f(x)$ is continuous on a closed interval $[a, b]$, then $f(x)$ has both a min and a max on $[a, b]$.
- $f(c)$ is a *local minimum* if $f(x) \geq f(c)$ for all x in some open interval around c. Local maxima are defined similarly.
- $x = c$ is a *critical point* of $f(x)$ if either $f'(c) = 0$ or $f'(c)$ does not exist.
- Fermat's Theorem: If $f(c)$ is a local min or max, then c is a critical point.
- To find the extreme values of a continuous function $f(x)$ on a closed interval $[a, b]$:

Step 1. Find the critical points of $f(x)$ in $[a, b]$.

Step 2. Calculate $f(x)$ at the critical points in $[a, b]$ and at the endpoints.

The min and max on $[a, b]$ are the smallest and largest among the values computed in Step 2.

• Rolle's Theorem: If $f(x)$ is continuous on $[a, b]$ and differentiable on (a, b), and if $f(a) = f(b)$, then there exists c between a and b such that $f'(c) = 0$.

4.2 EXERCISES

Preliminary Questions

1. What is the definition of a critical point?

In Questions 2 and 3, choose the correct conclusion.

2. If $f(x)$ is not continuous on $[0, 1]$, then

(a) $f(x)$ has no extreme values on $[0, 1]$.

(b) $f(x)$ might not have any extreme values on $[0, 1]$.

3. If $f(x)$ is continuous but has no critical points in $[0, 1]$, then

(a) $f(x)$ has no min or max on $[0, 1]$.

(b) Either $f(0)$ or $f(1)$ is the minimum value on $[0, 1]$.

4. Fermat's Theorem *does not* claim that if $f'(c) = 0$, then $f(c)$ is a local extreme value (this is false). What *does* Fermat's Theorem assert?

Exercises

1. The following questions refer to Figure 15.

(a) How many critical points does $f(x)$ have on $[0, 8]$?

(b) What is the maximum value of $f(x)$ on $[0, 8]$?

(c) What are the local maximum values of $f(x)$?

(d) Find a closed interval on which both the minimum and maximum values of $f(x)$ occur at critical points.

(e) Find an interval on which the minimum value occurs at an endpoint.

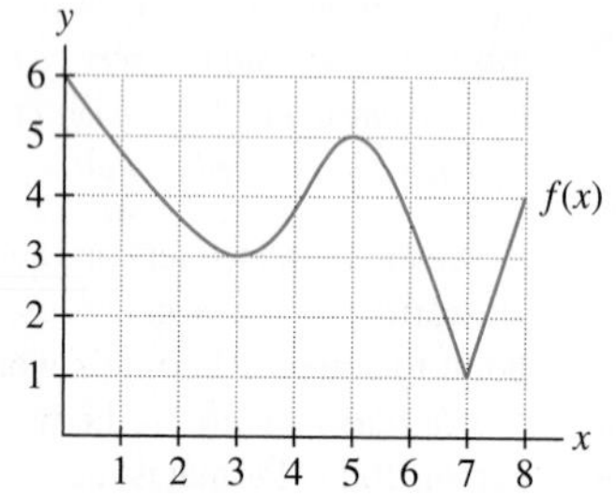

FIGURE 15

2. State whether $f(x) = x^{-1}$ (Figure 16) has a minimum or maximum value on the following intervals:

(a) $(0, 2)$ **(b)** $(1, 2)$ **(c)** $[1, 2]$

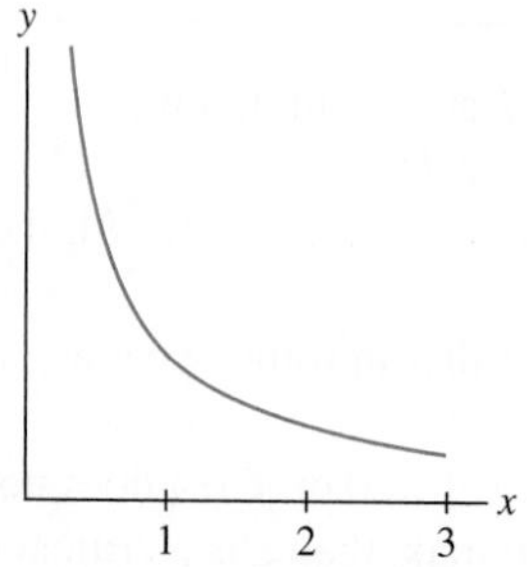

FIGURE 16 Graph of $f(x) = x^{-1}$.

In Exercises 3–16, find all critical points of the function.

3. $f(x) = x^2 - 2x + 4$

4. $f(x) = 7x - 2$

5. $f(x) = x^3 - \frac{9}{2}x^2 - 54x + 2$

6. $f(t) = 8t^3 - t^2$

7. $f(x) = x^{-1} - x^{-2}$

8. $g(z) = \dfrac{1}{z-1} - \dfrac{1}{z}$

9. $f(x) = \dfrac{x}{x^2 + 1}$

10. $f(x) = \dfrac{x^2}{x^2 - 4x + 8}$

11. $f(t) = t - 4\sqrt{t+1}$

12. $f(t) = 4t - \sqrt{t^2 + 1}$

13. $f(x) = x^2\sqrt{1 - x^2}$

14. $f(x) = x + |2x + 1|$

15. $g(\theta) = \sin^2 \theta$

16. $R(\theta) = \cos\theta + \sin^2\theta$

17. Let $f(x) = x^2 - 4x + 1$.

(a) Find the critical point c of $f(x)$ and compute $f(c)$.

(b) Compute the value of $f(x)$ at the endpoints of the interval $[0, 4]$.

(c) Determine the min and max of $f(x)$ on $[0, 4]$.

(d) Find the extreme values of $f(x)$ on $[0, 1]$.

18. Find the extreme values of $f(x) = 2x^3 - 9x^2 + 12x$ on $[0, 3]$ and $[0, 2]$.

19. Find the critical points of $f(x) = \sin x + \cos x$ and determine the extreme values on $\left[0, \frac{\pi}{2}\right]$.

20. Compute the critical points of $h(t) = (t^2 - 1)^{1/3}$. Check that your answer is consistent with Figure 17. Then find the extreme values of $h(t)$ on $[0, 1]$ and $[0, 2]$.

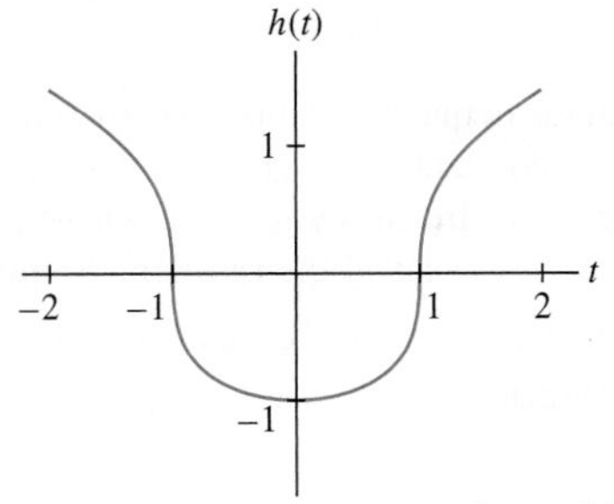

FIGURE 17 Graph of $h(t) = (t^2 - 1)^{1/3}$.

21. GU Plot $f(x) = 2\sqrt{x} - x$ on $[0, 4]$ and determine the maximum value graphically. Then verify your answer using calculus.

22. GU Plot $f(x) = 2x^3 - 9x^2 + 12x$ on $[0, 3]$ and locate the extreme values graphically. Then verify your answer using calculus.

23. CAS Approximate the critical points of $g(x) = x \cos x$ on $I = [0, 2\pi]$, and estimate the minimum value of $g(x)$ on I.

24. CAS Approximate the critical points of $g(x) = \tan^2 x - 5x$ on $I = \left(-\frac{\pi}{2}, \frac{\pi}{2}\right)$, and estimate the minimum value of $g(x)$ on I.

In Exercises 25–50, find the min and max of the function on the given interval by comparing values at the critical points and endpoints.

25. $y = 2x^2 + 4x + 5, \quad [-2, 2]$

26. $y = 2x^2 + 4x + 5, \quad [0, 2]$

27. $y = 6t - t^2, \quad [0, 5]$

28. $y = 6t - t^2, \quad [4, 6]$

29. $y = x^3 - 6x^2 + 8, \quad [1, 6]$

30. $y = x^3 + x^2 - x, \quad [-2, 2]$

31. $y = 2t^3 + 3t^2, \quad [1, 2]$

32. $y = x^3 - 12x^2 + 21x, \quad [0, 2]$

33. $y = z^5 - 80z, \quad [-3, 3]$

34. $y = 2x^5 + 5x^2, \quad [-2, 2]$

35. $y = \dfrac{x^2 + 1}{x - 4}, \quad [5, 6]$

36. $y = \dfrac{1 - x}{x^2 + 3x}, \quad [1, 4]$

37. $y = x - \dfrac{4x}{x + 1}, \quad [0, 3]$

38. $y = 2\sqrt{x^2 + 1} - x, \quad [0, 2]$

39. $y = (2 + x)\sqrt{2 + (2 - x)^2}, \quad [0, 2]$

40. $y = \sqrt{1 + x^2} - 2x, \quad [0, 1]$

41. $y = \sqrt{x + x^2} - 2\sqrt{x}, \quad [0, 4]$

42. $y = (t - t^2)^{1/3}, \quad [-1, 2]$

43. $y = \sin x \cos x, \quad \left[0, \frac{\pi}{2}\right]$

44. $y = x + \sin x, \quad [0, 2\pi]$

45. $y = \sqrt{2}\,\theta - \sec\theta, \quad \left[0, \frac{\pi}{3}\right]$

46. $y = \cos\theta + \sin\theta, \quad [0, 2\pi]$

47. $y = \theta - 2\sin\theta, \quad [0, 2\pi]$

48. $y = 4\sin^3\theta - 3\cos^2\theta, \quad [0, 2\pi]$

49. $y = \tan x - 2x, \quad [0, 1]$

50. $y = \sec^2 x - 2\tan x, \quad [-\pi/6, \pi/3]$

51. Let $f(\theta) = 2\sin 2\theta + \sin 4\theta$.

(a) Show that θ is a critical point if $\cos 4\theta = -\cos 2\theta$.

(b) Show, using a unit circle, that $\cos\theta_1 = -\cos\theta_2$ if and only if $\theta_1 = \pi \pm \theta_2 + 2\pi k$ for an integer k.

(c) Show that $\cos 4\theta = -\cos 2\theta$ if and only if $\theta = \frac{\pi}{2} + \pi k$ or $\theta = \frac{\pi}{6} + \left(\frac{\pi}{3}\right)k$.

(d) Find the six critical points of $f(\theta)$ on $[0, 2\pi]$ and find the extreme values of $f(\theta)$ on this interval.

(e) GU Check your results against a graph of $f(\theta)$.

52. GU Find the critical points of $f(x) = 2\cos 3x + 3\cos 2x$ in $[0, 2\pi]$. Check your answer against a graph of $f(x)$.

In Exercises 53–56, find the critical points and the extreme values on $[0, 4]$. *In Exercises 55 and 56, refer to Figure 18.*

53. $y = |x - 2|$

54. $y = |3x - 9|$

55. $y = |x^2 + 4x - 12|$

56. $y = |\cos x|$

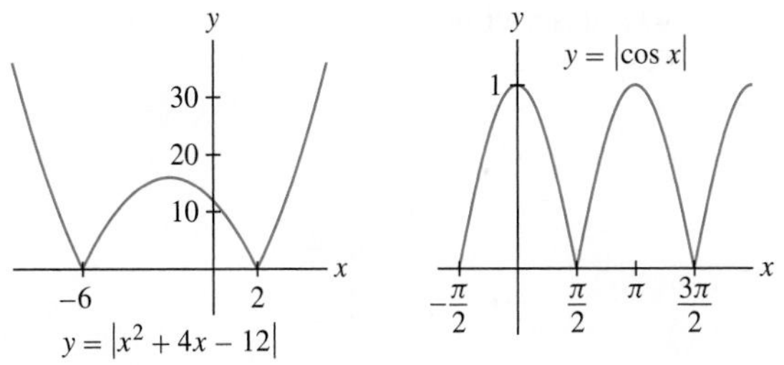

FIGURE 18

In Exercises 57–60, verify Rolle's Theorem for the given interval.

57. $f(x) = x + x^{-1}, \quad \left[\frac{1}{2}, 2\right]$

58. $f(x) = \sin x, \quad \left[\frac{\pi}{4}, \frac{3\pi}{4}\right]$

59. $f(x) = \dfrac{x^2}{8x - 15}, \quad [3, 5]$

60. $f(x) = \sin^2 x - \cos^2 x, \quad \left[\frac{\pi}{4}, \frac{3\pi}{4}\right]$

61. Prove that $f(x) = x^5 + 2x^3 + 4x - 12$ has precisely one real root.

62. Prove that $f(x) = x^3 + 3x^2 + 6x$ has precisely one real root.

63. Prove that $f(x) = x^4 + 5x^3 + 4x$ has no root c satisfying $c > 0$. *Hint:* Note that $x = 0$ is a root and apply Rolle's Theorem.

64. Prove that $c = 4$ is the largest root of $f(x) = x^4 - 8x^2 - 128$.

65. The position of a mass oscillating at the end of a spring is $s(t) = A\sin\omega t$, where A is the amplitude and ω is the angular frequency. Show that the speed $|v(t)|$ is at a maximum when the acceleration $a(t)$ is zero and that $|a(t)|$ is at a maximum when $v(t)$ is zero.

66. The concentration $C(t)$ (in mg/cm^3) of a drug in a patient's bloodstream after t hours is

$$C(t) = \frac{0.016t}{t^2 + 4t + 4}$$

Find the maximum concentration in the time interval $[0, 8]$ and the time at which it occurs.

67. In 1919, physicist Alfred Betz argued that the maximum efficiency of a wind turbine is around 59%. If wind enters a turbine with speed v_1 and exits with speed v_2, then the power extracted is the difference in kinetic energy per unit time:

$$P = \frac{1}{2}mv_1^2 - \frac{1}{2}mv_2^2 \quad \text{watts}$$

where m is the mass of wind flowing through the rotor per unit time (Figure 19). Betz assumed that $m = \rho A(v_1 + v_2)/2$, where ρ is the density of air and A is the area swept out by the rotor. Wind flowing undisturbed through the same area A would have mass per unit time $\rho A v_1$ and power $P_0 = \frac{1}{2}\rho A v_1^3$. The fraction of power extracted by the turbine is $F = P/P_0$.

(a) Show that F depends only on the ratio $r = v_2/v_1$ and is equal to $F(r) = \frac{1}{2}(1 - r^2)(1 + r)$, where $0 \le r \le 1$.

(b) Show that the maximum value of $F(r)$, called the **Betz Limit**, is $16/27 \approx 0.59$.

(c) Explain why Betz's formula for $F(r)$ is not meaningful for r close to zero. *Hint:* How much wind would pass through the turbine if v_2 were zero? Is this realistic?

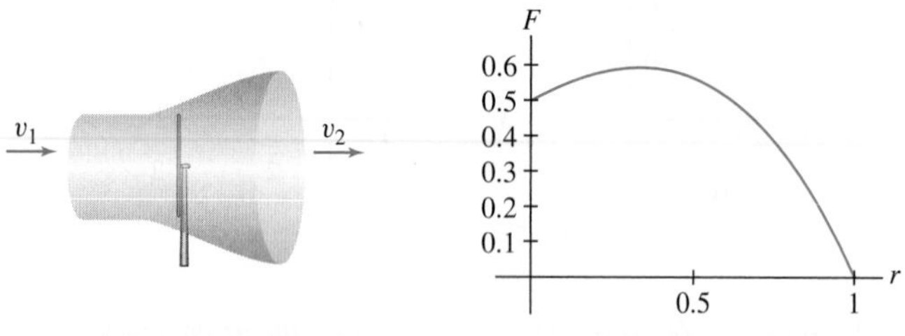

(A) Wind flowing through a turbine.

(B) F is the fraction of energy extracted by the turbine as a function of $r = v_2/v_1$.

FIGURE 19

68. GU The **Bohr radius** a_0 of the hydrogen atom is the value of r that minimizes the energy

$$E(r) = \frac{\hbar^2}{2mr^2} - \frac{e^2}{4\pi\epsilon_0 r}$$

where $\hbar$, m, e, and ϵ_0 are physical constants. Show that $a_0 = 4\pi\epsilon_0\hbar^2/(me^2)$. Assume that the minimum occurs at a critical point, as suggested by Figure 20.

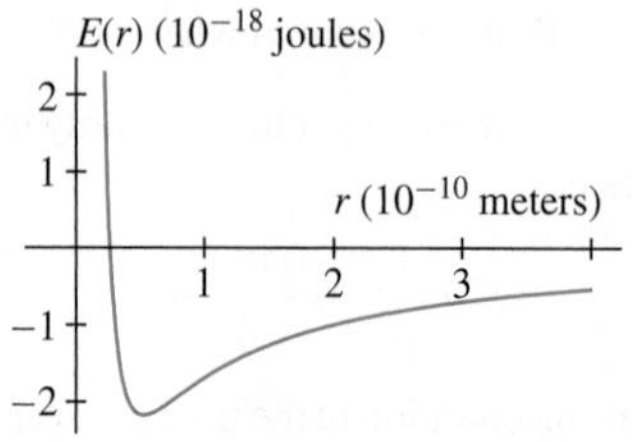

FIGURE 20

69. The response of a circuit or other oscillatory system to an input of frequency ω ("omega") is described by the function

$$\phi(\omega) = \frac{1}{\sqrt{(\omega_0^2 - \omega^2)^2 + 4D^2\omega^2}}$$

Both ω_0 (the natural frequency of the system) and D (the damping factor) are positive constants. The graph of ϕ is called a **resonance curve**, and the positive frequency $\omega_r > 0$, where ϕ takes its maximum value, if it exists, is called the **resonant frequency**. Show that $\omega_r = \sqrt{\omega_0^2 - 2D^2}$ if $0 < D < \omega_0/\sqrt{2}$ and that no resonant frequency exists otherwise (Figure 21).

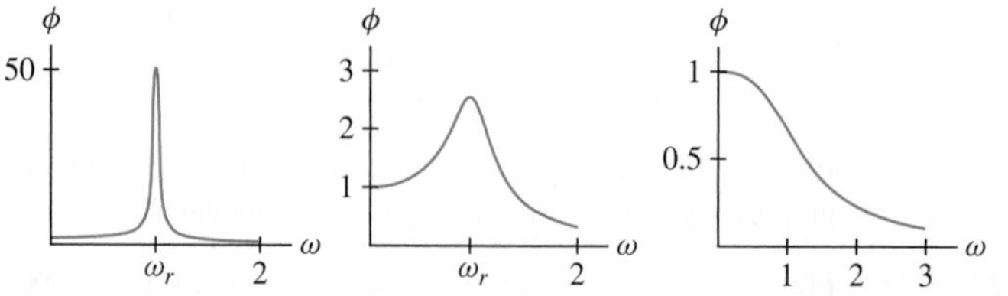

(A) $D = 0.01$ (B) $D = 0.2$ (C) $D = 0.75$ (no resonance)

FIGURE 21 Resonance curves with $\omega_0 = 1$.

70. Bees build honeycomb structures out of cells with a hexagonal base and three rhombus-shaped faces on top, as in Figure 22. We can show that the surface area of this cell is

$$A(\theta) = 6hs + \frac{3}{2}s^2(\sqrt{3}\csc\theta - \cot\theta)$$

with h, s, and θ as indicated in the figure. Remarkably, bees "know" which angle θ minimizes the surface area (and therefore requires the least amount of wax).

(a) Show that $\theta \approx 54.7°$ (assume h and s are constant). *Hint:* Find the critical point of $A(\theta)$ for $0 < \theta < \pi/2$.

(b) GU Confirm, by graphing $f(\theta) = \sqrt{3}\csc\theta - \cot\theta$, that the critical point indeed minimizes the surface area.

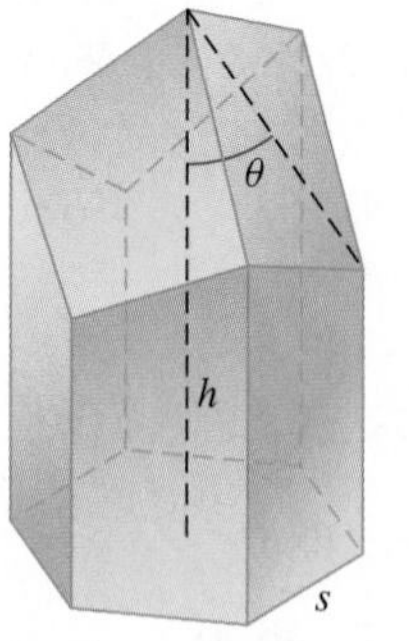

FIGURE 22 A cell in a honeycomb constructed by bees.

71. Find the maximum of $y = x^a - x^b$ on $[0, 1]$ where $0 < a < b$. In particular, find the maximum of $y = x^5 - x^{10}$ on $[0, 1]$.

In Exercises 72–74, plot the function using a graphing utility and find its critical points and extreme values on $[-5, 5]$.

72. GU $y = \dfrac{1}{1 + |x - 1|}$

73. GU $y = \dfrac{1}{1+|x-1|} + \dfrac{1}{1+|x-4|}$

74. GU $y = \dfrac{x}{|x^2-1|+|x^2-4|}$

75. **(a)** Use implicit differentiation to find the critical points on the curve $27x^2 = (x^2+y^2)^3$.

(b) GU Plot the curve and the horizontal tangent lines on the same set of axes.

76. Sketch the graph of a continuous function on $(0, 4)$ with a minimum value but no maximum value.

77. Sketch the graph of a continuous function on $(0, 4)$ having a local minimum but no absolute minimum.

78. Sketch the graph of a function on $[0, 4]$ having

(a) Two local maxima and one local minimum.

(b) An absolute minimum that occurs at an endpoint, and an absolute maximum that occurs at a critical point.

79. Sketch the graph of a function $f(x)$ on $[0, 4]$ with a discontinuity such that $f(x)$ has an absolute minimum but no absolute maximum.

80. A rainbow is produced by light rays that enter a raindrop (assumed spherical) and exit after being reflected internally as in Figure 23. The angle between the incoming and reflected rays is $\theta = 4r - 2i$, where the angle of incidence i and refraction r are related by Snell's Law $\sin i = n \sin r$ with $n \approx 1.33$ (the index of refraction for air and water).

(a) Use Snell's Law to show that $\dfrac{dr}{di} = \dfrac{\cos i}{n \cos r}$.

(b) Show that the maximum value θ_{max} of θ occurs when i satisfies $\cos i = \sqrt{\dfrac{n^2-1}{3}}$. *Hint:* Show that $\dfrac{d\theta}{di} = 0$ if $\cos i = \dfrac{n}{2}\cos r$. Then use Snell's Law to eliminate r.

(c) Show that $\theta_{max} \approx 59.58°$.

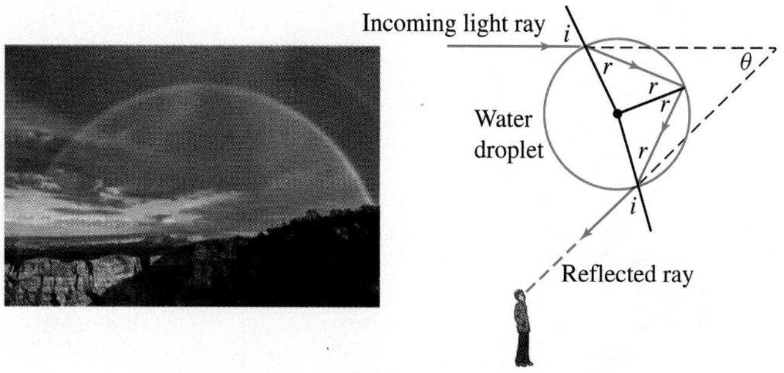

FIGURE 23

Further Insights and Challenges

81. Show that the extreme values of $f(x) = a \sin x + b \cos x$ are $\pm\sqrt{a^2+b^2}$.

82. Show, by considering its minimum, that $f(x) = x^2 - 2x + 3$ takes on only positive values. More generally, find the conditions on r and s under which the quadratic function $f(x) = x^2 + rx + s$ takes on only positive values. Give examples of r and s for which f takes on both positive and negative values.

83. Show that if the quadratic polynomial $f(x) = x^2 + rx + s$ takes on both positive and negative values, then its minimum value occurs at the midpoint between the two roots.

84. Generalize Exercise 83: Show that if the horizontal line $y = c$ intersects the graph of $f(x) = x^2 + rx + s$ at two points $(x_1, f(x_1))$ and $(x_2, f(x_2))$, then $f(x)$ takes its minimum value at the midpoint $M = \dfrac{x_1 + x_2}{2}$ (Figure 24).

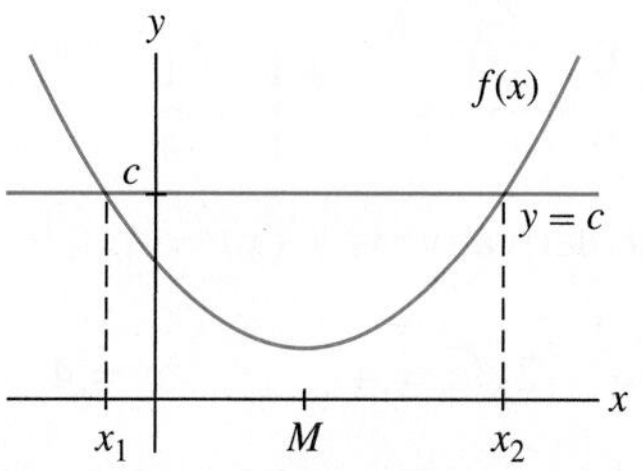

FIGURE 24

85. A cubic polynomial may have a local min and max, or it may have neither (Figure 25). Find conditions on the coefficients a and b of

$$f(x) = \frac{1}{3}x^3 + \frac{1}{2}ax^2 + bx + c$$

that ensure that f has neither a local min nor a local max. *Hint:* Apply Exercise 82 to $f'(x)$.

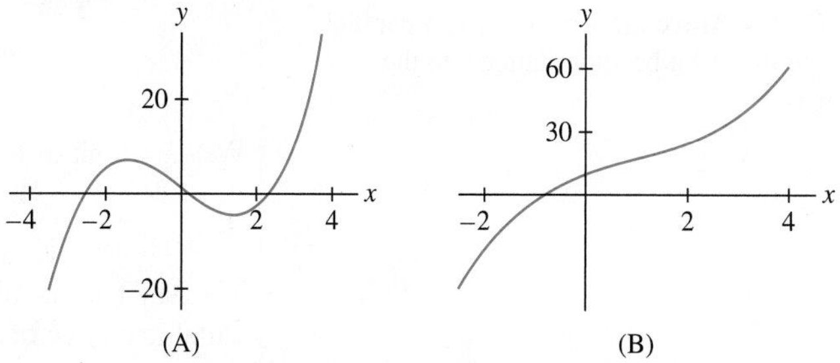

FIGURE 25 Cubic polynomials

86. Find the min and max of

$$f(x) = x^p(1-x)^q \quad \text{on } [0, 1],$$

where $p, q > 0$.

87. Prove that if f is continuous and $f(a)$ and $f(b)$ are local minima where $a < b$, then there exists a value c between a and b such that $f(c)$ is a local maximum. (*Hint:* Apply Theorem 1 to the interval $[a, b]$.) Show that continuity is a necessary hypothesis by sketching the graph of a function (necessarily discontinuous) with two local minima but no local maximum.

4.3 The Mean Value Theorem and Monotonicity

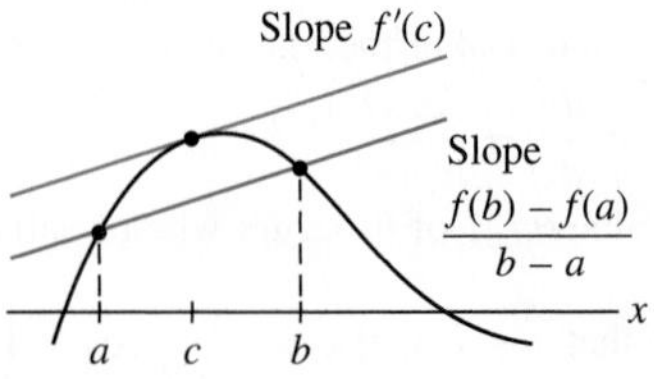

FIGURE 1 By the MVT, there exists at least one tangent line parallel to the secant line.

We have taken for granted that a function $f(x)$ is increasing if $f'(x)$ is positive and decreasing if $f'(x)$ is negative. In this section, we prove this rigorously using an important result called the Mean Value Theorem (MVT). Then we develop a method for "testing" critical points—that is, for determining whether they correspond to local minima or maxima.

The MVT says that a secant line between two points $(a, f(a))$ and $(b, f(b))$ on a graph is parallel to at least one tangent line in the interval (a, b) [Figure 1]. Because two lines are parallel if they have the same slope, what the MVT claims is that there exists a point c between a and b such that

$$\underbrace{f'(c)}_{\text{Slope of tangent line}} = \underbrace{\frac{f(b) - f(a)}{b - a}}_{\text{Slope of secant line}}$$

THEOREM 1 The Mean Value Theorem Assume that $f(x)$ is continuous on the closed interval $[a, b]$ and differentiable on (a, b). Then there exists at least one value c in (a, b) such that

$$f'(c) = \frac{f(b) - f(a)}{b - a}$$

Rolle's Theorem (Section 4.2) is the special case of the MVT in which $f(a) = f(b)$. In this case, the conclusion is that $f'(c) = 0$.

GRAPHICAL INSIGHT Imagine what happens when a secant line is moved parallel to itself. Eventually, it becomes a tangent line, as shown in Figure 2. This is the idea behind the MVT. We present a formal proof at the end of this section.

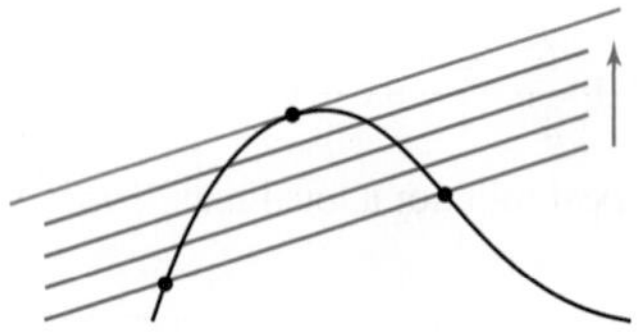

FIGURE 2 Move the secant line in a parallel fashion until it becomes tangent to the curve.

CONCEPTUAL INSIGHT The conclusion of the MVT can be rewritten as

$$f(b) - f(a) = f'(c)(b - a)$$

We can think of this as a variation on the Linear Approximation, which says

$$f(b) - f(a) \approx f'(a)(b - a).$$

The MVT turns this approximation into an equality by replacing $f'(a)$ with $f'(c)$ for a suitable choice of c in (a, b).

■ **EXAMPLE 1** Verify the MVT with $f(x) = \sqrt{x}$, $a = 1$, and $b = 9$.

Solution First, compute the slope of the secant line (Figure 3):

$$\frac{f(b) - f(a)}{b - a} = \frac{\sqrt{9} - \sqrt{1}}{9 - 1} = \frac{3 - 1}{9 - 1} = \frac{1}{4}$$

We must find c such that $f'(c) = 1/4$. The derivative is $f'(x) = \frac{1}{2}x^{-1/2}$, and

$$f'(c) = \frac{1}{2\sqrt{c}} = \frac{1}{4} \quad \Rightarrow \quad 2\sqrt{c} = 4 \quad \Rightarrow \quad c = 4$$

The value $c = 4$ lies in $(1, 9)$ and satisfies $f'(4) = \frac{1}{4}$. This verifies the MVT. ■

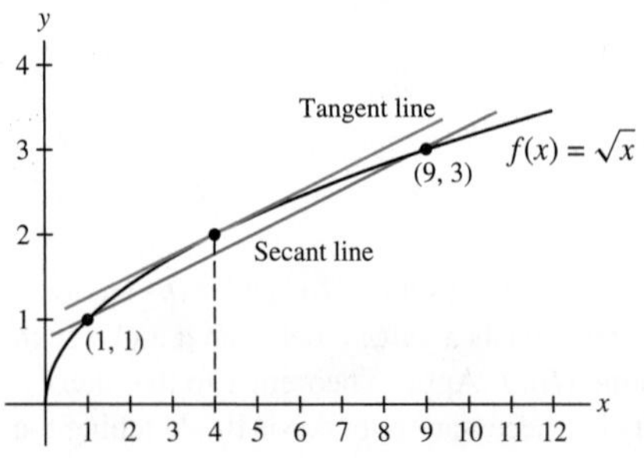

FIGURE 3 The tangent line at $c = 4$ is parallel to the secant line.

As a first application, we prove that a function with zero derivative is constant.

COROLLARY If $f(x)$ is differentiable and $f'(x) = 0$ for all $x \in (a, b)$, then $f(x)$ is constant on (a, b). In other words, $f(x) = C$ for some constant C.

Proof If a_1 and b_1 are any two distinct points in (a, b), then, by the MVT, there exists c between a_1 and b_1 such that

$$f(b_1) - f(a_1) = f'(c)(b_1 - a_1) = 0 \qquad (\text{since } f'(c) = 0)$$

Thus $f(b_1) = f(a_1)$. This says that $f(x)$ is constant on (a, b). ■

Increasing/Decreasing Behavior of Functions

We prove now that the sign of the derivative determines whether a function $f(x)$ is increasing or decreasing. Recall that $f(x)$ is

- **Increasing on (a, b)** if $f(x_1) < f(x_2)$ for all $x_1, x_2 \in (a, b)$ such that $x_1 < x_2$
- **Decreasing on (a, b)** if $f(x_1) > f(x_2)$ for all $x_1, x_2 \in (a, b)$ such that $x_1 < x_2$

We say that $f(x)$ is **monotonic** on (a, b) if it is either increasing or decreasing on (a, b).

We say that f is "nondecreasing" if

$$f(x_1) \le f(x_2) \quad \text{for} \quad x_1 \le x_2$$

"Nonincreasing" is defined similarly. In Theorem 2, if we assume that $f'(x) \ge 0$ (instead of > 0), then $f(x)$ is nondecreasing on (a, b). If $f'(x) \le 0$, then $f(x)$ is nonincreasing on (a, b).

THEOREM 2 The Sign of the Derivative Let f be a differentiable function on an open interval (a, b).

- If $f'(x) > 0$ for $x \in (a, b)$, then f is increasing on (a, b).
- If $f'(x) < 0$ for $x \in (a, b)$, then f is decreasing on (a, b).

Proof Suppose first that $f'(x) > 0$ for all $x \in (a, b)$. The MVT tells us that for any two points $x_1 < x_2$ in (a, b), there exists c between x_1 and x_2 such that

$$f(x_2) - f(x_1) = f'(c)(x_2 - x_1) > 0$$

The inequality holds because $f'(c)$ and $(x_2 - x_1)$ are both positive. Therefore, $f(x_2) > f(x_1)$, as required. The case $f'(x) < 0$ is similar. ■

GRAPHICAL INSIGHT Theorem 2 confirms our graphical intuition (Figure 4):

- $f'(x) > 0 \;\Rightarrow\;$ Tangent lines have positive slope $\;\Rightarrow\;$ f increasing
- $f'(x) < 0 \;\Rightarrow\;$ Tangent lines have negative slope $\;\Rightarrow\;$ f decreasing

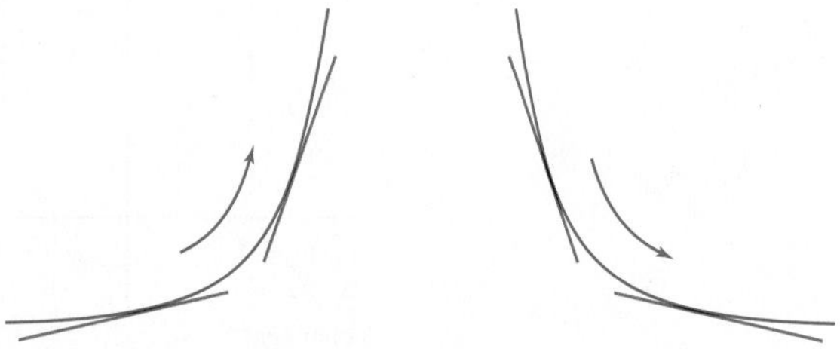

Increasing function: Tangent lines have positive slope.

Decreasing function: Tangent lines have negative slope.

FIGURE 4

■ **EXAMPLE 2** Show that the function $f(x) = 3x - \cos 2x$ is increasing.

Solution The derivative $f'(x) = 3 + 2\sin 2x$ satisfies $f(x) > 0$ for all x. Indeed, $\sin 2x \geq -1$, and thus $3 + 2\sin 2x \geq 3 - 2 = 1$. Therefore, $f(x)$ is an increasing function on the entire real line $(-\infty, \infty)$ [Figure 5]. ■

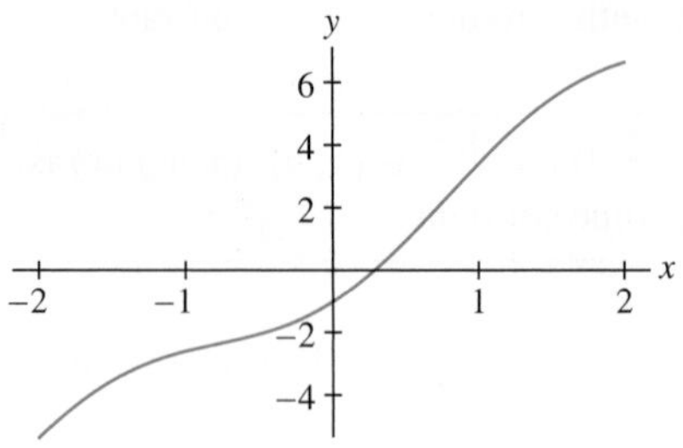

FIGURE 5

■ **EXAMPLE 3** Find the intervals on which $f(x) = x^2 - 2x - 3$ is monotonic.

Solution The derivative $f'(x) = 2x - 2 = 2(x - 1)$ is positive for $x > 1$ and negative for $x < 1$. By Theorem 2, f is decreasing on the interval $(-\infty, 1)$ and increasing on the interval $(1, \infty)$, as confirmed in Figure 6. ■

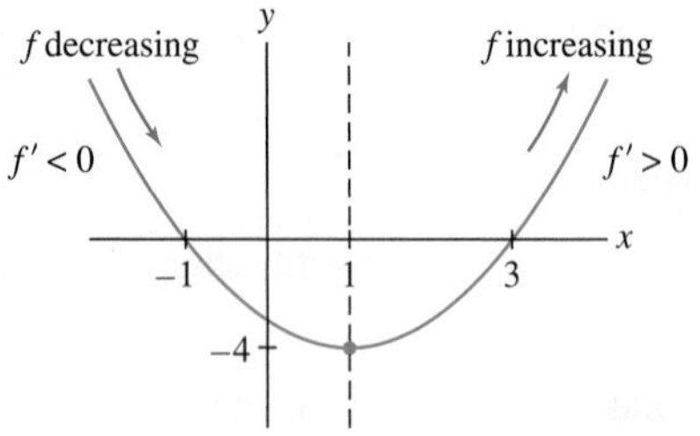

FIGURE 6 Graph of $f(x) = x^2 - 2x - 3$.

Testing Critical Points

There is a useful test for determining whether a critical point is a min or max (or neither) based on the *sign change* of the derivative $f'(x)$.

To explain the term "sign change," suppose that a function $F(x)$ satisfies $F(c) = 0$. We say that $F(x)$ *changes from positive to negative* at $x = c$ if $F(x) > 0$ to the left of c and $F(x) < 0$ to the right of c for x within a small open interval around c (Figure 7). A sign change from negative to positive is defined similarly. Observe in Figure 7 that $F(5) = 0$ but $F(x)$ does not change sign at $x = 5$.

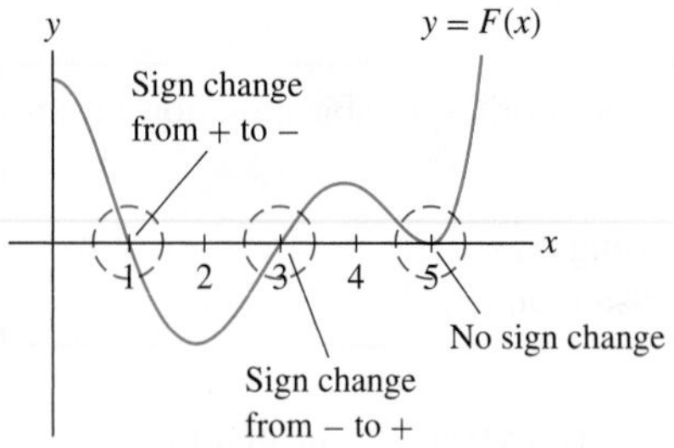

FIGURE 7

Now suppose that $f'(c) = 0$ and that $f'(x)$ changes sign at $x = c$, say from $+$ to $-$. Then $f(x)$ is increasing to the left of c and decreasing to the right, so $f(c)$ is a local maximum. Similarly, if $f'(x)$ changes sign from $-$ to $+$, then $f(c)$ is a local minimum. See Figure 8(A).

Figure 8(B) illustrates a case where $f'(c) = 0$ but f' does not change sign. In this case, $f'(x) > 0$ for all x near but not equal to c, so $f(x)$ is increasing and has neither a local min nor a local max at c.

> **THEOREM 3 First Derivative Test for Critical Points** Assume that $f(x)$ is differentiable and let c be a critical point of $f(x)$. Then
>
> - $f'(x)$ changes from $+$ to $-$ at c $\Rightarrow$ $f(c)$ is a local maximum.
> - $f'(x)$ changes from $-$ to $+$ at c $\Rightarrow$ $f(c)$ is a local minimum.

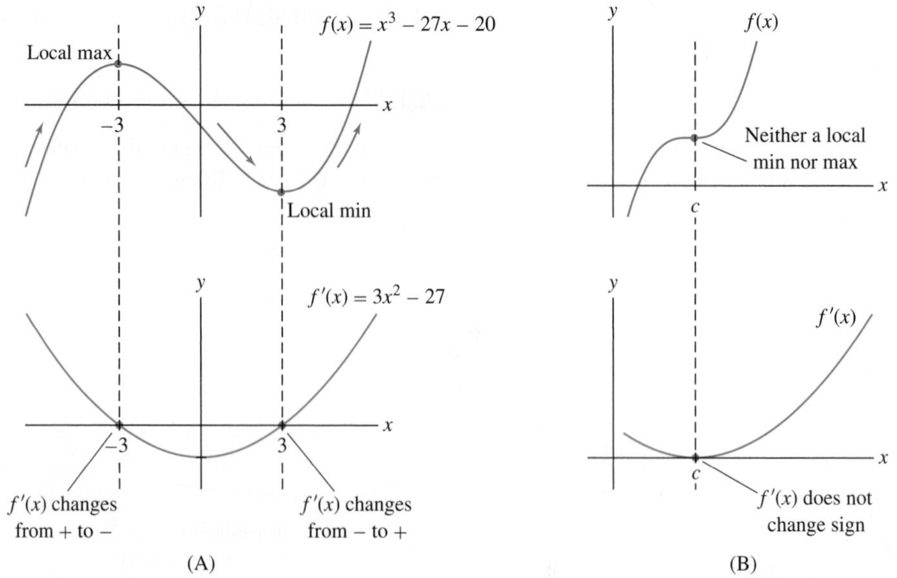

FIGURE 8

To carry out the First Derivative Test, we make a useful observation: $f'(x)$ can change sign at a critical point, but *it cannot change sign on the interval between two consecutive critical points* (one can prove this is true even if $f'(x)$ is not assumed to be continuous). So we can determine the sign of $f'(x)$ on an interval between consecutive critical points by evaluating $f'(x)$ at an any *test point* x_0 inside the interval. The sign of $f'(x_0)$ is the sign of $f'(x)$ on the entire interval.

■ **EXAMPLE 4** Analyze the critical points of $f(x) = x^3 - 27x - 20$.

Solution Our analysis will confirm the picture in Figure 8(A).

***Step 1.* Find the critical points.**
The roots of $f'(x) = 3x^2 - 27 = 3(x^2 - 9) = 0$ are $c = \pm 3$.

***Step 2.* Find the sign of f' on the intervals between the critical points.**
The critical points $c = \pm 3$ divide the real line into three intervals:

$$(-\infty, -3), \qquad (-3, 3), \qquad (3, \infty)$$

To determine the sign of f' on these intervals, we choose a test point inside each interval and evaluate. For example, in $(-\infty, -3)$ we choose $x = -4$. Because $f'(-4) = 21 > 0$, $f'(x)$ is positive on the entire interval $(-3, \infty)$. Similarly,

We chose the test points −4, 0, and 4 arbitrarily. To find the sign of $f'(x)$ on $(-\infty, -3)$, we could just as well have computed $f'(-5)$ or any other value of f' in the interval $(-\infty, -3)$.

$$\begin{aligned} f'(-4) = 21 > 0 \quad &\Rightarrow \quad f'(x) > 0 \quad \text{for all } x \in (-\infty, -3) \\ f'(0) = -27 < 0 \quad &\Rightarrow \quad f'(x) < 0 \quad \text{for all } x \in (-3, 3) \\ f'(4) = 21 > 0 \quad &\Rightarrow \quad f'(x) > 0 \quad \text{for all } x \in (3, \infty) \end{aligned}$$

This information is displayed in the following sign diagram:

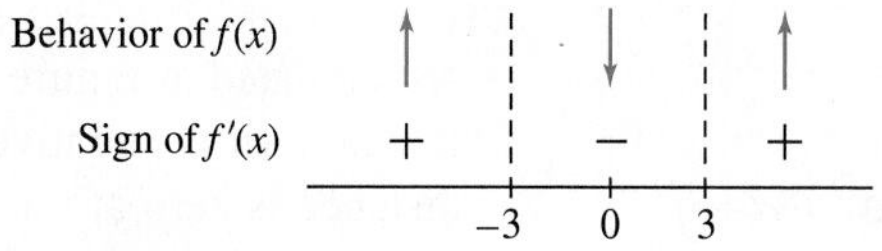

***Step 3.* Use the First Derivative Test.**

- $c = -3$: $f'(x)$ changes from + to − $\Rightarrow$ $f(-3)$ is a local max.
- $c = 3$: $f'(x)$ changes from − to + $\Rightarrow$ $f(3)$ is a local min. ■

■ **EXAMPLE 5** Analyze the critical points and the increase/decrease behavior of $f(x) = \cos^2 x + \sin x$ in $(0, \pi)$.

Solution First, find the critical points:

$$f'(x) = -2\cos x \sin x + \cos x = (\cos x)(1 - 2\sin x) = 0 \quad \Rightarrow \quad \cos x = 0 \text{ or } \sin x = \frac{1}{2}$$

The critical points are $\frac{\pi}{6}$, $\frac{\pi}{2}$, and $\frac{5\pi}{6}$. They divide $(0, \pi)$ into four intervals:

$$\left(0, \frac{\pi}{6}\right), \qquad \left(\frac{\pi}{6}, \frac{\pi}{2}\right), \qquad \left(\frac{\pi}{2}, \frac{5\pi}{6}\right), \qquad \left(\frac{5\pi}{6}, \pi\right)$$

We determine the sign of f' by evaluating f' at a test point inside each interval. Since $\frac{\pi}{6} \approx 0.52$, $\frac{\pi}{2} \approx 1.57$, $\frac{5\pi}{6} \approx 2.62$, and $\pi \approx 3.14$, we can use the following test points.

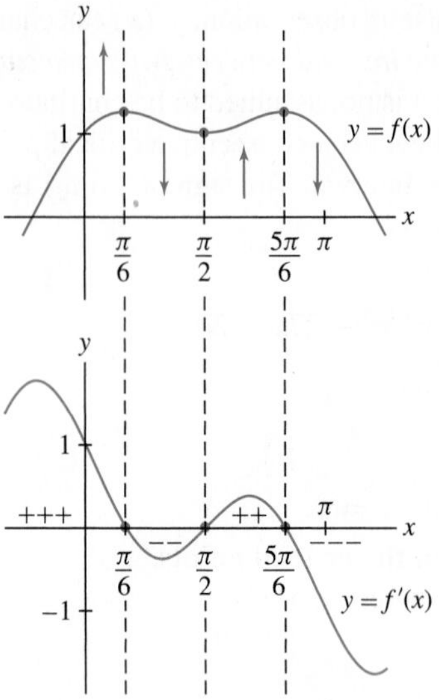

FIGURE 9 Graph of $f(x) = \cos^2 x + \sin x$ and its derivative.

Interval	Test Value	Sign of $f'(x)$	Behavior of $f(x)$
$(0, \frac{\pi}{6})$	$f'(0.5) \approx 0.04$	$+$	$\uparrow$
$(\frac{\pi}{6}, \frac{\pi}{2})$	$f'(1) \approx -0.37$	$-$	$\downarrow$
$(\frac{\pi}{2}, \frac{5\pi}{6})$	$f'(2) \approx 0.34$	$+$	$\uparrow$
$(\frac{5\pi}{6}, \pi)$	$f'(3) \approx -0.71$	$-$	$\downarrow$

Now apply the First Derivative Test:

- Local max at $c = \frac{\pi}{6}$ and $c = \frac{5\pi}{6}$ because f' changes from $+$ to $-$.
- Local min at $c = \frac{\pi}{2}$ because f' changes from $-$ to $+$.

The behavior of $f(x)$ and $f'(x)$ is reflected in the graphs in Figure 9. ■

■ **EXAMPLE 6 A Critical Point Without a Sign Transition** Analyze the critical points of $f(x) = \frac{1}{3}x^3 - x^2 + x$.

Solution The derivative is $f'(x) = x^2 - 2x + 1 = (x-1)^2$, so $c = 1$ is the only critical point. However, $(x-1)^2 \geq 0$, so $f'(x)$ does not change sign at $c = 1$, and $f(1)$ is neither a local min nor a local max (Figure 10). ■

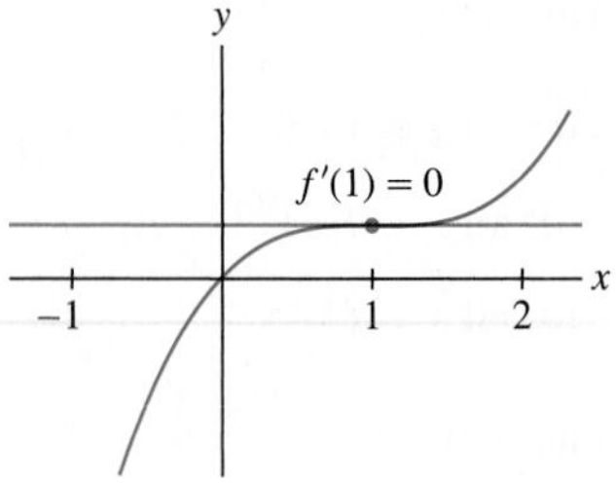

FIGURE 10 Graph of $f(x) = \frac{1}{3}x^3 - x^2 + x$.

Proof of the MVT

Let $m = \dfrac{f(b) - f(a)}{b - a}$ be the slope of the secant line joining $(a, f(a))$ and $(b, f(b))$. The secant line has equation $y = mx + r$ for some r (Figure 11). The value of r is not important, but you can check that $r = f(a) - ma$. Now consider the function

$$G(x) = f(x) - (mx + r)$$

As indicated in Figure 11, $G(x)$ is the vertical distance between the graph and the secant line at x (it is negative at points where the graph of f lies below the secant line). This distance is zero at the endpoints, and therefore $G(a) = G(b) = 0$. By Rolle's Theorem (Section 4.2), there exists a point c in (a, b) such that $G'(c) = 0$. But $G'(x) = f'(x) - m$, so $G'(c) = f'(c) - m = 0$, and $f'(c) = m$ as desired. ■

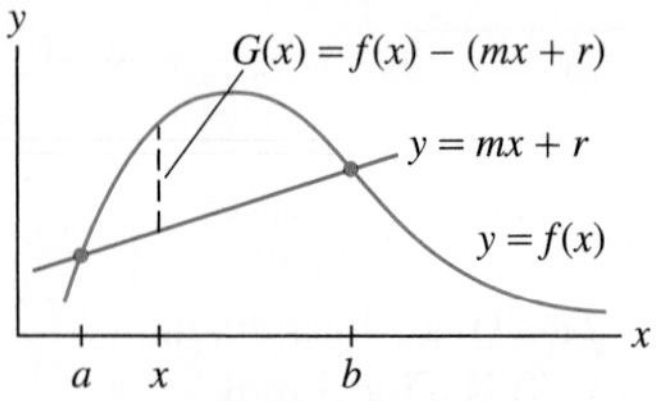

FIGURE 11 $G(x)$ is the vertical distance between the graph and the secant line.

4.3 SUMMARY

- The Mean Value Theorem (MVT): If $f(x)$ is continuous on $[a, b]$ and differentiable on (a, b), then there exists at least one value c in (a, b) such that

$$f'(c) = \frac{f(b) - f(a)}{b - a}$$

This conclusion can also be written

$$f(b) - f(a) = f'(c)(b - a)$$

- Important corollary of the MVT: If $f'(x) = 0$ for all $x \in (a, b)$, then $f(x)$ is constant on (a, b).
- The *sign* of $f'(x)$ determines whether $f(x)$ is increasing or decreasing:

$$f'(x) > 0 \text{ for } x \in (a, b) \quad \Rightarrow \quad f \text{ is increasing on } (a, b)$$

$$f'(x) < 0 \text{ for } x \in (a, b) \quad \Rightarrow \quad f \text{ is decreasing on } (a, b)$$

- The sign of $f'(x)$ can change only at the critical points, so $f(x)$ is *monotonic* (increasing or decreasing) on the intervals between the critical points.
- To find the sign of $f'(x)$ on the interval between two critical points, calculate the sign of $f'(x_0)$ at any test point x_0 in that interval.
- *First Derivative Test:* If $f(x)$ is differentiable and c is a critical point, then

Sign Change of f' at c	Type of Critical Point
From + to −	Local maximum
From − to +	Local minimum

4.3 EXERCISES

Preliminary Questions

1. For which value of m is the following statement correct? If $f(2) = 3$ and $f(4) = 9$, and $f(x)$ is differentiable, then f has a tangent line of slope m.

2. Assume f is differentiable. Which of the following statements does *not* follow from the MVT?

(a) If f has a secant line of slope 0, then f has a tangent line of slope 0.
(b) If $f(5) < f(9)$, then $f'(c) > 0$ for some $c \in (5, 9)$.
(c) If f has a tangent line of slope 0, then f has a secant line of slope 0.
(d) If $f'(x) > 0$ for all x, then every secant line has positive slope.

3. Can a function that takes on only negative values have a positive derivative? If so, sketch an example.

4. For $f(x)$ with derivative as in Figure 12:
(a) Is $f(c)$ a local minimum or maximum?
(b) Is $f(x)$ a decreasing function?

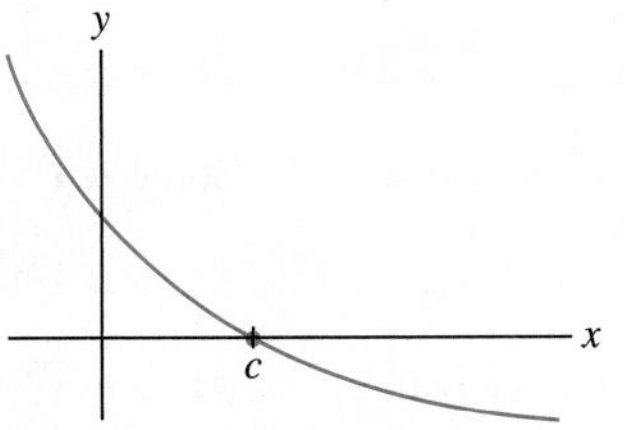

FIGURE 12 Graph of derivative $f'(x)$.

Exercises

In Exercises 1–8, find a point c satisfying the conclusion of the MVT for the given function and interval.

1. $y = x^{-1}$, $[2, 8]$

2. $y = \sqrt{x}$, $[9, 25]$

3. $y = \cos x - \sin x$, $[0, 2\pi]$

4. $y = \dfrac{x}{x+2}$, $[1, 4]$

5. $y = x^3$, $[-4, 5]$

6. $y = (x-1)(x-3)$, $[1, 3]$

7. $y = x \sin x$, $\left[-\frac{\pi}{2}, \frac{\pi}{2}\right]$

8. $y = x - \sin(\pi x)$, $[-1, 1]$

9. GU Let $f(x) = x^5 + x^2$. The secant line between $x = 0$ and $x = 1$ has slope 2 (check this), so by the MVT, $f'(c) = 2$ for some $c \in (0, 1)$. Plot $f(x)$ and the secant line on the same axes. Then plot $y = 2x + b$ for different values of b until the line becomes tangent to the graph of f. Zoom in on the point of tangency to estimate x-coordinate c of the point of tangency.

10. GU Plot the derivative of $f(x) = 3x^5 - 5x^3$. Describe its sign changes and use this to determine the local extreme values of $f(x)$. Then graph $f(x)$ to confirm your conclusions.

11. Determine the intervals on which $f'(x)$ is positive and negative, assuming that Figure 13 is the graph of $f(x)$.

12. Determine the intervals on which $f(x)$ is increasing or decreasing, assuming that Figure 13 is the graph of $f'(x)$.

13. State whether $f(2)$ and $f(4)$ are local minima or local maxima, assuming that Figure 13 is the graph of $f'(x)$.

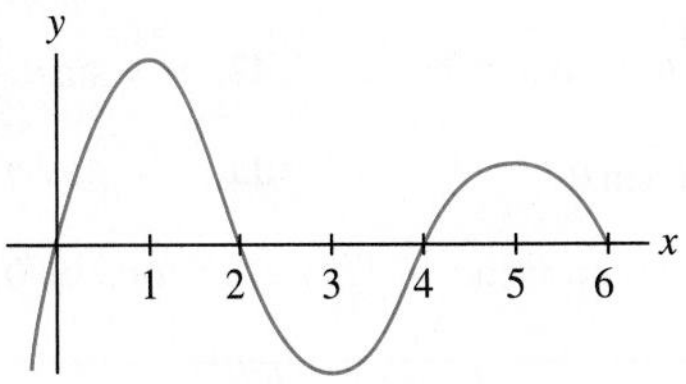

FIGURE 13

14. Figure 14 shows the graph of the derivative $f'(x)$ of a function $f(x)$. Find the critical points of $f(x)$ and determine whether they are local minima, local maxima, or neither.

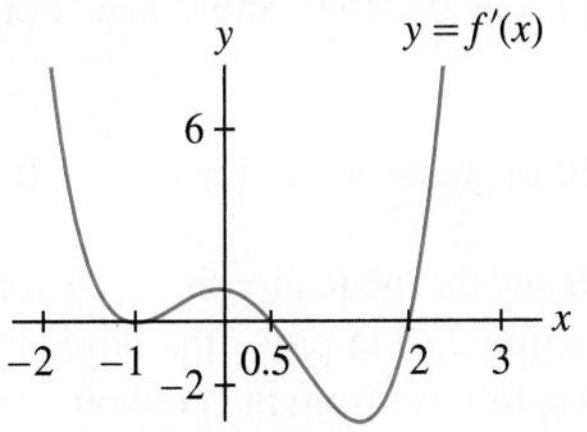

FIGURE 14

In Exercises 15–18, sketch the graph of a function $f(x)$ whose derivative $f'(x)$ has the given description.

15. $f'(x) > 0$ for $x > 3$ and $f'(x) < 0$ for $x < 3$

16. $f'(x) > 0$ for $x < 1$ and $f'(x) < 0$ for $x > 1$

17. $f'(x)$ is negative on $(1, 3)$ and positive everywhere else.

18. $f'(x)$ makes the sign transitions $+, -, +, -$.

In Exercises 19–22, find all critical points of f and use the First Derivative Test to determine whether they are local minima or maxima.

19. $f(x) = 4 + 6x - x^2$

20. $f(x) = x^3 - 12x - 4$

21. $f(x) = \dfrac{x^2}{x+1}$

22. $f(x) = x^3 + x^{-3}$

In Exercises 23–44, find the critical points and the intervals on which the function is increasing or decreasing. Use the First Derivative Test to determine whether the critical point is a local min or max (or neither).

23. $y = -x^2 + 7x - 17$

24. $y = 5x^2 + 6x - 4$

25. $y = x^3 - 12x^2$

26. $y = x(x-2)^3$

27. $y = 3x^4 + 8x^3 - 6x^2 - 24x$

28. $y = x^2 + (10 - x)^2$

29. $y = \frac{1}{3}x^3 + \frac{3}{2}x^2 + 2x + 4$

30. $y = x^4 + x^3$

31. $y = x^5 + x^3 + 1$

32. $y = x^5 + x^3 + x$

33. $y = x^4 - 4x^{3/2} \quad (x > 0)$

34. $y = x^{5/2} - x^2 \quad (x > 0)$

35. $y = x + x^{-1} \quad (x > 0)$

36. $y = x^{-2} - 4x^{-1} \quad (x > 0)$

37. $y = \dfrac{1}{x^2 + 1}$

38. $y = \dfrac{2x+1}{x^2+1}$

39. $y = \dfrac{x^3}{x^2+1}$

40. $y = \dfrac{x^3}{x^2-3}$

41. $y = \theta + \sin\theta + \cos\theta$

42. $y = \sin\theta + \sqrt{3}\cos\theta$

43. $y = \sin^2\theta + \sin\theta$

44. $y = \theta - 2\cos\theta, \quad [0, 2\pi]$

45. Find the minimum value of $f(x) = x^x$ for $x > 0$.

46. Show that $f(x) = x^2 + bx + c$ is decreasing on $\left(-\infty, -\frac{b}{2}\right)$ and increasing on $\left(-\frac{b}{2}, \infty\right)$.

47. Show that $f(x) = x^3 - 2x^2 + 2x$ is an increasing function. *Hint:* Find the minimum value of $f'(x)$.

48. Find conditions on a and b that ensure that $f(x) = x^3 + ax + b$ is increasing on $(-\infty, \infty)$.

49. GU Let $h(x) = \dfrac{x(x^2-1)}{x^2+1}$ and suppose that $f'(x) = h(x)$. Plot $h(x)$ and use the plot to describe the local extrema and the increasing/decreasing behavior of $f(x)$. Sketch a plausible graph for $f(x)$ itself.

50. Sam made two statements that Deborah found dubious.

(a) "The average velocity for my trip was 70 mph; at no point in time did my speedometer read 70 mph."

(b) "A policeman clocked me going 70 mph, but my speedometer never read 65 mph."

In each case, which theorem did Deborah apply to prove Sam's statement false: the Intermediate Value Theorem or the Mean Value Theorem? Explain.

51. Determine where $f(x) = (1{,}000 - x)^2 + x^2$ is decreasing. Use this to decide which is larger: $800^2 + 200^2$ or $600^2 + 400^2$.

52. Show that $f(x) = 1 - |x|$ satisfies the conclusion of the MVT on $[a, b]$ if both a and b are positive or negative, but not if $a < 0$ and $b > 0$.

53. Which values of c satisfy the conclusion of the MVT on the interval $[a, b]$ if $f(x)$ is a linear function?

54. Show that if $f(x)$ is any quadratic polynomial, then the midpoint $c = \dfrac{a+b}{2}$ satisfies the conclusion of the MVT on $[a, b]$ for any a and b.

55. Suppose that $f(0) = 2$ and $f'(x) \le 3$ for $x > 0$. Apply the MVT to the interval $[0, 4]$ to prove that $f(4) \le 14$. Prove more generally that $f(x) \le 2 + 3x$ for all $x > 0$.

56. Show that if $f(2) = -2$ and $f'(x) \ge 5$ for $x > 2$, then $f(4) \ge 8$.

57. Show that if $f(2) = 5$ and $f'(x) \ge 10$ for $x > 2$, then $f(x) \ge 10x - 15$ for all $x > 2$.

Further Insights and Challenges

58. Show that a cubic function $f(x) = x^3 + ax^2 + bx + c$ is increasing on $(-\infty, \infty)$ if $b > a^2/3$.

59. Prove that if $f(0) = g(0)$ and $f'(x) \le g'(x)$ for $x \ge 0$, then $f(x) \le g(x)$ for all $x \ge 0$. *Hint:* Show that $f(x) - g(x)$ is nonincreasing.

60. Use Exercise 59 to prove that $x \le \tan x$ for $0 \le x < \frac{\pi}{2}$.

61. Use Exercise 59 and the inequality $\sin x \le x$ for $x \ge 0$ (established in Theorem 3 of Section 2.6) to prove the following assertions for all $x \ge 0$ (each assertion follows from the previous one).

(a) $\cos x \ge 1 - \frac{1}{2}x^2$

(b) $\sin x \ge x - \frac{1}{6}x^3$

(c) $\cos x \le 1 - \frac{1}{2}x^2 + \frac{1}{24}x^4$

(d) Can you guess the next inequality in the series?

62. Suppose that $f(x)$ is a function such that $f(0) = 1$ and for all x, $f'(x) = f(x)$ and $f(x) > 0$ (in Chapter 7, we will see that $f(x)$ is the exponential function e^x). Prove that for all $x \ge 0$ (each assertion follows from the previous one),

(a) $f(x) \ge 1$

(b) $f(x) \ge 1 + x$

(c) $f(x) \ge 1 + x + \frac{1}{2}x^2$

Then prove by induction that for every whole number n and all $x \ge 0$,

$$f(x) \ge 1 + x + \frac{1}{2!}x^2 + \cdots + \frac{1}{n!}x^n$$

63. Assume that f'' exists and $f''(x) = 0$ for all x. Prove that $f(x) = mx + b$, where $m = f'(0)$ and $b = f(0)$.

64. Define $f(x) = x^3 \sin\left(\frac{1}{x}\right)$ for $x \neq 0$ and $f(0) = 0$.

(a) Show that $f'(x)$ is continuous at $x = 0$ and that $x = 0$ is a critical point of f.

(b) GU Examine the graphs of $f(x)$ and $f'(x)$. Can the First Derivative Test be applied?

(c) Show that $f(0)$ is neither a local min nor a local max.

65. Suppose that $f(x)$ satisfies the following equation (an example of a **differential equation**):

$$f''(x) = -f(x) \qquad (1)$$

(a) Show that $f(x)^2 + f'(x)^2 = f(0)^2 + f'(0)^2$ for all x. *Hint:* Show that the function on the left has zero derivative.

(b) Verify that $\sin x$ and $\cos x$ satisfy Eq. (1), and deduce that $\sin^2 x + \cos^2 x = 1$.

66. Suppose that functions f and g satisfy Eq. (1) and have the same initial values—that is, $f(0) = g(0)$ and $f'(0) = g'(0)$. Prove that $f(x) = g(x)$ for all x. *Hint:* Apply Exercise 65(a) to $f - g$.

67. Use Exercise 66 to prove: $f(x) = \sin x$ is the unique solution of Eq. (1) such that $f(0) = 0$ and $f'(0) = 1$; and $g(x) = \cos x$ is the unique solution such that $g(0) = 1$ and $g'(0) = 0$. This result can be used to develop all the properties of the trigonometric functions "analytically"—that is, without reference to triangles.

4.4 The Shape of a Graph

In the previous section, we studied the increasing/decreasing behavior of a function, as determined by the sign of the derivative. Another important property is concavity, which refers to the way the graph bends. Informally, a curve is *concave up* if it bends up and *concave down* if it bends down (Figure 1).

FIGURE 1

To analyze concavity in a precise fashion, let's examine how concavity is related to tangent lines and derivatives. Observe in Figure 2 that when $f(x)$ is concave up, $f'(x)$ is increasing (the slopes of the tangent lines increase as we move to the right). Similarly, when $f(x)$ is concave down, $f'(x)$ is decreasing. This suggests the following definition.

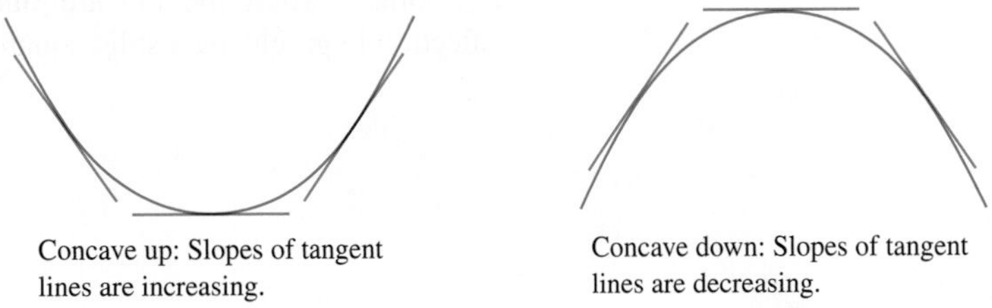

Concave up: Slopes of tangent lines are increasing.
Concave down: Slopes of tangent lines are decreasing.

FIGURE 2

DEFINITION Concavity Let $f(x)$ be a differentiable function on an open interval (a, b). Then

- f is **concave up** on (a, b) if $f'(x)$ is increasing on (a, b).
- f is **concave down** on (a, b) if $f'(x)$ is decreasing on (a, b).

■ **EXAMPLE 1 Concavity and Stock Prices** The stocks of two companies, A and B, went up in value, and both currently sell for \$75 (Figure 3). However, one is clearly a better investment than the other. Explain in terms of concavity.

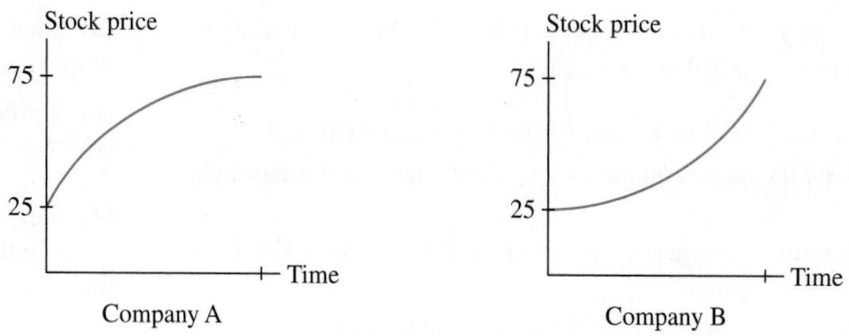

FIGURE 3

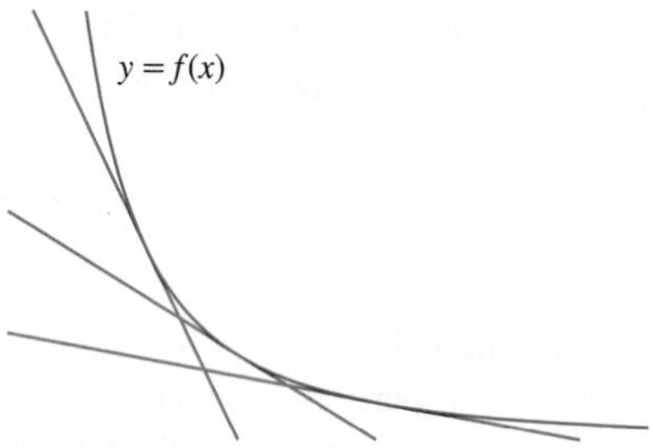

FIGURE 4 This function is decreasing. Its derivative is negative but increasing.

Solution The graph of Stock A is concave down, so its growth rate (first derivative) is declining as time goes on. The graph of Stock B is concave up, so its growth rate is increasing. If these trends continue, Stock B is the better investment. ■

GRAPHICAL INSIGHT Keep in mind that a function can decrease while its derivative increases. In Figure 4, the derivative $f'(x)$ is increasing. Although the tangent lines are getting less steep, their slopes are becoming *less negative*.

The concavity of a function is determined by the *sign* of its second derivative. Indeed, if $f''(x) > 0$, then $f'(x)$ is increasing and hence $f(x)$ is concave up. Similarly, if $f''(x) < 0$, then $f'(x)$ is decreasing and $f(x)$ is concave down.

THEOREM 1 Test for Concavity Assume that $f''(x)$ exists for all $x \in (a, b)$.

- If $f''(x) > 0$ for all $x \in (a, b)$, then f is concave up on (a, b).
- If $f''(x) < 0$ for all $x \in (a, b)$, then f is concave down on (a, b).

Of special interest are the points on the graph where the concavity changes. We say that $P = (c, f(c))$ is a **point of inflection** of $f(x)$ if the concavity changes from up to down or from down to up at $x = c$. Figure 5 shows a curve made up of two arcs—one is concave down and one is concave up (the word "arc" refers to a piece of a curve). The point P where the arcs are joined is a point of inflection. We will denote points of inflection in graphs by a solid square ■.

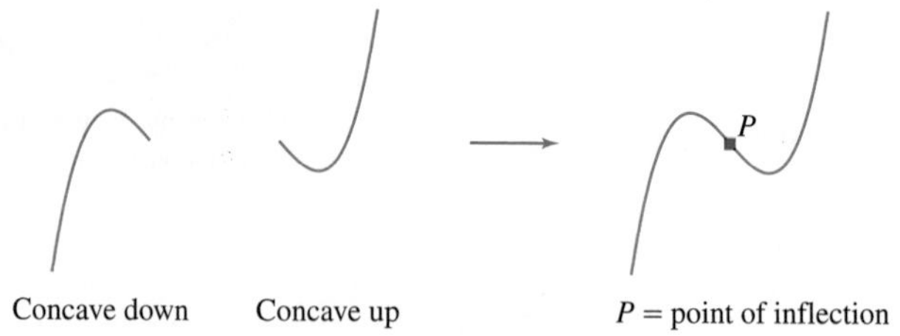

FIGURE 5

According to Theorem 1, the concavity of f is determined by the sign of f''. Therefore, a point of inflection is a point where $f''(x)$ changes sign.

THEOREM 2 Test for Inflection Points Assume that $f''(x)$ exists. If $f''(c) = 0$ and $f''(x)$ changes sign at $x = c$, then $f(x)$ has a point of inflection at $x = c$.

■ **EXAMPLE 2** Find the points of inflection of $f(x) = \cos x$ on $[0, 2\pi]$.

Solution We have $f''(x) = -\cos x$, and $f''(x) = 0$ for $x = \frac{\pi}{2}, \frac{3\pi}{2}$. Figure 6 shows that $f''(x)$ changes sign at $x = \frac{\pi}{2}$ and $\frac{3\pi}{2}$, so $f(x)$ has a point of inflection at both points. ■

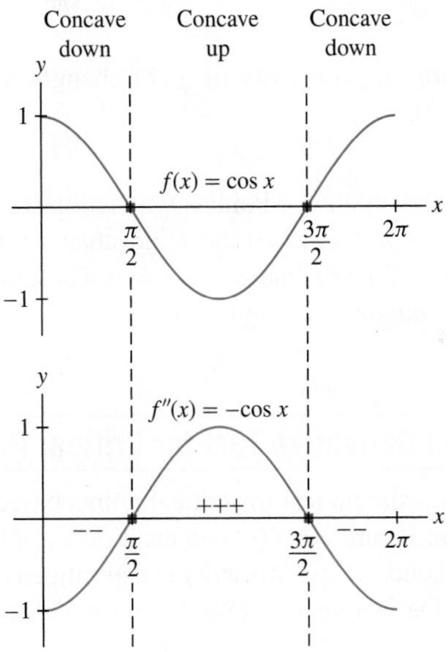

FIGURE 6

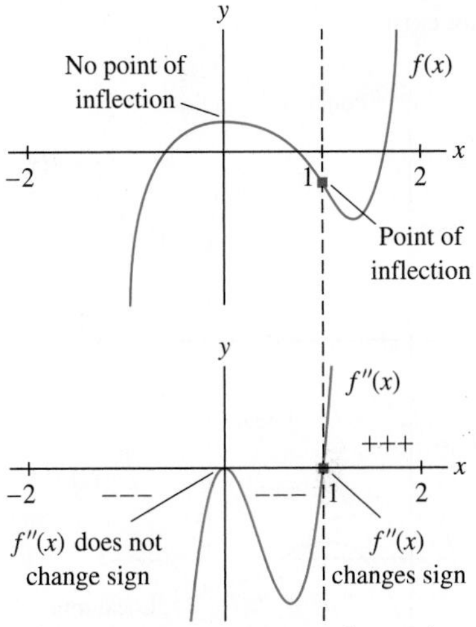

FIGURE 7 Graph of $f(x) = 3x^5 - 5x^4 + 1$ and its second derivative.

■ **EXAMPLE 3 Points of Inflection and Intervals of Concavity** Find the points of inflection and intervals of concavity of $f(x) = 3x^5 - 5x^4 + 1$.

Solution The first derivative is $f'(x) = 15x^4 - 20x^3$ and

$$f''(x) = 60x^3 - 60x^2 = 60x^2(x - 1)$$

The zeroes of $f''(x) = 60x^2(x - 1)$ are $x = 0, 1$. They divide the x-axis into three intervals: $(-\infty, 0)$, $(0, 1)$, and $(1, \infty)$. We determine the sign of $f''(x)$ and the concavity of f by computing "test values" within each interval (Figure 7):

Interval	Test Value	Sign of $f''(x)$	Behavior of $f(x)$
$(-\infty, 0)$	$f''(-1) = -120$	$-$	Concave down
$(0, 1)$	$f''\left(\frac{1}{2}\right) = -\frac{15}{2}$	$-$	Concave down
$(1, \infty)$	$f''(2) = 240$	$+$	Concave up

We can read off the points of inflection from this table:

- $c = 0$: no point of inflection, because $f''(x)$ does not change sign at 0.
- $c = 1$: point of inflection, because $f''(x)$ changes sign at 1. ■

Usually, we find the inflection points by solving $f''(x) = 0$. However, an inflection point can also occur at a point c where $f''(c)$ does not exist.

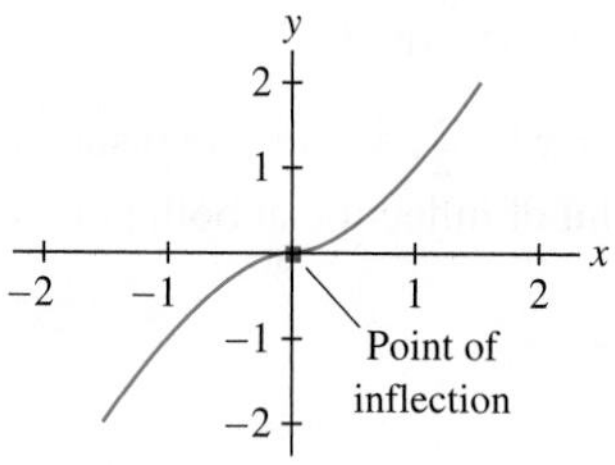

FIGURE 8 The concavity of $f(x) = x^{5/3}$ changes at $x = 0$ even though $f''(0)$ does not exist.

■ **EXAMPLE 4 A Case Where the Second Derivative Does Not Exist** Find the points of inflection of $f(x) = x^{5/3}$.

Solution In this case, $f'(x) = \frac{5}{3}x^{2/3}$ and $f''(x) = \frac{10}{9}x^{-1/3}$. Although $f''(0)$ does not exist, $f''(x)$ does change sign at $x = 0$:

$$f''(x) = \frac{10}{9x^{1/3}} = \begin{cases} > 0 & \text{for } x > 0 \\ < 0 & \text{for } x < 0 \end{cases}$$

Therefore, the concavity of $f(x)$ changes at $x = 0$, and $(0, 0)$ is a point of inflection (Figure 8). ■

GRAPHICAL INSIGHT Points of inflection are easy to spot on the graph of the first derivative $f'(x)$. If $f''(c) = 0$ and $f''(x)$ changes sign at $x = c$, then the increasing/decreasing behavior of $f'(x)$ changes at $x = c$. Thus, *inflection points of f occur where $f'(x)$ has a local min or max* (Figure 9).

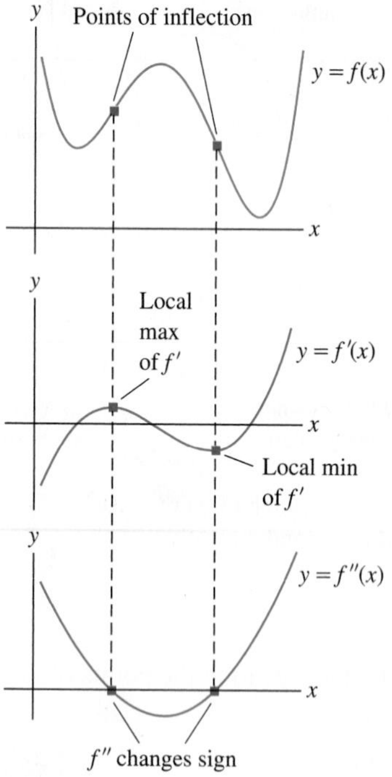

FIGURE 9

Second Derivative Test for Critical Points

There is a simple test for critical points based on concavity. Suppose that $f'(c) = 0$. As we see in Figure 10, $f(c)$ is a local max if $f(x)$ is concave down, and it is a local min if $f(x)$ is concave up. Concavity is determined by the sign of f'', so we obtain the following Second Derivative Test. (See Exercise 55 for a detailed proof.)

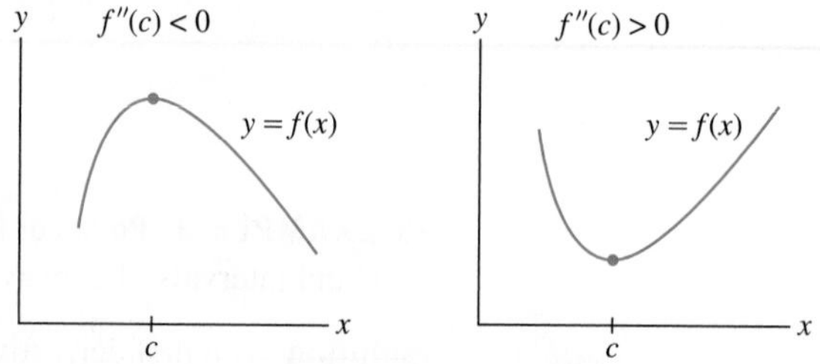

Concave down—local max   Concave up—local min

FIGURE 10 Concavity determines the type of the critical point.

> **THEOREM 3 Second Derivative Test** Let c be a critical point of $f(x)$. If $f''(c)$ exists, then
>
> - $f''(c) > 0 \Rightarrow f(c)$ is a local minimum
> - $f''(c) < 0 \Rightarrow f(c)$ is a local maximum
> - $f''(c) = 0 \Rightarrow$ inconclusive: $f(c)$ may be a local min, a local max, or neither

■ **EXAMPLE 5** Analyze the critical points of $f(x) = 2x^3 + 3x^2 - 12x$.

Solution First we find the critical points by solving

$$f'(x) = 6x^2 + 6x - 12 = 6(x^2 + x - 2) = 6(x + 2)(x - 1) = 0$$

The critical points are $c = -2, 1$ (Figure 11) and $f''(x) = 12x + 6$, so by the Second Derivative Test,

$$f''(-2) = -24 + 6 = -18 < 0 \quad \Rightarrow \quad f(-2) \text{ is a local max}$$

$$f''(1) = 12 + 6 = 18 > 0 \quad \Rightarrow \quad f(1) \text{ is a local min}$$ ■

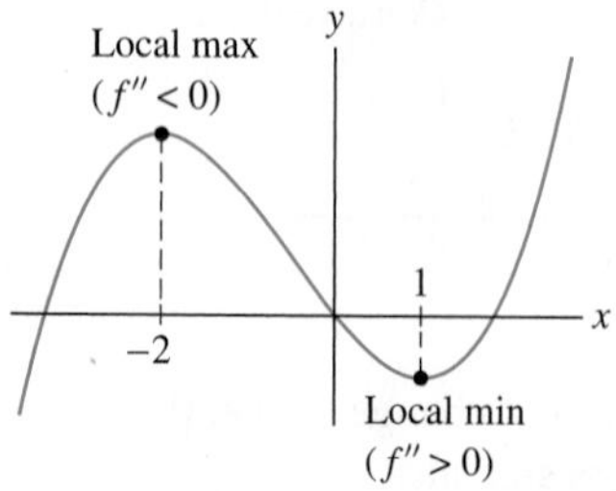

FIGURE 11 Graph of $f(x) = 2x^3 + 3x^2 - 12x$.

■ **EXAMPLE 6** **Second Derivative Test Inconclusive** Analyze the critical points of $f(x) = x^5 - 5x^4$.

Solution The first two derivatives are

$$f'(x) = 5x^4 - 20x^3 = 5x^3(x-4)$$
$$f''(x) = 20x^3 - 60x^2$$

The critical points are $c = 0, 4$, and the Second Derivative Test yields

$$f''(0) = 0 \quad \Rightarrow \quad \text{Second Derivative Test fails}$$
$$f''(4) = 320 > 0 \quad \Rightarrow \quad f(4) \text{ is a local min}$$

The Second Derivative Test fails at $c = 0$, so we fall back on the First Derivative Test. Choosing test points to the left and right of $c = 0$, we find

$$f'(-1) = 5 + 20 = 25 > 0 \quad \Rightarrow \quad f'(x) \text{ is positive on } (-\infty, 0)$$
$$f'(1) = 5 - 20 = -15 < 0 \quad \Rightarrow \quad f'(x) \text{ is negative on } (0, 4)$$

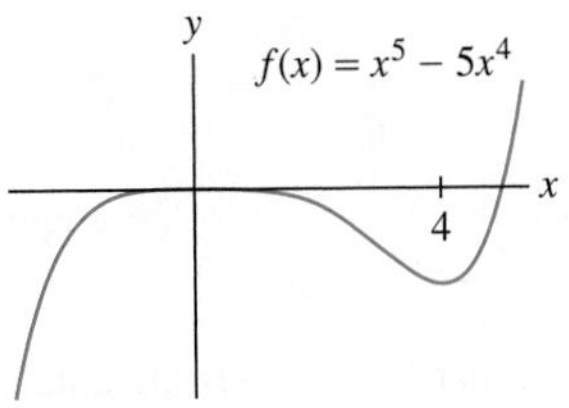

FIGURE 12 Graph of $f(x) = x^5 - 5x^4$.

Since $f'(x)$ changes from + to − at $c = 0$, $f(0)$ is a local max (Figure 12). ■

4.4 SUMMARY

- A differentiable function $f(x)$ is *concave up* on (a, b) if $f'(x)$ is increasing and *concave down* if $f'(x)$ is decreasing on (a, b).
- The signs of the first two derivatives provide the following information:

First Derivative			Second Derivative		
$f' > 0$	$\Rightarrow$	f is increasing	$f'' > 0$	$\Rightarrow$	f is concave up
$f' < 0$	$\Rightarrow$	f is decreasing	$f'' < 0$	$\Rightarrow$	f is concave down

- A *point of inflection* is a point where the concavity changes from concave up to concave down, or vice versa.
- If $f''(c) = 0$ and $f''(x)$ changes sign at c, then c is a point of inflection.
- Second Derivative Test: If $f'(c) = 0$ and $f''(c)$ exists, then
  - $f(c)$ is a local maximum if $f''(c) < 0$.
  - $f(c)$ is a local minimum if $f''(c) > 0$.
  - The test fails if $f''(c) = 0$.

If the test fails, use the First Derivative Test.

4.4 EXERCISES

Preliminary Questions

1. If f is concave up, then f' is (choose one):
(a) increasing **(b)** decreasing

2. What conclusion can you draw if $f'(c) = 0$ and $f''(c) < 0$?

3. True or False? If $f(c)$ is a local min, then $f''(c)$ must be positive.

4. True or False? If $f''(x)$ changes from + to − at $x = c$, then f has a point of inflection at $x = c$.

Exercises

1. Match the graphs in Figure 13 with the description:

(a) $f''(x) < 0$ for all x.
(b) $f''(x)$ goes from $+$ to $-$.
(c) $f''(x) > 0$ for all x.
(d) $f''(x)$ goes from $-$ to $+$.

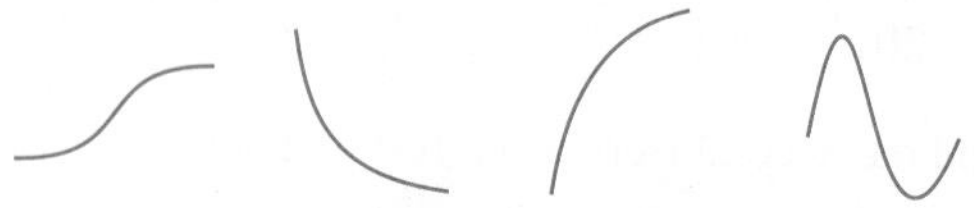

(A) (B) (C) (D)

FIGURE 13

2. Match each statement with a graph in Figure 14 that represents company profits as a function of time.

(a) The outlook is great: The growth rate keeps increasing.
(b) We're losing money, but not as quickly as before.
(c) We're losing money, and it's getting worse as time goes on.
(d) We're doing well, but our growth rate is leveling off.
(e) Business had been cooling off, but now it's picking up.
(f) Business had been picking up, but now it's cooling off.

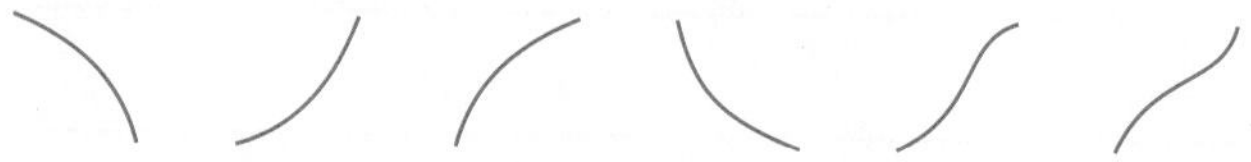

(i) (ii) (iii) (iv) (v) (vi)

FIGURE 14

In Exercises 3–14, determine the intervals on which the function is concave up or down and find the points of inflection.

3. $y = x^2 - 4x + 3$

4. $y = t^3 - 6t^2 + 4$

5. $y = 10x^3 - x^5$

6. $y = 5x^2 + x^4$

7. $y = \theta - 2\sin\theta, \quad [0, 2\pi]$

8. $y = \theta + \sin^2\theta, \quad [0, \pi]$

9. $y = x(x - 8\sqrt{x}) \quad (x \geq 0)$

10. $y = x^{7/2} - 35x^2$

11. $y = (x - 2)(1 - x^3)$

12. $y = x^{7/5}$

13. $y = \dfrac{1}{x^2 + 3}$

14. $y = \dfrac{x - 1}{x^2 + 8}$

15. The growth of a sunflower during the first 100 days after sprouting is modeled well by the *logistic curve* $y = h(t)$ shown in Figure 15. Estimate the growth rate at the point of inflection and explain its significance. Then make a rough sketch of the first and second derivatives of $h(t)$.

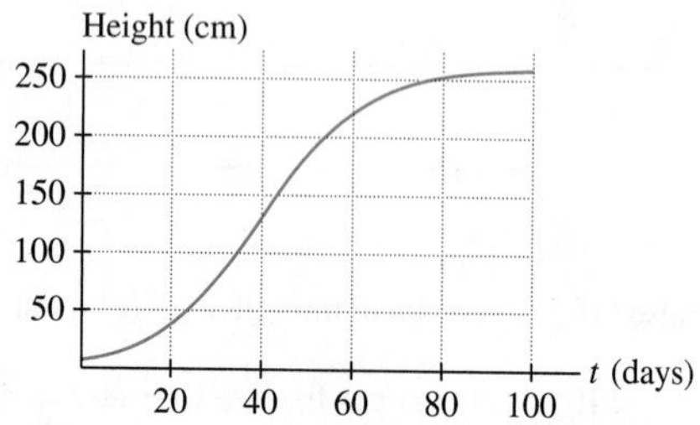

FIGURE 15

16. Assume that Figure 16 is the graph of $f(x)$. Where do the points of inflection of $f(x)$ occur, and on which interval is $f(x)$ concave down?

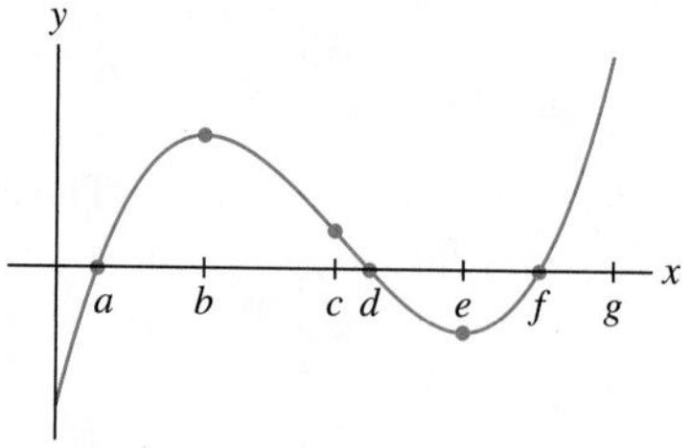

FIGURE 16

17. Repeat Exercise 16 but assume that Figure 16 is the graph of the *derivative* $f'(x)$.

18. Repeat Exercise 16 but assume that Figure 16 is the graph of the *second derivative* $f''(x)$.

19. Figure 17 shows the *derivative* $f'(x)$ on $[0, 1.2]$. Locate the points of inflection of $f(x)$ and the points where the local minima and maxima occur. Determine the intervals on which $f(x)$ has the following properties:

(a) Increasing
(b) Decreasing
(c) Concave up
(d) Concave down

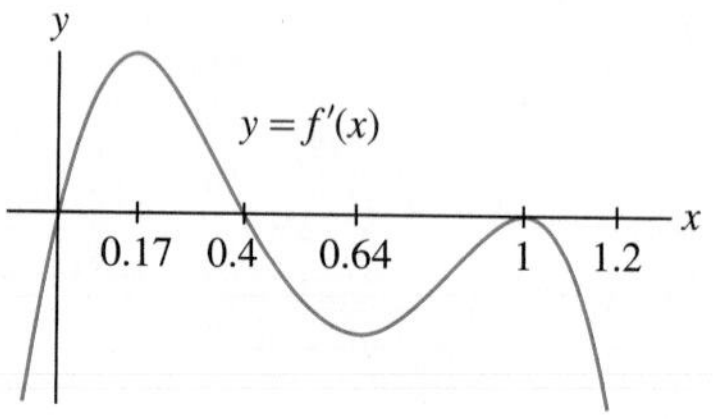

FIGURE 17

20. Leticia has been selling solar-powered laptop chargers through her website, with monthly sales as recorded below. In a report to investors, she states, "Sales reached a point of inflection when I started using pay-per-click advertising." In which month did that occur? Explain.

Month	1	2	3	4	5	6	7	8
Sales	2	30	50	60	90	150	230	340

In Exercises 21–34, find the critical points and apply the Second Derivative Test (or state that it fails).

21. $f(x) = x^3 - 12x^2 + 45x$

22. $f(x) = x^4 - 8x^2 + 1$

23. $f(x) = 3x^4 - 8x^3 + 6x^2$

24. $f(x) = x^5 - x^3$

25. $f(x) = \dfrac{x^2 - 8x}{x + 1}$

26. $f(x) = \dfrac{1}{x^2 - x + 2}$

27. $y = 6x^{3/2} - 4x^{1/2}$

28. $y = 9x^{7/3} - 21x^{1/2}$

29. $f(x) = x^3 + 48/x, \quad (0, \infty)$

30. $f(x) = x^4 + 128/x^2, \quad (0, \infty)$

31. $f(x) = \sin^2 x + \cos x, \quad [0, \pi]$

32. $y = \dfrac{1}{\sin x + 4}, \quad [0, 2\pi]$

33. $f(x) = 2 + \tan^2 x, \quad \left(-\frac{\pi}{2}, \frac{\pi}{2}\right)$

34. $f(x) = \sin x \cos^3 x, \quad [0, \pi]$

In Exercises 35–46, find the intervals on which f is concave up or down, the points of inflection, the critical points, and the local minima and maxima.

35. $f(x) = x^3 - 2x^2 + x$

36. $f(x) = x^2(x - 4)$

37. $f(t) = t^2 - t^3$

38. $f(x) = 2x^4 - 3x^2 + 2$

39. $f(x) = x^2 - 8x^{1/2} \quad (x \ge 0)$

40. $f(x) = x^{3/2} - 4x^{-1/2} \quad (x > 0)$

41. $f(x) = \dfrac{x}{x^2 + 27}$

42. $f(x) = \dfrac{1}{x^4 + 1}$

43. $f(\theta) = \theta + \sin\theta, \quad [0, 2\pi]$

44. $f(x) = \cos^2 x, \quad [0, \pi]$

45. $f(x) = \tan x, \quad \left(-\frac{\pi}{2}, \frac{\pi}{2}\right)$

46. $f(x) = \dfrac{x}{x^6 + 5}$

47. Sketch the graph of an increasing function such that $f''(x)$ changes from $+$ to $-$ at $x = 2$ and from $-$ to $+$ at $x = 4$. Do the same for a decreasing function.

In Exercises 48–50, sketch the graph of a function $f(x)$ satisfying all of the given conditions.

48. $f'(x) > 0$ and $f''(x) < 0$ for all x.

49. **(i)** $f'(x) > 0$ for all x, and
(ii) $f''(x) < 0$ for $x < 0$ and $f''(x) > 0$ for $x > 0$.

50. **(i)** $f'(x) < 0$ for $x < 0$ and $f'(x) > 0$ for $x > 0$, and
(ii) $f''(x) < 0$ for $|x| > 2$, and $f''(x) > 0$ for $|x| < 2$.

51. An infectious flu spreads slowly at the beginning of an epidemic. The infection process accelerates until a majority of the susceptible individuals are infected, at which point the process slows down.
(a) If $R(t)$ is the number of individuals infected at time t, describe the concavity of the graph of R near the beginning and end of the epidemic.
(b) Describe the status of the epidemic on the day that $R(t)$ has a point of inflection.

52. Water is pumped into a sphere at a constant rate (Figure 18). Let $h(t)$ be the water level at time t. Sketch the graph of $h(t)$ (approximately, but with the correct concavity). Where does the point of inflection occur?

53. Water is pumped into a sphere of radius R at a variable rate in such a way that the water level rises at a constant rate (Figure 18). Let $V(t)$ be the volume of water in the tank at time t. Sketch the graph $V(t)$ (approximately, but with the correct concavity). Where does the point of inflection occur?

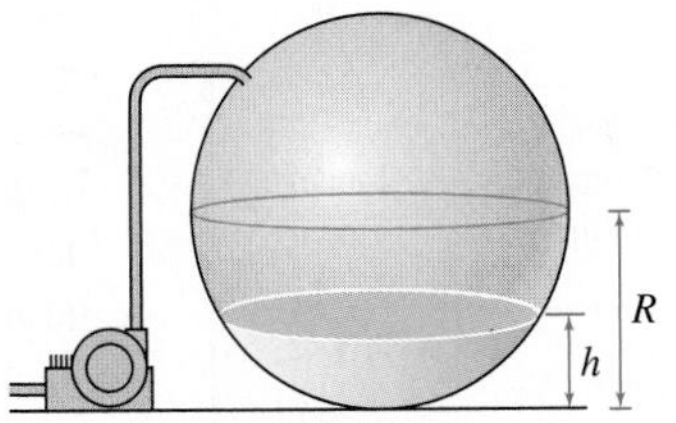

FIGURE 18

54. (Continuation of Exercise 53) If the sphere has radius R, the volume of water is $V = \pi\left(Rh^2 - \frac{1}{3}h^3\right)$ where h is the water level. Assume the level rises at a constant rate of 1 (that is, $h = t$).
(a) Find the inflection point of $V(t)$. Does this agree with your conclusion in Exercise 53?
(b) GU Plot $V(t)$ for $R = 1$.

Further Insights and Challenges

In Exercises 55–57, assume that $f(x)$ is differentiable.

55. Proof of the Second Derivative Test Let c be a critical point such that $f''(c) > 0$ (the case $f''(c) < 0$ is similar).
(a) Show that

$$f''(c) = \lim_{h \to 0} \frac{f'(c + h)}{h}$$

(b) Use (a) to show that there exists an open interval (a, b) containing c such that $f'(x) < 0$ if $a < x < c$ and $f'(x) > 0$ if $c < x < b$. Conclude that $f(c)$ is a local minimum.

56. Prove that if $f''(x)$ exists and $f''(x) > 0$ for all x, then the graph of $f(x)$ "sits above" its tangent lines.
(a) For any c, set $G(x) = f(x) - f'(c)(x - c) - f(c)$. It is sufficient to prove that $G(x) \ge 0$ for all c. Explain why with a sketch.
(b) Show that $G(c) = G'(c) = 0$ and $G''(x) > 0$ for all x. Conclude that $G'(x) < 0$ for $x < c$ and $G'(x) > 0$ for $x > c$. Then deduce, using the MVT, that $G(x) > G(c)$ for $x \ne c$.

57. Assume that $f''(x)$ exists and let c be a point of inflection of $f(x)$.
(a) Use the method of Exercise 56 to prove that the tangent line at $x = c$ *crosses the graph* (Figure 19). *Hint:* Show that $G(x)$ changes sign at $x = c$.
(b) GU Verify this conclusion for

$$f(x) = \frac{x}{3x^2 + 1}$$

by graphing $f(x)$ and the tangent line at each inflection point on the same set of axes.

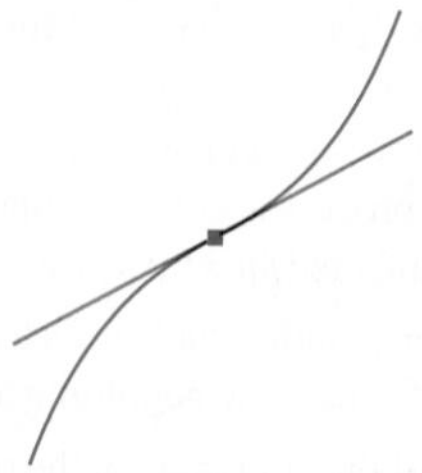

FIGURE 19 Tangent line crosses graph at point of inflection.

58. Let $C(x)$ be the cost of producing x units of a certain good. Assume that the graph of $C(x)$ is concave up.

(a) Show that the average cost $A(x) = C(x)/x$ is minimized at the production level x_0 such that average cost equals marginal cost—that is, $A(x_0) = C'(x_0)$.

(b) Show that the line through $(0, 0)$ and $(x_0, C(x_0))$ is tangent to the graph of $C(x)$.

59. Let $f(x)$ be a polynomial of degree $n \geq 2$. Show that $f(x)$ has at least one point of inflection if n is odd. Then give an example to show that $f(x)$ need not have a point of inflection if n is even.

60. Critical and Inflection Points If $f'(c) = 0$ and $f(c)$ is neither a local min nor a local max, must $x = c$ be a point of inflection? This is true for "reasonable" functions (including the functions studied in this text), but it is not true in general. Let

$$f(x) = \begin{cases} x^2 \sin \frac{1}{x} & \text{for } x \neq 0 \\ 0 & \text{for } x = 0 \end{cases}$$

(a) Use the limit definition of the derivative to show that $f'(0)$ exists and $f'(0) = 0$.

(b) Show that $f(0)$ is neither a local min nor a local max.

(c) Show that $f'(x)$ changes sign infinitely often near $x = 0$. Conclude that $x = 0$ is not a point of inflection.

4.5 Graph Sketching and Asymptotes

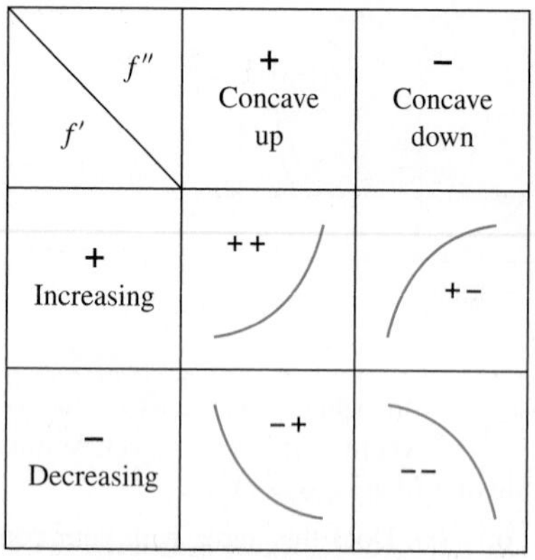

FIGURE 1 The four basic shapes.

In this section, our goal is to sketch graphs using the information provided by the first two derivatives f' and f''. We will see that a useful sketch can be produced without plotting a large number of points. Although nowadays almost all graphs are produced by computer (including, of course, the graphs in this textbook), sketching graphs by hand is a useful way of solidifying your understanding of the basic concepts in this chapter.

Most graphs are made up of smaller *arcs* that have one of the four basic shapes, corresponding to the four possible sign combinations of f' and f'' (Figure 1). Since f' and f'' can each have sign $+$ or $-$, the sign combinations are

$$++ \qquad +- \qquad -+ \qquad --$$

In this notation, the first sign refers to f' and the second sign to f''. For instance, $-+$ indicates that $f'(x) < 0$ and $f''(x) > 0$.

In graph sketching, we focus on the **transition points**, where the basic shape changes due to a sign change in either f' (local min or max) or f'' (point of inflection). In this section, local extrema are indicated by solid dots, and points of inflection are indicated by green solid squares (Figure 2).

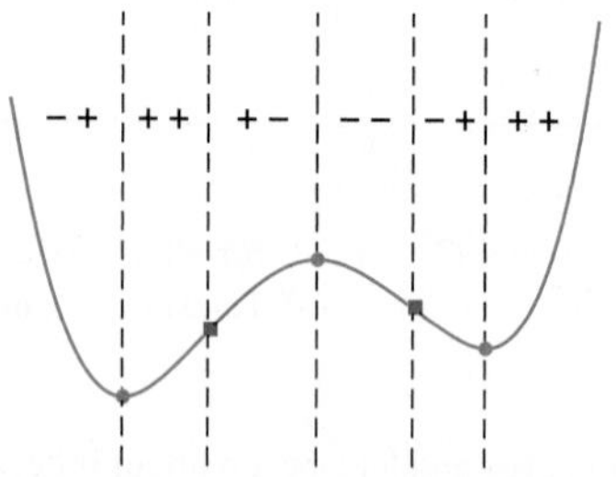

FIGURE 2 The graph of $f(x)$ with transition points and sign combinations of f' and f''.

In graph sketching, we must also pay attention to **asymptotic behavior**—that is, to the behavior of $f(x)$ as x approaches either $\pm\infty$ or a vertical asymptote.

The next three examples treat polynomials. Recall from Section 2.7 that the limits at infinity of a polynomial

$$f(x) = a_n x^n + a_{n-1} x^{n-1} + \cdots + a_1 x + a_0$$

(assuming that $a_n \neq 0$) are determined by

$$\lim_{x\to\infty} f(x) = a_n \lim_{x\to\pm\infty} x^n$$

In general, then, the graph of a polynomial "wiggles" up and down a finite number of times and then tends to positive or negative infinity (Figure 3).

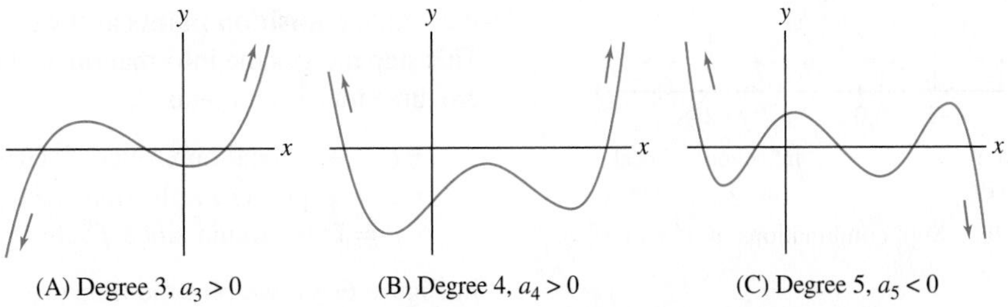

FIGURE 3 Graphs of polynomials.

■ **EXAMPLE 1 Quadratic Polynomial** Sketch the graph of $f(x) = x^2 - 4x + 3$.

Solution We have $f'(x) = 2x - 4 = 2(x - 2)$. We can see directly that $f'(x)$ is negative for $x < 2$ and positive for $x > 2$, but let's confirm this using test values, as in previous sections:

Interval	Test Value	Sign of f'
$(-\infty, 2)$	$f'(1) = -2$	$-$
$(2, \infty)$	$f'(3) = 2$	$+$

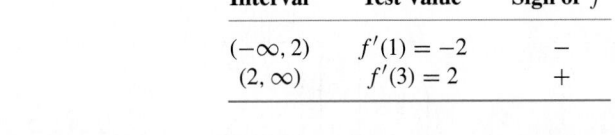

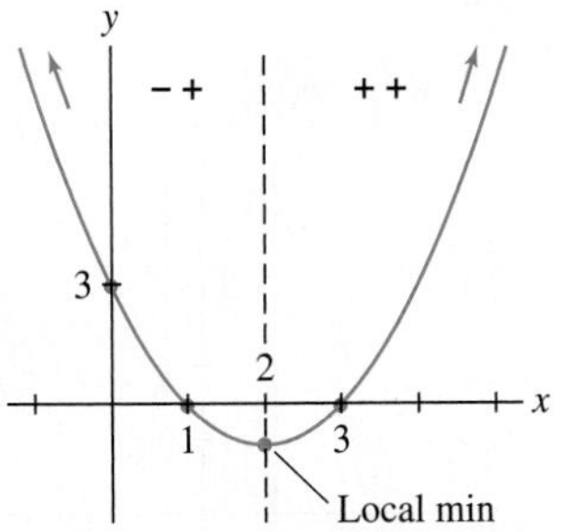

FIGURE 4 Graph of $f(x) = x^2 - 4x + 3$.

Furthermore, $f''(x) = 2$ is positive, so the graph is everywhere concave up. To sketch the graph, plot the local minimum $(2, -1)$, the y-intercept, and the roots $x = 1, 3$. Since the leading term of f is x^2, $f(x)$ tends to ∞ as $x \to \pm\infty$. This asymptotic behavior is noted by the arrows in Figure 4. ■

■ **EXAMPLE 2 Cubic Polynomial** Sketch the graph of $f(x) = \frac{1}{3}x^3 - \frac{1}{2}x^2 - 2x + 3$.

Solution

Step 1. **Determine the signs of f' and f''.**
First, solve for the critical points:

$$f'(x) = x^2 - x - 2 = (x + 1)(x - 2) = 0$$

The critical points $c = -1, 2$ divide the x-axis into three intervals $(-\infty, -1)$, $(-1, 2)$, and $(2, \infty)$, on which we determine the sign of f' by computing test values:

Interval	Test Value	Sign of f'
$(-\infty, -1)$	$f'(-2) = 4$	$+$
$(-1, 2)$	$f'(0) = -2$	$-$
$(2, \infty)$	$f'(3) = 4$	$+$

Next, solve $f''(x) = 2x - 1 = 0$. The solution is $c = \frac{1}{2}$ and we have

Interval	Test Value	Sign of f''
$(-\infty, \frac{1}{2})$	$f''(0) = -1$	$-$
$(\frac{1}{2}, \infty)$	$f''(1) = 1$	$+$

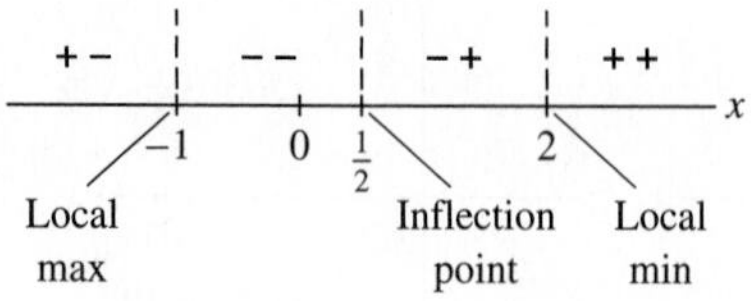

FIGURE 5 Sign combinations of f' and f''.

Step 2. **Note transition points and sign combinations.**
This step merges the information about f' and f'' in a sign diagram (Figure 5). There are three transition points:

- $c = -1$: local max since f' changes from + to − at $c = -1$.
- $c = \frac{1}{2}$: point of inflection since f'' changes sign at $c = \frac{1}{2}$.
- $c = 2$: local min since f' changes from − to + at $c = 2$.

In Figure 6(A), we plot the transition points and, for added accuracy, the y-intercept $f(0)$, using the values

$$f(-1) = \frac{25}{6}, \qquad f\left(\frac{1}{2}\right) = \frac{23}{12}, \qquad f(0) = 3, \qquad f(2) = -\frac{1}{3}$$

Step 3. **Draw arcs of appropriate shape and asymptotic behavior.**
The leading term of $f(x)$ is $\frac{1}{3}x^3$. Therefore, $\lim_{x\to\infty} f(x) = \infty$ and $\lim_{x\to-\infty} f(x) = -\infty$.

To create the sketch, it remains only to connect the transition points by arcs of the appropriate concavity and asymptotic behavior, as in Figure 6(B) and (C). ■

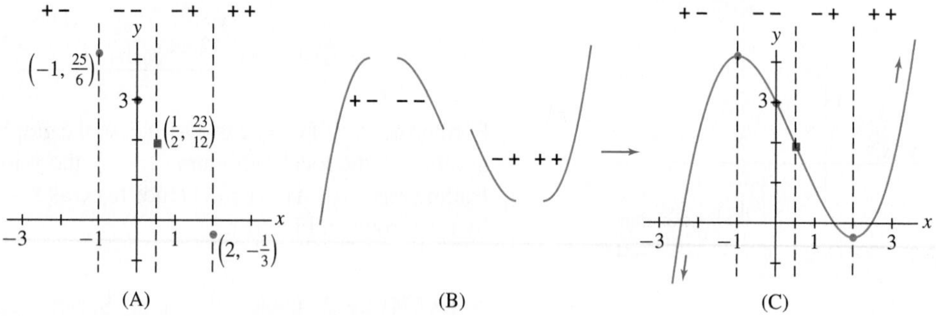

FIGURE 6 Graph of $f(x) = \frac{1}{3}x^3 - \frac{1}{2}x^2 - 2x + 3$.

■ **EXAMPLE 3** Sketch the graph of $f(x) = 3x^4 - 8x^3 + 6x^2 + 1$.

Solution

Step 1. **Determine the signs of f' and f''.**
First, solve for the transition points:

$$f'(x) = 12x^3 - 24x^2 + 12x = 12x(x-1)^2 = 0 \quad \Rightarrow \quad x = 0, 1$$

$$f''(x) = 36x^2 - 48x + 12 = 12(x-1)(3x-1) = 0 \quad \Rightarrow \quad x = \frac{1}{3}, 1$$

The signs of f' and f'' are recorded in the following tables.

Interval	Test Value	Sign of f'
$(-\infty, 0)$	$f'(-1) = -48$	−
$(0, 1)$	$f'(\frac{1}{2}) = \frac{3}{2}$	+
$(1, \infty)$	$f'(2) = 24$	+

Interval	Test Value	Sign of f''
$(-\infty, \frac{1}{3})$	$f''(0) = 12$	+
$(\frac{1}{3}, 1)$	$f''(\frac{1}{2}) = -3$	−
$(1, \infty)$	$f''(2) = 60$	+

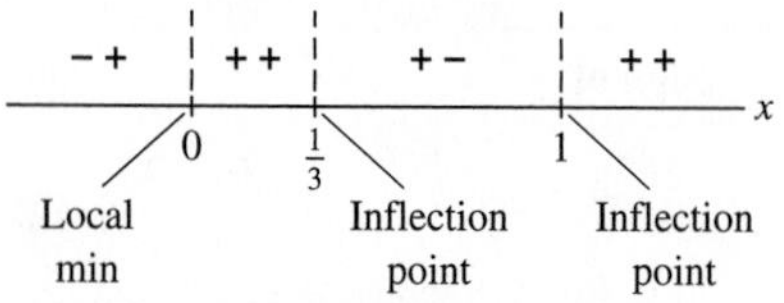

FIGURE 7

Step 2. **Note transition points and sign combinations.**
The transition points $c = 0, \frac{1}{3}, 1$ divide the x-axis into four intervals (Figure 7). The type of sign change determines the nature of the transition point:

- $c = 0$: local min since f' changes from − to + at $c = 0$.
- $c = \frac{1}{3}$: point of inflection since f'' changes sign at $c = \frac{1}{3}$.

- $c = 1$: neither a local min nor a local max since f' does not change sign, but it is a point of inflection since $f''(x)$ changes sign at $c = 1$.

We plot the transition points $c = 0, \frac{1}{3}, 1$ in Figure 8(A) using function values $f(0) = 1$, $f\left(\frac{1}{3}\right) = \frac{38}{27}$, and $f(1) = 2$.

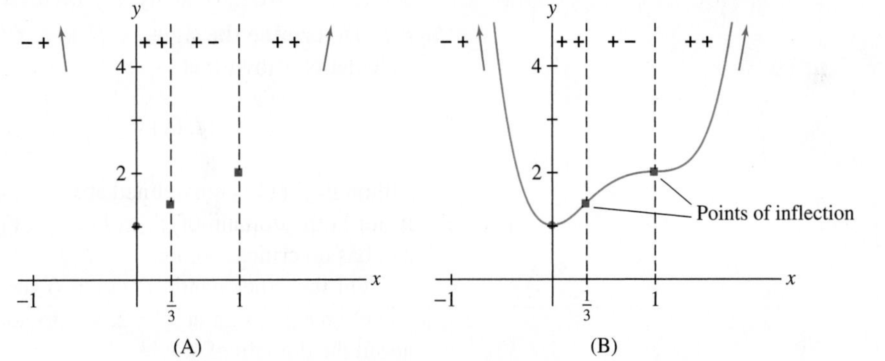

FIGURE 8 $f(x) = 3x^4 - 8x^3 + 6x^2 + 1$

Step 3. **Draw arcs of appropriate shape and asymptotic behavior.**
Before drawing the arcs, we note that $f(x)$ has leading term $3x^4$, so $f(x)$ tends to ∞ as $x \to \infty$ and as $x \to -\infty$. We obtain Figure 8(B). ■

■ **EXAMPLE 4 Trigonometric Function** Sketch $f(x) = \cos x + \frac{1}{2}x$ over $[0, \pi]$.

Solution First, we solve the transition points for x in $[0, \pi]$:

$$f'(x) = -\sin x + \frac{1}{2} = 0 \quad \Rightarrow \quad x = \frac{\pi}{6}, \frac{5\pi}{6}$$

$$f''(x) = -\cos x = 0 \quad \Rightarrow \quad x = \frac{\pi}{2}$$

The sign combinations are shown in the following tables.

Interval	Test Value	Sign of f'
$(0, \frac{\pi}{6})$	$f'(\frac{\pi}{12}) \approx 0.24$	+
$(\frac{\pi}{6}, \frac{5\pi}{6})$	$f'(\frac{\pi}{2}) = -\frac{1}{2}$	−
$(\frac{5\pi}{6}, \pi)$	$f'(\frac{11\pi}{12}) \approx 0.24$	+

Interval	Test Value	Sign of f''
$(0, \frac{\pi}{2})$	$f''(\frac{\pi}{4}) = -\frac{\sqrt{2}}{2}$	−
$(\frac{\pi}{2}, \pi)$	$f''(\frac{3\pi}{4}) = \frac{\sqrt{2}}{2}$	+

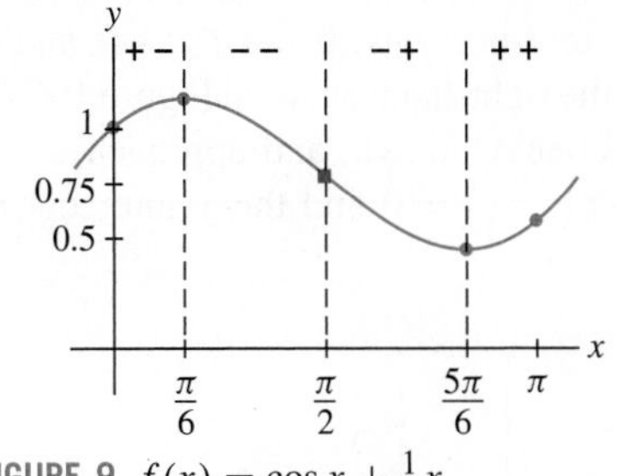

FIGURE 9 $f(x) = \cos x + \frac{1}{2}x$.

We record the sign changes and transition points in Figure 9 and sketch the graph using the values

$$f(0) = 1, \quad f\left(\frac{\pi}{6}\right) \approx 1.13, \quad f\left(\frac{\pi}{2}\right) \approx 0.79, \quad f\left(\frac{5\pi}{6}\right) \approx 0.44, \quad f(\pi) \approx 0.57$$ ■

The next two examples deal with horizontal and vertical asymptotes.

■ **EXAMPLE 5** Sketch the graph of $f(x) = \dfrac{3x + 2}{2x - 4}$.

Solution The function $f(x)$ is not defined for all x. This plays a role in our analysis so we add a Step 0 to our procedure.

***Step 0.* Determine the domain of f.**
Our function

$$f(x) = \frac{3x+2}{2x-4}$$

is not defined at $x = 2$. Therefore, the domain of f consists of the two intervals $(-\infty, 2)$ and $(2, \infty)$. We must analyze f on these intervals separately.

***Step 1.* Determine the signs of f' and f''.**
Calculation shows that

$$f'(x) = -\frac{4}{(x-2)^2}, \qquad f''(x) = \frac{8}{(x-2)^3}$$

Although $f'(x)$ is not defined at $x = 2$, we do not call it a critical point because $x = 2$ is not in the domain of f. In fact, $f'(x)$ is negative for $x \neq 2$, so $f(x)$ is decreasing and has no critical points.

On the other hand, $f''(x) > 0$ for $x > 2$ and $f''(x) < 0$ for $x < 2$. Although $f''(x)$ changes sign at $x = 2$, we do not call $x = 2$ a point of inflection because it is not in the domain of f.

***Step 2.* Note transition points and sign combinations.**
There are no transition points in the domain of f.

$(-\infty, 2)$	$f'(x) < 0$ and $f''(x) < 0$
$(2, \infty)$	$f'(x) < 0$ and $f''(x) > 0$

***Step 3.* Draw arcs of appropriate shape and asymptotic behavior.**
The following limits show that $y = \frac{3}{2}$ is a horizontal asymptote:

$$\lim_{x\to\pm\infty} \frac{3x+2}{2x-4} = \lim_{x\to\pm\infty} \frac{3+2x^{-1}}{2-4x^{-1}} = \frac{3}{2}$$

The line $x = 2$ is a vertical asymptote because $f(x)$ has infinite one-sided limits

$$\lim_{x\to 2-} \frac{3x+2}{2x-4} = -\infty, \qquad \lim_{x\to 2+} \frac{3x+2}{2x-4} = \infty$$

To verify this, note that for x near 2, the denominator $2x - 4$ is small negative if $x < 2$ and small positive if $x > 2$, whereas the numerator $3x + 4$ is positive.

Figure 10(A) summarizes the asymptotic behavior. What does the graph look like to the right of $x = 2$? It is decreasing and concave up since $f' < 0$ and $f'' > 0$, and it approaches the asymptotes. The only possibility is the right-hand curve in Figure 10(B). To the left of $x = 2$, the graph is decreasing, is concave down, and approaches the asymptotes. The x-intercept is $x = -\frac{2}{3}$ because $f\left(-\frac{2}{3}\right) = 0$ and the y-intercept is $y = f(0) = -\frac{1}{2}$. ■

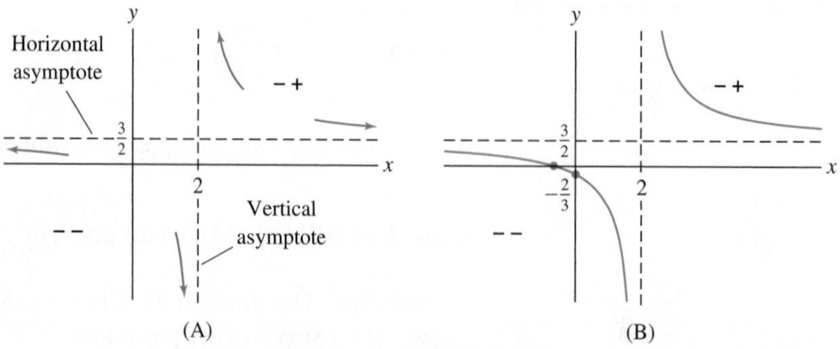

FIGURE 10 Graph of $y = \dfrac{3x+2}{2x-4}$.

■ **EXAMPLE 6** Sketch the graph of $f(x) = \dfrac{1}{x^2 - 1}$.

Solution The function $f(x)$ is defined for $x \neq \pm 1$. By calculation,

$$f'(x) = -\frac{2x}{(x^2 - 1)^2}, \qquad f''(x) = \frac{6x^2 + 2}{(x^2 - 1)^3}$$

For $x \neq \pm 1$, the denominator of $f'(x)$ is positive. Therefore, $f'(x)$ and x have opposite signs:

- $f'(x) > 0$ for $x < 0$, $f'(x) < 0$ for $x > 0$, $x = 0$ is a local max

The sign of $f''(x)$ is equal to the sign of $x^2 - 1$ because $6x^2 + 2$ is positive:

- $f''(x) > 0$ for $x < -1$ or $x > 1$ and $f''(x) < 0$ for $-1 < x < 1$

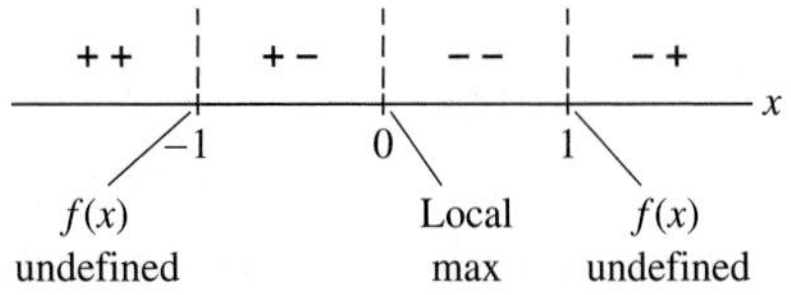

FIGURE 11

In this example,

$$f(x) = \frac{1}{x^2 - 1}$$

$$f'(x) = -\frac{2x}{(x^2 - 1)^2}$$

$$f''(x) = \frac{6x^2 + 2}{(x^2 - 1)^3}$$

Figure 11 summarizes the sign information.

The x-axis, $y = 0$, is a horizontal asymptote because

$$\lim_{x \to \infty} \frac{1}{x^2 - 1} = 0 \quad \text{and} \quad \lim_{x \to -\infty} \frac{1}{x^2 - 1} = 0$$

The lines $x = \pm 1$ are vertical asymptotes. To determine the one-sided limits, note that $f(x) < 0$ for $-1 < x < 1$ and $f(x) > 0$ for $|x| > 1$. Therefore, as $x \to \pm 1$, $f(x)$ approaches $-\infty$ from within the interval $(-1, 1)$, and it approaches ∞ from outside $(-1, 1)$ (Figure 12). We obtain the sketch in Figure 13.

Vertical Asymptote	Left-Hand Limit	Right-Hand Limit
$x = -1$	$\lim\limits_{x \to -1-} \dfrac{1}{x^2 - 1} = \infty$	$\lim\limits_{x \to -1+} \dfrac{1}{x^2 - 1} = -\infty$
$x = 1$	$\lim\limits_{x \to 1-} \dfrac{1}{x^2 - 1} = -\infty$	$\lim\limits_{x \to 1+} \dfrac{1}{x^2 - 1} = \infty$

■

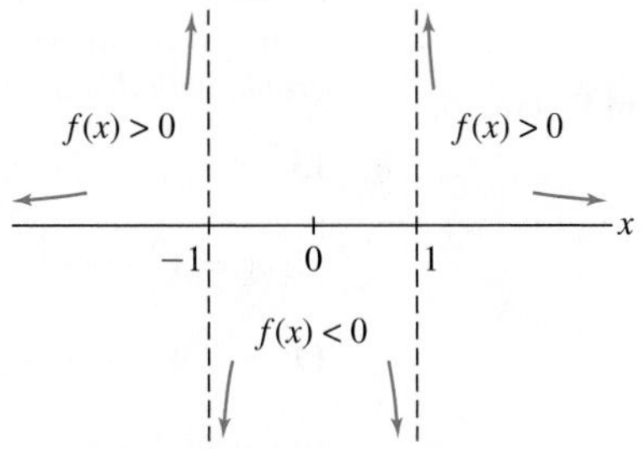

FIGURE 12 Behavior at vertical asymptotes.

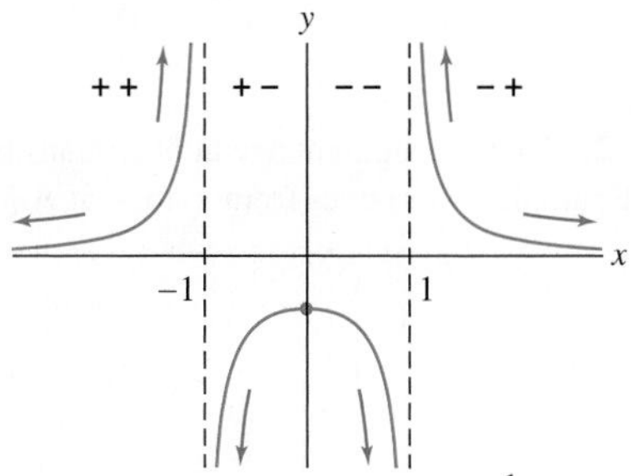

FIGURE 13 Graph of $y = \dfrac{1}{x^2 - 1}$.

4.5 SUMMARY

- Most graphs are made up of arcs that have one of the four basic shapes (Figure 14):

	Sign Combination	Curve Type
++	$f' > 0,\ f'' > 0$	Increasing and concave up
+−	$f' > 0,\ f'' < 0$	Increasing and concave down
−+	$f' < 0,\ f'' > 0$	Decreasing and concave up
−−	$f' < 0,\ f'' < 0$	Decreasing and concave down

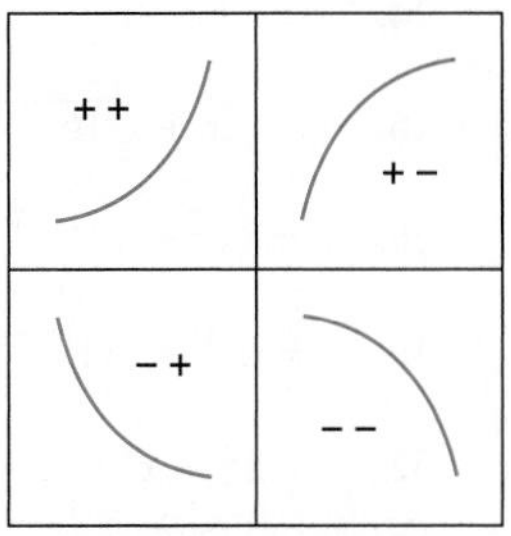

FIGURE 14 The four basic shapes.

- A *transition point* is a point in the domain of f at which either f' changes sign (local min or max) or f'' changes sign (point of inflection).
- It is convenient to break up the curve-sketching process into steps:

Step 0. Determine the domain of f.
Step 1. Determine the signs of f' and f''.
Step 2. Note transition points and sign combinations.
Step 3. Determine the asymptotic behavior of $f(x)$.
Step 4. Draw arcs of appropriate shape and asymptotic behavior.

4.5 EXERCISES

Preliminary Questions

1. Sketch an arc where f' and f'' have the sign combination $++$. Do the same for $-+$.

2. If the sign combination of f' and f'' changes from $++$ to $+-$ at $x = c$, then (choose the correct answer):

(a) $f(c)$ is a local min **(b)** $f(c)$ is a local max

(c) c is a point of inflection

3. The second derivative of the function $f(x) = (x-4)^{-1}$ is $f''(x) = 2(x-4)^{-3}$. Although $f''(x)$ changes sign at $x = 4$, $f(x)$ does not have a point of inflection at $x = 4$. Why not?

Exercises

1. Determine the sign combinations of f' and f'' for each interval A–G in Figure 15.

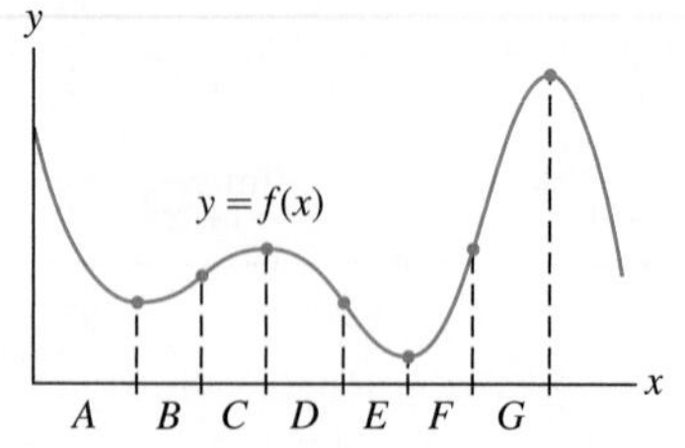

FIGURE 15

2. State the sign change at each transition point A–G in Figure 16. Example: $f'(x)$ goes from $+$ to $-$ at A.

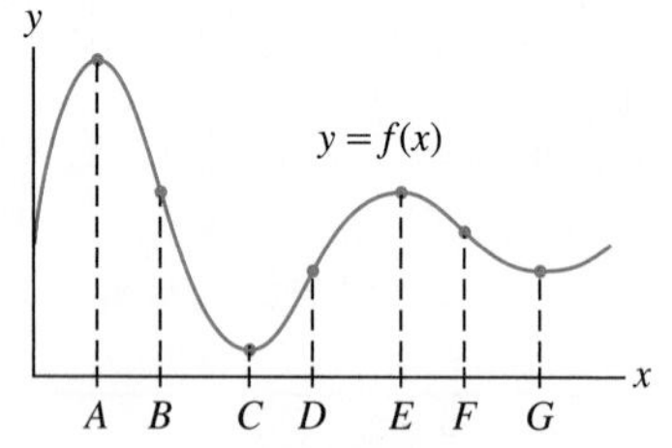

FIGURE 16

In Exercises 3–6, draw the graph of a function for which f' and f'' take on the given sign combinations.

3. $++$, $+-$, $--$

4. $+-$, $--$, $-+$

5. $-+$, $--$, $-+$

6. $-+$, $++$, $+-$

7. Sketch the graph of $y = x^2 - 5x + 4$.

8. Sketch the graph of $y = 12 - 5x - 2x^2$.

9. Sketch the graph of $f(x) = x^3 - 3x^2 + 2$. Include the zeros of $f(x)$, which are $x = 1$ and $1 \pm \sqrt{3}$ (approximately -0.73, 2.73).

10. Show that $f(x) = x^3 - 3x^2 + 6x$ has a point of inflection but no local extreme values. Sketch the graph.

11. Extend the sketch of the graph of $f(x) = \cos x + \frac{1}{2}x$ in Example 4 to the interval $[0, 5\pi]$.

12. Sketch the graphs of $y = x^{2/3}$ and $y = x^{4/3}$.

In Exercises 13–34, find the transition points, intervals of increase/decrease, concavity, and asymptotic behavior. Then sketch the graph, with this information indicated.

13. $y = x^3 + 24x^2$

14. $y = x^3 - 3x + 5$

15. $y = x^2 - 4x^3$

16. $y = \frac{1}{3}x^3 + x^2 + 3x$

17. $y = 4 - 2x^2 + \frac{1}{6}x^4$

18. $y = 7x^4 - 6x^2 + 1$

19. $y = x^5 + 5x$

20. $y = x^5 - 15x^3$

21. $y = x^4 - 3x^3 + 4x$

22. $y = x^2(x-4)^2$

23. $y = x^7 - 14x^6$

24. $y = x^6 - 9x^4$

25. $y = x - 4\sqrt{x}$

26. $y = \sqrt{x} + \sqrt{16 - x}$

27. $y = x(8 - x)^{1/3}$

28. $y = (x^2 - 4x)^{1/3}$

29. $y = (2x - x^2)^{1/3}$

30. $y = (x^3 - 3x)^{1/3}$

31. $y = x - x^{-1}$

32. $y = x^2 - x^{-2}$

33. $y = x^3 - 48/x^2$

34. $y = x^2 - x + x^{-1}$

35. Sketch the graph of $f(x) = 18(x-3)(x-1)^{2/3}$ using the formulas

$$f'(x) = \frac{30\left(x - \frac{9}{5}\right)}{(x-1)^{1/3}}, \qquad f''(x) = \frac{20\left(x - \frac{3}{5}\right)}{(x-1)^{4/3}}$$

36. Sketch the graph of $f(x) = \dfrac{x}{x^2+1}$ using the formulas

$$f'(x) = \frac{1-x^2}{(1+x^2)^2}, \qquad f''(x) = \frac{2x(x^2-3)}{(x^2+1)^3}$$

CAS *In Exercises 37–40, sketch the graph of the function, indicating all transition points. If necessary, use a graphing utility or computer algebra system to locate the transition points numerically.*

37. $y = x^3 - \dfrac{4}{x^2+1}$ **38.** $y = 12\sqrt{x^2+2x+4} - x^2$

39. $y = x^4 - 4x^2 + x + 1$

40. $y = 2\sqrt{x} - \sin x, \quad 0 \le x \le 2\pi$

In Exercises 41–46, sketch the graph over the given interval, with all transition points indicated.

41. $y = x + \sin x, \quad [0, 2\pi]$

42. $y = \sin x + \cos x, \quad [0, 2\pi]$

43. $y = 2\sin x - \cos^2 x, \quad [0, 2\pi]$ **44.** $y = \sin x + \frac{1}{2}x, \quad [0, 2\pi]$

45. $y = \sin x + \sqrt{3}\cos x, \quad [0, \pi]$

46. $y = \sin x - \frac{1}{2}\sin 2x, \quad [0, \pi]$

47. Are all sign transitions possible? Explain with a sketch why the transitions $++ \to -+$ and $-- \to +-$ do not occur if the function is differentiable. (See Exercise 76 for a proof.)

48. Suppose that f is twice differentiable satisfying (i) $f(0) = 1$, (ii) $f'(x) > 0$ for all $x \neq 0$, and (iii) $f''(x) < 0$ for $x < 0$ and $f''(x) > 0$ for $x > 0$. Let $g(x) = f(x^2)$.

(a) Sketch a possible graph of $f(x)$.

(b) Prove that $g(x)$ has no points of inflection and a unique local extreme value at $x = 0$. Sketch a possible graph of $g(x)$.

49. Which of the graphs in Figure 17 *cannot* be the graph of a polynomial? Explain.

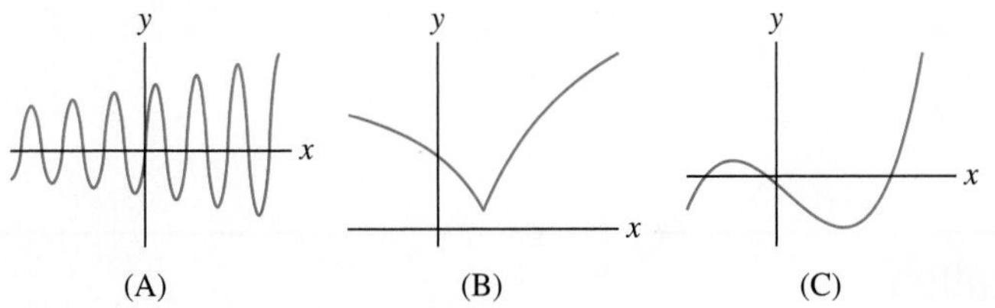

FIGURE 17

50. Which curve in Figure 18 is the graph of $f(x) = \dfrac{2x^4-1}{1+x^4}$? Explain on the basis of horizontal asymptotes.

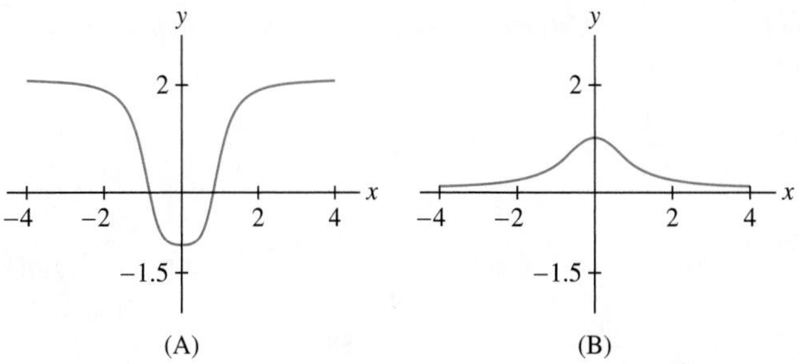

FIGURE 18

51. Match the graphs in Figure 19 with the two functions $y = \dfrac{3x}{x^2-1}$ and $y = \dfrac{3x^2}{x^2-1}$. Explain.

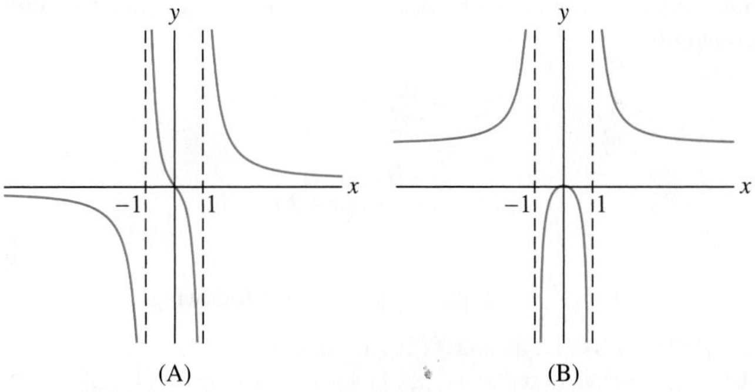

FIGURE 19

52. Match the functions with their graphs in Figure 20.

(a) $y = \dfrac{1}{x^2-1}$ **(b)** $y = \dfrac{x^2}{x^2+1}$

(c) $y = \dfrac{1}{x^2+1}$ **(d)** $y = \dfrac{x}{x^2-1}$

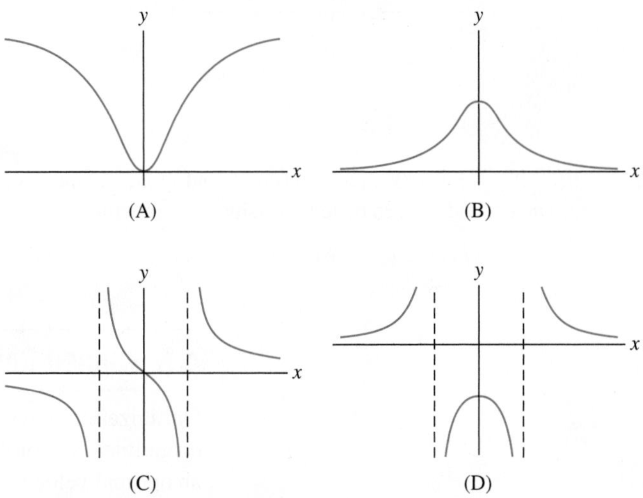

FIGURE 20

In Exercises 53–70, sketch the graph of the function. Indicate the transition points and asymptotes.

53. $y = \dfrac{1}{3x-1}$

54. $y = \dfrac{x-2}{x-3}$

55. $y = \dfrac{x+3}{x-2}$

56. $y = x + \dfrac{1}{x}$

57. $y = \dfrac{1}{x} + \dfrac{1}{x-1}$

58. $y = \dfrac{1}{x} - \dfrac{1}{x-1}$

59. $y = \dfrac{1}{x(x-2)}$

60. $y = \dfrac{x}{x^2-9}$

61. $y = \dfrac{1}{x^2-6x+8}$

62. $y = \dfrac{x^3+1}{x}$

63. $y = 1 - \dfrac{3}{x} + \dfrac{4}{x^3}$

64. $y = \dfrac{1}{x^2} + \dfrac{1}{(x-2)^2}$

65. $y = \dfrac{1}{x^2} - \dfrac{1}{(x-2)^2}$

66. $y = \dfrac{4}{x^2-9}$

67. $y = \dfrac{1}{(x^2+1)^2}$

68. $y = \dfrac{x^2}{(x^2-1)(x^2+1)}$

69. $y = \dfrac{1}{\sqrt{x^2+1}}$

70. $y = \dfrac{x}{\sqrt{x^2+1}}$

Further Insights and Challenges

In Exercises 71–75, we explore functions whose graphs approach a nonhorizontal line as $x \to \infty$. A line $y = ax + b$ is called a ***slant asymptote*** *if*

$$\lim_{x\to\infty} (f(x) - (ax+b)) = 0$$

or

$$\lim_{x\to-\infty} (f(x) - (ax+b)) = 0$$

71. Let $f(x) = \dfrac{x^2}{x-1}$ (Figure 21). Verify the following:

(a) $f(0)$ is a local max and $f(2)$ a local min.
(b) f is concave down on $(-\infty, 1)$ and concave up on $(1, \infty)$.
(c) $\lim_{x\to1-} f(x) = -\infty$ and $\lim_{x\to1+} f(x) = \infty$.
(d) $y = x + 1$ is a slant asymptote of $f(x)$ as $x \to \pm\infty$.
(e) The slant asymptote lies above the graph of $f(x)$ for $x < 1$ and below the graph for $x > 1$.

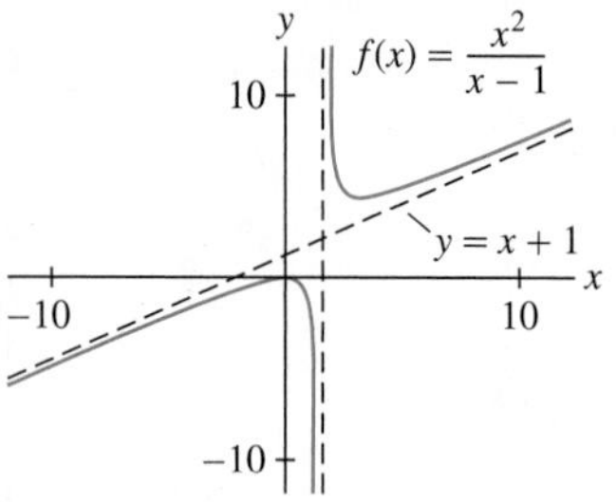

FIGURE 21

72. If $f(x) = P(x)/Q(x)$, where P and Q are polynomials of degrees $m+1$ and m, then by long division, we can write

$$f(x) = (ax+b) + P_1(x)/Q(x)$$

where P_1 is a polynomial of degree $< m$. Show that $y = ax + b$ is the slant asymptote of $f(x)$. Use this procedure to find the slant asymptotes of the following functions:

(a) $y = \dfrac{x^2}{x+2}$ **(b)** $y = \dfrac{x^3+x}{x^2+x+1}$

73. Sketch the graph of

$$f(x) = \frac{x^2}{x+1}.$$

Proceed as in the previous exercise to find the slant asymptote.

74. Show that $y = 3x$ is a slant asymptote for $f(x) = 3x + x^{-2}$. Determine whether $f(x)$ approaches the slant asymptote from above or below and make a sketch of the graph.

75. Sketch the graph of $f(x) = \dfrac{1-x^2}{2-x}$.

76. Assume that $f'(x)$ and $f''(x)$ exist for all x and let c be a critical point of $f(x)$. Show that $f(x)$ cannot make a transition from $++$ to $-+$ at $x = c$. *Hint:* Apply the MVT to $f'(x)$.

77. Assume that $f''(x)$ exists and $f''(x) > 0$ for all x. Show that $f(x)$ cannot be negative for all x. *Hint:* Show that $f'(b) \neq 0$ for some b and use the result of Exercise 56 in Section 4.4.

4.6 Applied Optimization

Optimization plays a role in a wide range of disciplines, including the physical sciences, economics, and biology. For example, scientists have studied how migrating birds choose an optimal velocity v that maximizes the distance they can travel without stopping, given the energy that can be stored as body fat (Figure 1).

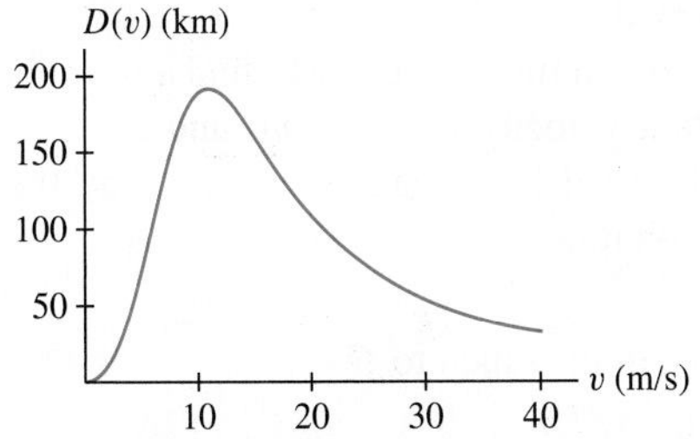

FIGURE 1 Physiology and aerodynamics are applied to obtain a plausible formula for bird migration distance $D(v)$ as a function of velocity v. The optimal velocity corresponds to the maximum point on the graph (see Exercise 56).

In many optimization problems, the first step is to write down the **objective function**. This is the function whose minimum or maximum we need. Once we find the objective function, we can apply the techniques developed in this chapter. Our first examples require optimization on a closed interval $[a, b]$. Let's recall the steps for finding extrema developed in Section 4.2:

(i) Find the critical points of $f(x)$ in $[a, b]$.
(ii) Evaluate $f(x)$ at the critical points and the endpoints a and b.
(iii) The largest and smallest values are the extreme values of $f(x)$ on $[a, b]$.

■ **EXAMPLE 1** A piece of wire of length L is bent into the shape of a rectangle (Figure 2). Which dimensions produce the rectangle of maximum area?

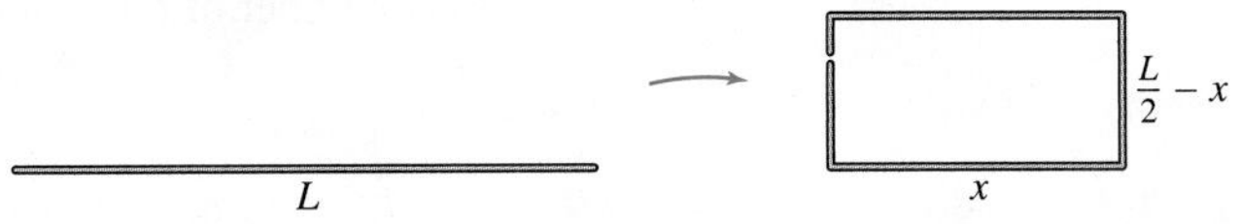

FIGURE 2

Solution The rectangle has area $A = xy$, where x and y are the lengths of the sides. Since A depends on two variables x and y, we cannot find the maximum until we eliminate one of the variables. We can do this because the variables are related: The rectangle has perimeter $L = 2x + 2y$, so $y = \frac{1}{2}L - x$. This allows us to rewrite the area in terms of x alone to obtain the objective function

An equation relating two or more variables in an optimization problem is called a "constraint equation." In Example 1, the constraint equation is

$$2x + 2y = L$$

$$A(x) = x\left(\frac{1}{2}L - x\right) = \frac{1}{2}Lx - x^2$$

On which interval does the optimization take place? The sides of the rectangle are nonnegative, so we require both $x \geq 0$ and $\frac{1}{2}L - x \geq 0$. Thus, $0 \leq x \leq \frac{1}{2}L$. Our problem is to maximize $A(x)$ on the closed interval $\left[0, \frac{1}{2}L\right]$.

We solve $A'(x) = \frac{1}{2}L - 2x = 0$ to obtain the critical point $x = \frac{1}{4}L$ and compare:

$$\text{Endpoints:} \quad A(0) = 0$$

$$A\left(\frac{1}{2}L\right) = \frac{1}{2}L\left(\frac{1}{2}L - \frac{1}{2}L\right) = 0$$

$$\text{Critical point:} \quad A\left(\frac{1}{4}L\right) = \left(\frac{1}{4}L\right)\left(\frac{1}{2}L - \frac{1}{4}L\right) = \frac{1}{16}L^2$$

The largest value occurs for $x = \frac{1}{4}L$, and in this case, $y = \frac{1}{2}L - \frac{1}{4}L = \frac{1}{4}L$. The rectangle of maximum area is the square of sides $x = y = \frac{1}{4}L$. ■

■ **EXAMPLE 2 Minimizing Travel Time** Your task is to build a road joining a ranch to a highway that enables drivers to reach the city in the shortest time (Figure 3). How should this be done if the speed limit is 60 km/h on the road and 110 km/h on the highway? The perpendicular distance from the ranch to the highway is 30 km, and the city is 50 km down the highway.

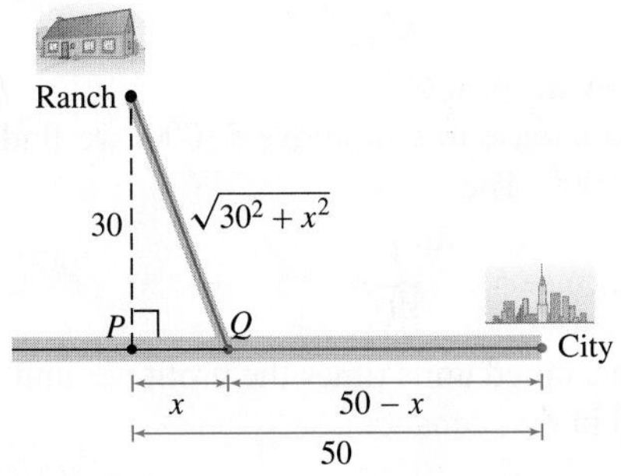

FIGURE 3

Solution This problem is more complicated than the previous one, so we'll analyze it in three steps. You can follow these steps to solve other optimization problems.

Step 1. **Choose variables.**

We need to determine the point Q where the road will join the highway. So let x be the distance from Q to the point P where the perpendicular joins the highway.

Step 2. **Find the objective function and the interval.**

Our objective function is the time $T(x)$ of the trip as a function of x. To find a formula for $T(x)$, recall that distance traveled at constant velocity v is $d = vt$, and the *time* required to travel a distance d is $t = d/v$. The road has length $\sqrt{30^2 + x^2}$ by the Pythagorean Theorem, so at velocity $v = 60$ km/h it takes

$$\frac{\sqrt{30^2 + x^2}}{60} \text{ hours to travel from the ranch to } Q$$

The strip of highway from Q to the city has length $50 - x$. At velocity $v = 110$ km/h, it takes

$$\frac{50 - x}{110} \text{ hours to travel from } Q \text{ to the city}$$

The total number of hours for the trip is

$$T(x) = \frac{\sqrt{30^2 + x^2}}{60} + \frac{50 - x}{110}$$

Our interval is $0 \le x \le 50$ because the road joins the highway somewhere between P and the city. So our task is to minimize $T(x)$ on $[0, 50]$ (Figure 4).

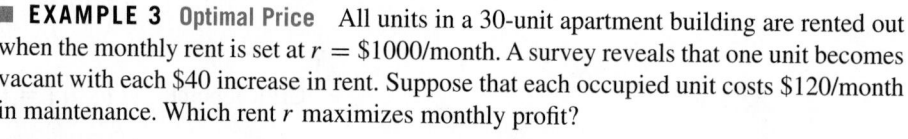

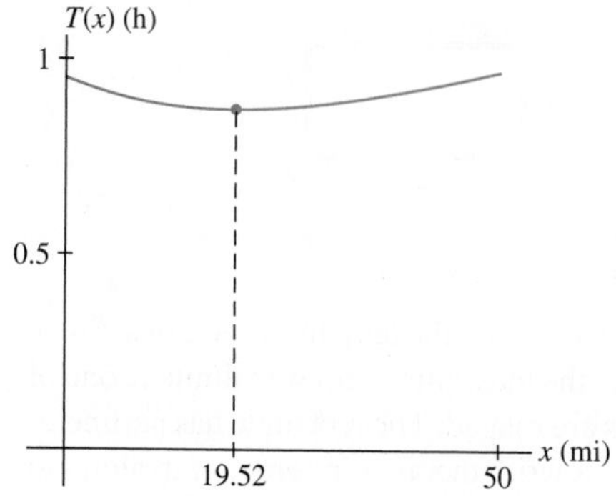

FIGURE 4 Graph of time of trip as function of x.

Step 3. **Optimize.**

Solve for the critical points:

$$T'(x) = \frac{x}{60\sqrt{30^2 + x^2}} - \frac{1}{110} = 0$$

$$110x = 60\sqrt{30^2 + x^2} \quad \Rightarrow \quad 11x = 6\sqrt{30^2 + x^2} \quad \Rightarrow$$

$$121x^2 = 36(30^2 + x^2) \quad \Rightarrow \quad 85x^2 = 32{,}400 \quad \Rightarrow \quad x = \sqrt{32{,}400/85} \approx 19.52$$

To find the minimum value of $T(x)$, we compare the values of $T(x)$ at the critical point and the endpoints of $[0, 50]$:

$$T(0) \approx 0.95 \text{ h}, \qquad T(19.52) \approx 0.87 \text{ h}, \qquad T(50) \approx 0.97 \text{ h}$$

We conclude that the travel time is minimized if the road joins the highway at a distance $x \approx 19.52$ km along the highway from P. ■

■ **EXAMPLE 3 Optimal Price** All units in a 30-unit apartment building are rented out when the monthly rent is set at $r = \$1000$/month. A survey reveals that one unit becomes vacant with each \$40 increase in rent. Suppose that each occupied unit costs \$120/month in maintenance. Which rent r maximizes monthly profit?

Solution

Step 1. **Choose variables.**

Our goal is to maximize the total monthly profit $P(r)$ as a function of rent r. It will depend on the number $N(r)$ of units occupied.

Step 2. **Find the objective function and the interval.**

Since one unit becomes vacant with each \$40 increase in rent above \$1000, we find that $(r - 1000)/40$ units are vacant when $r > 1000$. Therefore

$$N(r) = 30 - \frac{1}{40}(r - 1000) = 55 - \frac{1}{40}r$$

Total monthly profit is equal to the number of occupied units times the profit per unit, which is $r - 120$ (because each unit costs \$120 in maintenance), so

$$P(r) = N(r)(r - 120) = \left(55 - \frac{1}{40}r\right)(r - 120) = -6600 + 58r - \frac{1}{40}r^2$$

Which interval of r-values should we consider? There is no reason to lower the rent below $r = 1000$ because all units are already occupied when $r = 1000$. On the other hand, $N(r) = 0$ for $r = 40 \cdot 55 = 2200$. Therefore, zero units are occupied when $r = 2200$ and it makes sense to take $1000 \le r \le 2200$.

Step 3. **Optimize.**
Solve for the critical points:

$$P'(r) = 58 - \frac{1}{20}r = 0 \quad \Rightarrow \quad r = 1160$$

and compare values at the critical point and the endpoints:

$$P(1000) = 26{,}400, \qquad P(1160) = 27{,}040, \qquad P(2200) = 0$$

We conclude that the profit is maximized when the rent is set at $r = \$1160$. In this case, four units are left vacant. ■

Open Versus Closed Intervals

When we have to optimize over an open interval, there is no guarantee that a min or max exists (unlike the case of closed intervals) . However, if a min or max does exist, then it must occur at a critical point (because it is also a local min or max). Often, we can show that a min or max exists by examining $f(x)$ near the endpoints of the open interval. If $f(x)$ tends to infinity at the endpoints (as in Figure 6), then a minimum occurs at a critical point somewhere in the interval.

■ **EXAMPLE 4** Design a cylindrical can of volume 900 cm^3 so that it uses the least amount of metal (Figure 5). In other words, minimize the surface area of the can (including its top and bottom).

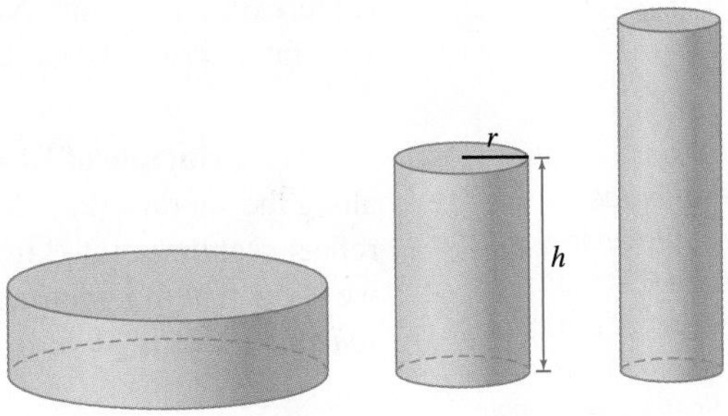

FIGURE 5 Cylinders with the same volume but different surface areas.

Solution

Step 1. **Choose variables.**
We must specify the can's radius and height. Therefore, let r be the radius and h the height. Let A be the surface area of the can.

Step 2. **Find the objective function and the interval.**
We compute A as a function of r and h:

$$A = \underbrace{\pi r^2}_{\text{Top}} + \underbrace{\pi r^2}_{\text{Bottom}} + \underbrace{2\pi r h}_{\text{Side}} = 2\pi r^2 + 2\pi r h$$

The can's volume is $V = \pi r^2 h$. Since we require that $V = 900$ cm^3, we have the constraint equation $\pi r^2 h = 900$. Thus $h = (900/\pi)r^{-2}$ and

$$A(r) = 2\pi r^2 + 2\pi r \left(\frac{900}{\pi r^2}\right) = 2\pi r^2 + \frac{1800}{r}$$

The radius r can take on any positive value, so we minimize $A(r)$ on $(0, \infty)$.

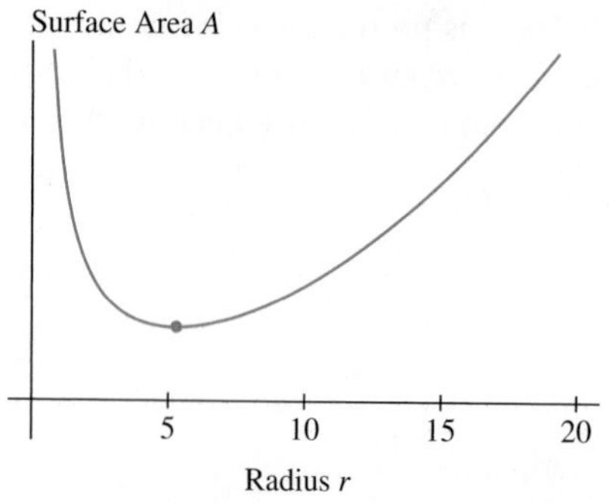

FIGURE 6 Surface area increases as r tends to 0 or ∞. The minimum value exists.

Step 3. **Optimize the function.**
Observe that $A(r)$ tends to infinity as r approaches the endpoints of $(0, \infty)$:

- $A(r) \to \infty$ as $r \to \infty$ (because of the r^2 term)
- $A(r) \to \infty$ as $r \to 0$ (because of the $1/r$ term)

Therefore $A(r)$ must take on a minimum value at a critical point in $(0, \infty)$ [Figure 6]. We solve in the usual way:

$$\frac{dA}{dr} = 4\pi r - \frac{1800}{r^2} = 0 \quad \Rightarrow \quad r^3 = \frac{450}{\pi} \quad \Rightarrow \quad r = \left(\frac{450}{\pi}\right)^{1/3} \approx 5.23 \text{ cm}$$

We also need to calculate the height:

$$h = \frac{900}{\pi r^2} = 2\left(\frac{450}{\pi}\right) r^{-2} = 2\left(\frac{450}{\pi}\right)\left(\frac{450}{\pi}\right)^{-2/3} = 2\left(\frac{450}{\pi}\right)^{1/3} \approx 10.46 \text{ cm}$$

Notice that the optimal dimensions satisfy $h = 2r$. In other words, the optimal can is as tall as it is wide. ■

■ **EXAMPLE 5 Optimization Problem with No Solution** Is it possible to design a cylinder of volume 900 cm^3 with the largest possible surface area?

Solution The answer is no. In the previous example, we showed that a cylinder of volume 900 and radius r has surface area

$$A(r) = 2\pi r^2 + \frac{1800}{r}$$

This function has no maximum value because it tends to infinity as $r \to 0$ or $r \to \infty$ (Figure 6). This means that a cylinder of fixed volume has a large surface area if it is either very fat and short (r large) or very tall and skinny (r small). ■

The Principle of Least Distance is also called ***Heron's Principle*** *after the mathematician Heron of Alexandria (c. 100 AD). See Exercise 69 for an elementary proof that does not use calculus and would have been known to Heron. Exercise 44 develops Snell's Law, a more general optical law based on the Principle of Least Time.*

The **Principle of Least Distance** states that a light beam reflected in a mirror travels along the shortest path. More precisely, a beam traveling from A to B, as in Figure 7, is reflected at the point P for which the path APB has minimum length. In the next example, we show that this minimum occurs when *the angle of incidence is equal to the angle of reflection*, that is, $\theta_1 = \theta_2$.

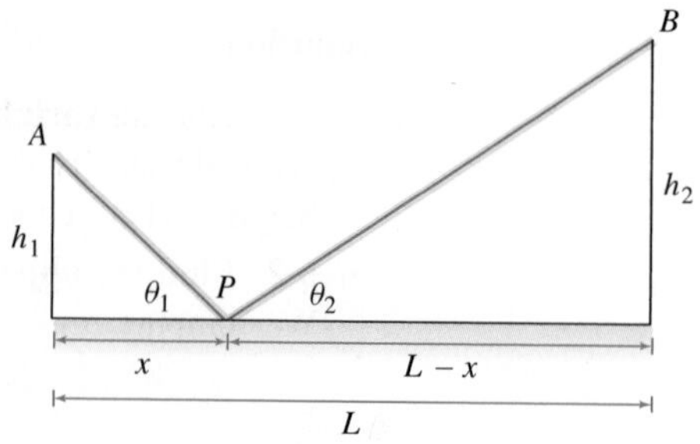

FIGURE 7 Reflection of a light beam in a mirror.

■ **EXAMPLE 6** Show that if P is the point for which the path APB in Figure 7 has minimal length, then $\theta_1 = \theta_2$.

Solution By the Pythagorean Theorem, the path APB has length

$$f(x) = AP + PB = \sqrt{x^2 + h_1^2} + \sqrt{(L - x)^2 + h_2^2}$$

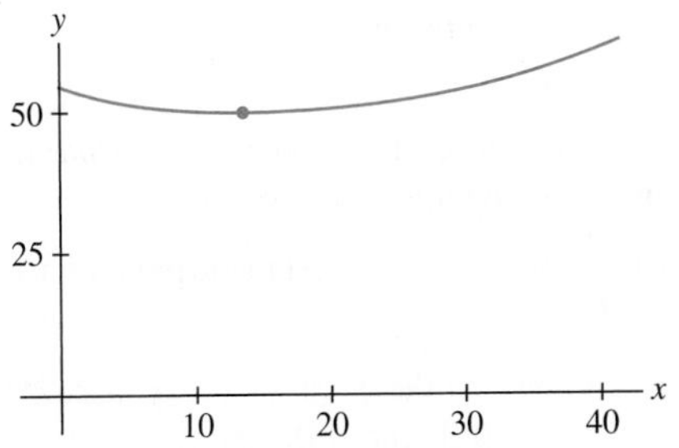

FIGURE 8 Graph of path length for $h_1 = 10$, $h_2 = 20$, $L = 40$.

with x, h_1, and h_2 as in the figure. The function $f(x)$ tends to infinity as x approaches $\pm\infty$ (that is, as P moves arbitrarily far to the right or left), so $f(x)$ takes on its minimum value at a critical point x such that (see Figure 8)

$$f'(x) = \frac{x}{\sqrt{x^2 + h_1^2}} - \frac{L - x}{\sqrt{(L - x)^2 + h_2^2}} = 0 \qquad \boxed{1}$$

It is not necessary to solve for x because our goal is not to find the critical point, but rather to show that $\theta_1 = \theta_2$. To do this, we rewrite Eq. (1) as

$$\underbrace{\frac{x}{\sqrt{x^2 + h_1^2}}}_{\cos\theta_1} = \underbrace{\frac{L - x}{\sqrt{(L - x)^2 + h_2^2}}}_{\cos\theta_2}$$

Referring to Figure 7, we see that this equation says $\cos\theta_1 = \cos\theta_2$, and since θ_1 and θ_2 lie between 0 and π, we conclude that $\theta_1 = \theta_2$ as claimed. ■

CONCEPTUAL INSIGHT The examples in this section were selected because they lead to optimization problems where the min or max occurs at a critical point. Often, the critical point represents the best compromise between "competing factors." In Example 3, we maximized profit by finding the best compromise between raising the rent and keeping the apartment units occupied. In Example 4, our solution minimizes surface area by finding the best compromise between height and width. In daily life, however, we often encounter endpoint rather than critical point solutions. For example, to run 10 meters in minimal time, you should run as fast as you can—the solution is not a critical point but rather an endpoint (your maximum speed).

4.6 SUMMARY

- There are usually three main steps in solving an applied optimization problem:

Step 1. Choose variables.
Determine which quantities are relevant, often by drawing a diagram, and assign appropriate variables.

Step 2. Find the objective function and the interval.
Restate as an optimization problem for a function f over an interval. If f depends on more than one variable, use a *constraint equation* to write f as a function of just one variable.

Step 3. Optimize the objective function.

- If the interval is open, f does not necessarily take on a minimum or maximum value. But if it does, these must occur at critical points within the interval. To determine if a min or max exists, analyze the behavior of f as x approaches the endpoints of the interval.

4.6 EXERCISES

Preliminary Questions

1. The problem is to find the right triangle of perimeter 10 whose area is as large as possible. What is the constraint equation relating the base b and height h of the triangle?

2. Describe a way of showing that a continuous function on an open interval (a, b) has a minimum value.

3. Is there a rectangle of area 100 of largest perimeter? Explain

Exercises

1. Find the dimensions x and y of the rectangle of maximum area that can be formed using 3 meters of wire.

(a) What is the constraint equation relating x and y?

(b) Find a formula for the area in terms of x alone.

(c) What is the interval of optimization? Is it open or closed?

(d) Solve the optimization problem.

2. Wire of length 12 m is divided into two pieces and each piece is bent into a square. How should this be done in order to minimize the sum of the areas of the two squares?

(a) Express the sum of the areas of the squares in terms of the lengths x and y of the two pieces.

(b) What is the constraint equation relating x and y?

(c) What is the interval of optimization? Is it open or closed?

(d) Solve the optimization problem.

3. Wire of length 12 m is divided into two pieces and the pieces are bend into a square and a circle. How should this be done in order to minimize the sum of their areas?

4. Find the positive number x such that the sum of x and its reciprocal is as small as possible. Does this problem require optimization over an open interval or a closed interval?

5. A flexible tube of length 4 m is bent into an L-shape. Where should the bend be made to minimize the distance between the two ends?

6. Find the dimensions of the box with square base with:

(a) Volume 12 and the minimal surface area.

(b) Surface area 20 and maximal volume.

7. A rancher will use 600 m of fencing to build a corral in the shape of a semicircle on top of a rectangle (Figure 9). Find the dimensions that maximize the area of the corral.

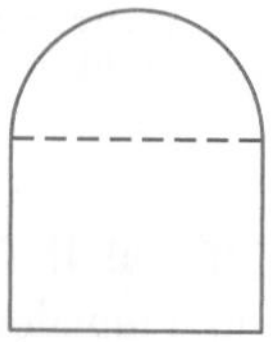

FIGURE 9

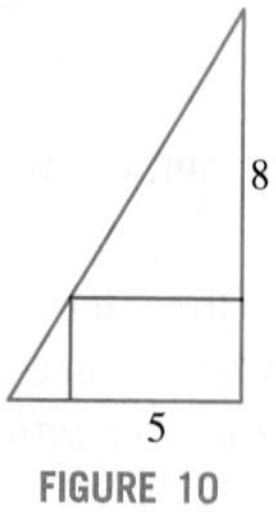

FIGURE 10

8. What is the maximum area of a rectangle inscribed in a right triangle with 5 and 8 as in Figure 10. The sides of the rectangle are parallel to the legs of the triangle.

9. Find the dimensions of the rectangle of maximum area that can be inscribed in a circle of radius $r = 4$ (Figure 11).

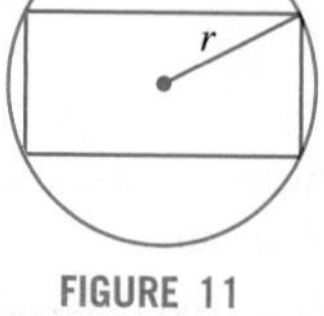

FIGURE 11

10. Find the dimensions x and y of the rectangle inscribed in a circle of radius r that maximizes the quantity xy^2.

11. Find the point on the line $y = x$ closest to the point $(1, 0)$. *Hint:* It is equivalent and easier to minimize the *square* of the distance.

12. Find the point P on the parabola $y = x^2$ closest to the point $(3, 0)$ (Figure 12).

13. Find the coordinates of the point on the graph of $y = x + 2x^{-1}$ closest to the origin in the region $x > 0$ (Figure 13).

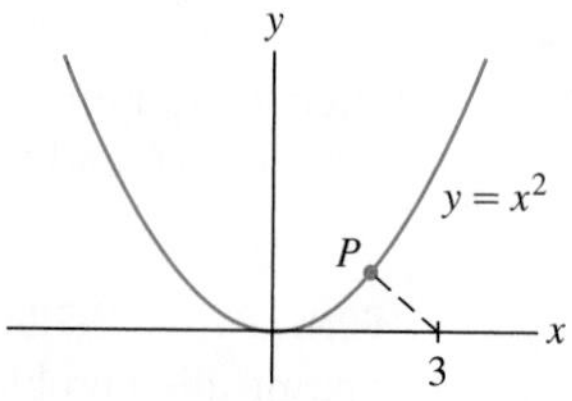

FIGURE 12

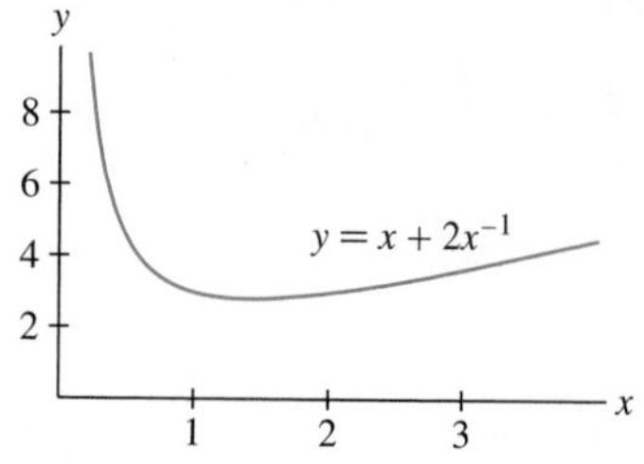

FIGURE 13

14. Problem of Tartaglia (1500–1557) Among all positive numbers a, b whose sum is 8, find those for which the product of the two numbers and their difference is largest.

15. Find the angle θ that maximizes the area of the isosceles triangle whose legs have length ℓ (Figure 14).

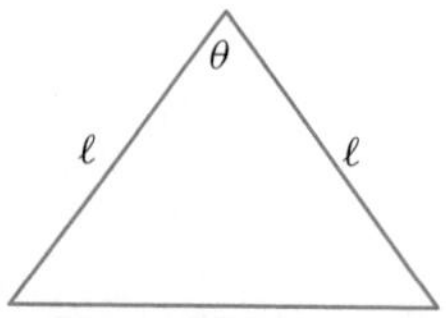

FIGURE 14

16. A right circular cone (Figure 15) has volume $V = \frac{\pi}{3}r^2h$ and surface area is $S = \pi r\sqrt{r^2 + h^2}$. Find the dimensions of the cone with surface area 1 and maximal volume.

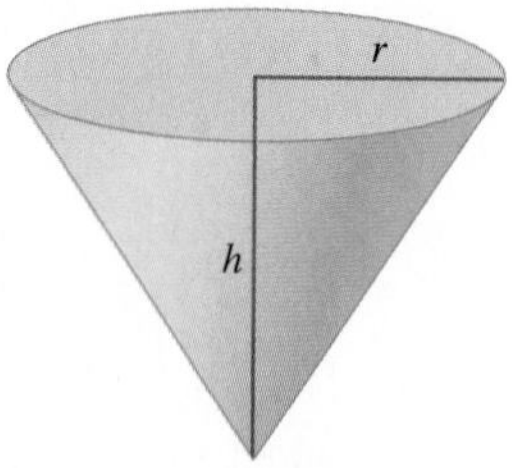

FIGURE 15

17. Find the area of the largest isosceles triangle that can be inscribed in a circle of radius r.

18. Find the radius and height of a cylindrical can of total surface area A whose volume is as large as possible. Does there exist a cylinder of surface area A and minimal total volume?

19. A poster of area 6000 cm^2 has blank margins of width 10 cm on the top and bottom and 6 cm on the sides. Find the dimensions that maximize the printed area.

20. According to postal regulations, a carton is classified as "oversized" if the sum of its height and girth (perimeter of its base) exceeds 108 in. Find the dimensions of a carton with square base that is not oversized and has maximum volume.

21. Kepler's Wine Barrel Problem In his work *Nova stereometria doliorum vinariorum* (New Solid Geometry of a Wine Barrel), published in 1615, astronomer Johannes Kepler stated and solved the following problem: Find the dimensions of the cylinder of largest volume that can be inscribed in a sphere of radius R. *Hint:* Show that an inscribed cylinder has volume $2\pi x(R^2 - x^2)$, where x is one-half the height of the cylinder.

22. Find the angle θ that maximizes the area of the trapezoid with a base of length 4 and sides of length 2, as in Figure 16.

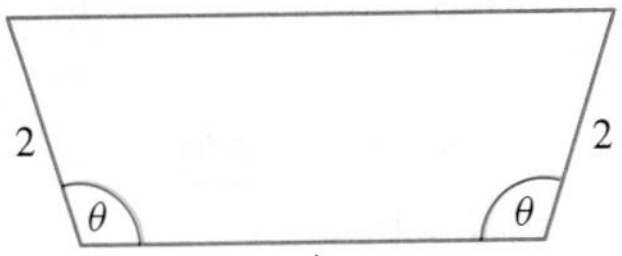

FIGURE 16

23. A landscape architect wishes to enclose a rectangular garden of area 1,000 m^2 on one side by a brick wall costing \$90/m and on the other three sides by a metal fence costing \$30/m. Which dimensions minimize the total cost?

24. The amount of light reaching a point at a distance r from a light source A of intensity I_A is I_A/r^2. Suppose that a second light source B of intensity $I_B = 4I_A$ is located 10 m from A. Find the point on the segment joining A and B where the total amount of light is at a minimum.

25. Find the maximum area of a rectangle inscribed in the region bounded by the graph of $y = \dfrac{4-x}{2+x}$ and the axes (Figure 17).

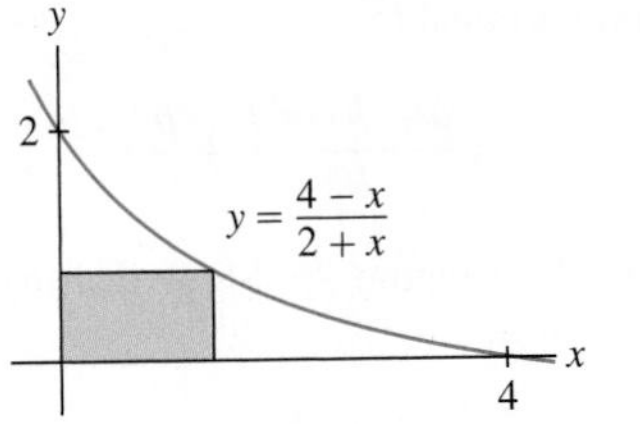

FIGURE 17

26. Find the maximum area of a triangle formed by the axes and a tangent line to the graph of $y = (x+1)^{-2}$ with $x > 0$.

27. Find the maximum area of a rectangle circumscribed around a rectangle of sides L and H. *Hint:* Express the area in terms of the angle θ (Figure 18).

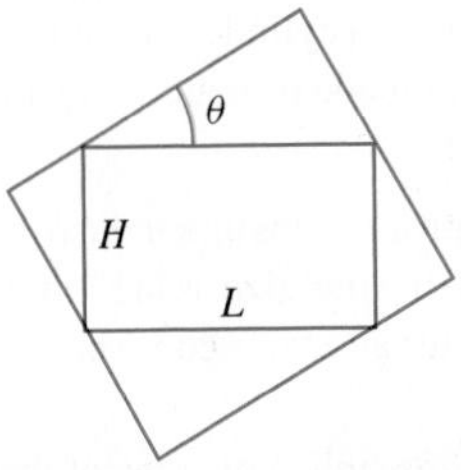

FIGURE 18

28. A contractor is engaged to build steps up the slope of a hill that has the shape of the graph of $y = x^2(120 - x)/6400$ for $0 \le x \le 80$ with x in meters (Figure 19). What is the maximum vertical rise of a stair if each stair has a horizontal length of one-third meter.

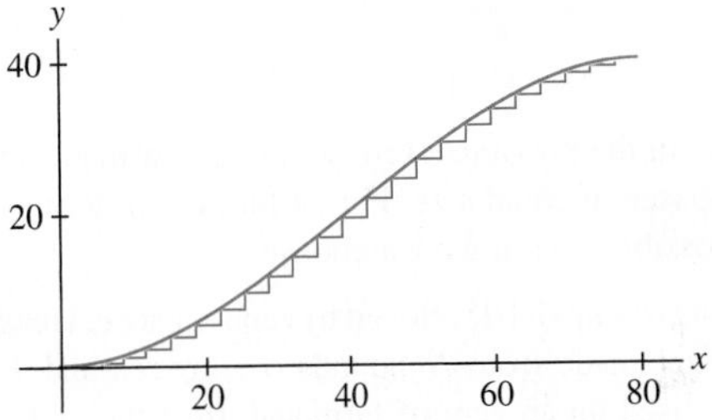

FIGURE 19

29. Find the equation of the line through $P = (4, 12)$ such that the triangle bounded by this line and the axes in the first quadrant has minimal area.

30. Let $P = (a, b)$ lie in the first quadrant. Find the slope of the line through P such that the triangle bounded by this line and the axes in the first quadrant has minimal area. Then show that P is the midpoint of the hypotenuse of this triangle.

31. Archimedes' Problem A spherical cap (Figure 20) of radius r and height h has volume $V = \pi h^2\left(r - \frac{1}{3}h\right)$ and surface area $S = 2\pi rh$. Prove that the hemisphere encloses the largest volume among all spherical caps of fixed surface area S.

32. Find the isosceles triangle of smallest area (Figure 21) that circumscribes a circle of radius 1 (from Thomas Simpson's *The Doctrine and Application of Fluxions*, a calculus text that appeared in 1750).

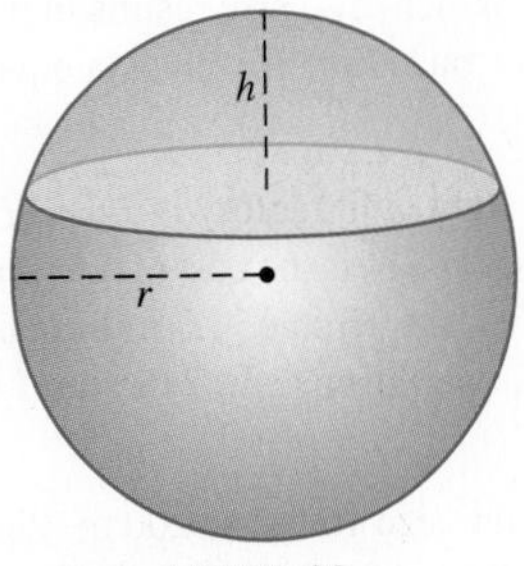

FIGURE 20

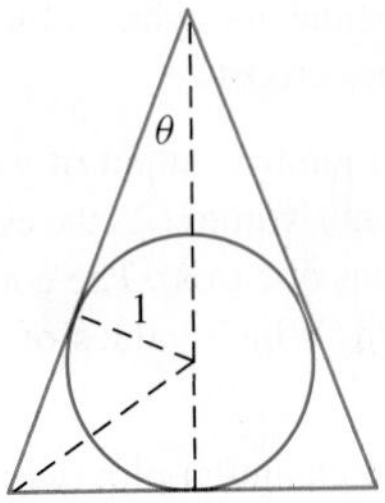

FIGURE 21

33. A box of volume 72 m^3 with square bottom and no top is constructed out of two different materials. The cost of the bottom is \$40/m^2 and the cost of the sides is \$30/m^2. Find the dimensions of the box that minimize total cost.

34. Find the dimensions of a cylinder of volume 1 m^3 of minimal cost if the top and bottom are made of material that costs twice as much as the material for the side.

35. Your task is to design a rectangular industrial warehouse consisting of three separate spaces of equal size as in Figure 22. The wall materials cost \$500 per linear meter and your company allocates \$2,400,000 for the project.

(a) Which dimensions maximize the area of the warehouse?

(b) What is the area of each compartment in this case?

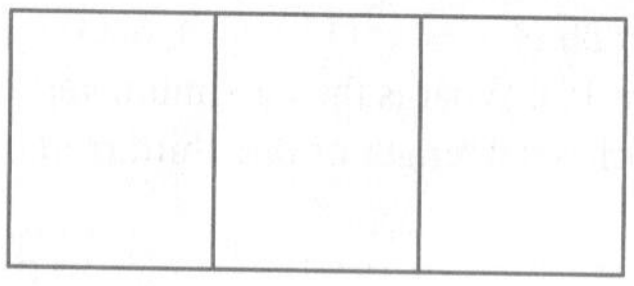

FIGURE 22

36. Suppose, in the previous exercise, that the warehouse consists of n separate spaces of equal size. Find a formula in terms of n for the maximum possible area of the warehouse.

37. According to a model developed by economists E. Heady and J. Pesek, if fertilizer made from N pounds of nitrogen and P pounds of phosphate is used on an acre of farmland, then the yield of corn (in bushels per acre) is

$$Y = 7.5 + 0.6N + 0.7P - 0.001N^2 - 0.002P^2 + 0.001NP$$

A farmer intends to spend \$30 per acre on fertilizer. If nitrogen costs 25 cents/lb and phosphate costs 20 cents/lb, which combination of N and L produces the highest yield of corn?

38. Experiments show that the quantities x of corn and y of soybean required to produce a hog of weight Q satisfy $Q = 0.5x^{1/2}y^{1/4}$. The unit of x, y, and Q is the cwt, an agricultural unit equal to 100 lbs. Find the values of x and y that minimize the cost of a hog of weight $Q = 2.5$ cwt if corn costs \$3/cwt and soy costs \$7/cwt.

39. All units in a 100-unit apartment building are rented out when the monthly rent is set at $r =$ \$900/month. Suppose that one unit becomes vacant with each \$10 increase in rent and that each occupied unit costs \$80/month in maintenance. Which rent r maximizes monthly profit?

40. An 8-billion-bushel corn crop brings a price of \$2.40/bu. A commodity broker uses the rule of thumb: If the crop is reduced by x percent, then the price increases by $10x$ cents. Which crop size results in maximum revenue and what is the price per bu? *Hint:* Revenue is equal to price times crop size.

41. The monthly output of a Spanish light bulb factory is $P = 2LK^2$ (in millions), where L is the cost of labor and K is the cost of equipment (in millions of euros). The company needs to produce 1.7 million units per month. Which values of L and K would minimize the total cost $L + K$?

42. The rectangular plot in Figure 23 has size 100 m × 200 m. Pipe is to be laid from A to a point P on side BC and from there to C. The cost of laying pipe along the side of the plot is \$45/m and the cost through the plot is \$80/m (since it is underground).

(a) Let $f(x)$ be the total cost, where x is the distance from P to B. Determine $f(x)$, but note that f is discontinuous at $x = 0$ (when $x = 0$, the cost of the entire pipe is \$45/ft).

(b) What is the most economical way to lay the pipe? What if the cost along the sides is \$65/m?

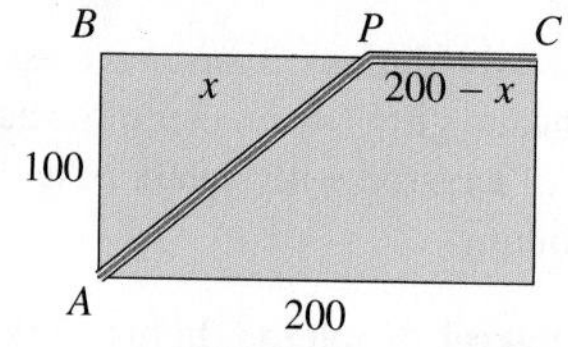

FIGURE 23

43. Brandon is on one side of a river that is 50 m wide and wants to reach a point 200 m downstream on the opposite side as quickly as possible by swimming diagonally across the river and then running the rest of the way. Find the best route if Brandon can swim at 1.5 m/s and run at 4 m/s.

44. Snell's Law When a light beam travels from a point A above a swimming pool to a point B below the water (Figure 24), it chooses the path that takes the *least time*. Let v_1 be the velocity of light in air and v_2 the velocity in water (it is known that $v_1 > v_2$). Prove Snell's Law of Refraction:

$$\frac{\sin\theta_1}{v_1} = \frac{\sin\theta_2}{v_2}$$

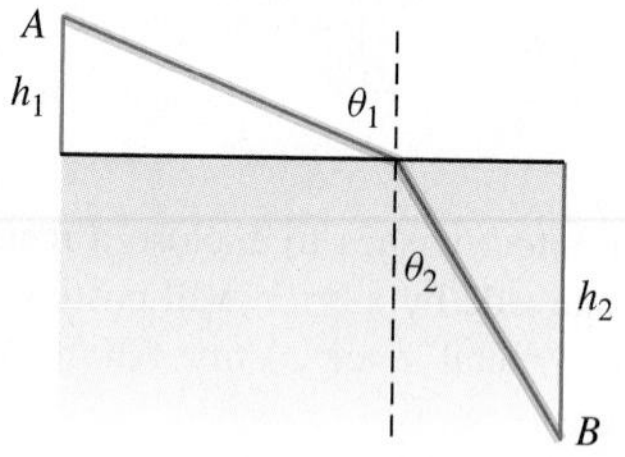

FIGURE 24

In Exercises 45–47, a box (with no top) is to be constructed from a piece of cardboard of sides A and B by cutting out squares of length h from the corners and folding up the sides (Figure 26).

45. Find the value of h that maximizes the volume of the box if $A = 15$ and $B = 24$. What are the dimensions of this box?

46. Vascular Branching A small blood vessel of radius r branches off at an angle θ from a larger vessel of radius R to supply blood along a path from A to B. According to Poiseuille's Law, the total resistance to blood flow is proportional to

$$T = \left(\frac{a - b\cot\theta}{R^4} + \frac{b\csc\theta}{r^4}\right)$$

where a and b are as in Figure 25. Show that the total resistance is minimized when $\cos\theta = (r/R)^4$.

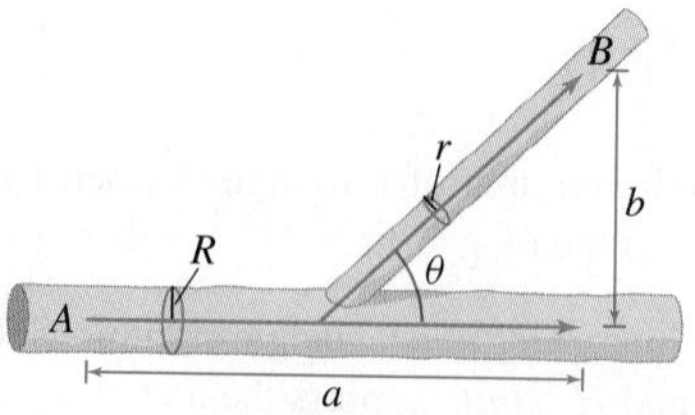

FIGURE 25

47. Which values of A and B maximize the volume of the box if $h = 10$ cm and $AB = 900$ cm.

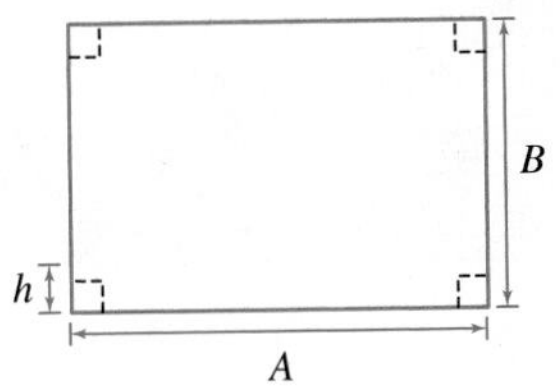

FIGURE 26

48. Given n numbers $x_1, \dots, x_n$, find the value of x minimizing the sum of the squares:

$$(x - x_1)^2 + (x - x_2)^2 + \cdots + (x - x_n)^2$$

First solve for $n = 2, 3$ and then try it for arbitrary n.

49. A billboard of height b is mounted on the side of a building with its bottom edge at a distance h from the street as in Figure 27. At what distance x should an observer stand from the wall to maximize the angle of observation θ?

50. Solve Exercise 49 again using geometry rather than calculus. There is a unique circle passing through points B and C which is tangent to the street. Let R be the point of tangency. Note that the two angles labeled ψ in Figure 27 are equal because they subtend equal arcs on the circle.

(a) Show that the maximum value of θ is $\theta = \psi$.

(b) Prove that this agrees with the answer to Exercise 49. *Hint:* Show that $\psi = \theta + \angle PBA$ where A is the intersection of the circle with PC.

(c) Show that $\angle QRB = \angle RCQ$ for the maximal angle ψ.

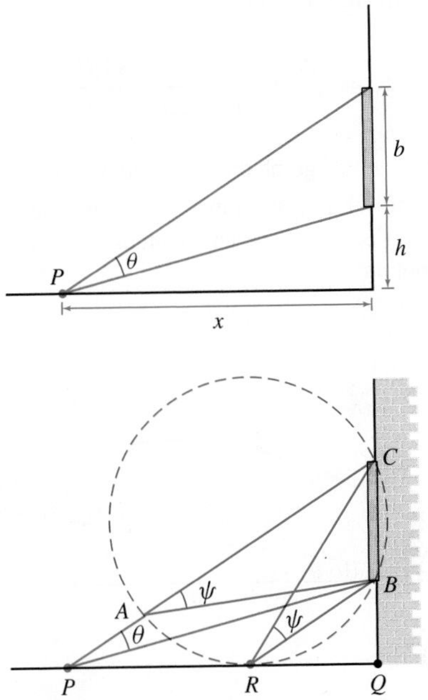

FIGURE 27

51. Optimal Delivery Schedule A gas station sells Q gallons of gasoline per year, which is delivered N times per year in equal shipments of Q/N gallons. The cost of each delivery is d dollars and the yearly storage costs are sQT, where T is the length of time (a fraction of a year) between shipments and s is a constant. Show that costs are minimized for $N = \sqrt{sQ/d}$. (*Hint:* $T = 1/N$.) Find the optimal number of deliveries if $Q = 2$ million gal, $d = \$8{,}000$, and $s = 30$ cents/gal-yr. Your answer should be a whole number, so compare costs for the two integer values of N nearest the optimal value.

52. Victor Klee's Endpoint Maximum Problem Given 40 meters of straight fence, your goal is to build a rectangular enclosure using 80 additional meters of fence that encompasses the greatest area. Let $A(x)$ be the area of the enclosure, with x as in Figure 28.

(a) Find the maximum value of $A(x)$.

(b) Which interval of x values is relevant to our problem? Find the maximum value of $A(x)$ on this interval.

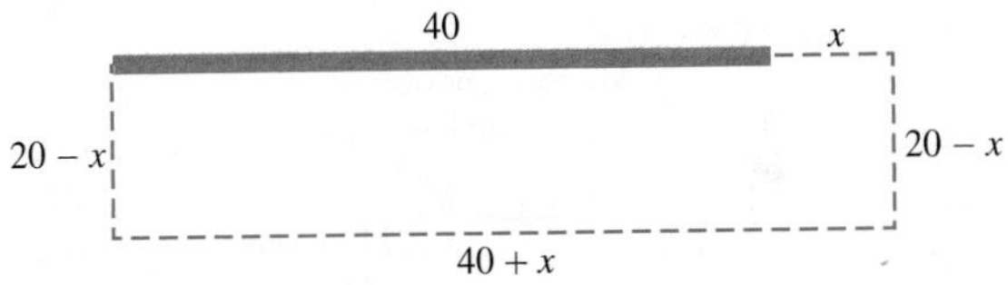

FIGURE 28

53. Let (a, b) be a fixed point in the first quadrant and let $S(d)$ be the sum of the distances from $(d, 0)$ to the points $(0, 0)$, (a, b), and $(a, -b)$.

(a) Find the value of d for which $S(d)$ is minimal. The answer depends on whether $b < \sqrt{3}a$ or $b \ge \sqrt{3}a$. *Hint:* Show that $d = 0$ when $b \ge \sqrt{3}a$.

(b) GU Let $a = 1$. Plot $S(d)$ for $b = 0.5, \sqrt{3}, 3$ and describe the position of the minimum.

54. The force F (in Newtons) required to move a box of mass m kg in motion by pulling on an attached rope (Figure 29) is

$$F(\theta) = \frac{fmg}{\cos\theta + f\sin\theta}$$

where θ is the angle between the rope and the horizontal, f is the coefficient of static friction, and $g = 9.8$ m/s^2. Find the angle θ that minimizes the required force F, assuming $f = 0.4$. *Hint:* Find the maximum value of $\cos\theta + f\sin\theta$.

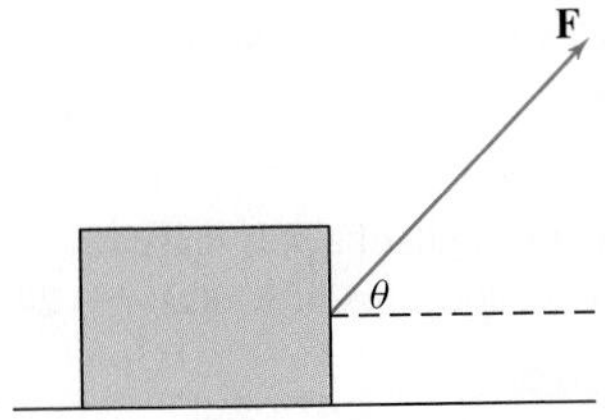

FIGURE 29

55. In the setting of Exercise 54, show that for any f the minimal force required is proportional to $1/\sqrt{1 + f^2}$.

56. Bird Migration Ornithologists have found that the power (in joules per second) consumed by a certain pigeon flying at velocity v m/s is described well by the function $P(v) = 17v^{-1} + 10^{-3}v^3$ J/s. Assume that the pigeon can store 5×10^4 J of usable energy as body fat.

(a) Show that at velocity v, a pigeon can fly a total distance of $D(v) = (5 \times 10^4)v/P(v)$ if it uses all of its stored energy.

(b) Find the velocity v_p that *minimizes* $P(v)$.

(c) Migrating birds are smart enough to fly at the velocity that maximizes distance traveled rather than minimizes power consumption. Show that the velocity v_d which maximizes $D(v)$ satisfies $P'(v_d) = P(v_d)/v_d$. Show that v_d is obtained graphically as the velocity coordinate of the point where a line through the origin is tangent to the graph of $P(v)$ (Figure 30).

(d) Find v_d and the maximum distance $D(v_d)$.

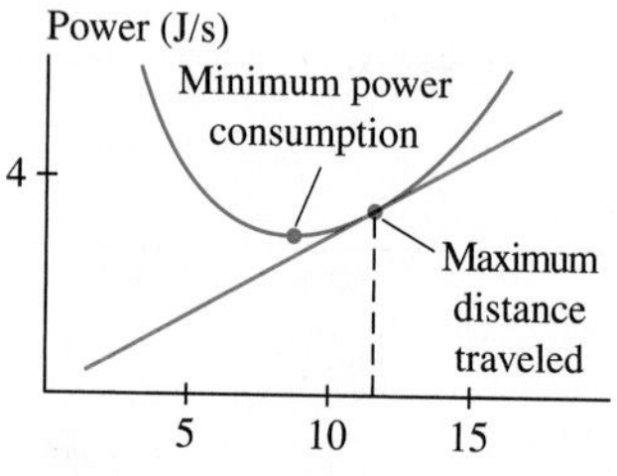

FIGURE 30

57. The problem is to put a "roof" of side s on an attic room of height h and width b. Find the smallest length s for which this is possible if $b = 27$ and $h = 8$ (Figure 31).

58. Redo Exercise 57 for arbitrary b and h.

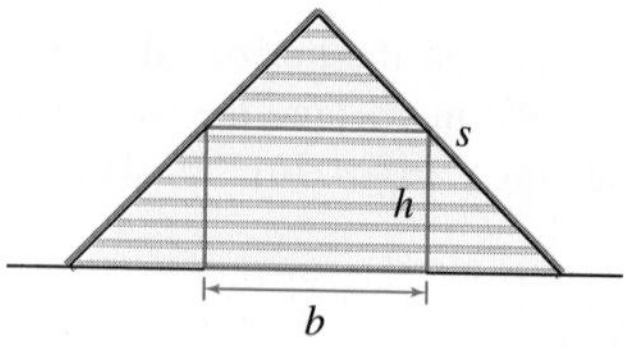

FIGURE 31

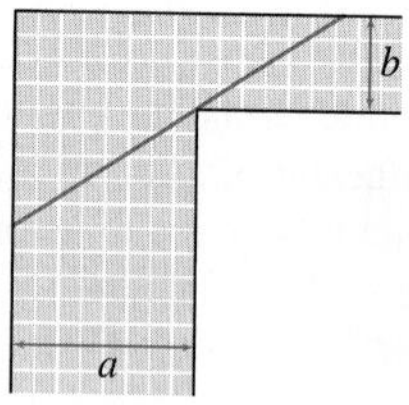

FIGURE 32

59. Find the maximum length of a pole that can be carried horizontally around a corner joining corridors of widths $a = 24$ and $b = 3$ (Figure 32).

60. Redo Exercise 59 for arbitrary widths a and b.

61. Find the minimum length ℓ of a beam that can clear a fence of height h and touch a wall located b ft behind the fence (Figure 33).

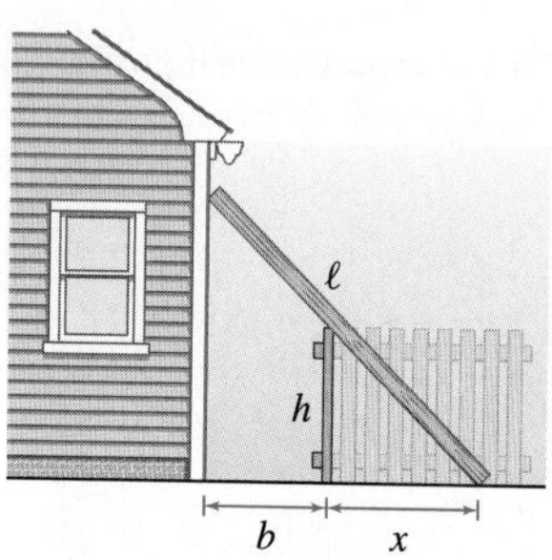

FIGURE 33

62. Which value of h maximizes the volume of the box if $A = B$?

63. A basketball player stands d feet from the basket. Let h and α be as in Figure 34. Using physics, one can show that if the player releases the ball at an angle θ, then the initial velocity required to make the ball go through the basket satisfies

$$v^2 = \frac{16d}{\cos^2\theta(\tan\theta - \tan\alpha)}$$

(a) Explain why this formula is meaningful only for $\alpha < \theta < \frac{\pi}{2}$. Why does v approach infinity at the endpoints of this interval?

(b) GU Take $\alpha = \frac{\pi}{6}$ and plot v^2 as a function of θ for $\frac{\pi}{6} < \theta < \frac{\pi}{2}$. Verify that the minimum occurs at $\theta = \frac{\pi}{3}$.

(c) Set $F(\theta) = \cos^2\theta(\tan\theta - \tan\alpha)$. Explain why v is minimized for θ such that $F(\theta)$ is maximized.

(d) Verify that $F'(\theta) = \cos(\alpha - 2\theta)\sec\alpha$ (you will need to use the addition formula for cosine) and show that the maximum value of $F(\theta)$ on $\left[\alpha, \frac{\pi}{2}\right]$ occurs at $\theta_0 = \frac{\alpha}{2} + \frac{\pi}{4}$.

(e) For a given α, the optimal angle for shooting the basket is θ_0 because it minimizes v^2 and therefore minimizes the energy required to make the shot (energy is proportional to v^2). Show that the velocity v_{opt} at the optimal angle θ_0 satisfies

$$v_{opt}^2 = \frac{32d\cos\alpha}{1 - \sin\alpha} = \frac{32\,d^2}{-h + \sqrt{d^2 + h^2}}$$

(f) GU Show with a graph that for fixed d (say, $d = 15$ ft, the distance of a free throw), v_{opt}^2 is an increasing function of h. Use this to explain why taller players have an advantage and why it can help to jump while shooting.

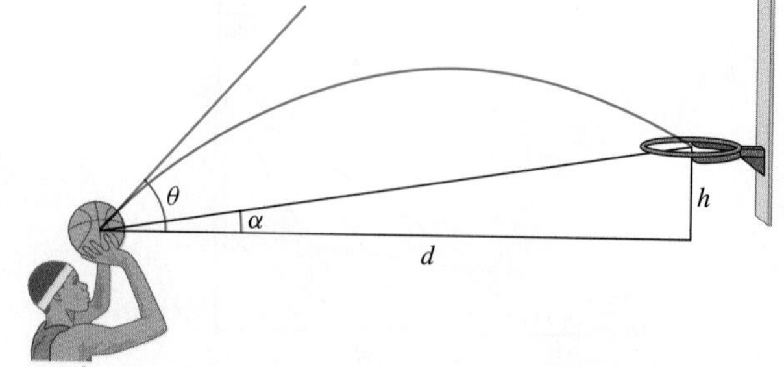

FIGURE 34

64. Three towns A, B, and C are to be joined by an underground fiber cable as illustrated in Figure 35(A). Assume that C is located directly below the midpoint of $\overline{AB}$. Find the junction point P that minimizes the total amount of cable used.

(a) First show that P must lie directly above C. *Hint:* Use the result of Example 6 to show that if the junction is placed at point Q in Figure 35(B), then we can reduce the cable length by moving Q horizontally over to the point P lying above C.

(b) With x as in Figure 35(A), let $f(x)$ be the total length of cable used. Show that $f(x)$ has a unique critical point c. Compute c and show that $0 \le c \le L$ if and only if $D \le 2\sqrt{3}\,L$.

(c) Find the minimum of $f(x)$ on $[0, L]$ in two cases: $D = 2$, $L = 4$ and $D = 8$, $L = 2$.

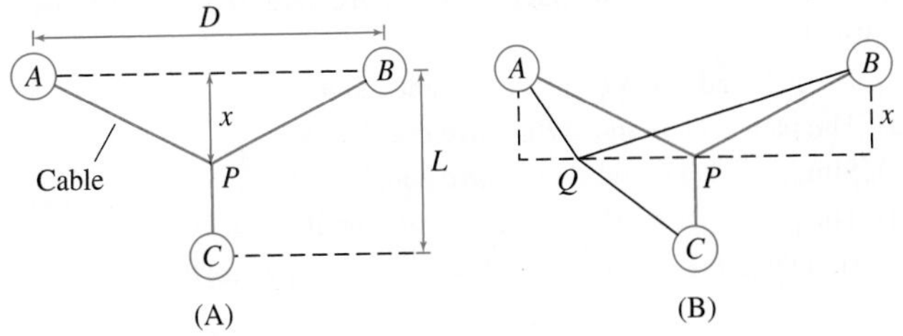

FIGURE 35

Further Insights and Challenges

65. Tom and Ali drive along a highway represented by the graph of $f(x)$ in Figure 36. During the trip, Ali views a billboard represented by the segment $\overline{BC}$ along the y-axis. Let Q be the y-intercept of the tangent line to $y = f(x)$. Show that θ is maximized at the value of x for which the angles $\angle QPB$ and $\angle QCP$ are equal. This generalizes Exercise 50 (c) (which corresponds to the case $f(x) = 0$). *Hints:*

(a) Show that $d\theta/dx$ is equal to

$$(b-c)\cdot\frac{(x^2 + (xf'(x))^2) - (b - (f(x) - xf'(x)))(c - (f(x) - xf'(x)))}{(x^2 + (b - f(x))^2)(x^2 + (c - f(x))^2)}$$

(b) Show that the y-coordinate of Q is $f(x) - xf'(x)$.

(c) Show that the condition $d\theta/dx = 0$ is equivalent to

$$PQ^2 = BQ \cdot CQ$$

(d) Conclude that $\triangle QPB$ and $\triangle QCP$ are similar triangles.

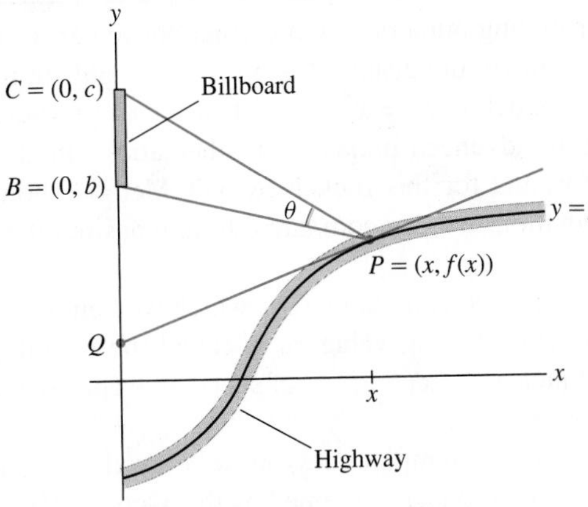

FIGURE 36

Seismic Prospecting *Exercises 66–68 are concerned with determining the thickness d of a layer of soil that lies on top of a rock formation. Geologists send two sound pulses from point A to point D separated by a distance s. The first pulse travels directly from A to D along the surface of the earth. The second pulse travels down to the rock formation, then along its surface, and then back up to D (path ABCD), as in Figure 37. The pulse travels with velocity v_1 in the soil and v_2 in the rock.*

66. **(a)** Show that the time required for the first pulse to travel from A to D is $t_1 = s/v_1$.

(b) Show that the time required for the second pulse is

$$t_2 = \frac{2d}{v_1}\sec\theta + \frac{s - 2d\tan\theta}{v_2}$$

provided that

$$\tan\theta \le \frac{s}{2d} \qquad \boxed{2}$$

(*Note:* If this inequality is not satisfied, then point B does not lie to the left of C.)

(c) Show that t_2 is minimized when $\sin\theta = v_1/v_2$.

67. In this exercise, assume that $v_2/v_1 \ge \sqrt{1 + 4(d/s)^2}$.

(a) Show that inequality (2) holds if $\sin\theta = v_1/v_2$.

(b) Show that the minimal time for the second pulse is

$$t_2 = \frac{2d}{v_1}(1 - k^2)^{1/2} + \frac{s}{v_2}$$

where $k = v_1/v_2$.

(c) Conclude that $\dfrac{t_2}{t_1} = \dfrac{2d(1-k^2)^{1/2}}{s} + k$.

68. Continue with the assumption of the previous exercise.

(a) Find the thickness of the soil layer, assuming that $v_1 = 0.7v_2$, $t_2/t_1 = 1.3$, and $s = 400$ m.

(b) The times t_1 and t_2 are measured experimentally. The equation in Exercise 67(c) shows that t_2/t_1 is a linear function of $1/s$. What might you conclude if experiments were formed for several values of s and the points $(1/s, t_2/t_1)$ did *not* lie on a straight line?

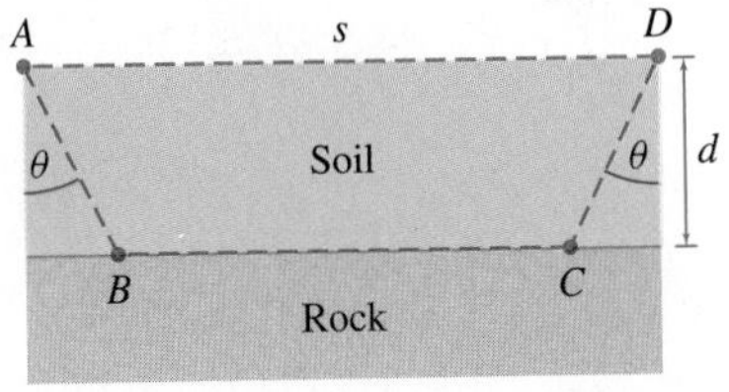

FIGURE 37

69. In this exercise we use Figure 38 to prove Heron's principle of Example 6 without calculus. By definition, C is the reflection of B across the line $\overline{MN}$ (so that $\overline{BC}$ is perpendicular to $\overline{MN}$ and $BN = CN$. Let P be the intersection of $\overline{AC}$ and $\overline{MN}$. Use geometry to justify:

(a) $\triangle PNB$ and $\triangle PNC$ are congruent and $\theta_1 = \theta_2$.

(b) The paths APB and APC have equal length.

(c) Similarly AQB and AQC have equal length.

(d) The path APC is shorter than AQC for all $Q \neq P$.

Conclude that the shortest path AQB occurs for $Q = P$.

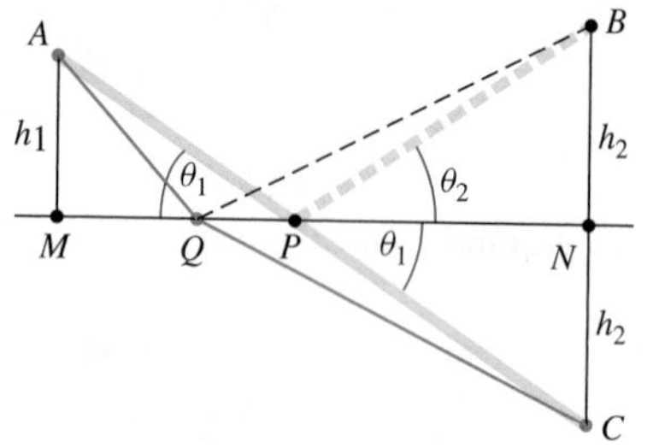

FIGURE 38

70. A jewelry designer plans to incorporate a component made of gold in the shape of a frustum of a cone of height 1 cm and fixed lower radius r (Figure 39). The upper radius x can take on any value between 0 and r. Note that $x = 0$ and $x = r$ correspond to a cone and cylinder, respectively. As a function of x, the surface area (not including the top and bottom) is $S(x) = \pi s(r + x)$, where s is the *slant height* as indicated in the figure. Which value of x yields the least expensive design [the minimum value of $S(x)$ for $0 \le x \le r$]?

(a) Show that $S(x) = \pi(r + x)\sqrt{1 + (r - x)^2}$.

(b) Show that if $r < \sqrt{2}$, then $S(x)$ is an increasing function. Conclude that the cone ($x = 0$) has minimal area in this case.

(c) Assume that $r > \sqrt{2}$. Show that $S(x)$ has two critical points $x_1 < x_2$ in $(0, r)$, and that $S(x_1)$ is a local maximum, and $S(x_2)$ is a local minimum.

(d) Conclude that the minimum occurs at $x = 0$ or x_2.

(e) Find the minimum in the cases $r = 1.5$ and $r = 2$.

(f) Challenge: Let $c = \sqrt{(5 + 3\sqrt{3})/4} \approx 1.597$. Prove that the minimum occurs at $x = 0$ (cone) if $\sqrt{2} < r < c$, but the minimum occurs at $x = x_2$ if $r > c$.

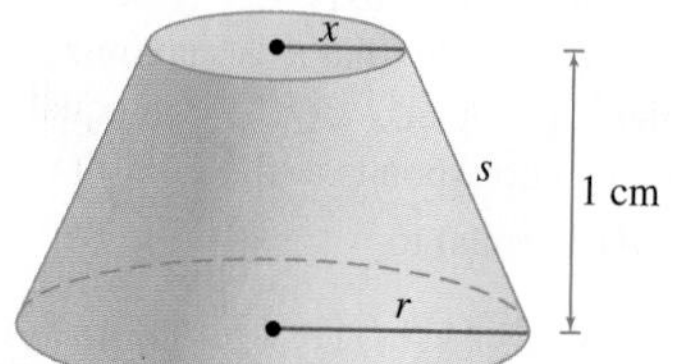

FIGURE 39 Frustum of height 1 cm.

4.7 Newton's Method

REMINDER A "zero" or "root" of a function $f(x)$ is a solution of the equation $f(x) = 0$.

Newton's Method is a procedure for finding numerical approximations to zeros of functions. Numerical approximations are important because it is often impossible to find the zeros exactly. For example, the polynomial $f(x) = x^5 - x - 1$ has one real root c (see Figure 1), but we can prove, using an advanced branch of mathematics called *Galois Theory*, that there is no algebraic formula for this root. Newton's Method shows that $c \approx 1.1673$, and with enough computation, we can compute c to any desired degree of accuracy.

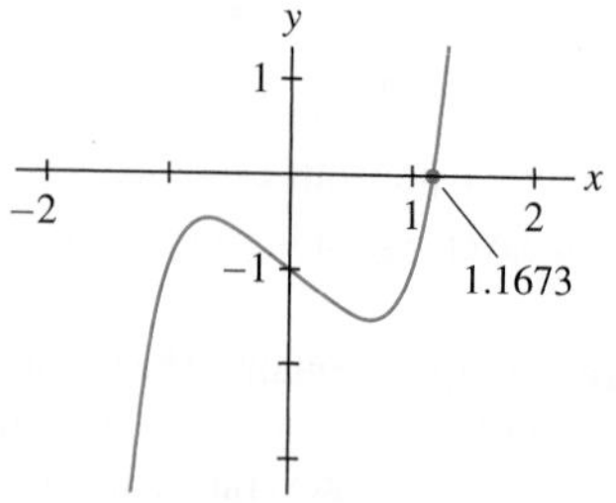

FIGURE 1 Graph of $y = x^5 - x - 1$. The value 1.1673 is a good numerical approximation to the root.

In Newton's Method, we begin by choosing a number x_0, which we believe is close to a root of the equation $f(x) = 0$. This starting value x_0 is called the **initial guess**. Newton's Method then produces a sequence $x_0, x_1, x_2, \ldots$ of successive approximations that, in favorable situations, converge to a root.

Figure 2 illustrates the procedure. Given an initial guess x_0, we draw the tangent line to the graph at $(x_0, f(x_0))$. The approximation x_1 is defined as the x-coordinate of the point where the tangent line intersects the x-axis. To produce the second approximation x_2 (also called the second iterate), we apply this procedure to x_1.

Let's derive a formula for x_1. The tangent line at $(x_0, f(x_0))$ has equation

$$y = f(x_0) + f'(x_0)(x - x_0)$$

The tangent line crosses the x-axis at x_1, where

$$y = f(x_0) + f'(x_0)(x_1 - x_0) = 0$$

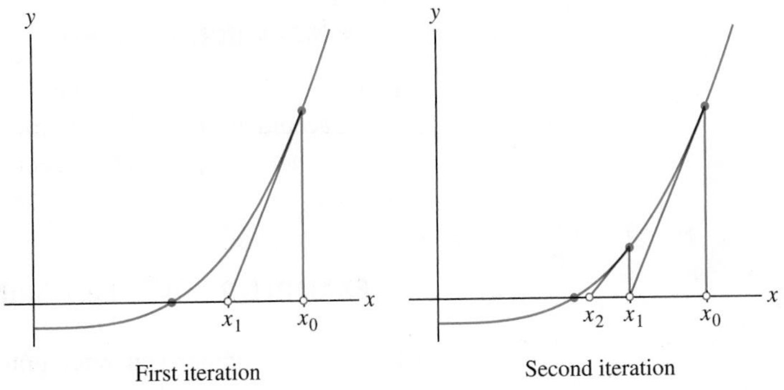

FIGURE 2 The sequence produced by iteration converges to a root.

If $f'(x_0) \neq 0$, we can solve for x_1 to obtain $x_1 - x_0 = -f(x_0)/f'(x_0)$, or

$$x_1 = x_0 - \frac{f(x_0)}{f'(x_0)}$$

The second iterate x_2 is obtained by applying this formula to x_1 instead of x_0:

$$x_2 = x_1 - \frac{f(x_1)}{f'(x_1)}$$

and so on. Notice in Figure 2 that x_1 is closer to the root than x_0 and that x_2 is closer still. This is typical: The successive approximations usually converge to the actual root. However, there are cases where Newton's Method fails (see Figure 4).

Newton's Method is an example of an iterative procedure. To "iterate" means to repeat, and in Newton's Method we use Eq. (1) repeatedly to produce the sequence of approximations.

Newton's Method To approximate a root of $f(x) = 0$:

Step 1. **Choose initial guess x_0 (close to the desired root if possible).**

Step 2. **Generate successive approximations $x_1, x_2, \dots,$ where**

$$x_{n+1} = x_n - \frac{f(x_n)}{f'(x_n)} \qquad \boxed{1}$$

■ **EXAMPLE 1** **Approximating $\sqrt{5}$** Calculate the first three approximations x_1, x_2, x_3 to a root of $f(x) = x^2 - 5$ using the initial guess $x_0 = 2$.

Solution We have $f'(x) = 2x$. Therefore,

$$x_1 = x_0 - \frac{f(x_0)}{f'(x_0)} = x_0 - \frac{x_0^2 - 5}{2x_0}$$

We compute the successive approximations as follows:

$$x_1 = x_0 - \frac{f(x_0)}{f'(x_0)} = 2 - \frac{2^2 - 5}{2 \cdot 2} = 2.25$$

$$x_2 = x_1 - \frac{f(x_1)}{f'(x_1)} = 2.25 - \frac{2.25^2 - 5}{2 \cdot 2.25} \approx 2.23611$$

$$x_3 = x_2 - \frac{f(x_2)}{f'(x_2)} = 2.23611 - \frac{2.23611^2 - 5}{2 \cdot 2.23611} \approx \mathbf{2.23606797}789$$

This sequence provides successive approximations to a root of $x^2 - 5 = 0$, namely

$$\sqrt{5} = \mathbf{2.236067977}499789696\ldots$$

Observe that x_3 is accurate to within an error of less than 10^{-9}. This is impressive accuracy for just three iterations of Newton's Method. ■

How Many Iterations Are Required?

How many iterations of Newton's Method are required to approximate a root to within a given accuracy? There is no definitive answer, but in practice, it is usually safe to assume that if x_n and x_{n+1} agree to m decimal places, then the approximation x_n is correct to these m places.

■ **EXAMPLE 2** GU Let c be the smallest positive solution of $\sin 3x = \cos x$.

(a) Use a computer-generated graph to choose an initial guess x_0 for c.

(b) Use Newton's Method to approximate c to within an error of at most 10^{-6}.

Solution

(a) A solution of $\sin 3x = \cos x$ is a zero of the function $f(x) = \sin 3x - \cos x$. Figure 3 shows that the smallest zero is approximately halfway between 0 and $\frac{\pi}{4}$. Because $\frac{\pi}{4} \approx 0.785$, a good initial guess is $x_0 = 0.4$.

(b) Since $f'(x) = 3\cos 3x + \sin x$, Eq. (1) yields the formula

$$x_{n+1} = x_n - \frac{\sin 3x_n - \cos x_n}{3\cos 3x_n + \sin x_n}$$

With $x_0 = 0.4$ as the initial guess, the first four iterates are

$$x_1 \approx \mathbf{0.392}5647447$$
$$x_2 \approx \mathbf{0.3926990}382$$
$$x_3 \approx \mathbf{0.3926990816987}196$$
$$x_4 \approx \mathbf{0.3926990816987}241$$

Stopping here, we can be fairly confident that x_4 approximates the smallest positive root c to at least twelve places. In fact, $c = \frac{\pi}{8}$ and x_4 is accurate to sixteen places. ■

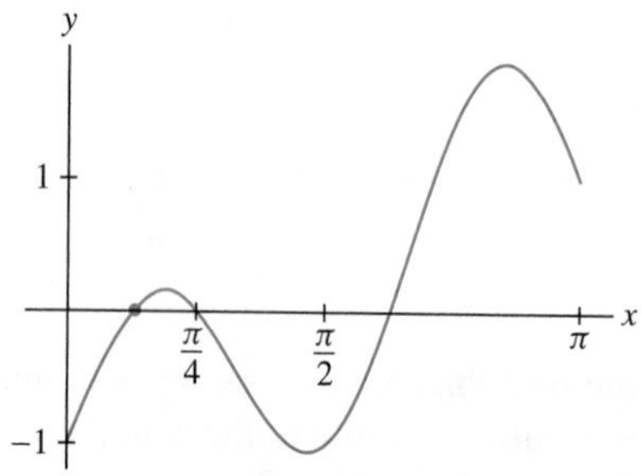

FIGURE 3 Graph of $f(x) = \sin 3x - \cos x$.

There is no single "correct" initial guess. In Example 2, we chose $x_0 = 0.4$, but another possible choice is $x_0 = 0$, leading to the sequence

$$x_1 \approx 0.3333333333$$
$$x_2 \approx 0.3864547725$$
$$x_3 \approx 0.3926082513$$
$$x_4 \approx 0.3926990816$$

You can check, however, that $x_0 = 1$ yields a sequence converging to $\frac{\pi}{4}$, which is the second positive solution of $\sin 3x = \cos x$.

Which Root Does Newton's Method Compute?

Sometimes, Newton's Method computes no root at all. In Figure 4, the iterates diverge to infinity. In practice, however, Newton's Method usually converges quickly, and if a particular choice of x_0 does not lead to a root, the best strategy is to try a different initial guess, consulting a graph if possible. If $f(x) = 0$ has more than one root, different initial guesses x_0 may lead to different roots.

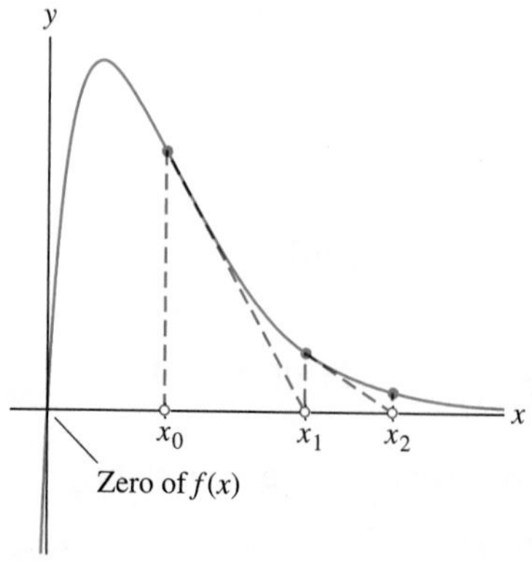

FIGURE 4 Function has only one zero but the sequence of Newton iterates goes off to infinity.

■ **EXAMPLE 3** Figure 5 shows that $f(x) = x^4 - 6x^2 + x + 5$ has four real roots.

(a) Show that with $x_0 = 0$, Newton's Method converges to the root near -2.

(b) Show that with $x_0 = -1$, Newton's Method converges to the root near -1.

Solution We have $f'(x) = 4x^3 - 12x + 1$ and

$$x_{n+1} = x_n - \frac{x_n^4 - 6x_n^2 + x_n + 5}{4x_n^3 - 12x_n + 1} = \frac{3x_n^4 - 6x_n^2 - 5}{4x_n^3 - 12x_n + 1}$$

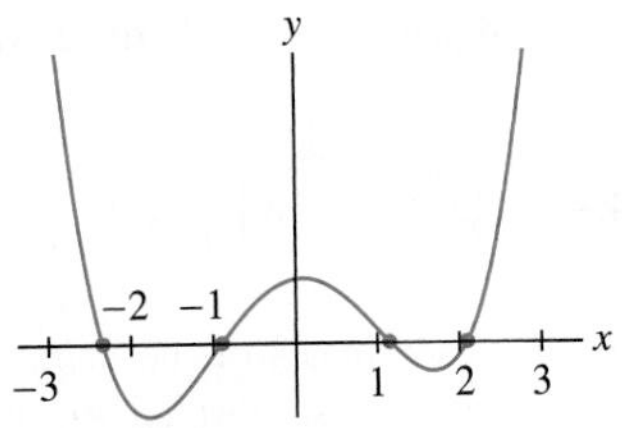

FIGURE 5 Graph of $f(x) = x^4 - 6x^2 + x + 5$.

(a) On the basis of Table 1, we can be confident that when $x_0 = 0$, Newton's Method converges to a root near -2.3. Notice in Figure 5 that this is not the closest root to x_0.

(b) Table 2 suggests that with $x_0 = -1$, Newton's Method converges to the root near -0.9. ■

TABLE 1

x_0	0
x_1	−5
x_2	−3.9179954
x_3	−3.1669480
x_4	−2.6871270
x_5	−2.4363303
x_6	−2.3572979
x_7	−**2.3**495000

TABLE 2

x_0	−1
x_1	−0.8888888888
x_2	−0.8882866140
x_3	−0.88828656234358
x_4	−0.**888286562343**575

4.7 SUMMARY

- Newton's Method: To find a sequence of numerical approximations to a root of $f(x)$, begin with an initial guess x_0. Then construct the sequence $x_0, x_1, x_2, \ldots$ using the formula

$$x_{n+1} = x_n - \frac{f(x_n)}{f'(x_n)}$$

You should choose the initial guess x_0 as close as possible to a root, possibly by referring to a graph. In favorable cases, the sequence converges rapidly to a root.

- If x_n and x_{n+1} agree to m decimal places, it is usually safe to assume that x_n agrees with a root to m decimal places.

4.7 EXERCISES

Preliminary Questions

1. How many iterations of Newton's Method are required to compute a root if $f(x)$ is a linear function?

2. What happens in Newton's Method if your initial guess happens to be a zero of f?

3. What happens in Newton's Method if your initial guess happens to be a local min or max of f?

4. Is the following a reasonable description of Newton's Method: "A root of the equation of the tangent line to $f(x)$ is used as an approximation to a root of $f(x)$ itself"? Explain.

Exercises

In this exercise set, all approximations should be carried out using Newton's Method.

In Exercises 1–6, apply Newton's Method to $f(x)$ and initial guess x_0 to calculate x_1, x_2, x_3.

1. $f(x) = x^2 - 6, \quad x_0 = 2$

2. $f(x) = x^2 - 3x + 1, \quad x_0 = 3$

3. $f(x) = x^3 - 10, \quad x_0 = 2$

4. $f(x) = x^3 + x + 1, \quad x_0 = -1$

5. $f(x) = \cos x - 4x, \quad x_0 = 1$

6. $f(x) = 1 - x\sin x, \quad x_0 = 7$

7. Use Figure 6 to choose an initial guess x_0 to the unique real root of $x^3 + 2x + 5 = 0$ and compute the first three Newton iterates.

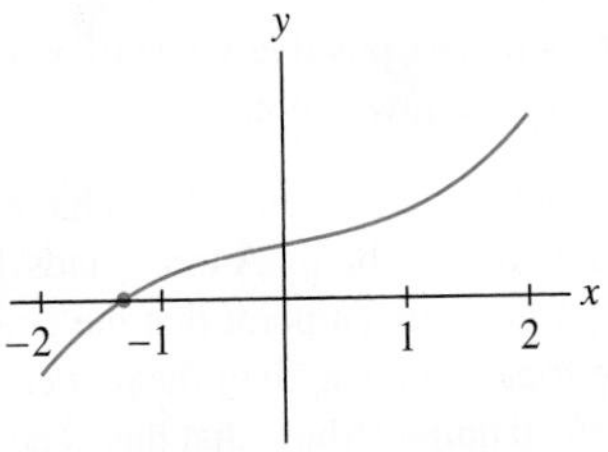

FIGURE 6 Graph of $y = x^3 + 2x + 5$.

8. Approximate a solution of $\sin x = \cos 2x$ in the interval $\left[0, \frac{\pi}{2}\right]$ to three decimal places. Then find the exact solution and compare with your approximation.

9. Approximate the point of intersection of the graphs $y = x^2 + 4 + 1/x$ and $y = 2/x^2$ to three decimal places (Figure 7).

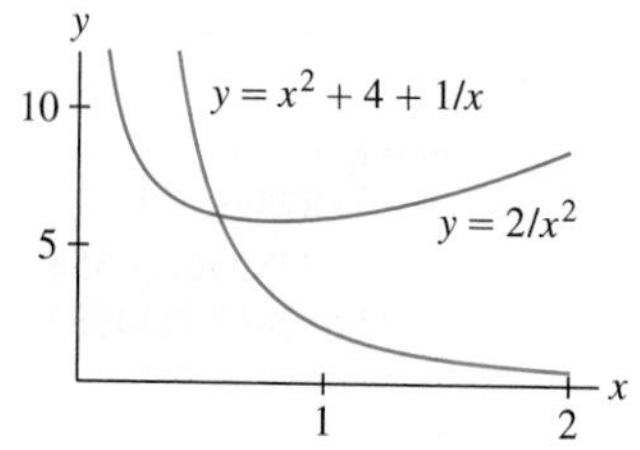

FIGURE 7

10. The first positive solution of $\sin x = 0$ is $x = \pi$. Use Newton's Method to calculate π to four decimal places.

In Exercises 11–14, approximate to three decimal places using Newton's Method and compare with the value from a calculator.

11. $\sqrt{11}$ **12.** $5^{1/3}$ **13.** $2^{7/3}$ **14.** $3^{-1/4}$

15. Approximate the largest positive root of $f(x) = x^4 - 6x^2 + x + 5$ to within an error of at most 10^{-4}. Refer to Figure 5.

GU *In Exercises 16–19, approximate the root specified to three decimal places using Newton's Method. Use a plot to choose an initial guess.*

16. Largest positive root of $f(x) = x^3 - 5x + 1$.

17. Negative root of $f(x) = x^5 - 20x + 10$.

18. Positive solution of $\sin\theta = 0.8\theta$.

19. Positive solution of $4\cos x = x^2$.

20. Let x_1, x_2 be the estimates to a root obtained by applying Newton's Method with $x_0 = 1$ to the function graphed in Figure 8. Estimate the numerical values of x_1 and x_2, and draw the tangent lines used to obtain them.

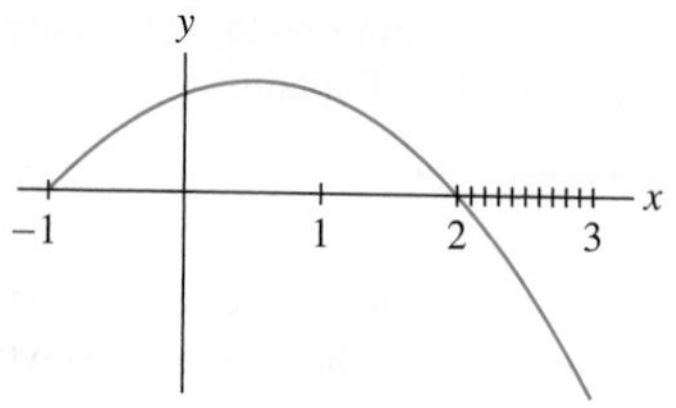

FIGURE 8

21. GU Find the smallest positive value of x at which $y = x$ and $y = \tan x$ intersect. *Hint:* Draw a plot.

22. In 1535, the mathematician Antonio Fior challenged his rival Niccolo Tartaglia to solve this problem: A tree stands 12 *braccia* high; it is broken into two parts at such a point that the height of the part left standing is the cube root of the length of the part cut away. What is the height of the part left standing? Show that this is equivalent to solving $x^3 + x = 12$ and find the height to three decimal places. Tartaglia, who had discovered the secret of the cubic equation, was able to determine the exact answer:

$$x = \left(\sqrt[3]{\sqrt{2{,}919} + 54} - \sqrt[3]{\sqrt{2{,}919} - 54}\right) \Big/ \sqrt[3]{9}$$

23. Find (to two decimal places) the coordinates of the point P in Figure 9 where the tangent line to $y = \cos x$ passes through the origin.

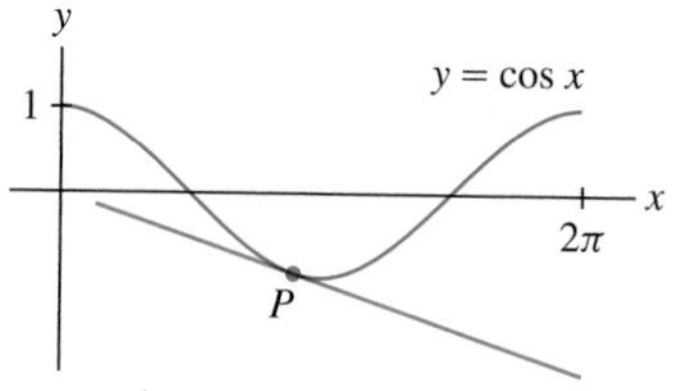

FIGURE 9

Newton's Method is often used to determine interest rates in financial calculations. In Exercises 24–26, r denotes a yearly interest rate expressed as a decimal (rather than as a percent).

24. If P dollars are deposited every month in an account earning interest at the yearly rate r, then the value S of the account after N years is

$$S = P\left(\frac{b^{12N+1} - b}{b - 1}\right) \qquad \text{where } b = 1 + \frac{r}{12}$$

You have decided to deposit $P = 100$ dollars per month.

(a) Determine S after 5 years if $r = 0.07$ (that is, 7%).

(b) Show that to save \$10,000 after 5 years, you must earn interest at a rate r determined bys the equation $b^{61} - 101b + 100 = 0$. Use Newton's Method to solve for b. Then find r. Note that $b = 1$ is a root, but you want the root satisfying $b > 1$.

25. If you borrow L dollars for N years at a yearly interest rate r, your monthly payment of P dollars is calculated using the equation

$$L = P\left(\frac{1 - b^{-12N}}{b - 1}\right) \qquad \text{where } b = 1 + \frac{r}{12}$$

(a) Find P if $L = \$5{,}000$, $N = 3$, and $r = 0.08$ (8%).

(b) You are offered a loan of $L = \$5{,}000$ to be paid back over 3 years with monthly payments of $P = \$200$. Use Newton's Method to compute b and find the implied interest rate r of this loan. *Hint:* Show that $(L/P)b^{12N+1} - (1 + L/P)b^{12N} + 1 = 0$.

26. If you deposit P dollars in a retirement fund every year for N years with the intention of then withdrawing Q dollars per year for M years, you must earn interest at a rate r satisfying $P(b^N - 1) = Q(1 - b^{-M})$, where $b = 1 + r$. Assume that \$2,000 is deposited each year for 30 years and the goal is to withdraw \$10,000 per year for 25 years. Use Newton's Method to compute b and then find r. Note that $b = 1$ is a root, but you want the root satisfying $b > 1$.

27. There is no simple formula for the position at time t of a planet P in its orbit (an ellipse) around the sun. Introduce the auxiliary circle and angle θ in Figure 10 (note that P determines θ because it is the central angle of point B on the circle). Let $a = OA$ and $e = OS/OA$ (the eccentricity of the orbit).

(a) Show that sector BSA has area $(a^2/2)(\theta - e\sin\theta)$.

(b) By Kepler's Second Law, the area of sector *BSA* is proportional to the time t elapsed since the planet passed point A, and because the circle has area πa^2, *BSA* has area $(\pi a^2)(t/T)$, where T is the period of the orbit. Deduce **Kepler's Equation**:

$$\frac{2\pi t}{T} = \theta - e\sin\theta$$

(c) The eccentricity of Mercury's orbit is approximately $e = 0.2$. Use Newton's Method to find θ after a quarter of Mercury's year has elapsed ($t = T/4$). Convert θ to degrees. Has Mercury covered more than a quarter of its orbit at $t = T/4$?

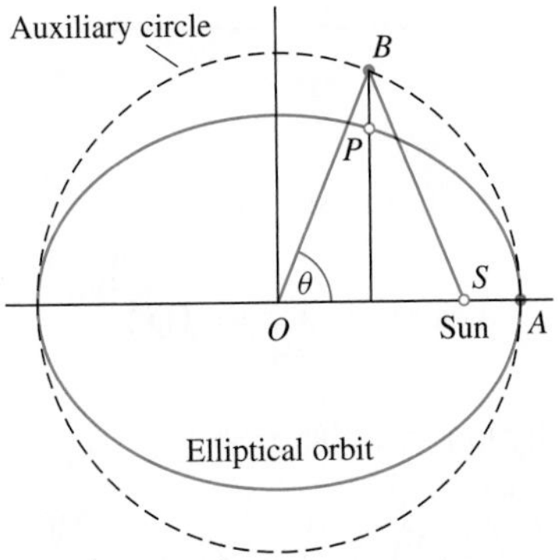

FIGURE 10

28. The roots of $f(x) = \frac{1}{3}x^3 - 4x + 1$ to three decimal places are -3.583, 0.251, and 3.332 (Figure 11). Determine the root to which Newton's Method converges for the initial choices $x_0 = 1.85$, 1.7, and 1.55. The answer shows that a small change in x_0 can have a significant effect on the outcome of Newton's Method.

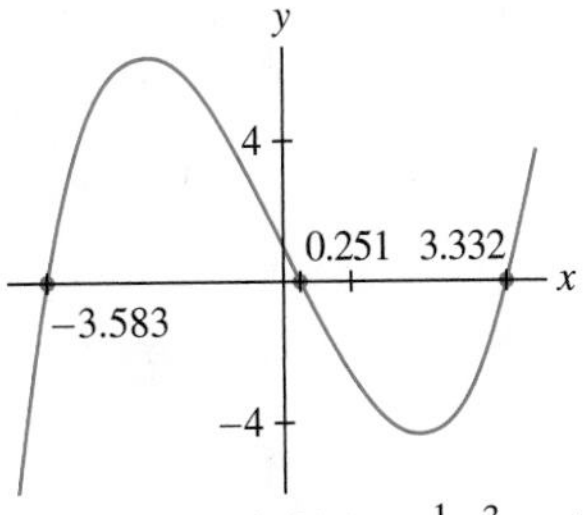

FIGURE 11 Graph of $f(x) = \frac{1}{3}x^3 - 4x + 1$.

29. What happens when you apply Newton's Method to find a zero of $f(x) = x^{1/3}$? Note that $x = 0$ is the only zero.

30. What happens when you apply Newton's Method to the equation $x^3 - 20x = 0$ with the unlucky initial guess $x_0 = 2$?

Further Insights and Challenges

31. Newton's Method can be used to compute reciprocals without performing division. Let $c > 0$ and set $f(x) = x^{-1} - c$.

(a) Show that $x - (f(x)/f'(x)) = 2x - cx^2$.

(b) Calculate the first three iterates of Newton's Method with $c = 10.3$ and the two initial guesses $x_0 = 0.1$ and $x_0 = 0.5$.

(c) Explain graphically why $x_0 = 0.5$ does not yield a sequence converging to $1/10.3$.

In Exercises 32 and 33, consider a metal rod of length L fastened at both ends. If you cut the rod and weld on an additional segment of length m, leaving the ends fixed, the rod will bow up into a circular arc of radius R (unknown), as indicated in Figure 12.

32. Let h be the maximum vertical displacement of the rod.

(a) Show that $L = 2R\sin\theta$ and conclude that

$$h = \frac{L(1-\cos\theta)}{2\sin\theta}$$

(b) Show that $L + m = 2R\theta$ and then prove

$$\frac{\sin\theta}{\theta} = \frac{L}{L+m} \qquad (2)$$

33. Let $L = 3$ and $m = 1$. Apply Newton's Method to Eq. (2) to estimate θ, and use this to estimate h.

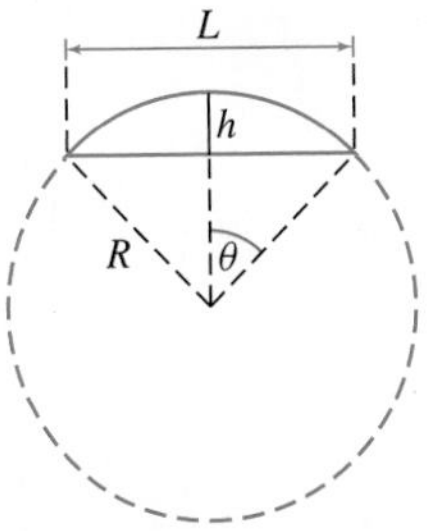

FIGURE 12 The bold circular arc has length $L + m$.

34. Quadratic Convergence to Square Roots Let $f(x) = x^2 - c$ and let $e_n = x_n - \sqrt{c}$ be the error in x_n.

(a) Show that $x_{n+1} = \frac{1}{2}(x_n + c/x_n)$ and $e_{n+1} = e_n^2/2x_n$.

(b) Show that if $x_0 > \sqrt{c}$, then $x_n > \sqrt{c}$ for all n. Explain graphically.

(c) Show that if $x_0 > \sqrt{c}$, then $e_{n+1} \le e_n^2/(2\sqrt{c})$.

4.8 Antiderivatives

In addition to finding derivatives, there is an important "inverse" problem: *Given the derivative, find the function itself.* For example, in physics we may know the velocity $v(t)$ (the derivative) and wish to compute the position $s(t)$ of an object. Since $s'(t) = v(t)$, this amounts to finding a function whose derivative is $v(t)$. A function $F(x)$ whose derivative is $f(x)$ is called an antiderivative of $f(x)$.

DEFINITION Antiderivatives A function $F(x)$ is an antiderivative of $f(x)$ on (a, b) if $F'(x) = f(x)$ for all $x \in (a, b)$.

Examples:

- $F(x) = -\cos x$ is an antiderivative of $f(x) = \sin x$ because

$$F'(x) = \frac{d}{dx}(-\cos x) = \sin x = f(x)$$

- $F(x) = \frac{1}{3}x^3$ is an antiderivative of $f(x) = x^2$ because

$$F'(x) = \frac{d}{dx}\left(\frac{1}{3}x^3\right) = x^2 = f(x)$$

One critical observation is that antiderivatives are not unique. We are free to add a constant C because the derivative of a constant is zero, and so, if $F'(x) = f(x)$, then $(F(x) + C)' = f(x)$. For example, each of the following is an antiderivative of x^2:

$$\frac{1}{3}x^3, \qquad \frac{1}{3}x^3 + 5, \qquad \frac{1}{3}x^3 - 4$$

Are there any antiderivatives of $f(x)$ other than those obtained by adding a constant to a given antiderivative $F(x)$? Our next theorem says that the answer is no if $f(x)$ is defined on an interval (a, b).

THEOREM 1 The General Antiderivative Let $F(x)$ be an antiderivative of $f(x)$ on (a, b). Then every other antiderivative on (a, b) is of the form $F(x) + C$ for some constant C.

Proof If $G(x)$ is a second antiderivative of $f(x)$, set $H(x) = G(x) - F(x)$. Then $H'(x) = G'(x) - F'(x) = f(x) - f(x) = 0$. By the Corollary in Section 4.3, $H(x)$ must be a constant—say, $H(x) = C$—and therefore $G(x) = F(x) + C$. ∎

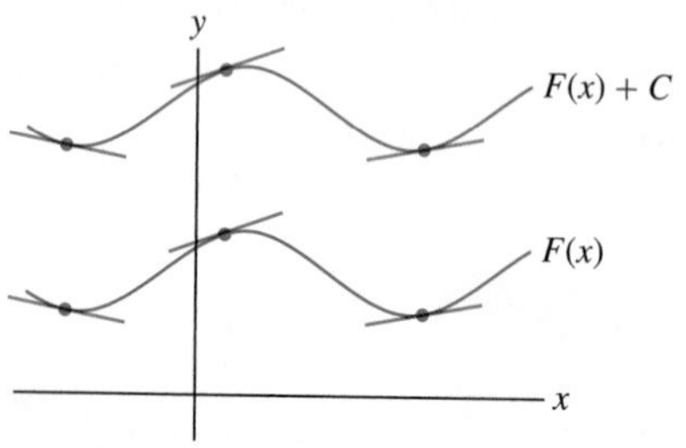

FIGURE 1 The tangent lines to the graphs of $y = F(x)$ and $y = F(x) + C$ are parallel.

GRAPHICAL INSIGHT The graph of $F(x) + C$ is obtained by shifting the graph of $F(x)$ vertically by C units. Since vertical shifting moves the tangent lines without changing their slopes, it makes sense that all of the functions $F(x) + C$ have the same derivative (Figure 1). Theorem 1 tells us that conversely, if two graphs have parallel tangent lines, then one graph is obtained from the other by a vertical shift.

We often describe the *general* antiderivative of a function in terms of an arbitrary constant C, as in the following example.

■ **EXAMPLE 1** Find two antiderivatives of $f(x) = \cos x$. Then determine the general antiderivative.

Solution The functions $F(x) = \sin x$ and $G(x) = \sin x + 2$ are both antiderivatives of $f(x)$. The general antiderivative is $F(x) = \sin x + C$, where C is any constant. ■

The process of finding an antiderivative is called **integration**. We will see why in Chapter 5, when we discuss the connection between antiderivatives and areas under curves given by the Fundamental Theorem of Calculus. Anticipating this result, we begin using the integral sign $\int$, the standard notation for antiderivatives.

The terms "antiderivative" and "indefinite integral" are used interchangeably. In some textbooks, an antiderivative is called a "primitive function."

NOTATION Indefinite Integral The notation

$$\int f(x)\,dx = F(x) + C \quad \text{means that} \quad F'(x) = f(x)$$

We say that $F(x) + C$ is the general antiderivative or **indefinite integral** of $f(x)$.

The function $f(x)$ appearing in the integral sign is called the **integrand**. The symbol dx is a *differential*. It is part of the integral notation and serves to indicate the independent variable. The constant C is called the *constant of integration*.

Some indefinite integrals can be evaluated by reversing the familiar derivative formulas. For example, we obtain the indefinite integral of x^n by reversing the Power Rule for derivatives.

There are no Product, Quotient, or Chain Rules for integrals. However, we will see that the Product Rule for derivatives leads to an important technique called Integration by Parts (Section 8.1) and the Chain Rule leads to the Substitution Method (Section 5.6).

THEOREM 2 Power Rule for Integrals

$$\int x^n\,dx = \frac{x^{n+1}}{n+1} + C \qquad \text{for } n \neq -1$$

Proof We just need to verify that $F(x) = \dfrac{x^{n+1}}{n+1}$ is an antiderivative of $f(x) = x^n$:

$$F'(x) = \frac{d}{dx}\left(\frac{x^{n+1}}{n+1} + C\right) = \frac{1}{n+1}((n+1)x^n) = x^n$$ ■

In words, the Power Rule for Integrals says that to integrate a power of x, "add one to the exponent and then divide by the new exponent." Here are some examples:

Notice that in integral notation, we treat dx as a movable variable, and thus we write $\int \frac{1}{x^9}dx$ as $\int \frac{dx}{x^9}$.

$$\int x^5\,dx = \frac{1}{6}x^6 + C, \qquad \int \frac{dx}{x^9} = -\frac{1}{8}x^{-8} + C, \qquad \int x^{3/5}\,dx = \frac{5}{8}x^{8/5} + C$$

The Power Rule is not valid for $n = -1$. In fact, for $n = -1$, we obtain the meaningless result

$$\int x^{-1}\,dx = \frac{x^{n+1}}{n+1} + C = \frac{x^0}{0} + C \qquad \text{(meaningless)}$$

It turns out that the antiderivative of $f(x) = x^{-1}$ is the natural logarithm. We will prove this in Section 7.3.

The indefinite integral obeys the usual linearity rules that allow us to integrate "term by term." These rules follow from the linearity rules for the derivative (see Exercise 79.)

THEOREM 3 Linearity of the Indefinite Integral

- **Sum Rule:** $\displaystyle\int (f(x)+g(x))\,dx = \int f(x)\,dx + \int g(x)\,dx$
- **Multiples Rule:** $\displaystyle\int cf(x)\,dx = c\int f(x)\,dx$

■ **EXAMPLE 2** Evaluate $\int (3x^4 - 5x^{2/3} + x^{-3})\,dx$.

Solution We integrate term by term and use the Power Rule:

$$\begin{aligned}\int (3x^4 - 5x^{2/3} + x^{-3})\,dx &= \int 3x^4\,dx - \int 5x^{2/3}\,dx + \int x^{-3}\,dx && \text{(Sum Rule)}\\ &= 3\int x^4\,dx - 5\int x^{2/3}\,dx + \int x^{-3}\,dx && \text{(Multiples Rule)}\\ &= 3\left(\frac{x^5}{5}\right) - 5\left(\frac{x^{5/3}}{5/3}\right) + \frac{x^{-2}}{-2} + C && \text{(Power Rule)}\\ &= \frac{3}{5}x^5 - 3x^{5/3} - \frac{1}{2}x^{-2} + C\end{aligned}$$

When we break up an indefinite integral into a sum of several integrals as in Example 2, it is not necessary to include a separate constant of integration for each integral.

To check the answer, we verify that the derivative is equal to the integrand:

$$\frac{d}{dx}\left(\frac{3}{5}x^5 - 3x^{5/3} - \frac{1}{2}x^{-2} + C\right) = 3x^4 - 5x^{2/3} + x^{-3}$$ ■

■ **EXAMPLE 3** Evaluate $\displaystyle\int \left(\frac{5}{x^2} - 3x^{-10}\right) dx$

Solution

$$\begin{aligned}\int \left(\frac{5}{x^2} - 3x^{-10}\right) dx &= 5\int \frac{dx}{x^2} - 3\int x^{-10}\,dx\\ &= 5\left(-x^{-1}\right) - 3\left(\frac{x^{-9}}{-9}\right) + C = -5x^{-1} + \frac{1}{3}x^{-9}\end{aligned}$$ ■

The differentiation formulas for the trigonometric functions give us the following integration formulas. Each formula can be checked by differentiation.

Basic Trigonometric Integrals

$$\int \sin x\,dx = -\cos x + C \qquad \int \cos x\,dx = \sin x + C$$

$$\int \sec^2 x\,dx = \tan x + C \qquad \int \csc^2 x\,dx = -\cot x + C$$

$$\int \sec x \tan x\,dx = \sec x + C \qquad \int \csc x \cot x\,dx = -\csc x + C$$

Similarly, for any constants b and k with $k \neq 0$, the formulas

$$\frac{d}{dx}\sin(kx+b) = k\cos(kx+b), \qquad \frac{d}{dx}\cos(kx+b) = -k\sin(kx+b)$$

translate to the following indefinite integral formulas:

$$\int \cos(kx+b)\,dx = \frac{1}{k}\sin(kx+b) + C$$

$$\int \sin(kx+b)\,dx = -\frac{1}{k}\cos(kx+b) + C$$

■ **EXAMPLE 4** Evaluate $\int \big(\sin(8t-3) + 20\cos 9t\big)\,dt$.

Solution

$$\int \big(\sin(8t-3) + 20\cos 9t\big)\,dt = \int \sin(8t-3)\,dt + 20\int \cos 9t\,dt$$

$$= -\frac{1}{8}\cos(8t-3) + \frac{20}{9}\sin 9t + C$$ ■

Initial Conditions

We can think of an antiderivative as a solution to the **differential equation**

$$\frac{dy}{dx} = f(x) \qquad \boxed{1}$$

In general, a differential equation is an equation relating an unknown function and its derivatives. The unknown in Eq. (1) is a function $y = F(x)$ whose derivative is $f(x)$; that is, $F(x)$ is an antiderivative of $f(x)$.

An initial condition is like the y-intercept of a line, which determines one particular line among all lines with the same slope. The graphs of the antiderivatives of $f(x)$ are all parallel (Figure 1), and the initial condition determines one of them.

Eq. (1) has infinitely many solutions (because the antiderivative is not unique), but we can specify a particular solution by imposing an **initial condition**—that is, by requiring that the solution satisfy $y(x_0) = y_0$ for some fixed values x_0 and y_0. A differential equation with an initial condition is called an **initial value problem**.

■ **EXAMPLE 5** Solve $\frac{dy}{dx} = 4x^7$ subject to the initial condition $y(0) = 4$.

Solution First, find the general antiderivative:

$$y(x) = \int 4x^7\,dx = \frac{1}{2}x^8 + C$$

Then choose C so that the initial condition is satisfied: $y(0) = 0 + C = 4$. This yields $C = 4$, and our solution is $y = \frac{1}{2}x^8 + 4$. ■

■ **EXAMPLE 6** Solve the initial value problem $\frac{dy}{dt} = \sin(\pi t)$, $y(2) = 2$.

Solution First find the general antiderivative:

$$y(t) = \int \sin(\pi t)\,dt = -\frac{1}{\pi}\cos(\pi t) + C$$

Then solve for C by evaluating at $t = 2$:

$$y(2) = -\frac{1}{\pi}\cos(2\pi) + C = 2 \quad \Rightarrow \quad C = 2 + \frac{1}{\pi}$$

The solution of the initial value problem is $y(t) = -\frac{1}{\pi}\cos(\pi t) + 2 + \frac{1}{\pi}$. ■

■ **EXAMPLE 7** A car traveling with velocity 24 m/s begins to slow down at time $t = 0$ with a constant deceleration of $a = -6$ m/s^2. Find **(a)** the velocity $v(t)$ at time t, and **(b)** the distance traveled before the car comes to a halt.

Solution **(a)** The derivative of velocity is acceleration, so *velocity is the antiderivative of acceleration*:

$$v(t) = \int a\,dt = \int (-6)\,dt = -6t + C$$

Relation between position, velocity, and acceleration:

$$s'(t) = v(t), \qquad s(t) = \int v(t)\,dt$$

$$v'(t) = a(t), \qquad v(t) = \int a(t)\,dt$$

The initial condition $v(0) = C = 24$ gives us $v(t) = -6t + 24$.

(b) Position is the antiderivative of velocity, so the car's position is

$$s(t) = \int v(t)\,dt = \int (-6t + 24)\,dt = -3t^2 + 24t + C_1$$

where C_1 is a constant. We are not told where the car is at $t = 0$, so let us set $s(0) = 0$ for convenience, getting $c_1 = 0$. With this choice, $s(t) = -3t^2 + 24t$. This is the distance traveled from time $t = 0$.

The car comes to a halt when its velocity is zero, so we solve

$$v(t) = -6t + 24 = 0 \quad \Rightarrow \quad t = 4 \text{ s}$$

The distance traveled before coming to a halt is $s(4) = -3(4^2) + 24(4) = 48$ m. ■

4.8 SUMMARY

- $F(x)$ is called an *antiderivative* of $f(x)$ if $F'(x) = f(x)$.
- Any two antiderivatives of $f(x)$ on an interval (a, b) differ by a constant.
- The general antiderivative is denoted by the indefinite integral

$$\int f(x)\,dx = F(x) + C$$

- Integration formulas:

$$\int x^n\,dx = \frac{x^{n+1}}{n+1} + C \qquad (n \neq -1)$$

$$\int \sin(kx + b)\,dx = -\frac{1}{k}\cos(kx + b) + C \qquad (k \neq 0)$$

$$\int \cos(kx + b)\,dx = \frac{1}{k}\sin(kx + b) + C \qquad (k \neq 0)$$

- To solve an initial value problem $\frac{dy}{dx} = f(x)$, $y(x_0) = y_0$, first find the general antiderivative $y = F(x) + C$. Then determine C using the initial condition $F(x_0) + C = y_0$.

4.8 EXERCISES

Preliminary Questions

1. Find an antiderivative of the function $f(x) = 0$.

2. Is there a difference between finding the general antiderivative of a function $f(x)$ and evaluating $\int f(x)\,dx$?

3. Jacques was told that $f(x)$ and $g(x)$ have the same derivative, and he wonders whether $f(x) = g(x)$. Does Jacques have sufficient information to answer his question?

4. Suppose that $F'(x) = f(x)$ and $G'(x) = g(x)$. Which of the following statements are true? Explain.

(a) If $f = g$, then $F = G$.
(b) If F and G differ by a constant, then $f = g$.
(c) If f and g differ by a constant, then $F = G$.

5. Is $y = x$ a solution of the following Initial Value Problem?

$$\frac{dy}{dx} = 1, \qquad y(0) = 1$$

Exercises

In Exercises 1–8, find the general antiderivative of $f(x)$ and check your answer by differentiating.

1. $f(x) = 18x^2$
2. $f(x) = x^{-3/5}$
3. $f(x) = 2x^4 - 24x^2 + 12x^{-1}$
4. $f(x) = 9x + 15x^{-2}$
5. $f(x) = 2\cos x - 9\sin x$
6. $f(x) = 4x^7 - 3\cos x$
7. $f(x) = \sin 2x + 12\cos 3x$
8. $f(x) = \sin(4 - 9x)$

9. Match functions (a)–(d) with their antiderivatives (i)–(iv).

(a) $f(x) = \sin x$	(i) $F(x) = \cos(1 - x)$
(b) $f(x) = x\sin(x^2)$	(ii) $F(x) = -\cos x$
(c) $f(x) = \sin(1 - x)$	(iii) $F(x) = -\frac{1}{2}\cos(x^2)$
(d) $f(x) = x\sin x$	(iv) $F(x) = \sin x - x\cos x$

In Exercises 10–39, evaluate the indefinite integral.

10. $\int (9x + 2)\,dx$
11. $\int (4 - 18x)\,dx$
12. $\int x^{-3}\,dx$
13. $\int t^{-6/11}\,dt$
14. $\int (5t^3 - t^{-3})\,dt$
15. $\int (18t^5 - 10t^4 - 28t)\,dt$
16. $\int 14s^{9/5}\,ds$
17. $\int (z^{-4/5} - z^{2/3} + z^{5/4})\,dz$
18. $\int \frac{3}{2}\,dx$
19. $\int \frac{1}{\sqrt[3]{x}}\,dx$
20. $\int \frac{dx}{x^{4/3}}$
21. $\int \frac{36\,dt}{t^3}$
22. $\int x(x^2 - 4)\,dx$
23. $\int (t^{1/2} + 1)(t + 1)\,dt$
24. $\int \frac{12 - z}{\sqrt{z}}\,dz$
25. $\int \frac{x^3 + 3x^2 - 4}{x^2}\,dx$
26. $\int \left(\frac{1}{3}\sin x - \frac{1}{4}\cos x\right)dx$
27. $\int 12\sec x\tan x\,dx$
28. $\int (\theta + \sec^2\theta)\,d\theta$
29. $\int (\csc t\cot t)\,dt$
30. $\int \sin(7x - 5)\,dx$
31. $\int \sec^2(7 - 3\theta)\,d\theta$
32. $\int (\theta - \cos(1 - \theta))\,d\theta$
33. $\int 25\sec^2(3z + 1)\,dz$
34. $\int (12\cos 4x + 9\sin 3x)\,dx$
35. $\int \sec 12t\tan 12t\,dt$
36. $\int 5\tan(4\theta + 3)\,d\theta$
37. $\int \sec 4x\,(3\sec 4x - 5\tan 4x)\,dx$
38. $\int \sec(x + 5)\tan(x + 5)\,dx$
39. $\int \left(\cos(3\theta) - \frac{1}{2}\sec^2\left(\frac{\theta}{4}\right)\right)d\theta$

40. In Figure 2, is graph (A) or graph (B) the graph of an antiderivative of $f(x)$?

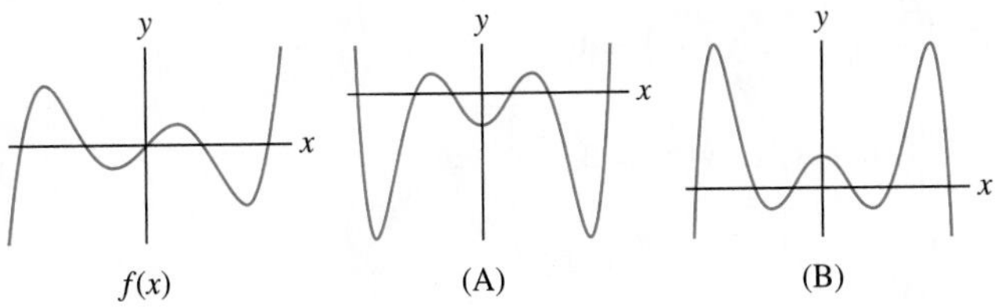

FIGURE 2

41. In Figure 3, which of graphs (A), (B), and (C) is *not* the graph of an antiderivative of $f(x)$? Explain.

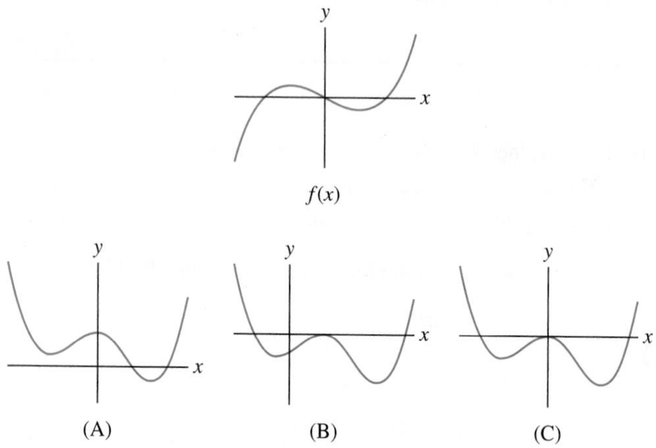

FIGURE 3

42. Show that $F(x) = \frac{1}{3}(x+13)^3$ is an antiderivative of $f(x) = (x+13)^2$.

In Exercises 43–46, verify by differentiation.

43. $\int (x+13)^6\,dx = \frac{1}{7}(x+13)^7 + C$

44. $\int (x+13)^{-5}\,dx = -\frac{1}{4}(x+13)^{-4} + C$

45. $\int (4x+13)^2\,dx = \frac{1}{12}(4x+13)^3 + C$

46. $\int (ax+b)^n\,dx = \frac{1}{a(n+1)}(ax+b)^{n+1} + C$

In Exercises 47–62, solve the initial value problem.

47. $\frac{dy}{dx} = x^3,\ y(0) = 4$

48. $\frac{dy}{dt} = 3 - 2t,\ y(0) = -5$

49. $\frac{dy}{dt} = 2t + 9t^2,\ y(1) = 2$

50. $\frac{dy}{dx} = 8x^3 + 3x^2,\ y(2) = 0$

51. $\frac{dy}{dt} = \sqrt{t},\ y(1) = 1$

52. $\frac{dz}{dt} = t^{-3/2},\ z(4) = -1$

53. $\frac{dy}{dx} = (3x+2)^3,\ y(0) = 1$

54. $\frac{dy}{dt} = (4t+3)^{-2},\ y(1) = 0$

55. $\frac{dy}{dx} = \sin x,\ y\left(\frac{\pi}{2}\right) = 1$

56. $\frac{dy}{dz} = \sin 2z,\ y\left(\frac{\pi}{4}\right) = 4$

57. $\frac{dy}{dx} = \cos 5x,\ y(\pi) = 3$

58. $\frac{dy}{dx} = \sec^2 3x,\ y\left(\frac{\pi}{4}\right) = 2$

59. $\frac{dy}{d\theta} = 6\sec 3\theta \tan 3\theta,\quad y\left(\frac{\pi}{12}\right) = -4$

60. $\frac{dy}{dt} = 4t - \sin 2t,\quad y(0) = 2$

61. $\frac{dy}{d\theta} = \cos\left(3\pi - \frac{1}{2}\theta\right),\quad y(3\pi) = 8$

62. $\frac{dy}{dx} = \frac{1}{x^2} - \csc^2 x,\quad y\left(\frac{\pi}{2}\right) = 0$

In Exercises 63–69, first find f' and then find f.

63. $f''(x) = 12x,\quad f'(0) = 1,\quad f(0) = 2$

64. $f''(x) = x^3 - 2x,\quad f'(1) = 0,\quad f(1) = 2$

65. $f''(x) = x^3 - 2x + 1,\quad f'(0) = 1,\quad f(0) = 0$

66. $f''(x) = x^3 - 2x + 1,\quad f'(1) = 0,\quad f(1) = 4$

67. $f''(t) = t^{-3/2},\quad f'(4) = 1,\quad f(4) = 4$

68. $f''(\theta) = \cos\theta,\quad f'\left(\frac{\pi}{2}\right) = 1,\quad f\left(\frac{\pi}{2}\right) = 6$

69. $f''(t) = t - \cos t,\quad f'(0) = 2,\quad f(0) = -2$

70. Show that $F(x) = \tan^2 x$ and $G(x) = \sec^2 x$ have the same derivative. What can you conclude about the relation between F and G? Verify this conclusion directly.

71. A particle located at the origin at $t = 1$ s moves along the x-axis with velocity $v(t) = (6t^2 - t)$ m/s. State the differential equation with initial condition satisfied by the position $s(t)$ of the particle, and find $s(t)$.

72. A particle moves along the x-axis with velocity $v(t) = (6t^2 - t)$ m/s. Find the particle's position $s(t)$ assuming that $s(2) = 4$.

73. A mass oscillates at the end of a spring. Let $s(t)$ be the displacement of the mass from the equilibrium position at time t. Assuming that the mass is located at the origin at $t = 0$ and has velocity $v(t) = \sin(\pi t/2)$ m/s, state the differential equation with initial condition satisfied by $s(t)$, and find $s(t)$.

74. Beginning at $t = 0$ with initial velocity 4 m/s, a particle moves in a straight line with acceleration $a(t) = 3t^{1/2}$ m/s^2. Find the distance traveled after 25 seconds.

75. A car traveling 25 m/s begins to decelerate at a constant rate of 4 m/s^2. After how many seconds does the car come to a stop and how far will the car have traveled before stopping?

76. At time $t = 1$ s, a particle is traveling at 72 m/s and begins to decelerate at the rate $a(t) = -t^{-1/2}$ until it stops. How far does the particle travel before stopping?

77. A 900-kg rocket is released from a space station. As it burns fuel, the rocket's mass decreases and its velocity increases. Let $v(m)$ be the velocity (in meters per second) as a function of mass m. Find the velocity when $m = 729$ if $dv/dm = -50m^{-1/2}$. Assume that $v(900) = 0$.

78. As water flows through a tube of radius $R = 10$ cm, the velocity v of an individual water particle depends only on its distance r from the center of the tube. The particles at the walls of the tube have zero velocity and $dv/dr = -0.06r$. Determine $v(r)$.

79. Verify the linearity properties of the indefinite integral stated in Theorem 3.

Further Insights and Challenges

80. Find constants c_1 and c_2 such that $F(x) = c_1 x \sin x + c_2 \cos x$ is an antiderivative of $f(x) = x \cos x$.

81. Find constants c_1 and c_2 such that $F(x) = c_1 x \cos x + c_2 \sin x$ is an antiderivative of $f(x) = x \sin x$.

82. Suppose that $F'(x) = f(x)$ and $G'(x) = g(x)$. Is it true that $F(x)G(x)$ is an antiderivative of $f(x)g(x)$? Confirm or provide a counterexample.

83. Suppose that $F'(x) = f(x)$.

(a) Show that $\frac{1}{2}F(2x)$ is an antiderivative of $f(2x)$.

(b) Find the general antiderivative of $f(kx)$ for $k \neq 0$.

84. Find an antiderivative for $f(x) = |x|$.

85. Using Theorem 1, prove that $F'(x) = f(x)$ where $f(x)$ is a polynomial of degree $n-1$, then $F(x)$ is a polynomial of degree n. Then prove that if $g(x)$ is any function such that $g^{(n)}(x) = 0$, then $g(x)$ is a polynomial of degree at most n.

86. The Power Rule for antiderivatives does not apply to $f(x) = x^{-1}$. Which of the graphs in Figure 4 could plausibly represent an antiderivative of $f(x) = x^{-1}$?

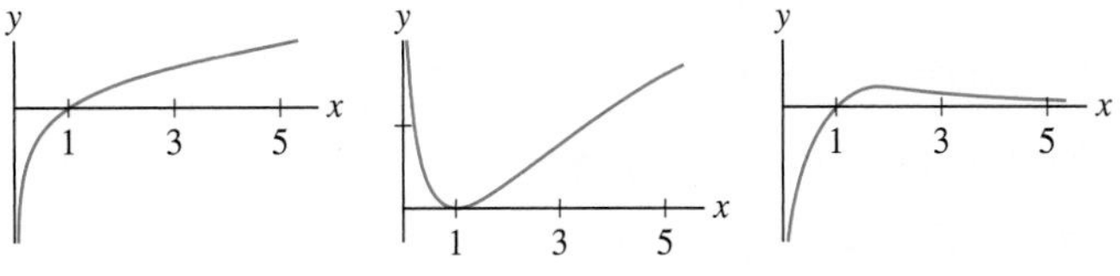

FIGURE 4

CHAPTER REVIEW EXERCISES

In Exercises 1–6, estimate using the Linear Approximation or linearization, and use a calculator to estimate the error.

1. $8.1^{1/3} - 2$

2. $\dfrac{1}{\sqrt{4.1}} - \dfrac{1}{2}$

3. $625^{1/4} - 624^{1/4}$

4. $\sqrt{101}$

5. $\dfrac{1}{1.02}$

6. $\sqrt[5]{33}$

In Exercises 7–12, find the linearization at the point indicated.

7. $y = \sqrt{x}, \quad a = 25$

8. $v(t) = 32t - 4t^2, \quad a = 2$

9. $A(r) = \frac{4}{3}\pi r^3, \quad a = 3$

10. $V(h) = 4h(2-h)(4-2h), \quad a = 1$

11. $P(\theta) = \sin(3\theta + \pi), \quad a = \dfrac{\pi}{3}$

12. $R(t) = \tan\left(\pi\left(t - \dfrac{1}{2}\right)\right), \quad a = \dfrac{1}{4}$

In Exercises 13–18, use the Linear Approximation.

13. The position of an object in linear motion at time t is $s(t) = 0.4t^2 + (t+1)^{-1}$. Estimate the distance traveled over the time interval $[4, 4.2]$.

14. A bond that pays $10,000 in 6 years is offered for sale at a price P. The percentage yield Y of the bond is

$$Y = 100\left(\left(\frac{10{,}000}{P}\right)^{1/6} - 1\right)$$

Verify that if $P = \$7{,}500$, then $Y = 4.91\%$. Estimate the drop in yield if the price rises to $7,700.

15. When a bus pass from Albuquerque to Los Alamos is priced at p dollars, a bus company takes in a monthly revenue of $R(p) = 1.5p - 0.01p^2$ (in thousands of dollars).

(a) Estimate ΔR if the price rises from $50 to $53.

(b) If $p = 80$, how will revenue be affected by a small increase in price? Explain using the Linear Approximation.

16. A store sells 80 MP4 players per week when the players are priced at $P = \$75$. Estimate the number N sold if P is raised to $80, assuming that $dN/dP = -4$. Estimate N if the price is lowered to $69.

17. The circumference of a sphere is measured at $C = 100$ cm. Estimate the maximum percentage error in V if the error in C is at most 3 cm.

18. Show that $\sqrt{a^2 + b} \approx a + \frac{b}{2a}$ if b is small. Use this to estimate $\sqrt{26}$ and find the error using a calculator.

19. Verify the MVT for $f(x) = x^{-1/3}$ on $[1, 8]$.

20. Show that $f(x) = 2x^3 + 2x + \sin x + 1$ has precisely one real root.

21. Verify the MVT for $f(x) = x + \frac{1}{x}$ on $[2, 5]$.

22. Suppose that $f(1) = 5$ and $f'(x) \geq 2$ for $x \geq 1$. Use the MVT to show that $f(8) \geq 19$.

23. Use the MVT to prove that if $f'(x) \leq 2$ for $x > 0$ and $f(0) = 4$, then $f(x) \leq 2x + 4$ for all $x \geq 0$.

24. A function $f(x)$ has derivative $f'(x) = \dfrac{1}{x^4 + 1}$. Where on the interval $[1, 4]$ does $f(x)$ take on its maximum value?

In Exercises 25–30, find the critical points and determine whether they are minima, maxima, or neither.

25. $f(x) = x^3 - 4x^2 + 4x$

26. $s(t) = t^4 - 8t^2$

27. $f(x) = x^2(x+2)^3$

28. $f(x) = x^{2/3}(1-x)$

29. $g(\theta) = \sin^2\theta + \theta$

30. $h(\theta) = 2\cos 2\theta + \cos 4\theta$

In Exercises 31–38, find the extreme values on the interval.

31. $f(x) = x(10 - x)$, $[-1, 3]$

32. $f(x) = 6x^4 - 4x^6$, $[-2, 2]$

33. $g(\theta) = \sin^2\theta - \cos\theta$, $[0, 2\pi]$

34. $R(t) = \dfrac{t}{t^2 + t + 1}$, $[0, 3]$

35. $f(x) = x^{2/3} - 2x^{1/3}$, $[-1, 3]$

36. $f(x) = x - \tan x$, $\left[-\frac{\pi}{4}, \frac{\pi}{4}\right]$

37. $f(x) = x - x^{3/2}$, $[0, 2]$

38. $f(x) = \sec x - \cos x$, $\left[-\frac{\pi}{4}, \frac{\pi}{4}\right]$

39. Find the critical points and extreme values of $f(x) = |x - 1| + |2x - 6|$ in $[0, 8]$.

40. Match the description of $f(x)$ with the graph of its *derivative* $f'(x)$ in Figure 1.

(a) $f(x)$ is increasing and concave up.

(b) $f(x)$ is decreasing and concave up.

(c) $f(x)$ is increasing and concave down.

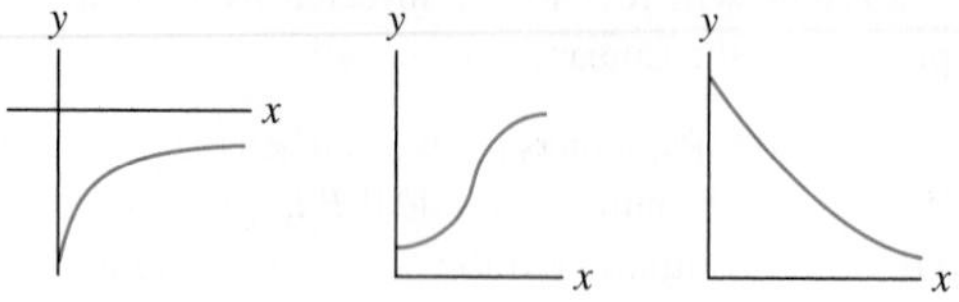

(i) (ii) (iii)

FIGURE 1 Graphs of the derivative.

In Exercises 41–46, find the points of inflection.

41. $y = x^3 - 4x^2 + 4x$

42. $y = x - 2\cos x$

43. $y = \dfrac{x^2}{x^2 + 4}$

44. $y = \dfrac{x}{(x^2 - 4)^{1/3}}$

45. $f(x) = \dfrac{x^3 - x}{x^2 + 1}$

46. $f(x) = \sin 2x - 4\cos x$

In Exercises 47–56, sketch the graph, noting the transition points and asymptotic behavior.

47. $y = 12x - 3x^2$

48. $y = 8x^2 - x^4$

49. $y = x^3 - 2x^2 + 3$

50. $y = 4x - x^{3/2}$

51. $y = \dfrac{x}{x^3 + 1}$

52. $y = \dfrac{x}{(x^2 - 4)^{2/3}}$

53. $y = \dfrac{1}{|x + 2| + 1}$

54. $y = \sqrt{2 - x^3}$

55. $y = \sqrt{3}\sin x - \cos x$ on $[0, 2\pi]$

56. $y = 2x - \tan x$ on $[0, 2\pi]$

57. Draw a curve $y = f(x)$ for which f' and f'' have signs as indicated in Figure 2.

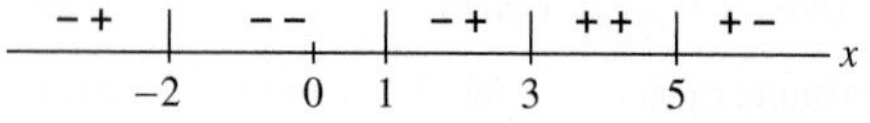

FIGURE 2

58. Find the dimensions of a cylindrical can with a bottom but no top of volume 4 m^3 that uses the least amount of metal.

59. A rectangular box of height h with square base of side b has volume $V = 4$ m^3. Two of the side faces are made of material costing $40/$m^2$. The remaining sides cost $20/$m^2$. Which values of b and h minimize the cost of the box?

60. The corn yield on a certain farm is

$$Y = -0.118x^2 + 8.5x + 12.9 \quad \text{(bushels per acre)}$$

where x is the number of corn plants per acre (in thousands). Assume that corn seed costs $1.25 (per thousand seeds) and that corn can be sold for $1.50/bushel. Let $P(x)$ be the profit (revenue minus the cost of seeds) at planting level x.

(a) Compute $P(x_0)$ for the value x_0 that maximizes yield Y.

(b) Find the maximum value of $P(x)$. Does maximum yield lead to maximum profit?

61. A quantity $N(t)$ satisfies $dN/dt = 2/t - 8/t^2$ for $t \geq 4$ (t in days). At which time is N increasing most rapidly?

62. A truck gets 10 miles per gallon of diesel fuel traveling along an interstate highway at 50 mph. This mileage decreases by 0.15 mpg for each mile per hour increase above 50 mph.

(a) If the truck driver is paid $30/hour and diesel fuel costs $P = \$3$/gal, which speed v between 50 and 70 mph will minimize the cost of a trip along the highway? Notice that the actual cost depends on the length of the trip, but the optimal speed does not.

(b) GU Plot cost as a function of v (choose the length arbitrarily) and verify your answer to part (a).

(c) GU Do you expect the optimal speed v to increase or decrease if fuel costs go down to $P = \$2$/gal? Plot the graphs of cost as a function of v for $P = 2$ and $P = 3$ on the same axis and verify your conclusion.

63. Find the maximum volume of a right-circular cone placed upside-down in a right-circular cone of radius $R = 3$ and height $H = 4$ as in Figure 3. A cone of radius r and height h has volume $\frac{1}{3}\pi r^2 h$.

64. Redo Exercise 63 for arbitrary R and H.

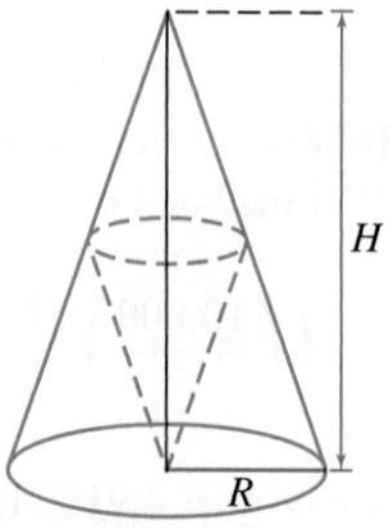

FIGURE 3

65. Show that the maximum area of a parallelogram $ADEF$ that is inscribed in a triangle ABC, as in Figure 4, is equal to one-half the area of $\triangle ABC$.

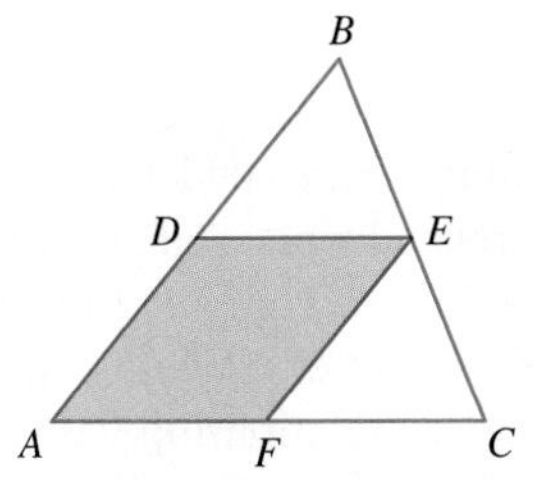

FIGURE 4

66. A box of volume 8 m^3 with a square top and bottom is constructed out of two types of metal. The metal for the top and bottom costs \$50/m^2 and the metal for the sides costs \$30/m^2. Find the dimensions of the box that minimize total cost.

67. Let $f(x)$ be a function whose graph does not pass through the x-axis and let $Q = (a, 0)$. Let $P = (x_0, f(x_0))$ be the point on the graph closest to Q (Figure 5). Prove that $\overline{PQ}$ is perpendicular to the tangent line to the graph of x_0. *Hint:* Find the minimum value of the *square* of the distance from $(x, f(x))$ to $(a, 0)$.

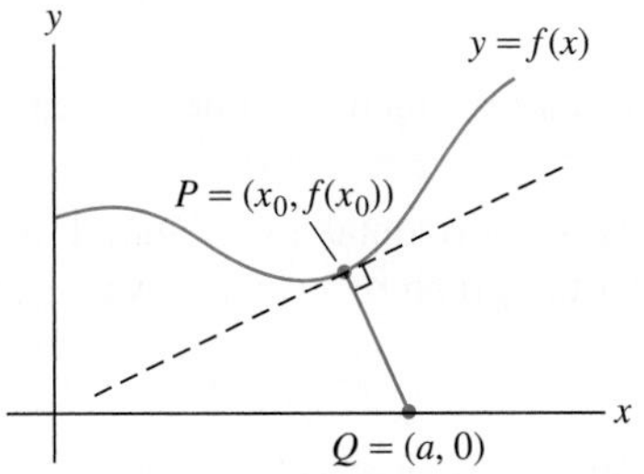

FIGURE 5

68. Take a circular piece of paper of radius R, remove a sector of angle θ (Figure 6), and fold the remaining piece into a cone-shaped cup. Which angle θ produces the cup of largest volume?

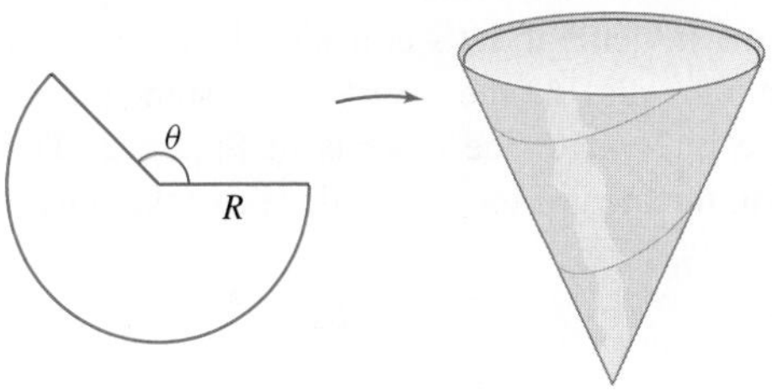

FIGURE 6

69. Use Newton's Method to estimate $\sqrt[3]{25}$ to four decimal places.

70. Use Newton's Method to find a root of $f(x) = x^2 - x - 1$ to four decimal places.

In Exercises 71–84, calculate the indefinite integral.

71. $\displaystyle\int (4x^3 - 2x^2)\,dx$

72. $\displaystyle\int x^{9/4}\,dx$

73. $\displaystyle\int \sin(\theta - 8)\,d\theta$

74. $\displaystyle\int \cos(5 - 7\theta)\,d\theta$

75. $\displaystyle\int (4t^{-3} - 12t^{-4})\,dt$

76. $\displaystyle\int (9t^{-2/3} + 4t^{7/3})\,dt$

77. $\displaystyle\int \sec^2 x\,dx$

78. $\displaystyle\int \tan 3\theta \sec 3\theta\,d\theta$

79. $\displaystyle\int (y + 2)^4\,dy$

80. $\displaystyle\int \frac{3x^3 - 9}{x^2}\,dx$

81. $\displaystyle\int (\cos\theta - \theta)\,d\theta$

82. $\displaystyle\int \sec^2(12 - 25\theta)\,d\theta$

83. $\displaystyle\int \frac{8\,dx}{x^3}$

84. $\displaystyle\int \sin(4x - 9)\,dx$

In Exercises 85–90, solve the differential equation with the given initial condition.

85. $\dfrac{dy}{dx} = 4x^3, \quad y(1) = 4$

86. $\dfrac{dy}{dt} = 3t^2 + \cos t, \quad y(0) = 12$

87. $\dfrac{dy}{dx} = x^{-1/2}, \quad y(1) = 1$

88. $\dfrac{dy}{dx} = \sec^2 x, \quad y(\frac{\pi}{4}) = 2$

89. $\dfrac{dy}{dt} = 1 + \pi \sin 3t, \quad y(\pi) = \pi$

90. $\dfrac{dy}{dt} = \cos 3\pi t + \sin 4\pi t, \quad y\left(\dfrac{1}{3}\right) = 0$

91. Find $f(t)$ if $f''(t) = 1 - 2t$, $f(0) = 2$, and $f'(0) = -1$.

92. At time $t = 0$, a driver begins decelerating at a constant rate of -10 m/s^2 and comes to a halt after traveling 500 m. Find the velocity at $t = 0$.

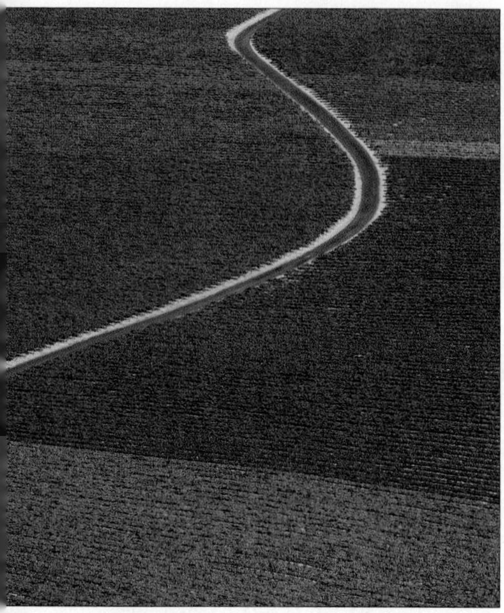

Integration solves an ancient mathematical problem—finding the area of an irregular region.

5 THE INTEGRAL

The basic problem in integral calculus is finding the area under a curve. You may wonder why calculus deals with two seemingly unrelated topics: tangent lines on the one hand and areas on the other. One reason is that both are computed using limits. A deeper connection is revealed by the Fundamental Theorem of Calculus, discussed in Sections 5.3 and 5.4. This theorem expresses the "inverse" relationship between integration and differentiation. It plays a truly fundamental role in nearly all applications of calculus, both theoretical and practical.

5.1 Approximating and Computing Area

Why might we be interested in the area under a graph? Consider an object moving in a straight line with *constant velocity* v (assumed positive). The distance traveled over a time interval $[t_1, t_2]$ is equal to $v\Delta t$ where $\Delta t = (t_2 - t_1)$ is the time elapsed. This is the well-known formula

$$\text{Distance traveled} = \overbrace{\text{velocity} \times \text{time elapsed}}^{v\Delta t} \qquad \boxed{1}$$

Because v is constant, the graph of velocity is a horizontal line (Figure 1) and $v\Delta t$ is equal to the area of the rectangular region under the graph of velocity over $[t_1, t_2]$. So we can write Eq. (1) as

$$\text{Distance traveled} = \text{area under the graph of velocity over } [t_1, t_2] \qquad \boxed{2}$$

There is, however, an important difference between these two equations: Eq. (1) makes sense only if velocity v is constant whereas Eq. (2) is correct *even if the velocity changes with time* (we will prove this in Section 5.5). Thus, the advantage of expressing distance traveled as an area is that it enables us to deal with much more general types of motion.

To see why Eq. (2) might be true in general, let's consider the case where velocity changes over time but is constant on intervals. In other words, we assume that the object's velocity changes abruptly from one interval to the next as in Figure 2. The distance traveled over each interval is equal to the area of the rectangle above that interval, so the

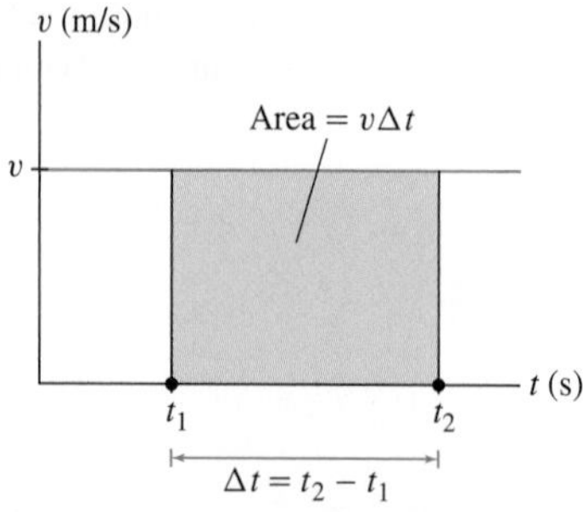

FIGURE 1 The rectangle has area $v\Delta t$, which is equal to the distance traveled.

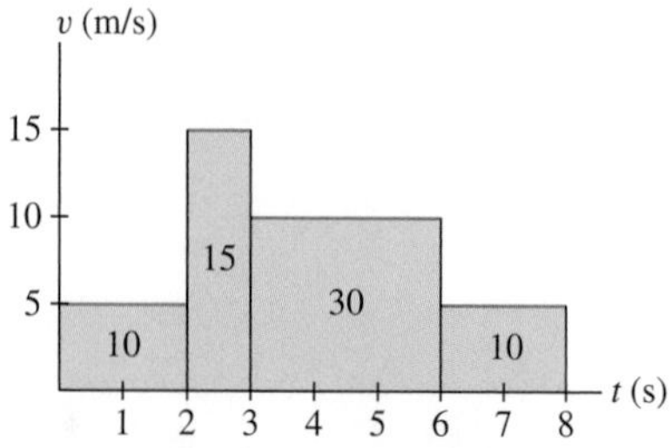

FIGURE 2 Distance traveled equals the sum of the areas of the rectangles.

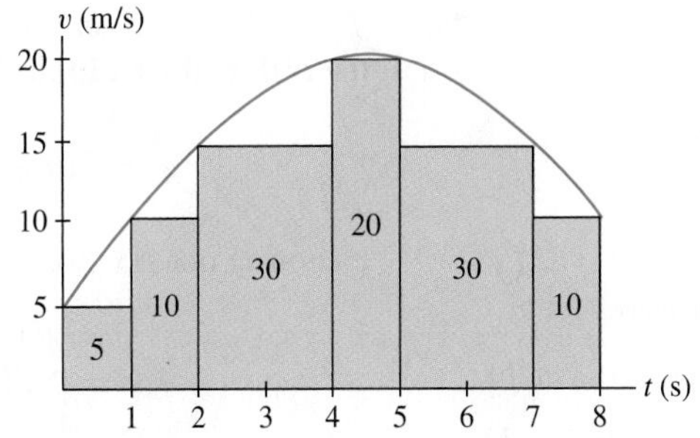

FIGURE 3 Distance traveled is equal to the area under the graph. It is *approximated* by the sum of the areas of the rectangles.

total distance traveled is the sum of the areas of the rectangles. In Figure 2,

$$\text{Distance traveled over } [0, 8] = \underbrace{10 + 15 + 30 + 10}_{\text{Sum of areas of rectangles}} = 65 \text{ m}$$

Our strategy when velocity changes continuously (Figure 3) is to *approximate* the area under the graph by sums of areas of rectangles and then pass to a limit. This idea leads to the concept of an integral.

Approximating Area by Rectangles

Our goal is to compute the area under the graph of a function $f(x)$. In this section, we assume that $f(x)$ is continuous and *positive*, so that the graph of $f(x)$ lies above the x-axis (Figure 4). The first step is to approximate the area using rectangles.

Recall the two-step procedure for finding the slope of the tangent line (the derivative): First approximate the slope using secant lines and then compute the limit of these approximations. In integral calculus, there are also two steps:

- *First, approximate the area under the graph using rectangles, and then*
- *Compute the exact area (the integral) as the limit of these approximations.*

To begin, choose a whole number N and divide $[a, b]$ into N subintervals of equal width, as in Figure 4(A). The full interval $[a, b]$ has width $b - a$, so each subinterval has width $\Delta x = (b - a)/N$. The right endpoints of the subintervals are

$$a + \Delta x, \ a + 2\Delta x, \ \ldots, \ a + (N - 1)\Delta x, \ a + N\Delta x$$

Note that the last right endpoint is b because $a + N\Delta x = a + N((b - a)/N) = b$. Next, as in Figure 4(B), construct, above each subinterval, a rectangle whose height is the value of $f(x)$ at the *right endpoint* of the subinterval.

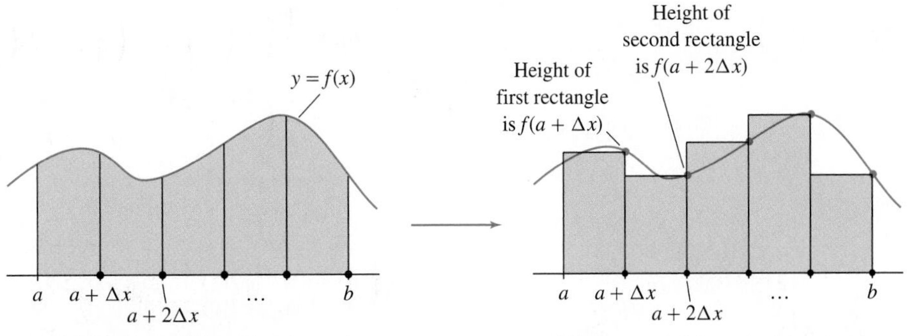

(A) Divide $[a, b]$ into N subintervals, each of width Δx

(B) Construct right-endpoint rectangles

FIGURE 4

The sum of the areas of these rectangles provides an *approximation* to the area under the graph. The first rectangle has base Δx and height $f(a + \Delta x)$, so its area is $f(a + \Delta x)\Delta x$. Similarly, the second rectangle has height $f(a + 2\Delta x)$ and area

$f(a + 2\Delta x)\,\Delta x$, etc. The sum of the areas of the rectangles is denoted R_N and is called the **Nth right-endpoint approximation**:

$$R_N = f(a + \Delta x)\Delta x + f(a + 2\Delta x)\Delta x + \cdots + f(a + N\Delta x)\Delta x$$

Factoring out Δx, we obtain the formula

$$R_N = \Delta x\,\big(f(a + \Delta x) + f(a + 2\Delta x) + \cdots + f(a + N\Delta x)\big)$$

In words: *R_N is equal to Δx times the sum of the function values at the right endpoints of the subintervals.*

To summarize,

a = *left endpoint of interval* $[a, b]$

b = *right endpoint of interval* $[a, b]$

N = *number of subintervals in* $[a, b]$

$$\Delta x = \frac{b - a}{N}$$

EXAMPLE 1 Calculate R_4 and R_6 for $f(x) = x^2$ on the interval $[1, 3]$.

Solution

***Step 1.* Determine Δx and the right endpoints.**
To calculate R_4, divide $[1, 3]$ into four subintervals of width $\Delta x = \frac{3-1}{4} = \frac{1}{2}$. The right endpoints are the numbers $a + j\Delta x = 1 + j\left(\frac{1}{2}\right)$ for $j = 1, 2, 3, 4$. They are spaced at intervals of $\frac{1}{2}$ beginning at $\frac{3}{2}$, so, as we see in Figure 5(A), the right endpoints are $\frac{3}{2}$, $\frac{4}{2}$, $\frac{5}{2}$, $\frac{6}{2}$.

***Step 2.* Calculate Δx times the sum of function values.**
R_4 is Δx times the sum of the function values at the right endpoints:

$$R_4 = \frac{1}{2}\left(f\left(\frac{3}{2}\right) + f\left(\frac{4}{2}\right) + f\left(\frac{5}{2}\right) + f\left(\frac{6}{2}\right)\right)$$

$$= \frac{1}{2}\left(\left(\frac{3}{2}\right)^2 + \left(\frac{4}{2}\right)^2 + \left(\frac{5}{2}\right)^2 + \left(\frac{6}{2}\right)^2\right) = \frac{43}{4} = 10.75$$

R_6 is similar: $\Delta x = \frac{3-1}{6} = \frac{1}{3}$, and the right endpoints are spaced at intervals of $\frac{1}{3}$ beginning at $\frac{4}{3}$ and ending at 3, as in Figure 5(B). Thus,

$$R_6 = \frac{1}{3}\left(f\left(\frac{4}{3}\right) + f\left(\frac{5}{3}\right) + f\left(\frac{6}{3}\right) + f\left(\frac{7}{3}\right) + f\left(\frac{8}{3}\right) + f\left(\frac{9}{3}\right)\right)$$

$$= \frac{1}{3}\left(\frac{16}{9} + \frac{25}{9} + \frac{36}{9} + \frac{49}{9} + \frac{64}{9} + \frac{81}{9}\right) = \frac{271}{27} \approx 10.037$$

(A) The approximation R_4

(B) The approximation R_6

FIGURE 5

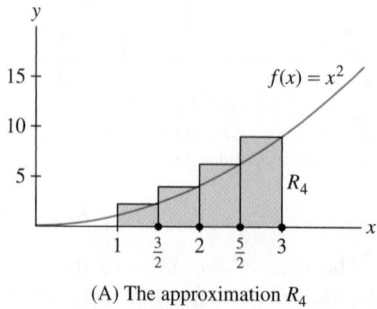

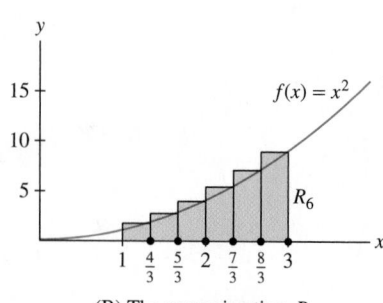

Summation Notation

Summation notation is a standard notation for writing sums in compact form. The sum of numbers $a_m, \ldots, a_n$ $(m \le n)$ is denoted

$$\sum_{j=m}^{n} a_j = a_m + a_{m+1} + \cdots + a_n$$

The Greek letter $\sum$ (capital sigma) stands for "sum," and the notation $\sum_{j=m}^{n}$ tells us to start the summation at $j = m$ and end it at $j = n$. For example,

$$\sum_{j=1}^{5} j^2 = 1^2 + 2^2 + 3^2 + 4^2 + 5^2 = 55$$

In this summation, the jth term is $a_j = j^2$. We refer to j^2 as the **general term**. The letter j is called the **summation index**. It is also referred to as a **dummy variable** because any other letter can be used instead. For example,

$$\sum_{k=4}^{6} (k^2 - 2k) = \overbrace{\big(4^2 - 2(4)\big)}^{k=4} + \overbrace{\big(5^2 - 2(5)\big)}^{k=5} + \overbrace{\big(6^2 - 2(6)\big)}^{k=6} = 47$$

$$\sum_{m=7}^{9} 1 = 1 + 1 + 1 = 3 \qquad \text{(because } a_7 = a_8 = a_9 = 1\text{)}$$

The usual commutative, associative, and distributive laws of addition give us the following rules for manipulating summations.

Linearity of Summations

- $\displaystyle\sum_{j=m}^{n} (a_j + b_j) = \sum_{j=m}^{n} a_j + \sum_{j=m}^{n} b_j$
- $\displaystyle\sum_{j=m}^{n} C a_j = C \sum_{j=m}^{n} a_j$ $\quad$ (C any constant)
- $\displaystyle\sum_{j=1}^{n} k = nk$ $\quad$ (k any constant and $n \ge 1$)

For example,

$$\sum_{j=3}^{5} (j^2 + j) = (3^2 + 3) + (4^2 + 4) + (5^2 + 5)$$

is equal to

$$\sum_{j=3}^{5} j^2 + \sum_{j=3}^{5} j = \big(3^2 + 4^2 + 5^2\big) + \big(3 + 4 + 5\big)$$

Linearity can be used to write a single summation as a sum of several summations. For example,

$$\sum_{k=0}^{100}(7k^2 - 4k + 9) = \sum_{k=0}^{100} 7k^2 + \sum_{k=0}^{100}(-4k) + \sum_{k=0}^{100} 9$$

$$= 7\sum_{k=0}^{100} k^2 - 4\sum_{k=0}^{100} k + 9\sum_{k=0}^{100} 1$$

It is convenient to use summation notation when working with area approximations. For example, R_N is a sum with general term $f(a + j\Delta x)$:

$$R_N = \Delta x\big[f(a + \Delta x) + f(a + 2\Delta x) + \cdots + f(a + N\Delta x)\big]$$

The summation extends from $j = 1$ to $j = N$, so we can write R_N concisely as

$$R_N = \Delta x \sum_{j=1}^{N} f(a + j\Delta x)$$

We shall make use of two other rectangular approximations to area: the left-endpoint and the midpoint approximations. Divide $[a, b]$ into N subintervals as before. In the **left-endpoint approximation** L_N, the heights of the rectangles are the values of $f(x)$ at the left endpoints [Figure 6(A)]. These left endpoints are

← REMINDER

$$\Delta x = \frac{b - a}{N}$$

$$a,\ a + \Delta x,\ a + 2\Delta x,\ \ldots,\ a + (N - 1)\Delta x$$

and the sum of the areas of the left-endpoint rectangles is

$$L_N = \Delta x\Big(f(a) + f(a + \Delta x) + f(a + 2\Delta x) + \cdots + f(a + (N - 1)\Delta x)\Big)$$

Note that both R_N and L_N have general term $f(a + j\Delta x)$, but the sum for L_N runs from $j = 0$ to $j = N - 1$ rather than from $j = 1$ to $j = N$:

$$L_N = \Delta x \sum_{j=0}^{N-1} f(a + j\Delta x)$$

In the **midpoint approximation** M_N, the heights of the rectangles are the values of $f(x)$ at the midpoints of the subintervals rather than at the endpoints. As we see in Figure 6(B), the midpoints are

$$a + \frac{1}{2}\Delta x,\ a + \frac{3}{2}\Delta x,\ \ldots,\ a + \left(N - \frac{1}{2}\right)\Delta x$$

The sum of the areas of the midpoint rectangles is

$$M_N = \Delta x\left(f\left(a + \frac{1}{2}\Delta x\right) + f\left(a + \frac{3}{2}\Delta x\right) + \cdots + f\left(a + \left(N - \frac{1}{2}\right)\Delta x\right)\right)$$

In summation notation,

$$M_N = \Delta x \sum_{j=1}^{N} f\left(a + \left(j - \frac{1}{2}\right)\Delta x\right)$$

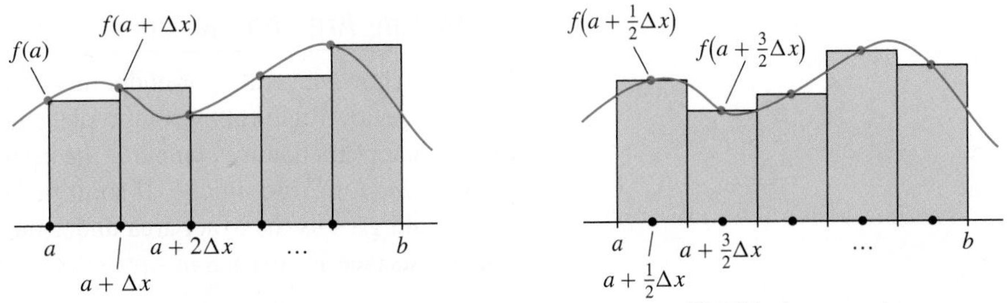

(A) Left-endpoint rectangles (B) Midpoint rectangles

FIGURE 6

■ **EXAMPLE 2** Calculate R_6, L_6, and M_6 for $f(x) = x^{-1}$ on $[2, 4]$.

Solution In this case, $\Delta x = (b-a)/N = (4-2)/6 = \frac{1}{3}$. The general term in the summation for R_6 and L_6 is

$$f(a + j\Delta x) = f\left(2 + j\left(\frac{1}{3}\right)\right) = \frac{1}{2+\frac{1}{3}j} = \frac{3}{6+j}$$

Therefore (Figure 7),

$$R_6 = \frac{1}{3}\sum_{j=1}^{6} f\left(2 + \left(\frac{1}{3}\right)j\right) = \frac{1}{3}\sum_{j=1}^{6} \frac{3}{6+j}$$

$$= \frac{1}{3}\left(\frac{3}{7} + \frac{3}{8} + \frac{3}{9} + \frac{3}{10} + \frac{3}{11} + \frac{3}{12}\right) \approx 0.653$$

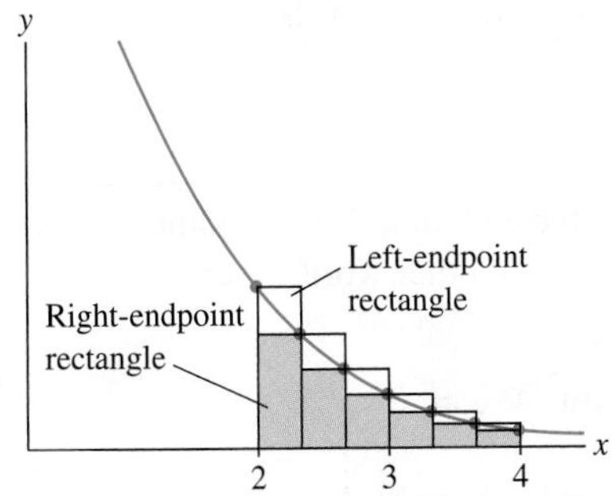

FIGURE 7 L_6 and R_6 for $f(x) = x^{-1}$ on $[2, 4]$.

In L_6, the sum begins at $j = 0$ and ends at $j = 5$:

$$L_6 = \frac{1}{3}\sum_{j=0}^{5} \frac{3}{6+j} = \frac{1}{3}\left(\frac{3}{6} + \frac{3}{7} + \frac{3}{8} + \frac{3}{9} + \frac{3}{10} + \frac{3}{11}\right) \approx 0.737$$

The general term in M_6 is

$$f\left(a + \left(j - \frac{1}{2}\right)\Delta x\right) = f\left(2 + \left(j - \frac{1}{2}\right)\frac{1}{3}\right) = \frac{1}{2 + \frac{j}{3} - \frac{1}{6}} = \frac{6}{12 + 2j - 1}$$

Summing up from $j = 1$ to 6, we obtain (Figure 8)

$$M_6 = \frac{1}{3}\sum_{j=1}^{6} f\left(2 + \left(j - \frac{1}{2}\right)\frac{1}{3}\right) = \frac{1}{3}\sum_{j=1}^{6} \frac{6}{12 + 2j - 1}$$

$$= \frac{1}{3}\left(\frac{6}{13} + \frac{6}{15} + \frac{6}{17} + \frac{6}{19} + \frac{6}{21} + \frac{6}{23}\right) \approx 0.692$$ ■

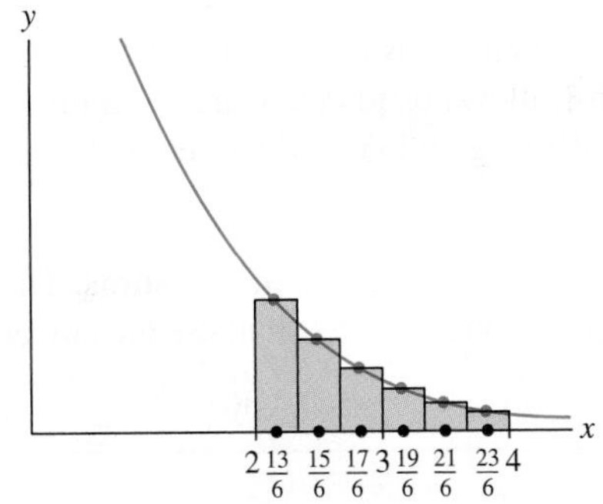

FIGURE 8 M_6 for $f(x) = x^{-1}$ on $[2, 4]$.

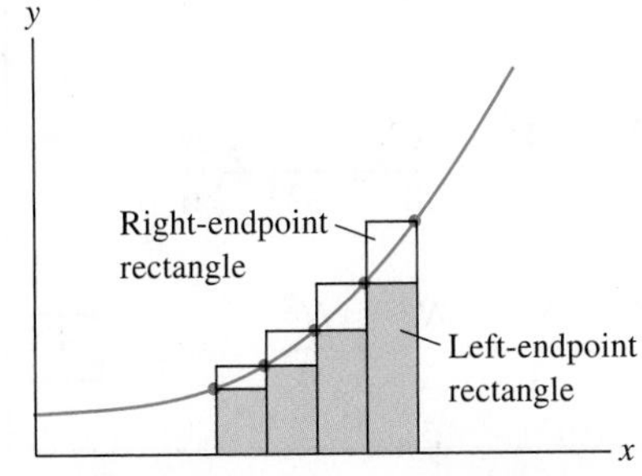

FIGURE 9 When $f(x)$ is increasing, the left-endpoint rectangles lie below the graph and right-endpoint rectangles lie above it.

GRAPHICAL INSIGHT Monotonic Functions Observe in Figure 7 that the left-endpoint rectangles for $f(x) = x^{-1}$ extend above the graph and the right-endpoint rectangles lie below it. The exact area A must lie between R_6 and L_6, and so, according to the previous example, $0.65 \le A \le 0.74$. More generally, *when $f(x)$ is monotonic (increasing or decreasing), the exact area lies between R_N and L_N* (Figure 9):

- $f(x)$ increasing $\Rightarrow$ $L_N \le$ area under graph $\le R_N$
- $f(x)$ decreasing $\Rightarrow$ $R_N \le$ area under graph $\le L_N$

Computing Area as the Limit of Approximations

Figure 10 shows several right-endpoint approximations. Notice that the *error*, corresponding to the yellow region above the graph, gets smaller as the number of rectangles increases. In fact, it appears that *we can make the error as small as we please by taking the number N of rectangles large enough.* If so, it makes sense to consider the limit as $N \to \infty$, as this should give us the exact area under the curve. The next theorem guarantees that the limit exists (see Theorem 8 in Appendix D for a proof and Exercise 85 for a special case).

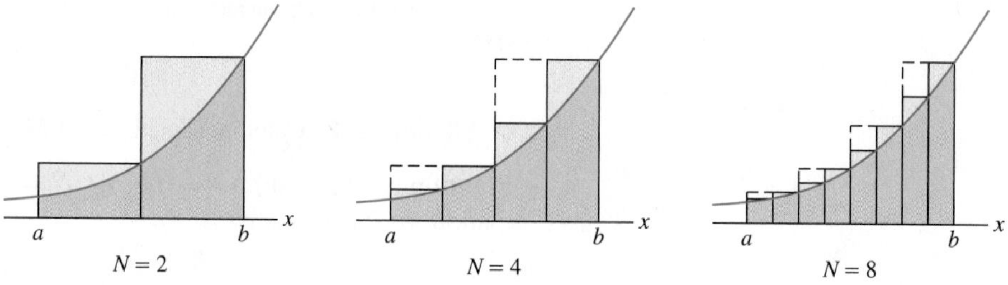

FIGURE 10 The error decreases as we use more rectangles.

In Theorem 1, it is not assumed that $f(x) \geq 0$. If $f(x)$ takes on negative values, the limit L no longer represents area under the graph, but we can interpret it as a "signed area," discussed in the next section.

THEOREM 1 If $f(x)$ is continuous on $[a, b]$, then the endpoint and midpoint approximations approach one and the same limit as $N \to \infty$. In other words, there is a value L such that

$$\lim_{N\to\infty} R_N = \lim_{N\to\infty} L_N = \lim_{N\to\infty} M_N = L$$

If $f(x) \geq 0$, we define the area under the graph over $[a, b]$ to be L.

CONCEPTUAL INSIGHT In calculus, limits are used to define basic quantities that otherwise would not have a precise meaning. Theorem 1 allows us to *define* area as a limit L in much the same way that we define the slope of a tangent line as the limit of slopes of secant lines.

The next three examples illustrate Theorem 1 using formulas for **power sums**. The kth power sum is the sum of the kth powers of the first N integers. We shall use the power sum formulas for $k = 1, 2, 3$.

A method for proving power sum formulas is developed in Exercises 40–43 of Section 1.3. Formulas (3)–(5) can also be verified using the method of induction.

Power Sums

$$\sum_{j=1}^{N} j = 1 + 2 + \cdots + N = \frac{N(N+1)}{2} = \frac{N^2}{2} + \frac{N}{2} \qquad (3)$$

$$\sum_{j=1}^{N} j^2 = 1^2 + 2^2 + \cdots + N^2 = \frac{N(N+1)(2N+1)}{6} = \frac{N^3}{3} + \frac{N^2}{2} + \frac{N}{6} \qquad (4)$$

$$\sum_{j=1}^{N} j^3 = 1^3 + 2^3 + \cdots + N^3 = \frac{N^2(N+1)^2}{4} = \frac{N^4}{4} + \frac{N^3}{2} + \frac{N^2}{4} \qquad (5)$$

For example, by Eq. (4),

$$\sum_{j=1}^{6} j^2 = 1^2 + 2^2 + 3^2 + 4^2 + 5^2 + 6^2 = \underbrace{\frac{6^3}{3} + \frac{6^2}{2} + \frac{6}{6}}_{\frac{N^3}{3} + \frac{N^2}{2} + \frac{N}{6} \text{ for } N=6} = 91$$

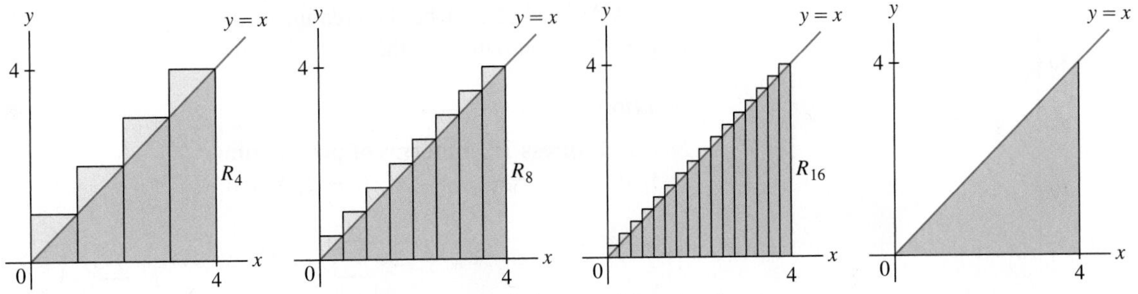

FIGURE 11 The right-endpoint approximations approach the area of the triangle.

As a first illustration, we compute the area of a right triangle "the hard way."

EXAMPLE 3 Find the area A under the graph of $f(x) = x$ over $[0, 4]$ in three ways:

(a) Using geometry **(b)** $\lim_{N\to\infty} R_N$ **(c)** $\lim_{N\to\infty} L_N$

Solution The region under the graph is a right triangle with base $b = 4$ and height $h = 4$ (Figure 11).

(a) By geometry, $A = \frac{1}{2}bh = \left(\frac{1}{2}\right)(4)(4) = 8$.

(b) We compute this area again as a limit. Since $\Delta x = (b - a)/N = 4/N$ and $f(x) = x$,

$$f(a + j\Delta x) = f\left(0 + j\left(\frac{4}{N}\right)\right) = \frac{4j}{N}$$

$$R_N = \Delta x \sum_{j=1}^{N} f\,(a + j\Delta x) = \frac{4}{N}\sum_{j=1}^{N}\frac{4j}{N} = \frac{16}{N^2}\sum_{j=1}^{N} j$$

In the last equality, we factored out $4/N$ from the sum. This is valid because $4/N$ is a constant that does not depend on j. Now use formula (3):

$$R_N = \frac{16}{N^2}\sum_{j=1}^{N} j = \frac{16}{N^2}\underbrace{\left(\frac{N(N+1)}{2}\right)}_{\text{Formula for power sum}} = \frac{8}{N^2}\left(N^2 + N\right) = 8 + \frac{8}{N}$$

The second term $8/N$ tends to zero as N approaches ∞, so

$$A = \lim_{N\to\infty} R_N = \lim_{N\to\infty}\left(8 + \frac{8}{N}\right) = 8$$

As expected, this limit yields the same value as the formula $\frac{1}{2}bh$.

(c) The left-endpoint approximation is similar, but the sum begins at $j = 0$ and ends at $j = N - 1$:

$$L_N = \frac{16}{N^2}\sum_{j=0}^{N-1} j = \frac{16}{N^2}\sum_{j=1}^{N-1} j = \frac{16}{N^2}\left(\frac{(N-1)N}{2}\right) = 8 - \frac{8}{N} \qquad \boxed{6}$$

Note in the second step that we replaced the sum beginning at $j = 0$ with a sum beginning at $j = 1$. This is valid because the term for $j = 0$ is zero and may be dropped. Again, we find that $A = \lim_{N\to\infty} L_N = \lim_{N\to\infty}(8 - 8/N) = 8$. ■

REMINDER

$$R_N = \Delta x \sum_{j=1}^{N} f(a + j\Delta x)$$

$$L_N = \Delta x \sum_{j=0}^{N-1} f(a + j\Delta x)$$

$$\Delta x = \frac{b - a}{N}$$

In Eq. (6), we apply the formula

$$\sum_{j=1}^{N} j = \frac{N(N+1)}{2}$$

with $N - 1$ in place of N:

$$\sum_{j=1}^{N-1} j = \frac{(N-1)N}{2}$$

In the next example, we compute the area under a curved graph. Unlike the previous example, it is not possible to compute this area directly using geometry.

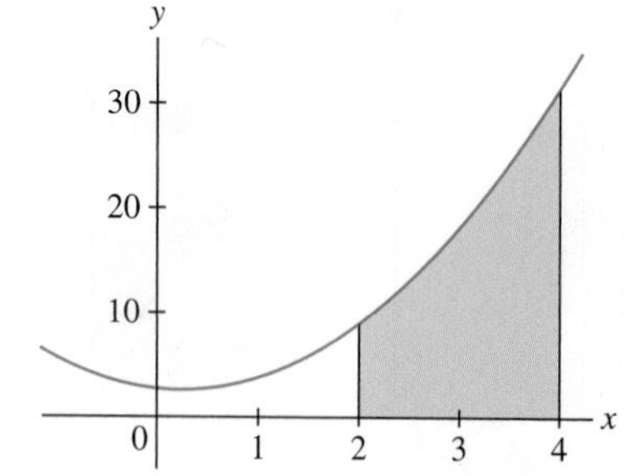

FIGURE 12 Area under the graph of $f(x) = 2x^2 - x + 3$ over [2, 4].

■ **EXAMPLE 4** Let A be the area under the graph of $f(x) = 2x^2 - x + 3$ over [2, 4] (Figure 12). Compute A as the limit $\lim_{N\to\infty} R_N$.

Solution

Step 1. **Express R_N in terms of power sums.**

In this case, $\Delta x = (4-2)/N = 2/N$ and

$$R_N = \Delta x \sum_{j=1}^{N} f(a + j\Delta x) = \frac{2}{N}\sum_{j=1}^{N} f\left(2 + \frac{2j}{N}\right)$$

Let's use algebra to simplify the general term. Since $f(x) = 2x^2 - x + 3$,

$$f\left(2+\frac{2j}{N}\right) = 2\left(2+\frac{2j}{N}\right)^2 - \left(2+\frac{2j}{N}\right) + 3$$

$$= 2\left(4 + \frac{8j}{N} + \frac{4j^2}{N^2}\right) - \left(2 + \frac{2j}{N}\right) + 3 = \frac{8}{N^2}j^2 + \frac{14}{N}j + 9$$

Now we can express R_N in terms of power sums:

$$R_N = \frac{2}{N}\sum_{j=1}^{N}\left(\frac{8}{N^2}j^2 + \frac{14}{N}j + 9\right) = \frac{2}{N}\sum_{j=1}^{N}\frac{8}{N^2}j^2 + \frac{2}{N}\sum_{j=1}^{N}\frac{14}{N}j + \frac{2}{N}\sum_{j=1}^{N}9$$

$$= \frac{16}{N^3}\sum_{j=1}^{N} j^2 + \frac{28}{N^2}\sum_{j=1}^{N} j + \frac{18}{N}\sum_{j=1}^{N} 1 \qquad \boxed{7}$$

Step 2. **Use the formulas for the power sums.**

Using formulas (3) and (4) for the power sums in Eq. (7), we obtain

$$R_N = \frac{16}{N^3}\left(\frac{N^3}{3} + \frac{N^2}{2} + \frac{N}{6}\right) + \frac{28}{N^2}\left(\frac{N^2}{2} + \frac{N}{2}\right) + \frac{18}{N}(N)$$

$$= \left(\frac{16}{3} + \frac{8}{N} + \frac{8}{3N^2}\right) + \left(14 + \frac{14}{N}\right) + 18$$

$$= \frac{112}{3} + \frac{22}{N} + \frac{8}{3N^2}$$

Step 3. **Calculate the limit.**

$$A = \lim_{N\to\infty} R_N = \lim_{N\to\infty}\left(\frac{112}{3} + \frac{22}{N} + \frac{8}{3N^2}\right) = \frac{112}{3}$$ ■

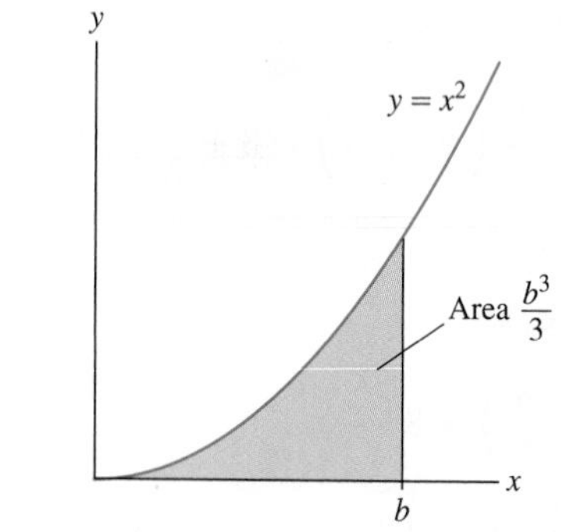

FIGURE 13

■ **EXAMPLE 5** Prove that for all $b > 0$, the area A under the graph of $f(x) = x^2$ over $[0, b]$ is equal to $b^3/3$, as indicated in Figure 13.

Solution We'll compute with R_N. We have $\Delta x = (b-0)/N = b/N$ and

$$R_N = \Delta x \sum_{j=1}^{N} f(0 + j\Delta x) = \frac{b}{N}\sum_{j=1}^{N}\left(0 + j\frac{b}{N}\right)^2 = \frac{b}{N}\sum_{j=1}^{N}\left(j^2\frac{b^2}{N^2}\right) = \frac{b^3}{N^3}\sum_{j=1}^{N} j^2$$

By the formula for the power sum recalled in the margin,

$$R_N = \frac{b^3}{N^3}\left(\frac{N^3}{3} + \frac{N^2}{2} + \frac{N}{6}\right) = \frac{b^3}{3} + \frac{b^3}{2N} + \frac{b^3}{6N^2}$$

$$A = \lim_{N\to\infty}\left(\frac{b^3}{3} + \frac{b^3}{2N} + \frac{b^3}{6N^2}\right) = \frac{b^3}{3}$$ ■

← *REMINDER* By Eq. (4)

$$\sum_{j=1}^{N} j^2 = \frac{N^3}{3} + \frac{N^2}{2} + \frac{N}{6}$$

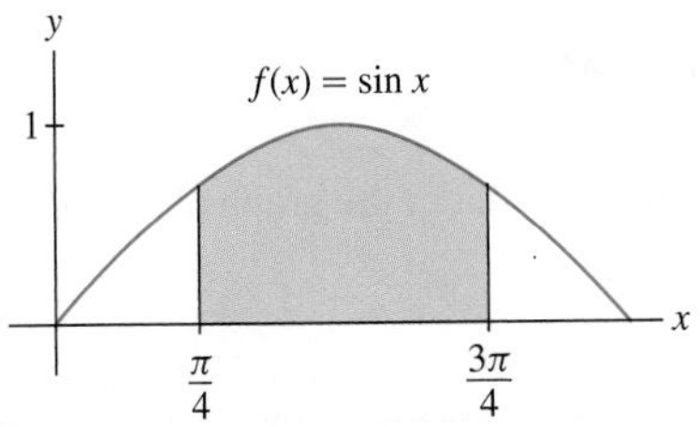

FIGURE 14 The area of this region is more difficult to compute as a limit of endpoint approximations.

The area under the graph of any polynomial can be calculated using power sum formulas as in the examples above. For other functions, the limit defining the area may be hard or impossible to evaluate directly. Consider $f(x) = \sin x$ on the interval $\left[\frac{\pi}{4}, \frac{3\pi}{4}\right]$. In this case (Figure 14), $\Delta x = (3\pi/4 - \pi/4)/N = \pi/(2N)$ and the area A is

$$A = \lim_{N\to\infty} R_N = \lim_{N\to\infty} \Delta x \sum_{j=1}^{N} f(a + j\Delta x) = \lim_{N\to\infty} \frac{\pi}{2N} \sum_{j=1}^{N} \sin\left(\frac{\pi}{4} + \frac{\pi j}{2N}\right)$$

With some work, we can show that the limit is equal to $A = \sqrt{2}$. However, in Section 5.3 we will see that it is much easier to apply the Fundamental Theorem of Calculus, which reduces area computations to the problem of finding antiderivatives.

HISTORICAL PERSPECTIVE

Jacob Bernoulli (1654–1705)

We used the formulas for the kth power sums for $k = 1, 2, 3$. Do similar formulas exist for all powers k? This problem was studied in the seventeenth century and eventually solved around 1690 by the great Swiss mathematician Jacob Bernoulli. Of this discovery, he wrote

> With the help of [these formulas] it took me less than half of a quarter of an hour to find that the 10th powers of the first 1000 numbers being added together will yield the sum
>
> 91409924241424243424241924242500

Bernoulli's formula has the general form

$$\sum_{j=1}^{n} j^k = \frac{1}{k+1} n^{k+1} + \frac{1}{2} n^k + \frac{k}{12} n^{k-1} + \cdots$$

The dots indicate terms involving smaller powers of n whose coefficients are expressed in terms of the so-called Bernoulli numbers. For example,

$$\sum_{j=1}^{n} j^4 = \frac{1}{5}n^5 + \frac{1}{2}n^4 + \frac{1}{3}n^3 - \frac{1}{30}n$$

These formulas are available on most computer algebra systems.

5.1 SUMMARY

- Approximations to the area under the graph of $f(x)$ over $[a, b]$ $\left(\Delta x = \dfrac{b-a}{N}\right)$:

$$R_N = \Delta x \sum_{j=1}^{N} f(a + j\Delta x) = \Delta x\big(f(a + \Delta x) + f(a + 2\Delta x) + \cdots + f(a + N\Delta x)\big)$$

$$L_N = \Delta x \sum_{j=0}^{N-1} f(a + j\Delta x) = \Delta x\big(f(a) + f(a + \Delta x) + \cdots + f(a + (N-1)\Delta x)\big)$$

$$M_N = \Delta x \sum_{j=1}^{N} f\left(a + \left(j - \frac{1}{2}\right)\Delta x\right)$$

$$= \Delta x\left(f\left(a + \frac{1}{2}\Delta x\right) + \cdots + f\left(a + \left(N - \frac{1}{2}\right)\Delta x\right)\right)$$

Power Sums

$$\sum_{j=1}^{N} j = \frac{N(N+1)}{2} = \frac{N^2}{2} + \frac{N}{2}$$

$$\sum_{j=1}^{N} j^2 = \frac{N(N+1)(2N+1)}{6} = \frac{N^3}{3} + \frac{N^2}{2} + \frac{N}{6}$$

$$\sum_{j=1}^{N} j^3 = \frac{N^2(N+1)^2}{4} = \frac{N^4}{4} + \frac{N^3}{2} + \frac{N^2}{4}$$

- If $f(x)$ is continuous on $[a, b]$, then the endpoint and midpoint approximations approach one and the same limit L:

$$\lim_{N\to\infty} R_N = \lim_{N\to\infty} L_N = \lim_{N\to\infty} M_N = L$$

- If $f(x) \geq 0$ on $[a, b]$, we take L as the definition of the area under the graph of $y = f(x)$ over $[a, b]$.

5.1 EXERCISES

Preliminary Questions

1. What are the right and left endpoints if $[2, 5]$ is divided into six subintervals?

2. The interval $[1, 5]$ is divided into eight subintervals.

(a) What is the left endpoint of the last subinterval?

(b) What are the right endpoints of the first two subintervals?

3. Which of the following pairs of sums are *not* equal?

(a) $\sum_{i=1}^{4} i$, $\sum_{\ell=1}^{4} \ell$ **(b)** $\sum_{j=1}^{4} j^2$, $\sum_{k=2}^{5} k^2$

(c) $\sum_{j=1}^{4} j$, $\sum_{i=2}^{5} (i-1)$ **(d)** $\sum_{i=1}^{4} i(i+1)$, $\sum_{j=2}^{5} (j-1)j$

4. Explain: $\sum_{j=1}^{100} j = \sum_{j=0}^{100} j$ but $\sum_{j=1}^{100} 1$ is not equal to $\sum_{j=0}^{100} 1$.

5. Explain why $L_{100} \geq R_{100}$ for $f(x) = x^{-2}$ on $[3, 7]$.

Exercises

1. Figure 15 shows the velocity of an object over a 3-min interval. Determine the distance traveled over the intervals $[0, 3]$ and $[1, 2.5]$ (remember to convert from km/h to km/min).

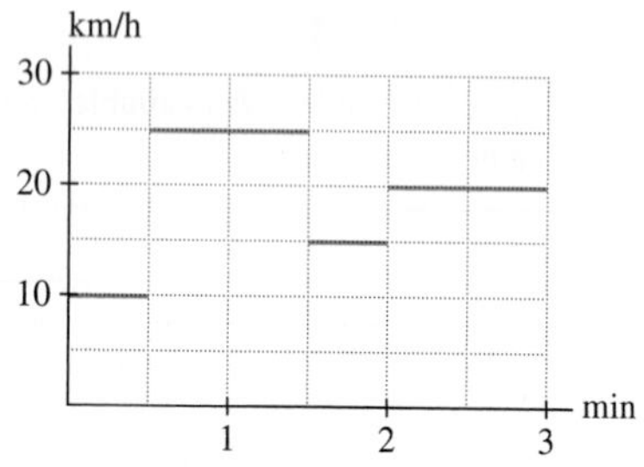

FIGURE 15

2. An ostrich (Figure 16) runs with velocity 20 km/h for 2 minutes, 12 km/h for 3 minutes, and 40 km/h for another minute. Compute the total distance traveled and indicate with a graph how this quantity can be interpreted as an area.

FIGURE 16 Ostriches can reach speeds as high as 70 km/h.

3. A rainstorm hit Portland, Maine, in October 1996, resulting in record rainfall. The rainfall rate $R(t)$ on October 21 is recorded, in centimeters per hour, in the following table, where t is the number of hours since midnight. Compute the total rainfall during this 24-hour period and indicate on a graph how this quantity can be interpreted as an area.

t (h)	0–2	2–4	4–9	9–12	12–20	20–24
$R(t)$ (cm)	0.5	0.3	1.0	2.5	1.5	0.6

4. The velocity of an object is $v(t) = 12t$ m/s. Use Eq. (2) and geometry to find the distance traveled over the time intervals $[0, 2]$ and $[2, 5]$.

5. Compute R_5 and L_5 over $[0, 1]$ using the following values.

x	0	0.2	0.4	0.6	0.8	1
$f(x)$	50	48	46	44	42	40

6. Compute R_6, L_6, and M_3 to estimate the distance traveled over $[0, 3]$ if the velocity at half-second intervals is as follows:

t (s)	0	0.5	1	1.5	2	2.5	3
v (m/s)	0	12	18	25	20	14	20

7. Let $f(x) = 2x + 3$.

(a) Compute R_6 and L_6 over $[0, 3]$.

(b) Use geometry to find the exact area A and compute the errors $|A - R_6|$ and $|A - L_6|$ in the approximations.

8. Repeat Exercise 7 for $f(x) = 20 - 3x$ over $[2, 4]$.

9. Calculate R_3 and L_3

$$\text{for } f(x) = x^2 - x + 4 \quad \text{over } [1, 4]$$

Then sketch the graph of f and the rectangles that make up each approximation. Is the area under the graph larger or smaller than R_3? Is it larger or smaller than L_3?

10. Let $f(x) = \sqrt{x^2 + 1}$ and $\Delta x = \frac{1}{3}$. Sketch the graph of $f(x)$ and draw the right-endpoint rectangles whose area is represented by the sum $\sum_{i=1}^{6} f(1 + i\Delta x)\Delta x$.

11. Estimate R_3, M_3, and L_6 over $[0, 1.5]$ for the function in Figure 17.

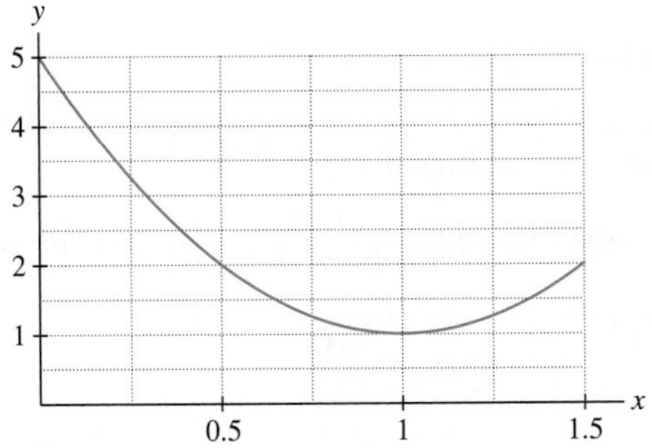

FIGURE 17

12. Calculate the area of the shaded rectangles in Figure 18. Which approximation do these rectangles represent?

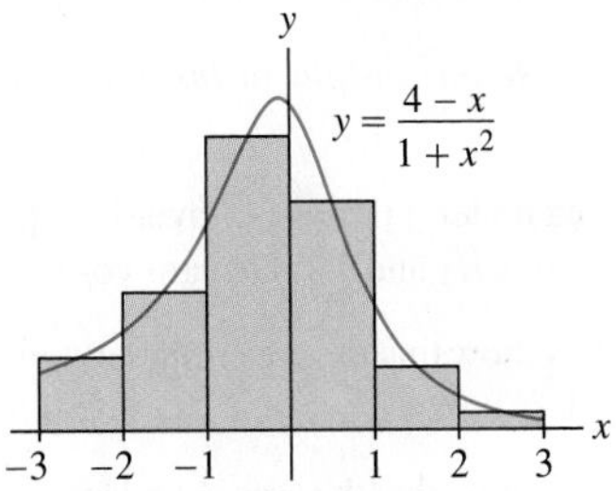

FIGURE 18

In Exercises 13–20, calculate the approximation for the given function and interval.

13. R_3, $f(x) = 7 - x$, $[3, 5]$

14. L_6, $f(x) = \sqrt{6x + 2}$, $[1, 3]$

15. M_6, $f(x) = 4x + 3$, $[5, 8]$

16. R_5, $f(x) = x^2 + x$, $[-1, 1]$

17. L_6, $f(x) = x^2 + 3|x|$, $[-2, 1]$

18. M_4, $f(x) = \sqrt{x}$, $[3, 5]$

19. L_4, $f(x) = \cos^2 x$, $\left[\frac{\pi}{6}, \frac{\pi}{2}\right]$

20. M_4, $f(x) = \frac{1}{x^2+1}$, $[1, 5]$

In Exercises 21–26, write the sum in summation notation.

21. $4^7 + 5^7 + 6^7 + 7^7 + 8^7$

22. $(2^2 + 2) + (3^2 + 3) + (4^2 + 4) + (5^2 + 5)$

23. $(2^2 + 2) + (2^3 + 2) + (2^4 + 2) + (2^5 + 2)$

24. $\sqrt{1 + 1^3} + \sqrt{2 + 2^3} + \cdots + \sqrt{n + n^3}$

25. $\dfrac{1}{2 \cdot 3} + \dfrac{2}{3 \cdot 4} + \cdots + \dfrac{n}{(n+1)(n+2)}$

26. $\sin(\pi) + \sin\left(\frac{\pi}{2}\right) + \sin\left(\frac{\pi}{3}\right) + \cdots + \sin\left(\frac{\pi}{n+1}\right)$

27. Calculate the sums:

(a) $\sum_{i=1}^{5} 9$ **(b)** $\sum_{i=0}^{5} 4$ **(c)** $\sum_{k=2}^{4} k^3$

28. Calculate the sums:

(a) $\sum_{j=3}^{4} \sin\left(j\frac{\pi}{2}\right)$ **(b)** $\sum_{k=3}^{5} \frac{1}{k-1}$ **(c)** $\sum_{j=0}^{2} 3^{j-1}$

29. Let $b_1 = 4$, $b_2 = 1$, $b_3 = 2$, and $b_4 = -4$. Calculate:

(a) $\sum_{i=2}^{4} b_i$ **(b)** $\sum_{j=1}^{2} (2^{b_j} - b_j)$ **(c)** $\sum_{k=1}^{3} kb_k$

30. Assume that $a_1 = -5$, $\sum_{i=1}^{10} a_i = 20$, and $\sum_{i=1}^{10} b_i = 7$. Calculate:

(a) $\sum_{i=1}^{10} (4a_i + 3)$ **(b)** $\sum_{i=2}^{10} a_i$ **(c)** $\sum_{i=1}^{10} (2a_i - 3b_i)$

31. Calculate $\sum_{j=101}^{200} j$. *Hint:* Write as a difference of two sums and use formula (3).

32. Calculate $\sum_{j=1}^{30} (2j + 1)^2$. *Hint:* Expand and use formulas (3)–(4).

In Exercises 33–40, use linearity and formulas (3)–(5) to rewrite and evaluate the sums.

33. $\sum_{j=1}^{20} 8j^3$

34. $\sum_{k=1}^{30} (4k - 3)$

35. $\sum_{n=51}^{150} n^2$

36. $\sum_{k=101}^{200} k^3$

37. $\sum_{j=0}^{50} j(j - 1)$

38. $\sum_{j=2}^{30} \left(6j + \frac{4j^2}{3}\right)$

39. $\sum_{m=1}^{30} (4 - m)^3$

40. $\sum_{m=1}^{20} \left(5 + \frac{3m}{2}\right)^2$

In Exercises 41–44, use formulas (3)–(5) to evaluate the limit.

41. $\lim_{N\to\infty} \sum_{i=1}^{N} \frac{i}{N^2}$

42. $\lim_{N\to\infty} \sum_{j=1}^{N} \frac{j^3}{N^4}$

43. $\lim_{N\to\infty} \sum_{i=1}^{N} \frac{i^2 - i + 1}{N^3}$

44. $\lim_{N\to\infty} \sum_{i=1}^{N} \left(\frac{i^3}{N^4} - \frac{20}{N}\right)$

In Exercises 45–50, calculate the limit for the given function and interval. Verify your answer by using geometry.

45. $\lim_{N\to\infty} R_N, \quad f(x) = 9x, \quad [0, 2]$

46. $\lim_{N\to\infty} R_N, \quad f(x) = 3x + 6, \quad [1, 4]$

47. $\lim_{N\to\infty} L_N, \quad f(x) = \frac{1}{2}x + 2, \quad [0, 4]$

48. $\lim_{N\to\infty} L_N, \quad f(x) = 4x - 2, \quad [1, 3]$

49. $\lim_{N\to\infty} M_N, \quad f(x) = x, \quad [0, 2]$

50. $\lim_{N\to\infty} M_N, \quad f(x) = 12 - 4x, \quad [2, 6]$

51. Show, for $f(x) = 3x^2 + 4x$ over $[0, 2]$, that

$$R_N = \frac{2}{N} \sum_{j=1}^{N} \left(\frac{24j^2}{N^2} + \frac{16j}{N}\right)$$

Then evaluate $\lim_{N\to\infty} R_N$.

52. Show, for $f(x) = 3x^3 - x^2$ over $[1, 5]$, that

$$R_N = \frac{4}{N} \sum_{j=1}^{N} \left(\frac{192j^3}{N^3} + \frac{128j^2}{N^2} + \frac{28j}{N} + 2\right)$$

Then evaluate $\lim_{N\to\infty} R_N$.

In Exercises 53–60, find a formula for R_N and compute the area under the graph as a limit.

53. $f(x) = x^2, \quad [0, 1]$

54. $f(x) = x^2, \quad [-1, 5]$

55. $f(x) = 6x^2 - 4, \quad [2, 5]$

56. $f(x) = x^2 + 7x, \quad [6, 11]$

57. $f(x) = x^3 - x, \quad [0, 2]$

58. $f(x) = 2x^3 + x^2, \quad [-2, 2]$

59. $f(x) = 2x + 1, \quad [a, b] \quad$ (a, b constants with $a < b$)

60. $f(x) = x^2, \quad [a, b] \quad$ (a, b constants with $a < b$)

In Exercises 61–64, describe the area represented by the limits.

61. $\lim_{N\to\infty} \frac{1}{N} \sum_{j=1}^{N} \left(\frac{j}{N}\right)^4$

62. $\lim_{N\to\infty} \frac{3}{N} \sum_{j=1}^{N} \left(2 + \frac{3j}{N}\right)^4$

63. $\lim_{N\to\infty} \frac{5}{N} \sum_{j=0}^{N-1} \left(-2 + 5\frac{j}{N}\right)^4$

64. $\lim_{N\to\infty} \frac{\pi}{2N} \sum_{j=1}^{N} \sin\left(\frac{\pi}{3} - \frac{\pi}{4N} + \frac{j\pi}{2N}\right)$

In Exercises 65–70, express the area under the graph as a limit using the approximation indicated (in summation notation), but do not evaluate.

65. $R_N, \quad f(x) = \sin x$ over $[0, \pi]$

66. $R_N, \quad f(x) = x^{-1}$ over $[1, 7]$

67. $L_N, \quad f(x) = \sqrt{2x + 1}$ over $[7, 11]$

68. $L_N, \quad f(x) = \cos x$ over $\left[\frac{\pi}{8}, \frac{\pi}{4}\right]$

69. $M_N, \quad f(x) = \tan x$ over $\left[\frac{1}{2}, 1\right]$

70. $M_N, \quad f(x) = x^{-2}$ over $[3, 5]$

71. Evaluate $\lim_{N\to\infty} \frac{1}{N} \sum_{j=1}^{N} \sqrt{1 - \left(\frac{j}{N}\right)^2}$ by interpreting it as the area of part of a familiar geometric figure.

In Exercises 72–74, let $f(x) = x^2$ and let R_N, L_N, and M_N be the approximations for the interval $[0, 1]$.

72. Show that $R_N = \frac{1}{3} + \frac{1}{2N} + \frac{1}{6N^2}$. Interpret the quantity $\frac{1}{2N} + \frac{1}{6N^2}$ as the area of a region.

73. Show that

$$L_N = \frac{1}{3} - \frac{1}{2N} + \frac{1}{6N^2}, \qquad M_N = \frac{1}{3} - \frac{1}{12N^2}$$

Then rank the three approximations R_N, L_N, and M_N in order of increasing accuracy (use Exercise 72).

74. For each of R_N, L_N, and M_N, find the smallest integer N for which the error is less than 0.001.

In Exercises 75–80, use the Graphical Insight on page 249 to obtain bounds on the area.

75. Let A be the area under $f(x) = \sqrt{x}$ over $[0, 1]$. Prove that $0.51 \le A \le 0.77$ by computing R_4 and L_4. Explain your reasoning.

76. Use R_5 and L_5 to show that the area A under $y = x^{-2}$ over $[10, 13]$ satisfies $0.0218 \le A \le 0.0244$.

77. Use R_4 and L_4 to show that the area A under the graph of $y = \sin x$ over $\left[0, \frac{\pi}{2}\right]$ satisfies $0.79 \le A \le 1.19$.

78. Show that the area A under $f(x) = x^{-1}$ over $[1, 8]$ satisfies

$$\frac{1}{2} + \frac{1}{3} + \frac{1}{4} + \frac{1}{5} + \frac{1}{6} + \frac{1}{7} + \frac{1}{8} \le A \le 1 + \frac{1}{2} + \frac{1}{3} + \frac{1}{4} + \frac{1}{5} + \frac{1}{6} + \frac{1}{7}$$

79. CAS Show that the area A under $y = x^{1/4}$ over $[0, 1]$ satisfies $L_N \le A \le R_N$ for all N. Use a computer algebra system to calculate L_N and R_N for $N = 100$ and 200, and determine A to two decimal places.

80. CAS Show that the area A under $y = 4/(x^2 + 1)$ over $[0, 1]$ satisfies $R_N \le A \le L_N$ for all N. Determine A to at least three decimal places using a computer algebra system. Can you guess the exact value of A?

Further Insights and Challenges

81. Although the accuracy of R_N generally improves as N increases, this need not be true for small values of N. Draw the graph of a positive continuous function $f(x)$ on an interval such that R_1 is closer than R_2 to the exact area under the graph. Can such a function be monotonic?

82. Draw the graph of a positive continuous function on an interval such that R_2 and L_2 are both smaller than the exact area under the graph. Can such a function be monotonic?

83. Explain graphically: *The endpoint approximations are less accurate when $f'(x)$ is large.*

84. Prove that for any function $f(x)$ on $[a, b]$,

$$R_N - L_N = \frac{b-a}{N}(f(b) - f(a)) \tag{8}$$

85. In this exercise, we prove that $\lim_{N\to\infty} R_N$ and $\lim_{N\to\infty} L_N$ exist and are equal if $f(x)$ is increasing [the case of $f(x)$ decreasing is similar]. We use the concept of a least upper bound discussed in Appendix B.

(a) Explain with a graph why $L_N \le R_M$ for all $N, M \ge 1$.

(b) By (a), the sequence $\{L_N\}$ is bounded, so it has a least upper bound L. By definition, L is the smallest number such that $L_N \le L$ for all N. Show that $L \le R_M$ for all M.

(c) According to (b), $L_N \le L \le R_N$ for all N. Use Eq. (8) to show that $\lim_{N\to\infty} L_N = L$ and $\lim_{N\to\infty} R_N = L$.

86. Use Eq. (8) to show that if $f(x)$ is positive and monotonic, then the area A under its graph over $[a, b]$ satisfies

$$|R_N - A| \le \frac{b-a}{N}|f(b) - f(a)| \tag{9}$$

In Exercises 87–88, use Eq. (9) to find a value of N such that $|R_N - A| < 10^{-4}$ for the given function and interval.

87. $f(x) = \sqrt{x}$, $[1, 4]$

88. $f(x) = \sqrt{9 - x^2}$, $[0, 3]$

89. Prove that if $f(x)$ is positive and monotonic, then M_N lies between R_N and L_N and is closer to the actual area under the graph than both R_N and L_N. *Hint:* In the case that $f(x)$ is increasing, Figure 19 shows that the part of the error in R_N due to the ith rectangle is the sum of the areas $A + B + D$, and for M_N it is $|B - E|$. On the other hand, $A \ge E$.

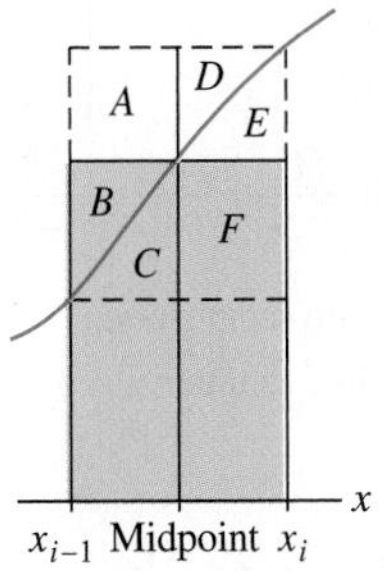

FIGURE 19

5.2 The Definite Integral

In the previous section, we saw that if $f(x)$ is continuous on an interval $[a, b]$, then the endpoint and midpoint approximations approach a common limit L as $N \to \infty$:

$$L = \lim_{N\to\infty} R_N = \lim_{N\to\infty} L_N = \lim_{N\to\infty} M_N \tag{1}$$

When $f(x) \ge 0$, L is the area under the graph of $f(x)$. In a moment, we will state formally that L is the *definite integral* of $f(x)$ over $[a, b]$. Before doing so, we introduce more general approximations called **Riemann sums**.

Recall that R_N, L_N, and M_N use rectangles of equal width Δx, whose heights are the values of $f(x)$ at the endpoints or midpoints of the subintervals. In Riemann sum approximations, we relax these requirements: The rectangles need not have equal width, and the height may be *any* value of $f(x)$ within the subinterval.

To specify a Riemann sum, we choose a partition and a set of sample points:

- **Partition** P of size N: a choice of points that divides $[a, b]$ into N subintervals.

$$P : a = x_0 < x_1 < x_2 < \cdots < x_N = b$$

- **Sample points** $C = \{c_1, \ldots, c_N\}$: c_i belongs to the subinterval $[x_{i-1}, x_i]$ for all i.

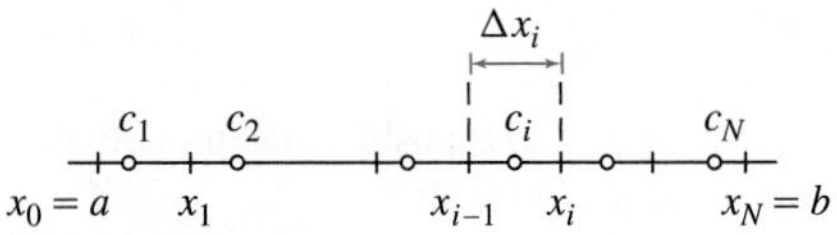

FIGURE 1 Partition of size N and set of sample points

See Figures 1 and 2(A). The length of the ith subinterval $[x_{i-1}, x_i]$ is

$$\boxed{\Delta x_i = x_i - x_{i-1}}$$

The **norm** of P, denoted $\|P\|$, is the maximum of the lengths Δx_i.

Given P and C, we construct the rectangle of height $f(c_i)$ and base Δx_i over each subinterval $[x_{i-1}, x_i]$, as in Figure 2(B). This rectangle has area $f(c_i)\Delta x_i$ if $f(c_i) \geq 0$. If $f(c_i) < 0$, the rectangle extends below the x-axis, and $f(c_i)\Delta x_i$ is the negative of its area. The Riemann sum is the sum

Keep in mind that R_N, L_N, and M_N are particular examples of Riemann sums in which $\Delta x_i = (b - a)/N$ for all i, and the sample points c_i are endpoints or midpoints.

$$R(f, P, C) = \sum_{i=1}^{N} f(c_i)\Delta x_i = f(c_1)\Delta x_1 + f(c_2)\Delta x_2 + \cdots + f(c_N)\Delta x_N \qquad \boxed{2}$$

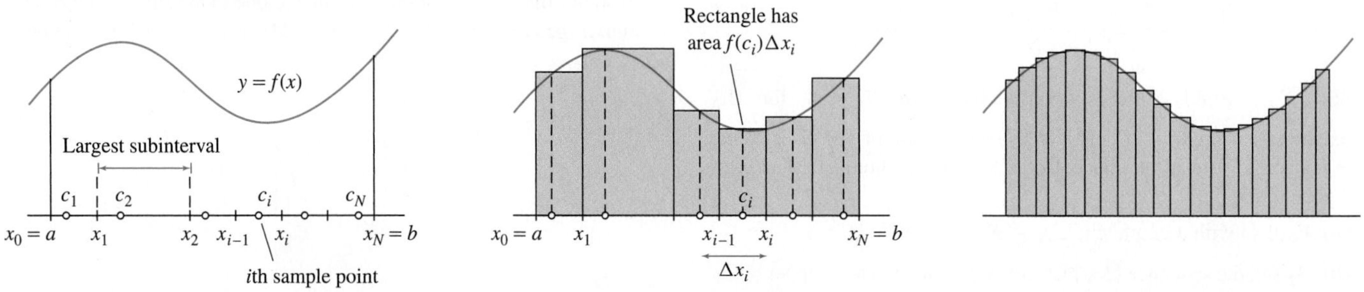

(A) Partition of $[a, b]$ into subintervals

(B) Construct rectangle above each subinterval of height $f(c_i)$

(C) Rectangles corresponding to a Riemann sum with $\|P\|$ small (a large number of rectangles)

FIGURE 2 Construction of $R(f, P, C)$.

■ **EXAMPLE 1** Calculate $R(f, P, C)$, where $f(x) = 8 + 12\sin x - 4x$ on $[0, 4]$,

$$P: x_0 = 0 < x_1 = 1 < x_2 = 1.8 < x_3 = 2.9 < x_4 = 4$$

$$C = \{0.4, 1.2, 2, 3.5\}$$

What is the norm $\|P\|$?

Solution The widths of the subintervals in the partition (Figure 3) are

$$\Delta x_1 = x_1 - x_0 = 1 - 0 = 1, \qquad \Delta x_2 = x_2 - x_1 = 1.8 - 1 = 0.8$$

$$\Delta x_3 = x_3 - x_2 = 2.9 - 1.8 = 1.1, \qquad \Delta x_4 = x_4 - x_3 = 4 - 2.9 = 1.1$$

The norm of the partition is $\|P\| = 1.1$ since the two longest subintervals have width 1.1. Using a calculator, we obtain

$$R(f, P, C) = f(0.4)\Delta x_1 + f(1.2)\Delta x_2 + f(2)\Delta x_3 + f(3.5)\Delta x_4$$

$$\approx 11.07(1) + 14.38(0.8) + 10.91(1.1) - 10.2(1.1) \approx 23.35 \qquad ■$$

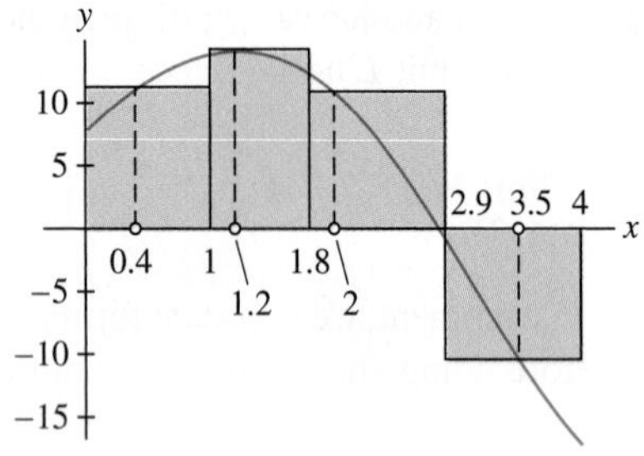

FIGURE 3 Rectangles defined by a Riemann sum for $f(x) = 8 + 12\sin x - 4x$.

Note in Figure 2(C) that as the norm $\|P\|$ tends to zero (meaning that the rectangles get thinner), the number of rectangles N tends to ∞ and they approximate the area under the graph more closely. This leads to the following definition: $f(x)$ is **integrable** over $[a, b]$ if *all* of the Riemann sums (not just the endpoint and midpoint approximations) approach one and the same limit L as $\|P\|$ tends to zero. Formally, we write

$$L = \lim_{\|P\| \to 0} R(f, P, C) = \lim_{\|P\| \to 0} \sum_{i=1}^{N} f(c_i)\Delta x_i \qquad \boxed{3}$$

if $|R(f, P, C) - L|$ gets arbitrarily small as the norm $\|P\|$ tends to zero, no matter how we choose the partition and sample points. The limit L is called the **definite integral** of $f(x)$ over $[a, b]$.

The notation $\int f(x)\,dx$ was introduced by Leibniz in 1686. The symbol $\int$ is an elongated S standing for "summation." The differential dx corresponds to the length Δx_i along the x-axis.

DEFINITION Definite Integral The definite integral of $f(x)$ over $[a, b]$, denoted by the integral sign, is the limit of Riemann sums:

$$\int_a^b f(x)\,dx = \lim_{\|P\|\to 0} R(f, P, C) = \lim_{\|P\|\to 0} \sum_{i=1}^{N} f(c_i)\Delta x_i$$

When this limit exists, we say that $f(x)$ is integrable over $[a, b]$.

One of the greatest mathematicians of the nineteenth century and perhaps second only to his teacher C. F. Gauss, Riemann transformed the fields of geometry, analysis, and number theory. Albert Einstein based his General Theory of Relativity *on Riemann's geometry. The "Riemann hypothesis" dealing with prime numbers is one of the great unsolved problems in present-day mathematics. The Clay Foundation has offered a \$1 million prize for its solution (http://www.claymath.org/millennium).*

Georg Friedrich Riemann (1826–1866)

The definite integral is often called, more simply, the *integral* of $f(x)$ over $[a, b]$. The process of computing integrals is called **integration**. The function $f(x)$ is called the **integrand**. The endpoints a and b of $[a, b]$ are called the **limits of integration**. Finally, we remark that any variable may be used as a variable of integration (this is a "dummy" variable). Thus, the following three integrals all denote the same quantity:

$$\int_a^b f(x)\,dx, \qquad \int_a^b f(t)\,dt, \qquad \int_a^b f(u)\,du$$

CONCEPTUAL INSIGHT Keep in mind that a Riemann sum $R(f, P, C)$ is nothing more than an approximation to area based on rectangles, and that $\int_a^b f(x)\,dx$ is the number we obtain in the limit as we take thinner and thinner rectangles.

However, general Riemann sums (with arbitrary partitions and sample points) are rarely used for computations. In practice, we use particular approximations such as M_N, or the Fundamental Theorem of Calculus, as we'll learn in the next section. If so, why bother introducing Riemann sums? The answer is that Riemann sums play a theoretical rather than a computational role. They are useful in proofs and for dealing rigorously with certain discontinuous functions. In later sections, Riemann sums are used to show that volumes and other quantities can be expressed as definite integrals.

The next theorem assures us that continuous functions (and even functions with finitely many jump discontinuities) are integrable (see Appendix D for a proof). In practice, we rely on this theorem rather than attempting to prove directly that a given function is integrable.

THEOREM 1 If $f(x)$ is continuous on $[a, b]$, or if $f(x)$ is continuous with at most finitely many jump discontinuities, then $f(x)$ is integrable over $[a, b]$.

Interpretation of the Definite Integral as Signed Area

When $f(x) \geq 0$, the definite integral defines the area under the graph. To interpret the integral when $f(x)$ takes on both positive and negative values, we define the notion of **signed area**, where regions below the x-axis are considered to have "negative area" (Figure 4); that is,

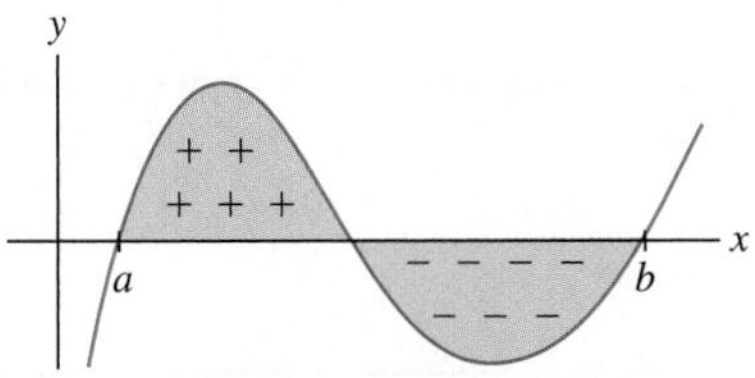

FIGURE 4 Signed area is the area above the x-axis minus the area below the x-axis.

Signed area of a region = (area above x-axis) − (area below x-axis)

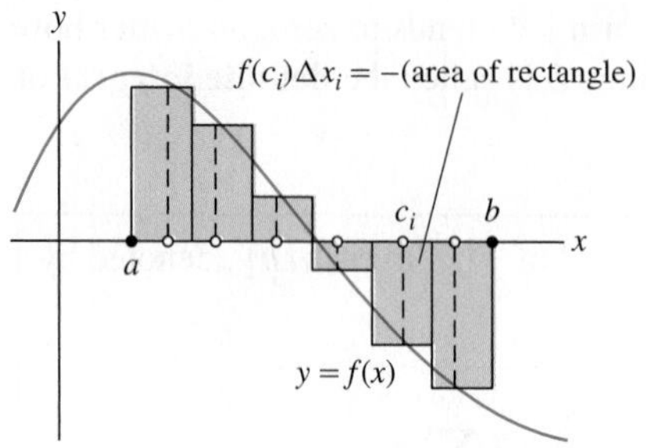

FIGURE 5

Now observe that a Riemann sum is equal to the signed area of the corresponding rectangles:

$$R(f, C, P) = f(c_1)\Delta x_1 + f(c_2)\Delta x_2 + \cdots + f(c_N)\Delta x_N$$

Indeed, if $f(c_i) < 0$, then the corresponding rectangle lies below the x-axis and has signed area $f(c_i)\Delta x_i$ (Figure 5). The limit of the Riemann sums is the signed area of the region between the graph and the x-axis:

$$\int_a^b f(x)\,dx = \text{signed area of region between the graph and } x\text{-axis over } [a, b]$$

■ **EXAMPLE 2 Signed Area** Calculate

$$\int_0^5 (3 - x)\,dx \qquad \text{and} \qquad \int_0^5 |3 - x|\,dx$$

Solution The region between $y = 3 - x$ and the x-axis consists of two triangles of areas $\frac{9}{2}$ and 2 [Figure 6(A)]. However, the second triangle lies below the x-axis, so it has signed area -2. In the graph of $y = |3 - x|$, both triangles lie above the x-axis [Figure 6(B)]. Therefore,

$$\int_0^5 (3 - x)\,dx = \frac{9}{2} - 2 = \frac{5}{2} \qquad \int_0^5 |3 - x|\,dx = \frac{9}{2} + 2 = \frac{13}{2}$$ ■

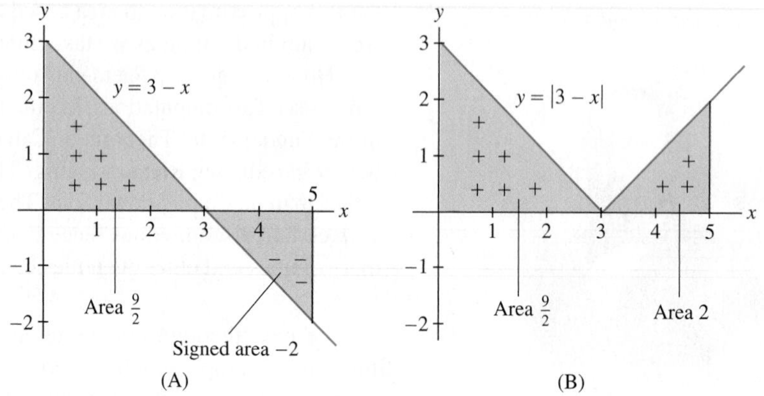

FIGURE 6

Properties of the Definite Integral

In the rest of this section, we discuss some basic properties of definite integrals. First, we note that the integral of a constant function $f(x) = C$ over $[a, b]$ is the signed area $C(b - a)$ of a rectangle (Figure 7).

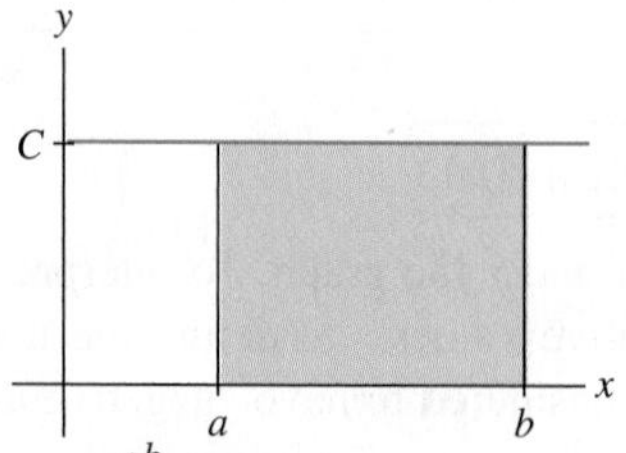

FIGURE 7 $\int_a^b C\,dx = C(b - a)$.

THEOREM 2 Integral of a Constant For any constant C,

$$\int_a^b C\,dx = C(b - a) \qquad \boxed{4}$$

Next, we state the linearity properties of the definite integral.

THEOREM 3 Linearity of the Definite Integral If f and g are integrable over $[a, b]$, then $f + g$ and Cf are integrable (for any constant C), and

- $$\int_a^b \big(f(x) + g(x)\big)\,dx = \int_a^b f(x)\,dx + \int_a^b g(x)\,dx$$
- $$\int_a^b Cf(x)\,dx = C\int_a^b f(x)\,dx$$

Proof These properties follow from the corresponding linearity properties of sums and limits. For example, Riemann sums are additive:

$$R(f+g, P, C) = \sum_{i=1}^{N} \big(f(c_i) + g(c_i)\big)\Delta x_i = \sum_{i=1}^{N} f(c_i)\Delta x_i + \sum_{i=1}^{N} g(c_i)\Delta x_i$$
$$= R(f, P, C) + R(g, P, C)$$

By the additivity of limits,

$$\int_a^b (f(x) + g(x))\,dx = \lim_{||P||\to 0} R(f+g, P, C)$$
$$= \lim_{||P||\to 0} R(f, P, C) + \lim_{||P||\to 0} R(g, P, C)$$
$$= \int_a^b f(x)\,dx + \int_a^b g(x)\,dx$$

The second property is proved similarly. ■

■ **EXAMPLE 3** Calculate $\int_0^3 (2x^2 - 5)\,dx$ using the formula

Eq. (5) was verified in Example 5 of Section 5.1.

$$\int_0^b x^2\,dx = \frac{b^3}{3} \qquad \boxed{5}$$

Solution

$$\int_0^3 (2x^2 - 5)\,dx = 2\int_0^3 x^2\,dx + \int_0^3 (-5)\,dx \qquad \text{(linearity)}$$
$$= 2\left(\frac{3^3}{3}\right) - 5(3 - 0) = 3 \qquad \text{[Eqs. (5) and (4)]}$$ ■

So far we have used the notation $\int_a^b f(x)\,dx$ with the understanding that $a < b$. It is convenient to define the definite integral for arbitrary a and b.

*According to Eq. (6), **the integral changes sign when the limits of integration are reversed**. Since we are free to define symbols as we please, why have we chosen to put the minus sign in Eq. (6)? Because it is only with this definition that the Fundamental Theorem of Calculus holds true.*

DEFINITION Reversing the Limits of Integration For $a < b$, we set

$$\int_b^a f(x)\,dx = -\int_a^b f(x)\,dx \qquad \boxed{6}$$

For example, by Eq. (5),

$$\int_5^0 x^2\,dx = -\int_0^5 x^2\,dx = -\frac{5^3}{3} = -\frac{125}{3}$$

When $a = b$, the interval $[a, b] = [a, a]$ has length zero and we define the definite integral to be zero:

$$\boxed{\int_a^a f(x)\,dx = 0}$$

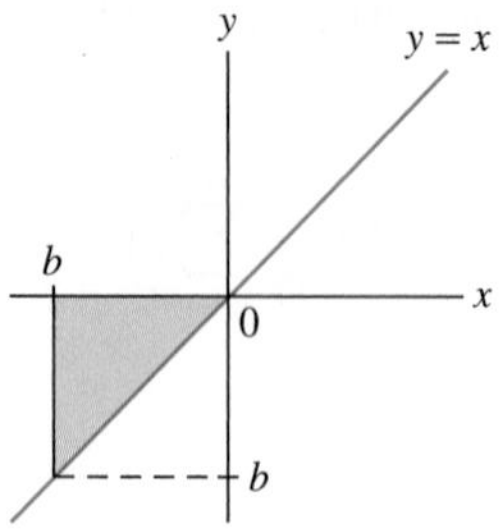

FIGURE 8 Here $b < 0$ and the signed area is $-\frac{1}{2}b^2$.

■ **EXAMPLE 4** Prove that, for all b (positive or negative),

$$\boxed{\int_0^b x\,dx = \frac{1}{2}b^2} \qquad \boxed{7}$$

Solution If $b > 0$, $\int_0^b x\,dx$ is the area $\frac{1}{2}b^2$ of a triangle of base b and height b. If $b < 0$, $\int_b^0 x\,dx$ is the signed area $-\frac{1}{2}b^2$ of the triangle in Figure 8, and Eq. (7) follows from the rule for reversing limits of integration:

$$\int_0^b x\,dx = -\int_b^0 x\,dx = -\left(-\frac{1}{2}b^2\right) = \frac{1}{2}b^2 \qquad ■$$

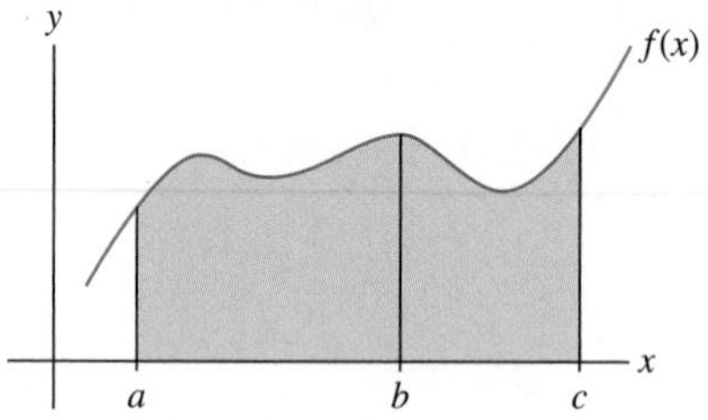

FIGURE 9 The area over $[a, c]$ is the *sum* of the areas over $[a, b]$ and $[b, c]$.

Definite integrals satisfy an important additivity property: If $f(x)$ is integrable and $a \le b \le c$ as in Figure 9, then the integral from a to c is equal to the integral from a to b *plus* the integral from b to c. We state this in the next theorem (a formal proof can be given using Riemann sums).

THEOREM 4 Additivity for Adjacent Intervals Let $a \le b \le c$, and assume that $f(x)$ is integrable. Then

$$\int_a^c f(x)\,dx = \int_a^b f(x)\,dx + \int_b^c f(x)\,dx$$

This theorem remains true as stated even if the condition $a \le b \le c$ is not satisfied (Exercise 87).

■ **EXAMPLE 5** Calculate $\displaystyle\int_4^7 x^2\,dx$.

Solution Before we can apply the formula $\int_0^b x^2 dx = b^3/3$ from Example 3, we must use the additivity property for adjacent intervals to write

$$\int_0^4 x^2\,dx + \int_4^7 x^2\,dx = \int_0^7 x^2\,dx$$

Now we can compute our integral as a difference:

$$\int_4^7 x^2\,dx = \int_0^7 x^2\,dx - \int_0^4 x^2\,dx = \left(\frac{1}{3}\right)7^3 - \left(\frac{1}{3}\right)4^3 = 93 \qquad ■$$

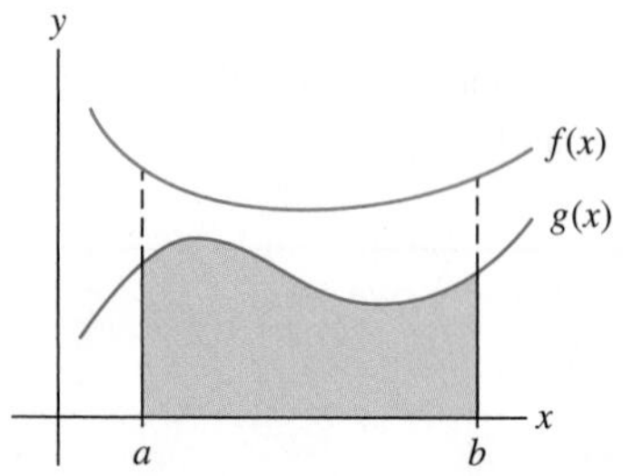

FIGURE 10 The integral of $f(x)$ is larger than the integral of $g(x)$.

Another basic property of the definite integral is that larger functions have larger integrals (Figure 10).

THEOREM 5 Comparison Theorem If f and g are integrable and $g(x) \le f(x)$ for x in $[a, b]$, then

$$\int_a^b g(x)\,dx \le \int_a^b f(x)\,dx$$

Proof If $g(x) \le f(x)$, then for any partition and choice of sample points, we have $g(c_i)\Delta x_i \le f(c_i)\Delta x_i$ for all i. Therefore, the Riemann sums satisfy

$$\sum_{i=1}^{N} g(c_i)\Delta x_i \le \sum_{i=1}^{N} f(c_i)\Delta x_i$$

Taking the limit as the norm $\|P\|$ tends to zero, we obtain

$$\int_a^b g(x)\,dx = \lim_{\|P\|\to 0} \sum_{i=1}^{N} g(c_i)\Delta x_i \le \lim_{\|P\|\to 0} \sum_{i=1}^{N} f(c_i)\Delta x_i = \int_a^b f(x)\,dx$$ ■

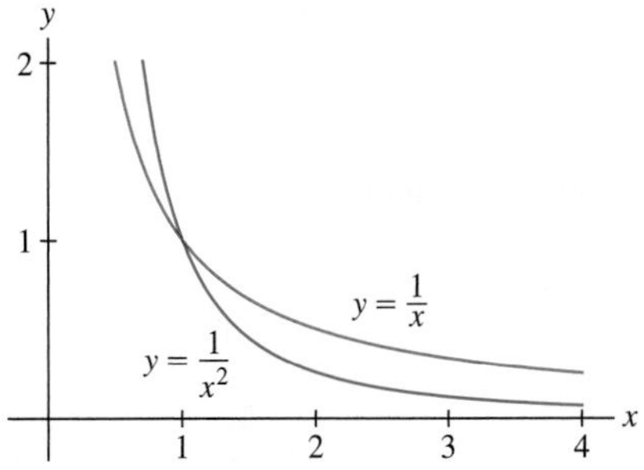

FIGURE 11

■ **EXAMPLE 6** Prove the inequality: $\displaystyle\int_1^4 \frac{1}{x^2}\,dx \le \int_1^4 \frac{1}{x}\,dx.$

Solution If $x \ge 1$, then $x^2 \ge x$, and $x^{-2} \le x^{-1}$ [Figure 11]. Therefore, the inequality follows from the Comparison Theorem, applied with $g(x) = x^{-2}$ and $f(x) = x^{-1}$. ■

Suppose there are numbers m and M such that $m \le f(x) \le M$ for x in $[a, b]$. We call m and M lower and upper bounds for $f(x)$ on $[a, b]$. By the Comparison Theorem,

$$\int_a^b m\,dx \le \int_a^b f(x)\,dx \le \int_a^b M\,dx$$

$$m(b - a) \le \int_a^b f(x)\,dx \le M(b - a) \qquad \boxed{8}$$

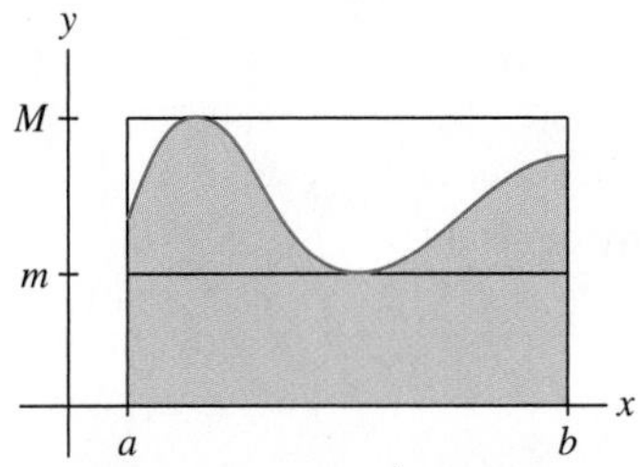

FIGURE 12 The integral $\int_a^b f(x)\,dx$ lies between the areas of the rectangles of heights m and M.

This says simply that the integral of $f(x)$ lies between the areas of two rectangles (Figure 12).

■ **EXAMPLE 7** Prove the inequalities: $\displaystyle\frac{3}{4} \le \int_{1/2}^{2} \frac{1}{x}\,dx \le 3.$

Solution Because $f(x) = x^{-1}$ is decreasing (Figure 13), its minimum value on $\left[\frac{1}{2}, 2\right]$ is $m = f(2) = \frac{1}{2}$ and its maximum value is $M = f\left(\frac{1}{2}\right) = 2$. By Eq. (8),

$$\underbrace{\frac{1}{2}\left(2 - \frac{1}{2}\right)}_{m(b-a)} = \frac{3}{4} \le \int_{1/2}^{2} \frac{1}{x}\,dx \le \underbrace{2\left(2 - \frac{1}{2}\right)}_{M(b-a)} = 3$$ ■

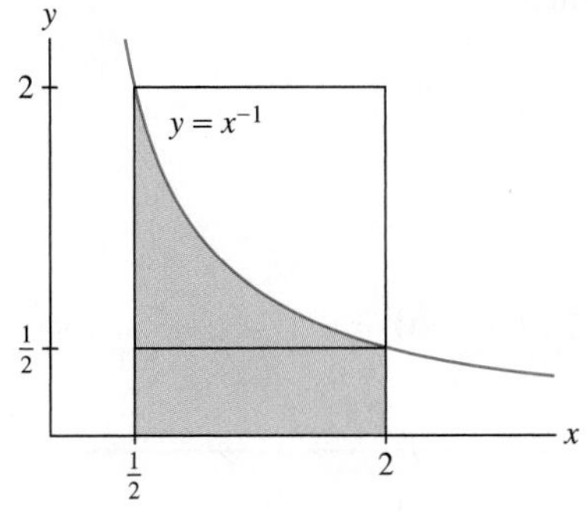

FIGURE 13

5.2 SUMMARY

- A Riemann sum $R(f, P, C)$ for the interval $[a, b]$ is defined by choosing a *partition*

$$P : a = x_0 < x_1 < x_2 < \cdots < x_N = b$$

and *sample points* $C = \{c_i\}$, where $c_i \in [x_{i-1}, x_i]$. Let $\Delta x_i = x_i - x_{i-1}$. Then

$$R(f, P, C) = \sum_{i=1}^{N} f(c_i)\Delta x_i$$

- The maximum of the widths Δx_i is called the norm $\|P\|$ of the partition.
- The *definite integral* is the limit of the Riemann sums (if it exists):

$$\int_a^b f(x)\,dx = \lim_{\|P\|\to 0} R(f, P, C)$$

We say that $f(x)$ is *integrable* over $[a, b]$ if the limit exists.
- Theorem: If $f(x)$ is continuous on $[a, b]$, then $f(x)$ is integrable over $[a, b]$.
- $\int_a^b f(x)\,dx =$ *signed area* of the region between the graph of $f(x)$ and the x-axis.
- Properties of definite integrals:

$$\int_a^b \big(f(x) + g(x)\big)\,dx = \int_a^b f(x)\,dx + \int_a^b g(x)\,dx$$

$$\int_a^b Cf(x)\,dx = C\int_a^b f(x)\,dx \quad \text{for any constant } C$$

$$\int_b^a f(x)\,dx = -\int_b^a f(x)\,dx$$

$$\int_a^a f(x)\,dx = 0$$

$$\int_a^b f(x)\,dx + \int_b^c f(x)\,dx = \int_a^c f(x)\,dx \quad \text{for all } a, b, c$$

- Formulas:

$$\int_a^b C\,dx = C(b-a) \quad (C \text{ any constant})$$

$$\int_0^b x\,dx = \frac{1}{2}b^2$$

$$\int_0^b x^2\,dx = \frac{1}{3}b^3$$

- Comparison Theorem: If $f(x) \le g(x)$ on $[a, b]$, then

$$\int_a^b f(x)\,dx \le \int_a^b g(x)\,dx$$

If $m \le f(x) \le M$ on $[a, b]$, then

$$m(b-a) \le \int_a^b f(x)\,dx \le M(b-a)$$

5.2 EXERCISES

Preliminary Questions

1. What is $\int_3^5 dx$ [the function is $f(x) = 1$]?

2. Let $I = \int_2^7 f(x)\,dx$, where $f(x)$ is continuous. State whether true or false:

(a) I is the area between the graph and the x-axis over $[2, 7]$.

(b) If $f(x) \ge 0$, then I is the area between the graph and the x-axis over $[2, 7]$.

(c) If $f(x) \le 0$, then $-I$ is the area between the graph of $f(x)$ and the x-axis over $[2, 7]$.

3. Explain graphically: $\int_0^{\pi} \cos x \, dx = 0$.

4. Which is negative, $\int_{-1}^{-5} 8 \, dx$ or $\int_{-5}^{-1} 8 \, dx$?

Exercises

In Exercises 1–10, draw a graph of the signed area represented by the integral and compute it using geometry.

1. $\int_{-3}^{3} 2x \, dx$

2. $\int_{-2}^{3} (2x + 4) \, dx$

3. $\int_{-2}^{1} (3x + 4) \, dx$

4. $\int_{-2}^{1} 4 \, dx$

5. $\int_{6}^{8} (7 - x) \, dx$

6. $\int_{\pi/2}^{3\pi/2} \sin x \, dx$

7. $\int_{0}^{5} \sqrt{25 - x^2} \, dx$

8. $\int_{-2}^{3} |x| \, dx$

9. $\int_{-2}^{2} (2 - |x|) \, dx$

10. $\int_{-2}^{5} (3 + x - 2|x|) \, dx$

11. Calculate $\int_0^{10} (8 - x) \, dx$ in two ways:

(a) As the limit $\lim_{N \to \infty} R_N$

(b) By sketching the relevant signed area and using geometry

12. Calculate $\int_{-1}^{4} (4x - 8) \, dx$ in two ways: As the limit $\lim_{N \to \infty} R_N$ and using geometry.

In Exercises 13 and 14, refer to Figure 14.

13. Evaluate: (a) $\int_0^2 f(x) \, dx$ (b) $\int_0^6 f(x) \, dx$

14. Evaluate: (a) $\int_1^4 f(x) \, dx$ (b) $\int_1^6 |f(x)| \, dx$

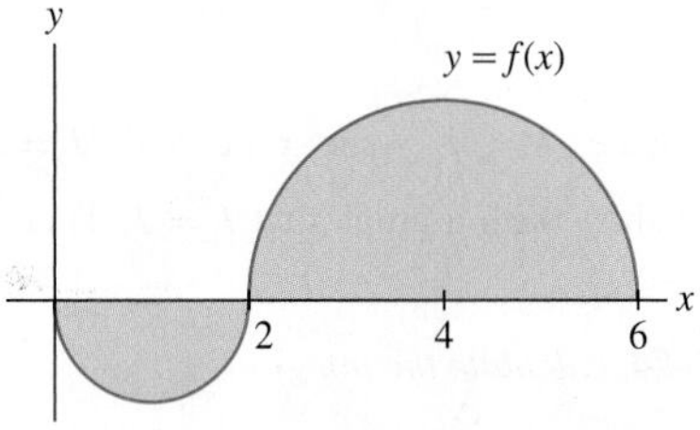

FIGURE 14 The two parts of the graph are semicircles.

In Exercises 15 and 16, refer to Figure 15.

15. Evaluate $\int_0^3 g(t) \, dt$ and $\int_3^5 g(t) \, dt$.

16. Find a, b, and c such that $\int_0^a g(t) \, dt$ and $\int_b^c g(t) \, dt$ are as large as possible.

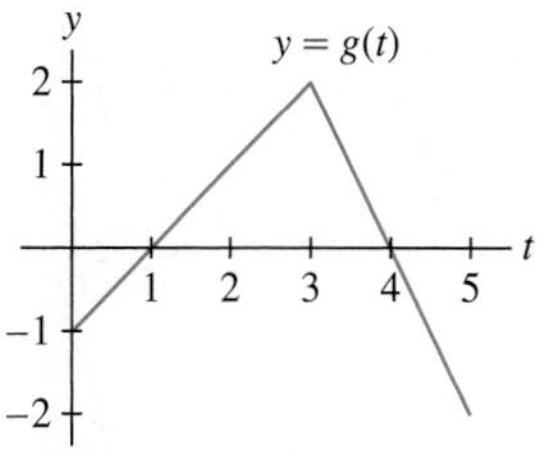

FIGURE 15

17. Describe the partition P and the set of sample points C for the Riemann sum shown in Figure 16. Compute the value of the Riemann sum.

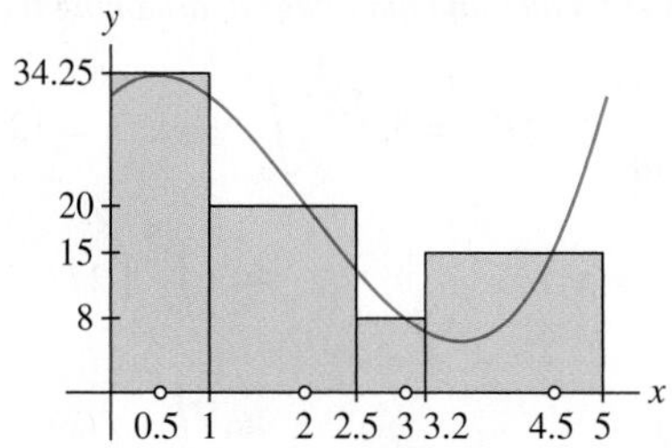

FIGURE 16

18. Compute $R(f, P, C)$ for $f(x) = x^2 + x$ for the partition P and the set of sample points C in Figure 16.

In Exercises 19–22, calculate the Riemann sum $R(f, P, C)$ for the given function, partition, and choice of sample points. Also, sketch the graph of f and the rectangles corresponding to $R(f, P, C)$.

19. $f(x) = x$, $P = \{1, 1.2, 1.5, 2\}$, $C = \{1.1, 1.4, 1.9\}$

20. $f(x) = 2x + 3$, $P = \{-4, -1, 1, 4, 8\}$, $C = \{-3, 0, 2, 5\}$

21. $f(x) = x^2 + x$, $P = \{2, 3, 4.5, 5\}$, $C = \{2, 3.5, 5\}$

22. $f(x) = \sin x$, $P = \left\{0, \frac{\pi}{6}, \frac{\pi}{3}, \frac{\pi}{2}\right\}$, $C = \{0.4, 0.7, 1.2\}$

In Exercises 23–28, sketch the signed area represented by the integral. Indicate the regions of positive and negative area.

23. $\int_0^5 (4x - x^2) \, dx$

24. $\int_{-\pi/4}^{\pi/4} \tan x \, dx$

25. $\int_{\pi}^{2\pi} \sin x \, dx$

26. $\int_0^{3\pi} \sin x \, dx$

27. $\int_0^6 (|12 - 4x| - 4) \, dx$

28. $\int_{-2}^{2} (t^2 - 1)(t^2 - 4) \, dx$

In Exercises 29–32, determine the sign of the integral without calculating it. Draw a graph if necessary.

29. $\int_{-2}^{1} x^4 \, dx$

30. $\int_{-2}^{1} x^3 \, dx$

31. GU $\int_0^{2\pi} x \sin x\, dx$

32. GU $\int_0^{2\pi} \frac{\sin x}{x}\, dx$

In Exercises 33–42, use properties of the integral and the formulas in the summary to calculate the integrals.

33. $\int_0^4 (6t - 3)\, dt$

34. $\int_{-3}^2 (4x + 7)\, dx$

35. $\int_0^9 x^2\, dx$

36. $\int_2^5 x^2\, dx$

37. $\int_0^1 (u^2 - 2u)\, du$

38. $\int_0^{1/2} (12y^2 + 6y)\, dy$

39. $\int_{-3}^1 (7t^2 + t + 1)\, dt$

40. $\int_{-3}^3 (9x - 4x^2)\, dx$

41. $\int_{-a}^1 (x^2 + x)\, dx$

42. $\int_a^{a^2} x^2\, dx$

In Exercises 43–47, calculate the integral, assuming that

$$\int_0^5 f(x)\, dx = 5, \qquad \int_0^5 g(x)\, dx = 12$$

43. $\int_0^5 (f(x) + g(x))\, dx$

44. $\int_0^5 \left(2f(x) - \frac{1}{3}g(x)\right) dx$

45. $\int_5^0 g(x)\, dx$

46. $\int_0^5 (f(x) - x)\, dx$

47. Is it possible to calculate $\int_0^5 g(x)f(x)\, dx$ from the information given?

48. Prove by computing the limit of right-endpoint approximations:

$$\int_0^b x^3\, dx = \frac{b^4}{4} \qquad (9)$$

In Exercises 49–54, evaluate the integral using the formulas in the summary and Eq. (9).

49. $\int_0^3 x^3\, dx$

50. $\int_1^3 x^3\, dx$

51. $\int_0^3 (x - x^3)\, dx$

52. $\int_0^1 (2x^3 - x + 4)\, dx$

53. $\int_0^1 (12x^3 + 24x^2 - 8x)\, dx$

54. $\int_{-2}^2 (2x^3 - 3x^2)\, dx$

In Exercises 55–58, calculate the integral, assuming that

$$\int_0^1 f(x)\, dx = 1, \qquad \int_0^2 f(x)\, dx = 4, \qquad \int_1^4 f(x)\, dx = 7$$

55. $\int_0^4 f(x)\, dx$

56. $\int_1^2 f(x)\, dx$

57. $\int_4^1 f(x)\, dx$

58. $\int_2^4 f(x)\, dx$

In Exercises 59–62, express each integral as a single integral.

59. $\int_0^3 f(x)\, dx + \int_3^7 f(x)\, dx$

60. $\int_2^9 f(x)\, dx - \int_4^9 f(x)\, dx$

61. $\int_2^9 f(x)\, dx - \int_2^5 f(x)\, dx$

62. $\int_7^3 f(x)\, dx + \int_3^9 f(x)\, dx$

In Exercises 63–66, calculate the integral, assuming that f is integrable and $\int_1^b f(x)\, dx = 1 - b^{-1}$ for all $b > 0$.

63. $\int_1^5 f(x)\, dx$

64. $\int_3^5 f(x)\, dx$

65. $\int_1^6 (3f(x) - 4)\, dx$

66. $\int_{1/2}^1 f(x)\, dx$

67. Explain the difference in graphical interpretation between $\int_a^b f(x)\, dx$ and $\int_a^b |f(x)|\, dx$.

68. Use the graphical interpretation of the definite integral to explain the inequality

$$\left| \int_a^b f(x)\, dx \right| \le \int_a^b |f(x)|\, dx$$

where $f(x)$ is continuous. Explain also why equality holds if and only if either $f(x) \ge 0$ for all x or $f(x) \le 0$ for all x.

69. Let $f(x) = x$. Find an interval $[a, b]$ such that

$$\left| \int_a^b f(x)\, dx \right| = \frac{1}{2} \quad \text{and} \quad \int_a^b |f(x)|\, dx = \frac{3}{2}$$

70. Evaluate $I = \int_0^{2\pi} \sin^2 x\, dx$ and $J = \int_0^{2\pi} \cos^2 x\, dx$ as follows. First show with a graph that $I = J$. Then prove that $I + J = 2\pi$.

In Exercises 71–74, calculate the integral.

71. $\int_0^6 |3 - x|\, dx$

72. $\int_1^3 |2x - 4|\, dx$

73. $\int_{-1}^1 |x^3|\, dx$

74. $\int_0^2 |x^2 - 1|\, dx$

75. Use the Comparison Theorem to show that

$$\int_0^1 x^5\, dx \le \int_0^1 x^4\, dx, \qquad \int_1^2 x^4\, dx \le \int_1^2 x^5\, dx$$

76. Prove that $\frac{1}{3} \le \int_4^6 \frac{1}{x}\,dx \le \frac{1}{2}$.

77. Prove that $0.0198 \le \int_{0.2}^{0.3} \sin x\,dx \le 0.0296$. *Hint:* Show that $0.198 \le \sin x \le 0.296$ for x in $[0.2, 0.3]$.

78. Prove that $0.277 \le \int_{\pi/8}^{\pi/4} \cos x\,dx \le 0.363$.

79. Prove that $0 \le \int_{\pi/4}^{\pi/2} \frac{\sin x}{x}\,dx \le \frac{\sqrt{2}}{2}$.

80. Find upper and lower bounds for $\int_0^1 \frac{dx}{\sqrt{5x^3+4}}$.

81. Suppose that $f(x) \le g(x)$ on $[a, b]$. By the Comparison Theorem, $\int_a^b f(x)\,dx \le \int_a^b g(x)\,dx$. Is it also true that $f'(x) \le g'(x)$ for $x \in [a, b]$? If not, give a counterexample.

82. State whether true or false. If false, sketch the graph of a counterexample.

(a) If $f(x) > 0$, then $\int_a^b f(x)\,dx > 0$.

(b) If $\int_a^b f(x)\,dx > 0$, then $f(x) > 0$.

Further Insights and Challenges

83. Explain graphically: If $f(x)$ is an odd function, then $\int_{-a}^{a} f(x)\,dx = 0$.

84. Compute $\int_{-1}^{1} \sin(\sin(x))(\sin^2(x) + 1)\,dx$.

85. Let k and b be positive. Show, by comparing the right-endpoint approximations, that

$$\int_0^b x^k\,dx = b^{k+1} \int_0^1 x^k\,dx$$

86. Suppose that f and g are continuous functions such that, *for all* a,

$$\int_{-a}^{a} f(x)\,dx = \int_{-a}^{a} g(x)\,dx$$

Give an *intuitive* argument showing that $f(0) = g(0)$. Explain your idea with a graph.

87. Theorem 4 remains true without the assumption $a \le b \le c$. Verify this for the cases $b < a < c$ and $c < a < b$.

5.3 The Fundamental Theorem of Calculus, Part I

The FTC was first stated clearly by Isaac Newton in 1666, although other mathematicians, including Newton's teacher Isaac Barrow, had discovered versions of it earlier.

The Fundamental Theorem of Calculus (FTC) reveals an unexpected connection between the two main operations of calculus: differentiation and integration. The theorem has two parts. Although they are closely related, we discuss them in separate sections to emphasize the different ways they are used.

To explain FTC I, recall a result from Example 5 of Section 5.2:

$$\int_4^7 x^2\,dx = \left(\frac{1}{3}\right)7^3 - \left(\frac{1}{3}\right)4^3 = 93$$

REMINDER
$F(x)$ is called an ***antiderivative*** of $f(x)$ if $F'(x) = f(x)$. We say also that $F(x)$ is an ***indefinite integral*** of $f(x)$, and we use the notation

$$\int f(x)\,dx = F(x) + C$$

Now observe that $F(x) = \frac{1}{3}x^3$ is an antiderivative of x^2, so we can write

$$\int_4^7 x^2\,dx = F(7) - F(4)$$

According to FTC I, this is no coincidence; this relation between the definite integral and the antiderivative holds in general.

THEOREM 1 The Fundamental Theorem of Calculus, Part I Assume that $f(x)$ is continuous on $[a, b]$. If $F(x)$ is an antiderivative of $f(x)$ on $[a, b]$, then

$$\int_a^b f(x)\,dx = F(b) - F(a) \qquad (1)$$

Proof The quantity $F(b) - F(a)$ is the total change in F (also called the "net change") over the interval $[a, b]$. Our task is to relate it to the integral of $F'(x) = f(x)$. There are two main steps.

Step 1. **Write total change as a sum of small changes.**
Given any partition P of $[a, b]$:

$$P : x_0 = a < x_1 < x_2 < \cdots < x_N = b$$

we can break up $F(b) - F(a)$ as a sum of changes over the intervals $[x_{i-1}, x_i]$:

$$F(b) - F(a) = \big(\cancel{F(x_1)} - F(a)\big) + \big(F(x_2) - \cancel{F(x_1)}\big) + \cdots + \big(F(b) - \cancel{F(x_{N-1})}\big)$$

On the right-hand side, $F(x_1)$ is canceled by $-F(x_1)$ in the second term, $F(x_2)$ is canceled by $-F(x_2)$, etc. (Figure 1). In summation notation,

$$F(b) - F(a) = \sum_{i=1}^{N} \big(F(x_i) - F(x_{i-1})\big) \qquad \boxed{2}$$

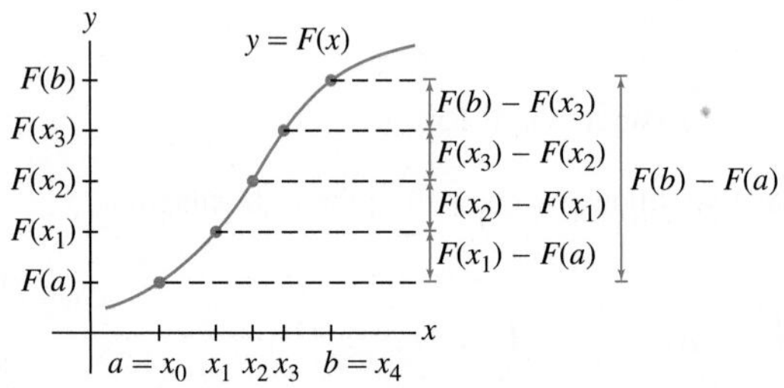

FIGURE 1 Note the cancelation when we write $F(b) - F(a)$ as a sum of small changes $F(x_i) - F(x_{i-1})$.

Step 2. **Interpret Eq. (2) as a Riemann sum.**
The Mean Value Theorem tells us that there is a point c_i^* in $[x_{i-1}, x_i]$ such that

$$F(x_i) - F(x_{i-1}) = F'(c_i^*)(x_i - x_{i-1}) = f(c_i^*)(x_i - x_{i-1}) = f(c_i^*)\,\Delta x_i$$

Therefore, Eq. (2) can be written

$$F(b) - F(a) = \sum_{i=1}^{N} f(c_i^*)\,\Delta x_i$$

This sum is the Riemann sum $R(f, P, C^*)$ with sample points $C^* = \{c_i^*\}$.

Now, $f(x)$ is integrable (Theorem 1, Section 5.2), so $R(f, P, C^*)$ approaches $\int_a^b f(x)\,dx$ as the norm $\|P\|$ tends to zero. On the other hand, $R(f, P, C^*)$ is *equal* to $F(b) - F(a)$ with our particular choice C^* of sample points. This proves the desired result:

$$F(b) - F(a) = \lim_{\|P\|\to 0} R(f, P, C^*) = \int_a^b f(x)\,dx \qquad \blacksquare$$

CONCEPTUAL INSIGHT **A Tale of Two Graphs** In the proof of FTC I, we used the MVT to write a small change in $F(x)$ in terms of the derivative $F'(x) = f(x)$:

$$F(x_i) - F(x_{i-1}) = f(c_i^*)\Delta x_i$$

But $f(c_i^*)\Delta x_i$ is the area of a thin rectangle that approximates a sliver of area under the graph of $f(x)$ (Figure 2). *This is the essence of the Fundamental Theorem*: the total change $F(b) - F(a)$ is equal to the sum of small changes $F(x_i) - F(x_{i-1})$, which in turn is equal to the sum of the areas of rectangles in a Riemann sum approximation for $f(x)$. We derive the Fundamental Theorem itself by taking the limit as the width of the rectangles tends to zero.

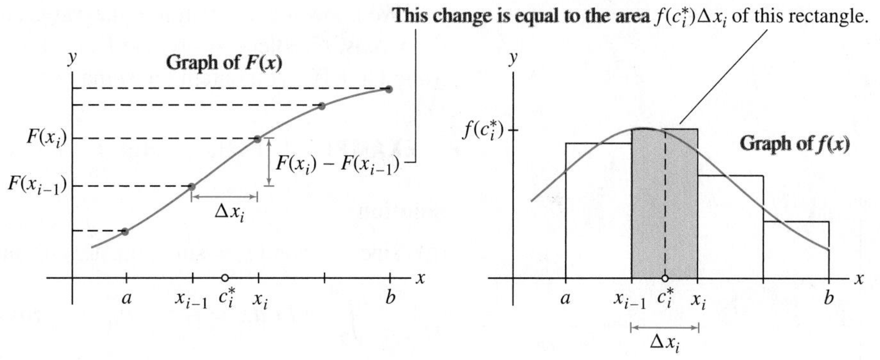

FIGURE 2

FTC I tells us that if we can find an antiderivative of $f(x)$, then we can compute the definite integral easily, without calculating any limits. It is for this reason that we use the integral sign $\int$ for both the definite integral $\int_a^b f(x)dx$ and the indefinite integral (antiderivative) $\int f(x)dx$.

Notation: $F(b) - F(a)$ is denoted $F(x)\big|_a^b$. In this notation, the FTC reads

$$\int_a^b f(x)\,dx = F(x)\big|_a^b$$

← REMINDER The Power Rule for Integrals (valid for $n \neq -1$) states:

$$\int x^n\,dx = \frac{x^{n+1}}{n+1} + C$$

■ **EXAMPLE 1** Calculate the area under the graph of $f(x) = x^3$ over $[2, 4]$.

Solution Since $F(x) = \frac{1}{4}x^4$ is an antiderivative of $f(x) = x^3$, FTC I gives us

$$\int_2^4 x^3\,dx = F(4) - F(2) = \frac{1}{4}x^4\Big|_2^4 = \frac{1}{4}4^4 - \frac{1}{4}2^4 = 60$$ ■

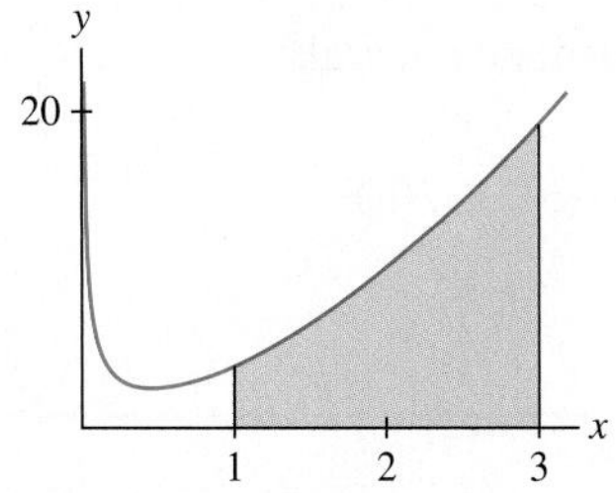

FIGURE 3 Region under the graph of $g(x) = x^{-3/4} + 3x^{5/3}$ over $[1, 3]$.

■ **EXAMPLE 2** Find the areaunder $g(x) = x^{-3/4} + 3x^{5/3}$ over $[1, 3]$

Solution The function $G(x) = 4x^{1/4} + \frac{9}{8}x^{8/3}$ is an antiderivative of $g(x)$. The area (Figure 3) is equal to

$$\int_1^3 (x^{-3/4} + 3x^{5/3})\,dx = G(x)\Big|_1^3 = \left(4x^{1/4} + \frac{9}{8}x^{8/3}\right)\Big|_1^3$$

$$= \left(4 \cdot 3^{1/4} + \frac{9}{8} \cdot 3^{8/3}\right) - \left(4 \cdot 1^{1/4} + \frac{9}{8} \cdot 1^{8/3}\right)$$

$$\approx 26.325 - 5.125 = 21.2$$ ■

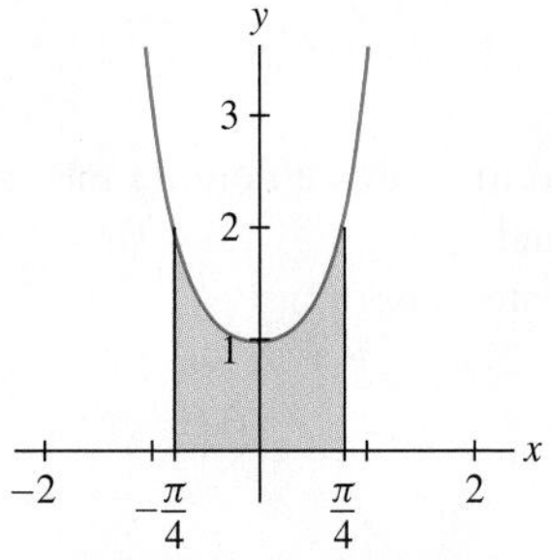

FIGURE 4 Graph of $y = \sec^2 x$.

■ **EXAMPLE 3** Calculate $\int_{-\pi/4}^{\pi/4} \sec^2 x\,dx$ and sketch the corresponding region.

Solution Figure 4 shows the region. Recall that $(\tan x)' = \sec^2 x$. Therefore,

$$\int_{-\pi/4}^{\pi/4} \sec^2 x\,dx = \tan x\Big|_{-\pi/4}^{\pi/4} = \tan\left(\frac{\pi}{4}\right) - \tan\left(-\frac{\pi}{4}\right) = 1 - (-1) = 2$$ ■

We know that the definite integral is equal to the *signed* area between the graph and the x-axis. Needless to say, the FTC "knows" this also: When you evaluate an integral using the FTC, you obtain the signed area.

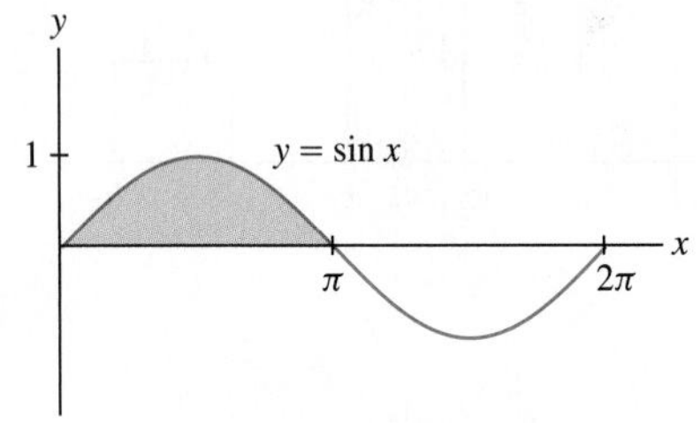

FIGURE 5 The area of one hump is 2. The signed area over $[0, 2\pi]$ is zero.

■ **EXAMPLE 4** Evaluate **(a)** $\int_0^{\pi} \sin x\, dx$ and **(b)** $\int_0^{2\pi} \sin x\, dx$.

Solution

(a) Since $(-\cos x)' = \sin x$, the area of one "hump" (Figure 5) is

$$\int_0^{\pi} \sin x\, dx = -\cos x\Big|_0^{\pi} = -\cos\pi - (-\cos 0) = -(-1) - (-1) = 2$$

(b) We expect the signed area over $[0, 2\pi]$ to be zero since the second hump lies below the x-axis, and, indeed,

$$\int_0^{2\pi} \sin x\, dx = -\cos x\Big|_0^{2\pi} = (-\cos(2\pi) - (-\cos 0)) = -1 - (-1) = 0$$ ■

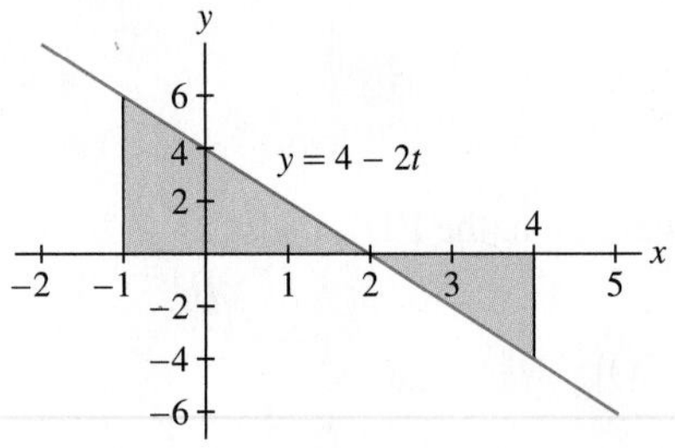

FIGURE 6

■ **EXAMPLE 5** Evaluate $\int_{-1}^{4} (4 - 2t)\, dt$.

Solution The function $F(x) = 4t - t^2$ is an antiderivative of $f(x) = 4 - 2t$, so the definite integral (the signed area under the graph in Figure 6) is

$$\int_{-1}^{4} (4 - 2t)\, dt = (4t - t^2)\Big|_{-1}^{4} = \left(4\cdot 4 - 4^2\right) - \left(4\cdot(-1) - (-1)^2\right) = 0 - (-5) = 5$$ ■

CONCEPTUAL INSIGHT Which Antiderivative? Antiderivatives are unique only to within an additive constant (Section 4.8). Does it matter which antiderivative is used in the FTC? The answer is no. If $F(x)$ and $G(x)$ are both antiderivatives of $f(x)$, then $F(x) = G(x) + C$ for some constant C, and

$$F(b) - F(a) = \underbrace{(G(b) + C) - (G(a) + C)}_{\text{The constant cancels}} = G(b) - G(a)$$

The two antiderivatives yield the same value for the definite integral:

$$\int_a^b f(x)\, dx = F(b) - F(a) = G(b) - G(a)$$

5.3 SUMMARY

- The Fundamental Theorem of Calculus, Part I, states that

$$\int_a^b f(x)\, dx = F(b) - F(a)$$

where $F(x)$ is an antiderivative of $f(x)$. FTC I is used to evaluate definite integrals in cases where we can find an antiderivative of the integrand.

- Basic antiderivative formulas for evaluating definite integrals:

$$\int x^n\, dx = \frac{x^{n+1}}{n+1} + C \qquad \text{for } n \neq -1$$

$$\int \sin x\, dx = -\cos x + C, \qquad \int \cos x\, dx = \sin x + C$$

$$\int \sec^2 x\,dx = \tan x + C, \qquad \int \csc^2 x\,dx = -\cot x + C$$

$$\int \sec x \tan x\,dx = \sec x + C, \qquad \int \csc x \cot x\,dx = -\csc x + C$$

5.3 EXERCISES

Preliminary Questions

1. Suppose that $F'(x) = f(x)$ and $F(0) = 3$, $F(2) = 7$.
(a) What is the area under $y = f(x)$ over $[0, 2]$ if $f(x) \geq 0$?
(b) What is the graphical interpretation of $F(2) - F(0)$ if $f(x)$ takes on both positive and negative values?

2. Suppose that $f(x)$ is a *negative* function with antiderivative F such that $F(1) = 7$ and $F(3) = 4$. What is the area (a positive number) between the x-axis and the graph of $f(x)$ over $[1, 3]$?

3. Are the following statements true or false? Explain.
(a) FTC I is valid only for positive functions.
(b) To use FTC I, you have to choose the right antiderivative.
(c) If you cannot find an antiderivative of $f(x)$, then the definite integral does not exist.

4. Evaluate $\int_2^9 f'(x)\,dx$ where $f(x)$ is differentiable and $f(2) = f(9) = 4$.

Exercises

In Exercises 1–4, sketch the region under the graph of the function and find its area using FTC I.

1. $f(x) = x^2$, $[0, 1]$
2. $f(x) = 2x - x^2$, $[0, 2]$
3. $f(x) = x^{-2}$, $[1, 2]$
4. $f(x) = \cos x$, $\left[0, \frac{\pi}{2}\right]$

In Exercises 5–34, evaluate the integral using FTC I.

5. $\int_3^6 x\,dx$
6. $\int_0^9 2\,dx$
7. $\int_{-3}^5 (3t - 4)\,dt$
8. $\int_2^4 (24 - 5u)\,du$
9. $\int_0^1 (4x - 9x^2)\,dx$
10. $\int_{-3}^2 u^2\,du$
11. $\int_0^2 (12x^5 + 3x^2 - 4x)\,dx$
12. $\int_{-2}^2 (10x^9 + 3x^5)\,dx$
13. $\int_3^0 (2t^3 - 6t^2)\,dt$
14. $\int_{-1}^1 (5u^4 + u^2 - u)\,du$
15. $\int_0^4 \sqrt{y}\,dy$
16. $\int_1^8 x^{4/3}\,dx$
17. $\int_{1/16}^1 t^{1/4}\,dt$
18. $\int_4^1 t^{5/2}\,dt$
19. $\int_1^3 \frac{dt}{t^2}$
20. $\int_1^4 x^{-4}\,dx$
21. $\int_{1/2}^1 \frac{8}{x^3}\,dx$
22. $\int_{-2}^{-1} \frac{1}{x^3}\,dx$
23. $\int_1^2 (x^2 - x^{-2})\,dx$
24. $\int_1^9 t^{-1/2}\,dt$
25. $\int_1^{27} \frac{t+1}{\sqrt{t}}\,dt$
26. $\int_{8/27}^1 \frac{10t^{4/3} - 8t^{1/3}}{t^2}\,dt$
27. $\int_{\pi/4}^{3\pi/4} \sin\theta\,d\theta$
28. $\int_{2\pi}^{4\pi} \sin x\,dx$
29. $\int_0^{\pi/2} \cos\left(\frac{1}{3}\theta\right)\,d\theta$
30. $\int_{\pi/4}^{5\pi/8} \cos 2x\,dx$
31. $\int_0^{\pi/6} \sec^2\left(3t - \frac{\pi}{6}\right)\,dt$
32. $\int_0^{\pi/6} \sec\theta\tan\theta\,d\theta$
33. $\int_{\pi/20}^{\pi/10} \csc 5x \cot 5x\,dx$
34. $\int_{\pi/28}^{\pi/14} \csc^2 7y\,dy$

In Exercises 35–40, write the integral as a sum of integrals without absolute values and evaluate.

35. $\int_{-2}^1 |x|\,dx$
36. $\int_0^5 |3 - x|\,dx$
37. $\int_{-2}^3 |x^3|\,dx$
38. $\int_0^3 |x^2 - 1|\,dx$
39. $\int_0^{\pi} |\cos x|\,dx$
40. $\int_0^5 |x^2 - 4x + 3|\,dx$

In Exercises 41–44, evaluate the integral in terms of the constants.

41. $\int_1^b x^3\,dx$
42. $\int_b^a x^4\,dx$

43. $\int_1^b x^5\,dx$

44. $\int_{-x}^{x} (t^3 + t)\,dt$

45. Calculate $\int_{-2}^{3} f(x)\,dx$, where

$$f(x) = \begin{cases} 12 - x^2 & \text{for } x \le 2 \\ x^3 & \text{for } x > 2 \end{cases}$$

46. Calculate $\int_0^{2\pi} f(x)\,dx$, where

$$f(x) = \begin{cases} \cos x & \text{for } x \le \pi \\ \cos x - \sin 2x & \text{for } x > \pi \end{cases}$$

47. Use FTC I to show that $\int_{-1}^{1} x^n\,dx = 0$ if n is an odd whole number. Explain graphically.

48. CAS Plot the function $f(x) = \sin 3x - x$. Find the positive root of $f(x)$ to three places and use it to find the area under the graph of $f(x)$ in the first quadrant.

49. Calculate $F(4)$ given that $F(1) = 3$ and $F'(x) = x^2$. *Hint:* Express $F(4) - F(1)$ as a definite integral.

50. Calculate $G(16)$, where $dG/dt = t^{-1/2}$ and $G(9) = -5$.

51. Does $\int_0^1 x^n\,dx$ get larger or smaller as n increases? Explain graphically.

52. Show that the area of the shaded parabolic arch in Figure 7 is equal to four-thirds the area of the triangle shown.

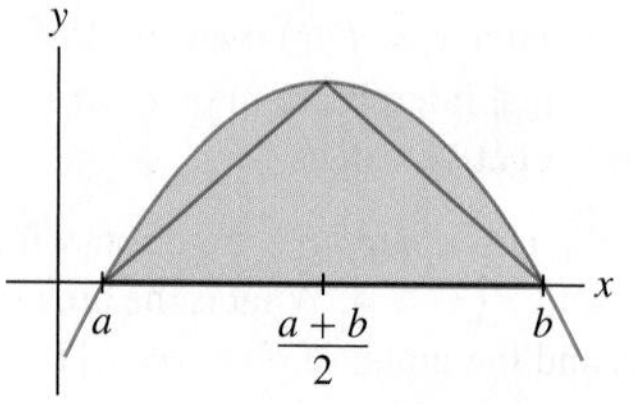

FIGURE 7 Graph of $y = (x - a)(b - x)$.

Further Insights and Challenges

53. Prove a famous result of Archimedes (generalizing Exercise 52): For $r < s$, the area of the shaded region in Figure 8 is equal to four-thirds the area of triangle $\triangle ACE$, where C is the point on the parabola at which the tangent line is parallel to secant line $\overline{AE}$.

(a) Show that C has x-coordinate $(r + s)/2$.

(b) Show that $ABDE$ has area $(s - r)^3/4$ by viewing it as a parallelogram of height $s - r$ and base of length $\overline{CF}$.

(c) Show that $\triangle ACE$ has area $(s - r)^3/8$ by observing that it has the same base and height as the parallelogram.

(d) Compute the shaded area as the area under the graph minus the area of a trapezoid, and prove Archimedes' result.

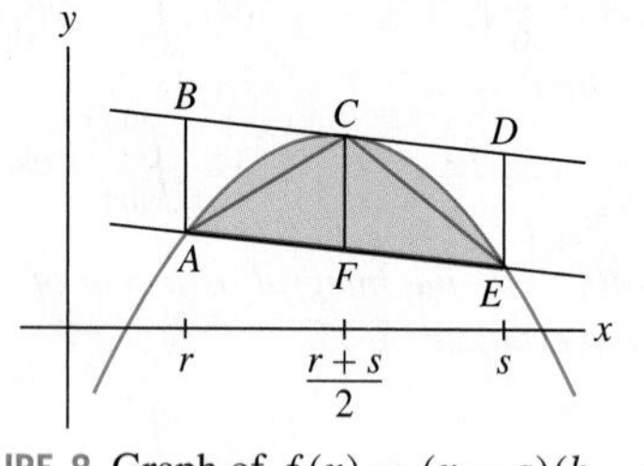

FIGURE 8 Graph of $f(x) = (x - a)(b - x)$.

54. (a) Apply the Comparison Theorem (Theorem 5 in Section 5.2) to the inequality $\sin x \le x$ (valid for $x \ge 0$) to prove that

$$1 - \frac{x^2}{2} \le \cos x \le 1$$

(b) Apply it again to prove that

$$x - \frac{x^3}{6} \le \sin x \le x \quad (\text{for } x \ge 0)$$

(c) Verify these inequalities for $x = 0.3$.

55. Use the method of Exercise 54 to prove that

$$1 - \frac{x^2}{2} \le \cos x \le 1 - \frac{x^2}{2} + \frac{x^4}{24}$$

$$x - \frac{x^3}{6} \le \sin x \le x - \frac{x^3}{6} + \frac{x^5}{120} \quad (\text{for } x \ge 0)$$

Verify these inequalities for $x = 0.1$. Why have we specified $x \ge 0$ for $\sin x$ but not for $\cos x$?

56. Calculate the next pair of inequalities for $\sin x$ and $\cos x$ by integrating the results of Exercise 55. Can you guess the general pattern?

57. Use FTC I to prove that if $|f'(x)| \le K$ for $x \in [a, b]$, then $|f(x) - f(a)| \le K|x - a|$ for $x \in [a, b]$.

58. (a) Use Exercise 57 to prove that $|\sin a - \sin b| \le |a - b|$ for all a, b.

(b) Let $f(x) = \sin(x + a) - \sin x$. Use part (a) to show that the graph of $f(x)$ lies between the horizontal lines $y = \pm a$.

(c) GU Plot $f(x)$ and the lines $y = \pm a$ to verify (b) for $a = 0.5$ and $a = 0.2$.

5.4 The Fundamental Theorem of Calculus, Part II

Part I of the Fundamental Theorem says that we can use antiderivatives to compute definite integrals. Part II turns this relationship around: It tells us that we can use the definite integral to *construct* antiderivatives.

To state Part II, we introduce the **area function** of f with lower limit a:

$A(x)$ is sometimes called the ***cumulative area function****. In the definition of $A(x)$, we use t as the variable of integration to avoid confusion with x, which is the upper limit of integration. In fact, t is a dummy variable and may be replaced by any other variable.*

$$A(x) = \int_a^x f(t)\,dt = \text{signed area from } a \text{ to } x$$

In essence, we turn the definite integral into a function by treating the upper limit x as a variable (Figure 1). Note that $A(a) = 0$ because $A(a) = \int_a^a f(t)\,dt = 0$.

In some cases we can find an explicit formula for $A(x)$ [Figure 2].

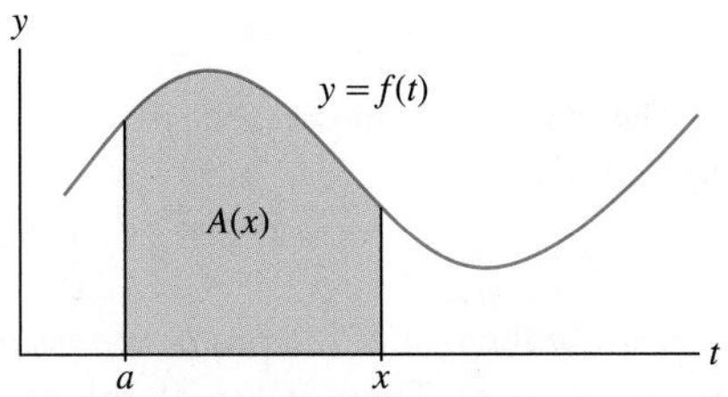

FIGURE 1 $A(x)$ is the area under the graph from a to x.

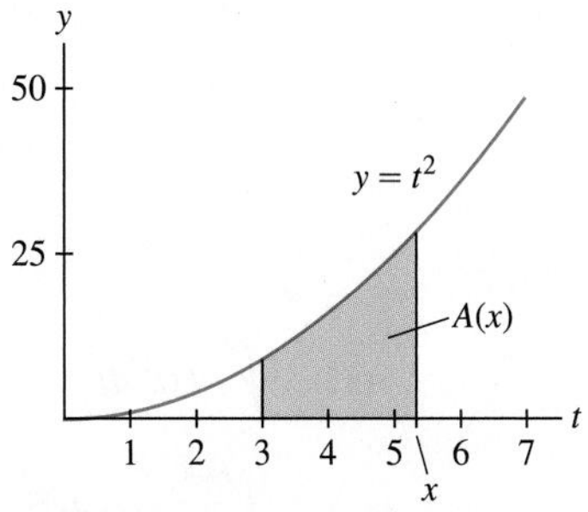

FIGURE 2 The area under $y = t^2$ from 3 to x is $A(x) = \frac{1}{3}x^3 - 9$.

■ **EXAMPLE 1** Find a formula for the area function $A(x) = \int_3^x t^2\,dt$.

Solution The function $F(t) = \frac{1}{3}t^3$ is an antiderivative for $f(t) = t^2$. By FTC I,

$$A(x) = \int_3^x t^2\,dt = F(x) - F(3) = \frac{1}{3}x^3 - \frac{1}{3}\cdot 3^3 = \frac{1}{3}x^3 - 9 \qquad ■$$

Note, in the previous example, that *the derivative of $A(x)$ is $f(x)$ itself:*

$$A'(x) = \frac{d}{dx}\left(\frac{1}{3}x^3 - 9\right) = x^2$$

FTC II states that this relation always holds: The derivative of the area function is equal to the original function.

THEOREM 1 Fundamental Theorem of Calculus, Part II Assume that $f(x)$ is continuous on an open interval I and let $a \in I$. Then the area function

$$A(x) = \int_a^x f(t)\,dt$$

is an antiderivative of $f(x)$ on I; that is, $A'(x) = f(x)$. Equivalently,

$$\frac{d}{dx}\int_a^x f(t)\,dt = f(x)$$

Furthermore, $A(x)$ satisfies the initial condition $A(a) = 0$.

Proof First, we use the additivity property of the definite integral to write the change in $A(x)$ over $[x, x+h]$ as an integral:

$$A(x+h) - A(x) = \int_a^{x+h} f(t)\,dt - \int_a^x f(t)\,dt = \int_x^{x+h} f(t)\,dt$$

In other words, $A(x+h) - A(x)$ is equal to the area of the thin sliver between the graph and the x-axis from x to $x+h$ in Figure 3.

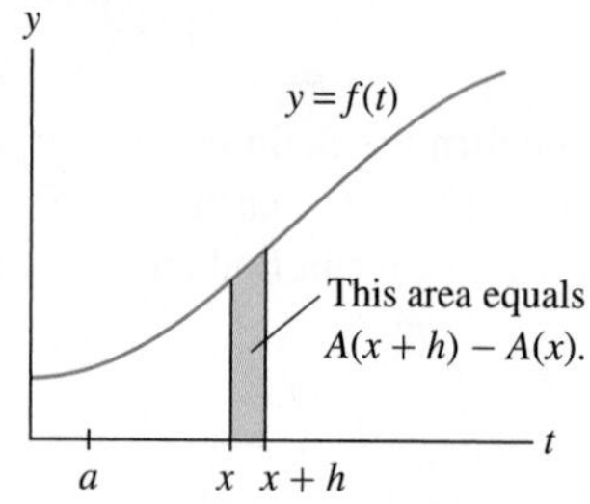

FIGURE 3 The area of the thin sliver equals $A(x+h) - A(x)$.

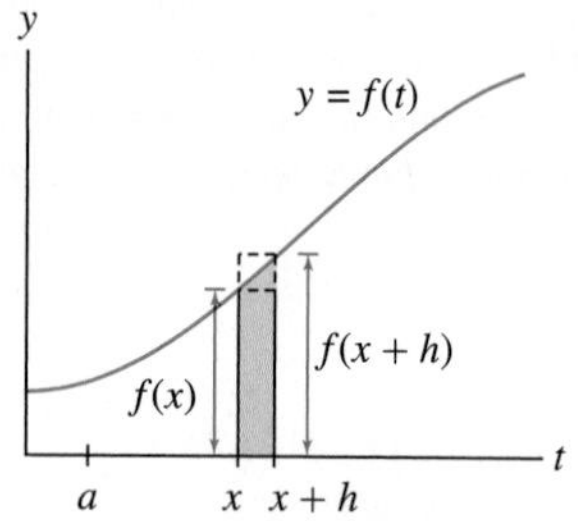

FIGURE 4 The shaded sliver lies between the rectangles of heights $f(x)$ and $f(x+h)$.

In this proof,

$$A(x) = \int_a^x f(t)\,dt$$

$$A(x+h) - A(x) = \int_x^{x+h} f(t)\,dt$$

$$A'(x) = \lim_{h\to 0} \frac{A(x+h) - A(x)}{h}$$

To simplify the rest of the proof, we assume that $f(x)$ is increasing (see Exercise 50 for the general case). Then, if $h > 0$, this thin sliver lies between the two rectangles of heights $f(x)$ and $f(x+h)$ in Figure 4, and we have

$$\underbrace{hf(x)}_{\text{Area of smaller rectangle}} \le \underbrace{A(x+h) - A(x)}_{\text{Area of sliver}} \le \underbrace{hf(x+h)}_{\text{Area of larger rectangle}}$$

Now divide by h to squeeze the difference quotient between $f(x)$ and $f(x+h)$:

$$f(x) \le \frac{A(x+h) - A(x)}{h} \le f(x+h)$$

We have $\lim_{h\to 0+} f(x+h) = f(x)$ because $f(x)$ is continuous, and $\lim_{h\to 0+} f(x) = f(x)$, so the Squeeze Theorem gives us

$$\lim_{h\to 0+} \frac{A(x+h) - A(x)}{h} = f(x) \qquad \boxed{1}$$

A similar argument shows that for $h < 0$,

$$f(x+h) \le \frac{A(x+h) - A(x)}{h} \le f(x)$$

Again, the Squeeze Theorem gives us

$$\lim_{h\to 0-} \frac{A(x+h) - A(x)}{h} = f(x) \qquad \boxed{2}$$

Equations (1) and (2) show that $A'(x)$ exists and $A'(x) = f(x)$. ■

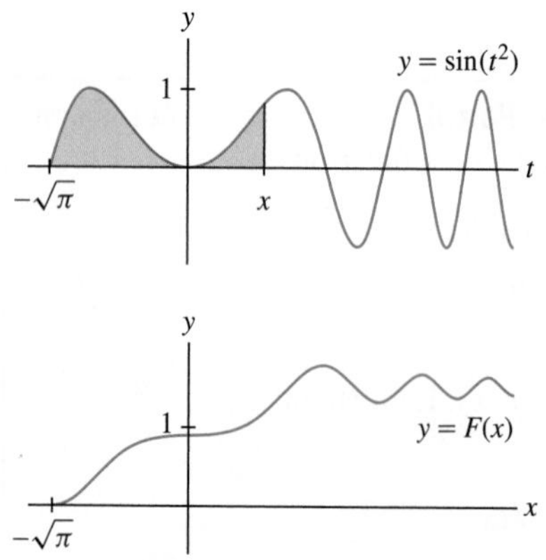

FIGURE 5 Computer-generated graph of $F(x) = \int_{-\sqrt{\pi}}^{x} \sin(t^2)\,dt$.

CONCEPTUAL INSIGHT Many applications (in the sciences, engineering, and statistics) involve functions for which there is no explicit formula. Often, however, these functions can be expressed as definite integrals (or as infinite series). This enables us to compute their values numerically and create plots using a computer algebra system. Figure 5 shows a computer-generated graph of an antiderivative of $f(x) = \sin(x^2)$, for which there is no explicit formula.

■ **EXAMPLE 2 Antiderivative as an Integral** Let $F(x)$ be the particular antiderivative of $f(x) = \sin(x^2)$ satisfying $F(-\sqrt{\pi}) = 0$. Express $F(x)$ as an integral.

Solution According to FTC II, the area function with lower limit $a = -\sqrt{\pi}$ is an antiderivative satisfying $F(-\sqrt{\pi}) = 0$:

$$F(x) = \int_{-\sqrt{\pi}}^{x} \sin(t^2)\,dt$$ ■

■ **EXAMPLE 3 Differentiating an Integral** Find the derivative of

$$A(x) = \int_{2}^{x} \sqrt{1+t^3}\,dt$$

and calculate $A'(2)$, $A'(3)$, and $A(2)$.

Solution By FTC II, $A'(x) = \sqrt{1+x^3}$. In particular,

$$A'(2) = \sqrt{1+2^3} = 3 \qquad \text{and} \qquad A'(3) = \sqrt{1+3^3} = \sqrt{28}$$

On the other hand, $A(2) = \displaystyle\int_{2}^{2} \sqrt{1+t^3}\,dt = 0$. ■

CONCEPTUAL INSIGHT The FTC shows that integration and differentiation are *inverse operations*. By FTC II, if you start with a continuous function $f(x)$ and form the integral $\int_a^x f(x)\,dx$, then you get back the original function by differentiating:

$$f(x) \overset{\text{Integrate}}{\longrightarrow} \int_a^x f(t)\,dt \overset{\text{Differentiate}}{\longrightarrow} \frac{d}{dx}\int_a^x f(t)\,dt = f(x)$$

On the other hand, by FTC I, if you differentiate first and then integrate, you also recover $f(x)$ [but only up to a constant $f(a)$]:

$$f(x) \overset{\text{Differentiate}}{\longrightarrow} f'(x) \overset{\text{Integrate}}{\longrightarrow} \int_a^x f'(t)\,dt = f(x) - f(a)$$

When the upper limit of the integral is a *function* of x rather than x itself, we use FTC II together with the Chain Rule to differentiate the integral.

■ **EXAMPLE 4 The FTC and the Chain Rule** Find the derivative of

$$G(x) = \int_{-2}^{x^2} \sin t\,dt$$

Solution FTC II does not apply directly because the upper limit is x^2 rather than x. It is necessary to recognize that $G(x)$ is a *composite function* with outer function $A(x) = \displaystyle\int_{-2}^{x} \sin t\,dt$:

$$G(x) = A(x^2) = \int_{-2}^{x^2} \sin t\,dt$$

FTC II tells us that $A'(x) = \sin x$, so by the Chain Rule,

$$G'(x) = A'(x^2)\cdot(x^2)' = \sin(x^2)\cdot(2x) = 2x\sin(x^2)$$

Alternatively, we may set $u = x^2$ and use the Chain Rule as follows:

$$\frac{dG}{dx} = \frac{d}{dx}\int_{-2}^{x^2} \sin t\,dt = \left(\frac{d}{du}\int_{-2}^{u} \sin t\,dt\right)\frac{du}{dx} = (\sin u)2x = 2x\sin(x^2)$$ ■

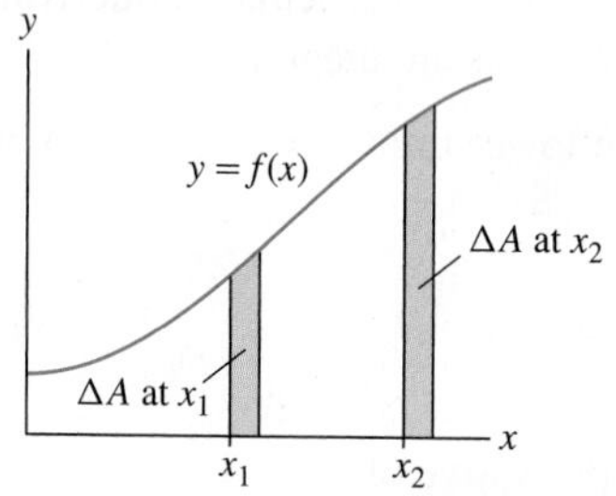

FIGURE 6 The change in area ΔA for a given Δx is larger when $f(x)$ is larger.

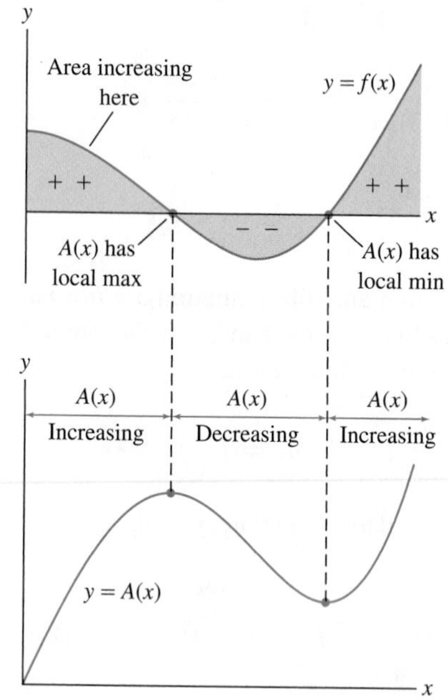

FIGURE 7 The sign of $f(x)$ determines the increasing/decreasing behavior of $A(x)$.

GRAPHICAL INSIGHT Another Tale of Two Graphs FTC II tells us that $A'(x) = f(x)$, or, in other words, $f(x)$ is the rate of change of $A(x)$. If we did not know this result, we might come to suspect it by comparing the graphs of $A(x)$ and $f(x)$. Consider the following:

- Figure 6 shows that the increase in area ΔA for a given Δx is larger at x_2 than at x_1 because $f(x_2) > f(x_1)$. So the size of $f(x)$ determines how quickly $A(x)$ changes, as we would expect if $A'(x) = f(x)$.
- Figure 7 shows that the sign of $f(x)$ determines whether $A(x)$ is increasing or decreasing. If $f(x) > 0$, then $A(x)$ is increasing because positive area is added as we move to the right. When $f(x)$ turns negative, $A(x)$ begins to decrease because we start adding negative area.
- $A(x)$ has a local max at points where $f(x)$ changes sign from + to − (the points where the area turns negative), and has a local min when $f(x)$ changes from − to +. This agrees with the First Derivative Test.

These observations show that $f(x)$ "behaves" like $A'(x)$, as claimed by FTC II.

5.4 SUMMARY

- The *area function* with lower limit a: $A(x) = \int_a^x f(t)\,dt$. It satisfies $A(a) = 0$.
- FTC II: $A'(x) = f(x)$, or, equivalently, $\frac{d}{dx}\int_a^x f(t)\,dt = f(x)$.
- FTC II shows that every continuous function has an antiderivative—namely, its area function (with any lower limit).
- To differentiate the function $G(x) = \int_a^{g(x)} f(t)\,dt$, write $G(x) = A(g(x))$, where $A(x) = \int_a^x f(t)\,dt$. Then use the Chain Rule:

$$G'(x) = A'(g(x))g'(x) = f(g(x))g'(x)$$

5.4 EXERCISES

Preliminary Questions

1. Let $G(x) = \int_4^x \sqrt{t^3+1}\,dt$.

(a) Is the FTC needed to calculate $G(4)$?

(b) Is the FTC needed to calculate $G'(4)$?

2. Which of the following is an antiderivative $F(x)$ of $f(x) = x^2$ satisfying $F(2) = 0$?

(a) $\int_2^x 2t\,dt$ **(b)** $\int_0^2 t^2\,dt$ **(c)** $\int_2^x t^2\,dt$

3. Does every continuous function have an antiderivative? Explain.

4. Let $G(x) = \int_4^{x^3} \sin t\,dt$. Which of the following statements are correct?

(a) $G(x)$ is the composite function $\sin(x^3)$.

(b) $G(x)$ is the composite function $A(x^3)$, where

$$A(x) = \int_4^x \sin(t)\,dt$$

(c) $G(x)$ is too complicated to differentiate.

(d) The Product Rule is used to differentiate $G(x)$.

(e) The Chain Rule is used to differentiate $G(x)$.

(f) $G'(x) = 3x^2 \sin(x^3)$.

Exercises

1. Write the area function of $f(x) = 2x + 4$ with lower limit $a = -2$ as an integral and find a formula for it.

2. Find a formula for the area function of $f(x) = 2x + 4$ with lower limit $a = 0$.

3. Let $G(x) = \int_1^x (t^2 - 2)\,dt$. Calculate $G(1)$, $G'(1)$ and $G'(2)$. Then find a formula for $G(x)$.

4. Find $F(0)$, $F'(0)$, and $F'(3)$, where $F(x) = \int_0^x \sqrt{t^2 + t}\,dt$.

5. Find $G(1)$, $G'(0)$, and $G'(\pi/4)$, where $G(x) = \int_1^x \tan t\,dt$.

6. Find $H(-2)$ and $H'(-2)$, where $H(x) = \int_{-2}^x \frac{du}{u^2 + 1}$.

In Exercises 7–16, find formulas for the functions represented by the integrals.

7. $\int_2^x u^4\,du$

8. $\int_2^x (12t^2 - 8t)\,dt$

9. $\int_0^x \sin u\,du$

10. $\int_{-\pi/4}^x \sec^2 \theta\,d\theta$

11. $\int_2^{\sqrt{x}} \frac{dt}{t^2}$

12. $\int_{\sin\theta}^4 (5t + 9)\,dt$

13. $\int_1^{x^2} t\,dt$

14. $\int_{x/2}^{x/4} \sec^2 u\,du$

15. $\int_{3\sqrt{x}}^{x^{3/2}} t^3\,dt$

16. $\int_{-2x}^x \sec^2 t\,dt$

In Exercises 17–20, express the antiderivative $F(x)$ of $f(x)$ satisfying the given initial condition as an integral.

17. $f(x) = \sqrt{x^3 + 1}$, $F(5) = 0$

18. $f(x) = \frac{x + 1}{x^2 + 9}$, $F(7) = 0$

19. $f(x) = \sec x$, $F(0) = 0$

20. $f(\theta) = \sin(\theta^3)$, $F(-\pi) = 0$

In Exercises 21–24, calculate the derivative.

21. $\frac{d}{dx} \int_0^x (t^5 - 9t^3)\,dt$

22. $\frac{d}{d\theta} \int_1^\theta \cot u\,du$

23. $\frac{d}{dt} \int_{100}^t \sec(5x - 9)\,dx$

24. $\frac{d}{ds} \int_{-2}^s \tan\left(\frac{1}{1 + u^2}\right) du$

25. Let $A(x) = \int_0^x f(t)\,dt$ for $f(x)$ in Figure 8.

(a) Calculate $A(2)$, $A(3)$, $A'(2)$, and $A'(3)$.

(b) Find formulas for $A(x)$ on $[0, 2]$ and $[2, 4]$ and sketch the graph of $A(x)$.

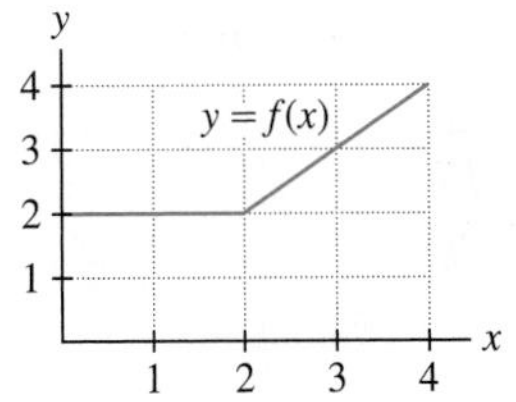

FIGURE 8

26. Make a rough sketch of the graph of $A(x) = \int_0^x g(t)\,dt$ for $g(x)$ in Figure 9.

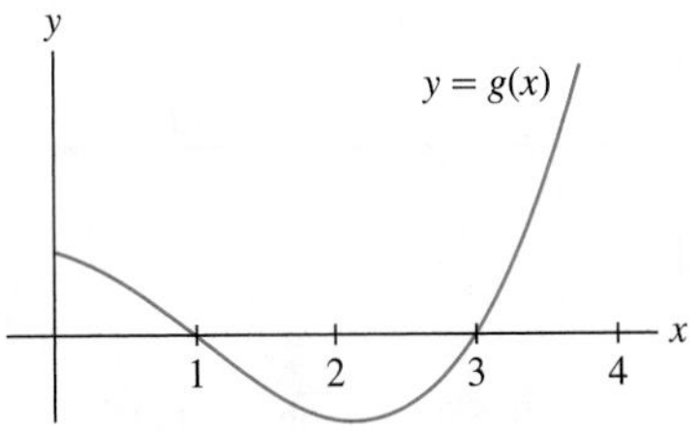

FIGURE 9

27. Verify: $\int_0^x |t|\,dt = \frac{1}{2}x|x|$. *Hint:* Consider $x \ge 0$ and $x \le 0$ separately.

28. Find $G'(1)$, where $G(x) = \int_0^{x^2} \sqrt{t^3 + 3}\,dt$.

In Exercises 29–34, calculate the derivative.

29. $\frac{d}{dx} \int_0^{x^2} \frac{t\,dt}{t + 1}$

30. $\frac{d}{dx} \int_1^{1/x} \cos^3 t\,dt$

31. $\frac{d}{ds} \int_{-6}^{\cos s} u^4\,du$

32. $\frac{d}{dx} \int_{x^2}^{x^4} \sqrt{t}\,dt$

Hint for Exercise 32: $F(x) = A(x^4) - A(x^2)$.

33. $\frac{d}{dx} \int_{\sqrt{x}}^{x^2} \tan t\,dt$

34. $\frac{d}{du} \int_{-u}^{3u} \sqrt{x^2 + 1}\,dx$

In Exercises 35–38, with $f(x)$ as in Figure 10 let

$$A(x) = \int_0^x f(t)\,dt \quad \text{and} \quad B(x) = \int_2^x f(t)\,dt.$$

35. Find the min and max of $A(x)$ on $[0, 6]$.

36. Find the min and max of $B(x)$ on $[0, 6]$.

37. Find formulas for $A(x)$ and $B(x)$ valid on $[2, 4]$.

38. Find formulas for $A(x)$ and $B(x)$ valid on $[4, 5]$.

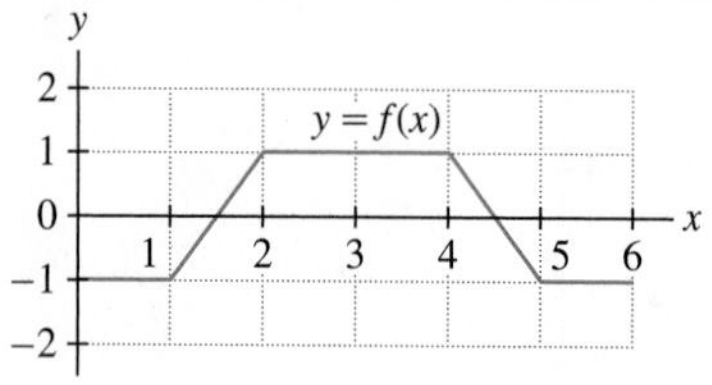

FIGURE 10

39. Let $A(x) = \int_0^x f(t)\,dt$, with $f(x)$ as in Figure 11.

(a) Does $A(x)$ have a local maximum at P?

(b) Where does $A(x)$ have a local minimum?

(c) Where does $A(x)$ have a local maximum?

(d) True or false? $A(x) < 0$ for all x in the interval shown.

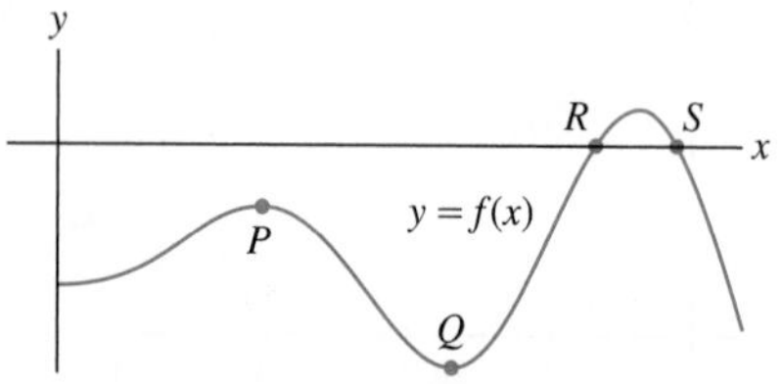

FIGURE 11 Graph of $f(x)$.

40. Determine $f(x)$, assuming that $\int_0^x f(t)\,dt = x^2 + x$.

41. Determine the function $g(x)$ and all values of c such that

$$\int_c^x g(t)\,dt = x^2 + x - 6$$

42. Find $a \le b$ such that $\int_a^b (x^2 - 9)\,dx$ has minimal value.

In Exercises 43–44, let $A(x) = \int_a^x f(t)\,dt$.

43. **Area Functions and Concavity** Explain why the following statements are true. Assume $f(x)$ is differentiable.

(a) If c is an inflection point of $A(x)$, then $f'(c) = 0$.

(b) $A(x)$ is concave up if $f(x)$ is increasing.

(c) $A(x)$ is concave down if $f(x)$ is decreasing.

44. Match the property of $A(x)$ with the corresponding property of the graph of $f(x)$. Assume $f(x)$ is differentiable.

Area function $A(x)$

(a) $A(x)$ is decreasing.

(b) $A(x)$ has a local maximum.

(c) $A(x)$ is concave up.

(d) $A(x)$ goes from concave up to concave down.

Graph of $f(x)$

(i) Lies below the x-axis.

(ii) Crosses the x-axis from positive to negative.

(iii) Has a local maximum.

(iv) $f(x)$ is increasing.

45. Let $A(x) = \int_0^x f(t)\,dt$, with $f(x)$ as in Figure 12. Determine:

(a) The intervals on which $A(x)$ is increasing and decreasing

(b) The values x where $A(x)$ has a local min or max

(c) The inflection points of $A(x)$

(d) The intervals where $A(x)$ is concave up or concave down

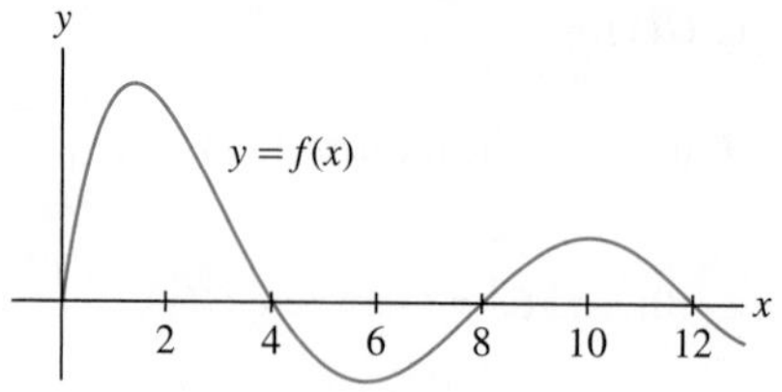

FIGURE 12

46. Let $f(x) = x^2 - 5x - 6$ and $F(x) = \int_0^x f(t)\,dt$.

(a) Find the critical points of $F(x)$ and determine whether they are local minima or local maxima.

(b) Find the points of inflection of $F(x)$ and determine whether the concavity changes from up to down or from down to up.

(c) GU Plot $f(x)$ and $F(x)$ on the same set of axes and confirm your answers to (a) and (b).

47. Sketch the graph of an increasing function $f(x)$ such that both $f'(x)$ and $A(x) = \int_0^x f(t)\,dt$ are decreasing.

48. Figure 13 shows the graph of $f(x) = x \sin x$. Let $F(x) = \int_0^x t \sin t\,dt$.

(a) Locate the local max and absolute max of $F(x)$ on $[0, 3\pi]$.

(b) Justify graphically: $F(x)$ has precisely one zero in $[\pi, 2\pi]$.

(c) How many zeros does $F(x)$ have in $[0, 3\pi]$?

(d) Find the inflection points of $F(x)$ on $[0, 3\pi]$. For each one, state whether the concavity changes from up to down or from down to up.

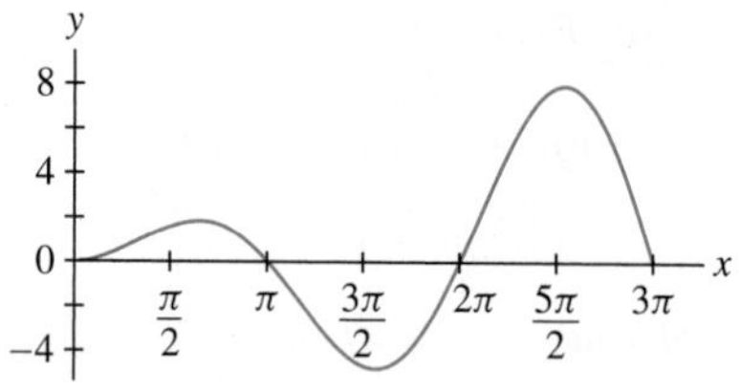

FIGURE 13 Graph of $f(x) = x \sin x$.

49. GU Find the smallest positive critical point of

$$F(x) = \int_0^x \cos(t^{3/2})\,dt$$

and determine whether it is a local min or max. Then find the smallest positive inflection point of $F(x)$ and use a graph of $y = \cos(x^{3/2})$ to determine whether the concavity changes from up to down or from down to up.

Further Insights and Challenges

50. Proof of FTC II The proof in the text assumes that $f(x)$ is increasing. To prove it for all continuous functions, let $m(h)$ and $M(h)$ denote the *minimum* and *maximum* of $f(t)$ on $[x, x+h]$ (Figure 14). The continuity of $f(x)$ implies that $\lim_{h\to 0} m(h) = \lim_{h\to 0} M(h) = f(x)$. Show that for $h > 0$,

$$hm(h) \le A(x+h) - A(x) \le hM(h)$$

For $h < 0$, the inequalities are reversed. Prove that $A'(x) = f(x)$.

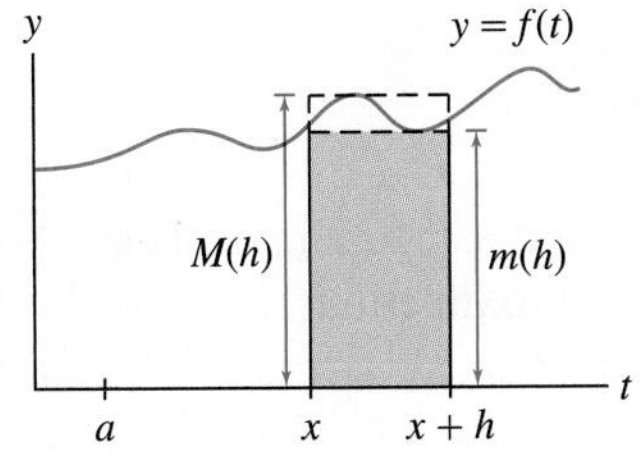

FIGURE 14 Graphical interpretation of $A(x+h) - A(x)$.

51. Proof of FTC I FTC I asserts that $\int_a^b f(t)\,dt = F(b) - F(a)$ if $F'(x) = f(x)$. Use FTC II to give a new proof of FTC I as follows. Set $A(x) = \int_a^x f(t)\,dt$.

(a) Show that $F(x) = A(x) + C$ for some constant.

(b) Show that $F(b) - F(a) = A(b) - A(a) = \displaystyle\int_a^b f(t)\,dt$.

52. Can Every Antiderivative Be Expressed as an Integral? The area function $\int_a^x f(t)\,dt$ is an antiderivative of $f(x)$ for every value of a. However, not all antiderivatives are obtained in this way. The general antiderivative of $f(x) = x$ is $F(x) = \frac{1}{2}x^2 + C$. Show that $F(x)$ is an area function if $C \le 0$ but not if $C > 0$.

53. Prove the formula

$$\frac{d}{dx}\int_{u(x)}^{v(x)} f(t)\,dt = f(v(x))v'(x) - f(u(x))u'(x)$$

54. Use the result of Exercise 53 to calculate

$$\frac{d}{dx}\int_{\ln x}^{e^x} \sin t\,dt$$

5.5 Net Change as the Integral of a Rate

So far we have focused on the area interpretation of the integral. In this section, we use the integral to compute net change.

Consider the following problem: Water flows into an empty bucket at a rate of $r(t)$ liters per second. How much water is in the bucket after 4 seconds? If the rate of water flow were *constant*—say, 1.5 liters/second—we would have

$$\text{Quantity of water} = \text{flow rate} \times \text{time elapsed} = (1.5)4 = 6 \text{ liters}$$

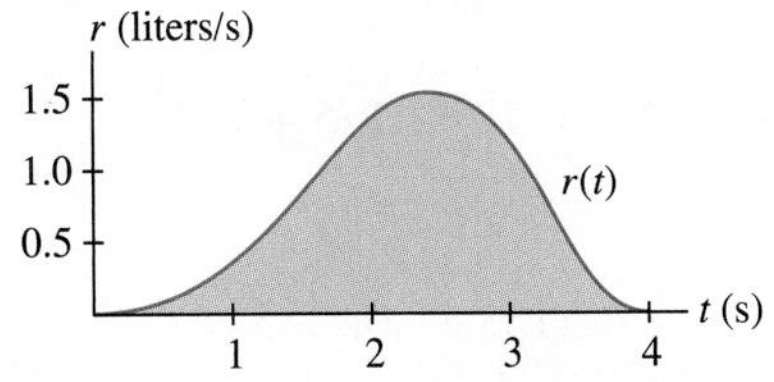

FIGURE 1 The quantity of water in the bucket is equal to the area under the graph of the flow rate $r(t)$.

Suppose, however, that the flow rate $r(t)$ varies as in Figure 1. Then *the quantity of water is equal to the area under the graph of* $r(t)$. To prove this, let $s(t)$ be the amount of water in the bucket at time t. Then $s'(t) = r(t)$ because $s'(t)$ is the rate at which the quantity of water is changing. Furthermore, $s(0) = 0$ because the bucket is initially empty. By FTC I,

$$\underbrace{\int_0^4 s'(t)\,dt}_{\text{Area under the graph of the flow rate}} = s(4) - s(0) = \underbrace{s(4)}_{\text{Water in bucket at } t=4}$$

More generally, $s(t_2) - s(t_1)$ is the **net change** in $s(t)$ over the interval $[t_1, t_2]$. FTC I yields the following result.

In Theorem 1, the variable t does not have to be a time variable.

THEOREM 1 Net Change as the Integral of a Rate The net change in $s(t)$ over an interval $[t_1, t_2]$ is given by the integral

$$\underbrace{\int_{t_1}^{t_2} s'(t)\,dt}_{\text{Integral of the rate of change}} = \underbrace{s(t_2) - s(t_1)}_{\text{Net change over } [t_1, t_2]}$$

EXAMPLE 1 Water leaks from a tank at a rate of $2 + 5t$ liters/hour, where t is the number of hours after 7 AM. How much water is lost between 9 and 11 AM?

Solution Let $s(t)$ be the quantity of water in the tank at time t. Then $s'(t) = -(2 + 5t)$, where the minus sign occurs because $s(t)$ is decreasing. Since 9 AM and 11 AM correspond to $t = 2$ and $t = 4$, respectively, the net change in $s(t)$ between 9 and 11 AM is

$$s(4) - s(2) = \int_2^4 s'(t)\,dt = -\int_2^4 (2 + 5t)\,dt$$

$$= -\left(2t + \frac{5}{2}t^2\right)\Big|_2^4 = (-48) - (-14) = -34 \text{ liters}$$

The tank lost 34 liters between 9 and 11 AM. ■

In the next example, we estimate an integral using numerical data. We shall compute the average of the left- and right-endpoint approximations, because this is usually more accurate than either endpoint approximation alone. (In Section 8.8, this average is called the Trapezoidal Approximation.)

EXAMPLE 2 Traffic Flow The number of cars per hour passing an observation point along a highway is called the traffic flow rate $q(t)$ (in cars per hour).

(a) Which quantity is represented by the integral $\displaystyle\int_{t_1}^{t_2} q(t)\,dt$?

(b) The flow rate is recorded at 15-min intervals between 7:00 and 9:00 AM. Estimate the number of cars using the highway during this 2-hour period.

t	7:00	7:15	7:30	7:45	8:00	8:15	8:30	8:45	9:00
$q(t)$	1044	1297	1478	1844	1451	1378	1155	802	542

Solution

(a) The integral $\int_{t_1}^{t_2} q(t)\,dt$ represents the total number of cars that passed the observation point during the time interval $[t_1, t_2]$.

(b) The data values are spaced at intervals of $\Delta t = 0.25$ hour. Thus,

$$L_N = 0.25\big(1044 + 1297 + 1478 + 1844 + 1451 + 1378 + 1155 + 802\big)$$
$$\approx 2612$$

$$R_N = 0.25\big(1297 + 1478 + 1844 + 1451 + 1378 + 1155 + 802 + 542\big)$$
$$\approx 2487$$

In Example 2, L_N is the sum of the values of $q(t)$ at the left endpoints

7:00, 7:15, ..., 8:45

and R_N is the sum of the values of $q(t)$ at the right endpoints

7:15, ..., 8:45, 9:00

We estimate the number of cars that passed the observation point between 7 and 9 AM by taking the average of R_N and L_N:

$$\int_7^9 q(t)\,dt \approx \frac{1}{2}(R_N + L_N) = \frac{1}{2}(2612 + 2487) \approx 2550$$

Approximately 2550 cars used the highway between 7 and 9 AM. ■

The Integral of Velocity

Let $s(t)$ be the position at time t of an object in linear motion. Then the object's velocity is $v(t) = s'(t)$, and the integral of $v(t)$ is equal to the *net change in position* or *displacement* over a time interval $[t_1, t_2]$:

$$\int_{t_1}^{t_2} v(t)\,dt = \int_{t_1}^{t_2} s'(t)\,dt = \underbrace{s(t_2) - s(t_1)}_{\text{Displacement or net change in position}}$$

We must distinguish between displacement and *distance traveled.* If you travel 10 km and return to your starting point, your displacement is zero but your distance traveled is 20 km. To compute distance traveled rather than displacement, integrate the *speed* $|v(t)|$.

THEOREM 2 The Integral of Velocity For an object in linear motion with velocity $v(t)$, then

$$\text{Displacement during } [t_1, t_2] = \int_{t_1}^{t_2} v(t)\,dt$$

$$\text{Distance traveled during } [t_1, t_2] = \int_{t_1}^{t_2} |v(t)|\,dt$$

EXAMPLE 3 A particle has velocity $v(t) = t^3 - 10t^2 + 24t$ m/s. Compute:

(a) Displacement over [0, 6] **(b)** Total distance traveled over [0, 6]

Indicate the particle's trajectory with a motion diagram.

Solution First, we compute the indefinite integral:

$$\int v(t)\,dt = \int (t^3 - 10t^2 + 24t)\,dt = \frac{1}{4}t^4 - \frac{10}{3}t^3 + 12t^2 + C$$

(a) The displacement over the time interval [0, 6] is

$$\int_0^6 v(t)\,dt = \left(\frac{1}{4}t^4 - \frac{10}{3}t^3 + 12t^2\right)\Big|_0^6 = 36 \text{ m}$$

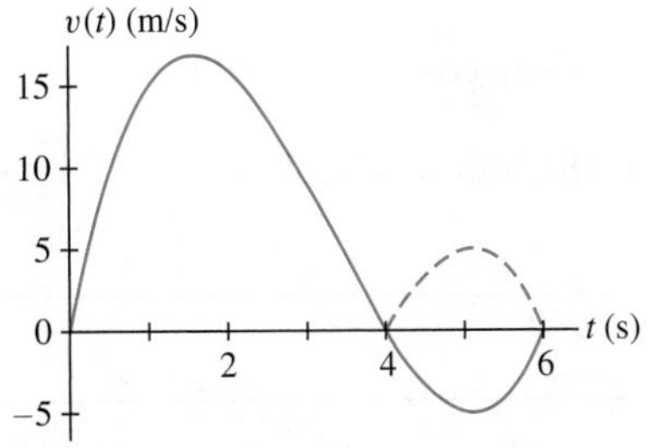

FIGURE 2 Graph of $v(t) = t^3 - 10t^2 + 24t$. Over [4, 6], the dashed curve is the graph of $|v(t)|$.

(b) The factorization $v(t) = t(t-4)(t-6)$ shows that $v(t)$ changes sign at $t = 4$. It is positive on [0, 4] and negative on [4, 6] as we see in Figure 2. Therefore, the total distance traveled is

$$\int_0^6 |v(t)|\,dt = \int_0^4 v(t)\,dt - \int_4^6 v(t)\,dt$$

We evaluate these two integrals separately:

$$[0, 4]: \quad \int_0^4 v(t)\,dt = \left(\frac{1}{4}t^4 - \frac{10}{3}t^3 + 12t^2\right)\Big|_0^4 = \frac{128}{3} \text{ m}$$

$$[4, 6]: \quad \int_4^6 v(t)\,dt = \left(\frac{1}{4}t^4 - \frac{10}{3}t^3 + 12t^2\right)\Big|_4^6 = -\frac{20}{3} \text{ m}$$

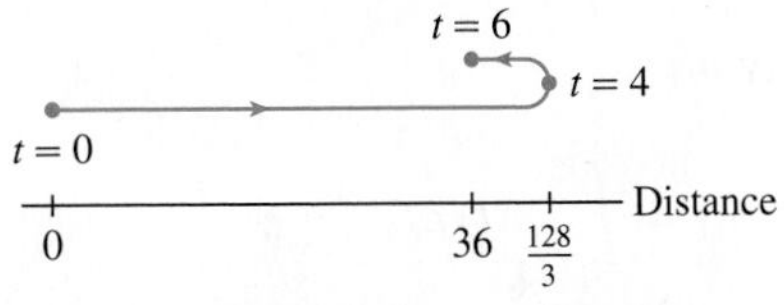

FIGURE 3 Path of the particle along a straight line.

The total distance traveled is $\frac{128}{3} + \frac{20}{3} = \frac{148}{3} = 49\frac{1}{3}$ m.

Figure 3 is a motion diagram indicating the particle's trajectory. The particle travels $\frac{128}{3}$ m during the first 4 s and then backtracks $\frac{20}{3}$ m over the next 2 s.

Total versus Marginal Cost

Consider the cost function $C(x)$ of a manufacturer (the dollar cost of producing x units of a particular product or commodity). The derivative $C'(x)$ is called the **marginal cost**. The cost of increasing production from a to b is the net change $C(b) - C(a)$, which is equal to the integral of the marginal cost:

In Section 3.4, we defined the marginal cost at production level x_0 as the cost

$$C(x_0 + 1) - C(x_0)$$

of producing one additional unit. Since this marginal cost is approximated well by the derivative $C'(x_0)$, we often refer to $C'(x)$ itself as the marginal cost.

$$\text{Cost of increasing production from } a \text{ to } b = \int_a^b C'(x)\,dx$$

■ **EXAMPLE 4** The marginal cost of producing x computer chips (in units of 1000) is $C'(x) = 300x^2 - 4000x + 40{,}000$ (dollars per thousand chips).

(a) Find the cost of increasing production from 10,000 to 15,000 chips.

(b) Determine the total cost of producing 15,000 chips, assuming that it costs \$30,000 to set up the manufacturing run [that is, $C(0) = 30{,}000$].

Solution

(a) The cost of increasing production from 10,000 ($x = 10$) to 15,000 ($x = 15$) is

$$\begin{aligned} C(15) - C(10) &= \int_{10}^{15} (300x^2 - 4000x + 40{,}000)\,dx \\ &= (100x^3 - 2000x^2 + 40{,}000x)\Big|_{10}^{15} \\ &= 487{,}500 - 300{,}000 = \$187{,}500 \end{aligned}$$

(b) The cost of increasing production from 0 to 15,000 chips is

$$\begin{aligned} C(15) - C(0) &= \int_{0}^{15} (300x^2 - 4000x + 40{,}000)\,dx \\ &= (100x^3 - 2000x^2 + 40{,}000x)\Big|_{0}^{15} = \$487{,}500 \end{aligned}$$

The total cost of producing 15,000 chips includes the set-up costs of \$30,000:

$$C(15) = C(0) + 487{,}500 = 30{,}000 + 487{,}500 = \$517{,}500$$ ■

5.5 SUMMARY

- Many applications are based on the following principle: *The net change in a quantity $s(t)$ is equal to the integral of its rate of change:*

$$\underbrace{s(t_2) - s(t_1)}_{\text{Net change over } [t_1, t_2]} = \int_{t_1}^{t_2} s'(t)\,dt$$

- For an object traveling in a straight line at velocity $v(t)$,

$$\text{Displacement during } [t_1, t_2] = \int_{t_1}^{t_2} v(t)\,dt$$

$$\text{Total distance traveled during } [t_1, t_2] = \int_{t_1}^{t_2} |v(t)|\,dt$$

- If $C(x)$ is the cost of producing x units of a commodity, then $C'(x)$ is the marginal cost and

$$\text{Cost of increasing production from } a \text{ to } b = \int_a^b C'(x)\,dx$$

5.5 EXERCISES

Preliminary Questions

1. A hot metal object is submerged in cold water. The rate at which the object cools (in degrees per minute) is a function $f(t)$ of time. Which quantity is represented by the integral $\int_0^T f(t)\,dt$?

2. A plane travels 560 km from Los Angeles to San Francisco in 1 hour. If the plane's velocity at time t is $v(t)$ km/h, what is the value of $\int_0^1 v(t)\,dt$?

3. Which of the following quantities would be naturally represented as derivatives and which as integrals?

(a) Velocity of a train

(b) Rainfall during a 6-month period

(c) Mileage per gallon of an automobile

(d) Increase in the U.S. population from 1990 to 2010

Exercises

1. Water flows into an empty reservoir at a rate of $3000 + 20t$ liters per hour. What is the quantity of water in the reservoir after 5 hours?

2. A population of insects increases at a rate of $200 + 10t + 0.25t^2$ insects per day. Find the insect population after 3 days, assuming that there are 35 insects at $t = 0$.

3. A survey shows that a mayoral candidate is gaining votes at a rate of $2000t + 1000$ votes per day, where t is the number of days since she announced her candidacy. How many supporters will the candidate have after 60 days, assuming that she had no supporters at $t = 0$?

4. A factory produces bicycles at a rate of $95 + 3t^2 - t$ bicycles per week. How many bicycles were produced from the beginning of week 2 to the end of week 3?

5. Find the displacement of a particle moving in a straight line with velocity $v(t) = 4t - 3$ m/s over the time interval [2, 5].

6. Find the displacement over the time interval [1, 6] of a helicopter whose (vertical) velocity at time t is $v(t) = 0.02t^2 + t$ m/s.

7. A cat falls from a tree (with zero initial velocity) at time $t = 0$. How far does the cat fall between $t = 0.5$ and $t = 1$ s? Use Galileo's formula $v(t) = -9.8t$ m/s.

8. A projectile is released with an initial (vertical) velocity of 100 m/s. Use the formula $v(t) = 100 - 9.8t$ for velocity to determine the distance traveled during the first 15 seconds.

In Exercises 9–12, a particle moves in a straight line with the given velocity (in m/s). Find the displacement and distance traveled over the time interval, and draw a motion diagram like Figure 3 (with distance and time labels).

9. $v(t) = 12 - 4t$, $[0, 5]$

10. $v(t) = 36 - 24t + 3t^2$, $[0, 10]$

11. $v(t) = t^{-2} - 1$, $[0.5, 2]$

12. $v(t) = \cos t$, $[0, 3\pi]$

13. Find the net change in velocity over [1, 4] of an object with $a(t) = 8t - t^2$ m/s^2.

14. Show that if acceleration is constant, then the change in velocity is proportional to the length of the time interval.

15. The traffic flow rate past a certain point on a highway is $q(t) = 3000 + 2000t - 300t^2$ (t in hours), where $t = 0$ is 8 AM. How many cars pass by in the time interval from 8 to 10 AM?

16. The marginal cost of producing x tablet computers is $C'(x) = 120 - 0.06x + 0.00001x^2$ What is the cost of producing 3000 units if the setup cost is \$90,000? If production is set at 3000 units, what is the cost of producing 200 additional units?

17. A small boutique produces wool sweaters at a marginal cost of $40 - 5[[x/5]]$ for $0 \le x \le 20$, where $[[x]]$ is the greatest integer function. Find the cost of producing 20 sweaters. Then compute the average cost of the first 10 sweaters and the last 10 sweaters.

18. The rate (in liters per minute) at which water drains from a tank is recorded at half-minute intervals. Compute the average of the left- and right-endpoint approximations to estimate the total amount of water drained during the first 3 minutes.

t (min)	0	0.5	1	1.5	2	2.5	3
r (l/min)	50	48	46	44	42	40	38

19. The velocity of a car is recorded at half-second intervals (in feet per second). Use the average of the left- and right-endpoint approximations to estimate the total distance traveled during the first 4 seconds.

t	0	0.5	1	1.5	2	2.5	3	3.5	4
$v(t)$	0	12	20	29	38	44	32	35	30

20. To model the effects of a **carbon tax** on CO_2 emissions, policymakers study the *marginal cost of abatement* $B(x)$, defined as the cost of increasing CO_2 reduction from x to $x + 1$ tons (in units of ten thousand tons—Figure 4). Which quantity is represented by the area under the curve over [0, 3] in Figure 4?

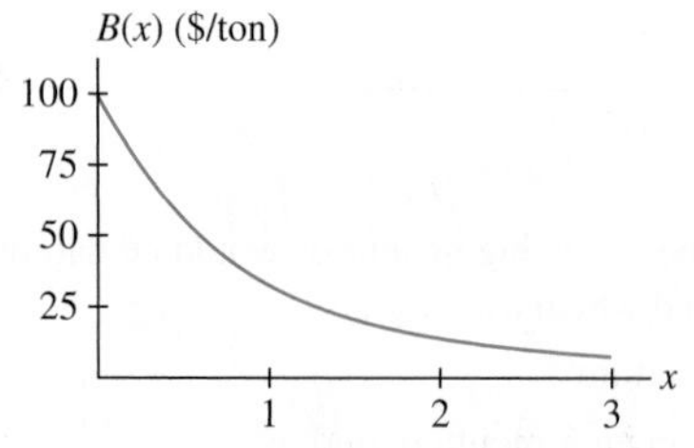

FIGURE 4 Marginal cost of abatement $B(x)$.

21. A megawatt of power is 10^6 W, or 3.6×10^9 J/hour. Which quantity is represented by the area under the graph in Figure 5? Estimate the energy (in joules) consumed during the period 4 PM to 8 PM.

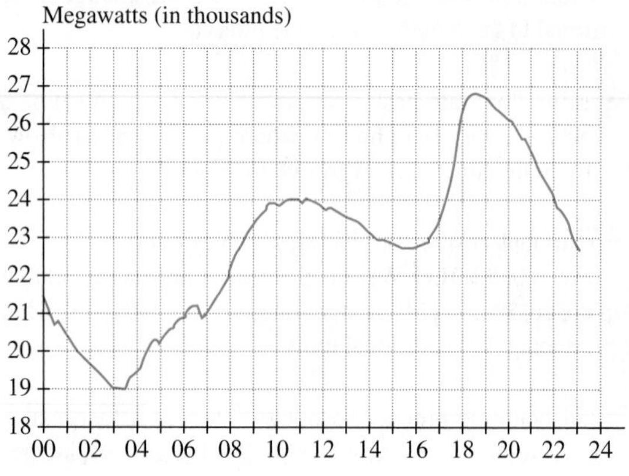

FIGURE 5 Power consumption over 1-day period in California (February 2010).

22. Figure 6 shows the migration rate $M(t)$ of Ireland in the period 1988–1998. This is the rate at which people (in thousands per year) move into or out of the country.

(a) Is the following integral positive or negative? What does this quantity represent?

$$\int_{1988}^{1998} M(t)\,dt$$

(b) Did migration in the period 1988–1998 result in a net influx of people into Ireland or a net outflow of people from Ireland?

(c) During which two years could the Irish prime minister announce, "We've hit an inflection point. We are still losing population, but the trend is now improving."

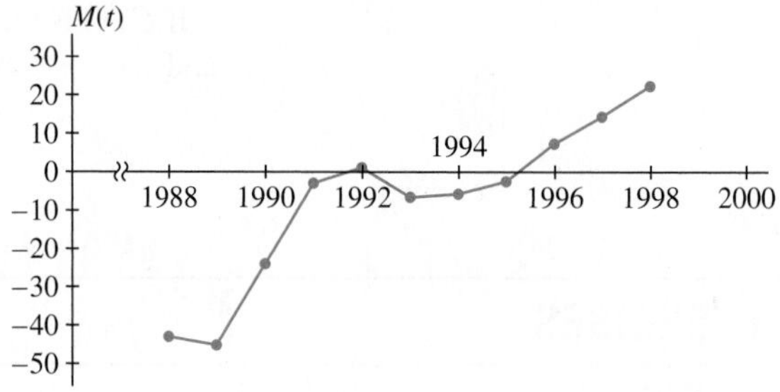

FIGURE 6 Irish migration rate (in thousands per year).

23. Let $N(d)$ be the number of asteroids of diameter $\le d$ kilometers. Data suggest that the diameters are distributed according to a piecewise power law:

$$N'(d) = \begin{cases} 1.9 \times 10^9 d^{-2.3} & \text{for } d < 70 \\ 2.6 \times 10^{12} d^{-4} & \text{for } d \ge 70 \end{cases}$$

(a) Compute the number of asteroids with diameter between 0.1 and 100 km.

(b) Using the approximation $N(d+1) - N(d) \approx N'(d)$, estimate the number of asteroids of diameter 50 km.

24. Heat Capacity The heat capacity $C(T)$ of a substance is the amount of energy (in joules) required to raise the temperature of 1 g by 1°C at temperature T.

(a) Explain why the energy required to raise the temperature from T_1 to T_2 is the area under the graph of $C(T)$ over $[T_1, T_2]$.

(b) How much energy is required to raise the temperature from 50 to 100°C if $C(T) = 6 + 0.2\sqrt{T}$?

25. Figure 7 shows the rate $R(t)$ of natural gas consumption (in billions of cubic feet per day) in the mid-Atlantic states (New York, New Jersey, Pennsylvania). Express the total quantity of natural gas consumed in 2009 as an integral (with respect to time t in days). Then estimate this quantity, given the following monthly values of $R(t)$:

3.18, 2.86, 2.39, 1.49, 1.08, 0.80,
1.01, 0.89, 0.89, 1.20, 1.64, 2.52

Keep in mind that the number of days in a month varies with the month.

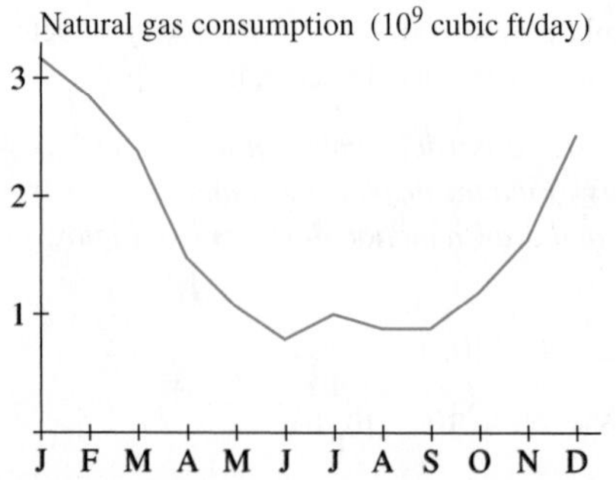

FIGURE 7 Natural gas consumption in 2009 in the mid-Atlantic states

26. Cardiac output is the rate R of volume of blood pumped by the heart per unit time (in liters per minute). Doctors measure R by injecting A mg of dye into a vein leading into the heart at $t = 0$ and recording the concentration $c(t)$ of dye (in milligrams per liter) pumped out at short regular time intervals (Figure 8).

(a) Explain: The quantity of dye pumped out in a small time interval $[t, t + \Delta t]$ is approximately $Rc(t)\Delta t$.

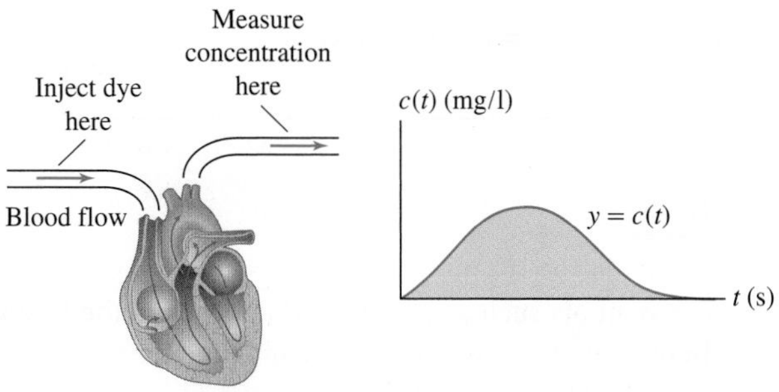

FIGURE 8

(b) Show that $A = R\int_0^T c(t)\,dt$, where T is large enough that all of the dye is pumped through the heart but not so large that the dye returns by recirculation.

(c) Assume $A = 5$ mg. Estimate R using the following values of $c(t)$ recorded at 1-second intervals from $t = 0$ to $t = 10$:

$$0,\quad 0.4,\quad 2.8,\quad 6.5,\quad 9.8,\quad 8.9,$$
$$6.1,\quad 4,\quad 2.3,\quad 1.1,\quad 0$$

Exercises 27 and 28: A study suggests that the extinction rate $r(t)$ of marine animal families during the Phanerozoic Eon can be modeled by the function $r(t) = 3130/(t + 262)$ for $0 \le t \le 544$, where t is time elapsed (in millions of years) since the beginning of the eon 544 million years ago. Thus, $t = 544$ refers to the present time, $t = 540$ is 4 million years ago, and so on.

27. Compute the average of R_N and L_N with $N = 5$ to estimate the total number of families that became extinct in the periods $100 \le t \le 150$ and $350 \le t \le 400$.

28. CAS Estimate the total number of extinct families from $t = 0$ to the present, using M_N with $N = 544$.

Further Insights and Challenges

29. Show that a particle, located at the origin at $t = 1$ and moving along the x-axis with velocity $v(t) = t^{-2}$, will never pass the point $x = 2$.

30. Show that a particle, located at the origin at $t = 1$ and moving along the x-axis with velocity $v(t) = t^{-1/2}$ moves arbitrarily far from the origin after sufficient time has elapsed.

5.6 Substitution Method

The term "integration" is used in two ways. It refers to:

- *The process of finding signed area (computing a definite integral), and also*
- *The process of finding an antiderivative (evaluating an indefinite integral).*

Integration (antidifferentiation) is generally more difficult than differentiation. There are no sure-fire methods, and many antiderivatives cannot be expressed in terms of elementary functions. However, there are a few important general techniques. One such technique is the **Substitution Method**, which uses the Chain Rule "in reverse."

Consider the integral $\int 2x\cos(x^2)\,dx$. We can evaluate it if we remember the Chain Rule calculation

$$\frac{d}{dx}\sin(x^2) = 2x\cos(x^2)$$

This tells us that $\sin(x^2)$ is an antiderivative of $2x\cos(x^2)$, and therefore,

$$\int \underbrace{2x}_{\substack{\text{Derivative of}\\ \text{inside function}}}\ \cos\underbrace{(x^2)}_{\substack{\text{Inside}\\ \text{function}}}\,dx = \sin(x^2) + C$$

A similar Chain Rule calculation shows that

$$\int \underbrace{(1 + 3x^2)}_{\substack{\text{Derivative of}\\ \text{inside function}}}\ \cos(\underbrace{x + x^3}_{\substack{\text{Inside}\\ \text{function}}})\,dx = \sin(x + x^3) + C$$

REMINDER A "composite function" is a function of the form $f(g(x))$. For convenience, we call $g(x)$ the inside function and $f(u)$ the outside function.

In both cases, the integrand is the product of a composite function and the derivative of the inside function. The Chain Rule does not help if the derivative of the inside function is missing. For instance, we cannot use the Chain Rule to compute $\int \cos(x + x^3)\,dx$ because the factor $(1 + 3x^2)$ does not appear.

In general, if $F'(u) = f(u)$, then by the Chain Rule,

$$\frac{d}{dx} F(u(x)) = F'(u(x))u'(x) = f(u(x))u'(x)$$

This translates into the following integration formula:

THEOREM 1 The Substitution Method If $F'(x) = f(x)$, then

$$\int f(u(x))u'(x)\, dx = F(u(x)) + C$$

Substitution Using Differentials

Before proceeding to the examples, we discuss the procedure for carrying out substitution using differentials. Differentials are symbols such as du or dx that occur in the Leibniz notations du/dx and $\int f(x)\, dx$. In our calculations, we shall manipulate them as though they are related by an equation in which the dx "cancels":

$$du = \frac{du}{dx}\, dx$$

Equivalently, du and dx are related by

$$du = u'(x)\, dx \qquad \boxed{1}$$

For example,

If $u = x^2$, then $du = 2x\, dx$

If $u = \cos(x^3)$, then $du = -3x^2 \sin(x^3)\, dx$

Now when the integrand has the form $f(u(x))\, u'(x)$, we can use Eq. (1) to rewrite the entire integral (including the dx term) in terms of u and its differential du:

The symbolic calculus of substitution using differentials was invented by Leibniz and is considered one of his most important achievements. It reduces the otherwise complicated process of transforming integrals to a convenient set of rules.

$$\int \underbrace{f(u(x))}_{f(u)}\ \underbrace{u'(x)\, dx}_{du} = \int f(u)\, du$$

This equation is called the **Change of Variables Formula**. It transforms an integral in the variable x into a (hopefully simpler) integral in the new variable u.

■ **EXAMPLE 1** Evaluate $\int 3x^2 \sin(x^3)\, dx$.

In substitution, the key step is to choose the appropriate inside function u.

Solution The integrand contains the composite function $\sin(x^3)$, so we set $u = x^3$. The differential $du = 3x^2\, dx$ also appears, so we can carry out the substitution:

$$\int 3x^2 \sin(x^3)\, dx = \int \underbrace{\sin(x^3)}_{\sin u}\ \underbrace{3x^2\, dx}_{du} = \int \sin u\, du$$

Now evaluate the integral in the u-variable and replace u by x^3 in the answer:

$$\int 3x^2 \sin(x^3)\, dx = \int \sin u\, du = -\cos u + C = -\cos(x^3) + C$$

Let's check our answer by differentiating:

$$\frac{d}{dx}(-\cos(x^3)) = \sin(x^3)\frac{d}{dx}x^3 = 3x^2\sin(x^3)$$ ■

■ **EXAMPLE 2 Multiplying du by a Constant** Evaluate $\int x(x^2+9)^5\,dx$.

Solution We let $u = x^2 + 9$ because the composite $u^5 = (x^2+9)^5$ appears in the integrand. The differential $du = 2x\,dx$ does not appear as is, but we can multiply by $\frac{1}{2}$ to obtain

$$\frac{1}{2}du = x\,dx \quad \Rightarrow \quad \frac{1}{2}u^5\,du = x(x^2+9)^5\,dx$$

Now we can apply substitution:

$$\int x(x^2+9)^5\,dx = \int \overbrace{(x^2+9)^5}^{u^5}\,\overbrace{x\,dx}^{\frac{1}{2}du} = \frac{1}{2}\int u^5\,du = \frac{1}{12}u^6 + C$$

Finally, we express the answer in terms of x by substituting $u = x^2 + 9$:

$$\int x(x^2+9)^5\,dx = \frac{1}{12}u^6 + C = \frac{1}{12}(x^2+9)^6 + C$$ ■

Substitution Method:

(1) Choose u and compute du.

(2) Rewrite the integral in terms of u and du, and evaluate.

(3) Express the final answer in terms of x.

■ **EXAMPLE 3** Evaluate $\displaystyle\int \frac{(x^2+2x)\,dx}{(x^3+3x^2+12)^6}$.

Solution The appearance of $(x^3+3x^2+12)^{-6}$ in the integrand suggests that we try $u = x^3 + 3x^2 + 12$. With this choice,

$$du = (3x^2+6x)\,dx = 3(x^2+2x)\,dx \quad \Rightarrow \quad \frac{1}{3}du = (x^2+2x)\,dx$$

$$\int \frac{(x^2+2x)\,dx}{(x^3+3x^2+12)^6} = \int \overbrace{(x^3+3x^2+12)^{-6}}^{u^{-6}}\,\overbrace{(x^2+2x)\,dx}^{\frac{1}{3}du}$$

$$= \frac{1}{3}\int u^{-6}\,du = \left(\frac{1}{3}\right)\left(\frac{u^{-5}}{-5}\right) + C$$

$$= -\frac{1}{15}(x^3+3x^2+12)^{-5} + C$$ ■

CONCEPTUAL INSIGHT An integration method that works for a given function may fail if we change the function even slightly. In the previous example, if we replace 2 by 2.1 and consider instead $\displaystyle\int \frac{(x^2+2.1x)\,dx}{(x^3+3x^2+12)^6}$, the Substitution Method does not work. The problem is that $(x^2+2.1x)\,dx$ is *not* a multiple of $du = (3x^2+6x)\,dx$.

■ **EXAMPLE 4** Evaluate $\int \sin(7\theta+5)\,d\theta$.

Solution Let $u = 7\theta + 5$. Then $du = 7\,d\theta$ and $\frac{1}{7}du = d\theta$. We obtain

$$\int \sin(7\theta+5)\,d\theta = \frac{1}{7}\int \sin u\,du = -\frac{1}{7}\cos u + C = -\frac{1}{7}\cos(7\theta+5) + C$$ ■

■ **EXAMPLE 5** Evaluate $\displaystyle\int \frac{\sin(t^{1/3})\,dt}{t^{2/3}}$.

Solution It makes sense to try $u = t^{1/3}$ because $du = \frac{1}{3}t^{-2/3}dt$, and thus the multiple $3\,du$ appears in the integrand. In other words,

$$u = t^{1/3}, \qquad \frac{dt}{t^{2/3}} = 3\,du$$

$$\int \frac{\sin(t^{1/3})\,dt}{t^{2/3}} = \int \sin(t^{1/3})\frac{dt}{t^{2/3}}$$

$$= \int \sin u\,(3\,du)$$

$$= -3\cos u + C = -3\cos(t^{1/3}) + C$$ ■

■ **EXAMPLE 6** **Additional Step Necessary** Evaluate $\displaystyle\int x\sqrt{5x+1}\,dx$.

Solution Since $\sqrt{5x+1}$ appears, we are tempted to set $u = 5x + 1$. Then

$$du = 5dx \quad \Rightarrow \quad \sqrt{5x+1}\,dx = \frac{1}{5}u^{1/2}\,du$$

Unfortunately, the integrand is not $\sqrt{5x+1}$ but $x\sqrt{5x+1}$. To take care of the extra factor of x, we solve $u = 5x + 1$ to obtain $x = \frac{1}{5}(u - 1)$. Then

$$x\sqrt{5x+1}\,dx = \left(\frac{1}{5}(u-1)\right)\frac{1}{5}u^{1/2}\,du = \frac{1}{25}(u-1)\,u^{1/2}\,du$$

$$\int x\sqrt{5x+1}\,dx = \frac{1}{25}\int (u-1)\,u^{1/2}\,du = \frac{1}{25}\int (u^{3/2} - u^{1/2})\,du$$

$$= \frac{1}{25}\left(\frac{2}{5}u^{5/2} - \frac{2}{3}u^{3/2}\right) + C$$

$$= \frac{2}{125}(5x+1)^{5/2} - \frac{2}{75}(5x+1)^{3/2} + C$$ ■

The substitution method does not always work, even when the integral looks relatively simple. For example, $\int \sin(x^2)\,dx$ cannot be evaluated explicitly by substitution, or any other method. With experience, you will learn to recognize when substitution is likely to be successful.

Change of Variables Formula for Definite Integrals

The Change of Variables Formula can be applied to definite integrals provided that the limits of integration are changed, as indicated in the next theorem.

Change of Variables Formula for Definite Integrals

$$\int_a^b f(u(x))u'(x)\,dx = \int_{u(a)}^{u(b)} f(u)\,du \qquad \boxed{2}$$

The new limits of integration with respect to the u-variable are $u(a)$ and $u(b)$. Think of it this way: As x varies from a to b, the variable $u = u(x)$ varies from $u(a)$ to $u(b)$.

Change of Variables for definite integrals:

$$\int_a^b f(u(x))u'(x)\,dx = \int_{u(a)}^{u(b)} f(u)\,du$$

Proof If $F(x)$ is an antiderivative of $f(x)$, then $F(u(x))$ is an antiderivative of $f(u(x))u'(x)$. FTC I shows that the two integrals are equal:

$$\int_a^b f(u(x))u'(x)\,dx = F(u(b)) - F(u(a))$$

$$\int_{u(a)}^{u(b)} f(u)\,du = F(u(b)) - F(u(a))$$ ■

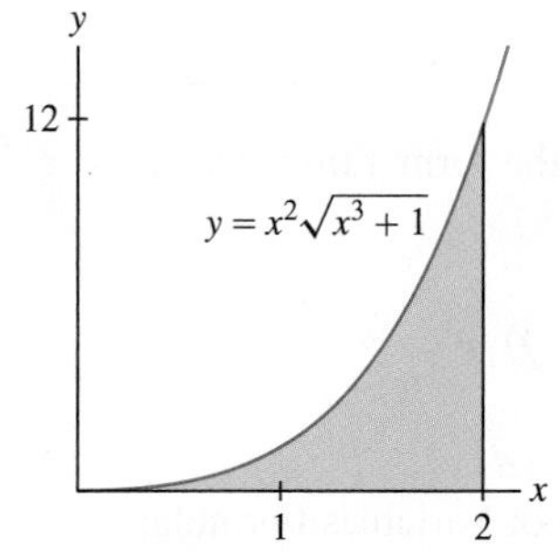

FIGURE 1 Region represented by $\int_0^2 x^2\sqrt{x^3+1}\,dx$.

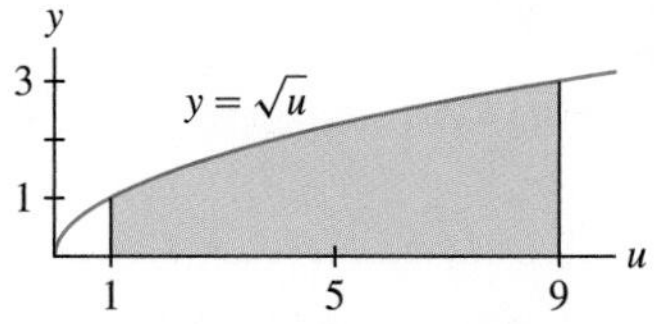

FIGURE 2 Region represented by $\int_1^9 \sqrt{u}\,du$.

■ **EXAMPLE 7** Evaluate $\int_0^2 x^2\sqrt{x^3+1}\,dx$.

Solution Use the substitution $u = x^3 + 1$, $du = 3x^2\,dx$:

$$x^2\sqrt{x^3+1}\,dx = \frac{1}{3}\sqrt{u}\,du$$

By Eq. (2), the new limits of integration

$$u(0) = 0^3 + 1 = 1 \qquad \text{and} \qquad u(2) = 2^3 + 1 = 9$$

Thus,

$$\int_0^2 x^2\sqrt{x^3+1}\,dx = \frac{1}{3}\int_1^9 \sqrt{u}\,du = \frac{2}{9}u^{3/2}\Big|_1^9 = \frac{52}{9}$$

This substitution shows that the area in Figure 1 is equal to one-third of the area in Figure 2 (but note that the figures are drawn to different scales). ■

In the previous example, we can avoid changing the limits of integration by evaluating the integral in terms of x.

$$\int x^2\sqrt{x^3+1}\,dx = \frac{1}{3}\int \sqrt{u}\,du = \frac{2}{9}u^{3/2} = \frac{2}{9}(x^3+1)^{3/2}$$

This leads to the same result: $\int_0^2 x^2\sqrt{x^3+1}\,dx = \frac{2}{9}(x^3+1)^{3/2}\Big|_0^2 = \frac{52}{9}$.

■ **EXAMPLE 8** Evaluate $\int_0^{\pi/4} \tan^3\theta \sec^2\theta\,d\theta$.

Solution The substitution $u = \tan\theta$ makes sense because $du = \sec^2\theta\,d\theta$ and therefore, $u^3\,du = \tan^3\theta\sec^2\theta\,d\theta$. The new limits of integration are

$$u(0) = \tan 0 = 0 \qquad \text{and} \qquad u\left(\frac{\pi}{4}\right) = \tan\left(\frac{\pi}{4}\right) = 1$$

Thus,

$$\int_0^{\pi/4} \tan^3\theta\sec^2\theta\,d\theta = \int_0^1 u^3\,du = \frac{u^4}{4}\Big|_0^1 = \frac{1}{4}$$ ■

■ **EXAMPLE 9** Calculate the area under the graph of $y = \dfrac{x}{(x^2+1)^2}$ over [1, 3].

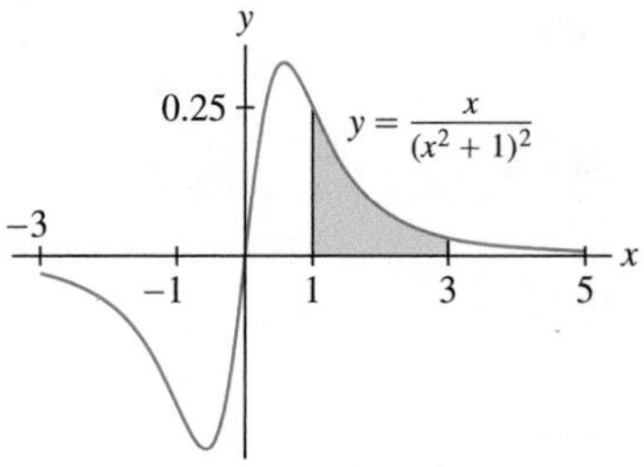

FIGURE 3 Area under the graph of $y = \dfrac{x}{(x^2+1)^2}$ over [1, 3].

Solution The area (Figure 3) is equal to the integral $\int_1^3 \dfrac{x}{(x^2+1)^2}\,dx$. We use the substitution

$$u = x^2 + 1, \qquad du = 2x\,dx, \qquad \frac{1}{2}\frac{du}{u^2} = \frac{x\,dx}{(x^2+1)^2}$$

The new limits of integration are $u(1) = 1^2 + 1 = 2$ and $u(3) = 3^2 + 1 = 10$, so

$$\int_1^3 \frac{x}{(x^2+1)^2}\,dx = \frac{1}{2}\int_2^{10}\frac{du}{u^2} = -\frac{1}{2}\frac{1}{u}\Big|_2^{10} = -\frac{1}{20} + \frac{1}{4} = \frac{1}{5}$$ ■

5.6 SUMMARY

- Try the Substitution Method when the integrand has the form $f(u(x))\,u'(x)$. If F is an antiderivative of f, then

$$\int f(u(x))\,u'(x)\,dx = F(u(x)) + C$$

- The differential of $u(x)$ is related to dx by $du = u'(x)\,dx$.
- The Substitution Method is expressed by the Change of Variables Formula:

$$\int f(u(x))\,u'(x)\,dx = \int f(u)\,du$$

- Change of Variables Formula for definite integrals:

$$\int_a^b f(u(x))\,u'(x)\,dx = \int_{u(a)}^{u(b)} f(u)\,du$$

5.6 EXERCISES

Preliminary Questions

1. Which of the following integrals is a candidate for the Substitution Method?

(a) $\int 5x^4 \sin(x^5)\,dx$ **(b)** $\int \sin^5 x \cos x\,dx$

(c) $\int x^5 \sin x\,dx$

2. Find an appropriate choice of u for evaluating the following integrals by substitution:

(a) $\int x(x^2+9)^4\,dx$ **(b)** $\int x^2 \sin(x^3)\,dx$

(c) $\int \sin x \cos^2 x\,dx$

3. Which of the following is equal to $\int_0^2 x^2(x^3+1)\,dx$ for a suitable substitution?

(a) $\frac{1}{3}\int_0^2 u\,du$ **(b)** $\int_0^9 u\,du$ **(c)** $\frac{1}{3}\int_1^9 u\,du$

Exercises

In Exercises 1–6, calculate du.

1. $u = x^3 - x^2$

2. $u = 2x^4 + 8x^{-1}$

3. $u = \cos(x^2)$

4. $u = \tan x$

5. $u = \sin^4\theta$

6. $u = \dfrac{t}{t+1}$

In Exercises 7–20, write the integral in terms of u and du. Then evaluate.

7. $\int (x-7)^3\,dx, \quad u = x - 7$

8. $\int (x+25)^{-2}\,dx, \quad u = x + 25$

9. $\int t\sqrt{t^2+1}\,dt, \quad u = t^2 + 1$

10. $\int (x^3+1)\cos(x^4+4x)\,dx, \quad u = x^4 + 4x$

11. $\int \dfrac{t^3}{(4-2t^4)^{11}}\,dt, \quad u = 4 - 2t^4$

12. $\int \sqrt{4x-1}\,dx, \quad u = 4x - 1$

13. $\int x(x+1)^9\,dx, \quad u = x + 1$

14. $\int x\sqrt{4x-1}\,dx, \quad u = 4x - 1$

15. $\int x^2\sqrt{x+1}\,dx, \quad u = x + 1$

16. $\int \sin(4\theta - 7)\,d\theta, \quad u = 4\theta - 7$

17. $\int \sin^2\theta \cos\theta\,d\theta, \quad u = \sin\theta$

18. $\int \sec^2 x \tan x\,dx, \quad u = \tan x$

19. $\int x \sec^2(x^2)\,dx, \quad u = x^2$

20. $\int \sec^2(\cos x)\sin x\,dx, \quad u = \cos x$

In Exercises 21–24, evaluate the integral in the form $a\sin(u(x)) + C$ for an appropriate choice of $u(x)$ and constant a.

21. $\int x^3\cos(x^4)\,dx$

22. $\int x^2\cos(x^3+1)\,dx$

23. $\int x^{1/2}\cos(x^{3/2})\,dx$

24. $\int \cos x\cos(\sin x)\,dx$

In Exercises 25–59, evaluate the indefinite integral.

25. $\int (4x+5)^9\,dx$

26. $\int \frac{dx}{(x-9)^5}$

27. $\int \frac{dt}{\sqrt{t+12}}$

28. $\int (9t+2)^{2/3}\,dt$

29. $\int \frac{x+1}{(x^2+2x)^3}\,dx$

30. $\int (x+1)(x^2+2x)^{3/4}\,dx$

31. $\int \frac{x}{\sqrt{x^2+9}}\,dx$

32. $\int \frac{2x^2+x}{(4x^3+3x^2)^2}\,dx$

33. $\int (3x^2+1)(x^3+x)^2\,dx$

34. $\int \frac{5x^4+2x}{(x^5+x^2)^3}\,dx$

35. $\int (3x+8)^{11}\,dx$

36. $\int x(3x+8)^{11}\,dx$

37. $\int x^2\sqrt{x^3+1}\,dx$

38. $\int x^5\sqrt{x^3+1}\,dx$

39. $\int \frac{dx}{(x+5)^3}$

40. $\int \frac{x\,dx}{(x+5)^{3/2}}$

41. $\int z^2(z^3+1)^{12}\,dz$

42. $\int (z^5+4z^2)(z^3+1)^{12}\,dz$

43. $\int (x+2)(x+1)^{1/4}\,dx$

44. $\int x^3(x^2-1)^{3/2}\,dx$

45. $\int \sin(8-3\theta)\,d\theta$

46. $\int \theta\sin(\theta^2)\,d\theta$

47. $\int \frac{\cos\sqrt{t}}{\sqrt{t}}\,dt$

48. $\int x^2\sin(x^3+1)\,dx$

49. $\int \frac{\sin x\cos x}{\sqrt{\sin x+1}}$

50. $\int \sin^8\theta\cos\theta\,d\theta$

51. $\int \sec^2 x\left(12\tan^3 x - 6\tan^2 x\right)dx$

52. $\int x^{-1/5}\sec\left(x^{4/5}\right)\tan\left(x^{4/5}\right)dx$

53. $\int \sec^2(4x+9)\,dx$

54. $\int \sec^2 x\tan^4 x\,dx$

55. $\int \frac{\sec^2(\sqrt{x})\,dx}{\sqrt{x}}$

56. $\int \frac{\cos 2x}{(1+\sin 2x)^2}\,dx$

57. $\int \sin 4x\sqrt{\cos 4x+1}\,dx$

58. $\int \cos x(3\sin x-1)\,dx$

59. $\int \sec\theta\tan\theta(\sec\theta-1)\,d\theta$

60. $\int \cos t\cos(\sin t)\,dt$

61. Evaluate $\int \frac{dx}{(1+\sqrt{x})^3}$ using $u = 1+\sqrt{x}$. *Hint:* Show that $dx = 2(u-1)du$.

62. Can They Both Be Right? Hannah uses the substitution $u = \tan x$ and Akiva uses $u = \sec x$ to evaluate $\int \tan x\sec^2 x\,dx$. Show that they obtain different answers, and explain the apparent contradiction.

63. Evaluate $\int \sin x\cos x\,dx$ using substitution in two different ways: first using $u = \sin x$ and then using $u = \cos x$. Reconcile the two different answers.

64. Some Choices Are Better Than Others Evaluate

$$\int \sin x\,\cos^2 x\,dx$$

twice. First use $u = \sin x$ to show that

$$\int \sin x\,\cos^2 x\,dx = \int u\sqrt{1-u^2}\,du$$

and evaluate the integral on the right by a further substitution. Then show that $u = \cos x$ is a better choice.

65. What are the new limits of integration if we apply the substitution $u = 3x+\pi$ to the integral $\int_0^\pi \sin(3x+\pi)\,dx$?

66. Which of the following is the result of applying the substitution $u = 4x-9$ to the integral $\int_2^8 (4x-9)^{20}\,dx$?

(a) $\int_2^8 u^{20}\,du$ **(b)** $\frac{1}{4}\int_2^8 u^{20}\,du$

(c) $4\int_{-1}^{23} u^{20}\,du$ **(d)** $\frac{1}{4}\int_{-1}^{23} u^{20}\,du$

In Exercises 67–78, use the Change-of-Variables Formula to evaluate the definite integral.

67. $\int_1^3 (x+2)^3\,dx$

68. $\int_1^6 \sqrt{x+3}\,dx$

69. $\int_0^1 \frac{x}{(x^2+1)^3}\,dx$

70. $\int_{-1}^2 \sqrt{5x+6}\,dx$

71. $\int_0^4 x\sqrt{x^2+9}\,dx$

72. $\int_1^2 \frac{4x+12}{(x^2+6x+1)^2}\,dx$

73. $\int_0^1 (x+1)(x^2+2x)^5\,dx$

74. $\int_{10}^{17} (x-9)^{-2/3}\,dx$

75. $\int_{-\pi/2}^{\pi/2} \frac{\cos x\,dx}{\sqrt{\sin x+1}}$

76. $\int_0^{\pi/6} \sec^2\left(2x-\frac{\pi}{6}\right)dx$

77. $\int_0^{\pi/2} \cos^3 x \sin x\, dx$

78. $\int_{\pi/3}^{\pi/2} \cot^2 \frac{x}{2} \csc^2 \frac{x}{2}\, dx$

79. Evaluate $\int_0^2 r\sqrt{5-\sqrt{4-r^2}}\, dr$.

80. Find numbers a and b such that

$$\int_a^b (u^2+1)\, du = \int_{-\pi/4}^{\pi/4} \sec^4 \theta\, d\theta$$

and evaluate. *Hint:* Use the identity $\sec^2\theta = \tan^2\theta + 1$.

In Exercises 81–82, use substitution to evaluate the integral in terms of $f(x)$.

81. $\int f(x)^3 f'(x)\, dx$

82. $\int \frac{f'(x)}{f(x)^2}\, dx$

83. Show that $\int_0^{\pi/6} f(\sin\theta)\, d\theta = \int_0^{1/2} f(u)\frac{1}{\sqrt{1-u^2}}\, du$.

84. Evaluate $\int_0^{\pi/2} \sin^n x \cos x\, dx$ for $n \ge 0$.

Further Insights and Challenges

85. Evaluate $I = \int_0^{\pi/2} \frac{d\theta}{1+\tan^{6{,}000}\theta}$. *Hint:* Use substitution to show that I is equal to $J = \int_0^{\pi/2} \frac{d\theta}{1+\cot^{6{,}000}\theta}$ and then check that $I + J = \int_0^{\pi/2} d\theta$.

86. Use the substitution $u = 1 + x^{1/n}$ to show that

$$\int \sqrt{1+x^{1/n}}\, dx = n\int u^{1/2}(u-1)^{n-1}\, du$$

Evaluate for $n = 2, 3$.

87. Use substitution to prove that $\int_{-a}^{a} f(x)\, dx = 0$ if f is an odd function.

88. Prove that $\int_a^b \frac{1}{x}\, dx = \int_1^{b/a} \frac{1}{x}\, dx$ for $a, b > 0$. Then show that the regions under the hyperbola over the intervals

$$[1, 2],\quad [2, 4],\quad [4, 8],\quad \ldots$$

all have the same area (Figure 4).

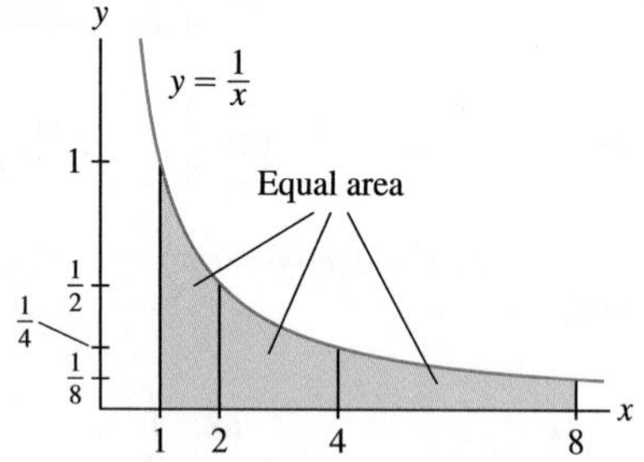

FIGURE 4 The area under $y = \frac{1}{x}$ over $[2^n, 2^{n+1}]$ is the same for all $n = 0, 1, 2, \ldots$.

89. Show that the two regions in Figure 5 have the same area. Then use the identity $\cos^2 u = \frac{1}{2}(1+\cos 2u)$ to compute the second area.

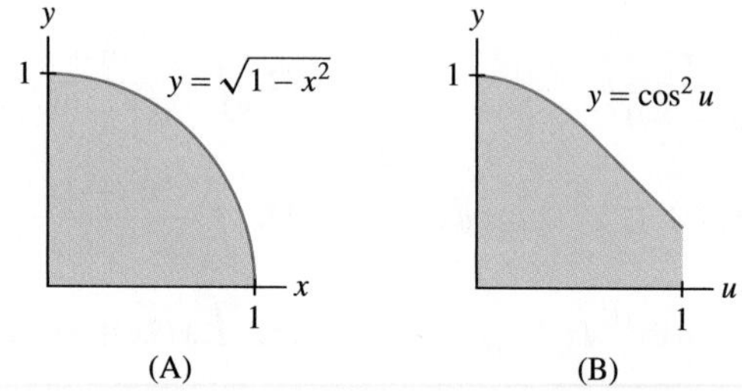

FIGURE 5

90. Area of an Ellipse Prove the formula $A = \pi ab$ for the area of the ellipse with equation (Figure 6)

$$\frac{x^2}{a^2} + \frac{y^2}{b^2} = 1$$

Hint: Use a change of variables to show that A is equal to ab times the area of the unit circle.

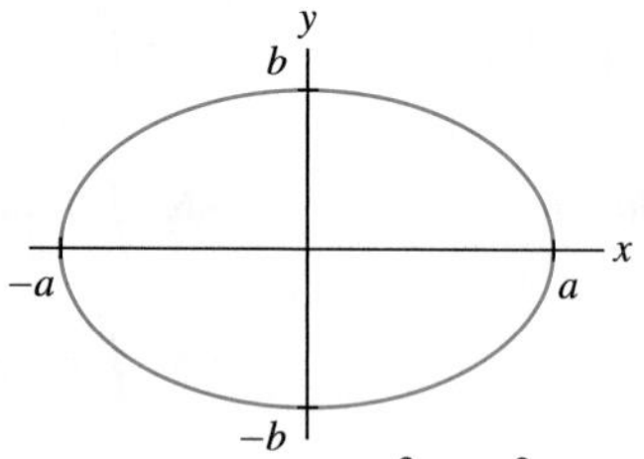

FIGURE 6 Graph of $\frac{x^2}{a^2} + \frac{y^2}{b^2} = 1$.

CHAPTER REVIEW EXERCISES

In Exercises 1–4, refer to the function $f(x)$ whose graph is shown in Figure 1.

1. Estimate L_4 and M_4 on $[0, 4]$.

2. Estimate R_4, L_4, and M_4 on $[1, 3]$.

3. Find an interval $[a, b]$ on which R_4 is larger than $\int_a^b f(x)\, dx$. Do the same for L_4.

4. Justify $\frac{3}{2} \le \int_1^2 f(x)\, dx \le \frac{9}{4}$.

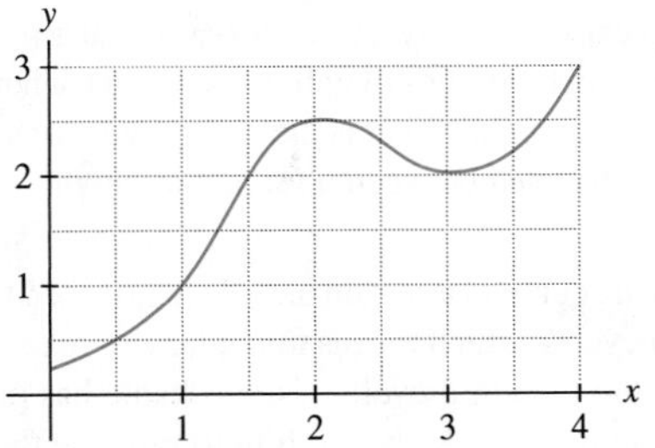

FIGURE 1

In Exercises 5–8, let $f(x) = x^2 + 3x$.

5. Calculate R_6, M_6, and L_6 for $f(x)$ on the interval [2, 5]. Sketch the graph of $f(x)$ and the corresponding rectangles for each approximation.

6. Use FTC I to evaluate $A(x) = \int_{-2}^{x} f(t)\,dt$.

7. Find a formula for R_N for $f(x)$ on [2, 5] and compute $\int_2^5 f(x)\,dx$ by taking the limit.

8. Find a formula for L_N for $f(x)$ on [0, 2] and compute $\int_0^2 f(x)\,dx$ by taking the limit.

9. Calculate R_5, M_5, and L_5 for $f(x) = (x^2+1)^{-1}$ on the interval [0, 1].

10. Let R_N be the Nth right-endpoint approximation for $f(x) = x^3$ on [0, 4] (Figure 2).

(a) Prove that $R_N = \dfrac{64(N+1)^2}{N^2}$.

(b) Prove that the area of the region within the right-endpoint rectangles above the graph is equal to

$$\frac{64(2N+1)}{N^2}$$

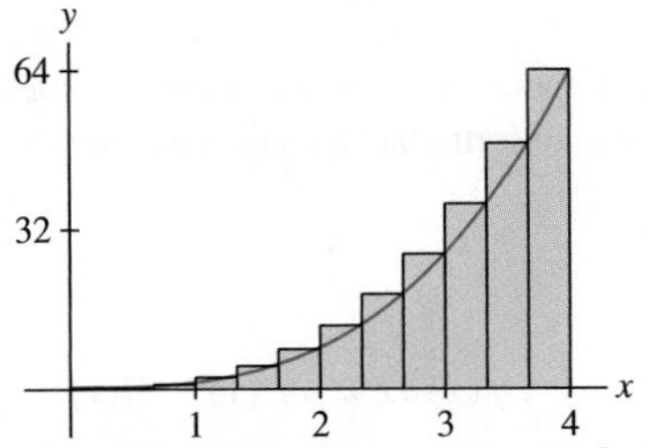

FIGURE 2 Approximation R_N for $f(x) = x^3$ on [0, 4].

11. Which approximation to the area is represented by the shaded rectangles in Figure 3? Compute R_5 and L_5.

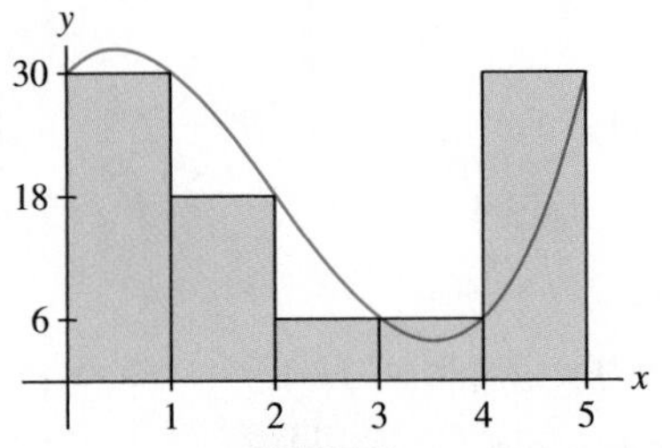

FIGURE 3

12. Calculate any two Riemann sums for $f(x) = x^2$ on the interval [2, 5], but choose partitions with at least five subintervals of unequal widths and intermediate points that are neither endpoints nor midpoints.

In Exercises 13–16, express the limit as an integral (or multiple of an integral) and evaluate.

13. $\displaystyle\lim_{N\to\infty} \frac{\pi}{6N} \sum_{j=1}^{N} \sin\left(\frac{\pi}{3} + \frac{\pi j}{6N}\right)$

14. $\displaystyle\lim_{N\to\infty} \frac{3}{N} \sum_{k=0}^{N-1} \left(10 + \frac{3k}{N}\right)$

15. $\displaystyle\lim_{N\to\infty} \frac{5}{N} \sum_{j=1}^{N} \sqrt{4 + 5j/N}$

16. $\displaystyle\lim_{N\to\infty} \frac{1^k + 2^k + \cdots + N^k}{N^{k+1}} \quad (k > 0)$

In Exercises 17–20, use the given substitution to evaluate the integral.

17. $\displaystyle\int_0^2 \frac{dt}{(4t+12)^2}$

18. $\displaystyle\int \frac{(x^2+1)\,dx}{(x^3+3x)^4}, \quad u = x^3 + 3x$

19. $\displaystyle\int_0^{\pi/6} \sin x \cos^4 x\,dx, \quad u = \cos x$

20. $\displaystyle\int \sec^2(2\theta)\tan(2\theta)\,d\theta, \quad u = \tan(2\theta)$

In Exercises 21–48, evaluate the integral.

21. $\displaystyle\int (20x^4 - 9x^3 - 2x)\,dx$

22. $\displaystyle\int_0^2 (12x^3 - 3x^2)\,dx$

23. $\displaystyle\int (2x^2 - 3x)^2\,dx$

24. $\displaystyle\int_0^1 (x^{7/3} - 2x^{1/4})\,dx$

25. $\displaystyle\int \frac{x^5 + 3x^4}{x^2}\,dx$

26. $\displaystyle\int_1^3 r^{-4}\,dr$

27. $\displaystyle\int_{-3}^{3} |x^2 - 4|\,dx$

28. $\displaystyle\int_{-2}^{4} |(x-1)(x-3)|\,dx$

29. $\displaystyle\int_1^3 [t]\,dt$

30. $\displaystyle\int_0^2 (t - [t])^2\,dt$

31. $\displaystyle\int (10t - 7)^{14}\,dt$

32. $\displaystyle\int_2^3 \sqrt{7y - 5}\,dy$

33. $\displaystyle\int \frac{(2x^3 + 3x)\,dx}{(3x^4 + 9x^2)^5}$

34. $\displaystyle\int_{-3}^{-1} \frac{x\,dx}{(x^2+5)^2}$

35. $\int_0^5 15x\sqrt{x+4}\,dx$

36. $\int t^2\sqrt{t+8}\,dt$

37. $\int_0^1 \cos\left(\frac{\pi}{3}(t+2)\right)\,dt$

38. $\int_{\pi/2}^{\pi} \sin\left(\frac{5\theta-\pi}{6}\right)\,d\theta$

39. $\int t^2\sec^2(9t^3+1)\,dt$

40. $\int \sin^2(3\theta)\cos(3\theta)\,d\theta$

41. $\int \csc^2(9-2\theta)\,d\theta$

42. $\int \sin\theta\sqrt{4-\cos\theta}\,d\theta$

43. $\int_0^{\pi/3} \frac{\sin\theta}{\cos^{2/3}\theta}\,d\theta$

44. $\int \frac{\sec^2 t\,dt}{(\tan t-1)^2}$

45. $\int y\sqrt{2y+3}\,dy$

46. $\int_1^8 t^2\sqrt{t+8}\,dt$

47. $\int_0^{\pi/2} \sec^2(\cos\theta)\sin\theta\,d\theta$

48. $\int_{-4}^{-2} \frac{12x\,dx}{(x^2+2)^3}$

49. Combine to write as a single integral:

$$\int_0^8 f(x)\,dx + \int_{-2}^0 f(x)\,dx + \int_8^6 f(x)\,dx$$

50. Let $A(x) = \int_0^x f(x)\,dx$, where $f(x)$ is the function shown in Figure 4. Identify the location of the local minima, the local maxima, and points of inflection of $A(x)$ on the interval $[0, E]$, as well as the intervals where $A(x)$ is increasing, decreasing, concave up, or concave down. Where does the absolute max of $A(x)$ occur?

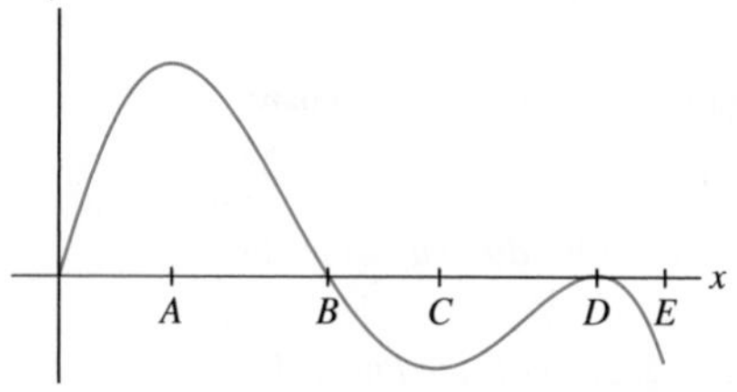

FIGURE 4

51. Find the inflection points of $A(x) = \int_3^x \frac{t\,dt}{t^2+1}$. However, do not evaluate $A(x)$ explicitly.

52. A particle starts at the origin at time $t = 0$ and moves with velocity $v(t)$ as shown in Figure 5.

(a) How many times does the particle return to the origin in the first 12 seconds?

(b) What is the particle's maximum distance from the origin?

(c) What is particle's maximum distance to the left of the origin?

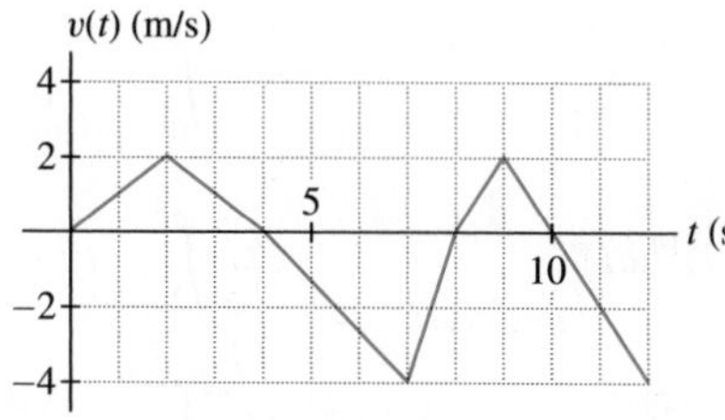

FIGURE 5

53. On a typical day, a city consumes water at the rate of $r(t) = 100 + 72t - 3t^2$ (in thousands of gallons per hour), where t is the number of hours past midnight. What is the daily water consumption? How much water is consumed between 6 PM and midnight?

54. The learning curve in a certain bicycle factory is $L(x) = 12x^{-1/5}$ (in hours per bicycle), which means that it takes a bike mechanic $L(n)$ hours to assemble the nth bicycle. If a mechanic has produced 24 bicycles, how long does it take her or him to produce the second batch of 12?

55. Cost engineers at NASA have the task of projecting the cost P of major space projects. It has been found that the cost C of developing a projection increases with P at the rate $dC/dP \approx 21P^{-0.65}$, where C is in thousands of dollars and P in millions of dollars. What is the cost of developing a projection for a project whose cost turns out to be $P = \$35$ million?

56. An astronomer estimates that in a certain constellation, the number of stars per magnitude m, per degree-squared of sky, is equal to $A(m) = 2.4 \times 10^{-6} m^{7.4}$ (fainter stars have higher magnitudes). Determine the total number of stars of magnitude between 6 and 15 in a one-degree-squared region of sky.

57. Evaluate $\int_{-8}^{8} \frac{x^{15}\,dx}{3+\cos^2 x}$, using the properties of odd functions.

58. Evaluate $\int_0^1 f(x)\,dx$, assuming that $f(x)$ is an even continuous function such that

$$\int_1^2 f(x)\,dx = 5, \qquad \int_{-2}^1 f(x)\,dx = 8$$

59. GU Plot the graph of $f(x) = \sin mx \sin nx$ on $[0, \pi]$ for the pairs $(m, n) = (2, 4), (3, 5)$ and in each case guess the value of $I = \int_0^\pi f(x)\,dx$. Experiment with a few more values (including two cases with $m = n$) and formulate a conjecture for when I is zero.

60. Show that

$$\int x\,f(x)\,dx = xF(x) - G(x)$$

where $F'(x) = f(x)$ and $G'(x) = F(x)$. Use this to evaluate $\int x\cos x\,dx$.

61. Prove

$$2 \le \int_1^2 2^x\,dx \le 4 \quad \text{and} \quad \frac{1}{9} \le \int_1^2 3^{-x}\,dx \le \frac{1}{3}$$

62. GU Plot the graph of $f(x) = x^{-2}\sin x$, and show that $0.2 \le \int_1^2 f(x)\,dx \le 0.9$.

63. Find upper and lower bounds for $\int_0^1 f(x)\,dx$, for $f(x)$ in Figure 6.

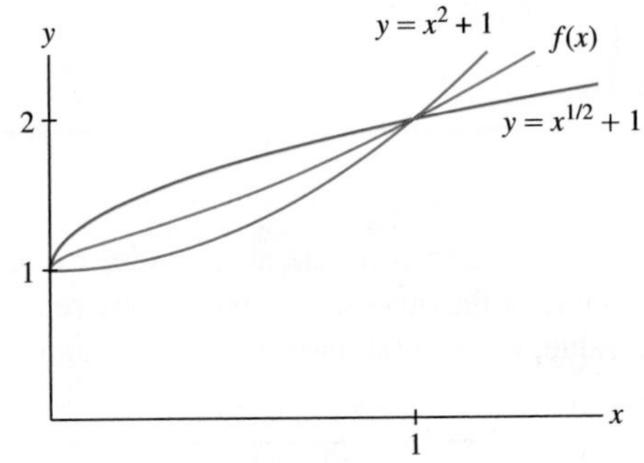

FIGURE 6

In Exercises 64–69, find the derivative.

64. $A'(x)$, where $A(x) = \int_3^x \sin(t^3)\,dt$

65. $A'(\pi)$, where $A(x) = \int_2^x \frac{\cos t}{1+t}\,dt$

66. $\frac{d}{dy}\int_{-2}^{y} 3^x\,dx$

67. $G'(x)$, where $G(x) = \int_{-2}^{\sin x} t^3\,dt$

68. $G'(2)$, where $G(x) = \int_0^{x^3} \sqrt{t+1}\,dt$

69. $H'(1)$, where $H(x) = \int_{4x^2}^{9} \frac{1}{t}\,dt$

70. Explain with a graph: If $f(x)$ is increasing and concave up on $[a, b]$, then L_N is more accurate than R_N. Which is more accurate if $f(x)$ is increasing and concave down?

71. Explain with a graph: If $f(x)$ is linear on $[a, b]$, then the $\int_a^b f(x)\,dx = \frac{1}{2}(R_N + L_N)$ for all N.

72. Let $f(x)$ be a positive increasing continuous function on $[a, b]$, where $0 \le a < b$ as in Figure 7. Show that the shaded region has area

$$I = bf(b) - af(a) - \int_a^b f(x)\,dx \qquad \boxed{1}$$

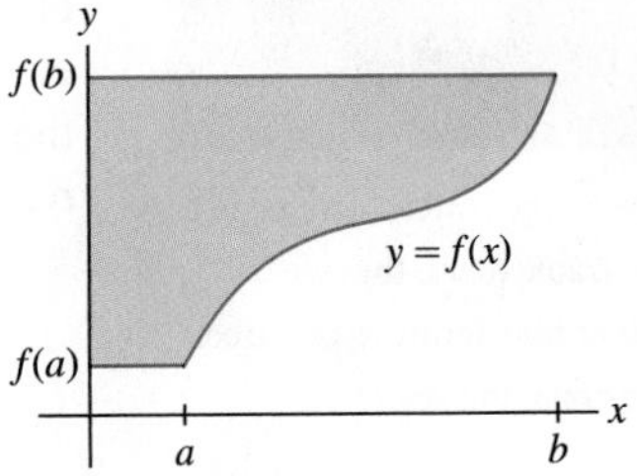

FIGURE 7

73. How can we interpret the quantity I in Eq. (1) if $a < b \le 0$? Explain with a graph.

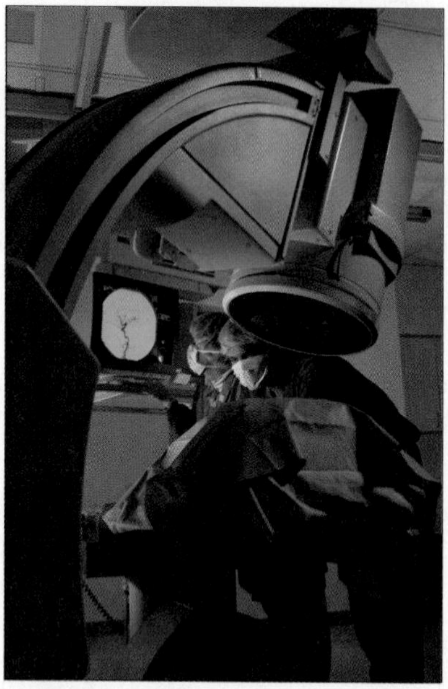

Magnetic Resonance Image (MRI) of veins in a patient's heart. MRI scanners use the mathematics of Fourier transforms to construct two and three-dimensional images.

6 APPLICATIONS OF THE INTEGRAL

In the previous chapter, we used the integral to compute areas under curves and net change. In this chapter, we discuss some of the other quantities that are represented by integrals, including volume, average value, work, total mass, population, and fluid flow.

6.1 Area Between Two Curves

Sometimes we are interested in the area between two curves. Figure 1 shows projected electric power generation in the U.S. through renewable resources (wind, solar, biofuels, etc.) under two scenarios: with and without government stimulus spending. The area of the shaded region between the two graphs represents the additional energy projected to result from stimulus spending.

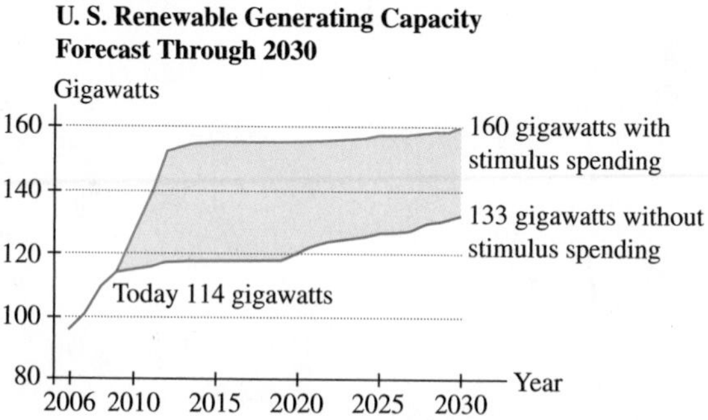

FIGURE 1 The area of the shaded region (which has units of *power* × *time*, or *energy*) represents the additional energy from renewable generating capacity projected to result from government stimulus spending in 2009–2010. *Source:* Energy Information Agency.

Now suppose that we are given two functions $y = f(x)$ and $y = g(x)$ such that $f(x) \geq g(x)$ for all x in an interval $[a, b]$. Then the graph of $f(x)$ lies above the graph of $g(x)$ [Figure 2], and the area between the graphs is equal to the integral of $f(x) - g(x)$:

$$\text{Area between the graphs} = \int_a^b f(x)\,dx - \int_a^b g(x)\,dx$$

$$= \int_a^b \big(f(x) - g(x)\big)\,dx \qquad \boxed{1}$$

Figure 2 illustrates this formula in the case that both graphs lie above the x-axis. We see that the region between the graphs is obtained by removing the region under $y = g(x)$ from the region under $y = f(x)$.

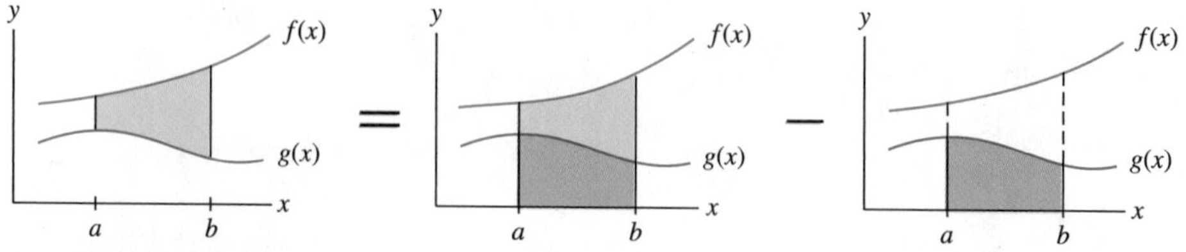

FIGURE 2 The area between the graphs is a difference of two areas.

■ **EXAMPLE 1** Find the area of the region between the graphs of the functions

$$f(x) = x^2 - 4x + 10, \qquad g(x) = 4x - x^2, \qquad 1 \le x \le 3$$

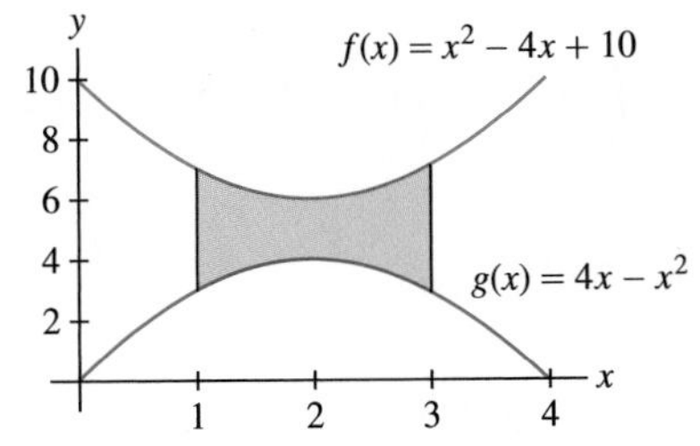

FIGURE 3

Solution First, we must determine which graph lies on top. Figure 3 shows that $f(x) \ge g(x)$, as we can verify directly by completing the square:

$$f(x) - g(x) = (x^2 - 4x + 10) - (4x - x^2) = 2x^2 - 8x + 10 = 2(x-2)^2 + 2 > 0$$

Therefore, by Eq. (1), the area between the graphs is

$$\int_1^3 \big(f(x) - g(x)\big)\,dx = \int_1^3 \left((x^2 - 4x + 10) - (4x - x^2)\right)dx$$

$$= \int_1^3 (2x^2 - 8x + 10)\,dx = \left(\frac{2}{3}x^3 - 4x^2 + 10x\right)\Bigg|_1^3 = 12 - \frac{20}{3} = \frac{16}{3} \qquad ■$$

Before continuing with more examples, we note that Eq. (1) remains valid whenever $f(x) \ge g(x)$, even if $f(x)$ and $g(x)$ are not assumed to be positive. Recall that the integral is a limit of Riemann sums:

$$\int_a^b \big(f(x) - g(x)\big)\,dx = \lim_{\|P\| \to 0} R(f - g, P, C) = \lim_{N \to \infty} \sum_{i=1}^{N} \big(f(c_i) - g(c_i)\big)\Delta x_i$$

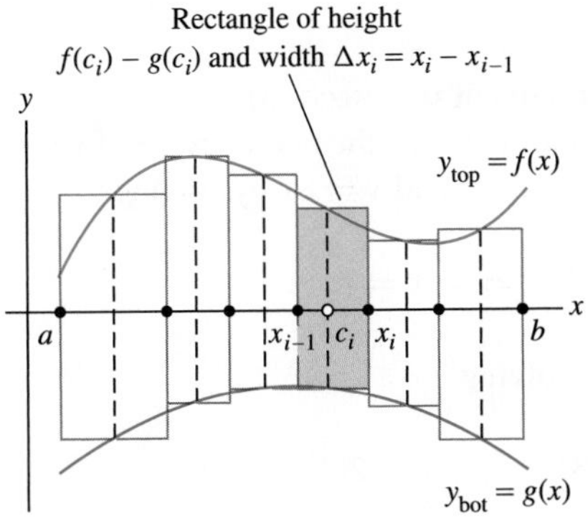

FIGURE 4 Riemann sum for $f(x) - g(x)$.

where $C = \{c_1, \dots, c_N\}$ is a set of sample points for a partition P of $[a, b]$ and $\Delta x_i = x_i - x_{i-1}$. The ith term in the sum is the area of a thin vertical rectangle (Figure 4):

$$\big(f(c_i) - g(c_i)\big)\Delta x_i = \text{height} \times \text{width}$$

Therefore, $R(f - g, P, C)$ is an approximation to the area between the graphs using thin vertical rectangles. As the norm $\|P\|$ tends to zero, the rectangles get thinner and the Riemann sum converges to the area between the graphs. Writing $y_{top} = f(x)$ for the upper graph and $y_{bot} = g(x)$ for the lower graph, we obtain

Keep in mind that $(y_{top} - y_{bot})$ is the height of a vertical slice of the region.

$$\text{Area between the graphs} = \int_a^b (y_{top} - y_{bot})\,dx = \int_a^b \big(f(x) - g(x)\big)\,dx \qquad \boxed{2}$$

■ **EXAMPLE 2** Find the area between the graphs of $f(x) = x^2 - 5x - 7$ and $g(x) = x - 12$ over $[-2, 5]$.

Solution First, we must determine which graph lies on top.

***Step 1.* Sketch the region (especially, find any points of intersection).**
We know that $y = f(x)$ is a parabola with y-intercept -7 and that $y = g(x)$ is a line with y-intercept -12 (Figure 5). To determine where the graphs intersect, we observe

$$f(x) - g(x) = (x^2 - 5x - 7) - (x - 12) = x^2 - 6x + 5 = (x-1)(x-5)$$

The graphs intersect where $(x - 1)(x - 5) = 0$, that is, at $x = 1$ and $x = 5$.

***Step 2.* Set up the integrals and evaluate.**
We also see that $f(x) - g(x) \le 0$ for $1 \le x < 5$, and thus

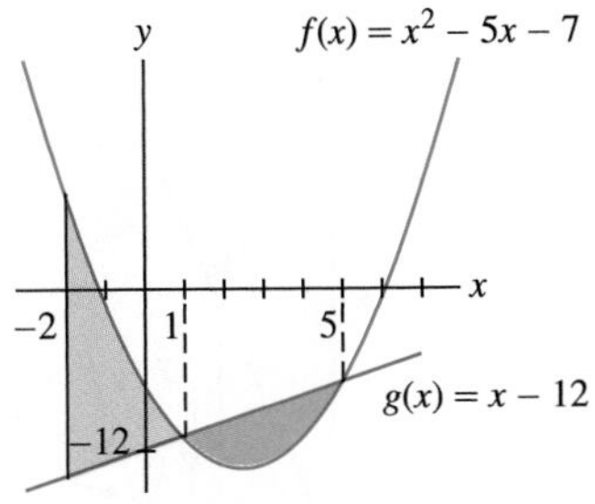

FIGURE 5

$$f(x) \ge g(x) \text{ on } [-2, 1] \qquad \text{and} \qquad g(x) \ge f(x) \text{ on } [1, 5]$$

In Example 2, we found the intersection points of $y = f(x)$ and $y = g(x)$ algebraically. For more complicated functions, it may be necessary to use a computer algebra system.

Therefore, we write the area as a sum of integrals over the two intervals:

$$\int_{-2}^{5} (y_{\text{top}} - y_{\text{bot}})\,dx = \int_{-2}^{1} \big(f(x) - g(x)\big)\,dx + \int_{1}^{5} \big(g(x) - f(x)\big)\,dx$$

$$= \int_{-2}^{1} \big((x^2 - 5x - 7) - (x - 12)\big)\,dx + \int_{1}^{5} \big((x - 12) - (x^2 - 5x - 7)\big)\,dx$$

$$= \int_{-2}^{1} (x^2 - 6x + 5)\,dx + \int_{1}^{5} (-x^2 + 6x - 5)\,dx$$

$$= \left(\frac{1}{3}x^3 - 3x^2 + 5x\right)\Bigg|_{-2}^{1} + \left(-\frac{1}{3}x^3 + 3x^2 - 5x\right)\Bigg|_{1}^{5}$$

$$= \left(\frac{7}{3} - \frac{(-74)}{3}\right) + \left(\frac{25}{3} - \frac{(-7)}{3}\right) = \frac{113}{3}$$

■

■ **EXAMPLE 3 Calculating Area by Dividing the Region** Find the area of the region bounded by the graphs of $y = 8/x^2$, $y = 8x$, and $y = x$.

Solution

Step 1. **Sketch the region (especially, find any points of intersection).**
The curve $y = 8/x^2$ cuts off a region in the sector between the two lines $y = 8x$ and $y = x$ (Figure 6). We find the intersection of $y = 8/x^2$ and $y = 8x$ by solving

$$\frac{8}{x^2} = 8x \quad \Rightarrow \quad x^3 = 1 \quad \Rightarrow \quad x = 1$$

and the intersection of $y = 8/x^2$ and $y = x$ by solving

$$\frac{8}{x^2} = x \quad \Rightarrow \quad x^3 = 8 \quad \Rightarrow \quad x = 2$$

Step 2. **Set up the integrals and evaluate.**
Figure 6 shows that $y_{\text{bot}} = x$, but y_{top} changes at $x = 1$ from $y_{\text{top}} = 8x$ to $y_{\text{top}} = 8/x^2$. Therefore, we break up the regions into two parts, A and B, and compute their areas separately.

$$\text{Area of } A = \int_{0}^{1} \big(y_{\text{top}} - y_{\text{bot}}\big)\,dx = \int_{0}^{1} (8x - x)\,dx = \int_{0}^{1} 7x\,dx = \frac{7}{2}x^2\Big|_{0}^{1} = \frac{7}{2}$$

$$\text{Area of } B = \int_{1}^{2} \big(y_{\text{top}} - y_{\text{bot}}\big)\,dx = \int_{1}^{2} \left(\frac{8}{x^2} - x\right)\,dx = \left(-\frac{8}{x} - \frac{1}{2}x^2\right)\Bigg|_{1}^{2} = \frac{5}{2}$$

The total area bounded by the curves is the sum $\frac{7}{2} + \frac{5}{2} = 6$. ■

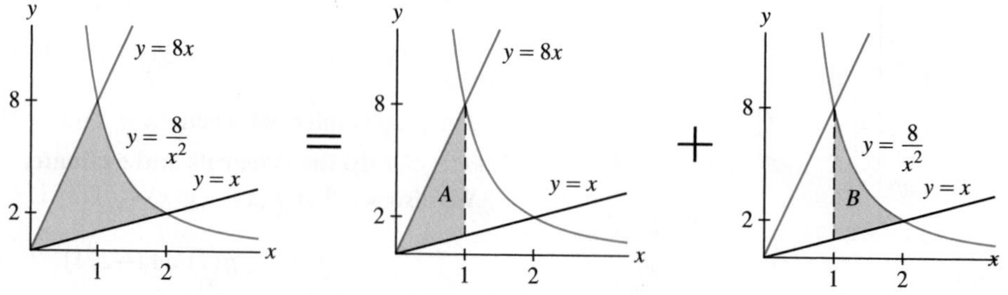

FIGURE 6 Area bounded by $y = 8/x^2$, $y = 8x$, and $y = x$ as a sum of two areas.

Integration Along the y-Axis

Suppose we are given x as a function of y, say, $x = g(y)$. What is the meaning of the integral $\int_c^d g(y)\,dy$? This integral can be interpreted as *signed area*, where regions to the *right* of the y-axis have positive area and regions to the *left* have negative area:

$$\int_c^d g(y)\,dy = \text{signed area between graph and } y\text{-axis for } c \le y \le d$$

In Figure 7(A), the part of the shaded region to the left of the y-axis has negative signed area. The signed area of the entire region is

$$\underbrace{\int_{-6}^{6} (y^2 - 9)\,dy}_{\substack{\text{Area to the right of } y\text{-axis minus} \\ \text{area to the left of } y\text{-axis}}} = \left(\frac{1}{3}y^3 - 9y\right)\Big|_{-6}^{6} = 36$$

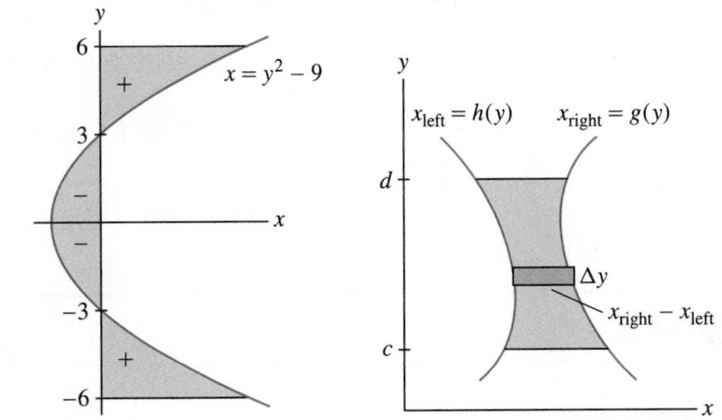

(A) Region between $x = y^2 - 9$ and the y-axis

(B) Region between $x = h(y)$ and $x = g(y)$

FIGURE 7

More generally, if $g(y) \ge h(y)$ as in Figure 7(B), then the graph of $x = g(y)$ lies to the right of the graph of $x = h(y)$. In this case, we write $x_{\text{right}} = g(y)$ and $x_{\text{left}} = h(y)$. The formula for area corresponding to Eq. (2) is

$$\text{Area between the graphs} = \int_c^d (x_{\text{right}} - x_{\text{left}})\,dy = \int_c^d \big(g(y) - h(y)\big)\,dy \qquad \boxed{3}$$

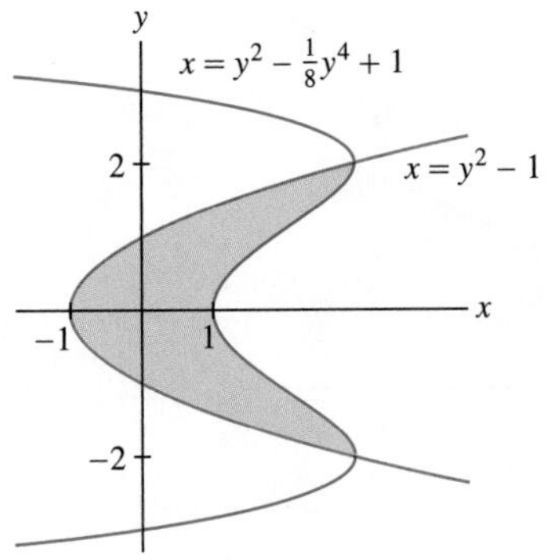

FIGURE 8

■ **EXAMPLE 4** Calculate the area enclosed by the graphs of $h(y) = y^2 - 1$ and $g(y) = y^2 - \frac{1}{8}y^4 + 1$.

Solution First, we find the points where the graphs intersect by solving $g(y) = h(y)$ for y:

$$y^2 - \frac{1}{8}y^4 + 1 = y^2 - 1 \quad \Rightarrow \quad \frac{1}{8}y^4 - 2 = 0 \quad \Rightarrow \quad y = \pm 2$$

Figure 8 shows that the enclosed region stretches from $y = -2$ to $y = 2$. On this interval, $g(y) \ge h(y)$. Therefore $x_{\text{right}} = g(y)$, $x_{\text{left}} = h(y)$, and

$$x_{\text{right}} - x_{\text{left}} = \left(y^2 - \frac{1}{8}y^4 + 1\right) - (y^2 - 1) = 2 - \frac{1}{8}y^4$$

It would be more difficult to calculate the area of the region in Figure 8 as an integral with respect to x because the curves are not graphs of functions of x.

The enclosed area is

$$\int_{-2}^{2} (x_{\text{right}} - x_{\text{left}})\, dy = \int_{-2}^{2} \left(2 - \frac{1}{8}y^4\right) dy = \left(2y - \frac{1}{40}y^5\right)\Big|_{-2}^{2}$$
$$= \frac{16}{5} - \left(-\frac{16}{5}\right) = \frac{32}{5}$$

6.1 SUMMARY

- If $f(x) \geq g(x)$ on $[a, b]$, then the area between the graphs is

$$\text{Area between the graphs} = \int_a^b \left(y_{\text{top}} - y_{\text{bot}}\right) dx = \int_a^b \left(f(x) - g(x)\right) dx$$

- To calculate the area between $y = f(x)$ and $y = g(x)$, sketch the region to find y_{top}. If necessary, find points of intersection by solving $f(x) = g(x)$.
- Integral along the y-axis: $\int_c^d g(y)\, dy$ is equal to the signed area between the graph and the y-axis for $c \leq y \leq d$. Area to the right of the y-axis is positive and area to the left is negative.
- If $g(y) \geq h(y)$ on $[c, d]$, then $x = g(y)$ lies to the right of $x = h(y)$ and

$$\text{Area between the graphs} = \int_c^d \left(x_{\text{right}} - x_{\text{left}}\right) dy = \int_c^d \left(g(y) - h(y)\right) dy$$

6.1 EXERCISES

Preliminary Questions

1. What is the area interpretation of $\int_a^b \left(f(x) - g(x)\right) dx$ if $f(x) \geq g(x)$?

2. Is $\int_a^b \left(f(x) - g(x)\right) dx$ still equal to the area between the graphs of f and g if $f(x) \geq 0$ but $g(x) \leq 0$?

3. Suppose that $f(x) \geq g(x)$ on $[0, 3]$ and $g(x) \geq f(x)$ on $[3, 5]$. Express the area between the graphs over $[0, 5]$ as a sum of integrals.

4. Suppose that the graph of $x = f(y)$ lies to the left of the y-axis. Is $\int_a^b f(y)\, dy$ positive or negative?

Exercises

1. Find the area of the region between $y = 3x^2 + 12$ and $y = 4x + 4$ over $[-3, 3]$ (Figure 9).

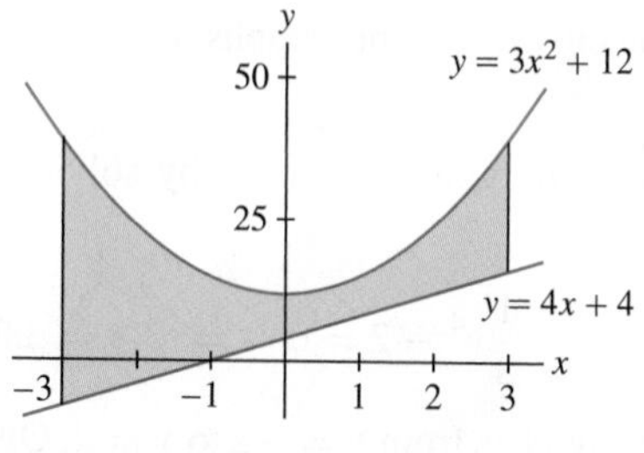

FIGURE 9

2. Find the area of the region between the graphs of $f(x) = 3x + 8$ and $g(x) = x^2 + 2x + 2$ over $[0, 2]$.

3. Find the area of the region enclosed by the graphs of $f(x) = x^2 + 2$ and $g(x) = 2x + 5$ (Figure 10).

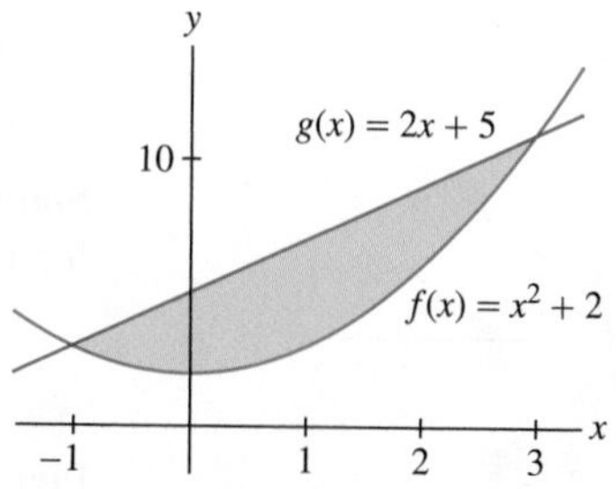

FIGURE 10

4. Find the area of the region enclosed by the graphs of $f(x) = x^3 - 10x$ and $g(x) = 6x$ (Figure 11).

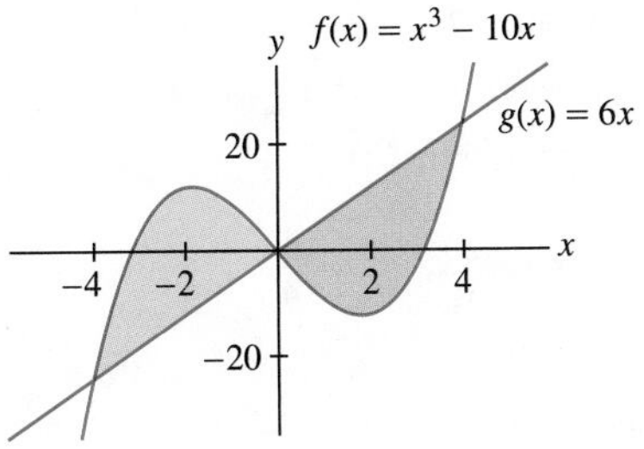

FIGURE 11

In Exercises 5 and 6, sketch the region between $y = \sin x$ and $y = \cos x$ over the interval and find its area.

5. $\left[\frac{\pi}{4}, \frac{\pi}{2}\right]$

6. $[0, \pi]$

In Exercises 7 and 8, let $f(x) = 20 + x - x^2$ and $g(x) = x^2 - 5x$.

7. Sketch the region enclosed by the graphs of $f(x)$ and $g(x)$ and compute its area.

8. Sketch the region between the graphs of $f(x)$ and $g(x)$ over $[4, 8]$ and compute its area as a sum of two integrals.

9. GU Find the points of intersection of $y = x(x^2 - 1)$ and $y = 1 - x^2$. Sketch the region enclosed by these curves over $[-1, 1]$ and compute its area.

10. GU Find the points of intersection of $y = x(4 - x)$ and $y = x^2(4 - x)$. Sketch the region enclosed by these curves over $[0, 4]$ and compute its area.

11. Sketch the region bounded by the line $y = 2$ and the graph of $y = \sec^2 x$ for $-\frac{\pi}{2} < x < \frac{\pi}{2}$ and find its area.

12. Sketch the region bounded by

$$y = \frac{x}{\sqrt{1 - x^2}} \quad \text{and} \quad y = -\frac{x}{\sqrt{1 - x^2}}$$

for $0 \le x \le 0.8$ and find its area.

In Exercises 13–16, find the area of the shaded region in Figures 12–15.

13.

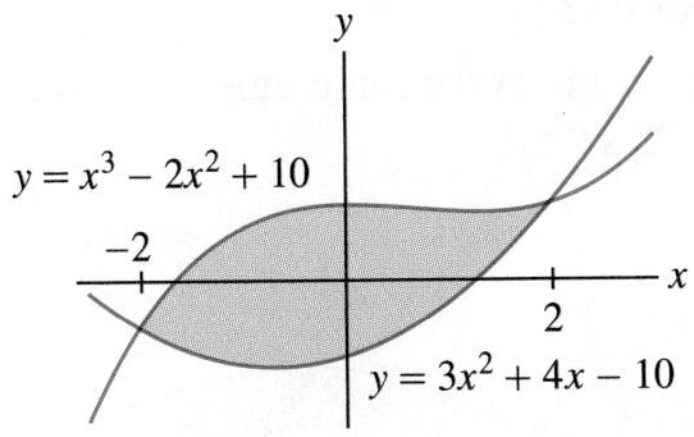

FIGURE 12

14.

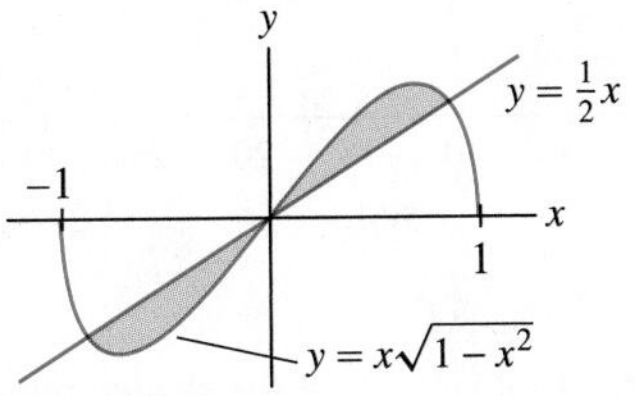

FIGURE 13

15.

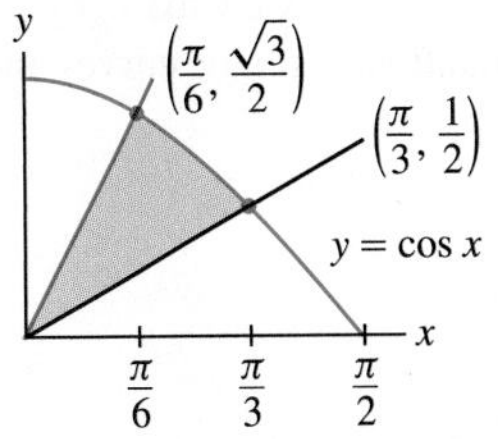

FIGURE 14

16.

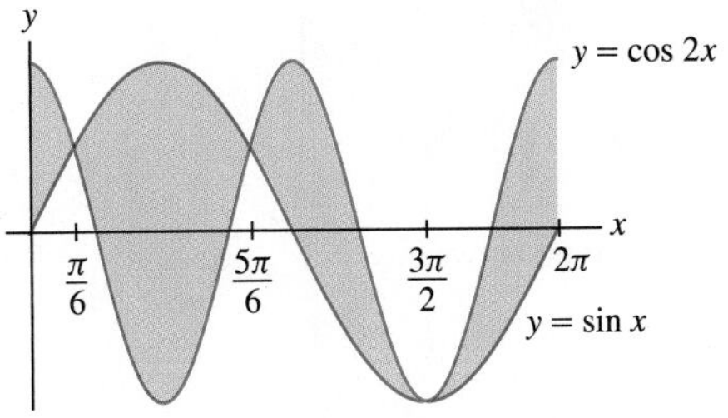

FIGURE 15

In Exercises 17 and 18, find the area between the graphs of $x = \sin y$ and $x = 1 - \cos y$ over the given interval (Figure 16).

17. $0 \le y \le \frac{\pi}{2}$

18. $-\frac{\pi}{2} \le y \le \frac{\pi}{2}$

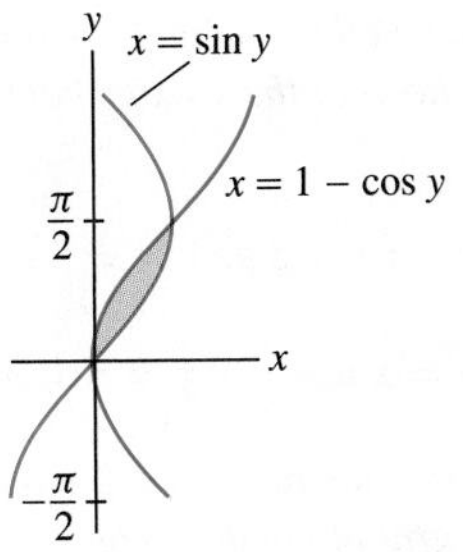

FIGURE 16

19. Find the area of the region lying to the right of $x = y^2 + 4y - 22$ and to the left of $x = 3y + 8$.

20. Find the area of the region lying to the right of $x = y^2 - 5$ and to the left of $x = 3 - y^2$.

21. Figure 17 shows the region enclosed by $x = y^3 - 26y + 10$ and $x = 40 - 6y^2 - y^3$. Match the equations with the curves and compute the area of the region.

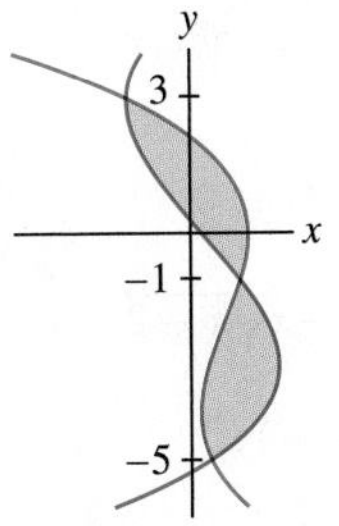

FIGURE 17

22. Figure 18 shows the region enclosed by $y = x^3 - 6x$ and $y = 8 - 3x^2$. Match the equations with the curves and compute the area of the region.

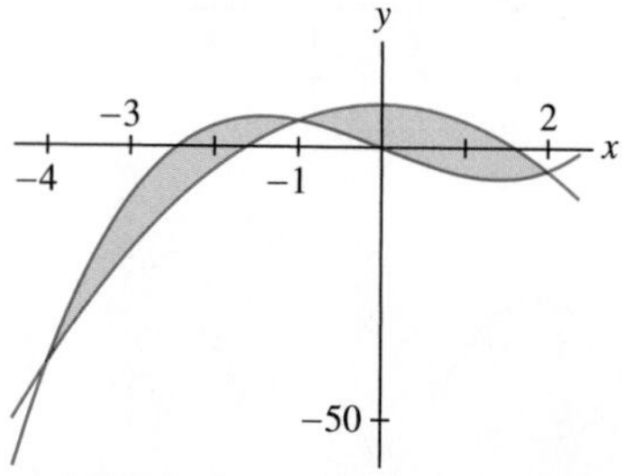

FIGURE 18 Region between $y = x^3 - 6x$ and $y = 8 - 3x^2$.

In Exercises 23 and 24, find the area enclosed by the graphs in two ways: by integrating along the x-axis and by integrating along the y-axis.

23. $x = 9 - y^2, \quad x = 5$

24. The *semicubical parabola* $y^2 = x^3$ and the line $x = 1$.

In Exercises 25 and 26, find the area of the region using the method (integration along either the x- or the y-axis) that requires you to evaluate just one integral.

25. Region between $y^2 = x + 5$ and $y^2 = 3 - x$

26. Region between $y = x$ and $x + y = 8$ over $[2, 3]$

In Exercises 27–44, sketch the region enclosed by the curves and compute its area as an integral along the x- or y-axis.

27. $y = 4 - x^2, \quad y = x^2 - 4$

28. $y = x^2 - 6, \quad y = 6 - x^3, \quad$ y-axis

29. $x + y = 4, \quad x - y = 0, \quad y + 3x = 4$

30. $y = 8 - 3x, \quad y = 6 - x, \quad y = 2$

31. $y = 8 - \sqrt{x}, \quad y = \sqrt{x}, \quad x = 0$

32. $y = |x^2 - 4|, \quad y = 5$

33. $x = |y|, \quad x = 1 - |y|$

34. $y = |x|, \quad y = x^2 - 6$

35. $x = y^3 - 18y, \quad y + 2x = 0$

36. $y = x\sqrt{x - 2}, \quad y = -x\sqrt{x - 2}, \quad x = 4$

37. $x = 2y, \quad x + 1 = (y - 1)^2$

38. $x + y = 1, \quad x^{1/2} + y^{1/2} = 1$

39. $y = \cos x, \quad y = \cos 2x, \quad x = 0, \quad x = \frac{2\pi}{3}$

40. $x = \tan x, \quad y = -\tan x, \quad x = \frac{\pi}{4}$

41. $y = \sin x, \quad y = \csc^2 x, \quad x = \frac{\pi}{4}$

42. $x = \sin y, \quad x = \frac{2}{\pi} y$

43. $y = \sin x, \quad y = x \sin(x^2), \quad 0 \le x \le 1$

44. $y = \frac{\sin(\sqrt{x})}{\sqrt{x}}, \quad y = 0, \quad \pi^2 \le x \le 9\pi^2$

45. CAS Plot

$$y = \frac{x}{\sqrt{x^2 + 1}} \quad \text{and} \quad y = (x - 1)^2$$

on the same set of axes. Use a computer algebra system to find the points of intersection numerically and compute the area between the curves.

46. Sketch a region whose area is represented by

$$\int_{-\sqrt{2}/2}^{\sqrt{2}/2} \left(\sqrt{1 - x^2} - |x|\right) dx$$

and evaluate using geometry.

47. Athletes 1 and 2 run along a straight track with velocities $v_1(t)$ and $v_2(t)$ (in m/s) as shown in Figure 19.

(a) Which of the following is represented by the area of the shaded region over [0, 10]?

i. The distance between athletes 1 and 2 at time $t = 10$ s.

ii. The difference in the distance traveled by the athletes over the time interval [0, 10].

(b) Does Figure 19 give us enough information to determine who is ahead at time $t = 10$ s?

(c) If the athletes begin at the same time and place, who is ahead at $t = 10$ s? At $t = 25$ s?

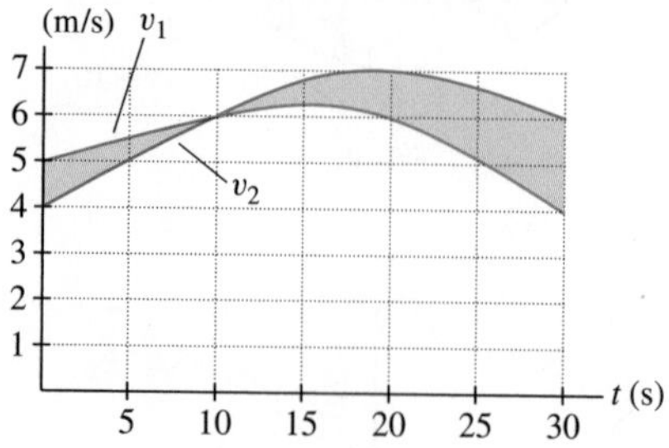

FIGURE 19

48. Express the area (not signed) of the shaded region in Figure 20 as a sum of three integrals involving $f(x)$ and $g(x)$.

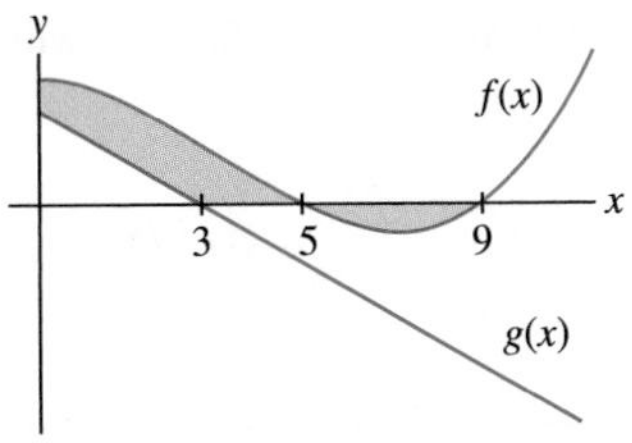

FIGURE 20

49. Find the area enclosed by the curves $y = c - x^2$ and $y = x^2 - c$ as a function of c. Find the value of c for which this area is equal to 1.

50. Set up (but do not evaluate) an integral that expresses the area between the circles $x^2 + y^2 = 2$ and $x^2 + (y-1)^2 = 1$.

51. Set up (but do not evaluate) an integral that expresses the area between the graphs of $y = (1+x^2)^{-1}$ and $y = x^2$.

52. CAS Find a numerical approximation to the area above $y = 1 - (x/\pi)$ and below $y = \sin x$ (find the points of intersection numerically).

53. CAS Find a numerical approximation to the area above $y = |x|$ and below $y = \cos x$.

54. CAS Use a computer algebra system to find a numerical approximation to the number c (besides zero) in $\left[0, \frac{\pi}{2}\right]$, where the curves $y = \sin x$ and $y = \tan^2 x$ intersect. Then find the area enclosed by the graphs over $[0, c]$.

55. The back of Jon's guitar (Figure 21) is 19 inches long. Jon measured the width at 1-in. intervals, beginning and ending $\frac{1}{2}$ in. from the ends, obtaining the results

6, 9, 10.25, 10.75, 10.75, 10.25, 9.75, 9.5, 10, 11.25,

12.75, 13.75, 14.25, 14.5, 14.5, 14, 13.25, 11.25, 9

Use the midpoint rule to estimate the area of the back.

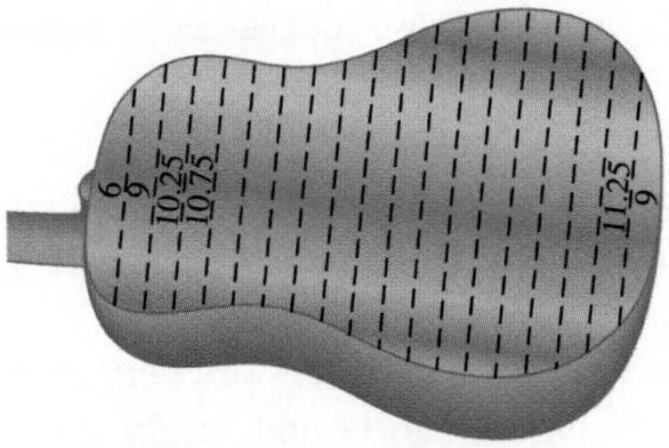

FIGURE 21 Back of guitar.

56. Referring to Figure 1 at the beginning of this section, estimate the projected number of additional joules produced in the years 2009–2030 as a result of government stimulus spending in 2009–2010. *Note:* One watt is equal to one joule per second, and one gigawatt is 10^9 watts.

Exercises 57 and 58 use the notation and results of Exercises 49–51 of Section 3.4. For a given country, $F(r)$ is the fraction of total income that goes to the bottom rth fraction of households. The graph of $y = F(r)$ is called the Lorenz curve.

57. Let A be the area between $y = r$ and $y = F(r)$ over the interval [0, 1] (Figure 22). The **Gini index** is the ratio $G = A/B$, where B is the area under $y = r$ over [0, 1].

(a) Show that $G = 2\int_0^1 (r - F(r))\,dr$.

(b) Calculate G if

$$F(r) = \begin{cases} \frac{1}{3}r & \text{for } 0 \le r \le \frac{1}{2} \\ \frac{5}{3}r - \frac{2}{3} & \text{for } \frac{1}{2} \le r \le 1 \end{cases}$$

(c) The Gini index is a measure of income distribution, with a lower value indicating a more equal distribution. Calculate G if $F(r) = r$ (in this case, all households have the same income by Exercise 51(b) of Section 3.4).

(d) What is G if all of the income goes to one household? *Hint:* In this extreme case, $F(r) = 0$ for $0 \le r < 1$.

58. Calculate the Gini index of the United States in the year 2001 from the Lorenz curve in Figure 22, which consists of segments joining the data points in the following table.

r	0	0.2	0.4	0.6	0.8	1
$F(r)$	0	0.035	0.123	0.269	0.499	1

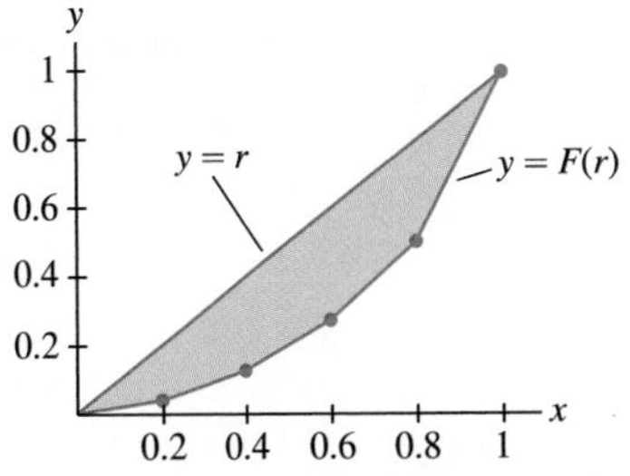

FIGURE 22 Lorenz Curve for U.S. in 2001.

Further Insights and Challenges

59. Find the line $y = mx$ that divides the area under the curve $y = x(1-x)$ over [0, 1] into two regions of equal area.

60. CAS Let c be the number such that the area under $y = \sin x$ over $[0, \pi]$ is divided in half by the line $y = cx$ (Figure 23). Find an equation for c and solve this equation *numerically* using a computer algebra system.

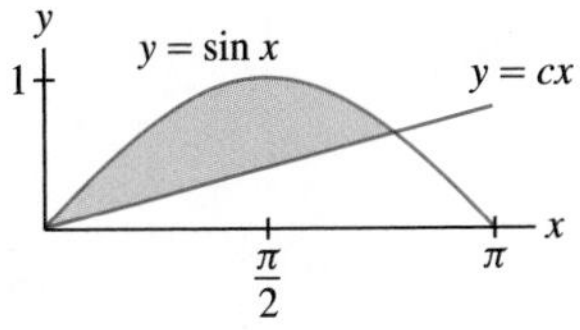

FIGURE 23

61. Explain geometrically (without calculation):

$$\int_0^1 x^n \, dx + \int_0^1 x^{1/n} \, dx = 1 \qquad \text{(for } n > 0\text{)}$$

62. Let $f(x)$ be an increasing function with inverse $g(x)$. Explain geometrically:

$$\int_0^a f(x)\, dx + \int_{f(0)}^{f(a)} g(x)\, dx = af(a)$$

6.2 Setting Up Integrals: Volume, Density, Average Value

Which quantities are represented by integrals? Roughly speaking, integrals represent quantities that are the "total amount" of something such as area, volume, or total mass. There is a two-step procedure for computing such quantities: (1) Approximate the quantity by a sum of N terms, and (2) Pass to the limit as $N \to \infty$ to obtain an integral. We'll use this procedure often in this and other sections.

Volume

The term "solid" or "solid body" refers to a solid three-dimensional object.

Our first example is the **volume** of a solid body. Before proceeding, let's recall that the volume of a *right cylinder* (Figure 1) is Ah, where A is the area of the base and h is the height, measured perpendicular to the base. Here we use the "right cylinder" in the general sense; the base does not have to be circular, but the sides are perpendicular to the base.

Suppose that the solid body extends from height $y = a$ to $y = b$ along the y-axis as in Figure 2. Let $A(y)$ be the area of the **horizontal cross section** at height y (the intersection of the solid with the horizontal plane at height y).

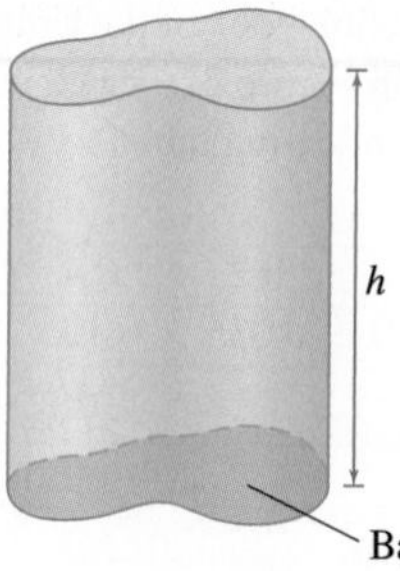

FIGURE 1 The volume of a right cylinder is Ah.

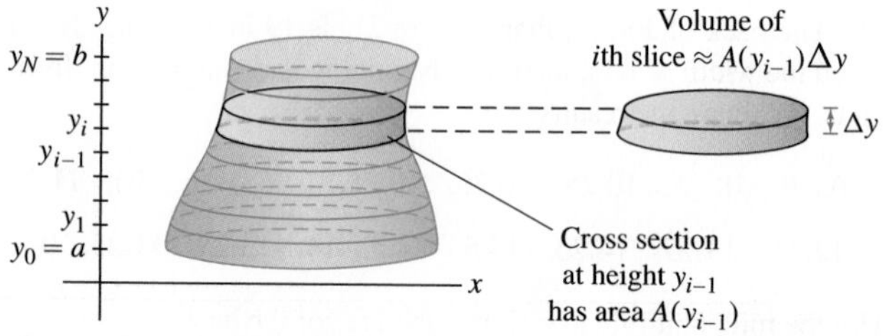

FIGURE 2 Divide the solid into thin horizontal slices. Each slice is nearly a right cylinder whose volume can be approximated as area times height.

To compute the volume V of the body, divide the body into N horizontal slices of thickness $\Delta y = (b - a)/N$. The ith slice extends from y_{i-1} to y_i, where $y_i = a + i\Delta y$. Let V_i be the volume of the slice.

If N is very large, then Δy is very small and the slices are very thin. In this case, the ith slice is nearly a right cylinder of base $A(y_{i-1})$ and height Δy, and therefore $V_i \approx A(y_{i-1})\Delta y$. Summing up, we obtain

$$V = \sum_{i=1}^{N} V_i \approx \sum_{i=1}^{N} A(y_{i-1})\Delta y$$

The sum on the right is a left-endpoint approximation to the integral $\int_a^b A(y)\, dy$. If we assume that $A(y)$ is a continuous function, then the approximation improves in accuracy and converges to the integral as $N \to \infty$. We conclude that *the volume of the solid is equal to the integral of its cross-sectional area.*

Volume as the Integral of Cross-Sectional Area Let $A(y)$ be the area of the horizontal cross section at height y of a solid body extending from $y = a$ to $y = b$. Then

$$\text{Volume of the solid body} = \int_a^b A(y)\,dy \qquad \boxed{1}$$

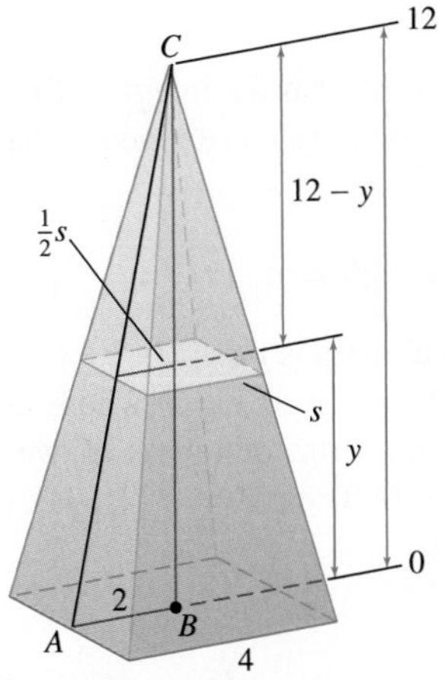

FIGURE 3 A horizontal cross section of the pyramid is a square.

■ **EXAMPLE 1 Volume of a Pyramid** Calculate the volume V of a pyramid of height 12 m whose base is a square of side 4 m.

Solution To use Eq. (1), we need a formula for the horizontal cross section $A(y)$.

***Step 1.* Find a formula for $A(y)$.**
Figure 3 shows that the horizontal cross section at height y is a square. To find the side s of this square, apply the law of similar triangles to $\triangle ABC$ and to the triangle of height $12 - y$ whose base of length $\frac{1}{2}s$ lies on the cross section:

$$\frac{\text{Base}}{\text{Height}} = \frac{2}{12} = \frac{\frac{1}{2}s}{12-y} \quad \Rightarrow \quad 2(12-y) = 6s$$

We find that $s = \frac{1}{3}(12 - y)$ and therefore $A(y) = s^2 = \frac{1}{9}(12 - y)^2$.

***Step 2.* Compute V as the integral of $A(y)$.**

$$V = \int_0^{12} A(y)\,dy = \int_0^{12} \frac{1}{9}(12-y)^2\,dy = -\frac{1}{27}(12-y)^3\Big|_0^{12} = 64 \text{ m}^3$$

This agrees with the result obtained using the formula $V = \frac{1}{3}Ah$ for the volume of a pyramid of base A and height h, since $\frac{1}{3}Ah = \frac{1}{3}(4^2)(12) = 64$. ■

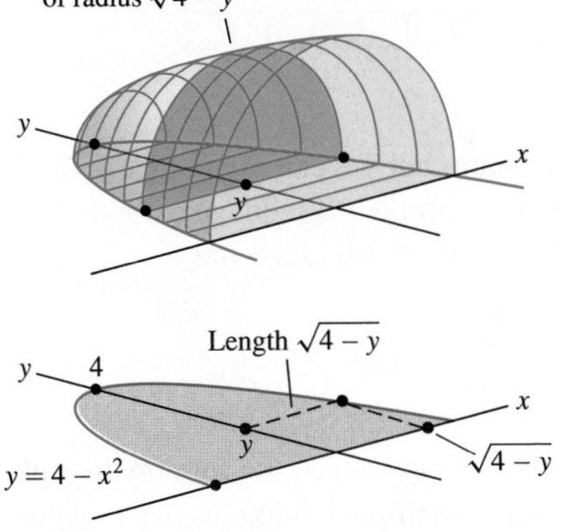

FIGURE 4

■ **EXAMPLE 2** Compute the volume V of the solid in Figure 4, whose base is the region between the inverted parabola $y = 4 - x^2$ and the x-axis, and whose vertical cross sections perpendicular to the y-axis are semicircles.

Solution To find a formula for the area $A(y)$ of the cross section, observe that $y = 4 - x^2$ can be written $x = \pm\sqrt{4-y}$. We see in Figure 4 that the cross section at y is a semicircle of radius $r = \sqrt{4-y}$. This semicircle has area $A(y) = \frac{1}{2}\pi r^2 = \frac{\pi}{2}(4 - y)$. Therefore

$$V = \int_0^4 A(y)\,dy = \frac{\pi}{2}\int_0^4 (4-y)\,dy = \frac{\pi}{2}\left(4y - \frac{1}{2}y^2\right)\Big|_0^4 = 4\pi \qquad ■$$

In the next example, we compute volume using vertical rather than horizontal cross sections. This leads to an integral with respect to x rather than y.

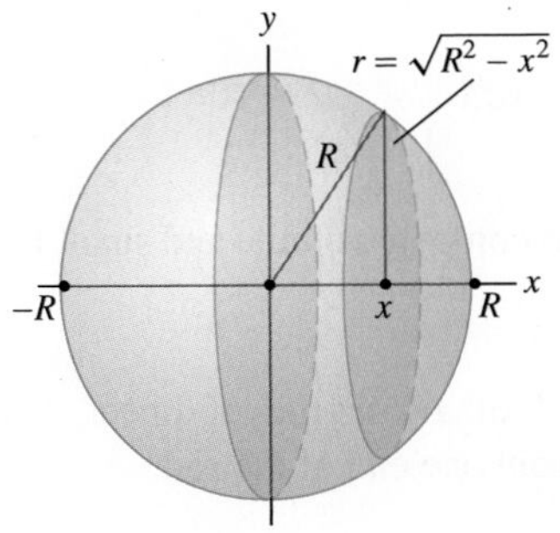

FIGURE 5

■ **EXAMPLE 3 Volume of a Sphere: Vertical Cross Sections** Compute the volume of a sphere of radius R.

Solution As we see in Figure 5, the vertical cross section of the sphere at x is a circle whose radius r satisfies $x^2 + r^2 = R^2$ or $r = \sqrt{R^2 - x^2}$. The area of the cross section is $A(x) = \pi r^2 = \pi(R^2 - x^2)$. Therefore, the sphere has volume

$$\int_{-R}^{R} \pi(R^2 - x^2)\,dx = \pi\left(R^2x - \frac{x^3}{3}\right)\Big|_{-R}^{R} = 2\left(\pi R^3 - \pi\frac{R^3}{3}\right) = \frac{4}{3}\pi R^3 \qquad ■$$

FIGURE 6 The two stacks of coins have equal cross-sections, hence equal volumes by Cavalieri's principle.

CONCEPTUAL INSIGHT **Cavalieri's principle** states: Solids with equal cross-sectional areas have equal volume. It is often illustrated convincingly with two stacks of coins (Figure 6). Our formula $V = \int_a^b A(y)\,dy$ includes Cavalieri's principle, because the volumes V are certainly equal if the cross-sectional areas $A(y)$ are equal.

Density

Next, we show that the total mass of an object can be expressed as the integral of its mass density. Consider a rod of length ℓ. The rod's **linear mass density** ρ is defined as the mass per unit length. If ρ is constant, then by definition,

$$\text{Total mass} = \text{linear mass density} \times \text{length} = \rho \cdot \ell \qquad \boxed{2}$$

The symbol ρ (lowercase Greek letter rho) is used often to denote density.

For example, if $\ell = 10$ cm and $\rho = 9$ g/cm, then the total mass is $\rho\ell = 9 \cdot 10 = 90$ g.

Now consider a rod extending along the x-axis from $x = a$ to $x = b$ whose density $\rho(x)$ is a continuous function of x, as in Figure 7. To compute the total mass M, we break up the rod into N small segments of length $\Delta x = (b - a)/N$. Then $M = \sum_{i=1}^{N} M_i$, where M_i is the mass of the ith segment.

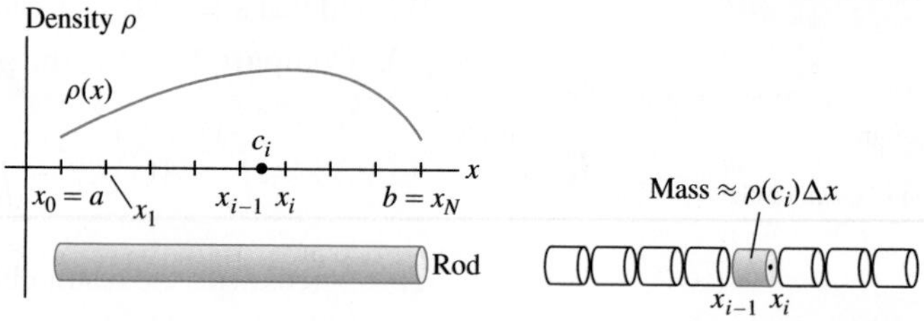

FIGURE 7 The total mass of the rod is equal to the area under the graph of mass density ρ.

We cannot use Eq. (2) because $\rho(x)$ is not constant, but we can argue that if Δx is small, then $\rho(x)$ is *nearly constant* along the ith segment. If the ith segment extends from x_{i-1} to x_i and c_i is any sample point in $[x_{i-1}, x_i]$, then $M_i \approx \rho(c_i)\Delta x$ and

$$\text{Total mass } M = \sum_{i=1}^{N} M_i \approx \sum_{i=1}^{N} \rho(c_i)\Delta x$$

As $N \to \infty$, the accuracy of the approximation improves. However, the sum on the right is a Riemann sum whose value approaches $\int_a^b \rho(x)\,dx$, and thus it makes sense to define *the total mass of a rod as the integral of its linear mass density*:

$$\boxed{\text{Total mass } M = \int_a^b \rho(x)\,dx} \qquad \boxed{3}$$

Note the similarity in the way we use thin slices to compute volume and small pieces to compute total mass.

■ **EXAMPLE 4** **Total Mass** Find the total mass M of a 2-m rod of linear density $\rho(x) = 1 + x(2 - x)$ kg/m, where x is the distance from one end of the rod.

Solution

$$M = \int_0^2 \rho(x)\,dx = \int_0^2 \big(1 + x(2 - x)\big)\,dx = \left(x + x^2 - \frac{1}{3}x^3\right)\Big|_0^2 = \frac{10}{3}\text{ kg} \qquad ■$$

In general, density is a function $\rho(x, y)$ that depends not just on the distance to the origin but also on the coordinates (x, y). Total mass or population is then computed using double integration, a topic in multivariable calculus.

In some situations, density is a function of distance to the origin. For example, in the study of urban populations, it might be assumed that the population density $\rho(r)$ (in people per square kilometer) depends only on the distance r from the center of a city. Such a density function is called a **radial density function.**

We now derive a formula for the total population P within a radius R of the city center assuming a radial density $\rho(r)$. First, divide the circle of radius R into N thin rings of equal width $\Delta r = R/N$ as in Figure 8.

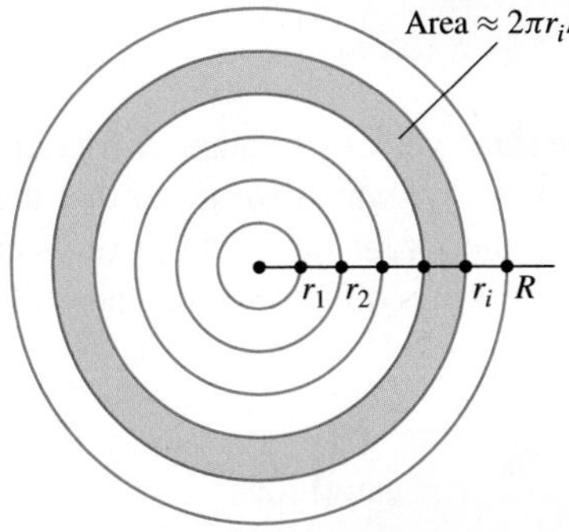

FIGURE 8 Dividing the circle of radius R into N thin rings of width $\Delta r = R/N$.

Let P_i be the population within the ith ring, so that $P = \sum_{i=1}^{N} P_i$. If the outer radius of the ith ring is r_i, then the circumference is $2\pi r_i$, and if Δr is small, the area of this ring is *approximately* $2\pi r_i \Delta r$ (outer circumference times width). Furthermore, the population density within the thin ring is nearly constant with value $\rho(r_i)$. With these approximations,

$$P_i \approx \underbrace{2\pi r_i \Delta r}_{\text{Area of ring}} \times \underbrace{\rho(r_i)}_{\text{Population density}} = 2\pi r_i \rho(r_i) \Delta r$$

$$P = \sum_{i=1}^{N} P_i \approx 2\pi \sum_{i=1}^{N} r_i \rho(r_i) \Delta r$$

This last sum is a right-endpoint approximation to the integral $2\pi \int_0^R r\rho(r)\,dr$. As N tends to ∞, the approximation improves in accuracy and the sum converges to the integral. Thus, for a population with a radial density function $\rho(r)$,

Remember that for a radial density function, the total population is obtained by integrating $2\pi r\rho(r)$ rather than $\rho(r)$.

$$\text{Population } P \text{ within a radius } R = 2\pi \int_0^R r\rho(r)\,dr \qquad \boxed{4}$$

■ **EXAMPLE 5 Computing Total Population** The population in a certain city has radial density function $\rho(r) = 15(1 + r^2)^{-1/2}$, where r is the distance from the city center in kilometers and ρ has units of thousands per square kilometer. How many people live in the ring between 10 and 30 km from the city center?

Solution The population P (in thousands) within the ring is

$$P = 2\pi \int_{10}^{30} r\left(15(1 + r^2)^{-1/2}\right) dr = 2\pi(15) \int_{10}^{30} \frac{r}{(1 + r^2)^{1/2}}\,dr$$

Now use the substitution $u = 1 + r^2$, $du = 2r\,dr$. The limits of integration become $u(10) = 101$ and $u(30) = 901$:

$$P = 30\pi \int_{101}^{901} u^{-1/2} \left(\frac{1}{2}\right) du = 30\pi u^{1/2}\Big|_{101}^{901} \approx 1881 \text{ thousand}$$

In other words, the population is approximately 1.9 million people. ■

Flow Rate

When fluid flows through a tube, the **flow rate** Q is the *volume per unit time* of fluid passing through the tube (Figure 9). The flow rate depends on the velocity of the fluid particles. If all particles of the fluid travel with the same velocity v (say, in units of cm^3/min), and the tube has radius R, then

$$\underbrace{\text{Flow rate } Q}_{\text{Volume per unit time}} = \text{cross-sectional area} \times \text{velocity} = \pi R^2 v \text{ cm}^3/\text{min}$$

Why is this formula true? Let's fix an observation point P in the tube and ask: Which fluid particles flow past P in a 1-min interval? A particle travels v centimeters each minute, so it flows past P during this minute if it is located not more than v centimeters to the left of P (assuming the fluid flows from left to right). Therefore, the column of fluid flowing past P in a 1-min interval is a cylinder of radius R, length v, and volume $\pi R^2 v$ (Figure 9).

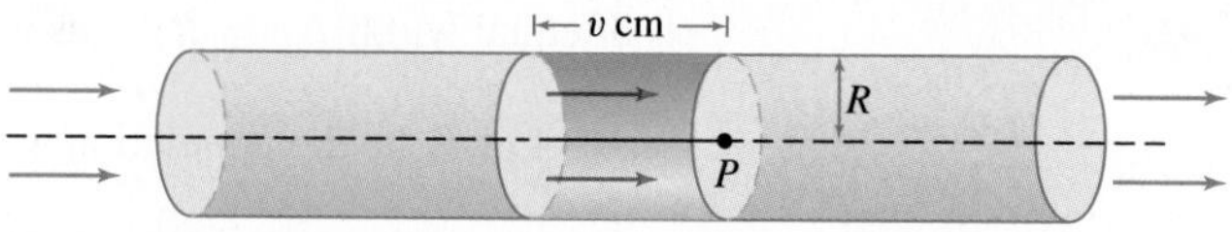

FIGURE 9 The column of fluid flowing past P in one unit of time is a cylinder of volume $\pi R^2 v$.

In reality, the fluid particles do not all travel at the same velocity because of friction. However, for a slowly moving fluid, the flow is **laminar**, by which we mean that the velocity $v(r)$ depends only on the distance r from the center of the tube. The particles at the center of the tube travel most quickly, and the velocity tapers off to zero near the walls of the tube (Figure 10).

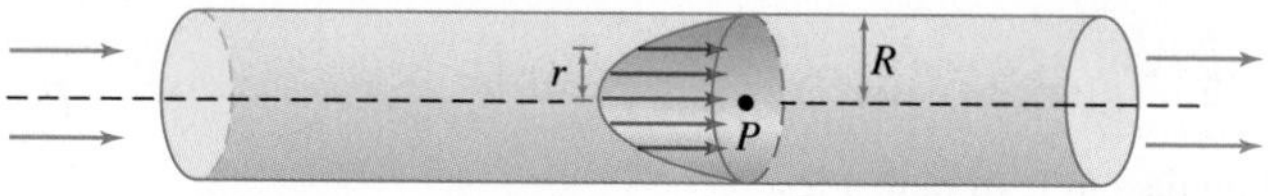

FIGURE 10 Laminar flow: Velocity of fluid increases toward the center of the tube.

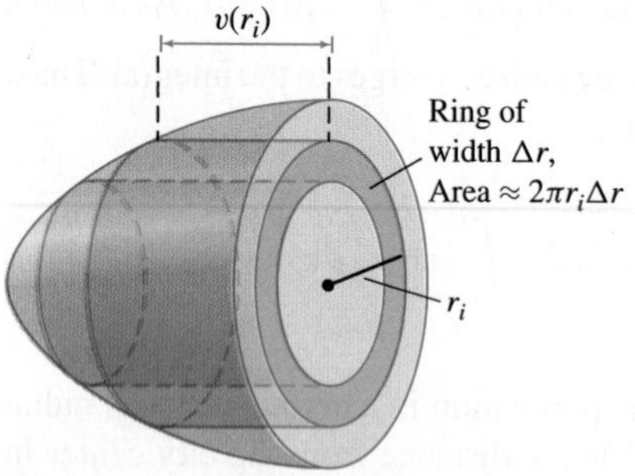

FIGURE 11 In a laminar flow, the fluid particles passing through a thin ring at distance r_i from the center all travel at nearly the same velocity $v(r_i)$.

If the flow is laminar, we can express the flow rate Q as an integral. We divide the circular cross-section of the tube into N thin concentric rings of width $\Delta r = R/N$ (Figure 11). The area of the ith ring is approximately $2\pi r_i \Delta r$ and the fluid particles flowing past this ring have velocity that is nearly constant with value $v(r_i)$. Therefore, we can approximate the flow rate Q_i through the ith ring by

$$Q_i \approx \text{cross-sectional area} \times \text{velocity} \approx (2\pi r_i \Delta r)v(r_i)$$

We obtain

$$Q = \sum_{i=1}^{N} Q_i \approx 2\pi \sum_{i=1}^{N} r_i v(r_i)\Delta r$$

The sum on the right is a right-endpoint approximation to the integral $2\pi \int_0^R r v(r)\,dr$. Once again, we let N tend to ∞ to obtain the formula

$$\text{Flow rate } Q = 2\pi \int_0^R r v(r)\,dr \qquad (5)$$

Note the similarity of this formula and its derivation to that of population with a radial density function.

The French physician Jean Poiseuille (1799–1869) discovered the law of laminar flow that cardiologists use to study blood flow in humans. Poiseuille's Law highlights the danger of cholesterol buildup in blood vessels: The flow rate through a blood vessel of radius R is proportional to R^4, so if R is reduced by one-half, the flow is reduced by a factor of 16.

■ **EXAMPLE 6 Laminar Flow** According to **Poiseuille's Law**, the velocity of blood flowing in a blood vessel of radius R cm is $v(r) = k(R^2 - r^2)$, where r is the distance from the center of the vessel (in centimeters) and k is a constant. Calculate the flow rate Q as function of R, assuming that $k = 0.5\ (\text{cm-s})^{-1}$.

Solution By Eq. (5),

$$Q = 2\pi \int_0^R (0.5)r(R^2 - r^2)\,dr = \pi \left(R^2\frac{r^2}{2} - \frac{r^4}{4}\right)\Bigg|_0^R = \frac{\pi}{4}R^4 \text{ cm}^3/\text{s}$$

Note that Q is proportional to R^4 (this is true for any value of k). ■

Average Value

As a final example, we discuss the *average value* of a function. Recall that the average of N numbers $a_1, a_2, \ldots, a_N$ is the sum divided by N:

$$\frac{a_1 + a_2 + \cdots + a_N}{N} = \frac{1}{N}\sum_{j=1}^{N} a_j$$

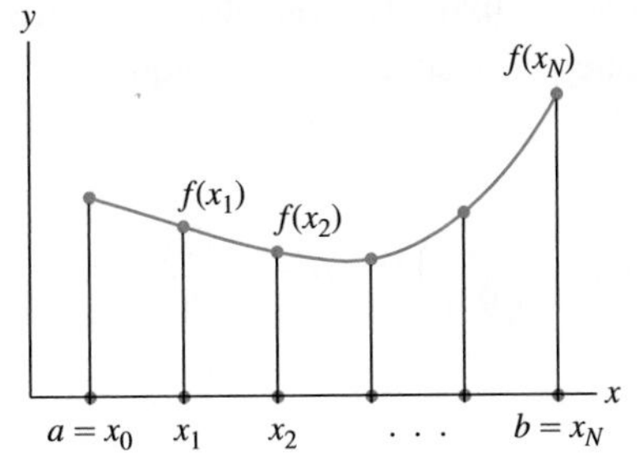

FIGURE 12 The average of the values of $f(x)$ at the points $x_1, x_2, \ldots, x_N$ is equal to $\dfrac{R_N}{b-a}$.

For example, the average of 18, 25, 22, and 31 is $\frac{1}{4}(18 + 25 + 22 + 31) = 24$.

We cannot define the average value of a function $f(x)$ on an interval $[a, b]$ as a sum because there are infinitely many values of x to consider. But recall the formula for the right-endpoint approximation R_N (Figure 12):

$$R_N = \frac{b-a}{N}\big(f(x_1) + f(x_2) + \cdots + f(x_N)\big)$$

where $x_i = a + i\left(\dfrac{b-a}{N}\right)$. We see that R_N divided by $(b - a)$ is equal to the average of the equally spaced function values $f(x_i)$:

$$\frac{1}{b-a}R_N = \underbrace{\frac{f(x_1) + f(x_2) + \cdots + f(x_N)}{N}}_{\text{Average of the function values}}$$

If N is large, it is reasonable to think of this quantity as an *approximation* to the average of $f(x)$ on $[a, b]$. Therefore, we define the average value itself as the limit:

$$\text{Average value} = \lim_{N\to\infty}\frac{1}{b-a}R_N(f) = \frac{1}{b-a}\int_a^b f(x)\,dx$$

DEFINITION Average Value The **average value** of an integrable function $f(x)$ on $[a, b]$ is the quantity

$$\text{Average value} = \frac{1}{b-a}\int_a^b f(x)\,dx \qquad \boxed{6}$$

The average value of a function is also called the **mean value**.

GRAPHICAL INSIGHT Think of the average value M of a function as the average height of its graph (Figure 13). The region under the graph has the same signed area as the rectangle of height M, because $\displaystyle\int_a^b f(x)\,dx = M(b-a)$.

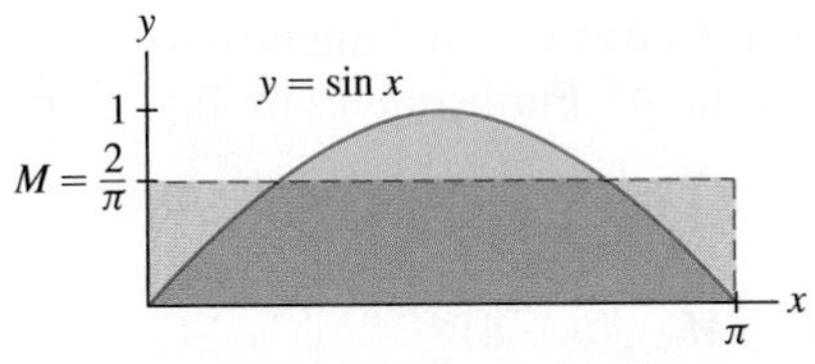

FIGURE 13 The area under the graph is equal to the area of the rectangle whose height is the average value M.

■ **EXAMPLE 7** Find the average value of $f(x) = \sin x$ on $[0, \pi]$.

Solution The average value of $\sin x$ on $[0, \pi]$ is

$$\frac{1}{\pi}\int_0^{\pi}\sin x\,dx = -\frac{1}{\pi}\cos x\Big|_0^{\pi} = \frac{1}{\pi}\big(-(-1) - (-1)\big) = \frac{2}{\pi} \approx 0.637$$

This answer is reasonable because $\sin x$ varies from 0 to 1 on the interval $[0, \pi]$ and the average 0.637 lies somewhere between the two extremes (Figure 13). ■

FIGURE 14 A bushbaby can jump as high as 2 meters (its center of mass rises more than five bodylengths). By contrast, Michael Jordan rises at most 0.6 body length when executing a slam dunk.

■ **EXAMPLE 8 Vertical Jump of a Bushbaby** The bushbaby (*Galago senegalensis*) is a small primate with remarkable jumping ability (Figure 14). Find the average speed during a jump if the initial vertical velocity is $v_0 = 600$ cm/s. Use Galileo's formula for the height $h(t) = v_0 t - \frac{1}{2}gt^2$ (in centimeters, where $g = 980$ cm/s^2).

Solution The bushbaby's height is $h(t) = v_0 t - \frac{1}{2}gt^2 = t\left(v_0 - \frac{1}{2}gt\right)$. The height is zero at $t = 0$ and at $t = 2v_0/g = \frac{1200}{980} = \frac{6}{4.9}$ s, when jump ends.

The bushbaby's velocity is $h'(t) = v_0 = gt = 600 - 980t$. The velocity is negative for $t > v_0/g = \frac{6}{9.8}$, so as we see in Figure 15, the integral of speed $|h'(t)|$ is equal to the sum of the areas of two triangles of base $\frac{6}{9.8}$ and height 600:

$$\int_0^{6/4.9} |600 - 980t|\, dt = \frac{1}{2}\left(\frac{6}{9.8}\right)(600) + \frac{1}{2}\left(\frac{6}{9.8}\right)(600) = \frac{3600}{9.8}$$

The average speed $\bar{s}$ is

$$\bar{s} = \frac{1}{\frac{6}{4.9}} \int_0^{6/4.9} |600 - 980t|\, dt = \frac{1}{\frac{6}{4.9}}\left(\frac{3600}{9.8}\right) = 300 \text{ cm/s}$$

■

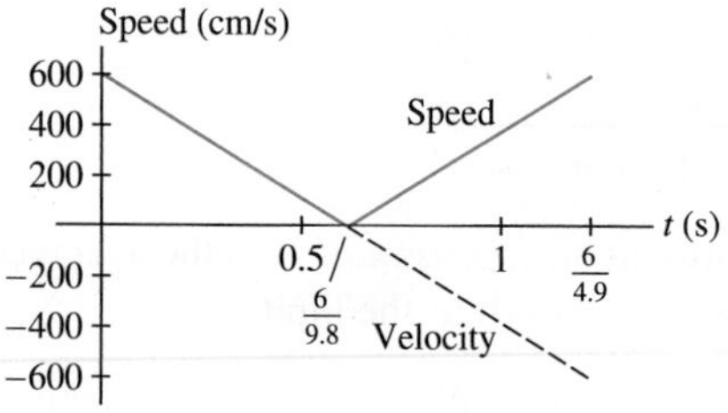

FIGURE 15 Graph of speed $|h'(t)| = |600 - 980t|$.

There is an important difference between the average of a list of numbers and the average value of a continuous function. If the average score on an exam is 84, then 84 lies between the highest and lowest scores, but it is possible that no student received a score of 84. By contrast, the Mean Value Theorem (MVT) for Integrals asserts that a continuous function always takes on its average value somewhere in the interval (Figure 16).

For example, the average of $f(x) = \sin x$ on $[0, \pi]$ is $2/\pi$ by Example 7. We have $f(c) = 2/\pi$ for $c = \sin^{-1}(2/\pi) \approx 0.69$. Since 0.69 lies in $[0, \pi]$, $f(x) = \sin x$ indeed takes on its average value at a point in the interval.

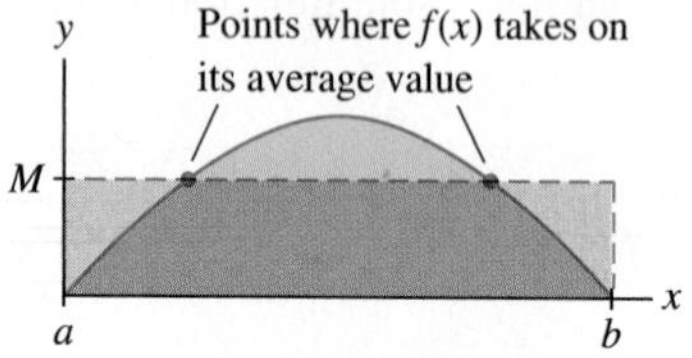

FIGURE 16 The function $f(x)$ takes on its average value M at the points where the upper edge of the rectangle intersects the graph.

THEOREM 1 Mean Value Theorem for Integrals If $f(x)$ is continuous on $[a, b]$, then there exists a value $c \in [a, b]$ such that

$$f(c) = \frac{1}{b-a}\int_a^b f(x)\, dx$$

Proof Let $M = \frac{1}{b-a}\int_a^b f(x)$ be the average value. Because $f(x)$ is continuous, we can apply Theorem 1 of Section 4.2 to conclude that f takes on a minimum value m_{min} and a maximum value M_{max} on the closed interval $[a, b]$. Furthermore, by Eq. (8) of Section 5.2,

$$m_{\text{min}}(b - a) \le \int_a^b f(x)\, dx \le M_{\text{max}}(b - a)$$

Dividing by $(b - a)$, we find

$$m_{\text{min}} \le M \le M_{\text{max}}$$

In other words, the average value M lies between m_{min} and M_{max}. The Intermediate Value Theorem guarantees that $f(x)$ takes on every value between its min and max, so $f(c) = M$ for some c in $[a, b]$. ■

6.2 SUMMARY

- Formulas

Volume	$V = \int_a^b A(y)\,dy$	$A(y) =$ cross-sectional area
Total Mass	$M = \int_a^b \rho(x)\,dx$	$\rho(x) =$ linear mass density
Total Population	$P = 2\pi \int_0^R r\rho(r)\,dr$	$\rho(r) =$ radial density
Flow Rate	$Q = 2\pi \int_0^R rv(r)\,dr$	$v(r) =$ velocity at radius r
Average value	$M = \frac{1}{b-a}\int_a^b f(x)\,dx$	$f(x)$ any continuous function

- The MVT for Integrals: If $f(x)$ is continuous on $[a, b]$ with average (or mean) value M, then $f(c) = M$ for some $c \in [a, b]$.

6.2 EXERCISES

Preliminary Questions

1. What is the average value of $f(x)$ on $[0, 4]$ if the area between the graph of $f(x)$ and the x-axis is equal to 12?

2. Find the volume of a solid extending from $y = 2$ to $y = 5$ if every cross section has area $A(y) = 5$.

3. What is the definition of flow rate?

4. Which assumption about fluid velocity did we use to compute the flow rate as an integral?

5. The average value of $f(x)$ on $[1, 4]$ is 5. Find $\int_1^4 f(x)\,dx$.

Exercises

1. Let V be the volume of a pyramid of height 20 whose base is a square of side 8.

(a) Use similar triangles as in Example 1 to find the area of the horizontal cross section at a height y.

(b) Calculate V by integrating the cross-sectional area.

2. Let V be the volume of a right circular cone of height 10 whose base is a circle of radius 4 [Figure 17(A)].

(a) Use similar triangles to find the area of a horizontal cross section at a height y.

(b) Calculate V by integrating the cross-sectional area.

3. Use the method of Exercise 2 to find the formula for the volume of a right circular cone of height h whose base is a circle of radius R [Figure 17(B)].

4. Calculate the volume of the ramp in Figure 18 in three ways by integrating the area of the cross sections:

(a) Perpendicular to the x-axis (rectangles).

(b) Perpendicular to the y-axis (triangles).

(c) Perpendicular to the z-axis (rectangles).

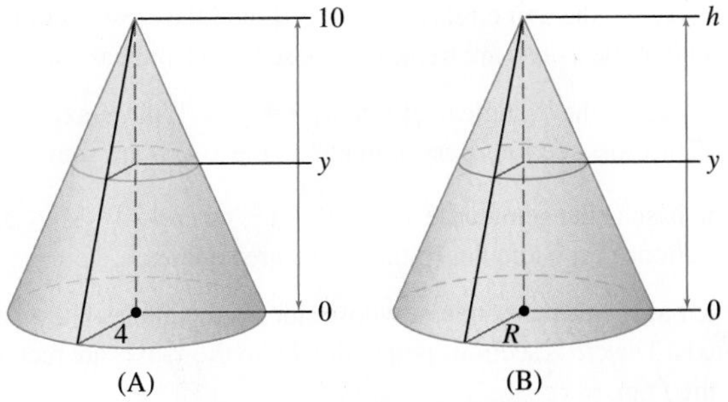

FIGURE 17 Right circular cones.

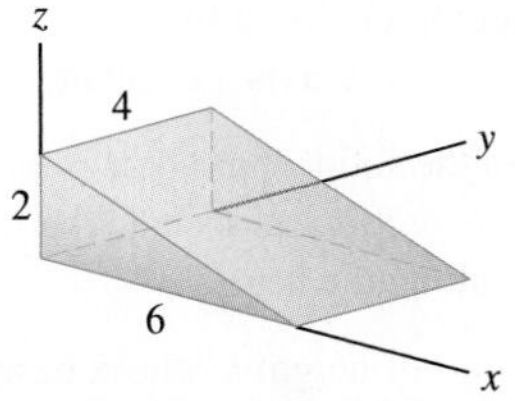

FIGURE 18 Ramp of length 6, width 4, and height 2.

5. Find the volume of liquid needed to fill a sphere of radius R to height h (Figure 19).

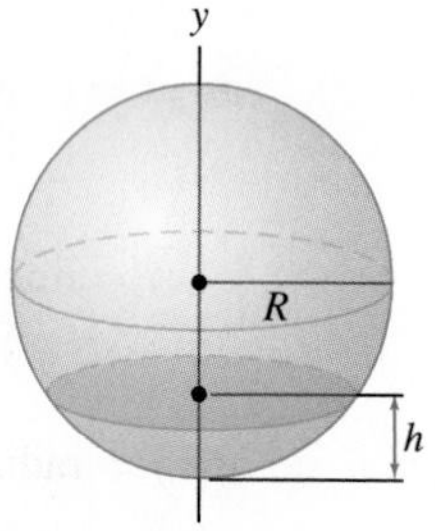

FIGURE 19 Sphere filled with liquid to height h.

6. Find the volume of the wedge in Figure 20(A) by integrating the area of vertical cross sections.

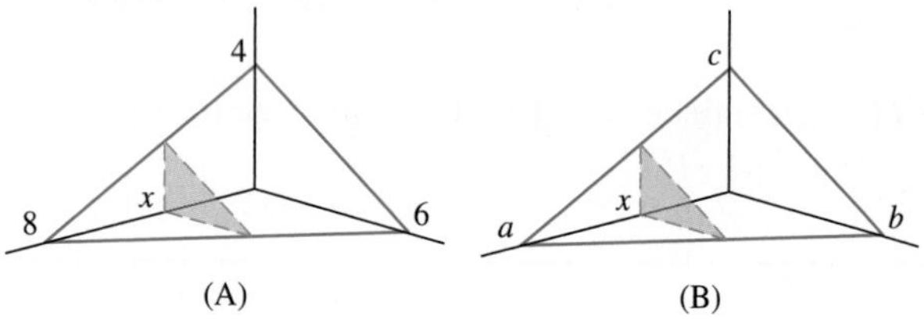

FIGURE 20

7. Derive a formula for the volume of the wedge in Figure 20(B) in terms of the constants a, b, and c.

8. Let B be the solid whose base is the unit circle $x^2 + y^2 = 1$ and whose vertical cross sections perpendicular to the x-axis are equilateral triangles. Show that the vertical cross sections have area $A(x) = \sqrt{3}(1 - x^2)$ and compute the volume of B.

In Exercises 9–14, find the volume of the solid with the given base and cross sections.

9. The base is the unit circle $x^2 + y^2 = 1$, and the cross sections perpendicular to the x-axis are triangles whose height and base are equal.

10. The base is the triangle enclosed by $x + y = 1$, the x-axis, and the y-axis. The cross sections perpendicular to the y-axis are semicircles.

11. The base is the semicircle $y = \sqrt{9 - x^2}$, where $-3 \le x \le 3$. The cross sections perpendicular to the x-axis are squares.

12. The base is a square, one of whose sides is the interval $[0, \ell]$ along the x-axis. The cross sections perpendicular to the x-axis are rectangles of height $f(x) = x^2$.

13. The base is the region enclosed by $y = x^2$ and $y = 3$. The cross sections perpendicular to the y-axis are squares.

14. The base is the region enclosed by $y = x^2$ and $y = 3$. The cross sections perpendicular to the y-axis are rectangles of height y^3.

15. Find the volume of the solid whose base is the region $|x| + |y| \le 1$ and whose vertical cross sections perpendicular to the y-axis are semicircles (with diameter along the base).

16. Show that a pyramid of height h whose base is an equilateral triangle of side s has volume $\frac{\sqrt{3}}{12}hs^2$.

17. The area of an ellipse is πab, where a and b are the lengths of the semimajor and semiminor axes (Figure 21). Compute the volume of a cone of height 12 whose base is an ellipse with semimajor axis $a = 6$ and semiminor axis $b = 4$.

18. Find the volume V of a *regular* tetrahedron (Figure 22) whose face is an equilateral triangle of side s. The tetrahedron has height $h = \sqrt{2/3}s$.

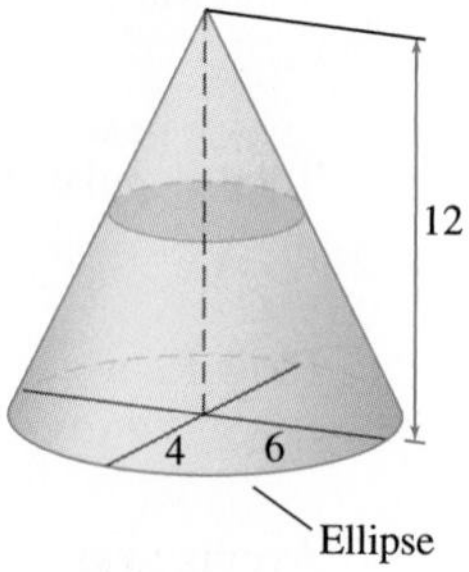

FIGURE 21

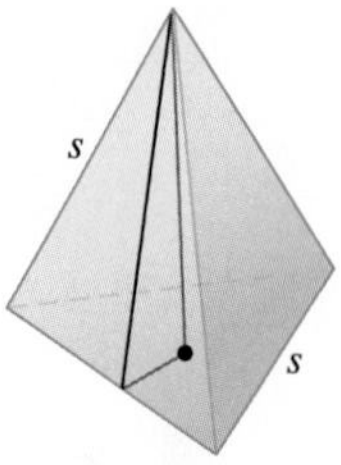

FIGURE 22

19. A frustum of a pyramid is a pyramid with its top cut off [Figure 23(A)]. Let V be the volume of a frustum of height h whose base is a square of side a and whose top is a square of side b with $a > b \ge 0$.

(a) Show that if the frustum were continued to a full pyramid, it would have height $ha/(a - b)$ [Figure 23(B)].

(b) Show that the cross section at height x is a square of side $(1/h)(a(h - x) + bx)$.

(c) Show that $V = \frac{1}{3}h(a^2 + ab + b^2)$. A papyrus dating to the year 1850 BCE indicates that Egyptian mathematicians had discovered this formula almost 4000 years ago.

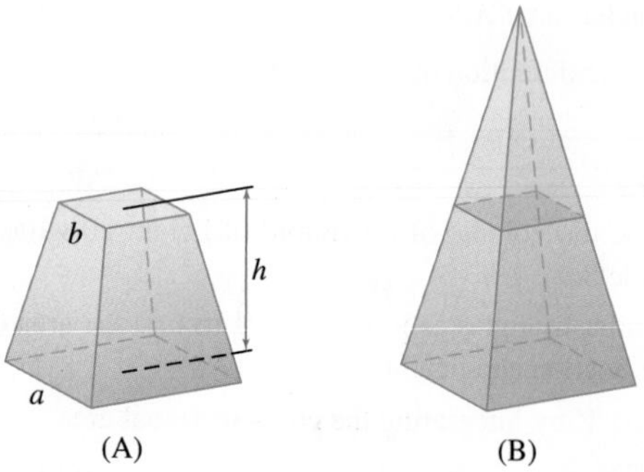

FIGURE 23

20. A plane inclined at an angle of 45° passes through a diameter of the base of a cylinder of radius r. Find the volume of the region within the cylinder and below the plane (Figure 24).

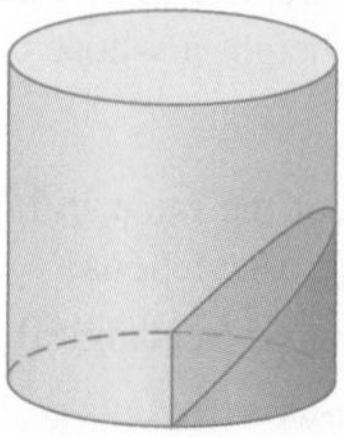

FIGURE 24

21. The solid S in Figure 25 is the intersection of two cylinders of radius r whose axes are perpendicular.

(a) The horizontal cross section of each cylinder at distance y from the central axis is a rectangular strip. Find the strip's width.

(b) Find the area of the horizontal cross section of S at distance y.

(c) Find the volume of S as a function of r.

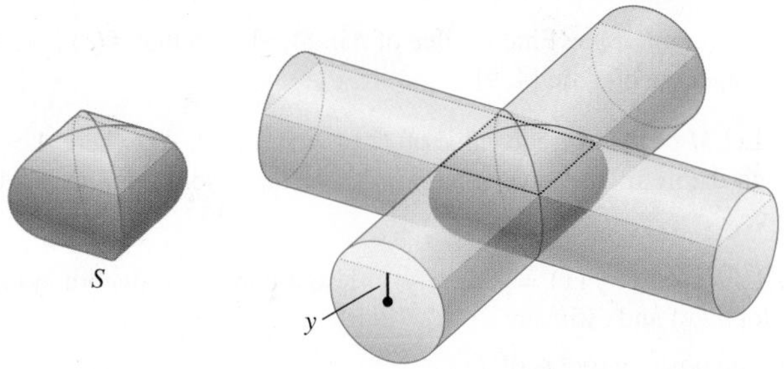

FIGURE 25 Two cylinders intersecting at right angles.

22. Let S be the intersection of two cylinders of radius r whose axes intersect at an angle θ. Find the volume of S as a function of r and θ.

23. Calculate the volume of a cylinder inclined at an angle $\theta = 30°$ with height 10 and base of radius 4 (Figure 26).

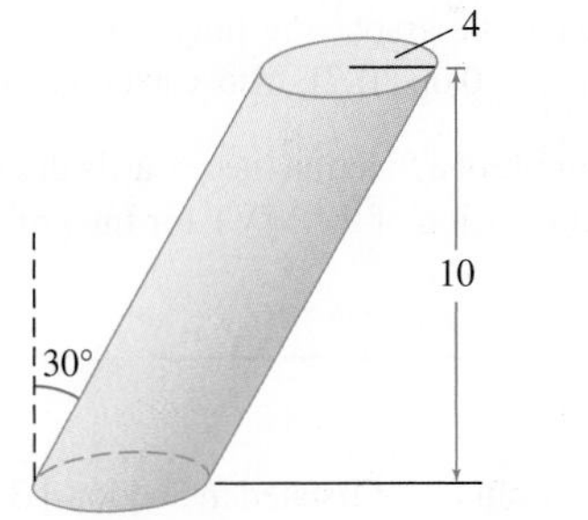

FIGURE 26 Cylinder inclined at an angle $\theta = 30°$.

24. The areas of cross sections of Lake Nogebow at 5-meter intervals are given in the table below. Figure 27 shows a contour map of the lake. Estimate the volume V of the lake by taking the average of the right- and left-endpoint approximations to the integral of cross-sectional area.

Depth (m)	0	5	10	15	20
Area (million m^2)	2.1	1.5	1.1	0.835	0.217

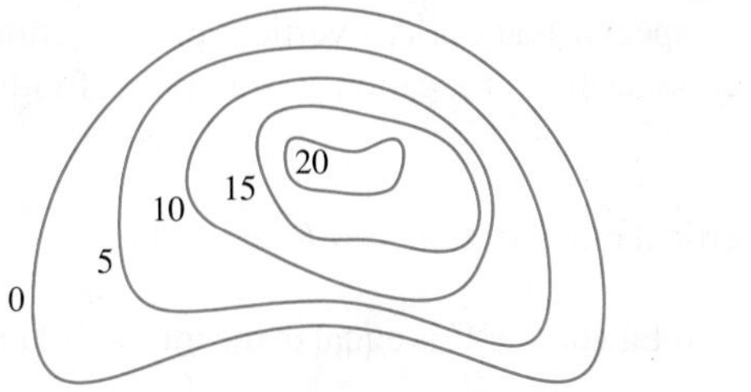

FIGURE 27 Depth contour map of Lake Nogebow.

25. Find the total mass of a 1-m rod whose linear density function is $\rho(x) = 10(x+1)^{-2}$ kg/m for $0 \le x \le 1$.

26. Find the total mass of a 2-m rod whose linear density function is $\rho(x) = 1 + 0.5\sin(\pi x)$ kg/m for $0 \le x \le 2$.

27. A mineral deposit along a strip of length 6 cm has density $s(x) = 0.01x(6-x)$ g/cm for $0 \le x \le 6$. Calculate the total mass of the deposit.

28. Charge is distributed along a glass tube of length 10 cm with linear charge density $\rho(x) = x(x^2+1)^{-2} \times 10^{-4}$ coulombs per centimeter for $0 \le x \le 10$. Calculate the total charge.

29. Calculate the population within a 10-mile radius of the city center if the radial population density is $\rho(r) = 4(1+r^2)^{1/3}$ (in thousands per square mile).

30. Odzala National Park in the Republic of the Congo has a high density of gorillas. Suppose that the radial population density is $\rho(r) = 52(1+r^2)^{-2}$ gorillas per square kilometer, where r is the distance from a grassy clearing with a source of water. Calculate the number of gorillas within a 5-km radius of the clearing.

31. Table 1 lists the population density (in people per square kilometer) as a function of distance r (in kilometers) from the center of a rural town. Estimate the total population within a 1.2-km radius of the center by taking the average of the left- and right-endpoint approximations.

TABLE 1 **Population Density**

r	$\rho(r)$	r	$\rho(r)$
0.0	125.0	0.8	56.2
0.2	102.3	1.0	46.0
0.4	83.8	1.2	37.6
0.6	68.6		

32. Find the total mass of a circular plate of radius 20 cm whose mass density is the radial function $\rho(r) = 0.03 + 0.01\cos(\pi r^2)$ g/cm^2.

33. The density of deer in a forest is the radial function $\rho(r) = 150(r^2+2)^{-2}$ deer per square kilometer, where r is the distance (in kilometers) to a small meadow. Calculate the number of deer in the region $2 \le r \le 5$ km.

34. Show that a circular plate of radius 2 cm with radial mass density $\rho(r) = \frac{4}{r}$ g/cm^2 has finite total mass, even though the density becomes infinite at the origin.

35. Find the flow rate through a tube of radius 4 cm, assuming that the velocity of fluid particles at a distance r cm from the center is $v(r) = (16 - r^2)$ cm/s.

36. The velocity of fluid particles flowing through a tube of radius 5 cm is $v(r) = (10 - 0.3r - 0.34r^2)$ cm/s, where r cm is the distance from the center. What quantity per second of fluid flows through the portion of the tube where $0 \le r \le 2$?

37. A solid rod of radius 1 cm is placed in a pipe of radius 3 cm so that their axes are aligned. Water flows through the pipe and around the rod. Find the flow rate if the velocity of the water is given by the radial function $v(r) = 0.5(r-1)(3-r)$ cm/s.

38. Let $v(r)$ be the velocity of blood in an arterial capillary of radius $R = 4 \times 10^{-5}$ m. Use Poiseuille's Law (Example 6) with $k = 10^6$ (m-s)$^{-1}$ to determine the velocity at the center of the capillary and the flow rate (use correct units).

In Exercises 39–48, calculate the average over the given interval.

39. $f(x) = x^3$, $[0, 4]$

40. $f(x) = x^3$, $[-1, 1]$

41. $f(x) = \cos x$, $\left[0, \frac{\pi}{6}\right]$

42. $f(x) = \sec^2 x$, $\left[\frac{\pi}{6}, \frac{\pi}{3}\right]$

43. $f(s) = s^{-2}$, $[2, 5]$

44. $f(x) = \dfrac{\sin(\pi/x)}{x^2}$, $[1, 2]$

45. $f(x) = 2x^3 - 6x^2$, $[-1, 3]$

46. $f(x) = \dfrac{x}{(x^2 + 16)^{3/2}}$, $[0, 3]$

47. $f(x) = x^n$ for $n \ge 0$, $[0, 1]$

48. $f(x) = \sin(nx)$, $[0, \pi]$

49. The temperature (in °C) at time t (in hours) in an art museum varies according to $T(t) = 20 + 5\cos\left(\frac{\pi}{12}t\right)$. Find the average over the time periods $[0, 24]$ and $[2, 6]$.

50. A ball thrown in the air vertically from ground level with initial velocity 18 m/s has height $h(t) = 18t - 9.8t^2$ at time t (in seconds). Find the average height and the average speed over the time interval extending from the ball's release to its return to ground level.

51. Find the average speed over the time interval $[1, 5]$ of a particle whose position at time t is $s(t) = t^3 - 6t^2$ m/s.

52. An object with zero initial velocity accelerates at a constant rate of 10 m/s^2. Find its average velocity during the first 15 seconds.

53. The acceleration of a particle is $a(t) = 60t - 4t^3$ m/s^2. Compute the average acceleration and the average speed over the time interval $[2, 6]$, assuming that the particle's initial velocity is zero.

54. What is the average area of the circles whose radii vary from 0 to R?

55. Let M be the average value of $f(x) = x^4$ on $[0, 3]$. Find a value of c in $[0, 3]$ such that $f(c) = M$.

56. Let $f(x) = \sqrt{x}$. Find a value of c in $[4, 9]$ such that $f(c)$ is equal to the average of f on $[4, 9]$.

57. Let M be the average value of $f(x) = x^3$ on $[0, A]$, where $A > 0$. Which theorem guarantees that $f(c) = M$ has a solution c in $[0, A]$? Find c.

58. CAS Let $f(x) = 2\sin x - x$. Use a computer algebra system to plot $f(x)$ and estimate:

(a) The positive root α of $f(x)$.

(b) The average value M of $f(x)$ on $[0, \alpha]$.

(c) A value $c \in [0, \alpha]$ such that $f(c) = M$.

59. Which of $f(x) = x\sin^2 x$ and $g(x) = x^2 \sin^2 x$ has a larger average value over $[0, 1]$? Over $[1, 2]$?

60. Find the average of $f(x) = ax + b$ over the interval $[-M, M]$, where a, b, and M are arbitrary constants.

61. Sketch the graph of a function $f(x)$ such that $f(x) \ge 0$ on $[0, 1]$ and $f(x) \le 0$ on $[1, 2]$, whose average on $[0, 2]$ is negative.

62. Give an example of a function (necessarily discontinuous) that does not satisfy the conclusion of the MVT for Integrals.

Further Insights and Challenges

63. An object is tossed into the air vertically from ground level with initial velocity v_0 ft/s at time $t = 0$. Find the average speed of the object over the time interval $[0, T]$, where T is the time the object returns to earth.

64. Review the MVT stated in Section 4.3 (Theorem 1, p. 194) and show how it can be used, together with the Fundamental Theorem of Calculus, to prove the MVT for Integrals.

6.3 Volumes of Revolution

We use the terms "revolve" and "rotate" interchangeably.

A **solid of revolution** is a solid obtained by rotating a region in the plane about an axis. The sphere and right circular cone are familiar examples of such solids. Each of these is "swept out" as a plane region revolves around an axis (Figure 1).

This method for computing the volume is referred to as the disk method because the vertical slices of the solid are circular disks.

Suppose that $f(x) \ge 0$ for $a \le x \le b$. The solid obtained by rotating the region under the graph about the x-axis has a special feature: All vertical cross sections are circles (Figure 2). In fact, the vertical cross section at location x is a circle of radius $R = f(x)$ and thus

$$\text{Area of the vertical cross section} = \pi R^2 = \pi f(x)^2$$

We know from Section 6.2 that the total volume V is equal to the integral of cross-sectional area. Therefore, $V = \displaystyle\int_a^b \pi f(x)^2\, dx$.

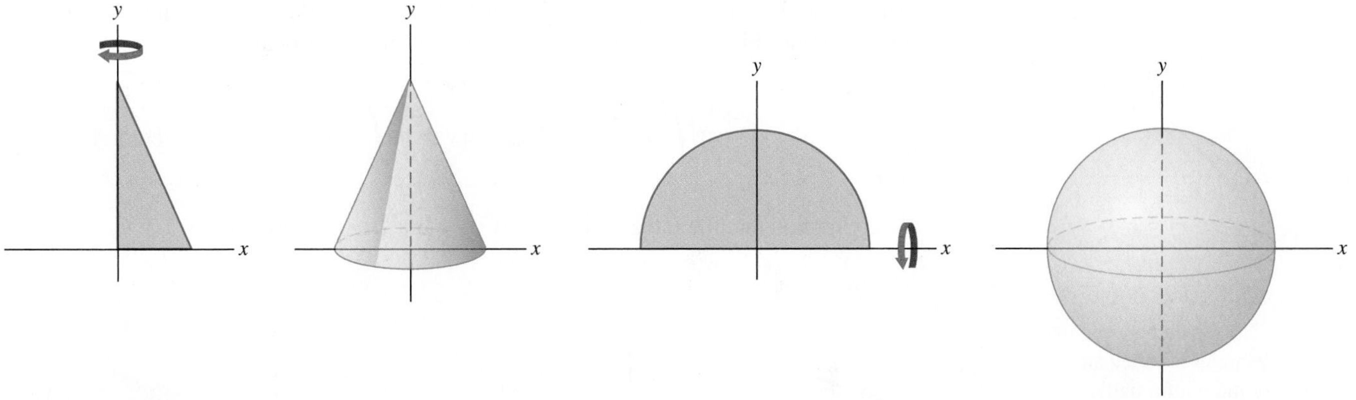

FIGURE 1 The right circular cone and the sphere are solids of revolution.

The cross sections of a solid of revolution are circles of radius $R = f(x)$ and area $\pi R^2 = \pi f(x)^2$. The volume, given by Eq. (1), is the integral of cross-sectional area.

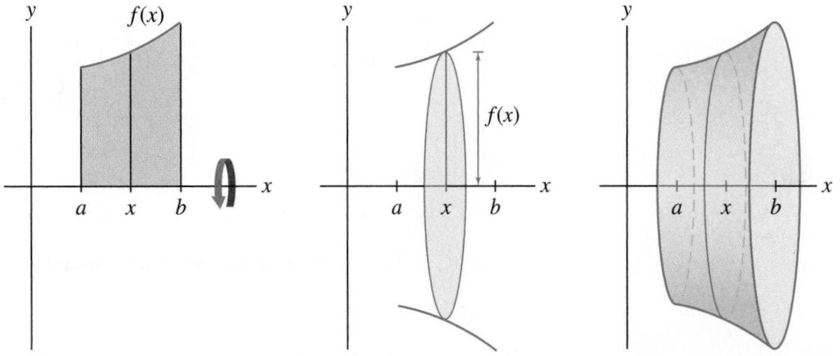

(A) (B) Cross section is a circle of radius $f(x)$. (C) Solid of revolution

FIGURE 2

Volume of Revolution: Disk Method If $f(x)$ is continuous and $f(x) \geq 0$ on $[a, b]$, then the solid obtained by rotating the region under the graph about the x-axis has volume [with $R = f(x)$]

$$V = \pi \int_a^b R^2\,dx = \pi \int_a^b f(x)^2\,dx \qquad \boxed{1}$$

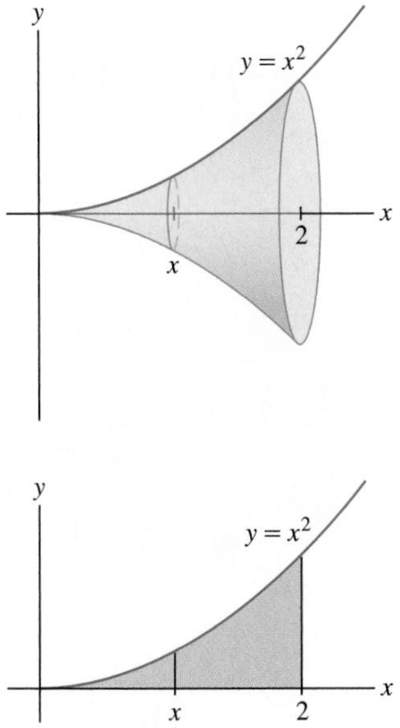

FIGURE 3 Region under $y = x^2$ rotated about the x-axis.

■ EXAMPLE 1 Calculate the volume V of the solid obtained by rotating the region under $y = x^2$ about the x-axis for $0 \leq x \leq 2$.

Solution The solid is shown in Figure 3. By Eq. (1) with $f(x) = x^2$, its volume is

$$V = \pi \int_0^2 R^2\,dx = \pi \int_0^2 (x^2)^2\,dx = \pi \int_0^2 x^4\,dx = \pi \left.\frac{x^5}{5}\right|_0^2 = \pi \frac{2^5}{5} = \frac{32}{5}\pi \qquad ■$$

There are some useful variations on the formula for a volume of revolution. First, consider the region *between* two curves $y = f(x)$ and $y = g(x)$, where $f(x) \geq g(x) \geq 0$ as in Figure 5(A). When this region is rotated about the x-axis, segment $\overline{AB}$ sweeps out the **washer** shown in Figure 5(B). The inner and outer radii of this washer (also called an annulus; see Figure 4) are

$$R_{\text{outer}} = f(x), \qquad R_{\text{inner}} = g(x)$$

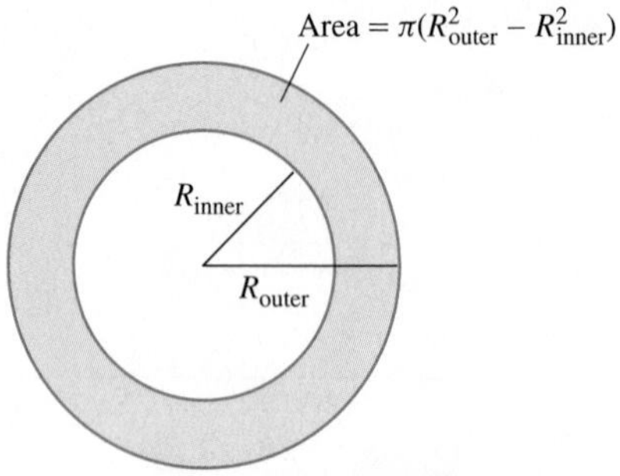

FIGURE 4 The region between two concentric circles is called an annulus, or more informally, a washer.

The washer has area $\pi R_{\text{outer}}^2 - \pi R_{\text{inner}}^2$ or $\pi(f(x)^2 - g(x)^2)$, and the volume of the solid of revolution [Figure 5(C)] is the integral of this cross-sectional area:

$$V = \pi \int_a^b \left(R_{\text{outer}}^2 - R_{\text{inner}}^2\right) dx = \pi \int_a^b \left(f(x)^2 - g(x)^2\right) dx \qquad \boxed{2}$$

Keep in mind that the integrand is $(f(x)^2 - g(x)^2)$, not $(f(x) - g(x))^2$.

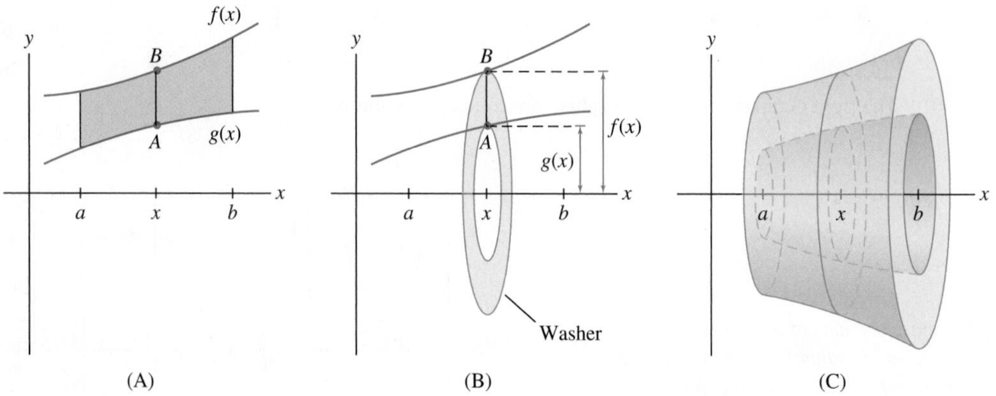

FIGURE 5 $\overline{AB}$ generates a washer when rotated about the x-axis.

■ **EXAMPLE 2 Region Between Two Curves** Find the volume V obtained by revolving the region between $y = x^2 + 4$ and $y = 2$ about the x-axis for $1 \le x \le 3$.

Solution The graph of $y = x^2 + 4$ lies above the graph of $y = 2$ (Figure 6). Therefore, $R_{\text{outer}} = x^2 + 4$ and $R_{\text{inner}} = 2$. By Eq. (2),

$$V = \pi \int_1^3 \left(R_{\text{outer}}^2 - R_{\text{inner}}^2\right) dx = \pi \int_1^3 \left((x^2 + 4)^2 - 2^2\right) dx$$

$$= \pi \int_1^3 \left(x^4 + 8x^2 + 12\right) dx = \pi \left(\frac{1}{5}x^5 + \frac{8}{3}x^3 + 12x\right)\Bigg|_1^3 = \frac{2126}{15}\pi \qquad ■$$

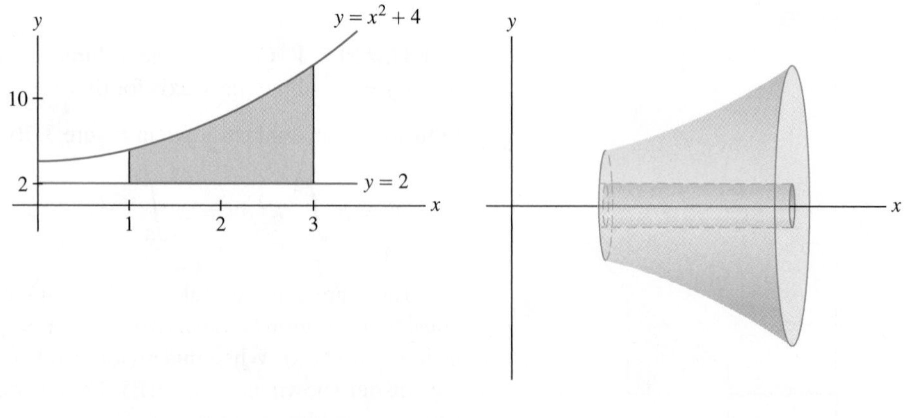

FIGURE 6 The area between $y = x^2 + 4$ and $y = 2$ over $[1, 3]$ rotated about the x-axis.

In the next example we calculate a volume of revolution about a horizontal axis parallel to the x-axis.

■ **EXAMPLE 3 Revolving About a Horizontal Axis** Find the volume V of the "wedding band" [Figure 7(C)] obtained by rotating the region between the graphs of $f(x) = x^2 + 2$ and $g(x) = 4 - x^2$ about the horizontal line $y = -3$.

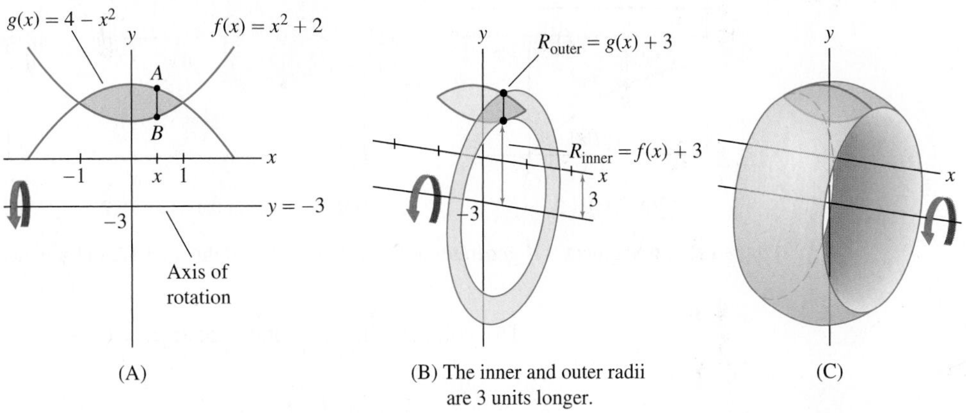

FIGURE 7

When you set up the integral for a volume of revolution, visualize the cross sections. These cross sections are washers (or disks) whose inner and outer radii depend on the axis of rotation.

Solution First, let's find the points of intersection of the two graphs by solving

$$f(x) = g(x) \quad \Rightarrow \quad x^2 + 2 = 4 - x^2 \quad \Rightarrow \quad x^2 = 1 \quad \Rightarrow \quad x = \pm 1$$

Figure 7(A) shows that $g(x) \geq f(x)$ for $-1 \leq x \leq 1$.

If we wanted to revolve about the x-axis, we would use Eq. (2). Since we want to revolve around $y = -3$, we must determine how the radii are affected. Figure 7(B) shows that when we rotate about $y = -3$, $\overline{AB}$ generates a washer whose outer and inner radii are both 3 units longer:

- $R_{\text{outer}} = g(x) - (-3) = (4 - x^2) + 3 = 7 - x^2$
- $R_{\text{inner}} = f(x) - (-3) = (x^2 + 2) + 3 = x^2 + 5$

The volume of revolution is equal to the integral of the area of this washer:

We get R_{outer} by subtracting $y = -3$ from $y = g(x)$ because vertical distance is the difference of the y-coordinates. Similarly, we subtract -3 from $f(x)$ to get R_{inner}.

$$\begin{aligned} V \text{ (about } y = -3) &= \pi \int_{-1}^{1} \left(R_{\text{outer}}^2 - R_{\text{inner}}^2\right) dx = \pi \int_{-1}^{1} \left((g(x)+3)^2 - (f(x)+3)^2\right) dx \\ &= \pi \int_{-1}^{1} \left((7 - x^2)^2 - (x^2 + 5)^2\right) dx \\ &= \pi \int_{-1}^{1} \left((49 - 14x^2 + x^4) - (x^4 + 10x^2 + 25)\right) dx \\ &= \pi \int_{-1}^{1} (24 - 24x^2)\, dx = \pi (24x - 8x^3)\Big|_{-1}^{1} = 32\pi \end{aligned}$$ ■

■ **EXAMPLE 4** Find the volume obtained by rotating the graphs of $f(x) = 9 - x^2$ and $y = 12$ for $0 \leq x \leq 3$ about

(a) the line $y = 12$ **(b)** the line $y = 15$.

Solution To set up the integrals, let's visualize the cross section. Is it a disk or a washer?

In Figure 8, the length of $\overline{AB}$ is $12 - f(x)$ rather than $f(x) - 12$ because the line $y = 12$ lies above the graph of $f(x)$.

(a) Figure 8(B) shows that $\overline{AB}$ rotated about $y = 12$ generates a *disk* of radius

$$R = \text{length of } \overline{AB} = 12 - f(x) = 12 - (9 - x^2) = 3 + x^2$$

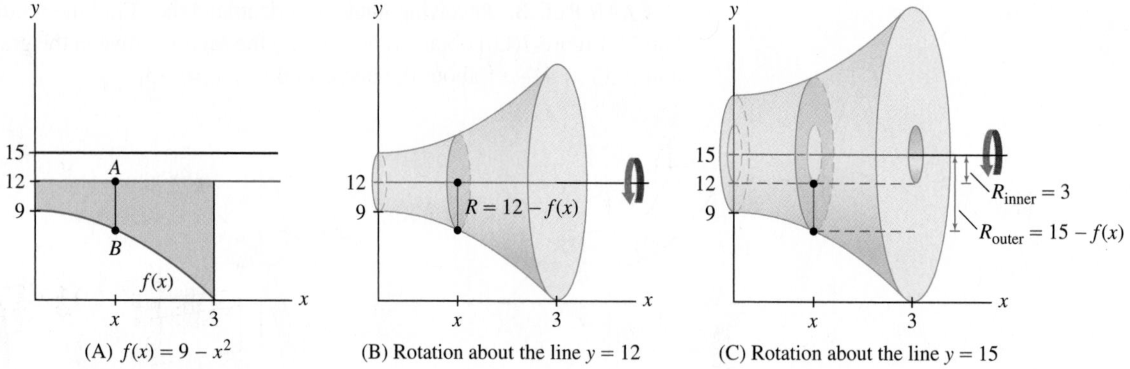

FIGURE 8 Segment $\overline{AB}$ generates a disk when rotated about $y = 12$, but it generates a washer when rotated about $y = 15$.

The volume when we rotate about $y = 12$ is

$$V = \pi \int_0^3 R^2\,dx = \pi \int_0^3 (3 + x^2)^2\,dx = \pi \int_0^3 (9 + 6x^2 + x^4)\,dx$$

$$= \pi \left(9x + 2x^3 + \frac{1}{5}x^5\right)\Big|_0^3 = \frac{648}{5}\pi$$

(b) Figure 8(C) shows that $\overline{AB}$ rotated about $y = 15$ generates a *washer*. The outer radius of this washer is the distance from B to the line $y = 15$:

$$R_{\text{outer}} = 15 - f(x) = 15 - (9 - x^2) = 6 + x^2$$

The inner radius is $R_{\text{inner}} = 3$, so the volume of revolution about $y = 15$ is

$$V = \pi \int_0^3 \left(R_{\text{outer}}^2 - R_{\text{inner}}^2\right) dx = \pi \int_0^3 \left((6 + x^2)^2 - 3^2\right) dx$$

$$= \pi \int_0^3 (36 + 12x^2 + x^4 - 9)\,dx$$

$$= \pi \left(27x + 4x^3 + \frac{1}{5}x^5\right)\Big|_0^3 = \frac{1188}{5}\pi \qquad ■$$

We can use the disk and washer methods for solids of revolution about vertical axes, but it is necessary to describe the graph as a function of y—that is, $x = g(y)$.

■ **EXAMPLE 5 Revolving About a Vertical Axis** Find the volume of the solid obtained by rotating the region under the graph of $f(x) = 9 - x^2$ for $0 \le x \le 3$ about the vertical axis $x = -2$.

Solution Figure 9 shows that $\overline{AB}$ sweeps out a horizontal washer when rotated about the vertical line $x = -2$. We are going to integrate with respect to y, so we need the inner and outer radii of this washer as functions of y. Solving for x in $y = 9 - x^2$, we obtain $x^2 = 9 - y$, or $x = \sqrt{9 - y}$ (since $x \ge 0$). Therefore,

$$R_{\text{outer}} = \sqrt{9 - y} + 2, \qquad R_{\text{inner}} = 2$$

$$R_{\text{outer}}^2 - R_{\text{inner}}^2 = \left(\sqrt{9 - y} + 2\right)^2 - 2^2 = (9 - y) + 4\sqrt{9 - y} + 4 - 4$$

$$= 9 - y + 4\sqrt{9 - y}$$

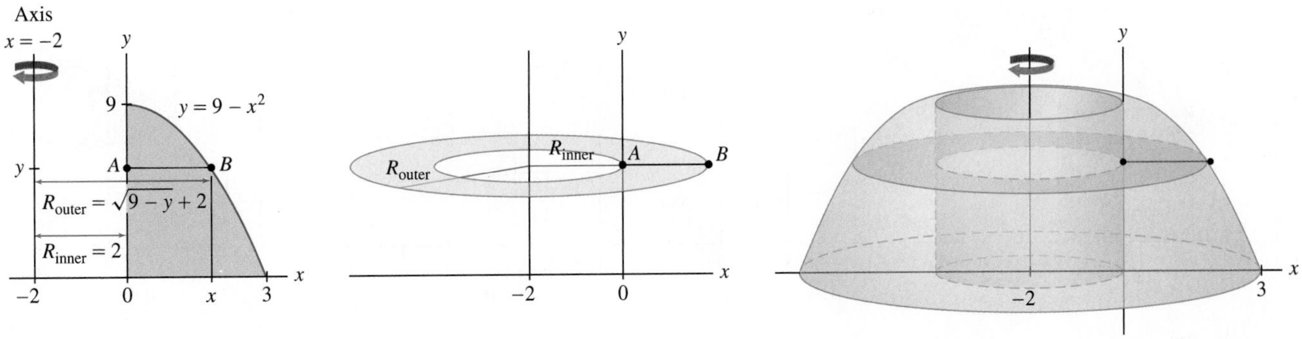

FIGURE 9

The region extends from $y = 0$ to $y = 9$ along the y-axis, so

$$V = \pi \int_0^9 \left(R_{\text{outer}}^2 - R_{\text{inner}}^2\right) dy = \pi \int_0^9 \left(9 - y + 4\sqrt{9 - y}\right) dy$$

$$= \pi \left(9y - \frac{1}{2}y^2 - \frac{8}{3}(9 - y)^{3/2}\right)\bigg|_0^9 = \frac{225}{2}\pi \quad ■$$

6.3 SUMMARY

- *Disk method* When you rotate the region between two graphs about an axis, the segments *perpendicular* to the axis generate disks or washers. The volume V of the solid of revolution is the integral of the areas of these disks or washers.
- Sketch the graphs to visualize the disks or washers.
- *Figure 10(A):* Region between $y = f(x)$ and the x-axis, rotated about the x-axis.
  - Vertical cross section: a circle of radius $R = f(x)$ and area $\pi R^2 = \pi f(x)^2$:

$$V = \pi \int_a^b R^2\, dx = \pi \int_a^b f(x)^2\, dx$$

- *Figure 10(B):* Region between $y = f(x)$ and $y = g(x)$, rotated about the x-axis.
  - Vertical cross section: a washer of outer radius $R_{\text{outer}} = f(x)$ and inner radius $R_{\text{inner}} = g(x)$:

$$V = \pi \int_a^b \left(R_{\text{outer}}^2 - R_{\text{inner}}^2\right) dx = \pi \int_a^b \left(f(x)^2 - g(x)^2\right) dx$$

- To rotate about a horizontal line $y = c$, modify the radii appropriately:
  - *Figure 10(C):* $c \geq f(x) \geq g(x)$:

$$R_{\text{outer}} = c - g(x), \qquad R_{\text{inner}} = c - f(x)$$

  - *Figure 10(D):* $f(x) \geq g(x) \geq c$:

$$R_{\text{outer}} = f(x) - c, \qquad R_{\text{inner}} = g(x) - c$$

- To rotate about a vertical line $x = c$, express R_{outer} and R_{inner} as functions of y and integrate along the y axis.

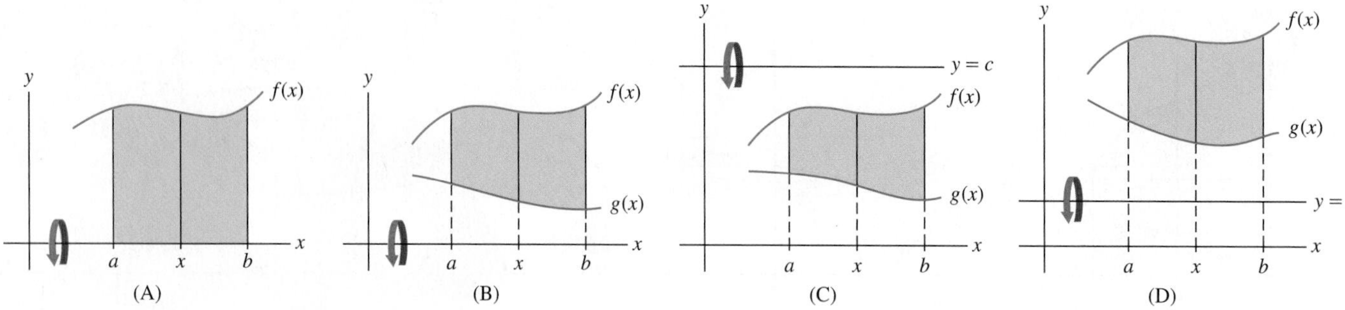

FIGURE 10

6.3 EXERCISES

Preliminary Questions

1. Which of the following is a solid of revolution?

(a) Sphere **(b)** Pyramid **(c)** Cylinder **(d)** Cube

2. True or false? When the region under a single graph is rotated about the x-axis, the cross sections of the solid perpendicular to the x-axis are circular disks.

3. True or false? When the region between two graphs is rotated about the x-axis, the cross sections to the solid perpendicular to the x-axis are circular disks.

4. Which of the following integrals expresses the volume obtained by rotating the area between $y = f(x)$ and $y = g(x)$ over $[a, b]$ around the x-axis? [Assume $f(x) \geq g(x) \geq 0$.]

(a) $\pi \int_a^b \left(f(x) - g(x)\right)^2 \, dx$

(b) $\pi \int_a^b \left(f(x)^2 - g(x)^2\right) dx$

Exercises

In Exercises 1–4, (a) sketch the solid obtained by revolving the region under the graph of $f(x)$ about the x-axis over the given interval, (b) describe the cross section perpendicular to the x-axis located at x, and (c) calculate the volume of the solid.

1. $f(x) = x + 1, \quad [0, 3]$

2. $f(x) = x^2, \quad [1, 3]$

3. $f(x) = \sqrt{x + 1}, \quad [1, 4]$

4. $f(x) = x^{-1}, \quad [1, 4]$

In Exercises 5–12, find the volume of revolution about the x-axis for the given function and interval.

5. $f(x) = x^2 - 3x, \quad [0, 3]$

6. $f(x) = \frac{1}{x^2}, \quad [1, 4]$

7. $f(x) = x^{5/3}, \quad [1, 8]$

8. $f(x) = 4 - x^2, \quad [0, 2]$

9. $f(x) = \frac{2}{x + 1}, \quad [1, 3]$

10. $f(x) = \sqrt{x^4 + 1}, \quad [1, 3]$

11. $f(x) = \csc x, \quad \left[\frac{\pi}{4}, \frac{3\pi}{4}\right]$

12. $f(x) = \sqrt{\cos x \sin x}, \quad \left[0, \frac{\pi}{2}\right]$

In Exercises 13 and 14, R is the shaded region in Figure 11.

13. Which of the integrands (i)–(iv) is used to compute the volume obtained by rotating region R about $y = -2$?

(i) $(f(x)^2 + 2^2) - (g(x)^2 + 2^2)$
(ii) $(f(x) + 2)^2 - (g(x) + 2)^2$
(iii) $(f(x)^2 - 2^2) - (g(x)^2 - 2^2)$
(iv) $(f(x) - 2)^2 - (g(x) - 2)^2$

14. Which of the integrands (i)–(iv) is used to compute the volume obtained by rotating R about $y = 9$?

(i) $(9 + f(x))^2 - (9 + g(x))^2$
(ii) $(9 + g(x))^2 - (9 + f(x))^2$
(iii) $(9 - f(x))^2 - (9 - g(x))^2$
(iv) $(9 - g(x))^2 - (9 - f(x))^2$

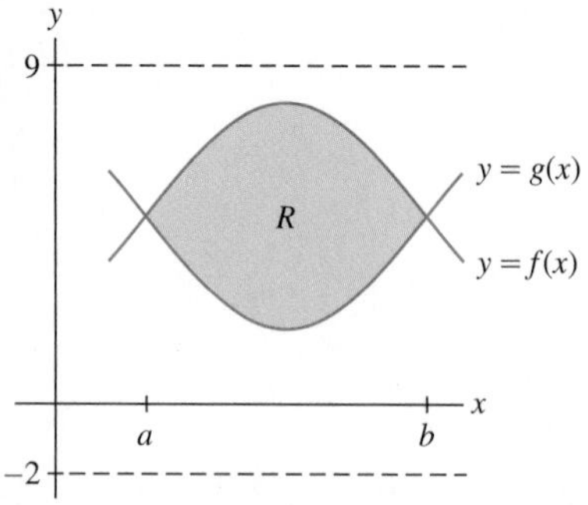

FIGURE 11

In Exercises 15–20, (a) sketch the region enclosed by the curves, (b) describe the cross section perpendicular to the x-axis located at x, and (c) find the volume of the solid obtained by rotating the region about the x-axis.

15. $y = x^2 + 2, \quad y = 10 - x^2$

16. $y = x^2, \quad y = 2x + 3$

17. $y = 16 - x, \quad y = 3x + 12, \quad x = -1$

18. $y = \frac{1}{x}, \quad y = \frac{5}{2} - x$

19. $y = \sec x, \quad y = 0, \quad x = -\frac{\pi}{4}, \quad x = \frac{\pi}{4}$

20. $y = \sec x, \quad y = 0, \quad x = 0, \quad x = \frac{\pi}{4}$

In Exercises 21–24, find the volume of the solid obtained by rotating the region enclosed by the graphs about the y-axis over the given interval.

21. $x = \sqrt{y}, \quad x = 0; \quad 1 \le y \le 4$

22. $x = \sqrt{\sin y}, \quad x = 0; \quad 0 \le y \le \pi$

23. $x = y^2, \quad x = \sqrt{y}$

24. $x = 4 - y, \quad x = 16 - y^2$

25. Rotation of the region in Figure 12 about the y-axis produces a solid with two types of different cross sections. Compute the volume as a sum of two integrals, one for $-12 \le y \le 4$ and one for $4 \le y \le 12$.

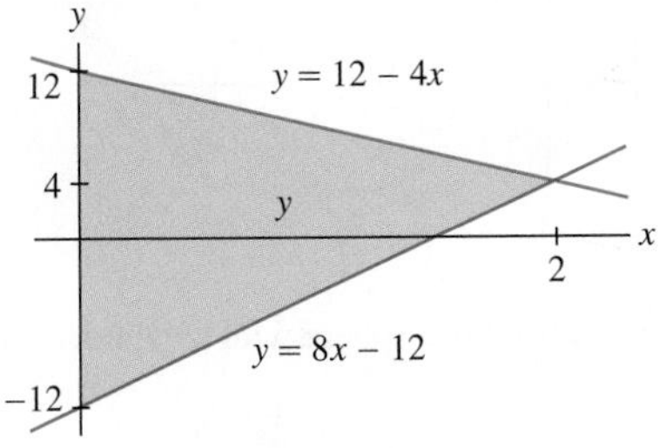

FIGURE 12

26. Let R be the region enclosed by $y = x^2 + 2$, $y = (x-2)^2$ and the axes $x = 0$ and $y = 0$. Compute the volume V obtained by rotating R about the x-axis. *Hint:* Express V as a sum of two integrals.

In Exercises 27–32, find the volume of the solid obtained by rotating region A in Figure 13 about the given axis.

27. x-axis

28. $y = -2$

29. $y = 2$

30. y-axis

31. $x = -3$

32. $x = 2$

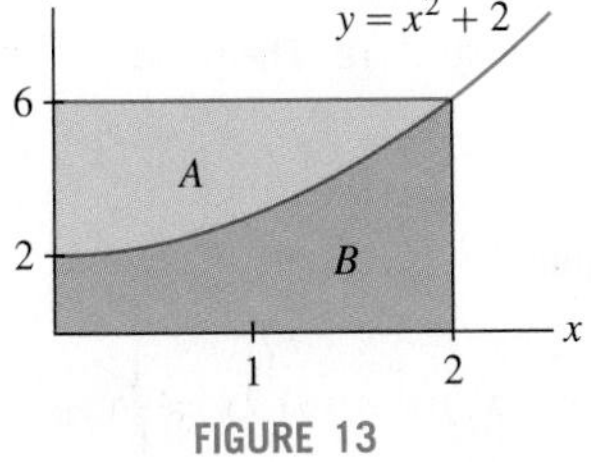

FIGURE 13

In Exercises 33–38, find the volume of the solid obtained by rotating region B in Figure 13 about the given axis.

33. x-axis

34. $y = -2$

35. $y = 6$

36. y-axis

Hint for Exercise 36: Express the volume as a sum of two integrals along the y-axis or use Exercise 30.

37. $x = 2$

38. $x = -3$

In Exercises 39–52, find the volume of the solid obtained by rotating the region enclosed by the graphs about the given axis.

39. $y = x^2, \quad y = 12 - x, \quad x = 0, \quad$ about $y = -2$

40. $y = x^2, \quad y = 12 - x, \quad x = 0, \quad$ about $y = 15$

41. $y = 16 - 2x, \quad y = 6, \quad x = 0, \quad$ about x-axis

42. $y = 32 - 2x, \quad y = 2 + 4x, \quad x = 0, \quad$ about y-axis

43. $y = \sec x, \quad y = 1 + \frac{3}{\pi}x, \quad$ about x-axis

44. $x = 2, \quad x = 3, \quad y = 16 - x^4, \quad y = 0, \quad$ about y-axis

45. $y = 2\sqrt{x}, \quad y = x, \quad$ about $x = -2$

46. $y = 2\sqrt{x}, \quad y = x, \quad$ about $y = 4$

47. $y = x^3, \quad y = x^{1/3}, \quad$ for $x \ge 0, \quad$ about y-axis

48. $y = x^2, \quad y = x^{1/2}, \quad$ about $x = -2$

49. $y = \frac{9}{x^2}, \quad y = 10 - x^2, \quad x \ge 0, \quad$ about $y = 12$

50. $y = \frac{9}{x^2}, \quad y = 10 - x^2, \quad x \ge 0, \quad$ about $x = -1$

51. $y = \frac{1}{x}, \quad y = \frac{5}{2} - x, \quad$ about y-axis

52. $y^2 = 4x, \quad y = x, \quad$ about $y = 8$

53. The bowl in Figure 14(A) is 21 cm high, obtained by rotating the curve in Figure 14(B) as indicated. Estimate the volume capacity of the bowl shown by taking the average of right- and left-endpoint approximations to the integral with $N = 7$.

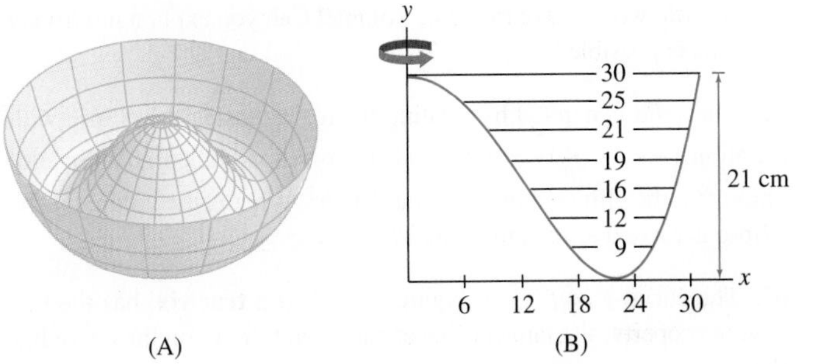

FIGURE 14

54. The region between the graphs of $f(x)$ and $g(x)$ over $[0, 1]$ is revolved about the line $y = -3$. Use the midpoint approximation with values from the following table to estimate the volume V of the resulting solid.

x	0.1	0.3	0.5	0.7	0.9
$f(x)$	8	7	6	7	8
$g(x)$	2	3.5	4	3.5	2

55. Find the volume of the cone obtained by rotating the region under the segment joining $(0, h)$ and $(r, 0)$ about the y-axis.

56. The **torus** (doughnut-shaped solid) in Figure 15 is obtained by rotating the circle $(x - a)^2 + y^2 = b^2$ around the y-axis (assume that $a > b$). Show that it has volume $2\pi^2 ab^2$. *Hint:* Evaluate the integral by interpreting it as the area of a circle.

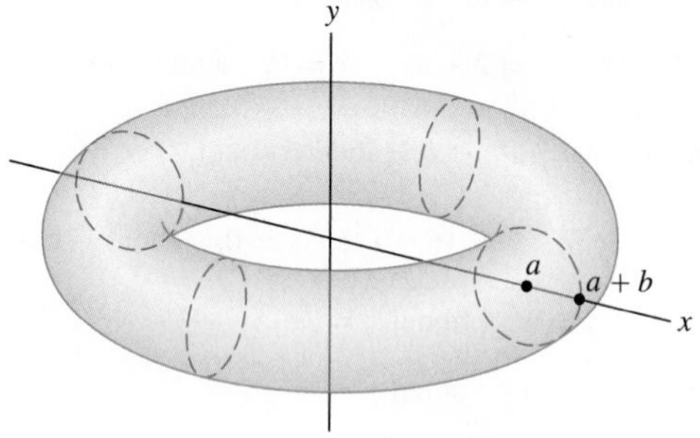

FIGURE 15 Torus obtained by rotating a circle about the y-axis.

57. GU Sketch the hypocycloid $x^{2/3} + y^{2/3} = 1$ and find the volume of the solid obtained by revolving it about the x-axis.

58. The solid generated by rotating the region between the branches of the hyperbola $y^2 - x^2 = 1$ about the x-axis is called a **hyperboloid** (Figure 16). Find the volume of the hyperboloid for $-a \le x \le a$.

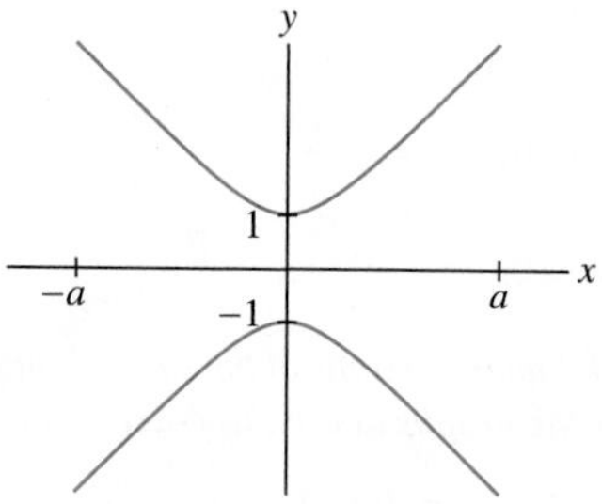

FIGURE 16 The hyperbola with equation $y^2 - x^2 = 1$.

59. A "bead" is formed by removing a cylinder of radius r from the center of a sphere of radius R (Figure 17). Find the volume of the bead with $r = 1$ and $R = 2$.

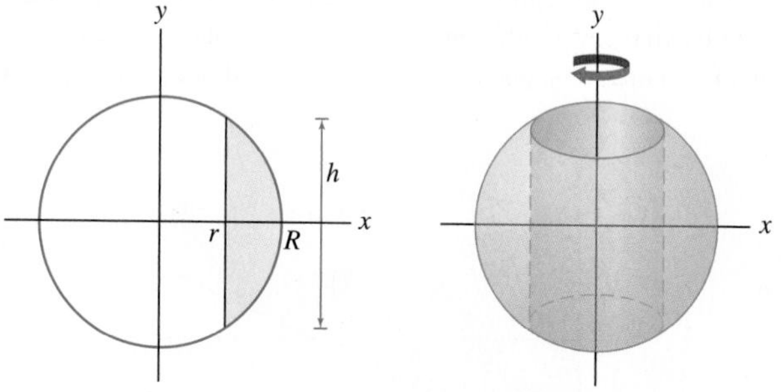

FIGURE 17 A bead is a sphere with a cylinder removed.

Further Insights and Challenges

60. Find the volume V of the bead (Figure 17) in terms of r and R. Then show that $V = \frac{\pi}{6}h^3$, where h is the height of the bead. This formula has a surprising consequence: Since V can be expressed in terms of h alone, it follows that two beads of height 1 cm, one formed from a sphere the size of an orange and the other from a sphere the size of the earth, would have the same volume! Can you explain intuitively how this is possible?

61. The solid generated by rotating the region inside the ellipse with equation $\left(\frac{x}{a}\right)^2 + \left(\frac{y}{b}\right)^2 = 1$ around the x-axis is called an **ellipsoid**. Show that the ellipsoid has volume $\frac{4}{3}\pi ab^2$. What is the volume if the ellipse is rotated around the y-axis?

62. The curve $y = f(x)$ in Figure 18, called a **tractrix**, has the following property: the tangent line at each point (x, y) on the curve has slope

$$\frac{dy}{dx} = \frac{-y}{\sqrt{1 - y^2}}$$

Let R be the shaded region under the graph of $0 \le x \le a$ in Figure 18. Compute the volume V of the solid obtained by revolving R around the x-axis in terms of the constant $c = f(a)$. *Hint:* Use the substitution $u = f(x)$ to show that

$$V = \pi \int_c^1 u\sqrt{1 - u^2}\, du$$

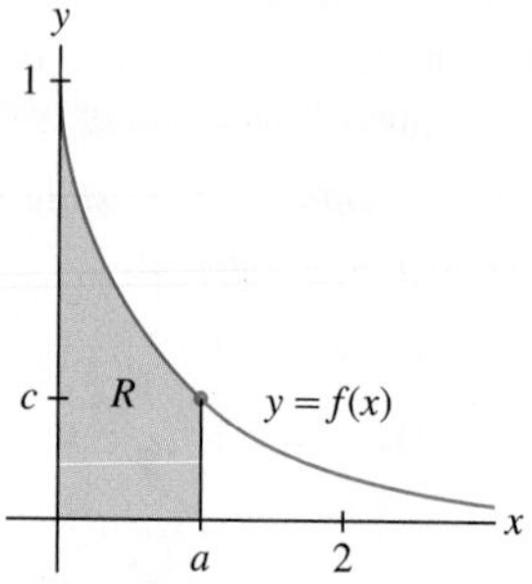

FIGURE 18 The tractrix.

63. Verify the formula

$$\int_{x_1}^{x_2} (x - x_1)(x - x_2)\, dx = \frac{1}{6}(x_1 - x_2)^3 \qquad \boxed{3}$$

Then prove that the solid obtained by rotating the shaded region in Figure 19 about the x-axis has volume $V = \frac{\pi}{6}BH^2$, with B and H as in the figure. *Hint:* Let x_1 and x_2 be the roots of $f(x) = ax + b - (mx + c)^2$, where $x_1 < x_2$. Show that

$$V = \pi \int_{x_1}^{x_2} f(x)\, dx$$

and use Eq. (3).

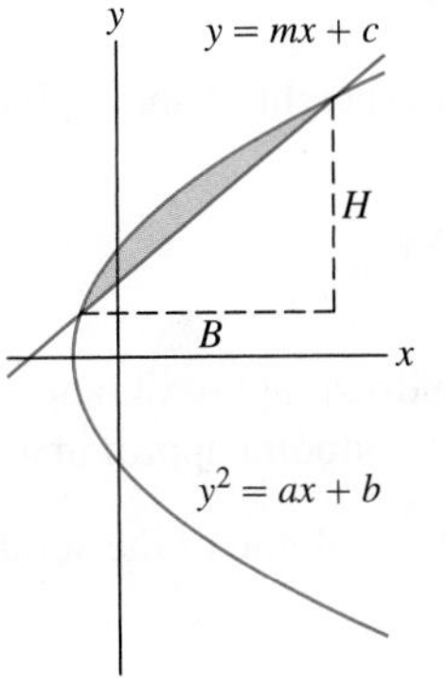

FIGURE 19 The line $y = mx + c$ intersects the parabola $y^2 = ax + b$ at two points above the x-axis.

64. Let R be the region in the unit circle lying above the cut with the line $y = mx + b$ (Figure 20). Assume the points where the line intersects the circle lie above the x-axis. Use the method of Exercise 63 to show that the solid obtained by rotating R about the x-axis has volume $V = \frac{\pi}{6}hd^2$, with h and d as in the figure.

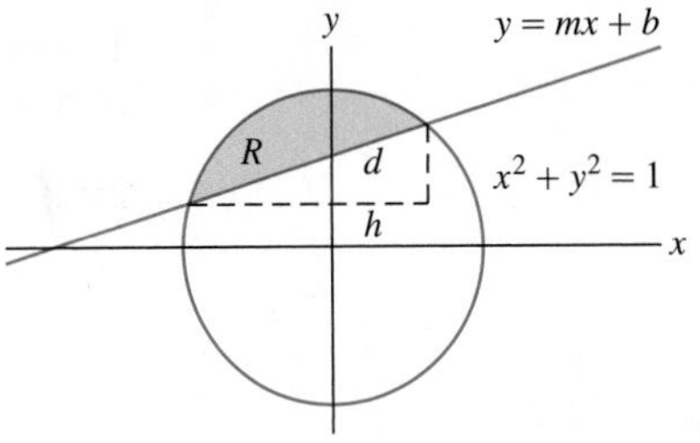

FIGURE 20

6.4 The Method of Cylindrical Shells

In the previous two sections, we computed volumes by integrating cross-sectional area. The **Shell Method**, based on cylindrical shells, is more convenient in some cases.

Consider a cylindrical shell (Figure 1) of height h, with outer radius R and inner radius r. Because the shell is obtained by removing a cylinder of radius r from the wider cylinder of radius R, it has volume

$$\pi R^2 h - \pi r^2 h = \pi h(R^2 - r^2) = \pi h(R + r)(R - r) = \pi h(R + r)\Delta r$$

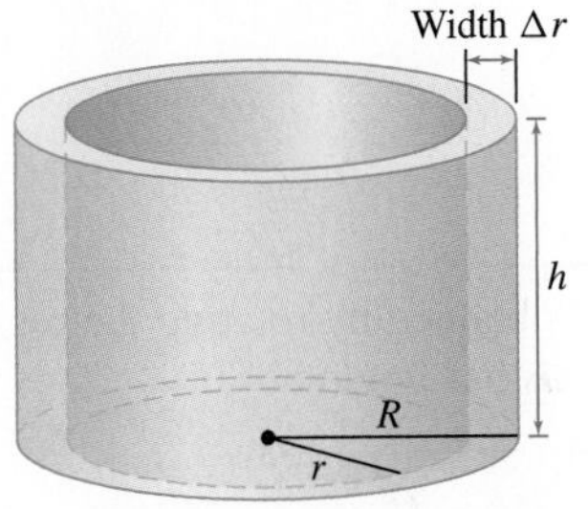

FIGURE 1 The volume of the cylindrical shell is approximately

$$2\pi R h \Delta r$$

where $\Delta r = R - r$.

where $\Delta r = R - r$ is the width of the shell. If the shell is very thin, then R and r are nearly equal and we may replace $(R + r)$ by $2R$ to obtain

$$\text{Volume of shell} \approx 2\pi R h \Delta r = 2\pi(\text{radius}) \times (\text{height of shell}) \times (\text{thickness}) \qquad \boxed{1}$$

Now, let us rotate the region under $y = f(x)$ from $x = a$ to $x = b$ about the y-axis as in Figure 2. The resulting solid can be divided into thin concentric shells. More precisely, we divide $[a, b]$ into N subintervals of length $\Delta x = (b - a)/N$ with endpoints $x_0, x_1, \dots, x_N$. When we rotate the thin strip of area above $[x_{i-1}, x_i]$ about the y-axis, we obtain a thin shell whose volume we denote by V_i. The volume of the solid is equal to the sum $V = \sum_{i=1}^{N} V_i$.

The top rim of the ith thin shell in Figure 2 is curved. However, when Δx is small, we can approximate this thin shell by the cylindrical shell (with flat rim) of height $f(x_i)$.

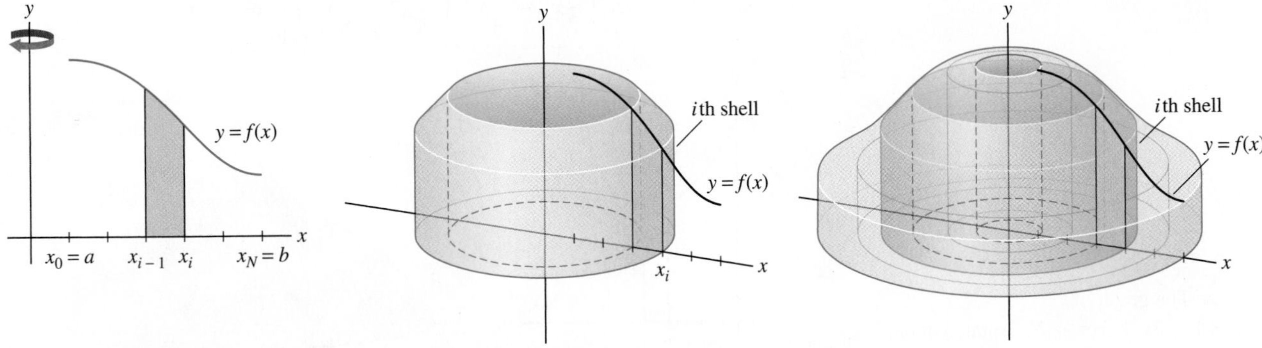

FIGURE 2 The shaded strip, when rotated about the y-axis, generates a "thin shell."

Then, using Eq. (1), we obtain

$$V_i \approx 2\pi x_i f(x_i)\Delta x = 2\pi(\text{radius})(\text{height of shell})(\text{thickness})$$

$$V = \sum_{i=1}^{N} V_i \approx 2\pi \sum_{i=1}^{N} x_i f(x_i)\Delta x$$

The sum on the right is the volume of a cylindrical approximation that converges to V as $N \to \infty$ (Figure 3). This sum is also a right-endpoint approximation that converges to $2\pi \int_a^b x f(x)\,dx$. Thus we obtain Eq. (2) for the volume of the solid.

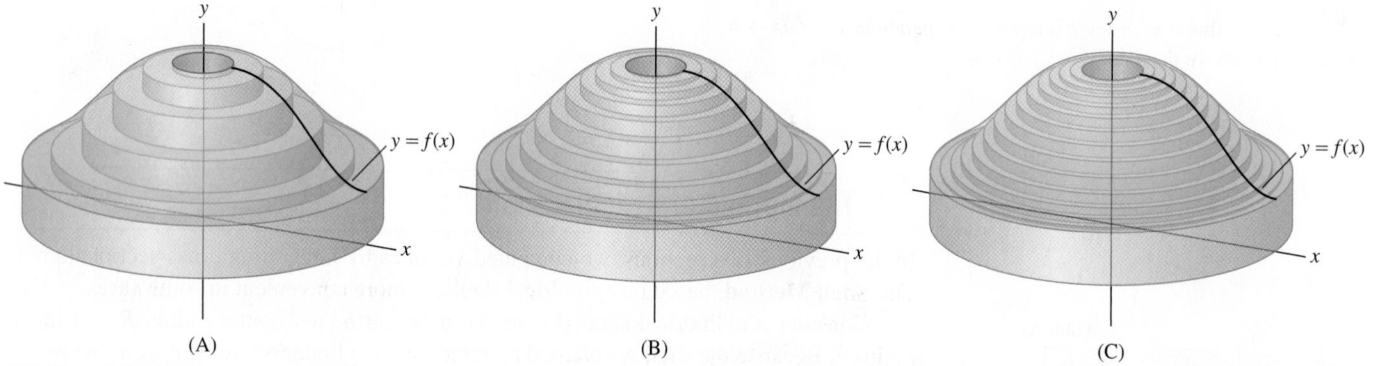

FIGURE 3 Cylindrical shell approximations as $N \to \infty$.

Note: In the Shell Method, we integrate with respect to x when the region is rotated about the y-axis.

Volume of Revolution: The Shell Method The solid obtained by rotating the region under $y = f(x)$ over the interval $[a, b]$ *about the y-axis* has volume

$$V = 2\pi \int_a^b (\text{radius})(\text{height of shell})\,dx = 2\pi \int_a^b x f(x)\,dx \qquad \boxed{2}$$

■ **EXAMPLE 1** Find the volume V of the solid obtained by rotating the region under the graph of $f(x) = 1 - 2x + 3x^2 - 2x^3$ over $[0, 1]$ about the y-axis.

Solution The solid is shown in Figure 4. By Eq. (2),

$$V = 2\pi \int_0^1 x f(x)\,dx = 2\pi \int_0^1 x(1 - 2x + 3x^2 - 2x^3)\,dx$$

$$= 2\pi \left(\frac{1}{2}x^2 - \frac{2}{3}x^3 + \frac{3}{4}x^4 - \frac{2}{5}x^5\right)\Bigg|_0^1 = \frac{11}{30}\pi \qquad ■$$

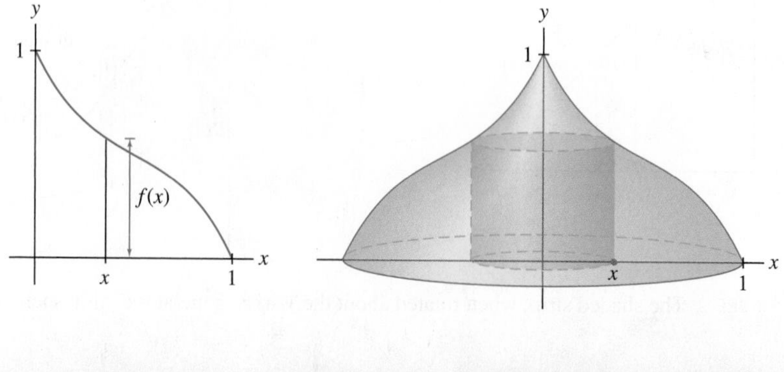

FIGURE 4 The graph of $f(x) = 1 - 2x + 3x^2 - 2x^3$ rotated about the y-axis.

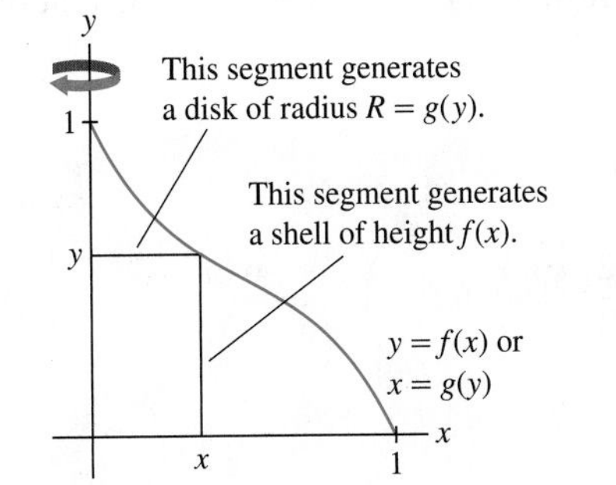

FIGURE 5 For rotation about the y-axis, the Shell Method uses $y = f(x)$ but the Disk Method requires the inverse function $x = g(y)$.

CONCEPTUAL INSIGHT Shells versus Disks and Washers Some volumes can be computed equally well using either the Shell Method or the Disk and Washer Method, but in Example 1, the Shell Method is much easier. To use the Disk Method, we would need to know the radius of the disk generated at height y because we're rotating about the y-axis (Figure 5). This would require finding the inverse $g(y) = f^{-1}(y)$. In general: Use the Shell Method if finding the shell height (which is *parallel* to the axis of rotation) is easier than finding the disk radius (which is *perpendicular* to the axis of rotation). Use the Disk Method when finding the disk radius is easier.

When we rotate the region between the graphs of two functions $f(x)$ and $g(x)$ satisfying $f(x) \geq g(x)$, the vertical segment at location x generates a cylindrical shell of radius x and height $f(x) - g(x)$ (Figure 6). Therefore, the volume is

$$V = 2\pi \int_a^b (\text{radius})(\text{Height of shell})\, dx = 2\pi \int_a^b x\big(f(x) - g(x)\big)\, dx \qquad \boxed{3}$$

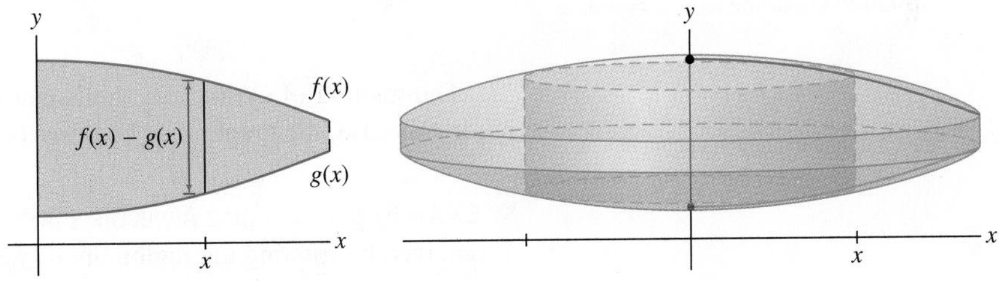

FIGURE 6 The vertical segment at location x generates a shell of radius x and height $f(x) - g(x)$.

■ **EXAMPLE 2 Region Between Two Curves** Find the volume V obtained by rotating the region enclosed by the graphs of $f(x) = x(5 - x)$ and $g(x) = 8 - x(5 - x)$ about the y-axis.

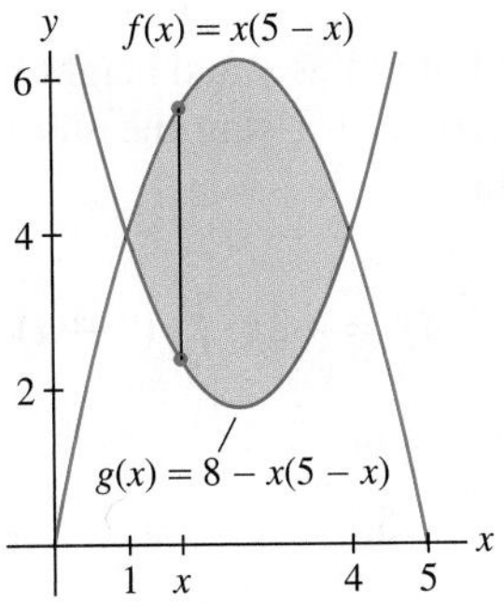

FIGURE 7

Solution First, find the points of intersection by solving $x(5 - x) = 8 - x(5 - x)$. We obtain $x^2 - 5x + 4 = (x - 1)(x - 4) = 0$, so the curves intersect at $x = 1, 4$. Sketching the graphs (Figure 7), we see that $f(x) \geq g(x)$ on the interval $[1, 4]$ and

$$\text{Height of shell} = f(x) - g(x) = x(5 - x) - \big(8 - x(5 - x)\big) = 10x - 2x^2 - 8$$

$$V = 2\pi \int_1^4 (\text{radius})(\text{height of shell})\, dx = 2\pi \int_1^4 x(10x - 2x^2 - 8)\, dx$$

$$= 2\pi \left(\frac{10}{3}x^3 - \frac{1}{2}x^4 - 4x^2\right)\Big|_1^4 = 2\pi \left(\frac{64}{3} - \left(-\frac{7}{6}\right)\right) = 45\pi \qquad ■$$

The reasoning in Example 3 shows that if we rotate the region under $y = f(x)$ over $[a, b]$ about the vertical line $x = c$, then the volume is

$$V = 2\pi \int_a^b (x - c) f(x)\, dx \quad \text{if } c \leq a$$

$$V = 2\pi \int_a^b (c - x) f(x)\, dx \quad \text{if } c \geq b$$

■ **EXAMPLE 3 Rotating About a Vertical Axis** Use the Shell Method to calculate the volume V obtained by rotating the region under the graph of $f(x) = x^{-1/2}$ over $[1, 4]$ about the axis $x = -3$.

Solution If we were rotating this region about the y-axis (that is, $x = 0$), we would use Eq. (3). To rotate it around the line $x = -3$, we must take into account that the radius of revolution is now 3 units longer.

Figure 8 shows that the radius of the shell is now $x - (-3) = x + 3$. The height of the shell is still $f(x) = x^{-1/2}$, so

$$V = 2\pi \int_1^4 (\text{radius})(\text{height of shell})\, dx$$

$$= 2\pi \int_1^4 (x+3)x^{-1/2}\, dx = 2\pi \left(\frac{2}{3}x^{3/2} + 6x^{1/2} \right) \Big|_1^4 = \frac{64\pi}{3}$$

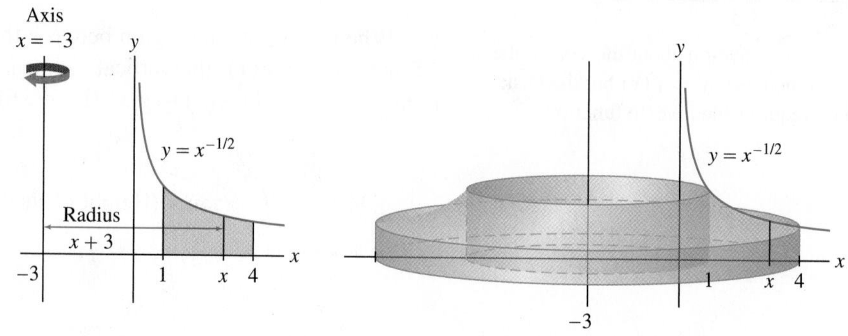

FIGURE 8 Rotation about the axis $x = -3$.

The method of cylindrical shells can be applied to rotations about horizontal axes, but in this case, the graph must be described in the form $x = g(y)$.

■ **EXAMPLE 4 Rotating About the x-Axis** Use the Shell Method to compute the volume V obtained by rotating the region under $y = 9 - x^2$ over $[0, 3]$ about the x-axis.

Solution When we rotate about the x-axis, the cylindrical shells are generated by horizontal segments and the Shell Method gives us an integral with respect to y. Therefore, we solve $y = 9 - x^2$ for x to obtain $x = \sqrt{9-y}$.

Segment $\overline{AB}$ in Figure 9 generates a cylindrical shell of radius y and height $\sqrt{9-y}$ (we use the term "height" even though the shell is horizontal). Using the substitution $u = 9 - y$, $du = -dy$ in the resulting integral, we obtain

← REMINDER After making the substitution $u = 9 - y$, the limits of integration must be changed. Since $u(0) = 9$ and $u(9) = 0$, we change $\int_0^9$ to $\int_9^0$.

$$V = 2\pi \int_0^9 (\text{radius})(\text{height of shell})\, dy = 2\pi \int_0^9 y\sqrt{9-y}\, dy = -2\pi \int_9^0 (9-u)\sqrt{u}\, du$$

$$= 2\pi \int_0^9 (9u^{1/2} - u^{3/2})\, du = 2\pi \left(6u^{3/2} - \frac{2}{5}u^{5/2} \right) \Big|_0^9 = \frac{648}{5}\pi$$

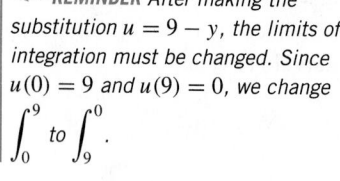

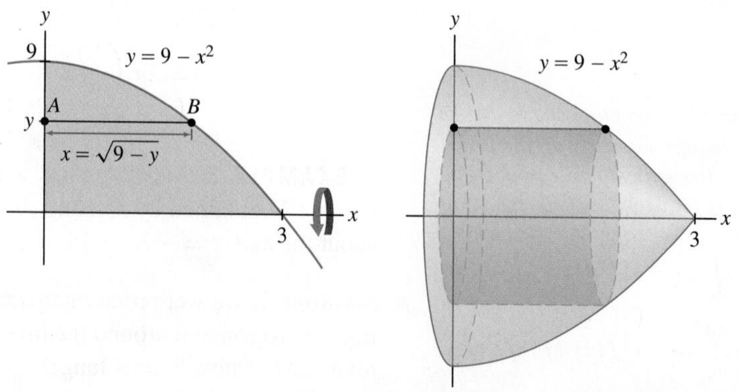

FIGURE 9 Shell generated by a horizontal segment in the region under the graph of $y = 9 - x^2$.

6.4 SUMMARY

- *Shell Method* When you rotate the region between two graphs about an axis, the segments *parallel* to the axis generate cylindrical shells [Figure 10(A)]. The volume V of the solid of revolution is the integral of the areas of these shells:

$$\text{area of shell} = 2\pi(\text{radius})(\text{height of shell})$$

- Sketch the graphs to visualize the shells.
- *Figure 10(B):* Region between $y = f(x)$ (with $f(x) \geq 0$) and the y-axis, rotated about the y-axis.

$$V = 2\pi \int_a^b (\text{radius})(\text{height of shell})\,dx = 2\pi \int_a^b x f(x)\,dx$$

- *Figure 10(C):* Region between $y = f(x)$ and $y = g(x)$ (with $f(x) \geq g(x) \geq 0$), rotated about the y-axis.

$$V = 2\pi \int_a^b (\text{radius})(\text{height of shell})\,dx = 2\pi \int_a^b x(f(x) - g(x))\,dx$$

- Rotation about a vertical axis $x = c$.
  - *Figure 10(D):* $c \leq a$, radius of shell is $(x - c)$:

$$V = 2\pi \int_a^b (x - c) f(x)\,dx$$

  - *Figure 10(E):* $c \geq a$, radius of shell is $(c - x)$:

$$V = 2\pi \int_a^b (c - x) f(x)\,dx$$

- Rotation about the x-axis using the Shell Method: Write the graph as $x = g(y)$:

$$V = 2\pi \int_c^d (\text{radius})(\text{height of shell})\,dy = 2\pi \int_c^d y g(y)\,dy$$

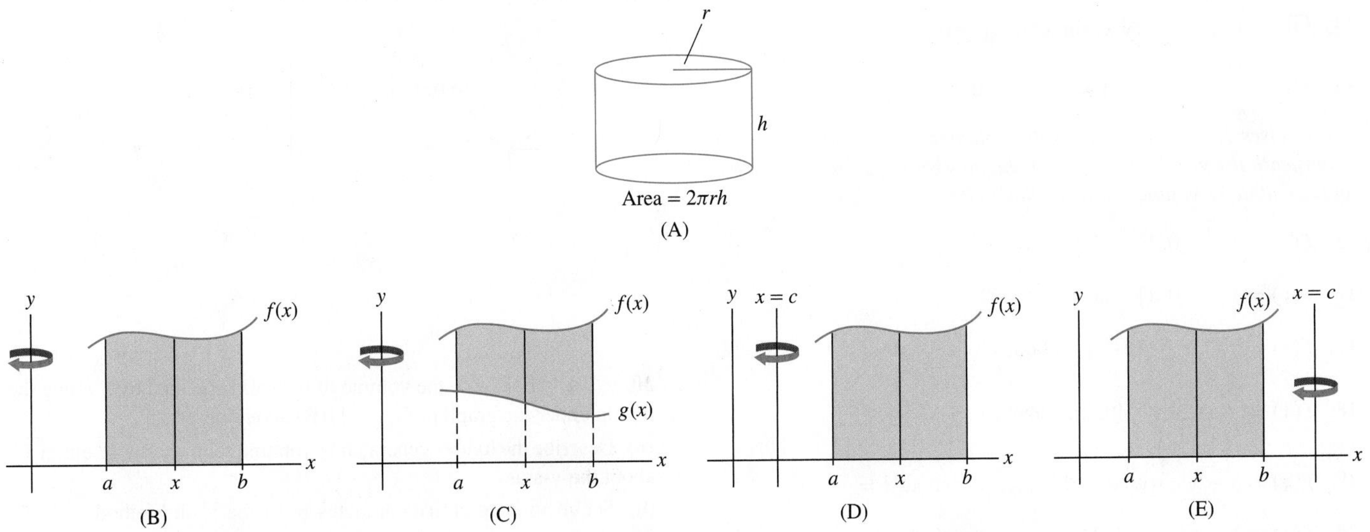

FIGURE 10

6.4 EXERCISES

Preliminary Questions

1. Consider the region $\mathcal{R}$ under the graph of the constant function $f(x) = h$ over the interval $[0, r]$. Give the height and the radius of the cylinder generated when $\mathcal{R}$ is rotated about:

(a) the x-axis **(b)** the y-axis

2. Let V be the volume of a solid of revolution about the y-axis.

(a) Does the Shell Method for computing V lead to an integral with respect to x or y?

(b) Does the Disk or Washer Method for computing V lead to an integral with respect to x or y?

Exercises

In Exercises 1–6, sketch the solid obtained by rotating the region underneath the graph of the function over the given interval about the y-axis, and find its volume.

1. $f(x) = x^3, \quad [0, 1]$

2. $f(x) = \sqrt{x}, \quad [0, 4]$

3. $f(x) = x^{-1}, \quad [1, 3]$

4. $f(x) = 4 - x^2, \quad [0, 2]$

5. $f(x) = \sqrt{x^2 + 9}, \quad [0, 3]$

6. $f(x) = \dfrac{x}{\sqrt{1 + x^3}}, \quad [1, 4]$

In Exercises 7–12, use the Shell Method to compute the volume obtained by rotating the region enclosed by the graphs as indicated, about the y-axis.

7. $y = 3x - 2, \quad y = 6 - x, \quad x = 0$

8. $y = \sqrt{x}, \quad y = x^2$

9. $y = x^2, \quad y = 8 - x^2, \quad x = 0, \quad \text{for } x \geq 0$

10. $y = 8 - x^3, \quad y = 8 - 4x, \quad \text{for } x \geq 0$

11. $y = (x^2 + 1)^{-2}, \quad y = 2 - (x^2 + 1)^{-2}, \quad x = 2$

12. $y = 1 - |x - 1|, \quad y = 0$

In Exercises 13 and 14, use a graphing utility to find the points of intersection of the curves numerically and then compute the volume of rotation of the enclosed region about the y-axis.

13. GU $y = \frac{1}{2}x^2, \quad y = \sin(x^2), \quad x \geq 0$

14. GU $y = 1 - x^4, \quad y = x, \quad x \geq 0$

In Exercises 15–20, sketch the solid obtained by rotating the region underneath the graph of f(x) over the interval about the given axis, and calculate its volume using the Shell Method.

15. $f(x) = x^3, \quad [0, 1], \quad \text{about } x = 2$

16. $f(x) = x^3, \quad [0, 1], \quad \text{about } x = -2$

17. $f(x) = x^{-4}, \quad [-3, -1], \quad \text{about } x = 4$

18. $f(x) = \dfrac{1}{\sqrt{x^2 + 1}}, \quad [0, 2], \quad \text{about } x = 0$

19. $f(x) = a - x \quad \text{with } a > 0, \quad [0, a], \quad \text{about } x = -1$

20. $f(x) = 1 - x^2, \quad [-1, 1], \quad x = c \quad \text{with } c > 1$

In Exercises 21–26, sketch the enclosed region and use the Shell Method to calculate the volume of rotation about the x-axis.

21. $x = y, \quad y = 0, \quad x = 1$

22. $x = \frac{1}{4}y + 1, \quad x = 3 - \frac{1}{4}y, \quad y = 0$

23. $x = y(4 - y), \quad y = 0$

24. $x = y(4 - y), \quad x = (y - 2)^2$

25. $y = 4 - x^2, \quad x = 0, \quad y = 0$

26. $y = x^{1/3} - 2, \quad y = 0, \quad x = 27$

27. Use both the Shell and Disk Methods to calculate the volume obtained by rotating the region under the graph of $f(x) = 8 - x^3$ for $0 \leq x \leq 2$ about:

(a) the x-axis **(b)** the y-axis

28. Sketch the solid of rotation about the y-axis for the region under the graph of the constant function $f(x) = c$ (where $c > 0$) for $0 \leq x \leq r$.

(a) Find the volume without using integration.

(b) Use the Shell Method to compute the volume.

29. The graph in Figure 11(A) can be described by both $y = f(x)$ and $x = h(y)$, where h is the inverse of f. Let V be the volume obtained by rotating the region under the graph about the y-axis.

(a) Describe the figures generated by rotating segments $\overline{AB}$ and $\overline{CB}$ about the y-axis.

(b) Set up integrals that compute V by the Shell and Disk Methods.

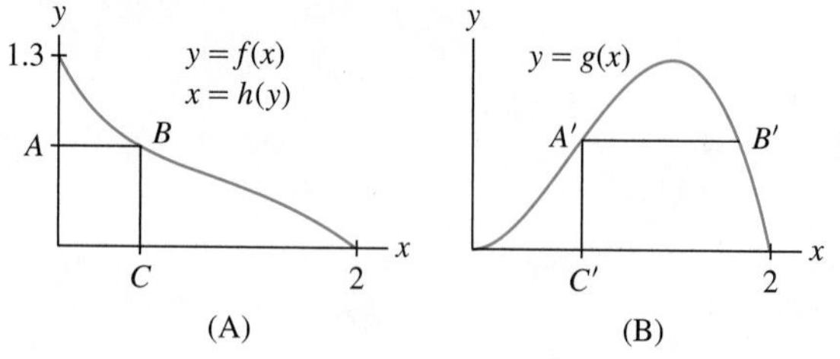

FIGURE 11

30. Let W be the volume of the solid obtained by rotating the region under the graph in Figure 11(B) about the y-axis.

(a) Describe the figures generated by rotating segments $\overline{A'B'}$ and $\overline{A'C'}$ about the y-axis.

(b) Set up an integral that computes W by the Shell Method.

(c) Explain the difficulty in computing W by the Washer Method.

31. Let R be the region under the graph of $y = 9 - x^2$ for $0 \le x \le 2$. Use the Shell Method to compute the volume of rotation of R about the x-axis as a sum of two integrals along the y-axis. *Hint:* The shells generated depend on whether $y \in [0, 5]$ or $y \in [5, 9]$.

32. Let R be the region under the graph of $y = 4x^{-1}$ for $1 \le y \le 4$. Use the Shell Method to compute the volume of rotation of R about the y-axis as a sum of two integrals along the x-axis.

In Exercises 33–38, use the Shell Method to find the volume obtained by rotating region A in Figure 12 about the given axis.

33. y-axis

34. $x = -3$

35. $x = 2$

36. x-axis

37. $y = -2$

38. $y = 6$

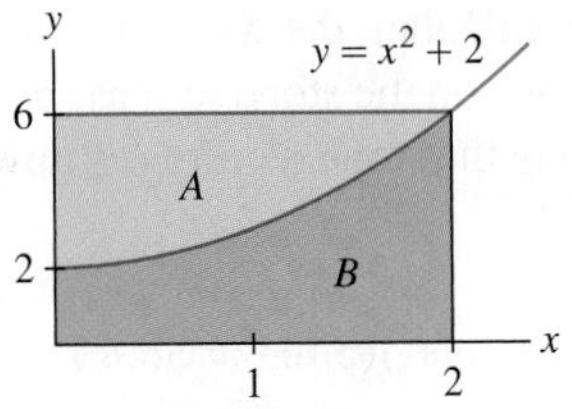

FIGURE 12

In Exercises 39–44, use the most convenient method (Disk or Shell Method) to find the volume obtained by rotating region B in Figure 12 about the given axis.

39. y-axis

40. $x = -3$

41. $x = 2$

42. x-axis

43. $y = -2$

44. $y = 8$

In Exercises 45–50, use the most convenient method (Disk or Shell Method) to find the given volume of rotation.

45. Region between $x = y(5 - y)$ and $x = 0$, rotated about the y-axis

46. Region between $x = y(5 - y)$ and $x = 0$, rotated about the x-axis

47. Region in Figure 13, rotated about the x-axis

48. Region in Figure 13, rotated about the y-axis

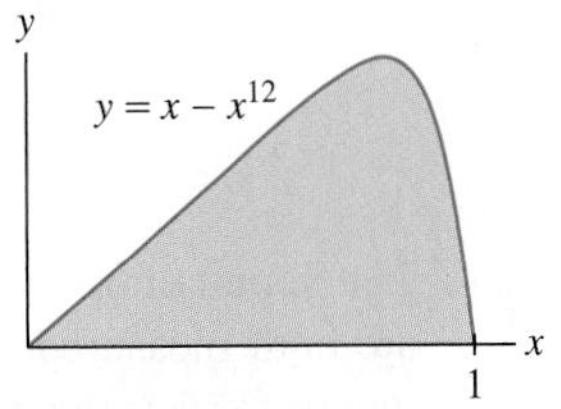

FIGURE 13

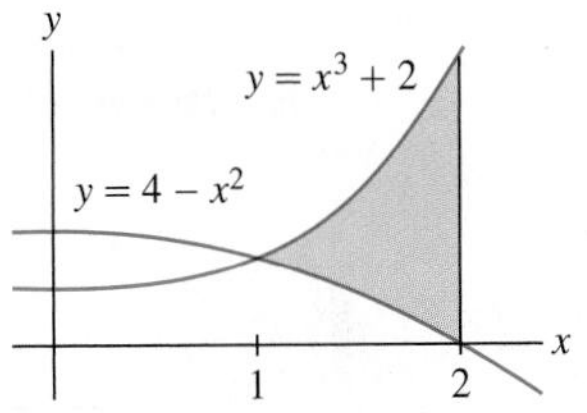

FIGURE 14

49. Region in Figure 14, rotated about $x = 4$

50. Region in Figure 14, rotated about $y = -2$

In Exercises 51–54, use the Shell Method to find the given volume of rotation.

51. A sphere of radius r

52. The "bead" formed by removing a cylinder of radius r from the center of a sphere of radius R (compare with Exercise 59 in Section 6.3)

53. The torus obtained by rotating the circle $(x - a)^2 + y^2 = b^2$ about the y-axis, where $a > b$ (compare with Exercise 56 in Section 6.3). *Hint:* Evaluate the integral by interpreting part of it as the area of a circle.

54. The "paraboloid" obtained by rotating the region between $y = x^2$ and $y = c$ $(c > 0)$ about the y-axis

Further Insights and Challenges

55. The surface area of a sphere of radius r is $4\pi r^2$. Use this to derive the formula for the volume V of a sphere of radius R in a new way.

(a) Show that the volume of a thin spherical shell of inner radius r and thickness Δr is approximately $4\pi r^2 \Delta r$.

(b) Approximate V by decomposing the sphere of radius R into N thin spherical shells of thickness $\Delta r = R/N$.

(c) Show that the approximation is a Riemann sum that converges to an integral. Evaluate the integral.

56. Show that the solid (an **ellipsoid**) obtained by rotating the region R in Figure 15 about the y-axis has volume $\frac{4}{3}\pi a^2 b$.

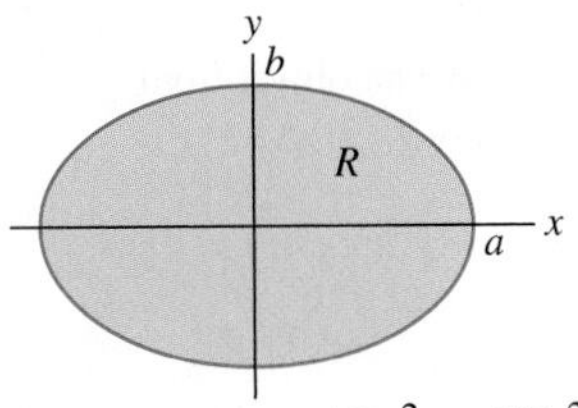

FIGURE 15 The ellipse $\left(\frac{x}{a}\right)^2 + \left(\frac{y}{b}\right)^2 = 1$.

57. The bell-shaped curve $y = f(x)$ in Figure 16 satisfies $dy/dx = -xy$. Use the Shell Method and the substitution $u = f(x)$ to show that the solid obtained by rotating the region R about the y-axis has volume $V = 2\pi(1 - c)$, where $c = f(a)$. Observe that as $c \to 0$, the region R becomes infinite but the volume V approaches 2π.

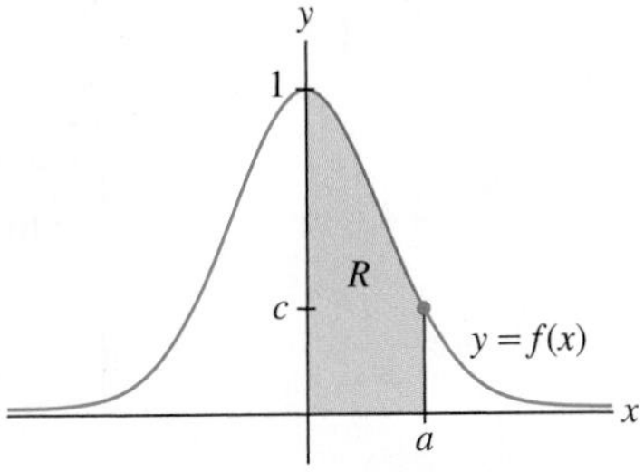

FIGURE 16 The bell-shaped curve.

6.5 Work and Energy

"For those who want some proof that physicists are human, the proof is in the idiocy of all the different units which they use for measuring energy."

—Richard Feynman,
The Character of Physical Law

All physical tasks, from running up a hill to turning on a computer, require an expenditure of energy. When a force is applied to an object to move it, the energy expended is called **work**. When a *constant* force F is applied to move the object a distance d in the direction of the force, the work W is defined as "force times distance" (Figure 1):

$$W = F \cdot d \qquad (1)$$

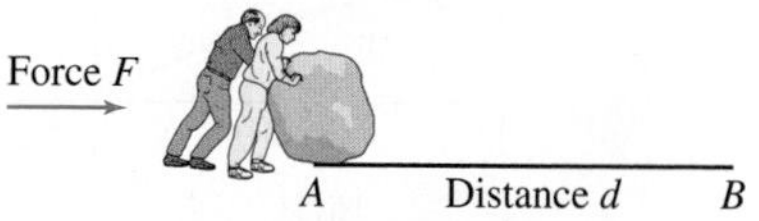

FIGURE 1 The work expended to move the object from A to B is $W = F \cdot d$.

The SI unit of force is the *newton* (abbreviated N), defined as 1 kg-m/s^2. Energy and work are both measured in units of the *joule* (J), equal to 1 N-m. In the British system, the unit of force is the pound, and both energy and work are measured in foot-pounds (ft-lb). Another unit of energy is the *calorie*. One ft-lb is approximately 0.738 J or 3.088 calories.

To become familiar with the units, let's calculate the work W required to lift a 2-kg stone 3 m above the ground. Gravity pulls down on the stone of mass m with a force equal to $-mg$, where $g = 9.8$ m/s^2. Therefore, lifting the stone requires an upward vertical force $F = mg$, and the work expended is

$$W = \underbrace{(mg)h}_{F \cdot d} = (2 \text{ kg})(9.8 \text{ m/s}^2)(3 \text{ m}) = 58.8 \text{ J}$$

The kilogram is a unit of mass, but the pound is a unit of force. Therefore, the factor g does not appear when work against gravity is computed in the British system. The work required to lift a 2-lb stone 3 ft is

$$W = \underbrace{(2 \text{ lb})(3 \text{ ft})}_{F \cdot d} = 6 \text{ ft-lb}$$

We are interested in the case where the force $F(x)$ varies as the object moves from a to b along the x-axis. Eq. (1) does not apply directly, but we can break up the task into a large number of smaller tasks for which Eq. (1) gives a good approximation. Divide $[a, b]$ into N subintervals of length $\Delta x = (b - a)/N$ as in Figure 2 and let W_i be the work required to move the object from x_{i-1} to x_i. If Δx is small, then the force $F(x)$ is nearly constant on the interval $[x_{i-1}, x_i]$ with value $F(x_i)$, so $W_i \approx F(x_i)\Delta x$. Summing the contributions, we obtain

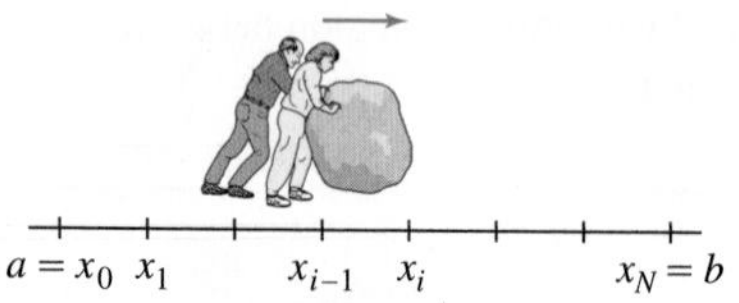

FIGURE 2 The work to move an object from x_{i-1} to x_i is approximately $F(x_i)\Delta x$.

$$W = \sum_{i=1}^{N} W_i \approx \underbrace{\sum_{i=1}^{N} F(x_i)\Delta x}_{\text{Right-endpoint approximation}}$$

The sum on the right is a right-endpoint approximation that converges to $\int_a^b F(x)\,dx$. This leads to the following definition.

DEFINITION Work The work performed in moving an object along the x-axis from a to b by applying a force of magnitude $F(x)$ is

$$W = \int_a^b F(x)\,dx \qquad (2)$$

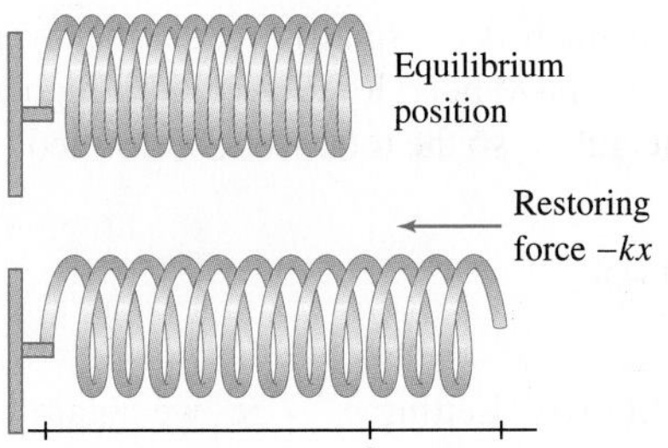

FIGURE 3 Hooke's Law.

Hooke's Law is named after the English scientist, inventor, and architect Robert Hooke (1635–1703), who made important discoveries in physics, astronomy, chemistry, and biology. He was a pioneer in the use of the microscope to study organisms. Unfortunately, Hooke was involved in several bitter disputes with other scientists, most notably with his contemporary Isaac Newton. Newton was furious when Hooke criticized his work on optics. Later, Hooke told Newton that he believed Kepler's Laws would follow from an inverse square law of gravitation, but Newton refused to acknowledge Hooke's contributions in his masterwork Principia. *Shortly before his death in 1955, Albert Einstein commented on Newton's behavior: "That, alas, is vanity. You find it in so many scientists... it has always hurt me to think that Galileo did not acknowledge the work of Kepler".*

One typical calculation involves finding the work required to stretch a spring. Assume that the free end of the spring has position $x = 0$ at equilibrium, when no force is acting (Figure 3). According to **Hooke's Law**, when the spring is stretched (or compressed) to position x, it exerts a restoring force of magnitude $-kx$ in the opposite direction, where k is the **spring constant**. If we want to stretch the spring further, we must apply a force $F(x) = kx$ to counteract the force exerted by the spring.

■ EXAMPLE 1 Hooke's Law Assuming a spring constant of $k = 400$ N/m, find the work required to

(a) Stretch the spring 10 cm beyond equilibrium.

(b) Compress the spring 2 cm more when it is already compressed 3 cm.

Solution A force $F(x) = 400x$ N is required to stretch the spring (with x in meters). Note that centimeters must be converted to meters.

(a) The work required to stretch the spring 10 cm (0.1 m) beyond equilibrium is

$$W = \int_0^{0.1} 400x\,dx = 200x^2\Big|_0^{0.1} = 2\text{ J}$$

(b) If the spring is at position $x = -3$ cm, then the work W required to compress it further to $x = -5$ cm is

$$W = \int_{-0.03}^{-0.05} 400x\,dx = 200x^2\Big|_{-0.03}^{-0.05} = 0.5 - 0.18 = 0.32\text{ J}$$

Observe that we integrate from the starting point $x = -0.03$ to the ending point $x = -0.05$ (even though the lower limit of the integral is larger than the upper limit in this case). ■

In the next two examples, we are not moving a single object through a fixed distance, so we cannot apply Eq. (2). Rather, each thin layer of the object is moved through a different distance. The work performed is computed by "summing" (i.e., *integrating*) the work performed on the thin layers.

On the earth's surface, work against gravity is equal to the force mg times the vertical distance through which the object is lifted. No work against gravity is done when an object is moved sideways.

■ EXAMPLE 2 Building a Cement Column Compute the work (against gravity) required to build a cement column of height 5 m and square base of side 2 m. Assume that cement has density 1500 kg/m^3.

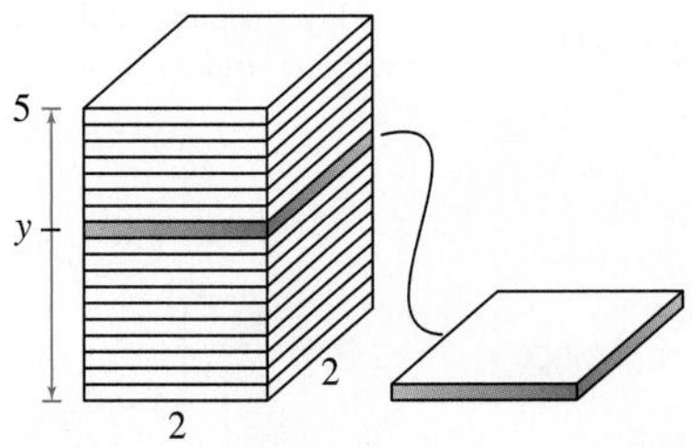

FIGURE 4 Total work is the sum of the work performed on each layer of the column.

Solution Think of the column as a stack of n thin layers of width $\Delta y = 5/n$. The work consists of lifting up these layers and placing them on the stack (Figure 4), but the work performed on a given layer depends on how high we lift it. First, let us compute the gravitational force on a thin layer of width Δy:

$$\begin{aligned} &\text{Volume of layer} = \text{area} \times \text{width} &&= 4\Delta y\text{ m}^3\\ &\text{Mass of layer} \quad = \text{density} \times \text{volume} &&= 1500 \cdot 4\Delta y\text{ kg}\\ &\text{Force on layer} \quad = g \times \text{mass} &&= 9.8 \cdot 1500 \cdot 4\Delta y = 58{,}800\,\Delta y\text{ N}\end{aligned}$$

The work performed in lifting this layer to height y is equal to the force times the distance y, which is $(58{,}800\Delta y)y$. Setting $L(y) = 58{,}800y$, we have

Work lifting layer to height $y \approx (58{,}800\Delta y)y = L(y)\Delta y$

This is only an approximation (although a very good one if Δy is small) because the layer has nonzero width and the cement particles at the top have been lifted a little bit higher than those at the bottom. The ith layer is lifted to height y_i, so the total work performed is

$$W \approx \sum_{i=1}^{n} L(y_i)\,\Delta y$$

This sum is a right-endpoint approximation to $\int_0^5 L(y)\,dy$. Letting $n \to \infty$, we obtain

$$W = \int_0^5 L(y)\,dy = \int_0^5 58{,}800y\,dy = 58{,}800\frac{y^2}{2}\bigg|_0^5 = 735{,}000 \text{ J}$$ ■

In Examples 2 and 3, the work performed on a thin layer is written

$$L(y)\Delta y$$

When we take the sum and let Δy approach zero, we obtain the integral of $L(y)$. Symbolically, the Δy "becomes" the dy of the integral. Note that

$$L(y) = g \times \textit{density} \times A(y) \times \textit{(vertical distance lifted)}$$

where $A(y)$ is the area of the cross section.

■ **EXAMPLE 3 Pumping Water out of a Tank** A spherical tank of radius R meters is filled with water. Calculate the work W performed (against gravity) in pumping out the water through a small hole at the top. The density of water is 1000 kg/m^3.

Solution The first step, as in the previous example, is to compute the work against gravity performed on a thin layer of water of width Δy. We place the origin of our coordinate system at the center of the sphere because this leads to a simple formula for the radius r of the cross section at height y (Figure 5).

***Step 1.* Compute work performed on a layer.**

Figure 5 shows that the cross section at height y is a circle of radius $r = \sqrt{R^2 - y^2}$ and area $A(y) = \pi r^2 = \pi(R^2 - y^2)$. A thin layer has volume $A(y)\Delta y$, and to lift it, we must exert a force against gravity equal to

$$\text{Force on layer} = g \times \overbrace{\text{density} \times A(y)\Delta y}^{\text{mass}} \approx (9.8)1000\pi(R^2 - y^2)\Delta y$$

The layer has to be lifted a vertical distance $R - y$, so

$$\text{Work on layer} \approx \overbrace{9800\pi(R^2 - y^2)\Delta y}^{\text{Force against gravity}} \times \overbrace{(R - y)}^{\text{Vertical distance lifted}} = L(y)\Delta y$$

where $L(y) = 9800\pi(R^3 - R^2y - Ry^2 + y^3)$.

***Step 2.* Compute total work.**

Now divide the sphere into N layers and let y_i be the height of the ith layer. The work performed on ith layer is approximately $L(y_i)\,\Delta y$, and therefore

$$W \approx \sum_{i=1}^{N} L(y_i)\,\Delta y$$

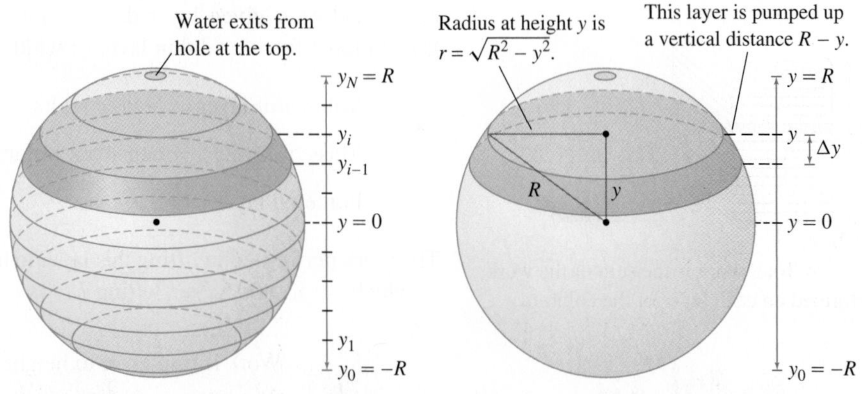

FIGURE 5 The sphere is divided into N thin layers.

This sum approaches the integral of $L(y)$ as $N \to \infty$ (that is, $\Delta y \to 0$), so

$$W = \int_{-R}^{R} L(y)\,dy = 9800\pi \int_{-R}^{R} (R^3 - R^2 y - R y^2 + y^3)\,dy$$

$$= 9800\pi \left(R^3 y - \frac{1}{2}R^2 y^2 - \frac{1}{3}R y^3 + \frac{1}{4}y^4 \right)\bigg|_{-R}^{R} = \frac{39{,}200\pi}{3} R^4 \text{ J}$$

Note that the integral extends from $-R$ to R because the y-coordinate along the sphere varies from $-R$ to R. ■

A liter of gasoline has an energy content of approximately 3.4×10^7 joules. The previous example shows that the work required to pump water out of a sphere of radius $R = 5$ meters is

$$W = \left(\frac{39{,}200\pi}{3} \right) 5^4 \approx 2.6 \times 10^7 \text{ J}$$

or the energy content of roughly three-fourths of a liter of gasoline.

6.5 SUMMARY

- Work performed to move an object:

$$\text{Constant force:} \quad W = F \cdot d, \qquad \text{Variable force:} \quad W = \int_a^b F(x)\,dx$$

- Hooke's Law: A spring stretched x units past equilibrium exerts a restoring force of magnitude $-kx$. A force $F(x) = kx$ is required to stretch the spring further.
- To compute work against gravity by decomposing an object into N thin layers of thickness Δy, express the work performed on a thin layer as $L(y)\Delta y$, where

$$L(y) = g \times \text{density} \times A(y) \times (\text{vertical distance lifted})$$

The total work performed is $W = \int_a^b L(y)\,dy$.

6.5 EXERCISES

Preliminary Questions

1. Why is integration needed to compute the work performed in stretching a spring?

2. Why is integration needed to compute the work performed in pumping water out of a tank but not to compute the work performed in lifting up the tank?

3. Which of the following represents the work required to stretch a spring (with spring constant k) a distance x beyond its equilibrium position: kx, $-kx$, $\frac{1}{2}mk^2$, $\frac{1}{2}kx^2$, or $\frac{1}{2}mx^2$?

Exercises

1. How much work is done raising a 4-kg mass to a height of 16 m above ground?

2. How much work is done raising a 4-lb mass to a height of 16 ft above ground?

In Exercises 3–6, compute the work (in joules) required to stretch or compress a spring as indicated, assuming a spring constant of $k = 800$ N/m.

3. Stretching from equilibrium to 12 cm past equilibrium

4. Compressing from equilibrium to 4 cm past equilibrium

5. Stretching from 5 cm to 15 cm past equilibrium

6. Compressing 4 cm more when it is already compressed 5 cm

7. If 5 J of work are needed to stretch a spring 10 cm beyond equilibrium, how much work is required to stretch it 15 cm beyond equilibrium?

8. To create images of samples at the molecular level, atomic force microscopes use silicon micro-cantilevers that obey Hooke's Law $F(x) = -kx$, where x is the distance through which the tip is deflected (Figure 6). Suppose that 10^{-17} J of work are required to deflect the tip a distance 10^{-8} m. Find the deflection if a force of 10^{-9} N is applied to the tip.

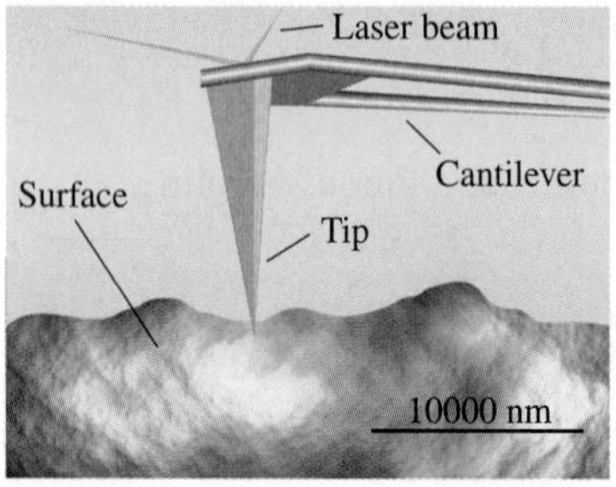

FIGURE 6

9. A spring obeys a force law $F(x) = -kx^{1.1}$ with $k = 100$ N/m. Find the work required to stretch a spring 0.3 m past equilibrium.

10. Show that the work required to stretch a spring from position a to position b is $\frac{1}{2}k(b^2 - a^2)$, where k is the spring constant. How do you interpret the negative work obtained when $|b| < |a|$?

In Exercises 11–14, use the method of Examples 2 and 3 to calculate the work against gravity required to build the structure out of a lightweight material of density 600 kg/m^3.

11. Box of height 3 m and square base of side 2 m

12. Cylindrical column of height 4 m and radius 0.8 m

13. Right circular cone of height 4 m and base of radius 1.2 m

14. Hemisphere of radius 0.8 m

15. Built around 2600 BCE, the Great Pyramid of Giza in Egypt (Figure 7) is 146 m high and has a square base of side 230 m. Find the work (against gravity) required to build the pyramid if the density of the stone is estimated at 2000 kg/m^3.

FIGURE 7 The Great Pyramid in Giza, Egypt.

16. Calculate the work (against gravity) required to build a box of height 3 m and square base of side 2 m out of material of variable density, assuming that the density at height y is $f(y) = 1000 - 100y$ kg/m^3.

In Exercises 17–22, calculate the work (in joules) required to pump all of the water out of a full tank. Distances are in meters, and the density of water is 1000 kg/m^3.

17. Rectangular tank in Figure 8; water exits from a small hole at the top.

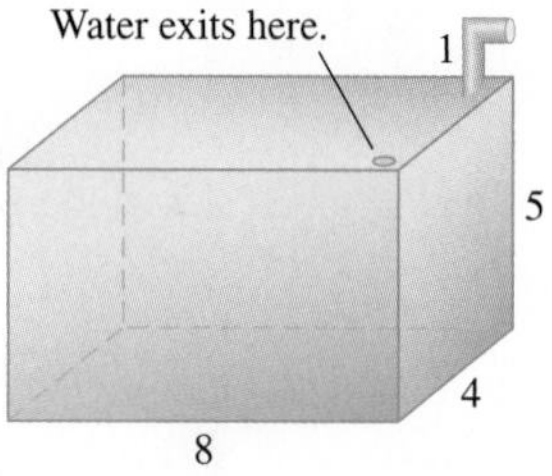

FIGURE 8

18. Rectangular tank in Figure 8; water exits through the spout.

19. Hemisphere in Figure 9; water exits through the spout.

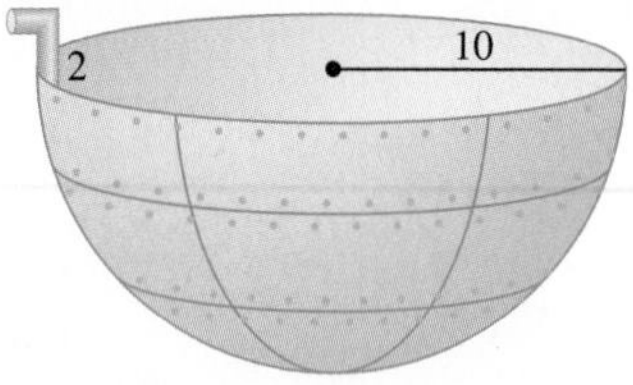

FIGURE 9

20. Conical tank in Figure 10; water exits through the spout.

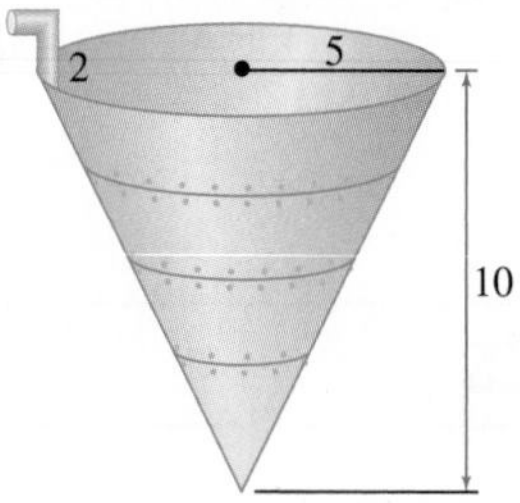

FIGURE 10

21. Horizontal cylinder in Figure 11; water exits from a small hole at the top. *Hint:* Evaluate the integral by interpreting part of it as the area of a circle.

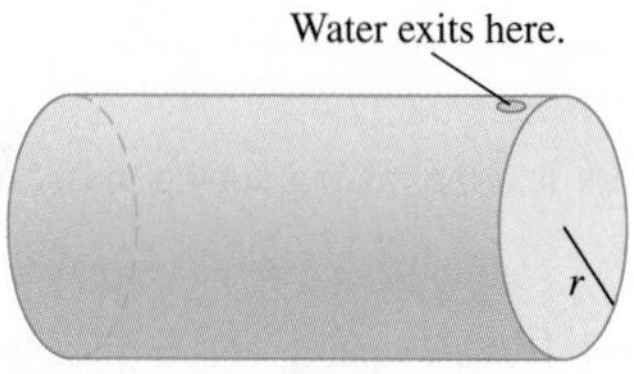

FIGURE 11

22. Trough in Figure 12; water exits by pouring over the sides.

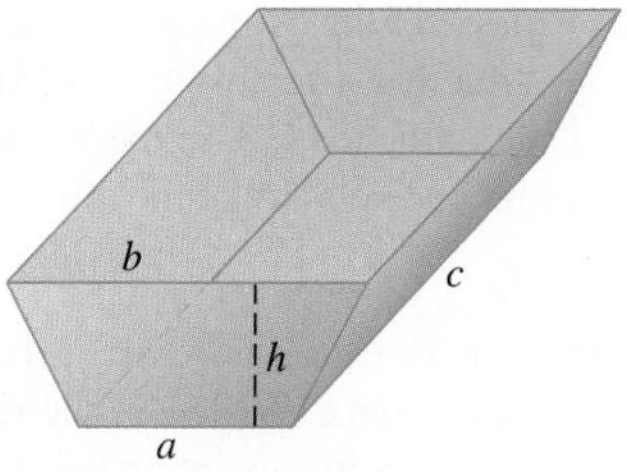

FIGURE 12

23. Find the work W required to empty the tank in Figure 8 through the hole at the top if the tank is half full of water.

24. Assume the tank in Figure 8 is full of water and let W be the work required to pump out half of the water through the hole at the top. Do you expect W to equal the work computed in Exercise 23? Explain and then compute W.

25. Assume the tank in Figure 10 is full. Find the work required to pump out half of the water. *Hint:* First, determine the level H at which the water remaining in the tank is equal to one-half the total capacity of the tank.

26. Assume that the tank in Figure 10 is full.

(a) Calculate the work $F(y)$ required to pump out water until the water level has reached level y.

(b) CAS Plot $F(y)$.

(c) What is the significance of $F'(y)$ as a rate of change?

(d) CAS If your goal is to pump out all of the water, at which water level y_0 will half of the work be done?

27. Calculate the work required to lift a 10-m chain over the side of a building (Figure 13) Assume that the chain has a density of 8 kg/m. *Hint:* Break up the chain into N segments, estimate the work performed on a segment, and compute the limit as $N \to \infty$ as an integral.

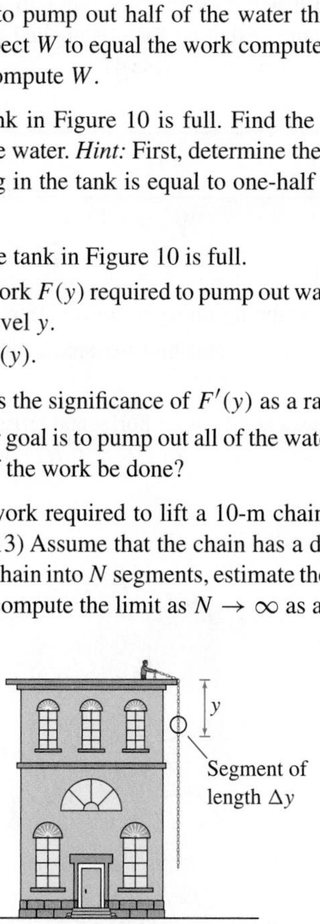

FIGURE 13 The small segment of the chain of length Δy located y meters from the top is lifted through a vertical distance y.

28. How much work is done lifting a 3-m chain over the side of a building if the chain has mass density 4 kg/m?

29. A 6-m chain has mass 18 kg. Find the work required to lift the chain over the side of a building.

30. A 10-m chain with mass density 4 kg/m is initially coiled on the ground. How much work is performed in lifting the chain so that it is fully extended (and one end touches the ground)?

31. How much work is done lifting a 12-m chain that has mass density 3 kg/m (initially coiled on the ground) so that its top end is 10 m above the ground?

32. A 500-kg wrecking ball hangs from a 12-m cable of density 15 kg/m attached to a crane. Calculate the work done if the crane lifts the ball from ground level to 12 m in the air by drawing in the cable.

33. Calculate the work required to lift a 3-m chain over the side of a building if the chain has variable density of $\rho(x) = x^2 - 3x + 10$ kg/m for $0 \le x \le 3$.

34. A 3-m chain with linear mass density $\rho(x) = 2x(4 - x)$ kg/m lies on the ground. Calculate the work required to lift the chain so that its bottom is 2 m above ground.

Exercises 35–37: The gravitational force between two objects of mass m and M, separated by a distance r, has magnitude GMm/r^2, where $G = 6.67 \times 10^{-11}\ \mathrm{m^3 kg^{-1} s^{-1}}$.

35. Show that if two objects of mass M and m are separated by a distance r_1, then the work required to increase the separation to a distance r_2 is equal to $W = GMm(r_1^{-1} - r_2^{-1})$.

36. Use the result of Exercise 35 to calculate the work required to place a 2000-kg satellite in an orbit 1200 km above the surface of the earth. Assume that the earth is a sphere of radius $R_e = 6.37 \times 10^6$ m and mass $M_e = 5.98 \times 10^{24}$ kg. Treat the satellite as a point mass.

37. Use the result of Exercise 35 to compute the work required to move a 1500-kg satellite from an orbit 1000 to an orbit 1500 km above the surface of the earth.

38. The pressure P and volume V of the gas in a cylinder of length 0.8 meters and radius 0.2 meters, with a movable piston, are related by $PV^{1.4} = k$, where k is a constant (Figure 14). When the piston is fully extended, the gas pressure is 2000 kilopascals (one kilopascal is 10^3 newtons per square meter).

(a) Calculate k.

(b) The force on the piston is PA, where A is the piston's area. Calculate the force as a function of the length x of the column of gas.

(c) Calculate the work required to compress the gas column from 1.5 m to 1.2 m.

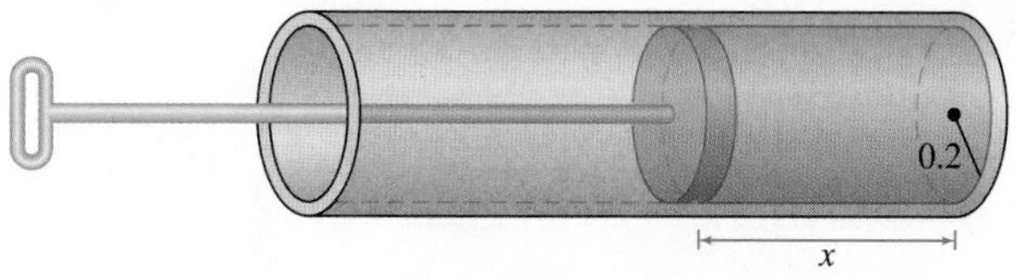

FIGURE 14 Gas in a cylinder with a piston.

Further Insights and Challenges

39. Work-Energy Theorem An object of mass m moves from x_1 to x_2 during the time interval $[t_1, t_2]$ due to a force $F(x)$ acting in the direction of motion. Let $x(t)$, $v(t)$, and $a(t)$ be the position, velocity, and acceleration at time t. The object's kinetic energy is $\text{KE} = \frac{1}{2}mv^2$.

(a) Use the change-of-variables formula to show that the work performed is equal to

$$W = \int_{x_1}^{x_2} F(x)\,dx = \int_{t_1}^{t_2} F(x(t))v(t)\,dt$$

(b) Use Newton's Second Law, $F(x(t)) = ma(t)$, to show that

$$\frac{d}{dt}\left(\frac{1}{2}mv(t)^2\right) = F(x(t))v(t)$$

(c) Use the FTC to prove the Work-Energy Theorem: The change in kinetic energy during the time interval $[t_1, t_2]$ is equal to the work performed.

40. A model train of mass 0.5 kg is placed at one end of a straight 3-m electric track. Assume that a force $F(x) = (3x - x^2)$ N acts on the train at distance x along the track. Use the Work-Energy Theorem (Exercise 39) to determine the velocity of the train when it reaches the end of the track.

41. With what initial velocity v_0 must we fire a rocket so it attains a maximum height r above the earth? *Hint:* Use the results of Exercises 35 and 39. As the rocket reaches its maximum height, its KE decreases from $\frac{1}{2}mv_0^2$ to zero.

42. With what initial velocity must we fire a rocket so it attains a maximum height of $r = 20$ km above the surface of the earth?

43. Calculate **escape velocity,** the minimum initial velocity of an object to ensure that it will continue traveling into space and never fall back to earth (assuming that no force is applied after takeoff). *Hint:* Take the limit as $r \to \infty$ in Exercise 41.

CHAPTER REVIEW EXERCISES

1. Compute the area of the region in Figure 1(A) enclosed by $y = 2 - x^2$ and $y = -2$.

2. Compute the area of the region in Figure 1(B) enclosed by $y = 2 - x^2$ and $y = x$.

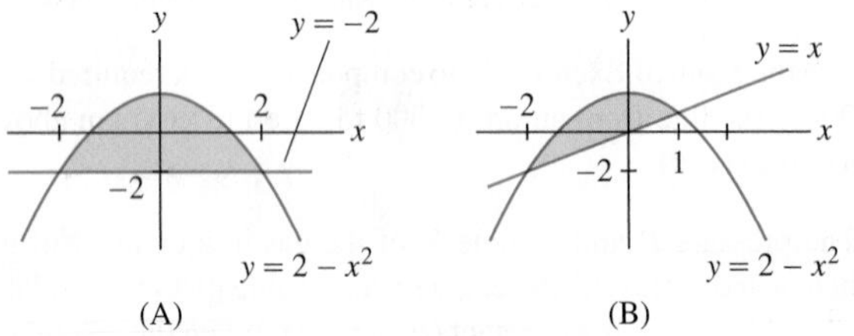

FIGURE 1

In Exercises 3–12, find the area of the region enclosed by the graphs of the functions.

3. $y = x^3 - 2x^2 + x, \quad y = x^2 - x$

4. $y = x^2 + 2x, \quad y = x^2 - 1, \quad h(x) = x^2 + x - 2$

5. $x = 4y, \quad x = 24 - 8y, \quad y = 0$

6. $x = y^2 - 9, \quad x = 15 - 2y$

7. $y = 4 - x^2, \quad y = 3x, \quad y = 4$

8. GU $x = \frac{1}{2}y, \quad x = y\sqrt{1 - y^2}, \quad 0 \le y \le 1$

9. $y = \sin x, \quad y = \cos x, \quad 0 \le x \le \frac{5\pi}{4}$

10. $f(x) = \sin x, \quad g(x) = \sin 2x, \quad \frac{\pi}{3} \le x \le \pi$

11. $y = \sec^2\left(\frac{\pi x}{4}\right), \quad y = \sec^2\left(\frac{\pi x}{8}\right), \quad 0 \le x \le 1$

12. $y = \frac{x}{\sqrt{x^2 + 1}}, \quad y = \frac{x}{\sqrt{x^2 + 4}}, \quad -1 \le x \le 1$

13. GU Use a graphing utility to locate the points of intersection of $y = x^2$ and $y = \cos x$, and find the area between the two curves (approximately).

14. Figure 2 shows a solid whose horizontal cross section at height y is a circle of radius $(1 + y)^{-2}$ for $0 \le y \le H$. Find the volume of the solid.

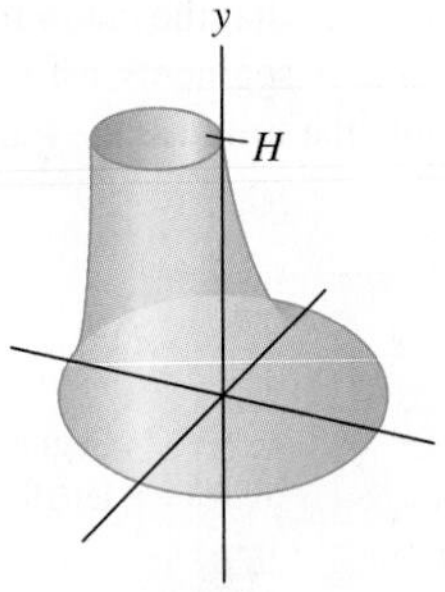

FIGURE 2

15. The base of a solid is the unit circle $x^2 + y^2 = 1$, and its cross sections perpendicular to the x-axis are rectangles of height 4. Find its volume.

16. The base of a solid is the triangle bounded by the axes and the line $2x + 3y = 12$, and its cross sections perpendicular to the y-axis have area $A(y) = (y + 2)$. Find its volume.

17. Find the total mass of a rod of length 1.2 m with linear density $\rho(x) = (1 + 2x + \frac{2}{9}x^3)$ kg/m.

18. Find the flow rate (in the correct units) through a pipe of diameter 6 cm if the velocity of fluid particles at a distance r from the center of the pipe is $v(r) = (3 - r)$ cm/s.

In Exercises 19–24, find the average value of the function over the interval.

19. $f(x) = x^3 - 2x + 2, \quad [-1, 2]$

20. $f(x) = |x|, \quad [-4, 4]$

21. $f(x) = (x + 1)(x^2 + 2x + 1)^{4/5}, \quad [0, 4]$

22. $f(x) = |x^2 - 1|, \quad [0, 4]$

23. $f(x) = \sqrt{9 - x^2}, \quad [0, 3]$ *Hint:* Use geometry to evaluate the integral.

24. $f(x) = x[x], \quad [0, 3]$, where $[x]$ is the greatest integer function.

25. Find $\int_2^5 g(t)\,dt$ if the average value of $g(t)$ on $[2, 5]$ is 9.

26. The average value of $R(x)$ over $[0, x]$ is equal to x for all x. Use the FTC to determine $R(x)$.

27. Use the Washer Method to find the volume obtained by rotating the region in Figure 3 about the x-axis.

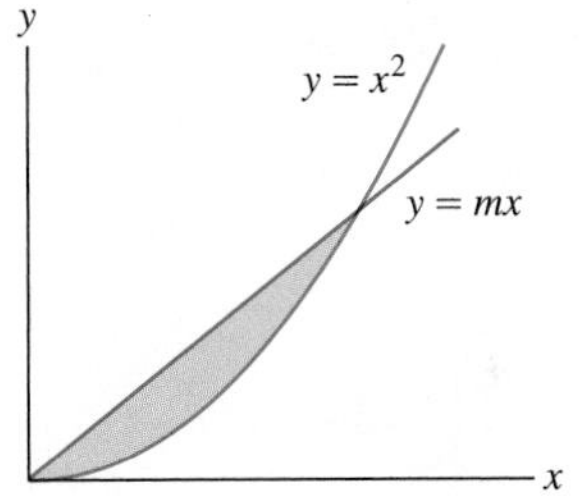

FIGURE 3

28. Use the Shell Method to find the volume obtained by rotating the region in Figure 3 about the x-axis.

In Exercises 29–40, use any method to find the volume of the solid obtained by rotating the region enclosed by the curves about the given axis.

29. $y = x^2 + 2, \quad y = x + 4, \quad x$-axis

30. $y = x^2 + 6, \quad y = 8x - 1, \quad y$-axis

31. $x = y^2 - 3, \quad x = 2y, \quad$ axis $y = 4$

32. $y = 2x, \quad y = 0, \quad x = 8, \quad$ axis $x = -3$

33. $y = x^2 - 1, \quad y = 2x - 1, \quad$ axis $x = -2$

34. $y = x^2 - 1, \quad y = 2x - 1, \quad$ axis $y = 4$

35. $y = -x^2 + 4x - 3, \quad y = 0, \quad$ axis $y = -1$

36. $y = -x^2 + 4x - 3, \quad y = 0, \quad$ axis $x = 4$

37. $x = 4y - y^3, \quad x = 0, \quad y \geq 0, \quad x$-axis

38. $y^2 = x^{-1}, \quad x = 1, \quad x = 3, \quad$ axis $y = -3$

39. $y = \cos(x^2), \quad y = 0, \quad 0 \leq x \leq \sqrt{\pi/2}, \quad y$-axis

40. $y = \sec x, \quad y = \csc x, \quad y = 0, \quad x = 0, \quad x = \frac{\pi}{2}, \quad x$-axis

In Exercises 41–44, find the volume obtained by rotating the region about the given axis. The regions refer to the graph of the hyperbola $y^2 - x^2 = 1$ in Figure 4.

41. The shaded region between the upper branch of the hyperbola and the x-axis for $-c \leq x \leq c$, about the x-axis.

42. The region between the upper branch of the hyperbola and the x-axis for $0 \leq x \leq c$, about the y-axis.

43. The region between the upper branch of the hyperbola and the line $y = x$ for $0 \leq x \leq c$, about the x-axis.

44. The region between the upper branch of the hyperbola and $y = 2$, about the y-axis.

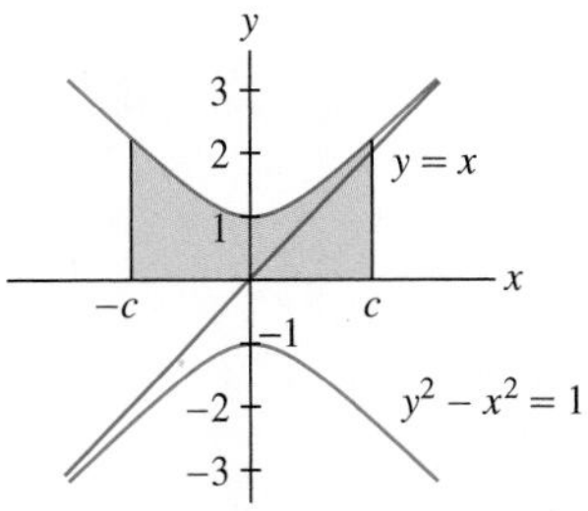

FIGURE 4

45. Let R be the intersection of the circles of radius 1 centered at $(1, 0)$ and $(0, 1)$. Express as an integral (but do not evaluate): **(a)** the area of R and **(b)** the volume of revolution of R about the x-axis.

46. Let $a > 0$. Show that the volume obtained when the region between $y = a\sqrt{x - ax^2}$ and the x-axis is rotated about the x-axis is independent of the constant a.

47. If 12 J of work are needed to stretch a spring 20 cm beyond equilibrium, how much work is required to compress it 6 cm beyond equilibrium?

48. A spring whose equilibrium length is 15 cm exerts a force of 50 N when it is stretched to 20 cm. Find the work required to stretch the spring from 22 to 24 cm.

49. If 18 ft-lb of work are needed to stretch a spring 1.5 ft beyond equilibrium, how far will the spring stretch if a 12-lb weight is attached to its end?

50. Let W be the work (against the sun's gravitational force) required to transport an 80-kg person from Earth to Mars when the two planets are aligned with the sun at their minimal distance of 55.7×10^6 km. Use Newton's Universal Law of Gravity (see Exercises 35–37 in Section 6.5) to express W as an integral and evaluate it. The sun has mass $M_s = 1.99 \times 10^{30}$ kg, and the distance from the sun to the earth is 149.6×10^6 km.

In Exercises 51 and 52, water is pumped into a spherical tank of radius 2 m *from a source located* 1 m *below a hole at the bottom (Figure 5). The density of water is* 1000 kg/m^3.

51. Calculate the work required to fill the tank.

52. Calculate the work $F(h)$ required to fill the tank to level h meters in the sphere.

53. A tank of mass 20 kg containing 100 kg of water (density 1000 kg/m^3) is raised vertically at a constant speed of 100 m/min for one minute, during which time it leaks water at a rate of 40 kg/min. Calculate the total work performed in raising the container.

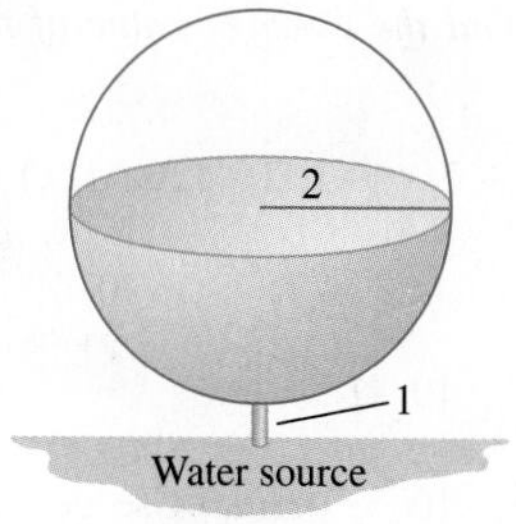

FIGURE 5

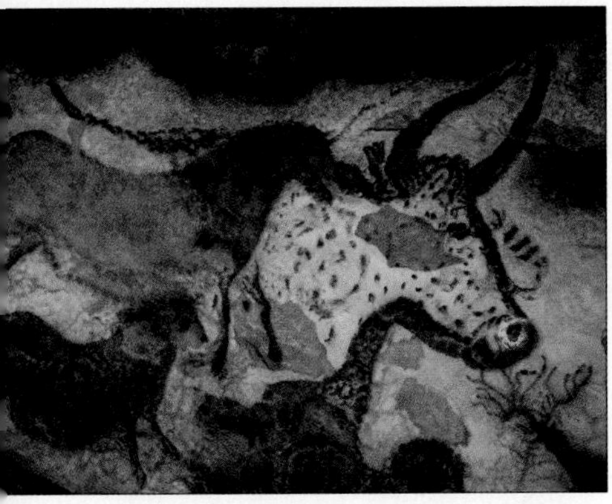

Detail of bison and other animals from a replica of the Lascaux cave mural.

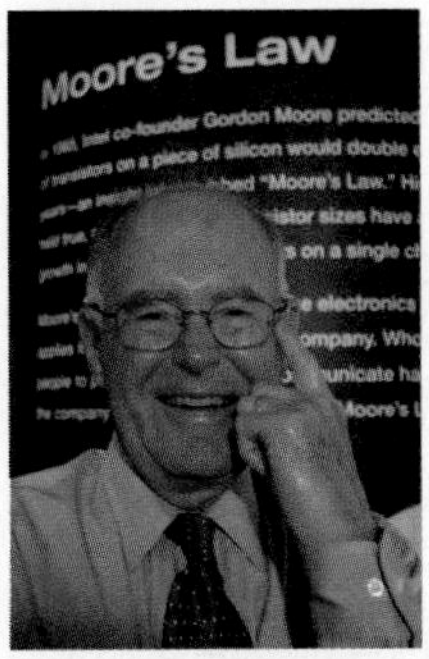

Gordon Moore (1929–). Moore, who later became chairman of Intel Corporation, predicted that in the decades following 1965, the number of transistors per integrated circuit would grow "exponentially." This prediction has held up for nearly five decades and may well continue for several more years. Moore has said, "Moore's Law is a term that got applied to a curve I plotted in the mid-sixties showing the increase in complexity of integrated circuits versus time. It's been expanded to include a lot more than that, and I'm happy to take credit for all of it."

7 EXPONENTIAL FUNCTIONS

This chapter is devoted to exponential functions and their applications. These functions are used to model a remarkably wide range of phenomena, such as radioactive decay, population growth, interest rates, atmospheric pressure, and the diffusion of molecules across a cell membrane. Calculus gives us insight into why exponential functions play an important role in so many different situations. The key, it turns out, is the relation between an exponential function and its derivative.

7.1 Derivative of $f(x) = b^x$ and the Number e

An **exponential function** is a function of the form $f(x) = b^x$, where $b > 0$ and $b \neq 1$. The number b is called the **base**. Some examples are 2^x, $(1.4)^x$, and 10^x. We exclude the case $b = 1$ because $f(x) = 1^x$ is a constant function. Calculators give good decimal approximations to values of exponential functions:

$$2^4 = 16, \quad 2^{-3} = 0.125, \quad (1.4)^3 = 2.744, \quad 10^{4.6} \approx 39{,}810.717$$

Three properties of exponential functions should be singled out from the start (see Figure 1 for the cases $b = 2$ and $b = \frac{1}{2}$):

- *Exponential functions are positive*: $b^x > 0$ for all x.
- The *range* of $f(x) = b^x$ is the set of all positive real numbers.
- $f(x) = b^x$ is increasing if $b > 1$ and decreasing if $0 < b < 1$.

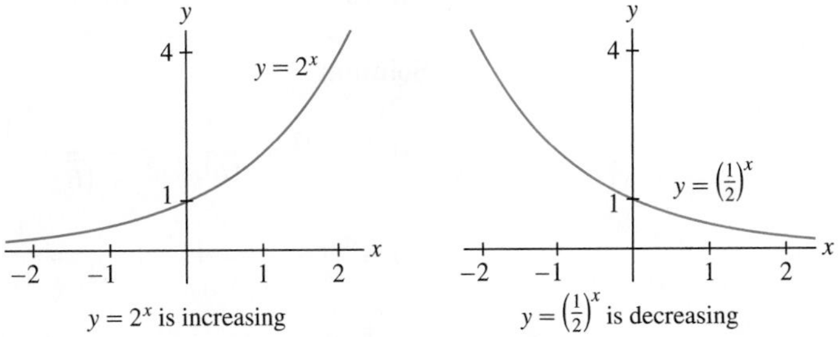

FIGURE 1

If $b > 1$, the exponential function $f(x) = b^x$ is not merely increasing but is, in a certain sense, rapidly increasing. Although the term "rapid increase" is perhaps subjective, the following precise statement is true: $f(x) = b^x$ increases more rapidly than the power function x^n for all n (we will prove this in Section 7.7). For example, Figure 2 shows that $f(x) = 3^x$ eventually overtakes and increases faster than the power functions x^3, x^4, and x^5. Table 1 compares 3^x and x^5.

We now review the laws of exponents. The most important law is

$$\boxed{b^x b^y = b^{x+y}}$$

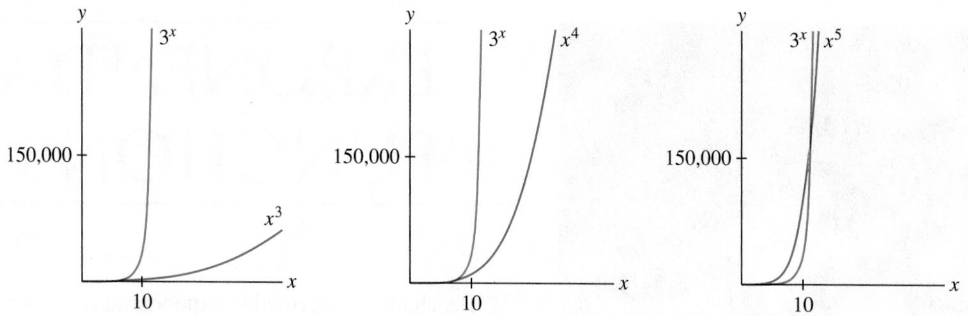

FIGURE 2 Comparison of 3^x and power functions.

In other words, *under multiplication, the exponents add*, provided that the bases are the same. This law does not apply to a product such as $3^2 \cdot 5^4$.

TABLE 1

x	x^5	3^x
1	1	3
5	3125	243
10	100,000	59,049
15	759,375	14,348,907
25	9,765,625	847,288,609,443

Be sure you are familiar with the laws of exponents. They are used throughout this text.

Laws of Exponents ($b > 0$)

	Rule	Example
Exponent zero	$b^0 = 1$	
Products	$b^x b^y = b^{x+y}$	$2^5 \cdot 2^3 = 2^{5+3} = 2^8$
Quotients	$\dfrac{b^x}{b^y} = b^{x-y}$	$\dfrac{4^7}{4^2} = 4^{7-2} = 4^5$
Negative exponents	$b^{-x} = \dfrac{1}{b^x}$	$3^{-4} = \dfrac{1}{3^4} = \dfrac{1}{81}$
Power to a power	$\left(b^x\right)^y = b^{xy}$	$\left(3^2\right)^4 = 3^{2(4)} = 3^8$
Roots	$b^{1/n} = \sqrt[n]{b}$	$5^{1/2} = \sqrt{5}$

EXAMPLE 1 Rewrite as a whole number or fraction:

(a) $16^{-1/2}$ **(b)** $27^{2/3}$ **(c)** $4^{16} \cdot 4^{-18}$ **(d)** $\dfrac{9^3}{3^7}$

Solution

(a) $16^{-1/2} = \dfrac{1}{16^{1/2}} = \dfrac{1}{\sqrt{16}} = \dfrac{1}{4}$

(b) $27^{2/3} = \left(27^{1/3}\right)^2 = 3^2 = 9$

(c) $4^{16} \cdot 4^{-18} = 4^{-2} = \dfrac{1}{4^2} = \dfrac{1}{16}$

(d) $\dfrac{9^3}{3^7} = \dfrac{\left(3^2\right)^3}{3^7} = \dfrac{3^6}{3^7} = 3^{-1} = \dfrac{1}{3}$

In the next example, we use the fact that $f(x) = b^x$ is one-to-one. In other words, if $b^x = b^y$, then $x = y$.

EXAMPLE 2 Solve for the unknown:

(a) $2^{3x+1} = 2^5$ **(b)** $b^3 = 5^6$ **(c)** $7^{t+1} = \left(\dfrac{1}{7}\right)^{2t}$

Solution

(a) If $2^{3x+1} = 2^5$, then $3x + 1 = 5$ and thus $x = \frac{4}{3}$.

(b) Raise both sides of $b^3 = 5^6$ to the $\frac{1}{3}$ power. By the "power to a power" rule,

$$b = \left(b^3\right)^{1/3} = \left(5^6\right)^{1/3} = 5^{6/3} = 5^2 = 25$$

(c) Since $\frac{1}{7} = 7^{-1}$, the right-hand side of the equation is $\left(\frac{1}{7}\right)^{2t} = (7^{-1})^{2t} = 7^{-2t}$. The equation becomes $7^{t+1} = 7^{-2t}$. Therefore, $t + 1 = -2t$, or $t = -\frac{1}{3}$. ■

Derivative of $f(x) = b^x$

At this point, it is natural to ask: What is the derivative of $f(x) = b^x$? Our rules of differentiation are of no help because b^x is neither a product, quotient, nor composite of functions with known derivatives. We must go back to the limit definition of the derivative. The difference quotient (for $h \neq 0$) is

$$\frac{f(x+h) - f(x)}{h} = \frac{b^{x+h} - b^x}{h} = \frac{b^x b^h - b^x}{h} = \frac{b^x(b^h - 1)}{h}$$

We shall take for granted that $f(x) = b^x$ is differentiable. Although the proof of this fact is somewhat technical, it is plausible because the graph of $y = b^x$ appears smooth and without corners.

Now take the limit as $h \to 0$. The factor b^x does not depend on h, so it may be taken outside the limit:

$$\frac{d}{dx}b^x = \lim_{h\to 0} \frac{b^{x+h} - b^x}{h} = \lim_{h\to 0} \frac{b^x(b^h - 1)}{h} = b^x \lim_{h\to 0}\left(\frac{b^h - 1}{h}\right)$$

This last limit (which exists because b^x is differentiable) does not depend on x. We denote its value by $m(b)$:

$$m(b) = \lim_{h\to 0}\left(\frac{b^h - 1}{h}\right)$$

What we have shown, then, is that *the derivative of b^x is proportional to b^x*:

$$\boxed{\frac{d}{dx}b^x = m(b)\, b^x} \qquad \boxed{1}$$

What is the factor $m(b)$? We cannot determine its exact value at this point (in Section 7.3, we will learn that $m(b)$ is equal to $\ln b$, the natural logarithm of b). To proceed further, let's investigate $m(b)$ numerically.

■ **EXAMPLE 3** Estimate $m(b)$ numerically for $b = 2, 2.5, 3$, and 10.

Solution We create a table of values of difference quotients to estimate $m(b)$:

h	$\frac{2^h - 1}{h}$	$\frac{(2.5)^h - 1}{h}$	$\frac{3^h - 1}{h}$	$\frac{10^h - 1}{h}$
0.01	0.69556	0.92050	1.10467	2.32930
0.001	0.69339	0.91671	1.09921	2.30524
0.0001	0.69317	0.91633	1.09867	2.30285
0.00001	0.69315	0.916295	1.09861	2.30261
	$m(2) \approx 0.69$	$m(2.5) \approx 0.92$	$m(3) \approx 1.10$	$m(10) \approx 2.30$

■

These computations suggest that $m(b)$ is an increasing function of b. In fact, it can be shown that $m(b)$ is both increasing and continuous as a function of b (we shall take these facts for granted). Then, since $m(2.5) \approx 0.92$ and $m(3) \approx 1.10$, there exists a unique number b *between* 2.5 and 3 such that $m(b) = 1$. This is the number e, whose value is approximately 2.718.

Using infinite series (Exercise 87 in Section 10.7), we can show that e is irrational and we can compute its value to any desired accuracy.

Whenever we refer to the *exponential function without specifying the base, the reference is to* $f(x) = e^x$. *In many books,* e^x *is denoted* $\exp(x)$.

Because e is defined by the property $m(e) = 1$, Eq. (1) tells us that $(e^x)' = e^x$. In other words, e^x *is equal to its own derivative.*

The Number e There is a unique positive real number e with the property:

$$\frac{d}{dx}e^x = e^x \qquad (2)$$

The number e is irrational, with approximate value $e \approx 2.718$.

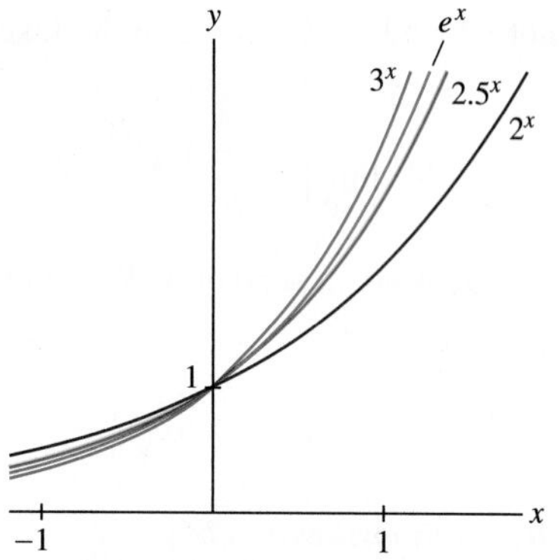

FIGURE 3 The tangent lines to $y = b^x$ at $x = 0$ grow steeper as b increases.

GRAPHICAL INSIGHT The graph of $f(x) = b^x$ passes through (0, 1) for all $b > 0$ because $b^0 = 1$. (Figure 3). The number $m(b)$ is simply the slope of the tangent line at $x = 0$:

$$\left.\frac{d}{dx}b^x\right|_{x=0} = m(b)\cdot b^0 = m(b)$$

These tangent lines become steeper as b increases and $b = e$ is the unique value for which the tangent line has slope 1. In Section 7.3, we will show that $m(b) = \ln b$, the natural logarithm of b.

CONCEPTUAL INSIGHT In some ways, the number e is "complicated". It has been computed to an accuracy of more than 100 billion digits, but it is irrational and it cannot be defined without using limits. To 20 places,

$$e = 2.71828182845904523536\ldots$$

However, the elegant formula $\frac{d}{dx}e^x = e^x$ shows that e is "simple" from the point of view of calculus and that e^x is simpler than the seemingly more natural exponential functions such as 2^x or 10^x.

Although written reference to the number π goes back more than 4000 years, mathematicians first became aware of the special role played by e in the seventeenth century. The notation e was introduced around 1730 by Leonhard Euler, who discovered many fundamental properties of this important number.

■ **EXAMPLE 4** Find the equation of the tangent line to the graph of $f(x) = 3e^x - 5x^2$ at $x = 2$.

Solution We compute both $f'(2)$ and $f(2)$:

$$f'(x) = \frac{d}{dx}(3e^x - 5x^2) = 3\frac{d}{dx}e^x - 5\frac{d}{dx}x^2 = 3e^x - 10x$$

$$f'(2) = 3e^2 - 10(2) \approx 2.17$$

$$f(2) = 3e^2 - 5(2^2) \approx 2.17$$

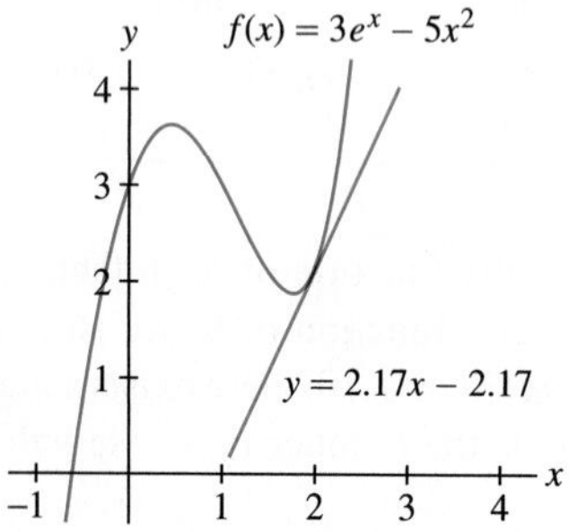

FIGURE 4

The equation of the tangent line is $y = f(2) + f'(2)(x - 2)$. Using these approximate values, we write the equation as (Figure 4)

$$y = 2.17 + 2.17(x - 2) \qquad \text{or} \qquad y = 2.17(x - 1)$$

■

■ **EXAMPLE 5** Calculate $f'(0)$, where $f(x) = e^x \cos x$.

Solution Use the Product Rule:

$$f'(x) = e^x \cdot (\cos x)' + \cos x \cdot (e^x)' = -e^x \sin x + \cos x \cdot e^x = e^x(\cos x - \sin x)$$

Then $f'(0) = e^0(1 - 0) = 1$. ■

To compute the derivative of a function of the form $e^{g(x)}$, write $e^{g(x)}$ as a composite $e^{g(x)} = f(g(x))$ where $f(u) = e^u$, and apply the Chain Rule:

$$\frac{d}{dx}\left(e^{g(x)}\right) = \left[f(g(x))\right]' = f'(g(x))g'(x) = e^{g(x)}g'(x) \quad [\text{since } f'(x) = e^x]$$

A special case is $(e^{kx+b})' = ke^{kx+b}$, where k and b are constants.

$$\frac{d}{dx}\left(e^{g(x)}\right) = g'(x)e^{g(x)}, \qquad \frac{d}{dx}\left(e^{kx+b}\right) = ke^{kx+b} \quad (k, b \text{ constants}) \tag{3}$$

■ **EXAMPLE 6** Differentiate:

(a) $f(x) = e^{9x-5}$ and **(b)** $f(x) = e^{\cos x}$

Solution Apply Eq. (3):

(a) $\dfrac{d}{dx}e^{9x-5} = 9e^{9x-5}$ and **(b)** $\dfrac{d}{dx}\left(e^{\cos x}\right) = -(\sin x)e^{\cos x}$ ■

■ **EXAMPLE 7 Graph Sketching Involving e^x** Sketch the graph of $f(x) = xe^x$ on the interval $[-4, 2]$.

Solution As usual, the first step is to solve for the critical points:

$$f'(x) = \frac{d}{dx}xe^x = xe^x + e^x = (x+1)e^x = 0$$

Since $e^x > 0$ for all x, the unique critical point is $x = -1$ and

$$f'(x) = \begin{cases} < 0 & \text{for } x < -1 \\ > 0 & \text{for } x > -1 \end{cases}$$

Thus, $f'(x)$ changes sign from $-$ to $+$ at $x = -1$ and $f(-1)$ is a local minimum. For the second derivative, we have

$$f''(x) = (x+1) \cdot (e^x)' + e^x \cdot (x+1)' = (x+1)e^x + e^x = (x+2)e^x$$

$$f''(x) = \begin{cases} < 0 & \text{for } x < -2 \\ > 0 & \text{for } x > -2 \end{cases}$$

Thus, $x = -2$ is a point of inflection, where the graph changes from concave down to concave up at $x = -2$. Figure 5 shows the graph with its local minimum and point of inflection. ■

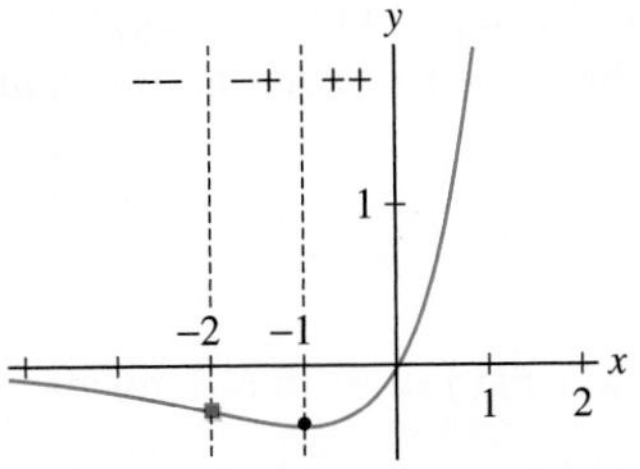

FIGURE 5 Graph of $f(x) = xe^x$. The sign combinations $--$, $-+$, $++$ indicate the signs of f' and f''.

Integrals Involving e^x

The formula $(e^x)' = e^x$ says that the function $f(x) = e^x$ is its own derivative. But this means $f(x) = e^x$ is also *its own antiderivative*. In other words,

$$\int e^x\,dx = e^x + C$$

More generally, for any constants b and k with $k \neq 0$,

$$\int e^{kx+b}\,dx = \frac{1}{k}\,e^{kx+b} + C$$

We verify this formula using substitution, or by noting that $\dfrac{d}{dx}\left(\dfrac{1}{k}\,e^{kx+b}\right) = e^{kx+b}$.

EXAMPLE 8 Evaluate:

(a) $\displaystyle\int e^{7x-5}\,dx$ **(b)** $\displaystyle\int xe^{2x^2}\,dx$ **(c)** $\displaystyle\int \frac{e^t}{1+2e^t+e^{2t}}\,dt$

Solution

(a) $\displaystyle\int e^{7x-5}\,dx = \frac{1}{7}e^{7x-5} + C.$

(b) Use the substitution $u = 2x^2$, $du = 4x\,dx$:

$$\int xe^{2x^2}\,dx = \frac{1}{4}\int e^u\,du = \frac{1}{4}e^u + C = \frac{1}{4}e^{2x^2} + C$$

(c) We have $1 + 2e^t + e^{2t} = (1+e^t)^2$. The substitution $u = e^t$, $du = e^t\,dt$ gives

$$\int \frac{e^t}{1+2e^t+e^{2t}}\,dt = \int \frac{du}{(1+u)^2} = -(1+u)^{-1} + C = -(1+e^t)^{-1} + C$$

CONCEPTUAL INSIGHT What precisely do we mean by b^x? We have taken for granted that b^x is meaningful for all real numbers x, but we never specified how b^x is defined when x is irrational. If n is a whole number, then b^n is simply the product $b \cdot b \cdots b$ (n times), and for any rational number $x = m/n$,

$$b^x = b^{m/n} = \left(b^{1/n}\right)^m = \left(\sqrt[n]{b}\right)^m$$

When x is irrational, this definition does not apply and b^x cannot be defined directly in terms of roots and powers of b. However, it makes sense to view $b^{m/n}$ as an approximation to b^x when m/n is a rational number close to x. For example, $3^{\sqrt{2}}$ should be approximately equal to $3^{1.4142} \approx 4.729$ because 1.4142 is a good rational approximation to $\sqrt{2}$. Formally, then, we may define b^x as a limit over rational numbers m/n approaching x:

$$b^x = \lim_{m/n \to x} b^{m/n}$$

It can be shown that this limit exists and that the function $f(x) = b^x$ thus defined is not only continuous but also differentiable.

7.1 SUMMARY

- $f(x) = b^x$ is the *exponential function* with base b (where $b > 0$ and $b \neq 1$).
- $f(x) = b^x$ is increasing if $b > 1$ and decreasing if $b < 1$.
- The derivative of b^x is proportional to b^x:

$$\frac{d}{dx}\,b^x = m(b)b^x$$

where $m(b) = \displaystyle\lim_{h\to 0} \frac{b^h - 1}{h}$.

- There is a unique number $e \approx 2.718$ with the property $m(e) = 1$, so that

$$\frac{d}{dx}e^x = e^x$$

- By the Chain Rule:

$$\frac{d}{dx}e^{f(x)} = f'(x)e^{f(x)}, \qquad \frac{d}{dx}e^{kx+b} = ke^{kx+b} \quad (k, b \text{ constants})$$

- $\int e^x\,dx = e^x + C.$
- $\int e^{kx+b}\,dx = \frac{1}{k}e^{kx+b} + C \quad (k, b \text{ constants with } k \neq 0).$

7.1 EXERCISES

Preliminary Questions

1. Which of the following equations is incorrect?

(a) $3^2 \cdot 3^5 = 3^7$ **(b)** $(\sqrt{5})^{4/3} = 5^{2/3}$

(c) $3^2 \cdot 2^3 = 1$ **(d)** $(2^{-2})^{-2} = 16$

2. What are the domain and range of $\ln x$? When is $\ln x$ negative?

3. To which of the following does the Power Rule apply?

(a) $f(x) = x^2$ **(b)** $f(x) = 2^e$ **(c)** $f(x) = x^e$

(d) $f(x) = e^x$ **(e)** $f(x) = x^x$

(f) $f(x) = x^{-4/5}$

4. For which values of b does b^x have a negative derivative?

5. For which values of b is the graph of $y = b^x$ concave up?

6. Which point lies on the graph of $y = b^x$ for all b?

7. Which of the following statements is not true?

(a) $(e^x)' = e^x$

(b) $\lim_{h\to 0} \frac{e^h - 1}{h} = 1$

(c) The tangent line to $y = e^x$ at $x = 0$ has slope e.

(d) The tangent line to $y = e^x$ at $x = 0$ has slope 1.

Exercises

1. Rewrite as a whole number (without using a calculator):

(a) 7^0 **(b)** $10^2(2^{-2} + 5^{-2})$

(c) $\frac{(4^3)^5}{(4^5)^3}$ **(d)** $27^{4/3}$

(e) $8^{-1/3} \cdot 8^{5/3}$ **(f)** $3 \cdot 4^{1/4} - 12 \cdot 2^{-3/2}$

2. Compute $(16^{-1/16})^4$.

In Exercises 3–10, solve for the unknown variable.

3. $9^{2x} = 9^8$ **4.** $e^{t^2} = e^{4t-3}$

5. $3^x = \left(\frac{1}{3}\right)^{x+1}$ **6.** $(\sqrt{5})^x = 125$

7. $4^{-x} = 2^{x+1}$ **8.** $b^4 = 10^{12}$

9. $k^{3/2} = 27$ **10.** $(b^2)^{x+1} = b^{-6}$

In Exercises 11–14, determine the limit.

11. $\lim_{x\to\infty} 4^x$ **12.** $\lim_{x\to\infty} 4^{-x}$

13. $\lim_{x\to\infty} \left(\frac{1}{4}\right)^{-x}$ **14.** $\lim_{x\to\infty} e^{x-x^2}$

In Exercises 15–18, find the equation of the tangent line at the point indicated.

15. $y = 4e^x, \quad x_0 = 0$ **16.** $y = e^{4x}, \quad x_0 = 0$

17. $y = e^{x+2}, \quad x_0 = -1$ **18.** $y = e^{x^2}, \quad x_0 = 1$

In Exercises 19–40, find the derivative.

19. $f(x) = 7e^{2x} + 3e^{4x}$ **20.** $f(x) = e^{-5x}$

21. $f(x) = e^{\pi x}$ **22.** $f(x) = e^3$

23. $f(x) = e^{-4x+9}$ **24.** $f(x) = 4e^{-x} + 7e^{-2x}$

25. $f(x) = \frac{e^{x^2}}{x}$ **26.** $f(x) = x^2e^{2x}$

27. $f(x) = (1 + e^x)^4$ **28.** $f(x) = (2e^{3x} + 2e^{-2x})^4$

29. $f(x) = e^{x^2+2x-3}$ **30.** $f(x) = e^{1/x}$

31. $f(x) = e^{\sin x}$ **32.** $f(x) = e^{(x^2+2x+3)^2}$

33. $f(\theta) = \sin(e^\theta)$ **34.** $f(t) = e^{\sqrt{t}}$

35. $f(t) = \frac{1}{1 - e^{-3t}}$ **36.** $f(t) = \cos(te^{-2t})$

37. $f(x) = \dfrac{e^x}{3x+1}$

38. $f(x) = \tan(e^{5-6x})$

39. $f(x) = \dfrac{e^{x+1}+x}{2e^x - 1}$

40. $f(x) = e^{e^x}$

In Exercises 41–46, calculate the derivative indicated.

41. $f''(x);\quad f(x) = e^{4x-3}$

42. $f'''(x);\quad f(x) = e^{12-3x}$

43. $\dfrac{d^2y}{dt^2};\quad y = e^t \sin t$

44. $\dfrac{d^2y}{dt^2};\quad y = e^{-2t}\sin 3t$

45. $\dfrac{d^2}{dt^2} e^{t-t^2}$

46. $\dfrac{d^3}{d\theta^3} \cos(e^\theta)$

In Exercises 47–52, find the critical points and determine whether they are local minima, maxima, or neither.

47. $f(x) = e^x - x$

48. $f(x) = x + e^{-x}$

49. $f(x) = \dfrac{e^x}{x}$ for $x > 0$

50. $f(x) = x^2 e^x$

51. $g(t) = \dfrac{e^t}{t^2+1}$

52. $g(t) = (t^3 - 2t)e^t$

In Exercises 53–58, find the critical points and points of inflection. Then sketch the graph.

53. $y = xe^{-x}$

54. $y = e^{-x} + e^x$

55. $y = e^{-x}\cos x$ on $\left[-\frac{\pi}{2}, \frac{\pi}{2}\right]$

56. $y = e^{-x^2}$

57. $y = e^x - x$

58. $y = x^2 e^{-x}$

59. Find $a > 0$ such that the tangent line to the graph of $f(x) = x^2 e^{-x}$ at $x = a$ passes through the origin (Figure 6).

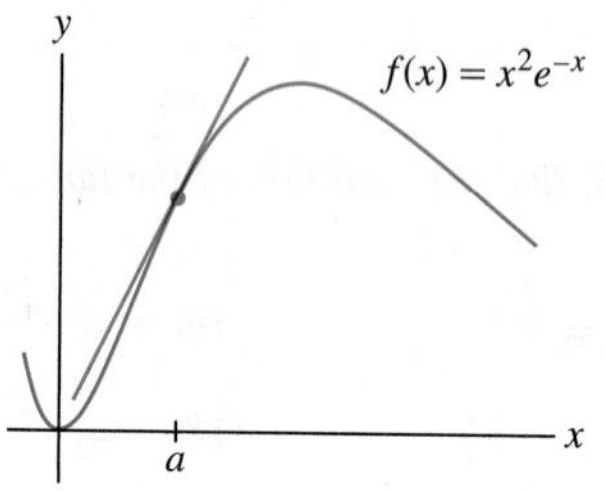

FIGURE 6

60. Use Newton's Method to find the two solutions of $e^x = 5x$ to three decimal places (Figure 7).

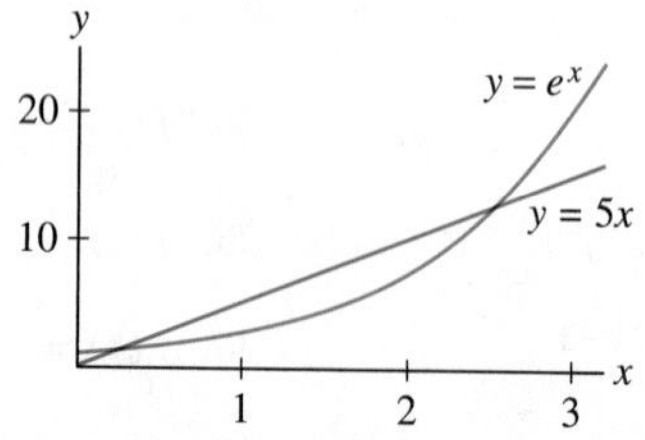

FIGURE 7 Graphs of e^x and $5x$.

61. Compute the linearization of $f(x) = e^{-2x}\sin x$ at $a = 0$.

62. Compute the linearization of $f(x) = xe^{6-3x}$ at $a = 2$.

63. Find the linearization of $f(x) = e^x$ at $a = 0$ and use it to estimate $e^{-0.1}$.

64. Use the linear approximation to estimate $f(1.03) - f(1)$ where $y = x^{1/3}e^{x-1}$.

65. A 2005 study by the Fisheries Research Services in Aberdeen Scotland showed that the average length of the species Clupea Harengus (Atlantic herring) as a function of age t (in years) can be modeled by $L(t) = 32(1 - e^{-0.37t})$ cm for $0 \le t \le 13$.

(a) How fast is the average length changing at age $t = 6$ yrs?

(b) At what age is the average length changing at a rate of 5 cm/yr?

(c) Calculate $L = \lim_{t\to\infty} L(t)$.

66. According to a 1999 study by Starkey and Scarnecchia, the average weight (kg) at age t (years) of channel catfish in the Lower Yellowstone River can be modeled by

$$W(t) = \left(3.46293 - 3.32173e^{-0.03456t}\right)^{3.4026}$$

Find the rate at which weight is changing at age $t = 10$.

67. The functions in Exercises 65 and 66 are examples of the **Von Bertalanffy growth function**

$$M(t) = \left(a + (b-a)e^{kmt}\right)^{1/m}$$

introduced in the 1930's by Austrian-born biologist Karl Ludwig Von Bertalanffy. Calculate $M'(0)$ in terms of the constants a, b, k, and m.

68. Find an approximation to $m(4)$ using the limit definition and estimate the slope of the tangent line to $y = 4^x$ at $x = 0$ and $x = 2$.

In Exercises 69–86, evaluate the integral.

69. $\displaystyle\int (e^x + 2)\,dx$

70. $\displaystyle\int e^{4x}\,dx$

71. $\displaystyle\int_0^1 e^{-3x}\,dx$

72. $\displaystyle\int_2^6 e^{-x/2}\,dx$

73. $\displaystyle\int_0^3 e^{1-6t}\,dt$

74. $\displaystyle\int_2^3 e^{4t-3}\,dt$

75. $\displaystyle\int (e^{4x} + 1)\,dx$

76. $\displaystyle\int (e^x + e^{-x})\,dx$

77. $\displaystyle\int_0^1 xe^{-x^2/2}\,dx$

78. $\displaystyle\int_0^2 ye^{3y^2}\,dy$

79. $\displaystyle\int e^t\sqrt{e^t+1}\,dt$

80. $\displaystyle\int (e^{-x} - 4x)\,dx$

81. $\displaystyle\int \frac{e^{2x} - e^{4x}}{e^x}\,dx$

82. $\displaystyle\int e^x\cos(e^x)\,dx$

83. $\displaystyle\int \frac{e^x}{\sqrt{e^x+1}}\,dx$

84. $\displaystyle\int e^x(e^{2x}+1)^3\,dx$

85. $\displaystyle\int \frac{e^{\sqrt{x}}\,dx}{\sqrt{x}}$

86. $\displaystyle\int x^{-2/3}e^{x^{1/3}}\,dx$

87. Find the area between $y = e^x$ and $y = e^{2x}$ over $[0, 1]$.

88. Find the area between $y = e^x$ and $y = e^{-x}$ over $[0, 2]$.

89. Find the area bounded by $y = e^2$, $y = e^x$, and $x = 0$.

90. Find the volume obtained by revolving $y = e^x$ about the x-axis for $0 \le x \le 1$.

91. Wind engineers have found that wind speed v (in m/s) at a given location follows a **Rayleigh distribution** of the type

$$W(v) = \frac{1}{32} v e^{-v^2/64}$$

This means that the probability that v lies between a and b is equal to the shaded area in Figure 8.

(a) Show that the probability that $v \in [0, b]$ is $1 - e^{-b^2/64}$.

(b) Calculate the probability that $v \in [2, 5]$.

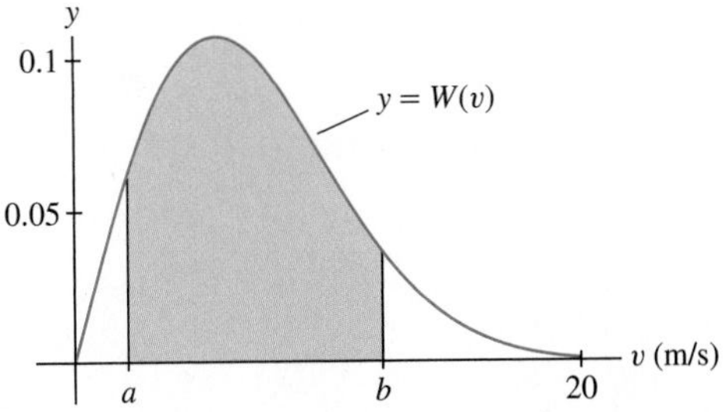

FIGURE 8 The shaded area is the probability that v lies between a and b.

92. The function $f(x) = e^x$ satisfies $f'(x) = f(x)$. Show that if $g(x)$ is another function satisfying $g'(x) = g(x)$, then $g(x) = Ce^x$ for some constant C. *Hint:* Compute the derivative of $g(x)e^{-x}$.

Further Insights and Challenges

93. Prove that $f(x) = e^x$ is not a polynomial function. *Hint:* Differentiation lowers the degree of a polynomial by 1.

94. Recall the following property of integrals: If $f(t) \ge g(t)$ for all $t \ge 0$, then for all $x \ge 0$,

$$\int_0^x f(t)\,dt \ge \int_0^x g(t)\,dt \qquad \boxed{4}$$

The inequality $e^t \ge 1$ holds for $t \ge 0$ because $e > 1$. Use (4) to prove that $e^x \ge 1 + x$ for $x \ge 0$. Then prove, by successive integration, the following inequalities (for $x \ge 0$):

$$e^x \ge 1 + x + \frac{1}{2}x^2, \qquad e^x \ge 1 + x + \frac{1}{2}x^2 + \frac{1}{6}x^3$$

95. Generalize Exercise 94; that is, use induction (if you are familiar with this method of proof) to prove that for all $n \ge 0$,

$$e^x \ge 1 + x + \frac{1}{2}x^2 + \frac{1}{6}x^3 + \cdots + \frac{1}{n!}x^n \quad (x \ge 0)$$

96. Use Exercise 94 to show that $\frac{e^x}{x^2} \ge \frac{x}{6}$ and conclude that $\lim_{x\to\infty} \frac{e^x}{x^2} = \infty$. Then use Exercise 95 to prove more generally that $\lim_{x\to\infty} \frac{e^x}{x^n} = \infty$ for all n.

97. Calculate the first three derivatives of $f(x) = xe^x$. Then guess the formula for $f^{(n)}(x)$ (use induction to prove it if you are familiar with this method of proof).

98. Consider the equation $e^x = \lambda x$, where λ is a constant.

(a) For which λ does it have a unique solution? For intuition, draw a graph of $y = e^x$ and the line $y = \lambda x$.

(b) For which λ does it have at least one solution?

99. Prove in two ways that the numbers $m(a)$ satisfy

$$m(ab) = m(a) + m(b)$$

(a) First method: Use the limit definition of m_b and

$$\frac{(ab)^h - 1}{h} = b^h\left(\frac{a^h - 1}{h}\right) + \frac{b^h - 1}{h}$$

(b) Second method: Apply the Product Rule to $a^x b^x = (ab)^x$.

7.2 Inverse Functions

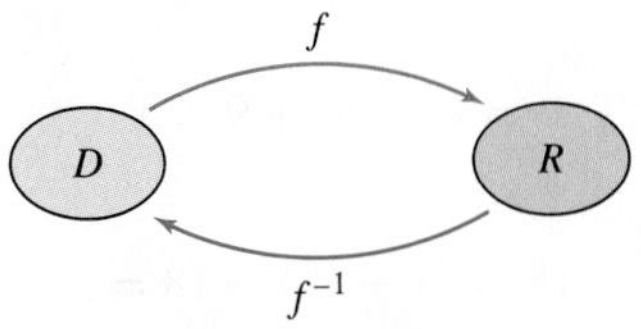

FIGURE 1 A function and its inverse.

In the next section, we will define logarithmic functions as inverses of exponential functions. But first, we review inverse functions and compute their derivatives.

The inverse of $f(x)$, denoted $f^{-1}(x)$, is the function that *reverses* the effect of $f(x)$ (Figure 1). For example, the inverse of $f(x) = x^3$ is the cube root function $f^{-1}(x) = x^{1/3}$. Given a table of function values for $f(x)$, we obtain a table for $f^{-1}(x)$ by interchanging the x and y columns:

Function	
x	$y = x^3$
−2	−8
−1	−1
0	0
1	1
2	8
3	27

(Interchange columns) $\Longrightarrow$

Inverse	
x	$y = x^{1/3}$
−8	−2
−1	−1
0	0
1	1
8	2
27	3

If we apply both f and f^{-1} to a number x in either order, we get back x. For instance,

$$\text{Apply } f \text{ and then } f^{-1}: \quad 2 \xrightarrow{(\text{Apply } x^3)} 8 \xrightarrow{(\text{Apply } x^{1/3})} 2$$

$$\text{Apply } f^{-1} \text{ and then } f: \quad 8 \xrightarrow{(\text{Apply } x^{1/3})} 2 \xrightarrow{(\text{Apply } x^3)} 8$$

This property is used in the formal definition of the inverse function.

← *REMINDER The "domain" is the set of numbers x such that $f(x)$ is defined (the set of allowable inputs), and the "range" is the set of all values $f(x)$ (the set of outputs).*

DEFINITION Inverse Let $f(x)$ have domain D and range R. If there is a function $g(x)$ with domain R such that

$$g(f(x)) = x \quad \text{for } x \in D \qquad \text{and} \qquad f(g(x)) = x \quad \text{for } x \in R$$

then $f(x)$ is said to be **invertible**. The function $g(x)$ is called the **inverse function** and is denoted $f^{-1}(x)$.

■ **EXAMPLE 1** Show that $f(x) = 2x - 18$ is invertible. What are the domain and range of $f^{-1}(x)$?

Solution We show that $f(x)$ is invertible by computing the inverse function in two steps.

Step 1. **Solve the equation $y = f(x)$ for x in terms of y.**

$$y = 2x - 18$$
$$y + 18 = 2x$$
$$x = \frac{1}{2}y + 9$$

This gives us the inverse as a function of the variable y: $f^{-1}(y) = \frac{1}{2}y + 9$.

Step 2. **Interchange variables.**

We usually prefer to write the inverse as a function of x, so we interchange the roles of x and y (Figure 2):

$$f^{-1}(x) = \frac{1}{2}x + 9$$

To check our calculation, let's verify that $f^{-1}(f(x)) = x$ and $f(f^{-1}(x)) = x$:

$$f^{-1}(f(x)) = f^{-1}(2x - 18) = \frac{1}{2}(2x - 18) + 9 = (x - 9) + 9 = x$$

$$f(f^{-1}(x)) = f\left(\frac{1}{2}x + 9\right) = 2\left(\frac{1}{2}x + 9\right) - 18 = (x + 18) - 18 = x$$

Because f^{-1} is a linear function, its domain and range are $\mathbf{R}$. ■

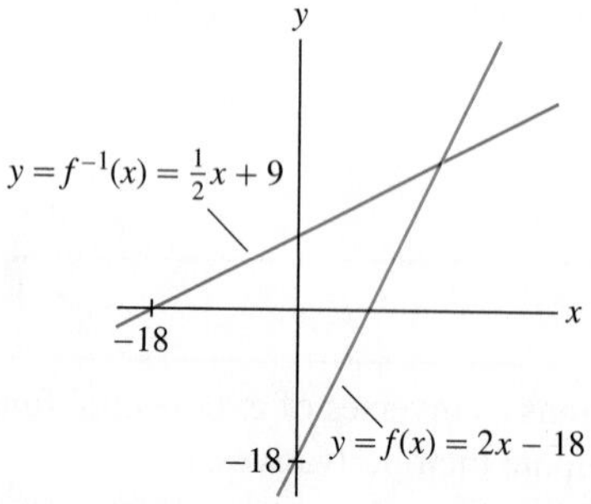

FIGURE 2

The inverse function, if it exists, is unique. However, some functions do not have an inverse. Consider $f(x) = x^2$. When we interchange the columns in a table of values (which should give us a table of values for f^{-1}), the resulting table does not define a function:

Function

x	$y = x^2$
−2	4
−1	**1**
0	0
1	**1**
2	4

(Interchange columns) $\Longrightarrow$

Inverse (?)

x	y
4	−2
1	−1
0	0
1	1
4	2

$f^{-1}(1)$ has two values: 1 and −1.

The problem is that every positive number occurs twice as an output of $f(x) = x^2$. For example, 1 occurs twice as an *output* in the first table and therefore occurs twice as an *input* in the second table. So the second table gives us two possible values for $f^{-1}(1)$, namely $f^{-1}(1) = 1$ and $f^{-1}(1) = -1$. Neither value satisfies the inverse property. For instance, if we set $f^{-1}(1) = 1$, then $f^{-1}(f(-1)) = f^{-1}(1) = 1$, but an inverse would have to satisfy $f^{-1}(f(-1)) = -1$.

Another standard term for one-to-one is ***injective****.*

So when does a function $f(x)$ have an inverse? The answer is: If $f(x)$ is **one-to-one**, which means that $f(x)$ takes on each value at most once (Figure 3). Here is the formal definition:

DEFINITION One-to-One Function A function $f(x)$ is one-to-one on a domain D if, for every value c, the equation $f(x) = c$ has at most one solution for $x \in D$. Or, equivalently, if for all $a, b \in D$,

$$f(a) \neq f(b) \qquad \text{unless} \qquad a = b$$

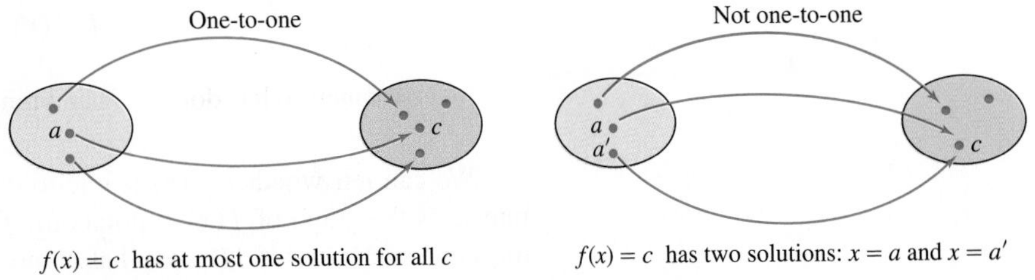

FIGURE 3 A one-to-one function takes on each value at most once.

Think of a function as a device for "labeling" members of the range by members of the domain. When f is one-to-one, this labeling is unique and f^{-1} maps each number in the range back to its label.

When $f(x)$ is one-to-one on its domain D, the inverse function $f^{-1}(x)$ exists and its domain is equal to the range R of f (Figure 4). Indeed, for every $c \in R$, there is precisely one element $a \in D$ such that $f(a) = c$ and we may define $f^{-1}(c) = a$. With this definition, $f(f^{-1}(c)) = f(a) = c$ and $f^{-1}(f(a)) = f^{-1}(c) = a$. This proves the following theorem.

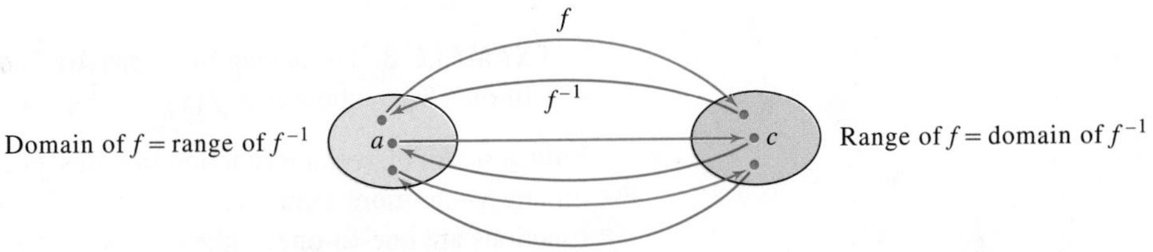

FIGURE 4 In passing from f to f^{-1}, the domain and range are interchanged.

THEOREM 1 Existence of Inverses The inverse function $f^{-1}(x)$ exists if and only if $f(x)$ is one-to-one on its domain D. Furthermore,

- Domain of f = range of f^{-1}.
- Range of f = domain of f^{-1}.

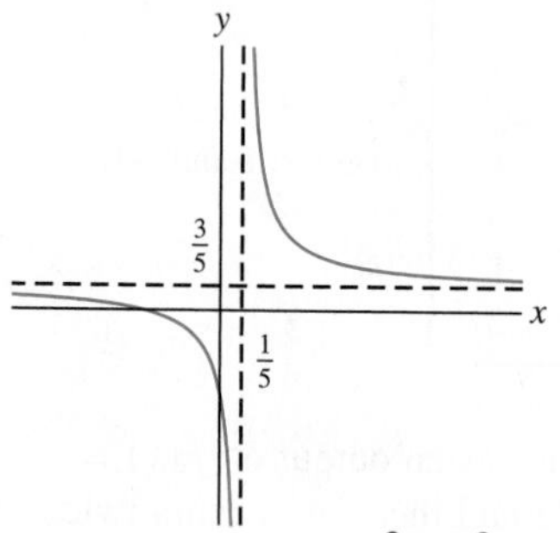

FIGURE 5 Graph of $f(x) = \dfrac{3x+2}{5x-1}$.

■ **EXAMPLE 2** Show that $f(x) = \dfrac{3x+2}{5x-1}$ is invertible. Determine the domain and range of f and f^{-1}.

Solution The domain of $f(x)$ is $D = \left\{x : x \neq \frac{1}{5}\right\}$ (Figure 5). Assume that $x \in D$, and let's solve $y = f(x)$ for x in terms of y:

$$y = \frac{3x+2}{5x-1}$$

$$y(5x-1) = 3x+2$$

$$5xy - y = 3x + 2$$

$$5xy - 3x = y + 2 \qquad \text{(gather terms involving } x\text{)}$$

$$x(5y-3) = y+2 \qquad \text{(factor out } x \text{ in order to solve for } x\text{)} \qquad \boxed{1}$$

$$x = \frac{y+2}{5y-3} \qquad \text{(divide by } 5y-3\text{)} \qquad \boxed{2}$$

Often, it is impossible to find a formula for the inverse because we cannot solve for x explicitly in the equation $y = f(x)$. For example, the function $f(x) = x + e^x$ has an inverse, but we must make do without an explicit formula for it.

The last step is valid if $5y - 3 \neq 0$—that is, if $y \neq \frac{3}{5}$. But note that $y = \frac{3}{5}$ is not in the range of $f(x)$. For if it were, Eq. (1) would yield the false equation $0 = \frac{3}{5} + 2$. Now Eq. (2) shows that for all $y \neq \frac{3}{5}$ there is a unique value x such that $f(x) = y$. Therefore, $f(x)$ is one-to-one on its domain. By Theorem 1, $f(x)$ is invertible. The range of $f(x)$ is $R = \left\{x : x \neq \frac{3}{5}\right\}$ and

$$f^{-1}(x) = \frac{x+2}{5x-3}.$$

The inverse function has domain R and range D. ■

We can tell whether $f(x)$ is one-to-one from its graph. The horizontal line $y = c$ intersects the graph of $f(x)$ at points $(a, f(a))$, where $f(a) = c$ (Figure 6). There is at most one such point if $f(x) = c$ has at most one solution. This gives us the

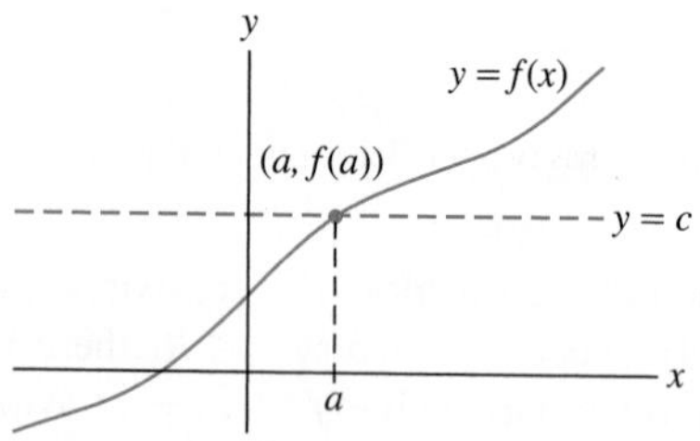

FIGURE 6 The line $y = c$ intersects the graph at points where $f(a) = c$.

Horizontal Line Test A function $f(x)$ is one-to-one if and only if every horizontal line intersects the graph of $f(x)$ in at most one point.

In Figure 7, we see that $f(x) = x^3$ passes the Horizontal Line Test and therefore is one-to-one, whereas $f(x) = x^2$ fails the test and is not one-to-one.

■ **EXAMPLE 3 Increasing Functions Are One-to-One** Show that increasing functions are one-to-one. Then show that $f(x) = x^5 + 4x + 3$ is one-to-one.

Solution An increasing function satisfies $f(a) < f(b)$ if $a < b$. Therefore f cannot take on any value more than once, and thus f is one-to-one. Note similarly that decreasing functions are one-to-one.

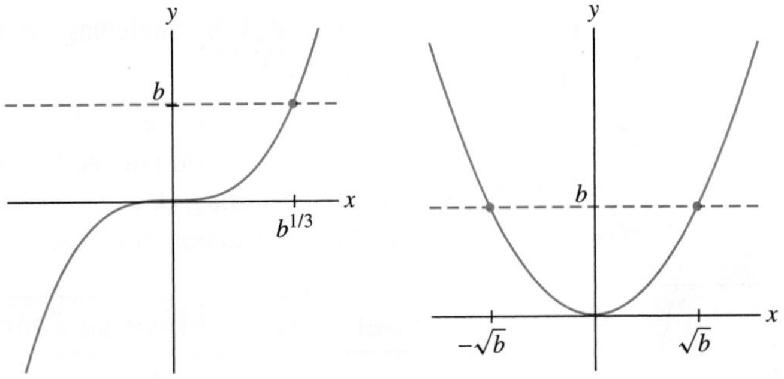

(A) $f(x) = x^3$ is one-to-one.

(B) $f(x) = x^2$ is not one-to-one.

FIGURE 7

Now observe that

- If n odd and $c > 0$, then cx^n is increasing.
- A sum of increasing functions is increasing.

Thus $x^5, 4x$, and hence the sum $x^5 + 4x$ are increasing. It follows that $f(x) = x^5 + 4x + 3$ is increasing and therefore one-to-one (Figure 8). ■

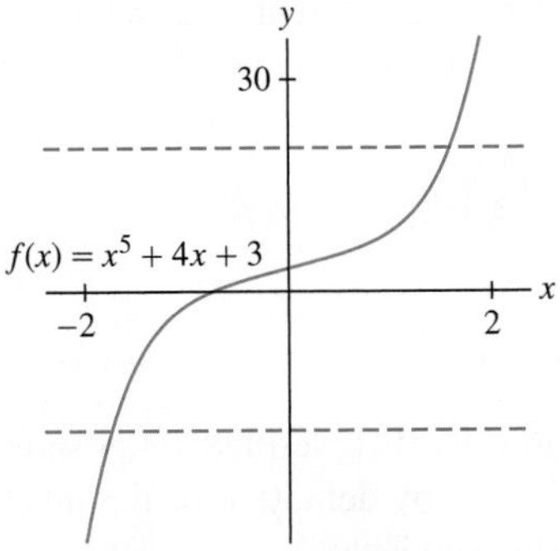

FIGURE 8 The increasing function $f(x) = x^5 + 4x + 3$ satisfies the Horizontal Line Test.

We can make a function one-to-one by restricting its domain suitably.

■ **EXAMPLE 4 Restricting the Domain** Find a domain on which $f(x) = x^2$ is one-to-one and determine its inverse on this domain.

Solution The function $f(x) = x^2$ is one-to-one on the domain $D = \{x : x \geq 0\}$, for if $a^2 = b^2$ where a and b are both nonnegative, then $a = b$ (Figure 9). The inverse of $f(x)$ on D is the positive square root $f^{-1}(x) = \sqrt{x}$. Alternatively, we may restrict $f(x)$ to the domain $\{x : x \leq 0\}$, on which the inverse function is $f^{-1}(x) = -\sqrt{x}$. ■

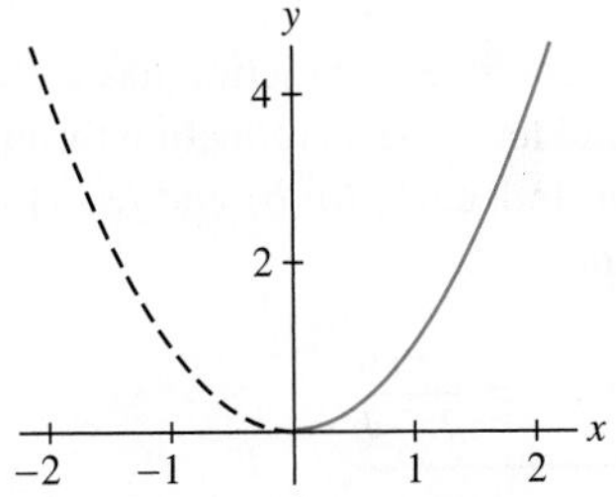

FIGURE 9 $f(x) = x^2$ satisfies the Horizontal Line Test on the domain $\{x : x \geq 0\}$.

Next we describe the graph of the inverse function. The **reflection** of a point (a, b) through the line $y = x$ is, by definition, the point (b, a) (Figure 10). Note that if the x- and y-axes are drawn to the same scale, then (a, b) and (b, a) are equidistant from the line $y = x$ and the segment joining them is perpendicular to $y = x$.

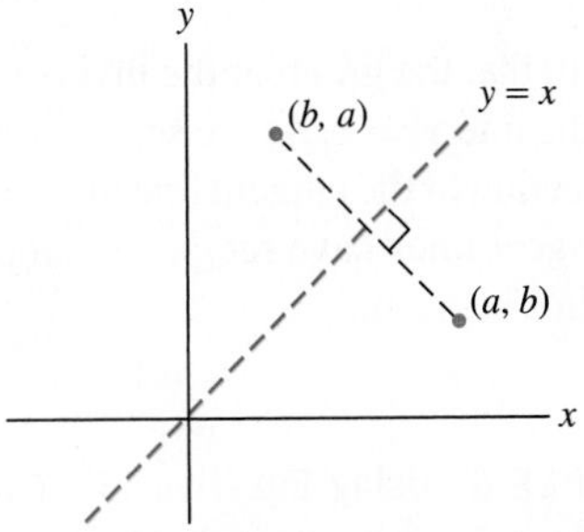

FIGURE 10 The reflection (a, b) through the line $y = x$ is the point (b, a).

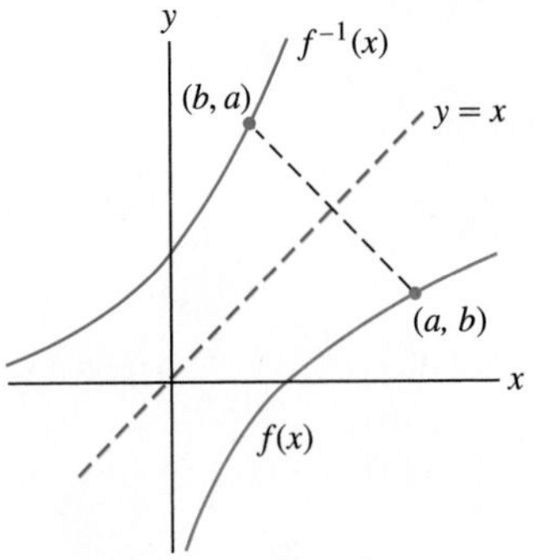

FIGURE 11 The graph of $f^{-1}(x)$ is the reflection of the graph of $f(x)$ through the line $y = x$.

The graph of f^{-1} is the reflection of the graph of f through $y = x$ (Figure 11). To check this, note that (a, b) lies on the graph of f if $f(a) = b$. But $f(a) = b$ if and only if $f^{-1}(b) = a$, and in this case, (b, a) lies on the graph of f^{-1}.

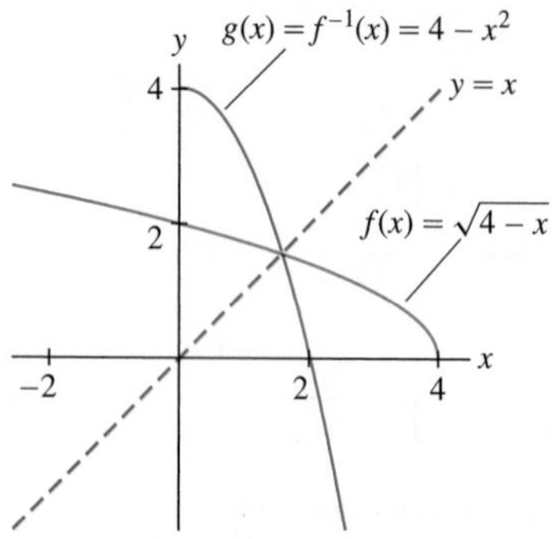

FIGURE 12 Graph of the inverse $g(x)$ of $f(x) = \sqrt{4 - x}$.

■ **EXAMPLE 5 Sketching the Graph of the Inverse** Sketch the graph of the inverse of $f(x) = \sqrt{4 - x}$.

Solution Let $g(x) = f^{-1}(x)$. Observe that $f(x)$ has domain $\{x : x \le 4\}$ and range $\{x : x \ge 0\}$. We do not need a formula for $g(x)$ to draw its graph. We simply reflect the graph of f through the line $y = x$ as in Figure 12. If desired, however, we can easily solve $y = \sqrt{4 - x}$ to obtain $x = 4 - y^2$ and thus $g(x) = 4 - x^2$ with domain $\{x : x \ge 0\}$. ■

Derivatives of Inverse Functions

Next, we derive a formula for the derivative of the inverse $f^{-1}(x)$. We will use this formula to differentiate logarithmic functions in Section 7.3.

THEOREM 2 Derivative of the Inverse Assume that $f(x)$ is differentiable and one-to-one with inverse $g(x) = f^{-1}(x)$. If b belongs to the domain of $g(x)$ and $f'(g(b)) \neq 0$, then $g'(b)$ exists and

$$g'(b) = \frac{1}{f'(g(b))} \qquad (3)$$

Proof The first claim, that $g(x)$ is differentiable if $f'(g(x)) \neq 0$, is verified in Appendix D (see Theorem 6). To prove Eq. (3), note that $f(g(x)) = x$ by definition of the inverse. Differentiate both sides of this equation, and apply the Chain Rule:

$$\frac{d}{dx} f(g(x)) = \frac{d}{dx} x \quad \Rightarrow \quad f'(g(x))g'(x) = 1 \quad \Rightarrow \quad g'(x) = \frac{1}{f'(g(x))}$$

Set $x = b$ to obtain Eq. (3). ■

GRAPHICAL INSIGHT The formula for the derivative of the inverse function has a clear graphical interpretation. Consider a line L of slope m and let L' be its reflection through $y = x$ as in Figure 13(A). Then the slope of L' is $1/m$. Indeed, if (a, b) and (c, d) are any two points on L, then (b, a) and (d, c) lie on L' and

$$\underbrace{\text{Slope of } L = \frac{d - b}{c - a}, \qquad \text{Slope of } L' = \frac{c - a}{d - b}}_{\text{Reciprocal slopes}}$$

Now recall that the graph of the inverse $g(x)$ is obtained by reflecting the graph of $f(x)$ through the line $y = x$. As we see in Figure 13(B), the tangent line to $y = g(x)$ at $x = b$ is the reflection of the tangent line to $y = f(x)$ at $x = a$ [where $b = f(a)$ and $a = g(b)$]. These tangent lines have reciprocal slopes, and thus $g'(b) = 1/f'(a) = 1/f'(g(b))$, as claimed in Theorem 2.

■ **EXAMPLE 6 Using Equation (3)** Calculate $g'(x)$, where $g(x)$ is the inverse of the function $f(x) = x^4 + 10$ on the domain $\{x : x \ge 0\}$.

Solution Solve $y = x^4 + 10$ for x to obtain $x = (y - 10)^{1/4}$. Thus $g(x) = (x - 10)^{1/4}$. Since $f'(x) = 4x^3$, we have $f'(g(x)) = 4g(x)^3$, and by Eq. (3),

$$g'(x) = \frac{1}{f'(g(x))} = \frac{1}{4g(x)^3} = \frac{1}{4(x - 10)^{3/4}} = \frac{1}{4}(x - 10)^{-3/4}$$

We obtain this same result by differentiating $g(x) = (x - 10)^{1/4}$ directly. ■

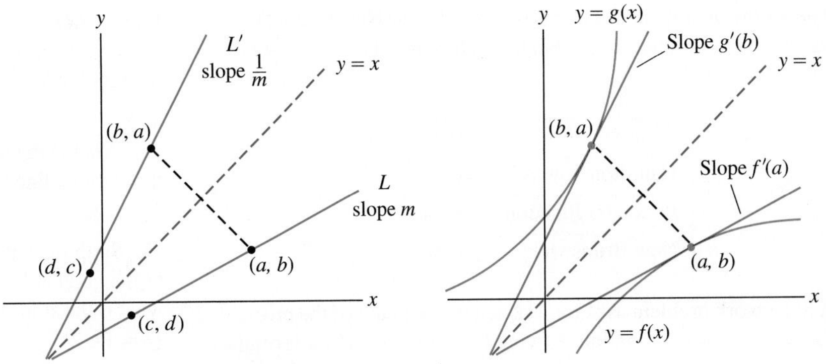

(A) If L has slope m, then its reflection L' has slope $1/m$.

(B) The tangent line to the inverse $y = g(x)$ is the reflection of the tangent line to $y = f(x)$.

FIGURE 13 Graphical illustration of the formula $g'(b) = 1/f'(g(b))$.

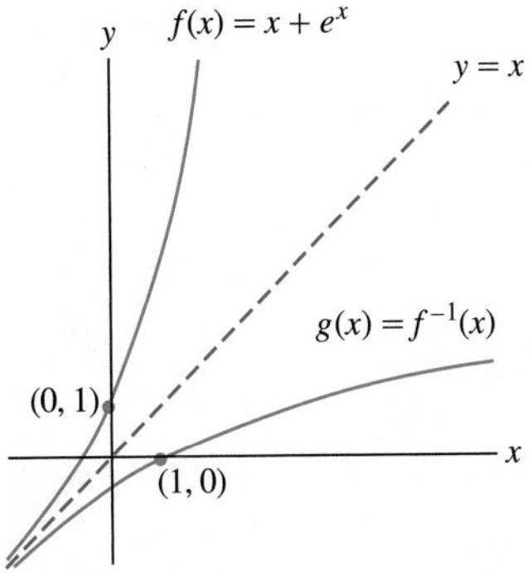

FIGURE 14 Graph of $f(x) = x + e^x$ and its inverse $g(x)$.

■ **EXAMPLE 7 Calculating $g'(x)$ Without Solving for $g(x)$** Calculate $g'(1)$, where $g(x)$ is the inverse of $f(x) = x + e^x$.

Solution In this case, we cannot solve for $g(x)$ explicitly, but a formula for $g(x)$ is not needed (Figure 14). All we need is the particular value $g(1)$, which we can find by solving $f(x) = 1$. By inspection, $x + e^x = 1$ has solution $x = 0$. Therefore, $f(0) = 1$ and, by definition of the inverse, $g(1) = 0$. Since $f'(x) = 1 + e^x$,

$$g'(1) = \frac{1}{f'(g(1))} = \frac{1}{f'(0)} = \frac{1}{1 + e^0} = \frac{1}{2}$$

■

7.2 SUMMARY

- A function $f(x)$ is *one-to-one* on a domain D if for every value c, the equation $f(x) = c$ has at most one solution for $x \in D$, or, equivalently, if for all $a, b \in D$, $f(a) \neq f(b)$ unless $a = b$.
- Let $f(x)$ have domain D and range R. The *inverse* $f^{-1}(x)$ (if it exists) is the unique function with domain R and range D satisfying $f(f^{-1}(x)) = x$ and $f^{-1}(f(x)) = x$.
- The inverse of $f(x)$ exists if and only if $f(x)$ is one-to-one on its domain.
- To find the inverse function, solve $y = f(x)$ for x in terms of y to obtain $x = g(y)$. The inverse is the function $g(x)$.
- *Horizontal Line Test:* $f(x)$ is one-to-one if and only if every horizontal line intersects the graph of $f(x)$ in at most one point.
- The graph of $f^{-1}(x)$ is obtained by reflecting the graph of $f(x)$ through the line $y = x$.
- Derivative of the inverse: If $f(x)$ is differentiable and one-to-one with inverse $g(x)$, then for x such that $f'(g(x)) \neq 0$,

$$g(x) = \frac{1}{f'(g(x))}$$

7.2 EXERCISES

Preliminary Questions

1. Which of the following satisfy $f^{-1}(x) = f(x)$?

(a) $f(x) = x$
(b) $f(x) = 1 - x$
(c) $f(x) = 1$
(d) $f(x) = \sqrt{x}$
(e) $f(x) = |x|$
(f) $f(x) = x^{-1}$

2. The graph of a function looks like the track of a roller coaster. Is the function one-to-one?

3. The function f maps teenagers in the United States to their last names. Explain why the inverse function f^{-1} does not exist.

4. The following fragment of a train schedule for the New Jersey Transit System defines a function f from towns to times. Is f one-to-one? What is $f^{-1}(6{:}27)$?

Trenton	6:21
Hamilton Township	6:27
Princeton Junction	6:34
New Brunswick	6:38

5. A homework problem asks for a sketch of the graph of the *inverse* of $f(x) = x + \cos x$. Frank, after trying but failing to find a formula for $f^{-1}(x)$, says it's impossible to graph the inverse. Bianca hands in an accurate sketch without solving for f^{-1}. How did Bianca complete the problem?

6. What is the slope of the line obtained by reflecting the line $y = \frac{x}{2}$ through the line $y = x$?

7. Suppose that $P = (2, 4)$ lies on the graph of $f(x)$ and that the slope of the tangent line through P is $m = 3$. Assuming that $f^{-1}(x)$ exists, what is the slope of the tangent line to the graph of $f^{-1}(x)$ at the point $Q = (4, 2)$?

Exercises

1. Show that $f(x) = 7x - 4$ is invertible and find its inverse.

2. Is $f(x) = x^2 + 2$ one-to-one? If not, describe a domain on which it is one-to-one.

3. What is the largest interval containing zero on which $f(x) = \sin x$ is one-to-one?

4. Show that $f(x) = \dfrac{x-2}{x+3}$ is invertible and find its inverse.

(a) What is the domain of $f(x)$? The range of $f^{-1}(x)$?

(b) What is the domain of $f^{-1}(x)$? The range of $f(x)$?

5. Verify that $f(x) = x^3 + 3$ and $g(x) = (x - 3)^{1/3}$ are inverses by showing that $f(g(x)) = x$ and $g(f(x)) = x$.

6. Repeat Exercise 5 for $f(t) = \dfrac{t+1}{t-1}$ and $g(t) = \dfrac{t+1}{t-1}$.

7. The escape velocity from a planet of radius R is $v(R) = \sqrt{\dfrac{2GM}{R}}$, where G is the universal gravitational constant and M is the mass. Find the inverse of $v(R)$ expressing R in terms of v.

In Exercises 8–15, find a domain on which f is one-to-one and a formula for the inverse of f restricted to this domain. Sketch the graphs of f and f^{-1}.

8. $f(x) = 3x - 2$

9. $f(x) = 4 - x$

10. $f(x) = \dfrac{1}{x+1}$

11. $f(x) = \dfrac{1}{7x-3}$

12. $f(s) = \dfrac{1}{s^2}$

13. $f(x) = \dfrac{1}{\sqrt{x^2+1}}$

14. $f(z) = z^3$

15. $f(x) = \sqrt{x^3 + 9}$

16. For each function shown in Figure 15, sketch the graph of the inverse (restrict the function's domain if necessary).

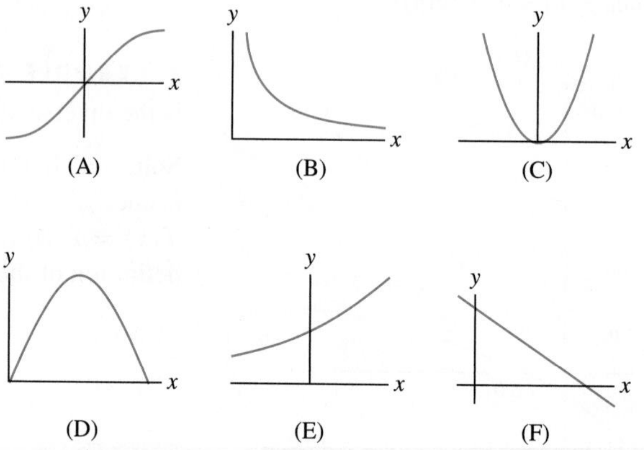

FIGURE 15

17. Which of the graphs in Figure 16 is the graph of a function satisfying $f^{-1} = f$?

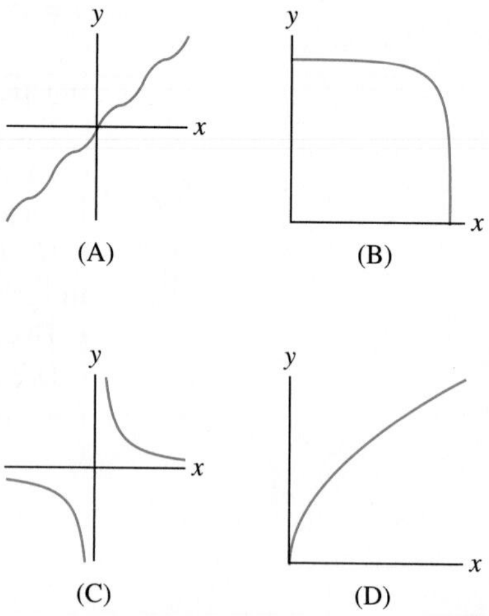

FIGURE 16

18. Let n be a nonzero integer. Find a domain on which $f(x) = (1 - x^n)^{1/n}$ coincides with its inverse. *Hint:* The answer depends on whether n is even or odd.

19. Let $f(x) = x^7 + x + 1$.

(a) Show that f^{-1} exists (but do not attempt to find it). *Hint:* Show that f is increasing.

(b) What is the domain of f^{-1}?

(c) Find $f^{-1}(3)$.

20. Show that $f(x) = (x^2 + 1)^{-1}$ is one-to-one on $(-\infty, 0]$, and find a formula for f^{-1} for this domain of f.

21. Let $f(x) = x^2 - 2x$. Determine a domain on which f^{-1} exists, and find a formula for f^{-1} for this domain of f.

22. Show that the inverse of $f(x) = e^{-x}$ exists (without finding it explicitly). What is the domain of f^{-1}?

23. Find the inverse $g(x)$ of $f(x) = \sqrt{x^2 + 9}$ with domain $x \geq 0$ and calculate $g'(x)$ in two ways: using Theorem 2 and by direct calculation.

24. Let $g(x)$ be the inverse of $f(x) = x^3 + 1$. Find a formula for $g(x)$ and calculate $g'(x)$ in two ways: using Theorem 2 and then by direct calculation.

In Exercises 25–30, use Theorem 2 to calculate $g'(x)$, where $g(x)$ is the inverse of $f(x)$.

25. $f(x) = 7x + 6$

26. $f(x) = \sqrt{3 - x}$

27. $f(x) = x^{-5}$

28. $f(x) = 4x^3 - 1$

29. $f(x) = \dfrac{x}{x + 1}$

30. $f(x) = 2 + x^{-1}$

31. Let $g(x)$ be the inverse of $f(x) = x^3 + 2x + 4$. Calculate $g(7)$ [without finding a formula for $g(x)$], and then calculate $g'(7)$.

32. Find $g'\left(-\frac{1}{2}\right)$, where $g(x)$ is the inverse of $f(x) = \dfrac{x^3}{x^2 + 1}$.

In Exercises 33–38, calculate $g(b)$ and $g'(b)$, where g is the inverse of f (in the given domain, if indicated).

33. $f(x) = x + \cos x, \quad b = 1$

34. $f(x) = 4x^3 - 2x, \quad b = -2$

35. $f(x) = \sqrt{x^2 + 6x} \quad$ for $x \geq 0, \quad b = 4$

36. $f(x) = \sqrt{x^2 + 6x} \quad$ for $x \leq -6, \quad b = 4$

37. $f(x) = \dfrac{1}{x + 1}, \quad b = \dfrac{1}{4}$

38. $f(x) = e^x, \quad b = e$

39. Let $f(x) = x^n$ and $g(x) = x^{1/n}$. Compute $g'(x)$ using Theorem 2 and check your answer using the Power Rule.

40. Show that $f(x) = \dfrac{1}{1 + x}$ and $g(x) = \dfrac{1 - x}{x}$ are inverses. Then compute $g'(x)$ directly and verify that $g'(x) = 1/f'(g(x))$.

41. Use graphical reasoning to determine if the following statements are true or false. If false, modify the statement to make it correct.

(a) If $f(x)$ is increasing, then $f^{-1}(x)$ is increasing.

(b) If $f(x)$ is decreasing, then $f^{-1}(x)$ is decreasing.

(c) If $f(x)$ is concave up, then $f^{-1}(x)$ is concave up.

(d) If $f(x)$ is concave down, then $f^{-1}(x)$ is concave down.

(e) Linear functions $f(x) = ax + b$ $(a \neq 0)$ are always one-to-one.

(f) Quadratic polynomials $f(x) = ax^2 + bx + c$ $(a \neq 0)$ are always one-to-one.

(g) $\sin x$ is not one-to-one.

Further Insights and Challenges

42. Show that if $f(x)$ is odd and $f^{-1}(x)$ exists, then $f^{-1}(x)$ is odd. Show, on the other hand, that an even function does not have an inverse.

43. Let g be the inverse of a function f satisfying $f'(x) = f(x)$. Show that $g'(x) = x^{-1}$. We will apply this in the next section to show that the inverse of $f(x) = e^x$ (the natural logarithm) is an antiderivative of x^{-1}.

FIGURE 1 Renato Solidum, director of the Philippine Institute of Volcanology and Seismology, checks the intensity of the October 8, 2004, Manila earthquake, which registered 6.2 on the Richter scale. The Richter scale is based on the logarithm (to base 10) of the amplitude of seismic waves. Each whole-number increase in Richter magnitude corresponds to a 10-fold increase in amplitude and around 31 times more energy.

7.3 Logarithms and Their Derivatives

Logarithm functions are inverses of exponential functions. More precisely, if $b > 0$ and $b \neq 1$, then the *logarithm to the base b*, denoted $\log_b x$, is the inverse of $f(x) = b^x$. By definition, $y = \log_b x$ if $b^y = x$.

$$b^{\log_b x} = x \qquad \text{and} \qquad \log_b(b^x) = x$$

Thus, $\log_b x$ is the number to which b must be raised in order to get x. For example,

$$\log_2(8) = 3 \quad \text{because} \quad 2^3 = 8$$

$$\log_{10}(1) = 0 \quad \text{because} \quad 10^0 = 1$$

$$\log_3\left(\frac{1}{9}\right) = -2 \quad \text{because} \quad 3^{-2} = \frac{1}{3^2} = \frac{1}{9}$$

The logarithm to the base e, denoted $\ln x$, plays a special role and is called the **natural logarithm**. We use a calculator to evaluate logarithms numerically. For example,

In this text, the natural logarithm is denoted $\ln x$. Other common notations are $\log x$ or $\text{Log}\, x$.

$$\ln 17 \approx 2.83321 \quad \text{because} \quad e^{2.83321} \approx 17$$

Recall that the domain of b^x is $\mathbf{R}$ and its range is the set of positive real numbers $\{x : x > 0\}$. Since the domain and range are reversed in the inverse function,

- The *domain* of $\log_b x$ is $\{x : x > 0\}$.
- The *range* of $\log_b x$ is the set of all real numbers $\mathbf{R}$.

If $b > 1$, then $\log_b x$ is positive for $x > 1$ and negative for $0 < x < 1$, and

$$\lim_{x \to 0+} \ln x = -\infty, \qquad \lim_{x \to \infty} \ln x = \infty$$

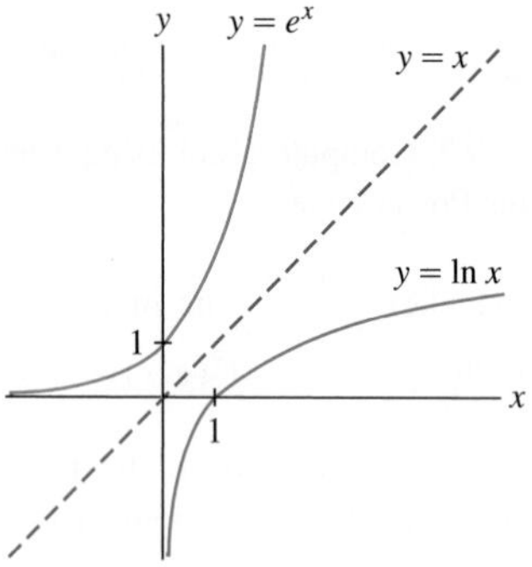

FIGURE 2 $y = \ln x$ is the inverse of $y = e^x$.

Figure 2 illustrates these facts for the base $b = e$. Keep in mind that the logarithm of a negative number does not exist. For example, $\log_{10}(-2)$ does not exist because $10^y = -2$ has no solution.

For each law of exponents, there is a corresponding law for logarithms. The rule $b^{x+y} = b^x b^y$ corresponds to the rule

$$\log_b(xy) = \log_b x + \log_b y$$

In words: *The log of a product is the sum of the logs.* To verify this rule, observe:

$$b^{\log_b(xy)} = xy = b^{\log_b x} \cdot b^{\log_b y} = b^{\log_b x + \log_b y}$$

It follows that the exponents $\log_b(xy)$ and $\log_b x + \log_b y$ are equal as claimed. The remaining logarithm laws are collected in the following table.

Laws of Logarithms

	Law	Example
Log of 1	$\log_b(1) = 0$	
Log of b	$\log_b(b) = 1$	
Products	$\log_b(xy) = \log_b x + \log_b y$	$\log_5(2 \cdot 3) = \log_5 2 + \log_5 3$
Quotients	$\log_b\left(\frac{x}{y}\right) = \log_b x - \log_b y$	$\log_2\left(\frac{3}{7}\right) = \log_2 3 - \log_2 7$
Reciprocals	$\log_b\left(\frac{1}{x}\right) = -\log_b x$	$\log_2\left(\frac{1}{7}\right) = -\log_2 7$
Powers (any n)	$\log_b(x^n) = n \log_b x$	$\log_{10}(8^2) = 2 \cdot \log_{10} 8$

We note also that all logarithm functions are proportional. More precisely, the following **change of base** formula holds (see Exercise 119):

$$\log_b x = \frac{\log_a x}{\log_a b}, \qquad \log_b x = \frac{\ln x}{\ln b} \tag{1}$$

EXAMPLE 1 Using the Logarithm Laws Evaluate:

(a) $\log_6 9 + \log_6 4$ **(b)** $\ln\left(\frac{1}{\sqrt{e}}\right)$ **(c)** $10 \log_b(b^3) - 4 \log_b(\sqrt{b})$

Solution

(a) $\log_6 9 + \log_6 4 = \log_6(9 \cdot 4) = \log_6(36) = \log_6(6^2) = 2$

(b) $\ln\left(\dfrac{1}{\sqrt{e}}\right) = \ln(e^{-1/2}) = -\dfrac{1}{2}\ln(e) = -\dfrac{1}{2}$

(c) $10\log_b(b^3) - 4\log_b(\sqrt{b}) = 10(3) - 4\log_b(b^{1/2}) = 30 - 4\left(\dfrac{1}{2}\right) = 28$ ■

■ **EXAMPLE 2 Solving an Exponential Equation** The bacteria population in a bottle at time t (in hours) has size $P(t) = 1000e^{0.35t}$. After how many hours will there be 5000 bacteria?

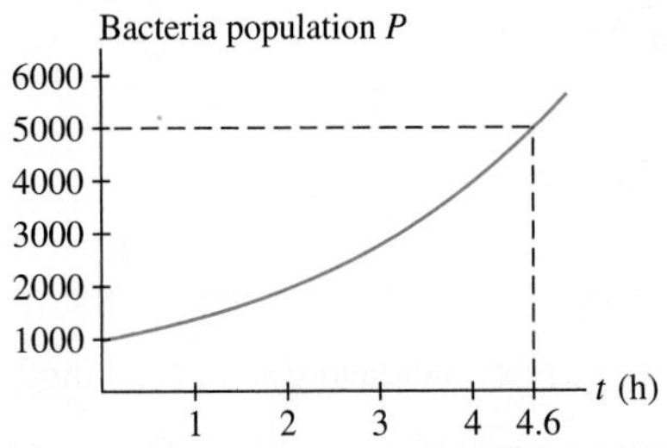

FIGURE 3 Bacteria population as a function of time.

Solution We must solve $P(t) = 1000e^{0.35t} = 5000$ for t (Figure 3):

$$e^{0.35t} = \frac{5000}{1000} = 5$$

$$\ln(e^{0.35t}) = \ln 5 \qquad \text{(take logarithms of both sides)}$$

$$0.35t = \ln 5 \approx 1.609 \qquad [\text{because } \ln(e^a) = a]$$

$$t \approx \frac{1.609}{0.35} \approx 4.6 \text{ hours}$$ ■

Calculus of Logarithms

In Section 7.1, we proved that for any base $b > 0$,

$$\frac{d}{dx}b^x = m(b)\,b^x, \qquad \text{where} \quad m(b) = \lim_{h\to 0}\frac{b^h - 1}{h}$$

However, we were not able to identify the factor $m(b)$ (other than to say that e is the unique number for which $m(e) = 1$). Now we can use the Chain Rule to prove that $m(b) = \ln b$. The key point is that every exponential function can be written in terms of e, namely, $b^x = (e^{\ln(b)})^x = e^{(\ln b)x}$. By the Chain Rule,

← REMINDER $\ln x$ *is the natural logarithm, that is,* $\ln x = \log_e x$.

$$\frac{d}{dx}b^x = \frac{d}{dx}e^{(\ln b)x} = (\ln b)e^{(\ln b)x} = (\ln b)b^x$$

THEOREM 1 Derivative of $f(x) = b^x$

$$\frac{d}{dx}b^x = (\ln b)b^x \qquad \text{for } b > 0 \qquad \boxed{2}$$

For example, $(10^x)' = (\ln 10)10^x$.

■ **EXAMPLE 3** Differentiate: **(a)** $f(x) = 4^{3x}$ and **(b)** $f(x) = 5^{x^2}$.

Solution

(a) The function $f(x) = 4^{3x}$ is a composite of 4^u and $u = 3x$:

$$\frac{d}{dx}4^{3x} = \left(\frac{d}{du}4^u\right)\frac{du}{dx} = (\ln 4)4^u(3x)' = (\ln 4)4^{3x}(3) = (3\ln 4)4^{3x}$$

(b) The function $f(x) = 5^{x^2}$ is a composite of 5^u and $u = x^2$:

$$\frac{d}{dx}5^{x^2} = \left(\frac{d}{du}5^u\right)\frac{du}{dx} = (\ln 5)5^u(x^2)' = (\ln 5)5^{x^2}(2x) = (2\ln 5)\,x\,5^{x^2}$$ ■

Next, we'll find the derivative of $\ln x$. Let $f(x) = e^x$ and $g(x) = \ln x$. Then $f(g(x)) = x$ and $g'(x) = 1/f'(g(x))$ because $g(x)$ is the inverse of $f(x)$. However, $f'(x) = f(x)$, so

$$\frac{d}{dx}\ln x = g'(x) = \frac{1}{f'(g(x))} = \frac{1}{f(g(x))} = \frac{1}{x}$$

The two most important calculus facts about exponentials and logs are

$$\frac{d}{dx}e^x = e^x, \qquad \frac{d}{dx}\ln x = \frac{1}{x}$$

THEOREM 2 Derivative of the Natural Logarithm

$$\frac{d}{dx}\ln x = \frac{1}{x} \quad \text{for } x > 0 \qquad (3)$$

■ **EXAMPLE 4** Describe the graph of $f(x) = \ln x$. Is $f(x)$ increasing or decreasing?

Solution The derivative $f'(x) = x^{-1}$ is positive on the domain $\{x : x > 0\}$, so $f(x) = \ln x$ is increasing. However, $f'(x) = x^{-1}$ is decreasing, so the graph of $f(x)$ is concave down and grows flatter as $x \to \infty$ (Figure 4). ■

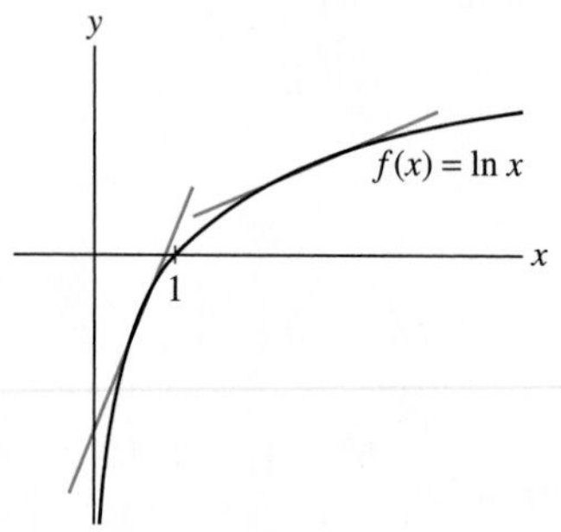

FIGURE 4 The tangent lines to $y = \ln x$ get flatter as $x \to \infty$.

■ **EXAMPLE 5** Differentiate: **(a)** $y = x\ln x$ and **(b)** $y = (\ln x)^2$.

Solution

(a) Use the Product Rule:

$$\frac{d}{dx}(x\ln x) = x\cdot(\ln x)' + (x)'\cdot\ln x = x\cdot\frac{1}{x} + \ln x = 1 + \ln x$$

(b) Use the General Power Rule,

$$\frac{d}{dx}(\ln x)^2 = 2\ln x\cdot\frac{d}{dx}\ln x = \frac{2\ln x}{x}$$ ■

In Section 3.2, we proved the Power Rule for whole number exponents. We can now prove it for all real exponents n and $x > 0$ by writing x^n as an exponential and using the Chain Rule:

$$x^n = (e^{\ln x})^n = e^{n\ln x}$$

$$\frac{d}{dx}x^n = \frac{d}{dx}e^{n\ln x} = \left(\frac{d}{dx}n\ln x\right)e^{n\ln x}$$

$$= \left(\frac{n}{x}\right)x^n = nx^{n-1}$$

There is a useful formula for the derivative of a composite function of the form $\ln(f(x))$. Let $u = f(x)$ and apply the Chain Rule:

$$\frac{d}{dx}\ln(f(x)) = \frac{d}{du}\ln(u)\frac{du}{dx} = \frac{1}{u}\cdot u' = \frac{1}{f(x)}f'(x)$$

$$\frac{d}{dx}\ln(f(x)) = \frac{f'(x)}{f(x)} \qquad (4)$$

■ **EXAMPLE 6** Differentiate: **(a)** $y = \ln(x^3+1)$ and **(b)** $y = \ln(\sqrt{\sin x})$.

Solution Use Eq. (4):

(a) $\displaystyle\frac{d}{dx}\ln(x^3+1) = \frac{(x^3+1)'}{x^3+1} = \frac{3x^2}{x^3+1}$.

(b) The algebra is simpler if we write $\ln(\sqrt{\sin x}) = \ln\left((\sin x)^{1/2}\right) = \frac{1}{2}\ln(\sin x)$:

$$\frac{d}{dx}\ln(\sqrt{\sin x}) = \frac{1}{2}\frac{d}{dx}\ln(\sin x) = \frac{1}{2}\frac{(\sin x)'}{\sin x} = \frac{1}{2}\frac{\cos x}{\sin x} = \frac{1}{2}\cot x$$ ■

EXAMPLE 7 Logarithm to Another Base Calculate $\frac{d}{dx}\log_{10} x$.

The "change of base" formula [Eq. (1)]

$$\log_b x = \frac{\ln x}{\ln b}$$

shows that for any base $b > 0$, $b \neq 1$:

$$\frac{d}{dx}\log_b x = \frac{1}{(\ln b)x}$$

Solution By the change of base formula recalled in the margin, $\log_{10} x = \frac{\ln x}{\ln 10}$, and therefore

$$\frac{d}{dx}\log_{10} x = \frac{d}{dx}\left(\frac{\ln x}{\ln 10}\right) = \frac{1}{\ln 10}\frac{d}{dx}\ln x = \frac{1}{(\ln 10)x}$$

The next example illustrates **logarithmic differentiation**. This technique saves work when the function is a product or quotient with several factors.

EXAMPLE 8 Logarithmic Differentiation Find the derivative of

$$f(x) = \frac{(x+1)^2(2x^2-3)}{\sqrt{x^2+1}}$$

Solution In logarithmic differentiation, we differentiate $\ln(f(x))$ rather than $f(x)$ itself. First, expand $\ln(f(x))$ using the logarithm rules:

$$\ln(f(x)) = \ln\left((x+1)^2\right) + \ln\left(2x^2-3\right) - \ln\left(\sqrt{x^2+1}\right)$$
$$= 2\ln(x+1) + \ln(2x^2-3) - \frac{1}{2}\ln(x^2+1)$$

Then use Eq. (4):

$$\frac{f'(x)}{f(x)} = \frac{d}{dx}\ln(f(x)) = 2\frac{d}{dx}\ln\left(x+1\right) + \frac{d}{dx}\ln\left(2x^2-3\right) - \frac{1}{2}\frac{d}{dx}\ln\left(x^2+1\right)$$

$$\frac{f'(x)}{f(x)} = 2\frac{1}{x+1} + \frac{4x}{2x^2-3} - \frac{1}{2}\frac{2x}{x^2+1}$$

Finally, multiply through by $f(x)$:

$$f'(x) = \left(\frac{(x+1)^2(2x^2-3)}{\sqrt{x^2+1}}\right)\left(\frac{2}{x+1} + \frac{4x}{2x^2-3} - \frac{x}{x^2+1}\right)$$

EXAMPLE 9 Differentiate (for $x > 0$): **(a)** $f(x) = x^x$ and **(b)** $g(x) = x^{\sin x}$.

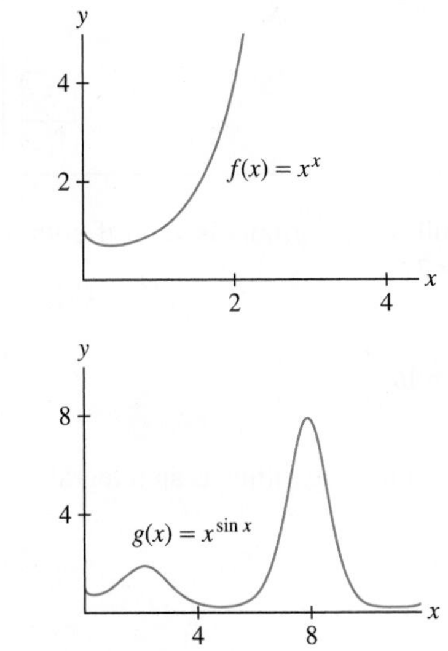

FIGURE 5 Graphs of $f(x) = x^x$ and $g(x) = x^{\sin x}$.

Solution The two problems are similar (Figure 5). We illustrate two different methods.

(a) Method 1: Use the identity $x = e^{\ln x}$ to write $f(x)$ as an exponential:

$$f(x) = x^x = (e^{\ln x})^x = e^{x\ln x}$$
$$f'(x) = (x\ln x)'e^{x\ln x} = (1+\ln x)e^{x\ln x} = (1+\ln x)x^x$$

(b) Method 2: Apply Eq. (4) to $\ln(g(x))$. Since $\ln(g(x)) = \ln(x^{\sin x}) = (\sin x)\ln x$,

$$\frac{g'(x)}{g(x)} = \frac{d}{dx}\ln(g(x)) = \frac{d}{dx}\Big((\sin x)\ln x\Big) = \frac{\sin x}{x} + (\cos x)\ln x$$

$$g'(x) = \left(\frac{\sin x}{x} + (\cos x)\ln x\right)g(x) = \left(\frac{\sin x}{x} + (\cos x)\ln x\right)x^{\sin x}$$

The Logarithm as an Integral

In Chapter 5, we noted that the Power Rule for Integrals is valid for all exponents $n \neq -1$:

$$\int x^n\,dx = \frac{x^{n+1}}{n+1} + C \qquad (n \neq -1)$$

This formula is not valid (or meaningful) for $n = -1$, so the question remained: *What is the antiderivative of* $y = x^{-1}$? We can now give the answer: the natural logarithm. Indeed, the formula $(\ln x)' = x^{-1}$ tells us that $\ln x$ is an antiderivative of $y = x^{-1}$ for $x > 0$:

$$\int \frac{dx}{x} = \ln x + C$$

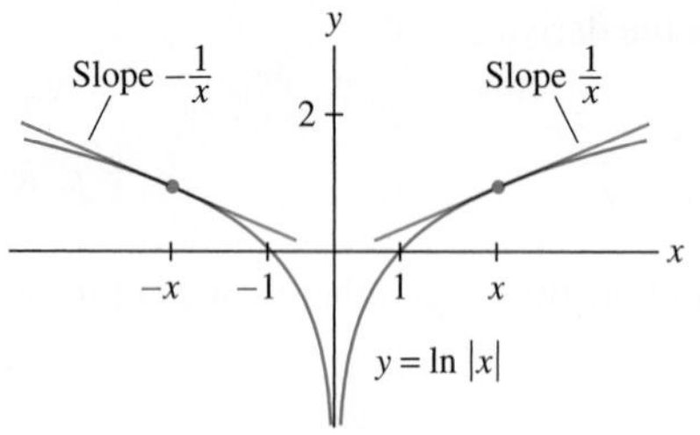

FIGURE 6

We would like to have an antiderivative of $y = \frac{1}{x}$ on its full domain, namely on the domain $\{x : x \neq 0\}$. To achieve this end, we extend $F(x)$ to an even function by setting $F(x) = \ln|x|$ (Figure 6). Then $F(x) = F(-x)$, and by the Chain Rule, $F'(x) = -F'(-x)$. For $x < 0$, we obtain

$$\frac{d}{dx}\ln|x| = F'(x) = -F'(-x) = -\frac{1}{-x} = \frac{1}{x}$$

This proves that $\frac{d}{dx}\ln|x| = \frac{1}{x}$ for all $x \neq 0$.

THEOREM 3 Antiderivative of $y = \frac{1}{x}$ The function $F(x) = \ln|x|$ is an antiderivative of $y = \frac{1}{x}$ in the domain $\{x : x \neq 0\}$, that is,

$$\int \frac{dx}{x} = \ln|x| + C \qquad \boxed{5}$$

By the Fundamental Theorem of Calculus, the following formula is valid if both a and b are either both positive or both negative (Figure 7):

$$\int_a^b \frac{dx}{x} = \ln|b| - \ln|a| = \ln\frac{b}{a} \qquad \boxed{6}$$

Setting $a = 1$ and $b = x$, we obtain a formula for the natural logarithm as an integral:

$$\ln x = \int_1^x \frac{dt}{t} \qquad \boxed{7}$$

■ **EXAMPLE 10 The Logarithm as an Antiderivative** Evaluate:

$$\textbf{(a)}\ \int_2^8 \frac{dx}{x} \quad \text{and} \quad \textbf{(b)}\ \int_{-4}^{-2} \frac{dx}{x}$$

Solution By Eq. (6),

$$\int_2^8 \frac{dx}{x} = \ln\frac{8}{2} = \ln 4 \approx 1.39 \quad \text{and} \quad \int_{-4}^{-2} \frac{dx}{x} = \ln\left(\frac{-2}{-4}\right) = \ln\frac{1}{2} \approx -0.69$$

The area represented by these integrals is shown in Figures 7(B) and (C). ■

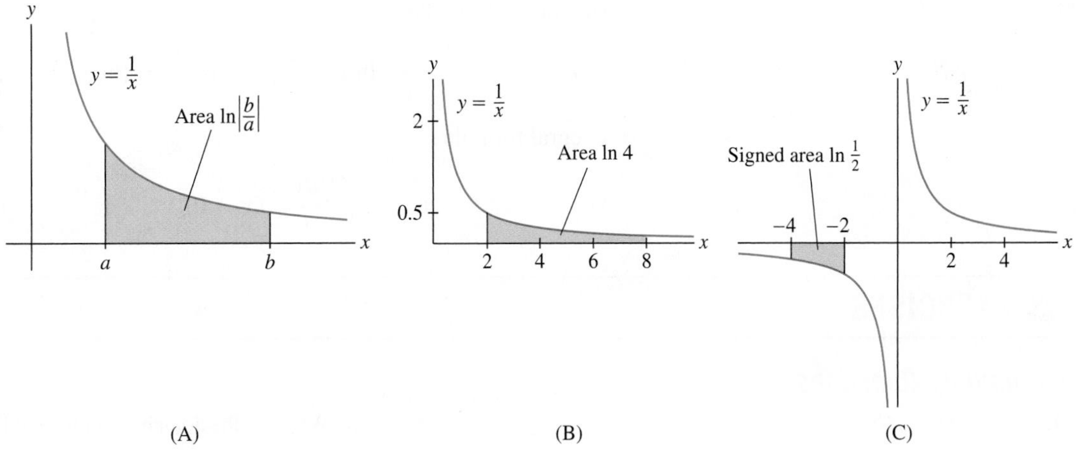

FIGURE 7

■ **EXAMPLE 11** Evaluate:

(a) $\displaystyle\int_1^3 \frac{x}{x^2+1}\,dx$ **(b)** $\displaystyle\int \tan x\,dx$

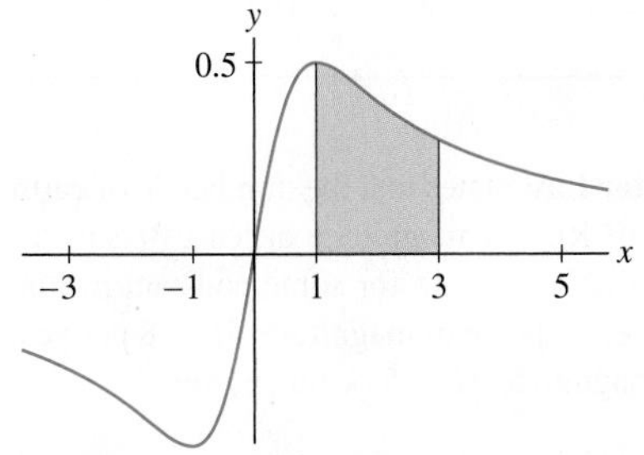

FIGURE 8 Area under the graph of $y = \dfrac{x}{x^2+1}$ over $[1, 3]$.

Solution

(a) Use the substitution $u = x^2 + 1$, $\frac{1}{2}du = x\,dx$. In the u-variable, the limits of the integral become $u(1) = 2$ and $u(3) = 10$. The integral is equal to the area shown in Figure 8:

$$\int_1^3 \frac{x}{x^2+1}\,dx = \frac{1}{2}\int_2^{10} \frac{du}{u} = \frac{1}{2}\ln|u|\Big|_2^{10} = \frac{1}{2}\ln 10 - \frac{1}{2}\ln 2 \approx 0.805$$

(b) Use the substitution $u = \cos x$, $du = -\sin x$:

$$\int \tan x\,dx = \int \frac{\sin x}{\cos x}\,dx = -\int \frac{du}{u} = -\ln|u| + C$$

$$= -\ln|\cos x| + C = \ln\left|\frac{1}{\cos x}\right| + C = \ln|\sec x| + C$$ ■

7.3 SUMMARY

- For $b > 0$ with $b \neq 1$, the *logarithm function* $\log_b x$ is the inverse of b^x;

$$x = b^y \quad \Leftrightarrow \quad y = \log_b x$$

- If $b > 1$, then $\log_b x$ is positive for $x > 1$ and negative for $0 < x < 1$, and

$$\lim_{x\to 0+} \ln x = -\infty, \qquad \lim_{x\to\infty} \ln x = \infty$$

- The *natural logarithm* is the logarithm to the base e and is denoted $\ln x$.
- Important logarithm laws:

(i) $\log_b(xy) = \log_b x + \log_b y$ **(ii)** $\log_b\left(\dfrac{x}{y}\right) = \log_b x - \log_b y$

(iii) $\log_b(x^n) = n\log_b x$ **(iv)** $\log_b 1 = 0$ and $\log_b b = 1$

- Derivative formulas:

$$(e^x)' = e^x, \qquad \frac{d}{dx}\ln x = \frac{1}{x}, \qquad (b^x)' = (\ln b)b^x, \qquad \frac{d}{dx}\log_b x = \frac{1}{(\ln b)\,x}$$

- Integral formulas:

$$\ln x = \int_1^x \frac{dt}{t} \quad (x > 0), \qquad \int \frac{dx}{x} = \ln|x| + C$$

7.3 EXERCISES

Preliminary Questions

1. Compute $\log_{b^2}(b^4)$.
2. When is $\ln x$ negative?
3. What is $\ln(-3)$? Explain.
4. Explain the phrase "The logarithm converts multiplication into addition."
5. What are the domain and range of $\ln x$?
6. Does x^{-1} have an antiderivative for $x < 0$? If so, describe one.
7. What is the slope of the tangent line to $y = 4^x$ at $x = 0$?
8. What is the rate of change of $y = \ln x$ at $x = 10$?

Exercises

In Exercises 1–16, calculate without using a calculator.

1. $\log_3 27$
2. $\log_5 \frac{1}{25}$
3. $\ln 1$
4. $\log_5(5^4)$
5. $\log_2(2^{5/3})$
6. $\log_2(8^{5/3})$
7. $\log_{64} 4$
8. $\log_7(49^2)$
9. $\log_8 2 + \log_4 2$
10. $\log_{25} 30 + \log_{25} \frac{5}{6}$
11. $\log_4 48 - \log_4 12$
12. $\ln(\sqrt{e}\cdot e^{7/5})$
13. $\ln(e^3) + \ln(e^4)$
14. $\log_2 \frac{4}{3} + \log_2 24$
15. $7^{\log_7(29)}$
16. $8^{3\log_8(2)}$

17. Write as the natural log of a single expression:
(a) $2\ln 5 + 3\ln 4$ **(b)** $5\ln(x^{1/2}) + \ln(9x)$

18. Solve for x: $\ln(x^2+1) - 3\ln x = \ln(2)$.

In Exercises 19–24, solve for the unknown.

19. $7e^{5t} = 100$
20. $6e^{-4t} = 2$
21. $2^{x^2-2x} = 8$
22. $e^{2t+1} = 9e^{1-t}$
23. $\ln(x^4) - \ln(x^2) = 2$
24. $\log_3 y + 3\log_3(y^2) = 14$

25. Show, by producing a counterexample, that $\ln(ab)$ is not equal to $(\ln a)(\ln b)$.

26. What is b if $(\log_b x)' = \dfrac{1}{3x}$?

27. The population of a city (in millions) at time t (years) is $P(t) = 2.4e^{0.06t}$, where $t = 0$ is the year 2000. When will the population double from its size at $t = 0$?

28. The **Gutenberg–Richter Law** states that the number N of earthquakes per year worldwide of Richter magnitude at least M satisfies an approximate relation $\log_{10} N = a - M$ for some constant a. Find a, assuming that there is one earthquake of magnitude $M \ge 8$ per year. How many earthquakes of magnitude $M \ge 5$ occur per year?

In Exercises 29–48, find the derivative.

29. $y = x\ln x$
30. $y = t\ln t - t$
31. $y = (\ln x)^2$
32. $y = \ln(x^5)$
33. $y = \ln(9x^2 - 8)$
34. $y = \ln(t5^t)$
35. $y = \ln(\sin t + 1)$
36. $y = x^2 \ln x$
37. $y = \dfrac{\ln x}{x}$
38. $y = e^{(\ln x)^2}$
39. $y = \ln(\ln x)$
40. $y = \ln(\cot x)$
41. $y = \left(\ln(\ln x)\right)^3$
42. $y = \ln\left((\ln x)^3\right)$
43. $y = \ln\left((x+1)(2x+9)\right)$
44. $y = \ln\left(\dfrac{x+1}{x^3+1}\right)$
45. $y = 11^x$
46. $y = 7^{4x-x^2}$
47. $y = \dfrac{2^x - 3^{-x}}{x}$
48. $y = 16^{\sin x}$

In Exercises 49–52, compute the derivative.

49. $f'(x)$, $f(x) = \log_2 x$
50. $f'(3)$, $f(x) = \log_5 x$
51. $\dfrac{d}{dt}\log_3(\sin t)$
52. $\dfrac{d}{dt}\log_{10}(t + 2^t)$

In Exercises 53–64, find an equation of the tangent line at the point indicated.

53. $f(x) = 6^x$, $x = 2$
54. $y = (\sqrt{2})^x$, $x = 8$

55. $s(t) = 3^{9t}, \quad t = 2$

56. $y = \pi^{5x-2}, \quad x = 1$

57. $f(x) = 5^{x^2-2x}, \quad x = 1$

58. $s(t) = \ln t, \quad t = 5$

59. $s(t) = \ln(8 - 4t), \quad t = 1$

60. $f(x) = \ln(x^2), \quad x = 4$

61. $R(z) = \log_5(2z^2 + 7), \quad z = 3$

62. $y = \ln(\sin x), \quad x = \dfrac{\pi}{4}$

63. $f(w) = \log_2 w, \quad w = \frac{1}{8}$

64. $y = \log_2(1 + 4x^{-1}), \quad x = 4$

In Exercises 65–72, find the derivative using logarithmic differentiation as in Example 8.

65. $y = (x + 5)(x + 9)$

66. $y = (3x + 5)(4x + 9)$

67. $y = (x - 1)(x - 12)(x + 7)$

68. $y = \dfrac{x(x + 1)^3}{(3x - 1)^2}$

69. $y = \dfrac{x(x^2 + 1)}{\sqrt{x + 1}}$

70. $y = (2x + 1)(4x^2)\sqrt{x - 9}$

71. $y = \sqrt{\dfrac{x(x + 2)}{(2x + 1)(3x + 2)}}$

72. $y = (x^3 + 1)(x^4 + 2)(x^5 + 3)^2$

In Exercises 73–78, find the derivative using either method of Example 9.

73. $f(x) = x^{3x}$

74. $f(x) = x^{\cos x}$

75. $f(x) = x^{e^x}$

76. $f(x) = x^{x^2}$

77. $f(x) = x^{3^x}$

78. $f(x) = e^{x^x}$

In Exercises 79–82, find the local extreme values in the domain $\{x : x > 0\}$ and use the Second Derivative Test to determine whether these values are local minima or maxima.

79. $g(x) = \dfrac{\ln x}{x}$

80. $g(x) = x \ln x$

81. $g(x) = \dfrac{\ln x}{x^3}$

82. $g(x) = x - \ln x$

In Exercises 83 and 84, find the local extreme values and points of inflection, and sketch the graph of $y = f(x)$ over the interval $[1, 4]$.

83. $f(x) = \dfrac{10 \ln x}{x^2}$

84. $f(x) = x^2 - 8 \ln x$

In Exercises 85–105, evaluate the indefinite integral, using substitution if necessary.

85. $\displaystyle\int \frac{7\,dx}{x}$

86. $\displaystyle\int \frac{dx}{x + 7}$

87. $\displaystyle\int \frac{dx}{2x + 4}$

88. $\displaystyle\int \frac{dx}{9x - 3}$

89. $\displaystyle\int \frac{t\,dt}{t^2 + 4}$

90. $\displaystyle\int \frac{x^2\,dx}{x^3 + 2}$

91. $\displaystyle\int \frac{(3x - 1)\,dx}{9 - 2x + 3x^2}$

92. $\displaystyle\int \tan(4x + 1)\,dx$

93. $\displaystyle\int \cot x\,dx$

94. $\displaystyle\int \frac{\cos x}{2 \sin x + 3}\,dx$

95. $\displaystyle\int \frac{\ln x}{x}\,dx$

96. $\displaystyle\int \frac{4 \ln x + 5}{x}\,dx$

97. $\displaystyle\int \frac{(\ln x)^2}{x}\,dx$

98. $\displaystyle\int \frac{dx}{x \ln x}$

99. $\displaystyle\int \frac{dx}{(4x - 1)\ln(8x - 2)}$

100. $\displaystyle\int \frac{\ln(\ln x)}{x \ln x}\,dx$

101. $\displaystyle\int \cot x \ln(\sin x)\,dx$

102. $\displaystyle\int 3^x\,dx$

103. $\displaystyle\int x 3^{x^2}\,dx$

104. $\displaystyle\int \cos x\, 3^{\sin x}\,dx$

105. $\displaystyle\int \left(\frac{1}{2}\right)^{3x+2} dx$

In Exercises 106–111, evaluate the definite integral.

106. $\displaystyle\int_1^2 \frac{1}{x}\,dx$

107. $\displaystyle\int_4^{12} \frac{1}{x}\,dx$

108. $\displaystyle\int_1^e \frac{1}{x}\,dx$

109. $\displaystyle\int_2^4 \frac{dt}{3t + 4}$

110. $\displaystyle\int_{-e^2}^{-e} \frac{1}{t}\,dt$

111. $\displaystyle\int_e^{e^2} \frac{1}{t \ln t}\,dt$

112. CAS Find a good numerical approximation to the coordinates of the point on the graph of $y = \ln x - x$ closest to the origin (Figure 9).

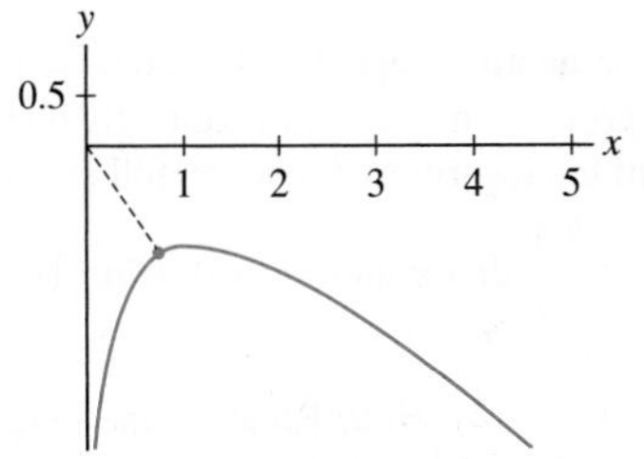

FIGURE 9 Graph of $y = \ln x - x$.

113. Find the minimum value of $f(x) = x^x$ for $x > 0$.

114. Use the formula $(\ln f(x))' = f'(x)/f(x)$ to show that $\ln x$ and $\ln(2x)$ have the same derivative. Is there a simpler explanation of this result?

115. According to one simplified model, the purchasing power of a dollar in the year $2000 + t$ is equal to $P(t) = 0.68(1.04)^{-t}$ (in 1983 dollars). Calculate the predicted rate of decline in purchasing power (in cents per year) in the year 2020.

116. The energy E (in joules) radiated as seismic waves by an earthquake of Richter magnitude M satisfies $\log_{10} E = 4.8 + 1.5M$.

(a) Show that when M increases by 1, the energy increases by a factor of approximately 31.5.

(b) Calculate dE/dM.

117. The Palermo Technical Impact Hazard Scale P is used to quantify the risk associated with the impact of an asteroid colliding with the earth:

$$P = \log_{10}\left(\frac{p_i E^{0.8}}{0.03T}\right)$$

where p_i is the probability of impact, T is the number of years until impact, and E is the energy of impact (in megatons of TNT). The risk is greater than a random event of similar magnitude if $P > 0$.

(a) Calculate dP/dT, assuming that $p_i = 2 \times 10^{-5}$ and $E = 2$ megatons.

(b) Use the derivative to estimate the change in P if T increases from 8 to 9 years.

Further Insights and Challenges

118. (a) Show that if f and g are differentiable, then

$$\frac{d}{dx}\ln(f(x)g(x)) = \frac{f'(x)}{f(x)} + \frac{g'(x)}{g(x)} \qquad \boxed{8}$$

(b) Give a new proof of the Product Rule by observing that the left-hand side of Eq. (8) is equal to $\dfrac{(f(x)g(x))'}{f(x)g(x)}$.

119. Prove the formula

$$\log_b x = \frac{\log_a x}{\log_a b}$$

for all positive numbers a, b with $a \neq 1$ and $b \neq 1$.

120. Prove the formula $\log_a b \log_b a = 1$ for all positive numbers a, b with $a \neq 1$ and $b \neq 1$.

Exercises 121–123 develop an elegant approach to the exponential and logarithm functions. Define a function $G(x)$ for $x > 0$:

$$G(x) = \int_1^x \frac{1}{t}\,dt$$

121. Defining $\ln x$ as an Integral This exercise proceeds as if we didn't know that $G(x) = \ln x$ and shows directly that $G(x)$ has all the basic properties of the logarithm. Prove the following statements.

(a) $\displaystyle\int_a^{ab} \frac{1}{t}\,dt = \int_1^b \frac{1}{t}\,dt$ for all $a, b > 0$. *Hint:* Use the substitution $u = t/a$.

(b) $G(ab) = G(a) + G(b)$. *Hint:* Break up the integral from 1 to ab into two integrals and use (a).

(c) $G(1) = 0$ and $G(a^{-1}) = -G(a)$ for $a > 0$.

(d) $G(a^n) = nG(a)$ for all $a > 0$ and integers n.

(e) $G(a^{1/n}) = \dfrac{1}{n}G(a)$ for all $a > 0$ and integers $n \neq 0$.

(f) $G(a^r) = rG(a)$ for all $a > 0$ and rational numbers r.

(g) $G(x)$ is increasing. *Hint:* Use FTC II.

(h) There exists a number a such that $G(a) > 1$. *Hint:* Show that $G(2) > 0$ and take $a = 2^m$ for $m > 1/G(2)$.

(i) $\lim_{x\to\infty} G(x) = \infty$ and $\lim_{x\to 0+} G(x) = -\infty$

(j) There exists a unique number E such that $G(E) = 1$.

(k) $G(E^r) = r$ for every rational number r.

122. Defining e^x Use Exercise 121 to prove the following statements.

(a) $G(x)$ has an inverse with domain **R** and range $\{x : x > 0\}$. Denote the inverse by $F(x)$.

(b) $F(x + y) = F(x)F(y)$ for all x, y. Hint: it suffices to show that $G(F(x)F(y)) = G(F(x + y))$.

(c) $F(r) = E^r$ for all numbers. In particular, $F(0) = 1$.

(d) $F'(x) = F(x)$. *Hint:* Use the formula for the derivative of an inverse function.

This shows that $E = e$ and that $F(x)$ is the function e^x as defined in the text.

123. Defining b^x Let $b > 0$ and let $f(x) = F(xG(b))$ with F as in Exercise 122. Use Exercise 121 (f) to prove that $f(r) = b^r$ for every rational number r. This gives us a way of defining b^x for irrational x, namely $b^x = f(x)$. With this definition, b^x is a differentiable function of x (because F is differentiable).

7.4 Exponential Growth and Decay

In this section, we explore some applications of the exponential function. Consider a quantity $P(t)$ that depends exponentially on time:

The constant k has units of "inverse time"; if t is measured in days, then k has units of $(\text{days})^{-1}$.

$$\boxed{P(t) = P_0 e^{kt}}$$

If $k > 0$, then $P(t)$ *grows exponentially* and k is called the *growth constant.* Note that P_0 is the initial size (the size at $t = 0$):

$$P(0) = P_0 e^{k\cdot 0} = P_0$$

We can also write $P(t) = P_0 b^t$ with $b = e^k$, because $b^t = (e^k)^t = e^{kt}$.

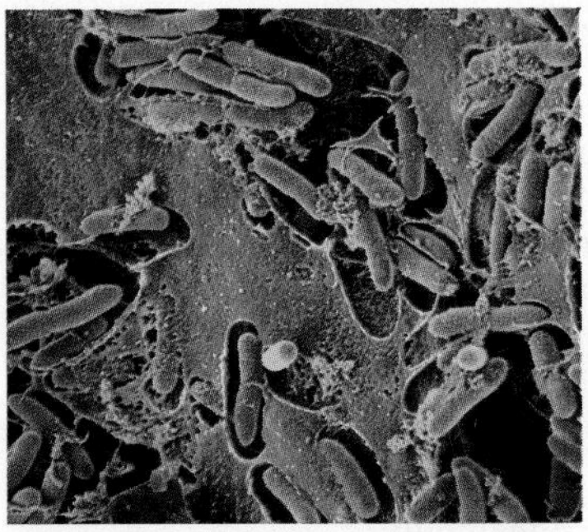

FIGURE 1 *E. coli* bacteria, found in the human intestine.

Exponential growth cannot continue over long periods of time. A colony starting with one E. coli *cell would grow to* 5×10^{89} *cells after 3 weeks—much more than the estimated number of atoms in the observable universe. In actual cell growth, the exponential phase is followed by a period in which growth slows and may decline.*

A quantity that decreases exponentially is said to have *exponential decay*. In this case, we write $P(t) = P_0 e^{-kt}$ with $k > 0$; k is then called the *decay constant.*

Population is a typical example of a quantity that grows exponentially, at least under suitable conditions. To understand why, consider a cell colony with initial population $P_0 = 100$ and assume that each cell divides into two cells after 1 hour. Then population $P(t)$ doubles with each passing hour:

$$P(0) = 100 \quad \text{(initial population)}$$
$$P(1) = 2(100) = 200 \quad \text{(population doubles)}$$
$$P(2) = 2(200) = 400 \quad \text{(population doubles again)}$$

After t hours, $P(t) = (100)2^t$.

■ **EXAMPLE 1** In the laboratory, the number of *Escherichia coli* bacteria (Figure 1) grows exponentially with growth constant of $k = 0.41$ (hours)$^{-1}$. Assume that 1000 bacteria are present at time $t = 0$.

(a) Find the formula for the number of bacteria $P(t)$ at time t.
(b) How large is the population after 5 hours?
(c) When will the population reach 10,000?

Solution The growth is exponential, so $P(t) = P_0 e^{kt}$.

(a) The initial size is $P_0 = 1000$ and $k = 0.41$, so $P(t) = 1000e^{0.41t}$ (t in hours).
(b) After 5 hours, $P(5) = 1000e^{0.41 \cdot 5} = 1000e^{2.05} \approx 7767.9$. Because the number of bacteria is a whole number, we round off the answer to 7768.
(c) The problem asks for the time t such that $P(t) = 10{,}000$, so we solve

$$1000e^{0.41t} = 10{,}000 \quad \Rightarrow \quad e^{0.41t} = \frac{10{,}000}{1000} = 10$$

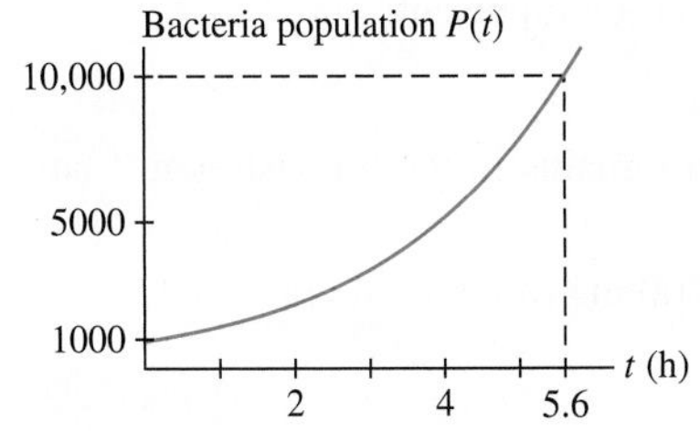

FIGURE 2 Growth of *E. coli* population.

Taking the logarithm of both sides, we obtain $\ln\left(e^{0.41t}\right) = \ln 10$, or

$$0.41t = \ln 10 \quad \Rightarrow \quad t = \frac{\ln 10}{0.41} \approx 5.62$$

Therefore, $P(t)$ reaches 10,000 after approximately 5 hours, 37 minutes (Figure 2). ■

The important role played by exponential functions is best understood in terms of the *differential equation* $y' = ky$. The function $y = P_0 e^{kt}$ satisfies this differential equation, as we can check directly:

$$y' = \frac{d}{dt}\left(P_0 e^{kt}\right) = kP_0 e^{kt} = ky$$

Theorem 1 goes further and asserts that the exponential functions are the *only* functions that satisfy this differential equation.

A differential equation is an equation relating a function $y = f(x)$ *to its derivative* y' *(or higher derivatives* y', y'', y''', ...*).*

THEOREM 1 If $y(t)$ is a differentiable function satisfying the differential equation

$$y' = ky$$

then $y(t) = P_0 e^{kt}$, where P_0 is the initial value $P_0 = y(0)$.

Proof Compute the derivative of ye^{-kt}. If $y' = ky$, then

$$\frac{d}{dt}\left(ye^{-kt}\right) = y'e^{-kt} - ke^{-kt}y = (ky)e^{-kt} - ke^{-kt}y = 0$$

Because the derivative is zero, $y(t)e^{-kt} = P_0$ for some constant P_0, and $y(t) = P_0e^{kt}$ as claimed. The initial value is $y(0) = P_0e^0 = P_0$. ■

FIGURE 3 Computer simulation of radioactive decay as a random process. The red squares are atoms that have not yet decayed. A fixed fraction of red squares turns white in each unit of time.

CONCEPTUAL INSIGHT Theorem 1 tells us that a process obeys an exponential law precisely when *its rate of change is proportional to the amount present.* This helps us understand why certain quantities grow or decay exponentially.

A population grows exponentially because each organism contributes to growth through reproduction, and thus the growth rate is proportional to the population size. However, this is true only under certain conditions. If the organisms interact—say, by competing for food or mates—then the growth rate may not be proportional to population size and we cannot expect exponential growth.

Similarly, experiments show that radioactive substances decay exponentially. This suggests that radioactive decay is a random process in which a fixed fraction of atoms, randomly chosen, decays per unit time (Figure 3). If exponential decay were not observed, we might suspect that the decay was influenced by some interaction between the atoms.

■ **EXAMPLE 2** Find all solutions of $y' = 3y$. Which solution satisfies $y(0) = 9$?

Solution The solutions to $y' = 3y$ are the functions $y(t) = Ce^{3t}$, where C is the initial value $C = y(0)$. The particular solution satisfying $y(0) = 9$ is $y(t) = 9e^{3t}$. ■

■ **EXAMPLE 3 Modeling Penicillin** Pharmacologists have shown that penicillin leaves a person's bloodstream at a rate proportional to the amount present.

(a) Express this statement as a differential equation.

(b) Find the decay constant if 50 mg of penicillin remains in the bloodstream 7 hours after an initial injection of 450 mg.

(c) Under the hypothesis of (b), at what time was 200 mg of penicillin present?

Solution

(a) Let $A(t)$ be the quantity of penicillin present in the bloodstream at time t. Since the rate at which penicillin leaves the bloodstream is proportional to $A(t)$,

$$A'(t) = -kA(t) \qquad \boxed{1}$$

where $k > 0$ because $A(t)$ is decreasing.

(b) Eq. (1) and the condition $A(0) = 450$ tell us that $A(t) = 450e^{-kt}$. The additional condition $A(7) = 50$ enables us to solve for k:

$$A(7) = 450e^{-7k} = 50 \quad \Rightarrow \quad e^{-7k} = \frac{1}{9} \quad \Rightarrow \quad -7k = \ln\frac{1}{9}$$

Thus, $k = -\frac{1}{7}\ln\frac{1}{9} \approx 0.31$.

(c) To find the time t at which 200 mg was present, we solve

$$A(t) = 450e^{-0.31t} = 200 \quad \Rightarrow \quad e^{-0.31t} = \frac{4}{9}$$

Therefore, $t = -\frac{1}{0.31}\ln\left(\frac{4}{9}\right) \approx 2.62$ hours (Figure 4). ■

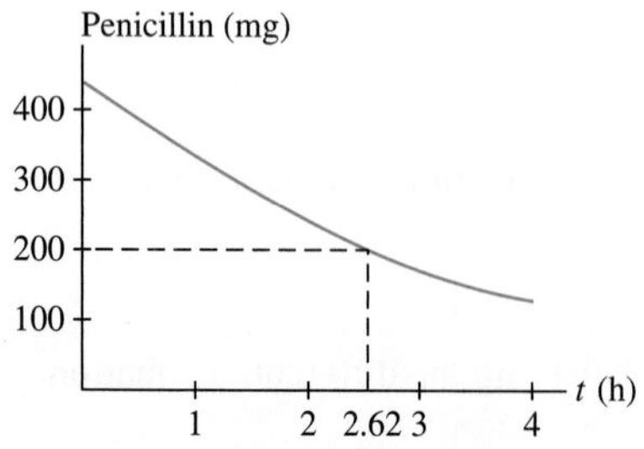

FIGURE 4 The quantity of penicillin in the bloodstream decays exponentially.

The constant k has units of time^{-1}, so the doubling time $T = (\ln 2)/k$ has units of time, as we should expect. A similar calculation shows that the tripling time is $(\ln 3)/k$, the quadrupling time is $(\ln 4)/k$, and, in general, the time to n-fold increase is $(\ln n)/k$.

Quantities that grow exponentially possess an important property: There is a doubling time T such that $P(t)$ doubles in size over every time interval of length T. To prove this, let $P(t) = P_0 e^{kt}$ and solve for T in the equation $P(t+T) = 2P(t)$.

$$P_0 e^{k(t+T)} = 2P_0 e^{kt}$$

$$e^{kt} e^{kT} = 2e^{kt}$$

$$e^{kT} = 2$$

We obtain $kT = \ln 2$ or $T = (\ln 2)/k$.

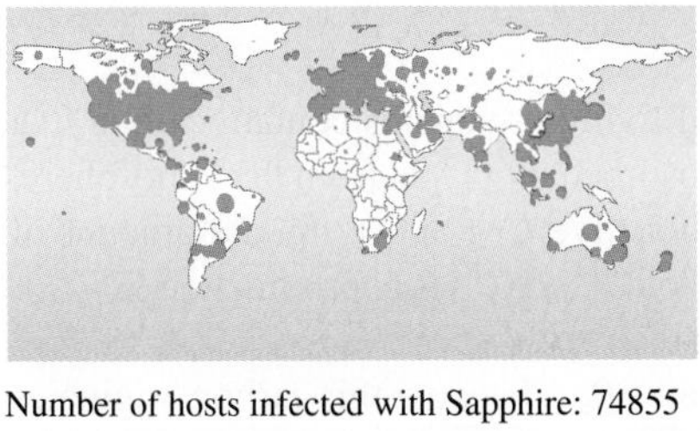

FIGURE 5 Spread of the Sapphire computer virus 30 minutes after release. The infected hosts spewed billions of copies of the virus into cyberspace, significantly slowing Internet traffic and interfering with businesses, flight schedules, and automated teller machines.

Doubling Time If $P(t) = P_0 e^{kt}$ with $k > 0$, then the doubling time of P is

$$\text{Doubling time} = \frac{\ln 2}{k}$$

■ **EXAMPLE 4 Spread of the Sapphire Worm** A computer virus nicknamed the *Sapphire Worm* spread throughout the Internet on January 25, 2003 (Figure 5). Studies suggest that during the first few minutes, the population of infected computer hosts increased exponentially with growth constant $k = 0.0815\ \text{s}^{-1}$.

(a) What was the doubling time of the virus?

(b) If the virus began in four computers, how many hosts were infected after 2 minutes? After 3 minutes?

Solution

(a) The doubling time is $(\ln 2)/0.0815 \approx 8.5$ seconds (Figure 6).

(b) If $P_0 = 4$, the number of infected hosts after t seconds is $P(t) = 4e^{0.0815t}$. After 2 minutes (120 seconds), the number of infected hosts is

$$P(120) = 4e^{0.0815(120)} \approx 70{,}700$$

After 3 minutes, the number would have been $P(180) = 4e^{0.0815(180)} \approx 9.4$ million. However, it is estimated that a total of around 75,000 hosts were infected, so the exponential phase of the virus could not have lasted much more than 2 minutes. ■

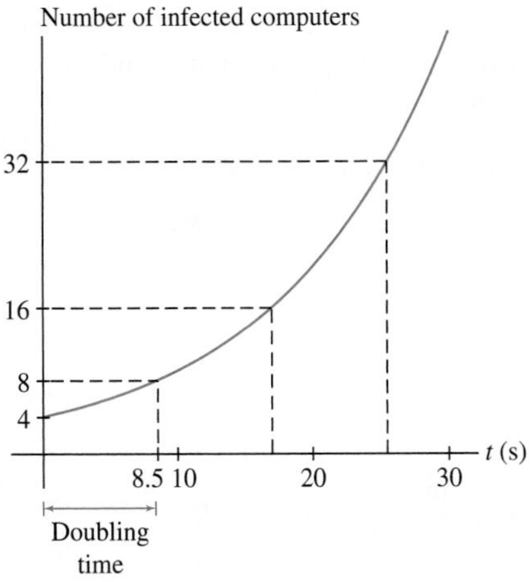

FIGURE 6 Doubling (from 4 to 8 to 16, etc.) occurs at equal time intervals.

In the situation of exponential decay $P(t) = P_0 e^{-kt}$, the **half-life** is the time it takes for the quantity to decrease by a factor of $\frac{1}{2}$. The calculation similar to that of doubling time above shows that

$$\text{Half-life} = \frac{\ln 2}{k}$$

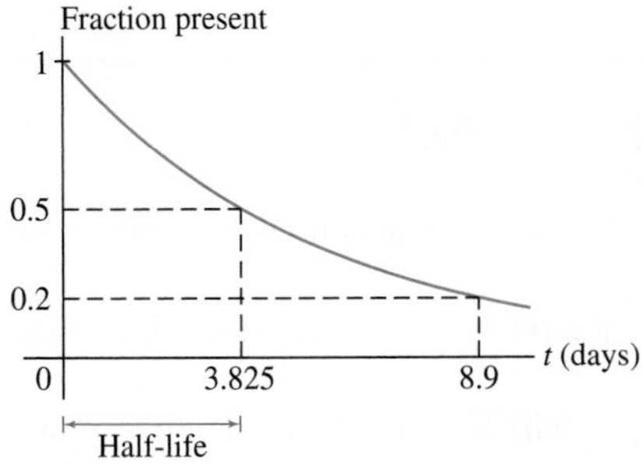

FIGURE 7 Fraction of radon-222 present at time t.

■ **EXAMPLE 5** The isotope radon-222 decays exponentially with a half-life of 3.825 days. How long will it take for 80% of the isotope to decay?

Solution By the equation for half-life, k equals $\ln 2$ divided by half-life:

$$k = \frac{\ln 2}{3.825} \approx 0.181$$

Therefore, the quantity of radon-222 at time t is $R(t) = R_0 e^{-0.181t}$, where R_0 is the amount present at $t = 0$ (Figure 7). When 80% has decayed, 20% remains, so we solve for t in the equation $R_0e^{-0.181t} = 0.2R_0$:

$$e^{-0.181t} = 0.2$$

$$-0.181t = \ln(0.2) \quad \Rightarrow \quad t = \frac{\ln(0.2)}{-0.181} \approx 8.9 \text{ days}$$

The quantity of radon-222 decreases by 80% after 8.9 days. ■

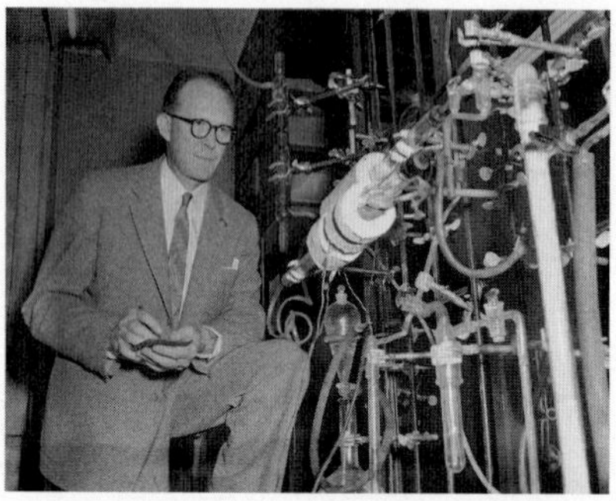

FIGURE 8 American chemist Willard Libby (1908–1980) developed the technique of carbon dating in 1946 to determine the age of fossils and was awarded the Nobel Prize in Chemistry for this work in 1960. Since then the technique has been refined considerably.

Carbon Dating

Carbon dating (Figure 8) relies on the fact that all living organisms contain carbon that enters the food chain through the carbon dioxide absorbed by plants from the atmosphere. Carbon in the atmosphere is made up of nonradioactive C^{12} and a minute amount of radioactive C^{14} that decays into nitrogen. The ratio of C^{14} to C^{12} is approximately $R_{\text{atm}} = 10^{-12}$.

The carbon in a living organism has the same ratio R_{atm} because this carbon originates in the atmosphere, but when the organism dies, its carbon is no longer replenished. The C^{14} begins to decay exponentially while the C^{12} remains unchanged. Therefore, the ratio of C^{14} to C^{12} in the organism decreases exponentially. By measuring this ratio, we can determine when the death occurred. The decay constant for C^{14} is $k = 0.000121 \text{ yr}^{-1}$, so

$$\text{Ratio of } C^{14} \text{ to } C^{12} \text{ after } t \text{ years} = R_{\text{atm}}e^{-0.000121t}$$

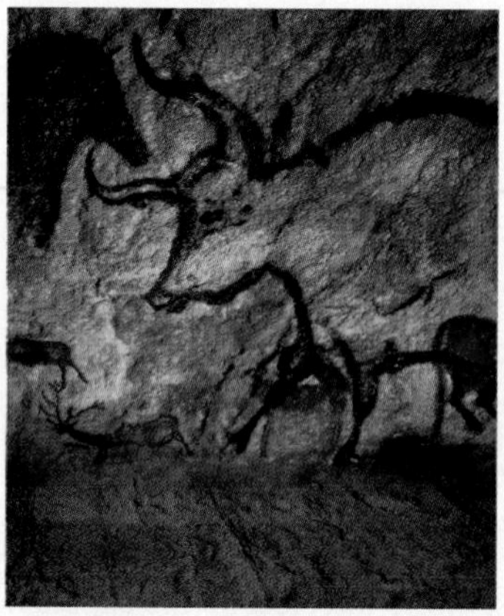

FIGURE 9 Replica of Lascaux cave painting of a bull and horse.

■ **EXAMPLE 6 Cave Paintings** In 1940, a remarkable gallery of prehistoric animal paintings was discovered in the Lascaux cave in Dordogne, France (Figure 9). A charcoal sample from the cave walls had a C^{14}-to-C^{12} ratio equal to 15% of that found in the atmosphere. Approximately how old are the paintings?

Solution The C^{14}-to-C^{12} ratio in the charcoal is now equal to $0.15R_{\text{atm}}$, so

$$R_{\text{atm}}e^{-0.000121t} = 0.15R_{\text{atm}}$$

where t is the age of the paintings. We solve for t:

$$e^{-0.000121t} = 0.15$$

$$-0.000121t = \ln(0.15) \quad \Rightarrow \quad t = \frac{\ln(0.15)}{0.000121} \approx 15{,}700$$

The cave paintings are approximately 16,000 years old (Figure 10). ■

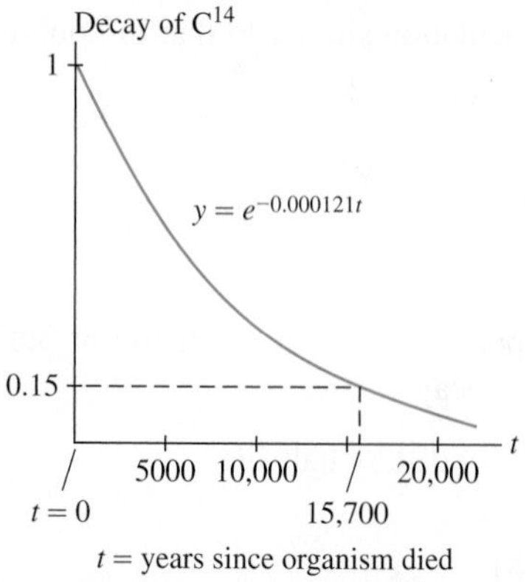

FIGURE 10 If only 15% of the C^{14} remains, the object is approximately 16,000 years old.

7.4 SUMMARY

- *Exponential growth* with growth constant $k > 0$: $P(t) = P_0e^{kt}$.
- *Exponential decay* with decay constant $k > 0$: $P(t) = P_0e^{-kt}$.
- The solutions of the differential equation $y' = ky$ are the exponential functions $y = Ce^{kt}$, where C is a constant.
- A quantity $P(t)$ grows exponentially if it grows at a rate proportional to its size—that is, if $P'(t) = kP(t)$.
- The *doubling time* for exponential growth and the *half-life* for exponential decay are both equal to $(\ln 2)/k$.
- For use in carbon dating: the decay constant of C^{14} is $k = 0.000121$.

7.4 EXERCISES

Preliminary Questions

1. Two quantities increase exponentially with growth constants $k = 1.2$ and $k = 3.4$, respectively. Which quantity doubles more rapidly?

2. A cell population grows exponentially beginning with one cell. Which takes longer: increasing from one to two cells or increasing from 15 million to 20 million cells?

3. Referring to his popular book *A Brief History of Time*, the renowned physicist Stephen Hawking said, "Someone told me that each equation I included in the book would halve its sales." Find a differential equation satisfied by the function $S(n)$, the number of copies sold if the book has n equations.

4. Carbon dating is based on the assumption that the ratio R of C^{14} to C^{12} in the atmosphere has been constant over the past 50,000 years. If R were actually smaller in the past than it is today, would the age estimates produced by carbon dating be too ancient or too recent?

Exercises

1. A certain population P of bacteria obeys the exponential growth law $P(t) = 2000e^{1.3t}$ (t in hours).

(a) How many bacteria are present initially?

(b) At what time will there be 10,000 bacteria?

2. A quantity P obeys the exponential growth law $P(t) = e^{5t}$ (t in years).

(a) At what time t is $P = 10$?

(b) What is the doubling time for P?

3. Write $f(t) = 5(7)^t$ in the form $f(t) = P_0 e^{kt}$ for some P_0 and k.

4. Write $f(t) = 9e^{1.4t}$ in the form $f(t) = P_0 b^t$ for some P_0 and b.

5. A certain RNA molecule replicates every 3 minutes. Find the differential equation for the number $N(t)$ of molecules present at time t (in minutes). How many molecules will be present after one hour if there is one molecule at $t = 0$?

6. A quantity P obeys the exponential growth law $P(t) = Ce^{kt}$ (t in years). Find the formula for $P(t)$, assuming that the doubling time is 7 years and $P(0) = 100$.

7. Find all solutions to the differential equation $y' = -5y$. Which solution satisfies the initial condition $y(0) = 3.4$?

8. Find the solution to $y' = \sqrt{2}y$ satisfying $y(0) = 20$.

9. Find the solution to $y' = 3y$ satisfying $y(2) = 1000$.

10. Find the function $y = f(t)$ that satisfies the differential equation $y' = -0.7y$ and the initial condition $y(0) = 10$.

11. The decay constant of cobalt-60 is 0.13 year^{-1}. Find its half-life.

12. The half-life radium-226 is 1622 years. Find its decay constant.

13. One of the world's smallest flowering plants, *Wolffia globosa* (Figure 11), has a doubling time of approximately 30 hours. Find the growth constant k and determine the initial population if the population grew to 1000 after 48 hours.

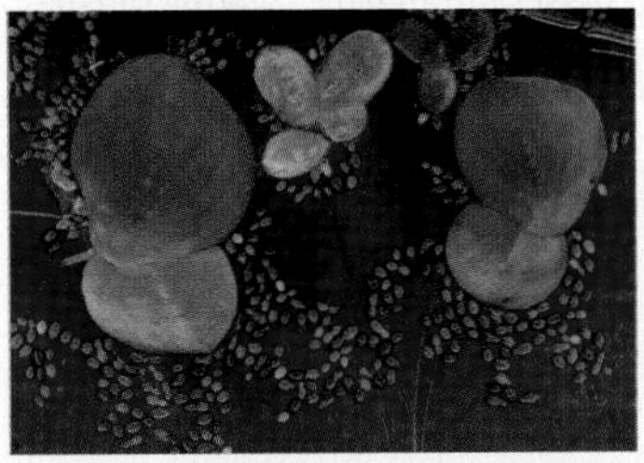

FIGURE 11 The tiny plants are *Wolffia*, with plant bodies smaller than the head of a pin.

14. A 10-kg quantity of a radioactive isotope decays to 3 kg after 17 years. Find the decay constant of the isotope.

15. The population of a city is $P(t) = 2 \cdot e^{0.06t}$ (in millions), where t is measured in years. Calculate the time it takes for the population to double, to triple, and to increase seven-fold.

16. What is the differential equation satisfied by $P(t)$, the number of infected computer hosts in Example 4? Over which time interval would $P(t)$ increase one hundred-fold?

17. The decay constant for a certain drug is $k = 0.35$ day^{-1}. Calculate the time it takes for the quantity present in the bloodstream to decrease by half, by one-third, and by one-tenth.

18. Light Intensity The intensity of light passing through an absorbing medium decreases exponentially with the distance traveled. Suppose the decay constant for a certain plastic block is $k = 4$ m^{-1}. How thick must the block be to reduce the intensity by a factor of one-third?

19. Assuming that population growth is approximately exponential, which of the following two sets of data is most likely to represent the population (in millions) of a city over a 5-year period?

Year	2000	2001	2002	2003	2004
Set I	3.14	3.36	3.60	3.85	4.11
Set II	3.14	3.24	3.54	4.04	4.74

20. The **atmospheric pressure** $P(h)$ (in kilopascals) at a height h meters above sea level satisfies a differential equation $P' = -kP$ for some positive constant k.

(a) Barometric measurements show that $P(0) = 101.3$ and $P(30,900) = 1.013$. What is the decay constant k?

(b) Determine the atmospheric pressure at $h = 500$.

21. Degrees in Physics One study suggests that from 1955 to 1970, the number of bachelor's degrees in physics awarded per year by U.S. universities grew exponentially, with growth constant $k = 0.1$.

(a) If exponential growth continues, how long will it take for the number of degrees awarded per year to increase 14-fold?

(b) If 2500 degrees were awarded in 1955, in which year were 10,000 degrees awarded?

22. The **Beer–Lambert Law** is used in spectroscopy to determine the molar absorptivity α or the concentration c of a compound dissolved in a solution at low concentrations (Figure 12). The law states that the intensity I of light as it passes through the solution satisfies $\ln(I/I_0) = \alpha c x$, where I_0 is the initial intensity and x is the distance traveled by the light. Show that I satisfies a differential equation $dI/dx = -kI$ for some constant k.

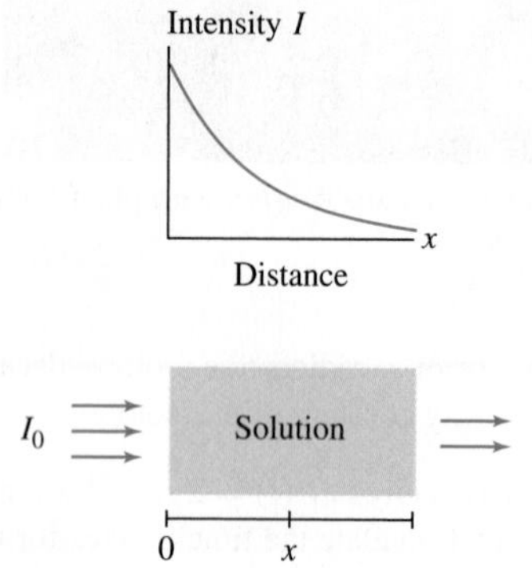

FIGURE 12 Light of intensity I_0 passing through a solution.

23. A sample of sheepskin parchment discovered by archaeologists had a C^{14}-to-C^{12} ratio equal to 40% of that found in the atmosphere. Approximately how old is the parchment?

24. Chauvet Caves In 1994, three French speleologists (geologists specializing in caves) discovered a cave in southern France containing prehistoric cave paintings. A C^{14} analysis carried out by archeologist Helene Valladas showed the paintings to be between 29,700 and 32,400 years old, much older than any previously known human art. Given that the C^{14}-to-C^{12} ratio of the atmosphere is $R = 10^{-12}$, what range of C^{14}-to-C^{12} ratios did Valladas find in the charcoal specimens?

25. A paleontologist discovers remains of animals that appear to have died at the onset of the Holocene ice age, between 10,000 and 12,000 years ago. What range of C^{14}-to-C^{12} ratio would the scientist expect to find in the animal remains?

26. Inversion of Sugar When cane sugar is dissolved in water, it converts to invert sugar over a period of several hours. The percentage $f(t)$ of unconverted cane sugar at time t (in hours) satisfies $f' = -0.2f$. What percentage of cane sugar remains after 5 hours? After 10 hours?

27. Continuing with Exercise 26, suppose that 50 grams of sugar are dissolved in a container of water. After how many hours will 20 grams of invert sugar be present?

28. Two bacteria colonies are cultivated in a laboratory. The first colony has a doubling time of 2 hours and the second a doubling time of 3 hours. Initially, the first colony contains 1000 bacteria and the second colony 3000 bacteria. At what time t will the sizes of the colonies be equal?

29. Moore's Law In 1965, Gordon Moore predicted that the number N of transistors on a microchip would increase exponentially.

(a) Does the table of data below confirm Moore's prediction for the period from 1971 to 2000? If so, estimate the growth constant k.

(b) CAS Plot the data in the table.

(c) Let $N(t)$ be the number of transistors t years after 1971. Find an approximate formula $N(t) \approx Ce^{kt}$, where t is the number of years after 1971.

(d) Estimate the doubling time in Moore's Law for the period from 1971 to 2000.

(e) How many transistors will a chip contain in 2015 if Moore's Law continues to hold?

(f) Can Moore have expected his prediction to hold indefinitely?

Processor	Year	No. Transistors
4004	1971	2250
8008	1972	2500
8080	1974	5000
8086	1978	29,000
286	1982	120,000
386 processor	1985	275,000
486 DX processor	1989	1,180,000
Pentium processor	1993	3,100,000
Pentium II processor	1997	7,500,000
Pentium III processor	1999	24,000,000
Pentium 4 processor	2000	42,000,000
Xeon processor	2008	1,900,000,000

30. Assume that in a certain country, the rate at which jobs are created is proportional to the number of people who already have jobs. If there are 15 million jobs at $t = 0$ and 15.1 million jobs 3 months later, how many jobs will there be after 2 years?

31. The only functions with a *constant* doubling time are the exponential functions P_0e^{kt} with $k > 0$. Show that the doubling time of linear function $f(t) = at + b$ at time t_0 is $t_0 + b/a$ (which increases with t_0). Compute the doubling times of $f(t) = 3t + 12$ at $t_0 = 10$ and $t_0 = 20$.

32. Verify that the half-life of a quantity that decays exponentially with decay constant k is equal to $(\ln 2)/k$.

33. **Drug Dosing Interval** Let $y(t)$ be the drug concentration (in mg/kg) in a patient's body at time t. The initial concentration is $y(0) = L$. Additional doses that increase the concentration by an amount d are administered at regular time intervals of length T. In between doses, $y(t)$ decays exponentially—that is, $y' = -ky$. Find the value of T (in terms of k and d) for which the the concentration varies between L and $L - d$ as in Figure 13.

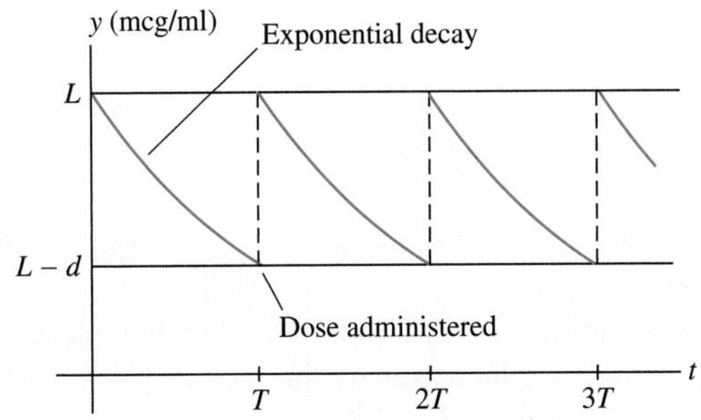

FIGURE 13 Drug concentration with periodic doses.

*Exercises 34 and 35: The **Gompertz differential equation***

$$\frac{dy}{dt} = ky \ln\left(\frac{y}{M}\right) \qquad \boxed{2}$$

(where M and k are constants) was introduced in 1825 by the English mathematician Benjamin Gompertz and is still used today to model aging and mortality.

34. Show that $y = Me^{ae^{kt}}$ satisfies Eq. (2) for any constant a.

35. To model mortality in a population of 200 laboratory rats, a scientist assumes that the number $P(t)$ of rats alive at time t (in months) satisfies Eq. (2) with $M = 204$ and $k = 0.15$ month^{-1} (Figure 14). Find $P(t)$ [note that $P(0) = 200$] and determine the population after 20 months.

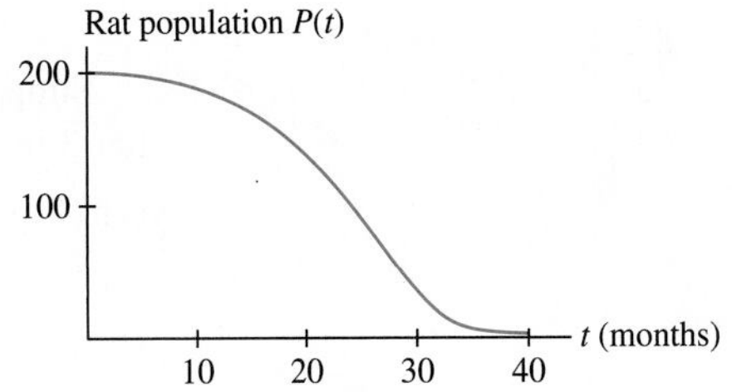

FIGURE 14

36. **Isotopes for Dating** Which of the following would be most suitable for dating extremely old rocks: carbon-14 (half-life 5570 years), lead-210 (half-life 22.26 years), or potassium-49 (half-life 1.3 billion years)? Explain why.

Further Insights and Challenges

37. Let $P = P(t)$ be a quantity that obeys an exponential growth law with growth constant k. Show that P increases m-fold after an interval of $(\ln m)/k$ years.

38. **Average Time of Decay** Physicists use the radioactive decay law $R = R_0 e^{-kt}$ to compute the average or *mean time M* until an atom decays. Let $F(t) = R/R_0 = e^{-kt}$ be the fraction of atoms that have survived to time t without decaying.

(a) Find the inverse function $t(F)$.

(b) By definition of $t(F)$, a fraction $1/N$ of atoms decays in the time interval

$$\left[t\left(\frac{j}{N}\right), t\left(\frac{j-1}{N}\right)\right]$$

Use this to justify the approximation $M \approx \dfrac{1}{N}\displaystyle\sum_{j=1}^{N} t\left(\frac{j}{N}\right)$. Then argue, by passing to the limit as $N \to \infty$, that $M = \int_0^1 t(F)\,dF$. Strictly speaking, this is an *improper integral* because $t(0)$ is infinite (it takes an infinite amount of time for all atoms to decay). Therefore, we define M as a limit

$$M = \lim_{c\to 0} \int_c^1 t(F)\,dF$$

(c) Verify the formula $\int \ln x\,dx = x\ln x - x$ by differentiation and use it to show that for $c > 0$,

$$M = \lim_{c\to 0}\left(\frac{1}{k} + \frac{1}{k}(c\ln c - c)\right)$$

(d) Verify numerically that $\lim_{c\to 0}(c - \ln c) = 0$ (we will prove this using L'Hôpital's Rule in Section 7.7). Use this to show that $M = 1/k$.

(e) What is the mean time to decay for radon (with a half-life of 3.825 days)?

7.5 Compound Interest and Present Value

Exponential functions are used extensively in financial calculations. Two basic applications are compound interest and present value.

Convention: Time t is measured in years and interest rates are given as yearly rates, either as a decimal or as a percentage. Thus, $r = 0.05$ corresponds to an interest rate of 5% per year.

When a sum of money P_0, called the **principal**, is deposited into an interest-bearing account, the amount or **balance** in the account at time t depends on two factors: the **interest rate** r and frequency with which interest is **compounded**. Interest paid out once a year at the end of the year is said to be *compounded annually*. The balance increases by the factor $(1 + r)$ after each year, leading to exponential growth:

	Principal	+	Interest	=	Balance
After 1 year	P_0	+	rP_0	=	$P_0(1+r)$
After 2 years	$P_0(1+r)$	+	$rP_0(1+r)$	=	$P_0(1+r)^2$
...		...			
After t years	$P_0(1+r)^{t-1}$	+	$rP_0(1+r)^{t-1}$	=	$P_0(1+r)^t$

Suppose that interest is paid out quarterly (every 3 months). Then the interest earned after 3 months is $\frac{r}{4}P_0$ dollars and the balance increases by the factor $\left(1+\frac{r}{4}\right)$. After one year (4 quarters), the balance increases to $P_0\left(1+\frac{r}{4}\right)^4$ and after t years,

$$\text{Balance after } t \text{ years} = P_0\left(1+\frac{r}{4}\right)^{4t}$$

For example, if $P_0 = 100$ and $r = 0.09$, then the balance after one year is

$$100\left(1+\frac{0.09}{12}\right)^{12} = 100(1.0075)^{12} \approx 100(1.09381) \approx 109.38$$

More generally,

Compound Interest If P_0 dollars are deposited into an account earning interest at an annual rate r, compounded M times yearly, then the value of the account after t years is

$$P(t) = P_0\left(1+\frac{r}{M}\right)^{Mt}$$

The factor $\left(1+\frac{r}{M}\right)^M$ is called the **yearly multiplier**.

TABLE 1 Compound Interest with Principal $P_0 = \$100$ and $r = 0.09$

	Principal after 1 Year
Annual	$100(1+0.09) = \$109$
Quarterly	$100(1+\frac{0.09}{4})^4 \approx \$109.31$
Monthly	$100(1+\frac{0.09}{12})^{12} \approx \$109.38$
Weekly	$100(1+\frac{0.09}{52})^{52} \approx \$109.41$
Daily	$100(1+\frac{0.09}{365})^{365} \approx \$109.42$

Table 1 shows the effect of more frequent compounding. What happens in the limit as M tends to infinity? This question is answered by the next theorem (a proof is given at the end of this section).

THEOREM 1 Limit Formula for e and e^x

$$e = \lim_{n\to\infty}\left(1+\frac{1}{n}\right)^n \quad \text{and} \quad e^x = \lim_{n\to\infty}\left(1+\frac{x}{n}\right)^n \quad \text{for all } x$$

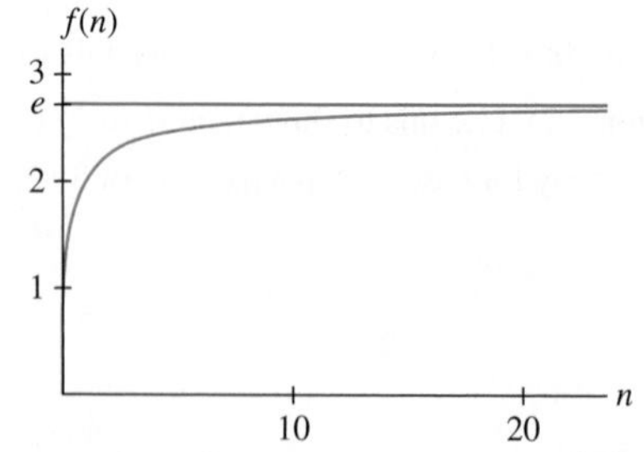

FIGURE 1 The function $f(n) = \left(1+\frac{1}{n}\right)^n$ approaches e as $n \to \infty$.

Figure 1 illustrates the first limit graphically. To compute the limit of the yearly multiplier as $M \to \infty$, we apply the second limit with $x = r$ and $n = M$:

$$\lim_{M\to\infty}\left(1+\frac{r}{M}\right)^M = e^r \qquad \boxed{1}$$

The multiplier after t years is $(e^r)^t = e^{rt}$. This leads to the following definition.

Continuously Compounded Interest If P_0 dollars are deposited into an account earning interest at an annual rate r, compounded continuously, then the value of the account after t years is

$$P(t) = P_0e^{rt}$$

■ **EXAMPLE 1** A principal of $P_0 = ¥\,100{,}000$ (Japanese yen) is deposited into an account paying 6% interest. Find the balance after 3 years if interest is compounded quarterly and if interest is compounded continuously.

Note: The mathematics of interest rates is the same for all currencies (dollars, euros, pesos, yen, etc.).

Solution After 3 years, the balance is

$$\text{Quarterly compounding:} \qquad 100{,}000\left(1+\frac{0.06}{4}\right)^{4(3)} \approx ¥\,119{,}562$$

$$\text{Continuous compounding:} \qquad 100{,}000e^{(0.06)3} \approx ¥\,119{,}722$$

■

Present Value

The concept of *present value* (PV) is used in business and finance to compare payments made at different times. Assume that there is an interest rate r (continuously compounded) at which an investor can lend or borrow money. By definition, the PV of P dollars to be received t years in the future is Pe^{-rt}:

In the financial world there are many different interest rates (federal funds rate, prime rate, LIBOR, etc.). We simplify the discussion by assuming that there is just one rate.

> The PV of P dollars received at time t is Pe^{-rt}.

What is the reasoning behind this definition? When you invest at the rate r for t years, your principal increases by the factor e^{rt}, so if you invest Pe^{-rt} dollars, your principal grows to $(Pe^{-rt})e^{rt} = P$ dollars at time t. The present value Pe^{-rt} is the amount you would have to invest *today* in order to have P dollars at time t.

■ **EXAMPLE 2** Is it better to receive \$2000 today or \$2200 in 2 years? Consider $r = 0.03$ and $r = 0.07$.

Solution We compare \$2000 today with the PV of \$2200 received in 2 years.

- If $r = 0.03$, the PV is $2200e^{-(0.03)2} \approx \$2071.88$. This is more than \$2000, so a payment of \$2200 in 2 years is preferable to a \$2000 payment today.
- If $r = 0.07$, the PV is $2200e^{-(0.07)2} \approx \$1912.59$. This PV is less than \$2000, so it is better to receive \$2000 today if $r = 0.07$. ■

■ **EXAMPLE 3 Deciding Whether to Invest** Chief Operating Officer Ryan Martinez must decide whether to upgrade his company's computer system. The upgrade costs \$400,000 and will save \$150,000 a year for each of the next 3 years. Is this a good investment if $r = 7\%$?

Solution Ryan must compare today's cost of the upgrade with the PV of the money saved. For simplicity, assume that the annual savings of \$150,000 is received as a lump sum at the end of each year.

If $r = 0.07$, the PV of the savings over 3 years is

$$150{,}000e^{-(0.07)} + 150{,}000e^{-(0.07)2} + 150{,}000e^{-(0.07)3} \approx \$391{,}850$$

The amount saved is *less* than the cost \$400,000, so the upgrade is not worthwhile. ■

An **income stream** is a sequence of periodic payments that continue over an interval of T years. Consider an investment that produces income at a rate of \$800/year for 5 years. A total of \$4000 is paid out over 5 years, but the PV of the income stream is less. For instance, if $r = 0.06$ and payments are made at the end of the year, then the PV is

$$800e^{-0.06} + 800e^{-(0.06)2} + 800e^{-(0.06)3} + 800e^{-(0.06)4} + 800e^{-(0.06)5} \approx \$3353.12$$

It is more convenient mathematically to assume that payments are made *continuously* at a rate of $R(t)$ dollars per year. We can then calculate PV as an integral. Divide the time interval $[0, T]$ into N subintervals of length $\Delta t = T/N$. If Δt is small, the amount paid out between time t and $t + \Delta t$ is approximately

$$\underbrace{R(t)}_{\text{Rate}} \times \underbrace{\Delta t}_{\text{Time interval}} = R(t)\Delta t$$

The PV of this payment is approximately $e^{-rt}R(t)\Delta t$. Setting $t_i = i\Delta t$, we obtain the approximation

$$\text{PV of income stream} \approx \sum_{i=1}^{N} e^{-rt_i} R(t_i)\Delta t$$

This is a Riemann sum whose value approaches $\int_0^T R(t)e^{-rt}\,dt$ as $\Delta t \to 0$.

In April 1720, Isaac Newton doubled his money by investing in the South Sea Company, an English company set up to conduct trade with the West Indies and South America. Having gained 7000 pounds, Newton invested a second time, but like many others, he did not realize that the company was built on fraud and manipulation. In what became known as the South Sea Bubble, the stock lost 80% of its value, and the famous scientist suffered a loss of 20,000 pounds.

PV of an Income Stream If the interest rate is r, the present value of an income stream paying out $R(t)$ dollars per year continuously for T years is

$$\text{PV} = \int_0^T R(t)e^{-rt}\,dt \qquad \boxed{2}$$

EXAMPLE 4 An investment pays out 800,000 Mexican pesos per year, continuously for 5 years. Find the PV of the investment for $r = 0.04$ and $r = 0.06$.

Solution In this case, $R(t) = 800{,}000$. If $r = 0.04$, the PV of the income stream is equal (in pesos) to

$$\int_0^5 800{,}000e^{-0.04t}\,dt = -800{,}000\frac{e^{-0.04t}}{0.04}\bigg|_0^5 \approx -16{,}374{,}615 - (-20{,}000{,}000)$$
$$= 3{,}625{,}385$$

If $r = 0.06$, the PV is equal (in pesos) to

$$\int_0^5 800{,}000e^{-0.06t}\,dt = -800{,}000\frac{e^{-0.06t}}{0.06}\bigg|_0^5 \approx -9{,}877{,}576 - (-13{,}333{,}333)$$
$$= 3{,}455{,}757$$

■

Proof of Theorem 1 Apply the formula $\ln b = \int_1^b t^{-1}\,dt$ with $b = 1 + 1/n$:

$$\ln\left(1 + \frac{1}{n}\right) = \int_1^{1+1/n} \frac{dt}{t}$$

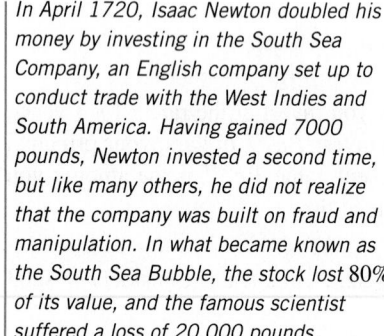

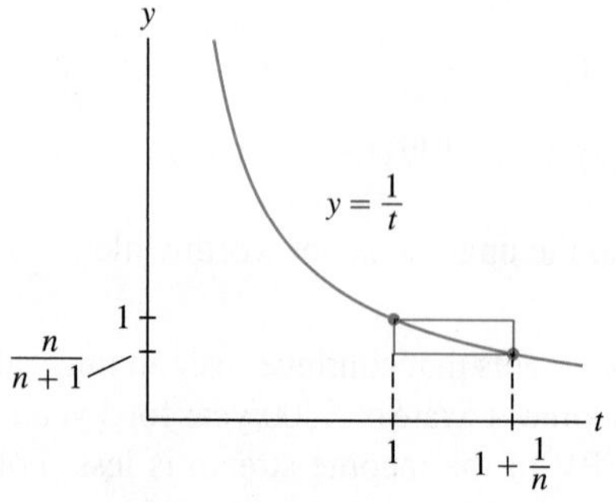

FIGURE 2

Figure 2 shows that the area represented by this integral lies between the areas of two rectangles of heights $n/(n+1)$ and 1, both of base $1/n$. These rectangles have areas $1/(n+1)$ and $1/n$, so

$$\frac{1}{n+1} \le \ln\left(1 + \frac{1}{n}\right) \le \frac{1}{n} \qquad \boxed{3}$$

Multiply through by n, using the rule $n \ln a = \ln a^n$:

$$\frac{n}{n+1} \le \ln\left(\left(1+\frac{1}{n}\right)^n\right) \le 1$$

Since $\lim_{n\to\infty} \frac{n}{n+1} = 1$, the middle quantity must approach 1 by the Squeeze Theorem:

$$\lim_{n\to\infty} \ln\left(\left(1+\frac{1}{n}\right)^n\right) = 1$$

Now we can apply e^x (because it is continuous) to obtain the desired result:

$$e^1 = e^{\lim_{n\to\infty} \ln\left(\left(1+\frac{1}{n}\right)^n\right)} = \lim_{n\to\infty} e^{\ln\left(\left(1+\frac{1}{n}\right)^n\right)} = \lim_{n\to\infty}\left(1+\frac{1}{n}\right)^n$$

See Exercise 27 for a proof of the more general formula $e^x = \lim_{n\to\infty}\left(1+\frac{x}{n}\right)^n$. ■

7.5 SUMMARY

- Interest rate r, compounded M times per year:

$$P(t) = P_0(1 + r/M)^{Mt}$$

- Interest rate r, compounded continuously: $P(t) = P_0 e^{rt}$.
- The *present value* (PV) of P dollars (or other currency), to be paid t years in the future, is Pe^{-rt}.
- Present value of an income stream paying $R(t)$ dollars per year continuously for T years:

$$\text{PV} = \int_0^T R(t)e^{-rt}\,dt$$

7.5 EXERCISES

Preliminary Questions

1. Which is preferable: an interest rate of 12% compounded quarterly, or an interest rate of 11% compounded continuously?

2. Find the yearly multiplier if $r = 9\%$ and interest is compounded (a) continuously and (b) quarterly.

3. The PV of N dollars received at time T is (choose the correct answer):

(a) The value at time T of N dollars invested today

(b) The amount you would have to invest today in order to receive N dollars at time T

4. In one year, you will be paid \$1. Will the PV increase or decrease if the interest rate goes up?

5. Xavier expects to receive a check for \$1000 one year from today. Explain using the concept of PV, whether he will be happy or sad to learn that the interest rate has just increased from 6% to 7%.

Exercises

1. Compute the balance after 10 years if \$2000 is deposited in an account paying 9% interest and interest is compounded (a) quarterly, (b) monthly, and (c) continuously.

2. Suppose \$500 is deposited into an account paying interest at a rate of 7%, continuously compounded. Find a formula for the value of the account at time t. What is the value of the account after 3 years?

3. A bank pays interest at a rate of 5%. What is the yearly multiplier if interest is compounded

(a) three times a year? **(b)** continuously?

4. How long will it take for \$4000 to double in value if it is deposited in an account bearing 7% interest, continuously compounded?

5. How much must one invest today in order to receive \$20,000 after 5 years if interest is compounded continuously at the rate $r = 9\%$?

6. An investment increases in value at a continuously compounded rate of 9%. How large must the initial investment be in order to build up a value of \$50,000 over a 7-year period?

7. Compute the PV of \$5000 received in 3 years if the interest rate is (a) 6% and (b) 11%. What is the PV in these two cases if the sum is instead received in 5 years?

8. Is it better to receive \$1000 today or \$1300 in 4 years? Consider $r = 0.08$ and $r = 0.03$.

9. Find the interest rate r if the PV of \$8000 to be received in 1 year is \$7300.

10. A company can earn additional profits of \$500,000/year for 5 years by investing \$2 million to upgrade its factory. Is the investment worthwhile if the interest rate is 6%? (Assume the savings are received as a lump sum at the end of each year.)

11. A new computer system costing \$25,000 will reduce labor costs by \$7,000/year for 5 years.

(a) Is it a good investment if $r = 8\%$?

(b) How much money will the company actually save?

12. After winning \$25 million in the state lottery, Jessica learns that she will receive five yearly payments of \$5 million beginning immediately.

(a) What is the PV of Jessica's prize if $r = 6\%$?

(b) How much more would the prize be worth if the entire amount were paid today?

13. Use Eq. (2) to compute the PV of an income stream paying out $R(t) = \$5000$/year continuously for 10 years, assuming $r = 0.05$.

14. Find the PV of an investment that pays out continuously at a rate of \$800/year for 5 years, assuming $r = 0.08$.

15. Find the PV of an income stream that pays out continuously at a rate $R(t) = \$5000e^{0.1t}$/year for 7 years, assuming $r = 0.05$.

16. A commercial property generates income at the rate $R(t)$. Suppose that $R(0) = \$70{,}000$/year and that $R(t)$ increases at a continuously compounded rate of 5%. Find the PV of the income generated in the first 4 years if $r = 6\%$.

17. Show that an investment that pays out R dollars per year continuously for T years has a PV of $R(1 - e^{-rT})/r$.

18. Explain this statement: If T is very large, then the PV of the income stream described in Exercise 17 is approximately R/r.

19. Suppose that $r = 0.06$. Use the result of Exercise 18 to estimate the payout rate R needed to produce an income stream whose PV is \$20,000, assuming that the stream continues for a large number of years.

20. Verify by differentiation:

$$\int te^{-rt}\,dt = -\frac{e^{-rt}(1+rt)}{r^2} + C \qquad \boxed{4}$$

Use Eq. (4) to compute the PV of an investment that pays out income continuously at a rate $R(t) = (5000 + 1000t)$ dollars per year for 5 years, assuming $r = 0.05$.

21. Use Eq. (4) to compute the PV of an investment that pays out income continuously at a rate $R(t) = (5000 + 1000t)e^{0.02t}$ dollars per year for 10 years, assuming $r = 0.08$.

22. **Banker's Rule of 70** If you earn an interest rate of R percent, continuously compounded, your money doubles after approximately $70/R$ years. For example, at $R = 5\%$, your money doubles after $70/5$ or 14 years. Use the concept of doubling time to justify the Banker's Rule. (*Note:* Sometimes, the rule $72/R$ is used. It is less accurate but easier to apply because 72 is divisible by more numbers than 70.)

In Exercises 23–26, calculate the limit.

23. $\displaystyle\lim_{n\to\infty}\left(1+\frac{1}{n}\right)^{6n}$

24. $\displaystyle\lim_{n\to\infty}\left(1+\frac{3}{n}\right)^{n}$

25. $\displaystyle\lim_{n\to\infty}\left(1+\frac{3}{n}\right)^{2n}$

26. $\displaystyle\lim_{n\to\infty}\left(1+\frac{1}{4n}\right)^{12n}$

Further Insights and Challenges

27. Modify the proof of the relation $e = \lim_{n\to\infty}\left(1+\frac{1}{n}\right)^n$ given in the text to prove $e^x = \lim_{n\to\infty}\left(1+\frac{x}{n}\right)^n$. *Hint:* Express $\ln(1 + xn^{-1})$ as an integral and estimate above and below by rectangles.

28. Prove that, for $n > 0$,

$$\left(1+\frac{1}{n}\right)^n \le e \le \left(1+\frac{1}{n}\right)^{n+1}$$

Hint: Take logarithms and use Eq. (3).

29. A bank pays interest at the rate r, compounded M times yearly. The **effective interest rate** r_e is the rate at which interest, if compounded annually, would have to be paid to produce the same yearly return.

(a) Find r_e if $r = 9\%$ compounded monthly.

(b) Show that $r_e = (1 + r/M)^M - 1$ and that $r_e = e^r - 1$ if interest is compounded continuously.

(c) Find r_e if $r = 11\%$ compounded continuously.

(d) Find the rate r that, compounded weekly, would yield an effective rate of 20%.

7.6 Models Involving $y' = k(y - b)$

We have seen that a quantity grows or decays exponentially if its *rate of change* is proportional to the amount present. This characteristic property is expressed by the differential equation $y' = ky$. We now study the closely related differential equation

*Every first-order, linear differential equation with **constant coefficients** can be written in the form of Eq. (1). This equation is used to model a variety of phenomena, such as the cooling process, free-fall with air resistance, and current in a circuit.*

$$\frac{dy}{dt} = k(y - b) \quad (1)$$

where k and b are constants and $k \neq 0$. This differential equation describes a quantity y whose *rate of change is proportional to the difference* $y - b$. The general solution is

$$y(t) = b + Ce^{kt} \quad (2)$$

To verify this, observe that $(y - b)' = y'$ since b is a constant, so Eq. (1) may be rewritten

$$\frac{d}{dt}(y - b) = k(y - b)$$

In other words, $y - b$ satisfies the differential equation of an exponential function and thus $y - b = Ce^{kt}$, or $y = b + Ce^{kt}$, as claimed.

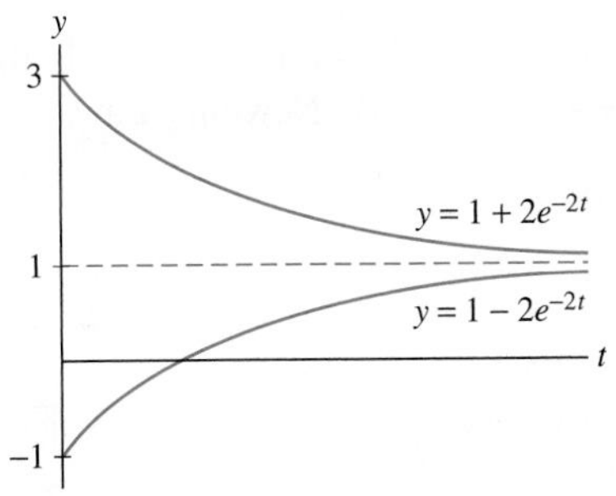

FIGURE 1 Two solutions to $y' = -2(y - 1)$ corresponding to $C = 2$ and $C = -2$.

GRAPHICAL INSIGHT The behavior of the solution $y(t)$ as $t \to \infty$ depends on whether C and k are positive or negative. When $k > 0$, e^{kt} tends to ∞ and therefore, $y(t)$ tends to ∞ if $C > 0$ and to $-\infty$ if $C < 0$. When $k < 0$, we usually rewrite the differential equation as $y' = -k(y - b)$ with $k > 0$. In this case, $y(t) = b + Ce^{-kt}$ and $y(t)$ approaches the horizontal asymptote $y = b$ since Ce^{-kt} tends to zero as $t \to \infty$ (Figure 1). However, $y(t)$ approaches the asymptote from above or below, depending on whether $C > 0$ or $C < 0$.

We now consider some applications of Eq. (1), beginning with Newton's Law of Cooling. Let $y(t)$ be the temperature of a hot object that is cooling off in an environment where the ambient temperature is T_0. Newton assumed that the *rate of cooling* is proportional to the temperature difference $y - T_0$. We express this hypothesis in a precise way by the differential equation

Newton's Law of Cooling implies that the object cools quickly when it is much hotter than its surroundings (when $y - T_0$ is large). The rate of cooling slows as y approaches T_0. When the object's initial temperature is less than T_0, y' is positive and Newton's Law models warming.

$$y' = -k(y - T_0) \qquad (T_0 = \text{ambient temperature})$$

The constant k, in units of $(\text{time})^{-1}$, is called the **cooling constant** and depends on the physical properties of the object.

■ **EXAMPLE 1 Newton's Law of Cooling** A hot metal bar with cooling constant $k = 2.1\ \text{min}^{-1}$ is submerged in a large tank of water held at temperature $T_0 = 10°\text{C}$. Let $y(t)$ be the bar's temperature at time t (in minutes).

(a) Find the differential equation satisfied by $y(t)$ and find its general solution.
(b) What is the bar's temperature after 1 min if its initial temperature was 180°C?
(c) What was the bar's initial temperature if it cooled to 80°C in 30 s?

Solution

← REMINDER The differential equation

$$\frac{dy}{dt} = k(y - b)$$

has general solution

$$y = b + Ce^{kt}$$

(a) Since $k = 2.1\ \text{min}^{-1}$, $y(t)$ (with t in minutes) satisfies

$$y' = -2.1(y - 10)$$

By Eq. (2), the general solution is $y(t) = 10 + Ce^{-2.1t}$ for some constant C.

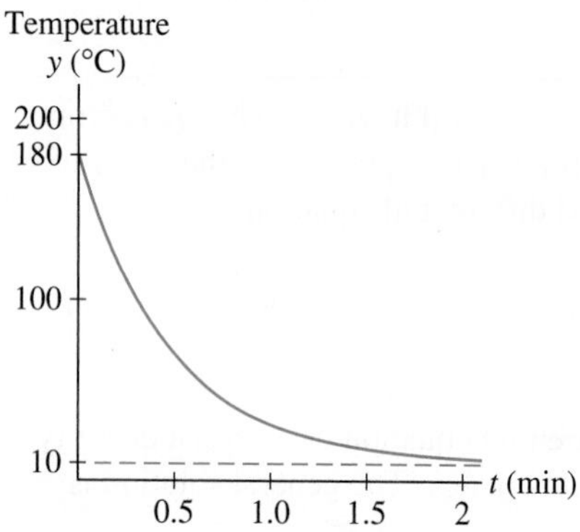

FIGURE 2 Temperature of metal bar as it cools.

(b) The general solution is $y(t) = 10 + Ce^{-2.1t}$. If the initial temperature was 180°C, then $y(0) = 10 + C = 180$. Thus, $C = 170$ and $y(t) = 10 + 170e^{-2.1t}$ (Figure 2). After 1 min,

$$y(1) = 10 + 170e^{-2.1(1)} \approx 30.8°\text{C}$$

(c) If the temperature after 30 s is 80°C, then $y(0.5) = 80$, and we have

$$10 + Ce^{-2.1(0.5)} = 80 \quad \Rightarrow \quad Ce^{-1.05} = 70 \quad \Rightarrow \quad C = 70e^{1.05} \approx 200$$

So in this case, $y(t) = 10 + 200e^{-2.1t}$ and the initial temperature was

$$y(0) = 10 + 200e^{-2.1(0)} = 10 + 200 = 210°\text{C}$$

The effect of air resistance depends on the physical situation. A high-speed bullet is affected differently than a skydiver. Our model is fairly realistic for a large object such as a skydiver falling from high altitudes.

The differential equation $y' = k(y - b)$ is also used to model free-fall when air resistance is taken into account. Assume that the force due to air resistance is proportional to the velocity v and acts opposite to the direction of the fall. We write this force as $-kv$, where $k > 0$. We take the upward direction to be positive, so $v < 0$ for a falling object and $-kv$ is an upward acting force.

The force due to gravity on a falling object of mass m is $-mg$, where g is the acceleration due to gravity, so the total force is $F = -mg - kv$. By Newton's Law,

$$F = ma = mv' \qquad (a = v' \text{ is the acceleration})$$

Thus $mv' = -mg - kv$, which can be written

In this model of free fall, k has units of mass per time, such as kg/s.

$$v' = -\frac{k}{m}\left(v + \frac{mg}{k}\right) \tag{3}$$

This equation has the form $v' = -k(v - b)$ with k replaced by k/m and $b = -mg/k$. By Eq. (2) the general solution is

$$v(t) = -\frac{mg}{k} + Ce^{-(k/m)t} \tag{4}$$

Since $Ce^{-(k/m)t}$ tends to zero as $t \to \infty$, $v(t)$ tends to a limiting terminal velocity:

$$\text{Terminal velocity} = \lim_{t\to\infty} v(t) = -\frac{mg}{k} \tag{5}$$

Without air resistance the velocity would increase indefinitely.

Skydiver in free fall.

EXAMPLE 2 An 80-kg skydiver steps out of an airplane.

(a) What is her terminal velocity if $k = 8$ kg/s?

(b) What is her velocity after 30 s?

Solution

(a) By Eq. (5), with $k = 8$ kg/s and $g = 9.8$ m/s^2, the terminal velocity is

$$-\frac{mg}{k} = -\frac{(80)9.8}{8} = -98 \text{ m/s}$$

(b) With t in seconds, we have, by Eq. (4),

$$v(t) = -98 + Ce^{-(k/m)t} = -98 + Ce^{-(8/80)t} = -98 + Ce^{-0.1t}$$

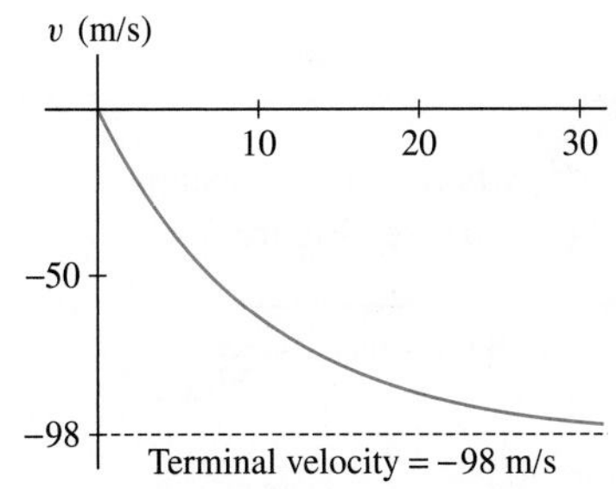

FIGURE 3 Velocity of 80-kg skydiver in free fall with air resistance ($k = 8$).

We assume that the skydiver leaves the airplane with no initial vertical velocity, so $v(0) = -98 + C = 0$, and $C = 98$. Thus we have $v(t) = -98(1 - e^{-0.1t})$ [Figure 3]. The skydiver's velocity after 30 s is

$$v(30) = -98(1 - e^{-0.1(30)}) \approx -93.1 \text{ m/s}$$ ■

An **annuity** is an investment in which a principal P_0 is placed in an account that earns interest (compounded continuously) at a rate r, and money is withdrawn at regular intervals. To model an annuity by a differential equation, we assume that the money is withdrawn continuously at a rate of N dollars per year. Let $P(t)$ be the balance in the annuity after t years. Then

Notice in Eq. (6) that $P'(t)$ is determined by the growth rate r and the withdrawal rate N. If no withdrawals occurred, $P(t)$ would grow with compound interest and would satisfy $P'(t) = rP(t)$.

$$\underbrace{P'(t)}_{\text{Rate of change}} = \underbrace{rP(t)}_{\text{Growth due to interest}} - \underbrace{N}_{\text{Withdrawal rate}} = r\left(P(t) - \frac{N}{r}\right) \qquad \boxed{6}$$

This equation has the form $y' = k(y - b)$ with $k = r$ and $b = N/r$, so by Eq. (2), the general solution is

$$P(t) = \frac{N}{r} + Ce^{rt} \qquad \boxed{7}$$

Since e^{rt} tends to infinity as $t \to \infty$, the balance $P(t)$ tends to ∞ if $C > 0$. If $C < 0$, then $P(t)$ tends to $-\infty$ (i.e., the annuity eventually runs out of money). If $C = 0$, then $P(t)$ remains constant with value N/r.

■ **EXAMPLE 3 Does an Annuity Pay Out Forever?** An annuity earns interest at the rate $r = 0.07$, and withdrawals are made continuously at a rate of $N = \$500$/year.

(a) When will the annuity run out of money if the initial deposit is $P(0) = \$5000$?

(b) Show that the balance increases indefinitely if $P(0) = \$9000$.

Solution We have $N/r = \frac{500}{0.07} \approx 7143$, so $P(t) = 7143 + Ce^{0.07t}$ by Eq. (7).

(a) If $P(0) = 5000 = 7143 + Ce^0$, then $C = -2143$ and

$$P(t) = 7143 - 2143e^{0.07t}$$

The account runs out of money when $P(t) = 7143 - 2143e^{0.07t} = 0$, or

$$e^{0.07t} = \frac{7143}{2143} \quad \Rightarrow \quad 0.07t = \ln\left(\frac{7143}{2143}\right) \approx 1.2$$

The annuity money runs out at time $t = \frac{1.2}{0.07} \approx 17$ years.

(b) If $P(0) = 9000 = 7143 + Ce^0$, then $C = 1857$ and

$$P(t) = 7143 + 1857e^{0.07t}$$

Since the coefficient $C = 1857$ is positive, the account never runs out of money. In fact, $P(t)$ increases indefinitely as $t \to \infty$. Figure 4 illustrates the two cases. ■

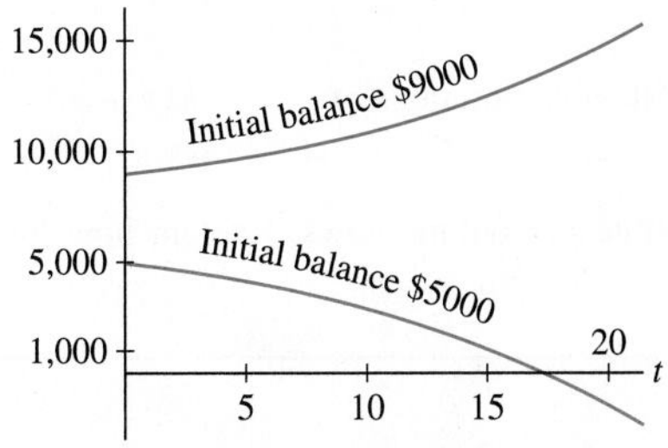

FIGURE 4 The balance in an annuity may increase indefinitely or decrease to zero (eventually becoming negative), depending on the size of initial deposit P_0.

7.6 SUMMARY

- The general solution of $y' = k(y - b)$ is $y = b + Ce^{kt}$, where C is a constant.
- The following tables describe the solutions to $y' = k(y - b)$ (see Figure 5).

Equation ($k > 0$)	**Solution**	**Behavior as** $t \to \infty$
$y' = k(y - b)$	$y(t) = b + Ce^{kt}$	$\lim_{t\to\infty} y(t) = \begin{cases} \infty & \text{if } C > 0 \\ -\infty & \text{if } C < 0 \end{cases}$
$y' = -k(y - b)$	$y(t) = b + Ce^{-kt}$	$\lim_{t\to\infty} y(t) = b$

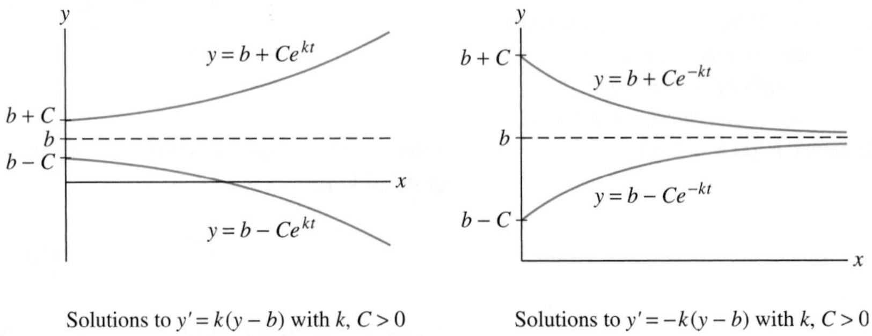

FIGURE 5

- Three applications:
  - Newton's law of cooling: $y' = -k(y - T_0)$, $y(t) =$ temperature of the object, $T_0 =$ ambient temperature, $k =$ cooling constant
  - Free-fall with air resistance: $v' = -\dfrac{k}{m}\left(v + \dfrac{mg}{k}\right)$, $v(t) =$ velocity, $m =$ mass, $k =$ air resistance constant, $g =$ acceleration due to gravity
  - Continuous annuity: $P' = r\left(P - \dfrac{N}{r}\right)$, $P(t) =$ balance in the annuity, $r =$ interest rate, $N =$ withdrawal rate

7.6 EXERCISES

Preliminary Questions

1. Write down a solution to $y' = 4(y - 5)$ that tends to $-\infty$ as $t \to \infty$.

2. Does $y' = -4(y - 5)$ have a solution that tends to ∞ as $t \to \infty$?

3. True or false? If $k > 0$, then all solutions of $y' = -k(y - b)$ approach the same limit as $t \to \infty$.

4. As an object cools, its rate of cooling slows. Explain how this follows from Newton's Law of Cooling.

Exercises

1. Find the general solution of

$$y' = 2(y - 10)$$

Then find the two solutions satisfying $y(0) = 25$ and $y(0) = 5$, and sketch their graphs.

2. Verify directly that $y = 12 + Ce^{-3t}$ satisfies

$$y' = -3(y - 12) \quad \text{for all } C$$

Then find the two solutions satisfying $y(0) = 20$ and $y(0) = 0$, and sketch their graphs.

3. Solve $y' = 4y + 24$ subject to $y(0) = 5$.

4. Solve $y' + 6y = 12$ subject to $y(2) = 10$.

In Exercises 5–12, use Newton's Law of Cooling.

5. A hot anvil with cooling constant $k = 0.02\ \text{s}^{-1}$ is submerged in a large pool of water whose temperature is 10°C. Let $y(t)$ be the anvil's temperature t seconds later.

(a) What is the differential equation satisfied by $y(t)$?

(b) Find a formula for $y(t)$, assuming the object's initial temperature is 100°C.

(c) How long does it take the object to cool down to 20°?

6. Frank's automobile engine runs at 100°C. On a day when the outside temperature is 21°C, he turns off the ignition and notes that five minutes later, the engine has cooled to 70°C.

(a) Determine the engine's cooling constant k.

(b) What is the formula for $y(t)$?

(c) When will the engine cool to 40°C?

7. At 10:30 AM, detectives discover a dead body in a room and measure its temperature at 26°C. One hour later, the body's temperature had dropped to 24.8°C. Determine the time of death (when the body temperature was a normal 37°C), assuming that the temperature in the room was held constant at 20°C.

8. A cup of coffee with cooling constant $k = 0.09\ \text{min}^{-1}$ is placed in a room at temperature 20°C.

(a) How fast is the coffee cooling (in degrees per minute) when its temperature is $T = 80$°C?

(b) Use the Linear Approximation to estimate the change in temperature over the next 6 s when $T = 80$°C.

(c) If the coffee is served at 90°C, how long will it take to reach an optimal drinking temperature of 65°C?

9. A cold metal bar at −30°C is submerged in a pool maintained at a temperature of 40°C. Half a minute later, the temperature of the bar is 20°C. How long will it take for the bar to attain a temperature of 30°C?

10. When a hot object is placed in a water bath whose temperature is 25°C, it cools from 100 to 50°C in 150 s. In another bath, the same cooling occurs in 120 s. Find the temperature of the second bath.

11. GU Objects A and B are placed in a warm bath at temperature $T_0 = 40$°C. Object A has initial temperature −20°C and cooling constant $k = 0.004\ \text{s}^{-1}$. Object B has initial temperature 0°C and cooling constant $k = 0.002\ \text{s}^{-1}$. Plot the temperatures of A and B for $0 \le t \le 1000$. After how many seconds will the objects have the same temperature?

12. In Newton's Law of Cooling, the constant $\tau = 1/k$ is called the "characteristic time." Show that τ is the time required for the temperature difference $(y - T_0)$ to decrease by the factor $e^{-1} \approx 0.37$. For example, if $y(0) = 100$°C and $T_0 = 0$°C, then the object cools to $100/e \approx 37$°C in time τ, to $100/e^2 \approx 13.5$°C in time 2τ, and so on.

In Exercises 13–16, use Eq. (3) as a model for free-fall with air resistance.

13. A 60-kg skydiver jumps out of an airplane. What is her terminal velocity, in meters per second, assuming that $k = 10$ kg/s for free-fall (no parachute)?

14. Find the terminal velocity of a skydiver of weight $w = 192$ lb if $k = 1.2$ lb-s/ft. How long does it take him to reach half of his terminal velocity if his initial velocity is zero? Mass and weight are related by $w = mg$, and Eq. (3) becomes $v' = -(kg/w)(v + w/k)$ with $g = 32$ ft/s^2.

15. A 80-kg skydiver jumps out of an airplane (with zero initial velocity). Assume that $k = 12$ kg/s with a closed parachute and $k = 70$ kg/s with an open parachute. What is the skydiver's velocity at $t = 25$ s if the parachute opens after 20 s of free fall?

16. Does a heavier or a lighter skydiver reach terminal velocity faster?

17. A continuous annuity with withdrawal rate $N = \$5000$/year and interest rate $r = 5\%$ is funded by an initial deposit of $P_0 = \$50{,}000$.

(a) What is the balance in the annuity after 10 years?

(b) When will the annuity run out of funds?

18. Show that a continuous annuity with withdrawal rate $N = \$5000$/year and interest rate $r = 8\%$, funded by an initial deposit of $P_0 = \$75{,}000$, never runs out of money.

19. Find the minimum initial deposit P_0 that will allow an annuity to pay out \$6000/year indefinitely if it earns interest at a rate of 5%.

20. Find the minimum initial deposit P_0 necessary to fund an annuity for 20 years if withdrawals are made at a rate of \$10,000/year and interest is earned at a rate of 7%.

21. An initial deposit of 100,000 euros are placed in an annuity with a French bank. What is the minimum interest rate the annuity must earn to allow withdrawals at a rate of 8000 euros/year to continue indefinitely?

22. Show that a continuous annuity never runs out of money if the initial balance is greater than or equal to N/r, where N is the withdrawal rate and r the interest rate.

23. Sam borrows \$10,000 from a bank at an interest rate of 9% and pays back the loan continuously at a rate of N dollars per year. Let $P(t)$ denote the amount still owed at time t.

(a) Explain why $P(t)$ satisfies the differential equation

$$y' = 0.09y - N$$

(b) How long will it take Sam to pay back the loan if $N = \$1200$?

(c) Will the loan ever be paid back if $N = \$800$?

24. April borrows \$18,000 at an interest rate of 5% to purchase a new automobile. At what rate (in dollars per year) must she pay back the loan, if the loan must be paid off in 5 years? *Hint:* Set up the differential equation as in Exercise 23).

25. Let $N(t)$ be the fraction of the population who have heard a given piece of news t hours after its initial release. According to one model, the rate $N'(t)$ at which the news spreads is equal to k times the fraction of the population that has not yet heard the news, for some constant $k > 0$.

(a) Determine the differential equation satisfied by $N(t)$.

(b) Find the solution of this differential equation with the initial condition $N(0) = 0$ in terms of k.

(c) Suppose that half of the population is aware of an earthquake 8 hours after it occurs. Use the model to calculate k and estimate the percentage that will know about the earthquake 12 hours after it occurs.

26. Current in a Circuit When the circuit in Figure 6 (which consists of a battery of V volts, a resistor of R ohms, and an inductor of L henries) is connected, the current $I(t)$ flowing in the circuit satisfies

$$L\frac{dI}{dt} + RI = V$$

with the initial condition $I(0) = 0$.

(a) Find a formula for $I(t)$ in terms of L, V, and R.

(b) Show that $\lim_{t\to\infty} I(t) = V/R$.

(c) Show that $I(t)$ reaches approximately 63% of its maximum value at the "characteristic time" $\tau = L/R$.

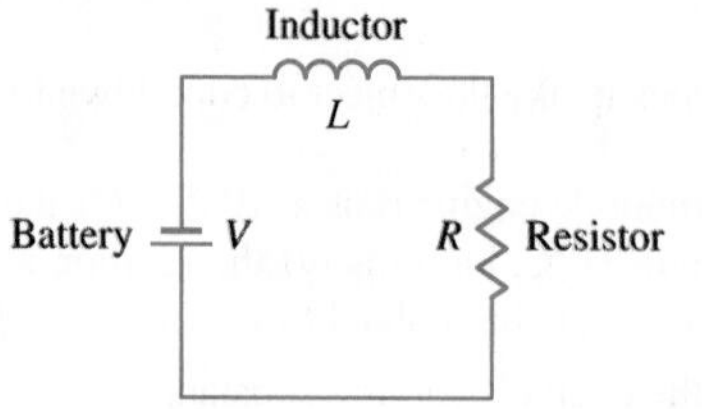

FIGURE 6 Current flow approaches the level $I_{max} = V/R$.

Further Insights and Challenges

27. Show that the cooling constant of an object can be determined from two temperature readings $y(t_1)$ and $y(t_2)$ at times $t_1 \neq t_2$ by the formula

$$k = \frac{1}{t_1 - t_2}\ln\left(\frac{y(t_2) - T_0}{y(t_1) - T_0}\right)$$

28. Show that by Newton's Law of Cooling, the time required to cool an object from temperature A to temperature B is

$$t = \frac{1}{k}\ln\left(\frac{A - T_0}{B - T_0}\right)$$

where T_0 is the ambient temperature.

29. Air Resistance A projectile of mass $m = 1$ travels straight up from ground level with initial velocity v_0. Suppose that the velocity v satisfies $v' = -g - kv$.

(a) Find a formula for $v(t)$.

(b) Show that the projectile's height $h(t)$ is given by

$$h(t) = C(1 - e^{-kt}) - \frac{g}{k}t$$

where $C = k^{-2}(g + kv_0)$.

(c) Show that the projectile reaches its maximum height at time $t_{max} = k^{-1}\ln(1 + kv_0/g)$.

(d) In the absence of air resistance, the maximum height is reached at time $t = v_0/g$. In view of this, explain why we should expect that

$$\lim_{k\to 0}\frac{\ln(1 + \frac{kv_0}{g})}{k} = \frac{v_0}{g} \qquad \boxed{8}$$

(e) Verify Eq. (8). *Hint:* Use Theorem 1 in Section 7.5 to show that $\lim_{k\to 0}\left(1 + \frac{kv_0}{g}\right)^{1/k} = e^{v_0/g}$.

7.7 L'Hôpital's Rule

L'Hôpital's Rule is named for the French mathematician Guillaume François Antoine Marquis de L'Hôpital (1661–1704), who wrote the first textbook on calculus in 1696. The name L'Hôpital is pronounced "Lo-pee-tal."

L'Hôpital's Rule is a valuable tool for computing certain limits that are otherwise difficult to evaluate, and also for determining "asymptotic behavior" (limits at infinity).

Consider the limit of a quotient

$$\lim_{x\to a}\frac{f(x)}{g(x)}$$

Roughly speaking, L'Hôpital's Rule states that *when $f(x)/g(x)$ has an indeterminate form of type $0/0$ or ∞/∞ at $x = a$, then we can replace $f(x)/g(x)$ by the quotient of the derivatives $f'(x)/g'(x)$.*

THEOREM 1 L'Hôpital's Rule Assume that $f(x)$ and $g(x)$ are differentiable on an open interval containing a and that

$$f(a) = g(a) = 0$$

Also assume that $g'(x) \neq 0$ (except possibly at a). Then

$$\lim_{x \to a} \frac{f(x)}{g(x)} = \lim_{x \to a} \frac{f'(x)}{g'(x)}$$

if the limit on the right exists or is infinite (∞ or $-\infty$). This conclusion also holds if $f(x)$ and $g(x)$ are differentiable for x near (but not equal to) a and

$$\lim_{x \to a} f(x) = \pm\infty \qquad \text{and} \qquad \lim_{x \to a} g(x) = \pm\infty$$

Furthermore, this rule if valid for one-sided limits.

CAUTION When using L'H^opital's Rule, be sure to take the derivative of the numerator and denominator separately:

$$\lim_{x \to a} \frac{f(x)}{g(x)} = \lim_{x \to a} \frac{f'(x)}{g'(x)}$$

Do not differentiate the quotient function $f(x)/g(x)$.

EXAMPLE 1 Use L'Hôpital's Rule to evaluate $\lim_{x \to 2} \frac{x^3 - 8}{x^4 + 2x - 20}$.

Solution Let $f(x) = x^3 - 8$ and $g(x) = x^4 + 2x - 20$. Both f and g are differentiable and $f(x)/g(x)$ is indeterminate of type $0/0$ at $a = 2$ because $f(2) = g(2) = 0$:

- Numerator: $f(2) = 2^3 - 1 = 0$
- Denominator: $g(1) = 2^4 + 2(2) - 20 = 0$

Furthermore, $g'(x) = 4x^3 + 2$ is nonzero near $x = 2$, so L'Hôpital's Rule applies. We may replace the numerator and denominator by their derivatives to obtain

$$\lim_{x \to 2} \frac{x^3 - 8}{x^4 + 2x - 2} = \lim_{x \to 2} \frac{3x^2}{4x^3 + 2} = \frac{3(2^2)}{4(2^3) + 2} = \frac{12}{34} = \frac{6}{17}$$ ■

EXAMPLE 2 Evaluate $\lim_{x \to 2} \frac{4 - x^2}{\sin \pi x}$.

Solution The quotient is indeterminate of type $0/0$ at $x = 2$:

- Numerator: $4 - x^2 = 4 - 2^2 = 0$
- Denominator: $\sin \pi x = \sin 2\pi = 0$

The other hypotheses (that f and g are differentiable and $g'(x) \neq 0$ for x near $a = 2$) are also satisfied, so we may apply L'Hôpital's Rule:

$$\lim_{x \to 2} \frac{4 - x^2}{\sin \pi x} = \lim_{x \to 2} \frac{(4 - x^2)'}{(\sin \pi x)'} = \lim_{x \to 2} \frac{-2x}{\pi \cos \pi x} = \frac{-2(2)}{\pi \cos 2\pi} = \frac{-4}{\pi}$$ ■

EXAMPLE 3 Evaluate $\lim_{x \to \pi/2} \frac{\cos^2 x}{1 - \sin x}$.

Solution Again, the quotient is indeterminate of type $0/0$ at $x = \frac{\pi}{2}$:

$$\cos^2\left(\frac{\pi}{2}\right) = 0, \qquad 1 - \sin \frac{\pi}{2} = 1 - 1 = 0$$

The other hypotheses are satisfied, so we may apply L'Hôpital's Rule:

$$\lim_{x \to \pi/2} \frac{\cos^2 x}{1 - \sin x} = \lim_{x \to \pi/2} \frac{(\cos^2 x)'}{(1 - \sin x)'} = \underbrace{\lim_{x \to \pi/2} \frac{-2\cos x \sin x}{-\cos x}}_{\text{L'Hôpital's Rule}} = \underbrace{\lim_{x \to \pi/2} (2 \sin x)}_{\text{Simplify}} = 2$$

Note that the quotient $\dfrac{-2\cos x\ \sin x}{-\cos x}$ is still indeterminate at $x = \pi/2$. We removed this indeterminacy by cancelling the factor $-\cos x$. ■

■ **EXAMPLE 4 The Form $0 \cdot \infty$** Evaluate $\lim\limits_{x\to 0+} x \ln x$.

Solution This limit is one-sided because $f(x) = x \ln x$ is not defined for $x \le 0$. Furthermore, as $x \to 0+$,

- x approaches 0
- $\ln x$ approaches $-\infty$

So $f(x)$ presents an indeterminate form of type $0 \cdot \infty$. To apply L'Hôpital's Rule we rewrite our function as $f(x) = (\ln x)/x^{-1}$ so that $f(x)$ presents an indeterminate form of type $-\infty/\infty$. Then L'Hôpital's Rule applies:

$$\lim_{x\to 0+} x\ln x = \lim_{x\to 0+} \frac{\ln x}{x^{-1}} = \underbrace{\lim_{x\to 0+} \frac{(\ln x)'}{(x^{-1})'} = \lim_{x\to 0+}\left(\frac{x^{-1}}{-x^{-2}}\right)}_{\text{L'Hôpital's Rule}} = \underbrace{\lim_{x\to 0+}(-x)}_{\text{Simplify}} = 0$$ ■

■ **EXAMPLE 5 Using L'Hôpital's Rule Twice** Evaluate $\lim\limits_{x\to 0} \dfrac{e^x - x - 1}{\cos x - 1}$.

Solution For $x = 0$, we have

$$e^x - x - 1 = e^0 - 0 - 1 = 0, \qquad \cos x - 1 = \cos 0 - 1 = 0$$

A first application of L'Hôpital's Rule gives

$$\lim_{x\to 0} \frac{e^x - x - 1}{\cos x - 1} = \lim_{x\to 0} \frac{(e^x - x - 1)'}{(\cos x - 1)'} = \lim_{x\to 0}\left(\frac{e^x - 1}{-\sin x}\right) = \lim_{x\to 0}\frac{1 - e^x}{\sin x}$$

This limit is again indeterminate of type $0/0$, so we apply L'Hôpital's Rule again:

$$\lim_{x\to 0} \frac{1 - e^x}{\sin x} = \lim_{x\to 0} \frac{-e^x}{\cos x} = \frac{-e^0}{\cos 0} = -1$$ ■

■ **EXAMPLE 6 Assumptions Matter** Can L'Hôpital's Rule be applied to $\lim\limits_{x\to 1} \dfrac{x^2+1}{2x+1}$?

Solution The answer is no. The function does *not* have an indeterminate form because

$$\left.\frac{x^2+1}{2x+1}\right|_{x=1} = \frac{1^2+1}{2\cdot 1 + 1} = \frac{2}{3}$$

However, the limit can be evaluated directly by substitution: $\lim\limits_{x\to 1} \dfrac{x^2+1}{2x+1} = \dfrac{2}{3}$. An incorrect application of L'Hôpital's Rule gives the wrong answer:

$$\lim_{x\to 1} \frac{(x^2+1)'}{(2x+1)'} = \lim_{x\to 1} \frac{2x}{2} = 1 \quad \text{(not equal to original limit)}$$ ■

EXAMPLE 7 The Form $\infty - \infty$ Evaluate $\lim_{x\to 0}\left(\frac{1}{\sin x} - \frac{1}{x}\right)$.

Solution Both $1/\sin x$ and $1/x$ become infinite at $x = 0$, so we have an indeterminate form of type $\infty - \infty$. We must rewrite the function as

$$\frac{1}{\sin x} - \frac{1}{x} = \frac{x - \sin x}{x \sin x}$$

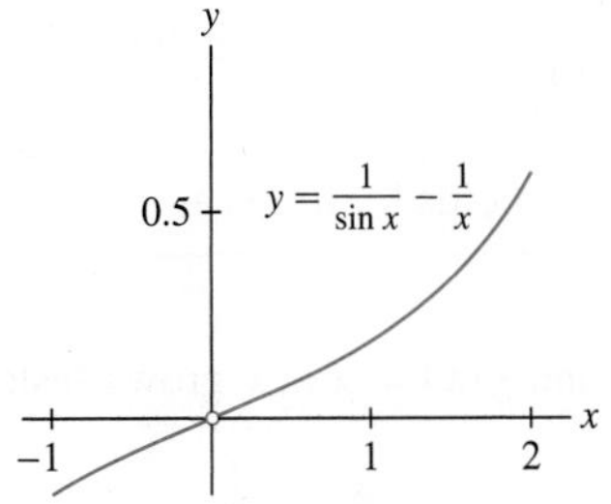

FIGURE 1 The graph confirms that $y = \frac{1}{\sin x} - \frac{1}{x}$ approaches 0 as $x \to 0$.

to obtain an indeterminate form of type 0/0. L'Hôpital's Rule yields (see Figure 1):

$$\lim_{x\to 0}\left(\frac{1}{\sin x} - \frac{1}{x}\right) = \lim_{x\to 0}\frac{x - \sin x}{x\sin x} = \underbrace{\lim_{x\to 0}\frac{1-\cos x}{x\cos x + \sin x}}_{\text{L'Hôpital's Rule}}$$

$$= \underbrace{\lim_{x\to 0}\frac{\sin x}{-x\sin x + 2\cos x}}_{\text{L'Hôpital's Rule needed again}} = \frac{0}{2} = 0 \quad \blacksquare$$

Limits of functions of the form $f(x)^{g(x)}$ can lead to the indeterminate forms 0^0, 1^∞, or ∞^0. In such cases, take the logarithm and then apply L'Hôpital's Rule.

EXAMPLE 8 The Form 0^0 Evaluate $\lim_{x\to 0+} x^x$.

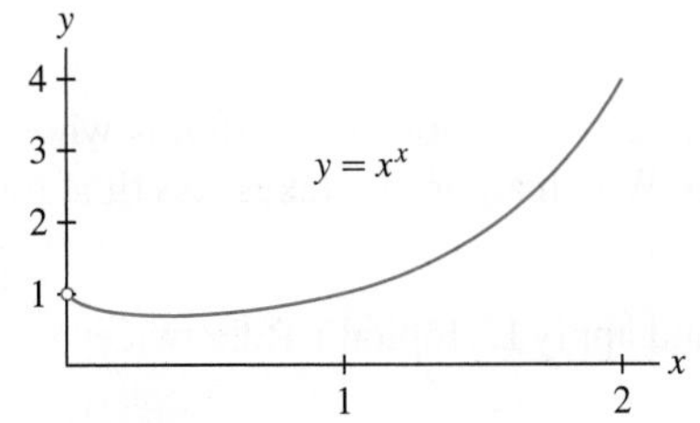

FIGURE 2 The function $y = x^x$ approaches 1 as $x \to 0+$.

Solution First, compute the limit of the logarithm $\ln x^x = x \ln x$:

$$\lim_{x\to 0+}\ln(x^x) = \lim_{x\to 0+} x\ln x = \lim_{x\to 0+}\frac{\ln x}{x^{-1}} = 0 \qquad \text{(by Example 4)}$$

Since $f(x) = e^x$ is continuous, we can exponentiate to obtain the desired limit (see Figure 2):

$$\lim_{x\to 0+} x^x = \lim_{x\to 0+} e^{\ln(x^x)} = e^{\lim_{x\to 0+}\ln(x^x)} = e^0 = 1 \quad \blacksquare$$

Comparing Growth of Functions

Sometimes, we are interested in determining which of two functions, $f(x)$ and $g(x)$, grows faster. For example, there are two standard computer algorithms for sorting data (alphabetizing, ordering according to rank, etc.): **Quick Sort** and **Bubble Sort**. The average time required to sort a list of size n has order of magnitude $n \ln n$ for Quick Sort and n^2 for Bubble Sort. Which algorithm is faster when the size n is large? Although n is a whole number, this problem amounts to comparing the growth of $f(x) = x \ln x$ and $g(x) = x^2$ as $x \to \infty$.

We say that $f(x)$ grows *faster* than $g(x)$ if

$$\lim_{x\to\infty}\frac{f(x)}{g(x)} = \infty \qquad \text{or, equivalently,} \qquad \lim_{x\to\infty}\frac{g(x)}{f(x)} = 0$$

To indicate that $f(x)$ grows faster than $g(x)$, we use the notation $g(x) \ll f(x)$. For example, $x \ll x^2$ because

$$\lim_{x\to\infty}\frac{x^2}{x} = \lim_{x\to\infty} x = \infty$$

To compare the growth of functions, we need a version of L'Hôpital's Rule that applies to limits at infinity.

THEOREM 2 L'Hôpital's Rule for Limits at Infinity Assume that $f(x)$ and $g(x)$ are differentiable in an interval (b, ∞) and that $g'(x) \neq 0$ for $x > b$. If $\lim_{x\to\infty} f(x)$ and $\lim_{x\to\infty} g(x)$ exist and either both are zero or both are infinite, then

$$\lim_{x\to\infty} \frac{f(x)}{g(x)} = \lim_{x\to\infty} \frac{f'(x)}{g'(x)}$$

provided that the limit on the right exists. A similar result holds for limits as $x \to -\infty$.

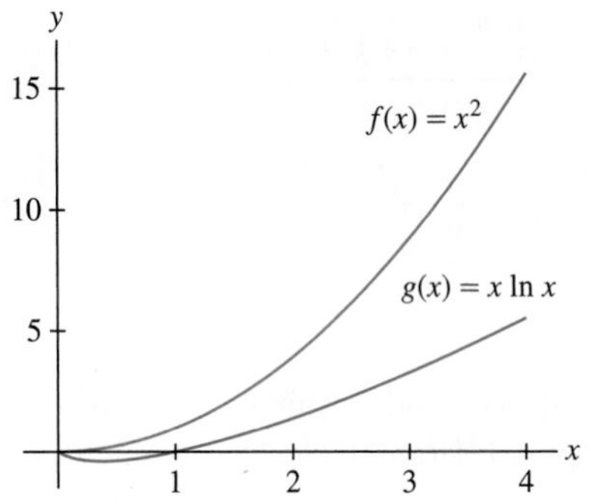

FIGURE 3

■ **EXAMPLE 9 The Form $\frac{\infty}{\infty}$** Which of $f(x) = x^2$ and $g(x) = x \ln x$ grows faster as $x \to \infty$?

Solution Both $f(x)$ and $g(x)$ approach infinity as $x \to \infty$, so L'Hôpital's Rule applies to the quotient:

$$\lim_{x\to\infty} \frac{f(x)}{g(x)} = \lim_{x\to\infty} \frac{x^2}{x \ln x} = \underbrace{\lim_{x\to\infty} \frac{x}{\ln x} = \lim_{x\to\infty} \frac{1}{x^{-1}}}_{\text{L'Hôpital's Rule}} = \lim_{x\to\infty} x = \infty$$

We conclude that $x \ln x \ll x^2$ (Figure 3). ■

■ **EXAMPLE 10** Jonathan is interested in comparing two computer algorithms whose average run times are approximately $(\ln n)^2$ and $\sqrt{n}$. Which algorithm takes less time for large values of n?

Solution Replace n by the continuous variable x and apply L'Hôpital's Rule twice:

$$\lim_{x\to\infty} \frac{\sqrt{x}}{(\ln x)^2} = \underbrace{\lim_{x\to\infty} \frac{\frac{1}{2}x^{-1/2}}{2x^{-1}\ln x}}_{\text{L'Hôpital's Rule}} = \underbrace{\lim_{x\to\infty} \frac{x^{1/2}}{4\ln x}}_{\text{Simplify}} = \underbrace{\lim_{x\to\infty} \frac{\frac{1}{2}x^{-1/2}}{4x^{-1}}}_{\text{L'Hôpital's Rule again}} = \underbrace{\lim_{x\to\infty} \frac{x^{1/2}}{8}}_{\text{Simplify}} = \infty$$

This shows that $(\ln x)^2 \ll \sqrt{x}$. We conclude that the algorithm whose average time is proportional to $(\ln n)^2$ takes less time for large n. ■

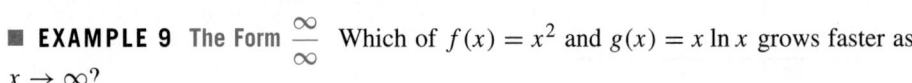

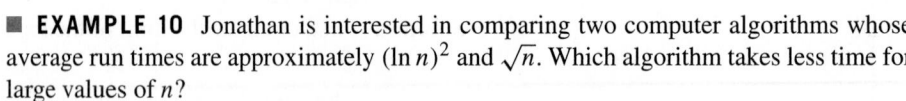

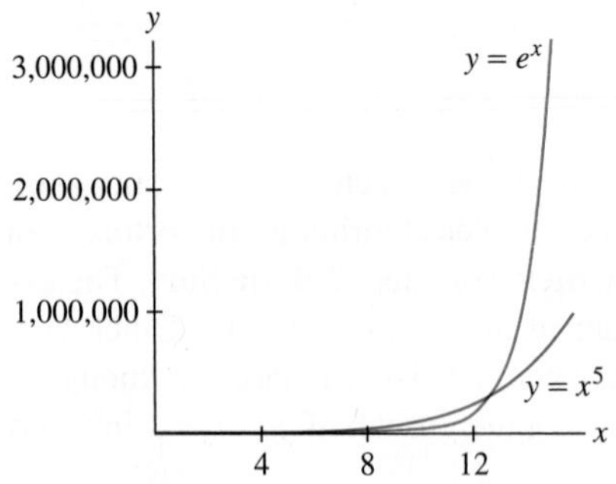

FIGURE 4 Graph illustrating that $x^5 \ll e^x$.

In Section 7.1, we asserted that exponential functions increase more rapidly than the power functions. We now prove this by showing that $x^n \ll e^x$ for every exponent n (Figure 4).

THEOREM 3 Growth of e^x

$$x^n \ll e^x \qquad \text{for every exponent } n$$

In other words, $\lim_{x\to\infty} \frac{e^x}{x^n} = \infty$ for all n.

Proof The theorem is true for $n = 0$ since $\lim_{x\to\infty} e^x = \infty$. We use L'Hôpital's Rule repeatedly to prove that e^x/x^n tends to ∞ for $n = 1, 2, 3 \ldots$. For example,

$$\lim_{x\to\infty} \frac{e^x}{x} = \lim_{x\to\infty} \frac{e^x}{1} = \lim_{x\to\infty} e^x = \infty$$

Then, having proved that $e^x/x \to \infty$, we use L'Hôpital's Rule again

$$\lim_{x\to\infty} \frac{e^x}{x^2} = \lim_{x\to\infty} \frac{e^x}{2x} = \frac{1}{2} \lim_{x\to\infty} \frac{e^x}{x} = \infty$$

Proceeding in this way, we prove the result for all whole numbers n. A more formal proof would use the principle of induction. Finally, if k is any exponent, choose any whole number n such that $n > k$. Then $e^x/x^n < e^x/x^k$ for $x > 1$, so e^x/x^k must also tend to infinity as $x \to \infty$. ■

Proof of L'Hôpital's Rule

A full proof of L'Hôpital's Rule, without simplifying assumptions, is presented in a supplement on the text's Comapanion Web Site.

We prove L'Hôpital's Rule here only in the first case of Theorem 1—namely, in the case that $f(a) = g(a) = 0$. We also assume that f' and g' are continuous at $x = a$ and that $g'(a) \neq 0$. Then $g(x) \neq g(a)$ for x near but not equal to a, and

$$\frac{f(x)}{g(x)} = \frac{f(x) - f(a)}{g(x) - g(a)} = \frac{\dfrac{f(x) - f(a)}{x - a}}{\dfrac{g(x) - g(a)}{x - a}}$$

By the Quotient Law for Limits and the definition of the derivative,

$$\lim_{x\to a} \frac{f(x)}{g(x)} = \frac{\lim\limits_{x\to a} \dfrac{f(x) - f(a)}{x - a}}{\lim\limits_{x\to a} \dfrac{g(x) - g(a)}{x - a}} = \frac{f'(a)}{g'(a)} = \lim_{x\to a} \frac{f'(x)}{g'(x)}$$ ■

7.7 SUMMARY

- *L'Hôpital's Rule:* Assume that f and g are differentiable near a and that

$$f(a) = g(a) = 0$$

Assume also that $g'(x) \neq 0$ (except possibly at a). Then

$$\lim_{x\to a} \frac{f(x)}{g(x)} = \lim_{x\to a} \frac{f'(x)}{g'(x)}$$

provided that the limit on the right exists or is infinite (∞ or $-\infty$).
- L'Hôpital's Rule also applies to limits as $x \to \infty$ or $x \to -\infty$.
- Limits involving the indeterminate forms 0^0, 1^∞, or ∞^0 can often be evaluated by first taking the logarithm and then applying L'Hôpital's Rule.
- In comparing the growth rates of functions, we say that $f(x)$ grows faster than $g(x)$, and we write $g << f$, if

$$\lim_{x\to\infty} \frac{f(x)}{g(x)} = \infty$$

7.7 EXERCISES

Preliminary Questions

1. What is wrong with applying L'Hôpital's Rule to $\lim\limits_{x\to 0} \dfrac{x^2 - 2x}{3x - 2}$?

2. Does L'Hôpital's Rule apply to $\lim\limits_{x\to a} f(x)g(x)$ if $f(x)$ and $g(x)$ both approach ∞ as $x \to a$?

Exercises

In Exercises 1–10, use L'Hôpital's Rule to evaluate the limit, or state that L'Hôpital's Rule does not apply.

1. $\lim_{x\to 3} \frac{2x^2-5x-3}{x-4}$

2. $\lim_{x\to -5} \frac{x^2-25}{5-4x-x^2}$

3. $\lim_{x\to 4} \frac{x^3-64}{x^2+16}$

4. $\lim_{x\to -1} \frac{x^4+2x+1}{x^5-2x-1}$

5. $\lim_{x\to 9} \frac{x^{1/2}+x-6}{x^{3/2}-27}$

6. $\lim_{x\to 3} \frac{\sqrt{x+1}-2}{x^3-7x-6}$

7. $\lim_{x\to 0} \frac{\sin 4x}{x^2+3x+1}$

8. $\lim_{x\to 0} \frac{x^3}{\sin x - x}$

9. $\lim_{x\to 0} \frac{\cos 2x - 1}{\sin 5x}$

10. $\lim_{x\to 0} \frac{\cos x - \sin^2 x}{\sin x}$

In Exercises 11–16, show that L'Hôpital's Rule is applicable to the limit as $x \to \pm\infty$ and evaluate.

11. $\lim_{x\to\infty} \frac{9x+4}{3-2x}$

12. $\lim_{x\to -\infty} x \sin\frac{1}{x}$

13. $\lim_{x\to\infty} \frac{\ln x}{x^{1/2}}$

14. $\lim_{x\to\infty} \frac{x}{e^x}$

15. $\lim_{x\to -\infty} \frac{\ln(x^4+1)}{x}$

16. $\lim_{x\to\infty} \frac{x^2}{e^x}$

In Exercises 17–50, evaluate the limit.

17. $\lim_{x\to 1} \frac{\sqrt{8+x}-3x^{1/3}}{x^2-3x+2}$

18. $\lim_{x\to 4} \left[\frac{1}{\sqrt{x}-2} - \frac{4}{x-4}\right]$

19. $\lim_{x\to -\infty} \frac{3x-2}{1-5x}$

20. $\lim_{x\to\infty} \frac{x^{2/3}+3x}{x^{5/3}-x}$

21. $\lim_{x\to -\infty} \frac{7x^2+4x}{9-3x^2}$

22. $\lim_{x\to\infty} \frac{3x^3+4x^2}{4x^3-7}$

23. $\lim_{x\to 1} \frac{(1+3x)^{1/2}-2}{(1+7x)^{1/3}-2}$

24. $\lim_{x\to 8} \frac{x^{5/3}-2x-16}{x^{1/3}-2}$

25. $\lim_{x\to 0} \frac{\sin 2x}{\sin 7x}$

26. $\lim_{x\to \pi/2} \frac{\tan 4x}{\tan 5x}$

27. $\lim_{x\to 0} \frac{\tan x}{x}$

28. $\lim_{x\to 0} \left(\cot x - \frac{1}{x}\right)$

29. $\lim_{x\to 0} \frac{\sin x - x\cos x}{x - \sin x}$

30. $\lim_{x\to \pi/2} \left(x - \frac{\pi}{2}\right)\tan x$

31. $\lim_{x\to 0} \frac{\cos(x+\frac{\pi}{2})}{\sin x}$

32. $\lim_{x\to 0} \frac{x^2}{1-\cos x}$

33. $\lim_{x\to \pi/2} \frac{\cos x}{\sin(2x)}$

34. $\lim_{x\to 0} \left(\frac{1}{x^2} - \csc^2 x\right)$

35. $\lim_{x\to \pi/2} (\sec x - \tan x)$

36. $\lim_{x\to 2} \frac{e^{x^2}-e^4}{x-2}$

37. $\lim_{x\to 1} \tan\left(\frac{\pi x}{2}\right)\ln x$

38. $\lim_{x\to 1} \frac{x(\ln x - 1)+1}{(x-1)\ln x}$

39. $\lim_{x\to 0} \frac{e^x-1}{\sin x}$

40. $\lim_{x\to 1} \frac{e^x-e}{\ln x}$

41. $\lim_{x\to 0} \frac{e^{2x}-1-x}{x^2}$

42. $\lim_{x\to\infty} \frac{e^{2x}-1-x}{x^2}$

43. $\lim_{t\to 0+} (\sin t)(\ln t)$

44. $\lim_{x\to\infty} e^{-x}(x^3-x^2+9)$

45. $\lim_{x\to 0} \frac{a^x-1}{x} \quad (a>0)$

46. $\lim_{x\to\infty} x^{1/x^2}$

47. $\lim_{x\to 1} (1+\ln x)^{1/(x-1)}$

48. $\lim_{x\to 0+} x^{\sin x}$

49. $\lim_{x\to 0} (\cos x)^{3/x^2}$

50. $\lim_{x\to\infty} \left(\frac{x}{x+1}\right)^x$

51. Evaluate $\lim_{x\to \pi/2} \frac{\cos mx}{\cos nx}$, where $m, n \neq 0$ are integers.

52. Evaluate $\lim_{x\to 1} \frac{x^m-1}{x^n-1}$ for any numbers $m, n \neq 0$.

53. Prove the following limit formula for e:

$$e = \lim_{x\to 0} (1+x)^{1/x}$$

Then find a value of x such that $|(1+x)^{1/x} - e| \le 0.001$.

54. GU Can L'Hôpital's Rule be applied to $\lim_{x\to 0+} x^{\sin(1/x)}$? Does a graphical or numerical investigation suggest that the limit exists?

55. Let $f(x) = x^{1/x}$ for $x > 0$.

(a) Calculate $\lim_{x\to 0+} f(x)$ and $\lim_{x\to\infty} f(x)$.

(b) Find the maximum value of $f(x)$, and determine the intervals on which $f(x)$ is increasing or decreasing.

56. (a) Use the results of Exercise 55 to prove that $x^{1/x} = c$ has a unique solution if $0 < c \le 1$ or $c = e^{1/e}$, two solutions if $1 < c < e^{1/e}$, and no solutions if $c > e^{1/e}$.

(b) GU Plot the graph of $f(x) = x^{1/x}$ and verify that it confirms the conclusions of (a).

57. Determine whether $f << g$ or $g << f$ (or neither) for the functions $f(x) = \log_{10} x$ and $g(x) = \ln x$.

58. Show that $(\ln x)^2 << \sqrt{x}$ and $(\ln x)^4 << x^{1/10}$.

59. Just as exponential functions are distinguished by their rapid rate of increase, the logarithm functions grow particularly slowly. Show that $\ln x << x^a$ for all $a > 0$.

60. Show that $(\ln x)^N << x^a$ for all N and all $a > 0$.

61. Determine whether $\sqrt{x} << e^{\sqrt{\ln x}}$ or $e^{\sqrt{\ln x}} << \sqrt{x}$. *Hint:* Use the substitution $u = \ln x$ instead of L'Hôpital's Rule.

62. Show that $\lim_{x\to\infty} x^n e^{-x} = 0$ for all whole numbers $n > 0$.

63. Assumptions Matter Let $f(x) = x(2 + \sin x)$ and $g(x) = x^2 + 1$.

(a) Show directly that $\lim_{x\to\infty} f(x)/g(x) = 0$.

(b) Show that $\lim_{x\to\infty} f(x) = \lim_{x\to\infty} g(x) = \infty$, but $\lim_{x\to\infty} f'(x)/g'(x)$ does not exist.

Do (a) and (b) contradict L'Hôpital's Rule? Explain.

64. Let $H(b) = \lim_{x\to\infty} \dfrac{\ln(1 + b^x)}{x}$ for $b > 0$.

(a) Show that $H(b) = \ln b$ if $b \geq 1$

(b) Determine $H(b)$ for $0 < b \leq 1$.

65. Let $G(b) = \lim_{x\to\infty} (1 + b^x)^{1/x}$.

(a) Use the result of Exercise 64 to evaluate $G(b)$ for all $b > 0$.

(b) GU Verify your result graphically by plotting $y = (1 + b^x)^{1/x}$ together with the horizontal line $y = G(b)$ for the values $b = 0.25, 0.5, 2, 3$.

66. Show that $\lim_{t\to\infty} t^k e^{-t^2} = 0$ for all k. *Hint:* Compare with $\lim_{t\to\infty} t^k e^{-t} = 0$.

In Exercises 67–69, let

$$f(x) = \begin{cases} e^{-1/x^2} & \text{for } x \neq 0 \\ 0 & \text{for } x = 0 \end{cases}$$

These exercises show that $f(x)$ has an unusual property: All of its derivatives at $x = 0$ exist and are equal to zero.

67. Show that $\lim_{x\to 0} \dfrac{f(x)}{x^k} = 0$ for all k. *Hint:* Let $t = x^{-1}$ and apply the result of Exercise 66.

68. Show that $f'(0)$ exists and is equal to zero. Also, verify that $f''(0)$ exists and is equal to zero.

69. Show that for $k \geq 1$ and $x \neq 0$,

$$f^{(k)}(x) = \frac{P(x)e^{-1/x^2}}{x^r}$$

for some polynomial $P(x)$ and some exponent $r \geq 1$. Use the result of Exercise 67 to show that $f^{(k)}(0)$ exists and is equal to zero for all $k \geq 1$.

70. (a) Verify for $r \neq 0$:

$$\int_0^T te^{rt}\,dt = \frac{e^{rT}(rT - 1) + 1}{r^2} \qquad \boxed{1}$$

Hint: For fixed r, let $F(T)$ be the value of the integral on the left. By FTC II, $F'(t) = te^{rt}$ and $F(0) = 0$. Show that the same is true of the function on the right.

(b) Use L'Hôpital's Rule to show that for fixed T, the limit as $r \to 0$ of the right-hand side of Eq. (1) is equal to the value of the integral for $r = 0$.

71. The formula $\int_1^x t^n\,dt = \dfrac{x^{n+1} - 1}{n + 1}$ is valid for $n \neq -1$. Use L'Hôpital's Rule to prove that

$$\lim_{n\to -1} \frac{x^{n+1} - 1}{n + 1} = \ln x$$

Use this to show that

$$\lim_{n\to -1} \int_1^x t^n\,dt = \int_1^x t^{-1}\,dt$$

Thus, the definite integral of x^{-1} is a limit of the definite integrals of x^n as n approaches -1.

Further Insights and Challenges

72. Show that L'Hôpital's Rule applies to $\lim_{x\to\infty} \dfrac{x}{\sqrt{x^2 + 1}}$ but that it does not help. Then evaluate the limit directly.

73. The Second Derivative Test for critical points fails if $f''(c) = 0$. This exercise develops a **Higher Derivative Test** based on the sign of the first nonzero derivative. Suppose that

$$f'(c) = f''(c) = \cdots = f^{(n-1)}(c) = 0, \quad \text{but} \quad f^{(n)}(c) \neq 0$$

(a) Show, by applying L'Hôpital's Rule n times, that

$$\lim_{x\to c} \frac{f(x) - f(c)}{(x - c)^n} = \frac{1}{n!} f^{(n)}(c)$$

where $n! = n(n - 1)(n - 2)\cdots(2)(1)$.

(b) Use (a) to show that if n is even, then $f(c)$ is a local minimum if $f^{(n)}(c) > 0$ and is a local maximum if $f^{(n)}(c) < 0$. *Hint:* If n is even, then $(x - c)^n > 0$ for $x \neq a$, so $f(x) - f(c)$ must be positive for x near c if $f^{(n)}(c) > 0$.

(c) Use (a) to show that if n is odd, then $f(c)$ is neither a local minimum nor a local maximum.

74. When a spring with natural frequency $\lambda/2\pi$ is driven with a sinusoidal force $\sin(\omega t)$ with $\omega \neq \lambda$, it oscillates according to

$$y(t) = \frac{1}{\lambda^2 - \omega^2}\big(\lambda \sin(\omega t) - \omega \sin(\lambda t)\big)$$

Let $y_0(t) = \lim_{\omega\to\lambda} y(t)$.

(a) Use L'Hôpital's Rule to determine $y_0(t)$.

(b) Show that $y_0(t)$ ceases to be periodic and that its amplitude $|y_0(t)|$ tends to ∞ as $t \to \infty$ (the system is said to be in **resonance**; eventually, the spring is stretched beyond its limits).

(c) CAS Plot $y(t)$ for $\lambda = 1$ and $\omega = 0.8, 0.9, 0.99$, and 0.999. Do the graphs confirm your conclusion in (b)?

75. We expended a lot of effort to evaluate $\lim_{x\to 0} \dfrac{\sin x}{x}$ in Chapter 2. Show that we could have evaluated it easily using L'Hôpital's Rule. Then explain why this method would involve *circular reasoning*.

76. By a fact from algebra, if f, g are polynomials such that $f(a) = g(a) = 0$, then there are polynomials f_1, g_1 such that

$$f(x) = (x-a)f_1(x), \qquad g(x) = (x-a)g_1(x)$$

Use this to verify L'Hôpital's Rule directly for $\lim_{x\to a} f(x)/g(x)$.

77. Patience Required Use L'Hôpital's Rule to evaluate and check your answers numerically:

(a) $\lim_{x\to 0+} \left(\dfrac{\sin x}{x}\right)^{1/x^2}$ **(b)** $\lim_{x\to 0} \left(\dfrac{1}{\sin^2 x} - \dfrac{1}{x^2}\right)$

78. In the following cases, check that $x = c$ is a critical point and use Exercise 73 to determine whether $f(c)$ is a local minimum or a local maximum.

(a) $f(x) = x^5 - 6x^4 + 14x^3 - 16x^2 + 9x + 12 \quad (c = 1)$

(b) $f(x) = x^6 - x^3 \quad (c = 0)$

7.8 Inverse Trigonometric Functions

In this section, we discuss the inverse trigonometric functions and their derivatives. Recall that an inverse $f^{-1}(x)$ exists if and only if the function $f(x)$ is one-to-one on its domain. However, the trigonometric functions are not one-to-one (because they are periodic). Therefore, to define their inverses, we shall restrict their domains so that the resulting functions are one-to-one.

Do not confuse the inverse $\sin^{-1} x$ with the reciprocal

$$(\sin x)^{-1} = \frac{1}{\sin x} = \csc x$$

The inverse functions $\sin^{-1} x, \cos^{-1} x, \ldots$ are often denoted $\arcsin x$, $\arccos x$, etc.

First consider the sine function. Figure 1 shows that $f(\theta) = \sin\theta$ is one-to-one on $\left[-\frac{\pi}{2}, \frac{\pi}{2}\right]$. With this interval as domain, the inverse is called the **arcsine function** and is denoted $\theta = \sin^{-1} x$ or $\theta = \arcsin x$. By definition,

$$\theta = \sin^{-1} x \text{ is the unique angle in } \left[-\frac{\pi}{2}, \frac{\pi}{2}\right] \text{ such that } \sin\theta = x$$

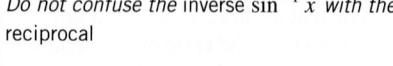

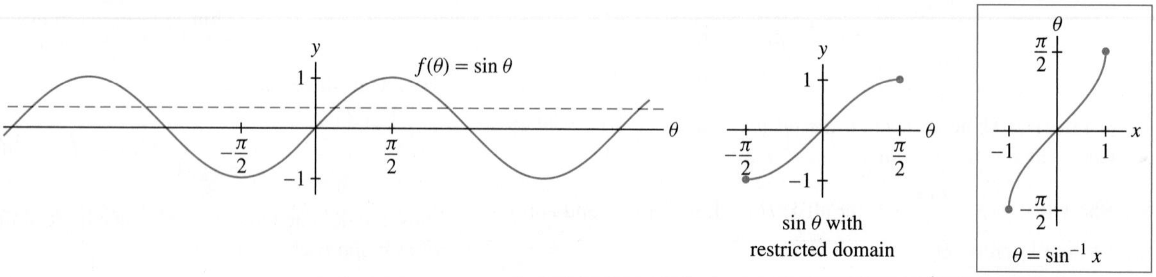

FIGURE 1

The range of $\sin x$ is $[-1, 1]$ and therefore $\sin^{-1} x$ has domain $[-1, 1]$. A table of values for the arcsine (Table 1) is obtained by reversing the columns in a table of values for $\sin x$.

Summary of inverse relation between the sine and arcsine functions:

$\sin(\sin^{-1} x) = x \quad$ for $-1 \le x \le 1$

$\sin^{-1}(\sin\theta) = \theta \quad$ for $-\frac{\pi}{2} \le \theta \le \frac{\pi}{2}$

TABLE 1

x	-1	$-\frac{\sqrt{3}}{2}$	$-\frac{\sqrt{2}}{2}$	$-\frac{1}{2}$	0	$\frac{1}{2}$	$\frac{\sqrt{2}}{2}$	$\frac{\sqrt{3}}{2}$	1
$\theta = \sin^{-1} x$	$-\frac{\pi}{2}$	$-\frac{\pi}{3}$	$-\frac{\pi}{4}$	$-\frac{\pi}{6}$	0	$\frac{\pi}{6}$	$\frac{\pi}{4}$	$\frac{\pi}{3}$	$\frac{\pi}{2}$

■ **EXAMPLE 1** **(a)** Show that $\sin^{-1}\left(\sin\left(\frac{\pi}{4}\right)\right) = \frac{\pi}{4}$.

(b) Explain why $\sin^{-1}\left(\sin\left(\dfrac{5\pi}{4}\right)\right) \ne \dfrac{5\pi}{4}$.

Solution The equation $\sin^{-1}(\sin\theta) = \theta$ is valid only if θ lies in $\left[-\frac{\pi}{2}, \frac{\pi}{2}\right]$.

(a) Since $\dfrac{\pi}{4}$ lies in the required interval, $\sin^{-1}\left(\sin\left(\frac{\pi}{4}\right)\right) = \dfrac{\pi}{4}$.

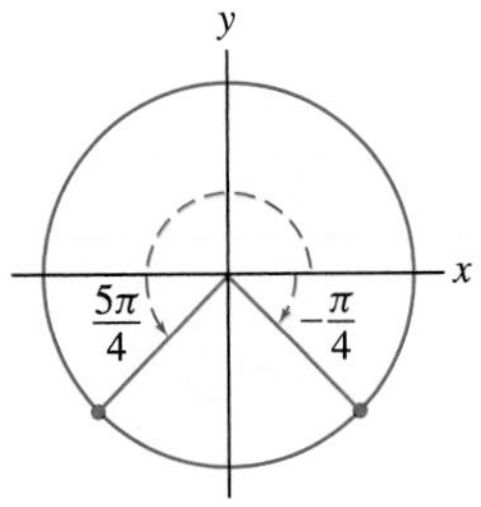

FIGURE 2 $\sin(\frac{5\pi}{4}) = \sin(-\frac{\pi}{4})$.

(b) Let $\theta = \sin^{-1}\left(\sin\left(\frac{5\pi}{4}\right)\right)$. By definition, θ is the angle in $\left[-\frac{\pi}{2}, \frac{\pi}{2}\right]$ such that $\sin\theta = \sin\left(\frac{5\pi}{4}\right)$. By the identity $\sin\theta = \sin(\pi - \theta)$ (Figure 2),

$$\sin\left(\frac{5\pi}{4}\right) = \sin\left(\pi - \frac{5\pi}{4}\right) = \sin\left(-\frac{\pi}{4}\right)$$

The angle $-\frac{\pi}{4}$ lies in the required interval, so $\theta = \sin^{-1}\left(\sin\left(\frac{5\pi}{4}\right)\right) = -\frac{\pi}{4}$. ■

Summary of inverse relation between the cosine and arccosine:

$\cos(\cos^{-1} x) = x$ for $-1 \le x \le 1$

$\cos^{-1}(\cos\theta) = \theta$ for $0 \le \theta \le \pi$

The cosine function is one-to-one on $[0, \pi]$ rather than $\left[-\frac{\pi}{2}, \frac{\pi}{2}\right]$ (Figure 3). With this domain, the inverse is called the **arccosine function** and is denoted $\theta = \cos^{-1} x$ or $\theta = \arccos x$. It has domain $[-1, 1]$. By definition,

$\theta = \cos^{-1} x$ is the unique angle in $[0, \pi]$ such that $\cos\theta = x$

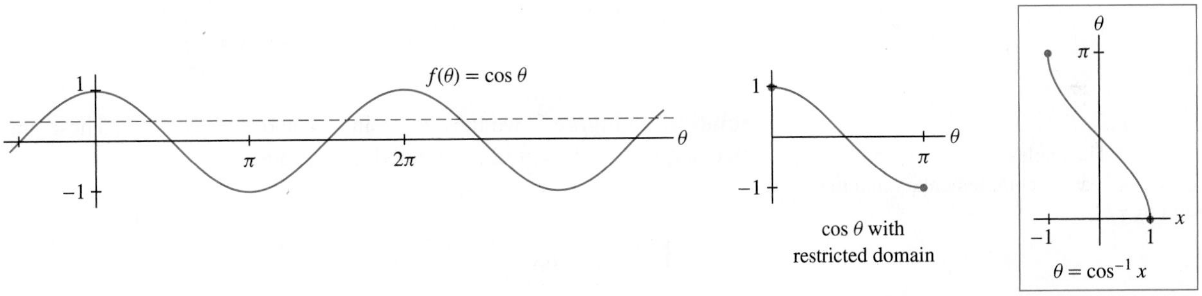

FIGURE 3

To compute the derivatives of the inverse trigonometric functions, we will need to simplify composite expressions such as $\cos(\sin^{-1} x)$ and $\tan(\sec^{-1} x)$. This can be done in two ways: by referring to the appropriate right triangle or by using trigonometric identities.

■ **EXAMPLE 2** Simplify $\cos(\sin^{-1} x)$ and $\tan(\sin^{-1} x)$.

Solution This problem asks for the values of $\cos\theta$ and $\tan\theta$ at the angle $\theta = \sin^{-1} x$. Consider a right triangle with hypotenuse of length 1 and angle θ such that $\sin\theta = x$ as in Figure 4. By the Pythagorean Theorem, the adjacent side has length $\sqrt{1-x^2}$. Now we can read off the values from Figure 4:

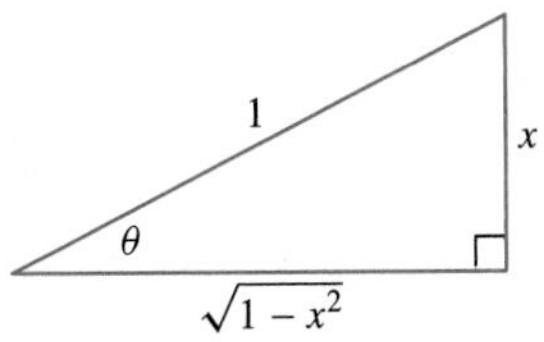

FIGURE 4 Right triangle constructed so that $\sin\theta = x$.

$$\cos(\sin^{-1} x) = \cos\theta = \frac{\text{adjacent}}{\text{hypotenuse}} = \sqrt{1-x^2}$$
$$\tan(\sin^{-1} x) = \tan\theta = \frac{\text{opposite}}{\text{adjacent}} = \frac{x}{\sqrt{1-x^2}} \qquad \boxed{1}$$

Alternatively, we may argue using trigonometric identities. Because $\sin\theta = x$,

$$\cos(\sin^{-1} x) = \cos\theta = \sqrt{1 - \sin^2\theta} = \sqrt{1-x^2}$$

We are justified in taking the positive square root because $\theta = \sin^{-1} x$ lies in $\left[-\frac{\pi}{2}, \frac{\pi}{2}\right]$ and $\cos\theta$ is positive in this interval. ■

In the next theorem, we compute the derivatives of arcsine and arccosine using the formula for the derivative of the inverse function recalled in the margin [Eq. (3)].

THEOREM 1 Derivatives of Arcsine and Arccosine

$$\frac{d}{dx}\sin^{-1}x = \frac{1}{\sqrt{1-x^2}}, \qquad \frac{d}{dx}\cos^{-1}x = -\frac{1}{\sqrt{1-x^2}} \qquad \boxed{2}$$

← *REMINDER If $g(x)$ is the inverse of $f(x)$, then*

$$g'(x) = \frac{1}{f'(g(x))} \qquad \boxed{3}$$

Proof Apply Eq. (3) in the margin to $f(x) = \sin x$ and $g(x) = \sin^{-1} x$:

$$\frac{d}{dx}\sin^{-1}x = \frac{1}{f'(\sin^{-1}x)} = \frac{1}{\cos(\sin^{-1}x)} = \frac{1}{\sqrt{1-x^2}}$$

In the last step, we use Eq. (1) from Example 2. The derivative of $\cos^{-1} x$ is similar (see Exercise 49 or the next example). ■

■ **EXAMPLE 3 Complementary Angles** The derivatives of $\sin^{-1} x$ and $\cos^{-1} x$ are equal up to a minus sign. Explain this by proving that

$$\sin^{-1}x + \cos^{-1}x = \frac{\pi}{2}$$

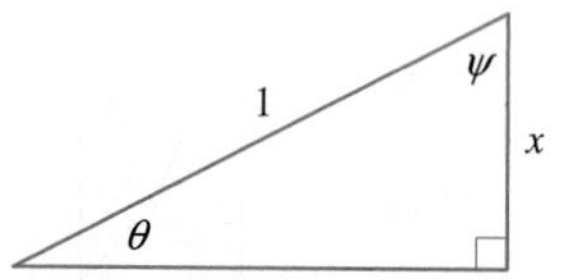

FIGURE 5 The angles $\theta = \sin^{-1} x$ and $\psi = \cos^{-1} x$ are complementary and thus sum to $\pi/2$.

Solution In Figure 5, we have $\theta = \sin^{-1} x$ and $\psi = \cos^{-1} x$. These angles are complementary, so $\theta + \psi = \pi/2$ as claimed. Therefore,

$$\frac{d}{dx}\cos^{-1}x = \frac{d}{dx}\left(\frac{\pi}{2} - \sin^{-1}x\right) = -\frac{d}{dx}\sin^{-1}x \qquad ■$$

■ **EXAMPLE 4** Calculate $f'\left(\frac{1}{2}\right)$, where $f(x) = \arcsin(x^2)$.

Solution Recall that $\arcsin x$ is another notation for $\sin^{-1} x$. By the Chain Rule,

$$\frac{d}{dx}\arcsin(x^2) = \frac{d}{dx}\sin^{-1}(x^2) = \frac{1}{\sqrt{1-x^4}}\frac{d}{dx}x^2 = \frac{2x}{\sqrt{1-x^4}}$$

$$f'\left(\frac{1}{2}\right) = \frac{2\left(\frac{1}{2}\right)}{\sqrt{1-\left(\frac{1}{2}\right)^4}} = \frac{1}{\sqrt{\frac{15}{16}}} = \frac{4}{\sqrt{15}} \qquad ■$$

We now address the remaining trigonometric functions. The function $f(\theta) = \tan\theta$ is one-to-one on $\left(-\frac{\pi}{2}, \frac{\pi}{2}\right)$, and $f(\theta) = \cot\theta$ is one-to-one on $(0, \pi)$ [see Figure 10 in Section 1.4]. We define their inverses by restricting them to these domains:

$\theta = \tan^{-1} x$ is the unique angle in $\left(-\frac{\pi}{2}, \frac{\pi}{2}\right)$ such that $\tan\theta = x$

$\theta = \cot^{-1} x$ is the unique angle in $(0, \pi)$ such that $\cot\theta = x$

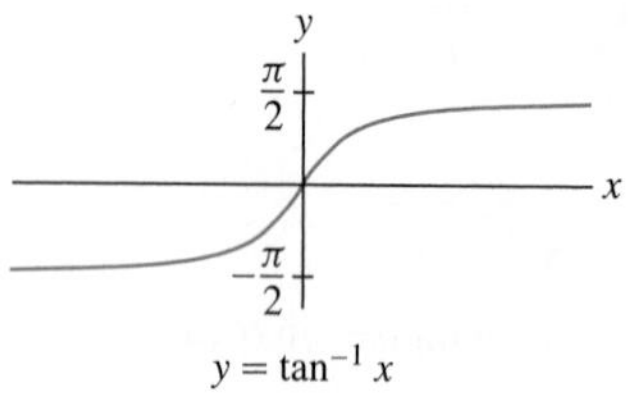

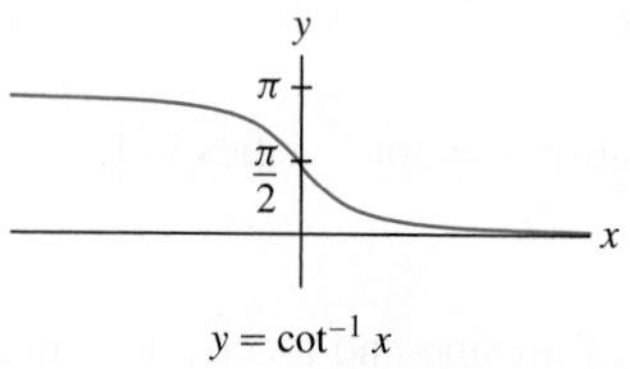

FIGURE 6

The range of both $\tan\theta$ and $\cot\theta$ is the set of all real numbers **R**. Therefore, $\theta = \tan^{-1} x$ and $\theta = \cot^{-1} x$ have domain **R** (Figure 6).

The function $f(\theta) = \sec\theta$ is not defined at $\theta = \frac{\pi}{2}$, but we see in Figure 7 that it is one-to-one on both $\left[0, \frac{\pi}{2}\right)$ and $\left(\frac{\pi}{2}, \pi\right]$. Similarly, $f(\theta) = \csc\theta$ is not defined at $\theta = 0$, but it is one-to-one on $\left[-\frac{\pi}{2}, 0\right)$ and $\left(0, \frac{\pi}{2}\right]$. We define the inverse functions as follows:

$$\theta = \sec^{-1} x \text{ is the unique angle in } \left[0, \frac{\pi}{2}\right) \cup \left(\frac{\pi}{2}, \pi\right] \text{ such that } \sec\theta = x$$

$$\theta = \csc^{-1} x \text{ is the unique angle in } \left[-\frac{\pi}{2}, 0\right) \cup \left(0, \frac{\pi}{2}\right] \text{ such that } \csc\theta = x$$

Figure 7 shows that the range of $f(\theta) = \sec\theta$ is the set of real numbers x such that $|x| \ge 1$. The same is true of $f(\theta) = \csc\theta$. It follows that both $\theta = \sec^{-1} x$ and $\theta = \csc^{-1} x$ have domain $\{x : |x| \ge 1\}$.

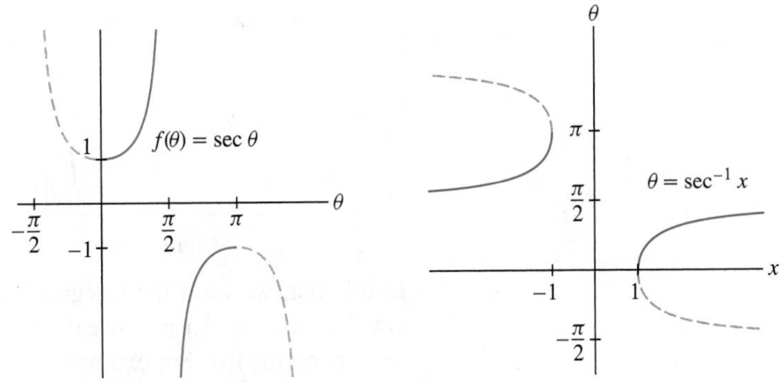

FIGURE 7 $f(\theta) = \sec\theta$ is one-to-one on the interval $[0, \pi]$ with $\pi/2$ removed.

THEOREM 2 Derivatives of Inverse Trigonometric Functions

$$\frac{d}{dx}\tan^{-1} x = \frac{1}{x^2+1}, \qquad \frac{d}{dx}\cot^{-1} x = -\frac{1}{x^2+1}$$

$$\frac{d}{dx}\sec^{-1} x = \frac{1}{|x|\sqrt{x^2-1}}, \qquad \frac{d}{dx}\csc^{-1} x = -\frac{1}{|x|\sqrt{x^2-1}}$$

The proofs of the formulas in Theorem 2 are similar to the proof of Theorem 1. See Exercises 50 and 52.

EXAMPLE 5 Calculate:

$$\textbf{(a)}\ \frac{d}{dx}\tan^{-1}(3x+1) \quad \text{and} \quad \textbf{(b)}\ \frac{d}{dx}\csc^{-1}(e^x+1)\bigg|_{x=0}$$

Solution

(a) Apply the Chain Rule using the formula $\frac{d}{du}\tan^{-1} u = \frac{1}{u^2+1}$:

$$\frac{d}{dx}\tan^{-1}(3x+1) = \frac{1}{(3x+1)^2+1}\frac{d}{dx}(3x+1) = \frac{3}{9x^2+6x+2}$$

(b) Apply the Chain Rule using the formula $\frac{d}{du}\csc^{-1} u = -\frac{1}{|u|\sqrt{u^2-1}}$:

$$\frac{d}{dx}\csc^{-1}(e^x+1) = -\frac{1}{|e^x+1|\sqrt{(e^x+1)^2-1}}\ \frac{d}{dx}(e^x+1)$$

$$= -\frac{e^x}{(e^x+1)\sqrt{e^{2x}+2e^x}}$$

We have replaced $|e^x + 1|$ by $e^x + 1$ because this quantity is positive. Now we have

$$\frac{d}{dx}\csc^{-1}(e^x+1)\bigg|_{x=0} = -\frac{e^0}{(e^0+1)\sqrt{e^0+2e^0}} = -\frac{1}{2\sqrt{3}}$$ ■

The formulas for the derivatives of the inverse trigonometric functions yield the following integration formulas.

Integral Formulas

$$\int \frac{dx}{\sqrt{1-x^2}} = \sin^{-1} x + C \qquad (4)$$

$$\int \frac{dx}{x^2+1} = \tan^{-1} x + C \qquad (5)$$

$$\int \frac{dx}{|x|\sqrt{x^2-1}} = \sec^{-1} x + C \qquad (6)$$

In this list, we omit the integral formulas corresponding to the derivatives of $\cos^{-1} x$, $\cot^{-1} x$, and $\csc^{-1} x$ because the resulting integrals differ only by a minus sign from those already on the list. For example,

$$\frac{d}{dx}\cos^{-1} x = -\frac{1}{\sqrt{1-x^2}}, \qquad \int \frac{dx}{\sqrt{1-x^2}} = -\cos^{-1} x + C$$

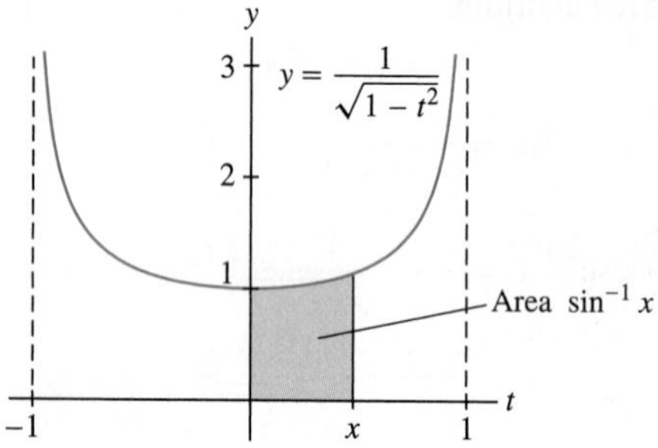

FIGURE 8

We can use these formulas to express the inverse trigonometric functions as definite integrals. For example, because $\sin^{-1} 0 = 0$, we have (Figure 8):

$$\sin^{-1} x = \int_0^x \frac{dt}{\sqrt{1-t^2}} \qquad \text{for } -1 < x < 1$$

■ **EXAMPLE 6** Evaluate $\displaystyle\int_0^1 \frac{dx}{x^2+1}$.

Solution This integral is the area of the region in Figure 9. By Eq. (5),

$$\int_0^1 \frac{dx}{x^2+1} = \tan^{-1} x\bigg|_0^1 = \tan^{-1} 1 - \tan^{-1} 0 = \frac{\pi}{4} - 0 = \frac{\pi}{4}$$ ■

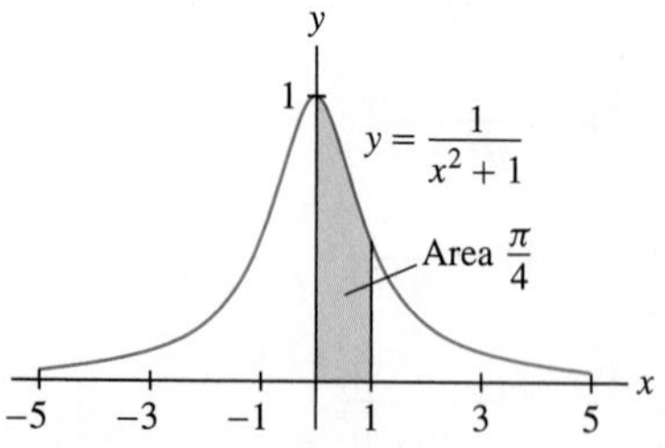

FIGURE 9

■ **EXAMPLE 7 Using Substitution** Evaluate $\displaystyle\int_{1/\sqrt{2}}^1 \frac{dx}{x\sqrt{4x^2-1}}$.

Solution Notice that $\sqrt{4x^2-1}$ can be written $\sqrt{(2x)^2-1}$, so it makes sense to try the substitution $u = 2x$, $du = 2\,dx$. Then

$$u^2 = 4x^2 \qquad \text{and} \qquad \sqrt{4x^2-1} = \sqrt{u^2-1}$$

The new limits of integration are $u\left(\frac{1}{\sqrt{2}}\right) = 2\left(\frac{1}{\sqrt{2}}\right) = \sqrt{2}$ and $u(1) = 2$. Now use Eq. (6). Note that we can ignore absolute value in Eq. (6) because the limits of integration are positive:

$$\int_{1/\sqrt{2}}^{1} \frac{dx}{x\sqrt{4x^2 - 1}} = \int_{\sqrt{2}}^{2} \frac{\frac{1}{2}\,du}{\frac{1}{2}u\sqrt{u^2 - 1}} = \int_{\sqrt{2}}^{2} \frac{du}{u\sqrt{u^2 - 1}}$$

$$= \sec^{-1} 2 - \sec^{-1}\sqrt{2}$$

$$= \frac{\pi}{3} - \frac{\pi}{4} = \frac{\pi}{12}$$

■

■ **EXAMPLE 8 Using Substitution** Evaluate $\displaystyle\int_{-3/4}^{0} \frac{dx}{\sqrt{9 - 16x^2}}$.

Solution Let us first rewrite the integrand:

$$\sqrt{9 - 16x^2} = \sqrt{9\left(1 - \frac{16x^2}{9}\right)} = 3\sqrt{1 - \left(\frac{4x}{3}\right)^2}$$

Thus it makes sense to use the substitution $u = \frac{4}{3}x$. Then $du = \frac{4}{3}\,dx$ and

$$x = \frac{3}{4}u, \qquad dx = \frac{3}{4}\,du, \qquad \sqrt{9 - 16x^2} = 3\sqrt{1 - u^2}$$

The new limits of integration are $u\left(-\frac{3}{4}\right) = -1$ and $u(0) = 0$. Eq. (4) gives us

$$\int_{-3/4}^{0} \frac{dx}{\sqrt{9 - 16x^2}} = \int_{-1}^{0} \frac{\frac{3}{4}\,du}{3\sqrt{1 - u^2}} = \frac{1}{4}\int_{-1}^{0} \frac{du}{\sqrt{1 - u^2}}$$

$$= \frac{1}{4}\left(\sin^{-1} 0 - \sin^{-1}(-1)\right) = -\frac{1}{4}\left(-\frac{\pi}{2}\right) = \frac{\pi}{8}$$

■

7.8 SUMMARY

- The *arcsine* and *arccosine* are defined for $-1 \le x \le 1$:

$$\theta = \sin^{-1} x \text{ is the unique angle in } \left[-\frac{\pi}{2}, \frac{\pi}{2}\right] \text{ such that } \sin\theta = x$$

$$\theta = \cos^{-1} x \text{ is the unique angle in } [0, \pi] \text{ such that } \cos\theta = x$$

- $\tan^{-1} x$ and $\cot^{-1} x$ are defined for all x:

$$\theta = \tan^{-1} x \text{ is the unique angle in } \left(-\frac{\pi}{2}, \frac{\pi}{2}\right) \text{ such that } \tan\theta = x$$

$$\theta = \cot^{-1} x \text{ is the unique angle in } (0, \pi) \text{ such that } \cot\theta = x$$

- $\sec^{-1} x$ and $\csc^{-1} x$ are defined for $|x| \ge 1$:

$$\theta = \sec^{-1} x \text{ is the unique angle in } \left[0, \frac{\pi}{2}\right) \cup \left(\frac{\pi}{2}, \pi\right] \text{ such that } \sec\theta = x$$

$$\theta = \csc^{-1} x \text{ is the unique angle in } \left[-\frac{\pi}{2}, 0\right) \cup \left(0, \frac{\pi}{2}\right] \text{ such that } \csc\theta = x$$

- Derivative formulas:

$$\frac{d}{dx}\sin^{-1}x = \frac{1}{\sqrt{1-x^2}}, \qquad \frac{d}{dx}\cos^{-1}x = -\frac{1}{\sqrt{1-x^2}}$$

$$\frac{d}{dx}\tan^{-1}x = \frac{1}{x^2+1}, \qquad \frac{d}{dx}\cot^{-1}x = -\frac{1}{x^2+1}$$

$$\frac{d}{dx}\sec^{-1}x = \frac{1}{|x|\sqrt{x^2-1}}, \qquad \frac{d}{dx}\csc^{-1}x = -\frac{1}{|x|\sqrt{x^2-1}}$$

- Integral formulas:

$$\int \frac{dx}{\sqrt{1-x^2}} = \sin^{-1}x + C$$

$$\int \frac{dx}{x^2+1} = \tan^{-1}x + C$$

$$\int \frac{dx}{|x|\sqrt{x^2-1}} = \sec^{-1}x + C$$

7.8 EXERCISES

Preliminary Questions

1. Which of the following quantities is undefined?

(a) $\sin^{-1}\left(-\frac{1}{2}\right)$ **(b)** $\cos^{-1}(2)$

(c) $\csc^{-1}\left(\frac{1}{2}\right)$ **(d)** $\csc^{-1}(2)$

2. Give an example of an angle θ such that $\cos^{-1}(\cos\theta) \neq \theta$. Does this contradict the definition of inverse function?

3. What is the geometric interpretation of the identity

$$\sin^{-1}x + \cos^{-1}x = \frac{\pi}{2}$$

What does this identity tell us about the derivatives of $\sin^{-1}x$ and $\cos^{-1}x$?

4. Find b such that $\int_0^b \frac{dx}{1+x^2} = \frac{\pi}{3}$.

5. Which relation between x and u yields $\sqrt{16+x^2} = 4\sqrt{1+u^2}$?

Exercises

In Exercises 1–6, evaluate without using a calculator.

1. $\cos^{-1}1$ **2.** $\sin^{-1}\frac{1}{2}$

3. $\cot^{-1}1$ **4.** $\sec^{-1}\frac{2}{\sqrt{3}}$

5. $\tan^{-1}\sqrt{3}$ **6.** $\sin^{-1}(-1)$

In Exercises 7–16, compute without using a calculator.

7. $\sin^{-1}\left(\sin\frac{\pi}{3}\right)$ **8.** $\sin^{-1}\left(\sin\frac{4\pi}{3}\right)$

9. $\cos^{-1}\left(\cos\frac{3\pi}{2}\right)$ **10.** $\sin^{-1}\left(\sin\left(-\frac{5\pi}{6}\right)\right)$

11. $\tan^{-1}\left(\tan\frac{3\pi}{4}\right)$ **12.** $\tan^{-1}(\tan\pi)$

13. $\sec^{-1}(\sec 3\pi)$ **14.** $\sec^{-1}\left(\sec\frac{3\pi}{2}\right)$

15. $\csc^{-1}(\csc(-\pi))$ **16.** $\cot^{-1}\left(\cot\left(-\frac{\pi}{4}\right)\right)$

In Exercises 17–20, simplify by referring to the appropriate triangle or trigonometric identity.

17. $\tan(\cos^{-1}x)$ **18.** $\cos(\tan^{-1}x)$

19. $\cot(\sec^{-1}x)$ **20.** $\cot(\sin^{-1}x)$

In Exercises 21–28, refer to the appropriate triangle or trigonometric identity to compute the given value.

21. $\cos\left(\sin^{-1}\frac{2}{3}\right)$ **22.** $\tan\left(\cos^{-1}\frac{2}{3}\right)$

23. $\tan(\sin^{-1}0.8)$ **24.** $\cos(\cot^{-1}1)$

25. $\cot(\csc^{-1}2)$ **26.** $\tan(\sec^{-1}(-2))$

27. $\cot(\tan^{-1}20)$ **28.** $\sin(\csc^{-1}20)$

In Exercises 29–32, compute the derivative at the point indicated without using a calculator.

29. $y = \sin^{-1} x, \quad x = \frac{3}{5}$

30. $y = \tan^{-1} x, \quad x = \frac{1}{2}$

31. $y = \sec^{-1} x, \quad x = 4$

32. $y = \arccos(4x), \quad x = \frac{1}{5}$

In Exercises 33–48, find the derivative.

33. $y = \sin^{-1}(7x)$

34. $y = \arctan\left(\frac{x}{3}\right)$

35. $y = \cos^{-1}(x^2)$

36. $y = \sec^{-1}(t+1)$

37. $y = x \tan^{-1} x$

38. $y = e^{\cos^{-1} x}$

39. $y = \arcsin(e^x)$

40. $y = \csc^{-1}(x^{-1})$

41. $y = \sqrt{1-t^2} + \sin^{-1} t$

42. $y = \tan^{-1}\left(\frac{1+t}{1-t}\right)$

43. $y = (\tan^{-1} x)^3$

44. $y = \frac{\cos^{-1} x}{\sin^{-1} x}$

45. $y = \cos^{-1} t^{-1} - \sec^{-1} t$

46. $y = \cos^{-1}(x + \sin^{-1} x)$

47. $y = \arccos(\ln x)$

48. $y = \ln(\arcsin x)$

49. Use Figure 10 to prove that $(\cos^{-1} x)' = -\frac{1}{\sqrt{1-x^2}}$.

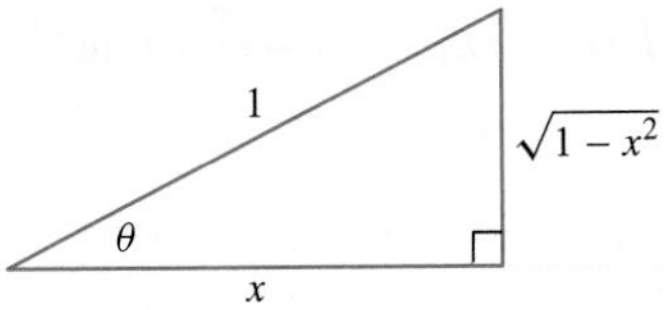

FIGURE 10 Right triangle with $\theta = \cos^{-1} x$.

50. Show that $(\tan^{-1} x)' = \cos^2(\tan^{-1} x)$ and then use Figure 11 to prove that $(\tan^{-1} x)' = (x^2+1)^{-1}$.

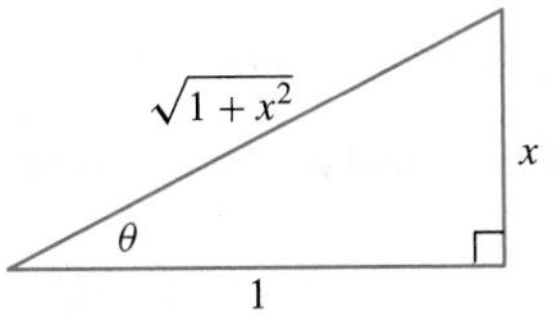

FIGURE 11 Right triangle with $\theta = \tan^{-1} x$.

51. Let $\theta = \sec^{-1} x$. Show that $\tan\theta = \sqrt{x^2-1}$ if $x \ge 1$ and that $\tan\theta = -\sqrt{x^2-1}$ if $x \le -1$. *Hint:* $\tan\theta \ge 0$ on $\left(0, \frac{\pi}{2}\right)$ and $\tan\theta \le 0$ on $\left(\frac{\pi}{2}, \pi\right)$.

52. Use Exercise 51 to verify the formula

$$(\sec^{-1} x)' = \frac{1}{|x|\sqrt{x^2-1}}$$

In Exercises 53–56, evaluate the definite integral.

53. $\int_{\tan 1}^{\tan 8} \frac{dx}{x^2+1}$

54. $\int_{2}^{7} \frac{x\,dx}{x^2+1}$

55. $\int_{0}^{1/2} \frac{dx}{\sqrt{1-x^2}}$

56. $\int_{-2}^{-2/\sqrt{3}} \frac{dx}{|x|\sqrt{x^2-1}}$

57. Use the substitution $u = x/3$ to prove

$$\int \frac{dx}{9+x^2} = \frac{1}{3}\tan^{-1}\frac{x}{3} + C$$

58. Use the substitution $u = 2x$ to evaluate $\int \frac{dx}{4x^2+1}$.

In Exercises 59–72, calculate the integral.

59. $\int_{0}^{3} \frac{dx}{x^2+3}$

60. $\int_{0}^{4} \frac{dt}{4t^2+9}$

61. $\int \frac{dt}{\sqrt{1-16t^2}}$

62. $\int_{-1}^{\sqrt{3}} \frac{dx}{\sqrt{4-25x^2}}$

63. $\int \frac{dt}{\sqrt{5-3t^2}}$

64. $\int_{1/4}^{1/2} \frac{dx}{x\sqrt{16x^2-1}}$

65. $\int \frac{dx}{x\sqrt{12x^2-3}}$

66. $\int \frac{x\,dx}{x^4+1}$

67. $\int \frac{dx}{x\sqrt{x^4-1}}$

68. $\int_{-1/2}^{0} \frac{(x+1)\,dx}{\sqrt{1-x^2}}$

69. $\int \frac{\ln(\cos^{-1} x)\,dx}{(\cos^{-1} x)\sqrt{1-x^2}}$

70. $\int \frac{\tan^{-1} x\,dx}{1+x^2}$

71. $\int_{1}^{\sqrt{3}} \frac{dx}{(\tan^{-1} x)(1+x^2)}$

72. $\int \frac{dx}{\sqrt{5^{2x}-1}}$

In Exercises 73–110, evaluate the integral using the methods covered in the text so far.

73. $\int y e^{y^2}\,dy$

74. $\int \frac{dx}{3x+5}$

75. $\int \frac{x\,dx}{\sqrt{4x^2+9}}$

76. $\int (x - x^{-2})^2\,dx$

77. $\int 7^{-x}\,dx$

78. $\int e^{9-12t}\,dt$

79. $\int \sec^2\theta \tan^7\theta\,d\theta$

80. $\int \frac{\cos(\ln t)\,dt}{t}$

81. $\int \frac{t\,dt}{\sqrt{7-t^2}}$

82. $\int 2^x e^{4x}\,dx$

83. $\int \frac{(3x+2)\,dx}{x^2+4}$

84. $\int \tan(4x+1)\,dx$

85. $\int \frac{dx}{\sqrt{1-16x^2}}$

86. $\int e^t\sqrt{e^t+1}\,dt$

87. $\int (e^{-x} - 4x)\,dx$

88. $\int (7 - e^{10x})\,dx$

89. $\int \frac{e^{2x} - e^{4x}}{e^x}\,dx$

90. $\int \frac{dx}{x\sqrt{25x^2-1}}$

91. $\displaystyle\int \frac{(x+5)\,dx}{\sqrt{4-x^2}}$

92. $\displaystyle\int (t+1)\sqrt{t+1}\,dt$

93. $\displaystyle\int e^x \cos(e^x)\,dx$

94. $\displaystyle\int \frac{e^x}{\sqrt{e^x+1}}\,dx$

95. $\displaystyle\int \frac{dx}{\sqrt{9-16x^2}}$

96. $\displaystyle\int \frac{dx}{(4x-1)\ln(8x-2)}$

97. $\displaystyle\int e^x(e^{2x}+1)^3\,dx$

98. $\displaystyle\int \frac{dx}{x(\ln x)^5}$

99. $\displaystyle\int \frac{x^2\,dx}{x^3+2}$

100. $\displaystyle\int \frac{(3x-1)\,dx}{9-2x+3x^2}$

101. $\displaystyle\int \cot x\,dx$

102. $\displaystyle\int \frac{\cos x}{2\sin x+3}\,dx$

103. $\displaystyle\int \frac{4\ln x+5}{x}\,dx$

104. $\displaystyle\int (\sec\theta\tan\theta)5^{\sec\theta}\,d\theta$

105. $\displaystyle\int x3^{x^2}\,dx$

106. $\displaystyle\int \frac{\ln(\ln x)}{x\ln x}\,dx$

107. $\displaystyle\int \cot x\ln(\sin x)\,dx$

108. $\displaystyle\int \frac{t\,dt}{\sqrt{1-t^4}}$

109. $\displaystyle\int t^2\sqrt{t-3}\,dt$

110. $\displaystyle\int \cos x\, 5^{-2\sin x}\,dx$

111. Use Figure 12 to prove

$$\int_0^x \sqrt{1-t^2}\,dt = \frac{1}{2}x\sqrt{1-x^2} + \frac{1}{2}\sin^{-1} x$$

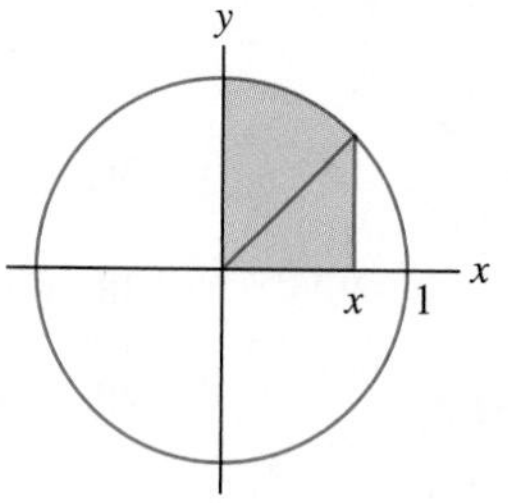

FIGURE 12

112. Use the substitution $u = \tan x$ to evaluate

$$\int \frac{dx}{1+\sin^2 x}$$

Hint: Show that

$$\frac{dx}{1+\sin^2 x} = \frac{du}{1+2u^2}$$

113. Prove:

$$\int \sin^{-1} t\,dt = \sqrt{1-t^2} + t\sin^{-1} t$$

Further Insights and Challenges

114. A cylindrical tank of radius R and length L lying horizontally as in Figure 13 is filled with oil to height h.

(a) Show that the volume $V(h)$ of oil in the tank as a function of height h is

$$V(h) = L\left(R^2\cos^{-1}\left(1-\frac{h}{R}\right) - (R-h)\sqrt{2hR-h^2}\right)$$

(b) Show that $\dfrac{dV}{dh} = 2L\sqrt{h(2R-h)}$.

(c) Suppose that $R = 2$ m and $L = 12$ m, and that the tank is filled at a constant rate of $1.5\ \text{m}^3/\text{min}$. How fast is the height h increasing when $h = 3$ m?

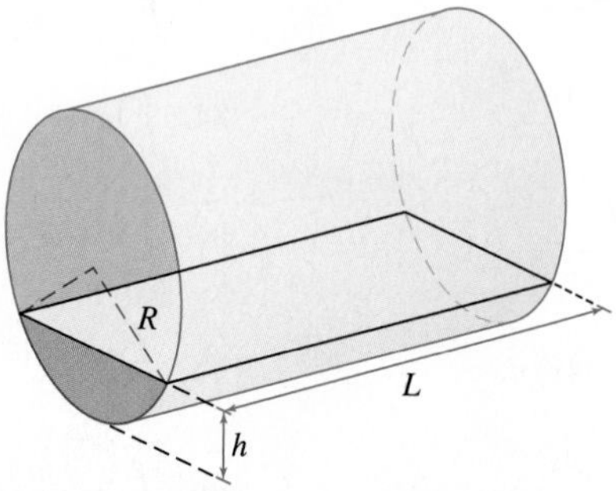

FIGURE 13 Oil in the tank has level h.

115. **(a)** Explain why the shaded region in Figure 14 has area $\int_0^{\ln a} e^y\,dy$.

(b) Prove the formula $\int_1^a \ln x\,dx = a\ln a - \int_0^{\ln a} e^y\,dy$.

(c) Conclude that $\int_1^a \ln x\,dx = a\ln a - a + 1$.

(d) Use the result of (a) to find an antiderivative of $\ln x$.

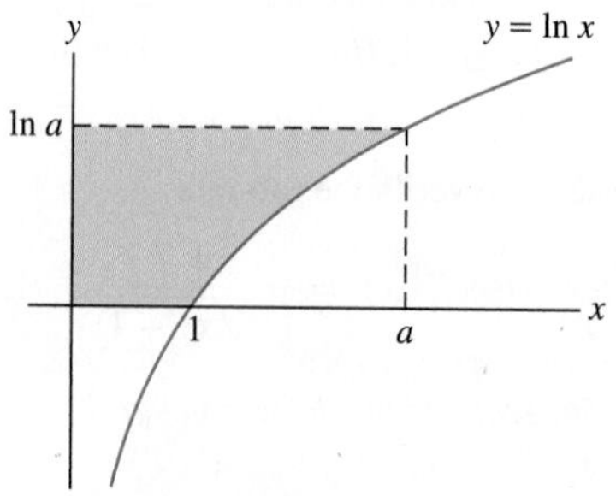

FIGURE 14

7.9 Hyperbolic Functions

The hyperbolic functions are certain special combinations of e^x and e^{-x} that play a role in engineering and physics (see Figure 1 for a real-life example). The hyperbolic sine and cosine, often called "cinch" and "cosh," are defined as follows:

$$\sinh x = \frac{e^x - e^{-x}}{2}, \qquad \cosh x = \frac{e^x + e^{-x}}{2}$$

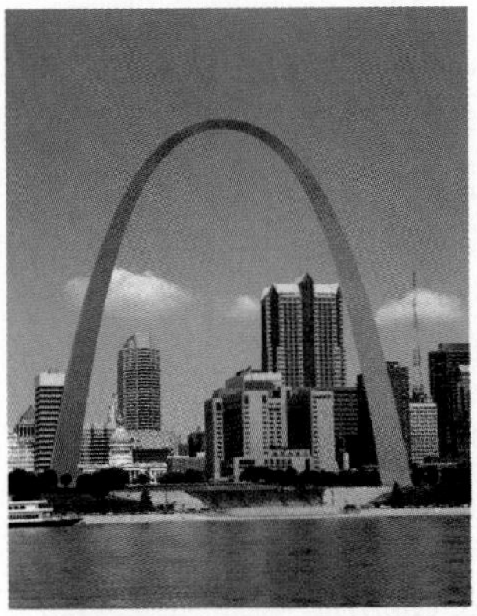

FIGURE 1 The St. Louis Arch has the shape of an inverted hyperbolic cosine.

As the terminology suggests, there are similarities between the hyperbolic and trigonometric functions. Here are some examples:

- **Parity:** The trigonometric functions and their hyperbolic analogs have the same parity. Thus, $\sin x$ and $\sinh x$ are both odd, and $\cos x$ and $\cosh x$ are both even (Figure 2):

$$\sinh(-x) = -\sinh x, \qquad \cosh(-x) = \cosh x$$

- **Identities:** The basic trigonometric identity $\sin^2 x + \cos^2 x = 1$ has a hyperbolic analog:

$$\cosh^2 x - \sinh^2 x = 1 \qquad \boxed{1}$$

The addition formulas satisfied by $\sin\theta$ and $\cos\theta$ also have hyperbolic analogs:

$$\sinh(x + y) = \sinh x \cosh y + \cosh x \sinh y$$
$$\cosh(x + y) = \cosh x \cosh y + \sinh x \sinh y$$

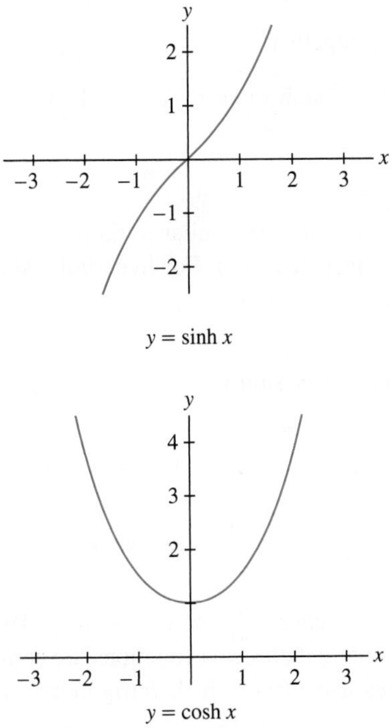

FIGURE 2 $y = \sinh x$ is an odd function, $y = \cosh x$ is an even function.

- **Hyperbola instead of the circle:** The identity $\sin^2 t + \cos^2 t = 1$ tells us that the point $(\cos t, \sin t)$ lies on the unit circle $x^2 + y^2 = 1$. Similarly, the identity $\cosh^2 t - \sinh^2 t = 1$ says that the point $(\cosh t, \sinh t)$ lies on the *hyperbola* $x^2 - y^2 = 1$ (Figure 3).

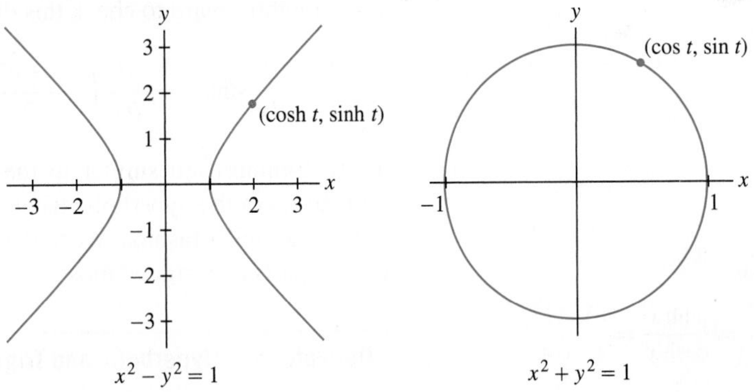

FIGURE 3

- **Other hyperbolic functions:** The hyperbolic tangent, cotangent, secant, and cosecant functions (see Figures 4 and 5) are defined like their trigonometric counterparts:

$$\tanh x = \frac{\sinh x}{\cosh x} = \frac{e^x - e^{-x}}{e^x + e^{-x}}, \qquad \operatorname{sech} x = \frac{1}{\cosh x} = \frac{2}{e^x + e^{-x}}$$

$$\coth x = \frac{\cosh x}{\sinh x} = \frac{e^x + e^{-x}}{e^x - e^{-x}}, \qquad \operatorname{csch} x = \frac{1}{\sinh x} = \frac{2}{e^x - e^{-x}}$$

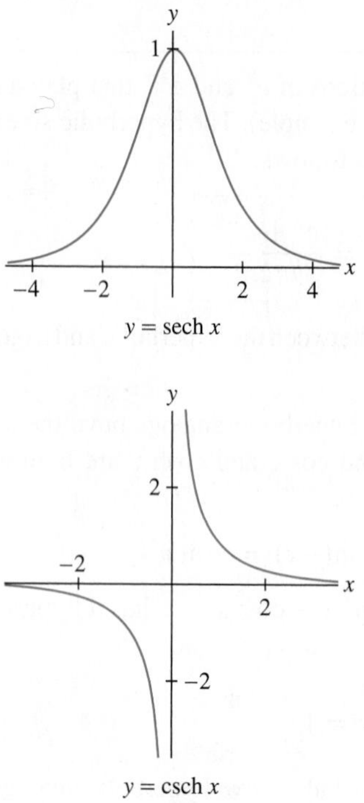

FIGURE 4 The hyperbolic secant and cosecant.

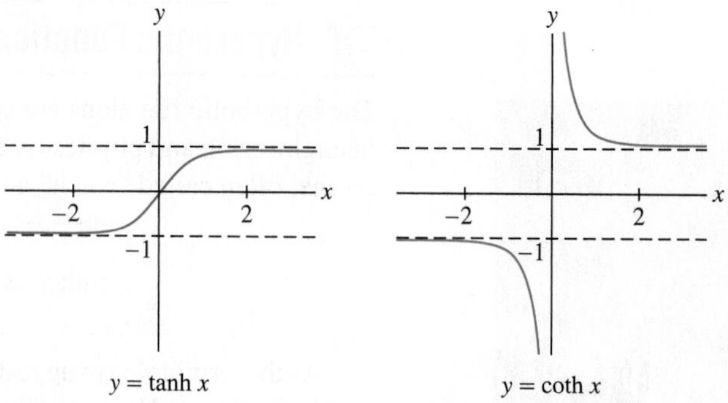

FIGURE 5 The hyperbolic tangent and cotangent.

■ **EXAMPLE 1 Verifying the Basic Identity** Verify Eq. (1).

Solution It follows from the definitions that

$$\cosh x + \sinh x = e^x, \qquad \cosh x - \sinh x = e^{-x}$$

We obtain Eq. (1) by multiplying these two equations together:

$$\cosh^2 x - \sinh^2 x = (\cosh x + \sinh x)(\cosh x - \sinh x) = e^x \cdot e^{-x} = 1$$ ■

Derivatives of Hyperbolic Functions

The formulas for the derivatives of the hyperbolic functions are similar to those for the corresponding trigonometric functions, differing at most by a sign. For hyperbolic sine and cosine we have

$$\frac{d}{dx}\sinh x = \cosh x, \qquad \frac{d}{dx}\cosh x = \sinh x$$

It is straightforward to check this directly. For example,

$$\frac{d}{dx}\sinh x = \frac{d}{dx}\left(\frac{e^x - e^{-x}}{2}\right) = \left(\frac{e^x - e^{-x}}{2}\right)' = \frac{e^x + e^{-x}}{2} = \cosh x$$

These formulas are similar to the formulas $\frac{d}{dx}\sin x = \cos x$, $\frac{d}{dx}\cos x = -\sin x$. The derivatives of the hyperbolic tangent, cotangent, secant, and cosecant functions are computed in a similar fashion. We find that their derivatives also differ from their trigonometric counterparts by a sign at most.

REMINDER

$$\tanh x = \frac{\sinh x}{\cosh x} = \frac{e^x - e^{-x}}{e^x + e^{-x}},$$

$$\operatorname{sech} x = \frac{1}{\cosh x} = \frac{2}{e^x + e^{-x}}$$

$$\coth x = \frac{\cosh x}{\sinh x} = \frac{e^x + e^{-x}}{e^x - e^{-x}},$$

$$\operatorname{csch} x = \frac{1}{\sinh x} = \frac{2}{e^x - e^{-x}}$$

Derivatives of Hyperbolic and Trigonometric Functions

$$\frac{d}{dx}\tanh x = \operatorname{sech}^2 x, \qquad \frac{d}{dx}\tan x = \sec^2 x$$

$$\frac{d}{dx}\coth x = -\operatorname{csch}^2 x, \qquad \frac{d}{dx}\cot x = -\csc^2 x$$

$$\frac{d}{dx}\operatorname{sech} x = -\operatorname{sech} x \tanh x, \qquad \frac{d}{dx}\sec x = \sec x \tan x$$

$$\frac{d}{dx}\operatorname{csch} x = -\operatorname{csch} x \coth x, \qquad \frac{d}{dx}\csc x = -\csc x \cot x$$

EXAMPLE 2 Verify: $\frac{d}{dx}\coth x = -\operatorname{csch}^2 x$.

Solution By the Quotient Rule and the identity $\cosh^2 x - \sinh^2 x = 1$,

$$\frac{d}{dx}\coth x = \left(\frac{\cosh x}{\sinh x}\right)' = \frac{(\sinh x)(\cosh x)' - (\cosh x)(\sinh x)'}{\sinh^2 x}$$

$$= \frac{\sinh^2 x - \cosh^2 x}{\sinh^2 x} = \frac{-1}{\sinh^2 x} = -\operatorname{csch}^2 x$$

■

EXAMPLE 3 Calculate **(a)** $\frac{d}{dx}\cosh(3x^2+1)$ and **(b)** $\frac{d}{dx}\sinh x \tanh x$.

Solution

(a) By the Chain Rule, $\frac{d}{dx}\cosh(3x^2+1) = 6x\sinh(3x^2+1)$

(b) By the Product Rule,

$$\frac{d}{dx}(\sinh x \tanh x) = \sinh x \operatorname{sech}^2 x + \tanh x \cosh x = \operatorname{sech} x \tanh x + \sinh x$$

■

The formulas for the derivatives of the hyperbolic functions are equivalent to the following integral formulas:

Integral Formulas

$$\int \sinh x\,dx = \cosh x + C, \qquad \int \cosh x\,dx = \sinh x + C$$

$$\int \operatorname{sech}^2 x\,dx = \tanh x + C, \qquad \int \operatorname{csch}^2 x\,dx = -\coth x + C$$

$$\int \operatorname{sech} x \tanh x\,dx = -\operatorname{sech} x + C, \qquad \int \operatorname{csch} x \coth x\,dx = -\operatorname{csch} x + C$$

EXAMPLE 4 Calculate $\int x\cosh(x^2)\,dx$.

Solution The substitution $u = x^2$, $du = 2x\,dx$ yields

$$\int x\cosh(x^2)\,dx = \frac{1}{2}\int \cosh u\,du = \frac{1}{2}\sinh u$$

$$= \frac{1}{2}\sinh(x^2) + C$$

■

Inverse Hyperbolic Functions

Each of the hyperbolic functions, except $y = \cosh x$ and $y = \operatorname{sech} x$, is one-to-one on its domain and therefore has a well-defined inverse. The functions $y = \cosh x$ and $y = \operatorname{sech} x$ are one-to-one on the restricted domain $\{x : x \ge 0\}$. Let $\cosh^{-1} x$ and $\operatorname{sech}^{-1} x$ denote the corresponding inverses. We have the following derivative formulas.

Inverse Hyperbolic Functions and Their Derivatives

Function	Domain	Derivative
$y = \sinh^{-1} x$	all x	$\frac{d}{dx}\sinh^{-1} x = \frac{1}{\sqrt{x^2+1}}$
$y = \cosh^{-1} x$	$x \ge 1$	$\frac{d}{dx}\cosh^{-1} x = \frac{1}{\sqrt{x^2-1}}$
$y = \tanh^{-1} x$	$\|x\| < 1$	$\frac{d}{dx}\tanh^{-1} x = \frac{1}{1-x^2}$
$y = \coth^{-1} x$	$\|x\| > 1$	$\frac{d}{dx}\coth^{-1} x = \frac{1}{1-x^2}$
$y = \operatorname{sech}^{-1} x$	$0 < x \le 1$	$\frac{d}{dx}\operatorname{sech}^{-1} x = -\frac{1}{x\sqrt{1-x^2}}$
$y = \operatorname{csch}^{-1} x$	$x \ne 0$	$\frac{d}{dx}\operatorname{csch}^{-1} x = -\frac{1}{\|x\|\sqrt{x^2+1}}$

Each derivative formula in this table can be written as an antiderivative formula. For instance,

$$\int \frac{dx}{\sqrt{x^2+1}} = \sinh^{-1} x + C, \qquad \int \frac{dx}{\sqrt{x^2-1}} = \cosh^{-1} x + C$$

■ **EXAMPLE 5** Verify the formula $\frac{d}{dx}\tanh^{-1} x = \frac{1}{1-x^2}$.

Solution Recall that if $g(x)$ is the inverse of $f(x)$, then $g'(x) = 1/f'(g(x))$. Applying this to $f(x) = \tanh x$, and using the formula $(\tanh x)' = \operatorname{sech}^2 x$, we have

$$\frac{d}{dx}\tanh^{-1} x = \frac{1}{\operatorname{sech}^2(\tanh^{-1} x)}$$

Now let $t = \tanh^{-1} x$. Then

$$\cosh^2 t - \sinh^2 t = 1 \qquad \text{(basic identity)}$$

$$1 - \tanh^2 t = \operatorname{sech}^2 t \qquad \text{(divide by } \cosh^2 t\text{)}$$

$$1 - x^2 = \operatorname{sech}^2(\tanh^{-1} x) \qquad \text{(because } x = \tanh t\text{)}$$

This gives the desired result:

$$\frac{d}{dx}\tanh^{-1} x = \frac{1}{\operatorname{sech}^2(\tanh^{-1} x)} = \frac{1}{1-x^2}$$

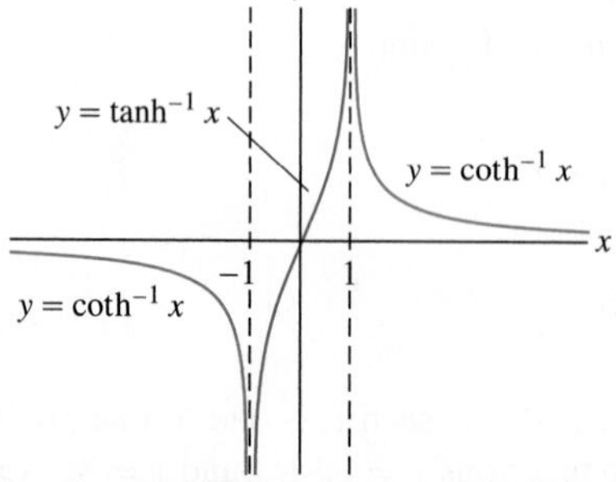

FIGURE 6 The functions $y = \tanh^{-1} x$ and $y = \coth^{-1} x$ have disjoint domains.

■ **EXAMPLE 6** The functions $y = \tanh^{-1} x$ and $y = \coth^{-1} x$ appear to have the same derivative. Does this imply that they differ by a constant?

Solution According to the table above, $y = \tanh^{-1} x$ and $y = \coth^{-1} x$ both have derivative $1/(1 - x^2)$. Although functions with the same derivative differ by a constant, this is the case only if they are defined on the same domain. The functions $y = \tanh^{-1} x$ and $y = \coth^{-1} x$ have disjoint domains and therefore do not differ by a constant (Figure 6). ■

■ **EXAMPLE 7** Evaluate:

$$\text{(a)} \int \frac{dx}{\sqrt{x^2-1}} \quad \text{and} \quad \text{(b)} \int_{0.2}^{0.6} \frac{x\,dx}{1-x^4}$$

Solution We use the table of derivatives on page 402.

(a) The second formula in the table is the derivative formula $\frac{d}{dx}\cosh^{-1} x = \frac{1}{\sqrt{x^2-1}}$. This yields the antiderivative formula:

$$\int \frac{dx}{\sqrt{x^2-1}} = \cosh^{-1} x + C$$

(b) First use a substitution $u = x^2$, $du = 2x\,dx$. The new limits of integration become $u = (0.2)^2 = 0.04$ and $u = (0.6)^2 = 0.36$, and we obtain

$$\int_{0.2}^{0.6} \frac{x\,dx}{1-x^4} = \int_{0.04}^{0.36} \frac{\frac{1}{2}\,du}{1-u^2} = \frac{1}{2}\int_{0.04}^{0.36} \frac{du}{1-u^2}$$

By the third and fourth formulas in the table, both $\tanh^{-1} u$ and $\coth^{-1} u$ are antiderivatives of $(1-u^2)^{-1}$. We must use $\tanh^{-1} u$ because the domain of $\tanh^{-1} u$ is $(-1, 1)$ and the limits of integration $u = 0.04$ and $u = 0.36$ are contained within this interval (if the limits of integration were contained in $|u| > 1$, we would use $\coth^{-1} u$). We obtain

$$\frac{1}{2}\int_{0.04}^{0.36} \frac{du}{1-u^2} = \frac{1}{2}\left(\tanh^{-1}(0.36) - \tanh^{-1}(0.04)\right) \approx 0.1684$$ ■

Einstein's Law of Velocity Addition

The hyperbolic tangent plays a role in the Special Theory of Relativity, developed by Albert Einstein in 1905. One consequence of this theory is that no object can travel faster than the speed of light $c \approx 3 \times 10^8$ m/s. Einstein realized that this contradicts a law stated by Galileo more than 250 years earlier, namely that *velocities add*. Imagine a train traveling at $u = 50$ m/s and a man walking down the aisle in the train at $v = 2$ m/s. According to Galileo, the man's velocity relative to the ground is $u + v = 52$ m/s. This agrees with our everyday experience. But now imagine an (unrealistic) rocket traveling away from the earth at $u = 2 \times 10^8$ m/s and suppose that the rocket fires a missile with velocity $v = 1.5 \times 10^8$ m/s (relative to the rocket). If Galileo's Law were correct, the velocity of the missile relative to the earth would be $u + v = 3.5 \times 10^8$ m/s, which exceeds Einstein's maximum speed limit of $c \approx 3 \times 10^8$ m/s.

Einstein's Law (2) reduces to Galileo's Law, $w = u + v$, when u and v are small relative to the velocity of light c. See Exercise 67 for another way of expressing (2).

However, Einstein's theory replaces Galileo's Law with a new law stating that the *inverse hyperbolic tangents of the velocities add*. More precisely, if u is the rocket's velocity relative to the earth and v is the missile's velocity relative to the rocket, then the velocity of the missile relative to the earth (Figure 7) is w, where

$$\tanh^{-1}\left(\frac{w}{c}\right) = \tanh^{-1}\left(\frac{u}{c}\right) + \tanh^{-1}\left(\frac{v}{c}\right) \qquad \boxed{2}$$

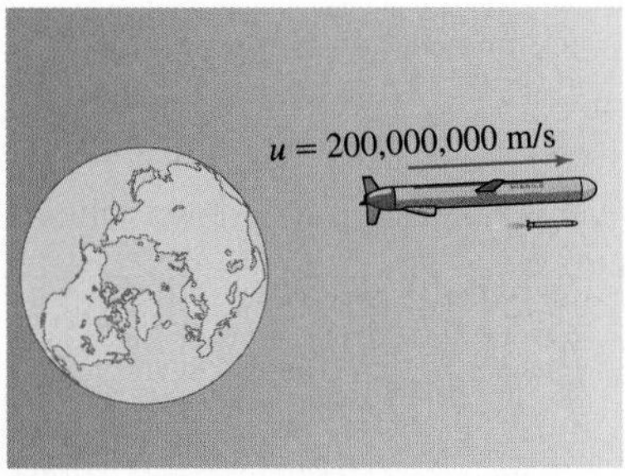

FIGURE 7 What is the missile's velocity relative to the earth?

■ **EXAMPLE 8** A rocket travels away from the earth at a velocity of 2×10^8 m/s. A missile is fired at a velocity of 1.5×10^8 m/s (relative to the rocket) away from the earth. Use Einstein's Law to find the velocity w of the missile relative to the earth.

Solution According to Eq. (2),

$$\tanh^{-1}\left(\frac{w}{c}\right) = \tanh^{-1}\left(\frac{2 \times 10^8}{3 \times 10^8}\right) + \tanh^{-1}\left(\frac{1.5 \times 10^8}{3 \times 10^8}\right) \approx 0.805 + 0.549 \approx 1.354$$

Therefore, $\frac{w}{c} \approx \tanh(1.354) \approx 0.875$, and $w \approx 0.875c \approx 2.6 \times 10^8$ m/s. This value obeys the Einstein speed limit of 3×10^8 m/s. ■

■ **EXAMPLE 9 Low Velocities** A plane traveling at 300 m/s fires a missile at a velocity of 200 m/s. Calculate the missile's velocity w relative to the earth (in mph) using both Einstein's Law and Galileo's Law.

Solution According to Einstein's law,

$$\tanh^{-1}\left(\frac{w}{c}\right) = \tanh^{-1}\left(\frac{300}{c}\right) + \tanh^{-1}\left(\frac{200}{c}\right)$$

$$w = c \cdot \tanh\left(\tanh^{-1}\left(\frac{300}{c}\right) + \tanh^{-1}\left(\frac{200}{c}\right)\right) \approx 499.99999999967$$

This is practically indistinguishable from the value $w = 300 + 200 = 500$ m/s, obtained using Galileo's Law. ■

Excursion: A Leap of Imagination

The terms "hyperbolic sine" and "hyperbolic cosine" suggest a connection between the hyperbolic and trigonometric functions. This excursion explores the source of this connection, which leads us to **complex numbers** and a famous formula of Euler (Figure 8).

Recall that $y = e^t$ satisfies the differential equation $y' = y$. In fact, we know that *every* solution is of the form $y = Ce^t$ for some constant C. Observe that both $y = e^t$ and $y = e^{-t}$ satisfy the **second-order differential equation**

This differential equation is called "second-order" because it involves the second derivative y''.

$$\boxed{y'' = y} \qquad (3)$$

Indeed, $(e^t)'' = e^t$ and $(e^{-t})'' = (-e^{-t})' = e^{-t}$. Furthermore, *every* solution of Eq. (3) has the form $y = Ae^t + Be^{-t}$ for some constants A and B.

Now let's see what happens when we change Eq. (3) by a minus sign:

$$\boxed{y'' = -y} \qquad (4)$$

In this case, $y = \sin t$ and $y = \cos t$ are solutions because

$$(\sin t)'' = (\cos t)' = -\sin t, \qquad (\cos t)'' = (-\sin t)' = -\cos t$$

And as before, every solution of Eq. (4) has the form

$$\boxed{y = A\cos t + B\sin t}$$

FIGURE 8 Leonhard Euler (1707–1783). Euler (pronounced "oil-er") ranks among the greatest mathematicians of all time. His work (printed in more than 70 volumes) contains fundamental contributions to almost every aspect of the mathematics and physics of his time. The French mathematician Pierre Simon de Laplace once declared: "Read Euler, he is our master in everything."

This might seem to be the end of the story. However, we can also write down solutions of Eq. (4) using the exponential functions $y = e^{it}$ and $y = e^{-it}$. Here

$$\boxed{i = \sqrt{-1}}$$

is the *imaginary* complex number satisfying $i^2 = -1$. Since i is not a real number, e^{it} is not defined without further explanation. But let's assume that e^{it} *can* be defined and that the usual rules of calculus apply:

$$(e^{it})' = ie^{it}$$

$$(e^{it})'' = (ie^{it})' = i^2e^{it} = -e^{it}$$

This shows that $y = e^{it}$ is a solution of $y'' = -y$, so there must exist constants A and B such that

$$e^{it} = A\cos t + B\sin t \qquad \boxed{5}$$

The constants are determined by initial conditions. First, set $t = 0$ in Eq. (5):

$$1 = e^{i0} = A\cos 0 + B\sin 0 = A$$

Then take the derivative of Eq. (5) and set $t = 0$:

$$ie^{it} = \frac{d}{dt}e^{it} = A\cos' t + B\sin' t = -A\sin t + B\cos t$$

$$i = ie^{i0} = -A\sin 0 + B\cos 0 = B$$

Thus $A = 1$ and $B = i$, and Eq. (5) yields **Euler's Formula**:

$$\boxed{e^{it} = \cos t + i\sin t}$$

Euler proved his formula using power series, which may be used to define e^{it} in a precise fashion. At $t = \pi$, Euler's Formula yields

$$\boxed{e^{i\pi} = -1}$$

Here we have a simple but surprising relation among the four important numbers e, i, π, and -1.

Euler's Formula also reveals the source of the analogy between hyperbolic and trigonometric functions. Let us calculate the hyperbolic cosine at $x = it$:

$$\cosh(it) = \frac{e^{it} + e^{-it}}{2} = \frac{\cos t + i\sin t}{2} + \frac{\cos(-t) + i\sin(-t)}{2} = \cos t$$

A similar calculation shows that $\sinh(it) = i\sin t$. In other words, the hyperbolic and trigonometric functions are not merely analogous—once we introduce complex numbers, we see that they are very nearly the same functions.

7.9 SUMMARY

- The hyperbolic sine and cosine:

$$\sinh x = \frac{e^x - e^{-x}}{2} \quad \text{(odd function)}, \qquad \cosh x = \frac{e^x + e^{-x}}{2} \quad \text{(even function)}$$

The remaining hyperbolic functions:

$$\tanh x = \frac{\sinh x}{\cosh x}, \qquad \coth x = \frac{\cosh x}{\sinh x}, \qquad \operatorname{sech} x = \frac{1}{\cosh x}, \qquad \operatorname{csch} x = \frac{1}{\sinh x}$$

- Basic identity: $\cosh^2 x - \sinh^2 x = 1$.

- Derivative and integral formulas:

$$\frac{d}{dx}\sinh x = \cosh x, \qquad \int \sinh x\,dx = \cosh x + C$$

$$\frac{d}{dx}\cosh x = \sinh x, \qquad \int \cosh x\,dx = \sinh x + C$$

$$\frac{d}{dx}\tanh x = \operatorname{sech}^2 x, \qquad \int \operatorname{sech}^2 x\,dx = \tanh x + C$$

$$\frac{d}{dx}\coth x = -\operatorname{csch}^2 x, \qquad \int \operatorname{csch}^2 x\,dx = -\coth x + C$$

$$\frac{d}{dx}\operatorname{sech} x = -\operatorname{sech} x \tanh x, \qquad \int \operatorname{sech} x \tanh x\,dx = -\operatorname{sech} x + C$$

$$\frac{d}{dx}\operatorname{csch} x = -\operatorname{csch} x \coth x, \qquad \int \operatorname{csch} x \coth x\,dx = -\operatorname{csch} x + C$$

- Inverse hyperbolic functions

$$\frac{d}{dx}\sinh^{-1} x = \frac{1}{\sqrt{x^2+1}}, \qquad \frac{d}{dx}\cosh^{-1} x = \frac{1}{\sqrt{x^2-1}} \quad (x > 1)$$

$$\frac{d}{dx}\tanh^{-1} x = \frac{1}{1-x^2} \quad (|x| < 1), \qquad \frac{d}{dx}\coth^{-1} x = \frac{1}{1-x^2} \quad (|x| > 1)$$

$$\frac{d}{dx}\operatorname{sech}^{-1} x = -\frac{1}{x\sqrt{1-x^2}} \quad (0 < x < 1), \qquad \frac{d}{dx}\operatorname{csch}^{-1} x = -\frac{1}{|x|\sqrt{x^2+1}} \quad (x \neq 0)$$

Each derivative formula in this table can be rewritten as an integral formula. For instance,

$$\int \frac{dx}{\sqrt{x^2+1}} = \sinh^{-1} x + C, \qquad \int \frac{dx}{\sqrt{x^2-1}} = \cosh^{-1} x + C$$

7.9 EXERCISES

Preliminary Questions

1. Which hyperbolic functions take on only positive values?

2. Which hyperbolic functions are increasing on their domains?

3. Describe three properties of hyperbolic functions that have trigonometric analogs.

4. What are $y^{(100)}$ and $y^{(101)}$ for $y = \cosh x$?

Exercises

1. Use a calculator to compute $\sinh x$ and $\cosh x$ for $x = -3, 0, 5$.

2. Compute $\sinh(\ln 5)$ and $\tanh(3\ln 5)$ without using a calculator.

3. For which values of x are $y = \sinh x$ and $y = \cosh x$ increasing and decreasing?

4. Show that $y = \tanh x$ is an odd function.

5. Refer to the graphs to explain why the equation $\sinh x = t$ has a unique solution for every t and why $\cosh x = t$ has two solutions for every $t > 1$.

6. Compute $\cosh x$ and $\tanh x$, assuming that $\sinh x = 0.8$.

7. Prove the addition formula for $\cosh x$.

8. Use the addition formulas to prove

$$\sinh(2x) = 2\cosh x \sinh x$$

$$\cosh(2x) = \cosh^2 x + \sinh^2 x$$

In Exercises 9–32, calculate the derivative.

9. $y = \sinh(9x)$

10. $y = \sinh(x^2)$

11. $y = \cosh^2(9 - 3t)$

12. $y = \tanh(t^2 + 1)$

13. $y = \sqrt{\cosh x + 1}$

14. $y = \sinh x \tanh x$

15. $y = \dfrac{\coth t}{1 + \tanh t}$

16. $y = (\ln(\cosh x))^5$

17. $y = \sinh(\ln x)$

18. $y = e^{\coth x}$

19. $y = \tanh(e^x)$

20. $y = \sinh(\cosh^3 x)$

21. $y = \operatorname{sech}(\sqrt{x})$

22. $y = \ln(\coth x)$

23. $y = \operatorname{sech} x \coth x$

24. $y = x^{\sinh x}$

25. $y = \cosh^{-1}(3x)$

26. $y = \tanh^{-1}(e^x + x^2)$

27. $y = (\sinh^{-1}(x^2))^3$

28. $y = (\operatorname{csch}^{-1} 3x)^4$

29. $y = e^{\cosh^{-1} x}$

30. $y = \sinh^{-1}(\sqrt{x^2 + 1})$

31. $y = \tanh^{-1}(\ln t)$

32. $y = \ln(\tanh^{-1} x)$

33. Show that for any constants M, k, and a, the function

$$y(t) = \frac{1}{2} M \left(1 + \tanh\left(\frac{k(t-a)}{2}\right)\right)$$

satisfies the **logistic equation:** $\dfrac{y'}{y} = k\left(1 - \dfrac{y}{M}\right)$.

34. Show that $V(x) = 2\ln(\tanh(x/2))$ satisfies the **Poisson-Boltzmann** equation $V''(x) = \sinh(V(x))$, which is used to describe electrostatic forces in certain molecules.

In Exercises 35–46, calculate the integral.

35. $\int \cosh(3x)\, dx$

36. $\int \sinh(x + 1)\, dx$

37. $\int x \sinh(x^2 + 1)\, dx$

38. $\int \sinh^2 x \cosh x\, dx$

39. $\int \operatorname{sech}^2(1 - 2x)\, dx$

40. $\int \tanh(3x) \operatorname{sech}(3x)\, dx$

41. $\int \tanh x \operatorname{sech}^2 x\, dx$

42. $\int \dfrac{\cosh x}{3 \sinh x + 4}\, dx$

43. $\int \tanh x\, dx$

44. $\int \dfrac{\cosh x}{\sinh x}\, dx$

45. $\int e^{-x} \sinh x\, dx$

46. $\int \dfrac{\cosh x}{\sinh^2 x}\, dx$

In Exercises 47–52, prove the formula.

47. $\dfrac{d}{dx} \tanh x = \operatorname{sech}^2 x$

48. $\dfrac{d}{dx} \operatorname{sech} x = -\operatorname{sech} x \tanh x$

49. $\cosh(\sinh^{-1} t) = \sqrt{t^2 + 1}$

50. $\sinh(\cosh^{-1} t) = \sqrt{t^2 - 1}$ for $t \ge 1$

51. $\dfrac{d}{dt} \sinh^{-1} t = \dfrac{1}{\sqrt{t^2 + 1}}$

52. $\dfrac{d}{dt} \cosh^{-1} t = \dfrac{1}{\sqrt{t^2 - 1}}$ for $t > 1$

In Exercises 53–60, calculate the integral in terms of inverse hyperbolic functions.

53. $\int_2^4 \dfrac{dx}{\sqrt{x^2 - 1}}$

54. $\int \dfrac{dx}{\sqrt{x^2 - 4}}$

55. $\int \dfrac{dx}{\sqrt{9 + x^2}}$

56. $\int \dfrac{dx}{\sqrt{1 + 9x^2}}$

57. $\int_{1/3}^{1/2} \dfrac{dx}{1 - x^2}$

58. $\int_0^1 \dfrac{dx}{\sqrt{1 + x^2}}$

59. $\int_2^{10} \dfrac{dx}{4x^2 - 1}$

60. $\int_{-3}^{-1} \dfrac{dx}{x\sqrt{x^2 + 16}}$

61. Prove that $\sinh^{-1} t = \ln(t + \sqrt{t^2 + 1})$. *Hint:* Let $t = \sinh x$. Prove that $\cosh x = \sqrt{t^2 + 1}$ and use the relation

$$\sinh x + \cosh x = e^x$$

62. Prove that $\cosh^{-1} t = \ln(t + \sqrt{t^2 - 1})$ for $t > 1$.

63. Prove that $\tanh^{-1} t = \dfrac{1}{2} \ln\left(\dfrac{1+t}{1-t}\right)$ for $|t| < 1$.

64. Use the substitution $u = \sinh x$ to prove

$$\int \operatorname{sech} x\, dx = \tan^{-1}(\sinh x) + C$$

65. An (imaginary) train moves along a track at velocity v. Bionica walks down the aisle of the train with velocity u in the direction of the train's motion. Compute the velocity w of Bionica relative to the ground using the laws of both Galileo and Einstein in the following cases.

(a) $v = 500$ m/s and $u = 10$ m/s. Is your calculator accurate enough to detect the difference between the two laws?

(b) $v = 10^7$ m/s and $u = 10^6$ m/s.

Further Insights and Challenges

66. Show that the linearization of the function $y = \tanh^{-1} x$ at $x = 0$ is $\tanh^{-1} x \approx x$. Use this to explain the following assertion: Einstein's Law of Velocity Addition [Eq. (2)] reduces to Galileo's Law if the velocities are small relative to the speed of light.

67. (a) Use the addition formulas for $\sinh x$ and $\cosh x$ to prove

$$\tanh(u + v) = \frac{\tanh u + \tanh v}{1 + \tanh u \tanh v}$$

(b) Use (a) to show that Einstein's Law of Velocity Addition [Eq. (2)] is equivalent to

$$w = \frac{u+v}{1+\frac{uv}{c^2}}$$

68. Prove that $\int_{-a}^{a} \cosh x \sinh x\, dx = 0$ for all a.

69. (a) Show that $y = \tanh t$ satisfies the differential equation $dy/dt = 1 - y^2$ with initial condition $y(0) = 0$.
(b) Show that for arbitrary constants A, B, the function

$$y = A \tanh(Bt)$$

satisfies

$$\frac{dy}{dt} = AB - \frac{B}{A}y^2, \qquad y(0) = 0$$

(c) Let $v(t)$ be the velocity of a falling object of mass m. For large velocities, air resistance is proportional to the square of velocity $v(t)^2$. If we choose coordinates so that $v(t) > 0$ for a falling object, then by Newton's Law of Motion, there is a constant $k > 0$ such that

$$\frac{dv}{dt} = g - \frac{k}{m}v^2$$

Solve for $v(t)$ by applying the result of (b) with $A = \sqrt{gm/k}$ and $B = \sqrt{gk/m}$.
(d) Calculate the terminal velocity $\lim_{t\to\infty} v(t)$.
(e) Find k if $m = 150$ lb and the terminal velocity is 100 mph.

In Exercises 70–72, a flexible chain of length L is suspended between two poles of equal height separated by a distance 2M (Figure 9). By Newton's laws, the chain describes a curve (called a ***catenary****) with equation $y = a\cosh(x/a) + C$. The constant C is arbitrary and a is the number such that $L = 2a\sinh(M/a)$. The sag s is the vertical distance from the highest to the lowest point on the chain.*

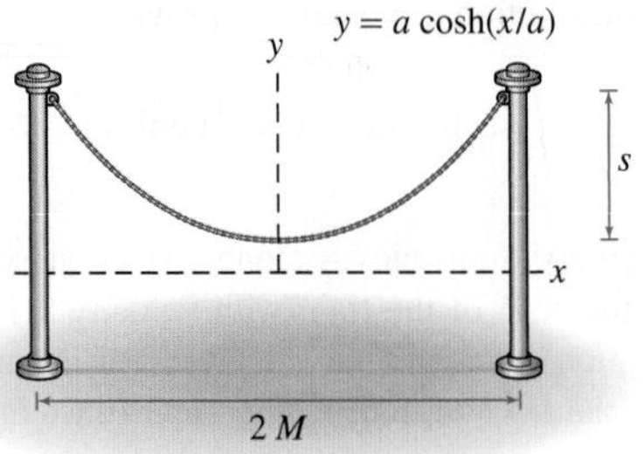

FIGURE 9 Chain hanging between two poles describes the curve $y = a\cosh(x/a)$.

70. GU Suppose that $L = 120$ and $M = 50$. Experiment with your calculator to find an approximate value of a satisfying $L = 2a\sinh(M/a)$ (for greater accuracy, use Newton's method or a computer algebra system).

71. Let M be a fixed constant. Show that the sag is given by $s = a\cosh(M/a) - a$.

(a) Calculate $\frac{ds}{da}$.

(b) Calculate da/dL by implicit differentiation using the relation $L = 2a\sinh(M/a)$.

(c) Use (a) and (b) and the Chain Rule to show that

$$\frac{ds}{dL} = \frac{ds}{da}\frac{da}{dL} = \frac{\cosh(M/a) - (M/a)\sinh(M/a) - 1}{2\sinh(M/a) - (2M/a)\cosh(M/a)} \qquad \boxed{6}$$

72. Assume that $M = 50$ and $L = 160$. In this case, a CAS can be used to show that $a \approx 28.46$.

(a) Use Eq. (6) and the Linear Approximation to estimate the increase in sag if L is increased from $L = 160$ to $L = 161$ and from $L = 160$ to $L = 165$.

(b) CAS If you have a CAS, compute $s(161) - s(160)$ and $s(165) - s(160)$ directly and compare with your estimates in (a).

73. Prove that every function $f(x)$ is the sum of an even function $f_+(x)$ and an odd function $f_-(x)$. [*Hint:* $f_\pm(x) = \frac{1}{2}(f(x) \pm f(-x))$.] Express $f(x) = 5e^x + 8e^{-x}$ in terms of $\cosh x$ and $\sinh x$.

74. Use the method of the previous problem to express

$$f(x) = 7e^{-3x} + 4e^{3x}$$

in terms of $\sinh(3x)$ and $\cosh(3x)$.

75. In the Excursion, we discussed the relations

$$\cosh(it) = \cos t \quad \text{and} \quad \sinh(it) = i \sin t$$

Use these relations to show that the identity $\cos^2 t + \sin^2 t = 1$ results from the identity $\cosh^2 x - \sinh^2 x = 1$ by setting $x = it$.

CHAPTER REVIEW EXERCISES

1. Match each quantity (a)–(d) with (i), (ii), or (iii) if possible, or state that no match exists.

(a) $2^a 3^b$ **(b)** $\frac{2^a}{3^b}$

(c) $(2^a)^b$ **(d)** $2^{a-b}3^{b-a}$

(i) 2^{ab} **(ii)** 6^{a+b} **(iii)** $\left(\frac{2}{3}\right)^{a-b}$

2. Match each quantity (a)–(d) with (i), (ii), or (iii) if possible, or state that no match exists.

(a) $\ln\left(\frac{a}{b}\right)$ **(b)** $\frac{\ln a}{\ln b}$

(c) $e^{\ln a - \ln b}$ (d) $(\ln a)(\ln b)$

(i) $\ln a + \ln b$ (ii) $\ln a - \ln b$ (iii) $\frac{a}{b}$

3. Which of the following is equal to $\frac{d}{dx}2^x$?

(a) 2^x (b) $(\ln 2)2^x$

(c) $x2^{x-1}$ (d) $\frac{1}{\ln 2}2^x$

4. Find the inverse of $f(x) = \sqrt{x^3 - 8}$ and determine its domain and range.

5. Find the inverse of $f(x) = \frac{x-2}{x-1}$ and determine its domain and range.

6. Find a domain on which $h(t) = (t-3)^2$ is one-to-one and determine the inverse on this domain.

7. Show that $g(x) = \frac{x}{x-1}$ is equal to its inverse on the domain $\{x : x \neq -1\}$.

8. Describe the graphical interpretation of the relation $g'(x) = 1/f'(g(x))$, where $f(x)$ and $g(x)$ are inverses of each other.

9. Suppose that $g(x)$ is the inverse of $f(x)$. Match the functions (a)–(d) with their inverses (i)–(iv).

(a) $f(x) + 1$ (b) $f(x+1)$

(c) $4f(x)$ (d) $f(4x)$

(i) $g(x)/4$ (ii) $g(x/4)$

(iii) $g(x-1)$ (iv) $g(x) - 1$

10. Find $g'(8)$ where $g(x)$ is the inverse of a differentiable function $f(x)$ such that $f(-1) = 8$ and $f'(-1) = 12$.

11. Suppose that $f(g(x)) = e^{x^2}$, where $g(1) = 2$ and $g'(1) = 4$. Find $f'(2)$.

12. Show that if $f(x)$ is a function satisfying $f'(x) = f(x)^2$, then its inverse $g(x)$ satisfies $g'(x) = x^{-2}$.

In Exercises 13–42, find the derivative.

13. $f(x) = 9e^{-4x}$ **14.** $f(x) = \ln(4x^2 + 1)$

15. $f(x) = \frac{e^{-x}}{x}$ **16.** $f(x) = \ln(x + e^x)$

17. $G(s) = (\ln(s))^2$ **18.** $G(s) = \ln(s^2)$

19. $g(t) = e^{4t - t^2}$ **20.** $g(t) = t^2 e^{1/t}$

21. $f(\theta) = \ln(\sin\theta)$ **22.** $f(\theta) = \sin(\ln\theta)$

23. $f(x) = \ln(e^x - 4x)$ **24.** $h(z) = \sec(z + \ln z)$

25. $f(x) = e^{x + \ln x}$ **26.** $f(x) = e^{\sin^2 x}$

27. $h(y) = 2^{1-y}$ **28.** $h(y) = \frac{1 + e^y}{1 - e^y}$

29. $f(x) = 7^{-2x}$ **30.** $g(x) = \tan^{-1}(\ln x)$

31. $G(s) = \cos^{-1}(s^{-1})$ **32.** $G(s) = \tan^{-1}(\sqrt{s})$

33. $f(x) = \ln(\csc^{-1} x)$ **34.** $f(x) = e^{\sec^{-1} x}$

35. $R(s) = s^{\ln s}$ **36.** $f(x) = (\cos^2 x)^{\cos x}$

37. $G(t) = (\sin^2 t)^t$ **38.** $h(t) = t^{(t^t)}$

39. $g(t) = \sinh(t^2)$ **40.** $h(y) = y\tanh(4y)$

41. $g(x) = \tanh^{-1}(e^x)$ **42.** $g(t) = \sqrt{t^2 - 1}\,\sinh^{-1} t$

43. The tangent line to the graph of $y = f(x)$ at $x = 4$ has equation $y = -2x + 12$. Find the equation of the tangent line to $y = g(x)$ at $x = 4$, where $g(x)$ is the inverse of $f(x)$.

In Exercises 44–46, let $f(x) = xe^{-x}$.

44. GU Plot $f(x)$ and use the zoom feature to find two solutions of $f(x) = 0.3$.

45. Show that $f(x)$ has an inverse on $[1, \infty)$. Let $g(x)$ be this inverse. Find the domain and range of $g(x)$ and compute $g'(2e^{-2})$.

46. Show that $f(x) = c$ has two solutions if $0 < c < e^{-1}$.

47. Find the local extrema of $f(x) = e^{2x} - 4e^x$.

48. Find the points of inflection of $f(x) = \ln(x^2 + 1)$ and determine whether the concavity changes from up to down or vice versa.

In Exercises 49–52, find the local extrema and points of inflection, and sketch the graph over the interval specified. Use L'Hôpital's Rule to determine the limits as $x \to 0+$ or $x \to \pm\infty$ if necessary.

49. $y = x\ln x, \quad x > 0$ **50.** $y = xe^{-x^2/2}$

51. $y = x(\ln x)^2, \quad x > 0$ **52.** $y = \tan^{-1}\left(\frac{x^2}{4}\right)$

In Exercises 53–58, use logarithmic differentiation to find the derivative.

53. $y = \frac{(x+1)^3}{(4x-2)^2}$ **54.** $y = \frac{(x+1)(x+2)^2}{(x+3)(x+4)}$

55. $y = e^{(x-1)^2}e^{(x-3)^2}$ **56.** $y = \frac{e^x \sin^{-1} x}{\ln x}$

57. $y = \frac{e^{3x}(x-2)^2}{(x+1)^2}$ **58.** $y = x^{\sqrt{x}}(x^{\ln x})$

59. **Image Processing** The intensity of a pixel in a digital image is measured by a number u between 0 and 1. Often, images can be enhanced by rescaling intensities (Figure 1), where pixels of intensity u are displayed with intensity $g(u)$ for a suitable function $g(u)$. One common choice is the **sigmoidal correction**, defined for constants a, b by

$$g(u) = \frac{f(u) - f(0)}{f(1) - f(0)} \quad \text{where} \quad f(u) = \left(1 + e^{b(a-u)}\right)^{-1}$$

Figure 2 shows that $g(u)$ reduces the intensity of low-intensity pixels (where $g(u) < u$) and increases the intensity of high-intensity pixels.

(a) Verify that $f'(u) > 0$ and use this to show that $g(u)$ increases from 0 to 1 for $0 \le u \le 1$.

(b) Where does $g(u)$ have a point of inflection?

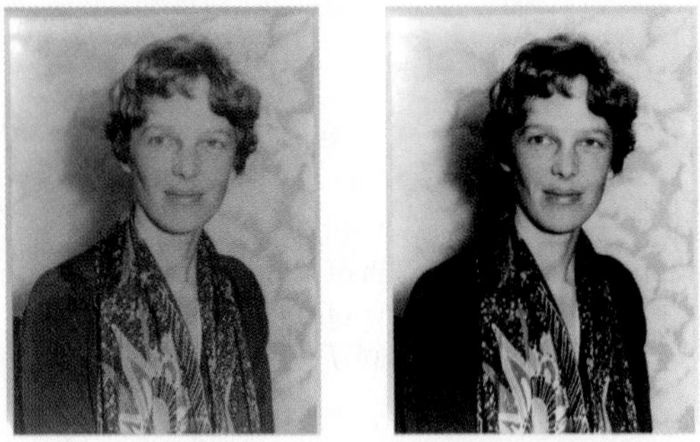

Original Sigmoidal correction

FIGURE 1

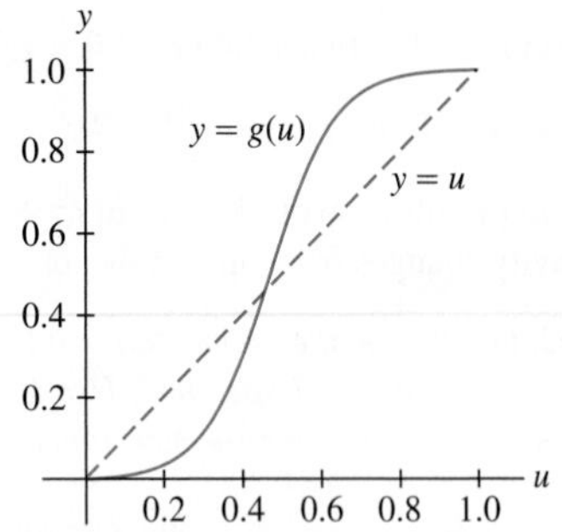

FIGURE 2 Sigmoidal correction with $a = 0.47$, $b = 12$.

60. Let $N(t)$ be the size of a tumor (in units of 10^6 cells) at time t (in days). According to the **Gompertz Model**, $dN/dt = N(a - b\ln N)$ where a, b are positive constants. Show that the maximum value of N is $e^{\frac{a}{b}}$ and that the tumor increases most rapidly when $N = e^{\frac{a}{b}-1}$.

In Exercises 61–66, use the given substitution to evaluate the integral.

61. $\displaystyle\int \frac{(\ln x)^2\,dx}{x}$, $u = \ln x$

62. $\displaystyle\int \frac{dx}{4x^2+9}$, $u = \frac{2x}{3}$

63. $\displaystyle\int \frac{dx}{\sqrt{e^{2x}-1}}$, $u = e^{-x}$

64. $\displaystyle\int \frac{\cos^{-1} t\,dt}{\sqrt{1-t^2}}$, $u = \cos^{-1} t$

65. $\displaystyle\int \frac{dt}{t(1+(\ln t)^2)}$, $u = \ln t$

66. $\displaystyle\int \frac{dt}{\cosh^2 t + \sinh^2 t}$, $u = \tanh t$

In Exercises 67–92, calculate the integral.

67. $\displaystyle\int e^{9-2x}\,dx$

68. $\displaystyle\int x^2 e^{x^3}\,dx$

69. $\displaystyle\int e^{-2x}\sin(e^{-2x})\,dx$

70. $\displaystyle\int \frac{\cos(\ln x)\,dx}{x}$

71. $\displaystyle\int_1^3 e^{4x-3}\,dx$

72. $\displaystyle\int \frac{dx}{x\sqrt{\ln x}}$

73. $\displaystyle\int_1^e \frac{\ln x\,dx}{x}$

74. $\displaystyle\int_0^{\ln 3} e^{x-e^x}\,dx$

75. $\displaystyle\int_{1/3}^{2/3} \frac{dx}{\sqrt{1-x^2}}$

76. $\displaystyle\int_4^{12} \frac{dx}{x\sqrt{x^2-1}}$

77. $\displaystyle\int_0^1 \cosh(2t)\,dt$

78. $\displaystyle\int_0^2 \frac{dt}{4t+12}$

79. $\displaystyle\int_0^3 \frac{x\,dx}{x^2+9}$

80. $\displaystyle\int_0^3 \frac{dx}{x^2+9}$

81. $\displaystyle\int \frac{x\,dx}{\sqrt{1-x^4}}$

82. $\displaystyle\int e^x 10^x\,dx$

83. $\displaystyle\int \frac{e^{-x}\,dx}{(e^{-x}+2)^3}$

84. $\displaystyle\int \sin\theta\cos\theta\, e^{\cos^2\theta+1}\,d\theta$

85. $\displaystyle\int_0^{\pi/6} \tan 2\theta\,d\theta$

86. $\displaystyle\int_{\pi/3}^{2\pi/3} \cot\left(\frac{1}{2}\theta\right)d\theta$

87. $\displaystyle\int \frac{\sin^{-1} x\,dx}{\sqrt{1-x^2}}$

88. $\displaystyle\int \tanh 5x\,dx$

89. $\displaystyle\int \sinh^3 x\cosh x\,dx$

90. $\displaystyle\int_0^1 \frac{dx}{25-x^2}$

91. $\displaystyle\int_0^4 \frac{dx}{2x^2+1}$

92. $\displaystyle\int_2^6 \frac{dx}{x\sqrt{x^2+12}}$

93. The isotope Thorium-234 has a half-life of 24.5 days.

(a) Find the differential equation satisfied by the amount $y(t)$ of Thorium-234 in a sample at time t.

(b) At $t = 0$, a sample contains 2 kg of Thorium-234. How much remains after 1 year?

94. The Oldest Snack Food In Bat Cave, New Mexico, archaeologists found ancient human remains, including cobs of popping corn, that had a C^{14} to C^{12} ratio equal to around 48% of that found in living matter. Estimate the age of the corn cobs.

95. The C^{14} to C^{12} ratio of a sample is proportional to the disintegration rate (number of beta particles emitted per minute) that is measured directly with a Geiger counter. The disintegration rate of carbon in a living organism is 15.3 beta particles/min per gram. Find the age of a sample that emits 9.5 beta particles/min per gram.

96. An investment pays out \$5000 at the end of the year for 3 years. Compute the PV, assuming an interest rate of 8%.

97. In a first-order chemical reaction, the quantity $y(t)$ of reactant at time t satisfies $y' = -ky$, where $k > 0$. The dependence of k on temperature T (in kelvins) is given by the **Arrhenius equation** $k =$

$Ae^{-E_a/(RT)}$, where E_a is the activation energy (J-mol^{-1}), $R = 8.314$ J-mol^{-1}-K^{-1}, and A is a constant. Assume that $A = 72 \times 10^{12}$ hour^{-1} and $E_a = 1.1 \times 10^5$. Calculate dk/dT for $T = 500$ and use the Linear Approximation to estimate the change in k if T is raised from 500 to 510 K.

98. Find the solutions to $y' = 4(y - 12)$ satisfying $y(0) = 20$ and $y(0) = 0$, and sketch their graphs.

99. Find the solutions to $y' = -2y + 8$ satisfying $y(0) = 3$ and $y(0) = 4$, and sketch their graphs.

100. Show that $y = \sin^{-1} x$ satisfies the differential equation $y' = \sec y$ with initial condition $y(0) = 0$.

101. What is the limit $\lim_{t\to\infty} y(t)$ if $y(t)$ is a solution of:

(a) $\dfrac{dy}{dt} = -4(y - 12)$?
(b) $\dfrac{dy}{dt} = 4(y - 12)$?
(c) $\dfrac{dy}{dt} = -4y - 12$?

102. Let A and B be constants. Prove that if $A > 0$, then all solutions of $\frac{dy}{dt} + Ay = B$ approach the same limit as $t \to \infty$.

103. An equipment upgrade costing \$1 million will save a company \$320,000 per year for 4 years. Is this a good investment if the interest rate is $r = 5\%$? What is the largest interest rate that would make the investment worthwhile? Assume that the savings are received as a lump sum at the end of each year.

104. Find the PV of an income stream paying out continuously at a rate of $5000e^{-0.1t}$ dollars per year for 5 years, assuming an interest rate of $r = 4\%$.

In Exercises 105–108, let $P(t)$ denote the balance at time t (years) of an annuity that earns 5% interest continuously compounded and pays out \$2000/year continuously.

105. Find the differential equation satisfied by $P(t)$.

106. Determine $P(2)$ if $P(0) = \$5000$.

107. When does the annuity run out of money if $P(0) = \$2000$?

108. What is the minimum initial balance that will allow the annuity to make payments indefinitely?

In Exercises 109–120, verify that L'Hôpital's Rule applies and evaluate the limit.

109. $\displaystyle\lim_{x\to 3} \frac{4x - 12}{x^2 - 5x + 6}$

110. $\displaystyle\lim_{x\to -2} \frac{x^3 + 2x^2 - x - 2}{x^4 + 2x^3 - 4x - 8}$

111. $\displaystyle\lim_{x\to 0+} x^{1/2} \ln x$

112. $\displaystyle\lim_{t\to\infty} \frac{\ln(e^t + 1)}{t}$

113. $\displaystyle\lim_{\theta\to 0} \frac{2\sin\theta - \sin 2\theta}{\sin\theta - \theta\cos\theta}$

114. $\displaystyle\lim_{x\to 0} \frac{\sqrt{4+x} - 2\sqrt[8]{1+x}}{x^2}$

115. $\displaystyle\lim_{t\to\infty} \frac{\ln(t+2)}{\log_2 t}$

116. $\displaystyle\lim_{x\to 0} \left(\frac{e^x}{e^x - 1} - \frac{1}{x}\right)$

117. $\displaystyle\lim_{y\to 0} \frac{\sin^{-1} y - y}{y^3}$

118. $\displaystyle\lim_{x\to 1} \frac{\sqrt{1 - x^2}}{\cos^{-1} x}$

119. $\displaystyle\lim_{x\to 0} \frac{\sinh(x^2)}{\cosh x - 1}$

120. $\displaystyle\lim_{x\to 0} \frac{\tanh x - \sinh x}{\sin x - x}$

121. Explain why L'Hôpital's Rule gives no information about $\displaystyle\lim_{x\to\infty} \frac{2x - \sin x}{3x + \cos 2x}$. Evaluate the limit by another method.

122. Let $f(x)$ be a differentiable function with inverse $g(x)$ such that $f(0) = 0$ and $f'(0) \neq 0$. Prove that

$$\lim_{x\to 0} \frac{f(x)}{g(x)} = f'(0)^2$$

123. Calculate the limit

$$\lim_{n\to\infty} \left(1 + \frac{4}{n}\right)^n$$

124. Calculate the limit

$$\lim_{n\to\infty} \left(1 + \frac{4}{n}\right)^{3n}$$

125. In this exercise, we prove that for all $x > 0$,

$$x - \frac{x^2}{2} \le \ln(1 + x) \le x \qquad \boxed{1}$$

(a) Show that $\ln(1 + x) = \displaystyle\int_0^x \frac{dt}{1+t}$ for $x > 0$.

(b) Verify that $1 - t \le \dfrac{1}{1+t} \le 1$ for all $t > 0$.

(c) Use (b) to prove Eq. (1).

(d) Verify Eq. (1) for $x = 0.5$, 0.1, and 0.01.

126. Let

$$F(x) = x\sqrt{x^2 - 1} - 2\int_1^x \sqrt{t^2 - 1}\, dt$$

Prove that $F(x)$ and $\cosh^{-1} x$ differ by a constant by showing that they have the same derivative. Then prove they are equal by evaluating both at $x = 1$.

In Exercises 127–130, let $gd(y) = \tan^{-1}(\sinh y)$ be the so-called ***gudermannian****, which arises in cartography. In a map of the earth constructed by Mercator projection, points located y radial units from the equator correspond to points on the globe of latitude $gd(y)$.*

127. Prove that $\dfrac{d}{dy} gd(y) = \operatorname{sech} y$.

128. Let $f(y) = 2\tan^{-1}(e^y) - \pi/2$. Prove that $gd(y) = f(y)$. *Hint:* Show that $gd'(y) = f'(y)$ and $f(0) = gd(0)$.

129. Show that $t(y) = \sinh^{-1}(\tan y)$ is the inverse of $gd(y)$ for $0 \le y < \pi/2$.

130. Verify that $t(y)$ in Exercise 129 satisfies $t'(y) = \sec y$ and find a value of a such that

$$t(y) = \int_a^y \frac{dt}{\cos t}$$

131. Use L'Hôpital's Rule to prove that for all $a > 0$ and $b > 0$,

$$\lim_{n\to\infty}\left(\frac{a^{1/n}+b^{1/n}}{2}\right)^n = \sqrt{ab}$$

132. Let

$$F(x) = \int_2^x \frac{dt}{\ln t} \quad \text{and} \quad G(x) = \frac{x}{\ln x}$$

Verify that L'Hôpital's Rule may be applied to the limit $L = \lim_{x\to\infty}\frac{F(x)}{G(x)}$ and evaluate L.

133. Let $f(x) = e^{-Ax^2/2}$, where $A > 0$. Given any n numbers $a_1, a_2, \ldots, a_n$, set

$$\Phi(x) = f(x-a_1)f(x-a_2)\cdots f(x-a_n)$$

(a) Assume $n = 2$ and prove that $\Phi(x)$ attains its maximum value at the average $x = \frac{1}{2}(a_1 + a_2)$. *Hint:* Show that $d/dx \ln(f(x)) = -Ax$ and calculate $\Phi'(x)$ using logarithmic differentiation.

(b) Show that for any n, $\Phi(x)$ attains its maximum value at $x = \frac{1}{n}(a_1 + a_2 + \cdots + a_n)$. This fact is related to the role of $f(x)$ (whose graph is a bell-shaped curve) in statistics.

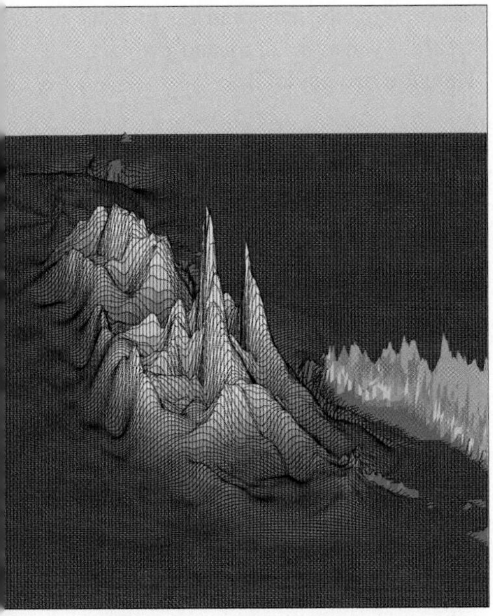

Computer simulation of the Indonesian tsunami of December 26, 2004 (8 minutes after the earthquake), created using models of wave motion based on advanced calculus by Steven Ward, University of California at Santa Cruz.

8 TECHNIQUES OF INTEGRATION

In Section 5.6 we introduced substitution, one of the most important techniques of integration. In this section, we develop a second fundamental technique, Integration by Parts, as well as several techniques for treating particular classes of functions such as trigonometric and rational functions. However, there is no surefire method, and in fact, many important antiderivatives cannot be expressed in elementary terms. Therefore, we discuss numerical integration in the last section. Every definite integral can be approximated numerically to any desired degree of accuracy.

8.1 Integration by Parts

The Integration by Parts formula is derived from the Product Rule:

$$\big(u(x)v(x)\big)' = u(x)v'(x) + u'(x)v(x)$$

According to this formula, $u(x)v(x)$ is an antiderivative of the right-hand side, so

$$u(x)v(x) = \int u(x)v'(x)\,dx + \int u'(x)v(x)\,dx$$

Moving the second integral on the right to the other side, we obtain:

The Integration by Parts formula is often written using differentials:

$$\int u\,dv = uv - \int v\,du$$

where $dv = v'(x)\,dx$ and $du = u'(x)\,dx$.

Integration by Parts Formula

$$\int u(x)v'(x)\,dx = u(x)v(x) - \int u'(x)v(x)\,dx \qquad \boxed{1}$$

Because the Integration by Parts formula applies to a product $u(x)v'(x)$, we should consider using it when the integrand is a product of two functions.

■ **EXAMPLE 1** Evaluate $\int x\cos x\,dx$.

Solution The integrand is a product, so we try writing $x\cos x = uv'$ with

$$u(x) = x, \qquad v'(x) = \cos x$$

In applying Eq. (1), any antiderivative $v(x)$ of $v'(x)$ may be used.

In this case, $u'(x) = 1$ and $v(x) = \sin x$. By the Integration by Parts formula,

$$\int \underbrace{x\cos x}_{uv'}\,dx = \underbrace{x\sin x}_{uv} - \int \underbrace{\sin x}_{u'v}\,dx = x\sin x + \cos x + C$$

Let's check the answer by taking the derivative:

$$\frac{d}{dx}(x\sin x + \cos x + C) = x\cos x + \sin x - \sin x = x\cos x$$ ■

The key step in Integration by Parts is deciding how to write the integrand as a product uv'. Keep in mind that Integration by Parts expresses $\int uv'\,dx$ in terms of uv and $\int u'v\,dx$. *This is useful if $u'v$ is easier to integrate than uv'.* Here are two guidelines:

- Choose u so that u' is "simpler" than u itself.
- Choose v' so that $v = \displaystyle\int v'\,dx$ can be evaluated.

EXAMPLE 2 Good Versus Bad Choices of u and v' Evaluate $\displaystyle\int xe^x\,dx$.

Solution Based on our guidelines, it makes sense to write $xe^x = uv'$ with

- $u = x$ (since $u' = 1$ is simpler)
- $v' = e^x$ (since we can evaluate $v = \displaystyle\int e^x\,dx = e^x + C$)

Integration by Parts gives us

$$\int xe^x\,dx = u(x)v(x) - \int u'(x)v(x)\,dx = xe^x - \int e^x\,dx = xe^x - e^x + C$$

Let's see what happens if we write $xe^x = uv'$ with $u = e^x$, $v' = x$. Then

$$u'(x) = e^x, \qquad v(x) = \int x\,dx = \frac{1}{2}x^2 + C$$

$$\int \underbrace{xe^x}_{uv'}\,dx = \underbrace{\frac{1}{2}x^2e^x}_{uv} - \int \underbrace{\frac{1}{2}x^2e^x}_{u'v}\,dx$$

This is a poor choice of u and v' because the integral on the right is more complicated than our original integral. ■

EXAMPLE 3 Integrating by Parts More Than Once Evaluate $\displaystyle\int x^2\cos x\,dx$.

In Example 3, it makes sense to take $u = x^2$ because Integration by Parts reduces the integration of $x^2\cos x$ to the integration of $2x\sin x$, which is easier.

Solution Apply Integration by Parts a first time with $u = x^2$ and $v' = \cos x$:

$$\int \underbrace{x^2\cos x}_{uv'}\,dx = \underbrace{x^2\sin x}_{uv} - \int \underbrace{2x\,\sin x}_{u'v}\,dx = x^2\sin x - 2\int x\sin x\,dx \qquad \boxed{2}$$

Now apply it again to the integral on the right, this time with $u = x$ and $v' = \sin x$:

$$\int \underbrace{x\sin x}_{uv'}\,dx = \underbrace{-x\cos x}_{uv} - \int \underbrace{(-\cos x)}_{u'v}\,dx = -x\cos x + \sin x + C$$

Using this result in Eq. (2), we obtain

$$\int x^2\cos x\,dx = x^2\sin x - 2\int x\sin x\,dx = x^2\sin x - 2(-x\cos x + \sin x) + C$$

$$= x^2\sin x + 2x\cos x - 2\sin x + C$$ ■

Integration by Parts applies to *definite integrals*:

$$\boxed{\int_a^b u(x)v'(x)\,dx = u(x)v(x)\Big|_a^b - \int_a^b u'(x)v(x)\,dx}$$

■ **EXAMPLE 4 Taking $v' = 1$** Evaluate $\int_1^3 \ln x \, dx$.

Surprisingly, the choice $v' = 1$ is effective in some cases. Using it as in Example 4, we find that

$$\int \ln x \, dx = x \ln x - x + C$$

This choice also works for the inverse trigonometric functions (see Exercise 6).

Solution The integrand is not a product, so at first glance, this integral does not look like a candidate for Integration by Parts. However, we are free to add a factor of 1 and write $\ln x = (\ln x) \cdot 1 = uv'$. Then

$$u = \ln x, \qquad v' = 1$$
$$u' = x^{-1}, \qquad v = x$$

$$\int_1^3 \underbrace{\ln x}_{uv'} \, dx = \underbrace{x \ln x}_{uv} \Big|_1^3 - \int_1^3 \underbrace{1}_{u'v} \, dx = (3 \ln 3 - 0) - 2 = 3 \ln 3 - 2$$ ■

■ **EXAMPLE 5 Going in a Circle?** Evaluate $\int e^x \cos x \, dx$.

In Example 5, the choice $u = e^x$, $v' = \cos x$ works equally well.

Solution There are two reasonable ways of writing $e^x \cos x$ as uv'. Let's try $u = \cos x$ and $v' = e^x$. Then

$$\int \underbrace{e^x \cos x}_{uv'} \, dx = \underbrace{e^x \cos x}_{uv} + \int \underbrace{e^x \sin x}_{-u'v} \, dx \tag{3}$$

Now use Integration by Parts to the integral on the right with $u = \sin x$ and $v' = e^x$:

$$\int e^x \sin x \, dx = e^x \sin x - \int e^x \cos x \, dx \tag{4}$$

Eq. (4) brings us back to our original integral of $e^x \cos x$, so it looks as if we're going in a circle. But we can substitute Eq. (4) in Eq. (3) and solve for the integral of $e^x \cos x$:

$$\int e^x \cos x \, dx = e^x \cos x + \int e^x \sin x \, dx = e^x \cos x + e^x \sin x - \int e^x \cos x \, dx$$

$$2 \int e^x \cos x \, dx = e^x \cos x + e^x \sin x + C$$

$$\int e^x \cos x \, dx = \frac{1}{2} e^x (\cos x + \sin x) + C$$ ■

A reduction formula (also called a ***recursive formula****) expresses the integral for a given value of n in terms of a similar integral for a smaller value of n. The desired integral is evaluated by applying the reduction formula repeatedly.*

Integration by Parts can be used to derive **reduction formulas** for integrals that depend on a positive integer n such as $\int x^n e^x \, dx$ or $\int \ln^n x \, dx$.

■ **EXAMPLE 6 A Reduction Formula** Derive the reduction formula

$$\boxed{\int x^n e^x \, dx = x^n e^x - n \int x^{n-1} e^x \, dx} \tag{5}$$

Then evaluate $\int x^3 e^x \, dx$.

In general, $\int x^n e^x \, dx = P_n(x) e^x + C$, where $P_n(x)$ is a polynomial of degree n (see Exercise 78).

Solution We apply Integration by Parts with $u = x^n$ and $v' = e^x$:

$$\int x^n e^x \, dx = uv - \int u'v \, dx = x^n e^x - n \int x^{n-1} e^x \, dx$$

To evaluate $\int x^3 e^x \, dx$, we'll need to use the reduction formula for $n = 3, 2, 1$:

$$\int x^3e^x\,dx = x^3e^x - 3\int x^2e^x\,dx$$
$$= x^3e^x - 3\left(x^2e^x - 2\int xe^x\,dx\right)$$
$$= x^3e^x - 3x^2e^x + 6\int xe^x\,dx$$
$$= x^3e^x - 3x^2e^x + 6\left(xe^x - \int e^x\,dx\right)$$
$$= x^3e^x - 3x^2e^x + 6xe^x - 6e^x + C$$
$$= (x^3 - 3x^2 + 6x - 6)e^x + C$$

■

8.1 SUMMARY

- *Integration by Parts* formula: $\int u(x)v'(x)\,dx = u(x)v(x) - \int u'(x)v(x)\,dx$.
- The key step is deciding how to write the integrand as a product uv'. Keep in mind that Integration by Parts is useful when $u'v$ is easier (or, at least, not more difficult) to integrate than uv'. Here are some guidelines:
  - Choose u so that u' is simpler than u itself.
  - Choose v' so that $v = \int v'\,dx$ can be evaluated.
  - Sometimes, $v' = 1$ is a good choice.

8.1 EXERCISES

Preliminary Questions

1. Which derivative rule is used to derive the Integration by Parts formula?

2. For each of the following integrals, state whether substitution or Integration by Parts should be used:

$$\int x\cos(x^2)\,dx, \qquad \int x\cos x\,dx, \qquad \int x^2e^x\,dx, \qquad \int xe^{x^2}\,dx$$

3. Why is $u = \cos x$, $v' = x$ a poor choice for evaluating $\int x\cos x\,dx$?

Exercises

In Exercises 1–6, evaluate the integral using the Integration by Parts formula with the given choice of u and v'.

1. $\int x\sin x\,dx$; $u = x$, $v' = \sin x$

2. $\int xe^{2x}\,dx$; $u = x$, $v' = e^{2x}$

3. $\int (2x+9)e^x\,dx$; $u = 2x+9$, $v' = e^x$

4. $\int x\cos 4x\,dx$; $u = x$, $v' = \cos 4x$

5. $\int x^3\ln x\,dx$; $u = \ln x$, $v' = x^3$

6. $\int \tan^{-1}x\,dx$; $u = \tan^{-1}x$, $v' = 1$

In Exercises 7–36, evaluate using Integration by Parts.

7. $\int (4x-3)e^{-x}\,dx$

8. $\int (2x+1)e^x\,dx$

9. $\int x\,e^{5x+2}\,dx$

10. $\int x^2e^x\,dx$

11. $\int x\cos 2x\,dx$

12. $\int x\sin(3-x)\,dx$

13. $\int x^2\sin x\,dx$

14. $\int x^2\cos 3x\,dx$

15. $\int e^{-x}\sin x\,dx$

16. $\int e^x\sin 2x\,dx$

17. $\int e^{-5x} \sin x \, dx$

18. $\int e^{3x} \cos 4x \, dx$

19. $\int x \ln x \, dx$

20. $\int \frac{\ln x}{x^2} \, dx$

21. $\int x^2 \ln x \, dx$

22. $\int x^{-5} \ln x \, dx$

23. $\int (\ln x)^2 \, dx$

24. $\int x(\ln x)^2 \, dx$

25. $\int x \sec^2 x \, dx$

26. $\int x \tan x \sec x \, dx$

27. $\int \cos^{-1} x \, dx$

28. $\int \sin^{-1} x \, dx$

29. $\int \sec^{-1} x \, dx$

30. $\int x 5^x \, dx$

31. $\int 3^x \cos x \, dx$

32. $\int x \sinh x \, dx$

33. $\int x^2 \cosh x \, dx$

34. $\int \cos x \cosh x \, dx$

35. $\int \tanh^{-1} 4x \, dx$

36. $\int \sinh^{-1} x \, dx$

In Exercises 37–38, evaluate using substitution and then Integration by Parts.

37. $\int e^{\sqrt{x}} \, dx$ *Hint:* Let $u = x^{1/2}$

38. $\int x^3 e^{x^2} \, dx$

In Exercises 39–48, evaluate using Integration by Parts, substitution, or both if necessary.

39. $\int x \cos 4x \, dx$

40. $\int \frac{\ln(\ln x) \, dx}{x}$

41. $\int \frac{x \, dx}{\sqrt{x+1}}$

42. $\int x^2 (x^3 + 9)^{15} \, dx$

43. $\int \cos x \ln(\sin x) \, dx$

44. $\int \sin \sqrt{x} \, dx$

45. $\int \sqrt{x} e^{\sqrt{x}} \, dx$

46. $\int \frac{\tan \sqrt{x} \, dx}{\sqrt{x}}$

47. $\int \frac{\ln(\ln x) \ln x \, dx}{x}$

48. $\int \sin(\ln x) \, dx$

In Exercises 49–54, compute the definite integral.

49. $\int_0^3 x e^{4x} \, dx$

50. $\int_0^{\pi/4} x \sin 2x \, dx$

51. $\int_1^2 x \ln x \, dx$

52. $\int_1^e \frac{\ln x \, dx}{x^2}$

53. $\int_0^{\pi} e^x \sin x \, dx$

54. $\int_0^1 \tan^{-1} x \, dx$

55. Use Eq. (5) to evaluate $\int x^4 e^x \, dx$.

56. Use substitution and then Eq. (5) to evaluate $\int x^4 e^{7x} \, dx$.

57. Find a reduction formula for $\int x^n e^{-x} \, dx$ similar to Eq. (5).

58. Evaluate $\int x^n \ln x \, dx$ for $n \neq -1$. Which method should be used to evaluate $\int x^{-1} \ln x \, dx$?

In Exercises 59–66, indicate a good method for evaluating the integral (but do not evaluate). Your choices are algebraic manipulation, substitution (specify u and du), and Integration by Parts (specify u and v′). If it appears that the techniques you have learned thus far are not sufficient, state this.

59. $\int \sqrt{x} \ln x \, dx$

60. $\int \frac{x^2 - \sqrt{x}}{2x} \, dx$

61. $\int \frac{x^3 \, dx}{\sqrt{4 - x^2}}$

62. $\int \frac{dx}{\sqrt{4 - x^2}}$

63. $\int \frac{x + 2}{x^2 + 4x + 3} \, dx$

64. $\int \frac{dx}{(x+2)(x^2 + 4x + 3)}$

65. $\int x \sin(3x + 4) \, dx$

66. $\int x \cos(9x^2) \, dx$

67. Evaluate $\int (\sin^{-1} x)^2 \, dx$. *Hint:* Use Integration by Parts first and then substitution.

68. Evaluate $\int \frac{(\ln x)^2 \, dx}{x^2}$. *Hint:* Use substitution first and then Integration by Parts.

69. Evaluate $\int x^7 \cos(x^4) \, dx$.

70. Find $f(x)$, assuming that

$$\int f(x) e^x \, dx = f(x) e^x - \int x^{-1} e^x \, dx$$

71. Find the volume of the solid obtained by revolving the region under $y = e^x$ for $0 \le x \le 2$ about the y-axis.

72. Find the area enclosed by $y = \ln x$ and $y = (\ln x)^2$.

73. Recall that the *present value* (PV) of an investment that pays out income continuously at a rate $R(t)$ for T years is $\int_0^T R(t) e^{-rt} \, dt$, where r is the interest rate. Find the PV if $R(t) = 5000 + 100t$ \$/year, $r = 0.05$ and $T = 10$ years.

74. Derive the reduction formula

$$\int (\ln x)^k \, dx = x(\ln x)^k - k \int (\ln x)^{k-1} \, dx \qquad \boxed{6}$$

75. Use Eq. (6) to calculate $\int (\ln x)^k \, dx$ for $k = 2, 3$.

76. Derive the reduction formulas

$$\int x^n \cos x \, dx = x^n \sin x - n \int x^{n-1} \sin x \, dx$$

$$\int x^n \sin x \, dx = -x^n \cos x + n \int x^{n-1} \cos x \, dx$$

77. Prove that $\int x b^x \, dx = b^x \left(\frac{x}{\ln b} - \frac{1}{\ln^2 b} \right) + C$.

78. Define $P_n(x)$ by

$$\int x^n e^x \, dx = P_n(x) e^x + C$$

Use Eq. (5) to prove that $P_n(x) = x^n - n P_{n-1}(x)$. Use this recursion relation to find $P_n(x)$ for $n = 1, 2, 3, 4$. Note that $P_0(x) = 1$.

Further Insights and Challenges

79. The Integration by Parts formula can be written

$$\int u(x)v(x)\,dx = u(x)V(x) - \int u'(x)V(x)\,dx \qquad \boxed{7}$$

where $V(x)$ satisfies $V'(x) = v(x)$.

(a) Show directly that the right-hand side of Eq. (7) does not change if $V(x)$ is replaced by $V(x) + C$, where C is a constant.

(b) Use $u = \tan^{-1} x$ and $v = x$ in Eq. (7) to calculate $\int x \tan^{-1} x\,dx$, but carry out the calculation twice: first with $V(x) = \frac{1}{2}x^2$ and then with $V(x) = \frac{1}{2}x^2 + \frac{1}{2}$. Which choice of $V(x)$ results in a simpler calculation?

80. Prove in two ways that

$$\int_0^a f(x)\,dx = af(a) - \int_0^a xf'(x)\,dx \qquad \boxed{8}$$

First use Integration by Parts. Then assume $f(x)$ is increasing. Use the substitution $u = f(x)$ to prove that $\int_0^a xf'(x)\,dx$ is equal to the area of the shaded region in Figure 1 and derive Eq. (8) a second time.

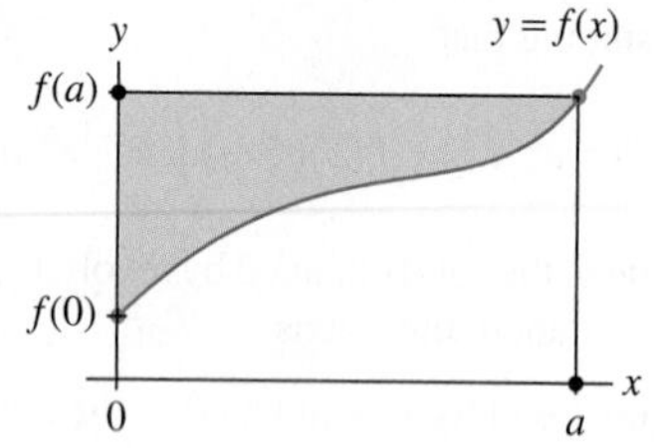

FIGURE 1

81. Assume that $f(0) = f(1) = 0$ and that f'' exists. Prove

$$\int_0^1 f''(x)f(x)\,dx = -\int_0^1 f'(x)^2\,dx \qquad \boxed{9}$$

Use this to prove that if $f(0) = f(1) = 0$ and $f''(x) = \lambda f(x)$ for some constant λ, then $\lambda < 0$. Can you think of a function satisfying these conditions for some λ?

82. Set $I(a, b) = \int_0^1 x^a(1 - x)^b\,dx$, where a, b are whole numbers.

(a) Use substitution to show that $I(a, b) = I(b, a)$.

(b) Show that $I(a, 0) = I(0, a) = \dfrac{1}{a + 1}$.

(c) Prove that for $a \ge 1$ and $b \ge 0$,

$$I(a, b) = \frac{a}{b + 1} I(a - 1, b + 1)$$

(d) Use (b) and (c) to calculate $I(1, 1)$ and $I(3, 2)$.

(e) Show that $I(a, b) = \dfrac{a!\,b!}{(a + b + 1)!}$.

83. Let $I_n = \int x^n \cos(x^2)\,dx$ and $J_n = \int x^n \sin(x^2)\,dx$.

(a) Find a reduction formula that expresses I_n in terms of J_{n-2}. *Hint:* Write $x^n \cos(x^2)$ as $x^{n-1}(x\cos(x^2))$.

(b) Use the result of (a) to show that I_n can be evaluated explicitly if n is odd.

(c) Evaluate I_3.

8.2 Trigonometric Integrals

Many trigonometric functions can be integrated by combining substitution and Integration by Parts with the appropriate trigonometric identities. First, consider

$$\int \sin^m x \cos^n x\,dx$$

where m, n are whole numbers. The easier case is when at least one of m, n is *odd*.

■ **EXAMPLE 1** **Odd Power of $\sin x$** Evaluate $\int \sin^3 x\,dx$.

Solution Because $\sin^3 x$ is an odd power, the identity $\sin^2 x = 1 - \cos^2 x$ allows us to split off a factor of $\sin x\,dx$:

$$\sin^3 x\,dx = \sin^2 x(\sin x\,dx) = (1 - \cos^2 x)\sin x\,dx$$

and use the substitution $u = \cos x$, $du = -\sin x\,dx$:

$$\int \sin^3 x\,dx = \int (1 - \cos^2 x)\sin x\,dx = -\int (1 - u^2)\,du$$

$$= \frac{u^3}{3} - u + C = \frac{\cos^3 x}{3} - \cos x + C$$ ■

Integrating $\sin^m x \cos^n x$

Case 1: $m = 2k + 1$ **odd**

Write $\sin^{2k+1} x$ *as* $(1 - \cos^2 x)^k \sin x$. *Then* $\int \sin^{2k+1} x \cos^n x\, dx$ *becomes*

$$\int \sin x (1 - \cos^2 x)^k \cos^n x\, dx$$

Substitute $u = \cos x$, $-du = \sin x\, dx$.

Case 2: $n = 2k + 1$ **odd**

Write $\cos^{2k+1} x$ *as* $(1 - \sin^2 x)^k \cos x$ *Then* $\int \sin^m x \cos^{2k+1} x\, dx$ *becomes*

$$\int \sin^m x (1 - \sin^2 x)^k \cos x\, dx$$

Substitute $u = \sin x$, $du = \cos x\, dx$.

Case 3: m, n **both even**

Use reduction formulas (1) or (2) as described below or use the method of Exercises 65–68.

The strategy of the previous example works when $\sin^m x$ appears with m odd. Similarly, if n is odd, write $\cos^n x$ as a power of $(1 - \sin^2 x)$ times $\cos x$.

■ **EXAMPLE 2** **Odd Power of $\sin x$ or $\cos x$** Evaluate $\int \sin^4 x \cos^5 x\, dx$.

Solution We take advantage of the fact that $\cos^5 x$ is an odd power to write

$$\sin^4 x \cos^5 x\, dx = \sin^4 x \cos^4 x (\cos x\, dx) = \sin^4 x (1 - \sin^2 x)^2 (\cos x\, dx)$$

This allows us to use the substitution $u = \sin x$, $du = \cos x\, dx$:

$$\int \sin^4 x \cos^5 x\, dx = \int (\sin^4 x)(1 - \sin^2 x)^2 \cos x\, dx$$

$$= \int u^4 (1 - u^2)^2\, du = \int (u^4 - 2u^6 + u^8)\, du$$

$$= \frac{u^5}{5} - \frac{2u^7}{7} + \frac{u^9}{9} + C = \frac{\sin^5 x}{5} - \frac{2\sin^7 x}{7} + \frac{\sin^9 x}{9} + C \quad ■$$

The following reduction formulas can be used to integrate $\sin^n x$ and $\cos^n x$ for any exponent n, even or odd (their proofs are left as exercises; see Exercise 64).

Reduction Formulas for Sine and Cosine

$$\int \sin^n x\, dx = -\frac{1}{n} \sin^{n-1} x \cos x + \frac{n-1}{n} \int \sin^{n-2} x\, dx \qquad \boxed{1}$$

$$\int \cos^n x\, dx = \frac{1}{n} \cos^{n-1} x \sin x + \frac{n-1}{n} \int \cos^{n-2} x\, dx \qquad \boxed{2}$$

■ **EXAMPLE 3** Evaluate $\int \sin^4 x\, dx$.

Solution Apply Eq. (1) with $n = 4$,

$$\int \sin^4 x\, dx = -\frac{1}{4} \sin^3 x \cos x + \frac{3}{4} \int \sin^2 x\, dx \qquad \boxed{3}$$

Then apply Eq. (1) again, with $n = 2$, to the integral on the right:

$$\int \sin^2 x\, dx = -\frac{1}{2} \sin x \cos x + \frac{1}{2} \int dx = -\frac{1}{2} \sin x \cos x + \frac{1}{2} x + C \qquad \boxed{4}$$

Using Eq. (4) in Eq. (3), we obtain

$$\int \sin^4 x\, dx = -\frac{1}{4} \sin^3 x \cos x - \frac{3}{8} \sin x \cos x + \frac{3}{8} x + C \quad ■$$

Trigonometric integrals can be expressed in many different ways because trigonometric functions satisfy a large number of identities. For example, a computer algebra system might evaluate the integral in the previous example as

$$\int \sin^4 x\, dx = \frac{1}{32}(x - 8 \sin 2x + \sin 4x) + C$$

You can check that this agrees with the result in Example 3 (Exercise 61).

More work is required to integrate $\sin^m x \cos^n x$ when both m and n are even. First of all, we have the following formulas, which are verified using the identities recalled in the margin.

REMINDER Useful Identities:

$$\sin^2 x = \frac{1}{2}(1 - \cos 2x)$$

$$\cos^2 x = \frac{1}{2}(1 + \cos 2x)$$

$$\sin 2x = 2 \sin x \cos x$$

$$\cos 2x = \cos^2 x - \sin^2 x$$

$$\int \sin^2 x\, dx = \frac{x}{2} - \frac{\sin 2x}{4} + C = \frac{x}{2} - \frac{1}{2} \sin x \cos x + C$$

$$\int \cos^2 x\, dx = \frac{x}{2} + \frac{\sin 2x}{4} + C = \frac{x}{2} + \frac{1}{2} \sin x \cos x + C$$

Here is a method for integrating $\sin^m x \cos^n x$ when both m and n are even. Another method is used in Exercises 65–68.

- If $m \le n$, use the identity $\sin^2 x = 1 - \cos^2 x$ to write
$$\int \sin^m x \cos^n x\, dx = \int (1 - \cos^2 x)^{m/2} \cos^n x\, dx$$
Expand the integral on the right to obtain a sum of integrals of powers of $\cos x$ and use reduction formula (2).
- If $m \ge n$, use the identity $\cos^2 x = 1 - \sin^2 x$ to write
$$\int \sin^m x \cos^n x\, dx = \int (\sin^m x)(1 - \sin^2 x)^{n/2}\, dx$$
Expand the integral on the right to obtain a sum of integrals of powers of $\sin x$, and again evaluate using reduction formula (1).

EXAMPLE 4 Even Powers of $\sin x$ and $\cos x$ Evaluate $\int \sin^2 x \cos^4 x\, dx$.

Solution Here $m = 2$ and $n = 4$. Since $m < n$, we replace $\sin^2 x$ by $1 - \cos^2 x$:

$$\int \sin^2 x \cos^4 x\, dx = \int (1 - \cos^2 x) \cos^4 x\, dx = \int \cos^4 x\, dx - \int \cos^6 x\, dx \qquad \boxed{5}$$

The reduction formula for $n = 6$ gives

$$\int \cos^6 x\, dx = \frac{1}{6} \cos^5 x \sin x + \frac{5}{6} \int \cos^4 x\, dx$$

Using this result in the right-hand side of Eq. (5), we obtain

$$\int \sin^2 x \cos^4 x\, dx = \int \cos^4 x\, dx - \left(\frac{1}{6} \cos^5 x \sin x + \frac{5}{6} \int \cos^4 x\, dx \right)$$
$$= -\frac{1}{6} \cos^5 x \sin x + \frac{1}{6} \int \cos^4 x\, dx$$

Next, we evaluate $\int \cos^4 x\, dx$ using the reduction formulas for $n = 4$ and $n = 2$:

$$\int \cos^4 x\, dx = \frac{1}{4} \cos^3 x \sin x + \frac{3}{4} \int \cos^2 x\, dx$$
$$= \frac{1}{4} \cos^3 x \sin x + \frac{3}{4} \left(\frac{1}{2} \cos x \sin x + \frac{1}{2} x \right) + C$$
$$= \frac{1}{4} \cos^3 x \sin x + \frac{3}{8} \cos x \sin x + \frac{3}{8} x + C$$

As we have noted, trigonometric integrals can be expressed in more than one way. According to Mathematica,

$$\int \sin^2 x \cos^4 x\, dx$$

$$= \tfrac{1}{16}x + \tfrac{1}{64}\sin 2x - \tfrac{1}{64}\sin 4x - \tfrac{1}{192}\sin 6x$$

Trigonometric identities show that this agrees with Eq. (6).

Altogether,

$$\int \sin^2 x \cos^4 x\, dx = -\frac{1}{6}\cos^5 x \sin x + \frac{1}{6}\left(\frac{1}{4}\cos^3 x \sin x + \frac{3}{8}\cos x \sin x + \frac{3}{8}x\right) + C$$

$$= -\frac{1}{6}\cos^5 x \sin x + \frac{1}{24}\cos^3 x \sin x + \frac{1}{16}\cos x \sin x + \frac{1}{16}x + C \qquad (6)$$

■

We turn now to the integrals of the remaining trigonometric functions.

■ **EXAMPLE 5 Integral of the Tangent and Secant** Derive the formulas

$$\int \tan x\, dx = \ln|\sec x| + C, \qquad \int \sec x\, dx = \ln|\sec x + \tan x| + C$$

The integral $\int \sec x\, dx$ was first computed numerically in the 1590s by the English mathematician Edward Wright, decades before the invention of calculus. Although he did not invent the concept of an integral, Wright realized that the sums that approximate the integral hold the key to understanding the Mercator map projection, of great importance in sea navigation because it enabled sailors to reach their destinations along lines of fixed compass direction. The formula for the integral was first proved by James Gregory in 1668.

Solution To integrate $\tan x$, use the substitution $u = \cos x$, $du = -\sin x\, dx$:

$$\int \tan x\, dx = \int \frac{\sin x}{\cos x}\, dx = -\int \frac{du}{u} = -\ln|u| + C = -\ln|\cos x| + C$$

$$= \ln\frac{1}{|\cos x|} + C = \ln|\sec x| + C$$

To integrate $\sec x$, we employ a clever substitution: $u = \sec x + \tan x$. Then

$$du = (\sec x \tan x + \sec^2 x)\, dx = (\sec x)\underbrace{(\tan x + \sec x)}_{u}\, dx = (\sec x)u\, dx$$

Thus $du = (\sec x)u\, dx$, and dividing by u gives $du/u = \sec x\, dx$. We obtain

$$\int \sec x\, dx = \int \frac{du}{u} = \ln|u| + C = \ln|\sec x + \tan x| + C$$

■

The table of integrals at the end of this section (page 423) contains a list of additional trigonometric integrals and reduction formulas.

■ **EXAMPLE 6 Using a Table of Integrals** Evaluate $\displaystyle\int_0^{\pi/4} \tan^3 x\, dx$.

Solution We use reduction formula (16) in the table with $k = 3$.

$$\int_0^{\pi/4} \tan^3 x\, dx = \frac{\tan^2 x}{2}\bigg|_0^{\pi/4} - \int_0^{\pi/4} \tan x\, dx = \left(\frac{1}{2}\tan^2 x - \ln|\sec x|\right)\bigg|_0^{\pi/4}$$

$$= \left(\frac{1}{2}\tan^2\frac{\pi}{4} - \ln\left|\sec\frac{\pi}{4}\right|\right) - \left(\frac{1}{2}\tan^2 0 - \ln|\sec 0|\right)$$

$$= \left(\frac{1}{2}(1)^2 - \ln\sqrt{2}\right) - \left(\frac{1}{2}0^2 - \ln|1|\right) = \frac{1}{2} - \ln\sqrt{2}$$

■

In the margin we describe a method for integrating $\tan^m x \sec^n x$.

■ **EXAMPLE 7** Evaluate $\int \tan^2 x \sec^3 x\, dx$.

Integrating $\tan^m x \sec^n x$

Case 1: $m = 2k + 1$ ***odd and*** $n \geq 1$

Use the identity $\tan^2 x = \sec^2 x - 1$ *to write* $\tan^{2k+1} x \sec^n x$ *as*

$$(\sec^2 x - 1)^k (\sec^{n-1} x)(\sec x \tan x)$$

Then substitute $u = \sec x$, $du = \sec x \tan x\, dx$ *to obtain an integral involving only powers of* u.

Case 2: $n = 2k$ ***even***

Use the identity $\sec^2 x = 1 + \tan^2 x$ *to write* $\tan^m x \sec^n x$ *as*

$$(\tan^m x)(1 + \tan^2 x)^{k-1} \sec^2 x$$

Then substitute $u = \tan x$, $du = \sec^2 x\, dx$ *to obtain an integral involving only powers of* u.

Case 3: m ***even and*** n ***odd***

Use the identity $\tan^2 x = \sec^2 x - 1$ *to write* $\tan^m x \sec^n x$ *as*

$$(\sec^2 x - 1)^{m/2} \sec^n x$$

Expand to obtain an integral involving only powers of $\sec x$ *and use the reduction formula (20).*

Solution Our integral is covered by Case 3 in the marginal note, because the integrand is $\tan^m x \sec^n x$, with $m = 2$ and $n = 3$.

The first step is to use the identity $\tan^2 x = \sec^2 x - 1$:

$$\int \tan^2 x \sec^3 x\, dx = \int (\sec^2 x - 1)\sec^3 x\, dx = \int \sec^5 x\, dx - \int \sec^3 x\, dx \qquad \boxed{7}$$

Next, use the reduction formula (20) in the table on page 423 with $m = 5$:

$$\int \sec^5 x\, dx = \frac{\tan x \sec^3 x}{4} + \frac{3}{4}\int \sec^3 x\, dx$$

Substitute this result in Eq. (7):

$$\begin{aligned}\int \tan^2 x \sec^3 x\, dx &= \left(\frac{\tan x \sec^3 x}{4} + \frac{3}{4}\int \sec^3 x\, dx\right) - \int \sec^3 x\, dx\\ &= \frac{1}{4}\tan x \sec^3 x - \frac{1}{4}\int \sec^3 x\, dx \qquad \boxed{8}\end{aligned}$$

and use the reduction formula (20) again with $m = 3$ and formula (19):

$$\begin{aligned}\int \sec^3 x\, dx &= \frac{\tan x \sec x}{2} + \frac{1}{2}\int \sec x\, dx\\ &= \frac{1}{2}\tan x \sec x + \frac{1}{2}\ln|\sec x + \tan x| + C\end{aligned}$$

Then Eq. (8) becomes

$$\begin{aligned}\int \tan^2 x \sec^3 x\, dx &= \frac{1}{4}\tan x \sec^3 x - \frac{1}{4}\left(\frac{1}{2}\tan x \sec x + \frac{1}{2}\ln|\sec x + \tan x|\right) + C\\ &= \frac{1}{4}\tan x \sec^3 x - \frac{1}{8}\tan x \sec x - \frac{1}{8}\ln|\sec x + \tan x| + C\end{aligned}$$ ■

Formulas (23)–(25) in the table describe the integrals of the products $\sin mx \sin nx$, $\cos mx \cos nx$, and $\sin mx \cos nx$. These integrals appear in the theory of Fourier Series, which is a fundamental technique used extensively in engineering and physics.

■ **EXAMPLE 8 Integral of $\sin mx \cos nx$** Evaluate $\int_0^\pi \sin 4x \cos 3x\, dx$.

Solution Apply reduction formula (24), with $m = 4$ and $n = 3$:

$$\begin{aligned}\int_0^\pi \sin 4x \cos 3x\, dx &= \left(-\frac{\cos(4-3)x}{2(4-3)} - \frac{\cos(4+3)x}{2(4+3)}\right)\Bigg|_0^\pi\\ &= \left(-\frac{\cos x}{2} - \frac{\cos 7x}{14}\right)\Bigg|_0^\pi\\ &= \left(\frac{1}{2} + \frac{1}{14}\right) - \left(-\frac{1}{2} - \frac{1}{14}\right) = \frac{8}{7}\end{aligned}$$ ■

TABLE OF TRIGONOMETRIC INTEGRALS

$$\int \sin^2 x\,dx = \frac{x}{2} - \frac{\sin 2x}{4} + C = \frac{x}{2} - \frac{1}{2}\sin x\cos x + C \tag{9}$$

$$\int \cos^2 x\,dx = \frac{x}{2} + \frac{\sin 2x}{4} + C = \frac{x}{2} + \frac{1}{2}\sin x\cos x + C \tag{10}$$

$$\int \sin^n x\,dx = -\frac{\sin^{n-1} x\cos x}{n} + \frac{n-1}{n}\int \sin^{n-2} x\,dx \tag{11}$$

$$\int \cos^n x\,dx = \frac{\cos^{n-1} x\sin x}{n} + \frac{n-1}{n}\int \cos^{n-2} x\,dx \tag{12}$$

$$\int \sin^m x\cos^n x\,dx = \frac{\sin^{m+1} x\cos^{n-1} x}{m+n} + \frac{n-1}{m+n}\int \sin^m x\cos^{n-2} x\,dx \tag{13}$$

$$\int \sin^m x\cos^n x\,dx = -\frac{\sin^{m-1} x\cos^{n+1} x}{m+n} + \frac{m-1}{m+n}\int \sin^{m-2} x\cos^n x\,dx \tag{14}$$

$$\int \tan x\,dx = \ln|\sec x| + C = -\ln|\cos x| + C \tag{15}$$

$$\int \tan^m x\,dx = \frac{\tan^{m-1} x}{m-1} - \int \tan^{m-2} x\,dx \tag{16}$$

$$\int \cot x\,dx = -\ln|\csc x| + C = \ln|\sin x| + C \tag{17}$$

$$\int \cot^m x\,dx = -\frac{\cot^{m-1} x}{m-1} - \int \cot^{m-2} x\,dx \tag{18}$$

$$\int \sec x\,dx = \ln\left|\sec x + \tan x\right| + C \tag{19}$$

$$\int \sec^m x\,dx = \frac{\tan x\sec^{m-2} x}{m-1} + \frac{m-2}{m-1}\int \sec^{m-2} x\,dx \tag{20}$$

$$\int \csc x\,dx = \ln\left|\csc x - \cot x\right| + C \tag{21}$$

$$\int \csc^m x\,dx = -\frac{\cot x\csc^{m-2} x}{m-1} + \frac{m-2}{m-1}\int \csc^{m-2} x\,dx \tag{22}$$

$$\int \sin mx\sin nx\,dx = \frac{\sin(m-n)x}{2(m-n)} - \frac{\sin(m+n)x}{2(m+n)} + C \quad (m \neq \pm n) \tag{23}$$

$$\int \sin mx\cos nx\,dx = -\frac{\cos(m-n)x}{2(m-n)} - \frac{\cos(m+n)x}{2(m+n)} + C \quad (m \neq \pm n) \tag{24}$$

$$\int \cos mx\cos nx\,dx = \frac{\sin(m-n)x}{2(m-n)} + \frac{\sin(m+n)x}{2(m+n)} + C \quad (m \neq \pm n) \tag{25}$$

8.2 SUMMARY

- To integrate an odd power of $\sin x$ times $\cos^n x$, write

$$\int \sin^{2k+1} x\cos^n x\,dx = \int (1-\cos^2 x)^k \cos^n x\sin x\,dx$$

Then use the substitution $u = \cos x$, $du = -\sin x\,dx$.

- To integrate an odd power of $\cos x$ times $\sin^m x$, write

$$\int \sin^m x \cos^{2k+1} x\,dx = \int (\sin^m x)(1 - \sin^2 x)^k \cos x\,dx$$

Then use the substitution $u = \sin x$, $du = \cos x\,dx$.
- If both $\sin x$ and $\cos x$ occur to an even power, write

$$\int \sin^m x \cos^n x\,dx = \int (1 - \cos^2 x)^{m/2} \cos^n x\,dx \quad (\text{if } m \le n)$$

$$\int \sin^m x \cos^n x\,dx = \int \sin^m x(1 - \sin^2 x)^{n/2}\,dx \quad (\text{if } m \ge n)$$

Expand the right-hand side to obtain a sum of powers of $\cos x$ or powers of $\sin x$. Then use the reduction formulas

$$\int \sin^n x\,dx = -\frac{1}{n}\sin^{n-1} x \cos x + \frac{n-1}{n}\int \sin^{n-2} x\,dx$$

$$\int \cos^n x\,dx = \frac{1}{n}\cos^{n-1} x \sin x + \frac{n-1}{n}\int \cos^{n-2} x\,dx$$

- The integral $\int \tan^m x \sec^n x\,dx$ can be evaluated by substitution. See the marginal note on page 422.

8.2 EXERCISES

Preliminary Questions

1. Describe the technique used to evaluate $\int \sin^5 x\,dx$.

2. Describe a way of evaluating $\int \sin^6 x\,dx$.

3. Are reduction formulas needed to evaluate $\int \sin^7 x \cos^2 x\,dx$? Why or why not?

4. Describe a way of evaluating $\int \sin^6 x \cos^2 x\,dx$.

5. Which integral requires more work to evaluate?

$$\int \sin^{798} x \cos x\,dx \quad \text{or} \quad \int \sin^4 x \cos^4 x\,dx$$

Explain your answer.

Exercises

In Exercises 1–6, use the method for odd powers to evaluate the integral.

1. $\int \cos^3 x\,dx$

2. $\int \sin^5 x\,dx$

3. $\int \sin^3 \theta \cos^2 \theta\,d\theta$

4. $\int \sin^5 x \cos x\,dx$

5. $\int \sin^3 t \cos^3 t\,dt$

6. $\int \sin^2 x \cos^5 x\,dx$

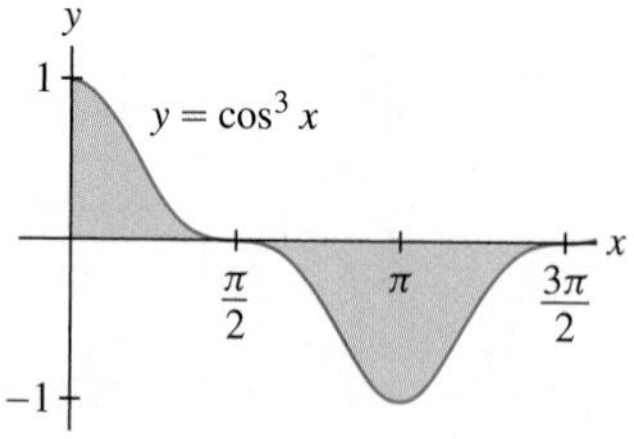

FIGURE 1 Graph of $y = \cos^3 x$.

7. Find the area of the shaded region in Figure 1.

8. Use the identity $\sin^2 x = 1 - \cos^2 x$ to write $\int \sin^2 x \cos^2 x\,dx$ as a sum of two integrals, and then evaluate using the reduction formula.

In Exercises 9–12, evaluate the integral using methods employed in Examples 3 and 4.

9. $\int \cos^4 y\,dy$

10. $\int \cos^2 \theta \sin^2 \theta\,d\theta$

11. $\int \sin^4 x \cos^2 x\,dx$

12. $\int \sin^2 x \cos^6 x\,dx$

In Exercises 13 and 14, evaluate using Eq. (13).

13. $\int \sin^3 x \cos^2 x\,dx$

14. $\int \sin^2 x \cos^4 x\,dx$

In Exercises 15–18, evaluate the integral using the method described on page 422 and the reduction formulas on page 423 as necessary.

15. $\int \tan^3 x \sec x\,dx$

16. $\int \tan^2 x \sec x\,dx$

17. $\int \tan^2 x \sec^4 x\,dx$

18. $\int \tan^8 x \sec^2 x\,dx$

In Exercises 19–22, evaluate using methods similar to those that apply to integral $\tan^m x \sec^n$.

19. $\int \cot^3 x\,dx$

20. $\int \sec^3 x\,dx$

21. $\int \cot^5 x \csc^2 x\,dx$

22. $\int \cot^4 x \csc x\,dx$

In Exercises 23–46, evaluate the integral.

23. $\int \cos^5 x \sin x\,dx$

24. $\int \cos^3(2-x)\sin(2-x)\,dx$

25. $\int \cos^4(3x+2)\,dx$

26. $\int \cos^7 3x\,dx$

27. $\int \cos^3(\pi\theta)\sin^4(\pi\theta)\,d\theta$

28. $\int \cos^{498} y \sin^3 y\,dy$

29. $\int \sin^4(3x)\,dx$

30. $\int \sin^2 x \cos^6 x\,dx$

31. $\int \csc^2(3-2x)\,dx$

32. $\int \csc^3 x\,dx$

33. $\int \tan x \sec^2 x\,dx$

34. $\int \tan^3\theta \sec^3\theta\,d\theta$

35. $\int \tan^5 x \sec^4 x\,dx$

36. $\int \tan^4 x \sec x\,dx$

37. $\int \tan^6 x \sec^4 x\,dx$

38. $\int \tan^2 x \sec^3 x\,dx$

39. $\int \cot^5 x \csc^5 x\,dx$

40. $\int \cot^2 x \csc^4 x\,dx$

41. $\int \sin 2x \cos 2x\,dx$

42. $\int \cos 4x \cos 6x\,dx$

43. $\int t\cos^3(t^2)\,dt$

44. $\int \frac{\tan^3(\ln t)}{t}\,dt$

45. $\int \cos^2(\sin t)\cos t\,dt$

46. $\int e^x \tan^2(e^x)\,dx$

In Exercises 47–60, evaluate the definite integral.

47. $\int_0^{2\pi} \sin^2 x\,dx$

48. $\int_0^{\pi/2} \cos^3 x\,dx$

49. $\int_0^{\pi/2} \sin^5 x\,dx$

50. $\int_0^{\pi/2} \sin^2 x \cos^3 x\,dx$

51. $\int_0^{\pi/4} \frac{dx}{\cos x}$

52. $\int_{\pi/4}^{\pi/2} \frac{dx}{\sin x}$

53. $\int_0^{\pi/3} \tan x\,dx$

54. $\int_0^{\pi/4} \tan^5 x\,dx$

55. $\int_{-\pi/4}^{\pi/4} \sec^4 x\,dx$

56. $\int_{\pi/4}^{3\pi/2} \cot^4 x \csc^2 x\,dx$

57. $\int_0^{\pi} \sin 3x \cos 4x\,dx$

58. $\int_0^{\pi} \sin x \sin 3x\,dx$

59. $\int_0^{\pi/6} \sin 2x \cos 4x\,dx$

60. $\int_0^{\pi/4} \sin 7x \cos 2x\,dx$

61. Use the identities for $\sin 2x$ and $\cos 2x$ on page 420 to verify that the following formulas are equivalent.

$$\int \sin^4 x\,dx = \frac{1}{32}(12x - 8\sin 2x + \sin 4x) + C$$

$$\int \sin^4 x\,dx = -\frac{1}{4}\sin^3 x\cos x - \frac{3}{8}\sin x\cos x + \frac{3}{8}x + C$$

62. Evaluate $\int \sin^2 x \cos^3 x\,dx$ using the method described in the text and verify that your result is equivalent to the following result produced by a computer algebra system.

$$\int \sin^2 x\cos^3 x\,dx = \frac{1}{30}(7 + 3\cos 2x)\sin^3 x + C$$

63. Find the volume of the solid obtained by revolving $y = \sin x$ for $0 \le x \le \pi$ about the x-axis.

64. Use Integration by Parts to prove Eqs. (1) and (2).

In Exercises 65–68, use the following alternative method for evaluating the integral $J = \int \sin^m x \cos^n x\,dx$ *when m and n are both even. Use the identities*

$$\sin^2 x = \frac{1}{2}(1 - \cos 2x), \qquad \cos^2 x = \frac{1}{2}(1 + \cos 2x)$$

to write $J = \frac{1}{4}\int (1-\cos 2x)^{m/2}(1+\cos 2x)^{n/2}\,dx$*, and expand the right-hand side as a sum of integrals involving smaller powers of sine and cosine in the variable* $2x$.

65. $\int \sin^2 x \cos^2 x\,dx$

66. $\int \cos^4 x\,dx$

67. $\int \sin^4 x \cos^2 x\,dx$

68. $\int \sin^6 x\,dx$

69. Prove the reduction formula

$$\int \tan^k x\,dx = \frac{\tan^{k-1} x}{k-1} - \int \tan^{k-2} x\,dx$$

Hint: $\tan^k x = (\sec^2 x - 1)\tan^{k-2} x$.

70. Use the substitution $u = \csc x - \cot x$ to evaluate $\int \csc x\,dx$ (see Example 5).

71. Let $I_m = \int_0^{\pi/2} \sin^m x\,dx$.

(a) Show that $I_0 = \frac{\pi}{2}$ and $I_1 = 1$.

(b) Prove that, for $m \ge 2$,

$$I_m = \frac{m-1}{m} I_{m-2}$$

(c) Use (a) and (b) to compute I_m for $m = 2, 3, 4, 5$.

72. Evaluate $\int_0^{\pi} \sin^2 mx\,dx$ for m an arbitrary integer.

73. Evaluate $\int \sin x \ln(\sin x)\,dx$. *Hint:* Use Integration by Parts as a first step.

74. Total Energy A 100-W light bulb has resistance $R = 144\ \Omega$ (ohms) when attached to household current, where the voltage varies as $V = V_0 \sin(2\pi f t)$ ($V_0 = 110$ V, $f = 60$ Hz). The energy (in joules) expended by the bulb over a period of T seconds is

$$U = \int_0^T P(t)\,dt$$

where $P = V^2/R$ (J/s) is the power. Compute U if the bulb remains on for 5 hours.

75. Let m, n be integers with $m \neq \pm n$. Use Eqs. (23)–(25) to prove the so-called **orthogonality relations** that play a basic role in the theory of Fourier Series (Figure 2):

$$\int_0^\pi \sin mx \sin nx\,dx = 0$$

$$\int_0^\pi \cos mx \cos nx\,dx = 0$$

$$\int_0^{2\pi} \sin mx \cos nx\,dx = 0$$

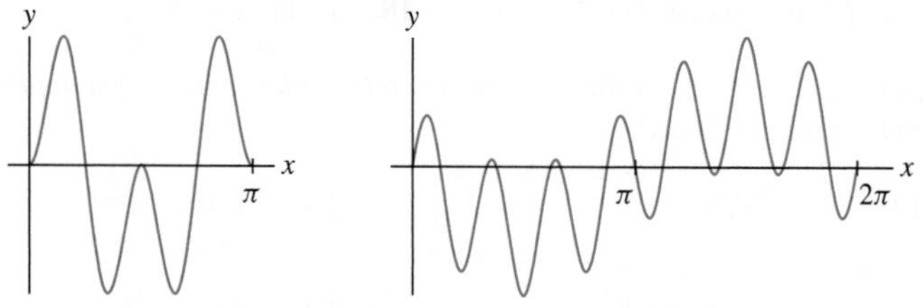

FIGURE 2 The integrals are zero by the orthogonality relations.

Further Insights and Challenges

76. Use the trigonometric identity

$$\sin mx \cos nx = \frac{1}{2}\big(\sin(m-n)x + \sin(m+n)x\big)$$

to prove Eq. (24) in the table of integrals on page 423.

77. Use Integration by Parts to prove that (for $m \neq 1$)

$$\int \sec^m x\,dx = \frac{\tan x \sec^{m-2} x}{m-1} + \frac{m-2}{m-1}\int \sec^{m-2} x\,dx$$

78. Set $I_m = \int_0^{\pi/2} \sin^m x\,dx$. Use Exercise 71 to prove that

$$I_{2m} = \frac{2m-1}{2m}\,\frac{2m-3}{2m-2}\cdots\frac{1}{2}\cdot\frac{\pi}{2}$$

$$I_{2m+1} = \frac{2m}{2m+1}\,\frac{2m-2}{2m-1}\cdots\frac{2}{3}$$

Conclude that

$$\frac{\pi}{2} = \frac{2\cdot 2}{1\cdot 3}\cdot\frac{4\cdot 4}{3\cdot 5}\cdots\frac{2m\cdot 2m}{(2m-1)(2m+1)}\,\frac{I_{2m}}{I_{2m+1}}$$

79. This is a continuation of Exercise 78.

(a) Prove that $I_{2m+1} \le I_{2m} \le I_{2m-1}$. *Hint:*

$$\sin^{2m+1} x \le \sin^{2m} x \le \sin^{2m-1} x \quad \text{for} \quad 0 \le x \le \tfrac{\pi}{2}$$

(b) Show that $\dfrac{I_{2m-1}}{I_{2m+1}} = 1 + \dfrac{1}{2m}$.

(c) Show that $1 \le \dfrac{I_{2m}}{I_{2m+1}} \le 1 + \dfrac{1}{2m}$.

(d) Prove that $\displaystyle\lim_{m\to\infty} \frac{I_{2m}}{I_{2m+1}} = 1$.

(e) Finally, deduce the infinite product for $\frac{\pi}{2}$ discovered by English mathematician John Wallis (1616–1703):

$$\frac{\pi}{2} = \lim_{m\to\infty} \frac{2}{1}\cdot\frac{2}{3}\cdot\frac{4}{3}\cdot\frac{4}{5}\cdots\frac{2m\cdot 2m}{(2m-1)(2m+1)}$$

8.3 Trigonometric Substitution

Our next goal is to integrate functions involving one of the square root expressions:

$$\sqrt{a^2 - x^2}, \qquad \sqrt{x^2 + a^2}, \qquad \sqrt{x^2 - a^2}$$

In each case, a substitution transforms the integral into a trigonometric integral.

■ **EXAMPLE 1** Evaluate $\displaystyle\int \sqrt{1 - x^2}\,dx$.

Solution

Step 1. **Substitute to eliminate the square root.**

The integrand is defined for $-1 \le x \le 1$, so we may set $x = \sin\theta$, where $-\frac{\pi}{2} \le \theta \le \frac{\pi}{2}$.

Because $\cos\theta \geq 0$ for such θ, we obtain the positive square root

$$\sqrt{1-x^2} = \sqrt{1-\sin^2\theta} = \sqrt{\cos^2\theta} = \cos\theta \qquad \boxed{1}$$

***Step 2.* Evaluate the trigonometric integral.**

Since $x = \sin\theta$, we have $dx = \cos\theta\,d\theta$, and $\sqrt{1-x^2}\,dx = \cos\theta(\cos\theta\,d\theta)$. Thus

$$\int \sqrt{1-x^2}\,dx = \int \cos^2\theta\,d\theta = \frac{1}{2}\theta + \frac{1}{2}\sin\theta\cos\theta + C$$

REMINDER

$$\int \cos^2\theta\,d\theta = \frac{1}{2}\theta + \frac{1}{2}\sin\theta\cos\theta + C$$

***Step 3.* Convert back to the original variable.**

It remains to express the answer in terms of x:

$$x = \sin\theta, \qquad \theta = \sin^{-1}x, \qquad \sqrt{1-x^2} = \cos\theta$$

$$\int \sqrt{1-x^2}\,dx = \frac{1}{2}\theta + \frac{1}{2}\sin\theta\cos\theta + C = \frac{1}{2}\sin^{-1}x + \frac{1}{2}x\sqrt{1-x^2} + C \qquad ■$$

Note: If $x = a\sin\theta$ and $a > 0$, then

$$a^2 - x^2 = a^2(1-\sin^2\theta) = a^2\cos^2\theta$$

For $-\frac{\pi}{2} \leq \theta \leq \frac{\pi}{2}$, $\cos\theta \geq 0$ and thus

$$\sqrt{a^2-x^2} = a\cos\theta$$

Integrals Involving $\sqrt{a^2-x^2}$ If $\sqrt{a^2-x^2}$ occurs in an integral where $a > 0$, try the substitution

$$x = a\sin\theta, \qquad dx = a\cos\theta\,d\theta, \qquad \sqrt{a^2-x^2} = a\cos\theta$$

The next example shows that trigonometric substitution can be used with integrands involving $(a^2-x^2)^{n/2}$, where n is any integer.

■ **EXAMPLE 2** **Integrand Involving $(a^2-x^2)^{3/2}$** Evaluate $\displaystyle\int \frac{x^2}{(4-x^2)^{3/2}}\,dx$.

Solution

***Step 1.* Substitute to eliminate the square root.**

In this case, $a = 2$ since $\sqrt{4-x^2} = \sqrt{2^2-x^2}$. Therefore, we use

$$x = 2\sin\theta, \qquad dx = 2\cos\theta\,d\theta, \qquad \sqrt{4-x^2} = 2\cos\theta$$

$$\int \frac{x^2}{(4-x^2)^{3/2}}\,dx = \int \frac{4\sin^2\theta}{2^3\cos^3\theta}\,2\cos\theta\,d\theta = \int \frac{\sin^2\theta}{\cos^2\theta}\,d\theta = \int \tan^2\theta\,d\theta$$

***Step 2.* Evaluate the trigonometric integral.**

Use the reduction formula in the marginal note with $m = 2$:

REMINDER

$$\int \tan^m x\,dx = \frac{\tan^{m-1}x}{m-1} - \int \tan^{m-2}x\,dx$$

$$\int \tan^2\theta\,d\theta = \tan\theta - \int d\theta = \tan\theta - \theta + C$$

We can also evaluate the integral using the identity $\tan^2\theta = \sec^2\theta - 1$.

***Step 3.* Convert back to the original variable.**

We must write $\tan\theta$ and θ in terms of x. By definition, $x = 2\sin\theta$, so

$$\sin\theta = \frac{x}{2}, \qquad \theta = \sin^{-1}\frac{x}{2}$$

To express $\tan\theta$ in terms of x, we use the right triangle in Figure 1. The angle θ satisfies $\sin\theta = \dfrac{x}{2}$ and

$$\tan\theta = \frac{\text{opposite}}{\text{adjacent}} = \frac{x}{\sqrt{4-x^2}}$$

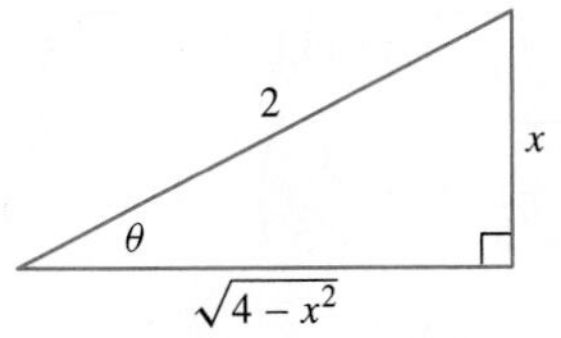

FIGURE 1 Right triangle with $\sin\theta = \frac{x}{2}$.

Thus we have

$$\int \frac{x^2}{(4 - x^2)^{3/2}}\,dx = \tan\theta - \theta + C = \frac{x}{\sqrt{4 - x^2}} - \sin^{-1}\frac{x}{2} + C$$ ■

When the integrand involves $\sqrt{x^2 + a^2}$, try the substitution $x = a\tan\theta$. Then

$$x^2 + a^2 = a^2\tan^2\theta + a^2 = a^2(1 + \tan^2\theta) = a^2\sec^2\theta$$

and thus $\sqrt{x^2 + a^2} = a\sec\theta$.

In the substitution $x = a\tan\theta$, we choose $-\frac{\pi}{2} < \theta < \frac{\pi}{2}$. Therefore, $a\sec\theta$ is the positive square root $\sqrt{x^2 + a^2}$.

Integrals Involving $\sqrt{x^2 + a^2}$ If $\sqrt{x^2 + a^2}$ occurs in an integral where $a > 0$, try the substitution

$$x = a\tan\theta, \qquad dx = a\sec^2\theta\,d\theta, \qquad \sqrt{x^2 + a^2} = a\sec\theta$$

■ **EXAMPLE 3** Evaluate $\int \sqrt{4x^2 + 20}\,dx$.

Solution First factor out a constant:

$$\int \sqrt{4x^2 + 20}\,dx = \int \sqrt{4(x^2 + 5)}\,dx = 2\int \sqrt{x^2 + 5}\,dx$$

Thus we have the form $\sqrt{x^2 + a^2}$ with $a = \sqrt{5}$.

***Step 1.* Substitute to eliminate the square root.**

$$x = \sqrt{5}\tan\theta, \qquad dx = \sqrt{5}\sec^2\theta\,d\theta, \qquad \sqrt{x^2 + 5} = \sqrt{5}\sec\theta$$

$$2\int \sqrt{x^2 + 5}\,dx = 2\int (\sqrt{5}\sec\theta)\sqrt{5}\sec^2\theta\,d\theta = 10\int \sec^3\theta\,d\theta$$

***Step 2.* Evaluate the trigonometric integral.**

Apply the reduction formula recalled in the margin with $m = 3$:

REMINDER

$$\int \sec^m x\,dx = \frac{\tan x\sec^{m-2}x}{m-1} + \frac{m-2}{m-1}\int \sec^{m-2}x\,dx$$

$$\int \sqrt{4x^2 + 20}\,dx = 10\int \sec^3\theta\,d\theta = 10\frac{\tan\theta\sec\theta}{2} + 10\left(\frac{1}{2}\right)\int \sec\theta\,dx$$

$$= 5\tan\theta\sec\theta + 5\ln(\sec\theta + \tan\theta) + C$$

Note: It is not necessary to write $\ln|\sec\theta + \tan\theta|$ with the absolute value because our substitution $x = \sqrt{5}\tan\theta$ assumes that $-\frac{\pi}{2} < \theta < \frac{\pi}{2}$, where $\sec\theta + \tan\theta > 0$.

***Step 3.* Convert back to the original variable.**

Since $x = \sqrt{5}\tan\theta$, we use the right triangle in Figure 2.

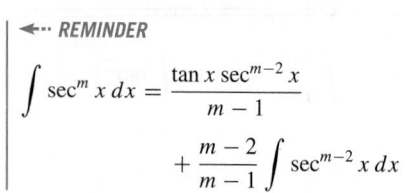

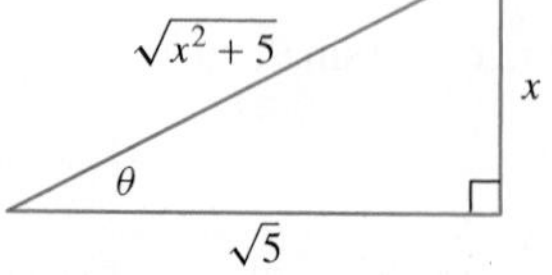

FIGURE 2

$$\tan\theta = \frac{\text{opposite}}{\text{adjacent}} = \frac{x}{\sqrt{5}}, \qquad \sec\theta = \frac{\text{hypotenuse}}{\text{adjacent}} = \frac{\sqrt{x^2 + 5}}{\sqrt{5}}$$

$$\int \sqrt{4x^2 + 20}\,dx = 5\frac{x}{\sqrt{5}}\frac{\sqrt{x^2 + 5}}{\sqrt{5}} + 5\ln\left(\frac{\sqrt{x^2 + 5}}{\sqrt{5}} + \frac{x}{\sqrt{5}}\right) + C$$

$$= x\sqrt{x^2 + 5} + 5\ln\left(\frac{\sqrt{x^2 + 5} + x}{\sqrt{5}}\right) + C$$

The logarithmic term can be rewritten as

$$5\ln\left(\frac{\sqrt{x^2+5}+x}{\sqrt{5}}\right)+C = 5\ln\left(\sqrt{x^2+5}+x\right) - \underbrace{5\ln\sqrt{5}+C}_{\text{Constant}}$$

Since the constant C is arbitrary, we can absorb $-5\ln\sqrt{5}$ into C and write

$$\int \sqrt{4x^2+20}\,dx = x\sqrt{x^2+5} + 5\ln\left(\sqrt{x^2+5}+x\right) + C$$ ■

Our last trigonometric substitution $x = a\sec\theta$ transforms $\sqrt{x^2-a^2}$ into $a\tan\theta$ because

$$x^2 - a^2 = a^2\sec^2\theta - a^2 = a^2(\sec^2\theta - 1) = a^2\tan^2\theta$$

In the substitution $x = a\sec\theta$, we choose $0 \le \theta < \frac{\pi}{2}$ if $x \ge a$ and $\pi \le \theta < \frac{3\pi}{2}$ if $x \le -a$. With these choices, $a\tan\theta$ is the positive square root $\sqrt{x^2-a^2}$.

Integrals Involving $\sqrt{x^2-a^2}$ If $\sqrt{x^2-a^2}$ occurs in an integral where $a > 0$, try the substitution

$$x = a\sec\theta, \qquad dx = a\sec\theta\tan\theta\,d\theta, \qquad \sqrt{x^2-a^2} = a\tan\theta$$

■ **EXAMPLE 4** Evaluate $\displaystyle\int \frac{dx}{x^2\sqrt{x^2-9}}$.

Solution In this case, make the substitution

$$x = 3\sec\theta, \qquad dx = 3\sec\theta\tan\theta\,d\theta, \qquad \sqrt{x^2-9} = 3\tan\theta$$

$$\int \frac{dx}{x^2\sqrt{x^2-9}} = \int \frac{3\sec\theta\tan\theta\,d\theta}{(9\sec^2\theta)(3\tan\theta)} = \frac{1}{9}\int \cos\theta\,d\theta = \frac{1}{9}\sin\theta + C$$

Since $x = 3\sec\theta$, we use the right triangle in Figure 3:

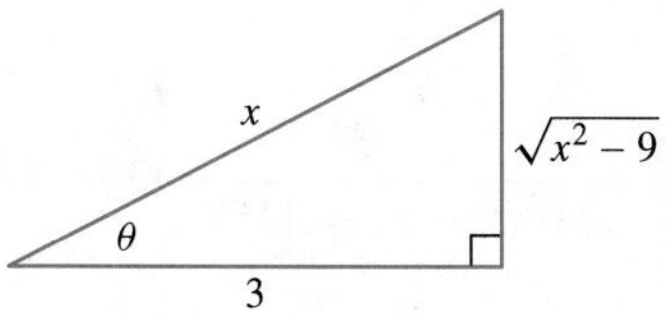

FIGURE 3

$$\sec\theta = \frac{\text{hypotenuse}}{\text{adjacent}} = \frac{x}{3}, \qquad \sin\theta = \frac{\text{opposite}}{\text{hypotenuse}} = \frac{\sqrt{x^2-9}}{x}$$

Therefore,

$$\int \frac{dx}{x^2\sqrt{x^2-9}} = \frac{1}{9}\sin\theta + C = \frac{\sqrt{x^2-9}}{9x} + C$$ ■

So far we have dealt with the expressions $\sqrt{x^2 \pm a^2}$ and $\sqrt{a^2-x^2}$. By completing the square (Section 1.2), we can treat the more general form $\sqrt{ax^2+bx+c}$.

■ **EXAMPLE 5 Completing the Square** Evaluate $\displaystyle\int \frac{dx}{(x^2-6x+11)^2}$.

Solution

***Step 1.* Complete the square.**

$$x^2 - 6x + 11 = (x^2 - 6x + 9) + 2 = \underbrace{(x-3)^2}_{u^2} + 2$$

Step 2. **Use substitution.**
Let $u = x - 3$, $du = dx$:

$$\int \frac{dx}{(x^2 - 6x + 11)^2} = \int \frac{du}{(u^2+2)^2} \qquad \boxed{2}$$

Step 3. **Trigonometric substitution.**
Evaluate the u-integral using trigonometric substitution:

REMINDER

$$\int \cos^2\theta\, d\theta = \frac{\theta}{2} + \frac{\sin\theta\cos\theta}{2} + C$$

$$u = \sqrt{2}\tan\theta, \qquad \sqrt{u^2+2} = \sqrt{2}\sec\theta, \qquad du = \sqrt{2}\sec^2\theta\, d\theta$$

$$\int \frac{du}{(u^2+2)^2} = \int \frac{\sqrt{2}\sec^2\theta\, d\theta}{4\sec^4\theta} = \frac{1}{2\sqrt{2}}\int \cos^2\theta\, d\theta$$

$$= \frac{1}{2\sqrt{2}}\left(\frac{\theta}{2} + \frac{\sin\theta\cos\theta}{2}\right) + C \qquad \boxed{3}$$

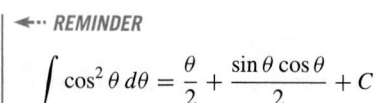

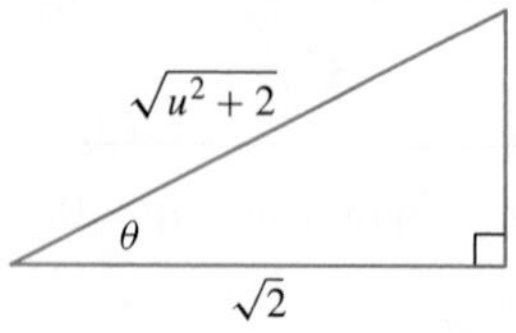

FIGURE 4

Since $\theta = \tan^{-1}\dfrac{u}{\sqrt{2}}$, we use the right triangle in Figure 4 to obtain

$$\sin\theta\cos\theta = \left(\frac{\text{opposite}}{\text{hypotenuse}}\right)\left(\frac{\text{adjacent}}{\text{hypotenuse}}\right) = \frac{u}{\sqrt{u^2+2}} \cdot \frac{\sqrt{2}}{\sqrt{u^2+2}} = \frac{\sqrt{2}u}{u^2+2}$$

Thus, Eq. (3) becomes

$$\int \frac{du}{(u^2+2)^2} = \frac{1}{4\sqrt{2}}\left(\tan^{-1}\frac{u}{\sqrt{2}} + \frac{\sqrt{2}u}{u^2+2}\right) + C$$

$$= \frac{1}{4\sqrt{2}}\tan^{-1}\frac{u}{\sqrt{2}} + \frac{u}{4(u^2+2)} + C \qquad \boxed{4}$$

Step 4. **Convert to the original variable.**
Since $u = x - 3$ and $u^2 + 2 = x^2 - 6x + 11$, Eq. (4) becomes

$$\int \frac{du}{(u^2+2)^2} = \frac{1}{4\sqrt{2}}\tan^{-1}\frac{x-3}{\sqrt{2}} + \frac{x-3}{4(x^2-6x+11)} + C$$

This is our final answer by Eq. (2):

$$\int \frac{dx}{(x^2-6x+11)^2} = \frac{1}{4\sqrt{2}}\tan^{-1}\frac{x-3}{\sqrt{2}} + \frac{x-3}{4(x^2-6x+11)} + C \qquad \blacksquare$$

8.3 SUMMARY

- Trigonometric substitution:

Square root form in integrand	Trigonometric substitution
$\sqrt{a^2 - x^2}$	$x = a\sin\theta,\quad dx = a\cos\theta\, d\theta,\quad \sqrt{a^2-x^2} = a\cos\theta$
$\sqrt{x^2 + a^2}$	$x = a\tan\theta,\quad dx = a\sec^2\theta\, d\theta,\quad \sqrt{x^2+a^2} = a\sec\theta$
$\sqrt{x^2 - a^2}$	$x = a\sec\theta,\quad dx = a\sec\theta\tan\theta\, d\theta,\quad \sqrt{x^2-a^2} = a\tan\theta$

Step 1. Substitute to eliminate the square root.
Step 2. Evaluate the trigonometric integral.
Step 3. Convert back to the original variable.

• The three trigonometric substitutions correspond to three right triangles (Figure 5) that we use to express the trigonometric functions of θ in terms of x.
• Integrands involving $\sqrt{x^2+bx+c}$ are treated by completing the square (see Example 5).

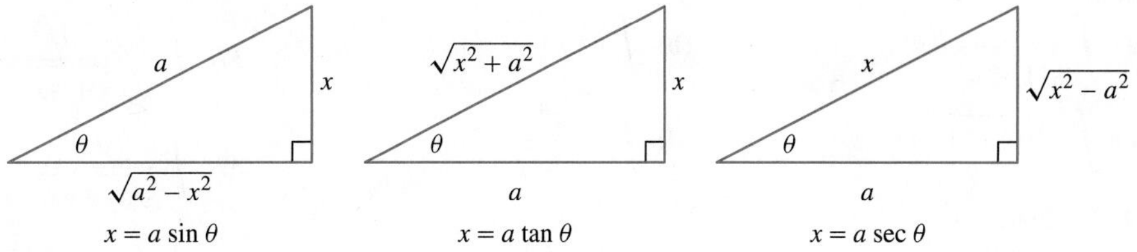

FIGURE 5

8.3 EXERCISES

Preliminary Questions

1. State the trigonometric substitution appropriate to the given integral:

(a) $\int \sqrt{9-x^2}\,dx$ **(b)** $\int x^2(x^2-16)^{3/2}\,dx$

(c) $\int x^2(x^2+16)^{3/2}\,dx$ **(d)** $\int (x^2-5)^{-2}\,dx$

2. Is trigonometric substitution needed to evaluate $\int x\sqrt{9-x^2}\,dx$?

3. Express $\sin 2\theta$ in terms of $x = \sin\theta$.

4. Draw a triangle that would be used together with the substitution $x = 3\sec\theta$.

Exercises

In Exercises 1–4, evaluate the integral by following the steps given.

1. $I = \int \frac{dx}{\sqrt{9-x^2}}$

(a) Show that the substitution $x = 3\sin\theta$ transforms I into $\int d\theta$, and evaluate I in terms of θ.
(b) Evaluate I in terms of x.

2. $I = \int \frac{dx}{x^2\sqrt{x^2-2}}$

(a) Show that the substitution $x = \sqrt{2}\sec\theta$ transforms the integral I into $\frac{1}{2}\int \cos\theta\,d\theta$, and evaluate I in terms of θ.
(b) Use a right triangle to show that with the above substitution, $\sin\theta = \sqrt{x^2-2}/x$.
(c) Evaluate I in terms of x.

3. $I = \int \frac{dx}{\sqrt{4x^2+9}}$

(a) Show that the substitution $x = \frac{3}{2}\tan\theta$ transforms I into $\frac{1}{2}\int \sec\theta\,d\theta$.
(b) Evaluate I in terms of θ (refer to the table of integrals on page 423 in Section 8.2 if necessary).
(c) Express I in terms of x.

4. $I = \int \frac{dx}{(x^2+4)^2}$

(a) Show that the substitution $x = 2\tan\theta$ transforms the integral I into $\frac{1}{8}\int \cos^2\theta\,d\theta$.
(b) Use the formula $\int \cos^2\theta\,d\theta = \frac{1}{2}\theta + \frac{1}{2}\sin\theta\cos\theta$ to evaluate I in terms of θ.
(c) Show that $\sin\theta = \frac{x}{\sqrt{x^2+4}}$ and $\cos\theta = \frac{2}{\sqrt{x^2+4}}$.
(d) Express I in terms of x.

In Exercises 5–10, use the indicated substitution to evaluate the integral.

5. $\int \sqrt{16-5x^2}\,dx, \quad x = \frac{4}{\sqrt{5}}\sin\theta$

6. $\int_0^{1/2} \frac{x^2}{\sqrt{1-x^2}}\,dx, \quad x = \sin\theta$

7. $\int \frac{dx}{x\sqrt{x^2-9}}, \quad x = 3\sec\theta$

8. $\int_{1/2}^{1} \frac{dx}{x^2\sqrt{x^2+4}}, \quad x = 2\tan\theta$

9. $\int \frac{dx}{(x^2-4)^{3/2}}, \quad x = 2\sec\theta$

10. $\int_0^1 \frac{dx}{(4+9x^2)^2}, \quad x = \frac{2}{3}\tan\theta$

11. Evaluate $\int \frac{x\,dx}{\sqrt{x^2-4}}$ in two ways: using the direct substitution $u = x^2-4$ and by trigonometric substitution.

12. Is the substitution $u = x^2 - 4$ effective for evaluating the integral $\int \frac{x^2\,dx}{\sqrt{x^2-4}}$? If not, evaluate using trigonometric substitution.

13. Evaluate using the substitution $u = 1 - x^2$ or trigonometric substitution.

(a) $\int \frac{x}{\sqrt{1-x^2}}\,dx$ **(b)** $\int x^2\sqrt{1-x^2}\,dx$

(c) $\int x^3\sqrt{1-x^2}\,dx$ **(d)** $\int \frac{x^4}{\sqrt{1-x^2}}\,dx$

14. Evaluate:

(a) $\int \frac{dt}{(t^2+1)^{3/2}}$ **(b)** $\int \frac{t\,dt}{(t^2+1)^{3/2}}$

In Exercises 15–32, evaluate using trigonometric substitution. Refer to the table of trigonometric integrals as necessary.

15. $\int \frac{x^2\,dx}{\sqrt{9-x^2}}$ **16.** $\int \frac{dt}{(16-t^2)^{3/2}}$

17. $\int \frac{dx}{x\sqrt{x^2+16}}$ **18.** $\int \sqrt{12+4t^2}\,dt$

19. $\int \frac{dx}{\sqrt{x^2-9}}$ **20.** $\int \frac{dt}{t^2\sqrt{t^2-25}}$

21. $\int \frac{dy}{y^2\sqrt{5-y^2}}$ **22.** $\int x^3\sqrt{9-x^2}\,dx$

23. $\int \frac{dx}{\sqrt{25x^2+2}}$ **24.** $\int \frac{dt}{(9t^2+4)^2}$

25. $\int \frac{dz}{z^3\sqrt{z^2-4}}$ **26.** $\int \frac{dy}{\sqrt{y^2-9}}$

27. $\int \frac{x^2\,dx}{(6x^2-49)^{1/2}}$ **28.** $\int \frac{dx}{(x^2-4)^2}$

29. $\int \frac{dt}{(t^2+9)^2}$ **30.** $\int \frac{dx}{(x^2+1)^3}$

31. $\int \frac{x^2\,dx}{(x^2-1)^{3/2}}$ **32.** $\int \frac{x^2\,dx}{(x^2+1)^{3/2}}$

33. Prove for $a > 0$:

$$\int \frac{dx}{x^2+a} = \frac{1}{\sqrt{a}}\tan^{-1}\frac{x}{\sqrt{a}} + C$$

34. Prove for $a > 0$:

$$\int \frac{dx}{(x^2+a)^2} = \frac{1}{2a}\left(\frac{x}{x^2+a} + \frac{1}{\sqrt{a}}\tan^{-1}\frac{x}{\sqrt{a}}\right) + C$$

35. Let $I = \int \frac{dx}{\sqrt{x^2-4x+8}}$.

(a) Complete the square to show that $x^2 - 4x + 8 = (x-2)^2 + 4$.

(b) Use the substitution $u = x - 2$ to show that $I = \int \frac{du}{\sqrt{u^2+2^2}}$. Evaluate the u-integral.

(c) Show that $I = \ln\left|\sqrt{(x-2)^2+4} + x - 2\right| + C$.

36. Evaluate $\int \frac{dx}{\sqrt{12x-x^2}}$. First complete the square to write $12x - x^2 = 36 - (x-6)^2$.

In Exercises 37–42, evaluate the integral by completing the square and using trigonometric substitution.

37. $\int \frac{dx}{\sqrt{x^2+4x+13}}$ **38.** $\int \frac{dx}{\sqrt{2+x-x^2}}$

39. $\int \frac{dx}{\sqrt{x+6x^2}}$ **40.** $\int \sqrt{x^2-4x+7}\,dx$

41. $\int \sqrt{x^2-4x+3}\,dx$ **42.** $\int \frac{dx}{(x^2+6x+6)^2}$

In Exercises 43–52, indicate a good method for evaluating the integral (but do not evaluate). Your choices are: substitution (specify u and du), Integration by Parts (specify u and v′), a trigonometric method, or trigonometric substitution (specify). If it appears that these techniques are not sufficient, state this.

43. $\int \frac{x\,dx}{\sqrt{12-6x-x^2}}$ **44.** $\int \sqrt{4x^2-1}\,dx$

45. $\int \sin^3 x\cos^3 x\,dx$ **46.** $\int x\sec^2 x\,dx$

47. $\int \frac{dx}{\sqrt{9-x^2}}$ **48.** $\int \sqrt{1-x^3}\,dx$

49. $\int \sin^{3/2} x\,dx$ **50.** $\int x^2\sqrt{x+1}\,dx$

51. $\int \frac{dx}{(x+1)(x+2)^3}$ **52.** $\int \frac{dx}{(x+12)^4}$

In Exercises 53–56, evaluate using Integration by Parts as a first step.

53. $\int \sec^{-1} x\,dx$ **54.** $\int \frac{\sin^{-1} x}{x^2}\,dx$

55. $\int \ln(x^2+1)\,dx$ **56.** $\int x^2\ln(x^2+1)\,dx$

57. Find the average height of a point on the semicircle $y = \sqrt{1-x^2}$ for $-1 \le x \le 1$.

58. Find the volume of the solid obtained by revolving the graph of $y = x\sqrt{1-x^2}$ over $[0, 1]$ about the y-axis.

59. Find the volume of the solid obtained by revolving the region between the graph of $y^2 - x^2 = 1$ and the line $y = 2$ about the line $y = 2$.

60. Find the volume of revolution for the region in Exercise 59, but revolve around $y = 3$.

61. Compute $\int \frac{dx}{x^2-1}$ in two ways and verify that the answers agree: first via trigonometric substitution and then using the identity

$$\frac{1}{x^2-1} = \frac{1}{2}\left(\frac{1}{x-1} - \frac{1}{x+1}\right)$$

62. CAS You want to divide an 18-inch pizza equally among three friends using vertical slices at $\pm x$ as in Figure 6. Find an equation satisfied by x and find the approximate value of x using a computer algebra system.

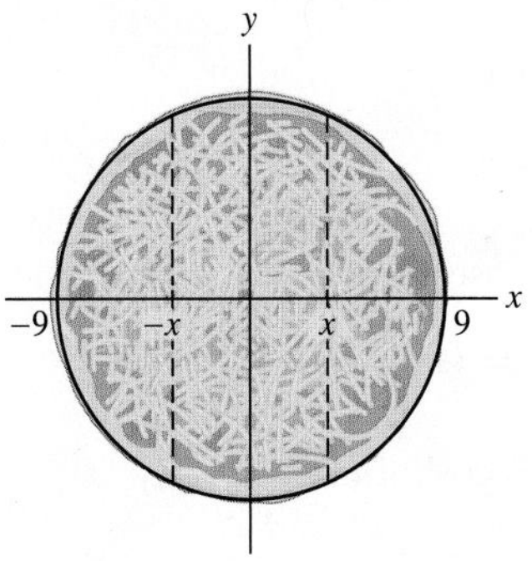

FIGURE 6 Dividing a pizza into three equal parts.

63. A charged wire creates an electric field at a point P located at a distance D from the wire (Figure 7). The component $E_\perp$ of the field perpendicular to the wire (in N/C) is

$$E_\perp = \int_{x_1}^{x_2} \frac{k\lambda D}{(x^2 + D^2)^{3/2}}\,dx$$

where λ is the charge density (coulombs per meter), $k = 8.99 \times 10^9$ N·m²/C² (Coulomb constant), and x_1, x_2 are as in the figure. Suppose that $\lambda = 6 \times 10^{-4}$ C/m, and $D = 3$ m. Find $E_\perp$ if (a) $x_1 = 0$ and $x_2 = 30$ m, and (b) $x_1 = -15$ m and $x_2 = 15$ m.

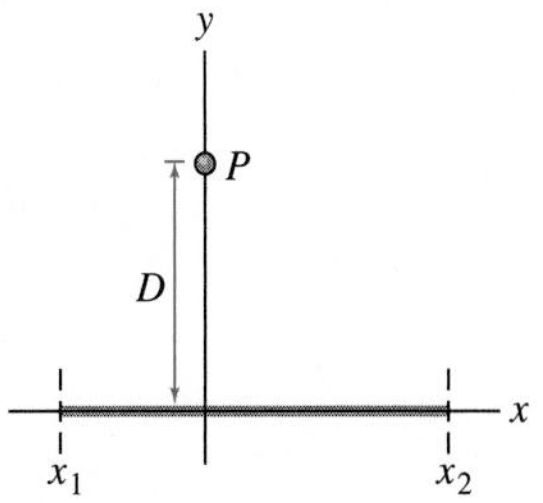

FIGURE 7

Further Insights and Challenges

64. Let $J_n = \displaystyle\int \frac{dx}{(x^2+1)^n}$. Use Integration by Parts to prove

$$J_{n+1} = \left(1 - \frac{1}{2n}\right) J_n + \left(\frac{1}{2n}\right) \frac{x}{(x^2+1)^n}$$

Then use this recursion relation to calculate J_2 and J_3.

65. Prove the formula

$$\int \sqrt{1 - x^2}\,dx = \frac{1}{2}\sin^{-1} x + \frac{1}{2} x\sqrt{1 - x^2} + C$$

using geometry by interpreting the integral as the area of part of the unit circle.

8.4 Integrals Involving Hyperbolic and Inverse Hyperbolic Functions

In Section 7.9, we noted the similarities between hyperbolic and trigonometric functions. We also saw that the formulas for their derivatives resemble each other, differing in at most a sign. Recall the following integral formulas.

← *REMINDER*

$$\sinh x = \frac{e^x - e^{-x}}{2} \qquad \cosh x = \frac{e^x + e^{-x}}{2}$$

$$\frac{d}{dx}\sinh x = \cosh x \qquad \frac{d}{dx}\cosh x = \sinh x$$

$$\frac{d}{dx}\tanh x = \operatorname{sech}^2 x$$

$$\frac{d}{dx}\coth x = -\operatorname{csch}^2 x$$

$$\frac{d}{dx}\operatorname{sech} x = -\operatorname{sech} x \tanh x$$

$$\frac{d}{dx}\operatorname{csch} x = -\operatorname{csch} x \coth x$$

Hyperbolic Integral Formulas

$$\int \sinh x\,dx = \cosh x + C, \qquad \int \cosh x\,dx = \sinh x + C$$

$$\int \operatorname{sech}^2 x\,dx = \tanh x + C, \qquad \int \operatorname{csch}^2 x\,dx = -\coth x + C$$

$$\int \operatorname{sech} x \tanh x\,dx = -\operatorname{sech} x + C, \qquad \int \operatorname{csch} x \coth x\,dx = -\operatorname{csch} x + C$$

■ **EXAMPLE 1** Calculate $\displaystyle\int x \operatorname{sech}^2(9 - x^2)\,dx$.

Solution Use the substitution $u = 9 - x^2$, $du = -2x\,dx$. Then $-\frac{1}{2}\,du = x\,dx$ and

$$\int x \operatorname{sech}^2(9 - x^2)\,dx = -\frac{1}{2}\int \operatorname{sech}^2 u\,du = -\frac{1}{2}\tanh u + C = -\frac{1}{2}\tanh(9 - x^2) + C$$

■

Hyperbolic Identities

$$\cosh^2 x - \sinh^2 x = 1,$$
$$\cosh^2 x = 1 + \sinh^2 x$$
$$\cosh^2 x = \tfrac{1}{2}(\cosh 2x + 1),$$
$$\sinh^2 x = \tfrac{1}{2}(\cosh 2x - 1)$$
$$\sinh 2x = 2\sinh x \cosh x,$$
$$\cosh 2x = \cosh^2 x + \sinh^2 x$$

The techniques for computing trigonometric integrals discussed in Section 8.2 apply with little change to hyperbolic integrals. In place of trigonometric identities, we use the corresponding hyperbolic identities (see margin).

■ **EXAMPLE 2 Powers of $\sinh x$ and $\cosh x$** Calculate:

(a) $\int \sinh^4 x \cosh^5 x\, dx$ **(b)** $\int \cosh^2 x\, dx$

Solution

(a) Since $\cosh x$ appears to an odd power, use $\cosh^2 x = 1 + \sinh^2 x$ to write

$$\cosh^5 x = \cosh^4 x \cdot \cosh x = (\sinh^2 x + 1)^2 \cosh x$$

Then use the substitution $u = \sinh x$, $du = \cosh x\, dx$:

$$\int \sinh^4 x \cosh^5 x\, dx = \int \underbrace{\sinh^4 x}_{u^4}\underbrace{(\sinh^2 x + 1)^2}_{(u^2+1)^2}\underbrace{\cosh x\, dx}_{du}$$
$$= \int u^4(u^2+1)^2\, du = \int (u^8 + 2u^6 + u^4)\, du$$
$$= \frac{u^9}{9} + \frac{2u^7}{7} + \frac{u^5}{5} + C = \frac{\sinh^9 x}{9} + \frac{2\sinh^7 x}{7} + \frac{\sinh^5 x}{5} + C$$

(b) Use the identity $\cosh^2 x = \dfrac{1}{2}(\cosh 2x + 1)$:

$$\int \cosh^2 x\, dx = \frac{1}{2}\int (\cosh 2x + 1)\, dx = \frac{1}{2}\left(\frac{\sinh 2x}{2} + x\right) + C$$
$$= \frac{1}{4}\sinh 2x + \frac{1}{2}x + C$$ ■

Hyperbolic substitution may be used as an alternative to trigonometric substitution to integrate functions involving the following square root expressions:

In trigonometric substitution, we treat $\sqrt{x^2+a^2}$ using the substitution $x = a\tan\theta$ and $\sqrt{x^2-a^2}$ using $x = a\sec\theta$. Identities can be used to show that the results coincide with those obtained from hyperbolic substitution (see Exercises 31–35).

Square root form	Hyperbolic substitution
$\sqrt{x^2+a^2}$	$x = a\sinh u,\ dx = a\cosh u,\ \sqrt{x^2+a^2} = a\cosh u$
$\sqrt{x^2-a^2}$	$x = a\cosh u,\ dx = a\sinh u,\ \sqrt{x^2-a^2} = a\sinh u$

■ **EXAMPLE 3 Hyperbolic Substitution** Calculate $\int \sqrt{x^2+16}\, dx$.

Solution

Step 1. **Substitute to eliminate the square root.**
Use the hyperbolic substitution $x = 4\sinh u$, $dx = 4\cosh u\, du$. Then

$$x^2 + 16 = 16(\sinh^2 u + 1) = (4\cosh u)^2$$

Furthermore, $4\cosh u > 0$, so $\sqrt{x^2+16} = 4\cosh u$ and thus,

$$\int \sqrt{x^2+16}\, dx = \int (4\cosh u) 4\cosh u\, du = 16\int \cosh^2 u\, du$$

Step 2. **Evaluate the hyperbolic integral.**
We evaluated the integral of $\cosh^2 u$ in Example 2(b):

$$\int \sqrt{x^2+16}\,dx = 16\int \cosh^2 u\,du = 16\left(\frac{1}{4}\sinh 2u + \frac{1}{2}u + C\right)$$

$$= 4\sinh 2u + 8u + C \qquad \boxed{1}$$

Step 3. **Convert back to original variable.**
To write the answer in terms of the original variable x, we note that

$$\sinh u = \frac{x}{4}, \qquad u = \sinh^{-1}\frac{x}{4}$$

Use the identities recalled in the margin to write

$$4\sinh 2u = 4(2\sinh u\cosh u) = 8\sinh u\sqrt{\sinh^2 u + 1}$$

$$= 8\left(\frac{x}{4}\right)\sqrt{\left(\frac{x}{4}\right)^2+1} = 2x\sqrt{\frac{x^2}{16}+1} = \frac{1}{2}x\sqrt{x^2+16}$$

REMINDER

$\sinh 2u = 2\sinh u\cosh u$

$\cosh u = \sqrt{\sinh^2 u + 1}$

Then Eq. (1) becomes

$$\int \sqrt{x^2+16}\,dx = 4\sinh 2u + 8u + C = \frac{1}{2}x\sqrt{x^2+16} + 8\sinh^{-1}\frac{x}{4} + C$$ ■

The next theorem states the integral formulas corresponding to the derivative formulas for the inverse hyperbolic functions recorded in Section 7.9. Each formula is valid on the domain where the integrand and inverse hyperbolic function are defined.

THEOREM 1 Integrals Involving Inverse Hyperbolic Functions

$$\int \frac{dx}{\sqrt{x^2+1}} = \sinh^{-1} x + C$$

$$\int \frac{dx}{\sqrt{x^2-1}} = \cosh^{-1} x + C \qquad (\text{for } x > 1)$$

$$\int \frac{dx}{1-x^2} = \tanh^{-1} x + C \qquad (\text{for } |x| < 1)$$

$$\int \frac{dx}{1-x^2} = \coth^{-1} x + C \qquad (\text{for } |x| > 1)$$

$$\int \frac{dx}{x\sqrt{1-x^2}} = -\operatorname{sech}^{-1} x + C \qquad (\text{for } 0 < x < 1)$$

$$\int \frac{dx}{|x|\sqrt{1+x^2}} = -\operatorname{csch}^{-1} x + C \qquad (\text{for } x \neq 0)$$

If your calculator does not provide values of inverse hyperbolic functions, you can use an online resource such as http://wolframalpha.com.

■ **EXAMPLE 4** Evaluate: **(a)** $\displaystyle\int_2^4 \frac{dx}{\sqrt{x^2-1}}$ and **(b)** $\displaystyle\int_{0.2}^{0.6} \frac{x\,dx}{1-x^4}$.

Solution

(a) By Theorem 1,

$$\int_2^4 \frac{dx}{\sqrt{x^2-1}} = \cosh^{-1} x\Big|_2^4 = \cosh^{-1} 4 - \cosh^{-1} 2 \approx 0.75$$

(b) First use the substitution $u = x^2$, $du = 2x\,du$. The new limits of integration become $u = (0.2)^2 = 0.04$ and $u = (0.6)^2 = 0.36$, so

$$\int_{0.2}^{0.6} \frac{x\,dx}{1-x^4} = \int_{0.04}^{0.36} \frac{\frac{1}{2}du}{1-u^2} = \frac{1}{2}\int_{0.04}^{0.36} \frac{du}{1-u^2}$$

By Theorem 1, both $\tanh^{-1} u$ and $\coth^{-1} u$ are antiderivatives of $f(u) = (1-u^2)^{-1}$. We use $\tanh^{-1} u$ because the interval of integration $[0.04, 0.36]$ is contained in the domain $(-1, 1)$ of $\tanh^{-1} u$. If the limits of integration were contained in $(1, \infty)$ or $(-\infty, -1)$, we would use $\coth^{-1} u$. The result is

$$\frac{1}{2}\int_{0.04}^{0.36} \frac{du}{1-u^2} = \frac{1}{2}\big(\tanh^{-1}(0.36) - \tanh^{-1}(0.04)\big) \approx 0.1684$$ ■

8.4 SUMMARY

- Integrals of hyperbolic functions:

$$\int \sinh x\,dx = \cosh x + C, \qquad \int \cosh x\,dx = \sinh x + C$$

$$\int \operatorname{sech}^2 x\,dx = \tanh x + C, \qquad \int \operatorname{csch}^2 x\,dx = -\coth x + C$$

$$\int \operatorname{sech} x \tanh x\,dx = -\operatorname{sech} x + C, \qquad \int \operatorname{csch} x \coth x\,dx = -\operatorname{csch} x + C$$

- Integrals involving inverse hyperbolic functions:

$$\int \frac{dx}{\sqrt{x^2+1}} = \sinh^{-1} x + C$$

$$\int \frac{dx}{\sqrt{x^2-1}} = \cosh^{-1} x + C \qquad (\text{for } x > 1)$$

$$\int \frac{dx}{1-x^2} = \tanh^{-1} x + C \qquad (\text{for } |x| < 1)$$

$$\int \frac{dx}{1-x^2} = \coth^{-1} x + C \qquad (\text{for } |x| > 1)$$

$$\int \frac{dx}{x\sqrt{1-x^2}} = -\operatorname{sech}^{-1} x + C \qquad (\text{for } 0 < x < 1)$$

$$\int \frac{dx}{|x|\sqrt{1+x^2}} = -\operatorname{csch}^{-1} x + C \qquad (\text{for } x \neq 0)$$

8.4 EXERCISES

Preliminary Questions

1. Which hyperbolic substitution can be used to evaluate the following integrals?

(a) $\displaystyle\int \frac{dx}{\sqrt{x^2+1}}$ (b) $\displaystyle\int \frac{dx}{\sqrt{x^2+9}}$ (c) $\displaystyle\int \frac{dx}{\sqrt{9x^2+1}}$

2. Which two of the hyperbolic integration formulas differ from their trigonometric counterparts by a minus sign?

3. Which antiderivative of $y = (1-x^2)^{-1}$ should we use to evaluate the integral $\displaystyle\int_3^5 (1-x^2)^{-1}\,dx$?

Exercises

In Exercises 1–16, calculate the integral.

1. $\int \cosh(3x)\,dx$

2. $\int \sinh(x+1)\,dx$

3. $\int x\sinh(x^2+1)\,dx$

4. $\int \sinh^2 x\cosh x\,dx$

5. $\int \operatorname{sech}^2(1-2x)\,dx$

6. $\int \tanh(3x)\operatorname{sech}(3x)\,dx$

7. $\int \tanh x\operatorname{sech}^2 x\,dx$

8. $\int \frac{\cosh x}{3\sinh x+4}\,dx$

9. $\int \tanh x\,dx$

10. $\int x\operatorname{csch}(x^2)\coth(x^2)\,dx$

11. $\int \frac{\cosh x}{\sinh x}\,dx$

12. $\int \frac{\cosh x}{\sinh^2 x}\,dx$

13. $\int \sinh^2(4x-9)\,dx$

14. $\int \sinh^3 x\cosh^6 x\,dx$

15. $\int \sinh^2 x\cosh^2 x\,dx$

16. $\int \tanh^3 x\,dx$

In Exercises 17–30, calculate the integral in terms of the inverse hyperbolic functions.

17. $\int \frac{dx}{\sqrt{x^2-1}}$

18. $\int \frac{dx}{\sqrt{9x^2-4}}$

19. $\int \frac{dx}{\sqrt{16+25x^2}}$

20. $\int \frac{dx}{\sqrt{1+3x^2}}$

21. $\int \sqrt{x^2-1}\,dx$

22. $\int \frac{x^2\,dx}{\sqrt{x^2+1}}$

23. $\int_{-1/2}^{1/2} \frac{dx}{1-x^2}$

24. $\int_4^5 \frac{dx}{1-x^2}$

25. $\int_0^1 \frac{dx}{\sqrt{1+x^2}}$

26. $\int_2^{10} \frac{dx}{4x^2-1}$

27. $\int_{-3}^{-1} \frac{dx}{x\sqrt{x^2+16}}$

28. $\int_{0.2}^{0.8} \frac{dx}{x\sqrt{1-x^2}}$

29. $\int \frac{\sqrt{x^2-1}\,dx}{x^2}$

30. $\int_1^9 \frac{dx}{x\sqrt{x^4+1}}$

31. Verify the formulas

$$\sinh^{-1} x = \ln|x+\sqrt{x^2+1}|$$

$$\cosh^{-1} x = \ln|x+\sqrt{x^2-1}| \qquad (\text{for } x \ge 1)$$

32. Verify that $\tanh^{-1} x = \frac{1}{2}\ln\left|\frac{1+x}{1-x}\right|$ for $|x|<1$.

33. Evaluate $\int \sqrt{x^2+16}\,dx$ using trigonometric substitution. Then use Exercise 31 to verify that your answer agrees with the answer in Example 3.

34. Evaluate $\int \sqrt{x^2-9}\,dx$ in two ways: using trigonometric substitution and using hyperbolic substitution. Then use Exercise 31 to verify that the two answers agree.

35. Prove the reduction formula for $n \ge 2$:

$$\int \cosh^n x\,dx = \frac{1}{n}\cosh^{n-1} x\sinh x + \frac{n-1}{n}\int \cosh^{n-2} x\,dx \qquad (2)$$

36. Use Eq. (2) to evaluate $\int \cosh^4 x\,dx$.

In Exercises 37–40, evaluate the integral.

37. $\int \frac{\tanh^{-1} x\,dx}{x^2-1}$

38. $\int \sinh^{-1} x\,dx$

39. $\int \tanh^{-1} x\,dx$

40. $\int x\tanh^{-1} x\,dx$

Further Insights and Challenges

41. Show that if $u = \tanh(x/2)$, then

$$\cosh x = \frac{1+u^2}{1-u^2}, \qquad \sinh x = \frac{2u}{1-u^2}, \qquad dx = \frac{2du}{1-u^2}$$

Hint: For the first relation, use the identities

$$\sinh^2\left(\frac{x}{2}\right) = \frac{1}{2}(\cosh x - 1), \qquad \cosh^2\left(\frac{x}{2}\right) = \frac{1}{2}(\cosh x + 1)$$

Exercises 42 and 43: evaluate using the substitution of Exercise 41.

42. $\int \operatorname{sech} x\,dx$

43. $\int \frac{dx}{1+\cosh x}$

44. Suppose that $y = f(x)$ satisfies $y'' = y$. Prove:

(a) $f(x)^2 - (f'(x))^2$ is constant.

(b) If $f(0) = f'(0) = 0$, then $f(x)$ is the zero function.

(c) $f(x) = f(0)\cosh x + f'(0)\sinh x$.

Exercises 45–48 refer to the function $gd(y) = \tan^{-1}(\sinh y)$, called the ***gudermannian****. In a map of the earth constructed by Mercator projection, points located y radial units from the equator correspond to points on the globe of latitude $gd(y)$.*

45. Prove that $\frac{d}{dy}gd(y) = \operatorname{sech} y$.

46. Let $f(y) = 2\tan^{-1}(e^y) - \pi/2$. Prove that $gd(y) = f(y)$. *Hint:* Show that $gd'(y) = f'(y)$ and $f(0) = g(0)$.

47. Let $t(y) = \sinh^{-1}(\tan y)$ Show that $t(y)$ is the inverse of $gd(y)$ for $0 \le y < \pi/2$.

48. Verify that $t(y)$ in Exercise 47 satisfies $t'(y) = \sec y$, and find a value of a such that

$$t(y) = \int_a^y \frac{dt}{\cos t}$$

49. The relations $\cosh(it) = \cos t$ and $\sinh(it) = i \sin t$ were discussed in the Excursion. Use these relations to show that the identity $\cos^2 t + \sin^2 t = 1$ results from setting $x = it$ in the identity $\cosh^2 x - \sinh^2 x = 1$.

8.5 The Method of Partial Fractions

The Method of Partial Fractions is used to integrate rational functions

$$f(x) = \frac{P(x)}{Q(x)}$$

where $P(x)$ and $Q(x)$ are polynomials. The idea is to write $f(x)$ as a sum of simpler rational functions that can be integrated directly. For example, in the simplest case we use the identity

$$\frac{1}{x^2 - 1} = \frac{\frac{1}{2}}{x - 1} - \frac{\frac{1}{2}}{x + 1}$$

to evaluate the integral

$$\int \frac{dx}{x^2 - 1} = \frac{1}{2}\int \frac{dx}{x - 1} - \frac{1}{2}\int \frac{dx}{x + 1} = \frac{1}{2}\ln|x - 1| - \frac{1}{2}\ln|x + 1|$$

It is a fact from algebra (known as the "Fundamental Theorem of Algebra") that every polynomial $Q(x)$ with real coefficients can be written as a product of linear and quadratic factors with real coefficients. However, it is not always possible to find these factors explicitly.

A rational function $P(x)/Q(x)$ is called **proper** if the degree of $P(x)$ [denoted $\deg(P)$] is *less than* the degree of $Q(x)$. For example,

$$\underbrace{\frac{x^2 - 3x + 7}{x^4 - 16}}_{\text{Proper}}, \qquad \underbrace{\frac{2x^2 + 7}{x - 5}, \qquad \frac{x - 2}{x - 5}}_{\text{Not proper}}$$

Suppose first that $P(x)/Q(x)$ is proper and that the denominator $Q(x)$ factors as a product of *distinct linear factors*. In other words,

$$\frac{P(x)}{Q(x)} = \frac{P(x)}{(x - a_1)(x - a_2)\cdots(x - a_n)}$$

Each distinct linear factor $(x - a)$ in the denominator contributes a term

$$\frac{A}{x - a}$$

to the partial fraction decomposition.

where the roots $a_1, a_2, \ldots, a_n$ are all distinct and $\deg(P) < n$. Then there is a **partial fraction decomposition**:

$$\frac{P(x)}{Q(x)} = \frac{A_1}{(x - a_1)} + \frac{A_2}{(x - a_2)} + \cdots + \frac{A_n}{(x - a_n)}$$

for suitable constants $A_1, \ldots, A_n$. For example,

$$\frac{5x^2 + x - 28}{(x + 1)(x - 2)(x - 3)} = -\frac{2}{x + 1} + \frac{2}{x - 2} + \frac{5}{x - 3}$$

Once we have found the partial fraction decomposition, we can integrate the individual terms.

■ EXAMPLE 1 Finding the Constants Evaluate $\displaystyle\int \frac{dx}{x^2 - 7x + 10}$.

Solution The denominator factors as $x^2 - 7x + 10 = (x - 2)(x - 5)$, so we look for a partial fraction decomposition:

$$\frac{1}{(x - 2)(x - 5)} = \frac{A}{x - 2} + \frac{B}{x - 5}$$

To find A and B, first multiply by $(x-2)(x-5)$ to clear denominators:

$$1 = (x-2)(x-5)\left(\frac{A}{x-2} + \frac{B}{x-5}\right)$$

$$1 = A(x-5) + B(x-2) \qquad \boxed{1}$$

This equation holds for all values of x (including $x = 2$ and $x = 5$, by continuity). We determine A by setting $x = 2$ (this makes the second term disappear):

$$1 = A(2-5) + \underbrace{B(2-2)}_{\text{This is zero}} = -3A \quad \Rightarrow \quad \boxed{A = -\frac{1}{3}}$$

Similarly, to calculate B, set $x = 5$ in Eq. (1):

$$1 = A(5-5) + B(5-2) = 3B \quad \Rightarrow \quad \boxed{B = \frac{1}{3}}$$

The resulting partial fraction decomposition is

$$\boxed{\frac{1}{(x-2)(x-5)} = \frac{-\frac{1}{3}}{x-2} + \frac{\frac{1}{3}}{x-5}}$$

The integration can now be carried out:

$$\int \frac{dx}{(x-2)(x-5)} = -\frac{1}{3}\int \frac{dx}{x-2} + \frac{1}{3}\int \frac{dx}{x-5}$$

$$= -\frac{1}{3}\ln|x-2| + \frac{1}{3}\ln|x-5| + C$$

■

■ **EXAMPLE 2** Evaluate $\displaystyle\int \frac{x^2+2}{(x-1)(2x-8)(x+2)}\,dx.$

Solution

Step 1. **Find the partial fraction decomposition.**

The decomposition has the form

$$\frac{x^2+2}{(x-1)(2x-8)(x+2)} = \frac{A}{x-1} + \frac{B}{2x-8} + \frac{C}{x+2} \qquad \boxed{2}$$

In Eq. (2), the linear factor $2x - 8$ does not have the form $(x - a)$ used previously, but the partial fraction decomposition can be carried out in the same way.

As before, multiply by $(x-1)(2x-8)(x+2)$ to clear denominators:

$$x^2 + 2 = A(2x-8)(x+2) + B(x-1)(x+2) + C(x-1)(2x-8) \qquad \boxed{3}$$

Since A goes with the factor $(x-1)$, we set $x = 1$ in Eq. (3) to compute A:

$$1^2 + 2 = A(2-8)(1+2) + \overbrace{B(1-1)(1+2) + C(1-1)(2-8)}^{\text{Zero}}$$

$$3 = -18A \quad \Rightarrow \quad \boxed{A = -\frac{1}{6}}$$

Similarly, 4 is the root of $2x - 8$, so we compute B by setting $x = 4$ in Eq. (3):

$$4^2 + 2 = A(8-8)(4+2) + B(4-1)(4+2) + C(4-1)(8-8)$$

$$18 = 18B \quad \Rightarrow \quad \boxed{B = 1}$$

Finally, C is determined by setting $x = -2$ in Eq. (3):

$$(-2)^2 + 2 = A(-4-8)(-2+2) + B(-2-1)(-2+2) + C(-2-1)(-4-8)$$

$$6 = 36C \quad \Rightarrow \quad \boxed{C = \frac{1}{6}}$$

The result is

$$\boxed{\frac{x^2+2}{(x-1)(2x-8)(x+2)} = -\frac{\frac{1}{6}}{x-1} + \frac{1}{2x-8} + \frac{\frac{1}{6}}{x+2}}$$

***Step 2.* Carry out the integration.**

$$\int \frac{x^2+2}{(x-1)(2x-8)(x+2)}\,dx = -\frac{1}{6}\int \frac{dx}{x-1} + \int \frac{dx}{2x-8} + \frac{1}{6}\int \frac{dx}{x+2}$$

$$= -\frac{1}{6}\ln|x-1| + \frac{1}{2}\ln|2x-8| + \frac{1}{6}\ln|x+2| + C$$

■

If $P(x)/Q(x)$ is not proper—that is, if $\deg(P) \ge \deg(Q)$—we use long division to write

$$\frac{P(x)}{Q(x)} = g(x) + \frac{R(x)}{Q(x)}$$

where $g(x)$ is a polynomial and $R(x)/Q(x)$ is proper. We may then integrate $P(x)/Q(x)$ using the partial fraction decomposition of $R(x)/Q(x)$.

Long division:

$$\begin{array}{r|l} & x \\ \hline x^2-4 & x^3 + 1 \\ & x^3 - 4x \\ \hline & \quad 4x + 1 \end{array}$$

The quotient $\frac{x^3+1}{x^2-4}$ *is equal to* x *with remainder* $4x+1$.

■ **EXAMPLE 3 Long Division Necessary** Evaluate $\displaystyle\int \frac{x^3+1}{x^2-4}\,dx$.

Solution Using long division, we write

$$\frac{x^3+1}{x^2-4} = x + \frac{4x+1}{x^2-4} = x + \frac{4x+1}{(x-2)(x+2)}$$

It is not hard to show that the second term has a partial fraction decomposition:

$$\frac{4x+1}{(x-2)(x+2)} = \frac{\frac{9}{4}}{x-2} + \frac{\frac{7}{4}}{x+2}$$

Therefore,

$$\int \frac{(x^3+1)\,dx}{x^2-4} = \int x\,dx + \frac{9}{4}\int \frac{dx}{x-2} + \frac{7}{4}\int \frac{dx}{x+2}$$

$$= \frac{1}{2}x^2 + \frac{9}{4}\ln|x-2| + \frac{7}{4}\ln|x+2| + C$$

■

Now suppose that the denominator has repeated linear factors:

$$\frac{P(x)}{Q(x)} = \frac{P(x)}{(x-a_1)^{M_1}(x-a_2)^{M_2}\cdots(x-a_n)^{M_n}}$$

Each factor $(x-a_i)^{M_i}$ contributes the following sum of terms to the partial fraction decomposition:

$$\frac{B_1}{(x-a_i)} + \frac{B_2}{(x-a_i)^2} + \cdots + \frac{B_{M_i}}{(x-a_i)^{M_i}}$$

■ **EXAMPLE 4 Repeated Linear Factors** Evaluate $\displaystyle\int \frac{3x-9}{(x-1)(x+2)^2}\,dx$.

Solution We are looking for a partial fraction decomposition of the form

$$\frac{3x-9}{(x-1)(x+2)^2} = \frac{A}{x-1} + \frac{B_1}{x+2} + \frac{B_2}{(x+2)^2}$$

Let's clear denominators to obtain

$$3x-9 = A(x+2)^2 + B_1(x-1)(x+2) + B_2(x-1) \qquad \boxed{4}$$

We compute A and B_2 by substituting in Eq. (4) in the usual way:

- Set $x=1$: This gives $-6 = 9A$, or $\boxed{A = -\frac{2}{3}}$.
- Set $x=-2$: This gives $-15 = -3B_2$, or $\boxed{B_2 = 5}$.

With these constants, Eq. (4) becomes

$$3x-9 = -\frac{2}{3}(x+2)^2 + B_1(x-1)(x+2) + 5(x-1) \qquad \boxed{5}$$

We cannot determine B_1 in the same way as A and B_2. Here are two ways to proceed.

- **First method (substitution):** There is no use substituting $x=1$ or $x=-2$ in Eq. (5) because the term involving B_1 drops out. But we are free to plug in any other value of x. Let's try $x=2$ in Eq. (5):

$$3(2)-9 = -\frac{2}{3}(2+2)^2 + B_1(2-1)(2+2) + 5(2-1)$$
$$-3 = -\frac{32}{3} + 4B_1 + 5$$
$$B_1 = \frac{1}{4}\left(-8 + \frac{32}{3}\right) = \frac{2}{3}$$

- **Second method (undetermined coefficients):** Expand the terms in Eq. (5):

$$3x-9 = -\frac{2}{3}(x^2+4x+4) + B_1(x^2+x-2) + 5(x-1)$$

The coefficients of the powers of x on each side of the equation must be equal. Since x^2 does not occur on the left-hand side, $0 = -\frac{2}{3} + B_1$, or $B_1 = \frac{2}{3}$.

Either way, we have shown that

$$\boxed{\frac{3x-9}{(x-1)(x+2)^2} = -\frac{\frac{2}{3}}{x-1} + \frac{\frac{2}{3}}{x+2} + \frac{5}{(x+2)^2}}$$

$$\int \frac{3x-9}{(x-1)(x+2)^2}\,dx = -\frac{2}{3}\int \frac{dx}{x-1} + \frac{2}{3}\int \frac{dx}{x+2} + 5\int \frac{dx}{(x+2)^2}$$
$$= -\frac{2}{3}\ln|x-1| + \frac{2}{3}\ln|x+2| - \frac{5}{x+2} + C$$ ■

Quadratic Factors

A quadratic polynomial $ax^2 + bx + c$ is called **irreducible** if it cannot be written as a product of two linear factors (without using complex numbers). A power of an irreducible quadratic factor $(ax^2 + bx + c)^M$ contributes a sum of the following type to a partial fraction decomposition:

$$\frac{A_1x + B_1}{ax^2 + bx + c} + \frac{A_2x + B_2}{(ax^2 + bx + c)^2} + \cdots + \frac{A_Mx + B_M}{(ax^2 + bx + c)^M}$$

For example,

$$\frac{4-x}{x(x^2+4x+2)^2} = \frac{1}{x} - \frac{x+4}{x^2+4x+2} - \frac{2x+9}{(x^2+4x+2)^2}$$

You may need to use trigonometric substitution to integrate these terms. In particular, the following result may be useful (see Exercise 33 in Section 8.3).

REMINDER If $b > 0$, then $x^2 + b$ is irreducible, but $x^2 - b$ is reducible because

$$x^2 - b = (x + \sqrt{b})(x - \sqrt{b})$$

$$\int \frac{dx}{x^2 + a} = \frac{1}{\sqrt{a}}\tan^{-1}\left(\frac{x}{\sqrt{a}}\right) + C \quad (\text{for } a > 0)$$ 6

■ **EXAMPLE 5** **Irreducible versus Reducible Quadratic Factors** Evaluate

(a) $\displaystyle\int \frac{18}{(x+3)(x^2+9)}\,dx$ **(b)** $\displaystyle\int \frac{18}{(x+3)(x^2-9)}\,dx$

Solution

(a) The quadratic factor $x^2 + 9$ is irreducible, so the partial fraction decomposition has the form

$$\frac{18}{(x+3)(x^2+9)} = \frac{A}{x+3} + \frac{Bx+C}{x^2+9}$$

Clear denominators to obtain

$$18 = A(x^2+9) + (Bx+C)(x+3)$$ 7

To find A, set $x = -3$:

$$18 = A\big((-3)^2 + 9\big) + 0 \quad \Rightarrow \quad A = 1$$

Then substitute $A = 1$ in Eq. (7) to obtain

$$18 = (x^2+9) + (Bx+C)(x+3) = (B+1)x^2 + (C+3B)x + (9+3C)$$

Equating coefficients, we get $B + 1 = 0$ and $9 + 3C = 18$. Hence (see margin):

$$B = -1, \qquad C = 3$$

In the second equality, we use

$$\int \frac{x\,dx}{x^2+9} = \frac{1}{2}\int \frac{du}{u} = \frac{1}{2}\ln(x^2+9) + C$$

and Eq. (6):

$$\int \frac{dx}{x^2+9} = \frac{1}{3}\tan^{-1}\frac{x}{3} + C$$

$$\begin{aligned}\int \frac{18\,dx}{(x+3)(x^2+9)} &= \int \frac{dx}{x+3} + \int \frac{(-x+3)\,dx}{x^2+9}\\ &= \int \frac{dx}{x+3} - \int \frac{x\,dx}{x^2+9} + \int \frac{3\,dx}{x^2+9}\\ &= \ln|x+3| - \frac{1}{2}\ln(x^2+9) + \tan^{-1}\frac{x}{3} + C\end{aligned}$$

(b) The polynomial $x^2 - 9$ is not irreducible because $x^2 - 9 = (x-3)(x+3)$. Therefore, the partial fraction decomposition has the form

$$\frac{18}{(x+3)(x^2-9)} = \frac{18}{(x+3)^2(x-3)} = \frac{A}{x-3} + \frac{B}{x+3} + \frac{C}{(x+3)^2}$$

Clear denominators:

$$18 = A(x+3)^2 + B(x+3)(x-3) + C(x-3)$$

For $x = 3$, this yields $18 = (6^2)A$, and for $x = -3$, this yields $18 = -6C$. Therefore,

$$A = \frac{1}{2}, \quad C = -3 \qquad \Rightarrow \qquad 18 = \frac{1}{2}(x+3)^2 + B(x+3)(x-3) - 3(x-3)$$

To solve for B, we can plug in any value of x other than ± 3. The choice $x = 2$ yields $18 = \frac{1}{2}(25) - 5B + 3$, or $B = -\frac{1}{2}$, and

$$\begin{aligned}\int \frac{18}{(x+3)(x^2-9)}\,dx &= \frac{1}{2}\int \frac{dx}{x-3} - \frac{1}{2}\int \frac{dx}{x+3} - 3\int \frac{dx}{(x+3)^2}\\ &= \frac{1}{2}\ln|x-3| - \frac{1}{2}\ln|x+3| + 3(x+3)^{-1} + C\end{aligned}$$

■

■ **EXAMPLE 6 Repeated Quadratic Factor** Evaluate $\displaystyle\int \frac{4-x}{x(x^2+2)^2}\,dx$.

Solution The partial fraction decomposition has the form

$$\frac{4-x}{x(x^2+2)^2} = \frac{A}{x} + \frac{Bx+C}{x^2+2} + \frac{Dx+E}{(x^2+2)^2}$$

Clear denominators by multiplying through by $x(x^2+2)^2$:

$$4 - x = A(x^2+2)^2 + (Bx+C)\big(x(x^2+2)\big) + (Dx+E)x \qquad \boxed{8}$$

We compute A directly by setting $x = 0$. Then Eq. (8) reduces to $4 = 4A$, or $A = 1$. We find the remaining coefficients by the method of undetermined coefficients. Set $A = 1$ in Eq. (8) and expand:

$$\begin{aligned}4 - x &= (x^4 + 4x^2 + 4) + (Bx^4 + 2Bx^2 + Cx^3 + 2C) + (Dx^2 + Ex)\\ &= (1+B)x^4 + Cx^3 + (4 + 2B + D)x^2 + Ex + 2C + 4\end{aligned}$$

Now equate the coefficients on the two sides of the equation:

$$\begin{aligned} 1 + B &= 0 && \text{(Coefficient of } x^4\text{)} \\ C &= 0 && \text{(Coefficient of } x^3\text{)} \\ 4 + 2B + D &= 0 && \text{(Coefficient of } x^2\text{)} \\ E &= -1 && \text{(Coefficient of } x\text{)} \\ 2C + 4 &= 4 && \text{(Constant term)} \end{aligned}$$

These equations yield $B = -1$, $C = 0$, $D = -2$, and $E = -1$. Thus,

$$\begin{aligned} \int \frac{(4-x)\,dx}{x(x^2+2)^2} &= \int \frac{dx}{x} - \int \frac{x\,dx}{x^2+2} - \int \frac{(2x+1)\,dx}{(x^2+2)^2} \\ &= \ln|x| - \frac{1}{2}\ln(x^2+2) - \int \frac{(2x+1)dx}{(x^2+2)^2} \end{aligned}$$

The middle integral was evaluated using the substitution $u = x^2 + 2$, $du = 2x\,dx$. The third integral breaks up as a sum:

$$\begin{aligned} \int \frac{(2x+1)\,dx}{(x^2+2)^2} &= \overbrace{\int \frac{2x\,dx}{(x^2+2)^2}}^{\text{Use substitution } u = x^2+2} + \int \frac{dx}{(x^2+2)^2} \\ &= -(x^2+2)^{-1} + \int \frac{dx}{(x^2+2)^2} \end{aligned} \qquad \boxed{9}$$

To evaluate the integral in Eq. (9), we use the trigonometric substitution

$$x = \sqrt{2}\tan\theta, \qquad dx = \sqrt{2}\sec^2\theta\,d\theta, \qquad x^2 + 2 = 2\tan^2\theta + 2 = 2\sec^2\theta$$

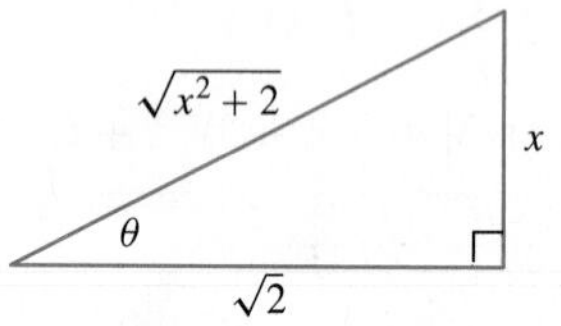

FIGURE 1

Referring to Figure 1, we obtain

$$\begin{aligned} \int \frac{dx}{(x^2+2)^2} &= \int \frac{\sqrt{2}\sec^2\theta\,d\theta}{(2\tan^2\theta+2)^2} = \int \frac{\sqrt{2}\sec^2\theta\,d\theta}{4\sec^4\theta} \\ &= \frac{\sqrt{2}}{4}\int \cos^2\theta\,d\theta = \frac{\sqrt{2}}{4}\left(\frac{1}{2}\theta + \frac{1}{2}\sin\theta\cos\theta\right) + C \\ &= \frac{\sqrt{2}}{8}\tan^{-1}\frac{x}{\sqrt{2}} + \frac{\sqrt{2}}{8}\frac{x}{\sqrt{x^2+2}}\frac{\sqrt{2}}{\sqrt{x^2+2}} + C \\ &= \frac{1}{4\sqrt{2}}\tan^{-1}\frac{x}{\sqrt{2}} + \frac{1}{4}\frac{x}{x^2+2} + C \end{aligned}$$

Collecting all the terms, we have

$$\int \frac{4-x}{x(x^2+2)^2}\,dx = \ln|x| - \frac{1}{2}\ln(x^2+2) + \frac{1-\frac{1}{4}x}{x^2+2} - \frac{1}{4\sqrt{2}}\tan^{-1}\frac{x}{\sqrt{2}} + C \qquad \blacksquare$$

CONCEPTUAL INSIGHT The examples in this section illustrate a general fact: The integral of a rational function can be expressed as a sum of rational functions, arctangents of linear or quadratic polynomials, and logarithms of linear or quadratic polynomials (provided that we can factor the denominator). Other types of functions, such as exponential and trigonometric functions, do not appear.

Using a Computer Algebra System

Finding partial fraction decompositions often requires laborious computation. Fortunately, most computer algebra systems have a command that produces partial fraction decompositions (with names such as "Apart" or "parfrac"). For example, the command

```
Apart[(x^2 − 2)/((x + 2)(x^2 + 4)^3)]
```

produces the partial fraction decomposition

$$\frac{x^2-2}{(x+2)(x^2+4)^3} = \frac{1}{256(2+x)} + \frac{3(x-2)}{4(4+x^2)^3} + \frac{2-x}{32(4+x^2)^2} + \frac{2-x}{256(4+x^2)}$$

However, a computer algebra system cannot produce a partial fraction decomposition in cases where $Q(x)$ cannot be factored explicitly.

The polynomial $x^5 + 2x + 2$ cannot be factored explicitly, so the command

```
Apart[1/(x^5 + 2x + 2)]
```

returns the useless response

$$\frac{1}{x^5+2x+2}$$

8.5 SUMMARY

Method of Partial Fractions: Assume first that $P(x)/Q(x)$ is a *proper* rational function [that is, $\deg(P) < \deg(Q)$] and that $Q(x)$ can be factored explicitly as a product of linear and irreducible quadratic terms.

- If $Q(x) = (x - a_1)(x - a_2)\cdots(x - a_n)$, where the roots a_j are distinct, then

$$\frac{P(x)}{(x-a_1)(x-a_2)\cdots(x-a_n)} = \frac{A_1}{x-a_1} + \frac{A_2}{x-a_2} + \cdots + \frac{A_n}{x-a_n}$$

To calculate the constants, clear denominators and substitute, in turn, the values $x = a_1$, $a_2, \ldots, a_n$.

- If $Q(x)$ is equal to a product of powers of linear factors $(x - a)^M$ and irreducible quadratic factors $(x^2 + b)^N$ with $b > 0$, then the partial fraction decomposition of $P(x)/Q(x)$ is a sum of terms of the following type:

$$(x-a)^M \quad \text{contributes} \quad \frac{A_1}{x-a} + \frac{A_2}{(x-a)^2} + \cdots + \frac{A_M}{(x-a)^M}$$

$$(x^2+b)^N \quad \text{contributes} \quad \frac{A_1x+B_1}{x^2+b} + \frac{A_2x+B_2}{(x^2+b)^2} + \cdots + \frac{A_Nx+B_N}{(x^2+b)^N}$$

Substitution and trigonometric substitution may be needed to integrate the terms corresponding to $(x^2 + b)^N$ (see Example 6).

- If $P(x)/Q(x)$ is improper, use long division (see Example 3).

8.5 EXERCISES

Preliminary Questions

1. Suppose that $\int f(x)\,dx = \ln x + \sqrt{x+1} + C$. Can $f(x)$ be a rational function? Explain.

2. Which of the following are *proper* rational functions?

(a) $\dfrac{x}{x-3}$ **(b)** $\dfrac{4}{9-x}$

(c) $\dfrac{x^2+12}{(x+2)(x+1)(x-3)}$ **(d)** $\dfrac{4x^3-7x}{(x-3)(2x+5)(9-x)}$

3. Which of the following quadratic polynomials are irreducible? To check, complete the square if necessary.

(a) $x^2 + 5$ **(b)** $x^2 - 5$

(c) $x^2 + 4x + 6$ **(d)** $x^2 + 4x + 2$

4. Let $P(x)/Q(x)$ be a proper rational function where $Q(x)$ factors as a product of distinct linear factors $(x - a_i)$. Then

$$\int \frac{P(x)\,dx}{Q(x)}$$

(choose the correct answer):

(a) is a sum of logarithmic terms $A_i \ln(x - a_i)$ for some constants A_i.

(b) may contain a term involving the arctangent.

Exercises

1. Match the rational functions (a)–(d) with the corresponding partial fraction decompositions (i)–(iv).

(a) $\dfrac{x^2+4x+12}{(x+2)(x^2+4)}$ **(b)** $\dfrac{2x^2+8x+24}{(x+2)^2(x^2+4)}$

(c) $\dfrac{x^2-4x+8}{(x-1)^2(x-2)^2}$ **(d)** $\dfrac{x^4-4x+8}{(x+2)(x^2+4)}$

(i) $x-2+\dfrac{4}{x+2}-\dfrac{4x-4}{x^2+4}$

(ii) $\dfrac{-8}{x-2}+\dfrac{4}{(x-2)^2}+\dfrac{8}{x-1}+\dfrac{5}{(x-1)^2}$

(iii) $\dfrac{1}{x+2}+\dfrac{2}{(x+2)^2}+\dfrac{-x+2}{x^2+4}$ **(iv)** $\dfrac{1}{x+2}+\dfrac{4}{x^2+4}$

2. Determine the constants A, B:

$$\frac{2x-3}{(x-3)(x-4)} = \frac{A}{x-3} + \frac{B}{x-4}$$

3. Clear denominators in the following partial fraction decomposition and determine the constant B (substitute a value of x or use the method of undetermined coefficients).

$$\frac{3x^2+11x+12}{(x+1)(x+3)^2} = \frac{1}{x+1} - \frac{B}{x+3} - \frac{3}{(x+3)^2}$$

4. Find the constants in the partial fraction decomposition

$$\frac{2x+4}{(x-2)(x^2+4)} = \frac{A}{x-2} + \frac{Bx+C}{x^2+4}$$

In Exercises 5–8, evaluate using long division first to write $f(x)$ as the sum of a polynomial and a proper rational function.

5. $\displaystyle\int \frac{x\,dx}{3x-4}$ **6.** $\displaystyle\int \frac{(x^2+2)\,dx}{x+3}$

7. $\displaystyle\int \frac{(x^3+2x^2+1)\,dx}{x+2}$ **8.** $\displaystyle\int \frac{(x^3+1)\,dx}{x^2+1}$

In Exercises 9–44, evaluate the integral.

9. $\displaystyle\int \frac{dx}{(x-2)(x-4)}$ **10.** $\displaystyle\int \frac{(x+3)\,dx}{x+4}$

11. $\displaystyle\int \frac{dx}{x(2x+1)}$ **12.** $\displaystyle\int \frac{(2x-1)\,dx}{x^2-5x+6}$

13. $\displaystyle\int \frac{x^2\,dx}{x^2+9}$ **14.** $\displaystyle\int \frac{dx}{(x-2)(x-3)(x+2)}$

15. $\displaystyle\int \frac{(x^2+3x-44)\,dx}{(x+3)(x+5)(3x-2)}$ **16.** $\displaystyle\int \frac{3\,dx}{(x+1)(x^2+x)}$

17. $\displaystyle\int \frac{(x^2+11x)\,dx}{(x-1)(x+1)^2}$ **18.** $\displaystyle\int \frac{(4x^2-21x)\,dx}{(x-3)^2(2x+3)}$

19. $\displaystyle\int \frac{dx}{(x-1)^2(x-2)^2}$ **20.** $\displaystyle\int \frac{(x^2-8x)\,dx}{(x+1)(x+4)^3}$

21. $\displaystyle\int \frac{8\,dx}{x(x+2)^3}$ **22.** $\displaystyle\int \frac{x^2\,dx}{x^2+3}$

23. $\displaystyle\int \frac{dx}{2x^2-3}$ **24.** $\displaystyle\int \frac{dx}{(x-4)^2(x-1)}$

25. $\displaystyle\int \frac{4x^2-20}{(2x+5)^3}\,dx$ **26.** $\displaystyle\int \frac{3x+6}{x^2(x-1)(x-3)}\,dx$

27. $\displaystyle\int \frac{dx}{x(x-1)^3}$ **28.** $\displaystyle\int \frac{(3x^2-2)\,dx}{x-4}$

29. $\displaystyle\int \frac{(x^2-x+1)\,dx}{x^2+x}$ **30.** $\displaystyle\int \frac{dx}{x(x^2+1)}$

31. $\displaystyle\int \frac{(3x^2-4x+5)\,dx}{(x-1)(x^2+1)}$ **32.** $\displaystyle\int \frac{x^2}{(x+1)(x^2+1)}\,dx$

33. $\displaystyle\int \frac{dx}{x(x^2+25)}$ **34.** $\displaystyle\int \frac{dx}{x^2(x^2+25)}$

35. $\displaystyle\int \frac{(6x^2+2)\,dx}{x^2+2x-3}$ **36.** $\displaystyle\int \frac{6x^2+7x-6}{(x^2-4)(x+2)}\,dx$

37. $\displaystyle\int \frac{10\,dx}{(x-1)^2(x^2+9)}$ **38.** $\displaystyle\int \frac{10\,dx}{(x+1)(x^2+9)^2}$

39. $\displaystyle\int \frac{dx}{x(x^2+8)^2}$ **40.** $\displaystyle\int \frac{100x\,dx}{(x-3)(x^2+1)^2}$

41. $\displaystyle\int \frac{dx}{(x+2)(x^2+4x+10)}$ **42.** $\displaystyle\int \frac{9\,dx}{(x+1)(x^2-2x+6)}$

43. $\displaystyle\int \frac{25\,dx}{x(x^2+2x+5)^2}$ **44.** $\displaystyle\int \frac{(x^2+3)\,dx}{(x^2+2x+3)^2}$

In Exercises 45–48, evaluate by using first substitution and then partial fractions if necessary.

45. $\displaystyle\int \frac{x\,dx}{x^4+1}$ **46.** $\displaystyle\int \frac{x\,dx}{(x+2)^4}$

47. $\displaystyle\int \frac{e^x\,dx}{e^{2x}-e^x}$ **48.** $\displaystyle\int \frac{\sec^2\theta\,d\theta}{\tan^2\theta-1}$

49. Evaluate $\displaystyle\int \frac{\sqrt{x}\,dx}{x-1}$. *Hint:* Use the substitution $u = \sqrt{x}$ (sometimes called a **rationalizing substitution**).

50. Evaluate $\displaystyle\int \frac{dx}{x^{1/2}-x^{1/3}}$.

51. Evaluate $\int \frac{dx}{x^2 - 1}$ in two ways: using partial fractions and using trigonometric substitution. Verify that the two answers agree.

52. GU Graph the equation $(x - 40)y^2 = 10x(x - 30)$ and find the volume of the solid obtained by revolving the region between the graph and the x-axis for $0 \le x \le 30$ around the x-axis.

In Exercises 53–66, evaluate the integral using the appropriate method or combination of methods covered thus far in the text.

53. $\int \frac{dx}{x^2\sqrt{4 - x^2}}$

54. $\int \frac{dx}{x(x-1)^2}$

55. $\int \cos^2 4x\, dx$

56. $\int x \sec^2 x\, dx$

57. $\int \frac{dx}{(x^2 + 9)^2}$

58. $\int \theta \sec^{-1} \theta\, d\theta$

59. $\int \tan^5 x \sec x\, dx$

60. $\int \frac{(3x^2 - 1)\, dx}{x(x^2 - 1)}$

61. $\int \ln(x^4 - 1)\, dx$

62. $\int \frac{x\, dx}{(x^2 - 1)^{3/2}}$

63. $\int \frac{x^2\, dx}{(x^2 - 1)^{3/2}}$

64. $\int \frac{(x + 1)\, dx}{(x^2 + 4x + 8)^2}$

65. $\int \frac{\sqrt{x}\, dx}{x^3 + 1}$

66. $\int \frac{x^{1/2}\, dx}{x^{1/3} + 1}$

67. Show that the substitution $\theta = 2\tan^{-1} t$ (Figure 2) yields the formulas

$$\cos\theta = \frac{1 - t^2}{1 + t^2}, \qquad \sin\theta = \frac{2t}{1 + t^2}, \qquad d\theta = \frac{2\, dt}{1 + t^2} \qquad \boxed{10}$$

This substitution transforms the integral of any rational function of $\cos\theta$ and $\sin\theta$ into an integral of a rational function of t (which can then be evaluated using partial fractions). Use it to evaluate $\int \frac{d\theta}{\cos\theta + \frac{3}{4}\sin\theta}$.

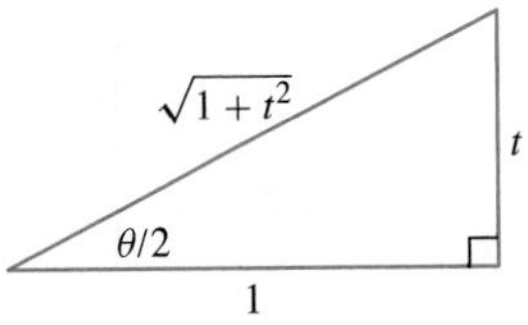

FIGURE 2

68. Use the substitution of Exercise 67 to evaluate $\int \frac{d\theta}{\cos\theta + \sin\theta}$.

Further Insights and Challenges

69. Prove the general formula

$$\int \frac{dx}{(x - a)(x - b)} = \frac{1}{a - b} \ln \frac{x - a}{x - b} + C$$

where a, b are constants such that $a \neq b$.

70. The method of partial fractions shows that

$$\int \frac{dx}{x^2 - 1} = \frac{1}{2} \ln |x - 1| - \frac{1}{2} \ln |x + 1| + C$$

The computer algebra system Mathematica evaluates this integral as $-\tanh^{-1} x$, where $\tanh^{-1} x$ is the inverse hyperbolic tangent function. Can you reconcile the two answers?

71. Suppose that $Q(x) = (x - a)(x - b)$, where $a \neq b$, and let $P(x)/Q(x)$ be a proper rational function so that

$$\frac{P(x)}{Q(x)} = \frac{A}{(x - a)} + \frac{B}{(x - b)}$$

(a) Show that $A = \frac{P(a)}{Q'(a)}$ and $B = \frac{P(b)}{Q'(b)}$.

(b) Use this result to find the partial fraction decomposition for $P(x) = 3x - 2$ and $Q(x) = x^2 - 4x - 12$.

72. Suppose that $Q(x) = (x - a_1)(x - a_2)\cdots(x - a_n)$, where the roots a_j are all distinct. Let $P(x)/Q(x)$ be a proper rational function so that

$$\frac{P(x)}{Q(x)} = \frac{A_1}{(x - a_1)} + \frac{A_2}{(x - a_2)} + \cdots + \frac{A_n}{(x - a_n)}$$

(a) Show that $A_j = \frac{P(a_j)}{Q'(a_j)}$ for $j = 1, \ldots, n$.

(b) Use this result to find the partial fraction decomposition for $P(x) = 2x^2 - 1$, $Q(x) = x^3 - 4x^2 + x + 6 = (x + 1)(x - 2)(x - 3)$.

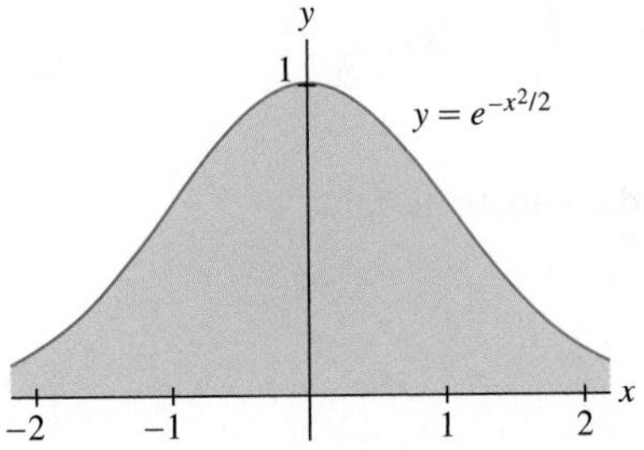

FIGURE 1 Bell-shaped curve. The region extends infinitely far in both directions, but the total area is finite.

8.6 Improper Integrals

The integrals we have studied so far represent signed areas of bounded regions. However, areas of unbounded regions (Figure 1) also arise in applications and are represented by **improper integrals**.

There are two ways an integral can be improper: (1) The interval of integration may be infinite, or (2) the integrand may tend to infinity. We deal first with improper integrals over infinite intervals. One or both endpoints may be infinite:

$$\int_{-\infty}^{a} f(x)\, dx, \qquad \int_{a}^{\infty} f(x)\, dx, \qquad \int_{-\infty}^{\infty} f(x)\, dx$$

How can an unbounded region have finite area? To answer this question, we must specify what we mean by the area of an unbounded region. Consider the area [Figure 2(A)] under the graph of $f(x) = e^{-x}$ over the finite interval $[0, R]$:

$$\int_0^R e^{-x}\, dx = -e^{-x}\Big|_0^R = -e^{-R} + e^0 = 1 - e^{-R}$$

As $R \to \infty$, this area approaches a finite value [Figure 2(B)]:

$$\int_0^\infty e^{-x}\, dx = \lim_{R \to \infty} \int_0^R e^{-x}\, dx = \lim_{R \to \infty}\left(1 - e^{-R}\right) = 1 \qquad \boxed{1}$$

It seems reasonable to take this limit as the *definition* of the area under the graph over the infinite interval $[0, \infty)$. Thus, the unbounded region in Figure 2(C) has area 1.

*The great British mathematician G. H. Hardy (1877–1947) observed that in calculus, we learn to ask, not " What is it?" but rather "How shall we **define** it?" We saw that tangent lines and areas under curves have no clear meaning until we define them precisely using limits. Here again, the key question is "How shall we define the area of an unbounded region?"*

(A) Finite region has area $1 - e^{-R}$ 　　 (B) Area approaches 1 as $R \to \infty$ 　　 (C) Unbounded region has area 1

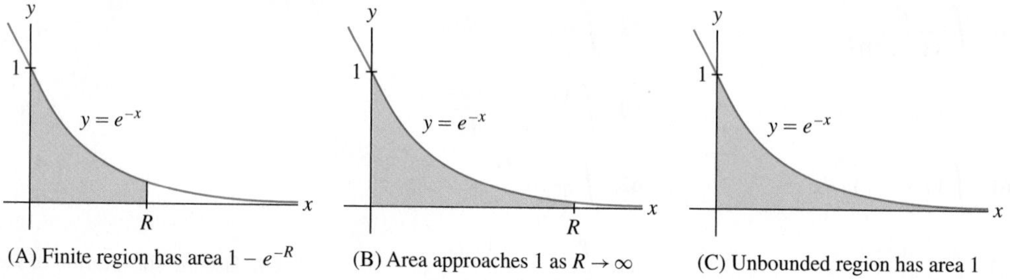

FIGURE 2

DEFINITION Improper Integral Fix a number a and assume that $f(x)$ is integrable over $[a, b]$ for all $b > a$. The *improper integral* of $f(x)$ over $[a, \infty)$ is defined as the following limit (if it exists):

$$\int_a^\infty f(x)\, dx = \lim_{R \to \infty} \int_a^R f(x)\, dx$$

We say that the improper integral *converges* if the limit exists (and is finite) and that it *diverges* if the limit does not exist.

Similarly, we define

$$\int_{-\infty}^a f(x)\, dx = \lim_{R \to -\infty} \int_R^a f(x)\, dx$$

A doubly infinite improper integral is defined as a sum (provided that both integrals on the right converge):

$$\int_{-\infty}^\infty f(x)\, dx = \int_{-\infty}^0 f(x)\, dx + \int_0^\infty f(x)\, dx \qquad \boxed{2}$$

■ **EXAMPLE 1** Show that $\displaystyle\int_2^\infty \frac{dx}{x^3}$ converges and compute its value.

Solution

Step 1. Integrate over a finite interval $[2, R]$.

$$\int_2^R \frac{dx}{x^3} = -\frac{1}{2}x^{-2}\Big|_2^R = -\frac{1}{2}\left(R^{-2}\right) + \frac{1}{2}\left(2^{-2}\right) = \frac{1}{8} - \frac{1}{2R^2}$$

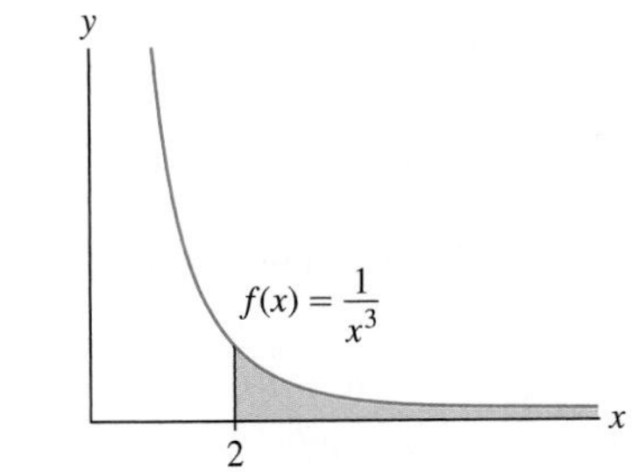

FIGURE 3 The area over $[2, \infty)$ is equal to $\frac{1}{8}$.

***Step 2.* Compute the limit as $R \to \infty$.**

$$\int_2^{\infty} \frac{dx}{x^3} = \lim_{R\to\infty} \int_2^R \frac{dx}{x^3} = \lim_{R\to\infty} \left(\frac{1}{8} - \frac{1}{2R^2}\right) = \frac{1}{8}$$

We conclude that the infinite shaded region in Figure 3 has area $\frac{1}{8}$. ■

■ **EXAMPLE 2** Determine whether $\int_{-\infty}^{-1} \frac{dx}{x}$ converges.

Solution First, we evaluate the definite integral over a finite interval $[R, -1]$ Since the lower limit of the integral is $-\infty$, we take $R < -1$:

$$\int_R^{-1} \frac{dx}{x} = \ln|x|\Big|_R^{-1} = \ln|-1| - \ln|R| = -\ln|R|$$

Then we compute the limit as $R \to -\infty$:

$$\lim_{R\to-\infty} \int_R^{-1} \frac{dx}{x} = \lim_{R\to-\infty} (-\ln|R|) = -\lim_{R\to-\infty} \ln|R| = -\infty$$

The limit is infinite, so the improper integral diverges. We conclude that the area of the unbounded region in Figure 4 is infinite. ■

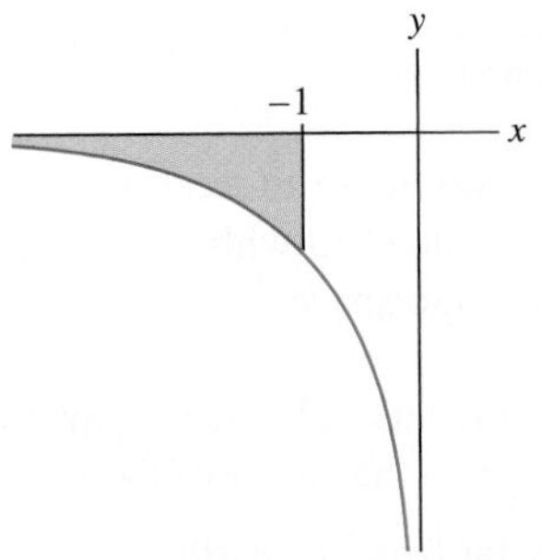

FIGURE 4 The integral of $f(x) = x^{-1}$ over $(-\infty, -1]$ is infinite.

CONCEPTUAL INSIGHT If you compare the unbounded shaded regions in Figures 3 and 4, you may wonder why one has finite area and the other has infinite area. Convergence of an improper integral depends on how rapidly the function $f(x)$ tends to zero as $x \to \infty$ (or $x \to -\infty$). Our calculations show that x^{-2} decreases rapidly enough for convergence, whereas x^{-1} does not.

An improper integral of a power function $f(x) = x^{-p}$ is called a ***p*-integral**. Note that $f(x) = x^{-p}$ decreases more rapidly as p gets larger. Interestingly, our next theorem shows that the exponent $p = -1$ is the dividing line between convergence and divergence.

p-integrals are particularly important because they are often used to determine the convergence or divergence of more complicated improper integrals by means of the Comparison Test (see Example 8).

THEOREM 1 The *p*-Integral over $[a, \infty)$ For $a > 0$,

$$\int_a^{\infty} \frac{dx}{x^p} = \begin{cases} \dfrac{a^{1-p}}{p-1} & \text{if } p > 1 \\ \text{diverges} & \text{if } p \le 1 \end{cases}$$

Proof Denote the p-integral by J. Then

$$J = \lim_{R\to\infty} \int_a^R x^{-p}\,dx = \lim_{R\to\infty} \frac{x^{1-p}}{1-p}\Big|_a^R = \lim_{R\to\infty} \left(\frac{R^{1-p}}{1-p} - \frac{a^{1-p}}{1-p}\right)$$

If $p > 1$, then $1 - p < 0$ and R^{1-p} tends to zero as $R \to \infty$. In this case, $J = \dfrac{a^{1-p}}{p-1}$. If $p < 1$, then $1 - p > 0$ and R^{1-p} tends to ∞. In this case, J diverges. If $p = 1$, then J diverges because $\lim_{R\to\infty} \int_a^R x^{-1}\,dx = \lim_{R\to\infty} (\ln R - \ln a) = \infty$. ■

Sometimes it is necessary to use L'Hôpital's Rule to determine the limits that arise in improper integrals.

■ **EXAMPLE 3 Using L'Hôpital's Rule** Calculate $\int_0^\infty xe^{-x}\,dx$.

Solution First, use Integration by Parts with $u = x$ and $v' = e^{-x}$:

$$\int xe^{-x}\,dx = -xe^{-x} + \int e^{-x}\,dx = -xe^{-x} - e^{-x} = -(x+1)e^{-x} + C$$

$$\int_0^R xe^{-x}\,dx = -(x+1)e^{-x}\Big|_0^R = -(R+1)e^{-R} + 1 = 1 - \frac{R+1}{e^R}$$

Then compute the improper integral as a limit using L'Hôpital's Rule:

$$\int_0^\infty xe^{-x}\,dx = 1 - \lim_{R\to\infty} \frac{R+1}{e^R} = 1 - \underbrace{\lim_{R\to\infty} \frac{1}{e^R}}_{\text{L'Hôpital's Rule}} = 1 - 0 = 1$$ ■

Improper integrals arise in applications when it makes sense to treat certain large quantities as if they were infinite. For example, an object launched with escape velocity never falls back to earth but rather, travels "infinitely far" into space.

In physics, we speak of moving an object "infinitely far away." In practice this means "very far away," but it is more convenient to work with an improper integral.

■ **EXAMPLE 4 Escape Velocity** The earth exerts a gravitational force of magnitude $F(r) = GM_em/r^2$ on an object of mass m at distance r from the center of the earth.

(a) Find the work required to move the object infinitely far from the earth.

(b) Calculate the escape velocity v_{esc} on the earth's surface.

← REMINDER The mass of the earth is

$$M_e \approx 5.98 \cdot 10^{24}\text{ kg}$$

The radius of the earth is

$$r_e \approx 6.37 \cdot 10^6\text{ m}$$

The universal gravitational constant is

$$G \approx 6.67 \cdot 10^{-11}\text{ N-m}^2/\text{kg}^2$$

A newton is 1 kg-m/s^2 *and a joule is* 1 N-m.

Solution This amounts to computing a p-integral with $p = 2$. Recall that work is the integral of force as a function of distance (Section 6.5).

(a) The work required to move an object from the earth's surface ($r = r_e$) to a distance R from the center is

$$\int_{r_e}^R \frac{GM_em}{r^2}\,dr = -\frac{GM_em}{r}\Big|_{r_e}^R = GM_em\left(\frac{1}{r_e} - \frac{1}{R}\right)\text{ joules}$$

The work moving the object "infinitely far away" is the improper integral

$$GM_em\int_{r_e}^\infty \frac{dr}{r^2} = \lim_{R\to\infty} GM_em\left(\frac{1}{r_e} - \frac{1}{R}\right) = \frac{GM_em}{r_e}\text{ joules}$$

(b) By the principle of Conservation of Energy, an object launched with velocity v_0 will escape the earth's gravitational field if its kinetic energy $\frac{1}{2}mv_0^2$ is at least as large as the work required to move the object to infinity—that is, if

$$\frac{1}{2}mv_0^2 \ge \frac{GM_em}{r_e} \quad\Rightarrow\quad v_0 \ge \left(\frac{2GM_e}{r_e}\right)^{1/2}$$

Escape velocity in miles per hour is approximately 25,000 mph.

Using the values recalled in the marginal note, we find that $v_0 \ge 11{,}200$ m/s. The minimal velocity is the escape velocity $v_{esc} = 11{,}200$ m/s. ■

In practice, the word "forever" means "a long but unspecified length of time." For example, if the investment pays out dividends for 100 years, then its present value is

$$\int_0^{100} 6000e^{-0.04t}\,dt \approx \$147{,}253$$

The improper integral ($150,000) gives a useful and convenient approximation to this value.

■ **EXAMPLE 5 Perpetual Annuity** An investment pays a dividend continuously at a rate of \$6000/year. Compute the present value of the income stream if the interest rate is 4% and the dividends continue forever.

Solution Recall from Section 7.5 that the present value (PV) after T years at interest rate $r = 0.04$ is $\int_0^T 6000e^{-0.04t}\,dt$. Over an infinite time interval,

$$PV = \int_0^\infty 6000e^{-0.04t}\,dt = \lim_{T\to\infty} \frac{6000e^{-0.04t}}{-0.04}\Big|_0^T = \frac{6000}{0.04} = \$150{,}000$$

Although an infinite number of dollars are paid out during the infinite time interval, their total present value is finite. ■

Infinite Discontinuities at the Endpoints

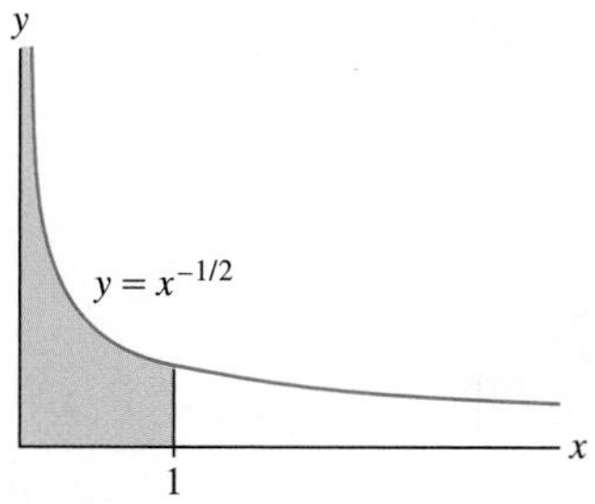

FIGURE 5 The infinite shaded region has area 2 by Example 2(a).

An integral over a finite interval $[a, b]$ is improper if the integrand becomes infinite at one or both of the endpoints of the interval. In this case, the region in question is unbounded in the vertical direction. For example, $\int_0^1 \frac{dx}{\sqrt{x}}$ is improper because the integrand $f(x) = x^{-1/2}$ tends to ∞ as $x \to 0+$ (Figure 5). Improper integrals of this type are defined as one-sided limits.

DEFINITION Integrands with Infinite Discontinuities If $f(x)$ is continuous on $[a, b)$ but discontinuous at $x = b$, we define

$$\int_a^b f(x)\,dx = \lim_{R\to b-} \int_a^R f(x)\,dx$$

Similarly, if $f(x)$ is continuous on $(a, b]$ but discontinuous at $x = a$,

$$\int_a^b f(x)\,dx = \lim_{R\to a+} \int_R^b f(x)\,dx$$

In both cases, we say that the improper integral converges if the limit exists and that it diverges otherwise.

■ **EXAMPLE 6** Calculate: **(a)** $\int_0^9 \frac{dx}{\sqrt{x}}$ and **(b)** $\int_0^{1/2} \frac{dx}{x}$.

Solution Both integrals are improper because the integrands have infinite discontinuities at $x = 0$. The first integral converges:

$$\int_0^9 \frac{dx}{\sqrt{x}} = \lim_{R\to 0+} \int_R^9 x^{-1/2}dx = \lim_{R\to 0+} 2x^{1/2}\Big|_R^9$$

$$= \lim_{R\to 0+} (6 - 2R^{1/2}) = 6$$

The second integral diverges:

$$\int_0^{1/2} \frac{dx}{x} = \lim_{R\to 0+} \int_R^{1/2} \frac{dx}{x} = \lim_{R\to 0+} \left(\ln\frac{1}{2} - \ln R\right)$$

$$= \ln\frac{1}{2} - \lim_{R\to 0+} \ln R = \infty$$ ■

The proof of the next theorem is similar to the proof of Theorem 1 (see Exercise 52).

Theorem 2 is valid for all exponents p. However, the integral is not improper if $p < 0$.

THEOREM 2 The p-Integral over $[0, a]$ For $a > 0$,

$$\int_0^a \frac{dx}{x^p} = \begin{cases} \dfrac{a^{1-p}}{1-p} & \text{if } p < 1 \\ \text{diverges} & \text{if } p \ge 1 \end{cases}$$

GRAPHICAL INSIGHT The p-integrals $\int_a^\infty x^{-p}\,dx$ and $\int_0^a x^{-p}\,dx$ have opposite behavior for $p \neq 1$. The first converges only for $p > 1$, and the second converges only for $p < 1$ (both diverge for $p = 1$). This is reflected in the graphs of $y = x^{-p}$ and $y = x^{-q}$, which switch places at $x = 1$ (Figure 6). We see that a large value of p helps $\int_a^\infty x^{-p}\,dx$ to converge but causes $\int_0^a x^{-p}\,dx$ to diverge.

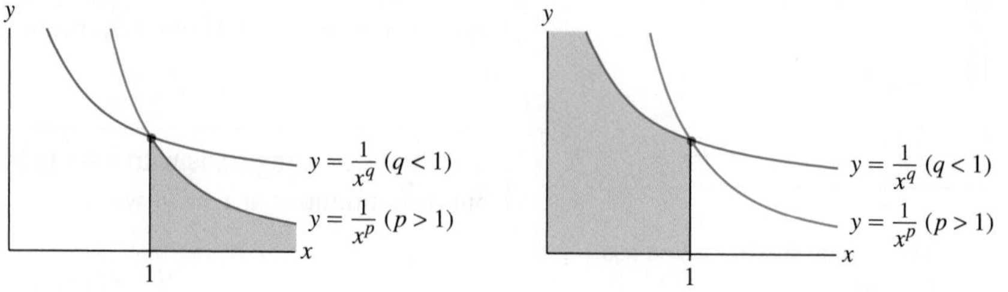

$\int_1^\infty \frac{dx}{x^p} < \infty$ but $\int_1^\infty \frac{dx}{x^q} = \infty$ $\qquad$ $\int_0^1 \frac{dx}{x^p} = \infty$ but $\int_0^1 \frac{dx}{x^q} < \infty$

FIGURE 6

In Section 9.1, we will compute the length of a curve as an integral. It turns out that the improper integral in our next example represents the length of one-quarter of a unit circle. Thus, we can expect its value to be $\frac{1}{4}(2\pi) = \pi/2$.

■ **EXAMPLE 7** Evaluate $\displaystyle\int_0^1 \frac{dx}{\sqrt{1-x^2}}$.

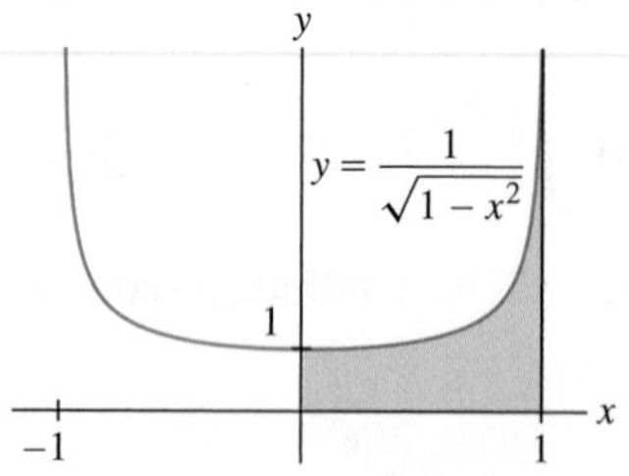

FIGURE 7 The infinite shaded region has area $\frac{\pi}{2}$.

Solution This integral is improper with an infinite discontinuity at $x = 1$ (Figure 7). Using the formula $\int dx/\sqrt{1-x^2} = \sin^{-1} x + C$, we find

$$\int_0^1 \frac{dx}{\sqrt{1-x^2}} = \lim_{R\to 1-} \int_0^R \frac{dx}{\sqrt{1-x^2}}$$

$$= \lim_{R\to 1-} (\sin^{-1} R - \sin^{-1} 0)$$

$$= \sin^{-1} 1 - \sin^{-1} 0 = \frac{\pi}{2} - 0 = \frac{\pi}{2}$$ ■

Comparing Integrals

Sometimes we are interested in determining whether an improper integral converges, even if we cannot find its exact value. For instance, the integral

$$\int_1^\infty \frac{e^{-x}}{x}\,dx$$

cannot be evaluated explicitly. However, if $x \geq 1$, then

$$0 \leq \frac{1}{x} \leq 1 \quad \Rightarrow \quad 0 \leq \frac{e^{-x}}{x} \leq e^{-x}$$

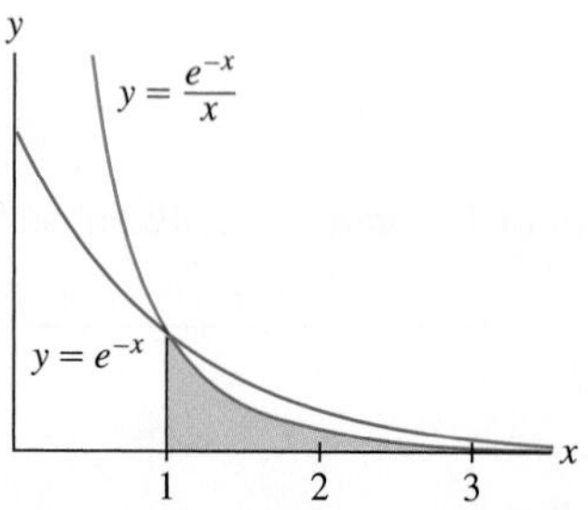

FIGURE 8 There is less area under $y = e^{-x}/x$ than $y = e^{-x}$ over the interval $[1, \infty)$.

In other words, the graph of $y = e^{-x}/x$ lies *underneath* the graph of $y = e^{-x}$ for $x \geq 1$ (Figure 8). Therefore

$$0 \leq \int_1^\infty \frac{e^{-x}}{x}\,dx \leq \underbrace{\int_1^\infty e^{-x}\,dx = e^{-1}}_{\text{Converges by direct computation}}$$

Since the larger integral converges, we can expect that the smaller integral also converges (and that its value is some positive number less than e^{-1}). This type of conclusion is stated in the next theorem. A proof is provided in a supplement on the text's Companion Web Site.

What the Comparison Test says (for nonnegative functions):

- *If the integral of the bigger function converges, then the integral of the smaller function also converges.*
- *If the integral of the smaller function diverges, then the integral of the larger function also diverges.*

THEOREM 3 Comparison Test for Improper Integrals
Assume that $f(x) \geq g(x) \geq 0$ for $x \geq a$.

- If $\int_a^\infty f(x)\,dx$ converges, then $\int_a^\infty g(x)\,dx$ also converges.
- If $\int_a^\infty g(x)\,dx$ diverges, then $\int_a^\infty f(x)\,dx$ also diverges.

The Comparison Test is also valid for improper integrals with infinite discontinuities at the endpoints.

■ **EXAMPLE 8** Show that $\displaystyle\int_1^\infty \frac{dx}{\sqrt{x^3+1}}$ converges.

Solution We cannot evaluate this integral, but we can use the Comparison Test. To show convergence, we must compare the integrand $(x^3+1)^{-1/2}$ with a *larger* function whose integral we can compute.

It makes sense to compare with $x^{-3/2}$ because $\sqrt{x^3} \leq \sqrt{x^3+1}$, and therefore

$$\frac{1}{\sqrt{x^3+1}} \leq \frac{1}{\sqrt{x^3}} = x^{-3/2}$$

The integral of the larger function converges, so the integral of the smaller function also converges:

$$\underbrace{\int_1^\infty \frac{dx}{x^{3/2}}}_{p\text{-integral with } p>1} \quad \text{converges} \quad \Rightarrow \quad \underbrace{\int_1^\infty \frac{dx}{\sqrt{x^3+1}}}_{\text{Integral of smaller function}} \quad \text{converges}$$ ■

■ **EXAMPLE 9 Choosing the Right Comparison** Does $\displaystyle\int_1^\infty \frac{dx}{\sqrt{x}+e^{3x}}$ converge?

Solution Since $\sqrt{x} \geq 0$, we have $\sqrt{x}+e^{3x} \geq e^{3x}$ and therefore

$$\frac{1}{\sqrt{x}+e^{3x}} \leq \frac{1}{e^{3x}}$$

Furthermore,

$$\int_1^\infty \frac{dx}{e^{3x}} = \lim_{R\to\infty} -\frac{1}{3}e^{-3x}\bigg|_1^R = \lim_{R\to\infty} \frac{1}{3}\left(e^{-3}-e^{-3R}\right) = \frac{1}{3}e^{-3} \quad \text{(converges)}$$

Our integral converges by the Comparison Test:

$$\underbrace{\int_1^\infty \frac{dx}{e^{3x}}}_{\text{Integral of larger function}} \quad \text{converges} \quad \Rightarrow \quad \underbrace{\int_1^\infty \frac{dx}{\sqrt{x}+e^{3x}}}_{\text{Integral of smaller function}} \quad \text{also converges}$$

Had we not been thinking, we might have tried to use the inequality

$$\frac{1}{\sqrt{x}+e^{3x}} \leq \frac{1}{\sqrt{x}}$$

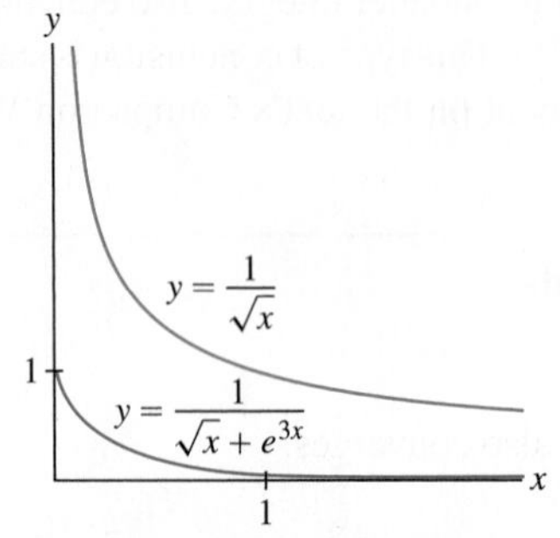

FIGURE 9 The divergence of a larger integral says nothing about the smaller integral.

However, $\int_1^\infty \frac{dx}{\sqrt{x}}$ diverges (p-integral with $p < 1$), and this says nothing about our smaller integral (Figure 9). ■

■ **EXAMPLE 10 Endpoint Discontinuity** Does $J = \int_0^{0.5} \frac{dx}{x^8 + x^2}$ converge?

Solution This integral has a discontinuity at $x = 0$. We might try the comparison

$$x^8 + x^2 > x^2 \quad \Rightarrow \quad \frac{1}{x^8 + x^2} < \frac{1}{x^2}$$

However, the p-integral $\int_0^{0.5} \frac{dx}{x^2}$ diverges, so this says nothing about our integral J, which is smaller. But notice that if $0 < x < 0.5$, then $x^8 < x^2$, and therefore

$$x^8 + x^2 < 2x^2 \quad \Rightarrow \quad \frac{1}{x^8 + x^2} > \frac{1}{2x^2}$$

Since $\int_0^{0.5} \frac{dx}{2x^2}$ diverges, the larger integral J also diverges. ■

8.6 SUMMARY

- An *improper integral* is defined as the limit of ordinary integrals:

$$\int_a^\infty f(x)\,dx = \lim_{R\to\infty} \int_a^R f(x)\,dx$$

The improper integral *converges* if this limit exists, and it *diverges* otherwise.

- If $f(x)$ is continuous on $[a, b)$ but discontinuous at $x = b$, then

$$\int_a^b f(x)\,dx = \lim_{R\to b-} \int_a^R f(x)\,dx$$

- An improper integral of x^{-p} is called a p-integral. For $a > 0$,

$p > 1$:	$\int_a^\infty \frac{dx}{x^p}$ converges	and	$\int_0^a \frac{dx}{x^p}$	diverges
$p < 1$:	$\int_a^\infty \frac{dx}{x^p}$ diverges	and	$\int_0^a \frac{dx}{x^p}$	converges
$p = 1$:	$\int_a^\infty \frac{dx}{x}$	and	$\int_0^a \frac{dx}{x}$	both diverge

- The Comparison Test: Assume that $f(x) \ge g(x) \ge 0$ for $x \ge a$. Then:

If $\int_a^\infty f(x)\,dx$ converges, then $\int_a^\infty g(x)\,dx$ converges.

If $\int_a^\infty g(x)\,dx$ diverges, then $\int_a^\infty f(x)\,dx$ diverges.

- Remember that the Comparison Test provides no information if the larger integral $\int_a^\infty f(x)\,dx$ diverges or the smaller integral $\int_a^\infty g(x)\,dx$ converges.
- The Comparison Test is also valid for improper integrals with infinite discontinuities at endpoints.

8.6 EXERCISES

Preliminary Questions

1. State whether the integral converges or diverges:

(a) $\int_1^\infty x^{-3}\,dx$ **(b)** $\int_0^1 x^{-3}\,dx$

(c) $\int_1^\infty x^{-2/3}\,dx$ **(d)** $\int_0^1 x^{-2/3}\,dx$

2. Is $\int_0^{\pi/2} \cot x\,dx$ an improper integral? Explain.

3. Find a value of $b > 0$ that makes $\int_0^b \frac{1}{x^2-4}\,dx$ an improper integral.

4. Which comparison would show that $\int_0^\infty \frac{dx}{x+e^x}$ converges?

5. Explain why it is not possible to draw any conclusions about the convergence of $\int_1^\infty \frac{e^{-x}}{x}\,dx$ by comparing with the integral $\int_1^\infty \frac{dx}{x}$.

Exercises

1. Which of the following integrals is improper? Explain your answer, but do not evaluate the integral.

(a) $\int_0^2 \frac{dx}{x^{1/3}}\,dx$ **(b)** $\int_1^\infty \frac{dx}{x^{0.2}}$ **(c)** $\int_{-1}^\infty e^{-x}\,dx$

(d) $\int_0^1 e^{-x}\,dx$ **(e)** $\int_0^{\pi/2} \sec x\,dx$ **(f)** $\int_0^\infty \sin x\,dx$

(g) $\int_0^1 \sin x\,dx$ **(h)** $\int_0^1 \frac{dx}{\sqrt{3-x^2}}$ **(i)** $\int_1^\infty \ln x\,dx$

(j) $\int_0^3 \ln x\,dx$

2. Let $f(x) = x^{-4/3}$.

(a) Evaluate $\int_1^R f(x)\,dx$.

(b) Evaluate $\int_1^\infty f(x)\,dx$ by computing the limit

$$\lim_{R\to\infty} \int_1^R f(x)\,dx$$

3. Prove that $\int_1^\infty x^{-2/3}\,dx$ diverges by showing that

$$\lim_{R\to\infty} \int_1^R x^{-2/3}\,dx = \infty$$

4. Determine whether $\int_0^3 \frac{dx}{(3-x)^{3/2}}$ converges by computing

$$\lim_{R\to 3-} \int_0^R \frac{dx}{(3-x)^{3/2}}$$

In Exercises 5–40, determine whether the improper integral converges and, if so, evaluate it.

5. $\int_1^\infty \frac{dx}{x^{19/20}}$

6. $\int_1^\infty \frac{dx}{x^{20/19}}$

7. $\int_{-\infty}^4 e^{0.0001t}\,dt$

8. $\int_{20}^\infty \frac{dt}{t}$

9. $\int_0^5 \frac{dx}{x^{20/19}}$

10. $\int_0^5 \frac{dx}{x^{19/20}}$

11. $\int_0^4 \frac{dx}{\sqrt{4-x}}$

12. $\int_5^6 \frac{dx}{(x-5)^{3/2}}$

13. $\int_2^\infty x^{-3}\,dx$

14. $\int_0^\infty \frac{dx}{(x+1)^3}$

15. $\int_{-3}^\infty \frac{dx}{(x+4)^{3/2}}$

16. $\int_2^\infty e^{-2x}\,dx$

17. $\int_0^1 \frac{dx}{x^{0.2}}$

18. $\int_2^\infty x^{-1/3}\,dx$

19. $\int_4^\infty e^{-3x}\,dx$

20. $\int_4^\infty e^{3x}\,dx$

21. $\int_{-\infty}^0 e^{3x}\,dx$

22. $\int_1^2 \frac{dx}{(x-1)^2}$

23. $\int_1^3 \frac{dx}{\sqrt{3-x}}$

24. $\int_{-2}^4 \frac{dx}{(x+2)^{1/3}}$

25. $\int_0^\infty \frac{dx}{1+x}$

26. $\int_{-\infty}^0 xe^{-x^2}\,dx$

27. $\int_0^\infty \frac{x\,dx}{(1+x^2)^2}$

28. $\int_3^6 \frac{x\,dx}{\sqrt{x-3}}$

29. $\int_0^\infty e^{-x} \cos x\, dx$

30. $\int_1^\infty xe^{-2x}\, dx$

31. $\int_0^3 \frac{dx}{\sqrt{9-x^2}}$

32. $\int_0^1 \frac{e^{\sqrt{x}}\, dx}{\sqrt{x}}$

33. $\int_1^\infty \frac{e^{\sqrt{x}}\, dx}{\sqrt{x}}$

34. $\int_0^{\pi/2} \sec\theta\, d\theta$

35. $\int_0^\infty \sin x\, dx$

36. $\int_0^{\pi/2} \tan x\, dx$

37. $\int_0^1 \ln x\, dx$

38. $\int_1^2 \frac{dx}{x \ln x}$

39. $\int_0^1 \frac{\ln x}{x^2}\, dx$

40. $\int_1^\infty \frac{\ln x}{x^2}\, dx$

41. Let $I = \int_4^\infty \frac{dx}{(x-2)(x-3)}$.

(a) Show that for $R > 4$,

$$\int_4^R \frac{dx}{(x-2)(x-3)} = \ln\left|\frac{R-3}{R-2}\right| - \ln\frac{1}{2}$$

(b) Then show that $I = \ln 2$.

42. Evaluate the integral $I = \int_1^\infty \frac{dx}{x(2x+5)}$.

43. Evaluate $I = \int_0^1 \frac{dx}{x(2x+5)}$ or state that it diverges.

44. Evaluate $I = \int_2^\infty \frac{dx}{(x+3)(x+1)^2}$ or state that it diverges.

In Exercises 45–48, determine whether the doubly infinite improper integral converges and, if so, evaluate it. Use definition (2).

45. $\int_{-\infty}^\infty \frac{x\, dx}{1+x^2}$

46. $\int_{-\infty}^\infty e^{-|x|}\, dx$

47. $\int_{-\infty}^\infty xe^{-x^2}\, dx$

48. $\int_{-\infty}^\infty \frac{dx}{(x^2+1)^{3/2}}$

49. Define $J = \int_{-1}^1 \frac{dx}{x^{1/3}}$ as the sum of the two improper integrals $\int_{-1}^0 \frac{dx}{x^{1/3}} + \int_0^1 \frac{dx}{x^{1/3}}$. Show that J converges and that $J = 0$.

50. Determine whether $J = \int_{-1}^1 \frac{dx}{x^2}$ (defined as in Exercise 49) converges.

51. For which values of a does $\int_0^\infty e^{ax}\, dx$ converge?

52. Show that $\int_0^1 \frac{dx}{x^p}$ converges if $p < 1$ and diverges if $p \ge 1$.

53. Sketch the region under the graph of $f(x) = \frac{1}{1+x^2}$ for $-\infty < x < \infty$, and show that its area is π.

54. Show that $\frac{1}{\sqrt{x^4+1}} \le \frac{1}{x^2}$ for all x, and use this to prove that $\int_1^\infty \frac{dx}{\sqrt{x^4+1}}$ converges.

55. Show that $\int_1^\infty \frac{dx}{x^3+4}$ converges by comparing with $\int_1^\infty x^{-3}\, dx$.

56. Show that $\int_2^\infty \frac{dx}{x^3-4}$ converges by comparing with $\int_2^\infty 2x^{-3}\, dx$.

57. Show that $0 \le e^{-x^2} \le e^{-x}$ for $x \ge 1$ (Figure 10). Use the Comparison Test to show that $\int_0^\infty e^{-x^2}\, dx$ converges. *Hint:* It suffices (why?) to make the comparison for $x \ge 1$ because

$$\int_0^\infty e^{-x^2}\, dx = \int_0^1 e^{-x^2}\, dx + \int_1^\infty e^{-x^2}\, dx$$

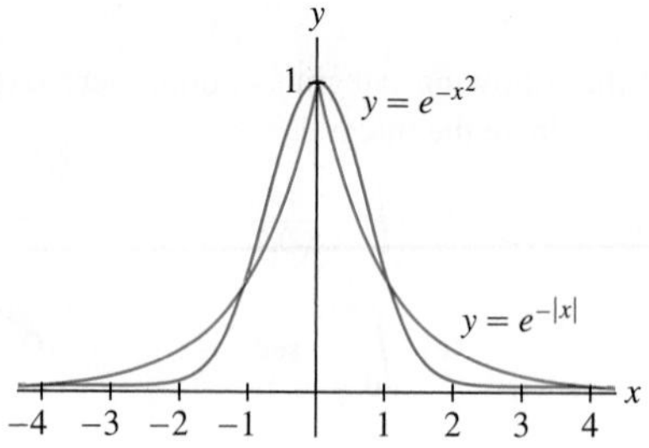

FIGURE 10 Comparison of $y = e^{-|x|}$ and $y = e^{-x^2}$.

58. Prove that $\int_{-\infty}^\infty e^{-x^2}\, dx$ converges by comparing with $\int_{-\infty}^\infty e^{-|x|}\, dx$ (Figure 10).

59. Show that $\int_1^\infty \frac{1-\sin x}{x^2}\, dx$ converges.

60. Let $a > 0$. Use L'Hôpital's Rule to prove that $\lim_{x\to\infty} \frac{x^a}{\ln x} = \infty$. Then:

(a) Show that $x^a > 2\ln x$ for all x sufficiently large.

(b) Show that $e^{-x^a} < x^{-2}$ for all x sufficiently large.

(c) Show that $\int_1^\infty e^{-x^a}\, dx$ converges.

In Exercises 61–74, use the Comparison Test to determine whether or not the integral converges.

61. $\int_1^\infty \frac{1}{\sqrt{x^5+2}}\, dx$

62. $\int_1^\infty \frac{dx}{(x^3+2x+4)^{1/2}}$

63. $\int_3^\infty \frac{dx}{\sqrt{x}-1}$

64. $\int_0^5 \frac{dx}{x^{1/3}+x^3}$

65. $\int_1^\infty e^{-(x+x^{-1})}\,dx$

66. $\int_0^1 \frac{|\sin x|}{\sqrt{x}}\,dx$

67. $\int_0^1 \frac{e^x}{x^2}\,dx$

68. $\int_1^\infty \frac{1}{x^4+e^x}\,dx$

69. $\int_0^1 \frac{1}{x^4+\sqrt{x}}\,dx$

70. $\int_1^\infty \frac{\ln x}{\sinh x}\,dx$

71. $\int_0^\infty \frac{dx}{\sqrt{x^{1/3}+x^3}}$

72. $\int_0^\infty \frac{dx}{(8x^2+x^4)^{1/3}}$

73. $\int_0^\infty \frac{dx}{(x+x^2)^{1/3}}$

74. $\int_0^\infty \frac{dx}{xe^x+x^2}$

Hint for Exercise 73: Show that for $x \ge 1$,

$$\frac{1}{(x+x^2)^{1/3}} \ge \frac{1}{2^{1/3}x^{2/3}}$$

Hint for Exercise 74: Show that for $0 \le x \le 1$,

$$\frac{1}{xe^x+x^2} \ge \frac{1}{(e+1)x}$$

75. Define $J = \int_0^\infty \frac{dx}{x^{1/2}(x+1)}$ as the sum of the two improper integrals

$$\int_0^1 \frac{dx}{x^{1/2}(x+1)} + \int_1^\infty \frac{dx}{x^{1/2}(x+1)}$$

Use the Comparison Test to show that J converges.

76. Determine whether $J = \int_0^\infty \frac{dx}{x^{3/2}(x+1)}$ (defined as in Exercise 75) converges.

77. An investment pays a dividend of \$250/year continuously forever. If the interest rate is 7%, what is the present value of the entire income stream generated by the investment?

78. An investment is expected to earn profits at a rate of $10{,}000e^{0.01t}$ dollars per year forever. Find the present value of the income stream if the interest rate is 4%.

79. Compute the present value of an investment that generates income at a rate of $5000te^{0.01t}$ dollars per year forever, assuming an interest rate of 6%.

80. Find the volume of the solid obtained by rotating the region below the graph of $y = e^{-x}$ about the x-axis for $0 \le x < \infty$.

81. The solid S obtained by rotating the region below the graph of $y = x^{-1}$ about the x-axis for $1 \le x < \infty$ is called **Gabriel's Horn** (Figure 11).

(a) Use the Disk Method (Section 6.3) to compute the volume of S. Note that the volume is finite even though S is an infinite region.

(b) It can be shown that the surface area of S is

$$A = 2\pi \int_1^\infty x^{-1}\sqrt{1+x^{-4}}\,dx$$

Show that A is infinite. If S were a container, you could fill its interior with a finite amount of paint, but you could not paint its surface with a finite amount of paint.

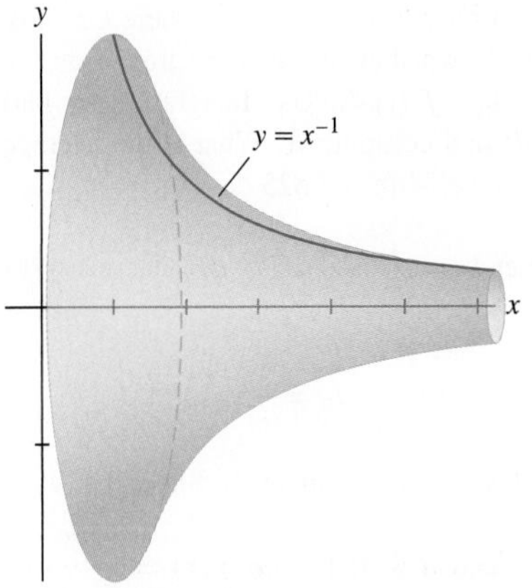

FIGURE 11

82. Compute the volume of the solid obtained by rotating the region below the graph of $y = e^{-|x|/2}$ about the x-axis for $-\infty < x < \infty$.

83. When a capacitor of capacitance C is charged by a source of voltage V, the power expended at time t is

$$P(t) = \frac{V^2}{R}(e^{-t/RC} - e^{-2t/RC})$$

where R is the resistance in the circuit. The total energy stored in the capacitor is

$$W = \int_0^\infty P(t)\,dt$$

Show that $W = \frac{1}{2}CV^2$.

84. For which integers p does $\int_0^{1/2} \frac{dx}{x(\ln x)^p}$ converge?

85. Conservation of Energy can be used to show that when a mass m oscillates at the end of a spring with spring constant k, the period of oscillation is

$$T = 4\sqrt{m}\int_0^{\sqrt{2E/k}} \frac{dx}{\sqrt{2E-kx^2}}$$

where E is the total energy of the mass. Show that this is an improper integral with value $T = 2\pi\sqrt{m/k}$.

In Exercises 86–89, the ***Laplace transform*** *of a function $f(x)$ is the function $Lf(s)$ of the variable s defined by the improper integral (if it converges):*

$$Lf(s) = \int_0^\infty f(x)e^{-sx}\,dx$$

Laplace transforms are widely used in physics and engineering.

86. Show that if $f(x) = C$, where C is a constant, then $Lf(s) = C/s$ for $s > 0$.

87. Show that if $f(x) = \sin\alpha x$, then $Lf(s) = \frac{\alpha}{s^2+\alpha^2}$.

88. Compute $Lf(s)$, where $f(x) = e^{\alpha x}$ and $s > \alpha$.

89. Compute $Lf(s)$, where $f(x) = \cos \alpha x$ and $s > 0$.

90. When a radioactive substance decays, the fraction of atoms present at time t is $f(t) = e^{-kt}$, where $k > 0$ is the decay constant. It can be shown that the *average* life of an atom (until it decays) is $A = -\int_0^\infty t f'(t)\,dt$. Use Integration by Parts to show that $A = \int_0^\infty f(t)\,dt$ and compute A. What is the average decay time of radon-222, whose half-life is 3.825 days?

91. Let $J_n = \displaystyle\int_0^\infty x^n e^{-\alpha x}\,dx$, where $n \geq 1$ is an integer and $\alpha > 0$. Prove that

$$J_n = \frac{n}{\alpha} J_{n-1}$$

and $J_0 = 1/\alpha$. Use this to compute J_4. Show that $J_n = n!/\alpha^{n+1}$.

92. Let $a > 0$ and $n > 1$. Define $f(x) = \dfrac{x^n}{e^{ax} - 1}$ for $x \neq 0$ and $f(0) = 0$.

(a) Use L'Hôpital's Rule to show that $f(x)$ is continuous at $x = 0$.

(b) Show that $\int_0^\infty f(x)\,dx$ converges. *Hint:* Show that $f(x) \leq 2x^n e^{-ax}$ if x is large enough. Then use the Comparison Test and Exercise 91.

93. According to **Planck's Radiation Law**, the amount of electromagnetic energy with frequency between ν and $\nu + \Delta\nu$ that is radiated by a so-called black body at temperature T is proportional to $F(\nu)\,\Delta\nu$, where

$$F(\nu) = \left(\frac{8\pi h}{c^3}\right) \frac{\nu^3}{e^{h\nu/kT} - 1}$$

where c, h, k are physical constants. Use Exercise 92 to show that the total radiated energy

$$E = \int_0^\infty F(\nu)\,d\nu$$

is finite. To derive his law, Planck introduced the quantum hypothesis in 1900, which marked the birth of quantum mechanics.

Further Insights and Challenges

94. Let $I = \displaystyle\int_0^1 x^p \ln x\,dx$.

(a) Show that I diverges for $p = -1$.

(b) Show that if $p \neq -1$, then

$$\int x^p \ln x\,dx = \frac{x^{p+1}}{p+1}\left(\ln x - \frac{1}{p+1}\right) + C$$

(c) Use L'Hôpital's Rule to show that I converges if $p > -1$ and diverges if $p < -1$.

95. Let

$$F(x) = \int_2^x \frac{dt}{\ln t} \quad \text{and} \quad G(x) = \frac{x}{\ln x}$$

Verify that L'Hôpital's Rule applies to the limit $L = \displaystyle\lim_{x\to\infty} \frac{F(x)}{G(x)}$ and evaluate L.

*In Exercises 96–98, an improper integral $I = \int_a^\infty f(x)\,dx$ is called **absolutely convergent** if $\int_a^\infty |f(x)|\,dx$ converges. It can be shown that if I is absolutely convergent, then it is convergent.*

96. Show that $\displaystyle\int_1^\infty \frac{\sin x}{x^2}\,dx$ is absolutely convergent.

97. Show that $\displaystyle\int_1^\infty e^{-x^2} \cos x\,dx$ is absolutely convergent.

98. Let $f(x) = \sin x/x$ and $I = \int_0^\infty f(x)\,dx$. We define $f(0) = 1$. Then $f(x)$ is continuous and I is not improper at $x = 0$.

(a) Show that

$$\int_1^R \frac{\sin x}{x}\,dx = -\frac{\cos x}{x}\bigg|_1^R - \int_1^R \frac{\cos x}{x^2}\,dx$$

(b) Show that $\int_1^\infty (\cos x/x^2)\,dx$ converges. Conclude that the limit as $R \to \infty$ of the integral in (a) exists and is finite.

(c) Show that I converges.

It is known that $I = \frac{\pi}{2}$. However, I is *not* absolutely convergent. The convergence depends on cancellation, as shown in Figure 12.

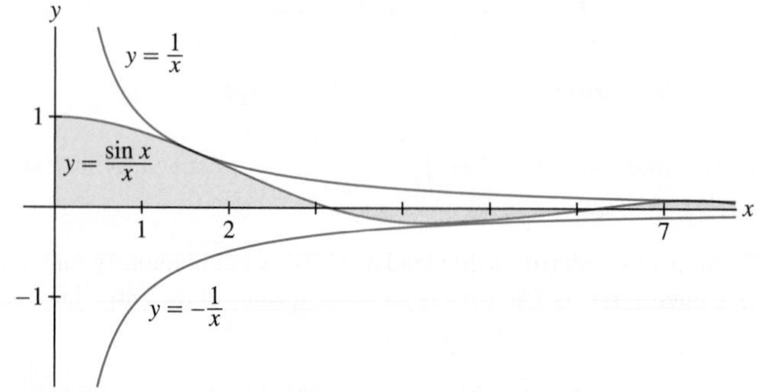

FIGURE 12 Convergence of $\int_1^\infty (\sin x/x)\,dx$ is due to the cancellation arising from the periodic change of sign.

99. The **gamma function**, which plays an important role in advanced applications, is defined for $n \geq 1$ by

$$\Gamma(n) = \int_0^\infty t^{n-1} e^{-t}\,dt$$

(a) Show that the integral defining $\Gamma(n)$ converges for $n \geq 1$ (it actually converges for all $n > 0$). *Hint:* Show that $t^{n-1}e^{-t} < t^{-2}$ for t sufficiently large.

(b) Show that $\Gamma(n+1) = n\Gamma(n)$ using Integration by Parts.

(c) Show that $\Gamma(n+1) = n!$ if $n \geq 1$ is an integer. *Hint:* Use (a) repeatedly. Thus, $\Gamma(n)$ provides a way of defining n-factorial when n is not an integer.

100. Use the results of Exercise 99 to show that the Laplace transform (see Exercises 86–89 above) of x^n is $\dfrac{n!}{s^{n+1}}$.

8.7 Probability and Integration

What is the probability that a customer will arrive at a fast-food restaurant in the next 45 seconds? Or of scoring above 90% on a standardized test? Probabilities such as these are described as areas under the graph of a function $p(x)$ called a **probability density function** (Figure 1). The methods of integration developed in this chapter are used extensively in the study of such functions.

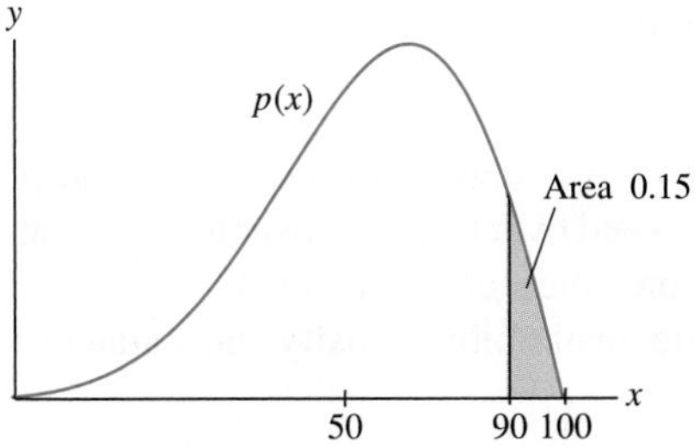

FIGURE 1 Probability density function for scores on an exam. The shaded region has area 0.15, so there is a 15% probability that a randomly chosen exam has a score above 90.

In probability theory, the quantity X that we are trying to predict (time to arrival, exam score, etc.) is called a **random variable**. The probability that X lies in a given range $[a, b]$ is denoted

$$P(a \le X \le b)$$

For example, the probability of a customer arriving within the next 30 to 45 seconds is denoted $P(30 \le X \le 45)$.

We say that X is a **continuous random variable** if there is a continuous probability density function $p(x)$ such that

$$P(a \le X \le b) = \int_a^b p(x)\,dx$$

We write $P(X \le b)$ for the probability that X is at most b, and $P(X \ge b)$ for the probability that X is at least b.

A probability density function $p(x)$ must satisfy two conditions. First, it must satisfy $p(x) \ge 0$ for all x, because a probability cannot be negative. Second,

$$\int_{-\infty}^{\infty} p(x) = 1 \qquad \boxed{1}$$

The integral represents $P(-\infty < X < \infty)$. It must equal 1 because it is certain (the probability is 1) that the value of X lies between $-\infty$ and ∞.

■ **EXAMPLE 1** Find a constant C for which $p(x) = \dfrac{C}{x^2+1}$ is a probability density function. Then compute $P(1 \le X \le 4)$.

Solution We must choose C so that Eq. (1) is satisfied. The improper integral is a sum of two integrals (see marginal note)

$$\int_{-\infty}^{\infty} p(x)\,dx = C\int_{-\infty}^{0} \frac{dx}{x^2+1} + C\int_0^{\infty} \frac{dx}{x^2+1} = C\frac{\pi}{2} + C\frac{\pi}{2} = C\pi$$

Therefore, Eq. (1) is satisfied if $C\pi = 1$ or $C = \pi^{-1}$. We have

$$P(1 < X < 4) = \int_1^4 p(x)\,dx = \int_1^4 \frac{\pi^{-1}\,dx}{x^2+1} = \pi^{-1}(\tan^{-1} 4 - \tan^{-1} 1) \approx 0.17$$

Therefore, X lies between 1 and 4 with probability 0.17, or 17% (Figure 2). ■

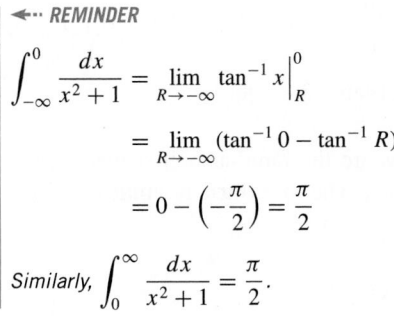

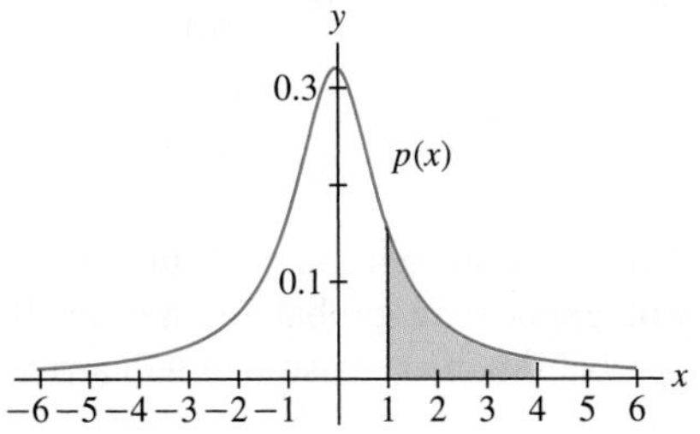

FIGURE 2 The probability density function $p(x) = \dfrac{1}{\pi(x^2+1)}$.

← REMINDER

$$\int_{-\infty}^{0} \frac{dx}{x^2+1} = \lim_{R\to-\infty} \tan^{-1} x\Big|_R^0 = \lim_{R\to-\infty} (\tan^{-1} 0 - \tan^{-1} R) = 0 - \left(-\frac{\pi}{2}\right) = \frac{\pi}{2}$$

Similarly, $\displaystyle\int_0^{\infty} \frac{dx}{x^2+1} = \frac{\pi}{2}$.

CONCEPTUAL INSIGHT If X is a continuous random variable, then the probability of X taking on any specific value a is zero because $\int_a^a p(x)\,dx = 0$. If so, what is the meaning of $p(a)$? We must think of it this way: the probability that X lies in a *small interval* $[a, a+\Delta x]$ is approximately $p(a)\Delta x$:

$$P(a \le X \le a+\Delta x) = \int_a^{a+\Delta x} p(x)\,dx \approx p(a)\Delta x$$

A probability density is similar to a linear mass density $\rho(x)$. The mass of a small segment $[a, a+\Delta x]$ is approximately $\rho(a)\Delta x$, but the mass of any particular point $x = a$ is zero.

The *mean* or *average value* of a random variable is the quantity

$$\mu = \mu(X) = \int_{-\infty}^{\infty} xp(x)\,dx \tag{2}$$

The symbol μ is a lowercase Greek letter mu. If $p(x)$ is defined on $[0, \infty)$ instead of $(-\infty, \infty)$, or on some other interval, then μ is computed by integrating over that interval. Similarly, in Eq. (1) we integrate over the interval on which $p(x)$ is defined.

In the next example, we consider the **exponential probability density** with parameter $r > 0$, defined on $[0, \infty)$ by

$$p(t) = \frac{1}{r}e^{-t/r}$$

This density function is often used to model "waiting times" between events that occur randomly. Exercise 10 asks you to verify that $p(t)$ satisfies Eq. (1).

■ **EXAMPLE 2 Mean of an Exponential Density** Let $r > 0$. Calculate the mean of the exponential probability density $p(t) = \frac{1}{r}e^{-t/r}$ on $[0, \infty)$.

Solution The mean is the integral of $tp(t)$ over $[0, \infty)$. Using Integration by Parts with $u = t/r$ and $v' = e^{-t/r}$, we have $u' = 1/r$, $v = -re^{-t/r}$, and

$$\int tp(t)\,dt = \int \left(\frac{t}{r}e^{-t/r}\right) dt = -te^{-t/r} + \int e^{-t/r}\,dt = -(r+t)e^{-t/r}$$

Thus (using that $re^{-R/r}$ and $Re^{-R/r}$ both tend to zero as $R \to \infty$ in the last step),

$$\mu = \int_0^{\infty} tp(t)\,dt = \int_0^{\infty} t\left(\frac{1}{r}e^{-t/r}\right) dt = \lim_{R\to\infty} -(r+t)e^{-t/r}\Big|_0^R$$

$$= \lim_{R\to\infty}\left(r - (r+R)\,e^{-R/r}\right) = r$$

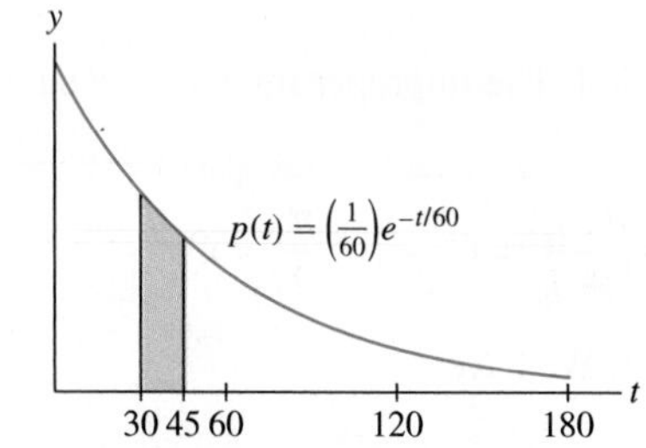

FIGURE 3 Customer arrivals have an exponential distribution.

■ **EXAMPLE 3 Waiting Time** The waiting time T between customer arrivals in a drive-through fast-food restaurant is a random variable with exponential probability density. If the average waiting time is 60 seconds, what is the probability that a customer will arrive within 30 to 45 seconds after another customer?

Solution If the average waiting time is 60 seconds, then $r = 60$ and $p(t) = \frac{1}{60}e^{-t/60}$ because the mean of $p(t)$ is r by the previous example. Therefore, the probability of waiting between 30 and 45 seconds for the next customer is

$$P(30 \le T \le 45) = \int_{30}^{45} \frac{1}{60}e^{-t/60} = -e^{-t/60}\Big|_{30}^{45} = -e^{-3/4} + e^{-1/2} \approx 0.134$$

This probability is the area of the shaded region in Figure 3. ■

The **normal density functions**, whose graphs are the familiar bell-shaped curves, appear in a surprisingly wide range of applications. The **standard normal** density is defined by

$$p(x) = \frac{1}{\sqrt{2\pi}}\,e^{-x^2/2} \tag{3}$$

We can prove that $p(x)$ satisfies Eq. (1) using multivariable calculus.

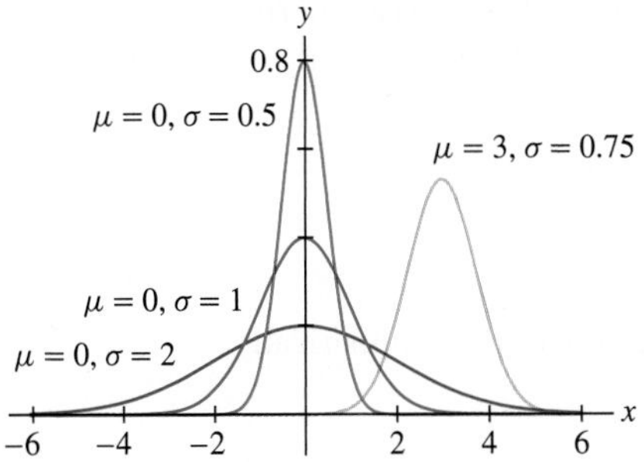

FIGURE 4 Normal density functions.

More generally, we define the normal density function with mean μ and standard deviation σ:

$$p(x) = \frac{1}{\sigma\sqrt{2\pi}} e^{-(x-\mu)^2/(2\sigma^2)}$$

The standard deviation σ measures the spread; for larger values of σ the graph is more spread out about the mean μ (Figure 4). The standard normal density in Eq. (3) has mean $\mu = 0$ and $\sigma = 1$. A random variable with a normal density function is said to have a **normal** or **Gaussian distribution**.

One difficulty is that normal density functions do not have elementary antiderivatives. As a result, we cannot evaluate the probabilities

$$P(a \le X \le b) = \frac{1}{\sigma\sqrt{2\pi}} \int_a^b e^{-(x-\mu)^2/(2\sigma^2)}\, dx$$

explicitly. However, the next theorem shows that these probabilities can all be expressed in terms of a single function called the **standard normal cumulative distribution function**:

$$F(z) = \frac{1}{\sqrt{2\pi}} \int_{-\infty}^{z} e^{-x^2/2}\, dx$$

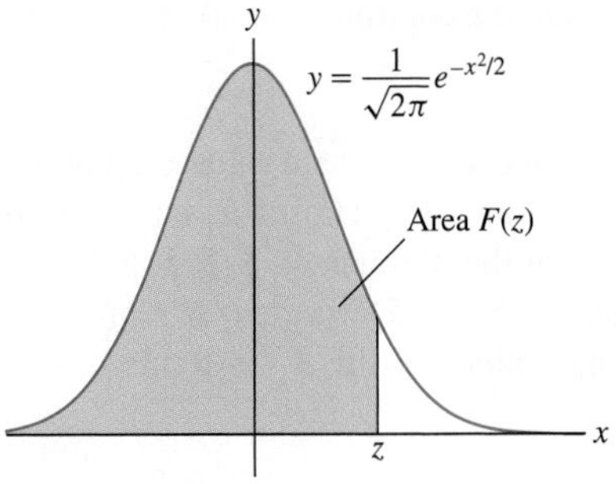

FIGURE 5 $F(z)$ is the area of the shaded region.

Observe that $F(z)$ is equal to the area under the graph over $(-\infty, z]$ in Figure 5. Numerical values of $F(z)$ are widely available on scientific calculators, on computer algebra systems, and online (search "standard cumulative normal distribution").

THEOREM 1 If X has a normal distribution with mean μ and standard deviation σ, then for all $a \le b$,

$$P(X \le b) = F\left(\frac{b-\mu}{\sigma}\right) \qquad \boxed{4}$$

$$P(a \le X \le b) = F\left(\frac{b-\mu}{\sigma}\right) - F\left(\frac{a-\mu}{\sigma}\right) \qquad \boxed{5}$$

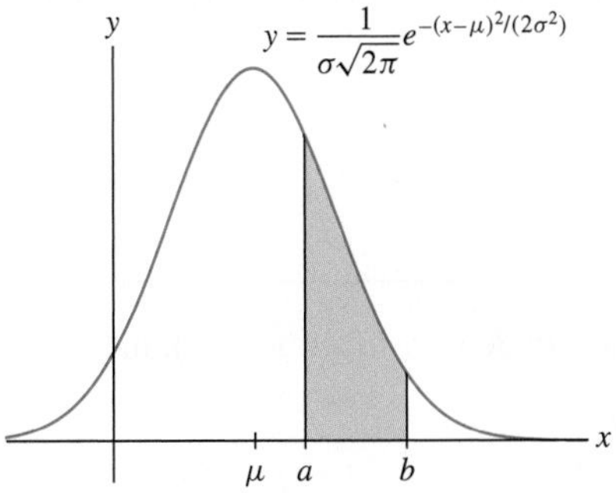

FIGURE 6 The shaded region has area $F\left(\frac{b-\mu}{\sigma}\right) - F\left(\frac{a-\mu}{\sigma}\right)$.

Proof We use two changes of variables, first $u = x - \mu$ and then $t = u/\sigma$:

$$P(X \le b) = \frac{1}{\sigma\sqrt{2\pi}} \int_{-\infty}^{b} e^{-(x-\mu)^2/(2\sigma^2)}\, dx = \frac{1}{\sigma\sqrt{2\pi}} \int_{-\infty}^{b-\mu} e^{-u^2/(2\sigma^2)}\, du$$

$$= \frac{1}{\sqrt{2\pi}} \int_{-\infty}^{(b-\mu)/\sigma} e^{-t^2/2}\, dt = F\left(\frac{b-\mu}{\sigma}\right)$$

This proves Eq. (4). Eq. (5) follows because $P(a \le X \le b)$ is the area under the graph between a and b, and this is equal to the area to the left of b minus the area to the left of a (Figure 6). ■

■ **EXAMPLE 4** Assume that the scores X on a standardized test are normally distributed with mean $\mu = 500$ and standard deviation $\sigma = 100$. Find the probability that a test chosen at random has score

(a) at most 600.

(b) between 450 and 650.

Solution We use a computer algebra system to evaluate $F(z)$ numerically.

(a) Apply Eq. (4) with $\mu = 500$ and $\sigma = 100$:

$$P(x \le 600) = F\left(\frac{600 - 500}{100}\right) = F(1) \approx 0.84$$

Thus, a randomly chosen score is 600 or less with a probability of 0.84, or 84%.

(b) Applying Eq. (5), we find that a randomly chosen score lies between 450 and 650 with a probability of 62.5%:

$$P(450 \le x \le 650) = F(1.5) - F(-0.5) \approx 0.933 - 0.308 = 0.625$$ ■

CONCEPTUAL INSIGHT Why have we defined the mean of a continuous random variable X as the integral $\mu = \int_{-\infty}^{\infty} xp(x)\,dx$?

Suppose first we are given N numbers $a_1, a_2, \dots, a_N$, and for each value x, let $N(x)$ be the number of times x occurs among the a_j. Then a randomly chosen a_j has value x with probability $p(x) = N(x)/N$. For example, given the numbers 4, 4, 5, 5, 5, 8, we have $N = 6$ and $N(5) = 3$. The probability of choosing a 5 is $p(5) = N(5)/N = \frac{3}{6} = \frac{1}{2}$. Now observe that we can write the mean (average value) of the a_j in terms of the probabilities $p(x)$:

$$\frac{a_1 + a_2 + \cdots + a_N}{N} = \frac{1}{N}\sum_x N(x)x = \sum_x xp(x)$$

For example,

$$\frac{4+4+5+5+5+8}{6} = \frac{1}{6}(2\cdot 4 + 3\cdot 5 + 1\cdot 8) = 4p(4) + 5p(5) + 8p(8)$$

In defining the mean of a continuous random variable X, we replace the sum $\sum_x xp(x)$ with the integral $\mu = \int_{-\infty}^{\infty} xp(x)\,dx$. This makes sense because the integral is the limit of sums $\sum x_i p(x_i)\Delta x$, and as we have seen, $p(x_i)\Delta x$ is the approximate probability that X lies in $[x_i, x_i + \Delta x]$.

8.7 SUMMARY

- If X is a continuous random variable with probability density function $p(x)$, then

$$P(a \le X \le b) = \int_a^b p(x)\,dx$$

- Probability densities satisfy two conditions: $p(x) \ge 0$ and $\int_{-\infty}^{\infty} p(x)\,dx = 1$.

- Mean (or average) value of X:

$$\mu = \int_{-\infty}^{\infty} xp(x)\,dx$$

- Exponential density function of mean r:

$$p(x) = \frac{1}{r}\,e^{-x/r}$$

- Normal density of mean μ and standard deviation σ:

$$p(x) = \frac{1}{\sigma\sqrt{2\pi}}\,e^{-(x-\mu)^2/(2\sigma^2)}$$

- Standard cumulative normal distribution function:

$$F(z) = \frac{1}{\sqrt{2\pi}}\int_{-\infty}^{z} e^{-t^2/2}\,dt$$

- If X has a normal distribution of mean μ and standard deviation σ, then

$$P(X \le b) = F\left(\frac{b-\mu}{\sigma}\right)$$

$$P(a \le X \le b) = F\left(\frac{b-\mu}{\sigma}\right) - F\left(\frac{a-\mu}{\sigma}\right)$$

8.7 EXERCISES

Preliminary Questions

1. The function $p(x) = \frac{1}{2}\cos x$ satisfies $\displaystyle\int_0^{\pi} p(x)\,dx = 1$. Is $p(x)$ a probability density function on $[0, \pi]$?

2. Estimate $P(2 \le X \le 2.1)$ assuming that the probability density function of X satisfies $p(2) = 0.2$.

3. Which exponential probability density has mean $\mu = \frac{1}{4}$?

Exercises

In Exercises 1–6, find a constant C such that p(x) is a probability density function on the given interval, and compute the probability indicated.

1. $p(x) = \dfrac{C}{(x+1)^3}$ on $[0, \infty)$; $P(0 \le X \le 1)$

2. $p(x) = Cx(4-x)$ on $[0, 4]$; $P(3 \le X \le 4)$

3. $p(x) = \dfrac{C}{\sqrt{1-x^2}}$ on $(-1, 1)$; $P\big(-\frac{1}{2} \le X \le \frac{1}{2}\big)$

4. $p(x) = \dfrac{Ce^{-x}}{1+e^{-2x}}$ on $(-\infty, \infty)$; $P(X \le -4)$

5. $p(x) = C\sqrt{1-x^2}$ on $(-1, 1)$; $P\big(-\frac{1}{2} \le X \le 1\big)$

6. $p(x) = Ce^{-x}e^{-e^{-x}}$ on $(-\infty, \infty)$; $P(-4 \le X \le 4)$
This function, called the **Gumbel density**, is used to model extreme events such as floods and earthquakes.

7. Verify that $p(x) = 3x^{-4}$ is a probability density function on $[1, \infty)$ and calculate its mean value.

8. Show that the density function $p(x) = \dfrac{2}{\pi(x^2+1)}$ on $[0, \infty)$ has infinite mean.

9. Verify that $p(t) = \frac{1}{50}e^{-t/50}$ satisfies the condition $\int_0^{\infty} p(t)\,dt = 1$.

10. Verify that for all $r > 0$, the exponential density function $p(t) = \frac{1}{r}e^{-t/r}$ satisfies the condition $\int_0^{\infty} p(t)\,dt = 1$.

11. The life X (in hours) of a battery in constant use is a random variable with exponential density. What is the probability that the battery will last more than 12 hours if the average life is 8 hours?

12. The time between incoming phone calls at a call center is a random variable with exponential density. There is a 50% probability of waiting 20 seconds or more between calls. What is the average time between calls?

13. The distance r between the electron and the nucleus in a hydrogen atom (in its lowest energy state) is a random variable with probability density $p(r) = 4a_0^{-3}r^2e^{-2r/a_0}$ for $r \ge 0$, where a_0 is the Bohr radius (Figure 7). Calculate the probability P that the electron is within one Bohr radius of the nucleus. The value of a_0 is approximately 5.29×10^{-11} m, but this value is not needed to compute P.

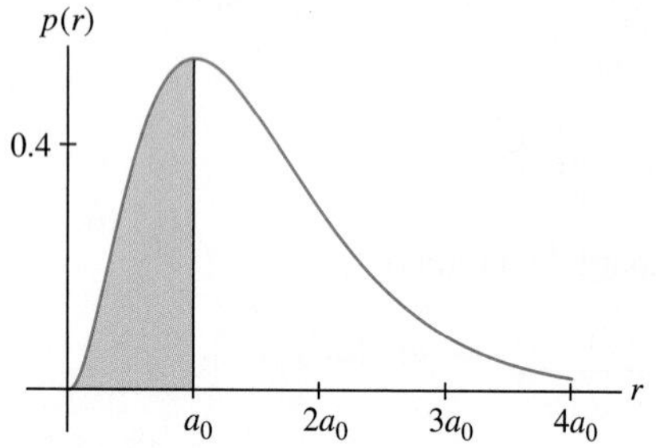

FIGURE 7 Probability density function $p(r) = 4a_0^{-3}r^2e^{-2r/a_0}$.

14. Show that the distance r between the electron and the nucleus in Exercise 13 has mean $\mu = 3a_0/2$.

In Exercises 15–21, $F(z)$ denotes the cumulative normal distribution function. Refer to a calculator, computer algebra system, or online resource to obtain values of $F(z)$.

15. Express the area of region A in Figure 8 in terms of $F(z)$ and compute its value.

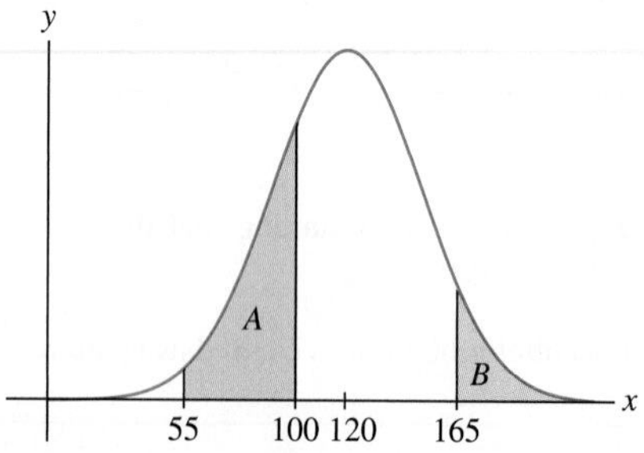

FIGURE 8 Normal density function with $\mu = 120$ and $\sigma = 30$.

16. Show that the area of region B in Figure 8 is equal to $1 - F(1.5)$ and compute its value. Verify numerically that this area is also equal to $F(-1.5)$ and explain why graphically.

17. Assume X has a standard normal distribution ($\mu = 0, \sigma = 1$). Find:

(a) $P(X \le 1.2)$ **(b)** $P(X \ge -0.4)$

18. Evaluate numerically: $\dfrac{1}{3\sqrt{2\pi}} \displaystyle\int_{14.5}^{\infty} e^{-(z-10)^2/18}\, dz$.

19. Use a graph to show that $F(-z) = 1 - F(z)$ for all z. Then show that if $p(x)$ is a normal density function with mean μ and standard deviation σ, then for all $r \ge 0$,

$$P(\mu - r\sigma \le X \le \mu + r\sigma) = 2F(r) - 1$$

20. The average September rainfall in Erie, Pennsylvania, is a random variable X with mean $\mu = 102$ mm. Assume that the amount of rainfall is normally distributed with standard deviation $\sigma = 48$.

(a) Express $P(128 \le X \le 150)$ in terms of $F(z)$ and compute its value numerically.

(b) Let P be the probability that September rainfall will be at least 120 mm. Express P as an integral of an appropriate density function and compute its value numerically.

21. A bottling company produces bottles of fruit juice that are filled, on average, with 32 ounces of juice. Due to random fluctuations in the machinery, the actual volume of juice is normally distributed with a standard deviation of 0.4 ounce. Let P be the probability of a bottle having less than 31 ounces. Express P as an integral of an appropriate density function and compute its value numerically.

22. According to **Maxwell's Distribution Law**, in a gas of molecular mass m, the speed v of a molecule in a gas at temperature T (kelvins) is a random variable with density

$$p(v) = 4\pi\left(\frac{m}{2\pi kT}\right)^{3/2} v^2 e^{-mv^2/(2kT)} \qquad (v \ge 0)$$

where k is Boltzmann's constant. Show that the average molecular speed is equal to $(8kT/\pi m)^{1/2}$. The average speed of oxygen molecules at room temperature is around 450 m/s.

In Exercises 23–26, calculate μ and σ, where σ is the ***standard deviation****, defined by*

$$\sigma^2 = \int_{-\infty}^{\infty} (x - \mu)^2\, p(x)\, dx$$

The smaller the value of σ, the more tightly clustered are the values of the random variable X about the mean μ.

23. $p(x) = \dfrac{5}{2x^{7/2}}$ on $[1, \infty)$

24. $p(x) = \dfrac{1}{\pi\sqrt{1 - x^2}}$ on $(-1, 1)$

25. $p(x) = \dfrac{1}{3}e^{-x/3}$ on $[0, \infty)$

26. $p(x) = \dfrac{1}{r}e^{-x/r}$ on $[0, \infty)$, where $r > 0$

Further Insights and Challenges

27. The time to decay of an atom in a radioactive substance is a random variable X. The law of radioactive decay states that if N atoms are present at time $t = 0$, then $Nf(t)$ atoms will be present at time t, where $f(t) = e^{-kt}$ ($k > 0$ is the decay constant). Explain the following statements:

(a) The fraction of atoms that decay in a small time interval $[t, t + \Delta t]$ is approximately $-f'(t)\Delta t$.

(b) The probability density function of X is $-f'(t)$.

(c) The average time to decay is $1/k$.

28. The half-life of radon-222, is 3.825 days. Use Exercise 27 to compute:

(a) The average time to decay of a radon-222 atom.

(b) The probability that a given atom will decay in the next 24 hours.

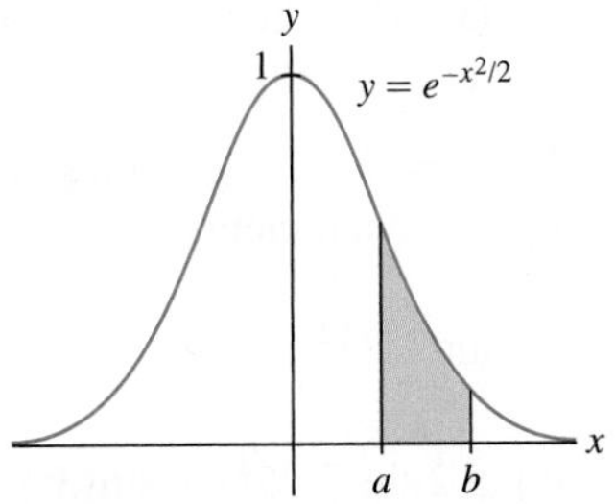

FIGURE 1 Areas under the bell-shaped curve are computed using numerical integration.

8.8 Numerical Integration

Numerical integration is the process of approximating a definite integral using well-chosen sums of function values. It is needed when we cannot find an antiderivative explicitly, as in the case of the Gaussian function $f(x) = e^{-x^2/2}$ (Figure 1).

To approximate the definite integral $\int_a^b f(x)\,dx$, we fix a whole number N and divide $[a, b]$ into N subintervals of length $\Delta x = (b - a)/N$. The endpoints of the subintervals (Figure 2) are

$$x_0 = a, \qquad x_1 = a + \Delta x, \qquad x_2 = a + 2\Delta x, \qquad \dots, \qquad x_N = b$$

We shall denote the values of $f(x)$ at these endpoints by y_j:

$$y_j = f(x_j) = f(a + j\Delta x)$$

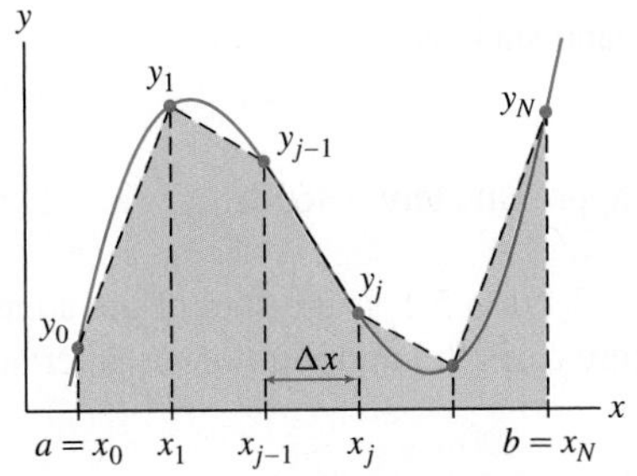

FIGURE 2 T_N approximates the area under the graph by trapezoids.

In particular, $y_0 = f(a)$ and $y_N = f(b)$.

The **Trapezoidal Rule** T_N approximates $\int_a^b f(x)\,dx$ by the area of the trapezoids obtained by joining the points $(x_0, y_0), (x_1, y_1), \dots, (x_N, y_N)$ with line segments as in Figure 2. The area of the jth trapezoid is $\frac{1}{2}\Delta x(y_{j-1} + y_j)$, and therefore,

$$\begin{aligned} T_N &= \frac{1}{2}\Delta x(y_0 + y_1) + \frac{1}{2}\Delta x(y_1 + y_2) + \cdots + \frac{1}{2}\Delta x(y_{N-1} + y_N) \\ &= \frac{1}{2}\Delta x\Big((y_0 + y_1) + (y_1 + y_2) + \cdots + (y_{N-1} + y_N)\Big) \end{aligned}$$

Note that each value y_j occurs twice except for y_0 and y_N, so we obtain

$$T_N = \frac{1}{2}\Delta x\Big(y_0 + 2y_1 + 2y_2 + \cdots + 2y_{N-1} + y_N\Big)$$

Trapezoidal Rule The Nth trapezoidal approximation to $\displaystyle\int_a^b f(x)\,dx$ is

$$T_N = \frac{1}{2}\Delta x\big(y_0 + 2y_1 + \cdots + 2y_{N-1} + y_N\big)$$

where $\Delta x = \dfrac{b - a}{N}$ and $y_j = f(a + j\,\Delta x)$.

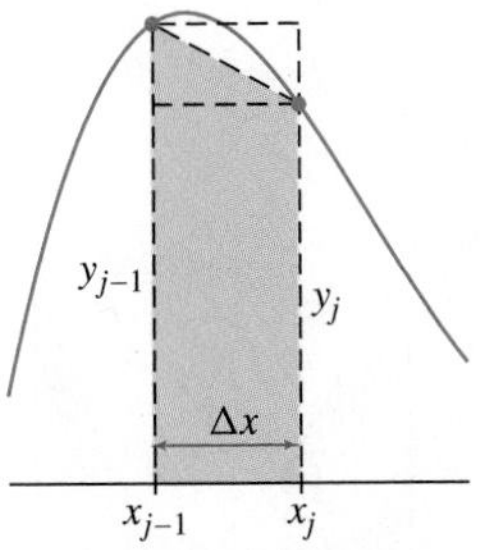

FIGURE 3 The shaded trapezoid has area $\frac{1}{2}\Delta x(y_{j-1} + y_j)$. This is the average of the areas of the left- and right-endpoint rectangles.

CONCEPTUAL INSIGHT We see in Figure 3 that the area of the jth trapezoid is equal to the average of the areas of the endpoint rectangles with heights y_{j-1} and y_j. It follows that T_N is equal to the average of the right- and left-endpoint approximations R_N and L_N introduced in Section 5.1:

$$T_N = \frac{1}{2}(R_N + L_N)$$

In general, this average is a better approximation than either R_N alone or L_N alone.

■ **EXAMPLE 1** CAS Calculate T_8 for the integral $\int_1^3 \sin(x^2)\,dx$. Then use a computer algebra system to calculate T_N for $N = 50, 100, 500, 1000$, and $10{,}000$.

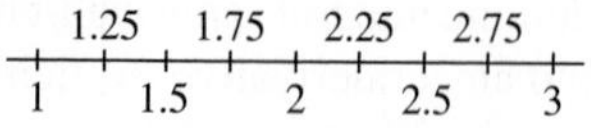

FIGURE 4 Division on [1, 3] into $N = 8$ subintervals.

Solution Divide [1, 3] into $N = 8$ subintervals of length $\Delta x = \frac{3-1}{8} = \frac{1}{4}$. Then sum the function values at the endpoints (Figure 4) with the appropriate coefficients:

$$T_8 = \frac{1}{2}\left(\frac{1}{4}\right)\Big[\sin(1^2) + 2\sin(1.25^2) + 2\sin(1.5^2) + 2\sin(1.75^2) + 2\sin(2^2) + 2\sin(2.25^2) + 2\sin(2.5^2) + 2\sin(2.75^2) + \sin(3^2)\Big]$$

$$\approx 0.4281$$

In general, $\Delta x = (3-1)/N = 2/N$ and $x_j = 1 + 2j/N$. In summation notation,

$$T_N = \frac{1}{2}\left(\frac{2}{N}\right)\Bigg[\sin(1^2) + 2\underbrace{\sum_{j=1}^{N-1} \sin\left(\left(1 + \frac{2j}{N}\right)^2\right)}_{\text{Sum of terms with coefficient 2}} + \sin(3^2)\Bigg]$$

We evaluate the inner sum on a CAS, using a command such as

```
Sum[Sin[(1 + 2j/N)^2], {j, 1, N − 1}]
```

The results in Table 1 suggest that $\int_1^3 \sin(x^2)\,dx$ is approximately 0.4633. ■

TABLE 1

N	T_N
50	0.4624205
100	0.4630759
500	**0.4632**855
1000	**0.4632**920
10,000	**0.4632**942

The midpoint approximation M_N, introduced in Section 5.1, is the sum of the areas of the rectangles of height $f(c_j)$ and base Δx, where c_j is the midpoint of the interval $[x_{j-1}, x_j]$ [Figure 6(A)].

Midpoint Rule The Nth midpoint approximation to $\int_a^b f(x)\,dx$ is

$$M_N = \Delta x\big(f(c_1) + f(c_2) + \cdots + f(c_N)\big)$$

where $\Delta x = \dfrac{b-a}{N}$ and $c_j = a + \left(j - \frac{1}{2}\right)\Delta x$ is the midpoint of $[x_{j-1}, x_j]$.

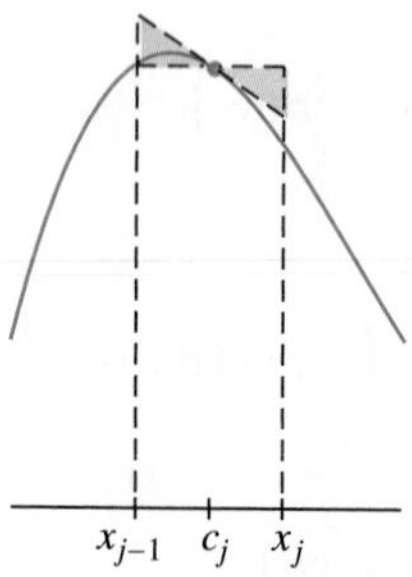

FIGURE 5 The rectangle and the trapezoid have the same area.

GRAPHICAL INSIGHT M_N has a second interpretation as the sum of the areas of tangential trapezoids—that is, trapezoids whose top edges are tangent to the graph of $f(x)$ at the midpoints c_j [Figure 6(B)]. The trapezoids have the same area as the rectangles because the top edge of the trapezoid passes through the midpoint of the top edge of the rectangle, as shown in Figure 5.

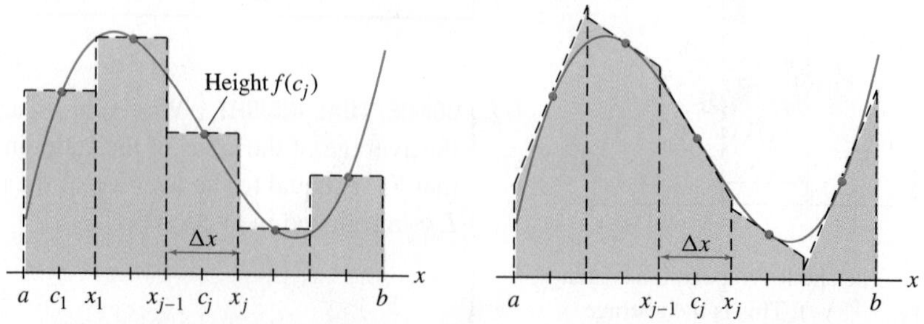

(A) M_N is the sum of the areas of the midpoint rectangles.

(B) M_N is also the sum of the areas of the tangential trapezoids.

FIGURE 6 Two interpretations of M_N.

Error Bounds

In applications, it is important to know the accuracy of a numerical approximation. We define the error in T_N and M_N by

$$\text{Error}(T_N) = \left| \int_a^b f(x)\,dx - T_N \right|, \qquad \text{Error}(M_N) = \left| \int_a^b f(x)\,dx - M_N \right|$$

According to the next theorem, the magnitudes of these errors are related to the size of the *second* derivative $f''(x)$. A proof of Theorem 1 is provided in a supplement on the text's Companion Web Site.

In the error bound, you can let K_2 be the maximum of $|f''(x)|$ on $[a, b]$, but if it is inconvenient to find this maximum exactly, take K_2 to be any number that is definitely larger than the maximum.

THEOREM 1 Error Bound for T_N and M_N Assume $f''(x)$ exists and is continuous. Let K_2 be a number such that $|f''(x)| \le K_2$ for all $x \in [a, b]$. Then

$$\text{Error}(T_N) \le \frac{K_2(b-a)^3}{12N^2}, \qquad \text{Error}(M_N) \le \frac{K_2(b-a)^3}{24N^2}$$

GRAPHICAL INSIGHT Note that the error bound for M_N is one-half of the error bound for T_N, suggesting that M_N is generally more accurate than T_N. Why do both error bounds depend on $f''(x)$? The second derivative measures concavity, so if $|f''(x)|$ is large, then the graph of f bends a lot and trapezoids do a poor job of approximating the region under the graph. Thus the errors in both T_N and M_N (which uses tangential trapezoids) are likely to be large (Figure 7).

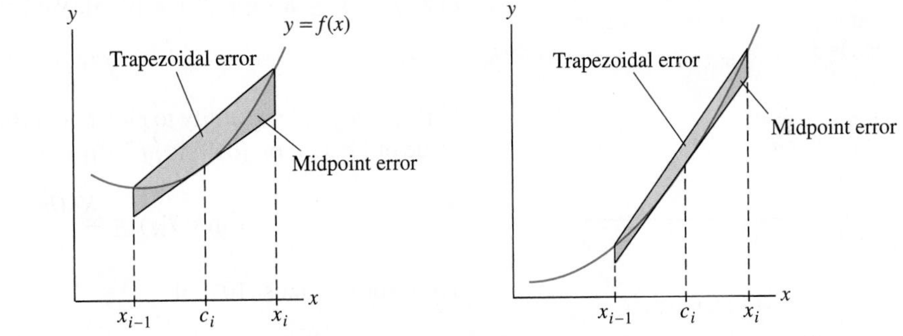

(A) $f''(x)$ is larger and the errors are larger. (B) $f''(x)$ is smaller and the errors are smaller.

FIGURE 7 T_N and M_N are more accurate when $|f''(x)|$ is small.

■ **EXAMPLE 2 Checking the Error Bound** Calculate T_6 and M_6 for $\int_1^4 \sqrt{x}\,dx$.

(a) Calculate the error bounds.

(b) Calculate the integral exactly and verify that the error bounds are satisfied.

Solution Divide [1, 4] into six subintervals of width $\Delta x = \frac{4-1}{6} = \frac{1}{2}$. Using the endpoints and midpoints shown in Figure 8, we obtain

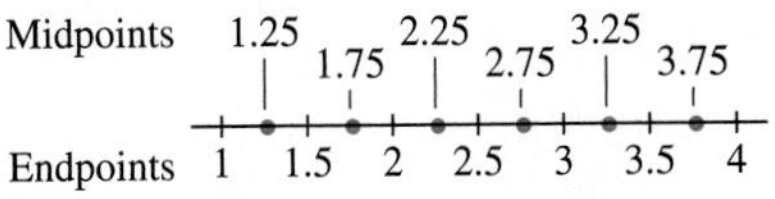

FIGURE 8 Interval [1, 4] divided into $N = 6$ subintervals.

$$T_6 = \frac{1}{2}\left(\frac{1}{2}\right)\left(\sqrt{1} + 2\sqrt{1.5} + 2\sqrt{2} + 2\sqrt{2.5} + 2\sqrt{3} + 2\sqrt{3.5} + \sqrt{4}\right) \approx 4.661488$$

$$M_6 = \frac{1}{2}\left(\sqrt{1.25} + \sqrt{1.75} + \sqrt{2.25} + \sqrt{2.75} + \sqrt{3.25} + \sqrt{3.75}\right) \approx 4.669245$$

(a) Let $f(x) = \sqrt{x}$. We must find a number K_2 such that $|f''(x)| \le K_2$ for $1 \le x \le 4$. We have $f''(x) = -\frac{1}{4}x^{-3/2}$. The absolute value $|f''(x)| = \frac{1}{4}x^{-3/2}$ is decreasing on $[1, 4]$, so its maximum occurs at $x = 1$ (Figure 9). Thus, we may take $K_2 = |f''(1)| = \frac{1}{4}$. By Theorem 1,

$$\text{Error}(T_6) \le \frac{K_2(b-a)^3}{12N^2} = \frac{\frac{1}{4}(4-1)^3}{12(6)^2} = \frac{1}{64} \approx 0.0156$$

$$\text{Error}(M_6) \le \frac{K_2(b-a)^3}{24N^2} = \frac{\frac{1}{4}(4-1)^3}{24(6)^2} = \frac{1}{128} \approx 0.0078$$

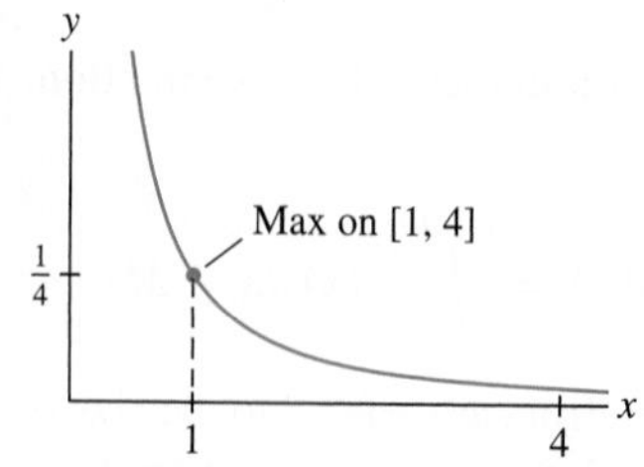

FIGURE 9 Graph of $y = |f''(x)| = \frac{1}{4}x^{-3/2}$ for $f(x) = \sqrt{x}$.

In Example 2, the error in T_6 is approximately twice as large as the error in M_6. In practice, this is often the case.

(b) The exact value is $\int_1^4 \sqrt{x}\,dx = \frac{2}{3}x^{3/2}\Big|_1^4 = \frac{14}{3}$, so the actual errors are

$$\text{Error}(T_6) \approx \left|\frac{14}{3} - 4.661488\right| \approx 0.00518 \quad \text{(less than error bound 0.0156)}$$

$$\text{Error}(M_6) \approx \left|\frac{14}{3} - 4.669245\right| \approx 0.00258 \quad \text{(less than error bound 0.0078)}$$

The actual errors are less than the error bound, so Theorem 1 is verified. ■

The error bound can be used to determine values of N that provide a given accuracy.

■ **EXAMPLE 3 Obtaining the Desired Accuracy** Find N such that T_N approximates $\int_0^3 e^{-x^2}\,dx$ with an error of at most 10^{-4}.

A quick way to find a value for K_2 is to plot $f''(x)$ using a graphing utility and find a bound for $|f''(x)|$ visually, as we do in Example 3.

Solution Let $f(x) = e^{-x^2}$. To apply the error bound, we must find a number K_2 such that $|f''(x)| \le K_2$ for all $x \in [0, 3]$. We have $f'(x) = -2xe^{-x^2}$ and

$$f''(x) = (4x^2 - 2)e^{-x^2}$$

Let's use a graphing utility to plot $f''(x)$ (Figure 10). The graph shows that the maximum value of $|f''(x)|$ on $[0, 3]$ is $|f''(0)| = |-2| = 2$, so we take $K_2 = 2$ in the error bound:

$$\text{Error}(T_N) \le \frac{K_2(b-a)^3}{12N^2} = \frac{2(3-0)^3}{12N^2} = \frac{9}{2N^2}$$

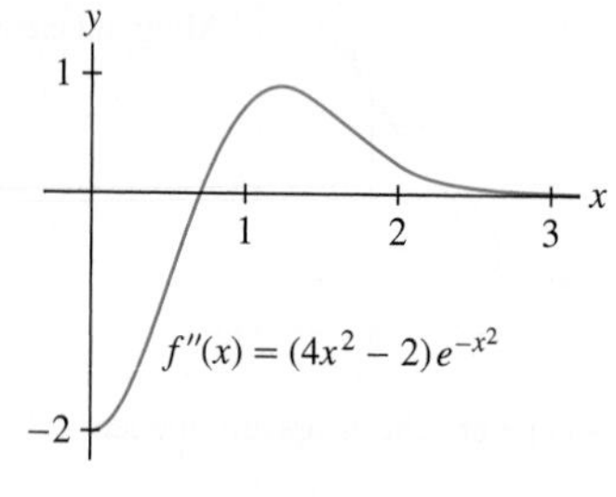

FIGURE 10 Graph of the second derivative of $f(x) = e^{-x^2}$.

The error is at most 10^{-4} if

$$\frac{9}{2N^2} \le 10^{-4} \quad \Rightarrow \quad N^2 \ge \frac{9 \times 10^4}{2} \quad \Rightarrow \quad N \ge \frac{300}{\sqrt{2}} \approx 212.1$$

We conclude that T_{213} has error at most 10^{-4}. We can confirm this using a computer algebra system. A CAS shows that $T_{213} \approx 0.886207$, whereas the value of the integral to nine places is 0.886207348. Thus the error is less than 10^{-6}. ■

Can we improve on the Trapezoidal and Midpoint Rules? One clue is that the exact value of the integral lies between T_N and M_N if $f(x)$ is concave up or down. In fact, we see geometrically (Figure 11) that

- $f(x)$ is concave down $\Rightarrow$ $T_N \le \displaystyle\int_a^b f(x)\,dx \le M_N$.

- $f(x)$ is concave up $\Rightarrow$ $M_N \le \displaystyle\int_a^b f(x)\,dx \le T_N$.

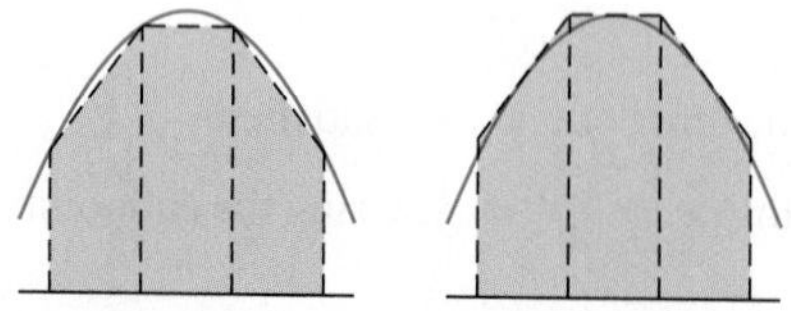

(A) Trapezoids used to compute T_N (B) Trapezoids used to compute M_N

FIGURE 11 If $f(x)$ is concave down, then T_N is smaller and M_N is larger than the integral.

This suggests that the errors in T_N and M_N may cancel partially if we take their average.

Simpson's Rule exploits this idea, but it takes into account that M_N is roughly twice as accurate as T_N. To minimize the error, Simpson's Rule S_N is defined as a **weighted average** that uses twice as much M_N as T_N. For N even, let

$$S_N = \frac{1}{3}T_{N/2} + \frac{2}{3}M_{N/2} \qquad \boxed{1}$$

To derive a formula for S_N, we divide $[a, b]$ into N subintervals as usual. Observe that the even-numbered endpoints divide $[a, b]$ into $N/2$ subintervals of length $2\Delta x$ (keep in mind that N is even):

$$[x_0, x_2], \quad [x_2, x_4], \quad \dots, \quad [x_{N-2}, x_N] \qquad \boxed{2}$$

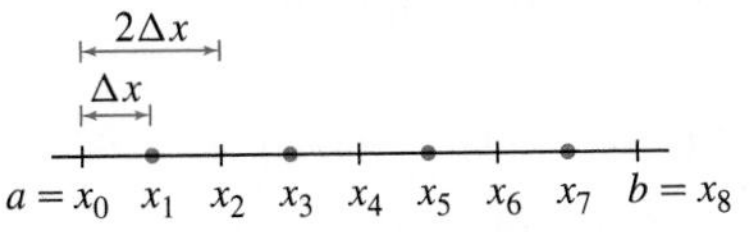

FIGURE 12 We compute S_8 using eight subintervals. The even endpoints are used for T_4, the odd endpoints for M_4, and $S_8 = \frac{1}{3}T_4 + \frac{2}{3}M_4$.

The *endpoints* of these intervals are $x_0, x_2, \dots, x_N$. They are used to compute $T_{N/2}$. The *midpoints* $x_1, x_3, \dots, x_{N-1}$ are used to compute $M_{N/2}$ (see Figure 12 for the case $N = 8$).

$$T_{N/2} = \frac{1}{2}(2\Delta x)\big(y_0 + 2y_2 + 2y_4 + \cdots + 2y_{N-2} + y_N\big)$$

$$M_{N/2} = 2\Delta x\big(y_1 + y_3 + y_5 + \cdots + y_{N-1}\big) = \Delta x\big(2y_1 + 2y_3 + 2y_5 + \cdots + 2y_{N-1}\big)$$

Thus,

$$S_N = \frac{1}{3}T_{N/2} + \frac{2}{3}M_{N/2} = \frac{1}{3}\Delta x\big(y_0 + 2y_2 + 2y_4 + \cdots + 2y_{N-2} + y_N\big) + \frac{1}{3}\Delta x\big(4y_1 + 4y_3 + 4y_5 + \cdots + 4y_{N-1}\big)$$

Pattern of coefficients in S_N:

$$1, 4, 2, 4, 2, 4, \dots, 4, 2, 4, 1$$

The intermediate coefficients alternate $4, 2, 4, 2, \dots, 2, 4$ (beginning and ending with 4).

Simpson's Rule For N even, the Nth approximation to $\int_a^b f(x)\,dx$ by Simpson's Rule is

$$S_N = \frac{1}{3}\Delta x\big[y_0 + 4y_1 + 2y_2 + \cdots + 4y_{N-3} + 2y_{N-2} + 4y_{N-1} + y_N\big] \qquad \boxed{3}$$

where $\Delta x = \dfrac{b-a}{N}$ and $y_j = f(a + j\,\Delta x)$.

CONCEPTUAL INSIGHT Both T_N and M_N give the exact value of the integral for all N when $f(x)$ is a linear function (Exercise 59). However, of all combinations of $T_{N/2}$ and $M_{N/2}$, only the particular combination $S_N = \frac{1}{3}T_{N/2} + \frac{2}{3}M_{N/2}$ gives the exact value for all quadratic polynomials (Exercises 60 and 61). In fact, S_N is also exact for all cubic polynomials (Exercise 62).

■ **EXAMPLE 4** Use Simpson's Rule with $N = 8$ to approximate $\int_2^4 \sqrt{1 + x^3}\,dx$.

Solution We have $\Delta x = \frac{4-2}{8} = \frac{1}{4}$. Figure 13 shows the endpoints and coefficients needed to compute S_8 using Eq. (3):

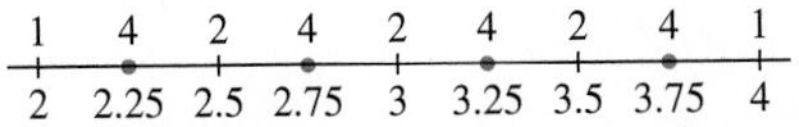

FIGURE 13 Coefficients for S_8 on $[2, 4]$ shown above the corresponding endpoint.

$$\frac{1}{3}\left(\frac{1}{4}\right)\Big[\sqrt{1+2^3} + 4\sqrt{1+2.25^3} + 2\sqrt{1+2.5^3} + 4\sqrt{1+2.75^3} + 2\sqrt{1+3^3} + 4\sqrt{1+3.25^3} + 2\sqrt{1+3.5^3} + 4\sqrt{1+3.75^3} + \sqrt{1+4^3}\Big]$$

$$\approx \frac{1}{12}\big[3 + 4(3.52003) + 2(4.07738) + 4(4.66871) + 2(5.2915) + 4(5.94375) + 2(6.62382) + 4(7.33037) + 8.06226\big] \approx 10.74159 \quad ■$$

The accuracy of Simpson's Rule is impressive. Using a computer algebra system, we find that the approximation in Example 4 has an error of less than 3×10^{-6}.

EXAMPLE 5 Estimating Integrals from Numerical Data The velocity (in km/h) of a Piper Cub aircraft traveling due west is recorded every minute during the first 10 minutes after takeoff. Use Simpson's Rule to estimate the distance traveled.

t (min)	0	1	2	3	4	5	6	7	8	9	10
$v(t)$ (km/h)	0	80	100	128	144	160	152	136	128	120	136

Solution The distance traveled is the integral of velocity. We convert from minutes to hours because velocity is given in km/h, and thus we apply Simpson's Rule, where the number of intervals is $N = 10$ and each interval has length $\Delta t = \frac{1}{60}$ hours:

$$S_{10} = \left(\frac{1}{3}\right)\left(\frac{1}{60}\right)\big(0 + 4(80) + 2(100) + 4(128) + 2(144) + 4(160) + 2(152) + 4(136) + 2(128) + 4(120) + 136\big) \approx 21.2 \text{ km}$$

The distance traveled is approximately 21.2 km (Figure 14). ■

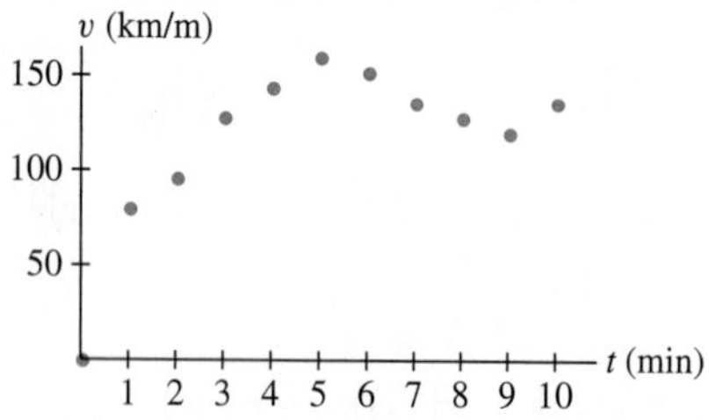

FIGURE 14 Velocity of a Piper Cub.

We now state (without proof) the error bound for Simpson's Rule. Set

$$\text{Error}(S_N) = \left| \int_a^b f(x) - S_N(f)\, dx \right|$$

The error involves the fourth derivative, which we assume exists and is continuous.

Although Simpson's Rule provides good approximations, more sophisticated techniques are implemented in computer algebra systems. These techniques are studied in the area of mathematics called numerical analysis.

THEOREM 2 Error Bound for S_N Let K_4 be a number such that $|f^{(4)}(x)| \le K_4$ for all $x \in [a, b]$. Then

$$\text{Error}(S_N) \le \frac{K_4(b-a)^5}{180N^4}$$

EXAMPLE 6 Calculate S_8 for $\int_1^3 \frac{1}{x}\, dx$. Then:

(a) Find a bound for the error in S_8.

(b) Find N such that S_N has an error of at most 10^{-6}.

Solution The width is $\Delta x = \frac{3-1}{8} = \frac{1}{4}$ and the endpoints in the partition of $[1, 3]$ are $1, 1.25, 1.5, \dots, 2.75, 3$. Using Eq. (3) with $f(x) = x^{-1}$, we obtain

$$S_8 = \frac{1}{3}\left(\frac{1}{4}\right)\left[\frac{1}{1} + \frac{4}{1.25} + \frac{2}{1.5} + \frac{4}{1.75} + \frac{2}{2} + \frac{4}{2.25} + \frac{2}{2.5} + \frac{4}{2.75} + \frac{1}{3}\right] \approx 1.09873$$

(a) The fourth derivative $f^{(4)}(x) = 24x^{-5}$ is decreasing, so the max of $|f^{(4)}(x)|$ on $[1, 3]$ is $|f^{(4)}(1)| = 24$. Therefore, we use the error bound with $K_4 = 24$:

$$\text{Error}(S_N) \le \frac{K_4(b-a)^5}{180N^4} = \frac{24(3-1)^5}{180N^4} = \frac{64}{15N^4}$$

$$\text{Error}(S_8) \le \frac{K_4(b-a)^5}{180(8)^4} = \frac{24(3-1)^5}{180(8^4)} \approx 0.001$$

Using a CAS, we find that

$$S_{46} \approx 1.09861241$$

$$\int_1^3 \frac{1}{x}\,dx = \ln 3 \approx 1.09861229$$

The error is indeed less than 10^{-6}.

(b) The error will be at most 10^{-6} if N satisfies

$$\text{Error}(S_N) = \frac{64}{15N^4} \le 10^{-6}$$

In other words,

$$N^4 \ge 10^6\left(\frac{64}{15}\right) \quad \text{or} \quad N \ge \left(\frac{10^6 \cdot 64}{15}\right)^{1/4} \approx 45.45$$

Thus, we may take $N = 46$ (see marginal comment). ■

GRAPHICAL INSIGHT Simpson's Rule has an interpretation in terms of parabolas (Figure 15). There is a unique parabola passing through the graph of $f(x)$ at the three points $x_{2j-2}, x_{2j-1}, x_{2j}$ [Figure 15(A)]. On the interval $[x_{2j-2}, x_{2j}]$, the area under the parabola approximates the area under the graph. Simpson's Rule S_N is equal to the sum of these parabolic approximations (see Exercises 60–61).

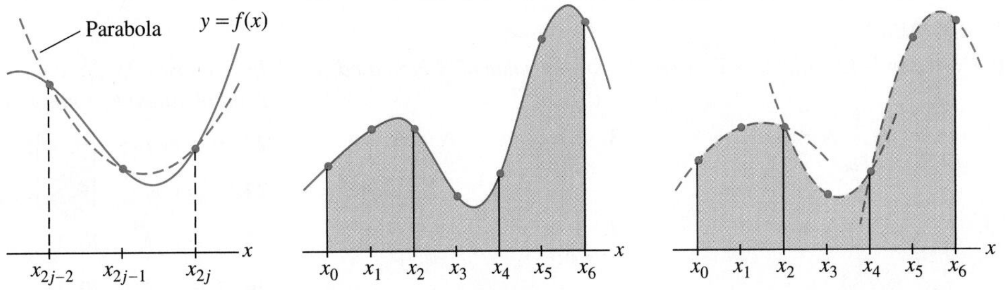

(A) Unique parabola through three points.

(B) Graph of $y = f(x)$

(C) Parabolic arcs used in Simpson's Rule.

FIGURE 15 Simpson's Rule approximates the graph by parabolic arcs.

8.8 SUMMARY

- We consider three numerical approximations to $\int_a^b f(x)\,dx$: the *Trapezoidal Rule* T_N, the *Midpoint Rule* M_N, and *Simpson's Rule* S_N (for N even).

$$T_N = \frac{1}{2}\Delta x\big(y_0 + 2y_1 + 2y_2 + \cdots + 2y_{N-1} + y_N\big)$$

$$M_N = \Delta x\big(f(c_1) + f(c_2) + \cdots + f(c_N)\big) \qquad \left(c_j = a + \left(j - \frac{1}{2}\right)\Delta x\right)$$

$$S_N = \frac{1}{3}\Delta x\big[y_0 + 4y_1 + 2y_2 + \cdots + 4y_{N-3} + 2y_{N-2} + 4y_{N-1} + y_N\big]$$

where $\Delta x = (b-a)/N$ and $y_j = f(a + j\,\Delta x)$.
- T_N is equal to the sum of the areas of the trapezoids obtained by connecting the points $(x_0, y_0), (x_1, y_1), \ldots, (x_N, y_N)$ with line segments.
- M_N has two geometric interpretations; it may be interpreted either as the sum of the areas of the midpoint rectangles or as the sum of the areas of the tangential trapezoids.
- S_N is equal to $\frac{1}{3}T_{N/2} + \frac{2}{3}M_{N/2}$.
- Error bounds:

$$\text{Error}(T_N) \le \frac{K_2(b-a)^3}{12N^2}, \qquad \text{Error}(M_N) \le \frac{K_2(b-a)^3}{24N^2}, \qquad \text{Error}(S_N) \le \frac{K_4(b-a)^5}{180N^4}$$

where K_2 is any number such that $|f''(x)| \le K_2$ for all $x \in [a, b]$ and K_4 is any number such that $|f^{(4)}(x)| \le K_4$ for all $x \in [a, b]$.

8.8 EXERCISES

Preliminary Questions

1. What are T_1 and T_2 for a function on $[0, 2]$ such that $f(0) = 3$, $f(1) = 4$, and $f(2) = 3$?

2. For which graph in Figure 16 will T_N *overestimate* the integral? What about M_N?

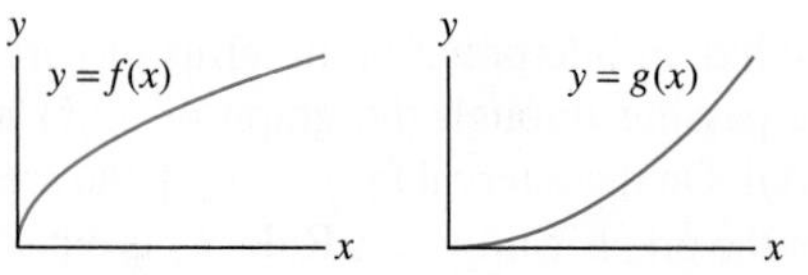

FIGURE 16

3. How large is the error when the Trapezoidal Rule is applied to a linear function? Explain graphically.

4. What is the maximum possible error if T_4 is used to approximate

$$\int_0^3 f(x)\,dx$$

where $|f''(x)| \le 2$ for all x.

5. What are the two graphical interpretations of the Midpoint Rule?

Exercises

In Exercises 1–12, calculate T_N and M_N for the value of N indicated.

1. $\displaystyle\int_0^2 x^2\,dx,\quad N = 4$

2. $\displaystyle\int_0^4 \sqrt{x}\,dx,\quad N = 4$

3. $\displaystyle\int_1^4 x^3\,dx,\quad N = 6$

4. $\displaystyle\int_1^2 \sqrt{x^4+1}\,dx,\quad N = 5$

5. $\displaystyle\int_1^4 \frac{dx}{x},\quad N = 6$

6. $\displaystyle\int_{-2}^{-1} \frac{dx}{x},\quad N = 5$

7. $\displaystyle\int_0^{\pi/2} \sqrt{\sin x}\,dx,\quad N = 6$

8. $\displaystyle\int_0^{\pi/4} \sec x\,dx,\quad N = 6$

9. $\displaystyle\int_1^2 \ln x\,dx,\quad N = 5$

10. $\displaystyle\int_2^3 \frac{dx}{\ln x},\quad N = 5$

11. $\displaystyle\int_0^1 e^{-x^2}\,dx,\quad N = 5$

12. $\displaystyle\int_{-2}^1 e^{x^2}\,dx,\quad N = 6$

In Exercises 13–22, calculate S_N given by Simpson's Rule for the value of N indicated.

13. $\displaystyle\int_0^4 \sqrt{x}\,dx,\quad N = 4$

14. $\displaystyle\int_3^5 (9 - x^2)\,dx,\quad N = 4$

15. $\displaystyle\int_0^3 \frac{dx}{x^4+1},\quad N = 6$

16. $\displaystyle\int_0^1 \cos(x^2)\,dx,\quad N = 6$

17. $\displaystyle\int_0^1 e^{-x^2}\,dx,\quad N = 4$

18. $\displaystyle\int_1^2 e^{-x}\,dx,\quad N = 6$

19. $\displaystyle\int_1^4 \ln x\,dx,\quad N = 8$

20. $\displaystyle\int_2^4 \sqrt{x^4+1}\,dx,\quad N = 8$

21. $\displaystyle\int_0^{\pi/4} \tan\theta\,d\theta,\quad N = 10$

22. $\displaystyle\int_0^2 (x^2+1)^{-1/3}\,dx,\quad N = 10$

In Exercises 23–26, calculate the approximation to the volume of the solid obtained by rotating the graph around the given axis.

23. $y = \cos x$; $\left[0, \frac{\pi}{2}\right]$; x-axis; M_8

24. $y = \cos x$; $\left[0, \frac{\pi}{2}\right]$; y-axis; S_8

25. $y = e^{-x^2}$; $[0, 1]$; x-axis; T_8

26. $y = e^{-x^2}$; $[0, 1]$; y-axis; S_8

27. An airplane's velocity is recorded at 5-min intervals during a 1-hour period with the following results, in miles per hour:

550, 575, 600, 580, 610, 640, 625,
595, 590, 620, 640, 640, 630

Use Simpson's Rule to estimate the distance traveled during the hour.

28. Use Simpson's Rule to determine the average temperature in a museum over a 3-hour period, if the temperatures (in degrees Celsius), recorded at 15-min intervals, are

21, 21.3, 21.5, 21.8, 21.6, 21.2, 20.8,
20.6, 20.9, 21.2, 21.1, 21.3, 21.2

29. **Tsunami Arrival Times** Scientists estimate the arrival times of tsunamis (seismic ocean waves) based on the point of origin P and ocean depths. The speed s of a tsunami in miles per hour is approximately $s = \sqrt{15d}$, where d is the ocean depth in feet.

(a) Let $f(x)$ be the ocean depth x miles from P (in the direction of the coast). Argue using Riemann sums that the time T required for the tsunami to travel M miles toward the coast is

$$T = \int_0^M \frac{dx}{\sqrt{15 f(x)}}$$

(b) Use Simpson's Rule to estimate T if $M = 1000$ and the ocean depths (in feet), measured at 100-mile intervals starting from P, are

13,000, 11,500, 10,500, 9000, 8500,
7000, 6000, 4400, 3800, 3200, 2000

30. Use S_8 to estimate $\displaystyle\int_0^{\pi/2} \frac{\sin x}{x}\,dx$, taking the value of $\dfrac{\sin x}{x}$ at $x = 0$ to be 1.

31. Calculate T_6 for the integral $I = \int_0^2 x^3\,dx$.

(a) Is T_6 too large or too small? Explain graphically.

(b) Show that $K_2 = |f''(2)|$ may be used in the error bound and find a bound for the error.

(c) Evaluate I and check that the actual error is less than the bound computed in (b).

32. Calculate M_4 for the integral $I = \int_0^1 x\sin(x^2)\,dx$.

(a) GU Use a plot of $f''(x)$ to show that $K_2 = 3.2$ may be used in the error bound and find a bound for the error.

(b) CAS Evaluate I numerically and check that the actual error is less than the bound computed in (a).

In Exercises 33–36, state whether T_N or M_N underestimates or overestimates the integral and find a bound for the error (but do not calculate T_N or M_N).

33. $\int_1^4 \frac{1}{x}\,dx, \quad T_{10}$

34. $\int_0^2 e^{-x/4}\,dx, \quad T_{20}$

35. $\int_1^4 \ln x\,dx, \quad M_{10}$

36. $\int_0^{\pi/4} \cos x, \quad M_{20}$

CAS *In Exercises 37–40, use the error bound to find a value of N for which Error(T_N) $\le 10^{-6}$. If you have a computer algebra system, calculate the corresponding approximation and confirm that the error satisfies the required bound.*

37. $\int_0^1 x^4\,dx$

38. $\int_0^3 (5x^4 - x^5)\,dx$

39. $\int_2^5 \frac{1}{x}\,dx$

40. $\int_0^3 e^{-x}\,dx$

41. Compute the error bound for the approximations T_{10} and M_{10} to $\int_0^3 (x^3+1)^{-1/2}\,dx$, using Figure 17 to determine a value of K_2. Then find a value of N such that the error in M_N is at most 10^{-6}.

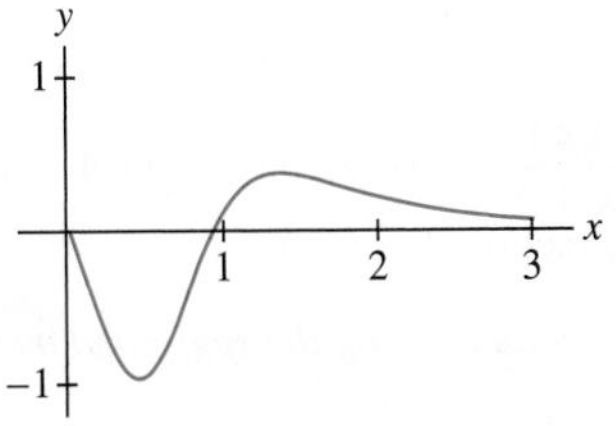

FIGURE 17 Graph of $f''(x)$, where $f(x) = (x^3+1)^{-1/2}$.

42. (a) Compute S_6 for the integral $I = \int_0^1 e^{-2x}\,dx$.

(b) Show that $K_4 = 16$ may be used in the error bound and compute the error bound.

(c) Evaluate I and check that the actual error is less than the bound for the error computed in (b).

43. Calculate S_8 for $\int_1^5 \ln x\,dx$ and calculate the error bound. Then find a value of N such that S_N has an error of at most 10^{-6}.

44. Find a bound for the error in the approximation S_{10} to $\int_0^3 e^{-x^2}\,dx$ (use Figure 18 to determine a value of K_4). Then find a value of N such that S_N has an error of at most 10^{-6}.

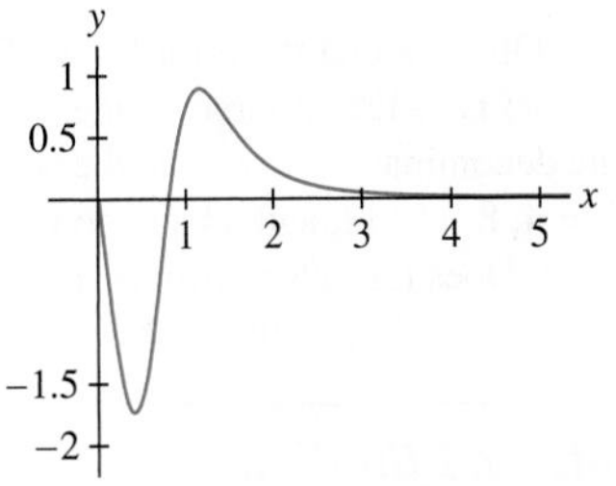

FIGURE 18 Graph of $f^{(4)}(x)$, where $f(x) = e^{-x^2}$.

45. CAS Use a computer algebra system to compute and graph $f^{(4)}(x)$ for $f(x) = \sqrt{1+x^4}$ and find a bound for the error in the approximation S_{40} to $\int_0^5 f(x)\,dx$.

46. CAS Use a computer algebra system to compute and graph $f^{(4)}(x)$ for $f(x) = \tan x - \sec x$ and find a bound for the error in the approximation S_{40} to $\int_0^{\pi/4} f(x)\,dx$.

In Exercises 47–50, use the error bound to find a value of N for which Error(S_N) $\le 10^{-9}$.

47. $\int_1^6 x^{4/3}\,dx$

48. $\int_0^4 xe^x\,dx$

49. $\int_0^1 e^{x^2}\,dx$

50. $\int_1^4 \sin(\ln x)\,dx$

51. CAS Show that $\int_0^1 \frac{dx}{1+x^2} = \frac{\pi}{4}$ [use Eq. (3) in Section 7.8].

(a) Use a computer algebra system to graph $f^{(4)}(x)$ for $f(x) = (1+x^2)^{-1}$ and find its maximum on [0, 1].

(b) Find a value of N such that S_N approximates the integral with an error of at most 10^{-6}. Calculate the corresponding approximation and confirm that you have computed $\frac{\pi}{4}$ to at least four places.

52. Let $J = \int_0^\infty e^{-x^2}\,dx$ and $J_N = \int_0^N e^{-x^2}\,dx$. Although e^{-x^2} has no elementary antiderivative, it is known that $J = \sqrt{\pi}/2$. Let T_N be the Nth trapezoidal approximation to J_N. Calculate T_4 and show that T_4 approximates J to three decimal places.

53. Let $f(x) = \sin(x^2)$ and $I = \int_0^1 f(x)\,dx$.

(a) Check that $f''(x) = 2\cos(x^2) - 4x^2\sin(x^2)$. Then show that $|f''(x)| \le 6$ for $x \in [0,1]$. *Hint:* Note that $|2\cos(x^2)| \le 2$ and $|4x^2\sin(x^2)| \le 4$ for $x \in [0,1]$.

(b) Show that Error(M_N) is at most $\frac{1}{4N^2}$.

(c) Find an N such that $|I - M_N| \le 10^{-3}$.

54. CAS The error bound for M_N is proportional to $1/N^2$, so the error bound decreases by $\frac{1}{4}$ if N is increased to $2N$. Compute the actual error in M_N for $\int_0^{\pi} \sin x \, dx$ for $N = 4, 8, 16, 32$, and 64. Does the actual error seem to decrease by $\frac{1}{4}$ as N is doubled?

55. CAS Observe that the error bound for T_N (which has 12 in the denominator) is twice as large as the error bound for M_N (which has 24 in the denominator). Compute the actual error in T_N for $\int_0^{\pi} \sin x \, dx$ for $N = 4, 8, 16, 32$, and 64 and compare with the calculations of Exercise 54. Does the actual error in T_N seem to be roughly twice as large as the error in M_N in this case?

56. CAS Explain why the error bound for S_N decreases by $\frac{1}{16}$ if N is increased to $2N$. Compute the actual error in S_N for $\int_0^{\pi} \sin x \, dx$ for $N = 4, 8, 16, 32$, and 64. Does the actual error seem to decrease by $\frac{1}{16}$ as N is doubled?

57. Verify that S_2 yields the exact value of $\displaystyle\int_0^1 (x - x^3)\, dx$.

58. Verify that S_2 yields the exact value of $\displaystyle\int_a^b (x - x^3)\, dx$ for all $a < b$.

Further Insights and Challenges

59. Show that if $f(x) = rx + s$ is a linear function (r, s constants), then $T_N = \displaystyle\int_a^b f(x)\, dx$ for all N and all endpoints a, b.

60. Show that if $f(x) = px^2 + qx + r$ is a quadratic polynomial, then $S_2 = \displaystyle\int_a^b f(x)\, dx$. In other words, show that

$$\int_a^b f(x)\, dx = \frac{b-a}{6}(y_0 + 4y_1 + y_2)$$

where $y_0 = f(a)$, $y_1 = f\left(\dfrac{a+b}{2}\right)$, and $y_2 = f(b)$. *Hint:* Show this first for $f(x) = 1, x, x^2$ and use linearity.

61. For N even, divide $[a, b]$ into N subintervals of width $\Delta x = \dfrac{b-a}{N}$. Set $x_j = a + j\,\Delta x$, $y_j = f(x_j)$, and

$$S_2^{2j} = \frac{b-a}{3N}(y_{2j} + 4y_{2j+1} + y_{2j+2})$$

(a) Show that S_N is the sum of the approximations on the intervals $[x_{2j}, x_{2j+2}]$—that is, $S_N = S_2^0 + S_2^2 + \cdots + S_2^{N-2}$.

(b) By Exercise 60, $S_2^{2j} = \displaystyle\int_{x_{2j}}^{x_{2j+2}} f(x)\, dx$ if $f(x)$ is a quadratic polynomial. Use (a) to show that S_N is exact *for all N* if $f(x)$ is a quadratic polynomial.

62. Show that S_2 also gives the exact value for $\displaystyle\int_a^b x^3\, dx$ and conclude, as in Exercise 61, that S_N is exact for all cubic polynomials. Show by counterexample that S_2 is not exact for integrals of x^4.

63. Use the error bound for S_N to obtain another proof that Simpson's Rule is exact for all cubic polynomials.

64. **Sometimes, Simpson's Rule Performs Poorly** Calculate M_{10} and S_{10} for the integral $\int_0^1 \sqrt{1 - x^2}\, dx$, whose value we know to be $\frac{\pi}{4}$ (one-quarter of the area of the unit circle).

(a) We usually expect S_N to be more accurate than M_N. Which of M_{10} and S_{10} is more accurate in this case?

(b) How do you explain the result of part (a)? *Hint:* The error bounds are not valid because $|f''(x)|$ and $|f^{(4)}(x)|$ tend to ∞ as $x \to 1$, but $|f^{(4)}(x)|$ goes to infinity faster.

CHAPTER REVIEW EXERCISES

1. Match the integrals (a)–(e) with their antiderivatives (i)–(v) on the basis of the general form (do not evaluate the integrals).

(a) $\displaystyle\int \frac{x\, dx}{x^2 - 4}$ **(b)** $\displaystyle\int \frac{(2x + 9)\, dx}{x^2 + 4}$

(c) $\displaystyle\int \sin^3 x \cos^2 x\, dx$ **(d)** $\displaystyle\int \frac{dx}{x\sqrt{16x^2 - 1}}$

(e) $\displaystyle\int \frac{16\, dx}{x(x-4)^2}$

(i) $\sec^{-1} 4x + C$

(ii) $\log|x| - \log|x - 4| - \dfrac{4}{x-4} + C$

(iii) $\dfrac{1}{30}(3\cos^5 x - 3\cos^3 x \sin^2 x - 7\cos^3 x) + C$

(iv) $\dfrac{9}{2}\tan^{-1}\dfrac{x}{2} + \ln(x^2 + 4) + C$ **(v)** $\sqrt{x^2 - 4} + C$

2. Evaluate $\displaystyle\int \frac{x\, dx}{x+2}$ in two ways: using substitution and using the Method of Partial Fractions.

In Exercises 3–12, evaluate using the suggested method.

3. $\displaystyle\int \cos^3\theta \sin^8\theta\, d\theta$ [write $\cos^3\theta$ as $\cos\theta(1 - \sin^2\theta)$]

4. $\displaystyle\int xe^{-12x}\, dx$ (Integration by Parts)

5. $\displaystyle\int \sec^3\theta \tan^4\theta\, d\theta$ (trigonometric identity, reduction formula)

6. $\displaystyle\int \frac{4x + 4}{(x-5)(x+3)}\, dx$ (partial fractions)

7. $\displaystyle\int \frac{dx}{x(x^2 - 1)^{3/2}}\, dx$ (trigonometric substitution)

8. $\int (1+x^2)^{-3/2}dx$ (trigonometric substitution)

9. $\int \frac{dx}{x^{3/2}+x^{1/2}}$ (substitution)

10. $\int \frac{dx}{x+x^{-1}}$ (rewrite integrand)

11. $\int x^{-2}\tan^{-1}x\,dx$ (Integration by Parts)

12. $\int \frac{dx}{x^2+4x-5}$ (complete the square, substitution, partial fractions)

In Exercises 13–64, evaluate using the appropriate method or combination of methods.

13. $\int_0^1 x^2 e^{4x}\,dx$

14. $\int \frac{x^2}{\sqrt{9-x^2}}\,dx$

15. $\int \cos^9 6\theta \sin^3 6\theta\,d\theta$

16. $\int \sec^2\theta \tan^4\theta\,d\theta$

17. $\int \frac{(6x+4)\,dx}{x^2-1}$

18. $\int_4^9 \frac{dt}{(t^2-1)^2}$

19. $\int \frac{d\theta}{\cos^4\theta}$

20. $\int \sin 2\theta \sin^2\theta\,d\theta$

21. $\int_0^1 \ln(4-2x)\,dx$

22. $\int (\ln(x+1))^2\,dx$

23. $\int \sin^5\theta\,d\theta$

24. $\int \cos^4(9x-2)\,dx$

25. $\int_0^{\pi/4} \sin 3x \cos 5x\,dx$

26. $\int \sin 2x \sec^2 x\,dx$

27. $\int \sqrt{\tan x}\,\sec^2 x\,dx$

28. $\int (\sec x + \tan x)^2\,dx$

29. $\int \sin^5\theta \cos^3\theta\,d\theta$

30. $\int \cot^3 x \csc x\,dx$

31. $\int \cot^2 x \csc^2 x\,dx$

32. $\int_{\pi/2}^{\pi} \cot^2\frac{\theta}{2}\,d\theta$

33. $\int_{\pi/4}^{\pi/2} \cot^2 x \csc^3 x\,dx$

34. $\int_4^6 \frac{dt}{(t-3)(t+4)}$

35. $\int \frac{dt}{(t-3)^2(t+4)}$

36. $\int \sqrt{x^2+9}\,dx$

37. $\int \frac{dx}{x\sqrt{x^2-4}}$

38. $\int_8^{27} \frac{dx}{x+x^{2/3}}$

39. $\int \frac{dx}{x^{3/2}+ax^{1/2}}$

40. $\int \frac{dx}{(x-b)^2+4}$

41. $\int \frac{(x^2-x)\,dx}{(x+2)^3}$

42. $\int \frac{(7x^2+x)\,dx}{(x-2)(2x+1)(x+1)}$

43. $\int \frac{16\,dx}{(x-2)^2(x^2+4)}$

44. $\int \frac{dx}{(x^2+25)^2}$

45. $\int \frac{dx}{x^2+8x+25}$

46. $\int \frac{dx}{x^2+8x+4}$

47. $\int \frac{(x^2-x)\,dx}{(x+2)^3}$

48. $\int_0^1 t^2\sqrt{1-t^2}\,dt$

49. $\int \frac{dx}{x^4\sqrt{x^2+4}}$

50. $\int \frac{dx}{(x^2+5)^{3/2}}$

51. $\int (x+1)e^{4-3x}\,dx$

52. $\int x^{-2}\tan^{-1}x\,dx$

53. $\int x^3\cos(x^2)\,dx$

54. $\int x^2(\ln x)^2\,dx$

55. $\int x\tanh^{-1}x\,dx$

56. $\int \frac{\tan^{-1}t\,dt}{1+t^2}$

57. $\int \ln(x^2+9)\,dx$

58. $\int (\sin x)(\cosh x)\,dx$

59. $\int_0^1 \cosh 2t\,dt$

60. $\int \sinh^3 x \cosh x\,dx$

61. $\int \coth^2(1-4t)\,dt$

62. $\int_{-0.3}^{0.3} \frac{dx}{1-x^2}$

63. $\int_0^{3\sqrt{3}/2} \frac{dx}{\sqrt{9-x^2}}$

64. $\int \frac{\sqrt{x^2+1}\,dx}{x^2}$

65. Use the substitution $u = \tanh t$ to evaluate $\int \frac{dt}{\cosh^2 t + \sinh^2 t}$.

66. Find the volume obtained by rotating the region enclosed by $y = \ln x$ and $y = (\ln x)^2$ about the y-axis.

67. Let $I_n = \int \frac{x^n\,dx}{x^2+1}$.

(a) Prove that $I_n = \frac{x^{n-1}}{n-1} - I_{n-2}$.

(b) Use (a) to calculate I_n for $0 \le n \le 5$.

(c) Show that, in general,

$$I_{2n+1} = \frac{x^{2n}}{2n} - \frac{x^{2n-2}}{2n-2} + \cdots + (-1)^{n-1}\frac{x^2}{2} + (-1)^n\frac{1}{2}\ln(x^2+1) + C$$

$$I_{2n} = \frac{x^{2n-1}}{2n-1} - \frac{x^{2n-3}}{2n-3} + \cdots + (-1)^{n-1}x + (-1)^n\tan^{-1}x + C$$

68. Let $J_n = \int x^n e^{-x^2/2}\,dx$.

(a) Show that $J_1 = -e^{-x^2/2}$.

(b) Prove that $J_n = -x^{n-1}e^{-x^2/2} + (n-1)J_{n-2}$.

(c) Use (a) and (b) to compute J_3 and J_5.

69. Compute $p(X \le 1)$, where X is a continuous random variable with probability density $p(x) = \dfrac{1}{\pi(x^2+1)}$.

70. Show that $p(x) = \frac{1}{4}e^{-t/2} + \frac{1}{6}e^{-t/3}$ is a probability density and find its mean.

71. Find a constant C such that $p(x) = Cx^3e^{-x^2}$ is a probability density and compute $p(0 \le X \le 1)$.

72. The interval between patient arrivals in an emergency room is a random variable with exponential density function $p(x) = 0.125e^{-0.125t}$ (t in minutes). What is the average time between patient arrivals? What is the probability of two patients arriving within 3 minutes of each other?

73. Calculate the following probabilities, assuming that X is normally distributed with mean $\mu = 40$ and $\sigma = 5$.

(a) $p(X \ge 45)$ **(b)** $p(0 \le X \le 40)$

74. According to kinetic theory, the molecules of ordinary matter are in constant random motion. The energy E of a molecule is a random variable with density function $p(E) = \frac{1}{kT}e^{-E/(kT)}$, where T is the temperature (in kelvins) and k is Boltzmann's constant. Compute the *mean* kinetic energy $\overline{E}$ in terms of k and T.

In Exercises 75–84, determine whether the improper integral converges and, if so, evaluate it.

75. $\displaystyle\int_0^\infty \frac{dx}{(x+2)^2}$ **76.** $\displaystyle\int_4^\infty \frac{dx}{x^{2/3}}$

77. $\displaystyle\int_0^4 \frac{dx}{x^{2/3}}$ **78.** $\displaystyle\int_9^\infty \frac{dx}{x^{12/5}}$

79. $\displaystyle\int_{-\infty}^0 \frac{dx}{x^2+1}$ **80.** $\displaystyle\int_{-\infty}^9 e^{4x}\,dx$

81. $\displaystyle\int_0^{\pi/2} \cot\theta\,d\theta$ **82.** $\displaystyle\int_1^\infty \frac{dx}{(x+2)(2x+3)}$

83. $\displaystyle\int_0^\infty (5+x)^{-1/3}\,dx$ **84.** $\displaystyle\int_2^5 (5-x)^{-1/3}\,dx$

In Exercises 85–90, use the Comparison Test to determine whether the improper integral converges or diverges.

85. $\displaystyle\int_8^\infty \frac{dx}{x^2-4}$ **86.** $\displaystyle\int_8^\infty (\sin^2 x)e^{-x}\,dx$

87. $\displaystyle\int_3^\infty \frac{dx}{x^4+\cos^2 x}$ **88.** $\displaystyle\int_1^\infty \frac{dx}{x^{1/3}+x^{2/3}}$

89. $\displaystyle\int_0^1 \frac{dx}{x^{1/3}+x^{2/3}}$ **90.** $\displaystyle\int_0^\infty e^{-x^3}\,dx$

91. Calculate the volume of the infinite solid obtained by rotating the region under $y = (x^2+1)^{-2}$ for $0 \le x < \infty$ about the y-axis.

92. Let R be the region under the graph of $y = (x+1)^{-1}$ for $0 \le x < \infty$. Which of the following quantities is finite?

(a) The area of R

(b) The volume of the solid obtained by rotating R about the x-axis

(c) The volume of the solid obtained by rotating R about the y-axis

93. Show that $\int_0^\infty x^n e^{-x^2}\,dx$ converges for all $n > 0$. *Hint:* First observe that $x^n e^{-x^2} < x^n e^{-x}$ for $x > 1$. Then show that $x^n e^{-x} < x^{-2}$ for x sufficiently large.

94. Compute the Laplace transform $Lf(s)$ of the function $f(x) = x$ for $s > 0$. See Exercises 86–89 in Section 8.6 for the definition of $Lf(s)$.

95. Compute the Laplace transform $Lf(s)$ of the function $f(x) = x^2e^{\alpha x}$ for $s > \alpha$.

96. Estimate $\displaystyle\int_2^5 f(x)\,dx$ by computing T_2, M_3, T_6, and S_6 for a function $f(x)$ taking on the values in the following table:

x	2	2.5	3	3.5	4	4.5	5
$f(x)$	$\frac{1}{2}$	2	1	0	$-\frac{3}{2}$	-4	-2

97. State whether the approximation M_N or T_N is larger or smaller than the integral.

(a) $\displaystyle\int_0^\pi \sin x\,dx$ **(b)** $\displaystyle\int_\pi^{2\pi} \sin x\,dx$

(c) $\displaystyle\int_1^8 \frac{dx}{x^2}$ **(d)** $\displaystyle\int_2^5 \ln x\,dx$

98. The rainfall rate (in inches per hour) was measured hourly during a 10-hour thunderstorm with the following results:

0, 0.41, 0.49, 0.32, 0.3, 0.23,
0.09, 0.08, 0.05, 0.11, 0.12

Use Simpson's Rule to estimate the total rainfall during the 10-hour period.

In Exercises 99–104, compute the given approximation to the integral.

99. $\displaystyle\int_0^1 e^{-x^2}\,dx,\quad M_5$ **100.** $\displaystyle\int_2^4 \sqrt{6t^3+1}\,dt,\quad T_3$

101. $\displaystyle\int_{\pi/4}^{\pi/2} \sqrt{\sin\theta}\,d\theta,\quad M_4$ **102.** $\displaystyle\int_1^4 \frac{dx}{x^3+1},\quad T_6$

103. $\displaystyle\int_0^1 e^{-x^2}\,dx,\quad S_4$ **104.** $\displaystyle\int_5^9 \cos(x^2)\,dx,\quad S_8$

105. The following table gives the area $A(h)$ of a horizontal cross section of a pond at depth h. Use the Trapezoidal Rule to estimate the volume V of the pond (Figure 1).

h (ft)	$A(h)$ (acres)	h (ft)	$A(h)$ (acres)
0	2.8	10	0.8
2	2.4	12	0.6
4	1.8	14	0.2
6	1.5	16	0.1
8	1.2	18	0

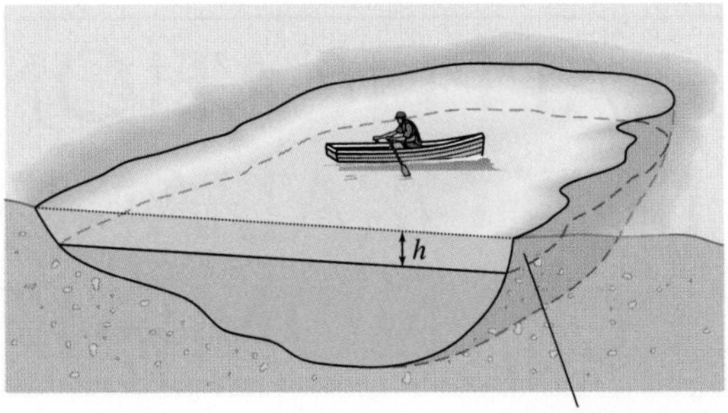

FIGURE 1

106. Suppose that the second derivative of the function $A(h)$ in Exercise 105 satisfies $|A''(h)| \leq 1.5$. Use the error bound to find the maximum possible error in your estimate of the volume V of the pond.

107. Find a bound for the error $\left| M_{16} - \int_1^3 x^3\,dx \right|$.

108. GU Let $f(x) = \sin(x^3)$. Find a bound for the error

$$\left| T_{24} - \int_0^{\pi/2} f(x)\,dx \right|$$

Hint: Find a bound K_2 for $|f''(x)|$ by plotting $f''(x)$ with a graphing utility.

109. Find a value of N such that

$$\left| M_N - \int_0^{\pi/4} \tan x\,dx \right| \leq 10^{-4}$$

110. Find a value of N such that S_N approximates $\int_2^5 x^{-1/4}\,dx$ with an error of at most 10^{-2} (but do not calculate S_N).

9 FURTHER APPLICATIONS OF THE INTEGRAL AND TAYLOR POLYNOMIALS

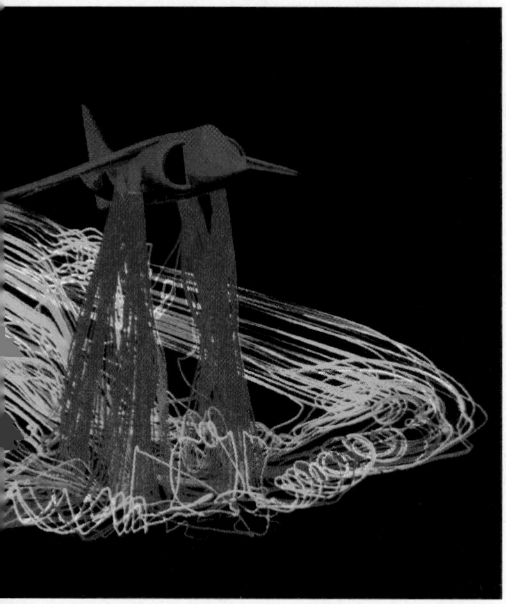

This NASA simulation, depicting streamlines of hot gas from the nozzles of a Harrier Jet during vertical takeoff, is based on a branch of mathematics called computational fluid dynamics.

The first three sections of this chapter develop some additional uses of integration, including two important physical applications. The last section introduces Taylor polynomials, the higher-order generalizations of the linear approximation. Taylor polynomials illustrate beautifully the power of calculus to yield valuable insight into functions.

9.1 Arc Length and Surface Area

We have seen that integrals are used to compute "total amounts" (such as distance traveled, total mass, total cost, etc.). Another such quantity is the length of a curve (also called **arc length**). We shall derive a formula for arc length using our standard procedure: approximation followed by passage to a limit.

Consider the graph of $y = f(x)$ over an interval $[a, b]$. Choose a partition P of $[a, b]$ into N subintervals with endpoints

$$P : a = x_0 < x_1 < \cdots < x_N = b$$

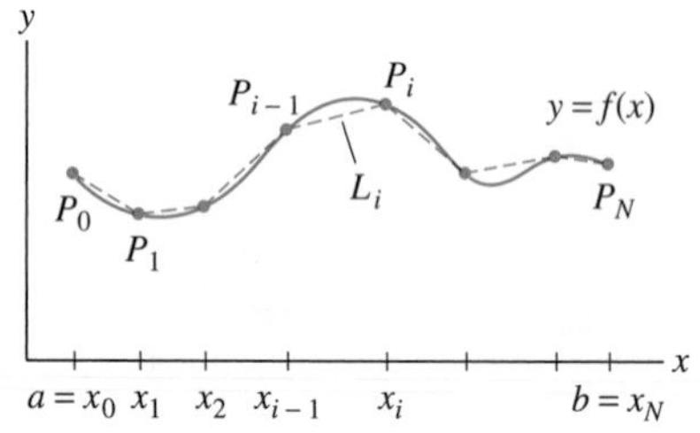

FIGURE 1 A polygonal approximation L to $y = f(x)$.

and let $P_i = (x_i, f(x_i))$ be the point on the graph above x_i. Now join these points by line segments $L_i = \overline{P_{i-1}P_i}$. The resulting curve L is called a **polygonal approximation** (Figure 1). The length of L, which we denote $|L|$, is the sum of the lengths $|L_i|$ of the segments:

$$|L| = |L_1| + |L_2| + \cdots + |L_N| = \sum_{i=1}^{N} |L_i|$$

The letter s is commonly used to denote arc length.

As may be expected, the polygonal approximations L approximate the curve more and more closely as the width of the partition decreases (Figure 2). Based on this idea, we define the arc length s of the graph to be the limit of the lengths $|L|$ as the width $\|P\|$ of the partition tends to zero:

$$\text{arc length } s = \lim_{\|P\|\to 0} \sum_{i=1}^{N} |L_i|$$

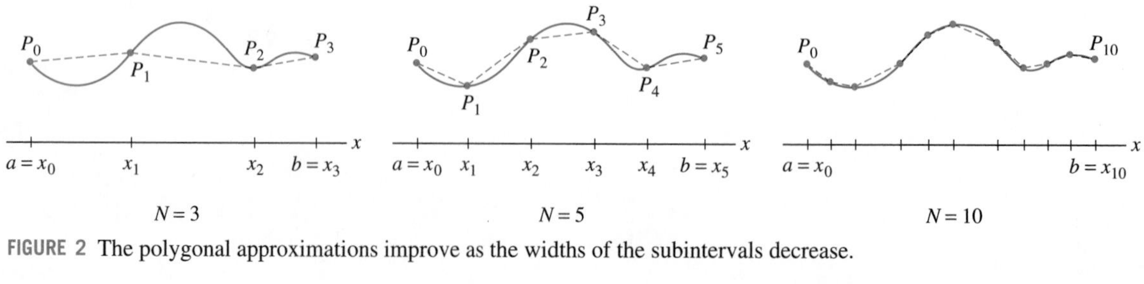

FIGURE 2 The polygonal approximations improve as the widths of the subintervals decrease.

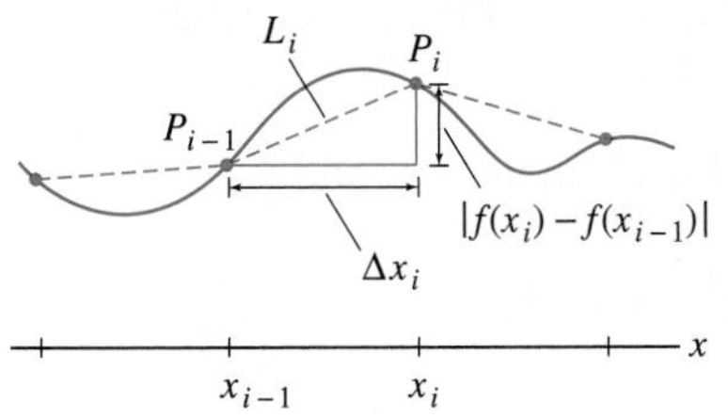

FIGURE 3

To compute the arc length s, we must express the limit of the polygonal approximations as an integral. Figure 3 shows that the segment L_i is the hypotenuse of a right triangle of base $\Delta x_i = x_i - x_{i-1}$ and height $|f(x_i) - f(x_{i-1})|$. By the Pythagorean Theorem,

$$|L_i| = \sqrt{\Delta x_i^2 + (f(x_i) - f(x_{i-1}))^2}$$

We shall assume that $f'(x)$ exists and is continuous. Then, by the Mean Value Theorem, there is a value c_i in $[x_{i-1}, x_i]$ such that

$$f(x_i) - f(x_{i-1}) = f'(c_i)(x_i - x_{i-1}) = f'(c_i)\Delta x_i$$

and therefore,

$$|L_i| = \sqrt{(\Delta x_i)^2 + (f'(c_i)\Delta x_i)^2} = \sqrt{(\Delta x_i)^2(1 + [f'(c_i)]^2)} = \sqrt{1 + [f'(c_i)]^2}\,\Delta x_i$$

We find that the length $|L|$ is a Riemann sum for the function $\sqrt{1 + [f'(x)]^2}$:

$$|L| = |L_1| + |L_2| + \cdots + |L_N| = \sum_{i=1}^{N} \sqrt{1 + [f'(c_i)]^2}\,\Delta x_i$$

REMINDER A Riemann sum *for the integral* $\int_a^b g(x)\,dx$ *is a sum*

$$\sum_{i=1}^{N} g(c_i)\Delta x_i$$

where $x_0, x_1, \ldots, x_N$ *is a partition of* $[a, b]$, $\Delta x_i = x_i - x_{i-1}$, *and* c_i *is any number in* $[x_{i-1}, x_i]$.

This function is continuous, and hence integrable, so the Riemann sums approach

$$\int_a^b \sqrt{1 + [f'(x)]^2}\,dx$$

as the norm (maximum of the widths Δx_i) of the partition tends to zero.

In Exercises 20–22, we verify that Eq. (1) correctly gives the lengths of line segments and circles.

THEOREM 1 Formula for Arc Length Assume that $f'(x)$ exists and is continuous on $[a, b]$. Then the arc length s of $y = f(x)$ over $[a, b]$ is equal to

$$s = \int_a^b \sqrt{1 + [f'(x)]^2}\,dx \qquad \boxed{1}$$

■ **EXAMPLE 1** Find the arc length s of the graph of $f(x) = \frac{1}{12}x^3 + x^{-1}$ over $[1, 3]$ (Figure 4).

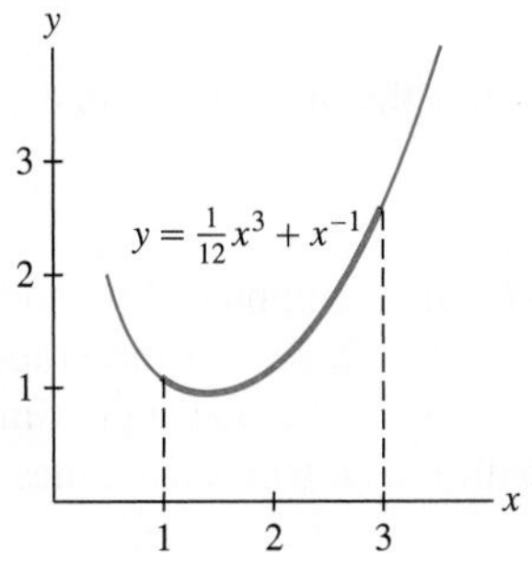

FIGURE 4 The arc length over [1, 3] is $\frac{17}{6}$.

Solution First, let's calculate $1 + f'(x)^2$. Since $f'(x) = \frac{1}{4}x^2 - x^{-2}$,

$$1 + f'(x)^2 = 1 + \left(\frac{1}{4}x^2 - x^{-2}\right)^2 = 1 + \left(\frac{1}{16}x^4 - \frac{1}{2} + x^{-4}\right)$$

$$= \frac{1}{16}x^4 + \frac{1}{2} + x^{-4} = \left(\frac{1}{4}x^2 + x^{-2}\right)^2$$

Fortunately, $1 + f'(x)^2$ is a square, so we can easily compute the arc length:

$$s = \int_1^3 \sqrt{1 + f'(x)^2}\,dx = \int_1^3 \left(\frac{1}{4}x^2 + x^{-2}\right)dx = \left(\frac{1}{12}x^3 - x^{-1}\right)\Bigg|_1^3$$

$$= \left(\frac{9}{4} - \frac{1}{3}\right) - \left(\frac{1}{12} - 1\right) = \frac{17}{6}$$ ■

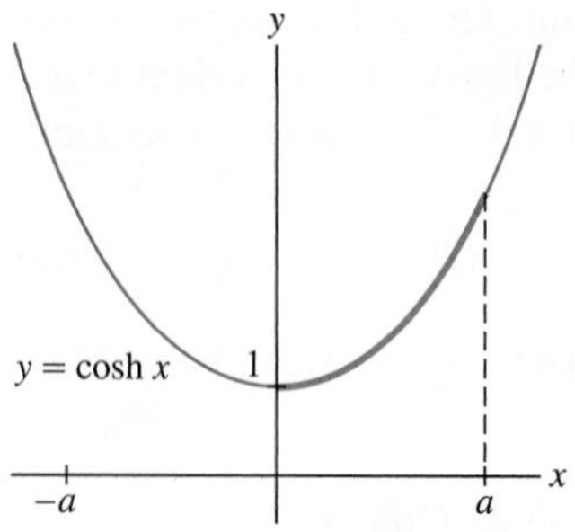

FIGURE 5

← REMINDER

$$\cosh x = \frac{1}{2}(e^x + e^{-x})$$

$$\sinh x = \frac{1}{2}(e^x - e^{-x})$$

$$\cosh^2 x - \sinh^2 x = 1 \qquad (2)$$

■ **EXAMPLE 2 Arc Length as a Function of the Upper Limit** Find the arc length $s(a)$ of $y = \cosh x$ over $[0, a]$ (Figure 5). Then find the arc length over $[0, 2]$.

Solution Recall that $y' = (\cosh x)' = \sinh x$. By Eq. (2) in the margin,

$$1 + (y')^2 = 1 + \sinh^2 x = \cosh^2 x$$

Because $\cosh x > 0$, we have $\sqrt{1 + (y')^2} = \cosh x$ and

$$s(a) = \int_0^a \sqrt{1 + (y')^2}\, dx = \int_0^a \cosh x\, dx = \sinh x \Big|_0^a = \sinh a$$

The arc length over $[0, 2]$ is $s(2) = \sinh 2 \approx 3.63$. ■

In Examples 1 and 2, the quantity $1 + f'(x)^2$ turned out to be a perfect square, and we were able to compute s exactly. Usually, $\sqrt{1 + f'(x)^2}$ does not have an elementary antiderivative and there is no explicit formula for the arc length. However, we can always approximate arc length using numerical integration.

■ **EXAMPLE 3 No Exact Formula for Arc Length** CAS Approximate the length s of $y = \sin x$ over $[0, \pi]$ using Simpson's Rule S_N with $N = 6$.

Solution We have $y' = \cos x$ and $\sqrt{1 + (y')^2} = \sqrt{1 + \cos^2 x}$. The arc length is

$$s = \int_0^\pi \sqrt{1 + \cos^2 x}\, dx$$

This integral cannot be evaluated explicitly, so we approximate it by applying Simpson's Rule (Section 8.8) to the integrand $g(x) = \sqrt{1 + \cos^2 x}$. Divide $[0, \pi]$ into $N = 6$ subintervals of width $\Delta x = \pi/6$. Then

$$S_6 = \frac{\Delta x}{3}\left(g(0) + 4g\left(\frac{\pi}{6}\right) + 2g\left(\frac{2\pi}{6}\right) + 4g\left(\frac{3\pi}{6}\right) + 2g\left(\frac{4\pi}{6}\right) + 4g\left(\frac{5\pi}{6}\right) + g(\pi)\right)$$

$$\approx \frac{\pi}{18}(1.4142 + 5.2915 + 2.2361 + 4 + 2.2361 + 5.2915 + 1.4142) \approx 3.82$$

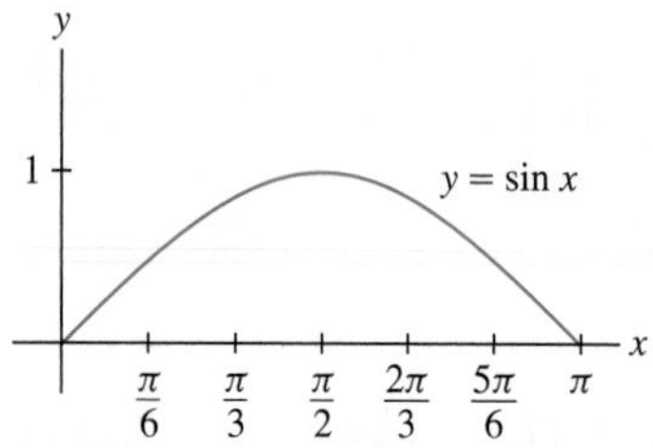

FIGURE 6 The arc length from 0 to π is approximately 3.82.

Thus $s \approx 3.82$ (Figure 6). A computer algebra system yields the more accurate approximation $s \approx 3.820198$. ■

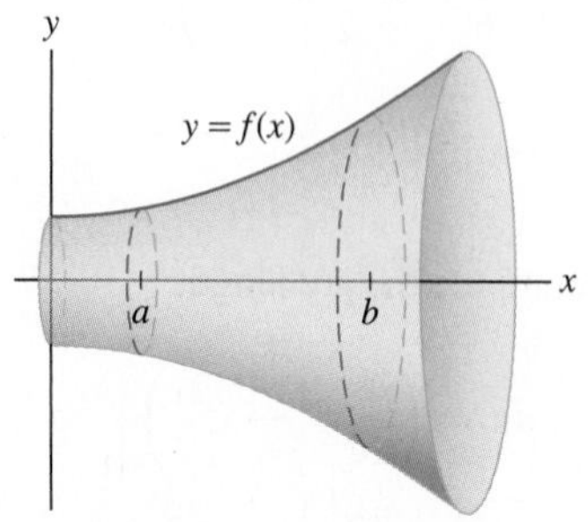

FIGURE 7 Surface obtained by revolving $y = f(x)$ about the x-axis.

The surface area S of a surface of revolution (Figure 7) can be computed by an integral that is similar to the arc length integral. Suppose that $f(x) \ge 0$, so that the graph lies above the x-axis. We can approximate the surface by rotating a polygonal approximation to $y = f(x)$ about the x-axis. The result is a surface built out of truncated cones (Figure 8).

The surface area of a truncated cone is equal to π times the sum of the left- and right-hand radii times the length of the slanted side. Using the notation from the derivation of the arc length formula above, we find that the surface area of the truncated cone along the subinterval $[x_{i-1}, x_i]$ is

$$\pi \underbrace{(f(x_{i-1}) + f(x_i))}_{\text{Sum of radii}} \underbrace{|\overline{P_{i-1}P_i}|}_{\text{Slant length}} = 2\pi \left(\frac{f(x_{i-1}) + f(x_i)}{2}\right)\sqrt{1 + f'(c_i)^2}\Delta x_i$$

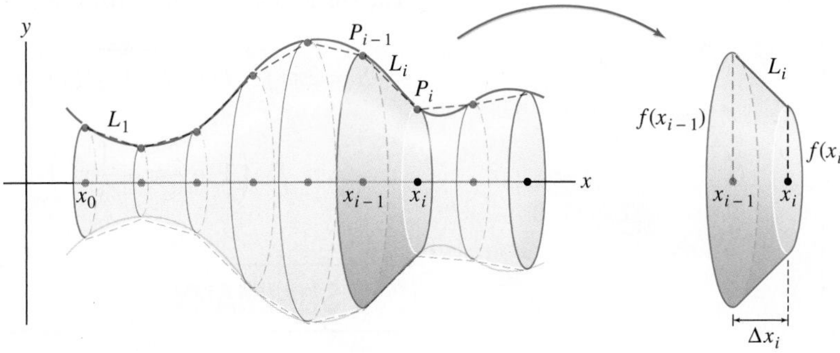

FIGURE 8 Rotating a polygonal approximation produces an approximation by truncated cones.

The surface area S is equal to the limit of the sums of the surface areas of the truncated cones as $N \to \infty$. We can show that the limit is not affected if we replace x_{i-1} and x_i by c_i. Therefore

$$S = 2\pi \lim_{N\to\infty} \sum_{i=1}^{N} f(c_i)\sqrt{1 + f'(c_i)^2}\,\Delta x_i$$

This is a limit of Riemann sums that converges to the integral in Eq. (3) below.

Area of a Surface of Revolution Assume that $f(x) \geq 0$ and that $f'(x)$ exists and is continuous on $[a, b]$. The surface area S of the surface obtained by rotating the graph of $f(x)$ about the x-axis for $a \leq x \leq b$ is equal to

$$S = 2\pi \int_a^b f(x)\sqrt{1 + f'(x)^2}\,dx \qquad (3)$$

EXAMPLE 4 Calculate the surface area of a sphere of radius R.

Solution The graph of $f(x) = \sqrt{R^2 - x^2}$ is a semicircle of radius R (Figure 9). We obtain a sphere by rotating it about the x-axis. We have

$$f'(x) = -\frac{x}{\sqrt{R^2 - x^2}}, \qquad 1 + f'(x)^2 = 1 + \frac{x^2}{R^2 - x^2} = \frac{R^2}{R^2 - x^2}$$

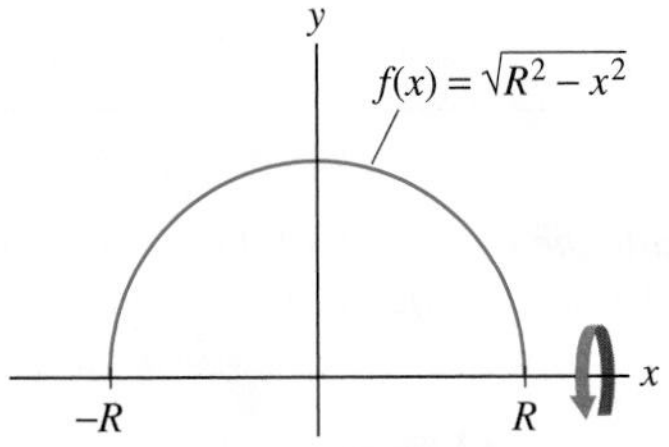

FIGURE 9 A sphere is obtained by revolving the semicircle about the x-axis.

The surface area integral gives us the usual formula for the surface area of a sphere:

$$S = 2\pi \int_{-R}^{R} f(x)\sqrt{1 + f'(x)^2}\,dx = 2\pi \int_{-R}^{R} \sqrt{R^2 - x^2}\,\frac{R}{\sqrt{R^2 - x^2}}\,dx$$

$$= 2\pi R \int_{-R}^{R} dx = 2\pi R(2R) = 4\pi R^2.$$

EXAMPLE 5 Find the surface area S of the surface obtained by rotating the graph of $y = x^{1/2} - \frac{1}{3}x^{3/2}$ about the x-axis for $1 \leq x \leq 3$.

Solution Let $f(x) = x^{1/2} - \frac{1}{3}x^{3/2}$. Then $f'(x) = \frac{1}{2}(x^{-1/2} - x^{1/2})$ and

$$1 + f'(x)^2 = 1 + \left(\frac{x^{-1/2} - x^{1/2}}{2}\right)^2 = 1 + \frac{x^{-1} - 2 + x}{4}$$

$$= \frac{x^{-1} + 2 + x}{4} = \left(\frac{x^{1/2} + x^{-1/2}}{2}\right)^2$$

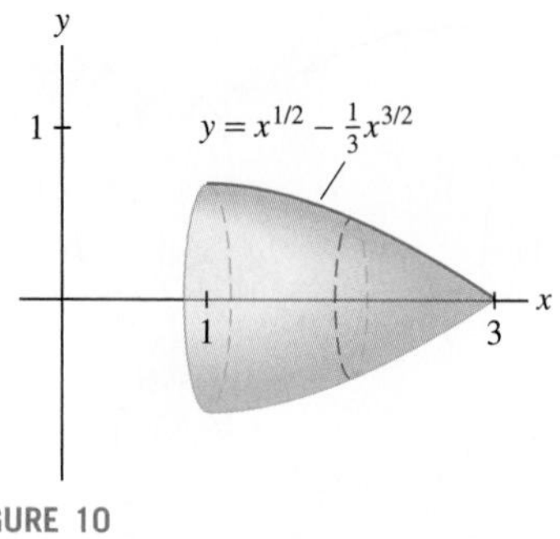

FIGURE 10

The surface area (Figure 10) is equal to

$$S = 2\pi \int_1^3 f(x)\sqrt{1 + f'(x)^2}\, dx = 2\pi \int_1^3 \left(x^{1/2} - \frac{1}{3}x^{3/2}\right)\left(\frac{x^{1/2} + x^{-1/2}}{2}\right) dx$$

$$= \pi \int_1^3 \left(1 + \frac{2}{3}x - \frac{1}{3}x^2\right) dx = \pi \left(x + \frac{1}{3}x^2 - \frac{1}{9}x^3\right)\Bigg|_1^3 = \frac{16\pi}{9}$$

■

9.1 SUMMARY

- The *arc length* of $y = f(x)$ over $[a, b]$ is

$$s = \int_a^b \sqrt{1 + f'(x)^2}\, dx$$

- Use numerical integration to approximate arc length when the arc length integral cannot be evaluated explicitly.
- Assume that $f(x) \geq 0$. The *surface area* of the surface obtained by rotating the graph of $f(x)$ about the x-axis for $a \leq x \leq b$ is

$$\text{Surface area} = 2\pi \int_a^b f(x)\sqrt{1 + f'(x)^2}\, dx$$

9.1 EXERCISES

Preliminary Questions

1. Which integral represents the length of the curve $y = \cos x$ between 0 and π?

$$\int_0^\pi \sqrt{1 + \cos^2 x}\, dx, \qquad \int_0^\pi \sqrt{1 + \sin^2 x}\, dx$$

2. Use the formula for arc length to show that for any constant C, the graphs $y = f(x)$ and $y = f(x) + C$ have the same length over every interval $[a, b]$. Explain geometrically.

3. Use the formula for arc length to show that the length of a graph over $[1, 4]$ cannot be less than 3.

Exercises

1. Express the arc length of the curve $y = x^4$ between $x = 2$ and $x = 6$ as an integral (but do not evaluate).

2. Express the arc length of the curve $y = \tan x$ for $0 \leq x \leq \frac{\pi}{4}$ as an integral (but do not evaluate).

3. Find the arc length of $y = \frac{1}{12}x^3 + x^{-1}$ for $1 \leq x \leq 2$. *Hint:* Show that $1 + (y')^2 = \left(\frac{1}{4}x^2 + x^{-2}\right)^2$.

4. Find the arc length of $y = \left(\frac{x}{2}\right)^4 + \frac{1}{2x^2}$ over $[1, 4]$. *Hint:* Show that $1 + (y')^2$ is a perfect square.

In Exercises 5–10, calculate the arc length over the given interval.

5. $y = 3x + 1$, $[0, 3]$

6. $y = 9 - 3x$, $[1, 3]$

7. $y = x^{3/2}$, $[1, 2]$

8. $y = \frac{1}{3}x^{3/2} - x^{1/2}$, $[2, 8]$

9. $y = \frac{1}{4}x^2 - \frac{1}{2}\ln x$, $[1, 2e]$

10. $y = \ln(\cos x)$, $\left[0, \frac{\pi}{4}\right]$

In Exercises 11–14, approximate the arc length of the curve over the interval using the Trapezoidal Rule T_N, the Midpoint Rule M_N, or Simpson's Rule S_N as indicated.

11. $y = \frac{1}{4}x^4$, $[1, 2]$, T_5

12. $y = \sin x$, $\left[0, \frac{\pi}{2}\right]$, M_8

13. $y = x^{-1}$, $[1, 2]$, S_8

14. $y = e^{-x^2}$, $[0, 2]$, S_8

15. Calculate the length of the astroid $x^{2/3} + y^{2/3} = 1$ (Figure 11).

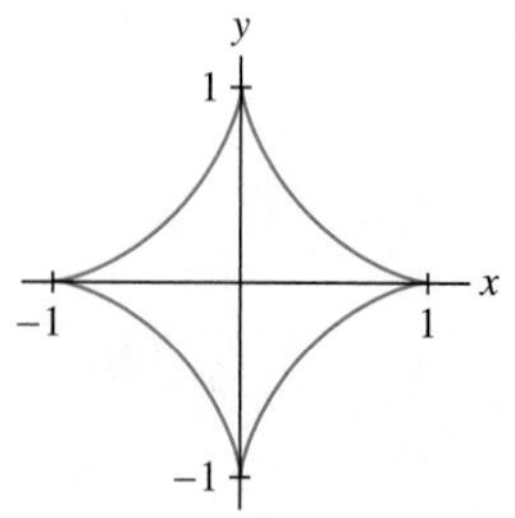

FIGURE 11 Graph of $x^{2/3} + y^{2/3} = 1$.

16. Show that the arc length of the astroid $x^{2/3} + y^{2/3} = a^{2/3}$ (for $a > 0$) is proportional to a.

17. Let $a, r > 0$. Show that the arc length of the curve $x^r + y^r = a^r$ for $0 \le x \le a$ is proportional to a.

18. Find the arc length of the curve shown in Figure 12.

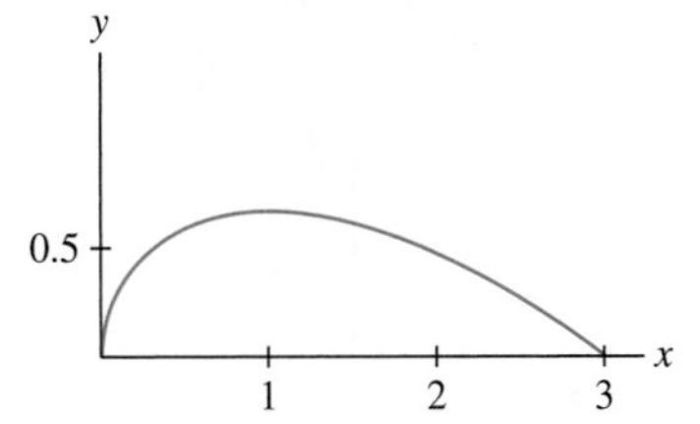

FIGURE 12 Graph of $9y^2 = x(x-3)^2$.

19. Find the value of a such that the arc length of the *catenary* $y = \cosh x$ for $-a \le x \le a$ equals 10.

20. Calculate the arc length of the graph of $f(x) = mx + r$ over $[a, b]$ in two ways: using the Pythagorean theorem (Figure 13) and using the arc length integral.

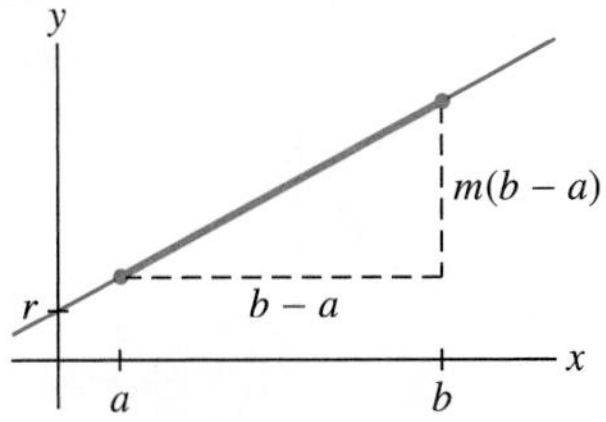

FIGURE 13

21. Show that the circumference of the unit circle is equal to

$$2\int_{-1}^{1} \frac{dx}{\sqrt{1-x^2}} \quad \text{(an improper integral)}$$

Evaluate, thus verifying that the circumference is 2π.

22. Generalize the result of Exercise 21 to show that the circumference of the circle of radius r is $2\pi r$.

23. Calculate the arc length of $y = x^2$ over $[0, a]$. *Hint:* Use trigonometric substitution. Evaluate for $a = 1$.

24. Express the arc length of $g(x) = \sqrt{x}$ over $[0, 1]$ as a definite integral. Then use the substitution $u = \sqrt{x}$ to show that this arc length is equal to the arc length of x^2 over $[0, 1]$ (but do not evaluate the integrals). Explain this result graphically.

25. Find the arc length of $y = e^x$ over $[0, a]$. *Hint:* Try the substitution $u = \sqrt{1 + e^{2x}}$ followed by partial fractions.

26. Show that the arc length of $y = \ln(f(x))$ for $a \le x \le b$ is

$$\int_a^b \frac{\sqrt{f(x)^2 + f'(x)^2}}{f(x)}\, dx$$

27. Use Eq. (4) to compute the arc length of $y = \ln(\sin x)$ for $\frac{\pi}{4} \le x \le \frac{\pi}{2}$.

28. Use Eq. (4) to compute the arc length of $y = \ln\left(\frac{e^x + 1}{e^x - 1}\right)$ over $[1, 3]$.

29. Show that if $0 \le f'(x) \le 1$ for all x, then the arc length of $y = f(x)$ over $[a, b]$ is at most $\sqrt{2}(b - a)$. Show that for $f(x) = x$, the arc length equals $\sqrt{2}(b - a)$.

30. Use the Comparison Theorem (Section 5.2) to prove that the arc length of $y = x^{4/3}$ over $[1, 2]$ is not less than $\frac{5}{3}$.

31. Approximate the arc length of one-quarter of the unit circle (which we know is $\frac{\pi}{2}$) by computing the length of the polygonal approximation with $N = 4$ segments (Figure 14).

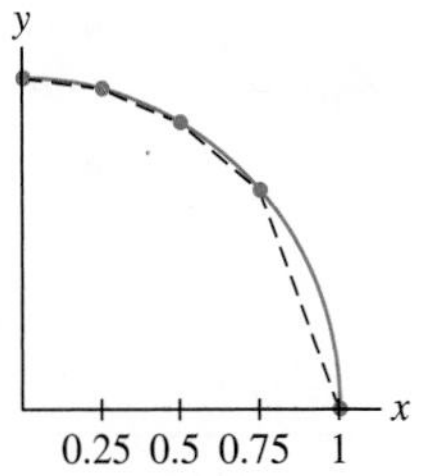

FIGURE 14 One-quarter of the unit circle

32. CAS A merchant intends to produce specialty carpets in the shape of the region in Figure 15, bounded by the axes and graph of $y = 1 - x^n$ (units in yards). Assume that material costs \$50/yd^2 and that it costs $50L$ dollars to cut the carpet, where L is the length of the curved side of the carpet. The carpet can be sold for $150A$ dollars, where A is the carpet's area. Using numerical integration with a computer algebra system, find the whole number n for which the merchant's profits are maximal.

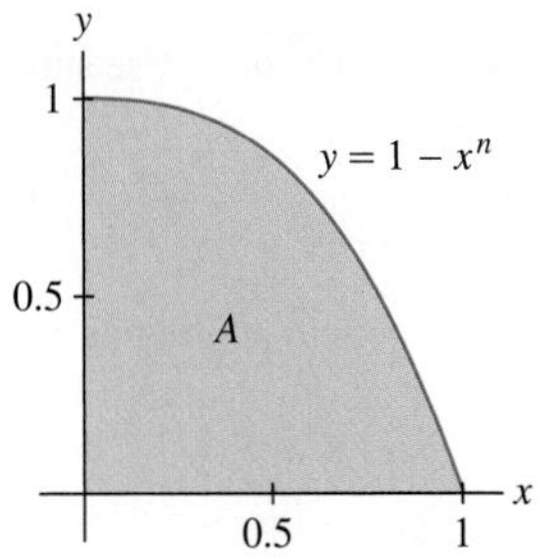

FIGURE 15

In Exercises 33–40, compute the surface area of revolution about the x-axis over the interval.

33. $y = x, \quad [0, 4]$

34. $y = 4x + 3, \quad [0, 1]$

35. $y = x^3, \quad [0, 2]$

36. $y = x^2, \quad [0, 4]$

37. $y = (4 - x^{2/3})^{3/2}, \quad [0, 8]$

38. $y = e^{-x}, \quad [0, 1]$

39. $y = \frac{1}{4}x^2 - \frac{1}{2}\ln x, \quad [1, e]$

40. $y = \sin x, \quad [0, \pi]$

CAS *In Exercises 41–44, use a computer algebra system to find the approximate surface area of the solid generated by rotating the curve about the x-axis.*

41. $y = x^{-1}$, $[1, 3]$

42. $y = x^4$, $[0, 1]$

43. $y = e^{-x^2/2}$, $[0, 2]$

44. $y = \tan x$, $\left[0, \frac{\pi}{4}\right]$

45. Find the area of the surface obtained by rotating $y = \cosh x$ over $[-\ln 2, \ln 2]$ around the x-axis.

46. Show that the surface area of a spherical cap of height h and radius R (Figure 16) has surface area $2\pi Rh$.

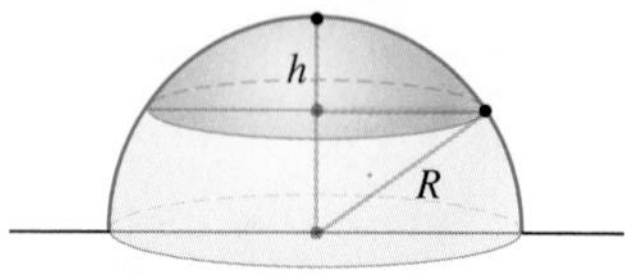

FIGURE 16

47. Find the surface area of the torus obtained by rotating the circle $x^2 + (y - b)^2 = r^2$ about the x-axis (Figure 17).

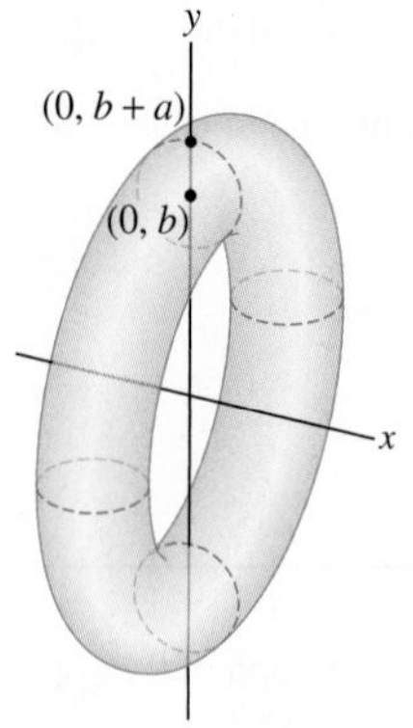

FIGURE 17 Torus obtained by rotating a circle about the x-axis.

48. Show that the surface area of a right circular cone of radius r and height h is $\pi r\sqrt{r^2 + h^2}$. *Hint:* Rotate a line $y = mx$ about the x-axis for $0 \le x \le h$, where m is determined suitably by the radius r.

Further Insights and Challenges

49. Find the surface area of the ellipsoid obtained by rotating the ellipse $\left(\frac{x}{a}\right)^2 + \left(\frac{y}{b}\right)^2 = 1$ about the x-axis.

50. Show that if the arc length of $f(x)$ over $[0, a]$ is proportional to a, then $f(x)$ must be a linear function.

51. CAS Let L be the arc length of the upper half of the ellipse with equation

$$y = \frac{b}{a}\sqrt{a^2 - x^2}$$

(Figure 18) and let $\eta = \sqrt{1 - (b^2/a^2)}$. Use substitution to show that

$$L = a\int_{-\pi/2}^{\pi/2} \sqrt{1 - \eta^2 \sin^2\theta}\, d\theta$$

Use a computer algebra system to approximate L for $a = 2$, $b = 1$.

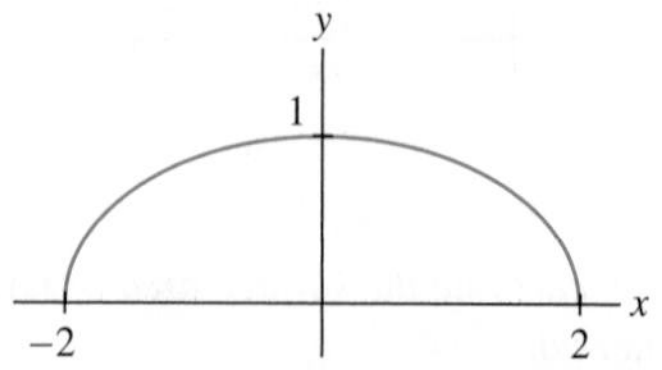

FIGURE 18 Graph of the ellipse $y = \frac{1}{2}\sqrt{4 - x^2}$.

52. Prove that the portion of a sphere of radius R seen by an observer located at a distance d above the North Pole has area $A = 2\pi dR^2/(d + R)$. *Hint:* According to Exercise 46, the cap has surface area is $2\pi Rh$. Show that $h = dR/(d + R)$ by applying the Pythagorean Theorem to the three right triangles in Figure 19.

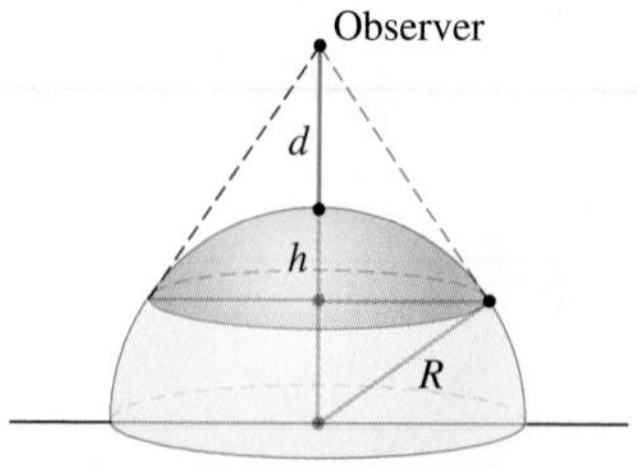

FIGURE 19 Spherical cap observed from a distance d above the North Pole.

53. Suppose that the observer in Exercise 52 moves off to infinity—that is, $d \to \infty$. What do you expect the limiting value of the observed area to be? Check your guess by calculating the limit using the formula for the area in the previous exercise.

54. Let M be the total mass of a metal rod in the shape of the curve $y = f(x)$ over $[a, b]$ whose mass density $\rho(x)$ varies as a function of x. Use Riemann sums to justify the formula

$$M = \int_a^b \rho(x)\sqrt{1 + f'(x)^2}\, dx$$

55. Let $f(x)$ be an increasing function on $[a, b]$ and let $g(x)$ be its inverse. Argue on the basis of arc length that the following equality holds:

$$\int_a^b \sqrt{1 + f'(x)^2}\, dx = \int_{f(a)}^{f(b)} \sqrt{1 + g'(y)^2}\, dy \qquad \boxed{5}$$

Then use the substitution $u = f(x)$ to prove Eq. (5).

9.2 Fluid Pressure and Force

Fluid force is the force on an object submerged in a fluid. Divers feel this force as they descend below the water surface (Figure 1). Our calculation of fluid force is based on two laws that determine the pressure exerted by a fluid:

- Fluid pressure p is proportional to depth.
- Fluid pressure does not act in a specific direction. Rather, a fluid exerts pressure on each side of an object in the perpendicular direction (Figure 2).

This second fact, known as Pascal's principle, points to an important difference between fluid pressure and the pressure exerted by one solid object on another.

FIGURE 1 Since water pressure is proportional to depth, divers breathe compressed air to equalize the pressure and avoid lung injury.

***Pressure**, by definition, is force per unit area.*

- *The SI unit of pressure is the pascal (Pa)* $(1\ Pa = 1\ \mathrm{N/m^2})$.
- *Mass density (mass per unit volume) is denoted* ρ *(Greek rho).*
- *The factor* ρg *is the density by weight, where* $g = 9.8\ \mathrm{m/s^2}$ *is the acceleration due to gravity.*

Fluid Pressure The pressure p at depth h in a fluid of mass density ρ is

$$p = \rho g h \qquad (1)$$

The pressure acts at each point on an object in the direction perpendicular to the object's surface at that point.

Our first example does not require integration because the pressure p is constant. In this case, the total force acting on a surface of area A is

$$\text{Force} = \text{pressure} \times \text{area} = pA$$

■ **EXAMPLE 1** Calculate the fluid force on the top and bottom of a box of dimensions $2 \times 2 \times 5$ m, submerged in a pool of water with its top 3 m below the water surface (Figure 2). The density of water is $\rho = 10^3\ \mathrm{kg/m^3}$.

Solution The top of the box is located at depth $h = 3$ m, so, by Eq. (1) with $g = 9.8$,

$$\text{Pressure on top} = \rho g h = 10^3(9.8)(3) = 29{,}400\ \text{Pa}$$

The top has area $A = 4\ \mathrm{m^2}$ and the pressure is constant, so

$$\text{Downward force on top} = pA = 10^3(9.8)(3) \times 4 = 117{,}600\ \text{N}$$

The bottom of the box is at depth $h = 8$ m, so the total force on the bottom is

$$\text{Upward force on bottom} = pA = 10^3(9.8)(8) \times 4 = 313{,}600\ \text{N}$$ ■

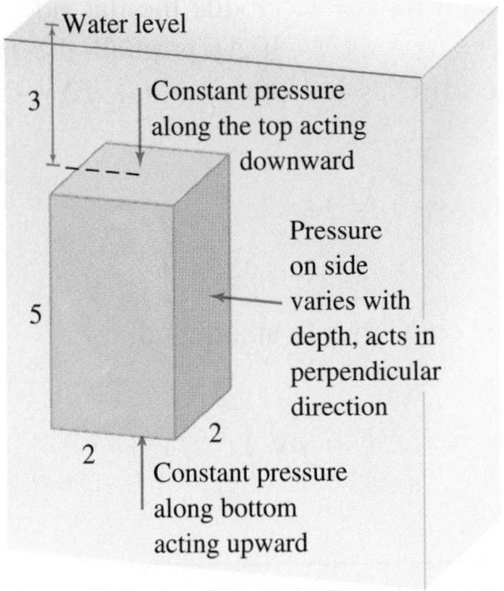

FIGURE 2 Fluid pressure acts on each side in the perpendicular direction.

In the next example, the pressure varies with depth, and it is necessary to calculate the force as an integral.

■ **EXAMPLE 2 Calculating Force Using Integration** Calculate the fluid force F on the side of the box in Example 1.

Solution Since the pressure varies with depth, we divide the side of the box into N thin horizontal strips (Figure 3). Let F_j be the force on the jth strip. The total force F is equal to the sum of the forces on the strips:

$$F = F_1 + F_2 + \cdots + F_N$$

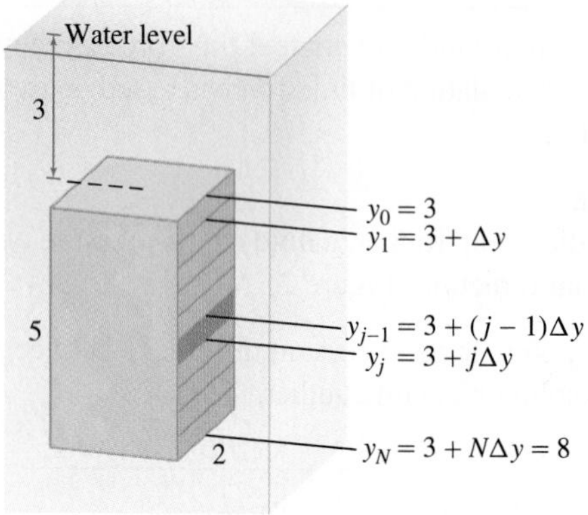

FIGURE 3 Each strip has area $2\Delta y$.

Step 1. **Approximate the force on a strip.**

We'll use the variable y to denote depth, where $y = 0$ at the water level and y is positive in the downward direction. Thus, a larger value of y denotes greater depth. Each strip is a rectangle of height $\Delta y = 5/N$ and length 2, so the area of a strip is $2\Delta y$. The bottom edge of the jth strip has depth $y_j = 3 + j\Delta y$.

If Δy is small, the pressure on the jth strip is nearly constant with value $\rho g y_j$ (because all points on the strip lie at nearly the same depth y_j), so we can approximate the force on the jth strip:

$$F_j \approx \underbrace{\rho g y_j}_{\text{Pressure}} \times \underbrace{(2\Delta y)}_{\text{Area}} = (\rho g)2y_j\Delta y$$

Step 2. **Approximate total force as a Riemann sum.**

$$F = F_1 + F_2 + \cdots + F_N \approx \rho g \sum_{j=1}^{N} 2y_j\Delta y$$

The sum on the right is a Riemann sum that converges to the integral $\rho g \int_3^8 2y\,dy$. The interval of integration is $[3, 8]$ because the box extends from $y = 3$ to $y = 8$ (the Riemann sum has been set up with $y_0 = 3$ and $y_N = 8$).

Step 3. **Evaluate total force as an integral.**

As Δy tends to zero, the Riemann sum approaches the integral, and we obtain

$$F = \rho g \int_3^8 2y\,dy = (\rho g)y^2\Big|_3^8 = (10^3)(9.8)(8^2 - 3^2) = 539{,}000 \text{ N}$$ ■

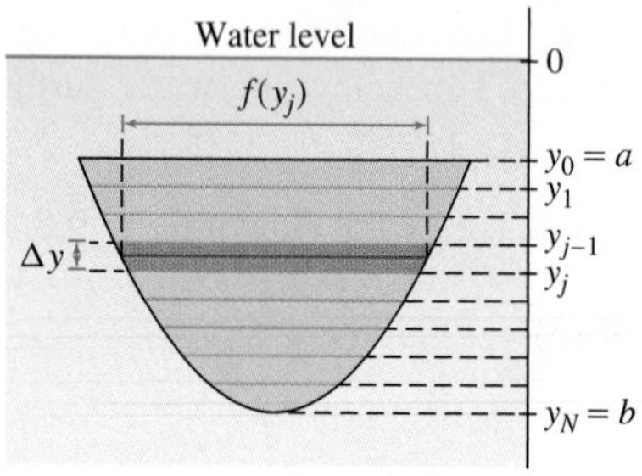

FIGURE 4 The area of the shaded strip is approximately $f(y_j)\,\Delta y$.

Now we'll add another complication: allowing the widths of the horizontal strips to vary with depth (Figure 4). Denote the width at depth y by $f(y)$:

$$f(y) = \text{width of the side at depth } y$$

As before, assume that the object extends from $y = a$ to $y = b$. Divide the flat side of the object into N horizontal strips of thickness $\Delta y = (b - a)/N$. If Δy is small, the jth strip is nearly rectangular of area $f(y)\Delta y$. Since the strip lies at depth $y_j = a + j\Delta y$, the force F_j on the jth strip can be approximated:

$$F_j \approx \underbrace{\rho g y_j}_{\text{Pressure}} \times \underbrace{f(y_j)\Delta y}_{\text{Area}} = (\rho g)y_j f(y_j)\Delta y$$

The force F is approximated by a Riemann sum that converges to an integral:

$$F = F_1 + \cdots + F_N \approx \rho g \sum_{j=1}^{N} y_j f(y_j)\Delta y \quad \Rightarrow \quad F = \rho g \int_a^b y f(y)\,dy$$

THEOREM 1 Fluid Force on a Flat Surface Submerged Vertically The fluid force F on a flat side of an object submerged vertically in a fluid is

$$F = \rho g \int_a^b y f(y)\,dy \qquad \boxed{2}$$

where $f(y)$ is the horizontal width of the side at depth y, and the object extends from depth $y = a$ to depth $y = b$.

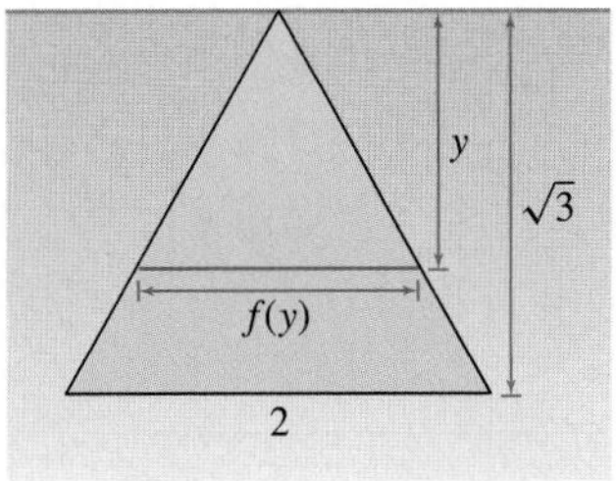

FIGURE 5 Triangular plate submerged in a tank of oil.

■ **EXAMPLE 3** Calculate the fluid force F on one side of an equilateral triangular plate of side 2 m submerged vertically in a tank of oil of mass density $\rho = 900\ \text{kg/m}^3$ (Figure 5).

Solution To use Eq. (2), we need to find the horizontal width $f(y)$ of the plate at depth y. An equilateral triangle of side $s = 2$ has height $\sqrt{3}s/2 = \sqrt{3}$. By similar triangles, $y/f(y) = \sqrt{3}/2$ and thus $f(y) = 2y/\sqrt{3}$. By Eq. (2),

$$F = \rho g \int_0^{\sqrt{3}} y\, f(y)\, dy = (900)(9.8) \int_0^{\sqrt{3}} \frac{2}{\sqrt{3}} y^2\, dy = \left(\frac{17{,}640}{\sqrt{3}}\right) \frac{y^3}{3}\bigg|_0^{\sqrt{3}} = 17{,}640\ \text{N}$$

■

The next example shows how to modify the force calculation when the side of the submerged object is inclined at an angle.

Hoover Dam, with recently completed Colorado river bridge

■ **EXAMPLE 4 Force on an Inclined Surface** The side of a dam is inclined at an angle of 45°. The dam has height 700 ft and width 1500 ft as in Figure 6. Calculate the force F on the dam if the reservoir is filled to the top of the dam. Water has density $w = 62.5$ lb/ft^3.

Solution The vertical height of the dam is 700 ft, so we divide the vertical axis from 0 to 700 into N subintervals of length $\Delta y = 700/N$. This divides the face of the dam into N strips as in Figure 6. By trigonometry, each strip has width equal to $\Delta y/\sin(45°) = \sqrt{2}\Delta y$. Therefore,

$$\text{Area of each strip} = \text{length} \times \text{width} = 1500(\sqrt{2}\,\Delta y)$$

As usual, we approximate the force F_j on the jth strip. The factor ρg is equal to weight per unit volume, so we use $w = 62.5$ lb/ft^3 in place of ρg:

$$F_j \approx \overbrace{wy_j}^{\text{Pressure}} \times \overbrace{1500\sqrt{2}\Delta y}^{\text{Area of strip}} = wy_j \times 1500\sqrt{2}\,\Delta y \ \ \text{lb}$$

$$F = \sum_{j=1}^{N} F_j \approx \sum_{j=1}^{N} wy_j\left(1500\sqrt{2}\,\Delta y\right) = 1500\sqrt{2}\, w \sum_{j=1}^{N} y_j \Delta y$$

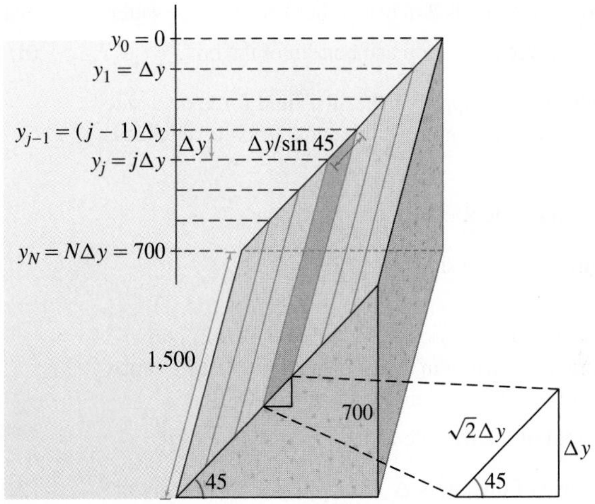

FIGURE 6

This is a Riemann sum for the integral $1500\sqrt{2}w \int_0^{700} y\,dy$. Therefore,

$$F = 1500\sqrt{2}w \int_0^{700} y\,dy = 1500\sqrt{2}(62.5)\frac{700^2}{2} \approx 3.25 \times 10^{10} \text{ lb}$$

9.2 SUMMARY

- If pressure is constant, then force = pressure × area.
- The fluid pressure at depth h is equal to $\rho g h$, where ρ is the fluid density (mass per unit volume) and $g = 9.8$ m/s^2 is the acceleration due to gravity. Fluid pressure acts on a surface in the direction perpendicular to the surface. Water has mass density 1000 kg/m^3.
- If an object is submerged vertically in a fluid and extends from depth $y = a$ to $y = b$, then the total fluid force on a side of the object is

$$F = \rho g \int_a^b y f(y)\,dy$$

where $f(y)$ is the horizontal width of the side at depth y.
- If fluid density is given as *weight* per unit volume, the factor g does not appear. Water has weight density 62.5 lb/ft^3.

9.2 EXERCISES

Preliminary Questions

1. How is pressure defined?

2. Fluid pressure is proportional to depth. What is the factor of proportionality?

3. When fluid force acts on the side of a submerged object, in which direction does it act?

4. Why is fluid pressure on a surface calculated using thin horizontal strips rather than thin vertical strips?

5. If a thin plate is submerged horizontally, then the fluid force on one side of the plate is equal to pressure times area. Is this true if the plate is submerged vertically?

Exercises

1. A box of height 6 m and square base of side 3 m is submerged in a pool of water. The top of the box is 2 m below the surface of the water.

(a) Calculate the fluid force on the top and bottom of the box.

(b) Write a Riemann sum that approximates the fluid force on a side of the box by dividing the side into N horizontal strips of thickness $\Delta y = 6/N$.

(c) To which integral does the Riemann sum converge?

(d) Compute the fluid force on a side of the box.

2. A plate in the shape of an isosceles triangle with base 1 m and height 2 m is submerged vertically in a tank of water so that its vertex touches the surface of the water (Figure 7).

(a) Show that the width of the triangle at depth y is $f(y) = \frac{1}{2}y$.

(b) Consider a thin strip of thickness Δy at depth y. Explain why the fluid force on a side of this strip is approximately equal to $\rho g \frac{1}{2} y^2 \Delta y$.

(c) Write an approximation for the total fluid force F on a side of the plate as a Riemann sum and indicate the integral to which it converges.

(d) Calculate F.

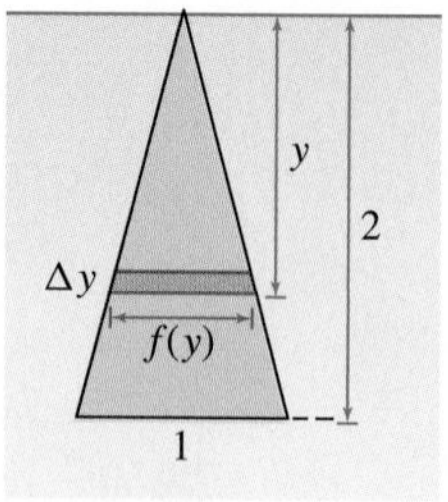

FIGURE 7

3. Repeat Exercise 2, but assume that the top of the triangle is located 3 m below the surface of the water.

4. The plate R in Figure 8, bounded by the parabola $y = x^2$ and $y = 1$, is submerged vertically in water (distance in meters).

(a) Show that the width of R at height y is $f(y) = 2\sqrt{y}$ and the fluid force on a side of a horizontal strip of thickness Δy at height y is approximately $(\rho g)2y^{1/2}(1 - y)\Delta y$.

(b) Write a Riemann sum that approximates the fluid force F on a side of R and use it to explain why

$$F = \rho g \int_0^1 2y^{1/2}(1 - y)\, dy$$

(c) Calculate F.

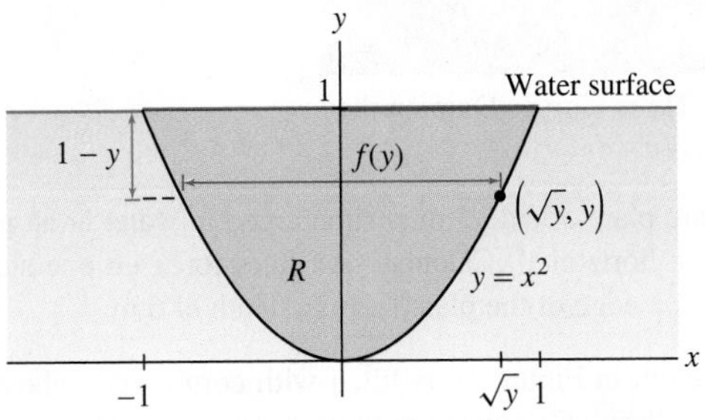

FIGURE 8

5. Let F be the fluid force on a side of a semicircular plate of radius r meters, submerged vertically in water so that its diameter is level with the water's surface (Figure 9).

(a) Show that the width of the plate at depth y is $2\sqrt{r^2 - y^2}$.

(b) Calculate F as a function of r using Eq. (2).

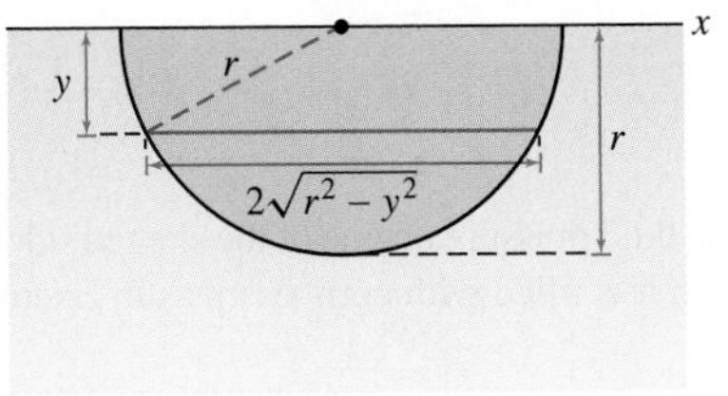

FIGURE 9

6. Calculate the force on one side of a circular plate with radius 2 m, submerged vertically in a tank of water so that the top of the circle is tangent to the water surface.

7. A semicircular plate of radius r meters, oriented as in Figure 9, is submerged in water so that its diameter is located at a depth of m meters. Calculate the fluid force on one side of the plate in terms of m and r.

8. A plate extending from depth $y = 2$ m to $y = 5$ m is submerged in a fluid of density $\rho = 850$ kg/m^3. The horizontal width of the plate at depth y is $f(y) = 2(1 + y^2)^{-1}$. Calculate the fluid force on one side of the plate.

9. Figure 10 shows the wall of a dam on a water reservoir. Use the Trapezoidal Rule and the width and depth measurements in the figure to estimate the fluid force on the wall.

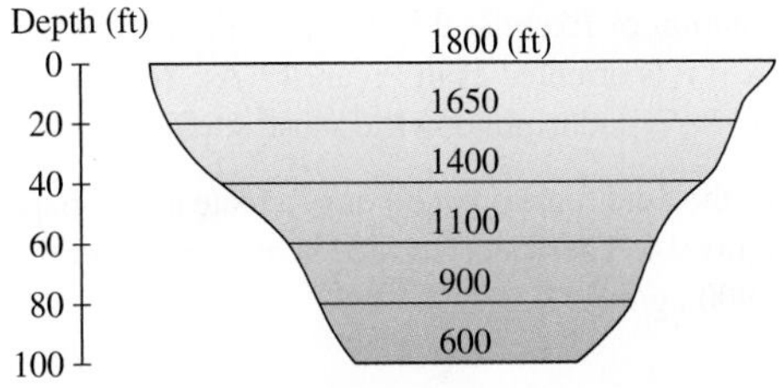

FIGURE 10

10. Calculate the fluid force on a side of the plate in Figure 11(A), submerged in water.

11. Calculate the fluid force on a side of the plate in Figure 11(B), submerged in a fluid of mass density $\rho = 800$ kg/m^3.

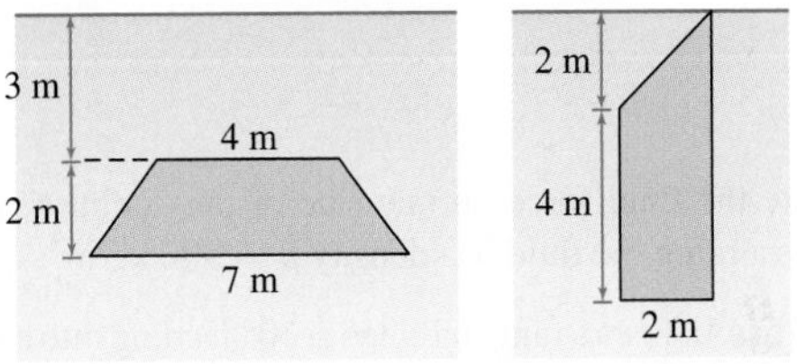

(A) (B)

FIGURE 11

12. Find the fluid force on the side of the plate in Figure 12, submerged in a fluid of density $\rho = 1200$ kg/m^3. The top of the place is level with the fluid surface. The edges of the plate are the curves $y = x^{1/3}$ and $y = -x^{1/3}$.

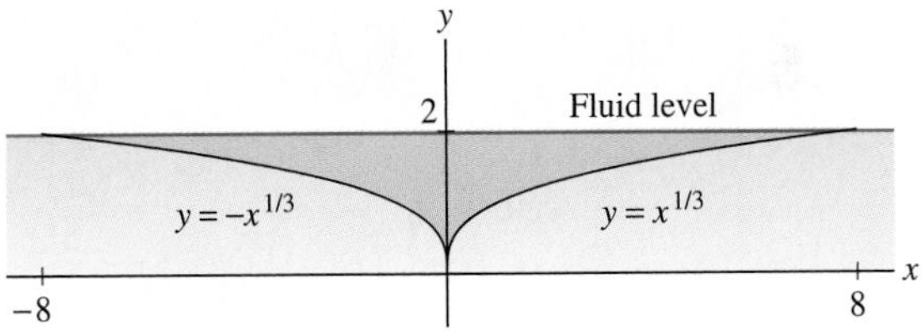

FIGURE 12

13. Let R be the plate in the shape of the region under $y = \sin x$ for $0 \le x \le \frac{\pi}{2}$ in Figure 13(A). Find the fluid force on a side of R if it is rotated counterclockwise by 90° and submerged in a fluid of density 1100 kg/m^3 with its top edge level with the surface of the fluid as in (B).

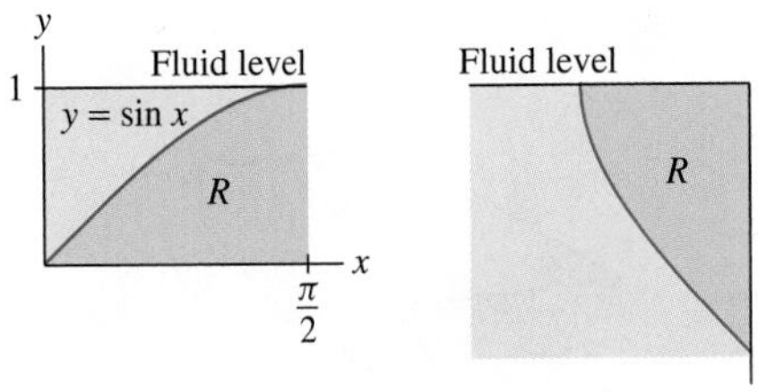

(A) (B)

FIGURE 13

14. In the notation of Exercise 13, calculate the fluid force on a side of the plate R if it is oriented as in Figure 13(A). You may need to use Integration by Parts and trigonometric substitution.

15. Calculate the fluid force on one side of a plate in the shape of region A shown Figure 14. The water surface is at $y = 1$, and the fluid has density $\rho = 900\ \text{kg/m}^3$.

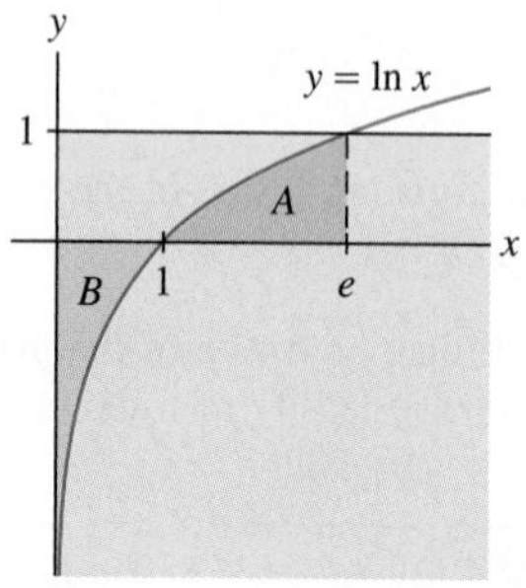

FIGURE 14

16. Calculate the fluid force on one side of the "infinite" plate B in Figure 14, assuming the fluid has density $\rho = 900\ \text{kg/m}^3$.

17. Figure 15(A) shows a ramp inclined at 30° leading into a swimming pool. Calculate the fluid force on the ramp.

18. Calculate the fluid force on one side of the plate (an isosceles triangle) shown in Figure 15(B).

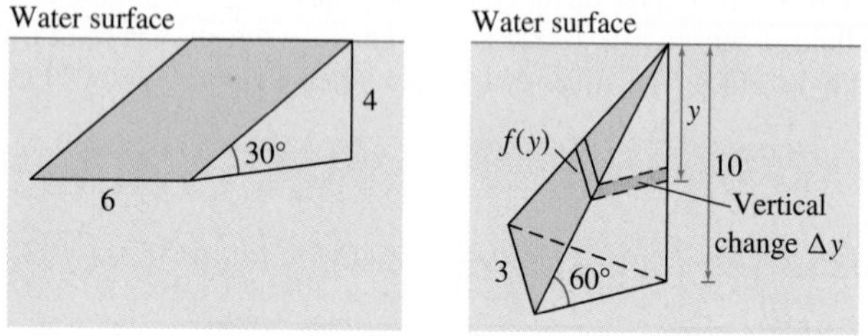

(A) (B)

FIGURE 15

19. The massive Three Gorges Dam on China's Yangtze River has height 185 m (Figure 16). Calculate the force on the dam, assuming that the dam is a trapezoid of base 2000 m and upper edge 3000 m, inclined at an angle of 55° to the horizontal (Figure 17).

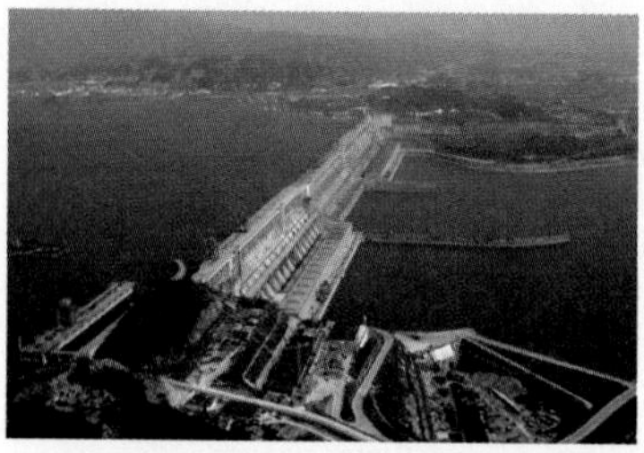

FIGURE 16 Three Gorges Dam on the Yangtze River

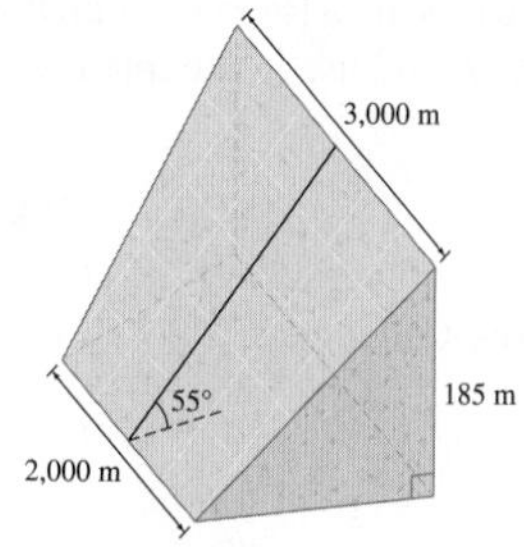

FIGURE 17

20. A square plate of side 3 m is submerged in water at an incline of 30° with the horizontal. Calculate the fluid force on one side of the plate if the top edge of the plate lies at a depth of 6 m.

21. The trough in Figure 18 is filled with corn syrup, whose weight density is 90 lb/ft^3. Calculate the force on the front side of the trough.

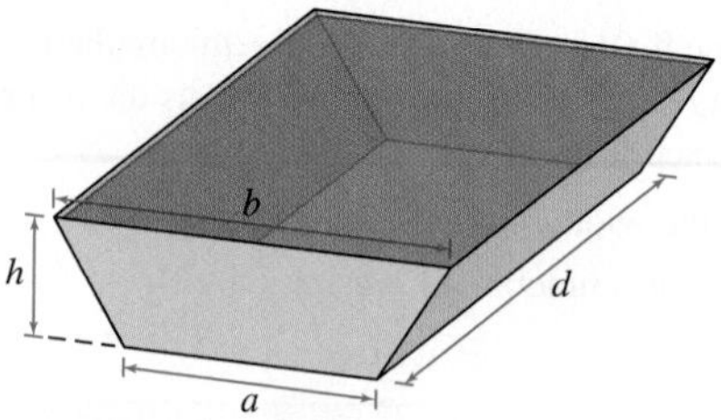

FIGURE 18

22. Calculate the fluid pressure on one of the slanted sides of the trough in Figure 18 when it is filled with corn syrup as in Exercise 21.

Further Insights and Challenges

23. The end of the trough in Figure 19 is an equilateral triangle of side 3. Assume that the trough is filled with water to height H. Calculate the fluid force on each side of the trough as a function of H and the length l of the trough.

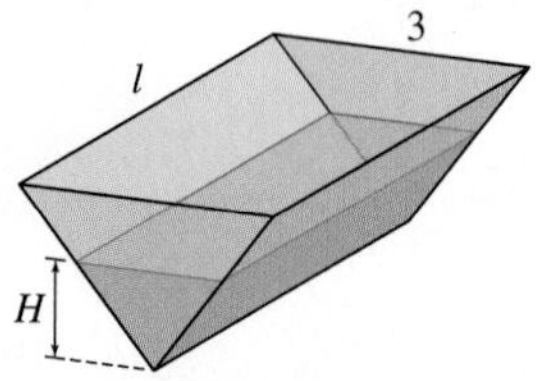

FIGURE 19

24. A rectangular plate of side ℓ is submerged vertically in a fluid of density w, with its top edge at depth h. Show that if the depth is increased by an amount Δh, then the force on a side of the plate increases by $wA\Delta h$, where A is the area of the plate.

25. Prove that the force on the side of a rectangular plate of area A submerged vertically in a fluid is equal to $p_0 A$, where p_0 is the fluid pressure at the center point of the rectangle.

26. If the density of a fluid varies with depth, then the pressure at depth y is a function $p(y)$ (which need not equal wy as in the case of constant density). Use Riemann sums to argue that the total force F on the flat side of a submerged object submerged vertically is $F = \int_a^b f(y)p(y)\,dy$, where $f(y)$ is the width of the side at depth y.

9.3 Center of Mass

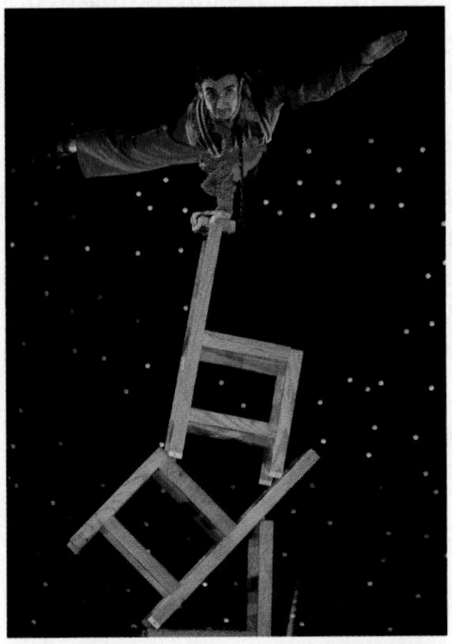

FIGURE 1 This acrobat with Cirque du Soleil must distribute his weight so that his arm provides support directly below his center of mass.

Every object has a balance point called the *center of mass* (Figure 1). When a rigid object such as a hammer is tossed in the air, it may rotate in a complicated fashion, but its center of mass follows the same simple parabolic trajectory as a stone tossed in the air. In this section we use integration to compute the center of mass of a thin plate (also called a **lamina**) of constant mass density ρ.

The center of mass (COM) is expressed in terms of quantities called **moments**. The moment of a single particle of mass m with respect to a line L is the product of the particle's mass m and its directed distance (positive or negative) to the line:

$$\text{Moment with respect to line } L = m \times \text{directed distance to } L$$

The particular moments with respect to the x- and y-axes are denoted M_x and M_y. For a particle located at the point (x, y) (Figure 2),

$$M_x = my \qquad \text{(mass times directed distance to } x\text{-axis)}$$

$$M_y = mx \qquad \text{(mass times directed distance to } y\text{-axis)}$$

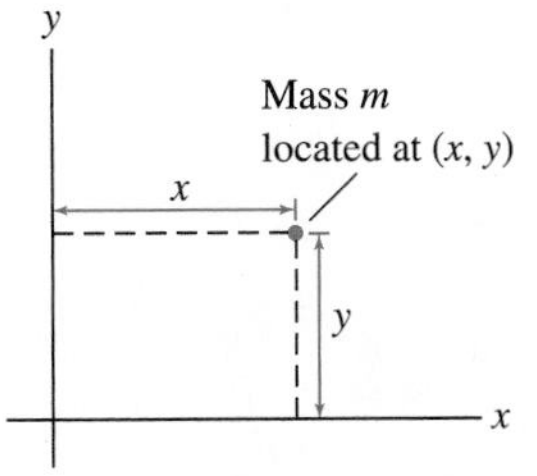

FIGURE 2

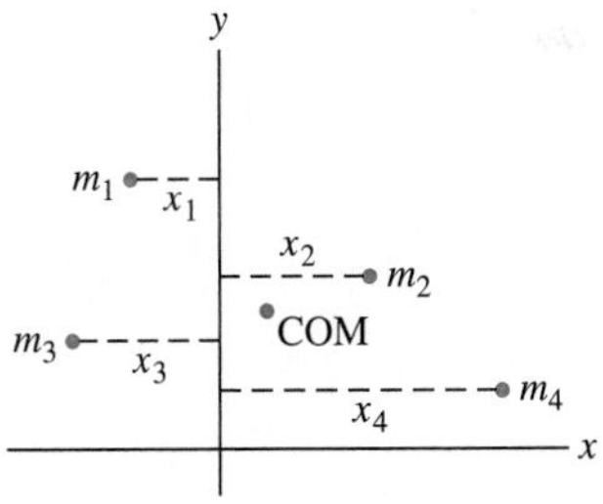

FIGURE 3

By definition, moments are additive: the moment of a system of n particles with coordinates (x_j, y_j) and mass m_j (Figure 3) is the sum

CAUTION The notation is potentially confusing: M_x is defined in terms of y-coordinates and M_y in terms of x-coordinates.

$$M_x = m_1 y_1 + m_2 y_2 + \cdots + m_n y_n$$

$$M_y = m_1 x_1 + m_2 x_2 + \cdots + m_n x_n$$

The **center of mass** (COM) is the point $P = (x_{\rm CM}, y_{\rm CM})$ with coordinates

$$x_{\rm CM} = \frac{M_y}{M}, \qquad y_{\rm CM} = \frac{M_x}{M}$$

where $M = m_1 + m_2 + \cdots + m_n$ is the total mass of the system.

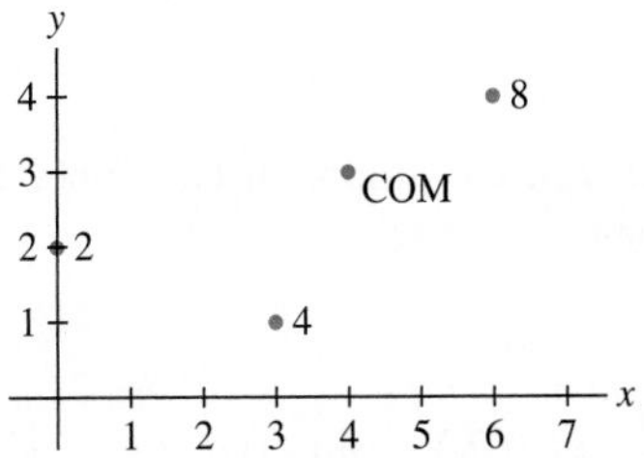

FIGURE 4 Centers of mass for Example 1.

■ **EXAMPLE 1** Find the COM of the system of three particles in Figure 4, having masses 2, 4, and 8 at locations (0, 2), (3, 1), and (6, 4).

Solution The total mass is $M = 2 + 4 + 8 = 14$ and the moments are

$$M_x = m_1 y_1 + m_2 y_2 + m_3 y_3 = 2 \cdot 2 + 4 \cdot 1 + 8 \cdot 4 = 40$$

$$M_y = m_1 x_1 + m_2 x_2 + m_3 x_3 = 2 \cdot 0 + 4 \cdot 3 + 8 \cdot 6 = 60$$

Therefore, $x_{\rm CM} = \frac{60}{14} = \frac{30}{7}$ and $y_{\rm CM} = \frac{40}{14} = \frac{20}{7}$. The COM is $\left(\frac{30}{7}, \frac{20}{7}\right)$. ■

Laminas (Thin Plates)

In this section, we restrict our attention to thin plates of constant mass density (also called "uniform density"). COM computations when mass density is not constant require multiple integration and are covered in Section 16.5.

Now consider a lamina (thin plate) of constant mass density ρ occupying the region under the graph of $f(x)$ over an interval $[a, b]$, where $f(x)$ is continuous and $f(x) \geq 0$ (Figure 5). In our calculations we will use the principle of *additivity of moments* mentioned above for point masses:

If a region is decomposed into smaller, non-overlapping regions, then the moment of the region is the sum of the moments of the smaller regions.

To compute the y-moment M_y, we begin as usual, by dividing $[a, b]$ into N subintervals of width $\Delta x = (b - a)/N$ and endpoints $x_j = a + j\Delta x$. This divides the lamina into N vertical strips (Figure 6). If Δx is small, the jth strip is nearly rectangular of area $f(x_j)\Delta x$ and mass $\rho f(x_j)\Delta x$. Since all points in the strip lie at approximately the same distance x_j from the y-axis, the moment $M_{y,j}$ of the jth strip is approximately

$$M_{y,j} \approx (\text{mass}) \times (\text{directed distance to } y\text{-axis}) = (\rho f(x_j)\Delta x)x_j$$

By additivity of moments,

$$M_y = \sum_{j=1}^{N} M_{y,j} \approx \rho \sum_{j=1}^{N} x_j f(x_j)\Delta x$$

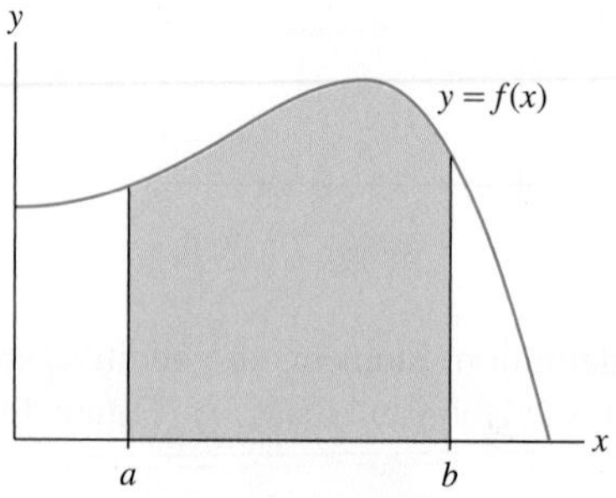

FIGURE 5 Lamina occupying the region under the graph of $f(x)$ over $[a, b]$.

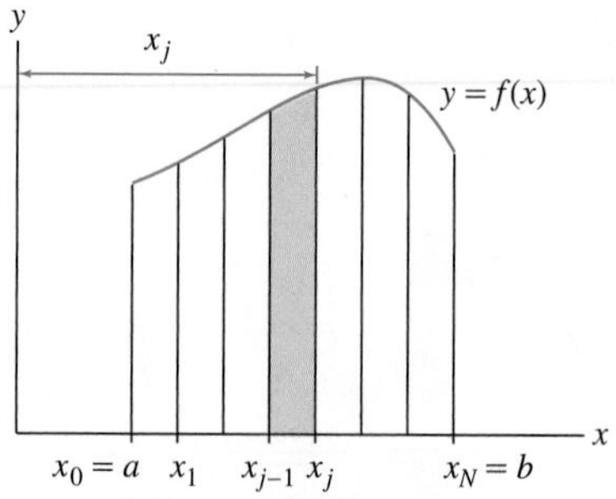

FIGURE 6 The shaded strip is nearly rectangular of area with $f(x_j)\Delta x$.

This is a Riemann sum whose value approaches $\rho \int_a^b x f(x)\,dx$ as $N \to \infty$, and thus

$$M_y = \rho \int_a^b x f(x)\,dx$$

More generally, if the lamina occupies the region *between* the graphs of two functions $f_1(x)$ and $f_2(x)$ over $[a, b]$, where $f_1(x) \geq f_2(x)$, then

$$M_y = \rho \int_a^b x(\text{length of vertical cut})\,dx = \rho \int_a^b x\big(f_1(x) - f_2(x)\big)\,dx \qquad \boxed{1}$$

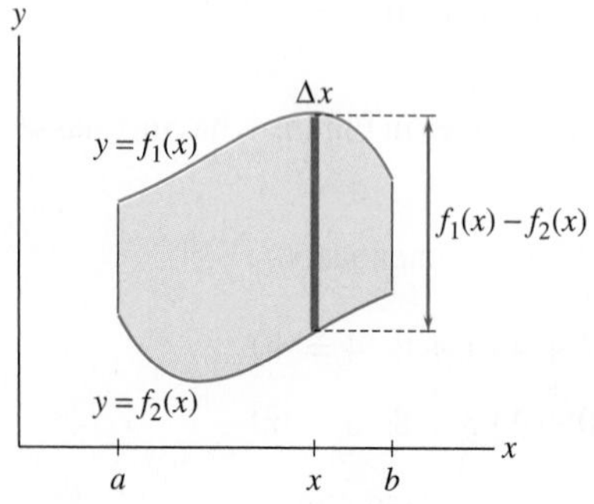

FIGURE 7

Think of the lamina as made up of vertical strips of length $f_1(x) - f_2(x)$ at distance x from the y-axis (Figure 7).

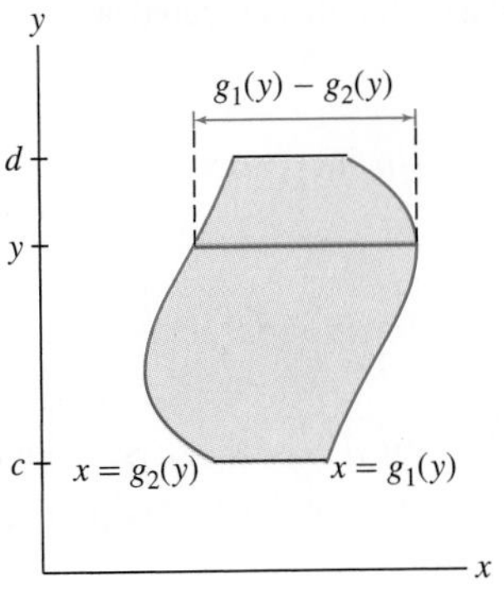

FIGURE 8

We can compute the x-moment by dividing the lamina into *horizontal* strips, but this requires us to describe the lamina as a region between two curves $x = g_1(y)$ and $x = g_2(y)$ with $g_1(y) \geq g_2(y)$ over an interval $[c, d]$ along the y-axis (Figure 8):

$$M_x = \rho \int_c^d y(\text{length of horizontal cut})\, dy = \rho \int_c^d y\big(g_1(y) - g_2(y)\big)\, dy \qquad \boxed{2}$$

The total mass of the lamina is $M = \rho A$, where A is the area of the lamina:

$$M = \rho A = \rho \int_a^b \big(f_1(x) - f_2(x)\big)\, dx \qquad \text{or} \qquad \rho \int_c^d \big(g_1(y) - g_2(y)\big)\, dy$$

The center-of-mass coordinates are the moments divided by the total mass:

$$x_{\text{CM}} = \frac{M_y}{M}, \qquad y_{\text{CM}} = \frac{M_x}{M}$$

■ **EXAMPLE 2** Find the moments and COM of the lamina of uniform density ρ occupying the region underneath the graph of $f(x) = x^2$ for $0 \leq x \leq 2$.

Solution First, compute M_y using Eq. (1):

$$M_y = \rho \int_0^2 x f(x)\, dx = \rho \int_0^2 x(x^2)\, dx = \rho \left.\frac{x^4}{4}\right|_0^2 = 4\rho$$

Then compute M_x using Eq. (2), describing the lamina as the region between $x = \sqrt{y}$ and $x = 2$ over the interval $[0, 4]$ along the y-axis (Figure 9). By Eq. (2),

$$M_x = \rho \int_0^4 y\big(g_1(y) - g_2(y)\big)\, dy = \rho \int_0^4 y(2 - \sqrt{y})\, dy$$

$$= \rho \left(y^2 - \frac{2}{5} y^{5/2}\right)\Bigg|_0^4 = \rho \left(16 - \frac{2}{5} \cdot 32\right) = \frac{16}{5}\rho$$

The plate has area $A = \int_0^2 x^2\, dx = \frac{8}{3}$ and total mass $M = \frac{8}{3}\rho$. Therefore,

$$x_{\text{CM}} = \frac{M_y}{M} = \frac{4\rho}{\frac{8}{3}\rho} = \frac{3}{2}, \qquad y_{\text{CM}} = \frac{M_x}{M} = \frac{\frac{16}{5}\rho}{\frac{8}{3}\rho} = \frac{6}{5}$$ ■

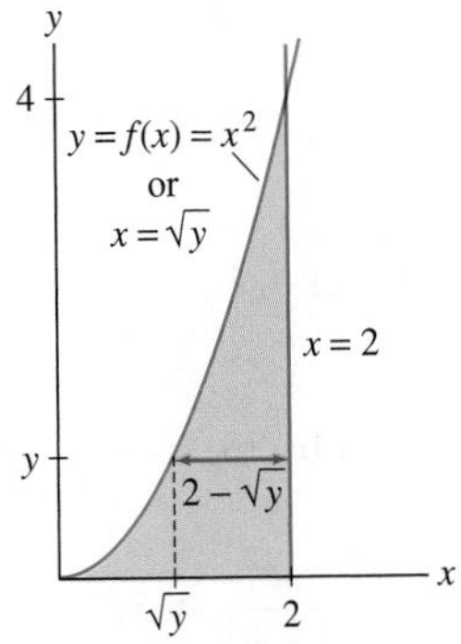

FIGURE 9 Lamina occupying the region under the graph of $f(x) = x^2$ over $[0, 2]$.

CONCEPTUAL INSIGHT The COM of a lamina of constant mass density ρ is also called the **centroid**. The centroid depends on the shape of the lamina, but not on its mass density because the factor ρ cancels in the ratios M_x/M and M_y/M. In particular, *in calculating the centroid, we can take* $\rho = 1$. When mass density is not constant, the COM depends on both shape and mass density. In this case, the COM is computed using multiple integration (Section 16.5).

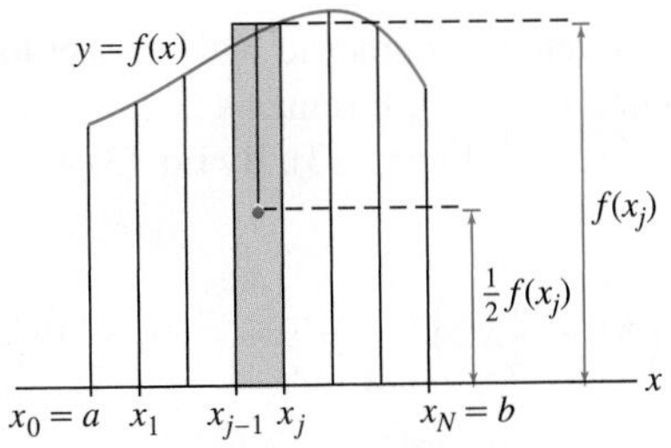

FIGURE 10 Because the shaded strip is nearly rectangular, its COM has an approximate height of $\frac{1}{2} f(x_j)$.

A drawback of Eq. (2) for M_x is that it requires integration along the y-axis. Fortunately, there is a second formula for M_x as an integral along the x-axis. As before, divide the region into N thin vertical strips of width Δx (see Figure 10). Let $M_{x,j}$ be the x-moment of the jth strip and let m_j be its mass. We can use the following trick to approximate $M_{x,j}$. The strip is nearly rectangular with height $f(x_j)$ and width Δx, so $m_j \approx \rho f(x_j)\, \Delta x$. Furthermore, $M_{x,j} = y_j m_j$, where y_j is the y-coordinate of the COM of the strip. However, $y_j \approx \frac{1}{2}f(x_j)$ because the COM of a rectangle is located at its center. Thus,

$$M_{x,j} = m_j y_j \approx \rho f(x_j)\Delta x \cdot \frac{1}{2}f(x_j) = \frac{1}{2}\rho f(x_j)^2\Delta x$$

$$M_x = \sum_{j=1}^{N} M_{x,j} \approx \frac{1}{2}\rho \sum_{j=1}^{N} f(x_j)^2\Delta x$$

This is a Riemann sum whose value approaches $\frac{1}{2}\rho\int_a^b f(x)^2\,dx$ as $N \to \infty$. The case of a region *between* the graphs of functions $f_1(x)$ and $f_2(x)$ where $f_1(x) \ge f_2(x) \ge 0$ is treated similarly, so we obtain the alternative formulas

$$M_x = \frac{1}{2}\rho\int_a^b f(x)^2\,dx \qquad \text{or} \qquad \frac{1}{2}\rho\int_a^b \left(f_1(x)^2 - f_2(x)^2\right)dx \tag{3}$$

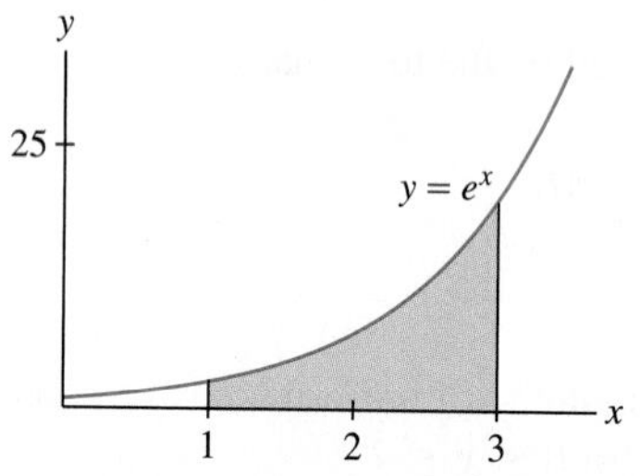

FIGURE 11 Region under the curve $y = e^x$ between $x = 1$ and $x = 3$.

■ **EXAMPLE 3** Find the centroid of the shaded region in Figure 11.

Solution The centroid does not depend on ρ, so we may set $\rho = 1$ and apply Eqs. (1) and (3) with $f(x) = e^x$:

$$M_x = \frac{1}{2}\int_1^3 f(x)^2\,dx = \frac{1}{2}\int_1^3 e^{2x}\,dx = \frac{1}{4}e^{2x}\Big|_1^3 = \frac{e^6 - e^2}{4}$$

Using Integration by Parts,

$$M_y = \int_1^3 xf(x)\,dx = \int_1^3 xe^x\,dx = (x-1)e^x\Big|_1^3 = 2e^3$$

The total mass is $M = \int_1^3 e^x\,dx = (e^3 - e)$. The centroid has coordinates

$$x_{CM} = \frac{M_y}{M} = \frac{2e^3}{e^3 - e} \approx 2.313, \qquad y_{CM} = \frac{M_x}{M} = \frac{e^6 - e^2}{4(e^3 - e)} \approx 5.701$$ ■

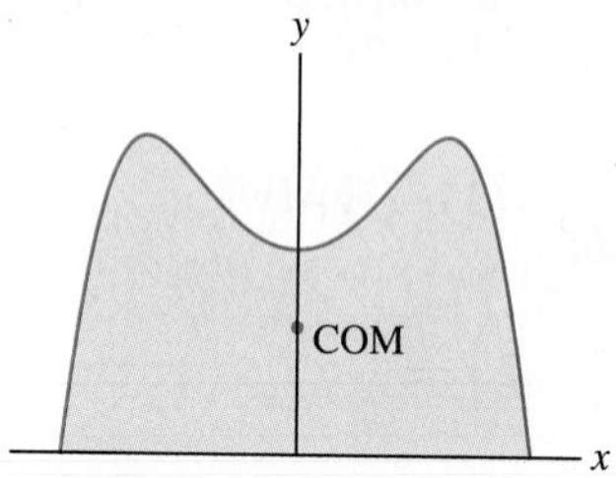

FIGURE 12 The COM of a symmetric plate lies on the axis of symmetry.

The symmetry properties of an object give information about its centroid (Figure 12). For instance, the centroid of a square or circular plate is located at its center. Here is a precise formulation (see Exercise 43).

THEOREM 1 Symmetry Principle If a lamina is symmetric with respect to a line, then its centroid lies on that line.

■ **EXAMPLE 4 Using Symmetry** Find the centroid of a semicircle of radius 3.

Solution Symmetry cuts our work in half. The semicircle is symmetric with respect to the y-axis, so the centroid lies on the y-axis, and hence $x_{CM} = 0$. It remains to calculate M_x and y_{CM}. The semicircle is the graph of $f(x) = \sqrt{9 - x^2}$ (Figure 13). By Eq. (3) with $\rho = 1$,

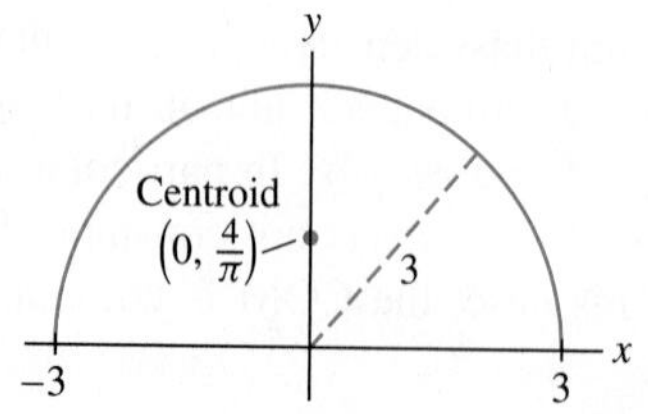

FIGURE 13 The semicircle $y = \sqrt{9 - x^2}$.

$$M_x = \frac{1}{2}\int_{-3}^{3} f(x)^2\,dx = \frac{1}{2}\int_{-3}^{3}(9 - x^2)\,dx = \frac{1}{2}\left(9x - \frac{1}{3}x^3\right)\Big|_{-3}^{3} = 9 - (-9) = 18$$

The semicircle has area (and mass) equal to $A = \frac{1}{2}\pi(3^2) = 9\pi/2$, so

$$y_{CM} = \frac{M_x}{M} = \frac{18}{9\pi/2} = \frac{4}{\pi} \approx 1.27$$ ■

■ **EXAMPLE 5** **Using Additivity and Symmetry** Find the centroid of the region R in Figure 14.

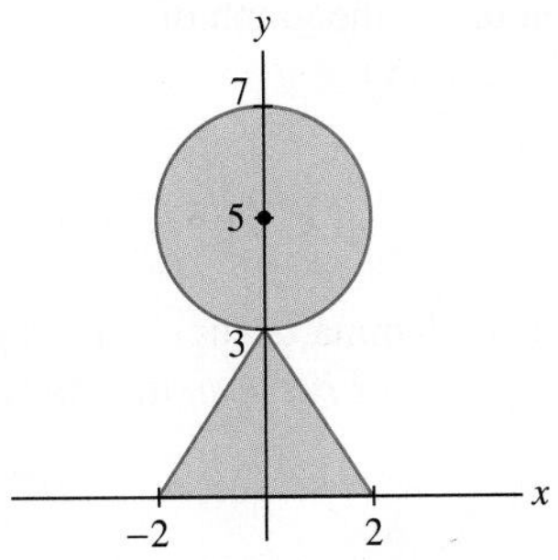

FIGURE 14 The moment of region R is the sum of the moments of the triangle and circle.

Solution We set $\rho = 1$ because we are computing a centroid. The region R is symmetric with respect to the y-axis, so we know in advance that $x_{CM} = 0$. To find y_{CM}, we compute the moment M_x.

Step 1. **Use additivity of moments.**

Let M_x^{triangle} and M_x^{circle} be the x-moments of the triangle and the circle. Then

$$M_x = M_x^{\text{triangle}} + M_x^{\text{circle}}$$

Step 2. **Moment of the circle.**

To save work, we use the fact that the centroid of the circle is located at the center $(0, 5)$ by symmetry. Thus $y_{CM}^{\text{circle}} = 5$ and we can solve for the moment:

$$y_{CM}^{\text{circle}} = \frac{M_x^{\text{circle}}}{M^{\text{circle}}} = \frac{M_x^{\text{circle}}}{4\pi} = 5 \quad \Rightarrow \quad M_x^{\text{circle}} = 20\pi$$

Here, the mass of the circle is its area $M^{\text{circle}} = \pi(2^2) = 4\pi$ (since $\rho = 1$).

Step 3. **Moment of a triangle.**

Let's compute M_x^{triangle} for an arbitrary triangle of height h and base b (Figure 15). Let $\ell(y)$ be the width of the triangle at height y. By similar triangles,

$$\frac{\ell(y)}{h - y} = \frac{b}{h} \quad \Rightarrow \quad \ell(y) = b - \frac{b}{h}y$$

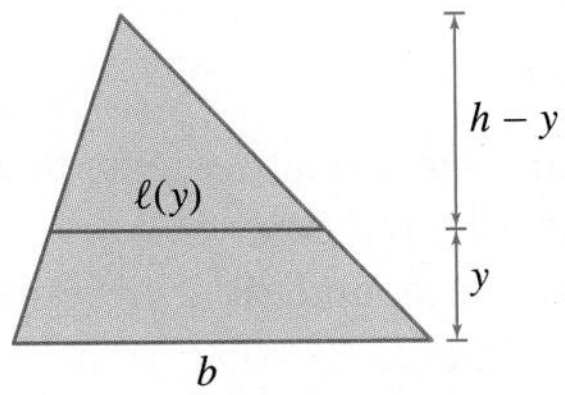

FIGURE 15 By similar triangles, $\dfrac{\ell(y)}{h-y} = \dfrac{b}{h}$.

By Eq. (2),

$$M_x^{\text{triangle}} = \int_0^h y\ell(y)\,dy = \int_0^h y\left(b - \frac{b}{h}y\right)dy = \left(\frac{by^2}{2} - \frac{by^3}{3h}\right)\bigg|_0^h = \frac{bh^2}{6}$$

In our case, $b = 4$, $h = 3$, and $M_x^{\text{triangle}} = \dfrac{4 \cdot 3^2}{6} = 6$.

Step 4. **Computation of y_{CM}.**

$$M_x = M_x^{\text{triangle}} + M_x^{\text{circle}} = 6 + 20\pi$$

The triangle has mass $\frac{1}{2} \cdot 4 \cdot 3 = 6$, and the circle has mass 4π, so R has mass $M = 6 + 4\pi$ and

$$y_{CM} = \frac{M_x}{M} = \frac{6 + 20\pi}{6 + 4\pi} \approx 3.71$$ ■

9.3 SUMMARY

- The *moments* of a system of particles of mass m_j located at (x_j, y_j) are

$$M_x = m_1y_1 + \cdots + m_ny_n, \qquad M_y = m_1x_1 + \cdots + m_nx_n$$

The *center of mass* (COM) has coordinates

$$x_{CM} = \frac{M_y}{M} \quad \text{and} \quad y_{CM} = \frac{M_x}{M}$$

where $M = m_1 + \cdots + m_n$.

- Lamina (thin plate) of constant mass density ρ (region under the graph of $f(x)$ where $f(x) \geq 0$, or *between* the graphs of $f_1(x)$ and $f_2(x)$ where $f_1(x) \geq f_2(x)$):

$$M_y = \rho \int_a^b x f(x)\,dx \qquad \text{or} \qquad \rho \int_a^b x\big(f_1(x) - f_2(x)\big)\,dx$$

- There are two ways to compute the x-moment M_x. If the lamina occupies the region between the graph of $x = g(y)$ and the y-axis where $g(y) \geq 0$, or *between* the graphs of $g_1(y)$ and $g_2(y)$ where $g_1(y) \geq g_2(y)$, then

$$M_x = \rho \int_c^d y g(y)\,dy \qquad \text{or} \qquad \rho \int_c^d y\big(g_1(y) - g_2(y)\big)\,dy$$

- Alternative (often more convenient) formula for M_x:

$$M_x = \frac{1}{2}\rho \int_a^b f(x)^2\,dx \qquad \text{or} \qquad \frac{1}{2}\rho \int_a^b \big(f_1(x)^2 - f_2(x)^2\big)\,dx$$

- The total mass of the lamina is $M = \rho \int_a^b \big(f_1(x) - f_2(x)\big)\,dx$. The coordinates of the center of mass (also called the *centroid*) are

$$x_{\mathrm{CM}} = \frac{M_y}{M}, \qquad y_{\mathrm{CM}} = \frac{M_x}{M}$$

- Additivity: If a region is decomposed into smaller non-overlapping regions, then the moment of the region is the sum of the moments of the smaller regions.
- Symmetry Principle: If a lamina of constant mass density is symmetric with respect to a given line, then the center of mass (centroid) lies on that line.

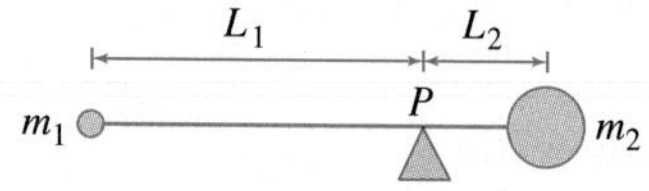

FIGURE 16 Archimedes' Law of the Lever: $m_1 L_1 = m_2 L_2$

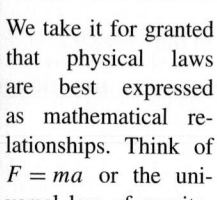

HISTORICAL PERSPECTIVE

We take it for granted that physical laws are best expressed as mathematical relationships. Think of $F = ma$ or the universal law of gravitation. However, the fundamental insight that mathematics could be used to formulate laws of nature (and not just for counting or measuring) developed gradually, beginning with the philosophers of ancient Greece and culminating some 2000 years later in the discoveries of Galileo and Newton. Archimedes (287–212 BCE) was one of the first scientists (perhaps *the* first) to formulate a precise physical law. Concerning the principle of the lever, Archimedes wrote, "Commensurable magnitudes balance at distances reciprocally proportional to their weight." In other words, if weights of mass m_1 and m_2 are placed on a weightless lever at distances L_1 and L_2 from the fulcrum P (Figure 16), then the lever will balance if $m_1/m_2 = L_2/L_1$, or

$$\boxed{m_1 L_1 = m_2 L_2}$$

In our terminology, what Archimedes had discovered was the center of mass P of the system of weights (see Exercises 41 and 42).

9.3 EXERCISES

Preliminary Questions

1. What are the x- and y-moments of a lamina whose center of mass is located at the origin?

2. A thin plate has mass 3. What is the x-moment of the plate if its center of mass has coordinates $(2, 7)$?

3. The center of mass of a lamina of total mass 5 has coordinates $(2, 1)$. What are the lamina's x- and y-moments?

4. Explain how the Symmetry Principle is used to conclude that the centroid of a rectangle is the center of the rectangle.

Exercises

1. Four particles are located at points (1, 1), (1, 2), (4, 0), (3, 1).

(a) Find the moments M_x and M_y and the center of mass of the system, assuming that the particles have equal mass m.

(b) Find the center of mass of the system, assuming the particles have masses 3, 2, 5, and 7, respectively.

2. Find the center of mass for the system of particles of masses 4, 2, 5, 1 located at (1, 2), (−3, 2), (2, −1), (4, 0).

3. Point masses of equal size are placed at the vertices of the triangle with coordinates $(a, 0)$, $(b, 0)$, and $(0, c)$. Show that the center of mass of the system of masses has coordinates $\left(\frac{1}{3}(a+b), \frac{1}{3}c\right)$.

4. Point masses of mass m_1, m_2, and m_3 are placed at the points (−1, 0), (3, 0), and (0, 4).

(a) Suppose that $m_1 = 6$. Find m_2 such that the center of mass lies on the y-axis.

(b) Suppose that $m_1 = 6$ and $m_2 = 4$. Find the value of m_3 such that $y_{CM} = 2$.

5. Sketch the lamina S of constant density $\rho = 3$ g/cm^2 occupying the region beneath the graph of $y = x^2$ for $0 \le x \le 3$.

(a) Use Eqs. (1) and (2) to compute M_x and M_y.

(b) Find the area and the center of mass of S.

6. Use Eqs. (1) and (3) to find the moments and center of mass of the lamina S of constant density $\rho = 2$ g/cm^2 occupying the region between $y = x^2$ and $y = 9x$ over [0, 3]. Sketch S, indicating the location of the center of mass.

7. Find the moments and center of mass of the lamina of uniform density ρ occupying the region underneath $y = x^3$ for $0 \le x \le 2$.

8. Calculate M_x (assuming $\rho = 1$) for the region underneath the graph of $y = 1 - x^2$ for $0 \le x \le 1$ in two ways, first using Eq. (2) and then using Eq. (3).

9. Let T be the triangular lamina in Figure 17.

(a) Show that the horizontal cut at height y has length $4 - \frac{2}{3}y$ and use Eq. (2) to compute M_x (with $\rho = 1$).

(b) Use the Symmetry Principle to show that $M_y = 0$ and find the center of mass.

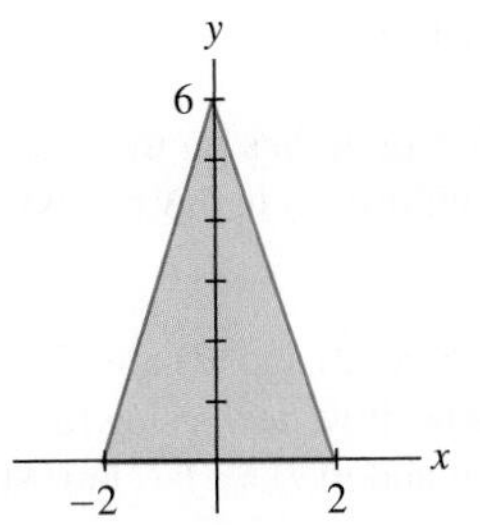

FIGURE 17 Isosceles triangle.

In Exercises 10–17, find the centroid of the region lying underneath the graph of the function over the given interval.

10. $f(x) = 6 - 2x$, [0, 3]

11. $f(x) = \sqrt{x}$, [1, 4]

12. $f(x) = x^3$, [0, 1]

13. $f(x) = 9 - x^2$, [0, 3]

14. $f(x) = (1 + x^2)^{-1/2}$, [0, 3]

15. $f(x) = e^{-x}$, [0, 4]

16. $f(x) = \ln x$, [1, 2]

17. $f(x) = \sin x$, $[0, \pi]$

18. Calculate the moments and center of mass of the lamina occupying the region between the curves $y = x$ and $y = x^2$ for $0 \le x \le 1$.

19. Sketch the region between $y = x + 4$ and $y = 2 - x$ for $0 \le x \le 2$. Using symmetry, explain why the centroid of the region lies on the line $y = 3$. Verify this by computing the moments and the centroid.

In Exercises 20–25, find the centroid of the region lying between the graphs of the functions over the given interval.

20. $y = x$, $y = \sqrt{x}$, [0, 1]

21. $y = x^2$, $y = \sqrt{x}$, [0, 1]

22. $y = x^{-1}$, $y = 2 - x$, [1, 2]

23. $y = e^x$, $y = 1$, [0, 1]

24. $y = \ln x$, $y = x - 1$, [1, 3]

25. $y = \sin x$, $y = \cos x$, $[0, \pi/4]$

26. Sketch the region enclosed by $y = x + 1$, and $y = (x-1)^2$, and find its centroid.

27. Sketch the region enclosed by $y = 0$, $y = (x+1)^3$, and $y = (1-x)^3$, and find its centroid.

In Exercises 28–32, find the centroid of the region.

28. Top half of the ellipse $\left(\frac{x}{2}\right)^2 + \left(\frac{y}{4}\right)^2 = 1$

29. Top half of the ellipse $\left(\frac{x}{a}\right)^2 + \left(\frac{y}{b}\right)^2 = 1$ for arbitrary $a, b > 0$

30. Semicircle of radius r with center at the origin

31. Quarter of the unit circle lying in the first quadrant

32. Triangular plate with vertices $(-c, 0)$, $(0, c)$, (a, b), where $a, b, c > 0$, and $b < c$

33. Find the centroid for the shaded region of the semicircle of radius r in Figure 18. What is the centroid when $r = 1$ and $h = \frac{1}{2}$? *Hint:* Use geometry rather than integration to show that the *area* of the region is $r^2 \sin^{-1}(\sqrt{1 - h^2/r^2}) - h\sqrt{r^2 - h^2}$.

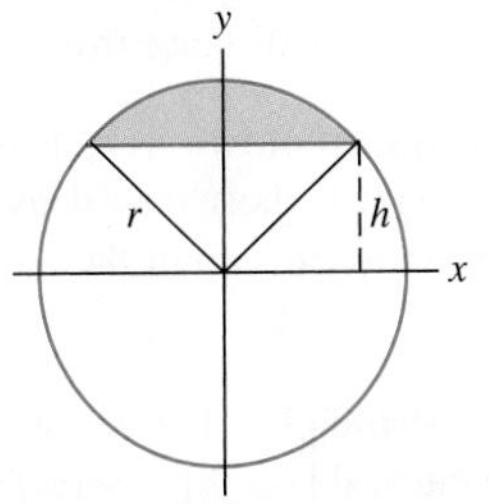

FIGURE 18

34. Sketch the region between $y = x^n$ and $y = x^m$ for $0 \le x \le 1$, where $m > n \ge 0$ and find the COM of the region. Find a pair (n, m) such that the COM lies outside the region.

In Exercises 35–37, use the additivity of moments to find the COM of the region.

35. Isosceles triangle of height 2 on top of a rectangle of base 4 and height 3 (Figure 19)

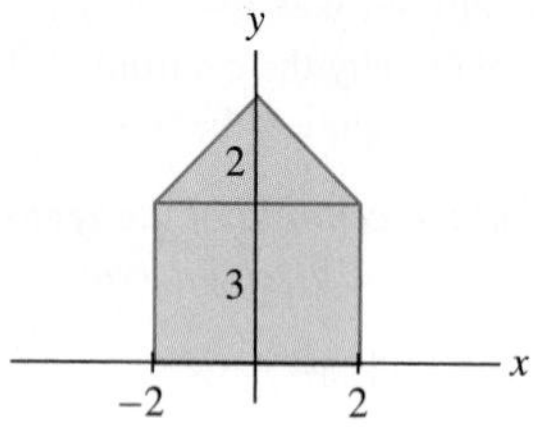

FIGURE 19

36. An ice cream cone consisting of a semicircle on top of an equilateral triangle of side 6 (Figure 20)

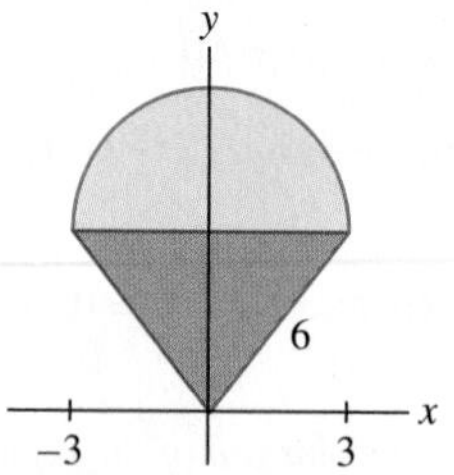

FIGURE 20

37. Three-quarters of the unit circle (remove the part in the fourth quadrant)

38. Let S be the lamina of mass density $\rho = 1$ obtained by removing a circle of radius r from the circle of radius $2r$ shown in Figure 21. Let M_x^S and M_y^S denote the moments of S. Similarly, let M_y^{big} and M_y^{small} be the y-moments of the larger and smaller circles.

(a) Use the Symmetry Principle to show that $M_x^S = 0$.

(b) Show that $M_y^S = M_y^{\text{big}} - M_y^{\text{small}}$ using the additivity of moments.

(c) Find M_y^{big} and M_y^{small} using the fact that the COM of a circle is its center. Then compute M_y^S using (b).

(d) Determine the COM of S.

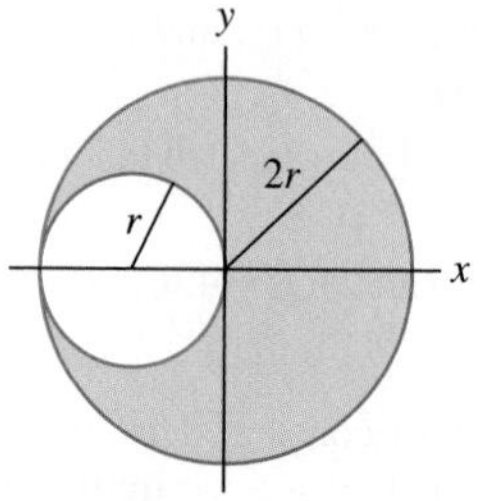

FIGURE 21

39. Find the COM of the laminas in Figure 22 obtained by removing squares of side 2 from a square of side 8.

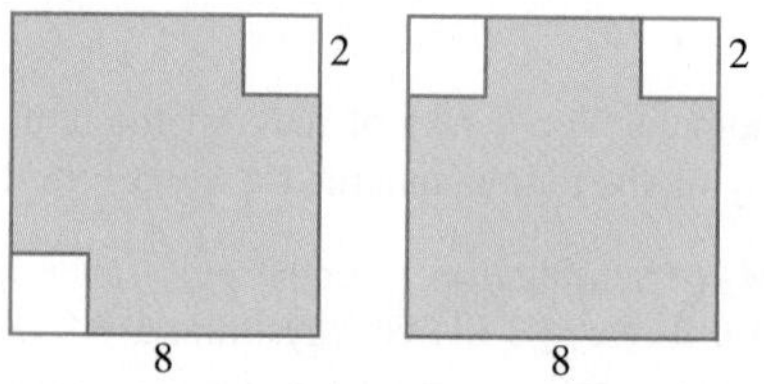

FIGURE 22

Further Insights and Challenges

40. A **median** of a triangle is a segment joining a vertex to the midpoint of the opposite side. Show that the centroid of a triangle lies on each of its medians, at a distance two-thirds down from the vertex. Then use this fact to prove that the three medians intersect at a single point. *Hint:* Simplify the calculation by assuming that one vertex lies at the origin and another on the x-axis.

41. Let P be the COM of a system of two weights with masses m_1 and m_2 separated by a distance d. Prove Archimedes' Law of the (weightless) Lever: P is the point on a line between the two weights such that $m_1 L_1 = m_2 L_2$, where L_j is the distance from mass j to P.

42. Find the COM of a system of two weights of masses m_1 and m_2 connected by a lever of length d whose mass density ρ is uniform. *Hint:* The moment of the system is the sum of the moments of the weights and the lever.

43.  **Symmetry Principle** Let $\mathcal{R}$ be the region under the graph of $f(x)$ over the interval $[-a, a]$, where $f(x) \ge 0$. Assume that $\mathcal{R}$ is symmetric with respect to the y-axis.

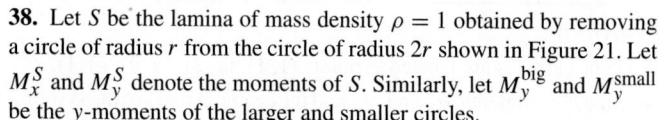

(a) Explain why $f(x)$ is even—that is, why $f(x) = f(-x)$.

(b) Show that $xf(x)$ is an *odd* function.

(c) Use (b) to prove that $M_y = 0$.

(d) Prove that the COM of $\mathcal{R}$ lies on the y-axis (a similar argument applies to symmetry with respect to the x-axis).

44. Prove directly that Eqs. (2) and (3) are equivalent in the following situation. Let $f(x)$ be a positive decreasing function on $[0, b]$ such that $f(b) = 0$. Set $d = f(0)$ and $g(y) = f^{-1}(y)$. Show that

$$\frac{1}{2}\int_0^b f(x)^2\,dx = \int_0^d y g(y)\,dy$$

Hint: First apply the substitution $y = f(x)$ to the integral on the left and observe that $dx = g'(y)\,dy$. Then apply Integration by Parts.

45. Let R be a lamina of uniform density submerged in a fluid of density w (Figure 23). Prove the following law: The fluid force on one side of R is equal to the area of R times the fluid pressure on the centroid. *Hint:* Let $g(y)$ be the horizontal width of R at depth y. Express both the fluid pressure [Eq. (2) in Section 9.2] and y-coordinate of the centroid in terms of $g(y)$.

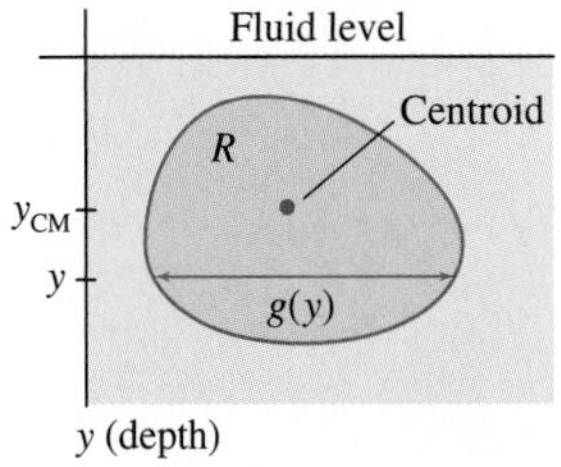

FIGURE 23

English mathematician Brook Taylor (1685–1731) made important contributions to calculus and physics, as well as to the theory of linear perspective used in drawing.

9.4 Taylor Polynomials

In Section 4.1, we used the linearization $L(x)$ to approximate a function $f(x)$ near a point $x = a$:

$$L(x) = f(a) + f'(a)(x - a)$$

We refer to $L(x)$ as a "first-order" approximation to $f(x)$ at $x = a$ because $f(x)$ and $L(x)$ have the same value and the same first derivative at $x = a$ (Figure 1):

$$L(a) = f(a), \qquad L'(a) = f'(a)$$

A first-order approximation is useful only in a small interval around $x = a$. In this section we learn how to achieve greater accuracy over larger intervals using the higher-order approximations (Figure 2).

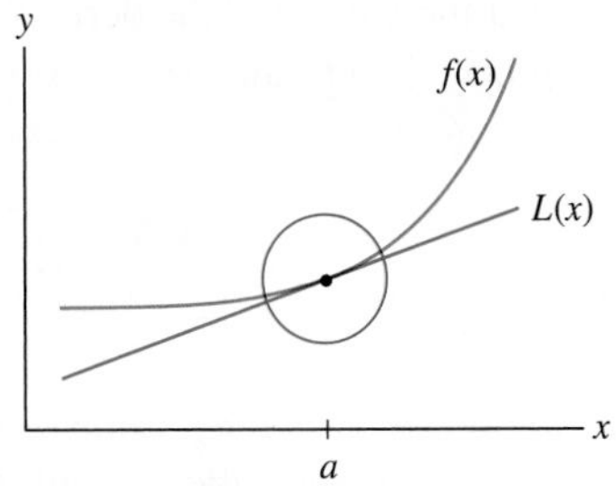

FIGURE 1 The linear approximation $L(x)$ is a first-order approximation to $f(x)$.

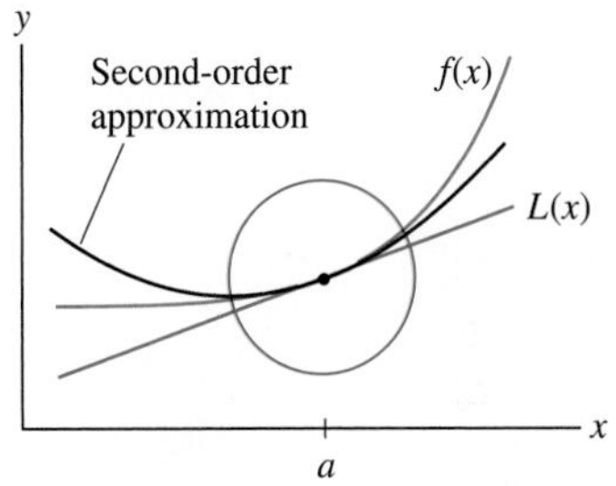

FIGURE 2 A second-order approximation is more accurate over a larger interval.

In what follows, assume that $f(x)$ is defined on an open interval I and that all higher derivatives $f^{(k)}(x)$ exist on I. Let $a \in I$. We say that two functions $f(x)$ and $g(x)$ **agree to order *n*** at $x = a$ if their derivatives up to order n at $x = a$ are equal:

$$f(a) = g(a), \quad f'(a) = g'(a), \quad f''(a) = g''(a), \quad \dots, \quad f^{(n)}(a) = g^{(n)}(a)$$

We also say that $g(x)$ "approximates $f(x)$ to order n" at $x = a$.

Define the nth **Taylor polynomial centered at** $x = a$ as follows:

$$T_n(x) = f(a) + \frac{f'(a)}{1!}(x - a) + \frac{f''(a)}{2!}(x - a)^2 + \cdots + \frac{f^{(n)}(a)}{n!}(x - a)^n$$

The first few Taylor polynomials are

$$T_0(x) = f(a)$$

$$T_1(x) = f(a) + f'(a)(x - a)$$

$$T_2(x) = f(a) + f'(a)(x-a) + \frac{1}{2}f''(a)(x-a)^2$$

$$T_3(x) = f(a) + f'(a)(x-a) + \frac{1}{2}f''(a)(x-a)^2 + \frac{1}{6}f'''(a)(x-a)^3$$

Note that $T_1(x)$ is the linearization of $f(x)$ at a. Note also that $T_n(x)$ is obtained from $T_{n-1}(x)$ by adding on a term of degree n:

$$T_n(x) = T_{n-1}(x) + \frac{f^{(n)}(a)}{n!}(x-a)^n$$

The next theorem justifies our definition of $T_n(x)$.

REMINDER k-factorial *is the number* $k! = k(k-1)(k-2)\cdots(2)(1)$. *Thus,*

$$1! = 1, \quad 2! = (2)1 = 2$$

$$3! = (3)(2)1 = 6$$

By convention, we define $0! = 1$.

THEOREM 1 The polynomial $T_n(x)$ centered at a agrees with $f(x)$ to order n at $x = a$, and it is the only polynomial of degree at most n with this property.

The verification of Theorem 1 is left to the exercises (Exercises 70–71), but we'll illustrate the idea by checking that $T_2(x)$ agrees with $f(x)$ to order $n = 2$.

$$T_2(x) = f(a) + f'(a)(x-a) + \frac{1}{2}f''(a)(x-a)^2, \qquad T_2(a) = f(a)$$

$$T_2'(x) = f'(a) + f''(a)(x-a), \qquad T_2'(a) = f'(a)$$

$$T_2''(x) = f''(a), \qquad T_2''(a) = f''(a)$$

This shows that the value and the derivatives of order up to $n = 2$ at $x = a$ are equal.

Before proceeding to the examples, we write $T_n(x)$ in summation notation:

$$T_n(x) = \sum_{j=0}^{n} \frac{f^{(j)}(a)}{j!}(x-a)^j$$

By convention, we regard $f(x)$ as the *zeroeth* derivative, and thus $f^{(0)}(x)$ is $f(x)$ itself. When $a = 0$, $T_n(x)$ is also called the nth **Maclaurin polynomial**.

EXAMPLE 1 Maclaurin Polynomials for e^x Plot the third and fourth Maclaurin polynomials for $f(x) = e^x$. Compare with the linear approximation.

Solution All higher derivatives coincide with $f(x)$ itself: $f^{(k)}(x) = e^x$. Therefore,

$$f(0) = f'(0) = f''(0) = f'''(0) = f^{(4)}(0) = e^0 = 1$$

The third Maclaurin polynomial (the case $a = 0$) is

$$T_3(x) = f(0) + f'(0)x + \frac{1}{2}f''(0)x^2 + \frac{1}{3!}f'''(0)x^3 = 1 + x + \frac{1}{2}x^2 + \frac{1}{6}x^3$$

We obtain $T_4(x)$ by adding the term of degree 4 to $T_3(x)$:

$$T_4(x) = T_3(x) + \frac{1}{4!}f^{(4)}(0)x^4 = 1 + x + \frac{1}{2}x^2 + \frac{1}{6}x^3 + \frac{1}{24}x^4$$

Figure 3 shows that T_3 and T_4 approximate $f(x) = e^x$ much more closely than the linear approximation $T_1(x)$ on an interval around $a = 0$. Higher-degree Taylor polynomials would provide even better approximations on larger intervals. ∎

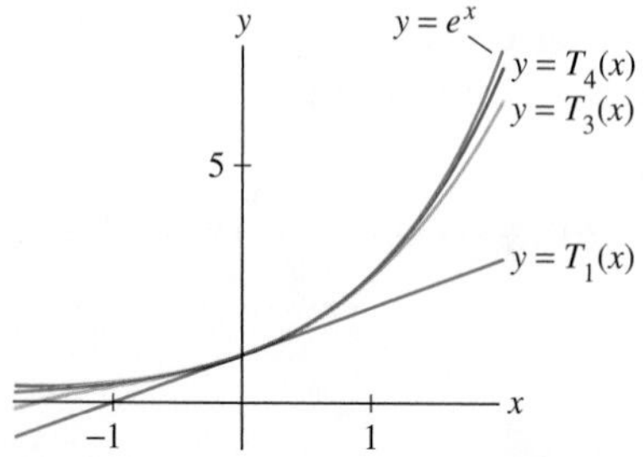

FIGURE 3 Maclaurin polynomials for $f(x) = e^x$.

■ **EXAMPLE 2 Computing Taylor Polynomials** Compute the Taylor polynomial $T_4(x)$ centered at $a = 3$ for $f(x) = \sqrt{x+1}$.

Solution First evaluate the derivatives up to degree 4 at $a = 3$:

$$\begin{aligned} f(x) &= (x+1)^{1/2}, & f(3) &= 2 \\ f'(x) &= \frac{1}{2}(x+1)^{-1/2}, & f'(3) &= \frac{1}{4} \\ f''(x) &= -\frac{1}{4}(x+1)^{-3/2}, & f''(3) &= -\frac{1}{32} \\ f'''(x) &= \frac{3}{8}(x+1)^{-5/2}, & f'''(3) &= \frac{3}{256} \\ f^{(4)}(x) &= -\frac{15}{16}(x+1)^{-7/2}, & f^{(4)}(3) &= -\frac{15}{2048} \end{aligned}$$

Then compute the coefficients $\dfrac{f^{(j)}(3)}{j!}$:

The first term $f(a)$ in the Taylor polynomial $T_n(x)$ is called the constant term.

$$\begin{aligned} \text{Constant term} &= f(3) = 2 \\ \text{Coefficient of } (x-3) &= f'(3) = \frac{1}{4} \\ \text{Coefficient of } (x-3)^2 &= \frac{f''(3)}{2!} = -\frac{1}{32}\cdot\frac{1}{2!} = -\frac{1}{64} \\ \text{Coefficient of } (x-3)^3 &= \frac{f'''(3)}{3!} = \frac{3}{256}\cdot\frac{1}{3!} = \frac{1}{512} \\ \text{Coefficient of } (x-3)^4 &= \frac{f^{(4)}(3)}{4!} = -\frac{15}{2048}\cdot\frac{1}{4!} = -\frac{5}{16{,}384} \end{aligned}$$

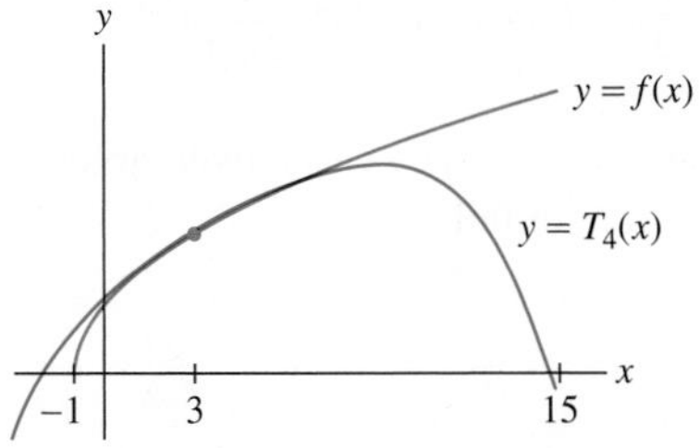

FIGURE 4 Graph of $f(x) = \sqrt{x+1}$ and $T_4(x)$ centered at $x = 3$.

The Taylor polynomial $T_4(x)$ centered at $a = 3$ is (see Figure 4):

$$T_4(x) = 2 + \frac{1}{4}(x-3) - \frac{1}{64}(x-3)^2 + \frac{1}{512}(x-3)^3 - \frac{5}{16{,}384}(x-3)^4$$ ■

■ **EXAMPLE 3 Finding a General Formula for T_n** Find the Taylor polynomials $T_n(x)$ of $f(x) = \ln x$ centered at $a = 1$.

Solution For $f(x) = \ln x$, the constant term of $T_n(x)$ at $a = 1$ is zero because $f(1) = \ln 1 = 0$. Next, we compute the derivatives:

After computing several derivatives of $f(x) = \ln x$, we begin to discern the pattern. For many functions of interest, however, the derivatives follow no simple pattern and there is no convenient formula for the general Taylor polynomial.

$$f'(x) = x^{-1}, \qquad f''(x) = -x^{-2}, \qquad f'''(x) = 2x^{-3}, \qquad f^{(4)}(x) = -3\cdot 2x^{-4}$$

Similarly, $f^{(5)}(x) = 4\cdot 3\cdot 2x^{-5}$. The general pattern is that $f^{(k)}(x)$ is a multiple of x^{-k}, with a coefficient $\pm(k-1)!$ that alternates in sign:

$$f^{(k)}(x) = (-1)^{k-1}(k-1)!\,x^{-k} \qquad \boxed{1}$$

The coefficient of $(x-1)^k$ in $T_n(x)$ is

$$\frac{f^{(k)}(1)}{k!} = \frac{(-1)^{k-1}(k-1)!}{k!} = \frac{(-1)^{k-1}}{k} \qquad (\text{for } k \geq 1)$$

Taylor polynomials for $\ln x$ *at* $a = 1$:

$T_1(x) = (x-1)$

$T_2(x) = (x-1) - \frac{1}{2}(x-1)^2$

$T_3(x) = (x-1) - \frac{1}{2}(x-1)^2 + \frac{1}{3}(x-1)^3$

Thus, the coefficients for $k \geq 1$ form a sequence $1, -\frac{1}{2}, \frac{1}{3}, -\frac{1}{4}, \dots$, and

$$T_n(x) = (x-1) - \frac{1}{2}(x-1)^2 + \frac{1}{3}(x-1)^3 - \cdots + (-1)^{n-1}\frac{1}{n}(x-1)^n$$ ■

■ **EXAMPLE 4 Cosine** Find the Maclaurin polynomials of $f(x) = \cos x$.

Solution The derivatives form a repeating pattern of period 4:

$$f(x) = \cos x, \quad f'(x) = -\sin x, \quad f''(x) = -\cos x, \quad f'''(x) = \sin x,$$
$$f^{(4)}(x) = \cos x, \quad f^{(5)}(x) = -\sin x, \quad \dots$$

In general, $f^{(j+4)}(x) = f^{(j)}(x)$. The derivatives at $x = 0$ also form a pattern:

$f(0)$	$f'(0)$	$f''(0)$	$f'''(0)$	$f^{(4)}(0)$	$f^{(5)}(0)$	$f^{(6)}(0)$	$f^{(7)}(0)$	...
1	0	−1	0	1	0	−1	0	...

Scottish mathematician Colin Maclaurin (1698–1746) was a professor in Edinburgh. Newton was so impressed by his work that he once offered to pay part of Maclaurin's salary.

Therefore, the coefficients of the odd powers x^{2k+1} are zero, and the coefficients of the even powers x^{2k} alternate in sign with value $(-1)^k/(2k)!$:

$$T_0(x) = T_1(x) = 1, \qquad T_2(x) = T_3(x) = 1 - \frac{1}{2!}x^2$$

$$T_4(x) = T_5(x) = 1 - \frac{x^2}{2} + \frac{x^4}{4!}$$

$$T_{2n}(x) = T_{2n+1}(x) = 1 - \frac{1}{2}x^2 + \frac{1}{4!}x^4 - \frac{1}{6!}x^6 + \cdots + (-1)^n\frac{1}{(2n)!}x^{2n}$$

Figure 5 shows that as n increases, $T_n(x)$ approximates $f(x) = \cos x$ well over larger and larger intervals, but outside this interval, the approximation fails. ■

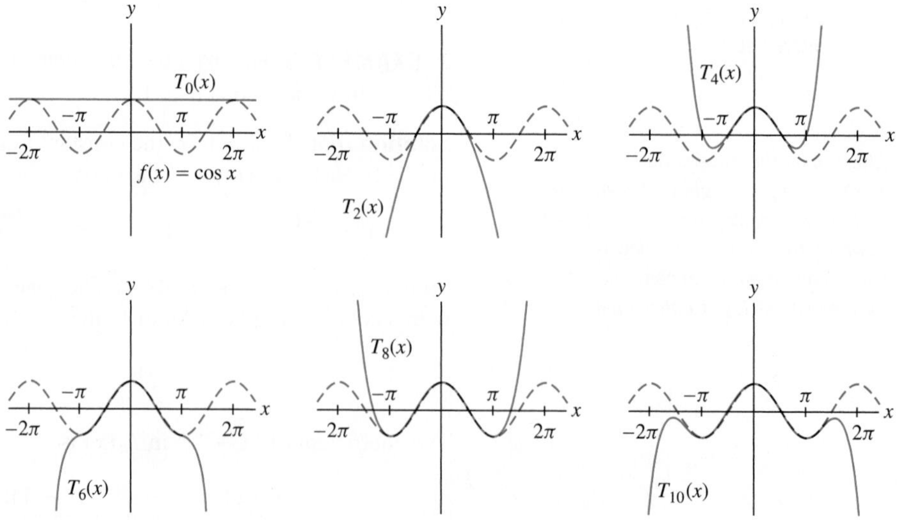

FIGURE 5 Maclaurin polynomials for $f(x) = \cos x$. The graph of $f(x)$ is shown as a dashed curve.

■ **EXAMPLE 5 How far is the horizon?** Valerie is at the beach, looking out over the ocean (Figure 6). How far can she see? Use Maclaurin polynomials to estimate the distance d, assuming that Valerie's eye level is $h = 1.7$ m above ground. What if she looks out from a window where her eye level is 20 m?

FIGURE 6 View from the beach

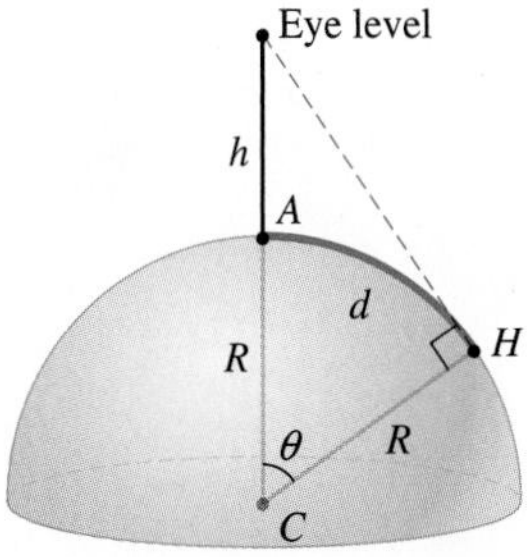

FIGURE 7 Valerie can see a distance $d = R\theta$, the length of arc AH.

Solution Let R be the radius of the earth. Figure 7 shows that Valerie can see a distance $d = R\theta$, the length of the circular arc AH in Figure 7. We have

$$\cos\theta = \frac{R}{R+h}$$

This calculation ignores the bending of light (called refraction) as it passes through the atmosphere. Refraction typically increases d by around 10%, although the actual effect is complex and varies with atmospheric temperature.

Our key observation is that θ is close to zero (both θ and h are much smaller than shown in the figure), so we lose very little accuracy if we replace $\cos\theta$ by its second Maclaurin polynomial $T_2(\theta) = 1 - \frac{1}{2}\theta^2$, as computed in Example 4:

$$1 - \frac{1}{2}\theta^2 \approx \frac{R}{R+h} \quad \Rightarrow \quad \theta^2 \approx 2 - \frac{2R}{R+h} \quad \Rightarrow \quad \theta \approx \sqrt{\frac{2h}{R+h}}$$

Furthermore, h is very small relative to R, so we may replace $R + h$ by R to obtain

$$d = R\theta \approx R\sqrt{\frac{2h}{R}} = \sqrt{2Rh}$$

The earth's radius is approximately $R \approx 6.37 \times 10^6$ m, so

$$d = \sqrt{2Rh} \approx \sqrt{2(6.37 \times 10^6)h} \approx 3569\sqrt{h} \text{ m}$$

In particular, we see that d is proportional to $\sqrt{h}$.

If Valerie's eye level is $h = 1.7$ m, then $d \approx 3569\sqrt{1.7} \approx 4653$ m, or roughly 4.7 km. If $h = 20$ m, then $d \approx 3569\sqrt{20} \approx 15.96$ m, or nearly 16 km.

The Error Bound

To use Taylor polynomials effectively, we need a way to estimate the size of the error. This is provided by the next theorem, which shows that the size of this error depends on the size of the $(n + 1)$st derivative.

A proof of Theorem 2 is presented at the end of this section.

THEOREM 2 Error Bound Assume that $f^{(n+1)}(x)$ exists and is continuous. Let K be a number such that $|f^{(n+1)}(u)| \le K$ for all u between a and x. Then

$$|f(x) - T_n(x)| \le K\frac{|x-a|^{n+1}}{(n+1)!}$$

where $T_n(x)$ is the nth Taylor polynomial centered at $x = a$.

■ **EXAMPLE 6 Using the Error Bound** Apply the error bound to

$$|\ln 1.2 - T_3(1.2)|$$

where $T_3(x)$ is the third Taylor polynomial for $f(x) = \ln x$ at $a = 1$. Check your result with a calculator.

Solution

***Step 1.* Find a value of K.**
To use the error bound with $n = 3$, we must find a value of K such that $|f^{(4)}(u)| \le K$ for all u between $a = 1$ and $x = 1.2$. As we computed in Example 3, $f^{(4)}(x) = -6x^{-4}$. The absolute value $|f^{(4)}(x)|$ is decreasing for $x > 0$, so its maximum value on $[1, 1.2]$ is $|f^{(4)}(1)| = 6$. Therefore, we may take $K = 6$.

***Step 2.* Apply the error bound.**

$$|\ln 1.2 - T_3(1.2)| \le K\frac{|x-a|^{n+1}}{(n+1)!} = 6\frac{|1.2-1|^4}{4!} \approx 0.0004$$

***Step 3.* Check the result.**
Recall from Example 3 that

$$T_3(x) = (x-1) - \frac{1}{2}(x-1)^2 + \frac{1}{3}(x-1)^3$$

The following values from a calculator confirm that the error is at most 0.0004:

$$|\ln 1.2 - T_3(1.2)| \approx |0.182667 - 0.182322| \approx 0.00035 < 0.0004$$

Observe in Figure 8 that $\ln x$ and $T_3(x)$ are indistinguishable near $x = 1.2$.

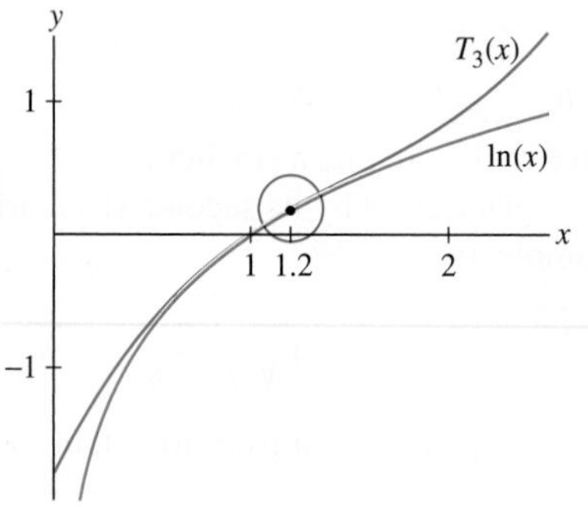

FIGURE 8 $\ln x$ and $T_3(x)$ are indistinguishable near $x = 1.2$.

■ **EXAMPLE 7 Approximating with a Given Accuracy** Let $T_n(x)$ be the nth Maclaurin polynomial for $f(x) = \cos x$. Find a value of n such that

$$|\cos 0.2 - T_n(0.2)| < 10^{-5}$$

Solution

***Step 1.* Find a value of K.**
Since $|f^{(n)}(x)|$ is $|\cos x|$ or $|\sin x|$, depending on whether n is even or odd, we have $|f^{(n)}(u)| \le 1$ for all u. Thus, we may apply the error bound with $K = 1$.

To use the error bound, it is not necessary to find the smallest possible value of K. In this example, we take $K = 1$. This works for all n, but for odd n we could have used the smaller value $K = \sin 0.2 \approx 0.2$.

***Step 2.* Find a value of n.**
The error bound gives us

$$|\cos 0.2 - T_n(0.2)| \le K\frac{|0.2-0|^{n+1}}{(n+1)!} = \frac{|0.2|^{n+1}}{(n+1)!}$$

To make the error less than 10^{-5}, we must choose n so that

$$\frac{|0.2|^{n+1}}{(n+1)!} < 10^{-5}$$

It's not possible to solve this inequality for n, but we can find a suitable n by checking several values:

n	2	3	4
$\frac{\|0.2\|^{n+1}}{(n+1)!}$	$\frac{0.2^3}{3!} \approx 0.0013$	$\frac{0.2^4}{4!} \approx 6.67 \times 10^{-5}$	$\frac{0.2^5}{5!} \approx 2.67 \times 10^{-6} < 10^{-5}$

We see that the error is less than 10^{-5} for $n = 4$. ■

The rest of this section is devoted to a proof of the error bound (Theorem 2). Define the ***n*th remainder**:

$$R_n(x) = f(x) - T_n(x)$$

The error in $T_n(x)$ is the absolute value $|R_n(x)|$. As a first step in proving the error bound, we show that $R_n(x)$ can be represented as an integral.

Taylor's Theorem Assume that $f^{(n+1)}(x)$ exists and is continuous. Then

$$R_n(x) = \frac{1}{n!}\int_a^x (x-u)^n f^{(n+1)}(u)\,du \qquad \boxed{2}$$

Proof Set

$$I_n(x) = \frac{1}{n!}\int_a^x (x-u)^n f^{(n+1)}(u)\,du$$

Our goal is to show that $R_n(x) = I_n(x)$. For $n = 0$, $R_0(x) = f(x) - f(a)$ and the desired result is just a restatement of the Fundamental Theorem of Calculus:

$$I_0(x) = \int_a^x f'(u)\,du = f(x) - f(a) = R_0(x)$$

Exercise 64 reviews this proof for the special case $n = 2$.

To prove the formula for $n > 0$, we apply Integration by Parts to $I_n(x)$ with

$$h(u) = \frac{1}{n!}(x-u)^n, \qquad g(u) = f^{(n)}(u)$$

Then $g'(u) = f^{(n+1)}(u)$, and so

$$\begin{aligned} I_n(x) &= \int_a^x h(u)\,g'(u)\,du = h(u)g(u)\Big|_a^x - \int_a^x h'(u)g(u)\,du \\ &= \frac{1}{n!}(x-u)^n f^{(n)}(u)\Big|_a^x - \frac{1}{n!}\int_a^x (-n)(x-u)^{n-1} f^{(n)}(u)\,du \\ &= -\frac{1}{n!}(x-a)^n f^{(n)}(a) + I_{n-1}(x) \end{aligned}$$

This can be rewritten as

$$I_{n-1}(x) = \frac{f^{(n)}(a)}{n!}(x-a)^n + I_n(x)$$

Now apply this relation n times, noting that $I_0(x) = f(x) - f(a)$:

$$\begin{aligned} f(x) &= f(a) + I_0(x) \\ &= f(a) + \frac{f'(a)}{1!}(x-a) + I_1(x) \\ &= f(a) + \frac{f'(a)}{1!}(x-a) + \frac{f''(a)}{2!}(x-a)^2 + I_2(x) \\ &\ \ \vdots \\ &= f(a) + \frac{f'(a)}{1!}(x-a) + \cdots + \frac{f^{(n)}(a)}{n!}(x-a)^n + I_n(x) \end{aligned}$$

This shows that $f(x) = T_n(x) + I_n(x)$ and hence $I_n(x) = R_n(x)$, as desired. ■

Now we can prove Theorem 2. Assume first that $x \geq a$. Then,

$$|f(x) - T_n(x)| = |R_n(x)| = \left| \frac{1}{n!} \int_a^x (x-u)^n f^{(n+1)}(u)\,du \right|$$

$$\leq \frac{1}{n!} \int_a^x \left| (x-u)^n f^{(n+1)}(u) \right| du \qquad \boxed{3}$$

$$\leq \frac{K}{n!} \int_a^x |x-u|^n\,du \qquad \boxed{4}$$

$$= \frac{K}{n!} \left. \frac{-(x-u)^{n+1}}{n+1} \right|_{u=a}^{x} = K \frac{|x-a|^{n+1}}{(n+1)!}$$

In Eq. (3), we use the inequality

$$\left| \int_a^b f(x)\,dx \right| \leq \int_a^b |f(x)|\,dx$$

which is valid for all integrable functions.

Note that the absolute value is not needed in Eq. (4) because $x - u \geq 0$ for $a \leq u \leq x$. If $x \leq a$, we must interchange the upper and lower limits of the integral in Eq. (3) and Eq. (4). ■

9.4 SUMMARY

- The nth *Taylor polynomial* centered at $x = a$ for the function $f(x)$ is

$$T_n(x) = f(a) + \frac{f'(a)}{1!}(x-a)^1 + \frac{f''(a)}{2!}(x-a)^2 + \cdots + \frac{f^{(n)}(a)}{n!}(x-a)^n$$

When $a = 0$, $T_n(x)$ is also called the nth *Maclaurin polynomial*.

- If $f^{(n+1)}(x)$ exists and is continuous, then we have the *error bound*

$$|T_n(x) - f(x)| \leq K \frac{|x-a|^{n+1}}{(n+1)!}$$

where K is a number such that $|f^{(n+1)}(u)| \leq K$ for all u between a and x.

- For reference, we include a table of standard Maclaurin and Taylor polynomials.

$f(x)$	a	Maclaurin or Taylor Polynomial
e^x	0	$T_n(x) = 1 + x + \frac{x^2}{2!} + \frac{x^3}{3!} + \cdots + \frac{x^n}{n!}$
$\sin x$	0	$T_{2n+1}(x) = T_{2n+2}(x) = x - \frac{x^3}{3!} + \cdots + (-1)^n \frac{x^{2n+1}}{(2n+1)!}$
$\cos x$	0	$T_{2n}(x) = T_{2n+1}(x) = 1 - \frac{x^2}{2!} + \frac{x^4}{4!} - \cdots + (-1)^n \frac{x^{2n}}{(2n)!}$
$\ln x$	1	$T_n(x) = (x-1) - \frac{1}{2}(x-1)^2 + \cdots + \frac{(-1)^{n-1}}{n}(x-1)^n$
$\frac{1}{1-x}$	0	$T_n(x) = 1 + x + x^2 + \cdots + x^n$

9.4 EXERCISES

Preliminary Questions

1. What is $T_3(x)$ centered at $a = 3$ for a function $f(x)$ such that $f(3) = 9$, $f'(3) = 8$, $f''(3) = 4$, and $f'''(3) = 12$?

2. The dashed graphs in Figure 9 are Taylor polynomials for a function $f(x)$. Which of the two is a Maclaurin polynomial?

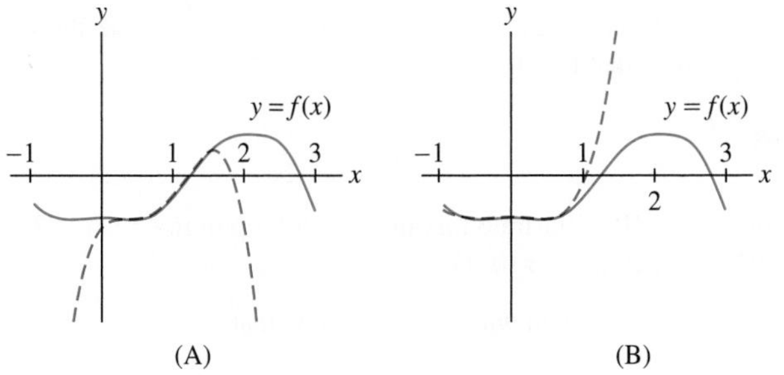

FIGURE 9

3. For which value of x does the Maclaurin polynomial $T_n(x)$ satisfy $T_n(x) = f(x)$, no matter what $f(x)$ is?

4. Let $T_n(x)$ be the Maclaurin polynomial of a function $f(x)$ satisfying $|f^{(4)}(x)| \le 1$ for all x. Which of the following statements follow from the error bound?

(a) $|T_4(2) - f(2)| \le \frac{2}{3}$

(b) $|T_3(2) - f(2)| \le \frac{2}{3}$

(c) $|T_3(2) - f(2)| \le \frac{1}{3}$

Exercises

In Exercises 1–14, calculate the Taylor polynomials $T_2(x)$ and $T_3(x)$ centered at $x = a$ for the given function and value of a.

1. $f(x) = \sin x, \quad a = 0$

2. $f(x) = \sin x, \quad a = \frac{\pi}{2}$

3. $f(x) = \frac{1}{1+x}, \quad a = 2$

4. $f(x) = \frac{1}{1+x^2}, \quad a = -1$

5. $f(x) = x^4 - 2x, \quad a = 3$

6. $f(x) = \frac{x^2+1}{x+1}, \quad a = -2$

7. $f(x) = \tan x, \quad a = 0$

8. $f(x) = \tan x, \quad a = \frac{\pi}{4}$

9. $f(x) = e^{-x} + e^{-2x}, \quad a = 0$

10. $f(x) = e^{2x}, \quad a = \ln 2$

11. $f(x) = x^2 e^{-x}, \quad a = 1$

12. $f(x) = \cosh 2x, \quad a = 0$

13. $f(x) = \frac{\ln x}{x}, \quad a = 1$

14. $f(x) = \ln(x+1), \quad a = 0$

15. Show that the nth Maclaurin polynomial for e^x is

$$T_n(x) = 1 + \frac{x}{1!} + \frac{x^2}{2!} + \cdots + \frac{x^n}{n!}$$

16. Show that the nth Taylor polynomial for $\frac{1}{x+1}$ at $a = 1$ is

$$T_n(x) = \frac{1}{2} - \frac{(x-1)}{4} + \frac{(x-1)^2}{8} + \cdots + (-1)^n \frac{(x-1)^n}{2^{n+1}}$$

17. Show that the Maclaurin polynomials for $\sin x$ are

$$T_{2n+1}(x) = T_{2n+2}(x) = x - \frac{x^3}{3!} + \frac{x^5}{5!} - \cdots + (-1)^n \frac{x^{2n+1}}{(2n+1)!}$$

18. Show that the Maclaurin polynomials for $\ln(1+x)$ are

$$T_n(x) = x - \frac{x^2}{2} + \frac{x^3}{3} + \cdots + (-1)^{n-1}\frac{x^n}{n}$$

In Exercises 19–24, find $T_n(x)$ at $x = a$ for all n.

19. $f(x) = \frac{1}{1+x}, \quad a = 0$

20. $f(x) = \frac{1}{x-1}, \quad a = 4$

21. $f(x) = e^x, \quad a = 1$

22. $f(x) = x^{-2}, \quad a = 2$

23. $f(x) = \cos x, \quad a = \frac{\pi}{4}$

24. $f(\theta) = \sin 3\theta, \quad a = 0$

In Exercises 25–28, find $T_2(x)$ and use a calculator to compute the error $|f(x) - T_2(x)|$ for the given values of a and x.

25. $y = e^x, \quad a = 0, \quad x = -0.5$

26. $y = \cos x, \quad a = 0, \quad x = \frac{\pi}{12}$

27. $y = x^{-2/3}, \quad a = 1, \quad x = 1.2$

28. $y = e^{\sin x}, \quad a = \frac{\pi}{2}, \quad x = 1.5$

29. GU Compute $T_3(x)$ for $f(x) = \sqrt{x}$ centered at $a = 1$. Then use a plot of the error $|f(x) - T_3(x)|$ to find a value $c > 1$ such that the error on the interval $[1, c]$ is at most 0.25.

30. CAS Plot $f(x) = 1/(1+x)$ together with the Taylor polynomials $T_n(x)$ at $a = 1$ for $1 \le n \le 4$ on the interval $[-2, 8]$ (be sure to limit the upper plot range).

(a) Over which interval does $T_4(x)$ appear to approximate $f(x)$ closely?

(b) What happens for $x < -1$?

(c) Use your computer algebra system to produce and plot T_{30} together with $f(x)$ on $[-2, 8]$. Over which interval does T_{30} appear to give a close approximation?

31. Let $T_3(x)$ be the Maclaurin polynomial of $f(x) = e^x$. Use the error bound to find the maximum possible value of $|f(1.1) - T_3(1.1)|$. Show that we can take $K = e^{1.1}$.

32. Let $T_2(x)$ be the Taylor polynomial of $f(x) = \sqrt{x}$ at $a = 4$. Apply the error bound to find the maximum possible value of the error $|f(3.9) - T_2(3.9)|$.

In Exercises 33–36, compute the Taylor polynomial indicated and use the error bound to find the maximum possible size of the error. Verify your result with a calculator.

33. $f(x) = \cos x, \quad a = 0; \quad |\cos 0.25 - T_5(0.25)|$

34. $f(x) = x^{11/2}, \quad a = 1; \quad |f(1.2) - T_4(1.2)|$

35. $f(x) = x^{-1/2}, \quad a = 4; \quad |f(4.3) - T_3(4.3)|$

36. $f(x) = \sqrt{1+x}, \quad a = 8; \quad |\sqrt{9.02} - T_3(8.02)|$

37. Calculate the Maclaurin polynomial $T_3(x)$ for $f(x) = \tan^{-1} x$. Compute $T_3(\frac{1}{2})$ and use the error bound to find a bound for the error $\left|\tan^{-1}\frac{1}{2} - T_3(\frac{1}{2})\right|$. Refer to the graph in Figure 10 to find an acceptable value of K. Verify your result by computing $\left|\tan^{-1}\frac{1}{2} - T_3(\frac{1}{2})\right|$ using a calculator.

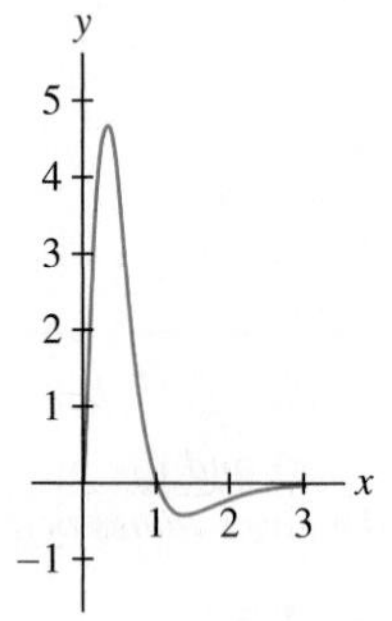

FIGURE 10 Graph of $f^{(4)}(x) = \dfrac{-24x(x^2-1)}{(x^2+1)^4}$, where $f(x) = \tan^{-1} x$.

38. Let $f(x) = \ln(x^3 - x + 1)$. The third Taylor polynomial at $a = 1$ is

$$T_3(x) = 2(x-1) + (x-1)^2 - \frac{7}{3}(x-1)^3$$

Find the maximum possible value of $|f(1.1) - T_3(1.1)|$, using the graph in Figure 11 to find an acceptable value of K. Verify your result by computing $|f(1.1) - T_3(1.1)|$ using a calculator.

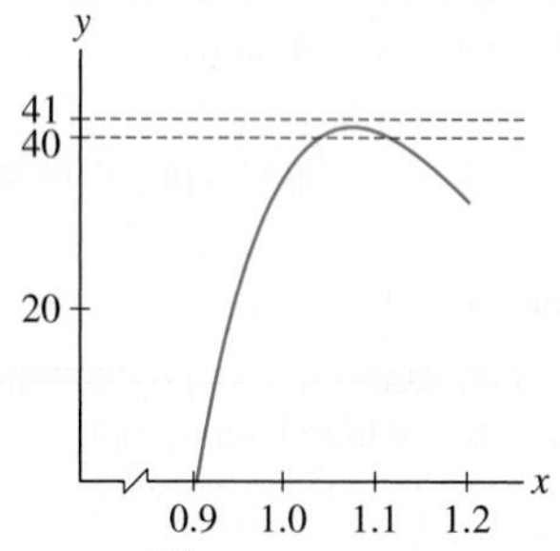

FIGURE 11 Graph of $f^{(4)}(x)$, where $f(x) = \ln(x^3 - x + 1)$.

39. GU Calculate the $T_3(x)$ at $a = 0.5$ for $f(x) = \cos(x^2)$, and use the error bound to find the maximum possible value of $|f(0.6) - T_2(0.6)|$. Plot $f^{(4)}(x)$ to find an acceptable value of K.

40. GU Calculate the Maclaurin polynomial $T_2(x)$ for $f(x) = \operatorname{sech} x$ and use the error bound to find the maximum possible value of $\left|f(\frac{1}{2}) - T_2(\frac{1}{2})\right|$. Plot $f'''(x)$ to find an acceptable value of K.

In Exercises 41–44, use the error bound to find a value of n for which the given inequality is satisfied. Then verify your result using a calculator.

41. $|\cos 0.1 - T_n(0.1)| \le 10^{-7}, \quad a = 0$

42. $|\ln 1.3 - T_n(1.3)| \le 10^{-4}, \quad a = 1$

43. $|\sqrt{1.3} - T_n(1.3)| \le 10^{-6}, \quad a = 1$

44. $|e^{-0.1} - T_n(-0.1)| \le 10^{-6}, \quad a = 0$

45. Let $f(x) = e^{-x}$ and $T_3(x) = 1 - x + \dfrac{x^2}{2} - \dfrac{x^3}{6}$. Use the error bound to show that for all $x \ge 0$,

$$|f(x) - T_3(x)| \le \frac{x^4}{24}$$

If you have a GU, illustrate this inequality by plotting $f(x) - T_3(x)$ and $x^4/24$ together over $[0, 1]$.

46. Use the error bound with $n = 4$ to show that

$$\left|\sin x - \left(x - \frac{x^3}{6}\right)\right| \le \frac{|x|^5}{120} \quad \text{(for all } x\text{)}$$

47. Let $T_n(x)$ be the Taylor polynomial for $f(x) = \ln x$ at $a = 1$, and let $c > 1$. Show that

$$|\ln c - T_n(c)| \le \frac{|c-1|^{n+1}}{n+1}$$

Then find a value of n such that $|\ln 1.5 - T_n(1.5)| \le 10^{-2}$.

48. Let $n \ge 1$. Show that if $|x|$ is small, then

$$(x+1)^{1/n} \approx 1 + \frac{x}{n} + \frac{1-n}{2n^2}x^2$$

Use this approximation with $n = 6$ to estimate $1.5^{1/6}$.

49. Verify that the third Maclaurin polynomial for $f(x) = e^x \sin x$ is equal to the product of the third Maclaurin polynomials of e^x and $\sin x$ (after discarding terms of degree greater than 3 in the product).

50. Find the fourth Maclaurin polynomial for $f(x) = \sin x \cos x$ by multiplying the fourth Maclaurin polynomials for $f(x) = \sin x$ and $f(x) = \cos x$.

51. Find the Maclaurin polynomials $T_n(x)$ for $f(x) = \cos(x^2)$. You may use the fact that $T_n(x)$ is equal to the sum of the terms up to degree n obtained by substituting x^2 for x in the nth Maclaurin polynomial of $\cos x$.

52. Find the Maclaurin polynomials of $1/(1+x^2)$ by substituting $-x^2$ for x in the Maclaurin polynomials of $1/(1-x)$.

53. Let $f(x) = 3x^3 + 2x^2 - x - 4$. Calculate $T_j(x)$ for $j = 1, 2, 3, 4, 5$ at both $a = 0$ and $a = 1$. Show that $T_3(x) = f(x)$ in both cases.

54. Let $T_n(x)$ be the nth Taylor polynomial at $x = a$ for a polynomial $f(x)$ of degree n. Based on the result of Exercise 53, guess the value of $|f(x) - T_n(x)|$. Prove that your guess is correct using the error bound.

55. Let $s(t)$ be the distance of a truck to an intersection. At time $t = 0$, the truck is 60 meters from the intersection, is traveling at a velocity of 24 m/s, and begins to slow down with an acceleration of $a = -3$ m/s^2. Determine the second Maclaurin polynomial of $s(t)$, and use it to estimate the truck's distance from the intersection after 4 s.

56. A bank owns a portfolio of bonds whose value $P(r)$ depends on the interest rate r (measured in percent; for example, $r = 5$ means a 5% interest rate). The bank's quantitative analyst determines that

$$P(5) = 100{,}000, \qquad \left.\frac{dP}{dr}\right|_{r=5} = -40{,}000, \qquad \left.\frac{d^2P}{dr^2}\right|_{r=5} = 50{,}000$$

In finance, this second derivative is called **bond convexity**. Find the second Taylor polynomial of $P(r)$ centered at $r = 5$ and use it to estimate the value of the portfolio if the interest rate moves to $r = 5.5\%$.

57. A narrow, negatively charged ring of radius R exerts a force on a positively charged particle P located at distance x above the center of the ring of magnitude

$$F(x) = -\frac{kx}{(x^2 + R^2)^{3/2}}$$

where $k > 0$ is a constant (Figure 12).

(a) Compute the third-degree Maclaurin polynomial for $F(x)$.

(b) Show that $F \approx -(k/R^3)x$ to second order. This shows that when x is small, $F(x)$ behaves like a restoring force similar to the force exerted by a spring.

(c) Show that $F(x) \approx -k/x^2$ when x is large by showing that

$$\lim_{x\to\infty} \frac{F(x)}{-k/x^2} = 1$$

Thus, $F(x)$ behaves like an inverse square law, and the charged ring looks like a point charge from far away.

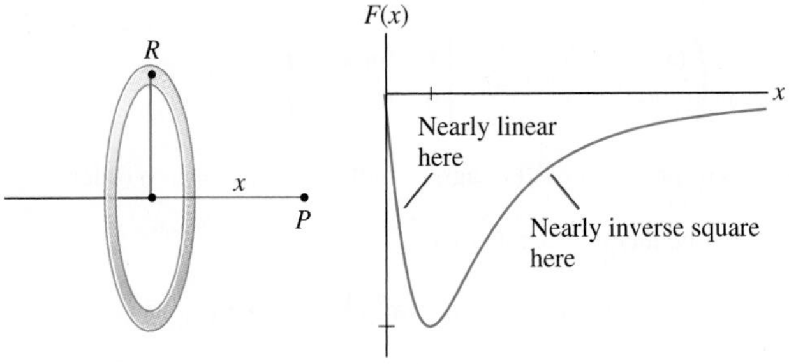

FIGURE 12

58. A light wave of wavelength λ travels from A to B by passing through an aperture (circular region) located in a plane that is perpendicular to $\overline{AB}$ (see Figure 13 for the notation). Let $f(r) = d' + h'$; that is, $f(r)$ is the distance $AC + CB$ as a function of r.

(a) Show that $f(r) = \sqrt{d^2 + r^2} + \sqrt{h^2 + r^2}$, and use the Maclaurin polynomial of order 2 to show that

$$f(r) \approx d + h + \frac{1}{2}\left(\frac{1}{d} + \frac{1}{h}\right) r^2$$

(b) The **Fresnel zones**, used to determine the optical disturbance at B, are the concentric bands bounded by the circles of radius R_n such that $f(R_n) = d + h + n\lambda/2$. Show that $R_n \approx \sqrt{n\lambda L}$, where $L = (d^{-1} + h^{-1})^{-1}$.

(c) Estimate the radii R_1 and R_{100} for blue light ($\lambda = 475 \times 10^{-7}$ cm) if $d = h = 100$ cm.

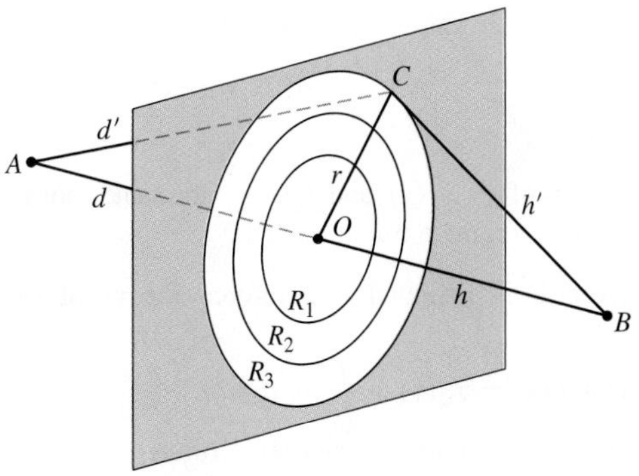

FIGURE 13 The Fresnel zones are the regions between the circles of radius R_n.

59. Referring to Figure 14, let a be the length of the chord $\overline{AC}$ of angle θ of the unit circle. Derive the following approximation for the excess of the arc over the chord.

$$\theta - a \approx \frac{\theta^3}{24}$$

Hint: Show that $\theta - a = \theta - 2\sin(\theta/2)$ and use the third Maclaurin polynomial as an approximation.

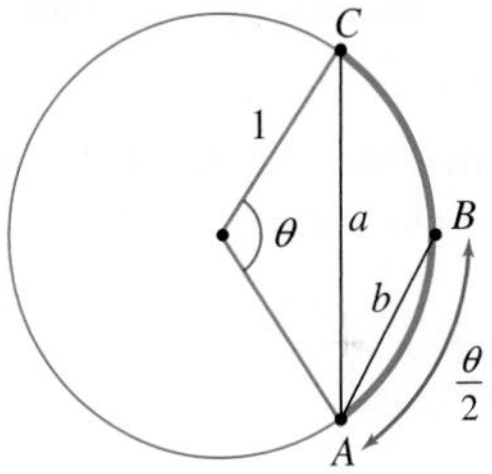

FIGURE 14 Unit circle.

60. To estimate the length θ of a circular arc of the unit circle, the seventeenth-century Dutch scientist Christian Huygens used the approximation $\theta \approx (8b - a)/3$, where a is the length of the chord $\overline{AC}$ of angle θ and b is length of the chord $\overline{AB}$ of angle $\theta/2$ (Figure 14).

(a) Prove that $a = 2\sin(\theta/2)$ and $b = 2\sin(\theta/4)$, and show that the Huygens approximation amounts to the approximation

$$\theta \approx \frac{16}{3}\sin\frac{\theta}{4} - \frac{2}{3}\sin\frac{\theta}{2}$$

(b) Compute the fifth Maclaurin polynomial of the function on the right.

(c) Use the error bound to show that the error in the Huygens approximation is less than $0.00022|\theta|^5$.

Further Insights and Challenges

61. Show that the nth Maclaurin polynomial of $f(x) = \arcsin x$ for n odd is

$$T_n(x) = x + \frac{1}{2}\frac{x^3}{3} + \frac{1\cdot 3}{2\cdot 4}\frac{x^5}{5} + \cdots + \frac{1\cdot 3\cdot 5\cdots(n-2)}{2\cdot 4\cdot 6\cdots(n-1)}\frac{x^n}{n}$$

62. Let $x \geq 0$ and assume that $f^{(n+1)}(t) \geq 0$ for $0 \leq t \leq x$. Use Taylor's Theorem to show that the nth Maclaurin polynomial $T_n(x)$ satisfies

$$T_n(x) \leq f(x) \quad \text{for all } x \geq 0$$

63. Use Exercise 62 to show that for $x \geq 0$ and all n,

$$e^x \geq 1 + x + \frac{x^2}{2!} + \cdots + \frac{x^n}{n!}$$

Sketch the graphs of e^x, $T_1(x)$, and $T_2(x)$ on the same coordinate axes. Does this inequality remain true for $x < 0$?

64. This exercise is intended to reinforce the proof of Taylor's Theorem.

(a) Show that $f(x) = T_0(x) + \int_a^x f'(u)\,du$.

(b) Use Integration by Parts to prove the formula

$$\int_a^x (x-u)f^{(2)}(u)\,du = -f'(a)(x-a) + \int_a^x f'(u)\,du$$

(c) Prove the case $n = 2$ of Taylor's Theorem:

$$f(x) = T_1(x) + \int_a^x (x-u)f^{(2)}(u)\,du.$$

In Exercises 65–69, we estimate integrals using Taylor polynomials. Exercise 66 is used to estimate the error.

65. Find the fourth Maclaurin polynomial $T_4(x)$ for $f(x) = e^{-x^2}$, and calculate $I = \int_0^{1/2} T_4(x)\,dx$ as an estimate $\int_0^{1/2} e^{-x^2}\,dx$. A CAS yields the value $I \approx 0.4794255$. How large is the error in your approximation? *Hint:* $T_4(x)$ is obtained by substituting $-x^2$ in the second Maclaurin polynomial for e^x.

66. Approximating Integrals Let $L > 0$. Show that if two functions $f(x)$ and $g(x)$ satisfy $|f(x) - g(x)| < L$ for all $x \in [a, b]$, then

$$\left| \int_a^b f(x)\,dx - \int_a^b g(x)\,dx \right| dx < L(b-a)$$

67. Let $T_4(x)$ be the fourth Maclaurin polynomial for $\cos x$.

(a) Show that $|\cos x - T_4(x)| \leq \left(\frac{1}{2}\right)^6/6!$ for all $x \in \left[0, \frac{1}{2}\right]$. *Hint:* $T_4(x) = T_5(x)$.

(b) Evaluate $\int_0^{1/2} T_4(x)\,dx$ as an approximation to $\int_0^{1/2} \cos x\,dx$. Use Exercise 66 to find a bound for the size of the error.

68. Let $Q(x) = 1 - x^2/6$. Use the error bound for $\sin x$ to show that

$$\left| \frac{\sin x}{x} - Q(x) \right| \leq \frac{|x|^4}{5!}$$

Then calculate $\int_0^1 Q(x)\,dx$ as an approximation to $\int_0^1 (\sin x/x)\,dx$ and find a bound for the error.

69. (a) Compute the sixth Maclaurin polynomial $T_6(x)$ for $\sin(x^2)$ by substituting x^2 in $P(x) = x - x^3/6$, the third Maclaurin polynomial for $\sin x$.

(b) Show that $|\sin(x^2) - T_6(x)| \leq \dfrac{|x|^{10}}{5!}$.

Hint: Substitute x^2 for x in the error bound for $|\sin x - P(x)|$, noting that $P(x)$ is also the fourth Maclaurin polynomial for $\sin x$.

(c) Use $T_6(x)$ to approximate $\int_0^{1/2} \sin(x^2)\,dx$ and find a bound for the error.

70. Prove by induction that for all k,

$$\frac{d^j}{dx^j}\left(\frac{(x-a)^k}{k!}\right) = \frac{k(k-1)\cdots(k-j+1)(x-a)^{k-j}}{k!}$$

$$\frac{d^j}{dx^j}\left(\frac{(x-a)^k}{k!}\right)\Bigg|_{x=a} = \begin{cases} 1 & \text{for } k = j \\ 0 & \text{for } k \neq j \end{cases}$$

Use this to prove that $T_n(x)$ agrees with $f(x)$ at $x = a$ to order n.

71. Let a be any number and let

$$P(x) = a_n x^n + a_{n-1}x^{n-1} + \cdots + a_1 + a_0$$

be a polynomial of degree n or less.

(a) Show that if $P^{(j)}(a) = 0$ for $j = 0, 1, \ldots, n$, then $P(x) = 0$, that is, $a_j = 0$ for all j. *Hint:* Use induction, noting that if the statement is true for degree $n - 1$, then $P'(x) = 0$.

(b) Prove that $T_n(x)$ is the only polynomial of degree n or less that agrees with $f(x)$ at $x = a$ to order n. *Hint:* If $Q(x)$ is another such polynomial, apply (a) to $P(x) = T_n(x) - Q(x)$.

CHAPTER REVIEW EXERCISES

In Exercises 1–4, calculate the arc length over the given interval.

1. $y = \dfrac{x^5}{10} + \dfrac{x^{-3}}{6}$, $[1, 2]$

2. $y = e^{x/2} + e^{-x/2}$, $[0, 2]$

3. $y = 4x - 2$, $[-2, 2]$

4. $y = x^{2/3}$, $[1, 8]$

5. Show that the arc length of $y = 2\sqrt{x}$ over $[0, a]$ is equal to $\sqrt{a(a+1)} + \ln(\sqrt{a} + \sqrt{a+1})$. *Hint:* Apply the substitution $x = \tan^2\theta$ to the arc length integral.

6. CAS Compute the trapezoidal approximation T_5 to the arc length s of $y = \tan x$ over $\left[0, \frac{\pi}{4}\right]$.

In Exercises 7–10, calculate the surface area of the solid obtained by rotating the curve over the given interval about the x-axis.

7. $y = x + 1$, $[0, 4]$

8. $y = \frac{2}{3}x^{3/4} - \frac{2}{5}x^{5/4}$, $[0, 1]$

9. $y = \frac{2}{3}x^{3/2} - \frac{1}{2}x^{1/2}$, $[1, 2]$

10. $y = \frac{1}{2}x^2$, $[0, 2]$

11. Compute the total surface area of the coin obtained by rotating the region in Figure 1 about the x-axis. The top and bottom parts of the region are semicircles with a radius of 1 mm.

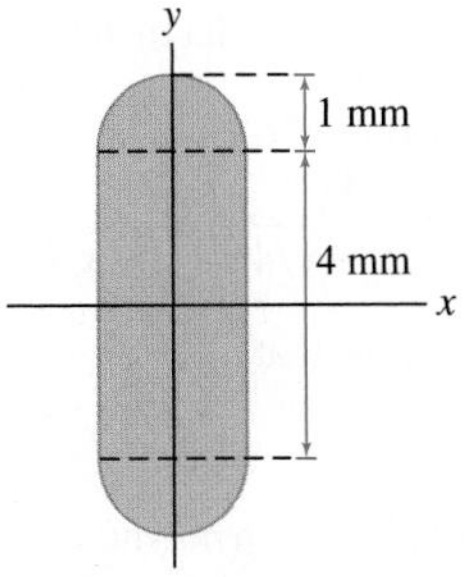

FIGURE 1

12. Calculate the fluid force on the side of a right triangle of height 3 m and base 2 m submerged in water vertically, with its upper vertex at the surface of the water.

13. Calculate the fluid force on the side of a right triangle of height 3 m and base 2 m submerged in water vertically, with its upper vertex located at a depth of 4 m.

14. A plate in the shape of the shaded region in Figure 2 is submerged in water. Calculate the fluid force on a side of the plate if the water surface is $y = 1$.

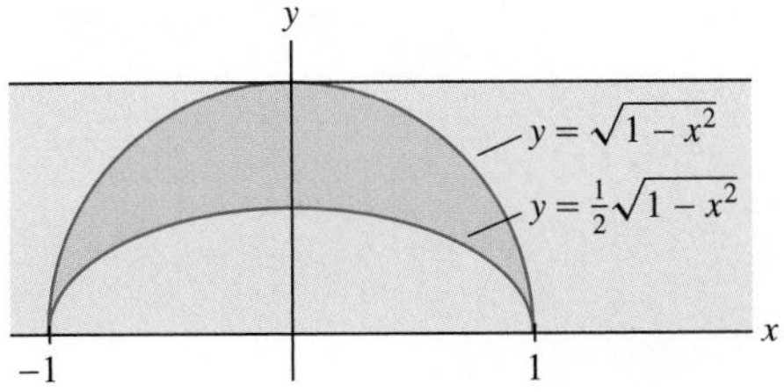

FIGURE 2

15. Figure 3 shows an object whose face is an equilateral triangle with 5-m sides. The object is 2 m thick and is submerged in water with its vertex 3 m below the water surface. Calculate the fluid force on both a triangular face and a slanted rectangular edge of the object.

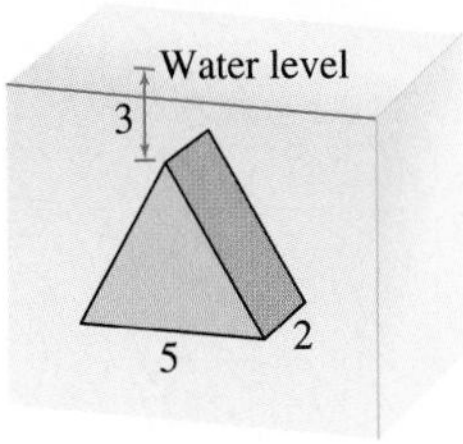

FIGURE 3

16. The end of a horizontal oil tank is an ellipse (Figure 4) with equation $(x/4)^2 + (y/3)^2 = 1$ (length in meters). Assume that the tank is filled with oil of density 900 kg/m^3.

(a) Calculate the total force F on the end of the tank when the tank is full.

(b) Would you expect the total force on the lower half of the tank to be greater than, less than, or equal to $\frac{1}{2}F$? Explain. Then compute the force on the lower half exactly and confirm (or refute) your expectation.

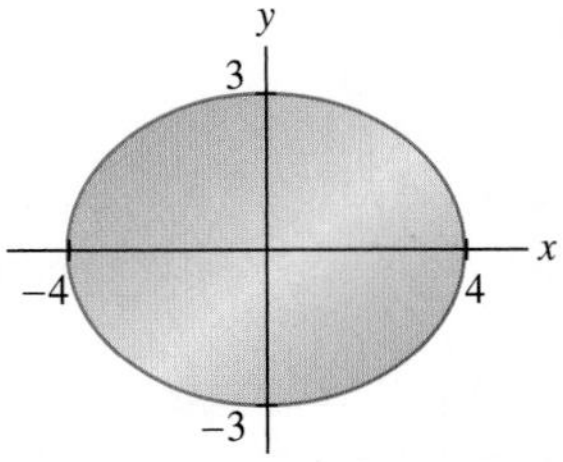

FIGURE 4

17. Calculate the moments and COM of the lamina occupying the region under $y = x(4 - x)$ for $0 \le x \le 4$, assuming a density of $\rho = 1200$ kg/m^3.

18. Sketch the region between $y = 4(x + 1)^{-1}$ and $y = 1$ for $0 \le x \le 3$, and find its centroid.

19. Find the centroid of the region between the semicircle $y = \sqrt{1 - x^2}$ and the top half of the ellipse $y = \frac{1}{2}\sqrt{1 - x^2}$ (Figure 2).

20. Find the centroid of the shaded region in Figure 5 bounded on the left by $x = 2y^2 - 2$ and on the right by a semicircle of radius 1. *Hint:* Use symmetry and additivity of moments.

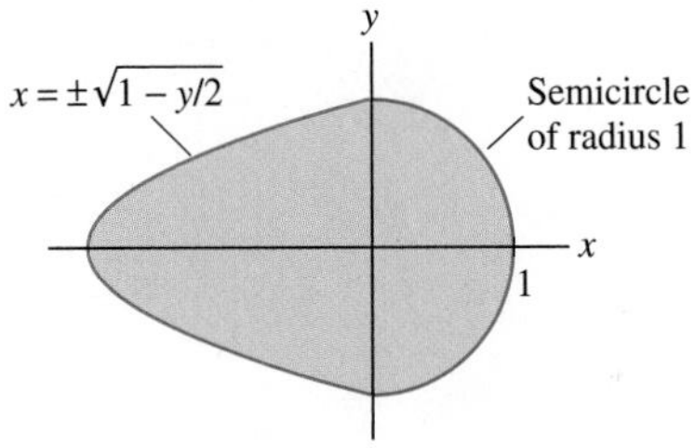

FIGURE 5

In Exercises 21–26, find the Taylor polynomial at $x = a$ for the given function.

21. $f(x) = x^3$, $T_3(x)$, $a = 1$

22. $f(x) = 3(x + 2)^3 - 5(x + 2)$, $T_3(x)$, $a = -2$

23. $f(x) = x\ln(x)$, $T_4(x)$, $a = 1$

24. $f(x) = (3x + 2)^{1/3}$, $T_3(x)$, $a = 2$

25. $f(x) = xe^{-x^2}$, $T_4(x)$, $a = 0$

26. $f(x) = \ln(\cos x)$, $T_3(x)$, $a = 0$

27. Find the nth Maclaurin polynomial for $f(x) = e^{3x}$.

28. Use the fifth Maclaurin polynomial of $f(x) = e^x$ to approximate $\sqrt{e}$. Use a calculator to determine the error.

29. Use the third Taylor polynomial of $f(x) = \tan^{-1} x$ at $a = 1$ to approximate $f(1.1)$. Use a calculator to determine the error.

30. Let $T_4(x)$ be the Taylor polynomial for $f(x) = \sqrt{x}$ at $a = 16$. Use the error bound to find the maximum possible size of $|f(17) - T_4(17)|$.

31. Find n such that $|e - T_n(1)| < 10^{-8}$, where $T_n(x)$ is the nth Maclaurin polynomial for $f(x) = e^x$.

32. Let $T_4(x)$ be the Taylor polynomial for $f(x) = x \ln x$ at $a = 1$ computed in Exercise 23. Use the error bound to find a bound for $|f(1.2) - T_4(1.2)|$.

33. Verify that $T_n(x) = 1 + x + x^2 + \cdots + x^n$ is the nth Maclaurin polynomial of $f(x) = 1/(1-x)$. Show using substitution that the nth Maclaurin polynomial for $f(x) = 1/(1 - x/4)$ is

$$T_n(x) = 1 + \frac{1}{4}x + \frac{1}{4^2}x^2 + \cdots + \frac{1}{4^n}x^n$$

What is the nth Maclaurin polynomial for $g(x) = \dfrac{1}{1+x}$?

34. Let $f(x) = \dfrac{5}{4 + 3x - x^2}$ and let a_k be the coefficient of x^k in the Maclaurin polynomial $T_n(x)$ of for $k \le n$.

(a) Show that $f(x) = \left(\dfrac{1/4}{1 - x/4} + \dfrac{1}{1+x}\right)$.

(b) Use Exercise 33 to show that $a_k = \dfrac{1}{4^{k+1}} + (-1)^k$.

(c) Compute $T_3(x)$.

35. Let $T_n(x)$ be the nth Maclaurin polynomial for the function $f(x) = \sin x + \sinh x$.

(a) Show that $T_5(x) = T_6(x) = T_7(x) = T_8(x)$.

(b) Show that $|f^n(x)| \le 1 + \cosh x$ for all n. *Hint:* Note that $|\sinh x| \le |\cosh x|$ for all x.

(c) Show that $|T_8(x) - f(x)| \le \dfrac{2.6}{9!}|x|^9$ for $-1 \le x \le 1$.

10 INTRODUCTION TO DIFFERENTIAL EQUATIONS

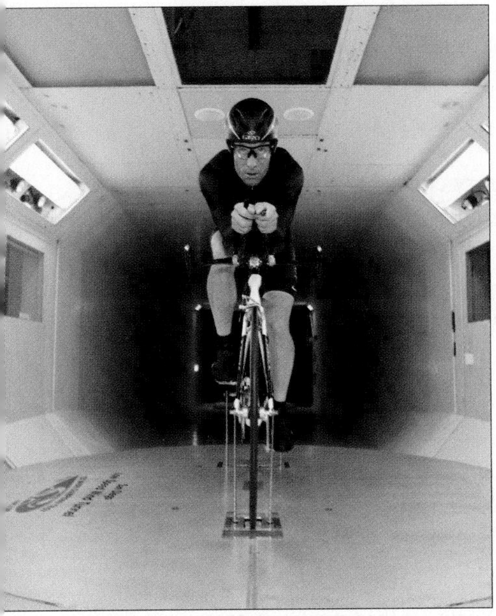

Tour de France champion Lance Armstrong testing a bicycle at the San Diego Air & Space Technology Low Speed Wind Tunnel in November 2008. Armstrong's clothing, helmet, posture, and hand position are also aerodynamically optimized.

Will this airplane fly?... How can we create an image of the interior of the human body using very weak X-rays?... What is a design of a bicycle frame that combines low weight with rigidity?... How much would the mean temperature of the earth increase if the amount of carbon dioxide in the atmosphere increased by 20 percent?

—An overview of applications of differential equations in *Computational Differential Equations*, K. Eriksson, D. Estep, P. Hansbo, and C. Johnson, Cambridge University Press, New York, 1996

Differential equations are among the most powerful tools we have for analyzing the world mathematically. They are used to formulate the fundamental laws of nature (from Newton's Laws to Maxwell's equations and the laws of quantum mechanics) and to model the most diverse physical phenomena. The quotation above lists just a few of the myriad applications. This chapter provides an introduction to some elementary techniques and applications of this important subject.

10.1 Solving Differential Equations

A differential equation is an equation that involves an unknown function $y = y(x)$ and its first or higher derivatives. A **solution** is a function $y = f(x)$ satisfying the given equation. As we have seen in previous chapters, solutions usually depend on one or more arbitrary constants (denoted A, B, and C in the following examples):

Differential equation	General solution
$y' = -2y$	$y = Ce^{-2x}$
$\dfrac{dy}{dt} = t$	$y = \dfrac{1}{2}t^2 + C$
$y'' + y = 0$	$y = A\sin x + B\cos x$

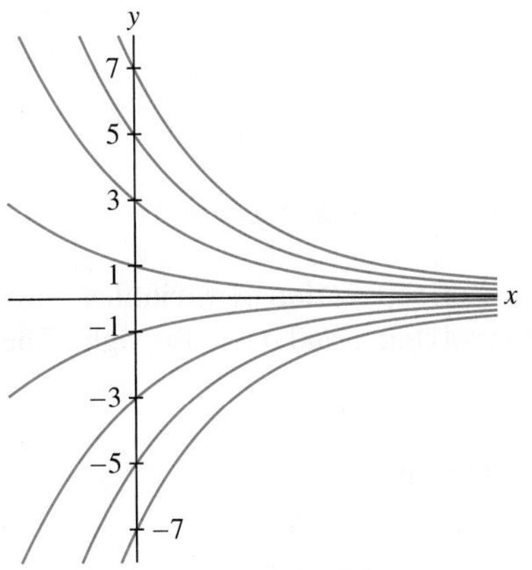

FIGURE 1 Family of solutions of $y' = -2y$.

An expression such as $y = Ce^{-2x}$ is called a **general solution**. For each value of C, we obtain a **particular solution**. The graphs of the solutions as C varies form a family of curves in the xy-plane (Figure 1).

The first step in any study of differential equations is to classify the equations according to various properties. The most important attributes of a differential equation are its order and whether or not it is linear.

The **order** of a differential equation is the order of the highest derivative appearing in the equation. The general solution of an equation of order n usually involves n arbitrary constants. For example,

$$y'' + y = 0$$

has order 2 and its general solution has two arbitrary constants A and B as listed above.

A differential equation is called **linear** if it can be written in the form

$$a_n(x)y^{(n)} + a_{n-1}(x)y^{(n-1)} + \cdots + a_1(x)y' + a_0(x)y = b(x)$$

The coefficients $a_j(x)$ and $b(x)$ can be arbitrary functions of x, but a linear equation cannot have terms such as y^3, yy', or $\sin y$.

Differential equation	Order	Linear or nonlinear
$x^2y' + e^x y = 4$	First-order	Linear
$x(y')^2 = y + x$	First-order	Nonlinear (because $(y')^2$ appears)
$y'' = (\sin x)y'$	Second-order	Linear
$y''' = x(\sin y)$	Third-order	Nonlinear (because $\sin y$ appears)

In this chapter we restrict our attention to first-order equations.

Separation of Variables

We are familiar with the simplest type of differential equation, namely $y' = f(x)$. A solution is simply an antiderivative of $f(x)$, so can write the general solution as

$$y = \int f(x)\,dx$$

A more general class of first-order equations that can be solved directly by integration are the **separable equations**, which have the form

$$\frac{dy}{dx} = f(x)g(y) \qquad \boxed{1}$$

For example,

- $\dfrac{dy}{dx} = (\sin x)y$ is separable.
- $\dfrac{dy}{dx} = x + y$ is not separable because $x + y$ is not a *product* $f(x)g(y)$.

In separation of variables, we manipulate dx and dy symbolically, just as in the Substitution Rule.

Separable equations are solved using the method of **separation of variables**: Move the terms involving y and dy to the left and those involving x and dx to the right. Then integrate both sides:

$$\frac{dy}{dx} = f(x)g(y) \qquad \text{(separable equation)}$$

$$\frac{dy}{g(y)} = f(x)\,dx \qquad \text{(separate the variables)}$$

$$\int \frac{dy}{g(y)} = \int f(x)\,dx \qquad \text{(integrate)}$$

If these integrals can be evaluated, we can try to solve for y as a function of x.

■ **EXAMPLE 1** Show that $y\dfrac{dy}{dx} - x = 0$ is separable but not linear. Then find the general solution and plot the family of solutions.

Solution This differential equation is nonlinear because it contains the term yy'. To show that it is separable, rewrite the equation:

$$y\frac{dy}{dx} - x = 0 \quad \Rightarrow \quad \frac{dy}{dx} = xy^{-1} \quad \text{(separable equation)}$$

Now use separation of variables:

$$y\,dy = x\,dx \qquad \text{(separate the variables)}$$

$$\int y\,dy = \int x\,dx \qquad \text{(integrate)}$$

$$\frac{1}{2}y^2 = \frac{1}{2}x^2 + C \qquad \boxed{2}$$

$$y = \pm\sqrt{x^2 + 2C} \qquad \text{(solve for } y\text{)}$$

Note that one constant of integration is sufficient in Eq. (2). An additional constant for the integral on the left is not needed.

Since C is arbitrary, we may replace $2C$ by C to obtain (Figure 2)

$$y = \pm\sqrt{x^2 + C}$$

Each choice of sign yields a solution.

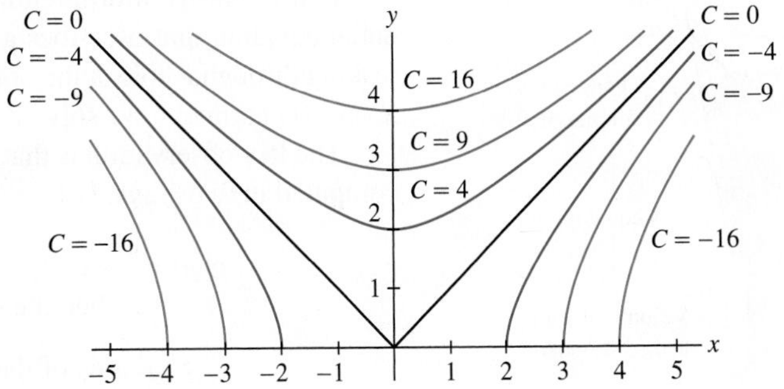

FIGURE 2 Solutions $y = \sqrt{x^2 + C}$ to $y\dfrac{dy}{dx} - x = 0$.

It is a good idea to verify that solutions you have found satisfy the differential equation. In our case, for the positive square root (the negative square root is similar), we have

$$\frac{dy}{dx} = \frac{d}{dx}\sqrt{x^2 + C} = \frac{x}{\sqrt{x^2 + C}}$$

$$y\frac{dy}{dx} = \sqrt{x^2 + C}\left(\frac{x}{\sqrt{x^2 + C}}\right) = x \quad \Rightarrow \quad y\frac{dy}{dx} - x = 0$$

This verifies that $y = \sqrt{x^2 + C}$ is a solution. ■

Most differential equations arising in applications have an existence and uniqueness *property: There exists one and only one solution satisfying a given initial condition. General existence and uniqueness theorems are discussed in textbooks on differential equations.*

Although it is useful to find general solutions, in applications we are usually interested in the solution that describes a particular physical situation. The general solution to a first-order equation generally depends on one arbitrary constant, so we can pick out a particular solution $y(x)$ by specifying the value $y(x_0)$ for some fixed x_0 (Figure 3). This specification is called an **initial condition.** A differential equation together with an initial condition is called an **initial value problem.**

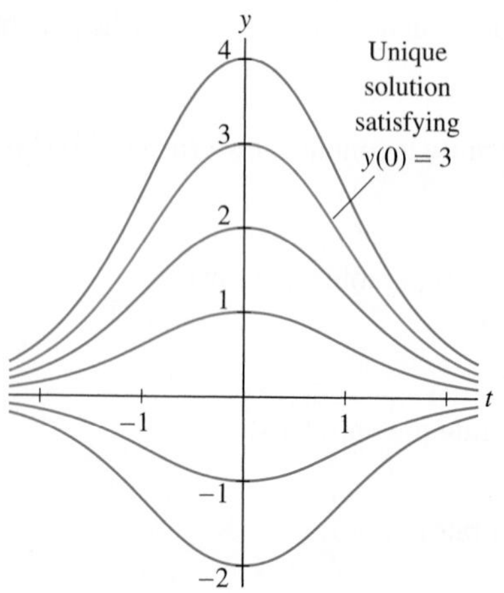

FIGURE 3 The initial condition $y(0) = 3$ determines one curve in the family of solutions to $y' = -ty$.

If we set $C = 0$ in Eq. (3), we obtain the solution $y = 0$. The separation of variables procedure did not directly yield this solution because we divided by y (and thus assumed implicitly that $y \neq 0$).

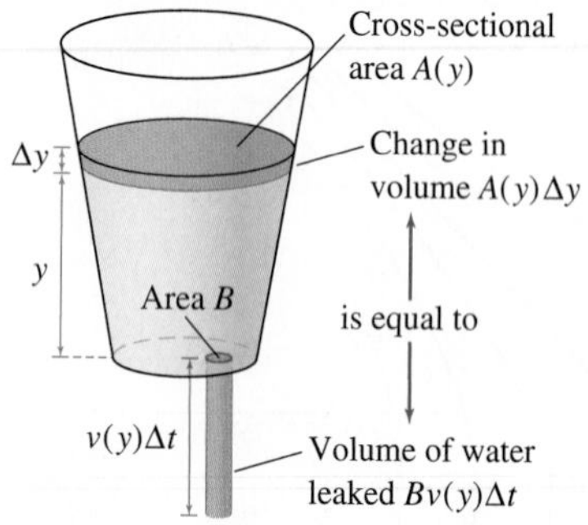

FIGURE 4 Water leaks out of a tank through a hole of area B at the bottom.

■ **EXAMPLE 2 Initial Value Problem** Solve the initial value problem

$$y' = -ty, \qquad y(0) = 3$$

Solution Use separation of variables to find the general solution:

$$\frac{dy}{dt} = -ty \quad \Rightarrow \quad \frac{dy}{y} = -t\,dt$$

$$\int \frac{dy}{y} = -\int t\,dt$$

$$\ln|y| = -\frac{1}{2}t^2 + C$$

$$|y| = e^{-t^2/2+C} = e^C e^{-t^2/2}$$

Thus, $y = \pm e^C e^{-t^2/2}$. Since C is arbitrary, e^C represents an arbitrary *positive number*, and $\pm e^C$ is an arbitrary nonzero number. We replace $\pm e^C$ by C and write the general solution as

$$y = Ce^{-t^2/2} \qquad \boxed{3}$$

Now use the initial condition $y(0) = Ce^{-0^2/2} = 3$. Thus, $C = 3$ and $y = 3e^{-t^2/2}$ is the solution to the initial value problem (Figure 3). ■

In the context of differential equations, the term "modeling" means finding a differential equation that describes a given physical situation. As an example, consider water leaking through a hole at the bottom of a tank (Figure 4). The problem is to find the water level $y(t)$ at time t. We solve it by showing that $y(t)$ satisfies a differential equation.

The key observation is that the water lost during the interval from t to $t + \Delta t$ can be computed in two ways. Let

$v(y)$ = velocity of the water flowing through the hole when the tank is filled to height y

B = area of the hole

$A(y)$ = area of horizontal cross section of the tank at height y

First, we observe that the water exiting through the hole during a time interval Δt forms a cylinder of base B and height $v(y)\Delta t$ (because the water travels a distance $v(y)\Delta t$—see Figure 4). The volume of this cylinder is approximately $Bv(y)\Delta t$ [approximately but not exactly, because $v(y)$ may not be constant]. Thus,

$$\text{Water lost between } t \text{ and } t + \Delta t \approx Bv(y)\Delta t$$

Second, we note that if the water level drops by an amount Δy during the interval Δt, then the volume of water lost is approximately $A(y)\Delta y$ (Figure 4). Therefore,

$$\text{Water lost between } t \text{ and } t + \Delta t \approx A(y)\,\Delta y$$

This is also an approximation because the cross-sectional area may not be constant. Comparing the two results, we obtain $A(y)\Delta y \approx Bv(y)\Delta t$, or

$$\frac{\Delta y}{\Delta t} \approx \frac{Bv(y)}{A(y)}$$

Like most if not all mathematical models, our model of water draining from a tank is at best an approximation. The differential equation (4) does not take into account viscosity (resistance of a fluid to flow). This can be remedied by using the differential equation

$$\frac{dy}{dt} = k\frac{Bv(y)}{A(y)}$$

where $k < 1$ is a viscosity constant. Furthermore, Torricelli's Law is valid only when the hole size B is small relative to the cross-sectional areas $A(y)$.

Now take the limit as $\Delta t \to 0$ to obtain the differential equation

$$\frac{dy}{dt} = \frac{Bv(y)}{A(y)} \qquad \boxed{4}$$

To use Eq. (4), we need to know the velocity of the water leaving the hole. This is given by **Torricelli's Law** ($g = 9.8$ m/s^2):

$$v(y) = -\sqrt{2gy} = -\sqrt{2(9.8)y} \approx -4.43\sqrt{y} \text{ m/s} \qquad \boxed{5}$$

EXAMPLE 3 Application of Torricelli's Law A cylindrical tank of height 4 m and radius 1 m is filled with water. Water drains through a square hole of side 2 cm in the bottom. Determine the water level $y(t)$ at time t (seconds). How long does it take for the tank to go from full to empty?

Solution We use units of centimeters.

***Step 1.* Write down and solve the differential equation.**
The horizontal cross section of the cylinder is a circle of radius $r = 100$ cm and area $A(y) = \pi r^2 = 10{,}000\pi$ cm^2 (Figure 5). The hole is a square of side 2 cm and area $B = 4$ cm^2. By Torricelli's Law [Eq. (5)], $v(y) = -44.3\sqrt{y}$ cm/s, so Eq. (4) becomes

$$\frac{dy}{dt} = \frac{Bv(y)}{A(y)} = -\frac{4(44.3\sqrt{y})}{10{,}000\pi} \approx -0.0056\sqrt{y} \qquad \boxed{6}$$

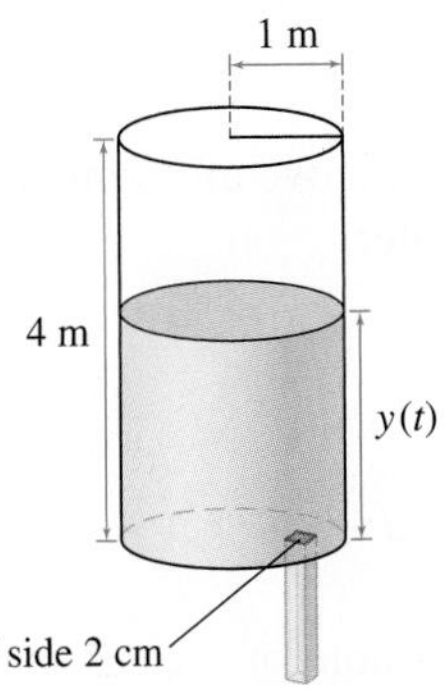

FIGURE 5

Solve using separation of variables:

$$\int \frac{dy}{\sqrt{y}} = -0.0056 \int dt$$

$$2y^{1/2} = -0.0056t + C \qquad \boxed{7}$$

$$y = \left(-0.0028t + \frac{1}{2}C\right)^2$$

Since C is arbitrary, we may replace $\frac{1}{2}C$ by C and write

$$y = (C - 0.0028t)^2$$

***Step 2.* Use the initial condition.**
The tank is full at $t = 0$, so we have the initial condition $y(0) = 400$ cm. Thus

$$y(0) = C^2 = 400 \quad \Rightarrow \quad C = \pm 20$$

Which sign is correct? You might think that both sign choices are possible, but notice that the water level y is a decreasing function of t, and the function $y = (C - 0.0028t)^2$ decreases to 0 only if C is positive. Alternatively, we can see directly from Eq. (7) that $C > 0$, because $2y^{1/2}$ is *nonnegative*. Thus,

$$y(t) = (20 - 0.0028t)^2$$

To determine the time t_e that it takes to empty the tank, we solve

$$y(t_e) = (20 - 0.0028t_e)^2 = 0 \quad \Rightarrow \quad t_e \approx 7142 \text{ s}$$

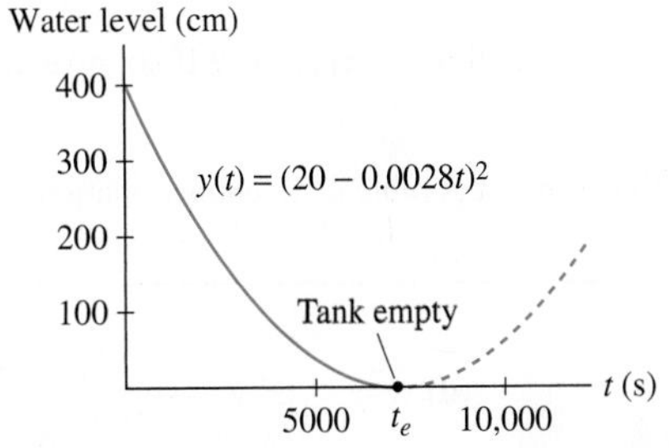

FIGURE 6

Thus, the tank is empty after 7142 s, or nearly two hours (Figure 6). ■

CONCEPTUAL INSIGHT The previous example highlights the need to analyze solutions to differential equations rather than relying on algebra alone. The algebra seemed to suggest that $C = \pm 20$, but further analysis showed that $C = -20$ does not yield a solution for $t \geq 0$. Note also that the function

$$y(t) = (20 - 0.0028t)^2$$

is a solution only for $t \leq t_e$—that is, until the tank is empty. This function cannot satisfy Eq. (6) for $t > t_e$ because its derivative is positive for $t > t_e$ (Figure 6), but solutions of Eq. (6) have nonpositive derivatives.

10.1 SUMMARY

- A differential equation has order n if $y^{(n)}$ is the highest-order derivative appearing in the equation.
- A differential equation is *linear* if it can be written as

$$a_n(x)y^{(n)} + a_{n-1}(x)y^{(n-1)} + \cdots + a_1(x)y' + a_0(x)y = b(x)$$

- *Separable first-order equation* $\dfrac{dy}{dx} = f(x)g(y)$
- *Separation of Variables* (for a separable equation): move all terms involving y to the left and all terms involving x to the right and integrate:

$$\frac{dy}{g(y)} = f(x)\,dx$$

$$\int \frac{dy}{g(y)} = \int f(x)\,dx$$

- Differential equation for water leaking through a hole of area B in a tank of cross-sectional areas $A(y)$:

$$\boxed{\frac{dy}{dt} = \frac{Bv(y)}{A(y)}}$$

Torricelli's Law: $v(y) = -\sqrt{2gy}$, where $g = 9.8$ m/s^2.

10.1 EXERCISES

Preliminary Questions

1. Determine the order of the following differential equations:

(a) $x^5 y' = 1$

(b) $(y')^3 + x = 1$

(c) $y''' + x^4 y' = 2$

(d) $\sin(y'') + x = y$

2. Is $y'' = \sin x$ a linear differential equation?

3. Give an example of a nonlinear differential equation of the form $y' = f(y)$.

4. Can a nonlinear differential equation be separable? If so, give an example.

5. Give an example of a linear, nonseparable differential equation.

Exercises

1. Which of the following differential equations are first-order?

(a) $y' = x^2$

(b) $y'' = y^2$

(c) $(y')^3 + yy' = \sin x$

(d) $x^2 y' - e^x y = \sin y$

(e) $y'' + 3y' = \dfrac{y}{x}$

(f) $yy' + x + y = 0$

2. Which of the equations in Exercise 1 are linear?

In Exercises 3–8, verify that the given function is a solution of the differential equation.

3. $y' - 8x = 0, \quad y = 4x^2$

4. $yy' + 4x = 0, \quad y = \sqrt{12 - 4x^2}$

5. $y' + 4xy = 0, \quad y = 25e^{-2x^2}$

6. $(x^2 - 1)y' + xy = 0, \quad y = 4(x^2 - 1)^{-1/2}$

7. $y'' - 2xy' + 8y = 0, \quad y = 4x^4 - 12x^2 + 3$

8. $y'' - 2y' + 5y = 0, \quad y = e^x \sin 2x$

9. Which of the following equations are separable? Write those that are separable in the form $y' = f(x)g(y)$ (but do not solve).

(a) $xy' - 9y^2 = 0$

(b) $\sqrt{4 - x^2}\, y' = e^{3y} \sin x$

(c) $y' = x^2 + y^2$

(d) $y' = 9 - y^2$

10. The following differential equations appear similar but have very different solutions.

$$\frac{dy}{dx} = x, \qquad \frac{dy}{dx} = y$$

Solve both subject to the initial condition $y(1) = 2$.

11. Consider the differential equation $y^3 y' - 9x^2 = 0$.

(a) Write it as $y^3\,dy = 9x^2\,dx$.

(b) Integrate both sides to obtain $\frac{1}{4}y^4 = 3x^3 + C$.

(c) Verify that $y = (12x^3 + C)^{1/4}$ is the general solution.

(d) Find the particular solution satisfying $y(1) = 2$.

12. Verify that $x^2 y' + e^{-y} = 0$ is separable.

(a) Write it as $e^y\,dy = -x^{-2}\,dx$.

(b) Integrate both sides to obtain $e^y = x^{-1} + C$.

(c) Verify that $y = \ln(x^{-1} + C)$ is the general solution.

(d) Find the particular solution satisfying $y(2) = 4$.

In Exercises 13–28, use separation of variables to find the general solution.

13. $y' + 4xy^2 = 0$

14. $y' + x^2 y = 0$

15. $\dfrac{dy}{dt} - 20t^4 e^{-y} = 0$

16. $t^3 y' + 4y^2 = 0$

17. $2y' + 5y = 4$

18. $\dfrac{dy}{dt} = 8\sqrt{y}$

19. $\sqrt{1 - x^2}\, y' = xy$

20. $y' = y^2(1 - x^2)$

21. $yy' = x$

22. $(\ln y)y' - ty = 0$

23. $\dfrac{dx}{dt} = (t + 1)(x^2 + 1)$

24. $(1 + x^2)y' = x^3 y$

25. $y' = x \sec y$

26. $\dfrac{dy}{d\theta} = \tan y$

27. $\dfrac{dy}{dt} = y \tan t$

28. $\dfrac{dx}{dt} = t \tan x$

In Exercises 29–42, solve the initial value problem.

29. $y' + 2y = 0, \quad y(\ln 5) = 3$

30. $y' - 3y + 12 = 0, \quad y(2) = 1$

31. $yy' = xe^{-y^2}, \quad y(0) = -2$

32. $y^2 \dfrac{dy}{dx} = x^{-3}, \quad y(1) = 0$

33. $y' = (x - 1)(y - 2), \quad y(2) = 4$

34. $y' = (x - 1)(y - 2), \quad y(2) = 2$

35. $y' = x(y^2 + 1), \quad y(0) = 0$

36. $(1 - t)\dfrac{dy}{dt} - y = 0, \quad y(2) = -4$

37. $\dfrac{dy}{dt} = ye^{-t}, \quad y(0) = 1$

38. $\dfrac{dy}{dt} = te^{-y}, \quad y(1) = 0$

39. $t^2 \dfrac{dy}{dt} - t = 1 + y + ty, \quad y(1) = 0$

40. $\sqrt{1 - x^2}\, y' = y^2 + 1, \quad y(0) = 0$

41. $y' = \tan y, \quad y(\ln 2) = \dfrac{\pi}{2}$

42. $y' = y^2 \sin x, \quad y(\pi) = 2$

43. Find all values of a such that $y = x^a$ is a solution of

$$y'' - 12x^{-2} y = 0$$

44. Find all values of a such that $y = e^{ax}$ is a solution of

$$y'' + 4y' - 12y = 0$$

In Exercises 45 and 46, let $y(t)$ be a solution of $(\cos y + 1)\dfrac{dy}{dt} = 2t$ such that $y(2) = 0$.

45. Show that $\sin y + y = t^2 + C$. We cannot solve for y as a function of t, but, assuming that $y(2) = 0$, find the values of t at which $y(t) = \pi$.

46. Assuming that $y(6) = \pi/3$, find an equation of the tangent line to the graph of $y(t)$ at $(6, \pi/3)$.

In Exercises 47–52, use Eq. (4) and Torricelli's Law [Eq. (5)].

47. Water leaks through a hole of area $0.002\ \text{m}^2$ at the bottom of a cylindrical tank that is filled with water and has height 3 m and a base of area $10\ \text{m}^2$. How long does it take (a) for half of the water to leak out and (b) for the tank to empty?

48. At $t = 0$, a conical tank of height 300 cm and top radius 100 cm [Figure 7(A)] is filled with water. Water leaks through a hole in the bottom of area $3\ \text{cm}^2$. Let $y(t)$ be the water level at time t.

(a) Show that the tank's cross-sectional area at height y is $A(y) = \frac{\pi}{9}y^2$.

(b) Find and solve the differential equation satisfied by $y(t)$

(c) How long does it take for the tank to empty?

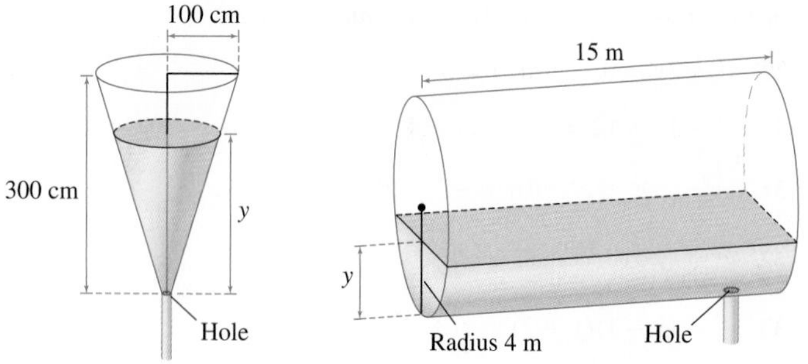

(A) Conical tank (B) Horizontal tank

FIGURE 7

49. The tank in Figure 7(B) is a cylinder of radius 4 m and height 15 m. Assume that the tank is half-filled with water and that water leaks through a hole in the bottom of area $B = 0.001\ \text{m}^2$. Determine the water level $y(t)$ and the time t_e when the tank is empty.

50. A tank has the shape of the parabola $y = x^2$, revolved around the y-axis. Water leaks from a hole of area $B = 0.0005\ \text{m}^2$ at the bottom of the tank. Let $y(t)$ be the water level at time t. How long does it take for the tank to empty if it is initially filled to height $y_0 = 1$ m.

51. A tank has the shape of the parabola $y = ax^2$ (where a is a constant) revolved around the y-axis. Water drains from a hole of area $B\ \text{m}^2$ at the bottom of the tank.

(a) Show that the water level at time t is

$$y(t) = \left(y_0^{3/2} - \frac{3aB\sqrt{2g}}{2\pi} t \right)^{2/3}$$

where y_0 is the water level at time $t = 0$.

(b) Show that if the total volume of water in the tank has volume V at time $t = 0$, then $y_0 = \sqrt{2aV/\pi}$. *Hint:* Compute the volume of the tank as a volume of rotation.

(c) Show that the tank is empty at time

$$t_e = \left(\frac{2}{3B\sqrt{g}} \right) \left(\frac{2\pi V^3}{a} \right)^{1/4}$$

We see that for fixed initial water volume V, the time t_e is proportional to $a^{-1/4}$. A large value of a corresponds to a tall thin tank. Such a tank drains more quickly than a short wide tank of the same initial volume.

52. A cylindrical tank filled with water has height h and a base of area A. Water leaks through a hole in the bottom of area B.

(a) Show that the time required for the tank to empty is proportional to $A\sqrt{h}/B$.

(b) Show that the emptying time is proportional to $Vh^{-1/2}$, where V is the volume of the tank.

(c) Two tanks have the same volume and a hole of the same size, but they have different heights and bases. Which tank empties first: the taller or the shorter tank?

53. Figure 8 shows a circuit consisting of a resistor of R ohms, a capacitor of C farads, and a battery of voltage V. When the circuit is completed, the amount of charge $q(t)$ (in coulombs) on the plates of the capacitor varies according to the differential equation (t in seconds)

$$R\frac{dq}{dt} + \frac{1}{C} q = V$$

where R, C, and V are constants.

(a) Solve for $q(t)$, assuming that $q(0) = 0$.

(b) Show that $\lim_{t\to\infty} q(t) = CV$.

(c) Show that the capacitor charges to approximately 63% of its final value CV after a time period of length $\tau = RC$ (τ is called the time constant of the capacitor).

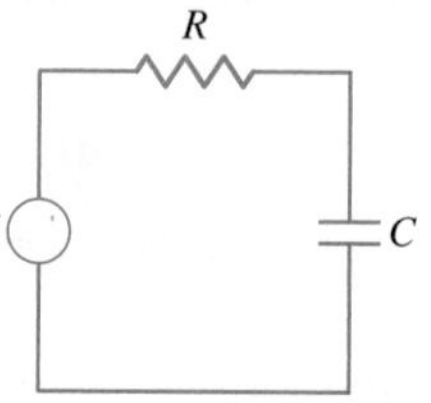

FIGURE 8 An RC circuit.

54. Assume in the circuit of Figure 8 that $R = 200\ \Omega$, $C = 0.02$ F, and $V = 12$ V. How many seconds does it take for the charge on the capacitor plates to reach half of its limiting value?

55. According to one hypothesis, the growth rate dV/dt of a cell's volume V is proportional to its surface area A. Since V has cubic units such as cm^3 and A has square units such as cm^2, we may assume roughly that $A \propto V^{2/3}$, and hence $dV/dt = kV^{2/3}$ for some constant k. If this hypothesis is correct, which dependence of volume on time would we expect to see (again, roughly speaking) in the laboratory?

(a) Linear **(b)** Quadratic **(c)** Cubic

56. We might also guess that the volume V of a melting snowball decreases at a rate proportional to its surface area. Argue as in Exercise 55 to find a differential equation satisfied by V. Suppose the snowball has volume 1000 cm^3 and that it loses half of its volume after 5 min. According to this model, when will the snowball disappear?

57. In general, $(fg)'$ is not equal to $f'g'$, but let $f(x) = e^{3x}$ and find a function $g(x)$ such that $(fg)' = f'g'$. Do the same for $f(x) = x$.

58. A boy standing at point B on a dock holds a rope of length ℓ attached to a boat at point A [Figure 9(A)]. As the boy walks along the dock, holding the rope taut, the boat moves along a curve called a **tractrix** (from the Latin *tractus*, meaning "to pull"). The segment from a point P on the curve to the x-axis along the tangent line has constant length ℓ. Let $y = f(x)$ be the equation of the tractrix.

(a) Show that $y^2 + (y/y')^2 = \ell^2$ and conclude $y' = -\dfrac{y}{\sqrt{\ell^2 - y^2}}$. Why must we choose the negative square root?

(b) Prove that the tractrix is the graph of

$$x = \ell \ln\left(\frac{\ell + \sqrt{\ell^2 - y^2}}{y} \right) - \sqrt{\ell^2 - y^2}$$

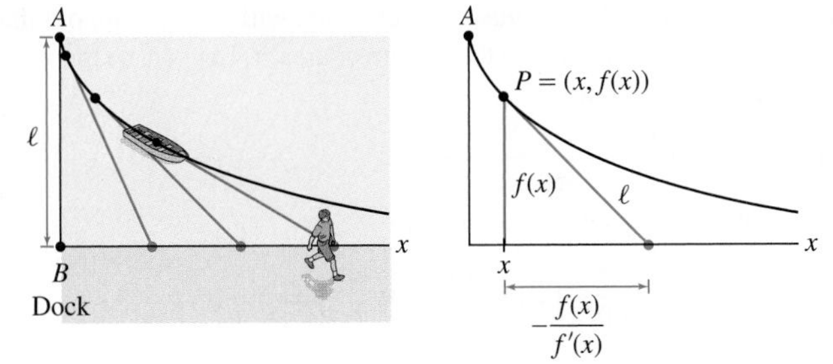

FIGURE 9

59. Show that the differential equations $y' = 3y/x$ and $y' = -x/3y$ define **orthogonal families** of curves; that is, the graphs of solutions to the first equation intersect the graphs of the solutions to the second equation in right angles (Figure 10). Find these curves explicitly.

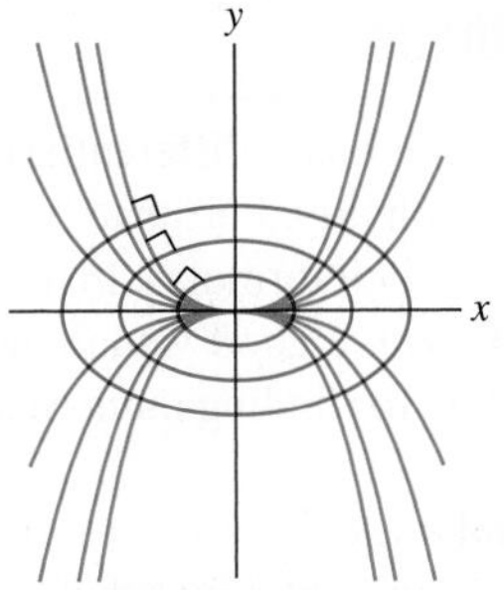

FIGURE 10 Two orthogonal families of curves.

60. Find the family of curves satisfying $y' = x/y$ and sketch several members of the family. Then find the differential equation for the orthogonal family (see Exercise 59), find its general solution, and add some members of this orthogonal family to your plot.

61. A 50-kg model rocket lifts off by expelling fuel at a rate of $k = 4.75$ kg/s for 10 s. The fuel leaves the end of the rocket with an exhaust velocity of $b = 100$ m/s. Let $m(t)$ be the mass of the rocket at time t. From the law of conservation of momentum, we find the following differential equation for the rocket's velocity $v(t)$ (in meters per second):

$$m(t)v'(t) = -9.8m(t) + b\frac{dm}{dt}$$

(a) Show that $m(t) = 50 - 4.75t$ kg.

(b) Solve for $v(t)$ and compute the rocket's velocity at rocket burnout (after 10 s).

62. Let $v(t)$ be the velocity of an object of mass m in free fall near the earth's surface. If we assume that air resistance is proportional to v^2, then v satisfies the differential equation $m\frac{dv}{dt} = -g + kv^2$ for some constant $k > 0$.

(a) Set $\alpha = (g/k)^{1/2}$ and rewrite the differential equation as

$$\frac{dv}{dt} = -\frac{k}{m}(\alpha^2 - v^2)$$

Then solve using separation of variables with initial condition $v(0) = 0$.

(b) Show that the terminal velocity $\lim_{t\to\infty} v(t)$ is equal to $-\alpha$.

63. If a bucket of water spins about a vertical axis with constant angular velocity ω (in radians per second), the water climbs up the side of the bucket until it reaches an equilibrium position (Figure 11). Two forces act on a particle located at a distance x from the vertical axis: the gravitational force $-mg$ acting downward and the force of the bucket on the particle (transmitted indirectly through the liquid) in the direction perpendicular to the surface of the water. These two forces must combine to supply a centripetal force $m\omega^2 x$, and this occurs if the diagonal of the rectangle in Figure 11 is normal to the water's surface (that is, perpendicular to the tangent line). Prove that if $y = f(x)$ is the equation of the curve obtained by taking a vertical cross section through the axis, then $-1/y' = -g/(\omega^2 x)$. Show that $y = f(x)$ is a parabola.

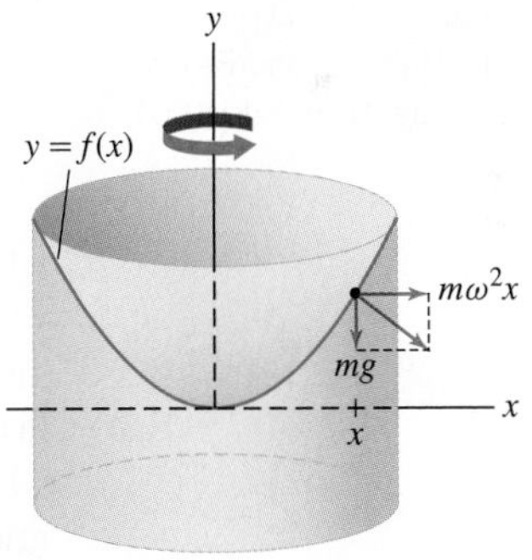

FIGURE 11

Further Insights and Challenges

64. In Section 6.2, we computed the volume V of a solid as the integral of cross-sectional area. Explain this formula in terms of differential equations. Let $V(y)$ be the volume of the solid up to height y, and let $A(y)$ be the cross-sectional area at height y as in Figure 12.

(a) Explain the following approximation for small Δy:

$$V(y + \Delta y) - V(y) \approx A(y)\,\Delta y \qquad \boxed{8}$$

(b) Use Eq. (8) to justify the differential equation $dV/dy = A(y)$. Then derive the formula

$$V = \int_a^b A(y)\,dy$$

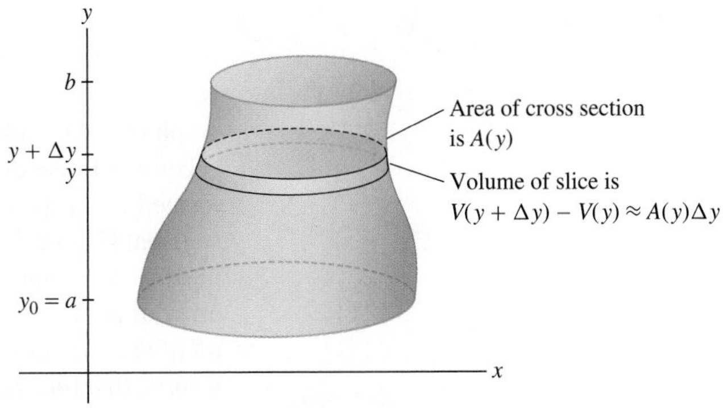

FIGURE 12

65. A basic theorem states that a *linear* differential equation of order n has a general solution that depends on n arbitrary constants. There are, however, nonlinear exceptions.

(a) Show that $(y')^2 + y^2 = 0$ is a first-order equation with only one solution $y = 0$.

(b) Show that $(y')^2 + y^2 + 1 = 0$ is a first-order equation with no solutions.

66. Show that $y = Ce^{rx}$ is a solution of $y'' + ay' + by = 0$ if and only if r is a root of $P(r) = r^2 + ar + b$. Then verify directly that $y = C_1e^{3x} + C_2e^{-x}$ is a solution of $y'' - 2y' - 3y = 0$ for any constants C_1, C_2.

67. A spherical tank of radius R is half-filled with water. Suppose that water leaks through a hole in the bottom of area B. Let $y(t)$ be the water level at time t (seconds).

(a) Show that $\dfrac{dy}{dt} = \dfrac{-8B\sqrt{y}}{\pi(2Ry - y^2)}$.

(b) Show that for some constant C,

$$\frac{\pi}{60B}\left(10Ry^{3/2} - 3y^{5/2}\right) = C - t$$

(c) Use the initial condition $y(0) = R$ to compute C, and show that $C = t_e$, the time at which the tank is empty.

(d) Show that t_e is proportional to $R^{5/2}$ and inversely proportional to B.

10.2 Graphical and Numerical Methods

"To imagine yourself subject to a differential equation, start somewhere. There you are tugged in some direction, so you move that way ... as you move, the tugging forces change, pulling you in a new direction; for your motion to solve the differential equation you must keep drifting with and responding to the ambient forces."

—From the introduction to *Differential Equations*, J. H. Hubbard and Beverly West, Springer-Verlag, New York, 1991

In the previous section, we focused on finding solutions to differential equations. However, most differential equations cannot be solved explicitly. Fortunately, there are techniques for analyzing the solutions that do not rely on explicit formulas. In this section, we discuss the method of slope fields, which provides us with a good visual understanding of first-order equations. We also discuss Euler's Method for finding numerical approximations to solutions.

We use t as the independent variable and write $\dot{y}$ for dy/dt. The notation $\dot{y}$, often used for time derivatives in physics and engineering, was introduced by Isaac Newton. A first-order differential equation can then be written in the form

$$\dot{y} = F(t, y) \tag{1}$$

where $F(t, y)$ is a function of t and y. For example, $dy/dt = ty$ becomes $\dot{y} = ty$.

It is useful to think of Eq. (1) as a set of instructions that "tells a solution" which direction to go in. Thus, a solution passing through a point (t, y) is "instructed" to continue in the direction of slope $F(t, y)$. To visualize this set of instructions, we draw a **slope field**, which is an array of small segments of slope $F(t, y)$ at points (t, y) lying on a rectangular grid in the plane.

To illustrate, let's return to the differential equation:

$$\dot{y} = -ty$$

In this case, $F(t, y) = -ty$. According to Example 2 of Section 10.1, the general solution is $y = Ce^{-t^2/2}$. Figure 1(A) shows segments of slope $-ty$ at points (t, y) along the graph of a particular solution $y(t)$. This particular solution passes through $(-1, 3)$, and according to the differential equation, $\dot{y}(-1) = -ty = -(-1)3 = 3$. Thus, the segment located at the point $(-1, 3)$ has slope 3. The graph of the solution is tangent to each segment [Figure 1(B)].

To sketch the slope field for $\dot{y} = -ty$, we draw small segments of slope $-ty$ at an array of points (t, y) in the plane, as in Figure 2(A). *The slope field allows us to visualize all of the solutions at a glance.* Starting at any point, we can sketch a solution by drawing a curve that runs tangent to the slope segments at each point [Figure 2(B)]. The graph of a solution is also called an **integral curve**.

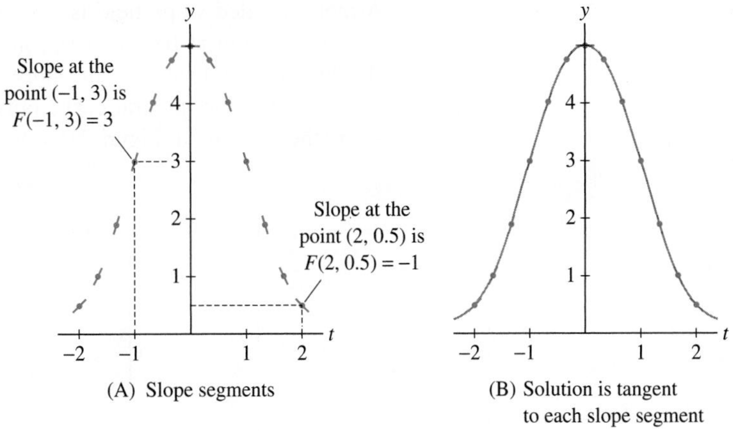

FIGURE 1 The solution of $\dot{y} = -ty$ satisfying $y(-1) = 3$.

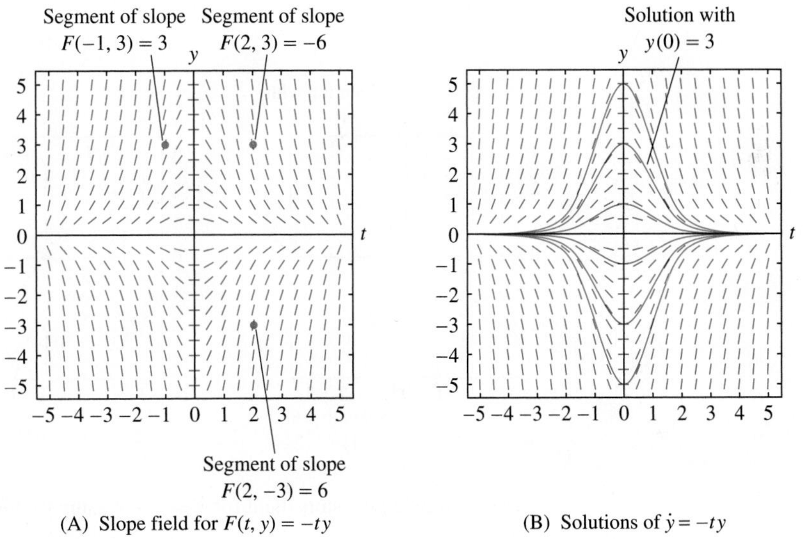

FIGURE 2 Slope field for $F(t, y) = -ty$.

■ **EXAMPLE 1** **Using Isoclines** Draw the slope field for

$$\dot{y} = y - t$$

and sketch the integral curves satisfying the initial conditions **(a)** $y(0) = 1$ and **(b)** $y(1) = -2$.

Solution A good way to sketch the slope field of $\dot{y} = F(t, y)$ is to choose several values c and identify the curve $F(t, y) = c$, called the **isocline** of slope c. The isocline is the curve consisting of all points where the slope field has slope c.

In our case, $F(t, y) = y - t$, so the isocline of fixed slope c has equation $y - t = c$, or $y = t + c$, which is a line. Consider the following values:

- $c = 0$: This isocline is $y - t = 0$, or $y = t$. We draw segments of slope $c = 0$ at points along the line $y = t$, as in Figure 3(A).
- $c = 1$: This isocline is $y - t = 1$, or $y = t + 1$. We draw segments of slope 1 at points along $y = t + 1$, as in Figure 3(B).
- $c = 2$: This isocline is $y - t = 2$, or $y = t + 2$. We draw segments of slope 2 at points along $y = t + 2$, as in Figure 3(C).
- $c = -1$: This isocline is $y - t = -1$, or $y = t - 1$ [Figure 3(C)].

A more detailed slope field is shown in Figure 3(D). To sketch the solution satisfying $y(0) = 1$, begin at the point $(t_0, y_0) = (0, 1)$ and draw the integral curve that follows the directions indicated by the slope field. Similarly, the graph of the solution satisfying $y(1) = -2$ is the integral curve obtained by starting at $(t_0, y_0) = (1, -2)$ and moving along the slope field. Figure 3(E) shows several other solutions (integral curves). ■

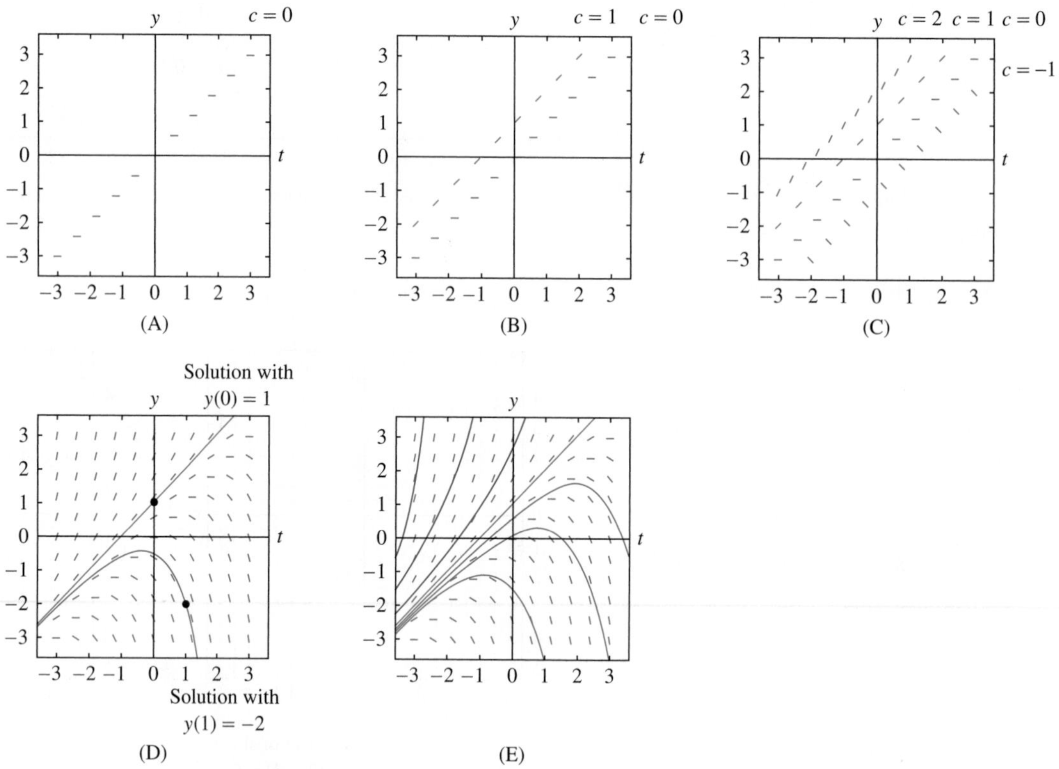

FIGURE 3 Drawing the slope field for $\dot{y} = y - t$ using isoclines.

> **GRAPHICAL INSIGHT** Slope fields often let us see the *asymptotic* behavior of solutions (as $t \to \infty$) at a glance. Figure 3(E) suggests that the asymptotic behavior depends on the initial value (the y-intercept): If $y(0) > 1$, then $y(t)$ tends to ∞, and if $y(0) < 1$, then $y(t)$ tends to $-\infty$. We can check this using the general solution $y(t) = 1 + t + Ce^t$, where $y(0) = 1 + C$. If $y(0) > 1$, then $C > 0$ and $y(t)$ tends to ∞, but if $y(0) < 1$, then $C < 0$ and $y(t)$ tends to $-\infty$. The solution $y = 1 + t$ with initial condition $y(0) = 1$ is the straight line shown in Figure 3(D).

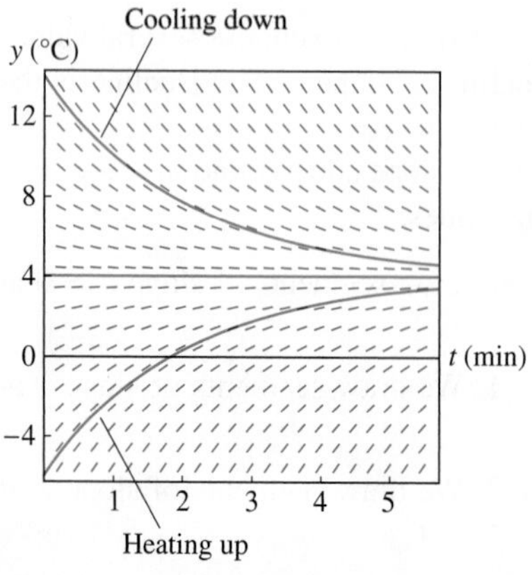

FIGURE 4 Slope field for $\dot{y} = -0.5(y - 4)$.

■ **EXAMPLE 2 Newton's Law of Cooling Revisited** The temperature $y(t)$ (°C) of an object placed in a refrigerator satisfies $\dot{y} = -0.5(y - 4)$ (t in minutes). Draw the slope field and describe the behavior of the solutions.

Solution The function $F(t, y) = -0.5(y - 4)$ depends only on y, so slopes of the segments in the slope field do not vary in the t-direction. The slope $F(t, y)$ is positive for $y < 4$ and negative for $y > 4$. More precisely, the slope at height y is $-0.5(y - 4) = -0.5y + 2$, so the segments grow steeper with positive slope as $y \to -\infty$, and they grow steeper with negative slope as $y \to \infty$ (Figure 4).

The slope field shows that if the initial temperature satisfies $y_0 > 4$, then $y(t)$ decreases to $y = 4$ as $t \to \infty$. In other words, the object cools down to 4°C when placed in the refrigerator. If $y_0 < 4$, then $y(t)$ increases to $y = 4$ as $t \to \infty$— the object warms up when placed in the refrigerator. If $y_0 = 4$, then y remains at 4°C for all time t. ■

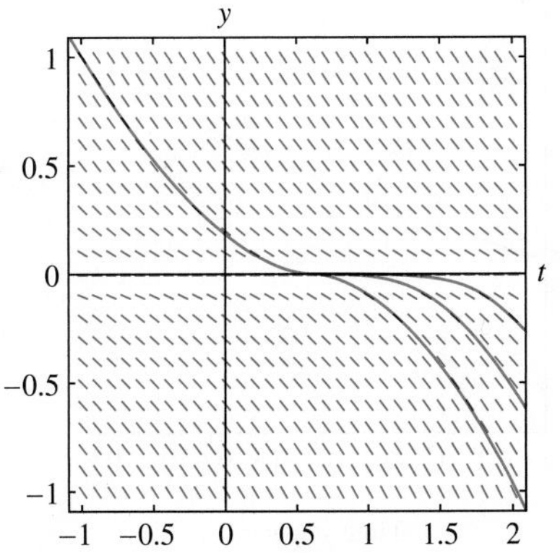

FIGURE 5 Overlapping integral curves for $\dot{y} = -\sqrt{|y|}$ (uniqueness fails for this differential equation).

CONCEPTUAL INSIGHT Most first-order equations arising in applications have a uniqueness property: There is precisely one solution $y(t)$ satisfying a given initial condition $y(t_0) = y_0$. Graphically, this means that precisely one integral curve (solution) passes through the point (t_0, y_0). Thus, when uniqueness holds, distinct integral curves never cross or overlap. Figure 5 shows the slope field of $\dot{y} = -\sqrt{|y|}$, where uniqueness fails. We can prove that once an integral curve touches the t-axis, it either remains on the t-axis or continues along the t-axis for a period of time before moving below the t-axis. Therefore, infinitely many integral curves pass through each point on the t-axis. However, the slope field does not show this clearly. This highlights again the need to analyze solutions rather than rely on visual impressions alone.

Euler's Method

Euler's Method produces numerical approximations to the solution of a first-order initial value problem:

$$\dot{y} = F(t, y), \qquad y(t_0) = y_0 \tag{2}$$

Euler's Method is the simplest method for solving initial value problems numerically, but it is not very efficient. Computer systems use more sophisticated schemes, making it possible to plot and analyze solutions to the complex systems of differential equations arising in areas such as weather prediction, aerodynamic modeling, and economic forecasting.

We begin by choosing a small number h, called the **time step**, and consider the sequence of times spaced at intervals of size h:

$$t_0, \qquad t_1 = t_0 + h, \qquad t_2 = t_0 + 2h, \qquad t_3 = t_0 + 3h, \qquad \dots$$

In general, $t_k = t_0 + kh$. Euler's Method consists of computing a sequence of values $y_1, y_2, y_3, \dots, y_n$ successively using the formula

$$\boxed{y_k = y_{k-1} + hF(t_{k-1}, y_{k-1})} \tag{3}$$

Starting with the initial value $y_0 = y(t_0)$, we compute $y_1 = y_0 + hF(t_0, y_0)$, etc. The value y_k is the Euler approximation to $y(t_k)$. We connect the points $P_k = (t_k, y_k)$ by segments to obtain an approximation to the graph of $y(t)$ (Figure 6).

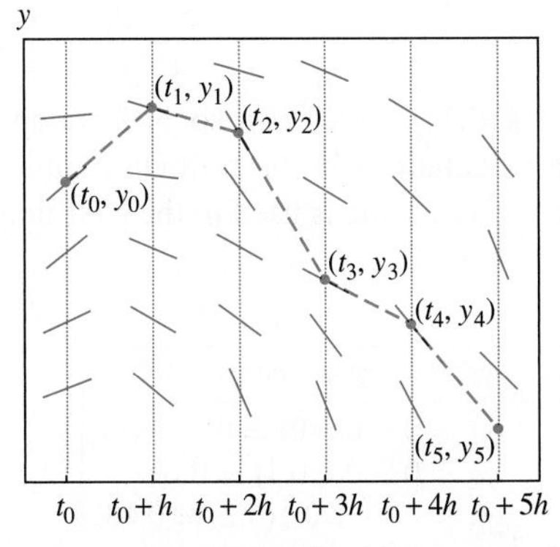

FIGURE 6 In Euler's Method, we move from one point to the next by traveling along the line indicated by the slope field.

GRAPHICAL INSIGHT The values y_k are defined so that the segment joining P_{k-1} to P_k has slope

$$\frac{y_k - y_{k-1}}{t_k - t_{k-1}} = \frac{(y_{k-1} + hF(t_{k-1}, y_{k-1})) - y_{k-1}}{h} = F(t_{k-1}, y_{k-1})$$

Thus, in Euler's method we move from P_{k-1} to P_k by traveling in the direction specified by the slope field at P_{k-1} for a time interval of length h (Figure 6).

■ **EXAMPLE 3** Use Euler's Method with time step $h = 0.2$ and $n = 4$ steps to approximate the solution of $\dot{y} = y - t^2$, $y(0) = 3$.

Solution Our initial value at $t_0 = 0$ is $y_0 = 3$. Since $h = 0.2$, the time values are $t_1 = 0.2$, $t_2 = 0.4$, $t_3 = 0.6$, and $t_4 = 0.8$. We use Eq. (3) with $F(t, y) = y - t^2$ to calculate

$$y_1 = y_0 + hF(t_0, y_0) = 3 + 0.2(3 - (0)^2) = 3.6$$

$$y_2 = y_1 + hF(t_1, y_1) = 3.6 + 0.2(3.6 - (0.2)^2) \approx 4.3$$

$$y_3 = y_2 + hF(t_2, y_2) = 4.3 + 0.2(4.3 - (0.4)^2) \approx 5.14$$

$$y_4 = y_3 + hF(t_3, y_3) = 5.14 + 0.2(5.14 - (0.6)^2) \approx 6.1$$

Figure 7(A) shows the exact solution $y(t) = 2 + 2t + t^2 + e^t$ together with a plot of the points (t_k, y_k) for $k = 0, 1, 2, 3, 4$ connected by line segments. ■

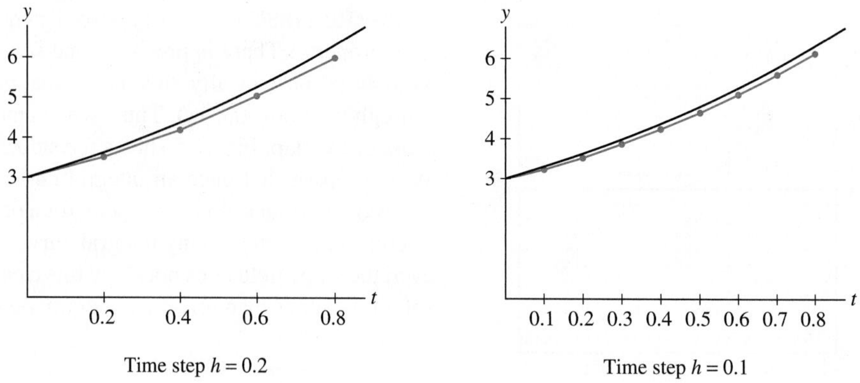

FIGURE 7 Euler's Method applied to $\dot{y} = y - t^2$, $y(0) = 3$.

CONCEPTUAL INSIGHT Figure 7(B) shows that the time step $h = 0.1$ gives a better approximation than $h = 0.2$. In general, the smaller the time step, the better the approximation. In fact, if we start at a point $(a, y(a))$ and use Euler's Method to approximate $(b, y(b))$ using N steps with $h = (b - a)/N$, then the error is roughly proportional to $1/N$ (provided that $F(t, y)$ is a well-behaved function). This is similar to the error size in the Nth left- and right-endpoint approximations to an integral. What this means, however, is that Euler's Method is quite inefficient; to cut the error in half, it is necessary to double the number of steps, and to achieve n-digit accuracy requires roughly 10^n steps. Fortunately, there are several methods that improve on Euler's Method in much the same way as the Midpoint Rule and Simpson's Rule improve on the endpoint approximations (see Exercises 22–27).

■ **EXAMPLE 4** CAS Let $y(t)$ be the solution of $\dot{y} = \sin t \cos y$, $y(0) = 0$.

(a) Use Euler's Method with time step $h = 0.1$ to approximate $y(0.5)$.

(b) Use a computer algebra system to implement Euler's Method with time steps $h = 0.01, 0.001$, and 0.0001 to approximate $y(0.5)$.

Solution

Euler's Method:

$$y_k = y_{k-1} + hF(t_{k-1}, y_{k-1})$$

(a) When $h = 0.1$, y_k is an approximation to $y(0 + k(0.1)) = y(0.1k)$, so y_5 is an approximation to $y(0.5)$. It is convenient to organize calculations in the following table. Note that the value y_{k+1} computed in the last column of each line is used in the next line to continue the process.

t_k	y_k	$F(t_k, y_k) = \sin t_k \cos y_k$	$y_{k+1} = y_k + hF(t_k, y_k)$
$t_0 = 0$	$y_0 = 0$	$(\sin 0)\cos 0 = 0$	$y_1 = 0 + 0.1(0) = 0$
$t_1 = 0.1$	$y_1 = 0$	$(\sin 0.1)\cos 0 \approx 0.1$	$y_2 \approx 0 + 0.1(0.1) = 0.01$
$t_2 = 0.2$	$y_2 \approx 0.01$	$(\sin 0.2)\cos(0.01) \approx 0.2$	$y_3 \approx 0.01 + 0.1(0.2) = 0.03$
$t_3 = 0.3$	$y_3 \approx 0.03$	$(\sin 0.3)\cos(0.03) \approx 0.3$	$y_4 \approx 0.03 + 0.1(0.3) = 0.06$
$t_4 = 0.4$	$y_4 \approx 0.06$	$(\sin 0.4)\cos(0.06) \approx 0.4$	$y_5 \approx 0.06 + 0.1(0.4) = 0.10$

A typical CAS command to implement Euler's Method with time step $h = 0.01$ reads as follows:

```
>> For[n = 0; y = 0, n < 50, n++,
>> y = y + (.01) * (Sin[.01 * n] * Cos[y])]
>> y
>> 0.119746
```

The command `For[...]` *updates the variable y successively through the values $y_1, y_2, \ldots, y_{50}$ according to Euler's Method.*

Thus, Euler's Method yields the approximation $y(0.5) \approx y_5 \approx 0.1$.

(b) When the number of steps is large, the calculations are too lengthy to do by hand, but they are easily carried out using a CAS. Note that for $h = 0.01$, the kth value y_k is an approximation to $y(0 + k(0.01)) = y(0.01k)$, and y_{50} gives an approximation to $y(0.5)$. Similarly, when $h = 0.001$, y_{500} is an approximation to $y(0.5)$, and when $h = 0.0001$, $y_{5,000}$ is an approximation to $y(0.5)$. Here are the results obtained using a CAS:

Time step $h = 0.01$	$y_{50} \approx 0.1197$
Time step $h = 0.001$	$y_{500} \approx 0.1219$
Time step $h = 0.0001$	$y_{5000} \approx 0.1221$

The values appear to converge and we may assume that $y(0.5) \approx 0.12$. However, we see here that Euler's Method converges quite slowly. ■

10.2 SUMMARY

- The *slope field* for a first-order differential equation $\dot{y} = F(t, y)$ is obtained by drawing small segments of slope $F(t, y)$ at points (t, y) lying on a rectangular grid in the plane.
- The graph of a solution (also called an *integral curve*) satisfying $y(t_0) = y_0$ is a curve through (t_0, y_0) that runs tangent to the segments of the slope field at each point.
- *Euler's Method*: to approximate a solution to $\dot{y} = F(t, y)$ with initial condition $y(t_0) = y_0$, fix a time step h and set $t_k = t_0 + kh$. Define $y_1, y_2, \ldots$ successively by the formula

$$y_k = y_{k-1} + hF(t_{k-1}, y_{k-1}) \qquad \boxed{4}$$

The values $y_0, y_1, y_2, \ldots$ are approximations to the values $y(t_0), y(t_1), y(t_2), \ldots$.

10.2 EXERCISES

Preliminary Questions

1. What is the slope of the segment in the slope field for $\dot{y} = ty + 1$ at the point $(2, 3)$?

2. What is the equation of the isocline of slope $c = 1$ for $\dot{y} = y^2 - t$?

3. For which of the following differential equations are the slopes at points on a vertical line $t = C$ all equal?

(a) $\dot{y} = \ln y$ **(b)** $\dot{y} = \ln t$

4. Let $y(t)$ be the solution to $\dot{y} = F(t, y)$ with $y(1) = 3$. How many iterations of Euler's Method are required to approximate $y(3)$ if the time step is $h = 0.1$?

Exercises

1. Figure 8 shows the slope field for $\dot{y} = \sin y \sin t$. Sketch the graphs of the solutions with initial conditions $y(0) = 1$ and $y(0) = -1$. Show that $y(t) = 0$ is a solution and add its graph to the plot.

2. Figure 9 shows the slope field for $\dot{y} = y^2 - t^2$. Sketch the integral curve passing through the point $(0, -1)$, the curve through $(0, 0)$, and the curve through $(0, 2)$. Is $y(t) = 0$ a solution?

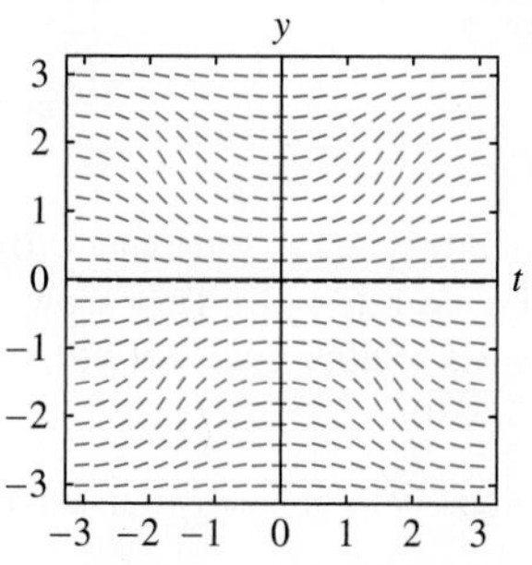

FIGURE 8 Slope field for $\dot{y} = \sin y \sin t$.

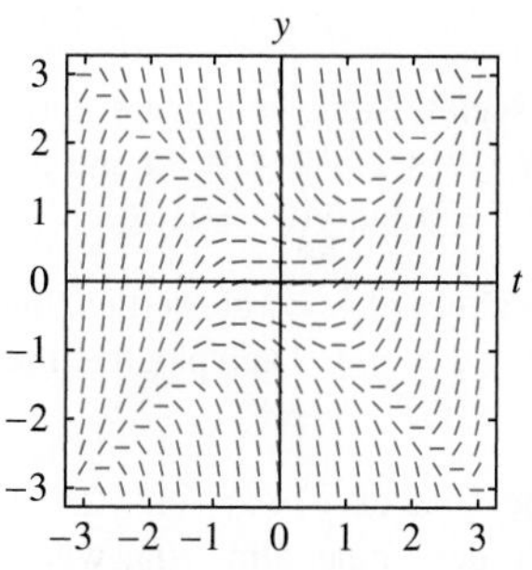

FIGURE 9 Slope field for $\dot{y} = y^2 - t^2$.

3. Show that $f(t) = \frac{1}{2}\left(t - \frac{1}{2}\right)$ is a solution to $\dot{y} = t - 2y$. Sketch the four solutions with $y(0) = \pm 0.5, \pm 1$ on the slope field in Figure 10. The slope field suggests that every solution approaches $f(t)$ as $t \to \infty$. Confirm this by showing that $y = f(t) + Ce^{-2t}$ is the general solution.

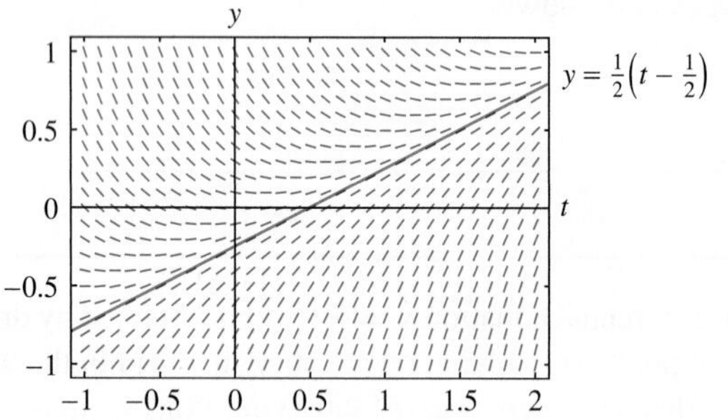

FIGURE 10 Slope field for $\dot{y} = t - 2y$.

4. One of the slope fields in Figures 11(A) and (B) is the slope field for $\dot{y} = t^2$. The other is for $\dot{y} = y^2$. Identify which is which. In each case, sketch the solutions with initial conditions $y(0) = 1$, $y(0) = 0$, and $y(0) = -1$.

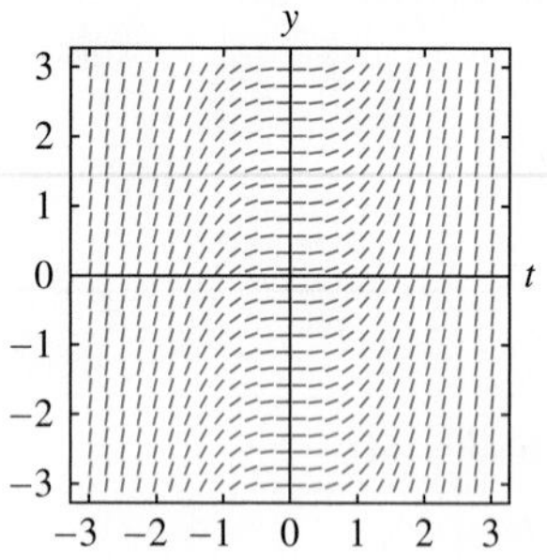

FIGURE 11(A)

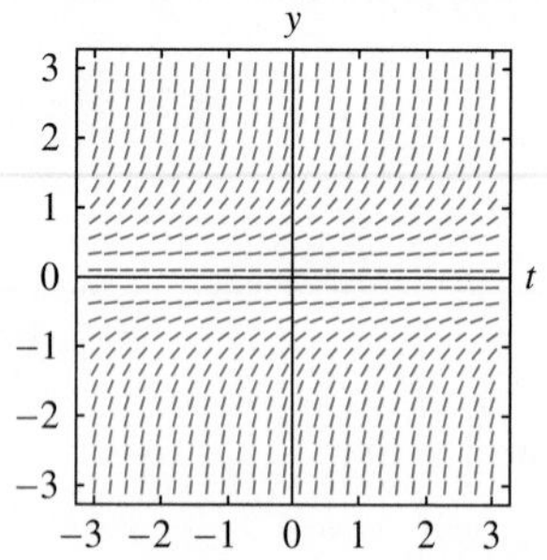

FIGURE 11(B)

5. Consider the differential equation $\dot{y} = t - y$.

(a) Sketch the slope field of the differential equation $\dot{y} = t - y$ in the range $-1 \le t \le 3$, $-1 \le y \le 3$. As an aid, observe that the isocline of slope c is the line $t - y = c$, so the segments have slope c at points on the line $y = t - c$.

(b) Show that $y = t - 1 + Ce^{-t}$ is a solution for all C. Since $\lim_{t\to\infty} e^{-t} = 0$, these solutions approach the particular solution $y = t - 1$ as $t \to \infty$. Explain how this behavior is reflected in your slope field.

6. Show that the isoclines of $\dot{y} = 1/y$ are horizontal lines. Sketch the slope field for $-2 \le t \le 2$, $-2 \le y \le 2$ and plot the solutions with initial conditions $y(0) = 0$ and $y(0) = 1$.

7. Show that the isoclines of $\dot{y} = t$ are vertical lines. Sketch the slope field for $-2 \le t \le 2$, $-2 \le y \le 2$ and plot the integral curves passing through $(0, -1)$ and $(0, 1)$.

8. Sketch the slope field of $\dot{y} = ty$ for $-2 \le t \le 2$, $-2 \le y \le 2$. Based on the sketch, determine $\lim_{t\to\infty} y(t)$, where $y(t)$ is a solution with $y(0) > 0$. What is $\lim_{t\to\infty} y(t)$ if $y(0) < 0$?

9. Match each differential equation with its slope field in Figures 12(A)–(F).

(i) $\dot{y} = -1$ **(ii)** $\dot{y} = \dfrac{y}{t}$ **(iii)** $\dot{y} = t^2 y$

(iv) $\dot{y} = ty^2$ **(v)** $\dot{y} = t^2 + y^2$ **(vi)** $\dot{y} = t$

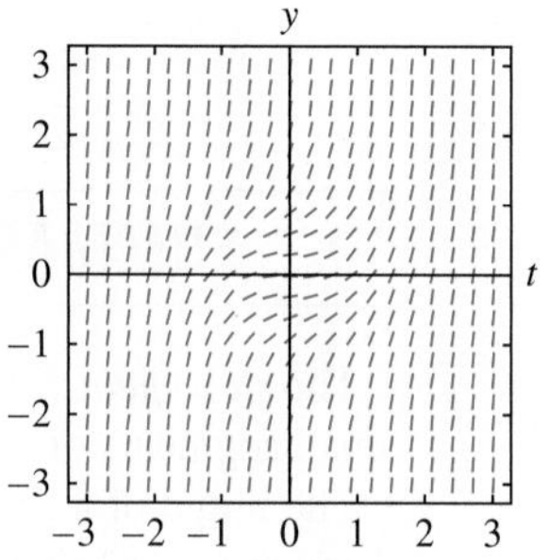

FIGURE 12(A)

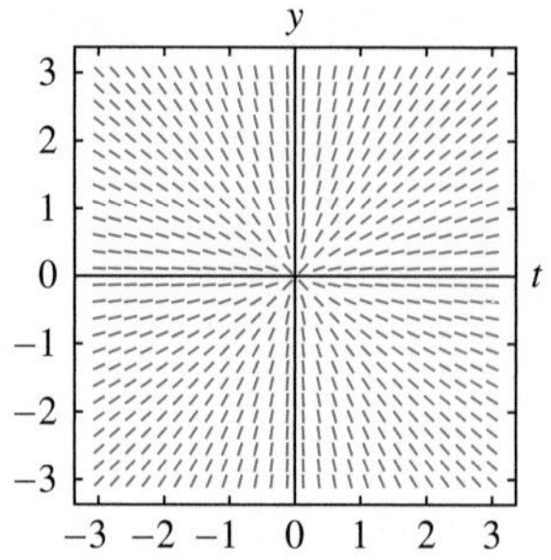

FIGURE 12(B)

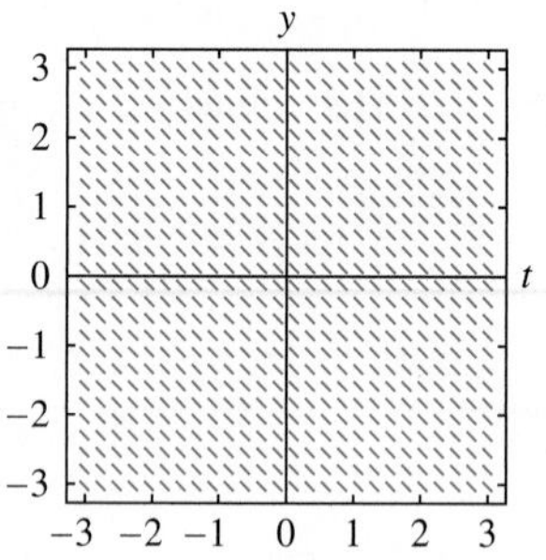

FIGURE 12(C)

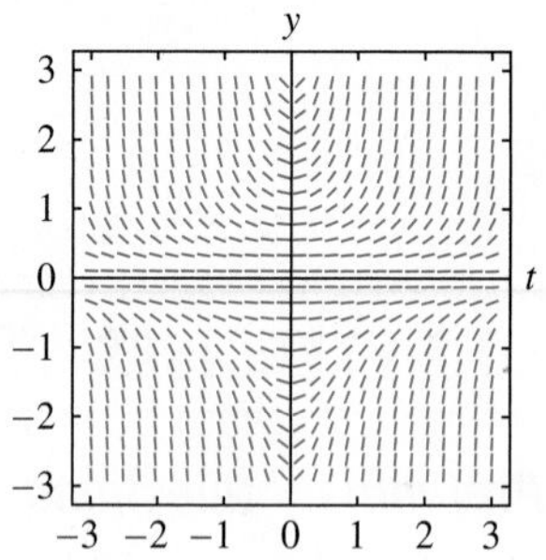

FIGURE 12(D)

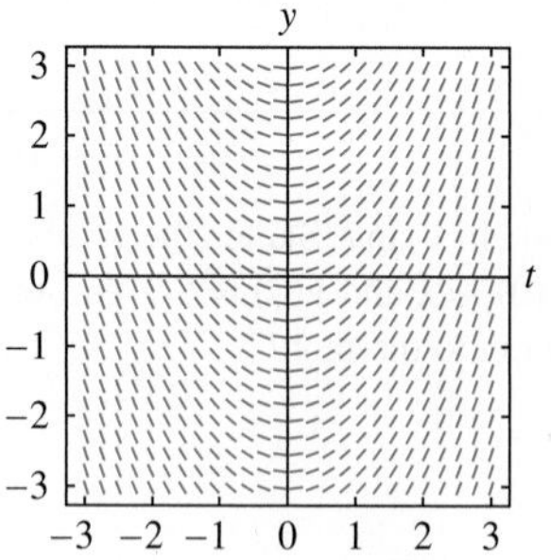

FIGURE 12(E)

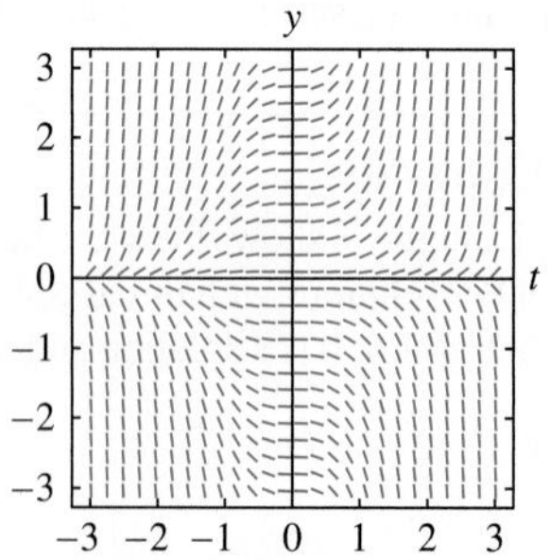

FIGURE 12(F)

10. Sketch the solution of $\dot{y} = ty^2$ satisfying $y(0) = 1$ in the appropriate slope field of Figure 12(A)–(F). Then show, using separation of variables, that if $y(t)$ is a solution such that $y(0) > 0$, then $y(t)$ tends to infinity as $t \to \sqrt{2/y(0)}$.

11. **(a)** Sketch the slope field of $\dot{y} = t/y$ in the region $-2 \le t \le 2$, $-2 \le y \le 2$.

(b) Check that $y = \pm\sqrt{t^2 + C}$ is the general solution.

(c) Sketch the solutions on the slope field with initial conditions $y(0) = 1$ and $y(0) = -1$.

12. Sketch the slope field of $\dot{y} = t^2 - y$ in the region $-3 \le t \le 3$, $-3 \le y \le 3$ and sketch the solutions satisfying $y(1) = 0$, $y(1) = 1$, and $y(1) = -1$.

13. Let $F(t, y) = t^2 - y$ and let $y(t)$ be the solution of $\dot{y} = F(t, y)$ satisfying $y(2) = 3$. Let $h = 0.1$ be the time step in Euler's Method, and set $y_0 = y(2) = 3$.

(a) Calculate $y_1 = y_0 + hF(2, 3)$.

(b) Calculate $y_2 = y_1 + hF(2.1, y_1)$.

(c) Calculate $y_3 = y_2 + hF(2.2, y_2)$ and continue computing y_4, y_5, and y_6.

(d) Find approximations to $y(2.2)$ and $y(2.5)$.

14. Let $y(t)$ be the solution to $\dot{y} = te^{-y}$ satisfying $y(0) = 0$.

(a) Use Euler's Method with time step $h = 0.1$ to approximate $y(0.1), y(0.2), \dots, y(0.5)$.

(b) Use separation of variables to find $y(t)$ exactly.

(c) Compute the errors in the approximations to $y(0.1)$ and $y(0.5)$.

In Exercises 15–20, use Euler's Method to approximate the given value of $y(t)$ with the time step h indicated.

15. $y(0.5)$; $\dot{y} = y + t$, $y(0) = 1$, $h = 0.1$

16. $y(0.7)$; $\dot{y} = 2y$, $y(0) = 3$, $h = 0.1$

17. $y(3.3)$; $\dot{y} = t^2 - y$, $y(3) = 1$, $h = 0.05$

18. $y(3)$; $\dot{y} = \sqrt{t + y}$, $y(2.7) = 5$, $h = 0.05$

19. $y(2)$; $\dot{y} = t \sin y$, $y(1) = 2$, $h = 0.2$

20. $y(5.2)$; $\dot{y} = t - \sec y$, $y(4) = -2$, $h = 0.2$

Further Insights and Challenges

21. If $f(t)$ is continuous on $[a, b]$, then the solution to $\dot{y} = f(t)$ with initial condition $y(a) = 0$ is $y(t) = \int_a^t f(u)\,du$. Show that Euler's Method with time step $h = (b - a)/N$ for N steps yields the Nth left-endpoint approximation to $y(b) = \int_a^b f(u)\,du$.

*Exercises 22–27: **Euler's Midpoint Method** is a variation on Euler's Method that is significantly more accurate in general. For time step h and initial value $y_0 = y(t_0)$, the values y_k are defined successively by*

$$y_k = y_{k-1} + hm_{k-1}$$

where $m_{k-1} = F\left(t_{k-1} + \frac{h}{2}, y_{k-1} + \frac{h}{2}F(t_{k-1}, y_{k-1})\right)$.

22. Apply both Euler's Method and the Euler Midpoint Method with $h = 0.1$ to estimate $y(1.5)$, where $y(t)$ satisfies $\dot{y} = y$ with $y(0) = 1$. Find $y(t)$ exactly and compute the errors in these two approximations.

In Exercises 23–26, use Euler's Midpoint Method with the time step indicated to approximate the given value of $y(t)$.

23. $y(0.5)$; $\dot{y} = y + t$, $y(0) = 1$, $h = 0.1$

24. $y(2)$; $\dot{y} = t^2 - y$, $y(1) = 3$, $h = 0.2$

25. $y(0.25)$; $\dot{y} = \cos(y + t)$, $y(0) = 1$, $h = 0.05$

26. $y(2.3)$; $\dot{y} = y + t^2$, $y(2) = 1$, $h = 0.05$

27. Assume that $f(t)$ is continuous on $[a, b]$. Show that Euler's Midpoint Method applied to $\dot{y} = f(t)$ with initial condition $y(a) = 0$ and time step $h = (b - a)/N$ for N steps yields the Nth midpoint approximation to

$$y(b) = \int_a^b f(u)\,du$$

10.3 The Logistic Equation

The logistic equation was first introduced in 1838 by the Belgian mathematician Pierre-François Verhulst (1804–1849). Based on the population of Belgium for three years (1815, 1830, and 1845), which was then between 4 and 4.5 million, Verhulst predicted that the population would never exceed 9.4 million. This prediction has held up reasonably well. Belgium's current population is around 10.4 million.

The simplest model of population growth is $dy/dt = ky$, according to which populations grow exponentially. This may be true over short periods of time, but it is clear that no population can increase without limit. Therefore, population biologists use a variety of other differential equations that take into account environmental limitations to growth such as food scarcity and competition between species. One widely used model is based on the **logistic differential equation**:

$$\frac{dy}{dt} = ky\left(1 - \frac{y}{A}\right) \qquad \boxed{1}$$

Here $k > 0$ is the growth constant, and $A > 0$ is a constant called the **carrying capacity**. Figure 1 shows a typical S-shaped solution of Eq. (1). As in the previous section, we also denote dy/dt by $\dot{y}$.

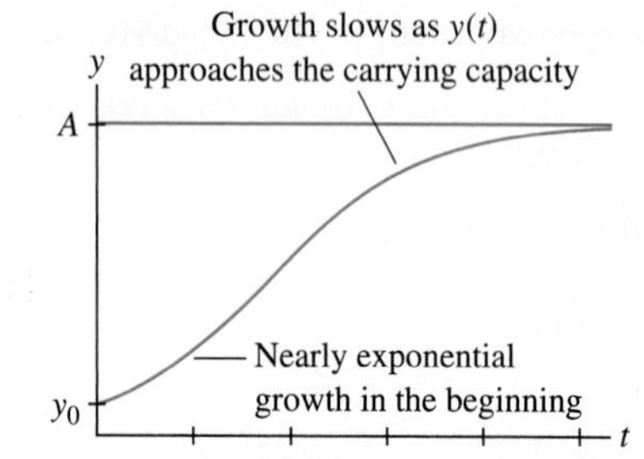

FIGURE 1 Solution of the logistic equation.

CONCEPTUAL INSIGHT The logistic equation $\dot{y} = ky(1 - y/A)$ differs from the exponential differential equation $\dot{y} = ky$ only by the additional factor $(1 - y/A)$. As long as y is small relative to A, this factor is close to 1 and can be ignored, yielding $\dot{y} \approx ky$. Thus, $y(t)$ grows nearly exponentially when the population is small (Figure 1). As $y(t)$ approaches A, the factor $(1 - y/A)$ tends to zero. This causes $\dot{y}$ to decrease and prevents $y(t)$ from exceeding the carrying capacity A.

The slope field in Figure 2 shows clearly that there are three families of solutions, depending on the initial value $y_0 = y(0)$.

- If $y_0 > A$, then $y(t)$ is decreasing and approaches A as $t \to \infty$.
- If $0 < y_0 < A$, then $y(t)$ is increasing and approaches A as $t \to \infty$.
- If $y_0 < 0$, then $y(t)$ is decreasing and $\lim_{t \to t_b-} y(t) = -\infty$ for some time t_b.

Solutions of the logistic equation with $y_0 < 0$ are not relevant to populations because a population cannot be negative (see Exercise 18).

Equation (1) also has two constant solutions: $y = 0$ and $y = A$. They correspond to the roots of $ky(1 - y/A) = 0$, and they satisfy Eq. (1) because $\dot{y} = 0$ when y is a constant. Constant solutions are called **equilibrium** or **steady-state** solutions. The equilibrium solution $y = A$ is a **stable equilibrium** because every solution with initial value y_0 close to A approaches the equilibrium $y = A$ as $t \to \infty$. By contrast, $y = 0$ is an **unstable equilibrium** because every nonequilibrium solution with initial value y_0 near $y = 0$ either increases to A or decreases to $-\infty$.

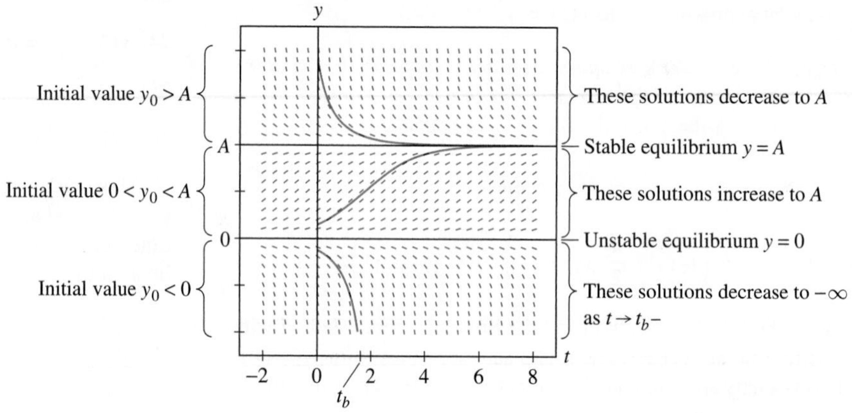

FIGURE 2 Slope field for $\frac{dy}{dt} = ky\left(1 - \frac{y}{A}\right)$.

Having described the solutions qualitatively, let us now find the nonequilibrium solutions explicitly using separation of variables. Assuming that $y \neq 0$ and $y \neq A$, we have

$$\frac{dy}{dt} = ky\left(1 - \frac{y}{A}\right)$$

$$\frac{dy}{y(1 - y/A)} = k\,dt$$

$$\int \left(\frac{1}{y} - \frac{1}{y - A}\right) dy = \int k\,dt \qquad \boxed{2}$$

$$\ln|y| - \ln|y - A| = kt + C$$

$$\left|\frac{y}{y - A}\right| = e^{kt+C} \quad \Rightarrow \quad \frac{y}{y - A} = \pm e^C e^{kt}$$

In Eq. (2), we use the the partial fraction decomposition

$$\frac{1}{y(1 - y/A)} = \frac{1}{y} - \frac{1}{y - A}$$

Since $\pm e^C$ takes on arbitrary nonzero values, we replace $\pm e^C$ with C (nonzero):

$$\frac{y}{y - A} = Ce^{kt} \qquad \boxed{3}$$

For $t = 0$, this gives a useful relation between C and the initial value $y_0 = y(0)$:

$$\frac{y_0}{y_0 - A} = C \qquad \boxed{4}$$

To solve for y, multiply each side of Eq. (3) by $(y - A)$:

$$y = (y - A)Ce^{kt}$$
$$y(1 - Ce^{kt}) = -ACe^{kt}$$
$$y = \frac{ACe^{kt}}{Ce^{kt} - 1}$$

As $C \neq 0$, we may divide by Ce^{kt} to obtain the general nonequilibrium solution:

$$\frac{dy}{dt} = ky\left(1 - \frac{y}{A}\right), \qquad y = \frac{A}{1 - e^{-kt}/C} \qquad \boxed{5}$$

■ **EXAMPLE 1** Solve $\dot{y} = 0.3y(4 - y)$ with initial condition $y(0) = 1$.

Solution Recall that $\dot{y} = \dfrac{dy}{dt}$. To apply Eq. (5), we must rewrite the equation in the form

$$\dot{y} = 1.2y\left(1 - \frac{y}{4}\right)$$

Thus, $k = 1.2$ and $A = 4$, and the general solution is

$$y = \frac{4}{1 - e^{-1.2t}/C}$$

There are two ways to find C. One way is to solve $y(0) = 1$ for C directly. An easier way is to use Eq. (4):

$$C = \frac{y_0}{y_0 - A} = \frac{1}{1 - 4} = -\frac{1}{3}$$

We find that the particular solution is $y = \dfrac{4}{1 + 3e^{-1.2t}}$ (Figure 3). ■

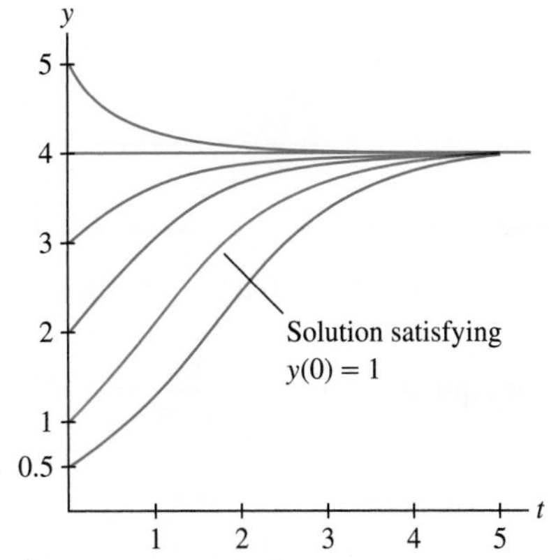

FIGURE 3 Several solutions of $\dot{y} = 0.3y(4 - y)$.

■ **EXAMPLE 2** **Deer Population** A deer population (Figure 4) grows logistically with growth constant $k = 0.4$ year^{-1} in a forest with a carrying capacity of 1000 deer.

(a) Find the deer population $P(t)$ if the initial population is $P_0 = 100$.

(b) How long does it take for the deer population to reach 500?

FIGURE 4

Solution The time unit is the year because the unit of k is year^{-1}.

(a) Since $k = 0.4$ and $A = 1000$, $P(t)$ satisfies the differential equation

$$\frac{dP}{dt} = 0.4P\left(1 - \frac{P}{1000}\right)$$

The logistic equation may be too simple to describe a real deer population accurately, but it serves as a starting point for more sophisticated models used by ecologists, population biologists, and forestry professionals.

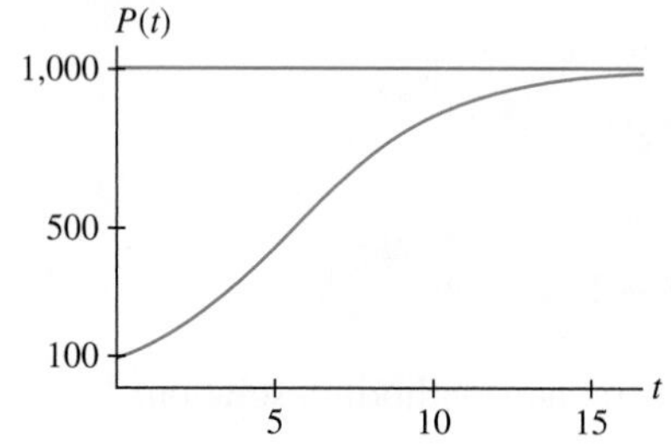

FIGURE 5 Deer population as a function of t (in years).

The general solution is given by Eq. (5):

$$P(t) = \frac{1000}{1 - e^{-0.4t}/C}$$ 6

Using Eq. (4) to compute C, we find (Figure 5)

$$C = \frac{P_0}{P_0 - A} = \frac{100}{100 - 1000} = -\frac{1}{9} \quad \Rightarrow \quad P(t) = \frac{1000}{1 + 9e^{-0.4t}}$$

(b) To find the time t when $P(t) = 500$, we could solve the equation

$$P(t) = \frac{1000}{1 + 9e^{-0.4t}} = 500$$

But it is easier to use Eq. (3):

$$\frac{P}{P - A} = Ce^{kt}$$

$$\frac{P}{P - 1000} = -\frac{1}{9}e^{0.4t}$$

Set $P = 500$ and solve for t:

$$-\frac{1}{9}e^{0.4t} = \frac{500}{500 - 1000} = -1 \quad \Rightarrow \quad e^{0.4t} = 9 \quad \Rightarrow \quad 0.4t = \ln 9$$

This gives $t = (\ln 9)/0.4 \approx 5.5$ years. ■

10.3 SUMMARY

- The *logistic equation* and its general nonequilibrium solution ($k > 0$ and $A > 0$):

$$\frac{dy}{dt} = ky\left(1 - \frac{y}{A}\right), \qquad y = \frac{A}{1 - e^{-kt}/C}, \qquad \text{or equivalently} \qquad \frac{y}{y - A} = Ce^{kt}$$

- Two equilibrium (constant) solutions:
  - $y = 0$ is an unstable equilibrium.
  - $y = A$ is a stable equilibrium.
- If the initial value $y_0 = y(0)$ satisfies $y_0 > 0$, then $y(t)$ approaches the stable equilibrium $y = A$; that is, $\lim_{t\to\infty} y(t) = A$.

10.3 EXERCISES

Preliminary Questions

1. Which of the following differential equations is a logistic differential equation?

(a) $\dot{y} = 2y(1 - y^2)$

(b) $\dot{y} = 2y\left(1 - \frac{y}{3}\right)$

(c) $\dot{y} = 2y\left(1 - \frac{t}{4}\right)$

(d) $\dot{y} = 2y(1 - 3y)$

2. Is the logistic equation a linear differential equation?

3. Is the logistic equation separable?

Exercises

1. Find the general solution of the logistic equation

$$\dot{y} = 3y\left(1 - \frac{y}{5}\right)$$

Then find the particular solution satisfying $y(0) = 2$.

2. Find the solution of $\dot{y} = 2y(3 - y)$, $y(0) = 10$.

3. Let $y(t)$ be a solution of $\dot{y} = 0.5y(1 - 0.5y)$ such that $y(0) = 4$. Determine $\lim_{t\to\infty} y(t)$ without finding $y(t)$ explicitly.

4. Let $y(t)$ be a solution of $\dot{y} = 5y(1 - y/5)$. State whether $y(t)$ is increasing, decreasing, or constant in the following cases:

(a) $y(0) = 2$ **(b)** $y(0) = 5$ **(c)** $y(0) = 8$

5. A population of squirrels lives in a forest with a carrying capacity of 2000. Assume logistic growth with growth constant $k = 0.6\ \text{yr}^{-1}$.

(a) Find a formula for the squirrel population $P(t)$, assuming an initial population of 500 squirrels.

(b) How long will it take for the squirrel population to double?

6. The population $P(t)$ of mosquito larvae growing in a tree hole increases according to the logistic equation with growth constant $k = 0.3\ \text{day}^{-1}$ and carrying capacity $A = 500$.

(a) Find a formula for the larvae population $P(t)$, assuming an initial population of $P_0 = 50$ larvae.

(b) After how many days will the larvae population reach 200?

7. Sunset Lake is stocked with 2000 rainbow trout, and after 1 year the population has grown to 4500. Assuming logistic growth with a carrying capacity of 20,000, find the growth constant k (specify the units) and determine when the population will increase to 10,000.

8. Spread of a Rumor A rumor spreads through a small town. Let $y(t)$ be the fraction of the population that has heard the rumor at time t and assume that the rate at which the rumor spreads is proportional to the product of the fraction y of the population that has heard the rumor and the fraction $1 - y$ that has not yet heard the rumor.

(a) Write down the differential equation satisfied by y in terms of a proportionality factor k.

(b) Find k (in units of day^{-1}), assuming that 10% of the population knows the rumor at $t = 0$ and 40% knows it at $t = 2$ days.

(c) Using the assumptions of part (b), determine when 75% of the population will know the rumor.

9. A rumor spreads through a school with 1000 students. At 8 AM, 80 students have heard the rumor, and by noon, half the school has heard it. Using the logistic model of Exercise 8, determine when 90% of the students will have heard the rumor.

10. GU A simpler model for the spread of a rumor assumes that the rate at which the rumor spreads is proportional (with factor k) to the fraction of the population that has not yet heard the rumor.

(a) Compute the solutions to this model and the model of Exercise 8 with the values $k = 0.9$ and $y_0 = 0.1$.

(b) Graph the two solutions on the same axis.

(c) Which model seems more realistic? Why?

11. Let $k = 1$ and $A = 1$ in the logistic equation.

(a) Find the solutions satisfying $y_1(0) = 10$ and $y_2(0) = -1$.

(b) Find the time t when $y_1(t) = 5$.

(c) When does $y_2(t)$ become infinite?

12. A tissue culture grows until it has a maximum area of $M\ \text{cm}^2$. The area $A(t)$ of the culture at time t may be modeled by the differential equation

$$\dot{A} = k\sqrt{A}\left(1 - \frac{A}{M}\right) \tag{7}$$

where k is a growth constant.

(a) Show that if we set $A = u^2$, then

$$\dot{u} = \frac{1}{2}k\left(1 - \frac{u^2}{M}\right)$$

Then find the general solution using separation of variables.

(b) Show that the general solution to Eq. (7) is

$$A(t) = M\left(\frac{Ce^{(k/\sqrt{M})t} - 1}{Ce^{(k/\sqrt{M})t} + 1}\right)^2$$

13. GU In the model of Exercise 12, let $A(t)$ be the area at time t (hours) of a growing tissue culture with initial size $A(0) = 1\ \text{cm}^2$, assuming that the maximum area is $M = 16\ \text{cm}^2$ and the growth constant is $k = 0.1$.

(a) Find a formula for $A(t)$. *Note:* The initial condition is satisfied for two values of the constant C. Choose the value of C for which $A(t)$ is increasing.

(b) Determine the area of the culture at $t = 10$ hours.

(c) GU Graph the solution using a graphing utility.

14. Show that if a tissue culture grows according to Eq. (7), then the growth rate reaches a maximum when $A = M/3$.

15. In 1751, Benjamin Franklin predicted that the U.S. population $P(t)$ would increase with growth constant $k = 0.028\ \text{year}^{-1}$. According to the census, the U.S. population was 5 million in 1800 and 76 million in 1900. Assuming logistic growth with $k = 0.028$, find the predicted carrying capacity for the U.S. population. *Hint:* Use Eqs. (3) and (4) to show that

$$\frac{P(t)}{P(t) - A} = \frac{P_0}{P_0 - A}e^{kt}$$

16. **Reverse Logistic Equation** Consider the following logistic equation (with $k, B > 0$):

$$\frac{dP}{dt} = -kP\left(1 - \frac{P}{B}\right) \tag{8}$$

(a) Sketch the slope field of this equation.

(b) The general solution is $P(t) = B/(1 - e^{kt}/C)$, where C is a nonzero constant. Show that $P(0) > B$ if $C > 1$ and $0 < P(0) < B$ if $C < 0$.

(c) Show that Eq. (8) models an "extinction–explosion" population. That is, $P(t)$ tends to zero if the initial population satisfies $0 < P(0) < B$, and it tends to ∞ after a finite amount of time if $P(0) > B$.

(d) Show that $P = 0$ is a stable equilibrium and $P = B$ an unstable equilibrium.

Further Insights and Challenges

In Exercises 17 and 18, let $y(t)$ be a solution of the logistic equation

$$\frac{dy}{dt} = ky\left(1 - \frac{y}{A}\right) \qquad \boxed{9}$$

where $A > 0$ and $k > 0$.

17. (a) Differentiate Eq. (9) with respect to t and use the Chain Rule to show that

$$\frac{d^2y}{dt^2} = k^2y\left(1 - \frac{y}{A}\right)\left(1 - \frac{2y}{A}\right)$$

(b) Show that $y(t)$ is concave up if $0 < y < A/2$ and concave down if $A/2 < y < A$.

(c) Show that if $0 < y(0) < A/2$, then $y(t)$ has a point of inflection at $y = A/2$ (Figure 6).

(d) Assume that $0 < y(0) < A/2$. Find the time t when $y(t)$ reaches the inflection point.

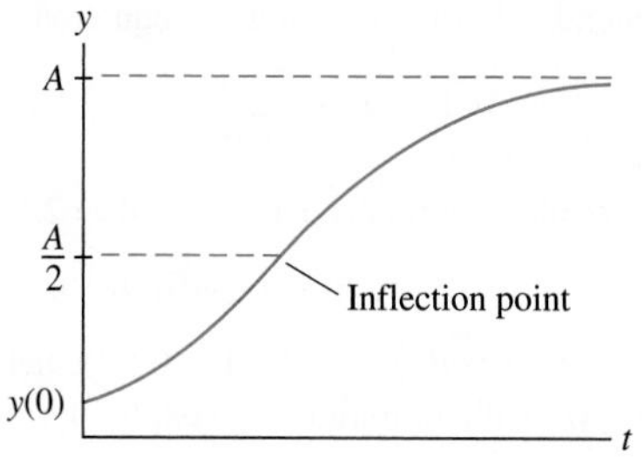

FIGURE 6 An inflection point occurs at $y = A/2$ in the logistic curve.

18. Let $y = \dfrac{A}{1 - e^{-kt}/C}$ be the general nonequilibrium Eq. (9). If $y(t)$ has a vertical asymptote at $t = t_b$, that is, if $\lim_{t \to t_b -} y(t) = \pm\infty$, we say that the solution "blows up" at $t = t_b$.

(a) Show that if $0 < y(0) < A$, then y does not blow up at any time t_b.

(b) Show that if $y(0) > A$, then y blows up at a time t_b, which is negative (and hence does not correspond to a real time).

(c) Show that y blows up at some positive time t_b if and only if $y(0) < 0$ (and hence does not correspond to a real population).

10.4 First-Order Linear Equations

This section introduces the method of "integrating factors" for solving first-order linear equations. Although we already have a method (separation of variables) for solving separable equations, this new method applies to all linear equations, whether separable or not (Figure 1).

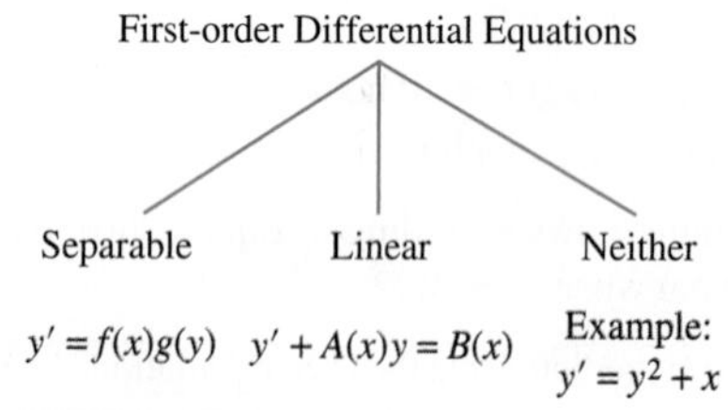

FIGURE 1

A first-order linear equation has the form $a(x)y' + b(x)y = c(x)$, where $a(x)$ is not the zero function. We divide by $a(x)$ and write the equation in the standard form

$$\boxed{y' + A(x)y = B(x)} \qquad \boxed{1}$$

Note that in this section, x is used as an independent variable (but t is used in Example 3 below). To solve Eq. (1), we shall multiply through by a function $\alpha(x)$, called an **integrating factor**, that turns the left-hand side into the derivative of $\alpha(x)y$:

$$\alpha(x)\big(y' + A(x)y\big) = \big(\alpha(x)y\big)' \qquad \boxed{2}$$

Suppose we can find a function $\alpha(x)$ satisfying Eq. (2). Then Eq. (1) yields

$$\alpha(x)\big(y' + A(x)y\big) = \alpha(x)B(x)$$

$$\big(\alpha(x)y\big)' = \alpha(x)B(x)$$

We can solve this equation by integration:

$$\alpha(x)y = \int \alpha(x)B(x)\,dx + C \qquad \text{or} \qquad y = \frac{1}{\alpha(x)}\left(\int \alpha(x)B(x)\,dx + C\right)$$

To find $\alpha(x)$, expand Eq. (2), using the Product Rule on the right-hand side:

$$\alpha(x)y' + \alpha(x)A(x)y = \alpha(x)y' + \alpha'(x)y \quad \Rightarrow \quad \alpha(x)A(x)y = \alpha'(x)y$$

Dividing by y, we obtain

$$\frac{d\alpha}{dx} = \alpha(x)A(x) \tag{3}$$

We solve this equation using separation of variables:

$$\frac{d\alpha}{\alpha} = A(x)\,dx \quad \Rightarrow \quad \int \frac{d\alpha}{\alpha} = \int A(x)\,dx$$

Therefore, $\ln|\alpha(x)| = \int A(x)\,dx$, and by exponentiation, $\alpha(x) = \pm e^{\int A(x)\,dx}$. Since we need just one solution of Eq. (3), we choose the positive solution.

In the formula for the integrating factor $\alpha(x)$, the integral $\int A(x)\,dx$ denotes any antiderivative of $A(x)$.

THEOREM 1 The general solution of $y' + A(x)y = B(x)$ is

$$y = \frac{1}{\alpha(x)}\left(\int \alpha(x)B(x)\,dx + C\right) \tag{4}$$

where $\alpha(x)$ is an integrating factor:

$$\alpha(x) = e^{\int A(x)\,dx} \tag{5}$$

■ **EXAMPLE 1** Solve $xy' - 3y = x^2$, $y(1) = 2$.

Solution First divide by x to put the equation in the form $y' + A(x)y = B(x)$:

$$y' - \frac{3}{x}\,y = x$$

Thus $A(x) = -3x^{-1}$ and $B(x) = x$.

***Step 1.* Find an integrating factor.**
In our case, $A(x) = -3x^{-1}$, and by Eq. (5),

$$\alpha(x) = e^{\int A(x)\,dx} = e^{\int(-3/x)\,dx} = e^{-3\ln x} = e^{\ln(x^{-3})} = x^{-3}$$

***Step 2.* Find the general solution.**
We have found $\alpha(x)$, so we can use Eq. (4) to write down the general solution:

$$y = \alpha(x)^{-1}\int \alpha(x)B(x)\,dx = x^3\left(\int x^{-3}\cdot x\,dx\right)$$

$$= x^3\left(\int x^{-2}\,dx\right) = x^3\left(-x^{-1} + C\right) \tag{6}$$

$$y = -x^2 + Cx^3$$

CAUTION We have to include the constant of integration C in Eq. (6), but note that in the general solution, C does not appear as an additive constant. The general solution is $y = -x^2 + Cx^3$. It is not correct to write $-x^2 + C$.

***Step 3.* Solve the initial value problem.**
Now solve for C using the initial condition $y(1) = 2$:

$$y(1) = -1^2 + C\cdot 1^3 = 2 \qquad \text{or} \qquad C = 3$$

Therefore, the solution of the initial value problem is $y = -x^2 + 3x^3$.

Finally, let's check that $y = -x^2 + 3x^3$ satisfies our equation $xy' - 3y = x^2$:

$$xy' - 3y = x(-2x + 9x^2) - 3(-x^2 + 3x^3)$$
$$= (-2x^2 + 9x^3) + (3x^2 - 9x^3) = x^2$$

■ **EXAMPLE 2** Solve the initial value problem: $y' + (1 - x^{-1})y = x^2$, $y(1) = 2$.

Summary: The general solution of $y' + A(x)y = B(x)$ *is*

$$y = \alpha(x)^{-1}\left(\int \alpha(x)B(x) + C\right)$$

where

$$\alpha(x) = e^{\int A(x)\,dx}$$

Solution This equation has the form $y' + A(x)y = B(x)$ with $A(x) = (1 - x^{-1})$. By Eq. (5), an integrating factor is

$$\alpha(x) = e^{\int (1-x^{-1})\,dx} = e^{x - \ln x} = e^x e^{\ln x^{-1}} = x^{-1}e^x$$

Using Eq. (4) with $B(x) = x^2$, we obtain the general solution:

$$y = \alpha(x)^{-1}\left(\int \alpha(x)B(x)\,dx + C\right) = xe^{-x}\left(\int (x^{-1}e^x)x^2\,dx + C\right)$$
$$= xe^{-x}\left(\int xe^x\,dx + C\right)$$

Integration by Parts shows that $\int xe^x\,dx = (x-1)e^x + C$, so we obtain

$$y = xe^{-x}\big((x-1)e^x + C\big) = x(x-1) + Cxe^{-x}$$

The initial condition $y(1) = 2$ gives

$$y(1) = 1(1-1) + Ce^{-1} = Ce^{-1} = 2 \quad \Rightarrow \quad C = 2e$$

The desired particular solution is

$$y = x(x-1) + (2e)xe^{-x} = x(x-1) + 2xe^{1-x}$$

CONCEPTUAL INSIGHT We have expressed the general solution of a first-order linear differential equation in terms of the integrals in Eqs. (4) and (5). Keep in mind, however, that it is not always possible to evaluate these integrals explicitly. For example, the general solution of $y' + xy = 1$ is

$$y = e^{-x^2/2}\left(\int e^{x^2/2}\,dx + C\right)$$

The integral $\int e^{x^2/2}\,dx$ cannot be evaluated in elementary terms. However, we can approximate the integral numerically and plot the solutions by computer (Figure 2).

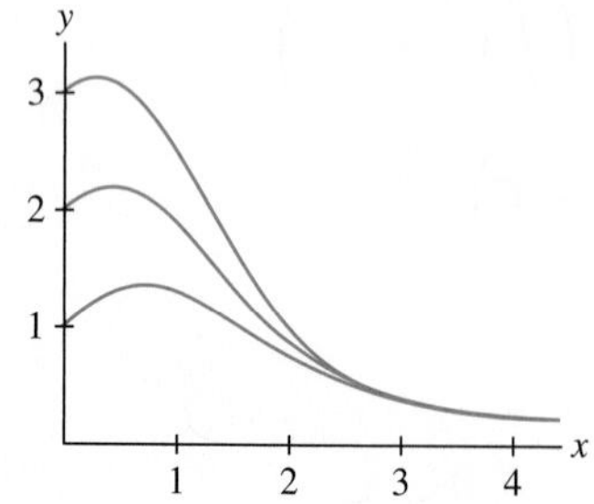

FIGURE 2 Solutions to $y' + xy = 1$ solved numerically and plotted by computer.

In the next example, we use a differential equation to model a "mixing problem," which has applications in biology, chemistry, and medicine.

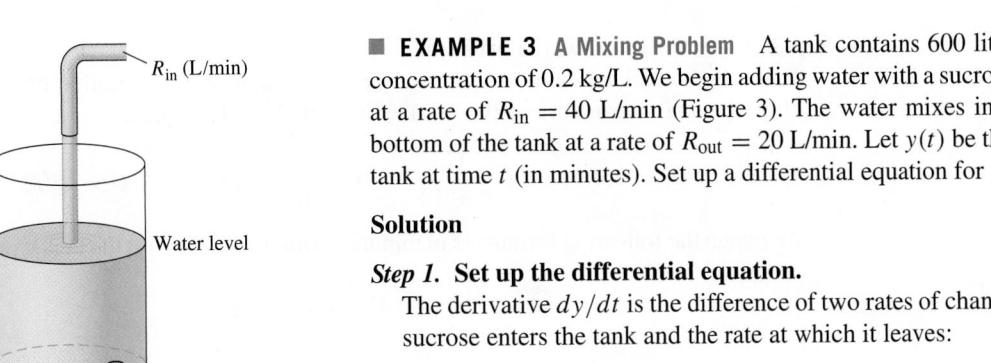

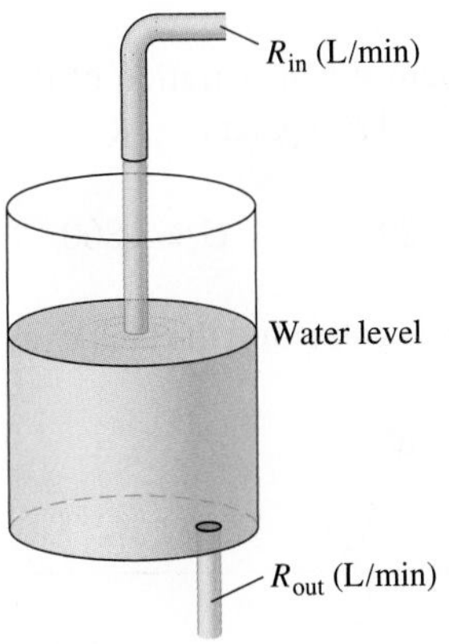

FIGURE 3

■ **EXAMPLE 3 A Mixing Problem** A tank contains 600 liters of water with a sucrose concentration of 0.2 kg/L. We begin adding water with a sucrose concentration of 0.1 kg/L at a rate of $R_{in} = 40$ L/min (Figure 3). The water mixes instantaneously and exits the bottom of the tank at a rate of $R_{out} = 20$ L/min. Let $y(t)$ be the quantity of sucrose in the tank at time t (in minutes). Set up a differential equation for $y(t)$ and solve for $y(t)$.

Solution

***Step 1.* Set up the differential equation.**

The derivative dy/dt is the difference of two rates of change, namely the rate at which sucrose enters the tank and the rate at which it leaves:

$$\frac{dy}{dt} = \text{sucrose rate in} - \text{sucrose rate out} \tag{7}$$

The rate at which sucrose enters the tank is

$$\text{Sucrose rate in} = \underbrace{(0.1 \text{ kg/L})(40 \text{ L/min})}_{\text{Concentration times water rate in}} = 4 \text{ kg/min}$$

Next, we compute the sucrose concentration in the tank at time t. Water flows in at 40 L/min and out at 20 L/min, so there is a net inflow of 20 L/min. The tank has 600 L at time $t = 0$, so it has $600 + 20t$ liters at time t, and

$$\text{Concentration at time } t = \frac{\text{kilograms of sucrose in tank}}{\text{liters of water in tank}} = \frac{y(t)}{600 + 20t} \text{ kg/L}$$

The rate at which sucrose leaves the tank is the product of the concentration and the rate at which water flows out:

$$\text{Sucrose rate out} = \underbrace{\left(\frac{y}{600 + 20t}\,\frac{\text{kg}}{\text{L}}\right)\left(20\,\frac{\text{L}}{\text{min}}\right)}_{\text{Concentration times water rate out}} = \frac{20y}{600 + 20t} = \frac{y}{t + 30} \text{ kg/min}$$

Now Eq. (7) gives us the differential equation

$$\frac{dy}{dt} = 4 - \frac{y}{t + 30} \tag{8}$$

***Step 2.* Find the general solution.**

We write Eq. (8) in standard form:

$$\frac{dy}{dt} + \underbrace{\frac{1}{t + 30}}_{A(t)}\, y = \underbrace{4}_{B(t)} \tag{9}$$

An integrating factor is

$$\alpha(t) = e^{\int A(t)\,dt} = e^{\int dt/(t+30)} = e^{\ln(t+30)} = t + 30$$

The general solution is

$$\begin{aligned} y(t) &= \alpha(t)^{-1}\left(\int \alpha(t)B(t)\,dt + C\right) \\ &= \frac{1}{t + 30}\left(\int (t + 30)(4)\,dt + C\right) \\ &= \frac{1}{t + 30}\left(2(t + 30)^2 + C\right) = 2t + 60 + \frac{C}{t + 30} \end{aligned}$$

Summary:

sucrose rate in = 4 kg/min

$$\text{sucrose rate out} = \frac{y}{t + 30} \text{ kg/min}$$

$$\frac{dy}{dt} = 4 - \frac{y}{t + 30}$$

$$\alpha(t) = t + 30$$

$$y(t) = 2t + 60 + \frac{C}{t + 30}$$

***Step 3.* Solve the initial value problem.**

At $t = 0$, the tank contains 600 L of water with a sucrose concentration of 0.2 kg/L. Thus, the total sucrose at $t = 0$ is $y(0) = (600)(0.2) = 120$ kg, and

$$y(0) = 2(0) + 60 + \frac{C}{0+30} = 60 + \frac{C}{30} = 120 \quad \Rightarrow \quad C = 1800$$

We obtain the following formula (t in minutes), which is valid until the tank overflows:

$$y(t) = 2t + 60 + \frac{1800}{t+30} \text{ kg sucrose}$$

■

10.4 SUMMARY

- A *first-order linear differential equation* can always be written in the form

$$y' + A(x)y = B(x)$$

- The general solution is

$$y = \alpha(x)^{-1}\left(\int \alpha(x)B(x)\,dx + C\right)$$

where $\alpha(x)$ is an *integrating factor*: $\alpha(x) = e^{\int A(x)\,dx}$.

10.4 EXERCISES

Preliminary Questions

1. Which of the following are first-order linear equations?

(a) $y' + x^2y = 1$ **(b)** $y' + xy^2 = 1$

(c) $x^5y' + y = e^x$ **(d)** $x^5y' + y = e^y$

2. If $\alpha(x)$ is an integrating factor for $y' + A(x)y = B(x)$, then $\alpha'(x)$ is equal to (choose the correct answer):

(a) $B(x)$ **(b)** $\alpha(x)A(x)$

(c) $\alpha(x)A'(x)$ **(d)** $\alpha(x)B(x)$

Exercises

1. Consider $y' + x^{-1}y = x^3$.

(a) Verify that $\alpha(x) = x$ is an integrating factor.

(b) Show that when multiplied by $\alpha(x)$, the differential equation can be written $(xy)' = x^4$.

(c) Conclude that xy is an antiderivative of x^4 and use this information to find the general solution.

(d) Find the particular solution satisfying $y(1) = 0$.

2. Consider $\frac{dy}{dt} + 2y = e^{-3t}$.

(a) Verify that $\alpha(t) = e^{2t}$ is an integrating factor.

(b) Use Eq. (4) to find the general solution.

(c) Find the particular solution with initial condition $y(0) = 1$.

3. Let $\alpha(x) = e^{x^2}$. Verify the identity

$$(\alpha(x)y)' = \alpha(x)(y' + 2xy)$$

and explain how it is used to find the general solution of

$$y' + 2xy = x$$

4. Find the solution of $y' - y = e^{2x}$, $y(0) = 1$.

In Exercises 5–18, find the general solution of the first-order linear differential equation.

5. $xy' + y = x$

6. $xy' - y = x^2 - x$

7. $3xy' - y = x^{-1}$

8. $y' + xy = x$

9. $y' + 3x^{-1}y = x + x^{-1}$

10. $y' + x^{-1}y = \cos(x^2)$

11. $xy' = y - x$

12. $xy' = x^{-2} - \frac{3y}{x}$

13. $y' + y = e^x$

14. $y' + (\sec x)y = \cos x$

15. $y' + (\tan x)y = \cos x$

16. $e^{2x}y' = 1 - e^x y$

17. $y' - (\ln x)y = x^x$

18. $y' + y = \cos x$

In Exercises 19–26, solve the initial value problem.

19. $y' + 3y = e^{2x}$, $y(0) = -1$

20. $xy' + y = e^x$, $y(1) = 3$

21. $y' + \dfrac{1}{x+1}y = x^{-2}, \quad y(1) = 2$

22. $y' + y = \sin x, \quad y(0) = 1$

23. $(\sin x)y' = (\cos x)y + 1, \quad y\left(\frac{\pi}{4}\right) = 0$

24. $y' + (\sec t)y = \sec t, \quad y\left(\frac{\pi}{4}\right) = 1$

25. $y' + (\tanh x)y = 1, \quad y(0) = 3$

26. $y' + \dfrac{x}{1+x^2}y = \dfrac{1}{(1+x^2)^{3/2}}, \quad y(1) = 0$

27. Find the general solution of $y' + ny = e^{mx}$ for all m, n. *Note:* The case $m = -n$ must be treated separately.

28. Find the general solution of $y' + ny = \cos x$ for all n.

In Exercises 29–32, a 1000 L *tank contains* 500 L *of water with a salt concentration of* 10 g/L*. Water with a salt concentration of* 50 g/L *flows into the tank at a rate of* 80 L/min*. The fluid mixes instantaneously and is pumped out at a specified rate R_{out}. Let $y(t)$ denote the quantity of salt in the tank at time t.*

29. Assume that $R_{out} = 40$ L/min.

(a) Set up and solve the differential equation for $y(t)$.

(b) What is the salt concentration when the tank overflows?

30. Find the salt concentration when the tank overflows, assuming that $R_{out} = 60$ L/min.

31. Find the limiting salt concentration as $t \to \infty$ assuming that $R_{out} = 80$ L/min.

32. Assuming that $R_{out} = 120$ L/min. Find $y(t)$. Then calculate the tank volume and the salt concentration at $t = 10$ minutes.

33. Water flows into a tank at the variable rate of $R_{in} = 20/(1+t)$ gal/min and out at the constant rate $R_{out} = 5$ gal/min. Let $V(t)$ be the volume of water in the tank at time t.

(a) Set up a differential equation for $V(t)$ and solve it with the initial condition $V(0) = 100$.

(b) Find the maximum value of V.

(c) CAS Plot $V(t)$ and estimate the time t when the tank is empty.

34. A stream feeds into a lake at a rate of 1000 m^3/day. The stream is polluted with a toxin whose concentration is 5 g/m^3. Assume that the lake has volume 10^6 m^3 and that water flows out of the lake at the same rate of 1000 m^3/day.

(a) Set up a differential equation for the concentration $c(t)$ of toxin in the lake and solve for $c(t)$, assuming that $c(0) = 0$. *Hint:* Find the differential equation for the quantity of toxin $y(t)$, and observe that $c(t) = y(t)/10^6$.

(b) What is the limiting concentration for large t?

In Exercises 35–38, consider a series circuit (Figure 4) consisting of a resistor of R ohms, an inductor of L henries, and a variable voltage source of $V(t)$ volts (time t in seconds). The current through the circuit $I(t)$ (in amperes) satisfies the differential equation

$$\frac{dI}{dt} + \frac{R}{L}I = \frac{1}{L}V(t) \qquad \boxed{10}$$

35. Find the solution to Eq. (10) with initial condition $I(0) = 0$, assuming that $R = 100\ \Omega$, $L = 5$ H, and $V(t)$ is constant with $V(t) = 10$ V.

36. Assume that $R = 110\ \Omega$, $L = 10$ H, and $V(t) = e^{-t}$.

(a) Solve Eq. (10) with initial condition $I(0) = 0$.

(b) Calculate t_m and $I(t_m)$, where t_m is the time at which $I(t)$ has a maximum value.

(c) GU Use a computer algebra system to sketch the graph of the solution for $0 \le t \le 3$.

37. Assume that $V(t) = V$ is constant and $I(0) = 0$.

(a) Solve for $I(t)$.

(b) Show that $\lim_{t\to\infty} I(t) = V/R$ and that $I(t)$ reaches approximately 63% of its limiting value after L/R seconds.

(c) How long does it take for $I(t)$ to reach 90% of its limiting value if $R = 500\ \Omega$, $L = 4$ H, and $V = 20$ V?

38. Solve for $I(t)$, assuming that $R = 500\ \Omega$, $L = 4$ H, and $V = 20\cos(80)$ V?

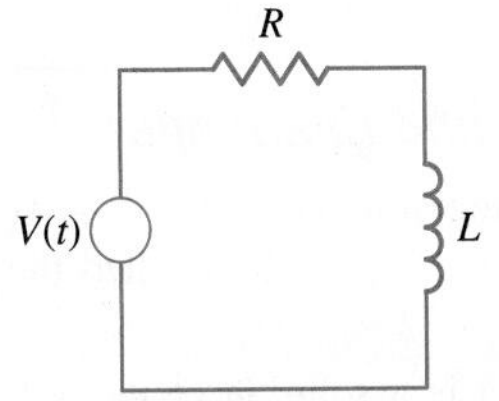

FIGURE 4 RL circuit.

39. Tank 1 in Figure 5 is filled with V_1 liters of water containing blue dye at an initial concentration of c_0 g/L. Water flows into the tank at a rate of R L/min, is mixed instantaneously with the dye solution, and flows out through the bottom at the same rate R. Let $c_1(t)$ be the dye concentration in the tank at time t.

(a) Explain why c_1 satisfies the differential equation $\dfrac{dc_1}{dt} = -\dfrac{R}{V_1}c_1$.

(b) Solve for $c_1(t)$ with $V_1 = 300$ L, $R = 50$, and $c_0 = 10$ g/L.

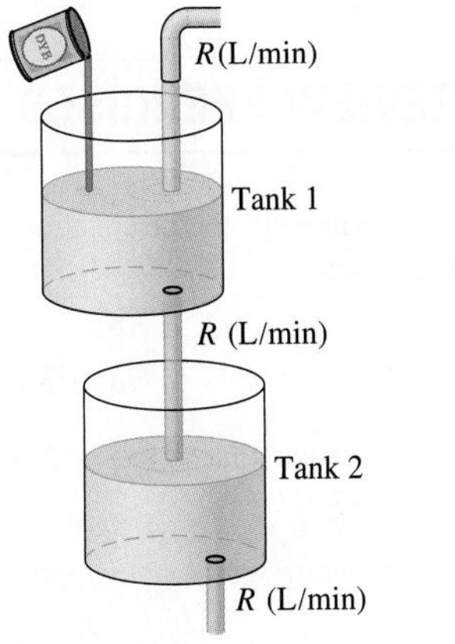

FIGURE 5

40. Continuing with the previous exercise, let Tank 2 be another tank filled with V_2 gal of water. Assume that the dye solution from Tank 1 empties into Tank 2 as in Figure 5, mixes instantaneously, and leaves Tank 2 at the same rate R. Let $c_2(t)$ be the dye concentration in Tank 2 at time t.

(a) Explain why c_2 satisfies the differential equation

$$\frac{dc_2}{dt} = \frac{R}{V_2}(c_1 - c_2)$$

(b) Use the solution to Exercise 39 to solve for $c_2(t)$ if $V_1 = 300$, $V_2 = 200$, $R = 50$, and $c_0 = 10$.

(c) Find the maximum concentration in Tank 2.

(d) GU Plot the solution.

41. Let a, b, r be constants. Show that

$$y = Ce^{-kt} + a + bk\left(\frac{k\sin rt - r\cos rt}{k^2 + r^2}\right)$$

is a general solution of

$$\frac{dy}{dt} = -k\big(y - a - b\sin rt\big)$$

42. Assume that the outside temperature varies as

$$T(t) = 15 + 5\sin(\pi t/12)$$

where $t = 0$ is 12 noon. A house is heated to 25°C at $t = 0$ and after that, its temperature $y(t)$ varies according to Newton's Law of Cooling (Figure 6):

$$\frac{dy}{dt} = -0.1\big(y(t) - T(t)\big)$$

Use Exercise 41 to solve for $y(t)$.

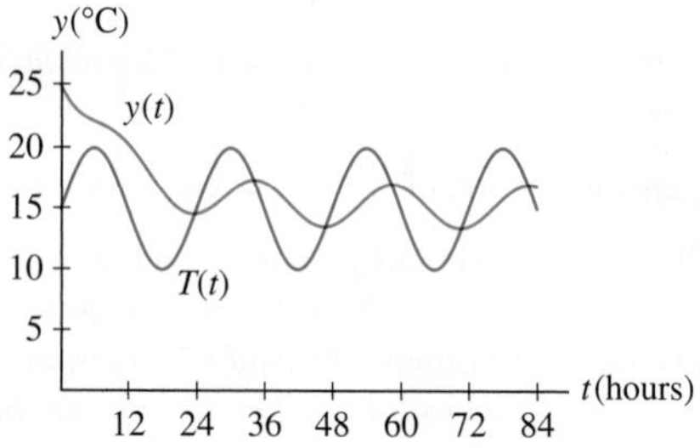

FIGURE 6 House temperature $y(t)$

Further Insights and Challenges

43. Let $\alpha(x)$ be an integrating factor for $y' + A(x)y = B(x)$. The differential equation $y' + A(x)y = 0$ is called the associated **homogeneous equation**.

(a) Show that $1/\alpha(x)$ is a solution of the associated homogeneous equation.

(b) Show that if $y = f(x)$ is a particular solution of $y' + A(x)y = B(x)$, then $f(x) + C/\alpha(x)$ is also a solution for any constant C.

44. Use the Fundamental Theorem of Calculus and the Product Rule to verify directly that for any x_0, the function

$$f(x) = \alpha(x)^{-1}\int_{x_0}^{x} \alpha(t)B(t)\,dt$$

is a solution of the initial value problem

$$y' + A(x)y = B(x), \qquad y(x_0) = 0$$

where $\alpha(x)$ is an integrating factor [a solution to Eq. (3)].

45. Transient Currents Suppose the circuit described by Eq. (10) is driven by a sinusoidal voltage source $V(t) = V\sin\omega t$ (where V and ω are constant).

(a) Show that

$$I(t) = \frac{V}{R^2 + L^2\omega^2}(R\sin\omega t - L\omega\cos\omega t) + Ce^{-(R/L)t}$$

(b) Let $Z = \sqrt{R^2 + L^2\omega^2}$. Choose θ so that $Z\cos\theta = R$ and $Z\sin\theta = L\omega$. Use the addition formula for the sine function to show that

$$I(t) = \frac{V}{Z}\sin(\omega t - \theta) + Ce^{-(R/L)t}$$

This shows that the current in the circuit varies sinusoidally apart from a DC term (called the **transient current** in electronics) that decreases exponentially.

CHAPTER REVIEW EXERCISES

1. Which of the following differential equations are linear? Determine the order of each equation.

(a) $y' = y^5 - 3x^4y$

(b) $y' = x^5 - 3x^4y$

(c) $y = y''' - 3x\sqrt{y}$

(d) $\sin x \cdot y'' = y - 1$

2. Find a value of c such that $y = x - 2 + e^{cx}$ is a solution of $2y' + y = x$.

In Exercises 3–6, solve using separation of variables.

3. $\dfrac{dy}{dt} = t^2y^{-3}$

4. $xyy' = 1 - x^2$

5. $x\dfrac{dy}{dx} - y = 1$

6. $y' = \dfrac{xy^2}{x^2 + 1}$

In Exercises 7–10, solve the initial value problem using separation of variables.

7. $y' = \cos^2 x, \quad y(0) = \dfrac{\pi}{4}$

8. $y' = \cos^2 y, \quad y(0) = \dfrac{\pi}{4}$

9. $y' = xy^2, \quad y(1) = 2$

10. $xyy' = 1, \quad y(3) = 2$

11. Figure 1 shows the slope field for $\dot{y} = \sin y + ty$. Sketch the graphs of the solutions with the initial conditions $y(0) = 1$, $y(0) = 0$, and $y(0) = -1$.

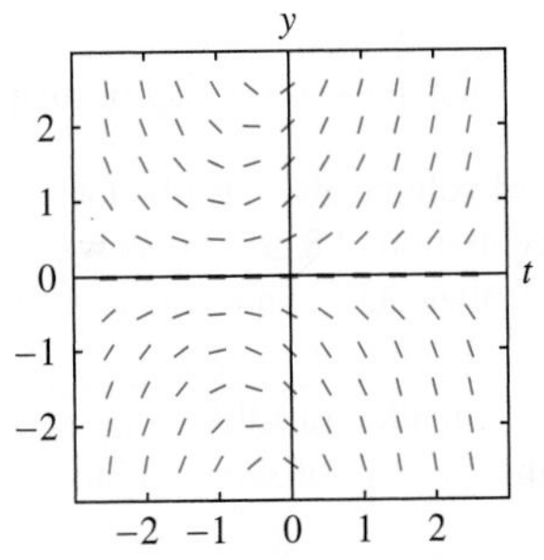

FIGURE 1

12. Which of the equations (i)–(iii) corresponds to the slope field in Figure 2?

(i) $\dot{y} = 1 - y^2$ **(ii)** $\dot{y} = 1 + y^2$ **(iii)** $\dot{y} = y^2$

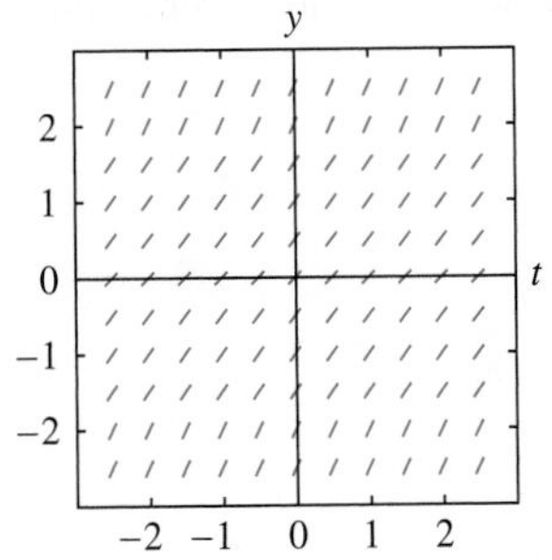

FIGURE 2

13. Let $y(t)$ be the solution to the differential equation with slope field as shown in Figure 2, satisfying $y(0) = 0$. Sketch the graph of $y(t)$. Then use your answer to Exercise 12 to solve for $y(t)$.

14. Let $y(t)$ be the solution of $4\dot{y} = y^2 + t$ satisfying $y(2) = 1$. Carry out Euler's Method with time step $h = 0.05$ for $n = 6$ steps.

15. Let $y(t)$ be the solution of $(x^3 + 1)\dot{y} = y$ satisfying $y(0) = 1$. Compute approximations to $y(0.1)$, $y(0.2)$, and $y(0.3)$ using Euler's Method with time step $h = 0.1$.

In Exercises 16–19, solve using the method of integrating factors.

16. $\dfrac{dy}{dt} = y + t^2, \quad y(0) = 4$

17. $\dfrac{dy}{dx} = \dfrac{y}{x} + x, \quad y(1) = 3$

18. $\dfrac{dy}{dt} = y - 3t, \quad y(-1) = 2$

19. $y' + 2y = 1 + e^{-x}, \quad y(0) = -4$

In Exercises 20–27, solve using the appropriate method.

20. $x^2 y' = x^2 + 1, \quad y(1) = 10$

21. $y' + (\tan x)y = \cos^2 x, \quad y(\pi) = 2$

22. $xy' = 2y + x - 1, \quad y\left(\frac{3}{2}\right) = 9$

23. $(y - 1)y' = t, \quad y(1) = -3$

24. $(\sqrt{y} + 1)y' = yte^{t^2}, \quad y(0) = 1$

25. $\dfrac{dw}{dx} = k\dfrac{1 + w^2}{x}, \quad w(1) = 1$

26. $y' + \dfrac{3y - 1}{t} = t + 2$

27. $y' + \dfrac{y}{x} = \sin x$

28. Find the solutions to $y' = 4(y - 12)$ satisfying $y(0) = 20$ and $y(0) = 0$, and sketch their graphs.

29. Find the solutions to $y' = -2y + 8$ satisfying $y(0) = 3$ and $y(0) = 4$, and sketch their graphs.

30. Show that $y = \sin^{-1} x$ satisfies the differential equation $y' = \sec y$ with initial condition $y(0) = 0$.

31. Find the solution $y = f(x)$ of $y' = \sqrt{y^2 - 1}$ satisfying the initial condition $y(0) = 1$.

32. State whether the differential equation can be solved using separation of variables, the method of integrating factors, both, or neither.

(a) $y' = y + x^2$ **(b)** $xy' = y + 1$

(c) $y' = y^2 + x^2$ **(d)** $xy' = y^2$

33. Let A and B be constants. Prove that if $A > 0$, then all solutions of $\frac{dy}{dt} + Ay = B$ approach the same limit as $t \to \infty$.

34. At time $t = 0$, a tank of height 5 m in the shape of an inverted pyramid whose cross section at the top is a square of side 2 m is filled with water. Water flows through a hole at the bottom of area 0.002 m^2. Use Torricelli's Law to determine the time required for the tank to empty.

35. The trough in Figure 3 (dimensions in centimeters) is filled with water. At time $t = 0$ (in seconds), water begins leaking through a hole at the bottom of area 4 cm^2. Let $y(t)$ be the water height at time t. Find a differential equation for $y(t)$ and solve it to determine when the water level decreases to 60 cm.

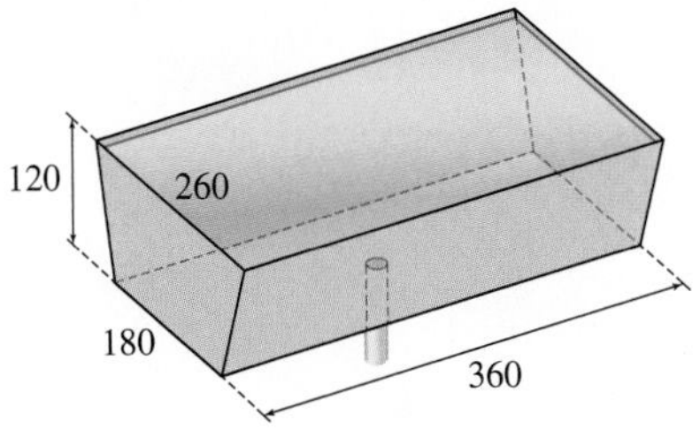

FIGURE 3

36. Find the solution of the logistic equation $\dot{y} = 0.4y(4 - y)$ satisfying $y(0) = 8$.

37. Let $y(t)$ be the solution of $\dot{y} = 0.3y(2 - y)$ with $y(0) = 1$. Determine $\lim_{t\to\infty} y(t)$ without solving for y explicitly.

38. Suppose that $y' = ky(1 - y/8)$ has a solution satisfying $y(0) = 12$ and $y(10) = 24$. Find k.

39. A lake has a carrying capacity of 1000 fish. Assume that the fish population grows logistically with growth constant $k = 0.2$ day^{-1}. How many days will it take for the population to reach 900 fish if the initial population is 20 fish?

40. A rabbit population on an island increases exponentially with growth rate $k = 0.12$ months^{-1}. When the population reaches 300 rabbits (say, at time $t = 0$), wolves begin eating the rabbits at a rate of r rabbits per month.

(a) Find a differential equation satisfied by the rabbit population $P(t)$.

(b) How large can r be without the rabbit population becoming extinct?

41. Show that $y = \sin(\tan^{-1} x + C)$ is the general solution of $y' = \sqrt{1 - y^2}/(1 + x^2)$. Then use the addition formula for the sine function to show that the general solution may be written

$$y = \frac{(\cos C)x + \sin C}{\sqrt{1 + x^2}}$$

42. A tank is filled with 300 liters of contaminated water containing 3 kg of toxin. Pure water is pumped in at a rate of 40 L/min, mixes instantaneously, and is then pumped out at the same rate. Let $y(t)$ be the quantity of toxin present in the tank at time t.

(a) Find a differential equation satisfied by $y(t)$.

(b) Solve for $y(t)$.

(c) Find the time at which there is 0.01 kg of toxin present.

43. At $t = 0$, a tank of volume 300 L is filled with 100 L of water containing salt at a concentration of 8 g/L. Fresh water flows in at a rate of 40 L/min, mixes instantaneously, and exits at the same rate. Let $c_1(t)$ be the salt concentration at time t.

(a) Find a differential equation satisfied by $c_1(t)$ *Hint:* Find the differential equation for the quantity of salt $y(t)$, and observe that $c_1(t) = y(t)/100$.

(b) Find the salt concentration $c_1(t)$ in the tank as a function of time.

44. The outflow of the tank in Exercise 43 is directed into a second tank containing V liters of fresh water where it mixes instantaneously and exits at the same rate of 40 L/min. Determine the salt concentration $c_2(t)$ in the second tank as a function of time in the following two cases:

(a) $V = 200$ **(b)** $V = 300$

In each case, determine the maximum concentration.

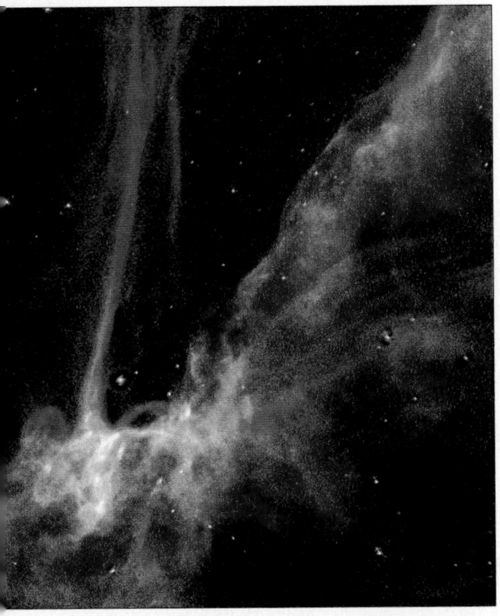

Our knowledge of what stars are made of is based on the study of absorption spectra, the sequences of wavelengths absorbed by gases in the star's atmosphere.

11 INFINITE SERIES

The theory of infinite series is a third branch of calculus, in addition to differential and integral calculus. Infinite series yield a new perspective on functions and on many interesting numbers. Two examples are the infinite series for the exponential function

$$e^x = 1 + x + \frac{x^2}{2!} + \frac{x^3}{3!} + \frac{x^4}{4!} + \cdots$$

and the Gregory–Leibniz series (see Exercise 53 in Section 2)

$$\frac{\pi}{4} = 1 - \frac{1}{3} + \frac{1}{5} - \frac{1}{7} + \frac{1}{9} - \cdots$$

The first shows that e^x can be expressed as an "infinite polynomial," and the second reveals that π is related to the reciprocals of the odd integers in an unexpected way. To make sense of infinite series, we need to define precisely what it means to add up infinitely many terms. Limits play a key role here, just as they do in differential and integral calculus.

11.1 Sequences

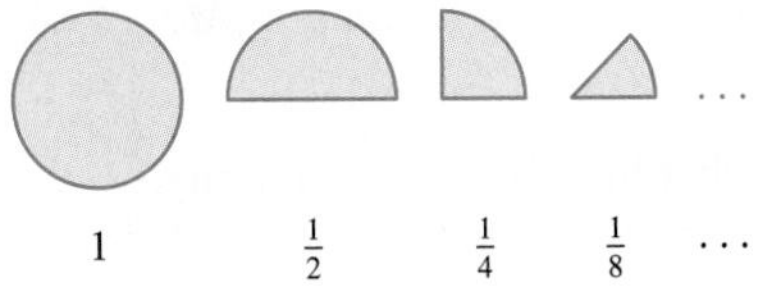

FIGURE 1

Sequences of numbers appear in diverse situations. If you divide a cake in half, and then divide the remaining half in half, and continue dividing in half indefinitely (Figure 1), then the fraction of cake remaining at each step forms the sequence

$$1, \quad \frac{1}{2}, \quad \frac{1}{4}, \quad \frac{1}{8}, \quad \ldots$$

This is the sequence of values of the function $f(n) = \dfrac{1}{2^n}$ for $n = 0, 1, 2, \ldots$.

Formally, a **sequence** is an ordered collection of numbers defined by a function $f(n)$ on a set of integers. The values $a_n = f(n)$ are called the **terms** of the sequence, and n is called the **index**. Informally, we think of a sequence $\{a_n\}$ as a list of terms:

$$a_1, \quad a_2, \quad a_3, \quad a_4, \quad \ldots$$

The sequence does not have to start at $n = 1$, but may start at $n = 0$, $n = 2$, or any other integer. When a_n is given by a formula, we refer to a_n as the **general term**.

The sequence b_n is the Balmer series of absorption wavelengths of the hydrogen atom in nanometers. It plays a key role in spectroscopy.

General term	Domain	Sequence
$a_n = 1 - \dfrac{1}{n}$	$n \ge 1$	$0, \ \frac{1}{2}, \ \frac{2}{3}, \ \frac{3}{4}, \ \frac{4}{5}, \ \ldots$
$a_n = (-1)^n n$	$n \ge 0$	$0, \ -1, \ 2, \ -3, \ 4, \ \ldots$
$b_n = \dfrac{364.5n^2}{n^2 - 4}$	$n \ge 3$	$656.1, \ 486, \ 433.9, \ 410.1, \ 396.9, \ \ldots$

The sequence in the next example is defined *recursively*. The first term is given and the nth term a_n is computed in terms of the preceding term a_{n-1}.

EXAMPLE 1 Recursive Sequence Compute a_2, a_3, a_4 for the sequence defined recursively by

$$a_1 = 1, \qquad a_n = \frac{1}{2}\left(a_{n-1} + \frac{2}{a_{n-1}}\right)$$

Solution

You may recognize the sequence in Example 1 as the sequence of approximations to $\sqrt{2} \approx 1.4142136$ produced by Newton's method with starting value $a_1 = 1$. As n tends to infinity, a_n approaches $\sqrt{2}$.

$$a_2 = \frac{1}{2}\left(a_1 + \frac{2}{a_1}\right) = \frac{1}{2}\left(1 + \frac{2}{1}\right) = \frac{3}{2} = 1.5$$

$$a_3 = \frac{1}{2}\left(a_2 + \frac{2}{a_2}\right) = \frac{1}{2}\left(\frac{3}{2} + \frac{2}{3/2}\right) = \frac{17}{12} \approx 1.4167$$

$$a_4 = \frac{1}{2}\left(a_3 + \frac{2}{a_3}\right) = \frac{1}{2}\left(\frac{17}{12} + \frac{2}{17/12}\right) = \frac{577}{408} \approx 1.414216$$ ■

Our main goal is to study convergence of sequences. A sequence $\{a_n\}$ converges to a limit L if $|a_n - L|$ becomes arbitrary small when n is sufficiently large. Here is the formal definition.

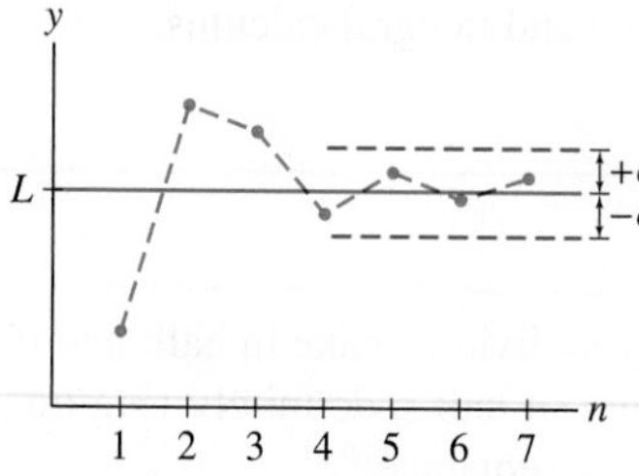

FIGURE 2 Plot of a sequence with limit L. For any ϵ, the dots eventually remain within an ϵ-band around L.

DEFINITION Limit of a Sequence We say that $\{a_n\}$ **converges** to a limit L, and we write

$$\lim_{n\to\infty} a_n = L \qquad \text{or} \qquad a_n \to L$$

if, for every $\epsilon > 0$, there is a number M such that $|a_n - L| < \epsilon$ for all $n > M$.

- If no limit exists, we say that $\{a_n\}$ diverges.
- If the terms increase without bound, we say that $\{a_n\}$ diverges to infinity.

If $\{a_n\}$ converges, then its limit L is unique. A good way to visualize the limit is to plot the points $(1, a_1), (2, a_2), (3, a_3), \ldots$, as in Figure 2. The sequence converges to L if, for every $\epsilon > 0$, the plotted points eventually remain within an ϵ-band around the horizontal line $y = L$. Figure 3 shows the plot of a sequence converging to $L = 1$. On the other hand, we can show that the sequence $a_n = \cos n$ in Figure 4 has no limit.

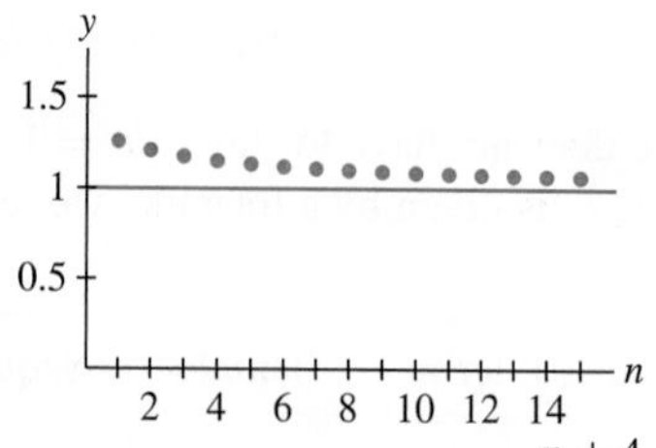

FIGURE 3 The sequence $a_n = \dfrac{n+4}{n+1}$.

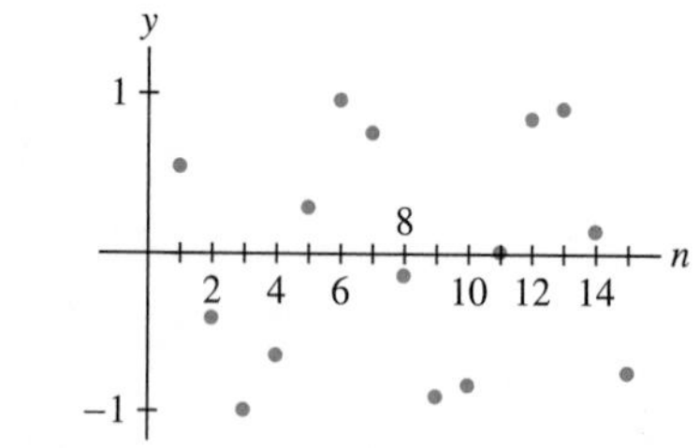

FIGURE 4 The sequence $a_n = \cos n$ has no limit.

EXAMPLE 2 Proving Convergence Let $a_n = \dfrac{n+4}{n+1}$. Prove formally that $\lim_{n\to\infty} a_n = 1$.

Solution The definition requires us to find, for every $\epsilon > 0$, a number M such that

$$|a_n - 1| < \epsilon \qquad \text{for all } n > M \tag{1}$$

We have

$$|a_n - 1| = \left|\frac{n+4}{n+1} - 1\right| = \frac{3}{n+1}$$

Therefore, $|a_n - 1| < \epsilon$ if

$$\frac{3}{n+1} < \epsilon \qquad \text{or} \qquad n > \frac{3}{\epsilon} - 1$$

In other words, Eq. (1) is valid with $M = \frac{3}{\epsilon} - 1$. This proves that $\lim_{n\to\infty} a_n = 1$. ■

Note the following two facts about sequences:

- The limit does not change if we change or drop finitely many terms of the sequence.
- If C is a constant and $a_n = C$ for all n sufficiently large, then $\lim_{n\to\infty} a_n = C$.

Many of the sequences we consider are defined by functions; that is, $a_n = f(n)$ for some function $f(x)$. For example,

$$a_n = \frac{n-1}{n} \quad \text{is defined by} \quad f(x) = \frac{x-1}{x}$$

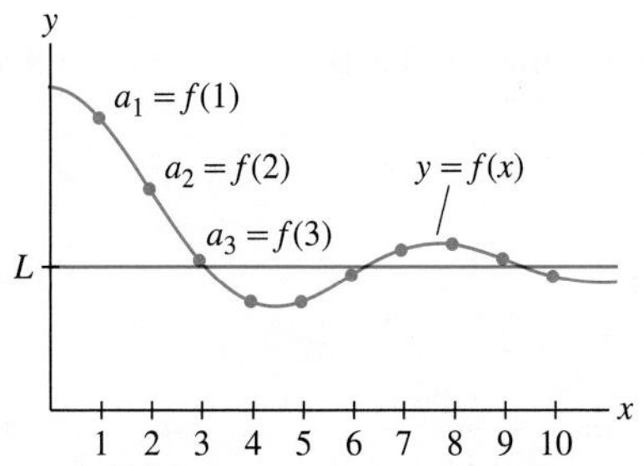

FIGURE 5 If $f(x)$ converges to L, then the sequence $a_n = f(n)$ also converges to L.

A fact we will use often is that if $f(x)$ approaches a limit L as $x \to \infty$, then the sequence $a_n = f(n)$ approaches the same limit L (Figure 5). Indeed, for all $\epsilon > 0$, we can find M so that $|f(x) - L| < \epsilon$ for all $x > M$. It follows automatically that $|f(n) - L| < \epsilon$ for all integers $n > M$.

THEOREM 1 Sequence Defined by a Function If $\lim_{x\to\infty} f(x)$ exists, then the sequence $a_n = f(n)$ converges to the same limit:

$$\lim_{n\to\infty} a_n = \lim_{x\to\infty} f(x)$$

■ **EXAMPLE 3** Find the limit of the sequence

$$\frac{2^2-2}{2^2}, \quad \frac{3^2-2}{3^2}, \quad \frac{4^2-2}{4^2}, \quad \frac{5^2-2}{5^2}, \quad \dots$$

Solution This is the sequence with general term

$$a_n = \frac{n^2-2}{n^2} = 1 - \frac{2}{n}$$

Therefore, we apply Theorem 1 with $f(x) = 1 - \frac{2}{x}$:

$$\lim_{n\to\infty} a_n = \lim_{x\to\infty}\left(1 - \frac{2}{x}\right) = 1 - \lim_{x\to\infty}\frac{2}{x} = 1 - 0 = 1$$ ■

■ **EXAMPLE 4** Calculate $\displaystyle\lim_{n\to\infty} \frac{n + \ln n}{n^2}$.

Solution Apply Theorem 1, using L'Hôpital's Rule in the second step:

$$\lim_{n\to\infty}\frac{n+\ln n}{n^2} = \lim_{x\to\infty}\frac{x+\ln x}{x^2} = \lim_{x\to\infty}\frac{1+(1/x)}{2x} = 0$$ ■

The limit of the Balmer wavelengths b_n in the next example plays a role in physics and chemistry because it determines the ionization energy of the hydrogen atom. Table 1 suggests that b_n approaches 364.5. Figure 6 shows the graph, and in Figure 7, the wavelengths are shown "crowding in" toward their limiting value.

TABLE 1
Balmer Wavelengths

n	b_n
3	656.1
4	486
5	433.9
6	410.1
7	396.9
10	379.7
20	368.2
40	365.4
60	364.9
80	364.7
100	364.6

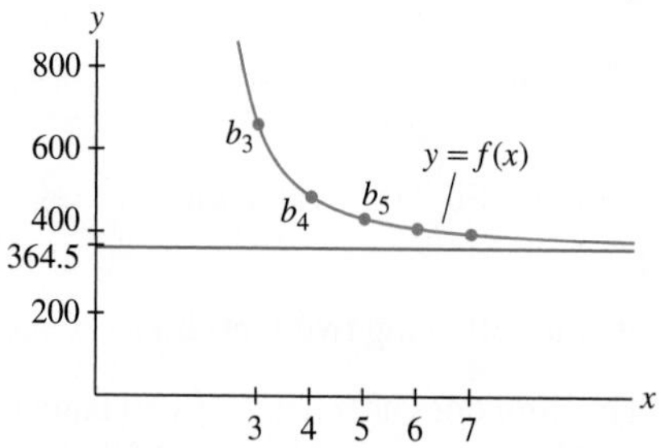

FIGURE 6 The sequence and the function approach the same limit.

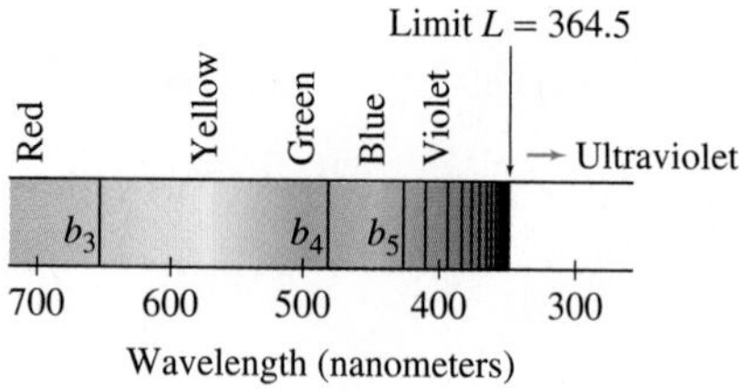

FIGURE 7

■ **EXAMPLE 5 Balmer Wavelengths** Calculate the limit of the Balmer wavelengths $b_n = \dfrac{364.5n^2}{n^2 - 4}$, where $n \geq 3$.

Solution Apply Theorem 1 with $f(x) = \dfrac{364.5x^2}{x^2 - 4}$:

$$\lim_{n\to\infty} b_n = \lim_{x\to\infty} \frac{364.5x^2}{x^2 - 4} = \lim_{x\to\infty} \frac{364.5}{1 - 4/x^2} = \frac{364.5}{\lim\limits_{x\to\infty}(1 - 4/x^2)} = 364.5$$ ■

A **geometric sequence** is a sequence $a_n = cr^n$, where c and r are nonzero constants. Each term is r times the previous term; that is, $a_n/a_{n-1} = r$. The number r is called the **common ratio**. For instance, if $r = 3$ and $c = 2$, we obtain the sequence (starting at $n = 0$)

$$2, \quad 2 \cdot 3, \quad 2 \cdot 3^2, \quad 2 \cdot 3^3, \quad 2 \cdot 3^4, \quad 2 \cdot 3^5, \quad \ldots$$

In the next example, we determine when a geometric series converges. Recall that $\{a_n\}$ **diverges to** ∞ if the terms a_n increase beyond all bounds (Figure 8); that is,

$$\lim_{n\to\infty} a_n = \infty \quad \text{if, for every number } N,\ a_n > N \text{ for all sufficiently large } n$$

We define $\lim\limits_{n\to\infty} a_n = -\infty$ similarly.

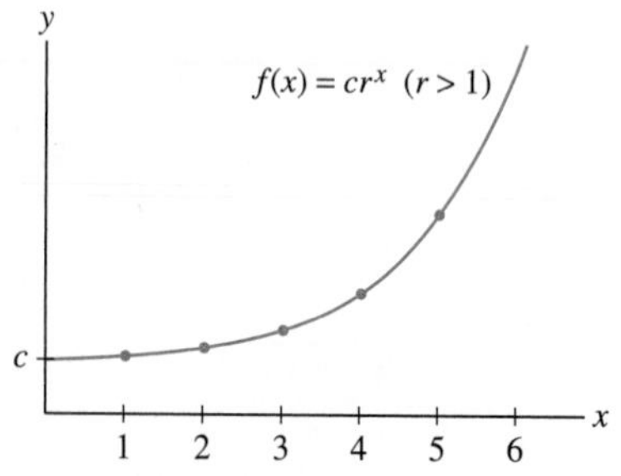

FIGURE 8 If $r > 1$, the geometric sequence $a_n = r^n$ diverges to ∞.

■ **EXAMPLE 6 Geometric Sequences with $r \geq 0$** Prove that for $r \geq 0$ and $c > 0$,

$$\lim_{n\to\infty} cr^n = \begin{cases} 0 & \text{if} \quad 0 \leq r < 1 \\ c & \text{if} \quad r = 1 \\ \infty & \text{if} \quad r > 1 \end{cases}$$

Solution Set $f(r) = cr^x$. If $0 \leq r < 1$, then (Figure 9)

$$\lim_{n\to\infty} cr^n = \lim_{x\to\infty} f(x) = c \lim_{x\to\infty} r^x = 0$$

If $r > 1$, then both $f(x)$ and the sequence $\{cr^n\}$ diverge to ∞ (because $c > 0$) (Figure 8). If $r = 1$, then $cr^n = c$ for all n, and the limit is c. ■

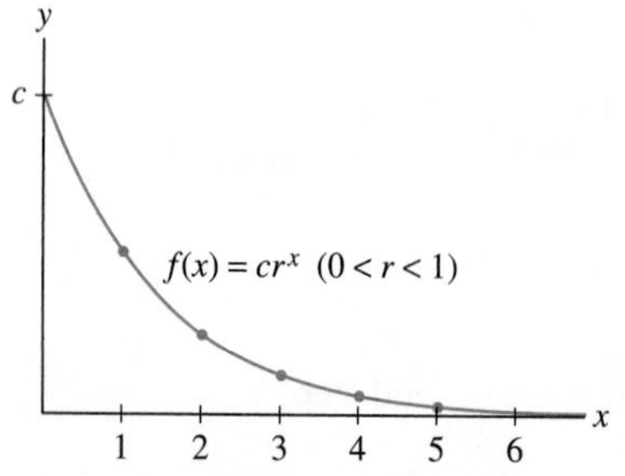

FIGURE 9 If $0 < r < 1$, the geometric sequence $a_n = r^n$ converges to 0.

The limit laws we have used for functions also apply to sequences and are proved in a similar fashion.

THEOREM 2 Limit Laws for Sequences Assume that $\{a_n\}$ and $\{b_n\}$ are convergent sequences with

$$\lim_{n\to\infty} a_n = L, \qquad \lim_{n\to\infty} b_n = M$$

Then:

(i) $\lim_{n\to\infty} (a_n \pm b_n) = \lim_{n\to\infty} a_n \pm \lim_{n\to\infty} b_n = L \pm M$

(ii) $\lim_{n\to\infty} a_n b_n = \left(\lim_{n\to\infty} a_n\right)\left(\lim_{n\to\infty} b_n\right) = LM$

(iii) $\lim_{n\to\infty} \dfrac{a_n}{b_n} = \dfrac{\lim_{n\to\infty} a_n}{\lim_{n\to\infty} b_n} = \dfrac{L}{M}$ if $M \neq 0$

(iv) $\lim_{n\to\infty} ca_n = c \lim_{n\to\infty} a_n = cL$ for any constant c

THEOREM 3 Squeeze Theorem for Sequences Let $\{a_n\}$, $\{b_n\}$, $\{c_n\}$ be sequences such that for some number M,

$$b_n \le a_n \le c_n \quad \text{for } n > M \qquad \text{and} \qquad \lim_{n\to\infty} b_n = \lim_{n\to\infty} c_n = L$$

Then $\lim_{n\to\infty} a_n = L$.

REMINDER $n!$ *(n-factorial) is the number*

$$n! = n(n-1)(n-2)\cdots 2\cdot 1$$

For example, $4! = 4\cdot 3\cdot 2\cdot 1 = 24$.

EXAMPLE 7 Show that if $\lim_{n\to\infty} |a_n| = 0$, then $\lim_{n\to\infty} a_n = 0$.

Solution We have

$$-|a_n| \le a_n \le |a_n|$$

By hypothesis, $\lim_{n\to\infty} |a_n| = 0$, and thus also $\lim_{n\to\infty} -|a_n| = -\lim_{n\to\infty} |a_n| = 0$. Therefore, we can apply the Squeeze Theorem to conclude that $\lim_{n\to\infty} a_n = 0$. ■

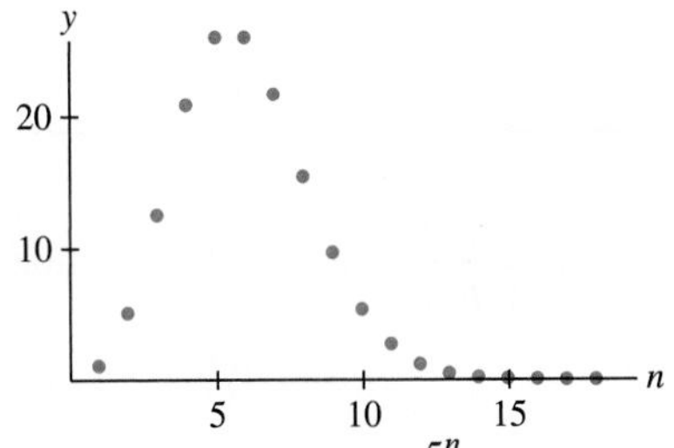

FIGURE 10 Graph of $a_n = \dfrac{5^n}{n!}$.

TABLE 2

n	$a_n = \dfrac{5^n}{n!}$
1	5
2	12.5
3	20.83
4	26.04
10	2.69
15	0.023
20	0.000039
50	2.92×10^{-30}

EXAMPLE 8 Geometric Sequences with $r < 0$ Prove that for $c \neq 0$,

$$\lim_{n\to\infty} cr^n = \begin{cases} 0 & \text{if} \quad -1 < r < 0 \\ \text{diverges} & \text{if} \quad r \le -1 \end{cases}$$

Solution If $-1 < r < 0$, then $0 < |r| < 1$ and $\lim_{n\to\infty} |cr^n| = 0$ by Example 6. Thus $\lim_{n\to\infty} cr^n = 0$ by Example 7. If $r = -1$, then the sequence $cr^n = (-1)^n c$ alternates in sign and does not approach a limit. The sequence also diverges if $r < -1$ because cr^n alternates in sign and $|cr^n|$ grows arbitrarily large. ■

As another application of the Squeeze Theorem, consider the sequence

$$a_n = \frac{5^n}{n!}$$

Both the numerator and the denominator grow without bound, so it is not clear in advance whether $\{a_n\}$ converges. Figure 10 and Table 2 suggest that a_n increases initially and then tends to zero. In the next example, we verify that $a_n = R^n/n!$ converges to zero for all R. This fact is used in the discussion of Taylor series in Section 11.7.

EXAMPLE 9 Prove that $\lim_{n\to\infty} \frac{R^n}{n!} = 0$ for all R.

Solution Assume first that $R > 0$ and let M be the positive integer such that

$$M \le R < M + 1$$

For $n > M$, we write $R^n/n!$ as a product of n factors:

$$\frac{R^n}{n!} = \underbrace{\left(\frac{R}{1}\frac{R}{2}\cdots\frac{R}{M}\right)}_{\text{Call this constant } C}\underbrace{\left(\frac{R}{M+1}\right)\left(\frac{R}{M+2}\right)\cdots\left(\frac{R}{n}\right)}_{\text{Each factor is less than 1}} \le C\left(\frac{R}{n}\right) \qquad \boxed{2}$$

The first M factors are ≥ 1 and the last $n - M$ factors are < 1. If we lump together the first M factors and call the product C, and drop all the remaining factors except the last factor R/n, we see that

$$0 \le \frac{R^n}{n!} \le \frac{CR}{n}$$

Since $CR/n \to 0$, the Squeeze Theorem gives us $\lim_{n\to\infty} R^n/n! = 0$ as claimed. If $R < 0$, the limit is also zero by Example 7 because $|R^n/n!|$ tends to zero. ■

Given a sequence $\{a_n\}$ and a function $f(x)$, we can form the new sequence $f(a_n)$. It is useful to know that if $f(x)$ is continuous and $a_n \to L$, then $f(a_n) \to f(L)$. A proof is given in Appendix D.

THEOREM 4 If $f(x)$ is continuous and $\lim_{n\to\infty} a_n = L$, then

$$\lim_{n\to\infty} f(a_n) = f\left(\lim_{n\to\infty} a_n\right) = f(L)$$

In other words, we may "bring a limit inside a continuous function."

EXAMPLE 10 Apply Theorem 4 to the sequence $a_n = \frac{3n}{n+1}$ and to the functions **(a)** $f(x) = e^x$ and **(b)** $g(x) = x^2$.

Solution Observe first that

$$L = \lim_{n\to\infty} a_n = \lim_{n\to\infty} \frac{3n}{n+1} = \lim_{n\to\infty} \frac{3}{1+n^{-1}} = 3$$

(a) With $f(x) = e^x$ we have $f(a_n) = e^{a_n} = e^{\frac{3n}{n+1}}$. According to Theorem 4,

$$\lim_{n\to\infty} e^{\frac{3n}{n+1}} = \lim_{n\to\infty} f(a_n) = f\left(\lim_{n\to\infty} a_n\right) = e^{\lim_{n\to\infty} \frac{3n}{n+1}} = e^3$$

(b) With $g(x) = x^2$ we have $g(a_n) = a_n^2$, and according to Theorem 4,

$$\lim_{n\to\infty} \left(\frac{3n}{n+1}\right)^2 = \lim_{n\to\infty} g(a_n) = g\left(\lim_{n\to\infty} a_n\right) = \left(\lim_{n\to\infty} \frac{3n}{n+1}\right)^2 = 3^2 = 9 \qquad ■$$

Of great importance for understanding convergence are the concepts of a bounded sequence and a monotonic sequence.

DEFINITION Bounded Sequences A sequence $\{a_n\}$ is:

- **Bounded from above** if there is a number M such that $a_n \le M$ for all n. The number M is called an *upper bound.*
- **Bounded from below** if there is a number m such that $a_n \ge m$ for all n. The number m is called a *lower bound.*

The sequence $\{a_n\}$ is called **bounded** if it is bounded from above and below. A sequence that is not bounded is called an **unbounded sequence**.

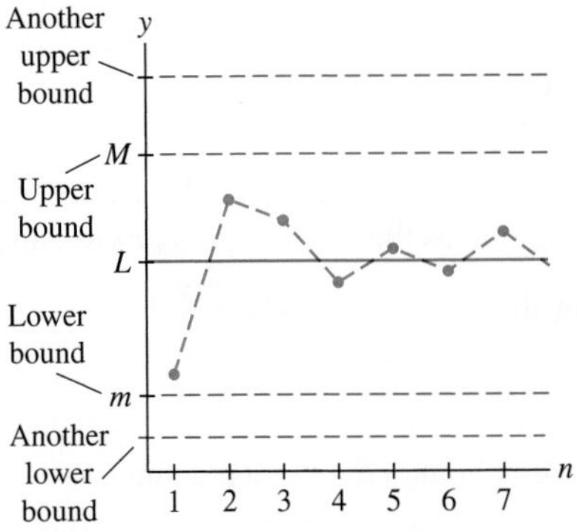

FIGURE 11 A convergent sequence is bounded.

Upper and lower bounds are not unique. If M is an upper bound, then any larger number is also an upper bound, and if m is a lower bound, then any smaller number is also a lower bound (Figure 11).

As we might expect, a convergent sequence $\{a_n\}$ is necessarily bounded because the terms a_n get closer and closer to the limit. This fact is recorded in the next theorem.

THEOREM 5 Convergent Sequences Are Bounded If $\{a_n\}$ converges, then $\{a_n\}$ is bounded.

Proof Let $L = \lim_{n\to\infty} a_n$. Then there exists $N > 0$ such that $|a_n - L| < 1$ for $n > N$. In other words,

$$L - 1 < a_n < L + 1 \qquad \text{for } n > N$$

If M is any number larger than $L + 1$ and also larger than the numbers $a_1, a_2, \dots, a_N$, then $a_n < M$ for all n. Thus, M is an upper bound. Similarly, any number m smaller than $L - 1$ and also smaller than the numbers $a_1, a_2, \dots, a_N$ is a lower bound. ■

There are two ways that a sequence $\{a_n\}$ can diverge. One way is by being unbounded. For example, the unbounded sequence $a_n = n$ diverges:

$$1, \quad 2, \quad 3, \quad 4, \quad 5, \quad 6, \quad \dots$$

However, a sequence can diverge even if it is bounded. This is the case with $a_n = (-1)^{n+1}$, whose terms a_n bounce back and forth but never settle down to approach a limit:

$$1, \quad -1, \quad 1, \quad -1, \quad 1, \quad -1, \quad \dots$$

There is no surefire method for determining whether a sequence $\{a_n\}$ converges, unless the sequence happens to be both bounded and **monotonic**. By definition, $\{a_n\}$ is monotonic if it is either increasing or decreasing:

- $\{a_n\}$ is *increasing* if $a_n < a_{n+1}$ for all n.
- $\{a_n\}$ is *decreasing* if $a_n > a_{n+1}$ for all n.

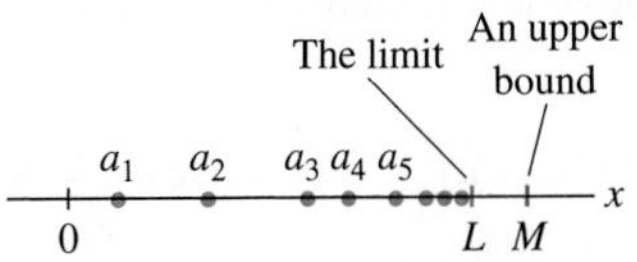

FIGURE 12 An increasing sequence with upper bound M approaches a limit L.

Intuitively, if $\{a_n\}$ is increasing and bounded above by M, then the terms must bunch up near some limiting value L that is not greater than M (Figure 12). See Appendix B for a proof of the next theorem.

THEOREM 6 Bounded Monotonic Sequences Converge

- If $\{a_n\}$ is increasing and $a_n \le M$, then $\{a_n\}$ converges and $\lim_{n\to\infty} a_n \le M$.
- If $\{a_n\}$ is decreasing and $a_n \ge m$, then $\{a_n\}$ converges and $\lim_{n\to\infty} a_n \ge m$.

EXAMPLE 11 Verify that $a_n = \sqrt{n+1} - \sqrt{n}$ is decreasing and bounded below. Does $\lim_{n\to\infty} a_n$ exist?

TABLE 3

$a_n = \sqrt{n+1} - \sqrt{n}$
$a_1 \approx 0.4142$
$a_2 \approx 0.3178$
$a_3 \approx 0.2679$
$a_4 \approx 0.2361$
$a_5 \approx 0.2134$
$a_6 \approx 0.1963$
$a_7 \approx 0.1827$
$a_8 \approx 0.1716$

Solution The function $f(x) = \sqrt{x+1} - \sqrt{x}$ is decreasing because its derivative is negative:

$$f'(x) = \frac{1}{2\sqrt{x+1}} - \frac{1}{2\sqrt{x}} < 0 \qquad \text{for } x > 0$$

It follows that $a_n = f(n)$ is decreasing (see Table 3). Furthermore, $a_n > 0$ for all n, so the sequence has lower bound $m = 0$. Theorem 6 guarantees that $L = \lim_{n\to\infty} a_n$ exists and $L \ge 0$. In fact, we can show that $L = 0$ by noting that $f(x) = 1/(\sqrt{x+1} + \sqrt{x})$ and hence $\lim_{x\to\infty} f(x) = 0$. ■

EXAMPLE 12 Show that the following sequence is bounded and increasing:

$$a_1 = \sqrt{2}, \quad a_2 = \sqrt{2\sqrt{2}}, \quad a_3 = \sqrt{2\sqrt{2\sqrt{2}}}, \quad \dots$$

Then prove that $L = \lim_{n\to\infty} a_n$ exists and compute its value.

Solution If we knew in advance that the limit L exists, we could find its value as follows. The idea is that L "contains a copy" of itself under the square root sign:

$$L = \sqrt{2\sqrt{2\sqrt{2\sqrt{2\cdots}}}} = \sqrt{2\left(\sqrt{2\sqrt{2\sqrt{2\cdots}}}\right)} = \sqrt{2L}$$

Thus $L^2 = 2L$, which implies that $L = 2$ or $L = 0$. We eliminate $L = 0$ because the terms a_n are positive and increasing (as shown below), so we must have $L = 2$ (see Table 4).

TABLE 4 Recursive Sequence $a_{n+1} = \sqrt{2a_n}$

a_1	1.4142
a_2	1.6818
a_3	1.8340
a_4	1.9152
a_5	1.9571
a_6	1.9785
a_7	1.9892
a_8	1.9946

This argument is phrased more formally by noting that the sequence is defined recursively by

$$a_1 = \sqrt{2}, \qquad a_{n+1} = \sqrt{2a_n}$$

If a_n converges to L, then the sequence $b_n = a_{n+1}$ also converges to L (because it is the same sequence, with terms shifted one to the left). Then, using Theorem 4, we would have

$$L = \lim_{n\to\infty} a_{n+1} = \lim_{n\to\infty} \sqrt{2a_n} = \sqrt{2 \lim_{n\to\infty} a_n} = \sqrt{2L}$$

However, none of this is valid unless we know in advance that the limit L exists. By Theorem 6, it suffices to show that $\{a_n\}$ is bounded above and increasing.

***Step 1.* Show that $\{a_n\}$ is bounded above.**

We claim that $M = 2$ is an upper bound. We certainly have $a_1 < 2$ because $a_1 = \sqrt{2} \approx 1.414$. On the other hand,

$$\text{if} \quad a_n < 2, \qquad \text{then} \quad a_{n+1} < 2 \qquad \boxed{3}$$

This is true because $a_{n+1} = \sqrt{2a_n} < \sqrt{2 \cdot 2} = 2$. Now, since $a_1 < 2$, we can apply (3) to conclude that $a_2 < 2$. Similarly, $a_2 < 2$ implies $a_3 < 2$, and so on for all n. Formally speaking, this is a proof by induction.

Step 2. **Show that $\{a_n\}$ is increasing.**
Since a_n is positive and $a_n < 2$, we have

$$a_{n+1} = \sqrt{2a_n} > \sqrt{a_n \cdot a_n} = a_n$$

This shows that $\{a_n\}$ is increasing. ■

We conclude that the limit L exists and hence $L = 2$.

11.1 SUMMARY

- A sequence $\{a_n\}$ *converges* to a limit L if, for every $\epsilon > 0$, there is a number M such that

$$|a_n - L| < \epsilon \qquad \text{for all } n > M$$

We write $\lim_{n\to\infty} a_n = L$ or $a_n \to L$.
- If no limit exists, we say that $\{a_n\}$ *diverges*.
- In particular, if the terms increase without bound, we say that $\{a_n\}$ diverges to infinity.
- If $a_n = f(n)$ and $\lim_{x\to\infty} f(x) = L$, then $\lim_{n\to\infty} a_n = L$.
- A *geometric sequence* is a sequence $a_n = cr^n$, where c and r are nonzero.
- The Basic Limit Laws and the Squeeze Theorem apply to sequences.
- If $f(x)$ is continuous and $\lim_{n\to\infty} a_n = L$, then $\lim_{n\to\infty} f(a_n) = f(L)$.
- A sequence $\{a_n\}$ is
  - *bounded above* by M if $a_n \le M$ for all n.
  - *bounded below* by m if $a_n \ge m$ for all n.

If $\{a_n\}$ is bounded above and below, $\{a_n\}$ is called *bounded*.
- A sequence $\{a_n\}$ is *monotonic* if it is increasing ($a_n < a_{n+1}$) or decreasing ($a_n > a_{n+1}$).
- Bounded monotonic sequences converge (Theorem 6).

11.1 EXERCISES

Preliminary Questions

1. What is a_4 for the sequence $a_n = n^2 - n$?

2. Which of the following sequences converge to zero?

(a) $\dfrac{n^2}{n^2+1}$ **(b)** 2^n **(c)** $\left(\dfrac{-1}{2}\right)^n$

3. Let a_n be the nth decimal approximation to $\sqrt{2}$. That is, $a_1 = 1$, $a_2 = 1.4$, $a_3 = 1.41$, etc. What is $\lim_{n\to\infty} a_n$?

4. Which of the following sequences is defined recursively?

(a) $a_n = \sqrt{4+n}$ **(b)** $b_n = \sqrt{4+b_{n-1}}$

5. Theorem 5 says that every convergent sequence is bounded. Determine if the following statements are true or false and if false, give a counterexample.

(a) If $\{a_n\}$ is bounded, then it converges.

(b) If $\{a_n\}$ is not bounded, then it diverges.

(c) If $\{a_n\}$ diverges, then it is not bounded.

Exercises

1. Match each sequence with its general term:

$a_1, a_2, a_3, a_4, \ldots$	General term
(a) $\frac{1}{2}, \frac{2}{3}, \frac{3}{4}, \frac{4}{5}, \ldots$	(i) $\cos \pi n$
(b) $-1, 1, -1, 1, \ldots$	(ii) $\dfrac{n!}{2^n}$
(c) $1, -1, 1, -1, \ldots$	(iii) $(-1)^{n+1}$
(d) $\frac{1}{2}, \frac{2}{4}, \frac{6}{8}, \frac{24}{16} \ldots$	(iv) $\dfrac{n}{n+1}$

2. Let $a_n = \dfrac{1}{2n-1}$ for $n = 1, 2, 3, \ldots$. Write out the first three terms of the following sequences.

(a) $b_n = a_{n+1}$ **(b)** $c_n = a_{n+3}$

(c) $d_n = a_n^2$ **(d)** $e_n = 2a_n - a_{n+1}$

In Exercises 3–12, calculate the first four terms of the sequence, starting with $n = 1$.

3. $c_n = \dfrac{3^n}{n!}$ **4.** $b_n = \dfrac{(2n-1)!}{n!}$

5. $a_1 = 2, \quad a_{n+1} = 2a_n^2 - 3$

6. $b_1 = 1, \quad b_n = b_{n-1} + \dfrac{1}{b_{n-1}}$

7. $b_n = 5 + \cos \pi n$ **8.** $c_n = (-1)^{2n+1}$

9. $c_n = 1 + \dfrac{1}{2} + \dfrac{1}{3} + \cdots + \dfrac{1}{n}$

10. $a_n = n + (n+1) + (n+2) + \cdots + (2n)$

11. $b_1 = 2, \quad b_2 = 3, \quad b_n = 2b_{n-1} + b_{n-2}$

12. $c_n = n$-place decimal approximation to e

13. Find a formula for the nth term of each sequence.

(a) $\dfrac{1}{1}, \dfrac{-1}{8}, \dfrac{1}{27}, \ldots$ **(b)** $\dfrac{2}{6}, \dfrac{3}{7}, \dfrac{4}{8}, \ldots$

14. Suppose that $\lim_{n\to\infty} a_n = 4$ and $\lim_{n\to\infty} b_n = 7$. Determine:

(a) $\lim_{n\to\infty} (a_n + b_n)$ **(b)** $\lim_{n\to\infty} a_n^3$

(c) $\lim_{n\to\infty} \cos(\pi b_n)$ **(d)** $\lim_{n\to\infty} (a_n^2 - 2a_n b_n)$

In Exercises 15–26, use Theorem 1 to determine the limit of the sequence or state that the sequence diverges.

15. $a_n = 12$ **16.** $a_n = 20 - \dfrac{4}{n^2}$

17. $b_n = \dfrac{5n-1}{12n+9}$ **18.** $a_n = \dfrac{4+n-3n^2}{4n^2+1}$

19. $c_n = -2^{-n}$ **20.** $z_n = \left(\dfrac{1}{3}\right)^n$

21. $c_n = 9^n$ **22.** $z_n = 10^{-1/n}$

23. $a_n = \dfrac{n}{\sqrt{n^2+1}}$ **24.** $a_n = \dfrac{n}{\sqrt{n^3+1}}$

25. $a_n = \ln\left(\dfrac{12n+2}{-9+4n}\right)$ **26.** $r_n = \ln n - \ln(n^2+1)$

In Exercises 27–30, use Theorem 4 to determine the limit of the sequence.

27. $a_n = \sqrt{4 + \dfrac{1}{n}}$ **28.** $a_n = e^{4n/(3n+9)}$

29. $a_n = \cos^{-1}\left(\dfrac{n^3}{2n^3+1}\right)$ **30.** $a_n = \tan^{-1}(e^{-n})$

31. Let $a_n = \dfrac{n}{n+1}$. Find a number M such that:

(a) $|a_n - 1| \le 0.001$ for $n \ge M$.

(b) $|a_n - 1| \le 0.00001$ for $n \ge M$.

Then use the limit definition to prove that $\lim_{n\to\infty} a_n = 1$.

32. Let $b_n = \left(\frac{1}{3}\right)^n$.

(a) Find a value of M such that $|b_n| \le 10^{-5}$ for $n \ge M$.

(b) Use the limit definition to prove that $\lim_{n\to\infty} b_n = 0$.

33. Use the limit definition to prove that $\lim_{n\to\infty} n^{-2} = 0$.

34. Use the limit definition to prove that $\lim_{n\to\infty} \dfrac{n}{n+n^{-1}} = 1$.

In Exercises 35–62, use the appropriate limit laws and theorems to determine the limit of the sequence or show that it diverges.

35. $a_n = 10 + \left(-\dfrac{1}{9}\right)^n$ **36.** $d_n = \sqrt{n+3} - \sqrt{n}$

37. $c_n = 1.01^n$ **38.** $b_n = e^{1-n^2}$

39. $a_n = 2^{1/n}$ **40.** $b_n = n^{1/n}$

41. $c_n = \dfrac{9^n}{n!}$ **42.** $a_n = \dfrac{8^{2n}}{n!}$

43. $a_n = \dfrac{3n^2+n+2}{2n^2-3}$ **44.** $a_n = \dfrac{\sqrt{n}}{\sqrt{n}+4}$

45. $a_n = \dfrac{\cos n}{n}$ **46.** $c_n = \dfrac{(-1)^n}{\sqrt{n}}$

47. $d_n = \ln 5^n - \ln n!$

48. $d_n = \ln(n^2+4) - \ln(n^2-1)$

49. $a_n = \left(2 + \dfrac{4}{n^2}\right)^{1/3}$ **50.** $b_n = \tan^{-1}\left(1 - \dfrac{2}{n}\right)$

51. $c_n = \ln\left(\dfrac{2n+1}{3n+4}\right)$ **52.** $c_n = \dfrac{n}{n+n^{1/n}}$

53. $y_n = \dfrac{e^n}{2^n}$ **54.** $a_n = \dfrac{n}{2^n}$

55. $y_n = \dfrac{e^n + (-3)^n}{5^n}$

56. $b_n = \dfrac{(-1)^n n^3 + 2^{-n}}{3n^3 + 4^{-n}}$

57. $a_n = n \sin \dfrac{\pi}{n}$

58. $b_n = \dfrac{n!}{\pi^n}$

59. $b_n = \dfrac{3 - 4^n}{2 + 7 \cdot 4^n}$

60. $a_n = \dfrac{3 - 4^n}{2 + 7 \cdot 3^n}$

61. $a_n = \left(1 + \dfrac{1}{n}\right)^n$

62. $a_n = \left(1 + \dfrac{1}{n^2}\right)^n$

In Exercises 63–66, find the limit of the sequence using L'Hôpital's Rule.

63. $a_n = \dfrac{(\ln n)^2}{n}$

64. $b_n = \sqrt{n} \ln \left(1 + \dfrac{1}{n}\right)$

65. $c_n = n\left(\sqrt{n^2 + 1} - n\right)$

66. $d_n = n^2\left(\sqrt[3]{n^3 + 1} - n\right)$

In Exercises 67–70, use the Squeeze Theorem to evaluate $\lim_{n\to\infty} a_n$ *by verifying the given inequality.*

67. $a_n = \dfrac{1}{\sqrt{n^4 + n^8}}, \quad \dfrac{1}{\sqrt{2n^4}} \le a_n \le \dfrac{1}{\sqrt{2n^2}}$

68. $c_n = \dfrac{1}{\sqrt{n^2 + 1}} + \dfrac{1}{\sqrt{n^2 + 2}} + \cdots + \dfrac{1}{\sqrt{n^2 + n}},$

$\dfrac{n}{\sqrt{n^2 + n}} \le c_n \le \dfrac{n}{\sqrt{n^2 + 1}}$

69. $a_n = (2^n + 3^n)^{1/n}, \quad 3 \le a_n \le (2 \cdot 3^n)^{1/n} = 2^{1/n} \cdot 3$

70. $a_n = (n + 10^n)^{1/n}, \quad 10 \le a_n \le (2 \cdot 10^n)^{1/n}$

71. Which of the following statements is equivalent to the assertion $\lim_{n\to\infty} a_n = L$? Explain.

(a) For every $\epsilon > 0$, the interval $(L - \epsilon, L + \epsilon)$ contains at least one element of the sequence $\{a_n\}$.

(b) For every $\epsilon > 0$, the interval $(L - \epsilon, L + \epsilon)$ contains all but at most finitely many elements of the sequence $\{a_n\}$.

72. Show that $a_n = \dfrac{1}{2n + 1}$ is decreasing.

73. Show that $a_n = \dfrac{3n^2}{n^2 + 2}$ is increasing. Find an upper bound.

74. Show that $a_n = \sqrt[3]{n + 1} - n$ is decreasing.

75. Give an example of a divergent sequence $\{a_n\}$ such that $\lim_{n\to\infty} |a_n|$ converges.

76. Give an example of *divergent* sequences $\{a_n\}$ and $\{b_n\}$ such that $\{a_n + b_n\}$ converges.

77. Using the limit definition, prove that if $\{a_n\}$ converges and $\{b_n\}$ diverges, then $\{a_n + b_n\}$ diverges.

78. Use the limit definition to prove that if $\{a_n\}$ is a convergent sequence of integers with limit L, then there exists a number M such that $a_n = L$ for all $n \ge M$.

79. Theorem 1 states that if $\lim_{x\to\infty} f(x) = L$, then the sequence $a_n = f(n)$ converges and $\lim_{n\to\infty} a_n = L$. Show that the *converse* is false. In other words, find a function $f(x)$ such that $a_n = f(n)$ converges but $\lim_{x\to\infty} f(x)$ does not exist.

80. Use the limit definition to prove that the limit does not change if a finite number of terms are added or removed from a convergent sequence.

81. Let $b_n = a_{n+1}$. Use the limit definition to prove that if $\{a_n\}$ converges, then $\{b_n\}$ also converges and $\lim_{n\to\infty} a_n = \lim_{n\to\infty} b_n$.

82. Let $\{a_n\}$ be a sequence such that $\lim_{n\to\infty} |a_n|$ exists and is nonzero. Show that $\lim_{n\to\infty} a_n$ exists if and only if there exists an integer M such that the sign of a_n does not change for $n > M$.

83. Proceed as in Example 12 to show that the sequence $\sqrt{3}$, $\sqrt{3\sqrt{3}}$, $\sqrt{3\sqrt{3\sqrt{3}}}, \dots$ is increasing and bounded above by $M = 3$. Then prove that the limit exists and find its value.

84. Let $\{a_n\}$ be the sequence defined recursively by

$$a_0 = 0, \qquad a_{n+1} = \sqrt{2 + a_n}$$

Thus, $a_1 = \sqrt{2}, \quad a_2 = \sqrt{2 + \sqrt{2}}, \quad a_3 = \sqrt{2 + \sqrt{2 + \sqrt{2}}}, \dots.$

(a) Show that if $a_n < 2$, then $a_{n+1} < 2$. Conclude by induction that $a_n < 2$ for all n.

(b) Show that if $a_n < 2$, then $a_n \le a_{n+1}$. Conclude by induction that $\{a_n\}$ is increasing.

(c) Use (a) and (b) to conclude that $L = \lim_{n\to\infty} a_n$ exists. Then compute L by showing that $L = \sqrt{2 + L}$.

Further Insights and Challenges

85. Show that $\lim_{n\to\infty} \sqrt[n]{n!} = \infty$. *Hint:* Verify that $n! \ge (n/2)^{n/2}$ by observing that half of the factors of $n!$ are greater than or equal to $n/2$.

86. Let $b_n = \dfrac{\sqrt[n]{n!}}{n}$.

(a) Show that $\ln b_n = \dfrac{1}{n} \sum_{k=1}^{n} \ln \dfrac{k}{n}$.

(b) Show that $\ln b_n$ converges to $\int_0^1 \ln x \, dx$, and conclude that $b_n \to e^{-1}$.

87. Given positive numbers $a_1 < b_1$, define two sequences recursively by

$$a_{n+1} = \sqrt{a_n b_n}, \qquad b_{n+1} = \frac{a_n + b_n}{2}$$

(a) Show that $a_n \le b_n$ for all n (Figure 13).

(b) Show that $\{a_n\}$ is increasing and $\{b_n\}$ is decreasing.

(c) Show that $b_{n+1} - a_{n+1} \le \dfrac{b_n - a_n}{2}$.

(d) Prove that both $\{a_n\}$ and $\{b_n\}$ converge and have the same limit. This limit, denoted $\mathrm{AGM}(a_1, b_1)$, is called the **arithmetic-geometric mean** of a_1 and b_1.

(e) Estimate $\mathrm{AGM}(1, \sqrt{2})$ to three decimal places.

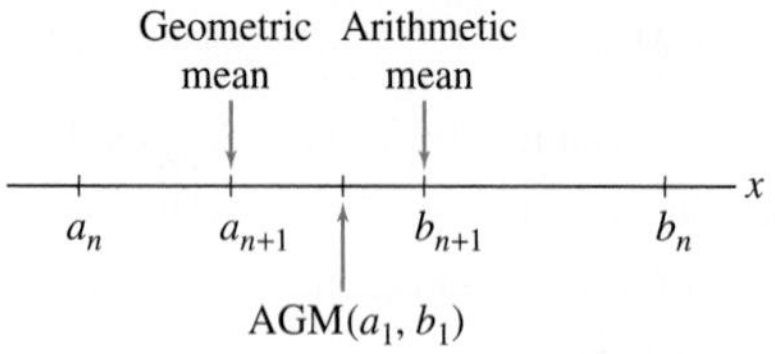

FIGURE 13

88. Let $c_n = \dfrac{1}{n} + \dfrac{1}{n+1} + \dfrac{1}{n+2} + \cdots + \dfrac{1}{2n}$.

(a) Calculate c_1, c_2, c_3, c_4.

(b) Use a comparison of rectangles with the area under $y = x^{-1}$ over the interval $[n, 2n]$ to prove that

$$\int_n^{2n} \frac{dx}{x} + \frac{1}{2n} \le c_n \le \int_n^{2n} \frac{dx}{x} + \frac{1}{n}$$

(c) Use the Squeeze Theorem to determine $\lim_{n\to\infty} c_n$.

89. Let $a_n = H_n - \ln n$, where H_n is the nth harmonic number

$$H_n = 1 + \frac{1}{2} + \frac{1}{3} + \cdots + \frac{1}{n}$$

(a) Show that $a_n \ge 0$ for $n \ge 1$. *Hint:* Show that $H_n \ge \displaystyle\int_1^{n+1} \frac{dx}{x}$.

(b) Show that $\{a_n\}$ is decreasing by interpreting $a_n - a_{n+1}$ as an area.

(c) Prove that $\lim_{n\to\infty} a_n$ exists.

This limit, denoted γ, is known as *Euler's Constant*. It appears in many areas of mathematics, including analysis and number theory, and has been calculated to more than 100 million decimal places, but it is still not known whether γ is an irrational number. The first 10 digits are $\gamma \approx 0.5772156649$.

11.2 Summing an Infinite Series

Many quantities that arise in applications cannot be computed exactly. We cannot write down an exact decimal expression for the number π or for values of the sine function such as sin 1. However, sometimes these quantities can be represented as infinite sums. For example, using Taylor series (Section 11.7), we can show that

$$\sin 1 = 1 - \frac{1}{3!} + \frac{1}{5!} - \frac{1}{7!} + \frac{1}{9!} - \frac{1}{11!} + \cdots \qquad \boxed{1}$$

Infinite sums of this type are called **infinite series**.

What precisely does Eq. (1) mean? It is impossible to add up infinitely many numbers, but what we can do is compute the **partial sums** S_N, defined as the finite sum of the terms up to and including Nth term. Here are the first five partial sums of the infinite series for sin 1:

$$S_1 = 1$$

$$S_2 = 1 - \frac{1}{3!} = 1 - \frac{1}{6} \approx 0.833$$

$$S_3 = 1 - \frac{1}{3!} + \frac{1}{5!} = 1 - \frac{1}{6} + \frac{1}{120} \approx 0.841667$$

$$S_4 = 1 - \frac{1}{6} + \frac{1}{120} - \frac{1}{5040} \approx 0.841468$$

$$S_5 = 1 - \frac{1}{6} + \frac{1}{120} - \frac{1}{5040} + \frac{1}{362{,}880} \approx \mathbf{0.8414709846}$$

Compare these values with the value obtained from a calculator:

$$\sin 1 \approx \mathbf{0.8414709848}079 \qquad \text{(calculator value)}$$

We see that S_5 differs from $\sin 1$ by less than 10^{-9}. This suggests that the partial sums converge to $\sin 1$, and in fact, we will prove that

$$\sin 1 = \lim_{N\to\infty} S_N$$

(Example 2 in Section 11.7). So although we cannot add up infinitely many numbers, it makes sense to *define* the sum of an infinite series as a limit of partial sums.

In general, an infinite series is an expression of the form

$$\sum_{n=1}^{\infty} a_n = a_1 + a_2 + a_3 + a_4 + \cdots$$

where $\{a_n\}$ is any sequence. For example,

Sequence	General term	Infinite series
$\frac{1}{3}, \frac{1}{9}, \frac{1}{27}, \dots$	$a_n = \frac{1}{3^n}$	$\sum_{n=1}^{\infty} \frac{1}{3^n} = \frac{1}{3} + \frac{1}{9} + \frac{1}{27} + \frac{1}{81} + \cdots$
$\frac{1}{1}, \frac{1}{4}, \frac{1}{9}, \frac{1}{16}, \dots$	$a_n = \frac{1}{n^2}$	$\sum_{n=1}^{\infty} \frac{1}{n^2} = \frac{1}{1} + \frac{1}{4} + \frac{1}{9} + \frac{1}{16} + \cdots$

- *Infinite series may begin with any index. For example,*

$$\sum_{n=3}^{\infty} \frac{1}{n} = \frac{1}{3} + \frac{1}{4} + \frac{1}{5} + \cdots$$

When it is not necessary to specify the starting point, we write simply $\sum a_n$.
- *Any letter may be used for the index. Thus, we may write* a_m, a_k, a_i, *etc.*

The Nth partial sum S_N is the finite sum of the terms up to and including a_N:

$$S_N = \sum_{n=1}^{N} a_n = a_1 + a_2 + a_3 + \cdots + a_N$$

If the series begins at k, then $S_N = \sum_{n=k}^{N} a_n$.

DEFINITION Convergence of an Infinite Series An infinite series $\sum_{n=k}^{\infty} a_n$ *converges* to the sum S if its partial sums converge to S:

$$\lim_{N\to\infty} S_N = S$$

In this case, we write $S = \sum_{n=k}^{\infty} a_n$.

- If the limit does not exist, we say that the infinite series diverges.
- If the limit is infinite, we say that the infinite series diverges to infinity.

We can investigate series numerically by computing several partial sums S_N. If the partial sums show a trend of convergence to some number S, then we have evidence (but not proof) that the series converges to S. The next example treats a **telescoping series**, where the partial sums are particularly easy to evaluate.

■ **EXAMPLE 1 Telescoping Series** Investigate numerically:

$$S = \sum_{n=1}^{\infty} \frac{1}{n(n+1)} = \frac{1}{1(2)} + \frac{1}{2(3)} + \frac{1}{3(4)} + \frac{1}{4(5)} + \cdots$$

Then compute the sum S using the identity:

$$\frac{1}{n(n+1)} = \frac{1}{n} - \frac{1}{n+1}$$

Solution The values of the partial sums listed in Table 1 suggest convergence to $S = 1$. To prove this, we observe that because of the identity, each partial sum collapses down to just two terms:

$$S_1 = \frac{1}{1(2)} = \frac{1}{1} - \frac{1}{2}$$

$$S_2 = \frac{1}{1(2)} + \frac{1}{2(3)} = \left(\frac{1}{1} - \frac{\cancel{1}}{\cancel{2}}\right) + \left(\frac{\cancel{1}}{\cancel{2}} - \frac{1}{3}\right) = 1 - \frac{1}{3}$$

$$S_3 = \frac{1}{1(2)} + \frac{1}{2(3)} + \frac{1}{3(4)} = \left(\frac{1}{1} - \frac{\cancel{1}}{\cancel{2}}\right) + \left(\frac{\cancel{1}}{\cancel{2}} - \frac{\cancel{1}}{\cancel{3}}\right) + \left(\frac{\cancel{1}}{\cancel{3}} - \frac{1}{4}\right) = 1 - \frac{1}{4}$$

In general,

$$S_N = \left(\frac{1}{1} - \frac{\cancel{1}}{\cancel{2}}\right) + \left(\frac{\cancel{1}}{\cancel{2}} - \frac{\cancel{1}}{\cancel{3}}\right) + \cdots + \left(\frac{\cancel{1}}{\cancel{N}} - \frac{1}{N+1}\right) = 1 - \frac{1}{N+1} \qquad \boxed{2}$$

The sum S is the limit of the partial sums:

$$S = \lim_{N \to \infty} S_N = \lim_{N \to \infty} \left(1 - \frac{1}{N+1}\right) = 1 \qquad ■$$

TABLE 1 **Partial Sums**
for $\displaystyle\sum_{n=1}^{\infty} \frac{1}{n(n+1)}$

N	S_N
10	0.90909
50	0.98039
100	0.990099
200	0.995025
300	0.996678

In most cases (apart from telescoping series and the geometric series introduced below), there is no simple formula like Eq. (2) for the partial sum S_N. Therefore, we shall develop techniques that do not rely on formulas for S_N.

It is important to keep in mind the difference between a sequence $\{a_n\}$ and an infinite series $\displaystyle\sum_{n=1}^{\infty} a_n$.

■ **EXAMPLE 2 Sequences versus Series** Discuss the difference between $\{a_n\}$ and $\displaystyle\sum_{n=1}^{\infty} a_n$, where $a_n = \frac{1}{n(n+1)}$.

Solution The sequence is the list of numbers $\frac{1}{1(2)}, \frac{1}{2(3)}, \frac{1}{3(4)}, \ldots$. This sequence converges to zero:

$$\lim_{n \to \infty} a_n = \lim_{n \to \infty} \frac{1}{n(n+1)} = 0$$

The infinite series is the *sum* of the numbers a_n, defined formally as the limit of the partial sums. This sum is not zero. In fact, the sum is equal to 1 by Example 1:

$$\sum_{n=1}^{\infty} a_n = \sum_{n=1}^{\infty} \frac{1}{n(n+1)} = \frac{1}{1(2)} + \frac{1}{2(3)} + \frac{1}{3(4)} + \cdots = 1 \qquad ■$$

Make sure you understand the difference between sequences and series.

• With a sequence, we consider the limit of the individual terms a_n.

• With a series, we are interested in the sum of the terms

$$a_1 + a_2 + a_3 + \cdots$$

which is defined as the limit of the partial sums.

The next theorem shows that infinite series may be added or subtracted like ordinary sums, *provided that the series converge.*

THEOREM 1 Linearity of Infinite Series If $\sum a_n$ and $\sum b_n$ converge, then $\sum (a_n \pm b_n)$ and $\sum c a_n$ also converge (c any constant), and

$$\sum a_n + \sum b_n = \sum (a_n + b_n)$$

$$\sum a_n - \sum b_n = \sum (a_n - b_n)$$

$$\sum c a_n = c \sum a_n \qquad (c \text{ any constant})$$

Proof These rules follow from the corresponding linearity rules for limits. For example,

$$\sum_{n=1}^{\infty} (a_n + b_n) = \lim_{N\to\infty} \sum_{n=1}^{N} (a_n + b_n) = \lim_{N\to\infty} \left(\sum_{n=1}^{N} a_n + \sum_{n=1}^{N} b_n \right)$$

$$= \lim_{N\to\infty} \sum_{n=1}^{N} a_n + \lim_{N\to\infty} \sum_{n=1}^{\infty} b_n = \sum_{n=1}^{\infty} a_n + \sum_{n=1}^{\infty} b_n \qquad \blacksquare$$

A main goal in this chapter is to develop techniques for determining whether a series converges or diverges. It is easy to give examples of series that diverge:

- $S = \sum_{n=1}^{\infty} 1$ diverges to infinity (the partial sums increase without bound):

$$S_1 = 1, \quad S_2 = 1 + 1 = 2, \quad S_3 = 1 + 1 + 1 = 3, \quad S_4 = 1 + 1 + 1 + 1 = 4, \quad \ldots$$

- $\sum_{n=1}^{\infty} (-1)^{n-1}$ diverges (the partial sums jump between 1 and 0):

$$S_1 = 1, \quad S_2 = 1 - 1 = 0, \quad S_3 = 1 - 1 + 1 = 1, \quad S_4 = 1 - 1 + 1 - 1 = 0, \quad \ldots$$

Next, we study the geometric series, which converge or diverge depending on the common ratio r.

A **geometric series** with common ratio $r \neq 0$ is a series defined by a geometric sequence cr^n, where $c \neq 0$. If the series begins at $n = 0$, then

$$S = \sum_{n=0}^{\infty} cr^n = c + cr + cr^2 + cr^3 + cr^4 + cr^5 + \cdots$$

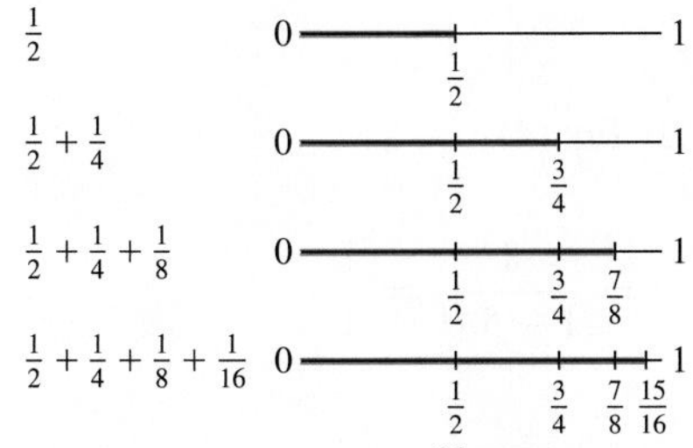

FIGURE 1 Partial sums of $\sum_{n=1}^{\infty} \frac{1}{2^n}$.

For $r = \frac{1}{2}$ and $c = 1$, we can visualize the geometric series starting at $n = 1$ (Figure 1):

$$S = \sum_{n=1}^{\infty} \frac{1}{2^n} = \frac{1}{2} + \frac{1}{4} + \frac{1}{8} + \frac{1}{16} + \cdots = 1$$

Adding up the terms corresponds to moving stepwise from 0 to 1, where each step is a move to the right by half of the remaining distance. Thus $S = 1$.

There is a simple device for computing the partial sums of a geometric series:

$$\begin{aligned} S_N &= c + cr + cr^2 + cr^3 + \cdots + cr^N \\ rS_N &= \qquad cr + cr^2 + cr^3 + \cdots + cr^N + cr^{N+1} \\ S_N - rS_N &= c - cr^{N+1} \\ S_N(1 - r) &= c(1 - r^{N+1}) \end{aligned}$$

If $r \neq 1$, we may divide by $(1 - r)$ to obtain

$$S_N = c + cr + cr^2 + cr^3 + \cdots + cr^N = \frac{c(1 - r^{N+1})}{1 - r} \tag{3}$$

This formula enables us to sum the geometric series.

Geometric series are important because they

- *arise often in applications.*
- *can be evaluated explicitly.*
- *are used to study other, nongeometric series (by comparison).*

THEOREM 2 Sum of a Geometric Series Let $c \neq 0$. If $|r| < 1$, then

$$\sum_{n=0}^{\infty} cr^n = c + cr + cr^2 + cr^3 + \cdots = \frac{c}{1 - r} \tag{4}$$

$$\sum_{n=M}^{\infty} cr^n = cr^M + cr^{M+1} + cr^{M+2} + cr^{M+3} + \cdots = \frac{cr^M}{1 - r} \tag{5}$$

If $|r| \geq 1$, then the geometric series diverges.

Proof If $r = 1$, then the series certainly diverges because the partial sums $S_N = Nc$ grow arbitrarily large. If $r \neq 1$, then Eq. (3) yields

$$S = \lim_{N\to\infty} S_N = \lim_{N\to\infty} \frac{c(1 - r^{N+1})}{1 - r} = \frac{c}{1 - r} - \frac{c}{1 - r} \lim_{N\to\infty} r^{N+1}$$

If $|r| < 1$, then $\lim_{N\to\infty} r^{N+1} = 0$ and we obtain Eq. (4). If $|r| \geq 1$ and $r \neq 1$, then $\lim_{N\to\infty} r^{N+1}$ does not exist and the geometric series diverges. Finally, if the series starts with cr^M rather than cr^0, then

$$S = cr^M + cr^{M+1} + cr^{M+2} + cr^{M+3} + \cdots = r^M \sum_{n=0}^{\infty} cr^n = \frac{cr^M}{1 - r} \quad ■$$

■ **EXAMPLE 3** Evaluate $\displaystyle\sum_{n=0}^{\infty} 5^{-n}$.

Solution This is a geometric series with $r = 5^{-1}$. By Eq. (4),

$$\sum_{n=0}^{\infty} 5^{-n} = 1 + \frac{1}{5} + \frac{1}{5^2} + \frac{1}{5^3} + \cdots = \frac{1}{1 - 5^{-1}} = \frac{5}{4} \quad ■$$

■ **EXAMPLE 4** Evaluate $\displaystyle\sum_{n=3}^{\infty} 7\left(-\frac{3}{4}\right)^n = 7\left(-\frac{3}{4}\right)^3 + 7\left(-\frac{3}{4}\right)^4 + 7\left(-\frac{3}{4}\right)^5 + \cdots$.

Solution This is a geometric series with $r = -\frac{3}{4}$ and $c = 7$, starting at $n = 3$. By Eq. (5),

$$\sum_{n=3}^{\infty} 7\left(-\frac{3}{4}\right)^n = \frac{7(-\frac{3}{4})^3}{1 - (-\frac{3}{4})} = -\frac{27}{16} \quad ■$$

■ **EXAMPLE 5** Evaluate $S = \sum_{n=0}^{\infty} \frac{2+3^n}{5^n}$.

Solution Write S as a sum of two geometric series. This is valid by Theorem 1 because both geometric series converge:

$$\sum_{n=0}^{\infty} \frac{2+3^n}{5^n} = \sum_{n=0}^{\infty} \frac{2}{5^n} + \sum_{n=0}^{\infty} \frac{3^n}{5^n} = \overbrace{2\sum_{n=0}^{\infty} \frac{1}{5^n} + \sum_{n=0}^{\infty} \left(\frac{3}{5}\right)^n}^{\text{Both geometric series converge}}$$

$$= 2 \cdot \frac{1}{1-\frac{1}{5}} + \frac{1}{1-\frac{3}{5}} = 5$$ ■

CONCEPTUAL INSIGHT Sometimes, the following *incorrect argument* is given for summing a geometric series:

$$S = \frac{1}{2} + \frac{1}{4} + \frac{1}{8} + \cdots$$

$$2S = 1 + \frac{1}{2} + \frac{1}{4} + \frac{1}{8} + \cdots = 1 + S$$

Thus, $2S = 1 + S$, or $S = 1$. The answer is correct, so why is the argument wrong? It is wrong because we do not know in advance that the series converges. Observe what happens when this argument is applied to a divergent series:

$$S = 1 + 2 + 4 + 8 + 16 + \cdots$$

$$2S = 2 + 4 + 8 + 16 + \cdots = S - 1$$

This would yield $2S = S - 1$, or $S = -1$, which is absurd because S diverges. We avoid such erroneous conclusions by carefully defining the sum of an infinite series as the limit of partial sums.

The infinite series $\sum_{k=1}^{\infty} 1$ diverges because the Nth partial sum $S_N = N$ diverges to infinity. It is less clear whether the following series converges or diverges:

$$\sum_{n=1}^{\infty} (-1)^{n+1} \frac{n}{n+1} = \frac{1}{2} - \frac{2}{3} + \frac{3}{4} - \frac{4}{5} + \frac{5}{6} - \cdots$$

We now introduce a useful test that allows us to conclude that this series diverges.

*The Divergence Test (also called the **nth-Term Test**) is often stated as follows:*

If $\sum_{n=1}^{\infty} a_n$ converges, then $\lim_{n\to\infty} a_n = 0$.

In practice, we use it to prove that a given series diverges.

THEOREM 3 Divergence Test If the nth term a_n does not converge to zero, then the series $\sum_{n=1}^{\infty} a_n$ diverges.

Proof First, note that $a_n = S_n - S_{n-1}$ because

$$S_n = (a_1 + a_2 + \cdots + a_{n-1}) + a_n = S_{n-1} + a_n$$

If $\sum_{n=1}^{\infty} a_n$ converges with sum S, then

$$\lim_{n\to\infty} a_n = \lim_{n\to\infty} (S_n - S_{n-1}) = \lim_{n\to\infty} S_n - \lim_{n\to\infty} S_{n-1} = S - S = 0$$

Therefore, if a_n does not converge to zero, $\sum_{n=1}^{\infty} a_n$ cannot converge. ■

■ **EXAMPLE 6** Prove the divergence of $S = \sum_{n=1}^{\infty} \frac{n}{4n+1}$.

Solution We have

$$\lim_{n\to\infty} a_n = \lim_{n\to\infty} \frac{n}{4n+1} = \lim_{n\to\infty} \frac{1}{4+1/n} = \frac{1}{4}$$

The nth term a_n does not converge to zero, so the series diverges by Theorem 3. ■

■ **EXAMPLE 7** Determine the convergence or divergence of

$$S = \sum_{n=1}^{\infty} (-1)^{n-1} \frac{n}{n+1} = \frac{1}{2} - \frac{2}{3} + \frac{3}{4} - \frac{4}{5} + \cdots$$

Solution The general term $a_n = (-1)^{n-1} \frac{n}{n+1}$ does not approach a limit. Indeed, $\frac{n}{n+1}$ tends to 1, so the odd terms a_{2n+1} tend to 1, and the even terms a_{2n} tend to -1. Because $\lim_{n\to\infty} a_n$ does not exist, the series S diverges by Theorem 3. ■

The Divergence Test tells only part of the story. If a_n does not tend to zero, then $\sum a_n$ certainly diverges. But what if a_n does tend to zero? In this case, the series may converge or it may diverge. In other words, $\lim_{n\to\infty} a_n = 0$ is a *necessary* condition of convergence, but it is *not sufficient*. As we show in the next example, it is possible for a series to diverge even though its terms tend to zero.

■ **EXAMPLE 8 Sequence Tends to Zero, yet the Series Diverges** Prove the divergence of

$$\sum_{n=1}^{\infty} \frac{1}{\sqrt{n}} = \frac{1}{\sqrt{1}} + \frac{1}{\sqrt{2}} + \frac{1}{\sqrt{3}} + \cdots$$

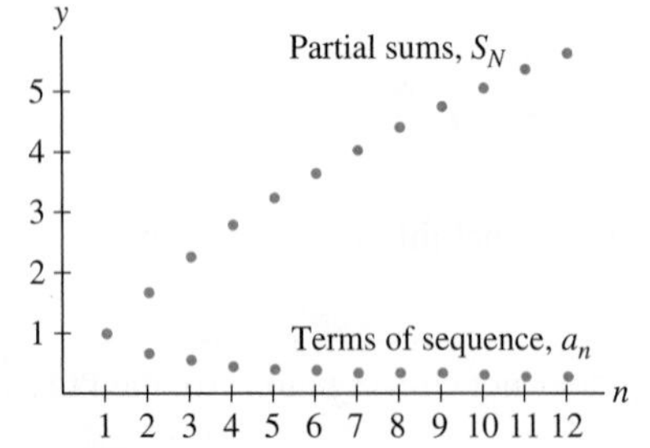

FIGURE 2 The partial sums of $\sum_{n=1}^{\infty} \frac{1}{\sqrt{n}}$ diverge even though the terms $a_n = 1/\sqrt{n}$ tend to zero.

Solution The general term $1/\sqrt{N}$ tends to zero. However, because each term in the sum S_N is greater than or equal to $1/\sqrt{N}$, we have

$$S_N = \overbrace{\frac{1}{\sqrt{1}} + \frac{1}{\sqrt{2}} + \cdots + \frac{1}{\sqrt{N}}}^{N \text{ terms}}$$
$$\geq \frac{1}{\sqrt{N}} + \frac{1}{\sqrt{N}} + \cdots + \frac{1}{\sqrt{N}}$$
$$= N\left(\frac{1}{\sqrt{N}}\right) = \sqrt{N}$$

This shows that $S_N \geq \sqrt{N}$. But $\sqrt{N}$ increases without bound (Figure 2). Therefore S_N also increases without bound. This proves that the series diverges. ■

Archimedes (287 BCE–212 BCE), who discovered the law of the lever, said "Give me a place to stand on, and I can move the earth" (quoted by Pappus of Alexandria c. AD 340).

HISTORICAL PERSPECTIVE

Geometric series were used as early as the third century BCE by Archimedes in a brilliant argument for determining the area S of a "parabolic segment" (shaded region in Figure 3). Given two points A and C on a parabola, there is a point B between A and C where the tangent line is parallel to $\overline{AC}$ (apparently, Archimedes knew the Mean Value Theorem more than 2000 years before the invention of calculus). Let T be the area of triangle $\triangle ABC$. Archimedes proved that if D is chosen in a similar fashion relative to $\overline{AB}$ and E is chosen relative to $\overline{BC}$, then

$$\frac{1}{4}T = \text{Area}(\triangle ADB) + \text{Area}(\triangle BEC) \qquad \boxed{6}$$

This construction of triangles can be continued. The next step would be to construct the four triangles on the segments $\overline{AD}$, $\overline{DB}$, $\overline{BE}$, $\overline{EC}$, of total area $\left(\frac{1}{4}\right)^2 T$. Then construct eight triangles of total area $\left(\frac{1}{4}\right)^3 T$, etc. In this way, we obtain infinitely many triangles that completely fill up the parabolic segment. By the formula for the sum of a geometric series,

$$S = T + \frac{1}{4}T + \frac{1}{16}T + \cdots = T\sum_{n=0}^{\infty}\frac{1}{4^n} = \frac{4}{3}T$$

For this and many other achievements, Archimedes is ranked together with Newton and Gauss as one of the greatest scientists of all time.

The modern study of infinite series began in the seventeenth century with Newton, Leibniz, and their contemporaries. The divergence of $\sum_{n=1}^{\infty} 1/n$ (called the **harmonic series**) was known to the medieval scholar Nicole d'Oresme (1323–1382), but his proof was lost for centuries, and the result was rediscovered on more than one occasion. It was also known that the sum of the reciprocal squares $\sum_{n=1}^{\infty} 1/n^2$ converges, and in the 1640s, the Italian Pietro Mengoli put forward the challenge of finding its sum. Despite the efforts of the best mathematicians of the day, including Leibniz and the Bernoulli brothers Jakob and Johann, the problem resisted solution for nearly a century. In 1735, the great master Leonhard Euler (at the time, 28 years old) astonished his contemporaries by proving that

$$\frac{1}{1^2} + \frac{1}{2^2} + \frac{1}{3^2} + \frac{1}{4^2} + \frac{1}{5^2} + \frac{1}{6^2} + \cdots = \frac{\pi^2}{6}$$

This formula, surprising in itself, plays a role in a variety of mathematical fields. A theorem from number theory states that two whole numbers, chosen randomly, have no common factor with probability $6/\pi^2 \approx 0.6$ (the reciprocal of Euler's result). On the other hand, Euler's result and its generalizations appear in the field of statistical mechanics.

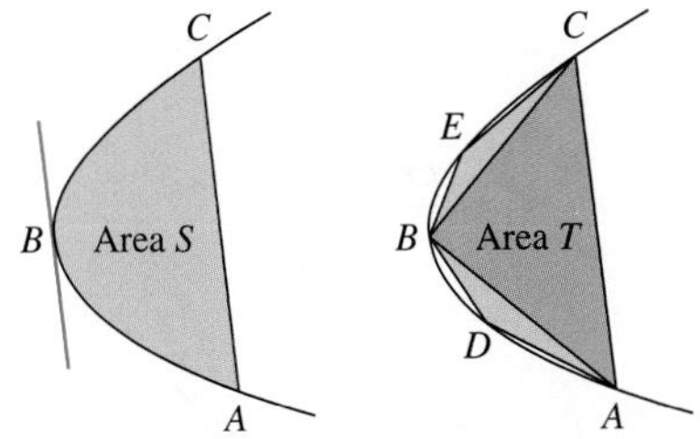

FIGURE 3 Archimedes showed that the area S of the parabolic segment is $\frac{4}{3}T$, where T is the area of $\triangle ABC$.

11.2 SUMMARY

- An *infinite series* is an expression

$$\sum_{n=1}^{\infty} a_n = a_1 + a_2 + a_3 + a_4 + \cdots$$

We call a_n the *general term* of the series. An infinite series can begin at $n = k$ for any integer k.

- The Nth *partial sum* is the finite sum of the terms up to and including the Nth term:

$$S_N = \sum_{n=1}^{N} a_n = a_1 + a_2 + a_3 + \cdots + a_N$$

- By definition, the sum of an infinite series is the limit $S = \lim_{N\to\infty} S_N$. If the limit exists, we say that the infinite series is *convergent* or *converges* to the sum S. If the limit does not exist, we say that the infinite series *diverges*.

- If the partial sums S_N increase without bound, we say that S diverges to infinity.
- *Divergence Test:* If $\{a_n\}$ does not tend to zero, then $\sum_{n=1}^{\infty} a_n$ diverges. However, a series may diverge even if its general term $\{a_n\}$ tends to zero.
- Partial sum of a geometric series:

$$c + cr + cr^2 + cr^3 + \cdots + cr^N = \frac{c(1 - r^{N+1})}{1 - r}$$

- *Geometric series*: If $|r| < 1$, then

$$\sum_{n=0}^{\infty} r^n = 1 + r + r^2 + r^3 + \cdots = \frac{1}{1-r}$$

$$\sum_{n=M}^{\infty} cr^n = cr^M + cr^{M+1} + cr^{M+2} + \cdots = \frac{cr^M}{1-r}$$

The geometric series diverges if $|r| \geq 1$.

11.2 EXERCISES

Preliminary Questions

1. What role do partial sums play in defining the sum of an infinite series?

2. What is the sum of the following infinite series?

$$\frac{1}{4} + \frac{1}{8} + \frac{1}{16} + \frac{1}{32} + \frac{1}{64} + \cdots$$

3. What happens if you apply the formula for the sum of a geometric series to the following series? Is the formula valid?

$$1 + 3 + 3^2 + 3^3 + 3^4 + \cdots$$

4. Arvind asserts that $\sum_{n=1}^{\infty} \frac{1}{n^2} = 0$ because $\frac{1}{n^2}$ tends to zero. Is this valid reasoning?

5. Colleen claims that $\sum_{n=1}^{\infty} \frac{1}{\sqrt{n}}$ converges because

$$\lim_{n\to\infty} \frac{1}{\sqrt{n}} = 0$$

Is this valid reasoning?

6. Find an N such that $S_N > 25$ for the series $\sum_{n=1}^{\infty} 2$.

7. Does there exist an N such that $S_N > 25$ for the series $\sum_{n=1}^{\infty} 2^{-n}$? Explain.

8. Give an example of a divergent infinite series whose general term tends to zero.

Exercises

1. Find a formula for the general term a_n (not the partial sum) of the infinite series.

(a) $\frac{1}{3} + \frac{1}{9} + \frac{1}{27} + \frac{1}{81} + \cdots$ **(b)** $\frac{1}{1} + \frac{5}{2} + \frac{25}{4} + \frac{125}{8} + \cdots$

(c) $\frac{1}{1} - \frac{2^2}{2 \cdot 1} + \frac{3^3}{3 \cdot 2 \cdot 1} - \frac{4^4}{4 \cdot 3 \cdot 2 \cdot 1} + \cdots$

(d) $\frac{2}{1^2+1} + \frac{1}{2^2+1} + \frac{2}{3^2+1} + \frac{1}{4^2+1} + \cdots$

2. Write in summation notation:

(a) $1 + \frac{1}{4} + \frac{1}{9} + \frac{1}{16} + \cdots$ **(b)** $\frac{1}{9} + \frac{1}{16} + \frac{1}{25} + \frac{1}{36} + \cdots$

(c) $1 - \frac{1}{3} + \frac{1}{5} - \frac{1}{7} + \cdots$

(d) $\frac{125}{9} + \frac{625}{16} + \frac{3125}{25} + \frac{15{,}625}{36} + \cdots$

In Exercises 3–6, compute the partial sums S_2, S_4, and S_6.

3. $1 + \frac{1}{2^2} + \frac{1}{3^2} + \frac{1}{4^2} + \cdots$ **4.** $\sum_{k=1}^{\infty} (-1)^k k^{-1}$

5. $\frac{1}{1 \cdot 2} + \frac{1}{2 \cdot 3} + \frac{1}{3 \cdot 4} + \cdots$ **6.** $\sum_{j=1}^{\infty} \frac{1}{j!}$

7. The series $S = 1 + (\frac{1}{5}) + (\frac{1}{5})^2 + (\frac{1}{5})^3 + \cdots$ converges to $\frac{5}{4}$. Calculate S_N for $N = 1, 2, \ldots$ until you find an S_N that approximates $\frac{5}{4}$ with an error less than 0.0001.

8. The series $S = \frac{1}{0!} - \frac{1}{1!} + \frac{1}{2!} - \frac{1}{3!} + \cdots$ is known to converge to e^{-1} (recall that $0! = 1$). Calculate S_N for $N = 1, 2, \ldots$ until you find an S_N that approximates e^{-1} with an error less than 0.001.

In Exercises 9 and 10, use a computer algebra system to compute S_{10}, S_{100}, S_{500}, and S_{1000} for the series. Do these values suggest convergence to the given value?

9. CAS

$$\frac{\pi - 3}{4} = \frac{1}{2 \cdot 3 \cdot 4} - \frac{1}{4 \cdot 5 \cdot 6} + \frac{1}{6 \cdot 7 \cdot 8} - \frac{1}{8 \cdot 9 \cdot 10} + \cdots$$

10. CAS

$$\frac{\pi^4}{90} = 1 + \frac{1}{2^4} + \frac{1}{3^4} + \frac{1}{4^4} + \cdots$$

11. Calculate S_3, S_4, and S_5 and then find the sum of the telescoping series

$$S = \sum_{n=1}^{\infty} \left(\frac{1}{n+1} - \frac{1}{n+2} \right)$$

12. Write $\sum_{n=3}^{\infty} \frac{1}{n(n-1)}$ as a telescoping series and find its sum.

13. Calculate S_3, S_4, and S_5 and then find the sum $S = \sum_{n=1}^{\infty} \frac{1}{4n^2 - 1}$ using the identity

$$\frac{1}{4n^2 - 1} = \frac{1}{2} \left(\frac{1}{2n - 1} - \frac{1}{2n + 1} \right)$$

14. Use partial fractions to rewrite $\sum_{n=1}^{\infty} \frac{1}{n(n+3)}$ as a telescoping series and find its sum.

15. Find the sum of $\frac{1}{1 \cdot 3} + \frac{1}{3 \cdot 5} + \frac{1}{5 \cdot 7} + \cdots$.

16. Find a formula for the partial sum S_N of $\sum_{n=1}^{\infty} (-1)^{n-1}$ and show that the series diverges.

In Exercises 17–22, use Theorem 3 to prove that the following series diverge.

17. $\sum_{n=1}^{\infty} \frac{n}{10n + 12}$

18. $\sum_{n=1}^{\infty} \frac{n}{\sqrt{n^2 + 1}}$

19. $\frac{0}{1} - \frac{1}{2} + \frac{2}{3} - \frac{3}{4} + \cdots$

20. $\sum_{n=1}^{\infty} (-1)^n n^2$

21. $\cos \frac{1}{2} + \cos \frac{1}{3} + \cos \frac{1}{4} + \cdots$

22. $\sum_{n=0}^{\infty} (\sqrt{4n^2 + 1} - n)$

In Exercises 23–36, use the formula for the sum of a geometric series to find the sum or state that the series diverges.

23. $\frac{1}{1} + \frac{1}{8} + \frac{1}{8^2} + \cdots$

24. $\frac{4^3}{5^3} + \frac{4^4}{5^4} + \frac{4^5}{5^5} + \cdots$

25. $\sum_{n=3}^{\infty} \left(\frac{3}{11} \right)^{-n}$

26. $\sum_{n=2}^{\infty} \frac{7 \cdot (-3)^n}{5^n}$

27. $\sum_{n=-4}^{\infty} \left(-\frac{4}{9} \right)^n$

28. $\sum_{n=0}^{\infty} \left(\frac{\pi}{e} \right)^n$

29. $\sum_{n=1}^{\infty} e^{-n}$

30. $\sum_{n=2}^{\infty} e^{3-2n}$

31. $\sum_{n=0}^{\infty} \frac{8 + 2^n}{5^n}$

32. $\sum_{n=0}^{\infty} \frac{3(-2)^n - 5^n}{8^n}$

33. $5 - \frac{5}{4} + \frac{5}{4^2} - \frac{5}{4^3} + \cdots$

34. $\frac{2^3}{7} + \frac{2^4}{7^2} + \frac{2^5}{7^3} + \frac{2^6}{7^4} + \cdots$

35. $\frac{7}{8} - \frac{49}{64} + \frac{343}{512} - \frac{2401}{4096} + \cdots$

36. $\frac{25}{9} + \frac{5}{3} + 1 + \frac{3}{5} + \frac{9}{25} + \frac{27}{125} + \cdots$

37. Which of the following are *not* geometric series?

(a) $\sum_{n=0}^{\infty} \frac{7^n}{29^n}$

(b) $\sum_{n=3}^{\infty} \frac{1}{n^4}$

(c) $\sum_{n=0}^{\infty} \frac{n^2}{2^n}$

(d) $\sum_{n=5}^{\infty} \pi^{-n}$

38. Use the method of Example 8 to show that $\sum_{k=1}^{\infty} \frac{1}{k^{1/3}}$ diverges.

39. Prove that if $\sum_{n=1}^{\infty} a_n$ converges and $\sum_{n=1}^{\infty} b_n$ diverges, then $\sum_{n=1}^{\infty} (a_n + b_n)$ diverges. *Hint:* If not, derive a contradiction by writing

$$\sum_{n=1}^{\infty} b_n = \sum_{n=1}^{\infty} (a_n + b_n) - \sum_{n=1}^{\infty} a_n$$

40. Prove the divergence of $\sum_{n=0}^{\infty} \frac{9^n + 2^n}{5^n}$.

41. Give a counterexample to show that each of the following statements is false.

(a) If the general term a_n tends to zero, then $\sum_{n=1}^{\infty} a_n = 0$.

(b) The Nth partial sum of the infinite series defined by $\{a_n\}$ is a_N.

(c) If a_n tends to zero, then $\sum_{n=1}^{\infty} a_n$ converges.

(d) If a_n tends to L, then $\sum_{n=1}^{\infty} a_n = L$.

42. Suppose that $S = \sum_{n=1}^{\infty} a_n$ is an infinite series with partial sum $S_N = 5 - \frac{2}{N^2}$.

(a) What are the values of $\sum_{n=1}^{10} a_n$ and $\sum_{n=5}^{16} a_n$?

(b) What is the value of a_3?

(c) Find a general formula for a_n.

(d) Find the sum $\sum_{n=1}^{\infty} a_n$.

43. Compute the total area of the (infinitely many) triangles in Figure 4.

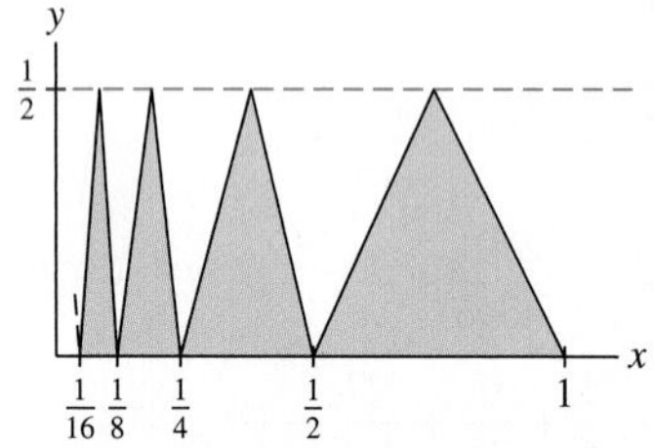

FIGURE 4

44. The winner of a lottery receives m dollars at the end of each year for N years. The present value (PV) of this prize in today's dollars is $\text{PV} = \sum_{i=1}^{N} m(1+r)^{-i}$, where r is the interest rate. Calculate PV if $m = \$50{,}000$, $r = 0.06$, and $N = 20$. What is PV if $N = \infty$?

45. Find the total length of the infinite zigzag path in Figure 5 (each zag occurs at an angle of $\frac{\pi}{4}$).

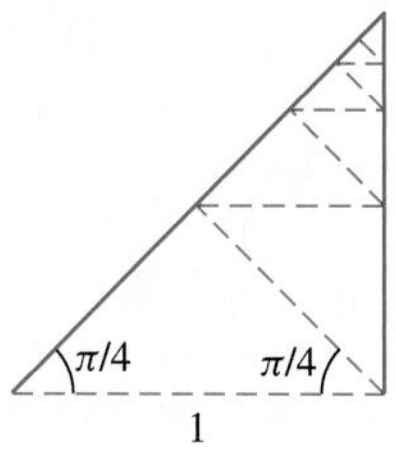

FIGURE 5

46. Evaluate $\sum_{n=1}^{\infty} \frac{1}{n(n+1)(n+2)}$. *Hint:* Find constants A, B, and C such that

$$\frac{1}{n(n+1)(n+2)} = \frac{A}{n} + \frac{B}{n+1} + \frac{C}{n+2}$$

47. Show that if a is a positive integer, then

$$\sum_{n=1}^{\infty} \frac{1}{n(n+a)} = \frac{1}{a}\left(1 + \frac{1}{2} + \cdots + \frac{1}{a}\right)$$

48. A ball dropped from a height of 10 ft begins to bounce. Each time it strikes the ground, it returns to two-thirds of its previous height. What is the total distance traveled by the ball if it bounces infinitely many times?

49. Let $\{b_n\}$ be a sequence and let $a_n = b_n - b_{n-1}$. Show that $\sum_{n=1}^{\infty} a_n$ converges if and only if $\lim_{n\to\infty} b_n$ exists.

50. Assumptions Matter Show, by giving counterexamples, that the assertions of Theorem 1 are not valid if the series $\sum_{n=0}^{\infty} a_n$ and $\sum_{n=0}^{\infty} b_n$ are not convergent.

Further Insights and Challenges

Exercises 51–53 use the formula

$$1 + r + r^2 + \cdots + r^{N-1} = \frac{1-r^N}{1-r} \qquad \boxed{7}$$

51. Professor George Andrews of Pennsylvania State University observed that we can use Eq. (7) to calculate the derivative of $f(x) = x^N$ (for $N \ge 0$). Assume that $a \neq 0$ and let $x = ra$. Show that

$$f'(a) = \lim_{x\to a} \frac{x^N - a^N}{x - a} = a^{N-1} \lim_{r\to 1} \frac{r^N - 1}{r - 1}$$

and evaluate the limit.

52. Pierre de Fermat used geometric series to compute the area under the graph of $f(x) = x^N$ over $[0, A]$. For $0 < r < 1$, let $F(r)$ be the sum of the areas of the infinitely many right-endpoint rectangles with endpoints Ar^n, as in Figure 6. As r tends to 1, the rectangles become narrower and $F(r)$ tends to the area under the graph.

(a) Show that $F(r) = A^{N+1}\frac{1-r}{1-r^{N+1}}$.

(b) Use Eq. (7) to evaluate $\int_0^A x^N\,dx = \lim_{r\to 1} F(r)$.

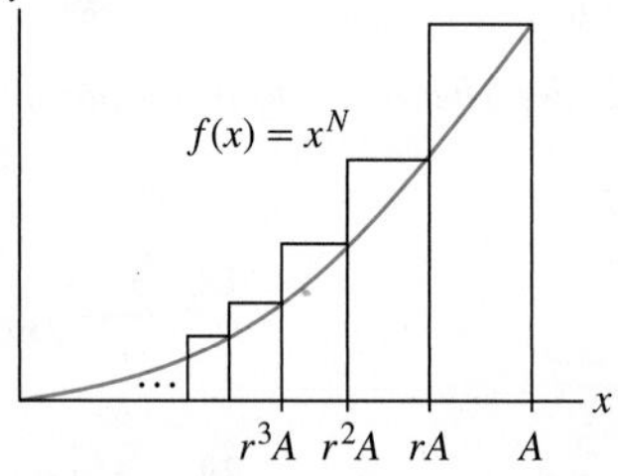

FIGURE 6

53. Verify the Gregory–Leibniz formula as follows.

(a) Set $r = -x^2$ in Eq. (7) and rearrange to show that

$$\frac{1}{1+x^2} = 1 - x^2 + x^4 - \cdots + (-1)^{N-1}x^{2N-2} + \frac{(-1)^N x^{2N}}{1+x^2}$$

(b) Show, by integrating over [0, 1], that

$$\frac{\pi}{4} = 1 - \frac{1}{3} + \frac{1}{5} - \frac{1}{7} + \cdots + \frac{(-1)^{N-1}}{2N-1} + (-1)^N \int_0^1 \frac{x^{2N}\,dx}{1+x^2}$$

(c) Use the Comparison Theorem for integrals to prove that

$$0 \le \int_0^1 \frac{x^{2N}\,dx}{1+x^2} \le \frac{1}{2N+1}$$

Hint: Observe that the integrand is $\le x^{2N}$.

(d) Prove that

$$\frac{\pi}{4} = 1 - \frac{1}{3} + \frac{1}{5} - \frac{1}{7} + \frac{1}{9} - \cdots$$

Hint: Use (b) and (c) to show that the partial sums S_N of satisfy $\left|S_N - \frac{\pi}{4}\right| \le \frac{1}{2N+1}$, and thereby conclude that $\lim_{N\to\infty} S_N = \frac{\pi}{4}$.

54. Cantor's Disappearing Table (following Larry Knop of Hamilton College) Take a table of length L (Figure 7). At stage 1, remove the section of length $L/4$ centered at the midpoint. Two sections remain, each with length less than $L/2$. At stage 2, remove sections of length $L/4^2$ from each of these two sections (this stage removes $L/8$ of the table). Now four sections remain, each of length less than $L/4$. At stage 3, remove the four central sections of length $L/4^3$, etc.

(a) Show that at the Nth stage, each remaining section has length less than $L/2^N$ and that the total amount of table removed is

$$L\left(\frac{1}{4} + \frac{1}{8} + \frac{1}{16} + \cdots + \frac{1}{2^{N+1}}\right)$$

(b) Show that in the limit as $N \to \infty$, precisely one-half of the table remains.

This result is curious, because there are no nonzero intervals of table left (at each stage, the remaining sections have a length less than $L/2^N$). So the table has "disappeared." However, we can place any object longer than $L/4$ on the table. It will not fall through because it will not fit through any of the removed sections.

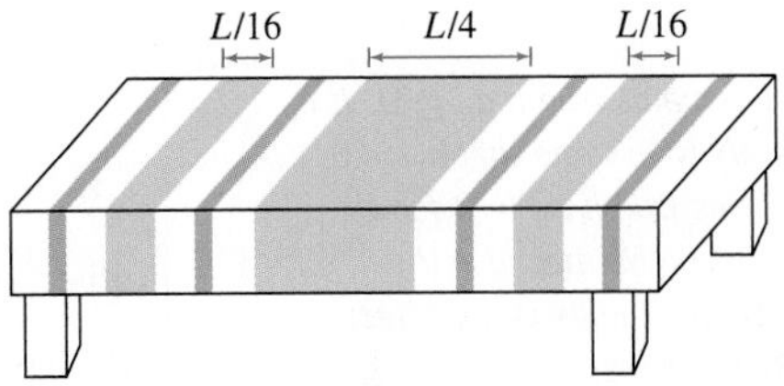

FIGURE 7

55. The **Koch snowflake** (described in 1904 by Swedish mathematician Helge von Koch) is an infinitely jagged "fractal" curve obtained as a limit of polygonal curves (it is continuous but has no tangent line at any point). Begin with an equilateral triangle (stage 0) and produce stage 1 by replacing each edge with four edges of one-third the length, arranged as in Figure 8. Continue the process: At the nth stage, replace each edge with four edges of one-third the length.

(a) Show that the perimeter P_n of the polygon at the nth stage satisfies $P_n = \frac{4}{3}P_{n-1}$. Prove that $\lim_{n\to\infty} P_n = \infty$. The snowflake has infinite length.

(b) Let A_0 be the area of the original equilateral triangle. Show that $(3)4^{n-1}$ new triangles are added at the nth stage, each with area $A_0/9^n$ (for $n \ge 1$). Show that the total area of the Koch snowflake is $\frac{8}{5}A_0$.

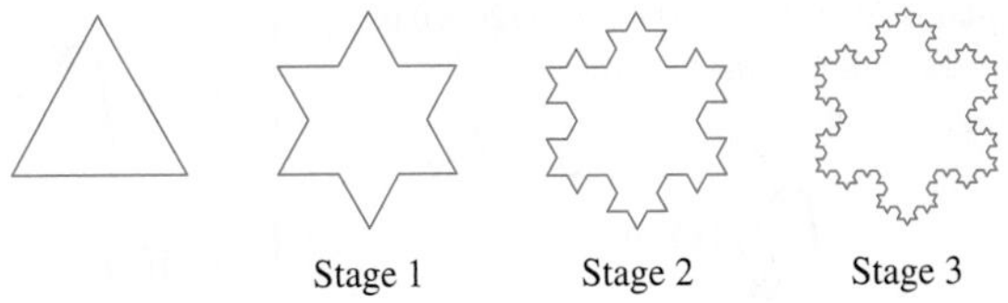

FIGURE 8

11.3 Convergence of Series with Positive Terms

The next three sections develop techniques for determining whether an infinite series converges or diverges. This is easier than finding the sum of an infinite series, which is possible only in special cases.

In this section, we consider **positive series** $\sum a_n$, where $a_n > 0$ for all n. We can visualize the terms of a positive series as rectangles of width 1 and height a_n (Figure 1). The partial sum

$$S_N = a_1 + a_2 + \cdots + a_N$$

is equal to the area of the first N rectangles.

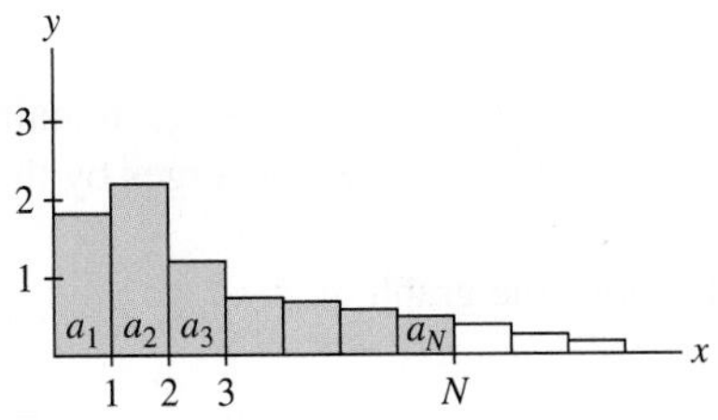

FIGURE 1 The partial sum S_N is the sum of the areas of the N shaded rectangles.

The key feature of positive series is that their partial sums form an increasing sequence:

$$S_N < S_{N+1}$$

for all N. This is because S_{N+1} is obtained from S_N by adding a positive number:

$$S_{N+1} = \left(a_1 + a_2 + \cdots + a_N\right) + a_{N+1} = S_N + \underbrace{a_{N+1}}_{\text{Positive}}$$

Recall that an increasing sequence converges if it is bounded above. Otherwise, it diverges (Theorem 6, Section 11.1). It follows that a positive series behaves in one of two ways (this is the dichotomy referred to in the next theorem).

- *Theorem 1 remains true if $a_n \geq 0$. It is not necessary to assume that $a_n > 0$.*
- *It also remains true if $a_n > 0$ for all $n \geq M$ for some M, because the convergence of a series is not affected by the first M terms.*

THEOREM 1 Dichotomy for Positive Series If $S = \sum_{n=1}^{\infty} a_n$ is a positive series, then either:

(i) The partial sums S_N are bounded above. In this case, S converges. Or,
(ii) The partial sums S_N are not bounded above. In this case, S diverges.

Assumptions Matter The dichotomy does not hold for nonpositive series. Consider

$$S = \sum_{n=1}^{\infty} (-1)^{n-1} = 1 - 1 + 1 - 1 + 1 - 1 + \cdots$$

The partial sums are bounded (because $S_N = 1$ or 0), but S diverges.

Our first application of Theorem 1 is the following Integral Test. It is extremely useful because integrals are easier to evaluate than series in most cases.

The Integral Test is valid for any series $\sum_{n=k}^{\infty} f(n)$, provided that for some $M > 0$, $f(x)$ is positive, decreasing, and continuous for $x \geq M$. The convergence of the series is determined by the convergence of

$$\int_M^{\infty} f(x)\,dx$$

THEOREM 2 Integral Test Let $a_n = f(n)$, where $f(x)$ is positive, decreasing, and continuous for $x \geq 1$.

(i) If $\int_1^{\infty} f(x)\,dx$ converges, then $\sum_{n=1}^{\infty} a_n$ converges.

(ii) If $\int_1^{\infty} f(x)\,dx$ diverges, then $\sum_{n=1}^{\infty} a_n$ diverges.

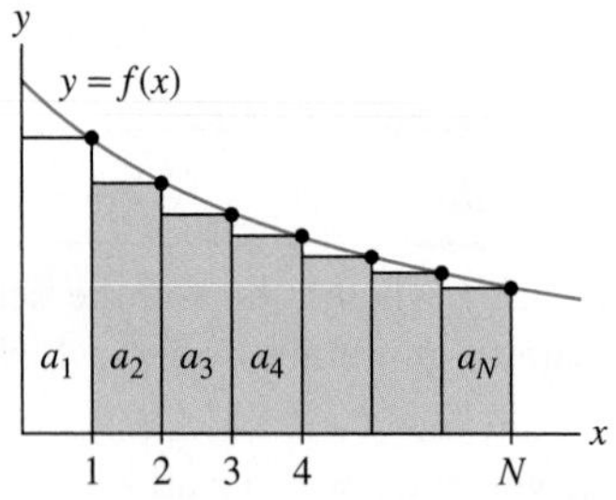

FIGURE 2

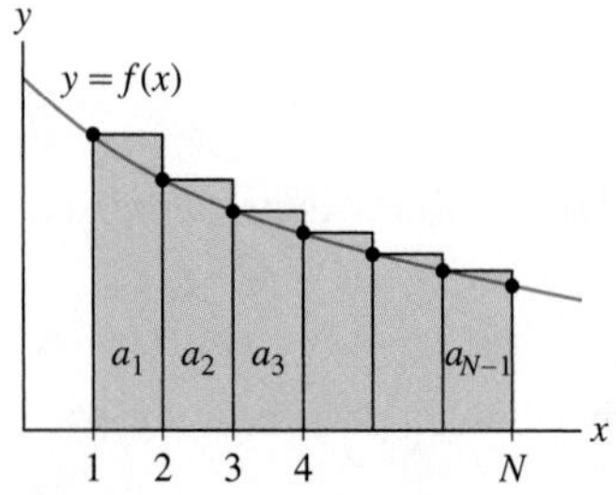

FIGURE 3

Proof Because $f(x)$ is decreasing, the shaded rectangles in Figure 2 lie below the graph of $f(x)$, and therefore for all N

$$\underbrace{a_2 + \cdots + a_N}_{\text{Area of shaded rectangles in Figure 2}} \leq \int_1^N f(x)\,dx \leq \int_1^{\infty} f(x)\,dx$$

If the improper integral on the right converges, then the sums $a_2 + \cdots + a_N$ remain bounded. In this case, S_N also remains bounded, and the infinite series converges by the Dichotomy Theorem (Theorem 1). This proves (i).

On the other hand, the rectangles in Figure 3 lie above the graph of $f(x)$, so

$$\int_1^N f(x)\,dx \leq \underbrace{a_1 + a_2 + \cdots + a_{N-1}}_{\text{Area of shaded rectangles in Figure 3}} \qquad \boxed{1}$$

If $\int_1^{\infty} f(x)\,dx$ diverges, then $\int_1^N f(x)\,dx$ tends to ∞, and Eq. (1) shows that S_N also tends to ∞. This proves (ii). ∎

■ **EXAMPLE 1 The Harmonic Series Diverges** Show that $\sum_{n=1}^{\infty} \frac{1}{n}$ diverges.

The infinite series

$$\sum_{n=1}^{\infty} \frac{1}{n}$$

is called the "harmonic series."

Solution Let $f(x) = \frac{1}{x}$. Then $f(n) = \frac{1}{n}$, and the Integral Test applies because f is positive, decreasing, and continuous for $x \geq 1$. The integral diverges:

$$\int_1^{\infty} \frac{dx}{x} = \lim_{R\to\infty} \int_1^R \frac{dx}{x} = \lim_{R\to\infty} \ln R = \infty$$

Therefore, the series $\sum_{n=1}^{\infty} \frac{1}{n}$ diverges. ■

■ **EXAMPLE 2** Does $\sum_{n=1}^{\infty} \frac{n}{(n^2+1)^2} = \frac{1}{2^2} + \frac{2}{5^2} + \frac{3}{10^2} + \cdots$ converge?

Solution The function $f(x) = \dfrac{x}{(x^2+1)^2}$ is positive and continuous for $x \geq 1$. It is decreasing because $f'(x)$ is negative:

$$f'(x) = \frac{1-3x^2}{(x^2+1)^3} < 0 \qquad \text{for } x \geq 1$$

Therefore, the Integral Test applies. Using the substitution $u = x^2 + 1$, $du = 2x\,dx$, we have

$$\int_1^{\infty} \frac{x}{(x^2+1)^2}\,dx = \lim_{R\to\infty} \int_1^R \frac{x}{(x^2+1)^2}\,dx = \lim_{R\to\infty} \frac{1}{2}\int_2^R \frac{du}{u^2}$$

$$= \lim_{R\to\infty} \frac{-1}{2u}\bigg|_2^R = \lim_{R\to\infty}\left(\frac{1}{4} - \frac{1}{2R}\right) = \frac{1}{4}$$

The integral converges. Therefore, $\sum_{n=1}^{\infty} \frac{n}{(n^2+1)^2}$ also converges. ■

The sum of the reciprocal powers n^{-p} is called a ***p*-series**.

> **THEOREM 3 Convergence of p-Series** The infinite series $\sum_{n=1}^{\infty} \frac{1}{n^p}$ converges if $p > 1$ and diverges otherwise.

Proof If $p \leq 0$, then the general term n^{-p} does not tend to zero, so the series diverges. If $p > 0$, then $f(x) = x^{-p}$ is positive and decreasing, so the Integral Test applies. According to Theorem 1 in Section 7.6,

$$\int_1^{\infty} \frac{1}{x^p}\,dx = \begin{cases} \dfrac{1}{p-1} & \text{if } p > 1 \\ \infty & \text{if } p \leq 1 \end{cases}$$

Therefore, $\sum_{n=1}^{\infty} \frac{1}{n^p}$ converges for $p > 1$ and diverges for $p \leq 1$. ■

Here are two examples of p-series:

$$p = \frac{1}{3}: \quad \sum_{n=1}^{\infty} \frac{1}{\sqrt[3]{n}} = \frac{1}{\sqrt[3]{1}} + \frac{1}{\sqrt[3]{2}} + \frac{1}{\sqrt[3]{3}} + \frac{1}{\sqrt[3]{4}} + \cdots = \infty \quad \text{diverges}$$

$$p = 2: \quad \sum_{n=1}^{\infty} \frac{1}{n^2} = \frac{1}{1} + \frac{1}{2^2} + \frac{1}{3^2} + \frac{1}{4^2} + \cdots \quad \text{converges}$$

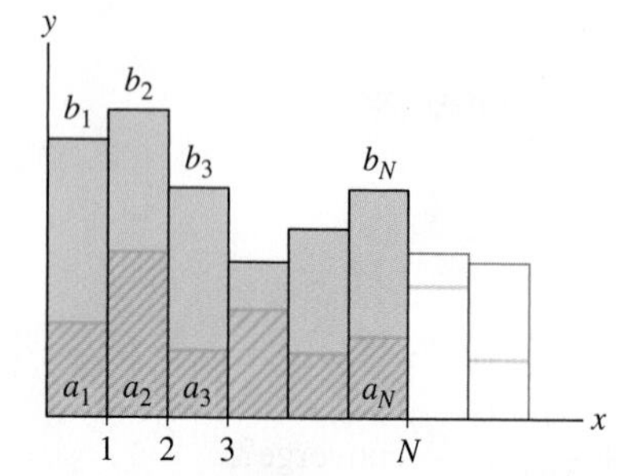

FIGURE 4 The series $\sum a_n$ is dominated by the series $\sum b_n$.

Another powerful method for determining convergence of positive series is comparison. Suppose that $0 \le a_n \le b_n$. Figure 4 suggests that if the larger sum $\sum b_n$ *converges*, then the smaller sum $\sum a_n$ also converges. Similarly, if the smaller sum *diverges*, then the larger sum also diverges.

THEOREM 4 Comparison Test
Assume that there exists $M > 0$ such that $0 \le a_n \le b_n$ for $n \ge M$.

(i) If $\sum_{n=1}^{\infty} b_n$ converges, then $\sum_{n=1}^{\infty} a_n$ also converges.

(ii) If $\sum_{n=1}^{\infty} a_n$ diverges, then $\sum_{n=1}^{\infty} b_n$ also diverges.

Proof We can assume, without loss of generality, that $M = 1$. If $S = \sum_{n=1}^{\infty} b_n$ converges, then the partial sums of $\sum_{n=1}^{\infty} a_n$ are bounded above by S because

$$a_1 + a_2 + \cdots + a_N \le b_1 + b_2 + \cdots + b_N \le \sum_{n=1}^{\infty} b_n = S \qquad \boxed{2}$$

Therefore, $\sum_{n=1}^{\infty} a_n$ converges by the Dichotomy Theorem (Theorem 1). This proves (i).

On the other hand, if $\sum_{n=1}^{\infty} a_n$ diverges, then $\sum_{n=1}^{\infty} b_n$ must also diverge. Otherwise we would have a contradiction to (i). ■

In words, the Comparison Test states that for positive series:

- *Convergence of the larger series forces convergence of the smaller series.*
- *Divergence of the smaller series forces divergence of the larger series.*

■ **EXAMPLE 3** Does $\sum_{n=1}^{\infty} \frac{1}{\sqrt{n}\, 3^n}$ converge?

Solution For $n \ge 1$, we have

$$\frac{1}{\sqrt{n}\, 3^n} \le \frac{1}{3^n}$$

The larger series $\sum_{n=1}^{\infty} \frac{1}{3^n}$ converges because it is a geometric series with $r = \frac{1}{3} < 1$. By the Comparison Test, the smaller series $\sum_{n=1}^{\infty} \frac{1}{\sqrt{n}\, 3^n}$ also converges. ■

EXAMPLE 4 Does $S = \sum_{n=2}^{\infty} \frac{1}{(n^2+3)^{1/3}}$ converge?

Solution Let us show that

$$\frac{1}{n} \leq \frac{1}{(n^2+3)^{1/3}} \qquad \text{for } n \geq 2$$

This inequality is equivalent to $(n^2+3) \leq n^3$, so we must show that

$$f(x) = x^3 - (x^2+3) \geq 0 \qquad \text{for } x \geq 2$$

The function $f(x)$ is increasing because its derivative $f'(x) = 3x\left(x - \frac{2}{3}\right)$ is positive for $x \geq 2$. Since $f(2) = 1$, it follows that $f(x) \geq 1$ for $x \geq 2$, and our original inequality follows. We know that the smaller harmonic series $\sum_{n=2}^{\infty} \frac{1}{n}$ diverges. Therefore, the larger series $\sum_{n=2}^{\infty} \frac{1}{(n^2+1)^{1/3}}$ also diverges. ■

EXAMPLE 5 Using the Comparison Correctly Study the convergence of

$$\sum_{n=2}^{\infty} \frac{1}{n(\ln n)^2}$$

Solution We might be tempted to compare $\sum_{n=2}^{\infty} \frac{1}{n(\ln n)^2}$ to the harmonic series $\sum_{n=2}^{\infty} \frac{1}{n}$ using the inequality (valid for $n \geq 3$)

$$\frac{1}{n(\ln n)^2} \leq \frac{1}{n}$$

However, $\sum_{n=2}^{\infty} \frac{1}{n}$ diverges, and this says nothing about the *smaller* series $\sum \frac{1}{n(\ln n)^2}$. Fortunately, the Integral Test can be used. The substitution $u = \ln x$ yields

$$\int_2^{\infty} \frac{dx}{x(\ln x)^2} = \int_{\ln 2}^{\infty} \frac{du}{u^2} = \lim_{R\to\infty}\left(\frac{1}{\ln 2} - \frac{1}{R}\right) = \frac{1}{\ln 2} < \infty$$

The Integral Test shows that $\sum_{n=2}^{\infty} \frac{1}{n(\ln n)^2}$ converges. ■

Suppose we wish to study the convergence of

$$S = \sum_{n=2}^{\infty} \frac{n^2}{n^4 - n - 1}$$

For large n, the general term is very close to $1/n^2$:

$$\frac{n^2}{n^4 - n - 1} = \frac{1}{n^2 - n^{-1} - n^{-2}} \approx \frac{1}{n^2}$$

Thus we might try to compare S with $\sum_{n=2}^{\infty} \frac{1}{n^2}$. Unfortunately, however, the inequality goes in the wrong direction:

$$\frac{n^2}{n^4 - n - 1} > \frac{n^2}{n^4} = \frac{1}{n^2}$$

Although the smaller series $\sum_{n=2}^{\infty} \frac{1}{n^2}$ converges, we cannot use the Comparison Theorem to say anything about our larger series. In this situation, the following variation of the Comparison Test can be used.

CAUTION *The Limit Comparison Test is not valid if the series are not positive. See Exercise 44 in Section 11.4.*

THEOREM 5 Limit Comparison Test Let $\{a_n\}$ and $\{b_n\}$ be *positive* sequences. Assume that the following limit exists:

$$L = \lim_{n\to\infty} \frac{a_n}{b_n}$$

- If $L > 0$, then $\sum a_n$ converges if and only if $\sum b_n$ converges.
- If $L = \infty$ and $\sum a_n$ converges, then $\sum b_n$ converges.
- If $L = 0$ and $\sum b_n$ converges, then $\sum a_n$ converges.

Proof Assume first that L is finite (possibly zero) and that $\sum b_n$ converges. Choose a positive number $R > L$. Then $0 \le a_n/b_n \le R$ for all n sufficiently large because a_n/b_n approaches L. Therefore $a_n \le Rb_n$. The series $\sum Rb_n$ converges because it is a multiple of the convergent series $\sum b_n$. Therefore $\sum a_n$ converges by the Comparison Test.

Next, suppose that L is nonzero (positive or infinite) and that $\sum a_n$ converges. Let $L^{-1} = \lim_{n\to\infty} b_n/a_n$. Then L^{-1} is finite and we can apply the result of the previous paragraph with the roles of $\{a_n\}$ and $\{b_n\}$ reversed to conclude that $\sum b_n$ converges. ■

CONCEPTUAL INSIGHT To remember the different cases of the Limit Comparison Test, you can think of it this way. If $L > 0$, then $a_n \approx Lb_n$ for large n. In other words, the series $\sum a_n$ and $\sum b_n$ are *roughly* multiples of each other, so one converges if and only if the other converges. If $L = \infty$, then a_n is much larger than b_n (for large n), so if $\sum a_n$ converges, $\sum b_n$ certainly converges. Finally, if $L = 0$, then b_n is much larger than a_n and the convergence of $\sum b_n$ yields the convergence of $\sum a_n$.

■ **EXAMPLE 6** Show that $\sum_{n=2}^{\infty} \frac{n^2}{n^4 - n - 1}$ converges.

Solution Let

$$a_n = \frac{n^2}{n^4 - n - 1} \quad \text{and} \quad b_n = \frac{1}{n^2}$$

We observed above that $a_n \approx b_n$ for large n. To apply the Limit Comparison Test, we observe that the limit L exists and $L > 0$:

$$L = \lim_{n\to\infty} \frac{a_n}{b_n} = \lim_{n\to\infty} \frac{n^2}{n^4 - n - 1} \cdot \frac{n^2}{1} = \lim_{n\to\infty} \frac{1}{1 - n^{-3} - n^{-4}} = 1$$

Since $\displaystyle\sum_{n=2}^{\infty} \frac{1}{n^2}$ converges, our series $\displaystyle\sum_{n=2}^{\infty} \frac{n^2}{n^4 - n - 1}$ also converges by Theorem 5. ■

■ **EXAMPLE 7** Determine whether $\displaystyle\sum_{n=3}^{\infty} \frac{1}{\sqrt{n^2+4}}$ converges.

Solution Apply the Limit Comparison Test with $a_n = \dfrac{1}{\sqrt{n^2+4}}$ and $b_n = \dfrac{1}{n}$. Then

$$L = \lim_{n\to\infty} \frac{a_n}{b_n} = \lim_{n\to\infty} \frac{n}{\sqrt{n^2+4}} = \lim_{n\to\infty} \frac{1}{\sqrt{1+4/n^2}} = 1$$

Since $\displaystyle\sum_{n=3}^{\infty} \frac{1}{n}$ diverges and $L > 0$, the series $\displaystyle\sum_{n=3}^{\infty} \frac{1}{\sqrt{n^2+4}}$ also diverges. ■

11.3 SUMMARY

- The partial sums S_N of a positive series $S = \sum a_n$ form an increasing sequence.
- *Dichotomy Theorem:* A positive series S converges if its partial sums S_N remain bounded. Otherwise, it diverges.
- *Integral Test:* Assume that f is positive, decreasing, and continuous for $x > M$. Set $a_n = f(n)$. If $\int_M^\infty f(x)\,dx$ converges, then $S = \sum a_n$ converges, and if $\int_M^\infty f(x)\,dx$ diverges, then $S = \sum a_n$ diverges.
- *p-Series:* The series $\displaystyle\sum_{n=1}^{\infty} \frac{1}{n^p}$ converges if $p > 1$ and diverges if $p \le 1$.
- *Comparison Test:* Assume there exists $M > 0$ such that $0 \le a_n \le b_n$ for all $n \ge M$. If $\sum b_n$ converges, then $\sum a_n$ converges, and if $\sum a_n$ diverges, then $\sum b_n$ diverges.
- *Limit Comparison Test:* Assume that $\{a_n\}$ and $\{b_n\}$ are positive and that the following limit exists:

$$L = \lim_{n\to\infty} \frac{a_n}{b_n}$$

  – If $L > 0$, then $\sum a_n$ converges if and only if $\sum b_n$ converges.

  – If $L = \infty$ and $\sum a_n$ converges, then $\sum b_n$ converges.

  – If $L = 0$ and $\sum b_n$ converges, then $\sum a_n$ converges.

11.3 EXERCISES

Preliminary Questions

1. Let $S = \sum_{n=1}^{\infty} a_n$. If the partial sums S_N are increasing, then (choose the correct conclusion):

(a) $\{a_n\}$ is an increasing sequence.
(b) $\{a_n\}$ is a positive sequence.

2. What are the hypotheses of the Integral Test?

3. Which test would you use to determine whether $\sum_{n=1}^{\infty} n^{-3.2}$ converges?

4. Which test would you use to determine whether $\sum_{n=1}^{\infty} \frac{1}{2^n + \sqrt{n}}$ converges?

5. Ralph hopes to investigate the convergence of $\sum_{n=1}^{\infty} \frac{e^{-n}}{n}$ by comparing it with $\sum_{n=1}^{\infty} \frac{1}{n}$. Is Ralph on the right track?

Exercises

In Exercises 1–14, use the Integral Test to determine whether the infinite series is convergent.

1. $\sum_{n=1}^{\infty} \frac{1}{n^4}$

2. $\sum_{n=1}^{\infty} \frac{1}{n+3}$

3. $\sum_{n=1}^{\infty} n^{-1/3}$

4. $\sum_{n=5}^{\infty} \frac{1}{\sqrt{n-4}}$

5. $\sum_{n=25}^{\infty} \frac{n^2}{(n^3+9)^{5/2}}$

6. $\sum_{n=1}^{\infty} \frac{n}{(n^2+1)^{3/5}}$

7. $\sum_{n=1}^{\infty} \frac{1}{n^2+1}$

8. $\sum_{n=4}^{\infty} \frac{1}{n^2-1}$

9. $\sum_{n=1}^{\infty} \frac{1}{n(n+1)}$

10. $\sum_{n=1}^{\infty} ne^{-n^2}$

11. $\sum_{n=2}^{\infty} \frac{1}{n(\ln n)^2}$

12. $\sum_{n=1}^{\infty} \frac{\ln n}{n^2}$

13. $\sum_{n=1}^{\infty} \frac{1}{2^{\ln n}}$

14. $\sum_{n=1}^{\infty} \frac{1}{3^{\ln n}}$

15. Show that $\sum_{n=1}^{\infty} \frac{1}{n^3+8n}$ converges by using the Comparison Test with $\sum_{n=1}^{\infty} n^{-3}$.

16. Show that $\sum_{n=2}^{\infty} \frac{1}{\sqrt{n^2-3}}$ diverges by comparing with $\sum_{n=2}^{\infty} n^{-1}$.

17. Let $S = \sum_{n=1}^{\infty} \frac{1}{n+\sqrt{n}}$. Verify that for $n \ge 1$,

$$\frac{1}{n+\sqrt{n}} \le \frac{1}{n}, \qquad \frac{1}{n+\sqrt{n}} \le \frac{1}{\sqrt{n}}$$

Can either inequality be used to show that S diverges? Show that $\frac{1}{n+\sqrt{n}} \ge \frac{1}{2n}$ and conclude that S diverges.

18. Which of the following inequalities can be used to study the convergence of $\sum_{n=2}^{\infty} \frac{1}{n^2+\sqrt{n}}$? Explain.

$$\frac{1}{n^2+\sqrt{n}} \le \frac{1}{\sqrt{n}}, \qquad \frac{1}{n^2+\sqrt{n}} \le \frac{1}{n^2}$$

In Exercises 19–30, use the Comparison Test to determine whether the infinite series is convergent.

19. $\sum_{n=1}^{\infty} \frac{1}{n2^n}$

20. $\sum_{n=1}^{\infty} \frac{n^3}{n^5+4n+1}$

21. $\sum_{n=1}^{\infty} \frac{1}{n^{1/3}+2^n}$

22. $\sum_{n=1}^{\infty} \frac{1}{\sqrt{n^3+2n-1}}$

23. $\sum_{m=1}^{\infty} \frac{4}{m!+4^m}$

24. $\sum_{n=4}^{\infty} \frac{\sqrt{n}}{n-3}$

25. $\sum_{k=1}^{\infty} \frac{\sin^2 k}{k^2}$

26. $\sum_{k=2}^{\infty} \frac{k^{1/3}}{k^{5/4}-k}$

27. $\sum_{n=1}^{\infty} \frac{2}{3^n+3^{-n}}$

28. $\sum_{k=1}^{\infty} 2^{-k^2}$

29. $\sum_{n=1}^{\infty} \frac{1}{(n+1)!}$

30. $\sum_{n=1}^{\infty} \frac{n!}{n^3}$

Exercise 31–36: For all $a > 0$ and $b > 1$, the inequalities

$$\ln n \le n^a, \qquad n^a < b^n$$

are true for n sufficiently large (this can be proved using L'Hopital's Rule). Use this, together with the Comparison Theorem, to determine whether the series converges or diverges.

31. $\sum_{n=1}^{\infty} \frac{\ln n}{n^3}$

32. $\sum_{m=2}^{\infty} \frac{1}{\ln m}$

33. $\sum_{n=1}^{\infty} \frac{(\ln n)^{100}}{n^{1.1}}$

34. $\sum_{n=1}^{\infty} \frac{1}{(\ln n)^{10}}$

35. $\sum_{n=1}^{\infty} \frac{n}{3^n}$

36. $\sum_{n=1}^{\infty} \frac{n^5}{2^n}$

37. Show that $\sum_{n=1}^{\infty} \sin \frac{1}{n^2}$ converges. *Hint:* Use the inequality $\sin x \le x$ for $x \ge 0$.

38. Does $\sum_{n=1}^{\infty} \frac{\sin(1/n)}{\ln n}$ converge?

In Exercises 39–48, use the Limit Comparison Test to prove convergence or divergence of the infinite series.

39. $\sum_{n=2}^{\infty} \frac{n^2}{n^4 - 1}$

40. $\sum_{n=2}^{\infty} \frac{1}{n^2 - \sqrt{n}}$

41. $\sum_{n=2}^{\infty} \frac{n}{\sqrt{n^3 + 1}}$

42. $\sum_{n=2}^{\infty} \frac{n^3}{\sqrt{n^7 + 2n^2 + 1}}$

43. $\sum_{n=3}^{\infty} \frac{3n + 5}{n(n-1)(n-2)}$

44. $\sum_{n=1}^{\infty} \frac{e^n + n}{e^{2n} - n^2}$

45. $\sum_{n=1}^{\infty} \frac{1}{\sqrt{n} + \ln n}$

46. $\sum_{n=1}^{\infty} \frac{\ln(n+4)}{n^{5/2}}$

47. $\sum_{n=1}^{\infty} \left(1 - \cos \frac{1}{n}\right)$ *Hint:* Compare with $\sum_{n=1}^{\infty} n^{-2}$.

48. $\sum_{n=1}^{\infty} (1 - 2^{-1/n})$ *Hint:* Compare with the harmonic series.

In Exercises 49–74, determine convergence or divergence using any method covered so far.

49. $\sum_{n=4}^{\infty} \frac{1}{n^2 - 9}$

50. $\sum_{n=1}^{\infty} \frac{\cos^2 n}{n^2}$

51. $\sum_{n=1}^{\infty} \frac{\sqrt{n}}{4n + 9}$

52. $\sum_{n=1}^{\infty} \frac{n - \cos n}{n^3}$

53. $\sum_{n=1}^{\infty} \frac{n^2 - n}{n^5 + n}$

54. $\sum_{n=1}^{\infty} \frac{1}{n^2 + \sin n}$

55. $\sum_{n=5}^{\infty} (4/5)^{-n}$

56. $\sum_{n=1}^{\infty} \frac{1}{3^{n^2}}$

57. $\sum_{n=2}^{\infty} \frac{1}{n^{3/2} \ln n}$

58. $\sum_{n=2}^{\infty} \frac{(\ln n)^{12}}{n^{9/8}}$

59. $\sum_{k=1}^{\infty} 4^{1/k}$

60. $\sum_{n=1}^{\infty} \frac{4^n}{5^n - 2n}$

61. $\sum_{n=2}^{\infty} \frac{1}{(\ln n)^4}$

62. $\sum_{n=1}^{\infty} \frac{2^n}{3^n - n}$

63. $\sum_{n=1}^{\infty} \frac{1}{n \ln n - n}$

64. $\sum_{n=1}^{\infty} \frac{1}{n(\ln n)^2 - n}$

65. $\sum_{n=1}^{\infty} \frac{1}{n^n}$

66. $\sum_{n=1}^{\infty} \frac{n^2 - 4n^{3/2}}{n^3}$

67. $\sum_{n=1}^{\infty} \frac{1 + (-1)^n}{n}$

68. $\sum_{n=1}^{\infty} \frac{2 + (-1)^n}{n^{3/2}}$

69. $\sum_{n=1}^{\infty} \sin \frac{1}{n}$

70. $\sum_{n=1}^{\infty} \frac{\sin(1/n)}{\sqrt{n}}$

71. $\sum_{n=1}^{\infty} \frac{2n + 1}{4^n}$

72. $\sum_{n=3}^{\infty} \frac{1}{e^{\sqrt{n}}}$

73. $\sum_{n=4}^{\infty} \frac{\ln n}{n^2 - 3n}$

74. $\sum_{n=1}^{\infty} \frac{1}{3^{\ln n}}$

75. $\sum_{n=2}^{\infty} \frac{1}{n^{1/2} \ln n}$

76. $\sum_{n=1}^{\infty} \frac{1}{n^{3/2} - \ln^4 n}$

77. $\sum_{n=1}^{\infty} \frac{4n^2 + 15n}{3n^4 - 5n^2 - 17}$

78. $\sum_{n=1}^{\infty} \frac{n}{4^{-n} + 5^{-n}}$

79. For which a does $\sum_{n=2}^{\infty} \frac{1}{n(\ln n)^a}$ converge?

80. For which a does $\sum_{n=2}^{\infty} \frac{1}{n^a \ln n}$ converge?

Approximating Infinite Sums *In Exercises 81–83, let $a_n = f(n)$, where $f(x)$ is a continuous, decreasing function such that $f(x) \ge 0$ and $\int_1^{\infty} f(x)\,dx$ converges.*

81. Show that

$$\int_1^{\infty} f(x)\,dx \le \sum_{n=1}^{\infty} a_n \le a_1 + \int_1^{\infty} f(x)\,dx \qquad \boxed{3}$$

82. CAS Using Eq. (3), show that

$$5 \le \sum_{n=1}^{\infty} \frac{1}{n^{1.2}} \le 6$$

This series converges slowly. Use a computer algebra system to verify that $S_N < 5$ for $N \le 43{,}128$ and $S_{43,129} \approx 5.00000021$.

83. Let $S = \sum_{n=1}^{\infty} a_n$. Arguing as in Exercise 81, show that

$$\sum_{n=1}^{M} a_n + \int_{M+1}^{\infty} f(x)\,dx \le S \le \sum_{n=1}^{M+1} a_n + \int_{M+1}^{\infty} f(x)\,dx \qquad \boxed{4}$$

Conclude that

$$0 \le S - \left(\sum_{n=1}^{M} a_n + \int_{M+1}^{\infty} f(x)\,dx\right) \le a_{M+1} \qquad \boxed{5}$$

This provides a method for approximating S with an error of at most a_{M+1}.

84. CAS Use Eq. (4) with $M = 43{,}129$ to prove that

$$5.5915810 \le \sum_{n=1}^{\infty} \frac{1}{n^{1.2}} \le 5.5915839$$

85. CAS Apply Eq. (4) with $M = 40{,}000$ to show that

$$1.644934066 \le \sum_{n=1}^{\infty} \frac{1}{n^2} \le 1.644934068$$

Is this consistent with Euler's result, according to which this infinite series has sum $\pi^2/6$?

86. CAS Using a CAS and Eq. (5), determine the value of $\sum_{n=1}^{\infty} n^{-6}$ to within an error less than 10^{-4}. Check that your result is consistent with that of Euler, who proved that the sum is equal to $\pi^6/945$.

87. CAS Using a CAS and Eq. (5), determine the value of $\sum_{n=1}^{\infty} n^{-5}$ to within an error less than 10^{-4}.

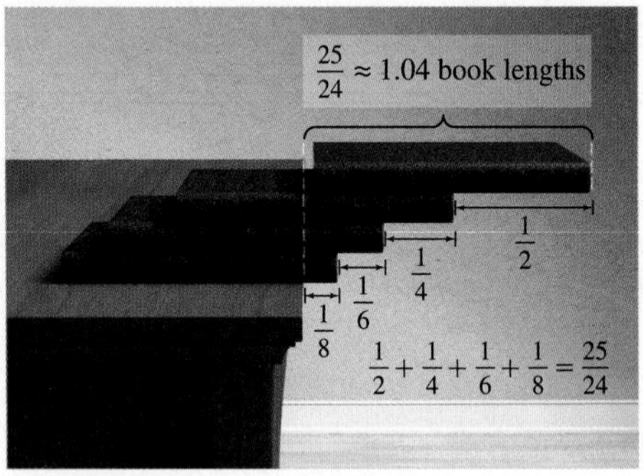

FIGURE 5

88. How far can a stack of identical books (of mass m and unit length) extend without tipping over? The stack will not tip over if the $(n+1)$st book is placed at the bottom of the stack with its right edge located at the center of mass of the first n books (Figure 5). Let c_n be the center of mass of the first n books, measured along the x-axis, where we take the positive x-axis to the left of the origin as in Figure 6. Recall that if an object of mass m_1 has center of mass at x_1 and a second object of m_2 has center of mass x_2, then the center of mass of the system has x-coordinate

$$\frac{m_1 x_1 + m_2 x_2}{m_1 + m_2}$$

(a) Show that if the $(n+1)$st book is placed with its right edge at c_n, then its center of mass is located at $c_n + \frac{1}{2}$.

(b) Consider the first n books as a single object of mass nm with center of mass at c_n and the $(n+1)$st book as a second object of mass m. Show that if the $(n+1)$st book is placed with its right edge at c_n, then $c_{n+1} = c_n + \dfrac{1}{2(n+1)}$.

(c) Prove that $\lim_{n\to\infty} c_n = \infty$. Thus, by using enough books, the stack can be extended as far as desired without tipping over.

89. The following argument proves the divergence of the harmonic series $S = \sum_{n=1}^{\infty} 1/n$ without using the Integral Test. Let

$$S_1 = 1 + \frac{1}{3} + \frac{1}{5} + \cdots, \qquad S_2 = \frac{1}{2} + \frac{1}{4} + \frac{1}{6} + \cdots$$

Show that if S converges, then

(a) S_1 and S_2 also converge and $S = S_1 + S_2$.

(b) $S_1 > S_2$ and $S_2 = \frac{1}{2}S$.

Observe that (b) contradicts (a), and conclude that S diverges.

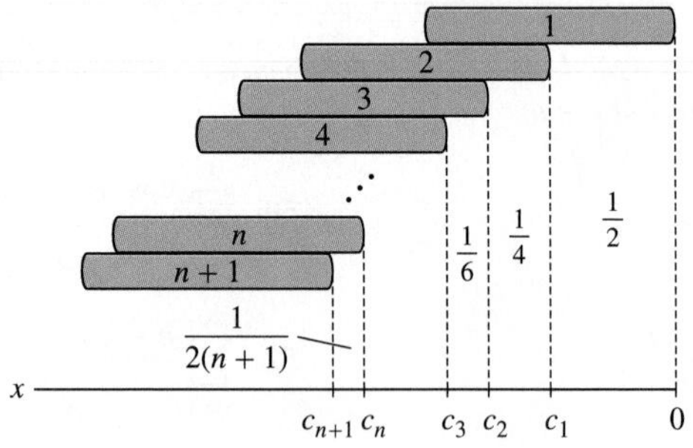

FIGURE 6

Further Insights and Challenges

90. Let $S = \sum_{n=2}^{\infty} a_n$, where $a_n = (\ln(\ln n))^{-\ln n}$.

(a) Show, by taking logarithms, that $a_n = n^{-\ln(\ln(\ln n))}$.

(b) Show that $\ln(\ln(\ln n)) \ge 2$ if $n > C$, where $C = e^{e^{e^2}}$.

(c) Show that S converges.

91. Kummer's Acceleration Method Suppose we wish to approximate $S = \sum_{n=1}^{\infty} 1/n^2$. There is a similar telescoping series whose value can be computed exactly (Example 1 in Section 11.2):

$$\sum_{n=1}^{\infty} \frac{1}{n(n+1)} = 1$$

(a) Verify that

$$S = \sum_{n=1}^{\infty} \frac{1}{n(n+1)} + \sum_{n=1}^{\infty} \left(\frac{1}{n^2} - \frac{1}{n(n+1)} \right)$$

Thus for M large,

$$S \approx 1 + \sum_{n=1}^{M} \frac{1}{n^2(n+1)} \qquad \boxed{6}$$

(b) Explain what has been gained. Why is Eq. (6) a better approximation to S than is $\sum_{n=1}^{M} 1/n^2$?

(c) CAS Compute

$$\sum_{n=1}^{1000} \frac{1}{n^2}, \qquad 1 + \sum_{n=1}^{100} \frac{1}{n^2(n+1)}$$

Which is a better approximation to S, whose exact value is $\pi^2/6$?

92. CAS The series $S = \sum_{k=1}^{\infty} k^{-3}$ has been computed to more than 100 million digits. The first 30 digits are

$$S = 1.202056903159594285399738161511$$

Approximate S using the Acceleration Method of Exercise 91 with $M = 100$ and auxiliary series $R = \sum_{n=1}^{\infty} (n(n+1)(n+2))^{-1}$. According to Exercise 46 in Section 11.2, R is a telescoping series with the sum $R = \frac{1}{4}$.

11.4 Absolute and Conditional Convergence

In the previous section, we studied positive series, but we still lack the tools to analyze series with both positive and negative terms. One of the keys to understanding such series is the concept of absolute convergence.

> **DEFINITION Absolute Convergence** The series $\sum a_n$ **converges absolutely** if $\sum |a_n|$ converges.

■ **EXAMPLE 1** Verify that the series

$$\sum_{n=1}^{\infty} \frac{(-1)^{n-1}}{n^2} = \frac{1}{1^2} - \frac{1}{2^2} + \frac{1}{3^2} - \frac{1}{4^2} + \cdots$$

converges absolutely.

Solution This series converges absolutely because the positive series (with absolute values) is a p-series with $p = 2 > 1$:

$$\sum_{n=1}^{\infty} \left| \frac{(-1)^{n-1}}{n^2} \right| = \frac{1}{1^2} + \frac{1}{2^2} + \frac{1}{3^2} + \frac{1}{4^2} + \cdots \qquad \text{(convergent } p\text{-series)}$$ ■

The next theorem tells us that if the series of absolute values converges, then the original series also converges.

> **THEOREM 1 Absolute Convergence Implies Convergence** If $\sum |a_n|$ converges, then $\sum a_n$ also converges.

Proof We have $-|a_n| \le a_n \le |a_n|$. By adding $|a_n|$ to all parts of the inequality, we get $0 \le |a_n| + a_n \le 2|a_n|$. If $\sum |a_n|$ converges, then $\sum 2|a_n|$ also converges, and therefore, $\sum (a_n + |a_n|)$ converges by the Comparison Test. Our original series converges because it is the difference of two convergent series:

$$\sum a_n = \sum (a_n + |a_n|) - \sum |a_n|$$ ■

■ **EXAMPLE 2** Verify that $S = \sum_{n=1}^{\infty} \frac{(-1)^{n-1}}{n^2}$ converges.

Solution We showed that S converges absolutely in Example 1. By Theorem 1, S itself converges. ■

■ **EXAMPLE 3** Does $S = \sum_{n=1}^{\infty} \frac{(-1)^{n-1}}{\sqrt{n}} = \frac{1}{\sqrt{1}} - \frac{1}{\sqrt{2}} + \frac{1}{\sqrt{3}} - \cdots$ converge absolutely?

Solution The positive series $\sum_{n=1}^{\infty} \frac{1}{\sqrt{n}}$ is a p-series with $p = \frac{1}{2}$. It diverges because $p < 1$. Therefore, S does not converge absolutely. ■

The series in the previous example does not converge *absolutely*, but we still do not know whether or not it converges. A series $\sum a_n$ may converge without converging absolutely. In this case, we say that $\sum a_n$ is conditionally convergent.

DEFINITION Conditional Convergence An infinite series $\sum a_n$ **converges conditionally** if $\sum a_n$ converges but $\sum |a_n|$ diverges.

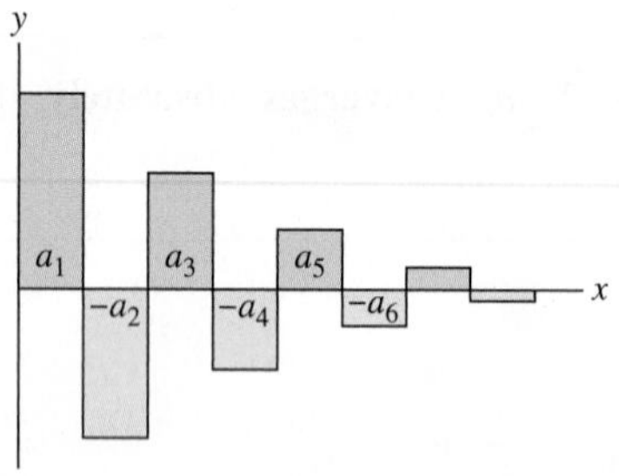

FIGURE 1 An alternating series with decreasing terms. The sum is the signed area, which is at most a_1.

If a series is not absolutely convergent, how can we determine whether it is conditionally convergent? This is often a more difficult question, because we cannot use the Integral Test or the Comparison Test (they apply only to positive series). However, convergence is guaranteed in the particular case of an **alternating series**

$$S = \sum_{n=1}^{\infty} (-1)^{n-1} a_n = a_1 - a_2 + a_3 - a_4 + \cdots$$

where the terms a_n are positive and decrease to zero (Figure 1).

THEOREM 2 Leibniz Test for Alternating Series Assume that $\{a_n\}$ is a positive sequence that is decreasing and converges to 0:

$$a_1 > a_2 > a_3 > a_4 > \cdots > 0, \qquad \lim_{n\to\infty} a_n = 0$$

Then the following alternating series converges:

$$S = \sum_{n=1}^{\infty} (-1)^{n-1} a_n = a_1 - a_2 + a_3 - a_4 + \cdots$$

Furthermore,

$$0 < S < a_1 \quad \text{and} \quad S_{2N} < S < S_{2N+1} \qquad N \ge 1$$

Assumptions Matter *The Leibniz Test is not valid if we drop the assumption that a_n is decreasing (see Exercise 35).*

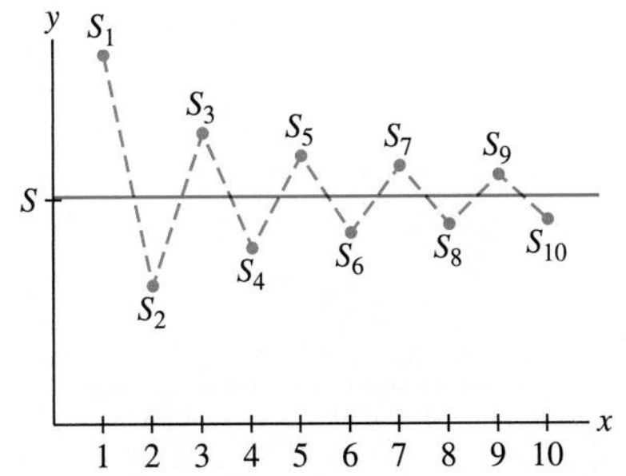

FIGURE 2 The partial sums of an alternating series zigzag above and below the limit. The odd partial sums decrease and the even partial sums increase.

Proof We will prove that the partial sums zigzag above and below the sum S as in Figure 2. Note first that the even partial sums are increasing. Indeed, the odd-numbered terms occur with a plus sign and thus, for example,

$$S_4 + a_5 - a_6 = S_6$$

But $a_5 - a_6 > 0$ because a_n is decreasing, and therefore $S_4 < S_6$. In general,

$$S_{2N} + (a_{2N+1} - a_{2N+2}) = S_{2N+2}$$

where $a_{2n+1} - a_{2N+2} > 0$. Thus $S_{2N} < S_{2N+2}$ and

$$0 < S_2 < S_4 < S_6 < \cdots$$

Similarly,

$$S_{2N-1} - (a_{2N} - a_{2N+1}) = S_{2N+1}$$

Therefore $S_{2N+1} < S_{2N-1}$, and the sequence of odd partial sums is decreasing:

$$\cdots < S_7 < S_5 < S_3 < S_1$$

Finally, $S_{2N} < S_{2N} + a_{2N+1} = S_{2N+1}$. The picture is as follows:

$$0 < S_2 < S_4 < S_6 < \quad \cdots \quad < S_7 < S_5 < S_3 < S_1$$

Now, because bounded monotonic sequences converge (Theorem 6 of Section 11.1), the even and odd partial sums approach limits that are sandwiched in the middle:

$$0 < S_2 < S_4 < \cdots < \lim_{N\to\infty} S_{2N} \le \lim_{N\to\infty} S_{2N+1} < \cdots < S_5 < S_3 < S_1 \tag{1}$$

These two limits must have a common value L because

$$\lim_{N\to\infty} S_{2N+1} - \lim_{N\to\infty} S_{2N} = \lim_{N\to\infty} (S_{2N+1} - S_{2N}) = \lim_{N\to\infty} a_{2N+1} = 0$$

Therefore, $\lim_{N\to\infty} S_N = L$ and the infinite series converges to $S = L$. From Eq. (1) we also see that $0 < S < S_1 = a_1$ and $S_{2N} < S < S_{2N+1}$ for all N as claimed. ■

The Leibniz Test is the only test for conditional convergence developed in this text. Other tests, such as Abel's Criterion and the Dirichlet Test, are discussed in textbooks on Analysis.

■ **EXAMPLE 4** Show that $S = \sum_{n=1}^{\infty} \frac{(-1)^{n-1}}{\sqrt{n}} = \frac{1}{\sqrt{1}} - \frac{1}{\sqrt{2}} + \frac{1}{\sqrt{3}} - \cdots$ converges conditionally and that $0 \le S \le 1$.

Solution The terms $a_n = 1/\sqrt{n}$ are positive and decreasing, and $\lim_{n\to\infty} a_n = 0$. Therefore, S converges by the Leibniz Test. Furthermore, $0 \le S \le 1$ because $a_1 = 1$. However, the positive series $\sum_{n=1}^{\infty} 1/\sqrt{n}$ diverges because it is a p-series with $p = \frac{1}{2} < 1$. Therefore, S is conditionally convergent but not absolutely convergent (Figure 3). ■

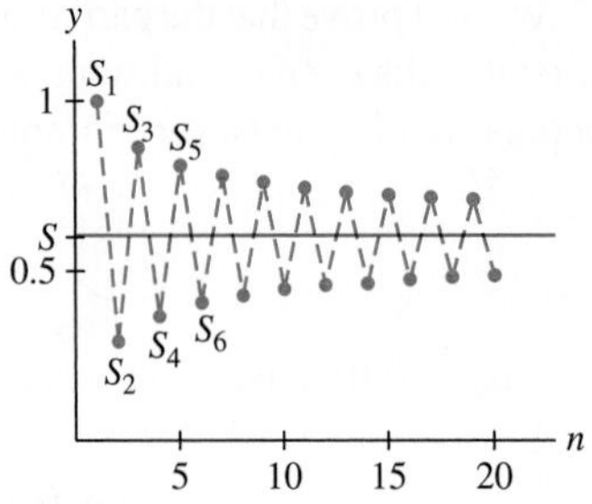

(A) Partial sums of $S = \sum_{n=1}^{\infty} (-1)^{n-1} \frac{1}{\sqrt{n}}$

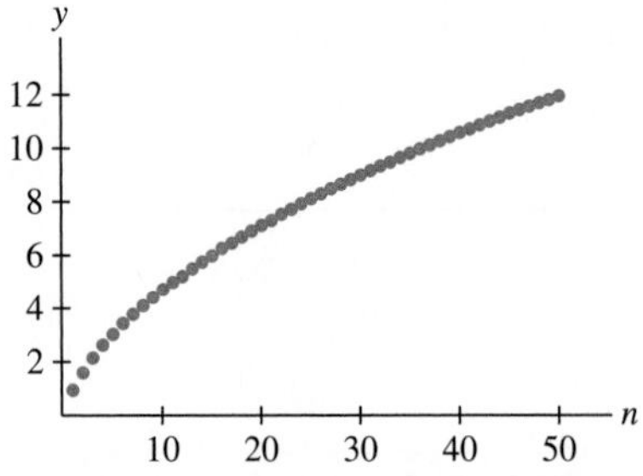

(B) Partial sums of $\sum_{n=1}^{\infty} \frac{1}{\sqrt{n}}$

FIGURE 3

The inequality $S_{2N} < S < S_{2N+1}$ in Theorem 2 gives us important information about the error; it tells us that $|S_N - S|$ is less than $|S_N - S_{N+1}| = a_{N+1}$ for all N.

THEOREM 3 Let $S = \sum_{n=1}^{\infty} (-1)^{n-1} a_n$, where $\{a_n\}$ is a positive decreasing sequence that converges to 0. Then

$$\left|S - S_N\right| < a_{N+1} \qquad (2)$$

In other words, *the error committed when we approximate S by S_N is less than the size of the first omitted term a_{N+1}.*

■ **EXAMPLE 5 Alternating Harmonic Series** Show that $S = \sum_{n=1}^{\infty} \frac{(-1)^{n-1}}{n}$ converges conditionally. Then:

(a) Show that $|S - S_6| < \frac{1}{7}$.

(b) Find an N such that S_N approximates S with an error less than 10^{-3}.

Solution The terms $a_n = 1/n$ are positive and decreasing, and $\lim_{n\to\infty} a_n = 0$. Therefore, S converges by the Leibniz Test. The harmonic series $\sum_{n=1}^{\infty} 1/n$ diverges, so S converges conditionally but not absolutely. Now, applying Eq. (2), we have

$$|S - S_N| < a_{N+1} = \frac{1}{N+1}$$

For $N = 6$, we obtain $|S - S_6| < a_7 = \frac{1}{7}$. We can make the error less than 10^{-3} by choosing N so that

$$\frac{1}{N+1} \le 10^{-3} \quad \Rightarrow \quad N + 1 \ge 10^3 \quad \Rightarrow \quad N \ge 999$$

Using a computer algebra system, we find that $S_{999} \approx 0.69365$. In Exercise 84 of Section 11.7, we will prove that $S = \ln 2 \approx 0.69314$, and thus we can verify that

$$|S - S_{999}| \approx |\ln 2 - 0.69365| \approx 0.0005 < 10^{-3}$$

■

CONCEPTUAL INSIGHT The convergence of an infinite series $\sum a_n$ depends on two factors: (1) how quickly a_n tends to zero, and (2) how much cancellation takes place among the terms. Consider

Harmonic series (diverges): $1+\frac{1}{2}+\frac{1}{3}+\frac{1}{4}+\frac{1}{5}+\cdots$

p-Series with $p=2$ (converges): $1+\frac{1}{2^2}+\frac{1}{3^2}+\frac{1}{4^2}+\frac{1}{5^2}+\cdots$

Alternating harmonic series (converges): $1-\frac{1}{2}+\frac{1}{3}-\frac{1}{4}+\frac{1}{5}-\cdots$

The harmonic series diverges because reciprocals $1/n$ do not tend to zero quickly enough. By contrast, the reciprocal squares $1/n^2$ tend to zero quickly enough for the p-series with $p=2$ to converge. The alternating harmonic series converges, but only due to the cancellation among the terms.

11.4 SUMMARY

- $\sum a_n$ *converges absolutely* if the positive series $\sum |a_n|$ converges.
- Absolute convergence implies convergence: If $\sum |a_n|$ converges, then $\sum a_n$ also converges.
- $\sum a_n$ *converges conditionally* if $\sum a_n$ converges but $\sum |a_n|$ diverges.
- *Leibniz Test:* If $\{a_n\}$ is positive and decreasing and $\lim_{n\to\infty} a_n = 0$, then the alternating series

$$S=\sum_{n=1}^{\infty}(-1)^{n-1}a_n = a_1 - a_2 + a_3 - a_4 + a_5 - \cdots$$

converges. Furthermore, $|S - S_N| < a_{N+1}$.

- We have developed two ways to handle nonpositive series: Show absolute convergence if possible, or use the Leibniz Test, if applicable.

11.4 EXERCISES

Preliminary Questions

1. Give an example of a series such that $\sum a_n$ converges but $\sum |a_n|$ diverges.

2. Which of the following statements is equivalent to Theorem 1?

(a) If $\sum_{n=0}^{\infty} |a_n|$ diverges, then $\sum_{n=0}^{\infty} a_n$ also diverges.

(b) If $\sum_{n=0}^{\infty} a_n$ diverges, then $\sum_{n=0}^{\infty} |a_n|$ also diverges.

(c) If $\sum_{n=0}^{\infty} a_n$ converges, then $\sum_{n=0}^{\infty} |a_n|$ also converges.

3. Lathika argues that $\sum_{n=1}^{\infty}(-1)^n\sqrt{n}$ is an alternating series and therefore converges. Is Lathika right?

4. Suppose that a_n is positive, decreasing, and tends to 0, and let $S=\sum_{n=1}^{\infty}(-1)^{n-1}a_n$. What can we say about $|S - S_{100}|$ if $a_{101} = 10^{-3}$? Is S larger or smaller than S_{100}?

Exercises

1. Show that

$$\sum_{n=0}^{\infty} \frac{(-1)^n}{2^n}$$

converges absolutely.

2. Show that the following series converges conditionally:

$$\sum_{n=1}^{\infty} (-1)^{n-1} \frac{1}{n^{2/3}} = \frac{1}{1^{2/3}} - \frac{1}{2^{2/3}} + \frac{1}{3^{2/3}} - \frac{1}{4^{2/3}} + \cdots$$

In Exercises 3–10, determine whether the series converges absolutely, conditionally, or not at all.

3. $\displaystyle\sum_{n=1}^{\infty} \frac{(-1)^{n-1}}{n^{1/3}}$

4. $\displaystyle\sum_{n=1}^{\infty} \frac{(-1)^n n^4}{n^3+1}$

5. $\displaystyle\sum_{n=0}^{\infty} \frac{(-1)^{n-1}}{(1.1)^n}$

6. $\displaystyle\sum_{n=1}^{\infty} \frac{\sin(\frac{\pi n}{4})}{n^2}$

7. $\displaystyle\sum_{n=2}^{\infty} \frac{(-1)^n}{n \ln n}$

8. $\displaystyle\sum_{n=1}^{\infty} \frac{(-1)^n}{1+\frac{1}{n}}$

9. $\displaystyle\sum_{n=2}^{\infty} \frac{\cos n\pi}{(\ln n)^2}$

10. $\displaystyle\sum_{n=1}^{\infty} \frac{\cos n}{2^n}$

11. Let $S = \displaystyle\sum_{n=1}^{\infty} (-1)^{n+1} \frac{1}{n^3}$.

(a) Calculate S_n for $1 \le n \le 10$.

(b) Use Eq. (2) to show that $0.9 \le S \le 0.902$.

12. Use Eq. (2) to approximate

$$\sum_{n=1}^{\infty} \frac{(-1)^{n+1}}{n!}$$

to four decimal places.

13. Approximate $\displaystyle\sum_{n=1}^{\infty} \frac{(-1)^{n+1}}{n^4}$ to three decimal places.

14. CAS Let

$$S = \sum_{n=1}^{\infty} (-1)^{n-1} \frac{n}{n^2+1}$$

Use a computer algebra system to calculate and plot the partial sums S_n for $1 \le n \le 100$. Observe that the partial sums zigzag above and below the limit.

In Exercises 15–16, find a value of N such that S_N approximates the series with an error of at most 10^{-5}. If you have a CAS, compute this value of S_N.

15. $\displaystyle\sum_{n=1}^{\infty} \frac{(-1)^{n+1}}{n(n+2)(n+3)}$

16. $\displaystyle\sum_{n=1}^{\infty} \frac{(-1)^{n+1} \ln n}{n!}$

In Exercises 17–32, determine convergence or divergence by any method.

17. $\displaystyle\sum_{n=0}^{\infty} 7^{-n}$

18. $\displaystyle\sum_{n=1}^{\infty} \frac{1}{n^{7.5}}$

19. $\displaystyle\sum_{n=1}^{\infty} \frac{1}{5^n - 3^n}$

20. $\displaystyle\sum_{n=2}^{\infty} \frac{n}{n^2 - n}$

21. $\displaystyle\sum_{n=1}^{\infty} \frac{1}{3n^4 + 12n}$

22. $\displaystyle\sum_{n=1}^{\infty} \frac{(-1)^n}{\sqrt{n^2+1}}$

23. $\displaystyle\sum_{n=1}^{\infty} \frac{1}{\sqrt{n^2+1}}$

24. $\displaystyle\sum_{n=0}^{\infty} \frac{(-1)^n n}{\sqrt{n^2+1}}$

25. $\displaystyle\sum_{n=1}^{\infty} \frac{3^n + (-2)^n}{5^n}$

26. $\displaystyle\sum_{n=1}^{\infty} \frac{(-1)^{n+1}}{(2n+1)!}$

27. $\displaystyle\sum_{n=1}^{\infty} (-1)^n n^2 e^{-n^3/3}$

28. $\displaystyle\sum_{n=1}^{\infty} n e^{-n^3/3}$

29. $\displaystyle\sum_{n=2}^{\infty} \frac{(-1)^n}{n^{1/2}(\ln n)^2}$

30. $\displaystyle\sum_{n=2}^{\infty} \frac{1}{n(\ln n)^{1/4}}$

31. $\displaystyle\sum_{n=1}^{\infty} \frac{\ln n}{n^{1.05}}$

32. $\displaystyle\sum_{n=2}^{\infty} \frac{1}{(\ln n)^2}$

33. Show that

$$S = \frac{1}{2} - \frac{1}{2} + \frac{1}{3} - \frac{1}{3} + \frac{1}{4} - \frac{1}{4} + \cdots$$

converges by computing the partial sums. Does it converge absolutely?

34. The Leibniz Test cannot be applied to

$$\frac{1}{2} - \frac{1}{3} + \frac{1}{2^2} - \frac{1}{3^2} + \frac{1}{2^3} - \frac{1}{3^3} + \cdots$$

Why not? Show that it converges by another method.

35. **Assumptions Matter** Show by counterexample that the Leibniz Test does not remain true if the sequence a_n tends to zero but is not assumed nonincreasing. *Hint:* Consider

$$R = \frac{1}{2} - \frac{1}{4} + \frac{1}{3} - \frac{1}{8} + \frac{1}{4} - \frac{1}{16} + \cdots + \left(\frac{1}{n} - \frac{1}{2^n}\right) + \cdots$$

36. Determine whether the following series converges conditionally:

$$1 - \frac{1}{3} + \frac{1}{2} - \frac{1}{5} + \frac{1}{3} - \frac{1}{7} + \frac{1}{4} - \frac{1}{9} + \frac{1}{5} - \frac{1}{11} + \cdots$$

37. Prove that if $\sum a_n$ converges absolutely, then $\sum a_n^2$ also converges. Then give an example where $\sum a_n$ is only conditionally convergent and $\sum a_n^2$ diverges.

Further Insights and Challenges

38. Prove the following variant of the Leibniz Test: If $\{a_n\}$ is a positive, decreasing sequence with $\lim_{n\to\infty} a_n = 0$, then the series

$$a_1 + a_2 - 2a_3 + a_4 + a_5 - 2a_6 + \cdots$$

converges. *Hint:* Show that S_{3N} is increasing and bounded by $a_1 + a_2$, and continue as in the proof of the Leibniz Test.

39. Use Exercise 38 to show that the following series converges:

$$S = \frac{1}{\ln 2} + \frac{1}{\ln 3} - \frac{2}{\ln 4} + \frac{1}{\ln 5} + \frac{1}{\ln 6} - \frac{2}{\ln 7} + \cdots$$

40. Prove the conditional convergence of

$$R = 1 + \frac{1}{2} + \frac{1}{3} - \frac{3}{4} + \frac{1}{5} + \frac{1}{6} + \frac{1}{7} - \frac{3}{8} + \cdots$$

41. Show that the following series diverges:

$$S = 1 + \frac{1}{2} + \frac{1}{3} - \frac{2}{4} + \frac{1}{5} + \frac{1}{6} + \frac{1}{7} - \frac{2}{8} + \cdots$$

Hint: Use the result of Exercise 40 to write S as the sum of a convergent series and a divergent series.

42. Prove that

$$\sum_{n=1}^{\infty} (-1)^{n+1} \frac{(\ln n)^a}{n}$$

converges for all exponents a. *Hint:* Show that $f(x) = (\ln x)^a/x$ is decreasing for x sufficiently large.

43. We say that $\{b_n\}$ is a rearrangement of $\{a_n\}$ if $\{b_n\}$ has the same terms as $\{a_n\}$ but occurring in a different order. Show that if $\{b_n\}$ is a rearrangement of $\{a_n\}$ and $S = \sum_{n=1}^{\infty} a_n$ converges absolutely, then $T = \sum_{n=1}^{\infty} b_n$ also converges absolutely. (This result does not hold if S is only conditionally convergent.) *Hint:* Prove that the partial sums $\sum_{n=1}^{N} |b_n|$ are bounded. It can be shown further that $S = T$.

44. Assumptions Matter In 1829, Lejeune Dirichlet pointed out that the great French mathematician Augustin Louis Cauchy made a mistake in a published paper by improperly assuming the Limit Comparison Test to be valid for nonpositive series. Here are Dirichlet's two series:

$$\sum_{n=1}^{\infty} \frac{(-1)^n}{\sqrt{n}}, \qquad \sum_{n=1}^{\infty} \frac{(-1)^n}{\sqrt{n}} \left(1 + \frac{(-1)^n}{\sqrt{n}}\right)$$

Explain how they provide a counterexample to the Limit Comparison Test when the series are not assumed to be positive.

11.5 The Ratio and Root Tests

Series such as

$$S = 1 + \frac{2}{1!} + \frac{2^2}{2!} + \frac{2^3}{3!} + \frac{2^4}{4!} + \cdots$$

arise in applications, but the convergence tests developed so far cannot be applied easily. Fortunately, the Ratio Test can be used for this and many other series.

THEOREM 1 Ratio Test Assume that the following limit exists:

$$\rho = \lim_{n\to\infty} \left|\frac{a_{n+1}}{a_n}\right|$$

The symbol ρ is a lowercase "rho," the seventeenth letter of the Greek alphabet.

(i) If $\rho < 1$, then $\sum a_n$ converges absolutely.

(ii) If $\rho > 1$, then $\sum a_n$ diverges.

(iii) If $\rho = 1$, the test is inconclusive (the series may converge or diverge).

Proof The idea is to compare with a geometric series. If $\rho < 1$, we may choose a number r such that $\rho < r < 1$. Since $|a_{n+1}/a_n|$ converges to ρ, there exists a number M such that $|a_{n+1}/a_n| < r$ for all $n \geq M$. Therefore,

$$|a_{M+1}| < r|a_M|$$
$$|a_{M+2}| < r|a_{M+1}| < r(r|a_M|) = r^2|a_M|$$
$$|a_{M+3}| < r|a_{M+2}| < r^3|a_M|$$

In general, $|a_{M+n}| < r^n|a_M|$, and thus,

$$\sum_{n=M}^{\infty} |a_n| = \sum_{n=0}^{\infty} |a_{M+n}| \leq \sum_{n=0}^{\infty} |a_M|\, r^n = |a_M| \sum_{n=0}^{\infty} r^n$$

The geometric series on the right converges because $0 < r < 1$, so $\sum_{n=M}^{\infty} |a_n|$ converges by the Comparison Test and thus $\sum a_n$ converges absolutely.

If $\rho > 1$, choose r such that $1 < r < \rho$. Then there exists a number M such that $|a_{n+1}/a_n| > r$ for all $n \geq M$. Arguing as before with the inequalities reversed, we find that $|a_{M+n}| \geq r^n|a_M|$. Since r^n tends to ∞, the terms a_{M+n} do not tend to zero, and consequently, $\sum a_n$ diverges. Finally, Example 4 below shows that both convergence and divergence are possible when $\rho = 1$, so the test is inconclusive in this case. ■

■ **EXAMPLE 1** Prove that $\displaystyle\sum_{n=1}^{\infty} \frac{2^n}{n!}$ converges.

Solution Compute the ratio and its limit with $a_n = \dfrac{2^n}{n!}$. Note that $(n+1)! = (n+1)n!$ and thus

$$\frac{a_{n+1}}{a_n} = \frac{2^{n+1}}{(n+1)!}\,\frac{n!}{2^n} = \frac{2^{n+1}}{2^n}\,\frac{n!}{(n+1)!} = \frac{2}{n+1}$$
$$\rho = \lim_{n\to\infty} \left|\frac{a_{n+1}}{a_n}\right| = \lim_{n\to\infty} \frac{2}{n+1} = 0$$

Since $\rho < 1$, the series $\displaystyle\sum_{n=1}^{\infty} \frac{2^n}{n!}$ converges by the Ratio Test. ■

■ **EXAMPLE 2** Does $\displaystyle\sum_{n=1}^{\infty} \frac{n^2}{2^n}$ converge?

Solution Apply the Ratio Test with $a_n = \dfrac{n^2}{2^n}$:

$$\left|\frac{a_{n+1}}{a_n}\right| = \frac{(n+1)^2}{2^{n+1}}\,\frac{2^n}{n^2} = \frac{1}{2}\left(\frac{n^2+2n+1}{n^2}\right) = \frac{1}{2}\left(1 + \frac{2}{n} + \frac{1}{n^2}\right)$$
$$\rho = \lim_{n\to\infty} \left|\frac{a_{n+1}}{a_n}\right| = \frac{1}{2}\lim_{n\to\infty}\left(1 + \frac{2}{n} + \frac{1}{n^2}\right) = \frac{1}{2}$$

Since $\rho < 1$, the series converges by the Ratio Test. ■

■ **EXAMPLE 3** Does $\displaystyle\sum_{n=0}^{\infty}(-1)^n \frac{n!}{1000^n}$ converge?

Solution This series diverges by the Ratio Test because $\rho > 1$:

$$\rho = \lim_{n\to\infty}\left|\frac{a_{n+1}}{a_n}\right| = \lim_{n\to\infty}\frac{(n+1)!}{1000^{n+1}}\,\frac{1000^n}{n!} = \lim_{n\to\infty}\frac{n+1}{1000} = \infty$$ ■

■ **EXAMPLE 4 Ratio Test Inconclusive** Show that both convergence and divergence are possible when $\rho = 1$ by considering $\displaystyle\sum_{n=1}^{\infty} n^2$ and $\displaystyle\sum_{n=1}^{\infty} n^{-2}$.

Solution For $a_n = n^2$, we have

$$\rho = \lim_{n\to\infty}\left|\frac{a_{n+1}}{a_n}\right| = \lim_{n\to\infty}\frac{(n+1)^2}{n^2} = \lim_{n\to\infty}\frac{n^2+2n+1}{n^2} = \lim_{n\to\infty}\left(1+\frac{2}{n}+\frac{1}{n^2}\right) = 1$$

On the other hand, for $b_n = n^{-2}$,

$$\rho = \lim_{n\to\infty}\left|\frac{b_{n+1}}{b_n}\right| = \lim_{n\to\infty}\left|\frac{a_n}{a_{n+1}}\right| = \frac{1}{\lim\limits_{n\to\infty}\left|\frac{a_{n+1}}{a_n}\right|} = 1$$

Thus, $\rho = 1$ in both cases, but $\displaystyle\sum_{n=1}^{\infty} n^2$ diverges and $\displaystyle\sum_{n=1}^{\infty} n^{-2}$ converges. This shows that both convergence and divergence are possible when $\rho = 1$. ■

Our next test is based on the limit of the nth roots $\sqrt[n]{a_n}$ rather than the ratios a_{n+1}/a_n. Its proof, like that of the Ratio Test, is based on a comparison with a geometric series (see Exercise 57).

THEOREM 2 Root Test Assume that the following limit exists:

$$L = \lim_{n\to\infty}\sqrt[n]{|a_n|}$$

(i) If $L < 1$, then $\sum a_n$ converges absolutely.

(ii) If $L > 1$, then $\sum a_n$ diverges.

(iii) If $L = 1$, the test is inconclusive (the series may converge or diverge).

■ **EXAMPLE 5** Does $\displaystyle\sum_{n=1}^{\infty}\left(\frac{n}{2n+3}\right)^n$ converge?

Solution We have $L = \lim\limits_{n\to\infty}\sqrt[n]{a_n} = \lim\limits_{n\to\infty}\dfrac{n}{2n+3} = \dfrac{1}{2}$. Since $L < 1$, the series converges by the Root Test. ■

11.5 SUMMARY

- *Ratio Test:* Assume that $\rho = \lim_{n\to\infty} \left|\frac{a_{n+1}}{a_n}\right|$ exists. Then $\sum a_n$
  - Converges absolutely if $\rho < 1$.
  - Diverges if $\rho > 1$.
  - Inconclusive if $\rho = 1$.
- *Root Test:* Assume that $L = \lim_{n\to\infty} \sqrt[n]{|a_n|}$ exists. Then $\sum a_n$
  - Converges absolutely if $L < 1$.
  - Diverges if $L > 1$.
  - Inconclusive if $L = 1$.

11.5 EXERCISES

Preliminary Questions

1. In the Ratio Test, is ρ equal to $\lim_{n\to\infty} \left|\frac{a_{n+1}}{a_n}\right|$ or $\lim_{n\to\infty} \left|\frac{a_n}{a_{n+1}}\right|$?

2. Is the Ratio Test conclusive for $\sum_{n=1}^{\infty} \frac{1}{2^n}$? Is it conclusive for $\sum_{n=1}^{\infty} \frac{1}{n}$?

3. Can the Ratio Test be used to show convergence if the series is only conditionally convergent?

Exercises

In Exercises 1–20, apply the Ratio Test to determine convergence or divergence, or state that the Ratio Test is inconclusive.

1. $\sum_{n=1}^{\infty} \frac{1}{5^n}$

2. $\sum_{n=1}^{\infty} \frac{(-1)^{n-1} n}{5^n}$

3. $\sum_{n=1}^{\infty} \frac{1}{n^n}$

4. $\sum_{n=0}^{\infty} \frac{3n+2}{5n^3+1}$

5. $\sum_{n=1}^{\infty} \frac{n}{n^2+1}$

6. $\sum_{n=1}^{\infty} \frac{2^n}{n}$

7. $\sum_{n=1}^{\infty} \frac{2^n}{n^{100}}$

8. $\sum_{n=1}^{\infty} \frac{n^3}{3^{n^2}}$

9. $\sum_{n=1}^{\infty} \frac{10^n}{2^{n^2}}$

10. $\sum_{n=1}^{\infty} \frac{e^n}{n!}$

11. $\sum_{n=1}^{\infty} \frac{e^n}{n^n}$

12. $\sum_{n=1}^{\infty} \frac{n^{40}}{n!}$

13. $\sum_{n=0}^{\infty} \frac{n!}{6^n}$

14. $\sum_{n=1}^{\infty} \frac{n!}{n^9}$

15. $\sum_{n=2}^{\infty} \frac{1}{n \ln n}$

16. $\sum_{n=1}^{\infty} \frac{1}{(2n)!}$

17. $\sum_{n=1}^{\infty} \frac{n^2}{(2n+1)!}$

18. $\sum_{n=1}^{\infty} \frac{(n!)^3}{(3n)!}$

19. $\sum_{n=2}^{\infty} \frac{1}{2^n+1}$

20. $\sum_{n=2}^{\infty} \frac{1}{\ln n}$

21. Show that $\sum_{n=1}^{\infty} n^k 3^{-n}$ converges for all exponents k.

22. Show that $\sum_{n=1}^{\infty} n^2 x^n$ converges if $|x| < 1$.

23. Show that $\sum_{n=1}^{\infty} 2^n x^n$ converges if $|x| < \frac{1}{2}$.

24. Show that $\sum_{n=1}^{\infty} \frac{r^n}{n!}$ converges for all r.

25. Show that $\sum_{n=1}^{\infty} \frac{r^n}{n}$ converges if $|r| < 1$.

26. Is there any value of k such that $\sum_{n=1}^{\infty} \frac{2^n}{n^k}$ converges?

27. Show that $\sum_{n=1}^{\infty} \frac{n!}{n^n}$ converges. *Hint:* Use $\lim_{n\to\infty} \left(1+\frac{1}{n}\right)^n = e$.

In Exercises 28–33, assume that $|a_{n+1}/a_n|$ converges to $\rho = \frac{1}{3}$. What can you say about the convergence of the given series?

28. $\sum_{n=1}^{\infty} na_n$

29. $\sum_{n=1}^{\infty} n^3 a_n$

30. $\sum_{n=1}^{\infty} 2^n a_n$

31. $\sum_{n=1}^{\infty} 3^n a_n$

32. $\sum_{n=1}^{\infty} 4^n a_n$

33. $\sum_{n=1}^{\infty} a_n^2$

34. Assume that $|a_{n+1}/a_n|$ converges to $\rho = 4$. Does $\sum_{n=1}^{\infty} a_n^{-1}$ converge (assume that $a_n \neq 0$ for all n)?

35. Is the Ratio Test conclusive for the p-series $\sum_{n=1}^{\infty} \frac{1}{n^p}$?

In Exercises 36–41, use the Root Test to determine convergence or divergence (or state that the test is inconclusive).

36. $\sum_{n=0}^{\infty} \frac{1}{10^n}$

37. $\sum_{n=1}^{\infty} \frac{1}{n^n}$

38. $\sum_{k=0}^{\infty} \left(\frac{k}{k+10}\right)^k$

39. $\sum_{k=0}^{\infty} \left(\frac{k}{3k+1}\right)^k$

40. $\sum_{n=1}^{\infty} \left(1+\frac{1}{n}\right)^{-n}$

41. $\sum_{n=4}^{\infty} \left(1+\frac{1}{n}\right)^{-n^2}$

42. Prove that $\sum_{n=1}^{\infty} \frac{2^{n^2}}{n!}$ diverges. *Hint:* Use $2^{n^2} = (2^n)^n$ and $n! \le n^n$.

In Exercises 43–56, determine convergence or divergence using any method covered in the text so far.

43. $\sum_{n=1}^{\infty} \frac{2^n + 4^n}{7^n}$

44. $\sum_{n=1}^{\infty} \frac{n^3}{n!}$

45. $\sum_{n=1}^{\infty} \frac{n^3}{5^n}$

46. $\sum_{n=2}^{\infty} \frac{1}{n(\ln n)^3}$

47. $\sum_{n=2}^{\infty} \frac{1}{\sqrt{n^3 - n^2}}$

48. $\sum_{n=1}^{\infty} \frac{n^2 + 4n}{3n^4 + 9}$

49. $\sum_{n=1}^{\infty} n^{-0.8}$

50. $\sum_{n=1}^{\infty} (0.8)^{-n} n^{-0.8}$

51. $\sum_{n=1}^{\infty} 4^{-2n+1}$

52. $\sum_{n=1}^{\infty} \frac{(-1)^{n-1}}{\sqrt{n}}$

53. $\sum_{n=1}^{\infty} \sin \frac{1}{n^2}$

54. $\sum_{n=1}^{\infty} (-1)^n \cos \frac{1}{n}$

55. $\sum_{n=1}^{\infty} \frac{(-2)^n}{\sqrt{n}}$

56. $\sum_{n=1}^{\infty} \left(\frac{n}{n+12}\right)^n$

Further Insights and Challenges

57. **Proof of the Root Test** Let $S = \sum_{n=0}^{\infty} a_n$ be a positive series, and assume that $L = \lim_{n\to\infty} \sqrt[n]{a_n}$ exists.

(a) Show that S converges if $L < 1$. *Hint:* Choose R with $L < R < 1$ and show that $a_n \le R^n$ for n sufficiently large. Then compare with the geometric series $\sum R^n$.

(b) Show that S diverges if $L > 1$.

58. Show that the Ratio Test does not apply, but verify convergence using the Comparison Test for the series

$$\frac{1}{2} + \frac{1}{3^2} + \frac{1}{2^3} + \frac{1}{3^4} + \frac{1}{2^5} + \cdots$$

59. Let $S = \sum_{n=1}^{\infty} \frac{c^n n!}{n^n}$, where c is a constant.

(a) Prove that S converges absolutely if $|c| < e$ and diverges if $|c| > e$.

(b) It is known that $\lim_{n\to\infty} \frac{e^n n!}{n^{n+1/2}} = \sqrt{2\pi}$. Verify this numerically.

(c) Use the Limit Comparison Test to prove that S diverges for $c = e$.

11.6 Power Series

A **power series** with center c is an infinite series

$$F(x) = \sum_{n=0}^{\infty} a_n(x-c)^n = a_0 + a_1(x-c) + a_2(x-c)^2 + a_3(x-c)^3 + \cdots$$

where x is a variable. For example,

$$F(x) = 1 + (x-2) + 2(x-2)^2 + 3(x-2)^3 + \cdots$$ 1

is a power series with center $c = 2$.

Many functions that arise in applications can be represented as power series. This includes not only the familiar trigonometric, exponential, logarithm, and root functions, but also the host of "special functions" of physics and engineering such as Bessel functions and elliptic functions.

A power series $F(x) = \sum_{n=0}^{\infty} a_n(x-c)^n$ converges for some values of x and may diverge for others. For example, if we set $x = \frac{9}{4}$ in the power series of Eq. (1), we obtain an infinite series that converges by the Ratio Test:

$$\begin{aligned} F\left(\frac{9}{4}\right) &= 1 + \left(\frac{9}{4}-2\right) + 2\left(\frac{9}{4}-2\right)^2 + 3\left(\frac{9}{4}-2\right)^3 + \cdots \\ &= 1 + \left(\frac{1}{4}\right) + 2\left(\frac{1}{4}\right)^2 + 3\left(\frac{1}{4}\right)^3 + \cdots \end{aligned}$$

On the other hand, the power series in Eq. (1) diverges for $x = 3$:

$$\begin{aligned} F(3) &= 1 + (3-2) + 2(3-2)^2 + 3(3-2)^3 + \cdots \\ &= 1 + 1 + 2 + 3 + \cdots \end{aligned}$$

There is a surprisingly simple way to describe the set of values x at which a power series $F(x)$ converges. According to our next theorem, either $F(x)$ converges absolutely for all values of x or there is a radius of convergence R such that

$F(x)$ converges absolutely when $|x-c| < R$ and diverges when $|x-c| > R$.

This means that $F(x)$ converges for x in an **interval of convergence** consisting of the open interval $(c-R, c+R)$ and possibly one or both of the endpoints $c-R$ and $c+R$ (Figure 1). Note that $F(x)$ automatically converges at $x = c$ because

$$F(c) = a_0 + a_1(c-c) + a_2(c-c)^2 + a_3(c-c)^3 + \cdots = a_0$$

We set $R = 0$ if $F(x)$ converges only for $x = c$, and we set $R = \infty$ if $F(x)$ converges for all values of x.

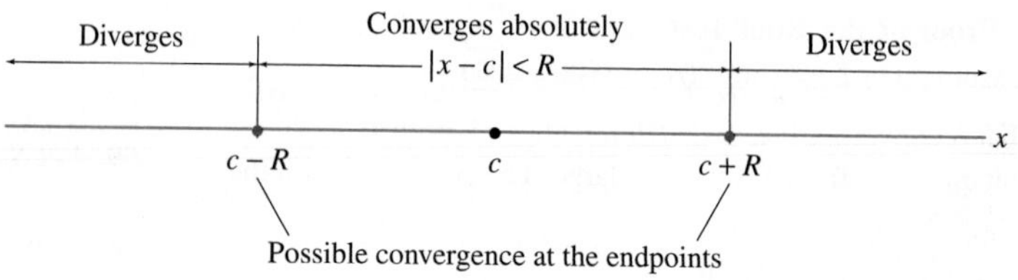

FIGURE 1 Interval of convergence of a power series.

THEOREM 1 Radius of Convergence Every power series

$$F(x) = \sum_{n=0}^{\infty} a_n(x-c)^n$$

has a radius of convergence R, which is either a nonnegative number ($R \ge 0$) or infinity ($R = \infty$). If R is finite, $F(x)$ converges absolutely when $|x-c| < R$ and diverges when $|x-c| > R$. If $R = \infty$, then $F(x)$ converges absolutely for all x.

Proof We assume that $c = 0$ to simplify the notation. If $F(x)$ converges only at $x = 0$, then $R = 0$. Otherwise, $F(x)$ converges for some nonzero value $x = B$. We claim that $F(x)$ must then converge absolutely for all $|x| < |B|$. To prove this, note that because $F(B) = \sum_{n=0}^{\infty} a_n B^n$ converges, the general term $a_n B^n$ tends to zero. In particular, there exists $M > 0$ such that $|a_n B^n| < M$ for all n. Therefore,

$$\sum_{n=0}^{\infty} |a_n x^n| = \sum_{n=0}^{\infty} |a_n B^n| \left|\frac{x}{B}\right|^n < M \sum_{n=0}^{\infty} \left|\frac{x}{B}\right|^n$$

If $|x| < |B|$, then $|x/B| < 1$ and the series on the right is a convergent geometric series. By the Comparison Test, the series on the left also converges. This proves that $F(x)$ converges absolutely if $|x| < |B|$.

Least Upper Bound Property: If S is a set of real numbers with an upper bound M (that is, $x \leq M$ for all $x \in S$), then S has a least upper bound L. See Appendix B.

Now let S be the set of numbers x such that $F(x)$ converges. Then S contains 0, and we have shown that if S contains a number $B \neq 0$, then S contains the open interval $(-|B|, |B|)$. If S is bounded, then S has a least upper bound $L > 0$ (see marginal note). In this case, there exist numbers $B \in S$ smaller than but arbitrarily close to L, and thus S contains $(-B, B)$ for all $0 < B < L$. It follows that S contains the open interval $(-L, L)$. The set S cannot contain any number x with $|x| > L$, but S may contain one or both of the endpoints $x = \pm L$. So in this case, $F(x)$ has radius of convergence $R = L$. If S is not bounded, then S contains intervals $(-B, B)$ for B arbitrarily large. In this case, S is the entire real line $\mathbf{R}$, and the radius of convergence is $R = \infty$. ■

From Theorem 1, we see that there are two steps in determining the interval of convergence of $F(x)$:

Step 1. Find the radius of convergence R (using the Ratio Test, in most cases).

Step 2. Check convergence at the endpoints (if $R \neq 0$ or ∞).

■ **EXAMPLE 1 Using the Ratio Test** Where does $F(x) = \displaystyle\sum_{n=0}^{\infty} \frac{x^n}{2^n}$ converge?

Solution

***Step 1.* Find the radius of convergence.**

Let $a_n = \dfrac{x^n}{2^n}$ and compute the ratio ρ of the Ratio Test:

$$\rho = \lim_{n\to\infty} \left|\frac{a_{n+1}}{a_n}\right| = \lim_{n\to\infty} \left|\frac{x^{n+1}}{2^{n+1}}\right| \cdot \left|\frac{2^n}{x^n}\right| = \lim_{n\to\infty} \frac{1}{2}|x| = \frac{1}{2}|x|$$

We find that

$$\rho < 1 \qquad \text{if} \quad \frac{1}{2}|x| < 1, \qquad \text{that is, if} \quad |x| < 2$$

Thus $F(x)$ converges if $|x| < 2$. Similarly, $\rho > 1$ if $\frac{1}{2}|x| > 1$, or $|x| > 2$. Thus $F(x)$ converges if $|x| > 2$. Therefore, the radius of convergence is $R = 2$.

***Step 2.* Check the endpoints.**

The Ratio Test is inconclusive for $x = \pm 2$, so we must check these cases directly:

$$F(2) = \sum_{n=0}^{\infty} \frac{2^n}{2^n} = 1 + 1 + 1 + 1 + 1 + 1 \cdots$$

$$F(-2) = \sum_{n=0}^{\infty} \frac{(-2)^n}{2^n} = 1 - 1 + 1 - 1 + 1 - 1 \cdots$$

Both series diverge. We conclude that $F(x)$ converges only for $|x| < 2$ (Figure 2). ■

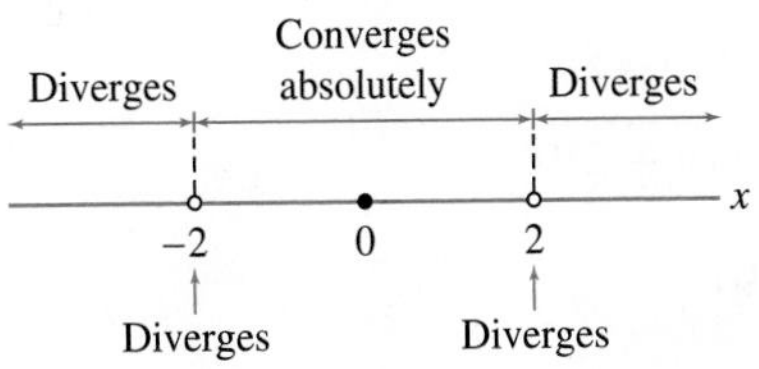

FIGURE 2 The power series $\displaystyle\sum_{n=0}^{\infty} \frac{x^n}{2^n}$ has interval of convergence $(-2, 2)$.

■ **EXAMPLE 2** Where does $F(x) = \sum_{n=1}^{\infty} \frac{(-1)^n}{4^n n}(x-5)^n$ converge?

Solution We compute ρ with $a_n = \frac{(-1)^n}{4^n n}(x-5)^n$:

$$\rho = \lim_{n\to\infty}\left|\frac{a_{n+1}}{a_n}\right| = \lim_{n\to\infty}\left|\frac{(x-5)^{n+1}}{4^{n+1}(n+1)}\,\frac{4^n n}{(x-5)^n}\right|$$
$$= |x-5| \lim_{n\to\infty}\left|\frac{n}{4(n+1)}\right|$$
$$= \frac{1}{4}|x-5|$$

We find that

$$\rho < 1 \qquad \text{if} \quad \frac{1}{4}|x-5| < 1, \qquad \text{that is, if} \quad |x-5| < 4$$

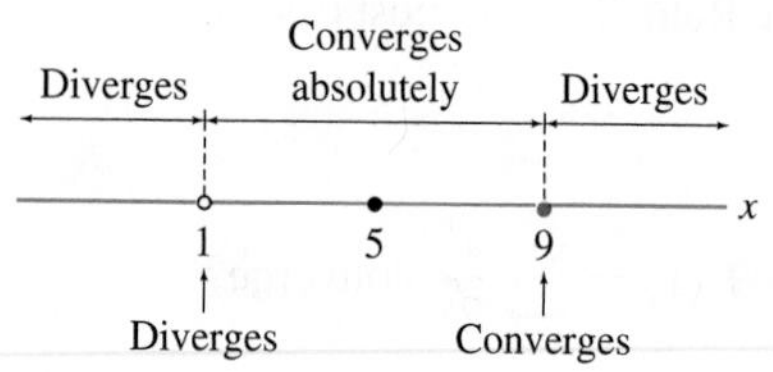

FIGURE 3 The power series

$$\sum_{n=1}^{\infty} \frac{(-1)^n}{4^n n}(x-5)^n$$

has interval of convergence $(1, 9]$.

Thus $F(x)$ converges absolutely on the open interval $(1, 9)$ of radius 4 with center $c = 5$. In other words, the radius of convergence is $R = 4$. Next, we check the endpoints:

$$x = 9: \quad \sum_{n=1}^{\infty} \frac{(-1)^n}{4^n n}(9-5)^n = \sum_{n=1}^{\infty} \frac{(-1)^n}{n} \qquad \text{converges (Leibniz Test)}$$

$$x = 1: \quad \sum_{n=1}^{\infty} \frac{(-1)^n}{4^n n}(-4)^n = \sum_{n=1}^{\infty} \frac{1}{n} \qquad \text{diverges (harmonic series)}$$

We conclude that $F(x)$ converges for x in the half-open interval $(1, 9]$ shown in Figure 3. ■

Some power series contain only even powers or only odd powers of x. The Ratio Test can still be used to find the radius of convergence.

■ **EXAMPLE 3 An Even Power Series** Where does $\sum_{n=0}^{\infty} \frac{x^{2n}}{(2n)!}$ converge?

Solution Although this power series has only even powers of x, we can still apply the Ratio Test with $a_n = x^{2n}/(2n)!$. We have

$$a_{n+1} = \frac{x^{2(n+1)}}{(2(n+1))!} = \frac{x^{2n+2}}{(2n+2)!}$$

Furthermore, $(2n+2)! = (2n+2)(2n+1)(2n)!$, so

$$\rho = \lim_{n\to\infty}\left|\frac{a_{n+1}}{a_n}\right| = \lim_{n\to\infty} \frac{x^{2n+2}}{(2n+2)!}\,\frac{(2n)!}{x^{2n}} = |x|^2 \lim_{n\to\infty} \frac{1}{(2n+2)(2n+1)} = 0$$

Thus $\rho = 0$ for all x, and $F(x)$ converges for all x. The radius of convergence is $R = \infty$. ■

When a function $f(x)$ is represented by a power series on an interval I, we refer to the power series expansion *of $f(x)$ on I.*

Geometric series are important examples of power series. Recall the formula $\sum_{n=0}^{\infty} r^n = 1/(1-r)$, valid for $|r| < 1$. Writing x in place of r, we obtain a power series expansion with radius of convergence $R = 1$:

$$\frac{1}{1-x} = \sum_{n=0}^{\infty} x^n \qquad \text{for } |x| < 1 \tag{2}$$

The next two examples show that we can modify this formula to find the power series expansions of other functions.

EXAMPLE 4 Geometric Series Prove that

$$\frac{1}{1-2x} = \sum_{n=0}^{\infty} 2^n x^n \qquad \text{for } |x| < \frac{1}{2}$$

Solution Substitute $2x$ for x in Eq. (2):

$$\frac{1}{1-2x} = \sum_{n=0}^{\infty} (2x)^n = \sum_{n=0}^{\infty} 2^n x^n \tag{3}$$

Expansion (2) is valid for $|x| < 1$, so Eq. (3) is valid for $|2x| < 1$, or $|x| < \frac{1}{2}$. ■

EXAMPLE 5 Find a power series expansion with center $c = 0$ for

$$f(x) = \frac{1}{2+x^2}$$

and find the interval of convergence.

Solution We need to rewrite $f(x)$ so we can use Eq. (2). We have

$$\frac{1}{2+x^2} = \frac{1}{2}\left(\frac{1}{1+\frac{1}{2}x^2}\right) = \frac{1}{2}\left(\frac{1}{1-\left(-\frac{1}{2}x^2\right)}\right) = \frac{1}{2}\left(\frac{1}{1-u}\right)$$

where $u = -\frac{1}{2}x^2$. Now substitute $u = -\frac{1}{2}x^2$ for x in Eq. (2) to obtain

$$f(x) = \frac{1}{2+x^2} = \frac{1}{2}\sum_{n=0}^{\infty}\left(-\frac{x^2}{2}\right)^n$$

$$= \sum_{n=0}^{\infty} \frac{(-1)^n x^{2n}}{2^{n+1}}$$

This expansion is valid if $|-x^2/2| < 1$, or $|x| < \sqrt{2}$. The interval of convergence is $(-\sqrt{2}, \sqrt{2})$. ■

Our next theorem tells us that within the interval of convergence, we can treat a power series as though it were a polynomial; that is, we can differentiate and integrate term by term.

THEOREM 2 Term-by-Term Differentiation and Integration Assume that

$$F(x) = \sum_{n=0}^{\infty} a_n(x-c)^n$$

has radius of convergence $R > 0$. Then $F(x)$ is differentiable on $(c - R, c + R)$ [or for all x if $R = \infty$]. Furthermore, we can integrate and differentiate term by term. For $x \in (c - R, c + R)$,

$$F'(x) = \sum_{n=1}^{\infty} na_n(x-c)^{n-1}$$

$$\int F(x)\,dx = A + \sum_{n=0}^{\infty} \frac{a_n}{n+1}(x-c)^{n+1} \qquad (A \text{ any constant})$$

These series have the same radius of convergence R.

The proof of Theorem 2 is somewhat technical and is omitted. See Exercise 66 for a proof that $F(x)$ is continuous.

■ **EXAMPLE 6 Differentiating a Power Series** Prove that for $-1 < x < 1$,

$$\frac{1}{(1-x)^2} = 1 + 2x + 3x^2 + 4x^3 + 5x^4 + \cdots$$

Solution The geometric series has radius of convergence $R = 1$:

$$\frac{1}{1-x} = 1 + x + x^2 + x^3 + x^4 + \cdots$$

By Theorem 2, we can differentiate term by term for $|x| < 1$ to obtain

$$\frac{d}{dx}\left(\frac{1}{1-x}\right) = \frac{d}{dx}(1 + x + x^2 + x^3 + x^4 + \cdots)$$

$$\frac{1}{(1-x)^2} = 1 + 2x + 3x^2 + 4x^3 + 5x^4 + \cdots$$ ■

Theorem 2 is a powerful tool in the study of power series.

■ **EXAMPLE 7 Power Series for Arctangent** Prove that for $-1 < x < 1$,

$$\tan^{-1} x = \sum_{n=0}^{\infty} \frac{(-1)^n x^{2n+1}}{2n+1} = x - \frac{x^3}{3} + \frac{x^5}{5} - \frac{x^7}{7} + \cdots \qquad \boxed{4}$$

Solution Recall that $\tan^{-1} x$ is an antiderivative of $(1 + x^2)^{-1}$. We obtain a power series expansion of this antiderivative by substituting $-x^2$ for x in the geometric series of Eq. (2):

$$\frac{1}{1+x^2} = 1 - x^2 + x^4 - x^6 + \cdots$$

This expansion is valid for $|x^2| < 1$—that is, for $|x| < 1$. By Theorem 2, we can integrate series term by term. The resulting expansion is also valid for $|x| < 1$:

$$\tan^{-1} x = \int \frac{dx}{1+x^2} = \int (1 - x^2 + x^4 - x^6 + \cdots)\,dx$$

$$= A + x - \frac{x^3}{3} + \frac{x^5}{5} - \frac{x^7}{7} + \cdots$$

Setting $x = 0$, we obtain $A = \tan^{-1} 0 = 0$. Thus Eq. (4) is valid for $-1 < x < 1$. ■

GRAPHICAL INSIGHT Let's examine the expansion of the previous example graphically. The partial sums of the power series for $f(x) = \tan^{-1} x$ are

$$S_N(x) = x - \frac{x^3}{3} + \frac{x^5}{5} - \frac{x^7}{7} + \cdots + (-1)^N \frac{x^{2N+1}}{2N+1}$$

For large N we can expect $S_N(x)$ to provide a good approximation to $f(x) = \tan^{-1} x$ on the interval $(-1, 1)$, where the power series expansion is valid. Figure 4 confirms this expectation: The graphs of $S_{50}(x)$ and $S_{51}(x)$ are nearly indistinguishable from the graph of $\tan^{-1} x$ on $(-1, 1)$. Thus we may use the partial sums to approximate the arctangent. For example, $\tan^{-1}(0.3)$ is approximated by

$$S_4(0.3) = 0.3 - \frac{(0.3)^3}{3} + \frac{(0.3)^5}{5} - \frac{(0.3)^7}{7} + \frac{(0.3)^9}{9} \approx 0.2914569$$

Since the power series is an alternating series, the error is less than the first omitted term:

$$|\tan^{-1}(0.3) - S_4(0.3)| < \frac{(0.3)^{11}}{11} \approx 1.61 \times 10^{-7}$$

The situation changes drastically in the region $|x| > 1$, where the power series diverges and the partial sums $S_N(x)$ deviate sharply from $\tan^{-1} x$.

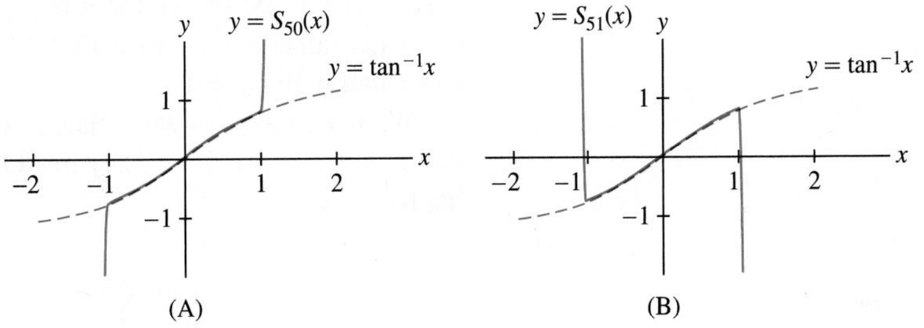

FIGURE 4 $S_{50}(x)$ and $S_{51}(x)$ are nearly indistinguishable from $\tan^{-1} x$ on $(-1, 1)$.

Power Series Solutions of Differential Equations

Power series are a basic tool in the study of differential equations. To illustrate, consider the differential equation with initial condition

$$y' = y, \qquad y(0) = 1$$

We know that $f(x) = e^x$ is the unique solution, but let's try to find a power series that satisfies this initial value problem. We have

$$F(x) = \sum_{n=0}^{\infty} a_n x^n = a_0 + a_1 x + a_2 x^2 + a_3 x^3 + \cdots$$

$$F'(x) = \sum_{n=0}^{\infty} n a_n x^{n-1} = a_1 + 2a_2 x + 3a_3 x^2 + 4a_4 x^3 + \cdots$$

Therefore, $F'(x) = F(x)$ if

$$a_0 = a_1, \quad a_1 = 2a_2, \quad a_2 = 3a_3, \quad a_3 = 4a_4, \quad \ldots$$

In other words, $F'(x) = F(x)$ if $a_{n-1} = na_n$, or

$$a_n = \frac{a_{n-1}}{n}$$

An equation of this type is called a *recursion relation*. It enables us to determine all of the coefficients a_n successively from the first coefficient a_0, which may be chosen arbitrarily. For example,

$$n = 1: \qquad a_1 = \frac{a_0}{1}$$

$$n = 2: \qquad a_2 = \frac{a_1}{2} = \frac{a_0}{2 \cdot 1} = \frac{a_0}{2!}$$

$$n = 3: \qquad a_3 = \frac{a_2}{3} = \frac{a_1}{3 \cdot 2} = \frac{a_0}{3 \cdot 2 \cdot 1} = \frac{a_0}{3!}$$

To obtain a general formula for a_n, apply the recursion relation n times:

$$a_n = \frac{a_{n-1}}{n} = \frac{a_{n-2}}{n(n-1)} = \frac{a_{n-3}}{n(n-1)(n-2)} = \cdots = \frac{a_0}{n!}$$

We conclude that

$$F(x) = a_0 \sum_{n=0}^{\infty} \frac{x^n}{n!}$$

In Example 3, we showed that this power series has radius of convergence $R = \infty$, so $y = F(x)$ satisfies $y' = y$ for all x. Moreover, $F(0) = a_0$, so the initial condition $y(0) = 1$ is satisfied with $a_0 = 1$.

What we have shown is that $f(x) = e^x$ and $F(x)$ with $a_0 = 1$ are both solutions of the initial value problem. They must be equal because the solution is unique. This proves that for all x,

$$e^x = \sum_{n=0}^{\infty} \frac{x^n}{n!} = 1 + x + \frac{x^2}{2!} + \frac{x^3}{3!} + \frac{x^4}{4!} + \cdots$$

In this example, we knew in advance that $y = e^x$ is a solution of $y' = y$, but suppose we are given a differential equation whose solution is unknown. We can try to find a solution in the form of a power series $F(x) = \sum_{n=0}^{\infty} a_n x^n$. In favorable cases, the differential equation leads to a recursion relation that enables us to determine the coefficients a_n.

The solution in Example 8 is called the "Bessel function of order 1." The Bessel function of order n is a solution of

$$x^2y'' + xy' + (x^2 - n^2)y = 0$$

These functions have applications in many areas of physics and engineering.

■ **EXAMPLE 8** Find a power series solution to the initial value problem

$$x^2y'' + xy' + (x^2 - 1)y = 0, \qquad y'(0) = 1 \qquad \boxed{5}$$

Solution Assume that Eq. (5) has a power series solution $F(x) = \sum_{n=0}^{\infty} a_n x^n$. Then

$$y' = F'(x) = \sum_{n=0}^{\infty} na_n x^{n-1} = a_1 + 2a_2x + 3a_3x^2 + \cdots$$

$$y'' = F''(x) = \sum_{n=0}^{\infty} n(n-1)a_n x^{n-2} = 2a_2 + 6a_3x + 12a_4x^2 + \cdots$$

Now substitute the series for y, y', and y'' into the differential equation (5) to determine the recursion relation satisfied by the coefficients a_n:

$$x^2y'' + xy' + (x^2 - 1)y$$
$$= x^2 \sum_{n=0}^{\infty} n(n-1)a_n x^{n-2} + x \sum_{n=0}^{\infty} n a_n x^{n-1} + (x^2 - 1) \sum_{n=0}^{\infty} a_n x^n$$
$$= \sum_{n=0}^{\infty} n(n-1)a_n x^n + \sum_{n=0}^{\infty} n a_n x^n - \sum_{n=0}^{\infty} a_n x^n + \sum_{n=0}^{\infty} a_n x^{n+2} \qquad \boxed{6}$$
$$= \sum_{n=0}^{\infty} (n^2 - 1)a_n x^n + \sum_{n=2}^{\infty} a_{n-2} x^n = 0$$

In Eq. (6), we combine the first three series into a single series using

$$n(n-1) + n - 1 = n^2 - 1$$

and we shift the fourth series to begin at $n = 2$ *rather than* $n = 0$.

The differential equation is satisfied if

$$\sum_{n=0}^{\infty} (n^2 - 1)a_n x^n = -\sum_{n=2}^{\infty} a_{n-2} x^n$$

The first few terms on each side of this equation are

$$-a_0 + 0 \cdot x + 3a_2x^2 + 8a_3x^3 + 15a_4x^4 + \cdots = 0 + 0 \cdot x - a_0x^2 - a_1x^3 - a_2x^4 - \cdots$$

Matching up the coefficients of x^n, we find that

$$-a_0 = 0, \qquad 3a_2 = -a_0, \qquad 8a_3 = -a_1, \qquad 15a_4 = -a_2 \qquad \boxed{7}$$

In general, $(n^2 - 1)a_n = -a_{n-2}$, and this yields the recursion relation

$$a_n = -\frac{a_{n-2}}{n^2 - 1} \quad \text{for } n \ge 2 \qquad \boxed{8}$$

Note that $a_0 = 0$ by Eq. (7). The recursion relation forces all of the even coefficients a_2, $a_4, a_6, \dots$ to be zero:

$$a_2 = \frac{a_0}{2^2 - 1} \text{ so } a_2 = 0, \qquad \text{and then} \qquad a_4 = \frac{a_2}{4^2 - 1} = 0 \text{ so } a_4 = 0, \qquad \text{etc.}$$

As for the odd coefficients, a_1 may be chosen arbitrarily. Because $F'(0) = a_1$, we set $a_1 = 1$ to obtain a solution $y = F(x)$ satisfying $F'(0) = 1$. Now apply Eq. (8):

$$n = 3: \qquad a_3 = -\frac{a_1}{3^2 - 1} = -\frac{1}{3^2 - 1}$$

$$n = 5: \qquad a_5 = -\frac{a_3}{5^2 - 1} = \frac{1}{(5^2 - 1)(3^2 - 1)}$$

$$n = 7: \qquad a_7 = -\frac{a_5}{7^2 - 1} = -\frac{1}{(7^2 - 1)(3^2 - 1)(5^2 - 1)}$$

This shows the general pattern of coefficients. To express the coefficients in a compact form, let $n = 2k + 1$. Then the denominator in the recursion relation (8) can be written

$$n^2 - 1 = (2k + 1)^2 - 1 = 4k^2 + 4k = 4k(k + 1)$$

and

$$a_{2k+1} = -\frac{a_{2k-1}}{4k(k + 1)}$$

Applying this recursion relation k times, we obtain the closed formula

$$a_{2k+1} = (-1)^k \left(\frac{1}{4k(k+1)}\right)\left(\frac{1}{4(k-1)k}\right)\cdots\left(\frac{1}{4(1)(2)}\right) = \frac{(-1)^k}{4^k\, k!\, (k+1)!}$$

Thus we obtain a power series representation of our solution:

$$F(x) = \sum_{k=0}^{\infty} \frac{(-1)^k}{4^k k!(k+1)!} x^{2k+1}$$

A straightforward application of the Ratio Test shows that $F(x)$ has an infinite radius of convergence. Therefore, $F(x)$ is a solution of the initial value problem for all x. ■

11.6 SUMMARY

- A *power series* is an infinite series of the form

$$F(x) = \sum_{n=0}^{\infty} a_n (x-c)^n$$

The constant c is called the *center* of $F(x)$.

- Every power series $F(x)$ has a *radius of convergence* R (Figure 5) such that
  - $F(x)$ converges absolutely for $|x-c| < R$ and diverges for $|x-c| > R$.
  - $F(x)$ may converge or diverge at the endpoints $c-R$ and $c+R$.

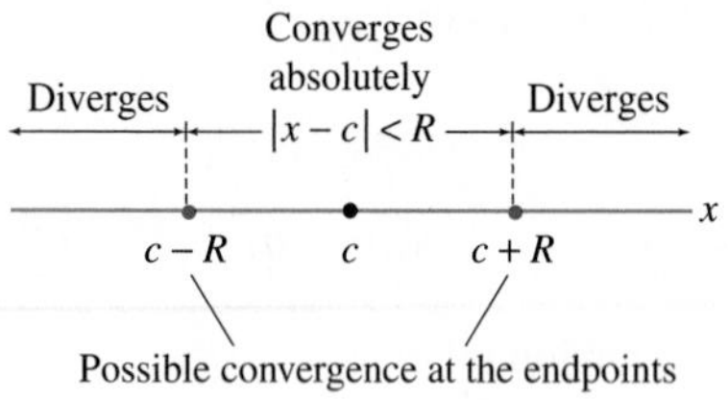

FIGURE 5 Interval of convergence of a power series.

We set $R = 0$ if $F(x)$ converges only for $x = c$ and $R = \infty$ if $F(x)$ converges for all x.

- The *interval of convergence* of $F(x)$ consists of the open interval $(c-R, c+R)$ and possibly one or both endpoints $c-R$ and $c+R$.
- In many cases, the Ratio Test can be used to find the radius of convergence R. It is necessary to check convergence at the endpoints separately.
- If $R > 0$, then $F(x)$ is differentiable on $(c-R, c+R)$ and

$$F'(x) = \sum_{n=1}^{\infty} n a_n (x-c)^{n-1}, \qquad \int F(x)\,dx = A + \sum_{n=0}^{\infty} \frac{a_n}{n+1}(x-c)^{n+1}$$

(A is any constant). These two power series have the same radius of convergence R.

- The expansion $\dfrac{1}{1-x} = \sum_{n=0}^{\infty} x^n$ is valid for $|x| < 1$. It can be used to derive expansions of other related functions by substitution, integration, or differentiation.

11.6 EXERCISES

Preliminary Questions

1. Suppose that $\sum a_n x^n$ converges for $x = 5$. Must it also converge for $x = 4$? What about $x = -3$?

2. Suppose that $\sum a_n (x-6)^n$ converges for $x = 10$. At which of the points (a)–(d) must it also converge?

(a) $x = 8$ **(b)** $x = 11$ **(c)** $x = 3$ **(d)** $x = 0$

3. What is the radius of convergence of $F(3x)$ if $F(x)$ is a power series with radius of convergence $R = 12$?

4. The power series $F(x) = \sum_{n=1}^{\infty} n x^n$ has radius of convergence $R = 1$. What is the power series expansion of $F'(x)$ and what is its radius of convergence?

Exercises

1. Use the Ratio Test to determine the radius of convergence R of $\sum_{n=0}^{\infty} \frac{x^n}{2^n}$. Does it converge at the endpoints $x = \pm R$?

2. Use the Ratio Test to show that $\sum_{n=1}^{\infty} \frac{x^n}{\sqrt{n}2^n}$ has radius of convergence $R = 2$. Then determine whether it converges at the endpoints $R = \pm 2$.

3. Show that the power series (a)–(c) have the same radius of convergence. Then show that (a) diverges at both endpoints, (b) converges at one endpoint but diverges at the other, and (c) converges at both endpoints.

(a) $\sum_{n=1}^{\infty} \frac{x^n}{3^n}$ **(b)** $\sum_{n=1}^{\infty} \frac{x^n}{n3^n}$ **(c)** $\sum_{n=1}^{\infty} \frac{x^n}{n^2 3^n}$

4. Repeat Exercise 3 for the following series:

(a) $\sum_{n=1}^{\infty} \frac{(x-5)^n}{9^n}$ **(b)** $\sum_{n=1}^{\infty} \frac{(x-5)^n}{n9^n}$ **(c)** $\sum_{n=1}^{\infty} \frac{(x-5)^n}{n^2 9^n}$

5. Show that $\sum_{n=0}^{\infty} n^n x^n$ diverges for all $x \neq 0$.

6. For which values of x does $\sum_{n=0}^{\infty} n! x^n$ converge?

7. Use the Ratio Test to show that $\sum_{n=0}^{\infty} \frac{x^{2n}}{3^n}$ has radius of convergence $R = \sqrt{3}$.

8. Show that $\sum_{n=0}^{\infty} \frac{x^{3n+1}}{64^n}$ has radius of convergence $R = 4$.

In Exercises 9–34, find the interval of convergence.

9. $\sum_{n=0}^{\infty} nx^n$

10. $\sum_{n=1}^{\infty} \frac{2^n}{n} x^n$

11. $\sum_{n=1}^{\infty} (-1)^n \frac{x^{2n+1}}{2^n n}$

12. $\sum_{n=0}^{\infty} (-1)^n \frac{n}{4^n} x^{2n}$

13. $\sum_{n=4}^{\infty} \frac{x^n}{n^5}$

14. $\sum_{n=8}^{\infty} n^7 x^n$

15. $\sum_{n=0}^{\infty} \frac{x^n}{(n!)^2}$

16. $\sum_{n=0}^{\infty} \frac{8^n}{n!} x^n$

17. $\sum_{n=0}^{\infty} \frac{(2n)!}{(n!)^2} x^n$

18. $\sum_{n=0}^{\infty} \frac{4^n}{(2n+1)!} x^{2n-1}$

19. $\sum_{n=0}^{\infty} \frac{(-1)^n x^n}{\sqrt{n^2+1}}$

20. $\sum_{n=0}^{\infty} \frac{x^n}{n^4+2}$

21. $\sum_{n=15}^{\infty} \frac{x^{2n+1}}{3n+1}$

22. $\sum_{n=1}^{\infty} \frac{x^n}{n - 4\ln n}$

23. $\sum_{n=2}^{\infty} \frac{x^n}{\ln n}$

24. $\sum_{n=2}^{\infty} \frac{x^{3n+2}}{\ln n}$

25. $\sum_{n=1}^{\infty} n(x-3)^n$

26. $\sum_{n=1}^{\infty} \frac{(-5)^n (x-3)^n}{n^2}$

27. $\sum_{n=1}^{\infty} (-1)^n n^5 (x-7)^n$

28. $\sum_{n=0}^{\infty} 27^n (x-1)^{3n+2}$

29. $\sum_{n=1}^{\infty} \frac{2^n}{3n} (x+3)^n$

30. $\sum_{n=0}^{\infty} \frac{(x-4)^n}{n!}$

31. $\sum_{n=0}^{\infty} \frac{(-5)^n}{n!} (x+10)^n$

32. $\sum_{n=10}^{\infty} n!\,(x+5)^n$

33. $\sum_{n=12}^{\infty} e^n (x-2)^n$

34. $\sum_{n=2}^{\infty} \frac{(x+4)^n}{(n \ln n)^2}$

In Exercises 35–40, use Eq. (2) to expand the function in a power series with center $c = 0$ and determine the interval of convergence.

35. $f(x) = \frac{1}{1-3x}$

36. $f(x) = \frac{1}{1+3x}$

37. $f(x) = \frac{1}{3-x}$

38. $f(x) = \frac{1}{4+3x}$

39. $f(x) = \frac{1}{1+x^2}$

40. $f(x) = \frac{1}{16+2x^3}$

41. Use the equalities

$$\frac{1}{1-x} = \frac{1}{-3-(x-4)} = \frac{-\frac{1}{3}}{1+\left(\frac{x-4}{3}\right)}$$

to show that for $|x-4| < 3$,

$$\frac{1}{1-x} = \sum_{n=0}^{\infty} (-1)^{n+1} \frac{(x-4)^n}{3^{n+1}}$$

42. Use the method of Exercise 41 to expand $1/(1-x)$ in power series with centers $c = 2$ and $c = -2$. Determine the interval of convergence.

43. Use the method of Exercise 41 to expand $1/(4-x)$ in a power series with center $c = 5$. Determine the interval of convergence.

44. Find a power series that converges only for x in $[2, 6)$.

45. Apply integration to the expansion

$$\frac{1}{1+x} = \sum_{n=0}^{\infty} (-1)^n x^n = 1 - x + x^2 - x^3 + \cdots$$

to prove that for $-1 < x < 1$,

$$\ln(1+x) = \sum_{n=1}^{\infty} \frac{(-1)^{n-1} x^n}{n} = x - \frac{x^2}{2} + \frac{x^3}{3} - \frac{x^4}{4} + \cdots$$

46. Use the result of Exercise 45 to prove that

$$\ln\frac{3}{2} = \frac{1}{2} - \frac{1}{2\cdot 2^2} + \frac{1}{3\cdot 2^3} - \frac{1}{4\cdot 2^4} + \cdots$$

Use your knowledge of alternating series to find an N such that the partial sum S_N approximates $\ln\frac{3}{2}$ to within an error of at most 10^{-3}. Confirm using a calculator to compute both S_N and $\ln\frac{3}{2}$.

47. Let $F(x) = (x+1)\ln(1+x) - x$.

(a) Apply integration to the result of Exercise 45 to prove that for $-1 < x < 1$,

$$F(x) = \sum_{n=1}^{\infty} (-1)^{n+1} \frac{x^{n+1}}{n(n+1)}$$

(b) Evaluate at $x = \frac{1}{2}$ to prove

$$\frac{3}{2}\ln\frac{3}{2} - \frac{1}{2} = \frac{1}{1\cdot 2\cdot 2^2} - \frac{1}{2\cdot 3\cdot 2^3} + \frac{1}{3\cdot 4\cdot 2^4} - \frac{1}{4\cdot 5\cdot 2^5} + \cdots$$

(c) Use a calculator to verify that the partial sum S_4 approximates the left-hand side with an error no greater than the term a_5 of the series.

48. Prove that for $|x| < 1$,

$$\int \frac{dx}{x^4+1} = x - \frac{x^5}{5} + \frac{x^9}{9} - \cdots$$

Use the first two terms to approximate $\int_0^{1/2} dx/(x^4+1)$ numerically. Use the fact that you have an alternating series to show that the error in this approximation is at most 0.00022.

49. Use the result of Example 7 to show that

$$F(x) = \frac{x^2}{1\cdot 2} - \frac{x^4}{3\cdot 4} + \frac{x^6}{5\cdot 6} - \frac{x^8}{7\cdot 8} + \cdots$$

is an antiderivative of $f(x) = \tan^{-1} x$ satisfying $F(0) = 0$. What is the radius of convergence of this power series?

50. Verify that function $F(x) = x\tan^{-1} x - \frac{1}{2}\log(x^2+1)$ is an antiderivative of $f(x) = \tan^{-1} x$ satisfying $F(0) = 0$. Then use the result of Exercise 49 with $x = \frac{\pi}{6}$ to show that

$$\frac{\pi}{6\sqrt{3}} - \frac{1}{2}\ln\frac{4}{3} = \frac{1}{1\cdot 2(3)} - \frac{1}{3\cdot 4(3^2)} + \frac{1}{5\cdot 6(3^3)} - \frac{1}{7\cdot 8(3^4)} + \cdots$$

Use a calculator to compare the value of the left-hand side with the partial sum S_4 of the series on the right.

51. Evaluate $\sum_{n=1}^{\infty} \frac{n}{2^n}$. *Hint:* Use differentiation to show that

$$(1-x)^{-2} = \sum_{n=1}^{\infty} nx^{n-1} \quad (\text{for } |x| < 1)$$

52. Use the power series for $(1+x^2)^{-1}$ and differentiation to prove that for $|x| < 1$,

$$\frac{2x}{(x^2+1)^2} = \sum_{n=1}^{\infty} (-1)^{n-1}(2n)x^{2n-1}$$

53. Show that the following series converges absolutely for $|x| < 1$ and compute its sum:

$$F(x) = 1 - x - x^2 + x^3 - x^4 - x^5 + x^6 - x^7 - x^8 + \cdots$$

Hint: Write $F(x)$ as a sum of three geometric series with common ratio x^3.

54. Show that for $|x| < 1$,

$$\frac{1+2x}{1+x+x^2} = 1 + x - 2x^2 + x^3 + x^4 - 2x^5 + x^6 + x^7 - 2x^8 + \cdots$$

Hint: Use the hint from Exercise 53.

55. Find all values of x such that $\sum_{n=1}^{\infty} \frac{x^{n^2}}{n!}$ converges.

56. Find all values of x such that the following series converges:

$$F(x) = 1 + 3x + x^2 + 27x^3 + x^4 + 243x^5 + \cdots$$

57. Find a power series $P(x) = \sum_{n=0}^{\infty} a_n x^n$ satisfying the differential equation $y' = -y$ with initial condition $y(0) = 1$. Then use Theorem 1 of Section 5.8 to conclude that $P(x) = e^{-x}$.

58. Let $C(x) = 1 - \frac{x^2}{2!} + \frac{x^4}{4!} - \frac{x^6}{6!} + \cdots$.

(a) Show that $C(x)$ has an infinite radius of convergence.

(b) Prove that $C(x)$ and $f(x) = \cos x$ are both solutions of $y'' = -y$ with initial conditions $y(0) = 1$, $y'(0) = 0$. This initial value problem has a unique solution, so we have $C(x) = \cos x$ for all x.

59. Use the power series for $y = e^x$ to show that

$$\frac{1}{e} = \frac{1}{2!} - \frac{1}{3!} + \frac{1}{4!} - \cdots$$

Use your knowledge of alternating series to find an N such that the partial sum S_N approximates e^{-1} to within an error of at most 10^{-3}. Confirm this using a calculator to compute both S_N and e^{-1}.

60. Let $P(x) = \sum_{n=0}^{\infty} a_n x^n$ be a power series solution to $y' = 2xy$ with initial condition $y(0) = 1$.

(a) Show that the odd coefficients a_{2k+1} are all zero.

(b) Prove that $a_{2k} = a_{2k-2}/k$ and use this result to determine the coefficients a_{2k}.

61. Find a power series $P(x)$ satisfying the differential equation

$$y'' - xy' + y = 0 \qquad \boxed{9}$$

with initial condition $y(0) = 1$, $y'(0) = 0$. What is the radius of convergence of the power series?

62. Find a power series satisfying Eq. (9) with initial condition $y(0) = 0$, $y'(0) = 1$.

63. Prove that

$$J_2(x) = \sum_{k=0}^{\infty} \frac{(-1)^k}{2^{2k+2}\, k!\,(k+3)!} x^{2k+2}$$

is a solution of the Bessel differential equation of order 2:

$$x^2 y'' + xy' + (x^2 - 4)y = 0$$

64. Why is it impossible to expand $f(x) = |x|$ as a power series that converges in an interval around $x = 0$? Explain using Theorem 2.

Further Insights and Challenges

65. Suppose that the coefficients of $F(x) = \sum_{n=0}^{\infty} a_n x^n$ are *periodic*; that is, for some whole number $M > 0$, we have $a_{M+n} = a_n$. Prove that $F(x)$ converges absolutely for $|x| < 1$ and that

$$F(x) = \frac{a_0 + a_1 x + \cdots + a_{M-1}x^{M-1}}{1 - x^M}$$

Hint: Use the hint for Exercise 53.

66. Continuity of Power Series Let $F(x) = \sum_{n=0}^{\infty} a_n x^n$ be a power series with radius of convergence $R > 0$.

(a) Prove the inequality

$$|x^n - y^n| \le n|x - y|(|x|^{n-1} + |y|^{n-1}) \qquad \boxed{10}$$

Hint: $x^n - y^n = (x - y)(x^{n-1} + x^{n-2}y + \cdots + y^{n-1})$.

(b) Choose R_1 with $0 < R_1 < R$. Show that the infinite series $M = \sum_{n=0}^{\infty} 2n|a_n|R_1^n$ converges. *Hint:* Show that $n|a_n|R_1^n < |a_n|x^n$ for all n sufficiently large if $R_1 < x < R$.

(c) Use Eq. (10) to show that if $|x| < R_1$ and $|y| < R_1$, then $|F(x) - F(y)| \le M|x - y|$.

(d) Prove that if $|x| < R$, then $F(x)$ is continuous at x. *Hint:* Choose R_1 such that $|x| < R_1 < R$. Show that if $\epsilon > 0$ is given, then $|F(x) - F(y)| \le \epsilon$ for all y such that $|x - y| < \delta$, where δ is any positive number that is less than ϵ/M and $R_1 - |x|$ (see Figure 6).

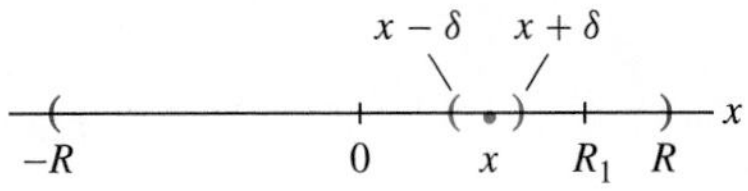

FIGURE 6 If $x > 0$, choose $\delta > 0$ less than ϵ/M and $R_1 - x$.

11.7 Taylor Series

In this section we develop general methods for finding power series representations. Suppose that $f(x)$ is represented by a power series centered at $x = c$ on an interval $(c - R, c + R)$ with $R > 0$:

$$f(x) = \sum_{n=0}^{\infty} a_n (x - c)^n = a_0 + a_1(x - c) + a_2(x - c)^2 + \cdots$$

According to Theorem 2 in Section 11.6, we can compute the derivatives of $f(x)$ by differentiating the series expansion term by term:

$$\begin{aligned}
f(x) &= a_0 + a_1(x - c) + a_2(x - c)^2 + a_3(x - c)^3 + \cdots \\
f'(x) &= a_1 + 2a_2(x - c) + 3a_3(x - c)^2 + 4a_4(x - c)^3 + \cdots \\
f''(x) &= 2a_2 + 2\cdot 3a_3(x - c) + 3\cdot 4a_4(x - c)^2 + 4\cdot 5a_5(x - c)^3 + \cdots \\
f'''(x) &= 2\cdot 3a_3 + 2\cdot 3\cdot 4a_4(x - 2) + 3\cdot 4\cdot 5a_5(x - 2)^2 + \cdots
\end{aligned}$$

In general,

$$f^{(k)}(x) = k!a_k + \Big(2\cdot 3\cdots(k+1)\Big)a_{k+1}(x - c) + \cdots$$

Setting $x = c$ in each of these series, we find that

$$f(c) = a_0, \quad f'(c) = a_1, \quad f''(c) = 2a_2, \quad f'''(c) = 2\cdot 3a_2, \quad \ldots, \quad f^{(k)}(c) = k!a_k, \quad \ldots$$

We see that a_k is the kth coefficient of the Taylor polynomial studied in Section 8.4:

$$a_k = \frac{f^{(k)}(c)}{k!} \qquad \boxed{1}$$

Therefore $f(x) = T(x)$, where $T(x)$ is the **Taylor series** of $f(x)$ centered at $x = c$:

$$T(x) = f(c) + f'(c)(x-c) + \frac{f''(c)}{2!}(x-c)^2 + \frac{f'''(c)}{3!}(x-c)^3 + \cdots$$

This proves the next theorem.

THEOREM 1 Taylor Series Expansion If $f(x)$ is represented by a power series centered at c in an interval $|x-c| < R$ with $R > 0$, then that power series is the Taylor series

$$T(x) = \sum_{n=0}^{\infty} \frac{f^{(n)}(c)}{n!}(x-c)^n$$

In the special case $c = 0$, $T(x)$ is also called the **Maclaurin series**:

$$f(x) = \sum_{n=0}^{\infty} \frac{f^{(n)}(0)}{n!}x^n = f(0) + f'(0)x + \frac{f''(0)}{2!}x^2 + \frac{f'''(0)}{3!}x^3 + \frac{f^{(4)}(0)}{4!}x^4 + \cdots$$

■ **EXAMPLE 1** Find the Taylor series for $f(x) = x^{-3}$ centered at $c = 1$.

Solution The derivatives of $f(x)$ are $f'(x) = -3x^{-4}$, $f''(x) = (-3)(-4)x^{-5}$, and in general,

$$f^{(n)}(x) = (-1)^n(3)(4)\cdots(n+2)x^{-3-n}$$

Note that $(3)(4)\cdots(n+2) = \frac{1}{2}(n+2)!$. Therefore,

$$f^{(n)}(1) = (-1)^n\frac{1}{2}(n+2)!$$

Noting that $(n+2)! = (n+2)(n+1)n!$, we write the coefficients of the Taylor series as:

$$a_n = \frac{f^{(n)}(1)}{n!} = \frac{(-1)^n\frac{1}{2}(n+2)!}{n!} = (-1)^n\frac{(n+2)(n+1)}{2}$$

The Taylor series for $f(x) = x^{-3}$ centered at $c = 1$ is

$$T(x) = 1 - 3(x-1) + 6(x-1)^2 - 10(x-1)^3 + \cdots$$

$$= \sum_{n=0}^{\infty}(-1)^n\frac{(n+2)(n+1)}{2}(x-1)^n \qquad ■$$

Theorem 1 tells us that if we want to represent a function $f(x)$ by a power series centered at c, then the only candidate for the job is the Taylor series:

$$T(x) = \sum_{n=0}^{\infty} \frac{f^{(n)}(c)}{n!}(x-c)^n$$

See Exercise 92 for an example where a Taylor series $T(x)$ converges but does not converge to $f(x)$.

However, *there is no guarantee that $T(x)$ converges to $f(x)$*, even if $T(x)$ converges. To study convergence, we consider the kth partial sum, which is the Taylor polynomial of degree k:

$$T_k(x) = f(c) + f'(c)(x-c) + \frac{f''(c)}{2!}(x-c)^2 + \cdots + \frac{f^{(k)}(c)}{k!}(x-c)^k$$

In Section 9.4, we defined the remainder

$$R_k(x) = f(x) - T_k(x)$$

Since $T(x)$ is the limit of the partial sums $T_k(x)$, we see that

The Taylor series converges to $f(x)$ if and only if $\lim_{k\to\infty} R_k(x) = 0$.

There is no general method for determining whether $R_k(x)$ tends to zero, but the following theorem can be applied in some important cases.

REMINDER $f(x)$ is called "infinitely differentiable" if $f^{(n)}(x)$ exists for all n.

THEOREM 2 Let $I = (c-R, c+R)$, where $R > 0$. Suppose there exists $K > 0$ such that all derivatives of f are bounded by K on I:

$$|f^{(k)}(x)| \le K \qquad \text{for all} \quad k \ge 0 \quad \text{and} \quad x \in I$$

Then $f(x)$ is represented by its Taylor series in I:

$$f(x) = \sum_{n=0}^{\infty} \frac{f^{(n)}(c)}{n!}(x-c)^n \qquad \text{for all} \quad x \in I$$

Proof According to the Error Bound for Taylor polynomials (Theorem 2 in Section 9.4),

$$|R_k(x)| = |f(x) - T_k(x)| \le K\frac{|x-c|^{k+1}}{(k+1)!}$$

If $x \in I$, then $|x-c| < R$ and

$$|R_k(x)| \le K\frac{R^{k+1}}{(k+1)!}$$

We showed in Example 9 of Section 11.1 that $R^k/k!$ tends to zero as $k \to \infty$. Therefore, $\lim_{k\to\infty} R_k(x) = 0$ for all $x \in (c-R, c+R)$, as required. ■

Taylor expansions were studied throughout the seventeenth and eighteenth centuries by Gregory, Leibniz, Newton, Maclaurin, Taylor, Euler, and others. These developments were anticipated by the great Hindu mathematician Madhava (c. 1340–1425), who discovered the expansions of sine and cosine and many other results two centuries earlier.

■ **EXAMPLE 2 Expansions of Sine and Cosine** Show that the following Maclaurin expansions are valid for all x.

$$\sin x = \sum_{n=0}^{\infty} (-1)^n \frac{x^{2n+1}}{(2n+1)!} = x - \frac{x^3}{3!} + \frac{x^5}{5!} - \frac{x^7}{7!} + \cdots$$

$$\cos x = \sum_{n=0}^{\infty} (-1)^n \frac{x^{2n}}{(2n)!} = 1 - \frac{x^2}{2!} + \frac{x^4}{4!} - \frac{x^6}{6!} + \cdots$$

Solution Recall that the derivatives of $f(x) = \sin x$ and their values at $x = 0$ form a repeating pattern of period 4:

$f(x)$	$f'(x)$	$f''(x)$	$f'''(x)$	$f^{(4)}(x)$	$\dots$
$\sin x$	$\cos x$	$-\sin x$	$-\cos x$	$\sin x$	$\dots$
0	1	0	-1	0	$\dots$

In other words, the even derivatives are zero and the odd derivatives alternate in sign: $f^{(2n+1)}(0) = (-1)^n$. Therefore, the nonzero Taylor coefficients for $\sin x$ are

$$a_{2n+1} = \frac{(-1)^n}{(2n+1)!}$$

For $f(x) = \cos x$, the situation is reversed. The odd derivatives are zero and the even derivatives alternate in sign: $f^{(2n)}(0) = (-1)^n \cos 0 = (-1)^n$. Therefore the nonzero Taylor coefficients for $\cos x$ are $a_{2n} = (-1)^n/(2n)!$.

We can apply Theorem 2 with $K = 1$ and any value of R because both sine and cosine satisfy $|f^{(n)}(x)| \le 1$ for all x and n. The conclusion is that the Taylor series converges to $f(x)$ for $|x| < R$. Since R is arbitrary, the Taylor expansions hold for all x. ■

■ **EXAMPLE 3** **Taylor Expansion of $f(x) = e^x$ at $x = c$** Find the Taylor series $T(x)$ of $f(x) = e^x$ at $x = c$.

Solution We have $f^{(n)}(c) = e^c$ for all x, and thus

$$T(x) = \sum_{n=0}^{\infty} \frac{e^c}{n!}(x-c)^n$$

Because e^x is increasing for all $R > 0$ we have $|f^{(k)}(x)| \le e^{c+R}$ for $x \in (c-R, c+R)$. Applying Theorem 2 with $K = e^{c+R}$, we conclude that $T(x)$ converges to $f(x)$ for all $x \in (c-R, c+R)$. Since R is arbitrary, the Taylor expansion holds for all x. For $c = 0$, we obtain the standard Maclaurin series

$$e^x = 1 + x + \frac{x^2}{2!} + \frac{x^3}{3!} + \cdots$$ ■

Shortcuts to Finding Taylor Series

There are several methods for generating new Taylor series from known ones. First of all, we can differentiate and integrate Taylor series term by term within its interval of convergence, by Theorem 2 of Section 11.6. We can also multiply two Taylor series or substitute one Taylor series into another (we omit the proofs of these facts).

In Example 4, we can also write the Maclaurin series as

$$\sum_{n=0}^{\infty} \frac{x^{n+2}}{n!}$$

■ **EXAMPLE 4** Find the Maclaurin series for $f(x) = x^2 e^x$.

Solution Multiply the known Maclaurin series for e^x by x^2.

$$x^2 e^x = x^2\left(1 + x + \frac{x^2}{2!} + \frac{x^3}{3!} + \frac{x^4}{4!} + \frac{x^5}{5!} + \cdots\right)$$

$$= x^2 + x^3 + \frac{x^4}{2!} + \frac{x^5}{3!} + \frac{x^6}{4!} + \frac{x^7}{5!} + \cdots = \sum_{n=2}^{\infty} \frac{x^n}{(n-2)!}$$ ■

■ **EXAMPLE 5 Substitution** Find the Maclaurin series for e^{-x^2}.

Solution Substitute $-x^2$ in the Maclaurin series for e^x.

$$e^{-x^2} = \sum_{n=0}^{\infty} \frac{(-x^2)^n}{n!} = \sum_{n=0}^{\infty} \frac{(-1)^n x^{2n}}{n!} = 1 - x^2 + \frac{x^4}{2!} - \frac{x^6}{3!} + \frac{x^8}{4!} - \cdots \qquad \boxed{2}$$

The Taylor expansion of e^x is valid for all x, so this expansion is also valid for all x. ■

■ **EXAMPLE 6 Integration** Find the Maclaurin series for $f(x) = \ln(1+x)$.

Solution We integrate the geometric series with common ratio $-x$ (valid for $|x| < 1$):

$$\frac{1}{1+x} = 1 - x + x^2 - x^3 + \cdots$$

$$\ln(1+x) = \int \frac{dx}{1+x} = x - \frac{x^2}{2} + \frac{x^3}{3} - \frac{x^4}{4} + \cdots = \sum_{n=1}^{\infty} (-1)^{n-1} \frac{x^n}{n}$$

The constant of integration on the right is zero because $\ln(1+x) = 0$ for $x = 0$. This expansion is valid for $|x| < 1$. It also holds for $x = 1$ (see Exercise 84). ■

In many cases, there is no convenient general formula for the Taylor coefficients, but we can still compute as many coefficients as desired.

■ **EXAMPLE 7 Multiplying Taylor Series** Write out the terms up to degree five in the Maclaurin series for $f(x) = e^x \cos x$.

Solution We multiply the fifth-order Taylor polynomials of e^x and $\cos x$ together, dropping the terms of degree greater than 5:

$$\left(1 + x + \frac{x^2}{2} + \frac{x^3}{6} + \frac{x^4}{24} + \frac{x^5}{120}\right)\left(1 - \frac{x^2}{2} + \frac{x^4}{24}\right)$$

Distributing the term on the left (and ignoring terms of degree greater than 5), we obtain

$$\left(1 + x + \frac{x^2}{2} + \frac{x^3}{6} + \frac{x^4}{24} + \frac{x^5}{120}\right) - \left(1 + x + \frac{x^2}{2} + \frac{x^3}{6}\right)\left(\frac{x^2}{2}\right) + (1+x)\left(\frac{x^4}{24}\right)$$

$$= \underbrace{1 + x - \frac{x^3}{3} - \frac{x^4}{6} - \frac{x^5}{30}}_{\text{Retain terms of degree } \le 5}$$

We conclude that the fifth Maclaurin polynomial for $f(x) = e^x \cos x$ is

$$T_5(x) = 1 + x - \frac{x^3}{3} - \frac{x^4}{6} - \frac{x^5}{30} \qquad ■$$

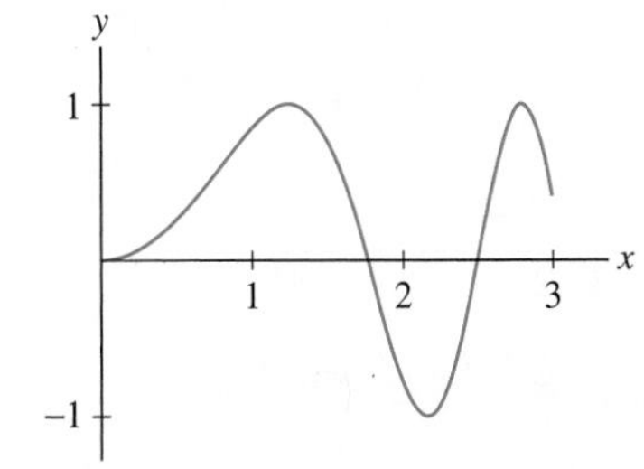

FIGURE 1 Graph of $T_{12}(x)$ for the power series expansion of the antiderivative

$$F(x) = \int_0^x \sin(t^2)\,dt$$

In the next example, we express the definite integral of $\sin(x^2)$ as an infinite series. This is useful because the integral cannot be evaluated explicitly. Figure 1 shows the graph of the Taylor polynomial $T_{12}(x)$ of the Taylor series expansion of the antiderivative.

■ **EXAMPLE 8** Let $J = \int_0^1 \sin(x^2)\,dx$.

(a) Express J as an infinite series.

(b) Determine J to within an error less than 10^{-4}.

Solution

(a) The Maclaurin expansion for $\sin x$ is valid for all x, so we have

$$\sin x = \sum_{n=0}^{\infty} \frac{(-1)^n}{(2n+1)!} x^{2n+1} \quad \Rightarrow \quad \sin(x^2) = \sum_{n=0}^{\infty} \frac{(-1)^n}{(2n+1)!} x^{4n+2}$$

We obtain an infinite series for J by integration:

$$J = \int_0^1 \sin(x^2)\,dx = \sum_{n=0}^{\infty} \frac{(-1)^n}{(2n+1)!} \int_0^1 x^{4n+2}\,dx = \sum_{n=0}^{\infty} \frac{(-1)^n}{(2n+1)!} \left(\frac{1}{4n+3}\right)$$

$$= \frac{1}{3} - \frac{1}{42} + \frac{1}{1320} - \frac{1}{75{,}600} + \cdots \qquad \boxed{3}$$

(b) The infinite series for J is an alternating series with decreasing terms, so the sum of the first N terms is accurate to within an error that is less than the $(N+1)$st term. The absolute value of the fourth term $1/75{,}600$ is smaller than 10^{-4} so we obtain the desired accuracy using the first three terms of the series for J:

$$J \approx \frac{1}{3} - \frac{1}{42} + \frac{1}{1320} \approx 0.31028$$

The error satisfies

$$\left| J - \left(\frac{1}{3} - \frac{1}{42} + \frac{1}{1320}\right) \right| < \frac{1}{75{,}600} \approx 1.3 \times 10^{-5}$$

The percentage error is less than 0.005% with just three terms. ■

Binomial Series

Isaac Newton discovered an important generalization of the Binomial Theorem around 1665. For any number a (integer or not) and integer $n \ge 0$, we define the **binomial coefficient**:

$$\binom{a}{n} = \frac{a(a-1)(a-2)\cdots(a-n+1)}{n!}, \qquad \binom{a}{0} = 1$$

For example,

$$\binom{6}{3} = \frac{6 \cdot 5 \cdot 4}{3 \cdot 2 \cdot 1} = 20, \qquad \binom{\frac{4}{3}}{3} = \frac{\frac{4}{3} \cdot \frac{1}{3} \cdot \left(-\frac{2}{3}\right)}{3 \cdot 2 \cdot 1} = -\frac{4}{81}$$

Let

$$f(x) = (1+x)^a$$

The **Binomial Theorem** of algebra (see Appendix C) states that for any whole number a,

$$(r+s)^a = r^a + \binom{a}{1} r^{a-1}s + \binom{a}{2} r^{a-2}s^2 + \cdots + \binom{a}{a-1} rs^{a-1} + s^a$$

Setting $r = 1$ and $s = x$, we obtain the expansion of $f(x)$:

$$(1+x)^a = 1 + \binom{a}{1}x + \binom{a}{2}x^2 + \cdots + \binom{a}{a-1}x^{a-1} + x^a$$

We derive Newton's generalization by computing the Maclaurin series of $f(x)$ without assuming that a is a whole number. Observe that the derivatives follow a pattern:

$$\begin{aligned} f(x) &= (1+x)^a & f(0) &= 1 \\ f'(x) &= a(1+x)^{a-1} & f'(0) &= a \\ f''(x) &= a(a-1)(1+x)^{a-2} & f''(0) &= a(a-1) \\ f'''(x) &= a(a-1)(a-2)(1+x)^{a-3} & f'''(0) &= a(a-1)(a-2) \end{aligned}$$

In general, $f^{(n)}(0) = a(a-1)(a-2)\cdots(a-n+1)$ and

$$\frac{f^{(n)}(0)}{n!} = \frac{a(a-1)(a-2)\cdots(a-n+1)}{n!} = \binom{a}{n}$$

Hence the Maclaurin series for $f(x) = (1+x)^a$ is the binomial series

When a is a whole number, $\binom{a}{n}$ is zero for $n > a$, and in this case, the binomial series breaks off at degree n. The binomial series is an infinite series when a is not a whole number.

$$\sum_{n=0}^{\infty} \binom{a}{n} x^n = 1 + ax + \frac{a(a-1)}{2!}x^2 + \frac{a(a-1)(a-2)}{3!}x^3 + \cdots + \binom{a}{n}x^n + \cdots$$

The Ratio Test shows that this series has radius of convergence $R = 1$ (Exercise 86) and an additional argument (developed in Exercise 87) shows that it converges to $(1+x)^a$ for $|x| < 1$.

THEOREM 3 The Binomial Series For any exponent a and for $|x| < 1$,

$$(1+x)^a = 1 + \frac{a}{1!}x + \frac{a(a-1)}{2!}x^2 + \frac{a(a-1)(a-2)}{3!}x^3 + \cdots + \binom{a}{n}x^n + \cdots$$

EXAMPLE 9 Find the terms through degree four in the Maclaurin expansion of

$$f(x) = (1+x)^{4/3}$$

Solution The binomial coefficients $\binom{a}{n}$ for $a = \frac{4}{3}$ for $0 < n < 4$ are

$$1, \quad \frac{\frac{4}{3}}{1!} = \frac{4}{3}, \quad \frac{\frac{4}{3}\left(\frac{1}{3}\right)}{2!} = \frac{2}{9}, \quad \frac{\frac{4}{3}\left(\frac{1}{3}\right)\left(-\frac{2}{3}\right)}{3!} = -\frac{4}{81}, \quad \frac{\frac{4}{3}\left(\frac{1}{3}\right)\left(-\frac{2}{3}\right)\left(-\frac{5}{3}\right)}{4!} = \frac{5}{243}$$

Therefore, $(1+x)^{4/3} \approx 1 + \frac{4}{3}x + \frac{2}{9}x^2 - \frac{4}{81}x^3 + \frac{5}{243}x^4 + \cdots$. ■

EXAMPLE 10 Find the Maclaurin series for

$$f(x) = \frac{1}{\sqrt{1-x^2}}$$

Solution First, let's find the coefficients in the binomial series for $(1+x)^{-1/2}$:

$$1, \quad \frac{-\frac{1}{2}}{1!} = -\frac{1}{2}, \quad \frac{-\frac{1}{2}\left(-\frac{3}{2}\right)}{1\cdot 2} = \frac{1\cdot 3}{2\cdot 4}, \quad \frac{-\frac{1}{2}\left(-\frac{3}{2}\right)\left(-\frac{5}{2}\right)}{1\cdot 2\cdot 3} = \frac{1\cdot 3\cdot 5}{2\cdot 4\cdot 6}$$

The general pattern is

$$\binom{-\frac{1}{2}}{n} = \frac{-\frac{1}{2}\left(-\frac{3}{2}\right)\left(-\frac{5}{2}\right)\cdots\left(-\frac{2n-1}{2}\right)}{1\cdot 2\cdot 3\cdots n} = (-1)^n \frac{1\cdot 3\cdot 5\cdots(2n-1)}{2\cdot 4\cdot 6\cdot 2n}$$

Thus, the following binomial expansion is valid for $|x| < 1$:

$$\frac{1}{\sqrt{1+x}} = 1 + \sum_{n=1}^{\infty} (-1)^n \frac{1 \cdot 3 \cdot 5 \cdots (2n-1)}{2 \cdot 4 \cdot 6 \cdots (2n)} x^n = 1 - \frac{1}{2}x + \frac{1 \cdot 3}{2 \cdot 4}x^2 - \cdots$$

If $|x| < 1$, then $|x|^2 < 1$, and we can substitute $-x^2$ for x to obtain

$$\frac{1}{\sqrt{1-x^2}} = 1 + \sum_{n=1}^{\infty} \frac{1 \cdot 3 \cdot 5 \cdots (2n-1)}{2 \cdot 4 \cdot 6 \cdots 2n} x^{2n} = 1 + \frac{1}{2}x^2 + \frac{1 \cdot 3}{2 \cdot 4}x^4 + \cdots \qquad \boxed{4}$$

■

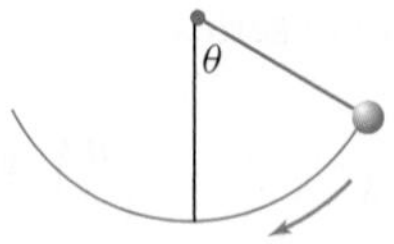

FIGURE 2 Pendulum released at an angle θ.

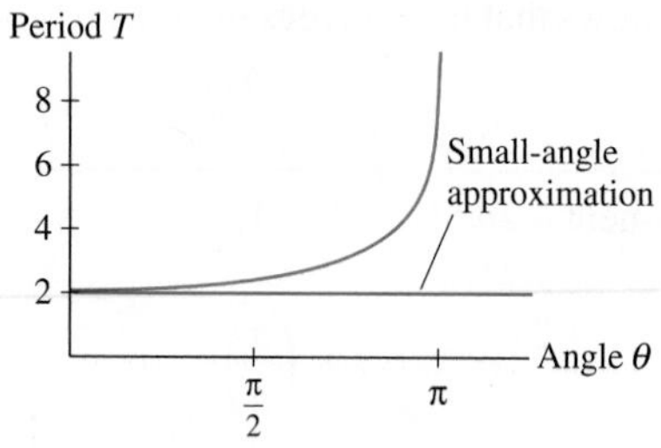

FIGURE 3 The period T of a 1-meter pendulum as a function of the angle θ at which it is released.

Taylor series are particularly useful for studying the so-called *special functions* (such as Bessel and hypergeometric functions) that appear in a wide range of physics and engineering applications. One example is the following **elliptic function of the first kind**, defined for $|k| < 1$:

$$E(k) = \int_0^{\pi/2} \frac{dt}{\sqrt{1 - k^2 \sin^2 t}}$$

This function is used in physics to compute the period T of pendulum of length L released from an angle θ (Figure 2). We can use the "small-angle approximation" $T \approx 2\pi\sqrt{L/g}$ when θ is small, but this approximation breaks down for large angles (Figure 3). The exact value of the period is $T = 4\sqrt{L/g}\,E(k)$, where $k = \sin \frac{1}{2}\theta$.

■ **EXAMPLE 11** **Elliptic Function** Find the Maclaurin series for $E(k)$ and estimate $E(k)$ for $k = \sin \frac{\pi}{6}$.

Solution Substitute $x = k \sin t$ in the Taylor expansion (4):

$$\frac{1}{\sqrt{1 - k^2 \sin^2 t}} = 1 + \frac{1}{2}k^2 \sin^2 t + \frac{1 \cdot 3}{2 \cdot 4}k^4 \sin^4 t + \frac{1 \cdot 3 \cdot 5}{2 \cdot 4 \cdot 6}k^6 \sin^6 t + \cdots$$

This expansion is valid because $|k| < 1$ and hence $|x| = |k \sin t| < 1$. Thus $E(k)$ is equal to

$$\int_0^{\pi/2} \frac{dt}{\sqrt{1 - k^2 \sin^2 t}} = \int_0^{\pi/2} dt + \sum_{n=1}^{\infty} \frac{1 \cdot 3 \cdots (2n-1)}{2 \cdot 4 \cdot (2n)} \left(\int_0^{\pi/2} \sin^{2n} t \, dt \right) k^{2n}$$

According to Exercise 78 in Section 8.2,

$$\int_0^{\pi/2} \sin^{2n} t \, dt = \left(\frac{1 \cdot 3 \cdots (2n-1)}{2 \cdot 4 \cdot (2n)} \right) \frac{\pi}{2}$$

This yields

$$E(k) = \frac{\pi}{2} + \frac{\pi}{2} \sum_{n=1}^{\infty} \left(\frac{1 \cdot 3 \cdots (2n-1)^2}{2 \cdot 4 \cdots (2n)} \right)^2 k^{2n}$$

We approximate $E(k)$ for $k = \sin\left(\frac{\pi}{6}\right) = \frac{1}{2}$ using the first five terms:

$$E\left(\frac{1}{2}\right) \approx \frac{\pi}{2}\left(1 + \left(\frac{1}{2}\right)^2\left(\frac{1}{2}\right)^2 + \left(\frac{1\cdot 3}{2\cdot 4}\right)^2\left(\frac{1}{2}\right)^4 + \left(\frac{1\cdot 3\cdot 5}{2\cdot 4\cdot 6}\right)^2\left(\frac{1}{2}\right)^6 + \left(\frac{1\cdot 3\cdot 5\cdot 7}{2\cdot 4\cdot 6\cdot 8}\right)^2\left(\frac{1}{2}\right)^8\right)$$

$$\approx 1.68517$$

The value given by a computer algebra system to seven places is $E\left(\frac{1}{2}\right) \approx 1.6856325$. ■

TABLE 1

Function $f(x)$	**Maclaurin series**	**Converges to $f(x)$ for**
e^x	$\sum_{n=0}^{\infty} \frac{x^n}{n!} = 1 + x + \frac{x^2}{2!} + \frac{x^3}{3!} + \frac{x^4}{4!} + \cdots$	All x
$\sin x$	$\sum_{n=0}^{\infty} \frac{(-1)^n x^{2n+1}}{(2n+1)!} = x - \frac{x^3}{3!} + \frac{x^5}{5!} - \frac{x^7}{7!} + \cdots$	All x
$\cos x$	$\sum_{n=0}^{\infty} \frac{(-1)^n x^{2n}}{(2n)!} = 1 - \frac{x^2}{2!} + \frac{x^4}{4!} - \frac{x^6}{6!} + \cdots$	All x
$\frac{1}{1-x}$	$\sum_{n=0}^{\infty} x^n = 1 + x + x^2 + x^3 + x^4 + \cdots$	$\lvert x\rvert < 1$
$\frac{1}{1+x}$	$\sum_{n=0}^{\infty} (-1)^n x^n = 1 - x + x^2 - x^3 + x^4 - \cdots$	$\lvert x\rvert < 1$
$\ln(1+x)$	$\sum_{n=1}^{\infty} \frac{(-1)^{n-1} x^n}{n} = x - \frac{x^2}{2} + \frac{x^3}{3} - \frac{x^4}{4} + \cdots$	$\lvert x\rvert < 1$ and $x = 1$
$\tan^{-1} x$	$\sum_{n=0}^{\infty} \frac{(-1)^n x^{2n+1}}{2n+1} = x - \frac{x^3}{3} + \frac{x^5}{5} - \frac{x^7}{7} + \cdots$	$\lvert x\rvert < 1$ and $x = 1$
$(1+x)^a$	$\sum_{n=0}^{\infty} \binom{a}{n} x^n = 1 + ax + \frac{a(a-1)}{2!}x^2 + \frac{a(a-1)(a-2)}{3!}x^3 + \cdots$	$\lvert x\rvert < 1$

11.7 SUMMARY

- *Taylor series* of $f(x)$ centered at $x = c$:

$$T(x) = \sum_{n=0}^{\infty} \frac{f^{(n)}(c)}{n!}(x-c)^n$$

The partial sum $T_k(x)$ is the kth Taylor polynomial.

- *Maclaurin series* ($c = 0$):

$$T(x) = \sum_{n=0}^{\infty} \frac{f^{(n)}(0)}{n!}x^n$$

• If $f(x)$ is represented by a power series $\sum_{n=0}^{\infty} a_n(x-c)^n$ for $|x-c| < R$ with $R > 0$, then this power series is necessarily the Taylor series centered at $x = c$.
• A function $f(x)$ is represented by its Taylor series $T(x)$ if and only if the remainder $R_k(x) = f(x) - T_k(x)$ tends to zero as $k \to \infty$.
• Let $I = (c - R, c + R)$ with $R > 0$. Suppose that there exists $K > 0$ such that $|f^{(k)}(x)| < K$ for all $x \in I$ and all k. Then $f(x)$ is represented by its Taylor series on I; that is, $f(x) = T(x)$ for $x \in I$.
• A good way to find the Taylor series of a function is to start with known Taylor series and apply one of the operations: differentiation, integration, multiplication, or substitution.
• For any exponent a, the binomial expansion is valid for $|x| < 1$:

$$(1+x)^a = 1 + ax + \frac{a(a-1)}{2!}x^2 + \frac{a(a-1)(a-2)}{3!}x^3 + \cdots + \binom{a}{n}x^n + \cdots$$

11.7 EXERCISES

Preliminary Questions

1. Determine $f(0)$ and $f'''(0)$ for a function $f(x)$ with Maclaurin series

$$T(x) = 3 + 2x + 12x^2 + 5x^3 + \cdots$$

2. Determine $f(-2)$ and $f^{(4)}(-2)$ for a function with Taylor series

$$T(x) = 3(x+2) + (x+2)^2 - 4(x+2)^3 + 2(x+2)^4 + \cdots$$

3. What is the easiest way to find the Maclaurin series for the function $f(x) = \sin(x^2)$?

4. Find the Taylor series for $f(x)$ centered at $c = 3$ if $f(3) = 4$ and $f'(x)$ has a Taylor expansion

$$f'(x) = \sum_{n=1}^{\infty} \frac{(x-3)^n}{n}$$

5. Let $T(x)$ be the Maclaurin series of $f(x)$. Which of the following guarantees that $f(2) = T(2)$?
(a) $T(x)$ converges for $x = 2$.
(b) The remainder $R_k(2)$ approaches a limit as $k \to \infty$.
(c) The remainder $R_k(2)$ approaches zero as $k \to \infty$.

Exercises

1. Write out the first four terms of the Maclaurin series of $f(x)$ if

$$f(0) = 2, \quad f'(0) = 3, \quad f''(0) = 4, \quad f'''(0) = 12$$

2. Write out the first four terms of the Taylor series of $f(x)$ centered at $c = 3$ if

$$f(3) = 1, \quad f'(3) = 2, \quad f''(3) = 12, \quad f'''(3) = 3$$

In Exercises 3–18, find the Maclaurin series and find the interval on which the expansion is valid.

3. $f(x) = \dfrac{1}{1-2x}$

4. $f(x) = \dfrac{x}{1-x^4}$

5. $f(x) = \cos 3x$

6. $f(x) = \sin(2x)$

7. $f(x) = \sin(x^2)$

8. $f(x) = e^{4x}$

9. $f(x) = \ln(1-x^2)$

10. $f(x) = (1-x)^{-1/2}$

11. $f(x) = \tan^{-1}(x^2)$

12. $f(x) = x^2e^{x^2}$

13. $f(x) = e^{x-2}$

14. $f(x) = \dfrac{1-\cos x}{x}$

15. $f(x) = \ln(1-5x)$

16. $f(x) = (x^2+2x)e^x$

17. $f(x) = \sinh x$

18. $f(x) = \cosh x$

In Exercises 19–28, find the terms through degree four of the Maclaurin series of $f(x)$. Use multiplication and substitution as necessary.

19. $f(x) = e^x \sin x$

20. $f(x) = e^x \ln(1-x)$

21. $f(x) = \dfrac{\sin x}{1-x}$

22. $f(x) = \dfrac{1}{1+\sin x}$

23. $f(x) = (1+x)^{1/4}$

24. $f(x) = (1+x)^{-3/2}$

25. $f(x) = e^x \tan^{-1} x$

26. $f(x) = \sin(x^3 - x)$

27. $f(x) = e^{\sin x}$

28. $f(x) = e^{(e^x)}$

In Exercises 29–38, find the Taylor series centered at c and find the interval on which the expansion is valid.

29. $f(x) = \dfrac{1}{x}, \quad c = 1$

30. $f(x) = e^{3x}, \quad c = -1$

31. $f(x) = \dfrac{1}{1-x}$, $c = 5$

32. $f(x) = \sin x$, $c = \dfrac{\pi}{2}$

33. $f(x) = x^4 + 3x - 1$, $c = 2$

34. $f(x) = x^4 + 3x - 1$, $c = 0$

35. $f(x) = \dfrac{1}{x^2}$, $c = 4$

36. $f(x) = \sqrt{x}$, $c = 4$

37. $f(x) = \dfrac{1}{1-x^2}$, $c = 3$

38. $f(x) = \dfrac{1}{3x-2}$, $c = -1$

39. Use the identity $\cos^2 x = \frac{1}{2}(1 + \cos 2x)$ to find the Maclaurin series for $\cos^2 x$.

40. Show that for $|x| < 1$,

$$\tanh^{-1} x = x + \frac{x^3}{3} + \frac{x^5}{5} + \cdots$$

Hint: Recall that $\dfrac{d}{dx} \tanh^{-1} x = \dfrac{1}{1-x^2}$.

41. Use the Maclaurin series for $\ln(1+x)$ and $\ln(1-x)$ to show that

$$\frac{1}{2} \ln\left(\frac{1+x}{1-x}\right) = x + \frac{x^3}{3} + \frac{x^5}{5} + \cdots$$

for $|x| < 1$. What can you conclude by comparing this result with that of Exercise 40?

42. Differentiate the Maclaurin series for $\dfrac{1}{1-x}$ twice to find the Maclaurin series of $\dfrac{1}{(1-x)^3}$.

43. Show, by integrating the Maclaurin series for $f(x) = \dfrac{1}{\sqrt{1-x^2}}$, that for $|x| < 1$,

$$\sin^{-1} x = x + \sum_{n=1}^{\infty} \frac{1 \cdot 3 \cdot 5 \cdots (2n-1)}{2 \cdot 4 \cdot 6 \cdots (2n)} \frac{x^{2n+1}}{2n+1}$$

44. Use the first five terms of the Maclaurin series in Exercise 43 to approximate $\sin^{-1} \frac{1}{2}$. Compare the result with the calculator value.

45. How many terms of the Maclaurin series of $f(x) = \ln(1+x)$ are needed to compute ln 1.2 to within an error of at most 0.0001? Make the computation and compare the result with the calculator value.

46. Show that

$$\pi - \frac{\pi^3}{3!} + \frac{\pi^5}{5!} - \frac{\pi^7}{7!} + \cdots$$

converges to zero. How many terms must be computed to get within 0.01 of zero?

47. Use the Maclaurin expansion for e^{-t^2} to express the function $F(x) = \int_0^x e^{-t^2}\,dt$ as an alternating power series in x (Figure 4).

(a) How many terms of the Maclaurin series are needed to approximate the integral for $x = 1$ to within an error of at most 0.001?

(b) CAS Carry out the computation and check your answer using a computer algebra system.

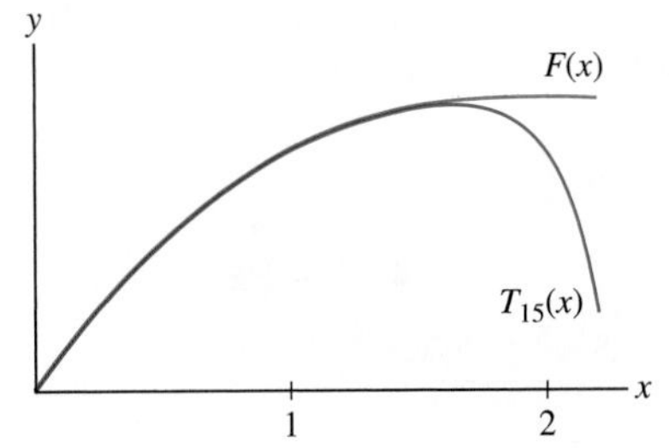

FIGURE 4 The Maclaurin polynomial $T_{15}(x)$ for $F(t) = \int_0^x e^{-t^2}\,dt$.

48. Let $F(x) = \displaystyle\int_0^x \frac{\sin t\,dt}{t}$. Show that

$$F(x) = x - \frac{x^3}{3 \cdot 3!} + \frac{x^5}{5 \cdot 5!} - \frac{x^7}{7 \cdot 7!} + \cdots$$

Evaluate $F(1)$ to three decimal places.

In Exercises 49–52, express the definite integral as an infinite series and find its value to within an error of at most 10^{-4}.

49. $\displaystyle\int_0^1 \cos(x^2)\,dx$

50. $\displaystyle\int_0^1 \tan^{-1}(x^2)\,dx$

51. $\displaystyle\int_0^1 e^{-x^3}\,dx$

52. $\displaystyle\int_0^1 \frac{dx}{\sqrt{x^4+1}}$

In Exercises 53–56, express the integral as an infinite series.

53. $\displaystyle\int_0^x \frac{1-\cos(t)}{t}\,dt$, for all x

54. $\displaystyle\int_0^x \frac{t - \sin t}{t}\,dt$, for all x

55. $\displaystyle\int_0^x \ln(1+t^2)\,dt$, for $|x| < 1$

56. $\displaystyle\int_0^x \frac{dt}{\sqrt{1-t^4}}$, for $|x| < 1$

57. Which function has Maclaurin series $\displaystyle\sum_{n=0}^{\infty} (-1)^n 2^n x^n$?

58. Which function has Maclaurin series

$$\sum_{k=0}^{\infty} \frac{(-1)^k}{3^{k+1}}(x-3)^k?$$

For which values of x is the expansion valid?

In Exercises 59–62, use Theorem 2 to prove that the $f(x)$ *is represented by its Maclaurin series on the interval* I.

59. $f(x) = \ln(1+x)$, $I = \left(-\frac{1}{2}, \frac{1}{2}\right)$

60. $f(x) = e^{-x}$, $I = (-c, c)$ for all $c > 0$

61. $f(x) = \sinh x$, $I = \mathbf{R}$ (see Exercise 17)

62. $f(x) = (1+x)^{100}$, $I = \mathbf{R}$

In Exercises 63–66, find the functions with the following Maclaurin series (refer to Table 1 on page 605).

63. $1 + x^3 + \dfrac{x^6}{2!} + \dfrac{x^9}{3!} + \dfrac{x^{12}}{4!} + \cdots$

64. $1 - 4x + 4^2x^2 - 4^3x^3 + 4^4x^4 - 4^5x^5 + \cdots$

65. $1 - \dfrac{5^3x^3}{3!} + \dfrac{5^5x^5}{5!} - \dfrac{5^7x^7}{7!} + \cdots$

66. $x^4 - \dfrac{x^{12}}{3} + \dfrac{x^{20}}{5} - \dfrac{x^{28}}{7} + \cdots$

In Exercises 67 and 68, let

$$f(x) = \frac{1}{(1-x)(1-2x)}$$

67. Find the Maclaurin series of $f(x)$ using the identity

$$f(x) = \frac{2}{1-2x} - \frac{1}{1-x}$$

68. Find the Taylor series for $f(x)$ at $c = 2$. *Hint:* Rewrite the identity of Exercise 67 as

$$f(x) = \frac{2}{-3 - 2(x-2)} - \frac{1}{-1-(x-2)}$$

69. When a voltage V is applied to a series circuit consisting of a resistor R and an inductor L, the current at time t is

$$I(t) = \left(\frac{V}{R}\right)\left(1 - e^{-Rt/L}\right)$$

Expand $I(t)$ in a Maclaurin series. Show that $I(t) \approx \dfrac{Vt}{L}$ for small t.

70. Use the result of Exercise 69 and your knowledge of alternating series to show that

$$\frac{Vt}{L}\left(1 - \frac{R}{2L}t\right) \le I(t) \le \frac{Vt}{L} \qquad \text{(for all } t\text{)}$$

71. Find the Maclaurin series for $f(x) = \cos(x^3)$ and use it to determine $f^{(6)}(0)$.

72. Find $f^{(7)}(0)$ and $f^{(8)}(0)$ for $f(x) = \tan^{-1} x$ using the Maclaurin series.

73. Use substitution to find the first three terms of the Maclaurin series for $f(x) = e^{x^{20}}$. How does the result show that $f^{(k)}(0) = 0$ for $1 \le k \le 19$?

74. Use the binomial series to find $f^{(8)}(0)$ for $f(x) = \sqrt{1 - x^2}$.

75. Does the Maclaurin series for $f(x) = (1 + x)^{3/4}$ converge to $f(x)$ at $x = 2$? Give numerical evidence to support your answer.

76. Explain the steps required to verify that the Maclaurin series for $f(x) = e^x$ converges to $f(x)$ for all x.

77. GU Let $f(x) = \sqrt{1 + x}$.

(a) Use a graphing calculator to compare the graph of f with the graphs of the first five Taylor polynomials for f. What do they suggest about the interval of convergence of the Taylor series?

(b) Investigate numerically whether or not the Taylor expansion for f is valid for $x = 1$ and $x = -1$.

78. Use the first five terms of the Maclaurin series for the elliptic function $E(k)$ to estimate the period T of a 1-meter pendulum released at an angle $\theta = \frac{\pi}{4}$ (see Example 11).

79. Use Example 11 and the approximation $\sin x \approx x$ to show that the period T of a pendulum released at an angle θ has the following second-order approximation:

$$T \approx 2\pi\sqrt{\frac{L}{g}}\left(1 + \frac{\theta^2}{16}\right)$$

In Exercises 80–83, find the Maclaurin series of the function and use it to calculate the limit.

80. $\displaystyle\lim_{x\to 0} \frac{\cos x - 1 + \frac{x^2}{2}}{x^4}$

81. $\displaystyle\lim_{x\to 0} \frac{\sin x - x + \frac{x^3}{6}}{x^5}$

82. $\displaystyle\lim_{x\to 0} \frac{\tan^{-1} x - x\cos x - \frac{1}{6}x^3}{x^5}$

83. $\displaystyle\lim_{x\to 0} \left(\frac{\sin(x^2)}{x^4} - \frac{\cos x}{x^2}\right)$

Further Insights and Challenges

84. In this exercise we show that the Maclaurin expansion of $f(x) = \ln(1 + x)$ is valid for $x = 1$.

(a) Show that for all $x \ne -1$,

$$\frac{1}{1+x} = \sum_{n=0}^{N} (-1)^n x^n + \frac{(-1)^{N+1}x^{N+1}}{1+x}$$

(b) Integrate from 0 to 1 to obtain

$$\ln 2 = \sum_{n=1}^{N} \frac{(-1)^{n-1}}{n} + (-1)^{N+1}\int_0^1 \frac{x^{N+1}\,dx}{1+x}$$

(c) Verify that the integral on the right tends to zero as $N \to \infty$ by showing that it is smaller than $\int_0^1 x^{N+1}dx$.

(d) Prove the formula

$$\ln 2 = 1 - \frac{1}{2} + \frac{1}{3} - \frac{1}{4} + \cdots$$

85. Let $g(t) = \dfrac{1}{1+t^2} - \dfrac{t}{1+t^2}$.

(a) Show that $\displaystyle\int_0^1 g(t)\,dt = \frac{\pi}{4} - \frac{1}{2}\ln 2$.

(b) Show that $g(t) = 1 - t - t^2 + t^3 - t^4 - t^5 + \cdots$.

(c) Evaluate $S = 1 - \frac{1}{2} - \frac{1}{3} + \frac{1}{4} - \frac{1}{5} - \frac{1}{6} + \cdots$.

In Exercises 86 and 87, we investigate the convergence of the binomial series

$$T_a(x) = \sum_{n=0}^{\infty} \binom{a}{n} x^n$$

86. Prove that $T_a(x)$ has radius of convergence $R = 1$ if a is not a whole number. What is the radius of convergence if a is a whole number?

87. By Exercise 86, $T_a(x)$ converges for $|x| < 1$, but we do not yet know whether $T_a(x) = (1+x)^a$.

(a) Verify the identity

$$a\binom{a}{n} = n\binom{a}{n} + (n+1)\binom{a}{n+1}$$

(b) Use (a) to show that $y = T_a(x)$ satisfies the differential equation $(1+x)y' = ay$ with initial condition $y(0) = 1$.

(c) Prove that $T_a(x) = (1+x)^a$ for $|x| < 1$ by showing that the derivative of the ratio $\dfrac{T_a(x)}{(1+x)^a}$ is zero.

88. The function $G(k) = \int_0^{\pi/2} \sqrt{1 - k^2 \sin^2 t}\, dt$ is called an **elliptic function of the second kind**. Prove that for $|k| < 1$,

$$G(k) = \frac{\pi}{2} - \frac{\pi}{2} \sum_{n=1}^{\infty} \left(\frac{1 \cdot 3 \cdots (2n-1)}{2 \cdots 4 \cdot (2n)} \right)^2 \frac{k^{2n}}{2n-1}$$

89. Assume that $a < b$ and let L be the arc length (circumference) of the ellipse $\left(\frac{x}{a}\right)^2 + \left(\frac{y}{b}\right)^2 = 1$ shown in Figure 5. There is no explicit formula for L, but it is known that $L = 4bG(k)$, with $G(k)$ as in Exercise 88 and $k = \sqrt{1 - a^2/b^2}$. Use the first three terms of the expansion of Exercise 88 to estimate L when $a = 4$ and $b = 5$.

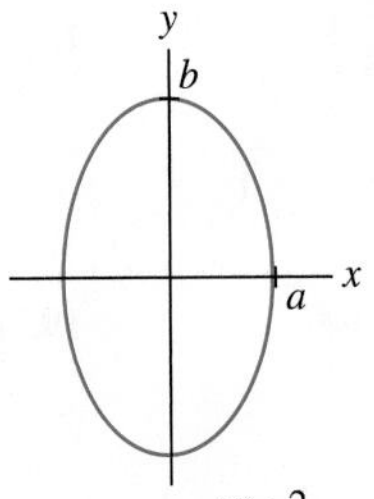

FIGURE 5 The ellipse $\left(\frac{x}{a}\right)^2 + \left(\frac{y}{b}\right)^2 = 1$.

90. Use Exercise 88 to prove that if $a < b$ and a/b is near 1 (a nearly circular ellipse), then

$$L \approx \frac{\pi}{2}\left(3b + \frac{a^2}{b}\right)$$

Hint: Use the first two terms of the series for $G(k)$.

91. Irrationality of *e* Prove that e is an irrational number using the following argument by contradiction. Suppose that $e = M/N$, where M, N are nonzero integers.

(a) Show that $M!\,e^{-1}$ is a whole number.

(b) Use the power series for e^x at $x = -1$ to show that there is an integer B such that $M!\,e^{-1}$ equals

$$B + (-1)^{M+1}\left(\frac{1}{M+1} - \frac{1}{(M+1)(M+2)} + \cdots\right)$$

(c) Use your knowledge of alternating series with decreasing terms to conclude that $0 < |M!\,e^{-1} - B| < 1$ and observe that this contradicts (a). Hence, e is not equal to M/N.

92. Use the result of Exercise 69 in Section 7.7 to show that the Maclaurin series of the function

$$f(x) = \begin{cases} e^{-1/x^2} & \text{for } x \neq 0 \\ 0 & \text{for } x = 0 \end{cases}$$

is $T(x) = 0$. This provides an example of a function $f(x)$ whose Maclaurin series converges but does not converge to $f(x)$ (except at $x = 0$).

CHAPTER REVIEW EXERCISES

1. Let $a_n = \dfrac{n-3}{n!}$ and $b_n = a_{n+3}$. Calculate the first three terms in each sequence.

(a) a_n^2

(b) b_n

(c) $a_n b_n$

(d) $2a_{n+1} - 3a_n$

2. Prove that $\lim\limits_{n\to\infty} \dfrac{2n-1}{3n+2} = \dfrac{2}{3}$ using the limit definition.

In Exercises 3–8, compute the limit (or state that it does not exist) assuming that $\lim\limits_{n\to\infty} a_n = 2$.

3. $\lim\limits_{n\to\infty} (5a_n - 2a_n^2)$

4. $\lim\limits_{n\to\infty} \dfrac{1}{a_n}$

5. $\lim\limits_{n\to\infty} e^{a_n}$

6. $\lim\limits_{n\to\infty} \cos(\pi a_n)$

7. $\lim\limits_{n\to\infty} (-1)^n a_n$

8. $\lim\limits_{n\to\infty} \dfrac{a_n + n}{a_n + n^2}$

In Exercises 9–22, determine the limit of the sequence or show that the sequence diverges.

9. $a_n = \sqrt{n+5} - \sqrt{n+2}$

10. $a_n = \dfrac{3n^3 - n}{1 - 2n^3}$

11. $a_n = 2^{1/n^2}$

12. $a_n = \dfrac{10^n}{n!}$

13. $b_m = 1 + (-1)^m$

14. $b_m = \dfrac{1 + (-1)^m}{m}$

15. $b_n = \tan^{-1}\left(\dfrac{n+2}{n+5}\right)$

16. $a_n = \dfrac{100^n}{n!} - \dfrac{3+\pi^n}{5^n}$

17. $b_n = \sqrt{n^2+n} - \sqrt{n^2+1}$

18. $c_n = \sqrt{n^2+n} - \sqrt{n^2-n}$

19. $b_m = \left(1+\dfrac{1}{m}\right)^{3m}$

20. $c_n = \left(1+\dfrac{3}{n}\right)^{n}$

21. $b_n = n\big(\ln(n+1) - \ln n\big)$

22. $c_n = \dfrac{\ln(n^2+1)}{\ln(n^3+1)}$

23. Use the Squeeze Theorem to show that $\lim\limits_{n\to\infty} \dfrac{\arctan(n^2)}{\sqrt{n}} = 0$.

24. Give an example of a divergent sequence $\{a_n\}$ such that $\{\sin a_n\}$ is convergent.

25. Calculate $\lim\limits_{n\to\infty} \dfrac{a_{n+1}}{a_n}$, where $a_n = \dfrac{1}{2}3^n - \dfrac{1}{3}2^n$.

26. Define $a_{n+1} = \sqrt{a_n + 6}$ with $a_1 = 2$.

(a) Compute a_n for $n = 2, 3, 4, 5$.

(b) Show that $\{a_n\}$ is increasing and is bounded by 3.

(c) Prove that $\lim\limits_{n\to\infty} a_n$ exists and find its value.

27. Calculate the partial sums S_4 and S_7 of the series $\displaystyle\sum_{n=1}^{\infty} \frac{n-2}{n^2+2n}$.

28. Find the sum $1 - \dfrac{1}{4} + \dfrac{1}{4^2} - \dfrac{1}{4^3} + \cdots$.

29. Find the sum $\dfrac{4}{9} + \dfrac{8}{27} + \dfrac{16}{81} + \dfrac{32}{243} + \cdots$.

30. Find the sum $\displaystyle\sum_{n=2}^{\infty} \left(\frac{2}{e}\right)^n$.

31. Find the sum $\displaystyle\sum_{n=-1}^{\infty} \frac{2^{n+3}}{3^n}$.

32. Show that $\displaystyle\sum_{n=1}^{\infty} (b - \tan^{-1} n^2)$ diverges if $b \neq \dfrac{\pi}{2}$.

33. Give an example of divergent series $\displaystyle\sum_{n=1}^{\infty} a_n$ and $\displaystyle\sum_{n=1}^{\infty} b_n$ such that $\displaystyle\sum_{n=1}^{\infty} (a_n + b_n) = 1$.

34. Let $S = \displaystyle\sum_{n=1}^{\infty} \left(\frac{1}{n} - \frac{1}{n+2}\right)$. Compute S_N for $N = 1, 2, 3, 4$. Find S by showing that

$$S_N = \frac{3}{2} - \frac{1}{N+1} - \frac{1}{N+2}$$

35. Evaluate $S = \displaystyle\sum_{n=3}^{\infty} \frac{1}{n(n+3)}$.

36. Find the total area of the infinitely many circles on the interval [0, 1] in Figure 1.

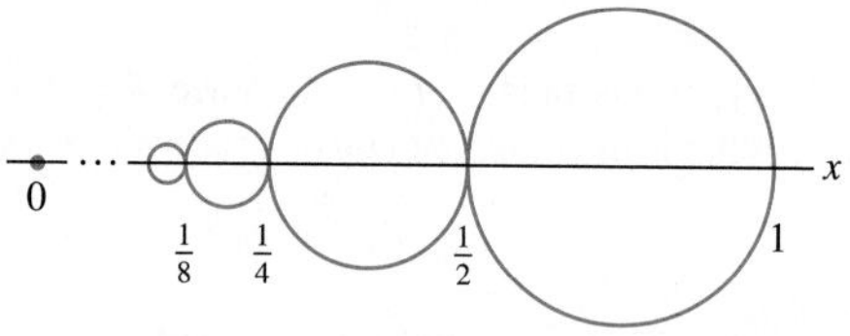

FIGURE 1

In Exercises 37–40, use the Integral Test to determine whether the infinite series converges.

37. $\displaystyle\sum_{n=1}^{\infty} \frac{n^2}{n^3+1}$

38. $\displaystyle\sum_{n=1}^{\infty} \frac{n^2}{(n^3+1)^{1.01}}$

39. $\displaystyle\sum_{n=1}^{\infty} \frac{1}{(n+2)(\ln(n+2))^3}$

40. $\displaystyle\sum_{n=1}^{\infty} \frac{n^3}{e^{n^4}}$

In Exercises 41–48, use the Comparison or Limit Comparison Test to determine whether the infinite series converges.

41. $\displaystyle\sum_{n=1}^{\infty} \frac{1}{(n+1)^2}$

42. $\displaystyle\sum_{n=1}^{\infty} \frac{1}{\sqrt{n}+n}$

43. $\displaystyle\sum_{n=2}^{\infty} \frac{n^2+1}{n^{3.5}-2}$

44. $\displaystyle\sum_{n=1}^{\infty} \frac{1}{n - \ln n}$

45. $\displaystyle\sum_{n=2}^{\infty} \frac{n}{\sqrt{n^5+5}}$

46. $\displaystyle\sum_{n=1}^{\infty} \frac{1}{3^n - 2^n}$

47. $\displaystyle\sum_{n=1}^{\infty} \frac{n^{10}+10^n}{n^{11}+11^n}$

48. $\displaystyle\sum_{n=1}^{\infty} \frac{n^{20}+21^n}{n^{21}+20^n}$

49. Determine the convergence of $\displaystyle\sum_{n=1}^{\infty} \frac{2^n+n}{3^n-2}$ using the Limit Comparison Test with $b_n = \left(\frac{2}{3}\right)^n$.

50. Determine the convergence of $\displaystyle\sum_{n=1}^{\infty} \frac{\ln n}{1.5^n}$ using the Limit Comparison Test with $b_n = \dfrac{1}{1.4^n}$.

51. Let $a_n = 1 - \sqrt{1 - \frac{1}{n}}$. Show that $\lim\limits_{n\to\infty} a_n = 0$ and that $\displaystyle\sum_{n=1}^{\infty} a_n$ diverges. *Hint:* Show that $a_n \geq \frac{1}{2n}$.

52. Determine whether $\displaystyle\sum_{n=2}^{\infty} \left(1 - \sqrt{1 - \frac{1}{n^2}}\right)$ converges.

53. Let $S = \displaystyle\sum_{n=1}^{\infty} \frac{n}{(n^2+1)^2}$.

(a) Show that S converges.

(b) CAS Use Eq. (4) in Exercise 83 of Section 11.3 with $M = 99$ to approximate S. What is the maximum size of the error?

In Exercises 54–57, determine whether the series converges absolutely. If it does not, determine whether it converges conditionally.

54. $\sum_{n=1}^{\infty} \frac{(-1)^n}{\sqrt[3]{n}+2n}$

55. $\sum_{n=1}^{\infty} \frac{(-1)^n}{n^{1.1}\ln(n+1)}$

56. $\sum_{n=1}^{\infty} \frac{\cos\left(\frac{\pi}{4}+\pi n\right)}{\sqrt{n}}$

57. $\sum_{n=1}^{\infty} \frac{\cos\left(\frac{\pi}{4}+2\pi n\right)}{\sqrt{n}}$

58. CAS Use a computer algebra system to approximate $\sum_{n=1}^{\infty} \frac{(-1)^n}{n^3+\sqrt{n}}$ to within an error of at most 10^{-5}.

59. Catalan's constant is defined by $K = \sum_{k=0}^{\infty} \frac{(-1)^k}{(2k+1)^2}$.

(a) How many terms of the series are needed to calculate K with an error of less than 10^{-6}?

(b) CAS Carry out the calculation.

60. Give an example of conditionally convergent series $\sum_{n=1}^{\infty} a_n$ and $\sum_{n=1}^{\infty} b_n$ such that $\sum_{n=1}^{\infty} (a_n + b_n)$ converges absolutely.

61. Let $\sum_{n=1}^{\infty} a_n$ be an absolutely convergent series. Determine whether the following series are convergent or divergent:

(a) $\sum_{n=1}^{\infty} \left(a_n + \frac{1}{n^2}\right)$

(b) $\sum_{n=1}^{\infty} (-1)^n a_n$

(c) $\sum_{n=1}^{\infty} \frac{1}{1+a_n^2}$

(d) $\sum_{n=1}^{\infty} \frac{|a_n|}{n}$

62. Let $\{a_n\}$ be a positive sequence such that $\lim_{n\to\infty} \sqrt[n]{a_n} = \frac{1}{2}$. Determine whether the following series converge or diverge:

(a) $\sum_{n=1}^{\infty} 2a_n$

(b) $\sum_{n=1}^{\infty} 3^n a_n$

(c) $\sum_{n=1}^{\infty} \sqrt{a_n}$

In Exercises 63–70, apply the Ratio Test to determine convergence or divergence, or state that the Ratio Test is inconclusive.

63. $\sum_{n=1}^{\infty} \frac{n^5}{5^n}$

64. $\sum_{n=1}^{\infty} \frac{\sqrt{n+1}}{n^8}$

65. $\sum_{n=1}^{\infty} \frac{1}{n2^n+n^3}$

66. $\sum_{n=1}^{\infty} \frac{n^4}{n!}$

67. $\sum_{n=1}^{\infty} \frac{2^{n^2}}{n!}$

68. $\sum_{n=4}^{\infty} \frac{\ln n}{n^{3/2}}$

69. $\sum_{n=1}^{\infty} \left(\frac{n}{2}\right)^n \frac{1}{n!}$

70. $\sum_{n=1}^{\infty} \left(\frac{n}{4}\right)^n \frac{1}{n!}$

In Exercises 71–74, apply the Root Test to determine convergence or divergence, or state that the Root Test is inconclusive.

71. $\sum_{n=1}^{\infty} \frac{1}{4^n}$

72. $\sum_{n=1}^{\infty} \left(\frac{2}{n}\right)^n$

73. $\sum_{n=1}^{\infty} \left(\frac{3}{4n}\right)^n$

74. $\sum_{n=1}^{\infty} \left(\cos\frac{1}{n}\right)^{n^3}$

In Exercises 75–92, determine convergence or divergence using any method covered in the text.

75. $\sum_{n=1}^{\infty} \left(\frac{2}{3}\right)^n$

76. $\sum_{n=1}^{\infty} \frac{\pi^{7n}}{e^{8n}}$

77. $\sum_{n=1}^{\infty} e^{-0.02n}$

78. $\sum_{n=1}^{\infty} ne^{-0.02n}$

79. $\sum_{n=1}^{\infty} \frac{(-1)^{n-1}}{\sqrt{n}+\sqrt{n+1}}$

80. $\sum_{n=10}^{\infty} \frac{1}{n(\ln n)^{3/2}}$

81. $\sum_{n=2}^{\infty} \frac{(-1)^n}{\ln n}$

82. $\sum_{n=1}^{\infty} \frac{e^n}{n!}$

83. $\sum_{n=1}^{\infty} \frac{1}{n\sqrt{n+\ln n}}$

84. $\sum_{n=1}^{\infty} \frac{1}{\sqrt[3]{n}(1+\sqrt{n})}$

85. $\sum_{n=1}^{\infty} \left(\frac{1}{\sqrt{n}} - \frac{1}{\sqrt{n+1}}\right)$

86. $\sum_{n=1}^{\infty} \left(\ln n - \ln(n+1)\right)$

87. $\sum_{n=1}^{\infty} \frac{1}{n+\sqrt{n}}$

88. $\sum_{n=2}^{\infty} \frac{\cos(\pi n)}{n^{2/3}}$

89. $\sum_{n=2}^{\infty} \frac{1}{n^{\ln n}}$

90. $\sum_{n=2}^{\infty} \frac{1}{\ln^3 n}$

91. $\sum_{n=1}^{\infty} \sin^2 \frac{\pi}{n}$

92. $\sum_{n=0}^{\infty} \frac{2^{2n}}{n!}$

In Exercises 93–98, find the interval of convergence of the power series.

93. $\sum_{n=0}^{\infty} \frac{2^n x^n}{n!}$

94. $\sum_{n=0}^{\infty} \frac{x^n}{n+1}$

95. $\sum_{n=0}^{\infty} \frac{n^6}{n^8+1}(x-3)^n$

96. $\sum_{n=0}^{\infty} nx^n$

97. $\sum_{n=0}^{\infty} (nx)^n$

98. $\sum_{n=0}^{\infty} \frac{(2x-3)^n}{n\ln n}$

99. Expand $f(x) = \frac{2}{4-3x}$ as a power series centered at $c = 0$. Determine the values of x for which the series converges.

100. Prove that

$$\sum_{n=0}^{\infty} ne^{-nx} = \frac{e^{-x}}{(1-e^{-x})^2}$$

Hint: Express the left-hand side as the derivative of a geometric series.

101. Let $F(x) = \sum_{k=0}^{\infty} \frac{x^{2k}}{2^k \cdot k!}$.

(a) Show that $F(x)$ has infinite radius of convergence.

(b) Show that $y = F(x)$ is a solution of

$$y'' = xy' + y, \qquad y(0) = 1, \qquad y'(0) = 0$$

(c) CAS Plot the partial sums S_N for $N = 1, 3, 5, 7$ on the same set of axes.

102. Find a power series $P(x) = \sum_{n=0}^{\infty} a_n x^n$ that satisfies the Laguerre differential equation

$$xy'' + (1-x)y' - y = 0$$

with initial condition satisfying $P(0) = 1$.

In Exercises 103–112, find the Taylor series centered at c.

103. $f(x) = e^{4x}, \quad c = 0$

104. $f(x) = e^{2x}, \quad c = -1$

105. $f(x) = x^4, \quad c = 2$

106. $f(x) = x^3 - x, \quad c = -2$

107. $f(x) = \sin x, \quad c = \pi$

108. $f(x) = e^{x-1}, \quad c = -1$

109. $f(x) = \frac{1}{1-2x}, \quad c = -2$

110. $f(x) = \frac{1}{(1-2x)^2}, \quad c = -2$

111. $f(x) = \ln \frac{x}{2}, \quad c = 2$

112. $f(x) = x \ln\left(1 + \frac{x}{2}\right), \quad c = 0$

In Exercises 113–116, find the first three terms of the Maclaurin series of $f(x)$ and use it to calculate $f^{(3)}(0)$.

113. $f(x) = (x^2 - x)e^{x^2}$

114. $f(x) = \tan^{-1}(x^2 - x)$

115. $f(x) = \frac{1}{1 + \tan x}$

116. $f(x) = (\sin x)\sqrt{1+x}$

117. Calculate $\frac{\pi}{2} - \frac{\pi^3}{2^3 3!} + \frac{\pi^5}{2^5 5!} - \frac{\pi^7}{2^7 7!} + \cdots$.

118. Find the Maclaurin series of the function $F(x) = \int_0^x \frac{e^t - 1}{t}\,dt$.

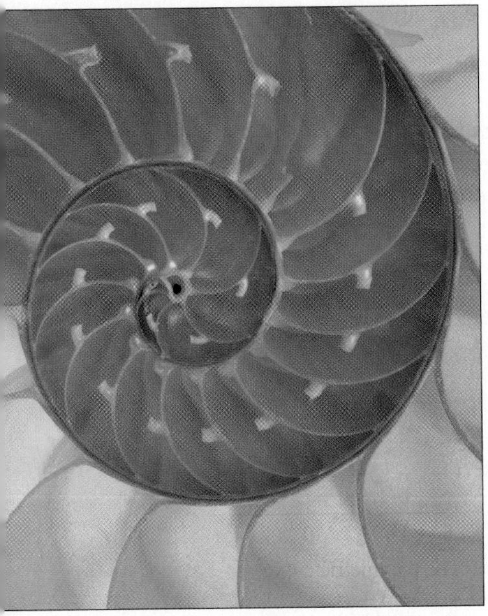

The beautiful shell of a chambered nautilus grows in the shape of an equiangular spiral, a curve described in polar coordinates by an equation $r = e^{a\theta}$.

12 PARAMETRIC EQUATIONS, POLAR COORDINATES, AND CONIC SECTIONS

This chapter introduces two important new tools. First, we consider parametric equations, which describe curves in a form that is particularly useful for analyzing motion and is indispensable in fields such as computer graphics and computer-aided design. We then study polar coordinates, an alternative to rectangular coordinates that simplifies computations in many applications. The chapter closes with a discussion of the conic sections (ellipses, hyperbolas, and parabolas).

12.1 Parametric Equations

We use the term "particle" when we treat an object as a moving point, ignoring its internal structure.

Imagine a particle moving along a curve $\mathcal{C}$ in the plane as in Figure 1. We can describe the particle's motion by specifying its coordinates as functions of time t:

$$x = f(t), \qquad y = g(t) \tag{1}$$

In other words, at time t, the particle is located at the point

$$c(t) = (f(t), g(t))$$

The equations (1) are called **parametric equations**, and $\mathcal{C}$ is called a **parametric curve**. We refer to $c(t)$ as a **parametrization** with **parameter** t.

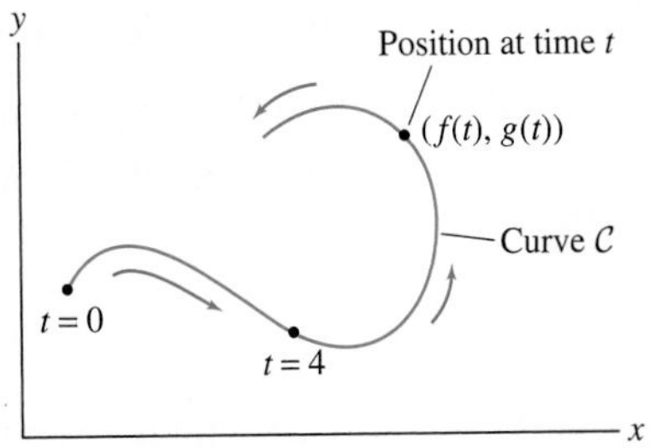

FIGURE 1 Particle moving along a curve $\mathcal{C}$ in the plane.

Because x and y are functions of t, we often write $c(t) = (x(t), y(t))$ instead of $(f(t), g(t))$. Of course, we are free to use any variable for the parameter (such as s or θ). In plots of parametric curves, the direction of motion is often indicated by an arrow as in Figure 1.

■ **EXAMPLE 1** Sketch the curve with parametric equations

$$x = 2t - 4, \qquad y = 3 + t^2 \tag{2}$$

Solution First compute the x- and y-coordinates for several values of t as in Table 1, and plot the corresponding points (x, y) as in Figure 2. Then join the points by a smooth curve, indicating the direction of motion with an arrow. ■

TABLE 1

t	$x = 2t - 4$	$y = 3 + t^2$
-2	-8	7
0	-4	3
2	0	7
4	4	19

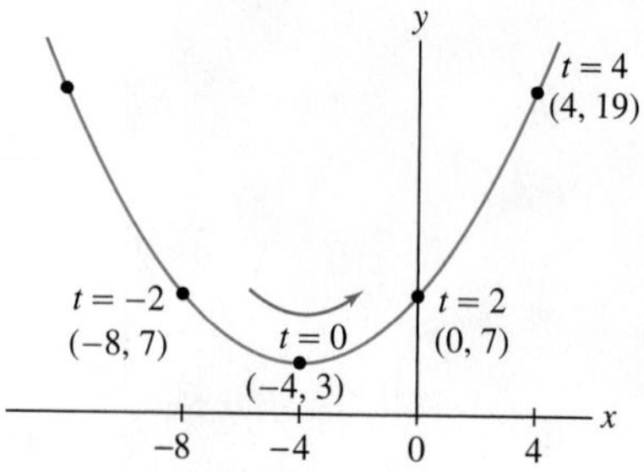

FIGURE 2 The parametric curve $x = 2t - 4$, $y = 3 + t^2$.

CONCEPTUAL INSIGHT The graph of a function $y = f(x)$ can always be parametrized in a simple way as

$$c(t) = (t, f(t))$$

For example, the parabola $y = x^2$ is parametrized by $c(t) = (t, t^2)$ and the curve $y = e^t$ by $c(t) = (t, e^t)$. An advantage of parametric equations is that they enable us to describe curves that are not graphs of functions. For example, the curve in Figure 3 is not of the form $y = f(x)$ but it can be expressed parametrically.

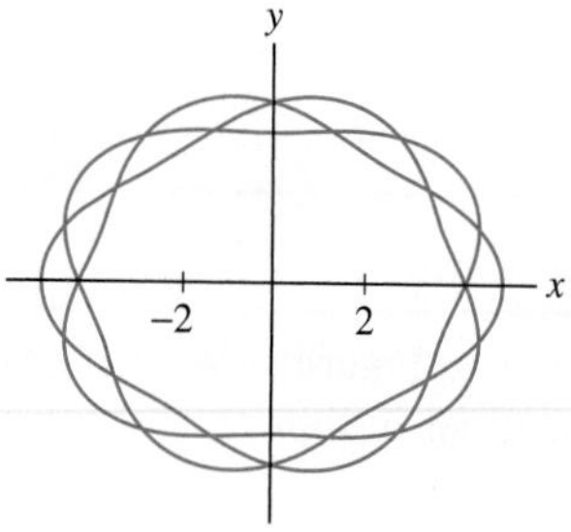

FIGURE 3 The parametric curve $x = 5\cos(3t)\cos\left(\frac{2}{3}\sin(5t)\right)$, $y = 4\sin(3t)\cos\left(\frac{2}{3}\sin(5t)\right)$.

As we have just noted, a parametric curve $c(t)$ need not be the graph of a function. If it is, however, it may be possible to find the function $f(x)$ by "eliminating the parameter" as in the next example.

■ **EXAMPLE 2 Eliminating the Parameter** Describe the parametric curve

$$c(t) = (2t - 4, 3 + t^2)$$

of the previous example in the form $y = f(x)$.

Solution We "eliminate the parameter" by solving for y as a function of x. First, express t in terms of x: Since $x = 2t - 4$, we have $t = \frac{1}{2}x + 2$. Then substitute

$$y = 3 + t^2 = 3 + \left(\frac{1}{2}x + 2\right)^2 = 7 + 2x + \frac{1}{4}x^2$$

Thus, $c(t)$ traces out the graph of $f(x) = 7 + 2x + \frac{1}{4}x^2$ shown in Figure 2. ■

■ **EXAMPLE 3** A bullet follows the trajectory

$$c(t) = (80t, 200t - 4.9t^2)$$

until it hits the ground, with t in seconds and distance in meters (Figure 4). Find:

(a) The bullet's height at $t = 5$ s.

(b) Its maximum height.

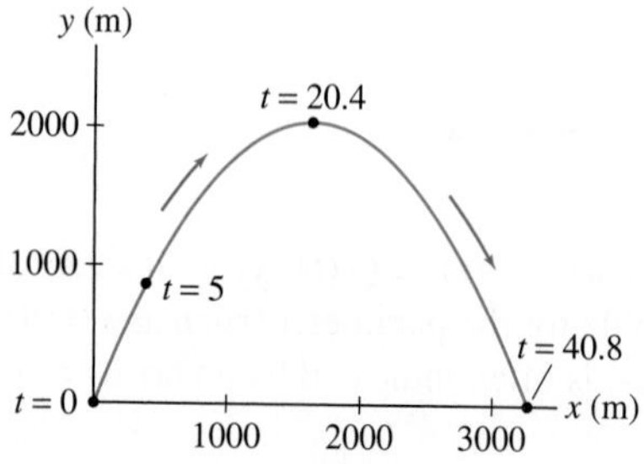

FIGURE 4 Trajectory of bullet.

CAUTION The graph of height versus time for an object tossed in the air is a parabola (by Galileo's formula). But keep in mind that Figure 4 is ***not*** *a graph of height versus time. It shows the actual path of the bullet (which has both a vertical and a horizontal displacement).*

Solution The height of the bullet at time t is $y(t) = 200t - 4.9t^2$.

(a) The height at $t = 5$ s is

$$y(5) = 200(5) - 4.9(5^2) = 877.5 \text{ m}$$

(b) The maximum height occurs at the critical point of $y(t)$:

$$y'(t) = \frac{d}{dt}(200t - 4.9t^2) = 200 - 9.8t = 0 \quad \Rightarrow \quad t = \frac{200}{9.8} \approx 20.4 \text{ s}$$

The bullet's maximum height is $y(20.4) = 200(20.4) - 4.9(20.4)^2 \approx 2041$ m. ■

We now discuss parametrizations of lines and circles. They will appear frequently in later chapters.

THEOREM 1 Parametrization of a Line

(a) The line through $P = (a, b)$ of slope m is parametrized by

$$x = a + rt, \qquad y = b + st \qquad -\infty < t < \infty \tag{3}$$

for any r and s (with $r \neq 0$) such that $m = s/r$.

(b) The line through $P = (a, b)$ and $Q = (c, d)$ has parametrization

$$x = a + t(c - a), \qquad y = b + t(d - b) \qquad -\infty < t < \infty \tag{4}$$

The segment from P to Q corresponds to $0 \le 1 \le t$.

Solution (a) Solve $x = a + rt$ for t in terms of x to obtain $t = (x - a)/r$. Then

$$y = b + st = b + s\left(\frac{x - a}{r}\right) = b + m(x - a) \qquad \text{or} \qquad y - b = m(x - a)$$

This is the equation of the line through $P = (a, b)$ of slope m. The choice $r = 1$ and $s = m$ yields the parametrization in Figure 5.

The parametrization in (b) defines a line that satisfies $(x(0), y(0)) = (a, b)$ and $(x(1), y(1)) = (c, d)$. Thus, it parametrizes the line through P and Q and traces the segment from P to Q as t varies from 0 to 1. ■

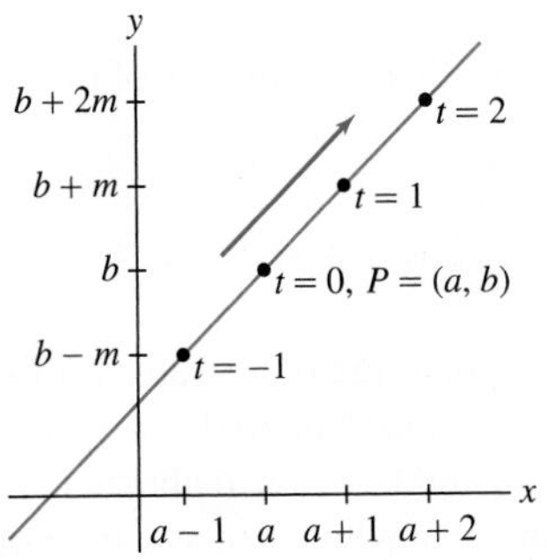

FIGURE 5 The line

$$y - a = m(x - b)$$

has parametrization

$$c(t) = (a + t, b + mt)$$

This corresponds to $r = 1$, $s = m$ in Eq. 3.

■ **EXAMPLE 4 Parametrization of a Line** Find parametric equations for the line through $P = (3, -1)$ of slope $m = 4$.

Solution We can parametrize the line by taking $r = 1$ and $s = 4$ in Eq. (3):

$$x = 3 + t, \qquad y = -1 + 4t$$

This is also written as $c(t) = (3 + t, -1 + 4t)$. Another parametrization of the line is $c(t) = (3 + 5t, -1 + 20t)$, corresponding to $r = 5$ and $s = 20$ in Eq. (3). ■

The circle of radius R centered at the origin has the parametrization

$$x = R\cos\theta, \qquad y = R\sin\theta$$

The parameter θ represents the angle corresponding to the point (x, y) on the circle (Figure 6). The circle is traversed once in the counterclockwise direction as θ varies over a half-open interval of length 2π such as $[0, 2\pi)$ or $[-\pi, \pi)$.

More generally, the circle of radius R with center (a, b) has parametrization (Figure 6)

$$x = a + R\cos\theta, \qquad y = b + R\sin\theta \tag{5}$$

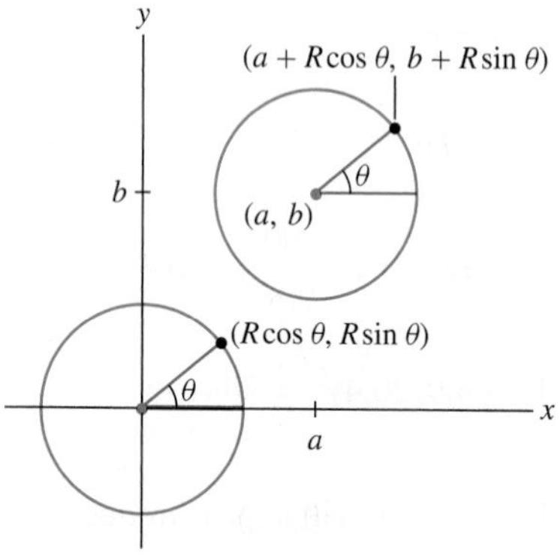

FIGURE 6 Parametrization of a circle of radius R with center (a, b).

As a check, let's verify that a point (x, y) given by Eq. (5) satisfies the equation of the circle of radius R centered at (a, b):

$$\begin{aligned}(x-a)^2 + (y-b)^2 &= (a + R\cos\theta - a)^2 + (b + R\sin\theta - b)^2 \\ &= R^2\cos^2\theta + R^2\sin^2\theta = R^2\end{aligned}$$

In general, to **translate** (meaning "to move") a parametric curve horizontally a units and vertically b units, replace $c(t) = (x(t), y(t))$ by $c(t) = (a + x(t), b + y(t))$.

Suppose we have a parametrization $c(t) = (x(t), y(t))$ where $x(t)$ is an even function and $y(t)$ is an odd function, that is, $x(-t) = x(t)$ and $y(-t) = -y(t)$. In this case, $c(-t)$ is the *reflection* of $c(t)$ across the x-axis:

$$c(-t) = (x(-t), y(-t)) = (x(t), -y(t))$$

The curve, therefore, is *symmetric* with respect to the x-axis. We apply this remark in the next example and in Example 7 below.

EXAMPLE 5 Parametrization of an Ellipse Verify that the ellipse with equation $\left(\frac{x}{a}\right)^2 + \left(\frac{y}{b}\right)^2 = 1$ is parametrized by

$$c(t) = (a\cos t,\ b\sin t) \qquad (\text{for } -\pi \le t < \pi)$$

Plot the case $a = 4$, $b = 2$.

Solution To verify that $c(t)$ parametrizes the ellipse, show that the equation of the ellipse is satisfied with $x = a\cos t$, $y = b\sin t$:

$$\left(\frac{x}{a}\right)^2 + \left(\frac{y}{b}\right)^2 = \left(\frac{a\cos t}{a}\right)^2 + \left(\frac{b\sin t}{b}\right)^2 = \cos^2 t + \sin^2 t = 1$$

TABLE 2

t	$x(t) = 4\cos t$	$y(t) = 2\sin t$
0	4	0
$\frac{\pi}{6}$	$2\sqrt{3}$	1
$\frac{\pi}{3}$	2	$\sqrt{3}$
$\frac{\pi}{2}$	0	2
$\frac{2\pi}{3}$	-2	$\sqrt{3}$
$\frac{5\pi}{6}$	$-2\sqrt{3}$	1
$\frac{\pi}{2}$	-4	0

To plot the case $a = 4, b = 2$, we connect the points corresponding to the t-values in Table 2 (see Figure 7). This gives us the top half of the ellipse corresponding to $0 \le t \le \pi$. Then we observe that $x(t) = 4\cos t$ is even and $y(t) = 2\sin t$ is odd. As noted above, this tells us that the bottom half of the ellipse is obtained by symmetry with respect to the x-axis. ■

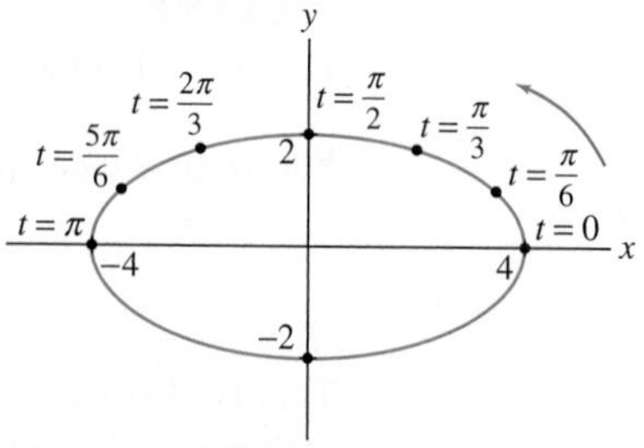

FIGURE 7 Ellipse with parametric equations $x = 4\cos t$, $y = 2\sin t$.

A parametric curve $c(t)$ is also called a **path**. This term emphasizes that $c(t)$ describes not just an underlying curve $\mathcal{C}$, but a particular way of moving along the curve.

CONCEPTUAL INSIGHT The parametric equations for the ellipse in Example 5 illustrate a key difference between the path $c(t)$ and its underlying curve $\mathcal{C}$. The curve $\mathcal{C}$ is an ellipse in the plane, whereas $c(t)$ describes a particular, counterclockwise motion of a particle along the ellipse. If we let t vary from 0 to 4π, then the particle goes around the ellipse twice.

A key feature of parametrizations is that they are not unique. In fact, every curve can be parametrized in infinitely many different ways. For instance, the parabola $y = x^2$ is parametrized not only by (t, t^2) but also by (t^3, t^6), or (t^5, t^{10}), and so on.

■ **EXAMPLE 6 Different Parametrizations of the Same Curve** Describe the motion of a particle moving along each of the following paths.

(a) $c_1(t) = (t^3, t^6)$ **(b)** $c_2(t) = (t^2, t^4)$ **(c)** $c_3(t) = (\cos t, \cos^2 t)$

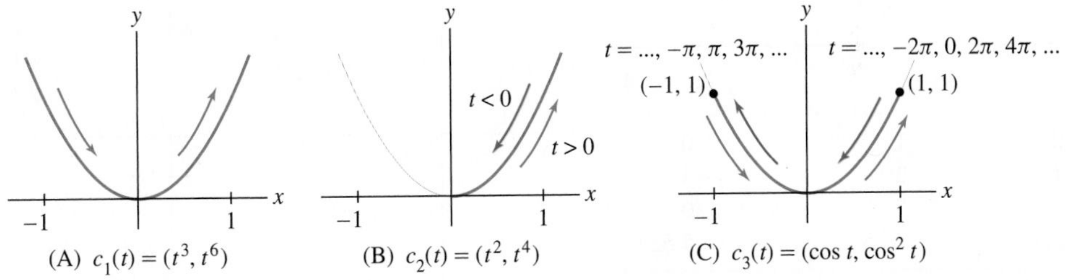

FIGURE 8 Three parametrizations of portions of the parabola.

Solution Each of these parametrizations satisfies $y = x^2$, so all three parametrize portions of the parabola $y = x^2$.

(a) As t varies from $-\infty$ to ∞, the function t^3 also varies from $-\infty$ to ∞. Therefore, $c_1(t) = (t^3, t^6)$ traces the entire parabola $y = x^2$, moving from left to right and passing through each point once [Figure 8(A)].

(b) Since $x = t^2 \geq 0$, the path $c_2(t) = (t^2, t^4)$ traces only the right half of the parabola. The particle comes in toward the origin as t varies from $-\infty$ to 0, and it goes back out to the right as t varies from 0 to ∞ [Figure 8(B)].

(c) As t varies from $-\infty$ and ∞, $\cos t$ oscillates between 1 and -1. Thus a particle following the path $c_3(t) = (\cos t, \cos^2 t)$ oscillates back and forth between the points $(1, 1)$ and $(-1, 1)$ on the parabola. [Figure 8(C)]. ■

■ **EXAMPLE 7 Using Symmetry to Sketch a Loop** Sketch the curve

$$c(t) = (t^2 + 1, t^3 - 4t)$$

Label the points corresponding to $t = 0, \pm 1, \pm 2, \pm 2.5$.

Solution

***Step 1.* Use symmetry.**

Observe that $x(t) = t^2 + 1$ is an even function and that $y(t) = t^3 - 4t$ is an odd function. As noted before Example 5, this tells us that $c(t)$ is symmetric with respect to the x-axis. Therefore, we will plot the curve for $t \geq 0$ and reflect across the x-axis to obtain the part for $t \leq 0$.

Step 2. **Analyze $x(t)$, $y(t)$ as functions of t.**
We have $x(t) = t^2 + 1$ and $y(t) = t^3 - 4t$. The x-coordinate $x(t) = t^2 + 1$ increases to ∞ as $t \to \infty$. To analyze the y-coordinate, we graph $y(t) = t^3 - 4t = t(t-2)(t+2)$ as a function of t (*not* as a function of x). Since $y(t)$ is the height above the x-axis, Figure 9(A) shows that

$$y(t) < 0 \quad \text{for} \quad 0 < t < 2, \quad \Rightarrow \quad \text{curve below } x\text{-axis}$$

$$y(t) > 0 \quad \text{for} \quad t > 2, \quad \Rightarrow \quad \text{curve above } x\text{-axis}$$

So the curve starts at $c(0) = (1, 0)$, dips below the x-axis and returns to the x-axis at $t = 2$. Both $x(t)$ and $y(t)$ tend to ∞ as $t \to \infty$. The curve is concave up because $y(t)$ increases more rapidly than $x(t)$.

Step 3. **Plot points and join by an arc.**
The points $c(0)$, $c(1)$, $c(2)$, $c(2.5)$ tabulated in Table 3 are plotted and joined by an arc to create the sketch for $t \geq 0$ as in Figure 9(B). The sketch is completed by reflecting across the x-axis as in Figure 9(C). ■

TABLE 3

t	$x = t^2 + 1$	$y = t^3 - 4t$
0	1	0
1	2	−3
2	5	0
2.5	7.25	5.625

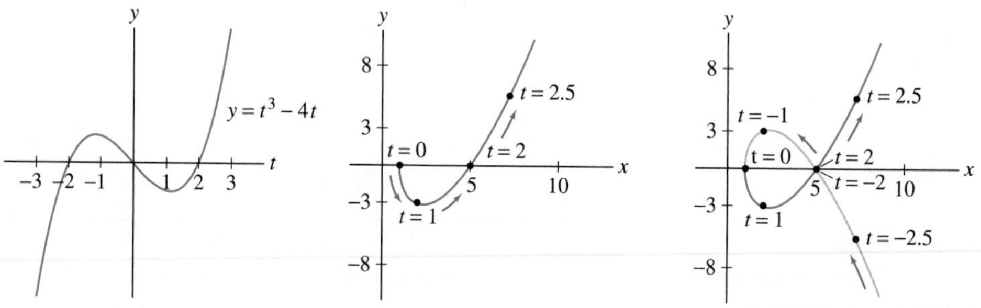

(A) Graph of y-coordinate $y(t) = t^3 - 4t$

(B) Graph for $t \geq 0$

(C) Complete sketch using symmetry.

FIGURE 9 The curve $c(t) = (t^2 + 1, t^3 - 4t)$.

A **cycloid** is a curve traced by a point on the circumference of a rolling wheel as in Figure 10. Cycloids are famous for their "brachistochrone property" (see the marginal note below).

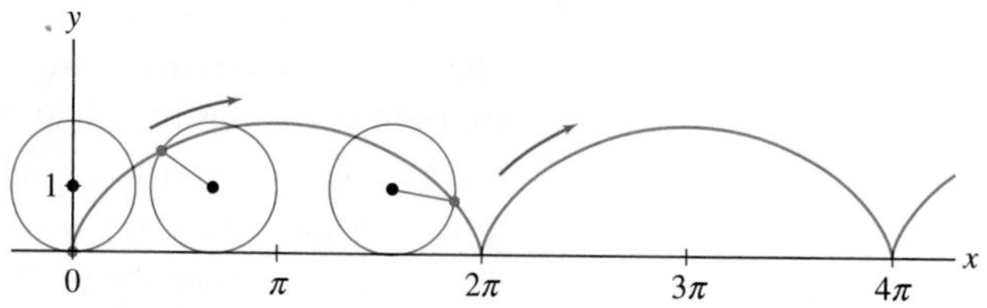

FIGURE 10 A cycloid.

A stellar cast of mathematicians (including Galileo, Pascal, Newton, Leibniz, Huygens, and Bernoulli) studied the cycloid and discovered many of its remarkable properties. A slide designed so that an object sliding down (without friction) reaches the bottom in the least time must have the shape of an inverted cycloid. This is the brachistochrone property, a term derived from the Greek brachistos, *"shortest," and* chronos, *"time."*

■ **EXAMPLE 8** **Parametrizing the Cycloid** Find parametric equations for the cycloid generated by a point P on the unit circle.

Solution The point P is located at the origin at $t = 0$. At time t, the circle has rolled t radians along the x axis and the center C of the circle then has coordinates $(t, 1)$ as in Figure 11(A). Figure 11(B) shows that we get from C to P by moving down $\cos t$ units and to the left $\sin t$ units, giving us the parametric equations

$$x(t) = t - \sin t, \qquad y(t) = 1 - \cos t \qquad \boxed{5}$$

■

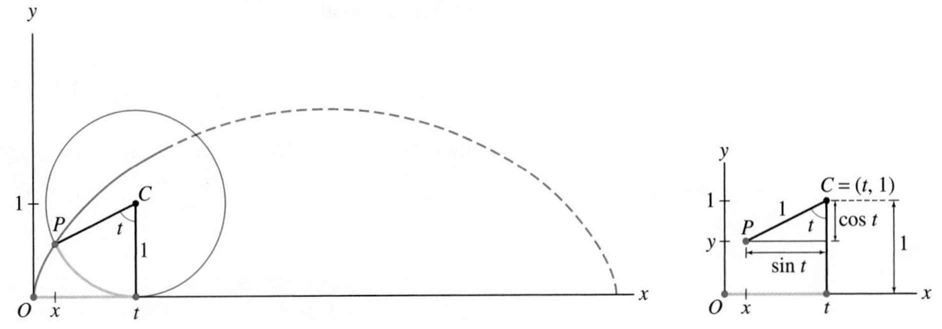

(A) Position of P at time t

(B) P has coordinates $x = t - \sin t$, $y = 1 - \cos t$

FIGURE 11

The argument in Example 8 shows in a similar fashion that the cycloid generated by a circle of radius R has parametric equations

$$x = Rt - R\sin t, \qquad y = R - R\cos t \tag{6}$$

Next, we address the problem of finding tangent lines to parametric curves. The slope of the tangent line is the derivative dy/dx, but we have to use the Chain Rule to compute it because y is not given explicitly as a function of x. Write $x = f(t)$, $y = g(t)$. Then, by the Chain Rule in Leibniz notation,

$$g'(t) = \frac{dy}{dt} = \frac{dy}{dx}\frac{dx}{dt} = \frac{dy}{dx}f'(t)$$

NOTATION In this section, we write $f'(t), x'(t), y'(t)$, and so on to denote the derivative with respect to t.

If $f'(t) \neq 0$, we can divide by $f'(t)$ to obtain

$$\frac{dy}{dx} = \frac{g'(t)}{f'(t)}$$

This calculation is valid if $f(t)$ and $g(t)$ are differentiable, $f'(t)$ is continuous, and $f'(t) \neq 0$. In this case, the inverse $t = f^{-1}(x)$ exists, and the composite $y = g(f^{-1}(x))$ is a differentiable function of x.

CAUTION Do not confuse dy/dx with the derivatives dx/dt and dy/dt, which are derivatives with respect to the parameter t. Only dy/dx is the slope of the tangent line.

THEOREM 2 Slope of the Tangent Line Let $c(t) = (x(t), y(t))$, where $x(t)$ and $y(t)$ are differentiable. Assume that $x'(t)$ is continuous and $x'(t) \neq 0$. Then

$$\frac{dy}{dx} = \frac{dy/dt}{dx/dt} = \frac{y'(t)}{x'(t)} \tag{7}$$

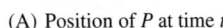

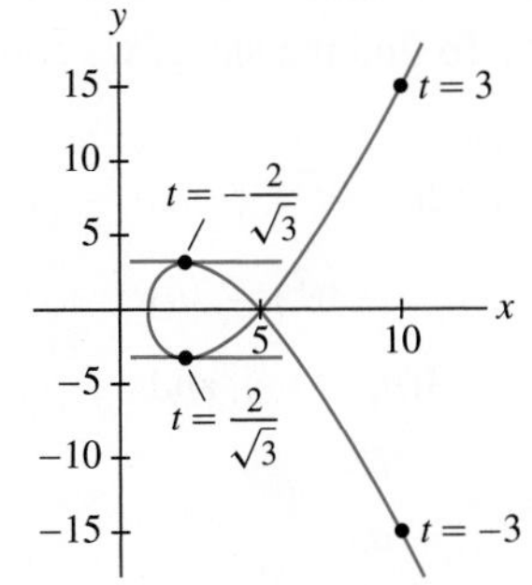

FIGURE 12 Horizontal tangent lines on $c(t) = (t^2 + 1, t^3 - 4t)$.

■ **EXAMPLE 9** Let $c(t) = (t^2 + 1, t^3 - 4t)$. Find:

(a) An equation of the tangent line at $t = 3$

(b) The points where the tangent is horizontal (Figure 12).

Solution We have

$$\frac{dy}{dx} = \frac{y'(t)}{x'(t)} = \frac{(t^3 - 4t)'}{(t^2 + 1)'} = \frac{3t^2 - 4}{2t}$$

(a) The slope at $t = 3$ is

$$\frac{dy}{dx} = \left.\frac{3t^2 - 4}{2t}\right|_{t=3} = \frac{3(3)^2 - 4}{2(3)} = \frac{23}{6}$$

Since $c(3) = (10, 15)$, the equation of the tangent line in point-slope form is

$$y - 15 = \frac{23}{6}(x - 10)$$

(b) The slope dy/dx is zero if $y'(t) = 0$ and $x'(t) \neq 0$. We have $y'(t) = 3t^2 - 4 = 0$ for $t = \pm 2/\sqrt{3}$ (and $x'(t) = 2t \neq 0$ for these values of t). Therefore, the tangent line is horizontal at the points

$$c\left(-\frac{2}{\sqrt{3}}\right) = \left(\frac{7}{3}, \frac{16}{3\sqrt{3}}\right), \qquad c\left(\frac{2}{\sqrt{3}}\right) = \left(\frac{7}{3}, -\frac{16}{3\sqrt{3}}\right)$$ ■

Bézier curves were invented in the 1960s by the French engineer Pierre Bézier (1910–1999), who worked for the Renault car company. They are based on the properties of Bernstein polynomials, introduced 50 years earlier by the Russian mathematician Sergei Bernstein to study the approximation of continuous functions by polynomials. Today, Bézier curves are used in standard graphics programs, such as Adobe Illustrator™ and Corel Draw™, and in the construction and storage of computer fonts such as TrueType™ and PostScript™ fonts.

Parametric curves are widely used in the field of computer graphics. A particularly important class of curves are **Bézier curves**, which we discuss here briefly in the cubic case. Given four "control points" (Figure 13):

$$P_0 = (a_0, b_0), \qquad P_1 = (a_1, b_1), \qquad P_2 = (a_2, b_2), \qquad P_3 = (a_3, b_3)$$

the Bézier curve $c(t) = (x(t), y(t))$ is defined for $0 \le t \le 1$ by

$$x(t) = a_0(1-t)^3 + 3a_1t(1-t)^2 + 3a_2t^2(1-t) + a_3t^3 \qquad (8)$$

$$y(t) = b_0(1-t)^3 + 3b_1t(1-t)^2 + 3b_2t^2(1-t) + b_3t^3 \qquad (9)$$

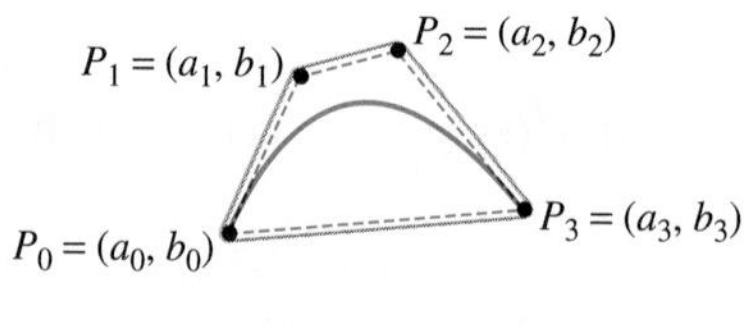

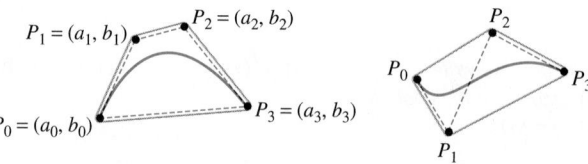

FIGURE 13 Cubic Bézier curves specified by four control points.

Note that $c(0) = (a_0, b_0)$ and $c(1) = (a_3, b_3)$, so the Bézier curve begins at P_0 and ends at P_3 (Figure 13). It can also be shown that the Bézier curve is contained within the quadrilateral (shown in blue) with vertices P_0, P_1, P_2, P_3. However, $c(t)$ does not pass through P_1 and P_2. Instead, these intermediate control points determine the slopes of the tangent lines at P_0 and P_3, as we show in the next example (also, see Exercises 65–68).

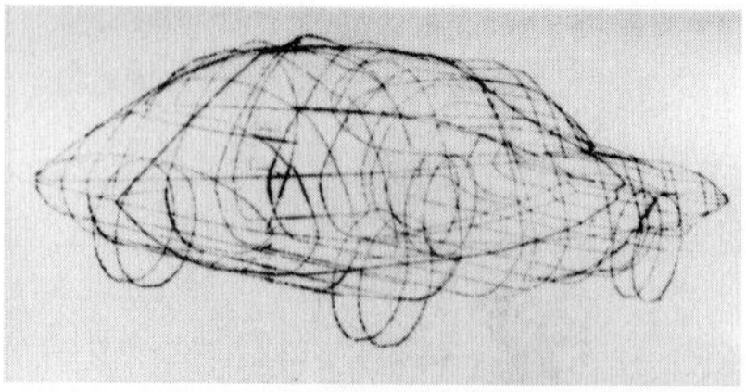

Hand sketch made in 1964 by Pierre Bézier for the French automobile manufacturer Renault.

■ **EXAMPLE 10** Show that the Bézier curve is tangent to segment $\overline{P_0P_1}$ at P_0.

Solution The Bézier curve passes through P_0 at $t = 0$, so we must show that the slope of the tangent line at $t = 0$ is equal to the slope of $\overline{P_0P_1}$. To find the slope, we compute the derivatives:

$$x'(t) = -3a_0(1-t)^2 + 3a_1(1 - 4t + 3t^2) + a_2(2t - 3t^2) + 3a_3t^2$$

$$y'(t) = -3b_0(1-t)^2 + 3b_1(1 - 4t + 3t^2) + b_2(2t - 3t^2) + 3b_3t^2$$

Evaluating at $t = 0$, we obtain $x'(0) = 3(a_1 - a_0)$, $y'(0) = 3(b_1 - b_0)$, and

$$\left.\frac{dy}{dx}\right|_{t=0} = \frac{y'(0)}{x'(0)} = \frac{3(b_1 - b_0)}{3(a_1 - a_0)} = \frac{b_1 - b_0}{a_1 - a_0}$$

This is equal to the slope of the line through $P_0 = (a_0, b_0)$ and $P_1 = (a_1, b_1)$ as claimed (provided that $a_1 \neq a_0$). ■

12.1 SUMMARY

- A *parametric curve* $c(t) = (f(t), g(t))$ describes the path of a particle moving along a curve as a function of the parameter t.
- Parametrizations are not unique: Every curve C can be parametrized in infinitely many ways. Furthermore, the path $c(t)$ may traverse all or part of C more than once.
- Slope of the tangent line at $c(t)$:

$$\frac{dy}{dx} = \frac{dy/dt}{dx/dt} = \frac{y'(t)}{x'(t)} \qquad \text{(valid if } x'(t) \neq 0\text{)}$$

- Do not confuse the slope of the tangent line dy/dx with the derivatives dy/dt and dx/dt, with respect to t.
- Standard parametrizations:
  - Line of slope $m = s/r$ through $P = (a, b)$: $c(t) = (a + rt, b + st)$.
  - Circle of radius R centered at $P = (a, b)$: $c(t) = (a + R\cos t, b + R\sin t)$.
  - Cycloid generated by a circle of radius R: $c(t) = (R(t - \sin t), R(1 - \cos t))$.

12.1 EXERCISES

Preliminary Questions

1. Describe the shape of the curve $x = 3\cos t$, $y = 3\sin t$.

2. How does $x = 4 + 3\cos t$, $y = 5 + 3\sin t$ differ from the curve in the previous question?

3. What is the maximum height of a particle whose path has parametric equations $x = t^9$, $y = 4 - t^2$?

4. Can the parametric curve $(t, \sin t)$ be represented as a graph $y = f(x)$? What about $(\sin t, t)$?

5. Match the derivatives with a verbal description:

(a) $\dfrac{dx}{dt}$ **(b)** $\dfrac{dy}{dt}$ **(c)** $\dfrac{dy}{dx}$

(i) Slope of the tangent line to the curve

(ii) Vertical rate of change with respect to time

(iii) Horizontal rate of change with respect to time

Exercises

1. Find the coordinates at times $t = 0, 2, 4$ of a particle following the path $x = 1 + t^3$, $y = 9 - 3t^2$.

2. Find the coordinates at $t = 0, \frac{\pi}{4}, \pi$ of a particle moving along the path $c(t) = (\cos 2t, \sin^2 t)$.

3. Show that the path traced by the bullet in Example 3 is a parabola by eliminating the parameter.

4. Use the table of values to sketch the parametric curve $(x(t), y(t))$, indicating the direction of motion.

t	−3	−2	−1	0	1	2	3
x	−15	0	3	0	−3	0	15
y	5	0	−3	−4	−3	0	5

5. Graph the parametric curves. Include arrows indicating the direction of motion.

(a) (t, t), $-\infty < t < \infty$

(b) $(\sin t, \sin t)$, $0 \le t \le 2\pi$

(c) (e^t, e^t), $-\infty < t < \infty$

(d) (t^3, t^3), $-1 \le t \le 1$

6. Give two different parametrizations of the line through $(4, 1)$ with slope 2.

In Exercises 7–14, express in the form $y = f(x)$ by eliminating the parameter.

7. $x = t + 3$, $y = 4t$

8. $x = t^{-1}$, $y = t^{-2}$

9. $x = t$, $y = \tan^{-1}(t^3 + e^t)$

10. $x = t^2$, $y = t^3 + 1$

11. $x = e^{-2t}$, $y = 6e^{4t}$

12. $x = 1 + t^{-1}$, $y = t^2$

13. $x = \ln t$, $y = 2 - t$

14. $x = \cos t$, $y = \tan t$

In Exercises 15–18, graph the curve and draw an arrow specifying the direction corresponding to motion.

15. $x = \frac{1}{2}t$, $y = 2t^2$

16. $x = 2 + 4t$, $y = 3 + 2t$

17. $x = \pi t$, $y = \sin t$

18. $x = t^2$, $y = t^3$

19. Match the parametrizations (a)–(d) below with their plots in Figure 14, and draw an arrow indicating the direction of motion.

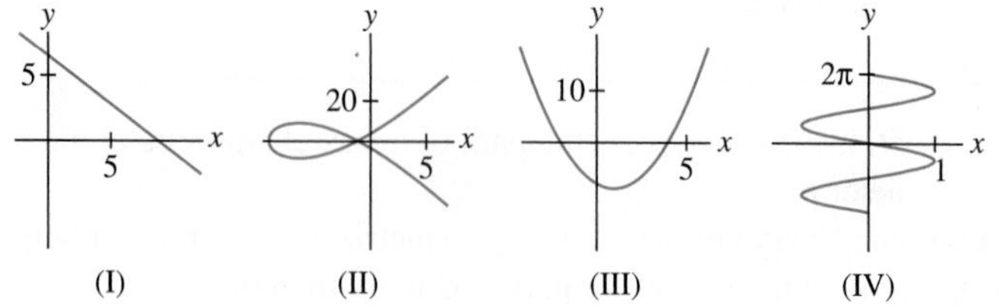

FIGURE 14

(a) $c(t) = (\sin t, -t)$

(b) $c(t) = (t^2 - 9, 8t - t^3)$

(c) $c(t) = (1 - t, t^2 - 9)$

(d) $c(t) = (4t + 2, 5 - 3t)$

20. A particle follows the trajectory

$$x(t) = \frac{1}{4}t^3 + 2t, \qquad y(t) = 20t - t^2$$

with t in seconds and distance in centimeters.

(a) What is the particle's maximum height?

(b) When does the particle hit the ground and how far from the origin does it land?

21. Find an interval of t-values such that $c(t) = (\cos t, \sin t)$ traces the lower half of the unit circle.

22. Find an interval of t-values such that $c(t) = (2t + 1, 4t - 5)$ parametrizes the segment from $(0, -7)$ to $(7, 7)$.

In Exercises 23–38, find parametric equations for the given curve.

23. $y = 9 - 4x$

24. $y = 8x^2 - 3x$

25. $4x - y^2 = 5$

26. $x^2 + y^2 = 49$

27. $(x + 9)^2 + (y - 4)^2 = 49$

28. $\left(\frac{x}{5}\right)^2 + \left(\frac{y}{12}\right)^2 = 1$

29. Line of slope 8 through $(-4, 9)$

30. Line through $(2, 5)$ perpendicular to $y = 3x$

31. Line through $(3, 1)$ and $(-5, 4)$

32. Line through $\left(\frac{1}{3}, \frac{1}{6}\right)$ and $\left(-\frac{7}{6}, \frac{5}{3}\right)$

33. Segment joining $(1, 1)$ and $(2, 3)$

34. Segment joining $(-3, 0)$ and $(0, 4)$

35. Circle of radius 4 with center $(3, 9)$

36. Ellipse of Exercise 28, with its center translated to $(7, 4)$

37. $y = x^2$, translated so that the minimum occurs at $(-4, -8)$

38. $y = \cos x$ translated so that a maximum occurs at $(3, 5)$

In Exercises 39–42, find a parametrization $c(t)$ of the curve satisfying the given condition.

39. $y = 3x - 4, \quad c(0) = (2, 2)$

40. $y = 3x - 4, \quad c(3) = (2, 2)$

41. $y = x^2, \quad c(0) = (3, 9)$

42. $x^2 + y^2 = 4, \quad c(0) = \left(\frac{1}{2}, \frac{\sqrt{3}}{2}\right)$

43. Describe $c(t) = (\sec t, \tan t)$ for $0 \le t < \frac{\pi}{2}$ in the form $y = f(x)$. Specify the domain of x.

44. Find a parametrization of the right branch $(x > 0)$ of the hyperbola

$$\left(\frac{x}{a}\right)^2 - \left(\frac{y}{b}\right)^2 = 1$$

using the functions $\cosh t$ and $\sinh t$. How can you parametrize the branch $x < 0$?

45. The graphs of $x(t)$ and $y(t)$ as functions of t are shown in Figure 15(A). Which of (I)–(III) is the plot of $c(t) = (x(t), y(t))$? Explain.

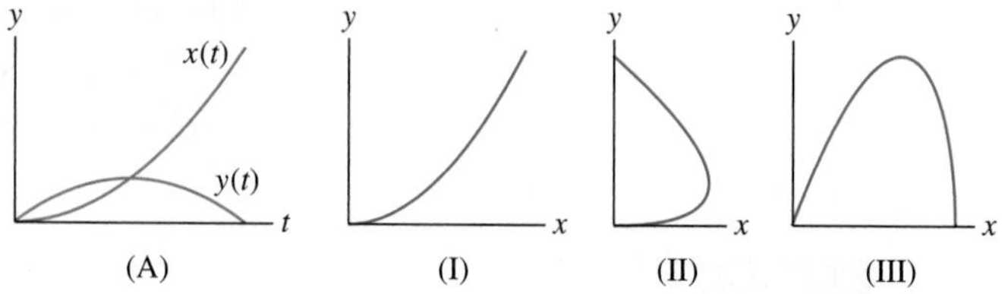

FIGURE 15

46. Which graph, (I) or (II), is the graph of $x(t)$ and which is the graph of $y(t)$ for the parametric curve in Figure 16(A)?

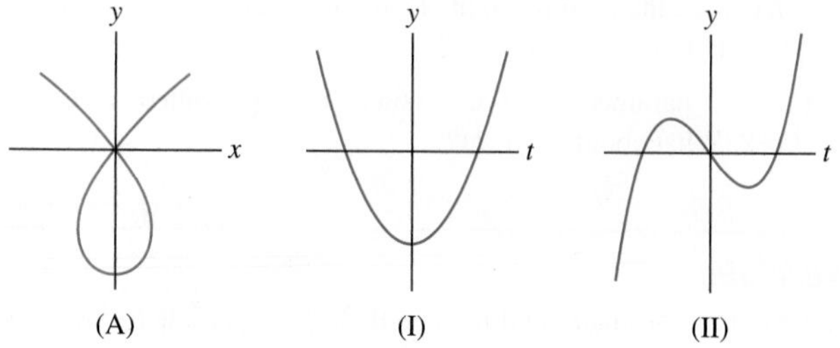

FIGURE 16

47. Sketch $c(t) = (t^3 - 4t, t^2)$ following the steps in Example 7.

48. Sketch $c(t) = (t^2 - 4t, 9 - t^2)$ for $-4 \le t \le 10$.

In Exercises 49–52, use Eq. (7) to find dy/dx at the given point.

49. $(t^3, t^2 - 1), \quad t = -4$

50. $(2t + 9, 7t - 9), \quad t = 1$

51. $(s^{-1} - 3s, s^3), \quad s = -1$

52. $(\sin 2\theta, \cos 3\theta), \quad \theta = \frac{\pi}{6}$

In Exercises 53–56, find an equation $y = f(x)$ for the parametric curve and compute dy/dx in two ways: using Eq. (7) and by differentiating $f(x)$.

53. $c(t) = (2t + 1, 1 - 9t)$

54. $c(t) = \left(\frac{1}{2}t, \frac{1}{4}t^2 - t\right)$

55. $x = s^3, \quad y = s^6 + s^{-3}$

56. $x = \cos\theta, \quad y = \cos\theta + \sin^2\theta$

57. Find the points on the curve $c(t) = (3t^2 - 2t, t^3 - 6t)$ where the tangent line has slope 3.

58. Find the equation of the tangent line to the cycloid generated by a circle of radius 4 at $t = \frac{\pi}{2}$.

In Exercises 59–62, let $c(t) = (t^2 - 9, t^2 - 8t)$ *(see Figure 17).*

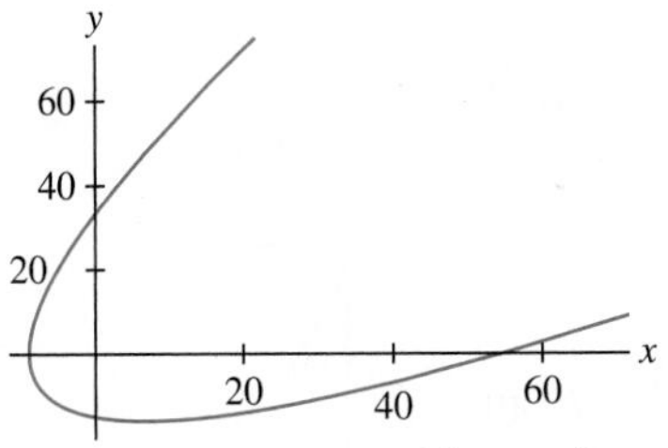

FIGURE 17 Plot of $c(t) = (t^2 - 9, t^2 - 8t)$.

59. Draw an arrow indicating the direction of motion, and determine the interval of t-values corresponding to the portion of the curve in each of the four quadrants.

60. Find the equation of the tangent line at $t = 4$.

61. Find the points where the tangent has slope $\frac{1}{2}$.

62. Find the points where the tangent is horizontal or vertical.

63. Let A and B be the points where the ray of angle θ intersects the two concentric circles of radii $r < R$ centered at the origin (Figure 18). Let P be the point of intersection of the horizontal line through A and the vertical line through B. Express the coordinates of P as a function of θ and describe the curve traced by P for $0 \le \theta \le 2\pi$.

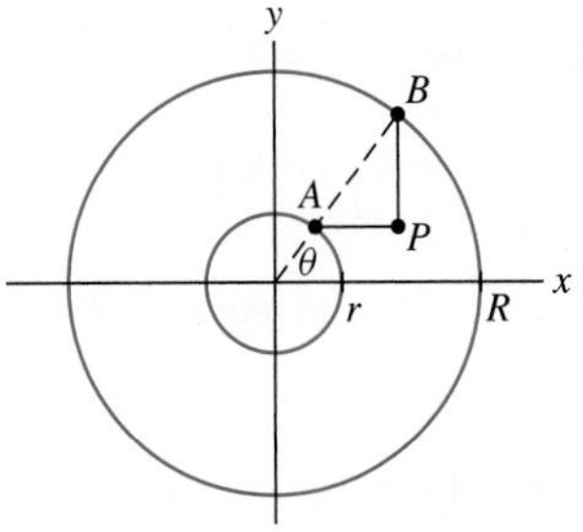

FIGURE 18

64. A 10-ft ladder slides down a wall as its bottom B is pulled away from the wall (Figure 19). Using the angle θ as parameter, find the parametric equations for the path followed by (a) the top of the ladder A, (b) the bottom of the ladder B, and (c) the point P located 4 ft from the top of the ladder. Show that P describes an ellipse.

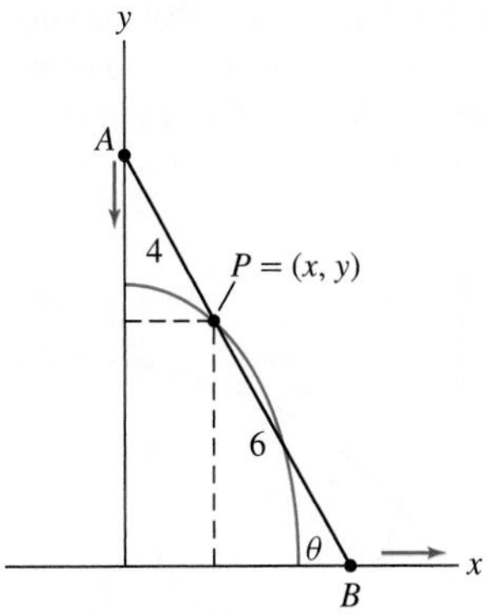

FIGURE 19

In Exercises 65–68, refer to the Bézier curve defined by Eqs. (8) and (9).

65. Show that the Bézier curve with control points

$$P_0 = (1, 4), \quad P_1 = (3, 12), \quad P_2 = (6, 15), \quad P_3 = (7, 4)$$

has parametrization

$$c(t) = (1 + 6t + 3t^2 - 3t^3, 4 + 24t - 15t^2 - 9t^3)$$

Verify that the slope at $t = 0$ is equal to the slope of the segment $\overline{P_0P_1}$.

66. Find an equation of the tangent line to the Bézier curve in Exercise 65 at $t = \frac{1}{3}$.

67. CAS Find and plot the Bézier curve $c(t)$ passing through the control points

$$P_0 = (3, 2), \quad P_1 = (0, 2), \quad P_2 = (5, 4), \quad P_3 = (2, 4)$$

68. Show that a cubic Bézier curve is tangent to the segment $\overline{P_2P_3}$ at P_3.

69. A bullet fired from a gun follows the trajectory

$$x = at, \qquad y = bt - 16t^2 \quad (a, b > 0)$$

Show that the bullet leaves the gun at an angle $\theta = \tan^{-1}\left(\frac{b}{a}\right)$ and lands at a distance $ab/16$ from the origin.

70. CAS Plot $c(t) = (t^3 - 4t, t^4 - 12t^2 + 48)$ for $-3 \le t \le 3$. Find the points where the tangent line is horizontal or vertical.

71. CAS Plot the astroid $x = \cos^3\theta$, $y = \sin^3\theta$ and find the equation of the tangent line at $\theta = \frac{\pi}{3}$.

72. Find the equation of the tangent line at $t = \frac{\pi}{4}$ to the cycloid generated by the unit circle with parametric equation (5).

73. Find the points with horizontal tangent line on the cycloid with parametric equation (5).

74. Property of the Cycloid Prove that the tangent line at a point P on the cycloid always passes through the top point on the rolling circle as indicated in Figure 20. Assume the generating circle of the cycloid has radius 1.

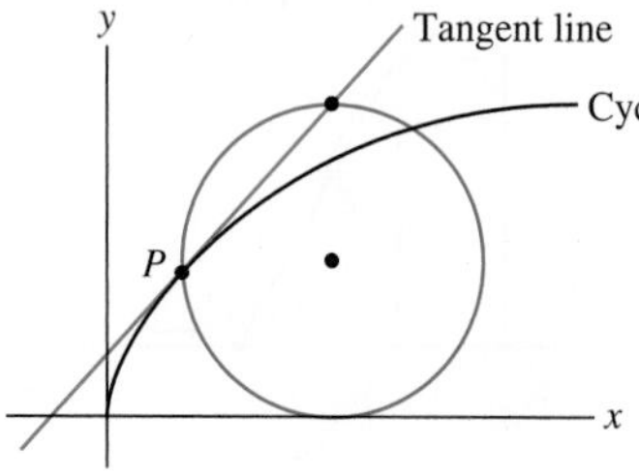

FIGURE 20

75. A *curtate cycloid* (Figure 21) is the curve traced by a point at a distance h from the center of a circle of radius R rolling along the x-axis where $h < R$. Show that this curve has parametric equations $x = Rt - h\sin t$, $y = R - h\cos t$.

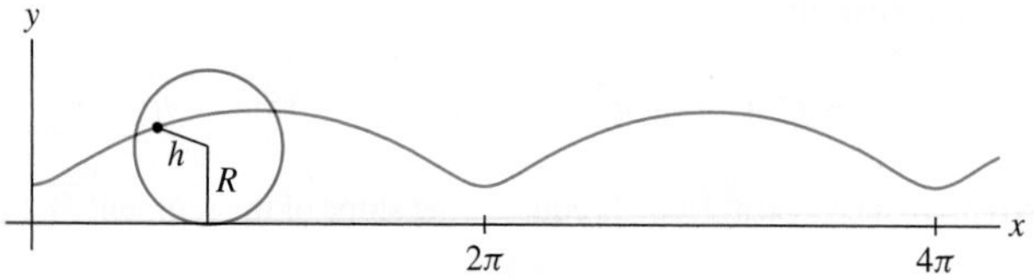

FIGURE 21 Curtate cycloid.

76. CAS Use a computer algebra system to explore what happens when $h > R$ in the parametric equations of Exercise 75. Describe the result.

77. Show that the line of slope t through $(-1, 0)$ intersects the unit circle in the point with coordinates

$$x = \frac{1-t^2}{t^2+1}, \qquad y = \frac{2t}{t^2+1} \qquad \boxed{10}$$

Conclude that these equations parametrize the unit circle with the point $(-1, 0)$ excluded (Figure 22). Show further that $t = y/(x+1)$.

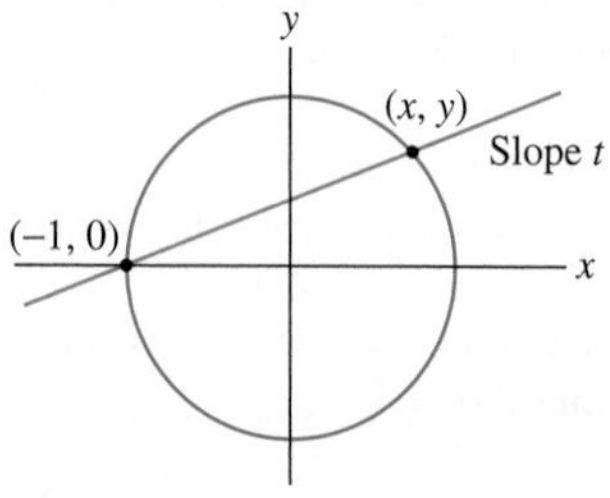

FIGURE 22 Unit circle.

78. The **folium of Descartes** is the curve with equation $x^3 + y^3 = 3axy$, where $a \neq 0$ is a constant (Figure 23).

(a) Show that the line $y = tx$ intersects the folium at the origin and at one other point P for all $t \neq -1, 0$. Express the coordinates of P in terms of t to obtain a parametrization of the folium. Indicate the direction of the parametrization on the graph.

(b) Describe the interval of t-values parametrizing the parts of the curve in quadrants I, II, and IV. Note that $t = -1$ is a point of discontinuity of the parametrization.

(c) Calculate dy/dx as a function of t and find the points with horizontal or vertical tangent.

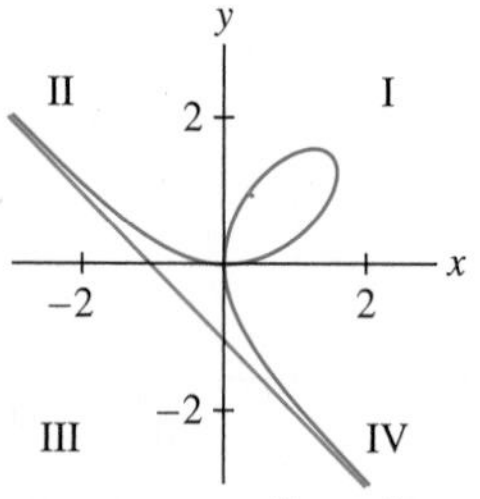

FIGURE 23 Folium $x^3 + y^3 = 3axy$.

79. Use the results of Exercise 78 to show that the asymptote of the folium is the line $x + y = -a$. *Hint:* Show that $\lim_{t \to -1} (x + y) = -a$.

80. Find a parametrization of $x^{2n+1} + y^{2n+1} = ax^n y^n$, where a and n are constants.

81. Second Derivative for a Parametrized Curve Given a parametrized curve $c(t) = (x(t), y(t))$, show that

$$\frac{d}{dt}\left(\frac{dy}{dx}\right) = \frac{x'(t)y''(t) - y'(t)x''(t)}{x'(t)^2}$$

Use this to prove the formula

$$\boxed{\frac{d^2y}{dx^2} = \frac{x'(t)y''(t) - y'(t)x''(t)}{x'(t)^3}} \qquad \boxed{11}$$

82. The second derivative of $y = x^2$ is $dy^2/d^2x = 2$. Verify that Eq. (11) applied to $c(t) = (t, t^2)$ yields $dy^2/d^2x = 2$. In fact, any parametrization may be used. Check that $c(t) = (t^3, t^6)$ and $c(t) = (\tan t, \tan^2 t)$ also yield $dy^2/d^2x = 2$.

In Exercises 83–86, use Eq. (11) to find d^2y/dx^2.

83. $x = t^3 + t^2$, $y = 7t^2 - 4$, $t = 2$

84. $x = s^{-1} + s$, $y = 4 - s^{-2}$, $s = 1$

85. $x = 8t + 9$, $y = 1 - 4t$, $t = -3$

86. $x = \cos\theta$, $y = \sin\theta$, $\theta = \frac{\pi}{4}$

87. Use Eq. (11) to find the t-intervals on which $c(t) = (t^2, t^3 - 4t)$ is concave up.

88. Use Eq. (11) to find the t-intervals on which $c(t) = (t^2, t^4 - 4t)$ is concave up.

89. Area Under a Parametrized Curve Let $c(t) = (x(t), y(t))$, where $y(t) > 0$ and $x'(t) > 0$ (Figure 24). Show that the area A under $c(t)$ for $t_0 \le t \le t_1$ is

$$A = \int_{t_0}^{t_1} y(t)x'(t)\,dt \qquad \boxed{12}$$

Hint: Because it is increasing, the function $x(t)$ has an inverse $t = g(x)$ and $c(t)$ is the graph of $y = y(g(x))$. Apply the change-of-variables formula to $A = \int_{x(t_0)}^{x(t_1)} y(g(x))\,dx$.

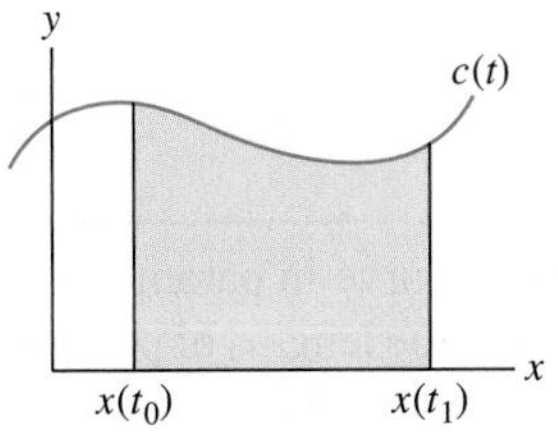

FIGURE 24

90. Calculate the area under $y = x^2$ over $[0, 1]$ using Eq. (12) with the parametrizations (t^3, t^6) and (t^2, t^4).

91. What does Eq. (12) say if $c(t) = (t, f(t))$?

92. Sketch the graph of $c(t) = (\ln t, 2 - t)$ for $1 \le t \le 2$ and compute the area under the graph using Eq. (12).

93. Galileo tried unsuccessfully to find the area under a cycloid. Around 1630, Gilles de Roberval proved that the area under one arch of the cycloid $c(t) = (Rt - R\sin t, R - R\cos t)$ generated by a circle of radius R is equal to three times the area of the circle (Figure 25). Verify Roberval's result using Eq. (12).

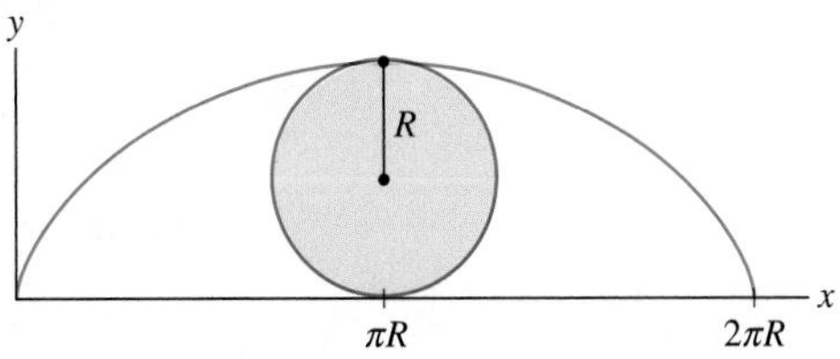

FIGURE 25 The area of one arch of the cycloid equals three times the area of the generating circle.

Further Insights and Challenges

94. Prove the following generalization of Exercise 93: For all $t > 0$, the area of the cycloidal sector OPC is equal to three times the area of the circular segment cut by the chord PC in Figure 26.

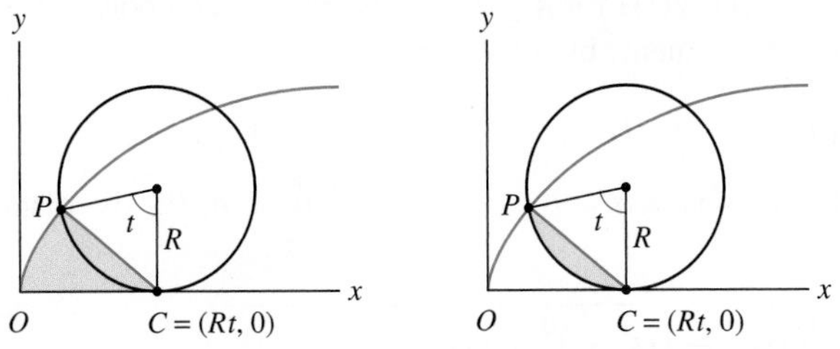

(A) Cycloidal sector OPC (B) Circular segment cut by the chord PC

FIGURE 26

95. Derive the formula for the slope of the tangent line to a parametric curve $c(t) = (x(t), y(t))$ using a method different from that presented in the text. Assume that $x'(t_0)$ and $y'(t_0)$ exist and that $x'(t_0) \ne 0$. Show that

$$\lim_{h\to 0} \frac{y(t_0 + h) - y(t_0)}{x(t_0 + h) - x(t_0)} = \frac{y'(t_0)}{x'(t_0)}$$

Then explain why this limit is equal to the slope dy/dx. Draw a diagram showing that the ratio in the limit is the slope of a secant line.

96. Verify that the **tractrix** curve $(\ell > 0)$

$$c(t) = \left(t - \ell \tanh \frac{t}{\ell}, \ell\,\text{sech}\,\frac{t}{\ell}\right)$$

has the following property: For all t, the segment from $c(t)$ to $(t, 0)$ is tangent to the curve and has length ℓ (Figure 27).

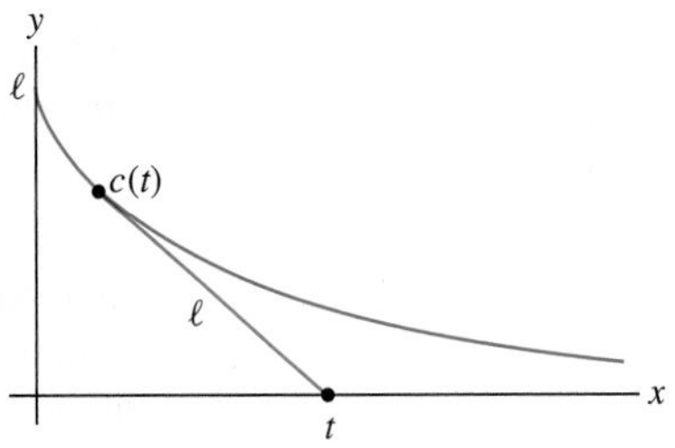

FIGURE 27 The tractrix $c(t) = \left(t - \ell \tanh \frac{t}{\ell}, \ell\,\text{sech}\,\frac{t}{\ell}\right)$.

97. In Exercise 54 of Section 9.1, we described the tractrix by the differential equation

$$\frac{dy}{dx} = -\frac{y}{\sqrt{\ell^2 - y^2}}$$

Show that the curve $c(t)$ identified as the tractrix in Exercise 96 satisfies this differential equation. Note that the derivative on the left is taken with respect to x, not t.

In Exercises 98 and 99, refer to Figure 28.

98. In the parametrization $c(t) = (a\cos t, b\sin t)$ of an ellipse, t is *not* an angular parameter unless $a = b$ (in which case the ellipse is a circle). However, t can be interpreted in terms of area: Show that if $c(t) = (x, y)$, then $t = (2/ab)A$, where A is the area of the shaded region in Figure 28. *Hint:* Use Eq. (12).

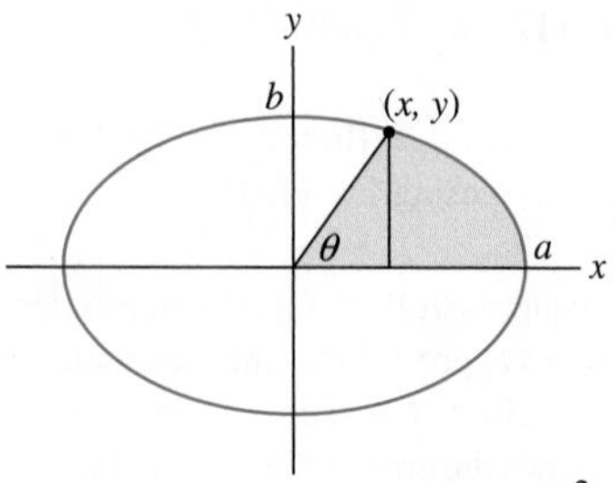

FIGURE 28 The parameter θ on the ellipse $\left(\frac{x}{a}\right)^2 + \left(\frac{y}{b}\right)^2 = 1$.

99. Show that the parametrization of the ellipse by the angle θ is

$$x = \frac{ab\cos\theta}{\sqrt{a^2\sin^2\theta + b^2\cos^2\theta}}$$

$$y = \frac{ab\sin\theta}{\sqrt{a^2\sin^2\theta + b^2\cos^2\theta}}$$

12.2 Arc Length and Speed

We now derive a formula for the arc length of a curve in parametric form. Recall that in Section 9.1, arc length was defined as the limit of the lengths of polygonal approximations (Figure 1).

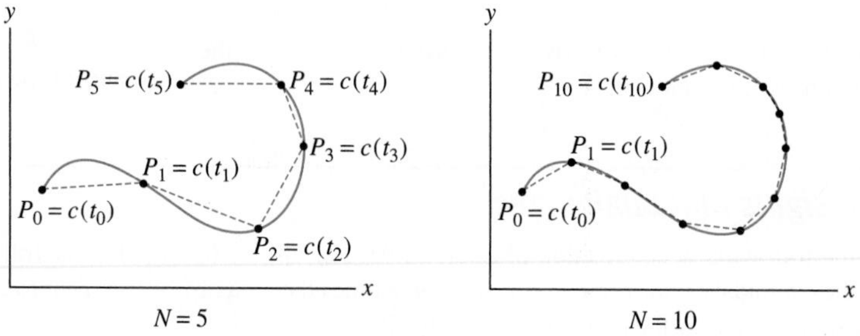

FIGURE 1 Polygonal approximations for $N = 5$ and $N = 10$.

Given a parametrization $c(t) = (x(t), y(t))$ for $a \le t \le b$, we construct a polygonal approximation L consisting of the N segments by joining points

$$P_0 = c(t_0), \quad P_1 = c(t_1), \quad \ldots, \quad P_N = c(t_N)$$

corresponding to a choice of values $t_0 = a < t_1 < t_2 < \cdots < t_N = b$. By the distance formula,

$$P_{i-1}P_i = \sqrt{\left(x(t_i) - x(t_{i-1})\right)^2 + \left(y(t_i) - y(t_{i-1})\right)^2} \qquad \boxed{1}$$

Now assume that $x(t)$ and $y(t)$ are differentiable. According to the Mean Value Theorem, there are values t_i^* and t_i^{**} in the interval $[t_{i-1}, t_i]$ such that

$$x(t_i) - x(t_{i-1}) = x'(t_i^*)\Delta t_i, \qquad y(t_i) - y(t_{i-1}) = y'(t_i^{**})\Delta t_i$$

where $\Delta t_i = t_i - t_{i-1}$, and therefore,

$$P_{i-1}P_i = \sqrt{x'(t_i^*)^2\Delta t_i^2 + y'(t_i^{**})^2\Delta t_i^2} = \sqrt{x'(t_i^*)^2 + y'(t_i^{**})^2}\,\Delta t_i$$

The length of the polygonal approximation L is equal to the sum

$$\sum_{i=1}^{N} P_{i-1}P_i = \sum_{i=1}^{N} \sqrt{x'(t_i^*)^2 + y'(t_i^{**})^2}\,\Delta t_i \qquad \boxed{2}$$

This is *nearly* a Riemann sum for the function $\sqrt{x'(t)^2 + y'(t)^2}$. It would be a true Riemann sum if the intermediate values t_i^* and t_i^{**} were equal. Although they are not necessarily equal, it can be shown (and we will take for granted) that if $x'(t)$ and $y'(t)$ are continuous, then the sum in Eq. (2) still approaches the integral as the widths Δt_i tend to 0. Thus,

$$s = \lim \sum_{i=1}^{N} P_{i-1}P_i = \int_a^b \sqrt{x'(t)^2 + y'(t)^2}\,dt$$

Because of the square root, the arc length integral cannot be evaluated explicitly except in special cases, but we can always approximate it numerically.

THEOREM 1 Arc Length Let $c(t) = (x(t), y(t))$, where $x'(t)$ and $y'(t)$ exist and are continuous. Then the arc length s of $c(t)$ for $a \le t \le b$ is equal to

$$s = \int_a^b \sqrt{x'(t)^2 + y'(t)^2}\,dt \qquad \boxed{3}$$

The graph of a function $y = f(x)$ has parametrization $c(t) = (t, f(t))$. In this case,

$$\sqrt{x'(t)^2 + y'(t)^2} = \sqrt{1 + f'(t)^2}$$

and Eq. (3) reduces to the arc length formula derived in Section 9.1.

As mentioned above, the arc length integral can be evaluated explicitly only in special cases. The circle and the cycloid are two such cases.

EXAMPLE 1 Use Eq. 3 to calculate the arc length of a circle of radius R.

Solution With the parametrization $x = R\cos\theta$, $y = R\sin\theta$,

$$x'(\theta)^2 + y'(\theta)^2 = (-R\sin\theta)^2 + (R\cos\theta)^2 = R^2(\sin^2\theta + \cos^2\theta) = R^2$$

We obtain the expected result:

$$s = \int_0^{2\pi} \sqrt{x'(\theta)^2 + y'(\theta)^2}\,d\theta = \int_0^{2\pi} R\,d\theta = 2\pi R$$

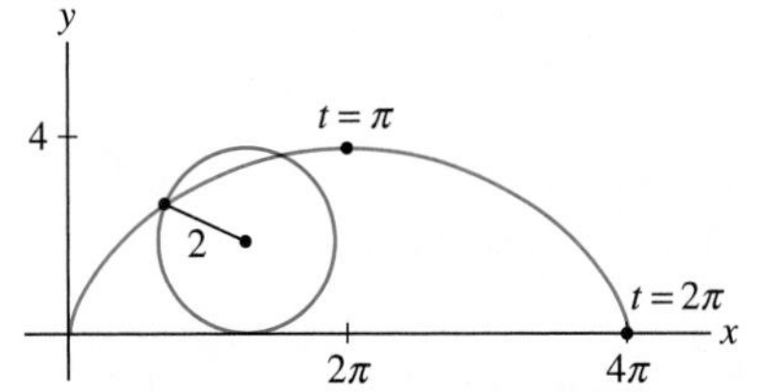

FIGURE 2 One arch of the cycloid generated by a circle of radius 2.

REMINDER

$$\frac{1-\cos t}{2} = \sin^2\frac{t}{2}$$

EXAMPLE 2 Length of the Cycloid Calculate the length s of one arch of the cycloid generated by a circle of radius $R = 2$ (Figure 2).

Solution We use the parametrization of the cycloid in Eq. (6) of Section 1:

$$x(t) = 2(t - \sin t), \qquad y(t) = 2(1 - \cos t)$$
$$x'(t) = 2(1 - \cos t), \qquad y'(t) = 2\sin t$$

Thus,

$$\begin{aligned} x'(t)^2 + y'(t)^2 &= 2^2(1-\cos t)^2 + 2^2\sin^2 t \\ &= 4 - 8\cos t + 4\cos^2 t + 4\sin^2 t \\ &= 8 - 8\cos t \\ &= 16\sin^2\frac{t}{2} \qquad \text{(Use the identity recalled in the margin.)} \end{aligned}$$

One arch of the cycloid is traced as t varies from 0 to 2π, and thus

$$s = \int_0^{2\pi} \sqrt{x'(t)^2 + y'(t)^2}\,dt = \int_0^{2\pi} 4\sin\frac{t}{2}\,dt = -8\cos\frac{t}{2}\Big|_0^{2\pi} = -8(-1) + 8 = 16$$

Note that because $\sin\frac{t}{2} \ge 0$ for $0 \le t \le 2\pi$, we did not need an absolute value when taking the square root of $16\sin^2\frac{t}{2}$.

In Chapter 13, we will discuss not just the speed but also the velocity *of a particle moving along a curved path. Velocity is "speed plus direction" and is represented by a vector.*

Now consider a particle moving along a path $c(t)$. The distance traveled by the particle over the time interval $[t_0, t]$ is given by the arc length integral:

$$s(t) = \int_{t_0}^{t} \sqrt{x'(u)^2 + y'(u)^2}\, du$$

On the other hand, speed is defined as the rate of change of distance traveled with respect to time, so by the Fundamental Theorem of Calculus,

$$\text{Speed} = \frac{ds}{dt} = \frac{d}{dt}\int_{t_0}^{t} \sqrt{x'(u)^2 + y'(u)^2}\, du = \sqrt{x'(t)^2 + y'(t)^2}$$

THEOREM 2 Speed Along a Parametrized Path The speed of $c(t) = (x(t), y(t))$ is

$$\text{Speed} = \frac{ds}{dt} = \sqrt{x'(t)^2 + y'(t)^2}$$

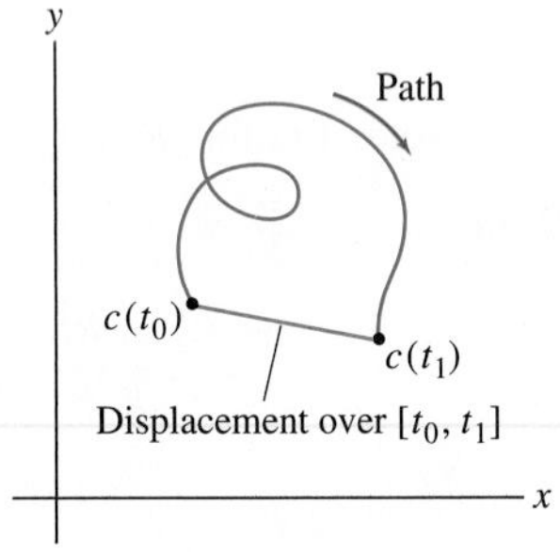

FIGURE 3 The distance along the path is greater than or equal to the displacement.

The next example illustrates the difference between distance traveled along a path and **displacement** (also called net change in position). The displacement along a path is the distance between the initial point $c(t_0)$ and the endpoint $c(t_1)$. The distance traveled is greater than the displacement unless the particle happens to move in a straight line (Figure 3).

EXAMPLE 3 A particle travels along the path $c(t) = (2t, 1 + t^{3/2})$. Find:

(a) The particle's speed at $t = 1$ (assume units of meters and minutes).

(b) The distance traveled s and displacement d during the interval $0 \le t \le 4$.

Solution We have

$$x'(t) = 2, \qquad y'(t) = \frac{3}{2}t^{1/2}$$

The speed at time t is

$$s'(t) = \sqrt{x'(t)^2 + y'(t)^2} = \sqrt{4 + \frac{9}{4}t} \quad \text{m/min}$$

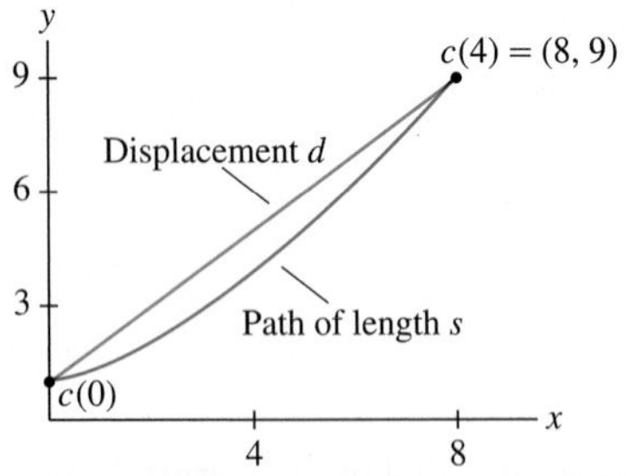

FIGURE 4 The path $c(t) = (2t, 1 + t^{3/2})$.

(a) The particle's speed at $t = 1$ is $s'(1) = \sqrt{4 + \frac{9}{4}} = 2.5$ m/min.

(b) The distance traveled in the first 4 min is

$$s = \int_0^4 \sqrt{4 + \frac{9}{4}t}\, dt = \frac{8}{27}\left(4 + \frac{9}{4}t\right)^{3/2}\Bigg|_0^4 = \frac{8}{27}\left(13^{3/2} - 8\right) \approx 11.52 \text{ m}$$

The displacement d is the distance from the initial point $c(0) = (0, 1)$ to the endpoint $c(4) = (8, 1 + 4^{3/2}) = (8, 9)$ (see Figure 4):

$$d = \sqrt{(8-0)^2 + (9-1)^2} = 8\sqrt{2} \approx 11.31 \text{ m}$$ ■

In physics, we often describe the path of a particle moving with constant speed along a circle of radius R in terms of a constant ω (lowercase Greek omega) as follows:

$$c(t) = (R\cos\omega t, R\sin\omega t)$$

The constant ω, called the *angular velocity*, is the rate of change with respect to time of the particle's angle θ (Figure 5).

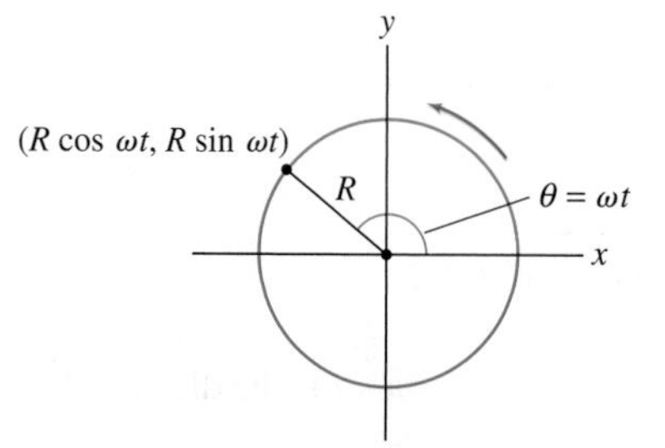

FIGURE 5 A particle moving on a circle of radius R with angular velocity ω has speed $|\omega R|$.

■ **EXAMPLE 4 Angular Velocity** Calculate the speed of the circular path of radius R and angular velocity ω. What is the speed if $R = 3$ m and $\omega = 4$ rad/s?

Solution We have $x = R\cos\omega t$ and $y = R\sin\omega t$, and

$$x'(t) = -\omega R\sin\omega t, \qquad y'(t) = \omega R\cos\omega t$$

The particle's speed is

$$\frac{ds}{dt} = \sqrt{x'(t)^2 + y'(t)^2} = \sqrt{(-\omega R\sin\omega t)^2 + (\omega R\cos\omega t)^2}$$

$$= \sqrt{\omega^2R^2(\sin^2\omega t + \cos^2\omega t)} = |\omega|R$$

Thus, the speed is constant with value $|\omega|R$. If $R = 3$ m and $\omega = 4$ rad/s, then the speed is $|\omega|R = 3(4) = 12$ m/s. ■

Consider the surface obtained by rotating a parametric curve $c(t) = (x(t), y(t))$ about the x-axis. The surface area is given by Eq. (4) in the next theorem. It can be derived in much the same way as the formula for a surface of revolution of a graph $y = f(x)$ in Section 9.1. In this theorem, we assume that $y(t) \ge 0$ so that the curve $c(t)$ lies above the x-axis, and that $x(t)$ is increasing so that the curve does not reverse direction.

THEOREM 3 Surface Area Let $c(t) = (x(t), y(t))$, where $y(t) \ge 0$, $x(t)$ is increasing, and $x'(t)$ and $y'(t)$ are continuous. Then the surface obtained by rotating $c(t)$ about the x-axis for $a \le t \le b$ has surface area

$$S = 2\pi\int_a^b y(t)\sqrt{x'(t)^2 + y'(t)^2}\,dt \qquad \boxed{4}$$

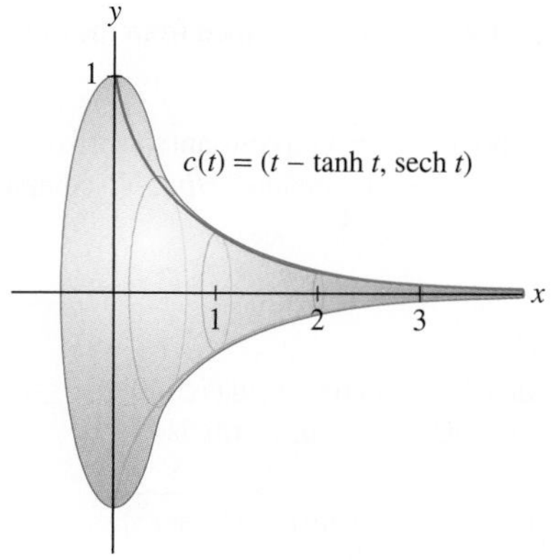

FIGURE 6 Surface generated by revolving the tractrix about the x-axis.

■ **EXAMPLE 5** Calculate the surface area of the surface obtained by rotating the *tractrix* $c(t) = (t - \tanh t, \operatorname{sech} t)$ about the x-axis for $0 \le t < \infty$.

Solution Note that the surface extends infinitely to the right (Figure 6). We have

$$x'(t) = \frac{d}{dt}(t - \tanh t) = 1 - \operatorname{sech}^2 t, \qquad y'(t) = \frac{d}{dt}\operatorname{sech} t = -\operatorname{sech} t\tanh t$$

Using the identities $1 - \operatorname{sech}^2 t = \tanh^2 t$ and $\operatorname{sech}^2 t = 1 - \tanh^2 t$, we obtain

$$x'(t)^2 + y'(t)^2 = (1 - \operatorname{sech}^2 t)^2 + (-\operatorname{sech} t\tanh t)^2$$

$$= (\tanh^2 t)^2 + (1 - \tanh^2 t)\tanh^2 t = \tanh^2 t$$

The surface area is given by an improper integral, which we evaluate using the integral formula recalled in the margin:

$$S = 2\pi\int_0^\infty \operatorname{sech} t\sqrt{\tanh^2 t}\,dt = 2\pi\int_0^\infty \operatorname{sech} t\tanh t\,dt = 2\pi\lim_{R\to\infty}\int_0^R \operatorname{sech} t\tanh t\,dt$$

$$= 2\pi\lim_{R\to\infty}(-\operatorname{sech} t)\Big|_0^R = 2\pi\lim_{R\to\infty}(\operatorname{sech} 0 - \operatorname{sech} R) = 2\pi\operatorname{sech} 0 = 2\pi$$

Here we use that $\operatorname{sech} R = \dfrac{1}{e^R + e^{-R}}$ tends to zero (because $e^R \to \infty$ while $e^{-R} \to 0$). ■

REMINDER

$$\operatorname{sech} t = \frac{1}{\cosh t} = \frac{2}{e^t + e^{-t}}$$

$$1 - \operatorname{sech}^2 t = \tanh^2 t$$

$$\frac{d}{dt}\tanh t = \operatorname{sech}^2 t$$

$$\frac{d}{dt}\operatorname{sech} t = -\operatorname{sech} t\tanh t$$

$$\int \operatorname{sech} t\tanh t\,dt = -\operatorname{sech} t + C$$

12.2 SUMMARY

- *Arc length* of $c(t) = (x(t), y(t))$ for $a \le t \le b$:

$$s = \text{arc length} = \int_a^b \sqrt{x'(t)^2 + y'(t)^2}\, dt$$

- The arc length is the distance along the path $c(t)$. The *displacement* is the distance from the starting point $c(a)$ to the endpoint $c(b)$.
- *Arc length integral:*

$$s(t) = \int_{t_0}^t \sqrt{x'(u)^2 + y'(u)^2}\, du$$

- Speed at time t:

$$\frac{ds}{dt} = \sqrt{x'(t)^2 + y'(t)^2}$$

- Surface area of the surface obtained by rotating $c(t) = (x(t), y(t))$ about the x-axis for $a \le t \le b$:

$$S = 2\pi \int_a^b y(t)\sqrt{x'(t)^2 + y'(t)^2}\, dt$$

12.2 EXERCISES

Preliminary Questions

1. What is the definition of arc length?

2. What is the interpretation of $\sqrt{x'(t)^2 + y'(t)^2}$ for a particle following the trajectory $(x(t), y(t))$?

3. A particle travels along a path from $(0, 0)$ to $(3, 4)$. What is the displacement? Can the distance traveled be determined from the information given?

4. A particle traverses the parabola $y = x^2$ with constant speed 3 cm/s. What is the distance traveled during the first minute? *Hint:* No computation is necessary.

Exercises

In Exercises 1–10, use Eq. (3) to find the length of the path over the given interval.

1. $(3t + 1, 9 - 4t), \quad 0 \le t \le 2$

2. $(1 + 2t, 2 + 4t), \quad 1 \le t \le 4$

3. $(2t^2, 3t^2 - 1), \quad 0 \le t \le 4$

4. $(3t, 4t^{3/2}), \quad 0 \le t \le 1$

5. $(3t^2, 4t^3), \quad 1 \le t \le 4$

6. $(t^3 + 1, t^2 - 3), \quad 0 \le t \le 1$

7. $(\sin 3t, \cos 3t), \quad 0 \le t \le \pi$

8. $(\sin\theta - \theta\cos\theta, \cos\theta + \theta\sin\theta), \quad 0 \le \theta \le 2$

In Exercises 9 and 10, use the identity

$$\frac{1 - \cos t}{2} = \sin^2 \frac{t}{2}$$

9. $(2\cos t - \cos 2t, 2\sin t - \sin 2t), \quad 0 \le t \le \frac{\pi}{2}$

10. $(5(\theta - \sin\theta), 5(1 - \cos\theta)), \quad 0 \le \theta \le 2\pi$

11. Show that one arch of a cycloid generated by a circle of radius R has length $8R$.

12. Find the length of the spiral $c(t) = (t\cos t, t\sin t)$ for $0 \le t \le 2\pi$ to three decimal places (Figure 7). *Hint:* Use the formula

$$\int \sqrt{1 + t^2}\, dt = \frac{1}{2}t\sqrt{1 + t^2} + \frac{1}{2}\ln\left(t + \sqrt{1 + t^2}\right)$$

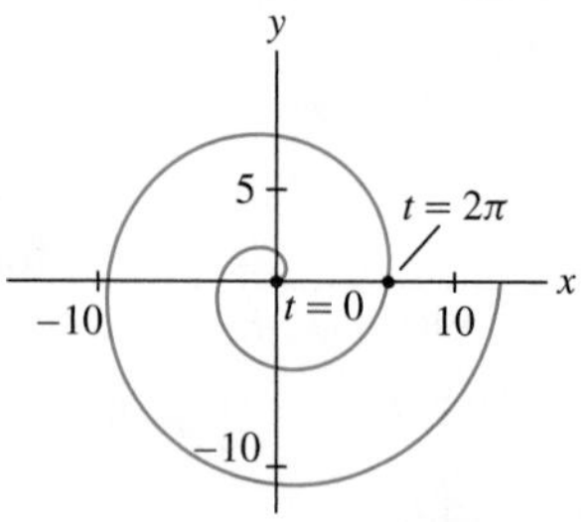

FIGURE 7 The spiral $c(t) = (t\cos t, t\sin t)$.

13. Find the length of the tractrix (see Figure 6)

$$c(t) = (t - \tanh(t), \operatorname{sech}(t)), \qquad 0 \le t \le A$$

14. CAS Find a numerical approximation to the length of $c(t) = (\cos 5t, \sin 3t)$ for $0 \le t \le 2\pi$ (Figure 8).

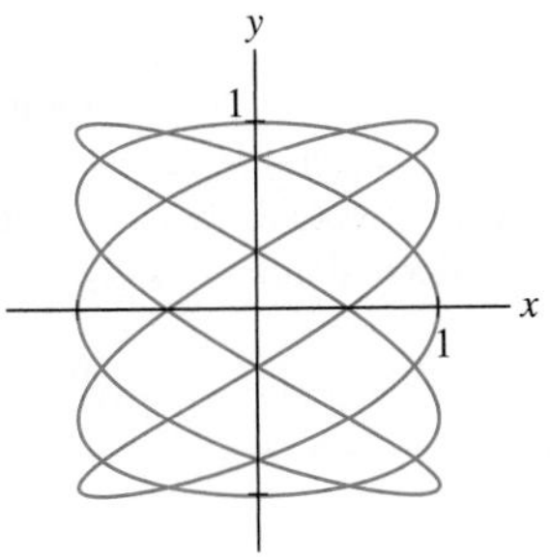

FIGURE 8

In Exercises 15–18, determine the speed s at time t (assume units of meters and seconds).

15. $(t^3, t^2), \quad t = 2$

16. $(3\sin 5t, 8\cos 5t), \quad t = \frac{\pi}{4}$

17. $(5t+1, 4t-3), \quad t = 9$

18. $(\ln(t^2+1), t^3), \quad t = 1$

19. Find the minimum speed of a particle with trajectory $c(t) = (t^3 - 4t, t^2 + 1)$ for $t \ge 0$. *Hint:* It is easier to find the minimum of the square of the speed.

20. Find the minimum speed of a particle with trajectory $c(t) = (t^3, t^{-2})$ for $t \ge 0.5$.

21. Find the speed of the cycloid $c(t) = (4t - 4\sin t, 4 - 4\cos t)$ at points where the tangent line is horizontal.

22. Calculate the arc length integral $s(t)$ for the *logarithmic spiral* $c(t) = (e^t \cos t, e^t \sin t)$.

CAS *In Exercises 23–26, plot the curve and use the Midpoint Rule with N = 10, 20, 30, and 50 to approximate its length.*

23. $c(t) = (\cos t, e^{\sin t}) \quad$ for $0 \le t \le 2\pi$

24. $c(t) = (t - \sin 2t, 1 - \cos 2t) \quad$ for $0 \le t \le 2\pi$

25. The ellipse $\left(\frac{x}{5}\right)^2 + \left(\frac{y}{3}\right)^2 = 1$

26. $x = \sin 2t, \quad y = \sin 3t \quad$ for $0 \le t \le 2\pi$

27. If you unwind thread from a stationary circular spool, keeping the thread taut at all times, then the endpoint traces a curve C called the **involute** of the circle (Figure 9). Observe that $\overline{PQ}$ has length $R\theta$. Show that C is parametrized by

$$c(\theta) = \big(R(\cos\theta + \theta\sin\theta), R(\sin\theta - \theta\cos\theta)\big)$$

Then find the length of the involute for $0 \le \theta \le 2\pi$.

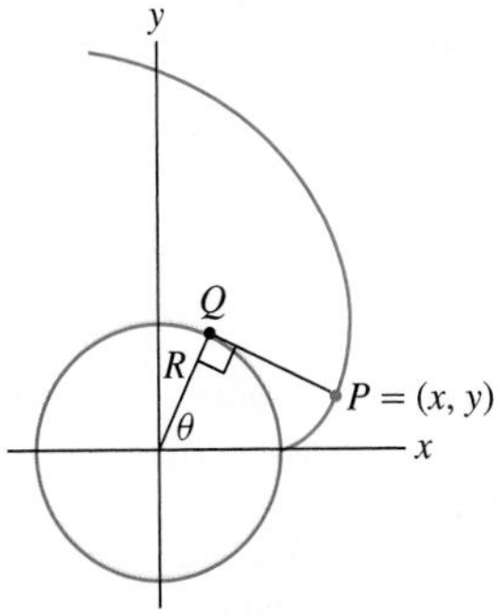

FIGURE 9 Involute of a circle.

28. Let $a > b$ and set

$$k = \sqrt{1 - \frac{b^2}{a^2}}$$

Use a parametric representation to show that the ellipse $\left(\frac{x}{a}\right)^2 + \left(\frac{y}{b}\right)^2 = 1$ has length $L = 4aG\left(\frac{\pi}{2}, k\right)$, where

$$G(\theta, k) = \int_0^\theta \sqrt{1 - k^2 \sin^2 t}\, dt$$

is the *elliptic integral of the second kind.*

In Exercises 29–32, use Eq. (4) to compute the surface area of the given surface.

29. The cone generated by revolving $c(t) = (t, mt)$ about the x-axis for $0 \le t \le A$

30. A sphere of radius R

31. The surface generated by revolving one arch of the cycloid $c(t) = (t - \sin t, 1 - \cos t)$ about the x-axis

32. The surface generated by revolving the astroid $c(t) = (\cos^3 t, \sin^3 t)$ about the x-axis for $0 \le t \le \frac{\pi}{2}$

Further Insights and Challenges

33. CAS Let $b(t)$ be the "Butterfly Curve":

$$x(t) = \sin t \left(e^{\cos t} - 2\cos 4t - \sin\left(\frac{t}{12}\right)^5\right)$$

$$y(t) = \cos t \left(e^{\cos t} - 2\cos 4t - \sin\left(\frac{t}{12}\right)^5\right)$$

(a) Use a computer algebra system to plot $b(t)$ and the speed $s'(t)$ for $0 \le t \le 12\pi$.

(b) Approximate the length $b(t)$ for $0 \le t \le 10\pi$.

34. CAS Let $a \ge b > 0$ and set $k = \dfrac{2\sqrt{ab}}{a-b}$. Show that the **trochoid**

$$x = at - b\sin t, \qquad y = a - b\cos t, \qquad 0 \le t \le T$$

has length $2(a-b)G\left(\frac{T}{2}, k\right)$ with $G(\theta, k)$ as in Exercise 28.

35. A satellite orbiting at a distance R from the center of the earth follows the circular path $x = R\cos\omega t$, $y = R\sin\omega t$.

(a) Show that the period T (the time of one revolution) is $T = 2\pi/\omega$.

(b) According to Newton's laws of motion and gravity,

$$x''(t) = -Gm_e \frac{x}{R^3}, \qquad y''(t) = -Gm_e \frac{y}{R^3}$$

where G is the universal gravitational constant and m_e is the mass of the earth. Prove that $R^3/T^2 = Gm_e/4\pi^2$. Thus, R^3/T^2 has the same value for all orbits (a special case of Kepler's Third Law).

36. The acceleration due to gravity on the surface of the earth is

$$g = \frac{Gm_e}{R_e^2} = 9.8 \text{ m/s}^2, \quad \text{where } R_e = 6378 \text{ km}$$

Use Exercise 35(b) to show that a satellite orbiting at the earth's surface would have period $T_e = 2\pi\sqrt{R_e/g} \approx 84.5$ min. Then estimate the distance R_m from the moon to the center of the earth. Assume that the period of the moon (sidereal month) is $T_m \approx 27.43$ days.

12.3 Polar Coordinates

Polar coordinates are appropriate when distance from the origin or angle plays a role. For example, the gravitational force exerted on a planet by the sun depends only on the distance r from the sun and is conveniently described in polar coordinates.

In polar coordinates, we label a point P by coordinates (r, θ), where r is the distance to the origin O and θ is the angle between $\overline{OP}$ and the positive x-axis (Figure 1). By convention, an angle is positive if the corresponding rotation is counterclockwise. We call r the **radial coordinate** and θ the **angular coordinate**.

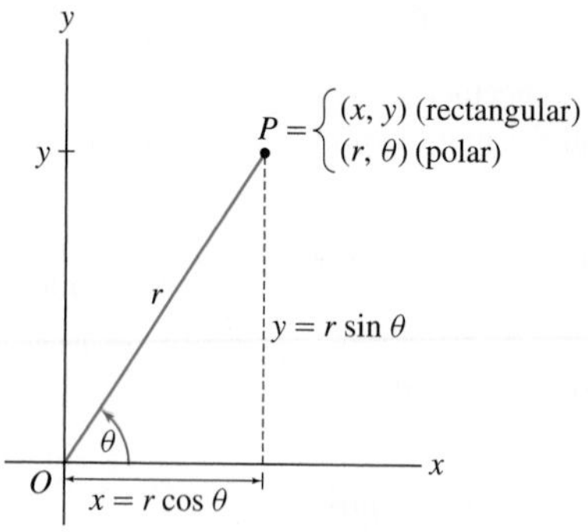

FIGURE 1

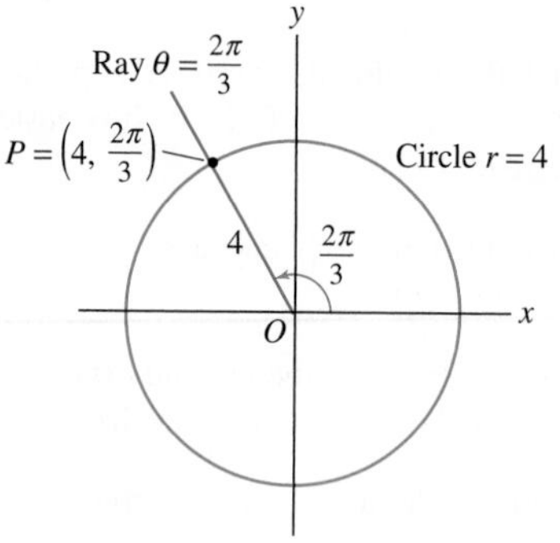

FIGURE 2

The point P in Figure 2 has polar coordinates $(r, \theta) = \left(4, \frac{2\pi}{3}\right)$. It is located at distance $r = 4$ from the origin (so it lies on the circle of radius 4), and it lies on the ray of angle $\theta = \frac{2\pi}{3}$.

Figure 3 shows the two families of **grid lines** in polar coordinates:

$$\text{Circle centered at } O \quad \longleftrightarrow \quad r = \text{constant}$$

$$\text{Ray starting at } O \quad \longleftrightarrow \quad \theta = \text{constant}$$

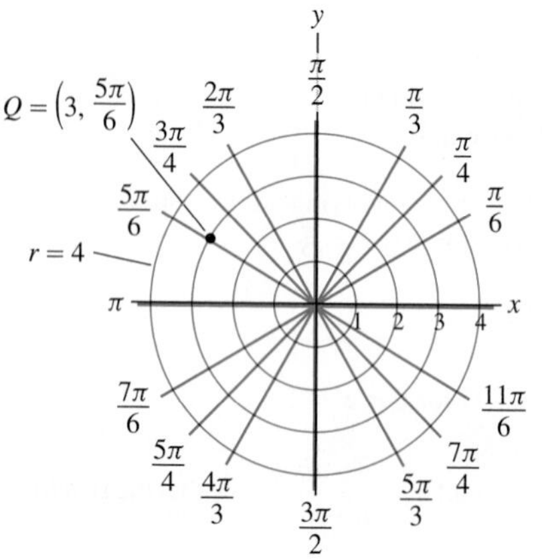

FIGURE 3 Grid lines in polar coordinates.

Every point in the plane other than the origin lies at the intersection of the two grid lines and these two grid lines determine its polar coordinates. For example, point Q in Figure 3 lies on the circle $r = 3$ and the ray $\theta = \frac{5\pi}{6}$, so $Q = \left(3, \frac{5\pi}{6}\right)$ in polar coordinates.

Figure 1 shows that polar and rectangular coordinates are related by the equations $x = r\cos\theta$ and $y = r\sin\theta$. On the other hand, $r^2 = x^2 + y^2$ by the distance formula, and $\tan\theta = y/x$ if $x \neq 0$. This yields the conversion formulas:

Polar to Rectangular	Rectangular to Polar
$x = r\cos\theta$	$r = \sqrt{x^2 + y^2}$
$y = r\sin\theta$	$\tan\theta = \dfrac{y}{x} \quad (x \neq 0)$

■ **EXAMPLE 1 From Polar to Rectangular Coordinates** Find the rectangular coordinates of point Q in Figure 3.

Solution The point $Q = (r, \theta) = \left(3, \frac{5\pi}{6}\right)$ has rectangular coordinates:

$$x = r\cos\theta = 3\cos\left(\frac{5\pi}{6}\right) = 3\left(-\frac{\sqrt{3}}{2}\right) = -\frac{3\sqrt{3}}{2}$$

$$y = r\sin\theta = 3\sin\left(\frac{5\pi}{6}\right) = 3\left(\frac{1}{2}\right) = \frac{3}{2}$$ ■

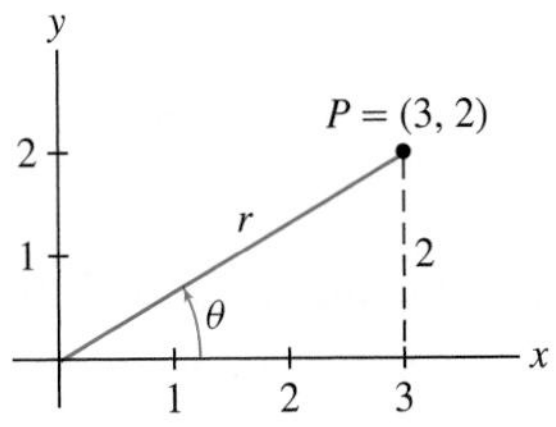

FIGURE 4 The polar coordinates of P satisfy $r = \sqrt{3^2 + 2^2}$ and $\tan\theta = \frac{2}{3}$.

■ **EXAMPLE 2 From Rectangular to Polar Coordinates** Find the polar coordinates of point P in Figure 4.

Solution Since $P = (x, y) = (3, 2)$,

$$r = \sqrt{x^2 + y^2} = \sqrt{3^2 + 2^2} = \sqrt{13} \approx 3.6$$

$$\tan\theta = \frac{y}{x} = \frac{2}{3}$$

and because P lies in the first quadrant,

$$\theta = \tan^{-1}\frac{y}{x} = \tan^{-1}\frac{2}{3} \approx 0.588$$

Thus, P has polar coordinates $(r, \theta) \approx (3.6, 0.588)$. ■

By definition,

$$-\frac{\pi}{2} < \tan^{-1} x < \frac{\pi}{2}$$

If $r > 0$, a coordinate θ of $P = (x, y)$ is

$$\theta = \begin{cases} \tan^{-1}\dfrac{y}{x} & \text{if } x > 0 \\ \tan^{-1}\dfrac{y}{x} + \pi & \text{if } x < 0 \\ \pm\dfrac{\pi}{2} & \text{if } x = 0 \end{cases}$$

A few remarks are in order before proceeding:

- The angular coordinate is not unique because (r, θ) and $(r, \theta + 2\pi n)$ *label the same point* for any integer n. For instance, point P in Figure 5 has radial coordinate $r = 2$, but its angular coordinate can be any one of $\frac{\pi}{2}, \frac{5\pi}{2}, \ldots$ or $-\frac{3\pi}{2}, -\frac{7\pi}{2}, \ldots$.
- The origin O has no well-defined angular coordinate, so we assign to O the polar coordinates $(0, \theta)$ for any angle θ.
- By convention, we allow *negative* radial coordinates. By definition, $(-r, \theta)$ is the reflection of (r, θ) through the origin (Figure 6). With this convention, $(-r, \theta)$ and $(r, \theta + \pi)$ represent the same point.
- We may specify unique polar coordinates for points other than the origin by placing restrictions on r and θ. We commonly choose $r > 0$ and $0 \le \theta < 2\pi$.

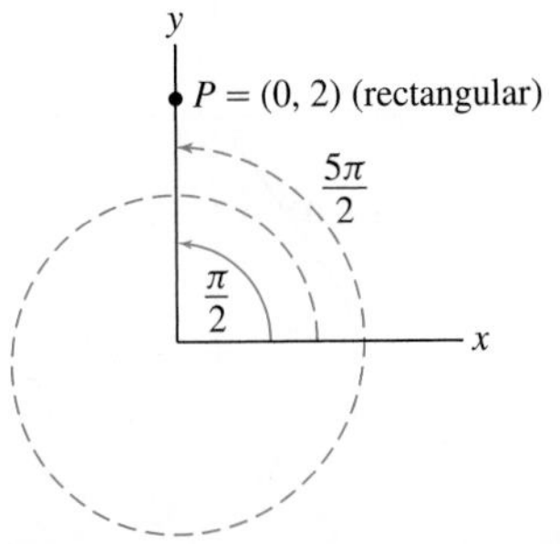

FIGURE 5 The angular coordinate of $P = (0, 2)$ is $\frac{\pi}{2}$ or any angle $\frac{\pi}{2} + 2\pi n$, where n is an integer.

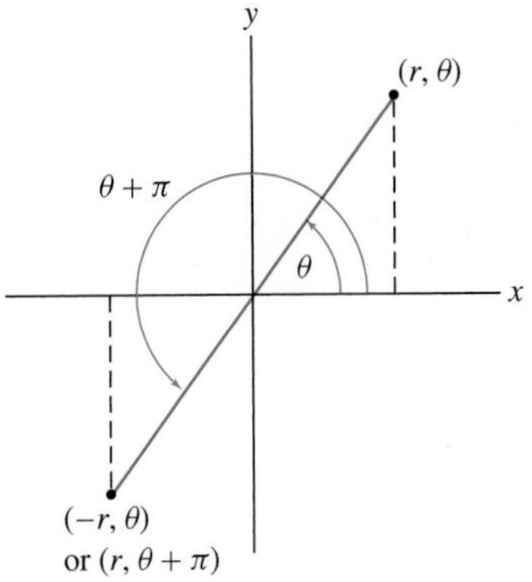

FIGURE 6 Relation between (r, θ) and $(-r, \theta)$.

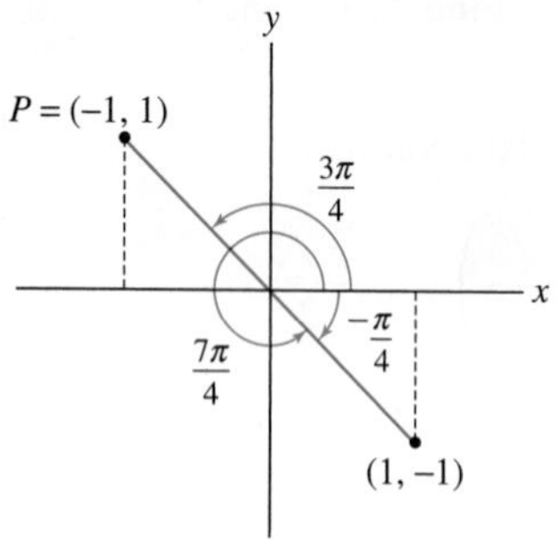

FIGURE 7

When determining the angular coordinate of a point $P = (x, y)$, remember that there are two angles between 0 and 2π satisfying $\tan\theta = y/x$. You must choose θ so that (r, θ) lies in the quadrant containing P and in the opposite quadrant (Figure 7).

EXAMPLE 3 Choosing θ Correctly Find two polar representations of $P = (-1, 1)$, one with $r > 0$ and one with $r < 0$.

Solution The point $P = (x, y) = (-1, 1)$ has polar coordinates (r, θ), where

$$r = \sqrt{(-1)^2 + 1^2} = \sqrt{2}, \qquad \tan\theta = \tan\frac{y}{x} = -1$$

However, θ is not given by

$$\tan^{-1}\frac{y}{x} = \tan^{-1}\left(\frac{1}{-1}\right) = -\frac{\pi}{4}$$

because $\theta = -\frac{\pi}{4}$ this would place P in the fourth quadrant (Figure 7). Since P is in the second quadrant, the correct angle is

$$\theta = \tan^{-1}\frac{y}{x} + \pi = -\frac{\pi}{4} + \pi = \frac{3\pi}{4}$$

If we wish to use the negative radial coordinate $r = -\sqrt{2}$, then the angle becomes $\theta = -\frac{\pi}{4}$ or $\frac{7\pi}{4}$. Thus,

$$P = \left(\sqrt{2}, \frac{3\pi}{4}\right) \quad \text{or} \quad \left(-\sqrt{2}, \frac{7\pi}{4}\right)$$

A curve is described in polar coordinates by an equation involving r and θ, which we call a **polar equation**. By convention, we allow solutions with $r < 0$.

A line through the origin O has the simple equation $\theta = \theta_0$, where θ_0 is the angle between the line and the x-axis (Figure 8). Indeed, the points with $\theta = \theta_0$ are (r, θ_0), where r is arbitrary (positive, negative, or zero).

EXAMPLE 4 Line Through the Origin Find the polar equation of the line through the origin of slope $\frac{3}{2}$ (Figure 9).

Solution A line of slope m makes an angle θ_0 with the x-axis, where $m = \tan\theta_0$. In our case, $\theta_0 = \tan^{-1}\frac{3}{2} \approx 0.98$. The equation of the line is $\theta = \tan^{-1}\frac{3}{2}$ or $\theta \approx 0.98$.

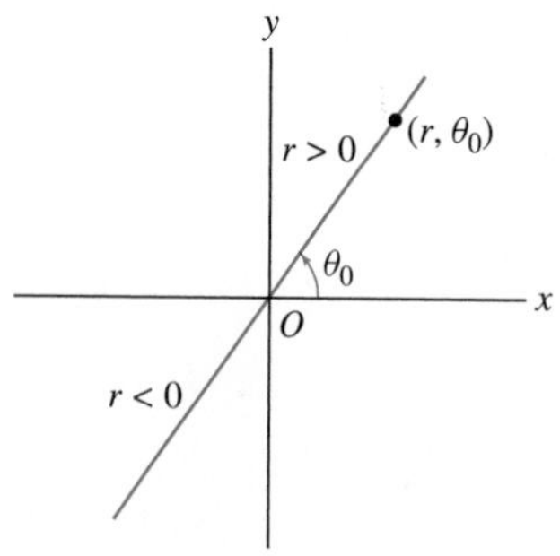

FIGURE 8 Lines through O with polar equation $\theta = \theta_0$.

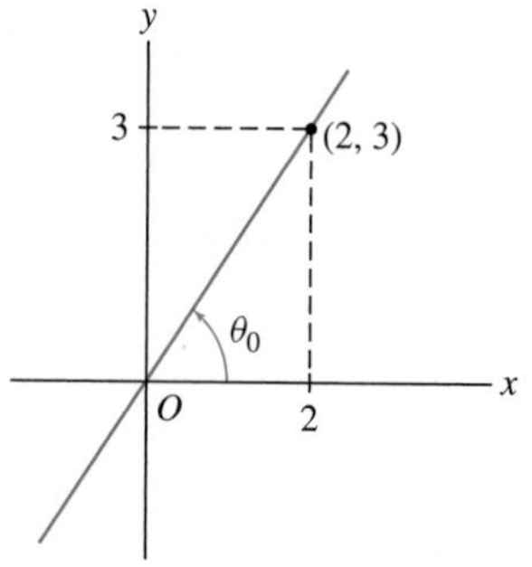

FIGURE 9 Line of slope $\frac{3}{2}$ through the origin.

To describe lines that do not pass through the origin, we note that any such line has a unique point P_0 that is *closest* to the origin. The next example shows how to write down the polar equation of the line in terms of P_0 (Figure 10).

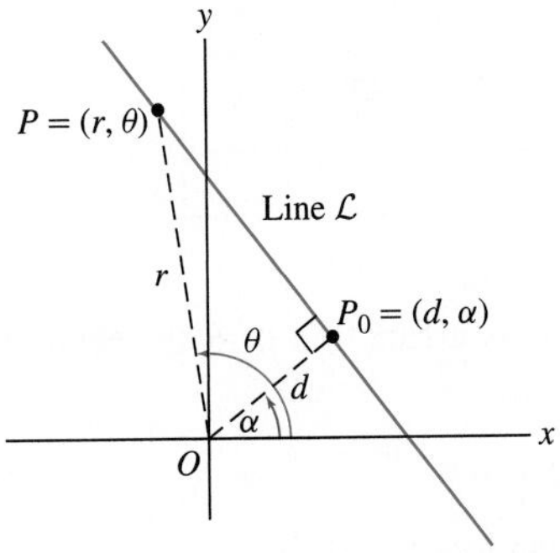

FIGURE 10 P_0 is the point on $\mathcal{L}$ closest to the origin.

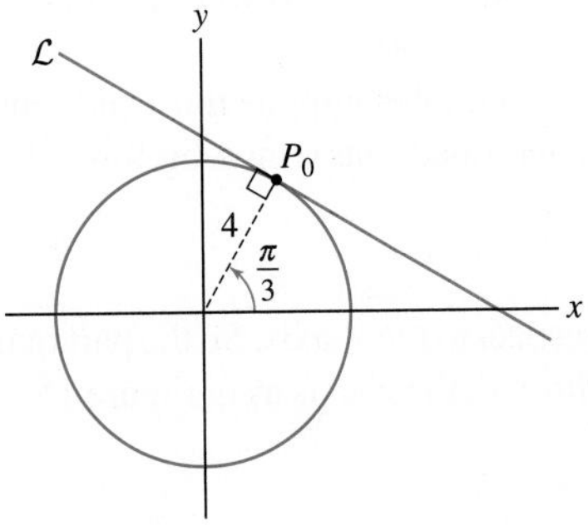

FIGURE 11 The tangent line has equation $r = 4\sec\left(\theta - \frac{\pi}{3}\right)$.

■ **EXAMPLE 5 Line Not Passing Through O** Show that

$$\boxed{r = d\sec(\theta - \alpha)} \qquad \boxed{1}$$

is the polar equation of the line $\mathcal{L}$ whose point closest to the origin is $P_0 = (d, \alpha)$.

Solution The point P_0 is obtained by dropping a perpendicular from the origin to $\mathcal{L}$ (Figure 10), and if $P = (r, \theta)$ is any point on $\mathcal{L}$ other than P_0, then $\triangle OPP_0$ is a right triangle. Therefore, $d/r = \cos(\theta - \alpha)$, or $r = d\sec(\theta - \alpha)$, as claimed. ■

■ **EXAMPLE 6** Find the polar equation of the line $\mathcal{L}$ tangent to the circle $r = 4$ at the point with polar coordinates $P_0 = \left(4, \frac{\pi}{3}\right)$.

Solution The point on $\mathcal{L}$ closest to the origin is P_0 itself (Figure 11). Therefore, we take $(d, \alpha) = \left(4, \frac{\pi}{3}\right)$ in Eq. (1) to obtain the equation $r = 4\sec\left(\theta - \frac{\pi}{3}\right)$. ■

Often, it is hard to guess the shape of a graph of a polar equation. In some cases, it is helpful rewrite the equation in rectangular coordinates.

■ **EXAMPLE 7 Converting to Rectangular Coordinates** Identify the curve with polar equation $r = 2a\cos\theta$ (a a constant).

Solution Multiply the equation by r to obtain $r^2 = 2ar\cos\theta$. Because $r^2 = x^2 + y^2$ and $x = r\cos\theta$, this equation becomes

$$x^2 + y^2 = 2ax \qquad \text{or} \qquad x^2 - 2ax + y^2 = 0$$

Then complete the square to obtain $(x - a)^2 + y^2 = a^2$. This is the equation of the circle of radius a and center $(a, 0)$ (Figure 12). ■

A similar calculation shows that the circle $x^2 + (y - a)^2 = a^2$ of radius a and center $(0, a)$ has polar equation $r = 2a\sin\theta$. In the next example, we make use of symmetry. Note that the points (r, θ) and $(r, -\theta)$ are symmetric with respect to the x-axis (Figure 13).

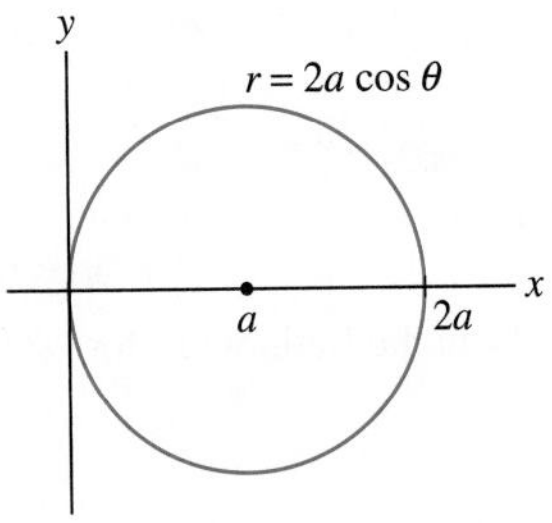

FIGURE 12

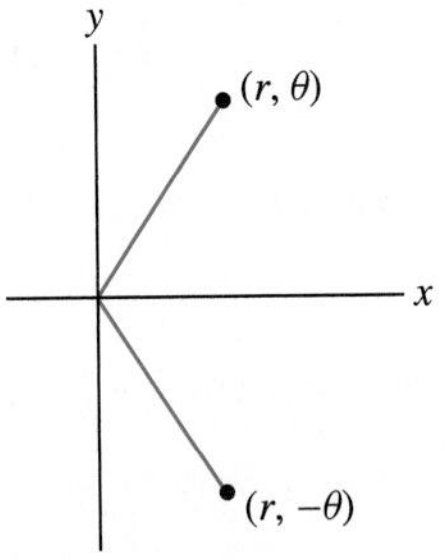

FIGURE 13 The points (r, θ) and $(r, -\theta)$ are symmetric with respect to the x-axis.

■ **EXAMPLE 8 Symmetry About the x-Axis** Sketch the *limaçon* curve $r = 2\cos\theta - 1$.

Solution Since $\cos\theta$ is periodic, it suffices to plot points for $-\pi \le \theta \le \pi$.

***Step 1.* Plot points.**
To get started, we plot points A–G on a grid and join them by a smooth curve (Figure 14).

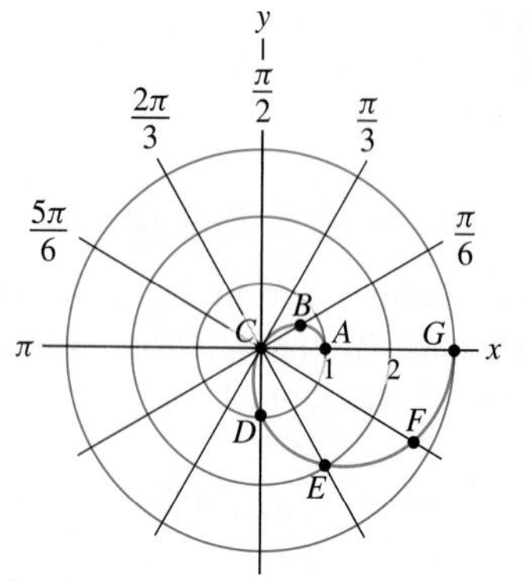

FIGURE 14 Plotting $r = 2\cos\theta - 1$ using a grid.

	A	B	C	D	E	F	G
θ	0	$\frac{\pi}{6}$	$\frac{\pi}{3}$	$\frac{\pi}{2}$	$\frac{2\pi}{3}$	$\frac{5\pi}{6}$	π
$r = 2\cos\theta - 1$	1	0.73	0	-1	-2	-2.73	-3

***Step 2.* Analyze r as a function of θ.**
For a better understanding, it is helpful to graph r as a function of θ in rectangular coordinates. Figure 15(A) shows that

As θ varies from 0 to $\frac{\pi}{3}$, $\quad r$ varies from 1 to 0.

As θ varies from $\frac{\pi}{3}$ to π, $\quad r$ is *negative* and varies from 0 to -3.

We conclude:

- The graph begins at point A in Figure 15(B) and moves in toward the origin as θ varies from 0 to $\frac{\pi}{3}$.
- Since r is negative for $\frac{\pi}{3} \le \theta \le \pi$, the curve continues into the third and fourth quadrants (rather than into the first and second quadrants), moving toward the point $G = (-3, \pi)$ in Figure 15(C).

***Step 3.* Use symmetry.**
Since $r(\theta) = r(-\theta)$, the curve is symmetric with respect to the x-axis. So the part of the curve with $-\pi \le \theta \le 0$ is obtained by reflection through the x-axis as in Figure 15(D). ■

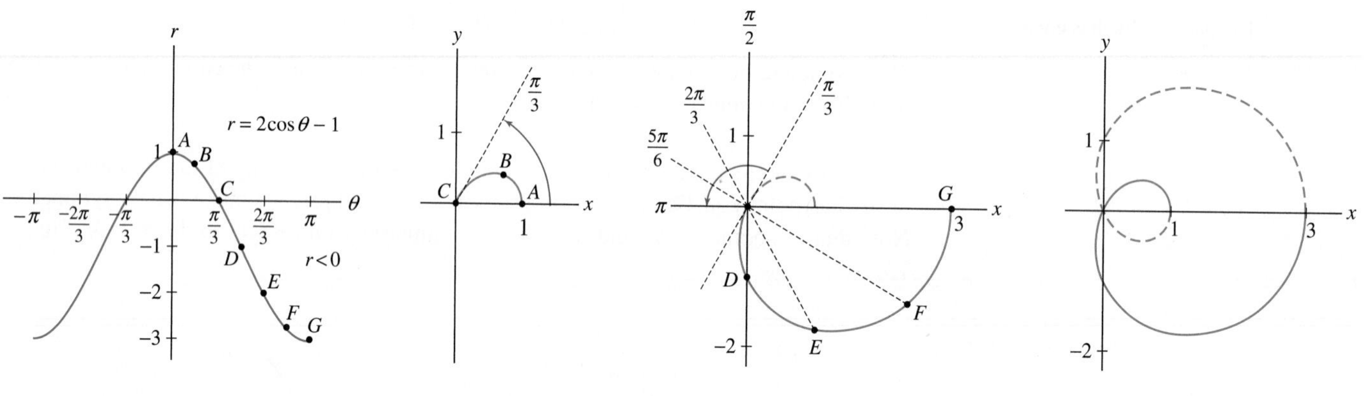

(A) Variation of r as a function of θ.

(B) θ varies from 0 to $\pi/3$; r varies from 1 to 0.

(C) θ varies from $\pi/3$ to π, but r is negative and varies from 0 to -3.

(D) The entire Limaçon.

FIGURE 15 The curve $r = 2\cos\theta - 1$ is called the *limaçon*, from the Latin word for "snail." It was first described in 1525 by the German artist Albrecht Dürer.

12.3 SUMMARY

- A point $P = (x, y)$ has polar coordinates (r, θ), where r is the distance to the origin and θ is the angle between the positive x-axis and the segment $\overline{OP}$, measured in the counterclockwise direction.

$$x = r\cos\theta \qquad r = \sqrt{x^2 + y^2}$$

$$y = r\sin\theta \qquad \tan\theta = \frac{y}{x} \quad (x \ne 0)$$

- The angular coordinate θ must be chosen so that (r, θ) lies in the proper quadrant. If $r > 0$, then

$$\theta = \begin{cases} \tan^{-1}\dfrac{y}{x} & \text{if } x > 0 \\ \tan^{-1}\dfrac{y}{x} + \pi & \text{if } x < 0 \\ \pm\dfrac{\pi}{2} & \text{if } x = 0 \end{cases}$$

- Nonuniqueness: (r, θ) and $(r, \theta + 2n\pi)$ represent the same point for all integers n. The origin O has polar coordinates $(0, \theta)$ for any θ.
- Negative radial coordinates: $(-r, \theta)$ and $(r, \theta + \pi)$ represent the same point.
- Polar equations:

Curve	Polar equation
Circle of radius R, center at the origin	$r = R$
Line through origin of slope $m = \tan\theta_0$	$\theta = \theta_0$
Line on which $P_0 = (d, \alpha)$ is the point closest to the origin	$r = d\sec(\theta - \alpha)$
Circle of radius a, center at $(a, 0)$ $(x-a)^2 + y^2 = a^2$	$r = 2a\cos\theta$
Circle of radius a, center at $(0, a)$ $x^2 + (y-a)^2 = a^2$	$r = 2a\sin\theta$

12.3 EXERCISES

Preliminary Questions

1. Points P and Q with the same radial coordinate (choose the correct answer):

(a) Lie on the same circle with the center at the origin.

(b) Lie on the same ray based at the origin.

2. Give two polar representations for the point $(x, y) = (0, 1)$, one with negative r and one with positive r.

3. Describe each of the following curves:

(a) $r = 2$ **(b)** $r^2 = 2$ **(c)** $r\cos\theta = 2$

4. If $f(-\theta) = f(\theta)$, then the curve $r = f(\theta)$ is symmetric with respect to the (choose the correct answer):

(a) x-axis **(b)** y-axis **(c)** origin

Exercises

1. Find polar coordinates for each of the seven points plotted in Figure 16.

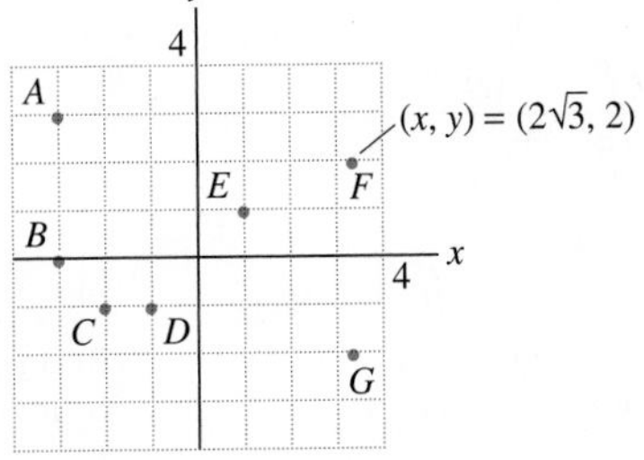

FIGURE 16

2. Plot the points with polar coordinates:

(a) $(2, \frac{\pi}{6})$ **(b)** $(4, \frac{3\pi}{4})$ **(c)** $(3, -\frac{\pi}{2})$ **(d)** $(0, \frac{\pi}{6})$

3. Convert from rectangular to polar coordinates.

(a) $(1, 0)$ **(b)** $(3, \sqrt{3})$ **(c)** $(-2, 2)$ **(d)** $(-1, \sqrt{3})$

4. Convert from rectangular to polar coordinates using a calculator (make sure your choice of θ gives the correct quadrant).

(a) $(2, 3)$ **(b)** $(4, -7)$ **(c)** $(-3, -8)$ **(d)** $(-5, 2)$

5. Convert from polar to rectangular coordinates:

(a) $(3, \frac{\pi}{6})$ **(b)** $(6, \frac{3\pi}{4})$ **(c)** $(0, \frac{\pi}{5})$ **(d)** $(5, -\frac{\pi}{2})$

6. Which of the following are possible polar coordinates for the point P with rectangular coordinates $(0, -2)$?

(a) $\left(2, \frac{\pi}{2}\right)$ **(b)** $\left(2, \frac{7\pi}{2}\right)$

(c) $\left(-2, -\frac{3\pi}{2}\right)$ **(d)** $\left(-2, \frac{7\pi}{2}\right)$

(e) $\left(-2, -\frac{\pi}{2}\right)$ **(f)** $\left(2, -\frac{7\pi}{2}\right)$

7. Describe each shaded sector in Figure 17 by inequalities in r and θ.

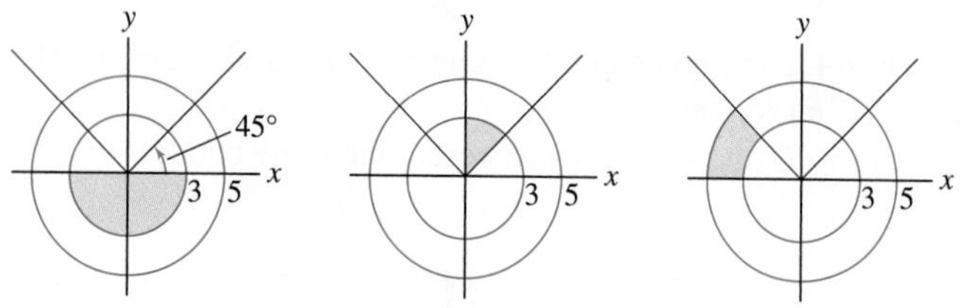

(A) (B) (C)

FIGURE 17

8. Find the equation in polar coordinates of the line through the origin with slope $\frac{1}{2}$.

9. What is the slope of the line $\theta = \frac{3\pi}{5}$?

10. Which of $r = 2\sec\theta$ and $r = 2\csc\theta$ defines a horizontal line?

In Exercises 11–16, convert to an equation in rectangular coordinates.

11. $r = 7$ **12.** $r = \sin\theta$

13. $r = 2\sin\theta$ **14.** $r = 2\csc\theta$

15. $r = \dfrac{1}{\cos\theta - \sin\theta}$ **16.** $r = \dfrac{1}{2 - \cos\theta}$

In Exercises 17–20, convert to an equation in polar coordinates.

17. $x^2 + y^2 = 5$ **18.** $x = 5$

19. $y = x^2$ **20.** $xy = 1$

21. Match each equation with its description.

(a) $r = 2$ **(i)** Vertical line
(b) $\theta = 2$ **(ii)** Horizontal line
(c) $r = 2\sec\theta$ **(iii)** Circle
(d) $r = 2\csc\theta$ **(iv)** Line through origin

22. Find the values of θ in the plot of $r = 4\cos\theta$ corresponding to points A, B, C, D in Figure 18. Then indicate the portion of the graph traced out as θ varies in the following intervals:

(a) $0 \le \theta \le \frac{\pi}{2}$ **(b)** $\frac{\pi}{2} \le \theta \le \pi$ **(c)** $\pi \le \theta \le \frac{3\pi}{2}$

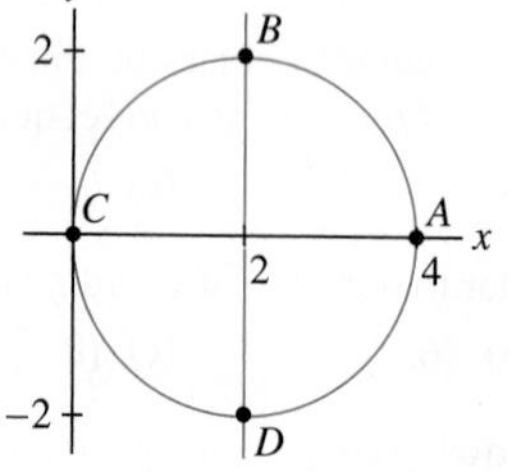

FIGURE 18 Plot of $r = 4\cos\theta$.

23. Suppose that $P = (x, y)$ has polar coordinates (r, θ). Find the polar coordinates for the points:

(a) $(x, -y)$ **(b)** $(-x, -y)$ **(c)** $(-x, y)$ **(d)** (y, x)

24. Match each equation in rectangular coordinates with its equation in polar coordinates.

(a) $x^2 + y^2 = 4$ **(i)** $r^2(1 - 2\sin^2\theta) = 4$
(b) $x^2 + (y-1)^2 = 1$ **(ii)** $r(\cos\theta + \sin\theta) = 4$
(c) $x^2 - y^2 = 4$ **(iii)** $r = 2\sin\theta$
(d) $x + y = 4$ **(iv)** $r = 2$

25. What are the polar equations of the lines parallel to the line $r\cos\left(\theta - \frac{\pi}{3}\right) = 1$?

26. Show that the circle with center at $\left(\frac{1}{2}, \frac{1}{2}\right)$ in Figure 19 has polar equation $r = \sin\theta + \cos\theta$ and find the values of θ between 0 and π corresponding to points A, B, C, and D.

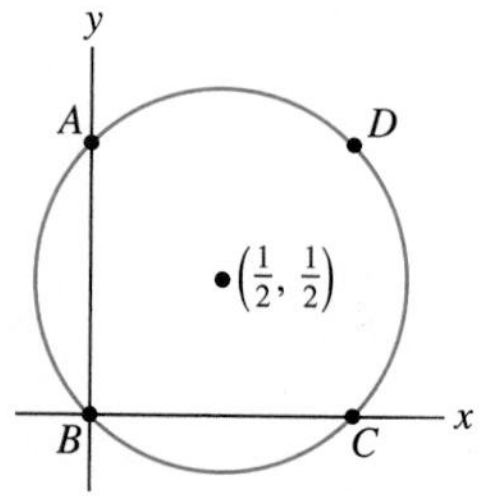

FIGURE 19 Plot of $r = \sin\theta + \cos\theta$.

27. Sketch the curve $r = \frac{1}{2}\theta$ (the spiral of Archimedes) for θ between 0 and 2π by plotting the points for $\theta = 0, \frac{\pi}{4}, \frac{\pi}{2}, \dots, 2\pi$.

28. Sketch $r = 3\cos\theta - 1$ (see Example 8).

29. Sketch the cardioid curve $r = 1 + \cos\theta$.

30. Show that the cardioid of Exercise 29 has equation

$$(x^2 + y^2 - x)^2 = x^2 + y^2$$

in rectangular coordinates.

31. Figure 20 displays the graphs of $r = \sin 2\theta$ in rectangular coordinates and in polar coordinates, where it is a "rose with four petals." Identify:

(a) The points in (B) corresponding to points A–I in (A).

(b) The parts of the curve in (B) corresponding to the angle intervals $\left[0, \frac{\pi}{2}\right]$, $\left[\frac{\pi}{2}, \pi\right]$, $\left[\pi, \frac{3\pi}{2}\right]$, and $\left[\frac{3\pi}{2}, 2\pi\right]$.

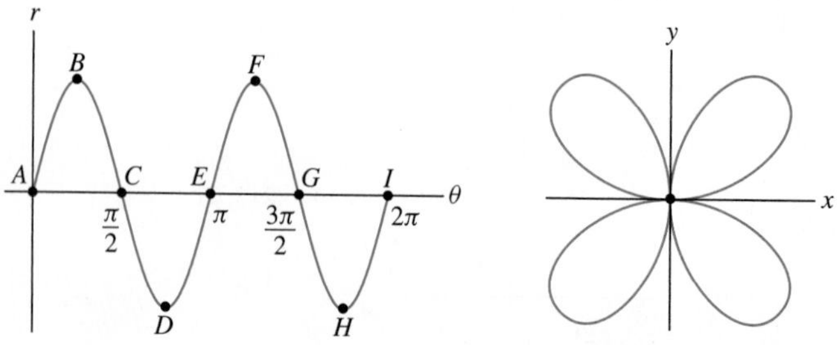

(A) Graph of r as a function of θ, where $r = \sin 2\theta$.

(B) Graph of $r = \sin 2\theta$ in polar coordinates.

FIGURE 20

32. Sketch the curve $r = \sin 3\theta$. First fill in the table of r-values below and plot the corresponding points of the curve. Notice that the three petals of the curve correspond to the angle intervals $[0, \frac{\pi}{3}]$, $[\frac{\pi}{3}, \frac{2\pi}{3}]$, and $[\frac{\pi}{3}, \pi]$. Then plot $r = \sin 3\theta$ in rectangular coordinates and label the points on this graph corresponding to (r, θ) in the table.

θ	0	$\frac{\pi}{12}$	$\frac{\pi}{6}$	$\frac{\pi}{4}$	$\frac{\pi}{3}$	$\frac{5\pi}{12}$	$\cdots$	$\frac{11\pi}{12}$	π
r									

33. CAS Plot the **cissoid** $r = 2\sin\theta\tan\theta$ and show that its equation in rectangular coordinates is

$$y^2 = \frac{x^3}{2 - x}$$

34. Prove that $r = 2a\cos\theta$ is the equation of the circle in Figure 21 using only the fact that a triangle inscribed in a circle with one side a diameter is a right triangle.

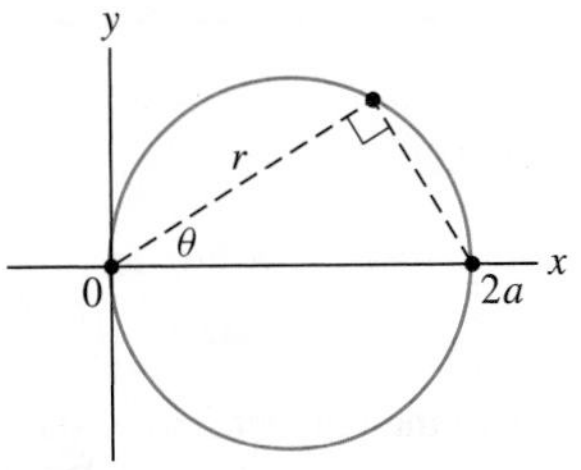

FIGURE 21

35. Show that

$$r = a\cos\theta + b\sin\theta$$

is the equation of a circle passing through the origin. Express the radius and center (in rectangular coordinates) in terms of a and b.

36. Use the previous exercise to write the equation of the circle of radius 5 and center (3, 4) in the form $r = a\cos\theta + b\sin\theta$.

37. Use the identity $\cos 2\theta = \cos^2\theta - \sin^2\theta$ to find a polar equation of the hyperbola $x^2 - y^2 = 1$.

38. Find an equation in rectangular coordinates for the curve $r^2 = \cos 2\theta$.

39. Show that $\cos 3\theta = \cos^3\theta - 3\cos\theta\sin^2\theta$ and use this identity to find an equation in rectangular coordinates for the curve $r = \cos 3\theta$.

40. Use the addition formula for the cosine to show that the line $\mathcal{L}$ with polar equation $r\cos(\theta - \alpha) = d$ has the equation in rectangular coordinates $(\cos\alpha)x + (\sin\alpha)y = d$. Show that $\mathcal{L}$ has slope $m = -\cot\alpha$ and y-intercept $d/\sin\alpha$.

In Exercises 41–44, find an equation in polar coordinates of the line $\mathcal{L}$ with the given description.

41. The point on $\mathcal{L}$ closest to the origin has polar coordinates $(2, \frac{\pi}{9})$.

42. The point on $\mathcal{L}$ closest to the origin has rectangular coordinates $(-2, 2)$.

43. $\mathcal{L}$ is tangent to the circle $r = 2\sqrt{10}$ at the point with rectangular coordinates $(-2, -6)$.

44. $\mathcal{L}$ has slope 3 and is tangent to the unit circle in the fourth quadrant.

45. Show that every line that does not pass through the origin has a polar equation of the form

$$r = \frac{b}{\sin\theta - a\cos\theta}$$

where $b \neq 0$.

46. By the Law of Cosines, the distance d between two points (Figure 22) with polar coordinates (r, θ) and (r_0, θ_0) is

$$d^2 = r^2 + r_0^2 - 2rr_0\cos(\theta - \theta_0)$$

Use this distance formula to show that

$$r^2 - 10r\cos\left(\theta - \frac{\pi}{4}\right) = 56$$

is the equation of the circle of radius 9 whose center has polar coordinates $(5, \frac{\pi}{4})$.

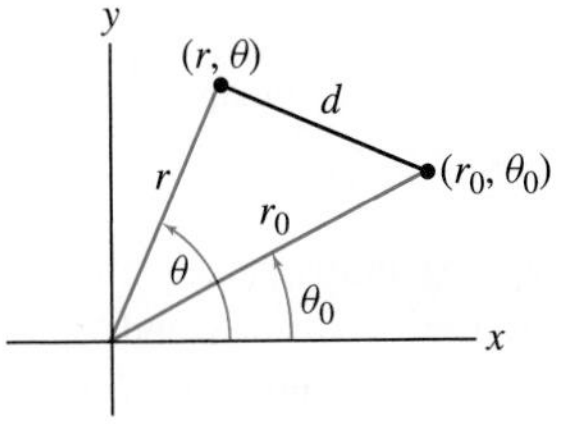

FIGURE 22

47. For $a > 0$, a **lemniscate curve** is the set of points P such that the product of the distances from P to $(a, 0)$ and $(-a, 0)$ is a^2. Show that the equation of the lemniscate is

$$(x^2 + y^2)^2 = 2a^2(x^2 - y^2)$$

Then find the equation in polar coordinates. To obtain the simplest form of the equation, use the identity $\cos 2\theta = \cos^2\theta - \sin^2\theta$. Plot the lemniscate for $a = 2$ if you have a computer algebra system.

48. Let c be a fixed constant. Explain the relationship between the graphs of:

(a) $y = f(x + c)$ and $y = f(x)$ (rectangular)
(b) $r = f(\theta + c)$ and $r = f(\theta)$ (polar)
(c) $y = f(x) + c$ and $y = f(x)$ (rectangular)
(d) $r = f(\theta) + c$ and $r = f(\theta)$ (polar)

49. The Derivative in Polar Coordinates Show that a polar curve $r = f(\theta)$ has parametric equations

$$x = f(\theta)\cos\theta, \qquad y = f(\theta)\sin\theta$$

Then apply Theorem 2 of Section 12.1 to prove

$$\frac{dy}{dx} = \frac{f(\theta)\cos\theta + f'(\theta)\sin\theta}{-f(\theta)\sin\theta + f'(\theta)\cos\theta} \qquad \boxed{2}$$

where $f'(\theta) = df/d\theta$.

50. Use Eq. (2) to find the slope of the tangent line to $r = \sin\theta$ at $\theta = \frac{\pi}{3}$.

51. Use Eq. (2) to find the slope of the tangent line to $r = \theta$ at $\theta = \frac{\pi}{2}$ and $\theta = \pi$.

52. Find the equation in rectangular coordinates of the tangent line to $r = 4\cos 3\theta$ at $\theta = \frac{\pi}{6}$.

53. Find the polar coordinates of the points on the lemniscate $r^2 = \cos 2t$ in Figure 23 where the tangent line is horizontal.

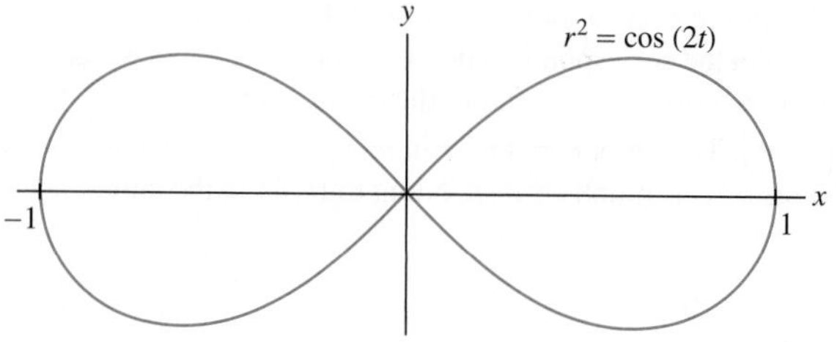

FIGURE 23

54. Find the polar coordinates of the points on the cardioid $r = 1 + \cos\theta$ where the tangent line is horizontal (see Figure 24).

55. Use Eq. (2) to show that for $r = \sin\theta + \cos\theta$,

$$\frac{dy}{dx} = \frac{\cos 2\theta + \sin 2\theta}{\cos 2\theta - \sin 2\theta}$$

Then calculate the slopes of the tangent lines at points A, B, C in Figure 19.

Further Insights and Challenges

56. Let $f(x)$ be a periodic function of period 2π—that is, $f(x) = f(x + 2\pi)$. Explain how this periodicity is reflected in the graph of:

(a) $y = f(x)$ in rectangular coordinates

(b) $r = f(\theta)$ in polar coordinates

57. GU Use a graphing utility to convince yourself that the polar equations $r = f_1(\theta) = 2\cos\theta - 1$ and $r = f_2(\theta) = 2\cos\theta + 1$ have the same graph. Then explain why. *Hint:* Show that the points $(f_1(\theta + \pi), \theta + \pi)$ and $(f_2(\theta), \theta)$ coincide.

58. CAS We investigate how the shape of the limaçon curve $r = b + \cos\theta$ depends on the constant b (see Figure 24).

(a) Argue as in Exercise 57 to show that the constants b and $-b$ yield the same curve.

(b) Plot the limaçon for $b = 0, 0.2, 0.5, 0.8, 1$ and describe how the curve changes.

(c) Plot the limaçon for 1.2, 1.5, 1.8, 2, 2.4 and describe how the curve changes.

(d) Use Eq. (2) to show that

$$\frac{dy}{dx} = -\left(\frac{b\cos\theta + \cos 2\theta}{b + 2\cos\theta}\right)\csc\theta$$

(e) Find the points where the tangent line is vertical. Note that there are three cases: $0 \le b < 2$, $b = 1$, and $b > 2$. Do the plots constructed in (b) and (c) reflect your results?

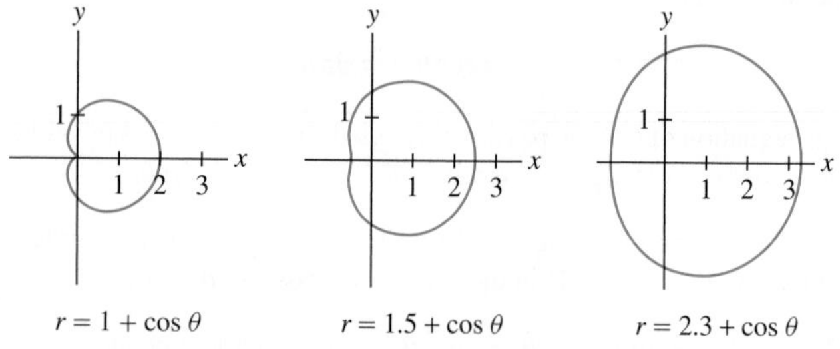

FIGURE 24

12.4 Area and Arc Length in Polar Coordinates

Integration in polar coordinates involves finding not the area *underneath* a curve but, rather, the area of a sector bounded by a curve as in Figure 1(A). Consider the region bounded by the curve $r = f(\theta)$ and the two rays $\theta = \alpha$ and $\theta = \beta$ with $\alpha < \beta$. To derive a formula for the area, divide the region into N narrow sectors of angle $\Delta\theta = (\beta - \alpha)/N$ corresponding to a partition of the interval $[\alpha, \beta]$:

$$\theta_0 = \alpha < \theta_1 < \theta_2 < \cdots < \theta_N = \beta$$

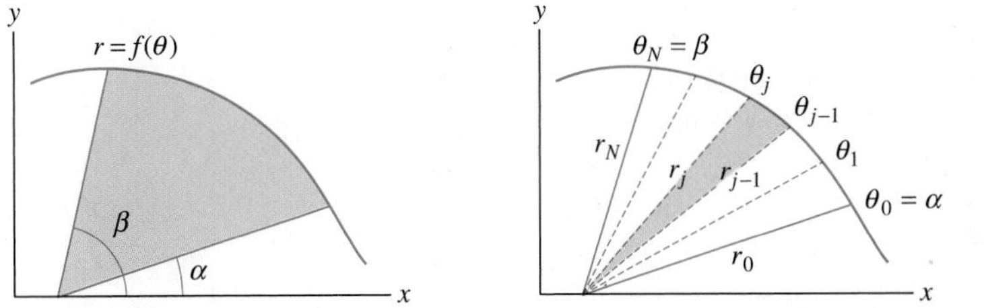

(A) Region $\alpha \le \theta \le \beta$ (B) Region divided into narrow sectors

FIGURE 1 Area bounded by the curve $r = f(\theta)$ and the two rays $\theta = \alpha$ and $\theta = \beta$.

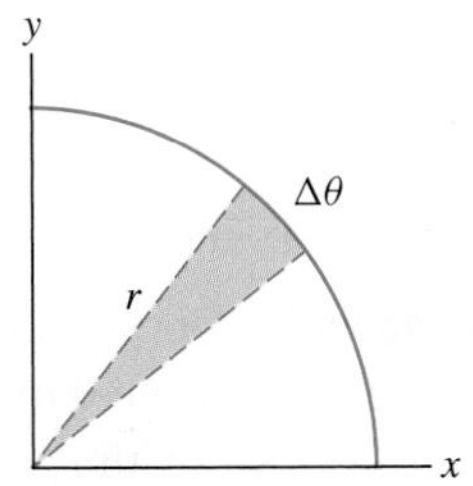

FIGURE 2 The area of a circular sector is *exactly* $\frac{1}{2}r^2\Delta\theta$.

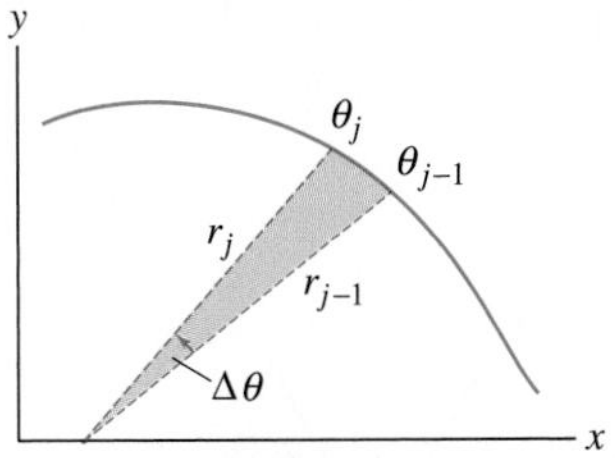

FIGURE 3 The area of the jth sector is *approximately* $\frac{1}{2}r_j^2\Delta\theta$.

Recall that a circular sector of angle $\Delta\theta$ and radius r has area $\frac{1}{2}r^2\Delta\theta$ (Figure 2). If $\Delta\theta$ is small, the jth narrow sector (Figure 3) is nearly a circular sector of radius $r_j = f(\theta_j)$, so its area is *approximately* $\frac{1}{2}r_j^2\Delta\theta$. The total area is approximated by the sum:

$$\text{Area of region} \approx \sum_{j=1}^{N} \frac{1}{2}r_j^2\Delta\theta = \frac{1}{2}\sum_{j=1}^{N} f(\theta_j)^2\Delta\theta \qquad \boxed{1}$$

This is a Riemann sum for the integral $\frac{1}{2}\int_\alpha^\beta f(\theta)^2\,d\theta$. If $f(\theta)$ is continuous, then the sum approaches the integral as $N \to \infty$, and we obtain the following formula.

THEOREM 1 Area in Polar Coordinates If $f(\theta)$ is a continuous function, then the area bounded by a curve in polar form $r = f(\theta)$ and the rays $\theta = \alpha$ and $\theta = \beta$ (with $\alpha < \beta$) is equal to

$$\frac{1}{2}\int_\alpha^\beta r^2\,d\theta = \frac{1}{2}\int_\alpha^\beta f(\theta)^2\,d\theta \qquad \boxed{2}$$

We know that $r = R$ defines a circle of radius R. By Eq. (2), the area is equal to $\frac{1}{2}\int_0^{2\pi} R^2\,d\theta = \frac{1}{2}R^2(2\pi) = \pi R^2$, as expected.

EXAMPLE 1 Use Theorem 1 to compute the area of the right semicircle with equation $r = 4\sin\theta$.

Solution The equation $r = 4\sin\theta$ defines a circle of radius 2 tangent to the x-axis at the origin. The right semicircle is "swept out" as θ varies from 0 to $\frac{\pi}{2}$ as in Figure 4(A). By Eq. (2), the area of the right semicircle is

REMINDER In Eq. (4), we use the identity

$$\sin^2\theta = \frac{1}{2}(1 - \cos 2\theta) \qquad \boxed{3}$$

$$\frac{1}{2}\int_0^{\pi/2} r^2\,d\theta = \frac{1}{2}\int_0^{\pi/2}(4\sin\theta)^2\,d\theta = 8\int_0^{\pi/2}\sin^2\theta\,d\theta \qquad \boxed{4}$$

$$= 8\int_0^{\pi/2}\frac{1}{2}(1 - \cos 2\theta)\,d\theta$$

$$= (4\theta - 2\sin 2\theta)\Big|_0^{\pi/2} = 4\left(\frac{\pi}{2}\right) - 0 = 2\pi$$

■

*CAUTION Keep in mind that the integral $\frac{1}{2}\int_\alpha^\beta r^2\,d\theta$ does **not** compute the area **under** a curve as in Figure 4(B), but rather computes the area "swept out" by a radial segment as θ varies from α to β, as in Figure 4(A).*

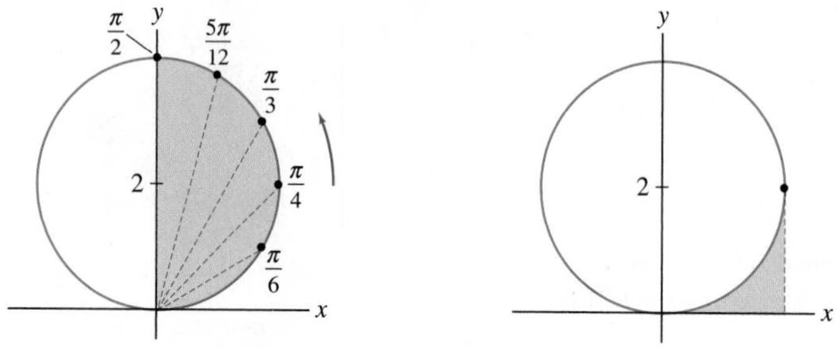

(A) The polar integral computes the area swept out by a radial segment.

(B) The ordinary integral in rectangular coordinates computes the area underneath a curve.

FIGURE 4

■ **EXAMPLE 2** Sketch $r = \sin 3\theta$ and compute the area of one "petal."

Solution To sketch the curve, we first graph $r = \sin 3\theta$ in rectangular coordinates. Figure 5 shows that the radius r varies from 0 to 1 and back to 0 as θ varies from 0 to $\frac{\pi}{3}$. This gives petal A in Figure 6. Petal B is traced as θ varies from $\frac{\pi}{3}$ to $\frac{2\pi}{3}$ (with $r \le 0$), and petal C is traced for $\frac{2\pi}{3} \le \theta \le \pi$. We find that the area of petal A (using Eq. (3) in the margin of the previous page to evaluate the integral) is equal to

$$\frac{1}{2}\int_0^{\pi/3} (\sin 3\theta)^2\,d\theta = \frac{1}{2}\int_0^{\pi/3}\left(\frac{1-\cos 6\theta}{2}\right)d\theta = \left(\frac{1}{4}\theta - \frac{1}{24}\sin 6\theta\right)\Bigg|_0^{\pi/3} = \frac{\pi}{12} \quad ■$$

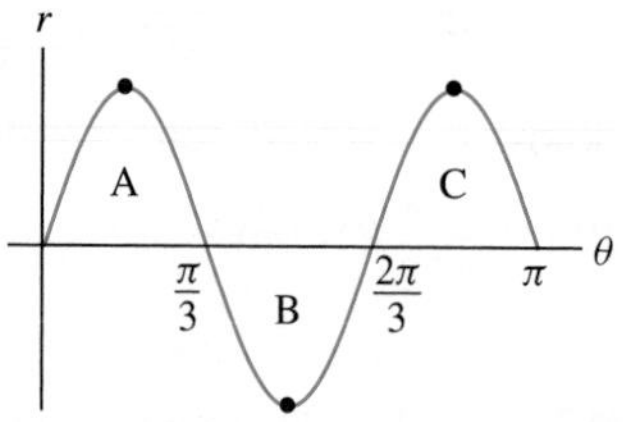

FIGURE 5 Graph of $r = \sin 3\theta$ as a function of θ.

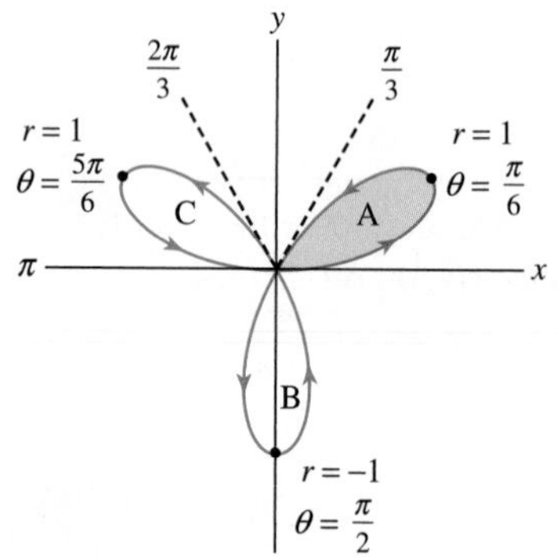

FIGURE 6 Graph of polar curve $r = \sin 3\theta$, a "rose with three petals."

The area between two polar curves $r = f_1(\theta)$ and $r = f_2(\theta)$ with $f_2(\theta) \ge f_1(\theta)$, for $\alpha \le \theta \le \beta$, is equal to (Figure 7):

$$\text{Area between two curves} = \frac{1}{2}\int_\alpha^\beta \left(f_2(\theta)^2 - f_1(\theta)^2\right)d\theta \qquad \boxed{5}$$

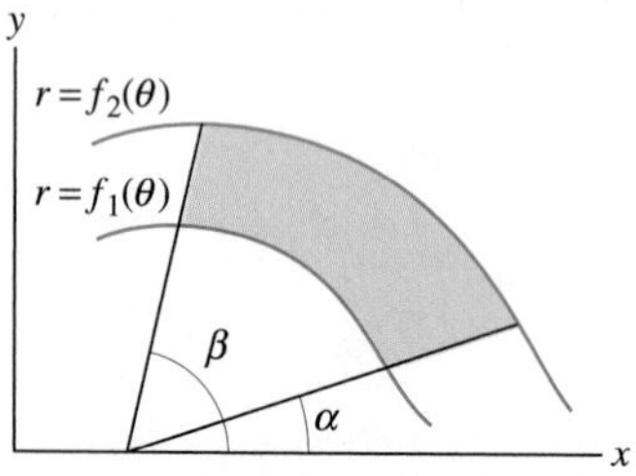

FIGURE 7 Area between two polar graphs in a sector.

■ **EXAMPLE 3 Area Between Two Curves** Find the area of the region inside the circle $r = 2\cos\theta$ but outside the circle $r = 1$ [Figure 8(A)].

Solution The two circles intersect at the points where $(r, 2\cos\theta) = (r, 1)$ or in other words, when $2\cos\theta = 1$. This yields $\cos\theta = \frac{1}{2}$, which has solutions $\theta = \pm\frac{\pi}{3}$.

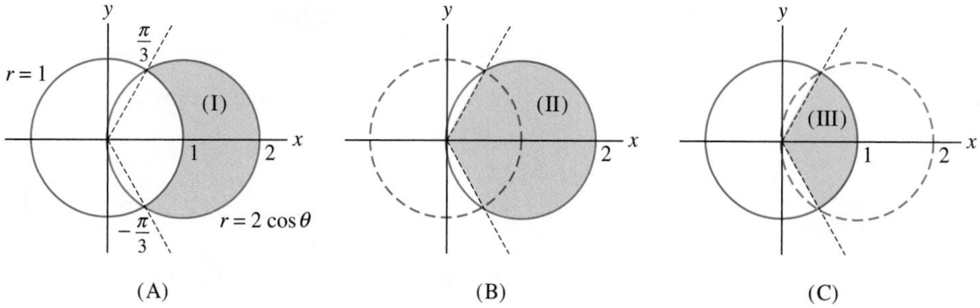

FIGURE 8 Region (I) is the difference of regions (II) and (III).

We see in Figure 8 that region (I) is the difference of regions (II) and (III) in Figures 8(B) and (C). Therefore,

$$\text{Area of (I)} = \text{area of (II)} - \text{area of (III)}$$
$$= \frac{1}{2}\int_{-\pi/3}^{\pi/3} (2\cos\theta)^2\,d\theta - \frac{1}{2}\int_{-\pi/3}^{\pi/3} (1)^2\,d\theta$$
$$= \frac{1}{2}\int_{-\pi/3}^{\pi/3} (4\cos^2\theta - 1)\,d\theta = \frac{1}{2}\int_{-\pi/3}^{\pi/3} (2\cos 2\theta + 1)\,d\theta \qquad \boxed{6}$$
$$= \frac{1}{2}(\sin 2\theta + \theta)\Big|_{-\pi/3}^{\pi/3} = \frac{\sqrt{3}}{2} + \frac{\pi}{3} \approx 1.91$$ ■

← REMINDER *In Eq. (6), we use the identity*

$$\cos^2\theta = \frac{1}{2}(1 + \cos 2\theta)$$

We close this section by deriving a formula for arc length in polar coordinates. Observe that a polar curve $r = f(\theta)$ has a parametrization with θ as a parameter:

$$x = r\cos\theta = f(\theta)\cos\theta, \qquad y = r\sin\theta = f(\theta)\sin\theta$$

Using a prime to denote the derivative with respect to θ, we have

$$x'(\theta) = \frac{dx}{d\theta} = -f(\theta)\sin\theta + f'(\theta)\cos\theta$$
$$y'(\theta) = \frac{dy}{d\theta} = f(\theta)\cos\theta + f'(\theta)\sin\theta$$

Recall from Section 12.2 that arc length is obtained by integrating $\sqrt{x'(\theta)^2 + y'(\theta)^2}$. Straightforward algebra shows that $x'(\theta)^2 + y'(\theta)^2 = f(\theta)^2 + f'(\theta)^2$, and thus

$$\text{Arc length } s = \int_\alpha^\beta \sqrt{f(\theta)^2 + f'(\theta)^2}\,d\theta \qquad \boxed{7}$$

■ **EXAMPLE 4** Find the total length of the circle $r = 2a\cos\theta$ for $a > 0$.

Solution In this case, $f(\theta) = 2a\cos\theta$ and

$$f(\theta)^2 + f'(\theta)^2 = 4a^2\cos^2\theta + 4a^2\sin^2\theta = 4a^2$$

The total length of this circle of radius a has the expected value:

$$\int_0^\pi \sqrt{f(\theta)^2 + f'(\theta)^2}\,d\theta = \int_0^\pi (2a)\,d\theta = 2\pi a$$

Note that the upper limit of integration is π rather than 2π because the entire circle is traced out as θ varies from 0 to π (see Figure 9). ■

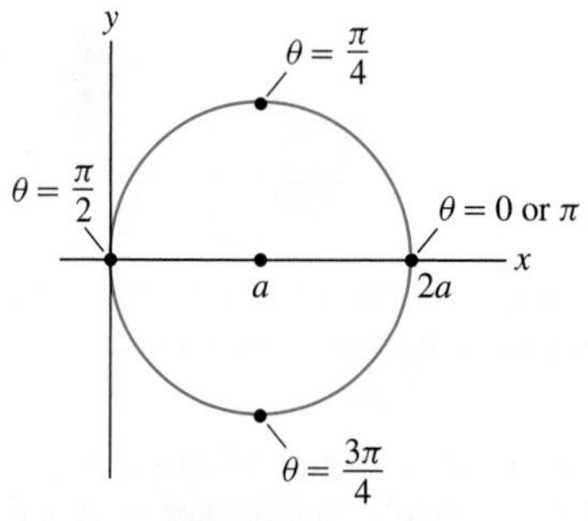

FIGURE 9 Graph of $r = 2a\cos\theta$.

12.4 SUMMARY

- Area of the sector bounded by a polar curve $r = f(\theta)$ and two rays $\theta = \alpha$ and $\theta = \beta$ (Figure 10):

$$\text{Area} = \frac{1}{2}\int_{\alpha}^{\beta} f(\theta)^2\, d\theta$$

- Area between $r = f_1(\theta)$ and $r = f_2(\theta)$, where $f_2(\theta) \ge f_1(\theta)$ (Figure 11):

$$\text{Area} = \frac{1}{2}\int_{\alpha}^{\beta} \left(f_2(\theta)^2 - f_1(\theta)^2\right) d\theta$$

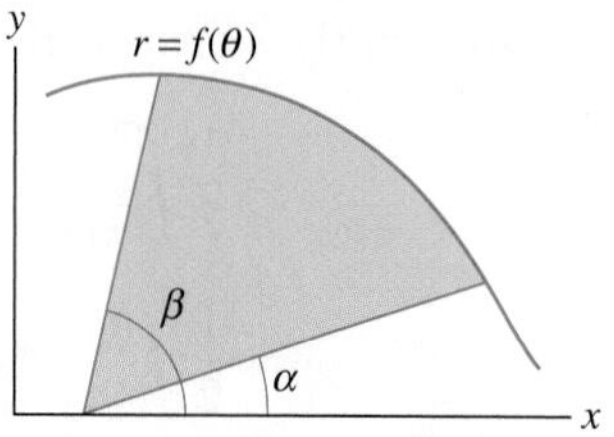

FIGURE 10 Region bounded by the polar curve $r = f(\theta)$ and the rays $\theta = \alpha$, $\theta = \beta$.

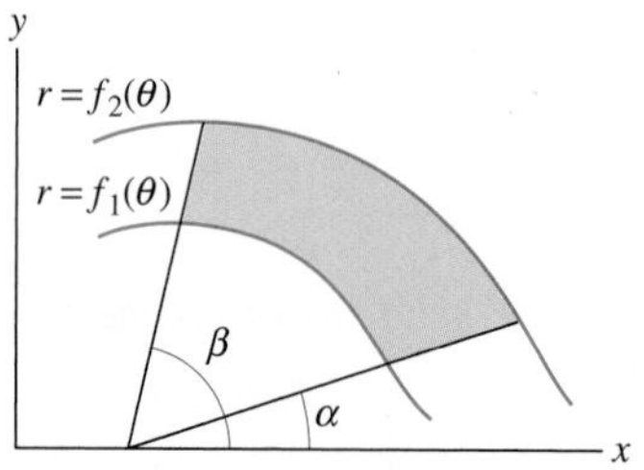

FIGURE 11 Region between two polar curves.

- Arc length of the polar curve $r = f(\theta)$ for $\alpha \le \theta \le \beta$:

$$\text{Arc length} = \int_{\alpha}^{\beta} \sqrt{f(\theta)^2 + f'(\theta)^2}\, d\theta$$

12.4 EXERCISES

Preliminary Questions

1. Polar coordinates are suited to finding the area (choose one):

(a) Under a curve between $x = a$ and $x = b$.

(b) Bounded by a curve and two rays through the origin.

2. Is the formula for area in polar coordinates valid if $f(\theta)$ takes negative values?

3. The horizontal line $y = 1$ has polar equation $r = \csc\theta$. Which area is represented by the integral $\frac{1}{2}\int_{\pi/6}^{\pi/2} \csc^2\theta\, d\theta$ (Figure 12)?

(a) $\square ABCD$ **(b)** $\triangle ABC$ **(c)** $\triangle ACD$

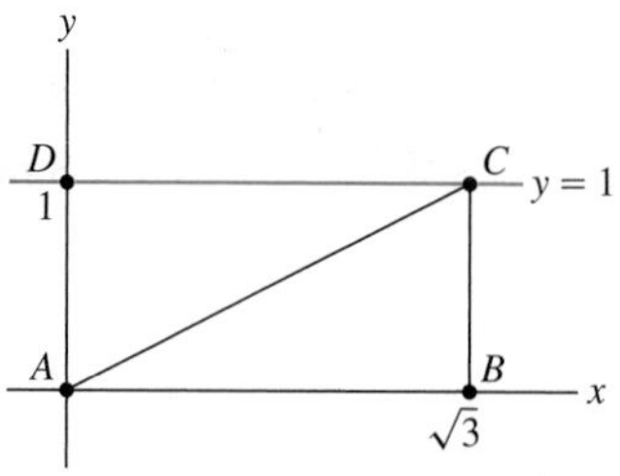

FIGURE 12

Exercises

1. Sketch the area bounded by the circle $r = 5$ and the rays $\theta = \frac{\pi}{2}$ and $\theta = \pi$, and compute its area as an integral in polar coordinates.

2. Sketch the region bounded by the line $r = \sec\theta$ and the rays $\theta = 0$ and $\theta = \frac{\pi}{3}$. Compute its area in two ways: as an integral in polar coordinates and using geometry.

3. Calculate the area of the circle $r = 4\sin\theta$ as an integral in polar coordinates (see Figure 4). Be careful to choose the correct limits of integration.

4. Find the area of the shaded triangle in Figure 13 as an integral in polar coordinates. Then find the rectangular coordinates of P and Q and compute the area via geometry.

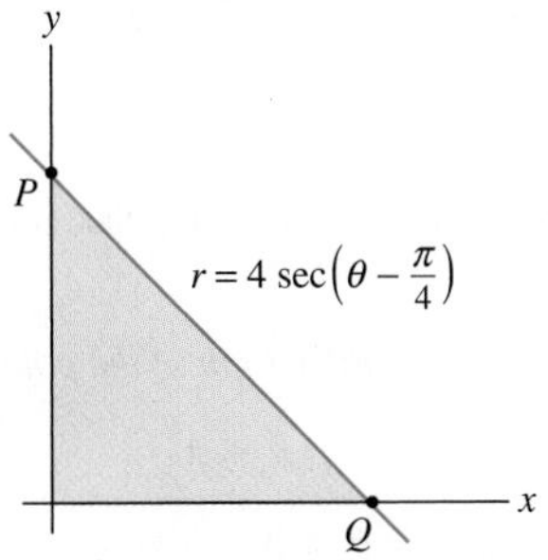

FIGURE 13

5. Find the area of the shaded region in Figure 14. Note that θ varies from 0 to $\frac{\pi}{2}$.

6. Which interval of θ-values corresponds to the the shaded region in Figure 15? Find the area of the region.

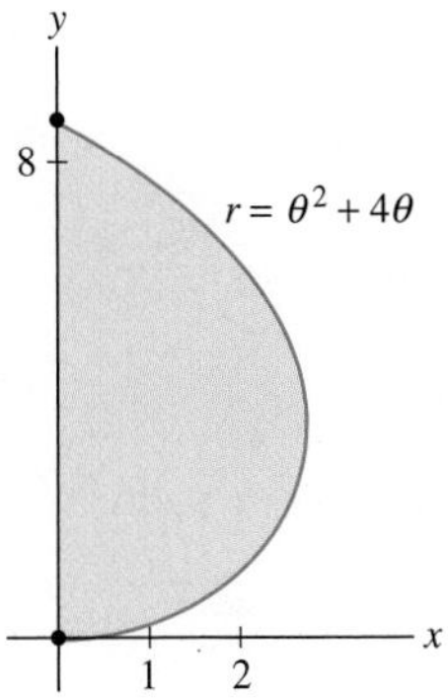

FIGURE 14

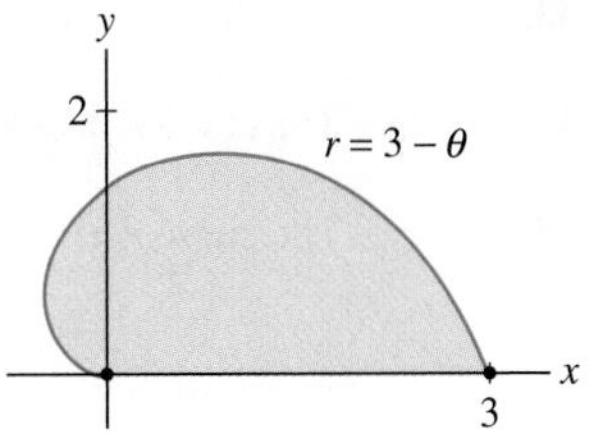

FIGURE 15

7. Find the total area enclosed by the cardioid in Figure 16.

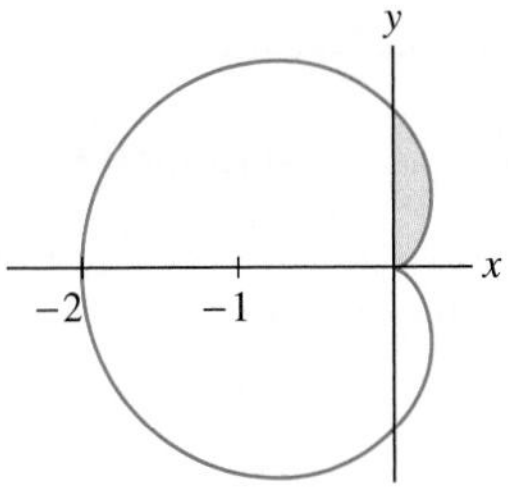

FIGURE 16 The cardioid $r = 1 - \cos\theta$.

8. Find the area of the shaded region in Figure 16.

9. Find the area of one leaf of the "four-petaled rose" $r = \sin 2\theta$ (Figure 17). Then prove that the total area of the rose is equal to one-half the area of the circumscribed circle.

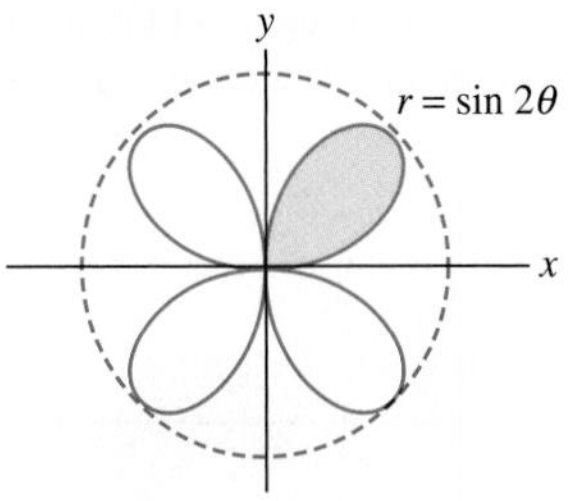

FIGURE 17 Four-petaled rose $r = \sin 2\theta$.

10. Find the area enclosed by one loop of the lemniscate with equation $r^2 = \cos 2\theta$ (Figure 18). Choose your limits of integration carefully.

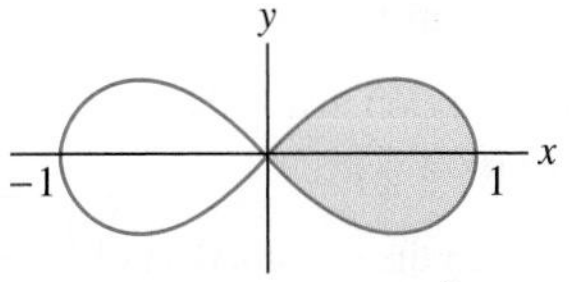

FIGURE 18 The lemniscate $r^2 = \cos 2\theta$.

11. Sketch the spiral $r = \theta$ for $0 \le \theta \le 2\pi$ and find the area bounded by the curve and the first quadrant.

12. Find the area of the intersection of the circles $r = \sin\theta$ and $r = \cos\theta$.

13. Find the area of region A in Figure 19.

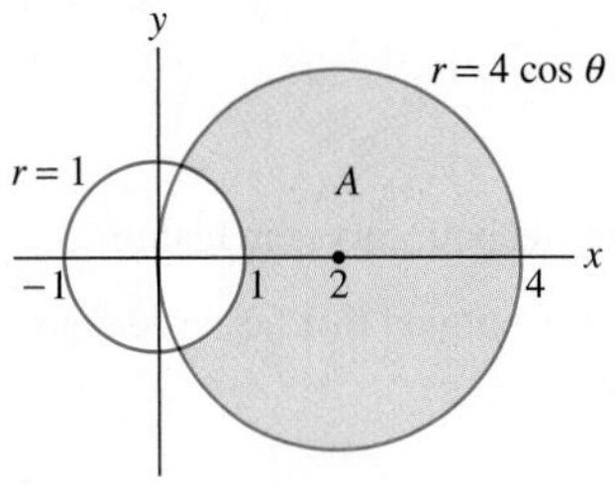

FIGURE 19

14. Find the area of the shaded region in Figure 20, enclosed by the circle $r = \frac{1}{2}$ and a petal of the curve $r = \cos 3\theta$. *Hint:* Compute the area of both the petal and the region inside the petal and outside the circle.

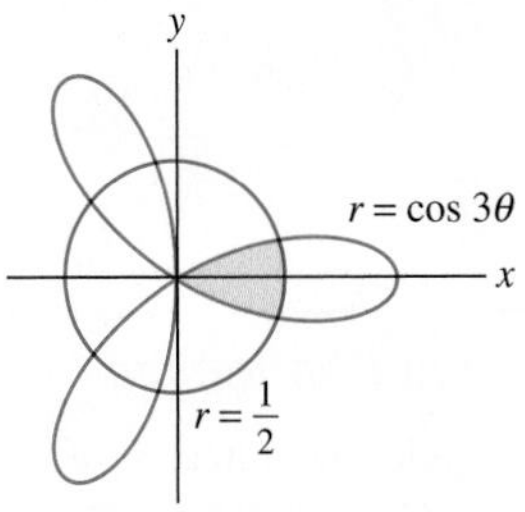

FIGURE 20

15. Find the area of the inner loop of the limaçon with polar equation $r = 2\cos\theta - 1$ (Figure 21).

16. Find the area of the shaded region in Figure 21 between the inner and outer loop of the limaçon $r = 2\cos\theta - 1$.

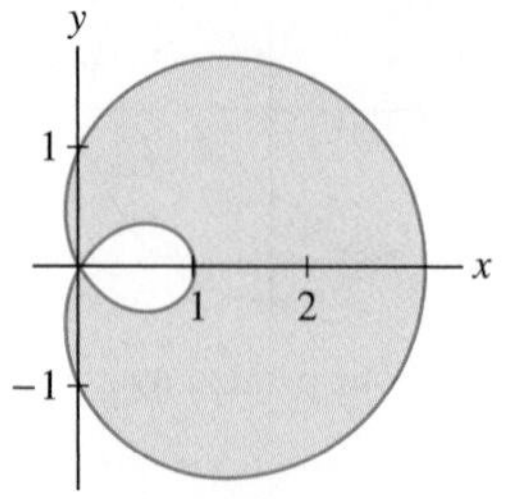

FIGURE 21 The limaçon $r = 2\cos\theta - 1$.

17. Find the area of the part of the circle $r = \sin\theta + \cos\theta$ in the fourth quadrant (see Exercise 26 in Section 12.3).

18. Find the area of the region inside the circle $r = 2\sin\left(\theta + \frac{\pi}{4}\right)$ and above the line $r = \sec\left(\theta - \frac{\pi}{4}\right)$.

19. Find the area between the two curves in Figure 22(A).

20. Find the area between the two curves in Figure 22(B).

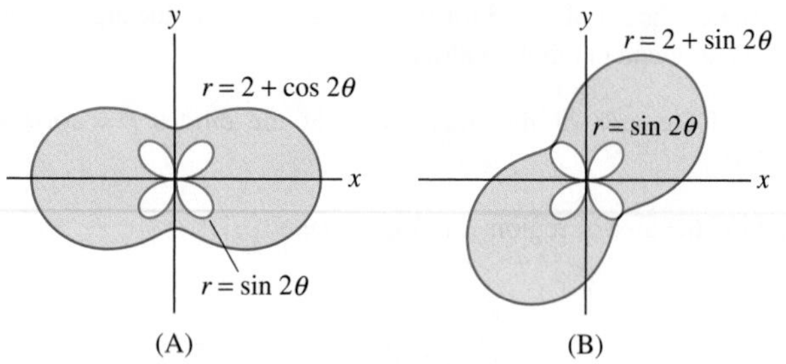

FIGURE 22

21. Find the area inside both curves in Figure 23.

22. Find the area of the region that lies inside one but not both of the curves in Figure 23.

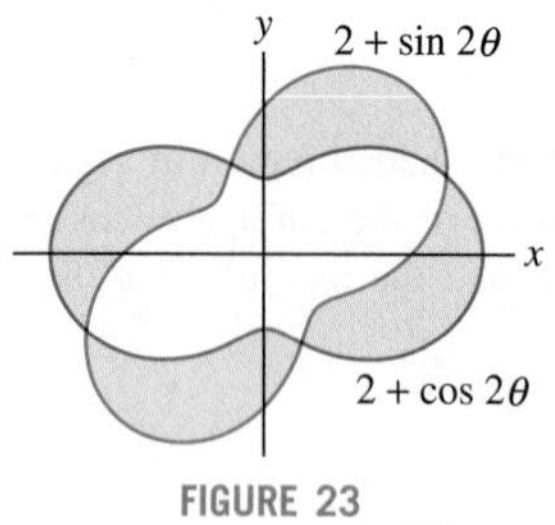

FIGURE 23

23. Calculate the total length of the circle $r = 4\sin\theta$ as an integral in polar coordinates.

24. Sketch the segment $r = \sec\theta$ for $0 \le \theta \le A$. Then compute its length in two ways: as an integral in polar coordinates and using trigonometry.

In Exercises 25–30, compute the length of the polar curve.

25. The length of $r = \theta^2$ for $0 \le \theta \le \pi$

26. The spiral $r = \theta$ for $0 \le \theta \le A$

27. The equiangular spiral $r = e^\theta$ for $0 \le \theta \le 2\pi$

28. The inner loop of $r = 2\cos\theta - 1$ in Figure 21

29. The cardioid $r = 1 - \cos\theta$ in Figure 16

30. $r = \cos^2\theta$

In Exercises 31 and 32, express the length of the curve as an integral but do not evaluate it.

31. $r = (2 - \cos\theta)^{-1}, \quad 0 \le \theta \le 2\pi$

32. $r = \sin^3 t, \quad 0 \le \theta \le 2\pi$

In Exercises 33–36, use a computer algebra system to calculate the total length to two decimal places.

33. CAS The three-petal rose $r = \cos 3\theta$ in Figure 20

34. CAS The curve $r = 2 + \sin 2\theta$ in Figure 23

35. CAS The curve $r = \theta\sin\theta$ in Figure 24 for $0 \le \theta \le 4\pi$

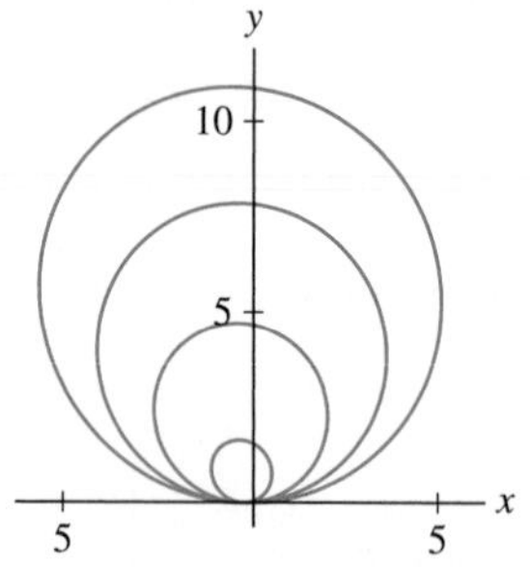

FIGURE 24 $r = \theta\sin\theta$ for $0 \le \theta \le 4\pi$.

36. CAS $r = \sqrt{\theta}, \quad 0 \le \theta \le 4\pi$

Further Insights and Challenges

37. Suppose that the polar coordinates of a moving particle at time t are $(r(t), \theta(t))$. Prove that the particle's speed is equal to $\sqrt{(dr/dt)^2 + r^2(d\theta/dt)^2}$.

38. Compute the speed at time $t = 1$ of a particle whose polar coordinates at time t are $r = t$, $\theta = t$ (use Exercise 37). What would the speed be if the particle's rectangular coordinates were $x = t$, $y = t$? Why is the speed increasing in one case and constant in the other?

12.5 Conic Sections

The conics were first studied by the ancient Greek mathematicians, beginning possibly with Menaechmus (c. 380–320 BCE) and including Archimedes (287–212 BCE) and Apollonius (c. 262–190 BCE).

Three familiar families of curves—ellipses, hyperbolas, and parabolas—appear throughout mathematics and its applications. They are called **conic sections** because they are obtained as the intersection of a cone with a suitable plane (Figure 1). Our goal in this section is to derive equations for the conic sections from their geometric definitions as curves in the plane.

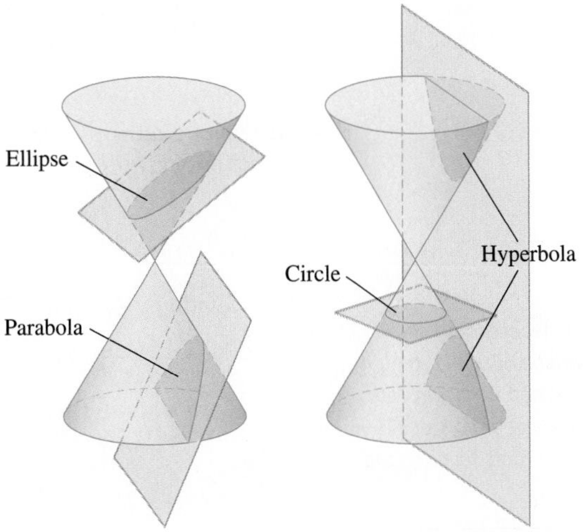

FIGURE 1 The conic sections are obtained by intersecting a plane and a cone.

An **ellipse** is an oval-shaped curve [Figure 2(A)] consisting of all points P such that the sum of the distances to two fixed points F_1 and F_2 is a constant $K > 0$:

$$PF_1 + PF_2 = K \qquad \boxed{1}$$

We assume always that K is greater than the distance F_1F_2 between the foci, because the ellipse reduces to the line segment $\overline{F_1F_2}$ if $K = F_1F_2$, and it has no points at all if $K < F_1F_2$.

The points F_1 and F_2 are called the **foci** (plural of "focus") of the ellipse. Note that if the foci coincide, then Eq. (1) reduces to $2PF_1 = K$ and we obtain a circle of radius $\frac{1}{2}K$ centered at F_1.

We use the following terminology:

- The midpoint of $\overline{F_1F_2}$ is the **center** of the ellipse.
- The line through the foci is the **focal axis**.
- The line through the center perpendicular to the focal axis is the **conjugate axis**.

The ellipse is said to be in **standard position** if the focal and conjugate axes are the x- and y-axes, as shown in Figure 2(B). In this case, the foci have coordinates $F_1 = (c, 0)$ and $F_2 = (-c, 0)$ for some $c > 0$. Let us prove that the equation of this ellipse has the particularly simple form

$$\left(\frac{x}{a}\right)^2 + \left(\frac{x}{b}\right)^2 = 1 \qquad \boxed{2}$$

where $a = K/2$ and $b = \sqrt{a^2 - c^2}$.

By the distance formula, $P = (x, y)$ lies on the ellipse in Figure 2(B) if

$$PF_1 + PF_2 = \sqrt{(x + c)^2 + y^2} + \sqrt{(x - c)^2 + y^2} = 2a \qquad \boxed{3}$$

Move the second term on the left over to the right and square both sides:

$$(x + c)^2 + y^2 = 4a^2 - 4a\sqrt{(x - c)^2 + y^2} + (x - c)^2 + y^2$$

$$4a\sqrt{(x - c)^2 + y^2} = 4a^2 + (x - c)^2 - (x + c)^2 = 4a^2 - 4cx$$

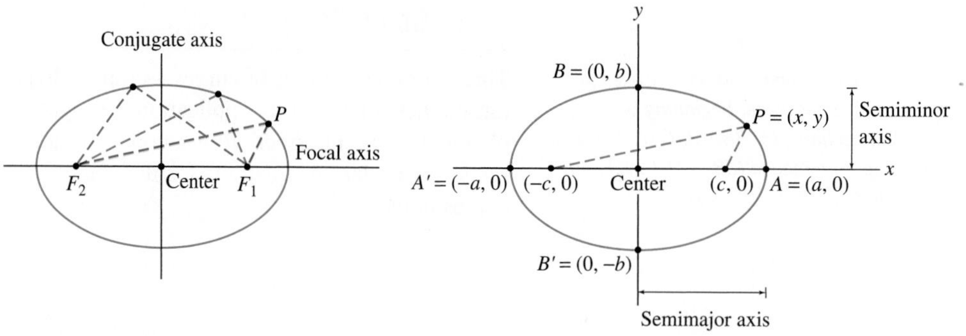

(A) The ellipse consists of all points P such that $PF_1 + PF_2 = K$.

(B) Ellipse in standard position: $\left(\frac{x}{a}\right)^2 + \left(\frac{y}{b}\right)^2 = 1$

FIGURE 2

Strictly speaking, it is necessary to show that if $P = (x, y)$ satisfies Eq. (4), then it also satisfies Eq. (3). When we begin with Eq. (4) and reverse the algebraic steps, the process of taking square roots leads to the relation

$$\sqrt{(x-c)^2 + y^2} \pm \sqrt{(x+c)^2 + y^2} = \pm 2a$$

However, this equation has no solutions unless both signs are positive because $a > c$.

Now divide by 4, square, and simplify:

$$a^2(x^2 - 2cx + c^2 + y^2) = a^4 - 2a^2cx + c^2x^2$$

$$(a^2 - c^2)x^2 + a^2y^2 = a^4 - a^2c^2 = a^2(a^2 - c^2)$$

$$\frac{x^2}{a^2} + \frac{y^2}{a^2 - c^2} = 1 \qquad \boxed{4}$$

This is Eq. (2) with $b^2 = a^2 - c^2$ as claimed.

The ellipse intersects the axes in four points A, A', B, B' called **vertices**. Vertices A and A' along the focal axis are called the **focal vertices**. Following common usage, the numbers a and b are referred to as the **semimajor axis** and the **semiminor axis** (even though they are numbers rather than axes).

THEOREM 1 Ellipse in Standard Position Let $a > b > 0$, and set $c = \sqrt{a^2 - b^2}$. The ellipse $PF_1 + PF_2 = 2a$ with foci $F_1 = (c, 0)$ and $F_2 = (-c, 0)$ has equation

$$\left(\frac{x}{a}\right)^2 + \left(\frac{y}{b}\right)^2 = 1 \qquad \boxed{5}$$

Furthermore, the ellipse has

- Semimajor axis a, semiminor axis b.
- Focal vertices $(\pm a, 0)$, minor vertices $(0, \pm b)$.

If $b > a > 0$, then Eq. (5) defines an ellipse with foci $(0, \pm c)$, where $c = \sqrt{b^2 - a^2}$.

EXAMPLE 1 Find the equation of the ellipse with foci $(\pm\sqrt{11}, 0)$ and semimajor axis $a = 6$. Then find the semiminor axis and sketch the graph.

Solution The foci are $(\pm c, 0)$ with $c = \sqrt{11}$, and the semimajor axis is $a = 6$, so we can use the relation $c = \sqrt{a^2 - b^2}$ to find b:

$$b^2 = a^2 - c^2 = 6^2 - (\sqrt{11})^2 = 25 \quad \Rightarrow \quad b = 5$$

Thus, the semiminor axis is $b = 5$ and the ellipse has equation $\left(\frac{x}{6}\right)^2 + \left(\frac{y}{5}\right)^2 = 1$. To sketch this ellipse, plot the vertices $(\pm 6, 0)$ and $(0, \pm 5)$ and connect them as in Figure 3. ■

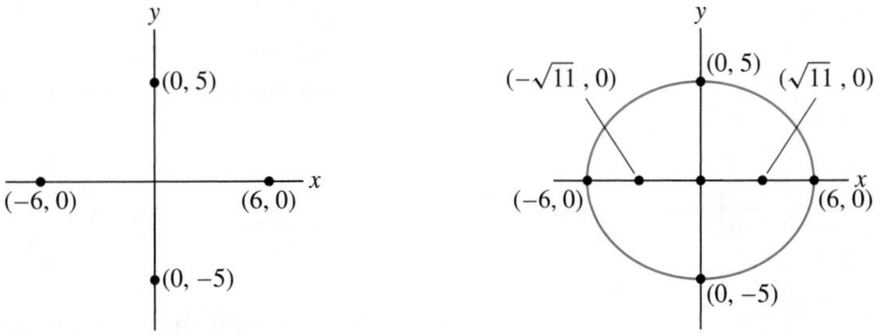

FIGURE 3

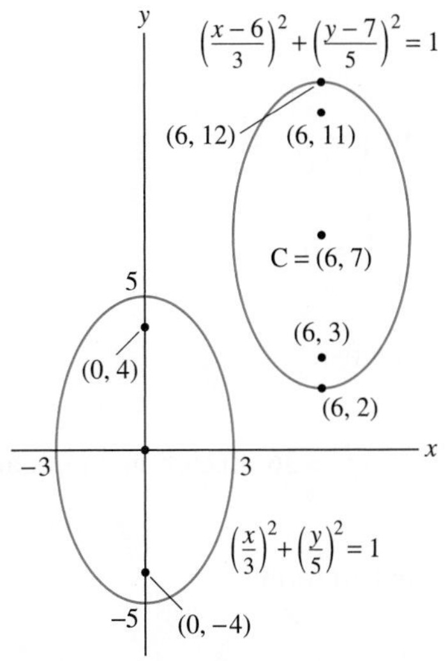

FIGURE 4 An ellipse with vertical major axis and its translate with center $C = (6, 7)$.

To write down the equation of an ellipse with axes parallel to the x- and y-axes and center translated to the point $C = (h, k)$, replace x by $x - h$ and y by $y - k$ in the equation (Figure 4):

$$\left(\frac{x-h}{a}\right)^2 + \left(\frac{y-k}{b}\right)^2 = 1$$

■ **EXAMPLE 2 Translating an Ellipse** Find an equation of the ellipse with center $C = (6, 7)$, vertical focal axis, semimajor axis 5, and semiminor axis 3. Where are the foci located?

Solution Since the focal axis is vertical, we have $a = 3$ and $b = 5$, so that $a < b$ (Figure 4). The ellipse centered at the origin would have equation $\left(\frac{x}{3}\right)^2 + \left(\frac{y}{5}\right)^2 = 1$. When the center is translated to $(h, k) = (6, 7)$, the equation becomes

$$\left(\frac{x-6}{3}\right)^2 + \left(\frac{y-7}{5}\right)^2 = 1$$

Furthermore, $c = \sqrt{b^2 - a^2} = \sqrt{5^2 - 3^2} = 4$, so the foci are located ± 4 vertical units above and below the center—that is, $F_1 = (6, 11)$ and $F_2 = (6, 3)$. ■

A **hyperbola** is the set of all points P such that the difference of the distances from P to two foci F_1 and F_2 is $\pm K$:

$$PF_1 - PF_2 = \pm K \qquad \boxed{6}$$

We assume that K is less than the distance F_1F_2 between the foci (the hyperbola has no points if $K > F_1F_2$). Note that a hyperbola consists of two branches corresponding to the choices of sign $\pm$ (Figure 5).

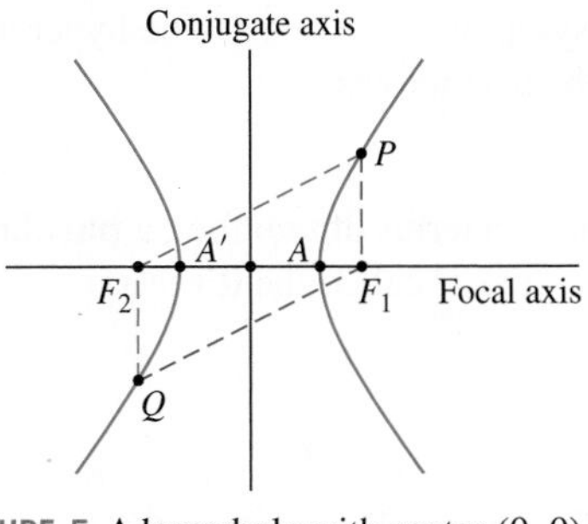

FIGURE 5 A hyperbola with center (0, 0).

As before, the midpoint of $\overline{F_1F_2}$ is the **center** of the hyperbola, the line through F_1 and F_2 is called the **focal axis**, and the line through the center perpendicular to the focal axis is called the **conjugate axis**. The **vertices** are the points where the focal axis intersects the hyperbola; they are labeled A and A' in Figure 5. The hyperbola is said to be in standard position when the focal and conjugate axes are the x- and y-axes as in Figure 6. The next theorem can be verified in much the same way as Theorem 1.

THEOREM 2 Hyperbola in Standard Position Let $a > 0$ and $b > 0$, and set $c = \sqrt{a^2 + b^2}$. The hyperbola $PF_1 - PF_2 = \pm 2a$ with foci $F_1 = (c, 0)$ and $F_2 = (-c, 0)$ has equation

$$\left(\frac{x}{a}\right)^2 - \left(\frac{y}{b}\right)^2 = 1 \qquad \boxed{7}$$

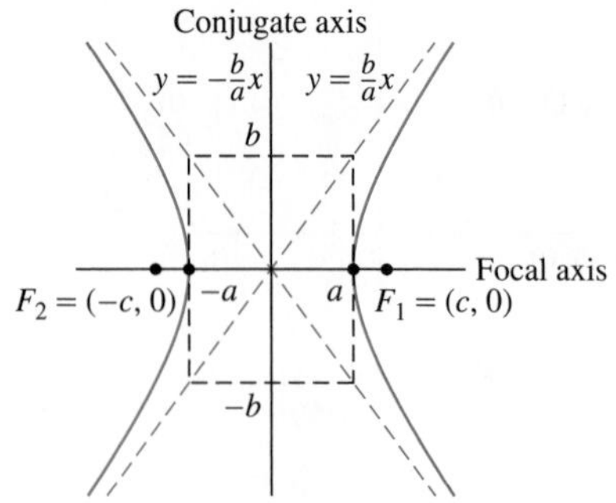

FIGURE 6 Hyperbola in standard position.

A hyperbola has two **asymptotes** $y = \pm\frac{b}{a}x$ which are, we claim, diagonals of the rectangle whose sides pass through $(\pm a, 0)$ and $(0, \pm b)$ as in Figure 6. To prove this, consider a point (x, y) on the hyperbola in the first quadrant. By Eq. (7),

$$y = \sqrt{\frac{b^2}{a^2}x^2 - b^2} = \frac{b}{a}\sqrt{x^2 - a^2}$$

The following limit shows that a point (x, y) on the hyperbola approaches the line $y = \frac{b}{a}x$ as $x \to \infty$:

$$\begin{aligned}\lim_{x\to\infty}\left(y - \frac{b}{a}x\right) &= \frac{b}{a}\lim_{x\to\infty}\left(\sqrt{x^2 - a^2} - x\right)\\ &= \frac{b}{a}\lim_{x\to\infty}\left(\sqrt{x^2 - a^2} - x\right)\left(\frac{\sqrt{x^2 - a^2} + x}{\sqrt{x^2 - a^2} + x}\right)\\ &= \frac{b}{a}\lim_{x\to\infty}\left(\frac{-a^2}{\sqrt{x^2 - a^2} + x}\right) = 0\end{aligned}$$

The asymptotic behavior in the remaining quadrants is similar.

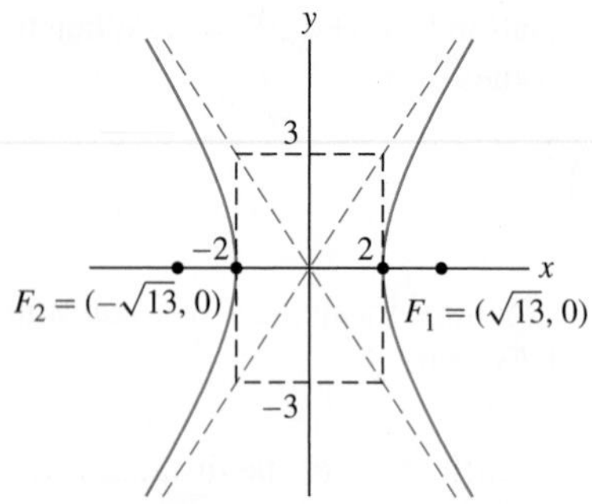

FIGURE 7 The hyperbola $9x^2 - 4y^2 = 36$.

■ **EXAMPLE 3** Find the foci of the hyperbola $9x^2 - 4y^2 = 36$. Sketch its graph and asymptotes.

Solution First divide by 36 to write the equation in standard form:

$$\frac{x^2}{4} - \frac{y^2}{9} = 1 \qquad \text{or} \qquad \left(\frac{x}{2}\right)^2 - \left(\frac{y}{3}\right)^2 = 1$$

Thus $a = 2$, $b = 3$, and $c = \sqrt{a^2 + b^2} = \sqrt{4 + 9} = \sqrt{13}$. The foci are

$$F_1 = (\sqrt{13}, 0), \qquad F_2 = (-\sqrt{13}, 0)$$

To sketch the graph, we draw the rectangle through the points $(\pm 2, 0)$ and $(0, \pm 3)$ as in Figure 7. The diagonals of the rectangle are the asymptotes $y = \pm\frac{3}{2}x$. The hyperbola passes through the vertices $(\pm 2, 0)$ and approaches the asymptotes. ■

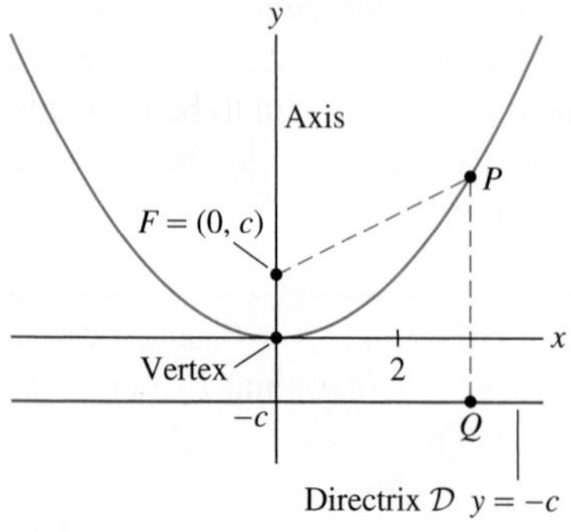

FIGURE 8 Parabola with focus $(0, c)$ and directrix $y = -c$.

Unlike the ellipse and hyperbola, which are defined in terms of two foci, a **parabola** is the set of points P equidistant from a focus F and a line $\mathcal{D}$ called the **directrix**:

$$PF = P\mathcal{D} \tag{8}$$

Here, when we speak of the *distance* from a point P to a line $\mathcal{D}$, we mean the distance from P to the point Q on $\mathcal{D}$ closest to P, obtained by dropping a perpendicular from P to $\mathcal{D}$ (Figure 8). We denote this distance by $P\mathcal{D}$.

The line through the focus F perpendicular to $\mathcal{D}$ is called the **axis** of the parabola. The **vertex** is the point where the parabola intersects its axis. We say that the parabola is in standard position if, for some c, the focus is $F = (0, c)$ and the directrix is $y = -c$, as shown in Figure 8. We verify in Exercise 73 that the vertex is then located at the origin and the equation of the parabola is $y = x^2/4c$. If $c < 0$, then the parabola opens downward.

THEOREM 3 Parabola in Standard Position Let $c \neq 0$. The parabola with focus $F = (0, c)$ and directrix $y = -c$ has equation

$$y = \frac{1}{4c}x^2 \tag{9}$$

The vertex is located at the origin. The parabola opens upward if $c > 0$ and downward if $c < 0$.

■ **EXAMPLE 4** The standard parabola with directrix $y = -2$ is translated so that its vertex is located at $(2, 8)$. Find its equation, directrix, and focus.

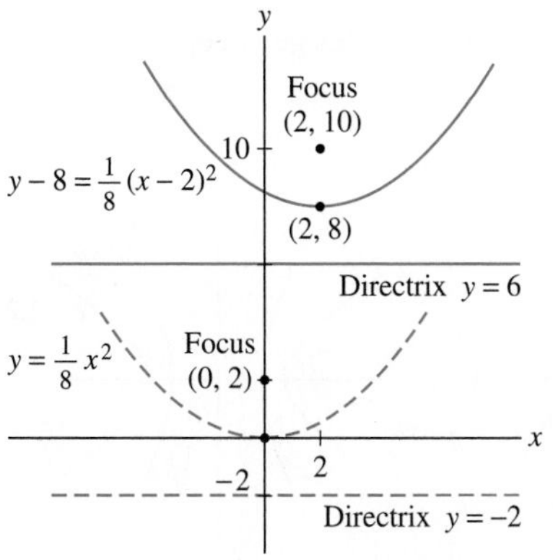

FIGURE 9 A parabola and its translate.

Solution By Eq. (9) with $c = 2$, the standard parabola with directrix $y = -2$ has equation $y = \frac{1}{8}x^2$ (Figure 9). The focus of this standard parabola is $(0, c) = (0, 2)$, which is two units above the vertex $(0, 0)$.

To obtain the equation when the parabola is translated with vertex at $(2, 8)$, we replace x by $x - 2$ and y by $y - 8$:

$$y - 8 = \frac{1}{8}(x-2)^2 \qquad \text{or} \qquad y = \frac{1}{8}x^2 - \frac{1}{2}x + \frac{17}{2}$$

The vertex has moved up 8 units, so the directrix also moves up 8 units to become $y = 6$. The new focus is two units above the new vertex $(2, 8)$, so the new focus is $(2, 10)$. ■

Eccentricity

Some ellipses are flatter than others, just as some hyperbolas are steeper. The "shape" of a conic section is measured by a number e called the **eccentricity**. For an ellipse or hyperbola,

$$e = \frac{\text{distance betweeen foci}}{\text{distance between vertices on focal axis}}$$

A parabola is defined to have eccentricity $e = 1$.

← REMINDER

Standard ellipse:

$$\left(\frac{x}{a}\right)^2 + \left(\frac{y}{b}\right)^2 = 1, \quad c = \sqrt{a^2 - b^2}$$

Standard hyperbola:

$$\left(\frac{x}{a}\right)^2 - \left(\frac{y}{b}\right)^2 = 1, \quad c = \sqrt{a^2 + b^2}$$

THEOREM 4 For ellipses and hyperbolas in standard position,

$$e = \frac{c}{a}$$

1. An ellipse has eccentricity $0 \le e < 1$.
2. A hyperbola has eccentricity $e > 1$.

Proof The foci are located at $(\pm c, 0)$ and the vertices are on the focal axis at $(\pm a, 0)$. Therefore,

$$e = \frac{\text{distance between foci}}{\text{distance between vertices on focal axis}} = \frac{2c}{2a} = \frac{c}{a}$$

For an ellipse, $c = \sqrt{a^2 - b^2}$ and so $e = c/a < 1$. For a hyperbola, $c = \sqrt{a^2 + b^2}$ and thus $e = c/a > 1$. ■

How does eccentricity determine the shape of a conic [Figure 10(A)]? Consider the ratio b/a of the semiminor axis to the semimajor axis of an ellipse. The ellipse is nearly circular if b/a is close to 1, whereas it is elongated and flat if b/a is small. Now

$$\frac{b}{a} = \frac{\sqrt{a^2 - c^2}}{a} = \sqrt{1 - \frac{c^2}{a^2}} = \sqrt{1 - e^2}$$

This shows that b/a gets smaller (and the ellipse get flatter) as $e \to 1$ [Figure 10(B)]. The "roundest" ellipse is the circle, with $e = 0$.

Similarly, for a hyperbola,

$$\frac{b}{a} = \sqrt{1 + e^2}$$

The ratios $\pm b/a$ are the slopes of the asymptotes, so the asymptotes get steeper as $e \to \infty$ [Figure 10(C)].

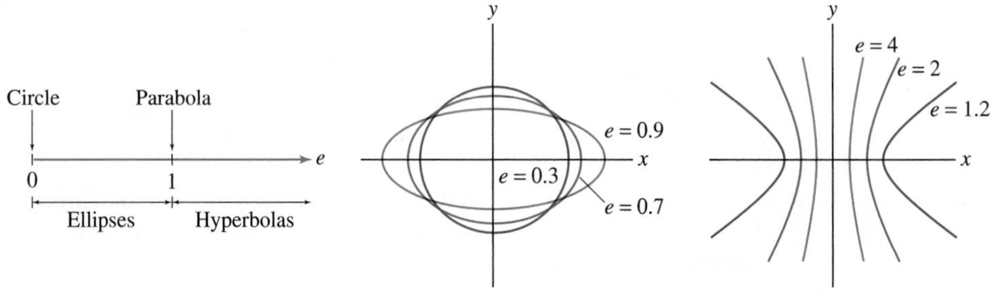

(A) Eccentricity e

(B) Ellipse flattens as $e \to 1$.

(C) Asymptotes of the hyperbola get steeper as $e \to \infty$.

FIGURE 10

CONCEPTUAL INSIGHT There is a more precise way to explain how eccentricity determines the shape of a conic. We can prove that if two conics C_1 and C_2 have same eccentricity e, then there is a change of scale that makes C_1 *congruent* to C_2. Changing the scale means changing the units along the x- and y-axes by a common positive factor. A curve scaled by a factor of 10 has the same shape but is ten times as large. This corresponds, for example, to changing units from centimeters to millimeters (smaller units make for a larger figure). By "congruent" we mean that after scaling, it is possible to move C_1 by a rigid motion (involving rotation and translation, but no stretching or bending) so that it lies directly on top of C_2.

All circles ($e = 0$) have the same shape because scaling by a factor $r > 0$ transforms a circle of radius R into a circle of radius rR. Similarly, any two parabolas ($e = 1$) become congruent after suitable scaling. However, an ellipse of eccentricity $e = 0.5$ cannot be made congruent to an ellipse of eccentricity $e = 0.8$ by scaling (see Exercise 74).

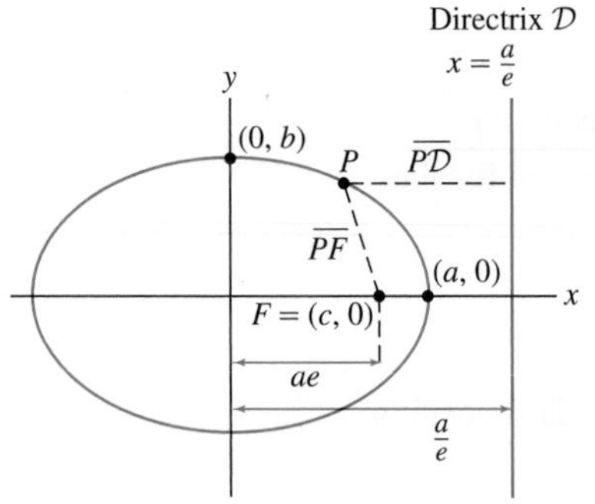

FIGURE 11 The ellipse consists of points P such that $PF = eP\mathcal{D}$.

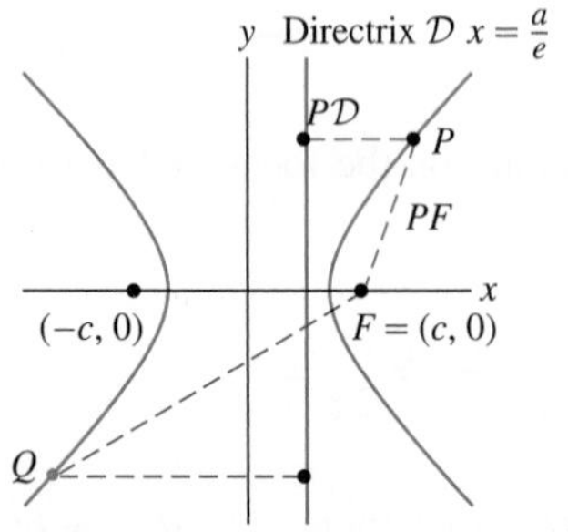

FIGURE 12 The hyperbola consists of points P such that $PF = eP\mathcal{D}$.

Eccentricity can be used to give a unified focus-directrix definition of the conic sections. Given a point F (the focus), a line $\mathcal{D}$ (the directrix), and a number $e > 0$, we consider the set of all points P such that

$$PF = eP\mathcal{D} \qquad \boxed{10}$$

For $e = 1$, this is our definition of a parabola. According to the next theorem, Eq. (10) defines a conic section of eccentricity e for all $e > 0$ (Figures 11 and 12). Note, however, that there is no focus-directrix definition for circles ($e = 0$).

THEOREM 5 Focus-Directrix Definition For all $e > 0$, the set of points satisfying Eq. (10) is a conic section of eccentricity e. Furthermore,

- **Ellipse:** Let $a > b > 0$ and $c = \sqrt{a^2 - b^2}$. The ellipse

$$\left(\frac{x}{a}\right)^2 + \left(\frac{y}{b}\right)^2 = 1$$

satisfies Eq. (10) with $F = (c, 0)$, $e = \frac{c}{a}$, and vertical directrix $x = \frac{a}{e}$.

- **Hyperbola:** Let $a, b > 0$ and $c = \sqrt{a^2 + b^2}$. The hyperbola

$$\left(\frac{x}{a}\right)^2 - \left(\frac{y}{b}\right)^2 = 1$$

satisfies Eq. (10) with $F = (c, 0)$, $e = \frac{c}{a}$, and vertical directrix $x = \frac{a}{e}$.

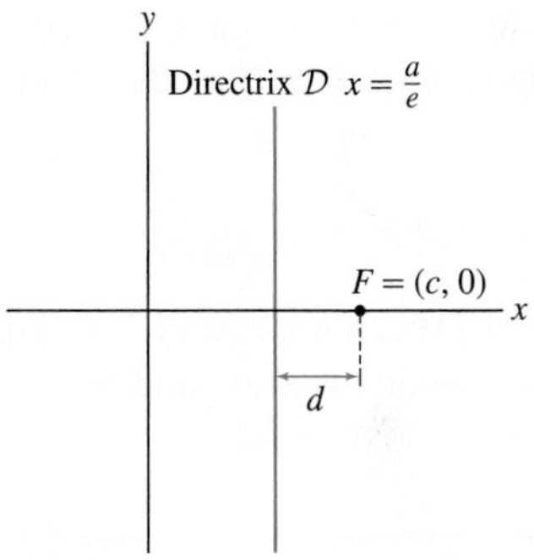

FIGURE 13

Proof Assume that $e > 1$ (the case $e < 1$ is similar, see Exercise 66). We may choose our coordinate axes so that the focus F lies on the x-axis and the directrix is vertical, lying to the left of F, as in Figure 13. Anticipating the final result, we let d be the distance from the focus F to the directrix $\mathcal{D}$ and set

$$c = \frac{d}{1 - e^{-2}}, \qquad a = \frac{c}{e}, \qquad b = \sqrt{c^2 - a^2}$$

Since we are free to shift the y-axis, let us choose the y-axis so that the focus has coordinates $F = (c, 0)$. Then the directrix is the line

$$\begin{aligned} x &= c - d = c - c(1 - e^{-2}) \\ &= c\,e^{-2} = \frac{a}{e} \end{aligned}$$

Now, the equation $PF = eP\mathcal{D}$ for a point $P = (x, y)$ may be written

$$\underbrace{\sqrt{(x - c)^2 + y^2}}_{PF} = e\underbrace{\sqrt{\left(x - (a/e)\right)^2}}_{P\mathcal{D}}$$

Algebraic manipulation yields

$$\begin{aligned} (x - c)^2 + y^2 &= e^2\left(x - (a/e)\right)^2 && \text{(square)} \\ x^2 - 2cx + c^2 + y^2 &= e^2x^2 - 2aex + a^2 && \\ x^2 - \cancel{2aex} + a^2e^2 + y^2 &= e^2x^2 - \cancel{2aex} + a^2 && \text{(use } c = ae\text{)} \\ (e^2 - 1)x^2 - y^2 &= a^2(e^2 - 1) && \text{(rearrange)} \\ \frac{x^2}{a^2} - \frac{y^2}{a^2(e^2 - 1)} &= 1 && \text{(divide)} \end{aligned}$$

This is the desired equation because $a^2(e^2 - 1) = c^2 - a^2 = b^2$. ■

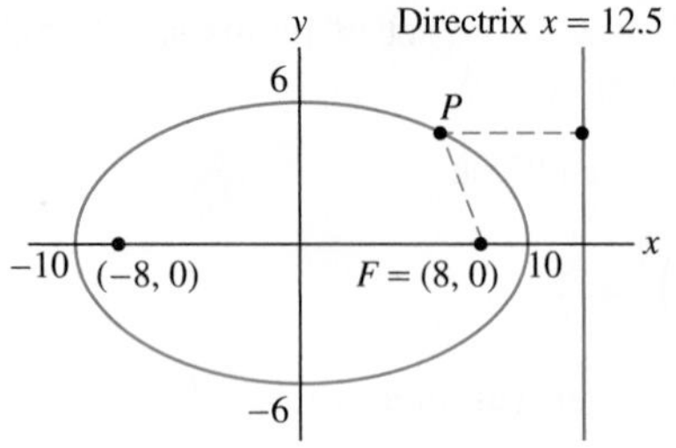

FIGURE 14 Ellipse of eccentricity $e = 0.8$ with focus at (8, 0).

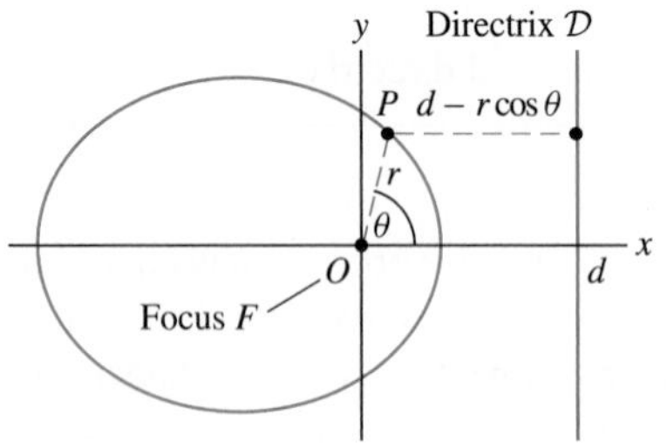

FIGURE 15 Focus-directrix definition of the ellipse in polar coordinates.

■ **EXAMPLE 5** Find the equation, foci, and directrix of the standard ellipse with eccentricity $e = 0.8$ and focal vertices $(\pm 10, 0)$.

Solution The vertices are $(\pm a, 0)$ with $a = 10$ (Figure 14). By Theorem 5,

$$c = ae = 10(0.8) = 8, \qquad b = \sqrt{a^2 - c^2} = \sqrt{10^2 - 8^2} = 6$$

Thus, our ellipse has equation

$$\left(\frac{x}{10}\right)^2 + \left(\frac{y}{6}\right)^2 = 1$$

The foci are $(\pm c, 0) = (\pm 8, 0)$ and the directrix is $x = \frac{a}{e} = \frac{10}{0.8} = 12.5$. ■

In Section 14.6, we discuss the famous law of Johannes Kepler stating that the orbit of a planet around the sun is an ellipse with one focus at the sun. In this discussion, we will need to write the equation of an ellipse in polar coordinates. To derive the polar equations of the conic sections, it is convenient to use the focus-directrix definition with focus F at the origin O and vertical line $x = d$ as directrix $\mathcal{D}$ (Figure 15). Note from the figure that if $P = (r, \theta)$, then

$$PF = r, \qquad P\mathcal{D} = d - r\cos\theta$$

Thus the focus-directrix equation of the ellipse $PF = eP\mathcal{D}$ becomes $r = e(d - r\cos\theta)$, or $r(1 + e\cos\theta) = ed$. This proves the following result, which is also valid for the hyperbola and parabola (see Exercise 67).

THEOREM 6 Polar Equation of a Conic Section The conic section of eccentricity $e > 0$ with focus at the origin and directrix $x = d$ has polar equation

$$r = \frac{ed}{1 + e\cos\theta} \qquad (11)$$

■ **EXAMPLE 6** Find the eccentricity, directrix, and focus of the conic section

$$r = \frac{24}{4 + 3\cos\theta}$$

Solution First, we write the equation in the standard form

$$r = \frac{24}{4 + 3\cos\theta} = \frac{6}{1 + \frac{3}{4}\cos\theta}$$

Comparing with Eq. (11), we see that $e = \frac{3}{4}$ and $ed = 6$. Therefore, $d = 8$. Since $e < 1$, the conic is an ellipse. By Theorem 6, the directrix is the line $x = 8$ and the focus is the origin. ■

Reflective Properties of Conic Sections

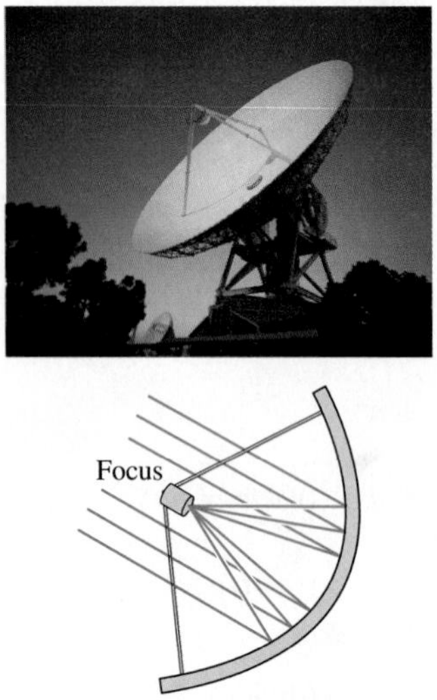

FIGURE 16 The paraboloid shape of this radio telescope directs the incoming signal to the focus.

The conic sections have numerous geometric properties. Especially important are the *reflective properties*, which are used in optics and communications (for example, in antenna and telescope design; Figure 16). We describe these properties here briefly without proof (but see Exercises 68–70 and Exercise 71 for proofs of the reflective property of ellipses).

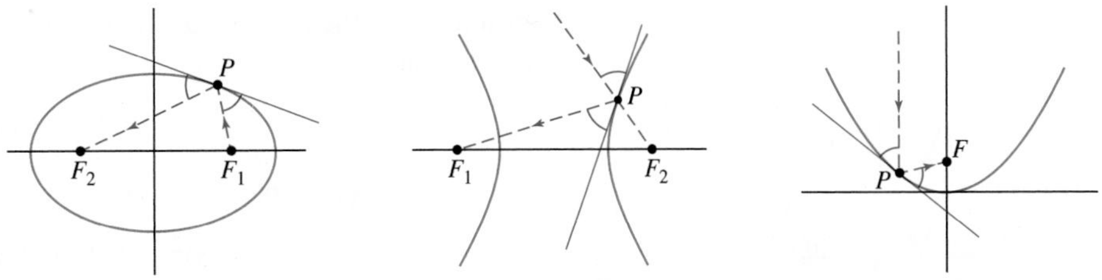

(A) Ellipse (B) Hyperbola (C) Parabola

FIGURE 17

FIGURE 18 The ellipsoidal dome of the National Statuary in the U.S. Capitol Building creates a "whisper chamber." Legend has it that John Quincy Adams would locate at one focus in order to eavesdrop on conversations taking place at the other focus.

- **Ellipse:** The segments F_1P and F_2P make equal angles with the tangent line at a point P on the ellipse. Therefore, a beam of light originating at focus F_1 is reflected off the ellipse toward the second focus F_2 [Figure 17(A)]. See also Figure 18.
- **Hyperbola:** The tangent line at a point P on the hyperbola bisects the angle formed by the segments F_1P and F_2P. Therefore, a beam of light directed toward F_2 is reflected off the hyperbola toward the second focus F_1 [Figure 17(B)].
- **Parabola:** The segment FP and the line through P parallel to the axis make equal angles with the tangent line at a point P on the parabola [Figure 17(C)]. Therefore, a beam of light approaching P from above in the axial direction is reflected off the parabola toward the focus F.

General Equations of Degree 2

The equations of the standard conic sections are special cases of the general equation of degree 2 in x and y:

$$ax^2 + bxy + cy^2 + dx + ey + f = 0 \qquad (12)$$

Here a, b, e, d, e, f are constants with a, b, c not all zero. It turns out that this general equation of degree 2 does not give rise to any new types of curves. Apart from certain "degenerate cases," Eq. (12) defines a conic section that is not necessarily in standard position: It need not be centered at the origin, and its focal and conjugate axes may be rotated relative to the coordinate axes. For example, the equation

$$6x^2 - 8xy + 8y^2 - 12x - 24y + 38 = 0$$

defines an ellipse with center at (3, 3) whose axes are rotated (Figure 19).

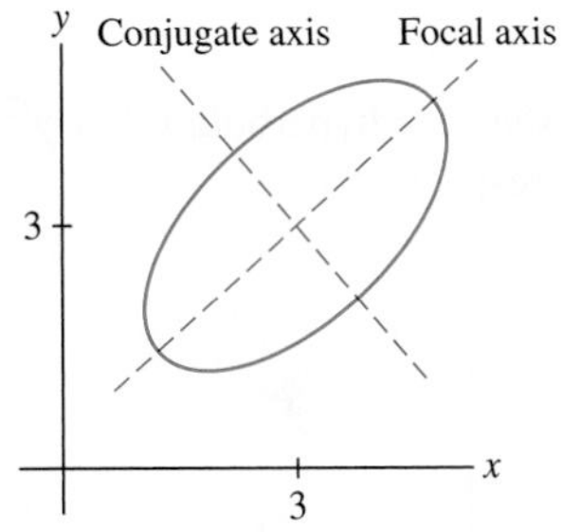

FIGURE 19 The ellipse with equation $6x^2 - 8xy + 8y^2 - 12x - 24y + 38 = 0$.

We say that Eq. (12) is **degenerate** if the set of solutions is a pair of intersecting lines, a pair of parallel lines, a single line, a point, or the empty set. For example:

- $x^2 - y^2 = 0$ defines a pair of intersecting lines $y = x$ and $y = -x$.
- $x^2 - x = 0$ defines a pair of parallel lines $x = 0$ and $x = 1$.
- $x^2 = 0$ defines a single line (the y-axis).
- $x^2 + y^2 = 0$ has just one solution (0, 0).
- $x^2 + y^2 = -1$ has no solutions.

Now assume that Eq. (12) is nondegenerate. The term bxy is called the *cross term*. When the cross term is zero (that is, when $b = 0$), we can "complete the square" to show that Eq. (12) defines a translate of the conic in standard position. In other words, the axes of the conic are parallel to the coordinate axes. This is illustrated in the next example.

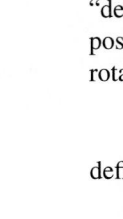

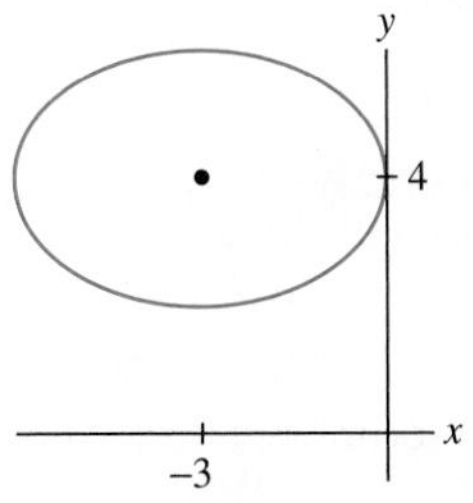

FIGURE 20 The ellipse with equation $4x^2 + 9y^2 + 24x - 72y + 144 = 0$.

EXAMPLE 7 Completing the Square Show that

$$4x^2 + 9y^2 + 24x - 72y + 144 = 0$$

defines a translate of a conic section in standard position (Figure 20).

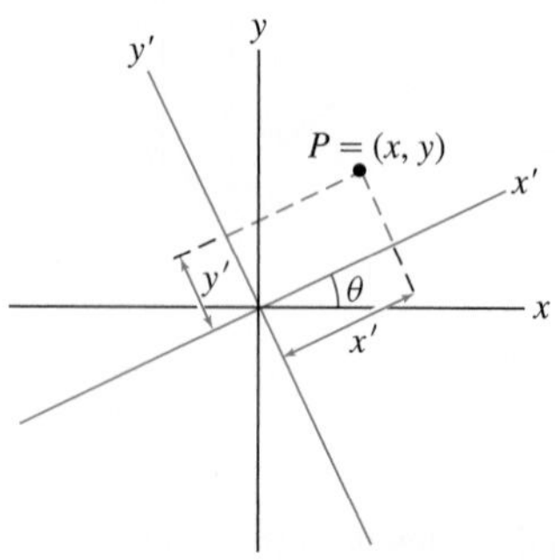

FIGURE 21

Solution Since there is no cross term, we may complete the square of the terms involving x and y separately:

$$4x^2 + 9y^2 + 24x - 72y + 144 = 0$$
$$4(x^2 + 6x + 9 - 9) + 9(y^2 - 8y + 16 - 16) + 144 = 0$$
$$4(x+3)^2 - 4(9) + 9(y-4)^2 - 9(16) + 144 = 0$$
$$4(x+3)^2 + 9(y-4)^2 = 36$$

Therefore, this quadratic equation can be rewritten in the form

$$\left(\frac{x+3}{3}\right)^2 + \left(\frac{y-4}{2}\right)^2 = 1$$ ■

When the cross term bxy is nonzero, Eq. (12) defines a conic whose axes are rotated relative to the coordinate axes. The marginal note describes how this may be verified in general. We illustrate with the following example.

If (x', y') are coordinates relative to axes rotated by an angle θ as in Figure 21, then

$$x = x'\cos\theta - y'\sin\theta \qquad \boxed{13}$$

$$y = x'\sin\theta + y'\cos\theta \qquad \boxed{14}$$

See Exercise 75. In Exercise 76, we show that the cross term disappears when Eq. (12) is rewritten in terms of x' and y' for the angle

$$\theta = \frac{1}{2}\cot^{-1}\frac{a-c}{b} \qquad \boxed{15}$$

■ **EXAMPLE 8** Show that $2xy = 1$ defines a conic section whose focal and conjugate axes are rotated relative to the coordinate axes.

Solution Figure 22(A) shows axes labeled x' and y' that are rotated by 45° relative to the standard coordinate axes. A point P with coordinates (x, y) may also be described by coordinates (x', y') relative to these rotated axes. Applying Eqs. (13) and (14) with $\theta = \frac{\pi}{4}$, we find that (x, y) and (x', y') are related by the formulas

$$x = \frac{x' - y'}{\sqrt{2}}, \qquad y = \frac{x' + y'}{\sqrt{2}}$$

Therefore, if $P = (x, y)$ lies on the hyperbola—that is, if $2xy = 1$—then

$$2xy = 2\left(\frac{x' - y'}{\sqrt{2}}\right)\left(\frac{x' + y'}{\sqrt{2}}\right) = x'^2 - y'^2 = 1$$

Thus, the coordinates (x', y') satisfy the equation of the standard hyperbola $x'^2 - y'^2 = 1$ whose focal and conjugate axes are the x'- and y'-axes, respectively. ■

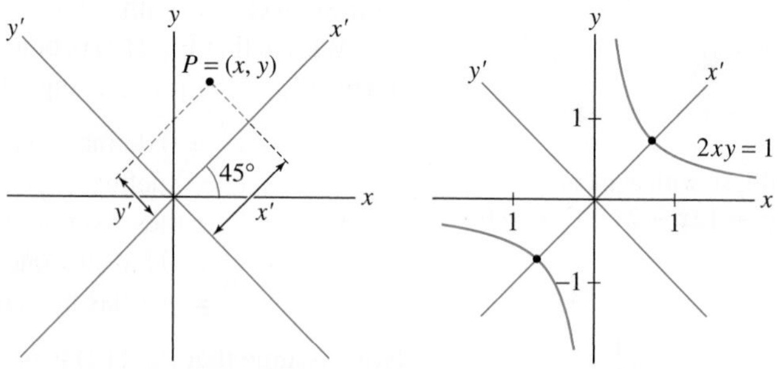

(A) The point $P = (x, y)$ may also be described by coordinates (x', y') relative to the rotated axis.

(B) The hyperbola $2xy = 1$ has the standard form $x'^2 - y'^2 = 1$ relative to the x', y' axes.

FIGURE 22 The x'- and y'-axes are rotated at a 45° angle relative to the x- and y-axes.

We conclude our discussion of conics by stating the Discriminant Test. Suppose that the equation

$$ax^2 + bxy + cy^2 + dx + ey + f = 0$$

is nondegenerate and thus defines a conic section. According to the Discriminant Test, the type of conic is determined by the **discriminant** D:

$$D = b^2 - 4ac$$

We have the following cases:

- $D < 0$: Ellipse or circle
- $D > 0$: Hyperbola
- $D = 0$: Parabola

For example, the discriminant of the equation $2xy = 1$ is

$$D = b^2 - 4ac = 2^2 - 0 = 4 > 0$$

According to the Discriminant Test, $2xy = 1$ defines a hyperbola. This agrees with our conclusion in Example 8.

12.5 SUMMARY

- An *ellipse* with foci F_1 and F_2 is the set of points P such that $PF_1 + PF_2 = K$, where K is a constant such that $K > F_1F_2$. The equation in standard position is

$$\left(\frac{x}{a}\right)^2 + \left(\frac{y}{b}\right)^2 = 1$$

The vertices of the ellipse are $(\pm a, 0)$ and $(0, \pm b)$.

	Focal axis	Foci	Focal vertices
$a > b$	x-axis	$(\pm c, 0)$ with $c = \sqrt{a^2 - b^2}$	$(\pm a, 0)$
$a < b$	y-axis	$(0, \pm c)$ with $c = \sqrt{b^2 - a^2}$	$(0, \pm b)$

Eccentricity: $e = \frac{c}{a}$ $(0 \le e < 1)$. Directrix: $x = \frac{a}{e}$ (if $a > b$).

- A *hyperbola* with foci F_1 and F_2 is the set of points P such that

$$PF_1 - PF_2 = \pm K$$

where K is a constant such that $0 < K < F_1F_2$. The equation in standard position is

$$\left(\frac{x}{a}\right)^2 - \left(\frac{y}{b}\right)^2 = 1$$

Focal axis	Foci	Vertices	Asymptotes
x-axis	$(\pm c, 0)$ with $c = \sqrt{a^2 + b^2}$	$(\pm a, 0)$	$y = \pm\frac{b}{a}x$

Eccentricity: $e = \frac{c}{a}$ $(e > 1)$. Directrix: $x = \frac{a}{e}$.

- A *parabola* with focus F and directrix $\mathcal{D}$ is the set of points P such that $PF = P\mathcal{D}$. The equation in standard position is

$$y = \frac{1}{4c}x^2$$

Focus $F = (0, c)$, directrix $y = -c$, and vertex at the origin $(0, 0)$.

- *Focus-directrix definition* of conic with focus F and directrix $\mathcal{D}$: $PF = eP\mathcal{D}$.
- To translate a conic section h units horizontally and k units vertically, replace x by $x - h$ and y by $y - k$ in the equation.
- Polar equation of conic of eccentricity $e > 0$, focus at the origin, directrix $x = d$:

$$r = \frac{ed}{1 + e\cos\theta}$$

12.5 EXERCISES

Preliminary Questions

1. Which of the following equations defines an ellipse? Which does not define a conic section?

(a) $4x^2 - 9y^2 = 12$ **(b)** $-4x + 9y^2 = 0$

(c) $4y^2 + 9x^2 = 12$ **(d)** $4x^3 + 9y^3 = 12$

2. For which conic sections do the vertices lie between the foci?

3. What are the foci of

$$\left(\frac{x}{a}\right)^2 + \left(\frac{y}{b}\right)^2 = 1 \quad \text{if } a < b?$$

4. What is the geometric interpretation of b/a in the equation of a hyperbola in standard position?

Exercises

In Exercises 1–6, find the vertices and foci of the conic section.

1. $\left(\frac{x}{9}\right)^2 + \left(\frac{y}{4}\right)^2 = 1$ **2.** $\frac{x^2}{9} + \frac{y^2}{4} = 1$

3. $\left(\frac{x}{4}\right)^2 - \left(\frac{y}{9}\right)^2 = 1$ **4.** $\frac{x^2}{4} - \frac{y^2}{9} = 36$

5. $\left(\frac{x-3}{7}\right)^2 - \left(\frac{y+1}{4}\right)^2 = 1$

6. $\left(\frac{x-3}{4}\right)^2 + \left(\frac{y+1}{7}\right)^2 = 1$

In Exercises 7–10, find the equation of the ellipse obtained by translating (as indicated) the ellipse

$$\left(\frac{x-8}{6}\right)^2 + \left(\frac{y+4}{3}\right)^2 = 1$$

7. Translated with center at the origin

8. Translated with center at $(-2, -12)$

9. Translated to the right six units

10. Translated down four units

In Exercises 11–14, find the equation of the given ellipse.

11. Vertices $(\pm 5, 0)$ and $(0, \pm 7)$

12. Foci $(\pm 6, 0)$ and focal vertices $(\pm 10, 0)$

13. Foci $(0, \pm 10)$ and eccentricity $e = \frac{3}{5}$

14. Vertices $(4, 0)$, $(28, 0)$ and eccentricity $e = \frac{2}{3}$

In Exercises 15–20, find the equation of the given hyperbola.

15. Vertices $(\pm 3, 0)$ and foci $(\pm 5, 0)$

16. Vertices $(\pm 3, 0)$ and asymptotes $y = \pm\frac{1}{2}x$

17. Foci $(\pm 4, 0)$ and eccentricity $e = 2$

18. Vertices $(0, \pm 6)$ and eccentricity $e = 3$

19. Vertices $(-3, 0)$, $(7, 0)$ and eccentricity $e = 3$

20. Vertices $(0, -6)$, $(0, 4)$ and foci $(0, -9)$, $(0, 7)$

In Exercises 21–28, find the equation of the parabola with the given properties.

21. Vertex $(0, 0)$, focus $\left(\frac{1}{12}, 0\right)$

22. Vertex $(0, 0)$, focus $(0, 2)$

23. Vertex $(0, 0)$, directrix $y = -5$

24. Vertex $(3, 4)$, directrix $y = -2$

25. Focus $(0, 4)$, directrix $y = -4$

26. Focus $(0, -4)$, directrix $y = 4$

27. Focus $(2, 0)$, directrix $x = -2$

28. Focus $(-2, 0)$, vertex $(2, 0)$

In Exercises 29–38, find the vertices, foci, center (if an ellipse or a hyperbola), and asymptotes (if a hyperbola).

29. $x^2 + 4y^2 = 16$ **30.** $4x^2 + y^2 = 16$

31. $\left(\frac{x-3}{4}\right)^2 - \left(\frac{y+5}{7}\right)^2 = 1$ **32.** $3x^2 - 27y^2 = 12$

33. $4x^2 - 3y^2 + 8x + 30y = 215$

34. $y = 4x^2$ **35.** $y = 4(x-4)^2$

36. $8y^2 + 6x^2 - 36x - 64y + 134 = 0$

37. $4x^2 + 25y^2 - 8x - 10y = 20$

38. $16x^2 + 25y^2 - 64x - 200y + 64 = 0$

In Exercises 39–42, use the Discriminant Test to determine the type of the conic section (in each case, the equation is nondegenerate). Plot the curve if you have a computer algebra system.

39. $4x^2 + 5xy + 7y^2 = 24$

40. $x^2 - 2xy + y^2 + 24x - 8 = 0$

41. $2x^2 - 8xy + 3y^2 - 4 = 0$

42. $2x^2 - 3xy + 5y^2 - 4 = 0$

43. Show that the "conic" $x^2 + 3y^2 - 6x + 12 + 23 = 0$ has no points.

44. For which values of a does the conic $3x^2 + 2y^2 - 16y + 12x = a$ have at least one point?

45. Show that $\dfrac{b}{a} = \sqrt{1 - e^2}$ for a standard ellipse of eccentricity e.

46. Show that the eccentricity of a hyperbola in standard position is $e = \sqrt{1 + m^2}$, where $\pm m$ are the slopes of the asymptotes.

47. Explain why the dots in Figure 23 lie on a parabola. Where are the focus and directrix located?

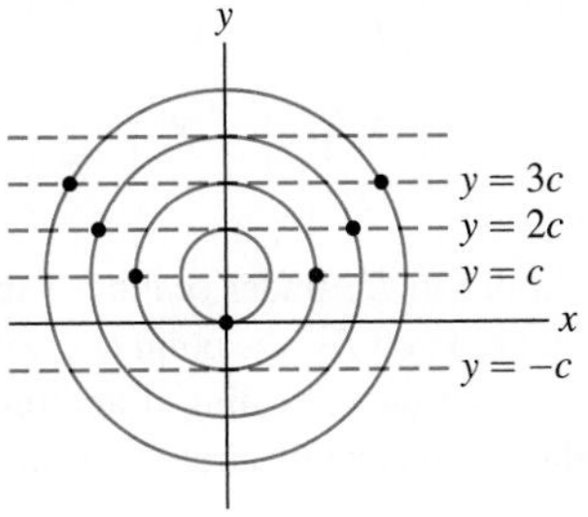

FIGURE 23

48. Find the equation of the ellipse consisting of points P such that $PF_1 + PF_2 = 12$, where $F_1 = (4, 0)$ and $F_2 = (-2, 0)$.

49. A **latus rectum** of a conic section is a chord through a focus parallel to the directrix. Find the area bounded by the parabola $y = x^2/(4c)$ and its latus rectum (refer to Figure 8).

50. Show that the tangent line at a point $P = (x_0, y_0)$ on the hyperbola $\left(\frac{x}{a}\right)^2 - \left(\frac{y}{b}\right)^2 = 1$ has equation

$$Ax - By = 1$$

where $A = \dfrac{x_0}{a^2}$ and $B = \dfrac{y_0}{b^2}$.

In Exercises 51–54, find the polar equation of the conic with the given eccentricity and directrix, and focus at the origin.

51. $e = \frac{1}{2}, \quad x = 3$

52. $e = \frac{1}{2}, \quad x = -3$

53. $e = 1, \quad x = 4$

54. $e = \frac{3}{2}, \quad x = -4$

In Exercises 55–58, identify the type of conic, the eccentricity, and the equation of the directrix.

55. $r = \dfrac{8}{1 + 4\cos\theta}$

56. $r = \dfrac{8}{4 + \cos\theta}$

57. $r = \dfrac{8}{4 + 3\cos\theta}$

58. $r = \dfrac{12}{4 + 3\cos\theta}$

59. Find a polar equation for the hyperbola with focus at the origin, directrix $x = -2$, and eccentricity $e = 1.2$.

60. Let C be the ellipse $r = de/(1 + e\cos\theta)$, where $e < 1$. Show that the x-coordinates of the points in Figure 24 are as follows:

Point	A	C	F_2	A'
x-coordinate	$\dfrac{de}{e+1}$	$-\dfrac{de^2}{1-e^2}$	$-\dfrac{2de^2}{1-e^2}$	$-\dfrac{de}{1-e}$

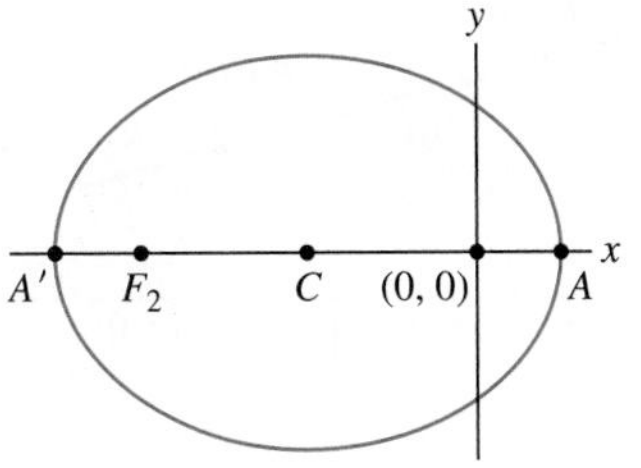

FIGURE 24

61. Find an equation in rectangular coordinates of the conic

$$r = \frac{16}{5 + 3\cos\theta}$$

Hint: Use the results of Exercise 60.

62. Let $e > 1$. Show that the vertices of the hyperbola $r = \dfrac{de}{1 + e\cos\theta}$ have x-coordinates $\dfrac{ed}{e+1}$ and $\dfrac{ed}{e-1}$.

63. Kepler's First Law states that planetary orbits are ellipses with the sun at one focus. The orbit of Pluto has eccentricity $e \approx 0.25$. Its **perihelion** (closest distance to the sun) is approximately 2.7 billion miles. Find the **aphelion** (farthest distance from the sun).

64. Kepler's Third Law states that the ratio $T/a^{3/2}$ is equal to a constant C for all planetary orbits around the sun, where T is the period (time for a complete orbit) and a is the semimajor axis.

(a) Compute C in units of days and kilometers, given that the semimajor axis of the earth's orbit is 150×10^6 km.

(b) Compute the period of Saturn's orbit, given that its semimajor axis is approximately 1.43×10^9 km.

(c) Saturn's orbit has eccentricity $e = 0.056$. Find the perihelion and aphelion of Saturn (see Exercise 63).

Further Insights and Challenges

65. Verify Theorem 2.

66. Verify Theorem 5 in the case $0 < e < 1$. *Hint:* Repeat the proof of Theorem 5, but set $c = d/(e^{-2} - 1)$.

67. Verify that if $e > 1$, then Eq. (11) defines a hyperbola of eccentricity e, with its focus at the origin and directrix at $x = d$.

Reflective Property of the Ellipse *In Exercises 68–70, we prove that the focal radii at a point on an ellipse make equal angles with the tangent line $\mathcal{L}$. Let $P = (x_0, y_0)$ be a point on the ellipse in Figure 25 with foci $F_1 = (-c, 0)$ and $F_2 = (c, 0)$, and eccentricity $e = c/a$.*

68. Show that the equation of the tangent line at P is $Ax + By = 1$, where $A = \dfrac{x_0}{a^2}$ and $B = \dfrac{y_0}{b^2}$.

69. Points R_1 and R_2 in Figure 25 are defined so that $\overline{F_1R_1}$ and $\overline{F_2R_2}$ are perpendicular to the tangent line.

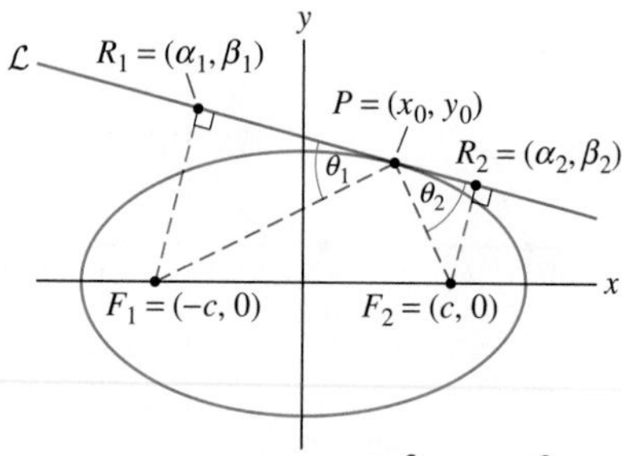

FIGURE 25 The ellipse $\left(\frac{x}{a}\right)^2 + \left(\frac{y}{b}\right)^2 = 1$.

(a) Show, with A and B as in Exercise 68, that

$$\frac{\alpha_1 + c}{\beta_1} = \frac{\alpha_2 - c}{\beta_2} = \frac{A}{B}$$

(b) Use (a) and the distance formula to show that

$$\frac{F_1R_1}{F_2R_2} = \frac{\beta_1}{\beta_2}$$

(c) Use (a) and the equation of the tangent line in Exercise 68 to show that

$$\beta_1 = \frac{B(1 + Ac)}{A^2 + B^2}, \qquad \beta_2 = \frac{B(1 - Ac)}{A^2 + B^2}$$

70. (a) Prove that $PF_1 = a + x_0 e$ and $PF_2 = a - x_0 e$. *Hint:* Show that $PF_1^2 - PF_2^2 = 4x_0 c$. Then use the defining property $PF_1 + PF_2 = 2a$ and the relation $e = c/a$.

(b) Verify that $\dfrac{F_1R_1}{PF_1} = \dfrac{F_2R_2}{PF_2}$.

(c) Show that $\sin\theta_1 = \sin\theta_2$. Conclude that $\theta_1 = \theta_2$.

71. Here is another proof of the Reflective Property.

(a) Figure 25 suggests that $\mathcal{L}$ is the unique line that intersects the ellipse only in the point P. Assuming this, prove that $QF_1 + QF_2 > PF_1 + PF_2$ for all points Q on the tangent line other than P.

(b) Use the Principle of Least Distance (Example 6 in Section 4.6) to prove that $\theta_1 = \theta_2$.

72. Show that the length QR in Figure 26 is independent of the point P.

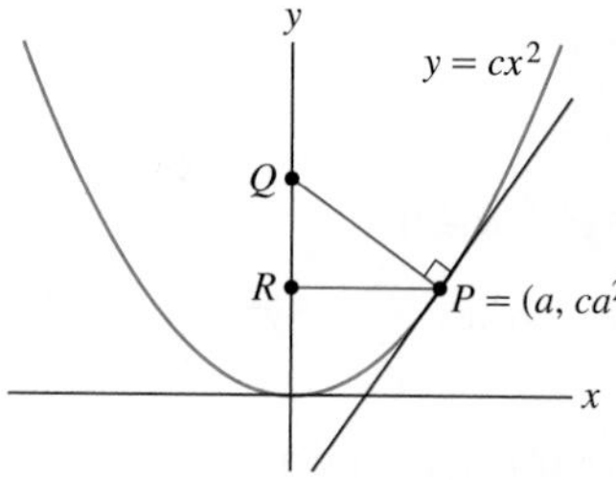

FIGURE 26

73. Show that $y = x^2/4c$ is the equation of a parabola with directrix $y = -c$, focus $(0, c)$, and the vertex at the origin, as stated in Theorem 3.

74. Consider two ellipses in standard position:

$$E_1: \quad \left(\frac{x}{a_1}\right)^2 + \left(\frac{y}{b_1}\right)^2 = 1$$

$$E_2: \quad \left(\frac{x}{a_2}\right)^2 + \left(\frac{y}{b_2}\right)^2 = 1$$

We say that E_1 is similar to E_2 under scaling if there exists a factor $r > 0$ such that for all (x, y) on E_1, the point (rx, ry) lies on E_2. Show that E_1 and E_2 are similar under scaling if and only if they have the same eccentricity. Show that any two circles are similar under scaling.

75. Derive Equations (13) and (14) in the text as follows. Write the coordinates of P with respect to the rotated axes in Figure 21 in polar form $x' = r\cos\alpha$, $y' = r\sin\alpha$. Explain why P has polar coordinates $(r, \alpha + \theta)$ with respect to the standard x and y-axes and derive (13) and (14) using the addition formulas for cosine and sine.

76. If we rewrite the general equation of degree 2 (Eq. 12) in terms of variables x' and y' that are related to x and y by Eqs. (13) and (14), we obtain a new equation of degree 2 in x' and y' of the same form but with different coefficients:

$$a'x^2 + b'xy + c'y^2 + d'x + e'y + f' = 0$$

(a) Show that $b' = b\cos 2\theta + (c - a)\sin 2\theta$.

(b) Show that if $b \neq 0$, then we obtain $b' = 0$ for

$$\theta = \frac{1}{2}\cot^{-1}\frac{a - c}{b}$$

This proves that it is always possible to eliminate the cross term bxy by rotating the axes through a suitable angle.

CHAPTER REVIEW EXERCISES

1. Which of the following curves pass through the point $(1, 4)$?

(a) $c(t) = (t^2, t + 3)$ **(b)** $c(t) = (t^2, t - 3)$

(c) $c(t) = (t^2, 3 - t)$ **(d)** $c(t) = (t - 3, t^2)$

2. Find parametric equations for the line through $P = (2, 5)$ perpendicular to the line $y = 4x - 3$.

3. Find parametric equations for the circle of radius 2 with center $(1, 1)$. Use the equations to find the points of intersection of the circle with the x- and y-axes.

4. Find a parametrization $c(t)$ of the line $y = 5 - 2x$ such that $c(0) = (2, 1)$.

5. Find a parametrization $c(\theta)$ of the unit circle such that $c(0) = (-1, 0)$.

6. Find a path $c(t)$ that traces the parabolic arc $y = x^2$ from $(0, 0)$ to $(3, 9)$ for $0 \le t \le 1$.

7. Find a path $c(t)$ that traces the line $y = 2x + 1$ from $(1, 3)$ to $(3, 7)$ for $0 \le t \le 1$.

8. Sketch the graph $c(t) = (1 + \cos t, \sin 2t)$ for $0 \le t \le 2\pi$ and draw arrows specifying the direction of motion.

In Exercises 9–12, express the parametric curve in the form $y = f(x)$.

9. $c(t) = (4t - 3, 10 - t)$ **10.** $c(t) = (t^3 + 1, t^2 - 4)$

11. $c(t) = \left(3 - \dfrac{2}{t}, t^3 + \dfrac{1}{t}\right)$ **12.** $x = \tan t, \quad y = \sec t$

In Exercises 13–16, calculate dy/dx at the point indicated.

13. $c(t) = (t^3 + t, t^2 - 1), \quad t = 3$

14. $c(\theta) = (\tan^2 \theta, \cos \theta), \quad \theta = \frac{\pi}{4}$

15. $c(t) = (e^t - 1, \sin t), \quad t = 20$

16. $c(t) = (\ln t, 3t^2 - t), \quad P = (0, 2)$

17. CAS Find the point on the cycloid $c(t) = (t - \sin t, 1 - \cos t)$ where the tangent line has slope $\frac{1}{2}$.

18. Find the points on $(t + \sin t, t - 2\sin t)$ where the tangent is vertical or horizontal.

19. Find the equation of the Bézier curve with control points

$$P_0 = (-1, -1), \quad P_1 = (-1, 1), \quad P_2 = (1, 1), \quad P_3(1, -1)$$

20. Find the speed at $t = \frac{\pi}{4}$ of a particle whose position at time t seconds is $c(t) = (\sin 4t, \cos 3t)$.

21. Find the speed (as a function of t) of a particle whose position at time t seconds is $c(t) = (\sin t + t, \cos t + t)$. What is the particle's maximal speed?

22. Find the length of $(3e^t - 3, 4e^t + 7)$ for $0 \le t \le 1$.

In Exercises 23 and 24, let $c(t) = (e^{-t}\cos t, e^{-t}\sin t)$.

23. Show that $c(t)$ for $0 \le t < \infty$ has finite length and calculate its value.

24. Find the first positive value of t_0 such that the tangent line to $c(t_0)$ is vertical, and calculate the speed at $t = t_0$.

25. CAS Plot $c(t) = (\sin 2t, 2\cos t)$ for $0 \le t \le \pi$. Express the length of the curve as a definite integral, and approximate it using a computer algebra system.

26. Convert the points $(x, y) = (1, -3), (3, -1)$ from rectangular to polar coordinates.

27. Convert the points $(r, \theta) = \left(1, \frac{\pi}{6}\right), \left(3, \frac{5\pi}{4}\right)$ from polar to rectangular coordinates.

28. Write $(x + y)^2 = xy + 6$ as an equation in polar coordinates.

29. Write $r = \dfrac{2\cos\theta}{\cos\theta - \sin\theta}$ as an equation in rectangular coordinates.

30. Show that $r = \dfrac{4}{7\cos\theta - \sin\theta}$ is the polar equation of a line.

31. GU Convert the equation

$$9(x^2 + y^2) = (x^2 + y^2 - 2y)^2$$

to polar coordinates, and plot it with a graphing utility.

32. Calculate the area of the circle $r = 3\sin\theta$ bounded by the rays $\theta = \frac{\pi}{3}$ and $\theta = \frac{2\pi}{3}$.

33. Calculate the area of one petal of $r = \sin 4\theta$ (see Figure 1).

34. The equation $r = \sin(n\theta)$, where $n \ge 2$ is even, is a "rose" of $2n$ petals (Figure 1). Compute the total area of the flower, and show that it does not depend on n.

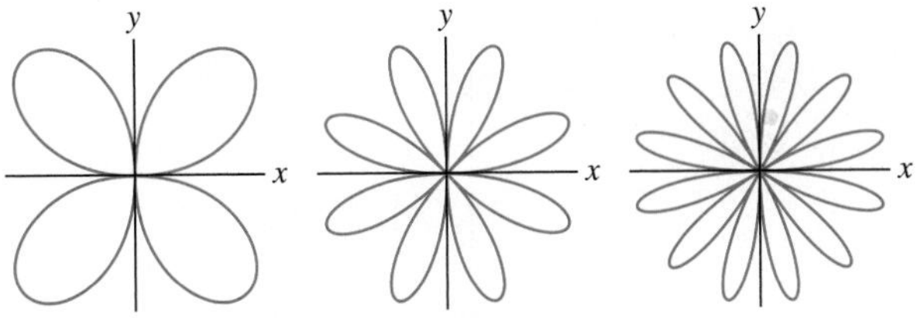

FIGURE 1 Plot of $r = \sin(n\theta)$.

35. Calculate the total area enclosed by the curve $r^2 = \cos\theta\, e^{\sin\theta}$ (Figure 2).

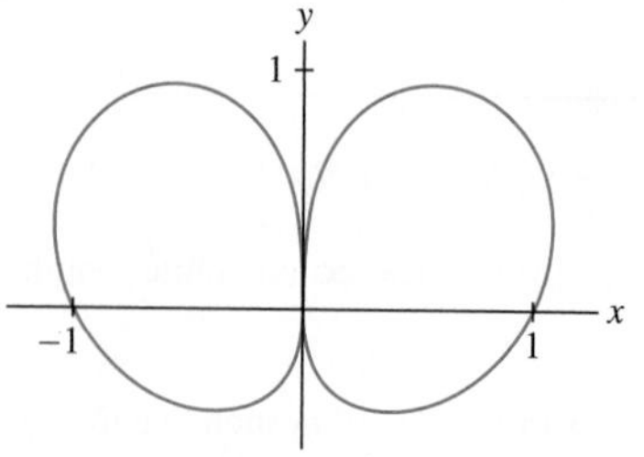

FIGURE 2 Graph of $r^2 = \cos\theta e^{\sin\theta}$.

36. Find the shaded area in Figure 3.

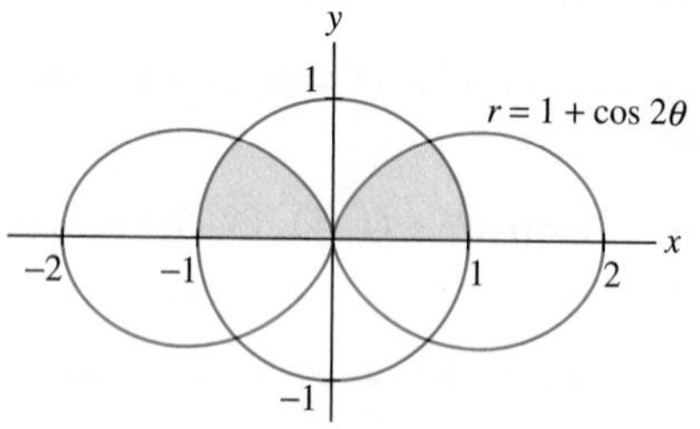

FIGURE 3

37. Find the area enclosed by the cardioid $r = a(1 + \cos\theta)$, where $a > 0$.

38. Calculate the length of the curve with polar equation $r = \theta$ in Figure 4.

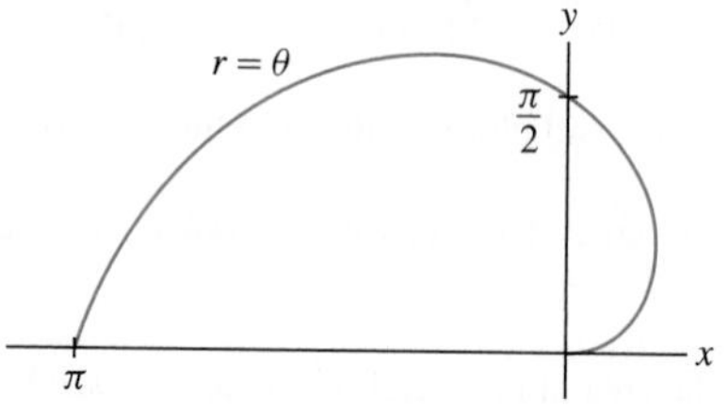

FIGURE 4

39. CAS Figure 5 shows the graph of $r = e^{0.5\theta}\sin\theta$ for $0 \le \theta \le 2\pi$. Use a computer algebra system to approximate the difference in length between the outer and inner loops.

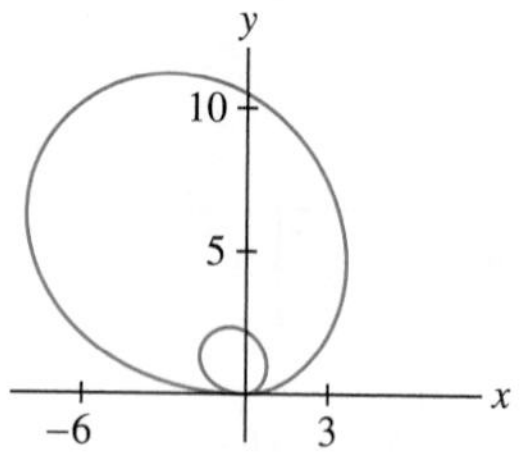

FIGURE 5

40. Show that $r = f_1(\theta)$ and $r = f_2(\theta)$ define the same curves in polar coordinates if $f_1(\theta) = -f_2(\theta + \pi)$. Use this to show that the following define the same conic section:

$$r = \frac{de}{1 - e\cos\theta}, \qquad r = \frac{-de}{1 + e\cos\theta}$$

In Exercises 41–44, identify the conic section. Find the vertices and foci.

41. $\left(\frac{x}{3}\right)^2 + \left(\frac{y}{2}\right)^2 = 1$

42. $x^2 - 2y^2 = 4$

43. $\left(2x + \frac{1}{2}y\right)^2 = 4 - (x - y)^2$

44. $(y - 3)^2 = 2x^2 - 1$

In Exercises 45–50, find the equation of the conic section indicated.

45. Ellipse with vertices $(\pm 8, 0)$ and foci $(\pm\sqrt{3}, 0)$

46. Ellipse with foci $(\pm 8, 0)$, eccentricity $\frac{1}{8}$

47. Hyperbola with vertices $(\pm 8, 0)$, asymptotes $y = \pm\frac{3}{4}x$

48. Hyperbola with foci $(2, 0)$ and $(10, 0)$, eccentricity $e = 4$

49. Parabola with focus $(8, 0)$, directrix $x = -8$

50. Parabola with vertex $(4, -1)$, directrix $x = 15$

51. Find the asymptotes of the hyperbola $3x^2 + 6x - y^2 - 10y = 1$.

52. Show that the "conic section" with equation $x^2 - 4x + y^2 + 5 = 0$ has no points.

53. Show that the relation $\frac{dy}{dx} = (e^2 - 1)\frac{x}{y}$ holds on a standard ellipse or hyperbola of eccentricity e.

54. The orbit of Jupiter is an ellipse with the sun at a focus. Find the eccentricity of the orbit if the perihelion (closest distance to the sun) equals 740×10^6 km and the aphelion (farthest distance from the sun) equals 816×10^6 km.

55. Refer to Figure 25 in Section 12.5. Prove that the product of the perpendicular distances F_1R_1 and F_2R_2 from the foci to a tangent line of an ellipse is equal to the square b^2 of the semiminor axes.

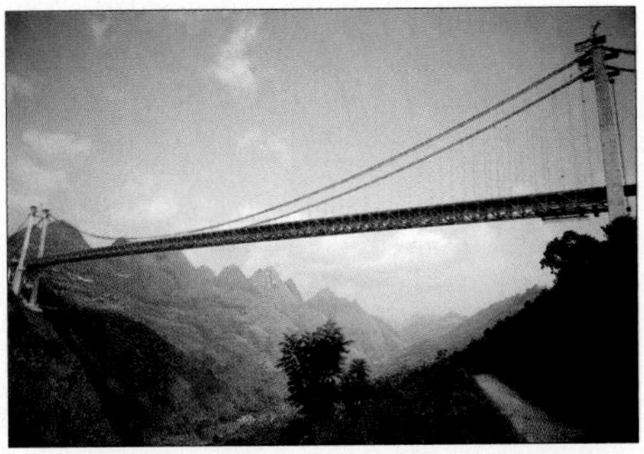

Recently completed Baling River Bridge in China's Guizhou province (2.25 km long and soaring 400 m above the Baling River). The tension in its cables and forces on its towers are described using vectors.

13 VECTOR GEOMETRY

Vectors play a role in nearly all areas of mathematics and its applications. In physical settings, they are used to represent quantities that have both magnitude and direction, such as velocity and force. They also appear in such diverse fields as computer graphics, economics, and statistics. This chapter develops the basic geometric and algebraic properties of vectors. Although no calculus is required, the concepts developed will be used throughout the remainder of the text.

13.1 Vectors in the Plane

A two-dimensional **vector v** is determined by two points in the plane: an initial point P (also called the "tail" or basepoint) and a terminal point Q (also called the "head"). We write

$$\mathbf{v} = \overrightarrow{PQ}$$

and we draw **v** as an arrow pointing from P to Q. This vector is said to be based at P. Figure 1(A) shows the vector with initial point $P = (2, 2)$ and terminal point $Q = (7, 5)$. The **length** or **magnitude** of **v**, denoted $\|\mathbf{v}\|$, is the distance from P to Q.

NOTATION In this text, vectors are represented by boldface lowercase letters such as **v**, **w**, **a**, **b**, *etc.*

The vector $\mathbf{v} = \overrightarrow{OR}$ pointing from the origin to a point R is called the **position vector** of R. Figure 1(B) shows the position vector of the point $R = (3, 5)$.

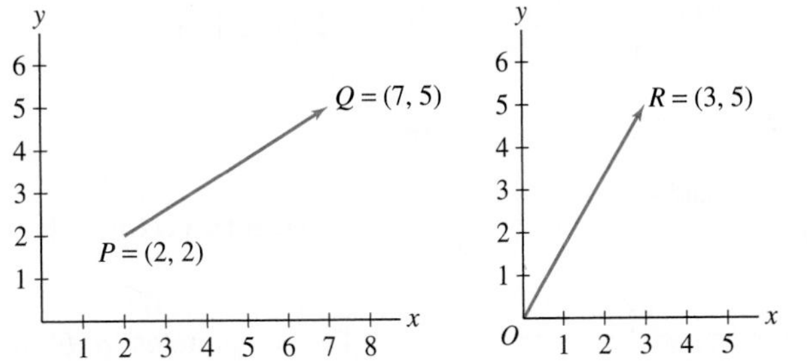

(A) The vector $\overrightarrow{PQ}$ (B) The position vector $\overrightarrow{OR}$

FIGURE 1

We now introduce some vector terminology.

- Two vectors **v** and **w** of nonzero length are called **parallel** if the lines through **v** and **w** are parallel. Parallel vectors point either in the same or in opposite directions [Figure 2(A)].
- A vector **v** is said to undergo a **translation** when it is moved parallel to itself without changing its length or direction. The resulting vector **w** is called a translate of **v** [Figure 2(B)]. Translates have the same length and direction but different basepoints.

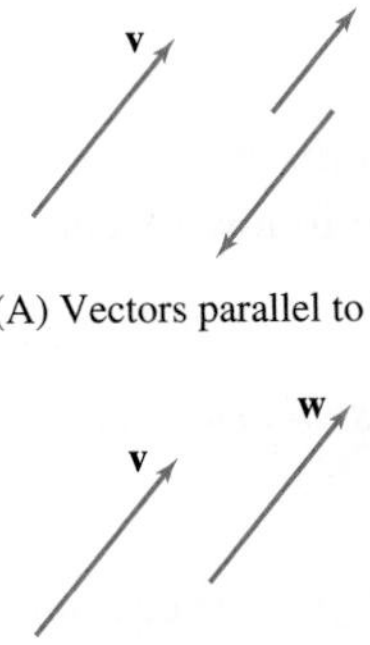

(A) Vectors parallel to **v**

(B) **w** is a translate of **v**

FIGURE 2

In many situations, it is convenient to treat vectors with the same length and direction as equivalent, even if they have different basepoints. With this in mind, we say that

- **v** and **w** are **equivalent** if **w** is a translate of **v** [Figure 3(A)].

Every vector can be translated so that its tail is at the origin [Figure 3(C)]. Therefore,

Every vector $\mathbf{v}$ *is equivalent to a unique vector* $\mathbf{v}_0$ *based at the origin.*

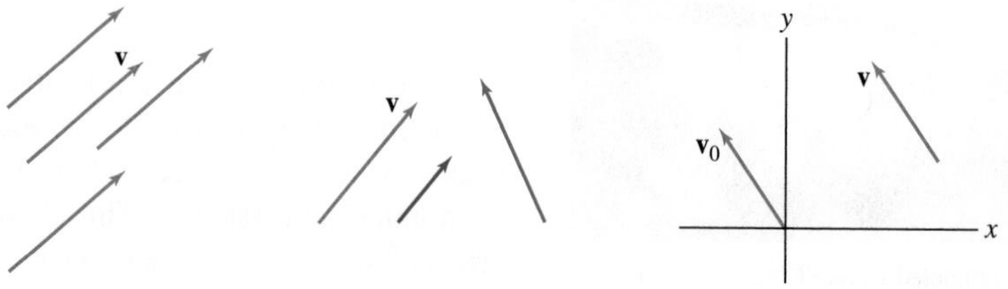

(A) Vectors equivalent to $\mathbf{v}$ (translates of $\mathbf{v}$) (B) Inequivalent vectors (C) $\mathbf{v}_0$ is the unique vector based at the origin and equivalent to $\mathbf{v}$.

FIGURE 3

To work algebraically, we define the components of a vector (Figure 4).

DEFINITION Components of a Vector The components of $\mathbf{v} = \overrightarrow{PQ}$, where $P = (a_1, b_1)$ and $Q = (a_2, b_2)$, are the quantities

$$a = a_2 - a_1 \quad (x\text{-component}), \qquad b = b_2 - b_1 \quad (y\text{-component})$$

The pair of components is denoted $\langle a, b \rangle$.

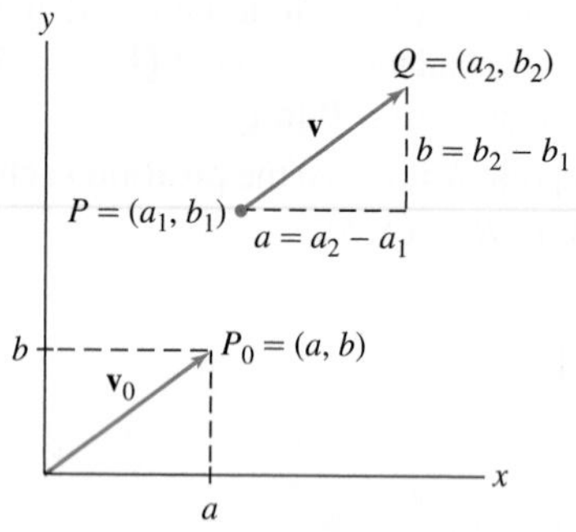

FIGURE 4 The vectors $\mathbf{v}$ and $\mathbf{v}_0$ have components $\langle a, b \rangle$.

- *In this text, "angle brackets" are used to distinguish between the vector* $\mathbf{v} = \langle a, b \rangle$ *and the point* $P = (a, b)$. *Some textbooks denote both* $\mathbf{v}$ *and* P *by* (a, b).
- *When referring to vectors, we use the terms "length" and "magnitude" interchangeably. The term "norm" is also commonly used.*

- The length of a vector in terms of its components (by the distance formula, see Figure 4) is

$$\|\mathbf{v}\| = \|\overrightarrow{PQ}\| = \sqrt{a^2 + b^2}$$

- The **zero vector** (whose head and tail coincide) is the vector $\mathbf{0} = \langle 0, 0 \rangle$ of length zero.

The components $\langle a, b \rangle$ determine the length and direction of $\mathbf{v}$, but not its basepoint. Therefore, *two vectors have the same components if and only if they are equivalent.* Nevertheless, the standard practice is to describe a vector by its components, and thus we write

$$\mathbf{v} = \langle a, b \rangle$$

Although this notation is ambiguous (because it does not specify the basepoint), it rarely causes confusion in practice. To further avoid confusion, the following convention will be in force for the remainder of the text:

We assume all vectors are based at the origin unless otherwise stated.

■ **EXAMPLE 1** Determine whether $\mathbf{v}_1 = \overrightarrow{P_1Q_1}$ and $\mathbf{v}_2 = \overrightarrow{P_2Q_2}$ are equivalent, where

$$P_1 = (3, 7), \quad Q_1 = (6, 5) \qquad \text{and} \qquad P_2 = (-1, 4), \quad Q_2 = (2, 1)$$

What is the magnitude of $\mathbf{v}_1$?

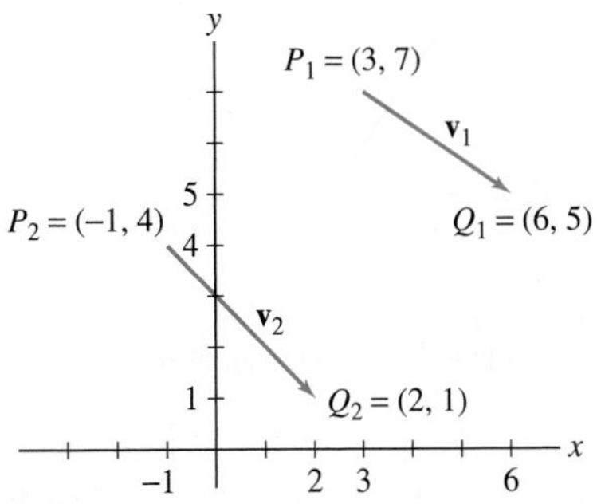

FIGURE 5

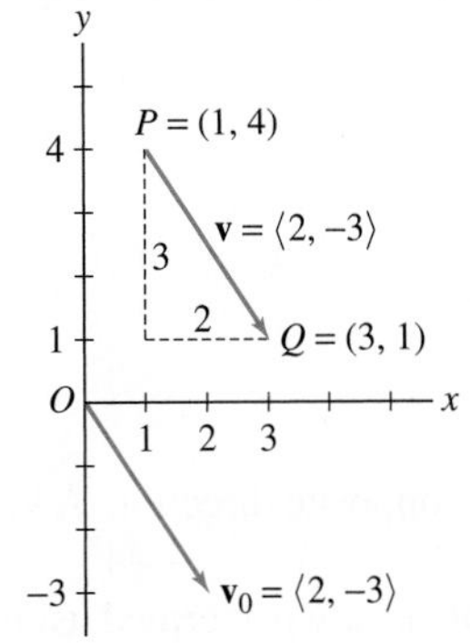

FIGURE 6 The vectors $\mathbf{v}$ and $\mathbf{v}_0$ have the same components but different basepoints.

Solution We can test for equivalence by computing the components (Figure 5):

$$\mathbf{v}_1 = \langle 6-3, 5-7 \rangle = \langle 3, -2 \rangle, \qquad \mathbf{v}_2 = \langle 2-(-1), 1-4 \rangle = \langle 3, -3 \rangle$$

The components of $\mathbf{v}_1$ and $\mathbf{v}_2$ are not the same, so $\mathbf{v}_1$ and $\mathbf{v}_2$ are not equivalent. Since $\mathbf{v}_1 = \langle 3, -2 \rangle$, its magnitude is

$$\|\mathbf{v}_1\| = \sqrt{3^2 + (-2)^2} = \sqrt{13}$$ ■

■ **EXAMPLE 2** Sketch the vector $\mathbf{v} = \langle 2, -3 \rangle$ based at $P = (1, 4)$ and the vector $\mathbf{v}_0$ equivalent to $\mathbf{v}$ based at the origin.

Solution The vector $\mathbf{v} = \langle 2, -3 \rangle$ based at $P = (1, 4)$ has terminal point $Q = (1+2, 4-3) = (3, 1)$, located two units to the right and three units down from P as shown in Figure 6. The vector $\mathbf{v}_0$ equivalent to $\mathbf{v}$ based at O has terminal point $(2, -3)$. ■

Vector Algebra

We now define two basic vector operations: vector addition and scalar multiplication.

The vector sum $\mathbf{v} + \mathbf{w}$ is defined when $\mathbf{v}$ and $\mathbf{w}$ have the same basepoint: Translate $\mathbf{w}$ to the equivalent vector $\mathbf{w}'$ whose tail coincides with the head of $\mathbf{v}$. The sum $\mathbf{v} + \mathbf{w}$ is the vector pointing from the tail of $\mathbf{v}$ to the head of $\mathbf{w}'$ [Figure 7(A)]. Alternatively, we can use the **Parallelogram Law**: $\mathbf{v} + \mathbf{w}$ is the vector pointing from the basepoint to the opposite vertex of the parallelogram formed by $\mathbf{v}$ and $\mathbf{w}$ [Figure 7(B)].

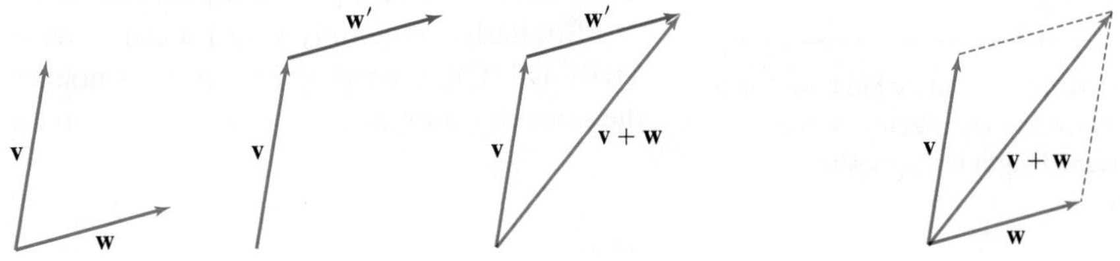

(A) The vector sum $\mathbf{v} + \mathbf{w}$

(B) Addition via the Parallelogram Law

FIGURE 7

To add several vectors $\mathbf{v}_1, \mathbf{v}_2, \ldots, \mathbf{v}_n$, translate the vectors to $\mathbf{v}_1 = \mathbf{v}_1', \mathbf{v}_2', \ldots, \mathbf{v}_n'$ so that they lie head to tail as in Figure 8. The vector sum $\mathbf{v} = \mathbf{v}_1 + \mathbf{v}_2 + \cdots + \mathbf{v}_n$ is the vector whose terminal point is the terminal point of $\mathbf{v}_n'$.

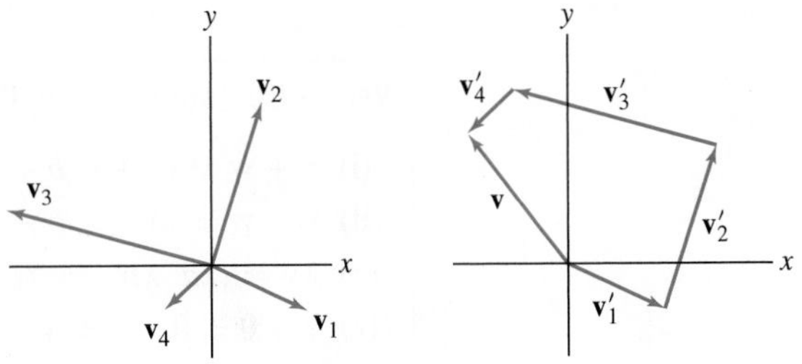

FIGURE 8 The sum $\mathbf{v} = \mathbf{v}_1 + \mathbf{v}_2 + \mathbf{v}_3 + \mathbf{v}_4$.

CAUTION Remember that the vector $\mathbf{v} - \mathbf{w}$ *points in the direction* from the tip of $\mathbf{w}$ to the tip of $\mathbf{v}$ *(not from the tip of* $\mathbf{v}$ *to the tip of* $\mathbf{w}$*).*

Vector subtraction $\mathbf{v} - \mathbf{w}$ is carried out by adding $-\mathbf{w}$ to $\mathbf{v}$ as in Figure 9(A). Or, more simply, draw the vector pointing from $\mathbf{w}$ to $\mathbf{v}$ as in Figure 9(B), and translate it back to the basepoint to obtain $\mathbf{v} - \mathbf{w}$.

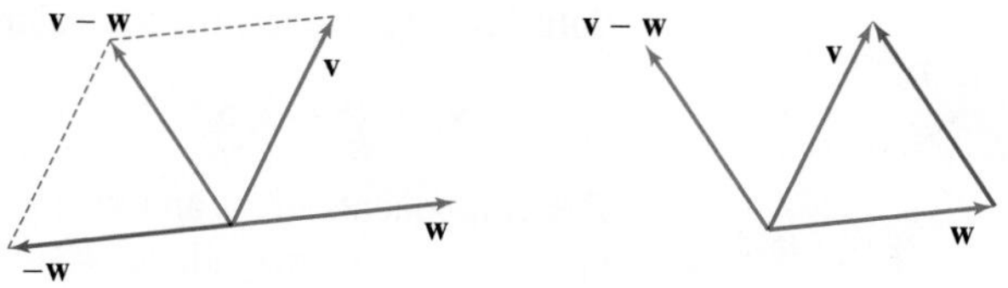

(A) **v** − **w** equals v plus (−**w**)

(B) More simply, **v** − **w** is the translate of the vector pointing from the tip of **w** to the tip of **v**.

FIGURE 9 Vector subtraction.

The term **scalar** is another word for "real number," and we often speak of scalar versus vector quantities. Thus, the number 8 is a scalar, while $\langle 8, 2\rangle$ is a vector. If λ is a scalar and $\mathbf{v}$ is a nonzero vector, the **scalar multiple** $\lambda\mathbf{v}$ is defined as follows (Figure 10):

- $\lambda\mathbf{v}$ has length $|\lambda|\ \|\mathbf{v}\|$.
- It points in the same direction as $\mathbf{v}$ if $\lambda > 0$.
- It points in the opposite direction if $\lambda < 0$.

NOTATION *λ (pronounced "lambda") is the eleventh letter in the Greek alphabet. We use the symbol λ often (but not exclusively) to denote a scalar.*

Note that $0\mathbf{v} = \mathbf{0}$ for all $\mathbf{v}$, and

$$\|\lambda\mathbf{v}\| = |\lambda|\ \|\mathbf{v}\|$$

In particular, $-\mathbf{v}$ has the same length as $\mathbf{v}$ but points in the opposite direction. A vector $\mathbf{w}$ is parallel to $\mathbf{v}$ if and only if $\mathbf{w} = \lambda\mathbf{v}$ for some nonzero scalar λ.

Vector addition and scalar multiplication operations are easily performed using components. To add or subtract two vectors $\mathbf{v}$ and $\mathbf{w}$, we add or subtract their components. This follows from the parallelogram law as indicated in Figure 11(A).

Similarly, to multiply $\mathbf{v}$ by a scalar λ, we multiply the components of $\mathbf{v}$ by λ [Figures 11(B) and (C)]. Indeed, if $\mathbf{v} = \langle a, b\rangle$ is nonzero, $\langle\lambda a, \lambda b\rangle$ has length $|\lambda|\ \|\mathbf{v}\|$. It points in the same direction as $\langle a, b\rangle$ if $\lambda > 0$, and in the opposite direction if $\lambda < 0$.

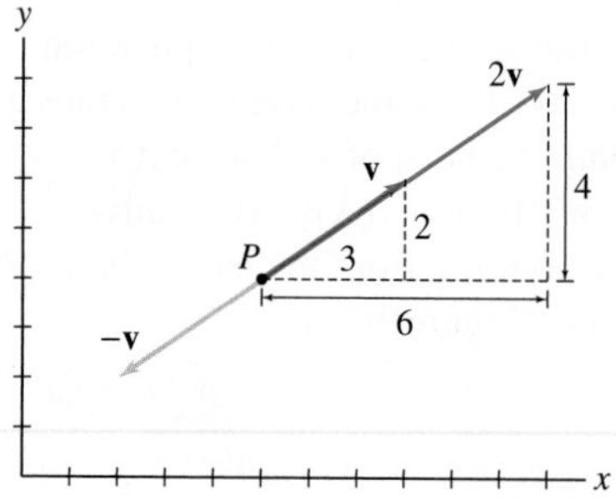

FIGURE 10 Vectors **v** and 2**v** are based at P but 2**v** is twice as long. Vectors **v** and −**v** have the same length but opposite directions.

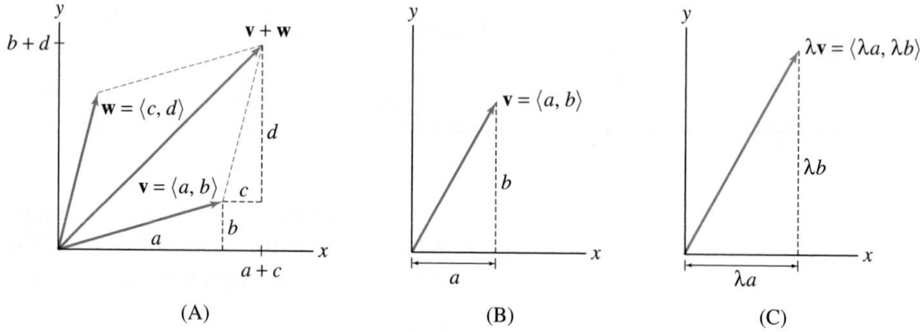

FIGURE 11 Vector operations using components.

Vector Operations Using Components If $\mathbf{v} = \langle a, b\rangle$ and $\mathbf{w} = \langle c, d\rangle$, then:

(i) $\mathbf{v} + \mathbf{w} = \langle a + c, b + d\rangle$

(ii) $\mathbf{v} - \mathbf{w} = \langle a - c, b - d\rangle$

(iii) $\lambda\mathbf{v} = \langle\lambda a, \lambda b\rangle$

(iv) $\mathbf{v} + \mathbf{0} = \mathbf{0} + \mathbf{v} = \mathbf{v}$

We also note that if $P = (a_1, b_1)$ and $Q = (a_2, b_2)$, then components of the vector $\mathbf{v} = \overrightarrow{PQ}$ are conveniently computed as the difference

$$\overrightarrow{PQ} = \overrightarrow{OQ} - \overrightarrow{OP} = \langle a_2, b_2\rangle - \langle a_1, b_1\rangle = \langle a_2 - a_1, b_2 - b_1\rangle$$

■ **EXAMPLE 3** For $\mathbf{v} = \langle 1, 4\rangle$, $\mathbf{w} = \langle 3, 2\rangle$, calculate

(a) $\mathbf{v} + \mathbf{w}$ **(b)** $5\mathbf{v}$

Solution

$$\mathbf{v} + \mathbf{w} = \langle 1, 4\rangle + \langle 3, 2\rangle = \langle 1+3, 4+2\rangle = \langle 4, 6\rangle$$
$$5\mathbf{v} = 5\,\langle 1, 4\rangle = \langle 5, 20\rangle$$

The vector sum is illustrated in Figure 12. ■

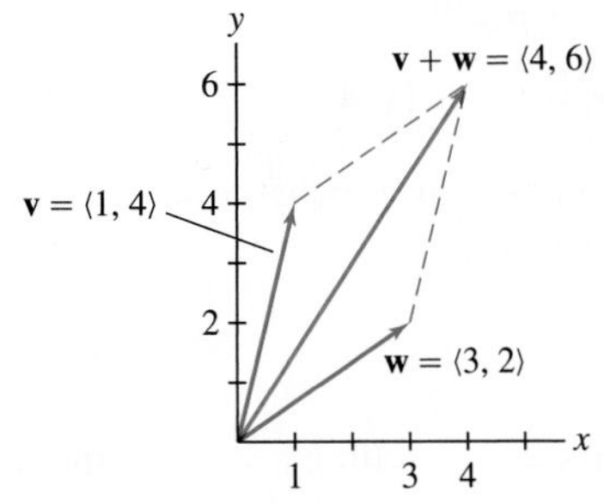

FIGURE 12

Vector operations obey the usual laws of algebra.

THEOREM 1 Basic Properties of Vector Algebra For all vectors $\mathbf{u}$, $\mathbf{v}$, $\mathbf{w}$ and for all scalars λ,

Commutative Law: $\mathbf{v} + \mathbf{w} = \mathbf{w} + \mathbf{v}$
Associative Law: $\mathbf{u} + (\mathbf{v} + \mathbf{w}) = (\mathbf{u} + \mathbf{v}) + \mathbf{w}$
Distributive Law for Scalars: $\lambda(\mathbf{v} + \mathbf{w}) = \lambda\mathbf{v} + \lambda\mathbf{w}$

These properties are verified easily using components. For example, we can check that vector addition is commutative:

$$\langle a, b\rangle + \langle c, d\rangle = \underbrace{\langle a+c, b+d\rangle = \langle c+a, d+b\rangle}_{\text{Commutativity of ordinary addition}} = \langle c, d\rangle + \langle a, b\rangle$$

A **linear combination** of vectors $\mathbf{v}$ and $\mathbf{w}$ is a vector

$$r\mathbf{v} + s\mathbf{w}$$

where r and s are scalars. If $\mathbf{v}$ and $\mathbf{w}$ are not parallel, then every vector $\mathbf{u}$ in the plane can be expressed as a linear combination $\mathbf{u} = r\mathbf{v} + s\mathbf{w}$ [Figure 13(A)]. The parallelogram $\mathcal{P}$ whose vertices are the origin and the terminal points of $\mathbf{v}$, $\mathbf{w}$ and $\mathbf{v} + \mathbf{w}$ is called the **parallelogram spanned** by $\mathbf{v}$ and $\mathbf{w}$ [Figure 13(B)]. It consists of the linear combinations $r\mathbf{v} + s\mathbf{w}$ with $0 \le r \le 1$ and $0 \le s \le 1$.

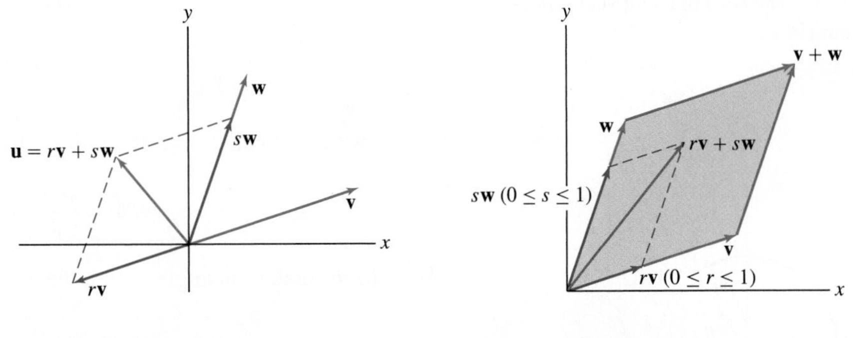

(A) The vector $\mathbf{u}$ can be expressed as a linear combination $\mathbf{u} = r\mathbf{v} + s\mathbf{w}$. In this figure, $r < 0$.

(B) The parallelogram $\mathcal{P}$ spanned by $\mathbf{v}$ and $\mathbf{w}$ consists of all linear combinations $r\mathbf{v} + s\mathbf{w}$ with $0 \le r, s \le 1$.

FIGURE 13

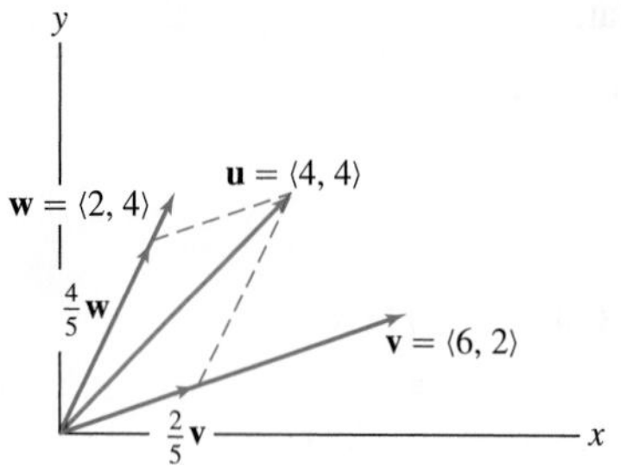

FIGURE 14

■ **EXAMPLE 4 Linear Combinations** Express the vector $\mathbf{u} = \langle 4, 4\rangle$ in Figure 14 as a linear combination of $\mathbf{v} = \langle 6, 2\rangle$ and $\mathbf{w} = \langle 2, 4\rangle$.

Solution We must find r and s such that $r\mathbf{v} + s\mathbf{w} = \langle 4, 4\rangle$, or

$$r\,\langle 6, 2\rangle + s\,\langle 2, 4\rangle = \langle 6r + 2s, 2r + 4s\rangle = \langle 4, 4\rangle$$

The components must be equal, so we have a system of two linear equations:

$$6r + 2s = 4$$

$$2r + 4s = 4$$

Subtracting the equations, we obtain $4r - 2s = 0$ or $s = 2r$. Setting $s = 2r$ in the first equation yields $6r + 4r = 4$ or $r = \frac{2}{5}$, and then $s = 2r = \frac{4}{5}$. Therefore,

$$\mathbf{u} = \langle 4, 4\rangle = \frac{2}{5}\,\langle 6, 2\rangle + \frac{4}{5}\,\langle 2, 4\rangle$$ ■

CONCEPTUAL INSIGHT In general, to write a vector $\mathbf{u} = \langle e, f\rangle$ as a linear combination of two other vectors $\mathbf{v} = \langle a, b\rangle$ and $\mathbf{w} = \langle c, d\rangle$, we have to solve a system of two linear equations in two unknowns r and s:

$$r\mathbf{v} + s\mathbf{w} = \mathbf{u} \quad \Leftrightarrow \quad r\,\langle a, b\rangle + s\,\langle c, d\rangle = \langle e, f\rangle \quad \Leftrightarrow \quad \begin{cases} ar + cs = e \\ br + ds = f \end{cases}$$

On the other hand, vectors give us a way of visualizing the system of equations geometrically. The solution is represented by a parallelogram as in Figure 14. This relation between vectors and systems of linear equations extends to any number of variables and is the starting point for the important subject of linear algebra.

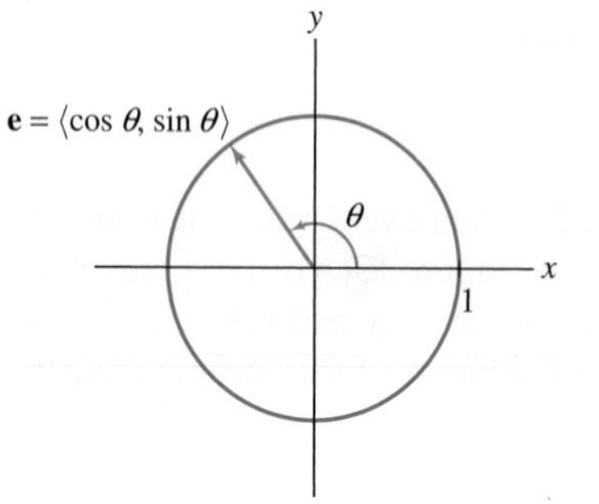

FIGURE 15 The head of a unit vector lies on the unit circle.

A vector of length 1 is called a **unit vector**. Unit vectors are often used to indicate direction, when it is not necessary to specify length. The head of a unit vector **e** based at the origin lies on the unit circle and has components

$$\mathbf{e} = \langle \cos\theta, \sin\theta\rangle$$

where θ is the angle between **e** and the positive x-axis (Figure 15).

We can always scale a nonzero vector $\mathbf{v} = \langle a, b\rangle$ to obtain a unit vector pointing in the same direction (Figure 16):

$$\boxed{\mathbf{e_v} = \frac{1}{\|\mathbf{v}\|}\mathbf{v}}$$

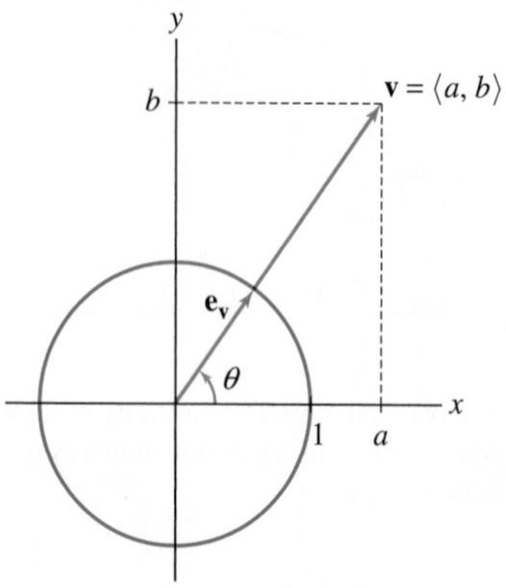

FIGURE 16 Unit vector in the direction of **v**.

Indeed, we can check that $\mathbf{e_v}$ is a unit vector as follows:

$$\|\mathbf{e_v}\| = \left\| \frac{1}{\|\mathbf{v}\|}\mathbf{v} \right\| = \frac{1}{\|\mathbf{v}\|}\|\mathbf{v}\| = 1$$

If $\mathbf{v} = \langle a, b\rangle$ makes an angle θ with the positive x-axis, then

$$\boxed{\mathbf{v} = \langle a, b\rangle = \|\mathbf{v}\|\mathbf{e_v} = \|\mathbf{v}\|\,\langle \cos\theta, \sin\theta\rangle} \qquad \boxed{1}$$

■ **EXAMPLE 5** Find the unit vector in the direction of $\mathbf{v} = \langle 3, 5\rangle$.

Solution $\|\mathbf{v}\| = \sqrt{3^2 + 5^2} = \sqrt{34}$, and thus $\mathbf{e_v} = \dfrac{1}{\sqrt{34}}\mathbf{v} = \left\langle \dfrac{3}{\sqrt{34}}, \dfrac{5}{\sqrt{34}} \right\rangle$. ■

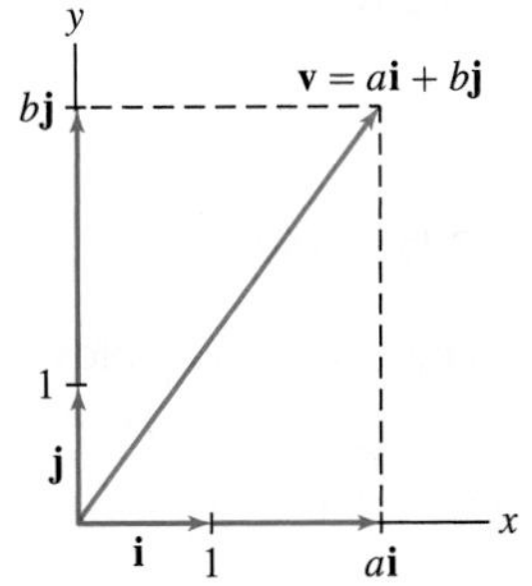

FIGURE 17

It is customary to introduce a special notation for the unit vectors in the direction of the positive x- and y-axes (Figure 17):

$$\mathbf{i} = \langle 1, 0 \rangle, \qquad \mathbf{j} = \langle 0, 1 \rangle$$

The vectors **i** and **j** are called the **standard basis vectors**. Every vector in the plane is a linear combination of **i** and **j** (Figure 17):

$$\mathbf{v} = \langle a, b \rangle = a\mathbf{i} + b\mathbf{j}$$

For example, $\langle 4, -2 \rangle = 4\mathbf{i} - 2\mathbf{j}$. Vector addition is performed by adding the **i** and **j** coefficients. For example,

$$(4\mathbf{i} - 2\mathbf{j}) + (5\mathbf{i} + 7\mathbf{j}) = (4+5)\mathbf{i} + (-2+7)\mathbf{j} = 9\mathbf{i} + 5\mathbf{j}$$

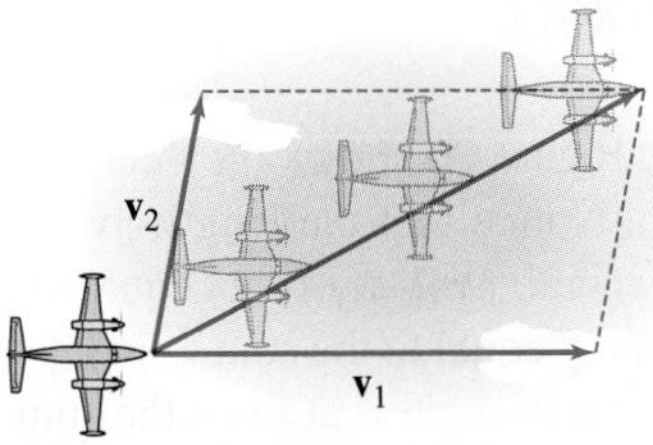

FIGURE 18 When an airplane traveling with velocity $\mathbf{v}_1$ encounters a wind of velocity $\mathbf{v}_2$, its resultant velocity is the vector sum $\mathbf{v}_1 + \mathbf{v}_2$.

CONCEPTUAL INSIGHT It is often said that quantities such as force and velocity are vectors because they have both magnitude and direction, but there is more to this statement than meets the eye. A vector quantity must obey the law of vector addition (Figure 18), so if we say that force is a vector, we are really claiming that forces add according to the Parallelogram Law. In other words, if forces $\mathbf{F}_1$ and $\mathbf{F}_2$ act on an object, then the resultant force is the **vector sum** $\mathbf{F}_1 + \mathbf{F}_2$. This is a physical fact that must be verified experimentally. It was well known to scientists and engineers long before the vector concept was introduced formally in the 1800s.

EXAMPLE 6 Find the forces on cables 1 and 2 in Figure 19(A).

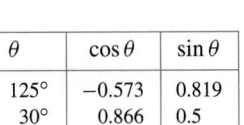

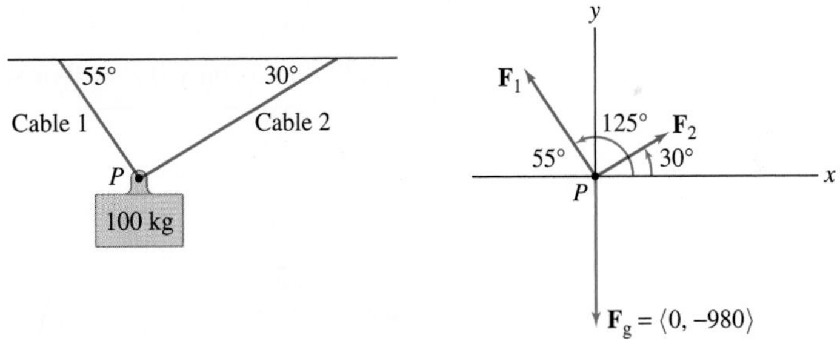

(A)

(B) Force diagram

FIGURE 19

Solution Three forces act on the point P in Figure 19(A): the force $\mathbf{F}_g$ due to gravity of $100\text{g} = 980$ newtons ($\text{g} = 9.8\ \text{m/s}^2$) acting vertically downward, and two unknown forces $\mathbf{F}_1$ and $\mathbf{F}_2$ acting through cables 1 and 2, as indicated in Figure 19(B).

Let $f_1 = \|\mathbf{F}_1\|$ and $f_2 = \|\mathbf{F}_2\|$. Because $\mathbf{F}_1$ makes an angle of 125° (the supplement of 55°) with the positive x-axis, and $\mathbf{F}_2$ makes an angle of 30°, we can use Eq. (1) and the table in the margin to write these vectors in component form:

θ	$\cos\theta$	$\sin\theta$
125°	−0.573	0.819
30°	0.866	0.5

$$\mathbf{F}_1 = f_1 \langle \cos 125°, \sin 125° \rangle \approx f_1 \langle -0.573, 0.819 \rangle$$

$$\mathbf{F}_2 = f_2 \langle \cos 30°, \sin 30° \rangle \approx f_2 \langle 0.866, 0.5 \rangle$$

$$\mathbf{F}_g = \langle 0, -980 \rangle$$

Now, the point P is not in motion, so the net force on P is zero:

$$\mathbf{F}_1 + \mathbf{F}_2 + \mathbf{F}_g = \mathbf{0}$$

$$f_1 \langle -0.573, 0.819\rangle + f_2 \langle 0.866, 0.5\rangle + \langle 0, -980\rangle = \langle 0, 0\rangle$$

This gives us two equations in two unknowns:

$$-0.573 f_1 + 0.866 f_2 = 0, \qquad 0.819 f_1 + 0.5 f_2 - 980 = 0$$

By the first equation, $f_2 = \left(\frac{0.573}{0.866}\right) f_1$. Substitution in the second equation yields

$$0.819 f_1 + 0.5\left(\frac{0.573}{0.866}\right) f_1 - 980 \approx 1.15 f_1 - 980 = 0$$

Therefore, the forces in newtons are

$$f_1 \approx \frac{980}{1.15} \approx 852 \text{ N} \qquad \text{and} \qquad f_2 \approx \left(\frac{0.573}{0.866}\right) 852 \approx 564 \text{ N}$$ ■

We close this section with the Triangle Inequality. Figure 20 shows the vector sum $\mathbf{v} + \mathbf{w}$ for three different vectors $\mathbf{w}$ of the same length. Notice that the length $\|\mathbf{v} + \mathbf{w}\|$ varies, depending on the angle between $\mathbf{v}$ and $\mathbf{w}$. So in general, $\|\mathbf{v} + \mathbf{w}\|$ is not equal to the sum $\|\mathbf{v}\| + \|\mathbf{w}\|$. What we can say is that $\|\mathbf{v} + \mathbf{w}\|$ is *at most* equal to the sum $\|\mathbf{v}\| + \|\mathbf{w}\|$. This corresponds to the fact that the length of one side of a triangle is at most the sum of the lengths of the other two sides. A formal proof may be given using the dot product (see Exercise 88 in Section 13.3).

THEOREM 2 Triangle Inequality For any two vectors $\mathbf{v}$ and $\mathbf{w}$,

$$\|\mathbf{v} + \mathbf{w}\| \le \|\mathbf{v}\| + \|\mathbf{w}\|$$

Equality holds only if $\mathbf{v} = \mathbf{0}$ or $\mathbf{w} = \mathbf{0}$, or if $\mathbf{w} = \lambda\mathbf{v}$, where $\lambda \ge 0$.

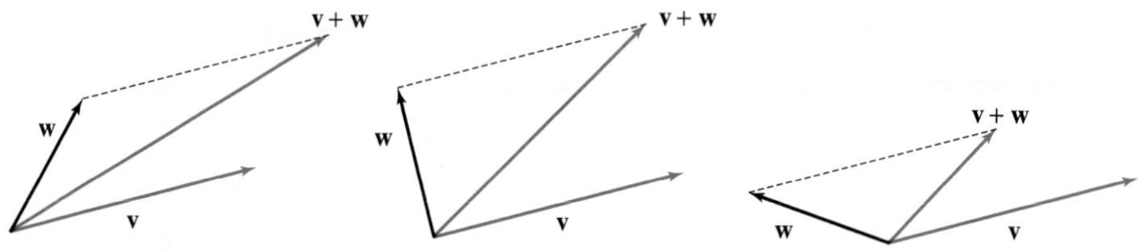

FIGURE 20 The length of $\mathbf{v} + \mathbf{w}$ depends on the angle between $\mathbf{v}$ and $\mathbf{w}$.

13.1 SUMMARY

- A *vector* $\mathbf{v} = \overrightarrow{PQ}$ is determined by a basepoint P (the "tail") and a terminal point Q (the "head").
- Components of $\mathbf{v} = \overrightarrow{PQ}$ where $P = (a_1, b_1)$ and $Q = (a_2, b_2)$:

$$\mathbf{v} = \langle a, b\rangle$$

with $a = a_2 - a_1$, $b = b_2 - b_1$.
- Length or magnitude: $\|\mathbf{v}\| = \sqrt{a^2 + b^2}$.
- The *length* $\|\mathbf{v}\|$ is the distance from P to Q.
- The *position vector* of $P_0 = (a, b)$ is the vector $\mathbf{v} = \langle a, b\rangle$ pointing from the origin O to P_0.

- Vectors $\mathbf{v}$ and $\mathbf{w}$ are *equivalent* if they are translates of each other: They have the same magnitude and direction, but possibly different basepoints. Two vectors are equivalent if and only if they have the same components.
- We assume all vectors are based at the origin unless otherwise indicated.
- The *zero vector* is the vector $\mathbf{0} = \langle 0, 0 \rangle$ of length 0.
- *Vector addition* is defined geometrically by the *Parallelogram Law*. In components,

$$\langle a_1, b_1 \rangle + \langle a_2, b_2 \rangle = \langle a_1 + a_2, b_1 + b_2 \rangle$$

- *Scalar multiplication*: $\lambda \mathbf{v}$ is the vector of length $|\lambda| \, \|\mathbf{v}\|$ in the same direction as $\mathbf{v}$ if $\lambda > 0$, and in the opposite direction if $\lambda < 0$. In components,

$$\lambda \langle a, b \rangle = \langle \lambda a, \lambda b \rangle$$

- Nonzero vectors $\mathbf{v}$ and $\mathbf{w}$ are *parallel* if $\mathbf{w} = \lambda \mathbf{v}$ for some scalar λ.
- Unit vector making an angle θ with the positive x-axis: $\mathbf{e} = \langle \cos\theta, \sin\theta \rangle$.
- Unit vector in the direction of $\mathbf{v} \neq \mathbf{0}$: $\mathbf{e_v} = \dfrac{1}{\|\mathbf{v}\|}\mathbf{v}$.
- If $\mathbf{v} = \langle a, b \rangle$ makes an angle θ with the positive x-axis, then

$$a = \|\mathbf{v}\| \cos\theta, \qquad b = \|\mathbf{v}\| \sin\theta, \qquad \mathbf{e_v} = \langle \cos\theta, \sin\theta \rangle$$

- *Standard basis vectors*: $\mathbf{i} = \langle 1, 0 \rangle$ and $\mathbf{j} = \langle 0, 1 \rangle$.
- Every vector $\mathbf{v} = \langle a, b \rangle$ is a linear combination $\mathbf{v} = a\mathbf{i} + b\mathbf{j}$.
- *Triangle Inequality*: $\|\mathbf{v} + \mathbf{w}\| \le \|\mathbf{v}\| + \|\mathbf{w}\|$.

13.1 EXERCISES

Preliminary Questions

1. Answer true or false. Every nonzero vector is:

(a) Equivalent to a vector based at the origin.

(b) Equivalent to a unit vector based at the origin.

(c) Parallel to a vector based at the origin.

(d) Parallel to a unit vector based at the origin.

2. What is the length of $-3\mathbf{a}$ if $\|\mathbf{a}\| = 5$?

3. Suppose that $\mathbf{v}$ has components $\langle 3, 1 \rangle$. How, if at all, do the components change if you translate $\mathbf{v}$ horizontally two units to the left?

4. What are the components of the zero vector based at $P = (3, 5)$?

5. True or false?

(a) The vectors $\mathbf{v}$ and $-2\mathbf{v}$ are parallel.

(b) The vectors $\mathbf{v}$ and $-2\mathbf{v}$ point in the same direction.

6. Explain the commutativity of vector addition in terms of the Parallelogram Law.

Exercises

1. Sketch the vectors $\mathbf{v}_1, \mathbf{v}_2, \mathbf{v}_3, \mathbf{v}_4$ with tail P and head Q, and compute their lengths. Are any two of these vectors equivalent?

	$\mathbf{v}_1$	$\mathbf{v}_2$	$\mathbf{v}_3$	$\mathbf{v}_4$
P	$(2, 4)$	$(-1, 3)$	$(-1, 3)$	$(4, 1)$
Q	$(4, 4)$	$(1, 3)$	$(2, 4)$	$(6, 3)$

2. Sketch the vector $\mathbf{b} = \langle 3, 4 \rangle$ based at $P = (-2, -1)$.

3. What is the terminal point of the vector $\mathbf{a} = \langle 1, 3 \rangle$ based at $P = (2, 2)$? Sketch $\mathbf{a}$ and the vector $\mathbf{a}_0$ based at the origin and equivalent to $\mathbf{a}$.

4. Let $\mathbf{v} = \overrightarrow{PQ}$, where $P = (1, 1)$ and $Q = (2, 2)$. What is the head of the vector $\mathbf{v}'$ equivalent to $\mathbf{v}$ based at $(2, 4)$? What is the head of the vector $\mathbf{v}_0$ equivalent to $\mathbf{v}$ based at the origin? Sketch $\mathbf{v}$, $\mathbf{v}_0$, and $\mathbf{v}'$.

In Exercises 5–8, find the components of $\overrightarrow{PQ}$.

5. $P = (3, 2), \quad Q = (2, 7)$

6. $P = (1, -4), \quad Q = (3, 5)$

7. $P = (3, 5), \quad Q = (1, -4)$

8. $P = (0, 2), \quad Q = (5, 0)$

In Exercises 9–14, calculate.

9. $\langle 2, 1 \rangle + \langle 3, 4 \rangle$

10. $\langle -4, 6 \rangle - \langle 3, -2 \rangle$

11. $5 \langle 6, 2 \rangle$

12. $4(\langle 1, 1 \rangle + \langle 3, 2 \rangle)$

13. $\left\langle -\frac{1}{2}, \frac{5}{3} \right\rangle + \left\langle 3, \frac{10}{3} \right\rangle$

14. $\langle \ln 2, e \rangle + \langle \ln 3, \pi \rangle$

15. Which of the vectors (A)–(C) in Figure 21 is equivalent to $\mathbf{v} - \mathbf{w}$?

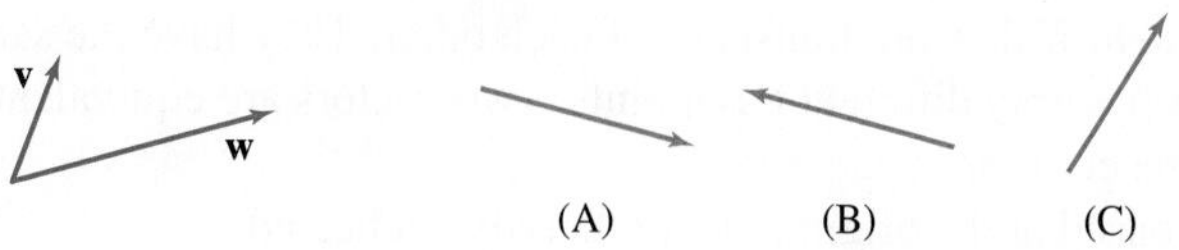

FIGURE 21

16. Sketch $\mathbf{v} + \mathbf{w}$ and $\mathbf{v} - \mathbf{w}$ for the vectors in Figure 22.

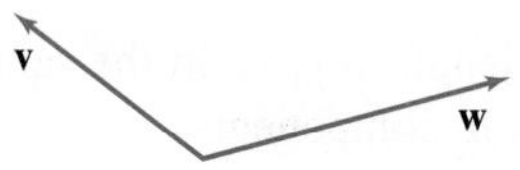

FIGURE 22

17. Sketch $2\mathbf{v}$, $-\mathbf{w}$, $\mathbf{v} + \mathbf{w}$, and $2\mathbf{v} - \mathbf{w}$ for the vectors in Figure 23.

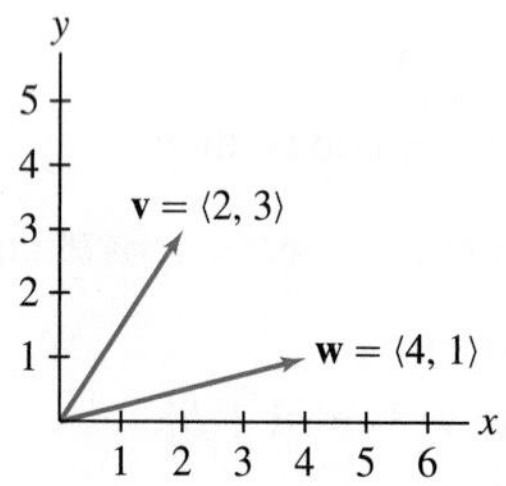

FIGURE 23

18. Sketch $\mathbf{v} = \langle 1, 3 \rangle$, $\mathbf{w} = \langle 2, -2 \rangle$, $\mathbf{v} + \mathbf{w}$, $\mathbf{v} - \mathbf{w}$.

19. Sketch $\mathbf{v} = \langle 0, 2 \rangle$, $\mathbf{w} = \langle -2, 4 \rangle$, $3\mathbf{v} + \mathbf{w}$, $2\mathbf{v} - 2\mathbf{w}$.

20. Sketch $\mathbf{v} = \langle -2, 1 \rangle$, $\mathbf{w} = \langle 2, 2 \rangle$, $\mathbf{v} + 2\mathbf{w}$, $\mathbf{v} - 2\mathbf{w}$.

21. Sketch the vector $\mathbf{v}$ such that $\mathbf{v} + \mathbf{v}_1 + \mathbf{v}_2 = \mathbf{0}$ for $\mathbf{v}_1$ and $\mathbf{v}_2$ in Figure 24(A).

22. Sketch the vector sum $\mathbf{v} = \mathbf{v}_1 + \mathbf{v}_2 + \mathbf{v}_3 + \mathbf{v}_4$ in Figure 24(B).

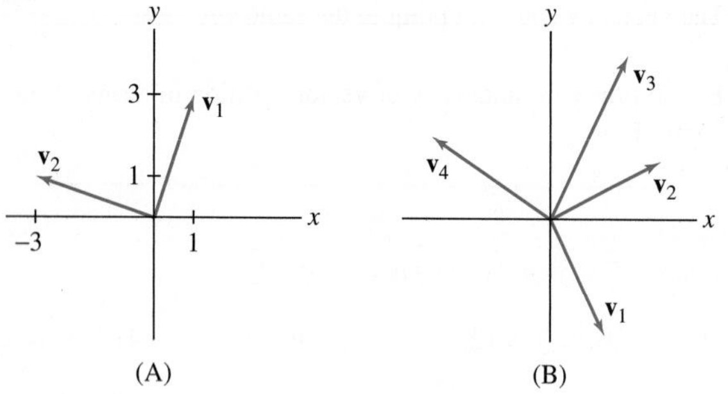

FIGURE 24

23. Let $\mathbf{v} = \overrightarrow{PQ}$, where $P = (-2, 5)$, $Q = (1, -2)$. Which of the following vectors with the given tails and heads are equivalent to $\mathbf{v}$?

(a) $(-3, 3)$, $(0, 4)$ **(b)** $(0, 0)$, $(3, -7)$
(c) $(-1, 2)$, $(2, -5)$ **(d)** $(4, -5)$, $(1, 4)$

24. Which of the following vectors are parallel to $\mathbf{v} = \langle 6, 9 \rangle$ and which point in the same direction?

(a) $\langle 12, 18 \rangle$ **(b)** $\langle 3, 2 \rangle$ **(c)** $\langle 2, 3 \rangle$
(d) $\langle -6, -9 \rangle$ **(e)** $\langle -24, -27 \rangle$ **(f)** $\langle -24, -36 \rangle$

In Exercises 25–28, sketch the vectors $\overrightarrow{AB}$ and $\overrightarrow{PQ}$, and determine whether they are equivalent.

25. $A = (1, 1)$, $B = (3, 7)$, $P = (4, -1)$, $Q = (6, 5)$

26. $A = (1, 4)$, $B = (-6, 3)$, $P = (1, 4)$, $Q = (6, 3)$

27. $A = (-3, 2)$, $B = (0, 0)$, $P = (0, 0)$, $Q = (3, -2)$

28. $A = (5, 8)$, $B = (1, 8)$, $P = (1, 8)$, $Q = (-3, 8)$

In Exercises 29–32, are $\overrightarrow{AB}$ and $\overrightarrow{PQ}$ parallel? And if so, do they point in the same direction?

29. $A = (1, 1)$, $B = (3, 4)$, $P = (1, 1)$, $Q = (7, 10)$

30. $A = (-3, 2)$, $B = (0, 0)$, $P = (0, 0)$, $Q = (3, 2)$

31. $A = (2, 2)$, $B = (-6, 3)$, $P = (9, 5)$, $Q = (17, 4)$

32. $A = (5, 8)$, $B = (2, 2)$, $P = (2, 2)$, $Q = (-3, 8)$

In Exercises 33–36, let $R = (-2, 7)$. Calculate the following.

33. The length of $\overrightarrow{OR}$

34. The components of $\mathbf{u} = \overrightarrow{PR}$, where $P = (1, 2)$

35. The point P such that $\overrightarrow{PR}$ has components $\langle -2, 7 \rangle$

36. The point Q such that $\overrightarrow{RQ}$ has components $\langle 8, -3 \rangle$

In Exercises 37–42, find the given vector.

37. Unit vector $\mathbf{e}_\mathbf{v}$ where $\mathbf{v} = \langle 3, 4 \rangle$

38. Unit vector $\mathbf{e}_\mathbf{w}$ where $\mathbf{w} = \langle 24, 7 \rangle$

39. Vector of length 4 in the direction of $\mathbf{u} = \langle -1, -1 \rangle$

40. Unit vector in the direction opposite to $\mathbf{v} = \langle -2, 4 \rangle$

41. Unit vector $\mathbf{e}$ making an angle of $\frac{4\pi}{7}$ with the x-axis

42. Vector $\mathbf{v}$ of length 2 making an angle of $30°$ with the x-axis

43. Find all scalars λ such that $\lambda \langle 2, 3 \rangle$ has length 1.

44. Find a vector $\mathbf{v}$ satisfying $3\mathbf{v} + \langle 5, 20 \rangle = \langle 11, 17 \rangle$.

45. What are the coordinates of the point P in the parallelogram in Figure 25(A)?

46. What are the coordinates a and b in the parallelogram in Figure 25(B)?

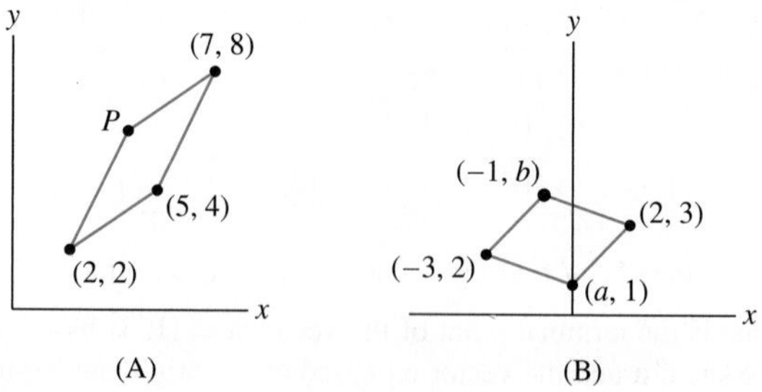

FIGURE 25

47. Let $\mathbf{v} = \overrightarrow{AB}$ and $\mathbf{w} = \overrightarrow{AC}$, where A, B, C are three distinct points in the plane. Match (a)–(d) with (i)–(iv). (*Hint:* Draw a picture.)

(a) $-\mathbf{w}$ **(b)** $-\mathbf{v}$ **(c)** $\mathbf{w}-\mathbf{v}$ **(d)** $\mathbf{v}-\mathbf{w}$

(i) $\overrightarrow{CB}$ **(ii)** $\overrightarrow{CA}$ **(iii)** $\overrightarrow{BC}$ **(iv)** $\overrightarrow{BA}$

48. Find the components and length of the following vectors:

(a) $4\mathbf{i}+3\mathbf{j}$ **(b)** $2\mathbf{i}-3\mathbf{j}$ **(c)** $\mathbf{i}+\mathbf{j}$ **(d)** $\mathbf{i}-3\mathbf{j}$

In Exercises 49–52, calculate the linear combination.

49. $3\mathbf{j}+(9\mathbf{i}+4\mathbf{j})$

50. $-\frac{3}{2}\mathbf{i}+5\left(\frac{1}{2}\mathbf{j}-\frac{1}{2}\mathbf{i}\right)$

51. $(3\mathbf{i}+\mathbf{j})-6\mathbf{j}+2(\mathbf{j}-4\mathbf{i})$

52. $3(3\mathbf{i}-4\mathbf{j})+5(\mathbf{i}+4\mathbf{j})$

53. For each of the position vectors $\mathbf{u}$ with endpoints A, B, and C in Figure 26, indicate with a diagram the multiples $r\mathbf{v}$ and $s\mathbf{w}$ such that $\mathbf{u}=r\mathbf{v}+s\mathbf{w}$. A sample is shown for $\mathbf{u}=\overrightarrow{OQ}$.

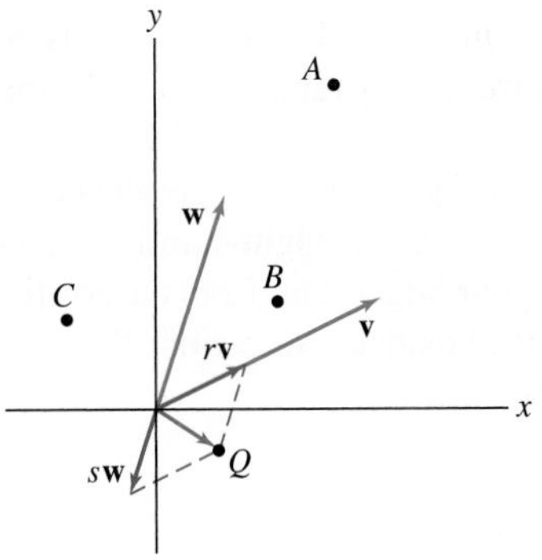

FIGURE 26

54. Sketch the parallelogram spanned by $\mathbf{v}=\langle 1,4\rangle$ and $\mathbf{w}=\langle 5,2\rangle$. Add the vector $\mathbf{u}=\langle 2,3\rangle$ to the sketch and express $\mathbf{u}$ as a linear combination of $\mathbf{v}$ and $\mathbf{w}$.

In Exercises 55 and 56, express $\mathbf{u}$ *as a linear combination* $\mathbf{u}=r\mathbf{v}+s\mathbf{w}$*. Then sketch* $\mathbf{u}$, $\mathbf{v}$, $\mathbf{w}$, *and the parallelogram formed by* $r\mathbf{v}$ *and* $s\mathbf{w}$.

55. $\mathbf{u}=\langle 3,-1\rangle$; $\mathbf{v}=\langle 2,1\rangle$, $\mathbf{w}=\langle 1,3\rangle$

56. $\mathbf{u}=\langle 6,-2\rangle$; $\mathbf{v}=\langle 1,1\rangle$, $\mathbf{w}=\langle 1,-1\rangle$

57. Calculate the magnitude of the force on cables 1 and 2 in Figure 27.

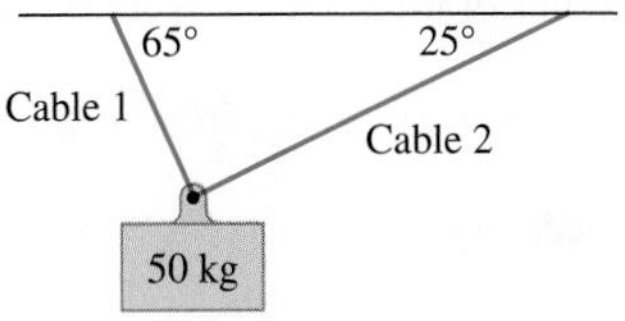

FIGURE 27

58. Determine the magnitude of the forces $\mathbf{F}_1$ and $\mathbf{F}_2$ in Figure 28, assuming that there is no net force on the object.

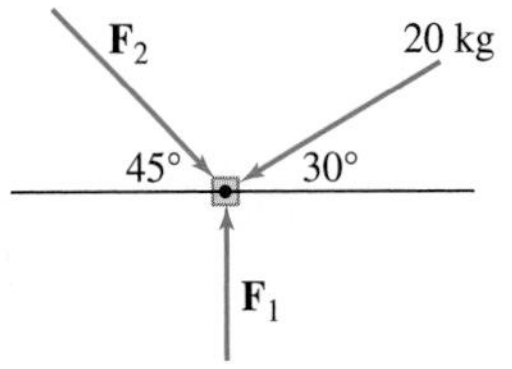

FIGURE 28

59. A plane flying due east at 200 km/h encounters a 40-km/h wind blowing in the north-east direction. The resultant velocity of the plane is the vector sum $\mathbf{v}=\mathbf{v}_1+\mathbf{v}_2$, where $\mathbf{v}_1$ is the velocity vector of the plane and $\mathbf{v}_2$ is the velocity vector of the wind (Figure 29). The angle between $\mathbf{v}_1$ and $\mathbf{v}_2$ is $\frac{\pi}{4}$. Determine the resultant *speed* of the plane (the length of the vector $\mathbf{v}$).

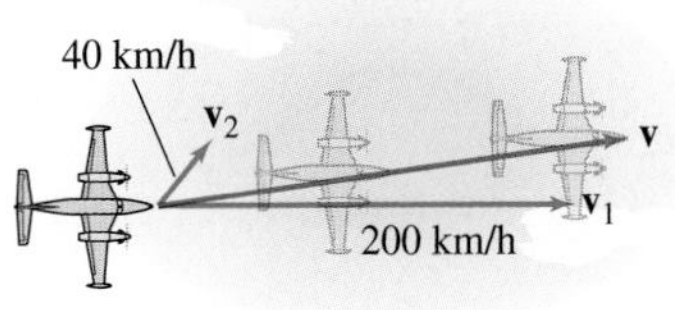

FIGURE 29

Further Insights and Challenges

In Exercises 60–62, refer to Figure 30, which shows a robotic arm consisting of two segments of lengths L_1 *and* L_2.

60. Find the components of the vector $\mathbf{r}=\overrightarrow{OP}$ in terms of θ_1 and θ_2.

61. Let $L_1=5$ and $L_2=3$. Find $\mathbf{r}$ for $\theta_1=\frac{\pi}{3}$, $\theta_2=\frac{\pi}{4}$.

62. Let $L_1=5$ and $L_2=3$. Show that the set of points reachable by the robotic arm with $\theta_1=\theta_2$ is an ellipse.

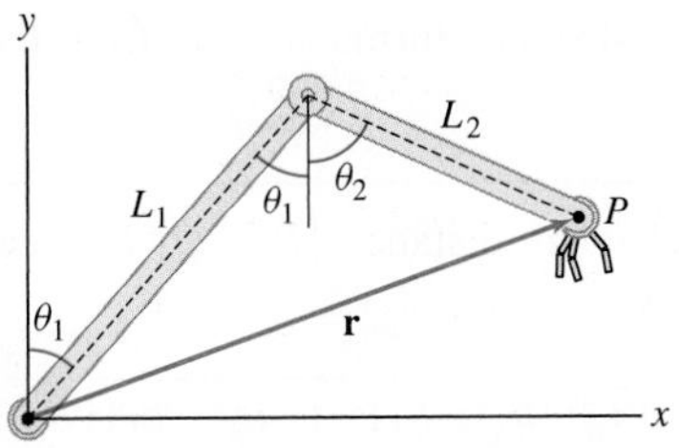

FIGURE 30

63. Use vectors to prove that the diagonals $\overline{AC}$ and $\overline{BD}$ of a parallelogram bisect each other (Figure 31). *Hint:* Observe that the midpoint of $\overline{BD}$ is the terminal point of $\mathbf{w}+\frac{1}{2}(\mathbf{v}-\mathbf{w})$.

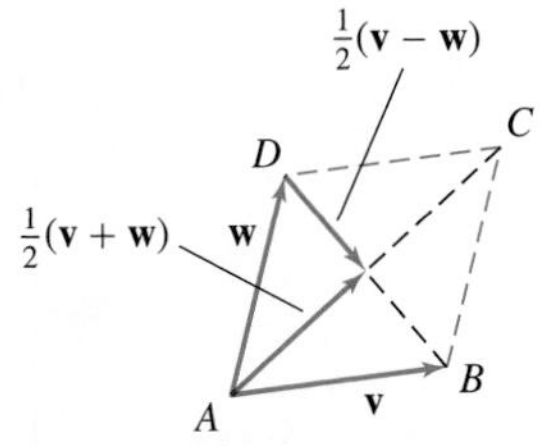

FIGURE 31

64. Use vectors to prove that the segments joining the midpoints of opposite sides of a quadrilateral bisect each other (Figure 32). *Hint:* Show that the midpoints of these segments are the terminal points of

$$\frac{1}{4}(2\mathbf{u}+\mathbf{v}+\mathbf{z}) \quad \text{and} \quad \frac{1}{4}(2\mathbf{v}+\mathbf{w}+\mathbf{u})$$

65. Prove that two vectors $\mathbf{v} = \langle a, b\rangle$ and $\mathbf{w} = \langle c, d\rangle$ are perpendicular if and only if

$$ac + bd = 0$$

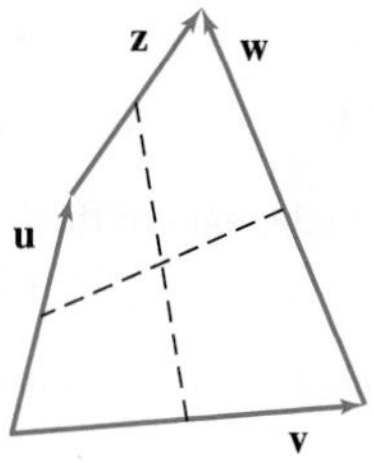

FIGURE 32

13.2 Vectors in Three Dimensions

This section extends the vector concepts introduced in the previous section to three-dimensional space. We begin with some introductory remarks about the three-dimensional coordinate system.

By convention, we label the axes as in Figure 1(A), where the positive sides of the axes are labeled x, y, and z. This labeling satisfies the **right-hand rule**, which means that when you position your right hand so that your fingers curl from the positive x-axis toward the positive y-axis, your thumb points in the positive z-direction. The axes in Figure 1(B) are not labeled according to the right-hand rule.

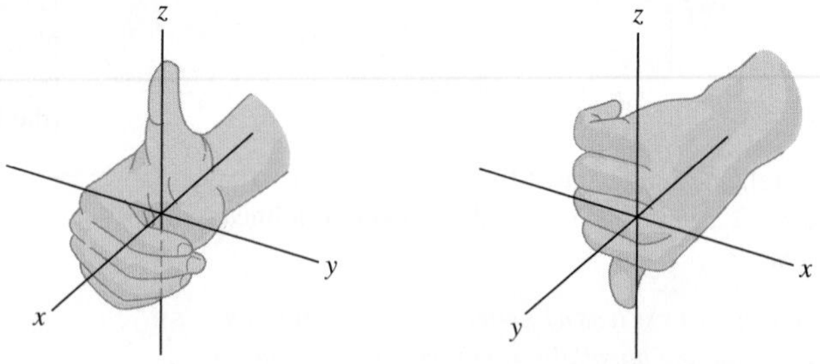

(A) Standard coordinate system (satisfies the right-hand rule)

(B) This coordinate system does not satisfy the right-hand rule because the thumb points in the negative z-direction.

FIGURE 1 The fingers of the right hand curl from the positive x-axis to the positive y-axis.

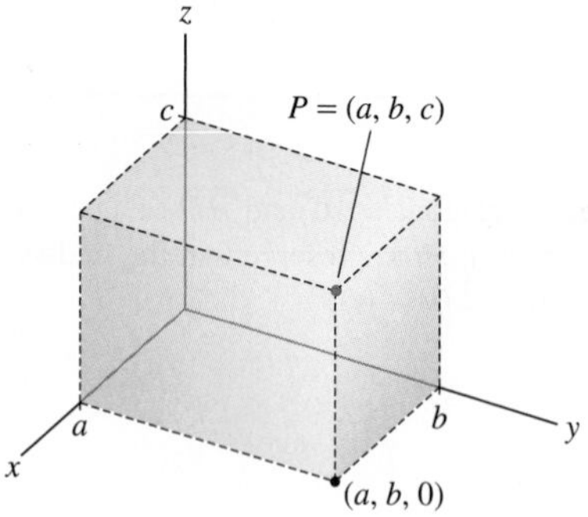

FIGURE 2

Each point in space has unique coordinates (a, b, c) relative to the axes (Figure 2). We denote the set of all triples (a, b, c) by $\mathbf{R}^3$. The **coordinate planes** in $\mathbf{R}^3$ are defined by setting one of the coordinates equal to zero (Figure 3). The xy-plane consists of the points $(a, b, 0)$ and is defined by the equation $z = 0$. Similarly, $x = 0$ defines the yz-plane consisting of the points $(0, b, c)$, and $y = 0$ defines the xz-plane consisting of the points $(a, 0, c)$. The coordinate planes divide $\mathbf{R}^3$ into eight **octants** (analogous to the four quadrants in the plane). Each octant corresponds to a possible combination of signs of the coordinates. The set of points (a, b, c) with $a, b, c > 0$ is called the **first octant**.

As in two dimensions, we derive the distance formula in $\mathbf{R}^3$ from the Pythagorean Theorem.

THEOREM 1 Distance Formula in $\mathbf{R}^3$ The distance $|P - Q|$ between the points $P = (a_1, b_1, c_1)$ and $Q = (a_2, b_2, c_2)$ is

$$|P - Q| = \sqrt{(a_2 - a_1)^2 + (b_2 - b_1)^2 + (c_2 - c_1)^2} \qquad \boxed{1}$$

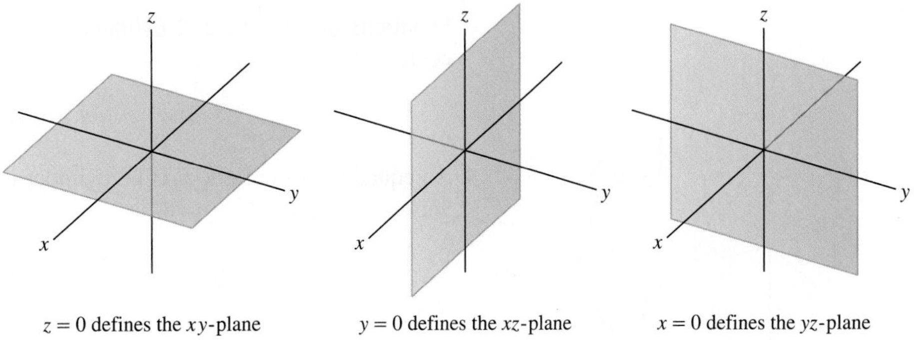
z
y
x
z = 0 defines the xy-plane
z
y
x
y = 0 defines the xz-plane
z
y
x
x = 0 defines the yz-plane

FIGURE 3

Proof First apply the distance formula in the plane to the points P and R (Figure 4):

$$|P - R|^2 = (a_2 - a_1)^2 + (b_2 - b_1)^2$$

Then observe that $\triangle PRQ$ is a right triangle [Figure 4(B)] and use the Pythagorean Theorem:

$$|P - Q|^2 = |P - R|^2 + |R - Q|^2 = (a_2 - a_1)^2 + (b_2 - b_1)^2 + (c_2 - c_1)^2 \quad \blacksquare$$

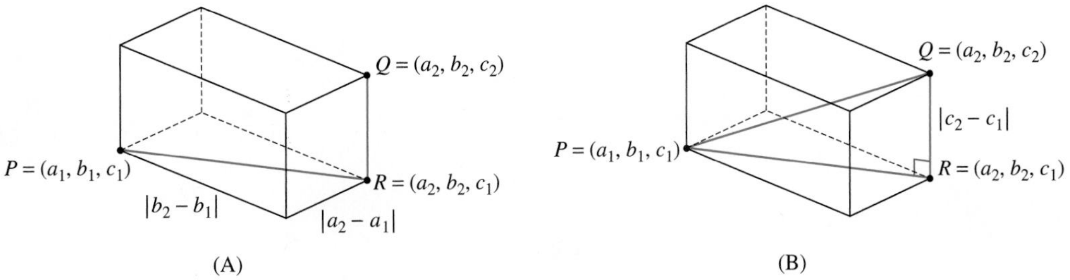

FIGURE 4 Compute $|P - Q|$ using the right triangle $\triangle PRQ$.

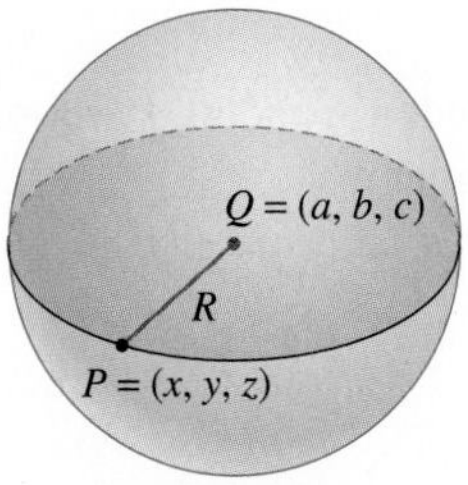

FIGURE 5 Sphere of radius R centered at (a, b, c).

The sphere of radius R with center $Q = (a, b, c)$ consists of all points $P = (x, y, z)$ located a distance R from Q (Figure 5). By the distance formula, the coordinates of $P = (x, y, z)$ must satisfy

$$\sqrt{(x - a)^2 + (y - b)^2 + (z - c)^2} = R$$

On squaring both sides, we obtain the standard equation of the sphere [Eq. (3) below].

Now consider the equation

$$(x - a)^2 + (y - b)^2 = R^2 \qquad \boxed{2}$$

In the xy-plane, Eq. (2) defines the circle of radius R with center (a, b). However, as an equation in $\mathbf{R}^3$, it defines the right circular cylinder of radius R whose central axis is the vertical line through $(a, b, 0)$ (Figure 6). Indeed, a point (x, y, z) satisfies Eq. (2) for any value of z if (x, y) lies on the circle. It is usually clear from the context which of the following is intended:

$$\text{Circle} = \{(x, y) : (x - a)^2 + (y - b)^2 = R^2\}$$

$$\text{Right circular cylinder} = \{(x, y, z) : (x - a)^2 + (y - b)^2 = R^2\}$$

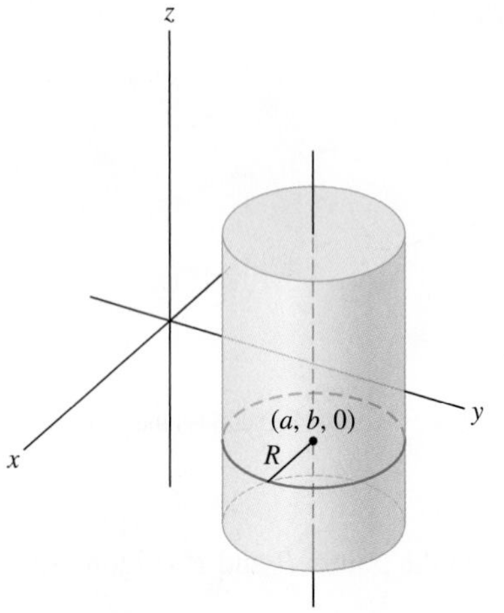

FIGURE 6 Right circular cylinder of radius R centered at $(a, b, 0)$.

Equations of Spheres and Cylinders An equation of the sphere in $\mathbf{R}^3$ of radius R centered at $Q = (a, b, c)$ is

$$(x - a)^2 + (y - b)^2 + (z - c)^2 = R^2 \qquad \boxed{3}$$

An equation of the right circular cylinder in $\mathbf{R}^3$ of radius R whose central axis is the vertical line through $(a, b, 0)$ is

$$(x - a)^2 + (y - b)^2 = R^2 \qquad \boxed{4}$$

■ **EXAMPLE 1** Describe the sets of points defined by the following conditions:

(a) $x^2 + y^2 + z^2 = 4, \quad y \geq 0$ **(b)** $(x - 3)^2 + (y - 2)^2 = 1, \quad z \geq -1$

Solution

(a) The equation $x^2 + y^2 + z^2 = 4$ defines a sphere of radius 2 centered at the origin. The inequality $y \geq 0$ holds for points lying on the positive side of the xz-plane. We obtain the right hemisphere of radius 2 illustrated in Figure 7(A).

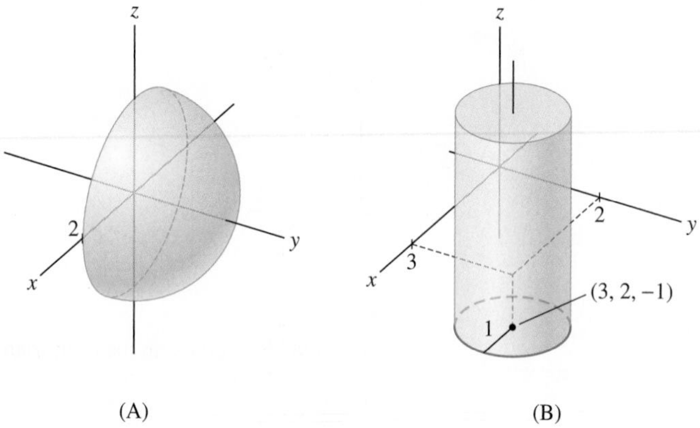

FIGURE 7 Hemisphere and upper cylinder.

(b) The equation $(x - 3)^2 + (y - 2)^2 = 1$ defines a cylinder of radius 1 whose central axis is the vertical line through $(3, 2, 0)$. The part of the cylinder where $z \geq -1$ is illustrated in Figure 7(B). ■

Vector Concepts

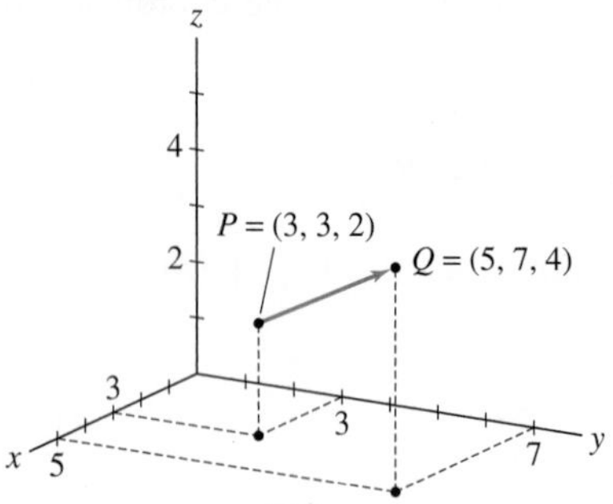

FIGURE 8 A vector $\overrightarrow{PQ}$ in 3-space.

As in the plane, a vector $\mathbf{v} = \overrightarrow{PQ}$ in $\mathbf{R}^3$ is determined by an initial point P and a terminal point Q (Figure 8). If $P = (a_1, b_1, c_1)$ and $Q = (a_2, b_2, c_2)$, then the **length** or **magnitude** of $\mathbf{v} = \overrightarrow{PQ}$, denoted $\|\mathbf{v}\|$, is the distance from P to Q:

$$\|\mathbf{v}\| = \|\overrightarrow{PQ}\| = \sqrt{(a_2 - a_1)^2 + (b_2 - b_1)^2 + (c_2 - c_1)^2}$$

The terminology and basic properties discussed in the previous section carry over to $\mathbf{R}^3$ with little change.

- A vector $\mathbf{v}$ is said to undergo a **translation** if it is moved without changing direction or magnitude.
- Two vectors $\mathbf{v}$ and $\mathbf{w}$ are **equivalent** if $\mathbf{w}$ is a translate of $\mathbf{v}$; that is, $\mathbf{v}$ and $\mathbf{w}$ have the same length and direction.

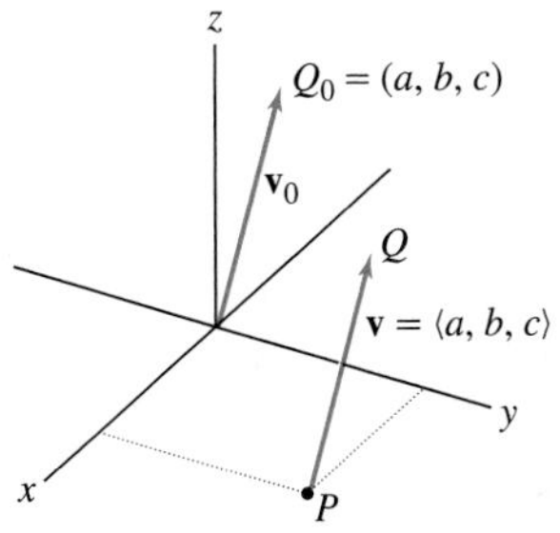

FIGURE 9 A vector **v** and its translate based at the origin.

Our basepoint convention remains in force: vectors are assumed to be based at the origin unless otherwise indicated.

- Two nonzero vectors **v** and **w** are **parallel** if $\mathbf{v} = \lambda\mathbf{w}$ for some scalar λ.
- The **position vector** of a point Q_0 is the vector $\mathbf{v}_0 = \overrightarrow{OQ_0}$ based at the origin (Figure 9).
- A vector $\mathbf{v} = \overrightarrow{PQ}$ with components $\langle a, b, c\rangle$ is equivalent to the vector $\mathbf{v}_0 = \overrightarrow{OQ_0}$ based at the origin with $Q_0 = (a, b, c)$ (Figure 9).
- The **components** of $\mathbf{v} = \overrightarrow{PQ}$, where $P = (a_1, b_1, c_1)$ and $Q = (a_2, b_2, c_2)$, are the differences $a = a_2 - a_1$, $b = b_2 - b_1$, $c = c_2 - c_1$; that is,

$$\mathbf{v} = \overrightarrow{PQ} = \overrightarrow{OQ} - \overrightarrow{OP} = \langle a_2, b_2, c_2\rangle - \langle a_1, b_1, c_1\rangle$$

For example, if $P = (3, -4, -4)$ and $Q = (2, 5, -1)$, then

$$\mathbf{v} = \overrightarrow{PQ} = \langle 2, 5, -1\rangle - \langle 3, -4, -4\rangle = \langle -1, 9, 3\rangle$$

- Two vectors are equivalent if and only if they have the same components.
- Vector addition and scalar multiplication are defined as in the two-dimensional case. Vector addition is defined by the Parallelogram Law (Figure 10).
- In terms of components, if $\mathbf{v} = \langle a_1, b_1, c_1\rangle$ and $\mathbf{w} = \langle a_2, b_2, c_2\rangle$, then

$$\lambda\mathbf{v} = \lambda\langle a_1, b_1, c_1\rangle = \langle \lambda a_1, \lambda b_1, \lambda c_1\rangle$$
$$\mathbf{v} + \mathbf{w} = \langle a_1, b_1, c_1\rangle + \langle a_2, b_2, c_2\rangle = \langle a_1 + a_2, b_1 + b_2, c_1 + c_2\rangle$$

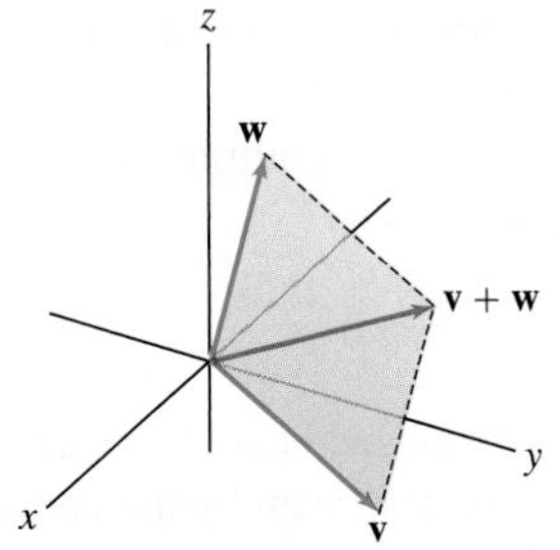

FIGURE 10 Vector addition is defined by the Parallelogram Law.

- Vector addition is commutative, is associative, and satisfies the distributive property with respect to scalar multiplication (Theorem 1 in Section 13.1).

■ **EXAMPLE 2 Vector Calculations** Calculate $\|\mathbf{v}\|$ and $6\mathbf{v} - \frac{1}{2}\mathbf{w}$, where $\mathbf{v} = \langle 3, -1, 2\rangle$ and $\mathbf{w} = \langle 4, 6, -8\rangle$.

Solution

$$\|\mathbf{v}\| = \sqrt{3^2 + (-1)^2 + 2^2} = \sqrt{14}$$

$$\begin{aligned} 6\mathbf{v} - \frac{1}{2}\mathbf{w} &= 6\langle 3, -1, 2\rangle - \frac{1}{2}\langle 4, 6, -8\rangle \\ &= \langle 18, -6, 12\rangle - \langle 2, 3, -4\rangle \\ &= \langle 16, -9, 16\rangle \end{aligned}$$ ■

The **standard basis vectors** in $\mathbf{R}^3$ are

$$\mathbf{i} = \langle 1, 0, 0\rangle, \qquad \mathbf{j} = \langle 0, 1, 0\rangle, \qquad \mathbf{k} = \langle 0, 0, 1\rangle$$

Every vector is a **linear combination** of the standard basis vectors (Figure 11):

$$\langle a, b, c\rangle = a\langle 1, 0, 0\rangle + b\langle 0, 1, 0\rangle + c\langle 0, 0, 1\rangle = a\mathbf{i} + b\mathbf{j} + c\mathbf{k}$$

For example, $\langle -9, -4, 17\rangle = -9\mathbf{i} - 4\mathbf{j} + 17\mathbf{k}$.

■ **EXAMPLE 3** Find the unit vector $\mathbf{e}_{\mathbf{v}}$ in the direction of $\mathbf{v} = 3\mathbf{i} + 2\mathbf{j} - 4\mathbf{k}$.

Solution Since $\|\mathbf{v}\| = \sqrt{3^2 + 2^2 + (-4)^2} = \sqrt{29}$,

$$\mathbf{e}_{\mathbf{v}} = \frac{1}{\|\mathbf{v}\|}\mathbf{v} = \frac{1}{\sqrt{29}}(3\mathbf{i} + 2\mathbf{j} - 4\mathbf{k}) = \left\langle \frac{3}{\sqrt{29}}, \frac{2}{\sqrt{29}}, \frac{-4}{\sqrt{29}} \right\rangle$$ ■

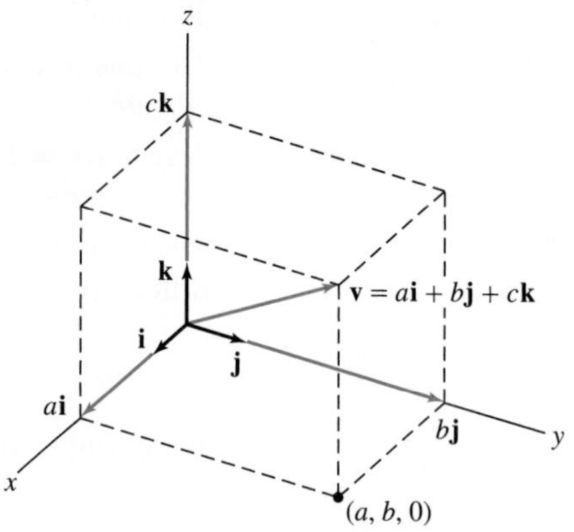

FIGURE 11 Writing $\mathbf{v} = \langle a, b, c\rangle$ as the sum $a\mathbf{i} + b\mathbf{j} + c\mathbf{k}$.

Parametric Equations of a Line

Although the basic vector concepts in two and three dimensions are essentially the same, there is an important difference in the way lines are described. A line in $\mathbf{R}^2$ is defined by a single linear equation such as $y = mx + b$. In $\mathbf{R}^3$, a single linear equation defines a plane rather than a line. Therefore, we describe lines in $\mathbf{R}^3$ in parametric form.

We note first that a line $\mathcal{L}_0$ through the origin consists of the multiples of a nonzero vector $\mathbf{v} = \langle a, b, c\rangle$, as in Figure 12(A). More precisely, set

$$\mathbf{r}_0 = t\mathbf{v} = \langle ta, tb, tc\rangle \qquad (-\infty < t < \infty)$$

Then the line $\mathcal{L}_0$ consists of the terminal points (ta, tb, tc) of the vectors $\mathbf{r}_0$ as t varies from $-\infty$ to ∞. The coordinates (x, y, z) of the points on the line are given by the parametric equations

$$x = at, \quad y = bt, \quad z = ct$$

Suppose, more generally, that we would like to parametrize the line $\mathcal{L}$ parallel to $\mathbf{v}$ but passing through a point $P_0 = (x_0, y_0, z_0)$ as in Figure 12(B). We must translate the line $t\mathbf{v}$ so that it passes through P_0. To do this, we add the position vector $\overrightarrow{OP_0}$ to the multiples $t\mathbf{v}$:

$$\mathbf{r}(t) = \overrightarrow{OP_0} + t\mathbf{v} = \langle x_0, y_0, z_0\rangle + t\langle a, b, c\rangle$$

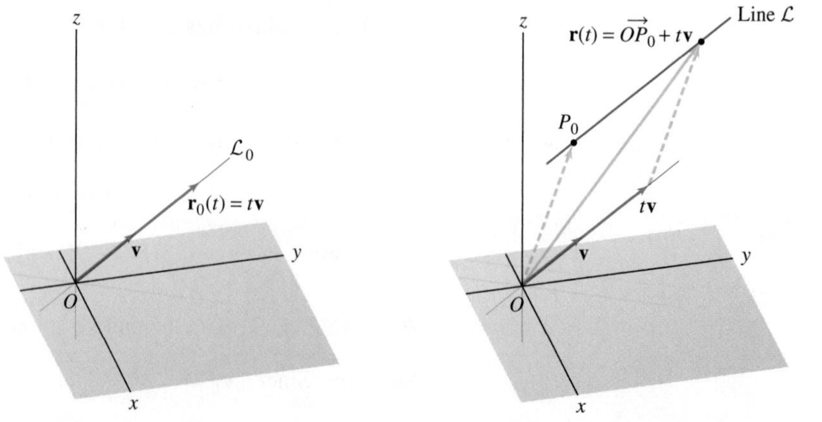

(A) Line through the origin (multiples of **v**).

(B) Line through P_0 in the direction of **v**.

FIGURE 12

The terminal point of $\mathbf{r}(t)$ traces out $\mathcal{L}$ as t varies from $-\infty$ to ∞. The vector $\mathbf{v}$ is called a **direction vector** for $\mathcal{L}$, and coordinates (x, y, z) of the points on the line $\mathcal{L}$ are given by the parametric equations

$$x = x_0 + at, \quad y = y_0 + bt, \quad z = z_0 + ct$$

Equation of a Line (Point-Direction Form) The line $\mathcal{L}$ through $P_0 = (x_0, y_0, z_0)$ in the direction of $\mathbf{v} = \langle a, b, c \rangle$ is described by

Vector parametrization:

$$\mathbf{r}(t) = \overrightarrow{OP_0} + t\mathbf{v} = \langle x_0, y_0, z_0 \rangle + t \langle a, b, c \rangle \qquad (5)$$

Parametric equations:

$$x = x_0 + at, \quad y = y_0 + bt, \quad z = z_0 + ct \qquad (6)$$

The parameter t varies for $-\infty < t < \infty$.

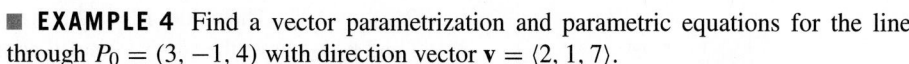

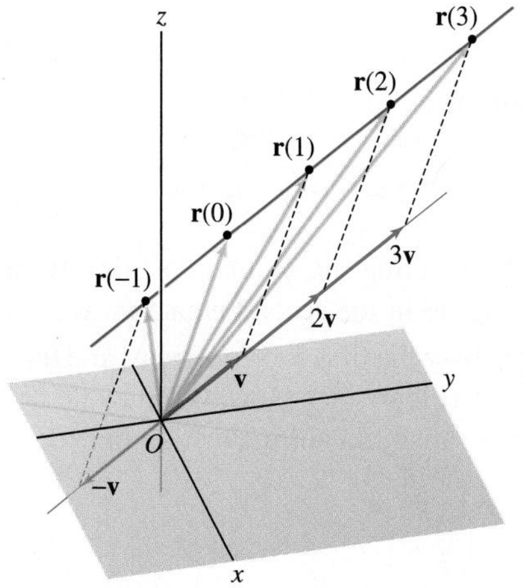

FIGURE 13 The terminal point of $\mathbf{r}(t)$ traces out a line as t varies from $-\infty$ to ∞.

Parametric equations specify the x, y, and z coordinates of a point on the line as a function of the parameter t. These are familiar from our discussion of parametric curves in the plane in Section 12.1. What is new here is the notion of a vector parametrization, the idea that $\mathbf{r}(t)$ describes a vector whose terminal point traces out a line as t varies from $-\infty$ to ∞ (Figure 13).

■ **EXAMPLE 4** Find a vector parametrization and parametric equations for the line through $P_0 = (3, -1, 4)$ with direction vector $\mathbf{v} = \langle 2, 1, 7 \rangle$.

Solution By Eq. (5), the following is a vector parametrization:

$$\mathbf{r}(t) = \underbrace{\langle 3, -1, 4 \rangle}_{\text{Coordinates of } P_0} + \underbrace{t \langle 2, 1, 7 \rangle}_{\text{Direction vector}} = \langle 3 + 2t, -1 + t, 4 + 7t \rangle$$

The corresponding parametric equations are $x = 3 + 2t$, $y = -1 + t$, $z = 4 + 7t$. ■

The parametrization of a line $\mathcal{L}$ is not unique. We are free to choose any point P_0 on $\mathcal{L}$ and we may replace the direction vector $\mathbf{v}$ by any nonzero scalar multiple $\lambda\mathbf{v}$. However, two lines in $\mathbf{R}^3$ coincide if they are parallel and pass through a common point, so we can always check whether two parametrizations describe the same line.

■ **EXAMPLE 5 Different Parametrizations of the Same Line** Show that

$$\mathbf{r}_1(t) = \langle 1, 1, 0 \rangle + t \langle -2, 1, 3 \rangle \quad \text{and} \quad \mathbf{r}_2(t) = \langle -3, 3, 6 \rangle + t \langle 4, -2, -6 \rangle$$

parametrize the same line.

Solution The line $\mathbf{r}_1$ has direction vector $\mathbf{v} = \langle -2, 1, 3 \rangle$, whereas $\mathbf{r}_2$ has direction vector $\mathbf{w} = \langle 4, -2, -6 \rangle$. These vectors are parallel because $\mathbf{w} = -2\mathbf{v}$. Therefore, the lines described by $\mathbf{r}_1$ and $\mathbf{r}_2$ are parallel. We must check that they have a point in common. Choose any point on $\mathbf{r}_1$, say $P = (1, 1, 0)$ [corresponding to $t = 0$]. This point lies on $\mathbf{r}_2$ if there is a value of t such that

$$\langle 1, 1, 0 \rangle = \langle -3, 3, 6 \rangle + t \langle 4, -2, -6 \rangle \qquad (7)$$

This yields three equations

$$1 = -3 + 4t, \qquad 1 = 3 - 2t, \qquad 0 = 6 - 6t$$

All three are satisfied with $t = 1$. Therefore P also lies on $\mathbf{r}_2$. We conclude that $\mathbf{r}_1$ and $\mathbf{r}_2$ parametrize the same line. If Eq. (7) had no solution, we would conclude that $\mathbf{r}_1$ and $\mathbf{r}_2$ are parallel but do not coincide. ■

EXAMPLE 6 **Intersection of Two Lines** Determine whether the following two lines intersect:

$$\mathbf{r}_1(t) = \langle 1, 0, 1\rangle + t\,\langle 3, 3, 5\rangle$$
$$\mathbf{r}_2(t) = \langle 3, 6, 1\rangle + t\,\langle 4, -2, 7\rangle$$

CAUTION We cannot assume in Eq. (8) that the parameter values t_1 and t_2 are equal. The point of intersection may correspond to different parameter values on the two lines.

Solution The two lines intersect if there exist parameter values t_1 and t_2 such that $\mathbf{r}_1(t_1) = \mathbf{r}_2(t_2)$—that is, if

$$\langle 1, 0, 1\rangle + t_1\,\langle 3, 3, 5\rangle = \langle 3, 6, 1\rangle + t_2\,\langle 4, -2, 7\rangle \qquad \boxed{8}$$

This is equivalent to three equations for the components:

$$x = 1 + 3t_1 = 3 + 4t_2, \qquad y = 3t_1 = 6 - 2t_2, \qquad z = 1 + 5t_1 = 1 + 7t_2 \qquad \boxed{9}$$

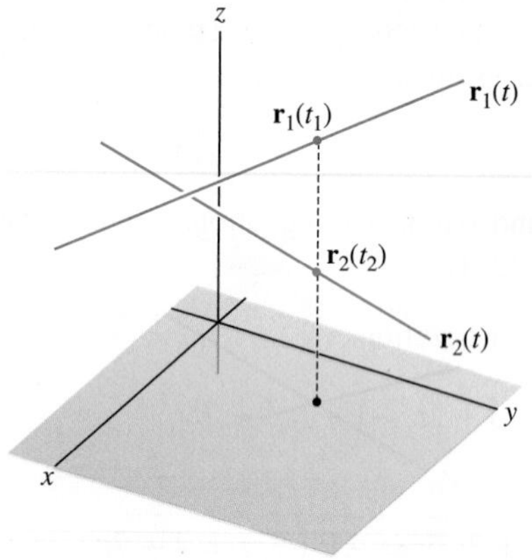

FIGURE 14 The lines $\mathbf{r}_1(t)$ and $\mathbf{r}_2(t)$ do not intersect, but the particular points $\mathbf{r}_1(t_1)$ and $\mathbf{r}_2(t_2)$ have the same x- and y-coordinates.

Let's solve the first two equations for t_1 and t_2. Subtracting the second equation from the first, we get $1 = 6t_2 - 3$ or $t_2 = \frac{2}{3}$. Using this value in the second equation, we get $t_1 = 2 - \frac{2}{3}t_2 = \frac{14}{9}$. The values $t_1 = \frac{14}{9}$ and $t_2 = \frac{2}{3}$ satisfy the first two equations, and thus $\mathbf{r}_1(t_1)$ and $\mathbf{r}_2(t_2)$ have the same x- and y-coordinates (Figure 14). However, they do not have the same z-coordinates because t_1 and t_2 do not satisfy the third equation in (9):

$$1 + 5\left(\frac{14}{9}\right) \neq 1 + 7\left(\frac{2}{3}\right)$$

Therefore, Eq. (8) has no solution and the lines do not intersect. ■

We can describe the line $\mathcal{L}$ passing through two points $P = (a_1, b_1, c_1)$ and $Q = (a_2, b_2, c_2)$ by the vector parametrization (Figure 15):

$$\mathbf{r}(t) = (1 - t)\,\overrightarrow{OP} + t\,\overrightarrow{OQ}$$

Why does $\mathbf{r}$ pass through P and Q? Because $\mathbf{r}(0) = \overrightarrow{OP}$ and $\mathbf{r}(1) = \overrightarrow{OQ}$. Thus $\mathbf{r}(t)$ traces the segment $\overline{PQ}$ joining P and Q as t varies from 0 to 1. Explicitly,

$$\mathbf{r}(t) = (1 - t)\,\langle a_1, b_1, c_1\rangle + t\,\langle a_2, b_2, c_2\rangle$$

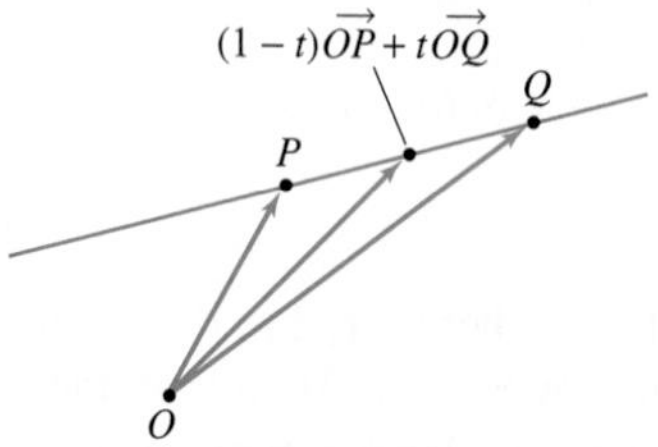

FIGURE 15 Line through two points P and Q.

The parametric equations are

$$x = a_1 + (a_2 - a_1)t, \qquad y = b_1 + (b_2 - b_1)t, \qquad z = c_1 + (c_2 - c_1)t$$

The **midpoint** of $\overline{PQ}$ corresponds to $t = \frac{1}{2}$:

$$\text{Midpoint of } \overline{PQ} = \left(\frac{a_1 + a_2}{2}, \frac{b_1 + b_2}{2}, \frac{c_1 + c_2}{2}\right)$$

Line through Two Points The line through $P = (a_1, b_1, c_1)$ and $Q = (a_2, b_2, c_2)$ is described by

Vector parametrization:

$$\mathbf{r}(t) = (1-t)\overrightarrow{OP} + t\overrightarrow{OQ} = (1-t)\langle a_1, b_1, c_1\rangle + t\langle a_2, b_2, c_2\rangle \qquad \boxed{10}$$

Parametric equations:

$$x = a_1 + (a_2 - a_1)t, \quad y = b_1 + (b_2 - b_1)t, \quad z = c_1 + (c_2 - c_1)t \qquad \boxed{11}$$

for $-\infty < t < \infty$. This parametrization traces the segment $\overline{PQ}$ from P to Q as t varies from 0 to 1.

■ **EXAMPLE 7** Parametrize the segment $\overline{PQ}$ where $P = (1, 0, 4)$ and $Q = (3, 2, 1)$. Find the midpoint of the segment.

Solution The line through $P = (1, 0, 4)$ and $Q = (3, 2, 1)$ has the parametrization

$$\mathbf{r}(t) = (1-t)\langle 1, 0, 4\rangle + t\langle 3, 2, 1\rangle = \langle 1+2t, 2t, 4-3t\rangle$$

The segment $\overline{PQ}$ is traced for $0 \le t \le 1$. The midpoint of $\overline{PQ}$ is the terminal point of the vector

$$\mathbf{r}\left(\frac{1}{2}\right) = \frac{1}{2}\langle 1, 0, 4\rangle + \frac{1}{2}\langle 3, 2, 1\rangle = \left\langle 2, 1, \frac{5}{2}\right\rangle$$ ■

In other words, the midpoint is $\left(2, 1, \frac{5}{2}\right)$.

13.2 SUMMARY

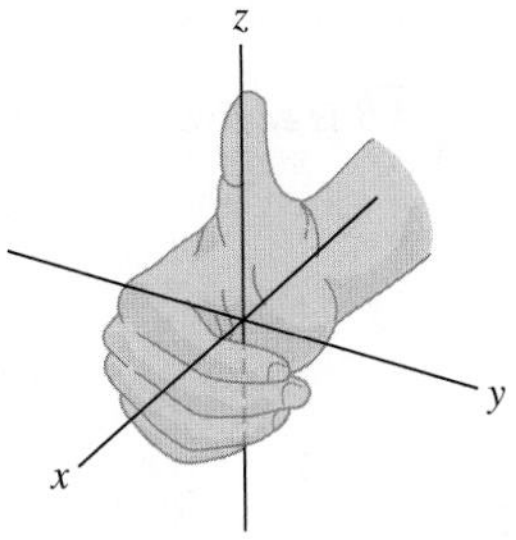

FIGURE 16

- The axes in $\mathbf{R}^3$ are labeled so that they satisfy the *right-hand rule*: When the fingers of your right hand curl from the positive x-axis toward the positive y-axis, your thumb points in the positive z-direction (Figure 16).

Sphere of radius R and center (a, b, c)	$(x-a)^2 + (y-b)^2 + (z-c)^2 = R^2$
Cylinder of radius R with vertical axis through $(a, b, 0)$	$(x-a)^2 + (y-b)^2 = R^2$

- The notation and terminology for vectors in the plane carry over to vectors in $\mathbf{R}^3$.
- The length (or magnitude) of $\mathbf{v} = \overrightarrow{PQ}$, where $P = (a_1, b_1, c_1)$ and $Q = (a_2, b_2, c_2)$, is

$$\|\mathbf{v}\| = \|\overrightarrow{PQ}\| = \sqrt{(a_2-a_1)^2 + (b_2-b_1)^2 + (c_2-c_1)^2}$$

- Equations for the line through $P_0 = (x_0, y_0, z_0)$ with direction vector $\mathbf{v} = \langle a, b, c\rangle$:

Vector parametrization: $\mathbf{r}(t) = \overrightarrow{OP_0} + t\mathbf{v} = \langle x_0, y_0, z_0\rangle + t\langle a, b, c\rangle$

Parametric equations: $x = x_0 + at, \quad y = y_0 + bt, \quad z = z_0 + ct$

- Equation of the line through $P = (a_1, b_1, c_1)$ and $Q = (a_2, b_2, c_2)$:

Vector parametrization: $\mathbf{r}(t) = (1-t)\langle a_1, b_1, c_1\rangle + t\langle a_2, b_2, c_2\rangle$

Parametric equations: $x = a_1 + (a_2-a_1)t, \quad y = b_1 + (b_2-b_1)t,$

$z = c_1 + (c_2 - c_1)t$

The segment $\overline{PQ}$ is parametrized by $\mathbf{r}(t)$ for $0 \le t \le 1$. The midpoint of $\overline{PQ}$ is the terminal point of $\mathbf{r}(\frac{1}{2})$, namely $\left(\frac{1}{2}(a_1 + a_2), \frac{1}{2}(b_1 + b_2), \frac{1}{2}(c_1 + c_2)\right)$.

13.2 EXERCISES

Preliminary Questions

1. What is the terminal point of the vector $\mathbf{v} = \langle 3, 2, 1 \rangle$ based at the point $P = (1, 1, 1)$?

2. What are the components of the vector $\mathbf{v} = \langle 3, 2, 1 \rangle$ based at the point $P = (1, 1, 1)$?

3. If $\mathbf{v} = -3\mathbf{w}$, then (choose the correct answer):

(a) $\mathbf{v}$ and $\mathbf{w}$ are parallel.

(b) $\mathbf{v}$ and $\mathbf{w}$ point in the same direction.

4. Which of the following is a direction vector for the line through $P = (3, 2, 1)$ and $Q = (1, 1, 1)$?

(a) $\langle 3, 2, 1 \rangle$ **(b)** $\langle 1, 1, 1 \rangle$ **(c)** $\langle 2, 1, 0 \rangle$

5. How many different direction vectors does a line have?

6. True or false? If $\mathbf{v}$ is a direction vector for a line $\mathcal{L}$, then $-\mathbf{v}$ is also a direction vector for $\mathcal{L}$.

Exercises

1. Sketch the vector $\mathbf{v} = \langle 1, 3, 2 \rangle$ and compute its length.

2. Let $\mathbf{v} = \overrightarrow{P_0Q_0}$, where $P_0 = (1, -2, 5)$ and $Q_0 = (0, 1, -4)$. Which of the following vectors (with tail P and head Q) are equivalent to $\mathbf{v}$?

	$\mathbf{v}_1$	$\mathbf{v}_2$	$\mathbf{v}_3$	$\mathbf{v}_4$
P	$(1, 2, 4)$	$(1, 5, 4)$	$(0, 0, 0)$	$(2, 4, 5)$
Q	$(0, 5, -5)$	$(0, -8, 13)$	$(-1, 3, -9)$	$(1, 7, 4)$

3. Sketch the vector $\mathbf{v} = \langle 1, 1, 0 \rangle$ based at $P = (0, 1, 1)$. Describe this vector in the form $\overrightarrow{PQ}$ for some point Q, and sketch the vector $\mathbf{v}_0$ based at the origin equivalent to $\mathbf{v}$.

4. Determine whether the coordinate systems (A)–(C) in Figure 17 satisfy the right-hand rule.

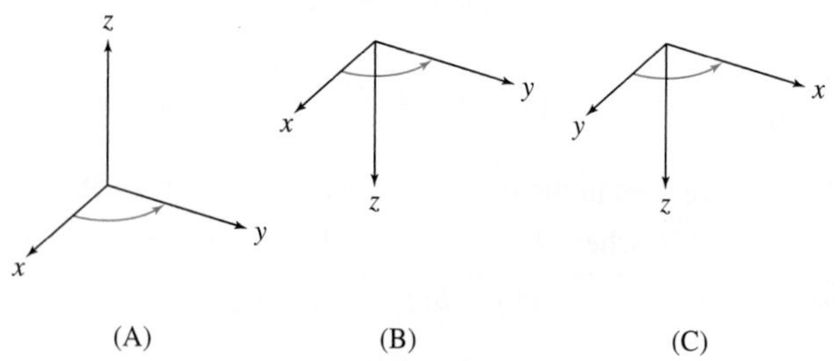

FIGURE 17

In Exercises 5–8, find the components of the vector $\overrightarrow{PQ}$.

5. $P = (1, 0, 1)$, $Q = (2, 1, 0)$

6. $P = (-3, -4, 2)$, $Q = (1, -4, 3)$

7. $P = (4, 6, 0)$, $Q = \left(-\frac{1}{2}, \frac{9}{2}, 1\right)$

8. $P = \left(-\frac{1}{2}, \frac{9}{2}, 1\right)$, $Q = (4, 6, 0)$

In Exercises 9–12, let $R = (1, 4, 3)$.

9. Calculate the length of $\overrightarrow{OR}$.

10. Find the point Q such that $\mathbf{v} = \overrightarrow{RQ}$ has components $\langle 4, 1, 1 \rangle$, and sketch $\mathbf{v}$.

11. Find the point P such that $\mathbf{w} = \overrightarrow{PR}$ has components $\langle 3, -2, 3 \rangle$, and sketch $\mathbf{w}$.

12. Find the components of $\mathbf{u} = \overrightarrow{PR}$, where $P = (1, 2, 2)$.

13. Let $\mathbf{v} = \langle 4, 8, 12 \rangle$. Which of the following vectors is parallel to $\mathbf{v}$? Which point in the same direction?

(a) $\langle 2, 4, 6 \rangle$ **(b)** $\langle -1, -2, 3 \rangle$

(c) $\langle -7, -14, -21 \rangle$ **(d)** $\langle 6, 10, 14 \rangle$

In Exercises 14–17, determine whether $\overrightarrow{AB}$ is equivalent to $\overrightarrow{PQ}$.

14. $A = (1, 1, 1)$ $B = (3, 3, 3)$
$P = (1, 4, 5)$ $Q = (3, 6, 7)$

15. $A = (1, 4, 1)$ $B = (-2, 2, 0)$
$P = (2, 5, 7)$ $Q = (-3, 2, 1)$

16. $A = (0, 0, 0)$ $B = (-4, 2, 3)$
$P = (4, -2, -3)$ $Q = (0, 0, 0)$

17. $A = (1, 1, 0)$ $B = (3, 3, 5)$
$P = (2, -9, 7)$ $Q = (4, -7, 13)$

In Exercises 18–23, calculate the linear combinations.

18. $5\langle 2, 2, -3 \rangle + 3\langle 1, 7, 2 \rangle$

19. $-2\langle 8, 11, 3 \rangle + 4\langle 2, 1, 1 \rangle$

20. $6(4\mathbf{j} + 2\mathbf{k}) - 3(2\mathbf{i} + 7\mathbf{k})$

21. $\frac{1}{2}\langle 4, -2, 8 \rangle - \frac{1}{3}\langle 12, 3, 3 \rangle$

22. $5(\mathbf{i} + 2\mathbf{j}) - 3(2\mathbf{j} + \mathbf{k}) + 7(2\mathbf{k} - \mathbf{i})$

23. $4\langle 6, -1, 1 \rangle - 2\langle 1, 0, -1 \rangle + 3\langle -2, 1, 1 \rangle$

In Exercises 24–27, find the given vector.

24. $\mathbf{e}_{\mathbf{v}}$, where $\mathbf{v} = \langle 1, 1, 2 \rangle$

25. $\mathbf{e_w}$, where $\mathbf{w} = \langle 4, -2, -1\rangle$

26. Unit vector in the direction of $\mathbf{u} = \langle 1, 0, 7\rangle$

27. Unit vector in the direction opposite to $\mathbf{v} = \langle -4, 4, 2\rangle$

28. Sketch the following vectors, and find their components and lengths.

(a) $4\mathbf{i} + 3\mathbf{j} - 2\mathbf{k}$ **(b)** $\mathbf{i} + \mathbf{j} + \mathbf{k}$

(c) $4\mathbf{j} + 3\mathbf{k}$ **(d)** $12\mathbf{i} + 8\mathbf{j} - \mathbf{k}$

In Exercises 29–36, find a vector parametrization for the line with the given description.

29. Passes through $P = (1, 2, -8)$, direction vector $\mathbf{v} = \langle 2, 1, 3\rangle$

30. Passes through $P = (4, 0, 8)$, direction vector $\mathbf{v} = \langle 1, 0, 1\rangle$

31. Passes through $P = (4, 0, 8)$, direction vector $\mathbf{v} = 7\mathbf{i} + 4\mathbf{k}$

32. Passes through O, direction vector $\mathbf{v} = \langle 3, -1, -4\rangle$

33. Passes through $(1, 1, 1)$ and $(3, -5, 2)$

34. Passes through $(-2, 0, -2)$ and $(4, 3, 7)$

35. Passes through O and $(4, 1, 1)$

36. Passes through $(1, 1, 1)$ parallel to the line through $(2, 0, -1)$ and $(4, 1, 3)$

In Exercises 37–40, find parametric equations for the lines with the given description.

37. Perpendicular to the xy-plane, passes through the origin

38. Perpendicular to the yz-plane, passes through $(0, 0, 2)$

39. Parallel to the line through $(1, 1, 0)$ and $(0, -1, -2)$, passes through $(0, 0, 4)$

40. Passes through $(1, -1, 0)$ and $(0, -1, 2)$

41. Which of the following is a parametrization of the line through $P = (4, 9, 8)$ perpendicular to the xz-plane (Figure 18)?

(a) $\mathbf{r}(t) = \langle 4, 9, 8\rangle + t\langle 1, 0, 1\rangle$ **(b)** $\mathbf{r}(t) = \langle 4, 9, 8\rangle + t\langle 0, 0, 1\rangle$

(c) $\mathbf{r}(t) = \langle 4, 9, 8\rangle + t\langle 0, 1, 0\rangle$ **(d)** $\mathbf{r}(t) = \langle 4, 9, 8\rangle + t\langle 1, 1, 0\rangle$

42. Find a parametrization of the line through $P = (4, 9, 8)$ perpendicular to the yz-plane.

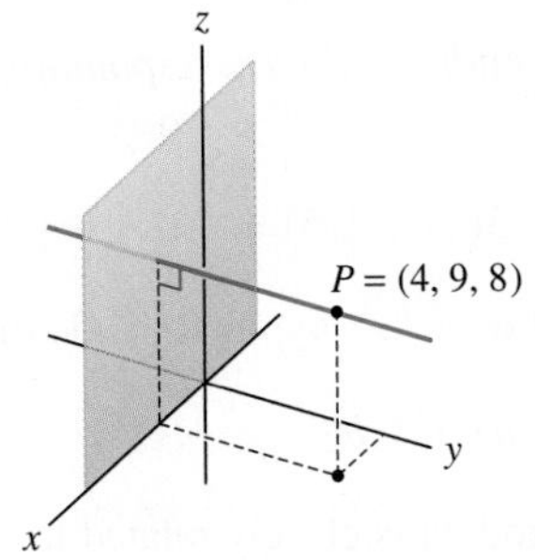

FIGURE 18

In Exercises 43–46, let $P = (2, 1, -1)$ and $Q = (4, 7, 7)$. Find the coordinates of each of the following.

43. The midpoint of $\overline{PQ}$

44. The point on $\overline{PQ}$ lying two-thirds of the way from P to Q

45. The point R such that Q is the midpoint of $\overline{PR}$

46. The two points on the line through $\overline{PQ}$ whose distance from P is twice its distance from Q

47. Show that $\mathbf{r}_1(t)$ and $\mathbf{r}_2(t)$ define the same line, where

$$\mathbf{r}_1(t) = \langle 3, -1, 4\rangle + t\langle 8, 12, -6\rangle$$

$$\mathbf{r}_2(t) = \langle 11, 11, -2\rangle + t\langle 4, 6, -3\rangle$$

Hint: Show that $\mathbf{r}_2$ passes through $(3, -1, 4)$ and that the direction vectors for $\mathbf{r}_1$ and $\mathbf{r}_2$ are parallel.

48. Show that $\mathbf{r}_1(t)$ and $\mathbf{r}_2(t)$ define the same line, where

$$\mathbf{r}_1(t) = t\langle 2, 1, 3\rangle, \qquad \mathbf{r}_2(t) = \langle -6, -3, -9\rangle + t\langle 8, 4, 12\rangle$$

49. Find two different vector parametrizations of the line through $P = (5, 5, 2)$ with direction vector $\mathbf{v} = \langle 0, -2, 1\rangle$.

50. Find the point of intersection of the lines $\mathbf{r}(t) = \langle 1, 0, 0\rangle + t\langle -3, 1, 0\rangle$ and $\mathbf{s}(t) = \langle 0, 1, 1\rangle + t\langle 2, 0, 1\rangle$.

51. Show that the lines $\mathbf{r}_1(t) = \langle -1, 2, 2\rangle + t\langle 4, -2, 1\rangle$ and $\mathbf{r}_2(t) = \langle 0, 1, 1\rangle + t\langle 2, 0, 1\rangle$ do not intersect.

52. Determine whether the lines $\mathbf{r}_1(t) = \langle 2, 1, 1\rangle + t\langle -4, 0, 1\rangle$ and $\mathbf{r}_2(s) = \langle -4, 1, 5\rangle + s\langle 2, 1, -2\rangle$ intersect, and if so, find the point of intersection.

53. Determine whether the lines $\mathbf{r}_1(t) = \langle 0, 1, 1\rangle + t\langle 1, 1, 2\rangle$ and $\mathbf{r}_2(s) = \langle 2, 0, 3\rangle + s\langle 1, 4, 4\rangle$ intersect, and if so, find the point of intersection.

54. Find the intersection of the lines $\mathbf{r}_1(t) = \langle -1, 1\rangle + t\langle 2, 4\rangle$ and $\mathbf{r}_2(s) = \langle 2, 1\rangle + s\langle -1, 6\rangle$ in $\mathbf{R}^2$.

55. Find the components of the vector $\mathbf{v}$ whose tail and head are the midpoints of segments $\overline{AC}$ and $\overline{BC}$ in Figure 19.

56. Find the components of the vector $\mathbf{w}$ whose tail is C and head is the midpoint of $\overline{AB}$ in Figure 19.

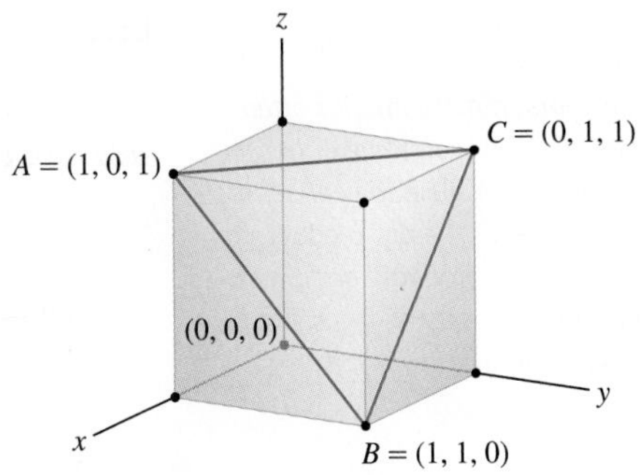

FIGURE 19

Further Insights and Challenges

In Exercises 57–63, we consider the equations of a line in ***symmetric form****, when $a \neq 0$, $b \neq 0$, $c \neq 0$.*

$$\frac{x - x_0}{a} = \frac{y - y_0}{b} = \frac{z - z_0}{c} \qquad 12$$

57. Let $\mathcal{L}$ be the line through $P_0 = (x_0, y_0, c_0)$ with direction vector $\mathbf{v} = \langle a, b, c \rangle$. Show that $\mathcal{L}$ is defined by the symmetric equations (12). *Hint:* Use the vector parametrization to show that every point on $\mathcal{L}$ satisfies (12).

58. Find the symmetric equations of the line through $P_0 = (-2, 3, 3)$ with direction vector $\mathbf{v} = \langle 2, 4, 3 \rangle$.

59. Find the symmetric equations of the line through $P = (1, 1, 2)$ and $Q = (-2, 4, 0)$.

60. Find the symmetric equations of the line

$$x = 3 + 2t, \quad y = 4 - 9t, \quad z = 12t$$

61. Find a vector parametrization for the line

$$\frac{x - 5}{9} = \frac{y + 3}{7} = z - 10$$

62. Find a vector parametrization for the line $\frac{x}{2} = \frac{y}{7} = \frac{z}{8}$.

63. Show that the line in the plane through (x_0, y_0) of slope m has symmetric equations

$$x - x_0 = \frac{y - y_0}{m}$$

64. A median of a triangle is a segment joining a vertex to the midpoint of the opposite side. Referring to Figure 20(A), prove that three medians of triangle ABC intersect at the terminal point P of the vector $\frac{1}{3}(\mathbf{u} + \mathbf{v} + \mathbf{w})$. The point P is the *centroid* of the triangle. *Hint:* Show, by parametrizing the segment $\overline{AA'}$, that P lies two-thirds of the way from A to A'. It will follow similarly that P lies on the other two medians.

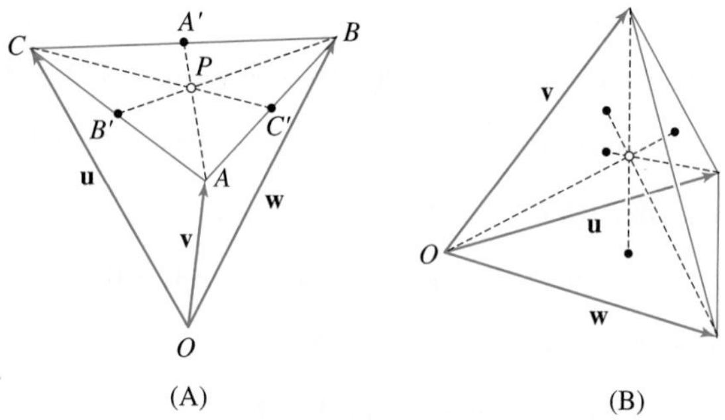

FIGURE 20

65. A median of a tetrahedron is a segment joining a vertex to the centroid of the opposite face. The tetrahedron in Figure 20(B) has vertices at the origin and at the terminal points of vectors **u**, **v**, and **w**. Show that the medians intersect at the terminal point of $\frac{1}{4}(\mathbf{u} + \mathbf{v} + \mathbf{w})$.

13.3 Dot Product and the Angle between Two Vectors

The dot product is one of the most important vector operations. It plays a role in nearly all aspects of multivariable calculus.

DEFINITION Dot Product The **dot product** $\mathbf{v} \cdot \mathbf{w}$ of two vectors

$$\mathbf{v} = \langle a_1, b_1, c_1 \rangle, \qquad \mathbf{w} = \langle a_2, b_2, c_2 \rangle$$

is the scalar defined by

$$\mathbf{v} \cdot \mathbf{w} = a_1 a_2 + b_1 b_2 + c_1 c_2$$

Important concepts in mathematics often have multiple names or notations either for historical reasons or because they arise in more than one context. The dot product *is also called the "scalar product" or "inner product" and in many texts, $\mathbf{v} \cdot \mathbf{w}$ is denoted $(\mathbf{v}, \mathbf{w})$ or $\langle \mathbf{v}, \mathbf{w} \rangle$.*

In words, to compute the dot product, *multiply the corresponding components and add*. For example,

$$\langle 2, 3, 1 \rangle \cdot \langle -4, 2, 5 \rangle = 2(-4) + 3(2) + 1(5) = -8 + 6 + 5 = 3$$

The dot product of vectors $\mathbf{v} = \langle a_1, b_1 \rangle$ and $\mathbf{w} = \langle a_2, b_2 \rangle$ in $\mathbf{R}^2$ is defined similarly:

$$\mathbf{v} \cdot \mathbf{w} = a_1 a_2 + b_1 b_2$$

We will see in a moment that the dot product is closely related to the angle between **v** and **w**. Before getting to this, we describe some elementary properties of dot products.

The dot product appears in a very wide range of applications. To rank how closely a Web document matches a search input at Google,

"We take the dot product of the vector of count-weights with the vector of type-weights to compute an IR score for the document."

From "The Anatomy of a Large-Scale Hypertextual Web Search Engine" by Google founders Sergey Brin and Lawrence Page.

First, the dot product is *commutative*: $\mathbf{v} \cdot \mathbf{w} = \mathbf{w} \cdot \mathbf{v}$, because the components can be multiplied in either order. Second, the dot product of a vector with itself is the square of the length: If $\mathbf{v} = \langle a, b, c \rangle$, then

$$\mathbf{v} \cdot \mathbf{v} = a \cdot a + b \cdot b + c \cdot c = a^2 + b^2 + c^2 = \|\mathbf{v}\|^2$$

The dot product also satisfies a Distributive Law and a scalar property as summarized in the next theorem (see Exercises 84 and 85).

THEOREM 1 Properties of the Dot Product

(i) $\mathbf{0} \cdot \mathbf{v} = \mathbf{v} \cdot \mathbf{0} = 0$

(ii) Commutativity: $\mathbf{v} \cdot \mathbf{w} = \mathbf{w} \cdot \mathbf{v}$

(iii) Pulling out scalars: $(\lambda \mathbf{v}) \cdot \mathbf{w} = \mathbf{v} \cdot (\lambda \mathbf{w}) = \lambda(\mathbf{v} \cdot \mathbf{w})$

(iv) Distributive Law: $\mathbf{u} \cdot (\mathbf{v} + \mathbf{w}) = \mathbf{u} \cdot \mathbf{v} + \mathbf{u} \cdot \mathbf{w}$

$(\mathbf{v} + \mathbf{w}) \cdot \mathbf{u} = \mathbf{v} \cdot \mathbf{u} + \mathbf{w} \cdot \mathbf{u}$

(v) Relation with length: $\mathbf{v} \cdot \mathbf{v} = \|\mathbf{v}\|^2$

EXAMPLE 1 Verify the Distributive Law $\mathbf{u} \cdot (\mathbf{v} + \mathbf{w}) = \mathbf{u} \cdot \mathbf{v} + \mathbf{u} \cdot \mathbf{w}$ for

$$\mathbf{u} = \langle 4, 3, 3 \rangle, \qquad \mathbf{v} = \langle 1, 2, 2 \rangle, \qquad \mathbf{w} = \langle 3, -2, 5 \rangle$$

Solution We compute both sides and check that they are equal:

$$\begin{aligned}
\mathbf{u} \cdot (\mathbf{v} + \mathbf{w}) &= \langle 4, 3, 3 \rangle \cdot \big(\langle 1, 2, 2 \rangle + \langle 3, -2, 5 \rangle \big) \\
&= \langle 4, 3, 3 \rangle \cdot \langle 4, 0, 7 \rangle = 4(4) + 3(0) + 3(7) = 37 \\
\mathbf{u} \cdot \mathbf{v} + \mathbf{u} \cdot \mathbf{w} &= \langle 4, 3, 3 \rangle \cdot \langle 1, 2, 2 \rangle + \langle 4, 3, 3 \rangle \cdot \langle 3, -2, 5 \rangle \\
&= \big(4(1) + 3(2) + 3(2)\big) + \big(4(3) + 3(-2) + 3(5)\big) \\
&= 16 + 21 = 37
\end{aligned}$$

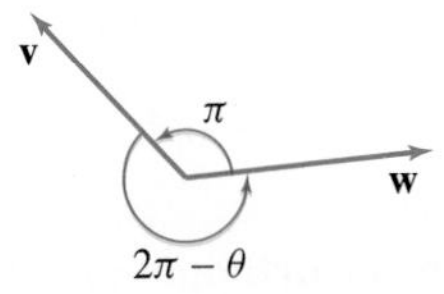

FIGURE 1 By convention, the angle θ between two vectors is chosen so that $0 \le \theta \le \pi$.

As mentioned above, the dot product $\mathbf{v} \cdot \mathbf{w}$ is related to the angle θ between $\mathbf{v}$ and $\mathbf{w}$. This angle θ is not uniquely defined because, as we see in Figure 1, both θ and $2\pi - \theta$ can serve as an angle between $\mathbf{v}$ and $\mathbf{w}$. Furthermore, any multiple of 2π may be added to θ. All of these angles have the same cosine, so it does not matter which angle we use in the next theorem. However, we shall adopt the following convention:

The angle between two vectors is chosen to satisfy $0 \le \theta \le \pi$.

THEOREM 2 Dot Product and the Angle Let θ be the angle between two nonzero vectors $\mathbf{v}$ and $\mathbf{w}$. Then

$$\mathbf{v} \cdot \mathbf{w} = \|\mathbf{v}\| \, \|\mathbf{w}\| \cos\theta \quad \text{or} \quad \cos\theta = \frac{\mathbf{v} \cdot \mathbf{w}}{\|\mathbf{v}\| \, \|\mathbf{w}\|} \qquad \boxed{1}$$

Proof According to the Law of Cosines, the three sides of a triangle satisfy (Figure 2)

$$c^2 = a^2 + b^2 - 2ab\cos\theta$$

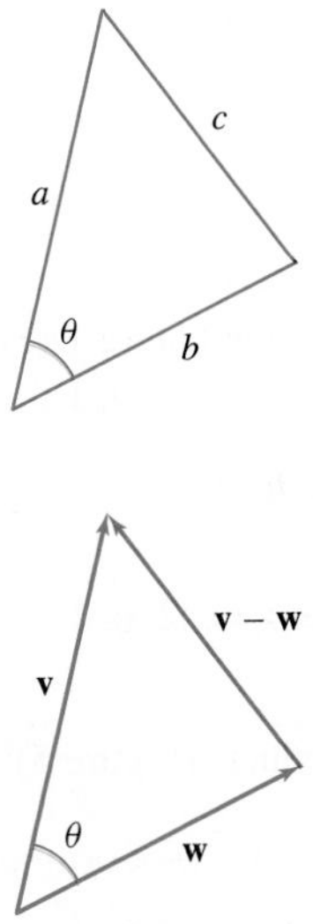

FIGURE 2

If two sides of the triangle are **v** and **w**, then the third side is **v** − **w**, as in the figure, and the Law of Cosines gives

$$\|\mathbf{v}-\mathbf{w}\|^2 = \|\mathbf{v}\|^2 + \|\mathbf{w}\|^2 - 2\cos\theta\,\|\mathbf{v}\|\,\|\mathbf{w}\| \tag{2}$$

Now, by property (v) of Theorem 1 and the Distributive Law,

$$\|\mathbf{v}-\mathbf{w}\|^2 = (\mathbf{v}-\mathbf{w})\cdot(\mathbf{v}-\mathbf{w}) = \mathbf{v}\cdot\mathbf{v} - 2\mathbf{v}\cdot\mathbf{w} + \mathbf{w}\cdot\mathbf{w}$$

$$= \|\mathbf{v}\|^2 + \|\mathbf{w}\|^2 - 2\mathbf{v}\cdot\mathbf{w} \tag{3}$$

Comparing Eq. (2) and Eq. (3), we obtain $-2\cos\theta\,\|\mathbf{v}\|\,\|\mathbf{w}\| = -2\mathbf{v}\cdot\mathbf{w}$, and Eq. (1) follows. ■

By definition of the arccosine, the angle $\theta = \cos^{-1} x$ is the angle in the interval $[0, \pi]$ satisfying $\cos\theta = x$. Thus, for nonzero vectors **v** and **w**, we have

$$\theta = \cos^{-1}\left(\frac{\mathbf{v}\cdot\mathbf{w}}{\|\mathbf{v}\|\,\|\mathbf{w}\|}\right)$$

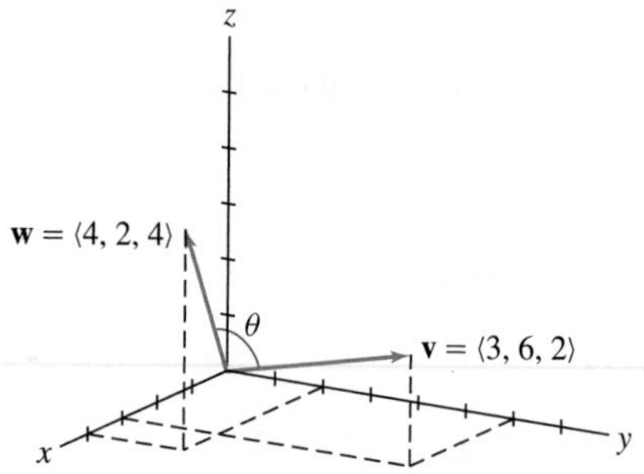

FIGURE 3

■ **EXAMPLE 2** Find the angle θ between $\mathbf{v} = \langle 3, 6, 2\rangle$ and $\mathbf{w} = \langle 4, 2, 4\rangle$.

Solution Compute $\cos\theta$ using the dot product:

$$\|\mathbf{v}\| = \sqrt{3^2+6^2+2^2} = \sqrt{49} = 7, \qquad \|\mathbf{w}\| = \sqrt{4^2+2^2+4^2} = \sqrt{36} = 6$$

$$\cos\theta = \frac{\mathbf{v}\cdot\mathbf{w}}{\|\mathbf{v}\|\|\mathbf{w}\|} = \frac{\langle 3,6,2\rangle\cdot\langle 4,2,4\rangle}{7\cdot 6} = \frac{3\cdot 4+6\cdot 2+2\cdot 4}{42} = \frac{32}{42} = \frac{16}{21}$$

The angle itself is $\theta = \cos^{-1}\left(\frac{16}{21}\right) \approx 0.705$ rad (Figure 3). ■

The terms "orthogonal" and "perpendicular" are synonymous and are used interchangeably, although we usually use "orthogonal" when dealing with vectors.

Two nonzero vectors **v** and **w** are called **perpendicular** or **orthogonal** if the angle between them is $\frac{\pi}{2}$. In this case we write $\mathbf{v} \perp \mathbf{w}$.

We can use the dot product to test whether **v** and **w** are orthogonal. Because an angle between 0 and π satisfies $\cos\theta = 0$ if and only if $\theta = \frac{\pi}{2}$, we see that

$$\mathbf{v}\cdot\mathbf{w} = \|\mathbf{v}\|\,\|\mathbf{w}\|\cos\theta = 0 \quad \Leftrightarrow \quad \theta = \frac{\pi}{2}$$

and thus

$$\mathbf{v} \perp \mathbf{w} \quad \text{if and only if} \quad \mathbf{v}\cdot\mathbf{w} = 0$$

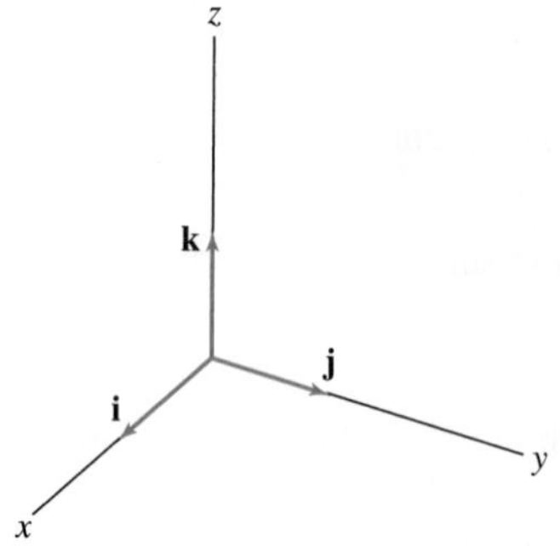

FIGURE 4 The standard basis vectors are mutually orthogonal and have length 1.

The standard basis vectors are mutually orthogonal and have length 1 (Figure 4). In terms of dot products, because $\mathbf{i} = \langle 1,0,0\rangle$, $\mathbf{j} = \langle 0,1,0\rangle$, and $\mathbf{k} = \langle 0,0,1\rangle$,

$$\mathbf{i}\cdot\mathbf{j} = \mathbf{i}\cdot\mathbf{k} = \mathbf{j}\cdot\mathbf{k} = 0, \qquad \mathbf{i}\cdot\mathbf{i} = \mathbf{j}\cdot\mathbf{j} = \mathbf{k}\cdot\mathbf{k} = 1$$

■ **EXAMPLE 3 Testing for Orthogonality** Determine whether $\mathbf{v} = \langle 2, 6, 1\rangle$ is orthogonal to $\mathbf{u} = \langle 2, -1, 1\rangle$ or $\mathbf{w} = \langle -4, 1, 2\rangle$.

Solution We test for orthogonality by computing the dot products (Figure 5):

$$\mathbf{v}\cdot\mathbf{u} = \langle 2,6,1\rangle\cdot\langle 2,-1,1\rangle = 2(2)+6(-1)+1(1) = -1 \quad \text{(not orthogonal)}$$

$$\mathbf{v}\cdot\mathbf{w} = \langle 2,6,1\rangle\cdot\langle -4,1,2\rangle = 2(-4)+6(1)+1(2) = 0 \quad \text{(orthogonal)}$$ ■

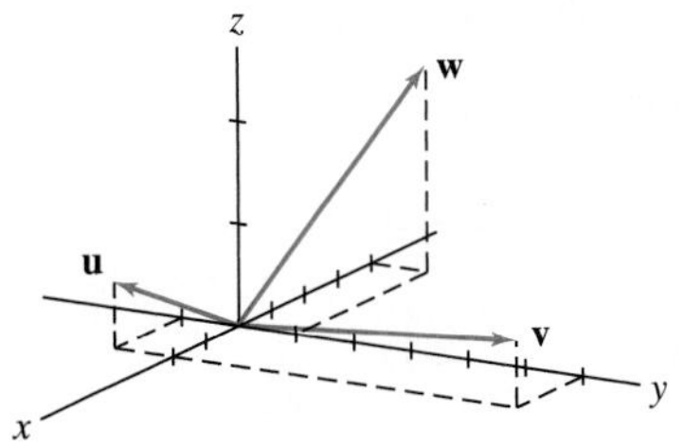

FIGURE 5 Vectors **v**, **w**, and **u** in Example 3.

■ **EXAMPLE 4 Testing for Obtuseness** Determine whether the angles between the vector $\mathbf{v} = \langle 3, 1, -2\rangle$ and the vectors $\mathbf{u} = \left\langle \frac{1}{2}, \frac{1}{2}, 5\right\rangle$ and $\mathbf{w} = \langle 4, -3, 0\rangle$ are obtuse.

Solution By definition, the angle θ between **v** and **u** is obtuse if $\frac{\pi}{2} < \theta \le \pi$, and this is the case if $\cos\theta < 0$. Since $\mathbf{v}\cdot\mathbf{u} = \|\mathbf{v}\|\,\|\mathbf{u}\|\cos\theta$ and the lengths $\|\mathbf{v}\|$ and $\|\mathbf{u}\|$ are positive, we see that $\cos\theta$ is negative if and only if $\mathbf{v}\cdot\mathbf{u}$ is negative. Thus,

The angle θ between **v** and **u** is obtuse if $\mathbf{v}\cdot\mathbf{u} < 0$.

We have

$$\mathbf{v}\cdot\mathbf{u} = \langle 3, 1, -2\rangle \cdot \left\langle \frac{1}{2}, \frac{1}{2}, 5\right\rangle = \frac{3}{2} + \frac{1}{2} - 10 = -8 < 0 \quad \text{(angle is obtuse)}$$

$$\mathbf{v}\cdot\mathbf{w} = \langle 3, 1, -2\rangle \cdot \langle 4, -3, 0\rangle = 12 - 3 + 0 = 9 > 0 \quad \text{(angle is acute)}$$ ■

■ **EXAMPLE 5 Using the Distributive Law** Calculate the dot product $\mathbf{v}\cdot\mathbf{w}$, where $\mathbf{v} = 4\mathbf{i} - 3\mathbf{j}$ and $\mathbf{w} = \mathbf{i} + 2\mathbf{j} + \mathbf{k}$.

Solution Use the Distributive Law and the orthogonality of **i**, **j**, and **k**:

$$\begin{aligned}\mathbf{v}\cdot\mathbf{w} &= (4\mathbf{i} - 3\mathbf{j})\cdot(\mathbf{i} + 2\mathbf{j} + \mathbf{k})\\ &= 4\mathbf{i}\cdot(\mathbf{i} + 2\mathbf{j} + \mathbf{k}) - 3\mathbf{j}\cdot(\mathbf{i} + 2\mathbf{j} + \mathbf{k})\\ &= 4\mathbf{i}\cdot\mathbf{i} - 3\mathbf{j}\cdot(2\mathbf{j}) = 4 - 6 = -2\end{aligned}$$ ■

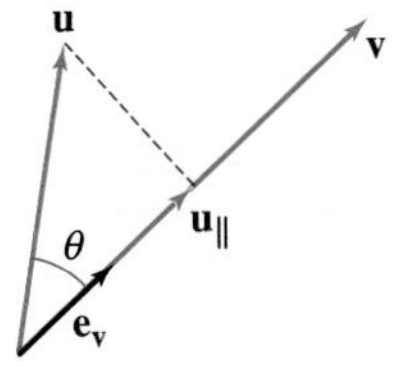

FIGURE 6 The projection $\mathbf{u}_{||}$ of **u** along **v** has length $\|\mathbf{u}\|\cos\theta$.

Another important use of the dot product is in finding the **projection** $\mathbf{u}_{||}$ of a vector **u** along a nonzero vector **v**. By definition, $\mathbf{u}_{||}$ is the vector obtained by dropping a perpendicular from **u** to the line through **v** as in Figures 6 and 7. In the next theorem, recall that $\mathbf{e_v} = \mathbf{v}/\|\mathbf{v}\|$ is the unit vector in the direction of **v**.

THEOREM 3 Projection Assume $\mathbf{v} \ne \mathbf{0}$. The **projection** of **u** along **v** is the vector

$$\mathbf{u}_{||} = (\mathbf{u}\cdot\mathbf{e_v})\mathbf{e_v} \quad \text{or} \quad \mathbf{u}_{||} = \left(\frac{\mathbf{u}\cdot\mathbf{v}}{\mathbf{v}\cdot\mathbf{v}}\right)\mathbf{v} \qquad \boxed{4}$$

The scalar $\mathbf{u}\cdot\mathbf{e_v}$ is called the **component** of **u** along **v**.

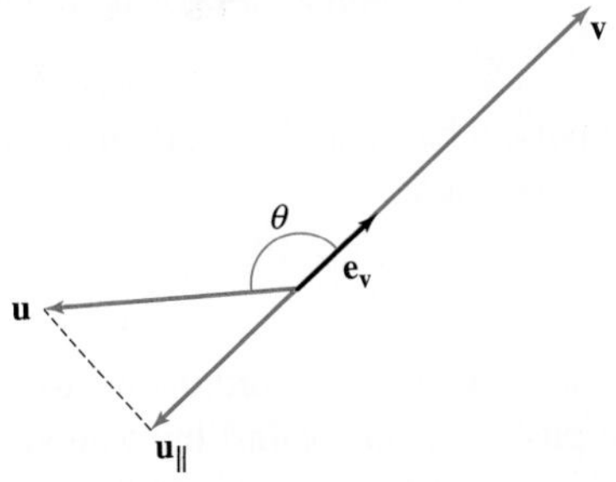

FIGURE 7 When θ is obtuse, $\mathbf{u}_{||}$ and $\mathbf{e_v}$ point in opposite directions.

Proof Referring to Figures 6 and 7, we see by trigonometry that $\mathbf{u}_{||}$ has length $\|\mathbf{u}\|\,|\cos\theta|$. If θ is acute, then $\mathbf{u}_{||}$ is a positive multiple of $\mathbf{e_v}$ and thus $\mathbf{u}_{||} = (\|\mathbf{u}\|\cos\theta)\mathbf{e_v}$ since $\cos\theta > 0$. Similarly, if θ is obtuse, then $\mathbf{u}_{||}$ is a negative multiple of $\mathbf{e_v}$ and $\mathbf{u}_{||} = (\|\mathbf{u}\|\cos\theta)\mathbf{e_v}$ since $\cos\theta < 0$. The first formula for $\mathbf{u}_{||}$ now follows because $\mathbf{u}\cdot\mathbf{e_v} = \|\mathbf{u}\|\,\|\mathbf{e_v}\|\cos\theta = \|\mathbf{u}\|\cos\theta$.

The second equality in Eq. (4) follows from the computation:

$$\begin{aligned}\mathbf{u}_{||} &= (\mathbf{u}\cdot\mathbf{e_v})\mathbf{e_v} = \left(\mathbf{u}\cdot\frac{\mathbf{v}}{\|\mathbf{v}\|}\right)\frac{\mathbf{v}}{\|\mathbf{v}\|}\\ &= \left(\frac{\mathbf{u}\cdot\mathbf{v}}{\|\mathbf{v}\|^2}\right)\mathbf{v} = \left(\frac{\mathbf{u}\cdot\mathbf{v}}{\mathbf{v}\cdot\mathbf{v}}\right)\mathbf{v}\end{aligned}$$ ■

■ **EXAMPLE 6** Find the projection of $\mathbf{u} = \langle 5, 1, -3\rangle$ along $\mathbf{v} = \langle 4, 4, 2\rangle$.

Solution It is convenient to use the second formula in Eq. (4):

$$\mathbf{u}\cdot\mathbf{v} = \langle 5, 1, -3\rangle \cdot \langle 4, 4, 2\rangle = 20 + 4 - 6 = 18, \qquad \mathbf{v}\cdot\mathbf{v} = 4^2 + 4^2 + 2^2 = 36$$

$$\mathbf{u}_{||} = \left(\frac{\mathbf{u}\cdot\mathbf{v}}{\mathbf{v}\cdot\mathbf{v}}\right)\mathbf{v} = \left(\frac{18}{36}\right)\langle 4, 4, 2\rangle = \langle 2, 2, 1\rangle$$ ■

We show now that if $\mathbf{v} \neq \mathbf{0}$, then every vector $\mathbf{u}$ can be written as the sum of the projection $\mathbf{u}_{||}$ and a vector $\mathbf{u}_{\perp}$ that is orthogonal to $\mathbf{v}$ (see Figure 8). In fact, if we set

$$\mathbf{u}_{\perp} = \mathbf{u} - \mathbf{u}_{||}$$

then we have

$$\boxed{\mathbf{u} = \mathbf{u}_{||} + \mathbf{u}_{\perp}} \qquad \boxed{5}$$

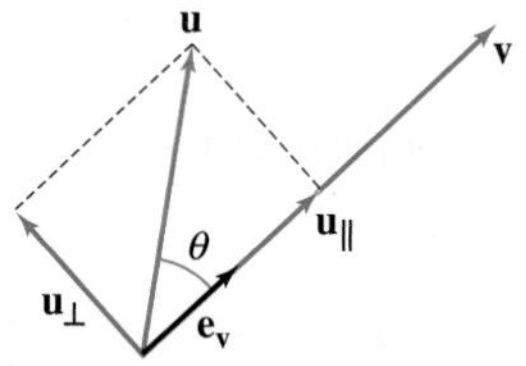

FIGURE 8 Decomposition of $\mathbf{u}$ as a sum $\mathbf{u} = \mathbf{u}_{||} + \mathbf{u}_{\perp}$ of vectors parallel and orthogonal to $\mathbf{v}$.

Eq. (5) is called the decomposition of $\mathbf{u}$ with respect to $\mathbf{v}$. We must verify, however, that $\mathbf{u}_{\perp}$ is orthogonal to $\mathbf{v}$. We do this by showing that the dot product is zero:

$$\mathbf{u}_{\perp}\cdot\mathbf{v} = (\mathbf{u} - \mathbf{u}_{||})\cdot\mathbf{v} = \left(\mathbf{u} - \left(\frac{\mathbf{u}\cdot\mathbf{v}}{\mathbf{v}\cdot\mathbf{v}}\right)\mathbf{v}\right)\cdot\mathbf{v} = \mathbf{u}\cdot\mathbf{v} - \left(\frac{\mathbf{u}\cdot\mathbf{v}}{\mathbf{v}\cdot\mathbf{v}}\right)(\mathbf{v}\cdot\mathbf{v}) = 0$$

■ **EXAMPLE 7** Find the decomposition of $\mathbf{u} = \langle 5, 1, -3\rangle$ with respect to $\mathbf{v} = \langle 4, 4, 2\rangle$.

Solution In Example 6 we showed that $\mathbf{u}_{||} = \langle 2, 2, 1\rangle$. The orthogonal vector is

$$\mathbf{u}_{\perp} = \mathbf{u} - \mathbf{u}_{||} = \langle 5, 1, -3\rangle - \langle 2, 2, 1\rangle = \langle 3, -1, -4\rangle$$

The decomposition of $\mathbf{u}$ with respect to $\mathbf{v}$ is

$$\mathbf{u} = \langle 5, 1, -3\rangle = \mathbf{u}_{||} + \mathbf{u}_{\perp} = \underbrace{\langle 2, 2, 1\rangle}_{\text{Projection along } \mathbf{v}} + \underbrace{\langle 3, -1, -4\rangle}_{\text{Orthogonal to } \mathbf{v}}$$ ■

The decomposition into parallel and orthogonal vectors is useful in many applications.

■ **EXAMPLE 8** What is the minimum force you must apply to pull a 20-kg wagon up a frictionless ramp inclined at an angle $\theta = 15°$?

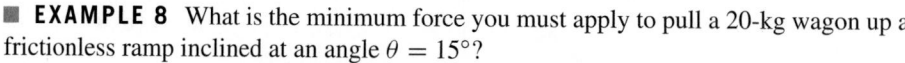

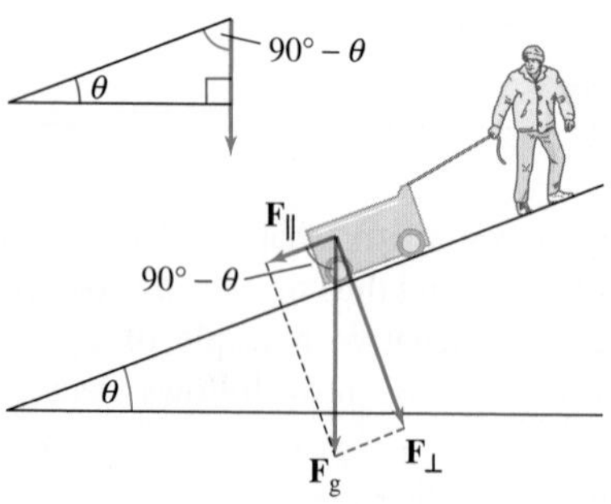

FIGURE 9 The angle between $\mathbf{F}_g$ and $\mathbf{F}_{||}$ is $90° - \theta$.

Solution Let $\mathbf{F}_g$ be the force on the wagon due to gravity. It has magnitude $20g$ newtons with $g = 9.8$. Referring to Figure 9, we decompose $\mathbf{F}_g$ as a sum

$$\mathbf{F}_g = \mathbf{F}_{||} + \mathbf{F}_{\perp}$$

where $\mathbf{F}_{||}$ is the projection along the ramp and $\mathbf{F}_{\perp}$ is the "normal force" orthogonal to the ramp. The normal force $\mathbf{F}_{\perp}$ is canceled by the ramp pushing back against the wagon in the normal direction, and thus (because there is no friction), you need only pull against $\mathbf{F}_{||}$.

Notice that the angle between $\mathbf{F}_g$ and the ramp is the complementary angle $90° - \theta$. Since $\mathbf{F}_{||}$ is parallel to the ramp, the angle between $\mathbf{F}_g$ and $\mathbf{F}_{||}$ is also $90° - \theta$, or 75°, and

$$\|\mathbf{F}_{||}\| = \|\mathbf{F}_g\|\cos(75°) \approx 20(9.8)(0.26) \approx 51 \text{ N}$$

Since gravity pulls the wagon down the ramp with a 51-newton force, it takes a minimum force of 51 newtons to pull the wagon up the ramp. ■

GRAPHICAL INSIGHT It seems that we are using the term "component" in two ways. We say that a vector $\mathbf{u} = \langle a, b \rangle$ has components a and b. On the other hand, $\mathbf{u} \cdot \mathbf{e}$ is called the component of **u** along the unit vector **e**.

In fact, these two notions of component are not different. The components a and b are the dot products of **u** with the standard unit vectors:

$$\mathbf{u} \cdot \mathbf{i} = \langle a, b \rangle \cdot \langle 1, 0 \rangle = a$$

$$\mathbf{u} \cdot \mathbf{j} = \langle a, b \rangle \cdot \langle 0, 1 \rangle = b$$

and we have the decomposition [Figure 10(A)]

$$\mathbf{u} = a\mathbf{i} + b\mathbf{j}$$

But any two orthogonal unit vectors **e** and **f** give rise to a rotated coordinate system, and we see in Figure 10(B) that

$$\mathbf{u} = (\mathbf{u} \cdot \mathbf{e})\mathbf{e} + (\mathbf{u} \cdot \mathbf{f})\mathbf{f}$$

In other words, $\mathbf{u} \cdot \mathbf{e}$ and $\mathbf{u} \cdot \mathbf{f}$ really are the components when we express **u** relative to the rotated system.

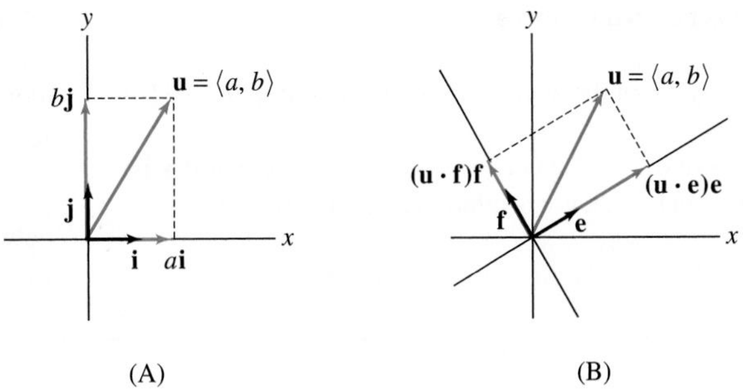

FIGURE 10

13.3 SUMMARY

- The *dot product* of $\mathbf{v} = \langle a_1, b_1, c_1 \rangle$ and $\mathbf{w} = \langle a_2, b_2, c_2 \rangle$ is

$$\mathbf{v} \cdot \mathbf{w} = a_1 a_2 + b_1 b_2 + c_1 c_2$$

- Basic Properties:
  - Commutativity: $\mathbf{v} \cdot \mathbf{w} = \mathbf{w} \cdot \mathbf{v}$
  - Pulling out scalars: $(\lambda \mathbf{v}) \cdot \mathbf{w} = \mathbf{v} \cdot (\lambda \mathbf{w}) = \lambda(\mathbf{v} \cdot \mathbf{w})$
  - Distributive Law: $\mathbf{u} \cdot (\mathbf{v} + \mathbf{w}) = \mathbf{u} \cdot \mathbf{v} + \mathbf{u} \cdot \mathbf{w}$

    $(\mathbf{v} + \mathbf{w}) \cdot \mathbf{u} = \mathbf{v} \cdot \mathbf{u} + \mathbf{w} \cdot \mathbf{u}$
  - $\mathbf{v} \cdot \mathbf{v} = \|\mathbf{v}\|^2$
  - $\mathbf{v} \cdot \mathbf{w} = \|\mathbf{v}\| \, \|\mathbf{w}\| \cos\theta$ where θ is the angle between **v** and **w**.
- By convention, the angle θ is chosen to satisfy $0 \le \theta \le \pi$.
- Test for orthogonality: $\mathbf{v} \perp \mathbf{w}$ if and only if $\mathbf{v} \cdot \mathbf{w} = 0$.

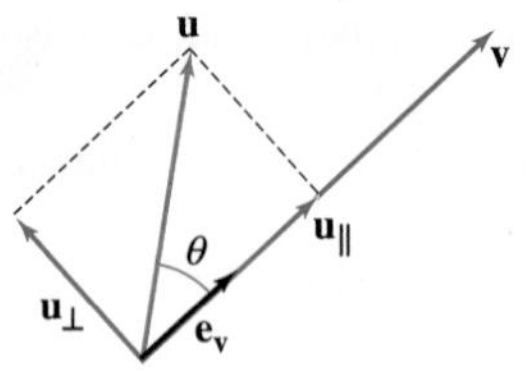

FIGURE 11

- The angle between $\mathbf{v}$ and $\mathbf{w}$ is acute if $\mathbf{v}\cdot\mathbf{w} > 0$ and obtuse if $\mathbf{v}\cdot\mathbf{w} < 0$.
- Assume $\mathbf{v} \neq \mathbf{0}$. Every vector $\mathbf{u}$ has a decomposition $\mathbf{u} = \mathbf{u}_{||} + \mathbf{u}_\perp$, where $\mathbf{u}_{||}$ is parallel to $\mathbf{v}$, and $\mathbf{u}_\perp$ is orthogonal to $\mathbf{v}$ (see Figure 11). The vector $\mathbf{u}_{||}$ is called the *projection* of $\mathbf{u}$ along $\mathbf{v}$.
- Let $\mathbf{e}_\mathbf{v} = \dfrac{\mathbf{v}}{\|\mathbf{v}\|}$. Then

$$\mathbf{u}_{||} = (\mathbf{u}\cdot\mathbf{e}_\mathbf{v})\mathbf{e}_\mathbf{v} = \left(\frac{\mathbf{u}\cdot\mathbf{v}}{\mathbf{v}\cdot\mathbf{v}}\right)\mathbf{v}, \qquad \mathbf{u}_\perp = \mathbf{u} - \mathbf{u}_{||}$$

- The coefficient $\mathbf{u}\cdot\mathbf{e}_\mathbf{v}$ is called the *component* of $\mathbf{u}$ along $\mathbf{v}$:

$$\text{Component of } \mathbf{u} \text{ along } \mathbf{v} = \mathbf{u}\cdot\mathbf{e}_\mathbf{v} = \|\mathbf{u}\|\cos\theta$$

13.3 EXERCISES

Preliminary Questions

1. Is the dot product of two vectors a scalar or a vector?

2. What can you say about the angle between **a** and **b** if $\mathbf{a}\cdot\mathbf{b} < 0$?

3. Which property of dot products allows us to conclude that if **v** is orthogonal to both **u** and **w**, then **v** is orthogonal to $\mathbf{u}+\mathbf{w}$?

4. Which is the projection of **v** along **v**: (a) **v** or (b) $\mathbf{e}_\mathbf{v}$?

5. Let $\mathbf{u}_{||}$ be the projection of **u** along **v**. Which of the following is the projection **u** along the vector 2**v** and which is the projection of 2**u** along **v**?

(a) $\frac{1}{2}\mathbf{u}_{||}$ **(b)** $\mathbf{u}_{||}$ **(c)** $2\mathbf{u}_{||}$

6. Which of the following is equal to $\cos\theta$, where θ is the angle between **u** and **v**?

(a) $\mathbf{u}\cdot\mathbf{v}$ **(b)** $\mathbf{u}\cdot\mathbf{e}_\mathbf{v}$ **(c)** $\mathbf{e}_\mathbf{u}\cdot\mathbf{e}_\mathbf{v}$

Exercises

In Exercises 1–12, compute the dot product.

1. $\langle 1, 2, 1\rangle \cdot \langle 4, 3, 5\rangle$

2. $\langle 3, -2, 2\rangle \cdot \langle 1, 0, 1\rangle$

3. $\langle 0, 1, 0\rangle \cdot \langle 7, 41, -3\rangle$

4. $\langle 1, 1, 1\rangle \cdot \langle 6, 4, 2\rangle$

5. $\langle 3, 1\rangle \cdot \langle 4, -7\rangle$

6. $\langle \frac{1}{6}, \frac{1}{2}\rangle \cdot \langle 3, \frac{1}{2}\rangle$

7. $\mathbf{k}\cdot\mathbf{j}$

8. $\mathbf{k}\cdot\mathbf{k}$

9. $(\mathbf{i}+\mathbf{j})\cdot(\mathbf{j}+\mathbf{k})$

10. $(3\mathbf{j}+2\mathbf{k})\cdot(\mathbf{i}-4\mathbf{k})$

11. $(\mathbf{i}+\mathbf{j}+\mathbf{k})\cdot(3\mathbf{i}+2\mathbf{j}-5\mathbf{k})$

12. $(-\mathbf{k})\cdot(\mathbf{i}-2\mathbf{j}+7\mathbf{k})$

In Exercises 13–18, determine whether the two vectors are orthogonal and, if not, whether the angle between them is acute or obtuse.

13. $\langle 1, 1, 1\rangle$, $\langle 1, -2, -2\rangle$

14. $\langle 0, 2, 4\rangle$, $\langle -5, 0, 0\rangle$

15. $\langle 1, 2, 1\rangle$, $\langle 7, -3, -1\rangle$

16. $\langle 0, 2, 4\rangle$, $\langle 3, 1, 0\rangle$

17. $\langle \frac{12}{5}, -\frac{4}{5}\rangle$, $\langle \frac{1}{2}, -\frac{7}{4}\rangle$

18. $\langle 12, 6\rangle$, $\langle 2, -4\rangle$

In Exercises 19–22, find the cosine of the angle between the vectors.

19. $\langle 0, 3, 1\rangle$, $\langle 4, 0, 0\rangle$

20. $\langle 1, 1, 1\rangle$, $\langle 2, -1, 2\rangle$

21. $\mathbf{i}+\mathbf{j}$, $\mathbf{j}+2\mathbf{k}$

22. $3\mathbf{i}+\mathbf{k}$, $\mathbf{i}+\mathbf{j}+\mathbf{k}$

In Exercises 23–28, find the angle between the vectors. Use a calculator if necessary.

23. $\langle 2, \sqrt{2}\rangle$, $\langle 1+\sqrt{2}, 1-\sqrt{2}\rangle$

24. $\langle 5, \sqrt{3}\rangle$, $\langle \sqrt{3}, 2\rangle$

25. $\langle 1, 1, 1\rangle$, $\langle 1, 0, 1\rangle$

26. $\langle 3, 1, 1\rangle$, $\langle 2, -4, 2\rangle$

27. $\langle 0, 1, 1\rangle$, $\langle 1, -1, 0\rangle$

28. $\langle 1, 1, -1\rangle$, $\langle 1, -2, -1\rangle$

29. Find all values of b for which the vectors are orthogonal.

(a) $\langle b, 3, 2\rangle$, $\langle 1, b, 1\rangle$ **(b)** $\langle 4, -2, 7\rangle$, $\langle b^2, b, 0\rangle$

30. Find a vector that is orthogonal to $\langle -1, 2, 2\rangle$.

31. Find two vectors that are not multiples of each other and are both orthogonal to $\langle 2, 0, -3\rangle$.

32. Find a vector that is orthogonal to $\mathbf{v} = \langle 1, 2, 1\rangle$ but not to $\mathbf{w} = \langle 1, 0, -1\rangle$.

33. Find $\mathbf{v} \cdot \mathbf{e}$ where $\|\mathbf{v}\| = 3$, $\mathbf{e}$ is a unit vector, and the angle between $\mathbf{e}$ and $\mathbf{v}$ is $\frac{2\pi}{3}$.

34. Assume that $\mathbf{v}$ lies in the yz-plane. Which of the following dot products is equal to zero for all choices of $\mathbf{v}$?

(a) $\mathbf{v} \cdot \langle 0, 2, 1\rangle$ **(b)** $\mathbf{v} \cdot \mathbf{k}$

(c) $\mathbf{v} \cdot \langle -3, 0, 0\rangle$ **(d)** $\mathbf{v} \cdot \mathbf{j}$

In Exercises 35–38, simplify the expression.

35. $(\mathbf{v} - \mathbf{w}) \cdot \mathbf{v} + \mathbf{v} \cdot \mathbf{w}$

36. $(\mathbf{v} + \mathbf{w}) \cdot (\mathbf{v} + \mathbf{w}) - 2\mathbf{v} \cdot \mathbf{w}$

37. $(\mathbf{v} + \mathbf{w}) \cdot \mathbf{v} - (\mathbf{v} + \mathbf{w}) \cdot \mathbf{w}$ **38.** $(\mathbf{v} + \mathbf{w}) \cdot \mathbf{v} - (\mathbf{v} - \mathbf{w}) \cdot \mathbf{w}$

In Exercises 39–42, use the properties of the dot product to evaluate the expression, assuming that $\mathbf{u} \cdot \mathbf{v} = 2$, $\|\mathbf{u}\| = 1$, *and* $\|\mathbf{v}\| = 3$.

39. $\mathbf{u} \cdot (4\mathbf{v})$ **40.** $(\mathbf{u} + \mathbf{v}) \cdot \mathbf{v}$

41. $2\mathbf{u} \cdot (3\mathbf{u} - \mathbf{v})$ **42.** $(\mathbf{u} + \mathbf{v}) \cdot (\mathbf{u} - \mathbf{v})$

43. Find the angle between $\mathbf{v}$ and $\mathbf{w}$ if $\mathbf{v} \cdot \mathbf{w} = -\|\mathbf{v}\|\ \|\mathbf{w}\|$.

44. Find the angle between $\mathbf{v}$ and $\mathbf{w}$ if $\mathbf{v} \cdot \mathbf{w} = \frac{1}{2}\|\mathbf{v}\|\ \|\mathbf{w}\|$.

45. Assume that $\|\mathbf{v}\| = 3$, $\|\mathbf{w}\| = 5$ and that the angle between $\mathbf{v}$ and $\mathbf{w}$ is $\theta = \frac{\pi}{3}$.

(a) Use the relation $\|\mathbf{v} + \mathbf{w}\|^2 = (\mathbf{v} + \mathbf{w}) \cdot (\mathbf{v} + \mathbf{w})$ to show that $\|\mathbf{v} + \mathbf{w}\|^2 = 3^2 + 5^2 + 2\mathbf{v} \cdot \mathbf{w}$.

(b) Find $\|\mathbf{v} + \mathbf{w}\|$.

46. Assume that $\|\mathbf{v}\| = 2$, $\|\mathbf{w}\| = 3$, and the angle between $\mathbf{v}$ and $\mathbf{w}$ is $120°$. Determine:

(a) $\mathbf{v} \cdot \mathbf{w}$ **(b)** $\|2\mathbf{v} + \mathbf{w}\|$ **(c)** $\|2\mathbf{v} - 3\mathbf{w}\|$

47. Show that if $\mathbf{e}$ and $\mathbf{f}$ are unit vectors such that $\|\mathbf{e} + \mathbf{f}\| = \frac{3}{2}$, then $\|\mathbf{e} - \mathbf{f}\| = \frac{\sqrt{7}}{2}$. *Hint:* Show that $\mathbf{e} \cdot \mathbf{f} = \frac{1}{8}$.

48. Find $\|2\mathbf{e} - 3\mathbf{f}\|$ assuming that $\mathbf{e}$ and $\mathbf{f}$ are unit vectors such that $\|\mathbf{e} + \mathbf{f}\| = \sqrt{3/2}$.

49. Find the angle θ in the triangle in Figure 12.

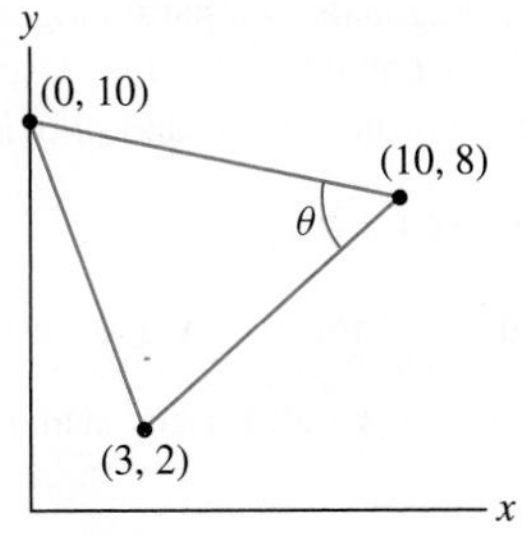

FIGURE 12

50. Find all three angles in the triangle in Figure 13.

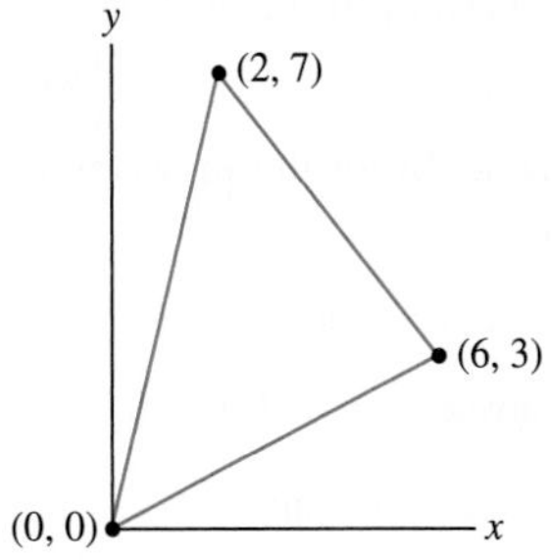

FIGURE 13

In Exercises 51–58, find the projection of $\mathbf{u}$ *along* $\mathbf{v}$.

51. $\mathbf{u} = \langle 2, 5\rangle, \quad \mathbf{v} = \langle 1, 1\rangle$ **52.** $\mathbf{u} = \langle 2, -3\rangle, \quad \mathbf{v} = \langle 1, 2\rangle$

53. $\mathbf{u} = \langle -1, 2, 0\rangle, \quad \mathbf{v} = \langle 2, 0, 1\rangle$

54. $\mathbf{u} = \langle 1, 1, 1\rangle, \quad \mathbf{v} = \langle 1, 1, 0\rangle$

55. $\mathbf{u} = 5\mathbf{i} + 7\mathbf{j} - 4\mathbf{k}, \quad \mathbf{v} = \mathbf{k}$ **56.** $\mathbf{u} = \mathbf{i} + 29\mathbf{k}, \quad \mathbf{v} = \mathbf{j}$

57. $\mathbf{u} = \langle a, b, c\rangle, \quad \mathbf{v} = \mathbf{i}$ **58.** $\mathbf{u} = \langle a, a, b\rangle, \quad \mathbf{v} = \mathbf{i} - \mathbf{j}$

In Exercises 59 and 60, compute the component of $\mathbf{u}$ *along* $\mathbf{v}$.

59. $\mathbf{u} = \langle 3, 2, 1\rangle, \quad \mathbf{v} = \langle 1, 0, 1\rangle$

60. $\mathbf{u} = \langle 3, 0, 9\rangle, \quad \mathbf{v} = \langle 1, 2, 2\rangle$

61. Find the length of $\overline{OP}$ in Figure 14.

62. Find $\|\mathbf{u}_\perp\|$ in Figure 14.

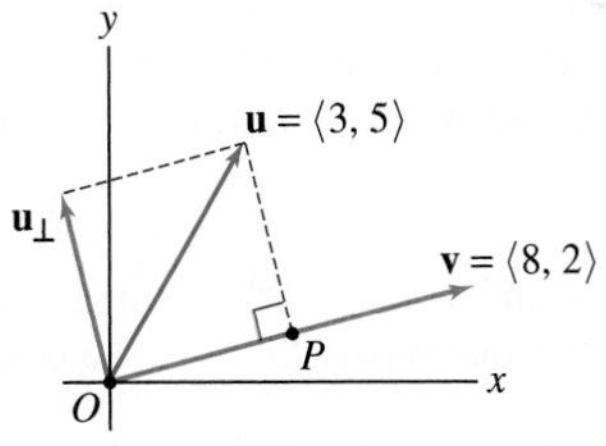

FIGURE 14

In Exercises 63–68, find the decomposition $\mathbf{a} = \mathbf{a}_{||} + \mathbf{a}_\perp$ *with respect to* $\mathbf{b}$.

63. $\mathbf{a} = \langle 1, 0\rangle, \quad \mathbf{b} = \langle 1, 1\rangle$

64. $\mathbf{a} = \langle 2, -3\rangle, \quad \mathbf{b} = \langle 5, 0\rangle$

65. $\mathbf{a} = \langle 4, -1, 0\rangle, \quad \mathbf{b} = \langle 0, 1, 1\rangle$

66. $\mathbf{a} = \langle 4, -1, 5\rangle, \quad \mathbf{b} = \langle 2, 1, 1\rangle$

67. $\mathbf{a} = \langle x, y\rangle, \quad \mathbf{b} = \langle 1, -1\rangle$

68. $\mathbf{a} = \langle x, y, z\rangle, \quad \mathbf{b} = \langle 1, 1, 1\rangle$

69. Let $\mathbf{e}_\theta = \langle \cos\theta, \sin\theta\rangle$. Show that $\mathbf{e}_\theta \cdot \mathbf{e}_\psi = \cos(\theta - \psi)$ for any two angles θ and ψ.

70. Let $\mathbf{v}$ and $\mathbf{w}$ be vectors in the plane.

(a) Use Theorem 2 to explain why the dot product $\mathbf{v} \cdot \mathbf{w}$ does not change if both $\mathbf{v}$ and $\mathbf{w}$ are rotated by the same angle θ.

(b) Sketch the vectors $\mathbf{e}_1 = \langle 1, 0 \rangle$ and $\mathbf{e}_2 = \left\langle \frac{\sqrt{2}}{2}, \frac{\sqrt{2}}{2} \right\rangle$, and determine the vectors $\mathbf{e}_1', \mathbf{e}_2'$ obtained by rotating $\mathbf{e}_1, \mathbf{e}_2$ through an angle $\frac{\pi}{4}$. Verify that $\mathbf{e}_1 \cdot \mathbf{e}_2 = \mathbf{e}_1' \cdot \mathbf{e}_2'$.

In Exercises 71–74, refer to Figure 15.

71. Find the angle between $\overline{AB}$ and $\overline{AC}$.

72. Find the angle between $\overline{AB}$ and $\overline{AD}$.

73. Calculate the projection of $\overrightarrow{AC}$ along $\overrightarrow{AD}$.

74. Calculate the projection of $\overrightarrow{AD}$ along $\overrightarrow{AB}$.

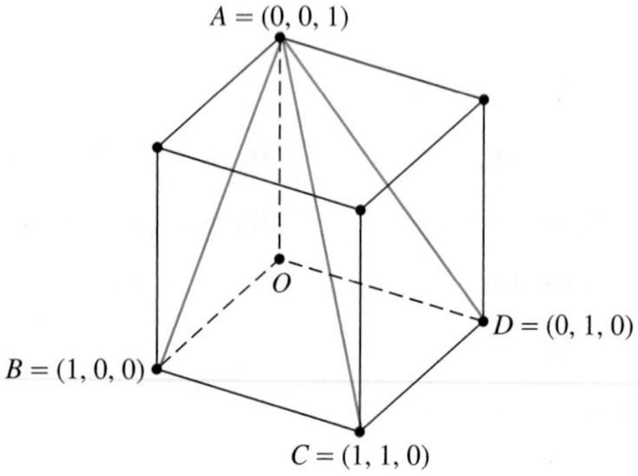

FIGURE 15 Unit cube in $\mathbf{R}^3$.

75. Let $\mathbf{v}$ and $\mathbf{w}$ be nonzero vectors and set $\mathbf{u} = \mathbf{e_v} + \mathbf{e_w}$. Use the dot product to show that the angle between $\mathbf{u}$ and $\mathbf{v}$ is equal to the angle between $\mathbf{u}$ and $\mathbf{w}$. Explain this result geometrically with a diagram.

76. Let $\mathbf{v}$, $\mathbf{w}$, and $\mathbf{a}$ be nonzero vectors such that $\mathbf{v} \cdot \mathbf{a} = \mathbf{w} \cdot \mathbf{a}$. Is it true that $\mathbf{v} = \mathbf{w}$? Either prove this or give a counterexample.

77. Calculate the force (in newtons) required to push a 40-kg wagon up a 10° incline (Figure 16).

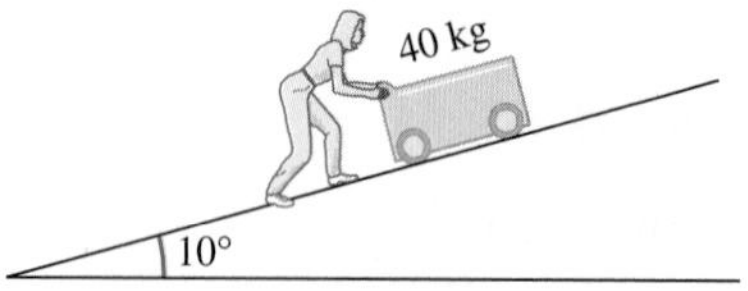

FIGURE 16

78. A force $\mathbf{F}$ is applied to each of two ropes (of negligible weight) attached to opposite ends of a 40-kg wagon and making an angle of 35° with the horizontal (Figure 17). What is the maximum magnitude of $\mathbf{F}$ (in newtons) that can be applied without lifting the wagon off the ground?

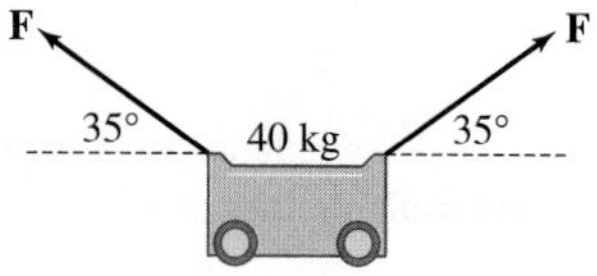

FIGURE 17

79. A light beam travels along the ray determined by a unit vector $\mathbf{L}$, strikes a flat surface at point P, and is reflected along the ray determined by a unit vector $\mathbf{R}$, where $\theta_1 = \theta_2$ (Figure 18). Show that if $\mathbf{N}$ is the unit vector orthogonal to the surface, then

$$\mathbf{R} = 2(\mathbf{L} \cdot \mathbf{N})\mathbf{N} - \mathbf{L}$$

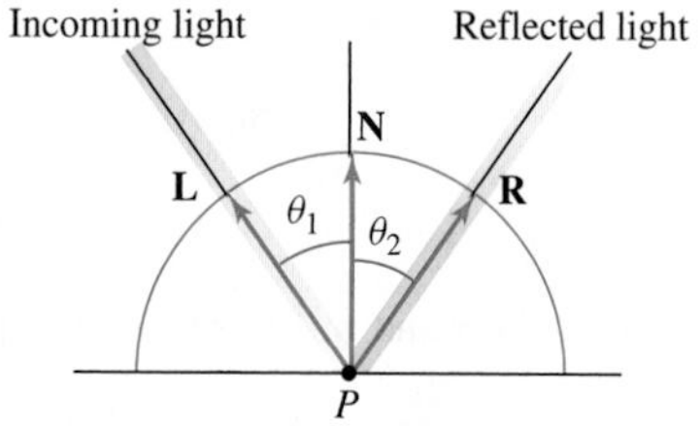

FIGURE 18

80. Let P and Q be antipodal (opposite) points on a sphere of radius r centered at the origin and let R be a third point on the sphere (Figure 19). Prove that $\overline{PR}$ and $\overline{QR}$ are orthogonal.

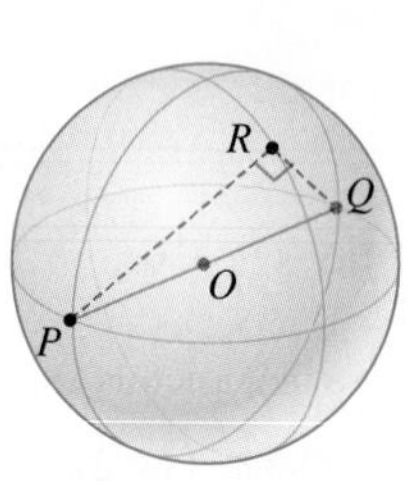

FIGURE 19

81. Prove that $\|\mathbf{v} + \mathbf{w}\|^2 - \|\mathbf{v} - \mathbf{w}\|^2 = 4\mathbf{v} \cdot \mathbf{w}$.

82. Use Exercise 81 to show that $\mathbf{v}$ and $\mathbf{w}$ are orthogonal if and only if $\|\mathbf{v} - \mathbf{w}\| = \|\mathbf{v} + \mathbf{w}\|$.

83. Show that the two diagonals of a parallelogram are perpendicular if and only if its sides have equal length. *Hint:* Use Exercise 82 to show that $\mathbf{v} - \mathbf{w}$ and $\mathbf{v} + \mathbf{w}$ are orthogonal if and only if $\|\mathbf{v}\| = \|\mathbf{w}\|$.

84. Verify the Distributive Law:

$$\mathbf{u} \cdot (\mathbf{v} + \mathbf{w}) = \mathbf{u} \cdot \mathbf{v} + \mathbf{u} \cdot \mathbf{w}$$

85. Verify that $(\lambda \mathbf{v}) \cdot \mathbf{w} = \lambda(\mathbf{v} \cdot \mathbf{w})$ for any scalar λ.

Further Insights and Challenges

86. Prove the Law of Cosines, $c^2 = a^2 + b^2 - 2ab\cos\theta$, by referring to Figure 20. *Hint:* Consider the right triangle $\triangle PQR$.

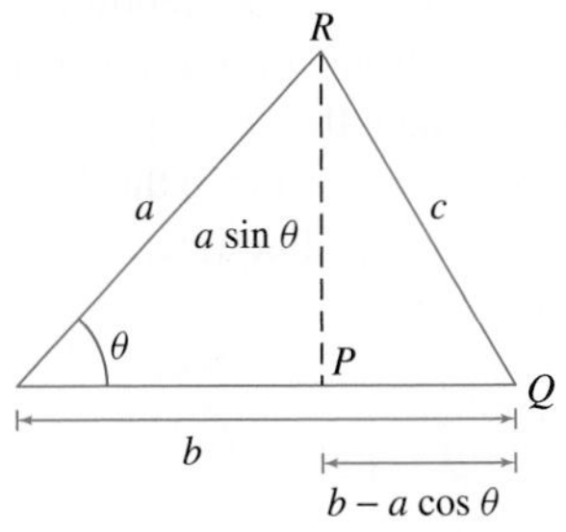

FIGURE 20

87. In this exercise, we prove the Cauchy–Schwarz inequality: If $\mathbf{v}$ and $\mathbf{w}$ are any two vectors, then

$$|\mathbf{v}\cdot\mathbf{w}| \le \|\mathbf{v}\|\,\|\mathbf{w}\| \qquad \boxed{6}$$

(a) Let $f(x) = \|x\mathbf{v}+\mathbf{w}\|^2$ for x a scalar. Show that $f(x) = ax^2 + bx + c$, where $a = \|\mathbf{v}\|^2$, $b = 2\mathbf{v}\cdot\mathbf{w}$, and $c = \|\mathbf{w}\|^2$.
(b) Conclude that $b^2 - 4ac \le 0$. *Hint:* Observe that $f(x) \ge 0$ for all x.

88. Use (6) to prove the Triangle Inequality

$$\|\mathbf{v}+\mathbf{w}\| \le \|\mathbf{v}\| + \|\mathbf{w}\|$$

Hint: First use the Triangle Inequality for numbers to prove

$$|(\mathbf{v}+\mathbf{w})\cdot(\mathbf{v}+\mathbf{w})| \le |(\mathbf{v}+\mathbf{w})\cdot\mathbf{v}| + |(\mathbf{v}+\mathbf{w})\cdot\mathbf{w}|$$

89. This exercise gives another proof of the relation between the dot product and the angle θ between two vectors $\mathbf{v} = \langle a_1, b_1\rangle$ and $\mathbf{w} = \langle a_2, b_2\rangle$ in the plane. Observe that $\mathbf{v} = \|\mathbf{v}\|\langle\cos\theta_1, \sin\theta_1\rangle$ and $\mathbf{w} = \|\mathbf{w}\|\langle\cos\theta_2, \sin\theta_2\rangle$, with θ_1 and θ_2 as in Figure 21. Then use the addition formula for the cosine to show that

$$\mathbf{v}\cdot\mathbf{w} = \|\mathbf{v}\|\,\|\mathbf{w}\|\cos\theta$$

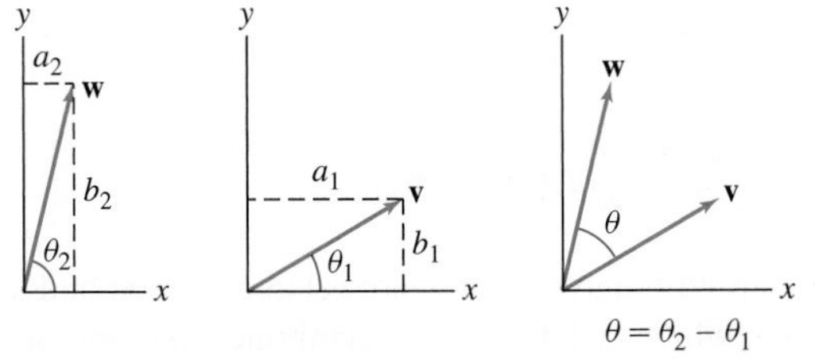

FIGURE 21

90. Let $\mathbf{v} = \langle x, y\rangle$ and

$$\mathbf{v}_\theta = \langle x\cos\theta + y\sin\theta, -x\sin\theta + y\cos\theta\rangle$$

Prove that the angle between $\mathbf{v}$ and $\mathbf{v}_\theta$ is θ.

91. Let $\mathbf{v}$ be a nonzero vector. The angles α, β, γ between $\mathbf{v}$ and the unit vectors $\mathbf{i}, \mathbf{j}, \mathbf{k}$ are called the direction angles of $\mathbf{v}$ (Figure 22). The cosines of these angles are called the **direction cosines** of $\mathbf{v}$. Prove that

$$\cos^2\alpha + \cos^2\beta + \cos^2\gamma = 1$$

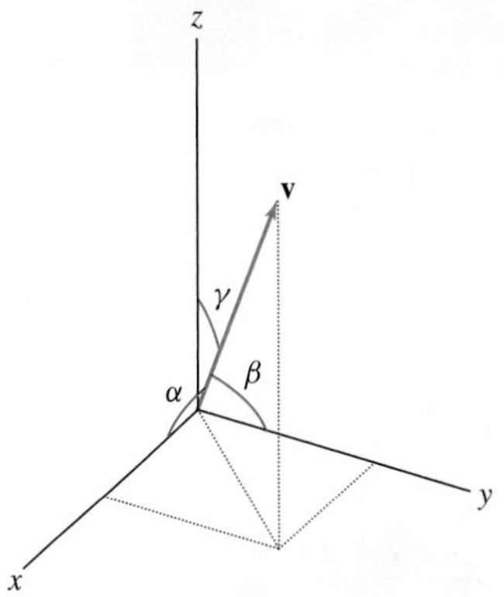

FIGURE 22 Direction angles of $\mathbf{v}$.

92. Find the direction cosines of $\mathbf{v} = \langle 3, 6, -2\rangle$.

93. The set of all points $X = (x, y, z)$ equidistant from two points P, Q in $\mathbf{R}^3$ is a plane (Figure 23). Show that X lies on this plane if

$$\overrightarrow{PQ}\cdot\overrightarrow{OX} = \frac{1}{2}\left(\|\overrightarrow{OQ}\|^2 - \|\overrightarrow{OP}\|^2\right) \qquad \boxed{7}$$

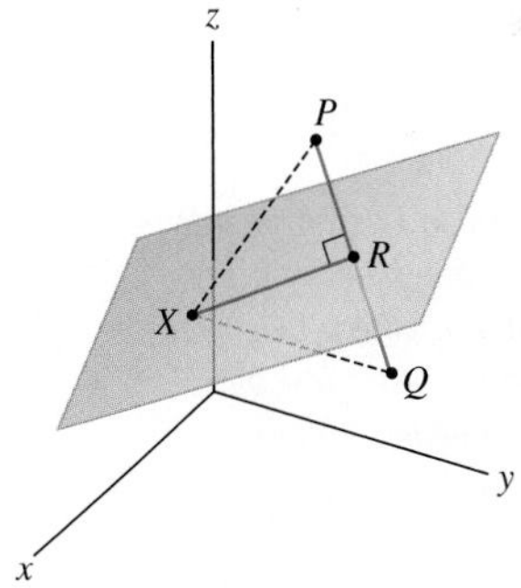

FIGURE 23

Hint: If R is the midpoint of $\overline{PQ}$, then X is equidistant from P and Q if and only if $\overrightarrow{XR}$ is orthogonal to $\overrightarrow{PQ}$.

94. Sketch the plane consisting of all points $X = (x, y, z)$ equidistant from the points $P = (0, 1, 0)$ and $Q = (0, 0, 1)$. Use Eq. (7) to show that X lies on this plane if and only if $y = z$.

95. Use Eq. (7) to find the equation of the plane consisting of all points $X = (x, y, z)$ equidistant from $P = (2, 1, 1)$ and $Q = (1, 0, 2)$.

13.4 The Cross Product

FIGURE 1 The spiral paths of charged particles in a bubble chamber in the presence of a magnetic field are described using cross products.

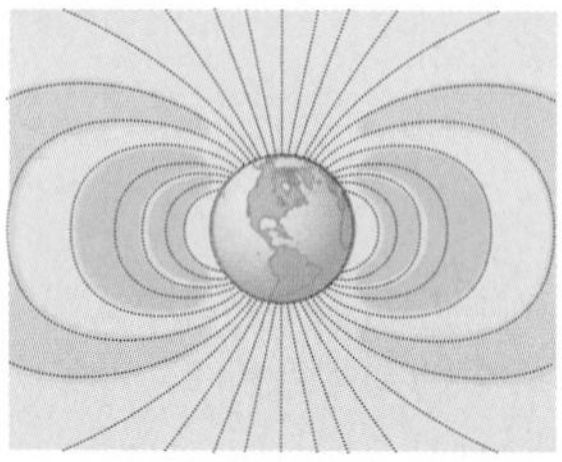

FIGURE 2 The Van Allen radiation belts, located thousands of miles above the earth's surface, are made up of streams of protons and electrons that oscillate back and forth in helical paths between two "magnetic mirrors" set up by the earth's magnetic field. This helical motion is explained by the "cross-product" nature of magnetic forces.

This section introduces the **cross product** $\mathbf{v} \times \mathbf{w}$ of two vectors $\mathbf{v}$ and $\mathbf{w}$. The cross product (sometimes called the **vector product**) is used in physics and engineering to describe quantities involving rotation, such as torque and angular momentum. In electromagnetic theory, magnetic forces are described using cross products (Figures 1 and 2).

Unlike the dot product $\mathbf{v} \cdot \mathbf{w}$ (which is a scalar), the cross product $\mathbf{v} \times \mathbf{w}$ is again a vector. It is defined using determinants, which we now define in the 2×2 and 3×3 cases. A 2×2 determinant is a number formed from an array of numbers with two rows and two columns (called a **matrix**) according to the formula

$$\begin{vmatrix} a & b \\ c & d \end{vmatrix} = ad - bc \tag{1}$$

Note that the determinant is the difference of the diagonal products. For example,

$$\begin{vmatrix} 3 & 2 \\ \frac{1}{2} & 4 \end{vmatrix} = \begin{vmatrix} 3 & 2 \\ \frac{1}{2} & 4 \end{vmatrix} - \begin{vmatrix} 3 & 2 \\ \frac{1}{2} & 4 \end{vmatrix} = 3 \cdot 4 - 2 \cdot \frac{1}{2} = 11$$

The determinant of a 3×3 matrix is defined by the formula

$$\begin{vmatrix} a_1 & b_1 & c_1 \\ a_2 & b_2 & c_2 \\ a_3 & b_3 & c_3 \end{vmatrix} = a_1 \underbrace{\begin{vmatrix} b_2 & c_2 \\ b_3 & c_3 \end{vmatrix}}_{(1,\,1)\text{-minor}} - b_1 \underbrace{\begin{vmatrix} a_2 & c_2 \\ a_3 & c_3 \end{vmatrix}}_{(1,\,2)\text{-minor}} + c_1 \underbrace{\begin{vmatrix} a_2 & b_2 \\ a_3 & b_3 \end{vmatrix}}_{(1,\,3)\text{-minor}} \tag{2}$$

This formula expresses the 3×3 determinant in terms of 2×2 determinants called **minors**. The minors are obtained by crossing out the first row and one of the three columns of the 3×3 matrix. For example, the minor labeled $(1, 2)$ above is obtained as follows:

$$\underbrace{\begin{vmatrix} a_1 & b_1 & c_1 \\ a_2 & b_2 & c_2 \\ a_3 & b_3 & c_3 \end{vmatrix}}_{\text{Cross out row 1 and column 2}} \quad \text{to obtain the } (1,2)\text{-minor} \quad \underbrace{\begin{vmatrix} a_2 & c_2 \\ a_3 & c_3 \end{vmatrix}}_{(1,2)\text{-minor}}$$

The theory of matrices and determinants is part of linear algebra, *a subject of great importance throughout mathematics. In this section, we discuss just a few basic definitions and facts needed for our treatment of multivariable calculus.*

■ **EXAMPLE 1 A 3 × 3 Determinant** Calculate $\begin{vmatrix} 2 & 4 & 3 \\ 0 & 1 & -7 \\ -1 & 5 & 3 \end{vmatrix}$.

Solution

$$\begin{vmatrix} \textcircled{2} & \textcircled{4} & \textcircled{3} \\ 0 & 1 & -7 \\ -1 & 5 & 3 \end{vmatrix} = \textcircled{2} \begin{vmatrix} 1 & -7 \\ 5 & 3 \end{vmatrix} - \textcircled{4} \begin{vmatrix} 0 & -7 \\ -1 & 3 \end{vmatrix} + \textcircled{3} \begin{vmatrix} 0 & 1 \\ -1 & 5 \end{vmatrix}$$

$$= 2(38) - 4(-7) + 3(1) = 107$$ ■

Later in this section we will see how determinants are related to area and volume. First, we introduce the cross product, which is defined as a "symbolic" determinant whose first row has the vector entries $\mathbf{i}, \mathbf{j}, \mathbf{k}$.

DEFINITION The Cross Product The cross product of vectors $\mathbf{v} = \langle a_1, b_1, c_1 \rangle$ and $\mathbf{w} = \langle a_2, b_2, c_2 \rangle$ is the vector

$$\mathbf{v} \times \mathbf{w} = \begin{vmatrix} \mathbf{i} & \mathbf{j} & \mathbf{k} \\ a_1 & b_1 & c_1 \\ a_2 & b_2 & c_2 \end{vmatrix} = \begin{vmatrix} b_1 & c_1 \\ b_2 & c_2 \end{vmatrix} \mathbf{i} - \begin{vmatrix} a_1 & c_1 \\ a_2 & c_2 \end{vmatrix} \mathbf{j} + \begin{vmatrix} a_1 & b_1 \\ a_2 & b_2 \end{vmatrix} \mathbf{k} \tag{3}$$

CAUTION Note in Eq. (3) that the middle term comes with a minus sign.

■ **EXAMPLE 2** Calculate $\mathbf{v} \times \mathbf{w}$, where $\mathbf{v} = \langle -2, 1, 4 \rangle$ and $\mathbf{w} = \langle 3, 2, 5 \rangle$.

Solution

$$\mathbf{v} \times \mathbf{w} = \begin{vmatrix} \mathbf{i} & \mathbf{j} & \mathbf{k} \\ -2 & 1 & 4 \\ 3 & 2 & 5 \end{vmatrix} = \begin{vmatrix} 1 & 4 \\ 2 & 5 \end{vmatrix} \mathbf{i} - \begin{vmatrix} -2 & 4 \\ 3 & 5 \end{vmatrix} \mathbf{j} + \begin{vmatrix} -2 & 1 \\ 3 & 2 \end{vmatrix} \mathbf{k}$$

$$= (-3)\mathbf{i} - (-22)\mathbf{j} + (-7)\mathbf{k} = \langle -3, 22, -7 \rangle$$ ■

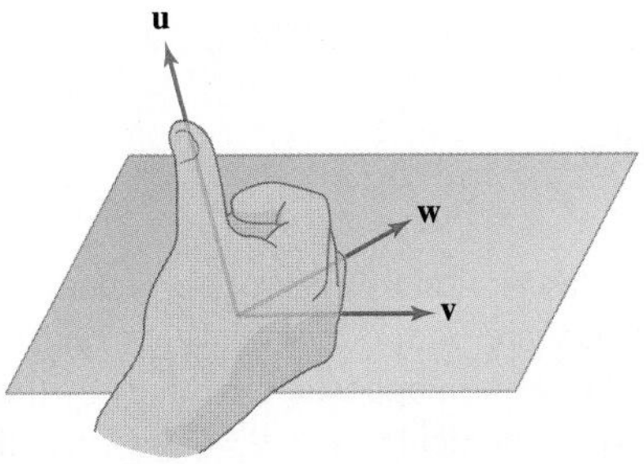

FIGURE 3 $\{\mathbf{v}, \mathbf{w}, \mathbf{u}\}$ forms a right-handed system.

Formula (3) gives no hint of the geometric meaning of the cross product. However, there is a simple way to visualize the vector $\mathbf{v} \times \mathbf{w}$ using the **right-hand rule**. Suppose that $\mathbf{v}$, $\mathbf{w}$, and $\mathbf{u}$ are nonzero vectors that do not all lie in a plane. We say that $\{\mathbf{v}, \mathbf{w}, \mathbf{u}\}$ forms a **right-handed system** if the direction of $\mathbf{u}$ is determined by the right-hand rule: *When the fingers of your right hand curl from* $\mathbf{v}$ *to* $\mathbf{w}$, *your thumb points to the same side of the plane spanned by* $\mathbf{v}$ *and* $\mathbf{w}$ *as* $\mathbf{u}$ (Figure 3). The following theorem is proved at the end of this section.

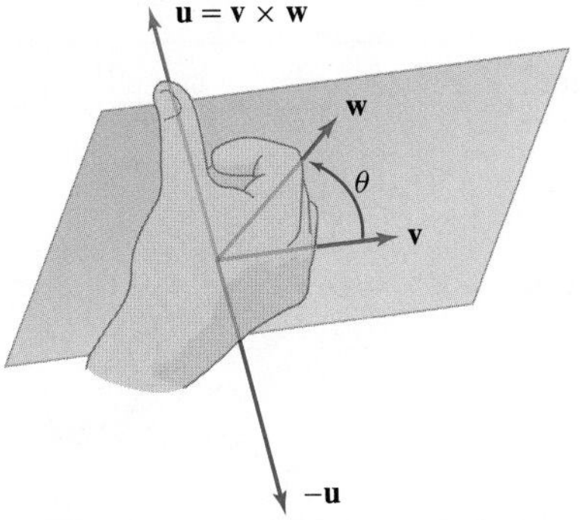

FIGURE 4 There are two vectors orthogonal to $\mathbf{v}$ and $\mathbf{w}$ with length $\|\mathbf{v}\|\,\|\mathbf{w}\|\sin\theta$. The right-hand rule determines which is $\mathbf{v} \times \mathbf{w}$.

THEOREM 1 Geometric Description of the Cross Product The cross product $\mathbf{v} \times \mathbf{w}$ is the unique vector with the following three properties:

(i) $\mathbf{v} \times \mathbf{w}$ is orthogonal to $\mathbf{v}$ and $\mathbf{w}$.

(ii) $\mathbf{v} \times \mathbf{w}$ has length $\|\mathbf{v}\|\,\|\mathbf{w}\|\sin\theta$ (θ is the angle between $\mathbf{v}$ and $\mathbf{w}$, $0 \le \theta \le \pi$).

(iii) $\{\mathbf{v}, \mathbf{w}, \mathbf{v} \times \mathbf{w}\}$ forms a right-handed system.

How do the three properties in Theorem 1 determine $\mathbf{v} \times \mathbf{w}$? By property (i), $\mathbf{v} \times \mathbf{w}$ lies on the line orthogonal to $\mathbf{v}$ and $\mathbf{w}$. By property (ii), $\mathbf{v} \times \mathbf{w}$ is one of the two vectors on this line of length $\|\mathbf{v}\|\,\|\mathbf{w}\|\sin\theta$. Finally, property (iii) tells us which of these two vectors is $\mathbf{v} \times \mathbf{w}$—namely, the vector for which $\{\mathbf{v}, \mathbf{w}, \mathbf{u}\}$ is right-handed (Figure 4).

■ **EXAMPLE 3** Let $\mathbf{v} = \langle 2, 0, 0 \rangle$ and $\mathbf{w} = \langle 0, 1, 1 \rangle$. Determine $\mathbf{u} = \mathbf{v} \times \mathbf{w}$ using the geometric properties of the cross product rather than Eq. (3).

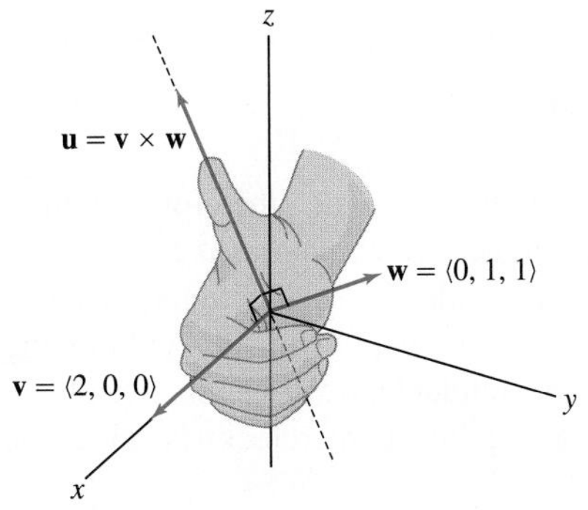

FIGURE 5 The direction of $\mathbf{u} = \mathbf{v} \times \mathbf{w}$ is determined by the right-hand rule. Thus, $\mathbf{u}$ has a positive z-component.

Solution We use Theorem 1. First, by Property (i), $\mathbf{u} = \mathbf{v} \times \mathbf{w}$ is orthogonal to $\mathbf{v}$ and $\mathbf{w}$. Since $\mathbf{v}$ lies along the x-axis, $\mathbf{u}$ must lie in the yz-plane (Figure 5). In other words, $\mathbf{u} = \langle 0, b, c \rangle$. But $\mathbf{u}$ is also orthogonal to $\mathbf{w} = \langle 0, 1, 1 \rangle$, so $\mathbf{u} \cdot \mathbf{w} = b + c = 0$ and thus $\mathbf{u} = \langle 0, b, -b \rangle$.

Next, direct computation shows that $\|\mathbf{v}\| = 2$ and $\|\mathbf{w}\| = \sqrt{2}$. Furthermore, the angle between $\mathbf{v}$ and $\mathbf{w}$ is $\theta = \frac{\pi}{2}$ since $\mathbf{v} \cdot \mathbf{w} = 0$. By property (ii),

$$\|\mathbf{u}\| = \sqrt{b^2 + (-b)^2} = |b|\sqrt{2} \qquad \text{is equal to} \qquad \|\mathbf{v}\|\,\|\mathbf{w}\|\sin\frac{\pi}{2} = 2\sqrt{2}$$

Therefore, $|b| = 2$ and $b = \pm 2$. Finally, property (iii) tells us that $\mathbf{u}$ points in the positive z-direction (Figure 5). Thus, $b = -2$ and $\mathbf{u} = \langle 0, -2, 2 \rangle$. You can verify that the formula for the cross product yields the same answer. ■

One of the most striking properties of the cross product is that it is *anticommutative*. Reversing the order changes the sign:

$$\mathbf{w} \times \mathbf{v} = -\mathbf{v} \times \mathbf{w} \qquad (4)$$

We verify this using Eq. (3): when we interchange **v** and **w**, each of the 2 × 2 determinants changes sign. For example,

$$\begin{vmatrix} a_1 & b_1 \\ a_2 & b_2 \end{vmatrix} = a_1b_2 - b_1a_2$$

$$= -(b_1a_2 - a_1b_2) = -\begin{vmatrix} a_2 & b_2 \\ a_1 & b_1 \end{vmatrix}$$

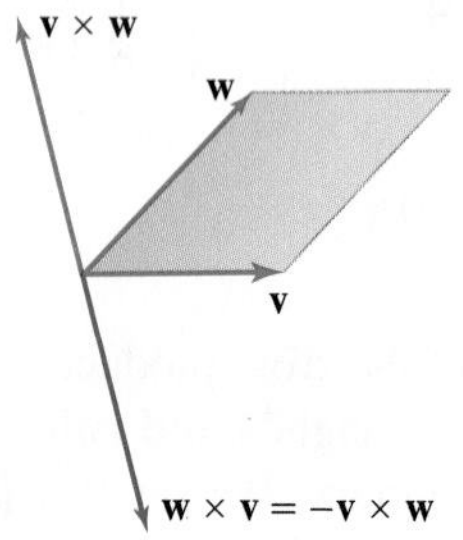

FIGURE 6

Anticommutativity also follows from the geometric description of the cross product. By properties (i) and (ii) in Theorem 1, $\mathbf{v} \times \mathbf{w}$ and $\mathbf{w} \times \mathbf{v}$ are both orthogonal to **v** and **w** and have the same length. However, $\mathbf{v} \times \mathbf{w}$ and $\mathbf{w} \times \mathbf{v}$ point in opposite directions by the right-hand rule, and thus $\mathbf{v} \times \mathbf{w} = -\mathbf{w} \times \mathbf{v}$ (Figure 6). In particular, $\mathbf{v} \times \mathbf{v} = -\mathbf{v} \times \mathbf{v}$ and hence $\mathbf{v} \times \mathbf{v} = \mathbf{0}$.

The next theorem lists some further properties of cross products (the proofs are given as Exercises 45–48).

Note an important distinction between the dot product and cross product of a vector with itself:

$$\mathbf{v} \times \mathbf{v} = \mathbf{0}$$
$$\mathbf{v} \cdot \mathbf{v} = \|\mathbf{v}\|^2$$

THEOREM 2 Basic Properties of the Cross Product

(i) $\mathbf{w} \times \mathbf{v} = -\mathbf{v} \times \mathbf{w}$

(ii) $\mathbf{v} \times \mathbf{v} = \mathbf{0}$

(iii) $\mathbf{v} \times \mathbf{w} = \mathbf{0}$ if and only if $\mathbf{w} = \lambda\mathbf{v}$ for some scalar λ or $\mathbf{v} = \mathbf{0}$.

(iv) $(\lambda\mathbf{v}) \times \mathbf{w} = \mathbf{v} \times (\lambda\mathbf{w}) = \lambda(\mathbf{v} \times \mathbf{w})$

(v) $(\mathbf{u} + \mathbf{v}) \times \mathbf{w} = \mathbf{u} \times \mathbf{w} + \mathbf{v} \times \mathbf{w}$

$\mathbf{u} \times (\mathbf{v} + \mathbf{w}) = \mathbf{u} \times \mathbf{v} + \mathbf{u} \times \mathbf{w}$

The cross product of any two of the standard basis vectors **i**, **j**, and **k** is equal to the third, possibly with a minus sign. More precisely (see Exercise 49),

$$\mathbf{i} \times \mathbf{j} = \mathbf{k}, \qquad \mathbf{j} \times \mathbf{k} = \mathbf{i}, \qquad \mathbf{k} \times \mathbf{i} = \mathbf{j} \tag{5}$$
$$\mathbf{i} \times \mathbf{i} = \mathbf{j} \times \mathbf{j} = \mathbf{k} \times \mathbf{k} = \mathbf{0}$$

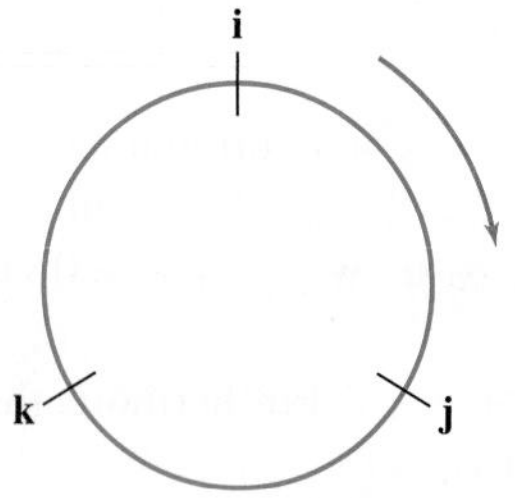

FIGURE 7 Circle for computing the cross products of the basis vectors.

Since the cross product is anticommutative, minus signs occur when the cross products are taken in the opposite order. An easy way to remember these relations is to draw **i**, **j**, and **k** in a circle as in Figure 7. Go around the circle in the clockwise direction (starting at any point) and you obtain one of the relations (5). For example, starting at **i** and moving clockwise yields $\mathbf{i} \times \mathbf{j} = \mathbf{k}$. If you go around in the counterclockwise direction, you obtain the relations with a minus sign. Thus, starting at **k** and going counterclockwise gives the relation $\mathbf{k} \times \mathbf{j} = -\mathbf{i}$.

■ **EXAMPLE 4 Using the ijk Relations** Compute $(2\mathbf{i} + \mathbf{k}) \times (3\mathbf{j} + 5\mathbf{k})$.

Solution We use the Distributive Law for cross products:

$$(2\mathbf{i} + \mathbf{k}) \times (3\mathbf{j} + 5\mathbf{k}) = (2\mathbf{i}) \times (3\mathbf{j}) + (2\mathbf{i}) \times (5\mathbf{k}) + \mathbf{k} \times (3\mathbf{j}) + \mathbf{k} \times (5\mathbf{k})$$

$$= 6(\mathbf{i} \times \mathbf{j}) + 10(\mathbf{i} \times \mathbf{k}) + 3(\mathbf{k} \times \mathbf{j}) + 5(\mathbf{k} \times \mathbf{k})$$

$$= 6\mathbf{k} - 10\mathbf{j} - 3\mathbf{i} + 5(\mathbf{0}) = -3\mathbf{i} - 10\mathbf{j} + 6\mathbf{k}$$ ■

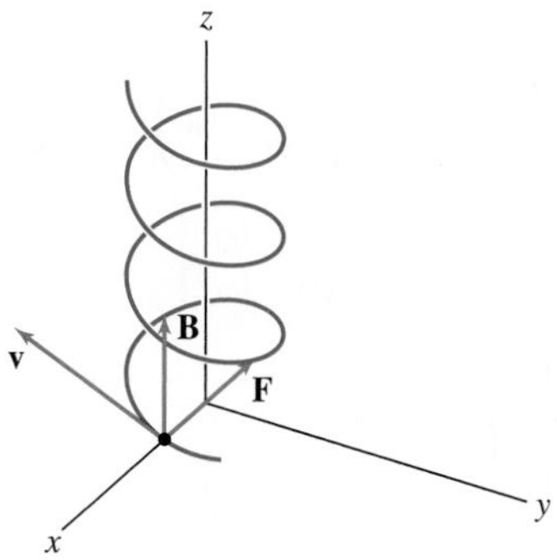

FIGURE 8 A proton in a uniform magnetic field travels in a helical path.

■ **EXAMPLE 5 Velocity in a Magnetic Field** The force $\mathbf{F}$ on a proton moving at velocity $\mathbf{v}$ m/s in a uniform magnetic field $\mathbf{B}$ (in teslas) is $\mathbf{F} = q(\mathbf{v} \times \mathbf{B})$ in newtons, where $q = 1.6 \times 10^{-19}$ coulombs (Figure 8). Calculate $\mathbf{F}$ if $\mathbf{B} = 0.0004\mathbf{k}$ T and $\mathbf{v}$ has magnitude 10^6 m/s in the direction $-\mathbf{j} + \mathbf{k}$.

Solution The vector $-\mathbf{j} + \mathbf{k}$ has length $\sqrt{2}$, and since $\mathbf{v}$ has magnitude 10^6,

$$\mathbf{v} = 10^6 \left(\frac{-\mathbf{j} + \mathbf{k}}{\sqrt{2}} \right)$$

Therefore, the force (in newtons) is

$$\mathbf{F} = q(\mathbf{v} \times \mathbf{B}) = 10^6 q \left(\frac{-\mathbf{j} + \mathbf{k}}{\sqrt{2}} \right) \times (0.0004\mathbf{k}) = \frac{400q}{\sqrt{2}} ((-\mathbf{j} + \mathbf{k}) \times \mathbf{k})$$

$$= -\frac{400q}{\sqrt{2}} \mathbf{i} = \frac{-400(1.6 \times 10^{-19})}{\sqrt{2}} \mathbf{i} \approx -(4.5 \times 10^{-17})\mathbf{i}$$ ■

Cross Products, Area, and Volume

Cross products and determinants are closely related to area and volume. Consider the parallelogram $\mathcal{P}$ spanned by nonzero vectors $\mathbf{v}$ and $\mathbf{w}$ with a common basepoint. In Figure 9(A), we see that $\mathcal{P}$ has base $b = \|\mathbf{v}\|$ and height $h = \|\mathbf{w}\| \sin\theta$, where θ is the angle between $\mathbf{v}$ and $\mathbf{w}$. Therefore, $\mathcal{P}$ has area $A = bh = \|\mathbf{v}\|\,\|\mathbf{w}\| \sin\theta = \|\mathbf{v} \times \mathbf{w}\|$.

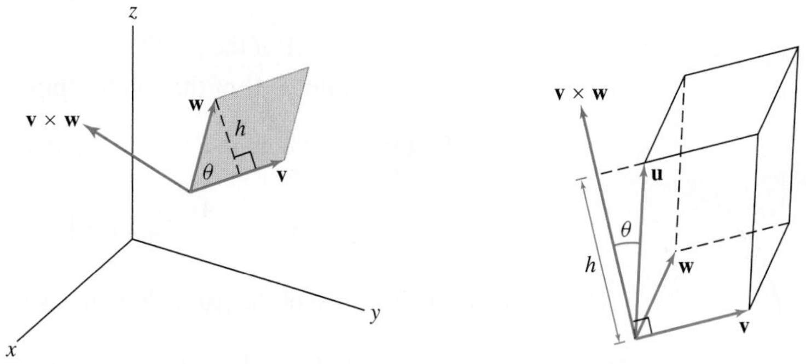

(A) The area of the parallelogram $\mathcal{P}$ is $\|\mathbf{v} \times \mathbf{w}\| = \|\mathbf{v}\|\ \|\mathbf{w}\|\ \sin\theta$.

(B) The volume of the parallelpiped $\mathbf{P}$ is $|\mathbf{u} \cdot (\mathbf{v} \times \mathbf{w})|$.

FIGURE 9

A "parallelepiped" is the solid spanned by three vectors. Each face is a parallelogram.

Next, consider the **parallelepiped P** spanned by three nonzero vectors $\mathbf{u}$, $\mathbf{v}$, $\mathbf{w}$ in $\mathbf{R}^3$ [the three-dimensional prism in Figure 9(B)]. The base of $\mathbf{P}$ is the parallelogram spanned by $\mathbf{v}$ and $\mathbf{w}$, so the area of the base is $\|\mathbf{v} \times \mathbf{w}\|$. The height of $\mathbf{P}$ is $h = \|\mathbf{u}\| \cdot |\cos\theta|$, where θ is the angle between $\mathbf{u}$ and $\mathbf{v} \times \mathbf{w}$. Therefore,

$$\text{Volume of } \mathbf{P} = (\text{area of base})(\text{height}) = \|\mathbf{v} \times \mathbf{w}\| \cdot \|\mathbf{u}\| \cdot |\cos\theta|$$

But $\|\mathbf{v} \times \mathbf{w}\|\ \|\mathbf{u}\| \cos\theta$ is equal to the dot product of $\mathbf{v} \times \mathbf{w}$ and $\mathbf{u}$. This proves the formula

$$\text{Volume of } \mathbf{P} = |\mathbf{u} \cdot (\mathbf{v} \times \mathbf{w})|$$

The quantity $\mathbf{u} \cdot (\mathbf{v} \times \mathbf{w})$, called the **vector triple product**, can be expressed as a determinant. Let

$$\mathbf{u} = \langle a_1, b_1, c_1 \rangle, \qquad \mathbf{v} = \langle a_2, b_2, c_2 \rangle, \qquad \mathbf{w} = \langle a_3, b_3, c_3 \rangle$$

We use the following notation for the determinant of the matrix whose rows are the vectors $\mathbf{v}, \mathbf{w}, \mathbf{u}$*:*

$$\det\begin{pmatrix} \mathbf{u} \\ \mathbf{v} \\ \mathbf{w} \end{pmatrix} = \begin{vmatrix} a_1 & b_1 & c_1 \\ a_2 & b_2 & c_2 \\ a_3 & b_3 & c_3 \end{vmatrix}$$

It is awkward to write the absolute value of a determinant in the notation on the right, but we may denote it

$$\left|\det\begin{pmatrix} \mathbf{u} \\ \mathbf{v} \\ \mathbf{w} \end{pmatrix}\right|$$

Then

$$\mathbf{u}\cdot(\mathbf{v}\times\mathbf{w}) = \mathbf{u}\cdot\left(\begin{vmatrix} b_2 & c_2 \\ b_3 & c_3 \end{vmatrix}\mathbf{i} - \begin{vmatrix} a_2 & c_2 \\ a_3 & c_3 \end{vmatrix}\mathbf{j} + \begin{vmatrix} a_2 & b_2 \\ a_3 & b_3 \end{vmatrix}\mathbf{k}\right)$$

$$= a_1\begin{vmatrix} b_2 & c_2 \\ b_3 & c_3 \end{vmatrix} - b_1\begin{vmatrix} a_2 & c_2 \\ a_3 & c_3 \end{vmatrix} + c_1\begin{vmatrix} a_2 & b_2 \\ a_3 & b_3 \end{vmatrix}$$

$$= \begin{vmatrix} a_1 & b_1 & c_1 \\ a_2 & b_2 & c_2 \\ a_3 & b_3 & c_3 \end{vmatrix} = \det\begin{pmatrix} \mathbf{u} \\ \mathbf{v} \\ \mathbf{w} \end{pmatrix} \qquad \boxed{6}$$

We obtain the following formulas for area and volume.

THEOREM 3 Area and Volume via Cross Products and Determinants Let $\mathbf{u}, \mathbf{v}, \mathbf{w}$ be nonzero vectors in $\mathbf{R}^3$. Then

(i) The parallelogram $\mathcal{P}$ spanned by $\mathbf{v}$ and $\mathbf{w}$ has area $A = \|\mathbf{v}\times\mathbf{w}\|$.

(ii) The parallelepiped $\mathbf{P}$ spanned by $\mathbf{u}$, $\mathbf{v}$, and $\mathbf{w}$ has volume

$$V = |\mathbf{u}\cdot(\mathbf{v}\times\mathbf{w})| = \left|\det\begin{pmatrix} \mathbf{u} \\ \mathbf{v} \\ \mathbf{w} \end{pmatrix}\right| \qquad \boxed{7}$$

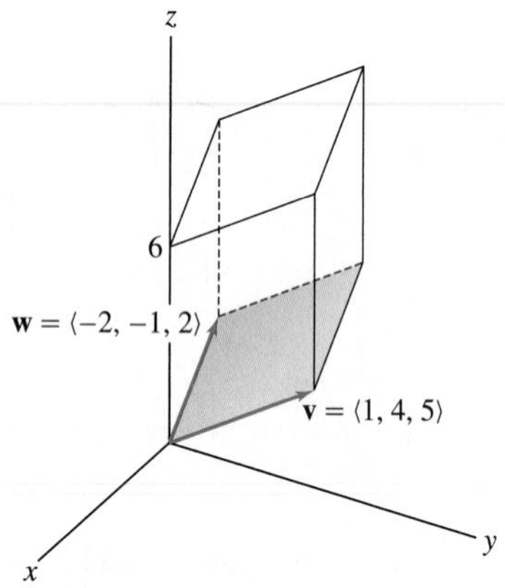

FIGURE 10

■ **EXAMPLE 6** Let $\mathbf{v} = \langle 1, 4, 5\rangle$ and $\mathbf{w} = \langle -2, -1, 2\rangle$. Calculate:

(a) The area A of the parallelogram spanned by $\mathbf{v}$ and $\mathbf{w}$

(b) The volume V of the parallelepiped in Figure 10

Solution We compute the cross product and apply Theorem 3:

$$\mathbf{v}\times\mathbf{w} = \begin{vmatrix} 4 & 5 \\ -1 & 2 \end{vmatrix}\mathbf{i} - \begin{vmatrix} 1 & 5 \\ -2 & 2 \end{vmatrix}\mathbf{j} + \begin{vmatrix} 1 & 4 \\ -2 & -1 \end{vmatrix}\mathbf{k} = \langle 13, -12, 7\rangle$$

(a) The area of the parallelogram spanned by $\mathbf{v}$ and $\mathbf{w}$ is

$$A = \|\mathbf{v}\times\mathbf{w}\| = \sqrt{13^2 + (-12)^2 + 7^2} = \sqrt{362} \approx 19$$

(b) The vertical leg of the parallelepiped is the vector $6\mathbf{k}$, so by Eq. (7),

$$V = |(6\mathbf{k})\cdot(\mathbf{v}\times\mathbf{w})| = |\langle 0, 0, 6\rangle\cdot\langle 13, -12, 7\rangle| = 6(7) = 42$$ ■

We can compute the area A of the parallelogram spanned by vectors $\mathbf{v} = \langle a, b\rangle$ and $\mathbf{w} = \langle c, d\rangle$ by regarding $\mathbf{v}$ and $\mathbf{w}$ as vectors in $\mathbf{R}^3$ with zero component in the z-direction (Figure 11). Thus, we write $\mathbf{v} = \langle a, b, 0\rangle$ and $\mathbf{w} = \langle c, d, 0\rangle$. The cross product $\mathbf{v}\times\mathbf{w}$ is a vector pointing in the z-direction:

$$\mathbf{v}\times\mathbf{w} = \begin{vmatrix} \mathbf{i} & \mathbf{j} & \mathbf{k} \\ a & b & 0 \\ c & d & 0 \end{vmatrix} = \begin{vmatrix} b & 0 \\ d & 0 \end{vmatrix}\mathbf{i} - \begin{vmatrix} a & 0 \\ c & 0 \end{vmatrix}\mathbf{j} + \begin{vmatrix} a & b \\ c & d \end{vmatrix}\mathbf{k} = \begin{vmatrix} a & b \\ c & d \end{vmatrix}\mathbf{k}$$

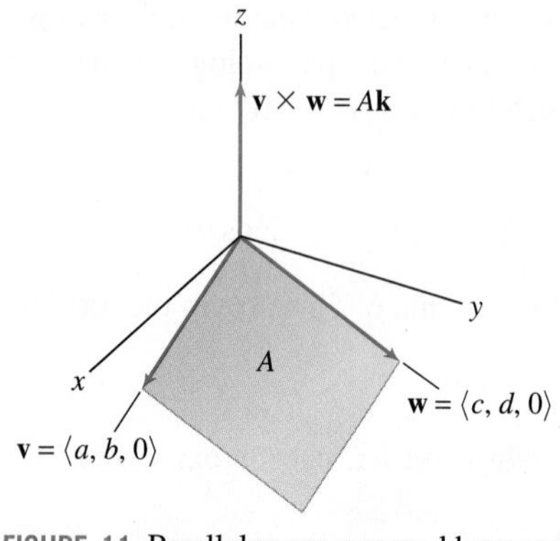

FIGURE 11 Parallelogram spanned by $\mathbf{v}$ and $\mathbf{w}$ in the xy-plane.

By Theorem 3, the parallelogram spanned by $\mathbf{v}$ and $\mathbf{w}$ has area $A = \|\mathbf{v}\times\mathbf{w}\|$, and thus,

$$\boxed{A = \left|\det\begin{pmatrix} \mathbf{v} \\ \mathbf{w} \end{pmatrix}\right| = \left|\det\begin{pmatrix} a & b \\ c & d \end{pmatrix}\right|} \qquad \boxed{8}$$

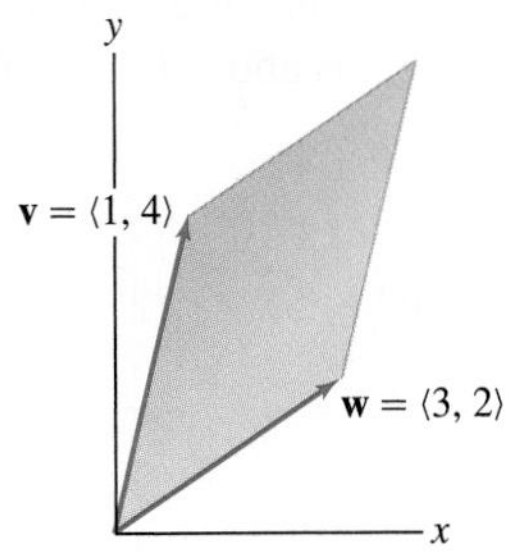

FIGURE 12

■ **EXAMPLE 7** Compute the area A of the parallelogram in Figure 12.

Solution We have $\begin{vmatrix} \mathbf{v} \\ \mathbf{w} \end{vmatrix} = \begin{vmatrix} 1 & 4 \\ 3 & 2 \end{vmatrix} = 1 \cdot 2 - 3 \cdot 4 = -10$. The area is the absolute value $A = |-10| = 10$. ■

Proofs of Cross-Product Properties

We now derive the properties of the cross product listed in Theorem 1. Let

$$\mathbf{v} = \langle a_1, b_1, c_1 \rangle, \qquad \mathbf{w} = \langle a_2, b_2, c_2 \rangle$$

We prove that $\mathbf{v} \times \mathbf{w}$ is orthogonal to $\mathbf{v}$ by showing that $\mathbf{v} \cdot (\mathbf{v} \times \mathbf{w}) = 0$. By Eq. (6),

$$\mathbf{v} \cdot (\mathbf{v} \times \mathbf{w}) = \det \begin{pmatrix} \mathbf{v} \\ \mathbf{v} \\ \mathbf{w} \end{pmatrix} = a_1 \begin{vmatrix} b_1 & c_1 \\ b_2 & c_2 \end{vmatrix} - b_1 \begin{vmatrix} a_1 & c_1 \\ a_2 & c_2 \end{vmatrix} + c_1 \begin{vmatrix} a_1 & b_1 \\ a_2 & b_2 \end{vmatrix} \qquad \boxed{9}$$

Straightforward algebra (left to the reader) shows that the right-hand side of Eq. (9) is equal to zero. This shows that $\mathbf{v} \cdot (\mathbf{v} \times \mathbf{w}) = 0$ and thus $\mathbf{v} \times \mathbf{w}$ is orthogonal to $\mathbf{v}$ as claimed. Interchanging the roles of $\mathbf{v}$ and $\mathbf{w}$, we conclude also that $\mathbf{w} \times \mathbf{v}$ is orthogonal to $\mathbf{w}$, and since $\mathbf{v} \times \mathbf{w} = -\mathbf{w} \times \mathbf{v}$, it follows that $\mathbf{v} \times \mathbf{w}$ is orthogonal to $\mathbf{w}$. This proves part (i) of Theorem 1. To prove (ii), we shall use the following identity:

$$\boxed{\|\mathbf{v} \times \mathbf{w}\|^2 = \|\mathbf{v}\|^2 \|\mathbf{w}\|^2 - (\mathbf{v} \cdot \mathbf{w})^2} \qquad \boxed{10}$$

To verify this identity, we compute $\|\mathbf{v} \times \mathbf{w}\|^2$ as the sum of the squares of the components of $\mathbf{v} \times \mathbf{w}$:

$$\|\mathbf{v} \times \mathbf{w}\|^2 = \begin{vmatrix} b_1 & c_1 \\ b_2 & c_2 \end{vmatrix}^2 + \begin{vmatrix} a_1 & c_1 \\ a_2 & c_2 \end{vmatrix}^2 + \begin{vmatrix} a_1 & b_1 \\ a_2 & b_2 \end{vmatrix}^2$$

$$= (b_1c_2 - c_1b_2)^2 + (a_1c_2 - c_1a_2)^2 + (a_1b_2 - b_1a_2)^2 \qquad \boxed{11}$$

On the other hand, by definition,

$$\|\mathbf{v}\|^2 \|\mathbf{w}\|^2 - (\mathbf{v} \cdot \mathbf{w})^2 = \left(a_1^2 + b_1^2 + c_1^2\right)\left(a_2^2 + b_2^2 + c_2^2\right) - (a_1a_2 + b_1b_2 + c_1c_2)^2 \qquad \boxed{12}$$

Again, algebra (left to the reader) shows that Eq. (11) is equal to Eq. (12).

Now let θ be the angle between $\mathbf{v}$ and $\mathbf{w}$. By Eq. (10),

$$\|\mathbf{v} \times \mathbf{w}\|^2 = \|\mathbf{v}\|^2 \|\mathbf{w}\|^2 - (\mathbf{v} \cdot \mathbf{w})^2 = \|\mathbf{v}\|^2 \|\mathbf{w}\|^2 - \|\mathbf{v}\|^2 \|\mathbf{w}\|^2 \cos^2 \theta$$

$$= \|\mathbf{v}\|^2 \|\mathbf{w}\|^2 (1 - \cos^2 \theta) = \|\mathbf{v}\|^2 \|\mathbf{w}\|^2 \sin^2 \theta$$

Therefore, $\|\mathbf{v} \times \mathbf{w}\| = \|\mathbf{v}\| \|\mathbf{w}\| \sin \theta$. Note that $\sin \theta \geq 0$ since, by convention, θ lies between 0 and π. This proves (ii).

Part (iii) of Theorem 1 asserts that $\{\mathbf{v}, \mathbf{w}, \mathbf{v} \times \mathbf{w}\}$ is a right-handed system. This is a more subtle property that cannot be verified by algebra alone. We must rely on the following relation between right-handedness and the sign of the determinant, which can be established using the continuity of determinants:

$$\det \begin{pmatrix} \mathbf{u} \\ \mathbf{v} \\ \mathbf{w} \end{pmatrix} > 0 \quad \text{if and only if } \{\mathbf{u}, \mathbf{v}, \mathbf{w}\} \text{ is a right-handed system}$$

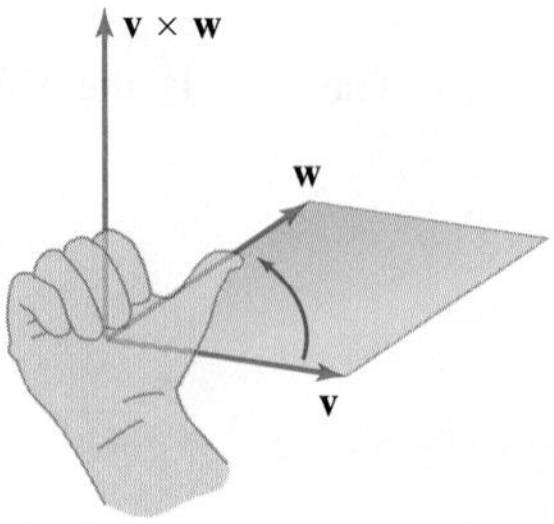

FIGURE 13 Both $\{\mathbf{v} \times \mathbf{w}, \mathbf{v}, \mathbf{w}\}$ and $\{\mathbf{v}, \mathbf{w}, \mathbf{v} \times \mathbf{w}\}$ are right-handed.

Furthermore, it can be checked directly from Eq. (2) that the determinant does not change when we replace $\{\mathbf{u}, \mathbf{v}, \mathbf{w}\}$ by $\{\mathbf{v}, \mathbf{w}, \mathbf{u}\}$ (or $\{\mathbf{w}, \mathbf{u}, \mathbf{v}\}$). Granting this and using Eq. (6), we obtain

$$\det\begin{pmatrix} \mathbf{v} \\ \mathbf{w} \\ \mathbf{v} \times \mathbf{w} \end{pmatrix} = \det\begin{pmatrix} \mathbf{v} \times \mathbf{w} \\ \mathbf{v} \\ \mathbf{w} \end{pmatrix} = (\mathbf{v} \times \mathbf{w}) \cdot (\mathbf{v} \times \mathbf{w}) = \|\mathbf{v} \times \mathbf{w}\|^2 > 0$$

Therefore $\{\mathbf{v}, \mathbf{w}, \mathbf{v} \times \mathbf{w}\}$ is right-handed as claimed (Figure 13).

13.4 SUMMARY

- Determinants of sizes 2×2 and 3×3:

$$\begin{vmatrix} a_{11} & a_{12} \\ a_{21} & a_{22} \end{vmatrix} = a_{11}a_{22} - a_{12}a_{21}$$

$$\begin{vmatrix} a_{11} & a_{12} & a_{13} \\ a_{21} & a_{22} & a_{23} \\ a_{31} & a_{32} & a_{33} \end{vmatrix} = a_{11}\begin{vmatrix} a_{22} & a_{23} \\ a_{32} & a_{33} \end{vmatrix} - a_{12}\begin{vmatrix} a_{21} & a_{23} \\ a_{31} & a_{33} \end{vmatrix} + a_{13}\begin{vmatrix} a_{21} & a_{22} \\ a_{31} & a_{32} \end{vmatrix}$$

- The *cross product* of $\mathbf{v} = \langle a_1, b_1, c_1 \rangle$ and $\mathbf{w} = \langle a_2, b_2, c_2 \rangle$ is the symbolic determinant

$$\mathbf{v} \times \mathbf{w} = \begin{vmatrix} \mathbf{i} & \mathbf{j} & \mathbf{k} \\ a_1 & b_1 & c_1 \\ a_2 & b_2 & c_2 \end{vmatrix} = \begin{vmatrix} b_1 & c_1 \\ b_2 & c_2 \end{vmatrix}\mathbf{i} - \begin{vmatrix} a_1 & c_1 \\ a_2 & c_2 \end{vmatrix}\mathbf{j} + \begin{vmatrix} a_1 & b_1 \\ a_2 & b_2 \end{vmatrix}\mathbf{k}$$

- The cross product $\mathbf{v} \times \mathbf{w}$ is the unique vector with the following three properties:

  (i) $\mathbf{v} \times \mathbf{w}$ is orthogonal to $\mathbf{v}$ and $\mathbf{w}$.
  (ii) $\mathbf{v} \times \mathbf{w}$ has length $\|\mathbf{v}\|\,\|\mathbf{w}\|\sin\theta$ (θ is the angle between $\mathbf{v}$ and $\mathbf{w}$, $0 \le \theta \le \pi$).
  (iii) $\{\mathbf{v}, \mathbf{w}, \mathbf{v} \times \mathbf{w}\}$ is a right-handed system.

- Properties of the cross product:

  (i) $\mathbf{w} \times \mathbf{v} = -\mathbf{v} \times \mathbf{w}$
  (ii) $\mathbf{v} \times \mathbf{w} = \mathbf{0}$ if and only if $\mathbf{w} = \lambda\mathbf{v}$ for some scalar or $\mathbf{v} = \mathbf{0}$
  (iii) $(\lambda\mathbf{v}) \times \mathbf{w} = \mathbf{v} \times (\lambda\mathbf{w}) = \lambda(\mathbf{v} \times \mathbf{w})$
  (iv) $(\mathbf{u} + \mathbf{v}) \times \mathbf{w} = \mathbf{u} \times \mathbf{w} + \mathbf{v} \times \mathbf{w}$

  $\mathbf{v} \times (\mathbf{u} + \mathbf{w}) = \mathbf{v} \times \mathbf{u} + \mathbf{v} \times \mathbf{w}$

- Cross products of standard basis vectors (Figure 14):

$$\mathbf{i} \times \mathbf{j} = \mathbf{k}, \qquad \mathbf{j} \times \mathbf{k} = \mathbf{i}, \qquad \mathbf{k} \times \mathbf{i} = \mathbf{j}$$

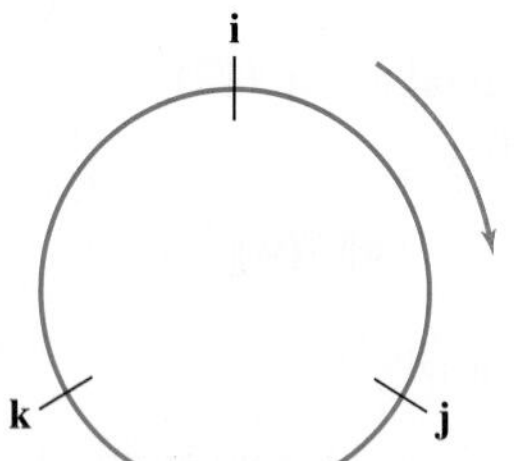

FIGURE 14 Circle for computing the cross products of the basis vectors.

- The parallelogram spanned by $\mathbf{v}$ and $\mathbf{w}$ has area $\|\mathbf{v} \times \mathbf{w}\|$.
- Cross-product identity: $\|\mathbf{v} \times \mathbf{w}\|^2 = \|\mathbf{v}\|^2\|\mathbf{w}\|^2 - (\mathbf{v} \cdot \mathbf{w})^2$.
- The *vector triple product* is defined by $\mathbf{u} \cdot (\mathbf{v} \times \mathbf{w})$. We have

$$\mathbf{u} \cdot (\mathbf{v} \times \mathbf{w}) = \det\begin{pmatrix} \mathbf{u} \\ \mathbf{v} \\ \mathbf{w} \end{pmatrix}$$

- The parallelepiped spanned by $\mathbf{u}$, $\mathbf{v}$, and $\mathbf{w}$ has volume $|\mathbf{u} \cdot (\mathbf{v} \times \mathbf{w})|$.

13.4 EXERCISES

Preliminary Questions

1. What is the (1, 3) minor of the matrix $\begin{vmatrix} 3 & 4 & 2 \\ -5 & -1 & 1 \\ 4 & 0 & 3 \end{vmatrix}$?

2. The angle between two unit vectors **e** and **f** is $\frac{\pi}{6}$. What is the length of $\mathbf{e} \times \mathbf{f}$?

3. What is $\mathbf{u} \times \mathbf{w}$, assuming that $\mathbf{w} \times \mathbf{u} = \langle 2, 2, 1 \rangle$?

4. Find the cross product without using the formula:
(a) $\langle 4, 8, 2 \rangle \times \langle 4, 8, 2 \rangle$ **(b)** $\langle 4, 8, 2 \rangle \times \langle 2, 4, 1 \rangle$

5. What are $\mathbf{i} \times \mathbf{j}$ and $\mathbf{i} \times \mathbf{k}$?

6. When is the cross product $\mathbf{v} \times \mathbf{w}$ equal to zero?

Exercises

In Exercises 1–4, calculate the 2×2 *determinant.*

1. $\begin{vmatrix} 1 & 2 \\ 4 & 3 \end{vmatrix}$

2. $\begin{vmatrix} \frac{2}{3} & \frac{1}{6} \\ -5 & 2 \end{vmatrix}$

3. $\begin{vmatrix} -6 & 9 \\ 1 & 1 \end{vmatrix}$

4. $\begin{vmatrix} 9 & 25 \\ 5 & 14 \end{vmatrix}$

In Exercises 5–8, calculate the 3×3 *determinant.*

5. $\begin{vmatrix} 1 & 2 & 1 \\ 4 & -3 & 0 \\ 1 & 0 & 1 \end{vmatrix}$

6. $\begin{vmatrix} 1 & 0 & 1 \\ -2 & 0 & 3 \\ 1 & 3 & -1 \end{vmatrix}$

7. $\begin{vmatrix} 1 & 2 & 3 \\ 2 & 4 & 6 \\ -3 & -4 & 2 \end{vmatrix}$

8. $\begin{vmatrix} 1 & 0 & 0 \\ 0 & 0 & -1 \\ 0 & 1 & 0 \end{vmatrix}$

In Exercises 9–12, calculate $\mathbf{v} \times \mathbf{w}$.

9. $\mathbf{v} = \langle 1, 2, 1 \rangle$, $\mathbf{w} = \langle 3, 1, 1 \rangle$

10. $\mathbf{v} = \langle 2, 0, 0 \rangle$, $\mathbf{w} = \langle -1, 0, 1 \rangle$

11. $\mathbf{v} = \langle \frac{2}{3}, 1, \frac{1}{2} \rangle$, $\mathbf{w} = \langle 4, -6, 3 \rangle$

12. $\mathbf{v} = \langle 1, 1, 0 \rangle$, $\mathbf{w} = \langle 0, 1, 1 \rangle$

In Exercises 13–16, use the relations in Eq. (5) to calculate the cross product.

13. $(\mathbf{i} + \mathbf{j}) \times \mathbf{k}$

14. $(\mathbf{j} - \mathbf{k}) \times (\mathbf{j} + \mathbf{k})$

15. $(\mathbf{i} - 3\mathbf{j} + 2\mathbf{k}) \times (\mathbf{j} - \mathbf{k})$

16. $(2\mathbf{i} - 3\mathbf{j} + 4\mathbf{k}) \times (\mathbf{i} + \mathbf{j} - 7\mathbf{k})$

In Exercises 17–22, calculate the cross product assuming that

$$\mathbf{u} \times \mathbf{v} = \langle 1, 1, 0 \rangle, \quad \mathbf{u} \times \mathbf{w} = \langle 0, 3, 1 \rangle, \quad \mathbf{v} \times \mathbf{w} = \langle 2, -1, 1 \rangle$$

17. $\mathbf{v} \times \mathbf{u}$

18. $\mathbf{v} \times (\mathbf{u} + \mathbf{v})$

19. $\mathbf{w} \times (\mathbf{u} + \mathbf{v})$

20. $(3\mathbf{u} + 4\mathbf{w}) \times \mathbf{w}$

21. $(\mathbf{u} - 2\mathbf{v}) \times (\mathbf{u} + 2\mathbf{v})$

22. $(\mathbf{v} + \mathbf{w}) \times (3\mathbf{u} + 2\mathbf{v})$

23. Let $\mathbf{v} = \langle a, b, c \rangle$. Calculate $\mathbf{v} \times \mathbf{i}$, $\mathbf{v} \times \mathbf{j}$, and $\mathbf{v} \times \mathbf{k}$.

24. Find $\mathbf{v} \times \mathbf{w}$, where **v** and **w** are vectors of length 3 in the xz-plane, oriented as in Figure 15, and $\theta = \frac{\pi}{6}$.

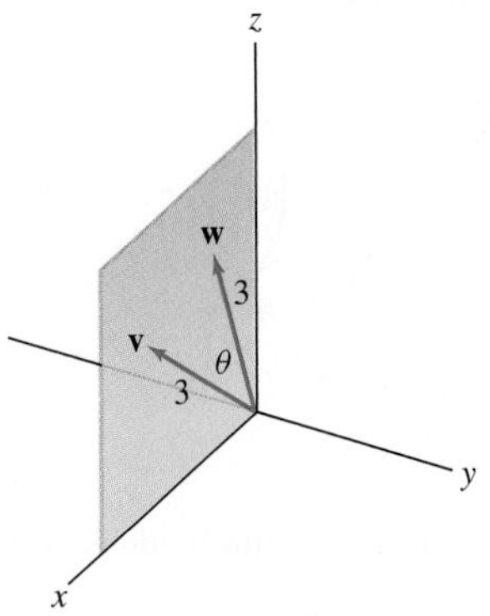

FIGURE 15

In Exercises 25 and 26, refer to Figure 16.

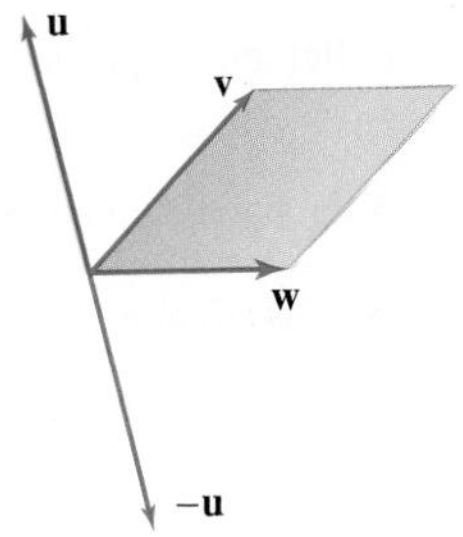

FIGURE 16

25. Which of **u** and $-\mathbf{u}$ is equal to $\mathbf{v} \times \mathbf{w}$?

26. Which of the following form a right-handed system?
(a) $\{\mathbf{v}, \mathbf{w}, \mathbf{u}\}$ **(b)** $\{\mathbf{w}, \mathbf{v}, \mathbf{u}\}$ **(c)** $\{\mathbf{v}, \mathbf{u}, \mathbf{w}\}$
(d) $\{\mathbf{u}, \mathbf{v}, \mathbf{w}\}$ **(e)** $\{\mathbf{w}, \mathbf{v}, -\mathbf{u}\}$ **(f)** $\{\mathbf{v}, -\mathbf{u}, \mathbf{w}\}$

27. Let $\mathbf{v} = \langle 3, 0, 0 \rangle$ and $\mathbf{w} = \langle 0, 1, -1 \rangle$. Determine $\mathbf{u} = \mathbf{v} \times \mathbf{w}$ using the geometric properties of the cross product rather than the formula.

28. What are the possible angles θ between two unit vectors **e** and **f** if $\|\mathbf{e} \times \mathbf{f}\| = \frac{1}{2}$?

29. Show that if **v** and **w** lie in the yz-plane, then $\mathbf{v} \times \mathbf{w}$ is a multiple of **i**.

30. Find the two unit vectors orthogonal to both $\mathbf{a} = \langle 3, 1, 1 \rangle$ and $\mathbf{b} = \langle -1, 2, 1 \rangle$.

31. Let **e** and **e**′ be unit vectors in $\mathbf{R}^3$ such that $\mathbf{e} \perp \mathbf{e}'$. Use the geometric properties of the cross product to compute $\mathbf{e} \times (\mathbf{e}' \times \mathbf{e})$.

32. Calculate the force **F** on an electron (charge $q = -1.6 \times 10^{-19}$ C) moving with velocity 10^5 m/s in the direction **i** in a uniform magnetic field **B**, where $\mathbf{B} = 0.0004\mathbf{i} + 0.0001\mathbf{j}$ teslas (see Example 5).

33. An electron moving with velocity **v** in the plane experiences a force $\mathbf{F} = q(\mathbf{v} \times \mathbf{B})$, where q is the charge on the electron and **B** is a uniform magnetic field pointing directly out of the page. Which of the two vectors $\mathbf{F}_1$ or $\mathbf{F}_2$ in Figure 17 represents the force on the electron? Remember that q is negative.

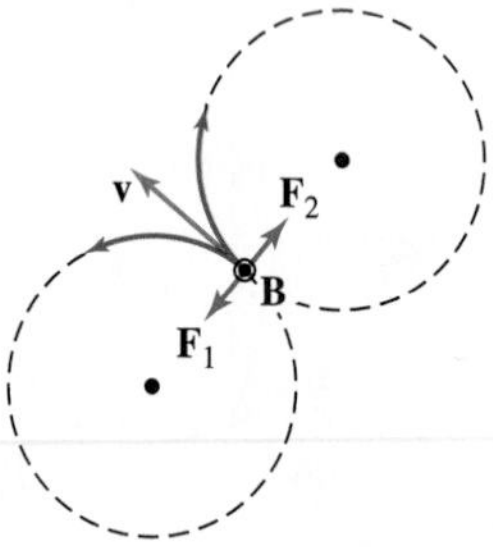

FIGURE 17 The magnetic field vector **B** points directly out of the page.

34. Calculate the scalar triple product $\mathbf{u} \cdot (\mathbf{v} \times \mathbf{w})$, where $\mathbf{u} = \langle 1, 1, 0 \rangle$, $\mathbf{v} = \langle 3, -2, 2 \rangle$, and $\mathbf{w} = \langle 4, -1, 2 \rangle$.

35. Verify identity (10) for vectors $\mathbf{v} = \langle 3, -2, 2 \rangle$ and $\mathbf{w} = \langle 4, -1, 2 \rangle$.

36. Find the volume of the parallelepiped spanned by **u**, **v**, and **w** in Figure 18.

37. Find the area of the parallelogram spanned by **v** and **w** in Figure 18.

38. Calculate the volume of the parallelepiped spanned by

$$\mathbf{u} = \langle 2, 2, 1 \rangle, \qquad \mathbf{v} = \langle 1, 0, 3 \rangle, \qquad \mathbf{w} = \langle 0, -4, 0 \rangle$$

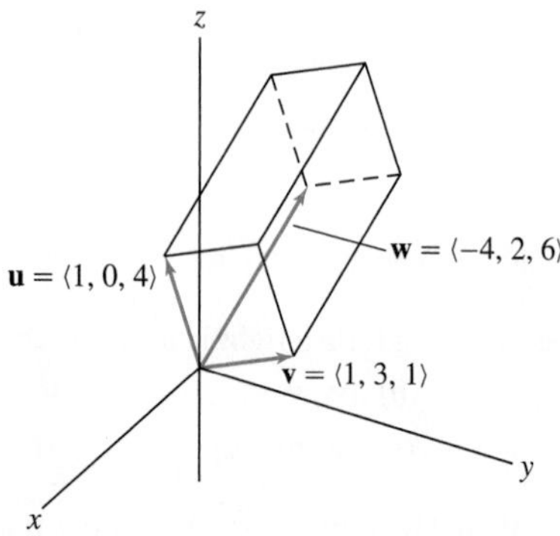

FIGURE 18

39. Sketch and compute the volume of the parallelepiped spanned by

$$\mathbf{u} = \langle 1, 0, 0 \rangle, \qquad \mathbf{v} = \langle 0, 2, 0 \rangle, \qquad \mathbf{w} = \langle 1, 1, 2 \rangle$$

40. Sketch the parallelogram spanned by $\mathbf{u} = \langle 1, 1, 1 \rangle$ and $\mathbf{v} = \langle 0, 0, 4 \rangle$, and compute its area.

41. Calculate the area of the parallelogram spanned by $\mathbf{u} = \langle 1, 0, 3 \rangle$ and $\mathbf{v} = \langle 2, 1, 1 \rangle$.

42. Find the area of the parallelogram determined by the vectors $\langle a, 0, 0 \rangle$ and $\langle 0, b, c \rangle$.

43. Sketch the triangle with vertices at the origin O, $P = (3, 3, 0)$, and $Q = (0, 3, 3)$, and compute its area using cross products.

44. Use the cross product to find the area of the triangle with vertices $P = (1, 1, 5)$, $Q = (3, 4, 3)$, and $R = (1, 5, 7)$ (Figure 19).

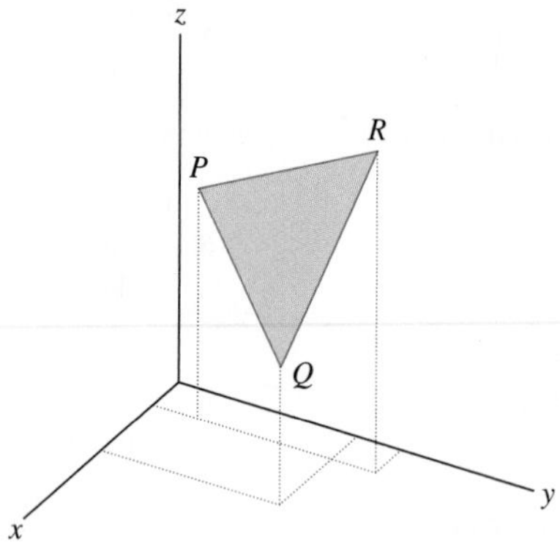

FIGURE 19

In Exercises 45–47, verify the identity using the formula for the cross product.

45. $\mathbf{v} \times \mathbf{w} = -\mathbf{w} \times \mathbf{v}$

46. $(\lambda \mathbf{v}) \times \mathbf{w} = \lambda(\mathbf{v} \times \mathbf{w})$ (λ a scalar)

47. $(\mathbf{u} + \mathbf{v}) \times \mathbf{w} = \mathbf{u} \times \mathbf{w} + \mathbf{v} \times \mathbf{w}$

48. Use the geometric description in Theorem 1 to prove Theorem 2 (iii): $\mathbf{v} \times \mathbf{w} = \mathbf{0}$ if and only if $\mathbf{w} = \lambda \mathbf{v}$ for some scalar λ or $\mathbf{v} = \mathbf{0}$.

49. Verify the relations (5).

50. Show that

$$(\mathbf{i} \times \mathbf{j}) \times \mathbf{j} \neq \mathbf{i} \times (\mathbf{j} \times \mathbf{j})$$

Conclude that the Associative Law does not hold for cross products.

51. The components of the cross product have a geometric interpretation. Show that the absolute value of the **k**-component of $\mathbf{v} \times \mathbf{w}$ is equal to the area of the parallelogram spanned by the projections $\mathbf{v}_0$ and $\mathbf{w}_0$ onto the xy-plane (Figure 20).

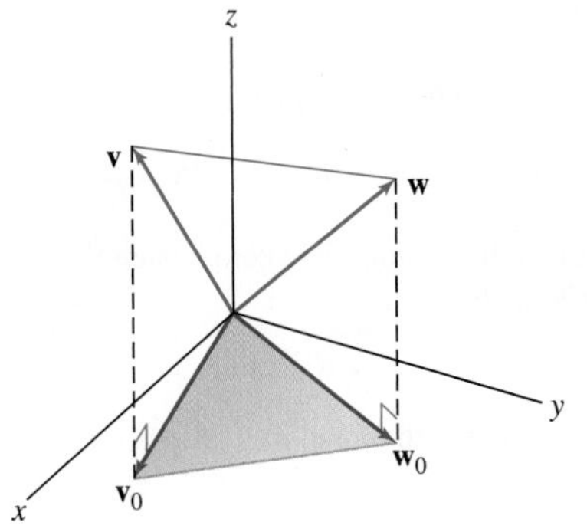

FIGURE 20

52. Formulate and prove analogs of the result in Exercise 51 for the **i**- and **j**-components of $\mathbf{v} \times \mathbf{w}$.

53. Show that three points P, Q, R are collinear (lie on a line) if and only if $\overrightarrow{PQ} \times \overrightarrow{PR} = \mathbf{0}$.

54. Use the result of Exercise 53 to determine whether the points P, Q, and R are collinear, and if not, find a vector normal to the plane containing them.

(a) $P = (2, 1, 0)$, $Q = (1, 5, 2)$, $R = (-1, 13, 6)$

(b) $P = (2, 1, 0)$, $Q = (-3, 21, 10)$, $R = (5, -2, 9)$

(c) $P = (1, 1, 0)$, $Q = (1, -2, -1)$, $R = (3, 2, -4)$

55. Solve the equation $\langle 1, 1, 1 \rangle \times \mathbf{X} = \langle 1, -1, 0 \rangle$, where $\mathbf{X} = \langle x, y, z \rangle$. *Note:* There are infinitely many solutions.

56. Explain geometrically why $\langle 1, 1, 1 \rangle \times \mathbf{X} = \langle 1, 0, 0 \rangle$ has no solution, where $\mathbf{X} = \langle x, y, z \rangle$.

57. Let $\mathbf{X} = \langle x, y, z \rangle$. Show that $\mathbf{i} \times \mathbf{X} = \mathbf{v}$ has a solution if and only if $\mathbf{v}$ is contained in the yz-plane (the **i**-component is zero).

58. Suppose that vectors **u**, **v**, and **w** are mutually orthogonal—that is, $\mathbf{u} \perp \mathbf{v}$, $\mathbf{u} \perp \mathbf{w}$, and $\mathbf{v} \perp \mathbf{w}$. Prove that $(\mathbf{u} \times \mathbf{v}) \times \mathbf{w} = \mathbf{0}$ and $\mathbf{u} \times (\mathbf{v} \times \mathbf{w}) = \mathbf{0}$.

In Exercises 59–62: The ***torque*** *about the origin O due to a force* **F** *acting on an object with position vector* **r** *is the vector quantity* $\tau = \mathbf{r} \times \mathbf{F}$. *If several forces* $\mathbf{F}_j$ *act at positions* $\mathbf{r}_j$, *then the net torque (units: N-m or lb-ft) is the sum*

$$\tau = \sum \mathbf{r}_j \times \mathbf{F}_j$$

Torque measures how much the force causes the object to rotate. By Newton's Laws, τ *is equal to the rate of change of angular momentum.*

59. Calculate the torque τ about O acting at the point P on the mechanical arm in Figure 21(A), assuming that a 25-N force acts as indicated. Ignore the weight of the arm itself.

60. Calculate the net torque about O at P, assuming that a 30-kg mass is attached at P [Figure 21(B)]. The force $\mathbf{F}_g$ due to gravity on a mass m has magnitude $9.8m$ m/s^2 in the downward direction.

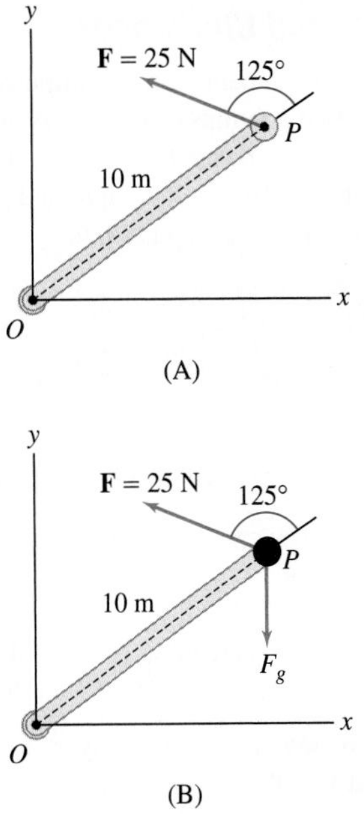

FIGURE 21

61. Let τ be the net torque about O acting on the robotic arm of Figure 22. Assume that the two segments of the arms have mass m_1 and m_2 (in kg) and that a weight of m_3 kg is located at the endpoint P. In calculating the torque, we may assume that the entire mass of each arm segment lies at the midpoint of the arm (its center of mass). Show that the position vectors of the masses m_1, m_2, and m_3 are

$$\mathbf{r}_1 = \frac{1}{2}L_1(\sin\theta_1\mathbf{i} + \cos\theta_1\mathbf{j})$$

$$\mathbf{r}_2 = L_1(\sin\theta_1\mathbf{i} + \cos\theta_1\mathbf{j}) + \frac{1}{2}L_2(\sin\theta_2\mathbf{i} - \cos\theta_2\mathbf{j})$$

$$\mathbf{r}_3 = L_1(\sin\theta_1\mathbf{i} + \cos\theta_1\mathbf{j}) + L_2(\sin\theta_2\mathbf{i} - \cos\theta_2\mathbf{j})$$

Then show that

$$\tau = -\left(L_1\left(\frac{1}{2}m_1 + m_2 + m_3\right)\sin\theta_1 + L_2\left(\frac{1}{2}m_2 + m_3\right)\sin\theta_2\right)\mathbf{k}$$

To simplify the computation, note that all three gravitational forces act in the $-\mathbf{j}$ direction, so the **j**-components of the position vectors $\mathbf{r}_i$ do not contribute to the torque.

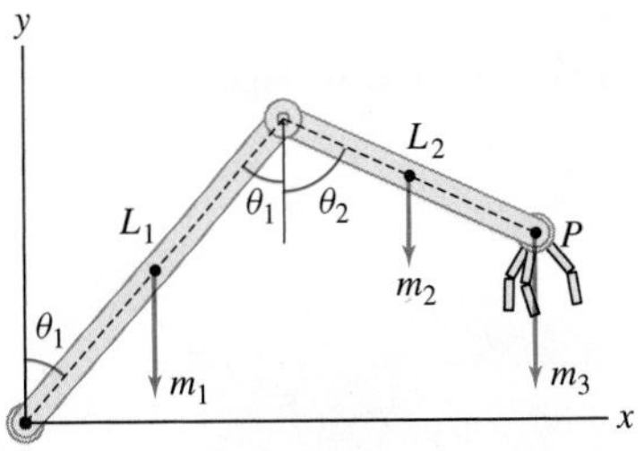

FIGURE 22

62. Continuing with Exercise 61, suppose that $L_1 = 5$ ft, $L_2 = 3$ ft, $m_1 = 30$ lb, $m_2 = 20$ lb, and $m_3 = 50$ lb. If the angles θ_1, θ_2 are equal (say, to θ), what is the maximum allowable value of θ if we assume that the robotic arm can sustain a maximum torque of 400 ft-lb?

Further Insights and Challenges

63. Show that 3×3 determinants can be computed using the **diagonal rule**: Repeat the first two columns of the matrix and form the products of the numbers along the six diagonals indicated. Then add the products for the diagonals that slant from left to right and subtract the products for the diagonals that slant from right to left.

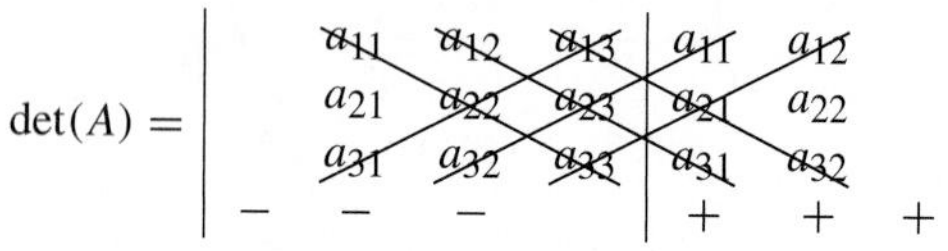

$$= a_{11}a_{22}a_{33} + a_{12}a_{23}a_{31} + a_{13}a_{21}a_{32} - a_{13}a_{22}a_{31} - a_{11}a_{23}a_{32} - a_{12}a_{21}a_{33}$$

64. Use the diagonal rule to calculate $\begin{vmatrix} 2 & 4 & 3 \\ 0 & 1 & -7 \\ -1 & 5 & 3 \end{vmatrix}$.

65. Prove that $\mathbf{v} \times \mathbf{w} = \mathbf{v} \times \mathbf{u}$ if and only if $\mathbf{u} = \mathbf{w} + \lambda\mathbf{v}$ for some scalar λ. Assume that $\mathbf{v} \neq \mathbf{0}$.

66. Use Eq. (10) to prove the Cauchy–Schwarz inequality:

$$|\mathbf{v} \cdot \mathbf{w}| \le \|\mathbf{v}\| \, \|\mathbf{w}\|$$

Show that equality holds if and only if $\mathbf{w}$ is a multiple of $\mathbf{v}$ or at least one of $\mathbf{v}$ and $\mathbf{w}$ is zero.

67. Show that if $\mathbf{u}$, $\mathbf{v}$, and $\mathbf{w}$ are nonzero vectors and $(\mathbf{u} \times \mathbf{v}) \times \mathbf{w} = \mathbf{0}$, then either (i) $\mathbf{u}$ and $\mathbf{v}$ are parallel, or (ii) $\mathbf{w}$ is orthogonal to $\mathbf{u}$ and $\mathbf{v}$.

68. Suppose that $\mathbf{u}$, $\mathbf{v}$, $\mathbf{w}$ are nonzero and

$$(\mathbf{u} \times \mathbf{v}) \times \mathbf{w} = \mathbf{u} \times (\mathbf{v} \times \mathbf{w}) = \mathbf{0}$$

Show that $\mathbf{u}$, $\mathbf{v}$, and $\mathbf{w}$ are either mutually parallel or mutually perpendicular. *Hint:* Use Exercise 67.

69. Let $\mathbf{a}$, $\mathbf{b}$, $\mathbf{c}$ be nonzero vectors, and set

$$\mathbf{v} = \mathbf{a} \times (\mathbf{b} \times \mathbf{c}), \qquad \mathbf{w} = (\mathbf{a} \cdot \mathbf{c})\mathbf{b} - (\mathbf{a} \cdot \mathbf{b})\mathbf{c}$$

(a) Prove that

(i) $\mathbf{v}$ lies in the plane spanned by $\mathbf{b}$ and $\mathbf{c}$.

(ii) $\mathbf{v}$ is orthogonal to $\mathbf{a}$.

(b) Prove that $\mathbf{w}$ also satisfies (i) and (ii). Conclude that $\mathbf{v}$ and $\mathbf{w}$ are parallel.

(c) Show algebraically that $\mathbf{v} = \mathbf{w}$ (Figure 23).

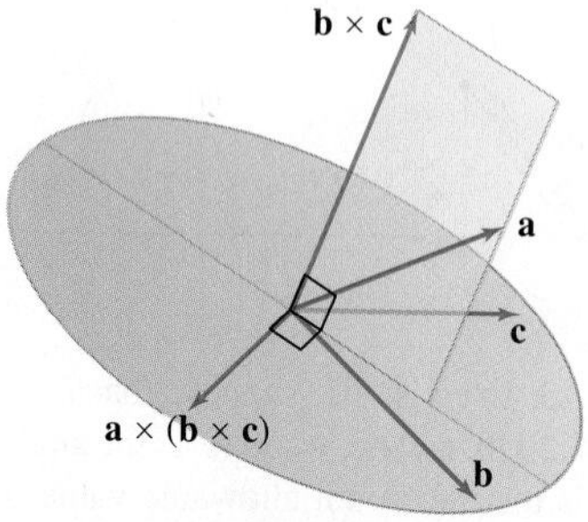

FIGURE 23

70. Use Exercise 69 to prove the identity

$$(\mathbf{a} \times \mathbf{b}) \times \mathbf{c} - \mathbf{a} \times (\mathbf{b} \times \mathbf{c}) = (\mathbf{a} \cdot \mathbf{b})\mathbf{c} - (\mathbf{b} \cdot \mathbf{c})\mathbf{a}$$

71. Show that if $\mathbf{a}$, $\mathbf{b}$ are nonzero vectors such that $\mathbf{a} \perp \mathbf{b}$, then there exists a vector $\mathbf{X}$ such that

$$\mathbf{a} \times \mathbf{X} = \mathbf{b} \qquad \boxed{13}$$

Hint: Show that if $\mathbf{X}$ is orthogonal to $\mathbf{b}$ and is not a multiple of $\mathbf{a}$, then $\mathbf{a} \times \mathbf{X}$ is a multiple of $\mathbf{b}$.

72. Show that if $\mathbf{a}$, $\mathbf{b}$ are nonzero vectors such that $\mathbf{a} \perp \mathbf{b}$, then the set of all solutions of Eq. (13) is a line with $\mathbf{a}$ as direction vector. *Hint:* Let $\mathbf{X}_0$ be any solution (which exists by Exercise 71), and show that every other solution is of the form $\mathbf{X}_0 + \lambda\mathbf{a}$ for some scalar λ.

73. Assume that $\mathbf{v}$ and $\mathbf{w}$ lie in the first quadrant in $\mathbf{R}^2$ as in Figure 24. Use geometry to prove that the area of the parallelogram is equal to $\det\begin{pmatrix} \mathbf{v} \\ \mathbf{w} \end{pmatrix}$.

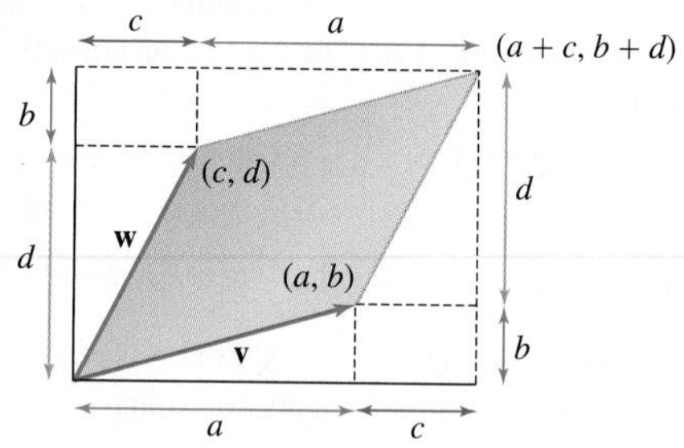

FIGURE 24

74. Consider the tetrahedron spanned by vectors $\mathbf{a}$, $\mathbf{b}$, and $\mathbf{c}$ as in Figure 25(A). Let A, B, C be the faces containing the origin O, and let D be the fourth face opposite O. For each face F, let $\mathbf{v}_F$ be the vector normal to the face, pointing outside the tetrahedron, of magnitude equal to twice the area of F. Prove the relations

$$\mathbf{v}_A + \mathbf{v}_B + \mathbf{v}_C = \mathbf{a} \times \mathbf{b} + \mathbf{b} \times \mathbf{c} + \mathbf{c} \times \mathbf{a}$$

$$\mathbf{v}_A + \mathbf{v}_B + \mathbf{v}_C + \mathbf{v}_D = 0$$

Hint: Show that $\mathbf{v}_D = (\mathbf{c} - \mathbf{b}) \times (\mathbf{b} - \mathbf{a})$.

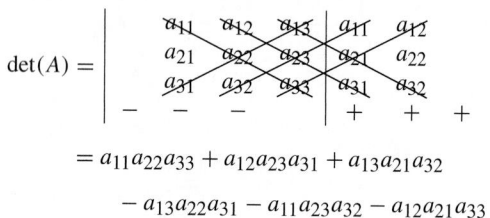

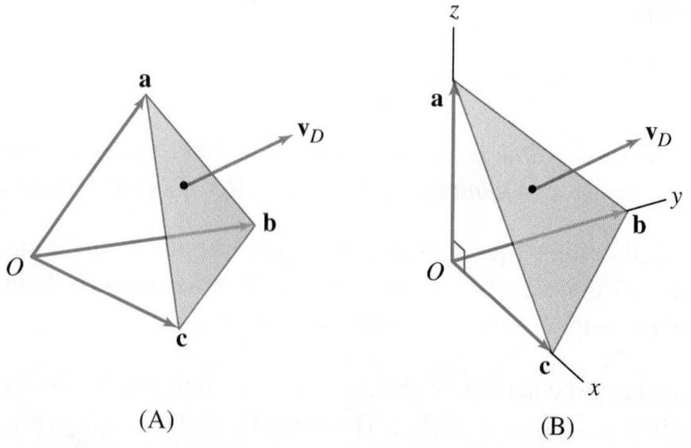

FIGURE 25 The vector $\mathbf{v}_D$ is perpendicular to the face.

75. In the notation of Exercise 74, suppose that **a**, **b**, **c** are mutually perpendicular as in Figure 25(B). Let S_F be the area of face F. Prove the following three-dimensional version of the Pythagorean Theorem:

$$S_A^2 + S_B^2 + S_C^2 = S_D^2$$

13.5 Planes in Three-Space

A linear equation $ax + by = c$ in two variables defines a line in $\mathbf{R}^2$. In this section we show that a linear equation $ax + by + cz = d$ in three variables defines a plane in $\mathbf{R}^3$.

Consider a plane $\mathcal{P}$ that passes through a point $P_0 = (x_0, y_0, z_0)$. We can determine $\mathcal{P}$ completely by specifying a nonzero vector $\mathbf{n} = \langle a, b, c\rangle$ that is orthogonal to $\mathcal{P}$. Such a vector is called a **normal vector**. Basing $\mathbf{n}$ at P_0 as in Figure 1, we see that a point $P = (x, y, z)$ lies on $\mathcal{P}$ precisely when $\overrightarrow{P_0P}$ is orthogonal to $\mathbf{n}$. Therefore, P lies on the plane if

The term "normal" is another word for "orthogonal" or "perpendicular."

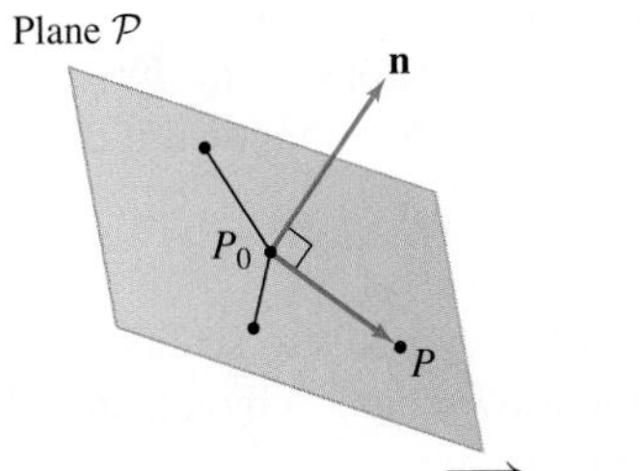

FIGURE 1 A point P lies on $\mathcal{P}$ if $\overrightarrow{P_0P} \perp \mathbf{n}$.

$$\mathbf{n} \cdot \overrightarrow{P_0P} = 0 \qquad \boxed{1}$$

In components, $\overrightarrow{P_0P} = \langle x - x_0, y - y_0, z - z_0\rangle$, so Eq. (1) reads

$$\langle a, b, c\rangle \cdot \langle x - x_0, y - y_0, z - z_0\rangle = 0$$

This gives us the following equation for the plane:

$$a(x - x_0) + b(y - y_0) + c(z - z_0) = 0$$

This can also be written

$$ax + by + cz = ax_0 + by_0 + cz_0 \quad \text{or} \quad \mathbf{n} \cdot \overrightarrow{OP} = \mathbf{n} \cdot \overrightarrow{OP_0} \qquad \boxed{2}$$

When we set $d = ax_0 + by_0 + cz_0 = \mathbf{n} \cdot \overrightarrow{OP_0}$, Eq. (2) becomes $\mathbf{n} \cdot \langle x, y, z\rangle = d$, or

$$ax + by + cz = d$$

THEOREM 1 Equation of a Plane Plane through $P_0 = (x_0, y_0, z_0)$ with normal vector $\mathbf{n} = \langle a, b, c\rangle$:

Vector form: $$\mathbf{n} \cdot \langle x, y, z\rangle = d \qquad \boxed{3}$$

Scalar forms: $$a(x - x_0) + b(y - y_0) + c(z - z_0) = 0 \qquad \boxed{4}$$

$$ax + by + cz = d \qquad \boxed{5}$$

where $d = \mathbf{n} \cdot \langle x_0, y_0, z_0\rangle = ax_0 + by_0 + cz_0$.

To show how this works in a simple case, consider the plane $\mathcal{P}$ through $P_0 = (1, 2, 0)$ with normal vector $\mathbf{n} = \langle 0, 0, 3\rangle$ (Figure 2). Because $\mathbf{n}$ points in the z-direction, $\mathcal{P}$ must be parallel to the xy-plane. On the other hand, P_0 lies on the xy-plane, so $\mathcal{P}$ must be the xy-plane itself. This is precisely what Eq. (3) gives us:

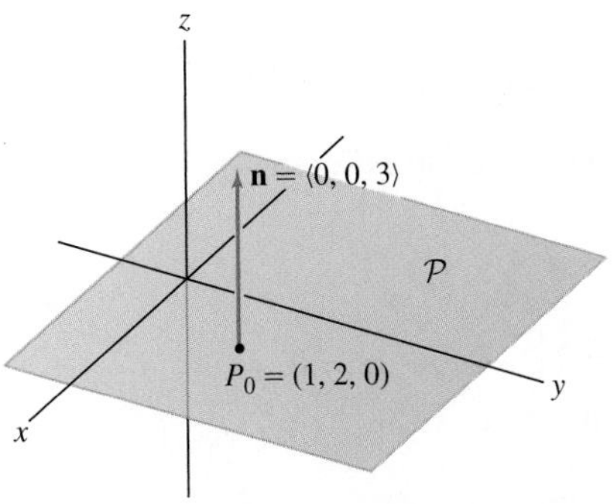

FIGURE 2 The plane with normal vector $\mathbf{n} = \langle 0, 0, 3\rangle$ passing through $P_0 = (1, 2, 0)$ is the xy-plane.

$$\mathbf{n} \cdot \langle x, y, z\rangle = \mathbf{n} \cdot \langle 1, 2, 0\rangle$$

$$\langle 0, 0, 3\rangle \cdot \langle x, y, z\rangle = \langle 0, 0, 3\rangle \cdot \langle 1, 2, 0\rangle$$

$$3z = 0 \quad \text{or} \quad z = 0$$

In other words, $\mathcal{P}$ has equation $z = 0$, so $\mathcal{P}$ is the xy-plane.

■ **EXAMPLE 1** Find an equation of the plane through $P_0 = (3, 1, 0)$ with normal vector $\mathbf{n} = \langle 3, 2, -5 \rangle$.

Solution Using Eq. (4), we obtain

$$3(x - 3) + 2(y - 1) - 5z = 0$$

Alternatively, we can compute

$$d = \mathbf{n} \cdot \overrightarrow{OP_0} = \langle 3, 2, -5 \rangle \cdot \langle 3, 1, 0 \rangle = 11$$

and write the equation as $\langle 3, 2, -5 \rangle \cdot \langle x, y, z \rangle = 11$, or $3x + 2y - 5z = 11$. ■

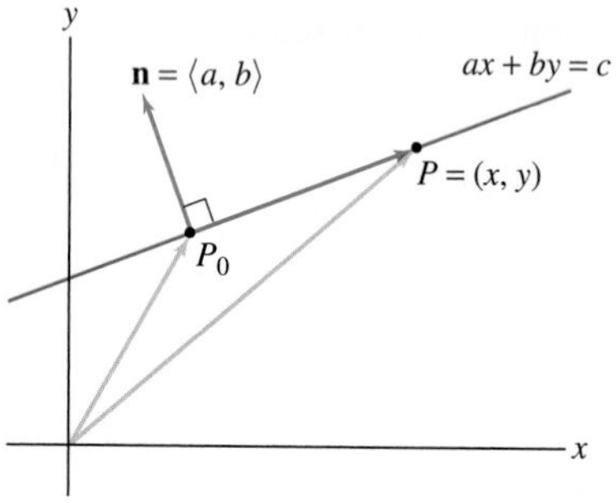

FIGURE 3 A line with normal vector **n**.

CONCEPTUAL INSIGHT Keep in mind that the components of a normal vector are "lurking" inside the equation $ax + by + cz = d$, because $\mathbf{n} = \langle a, b, c \rangle$. The same is true for lines in $\mathbf{R}^2$. The line $ax + by = c$ in Figure 3 has normal vector $\mathbf{n} = \langle a, b \rangle$ because the line has slope $-a/b$ and the vector $\mathbf{n}$ has slope b/a (lines are orthogonal if the product of their slopes is -1).

Note that if $\mathbf{n}$ is normal to a plane $\mathcal{P}$, then so is every nonzero scalar multiple λn. When we use $\lambda\mathbf{n}$ instead of $\mathbf{n}$, the resulting equation for $\mathcal{P}$ changes by a factor of λ. For example, the following two equations define the same plane:

$$x + y + z = 1, \qquad 4x + 4y + 4z = 4$$

The first equation uses the normal $\langle 1, 1, 1 \rangle$, and the second uses the normal $\langle 4, 4, 4 \rangle$.

On the other hand, two planes $\mathcal{P}$ and $\mathcal{P}'$ are parallel if they have a common normal vector. The following planes are parallel because each is normal to $\mathbf{n} = \langle 1, 1, 1 \rangle$:

$$x + y + z = 1, \qquad x + y + z = 2, \qquad 4x + 4y + 4z = 7$$

In general, a family of parallel planes is obtained by choosing a normal vector $\mathbf{n} = \langle a, b, c \rangle$ and varying the constant d in the equation

$$ax + by + cz = d$$

The unique plane in this family through the origin has equation $ax + by + cz = 0$.

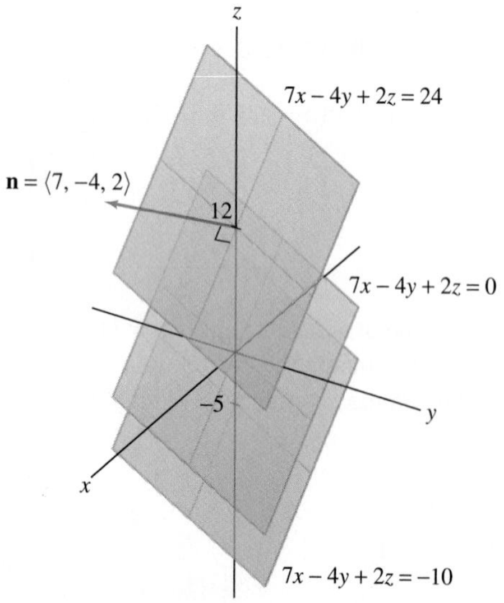

FIGURE 4 Parallel planes with normal vector $\mathbf{n} = \langle 7, -4, 2 \rangle$.

■ **EXAMPLE 2 Parallel Planes** Let $\mathcal{P}$ have equation $7x - 4y + 2z = -10$. Find an equation of the plane parallel to $\mathcal{P}$ passing through

(a) The origin. **(b)** $Q = (2, -1, 3)$.

Solution The planes parallel to $\mathcal{P}$ have an equation of the form (Figure 4)

$$7x - 4y + 2z = d \qquad \boxed{6}$$

(a) For $d = 0$, we get the plane through the origin: $7x - 4y + 2z = 0$.

(b) The point $Q = (2, -1, 3)$ satisfies Eq. (6) with

$$d = 7(2) - 4(-1) + 2(3) = 24$$

Therefore, the plane parallel to $\mathcal{P}$ through Q has equation $7x - 4y + 2z = 24$. ■

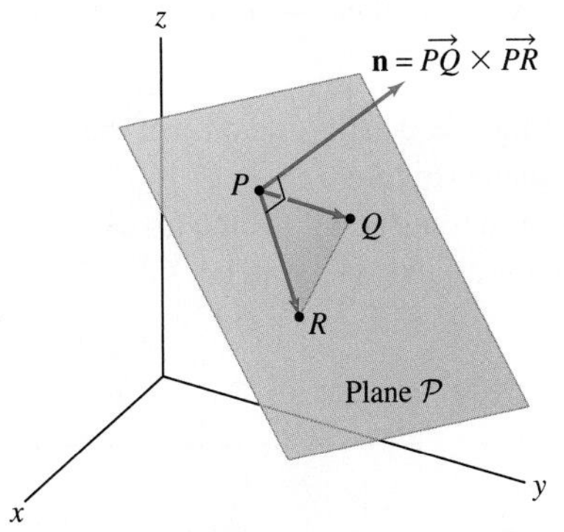

FIGURE 5 Three points P, Q, and R determine a plane (assuming they do not lie in a straight line).

Points that lie on a line are called **collinear**. If we are given three points P, Q, and R that are not collinear, then there is just one plane passing through P, Q, and R (Figure 5). The next example shows how to find an equation of this plane.

■ **EXAMPLE 3 The Plane Determined by Three Points** Find an equation of the plane $\mathcal{P}$ determined by the points

$$P = (1, 0, -1), \qquad Q = (2, 2, 1), \qquad R = (4, 1, 2)$$

Solution

***Step 1.* Find a normal vector.**

The vectors $\overrightarrow{PQ}$ and $\overrightarrow{PR}$ lie in the plane $\mathcal{P}$, so their cross product is normal to $\mathcal{P}$:

$$\overrightarrow{PQ} = \langle 2, 2, 1\rangle - \langle 1, 0, -1\rangle = \langle 1, 2, 2\rangle$$

$$\overrightarrow{PR} = \langle 4, 1, 2\rangle - \langle 1, 0, -1\rangle = \langle 3, 1, 3\rangle$$

$$\mathbf{n} = \overrightarrow{PQ} \times \overrightarrow{PR} = \begin{vmatrix} \mathbf{i} & \mathbf{j} & \mathbf{k} \\ 1 & 2 & 2 \\ 3 & 1 & 3 \end{vmatrix} = 4\mathbf{i} + 3\mathbf{j} - 5\mathbf{k} = \langle 4, 3, -5\rangle$$

In Example 3, we could just as well have used the vectors $\overrightarrow{QP}$ and $\overrightarrow{QR}$ (or $\overrightarrow{RP}$ and $\overrightarrow{RQ}$) to find a normal vector $\mathbf{n}$.

By Eq. (5), $\mathcal{P}$ has equation $4x + 3y - 5z = d$ for some d.

***Step 2.* Choose a point on the plane and compute d.**

Now choose any one of the three points—say, $P = (1, 0, -1)$—and compute

$$d = \mathbf{n} \cdot \overrightarrow{OP} = \langle 4, 3, -5\rangle \cdot \langle 1, 0, -1\rangle = 9$$

We conclude that $\mathcal{P}$ has equation $4x + 3y - 5z = 9$. ■

CAUTION When you find a normal vector to the plane containing points P, Q, R, be sure to compute a cross product such as $\overrightarrow{PQ} \times \overrightarrow{PR}$. A common mistake is to use a cross product such as $\overrightarrow{OP} \times \overrightarrow{OQ}$ or $\overrightarrow{OP} \times \overrightarrow{OR}$, which need not be normal to the plane.

■ **EXAMPLE 4 Intersection of a Plane and a Line** Find the point P where the plane $3x - 9y + 2z = 7$ and the line $\mathbf{r}(t) = \langle 1, 2, 1\rangle + t\langle -2, 0, 1\rangle$ intersect.

Solution The line has parametric equations

$$x = 1 - 2t, \qquad y = 2, \qquad z = 1 + t$$

Substitute in the equation of the plane and solve for t:

$$3x - 9y + 2z = 3(1 - 2t) - 9(2) + 2(1 + t) = 7$$

Simplification yields $-4t - 13 = 7$ or $t = -5$. Therefore, P has coordinates

$$x = 1 - 2(-5) = 11, \qquad y = 2, \qquad z = 1 + (-5) = -4$$

The plane and line intersect at the point $P = (11, 2, -4)$. ■

The intersection of a plane $\mathcal{P}$ with a coordinate plane or a plane parallel to a coordinate plane is called a **trace**. The trace is a line unless $\mathcal{P}$ is parallel to the coordinate plane (in which case the trace is empty or is $\mathcal{P}$ itself).

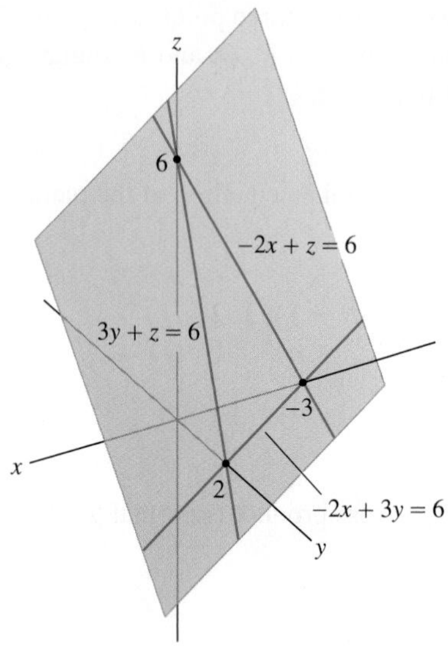

FIGURE 6 The three blue lines are the traces of the plane $-2x + 3y + z = 6$ in the coordinate planes.

EXAMPLE 5 Traces of the Plane Find the traces of the plane $-2x + 3y + z = 6$ in the coordinate planes.

Solution We obtain the trace in the xy-plane by setting $z = 0$ in the equation of the plane. Thus, the trace is the line $-2x + 3y = 6$ in the xy-plane (Figure 6).

Similarly, the trace in the xz-plane is obtained by setting $y = 0$, which gives the line $-2x + z = 6$ in the xz-plane. Finally, the trace in the yz-plane is $3y + z = 6$. ■

13.5 SUMMARY

- Equation of plane through $P_0 = (x_0, y_0, z_0)$ with normal vector $\mathbf{n} = \langle a, b, c\rangle$:

Vector form: $\mathbf{n} \cdot \langle x, y, z\rangle = d$

Scalar forms: $a(x - x_0) + b(y - y_0) + c(z - z_0) = 0$

$ax + by + cz = d$

where $d = \mathbf{n} \cdot \langle x_0, y_0, z_0\rangle = ax_0 + by_0 + cz_0$.

- The family of parallel planes with given normal vector $\mathbf{n} = \langle a, b, c\rangle$ consists of all planes with equation $ax + by + cz = d$ for some d.
- The plane through three points P, Q, R that are not collinear:
  - $\mathbf{n} = \overrightarrow{PQ} \times \overrightarrow{PR}$
  - $d = \mathbf{n} \cdot \langle x_0, y_0, z_0\rangle$, where $P = (x_0, y_0, z_0)$
- The intersection of a plane $\mathcal{P}$ with a coordinate plane or a plane parallel to a coordinate plane is called a *trace*. The trace in the yz-plane is obtained by setting $x = 0$ in the equation of the plane (and similarly for the traces in the xz- and xy-planes).

13.5 EXERCISES

Preliminary Questions

1. What is the equation of the plane parallel to $3x + 4y - z = 5$ passing through the origin?

2. The vector **k** is normal to which of the following planes?

(a) $x = 1$ **(b)** $y = 1$ **(c)** $z = 1$

3. Which of the following planes is not parallel to the plane $x + y + z = 1$?

(a) $2x + 2y + 2z = 1$ **(b)** $x + y + z = 3$

(c) $x - y + z = 0$

4. To which coordinate plane is the plane $y = 1$ parallel?

5. Which of the following planes contains the z-axis?

(a) $z = 1$ **(b)** $x + y = 1$ **(c)** $x + y = 0$

6. Suppose that a plane $\mathcal{P}$ with normal vector **n** and a line $\mathcal{L}$ with direction vector **v** both pass through the origin and that $\mathbf{n} \cdot \mathbf{v} = 0$. Which of the following statements is correct?

(a) $\mathcal{L}$ is contained in $\mathcal{P}$.

(b) $\mathcal{L}$ is orthogonal to $\mathcal{P}$.

Exercises

In Exercises 1–8, write the equation of the plane with normal vector **n** *passing through the given point in each of the three forms (one vector form and two scalar forms).*

1. $\mathbf{n} = \langle 1, 3, 2\rangle$, $(4, -1, 1)$

2. $\mathbf{n} = \langle -1, 2, 1\rangle$, $(3, 1, 9)$

3. $\mathbf{n} = \langle -1, 2, 1\rangle$, $(4, 1, 5)$

4. $\mathbf{n} = \langle 2, -4, 1\rangle$, $\left(\frac{1}{3}, \frac{2}{3}, 1\right)$

5. $\mathbf{n} = \mathbf{i}$, $(3, 1, -9)$

6. $\mathbf{n} = \mathbf{j}$, $\left(-5, \frac{1}{2}, \frac{1}{2}\right)$

7. $\mathbf{n} = \mathbf{k}$, $(6, 7, 2)$

8. $\mathbf{n} = \mathbf{i} - \mathbf{k}$, $(4, 2, -8)$

9. Write down the equation of any plane through the origin.

10. Write down the equations of any two distinct planes with normal vector $\mathbf{n} = \langle 3, 2, 1 \rangle$ that do not pass through the origin.

11. Which of the following statements are true of a plane that is parallel to the yz-plane?

(a) $\mathbf{n} = \langle 0, 0, 1 \rangle$ is a normal vector.

(b) $\mathbf{n} = \langle 1, 0, 0 \rangle$ is a normal vector.

(c) The equation has the form $ay + bz = d$

(d) The equation has the form $x = d$

12. Find a normal vector $\mathbf{n}$ and an equation for the planes in Figures 7(A)–(C).

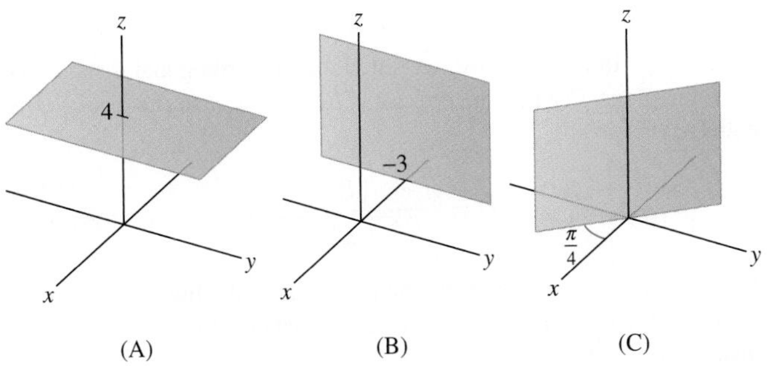

FIGURE 7

In Exercises 13–16, find a vector normal to the plane with the given equation.

13. $9x - 4y - 11z = 2$

14. $x - z = 0$

15. $3(x - 4) - 8(y - 1) + 11z = 0$

16. $x = 1$

In Exercises 17–20, find an equation of the plane passing through the three points given.

17. $P = (2, -1, 4)$, $Q = (1, 1, 1)$, $R = (3, 1, -2)$

18. $P = (5, 1, 1)$, $Q = (1, 1, 2)$, $R = (2, 1, 1)$

19. $P = (1, 0, 0)$, $Q = (0, 1, 1)$, $R = (2, 0, 1)$

20. $P = (2, 0, 0)$, $Q = (0, 4, 0)$, $R = (0, 0, 2)$

In Exercises 21–28, find the equation of the plane with the given description.

21. Passes through O and is parallel to $4x - 9y + z = 3$

22. Passes through $(4, 1, 9)$ and is parallel to $x + y + z = 3$

23. Passes through $(4, 1, 9)$ and is parallel to $x = 3$

24. Passes through $P = (3, 5, -9)$ and is parallel to the xz-plane

25. Passes through $(-2, -3, 5)$ and has normal vector $\mathbf{i} + \mathbf{k}$

26. Contains the lines $\mathbf{r}_1(t) = \langle t, 2t, 3t \rangle$ and $\mathbf{r}_2(t) = \langle 3t, t, 8t \rangle$

27. Contains the lines $\mathbf{r}_1(t) = \langle 2, 1, 0 \rangle + \langle t, 2t, 3t \rangle$ and $\mathbf{r}_2(t) = \langle 2, 1, 0 \rangle + \langle 3t, t, 8t \rangle$

28. Contains $P = (-1, 0, 1)$ and $\mathbf{r}(t) = \langle t + 1, 2t, 3t - 1 \rangle$

29. Are the planes $\frac{1}{2}x + 2x - y = 5$ and $3x + 12x - 6y = 1$ parallel?

30. Let a, b, c be constants. Which two of the following equations define the plane passing through $(a, 0, 0)$, $(0, b, 0)$, $(0, 0, c)$?

(a) $ax + by + cz = 1$

(b) $bcx + acy + abz = abc$

(c) $bx + cy + az = 1$

(d) $\dfrac{x}{a} + \dfrac{y}{b} + \dfrac{z}{c} = 1$

31. Find an equation of the plane $\mathcal{P}$ in Figure 8.

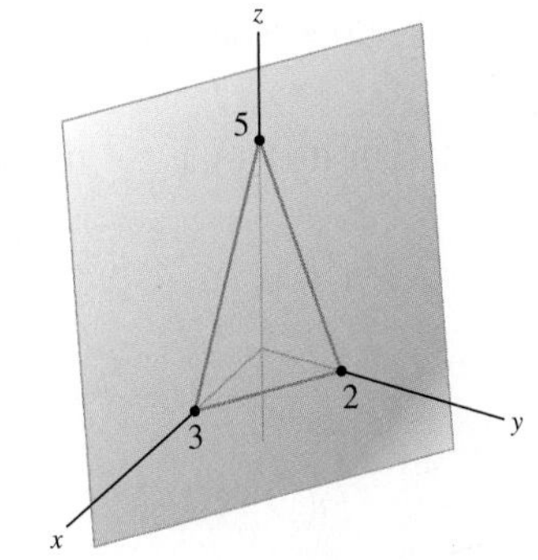

FIGURE 8

32. Verify that the plane $x - y + 5z = 10$ and the line $\mathbf{r}(t) = \langle 1, 0, 1 \rangle + t \langle -2, 1, 1 \rangle$ intersect at $P = (-3, 2, 3)$.

In Exercises 33–36, find the intersection of the line and the plane.

33. $x + y + z = 14$, $\mathbf{r}(t) = \langle 1, 1, 0 \rangle + t \langle 0, 2, 4 \rangle$

34. $2x + y = 3$, $\mathbf{r}(t) = \langle 2, -1, -1 \rangle + t \langle 1, 2, -4 \rangle$

35. $z = 12$, $\mathbf{r}(t) = t \langle -6, 9, 36 \rangle$

36. $x - z = 6$, $\mathbf{r}(t) = \langle 1, 0, -1 \rangle + t \langle 4, 9, 2 \rangle$

In Exercises 37–42, find the trace of the plane in the given coordinate plane.

37. $3x - 9y + 4z = 5$, yz

38. $3x - 9y + 4z = 5$, xz

39. $3x + 4z = -2$, xy

40. $3x + 4z = -2$, xz

41. $-x + y = 4$, xz

42. $-x + y = 4$, yz

43. Does the plane $x = 5$ have a trace in the yz-plane? Explain.

44. Give equations for two distinct planes whose trace in the xy-plane has equation $4x + 3y = 8$.

45. Give equations for two distinct planes whose trace in the yz-plane has equation $y = 4z$.

46. Find parametric equations for the line through $P_0 = (3, -1, 1)$ perpendicular to the plane $3x + 5y - 7z = 29$.

47. Find all planes in $\mathbf{R}^3$ whose intersection with the xz-plane is the line with equation $3x + 2z = 5$.

48. Find all planes in $\mathbf{R}^3$ whose intersection with the xy-plane is the line $\mathbf{r}(t) = t\,\langle 2, 1, 0\rangle$.

In Exercises 49–54, compute the angle between the two planes, defined as the angle θ (between 0 and π) between their normal vectors (Figure 9).

49. Planes with normals $\mathbf{n}_1 = \langle 1, 0, 1\rangle$, $\mathbf{n}_2 = \langle -1, 1, 1\rangle$

50. Planes with normals $\mathbf{n}_1 = \langle 1, 2, 1\rangle$, $\mathbf{n}_2 = \langle 4, 1, 3\rangle$

51. $2x + 3y + 7z = 2$ and $4x - 2y + 2z = 4$

52. $x - 3y + z = 3$ and $2x - 3z = 4$

53. $3(x - 1) - 5y + 2(z - 12) = 0$ and the plane with normal $\mathbf{n} = \langle 1, 0, 1\rangle$

54. The plane through $(1, 0, 0)$, $(0, 1, 0)$, and $(0, 0, 1)$ and the yz-plane

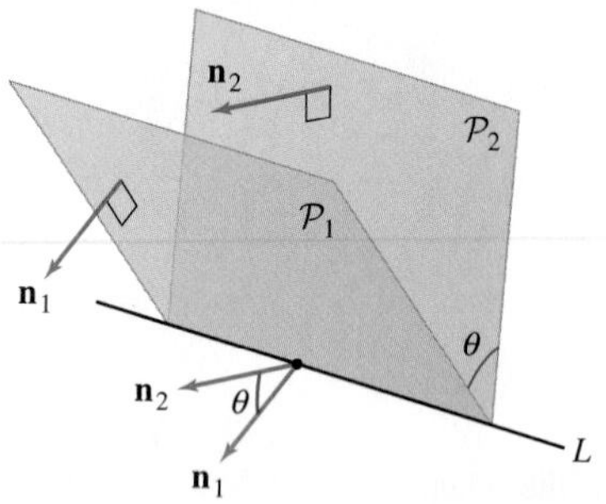

FIGURE 9 By definition, the angle between two planes is the angle between their normal vectors.

55. Find an equation of a plane making an angle of $\frac{\pi}{2}$ with the plane $3x + y - 4z = 2$.

56. Let $\mathcal{P}_1$ and $\mathcal{P}_2$ be planes with normal vectors $\mathbf{n}_1$ and $\mathbf{n}_2$. Assume that the planes are not parallel, and let $\mathcal{L}$ be their intersection (a line). Show that $\mathbf{n}_1 \times \mathbf{n}_2$ is a direction vector for $\mathcal{L}$.

57. Find a plane that is perpendicular to the two planes $x + y = 3$ and $x + 2y - z = 4$.

58. Let $\mathcal{L}$ be the intersection of the planes $x + y + z = 1$ and $x + 2y + 3z = 1$. Use Exercise 56 to find a direction vector for $\mathcal{L}$. Then find a point P on $\mathcal{L}$ by *inspection*, and write down the parametric equations for $\mathcal{L}$.

59. Let $\mathcal{L}$ denote the intersection of the planes $x - y - z = 1$ and $2x + 3y + z = 2$. Find parametric equations for the line $\mathcal{L}$. *Hint:* To find a point on $\mathcal{L}$, substitute an arbitrary value for z (say, $z = 2$) and then solve the resulting pair of equations for x and y.

60. Find parametric equations for the intersection of the planes $2x + y - 3z = 0$ and $x + y = 1$.

61. Two vectors $\mathbf{v}$ and $\mathbf{w}$, each of length 12, lie in the plane $x + 2y - 2z = 0$. The angle between $\mathbf{v}$ and $\mathbf{w}$ is $\pi/6$. This information determines $\mathbf{v} \times \mathbf{w}$ up to a sign ± 1. What are the two possible values of $\mathbf{v} \times \mathbf{w}$?

62. The plane

$$\frac{x}{2} + \frac{y}{4} + \frac{z}{3} = 1$$

intersects the x-, y-, and z-axes in points P, Q, and R. Find the area of the triangle $\triangle PQR$.

63. In this exercise, we show that the orthogonal distance D from the plane $\mathcal{P}$ with equation $ax + by + cz = d$ to the origin O is equal to (Figure 10)

$$D = \frac{|d|}{\sqrt{a^2 + b^2 + c^2}}$$

Let $\mathbf{n} = \langle a, b, c\rangle$, and let P be the point where the line through $\mathbf{n}$ intersects $\mathcal{P}$. By definition, the orthogonal distance from $\mathcal{P}$ to O is the distance from P to O.

(a) Show that P is the terminal point of $\mathbf{v} = \left(\frac{d}{\mathbf{n} \cdot \mathbf{n}}\right)\mathbf{n}$.

(b) Show that the distance from P to O is D.

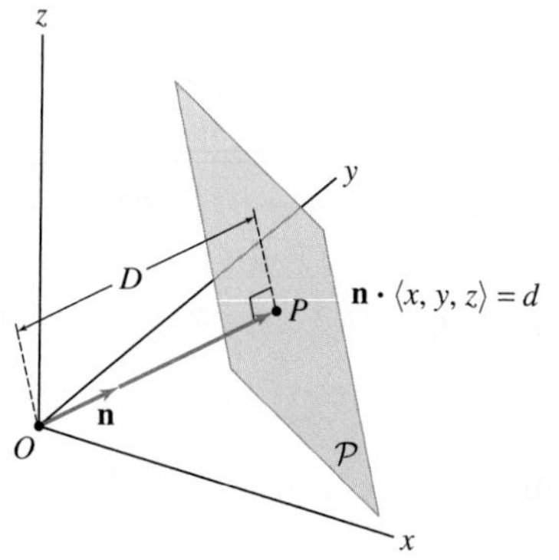

FIGURE 10

64. Use Exercise 63 to compute the orthogonal distance from the plane $x + 2y + 3z = 5$ to the origin.

Further Insights and Challenges

In Exercises 65 and 66, let $\mathcal{P}$ be a plane with equation

$$ax + by + cz = d$$

and normal vector $\mathbf{n} = \langle a, b, c\rangle$. For any point Q, there is a unique point P on $\mathcal{P}$ that is closest to Q, and is such that $\overline{PQ}$ is orthogonal to $\mathcal{P}$ (Figure 11).

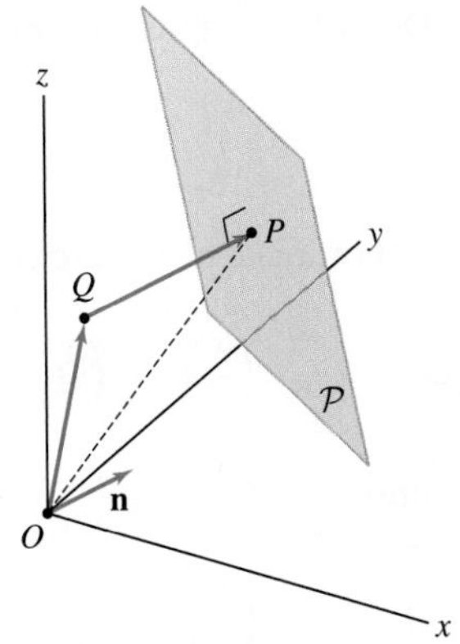

FIGURE 11

65. Show that the point P on $\mathcal{P}$ closest to Q is determined by the equation

$$\overrightarrow{OP} = \overrightarrow{OQ} + \left(\frac{d - \overrightarrow{OQ} \cdot \mathbf{n}}{\mathbf{n} \cdot \mathbf{n}}\right)\mathbf{n} \qquad \boxed{7}$$

66. By definition, the distance from a point $Q = (x_1, y_1, z_1)$ to the plane $\mathcal{P}$ is $\|QP\|$ where P is the point on $\mathcal{P}$ that is closest to Q. Prove:

$$\text{Distance from } Q \text{ to } \mathcal{P} = \frac{|ax_1 + by_1 + cz_1 - d|}{\|\mathbf{n}\|} \qquad \boxed{8}$$

67. Use Eq. (7) to find the point P nearest to $Q = (2, 1, 2)$ on the plane $x + y + z = 1$.

68. Find the point P nearest to $Q = (-1, 3, -1)$ on the plane

$$x - 4z = 2$$

69. Use Eq. (8) to find the distance from $Q = (1, 1, 1)$ to the plane $2x + y + 5z = 2$.

70. Find the distance from $Q = (1, 2, 2)$ to the plane $\mathbf{n} \cdot \langle x, y, z \rangle = 3$, where $\mathbf{n} = \langle \frac{3}{5}, \frac{4}{5}, 0 \rangle$.

71. What is the distance from $Q = (a, b, c)$ to the plane $x = 0$? Visualize your answer geometrically and explain without computation. Then verify that Eq. (8) yields the same answer.

72. The equation of a plane $\mathbf{n} \cdot \langle x, y, z \rangle = d$ is said to be in **normal form** if $\mathbf{n}$ is a unit vector. Show that in this case, $|d|$ is the distance from the plane to the origin. Write the equation of the plane $4x - 2y + 4z = 24$ in normal form.

13.6 A Survey of Quadric Surfaces

Quadric surfaces are the surface analogs of conic sections. Recall that a conic section is a curve in $\mathbf{R}^2$ defined by a quadratic equation in two variables. A quadric surface is defined by a quadratic equation in *three* variables:

To ensure that Eq. (1) is genuinely quadratic, we assume that the degree-2 coefficients A, B, C, D, E, F are not all zero.

$$Ax^2 + By^2 + Cz^2 + Dxy + Eyz + Fzx + ax + by + cz + d = 0 \qquad \boxed{1}$$

Like conic sections, quadric surfaces are classified into a small number of types. When the coordinate axes are chosen to coincide with the axes of the quadric, the equation of the quadric has a simple form. The quadric is then said to be in **standard position.** In standard position, the coefficients D, E, F are all zero and the linear part $(ax + by + cz + d)$ also reduces to just one term. In this short survey of quadric surfaces, we restrict our attention to quadrics in standard position.

The surface analogs of ellipses are the egg-shaped **ellipsoids** (Figure 1). In standard form, an ellipsoid has the equation

Ellipsoid $\quad \left(\frac{x}{a}\right)^2 + \left(\frac{y}{b}\right)^2 + \left(\frac{z}{c}\right)^2 = 1$

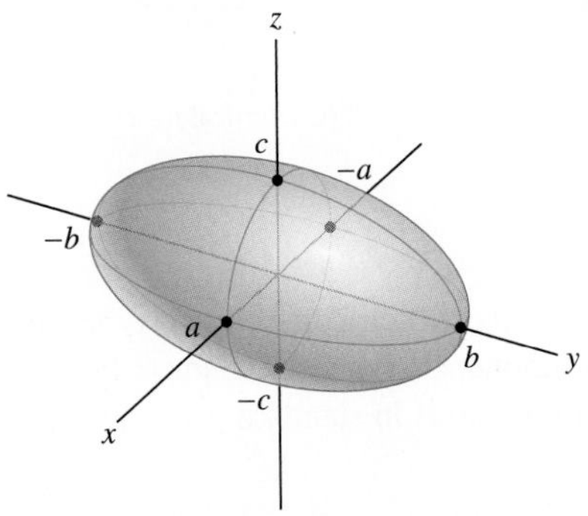

FIGURE 1 Ellipsoid with equation $\left(\frac{x}{a}\right)^2 + \left(\frac{y}{b}\right)^2 + \left(\frac{z}{c}\right)^2 = 1$.

For $a = b = c$, this equation is equivalent to $x^2 + y^2 + z^2 = a^2$ and the ellipsoid is a sphere of radius a.

Surfaces are often represented graphically by a mesh of curves called **traces**, obtained by intersecting the surface with planes parallel to one of the coordinate planes (Figure 2). Algebraically, this corresponds to **freezing** one of the three variables (holding it constant). For example, the intersection of the horizontal plane $z = z_0$ with the surface is a horizontal trace curve.

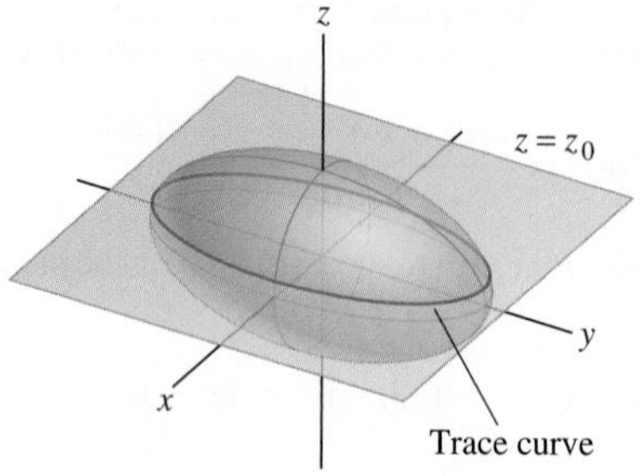

FIGURE 2 The intersection of the plane $z = z_0$ with an ellipsoid is an ellipse.

EXAMPLE 1 The Traces of an Ellipsoid Describe the traces of the ellipsoid

$$\left(\frac{x}{5}\right)^2 + \left(\frac{y}{7}\right)^2 + \left(\frac{z}{9}\right)^2 = 1$$

Solution First we observe that the traces in the coordinate planes are ellipses (Figure 3A):

xy-trace (set $z = 0$, blue in figure): $\left(\frac{x}{5}\right)^2 + \left(\frac{y}{7}\right)^2 = 1$

yz-trace (set $x = 0$, green in figure): $\left(\frac{y}{7}\right)^2 + \left(\frac{z}{9}\right)^2 = 1$

xz-trace (set $y = 0$, red in figure): $\left(\frac{x}{5}\right)^2 + \left(\frac{z}{9}\right)^2 = 1$

In fact, all the traces of an ellipsoid are ellipses. For example, the horizontal trace defined by setting $z = z_0$ is the ellipse [Figure 3(B)]

$$\text{Trace at height } z_0\text{:} \quad \left(\frac{x}{5}\right)^2 + \left(\frac{y}{7}\right)^2 + \left(\frac{z_0}{9}\right)^2 = 1 \quad \text{or} \quad \frac{x^2}{25} + \frac{y^2}{49} = \underbrace{1 - \frac{z_0^2}{81}}_{\text{A constant}}$$

The trace at height $z_0 = 9$ is the single point $(0, 0, 9)$ because $x^2/25 + y^2/49 = 0$ has only one solution: $x = 0$, $y = 0$. Similarly, for $z_0 = -9$ the trace is the point $(0, 0, -9)$. If $|z_0| > 9$, then $1 - z_0^2/81 < 0$ and the plane lies above or below the ellipsoid. The trace has no points in this case. The traces in the vertical planes $x = x_0$ and $y = y_0$ have a similar description [Figure 3(C)]. ■

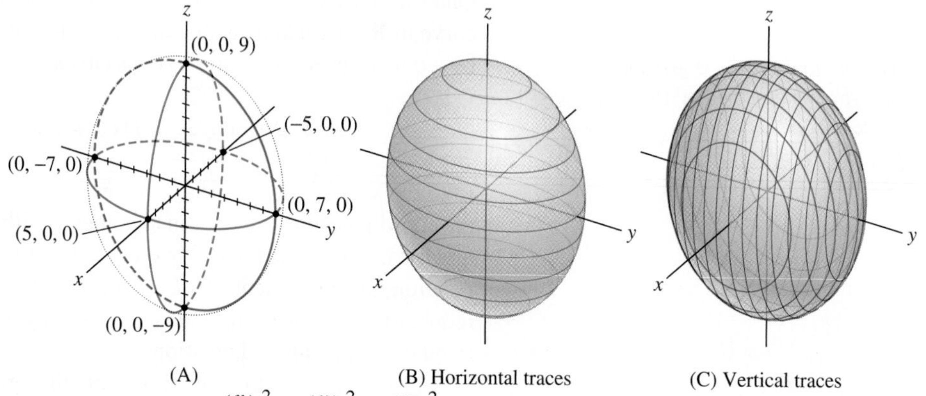

FIGURE 3 The ellipsoid $\left(\frac{x}{5}\right)^2 + \left(\frac{y}{7}\right)^2 + \left(\frac{z}{9}\right)^2 = 1$.

The analogs of the hyperbolas are the **hyperboloids**, which come in two types, depending on whether the surface has one or two components. We refer to these types as hyperboloids of one or two sheets (Figure 4). Their equations in standard position are

Hyperboloids

One Sheet: $\left(\frac{x}{a}\right)^2 + \left(\frac{y}{b}\right)^2 = \left(\frac{z}{c}\right)^2 + 1$

Two Sheets: $\left(\frac{x}{a}\right)^2 + \left(\frac{y}{b}\right)^2 = \left(\frac{z}{c}\right)^2 - 1$ (2)

Notice that a hyperboloid of two sheets does not contain any points whose z-coordinate satisfies $-c < z < c$ because the right-hand side $\left(\frac{z}{c}\right)^2 - 1$ is then negative, but the left-hand side of the equation is greater than or equal to zero.

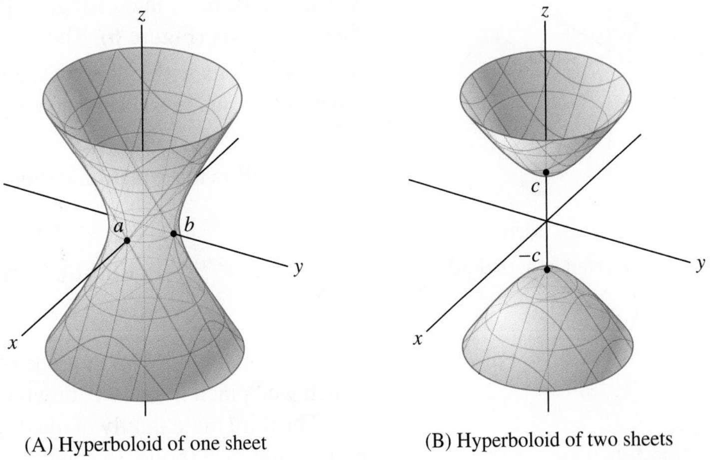

FIGURE 4 Hyperboloids of one and two sheets.

■ **EXAMPLE 2 The Traces of a Hyperboloid of One Sheet** Determine the traces of the hyperboloid $\left(\frac{x}{2}\right)^2 + \left(\frac{y}{3}\right)^2 = \left(\frac{z}{4}\right)^2 + 1$.

Solution The horizontal traces are ellipses and the vertical traces (parallel to both the yz-plane and the xz-plane) are hyperbolas (Figure 5):

Trace $z = z_0$ (ellipse, blue in figure): $\left(\frac{x}{2}\right)^2 + \left(\frac{y}{3}\right)^2 = \left(\frac{z_0}{4}\right)^2 + 1$

Trace $x = x_0$ (hyperbola, green in figure): $\left(\frac{y}{3}\right)^2 - \left(\frac{z}{4}\right)^2 = 1 - \left(\frac{x_0}{2}\right)^2$

Trace $y = y_0$ (hyperbola, red in figure): $\left(\frac{x}{2}\right)^2 - \left(\frac{z}{4}\right)^2 = 1 - \left(\frac{y_0}{3}\right)^2$ ■

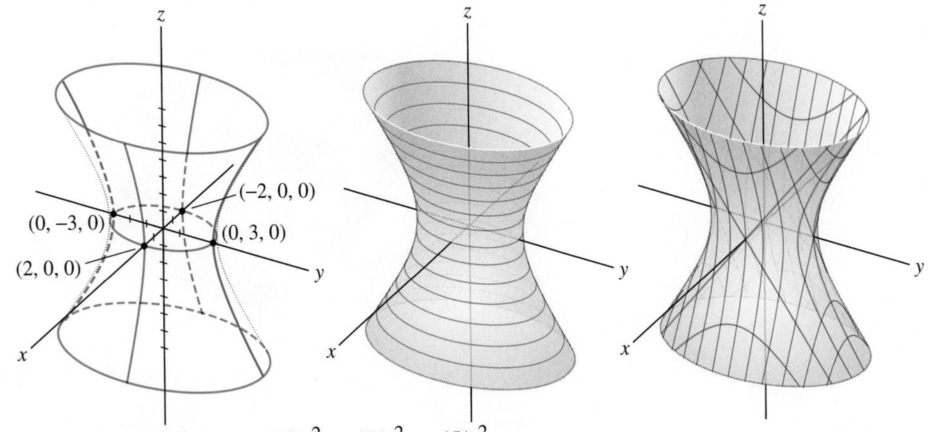

FIGURE 5 The hyperboloid $\left(\frac{x}{2}\right)^2 + \left(\frac{y}{3}\right)^2 = \left(\frac{z}{4}\right)^2 + 1$.

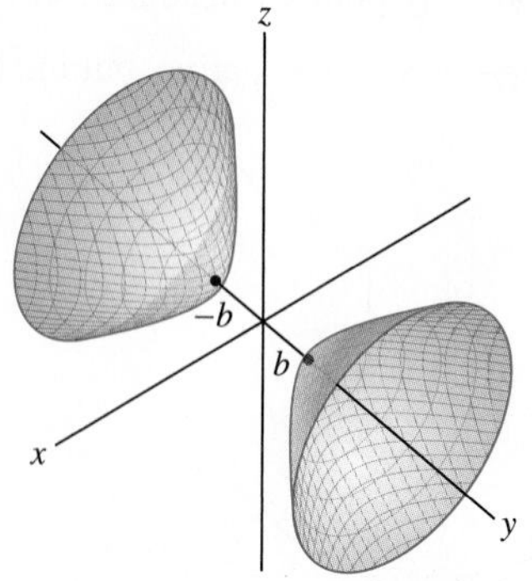

Hyperboloid of two sheets

FIGURE 6 The two-sheeted hyperboloid $\left(\frac{x}{a}\right)^2 + \left(\frac{z}{c}\right)^2 = \left(\frac{y}{b}\right)^2 - 1$.

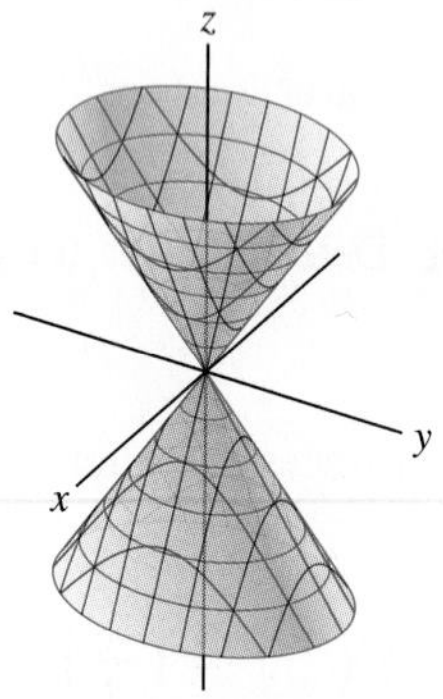

FIGURE 7 Elliptic cone $\left(\frac{x}{a}\right)^2 + \left(\frac{y}{b}\right)^2 = \left(\frac{z}{c}\right)^2$.

EXAMPLE 3 Hyperboloid of Two Sheets Symmetric about the *y*-axis Show that $\left(\frac{x}{a}\right)^2 + \left(\frac{z}{c}\right)^2 = \left(\frac{y}{b}\right)^2 - 1$ has no points for $-b < y < b$.

Solution This equation does not have the same form as Eq. (3) because the variables y and z have been interchanged. This hyperboloid is symmetric about the y-axis rather than the z-axis (Figure 6). The left-hand side of the equation is always ≥ 0. Thus, there are no solutions with $|y| < b$ because the right-hand side is $\left(\frac{y}{b}\right)^2 - 1 < 0$. Therefore, the hyperboloid has two sheets, corresponding to $y \geq b$ and $y \leq -b$. ■

The following equation defines an **elliptic cone** (Figure 7):

Elliptic Cone: $$\left(\frac{x}{a}\right)^2 + \left(\frac{y}{b}\right)^2 = \left(\frac{z}{c}\right)^2$$

An elliptic cone may be thought of as a limiting case of a hyperboloid of one sheet in which we "pinch the waist" down to a point.

The third main family of quadric surfaces are the **paraboloids**. There are two types—elliptic and hyperbolic. In standard position, their equations are

Paraboloids

Elliptic: $$z = \left(\frac{x}{a}\right)^2 + \left(\frac{y}{b}\right)^2$$

Hyperbolic: $$z = \left(\frac{x}{a}\right)^2 - \left(\frac{y}{b}\right)^2 \qquad (3)$$

Let's compare their traces (Figure 8):

	Elliptic paraboloid	Hyperbolic paraboloid
Horizontal traces	ellipses	hyperbolas
Vertical traces	upward parabolas	upward and downward parabolas

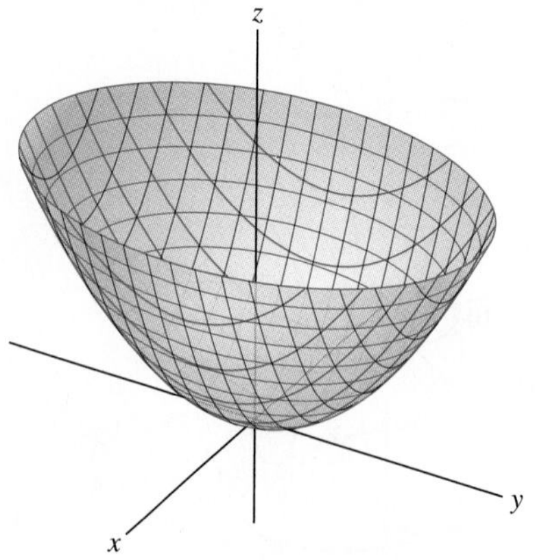

(A) Elliptic paraboloid $z = \left(\frac{x}{2}\right)^2 + \left(\frac{y}{3}\right)^2$

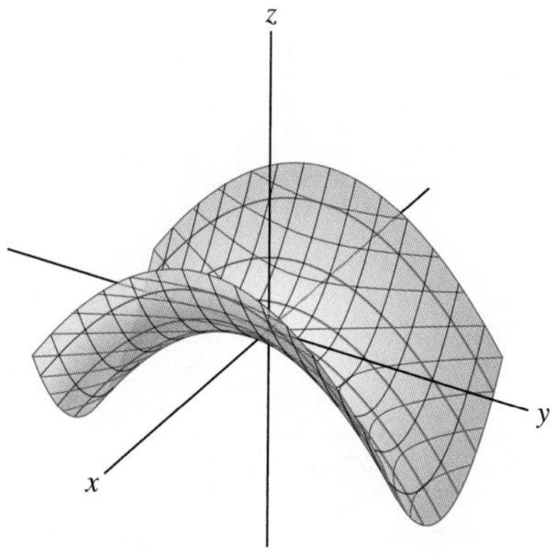

(B) Hyperbolic paraboloid $z = \left(\frac{x}{2}\right)^2 - \left(\frac{y}{3}\right)^2$

FIGURE 8

Notice, for example, that for the hyperbolic paraboloid, the vertical traces $x = x_0$ are downward parabolas (green in the figure)

$$\underbrace{z = -\left(\frac{y}{b}\right)^2 + \left(\frac{x_0}{a}\right)^2}_{\text{Trace } x = x_0 \text{ of hyperbolic paraboloid}}$$

whereas the vertical traces $y = y_0$ are upward parabolas (red in the figure)

$$\underbrace{z = \left(\frac{x}{a}\right)^2 - \left(\frac{y_0}{b}\right)^2}_{\text{Trace } y = y_0 \text{ of hyperbolic paraboloid}}$$

Paraboloids play an important role in the optimization of functions of two variables. The elliptic paraboloid in Figure 8 has a local minimum at the origin. The hyperbolic paraboloid is a "saddle shape" at the origin, which is an analog for surfaces of a point of inflection.

■ **EXAMPLE 4 Alternative Form of a Hyperbolic Paraboloid** Show that $z = 4xy$ is a hyperbolic paraboloid by writing the equation in terms of the variables $u = x + y$ and $v = x - y$.

Solution Note that $u + v = 2x$ and $u - v = 2y$. Therefore,

$$4xy = (u + v)(u - v) = u^2 - v^2$$

and thus the equation takes the form $z = u^2 - v^2$ in the coordinates $\{u, v, z\}$. The coordinates $\{u, v, z\}$ are obtained by rotating the coordinates $\{x, y, z\}$ by 45° about the z-axis (Figure 9). ■

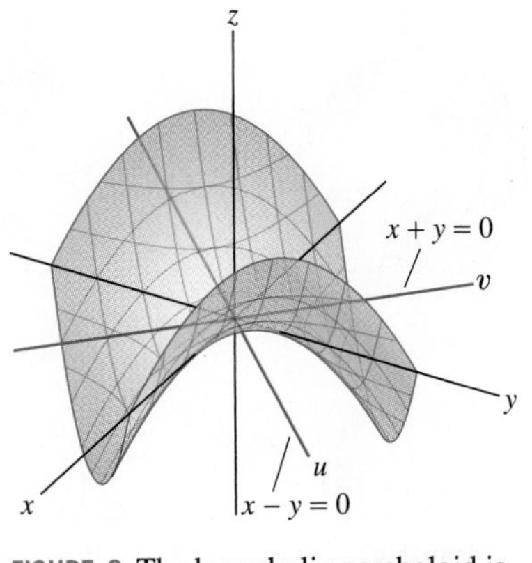

FIGURE 9 The hyperbolic paraboloid is defined by $z = 4xy$ or $z = u^2 - v^2$.

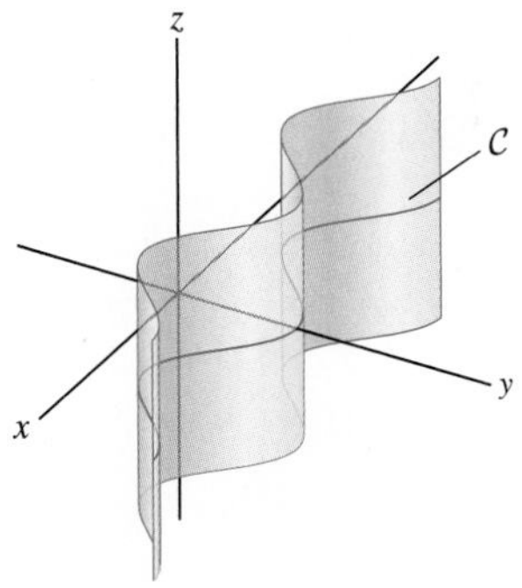

FIGURE 10 The cylinder with base $\mathcal{C}$.

Further examples of quadric surfaces are the **quadratic cylinders**. We use the term *cylinder* in the following sense: Given a curve $\mathcal{C}$ in the xy-plane, the cylinder with base $\mathcal{C}$ is the surface consisting of all vertical lines passing through $\mathcal{C}$ (Figure 10). Equations of cylinders involve just the two variables x and y. The equation $x^2 + y^2 = r^2$ defines a circular cylinder of radius r with the z-axis as central axis. Figure 11 shows a circular cylinder and three other types of quadratic cylinders.

The ellipsoids, hyperboloids, paraboloids, and quadratic cylinders are called **nondegenerate** quadric surfaces. There are also a certain number of "degenerate" quadric surfaces. For example, $x^2 + y^2 + z^2 = 0$ is a quadric that reduces to a single point (0, 0, 0), and $(x + y + z)^2 = 1$ reduces to the union of the two planes $x + y + z = \pm 1$.

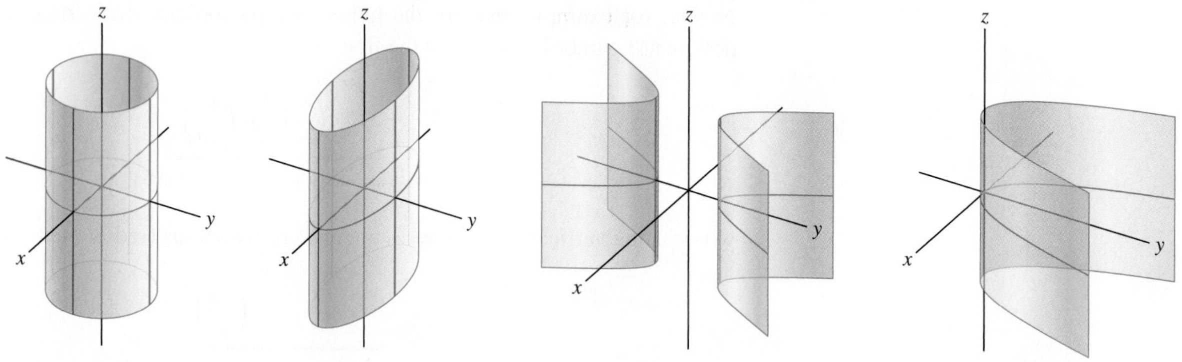

$x^2 + y^2 = r^2$
Right-Circular cylinder of radius r

$\left(\frac{x}{a}\right)^2 + \left(\frac{y}{b}\right)^2 = 1$
Elliptic cylinder

$\left(\frac{x}{a}\right)^2 - \left(\frac{y}{b}\right)^2 = 1$
Hyperbolic cylinder

$y = ax^2$
Parabolic cylinder

FIGURE 11

13.6 SUMMARY

- A *quadric surface* is a surface defined by a quadratic equation in three variables in which the coefficients A–F are not all zero:

$$Ax^2 + By^2 + Cz^2 + Dxy + Eyz + Fzx + ax + by + cz + d = 0$$

- Quadric surfaces in standard position:

Ellipsoid

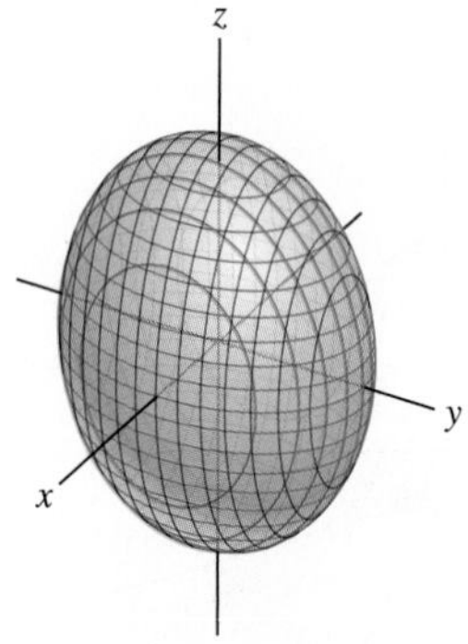

$\left(\frac{x}{a}\right)^2 + \left(\frac{y}{b}\right)^2 + \left(\frac{z}{c}\right)^2 = 1$

Hyperboloid (one sheet)

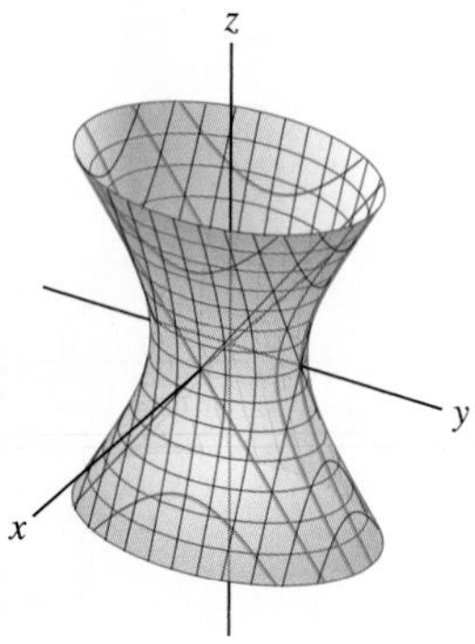

$\left(\frac{x}{a}\right)^2 + \left(\frac{y}{b}\right)^2 = \left(\frac{z}{c}\right)^2 + 1$

Hyperboloid (two sheets)

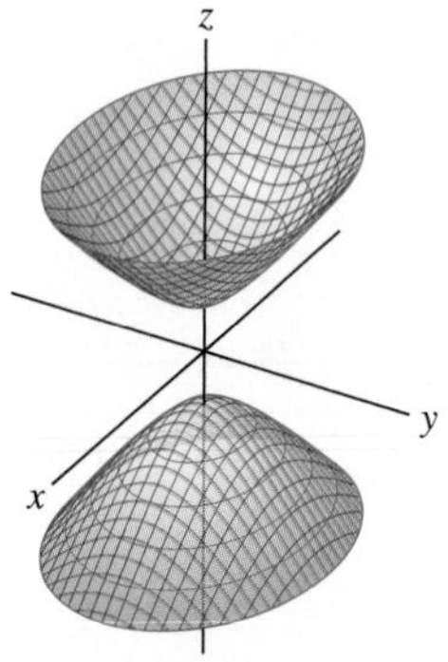

$\left(\frac{x}{a}\right)^2 + \left(\frac{y}{b}\right)^2 = \left(\frac{z}{c}\right)^2 - 1$

Paraboloid (elliptic)

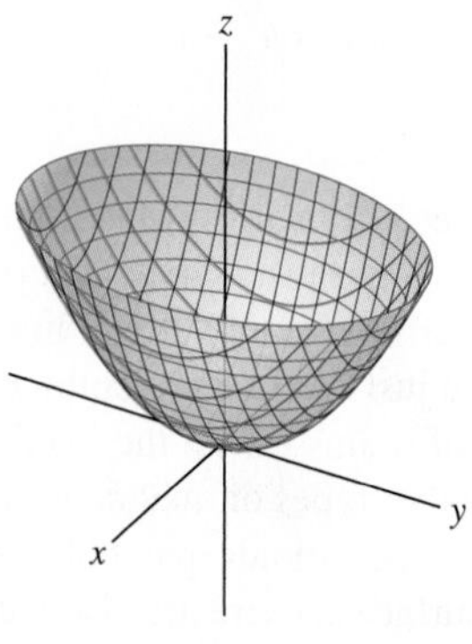

$z = \left(\frac{x}{a}\right)^2 + \left(\frac{y}{b}\right)^2$

Paraboloid (hyperbolic)

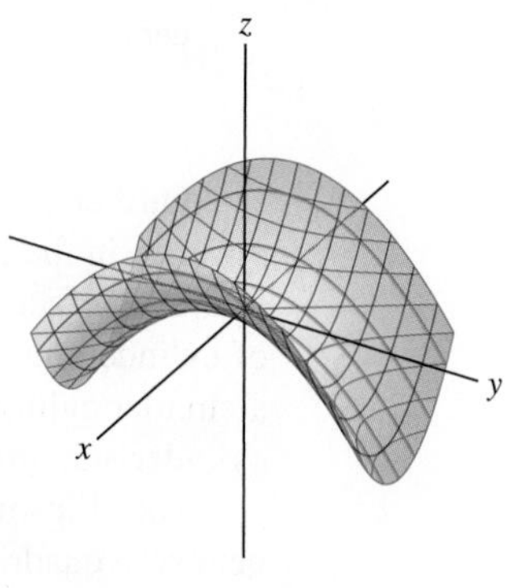

$z = \left(\frac{x}{a}\right)^2 - \left(\frac{y}{b}\right)^2$

Cone (elliptic)

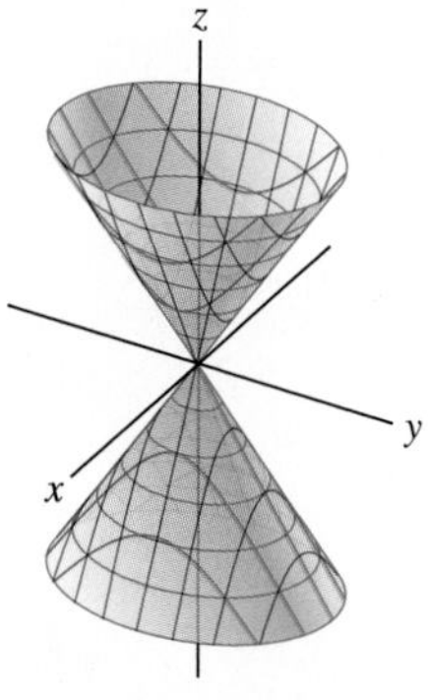

$\left(\frac{x}{a}\right)^2 + \left(\frac{y}{b}\right)^2 = \left(\frac{z}{c}\right)^2$

- A (vertical) cylinder is a surface consisting of all vertical lines passing through a curve (called the base) in the xy-plane. A quadratic cylinder is a cylinder whose base is a conic section. There are three types:

Elliptic cylinder

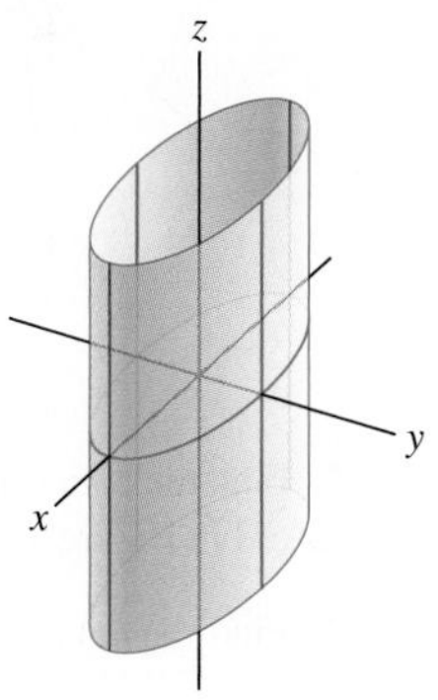

$\left(\frac{x}{a}\right)^2 + \left(\frac{y}{b}\right)^2 = 1$

Hyperbolic cylinder

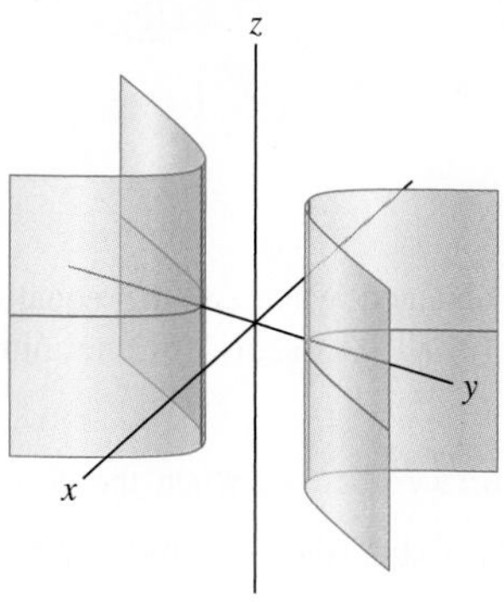

$\left(\frac{x}{a}\right)^2 - \left(\frac{y}{b}\right)^2 = 1$

Parabolic cylinder

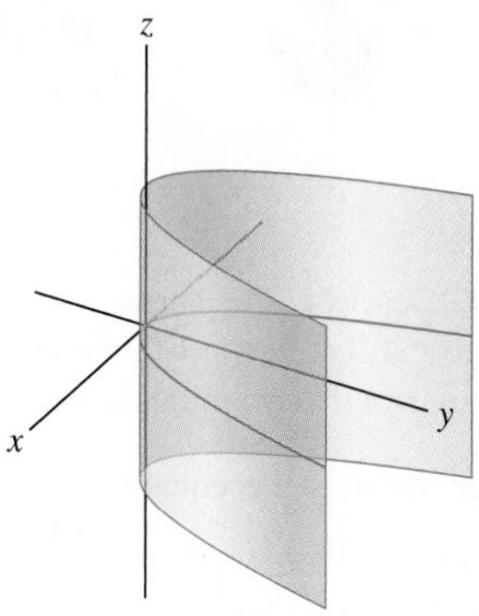

$y = ax^2$

13.6 EXERCISES

Preliminary Questions

1. True or false? All traces of an ellipsoid are ellipses.

2. True or false? All traces of a hyperboloid are hyperbolas.

3. Which quadric surfaces have both hyperbolas and parabolas as traces?

4. Is there any quadric surface whose traces are all parabolas?

5. A surface is called **bounded** if there exists $M > 0$ such that every point on the surface lies at a distance of at most M from the origin. Which of the quadric surfaces are bounded?

6. What is the definition of a parabolic cylinder?

Exercises

In Exercises 1–6, state whether the given equation defines an ellipsoid or hyperboloid, and if a hyperboloid, whether it is of one or two sheets.

1. $\left(\frac{x}{2}\right)^2 + \left(\frac{y}{3}\right)^2 + \left(\frac{z}{5}\right)^2 = 1$

2. $\left(\frac{x}{5}\right)^2 + \left(\frac{y}{5}\right)^2 - \left(\frac{z}{7}\right)^2 = 1$

3. $x^2 + 3y^2 + 9z^2 = 1$

4. $-\left(\frac{x}{2}\right)^2 - \left(\frac{y}{3}\right)^2 + \left(\frac{z}{5}\right)^2 = 1$

5. $x^2 - 3y^2 + 9z^2 = 1$

6. $x^2 - 3y^2 - 9z^2 = 1$

In Exercises 7–12, state whether the given equation defines an elliptic paraboloid, a hyperbolic paraboloid, or an elliptic cone.

7. $z = \left(\frac{x}{4}\right)^2 + \left(\frac{y}{3}\right)^2$

8. $z^2 = \left(\frac{x}{4}\right)^2 + \left(\frac{y}{3}\right)^2$

9. $z = \left(\frac{x}{9}\right)^2 - \left(\frac{y}{12}\right)^2$

10. $4z = 9x^2 + 5y^2$

11. $3x^2 - 7y^2 = z$

12. $3x^2 + 7y^2 = 14z^2$

In Exercises 13–20, state the type of the quadric surface and describe the trace obtained by intersecting with the given plane.

13. $x^2 + \left(\frac{y}{4}\right)^2 + z^2 = 1, \quad y = 0$

14. $x^2 + \left(\frac{y}{4}\right)^2 + z^2 = 1, \quad y = 5$

15. $x^2 + \left(\frac{y}{4}\right)^2 + z^2 = 1, \quad z = \frac{1}{4}$

16. $\left(\frac{x}{2}\right)^2 + \left(\frac{y}{5}\right)^2 - 5z^2 = 1, \quad x = 0$

17. $\left(\frac{x}{3}\right)^2 + \left(\frac{y}{5}\right)^2 - 5z^2 = 1, \quad y = 1$

18. $4x^2 + \left(\frac{y}{3}\right)^2 - 2z^2 = -1, \quad z = 1$

19. $y = 3x^2, \quad z = 27$

20. $y = 3x^2, \quad y = 27$

21. Match each of the ellipsoids in Figure 12 with the correct equation:

(a) $x^2 + 4y^2 + 4z^2 = 16$

(b) $4x^2 + y^2 + 4z^2 = 16$

(c) $4x^2 + 4y^2 + z^2 = 16$

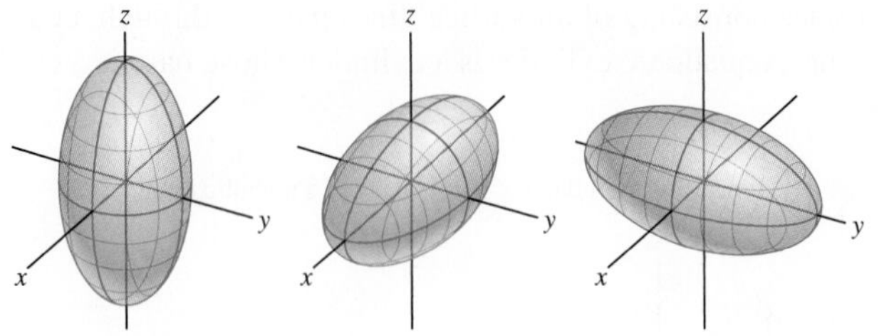

FIGURE 12

22. Describe the surface that is obtained when, in the equation $\pm 8x^2 \pm 3y^2 \pm z^2 = 1$, we choose (a) all plus signs, (b) one minus sign, and (c) two minus signs.

23. What is the equation of the surface obtained when the elliptic paraboloid $z = \left(\frac{x}{2}\right)^2 + \left(\frac{y}{4}\right)^2$ is rotated about the x-axis by $90°$? Refer to Figure 13.

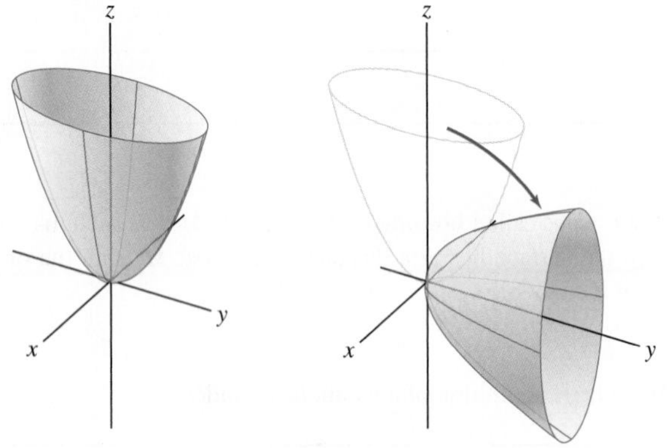

FIGURE 13

24. Describe the intersection of the horizontal plane $z = h$ and the hyperboloid $-x^2 - 4y^2 + 4z^2 = 1$. For which values of h is the intersection empty?

In Exercises 25–30, sketch the given surface.

25. $x^2 + y^2 - z^2 = 1$

26. $\left(\frac{x}{4}\right)^2 + \left(\frac{y}{8}\right)^2 + \left(\frac{z}{12}\right)^2 = 1$

27. $z = \left(\frac{x}{4}\right)^2 + \left(\frac{y}{8}\right)^2$

28. $z = \left(\frac{x}{4}\right)^2 - \left(\frac{y}{8}\right)^2$

29. $z^2 = \left(\frac{x}{4}\right)^2 + \left(\frac{y}{8}\right)^2$

30. $z = -x^2$

31. Find the equation of the ellipsoid passing through the points marked in Figure 14(A).

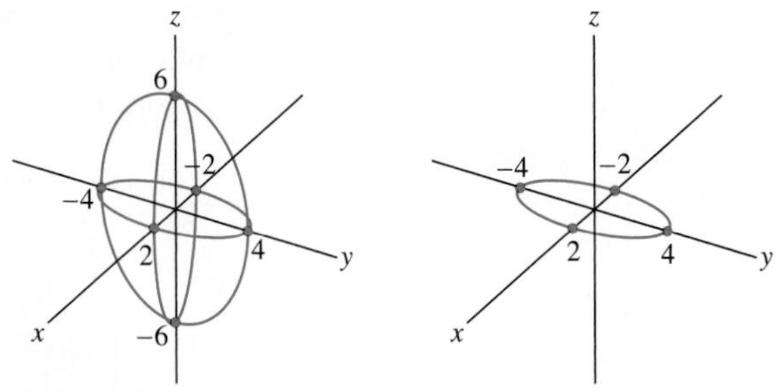

(A) (B)

FIGURE 14

32. Find the equation of the elliptic cylinder passing through the points marked in Figure 14(B).

33. Find the equation of the hyperboloid shown in Figure 15(A).

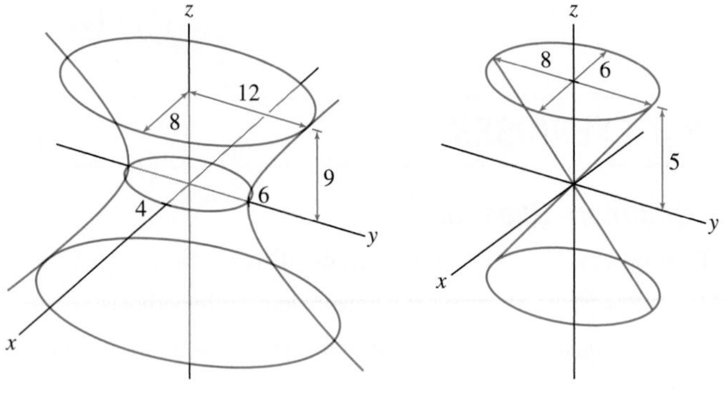

(A) (B)

FIGURE 15

34. Find the equation of the quadric surface shown in Figure 15(B).

35. Determine the vertical traces of elliptic and parabolic cylinders in standard form.

36. What is the equation of a hyperboloid of one or two sheets in standard form if every horizontal trace is a circle?

37. Let $\mathcal{C}$ be an ellipse in a horizonal plane lying above the xy-plane. Which type of quadric surface is made up of all lines passing through the origin and a point on $\mathcal{C}$?

38. The eccentricity of a conic section is defined in Section 11.5. Show that the horizontal traces of the ellipsoid

$$\left(\frac{x}{a}\right)^2 + \left(\frac{y}{b}\right)^2 + \left(\frac{z}{c}\right)^2 = 1$$

are ellipses of the same eccentricity (apart from the traces at height $h = \pm c$, which reduce to a single point). Find the eccentricity.

Further Insights and Challenges

39. Let $\mathcal{S}$ be the hyperboloid $x^2 + y^2 = z^2 + 1$ and let $P = (\alpha, \beta, 0)$ be a point on $\mathcal{S}$ in the (x, y)-plane. Show that there are precisely two lines through P entirely contained in $\mathcal{S}$ (Figure 16). *Hint:* Consider the line $\mathbf{r}(t) = \langle \alpha + at, \beta + bt, t \rangle$ through P. Show that $\mathbf{r}(t)$ is contained in $\mathcal{S}$ if (a, b) is one of the two points on the unit circle obtained by rotating (α, β) through $\pm\frac{\pi}{2}$. This proves that a hyperboloid of one sheet is a **doubly ruled surface**, which means that it can be swept out by moving a line in space in two different ways.

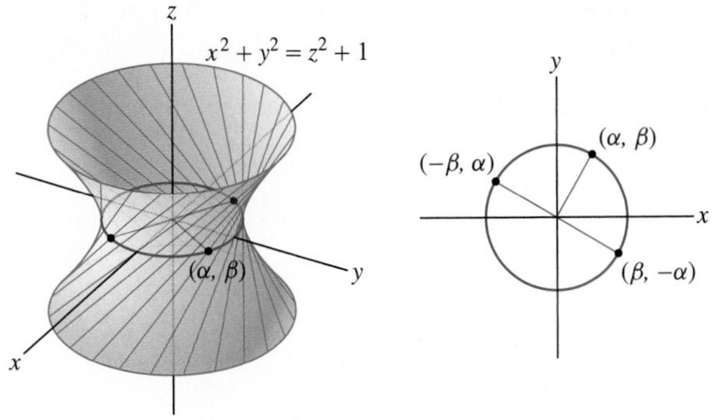

FIGURE 16

In Exercises 40 and 41, let $\mathcal{C}$ be a curve in $\mathbf{R}^3$ not passing through the origin. The cone on $\mathcal{C}$ is the surface consisting of all lines passing through the origin and a point on $\mathcal{C}$ [Figure 17(A)].

40. Show that the elliptic cone $\left(\frac{z}{c}\right)^2 = \left(\frac{x}{a}\right)^2 + \left(\frac{y}{b}\right)^2$ is, in fact, a cone on the ellipse $\mathcal{C}$ consisting of all points (x, y, c) such that $\left(\frac{x}{a}\right)^2 + \left(\frac{y}{b}\right)^2 = 1$.

41. Let a and c be nonzero constants and let $\mathcal{C}$ be the parabola at height c consisting of all points (x, ax^2, c) [Figure 17(B)]. Let $\mathcal{S}$ be the cone consisting of all lines passing through the origin and a point on $\mathcal{C}$. This exercise shows that $\mathcal{S}$ is also an elliptic cone.

(a) Show that $\mathcal{S}$ has equation $yz = acx^2$.

(b) Show that under the change of variables $y = u + v$ and $z = u - v$, this equation becomes $acx^2 = u^2 - v^2$ or $u^2 = acx^2 + v^2$ (the equation of an elliptic cone in the variables x, v, u).

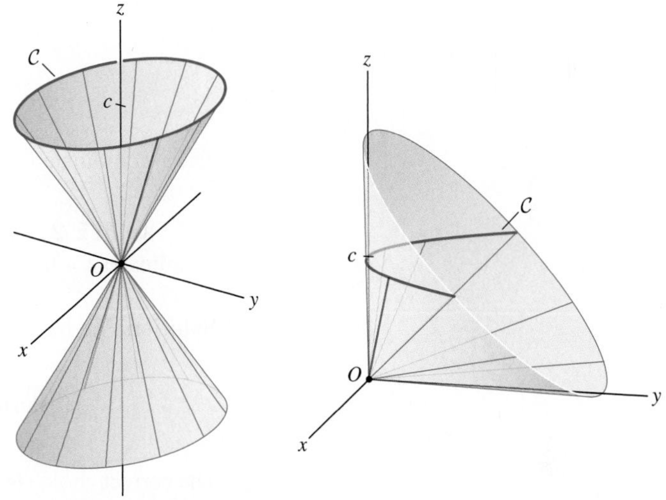

(A) Cone on ellipse $\mathcal{C}$

(B) Cone on parabola $\mathcal{C}$ (half of cone shown)

FIGURE 17

13.7 Cylindrical and Spherical Coordinates

This section introduces two generalizations of polar coordinates to $\mathbf{R}^3$: cylindrical and spherical coordinates. These coordinate systems are commonly used in problems having symmetry about an axis or rotational symmetry. For example, the magnetic field generated by a current flowing in a long, straight wire is conveniently expressed in cylindrical coordinates (Figure 1). We will also see the benefits of cylindrical and spherical coordinates when we study change of variables for multiple integrals.

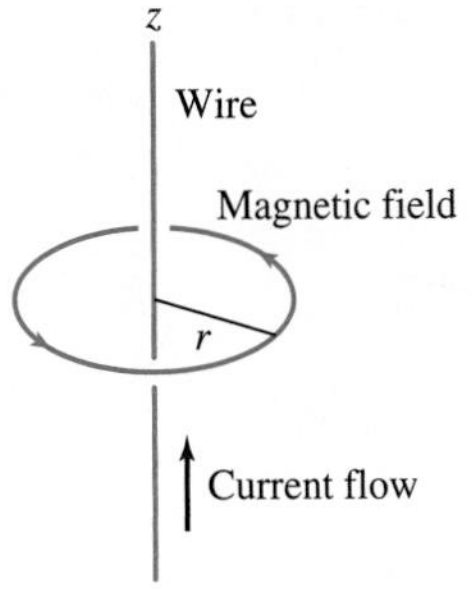

FIGURE 1 The magnetic field generated by a current flowing in a long, straight wire is conveniently expressed in cylindrical coordinates.

Cylindrical Coordinates

In cylindrical coordinates, we replace the x- and y-coordinates of a point $P = (x, y, z)$ by polar coordinates. Thus, the **cylindrical coordinates** of P are (r, θ, z), where (r, θ) are polar coordinates of the projection $Q = (x, y, 0)$ of P onto the xy-plane (Figure 2). Note that the points at fixed distance r from the z-axis make up a cylinder, hence the name cylindrical coordinates.

We convert between rectangular and cylindrical coordinates using the rectangular-polar formulas of Section 12.3. In cylindrical coordinates, we usually assume $r \geq 0$.

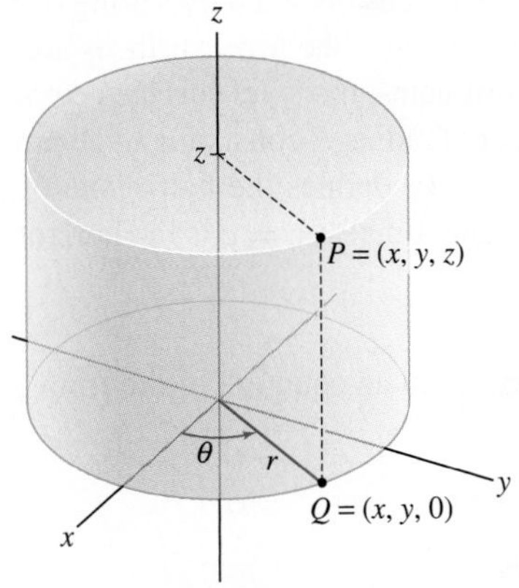

FIGURE 2 P has cylindrical coordinates (r, θ, z).

Cylindrical to rectangular	Rectangular to cylindrical
$x = r\cos\theta$	$r = \sqrt{x^2 + y^2}$
$y = r\sin\theta$	$\tan\theta = \dfrac{y}{x}$
$z = z$	$z = z$

■ **EXAMPLE 1** **Converting from Cylindrical to Rectangular Coordinates** Find the rectangular coordinates of the point P with cylindrical coordinates $(r, \theta, z) = \left(2, \frac{3\pi}{4}, 5\right)$.

Solution Converting to rectangular coordinates is straightforward (Figure 3):

$$x = r\cos\theta = 2\cos\frac{3\pi}{4} = 2\left(-\frac{\sqrt{2}}{2}\right) = -\sqrt{2}$$

$$y = r\sin\theta = 2\sin\frac{3\pi}{4} = 2\left(\frac{\sqrt{2}}{2}\right) = \sqrt{2}$$

The z-coordinate is unchanged, so $(x, y, z) = (-\sqrt{2}, \sqrt{2}, 5)$. ■

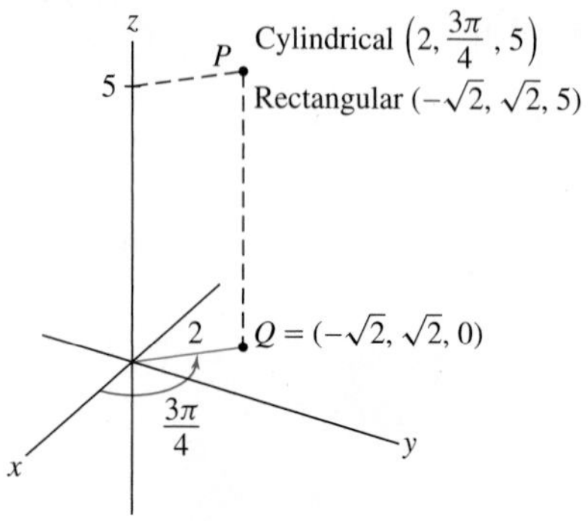

FIGURE 3

■ **EXAMPLE 2** **Converting from Rectangular to Cylindrical Coordinates** Find cylindrical coordinates for the point with rectangular coordinates $(x, y, z) = (-3\sqrt{3}, -3, 5)$.

Solution We have $r = \sqrt{x^2 + y^2} = \sqrt{(-3\sqrt{3})^2 + (-3)^2} = 6$. The angle θ satisfies

$$\tan\theta = \frac{y}{x} = \frac{-3}{-3\sqrt{3}} = \frac{1}{\sqrt{3}} \quad \Rightarrow \quad \theta = \frac{\pi}{6} \quad \text{or} \quad \frac{7\pi}{6}$$

The correct choice is $\theta = \frac{7\pi}{6}$ because the projection $Q = (-3\sqrt{3}, -3, 0)$ lies in the third quadrant (Figure 4). The cylindrical coordinates are $(r, \theta, z) = \left(6, \frac{7\pi}{6}, 5\right)$. ■

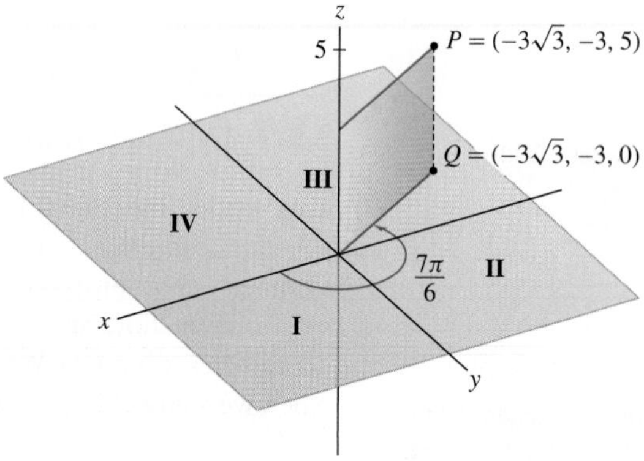

FIGURE 4 The projection Q lies in the third quadrant. Therefore, $\theta = \frac{7\pi}{6}$.

The **level surfaces** of a coordinate system are the surfaces obtained by setting one of the coordinates equal to a constant. In rectangular coordinates, the level surfaces are the planes $x = x_0$, $y = y_0$, and $z = z_0$. In cylindrical coordinates, the level surfaces come in three types (Figure 5). The surface $r = R$ is the cylinder of radius R consisting of all points located a distance R from the z-axis. The equation $\theta = \theta_0$ defines the half-plane of all points that project onto the ray $\theta = \theta_0$ in the (x, y)-plane. Finally, $z = c$ is the horizontal plane at height c.

Level Surfaces in Cylindrical Coordinates

$r = R$	*Cylinder of radius R with the z-axis as axis of symmetry*
$\theta = \theta_0$	*Half-plane through the z-axis making an angle θ_0 with the xz-plane*
$z = c$	*Horizontal plane at height c*

■ **EXAMPLE 3** **Equations in Cylindrical Coordinates** Find an equation of the form $z = f(r, \theta)$ for the surfaces

(a) $x^2 + y^2 + z^2 = 9$ **(b)** $x + y + z = 1$

Solution We use the formulas

$$x^2 + y^2 = r^2, \qquad x = r\cos\theta, \qquad y = r\sin\theta$$

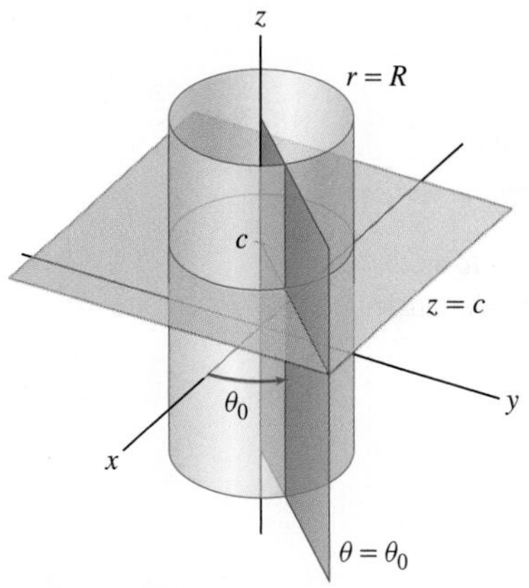

FIGURE 5 Level surfaces in cylindrical coordinates.

(a) The equation $x^2 + y^2 + z^2 = 9$ becomes $r^2 + z^2 = 9$, or $z = \pm\sqrt{9 - r^2}$. This is a sphere of radius 3.

(b) The plane $x + y + z = 1$ becomes

$$z = 1 - x - y = 1 - r\cos\theta - r\sin\theta \qquad \text{or} \qquad z = 1 - r(\cos\theta + \sin\theta)$$ ■

Spherical Coordinates

Spherical coordinates make use of the fact that a point P on a sphere of radius ρ is determined by two angular coordinates θ and ϕ (Figure 6):

- θ is the polar angle of the projection Q of P onto the xy-plane.
- ϕ is the **angle of declination**, which measures how much the ray through P declines from the vertical.

Thus P is determined by the triple (ρ, θ, ϕ), which are called **spherical coordinates**.

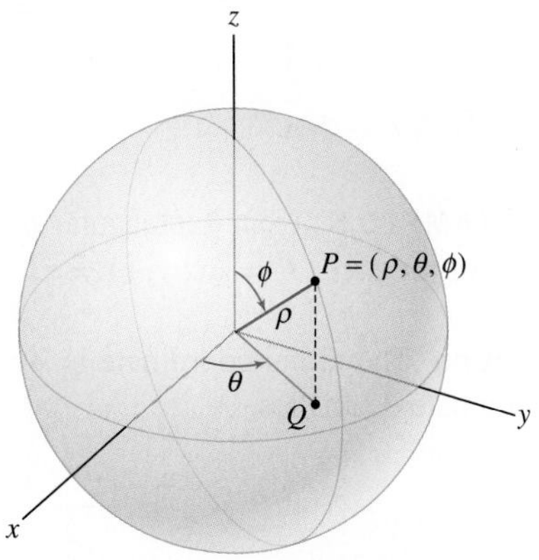

FIGURE 6 Spherical coordinates (ρ, θ, ϕ).

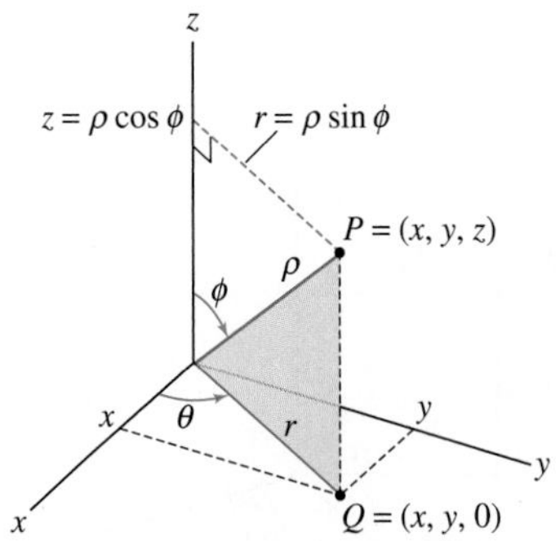

FIGURE 7

- The symbol ϕ (usually pronounced "fee," but sometimes pronounced "fie") is the twenty-first letter of the Greek alphabet.
- We use ρ for the radial coordinate, although r is also used to denote distance from the origin in other contexts.

Suppose that $P = (x, y, z)$ in rectangular coordinates. Since ρ is the distance from P to the origin,

$$\rho = \sqrt{x^2 + y^2 + z^2}$$

On the other hand, we see in Figure 7 that

$$\tan\theta = \frac{y}{x}, \qquad \cos\phi = \frac{z}{\rho}$$

Spherical Coordinates

ρ = distance from origin

θ = polar angle in the xy-plane

ϕ = angle of declination from the vertical

In some textbooks, θ is referred to as the azimuthal angle *and ϕ as the* polar angle.

The radial coordinate r of $Q = (x, y, 0)$ is $r = \rho \sin \phi$, and therefore,

$$x = r\cos\theta = \rho\cos\theta\sin\phi, \qquad y = r\sin\theta = \rho\sin\theta\sin\phi, \qquad z = \rho\cos\phi$$

Spherical to rectangular	Rectangular to spherical
$x = \rho\cos\theta\sin\phi$	$\rho = \sqrt{x^2 + y^2 + z^2}$
$y = \rho\sin\theta\sin\phi$	$\tan\theta = \dfrac{y}{x}$
$z = \rho\cos\phi$	$\cos\phi = \dfrac{z}{\rho}$

EXAMPLE 4 From Spherical to Rectangular Coordinates Find the rectangular coordinates of $P = (\rho, \theta, \phi) = \left(3, \frac{\pi}{3}, \frac{\pi}{4}\right)$, and find the radial coordinate r of its projection Q onto the xy-plane.

Solution By the formulas above,

$$x = \rho\cos\theta\sin\phi = 3\cos\frac{\pi}{3}\sin\frac{\pi}{4} = 3\left(\frac{1}{2}\right)\frac{\sqrt{2}}{2} = \frac{3\sqrt{2}}{4}$$

$$y = \rho\sin\theta\sin\phi = 3\sin\frac{\pi}{3}\sin\frac{\pi}{4} = 3\left(\frac{\sqrt{3}}{2}\right)\frac{\sqrt{2}}{2} = \frac{3\sqrt{6}}{4}$$

$$z = \rho\cos\phi = 3\cos\frac{\pi}{4} = 3\frac{\sqrt{2}}{2} = \frac{3\sqrt{2}}{2}$$

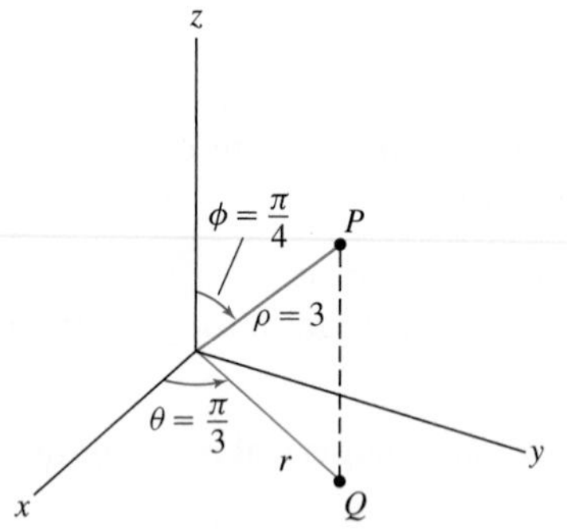

FIGURE 8 Point with spherical coordinates $\left(3, \frac{\pi}{3}, \frac{\pi}{4}\right)$.

Now consider the projection $Q = (x, y, 0) = \left(\frac{3\sqrt{2}}{4}, \frac{3\sqrt{6}}{4}, 0\right)$ (Figure 8). The radial coordinate r of Q satisfies

$$r^2 = x^2 + y^2 = \left(\frac{3\sqrt{2}}{4}\right)^2 + \left(\frac{3\sqrt{6}}{4}\right)^2 = \frac{9}{2}$$

Therefore, $r = 3/\sqrt{2}$. ■

EXAMPLE 5 From Rectangular to Spherical Coordinates Find the spherical coordinates of the point $P = (x, y, z) = (2, -2\sqrt{3}, 3)$.

Solution The radial coordinate is $\rho = \sqrt{2^2 + (-2\sqrt{3})^2 + 3^2} = \sqrt{25} = 5$. The angular coordinate θ satisfies

$$\tan\theta = \frac{y}{x} = \frac{-2\sqrt{3}}{2} = -\sqrt{3} \quad\Rightarrow\quad \theta = \frac{2\pi}{3} \text{ or } \frac{5\pi}{3}$$

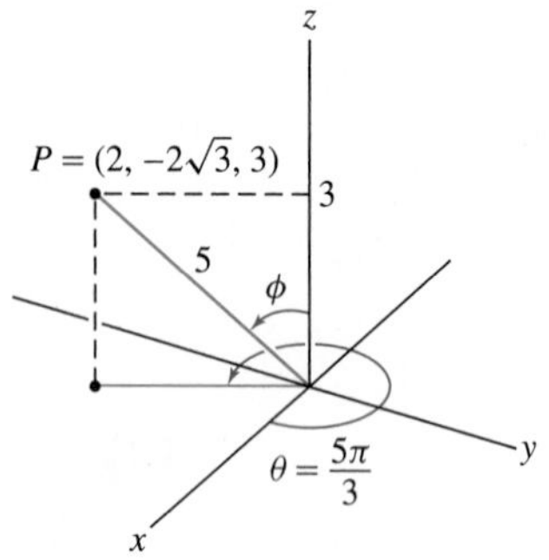

FIGURE 9 Point with rectangular coordinates $(2, -2\sqrt{3}, 3)$.

Since the point $(x, y) = (2, -2\sqrt{3})$ lies in the fourth quadrant, the correct choice is $\theta = \frac{5\pi}{3}$ (Figure 9). Finally, $\cos\phi = \frac{z}{\rho} = \frac{3}{5}$ and so $\phi = \cos^{-1}\frac{3}{5} \approx 0.93$. Therefore, P has spherical coordinates $\left(5, \frac{5\pi}{3}, 0.93\right)$. ■

Figure 10 shows the three types of level surfaces in spherical coordinates. Notice that if $\phi \neq 0, \frac{\pi}{2}$ or π, then the level surface $\phi = \phi_0$ is the right circular cone consisting of points P such that $\overline{OP}$ makes an angle ϕ_0 with the z-axis. There are three exceptional cases: $\phi = \frac{\pi}{2}$ defines the xy-plane, $\phi = 0$ is the positive z-axis, and $\phi = \pi$ is the negative z-axis.

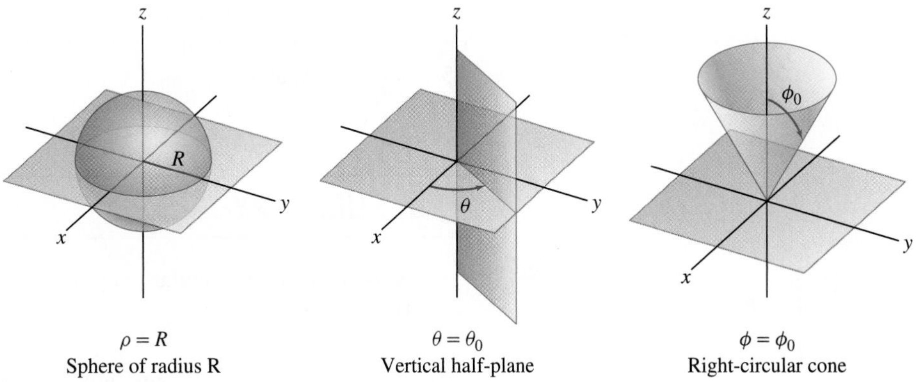

FIGURE 10

■ **EXAMPLE 6** **Finding an Equation in Spherical Coordinates** Find an equation of the form $\rho = f(\theta, \phi)$ for the following surfaces:

(a) $x^2 + y^2 + z^2 = 9$ **(b)** $z = x^2 - y^2$

Solution

(a) The equation $x^2 + y^2 + z^2 = 9$ defines the sphere of radius 3 centered at the origin. Since $\rho^2 = x^2 + y^2 + z^2$, the equation in spherical coordinates is $\rho = 3$.

(b) To convert $z = x^2 - y^2$ to spherical coordinates, we substitute the formulas for x, y, and z in terms of ρ, θ, and ϕ:

$$\overbrace{\rho\cos\phi}^{z} = \overbrace{(\rho\cos\theta\sin\phi)^2}^{x^2} - \overbrace{(\rho\sin\theta\sin\phi)^2}^{y^2}$$

$$\cos\phi = \rho\sin^2\phi(\cos^2\theta - \sin^2\theta) \qquad \text{(divide by } \rho \text{ and factor)}$$

$$\cos\phi = \rho\sin^2\phi\cos 2\theta \qquad \text{(since } \cos^2\theta - \sin^2\theta = \cos 2\theta\text{)}$$

Solving for ρ, we obtain $\rho = \dfrac{\cos\phi}{\sin^2\phi\cos 2\theta}$. ■

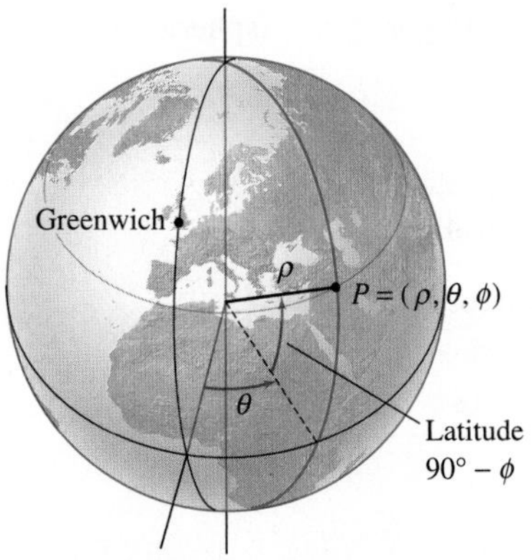

FIGURE 11 Longitude and latitude provide spherical coordinates on the surface of the earth.

The angular coordinates (θ, ϕ) on a sphere of fixed radius are closely related to the longitude-latitude system used to identify points on the surface of the earth (Figure 11). By convention, in this system we use degrees rather than radians.

- A **longitude** is a half-circle stretching from the North to the South Pole (Figure 12). The axes are chosen so that $\theta = 0$ passes through Greenwich, England (this longitude is called the *prime meridian*). We designate the longitude by an angle between 0 and 180° together with a label E or W, according to whether it lies to the east or west of the prime meridian.
- The set of points on the sphere satisfying $\phi = \phi_0$ is a horizontal circle called a **latitude**. We measure latitudes from the equator and use the label N or S to specify the Northern or Southern Hemisphere. Thus, in the upper hemisphere $0 \le \phi_0 \le 90°$, a spherical coordinate ϕ_0 corresponds to the latitude $(90° - \phi_0)$ N. In the lower hemisphere $90° \le \phi_0 \le 180°$, ϕ_0 corresponds to the latitude $(\phi_0 - 90°)$ S.

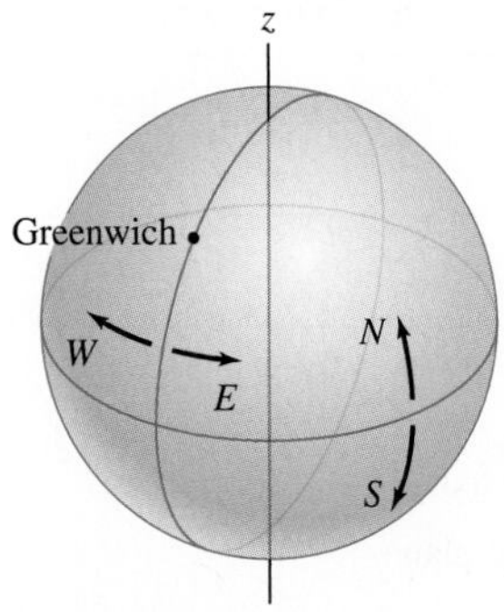

FIGURE 12 Latitude is measured from the equator and is labeled N (north) in the upper hemisphere, and S (south) in the lower hemisphere.

■ **EXAMPLE 7** **Spherical Coordinates via Longitude and Latitude** Find the angles (θ, ϕ) for Nairobi (1.17° S, 36.48° E) and Ottawa (45.27° N, 75.42° W).

Solution For Nairobi, $\theta = 36.48°$ since the longitude lies to the east of Greenwich. Nairobi's latitude is south of the equator, so $1.17 = \phi_0 - 90$ and $\phi_0 = 91.17°$.

For Ottawa, we have $\theta = 360 - 75.42 = 284.58°$ because 75.42° W refers to 75.42 degrees in the negative θ direction. Since the latitude of Ottawa is north of the equator, $45.27 = 90 - \phi_0$ and $\phi_0 = 44.73°$. ■

13.7 SUMMARY

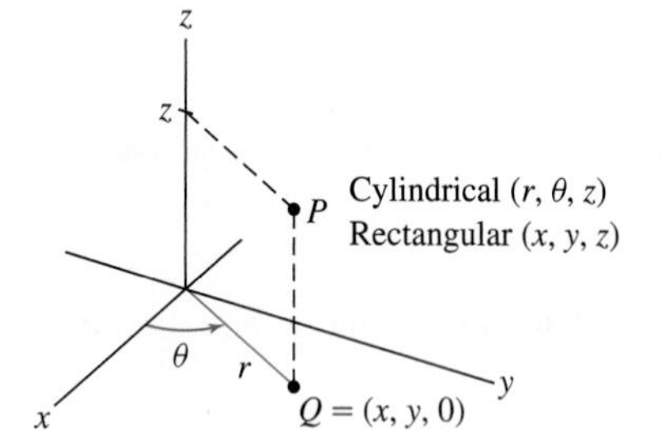

FIGURE 13 Cylindrical coordinates (r, θ, z).

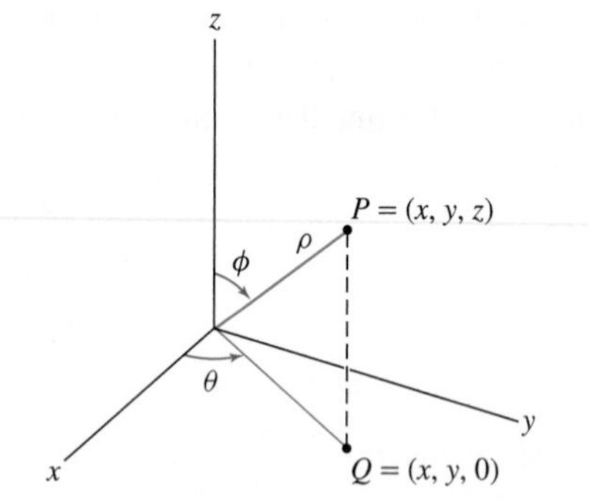

FIGURE 14 Spherical coordinates (ρ, θ, ϕ).

- Conversion from rectangular to cylindrical (Figure 13) and spherical coordinates (Figure 14):

Cylindrical	**Spherical**
$r = \sqrt{x^2 + y^2}$	$\rho = \sqrt{x^2 + y^2 + z^2}$
$\tan\theta = \frac{y}{x}$	$\tan\theta = \frac{y}{x}$
$z = z$	$\cos\phi = \frac{z}{\rho}$

The angles are chosen so that

$$0 \le \theta < 2\pi \quad \text{(cylindrical or spherical)}, \qquad 0 \le \phi \le \pi \quad \text{(spherical)}$$

- Conversion to rectangular coordinates:

Cylindrical (r, θ, z)	**Spherical** (ρ, θ, ϕ)
$x = r\cos\theta$	$x = \rho\cos\theta\sin\phi$
$y = r\sin\theta$	$y = \rho\sin\theta\sin\phi$
$z = z$	$z = \rho\cos\phi$

- Level surfaces:

Cylindrical		**Spherical**	
$r = R$:	Cylinder of radius R	$\rho = R$:	Sphere of radius R
$\theta = \theta_0$:	Vertical half-plane	$\theta = \theta_0$:	Vertical half-plane
$z = c$:	Horizontal plane	$\phi = \phi_0$:	Right-circular cone

13.7 EXERCISES

Preliminary Questions

1. Describe the surfaces $r = R$ in cylindrical coordinates and $\rho = R$ in spherical coordinates.

2. Which statement about cylindrical coordinates is correct?

(a) If $\theta = 0$, then P lies on the z-axis.

(b) If $\theta = 0$, then P lies in the xz-plane.

3. Which statement about spherical coordinates is correct?

(a) If $\phi = 0$, then P lies on the z-axis.

(b) If $\phi = 0$, then P lies in the xy-plane.

4. The level surface $\phi = \phi_0$ in spherical coordinates, usually a cone, reduces to a half-line for two values of ϕ_0. Which two values?

5. For which value of ϕ_0 is $\phi = \phi_0$ a plane? Which plane?

Exercises

In Exercises 1–4, convert from cylindrical to rectangular coordinates.

1. $(4, \pi, 4)$
2. $\left(2, \frac{\pi}{3}, -8\right)$

3. $\left(0, \frac{\pi}{5}, \frac{1}{2}\right)$
4. $\left(1, \frac{\pi}{2}, -2\right)$

In Exercises 5–10, convert from rectangular to cylindrical coordinates.

5. $(1, -1, 1)$
6. $(2, 2, 1)$
7. $(1, \sqrt{3}, 7)$

8. $\left(\frac{3}{2}, \frac{3\sqrt{3}}{2}, 9\right)$
9. $\left(\frac{5}{\sqrt{2}}, \frac{5}{\sqrt{2}}, 2\right)$
10. $(3, 3\sqrt{3}, 2)$

In Exercises 11–16, describe the set in cylindrical coordinates.

11. $x^2 + y^2 \le 1$
12. $x^2 + y^2 + z^2 \le 1$

13. $y^2 + z^2 \le 4, \quad x = 0$

14. $x^2 + y^2 + z^2 = 4, \quad x \ge 0, \quad y \ge 0, \quad z \ge 0$

15. $x^2 + y^2 \le 9, \quad x \ge y$
16. $y^2 + z^2 \le 9, \quad x \ge y$

In Exercises 17–24, sketch the set (described in cylindrical coordinates).

17. $r = 4$
18. $\theta = \frac{\pi}{3}$

19. $z = -2$
20. $r = 2, \quad z = 3$

21. $1 \le r \le 3, \quad 0 \le z \le 4$

22. $1 \le r \le 3, \quad 0 \le \theta \le \frac{\pi}{2}, \quad 0 \le z \le 4$

23. $z^2 + r^2 \le 4$

24. $r \le 3, \quad \pi \le \theta \le \frac{3\pi}{2}, \quad z = 4$

In Exercises 25–30, find an equation of the form $r = f(\theta, z)$ in cylindrical coordinates for the following surfaces.

25. $z = x + y$
26. $x^2 + y^2 + z^2 = 4$

27. $\frac{x^2}{yz} = 1$
28. $x^2 - y^2 = 4$

29. $x^2 + y^2 = 4$
30. $z = 3xy$

In Exercises 31–36, convert from spherical to rectangular coordinates.

31. $\left(3, 0, \frac{\pi}{2}\right)$
32. $\left(2, \frac{\pi}{4}, \frac{\pi}{3}\right)$
33. $(3, \pi, 0)$

34. $\left(5, \frac{3\pi}{4}, \frac{\pi}{4}\right)$
35. $\left(6, \frac{\pi}{6}, \frac{5\pi}{6}\right)$
36. $(0.5, 3.7, 2)$

In Exercises 37–42, convert from rectangular to spherical coordinates.

37. $(\sqrt{3}, 0, 1)$
38. $\left(\frac{\sqrt{3}}{2}, \frac{3}{2}, 1\right)$

39. $(1, 1, 1)$
40. $(1, -1, 1)$

41. $\left(\frac{1}{2}, \frac{\sqrt{3}}{2}, \sqrt{3}\right)$
42. $\left(\frac{\sqrt{2}}{2}, \frac{\sqrt{2}}{2}, \sqrt{3}\right)$

In Exercises 43 and 44, convert from cylindrical to spherical coordinates.

43. $(2, 0, 2)$
44. $(3, \pi, \sqrt{3})$

In Exercises 45 and 46, convert from spherical to cylindrical coordinates.

45. $\left(4, 0, \frac{\pi}{4}\right)$
46. $\left(2, \frac{\pi}{3}, \frac{\pi}{6}\right)$

In Exercises 47–52, describe the given set in spherical coordinates.

47. $x^2 + y^2 + z^2 \le 1$

48. $x^2 + y^2 + z^2 = 1, \quad z \ge 0$

49. $x^2 + y^2 + z^2 = 1, \quad x \ge 0, \quad y \ge 0, \quad z \ge 0$

50. $x^2 + y^2 + z^2 \le 1, \quad x = y, \quad x \ge 0, \quad y \ge 0$

51. $y^2 + z^2 \le 4, \quad x = 0$

52. $x^2 + y^2 = 3z^2$

In Exercises 53–60, sketch the set of points (described in spherical coordinates).

53. $\rho = 4$
54. $\phi = \frac{\pi}{4}$

55. $\rho = 2, \quad \theta = \frac{\pi}{4}$
56. $\rho = 2, \quad \phi = \frac{\pi}{4}$

57. $\rho = 2, \quad 0 \le \phi \le \frac{\pi}{2}$
58. $\theta = \frac{\pi}{2}, \quad \phi = \frac{\pi}{4}, \quad \rho \ge 1$

59. $\rho \le 2, \quad 0 \le \theta \le \frac{\pi}{2}, \quad \frac{\pi}{2} \le \phi \le \pi$

60. $\rho = 1, \quad \frac{\pi}{3} \le \phi \le \frac{2\pi}{3}$

In Exercises 61–66, find an equation of the form $\rho = f(\theta, \phi)$ in spherical coordinates for the following surfaces.

61. $z = 2$
62. $z^2 = 3(x^2 + y^2)$
63. $x = z^2$

64. $z = x^2 + y^2$
65. $x^2 - y^2 = 4$
66. $xy = z$

67. Which of (a)–(c) is the equation of the cylinder of radius R in spherical coordinates? Refer to Figure 15.
(a) $R\rho = \sin\phi$
(b) $\rho \sin\phi = R$
(c) $\rho = R \sin\phi$

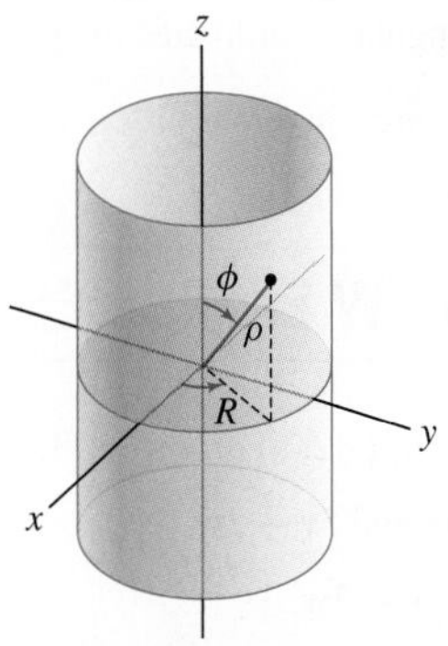

FIGURE 15

68. Let $P_1 = (1, -\sqrt{3}, 5)$ and $P_2 = (-1, \sqrt{3}, 5)$ in rectangular coordinates. In which quadrants do the projections of P_1 and P_2 onto the xy-plane lie? Find the polar angle θ of each point.

69. Find the spherical angles (θ, ϕ) for Helsinki, Finland (60.1° N, 25.0° E) and Sao Paulo, Brazil (23.52° S, 46.52° W).

70. Find the longitude and latitude for the points on the globe with angular coordinates $(\theta, \phi) = (\pi/8, 7\pi/12)$ and $(4, 2)$.

71. Consider a rectangular coordinate system with origin at the center of the earth, z-axis through the North Pole, and x-axis through the prime meridian. Find the rectangular coordinates of Sydney, Australia (34° S, 151° E), and Bogotá, Colombia (4° 32′ N, 74° 15′ W). A minute is $1/60°$. Assume that the earth is a sphere of radius $R = 6370$ km.

72. Find the equation in rectangular coordinates of the quadric surface consisting of the two cones $\phi = \frac{\pi}{4}$ and $\phi = \frac{3\pi}{4}$.

73. Find an equation of the form $z = f(r, \theta)$ in cylindrical coordinates for $z^2 = x^2 - y^2$.

74. Show that $\rho = 2\cos\phi$ is the equation of a sphere with its center on the z-axis. Find its radius and center.

75. Explain the following statement: If the equation of a surface in cylindrical or spherical coordinates does not involve the coordinate θ, then the surface is rotationally symmetric with respect to the z-axis.

76. CAS Plot the surface $\rho = 1 - \cos\phi$. Then plot the trace of S in the xz-plane and explain why S is obtained by rotating this trace.

77. Find equations $r = g(\theta, z)$ (cylindrical) and $\rho = f(\theta, \phi)$ (spherical) for the hyperboloid $x^2 + y^2 = z^2 + 1$ (Figure 16). Do there exist points on the hyperboloid with $\phi = 0$ or π? Which values of ϕ occur for points on the hyperboloid?

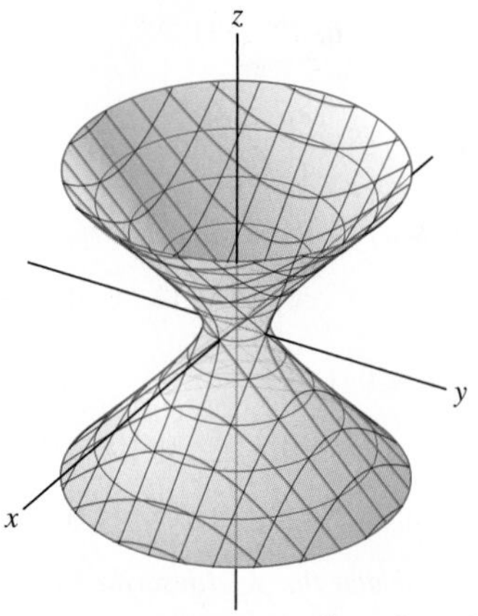

FIGURE 16 The hyperboloid $x^2 + y^2 = z^2 + 1$.

Further Insights and Challenges

In Exercises 78–82, a **great circle** *on a sphere S with center O is a circle obtained by intersecting S with a plane that passes through O (Figure 17). If P and Q are not antipodal (on opposite sides), there is a unique great circle through P and Q on S (intersect S with the plane through O, P, and Q). The geodesic distance from P to Q is defined as the length of the smaller of the two circular arcs of this great circle.*

78. Show that the geodesic distance from P to Q is equal to $R\psi$, where ψ is the *central angle* between P and Q (the angle between the vectors $\mathbf{v} = \overrightarrow{OP}$ and $\mathbf{u} = \overrightarrow{OQ}$).

79. Show that the geodesic distance from $Q = (a, b, c)$ to the North Pole $P = (0, 0, R)$ is equal to $R\cos^{-1}\left(\frac{c}{R}\right)$.

80. The coordinates of Los Angeles are 34° N and 118° W. Find the geodesic distance from the North Pole to Los Angeles, assuming that the earth is a sphere of radius $R = 6370$ km.

81. Show that the central angle ψ between points P and Q on a sphere (of any radius) with angular coordinates (θ, ϕ) and (θ', ϕ') is equal to

$$\psi = \cos^{-1}\left(\sin\phi\sin\phi'\cos(\theta - \theta') + \cos\phi\cos\phi'\right)$$

Hint: Compute the dot product of $\overrightarrow{OP}$ and $\overrightarrow{OQ}$. Check this formula by computing the geodesic distance between the North and South Poles.

82. Use Exercise 81 to find the geodesic distance between Los Angeles (34° N, 118° W) and Bombay (19° N, 72.8° E).

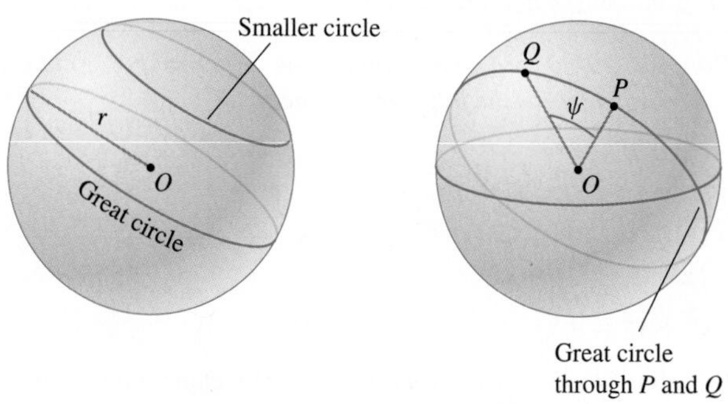

FIGURE 17

CHAPTER REVIEW EXERCISES

In Exercises 1–6, let $\mathbf{v} = \langle -2, 5 \rangle$ *and* $\mathbf{w} = \langle 3, -2 \rangle$.

1. Calculate $5\mathbf{w} - 3\mathbf{v}$ and $5\mathbf{v} - 3\mathbf{w}$.

2. Sketch $\mathbf{v}$, $\mathbf{w}$, and $2\mathbf{v} - 3\mathbf{w}$.

3. Find the unit vector in the direction of $\mathbf{v}$.

4. Find the length of $\mathbf{v} + \mathbf{w}$.

5. Express $\mathbf{i}$ as a linear combination $r\mathbf{v} + s\mathbf{w}$.

6. Find a scalar α such that $\|\mathbf{v} + \alpha\mathbf{w}\| = 6$.

7. If $P = (1, 4)$ and $Q = (-3, 5)$, what are the components of $\overrightarrow{PQ}$? What is the length of $\overrightarrow{PQ}$?

8. Let $A = (2, -1)$, $B = (1, 4)$, and $P = (2, 3)$. Find the point Q such that $\overrightarrow{PQ}$ is equivalent to $\overrightarrow{AB}$. Sketch $\overrightarrow{PQ}$ and $\overrightarrow{AB}$.

9. Find the vector with length 3 making an angle of $\frac{7\pi}{4}$ with the positive x-axis.

10. Calculate $3(\mathbf{i} - 2\mathbf{j}) - 6(\mathbf{i} + 6\mathbf{j})$.

11. Find the value of β for which $\mathbf{w} = \langle -2, \beta \rangle$ is parallel to $\mathbf{v} = \langle 4, -3 \rangle$.

12. Let $P = (1, 4, -3)$.

(a) Find the point Q such that $\overrightarrow{PQ}$ is equivalent to $\langle 3, -1, 5 \rangle$.

(b) Find a unit vector $\mathbf{e}$ equivalent to $\overrightarrow{PQ}$.

13. Let $\mathbf{w} = \langle 2, -2, 1 \rangle$ and $\mathbf{v} = \langle 4, 5, -4 \rangle$. Solve for $\mathbf{u}$ if $\mathbf{v} + 5\mathbf{u} = 3\mathbf{w} - \mathbf{u}$.

14. Let $\mathbf{v} = 3\mathbf{i} - \mathbf{j} + 4\mathbf{k}$. Find the length of $\mathbf{v}$ and the vector $2\mathbf{v} + 3(4\mathbf{i} - \mathbf{k})$.

15. Find a parametrization $\mathbf{r}_1(t)$ of the line passing through $(1, 4, 5)$ and $(-2, 3, -1)$. Then find a parametrization $\mathbf{r}_2(t)$ of the line parallel to $\mathbf{r}_1$ passing through $(1, 0, 0)$.

16. Let $\mathbf{r}_1(t) = \mathbf{v}_1 + t\mathbf{w}_1$ and $\mathbf{r}_2(t) = \mathbf{v}_2 + t\mathbf{w}_2$ be parametrizations of lines $\mathcal{L}_1$ and $\mathcal{L}_2$. For each statement (a)–(e), provide a proof if the statement is true and a counterexample if it is false.

(a) If $\mathcal{L}_1 = \mathcal{L}_2$, then $\mathbf{v}_1 = \mathbf{v}_2$ and $\mathbf{w}_1 = \mathbf{w}_2$.

(b) If $\mathcal{L}_1 = \mathcal{L}_2$ and $\mathbf{v}_1 = \mathbf{v}_2$, then $\mathbf{w}_1 = \mathbf{w}_2$.

(c) If $\mathcal{L}_1 = \mathcal{L}_2$ and $\mathbf{w}_1 = \mathbf{w}_2$, then $\mathbf{v}_1 = \mathbf{v}_2$.

(d) If $\mathcal{L}_1$ is parallel to $\mathcal{L}_2$, then $\mathbf{w}_1 = \mathbf{w}_2$.

(e) If $\mathcal{L}_1$ is parallel to $\mathcal{L}_2$, then $\mathbf{w}_1 = \lambda\mathbf{w}_2$ for some scalar λ.

17. Find a and b such that the lines $\mathbf{r}_1 = \langle 1, 2, 1 \rangle + t\langle 1, -1, 1 \rangle$ and $\mathbf{r}_2 = \langle 3, -1, 1 \rangle + t\langle a, b, -2 \rangle$ are parallel.

18. Find a such that the lines $\mathbf{r}_1 = \langle 1, 2, 1 \rangle + t\langle 1, -1, 1 \rangle$ and $\mathbf{r}_2 = \langle 3, -1, 1 \rangle + t\langle a, 4, -2 \rangle$ intersect.

19. Sketch the vector sum $\mathbf{v} = \mathbf{v}_1 - \mathbf{v}_2 + \mathbf{v}_3$ for the vectors in Figure 1(A).

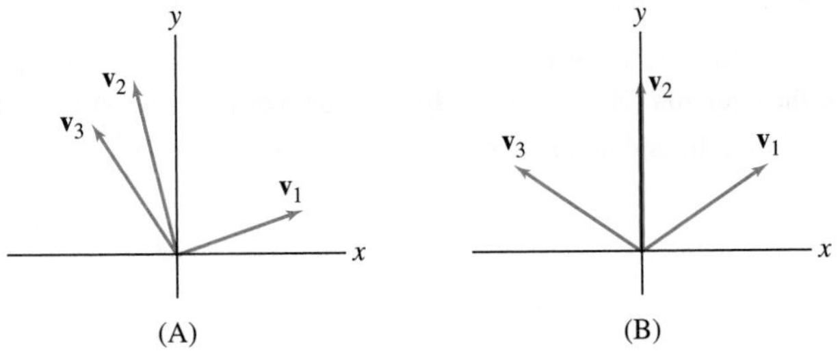

FIGURE 1

20. Sketch the sums $\mathbf{v}_1 + \mathbf{v}_2 + \mathbf{v}_3$, $\mathbf{v}_1 + 2\mathbf{v}_2$, and $\mathbf{v}_2 - \mathbf{v}_3$ for the vectors in Figure 1(B).

In Exercises 21–26, let $\mathbf{v} = \langle 1, 3, -2 \rangle$ *and* $\mathbf{w} = \langle 2, -1, 4 \rangle$.

21. Compute $\mathbf{v} \cdot \mathbf{w}$.

22. Compute the angle between $\mathbf{v}$ and $\mathbf{w}$.

23. Compute $\mathbf{v} \times \mathbf{w}$.

24. Find the area of the parallelogram spanned by $\mathbf{v}$ and $\mathbf{w}$.

25. Find the volume of the parallelepiped spanned by $\mathbf{v}$, $\mathbf{w}$, and $\mathbf{u} = \langle 1, 2, 6 \rangle$.

26. Find all the vectors orthogonal to both $\mathbf{v}$ and $\mathbf{w}$.

27. Use vectors to prove that the line connecting the midpoints of two sides of a triangle is parallel to the third side.

28. Let $\mathbf{v} = \langle 1, -1, 3 \rangle$ and $\mathbf{w} = \langle 4, -2, 1 \rangle$.

(a) Find the decomposition $\mathbf{v} = \mathbf{v}_{\parallel} + \mathbf{v}_{\perp}$ with respect to $\mathbf{w}$.

(b) Find the decomposition $\mathbf{w} = \mathbf{w}_{\parallel} + \mathbf{w}_{\perp}$ with respect to $\mathbf{v}$.

29. Calculate the component of $\mathbf{v} = \langle -2, \frac{1}{2}, 3 \rangle$ along $\mathbf{w} = \langle 1, 2, 2 \rangle$.

30. Calculate the magnitude of the forces on the two ropes in Figure 2.

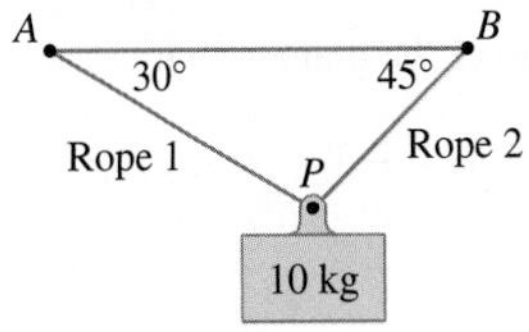

FIGURE 2

31. A 50-kg wagon is pulled to the right by a force $\mathbf{F}_1$ making an angle of 30° with the ground. At the same time the wagon is pulled to the left by a horizontal force $\mathbf{F}_2$.

(a) Find the magnitude of $\mathbf{F}_1$ in terms of the magnitude of $\mathbf{F}_2$ if the wagon does not move.

(b) What is the maximal magnitude of $\mathbf{F}_1$ that can be applied to the wagon without lifting it?

32. Let $\mathbf{v}$, $\mathbf{w}$, and $\mathbf{u}$ be the vectors in $\mathbf{R}^3$. Which of the following is a scalar?

(a) $\mathbf{v} \times (\mathbf{u} + \mathbf{w})$

(b) $(\mathbf{u} + \mathbf{w}) \cdot (\mathbf{v} \times \mathbf{w})$

(c) $(\mathbf{u} \times \mathbf{w}) + (\mathbf{w} - \mathbf{v})$

In Exercises 33–36, let $\mathbf{v} = \langle 1, 2, 4 \rangle$, $\mathbf{u} = \langle 6, -1, 2 \rangle$, *and* $\mathbf{w} = \langle 1, 0, -3 \rangle$. *Calculate the given quantity.*

33. $\mathbf{v} \times \mathbf{w}$

34. $\mathbf{w} \times \mathbf{u}$

35. $\det \begin{pmatrix} \mathbf{u} \\ \mathbf{v} \\ \mathbf{w} \end{pmatrix}$

36. $\mathbf{v} \cdot (\mathbf{u} \times \mathbf{w})$

37. Use the cross product to find the area of the triangle whose vertices are $(1, 3, -1)$, $(2, -1, 3)$, and $(4, 1, 1)$.

38. Calculate $\|\mathbf{v} \times \mathbf{w}\|$ if $\|\mathbf{v}\| = 2$, $\mathbf{v} \cdot \mathbf{w} = 3$, and the angle between $\mathbf{v}$ and $\mathbf{w}$ is $\frac{\pi}{6}$.

39. Show that if the vectors $\mathbf{v}$, $\mathbf{w}$ are orthogonal, then $\|\mathbf{v} + \mathbf{w}\|^2 = \|\mathbf{v}\|^2 + \|\mathbf{w}\|^2$.

40. Find the angle between $\mathbf{v}$ and $\mathbf{w}$ if $\|\mathbf{v}+\mathbf{w}\| = \|\mathbf{v}\| = \|\mathbf{w}\|$.

41. Find $\|\mathbf{e} - 4\mathbf{f}\|$, assuming that $\mathbf{e}$ and $\mathbf{f}$ are unit vectors such that $\|\mathbf{e}+\mathbf{f}\| = \sqrt{3}$.

42. Find the area of the parallelogram spanned by vectors $\mathbf{v}$ and $\mathbf{w}$ such that $\|\mathbf{v}\| = \|\mathbf{w}\| = 2$ and $\mathbf{v}\cdot\mathbf{w} = 1$.

43. Show that the equation $\langle 1, 2, 3\rangle \times \mathbf{v} = \langle -1, 2, a\rangle$ has no solution for $a \neq -1$.

44. Prove with a diagram the following: If $\mathbf{e}$ is a unit vector orthogonal to $\mathbf{v}$, then $\mathbf{e}\times(\mathbf{v}\times\mathbf{e}) = (\mathbf{e}\times\mathbf{v})\times\mathbf{e} = \mathbf{v}$.

45. Use the identity

$$\mathbf{u}\times(\mathbf{v}\times\mathbf{w}) = (\mathbf{u}\cdot\mathbf{w})\,\mathbf{v} - (\mathbf{u}\cdot\mathbf{v})\,\mathbf{w}$$

to prove that

$$\mathbf{u}\times(\mathbf{v}\times\mathbf{w}) + \mathbf{v}\times(\mathbf{w}\times\mathbf{u}) + \mathbf{w}\times(\mathbf{u}\times\mathbf{v}) = \mathbf{0}$$

46. Find an equation of the plane through $(1, -3, 5)$ with normal vector $\mathbf{n} = \langle 2, 1, -4\rangle$.

47. Write the equation of the plane $\mathcal{P}$ with vector equation

$$\langle 1, 4, -3\rangle \cdot \langle x, y, z\rangle = 7$$

in the form

$$a\,(x - x_0) + b\,(y - y_0) + c\,(z - z_0) = 0$$

Hint: You must find a point $P = (x_0, y_0, z_0)$ on $\mathcal{P}$.

48. Find all the planes parallel to the plane passing through the points $(1, 2, 3)$, $(1, 2, 7)$, and $(1, 1, -3)$.

49. Find the plane through $P = (4, -1, 9)$ containing the line $\mathbf{r}(t) = \langle 1, 4, -3\rangle + t\langle 2, 1, 1\rangle$.

50. Find the intersection of the line $\mathbf{r}(t) = \langle 3t + 2, 1, -7t\rangle$ and the plane $2x - 3y + z = 5$.

51. Find the trace of the plane $3x - 2y + 5z = 4$ in the xy-plane.

52. Find the intersection of the planes $x + y + z = 1$ and $3x - 2y + z = 5$.

In Exercises 53–58, determine the type of the quadric surface.

53. $\left(\frac{x}{3}\right)^2 + \left(\frac{y}{4}\right)^2 + 2z^2 = 1$

54. $\left(\frac{x}{3}\right)^2 - \left(\frac{y}{4}\right)^2 + 2z^2 = 1$

55. $\left(\frac{x}{3}\right)^2 + \left(\frac{y}{4}\right)^2 - 2z = 0$

56. $\left(\frac{x}{3}\right)^2 - \left(\frac{y}{4}\right)^2 - 2z = 0$

57. $\left(\frac{x}{3}\right)^2 - \left(\frac{y}{4}\right)^2 - 2z^2 = 0$

58. $\left(\frac{x}{3}\right)^2 - \left(\frac{y}{4}\right)^2 - 2z^2 = 1$

59. Determine the type of the quadric surface $ax^2 + by^2 - z^2 = 1$ if:

(a) $a < 0, \quad b < 0$

(b) $a > 0, \quad b > 0$

(c) $a > 0, \quad b < 0$

60. Describe the traces of the surface

$$\left(\frac{x}{2}\right)^2 - y^2 + \left(\frac{z}{2}\right)^2 = 1$$

in the three coordinate planes.

61. Convert $(x, y, z) = (3, 4, -1)$ from rectangular to cylindrical and spherical coordinates.

62. Convert $(r, \theta, z) = \left(3, \frac{\pi}{6}, 4\right)$ from cylindrical to spherical coordinates.

63. Convert the point $(\rho, \theta, \phi) = \left(3, \frac{\pi}{6}, \frac{\pi}{3}\right)$ from spherical to cylindrical coordinates.

64. Describe the set of all points $P = (x, y, z)$ satisfying $x^2 + y^2 \le 4$ in both cylindrical and spherical coordinates.

65. Sketch the graph of the cylindrical equation $z = 2r\cos\theta$ and write the equation in rectangular coordinates.

66. Write the surface $x^2 + y^2 - z^2 = 2\,(x + y)$ as an equation $r = f\,(\theta, z)$ in cylindrical coordinates.

67. Show that the cylindrical equation

$$r^2(1 - 2\sin^2\theta) + z^2 = 1$$

is a hyperboloid of one sheet.

68. Sketch the graph of the spherical equation $\rho = 2\cos\theta\sin\phi$ and write the equation in rectangular coordinates.

69. Describe how the surface with spherical equation

$$\rho^2(1 + A\cos^2\phi) = 1$$

depends on the constant A.

70. Show that the spherical equation $\cot\phi = 2\cos\theta + \sin\theta$ defines a plane through the origin (with the origin excluded). Find a normal vector to this plane.

71. Let c be a scalar, let $\mathbf{a}$ and $\mathbf{b}$ be vectors, and let $\mathbf{X} = \langle x, y, z\rangle$. Show that the equation $(\mathbf{X} - \mathbf{a})\cdot(\mathbf{X} - \mathbf{b}) = c^2$ defines a sphere with center $\mathbf{m} = \frac{1}{2}(\mathbf{a} + \mathbf{b})$ and radius R, where $R^2 = c^2 + \left\|\frac{1}{2}(\mathbf{a} - \mathbf{b})\right\|^2$.

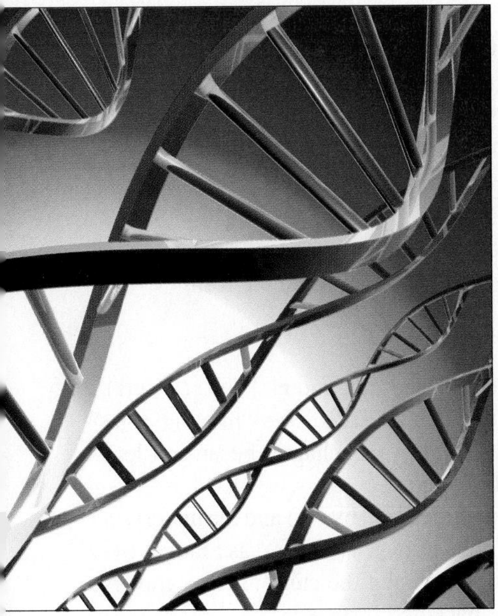

DNA polymers form helical curves whose spatial orientation influences their biochemical properties.

14 CALCULUS OF VECTOR-VALUED FUNCTIONS

In this chapter, we study vector-valued functions and their derivatives, and we use them to analyze curves and motion in three-space. Although many techniques from single-variable calculus carry over to the vector setting, there is an important new aspect to the derivative. A real-valued function $f(x)$ can change in just one of two ways: It can increase or decrease. By contrast, a vector-valued function can change not just in magnitude but also in direction, and the rate of change is not a single number but is itself a vector. To develop these new concepts, we begin with an introduction to vector-valued functions.

14.1 Vector-Valued Functions

Consider a particle moving in $\mathbf{R}^3$ whose coordinates at time t are $(x(t), y(t), z(t))$. It is convenient to represent the particle's path by the **vector-valued function**

$$\mathbf{r}(t) = \langle x(t), y(t), z(t)\rangle = x(t)\mathbf{i} + y(t)\mathbf{j} + z(t)\mathbf{k} \qquad \boxed{1}$$

Functions $f(x)$ (with real number values) are often called scalar-valued *to distinguish them from vector-valued functions.*

Think of $\mathbf{r}(t)$ as a moving vector that points from the origin to the position of the particle at time t (Figure 1).

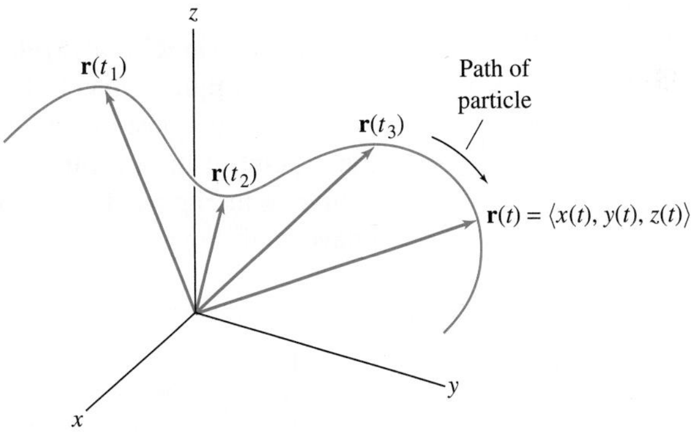

FIGURE 1

The parameter is often called t (for time), but we are free to use any other variable such as s or θ. It is best to avoid writing $\mathbf{r}(x)$ or $\mathbf{r}(y)$ to prevent confusion with the x- and y-components of $\mathbf{r}$.

More generally, a vector-valued function is any function $\mathbf{r}(t)$ of the form in Eq. (1) whose domain $\mathcal{D}$ is a set of real numbers and whose range is a set of position vectors. The variable t is called a **parameter**, and the functions $x(t), y(t), z(t)$ are called the **components** or **coordinate functions**. We usually take as domain the set of all values of t for which $\mathbf{r}(t)$ is defined—that is, all values of t that belong to the domains of all three coordinate functions $x(t), y(t), z(t)$. For example,

$$\mathbf{r}(t) = \langle t^2, e^t, 4 - 7t\rangle, \qquad \text{domain } \mathcal{D} = \mathbf{R}$$

$$\mathbf{r}(s) = \langle \sqrt{s}, e^s, s^{-1}\rangle, \qquad \text{domain } \mathcal{D} = \{s \in \mathbf{R} : s > 0\}$$

The terminal point of a vector-valued function $\mathbf{r}(t)$ traces a path in $\mathbf{R}^3$ as t varies. We refer to $\mathbf{r}(t)$ either as a path or as a **vector parametrization** of a path. We shall assume throughout this chapter that the components of $\mathbf{r}(t)$ have continuous derivatives.

We have already studied special cases of vector parametrizations. In Chapter 13, we described lines in $\mathbf{R}^3$ using vector parametrizations. Recall that

$$\mathbf{r}(t) = \langle x_0, y_0, z_0 \rangle + t\mathbf{v} = \langle x_0 + ta, y_0 + tb, z_0 + tc \rangle$$

parametrizes the line through $P = (x_0, y_0, z_0)$ in the direction of the vector $\mathbf{v} = \langle a, b, c \rangle$.

In Chapter 12, we studied parametrized curves in the plane $\mathbf{R}^2$ in the form

$$c(t) = (x(t), y(t))$$

Such a curve is described equally well by the vector-valued function $\mathbf{r}(t) = \langle x(t), y(t) \rangle$. The difference lies only in whether we visualize the path as traced by a "moving point" $c(t)$ or a "moving vector" $\mathbf{r}(t)$. The vector form is used in this chapter because it leads most naturally to the definition of vector-valued derivatives.

It is important to distinguish between the path parametrized by $\mathbf{r}(t)$ and the underlying curve $\mathcal{C}$ traced by $\mathbf{r}(t)$. The curve $\mathcal{C}$ is the set of all points $(x(t), y(t), z(t))$ as t ranges over the domain of $\mathbf{r}(t)$. The path is a particular way of traversing the curve; it may traverse the curve several times, reverse direction, or move back and forth, etc.

EXAMPLE 1 The Path versus the Curve Describe the path

$$\mathbf{r}(t) = \langle \cos t, \sin t, 1 \rangle, \qquad -\infty < t < \infty$$

How are the path and the curve $\mathcal{C}$ traced by $\mathbf{r}(t)$ different?

Solution As t varies from $-\infty$ to ∞, the endpoint of the vector $\mathbf{r}(t)$ moves around a unit circle at height $z = 1$ infinitely many times in the counterclockwise direction when viewed from above (Figure 2). The underlying curve $\mathcal{C}$ traced by $\mathbf{r}(t)$ is the circle itself. ■

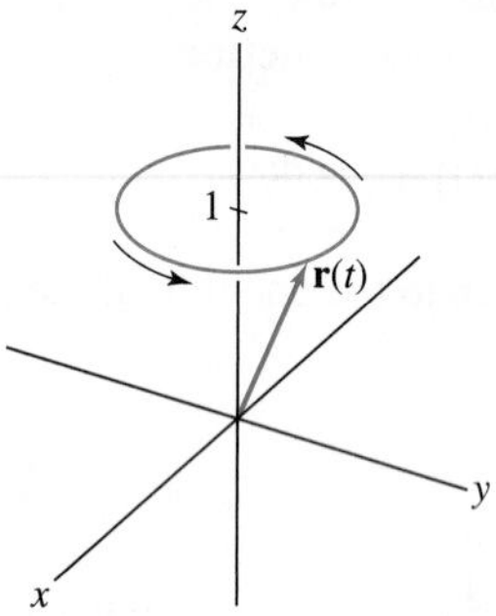

FIGURE 2 Plot of $\mathbf{r}(t) = \langle \cos t, \sin t, 1 \rangle$.

A curve in $\mathbf{R}^3$ is also referred to as a **space curve** (as opposed to a curve in $\mathbf{R}^2$, which is called a **plane curve**). Space curves can be quite complicated and difficult to sketch by hand. The most effective way to visualize a space curve is to plot it from different viewpoints using a computer (Figure 3). As an aid to visualization, we plot a "thickened" curve as in Figures 3 and 5, but keep in mind that space curves are one-dimensional and have no thickness.

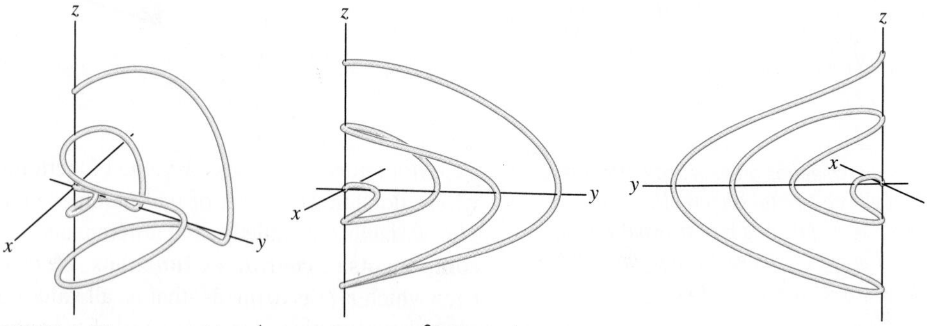

FIGURE 3 The curve $\mathbf{r}(t) = \langle t \sin 2t \cos t, t \sin^2 t, t \cos t \rangle$ for $0 \le t \le 4\pi$, seen from three different viewpoints.

The projections onto the coordinate planes are another aid in visualizing space curves. The projection of a path $\mathbf{r}(t) = \langle x(t), y(t), z(t) \rangle$ onto the xy-plane is the path $\mathbf{p}(t) = \langle x(t), y(t), 0 \rangle$ (Figure 4). Similarly, the projections onto the yz- and xz-planes are the paths $\langle 0, y(t), z(t) \rangle$ and $\langle x(t), 0, z(t) \rangle$, respectively.

EXAMPLE 2 Helix Describe the curve traced by $\mathbf{r}(t) = \langle -\sin t, \cos t, t\rangle$ for $t \geq 0$ in terms of its projections onto the coordinate planes.

Solution The projections are as follows (Figure 4):

- xy-plane (set $z = 0$): the path $\mathbf{p}(t) = \langle -\sin t, \cos t, 0\rangle$, which describes a point moving counterclockwise around the unit circle starting at $\mathbf{p}(0) = (0, 1, 0)$.
- xz-plane (set $y = 0$): the path $\langle -\sin t, 0, t\rangle$, which is a wave in the z-direction.
- yz-plane (set $x = 0$): the path $\langle 0, \cos t, t\rangle$, which is a wave in the z-direction.

The function $\mathbf{r}(t)$ describes a point moving above the unit circle in the xy-plane while its height $z = t$ increases linearly, resulting in the helix of Figure 4. ■

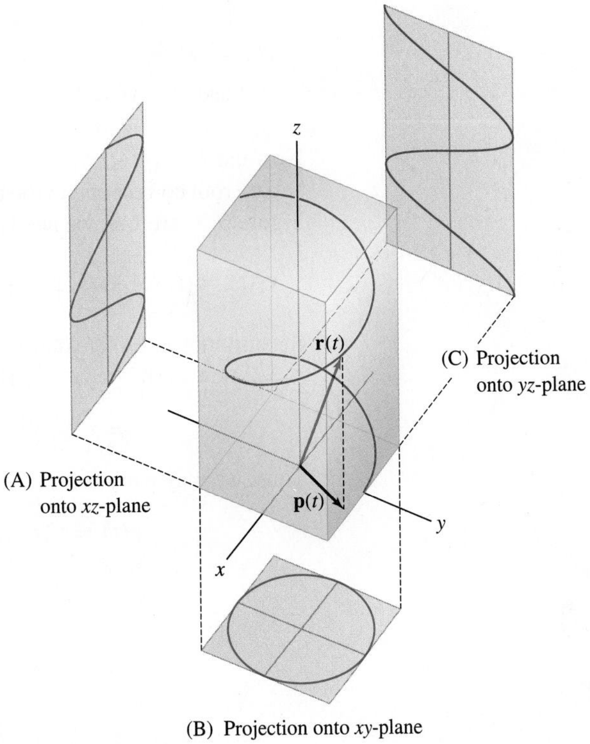

FIGURE 4 Projections of the helix $\mathbf{r}(t) = \langle -\sin t, \cos t, t\rangle$.

Every curve can be parametrized in infinitely many ways (because there are infinitely many ways that a point can traverse a curve as a function of time). The next example describes two very different parametrizations of the same curve.

EXAMPLE 3 Parametrizing the Intersection of Surfaces Parametrize the curve $\mathcal{C}$ obtained as the intersection of the surfaces $x^2 - y^2 = z - 1$ and $x^2 + y^2 = 4$ (Figure 5).

Solution We have to express the coordinates (x, y, z) of a point on the curve as functions of a parameter t. Here are two ways of doing this.

First method: Solve the given equations for y and z in terms of x. First, solve for y:

$$x^2 + y^2 = 4 \quad \Rightarrow \quad y^2 = 4 - x^2 \quad \Rightarrow \quad y = \pm\sqrt{4 - x^2}$$

The equation $x^2 - y^2 = z - 1$ can be written $z = x^2 - y^2 + 1$. Thus, we can substitute $y^2 = 4 - x^2$ to solve for z:

$$z = x^2 - y^2 + 1 = x^2 - (4 - x^2) + 1 = 2x^2 - 3$$

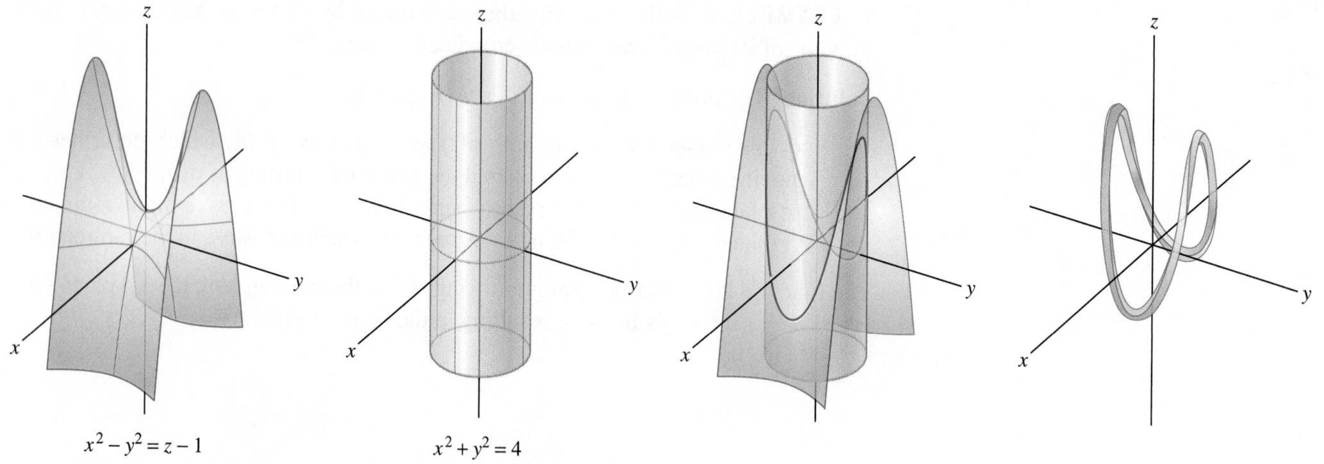

FIGURE 5 Intersection of surfaces $x^2 - y^2 = z - 1$ and $x^2 + y^2 = 4$.

Now use $t = x$ as the parameter. Then $y = \pm\sqrt{4 - t^2}$, $z = 2t^2 - 3$. The two signs of the square root correspond to the two halves of the curve where $y > 0$ and $y < 0$, as shown in Figure 6. Therefore, we need two vector-valued functions to parametrize the entire curve:

$$\mathbf{r}_1(t) = \left\langle t, \sqrt{4 - t^2}, 2t^2 - 3 \right\rangle, \quad \mathbf{r}_2(t) = \left\langle t, -\sqrt{4 - t^2}, 2t^2 - 3 \right\rangle, \qquad -2 \le t \le 2$$

Second method: Note that $x^2 + y^2 = 4$ has a trigonometric parametrization: $x = 2\cos t$, $y = 2\sin t$ for $0 \le t < 2\pi$. The equation $x^2 - y^2 = z - 1$ gives us

$$z = x^2 - y^2 + 1 = 4\cos^2 t - 4\sin^2 t + 1 = 4\cos 2t + 1$$

Thus, we may parametrize the entire curve by a single vector-valued function:

$$\mathbf{r}(t) = \langle 2\cos t, 2\sin t, 4\cos 2t + 1 \rangle, \qquad 0 \le t < 2\pi$$

■

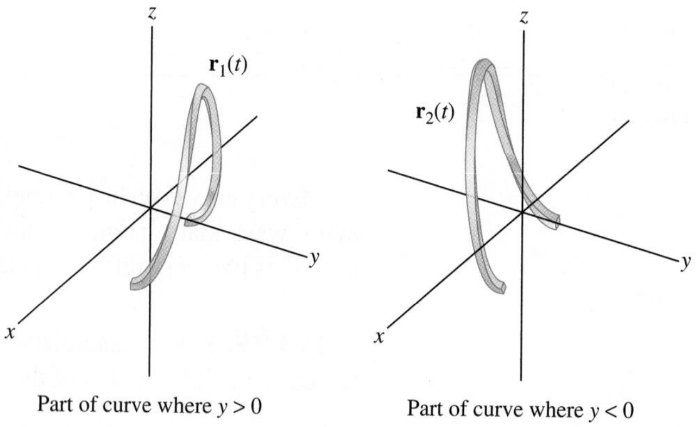

FIGURE 6 Two halves of the curve of intersection in Example 3.

■ **EXAMPLE 4** Parametrize the circle of radius 3 with center $P = (2, 6, 8)$ located in a plane:

(a) Parallel to the xy-plane **(b)** Parallel to the xz-plane

Solution **(a)** A circle of radius R in the xy-plane centered at the origin has parametrization $\langle R\cos t, R\sin t \rangle$. To place the circle in a three-dimensional coordinate system, we use the parametrization $\langle R\cos t, R\sin t, 0 \rangle$.

Thus, the circle of radius 3 centered at (0, 0, 0) has parametrization $\langle 3\cos t, 3\sin t, 0\rangle$. To move this circle in a parallel fashion so that its center lies at $P = (2, 6, 8)$, we translate by the vector $\langle 2, 6, 8\rangle$:

$$\mathbf{r}_1(t) = \langle 2, 6, 8\rangle + \langle 3\cos t, 3\sin t, 0\rangle = \langle 2 + 3\cos t, 6 + 3\sin t, 8\rangle$$

(b) The parametrization $\langle 3\cos t, 0, 3\sin t\rangle$ gives us a circle of radius 3 centered at the origin in the xz-plane. To move the circle in a parallel fashion so that its center lies at (2, 6, 8), we translate by the vector $\langle 2, 6, 8\rangle$:

$$\mathbf{r}_2(t) = \langle 2, 6, 8\rangle + \langle 3\cos t, 0, 3\sin t\rangle = \langle 2 + 3\cos t, 6, 8 + 3\sin t\rangle$$

These two circles are shown in Figure 7. ■

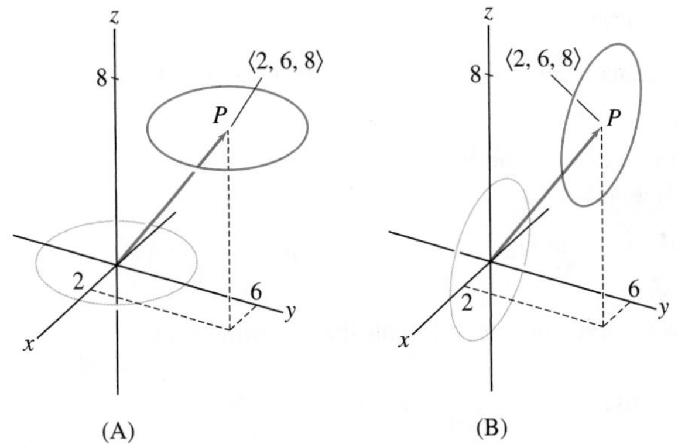

FIGURE 7 Horizontal and vertical circles of radius 3 and center $P = (2, 6, 8)$ obtained by translation.

14.1 SUMMARY

- A *vector-valued function* is a function of the form

$$\mathbf{r}(t) = \langle x(t), y(t), z(t)\rangle = x(t)\mathbf{i} + y(t)\mathbf{j} + z(t)\mathbf{k}$$

- We often think of t as time and $\mathbf{r}(t)$ as a "moving vector" whose terminal point traces out a path as a function of time. We refer to $\mathbf{r}(t)$ as a *vector parametrization* of the path, or simply as a "path."
- The underlying curve $\mathcal{C}$ traced by $\mathbf{r}(t)$ is the set of all points $(x(t), y(t), z(t))$ in $\mathbf{R}^3$ for t in the domain of $\mathbf{r}(t)$. A curve in $\mathbf{R}^3$ is also called a *space curve*.
- Every curve $\mathcal{C}$ can be parametrized in infinitely many ways.
- The projection of $\mathbf{r}(t)$ onto the xy-plane is the curve traced by $\langle x(t), y(t), 0\rangle$. The projection onto the xz-plane is $\langle x(t), 0, z(t)\rangle$, and the projection onto the yz-plane is $\langle 0, y(t), z(t)\rangle$.

14.1 EXERCISES

Preliminary Questions

1. Which one of the following does *not* parametrize a line?

(a) $\mathbf{r}_1(t) = \langle 8 - t, 2t, 3t\rangle$

(b) $\mathbf{r}_2(t) = t^3\mathbf{i} - 7t^3\mathbf{j} + t^3\mathbf{k}$

(c) $\mathbf{r}_3(t) = \langle 8 - 4t^3, 2 + 5t^2, 9t^3\rangle$

2. What is the projection of $\mathbf{r}(t) = t\mathbf{i} + t^4\mathbf{j} + e^t\mathbf{k}$ onto the xz-plane?

3. Which projection of $\langle \cos t, \cos 2t, \sin t\rangle$ is a circle?

4. What is the center of the circle with parametrization

$$\mathbf{r}(t) = (-2 + \cos t)\mathbf{i} + 2\mathbf{j} + (3 - \sin t)\mathbf{k}?$$

5. How do the paths $\mathbf{r}_1(t) = \langle \cos t, \sin t \rangle$ and $\mathbf{r}_2(t) = \langle \sin t, \cos t \rangle$ around the unit circle differ?

6. Which three of the following vector-valued functions parametrize the same space curve?

(a) $(-2 + \cos t)\mathbf{i} + 9\mathbf{j} + (3 - \sin t)\mathbf{k}$

(b) $(2 + \cos t)\mathbf{i} - 9\mathbf{j} + (-3 - \sin t)\mathbf{k}$

(c) $(-2 + \cos 3t)\mathbf{i} + 9\mathbf{j} + (3 - \sin 3t)\mathbf{k}$

(d) $(-2 - \cos t)\mathbf{i} + 9\mathbf{j} + (3 + \sin t)\mathbf{k}$

(e) $(2 + \cos t)\mathbf{i} + 9\mathbf{j} + (3 + \sin t)\mathbf{k}$

Exercises

1. What is the domain of $\mathbf{r}(t) = e^t\mathbf{i} + \dfrac{1}{t}\mathbf{j} + (t+1)^{-3}\mathbf{k}$?

2. What is the domain of $\mathbf{r}(s) = e^s\mathbf{i} + \sqrt{s}\mathbf{j} + \cos s\mathbf{k}$?

3. Evaluate $\mathbf{r}(2)$ and $\mathbf{r}(-1)$ for $\mathbf{r}(t) = \left\langle \sin \frac{\pi}{2}t, t^2, (t^2+1)^{-1} \right\rangle$.

4. Does either of $P = (4, 11, 20)$ or $Q = (-1, 6, 16)$ lie on the path $\mathbf{r}(t) = \langle 1 + t, 2 + t^2, t^4 \rangle$?

5. Find a vector parametrization of the line through $P = (3, -5, 7)$ in the direction $\mathbf{v} = \langle 3, 0, 1 \rangle$.

6. Find a direction vector for the line with parametrization $\mathbf{r}(t) = (4 - t)\mathbf{i} + (2 + 5t)\mathbf{j} + \frac{1}{2}t\mathbf{k}$.

7. Match the space curves in Figure 8 with their projections onto the xy-plane in Figure 9.

8. Match the space curves in Figure 8 with the following vector-valued functions:

(a) $\mathbf{r}_1(t) = \langle \cos 2t, \cos t, \sin t \rangle$ **(b)** $\mathbf{r}_2(t) = \langle t, \cos 2t, \sin 2t \rangle$

(c) $\mathbf{r}_3(t) = \langle 1, t, t \rangle$

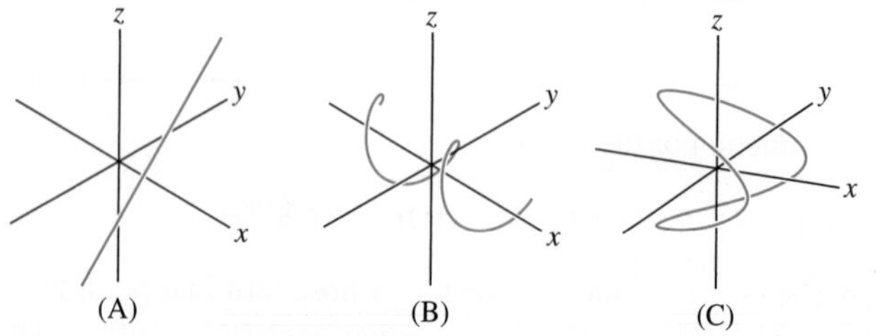

FIGURE 8

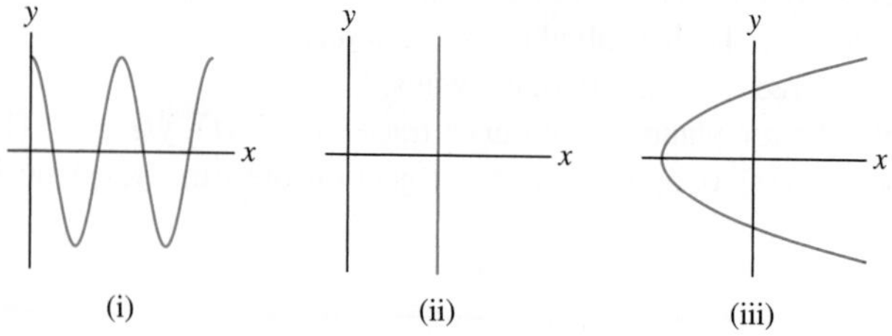

FIGURE 9

9. Match the vector-valued functions (a)–(f) with the space curves (i)–(vi) in Figure 10.

(a) $\mathbf{r}(t) = \langle t + 15, e^{0.08t}\cos t, e^{0.08t}\sin t \rangle$

(b) $\mathbf{r}(t) = \langle \cos t, \sin t, \sin 12t \rangle$ **(c)** $\mathbf{r}(t) = \left\langle t, t, \dfrac{25t}{1+t^2} \right\rangle$

(d) $\mathbf{r}(t) = \langle \cos^3 t, \sin^3 t, \sin 2t \rangle$ **(e)** $\mathbf{r}(t) = \langle t, t^2, 2t \rangle$

(f) $\mathbf{r}(t) = \langle \cos t, \sin t, \cos t \sin 12t \rangle$

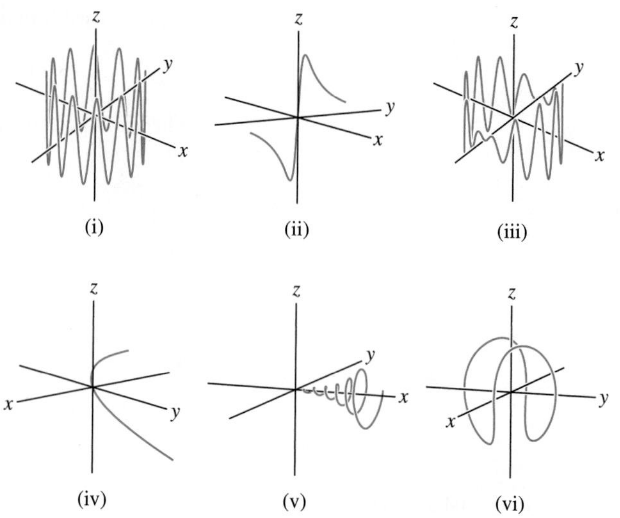

FIGURE 10

10. Which of the following curves have the same projection onto the xy-plane?

(a) $\mathbf{r}_1(t) = \langle t, t^2, e^t \rangle$ **(b)** $\mathbf{r}_2(t) = \langle e^t, t^2, t \rangle$

(c) $\mathbf{r}_3(t) = \langle t, t^2, \cos t \rangle$

11. Match the space curves (A)–(C) in Figure 11 with their projections (i)–(iii) onto the xy-plane.

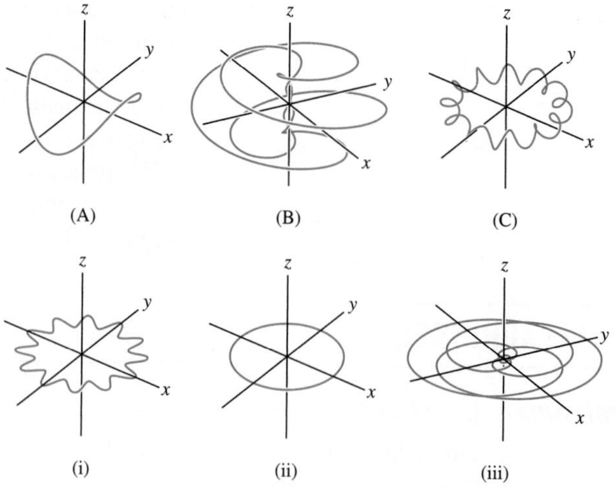

FIGURE 11

12. Describe the projections of the circle $\mathbf{r}(t) = \langle \sin t, 0, 4 + \cos t \rangle$ onto the coordinate planes.

In Exercises 13–16, the function $\mathbf{r}(t)$ *traces a circle. Determine the radius, center, and plane containing the circle.*

13. $\mathbf{r}(t) = (9\cos t)\mathbf{i} + (9\sin t)\mathbf{j}$

14. $\mathbf{r}(t) = 7\mathbf{i} + (12\cos t)\mathbf{j} + (12\sin t)\mathbf{k}$

15. $\mathbf{r}(t) = \langle \sin t, 0, 4 + \cos t \rangle$

16. $\mathbf{r}(t) = \langle 6 + 3\sin t, 9, 4 + 3\cos t \rangle$

17. Let $\mathcal{C}$ be the curve $\mathbf{r}(t) = \langle t\cos t, t\sin t, t \rangle$.

(a) Show that $\mathcal{C}$ lies on the cone $x^2 + y^2 = z^2$.

(b) Sketch the cone and make a rough sketch of $\mathcal{C}$ on the cone.

18. CAS Use a computer algebra system to plot the projections onto the xy- and xz-planes of the curve $\mathbf{r}(t) = \langle t\cos t, t\sin t, t \rangle$ in Exercise 17.

In Exercises 19 and 20, let

$$\mathbf{r}(t) = \langle \sin t, \cos t, \sin t \cos 2t \rangle$$

as shown in Figure 12.

19. Find the points where $\mathbf{r}(t)$ intersects the xy-plane.

20. Show that the projection of $\mathbf{r}(t)$ onto the xz-plane is the curve

$$z = x - 2x^3 \quad \text{for} \quad -1 \le x \le 1$$

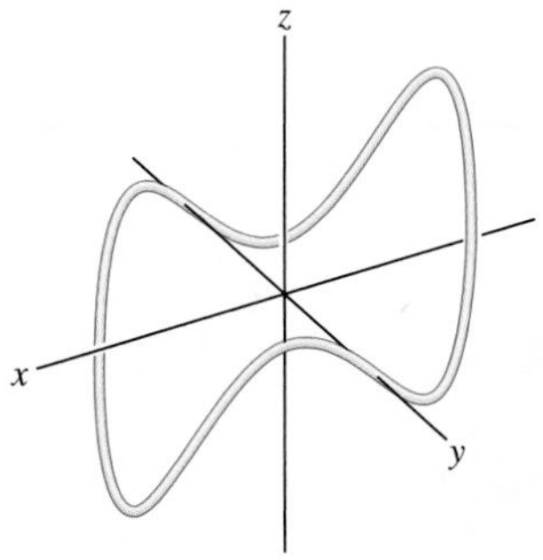

FIGURE 12

21. Parametrize the intersection of the surfaces

$$y^2 - z^2 = x - 2, \qquad y^2 + z^2 = 9$$

using $t = y$ as the parameter (two vector functions are needed as in Example 3).

22. Find a parametrization of the curve in Exercise 21 using trigonometric functions.

23. Viviani's Curve $\mathcal{C}$ is the intersection of the surfaces (Figure 13)

$$x^2 + y^2 = z^2, \qquad y = z^2$$

(a) Parametrize each of the two parts of $\mathcal{C}$ corresponding to $x \ge 0$ and $x \le 0$, taking $t = z$ as parameter.

(b) Describe the projection of $\mathcal{C}$ onto the xy-plane.

(c) Show that $\mathcal{C}$ lies on the sphere of radius 1 with center $(0, 1, 0)$. This curve looks like a figure eight lying on a sphere [Figure 13(B)].

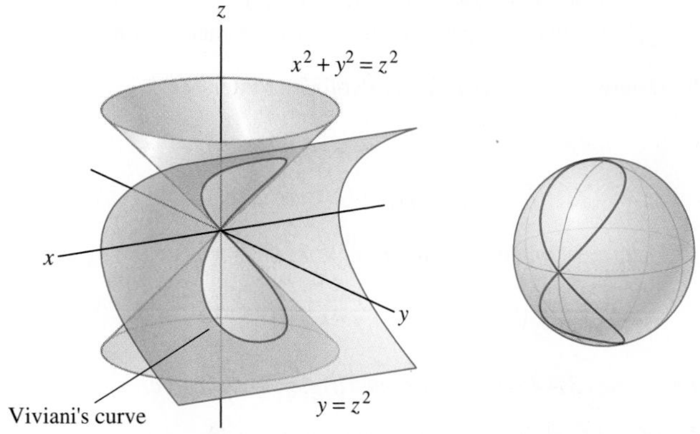

(A)

(B) Viviani's curve viewed from the negative y-axis

FIGURE 13 Viviani's curve is the intersection of the surfaces $x^2 + y^2 = z^2$ and $y = z^2$.

24. Show that any point on $x^2 + y^2 = z^2$ can be written in the form $(z\cos\theta, z\sin\theta, z)$ for some θ. Use this to find a parametrization of Viviani's curve (Exercise 23) with θ as parameter.

25. Use sine and cosine to parametrize the intersection of the cylinders $x^2 + y^2 = 1$ and $x^2 + z^2 = 1$ (use two vector-valued functions). Then describe the projections of this curve onto the three coordinate planes.

26. Use hyperbolic functions to parametrize the intersection of the surfaces $x^2 - y^2 = 4$ and $z = xy$.

27. Use sine and cosine to parametrize the intersection of the surfaces $x^2 + y^2 = 1$ and $z = 4x^2$ (Figure 14).

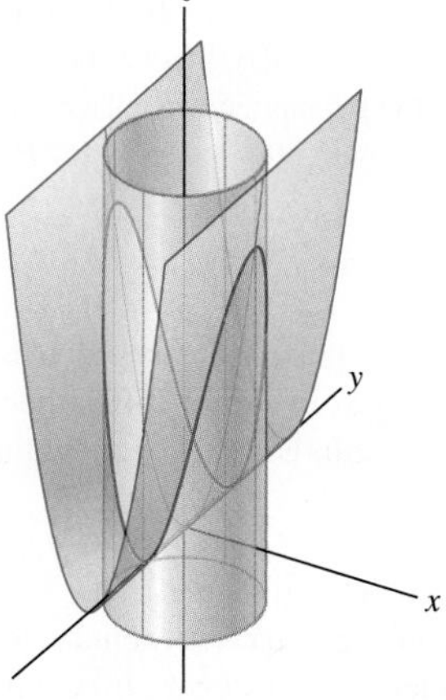

FIGURE 14 Intersection of the surfaces $x^2 + y^2 = 1$ and $z = 4x^2$.

In Exercises 28–30, two paths $\mathbf{r}_1(t)$ and $\mathbf{r}_2(t)$ intersect if there is a point P lying on both curves. We say that $\mathbf{r}_1(t)$ and $\mathbf{r}_2(t)$ collide if $\mathbf{r}_1(t_0) = \mathbf{r}_2(t_0)$ at some time t_0.

28. Which of the following statements are true?

(a) If $\mathbf{r}_1$ and $\mathbf{r}_2$ intersect, then they collide.

(b) If $\mathbf{r}_1$ and $\mathbf{r}_2$ collide, then they intersect.

(c) Intersection depends only on the underlying curves traced by $\mathbf{r}_1$ and $\mathbf{r}_2$, but collision depends on the actual parametrizations.

29. Determine whether $\mathbf{r}_1$ and $\mathbf{r}_2$ collide or intersect:

$$\mathbf{r}_1(t) = \langle t^2 + 3, t + 1, 6t^{-1} \rangle$$

$$\mathbf{r}_2(t) = \langle 4t, 2t - 2, t^2 - 7 \rangle$$

30. Determine whether $\mathbf{r}_1$ and $\mathbf{r}_2$ collide or intersect:

$$\mathbf{r}_1(t) = \langle t, t^2, t^3 \rangle, \qquad \mathbf{r}_2(t) = \langle 4t + 6, 4t^2, 7 - t \rangle$$

In Exercises 31–40, find a parametrization of the curve.

31. The vertical line passing through the point $(3, 2, 0)$

32. The line passing through $(1, 0, 4)$ and $(4, 1, 2)$

33. The line through the origin whose projection on the xy-plane is a line of slope 3 and whose projection on the yz-plane is a line of slope 5 (i.e., $\Delta z/\Delta y = 5$)

34. The horizontal circle of radius 1 with center $(2, -1, 4)$

35. The circle of radius 2 with center $(1, 2, 5)$ in a plane parallel to the yz-plane

36. The ellipse $\left(\frac{x}{2}\right)^2 + \left(\frac{y}{3}\right)^2 = 1$ in the xy-plane, translated to have center $(9, -4, 0)$

37. The intersection of the plane $y = \frac{1}{2}$ with the sphere $x^2 + y^2 + z^2 = 1$

38. The intersection of the surfaces

$$z = x^2 - y^2 \quad \text{and} \quad z = x^2 + xy - 1$$

39. The ellipse $\left(\frac{x}{2}\right)^2 + \left(\frac{z}{3}\right)^2 = 1$ in the xz-plane, translated to have center $(3, 1, 5)$ [Figure 15(A)]

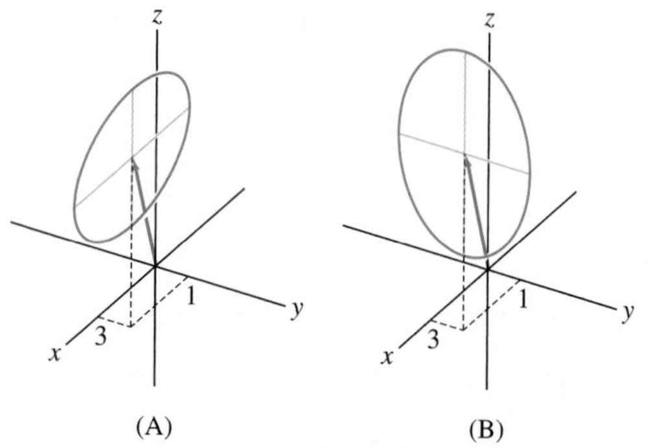

FIGURE 15 The ellipses described in Exercises 39 and 40.

40. The ellipse $\left(\frac{y}{2}\right)^2 + \left(\frac{z}{3}\right)^2 = 1$, translated to have center $(3, 1, 5)$ [Figure 15(B)]

Further Insights and Challenges

41. Sketch the curve parametrized by $\mathbf{r}(t) = \langle |t| + t, |t| - t \rangle$.

42. Find the maximum height above the xy-plane of a point on $\mathbf{r}(t) = \langle e^t, \sin t, t(4 - t) \rangle$.

43. Let C be the curve obtained by intersecting a cylinder of radius r and a plane. Insert two spheres of radius r into the cylinder above and below the plane, and let F_1 and F_2 be the points where the plane is tangent to the spheres [Figure 16(A)]. Let K be the vertical distance between the equators of the two spheres. Rediscover Archimedes's proof that C is an ellipse by showing that every point P on C satisfies

$$PF_1 + PF_2 = K \qquad \boxed{2}$$

Hint: If two lines through a point P are tangent to a sphere and intersect the sphere at Q_1 and Q_2 as in Figure 16(B), then the segments $\overline{PQ_1}$ and $\overline{PQ_2}$ have equal length. Use this to show that $PF_1 = PR_1$ and $PF_2 = PR_2$.

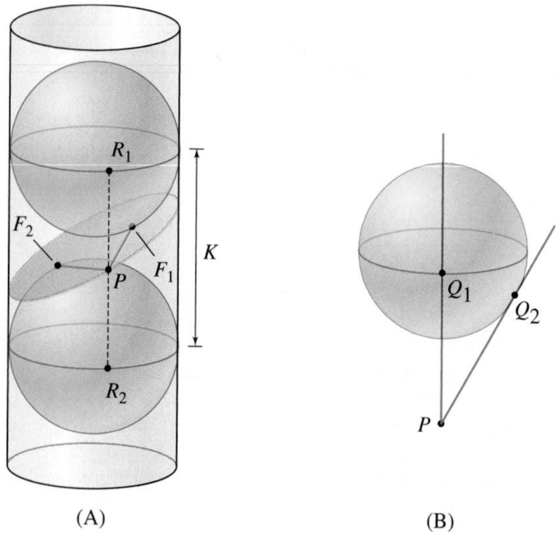

FIGURE 16

44. Assume that the cylinder in Figure 16 has equation $x^2 + y^2 = r^2$ and the plane has equation $z = ax + by$. Find a vector parametrization $\mathbf{r}(t)$ of the curve of intersection using the trigonometric functions $\cos t$ and $\sin t$.

45. CAS Now reprove the result of Exercise 43 using vector geometry. Assume that the cylinder has equation $x^2 + y^2 = r^2$ and the plane has equation $z = ax + by$.

(a) Show that the upper and lower spheres in Figure 16 have centers

$$C_1 = \left(0, 0, r\sqrt{a^2+b^2+1}\right)$$

$$C_2 = \left(0, 0, -r\sqrt{a^2+b^2+1}\right)$$

(b) Show that the points where the plane is tangent to the sphere are

$$F_1 = \frac{r}{\sqrt{a^2+b^2+1}}\left(a, b, a^2+b^2\right)$$

$$F_2 = \frac{-r}{\sqrt{a^2+b^2+1}}\left(a, b, a^2+b^2\right)$$

Hint: Show that $\overline{C_1F_1}$ and $\overline{C_2F_2}$ have length r and are orthogonal to the plane.

(c) Verify, with the aid of a computer algebra system, that Eq. (2) holds with

$$K = 2r\sqrt{a^2+b^2+1}$$

To simplify the algebra, observe that since a and b are arbitrary, it suffices to verify Eq. (2) for the point $P = (r, 0, ar)$.

14.2 Calculus of Vector-Valued Functions

In this section, we extend differentiation and integration to vector-valued functions. This is straightforward because the techniques of single-variable calculus carry over with little change. What is new and important, however, is the geometric interpretation of the derivative as a tangent vector. We describe this later in the section.

The first step is to define limits of vector-valued functions.

DEFINITION Limit of a Vector-Valued Function A vector-valued function $\mathbf{r}(t)$ approaches the limit $\mathbf{u}$ (a vector) as t approaches t_0 if $\lim_{t\to t_0} \|\mathbf{r}(t) - \mathbf{u}\| = 0$. In this case, we write

$$\lim_{t\to t_0} \mathbf{r}(t) = \mathbf{u}$$

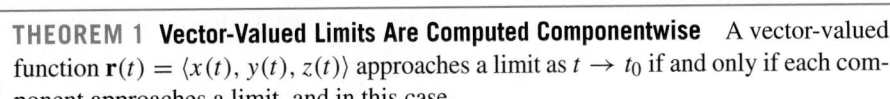

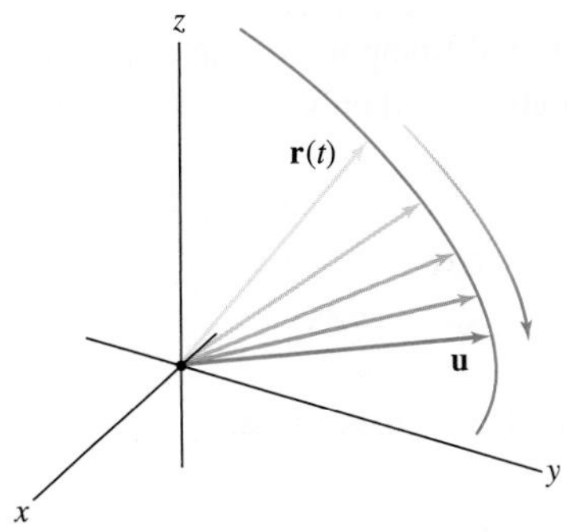

FIGURE 1 The vector-valued function $\mathbf{r}(t)$ approaches $\mathbf{u}$ as $t \to t_0$.

We can visualize the limit of a vector-valued function as a vector $\mathbf{r}(t)$ "moving" toward the limit vector $\mathbf{u}$ (Figure 1). According to the next theorem, vector limits may be computed componentwise.

THEOREM 1 Vector-Valued Limits Are Computed Componentwise A vector-valued function $\mathbf{r}(t) = \langle x(t), y(t), z(t)\rangle$ approaches a limit as $t \to t_0$ if and only if each component approaches a limit, and in this case,

$$\lim_{t\to t_0} \mathbf{r}(t) = \left\langle \lim_{t\to t_0} x(t), \lim_{t\to t_0} y(t), \lim_{t\to t_0} z(t)\right\rangle \qquad \boxed{1}$$

The Limit Laws of scalar functions remain valid in the vector-valued case. They are verified by applying the Limit Laws to the components.

Proof Let $\mathbf{u} = \langle a, b, c\rangle$ and consider the square of the length

$$\|\mathbf{r}(t) - \mathbf{u}\|^2 = (x(t) - a)^2 + (y(t) - b)^2 + (z(t) - c)^2 \qquad \boxed{2}$$

The term on the left approaches zero if and only if each term on the right approaches zero (because these terms are nonnegative). It follows that $\|\mathbf{r}(t) - \mathbf{u}\|$ approaches zero if and only if $|x(t) - a|$, $|y(t) - b|$, and $|z(t) - c|$ tend to zero. Therefore, $\mathbf{r}(t)$ approaches a limit $\mathbf{u}$ as $t \to t_0$ if and only if $x(t)$, $y(t)$, and $z(t)$ converge to the components a, b, and c. ■

■ **EXAMPLE 1** Calculate $\lim_{t\to 3} \mathbf{r}(t)$, where $\mathbf{r}(t) = \langle t^2, 1-t, t^{-1}\rangle$.

Solution By Theorem 1,

$$\lim_{t\to 3} \mathbf{r}(t) = \lim_{t\to 3} \langle t^2, 1-t, t^{-1}\rangle = \left\langle \lim_{t\to 3} t^2, \lim_{t\to 3}(1-t), \lim_{t\to 3} t^{-1}\right\rangle = \left\langle 9, -2, \frac{1}{3}\right\rangle$$ ■

Continuity of vector-valued functions is defined in the same way as in the scalar case. A vector-valued function $\mathbf{r}(t) = \langle x(t), y(t), z(t)\rangle$ is **continuous** at t_0 if

$$\lim_{t\to t_0} \mathbf{r}(t) = \mathbf{r}(t_0)$$

By Theorem 1, $\mathbf{r}(t)$ is continuous at t_0 if and only if the components $x(t)$, $y(t)$, $z(t)$ are continuous at t_0.

We define the derivative of $\mathbf{r}(t)$ as the limit of the difference quotient:

$$\mathbf{r}'(t) = \frac{d}{dt}\mathbf{r}(t) = \lim_{h\to 0} \frac{\mathbf{r}(t+h) - \mathbf{r}(t)}{h} \qquad \boxed{3}$$

In Leibniz notation, the derivative is written $d\mathbf{r}/dt$.

We say that $\mathbf{r}(t)$ is *differentiable* at t if the limit in Eq. (3) exists. Notice that the components of the difference quotient are difference quotients:

$$\lim_{h\to 0} \frac{\mathbf{r}(t+h) - \mathbf{r}(t)}{h} = \lim_{h\to 0} \left\langle \frac{x(t+h)-x(t)}{h}, \frac{y(t+h)-y(t)}{h}, \frac{z(t+h)-z(t)}{h} \right\rangle$$

and by Theorem 1, $\mathbf{r}(t)$ is differentiable if and only if the components are differentiable. In this case, $\mathbf{r}'(t)$ is equal to the vector of derivatives $\langle x'(t), y'(t), z'(t)\rangle$.

By Theorems 1 and 2, vector-valued limits and derivatives are computed "componentwise," so they are not more difficult to compute than ordinary limits and derivatives.

THEOREM 2 Vector-Valued Derivatives Are Computed Componentwise A vector-valued function $\mathbf{r}(t) = \langle x(t), y(t), z(t)\rangle$ is differentiable if and only if each component is differentiable. In this case,

$$\mathbf{r}'(t) = \frac{d}{dt}\mathbf{r}(t) = \langle x'(t), y'(t), z'(t)\rangle$$

Here are some vector-valued derivatives, computed componentwise:

$$\frac{d}{dt}\langle t^2, t^3, \sin t\rangle = \langle 2t, 3t^2, \cos t\rangle, \qquad \frac{d}{dt}\langle \cos t, -1, e^{2t}\rangle = \langle -\sin t, 0, 2e^{2t}\rangle$$

Higher-order derivatives are defined by repeated differentiation:

$$\mathbf{r}''(t) = \frac{d}{dt}\mathbf{r}'(t), \quad \mathbf{r}'''(t) = \frac{d}{dt}\mathbf{r}''(t), \quad \dots$$

EXAMPLE 2 Calculate $\mathbf{r}''(3)$, where $\mathbf{r}(t) = \langle \ln t, t, t^2\rangle$.

Solution We perform the differentiation componentwise:

$$\mathbf{r}'(t) = \frac{d}{dt}\langle \ln t, t, t^2\rangle = \langle t^{-1}, 1, 2t\rangle$$

$$\mathbf{r}''(t) = \frac{d}{dt}\langle t^{-1}, 1, 2t\rangle = \langle -t^{-2}, 0, 2\rangle$$

Therefore, $\mathbf{r}''(3) = \langle -\frac{1}{9}, 0, 2\rangle$. ∎

The differentiation rules of single-variable calculus carry over to the vector setting.

Differentiation Rules Assume that $\mathbf{r}(t)$, $\mathbf{r}_1(t)$, and $\mathbf{r}_2(t)$ are differentiable. Then

- **Sum Rule:** $(\mathbf{r}_1(t) + \mathbf{r}_2(t))' = \mathbf{r}_1'(t) + \mathbf{r}_2'(t)$
- **Constant Multiple Rule:** For any constant c, $(c\,\mathbf{r}(t))' = c\,\mathbf{r}'(t)$.
- **Product Rule:** For any differentiable scalar-valued function $f(t)$,

$$\frac{d}{dt}\big(f(t)\mathbf{r}(t)\big) = f(t)\mathbf{r}'(t) + f'(t)\mathbf{r}(t)$$

- **Chain Rule:** For any differentiable scalar-valued function $g(t)$,

$$\frac{d}{dt}\mathbf{r}(g(t)) = g'(t)\mathbf{r}'(g(t))$$

Proof Each rule is proved by applying the differentiation rules to the components. For example, to prove the Product Rule (we consider vector-valued functions in the plane, to keep the notation simple), we write

$$f(t)\mathbf{r}(t) = f(t)\,\langle x(t), y(t)\rangle = \langle f(t)x(t), f(t)y(t)\rangle$$

Now apply the Product Rule to each component:

$$\begin{aligned}\frac{d}{dt}f(t)\mathbf{r}(t) &= \left\langle \frac{d}{dt}f(t)x(t), \frac{d}{dt}f(t)y(t)\right\rangle \\ &= \big\langle f'(t)x(t) + f(t)x'(t), f'(t)y(t) + f(t)y'(t)\big\rangle \\ &= \big\langle f'(t)x(t), f'(t)y(t)\big\rangle + \big\langle f(t)x'(t), f(t)y'(t)\big\rangle \\ &= f'(t)\,\langle x(t), y(t)\rangle + f(t)\big\langle x'(t), y'(t)\big\rangle = f'(t)\mathbf{r}(t) + f(t)\mathbf{r}'(t)\end{aligned}$$

The remaining proofs are left as exercises (Exercises 69–70). ■

■ **EXAMPLE 3** Let $\mathbf{r}(t) = \langle t^2, 5t, 1\rangle$ and $f(t) = e^{3t}$. Calculate:

(a) $\dfrac{d}{dt}f(t)\mathbf{r}(t)$ **(b)** $\dfrac{d}{dt}\mathbf{r}(f(t))$

Solution We have $\mathbf{r}'(t) = \langle 2t, 5, 0\rangle$ and $f'(t) = 3e^{3t}$.

(a) By the Product Rule,

$$\begin{aligned}\frac{d}{dt}f(t)\mathbf{r}(t) &= f(t)\mathbf{r}'(t) + f'(t)\mathbf{r}(t) = e^{3t}\langle 2t, 5, 0\rangle + 3e^{3t}\langle t^2, 5t, 1\rangle \\ &= \big\langle (3t^2 + 2t)e^{3t}, (15t + 5)e^{3t}, 3e^{3t}\big\rangle\end{aligned}$$

(b) By the Chain Rule,

$$\frac{d}{dt}\mathbf{r}(f(t)) = f'(t)\mathbf{r}'(f(t)) = 3e^{3t}\mathbf{r}'(e^{3t}) = 3e^{3t}\langle 2e^{3t}, 5, 0\rangle = \langle 6e^{6t}, 15e^{3t}, 0\rangle$$ ■

There are three different Product Rules for vector-valued functions. In addition to the rule for the product of a scalar function $f(t)$ and a vector-valued function $\mathbf{r}(t)$ stated above, there are Product Rules for the dot and cross products. These rules are very important in applications, as we will see.

CAUTION Order is important in the Product Rule for cross products. The first term in Eq. (5) must be written as

$$\mathbf{r}_1(t) \times \mathbf{r}_2'(t)$$

not $\mathbf{r}_2'(t) \times \mathbf{r}_1(t)$. Similarly, the second term is $\mathbf{r}_1'(t) \times \mathbf{r}_2(t)$. Why is order not a concern for dot products?

THEOREM 3 Product Rule for Dot and Cross Products Assume that $\mathbf{r}_1(t)$ and $\mathbf{r}_2(t)$ are differentiable. Then

Dot Products: $$\frac{d}{dt}\big(\mathbf{r}_1(t) \cdot \mathbf{r}_2(t)\big) = \mathbf{r}_1(t) \cdot \mathbf{r}_2'(t) + \mathbf{r}_1'(t) \cdot \mathbf{r}_2(t) \qquad (4)$$

Cross Products: $$\frac{d}{dt}\big(\mathbf{r}_1(t) \times \mathbf{r}_2(t)\big) = \big[\mathbf{r}_1(t) \times \mathbf{r}_2'(t)\big] + \big[\mathbf{r}_1'(t) \times \mathbf{r}_2(t)\big] \qquad (5)$$

Proof We verify Eq. (4) for vector-valued functions in the plane. If $\mathbf{r}_1(t) = \langle x_1(t), y_1(t)\rangle$ and $\mathbf{r}_2(t) = \langle x_2(t), y_2(t)\rangle$, then

$$\begin{aligned}
\frac{d}{dt}\big(\mathbf{r}_1(t) \cdot \mathbf{r}_2(t)\big) &= \frac{d}{dt}\big(x_1(t)x_2(t) + y_1(t)y_2(t)\big) \\
&= x_1(t)x_2'(t) + x_1'(t)x_2(t) + y_1(t)y_2'(t) + y_1'(t)y_2(t) \\
&= \big(x_1(t)x_2'(t) + y_1(t)y_2'(t)\big) + \big(x_1'(t)x_2(t) + y_1'(t)y_2(t)\big) \\
&= \mathbf{r}_1(t) \cdot \mathbf{r}_2'(t) + \mathbf{r}_1'(t) \cdot \mathbf{r}_2(t)
\end{aligned}$$

The proof of Eq. (5) is left as an exercise (Exercise 71). ■

In the next example and throughout this chapter, *all vector-valued functions are assumed differentiable, unless otherwise stated.*

■ **EXAMPLE 4** Prove the formula $\frac{d}{dt}\big(\mathbf{r}(t) \times \mathbf{r}'(t)\big) = \mathbf{r}(t) \times \mathbf{r}''(t)$.

Solution By the Product Formula for cross products,

$$\frac{d}{dt}\big(\mathbf{r}(t) \times \mathbf{r}'(t)\big) = \mathbf{r}(t) \times \mathbf{r}''(t) + \underbrace{\mathbf{r}'(t) \times \mathbf{r}'(t)}_{\text{Equals } \mathbf{0}} = \mathbf{r}(t) \times \mathbf{r}''(t)$$

Here, $\mathbf{r}'(t) \times \mathbf{r}'(t) = \mathbf{0}$ because the cross product of a vector with itself is zero. ■

The Derivative as a Tangent Vector

The derivative vector $\mathbf{r}'(t_0)$ has an important geometric property: It points in the direction tangent to the path traced by $\mathbf{r}(t)$ at $t = t_0$.

To understand why, consider the difference quotient, where $\Delta\mathbf{r} = \mathbf{r}(t_0 + h) - \mathbf{r}(t_0)$ and $\Delta t = h$ with $h \neq 0$:

$$\frac{\Delta\mathbf{r}}{\Delta t} = \frac{\mathbf{r}(t_0 + h) - \mathbf{r}(t_0)}{h} \qquad (6)$$

Although it has been our convention to regard all vectors as based at the origin, the tangent vector $\mathbf{r}'(t)$ is an exception; we visualize it as a vector based at the terminal point of $\mathbf{r}(t)$. This makes sense because $\mathbf{r}'(t)$ then appears as a vector tangent to the curve (Figure 3).

The vector $\Delta\mathbf{r}$ points from the head of $\mathbf{r}(t_0)$ to the head of $\mathbf{r}(t_0 + h)$ as in Figure 2(A). The difference quotient $\Delta\mathbf{r}/\Delta t$ is a scalar multiple of $\Delta\mathbf{r}$ and therefore points in the same direction [Figure 2(B)].

As $h = \Delta t$ tends to zero, $\Delta\mathbf{r}$ also tends to zero but the quotient $\Delta\mathbf{r}/\Delta t$ approaches a vector $\mathbf{r}'(t_0)$, which, if nonzero, points in the direction tangent to the curve. Figure 3 illustrates the limiting process. We refer to $\mathbf{r}'(t_0)$ as the **tangent vector** or the **velocity vector** at $\mathbf{r}(t_0)$.

The tangent vector $\mathbf{r}'(t_0)$ (if it is nonzero) is a direction vector for the tangent line to the curve. Therefore, the tangent line has vector parametrization:

$$\text{Tangent line at } \mathbf{r}(t_0): \qquad \mathbf{L}(t) = \mathbf{r}(t_0) + t\,\mathbf{r}'(t_0) \qquad (7)$$

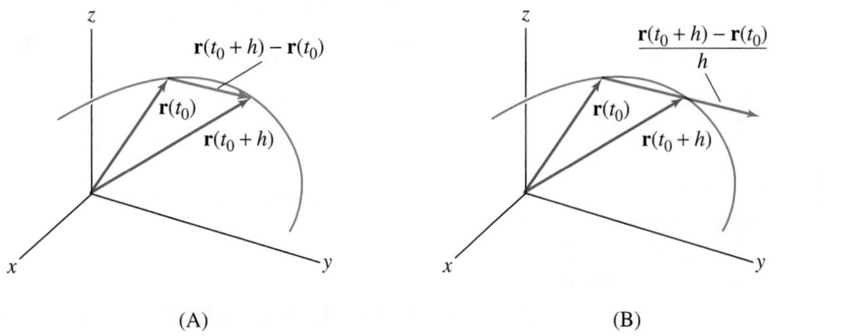

FIGURE 2 The difference quotient points in the direction of $\Delta\mathbf{r} = \mathbf{r}(t_0+h) - \mathbf{r}(t_0)$.

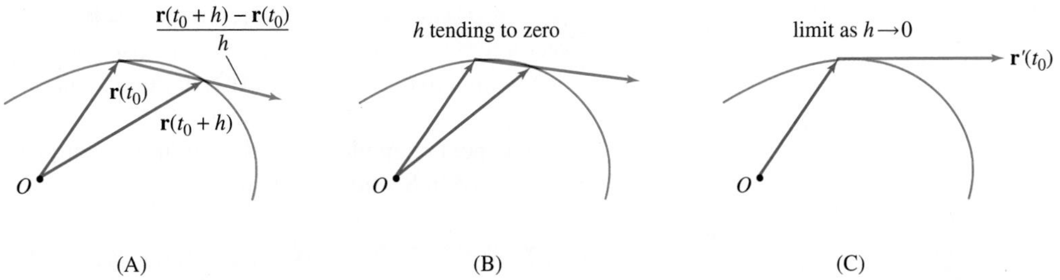

FIGURE 3 The difference quotient converges to a vector $\mathbf{r}'(t_0)$, tangent to the curve.

EXAMPLE 5 Plotting Tangent Vectors CAS Plot $\mathbf{r}(t) = \langle \cos t, \sin t, 4\cos^2 t\rangle$ together with its tangent vectors at $t = \frac{\pi}{4}$ and $\frac{3\pi}{2}$. Find a parametrization of the tangent line at $t = \frac{\pi}{4}$.

Solution The derivative is $\mathbf{r}'(t) = \langle -\sin t, \cos t, -8\cos t \sin t\rangle$, and thus the tangent vectors at $t = \frac{\pi}{4}$ and $\frac{3\pi}{2}$ are

$$\mathbf{r}'\left(\frac{\pi}{4}\right) = \left\langle -\frac{\sqrt{2}}{2}, \frac{\sqrt{2}}{2}, -4\right\rangle, \qquad \mathbf{r}'\left(\frac{3\pi}{2}\right) = \langle 1, 0, 0\rangle$$

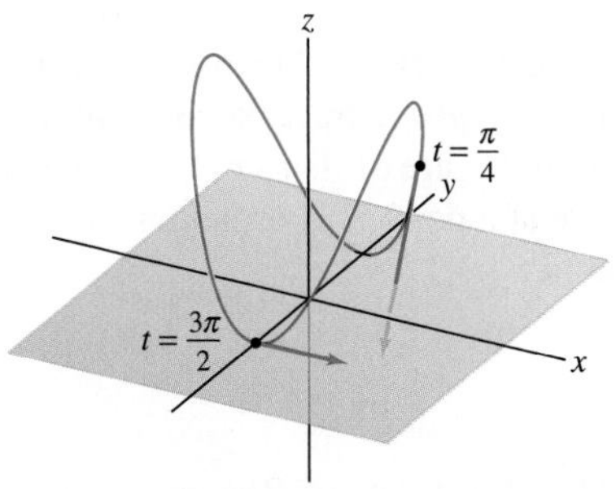

FIGURE 4 Tangent vectors to

$$\mathbf{r}(t) = \langle \cos t, \sin t, 4\cos^2 t\rangle$$

at $t = \frac{\pi}{4}$ and $\frac{3\pi}{2}$.

Figure 4 shows a plot of $\mathbf{r}(t)$ with $\mathbf{r}'(\frac{\pi}{4})$ based at $\mathbf{r}(\frac{\pi}{4})$ and $\mathbf{r}'(\frac{3\pi}{2})$ based at $\mathbf{r}(\frac{3\pi}{2})$.

At $t = \frac{\pi}{4}$, $\mathbf{r}(\frac{\pi}{4}) = \left\langle \frac{\sqrt{2}}{2}, \frac{\sqrt{2}}{2}, 2\right\rangle$ and thus the tangent line is parametrized by

$$\mathbf{L}(t) = \mathbf{r}\left(\frac{\pi}{4}\right) + t\,\mathbf{r}'\left(\frac{\pi}{4}\right) = \left\langle \frac{\sqrt{2}}{2}, \frac{\sqrt{2}}{2}, 2\right\rangle + t\left\langle -\frac{\sqrt{2}}{2}, \frac{\sqrt{2}}{2}, -4\right\rangle$$ ■

There are some important differences between vector- and scalar-valued derivatives. The tangent line to a plane curve $y = f(x)$ is horizontal at x_0 if $f'(x_0) = 0$. But in a vector parametrization, the tangent vector $\mathbf{r}'(t_0) = \langle x'(t_0), y'(t_0)\rangle$ is horizontal and nonzero if $y'(t_0) = 0$ but $x'(t_0) \neq 0$.

EXAMPLE 6 Horizontal Tangent Vectors on the Cycloid The function

$$\mathbf{r}(t) = \langle t - \sin t, 1 - \cos t\rangle$$

traces a cycloid. Find the points where:

(a) $\mathbf{r}'(t)$ is horizontal and nonzero. **(b)** $\mathbf{r}'(t)$ is the zero vector.

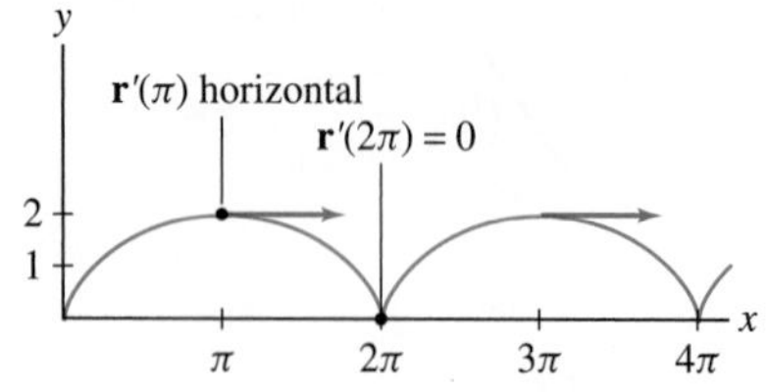

FIGURE 5 Points on the cycloid $r(t) = \langle t - \sin t, 1 - \cos t\rangle$ where the tangent vector is horizontal.

Solution The tangent vector is $\mathbf{r}'(t) = \langle 1 - \cos t, \sin t\rangle$. The y-component of $\mathbf{r}'(t)$ is zero if $\sin t = 0$—that is, if $t = 0, \pi, 2\pi, \dots$. We have

$$\mathbf{r}(0) = \langle 0, 0\rangle, \quad \mathbf{r}'(0) = \langle 1 - \cos 0, \sin 0\rangle = \langle 0, 0\rangle \quad \text{(zero vector)}$$

$$\mathbf{r}(\pi) = \langle \pi, 2\rangle, \quad \mathbf{r}'(\pi) = \langle 1 - \cos \pi, \sin \pi\rangle = \langle 2, 0\rangle \quad \text{(horizontal)}$$

By periodicity, we conclude that $\mathbf{r}'(t)$ is nonzero and horizontal for $t = \pi, 3\pi, 5\pi, \dots$ and $\mathbf{r}'(t) = \mathbf{0}$ for $t = 0, 2\pi, 4\pi, \dots$ (Figure 5). ■

CONCEPTUAL INSIGHT The cycloid in Figure 5 has sharp points called **cusps** at points where $x = 0, 2\pi, 4\pi, \dots$. If we represent the cycloid as the graph of a function $y = f(x)$, then $f'(x)$ does not exist at these points. By contrast, the vector derivative $\mathbf{r}'(t) = \langle 1 - \cos t, \sin t\rangle$ exists *for all* t, but $\mathbf{r}'(t) = \mathbf{0}$ at the cusps. In general, $\mathbf{r}'(t)$ is a direction vector for the tangent line whenever it exists, but we get no information about the tangent line (which may or may not exist) at points where $\mathbf{r}'(t) = \mathbf{0}$.

The next example establishes an important property of vector-valued functions that will be used in Sections 14.4–14.6.

■ **EXAMPLE 7 Orthogonality of r and r′ When r Has Constant Length** Prove that if $\mathbf{r}(t)$ has constant length, then $\mathbf{r}(t)$ is orthogonal to $\mathbf{r}'(t)$.

Solution By the Product Rule for Dot Products,

$$\frac{d}{dt}\|\mathbf{r}(t)\|^2 = \frac{d}{dt}\big(\mathbf{r}(t)\cdot\mathbf{r}(t)\big) = \mathbf{r}(t)\cdot\mathbf{r}'(t) + \mathbf{r}'(t)\cdot\mathbf{r}(t) = 2\mathbf{r}(t)\cdot\mathbf{r}'(t)$$

This derivative is zero because $\|\mathbf{r}(t)\|$ is constant. Therefore $\mathbf{r}(t)\cdot\mathbf{r}'(t) = 0$, and $\mathbf{r}(t)$ is orthogonal to $\mathbf{r}'(t)$ [or $\mathbf{r}'(t) = \mathbf{0}$]. ■

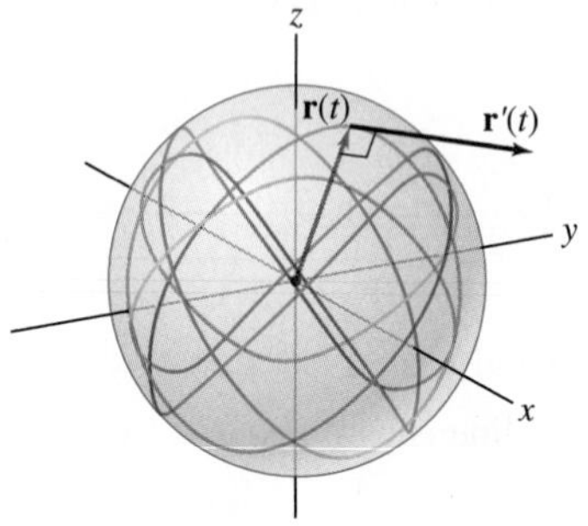

FIGURE 6

GRAPHICAL INSIGHT The result of Example 7 has a geometric explanation. A vector parametrization $\mathbf{r}(t)$ consisting of vectors of constant length R traces a curve on the surface of a sphere of radius R with center at the origin (Figure 6). Thus $\mathbf{r}'(t)$ is tangent to this sphere. But any line that is tangent to a sphere at a point P is orthogonal to the radial vector through P, and thus $\mathbf{r}(t)$ is orthogonal to $\mathbf{r}'(t)$.

Vector-Valued Integration

The integral of a vector-valued function can be defined in terms of Riemann sums as in Chapter 5. We will define it more simply via componentwise integration (the two definitions are equivalent). In other words,

$$\int_a^b \mathbf{r}(t)\,dt = \left\langle \int_a^b x(t)\,dt, \int_a^b y(t)\,dt, \int_a^b z(t)\,dt \right\rangle$$

The integral exists if each of the components $x(t)$, $y(t)$, $z(t)$ is integrable. For example,

$$\int_0^\pi \langle 1, t, \sin t\rangle\,dt = \left\langle \int_0^\pi 1\,dt, \int_0^\pi t\,dt, \int_0^\pi \sin t\,dt \right\rangle = \left\langle \pi, \frac{1}{2}\pi^2, 2 \right\rangle$$

Vector-valued integrals obey the same linearity rules as scalar-valued integrals (see Exercise 72).

An **antiderivative** of $\mathbf{r}(t)$ is a vector-valued function $\mathbf{R}(t)$ such that $\mathbf{R}'(t) = \mathbf{r}(t)$. In the single-variable case, two functions $f_1(x)$ and $f_2(x)$ with the same derivative differ by a constant. Similarly, two vector-valued functions with the same derivative differ by a *constant vector* (i.e., a vector that does not depend on t). This is proved by applying the scalar result to each component of $\mathbf{r}(t)$.

THEOREM 4 If $\mathbf{R}_1(t)$ and $\mathbf{R}_2(t)$ are differentiable and $\mathbf{R}_1'(t) = \mathbf{R}_2'(t)$, then

$$\mathbf{R}_1(t) = \mathbf{R}_2(t) + \mathbf{c}$$

for some constant vector $\mathbf{c}$.

The general antiderivative of $\mathbf{r}(t)$ is written

$$\int \mathbf{r}(t)\,dt = \mathbf{R}(t) + \mathbf{c}$$

where $\mathbf{c} = \langle c_1, c_2, c_3 \rangle$ is an arbitrary constant vector. For example,

$$\int \langle 1, t, \sin t \rangle\,dt = \left\langle t, \frac{1}{2}t^2, -\cos t \right\rangle + \mathbf{c} = \left\langle t + c_1, \frac{1}{2}t^2 + c_2, -\cos t + c_3 \right\rangle$$

Fundamental Theorem of Calculus for Vector-Valued Functions If $\mathbf{r}(t)$ is continuous on $[a, b]$, and $\mathbf{R}(t)$ is an antiderivative of $\mathbf{r}(t)$, then

$$\int_a^b \mathbf{r}(t)\,dt = \mathbf{R}(b) - \mathbf{R}(a)$$

EXAMPLE 8 Finding Position via Vector-Valued Differential Equations The path of a particle satisfies

$$\frac{d\mathbf{r}}{dt} = \left\langle 1 - 6\sin 3t, \frac{1}{5}t \right\rangle$$

Find the particle's location at $t = 4$ if $\mathbf{r}(0) = \langle 4, 1 \rangle$.

Solution The general solution is obtained by integration:

$$\mathbf{r}(t) = \int \left\langle 1 - 6\sin 3t, \frac{1}{5}t \right\rangle dt = \left\langle t + 2\cos 3t, \frac{1}{10}t^2 \right\rangle + \mathbf{c}$$

The initial condition $\mathbf{r}(0) = \langle 4, 1 \rangle$ gives us

$$\mathbf{r}(0) = \langle 2, 0 \rangle + \mathbf{c} = \langle 4, 1 \rangle$$

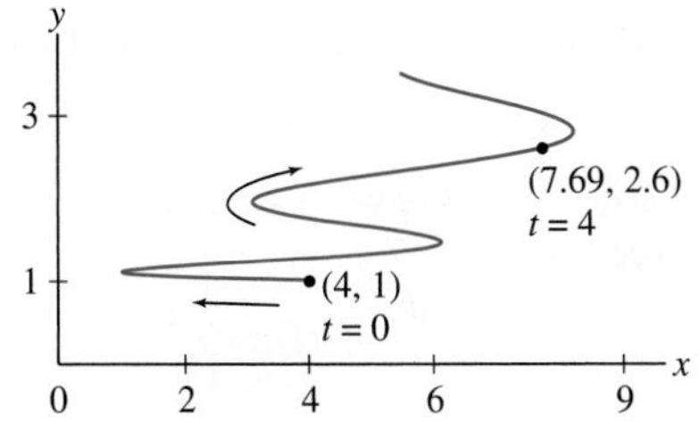

FIGURE 7 Particle path
$\mathbf{r}(t) = \langle t + 2\cos 3t + 2, \frac{1}{10}t^2 + 1 \rangle$

Therefore, $\mathbf{c} = \langle 2, 1 \rangle$ and (Figure 7)

$$\mathbf{r}(t) = \left\langle t + 2\cos 3t, \frac{1}{10}t^2 \right\rangle + \langle 2, 1 \rangle = \left\langle t + 2\cos 3t + 2, \frac{1}{10}t^2 + 1 \right\rangle$$

The particle's position at $t = 4$ is

$$\mathbf{r}(4) = \left\langle 4 + 2\cos 12 + 2, \frac{1}{10}(4^2) + 1 \right\rangle \approx \langle 7.69, 2.6 \rangle$$ ■

14.2 SUMMARY

- Limits, differentiation, and integration of vector-valued functions are performed componentwise.
- Differentation rules:
  - Sum Rule: $(\mathbf{r}_1(t) + \mathbf{r}_2(t))' = \mathbf{r}_1'(t) + \mathbf{r}_2'(t)$

– Constant Multiple Rule: $(c\,\mathbf{r}(t))' = c\,\mathbf{r}'(t)$

– Chain Rule: $\frac{d}{dt}\mathbf{r}(g(t)) = g'(t)\mathbf{r}'(g(t))$

- Product Rules:

Scalar times vector: $\frac{d}{dt}\big(f(t)\mathbf{r}(t)\big) = f(t)\mathbf{r}'(t) + f'(t)\mathbf{r}(t)$

Dot product: $\frac{d}{dt}\big(\mathbf{r}_1(t)\cdot\mathbf{r}_2(t)\big) = \mathbf{r}_1(t)\cdot\mathbf{r}_2'(t) + \mathbf{r}_1'(t)\cdot\mathbf{r}_2(t)$

Cross product: $\frac{d}{dt}\big(\mathbf{r}_1(t)\times\mathbf{r}_2(t)\big) = \big[\mathbf{r}_1(t)\times\mathbf{r}_2'(t)\big] + \big[\mathbf{r}_1'(t)\times\mathbf{r}_2(t)\big]$

- The derivative $\mathbf{r}'(t_0)$ is called the *tangent vector* or *velocity vector.*
- If $\mathbf{r}'(t_0)$ is nonzero, then it points in the direction tangent to the curve at $\mathbf{r}(t_0)$. The tangent line has vector parametrization

$$\mathbf{L}(t) = \mathbf{r}(t_0) + t\mathbf{r}'(t_0)$$

- If $\mathbf{R}_1'(t) = \mathbf{R}_2'(t)$, then $\mathbf{R}_1(t) = \mathbf{R}_2(t) + \mathbf{c}$ for some constant vector $\mathbf{c}$.
- The Fundamental Theorem for vector-valued functions: If $\mathbf{r}(t)$ is continuous and $\mathbf{R}(t)$ is an antiderivative of $\mathbf{r}(t)$, then

$$\int_a^b \mathbf{r}(t)\,dt = \mathbf{R}(b) - \mathbf{R}(a)$$

14.2 EXERCISES

Preliminary Questions

1. State the three forms of the Product Rule for vector-valued functions.

In Questions 2–6, indicate whether the statement is true or false, and if it is false, provide a correct statement.

2. The derivative of a vector-valued function is defined as the limit of the difference quotient, just as in the scalar-valued case.

3. There are two Chain Rules for vector-valued functions: one for the composite of two vector-valued functions and one for the composite of a vector-valued and a scalar-valued function.

4. The terms "velocity vector" and "tangent vector" for a path $\mathbf{r}(t)$ mean one and the same thing.

5. The derivative of a vector-valued function is the slope of the tangent line, just as in the scalar case.

6. The derivative of the cross product is the cross product of the derivatives.

7. State whether the following derivatives of vector-valued functions $\mathbf{r}_1(t)$ and $\mathbf{r}_2(t)$ are scalars or vectors:

(a) $\frac{d}{dt}\mathbf{r}_1(t)$ **(b)** $\frac{d}{dt}\big(\mathbf{r}_1(t)\cdot\mathbf{r}_2(t)\big)$

(c) $\frac{d}{dt}\big(\mathbf{r}_1(t)\times\mathbf{r}_2(t)\big)$

Exercises

In Exercises 1–6, evaluate the limit.

1. $\lim_{t\to 3}\left\langle t^2, 4t, \frac{1}{t}\right\rangle$

2. $\lim_{t\to\pi} \sin 2t\mathbf{i} + \cos t\mathbf{j} + \tan 4t\mathbf{k}$

3. $\lim_{t\to 0} e^{2t}\mathbf{i} + \ln(t+1)\mathbf{j} + 4\mathbf{k}$

4. $\lim_{t\to 0}\left\langle \frac{1}{t+1}, \frac{e^t-1}{t}, 4t\right\rangle$

5. Evaluate $\lim_{h\to 0}\frac{\mathbf{r}(t+h)-\mathbf{r}(t)}{h}$ for $\mathbf{r}(t) = \left\langle t^{-1}, \sin t, 4\right\rangle$.

6. Evaluate $\lim_{t\to 0}\frac{\mathbf{r}(t)}{t}$ for $\mathbf{r}(t) = \langle \sin t, 1-\cos t, -2t\rangle$.

In Exercises 7–12, compute the derivative.

7. $\mathbf{r}(t) = \left\langle t, t^2, t^3\right\rangle$

8. $\mathbf{r}(t) = \left\langle 7-t, 4\sqrt{t}, 8\right\rangle$

9. $\mathbf{r}(s) = \left\langle e^{3s}, e^{-s}, s^4\right\rangle$

10. $\mathbf{b}(t) = \left\langle e^{3t-4}, e^{6-t}, (t+1)^{-1}\right\rangle$ **11.** $\mathbf{c}(t) = t^{-1}\mathbf{i} - e^{2t}\mathbf{k}$

12. $\mathbf{a}(\theta) = (\cos 3\theta)\mathbf{i} + (\sin^2 \theta)\mathbf{j} + (\tan \theta)\mathbf{k}$

13. Calculate $\mathbf{r}'(t)$ and $\mathbf{r}''(t)$ for $\mathbf{r}(t) = \langle t, t^2, t^3 \rangle$.

14. Sketch the curve $\mathbf{r}(t) = \langle 1 - t^2, t \rangle$ for $-1 \le t \le 1$. Compute the tangent vector at $t = 1$ and add it to the sketch.

15. Sketch the curve $\mathbf{r}_1(t) = \langle t, t^2 \rangle$ together with its tangent vector at $t = 1$. Then do the same for $\mathbf{r}_2(t) = \langle t^3, t^6 \rangle$.

16. Sketch the cycloid $\mathbf{r}(t) = \langle t - \sin t, 1 - \cos t \rangle$ together with its tangent vectors at $t = \frac{\pi}{3}$ and $\frac{3\pi}{4}$.

In Exercises 17–20, evaluate the derivative by using the appropriate Product Rule, where

$$\mathbf{r}_1(t) = \langle t^2, t^3, t \rangle, \qquad \mathbf{r}_2(t) = \langle e^{3t}, e^{2t}, e^t \rangle$$

17. $\dfrac{d}{dt}(\mathbf{r}_1(t) \cdot \mathbf{r}_2(t))$

18. $\dfrac{d}{dt}(t^4 \mathbf{r}_1(t))$

19. $\dfrac{d}{dt}(\mathbf{r}_1(t) \times \mathbf{r}_2(t))$

20. $\dfrac{d}{dt}(\mathbf{r}(t) \cdot \mathbf{r}_1(t))\Big|_{t=2}$, assuming that

$$\mathbf{r}(2) = \langle 2, 1, 0 \rangle, \qquad \mathbf{r}'(2) = \langle 1, 4, 3 \rangle$$

In Exercises 21 and 22, let

$$\mathbf{r}_1(t) = \langle t^2, 1, 2t \rangle, \qquad \mathbf{r}_2(t) = \langle 1, 2, e^t \rangle$$

21. Compute $\dfrac{d}{dt}\mathbf{r}_1(t) \cdot \mathbf{r}_2(t)\Big|_{t=1}$ in two ways:

(a) Calculate $\mathbf{r}_1(t) \cdot \mathbf{r}_2(t)$ and differentiate.

(b) Use the Product Rule.

22. Compute $\dfrac{d}{dt}\mathbf{r}_1(t) \times \mathbf{r}_2(t)\Big|_{t=1}$ in two ways:

(a) Calculate $\mathbf{r}_1(t) \times \mathbf{r}_2(t)$ and differentiate.

(b) Use the Product Rule.

In Exercises 23–26, evaluate $\dfrac{d}{dt}\mathbf{r}(g(t))$ *using the Chain Rule.*

23. $\mathbf{r}(t) = \langle t^2, 1 - t \rangle, \quad g(t) = e^t$

24. $\mathbf{r}(t) = \langle t^2, t^3 \rangle, \quad g(t) = \sin t$

25. $\mathbf{r}(t) = \langle e^t, e^{2t}, 4 \rangle, \quad g(t) = 4t + 9$

26. $\mathbf{r}(t) = \langle 4 \sin 2t, 6 \cos 2t \rangle, \quad g(t) = t^2$

27. Let $\mathbf{r}(t) = \langle t^2, 1 - t, 4t \rangle$. Calculate the derivative of $\mathbf{r}(t) \cdot \mathbf{a}(t)$ at $t = 2$, assuming that $\mathbf{a}(2) = \langle 1, 3, 3 \rangle$ and $\mathbf{a}'(2) = \langle -1, 4, 1 \rangle$.

28. Let $\mathbf{v}(s) = s^2\mathbf{i} + 2s\mathbf{j} + 9s^{-2}\mathbf{k}$. Evaluate $\dfrac{d}{ds}\mathbf{v}(g(s))$ at $s = 4$, assuming that $g(4) = 3$ and $g'(4) = -9$.

In Exercises 29–34, find a parametrization of the tangent line at the point indicated.

29. $\mathbf{r}(t) = \langle t^2, t^4 \rangle, \quad t = -2$

30. $\mathbf{r}(t) = \langle \cos 2t, \sin 3t \rangle, \quad t = \frac{\pi}{4}$

31. $\mathbf{r}(t) = \langle 1 - t^2, 5t, 2t^3 \rangle, \quad t = 2$

32. $\mathbf{r}(t) = \langle 4t, 5t, 9t \rangle, \quad t = -4$

33. $\mathbf{r}(s) = 4s^{-1}\mathbf{i} - \frac{8}{3}s^{-3}\mathbf{k}, \quad s = 2$

34. $\mathbf{r}(s) = (\ln s)\mathbf{i} + s^{-1}\mathbf{j} + 9s\mathbf{k}, \quad s = 1$

35. Use Example 4 to calculate $\dfrac{d}{dt}(\mathbf{r} \times \mathbf{r}')$, where $\mathbf{r}(t) = \langle t, t^2, e^t \rangle$.

36. Let $\mathbf{r}(t) = \langle 3 \cos t, 5 \sin t, 4 \cos t \rangle$. Show that $\|\mathbf{r}(t)\|$ is constant and conclude, using Example 7, that $\mathbf{r}(t)$ and $\mathbf{r}'(t)$ are orthogonal. Then compute $\mathbf{r}'(t)$ and verify directly that $\mathbf{r}'(t)$ is orthogonal to $\mathbf{r}(t)$.

37. Show that the *derivative of the norm* is not equal to the *norm of the derivative* by verifying that $\|\mathbf{r}(t)\|' \ne \|\mathbf{r}(t)'\|$ for $\mathbf{r}(t) = \langle t, 1, 1 \rangle$.

38. Show that $\dfrac{d}{dt}(\mathbf{a} \times \mathbf{r}) = \mathbf{a} \times \mathbf{r}'$ for any constant vector $\mathbf{a}$.

In Exercises 39–46, evaluate the integrals.

39. $\displaystyle\int_{-1}^{3} \langle 8t^2 - t, 6t^3 + t \rangle\, dt$

40. $\displaystyle\int_0^1 \left\langle \frac{1}{1 + s^2}, \frac{s}{1 + s^2} \right\rangle ds$

41. $\displaystyle\int_{-2}^{2} (u^3\mathbf{i} + u^5\mathbf{j})\, du$

42. $\displaystyle\int_0^1 \left(te^{-t^2}\mathbf{i} + t \ln(t^2 + 1)\mathbf{j} \right) dt$

43. $\displaystyle\int_0^1 \langle 2t, 4t, -\cos 3t \rangle\, dt$

44. $\displaystyle\int_{1/2}^{1} \left\langle \frac{1}{u^2}, \frac{1}{u^4}, \frac{1}{u^5} \right\rangle du$

45. $\displaystyle\int_1^4 (t^{-1}\mathbf{i} + 4\sqrt{t}\,\mathbf{j} - 8t^{3/2}\mathbf{k})\, dt$

46. $\displaystyle\int_0^t (3s\mathbf{i} + 6s^2\mathbf{j} + 9\mathbf{k})\, ds$

In Exercises 47–54, find both the general solution of the differential equation and the solution with the given initial condition.

47. $\dfrac{d\mathbf{r}}{dt} = \langle 1 - 2t, 4t \rangle, \quad \mathbf{r}(0) = \langle 3, 1 \rangle$

48. $\mathbf{r}'(t) = \mathbf{i} - \mathbf{j}, \quad \mathbf{r}(0) = 2\mathbf{i} + 3\mathbf{k}$

49. $\mathbf{r}'(t) = t^2\mathbf{i} + 5t\mathbf{j} + \mathbf{k}, \quad \mathbf{r}(1) = \mathbf{j} + 2\mathbf{k}$

50. $\mathbf{r}'(t) = \langle \sin 3t, \sin 3t, t \rangle, \quad \mathbf{r}(\frac{\pi}{2}) = \left\langle 2, 4, \dfrac{\pi^2}{4} \right\rangle$

51. $\mathbf{r}''(t) = 16\mathbf{k}, \quad \mathbf{r}(0) = \langle 1, 0, 0 \rangle, \quad \mathbf{r}'(0) = \langle 0, 1, 0 \rangle$

52. $\mathbf{r}''(t) = \langle e^{2t-2}, t^2 - 1, 1 \rangle, \quad \mathbf{r}(1) = \langle 0, 0, 1 \rangle,$
$\mathbf{r}'(1) = \langle 2, 0, 0 \rangle$

53. $\mathbf{r}''(t) = \langle 0, 2, 0 \rangle, \quad \mathbf{r}(3) = \langle 1, 1, 0 \rangle,$
$\mathbf{r}'(3) = \langle 0, 0, 1 \rangle$

54. $\mathbf{r}''(t) = \langle e^t, \sin t, \cos t \rangle, \quad \mathbf{r}(0) = \langle 1, 0, 1 \rangle,$
$\mathbf{r}'(0) = \langle 0, 2, 2 \rangle$

55. Find the location at $t = 3$ of a particle whose path (Figure 8) satisfies

$$\frac{d\mathbf{r}}{dt} = \left\langle 2t - \frac{1}{(t + 1)^2}, 2t - 4 \right\rangle, \qquad \mathbf{r}(0) = \langle 3, 8 \rangle$$

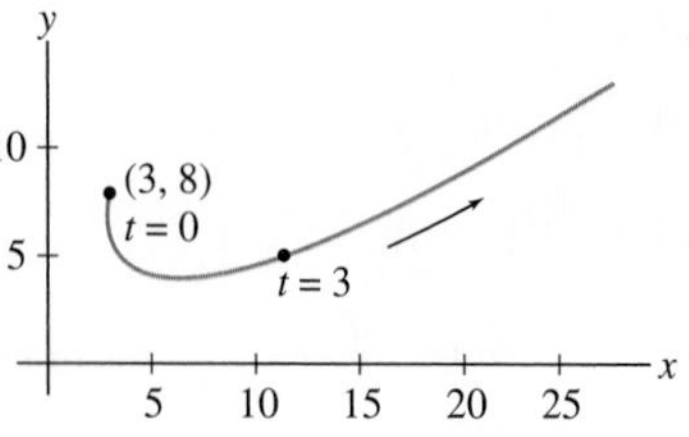

FIGURE 8 Particle path.

56. Find the location and velocity at $t = 4$ of a particle whose path satisfies

$$\frac{d\mathbf{r}}{dt} = \left\langle 2t^{-1/2}, 6, 8t \right\rangle, \qquad \mathbf{r}(1) = \langle 4, 9, 2 \rangle$$

57. A fighter plane, which can shoot a laser beam straight ahead, travels along the path $\mathbf{r}(t) = \left\langle 5 - t, 21 - t^2, 3 - t^3/27 \right\rangle$. Show that there is precisely one time t at which the pilot can hit a target located at the origin.

58. The fighter plane of Exercise 57 travels along the path $\mathbf{r}(t) = \left\langle t - t^3, 12 - t^2, 3 - t \right\rangle$. Show that the pilot cannot hit any target on the x-axis.

59. Find all solutions to $\mathbf{r}'(t) = \mathbf{v}$ with initial condition $\mathbf{r}(1) = \mathbf{w}$, where $\mathbf{v}$ and $\mathbf{w}$ are constant vectors in $\mathbf{R}^3$.

60. Let $\mathbf{u}$ be a constant vector in $\mathbf{R}^3$. Find the solution of the equation $\mathbf{r}'(t) = (\sin t)\mathbf{u}$ satisfying $\mathbf{r}'(0) = \mathbf{0}$.

61. Find all solutions to $\mathbf{r}'(t) = 2\mathbf{r}(t)$ where $\mathbf{r}(t)$ is a vector-valued function in three-space.

62. Show that $\mathbf{w}(t) = \langle \sin(3t + 4), \sin(3t - 2), \cos 3t \rangle$ satisfies the differential equation $\mathbf{w}''(t) = -9\mathbf{w}(t)$.

63. Prove that the **Bernoulli spiral** (Figure 9) with parametrization $\mathbf{r}(t) = \left\langle e^t \cos 4t, e^t \sin 4t \right\rangle$ has the property that the angle ψ between the position vector and the tangent vector is constant. Find the angle ψ in degrees.

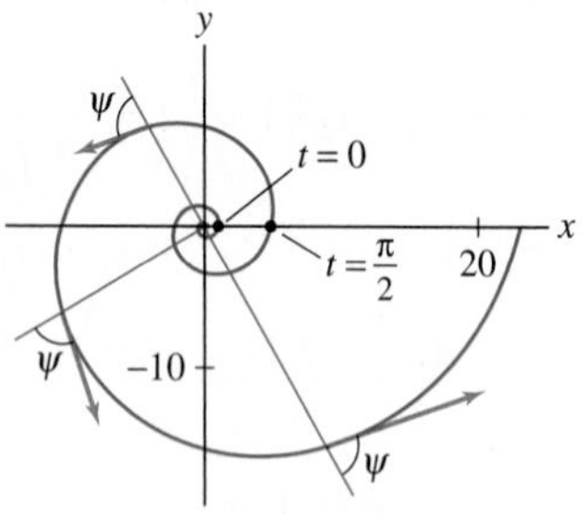

FIGURE 9 Bernoulli spiral.

64. A curve in polar form $r = f(\theta)$ has parametrization

$$\mathbf{r}(\theta) = f(\theta) \langle \cos\theta, \sin\theta \rangle$$

Let ψ be the angle between the radial and tangent vectors (Figure 10). Prove that

$$\tan\psi = \frac{r}{dr/d\theta} = \frac{f(\theta)}{f'(\theta)}$$

Hint: Compute $\mathbf{r}(\theta) \times \mathbf{r}'(\theta)$ and $\mathbf{r}(\theta) \cdot \mathbf{r}'(\theta)$.

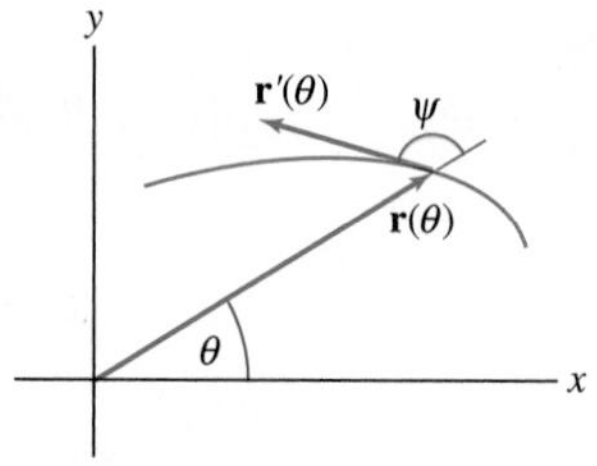

FIGURE 10 Curve with polar parametrization $\mathbf{r}(\theta) = f(\theta) \langle \cos\theta, \sin\theta \rangle$.

65. Prove that if $\|\mathbf{r}(t)\|$ takes on a local minimum or maximum value at t_0, then $\mathbf{r}(t_0)$ is orthogonal to $\mathbf{r}'(t_0)$. Explain how this result is related to Figure 11. *Hint:* Observe that if $\|\mathbf{r}(t_0)\|$ is a minimum, then $\mathbf{r}(t)$ is tangent at t_0 to the sphere of radius $\|\mathbf{r}(t_0)\|$ centered at the origin.

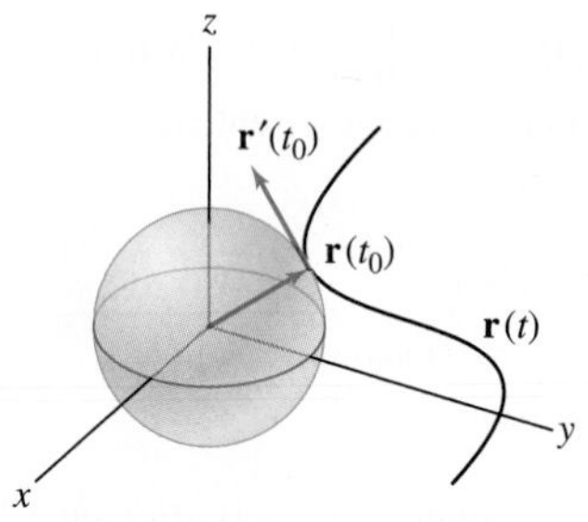

FIGURE 11

66. Newton's Second Law of Motion in vector form states that $\mathbf{F} = \frac{d\mathbf{p}}{dt}$ where $\mathbf{F}$ is the force acting on an object of mass m and $\mathbf{p} = m\mathbf{r}'(t)$ is the object's momentum. The analogs of force and momentum for rotational motion are the **torque** $\boldsymbol{\tau} = \mathbf{r} \times \mathbf{F}$ and **angular momentum**

$$\mathbf{J} = \mathbf{r}(t) \times \mathbf{p}(t)$$

Use the Second Law to prove that $\boldsymbol{\tau} = \frac{d\mathbf{J}}{dt}$.

Further Insights and Challenges

67. Let $\mathbf{r}(t) = \langle x(t), y(t) \rangle$ trace a plane curve $\mathcal{C}$. Assume that $x'(t_0) \neq 0$. Show that the slope of the tangent vector $\mathbf{r}'(t_0)$ is equal to the slope dy/dx of the curve at $\mathbf{r}(t_0)$.

68. Prove that $\frac{d}{dt}(\mathbf{r} \cdot (\mathbf{r}' \times \mathbf{r}'')) = \mathbf{r} \cdot (\mathbf{r}' \times \mathbf{r}''')$.

69. Verify the Sum and Product Rules for derivatives of vector-valued functions.

70. Verify the Chain Rule for vector-valued functions.

71. Verify the Product Rule for cross products [Eq. (5)].

72. Verify the linearity properties

$$\int c\mathbf{r}(t)\,dt = c\int \mathbf{r}(t)\,dt \qquad (c \text{ any constant})$$

$$\int (\mathbf{r}_1(t) + \mathbf{r}_2(t))\,dt = \int \mathbf{r}_1(t)\,dt + \int \mathbf{r}_2(t)\,dt$$

73. Prove the Substitution Rule (where $g(t)$ is a differentiable scalar function):

$$\int_a^b \mathbf{r}(g(t))g'(t)\,dt = \int_{g^{-1}(a)}^{g^{-1}(b)} \mathbf{r}(u)\,du$$

74. Prove that if $\|\mathbf{r}(t)\| \le K$ for $t \in [a,b]$, then

$$\left\| \int_a^b \mathbf{r}(t)\,dt \right\| \le K(b-a)$$

14.3 Arc Length and Speed

REMINDER The length of a path or curve is referred to as the arc length.

In Section 12.2, we derived a formula for the arc length of a plane curve given in parametric form. This discussion applies to paths in three-space with only minor changes.

Recall that arc length is defined as the limit of the lengths of polygonal approximations. To produce a polygonal approximation to a path

$$\mathbf{r}(t) = \langle x(t), y(t), z(t)\rangle, \qquad a \le t \le b$$

we choose a partition $a = t_0 < t_1 < t_2 < \cdots < t_N = b$ and join the terminal points of the vectors $\mathbf{r}(t_j)$ by segments, as in Figure 1. As in Section 12.2, we find that if $\mathbf{r}'(t)$ exists and is continuous on $[a,b]$, then the lengths of the polygonal approximations approach a limit L as the maximum of the widths $|t_j - t_{j-1}|$ tends to zero. This limit is the length s of the path which is computed by the integral in the next theorem.

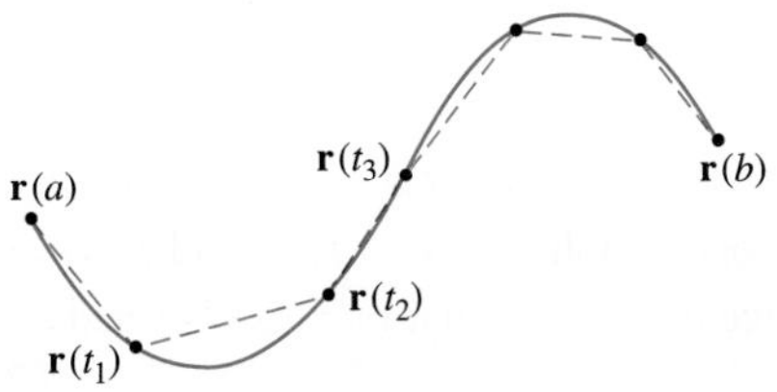

FIGURE 1 Polygonal approximation to the arc $\mathbf{r}(t)$ for $a \le t \le b$.

Keep in mind that the length s in Eq. (1) is the distance traveled by a particle following the path $\mathbf{r}(t)$. The path length s is not equal to the length of the underlying curve unless $\mathbf{r}(t)$ traverses the curve only once without reversing direction.

THEOREM 1 Length of a Path Assume that $\mathbf{r}(t)$ is differentiable and that $\mathbf{r}'(t)$ is continuous on $[a,b]$. Then the length s of the path $\mathbf{r}(t)$ for $a \le t \le b$ is equal to

$$s = \int_a^b \|\mathbf{r}'(t)\|\,dt = \int_a^b \sqrt{x'(t)^2 + y'(t)^2 + z'(t)^2}\,dt \qquad \boxed{1}$$

■ **EXAMPLE 1** Find the arc length s of $\mathbf{r}(t) = \langle \cos 3t, \sin 3t, 3t\rangle$ for $0 \le t \le 2\pi$.

Solution The derivative is $\mathbf{r}'(t) = \langle -3\sin 3t, 3\cos 3t, 3\rangle$, and

$$\|\mathbf{r}'(t)\|^2 = 9\sin^2 3t + 9\cos^2 3t + 9 = 9(\sin^2 3t + \cos^2 3t) + 9 = 18$$

Therefore, $s = \int_0^{2\pi} \|\mathbf{r}'(t)\|\,dt = \int_0^{2\pi} \sqrt{18}\,dt = 6\sqrt{2}\pi$. ■

Speed, by definition, is the rate of change of distance traveled with respect to time t. To calculate the speed, we define the **arc length function**:

$$s(t) = \int_a^t \|\mathbf{r}'(u)\|\,du$$

Thus $s(t)$ is the distance traveled during the time interval $[a,t]$. By the Fundamental Theorem of Calculus,

$$\text{Speed at time } t = \frac{ds}{dt} = \|\mathbf{r}'(t)\|$$

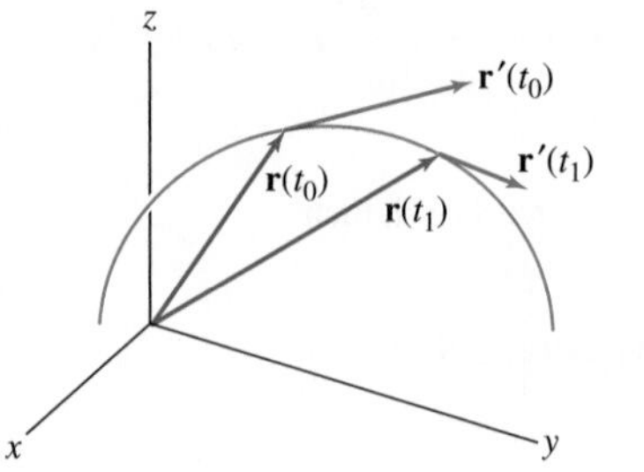

FIGURE 2 The velocity vector is longer at t_0 than at t_1, indicating that the particle is moving faster at t_0.

Now we can see why $\mathbf{r}'(t)$ is known as the velocity vector (and also as the tangent vector). It points in the direction of motion, and its magnitude is the speed (Figure 2). We often denote the velocity vector by $\mathbf{v}(t)$ and the speed by $v(t)$:

$$\mathbf{v}(t) = \mathbf{r}'(t), \qquad v(t) = \|\mathbf{v}(t)\|$$

■ **EXAMPLE 2** Find the speed at time $t = 2$ s of a particle whose position vector is

$$\mathbf{r}(t) = t^3\mathbf{i} - e^t\mathbf{j} + 4t\mathbf{k}$$

Solution The velocity vector is $\mathbf{v}(t) = \mathbf{r}'(t) = 3t^2\mathbf{i} - e^t\mathbf{j} + 4\mathbf{k}$, and at $t = 2$,

$$\mathbf{v}(2) = 12\mathbf{i} - e^2\mathbf{j} + 4\mathbf{k}$$

The particle's speed is $v(2) = \|\mathbf{v}(2)\| = \sqrt{12^2 + (-e^2)^2 + 4^2} \approx 14.65$ ft/s. ■

Arc Length Parametrization

Keep in mind that a parametrization $\mathbf{r}(t)$ describes not just a curve, but also how a particle traverses the curve, possibly speeding up, slowing down, or reversing direction along the way. Changing the parametrization amounts to describing a different way of traversing the same underlying curve.

We have seen that parametrizations are not unique. For example, $\mathbf{r}_1(t) = \langle t, t^2\rangle$ and $\mathbf{r}_2(s) = \langle s^3, s^6\rangle$ both parametrize the parabola $y = x^2$. Notice in this case that $\mathbf{r}_2(s)$ is obtained by substituting $t = s^3$ in $\mathbf{r}_1(t)$.

In general, we obtain a new parametrization by making a substitution $t = g(s)$—that is, by replacing $\mathbf{r}(t)$ with $\mathbf{r}_1(s) = \mathbf{r}(g(s))$ [Figure 3]. If $t = g(s)$ increases from a to b as s varies from c to d, then the path $\mathbf{r}(t)$ for $a \le t \le b$ is also parametrized by $\mathbf{r}_1(s)$ for $c \le s \le d$.

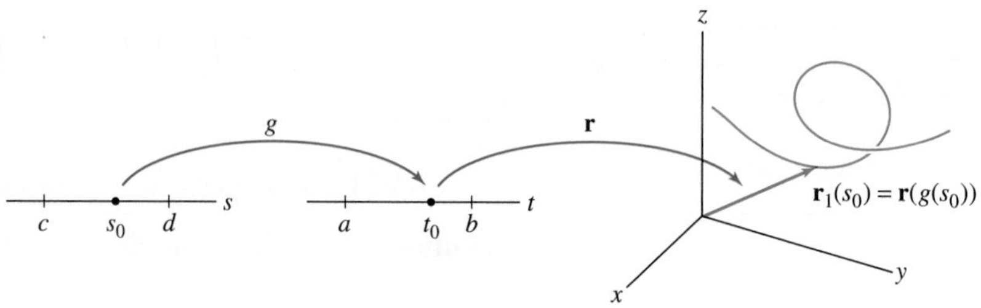

FIGURE 3 The path is parametrized by $\mathbf{r}(t)$ and by $\mathbf{r}_1(s) = \mathbf{r}(g(s))$.

■ **EXAMPLE 3** Parametrize the path $\mathbf{r}(t) = (t^2, \sin t, t)$ for $3 \le t \le 9$ using the parameter s, where $t = g(s) = e^s$.

Solution Substituting $t = e^s$ in $\mathbf{r}(t)$, we obtain the parametrization

$$\mathbf{r}_1(s) = \mathbf{r}(g(s)) = \langle e^{2s}, \sin e^s, e^s\rangle$$

Because $s = \ln t$, the parameter t varies from 3 to 9 as s varies from $\ln 3$ to $\ln 9$. Therefore, the path is parametrized by $\mathbf{r}_1(s)$ for $\ln 3 \le s \le \ln 9$. ■

One way of parametrizing a path is to choose a starting point and "walk along the path" at unit speed (say, 1 m/s). A parametrization of this type is called an **arc length parametrization** [Figure 4(A)]. It is defined by the property that the speed has constant value 1:

$$\|\mathbf{r}'(t)\| = 1 \qquad \text{for all } t$$

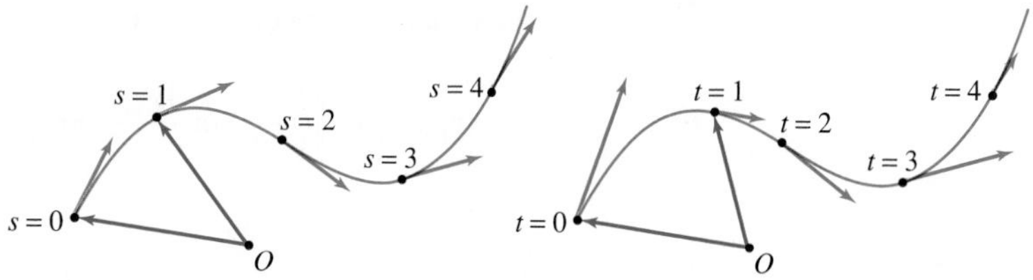

(A) An arc length parametrization: All tangent vectors have length 1, so speed is 1.

(B) Not an arc length parametrization: lengths of tangent vectors vary, so the speed is changing.

FIGURE 4

Arc length parametrizations are also called ***unit speed parametrizations****. We will use arc length parametrizations to define curvature in Section 14.4.*

In an arc length parametrization, the distance traveled over any time interval $[a, b]$ is equal to the length of the interval:

$$\text{Distance traveled over } [a, b] = \int_a^b \|\mathbf{r}'(t)\|\,dt = \int_a^b 1\,dt = b - a$$

The letter s is often used as the parameter in an arc length parametrization.

To find an arc length parametrization, start with any parametrization $\mathbf{r}(t)$ such that $\mathbf{r}'(t) \neq \mathbf{0}$ for all t, and form the arc length integral

$$s(t) = \int_0^t \|\mathbf{r}'(u)\|\,du$$

REMINDER By Theorem 2 in Section 7.2, if $g(x)$ is the inverse of $f(x)$, then

$$g'(x) = \frac{1}{f'(g(x))}$$

Because $\|\mathbf{r}'(t)\| \neq 0$, $s(t)$ is an increasing function and therefore has an inverse $t = g(s)$. By the formula for the derivative of an inverse (and since $s'(t) = \|\mathbf{r}'(t)\|$),

$$g'(s) = \frac{1}{s'(g(s))} = \frac{1}{\|\mathbf{r}'(g(s))\|}$$

Now we can show that the parametrization

$$\mathbf{r}_1(s) = \mathbf{r}(g(s))$$

is an arc length parametrization. Indeed, by the Chain Rule,

$$\|\mathbf{r}_1'(s)\| = \|\mathbf{r}'(g(s))g'(s)\| = \|\mathbf{r}'(g(s))\|\,\frac{1}{\|\mathbf{r}'(g(s))\|} = 1$$

In most cases we cannot evaluate the arc length integral $s(t)$ explicitly, and we cannot find a formula for its inverse $g(s)$ either. So although arc length parametrizations exist in general, we can find them explicitly only in special cases.

■ **EXAMPLE 4 Finding an Arc Length Parametrization** Find the arc length parametrization of the helix $\mathbf{r}(t) = \langle \cos 4t, \sin 4t, 3t\rangle$.

Solution First, we evaluate the arc length function

$$\|\mathbf{r}'(t)\| = \|\langle -4\sin 4t, 4\cos t, 3\rangle\| = \sqrt{16\sin^2 4t + 16\cos^2 4t + 3^2} = 5$$

$$s(t) = \int_0^t \|\mathbf{r}'(t)\|\,dt = \int_0^t 5\,dt = 5t$$

Then we observe that the inverse of $s(t) = 5t$ is $t = s/5$; that is, $g(s) = s/5$. As shown above, an arc length parametrization is

$$\mathbf{r}_1(s) = \mathbf{r}(g(s)) = \mathbf{r}\left(\frac{s}{5}\right) = \left\langle \cos\frac{4s}{5}, \sin\frac{4s}{5}, \frac{3s}{5}\right\rangle$$

As a check, let's verify that $\mathbf{r}_1(s)$ has unit speed:

$$\|\mathbf{r}_1'(s)\| = \left\|\left\langle -\frac{4}{5}\sin\frac{4s}{5}, \frac{4}{5}\cos\frac{4s}{5}, \frac{3}{5}\right\rangle\right\| = \sqrt{\frac{16}{25}\sin^2\frac{4s}{5} + \frac{16}{25}\cos^2\frac{4s}{5} + \frac{9}{25}} = 1 \quad \blacksquare$$

14.3 SUMMARY

- The length s of a path $\mathbf{r}(t) = \langle x(t), y(t), z(t)\rangle$ for $a \le t \le b$ is

$$s = \int_a^b \|\mathbf{r}'(t)\|\,dt = \int_a^b \sqrt{x'(t)^2 + y'(t)^2 + z'(t)^2}\,dt$$

- Arc length function: $s(t) = \int_a^t \|\mathbf{r}'(u)\|\,du$
- Speed is the derivative of distance traveled with respect to time:

$$v(t) = \frac{ds}{dt} = \|\mathbf{r}'(t)\|$$

- The velocity vector $\mathbf{v}(t) = \mathbf{r}'(t)$ points in the direction of motion [provided that $\mathbf{r}'(t) \neq \mathbf{0}$] and its magnitude $v(t) = \|\mathbf{r}'(t)\|$ is the object's speed.
- We say that $\mathbf{r}(s)$ is an *arc length parametrization* if $\|\mathbf{r}'(s)\| = 1$ for all s. In this case, the length of the path for $a \le s \le b$ is $b - a$.
- If $\mathbf{r}(t)$ is any parametrization such that $\mathbf{r}'(t) \neq \mathbf{0}$ for all t, then

$$\mathbf{r}_1(s) = \mathbf{r}(g(s))$$

is an arc length parametrization, where $t = g(s)$ is the inverse of the arc length function.

14.3 EXERCISES

Preliminary Questions

1. At a given instant, a car on a roller coaster has velocity vector $\mathbf{r}' = \langle 25, -35, 10\rangle$ (in miles per hour). What would the velocity vector be if the speed were doubled? What would it be if the car's direction were reversed but its speed remained unchanged?

2. Two cars travel in the same direction along the same roller coaster (at different times). Which of the following statements about their velocity vectors at a given point P on the roller coaster is/are true?

(a) The velocity vectors are identical.

(b) The velocity vectors point in the same direction but may have different lengths.

(c) The velocity vectors may point in opposite directions.

3. A mosquito flies along a parabola with speed $v(t) = t^2$. Let $L(t)$ be the total distance traveled at time t.

(a) How fast is $L(t)$ changing at $t = 2$?

(b) Is $L(t)$ equal to the mosquito's distance from the origin?

4. What is the length of the path traced by $\mathbf{r}(t)$ for $4 \le t \le 10$ if $\mathbf{r}(t)$ is an arc length parametrization?

Exercises

In Exercises 1–6, compute the length of the curve over the given interval.

1. $\mathbf{r}(t) = \langle 3t, 4t - 3, 6t + 1\rangle, \quad 0 \le t \le 3$

2. $\mathbf{r}(t) = 2t\mathbf{i} - 3t\mathbf{k}, \quad 11 \le t \le 15$

3. $\mathbf{r}(t) = \langle 2t, \ln t, t^2\rangle, \quad 1 \le t \le 4$

4. $\mathbf{r}(t) = \langle 2t^2 + 1, 2t^2 - 1, t^3\rangle, \quad 0 \le t \le 2$

5. $\mathbf{r}(t) = \langle t\cos t, t\sin t, 3t\rangle, \quad 0 \le t \le 2\pi$

6. $\mathbf{r}(t) = t\mathbf{i} + 2t\mathbf{j} + (t^2 - 3)\mathbf{k}, \quad 0 \le t \le 2$. Use the formula:

$$\int \sqrt{t^2 + a^2}\,dt = \frac{1}{2}t\sqrt{t^2 + a^2} + \frac{1}{2}a^2\ln\left(t + \sqrt{t^2 + a^2}\right)$$

In Exercises 7 and 8, compute the arc length function $s(t) = \int_a^t \|\mathbf{r}'(u)\|\, du$ *for the given value of a.*

7. $\mathbf{r}(t) = \langle t^2, 2t^2, t^3 \rangle, \quad a = 0$

8. $\mathbf{r}(t) = \langle 4t^{1/2}, \ln t, 2t \rangle, \quad a = 1$

In Exercises 9–12, find the speed at the given value of t.

9. $\mathbf{r}(t) = \langle 2t+3, 4t-3, 5-t \rangle, \quad t = 4$

10. $\mathbf{r}(t) = \langle e^{t-3}, 12, 3t^{-1} \rangle, \quad t = 3$

11. $\mathbf{r}(t) = \langle \sin 3t, \cos 4t, \cos 5t \rangle, \quad t = \frac{\pi}{2}$

12. $\mathbf{r}(t) = \langle \cosh t, \sinh t, t \rangle, \quad t = 0$

13. What is the velocity vector of a particle traveling to the right along the hyperbola $y = x^{-1}$ with constant speed 5 cm/s when the particle's location is $(2, \frac{1}{2})$?

14. A bee with velocity vector $\mathbf{r}'(t)$ starts out at the origin at $t = 0$ and flies around for T seconds. Where is the bee located at time T if $\int_0^T \mathbf{r}'(u)\, du = \mathbf{0}$? What does the quantity $\int_0^T \|\mathbf{r}'(u)\|\, du$ represent?

15. Let

$$\mathbf{r}(t) = \left\langle R\cos\left(\frac{2\pi N t}{h}\right), R\sin\left(\frac{2\pi N t}{h}\right), t \right\rangle, \qquad 0 \le t \le h$$

(a) Show that $\mathbf{r}(t)$ parametrizes a helix of radius R and height h making N complete turns.

(b) Guess which of the two springs in Figure 5 uses more wire.

(c) Compute the lengths of the two springs and compare.

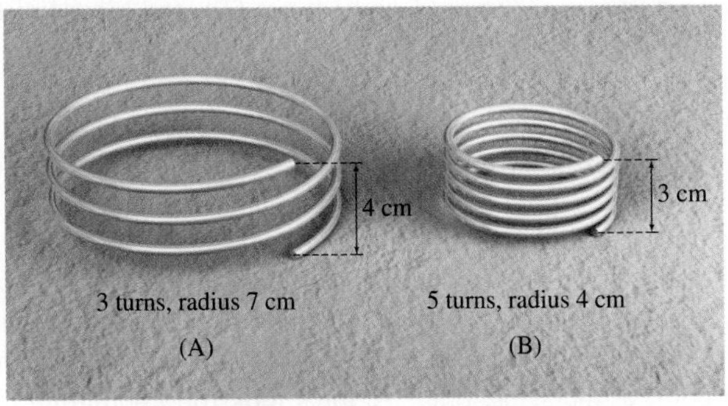

FIGURE 5 Which spring uses more wire?

16. Use Exercise 15 to find a general formula for the length of a helix of radius R and height h that makes N complete turns.

17. The cycloid generated by the unit circle has parametrization

$$\mathbf{r}(t) = \langle t - \sin t, 1 - \cos t \rangle$$

(a) Find the value of t in $[0, 2\pi]$ where the speed is at a maximum.

(b) Show that one arch of the cycloid has length 8. Recall the identity $\sin^2(t/2) = (1 - \cos t)/2$.

18. Which of the following is an arc length parametrization of a circle of radius 4 centered at the origin?

(a) $\mathbf{r}_1(t) = \langle 4\sin t, 4\cos t \rangle$

(b) $\mathbf{r}_2(t) = \langle 4\sin 4t, 4\cos 4t \rangle$

(c) $\mathbf{r}_3(t) = \langle 4\sin \frac{t}{4}, 4\cos \frac{t}{4} \rangle$

19. Let $\mathbf{r}(t) = \langle 3t+1, 4t-5, 2t \rangle$.

(a) Evaluate the arc length integral $s(t) = \int_0^t \|\mathbf{r}'(u)\|\, du$.

(b) Find the inverse $g(s)$ of $s(t)$.

(c) Verify that $\mathbf{r}_1(s) = \mathbf{r}(g(s))$ is an arc length parametrization.

20. Find an arc length parametrization of the line $y = 4x + 9$.

21. Let $\mathbf{r}(t) = \mathbf{w} + t\mathbf{v}$ be the parametrization of a line.

(a) Show that the arc length function $s(t) = \int_0^t \|\mathbf{r}'(u)\|\, du$ is given by $s(t) = t\|\mathbf{v}\|$. This shows that $\mathbf{r}(t)$ is an arc length parametrizaton if and only if $\mathbf{v}$ is a unit vector.

(b) Find an arc length parametrization of the line with $\mathbf{w} = \langle 1, 2, 3 \rangle$ and $\mathbf{v} = \langle 3, 4, 5 \rangle$.

22. Find an arc length parametrization of the circle in the plane $z = 9$ with radius 4 and center $(1, 4, 9)$.

23. Find a path that traces the circle in the plane $y = 10$ with radius 4 and center $(2, 10, -3)$ with constant speed 8.

24. Find an arc length parametrization of $\mathbf{r}(t) = \langle e^t \sin t, e^t \cos t, e^t \rangle$.

25. Find an arc length parametrization of $\mathbf{r}(t) = \langle t^2, t^3 \rangle$.

26. Find an arc length parametrization of the cycloid with parametrization $\mathbf{r}(t) = \langle t - \sin t, 1 - \cos t \rangle$.

27. Find an arc length parametrization of the line $y = mx$ for an arbitrary slope m.

28. Express the arc length L of $y = x^3$ for $0 \le x \le 8$ as an integral in two ways, using the parametrizations $\mathbf{r}_1(t) = \langle t, t^3 \rangle$ and $\mathbf{r}_2(t) = \langle t^3, t^9 \rangle$. Do not evaluate the integrals, but use substitution to show that they yield the same result.

29. The curve known as the **Bernoulli spiral** (Figure 6) has parametrization $\mathbf{r}(t) = \langle e^t \cos 4t, e^t \sin 4t \rangle$.

(a) Evaluate $s(t) = \int_{-\infty}^t \|\mathbf{r}'(u)\|\, du$. It is convenient to take lower limit $-\infty$ because $\mathbf{r}(-\infty) = \langle 0, 0 \rangle$.

(b) Use (a) to find an arc length parametrization of $\mathbf{r}(t)$.

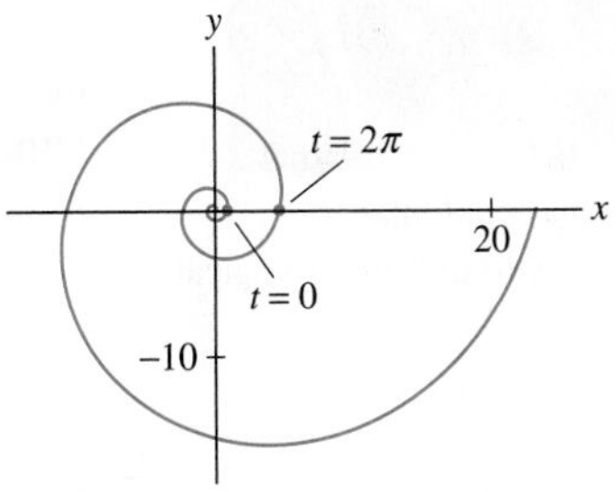

FIGURE 6 Bernoulli spiral.

Further Insights and Challenges

30. Prove that the length of a curve as computed using the arc length integral does not depend on its parametrization. More precisely, let $\mathcal{C}$ be the curve traced by $\mathbf{r}(t)$ for $a \le t \le b$. Let $f(s)$ be a differentiable function such that $f'(s) > 0$ and that $f(c) = a$ and $f(d) = b$. Then $\mathbf{r}_1(s) = \mathbf{r}(f(s))$ parametrizes $\mathcal{C}$ for $c \le s \le d$. Verify that

$$\int_a^b \|\mathbf{r}'(t)\|\,dt = \int_c^d \|\mathbf{r}_1'(s)\|\,ds$$

31. The unit circle with the point $(-1, 0)$ removed has parametrization (see Exercise 73 in Section 12.1)

$$\mathbf{r}(t) = \left\langle \frac{1-t^2}{1+t^2}, \frac{2t}{1+t^2} \right\rangle, \qquad -\infty < t < \infty$$

Use this parametrization to compute the length of the unit circle as an improper integral. *Hint:* The expression for $\|\mathbf{r}'(t)\|$ simplifies.

32. The involute of a circle, traced by a point at the end of a thread unwinding from a circular spool of radius R, has parametrization (see Exercise 26 in Section 13.2)

$$\mathbf{r}(\theta) = \big\langle R(\cos\theta + \theta\sin\theta), R(\sin\theta - \theta\cos\theta)\big\rangle$$

Find an arc length parametrization of the involute.

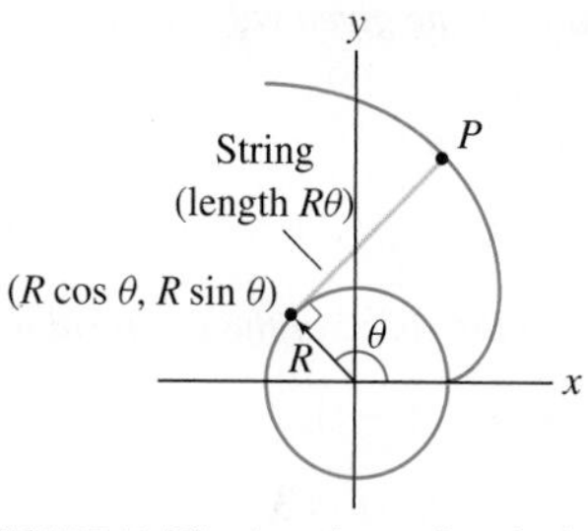

FIGURE 7 The involute of a circle.

33. The curve $\mathbf{r}(t) = \langle t - \tanh t, \operatorname{sech} t\rangle$ is called a **tractrix** (see Exercise 92 in Section 12.1).

(a) Show that $s(t) = \int_0^t \|\mathbf{r}'(u)\|\,du$ is equal to $s(t) = \ln(\cosh t)$.

(b) Show that $t = g(s) = \ln(e^s + \sqrt{e^{2s} - 1})$ is an inverse of $s(t)$ and verify that

$$\mathbf{r}_1(s) = \left\langle \tanh^{-1}\left(\sqrt{1 - e^{-2s}}\right) - \sqrt{1 - e^{-2s}}, e^{-s}\right\rangle$$

is an arc length parametrization of the tractrix.

14.4 Curvature

FIGURE 1 Curvature is a key ingredient in roller coaster design.

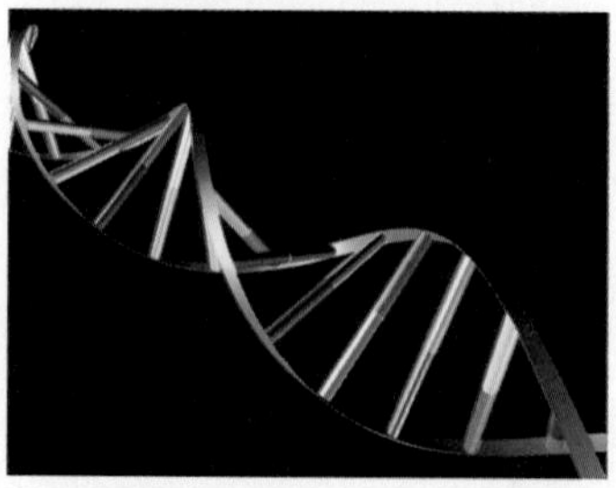

FIGURE 2 Biochemists study the effect of the curvature of DNA strands on biological processes.

Curvature is a measure of how much a curve bends. It is used to study geometric properties of curves and motion along curves, and has applications in diverse areas such as roller coaster design (Figure 1), optics, eye surgery (see Exercise 60), and biochemistry (Figure 2).

In Chapter 4, we used the second derivative $f''(x)$ to measure the bending or concavity of the graph of $y = f(x)$, so it might seem natural to take $f''(x)$ as our definition of curvature. However, there are two reasons why this proposed definition will not work. First, $f''(x)$ makes sense only for a graph $y = f(x)$ in the plane, and our goal is to define curvature for curves in three-space. A more serious problem is that $f''(x)$ does not truly capture the intrinsic curvature of a curve. A circle, for example, is symmetric, so its curvature ought to be the same at every point (Figure 3). But the upper semicircle is the graph of $f(x) = (1 - x^2)^{1/2}$ and the second derivative $f''(x) = -(1 - x^2)^{-3/2}$ does not have the same value at each point of the semicircle. We must look for a definition that depends only on the curve itself and not how it is oriented relative to the axes.

Consider a path with parametrization $\mathbf{r}(t) = \langle x(t), y(t), z(t)\rangle$. We assume that $\mathbf{r}'(t) \ne \mathbf{0}$ for all t in the domain of $\mathbf{r}(t)$. A parametrization with this property is called **regular**. At every point P along the path there is a **unit tangent vector** $\mathbf{T} = \mathbf{T}_P$ that points in the direction of motion of the parametrization. We write $\mathbf{T}(t)$ for the unit tangent vector at the terminal point of $\mathbf{r}(t)$:

$$\text{Unit tangent vector} = \mathbf{T}(t) = \frac{\mathbf{r}'(t)}{\|\mathbf{r}'(t)\|}$$

For example, if $\mathbf{r}(t) = \langle t, t^2, t^3\rangle$, then $\mathbf{r}'(t) = \langle 1, 2t, 3t^2\rangle$, and the unit tangent vector at $P = (1, 1, 1)$, which is the terminal point of $\mathbf{r}(1) = \langle 1, 1, 1\rangle$, is

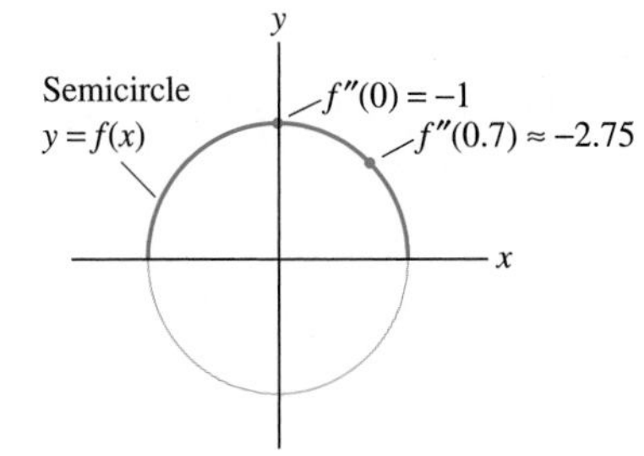

FIGURE 3 The second derivative of $f(x) = \sqrt{1 - x^2}$ does not capture the curvature of the circle, which by symmetry should be the same at all points.

$$\mathbf{T}_P = \frac{\langle 1, 2, 3\rangle}{\|\langle 1, 2, 3\rangle\|} = \frac{\langle 1, 2, 3\rangle}{\sqrt{1^2 + 2^2 + 3^2}} = \left\langle \frac{1}{\sqrt{14}}, \frac{2}{\sqrt{14}}, \frac{3}{\sqrt{14}} \right\rangle$$

If we choose another parametrization, say $\mathbf{r}_1(s)$, then we can also view $\mathbf{T}$ as function of s: $\mathbf{T}(s)$ is the unit tangent vector at the terminal point of $\mathbf{r}_1(s)$.

Now imagine walking along a path and observing how the unit tangent vector $\mathbf{T}$ changes direction (Figure 4). A change in $\mathbf{T}$ indicates that the path is bending, and the more rapidly $\mathbf{T}$ changes, the more the path bends. Thus, $\left\|\frac{d\mathbf{T}}{dt}\right\|$ would seem to be a good measure of curvature. However, $\left\|\frac{d\mathbf{T}}{dt}\right\|$ depends on how fast you walk (when you walk faster, the unit tangent vector changes more quickly). Therefore, we assume that you walk at unit speed. In other words, curvature is the magnitude $\kappa(s) = \left\|\frac{d\mathbf{T}}{ds}\right\|$, where s is the parameter of an arc length parametrization. Recall that $\mathbf{r}(s)$ is an arc length parametrization if $\|\mathbf{r}(s)\| = 1$ for all s.

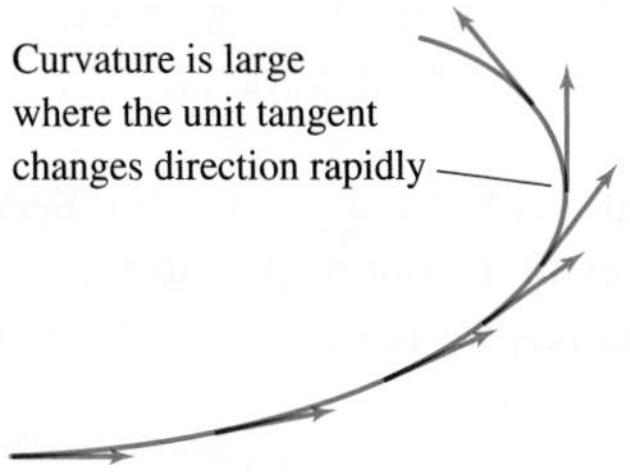

FIGURE 4 The unit tangent vector varies in direction but not in length.

DEFINITION Curvature Let $\mathbf{r}(s)$ be an arc length parametrization and $\mathbf{T}$ the unit tangent vector. The **curvature** at $\mathbf{r}(s)$ is the quantity (denoted by a lowercase Greek letter "kappa")

$$\kappa(s) = \left\|\frac{d\mathbf{T}}{ds}\right\| \qquad \boxed{1}$$

Our first two examples illustrate curvature in the case of lines and circles.

■ **EXAMPLE 1 A Line Has Zero Curvature** Compute the curvature at each point on the line $\mathbf{r}(t) = \langle x_0, y_0, z_0\rangle + t\mathbf{u}$, where $\|\mathbf{u}\| = 1$.

Solution First, we note that because $\mathbf{u}$ is a unit vector, $\mathbf{r}(t)$ is an arc length parametrization. Indeed, $\mathbf{r}'(t) = \mathbf{u}$ and thus $\|\mathbf{r}'(t)\| = \|\mathbf{u}\| = 1$. Thus we have $\mathbf{T}(t) = \mathbf{r}'(t)/\|\mathbf{r}'(t)\| = \mathbf{r}'(t)$ and hence $\mathbf{T}'(t) = \mathbf{r}''(t) = \mathbf{0}$ (because $\mathbf{r}'(t) = \mathbf{u}$ is constant). As expected, the curvature is zero at all points on a line:

$$\kappa(t) = \left\|\frac{d\mathbf{T}}{dt}\right\| = \|\mathbf{r}''(t)\| = 0 \qquad ■$$

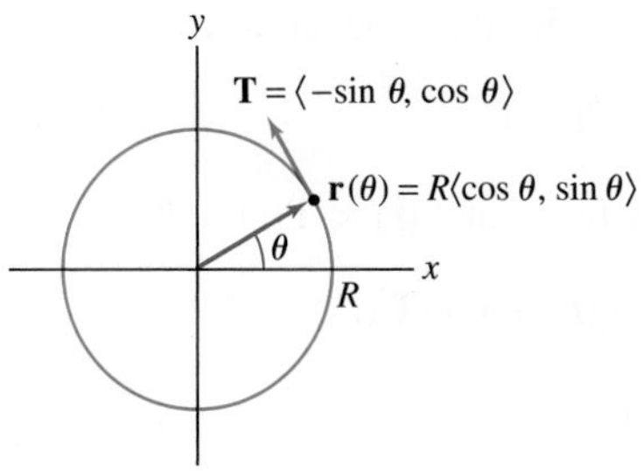

FIGURE 5 The unit tangent vector at a point on a circle of radius R.

■ **EXAMPLE 2 The Curvature of a Circle of Radius R Is $1/R$** Compute the curvature of a circle of radius R.

Solution Assume the circle is centered at the origin, so that it has parametrization $\mathbf{r}(\theta) = \langle R\cos\theta, R\sin\theta\rangle$ (Figure 5). This is not an arc length parametrization if $R \neq 1$. To find an arc length parametrization, we compute the arc length function:

$$s(\theta) = \int_0^\theta \|\mathbf{r}'(u)\|\, du = \int_0^\theta R\, du = R\theta$$

Thus $s = R\theta$, and the inverse of the arc length function is $\theta = g(s) = s/R$. In Section 14.3, we showed that $\mathbf{r}_1(s) = \mathbf{r}(g(s))$ is an arc length parametrization. In our case, we obtain

$$\mathbf{r}_1(s) = \mathbf{r}(g(s)) = \mathbf{r}\left(\frac{s}{R}\right) = \left\langle R\cos\frac{s}{R}, R\sin\frac{s}{R}\right\rangle$$

Example 2 shows that a circle of large radius R has small curvature $1/R$. This makes sense because your direction of motion changes slowly when you walk at unit speed along a circle of large radius.

The unit tangent vector and its derivative are

$$\mathbf{T}(s) = \frac{d\mathbf{r}_1}{ds} = \frac{d}{ds}\left\langle R\cos\frac{s}{R}, R\sin\frac{s}{R}\right\rangle = \left\langle -\sin\frac{s}{R}, \cos\frac{s}{R}\right\rangle$$

$$\frac{d\mathbf{T}}{ds} = -\frac{1}{R}\left\langle \cos\frac{s}{R}, \sin\frac{s}{R}\right\rangle$$

By definition of curvature,

$$\kappa(s) = \left\|\frac{d\mathbf{T}}{ds}\right\| = \frac{1}{R}\left\|\left\langle \cos\frac{s}{R}, \sin\frac{s}{R}\right\rangle\right\| = \frac{1}{R}$$

This shows that the curvature is $1/R$ at all points on the circle. ■

← REMINDER *To prove that $\mathbf{T}(t)$ and $\mathbf{T}'(t)$ are orthogonal, note that $\mathbf{T}(t)$ is a unit vector, so $\mathbf{T}(t)\cdot\mathbf{T}(t) = 1$. Differentiate using the Product Rule for Dot Products:*

$$\frac{d}{dt}\mathbf{T}(t)\cdot\mathbf{T}(t) = 2\mathbf{T}(t)\cdot\mathbf{T}'(t) = 0$$

This shows that $\mathbf{T}(t)\cdot\mathbf{T}'(t) = 0$

In practice, it is often impossible to find an arc length parametrization explicitly. Fortunately, we can compute curvature using any regular parametrization $\mathbf{r}(t)$. To derive a formula, we need the following two results.

First is the fact that $\mathbf{T}(t)$ and $\mathbf{T}'(t)$ are orthogonal (see the marginal note). Second, arc length s is function $s(t)$ of time t, so the derivatives of $\mathbf{T}$ with respect to t and s are related by the Chain Rule. Denoting the derivative with respect to t by a prime, we have

$$\mathbf{T}'(t) = \frac{d\mathbf{T}}{dt} = \frac{d\mathbf{T}}{ds}\frac{ds}{dt} = v(t)\frac{d\mathbf{T}}{ds}$$

where $v(t) = \dfrac{ds}{dt} = \|\mathbf{r}'(t)\|$ is the speed of $\mathbf{r}(t)$. Since curvature is the magnitude $\left\|\dfrac{d\mathbf{T}}{ds}\right\|$, we obtain

$$\|\mathbf{T}'(t)\| = v(t)\kappa(t) \qquad \boxed{2}$$

.

To apply Eq. (3) to plane curves, replace $\mathbf{r}(t) = \langle x(t), y(t)\rangle$ by $\mathbf{r}(t) = \langle x(t), y(t), 0\rangle$ and compute the cross product.

THEOREM 1 Formula for Curvature If $\mathbf{r}(t)$ is a regular parametrization, then the curvature at $\mathbf{r}(t)$ is

$$\kappa(t) = \frac{\|\mathbf{r}'(t)\times\mathbf{r}''(t)\|}{\|\mathbf{r}'(t)\|^3} \qquad \boxed{3}$$

Proof Since $v(t) = \|\mathbf{r}'(t)\|$, we have $\mathbf{r}'(t) = v(t)\mathbf{T}(t)$. By the Product Rule,

$$\mathbf{r}''(t) = v'(t)\mathbf{T}(t) + v(t)\mathbf{T}'(t)$$

Now compute the following cross product, using the fact that $\mathbf{T}(t)\times\mathbf{T}(t) = \mathbf{0}$:

$$\begin{aligned}\mathbf{r}'(t)\times\mathbf{r}''(t) &= v(t)\mathbf{T}(t)\times\big(v'(t)\mathbf{T}(t) + v(t)\mathbf{T}'(t)\big)\\ &= v(t)^2\mathbf{T}(t)\times\mathbf{T}'(t)\end{aligned} \qquad \boxed{4}$$

Because $\mathbf{T}(t)$ and $\mathbf{T}'(t)$ are orthogonal,

$$\|\mathbf{T}(t)\times\mathbf{T}'(t)\| = \|\mathbf{T}(t)\|\,\|\mathbf{T}'(t)\|\sin\frac{\pi}{2} = \|\mathbf{T}'(t)\|$$

← REMINDER *By Theorem 1 in Section 13.4,*

$$\|\mathbf{v}\times\mathbf{w}\| = \|\mathbf{v}\|\,\|\mathbf{w}\|\sin\theta$$

where θ is the angle between $\mathbf{v}$ and $\mathbf{w}$.

Eq. (4) yields $\|\mathbf{r}'(t)\times\mathbf{r}''(t)\| = v(t)^2\|\mathbf{T}'(t)\|$. Using Eq. (2), we obtain

$$\|\mathbf{r}'(t)\times\mathbf{r}''(t)\| = v(t)^2\|\mathbf{T}'(t)\| = v(t)^3\kappa(t) = \|\mathbf{r}'(t)\|^3\kappa(t)$$

This yields the desired formula. ■

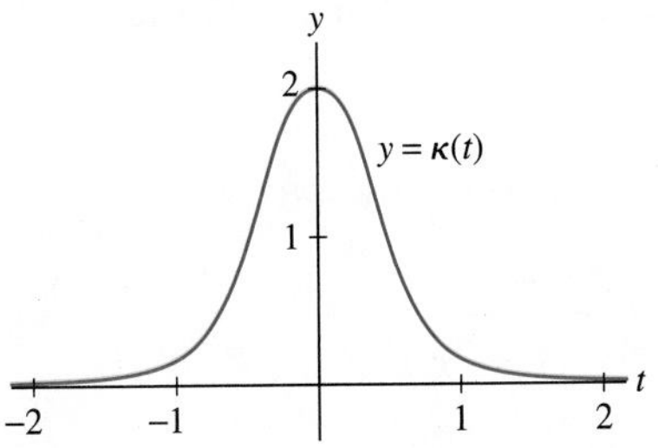

FIGURE 6 Graph of the curvature $\kappa(t)$ of the twisted cubic $\mathbf{r}(t) = \langle t, t^2, t^3 \rangle$.

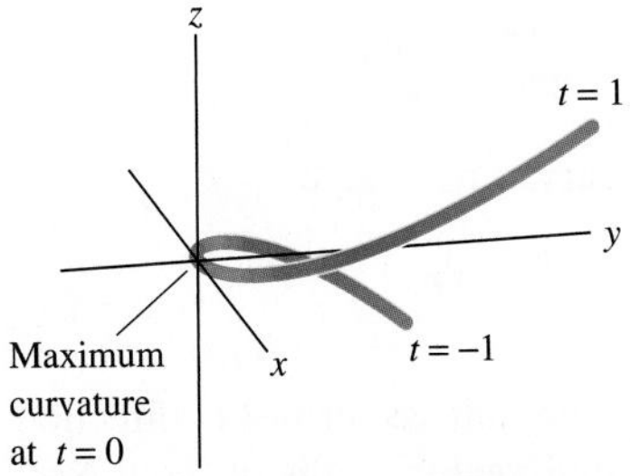

FIGURE 7 Graph of the twisted cubic $\mathbf{r}(t) = \langle t, t^2, t^3 \rangle$ colored by curvature.

■ **EXAMPLE 3 Twisted Cubic Curve** CAS Calculate the curvature $\kappa(t)$ of the twisted cubic $\mathbf{r}(t) = \langle t, t^2, t^3 \rangle$. Then plot the graph of $\kappa(t)$ and determine where the curvature is largest.

Solution The derivatives are

$$\mathbf{r}'(t) = \langle 1, 2t, 3t^2 \rangle, \qquad \mathbf{r}''(t) = \langle 0, 2, 6t \rangle$$

The parametrization is regular because $\mathbf{r}'(t) \neq \mathbf{0}$ for all t, so we may use Eq. (3):

$$\mathbf{r}'(t) \times \mathbf{r}''(t) = \begin{vmatrix} \mathbf{i} & \mathbf{j} & \mathbf{k} \\ 1 & 2t & 3t^2 \\ 0 & 2 & 6t \end{vmatrix} = 6t^2\mathbf{i} - 6t\mathbf{j} + 2\mathbf{k}$$

$$\kappa(t) = \frac{\|\mathbf{r}'(t) \times \mathbf{r}''(t)\|}{\|\mathbf{r}'(t)\|^3} = \frac{\sqrt{36t^4 + 36t^2 + 4}}{(1 + 4t^2 + 9t^4)^{3/2}}$$

The graph of $\kappa(t)$ in Figure 6 shows that the curvature is largest at $t = 0$. The curve $\mathbf{r}(t)$ is illustrated in Figure 7. The plot is colored by curvature, with large curvature represented in blue, small curvature in green. ■

In the second paragraph of this section, we pointed out that the curvature of a graph $y = f(x)$ must involve more than just the second derivative $f''(x)$. We now show that the curvature can be expressed in terms of both $f''(x)$ and $f'(x)$.

THEOREM 2 Curvature of a Graph in the Plane The curvature at the point $(x, f(x))$ on the graph of $y = f(x)$ is equal to

$$\kappa(x) = \frac{|f''(x)|}{\left(1 + f'(x)^2\right)^{3/2}} \qquad \boxed{5}$$

Proof The curve $y = f(x)$ has parametrization $\mathbf{r}(x) = \langle x, f(x) \rangle$. Therefore, $\mathbf{r}'(x) = \langle 1, f'(x) \rangle$ and $\mathbf{r}''(x) = \langle 0, f''(x) \rangle$. To apply Theorem 1, we treat $\mathbf{r}'(x)$ and $\mathbf{r}''(x)$ as vectors in $\mathbf{R}^3$ with z-component equal to zero. Then

$$\mathbf{r}'(x) \times \mathbf{r}''(x) = \begin{vmatrix} \mathbf{i} & \mathbf{j} & \mathbf{k} \\ 1 & f'(x) & 0 \\ 0 & f''(x) & 0 \end{vmatrix} = f''(x)\mathbf{k}$$

Since $\|\mathbf{r}'(x)\| = \|\langle 1, f'(x) \rangle\| = (1 + f'(x)^2)^{1/2}$, Eq. (3) yields

$$\kappa(x) = \frac{\|\mathbf{r}'(x) \times \mathbf{r}''(x)\|}{\|\mathbf{r}'(x)\|^3} = \frac{|f''(x)|}{\left(1 + f'(x)^2\right)^{3/2}}$$ ■

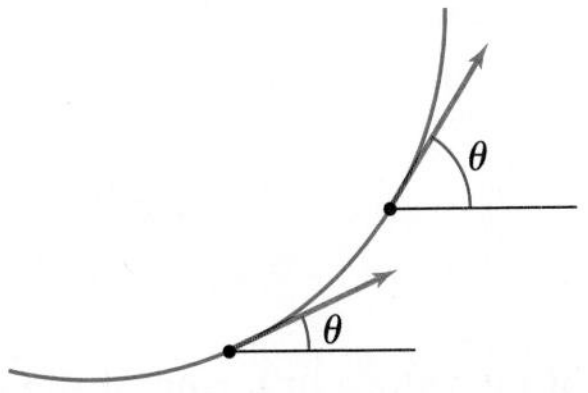

FIGURE 8 The angle θ changes as the curve bends.

CONCEPTUAL INSIGHT Curvature for plane curves has a geometric interpretation in terms of the angle of inclination, defined as the angle θ between the tangent vector and the horizonal (Figure 8). The angle θ changes as the curve bends, and we can show that the curvature κ is the rate of change of θ as you walk along the curve at unit speed (see Exercise 61).

EXAMPLE 4 Compute the curvature of $f(x) = x^3 - 3x^2 + 4$ at $x = 0, 1, 2, 3$.

Solution We apply Eq. (5):

$$f'(x) = 3x^2 - 6x = 3x(x-2), \qquad f''(x) = 6x - 6$$

$$\kappa(x) = \frac{|f''(x)|}{\left(1 + f'(x)^2\right)^{3/2}} = \frac{|6x - 6|}{\left(1 + 9x^2(x-2)^2\right)^{3/2}}$$

We obtain the following values:

$$\kappa(0) = \frac{6}{(1+0)^{3/2}} = 6, \qquad \kappa(1) = \frac{0}{(1+9)^{3/2}} = 0$$

$$\kappa(2) = \frac{6}{(1+0)^{3/2}} = 6, \qquad \kappa(3) = \frac{12}{82^{3/2}} \approx 0.016$$

Figure 9 shows that the graph bends more where the curvature is large. ■

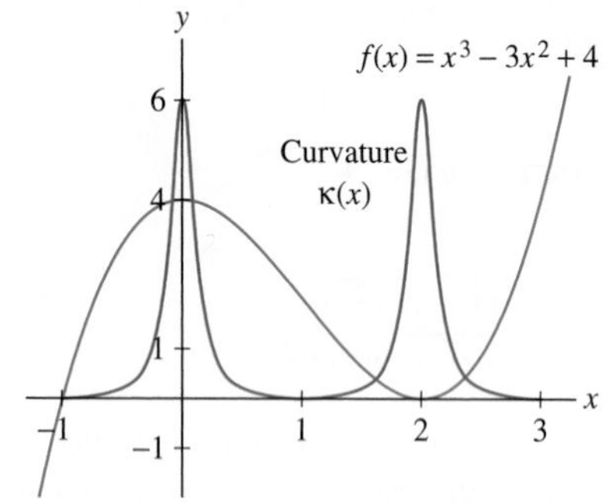

FIGURE 9 Graph of $f(x) = x^3 - 3x^2 + 4$ and the curvature $\kappa(x)$.

Unit Normal Vector

We noted above that $\mathbf{T}'(t)$ and $\mathbf{T}(t)$ are orthogonal. The unit vector in the direction of $\mathbf{T}'(t)$, assuming it is nonzero, is called the **unit normal vector** and is denoted $\mathbf{N}(t)$ or simply $\mathbf{N}$:

$$\text{Unit normal vector} = \mathbf{N}(t) = \frac{\mathbf{T}'(t)}{\|\mathbf{T}'(t)\|} \qquad \boxed{6}$$

Furthermore, $\|\mathbf{T}'(t)\| = v(t)\kappa(t)$ by Eq. (2), so we have

$$\mathbf{T}'(t) = v(t)\kappa(t)\mathbf{N}(t) \qquad \boxed{7}$$

Intuitively, $\mathbf{N}$ points the direction in which the curve is turning (see Figure 11). This is particularly clear for a plane curve. In this case, there are two unit vectors orthogonal to $\mathbf{T}$ (Figure 10), and of these two, $\mathbf{N}$ is the vector that points to the "inside" of the curve.

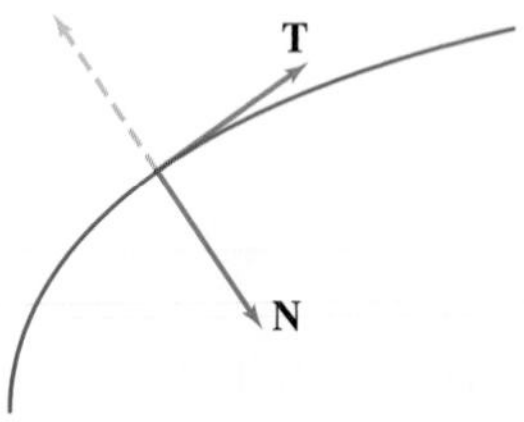

FIGURE 10 For a plane curve, the unit normal vector points in the direction of bending.

EXAMPLE 5 Unit Normal to a Helix Find the unit normal vector at $t = \frac{\pi}{4}$ to the helix $\mathbf{r}(t) = \langle \cos t, \sin t, t \rangle$.

Solution The tangent vector $\mathbf{r}'(t) = \langle -\sin t, \cos t, 1 \rangle$ has length $\sqrt{2}$, so

$$\mathbf{T}(t) = \frac{\mathbf{r}'(t)}{\|\mathbf{r}'(t)\|} = \frac{1}{\sqrt{2}} \langle -\sin t, \cos t, 1 \rangle$$

$$\mathbf{T}'(t) = \frac{1}{\sqrt{2}} \langle -\cos t, -\sin t, 0 \rangle$$

$$\mathbf{N}(t) = \frac{\mathbf{T}'(t)}{\|\mathbf{T}'(t)\|} = \langle -\cos t, -\sin t, 0 \rangle$$

Hence, $\mathbf{N}\left(\frac{\pi}{4}\right) = \left\langle -\frac{\sqrt{2}}{2}, -\frac{\sqrt{2}}{2}, 0 \right\rangle$ (Figure 11). ■

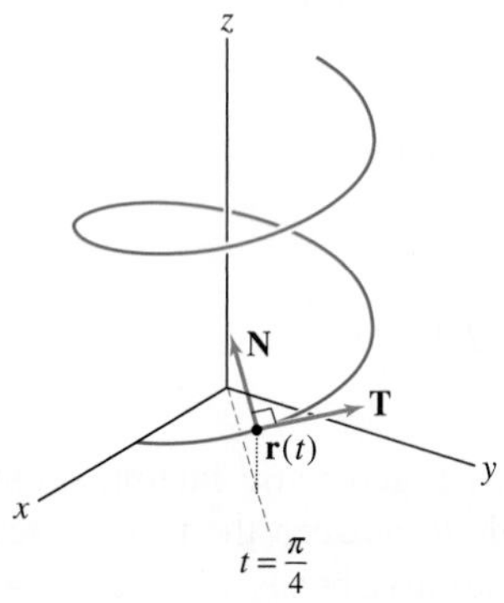

FIGURE 11 Unit tangent and unit normal vectors at $t = \frac{\pi}{4}$ on the helix in Example 5.

We conclude by describing another interpretation of curvature in terms of the osculating or "best-fitting circle" circle. Suppose that P is a point on a plane curve $\mathcal{C}$ where the curvature κ_P is nonzero. The **osculating circle**, denoted Osc_P, is the circle of radius

$R = 1/\kappa_P$ through P whose center Q lies in the direction of the unit normal $\mathbf{N}$ (Figure 12). In other words, the center Q is determined by

$$\overrightarrow{OQ} = \mathbf{r}(t_0) + \kappa_P^{-1}\mathbf{N} = \mathbf{r}(t_0) + R\mathbf{N} \qquad \boxed{8}$$

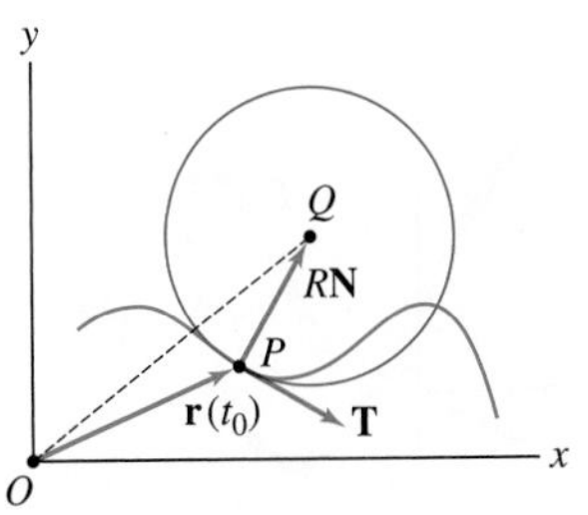

FIGURE 12 The center Q of the osculating circle at P lies at a distance $R = \kappa_P^{-1}$ from P in the normal direction.

Among all circles tangent to $\mathcal{C}$ at P, Osc_P "best fits" the curve (Figure 13; see also Exercise 71). We refer to $R = 1/\kappa_P$ as the **radius of curvature** at P. The center Q of Osc_P is called the **center of curvature** at P.

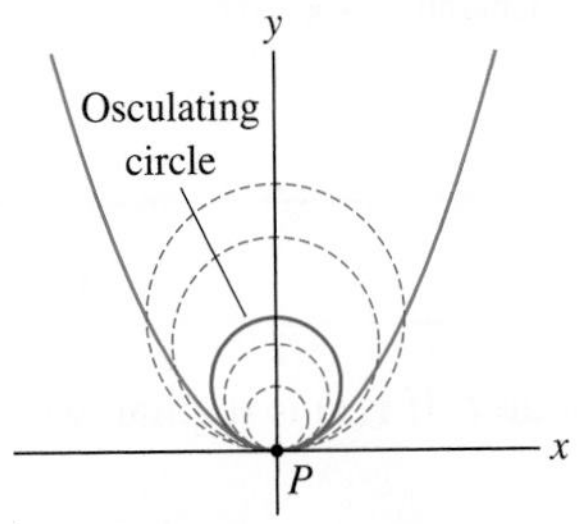

FIGURE 13 Among all circles tangent to the curve at P, the osculating circle is the "best fit" to the curve.

EXAMPLE 6 Parametrize the osculating circle to $y = x^2$ at $x = \frac{1}{2}$.

Solution Let $f(x) = x^2$. We use the parametrization

$$\mathbf{r}(x) = \langle x, f(x)\rangle = \langle x, x^2\rangle$$

and proceed by the following steps.

***Step 1.* Find the radius.**
Apply Eq. (5) to $f(x) = x^2$ to compute the curvature:

$$\kappa(x) = \frac{|f''(x)|}{(1 + f'(x)^2)^{3/2}} = \frac{2}{(1+4x^2)^{3/2}}, \qquad \kappa\left(\frac{1}{2}\right) = \frac{2}{2^{3/2}} = \frac{1}{\sqrt{2}}$$

The osculating circle has radius $R = 1/\kappa\left(\frac{1}{2}\right) = \sqrt{2}$.

***Step 2.* Find N at $t = \frac{1}{2}$.**
For a plane curve, there is an easy way to find $\mathbf{N}$ without computing $\mathbf{T}'$. The tangent vector is $\mathbf{r}'(x) = \langle 1, 2x\rangle$, and we know that $\langle 2x, -1\rangle$ is orthogonal to $\mathbf{r}'(x)$ (because their dot product is zero). Therefore, $\mathbf{N}(x)$ is the unit vector in one of the two directions $\pm\langle 2x, -1\rangle$. Figure 14 shows that the unit normal vector points in the positive y-direction (the direction of bending). Therefore,

$$\mathbf{N}(x) = \frac{\langle -2x, 1\rangle}{\|\langle -2x, 1\rangle\|} = \frac{\langle -2x, 1\rangle}{\sqrt{1+4x^2}}, \qquad \mathbf{N}\left(\frac{1}{2}\right) = \frac{1}{\sqrt{2}}\langle -1, 1\rangle$$

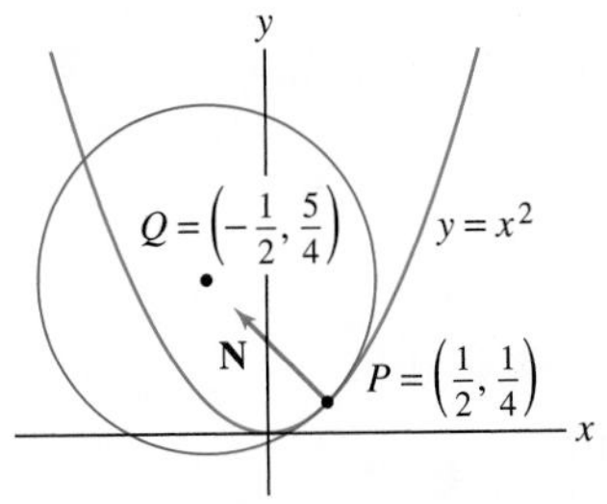

FIGURE 14 The osculating circle to $y = x^2$ at $x = \frac{1}{2}$ has center Q and radius $R = \sqrt{2}$.

***Step 3.* Find the center Q.**
Apply Eq. (8) with $t_0 = \frac{1}{2}$:

$$\overrightarrow{OQ} = \mathbf{r}\left(\frac{1}{2}\right) + \kappa\left(\frac{1}{2}\right)^{-1}\mathbf{N}\left(\frac{1}{2}\right) = \left\langle \frac{1}{2}, \frac{1}{4}\right\rangle + \sqrt{2}\left(\frac{\langle -1, 1\rangle}{\sqrt{2}}\right) = \left\langle -\frac{1}{2}, \frac{5}{4}\right\rangle$$

***Step 4.* Parametrize the osculating circle.**
The osculating circle has radius $R = \sqrt{2}$, so it has parametrization

$$\mathbf{c}(t) = \underbrace{\left\langle -\frac{1}{2}, \frac{5}{4}\right\rangle}_{\text{Center}} + \sqrt{2}\,\langle \cos t, \sin t\rangle \qquad \blacksquare$$

If a curve $\mathcal{C}$ lies in a plane, then this plane is the osculating plane. For a general curve in three-space, the osculating plane varies from point to point.

To define the osculating circle at a point P on a space curve $\mathcal{C}$, we must first specify the plane in which the circle lies. The **osculating plane** is the plane through P determined by the unit tangent $\mathbf{T}_P$ and the unit normal $\mathbf{N}_P$ at P (we assume that $\mathbf{T}' \neq 0$, so $\mathbf{N}$ is defined). Intuitively, the osculating plane is the plane that "most nearly" contains the curve $\mathcal{C}$ near P (see Figure 15). The osculating circle is the circle in the osculating plane through P of radius $R = 1/\kappa_P$ whose center is located in the normal direction N_P from P. Equation (8) remains valid for space curves.

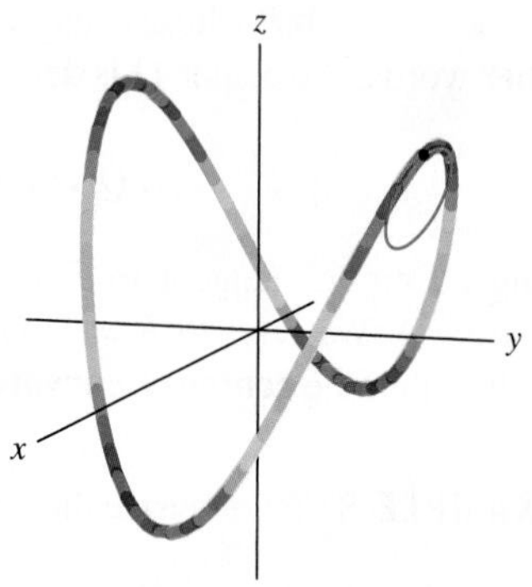

(A) Osculating circle at $t = \frac{\pi}{4}$
Curvature is $\kappa = 4.12$

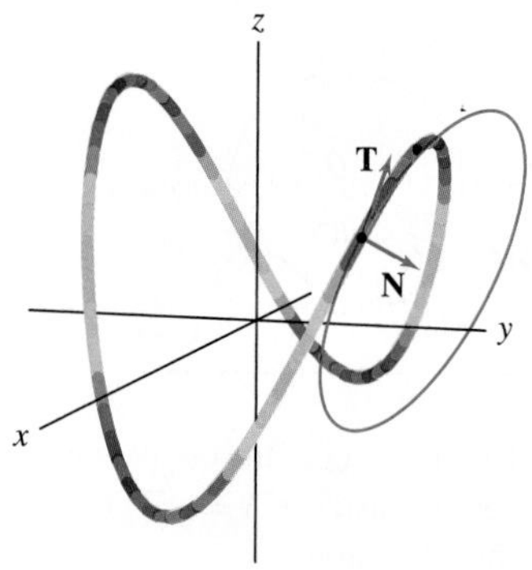

(B) Osculating circle at $t = \frac{\pi}{8}$
Curvature is $\kappa = 0.64$

FIGURE 15 Osculating circles to $\mathbf{r}(t) = \langle \cos t, \sin t, \sin 2t \rangle$.

14.4 SUMMARY

- A parametrization $\mathbf{r}(t)$ is called *regular* if $\mathbf{r}'(t) \neq \mathbf{0}$ for all t. If $\mathbf{r}(t)$ is regular, we define the *unit tangent vector* $\mathbf{T}(t) = \dfrac{\mathbf{r}'(t)}{\|\mathbf{r}'(t)\|}$.
- *Curvature* is defined by $\kappa(s) = \left\| \dfrac{d\mathbf{T}}{ds} \right\|$, where $\mathbf{r}(s)$ is an arc length parametrization.
- In practice, we compute curvature using the following formula, which is valid for arbitrary regular parametrizations:

$$\kappa(t) = \frac{\|\mathbf{r}'(t) \times \mathbf{r}''(t)\|}{\|\mathbf{r}'(t)\|^3}$$

- The curvature at a point on a graph $y = f(x)$ in the plane is

$$\kappa(x) = \frac{|f''(x)|}{\left(1 + f'(x)^2\right)^{3/2}}$$

- If $\|\mathbf{T}'(t)\| \neq 0$, we define the *unit normal vector* $\mathbf{N}(t) = \dfrac{\mathbf{T}'(t)}{\|\mathbf{T}'(t)\|}$.
- $\mathbf{T}'(t) = \kappa(t) v(t) \mathbf{N}(t)$
- The *osculating plane* at a point P on a curve $\mathcal{C}$ is the plane through P determined by the vectors $\mathbf{T}_P$ and $\mathbf{N}_P$. It is defined only if the curvature κ_P at P is nonzero.
- The *osculating circle* Osc_P is the circle in the osculating plane through P of radius $R = 1/\kappa_P$ whose center Q lies in the normal direction $\mathbf{N}_P$:

$$\overrightarrow{OQ} = \mathbf{r}(t_0) + \kappa_P^{-1} \mathbf{N}_P = \mathbf{r}(t_0) + R\mathbf{N}_P$$

The center of Osc_P is called the *center of curvature* and R is called the *radius of curvature*.

14.4 EXERCISES

Preliminary Questions

1. What is the unit tangent vector of a line with direction vector $\mathbf{v} = \langle 2, 1, -2 \rangle$?

2. What is the curvature of a circle of radius 4?

3. Which has larger curvature, a circle of radius 2 or a circle of radius 4?

4. What is the curvature of $\mathbf{r}(t) = \langle 2 + 3t, 7t, 5 - t \rangle$?

5. What is the curvature at a point where $\mathbf{T}'(s) = \langle 1, 2, 3\rangle$ in an arc length parametrization $\mathbf{r}(s)$?

6. What is the radius of curvature of a circle of radius 4?

7. What is the radius of curvature at P if $\kappa_P = 9$?

Exercises

In Exercises 1–6, calculate $\mathbf{r}'(t)$ and $\mathbf{T}(t)$, and evaluate $\mathbf{T}(1)$.

1. $\mathbf{r}(t) = \langle 4t^2, 9t\rangle$

2. $\mathbf{r}(t) = \langle e^t, t^2\rangle$

3. $\mathbf{r}(t) = \langle 3+4t, 3-5t, 9t\rangle$

4. $\mathbf{r}(t) = \langle 1+2t, t^2, 3-t^2\rangle$

5. $\mathbf{r}(t) = \langle \cos \pi t, \sin \pi t, t\rangle$

6. $\mathbf{r}(t) = \langle e^t, e^{-t}, t^2\rangle$

In Exercises 7–10, use Eq. (3) to calculate the curvature function $\kappa(t)$.

7. $\mathbf{r}(t) = \langle 1, e^t, t\rangle$

8. $\mathbf{r}(t) = \langle 4\cos t, t, 4\sin t\rangle$

9. $\mathbf{r}(t) = \langle 4t+1, 4t-3, 2t\rangle$

10. $\mathbf{r}(t) = \langle t^{-1}, 1, t\rangle$

In Exercises 11–14, use Eq. (3) to evaluate the curvature at the given point.

11. $\mathbf{r}(t) = \langle 1/t, 1/t^2, t^2\rangle, \quad t = -1$

12. $\mathbf{r}(t) = \langle 3-t, e^{t-4}, 8t-t^2\rangle, \quad t = 4$

13. $\mathbf{r}(t) = \langle \cos t, \sin t, t^2\rangle, \quad t = \frac{\pi}{2}$

14. $\mathbf{r}(t) = \langle \cosh t, \sinh t, t\rangle, \quad t = 0$

In Exercises 15–18, find the curvature of the plane curve at the point indicated.

15. $y = e^t, \quad t = 3$

16. $y = \cos x, \quad x = 0$

17. $y = t^4, \quad t = 2$

18. $y = t^n, \quad t = 1$

19. Find the curvature of $\mathbf{r}(t) = \langle 2\sin t, \cos 3t, t\rangle$ at $t = \frac{\pi}{3}$ and $t = \frac{\pi}{2}$ (Figure 16).

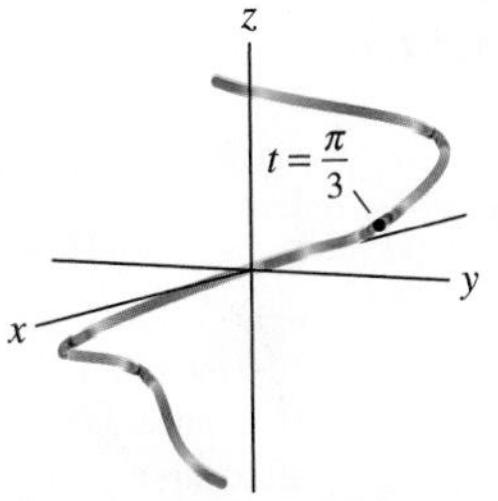

FIGURE 16 The curve $\mathbf{r}(t) = \langle 2\sin t, \cos 3t, t\rangle$.

20. GU Find the curvature function $\kappa(x)$ for $y = \sin x$. Use a computer algebra system to plot $\kappa(x)$ for $0 \le x \le 2\pi$. Prove that the curvature takes its maximum at $x = \frac{\pi}{2}$ and $\frac{3\pi}{2}$. *Hint:* As a shortcut to finding the max, observe that the maximum of the numerator and the minimum of the denominator of $\kappa(x)$ occur at the same points.

21. Show that the tractrix $\mathbf{r}(t) = \langle t - \tanh t, \operatorname{sech} t\rangle$ has the curvature function $\kappa(t) = \operatorname{sech} t$.

22. Show that curvature at an inflection point of a plane curve $y = f(x)$ is zero.

23. Find the value of α such that the curvature of $y = e^{\alpha x}$ at $x = 0$ is as large as possible.

24. Find the point of maximum curvature on $y = e^x$.

25. Show that the curvature function of the parametrization $\mathbf{r}(t) = \langle a\cos t, b\sin t\rangle$ of the ellipse $\left(\frac{x}{a}\right)^2 + \left(\frac{y}{b}\right)^2 = 1$ is

$$\kappa(t) = \frac{ab}{(b^2\cos^2 t + a^2 \sin^2 t)^{3/2}} \qquad \boxed{9}$$

26. Use a sketch to predict where the points of minimal and maximal curvature occur on an ellipse. Then use Eq. (9) to confirm or refute your prediction.

27. In the notation of Exercise 25, assume that $a \ge b$. Show that $b/a^2 \le \kappa(t) \le a/b^2$ for all t.

28. Use Eq. (3) to prove that for a plane curve $\mathbf{r}(t) = \langle x(t), y(t)\rangle$,

$$\kappa(t) = \frac{|x'(t)y''(t) - x''(t)y'(t)|}{\left(x'(t)^2 + y'(t)^2\right)^{3/2}} \qquad \boxed{10}$$

In Exercises 29–32, use Eq. (10) to compute the curvature at the given point.

29. $\langle t^2, t^3\rangle, \quad t = 2$

30. $\langle \cosh s, s\rangle, \quad s = 0$

31. $\langle t\cos t, \sin t\rangle, \quad t = \pi$

32. $\langle \sin 3s, 2\sin 4s\rangle, \quad s = \frac{\pi}{2}$

33. Let $s(t) = \int_{-\infty}^{t} \|\mathbf{r}'(u)\|\, du$ for the Bernoulli spiral $\mathbf{r}(t) = \langle e^t \cos 4t, e^t \sin 4t\rangle$ (see Exercise 29 in Section 14.3). Show that the radius of curvature is proportional to $s(t)$.

34. The **Cornu spiral** is the plane curve $\mathbf{r}(t) = \langle x(t), y(t)\rangle$, where

$$x(t) = \int_0^t \sin \frac{u^2}{2}\, du, \qquad y(t) = \int_0^t \cos \frac{u^2}{2}\, du$$

Verify that $\kappa(t) = |t|$. Since the curvature increases linearly, the Cornu spiral is used in highway design to create transitions between straight and curved road segments (Figure 17).

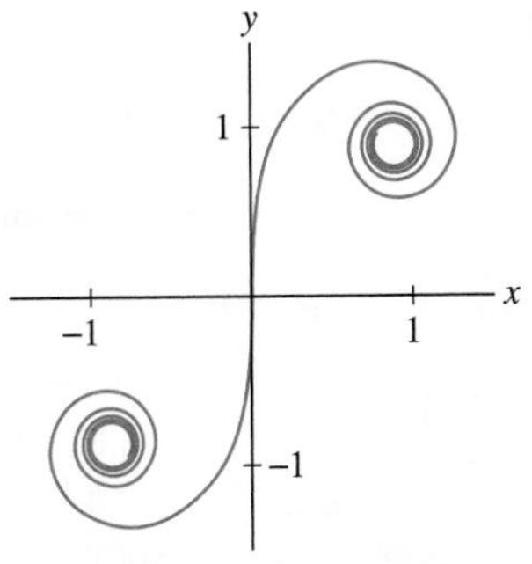

FIGURE 17 Cornu spiral.

35. CAS Plot and compute the curvature $\kappa(t)$ of the **clothoid** $\mathbf{r}(t) = \langle x(t), y(t)\rangle$, where

$$x(t) = \int_0^t \sin\frac{u^3}{3}\,du, \qquad y(t) = \int_0^t \cos\frac{u^3}{3}\,du$$

36. Find the unit normal vector $\mathbf{N}(\theta)$ to $\mathbf{r}(\theta) = R\langle\cos\theta, \sin\theta\rangle$, the circle of radius R. Does $\mathbf{N}(\theta)$ point inside or outside the circle? Draw $\mathbf{N}(\theta)$ at $\theta = \frac{\pi}{4}$ with $R = 4$.

37. Find the unit normal vector $\mathbf{N}(t)$ to $\mathbf{r}(t) = \langle 4, \sin 2t, \cos 2t\rangle$.

38. Sketch the graph of $\mathbf{r}(t) = \langle t, t^3\rangle$. Since $\mathbf{r}'(t) = \langle 1, 3t^2\rangle$, the unit normal $\mathbf{N}(t)$ points in one of the two directions $\pm\langle -3t^2, 1\rangle$. Which sign is correct at $t = 1$? Which is correct at $t = -1$?

39. Find the normal vectors to $\mathbf{r}(t) = \langle t, \cos t\rangle$ at $t = \frac{\pi}{4}$ and $t = \frac{3\pi}{4}$.

40. Find the unit normal to the Cornu spiral (Exercise 34) at $t = \sqrt{\pi}$.

41. Find the unit normal to the clothoid (Exercise 35) at $t = \pi^{1/3}$.

42. Method for Computing N Let $v(t) = \|\mathbf{r}'(t)\|$. Show that

$$\mathbf{N}(t) = \frac{v(t)\mathbf{r}''(t) - v'(t)\mathbf{r}'(t)}{\|v(t)\mathbf{r}''(t) - v'(t)\mathbf{r}'(t)\|} \qquad \boxed{11}$$

Hint: **N** is the unit vector in the direction $\mathbf{T}'(t)$. Differentiate $\mathbf{T}(t) = \mathbf{r}'(t)/v(t)$ to show that $v(t)\mathbf{r}''(t) - v'(t)\mathbf{r}'(t)$ is a positive multiple of $\mathbf{T}'(t)$.

In Exercises 43–48, use Eq. (11) to find **N** *at the point indicated.*

43. $\langle t^2, t^3\rangle$, $t = 1$

44. $\langle t - \sin t, 1 - \cos t\rangle$, $t = \pi$

45. $\langle t^2/2, t^3/3, t\rangle$, $t = 1$

46. $\langle t^{-1}, t, t^2\rangle$, $t = -1$

47. $\langle t, e^t, t\rangle$, $t = 0$

48. $\langle \cosh t, \sinh t, t^2\rangle$, $t = 0$

49. Let $f(x) = x^2$. Show that the center of the osculating circle at (x_0, x_0^2) is given by $\left(-4x_0^3, \frac{1}{2} + 3x_0^2\right)$.

50. Use Eq. (8) to find the center of curvature to $\mathbf{r}(t) = \langle t^2, t^3\rangle$ at $t = 1$.

In Exercises 51–58, find a parametrization of the osculating circle at the point indicated.

51. $\mathbf{r}(t) = \langle\cos t, \sin t\rangle$, $t = \frac{\pi}{4}$

52. $\mathbf{r}(t) = \langle\sin t, \cos t\rangle$, $t = 0$

53. $y = x^2$, $x = 1$

54. $y = \sin x$, $x = \frac{\pi}{2}$

55. $\langle t - \sin t, 1 - \cos t\rangle$, $t = \pi$

56. $\mathbf{r}(t) = \langle t^2/2, t^3/3, t\rangle$, $t = 0$

57. $\mathbf{r}(t) = \langle\cos t, \sin t, t\rangle$, $t = 0$

58. $\mathbf{r}(t) = \langle\cosh t, \sinh t, t\rangle$, $t = 0$

59. Figure 18 shows the graph of the half-ellipse $y = \pm\sqrt{2rx - px^2}$, where r and p are positive constants. Show that the radius of curvature at the origin is equal to r. *Hint:* One way of proceeding is to write the ellipse in the form of Exercise 25 and apply Eq. (9).

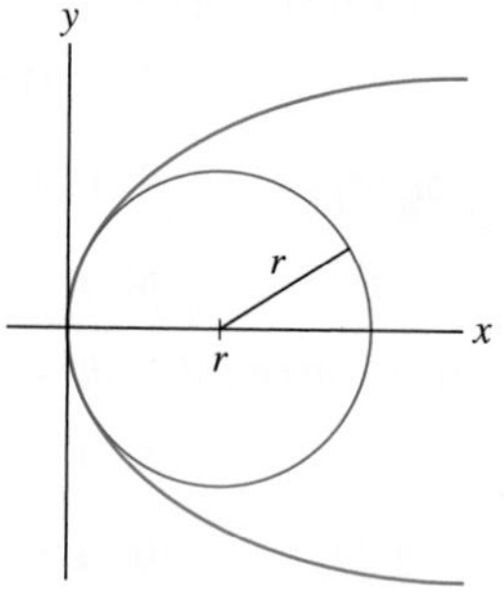

FIGURE 18 The curve $y = \pm\sqrt{2rx - px^2}$ and the osculating circle at the origin.

60. In a recent study of laser eye surgery by Gatinel, Hoang-Xuan, and Azar, a vertical cross section of the cornea is modeled by the half-ellipse of Exercise 59. Show that the half-ellipse can be written in the form $x = f(y)$, where $f(y) = p^{-1}\left(r - \sqrt{r^2 - py^2}\right)$. During surgery, tissue is removed to a depth $t(y)$ at height y for $-S \le y \le S$, where $t(y)$ is given by Munnerlyn's equation (for some $R > r$):

$$t(y) = \sqrt{R^2 - S^2} - \sqrt{R^2 - y^2} - \sqrt{r^2 - S^2} + \sqrt{r^2 - y^2}$$

After surgery, the cross section of the cornea has the shape $x = f(y) + t(y)$ (Figure 19). Show that after surgery, the radius of curvature at the point P (where $y = 0$) is R.

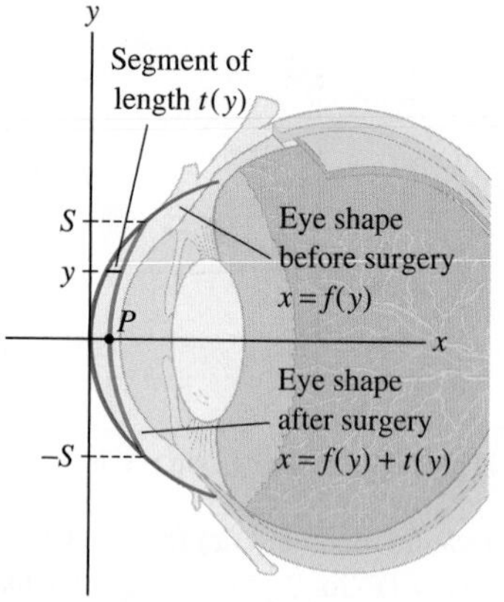

FIGURE 19 Contour of cornea before and after surgery.

61. The **angle of inclination** at a point P on a plane curve is the angle θ between the unit tangent vector **T** and the x-axis (Figure 20). Assume that $\mathbf{r}(s)$ is a arc length parametrization, and let $\theta = \theta(s)$ be the angle of inclination at $\mathbf{r}(s)$. Prove that

$$\kappa(s) = \left|\frac{d\theta}{ds}\right| \qquad \boxed{12}$$

Hint: Observe that $\mathbf{T}(s) = \langle\cos\theta(s), \sin\theta(s)\rangle$.

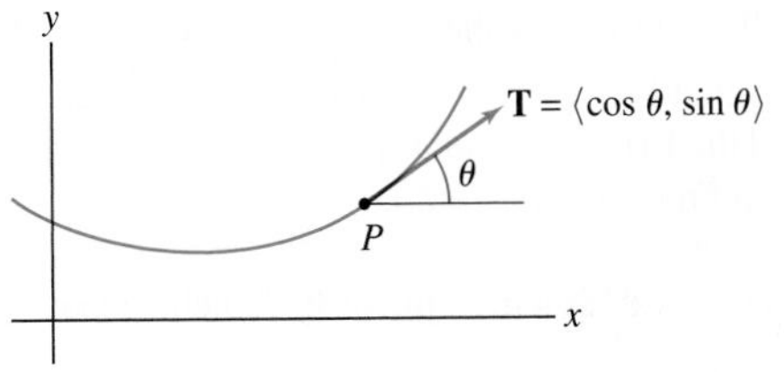

FIGURE 20 The curvature at P is the quantity $|d\theta/ds|$.

62. A particle moves along the path $y = x^3$ with unit speed. How fast is the tangent turning (i.e., how fast is the angle of inclination changing) when the particle passes through the point $(2, 8)$?

63. Let $\theta(x)$ be the angle of inclination at a point on the graph $y = f(x)$ (see Exercise 61).

(a) Use the relation $f'(x) = \tan\theta$ to prove that $\dfrac{d\theta}{dx} = \dfrac{f''(x)}{(1 + f'(x)^2)}$.

(b) Use the arc length integral to show that $\dfrac{ds}{dx} = \sqrt{1 + f'(x)^2}$.

(c) Now give a proof of Eq. (5) using Eq. (12).

64. Use the parametrization $\mathbf{r}(\theta) = \langle f(\theta)\cos\theta, f(\theta)\sin\theta\rangle$ to show that a curve $r = f(\theta)$ in polar coordinates has curvature

$$\kappa(\theta) = \frac{|f(\theta)^2 + 2f'(\theta)^2 - 2f(\theta)f''(\theta)|}{\left(f(\theta)^2 + f'(\theta)^2\right)^{3/2}} \qquad \boxed{13}$$

In Exercises 65–67, use Eq. (13) to find the curvature of the curve given in polar form.

65. $f(\theta) = 2\cos\theta$ **66.** $f(\theta) = \theta$ **67.** $f(\theta) = e^{\theta}$

68. Use Eq. (13) to find the curvature of the general Bernoulli spiral $r = ae^{b\theta}$ in polar form (a and b are constants).

69. Show that both $\mathbf{r}'(t)$ and $\mathbf{r}''(t)$ lie in the osculating plane for a vector function $\mathbf{r}(t)$. *Hint:* Differentiate $\mathbf{r}'(t) = v(t)\mathbf{T}(t)$.

70. Show that

$$\gamma(s) = \mathbf{r}(t_0) + \frac{1}{\kappa}\mathbf{N} + \frac{1}{\kappa}\left((\sin\kappa s)\mathbf{T} - (\cos\kappa s)\mathbf{N}\right)$$

is an arc length parametrization of the osculating circle at $\mathbf{r}(t_0)$.

71. Two vector-valued functions $\mathbf{r}_1(s)$ and $\mathbf{r}_2(s)$ are said to *agree to order 2* at s_0 if

$$\mathbf{r}_1(s_0) = \mathbf{r}_2(s_0), \quad \mathbf{r}_1'(s_0) = \mathbf{r}_2'(s_0), \quad \mathbf{r}_1''(s_0) = \mathbf{r}_2''(s_0)$$

Let $\mathbf{r}(s)$ be an arc length parametrization of a path C, and let P be the terminal point of $\mathbf{r}(0)$. Let $\gamma(s)$ be the arc length parametrization of the osculating circle given in Exercise 70. Show that $\mathbf{r}(s)$ and $\gamma(s)$ agree to order 2 at $s = 0$ (in fact, the osculating circle is the unique circle that approximates C to order 2 at P).

72. Let $\mathbf{r}(t) = \langle x(t), y(t), z(t)\rangle$ be a path with curvature $\kappa(t)$ and define the scaled path $\mathbf{r}_1(t) = \langle \lambda x(t), \lambda y(t), \lambda z(t)\rangle$, where $\lambda \neq 0$ is a constant. Prove that curvature varies inversely with the scale factor. That is, prove that the curvature $\kappa_1(t)$ of $\mathbf{r}_1(t)$ is $\kappa_1(t) = \lambda^{-1}\kappa(t)$. This explains why the curvature of a circle of radius R is proportional to $1/R$ (in fact, it is equal to $1/R$). *Hint:* Use Eq. (3).

Further Insights and Challenges

73. Show that the curvature of Viviani's curve, given by $\mathbf{r}(t) = \langle 1 + \cos t, \sin t, 2\sin(t/2)\rangle$, is

$$\kappa(t) = \frac{\sqrt{13 + 3\cos t}}{(3 + \cos t)^{3/2}}$$

74. Let $\mathbf{r}(s)$ be an arc length parametrization of a closed curve C of length L. We call C an **oval** if $d\theta/ds > 0$ (see Exercise 61). Observe that $-\mathbf{N}$ points to the *outside* of C. For $k > 0$, the curve C_1 defined by $\mathbf{r}_1(s) = \mathbf{r}(s) - k\mathbf{N}$ is called the expansion of $c(s)$ in the normal direction.

(a) Show that $\|\mathbf{r}_1'(s)\| = \|\mathbf{r}'(s)\| + k\kappa(s)$.

(b) As P moves around the oval counterclockwise, θ increases by 2π [Figure 21(A)]. Use this and a change of variables to prove that $\int_0^L \kappa(s)\,ds = 2\pi$.

(c) Show that C_1 has length $L + 2\pi k$.

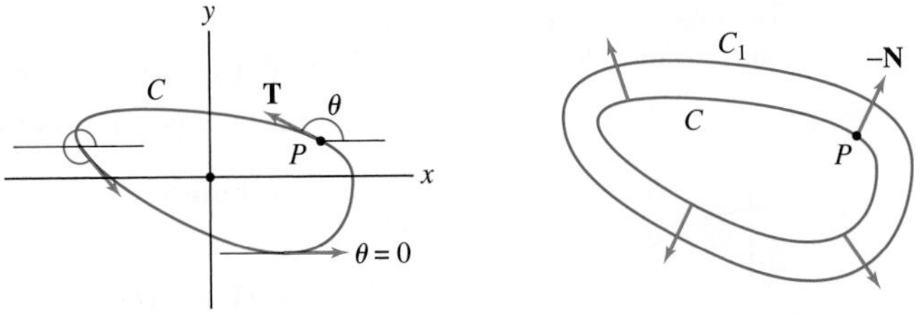

(A) An oval

(B) C_1 is the expansion of C in the normal direction.

FIGURE 21 As P moves around the oval, θ increases by 2π.

In Exercises 75–82, let $\mathbf{B}$ *denote the* **binormal vector** *at a point on a space curve* C*, defined by* $\mathbf{B} = \mathbf{T} \times \mathbf{N}$.

75. Show that $\mathbf{B}$ is a unit vector.

76. Follow steps (a)–(c) to prove that there is a number τ (lowercase Greek "tau") called the **torsion** such that

$$\frac{d\mathbf{B}}{ds} = -\tau\mathbf{N} \qquad \boxed{14}$$

(a) Show that $\dfrac{d\mathbf{B}}{ds} = \mathbf{T} \times \dfrac{d\mathbf{N}}{ds}$ and conclude that $d\mathbf{B}/ds$ is orthogonal to $\mathbf{T}$.

(b) Differentiate $\mathbf{B} \cdot \mathbf{B} = 1$ with respect to s to show that $d\mathbf{B}/ds$ is orthogonal to $\mathbf{B}$.

(c) Conclude that $d\mathbf{B}/ds$ is a multiple of $\mathbf{N}$.

77. Show that if C is contained in a plane $\mathcal{P}$, then $\mathbf{B}$ is a unit vector normal to $\mathcal{P}$. Conclude that $\tau = 0$ for a plane curve.

78. Torsion means "twisting." Is this an appropriate name for τ? Explain by interpreting τ geometrically.

79. Use the identity

$$\mathbf{a} \times (\mathbf{b} \times \mathbf{c}) = (\mathbf{a} \cdot \mathbf{c})\mathbf{b} - (\mathbf{a} \cdot \mathbf{b})\mathbf{c}$$

to prove

$$\mathbf{N} \times \mathbf{B} = \mathbf{T}, \qquad \mathbf{B} \times \mathbf{T} = \mathbf{N} \tag{15}$$

80. Follow steps (a)–(b) to prove

$$\frac{d\mathbf{N}}{ds} = -\kappa\mathbf{T} + \tau\mathbf{B} \tag{16}$$

(a) Show that $d\mathbf{N}/ds$ is orthogonal to $\mathbf{N}$. Conclude that $d\mathbf{N}/ds$ lies in the plane spanned by $\mathbf{T}$ and $\mathbf{B}$, and hence, $d\mathbf{N}/ds = a\mathbf{T} + b\mathbf{B}$ for some scalars a, b.

(b) Use $\mathbf{N} \cdot \mathbf{T} = 0$ to show that $\mathbf{T} \cdot \dfrac{d\mathbf{N}}{ds} = -\mathbf{N} \cdot \dfrac{d\mathbf{T}}{ds}$ and compute a. Compute b similarly. Equations (14) and (16) together with $d\mathbf{T}/dt = \kappa\mathbf{N}$ are called the **Frenet formulas** and were discovered by the French geometer Jean Frenet (1816–1900).

81. Show that $\mathbf{r}' \times \mathbf{r}''$ is a multiple of $\mathbf{B}$. Conclude that

$$\mathbf{B} = \frac{\mathbf{r}' \times \mathbf{r}''}{\|\mathbf{r}' \times \mathbf{r}''\|} \tag{17}$$

82. The vector $\mathbf{N}$ can be computed using $\mathbf{N} = \mathbf{B} \times \mathbf{T}$ [Eq. (15)] with $\mathbf{B}$, as in Eq. (17). Use this method to find $\mathbf{N}$ in the following cases:

(a) $\mathbf{r}(t) = \langle \cos t, t, t^2 \rangle$ at $t = 0$

(b) $\mathbf{r}(t) = \langle t^2, t^{-1}, t \rangle$ at $t = 1$

14.5 Motion in Three-Space

In this section, we study the motion of a particle traveling along a path $\mathbf{r}(t)$. Recall that the velocity vector is the derivative

$$\mathbf{v}(t) = \mathbf{r}'(t) = \lim_{h \to 0} \frac{\mathbf{r}(t+h) - \mathbf{r}(t)}{h}$$

FIGURE 1 The flight of the space shuttle is analyzed using vector calculus.

As we have seen, $\mathbf{v}(t)$ points in the direction of motion (if it is nonzero), and its magnitude $v(t) = \|\mathbf{v}(t)\|$ is the particle's speed. The **acceleration vector** is the second derivative $\mathbf{r}''(t)$, which we shall denote $\mathbf{a}(t)$. In summary,

$$\mathbf{v}(t) = \mathbf{r}'(t), \qquad v(t) = \|\mathbf{v}(t)\|, \qquad \mathbf{a}(t) = \mathbf{r}''(t)$$

■ **EXAMPLE 1** Calculate and plot the velocity and acceleration vectors at $t = 1$ of $\mathbf{r}(t) = \langle \sin 2t, -\cos 2t, \sqrt{t+1} \rangle$. Then find the speed at $t = 1$ (Figure 2).

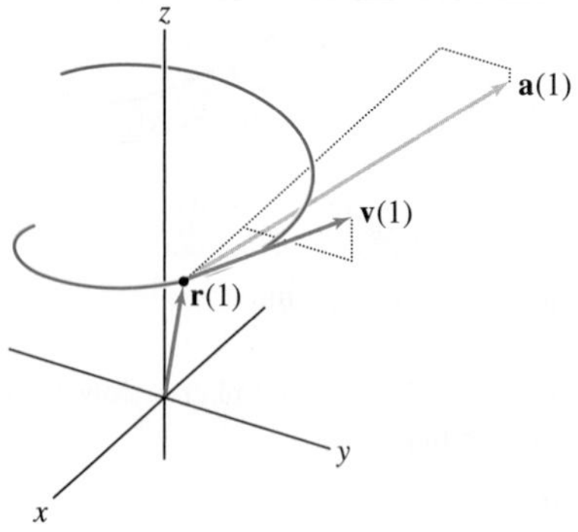

FIGURE 2

Solution

$$\mathbf{v}(t) = \mathbf{r}'(t) = \left\langle 2\cos 2t, 2\sin 2t, \frac{1}{2}(t+1)^{-1/2} \right\rangle, \qquad \mathbf{v}(1) \approx \langle -0.83, 0.84, 0.35 \rangle$$

$$\mathbf{a}(t) = \mathbf{r}''(t) = \left\langle -4\sin 2t, 4\cos 2t, -\frac{1}{4}(t+1)^{-3/2} \right\rangle, \qquad \mathbf{a}(1) \approx \langle -3.64, 0.54, -0.089 \rangle$$

The speed at $t = 1$ is

$$\|\mathbf{v}(1)\| \approx \sqrt{(-0.83)^2 + (0.84)^2 + (0.35)^2} \approx 1.23$$ ■

If an object's acceleration is given, we can solve for $\mathbf{v}(t)$ and $\mathbf{r}(t)$ by integrating twice:

$$\mathbf{v}(t) = \int \mathbf{a}(t)\,dt + \mathbf{v}_0$$

$$\mathbf{r}(t) = \int_0^t \mathbf{v}(t)\,dt + \mathbf{r}_0$$

with $\mathbf{v}_0$ and $\mathbf{r}_0$ determined by initial conditions.

■ **EXAMPLE 2** Find $\mathbf{r}(t)$ if

$$\mathbf{a}(t) = 2\mathbf{i} + 12t\mathbf{j}, \qquad \mathbf{v}(0) = 7\mathbf{i}, \qquad \mathbf{r}(0) = 2\mathbf{i} + 9\mathbf{k}$$

Solution We have

$$\mathbf{v}(t) = \int \mathbf{a}(t)\,dt + \mathbf{v}_0 = 2t\mathbf{i} + 6t^2\mathbf{j} + \mathbf{v}_0$$

The initial condition $\mathbf{v}(0) = \mathbf{v}_0 = 7\mathbf{i}$ gives us $\mathbf{v}(t) = 2t\mathbf{i} + 6t^2\mathbf{j} + 7\mathbf{i}$. Then we have

$$\mathbf{r}(t) = \int \mathbf{v}(t)\,dt + \mathbf{r}_0 = t^2\mathbf{i} + 2t^3\mathbf{j} + 7t\mathbf{i} + \mathbf{r}_0$$

The initial condition $\mathbf{r}(0) = \mathbf{r}_0 = 2\mathbf{i} + 9\mathbf{k}$ yields

$$\mathbf{r}(t) = t^2\mathbf{i} + 2t^3\mathbf{j} + 7t\mathbf{i} + (2\mathbf{i} + 9\mathbf{k}) = (t^2 + 7t + 2)\mathbf{i} + 2t^3\mathbf{j} + 9\mathbf{k}$$ ■

Newton's Second Law of Motion is often stated in the scalar form $F = ma$, but a more general statement is the vector law $\mathbf{F} = m\mathbf{a}$, where $\mathbf{F}$ is the net force vector acting on the object and $\mathbf{a}$ is the acceleration vector. When the force varies from position to position, we write $\mathbf{F}(\mathbf{r}(t))$ for the force acting on a particle with position vector $\mathbf{r}(t)$ at time t. Then Newton's Second Law reads

$$\mathbf{F}(\mathbf{r}(t)) = m\mathbf{a}(t) \qquad \text{or} \qquad \mathbf{F}(\mathbf{r}(t)) = m\mathbf{r}''(t) \tag{1}$$

■ **EXAMPLE 3** A bullet is fired from the ground at an angle of 60° above the horizontal. What initial speed v_0 must the bullet have in order to hit a point 150 m high on a tower located 250 m away (ignoring air resistance)?

Solution Place the gun at the origin, and let $\mathbf{r}(t)$ be the position vector of the bullet (Figure 3).

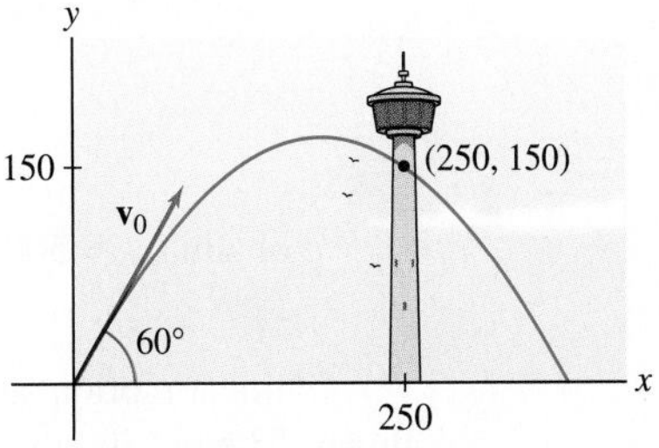

FIGURE 3 Trajectory of the bullet.

Step 1. **Use Newton's Law.**

Gravity exerts a downward force of magnitude mg, where m is the mass of the bullet and $g = 9.8$ m/s^2. In vector form,

$$\mathbf{F} = \langle 0, -mg\rangle = m\langle 0, -g\rangle$$

Newton's Second Law $\mathbf{F} = m\mathbf{r}''(t)$ yields $m\langle 0, -g\rangle = m\mathbf{r}''(t)$ or $\mathbf{r}''(t) = \langle 0, -g\rangle$. We determine $\mathbf{r}(t)$ by integrating twice:

$$\mathbf{r}'(t) = \int_0^t \mathbf{r}''(u)\,du = \int_0^t \langle 0, -g\rangle\,du = \langle 0, -gt\rangle + \mathbf{v}_0$$

$$\mathbf{r}(t) = \int_0^t \mathbf{r}'(u)\,du = \int_0^t \left(\langle 0, -gu\rangle + \mathbf{v}_0\right)du = \left\langle 0, -\frac{1}{2}gt^2\right\rangle + t\mathbf{v}_0 + \mathbf{r}_0$$

Step 2. **Use the initial conditions.**

By our choice of coordinates, $\mathbf{r}_0 = \mathbf{0}$. The initial velocity $\mathbf{v}_0$ has unknown magnitude v_0, but we know that it points in the direction of the unit vector $\langle \cos 60°, \sin 60° \rangle$. Therefore,

$$\mathbf{v}_0 = v_0 \langle \cos 60°, \sin 60° \rangle = v_0 \left\langle \frac{1}{2}, \frac{\sqrt{3}}{2} \right\rangle$$

$$\mathbf{r}(t) = \left\langle 0, -\frac{1}{2}gt^2 \right\rangle + t v_0 \left\langle \frac{1}{2}, \frac{\sqrt{3}}{2} \right\rangle$$

Step 3. **Solve for v_0.**

The bullet hits the point $\langle 250, 150 \rangle$ on the tower if there exists a time t such that $\mathbf{r}(t) = \langle 250, 150 \rangle$; that is,

$$\left\langle 0, -\frac{1}{2}gt^2 \right\rangle + t v_0 \left\langle \frac{1}{2}, \frac{\sqrt{3}}{2} \right\rangle = \langle 250, 150 \rangle$$

Equating components, we obtain

$$\frac{1}{2} t v_0 = 250, \qquad -\frac{1}{2}gt^2 + \frac{\sqrt{3}}{2} t v_0 = 150$$

The first equation yields $t = 500/v_0$. Now substitute in the second equation and solve, using $g = 9.8$:

$$-4.9 \left(\frac{500}{v_0} \right)^2 + \frac{\sqrt{3}}{2} \left(\frac{500}{v_0} \right) v_0 = 150$$

$$\left(\frac{500}{v_0} \right)^2 = \frac{250\sqrt{3} - 150}{4.9}$$

$$\left(\frac{v_0}{500} \right)^2 = \frac{4.9}{250\sqrt{3} - 150} \approx 0.0173$$

We obtain $v_0 \approx 500\sqrt{0.0173} \approx 66$ m/s. ■

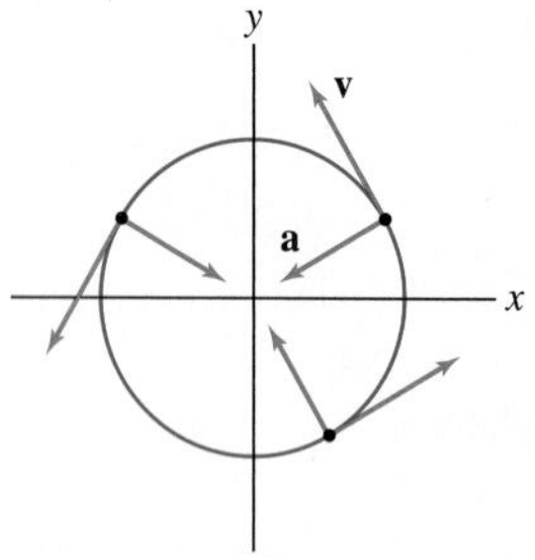

FIGURE 4 In uniform circular motion, **v** has constant length but turns continuously. The acceleration **a** is centripetal, pointing toward the center of the circle.

In linear motion, acceleration is the rate at which an object is speeding up or slowing down. The acceleration is zero if the speed is constant. By contrast, in two or three dimensions, the acceleration can be nonzero even when the object's speed is constant. This happens when $v(t) = \|\mathbf{v}(t)\|$ is constant but the *direction* of $\mathbf{v}(t)$ is changing. The simplest example is **uniform circular motion**, in which an object travels in a circular path at constant speed (Figure 4).

■ **EXAMPLE 4** **Uniform Circular Motion** Find $\mathbf{a}(t)$ and $\|\mathbf{a}(t)\|$ for motion around a circle of radius R with constant speed v.

Solution Assume that the particle follows the circular path $\mathbf{r}(t) = R\langle \cos \omega t, \sin \omega t \rangle$ for some constant ω. Then the velocity and speed of the particle are

$$\mathbf{v}(t) = R\omega \langle -\sin \omega t, \cos \omega t \rangle, \qquad v = \|\mathbf{v}(t)\| = R|\omega|$$

The constant ω (lowercase Greek "omega") is called the angular speed *because the particle's angle along the circle changes at a rate of ω radians per unit time.*

Thus $|\omega| = v/R$, and accordingly,

$$\mathbf{a}(t) = \mathbf{v}'(t) = -R\omega^2\langle \cos\omega t, \sin\omega t\rangle, \qquad \|\mathbf{a}(t)\| = R\omega^2 = R\left(\frac{v}{R}\right)^2 = \frac{v^2}{R}$$

The vector $\mathbf{a}(t)$ is called the **centripetal acceleration**: It has length v^2/R and points in toward the origin [because $\mathbf{a}(t)$ is a negative multiple of the position vector $\mathbf{r}(t)$], as in Figure 4. ■

Understanding the Acceleration Vector

We have noted that $\mathbf{v}(t)$ can change in two ways: in magnitude and in direction. To understand how the acceleration vector $\mathbf{a}(t)$ "encodes" both types of change, we decompose $\mathbf{a}(t)$ into a sum of tangential and normal components.

Recall the definition of unit tangent and unit normal vectors:

$$\mathbf{T}(t) = \frac{\mathbf{v}(t)}{\|\mathbf{v}(t)\|}, \qquad \mathbf{N}(t) = \frac{\mathbf{T}'(t)}{\|\mathbf{T}'(t)\|}$$

Thus, $\mathbf{v}(t) = v(t)\mathbf{T}(t)$, where $v(t) = \|\mathbf{v}(t)\|$, so by the Product Rule,

$$\mathbf{a}(t) = \frac{d\mathbf{v}}{dt} = \frac{d}{dt}v(t)\mathbf{T}(t) = v'(t)\mathbf{T}(t) + v(t)\mathbf{T}'(t)$$

Furthermore, $\mathbf{T}'(t) = v(t)\kappa(t)\mathbf{N}(t)$ by Eq. (7) of Section 14.4, where $\kappa(t)$ is the curvature. Thus we can write

$$\boxed{\mathbf{a} = a_{\mathbf{T}}\mathbf{T} + a_{\mathbf{N}}\mathbf{N}, \qquad a_{\mathbf{T}} = v'(t), \qquad a_{\mathbf{N}} = \kappa(t)v(t)^2} \qquad (2)$$

The coefficient $a_{\mathbf{T}}(t)$ is called the **tangential component** and $a_{\mathbf{N}}(t)$ the **normal component** of acceleration (Figure 5).

When you make a left turn in an automobile at constant speed, your tangential acceleration is zero (because $v'(t) = 0$) and you will not be pushed back against your seat. But the car seat (via friction) pushes you to the left toward the car door, causing you to accelerate in the normal direction. Due to inertia, you feel as if you are being pushed to the right toward the passenger's seat. This force is proportional to κv^2, so a sharp turn (large κ) or high speed (large v) produces a strong normal force.

The normal component $a_{\mathbf{N}}$ is often called the ***centripetal acceleration****, especially in the case of circular motion where it is directed toward the center of the circle.*

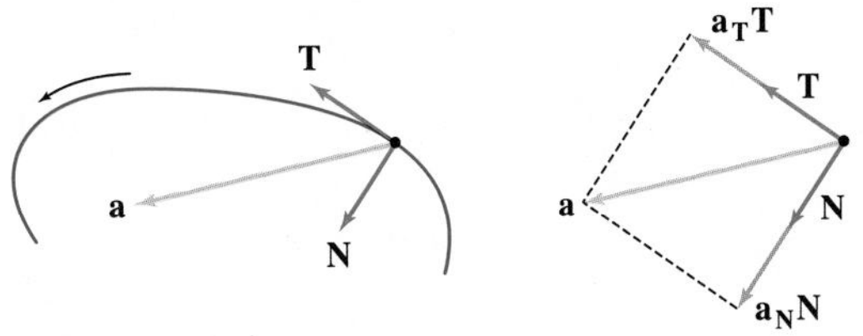

FIGURE 5 Decomposition of **a** into tangential and normal components.

CONCEPTUAL INSIGHT The tangential component $a_{\mathbf{T}} = v'(t)$ is the rate at which *speed* $v(t)$ changes, whereas the normal component $a_{\mathbf{N}} = \kappa(t)v(t)^2$ describes the change in $\mathbf{v}$ due to a change in *direction*. These interpretations become clear once we consider the following extreme cases:

- A particle travels in a straight line. Then direction does not change [$\kappa(t) = 0$] and $\mathbf{a}(t) = v'(t)\mathbf{T}$ is parallel to the direction of motion.
- A particle travels with constant speed along a curved path. Then $v'(t) = 0$ and the acceleration vector $\mathbf{a}(t) = \kappa(t)v(t)^2\mathbf{N}$ is normal to the direction of motion.

General motion combines both tangential and normal acceleration.

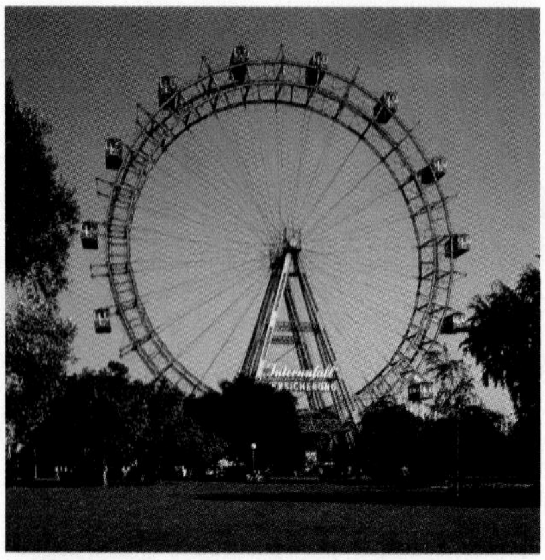

FIGURE 6 The Giant Ferris Wheel in Vienna, Austria, erected in 1897 to celebrate the 50th anniversary of the coronation of Emperor Franz Joseph I.

EXAMPLE 5 The Giant Ferris Wheel in Vienna has radius $R = 30$ m (Figure 6). Assume that at time $t = t_0$, the wheel rotates counterclockwise with a speed of 40 m/min and is slowing at a rate of 15 m/min^2. Find the acceleration vector **a** for a person seated in a car at the lowest point of the wheel.

Solution At the bottom of the wheel, $\mathbf{T} = \langle 1, 0\rangle$ and $\mathbf{N} = \langle 0, 1\rangle$. We are told that $a_{\mathbf{T}} = v' = -15$ at time t_0. The curvature of the wheel is $\kappa = 1/R = 1/30$, so the normal component is $a_{\mathbf{N}} = \kappa v^2 = v^2/R = (40)^2/30 \approx 53.3$. Therefore (Figure 7),

$$\mathbf{a} \approx -15\mathbf{T} + 53.3\mathbf{N} = \langle -15, 53.3\rangle \text{ m/min}^2$$

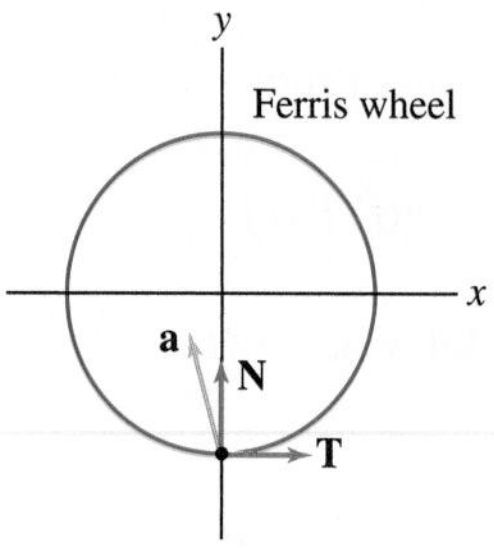

FIGURE 7

The following theorem provides useful formulas for the tangential and normal components.

THEOREM 1 Tangential and Normal Components of Acceleration In the decomposition $\mathbf{a} = a_{\mathbf{T}}\mathbf{T} + a_{\mathbf{N}}\mathbf{N}$, we have

$$a_{\mathbf{T}} = \mathbf{a}\cdot\mathbf{T} = \frac{\mathbf{a}\cdot\mathbf{v}}{\|\mathbf{v}\|}, \qquad a_{\mathbf{N}} = \mathbf{a}\cdot\mathbf{N} = \sqrt{\|\mathbf{a}\|^2 - |a_{\mathbf{T}}|^2} \qquad \boxed{3}$$

and

$$a_{\mathbf{T}}\mathbf{T} = \left(\frac{\mathbf{a}\cdot\mathbf{v}}{\mathbf{v}\cdot\mathbf{v}}\right)\mathbf{v}, \qquad a_{\mathbf{N}}\mathbf{N} = \mathbf{a} - a_{\mathbf{T}}\mathbf{T} = \mathbf{a} - \left(\frac{\mathbf{a}\cdot\mathbf{v}}{\mathbf{v}\cdot\mathbf{v}}\right)\mathbf{v} \qquad \boxed{4}$$

Proof We have $\mathbf{T}\cdot\mathbf{T} = 1$ and $\mathbf{N}\cdot\mathbf{T} = 0$. Thus

$$\mathbf{a}\cdot\mathbf{T} = (a_{\mathbf{T}}\mathbf{T} + a_{\mathbf{N}}\mathbf{N})\cdot\mathbf{T} = a_{\mathbf{T}}$$

$$\mathbf{a}\cdot\mathbf{N} = (a_{\mathbf{T}}\mathbf{T} + a_{\mathbf{N}}\mathbf{N})\cdot\mathbf{N} = a_{\mathbf{N}}$$

and since $\mathbf{T} = \dfrac{\mathbf{v}}{\|\mathbf{v}\|}$, we have

$$a_{\mathbf{T}}\mathbf{T} = (\mathbf{a}\cdot\mathbf{T})\mathbf{T} = \left(\frac{\mathbf{a}\cdot\mathbf{v}}{\|\mathbf{v}\|}\right)\frac{\mathbf{v}}{\|\mathbf{v}\|} = \left(\frac{\mathbf{a}\cdot\mathbf{v}}{\mathbf{v}\cdot\mathbf{v}}\right)\mathbf{v}$$

and

$$a_{\mathbf{N}}\mathbf{N} = \mathbf{a} - a_{\mathbf{T}}\mathbf{T} = \mathbf{a} - \left(\frac{\mathbf{a}\cdot\mathbf{v}}{\|\mathbf{v}\|}\right)\mathbf{v}$$

Finally, the vectors $a_{\mathbf{T}}\mathbf{T}$ and $a_{\mathbf{N}}\mathbf{N}$ are the sides of a right triangle with hypotenuse **a** as in Figure 5, so by the Pythagorean Theorem,

$$\|\mathbf{a}\|^2 = |a_{\mathbf{T}}|^2 + |a_{\mathbf{N}}|^2 \quad\Rightarrow\quad a_{\mathbf{N}} = \sqrt{\|\mathbf{a}\|^2 - |a_{\mathbf{T}}|^2}$$

Keep in mind that $a_{\mathbf{N}} \ge 0$ but $a_{\mathbf{T}}$ is positive or negative, depending on whether the object is speeding up or slowing down.

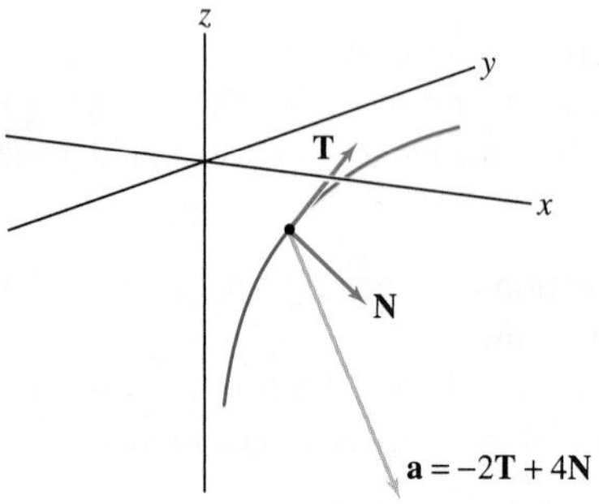

FIGURE 8 The vectors **T**, **N**, and **a** at $t = \frac{1}{2}$ on the curve $\mathbf{r}(t) = \langle t^2, 2t, \ln t\rangle$.

EXAMPLE 6 Decompose the acceleration vector **a** of $\mathbf{r}(t) = \langle t^2, 2t, \ln t\rangle$ into tangential and normal components at $t = \frac{1}{2}$ (Figure 8).

Solution First, we compute the tangential components **T** and $a_{\mathbf{T}}$. We have

$$\mathbf{v}(t) = \mathbf{r}'(t) = \langle 2t, 2, t^{-1}\rangle, \qquad \mathbf{a}(t) = \mathbf{r}''(t) = \langle 2, 0, -t^{-2}\rangle$$

At $t = \frac{1}{2}$,

$$\mathbf{v} = \mathbf{r}'\left(\frac{1}{2}\right) = \left\langle 2\left(\frac{1}{2}\right), 2, \left(\frac{1}{2}\right)^{-1}\right\rangle = \langle 1, 2, 2\rangle$$

$$\mathbf{a} = \mathbf{r}''\left(\frac{1}{2}\right) = \left\langle 2, 0, -\left(\frac{1}{2}\right)^{-2}\right\rangle = \langle 2, 0, -4\rangle$$

Thus

$$\mathbf{T} = \frac{\mathbf{v}}{\|\mathbf{v}\|} = \frac{\langle 1, 2, 2\rangle}{\sqrt{1^2 + 2^2 + 2^2}} = \left\langle \frac{1}{3}, \frac{2}{3}, \frac{2}{3}\right\rangle$$

and by Eq. (3),

$$a_{\mathbf{T}} = \mathbf{a} \cdot \mathbf{T} = \langle 2, 0, -4\rangle \cdot \left\langle \frac{1}{3}, \frac{2}{3}, \frac{2}{3}\right\rangle = -2$$

Next, we use Eq. (4):

$$a_{\mathbf{N}}\mathbf{N} = \mathbf{a} - a_{\mathbf{T}}\mathbf{T} = \langle 2, 0, -4\rangle - (-2)\left\langle \frac{1}{3}, \frac{2}{3}, \frac{2}{3}\right\rangle = \left\langle \frac{8}{3}, \frac{4}{3}, -\frac{8}{3}\right\rangle$$

Summary of steps in Example 6:

$$\mathbf{T} = \frac{\mathbf{v}}{\|\mathbf{v}\|}$$

$$a_{\mathbf{T}} = \mathbf{a} \cdot \mathbf{T}$$

$$a_{\mathbf{N}}\,\mathbf{N} = \mathbf{a} - a_{\mathbf{T}}\mathbf{T}$$

$$a_{\mathbf{N}} = \|a_{\mathbf{N}}\,\mathbf{N}\|$$

$$\mathbf{N} = \frac{a_{\mathbf{N}}\,\mathbf{N}}{a_{\mathbf{N}}}$$

This vector has length

$$a_{\mathbf{N}} = \|a_{\mathbf{N}}\mathbf{N}\| = \sqrt{\frac{64}{9} + \frac{16}{9} + \frac{64}{9}} = 4$$

and thus

$$\mathbf{N} = \frac{a_{\mathbf{N}}\mathbf{N}}{a_{\mathbf{N}}} = \frac{\left\langle \frac{8}{3}, \frac{4}{3}, -\frac{8}{3}\right\rangle}{4} = \left\langle \frac{2}{3}, \frac{1}{3}, -\frac{2}{3}\right\rangle$$

Finally, we obtain the decomposition

$$\mathbf{a} = \langle 2, 0, -4\rangle = a_{\mathbf{T}}\mathbf{T} + a_{\mathbf{N}}\mathbf{N} = -2\mathbf{T} + 4\mathbf{N}$$ ■

REMINDER

- *By Eq. (3),* $v' = a_{\mathbf{T}} = \mathbf{a} \cdot \mathbf{T}$
- $\mathbf{v} \cdot \mathbf{w} = \|\mathbf{v}\|\,\|\mathbf{w}\| \cos\theta$

where θ *is the angle between* $\mathbf{v}$ *and* $\mathbf{w}$.

■ **EXAMPLE 7 Nonuniform Circular Motion** Figure 9 shows the acceleration vectors of three particles moving *counterclockwise* around a circle. In each case, state whether the particle's speed v is increasing, decreasing, or momentarily constant.

Solution The rate of change of speed depends on the angle θ between **a** and **T**:

$$v' = a_{\mathbf{T}} = \mathbf{a} \cdot \mathbf{T} = \|\mathbf{a}\|\,\|\mathbf{T}\| \cos\theta = \|\mathbf{a}\| \cos\theta$$

- In (A), θ is obtuse so $\cos\theta < 0$ and $v' < 0$. The particle's speed is decreasing.
- In (B), $\theta = \frac{\pi}{2}$ so $\cos\theta = 0$ and $v' = 0$. The particle's speed is momentarily constant.
- In (C), θ is acute so $\cos\theta > 0$ and $v' > 0$. The particle's speed is increasing. ■

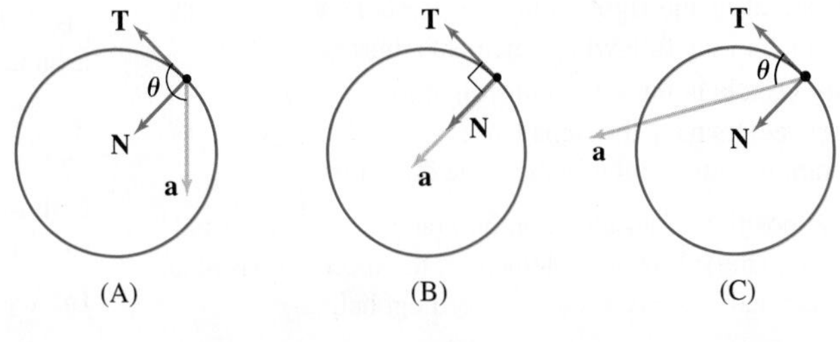

FIGURE 9 Acceleration vectors of particles moving counterclockwise (in the direction of **T**) around a circle.

EXAMPLE 8 Find the curvature $\kappa\left(\frac{1}{2}\right)$ for the path $\mathbf{r}(t) = \langle t^2, 2t, \ln t\rangle$ in Example 6.

Solution By Eq. (2), the normal component is

$$a_{\mathbf{N}} = \kappa v^2$$

In Example 6 we showed that $a_{\mathbf{N}} = 4$ and $\mathbf{v} = \langle 1, 2, 2\rangle$ at $t = \frac{1}{2}$. Therefore, $v^2 = \mathbf{v}\cdot\mathbf{v} = 9$ and the curvature is $\kappa\left(\frac{1}{2}\right) = a_{\mathbf{N}}/v^2 = \frac{4}{9}$. ■

14.5 SUMMARY

- For an object whose path is described by a vector-valued function $\mathbf{r}(t)$,

$$\mathbf{v}(t) = \mathbf{r}'(t), \qquad v(t) = \|\mathbf{v}(t)\|, \qquad \mathbf{a}(t) = \mathbf{r}''(t)$$

- The *velocity vector* $\mathbf{v}(t)$ points in the direction of motion. Its length $v(t) = \|\mathbf{v}(t)\|$ is the object's speed.
- The *acceleration vector* $\mathbf{a}$ is the sum of a tangential component (reflecting change in speed) and a normal component (reflecting change in direction):

$$\mathbf{a}(t) = a_{\mathbf{T}}(t)\mathbf{T}(t) + a_{\mathbf{N}}(t)\mathbf{N}(t)$$

Unit tangent vector	$\mathbf{T}(t) = \dfrac{\mathbf{v}(t)}{\|\mathbf{v}(t)\|}$
Unit normal vector	$\mathbf{N}(t) = \dfrac{\mathbf{T}'(t)}{\|\mathbf{T}'(t)\|}$
Tangential component	$a_{\mathbf{T}} = v'(t) = \mathbf{a}\cdot\mathbf{T} = \dfrac{\mathbf{a}\cdot\mathbf{v}}{\|\mathbf{v}\|}$
	$a_{\mathbf{T}}\mathbf{T} = \left(\dfrac{\mathbf{a}\cdot\mathbf{v}}{\mathbf{v}\cdot\mathbf{v}}\right)\mathbf{v}$
Normal component	$a_{\mathbf{N}} = \kappa(t)v(t)^2 = \sqrt{\|\mathbf{a}\|^2 - \lvert a_{\mathbf{T}}\rvert^2}$
	$a_{\mathbf{N}}\mathbf{N} = \mathbf{a} - a_{\mathbf{T}}\mathbf{T} = \mathbf{a} - \left(\dfrac{\mathbf{a}\cdot\mathbf{v}}{\mathbf{v}\cdot\mathbf{v}}\right)\mathbf{v}$

14.5 EXERCISES

Preliminary Questions

1. If a particle travels with constant speed, must its acceleration vector be zero? Explain.

2. For a particle in uniform circular motion around a circle, which of the vectors $\mathbf{v}(t)$ or $\mathbf{a}(t)$ always points toward the center of the circle?

3. Two objects travel to the right along the parabola $y = x^2$ with nonzero speed. Which of the following statements must be true?

(a) Their velocity vectors point in the same direction.
(b) Their velocity vectors have the same length.
(c) Their acceleration vectors point in the same direction.

4. Use the decomposition of acceleration into tangential and normal components to explain the following statement: If the speed is constant, then the acceleration and velocity vectors are orthogonal.

5. If a particle travels along a straight line, then the acceleration and velocity vectors are (choose the correct description):

(a) Orthogonal **(b)** Parallel

6. What is the length of the acceleration vector of a particle traveling around a circle of radius 2 cm with constant velocity 4 cm/s?

7. Two cars are racing around a circular track. If, at a certain moment, both of their speedometers read 110 mph. then the two cars have the same (choose one):

(a) $a_{\mathbf{T}}$ **(b)** $a_{\mathbf{N}}$

Exercises

1. Use the table below to calculate the difference quotients $\dfrac{\mathbf{r}(1+h)-\mathbf{r}(1)}{h}$ for $h=-0.2, -0.1, 0.1, 0.2$. Then estimate the velocity and speed at $t=1$.

$\mathbf{r}(0.8)$	$\langle 1.557, 2.459, -1.970\rangle$
$\mathbf{r}(0.9)$	$\langle 1.559, 2.634, -1.740\rangle$
$\mathbf{r}(1)$	$\langle 1.540, 2.841, -1.443\rangle$
$\mathbf{r}(1.1)$	$\langle 1.499, 3.078, -1.035\rangle$
$\mathbf{r}(1.2)$	$\langle 1.435, 3.342, -0.428\rangle$

2. Draw the vectors $\mathbf{r}(2+h)-\mathbf{r}(2)$ and $\dfrac{\mathbf{r}(2+h)-\mathbf{r}(2)}{h}$ for $h=0.5$ for the path in Figure 10. Draw $\mathbf{v}(2)$ (using a rough estimate for its length).

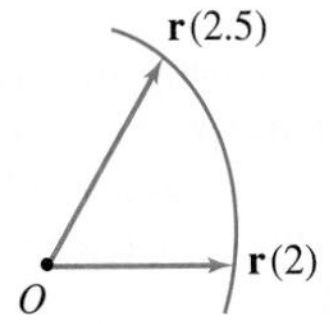

FIGURE 10

In Exercises 3–6, calculate the velocity and acceleration vectors and the speed at the time indicated.

3. $\mathbf{r}(t)=\langle t^3, 1-t, 4t^2\rangle, \quad t=1$

4. $\mathbf{r}(t)=e^t\mathbf{j}-\cos(2t)\mathbf{k}, \quad t=0$

5. $\mathbf{r}(\theta)=\langle\sin\theta, \cos\theta, \cos 3\theta\rangle, \quad \theta=\frac{\pi}{3}$

6. $\mathbf{r}(s)=\left\langle \dfrac{1}{1+s^2}, \dfrac{s}{1+s^2}\right\rangle, \quad s=2$

7. Find $\mathbf{a}(t)$ for a particle moving around a circle of radius 8 cm at a constant speed of $v=4$ cm/s (see Example 4). Draw the path and acceleration vector at $t=\frac{\pi}{4}$.

8. Sketch the path $\mathbf{r}(t)=\langle 1-t^2, 1-t\rangle$ for $-2\le t\le 2$, indicating the direction of motion. Draw the velocity and acceleration vectors at $t=0$ and $t=1$.

9. Sketch the path $\mathbf{r}(t)=\langle t^2, t^3\rangle$ together with the velocity and acceleration vectors at $t=1$.

10. The paths $\mathbf{r}(t)=\langle t^2, t^3\rangle$ and $\mathbf{r}_1(t)=\langle t^4, t^6\rangle$ trace the same curve, and $\mathbf{r}_1(1)=\mathbf{r}(1)$. Do you expect either the velocity vectors or the acceleration vectors of these paths at $t=1$ to point in the same direction? Compute these vectors and draw them on a single plot of the curve.

In Exercises 11–14, find $\mathbf{v}(t)$ given $\mathbf{a}(t)$ and the initial velocity.

11. $\mathbf{a}(t)=\langle t, 4\rangle, \quad \mathbf{v}(0)=\langle\frac{1}{3}, -2\rangle$

12. $\mathbf{a}(t)=\langle e^t, 0, t+1\rangle, \quad \mathbf{v}(0)=\langle 1, -3, \sqrt{2}\rangle$

13. $\mathbf{a}(t)=\mathbf{k}, \quad \mathbf{v}(0)=\mathbf{i}$

14. $\mathbf{a}(t)=t^2\mathbf{k}, \quad \mathbf{v}(0)=\mathbf{i}-\mathbf{j}$

In Exercises 15–18, find $\mathbf{r}(t)$ and $\mathbf{v}(t)$ given $\mathbf{a}(t)$ and the initial velocity and position.

15. $\mathbf{a}(t)=\langle t, 4\rangle, \quad \mathbf{v}(0)=\langle 3, -2\rangle, \quad \mathbf{r}(0)=\langle 0, 0\rangle$

16. $\mathbf{a}(t)=\langle e^t, 2t, t+1\rangle, \quad \mathbf{v}(0)=\langle 1, 0, 1\rangle, \quad \mathbf{r}(0)=\langle 2, 1, 1\rangle$

17. $\mathbf{a}(t)=t\mathbf{k}, \quad \mathbf{v}(0)=\mathbf{i}, \quad \mathbf{r}(0)=\mathbf{j}$

18. $\mathbf{a}(t)=\cos t\mathbf{k}, \quad \mathbf{v}(0)=\mathbf{i}-\mathbf{j}, \quad \mathbf{r}(0)=\mathbf{i}$

In Exercises 19–24, recall that $g=9.8$ m/s^2 is the acceleration due to gravity on the earth's surface.

19. A bullet is fired from the ground at an angle of 45°. What initial speed must the bullet have in order to hit the top of a 120-m tower located 180 m away?

20. Find the initial velocity vector $\mathbf{v}_0$ of a projectile released with initial speed 100 m/s that reaches a maximum height of 300 m.

21. Show that a projectile fired at an angle θ with initial speed v_0 travels a total distance $(v_0^2/g)\sin 2\theta$ before hitting the ground. Conclude that the maximum distance (for a given v_0) is attained for $\theta=45°$.

22. One player throws a baseball to another player standing 25 m away with initial speed 18 m/s. Use the result of Exercise 21 to find two angles θ at which the ball can be released. Which angle gets the ball there faster?

23. A bullet is fired at an angle $\theta=\frac{\pi}{4}$ at a tower located $d=600$ m away, with initial speed $v_0=120$ m/s. Find the height H at which the bullet hits the tower.

24. Show that a bullet fired at an angle θ will hit the top of an h-meter tower located d meters away if its initial speed is

$$v_0=\frac{\sqrt{g/2}\,d\sec\theta}{\sqrt{d\tan\theta-h}}$$

25. A constant force $\mathbf{F}=\langle 5, 2\rangle$ (in newtons) acts on a 10-kg mass. Find the position of the mass at $t=10$ s if it is located at the origin at $t=0$ and has initial velocity $\mathbf{v}_0=\langle 2, -3\rangle$ (in meters per second).

26. A force $\mathbf{F}=\langle 24t, 16-8t\rangle$ (in newtons) acts on a 4-kg mass. Find the position of the mass at $t=3$ s if it is located at $(10, 12)$ at $t=0$ and has zero initial velocity.

27. A particle follows a path $\mathbf{r}(t)$ for $0\le t\le T$, beginning at the origin O. The vector $\overline{\mathbf{v}}=\dfrac{1}{T}\displaystyle\int_0^T \mathbf{r}'(t)\,dt$ is called the **average velocity** vector. Suppose that $\overline{\mathbf{v}}=\mathbf{0}$. Answer and explain the following:

(a) Where is the particle located at time T if $\overline{\mathbf{v}}=\mathbf{0}$?

(b) Is the particle's average speed necessarily equal to zero?

28. At a certain moment, a moving particle has velocity $\mathbf{v}=\langle 2, 2, -1\rangle$ and $\mathbf{a}=\langle 0, 4, 3\rangle$. Find $\mathbf{T}$, $\mathbf{N}$, and the decomposition of $\mathbf{a}$ into tangential and normal components.

29. At a certain moment, a particle moving along a path has velocity $\mathbf{v}=\langle 12, 20, 20\rangle$ and acceleration $\mathbf{a}=\langle 2, 1, -3\rangle$. Is the particle speeding up or slowing down?

In Exercises 30–33, use Eq. (3) to find the coefficients $a_{\mathbf{T}}$ and $a_{\mathbf{N}}$ as a function of t (or at the specified value of t).

30. $\mathbf{r}(t) = \langle t^2, t^3 \rangle$

31. $\mathbf{r}(t) = \langle t, \cos t, \sin t \rangle$

32. $\mathbf{r}(t) = \langle t^{-1}, \ln t, t^2 \rangle, \quad t = 1$

33. $\mathbf{r}(t) = \langle e^{2t}, t, e^{-t} \rangle, \quad t = 0$

In Exercise 34–41, find the decomposition of $\mathbf{a}(t)$ into tangential and normal components at the point indicated, as in Example 6.

34. $\mathbf{r}(t) = \langle e^t, 1 - t \rangle, \quad t = 0$

35. $\mathbf{r}(t) = \left\langle \frac{1}{3}t^3, 1 - 3t \right\rangle, \quad t = -1$

36. $\mathbf{r}(t) = \left\langle t, \frac{1}{2}t^2, \frac{1}{6}t^3 \right\rangle, \quad t = 1$

37. $\mathbf{r}(t) = \left\langle t, \frac{1}{2}t^2, \frac{1}{6}t^3 \right\rangle, \quad t = 4$

38. $\mathbf{r}(t) = \langle 4 - t, t + 1, t^2 \rangle, \quad t = 2$

39. $\mathbf{r}(t) = \langle t, e^t, te^t \rangle, \quad t = 0$

40. $\mathbf{r}(\theta) = \langle \cos\theta, \sin\theta, \theta \rangle, \quad \theta = 0$

41. $\mathbf{r}(t) = \langle t, \cos t, t \sin t \rangle, \quad t = \frac{\pi}{2}$

42. Let $\mathbf{r}(t) = \langle t^2, 4t - 3 \rangle$. Find $\mathbf{T}(t)$ and $\mathbf{N}(t)$, and show that the decomposition of $\mathbf{a}(t)$ into tangential and normal components is

$$\mathbf{a}(t) = \left(\frac{2t}{\sqrt{t^2 + 4}} \right) \mathbf{T} + \left(\frac{4}{\sqrt{t^2 + 4}} \right) \mathbf{N}$$

43. Find the components $a_{\mathbf{T}}$ and $a_{\mathbf{N}}$ of the acceleration vector of a particle moving along a circular path of radius $R = 100$ cm with constant velocity $v_0 = 5$ cm/s.

44. In the notation of Example 5, find the acceleration vector for a person seated in a car at (a) the highest point of the Ferris wheel and (b) the two points level with the center of the wheel.

45. Suppose that the Ferris wheel in Example 5 is rotating clockwise and that the point P at angle 45° has acceleration vector $\mathbf{a} = \langle 0, -50 \rangle$ m/min^2 pointing down, as in Figure 11. Determine the speed and tangential acceleration of the Ferris wheel.

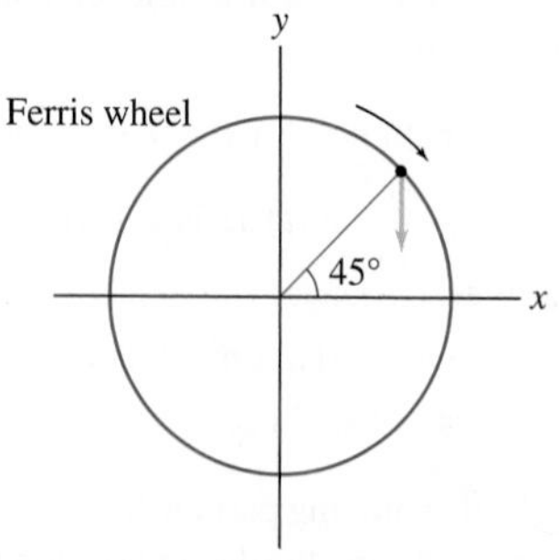

FIGURE 11

46. At time t_0, a moving particle has velocity vector $\mathbf{v} = 2\mathbf{i}$ and acceleration vector $\mathbf{a} = 3\mathbf{i} + 18\mathbf{k}$. Determine the curvature $\kappa(t_0)$ of the particle's path at time t_0.

47. A space shuttle orbits the earth at an altitude 400 km above the earth's surface, with constant speed $v = 28{,}000$ km/h. Find the magnitude of the shuttle's acceleration (in km/h^2), assuming that the radius of the earth is 6378 km (Figure 12).

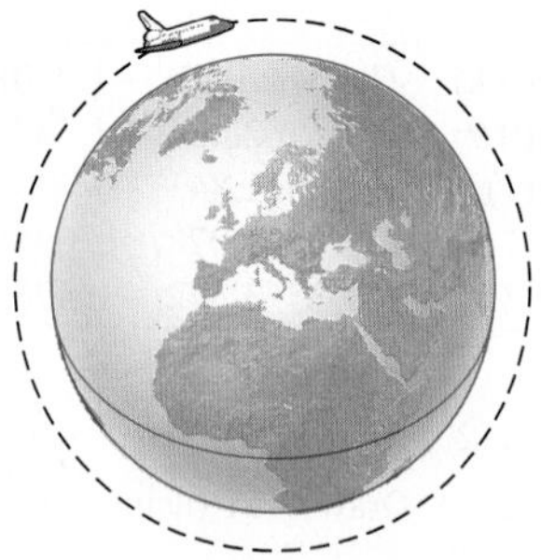

FIGURE 12 Space shuttle orbit.

48. A car proceeds along a circular path of radius $R = 300$ m centered at the origin. Starting at rest, its speed increases at a rate of t m/s^2. Find the acceleration vector $\mathbf{a}$ at time $t = 3$ s and determine its decomposition into normal and tangential components.

49. A runner runs along the helix $\mathbf{r}(t) = \langle \cos t, \sin t, t \rangle$. When he is at position $\mathbf{r}\left(\frac{\pi}{2}\right)$, his speed is 3 m/s and he is accelerating at a rate of $\frac{1}{2}$ m/s^2. Find his acceleration vector $\mathbf{a}$ at this moment. *Note:* The runner's acceleration vector does not coincide with the acceleration vector of $\mathbf{r}(t)$.

50. Explain why the vector $\mathbf{w}$ in Figure 13 cannot be the acceleration vector of a particle moving along the circle. *Hint:* Consider the sign of $\mathbf{w} \cdot \mathbf{N}$.

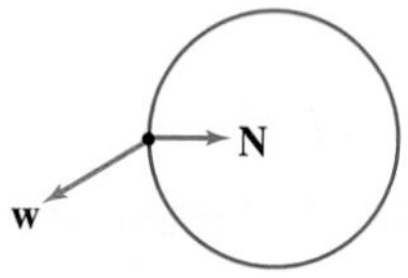

FIGURE 13

51. Figure 14 shows acceleration vectors of a particle moving clockwise around a circle. In each case, state whether the particle is speeding up, slowing down, or momentarily at constant speed. Explain.

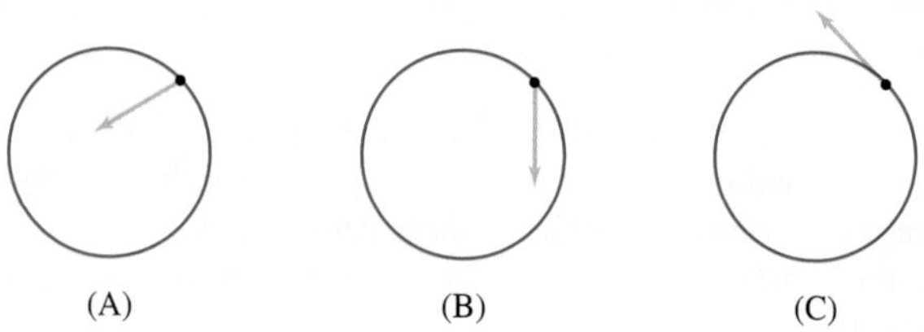

FIGURE 14

52. Prove that $a_{\mathbf{N}} = \dfrac{\|\mathbf{a} \times \mathbf{v}\|}{\|\mathbf{v}\|}$.

53. Suppose that $\mathbf{r} = \mathbf{r}(t)$ lies on a sphere of radius R for all t. Let $\mathbf{J} = \mathbf{r} \times \mathbf{r}'$. Show that $\mathbf{r}' = (\mathbf{J} \times \mathbf{r})/\|\mathbf{r}\|^2$. *Hint:* Observe that $\mathbf{r}$ and $\mathbf{r}'$ are perpendicular.

Further Insights and Challenges

54. The orbit of a planet is an ellipse with the sun at one focus. The sun's gravitational force acts along the radial line from the planet to the sun (the dashed lines in Figure 15), and by Newton's Second Law, the acceleration vector points in the same direction. Assuming that the orbit has positive eccentricity (the orbit is not a circle), explain why the planet must slow down in the upper half of the orbit (as it moves away from the sun) and speed up in the lower half. Kepler's Second Law, discussed in the next section, is a precise version of this qualitative conclusion. *Hint:* Consider the decomposition of **a** into normal and tangential components.

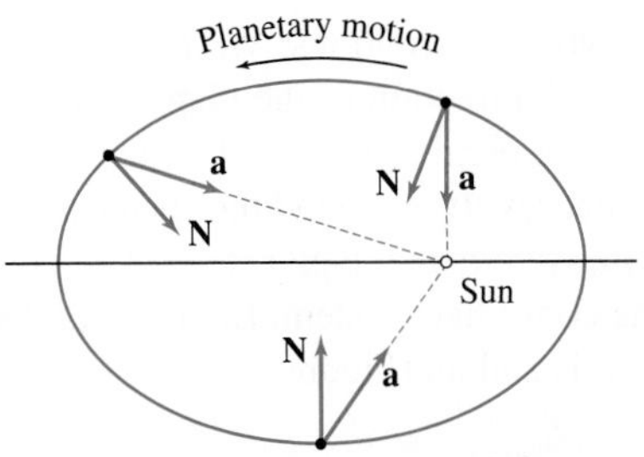

FIGURE 15 Elliptical orbit of a planet around the sun.

In Exercises 55–59, we consider an automobile of mass m traveling along a curved but level road. To avoid skidding, the road must supply a frictional force $\mathbf{F} = m\mathbf{a}$, where $\mathbf{a}$ is the car's acceleration vector. The maximum magnitude of the frictional force is μmg, where μ is the coefficient of friction and $g = 9.8$ m/s^2. Let v be the car's speed in meters per second.

55. Show that the car will not skid if the curvature κ of the road is such that (with $R = 1/\kappa$)

$$(v')^2 + \left(\frac{v^2}{R}\right)^2 \le (\mu g)^2 \qquad \boxed{5}$$

Note that braking ($v' < 0$) and speeding up ($v' > 0$) contribute equally to skidding.

56. Suppose that the maximum radius of curvature along a curved highway is $R = 180$ m. How fast can an automobile travel (at constant speed) along the highway without skidding if the coefficient of friction is $\mu = 0.5$?

57. Beginning at rest, an automobile drives around a circular track of radius $R = 300$ m, accelerating at a rate of 0.3 m/s^2. After how many seconds will the car begin to skid if the coefficient of friction is $\mu = 0.6$?

58. You want to reverse your direction in the shortest possible time by driving around a semicircular bend (Figure 16). If you travel at the maximum possible *constant speed* v that will not cause skidding, is it faster to hug the inside curve (radius r) or the outside curb (radius R)? *Hint:* Use Eq. (5) to show that at maximum speed, the time required to drive around the semicircle is proportional to the square root of the radius.

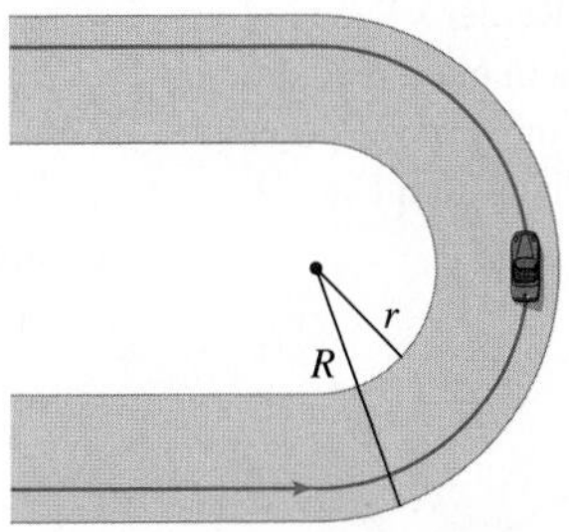

FIGURE 16 Car going around the bend.

59. What is the smallest radius R about which an automobile can turn without skidding at 100 km/h if $\mu = 0.75$ (a typical value)?

14.6 Planetary Motion According to Kepler and Newton

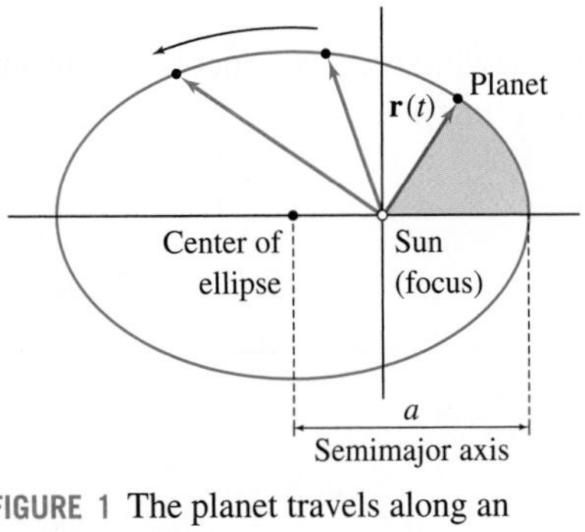

FIGURE 1 The planet travels along an ellipse with the sun at one focus.

In this section, we derive Kepler's laws of planetary motion, a feat first accomplished by Isaac Newton and published by him in 1687. No event was more emblematic of the scientific revolution. It demonstrated the power of mathematics to make the natural world comprehensible and it led succeeding generations of scientists to seek and discover mathematical laws governing other phenomena, such as electricity and magnetism, thermodynamics, and atomic processes.

According to Kepler, the planetary orbits are ellipses with the sun at one focus. Furthermore, if we imagine a radial vector $\mathbf{r}(t)$ pointing from the sun to the planet, as in Figure 1, then this radial vector sweeps out area at a constant rate or, as Kepler stated in his Second Law, the radial vector sweeps out equal areas in equal times (Figure 2). Kepler's Third Law determines the **period** T of the orbit, defined as the time required to complete one full revolution. These laws are valid not just for planets orbiting the sun, but for any body orbiting about another body according to the inverse-square law of gravitation.

Kepler's version of the Third Law stated only that T^2 is proportional to a^3. Newton discovered that the constant of proportionality is equal to $4\pi^2/(GM)$, and he observed that if you can measure T and a through observation, then you can use the Third Law to solve for the mass M. This method is used by astronomers to find the masses of the planets (by measuring T and a for moons revolving around the planet) as well as the masses of binary stars and galaxies. See Exercises 2–5.

Kepler's Three Laws

(i) Law of Ellipses: The orbit of a planet is an ellipse with the sun at one focus.

(ii) Law of Equal Area in Equal Time: The position vector pointing from the sun to the planet sweeps out equal areas in equal times.

(iii) Law of the Period of Motion: $T^2 = \left(\frac{4\pi^2}{GM}\right)a^3$, where

- a is the semimajor axis of the ellipse (Figure 1).
- G is the universal gravitational constant: $6.673 \times 10^{-11}\ \text{m}^3\,\text{kg}^{-1}\,\text{s}^{-2}$.
- M is the mass of the sun, approximately 1.989×10^{30} kg.

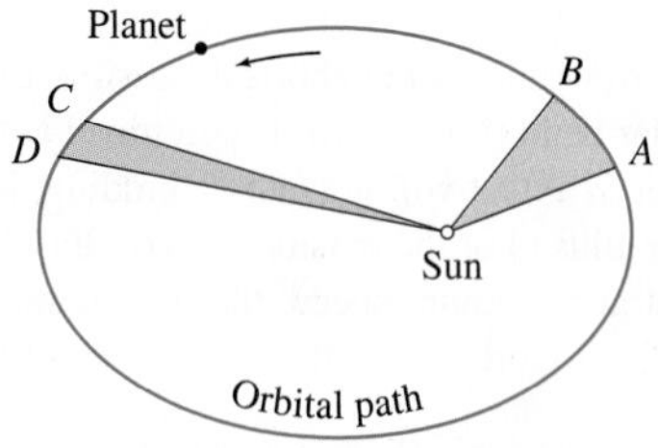

FIGURE 2 The two shaded regions have equal areas, and by Kepler's Second Law, the planet sweeps them out in equal times. To do so, the planet must travel faster going from A to B than from C to D.

Our derivation makes a few simplifying assumptions. We treat the sun and planet as point masses and ignore the gravitational attraction of the planets on each other. And although both the sun and the planet revolve around their mutual center of mass, we ignore the sun's motion and assume that the planet revolves around the center of the sun. This is justified because the sun is much more massive than the planet.

We place the sun at the origin of the coordinate system. Let $\mathbf{r} = \mathbf{r}(t)$ be the position vector of a planet of mass m, as in Figure 1, and let (Figure 3)

$$\mathbf{e}_r = \frac{\mathbf{r}(t)}{\|\mathbf{r}(t)\|}$$

be the unit radial vector at time t ($\mathbf{e}_r$ is the unit vector that points to the planet as it moves around the sun). By Newton's Universal Law of Gravitation (the inverse-square law), the sun attracts the planet with a gravitational force

$$\mathbf{F}(\mathbf{r}(t)) = -\left(\frac{km}{\|\mathbf{r}(t)\|^2}\right)\mathbf{e}_r$$

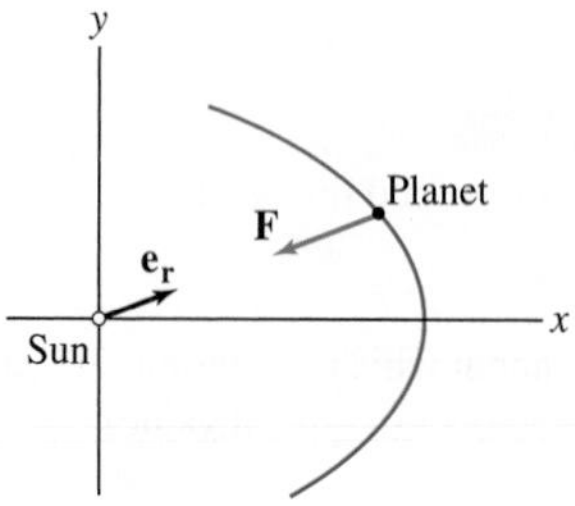

FIGURE 3 The gravitational force $\mathbf{F}$, directed from the planet to the sun, is a negative multiple of $\mathbf{e}_r$.

where $k = GM$ (Figure 3). Combining the Law of Gravitation with Newton's Second Law of Motion $\mathbf{F}(\mathbf{r}(t)) = m\mathbf{r}''(t)$, we obtain

$$\mathbf{r}''(t) = -\frac{k}{\|\mathbf{r}(t)\|^2}\mathbf{e}_r \qquad \boxed{1}$$

Kepler's Laws are a consequence of this *differential equation*.

Kepler's Second Law

The key to Kepler's Second Law is the fact that the following cross product is a constant vector (even though both $\mathbf{r}(t)$ and $\mathbf{r}'(t)$ are changing in time):

$$\mathbf{J} = \mathbf{r}(t) \times \mathbf{r}'(t)$$

*In physics, $m\mathbf{J}$ is called the **angular momentum** vector. In situations where $\mathbf{J}$ is constant, we say that angular momentum is* conserved. *This conservation law is valid whenever the force acts in the radial direction.*

THEOREM 1 The vector $\mathbf{J}$ is constant—that is,

$$\frac{d\mathbf{J}}{dt} = \mathbf{0} \qquad \boxed{2}$$

Proof By the Product Rule for cross products (Theorem 3 in Section 14.2)

$$\frac{d\mathbf{J}}{dt} = \frac{d}{dt}\big(\mathbf{r}(t) \times \mathbf{r}'(t)\big) = \mathbf{r}(t) \times \mathbf{r}''(t) + \mathbf{r}'(t) \times \mathbf{r}'(t)$$

The cross product of parallel vectors is zero, so the second term is certainly zero. The first term is also zero because $\mathbf{r}''(t)$ is a multiple of $\mathbf{e}_r$ by Eq. (1), and hence also of $\mathbf{r}(t)$. ■

REMINDER

- $\mathbf{a} \times \mathbf{b}$ *is orthogonal to both* $\mathbf{a}$ *and* $\mathbf{b}$
- $\mathbf{a} \times \mathbf{b} = \mathbf{0}$ *if* $\mathbf{a}$ *and* $\mathbf{b}$ *are parallel, that is, one is a multiple of the other.*

How can we use Eq. (2)? First of all, the cross product $\mathbf{J}$ is orthogonal to both $\mathbf{r}(t)$ and $\mathbf{r}'(t)$. Because $\mathbf{J}$ is constant, $\mathbf{r}(t)$ and $\mathbf{r}'(t)$ are confined to the fixed plane orthogonal to $\mathbf{J}$. This proves that the *motion of a planet around the sun takes place in a plane.*

We can choose coordinates so that the sun is at the origin and the planet moves in the counterclockwise direction (Figure 4). Let (r, θ) be the polar coordinates of the planet, where $r = r(t)$ and $\theta = \theta(t)$ are functions of time. Note that $r(t) = \|\mathbf{r}(t)\|$.

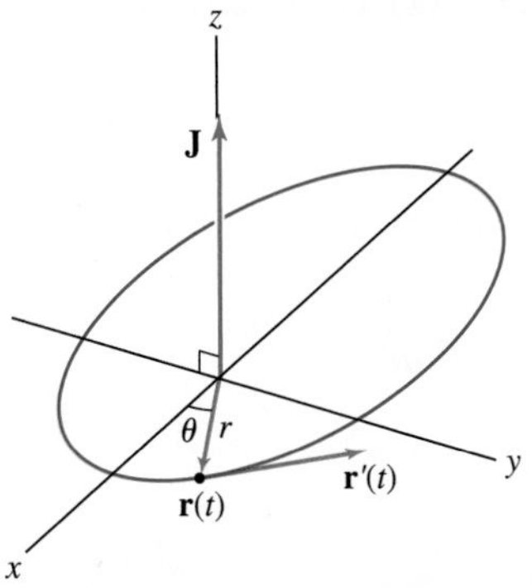

FIGURE 4 The orbit is contained in the plane orthogonal to $\mathbf{J}$. Of course, we have not yet shown that the orbit is an ellipse.

Recall from Section 12.4 (Theorem 1) that the area swept out by the planet's radial vector is

$$A = \frac{1}{2}\int_0^{\theta} r^2\, d\theta$$

Kepler's Second Law states that this area is swept out at a constant rate. But this rate is simply dA/dt. By the Fundamental Theorem of Calculus, $\dfrac{dA}{d\theta} = \dfrac{1}{2}r^2$, and by the Chain Rule,

$$\frac{dA}{dt} = \frac{dA}{d\theta}\frac{d\theta}{dt} = \frac{1}{2}\theta'(t)r(t)^2 = \frac{1}{2}r(t)^2\theta'(t)$$

Thus, Kepler's Second Law follows from the next theorem, which tells us that dA/dt has the constant value $\frac{1}{2}\|\mathbf{J}\|$.

THEOREM 2 Let $J = \|\mathbf{J}\|$ ($\mathbf{J}$ is constant by Theorem 1). Then

$$r(t)^2\theta'(t) = J \qquad (3)$$

Proof We note that in polar coordinates, $\mathbf{e}_r = \langle\cos\theta, \sin\theta\rangle$. We also define the unit vector $\mathbf{e}_\theta = \langle -\sin\theta, \cos\theta\rangle$ that is orthogonal to $\mathbf{e}_r$ (Figure 5). In summary,

$$r(t) = \|\mathbf{r}(t)\|, \quad \mathbf{e}_r = \langle\cos\theta, \sin\theta\rangle, \quad \mathbf{e}_\theta = \langle -\sin\theta, \cos\theta\rangle, \quad \mathbf{e}_r \cdot \mathbf{e}_\theta = 0$$

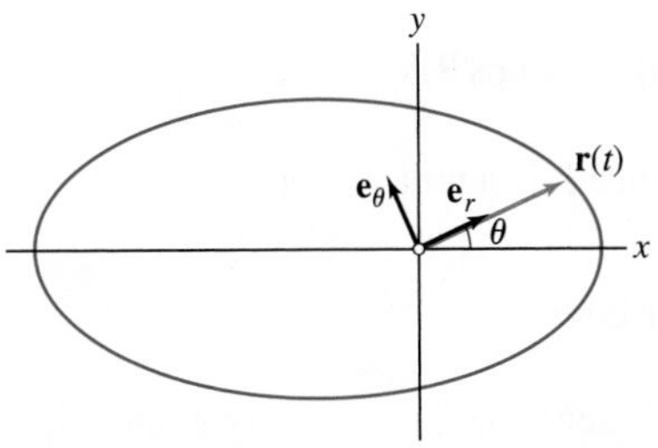

FIGURE 5 The unit vectors $\mathbf{e}_r$ and $\mathbf{e}_\theta$ are orthogonal, and rotate around the origin along with the planet.

We see directly that the derivatives of $\mathbf{e}_r$ and $\mathbf{e}_\theta$ with respect to θ are

$$\frac{d}{d\theta}\mathbf{e}_r = \mathbf{e}_\theta, \qquad \frac{d}{d\theta}\mathbf{e}_\theta = -\mathbf{e}_r \qquad (4)$$

The time derivative of $\mathbf{e}_r$ is computed using the Chain Rule:

$$\mathbf{e}_r' = \left(\frac{d\theta}{dt}\right)\left(\frac{d}{d\theta}\mathbf{e}_r\right) = \theta'(t)\,\mathbf{e}_\theta \qquad (5)$$

Now apply the Product Rule to $\mathbf{r} = r\mathbf{e}_r$:

$$\mathbf{r}' = \frac{d}{dt}r\mathbf{e}_r = r'\mathbf{e}_r + r\mathbf{e}_r' = r'\mathbf{e}_r + r\theta'\mathbf{e}_\theta$$

Using $\mathbf{e}_r \times \mathbf{e}_r = \mathbf{0}$, we obtain

$$\mathbf{J} = \mathbf{r} \times \mathbf{r}' = r\mathbf{e}_r \times (r'\mathbf{e}_r + r\theta'\mathbf{e}_\theta) = r^2\theta'(\mathbf{e}_r \times \mathbf{e}_\theta)$$

To compute cross products of vectors in the plane, such as $\mathbf{r}$, $\mathbf{e}_r$, and $\mathbf{e}_\theta$, we treat them as vectors in three-space with z-component equal to zero. The cross product is then a multiple of $\mathbf{k}$.

It is straightforward to check that $\mathbf{e}_r \times \mathbf{e}_\theta = \mathbf{k}$, and since $\mathbf{k}$ is a unit vector, $J = \|\mathbf{J}\| = |r^2\theta'|$. However, $\theta' > 0$ because the planet moves in the counterclockwise direction, so $J = r^2\theta'$. This proves Theorem 2. ■

Proof of the Law of Ellipses

← REMINDER Eq. (1) states:

$$\mathbf{r}''(t) = -\frac{k}{r(t)^2}\mathbf{e}_r$$

where $r(t) = \|\mathbf{r}(t)\|$.

Let $\mathbf{v} = \mathbf{r}'(t)$ be the velocity vector. Then $\mathbf{r}'' = \mathbf{v}'$ and Eq. (1) may be written

$$\frac{d\mathbf{v}}{dt} = -\frac{k}{r(t)^2}\mathbf{e}_r \qquad \boxed{6}$$

On the other hand, by the Chain Rule and the relation $r(t)^2\theta'(t) = J$ of Eq. (3),

$$\frac{d\mathbf{v}}{dt} = \frac{d\theta}{dt}\frac{d\mathbf{v}}{d\theta} = \theta'(t)\frac{d\mathbf{v}}{d\theta} = \frac{J}{r(t)^2}\frac{d\mathbf{v}}{d\theta}$$

Together with Eq. (6), this yields $J\dfrac{d\mathbf{v}}{d\theta} = -k\mathbf{e}_r$, or

$$\frac{d\mathbf{v}}{d\theta} = -\frac{k}{J}\mathbf{e}_r = -\frac{k}{J}\langle\cos\theta, \sin\theta\rangle$$

This is a first-order differential equation that no longer involves time t. We can solve it by integration:

$$\mathbf{v} = -\frac{k}{J}\int\langle\cos\theta, \sin\theta\rangle\,d\theta = \frac{k}{J}\langle-\sin\theta, \cos\theta\rangle + \mathbf{c} = \frac{k}{J}\mathbf{e}_\theta + \mathbf{c} \qquad \boxed{7}$$

where $\mathbf{c}$ is an arbitrary constant vector.

We are still free to rotate our coordinate system in the plane of motion, so we may assume that $\mathbf{c}$ points along the y-axis. We can then write $\mathbf{c} = \langle 0, (k/J)e\rangle$ for some constant e. We finish the proof by computing $\mathbf{J} = \mathbf{r} \times \mathbf{v}$:

$$\mathbf{J} = \mathbf{r} \times \mathbf{v} = r\mathbf{e}_r \times \left(\frac{k}{J}\mathbf{e}_\theta + \mathbf{c}\right) = \frac{k}{J}r\big(\mathbf{e}_r \times \mathbf{e}_\theta + \mathbf{e}_r \times \langle 0, e\rangle\big)$$

Direct calculation yields

$$\mathbf{e}_r \times \mathbf{e}_\theta = \mathbf{k}, \qquad \mathbf{e}_r \times \langle 0, e\rangle = (e\cos\theta)\mathbf{k}$$

so our equation becomes $\mathbf{J} = \dfrac{k}{J}r(1 + e\cos\theta)\mathbf{k}$. Since $\mathbf{k}$ is a unit vector,

$$J = \|\mathbf{J}\| = \frac{k}{J}r\big(1 + e\cos\theta\big)$$

← REMINDER The equation of a conic section in polar coordinates is discussed in Section 12.5.

Solving for r, we obtain the polar equation of a conic section of eccentricity e (an ellipse, parabola, or hyperbola):

$$r = \frac{J^2/k}{1 + e\cos\theta}$$

This result shows that if a planet travels around the sun in a bounded orbit, then the orbit must be an ellipse. There are also "open orbits" that are either parabolic and hyperbolic. They describe comets that pass by the sun and then continue into space, never to return. In our derivation, we assumed implicitly that $\mathbf{J} \neq \mathbf{0}$. If $\mathbf{J} = \mathbf{0}$, then $\theta'(t) = 0$. In this case, the orbit is a straight line, and the planet falls directly into the sun.

Kepler's Third Law is verified in Exercises 23 and 24.

CONCEPTUAL INSIGHT We exploited the fact that **J** is constant to prove the law of ellipses without ever finding a formula for the position vector $\mathbf{r}(t)$ of the planet as a function of time t. In fact, $\mathbf{r}(t)$ cannot be expressed in terms of elementary functions. This illustrates an important principle: Sometimes it is possible to describe solutions of a differential equation even if we cannot write them down explicitly.

The Hubble Space Telescope produced this image of the Antenna galaxies, a pair of spiral galaxies that began to collide hundreds of millions of years ago.

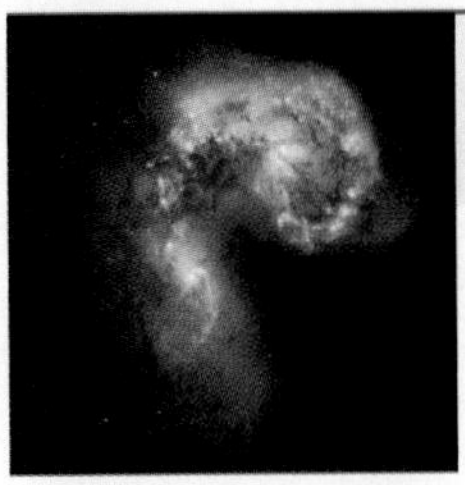

HISTORICAL PERSPECTIVE

The astronomers of the ancient world (Babylon, Egypt, and Greece) mapped out the nighttime sky with impressive accuracy, but their models of planetary motion were based on the erroneous assumption that the planets revolve around the earth. Although the Greek astronomer Aristarchus (310–230 BCE) had suggested that the earth revolves around the sun, this idea was rejected and forgotten for nearly eighteen centuries, until the Polish astronomer Nicolaus Copernicus (1473–1543) introduced a revolutionary set of ideas about the solar system, including the hypothesis that the planets revolve around the sun. Copernicus paved the way for the next generation, most notably Tycho Brahe (1546–1601), Galileo Galilei (1564–1642), and Johannes Kepler (1571–1630).

The German astronomer Johannes Kepler was the son of a mercenary soldier who apparently left his family when Johannes was 5 and may have died at war. He was raised by his mother in his grandfather's inn. Kepler's mathematical brilliance earned him a scholarship at the University of Tübingen and at age of 29, he went to work for the Danish astronomer Tycho Brahe, who had compiled the most complete and accurate data on planetary motion then available. When Brahe died in 1601, Kepler succeeded him as "Imperial Mathematician" to the Holy Roman Emperor, and in 1609 he formulated the first two of his laws of planetary motion in a work entitled *Astronomia Nova* (*New Astronomy*).

In the centuries since Kepler's death, as observational data improved, astronomers found that the planetary orbits are not exactly elliptical. Furthermore, the perihelion (the point on the orbit closest to the sun) shifts slowly over time (Figure 6). Most of these deviations can be explained by the mutual pull of the planets, but the perihelion shift of Mercury is larger than can be accounted for by Newton's Laws. On November 18, 1915, Albert Einstein made a discovery about which he later wrote to a friend, "I was beside myself with ecstasy for days." He had been working for a decade on his famous **General Theory of Relativity**, a theory that would replace Newton's law of gravitation with a new set of much more complicated equations called the Einstein Field Equations. On that 18th of November, Einstein showed that Mercury's perihelion shift was accurately explained by his new theory. At the time, this was the only substantial piece of evidence that the General Theory of Relativity was correct.

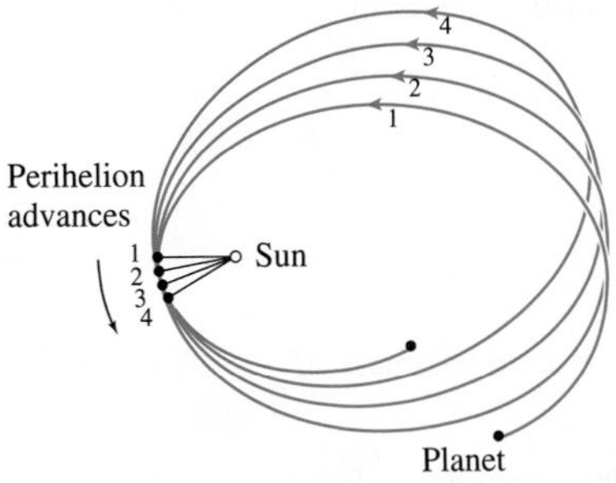

FIGURE 6 The perihelion of an orbit shifts slowly over time. For Mercury, the semimajor axis makes a full revolution approximately once every 24,000 years.

Constants:

- *Gravitational constant:*

  $$G \approx 6.673 \times 10^{-11}\ \mathrm{m}^3\,\mathrm{kg}^{-1}\,\mathrm{s}^{-2}$$

- *Mass of the sun:*

  $$M \approx 1.989 \times 10^{30}\ \mathrm{kg}$$

- $k = GM \approx 1.327 \times 10^{20}$

14.6 SUMMARY

- Kepler's three laws of planetary motion:
  - Law of Ellipses
  - Law of Equal Area in Equal Time
  - Law of the Period $T^2 = \left(\dfrac{4\pi^2}{GM}\right) a^3$, where T is the period (time to complete one full revolution) and a is the semimajor axis (Figure 7).
- According to Newton's Universal Law of Gravitation and Second Law of Motion, the position vector $\mathbf{r}(t)$ of a planet satisfies the differential equation

$$\mathbf{r}''(t) = -\frac{k}{r(t)^2}\mathbf{e}_r, \qquad \text{where } r(t) = \|\mathbf{r}(t)\|, \quad \mathbf{e}_r = \frac{\mathbf{r}(t)}{\|\mathbf{r}(t)\|}$$

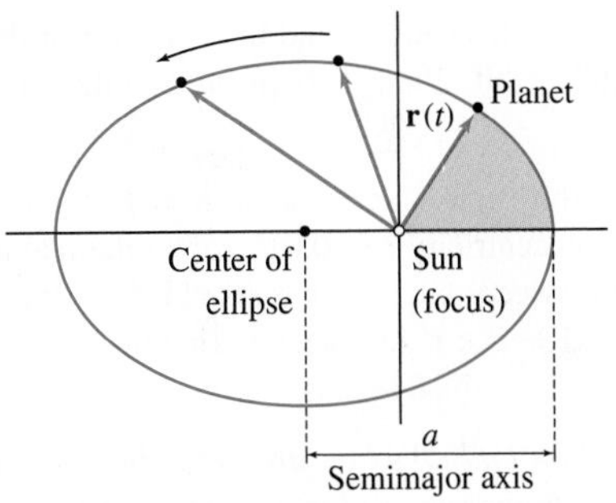

FIGURE 7 Planetary orbit.

- Properties of $\mathbf{J} = \mathbf{r}(t) \times \mathbf{r}'(t)$:
  - $\mathbf{J}$ is a constant of planetary motion.
  - Let $J = \|\mathbf{J}\|$. Then $J = r(t)^2\theta'(t)$.
  - The planet sweeps out area at the rate $\dfrac{dA}{dt} = \dfrac{1}{2}J$.
- A planetary orbit has polar equation $r = \dfrac{J^2/k}{1 + e\cos\theta}$, where e is the eccentricity of the orbit.

14.6 EXERCISES

Preliminary Questions

1. Describe the relation between the vector $\mathbf{J} = \mathbf{r} \times \mathbf{r}'$ and the rate at which the radial vector sweeps out area.

2. Equation (1) shows that $\mathbf{r}''$ is proportional to $\mathbf{r}$. Explain how this fact is used to prove Kepler's Second Law.

3. How is the period T affected if the semimajor axis a is increased four-fold?

Exercises

1. Kepler's Third Law states that T^2/a^3 has the same value for each planetary orbit. Do the data in the following table support this conclusion? Estimate the length of Jupiter's period, assuming that $a = 77.8 \times 10^{10}$ m.

Planet	Mercury	Venus	Earth	Mars
a (10^{10} m)	5.79	10.8	15.0	22.8
T (years)	0.241	0.615	1.00	1.88

2. Finding the Mass of a Star Using Kepler's Third Law, show that if a planet revolves around a star with period T and semimajor axis a, then the mass of the star is $M = \left(\dfrac{4\pi^2}{G}\right)\left(\dfrac{a^3}{T^2}\right)$.

3. Ganymede, one of Jupiter's moons discovered by Galileo, has an orbital period of 7.154 days and a semimajor axis of 1.07×10^9 m. Use Exercise 2 to estimate the mass of Jupiter.

4. An astronomer observes a planet orbiting a star with a period of 9.5 years and a semimajor axis of 3×10^8 km. Find the mass of the star using Exercise 2.

5. Mass of the Milky Way The sun revolves around the center of mass of the Milky Way galaxy in an orbit that is approximately circular, of radius $a \approx 2.8 \times 10^{17}$ km and velocity $v \approx 250$ km/s. Use the result of Exercise 2 to estimate the mass of the portion of the Milky Way inside the sun's orbit (place all of this mass at the center of the orbit).

6. A satellite orbiting above the equator of the earth is **geosynchronous** if the period is $T = 24$ hours (in this case, the satellite stays over a fixed point on the equator). Use Kepler's Third Law to show that in a circular geosynchronous orbit, the distance from the center of the earth is $R \approx 42{,}246$ km. Then compute the altitude h of the orbit above the earth's surface. The earth has mass $M \approx 5.974 \times 10^{24}$ kg and radius $R \approx 6371$ km.

7. Show that a planet in a circular orbit travels at constant speed. *Hint:* Use that $\mathbf{J}$ is constant and that $\mathbf{r}(t)$ is orthogonal to $\mathbf{r}'(t)$ for a circular orbit.

8. Verify that the circular orbit

$$\mathbf{r}(t) = \langle R\cos\omega t, R\sin\omega t\rangle$$

satisfies the differential equation, Eq. (1), provided that $\omega^2 = kR^{-3}$. Then deduce Kepler's Third Law $T^2 = \left(\dfrac{4\pi^2}{k}\right)R^3$ for this orbit.

9. Prove that if a planetary orbit is circular of radius R, then $vT = 2\pi R$, where v is the planet's speed (constant by Exercise 7) and T is the period. Then use Kepler's Third Law to prove that $v = \sqrt{\dfrac{k}{R}}$.

10. Find the velocity of a satellite in geosynchronous orbit about the earth. *Hint:* Use Exercises 6 and 9.

11. A communications satellite orbiting the earth has initial position $\mathbf{r} = \langle 29{,}000, 20{,}000, 0\rangle$ (in km) and initial velocity $\mathbf{r}' = \langle 1, 1, 1\rangle$ (in km/s), where the origin is the earth's center. Find the equation of the plane containing the satellite's orbit. *Hint:* This plane is orthogonal to $\mathbf{J}$.

12. Assume that the earth's orbit is circular of radius $R = 150 \times 10^6$ km (it is nearly circular with eccentricity $e = 0.017$). Find the rate at which the earth's radial vector sweeps out area in units of km^2/s. What is the magnitude of the vector $\mathbf{J} = \mathbf{r} \times \mathbf{r}'$ for the earth (in units of km^2 per second)?

Exercises 13–19: The perihelion and aphelion are the points on the orbit closest to and farthest from the sun, respectively (Figure 8). The distance from the sun at the perihelion is denoted $r_{\rm per}$ and the speed *at this point is denoted v_{per}. Similarly, we write r_{ap} and v_{ap} for the distance and speed at the aphelion. The semimajor axis is denoted a.*

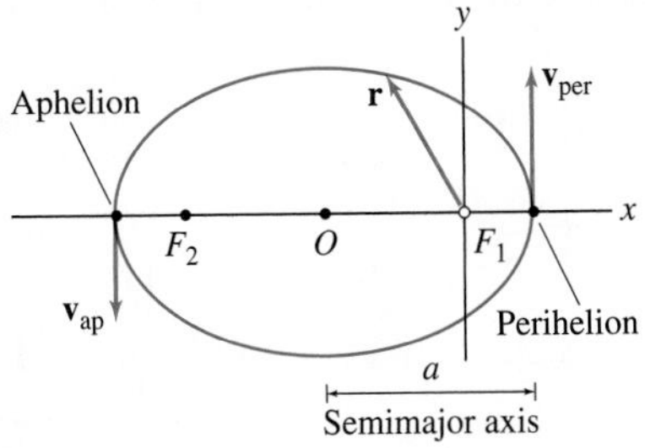

FIGURE 8 $\mathbf{r}$ and $\mathbf{v} = \mathbf{r}'$ are perpendicular at the perihelion and aphelion.

13. Use the polar equation of an ellipse

$$r = \frac{p}{1 + e\cos\theta}$$

to show that $r_{per} = a(1-e)$ and $r_{ap} = a(1+e)$. *Hint:* Use the fact that $r_{per} + r_{ap} = 2a$.

14. Use the result of Exercise 13 to prove the formulas

$$e = \frac{r_{ap} - r_{per}}{r_{ap} + r_{per}}, \qquad p = \frac{2r_{ap}r_{per}}{r_{ap} + r_{per}}$$

15. Use the fact that $\mathbf{J} = \mathbf{r} \times \mathbf{r}'$ is constant to prove

$$v_{per}(1-e) = v_{ap}(1+e)$$

Hint: $\mathbf{r}$ is perpendicular to $\mathbf{r}'$ at the perihelion and aphelion.

16. Compute r_{per} and r_{ap} for the orbit of Mercury, which has eccentricity $e = 0.244$ (see the table in Exercise 1 for the semimajor axis).

17. Conservation of Energy The total mechanical energy (kinetic energy plus potential energy) of a planet of mass m orbiting a sun of mass M with position $\mathbf{r}$ and speed $v = \|\mathbf{r}'\|$ is

$$E = \frac{1}{2}mv^2 - \frac{GMm}{\|\mathbf{r}\|} \qquad \boxed{8}$$

(a) Prove the equations

$$\frac{d}{dt}\frac{1}{2}mv^2 = \mathbf{v}\cdot(m\mathbf{a}), \qquad \frac{d}{dt}\frac{GMm}{\|\mathbf{r}\|} = \mathbf{v}\cdot\left(-\frac{GMm}{\|\mathbf{r}\|^3}\mathbf{r}\right)$$

(b) Then use Newton's Law $\mathbf{F} = m\mathbf{a}$ and Eq. (1) to prove that energy is conserved—that is, $\frac{dE}{dt} = 0$.

18. Show that the total energy [Eq. (8)] of a planet in a circular orbit of radius R is $E = -\frac{GMm}{2R}$. *Hint:* Use Exercise 9.

19. Prove that $v_{per} = \sqrt{\left(\frac{GM}{a}\right)\frac{1+e}{1-e}}$ as follows:

(a) Use Conservation of Energy (Exercise 17) to show that

$$v_{per}^2 - v_{ap}^2 = 2GM\left(r_{per}^{-1} - r_{ap}^{-1}\right)$$

(b) Show that $r_{per}^{-1} - r_{ap}^{-1} = \frac{2e}{a(1-e^2)}$ using Exercise 13.

(c) Show that $v_{per}^2 - v_{ap}^2 = 4\frac{e}{(1+e)^2}v_{per}^2$ using Exercise 15. Then solve for v_{per} using (a) and (b).

20. Show that a planet in an elliptical orbit has total mechanical energy $E = -\frac{GMm}{2a}$, where a is the semimajor axis. *Hint:* Use Exercise 19 to compute the total energy at the perihelion.

21. Prove that $v^2 = GM\left(\frac{2}{r} - \frac{1}{a}\right)$ at any point on an elliptical orbit, where $r = \|\mathbf{r}\|$, v is the velocity, and a is the semimajor axis of the orbit.

22. Two space shuttles A and B orbit the earth along the solid trajectory in Figure 9. Hoping to catch up to B, the pilot of A applies a forward thrust to increase her shuttle's kinetic energy. Use Exercise 20 to show that shuttle A will move off into a larger orbit as shown in the figure. Then use Kepler's Third Law to show that A's orbital period T will increase (and she will fall farther and farther behind B)!

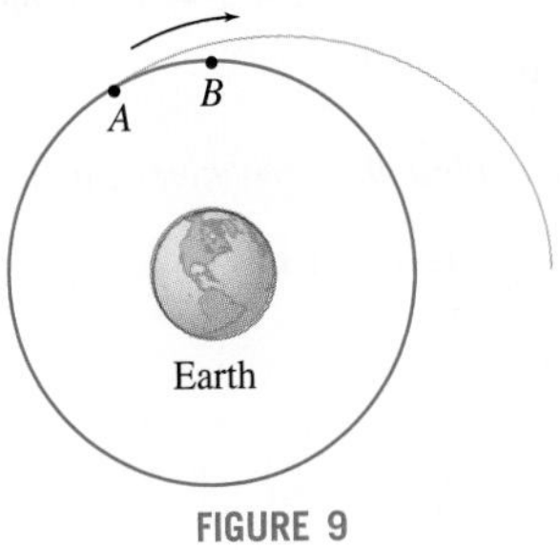

FIGURE 9

Further Insights and Challenges

Exercises 23 and 24 prove Kepler's Third Law. Figure 10 shows an elliptical orbit with polar equation

$$r = \frac{p}{1 + e\cos\theta}$$

where $p = J^2/k$. The origin of the polar coordinates is at F_1. Let a and b be the semimajor and semiminor axes, respectively.

23. This exercise shows that $b = \sqrt{pa}$.

(a) Show that $CF_1 = ae$. *Hint:* $r_{per} = a(1-e)$ by Exercise 13.

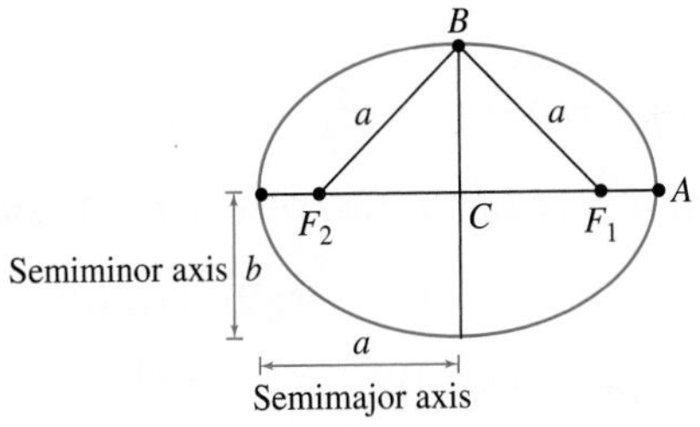

FIGURE 10

(b) Show that $a = \dfrac{p}{1 - e^2}$.

(c) Show that $F_1A + F_2A = 2a$. Conclude that $F_1B + F_2B = 2a$ and hence $F_1B = F_2B = a$.

(d) Use the Pythagorean Theorem to prove that $b = \sqrt{pa}$.

24. The area A of the ellipse is $A = \pi ab$.

(a) Prove, using Kepler's First Law, that $A = \frac{1}{2}JT$, where T is the period of the orbit.

(b) Use Exercise 23 to show that $A = (\pi\sqrt{p})a^{3/2}$.

(c) Deduce Kepler's Third Law: $T^2 = \dfrac{4\pi^2}{GM}a^3$.

25. According to Eq. (7) the velocity vector of a planet as a function of the angle θ is

$$\mathbf{v}(\theta) = \frac{k}{J}\mathbf{e}_\theta + \mathbf{c}$$

Use this to explain the following statement: As a planet revolves around the sun, its velocity vector traces out a circle of radius k/J with center at the terminal point of $\mathbf{c}$ (Figure 11). This beautiful but hidden property of orbits was discovered by William Rowan Hamilton in 1847.

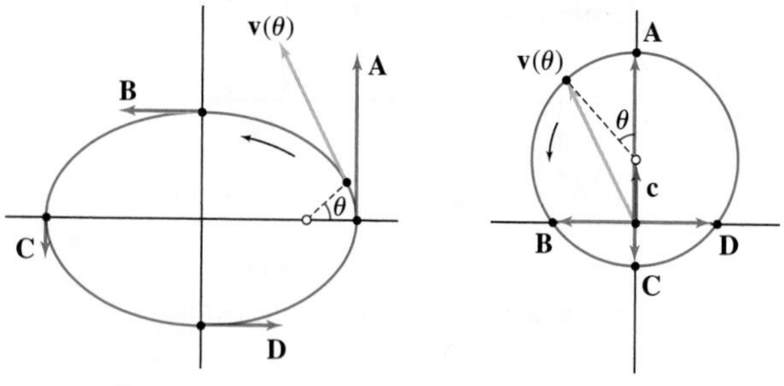

FIGURE 11 The velocity vector traces out a circle as the planet travels along its orbit.

CHAPTER REVIEW EXERCISES

1. Determine the domains of the vector-valued functions.

(a) $\mathbf{r}_1(t) = \langle t^{-1}, (t+1)^{-1}, \sin^{-1} t\rangle$

(b) $\mathbf{r}_2(t) = \langle \sqrt{8 - t^3}, \ln t, e^{\sqrt{t}}\rangle$

2. Sketch the paths $\mathbf{r}_1(\theta) = \langle \theta, \cos\theta\rangle$ and $\mathbf{r}_2(\theta) = \langle \cos\theta, \theta\rangle$ in the xy-plane.

3. Find a vector parametrization of the intersection of the surfaces $x^2 + y^4 + 2z^3 = 6$ and $x = y^2$ in $\mathbf{R}^3$.

4. Find a vector parametrization using trigonometric functions of the intersection of the plane $x + y + z = 1$ and the elliptical cylinder $\left(\frac{y}{3}\right)^2 + \left(\frac{z}{8}\right)^2 = 1$ in $\mathbf{R}^3$.

In Exercises 5–10, calculate the derivative indicated.

5. $\mathbf{r}'(t)$, $\quad \mathbf{r}(t) = \langle 1 - t, t^{-2}, \ln t\rangle$

6. $\mathbf{r}'''(t)$, $\quad \mathbf{r}(t) = \langle t^3, 4t^2, 7t\rangle$

7. $\mathbf{r}'(0)$, $\quad \mathbf{r}(t) = \langle e^{2t}, e^{-4t^2}, e^{6t}\rangle$

8. $\mathbf{r}''(-3)$, $\quad \mathbf{r}(t) = \langle t^{-2}, (t+1)^{-1}, t^3 - t\rangle$

9. $\dfrac{d}{dt}e^t\langle 1, t, t^2\rangle$

10. $\dfrac{d}{d\theta}\mathbf{r}(\cos\theta)$, $\quad \mathbf{r}(s) = \langle s, 2s, s^2\rangle$

In Exercises 11–14, calculate the derivative at $t = 3$, assuming that

$$\mathbf{r}_1(3) = \langle 1, 1, 0\rangle, \qquad \mathbf{r}_2(3) = \langle 1, 1, 0\rangle$$
$$\mathbf{r}_1'(3) = \langle 0, 0, 1\rangle, \qquad \mathbf{r}_2'(3) = \langle 0, 2, 4\rangle$$

11. $\dfrac{d}{dt}(6\mathbf{r}_1(t) - 4 \cdot \mathbf{r}_2(t))$ $\qquad$ **12.** $\dfrac{d}{dt}\left(e^t\mathbf{r}_2(t)\right)$

13. $\dfrac{d}{dt}(\mathbf{r}_1(t) \cdot \mathbf{r}_2(t))$ $\qquad$ **14.** $\dfrac{d}{dt}(\mathbf{r}_1(t) \times \mathbf{r}_2(t))$

15. Calculate $\displaystyle\int_0^3 \langle 4t + 3, t^2, -4t^3\rangle\, dt$.

16. Calculate $\displaystyle\int_0^\pi \langle \sin\theta, \theta, \cos 2\theta\rangle\, d\theta$.

17. A particle located at $(1, 1, 0)$ at time $t = 0$ follows a path whose velocity vector is $\mathbf{v}(t) = \langle 1, t, 2t^2\rangle$. Find the particle's location at $t = 2$.

18. Find the vector-valued function $\mathbf{r}(t) = \langle x(t), y(t)\rangle$ in $\mathbf{R}^2$ satisfying $\mathbf{r}'(t) = -\mathbf{r}(t)$ with initial conditions $\mathbf{r}(0) = \langle 1, 2\rangle$.

19. Calculate $\mathbf{r}(t)$ assuming that

$$\mathbf{r}''(t) = \left\langle 4 - 16t, 12t^2 - t\right\rangle, \qquad \mathbf{r}'(0) = \langle 1, 0\rangle, \qquad \mathbf{r}(0) = \langle 0, 1\rangle$$

20. Solve $\mathbf{r}''(t) = \left\langle t^2 - 1, t + 1, t^3\right\rangle$ subject to the initial conditions $\mathbf{r}(0) = \langle 1, 0, 0\rangle$ and $\mathbf{r}'(0) = \langle -1, 1, 0\rangle$

21. Compute the length of the path

$$\mathbf{r}(t) = \langle \sin 2t, \cos 2t, 3t - 1\rangle \quad \text{for } 1 \le t \le 3$$

22. CAS Express the length of the path $\mathbf{r}(t) = \langle \ln t, t, e^t\rangle$ for $1 \le t \le 2$ as a definite integral, and use a computer algebra system to find its value to two decimal places.

23. Find an arc length parametrization of a helix of height 20 cm that makes four full rotations over a circle of radius 5 cm.

24. Find the minimum speed of a particle with trajectory $\mathbf{r}(t) = \langle t, e^{t-3}, e^{4-t}\rangle$.

25. A projectile fired at an angle of 60° lands 400 m away. What was its initial speed?

26. A specially trained mouse runs counterclockwise in a circle of radius 0.6 m on the floor of an elevator with speed 0.3 m/s while the elevator ascends from ground level (along the z-axis) at a speed of 12 m/s. Find the mouse's acceleration vector as a function of time. Assume that the circle is centered at the origin of the xy-plane and the mouse is at $(2, 0, 0)$ at $t = 0$.

27. During a short time interval $[0.5, 1.5]$, the path of an unmanned spy plane is described by

$$\mathbf{r}(t) = \left\langle -\frac{100}{t^2}, 7 - t, 40 - t^2 \right\rangle$$

A laser is fired (in the tangential direction) toward the yz-plane at time $t = 1$. Which point in the yz-plane does the laser beam hit?

28. A force $\mathbf{F} = \langle 12t + 4, 8 - 24t \rangle$ (in newtons) acts on a 2-kg mass. Find the position of the mass at $t = 2$ s if it is located at $(4, 6)$ at $t = 0$ and has initial velocity $\langle 2, 3 \rangle$ in m/s.

29. Find the unit tangent vector to $\mathbf{r}(t) = \langle \sin t, t, \cos t \rangle$ at $t = \pi$.

30. Find the unit tangent vector to $\mathbf{r}(t) = \langle t^2, \tan^{-1} t, t \rangle$ at $t = 1$.

31. Calculate $\kappa(1)$ for $\mathbf{r}(t) = \langle \ln t, t \rangle$.

32. Calculate $\kappa\left(\frac{\pi}{4}\right)$ for $\mathbf{r}(t) = \langle \tan t, \sec t, \cos t \rangle$.

In Exercises 33 and 34, write the acceleration vector $\mathbf{a}$ *at the point indicated as a sum of tangential and normal components.*

33. $\mathbf{r}(\theta) = \langle \cos\theta, \sin 2\theta \rangle, \quad \theta = \frac{\pi}{4}$

34. $\mathbf{r}(t) = \langle t^2, 2t - t^2, t \rangle, \quad t = 2$

35. At a certain time t_0, the path of a moving particle is tangent to the y-axis. The particle's speed at time t_0 is 4 m/s, and its acceleration vector is $\mathbf{a} = \langle 5, 4, 12 \rangle$. Determine the curvature of the path at t_0.

36. Parametrize the osculating circle to $y = x^2 - x^3$ at $x = 1$.

37. Parametrize the osculating circle to $y = \sqrt{x}$ at $x = 4$.

38. If a planet has zero mass ($m = 0$), then Newton's laws of motion reduce to $\mathbf{r}''(t) = \mathbf{0}$ and the orbit is a straight line $\mathbf{r}(t) = \mathbf{r}_0 + t\mathbf{v}_0$, where $\mathbf{r}_0 = \mathbf{r}(0)$ and $\mathbf{v}_0 = \mathbf{r}'(0)$ (Figure 1). Show that the area swept out by the radial vector at time t is $A(t) = \frac{1}{2}\|\mathbf{r}_0 \times \mathbf{v}_0\| t$ and thus Kepler's Second Law continues to hold (the rate is constant).

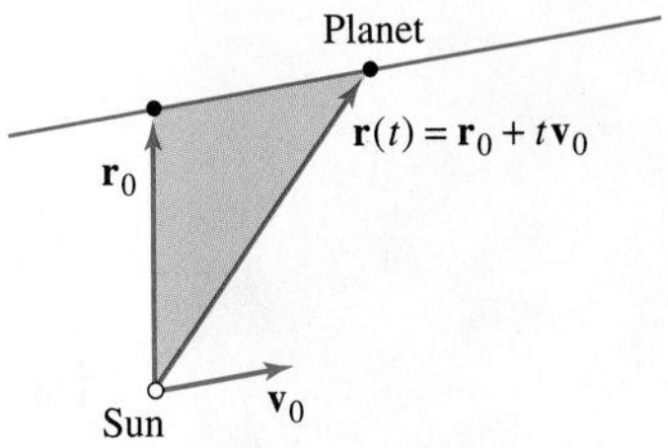

FIGURE 1

39. Suppose the orbit of a planet is an ellipse of eccentricity $e = c/a$ and period T (Figure 2). Use Kepler's Second Law to show that the time required to travel from A' to B' is equal to

$$\left(\frac{1}{4} + \frac{e}{2\pi}\right) T$$

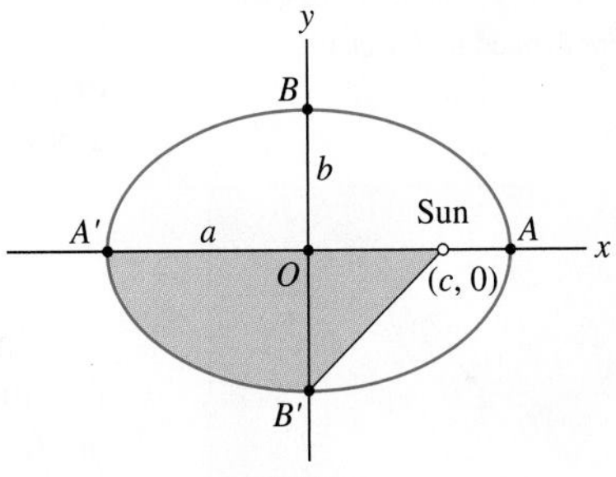

FIGURE 2

40. The period of Mercury is approximately 88 days, and its orbit has eccentricity 0.205. How much longer does it take Mercury to travel from A' to B' than from B' to A (Figure 2)?

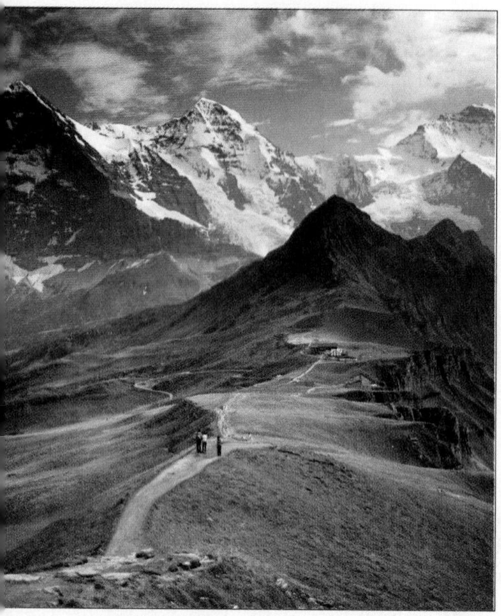

The famous triple peaks Eiger, Monch, and Jungfrau in the Swiss alps. The steepness at a point in a mountain range is measured by the gradient, a concept defined in this chapter.

15 DIFFERENTIATION IN SEVERAL VARIABLES

In this chapter we extend the concepts and techniques of differential calculus to functions of several variables. As we will see, a function f that depends on two or more variables has not just one derivative but rather a set of *partial derivatives*, one for each variable. The partial derivatives are the components of the gradient vector, which provides valuable insight into the function's behavior. In the last two sections, we apply the tools we have developed to optimization in several variables.

15.1 Functions of Two or More Variables

A familiar example of a function of two variables is the area A of a rectangle, equal to the product xy of the base x and height y. We write

$$A(x, y) = xy$$

or $A = f(x, y)$, where $f(x, y) = xy$. An example in three variables is the distance from a point $P = (x, y, z)$ to the origin:

$$g(x, y, z) = \sqrt{x^2 + y^2 + z^2}$$

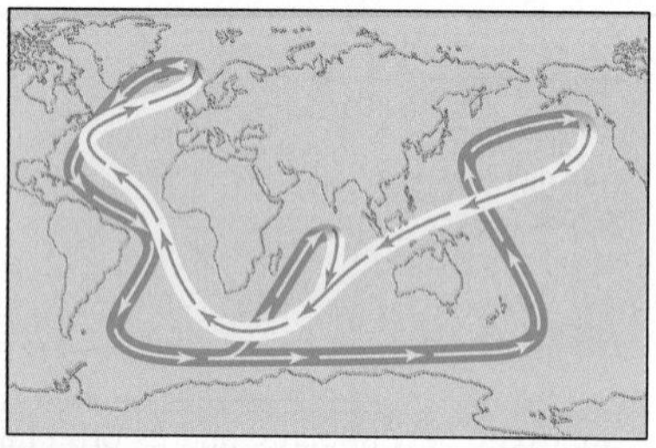

FIGURE 1 The global climate is influenced by the ocean "conveyer belt," a system of deep currents driven by variations in seawater density.

An important but less familiar example is the density of seawater, denoted ρ, which is a function $\rho(S, T)$ of salinity S and temperature T (Figure 1). Although there is no simple formula for $\rho(S, T)$, scientists determine function values experimentally (Figure 2). According to Table 1, if $S = 32$ (in parts per thousand) and $T = 10°C$, then

$$\rho(32, 10) = 1.0246 \text{ kg/m}^3$$

FIGURE 2 A Conductivity-Temperature-Depth (CDT) instrument is used to measure seawater variables such as density, temperature, pressure, and salinity.

TABLE 1 Seawater Density ρ (kg/m^3) as a Function of Temperature and Salinity.

	Salinity (ppt)		
°C	32	32.5	33
5	1.0253	1.0257	1.0261
10	1.0246	1.0250	1.0254
15	1.0237	1.0240	1.0244
20	1.0224	1.0229	1.0232

A function of n variables is a function $f(x_1, \dots, x_n)$ that assigns a real number to each n-tuple $(x_1, \dots, x_n)$ in a domain in $\mathbf{R}^n$. Sometimes we write $f(P)$ for the value of f at a point $P = (x_1, \dots, x_n)$. When f is defined by a formula, we usually take as domain the set of all n-tuples for which $f(x_1, \dots, x_n)$ is defined. The range of f is the set of all values $f(x_1, \dots, x_n)$ for $(x_1, \dots, x_n)$ in the domain. Since we focus on functions of two or three variables, we shall often use the variables x, y, and z (rather than x_1, x_2, x_3).

EXAMPLE 1 Sketch the domains of

(a) $f(x, y) = \sqrt{9 - x^2 - y}$ **(b)** $g(x, y, z) = x\sqrt{y} + \ln(z - 1)$

What are the ranges of these functions?

Solution

(a) $f(x, y) = \sqrt{9 - x^2 - y}$ is defined only when $9 - x^2 - y \geq 0$, or $y \leq 9 - x^2$. Thus the domain consists of all points (x, y) lying below the parabola $y = 9 - x^2$ [Figure 3(A)]:

$$\mathcal{D} = \{(x, y) : y \leq 9 - x^2\}$$

To determine the range, note that f is a nonnegative function and that $f(0, y) = \sqrt{9 - y}$. Since $9 - y$ can be any positive number, $f(0, y)$ takes on all nonnegative values. Therefore the range of f is the infinite interval $[0, \infty)$.

(b) $g(x, y, z) = x\sqrt{y} + \ln(z - 1)$ is defined only when both $\sqrt{y}$ and $\ln(z - 1)$ are defined. We must require that $y \geq 0$ and $z > 1$, so the domain is $\{(x, y, z) : y \geq 0, z > 1\}$ [Figure 3(B)]. The range of g is the entire real line $\mathbf{R}$. Indeed, for the particular choices $y = 1$ and $z = 2$, we have $g(x, 1, 2) = x\sqrt{1} + \ln 1 = x$, and since x is arbitrary, we see that g takes on all values. ■

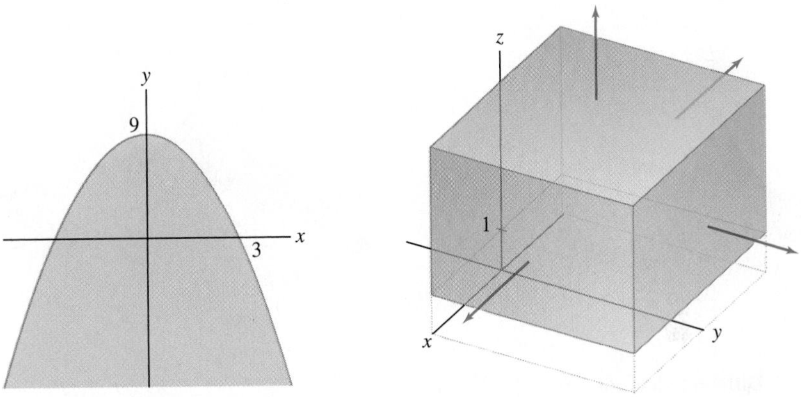

(A) The domain of $f(x, y) = \sqrt{9 - x^2 - y}$ is the set of all points lying below the parabola $y = 9 - x^2$.

(B) Domain of $g(x, y, z) = x\sqrt{y} + \ln(z - 1)$ is the set of points with $y \geq 0$ and $z > 1$. The domain continues out to infinity in the directions indicated by the arrows.

FIGURE 3

Graphing Functions of Two Variables

In single-variable calculus, we use graphs to visualize the important features of a function. Graphs play a similar role for functions of two variables. The graph of $f(x, y)$ consists of all points $(a, b, f(a, b))$ in $\mathbf{R}^3$ for (a, b) in the domain $\mathcal{D}$ of f. Assuming that f is continuous (as defined in the next section), the graph is a surface whose *height* above or below the xy-plane at (a, b) is the function value $f(a, b)$ [Figure 4]. We often write $z = f(x, y)$ to stress that the z-coordinate of a point on the graph is a function of x and y.

EXAMPLE 2 Sketch the graph of $f(x, y) = 2x^2 + 5y^2$.

Solution The graph is a paraboloid (Figure 5), which we saw in Section 13.6. We sketch the graph using the fact that the horizontal cross section (called the horizontal "trace" below) at height z is the ellipse $2x^2 + 5y^2 = z$. ■

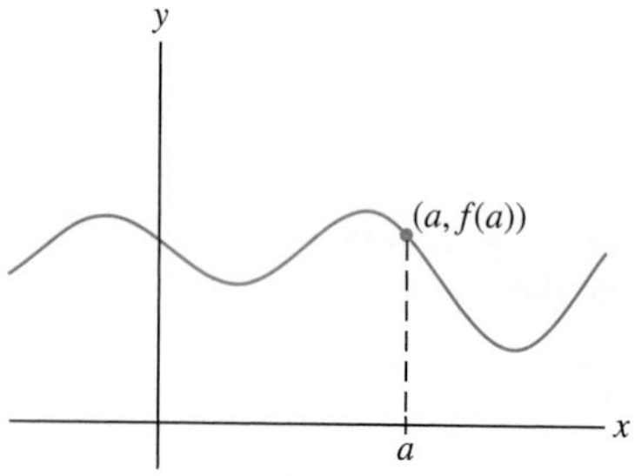

(A) Graph of $y = f(x)$

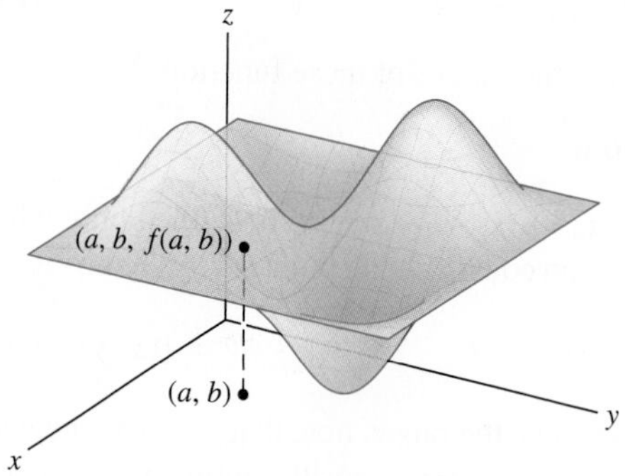

(B) Graph of $z = f(x, y)$

FIGURE 4

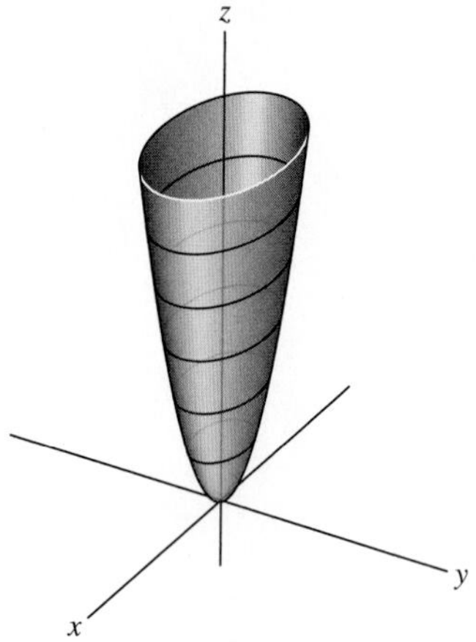

FIGURE 5 Graph of $f(x, y) = 2x^2 + 5y^2$

Plotting more complicated graphs by hand can be difficult. Fortunately, computer algebra systems eliminate the labor and greatly enhance our ability to explore functions graphically. Graphs can be rotated and viewed from different perspectives (Figure 6).

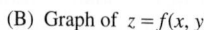

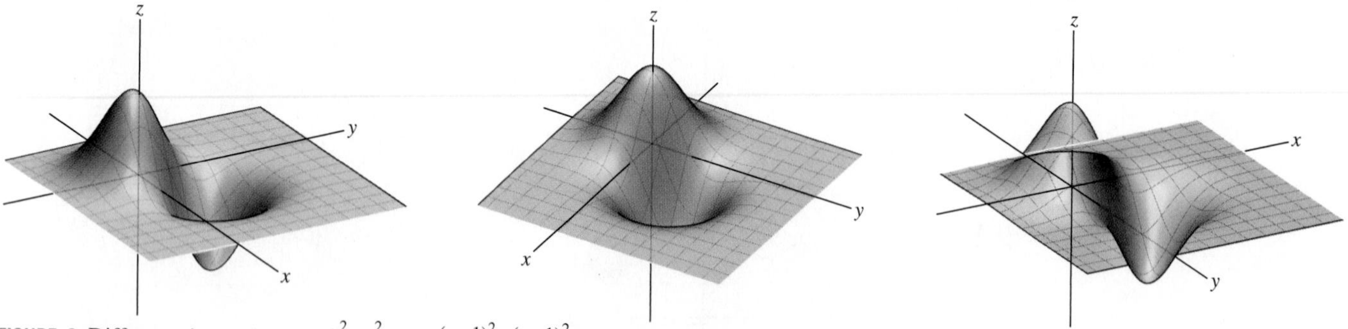

FIGURE 6 Different views of $z = e^{-x^2-y^2} - e^{-(x-1)^2-(y-1)^2}$

Traces and Level Curves

One way of analyzing the graph of a function $f(x, y)$ is to freeze the x-coordinate by setting $x = a$ and examine the resulting curve $z = f(a, y)$. Similarly, we may set $y = b$ and consider the curve $z = f(x, b)$. Curves of this type are called **vertical traces**. They are obtained by intersecting the graph with planes parallel to a vertical coordinate plane (Figure 7):

- **Vertical trace in the plane $x = a$:** Intersection of the graph with the vertical plane $x = a$, consisting of all points $(a, y, f(a, y))$.
- **Vertical trace in the plane $y = b$:** Intersection of the graph with the vertical plane $y = b$, consisting of all points $(x, b, f(x, b))$.

■ **EXAMPLE 3** Describe the vertical traces of $f(x, y) = x \sin y$.

Solution When we freeze the x-coordinate by setting $x = a$, we obtain the trace curve $z = a \sin y$ (see Figure 8). This is a sine curve located in the plane $x = a$. When we set $y = b$, we obtain a line $z = (\sin b)y$ of slope $\sin b$, located in the plane $y = b$. ■

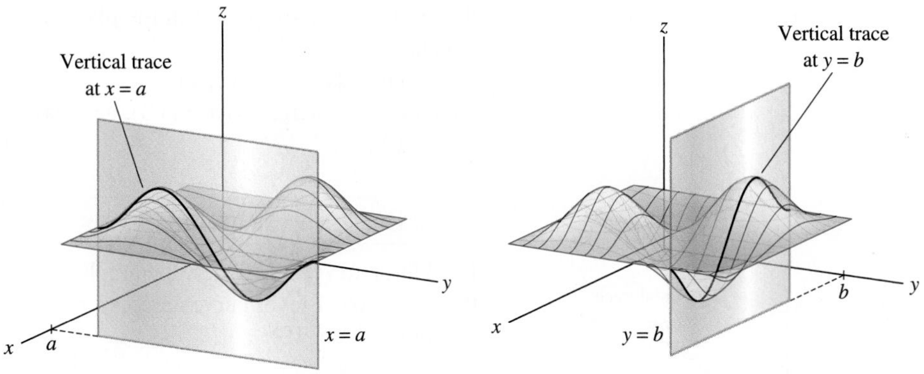

(A) Vertical traces parallel to yz-plane

(B) Vertical traces parallel to xz-plane

FIGURE 7

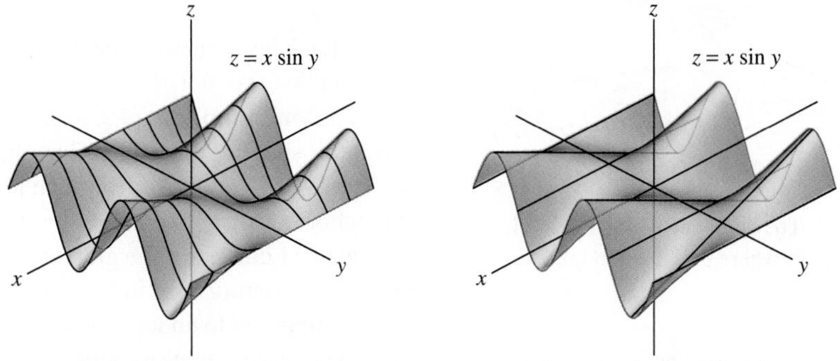

(A) The traces in the planes $x = a$ are the curves $z = a(\sin y)$.

(B) The traces in the planes $y = b$ are the lines $z = (\sin b)y$.

FIGURE 8 Vertical traces of $f(x, y) = x \sin y$.

■ **EXAMPLE 4 Identifying Features of a Graph** Match the graphs in Figure 9 with the following functions:

(i) $f(x, y) = x - y^2$ (ii) $g(x, y) = x^2 - y$

Solution Let's compare vertical traces. The vertical trace of $f(x, y) = x - y^2$ in the plane $x = a$ is a *downward* parabola $z = a - y^2$. This matches (B). On the other hand,

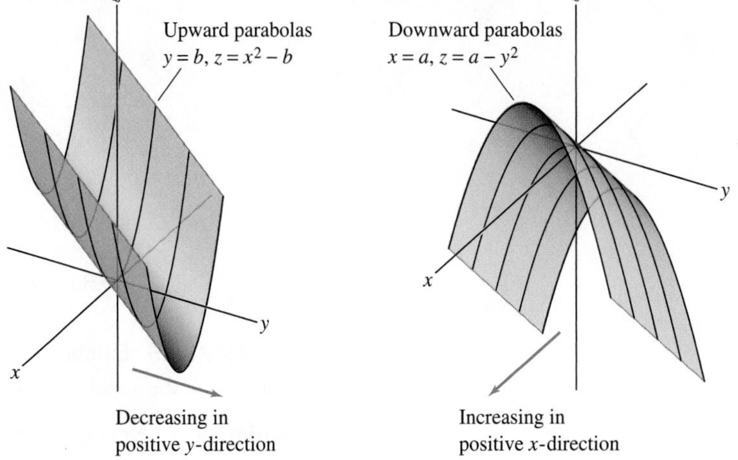

FIGURE 9 (A) (B)

the vertical trace of $g(x, y)$ in the plane $y = b$ is an *upward* parabola $z = x^2 - b$. This matches (A).

Notice also that $f(x, y) = x - y^2$ is an increasing function of x (that is, $f(x, y)$ increases as x increases) as in (B), whereas $g(x, y) = x^2 - y$ is a decreasing function of y as in (A). ■

Level Curves and Contour Maps

In addition to vertical traces, the graph of $f(x, y)$ has horizontal traces. These traces and their associated level curves are especially important in analyzing the behavior of the function (Figure 10):

- **Horizontal trace at height c:** Intersection of the graph with the horizontal plane $z = c$, consisting of the points $(x, y, f(x, y))$ such that $f(x, y) = c$.
- **Level curve:** The curve $f(x, y) = c$ in the xy-plane.

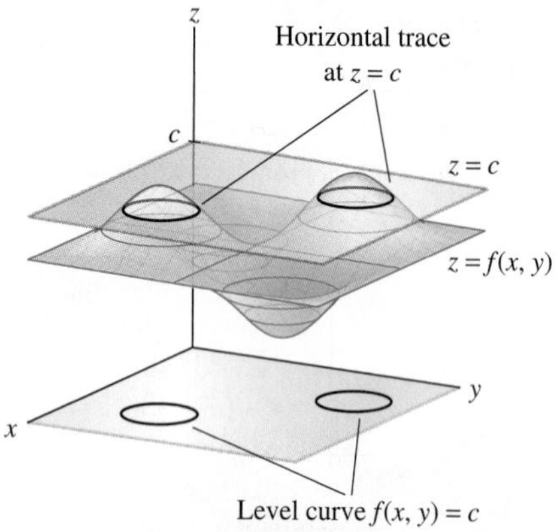

FIGURE 10 The level curve consists of all points (x, y) where the function takes on the value c.

*On contour maps level curves are often referred to as **contour lines**.*

Thus the level curve consists of all points (x, y) in the plane where the function takes the value c. Each level curve is the projection onto the xy-plane of the horizontal trace on the graph that lies above it.

A **contour map** is a plot in the xy-plane that shows the level curves $f(x, y) = c$ for equally spaced values of c. The interval m between the values is called the **contour interval**. When you move from one level curve to next, the value of $f(x, y)$ (and hence the height of the graph) changes by $\pm m$.

Figure 11 compares the graph of a function $f(x, y)$ in (A) and its horizontal traces in (B) with the contour map in (C). The contour map in (C) has contour interval $m = 100$.

It is important to understand how the contour map indicates the steepness of the graph. If the level curves are close together, then a small move from one level curve to the next in the xy-plane leads to a large change in height. In other words, *the level curves are close together if the graph is steep* (Figure 11). Similarly, the graph is flatter when the level curves are farther apart.

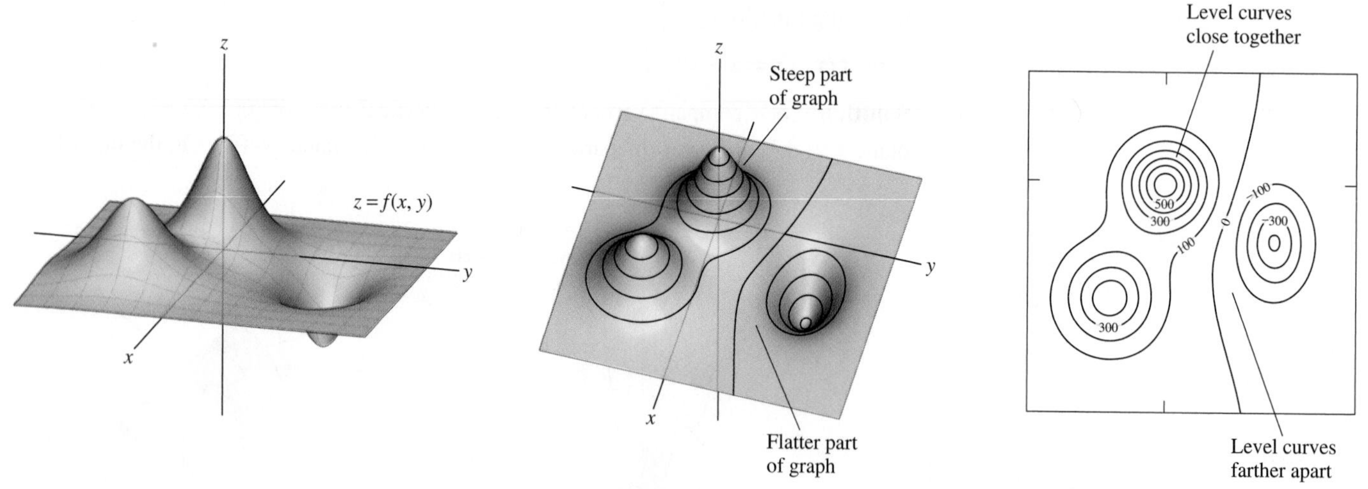

(A) (B) Horizontal traces (C) Contour map

FIGURE 11

■ **EXAMPLE 5 Elliptic Paraboloid** Sketch the contour map of $f(x, y) = x^2 + 3y^2$ and comment on the spacing of the contour curves.

Solution The level curves have equation $f(x, y) = c$, or

$$x^2 + 3y^2 = c$$

- For $c > 0$, the level curve is an ellipse.
- For $c = 0$, the level curve is just the point $(0, 0)$ because $x^2 + 3y^2 = 0$ only for $(x, y) = (0, 0)$.
- The level curve is empty if $c < 0$ because $f(x, y)$ is never negative.

The graph of $f(x, y)$ is an elliptic paraboloid (Figure 12). As we move away from the origin, $f(x, y)$ increases more rapidly. The graph gets steeper, and the level curves get closer together. ∎

EXAMPLE 6 Hyperbolic Paraboloid Sketch the contour map of $g(x, y) = x^2 - 3y^2$.

Solution The level curves have equation $g(x, y) = c$, or

$$x^2 - 3y^2 = c$$

- For $c \neq 0$, the level curve is the hyperbola $x^2 - 3y^2 = c$.
- For $c = 0$, the level curve consists of the two lines $x = \pm\sqrt{3}y$ because the equation $g(x, y) = 0$ factors as follows:

$$x^2 - 3y^2 = 0 = (x - \sqrt{3}y)(x + \sqrt{3}y) = 0$$

The graph of $g(x, y)$ is a hyperbolic paraboloid (Figure 13). When you stand at the origin, $g(x, y)$ increases as you move along the x-axis in either direction and decreases as you move along the y-axis in either direction. Furthermore, the graph gets steeper as you move out from the origin, so the level curves get closer together. ∎

REMINDER The hyperbolic paraboloid in Figure 13 is often called a "saddle" or "saddle-shaped surface."

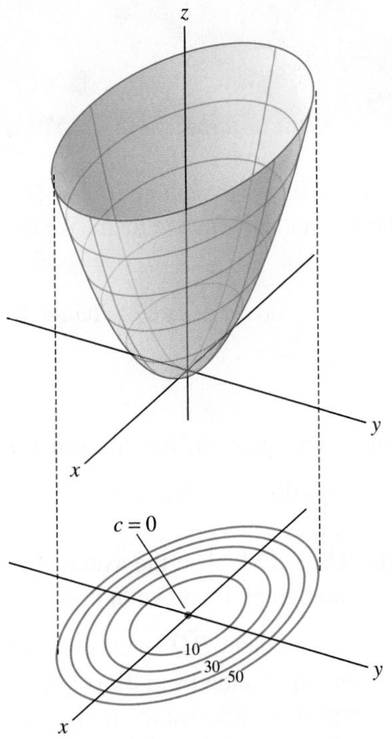

FIGURE 12 $f(x, y) = x^2 + 3y^2$. Contour interval $m = 10$.

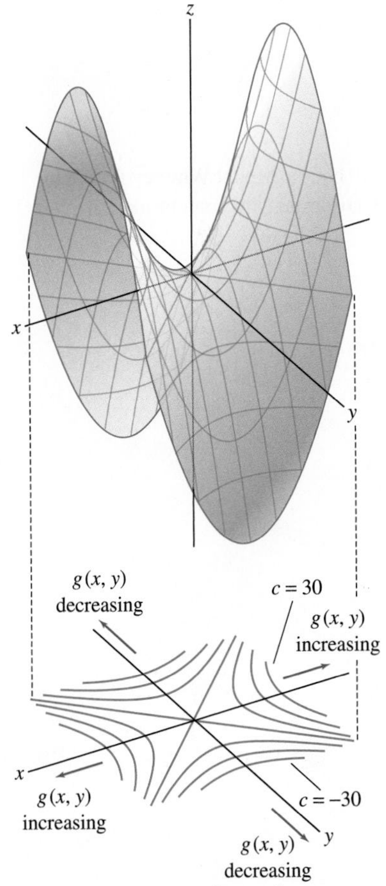

FIGURE 13 $g(x, y) = x^2 - 3y^2$. Contour interval $m = 10$.

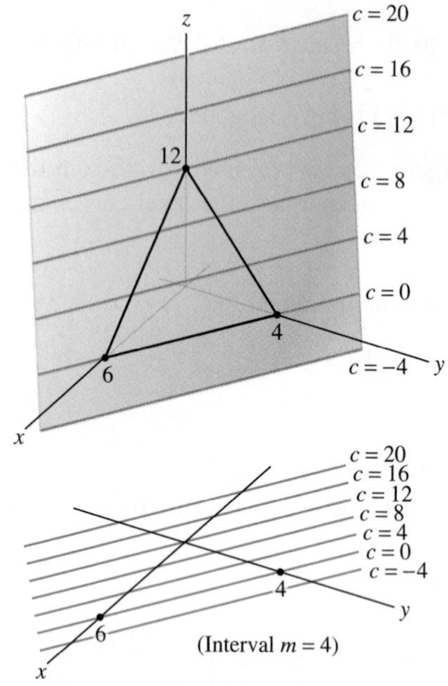

FIGURE 14 Graph and contour map of $f(x, y) = 12 - 2x - 3y$.

EXAMPLE 7 **Contour Map of a Linear Function** Sketch the graph of $f(x, y) = 12 - 2x - 3y$ and the associated contour map with contour interval $m = 4$.

Solution To plot the graph, which is a plane, we find the intercepts with the axes (Figure 14). The graph intercepts the z-axis at $z = f(0, 0) = 12$. To find the x-intercept, we set $y = z = 0$ to obtain $12 - 2x - 3(0) = 0$, or $x = 6$. Similarly, solving $12 - 3y = 0$ gives y-intercept $y = 4$. The graph is the plane determined by the three intercepts.

In general, the level curves of a linear function $f(x, y) = qx + ry + s$ are the lines with equation $qx + ry + s = c$. Therefore, *the contour map of a linear function consists of equally spaced parallel lines*. In our case, the level curves are the lines $12 - 2x - 3y = c$, or $2x + 3y = 12 - c$ (Figure 14). ■

How can we measure steepness quantitatively? Let's imagine the surface $z = f(x, y)$ as a mountain range. In fact, contour maps (also called topographical maps) are used extensively to describe terrain (Figure 15). We place the xy-plane at sea level, so that $f(a, b)$ is the height (also called altitude or elevation) of the mountain above sea level at the point (a, b) in the plane.

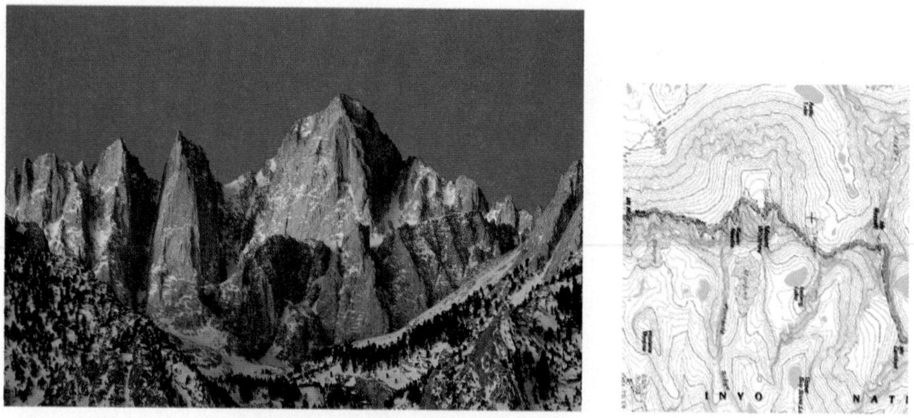

FIGURE 15 Mount Whitney Range in California, with contour map.

Figure 16 shows two points P and Q in the xy-plane, together with the points $\tilde{P}$ and $\tilde{Q}$ on the graph that lie above them. We define the **average rate of change**:

$$\text{Average rate of change from } P \text{ to } Q = \frac{\Delta \text{ altitude}}{\Delta \text{ horizontal}}$$

where

Δ altitude = change in the height from $\tilde{P}$ and $\tilde{Q}$

Δ horizontal = distance from P to Q

EXAMPLE 8 Calculate the average rate of change of $f(x, y)$ from P to Q for the function whose graph is shown in Figure 16.

Solution The segment $\overline{PQ}$ spans three level curves and the contour interval is 0.8 km, so the change in altitude from $\tilde{P}$ to $\tilde{Q}$ is $3(0.8) = 2.4$ km. From the horizontal scale of the contour map, we see that the horizontal distance PQ is 2 km, so

$$\text{Average rate of change from } P \text{ to } Q = \frac{\Delta \text{ altitude}}{\Delta \text{ horizontal}} = \frac{2.4}{2} = 1.2$$

On average, your altitude gain is 1.2 times your horizontal distance traveled as you climb from $\tilde{P}$ to $\tilde{Q}$. ■

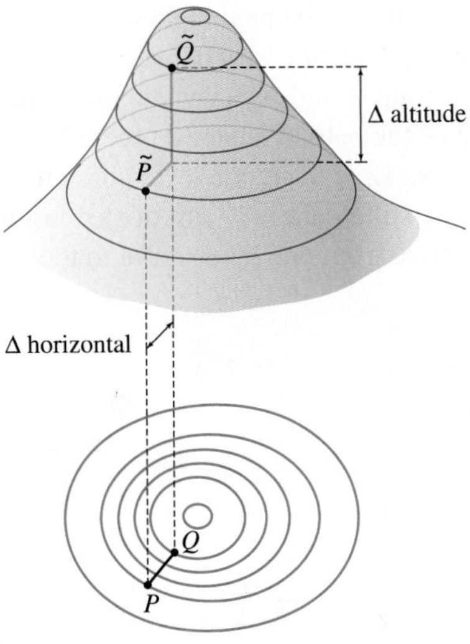

Contour interval: 0.8 km
Horizontal scale: 2 km

FIGURE 16

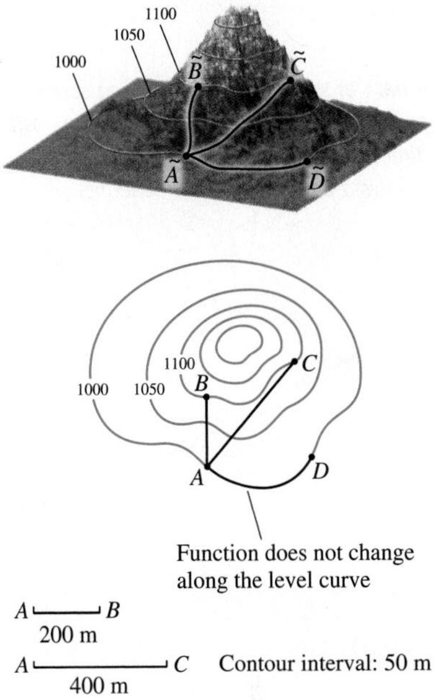

Contour interval: 50 m

FIGURE 17

CONCEPTUAL INSIGHT We will discuss the idea that rates of change depend on direction when we come to directional derivatives in Section 15.5. In single-variable calculus, we measure the rate of change by the derivative $f'(a)$. In the multivariable case, there is no single rate of change because the change in $f(x, y)$ depends on the direction: The rate is zero along a level curve (because $f(x, y)$ is constant along level curves), and the rate is nonzero in directions pointing from one level curve to the next (Figure 17).

■ **EXAMPLE 9 Average Rate of Change Depends on Direction** Compute the average rate of change from A to the points B, C, and D in Figure 17.

Solution The contour interval in Figure 17 is $m = 50$ m. Segments $\overline{AB}$ and $\overline{AC}$ both span two level curves, so the change in altitude is 100 m in both cases. The horizontal scale shows that AB corresponds to a horizontal change of 200 m, and $\overline{AC}$ corresponds to a horizontal change of 400 m. On the other hand, there is no change in altitude from A to D. Therefore:

$$\text{Average rate of change from } A \text{ to } B = \frac{\Delta \text{ altitude}}{\Delta \text{ horizontal}} = \frac{100}{200} = 0.5$$

$$\text{Average rate of change from } A \text{ to } C = \frac{\Delta \text{ altitude}}{\Delta \text{ horizontal}} = \frac{100}{400} = 0.25$$

$$\text{Average rate of change from } A \text{ to } D = \frac{\Delta \text{ altitude}}{\Delta \text{ horizontal}} = 0$$ ■

We see here explicitly that the average rate varies according to the direction.

When we walk up a mountain, the incline at each moment depends on the path we choose. If we walk "around" the mountain, our altitude does not change at all. On the other hand, at each point there is a *steepest* direction in which the altitude increases most rapidly.

A path of steepest descent *is the same as a path of steepest ascent but in the opposite direction. Water flowing down a mountain follows a path of steepest descent.*

On a contour map, the steepest direction is approximately the direction that takes us to the closest point on the next highest level curve [Figure 18(A)]. We say "approximately" because the terrain may vary between level curves. A **path of steepest ascent** is a path that begins at a point P and, everywhere along the way, points in the steepest direction. We can approximate the path of steepest ascent by drawing a sequence of segments that move as directly as possible from one level curve to the next. Figure 18(B) shows two paths from P to Q. The solid path is a path of steepest ascent, but the dashed path is not, because it does not move from one level curve to the next along the shortest possible segment.

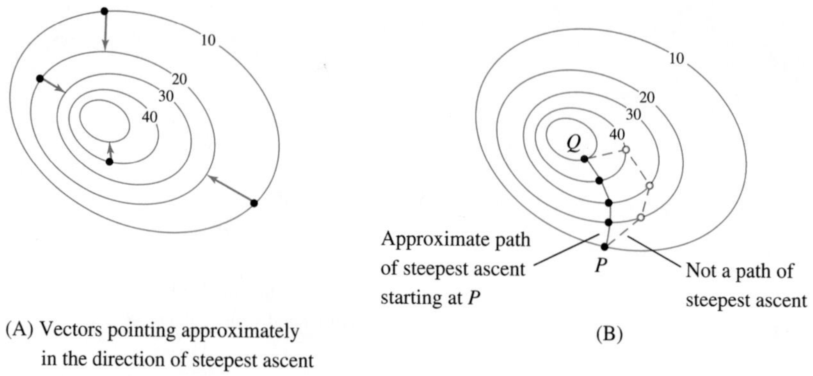

FIGURE 18

More Than Two Variables

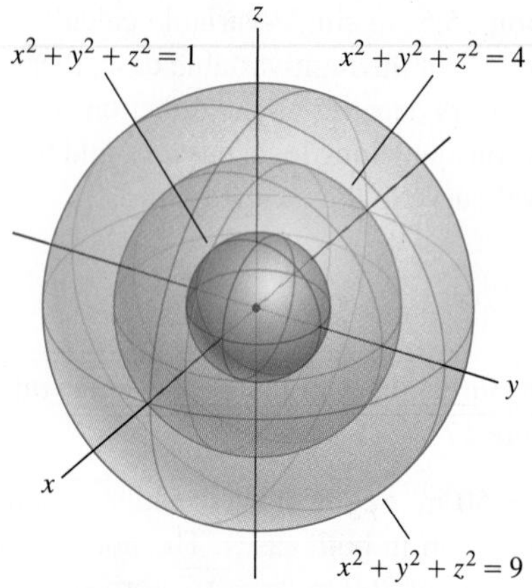

FIGURE 19 The level surfaces of $f(x, y, z) = x^2 + y^2 + z^2$ are spheres.

It is not possible to draw the graph of a function of more than two variables. The graph of a function $f(x, y, z)$ would consist of the set of points $(x, y, z, f(x, y, z))$ in four-dimensional space $\mathbf{R}^4$. However, it is possible to draw the **level surfaces** of a function of three variables $f(x, y, z)$. These are the surfaces with equation $f(x, y, z) = c$. For example, the level surfaces of

$$f(x, y, z) = x^2 + y^2 + z^2$$

are the spheres with equation $x^2 + y^2 + z^2 = c$ (Figure 19). For functions of four or more variables, we can no longer visualize the graph or the level surfaces. We must rely on intuition developed through the study of functions of two and three variables.

■ **EXAMPLE 10** Describe the level surfaces of $g(x, y, z) = x^2 + y^2 - z^2$.

Solution The level surface for $c = 0$ is the cone $x^2 + y^2 - z^2 = 0$. For $c \neq 0$, the level surfaces are the hyperboloids $x^2 + y^2 - z^2 = c$. The hyperboloid has one sheet if $c > 0$ and two sheets if $c < 0$ (Figure 20). ■

15.1 SUMMARY

- The domain $\mathcal{D}$ of a function $f(x_1, \dots, x_n)$ of n variables is the set of n-tuples $(a_1, \dots, a_n)$ in $\mathbf{R}^n$ for which $f(a_1, \dots, a_n)$ is defined. The range of f is the set of values taken by f.
- The graph of a continuous real-valued function $f(x, y)$ is the surface in $\mathbf{R}^3$ consisting of the points $(a, b, f(a, b))$ for (a, b) in the domain $\mathcal{D}$ of f.
- A *vertical trace* is a curve obtained by intersecting the graph with a vertical plane $x = a$ or $y = b$.

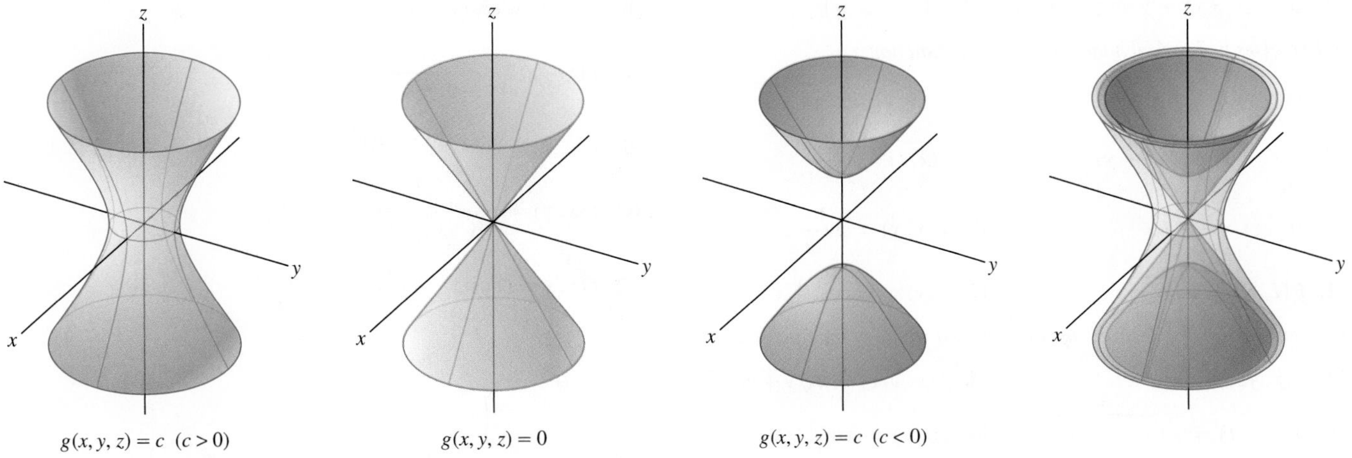

FIGURE 20 Level surfaces of $g(x, y, z) = x^2 + y^2 - z^2$.

- A *level curve* is a curve in the xy-plane defined by an equation $f(x, y) = c$. The level curve $f(x, y) = c$ is the projection onto the xy-plane of the horizontal trace curve, obtained by intersecting the graph with the horizontal plane $z = c$.
- A *contour map* shows the level curves $f(x, y) = c$ for equally spaced values of c. The spacing m is called the *contour interval*.
- When reading a contour map, keep in mind:
  - Your altitude does not change when you hike along a level curve.
  - Your altitude increases or decreases by m (the contour interval) when you hike from one level curve to the next.
- The spacing of the level curves indicates steepness: They are closer together where the graph is steeper.
- The *average rate of change* from P to Q is the ratio $\dfrac{\Delta\text{altitude}}{\Delta\text{horizontal}}$.
- A direction of steepest ascent at a point P is a direction along which $f(x, y)$ increases most rapidly. The steepest direction is obtained (approximately) by drawing the segment from P to the nearest point on the next level curve.

15.1 EXERCISES

Preliminary Questions

1. What is the difference between a horizontal trace and a level curve? How are they related?

2. Describe the trace of $f(x, y) = x^2 - \sin(x^3 y)$ in the xz-plane.

3. Is it possible for two different level curves of a function to intersect? Explain.

4. Describe the contour map of $f(x, y) = x$ with contour interval 1.

5. How will the contour maps of

$$f(x, y) = x \quad \text{and} \quad g(x, y) = 2x$$

with contour interval 1 look different?

Exercises

In Exercises 1–4, evaluate the function at the specified points.

1. $f(x, y) = x + yx^3$, $(2, 2)$, $(-1, 4)$

2. $g(x, y) = \dfrac{y}{x^2 + y^2}$, $(1, 3)$, $(3, -2)$

3. $h(x, y, z) = xyz^{-2}$, $(3, 8, 2)$, $(3, -2, -6)$

4. $Q(y, z) = y^2 + y \sin z, \quad (y, z) = \left(2, \frac{\pi}{2}\right), \left(-2, \frac{\pi}{6}\right)$

In Exercises 5–12, sketch the domain of the function.

5. $f(x, y) = 12x - 5y$

6. $f(x, y) = \sqrt{81 - x^2}$

7. $f(x, y) = \ln(4x^2 - y)$

8. $h(x, t) = \dfrac{1}{x + t}$

9. $g(y, z) = \dfrac{1}{z + y^2}$

10. $f(x, y) = \sin \dfrac{y}{x}$

11. $F(I, R) = \sqrt{IR}$

12. $f(x, y) = \cos^{-1}(x + y)$

In Exercises 13–16, describe the domain and range of the function.

13. $f(x, y, z) = xz + e^y$

14. $f(x, y, z) = x\sqrt{y + z}e^{z/x}$

15. $P(r, s, t) = \sqrt{16 - r^2 s^2 t^2}$

16. $g(r, s) = \cos^{-1}(rs)$

17. Match graphs (A) and (B) in Figure 21 with the functions

(i) $f(x, y) = -x + y^2$ (ii) $g(x, y) = x + y^2$

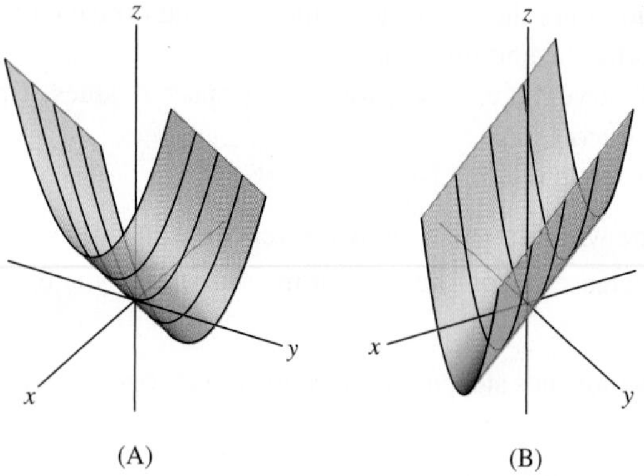

FIGURE 21

18. Match each of graphs (A) and (B) in Figure 22 with one of the following functions:

(i) $f(x, y) = (\cos x)(\cos y)$

(ii) $g(x, y) = \cos(x^2 + y^2)$

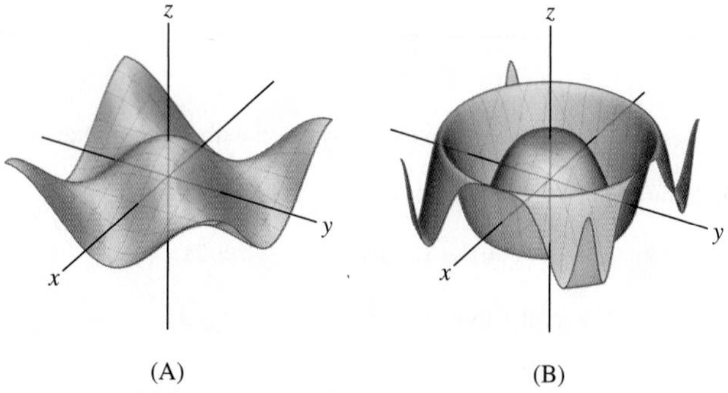

FIGURE 22

19. Match the functions (a)–(f) with their graphs (A)–(F) in Figure 23.

(a) $f(x, y) = |x| + |y|$

(b) $f(x, y) = \cos(x - y)$

(c) $f(x, y) = \dfrac{-1}{1 + 9x^2 + y^2}$

(d) $f(x, y) = \cos(y^2)e^{-0.1(x^2+y^2)}$

(e) $f(x, y) = \dfrac{-1}{1 + 9x^2 + 9y^2}$

(f) $f(x, y) = \cos(x^2 + y^2)e^{-0.1(x^2+y^2)}$

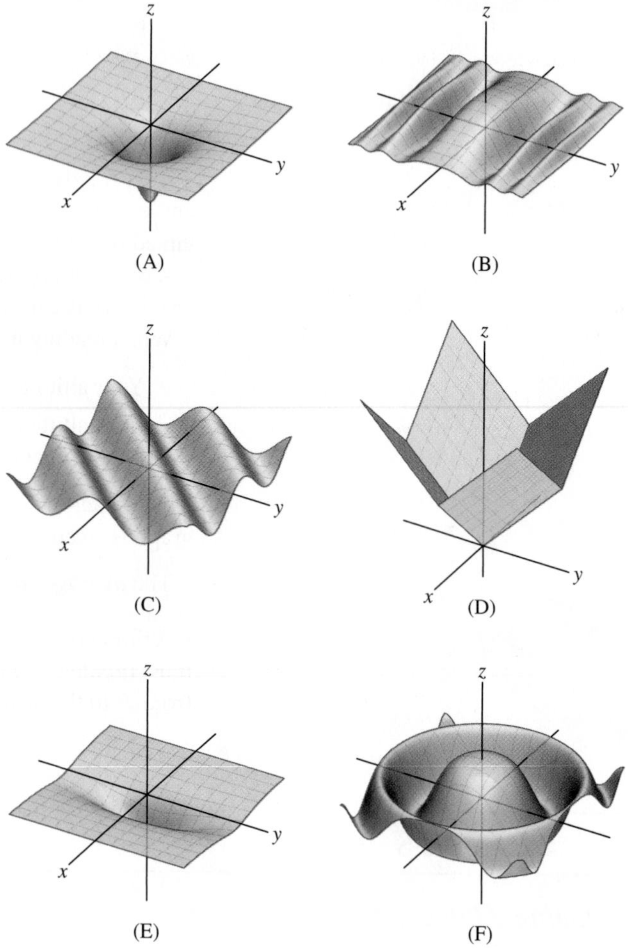

FIGURE 23

20. Match the functions (a)–(d) with their contour maps (A)–(D) in Figure 24.

(a) $f(x, y) = 3x + 4y$

(b) $g(x, y) = x^3 - y$

(c) $h(x, y) = 4x - 3y$

(d) $k(x, y) = x^2 - y$

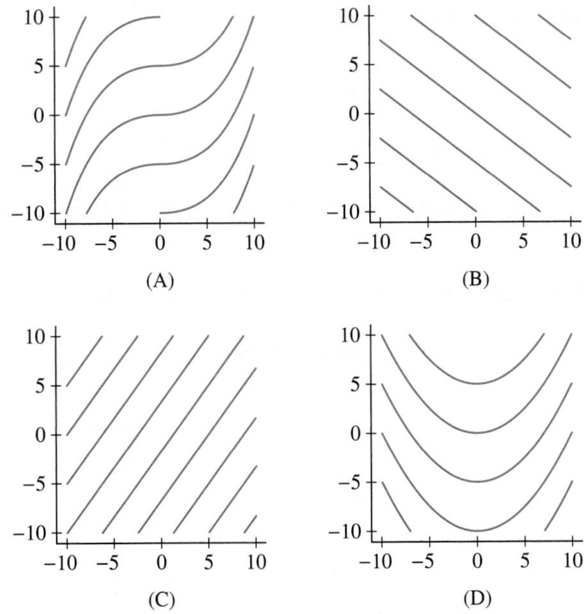

FIGURE 24

In Exercises 21–26, sketch the graph and describe the vertical and horizontal traces.

21. $f(x, y) = 12 - 3x - 4y$

22. $f(x, y) = \sqrt{4 - x^2 - y^2}$

23. $f(x, y) = x^2 + 4y^2$

24. $f(x, y) = y^2$

25. $f(x, y) = \sin(x - y)$

26. $f(x, y) = \dfrac{1}{x^2 + y^2 + 1}$

27. Sketch contour maps of $f(x, y) = x + y$ with contour intervals $m = 1$ and 2.

28. Sketch the contour map of $f(x, y) = x^2 + y^2$ with level curves $c = 0, 4, 8, 12, 16$.

In Exercises 29–36, draw a contour map of $f(x, y)$ with an appropriate contour interval, showing at least six level curves.

29. $f(x, y) = x^2 - y$

30. $f(x, y) = \dfrac{y}{x^2}$

31. $f(x, y) = \dfrac{y}{x}$

32. $f(x, y) = xy$

33. $f(x, y) = x^2 + 4y^2$

34. $f(x, y) = x + 2y - 1$

35. $f(x, y) = x^2$

36. $f(x, y) = 3x^2 - y^2$

37. Find the linear function whose contour map (with contour interval $m = 6$) is shown in Figure 25. What is the linear function if $m = 3$ (and the curve labeled $c = 6$ is relabeled $c = 3$)?

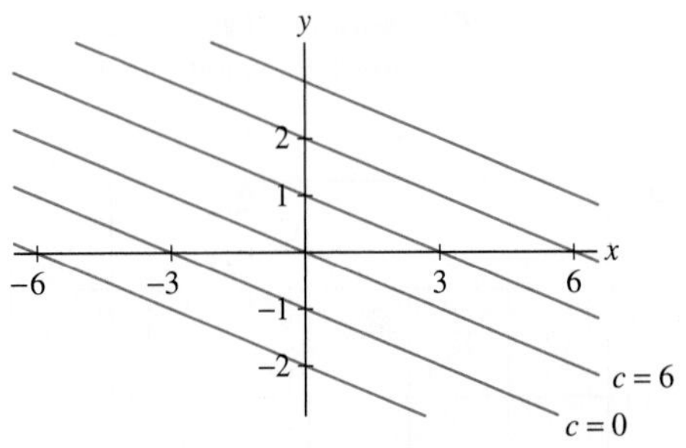

FIGURE 25 Contour map with contour interval $m = 6$

38. Use the contour map in Figure 26 to calculate the average rate of change:

(a) From A to B. **(b)** From A to C.

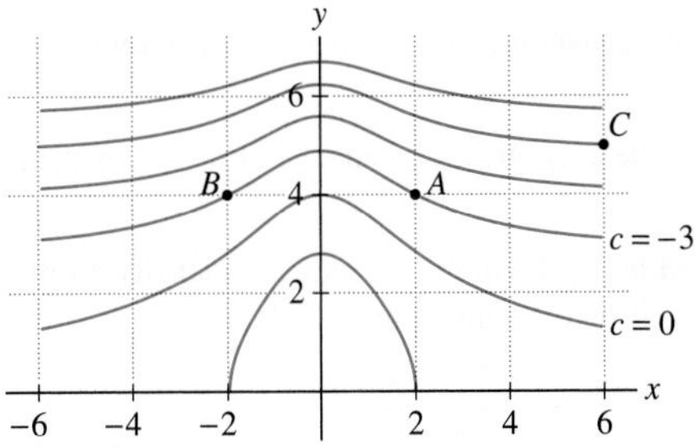

FIGURE 26

39. Referring to Figure 27, answer the following questions:

(a) At which of (A)–(C) is pressure increasing in the northern direction?

(b) At which of (A)–(C) is temperature increasing in the easterly direction?

(c) In which direction at (B) is temperature increasing most rapidly?

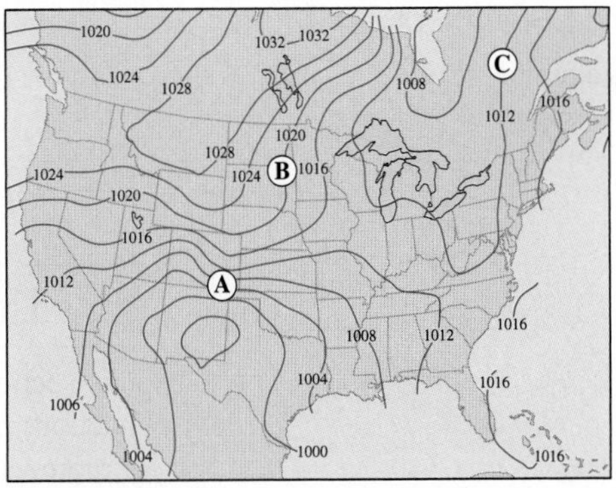

FIGURE 27 Atmospheric Pressure (in millibars) over the continental U.S. on March 26, 2009

In Exercises 40–43, $\rho(S, T)$ is seawater density (kg/m^3) *as a function of salinity S* (ppt) *and temperature T* (°C). *Refer to the contour map in Figure 28.*

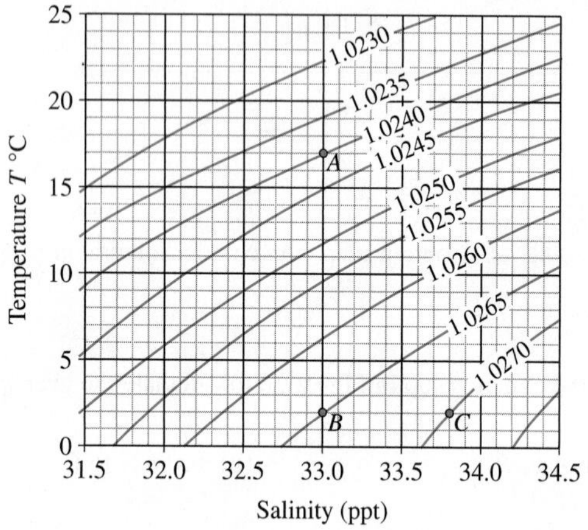

FIGURE 28 Contour map of seawater density $\rho(S, T)$ (kg/m^3).

40. Calculate the average rate of change of ρ with respect to T from B to A.

41. Calculate the average rate of change of ρ with respect to S from B to C.

42. At a fixed level of salinity, is seawater density an increasing or a decreasing function of temperature?

43. Does water density appear to be more sensitive to a change in temperature at point A or point B?

In Exercises 44–47, refer to Figure 29.

44. Find the change in elevation from A and B.

45. Estimate the average rate of change from A and B and from A to C.

46. Estimate the average rate of change from A to points i, ii, and iii.

47. Sketch the path of steepest ascent beginning at D.

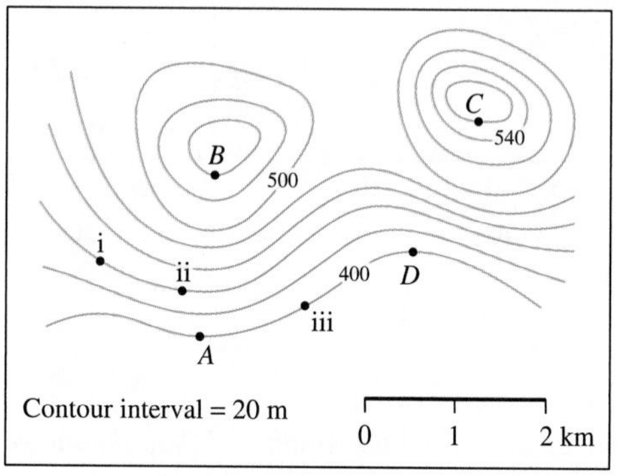

FIGURE 29

Further Insights and Challenges

48. The function $f(x, t) = t^{-1/2}e^{-x^2/t}$, whose graph is shown in Figure 30, models the temperature along a metal bar after an intense burst of heat is applied at its center point.

(a) Sketch the vertical traces at times $t = 1, 2, 3$. What do these traces tell us about the way heat diffuses through the bar?

(b) Sketch the vertical traces $x = c$ for $c = \pm 0.2, \pm 0.4$. Describe how temperature varies in time at points near the center.

49. Let

$$f(x, y) = \frac{x}{\sqrt{x^2 + y^2}} \quad \text{for } (x, y) \neq (0, 0)$$

Write f as a function $f(r, \theta)$ in polar coordinates, and use this to find the level curves of f.

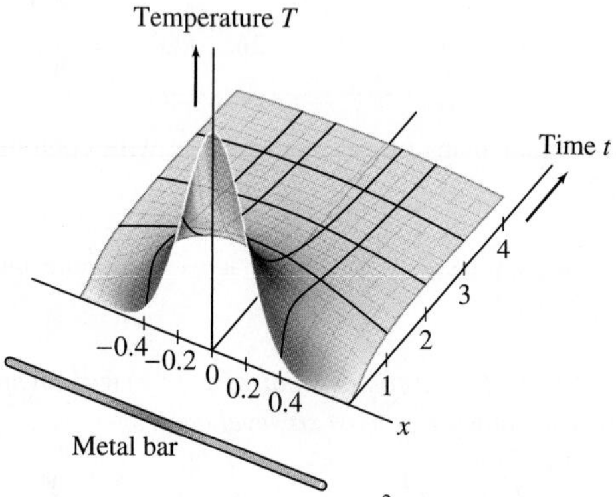

FIGURE 30 Graph of $f(x, t) = t^{-1/2}e^{-x^2/t}$ beginning shortly after $t = 0$.

15.2 Limits and Continuity in Several Variables

This section develops limits and continuity in the multivariable setting. We focus on functions of two variables, but similar definitions and results apply to functions of three or more variables.

Recall that a number x is close to a if the distance $|x - a|$ is small. In the plane, one point (x, y) is close to another point $P = (a, b)$ if the distance between them is small.

To express this precisely, we define the **open disk** of radius r and center $P = (a, b)$ (Figure 1):

$$D(P, r) = \{(x, y) \in \mathbf{R}^2 : (x - a)^2 + (y - b)^2 < r^2\}$$

The **open punctured disk** $D^*(P, r)$ is the disk $D(P, r)$ with its center point P removed. Thus $D^*(P, r)$ consists of all points whose distance to P is less than r, other than P itself.

Now assume that $f(x, y)$ is **defined near** P but not necessarily at P itself. In other words, $f(x, y)$ is defined for all (x, y) in some punctured disk $D^*(P, r)$ with $r > 0$. We say that $f(x, y)$ approaches the limit L as (x, y) approaches $P = (a, b)$ if $|f(x, y) - L|$ becomes arbitrarily small for (x, y) in a sufficiently small punctured disk centered at P [Figure 2(C)]. In this case, we write

$$\lim_{(x,y)\to P} f(x, y) = \lim_{(x,y)\to(a,b)} f(x, y) = L$$

Here is the formal definition.

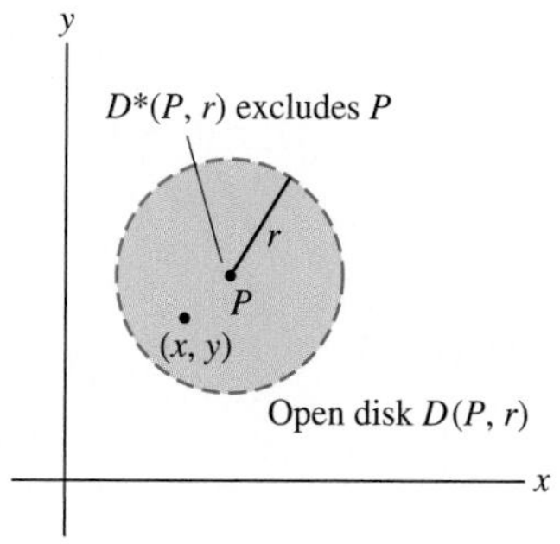

FIGURE 1 The open disk $D(P, r)$ consists of points (x, y) at distance $< r$ from P. It does not include the boundary circle.

DEFINITION Limit Assume that $f(x, y)$ is defined near $P = (a, b)$. Then

$$\lim_{(x,y)\to P} f(x, y) = L$$

if, for any $\epsilon > 0$, there exists $\delta > 0$ such that

$$|f(x, y) - L| < \epsilon \quad \text{for all} \quad (x, y) \in D^*(P, \delta)$$

This is similar to the limit definition in one variable, but there is an important difference. In a one-variable limit, we require that $f(x)$ tend to L as x approaches a from the left or right [Figure 2(A)]. In a multivariable limit, $f(x, y)$ must tend to L no matter how (x, y) approaches P [Figure 2(B)].

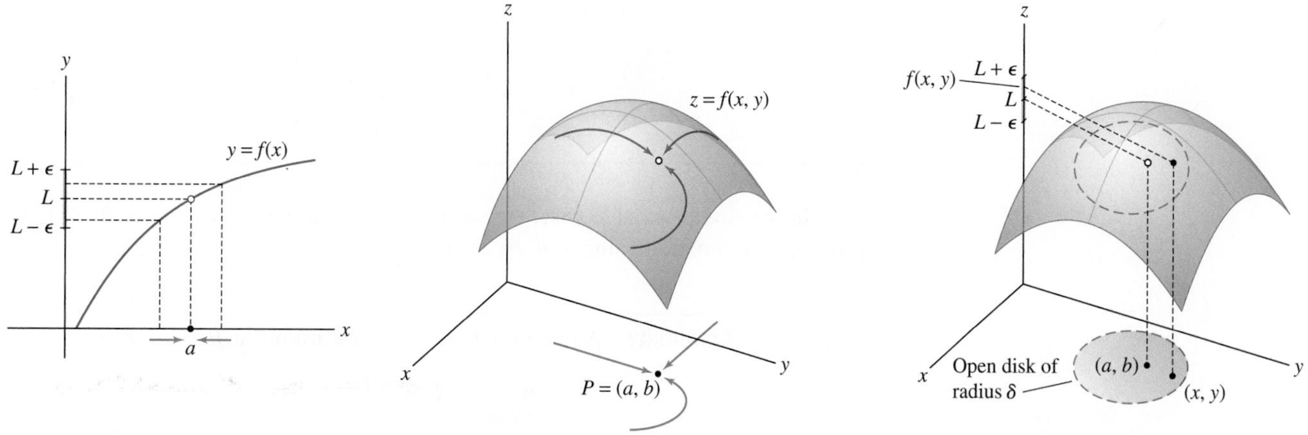

(A) In one variable, we can approach a from only two possible directions.

(B) In two variables, (x, y) can approach $P = (a, b)$ along any direction or path.

(C) $|f(x, y) - L| < \epsilon$ for all (x, y) inside the disk

FIGURE 2

EXAMPLE 1 Show that **(a)** $\lim_{(x,y)\to(a,b)} x = a$ and **(b)** $\lim_{(x,y)\to(a,b)} y = b$.

Solution Let $P = (a, b)$. To verify (a), let $f(x, y) = x$ and $L = a$. We must show that for any $\epsilon > 0$, we can find $\delta > 0$ such that

$$|f(x, y) - L| = |x - a| < \epsilon \quad \text{for all} \quad (x, y) \in D^*(P, \delta) \qquad \boxed{1}$$

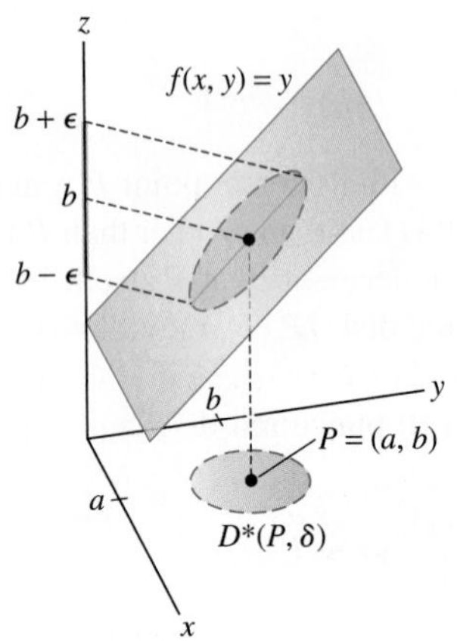

FIGURE 3 We have $|f(x, y) - b| < \epsilon$ if $|y - b| < \delta$ for $\delta = \epsilon$. Therefore,

$$\lim_{(x,y)\to(a,b)} y = b$$

In fact, we can choose $\delta = \epsilon$, for if $(x, y) \in D^*(P, \epsilon)$, then

$$(x-a)^2 + (y-b)^2 < \epsilon^2 \quad \Rightarrow \quad (x-a)^2 < \epsilon^2 \quad \Rightarrow \quad |x-a| < \epsilon$$

In other words, for any $\epsilon > 0$,

$$|x-a| < \epsilon \quad \text{for all} \quad (x, y) \in D^*(P, \epsilon)$$

This proves (a). The limit (b) is similar (see Figure 3). ■

The following theorem lists the basic laws for limits. We omit the proofs, which are similar to the proofs of the single-variable Limit Laws.

THEOREM 1 Limit Laws Assume that $\lim_{(x,y)\to P} f(x, y)$ and $\lim_{(x,y)\to P} g(x, y)$ exist. Then:

(i) Sum Law:

$$\lim_{(x,y)\to P} (f(x, y) + g(x, y)) = \lim_{(x,y)\to P} f(x, y) + \lim_{(x,y)\to P} g(x, y)$$

(ii) Constant Multiple Law: For any number k,

$$\lim_{(x,y)\to P} kf(x, y) = k \lim_{(x,y)\to P} f(x, y)$$

(iii) Product Law:

$$\lim_{(x,y)\to P} f(x, y)\, g(x, y) = \left(\lim_{(x,y)\to P} f(x, y)\right)\left(\lim_{(x,y)\to P} g(x, y)\right)$$

(iv) Quotient Law: If $\lim_{(x,y)\to P} g(x, y) \neq 0$, then

$$\lim_{(x,y)\to P} \frac{f(x, y)}{g(x, y)} = \frac{\lim_{(x,y)\to P} f(x, y)}{\lim_{(x,y)\to P} g(x, y)}$$

As in the single-variable case, we say that f is continuous at $P = (a, b)$ if $f(x, y)$ approaches the function value $f(a, b)$ as $(x, y) \to (a, b)$.

DEFINITION Continuity A function $f(x, y)$ is **continuous** at $P = (a, b)$ if

$$\lim_{(x,y)\to(a,b)} f(x, y) = f(a, b)$$

We say that f is continuous if it is continuous at each point (a, b) in its domain.

The Limit Laws tell us that all sums, multiples, and products of continuous functions are continuous. When we apply them to $f(x, y) = x$ and $g(x, y) = y$, which are continuous by Example 1, we find that the power functions $f(x, y) = x^m y^n$ are continuous for all whole numbers m, n and that all polynomials are continuous. Furthermore, a rational function $h(x, y)/g(x, y)$, where h and g are polynomials, is continuous at all points (a, b) where $g(a, b) \neq 0$. As in the single-variable case, we can evaluate limits of continuous functions using substitution.

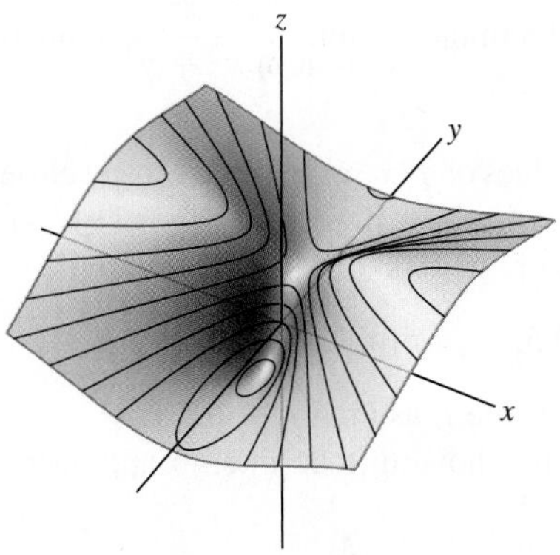

FIGURE 4 Top view of the graph $f(x, y) = \dfrac{3x + y}{x^2 + y^2 + 1}$.

■ **EXAMPLE 2 Evaluating Limits by Substitution** Show that

$$f(x, y) = \frac{3x + y}{x^2 + y^2 + 1}$$

is continuous (Figure 4). Then evaluate $\lim_{(x,y)\to(1,2)} f(x, y)$.

Solution The function $f(x, y)$ is continuous at all points (a, b) because it is a rational function whose denominator $Q(x, y) = x^2 + y^2 + 1$ is never zero. Therefore, we can evaluate the limit by substitution:

$$\lim_{(x,y)\to(1,2)} \frac{3x + y}{x^2 + y^2 + 1} = \frac{3(1) + 2}{1^2 + 2^2 + 1} = \frac{5}{6}$$ ■

If $f(x, y)$ is a product $f(x, y) = h(x)g(y)$, where $h(x)$ and $g(y)$ are continuous, then the limit is a product of limits by the Product Law:

$$\lim_{(x,y)\to(a,b)} f(x, y) = \lim_{(x,y)\to(a,b)} h(x)g(y) = \left(\lim_{x\to a} h(x)\right)\left(\lim_{y\to b} g(y)\right)$$

■ **EXAMPLE 3 Product Functions** Evaluate $\lim_{(x,y)\to(3,0)} x^3 \dfrac{\sin y}{y}$.

Solution The limit is equal to a product of limits:

$$\lim_{(x,y)\to(3,0)} x^3 \frac{\sin y}{y} = \left(\lim_{x\to 3} x^3\right)\left(\lim_{y\to 0} \frac{\sin y}{y}\right) = (3^3)(1) = 27$$ ■

Composition is another important way to build functions. If $f(x, y)$ is a function of two variables and $G(u)$ a function of one variable, then the composite $G \circ f$ is the function $G(f(x, y))$. According to the next theorem, a composite of continuous functions is again continuous.

THEOREM 2 A Composite of Continuous Functions Is Continuous If $f(x, y)$ is continuous at (a, b) and $G(u)$ is continuous at $c = f(a, b)$, then the composite function $G(f(x, y))$ is continuous at (a, b).

■ **EXAMPLE 4** Write $H(x, y) = e^{-x^2+2y}$ as a composite function and evaluate

$$\lim_{(x,y)\to(1,2)} H(x, y)$$

Solution We have $H(x, y) = G \circ f$, where $G(u) = e^u$ and $f(x, y) = -x^2 + 2y$. Both f and G are continuous, so H is also continuous and

$$\lim_{(x,y)\to(1,2)} H(x, y) = \lim_{(x,y)\to(1,2)} e^{-x^2+2y} = e^{-(1)^2+2(2)} = e^3$$ ■

We know that if a limit $\lim_{(x,y)\to(a,b)} f(x, y)$ exists and equals L, then $f(x, y)$ tends to L as (x, y) approaches (a, b) along any path. In the next example, we prove that a limit *does not exist* by showing that $f(x, y)$ approaches *different limits* along lines through the origin.

EXAMPLE 5 **Showing a Limit Does Not Exist** Examine $\displaystyle\lim_{(x,y)\to(0,0)} \frac{x^2}{x^2+y^2}$ numerically. Then prove that the limit does not exist.

Solution If the limit existed, we would expect the values of $f(x, y)$ in Table 1 to get closer to a limiting value L as (x, y) gets close to $(0, 0)$. But the table suggests that $f(x, y)$ takes on all values between 0 and 1, no matter how close (x, y) gets to $(0, 0)$. For example,

$$f(0.1, 0) = 1, \qquad f(0.1, 0.1) = 0.5, \qquad f(0, 0.1) = 0$$

Thus, $f(x, y)$ does not seem to approach any fixed value L as $(x, y) \to (0, 0)$.

Now let's prove that the limit does not exist by showing that $f(x, y)$ approaches different limits along the x- and y-axes (Figure 5):

$$\text{Limit along } x\text{-axis:} \qquad \lim_{x\to 0} f(x, 0) = \lim_{x\to 0} \frac{x^2}{x^2+0^2} = \lim_{x\to 0} 1 = 1$$

$$\text{Limit along } y\text{-axis:} \qquad \lim_{y\to 0} f(0, y) = \lim_{y\to 0} \frac{0^2}{0^2+y^2} = \lim_{y\to 0} 0 = 0$$

These two limits are different and hence $\displaystyle\lim_{(x,y)\to(0,0)} f(x, y)$ does not exist. ■

TABLE 1 **Values of** $f(x, y) = \dfrac{x^2}{x^2+y^2}$

y \ x	−0.5	−0.4	−0.3	−0.2	−0.1	0	0.1	0.2	0.3	0.4	0.5
0.5	**0.5**	0.39	0.265	0.138	0.038	**0**	0.038	0.138	0.265	0.39	**0.5**
0.4	0.61	**0.5**	0.36	0.2	0.059	**0**	0.059	0.2	0.36	**0.5**	0.61
0.3	0.735	0.64	**0.5**	0.308	0.1	**0**	0.1	0.308	**0.5**	0.64	0.735
0.2	0.862	0.8	0.692	**0.5**	0.2	**0**	0.2	**0.5**	0.692	0.8	0.862
0.1	0.962	0.941	0.9	0.8	**0.5**	**0**	**0.5**	0.8	0.9	0.941	0.962
0	**1**	**1**	**1**	**1**	**1**		**1**	**1**	**1**	**1**	**1**
−0.1	0.962	0.941	0.9	0.8	**0.5**	**0**	**0.5**	0.8	0.9	0.941	0.962
−0.2	0.862	0.8	0.692	**0.5**	0.2	**0**	0.2	**0.5**	0.692	0.8	0.862
−0.3	0.735	0.640	**0.5**	0.308	0.1	**0**	0.1	0.308	**0.5**	0.640	0.735
−0.4	0.610	**0.5**	0.360	0.2	0.059	**0**	0.059	0.2	0.36	**0.5**	0.61
−0.5	**0.5**	0.39	0.265	0.138	0.038	**0**	0.038	0.138	0.265	0.390	**0.5**

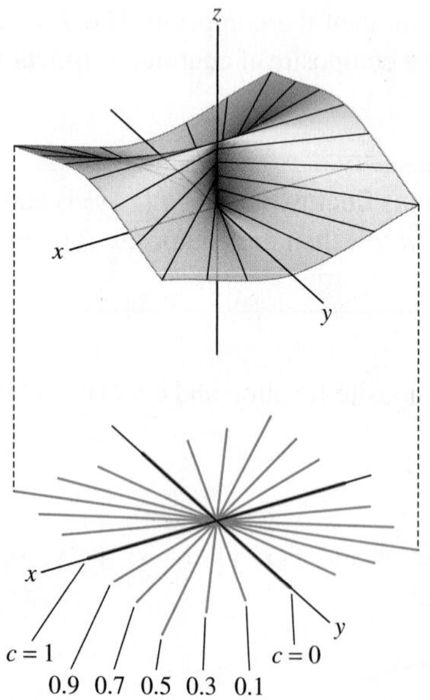

FIGURE 5 Graph of $f(x, y) = \dfrac{x^2}{x^2+y^2}$.

GRAPHICAL INSIGHT The contour map in Figure 5 shows clearly that the function $f(x, y) = x^2/(x^2 + y^2)$ does not approach a limit as (x, y) approaches $(0, 0)$. For nonzero c, the level curve $f(x, y) = c$ is the line $y = mx$ through the origin (with the origin deleted) where $c = (m^2 + 1)^{-1}$:

$$f(x, mx) = \frac{x^2}{x^2 + (mx)^2} = \frac{1}{m^2+1} \qquad (\text{for } x \neq 0)$$

The level curve $f(x, y) = 0$ is the y-axis (with the origin deleted). As the slope m varies, f takes on all values between 0 and 1 in every disk around the origin $(0, 0)$, no matter how small, so f cannot approach a limit.

As we know, there is no single method for computing limits that always works. The next example illustrates two different approaches to evaluating a limit in a case where substitution cannot be used.

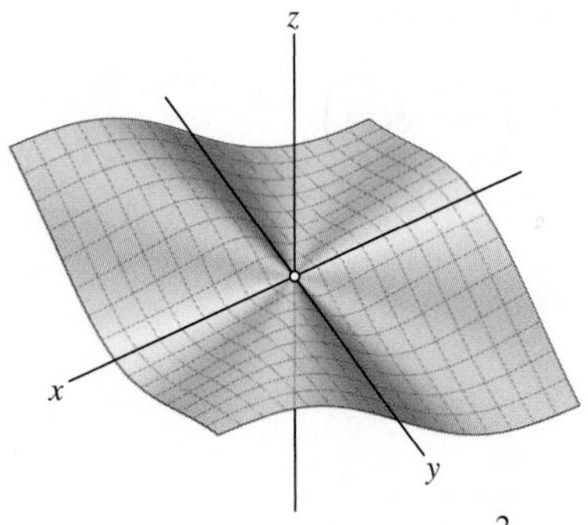

FIGURE 6 Graph of $f(x, y) = \dfrac{xy^2}{x^2 + y^2}$.

EXAMPLE 6 Two Methods for Verifying a Limit Calculate $\lim_{(x,y)\to(0,0)} f(x, y)$ where $f(x, y)$ is defined for $(x, y) \neq (0, 0)$ by (Figure 6)

$$f(x, y) = \frac{xy^2}{x^2 + y^2}$$

Solution

First Method For $(x, y) \neq (0, 0)$, we have

$$0 \le \left| \frac{y^2}{x^2 + y^2} \right| \le 1$$

because the numerator is not greater than the denominator. Multiply by $|x|$:

$$0 \le \left| \frac{xy^2}{x^2 + y^2} \right| \le |x|$$

and use the Squeeze Theorem (which is valid for limits in several variables):

$$0 \le \lim_{(x,y)\to(0,0)} \left| \frac{xy^2}{x^2 + y^2} \right| \le \lim_{(x,y)\to(0,0)} |x|$$

Because $\lim_{(x,y)\to(0,0)} |x| = 0$, we conclude that $\lim_{(x,y)\to(0,0)} f(x, y) = 0$ as desired.

Second Method Use polar coordinates:

$$x = r\cos\theta, \qquad y = r\sin\theta$$

Then $x^2 + y^2 = r^2$ and for $r \neq 0$,

$$0 \le \left| \frac{xy^2}{x^2 + y^2} \right| = \left| \frac{(r\cos\theta)(r\sin\theta)^2}{r^2} \right| = r|\cos\theta \sin^2\theta| \le r$$

As (x, y) approaches $(0, 0)$, the variable r also approaches 0, so again, the desired conclusion follows from the Squeeze Theorem:

$$0 \le \lim_{(x,y)\to(0,0)} \left| \frac{xy^2}{x^2 + y^2} \right| \le \lim_{r\to 0} r = 0$$ ■

15.2 SUMMARY

- The *open disk* of radius r centered at $P = (a, b)$ is defined by

$$D(P, r) = \{(x, y) \in \mathbf{R}^2 : (x - a)^2 + (y - b)^2 < r^2\}$$

The *punctured disk* $D^*(P, r)$ is $D(P, r)$ with P removed.

- Suppose that $f(x, y)$ is defined near $P = (a, b)$. Then

$$\lim_{(x,y)\to(a,b)} f(x, y) = L$$

if, for any $\epsilon > 0$, there exists $\delta > 0$ such that

$$|f(x, y) - L| < \epsilon \quad \text{for all} \quad (x, y) \in D^*(P, \delta)$$

- The limit of a product $f(x, y) = h(x)g(y)$ is a product of limits:

$$\lim_{(x,y)\to(a,b)} f(x, y) = \left(\lim_{x\to a} h(x)\right)\left(\lim_{y\to b} g(y)\right)$$

- A function $f(x, y)$ is *continuous* at $P = (a, b)$ if

$$\lim_{(x,y)\to(a,b)} f(x, y) = f(a, b)$$

15.2 EXERCISES

Preliminary Questions

1. What is the difference between $D(P, r)$ and $D^*(P, r)$?

2. Suppose that $f(x, y)$ is continuous at $(2, 3)$ and that $f(2, y) = y^3$ for $y \neq 3$. What is the value $f(2, 3)$?

3. Suppose that $Q(x, y)$ is a function such that $1/Q(x, y)$ is continuous for all (x, y). Which of the following statements are true?

(a) $Q(x, y)$ is continuous for all (x, y).

(b) $Q(x, y)$ is continuous for $(x, y) \neq (0, 0)$.

(c) $Q(x, y) \neq 0$ for all (x, y).

4. Suppose that $f(x, 0) = 3$ for all $x \neq 0$ and $f(0, y) = 5$ for all $y \neq 0$. What can you conclude about $\lim_{(x,y)\to(0,0)} f(x, y)$?

Exercises

In Exercises 1–8, evaluate the limit using continuity

1. $\lim_{(x,y)\to(1,2)} (x^2 + y)$

2. $\lim_{(x,y)\to(\frac{4}{9},\frac{2}{9})} \frac{x}{y}$

3. $\lim_{(x,y)\to(2,-1)} (xy - 3x^2y^3)$

4. $\lim_{(x,y)\to(-2,1)} \frac{2x^2}{4x + y}$

5. $\lim_{(x,y)\to(\frac{\pi}{4},0)} \tan x \cos y$

6. $\lim_{(x,y)\to(2,3)} \tan^{-1}(x^2 - y)$

7. $\lim_{(x,y)\to(1,1)} \frac{e^{x^2} - e^{-y^2}}{x + y}$

8. $\lim_{(x,y)\to(1,0)} \ln(x - y)$

In Exercises 9–12, assume that

$$\lim_{(x,y)\to(2,5)} f(x, y) = 3, \qquad \lim_{(x,y)\to(2,5)} g(x, y) = 7$$

9. $\lim_{(x,y)\to(2,5)} \big(g(x, y) - 2f(x, y)\big)$

10. $\lim_{(x,y)\to(2,5)} f(x, y)^2 g(x, y)$

11. $\lim_{(x,y)\to(2,5)} e^{f(x,y)^2 - g(x,y)}$

12. $\lim_{(x,y)\to(2,5)} \frac{f(x, y)}{f(x, y) + g(x, y)}$

13. Does $\lim_{(x,y)\to(0,0)} \frac{y^2}{x^2 + y^2}$ exist? Explain.

14. Let $f(x, y) = xy/(x^2 + y^2)$. Show that $f(x, y)$ approaches zero along the x- and y-axes. Then prove that $\lim_{(x,y)\to(0,0)} f(x, y)$ does not exist by showing that the limit along the line $y = x$ is nonzero.

15. Prove that

$$\lim_{(x,y)\to(0,0)} \frac{x}{x^2 + y^2}$$

does not exist by considering the limit along the x-axis.

16. Let $f(x, y) = x^3/(x^2 + y^2)$ and $g(x, y) = x^2/(x^2 + y^2)$. Using polar coordinates, prove that

$$\lim_{(x,y)\to(0,0)} f(x, y) = 0$$

and that $\lim_{(x,y)\to(0,0)} g(x, y)$ does not exist. *Hint:* Show that $g(x, y) = \cos^2\theta$ and observe that $\cos\theta$ can take on any value between -1 and 1 as $(x, y) \to (0, 0)$.

17. Use the Squeeze Theorem to evaluate

$$\lim_{(x,y)\to(4,0)} (x^2 - 16) \cos\left(\frac{1}{(x - 4)^2 + y^2}\right)$$

18. Evaluate $\lim_{(x,y)\to(0,0)} \tan x \sin\left(\frac{1}{|x| + |y|}\right)$.

In Exercises 19–32, evaluate the limit or determine that it does not exist.

19. $\lim_{(z,w)\to(-2,1)} \frac{z^4 \cos(\pi w)}{e^{z+w}}$

20. $\lim_{(z,w)\to(-1,2)} (z^2 w - 9z)$

21. $\lim_{(x,y)\to(4,2)} \frac{y - 2}{\sqrt{x^2 - 4}}$

22. $\lim_{(x,y)\to(0,0)} \frac{x^2 + y^2}{1 + y^2}$

23. $\lim_{(x,y)\to(3,4)} \frac{1}{\sqrt{x^2 + y^2}}$

24. $\lim_{(x,y)\to(0,0)} \frac{xy}{\sqrt{x^2 + y^2}}$

25. $\lim_{(x,y)\to(1,-3)} e^{x-y}\ln(x-y)$

26. $\lim_{(x,y)\to(0,0)} \frac{|x|}{|x|+|y|}$

27. $\lim_{(x,y)\to(-3,-2)} (x^2y^3+4xy)$

28. $\lim_{(x,y)\to(2,1)} e^{x^2-y^2}$

29. $\lim_{(x,y)\to(0,0)} \tan(x^2+y^2)\tan^{-1}\left(\frac{1}{x^2+y^2}\right)$

30. $\lim_{(x,y)\to(0,0)} (x+y+2)e^{-1/(x^2+y^2)}$

31. $\lim_{(x,y)\to(0,0)} \frac{x^2+y^2}{\sqrt{x^2+y^2+1}-1}$

32. $\lim_{(x,y)\to(1,1)} \frac{x^2+y^2-2}{|x-1|+|y-1|}$

Hint: Rewrite the limit in terms of $u = x-1$ and $v = y-1$.

33. Let $f(x,y) = \frac{x^3+y^3}{x^2+y^2}$.

(a) Show that

$$|x^3| \le |x|(x^2+y^2), \quad |y^3| \le |y|(x^2+y^2)$$

(b) Show that $|f(x,y)| \le |x|+|y|$.

(c) Use the Squeeze Theorem to prove that $\lim_{(x,y)\to(0,0)} f(x,y) = 0$.

34. Let $a, b \ge 0$. Show that $\lim_{(x,y)\to(0,0)} \frac{x^ay^b}{x^2+y^2} = 0$ if $a+b>2$ and that the limit does not exist if $a+b \le 2$.

35. Figure 7 shows the contour maps of two functions. Explain why the limit $\lim_{(x,y)\to P} f(x,y)$ does not exist. Does $\lim_{(x,y)\to Q} g(x,y)$ appear to exist in (B)? If so, what is its limit?

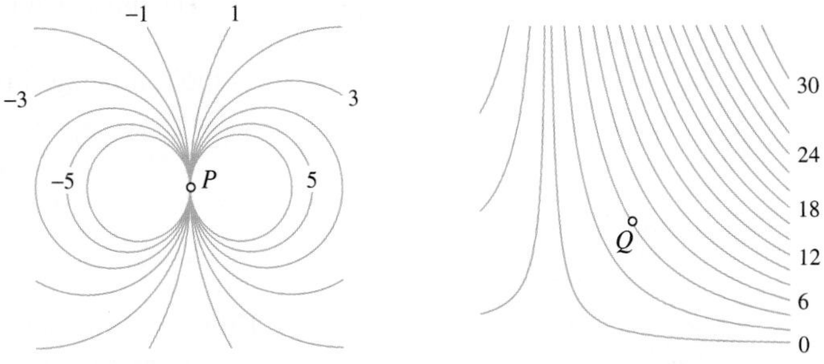

(A) Contour map of $f(x, y)$ (B) Contour map of $g(x, y)$

FIGURE 7

Further Insights and Challenges

36. Evaluate $\lim_{(x,y)\to(0,2)} (1+x)^{y/x}$.

37. Is the following function continuous?

$$f(x,y) = \begin{cases} x^2+y^2 & \text{if } x^2+y^2<1 \\ 1 & \text{if } x^2+y^2 \ge 1 \end{cases}$$

38. CAS The function $f(x,y) = \sin(xy)/xy$ is defined for $xy \ne 0$.

(a) Is it possible to extend the domain of $f(x,y)$ to all of $\mathbf{R}^2$ so that the result is a continuous function?

(b) Use a computer algebra system to plot $f(x,y)$. Does the result support your conclusion in (a)?

39. Prove that the function

$$f(x,y) = \begin{cases} \dfrac{(2^x-1)(\sin y)}{xy} & \text{if } xy \ne 0 \\ \ln 2 & \text{if } xy = 0 \end{cases}$$

is continuous at $(0,0)$.

40. Prove that if $f(x)$ is continuous at $x=a$ and $g(y)$ is continuous at $y=b$, then $F(x,y) = f(x)g(y)$ is continuous at (a,b).

41. The function $f(x,y) = x^2y/(x^4+y^2)$ provides an interesting example where the limit as $(x,y)\to(0,0)$ does not exist, even though the limit along every line $y=mx$ exists and is zero (Figure 8).

(a) Show that the limit along any line $y=mx$ exists and is equal to 0.

(b) Calculate $f(x,y)$ at the points $(10^{-1}, 10^{-2})$, $(10^{-5}, 10^{-10})$, $(10^{-20}, 10^{-40})$. Do not use a calculator.

(c) Show that $\lim_{(x,y)\to(0,0)} f(x,y)$ does not exist. *Hint:* Compute the limit along the parabola $y=x^2$.

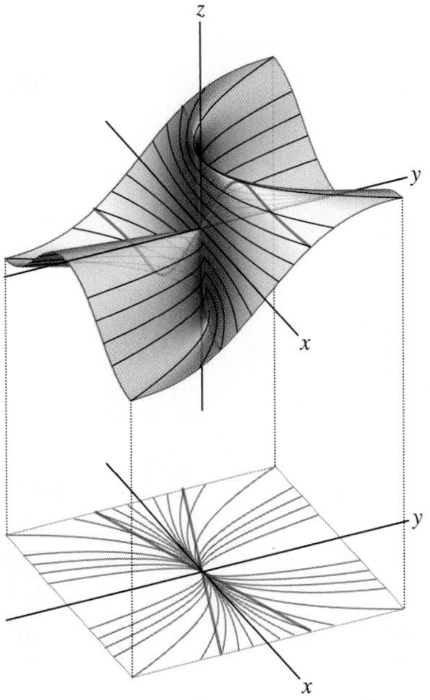

FIGURE 8 Graph of $f(x,y) = \frac{x^2y}{x^4+y^2}$.

15.3 Partial Derivatives

We have stressed that a function f of two or more variables does not have a unique rate of change because each variable may affect f in different ways. For example, the current I in a circuit is a function of both voltage V and resistance R given by Ohm's Law:

$$I(V, R) = \frac{V}{R}$$

The current I is *increasing* as a function of V but *decreasing* as a function of R.

The **partial derivatives** are the rates of change with respect to each variable separately. A function $f(x, y)$ of two variables has two partial derivatives, denoted f_x and f_y, defined by the following limits (if they exist):

$$f_x(a, b) = \lim_{h \to 0} \frac{f(a+h, b) - f(a, b)}{h}, \qquad f_y(a, b) = \lim_{k \to 0} \frac{f(a, b+k) - f(a, b)}{k}$$

Thus, f_x is the derivative of $f(x, b)$ as a function of x alone, and f_y is the derivative at $f(a, y)$ as a function of y alone. The Leibniz notation for partial derivatives is

The partial derivative symbol ∂ is a rounded "d." The symbols $\partial f/\partial x$ and $\partial f/\partial y$ are read as follows: "dee-eff dee-ex" and "dee-eff dee-why."

$$\frac{\partial f}{\partial x} = f_x, \qquad \frac{\partial f}{\partial y} = f_y$$

$$\left.\frac{\partial f}{\partial x}\right|_{(a,b)} = f_x(a, b), \qquad \left.\frac{\partial f}{\partial y}\right|_{(a,b)} = f_y(a, b)$$

If $z = f(x, y)$, then we also write $\partial z/\partial x$ and $\partial z/\partial y$.

Partial derivatives are computed just like ordinary derivatives in one variable with this difference: To compute f_x, treat y as a constant, and to compute f_y, treat x as a constant.

■ **EXAMPLE 1** Compute the partial derivatives of $f(x, y) = x^2y^5$.

Solution

$$\frac{\partial f}{\partial x} = \underbrace{\frac{\partial}{\partial x}(x^2y^5) = y^5\frac{\partial}{\partial x}(x^2)}_{\text{Treat } y^5 \text{ as a constant}} = y^5(2x) = 2xy^5$$

$$\frac{\partial f}{\partial y} = \underbrace{\frac{\partial}{\partial y}(x^2y^5) = x^2\frac{\partial}{\partial x}(y^5)}_{\text{Treat } x^2 \text{ as a constant}} = x^2(5y^4) = 5x^2y^4 \qquad ■$$

GRAPHICAL INSIGHT The partial derivatives at $P = (a, b)$ are the slopes of the tangent lines to the vertical trace curves through the point $(a, b, f(a, b))$ in Figure 1(A). To compute $f_x(a, b)$, we set $y = b$ and differentiate in the x-direction. This gives us the slope of the tangent line to the trace curve in the plane $y = b$ [Figure 1(B)]. Similarly, $f_y(a, b)$ is the slope of the trace curve in the vertical plane $x = a$ [Figure 1(C)].

The differentiation rules from calculus of one variable (the Product, Quotient, and Chain Rules) are valid for partial derivatives.

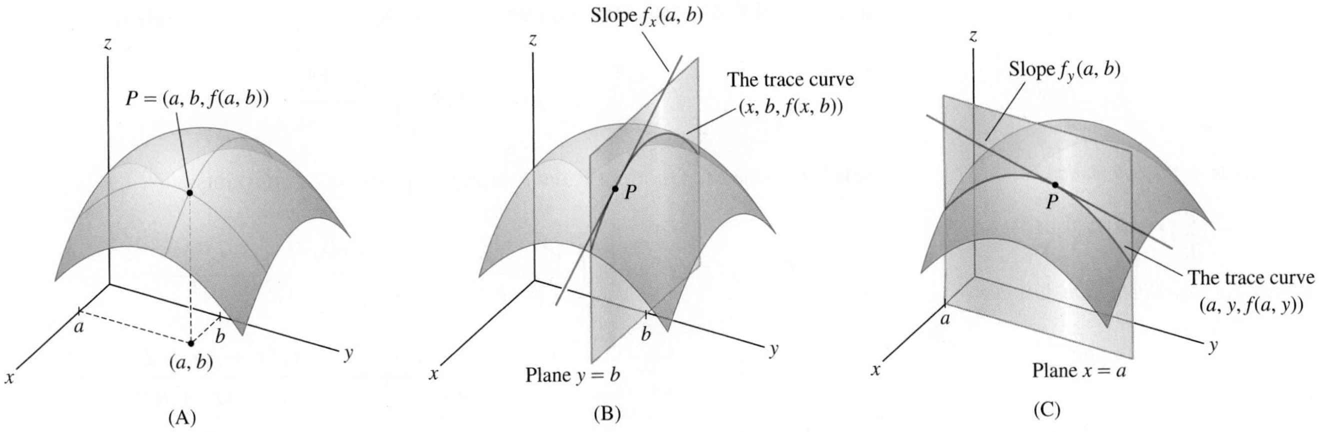

FIGURE 1 The partial derivatives are the slopes of the vertical trace curves.

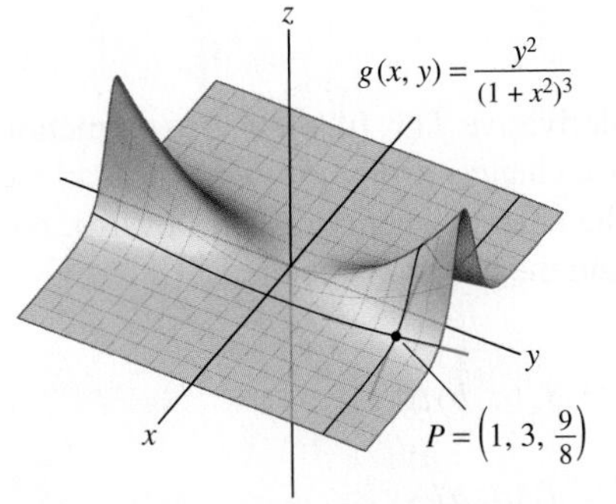

FIGURE 2 The slopes of the tangent lines to the trace curves are $g_x(1, 3)$ and $g_y(1, 3)$.

CAUTION It is not necessary to use the Quotient Rule to compute the partial derivative in Eq. (1). The denominator does not depend on y, so we treat it as a constant when differentiating with respect to y.

■ **EXAMPLE 2** Calculate $g_x(1, 3)$ and $g_y(1, 3)$, where $g(x, y) = \dfrac{y^2}{(1+x^2)^3}$.

Solution To calculate g_x, treat y (and therefore y^2) as a constant:

$$g_x(x, y) = \frac{\partial}{\partial x}\frac{y^2}{(1+x^2)^3} = y^2\frac{\partial}{\partial x}(1+x^2)^{-3} = \frac{-6xy^2}{(1+x^2)^4}$$

$$g_x(1, 3) = \frac{-6(1)3^2}{(1+1^2)^4} = -\frac{27}{8}$$

To calculate g_y, treat x (and therefore $1 + x^2$) as a constant:

$$g_y(x, y) = \frac{\partial}{\partial y}\frac{y^2}{(1+x^2)^3} = \frac{1}{(1+x^2)^3}\frac{\partial}{\partial y}y^2 = \frac{2y}{(1+x^2)^3} \qquad \boxed{1}$$

$$g_y(1, 3) = \frac{2(3)}{(1+1^2)^3} = \frac{3}{4}$$

These partial derivatives are the slopes of the trace curves through the point $\left(1, 3, \frac{9}{8}\right)$ shown in Figure 2. ■

We use the Chain Rule to compute partial derivatives of a composite function $f(x, y) = F(g(x, y))$, where $F(u)$ is a function of one variable and $u = g(x, y)$:

$$\frac{\partial f}{\partial x} = \frac{dF}{du}\frac{\partial u}{\partial x}, \qquad \frac{\partial f}{\partial y} = \frac{dF}{du}\frac{\partial u}{\partial y}$$

■ **EXAMPLE 3 Chain Rule for Partial Derivatives** Compute $\dfrac{\partial}{\partial x}\sin(x^2y^5)$.

Solution Write $\sin(x^2y^5) = F(u)$, where $F(u) = \sin u$ and $u = x^2y^5$. Then we have $\dfrac{dF}{du} = \cos u$ and the Chain Rule give us

$$\underbrace{\frac{\partial}{\partial x}\sin(x^2y^5) = \frac{dF}{du}\frac{\partial u}{\partial x} = \cos(x^2y^5)\frac{\partial}{\partial x}x^2y^5}_{\text{Chain Rule}} = 2xy^5\cos(x^2y^5)$$ ■

Partial derivatives are defined for functions of any number of variables. We compute the partial derivative with respect to any one of the variables by holding the remaining variables constant.

EXAMPLE 4 More Than Two Variables Calculate $f_z(0, 0, 1, 1)$, where

$$f(x, y, z, w) = \frac{e^{xz+y}}{z^2 + w}$$

In Example 4, the calculation

$$\frac{\partial}{\partial z} e^{xz+y} = xe^{xz+y}$$

follows from the Chain Rule, just like

$$\frac{d}{dz} e^{az+b} = ae^{az+b}$$

Solution Use the Quotient Rule, treating x, y, and w as constants:

$$f_z(x, y, z, w) = \frac{\partial}{\partial z}\left(\frac{e^{xz+y}}{z^2 + w}\right) = \frac{(z^2 + w)\frac{\partial}{\partial z} e^{xz+y} - e^{xz+y}\frac{\partial}{\partial z}(z^2 + w)}{(z^2 + w)^2}$$

$$= \frac{(z^2 + w)xe^{xz+y} - 2ze^{xz+y}}{(z^2 + w)^2} = \frac{(z^2x + wx - 2z)e^{xz+y}}{(z^2 + w)^2}$$

$$f_z(0, 0, 1, 1) = \frac{-2e^0}{(1^2 + 1)^2} = -\frac{1}{2}$$

Because the partial derivative $f_x(a, b)$ is the derivative $f(x, b)$, viewed as a function of x alone, we can estimate the change Δf when x changes from a to $a + \Delta x$ as in the single-variable case. Similarly, we can estimate the change when y changes by Δy. For small Δx and Δy (just how small depends on f and the accuracy required):

$$f(a + \Delta x, b) - f(a, b) \approx f_x(a, b)\Delta x$$

$$f(a, b + \Delta y) - f(a, b) \approx f_y(a, b)\Delta y$$

This applies to functions f in any number of variables. For example, $\Delta f \approx f_w \Delta w$ if one of the variables w changes by Δw and all other variables remain fixed.

EXAMPLE 5 Testing Microchips A **ball grid array** (BGA) is a microchip joined to a circuit board by small solder balls of radius R mm separated by a distance L mm (Figure 3). Manufacturers test the reliability of BGAs by subjecting them to repeated cycles in which the temperature is varied from 0°C to 100°C over a 40-min period. According to one model, the average number N of cycles before the chip fails is

$$N = \left(\frac{2200R}{Ld}\right)^{1.9}$$

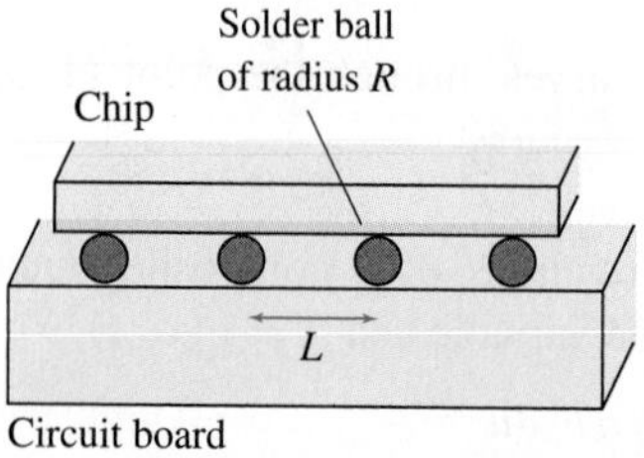

FIGURE 3 A BGA package. Temperature variations strain the BGA and may cause it to fail because the chip and board expand at different rates.

where d is the difference between the coefficients of expansion of the chip and the board. Estimate the change ΔN when $R = 0.12$, $d = 10$, and L is increased from 0.4 to 0.42.

Solution We use the approximation

$$\Delta N \approx \frac{\partial N}{\partial L} \Delta L$$

with $\Delta L = 0.42 - 0.4 = 0.02$. Since R and d are constant, the partial derivative is

$$\frac{\partial N}{\partial L} = \frac{\partial}{\partial L}\left(\frac{2200R}{Ld}\right)^{1.9} = \left(\frac{2200R}{d}\right)^{1.9} \frac{\partial}{\partial L} L^{-1.9} = -1.9\left(\frac{2200R}{d}\right)^{1.9} L^{-2.9}$$

Now evaluate at $L = 0.4$, $R = 0.12$, and $d = 10$:

$$\left.\frac{\partial N}{\partial L}\right|_{(L,R,d)=(0.4,0.12,10)} = -1.9\left(\frac{2200(0.12)}{10}\right)^{1.9}(0.4)^{-2.9} \approx -13{,}609$$

The decrease in the average number of cycles before a chip fails is

$$\Delta N \approx \frac{\partial N}{\partial L}\Delta L = -13{,}609(0.02) \approx -272 \text{ cycles}$$ ■

In the next example, we estimate a partial derivative numerically. Since f_x and f_y are limits of difference quotients, we have the following approximations when h and k are "small":

$$f_x(a,b) \approx \frac{\Delta f}{\Delta x} = \frac{f(a+h,b) - f(a,b)}{h}$$

$$f_y(a,b) \approx \frac{\Delta f}{\Delta y} = \frac{f(a,b+k) - f(a,b)}{k}$$

A similar approximation is valid in any number of variables.

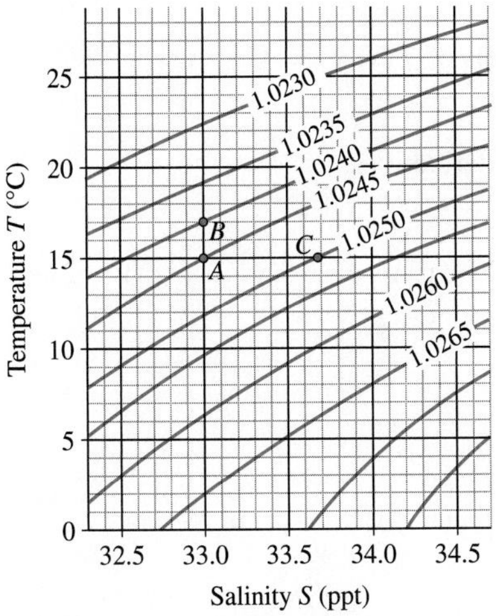

FIGURE 4 Contour map of seawater density as a function of temperature and salinity.

For greater accuracy, we can estimate $f_x(a,b)$ by taking the average of the difference quotients for Δx and $-\Delta x$. A similar remark applies to $f_y(a,b)$.

■ **EXAMPLE 6 Estimating Partial Derivatives Using Contour Maps** Seawater density ρ (kg/m^3) depends on salinity S (ppt) and the temperature T (°C). Use Figure 4 to estimate $\partial\rho/\partial T$ and $\partial\rho/\partial S$ at A.

Solution Point A has coordinates $(S, T) = (33, 15)$ and lies on the level curve $\rho = 1.0245$. We estimate $\partial\rho/\partial T$ at A in two steps.

Step 1. **Move vertically from A.**
Since T varies in the vertical direction, we move up vertically from point A to point B on the next higher level curve, where $\rho = 1.0240$. Point B has coordinates $(S, T) = (33, 17)$. Note that in moving from A to B, we have kept S constant because both points have salinity $S = 33$.

Step 2. **Compute the difference quotient.**

$$\Delta\rho = 1.0240 - 1.0245 = -0.0005 \text{ kg/m}^3$$

$$\Delta T = 17 - 15 = 2°\text{C}$$

This gives us the approximation

$$\left.\frac{\partial\rho}{\partial T}\right|_A \approx \frac{\Delta\rho}{\Delta T} = \frac{-0.0005}{2} = -0.00025 \text{ kg-m}^{-3}/°\text{C}$$

We estimate $\partial\rho/\partial S$ in a similar way, by moving to the right horizontally to point C with coordinates $(S, T) \approx (33.7, 15)$, where $\rho = 1.0250$:

$$\left.\frac{\partial\rho}{\partial S}\right|_A \approx \frac{\Delta\rho}{\Delta S} = \frac{1.0250 - 1.0245}{33.7 - 33} = \frac{0.0005}{0.7} \approx 0.0007 \text{ kg-m}^{-3}/\text{ppt}$$ ■

Higher-Order Partial Derivatives

The higher-order partial derivatives are the derivatives of derivatives. The *second-order* partial derivatives of f are the partial derivatives of f_x and f_y. We write f_{xx} for the x-derivative of f_x and f_{yy} for the y-derivative of f_y:

$$f_{xx} = \frac{\partial}{\partial x}\left(\frac{\partial f}{\partial x}\right), \qquad f_{yy} = \frac{\partial}{\partial y}\left(\frac{\partial f}{\partial y}\right)$$

We also have the *mixed partials*:

$$f_{xy} = \frac{\partial}{\partial y}\left(\frac{\partial f}{\partial x}\right), \qquad f_{yx} = \frac{\partial}{\partial x}\left(\frac{\partial f}{\partial y}\right)$$

The process can be continued. For example, f_{xyx} is the x-derivative of f_{xy}, and f_{xyy} is the y-derivative of f_{xy} (perform the differentiation in the order of the subscripts from left to right). The Leibniz notation for higher-order partial derivatives is

$$f_{xx} = \frac{\partial^2 f}{\partial x^2}, \qquad f_{xy} = \frac{\partial^2 f}{\partial y \partial x}, \qquad f_{yx} = \frac{\partial^2 f}{\partial x \partial y}, \qquad f_{yy} = \frac{\partial^2 f}{\partial y^2}$$

Higher partial derivatives are defined for functions of three or more variables in a similar manner.

EXAMPLE 7 Calculate the second-order partials of $f(x, y) = x^3 + y^2 e^x$.

Solution First, we compute the first-order partial derivatives:

$$f_x(x, y) = \frac{\partial}{\partial x}(x^3 + y^2 e^x) = 3x^2 + y^2 e^x, \qquad f_y(x, y) = \frac{\partial}{\partial y}(x^3 + y^2 e^x) = 2ye^x$$

Then we can compute the second-order derivatives:

$$\begin{aligned} f_{xx}(x, y) &= \frac{\partial}{\partial x} f_x = \frac{\partial}{\partial x}(3x^2 + y^2 e^x) & \qquad f_{yy}(x, y) &= \frac{\partial}{\partial y} f_y = \frac{\partial}{\partial y} 2ye^x \\ &= 6x + y^2 e^x, & &= 2e^x \\ f_{xy}(x, y) &= \frac{\partial f_x}{\partial y} = \frac{\partial}{\partial y}(3x^2 + y^2 e^x) & \qquad f_{yx}(x, y) &= \frac{\partial f_y}{\partial x} = \frac{\partial}{\partial x} 2ye^x \\ &= 2ye^x, & &= 2ye^x \end{aligned}$$

■

Remember how the subscripts are used in partial derivatives. The notation f_{xyy} means "first differentiate with respect to x and then differentiate twice with respect to y."

EXAMPLE 8 Calculate f_{xyy} for $f(x, y) = x^3 + y^2 e^x$.

Solution By the previous example, $f_{xy} = 2ye^x$. Therefore,

$$f_{xyy} = \frac{\partial}{\partial y} f_{xy} = \frac{\partial}{\partial y} 2ye^x = 2e^x$$

■

Observe in Example 7 that f_{xy} and f_{yx} are both equal to $2ye^x$. It is a pleasant circumstance that the equality $f_{xy} = f_{yx}$ holds in general, provided that the mixed partials are continuous. See Appendix D for a proof of the following theorem named for the French mathematician Alexis Clairaut (Figure 5).

The hypothesis of Clairaut's Theorem, that f_{xy} and f_{yx} are continuous, is almost always satisfied in practice, but see Exercise 84 for an example where the mixed partials are not equal.

THEOREM 1 Clairaut's Theorem: Equality of Mixed Partials If f_{xy} and f_{yx} are both continuous functions on a disk D, then $f_{xy}(a, b) = f_{yx}(a, b)$ for all $(a, b) \in D$. In other words,

$$\frac{\partial^2 f}{\partial x\, \partial y} = \frac{\partial^2 f}{\partial y\, \partial x}$$

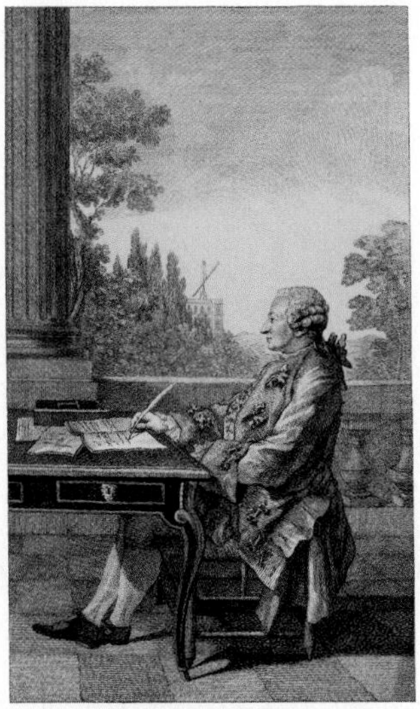

FIGURE 5 Alexis Clairaut (1713–1765) was a brilliant French mathematician who presented his first paper to the Paris Academy of Sciences at the age of 13. In 1752, Clairaut won a prize for an essay on lunar motion that Euler praised (surely an exaggeration) as "the most important and profound discovery that has ever been made in mathematics."

■ **EXAMPLE 9** Check that $\dfrac{\partial^2 W}{\partial U \partial T} = \dfrac{\partial^2 W}{\partial T \partial U}$ for $W = e^{U/T}$.

Solution We compute both derivatives and observe that they are equal:

$$\frac{\partial W}{\partial T} = e^{U/T}\frac{\partial}{\partial T}\left(\frac{U}{T}\right) = -UT^{-2}e^{U/T}, \qquad \frac{\partial W}{\partial U} = e^{U/T}\frac{\partial}{\partial U}\left(\frac{U}{T}\right) = T^{-1}e^{U/T}$$

$$\frac{\partial}{\partial U}\frac{\partial W}{\partial T} = -T^{-2}e^{U/T} - UT^{-3}e^{U/T}, \qquad \frac{\partial}{\partial T}\frac{\partial W}{\partial U} = -T^{-2}e^{U/T} - UT^{-3}e^{U/T}$$

■

Although Clairaut's Theorem is stated for f_{xy} and f_{yx}, it implies more generally that partial differentiation may be carried out in any order, provided that the derivatives in question are continuous (see Exercise 75). For example, we can compute f_{xyxy} by differentiating f twice with respect to x and twice with respect to y, in any order. Thus,

$$f_{xyxy} = f_{xxyy} = f_{yyxx} = f_{yxyx} = f_{xyyx} = f_{yxxy}$$

■ **EXAMPLE 10 Choosing the Order Wisely** Calculate the partial derivative g_{zzwx}, where $g(x, y, z, w) = x^3w^2z^2 + \sin\left(\dfrac{xy}{z^2}\right)$.

Solution Let's take advantage of the fact that the derivatives may be calculated in any order. If we differentiate with respect to w first, the second term disappears because it does not depend on w:

$$g_w = \frac{\partial}{\partial w}\left(x^3w^2z^2 + \sin\left(\frac{xy}{z^2}\right)\right) = 2x^3wz^2$$

Next, differentiate twice with respect to z and once with respect to x:

$$g_{wz} = \frac{\partial}{\partial z}2x^3wz^2 = 4x^3wz$$

$$g_{wzz} = \frac{\partial}{\partial z}4x^3wz = 4x^3w$$

$$g_{wzzx} = \frac{\partial}{\partial x}4x^3w = 12x^2w$$

We conclude that $g_{zzwx} = g_{wzzx} = 12x^2w$. ■

A **partial differential equation** (PDE) is a differential equation involving functions of several variables and their partial derivatives. The heat equation in the next example is a PDE that models temperature as heat spreads through an object. There are infinitely many solutions, but the particular function in the example describes temperature at times $t > 0$ along a metal rod when the center point is given a burst of heat at $t = 0$ (Figure 6).

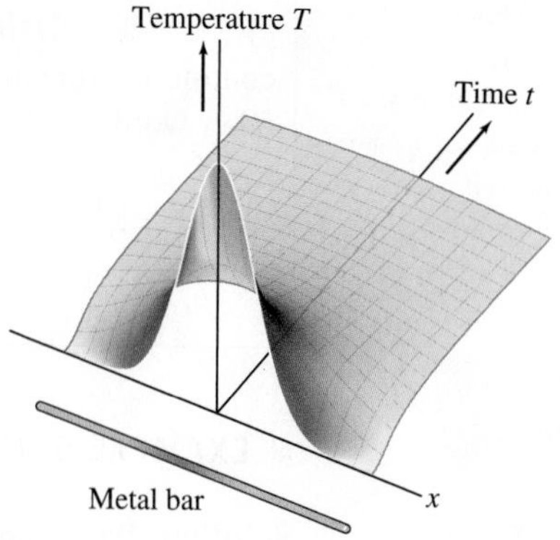

FIGURE 6 The plot of

$$u(x,t) = \frac{1}{2\sqrt{\pi t}} e^{-(x^2/4t)}$$

illustrates the diffusion of a burst of heat over time.

■ **EXAMPLE 11 The Heat Equation** Show that $u(x,t) = \dfrac{1}{2\sqrt{\pi t}} e^{-(x^2/4t)}$, defined for $t > 0$, satisfies the heat equation

$$\frac{\partial u}{\partial t} = \frac{\partial^2 u}{\partial x^2} \qquad \boxed{2}$$

Solution First, compute $\dfrac{\partial^2 u}{\partial x^2}$:

$$\frac{\partial u}{\partial x} = \frac{\partial}{\partial x} \frac{1}{2\sqrt{\pi}} t^{-1/2} e^{-(x^2/4t)} = -\frac{1}{4\sqrt{\pi}} x t^{-3/2} e^{-(x^2/4t)}$$

$$\frac{\partial^2 u}{\partial x^2} = \frac{\partial}{\partial x}\left(-\frac{1}{4\sqrt{\pi}} x t^{-3/2} e^{-(x^2/4t)}\right) = -\frac{1}{4\sqrt{\pi}} t^{-3/2} e^{-(x^2/4t)} + \frac{1}{8\sqrt{\pi}} x^2 t^{-5/2} e^{-(x^2/4t)}$$

Then compute $\partial u/\partial t$ and observe that it equals $\partial^2 u/\partial x^2$ as required:

$$\frac{\partial u}{\partial t} = \frac{\partial}{\partial t}\left(\frac{1}{2\sqrt{\pi}} t^{-1/2} e^{-(x^2/4t)}\right) = -\frac{1}{4\sqrt{\pi}} t^{-3/2} e^{-(x^2/4t)} + \frac{1}{8\sqrt{\pi}} x^2 t^{-5/2} e^{-(x^2/4t)} \quad ■$$

15.3 SUMMARY

- The partial derivatives of $f(x,y)$ are defined as the limits

$$f_x(a,b) = \left.\frac{\partial f}{\partial x}\right|_{(a,b)} = \lim_{h\to 0} \frac{f(a+h,b) - f(a,b)}{h}$$

$$f_y(a,b) = \left.\frac{\partial f}{\partial y}\right|_{(a,b)} = \lim_{k\to 0} \frac{f(a,b+k) - f(a,b)}{k}$$

- Compute f_x by holding y constant, and compute f_y by holding x constant.
- $f_x(a,b)$ is the slope at $x = a$ of the tangent line to the trace curve $z = f(x,b)$. Similarly, $f_y(a,b)$ is the slope at $y = b$ of the tangent line to the trace curve $z = f(a,y)$.
- For small changes Δx and Δy,

$$f(a+\Delta x, b) - f(a,b) \approx f_x(a,b)\Delta x$$

$$f(a, b+\Delta y) - f(a,b) \approx f_y(a,b)\Delta y$$

More generally, if f is a function of n variables and w is one of the variables, then $\Delta f \approx f_w \Delta w$ if w changes by Δw and all other variables remain fixed.

- The second-order partial derivatives are

$$\frac{\partial^2}{\partial x^2} f = f_{xx}, \qquad \frac{\partial^2}{\partial y\,\partial x} f = f_{xy}, \qquad \frac{\partial^2}{\partial x\,\partial y} f = f_{yx}, \qquad \frac{\partial^2}{\partial y^2} f = f_{yy}$$

- Clairaut's Theorem states that mixed partials are equal—that is, $f_{xy} = f_{yx}$ provided that f_{xy} and f_{yx} are continuous.
- More generally, higher-order partial derivatives may be computed in any order. For example, $f_{xyyz} = f_{yxzy}$ if f is a function of x, y, z whose fourth-order partial derivatives are continuous.

HISTORICAL PERSPECTIVE

Joseph Fourier (1768–1830)

Adolf Fick (1829–1901)

The general heat equation, of which Eq. (2) is a special case, was first introduced in 1807 by French mathematician Jean Baptiste Joseph Fourier. As a young man, Fourier was unsure whether to enter the priesthood or pursue mathematics, but he must have been very ambitious. He wrote in a letter, "Yesterday was my 21st birthday, at that age Newton and Pascal had already acquired many claims to immortality." In his twenties, Fourier got involved in the French Revolution and was imprisoned briefly in 1794 over an incident involving different factions. In 1798, he was summoned, along with more than 150 other scientists, to join Napoleon on his unsuccessful campaign in Egypt.

Fourier's true impact, however, lay in his mathematical contributions. The heat equation is applied throughout the physical sciences and engineering, from the study of heat flow through the earth's oceans and atmosphere to the use of heat probes to destroy tumors and treat heart disease.

Fourier also introduced a striking new technique—known as the **Fourier transform**—for solving his equation, based on the idea that a periodic function can be expressed as a (possibly infinite) sum of sines and cosines. Leading mathematicians of the day, including Lagrange and Laplace, initially raised objections because this technique was not easy to justify rigorously. Nevertheless, the Fourier transform turned out to be one of the most important mathematical discoveries of the nineteenth century. A Web search on the term "Fourier transform" reveals its vast range of modern applications.

In 1855, the German physiologist Adolf Fick showed that the heat equation describes not only heat conduction but also a wide range of diffusion processes, such as osmosis, ion transport at the cellular level, and the motion of pollutants through air or water. The heat equation thus became a basic tool in chemistry, molecular biology, and environmental science, where it is often called **Fick's Second Law**.

15.3 EXERCISES

Preliminary Questions

1. Patricia derived the following *incorrect* formula by misapplying the Product Rule:

$$\frac{\partial}{\partial x}(x^2y^2) = x^2(2y) + y^2(2x)$$

What was her mistake and what is the correct calculation?

2. Explain why it is not necessary to use the Quotient Rule to compute $\frac{\partial}{\partial x}\left(\frac{x+y}{y+1}\right)$. Should the Quotient Rule be used to compute $\frac{\partial}{\partial y}\left(\frac{x+y}{y+1}\right)$?

3. Which of the following partial derivatives should be evaluated without using the Quotient Rule?

(a) $\frac{\partial}{\partial x}\frac{xy}{y^2+1}$ **(b)** $\frac{\partial}{\partial y}\frac{xy}{y^2+1}$ **(c)** $\frac{\partial}{\partial x}\frac{y^2}{y^2+1}$

4. What is f_x, where $f(x, y, z) = (\sin yz)e^{z^3-z^{-1}\sqrt{y}}$?

5. Assuming the hypotheses of Clairaut's Theorem are satisfied, which of the following partial derivatives are equal to f_{xxy}?

(a) f_{xyx} **(b)** f_{yyx} **(c)** f_{xyy} **(d)** f_{yxx}

Exercises

1. Use the limit definition of the partial derivative to verify the formulas

$$\frac{\partial}{\partial x}xy^2 = y^2, \qquad \frac{\partial}{\partial y}xy^2 = 2xy$$

2. Use the Product Rule to compute $\frac{\partial}{\partial y}(x^2+y)(x+y^4)$.

3. Use the Quotient Rule to compute $\frac{\partial}{\partial y}\frac{y}{x+y}$.

4. Use the Chain Rule to compute $\frac{\partial}{\partial u}\ln(u^2+uv)$.

5. Calculate $f_z(2, 3, 1)$, where $f(x, y, z) = xyz$.

6. Explain the relation between the following two formulas (c is a constant).

$$\frac{d}{dx}\sin(cx) = c\cos(cx), \qquad \frac{\partial}{\partial x}\sin(xy) = y\cos(xy)$$

7. The plane $y = 1$ intersects the surface $z = x^4 + 6xy - y^4$ in a certain curve. Find the slope of the tangent line to this curve at the point $P = (1, 1, 6)$.

8. Determine whether the partial derivatives $\partial f/\partial x$ and $\partial f/\partial y$ are positive or negative at the point P on the graph in Figure 7.

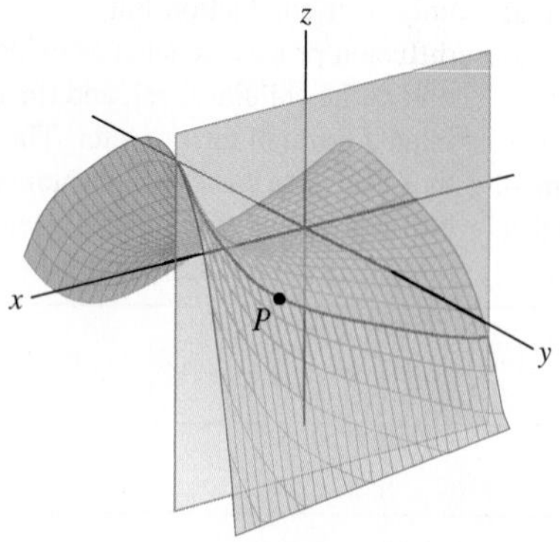

FIGURE 7

In Exercises 9–12, refer to Figure 8.

9. Estimate f_x and f_y at point A.

10. Is f_x positive or negative at B?

11. Starting at point B, in which compass direction (N, NE, SW, etc.) does f increase most rapidly?

12. At which of A, B, or C is f_y smallest?

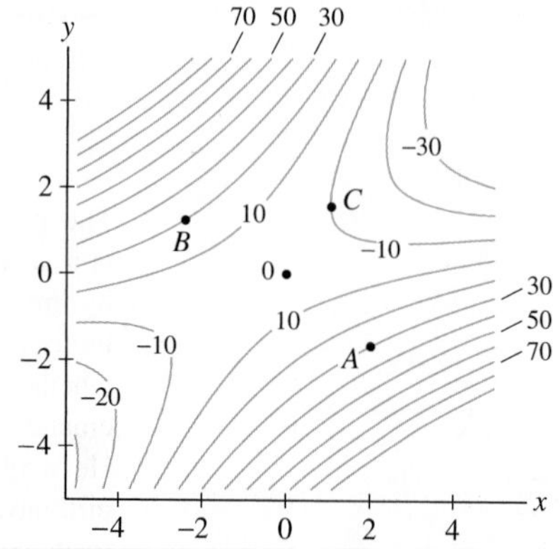

FIGURE 8 Contour map of $f(x, y)$.

In Exercises 13–40, compute the first-order partial derivatives.

13. $z = x^2 + y^2$

14. $z = x^4y^3$

15. $z = x^4y + xy^{-2}$

16. $V = \pi r^2 h$

17. $z = \frac{x}{y}$

18. $z = \frac{x}{x-y}$

19. $z = \sqrt{9-x^2-y^2}$

20. $z = \frac{x}{\sqrt{x^2+y^2}}$

21. $z = (\sin x)(\sin y)$

22. $z = \sin(u^2 v)$

23. $z = \tan\frac{x}{y}$

24. $S = \tan^{-1}(wz)$

25. $z = \ln(x^2+y^2)$

26. $A = \sin(4\theta - 9t)$

27. $W = e^{r+s}$

28. $Q = re^{\theta}$

29. $z = e^{xy}$

30. $R = e^{-v^2/k}$

31. $z = e^{-x^2-y^2}$

32. $P = e^{\sqrt{y^2+z^2}}$

33. $U = \frac{e^{-rt}}{r}$

34. $z = y^x$

35. $z = \sinh(x^2 y)$

36. $z = \cosh(t - \cos x)$

37. $w = xy^2z^3$

38. $w = \frac{x}{y+z}$

39. $Q = \frac{L}{M}e^{-Lt/M}$

40. $w = \frac{x}{(x^2 + y^2 + z^2)^{3/2}}$

In Exercises 41–44, compute the given partial derivatives.

41. $f(x, y) = 3x^2y + 4x^3y^2 - 7xy^5, \quad f_x(1, 2)$

42. $f(x, y) = \sin(x^2 - y), \quad f_y(0, \pi)$

43. $g(u, v) = u\ln(u + v), \quad g_u(1, 2)$

44. $h(x, z) = e^{xz - x^2z^3}, \quad h_z(3, 0)$

Exercises 45 and 46 refer to Example 5.

45. Calculate N for $L = 0.4$, $R = 0.12$, and $d = 10$, and use the linear approximation to estimate ΔN if d is increased from 10 to 10.4.

46. Estimate ΔN if $(L, R, d) = (0.5, 0.15, 8)$ and R is increased from 0.15 to 0.17.

47. The **heat index** I is a measure of how hot it feels when the relative humidity is H (as a percentage) and the actual air temperature is T (in degrees Fahrenheit). An approximate formula for the heat index that is valid for (T, H) near (90, 40) is

$$I(T, H) = 45.33 + 0.6845T + 5.758H - 0.00365T^2 - 0.1565HT + 0.001HT^2$$

(a) Calculate I at $(T, H) = (95, 50)$.

(b) Which partial derivative tells us the increase in I per degree increase in T when $(T, H) = (95, 50)$. Calculate this partial derivative.

48. The **wind-chill temperature** W measures how cold people feel (based on the rate of heat loss from exposed skin) when the outside temperature is T°C (with $T \le 10$) and wind velocity is v m/s (with $v \ge 2$):

$$W = 13.1267 + 0.6215T - 13.947v^{0.16} + 0.486Tv^{0.16}$$

Calculate $\partial W/\partial v$ at $(T, v) = (-10, 15)$ and use this value to estimate ΔW if $\Delta v = 2$.

49. The volume of a right-circular cone of radius r and height h is $V = \frac{\pi}{3}r^2h$. Suppose that $r = h = 12$ cm. What leads to a greater increase in V, a 1-cm increase in r or a 1-cm increase in h? Argue using partial derivatives.

50. Use the linear approximation to estimate the percentage change in volume of a right-circular cone of radius $r = 40$ cm if the height is increased from 40 to 41 cm.

51. Calculate $\partial W/\partial E$ and $\partial W/\partial T$, where $W = e^{-E/kT}$, where k is a constant.

52. Calculate $\partial P/\partial T$ and $\partial P/\partial V$, where pressure P, volume V, and temperature T are related by the ideal gas law, $PV = nRT$ (R and n are constants).

53. Use the contour map of $f(x, y)$ in Figure 9 to explain the following statements.

(a) f_y is larger at P than at Q, and f_x is smaller (more negative) at P than at Q.

(b) $f_x(x, y)$ is decreasing as a function of y; that is, for any fixed value $x = a$, $f_x(a, y)$ is decreasing in y.

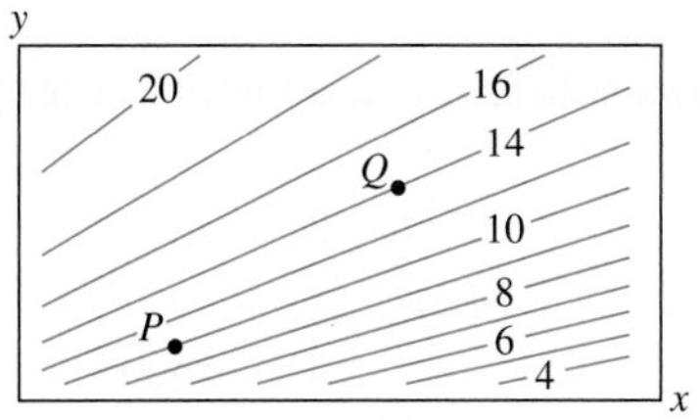

FIGURE 9 Contour interval 2.

54. Estimate the partial derivatives at P of the function whose contour map is shown in Figure 10.

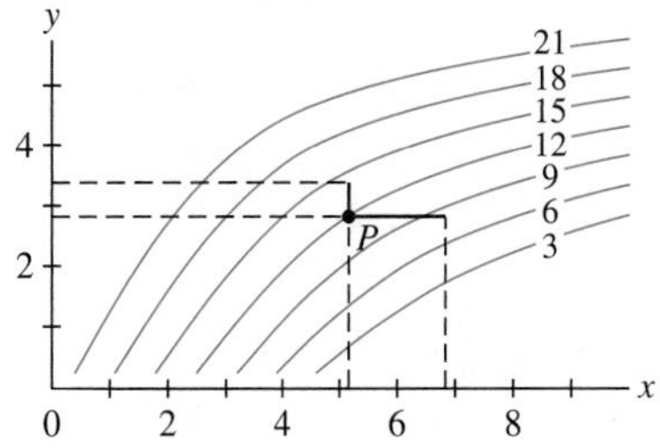

FIGURE 10

55. Over most of the earth, a magnetic compass does not point to true (geographic) north; instead, it points at some angle east or west of true north. The angle D between magnetic north and true north is called the **magnetic declination**. Use Figure 11 to determine which of the following statements is true.

(a) $\left.\frac{\partial D}{\partial y}\right|_A > \left.\frac{\partial D}{\partial y}\right|_B$ **(b)** $\left.\frac{\partial D}{\partial x}\right|_C > 0$ **(c)** $\left.\frac{\partial D}{\partial y}\right|_C > 0$

Note that the horizontal axis increases from right to left because of the way longitude is measured.

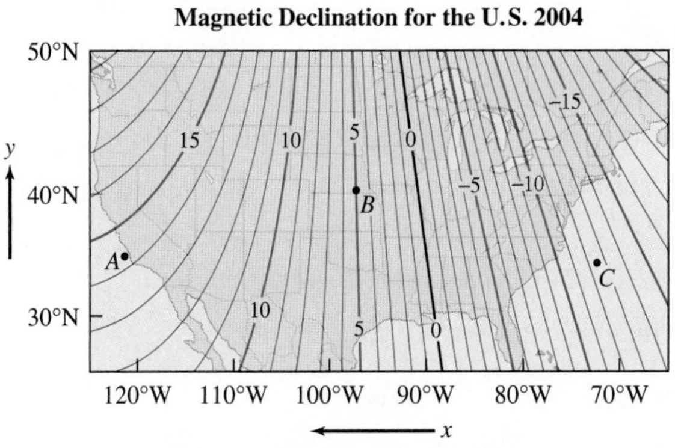

FIGURE 11 Contour interval 1°.

56. Refer to Table 1.

(a) Estimate $\partial\rho/\partial T$ and $\partial\rho/\partial S$ at the points $(S, T) = (34, 2)$ and $(35, 10)$ by computing the average of left-hand and right-hand difference quotients.

(b) For fixed salinity $S = 33$, is ρ concave up or concave down as a function of T? *Hint:* Determine whether the quotients $\Delta\rho/\Delta T$ are increasing or decreasing. What can you conclude about the sign of $\partial^2\rho/\partial T^2$?

TABLE 1 Seawater Density ρ as a Function of Temperature T and Salinity S

T \ S	30	31	32	33	34	35	36
12	22.75	23.51	24.27	25.07	25.82	26.6	27.36
10	23.07	23.85	24.62	25.42	26.17	26.99	27.73
8	23.36	24.15	24.93	25.73	26.5	27.28	29.09
6	23.62	24.44	25.22	26	26.77	27.55	28.35
4	23.85	24.62	25.42	26.23	27	27.8	28.61
2	24	24.78	25.61	26.38	27.18	28.01	28.78
0	24.11	24.92	25.72	26.5	27.34	28.12	28.91

In Exercises 57–62, compute the derivatives indicated.

57. $f(x, y) = 3x^2y - 6xy^4$, $\dfrac{\partial^2 f}{\partial x^2}$ and $\dfrac{\partial^2 f}{\partial y^2}$

58. $g(x, y) = \dfrac{xy}{x - y}$, $\dfrac{\partial^2 g}{\partial x\,\partial y}$

59. $h(u, v) = \dfrac{u}{u + 4v}$, $h_{vv}(u, v)$

60. $h(x, y) = \ln(x^3 + y^3)$, $h_{xy}(x, y)$

61. $f(x, y) = x\ln(y^2)$, $f_{yy}(2, 3)$

62. $g(x, y) = xe^{-xy}$, $g_{xy}(-3, 2)$

63. Compute f_{xyxzy} for

$$f(x, y, z) = y\sin(xz)\sin(x + z) + (x + z^2)\tan y + x\tan\left(\frac{z + z^{-1}}{y - y^{-1}}\right)$$

Hint: Use a well-chosen order of differentiation on each term.

64. Let

$$f(x, y, u, v) = \frac{x^2 + e^y v}{3y^2 + \ln(2 + u^2)}$$

What is the fastest way to show that $f_{uvxyvu}(x, y, u, v) = 0$ for all (x, y, u, v)?

In Exercises 65–72, compute the derivative indicated.

65. $f(u, v) = \cos(u + v^2)$, f_{uuv}

66. $g(x, y, z) = x^4y^5z^6$, g_{xxyz}

67. $F(r, s, t) = r(s^2 + t^2)$, F_{rst}

68. $u(x, t) = t^{-1/2}e^{-(x^2/4t)}$, u_{xx}

69. $F(\theta, u, v) = \sinh(uv + \theta^2)$, $F_{uu\theta}$

70. $R(u, v, w) = \dfrac{u}{v + w}$, R_{uvw}

71. $g(x, y, z) = \sqrt{x^2 + y^2 + z^2}$, g_{xyz}

72. $u(x, t) = \operatorname{sech}^2(x - t)$, u_{xxx}

73. Find a function such that $\dfrac{\partial f}{\partial x} = 2xy$ and $\dfrac{\partial f}{\partial y} = x^2$.

74. Prove that there does not exist any function $f(x, y)$ such that $\dfrac{\partial f}{\partial x} = xy$ and $\dfrac{\partial f}{\partial y} = x^2$. *Hint:* Show that f cannot satisfy Clairaut's Theorem.

75. Assume that f_{xy} and f_{yx} are continuous and that f_{yxx} exists. Show that f_{xyx} also exists and that $f_{yxx} = f_{xyx}$.

76. Show that $u(x, t) = \sin(nx)\,e^{-n^2t}$ satisfies the heat equation for any constant n:

$$\frac{\partial u}{\partial t} = \frac{\partial^2 u}{\partial x^2} \qquad \boxed{3}$$

77. Find all values of A and B such that $f(x, t) = e^{Ax+Bt}$ satisfies Eq. (3).

78. The function

$$f(x, t) = \frac{1}{2\sqrt{\pi t}}e^{-x^2/4t}$$

describes the temperature profile along a metal rod at time $t > 0$ when a burst of heat is applied at the origin (see Example 11). A small bug sitting on the rod at distance x from the origin feels the temperature rise and fall as heat diffuses through the bar. Show that the bug feels the maximum temperature at time $t = \frac{1}{2}x^2$.

In Exercises 79–82, the ***Laplace operator*** Δ *is defined by* $\Delta f = f_{xx} + f_{yy}$. *A function* $u(x, y)$ *satisfying the Laplace equation* $\Delta u = 0$ *is called* ***harmonic***.

79. Show that the following functions are harmonic:

(a) $u(x, y) = x$ **(b)** $u(x, y) = e^x\cos y$

(c) $u(x, y) = \tan^{-1}\dfrac{y}{x}$ **(d)** $u(x, y) = \ln(x^2 + y^2)$

80. Find all harmonic polynomials $u(x, y)$ of degree three, that is, $u(x, y) = ax^3 + bx^2y + cxy^2 + dy^3$.

81. Show that if $u(x, y)$ is harmonic, then the partial derivatives $\partial u/\partial x$ and $\partial u/\partial y$ are harmonic.

82. Find all constants a, b such that $u(x, y) = \cos(ax)e^{by}$ is harmonic.

83. Show that $u(x, t) = \operatorname{sech}^2(x - t)$ satisfies the **Korteweg–deVries equation** (which arises in the study of water waves):

$$4u_t + u_{xxx} + 12uu_x = 0$$

Further Insights and Challenges

84. Assumptions Matter This exercise shows that the hypotheses of Clairaut's Theorem are needed. Let

$$f(x, y) = xy\frac{x^2 - y^2}{x^2 + y^2}$$

for $(x, y) \neq (0, 0)$ and $f(0, 0) = 0$.

(a) Verify for $(x, y) \neq (0, 0)$:

$$f_x(x, y) = \frac{y(x^4 + 4x^2y^2 - y^4)}{(x^2 + y^2)^2}$$

$$f_y(x, y) = \frac{x(x^4 - 4x^2y^2 - y^4)}{(x^2 + y^2)^2}$$

(b) Use the limit definition of the partial derivative to show that $f_x(0, 0) = f_y(0, 0) = 0$ and that $f_{yx}(0, 0)$ and $f_{xy}(0, 0)$ both exist but are not equal.

(c) Show that for $(x, y) \neq (0, 0)$:

$$f_{xy}(x, y) = f_{yx}(x, y) = \frac{x^6 + 9x^4y^2 - 9x^2y^4 - y^6}{(x^2 + y^2)^3}$$

Show that f_{xy} is not continuous at $(0, 0)$. *Hint:* Show that $\lim_{h\to 0} f_{xy}(h, 0) \neq \lim_{h\to 0} f_{xy}(0, h)$.

(d) Explain why the result of part (b) does not contradict Clairaut's Theorem.

15.4 Differentiability and Tangent Planes

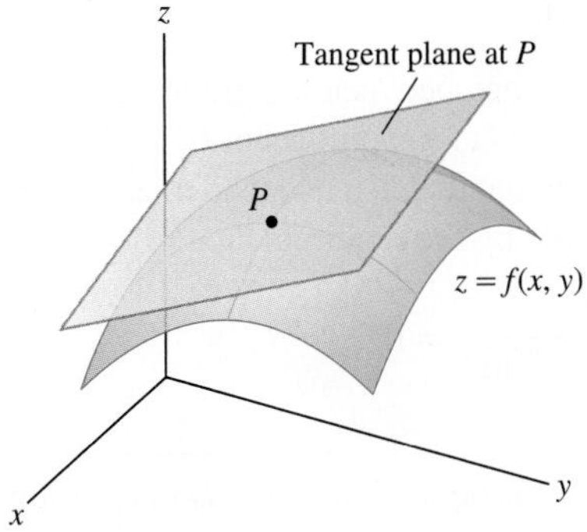

FIGURE 1 Tangent plane to the graph of $z = f(x, y)$.

In this section, we generalize two basic concepts from single-variable calculus: differentiability and the tangent line. The tangent line becomes the *tangent plane* for functions of two variables (Figure 1).

Intuitively, we would like to say that a continuous function $f(x, y)$ is differentiable if it is **locally linear**—that is, if its graph looks flatter and flatter as we zoom in on a point $P = (a, b, f(a, b))$ and eventually becomes indistinguishable from the tangent plane (Figure 2).

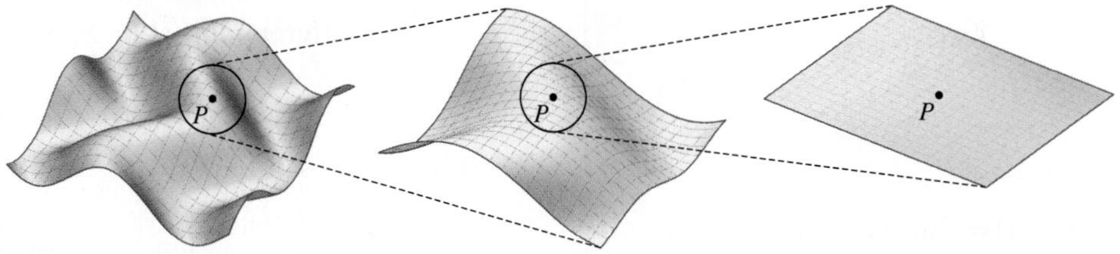

FIGURE 2 The graph looks flatter and flatter as we zoom in on a point P.

We can show that if the tangent plane at $P = (a, b, f(a, b))$ exists, then its equation must be $z = L(x, y)$, where $L(x, y)$ is the **linearization** at (a, b), defined by

$$L(x, y) = f(a, b) + f_x(a, b)(x - a) + f_y(a, b)(y - b)$$

Why must this be the tangent plane? Because it is the unique plane containing the tangent lines to the two vertical trace curves through P [Figure 3(A)]. Indeed, when we set $y = b$ in $z = L(x, y)$, the term $f_y(a, b)(y - b)$ drops out and we are left with the equation of the tangent line to the vertical trace $z = f(x, b)$ at P:

$$z = L(x, b) = f(a, b) + f_x(a, b)(x - a)$$

Similarly, $z = L(a, y)$ is the tangent line to the vertical trace $z = f(a, y)$ at P.

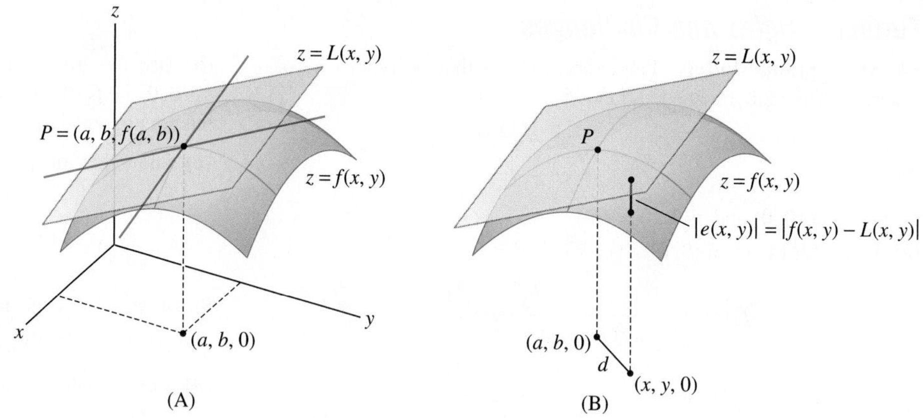

FIGURE 3

Before we can say that the tangent plane exists, however, we must impose a condition on $f(x, y)$ guaranteeing that the graph looks flat as we zoom in on P. Set

$$e(x, y) = f(x, y) - L(x, y)$$

As we see in Figure 3(B), $|e(x, y)|$ is the vertical distance between the graph of $f(x, y)$ and the plane $z = L(x, y)$. This distance tends to zero as (x, y) approaches (a, b) because $f(x, y)$ is continuous. To be locally linear, we require that the distance tend to zero *faster* than the distance from (x, y) to (a, b). We express this by the requirement

$$\lim_{(x,y)\to(a,b)} \frac{e(x, y)}{\sqrt{(x-a)^2 + (y-b)^2}} = 0$$

DEFINITION Differentiability Assume that $f(x, y)$ is defined in a disk D containing (a, b) and that $f_x(a, b)$ and $f_y(a, b)$ exist.

- $f(x, y)$ is **differentiable** at (a, b) if it is **locally linear**—that is, if

$$f(x, y) = L(x, y) + e(x, y) \qquad \boxed{1}$$

where $e(x, y)$ satisfies

$$\lim_{(x,y)\to(a,b)} \frac{e(x, y)}{\sqrt{(x-a)^2 + (y-b)^2}} = 0$$

- In this case, the **tangent plane** to the graph at $(a, b, f(a, b))$ is the plane with equation $z = L(x, y)$. Explicitly,

$$z = f(a, b) + f_x(a, b)(x - a) + f_y(a, b)(y - b) \qquad \boxed{2}$$

REMINDER

$$L(x, y) = f(a, b) + f_x(a, b)(x - a) + f_y(a, b)(y - b)$$

If $f(x, y)$ is differentiable at all points in a domain $\mathcal{D}$, we say that $f(x, y)$ is differentiable on $\mathcal{D}$.

It is cumbersome to check the local linearity condition directly (see Exercise 41), but fortunately, this is rarely necessary. The following theorem provides a criterion for differentiability that is easy to apply. It assures us that most functions arising in practice are differentiable on their domains. See Appendix D for a proof.

The definition of differentiability extends to functions of n-variables, and Theorem 1 holds in this setting: If all of the partial derivatives of $f(x_1, \ldots, x_n)$ exist and are continuous on an open domain $\mathcal{D}$, then $f(x_1, \ldots, x_n)$ is differentiable on $\mathcal{D}$.

THEOREM 1 Criterion for Differentiability If $f_x(x, y)$ and $f_y(x, y)$ exist and are continuous on an open disk D, then $f(x, y)$ is differentiable on D.

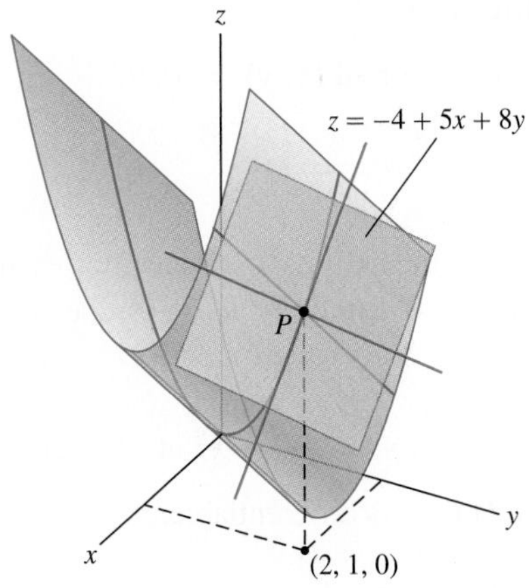

FIGURE 4 Graph of $f(x, y) = 5x + 4y^2$ and the tangent plane at $P = (2, 1, 14)$.

■ **EXAMPLE 1** Show that $f(x, y) = 5x + 4y^2$ is differentiable (Figure 4). Find the equation of the tangent plane at $(a, b) = (2, 1)$.

Solution The partial derivatives exist and are continuous functions:

$$f(x, y) = 5x + 4y^2, \qquad f_x(x, y) = 5, \qquad f_y(x, y) = 8y$$

Therefore, $f(x, y)$ is differentiable for all (x, y) by Theorem 1. To find the tangent plane, we evaluate the partial derivatives at (2, 1):

$$f(2, 1) = 14, \qquad f_x(2, 1) = 5, \qquad f_y(2, 1) = 8$$

The linearization at (2, 1) is

$$L(x, y) = \underbrace{14 + 5(x - 2) + 8(y - 1)}_{f(a,b)+f_x(a,b)(x-a)+f_y(a,b)(y-b)} = -4 + 5x + 8y$$

The tangent plane through $P = (2, 1, 14)$ has equation $z = -4 + 5x + 8y$. ■

Local linearity is used in the next section to prove the Chain Rule for Paths, upon which the fundamental properties of the gradient are based.

Assumptions Matter Local linearity plays a key role, and although most reasonable functions are locally linear, the mere existence of the partial derivatives does not guarantee local linearity. This is in contrast to the one-variable case, where $f(x)$ is automatically locally linear at $x = a$ if $f'(a)$ exists (Exercise 44).

The function $g(x, y)$ in Figure 5(A) shows what can go wrong. The graph contains the x- and y-axes—in other words, $g(x, y) = 0$ if x or y is zero—and therefore, the partial derivatives $g_x(0, 0)$ and $g_y(0, 0)$ are both zero. The tangent plane at the origin (0, 0), if it existed, would have to be the xy-plane. However, Figure 5(B) shows that the graph also contains lines through the origin that do not lie in the xy-plane (in fact, the graph is composed entirely of lines through the origin). As we zoom in on the origin, these lines remain at an angle to the xy-plane, and the surface does not get any flatter. Thus $g(x, y)$ cannot be locally linear at (0, 0), and the tangent plane does not exist. In particular, $g(x, y)$ cannot satisfy the assumptions of Theorem 1, so the partial derivatives $g_x(x, y)$ and $g_y(x, y)$ cannot be continuous at the origin (see Exercise 45 for details).

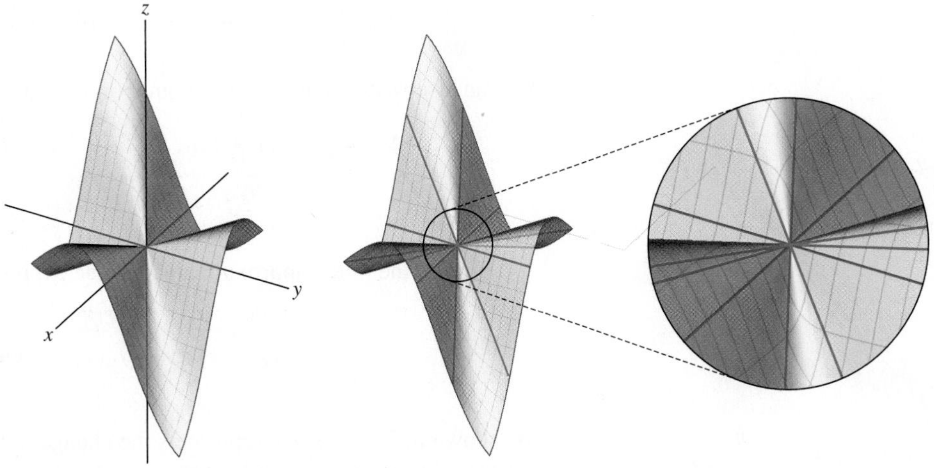

(A) The horizontal trace at $z = 0$ consists of the x and y axes.

(B) But the graph also contains non-horizontal lines through the origin.

(C) So the graph does not appear any flatter as we zoom in on the origin.

FIGURE 5 Graphs of $g(x, y) = \dfrac{2xy(x + y)}{x^2 + y^2}$.

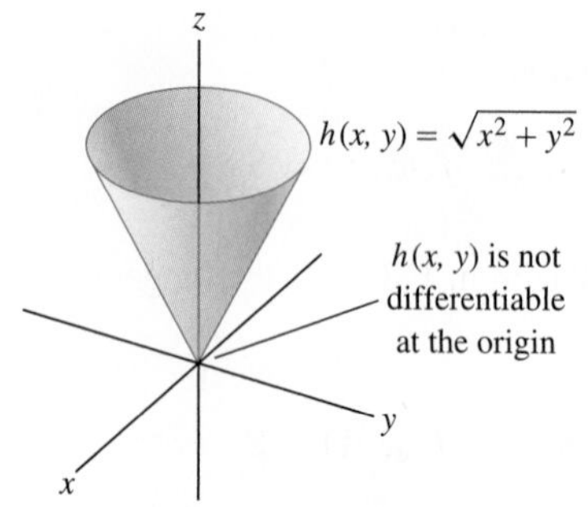

FIGURE 6 The function $h(x, y) = \sqrt{x^2 + y^2}$ is differentiable except at the origin.

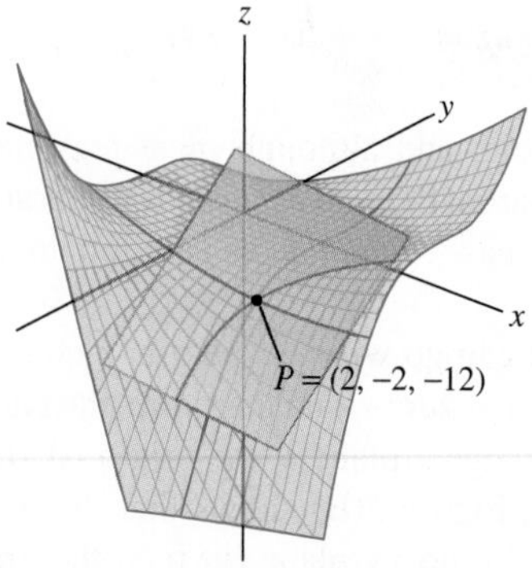

FIGURE 7 Tangent plane to the surface $f(x, y) = xy^3 + x^2$ passing through $P = (2, -2, -12)$.

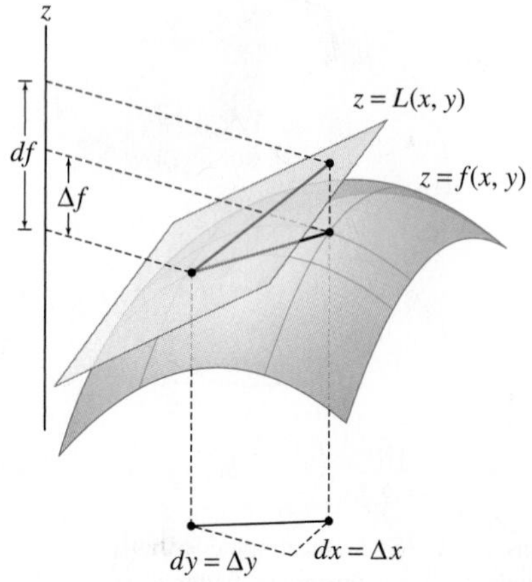

FIGURE 8 The quantity df is the change in height of the tangent plane.

■ **EXAMPLE 2** Where is $h(x, y) = \sqrt{x^2 + y^2}$ differentiable?

Solution The partial derivatives exist and are continuous for all $(x, y) \neq (0, 0)$:

$$h_x(x, y) = \frac{x}{\sqrt{x^2 + y^2}}, \qquad h_y(x, y) = \frac{y}{\sqrt{x^2 + y^2}}$$

However, the partial derivatives do not exist at $(0, 0)$. Indeed, $h_x(0, 0)$ does not exist because $h(x, 0) = \sqrt{x^2} = |x|$ is not differentiable at $x = 0$. Similarly, $h_y(0, 0)$ does not exist. By Theorem 1, $h(x, y)$ is differentiable except at $(0, 0)$ (Figure 6). ■

■ **EXAMPLE 3** Find a tangent plane of the graph of $f(x, y) = xy^3 + x^2$ at $(2, -2)$.

Solution The partial derivatives are continuous, so $f(x, y)$ is differentiable:

$$f_x(x, y) = y^3 + 2x, \qquad f_x(2, -2) = -4$$

$$f_y(x, y) = 3xy^2, \qquad f_y(2, -2) = 24$$

Since $f(2, -2) = -12$, the tangent plane through $(2, -2, -12)$ has equation

$$z = -12 - 4(x - 2) + 24(y + 2)$$

This can be rewritten as $z = 44 - 4x + 24y$ (Figure 7). ■

Linear Approximation and Differentials

By definition, if $f(x, y)$ is differentiable at (a, b), then it is locally linear and the **linear approximation** is

$$f(x, y) \approx L(x, y) \qquad \text{for } (x, y) \text{ near } (a, b)$$

where

$$L(x, y) = f(a, b) + f_x(a, b)(x - a) + f_y(a, b)(y - b)$$

We shall rewrite this in several useful ways. First, set $x = a + h$ and $y = b + k$. Then

$$f(a + h, b + k) \approx f(a, b) + f_x(a, b)h + f_y(a, b)k \qquad \boxed{3}$$

We can also write the linear approximation in terms of the *change in f*:

$$\Delta f = f(x, y) - f(a, b), \qquad \Delta x = x - a, \qquad \Delta y = y - b$$

$$\Delta f \approx f_x(a, b)\Delta x + f_y(a, b)\Delta y \qquad \boxed{4}$$

Finally, the linear approximation is often expressed in terms of **differentials**:

$$df = f_x(x, y)\,dx + f_y(x, y)\,dy = \frac{\partial f}{\partial x}dx + \frac{\partial f}{\partial y}dy$$

As shown in Figure 8, df represents the change in height of the tangent plane for given changes dx and dy in x and y (when we work with differentials, we call them dx and dy instead of Δx and Δy), whereas Δf is the change in the function itself. The linear approximation tells us that the two changes are approximately equal:

$$\Delta f \approx df$$

These approximations apply in any number of variables. In three variables,

$$f(a+h, b+k, c+\ell) \approx f(a,b,c) + f_x(a,b,c)h + f_y(a,b,c)k + f_z(a,b,c)\ell$$

or in terms of differentials, $\Delta f \approx df$, where

$$df = f_x(x,y,z)\,dx + f_y(x,y,z)\,dy + f_z(x,y,z)\,dz$$

EXAMPLE 4 Use the linear approximation to estimate

$$(3.99)^3(1.01)^4(1.98)^{-1}$$

Then use a calculator to find the percentage error.

REMINDER The percentage error is equal to

$$\left|\frac{\text{error}}{\text{actual value}}\right| \times 100\%$$

Solution Think of $(3.99)^3(1.01)^4(1.98)^{-1}$ as a value of $f(x,y,z) = x^3y^4z^{-1}$:

$$f(3.99, 1.01, 1.98) = (3.99)^3(1.01)^4(1.98)^{-1}$$

Then it makes sense to use the linear approximation at (4, 1, 2):

$$\begin{aligned} f(x,y,z) &= x^3y^4z^{-1}, & f(4,1,2) &= (4^3)(1^4)(2^{-1}) = 32 \\ f_x(x,y,z) &= 3x^2y^4z^{-1}, & f_x(4,1,2) &= 24 \\ f_y(x,y,z) &= 4x^3y^3z^{-1}, & f_y(4,1,2) &= 128 \\ f_z(x,y,z) &= -x^3y^4z^{-2}, & f_z(4,1,2) &= -16 \end{aligned}$$

The linear approximation in three variables stated above, with $a = 4, b = 1, c = 2$, gives us

$$\underbrace{(4+h)^3(1+k)^4(2+\ell)^{-1}}_{f(4+h,1+k,2+\ell)} \approx 32 + 24h + 128k - 16\ell$$

For $h = -0.01$, $k = 0.01$, and $\ell = -0.02$, we obtain the desired estimate

$$(3.99)^3(1.01)^4(1.98)^{-1} \approx 32 + 24(-0.01) + 128(0.01) - 16(-0.02) = 33.36$$

The calculator value is $(3.99)^3(1.01)^4(1.98)^{-1} \approx 33.384$, so the error in our estimate is less than 0.025. The percentage error is

$$\text{Percentage error} \approx \frac{|33.384 - 33.36|}{33.384} \times 100 \approx 0.075\%$$

■

EXAMPLE 5 Body Mass Index A person's BMI is $I = W/H^2$, where W is the body weight (in kilograms) and H is the body height (in meters). Estimate the change in a child's BMI if (W, H) changes from (40, 1.45) to (41.5, 1.47).

BMI is one factor used to assess the risk of certain diseases such as diabetes and high blood pressure. The range $18.5 \le I \le 24.9$ is considered normal for adults over 20 years of age.

Solution

***Step 1.* Compute the differential at $(W, H) = (40, 1.45)$.**

$$\frac{\partial I}{\partial W} = \frac{\partial}{\partial W}\left(\frac{W}{H^2}\right) = \frac{1}{H^2}, \qquad \frac{\partial I}{\partial H} = \frac{\partial}{\partial H}\left(\frac{W}{H^2}\right) = -\frac{2W}{H^3}$$

At $(W, H) = (40, 1.45)$, we have

$$\left.\frac{\partial I}{\partial W}\right|_{(40,1.45)} = \frac{1}{1.45^2} \approx 0.48, \qquad \left.\frac{\partial I}{\partial H}\right|_{(40,1.45)} = -\frac{2(40)}{1.45^3} \approx -26.24$$

Therefore, the differential at (40, 1.45) is

$$dI \approx 0.48\,dW - 26.24\,dH$$

Step 2. **Estimate the change.**
We have shown that the differential dI at (40, 1.45) is $0.48\,dW - 26.24\,dH$. If (W, H) changes from (40, 1.45) to (41.5, 1.47), then

$$dW = 41.5 - 40 = 1.5, \qquad dH = 1.47 - 1.45 = 0.02$$

Therefore,

$$\Delta I \approx dI = 0.48\,dW - 26.24\,dH = 0.48(1.5) - 26.24(0.02) \approx 0.2$$

We find that BMI increases by approximately 0.2. ■

15.4 SUMMARY

- The *linearization* in two and three variables:

$$L(x, y) = f(a, b) + f_x(a, b)(x - a) + f_y(a, b)(y - b)$$
$$L(x, y, z) = f(a, b, c) + f_x(a, b, c)(x - a) + f_y(a, b, c)(y - b) + f_z(a, b, c)(z - c)$$

- $f(x, y)$ is *differentiable* at (a, b) if $f_x(a, b)$ and $f_y(a, b)$ exist and

$$f(x, y) = L(x, y) + e(x, y)$$

where $e(x, y)$ is a function such that

$$\lim_{(x,y)\to(a,b)} \frac{e(x, y)}{\sqrt{(x - a)^2 + (y - b)^2}} = 0$$

- Result used in practice: *If $f_x(x, y)$ and $f_y(x, y)$ exist and are continuous in a disk D containing (a, b), then $f(x, y)$ is differentiable at (a, b).*
- Equation of the tangent plane to $z = f(x, y)$ at (a, b):

$$z = f(a, b) + f_x(a, b)(x - a) + f_y(a, b)(y - b)$$

- Equivalent forms of the linear approximation:

$$f(x, y) \approx f(a, b) + f_x(a, b)(x - a) + f_y(a, b)(y - b)$$
$$f(a + h, b + k) \approx f(a, b) + f_x(a, b)h + f_y(a, b)k$$
$$\Delta f \approx f_x(a, b)\,\Delta x + f_y(a, b)\,\Delta y$$

- In differential form, $\Delta f \approx df$, where

$$df = f_x(x, y)\,dx + f_y(x, y)\,dy = \frac{\partial f}{\partial x}dx + \frac{\partial f}{\partial y}dy$$

$$df = f_x(x, y, z)\,dx + f_y(x, y, z)\,dy + f_z(x, y, z)\,dz = \frac{\partial f}{\partial x}dx + \frac{\partial f}{\partial y}dy + \frac{\partial f}{\partial z}dz$$

15.4 EXERCISES

Preliminary Questions

1. How is the linearization of $f(x, y)$ at (a, b) defined?

2. Define local linearity for functions of two variables.

In Exercises 3–5, assume that

$$f(2, 3) = 8, \qquad f_x(2, 3) = 5, \qquad f_y(2, 3) = 7$$

3. Which of (a)–(b) is the linearization of f at (2, 3)?

(a) $L(x, y) = 8 + 5x + 7y$

(b) $L(x, y) = 8 + 5(x - 2) + 7(y - 3)$

4. Estimate $f(2, 3.1)$.

5. Estimate Δf at (2, 3) if $\Delta x = -0.3$ and $\Delta y = 0.2$.

6. Which theorem allows us to conclude that $f(x, y) = x^3y^8$ is differentiable?

Exercises

1. Use Eq. (2) to find an equation of the tangent plane to the graph of $f(x, y) = 2x^2 - 4xy^2$ at $(-1, 2)$.

2. Find the equation of the tangent plane in Figure 9. The point of tangency is $(a, b) = (1, 0.8, 0.34)$.

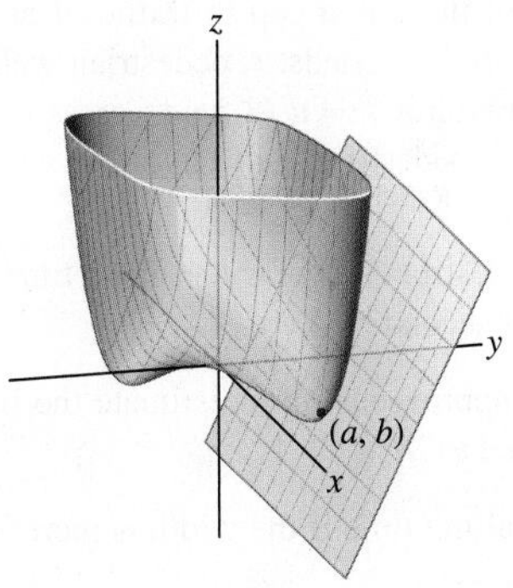

FIGURE 9 Graph of $f(x, y) = 0.2x^4 + y^6 - xy$.

In Exercises 3–10, find an equation of the tangent plane at the given point.

3. $f(x, y) = x^2y + xy^3$, $(2, 1)$

4. $f(x, y) = \dfrac{x}{\sqrt{y}}$, $(4, 4)$

5. $f(x, y) = x^2 + y^{-2}$, $(4, 1)$

6. $G(u, w) = \sin(uw)$, $\left(\frac{\pi}{6}, 1\right)$

7. $F(r, s) = r^2s^{-1/2} + s^{-3}$, $(2, 1)$

8. $g(x, y) = e^{x/y}$, $(2, 1)$

9. $f(x, y) = \operatorname{sech}(x - y)$, $(\ln 4, \ln 2)$

10. $f(x, y) = \ln(4x^2 - y^2)$, $(1, 1)$

11. Find the points on the graph of $z = 3x^2 - 4y^2$ at which the vector $\mathbf{n} = \langle 3, 2, 2\rangle$ is normal to the tangent plane.

12. Find the points on the graph of $z = xy^3 + 8y^{-1}$ where the tangent plane is parallel to $2x + 7y + 2z = 0$.

13. Find the linearization $L(x, y)$ of $f(x, y) = x^2y^3$ at $(a, b) = (2, 1)$. Use it to estimate $f(2.01, 1.02)$ and $f(1.97, 1.01)$ and compare with values obtained using a calculator.

14. Write the linear approximation to $f(x, y) = x(1 + y)^{-1}$ at $(a, b) = (8, 1)$ in the form

$$f(a + h, b + k) \approx f(a, b) + f_x(a, b)h + f_y(a, b)k$$

Use it to estimate $\frac{7.98}{2.02}$ and compare with the value obtained using a calculator.

15. Let $f(x, y) = x^3y^{-4}$. Use Eq. (4) to estimate the change

$$\Delta f = f(2.03, 0.9) - f(2, 1)$$

16. Use the linear approximation to $f(x, y) = \sqrt{x/y}$ at (9, 4) to estimate $\sqrt{9.1/3.9}$.

17. Use the linear approximation of $f(x, y) = e^{x^2+y}$ at (0, 0) to estimate $f(0.01, -0.02)$. Compare with the value obtained using a calculator.

18. Let $f(x, y) = x^2/(y^2 + 1)$. Use the linear approximation at an appropriate point (a, b) to estimate $f(4.01, 0.98)$.

19. Find the linearization of $f(x, y, z) = z\sqrt{x + y}$ at (8, 4, 5).

20. Find the linearization to $f(x, y, z) = xy/z$ at the point (2, 1, 2). Use it to estimate $f(2.05, 0.9, 2.01)$ and compare with the value obtained from a calculator.

21. Estimate $f(2.1, 3.8)$ assuming that

$$f(2, 4) = 5, \qquad f_x(2, 4) = 0.3, \qquad f_y(2, 4) = -0.2$$

22. Estimate $f(1.02, 0.01, -0.03)$ assuming that

$$f(1, 0, 0) = -3, \qquad f_x(1, 0, 0) = -2,$$
$$f_y(1, 0, 0) = 4, \qquad f_z(1, 0, 0) = 2$$

In Exercises 23–28, use the linear approximation to estimate the value. Compare with the value given by a calculator.

23. $(2.01)^3(1.02)^2$

24. $\dfrac{4.1}{7.9}$

25. $\sqrt{3.01^2 + 3.99^2}$

26. $\dfrac{0.98^2}{2.01^3 + 1}$

27. $\sqrt{(1.9)(2.02)(4.05)}$

28. $\dfrac{8.01}{\sqrt{(1.99)(2.01)}}$

29. Find an equation of the tangent plane to $z = f(x, y)$ at $P = (1, 2, 10)$ assuming that

$$f(1, 2) = 10, \qquad f(1.1, 2.01) = 10.3, \qquad f(1.04, 2.1) = 9.7$$

30. Suppose that the plane tangent to $z = f(x, y)$ at $(-2, 3, 4)$ has equation $4x + 2y + z = 2$. Estimate $f(-2.1, 3.1)$.

In Exercises 31–34, let $I = W/H^2$ denote the BMI described in Example 5.

31. A boy has weight $W = 34$ kg and height $H = 1.3$ m. Use the linear approximation to estimate the change in I if (W, H) changes to (36, 1.32).

32. Suppose that $(W, H) = (34, 1.3)$. Use the linear approximation to estimate the increase in H required to keep I constant if W increases to 35.

33. (a) Show that $\Delta I \approx 0$ if $\Delta H/\Delta W \approx H/2W$.
(b) Suppose that $(W, H) = (25, 1.1)$. What increase in H will leave I (approximately) constant if W is increased by 1 kg?

34. Estimate the change in height that will decrease I by 1 if $(W, H) = (25, 1.1)$, assuming that W remains constant.

35. A cylinder of radius r and height h has volume $V = \pi r^2 h$.
(a) Use the linear approximation to show that

$$\frac{\Delta V}{V} \approx \frac{2\Delta r}{r} + \frac{\Delta h}{h}$$

(b) Estimate the percentage increase in V if r and h are each increased by 2%.
(c) The volume of a certain cylinder V is determined by measuring r and h. Which will lead to a greater error in V: a 1% error in r or a 1% error in h?

36. Use the linear approximation to show that if $I = x^a y^b$, then

$$\frac{\Delta I}{I} \approx a\frac{\Delta x}{x} + b\frac{\Delta y}{y}$$

37. The monthly payment for a home loan is given by a function $f(P, r, N)$, where P is the principal (initial size of the loan), r the interest rate, and N is the length of the loan in months. Interest rates are expressed as a decimal: A 6% interest rate is denoted by $r = 0.06$. If $P = \$100{,}000$, $r = 0.06$, and $N = 240$ (a 20-year loan), then the monthly payment is $f(100{,}000, 0.06, 240) = 716.43$. Furthermore, at these values, we have

$$\frac{\partial f}{\partial P} = 0.0071, \qquad \frac{\partial f}{\partial r} = 5769, \qquad \frac{\partial f}{\partial N} = -1.5467$$

Estimate:

(a) The change in monthly payment per \$1000 increase in loan principal.

(b) The change in monthly payment if the interest rate increases to $r = 6.5\%$ and $r = 7\%$.

(c) The change in monthly payment if the length of the loan increases to 24 years.

38. Automobile traffic passes a point P on a road of width w ft at an average rate of R vehicles per second. Although the arrival of automobiles is irregular, traffic engineers have found that the average waiting time T until there is a gap in traffic of at least t seconds is approximately $T = te^{Rt}$ seconds. A pedestrian walking at a speed of 3.5 ft/s (5.1 mph) requires $t = w/3.5$ s to cross the road. Therefore, the average time the pedestrian will have to wait before crossing is $f(w, R) = (w/3.5)e^{wR/3.5}$ s.

(a) What is the pedestrian's average waiting time if $w = 25$ ft and $R = 0.2$ vehicle per second?

(b) Use the linear approximation to estimate the increase in waiting time if w is increased to 27 ft.

(c) Estimate the waiting time if the width is increased to 27 ft and R decreases to 0.18.

(d) What is the rate of increase in waiting time per 1-ft increase in width when $w = 30$ ft and $R = 0.3$ vehicle per second?

39. The volume V of a right-circular cylinder is computed using the values 3.5 m for diameter and 6.2 m for height. Use the linear approximation to estimate the maximum error in V if each of these values has a possible error of at most 5%. Recall that $V = \frac{1}{3}\pi r^2 h$.

Further Insights and Challenges

40. Show that if $f(x, y)$ is differentiable at (a, b), then the function of one variable $f(x, b)$ is differentiable at $x = a$. Use this to prove that $f(x, y) = \sqrt{x^2 + y^2}$ is *not* differentiable at $(0, 0)$.

41. This exercise shows directly (without using Theorem 1) that the function $f(x, y) = 5x + 4y^2$ from Example 1 is locally linear at $(a, b) = (2, 1)$.
(a) Show that $f(x, y) = L(x, y) + e(x, y)$ with $e(x, y) = 4(y - 1)^2$.
(b) Show that

$$0 \le \frac{e(x, y)}{\sqrt{(x - 2)^2 + (y - 1)^2}} \le 4|y - 1|$$

(c) Verify that $f(x, y)$ is locally linear.

42. Show directly, as in Exercise 41, that $f(x, y) = xy^2$ is differentiable at $(0, 2)$.

43. Differentiability Implies Continuity Use the definition of differentiability to prove that if f is differentiable at (a, b), then f is continuous at (a, b).

44. Let $f(x)$ be a function of one variable defined near $x = a$. Given a number M, set

$$L(x) = f(a) + M(x - a), \qquad e(x) = f(x) - L(x)$$

Thus $f(x) = L(x) + e(x)$. We say that f is locally linear at $x = a$ if M can be chosen so that $\lim_{x \to a} \frac{e(x)}{|x - a|} = 0$.

(a) Show that if $f(x)$ is differentiable at $x = a$, then $f(x)$ is locally linear with $M = f'(a)$.

(b) Show conversely that if f is locally linear at $x = a$, then $f(x)$ is differentiable and $M = f'(a)$.

45. Assumptions Matter Define $g(x, y) = 2xy(x + y)/(x^2 + y^2)$ for $(x, y) \neq 0$ and $g(0, 0) = 0$. In this exercise, we show that $g(x, y)$ is continuous at $(0, 0)$ and that $g_x(0, 0)$ and $g_y(0, 0)$ exist, but $g(x, y)$ is not differentiable at $(0, 0)$.

(a) Show using polar coordinates that $g(x, y)$ is continuous at $(0, 0)$.

(b) Use the limit definitions to show that $g_x(0, 0)$ and $g_y(0, 0)$ exist and that both are equal to zero.

(c) Show that the linearization of $g(x, y)$ at $(0, 0)$ is $L(x, y) = 0$.

(d) Show that if $g(x, y)$ were locally linear at $(0, 0)$, we would have $\lim_{h \to 0} \frac{g(h, h)}{h} = 0$. Then observe that this is not the case because $g(h, h) = 2h$. This shows that $g(x, y)$ is not locally linear at $(0, 0)$ and, hence, not differentiable at $(0, 0)$.

15.5 The Gradient and Directional Derivatives

We have seen that the rate of change of a function f of several variables depends on a choice of direction. Since directions are indicated by vectors, it is natural to use vectors to describe the derivative of f in a specified direction.

To do this, we introduce the **gradient** ∇f_P, which is the vector whose components are the partial derivatives of f at P.

The gradient of a function of n variables is the vector

$$\nabla f = \left\langle \frac{\partial f}{\partial x_1}, \frac{\partial f}{\partial x_2}, \ldots, \frac{\partial f}{\partial x_n} \right\rangle$$

DEFINITION The Gradient The gradient of a function $f(x, y)$ at a point $P = (a, b)$ is the vector

$$\nabla f_P = \langle f_x(a, b), f_y(a, b) \rangle$$

In three variables, if $P = (a, b, c)$,

$$\nabla f_P = \langle f_x(a, b, c), f_y(a, b, c), f_z(a, b, c) \rangle$$

The symbol ∇, called "del," is an upside-down Greek delta. It was popularized by the Scottish physicist P. G. Tait (1831–1901), who called the symbol "nabla," because of its resemblance to an ancient Assyrian harp. The great physicist James Clerk Maxwell was reluctant to adopt this term and would refer to the gradient simply as the "slope." He wrote jokingly to his friend Tait in 1871, "Still harping on that nabla?"

We also write $\nabla f_{(a,b)}$ or $\nabla f(a, b)$ for the gradient. Sometimes, we omit reference to the point P and write

$$\nabla f = \left\langle \frac{\partial f}{\partial x}, \frac{\partial f}{\partial y} \right\rangle \quad \text{or} \quad \nabla f = \left\langle \frac{\partial f}{\partial x}, \frac{\partial f}{\partial y}, \frac{\partial f}{\partial z} \right\rangle$$

The gradient ∇f "assigns" a vector ∇f_P to each point in the domain of f, as in Figure 1.

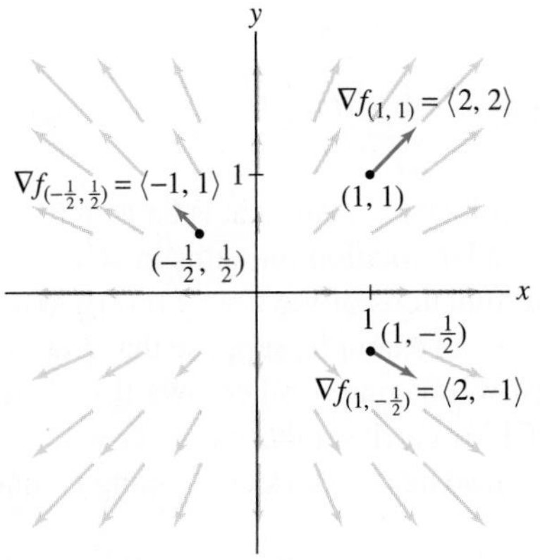

FIGURE 1 Gradient vectors of $f(x, y) = x^2 + y^2$ at several points (vectors not drawn to scale).

■ **EXAMPLE 1 Drawing Gradient Vectors** Let $f(x, y) = x^2 + y^2$. Calculate the gradient ∇f, draw several gradient vectors, and compute ∇f_P at $P = (1, 1)$.

Solution The partial derivatives are $f_x(x, y) = 2x$ and $f_y(x, y) = 2y$, so

$$\nabla f = \langle 2x, 2y \rangle$$

The gradient attaches the vector $\langle 2x, 2y\rangle$ to the point (x, y). As we see in Figure 1, these vectors point away from the origin. At the particular point $(1, 1)$,

$$\nabla f_P = \nabla f(1, 1) = \langle 2, 2 \rangle$$ ■

■ **EXAMPLE 2 Gradient in Three Variables** Calculate $\nabla f_{(3,-2,4)}$, where

$$f(x, y, z) = ze^{2x+3y}$$

Solution The partial derivatives and the gradient are

$$\frac{\partial f}{\partial x} = 2ze^{2x+3y}, \qquad \frac{\partial f}{\partial y} = 3ze^{2x+3y}, \qquad \frac{\partial f}{\partial z} = e^{2x+3y}$$

$$\nabla f = \langle 2ze^{2x+3y}, 3ze^{2x+3y}, e^{2x+3y} \rangle$$

Therefore, $\nabla f_{(3,-2,4)} = \langle 2 \cdot 4e^0, 3 \cdot 4e^0, e^0 \rangle = \langle 8, 12, 1 \rangle$. ■

The following theorem lists some useful properties of the gradient. The proofs are left as exercises (see Exercises 62–64).

THEOREM 1 Properties of the Gradient If $f(x, y, z)$ and $g(x, y, z)$ are differentiable and c is a constant, then

(i) $\nabla(f + g) = \nabla f + \nabla g$
(ii) $\nabla(cf) = c\nabla f$
(iii) Product Rule for Gradients: $\nabla(fg) = f\nabla g + g\nabla f$
(iv) Chain Rule for Gradients: If $F(t)$ is a differentiable function of one variable, then

$$\nabla(F(f(x, y, z))) = F'(f(x, y, z))\nabla f \qquad \boxed{1}$$

■ **EXAMPLE 3 Using the Chain Rule for Gradients** Find the gradient of

$$g(x, y, z) = (x^2 + y^2 + z^2)^8$$

Solution The function g is a composite $g(x, y, z) = F(f(x, y, z))$ with $F(t) = t^8$ and $f(x, y, z) = x^2 + y^2 + z^2$ and apply Eq. (1):

$$\nabla g = \nabla\big((x^2 + y^2 + z^2)^8\big) = 8(x^2 + y^2 + z^2)^7 \nabla(x^2 + y^2 + z^2)$$
$$= 8(x^2 + y^2 + z^2)^7 \langle 2x, 2y, 2z\rangle$$
$$= 16(x^2 + y^2 + z^2)^7 \langle x, y, z\rangle \qquad ■$$

The Chain Rule for Paths

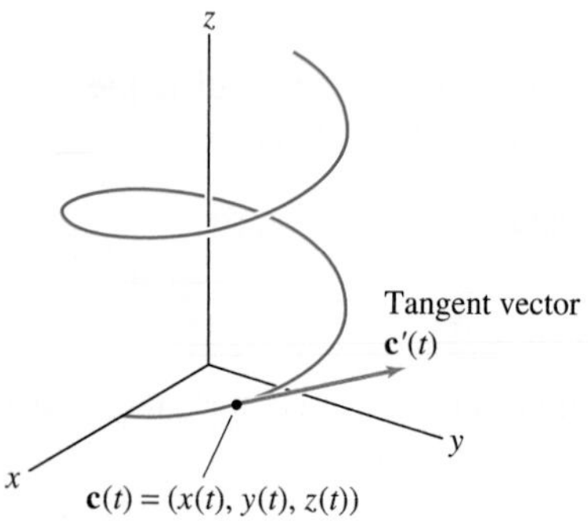

FIGURE 2 Tangent vector $\mathbf{c}'(t)$ to a path $\mathbf{c}(t) = (x(t), y(t), z(t))$.

Our first application of the gradient is the Chain Rule for Paths. In Chapter 14, we represented a path in $\mathbf{R}^3$ by a vector-valued function $\mathbf{r}(t) = \langle x(t), y(t), z(t)\rangle$. It is convenient to use a slightly different notation in this chapter.

A path will be represented by a function $\mathbf{c}(t) = (x(t), y(t), z(t))$. We think of $\mathbf{c}(t)$ as a moving *point* rather than as a moving *vector* (Figure 2). By definition, $\mathbf{c}'(t)$ is the vector of derivatives as before:

$$\mathbf{c}(t) = (x(t), y(t), z(t)), \qquad \mathbf{c}'(t) = \big\langle x'(t), y'(t), z'(t)\big\rangle$$

Recall from Section 14.2 that $\mathbf{c}'(t)$ is the tangent or "velocity" vector that is tangent to the path and points in the direction of motion. We use similar notation for paths in $\mathbf{R}^2$.

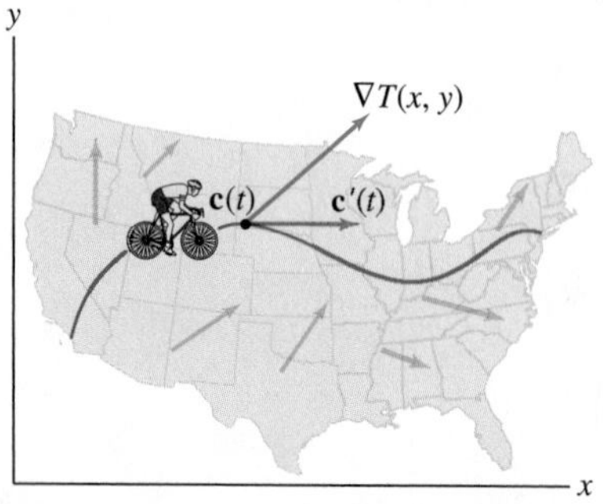

FIGURE 3 Chloe's temperature changes at the rate $\nabla T_{\mathbf{c}(t)} \cdot \mathbf{c}'(t)$.

The Chain Rule for Paths deals with composite functions of the type $f(\mathbf{c}(t))$. What is the idea behind a composite function of this type? As an example, suppose that $T(x, y)$ is the temperature at location (x, y) (Figure 3). Now imagine a biker—we'll call her Chloe—riding along a path $\mathbf{c}(t)$. We suppose that Chloe carries a thermometer with her and checks it as she rides. Her location at time t is $\mathbf{c}(t)$, so her temperature reading at time t is the composite function

$$T(\mathbf{c}(t)) = \text{Chloe's temperature at time } t$$

The temperature reading varies as Chloe's location changes, and the rate at which it changes is the derivative

$$\frac{d}{dt}T(\mathbf{c}(t))$$

The Chain Rule for Paths tells us that this derivative is simply the dot product of the temperature gradient ∇T evaluated at $\mathbf{c}(t)$ and Chloe's velocity vector $\mathbf{c}'(t)$.

CAUTION *Do not confuse the Chain Rule for Paths with the more elementary Chain Rule for Gradients stated in Theorem 1 above.*

THEOREM 2 Chain Rule for Paths If f and $\mathbf{c}(t)$ are differentiable, then

$$\frac{d}{dt} f(\mathbf{c}(t)) = \nabla f_{\mathbf{c}(t)} \cdot \mathbf{c}'(t)$$

Explicitly, in the case of two variables, if $\mathbf{c}(t) = (x(t), y(t))$, then

$$\frac{d}{dt} f(\mathbf{c}(t)) = \left\langle \frac{\partial f}{\partial x}, \frac{\partial f}{\partial y} \right\rangle \cdot \left\langle x'(t), y'(t) \right\rangle = \frac{\partial f}{\partial x}\frac{dx}{dt} + \frac{\partial f}{\partial y}\frac{dy}{dt}$$

Proof By definition,

$$\frac{d}{dt} f(\mathbf{c}(t)) = \lim_{h\to 0} \frac{f(x(t+h), y(t+h)) - f(x(t), y(t))}{h}$$

To calculate this derivative, set

$$\Delta f = f(x(t+h), y(t+h)) - f(x(t), y(t))$$

$$\Delta x = x(t+h) - x(t), \qquad \Delta y = y(t+h) - y(t)$$

The proof is based on the local linearity of f. As in Section 15.4, we write

$$\Delta f = f_x(x(t), y(t))\Delta x + f_y(x(t), y(t))\Delta y + e(x(t+h), y(t+h))$$

Now set $h = \Delta t$ and divide by Δt:

$$\frac{\Delta f}{\Delta t} = f_x(x(t), y(t))\frac{\Delta x}{\Delta t} + f_y(x(t), y(t))\frac{\Delta y}{\Delta t} + \frac{e(x(t+\Delta t), y(t+\Delta t))}{\Delta t}$$

Suppose for a moment that the last term tends to zero as $\Delta t \to 0$. Then we obtain the desired result:

$$\begin{aligned}
\frac{d}{dt} f(\mathbf{c}(t)) &= \lim_{\Delta t\to 0} \frac{\Delta f}{\Delta t} \\
&= f_x(x(t), y(t)) \lim_{\Delta t\to 0} \frac{\Delta x}{\Delta t} + f_y(x(t), y(t)) \lim_{\Delta t\to 0} \frac{\Delta y}{\Delta t} \\
&= f_x(x(t), y(t))\frac{dx}{dt} + f_y(x(t), y(t))\frac{dy}{dt} \\
&= \nabla f_{\mathbf{c}(t)} \cdot \mathbf{c}'(t)
\end{aligned}$$

We verify that the last term tends to zero as follows:

$$\lim_{\Delta t\to 0} \frac{e(x(t+\Delta t), y(t+\Delta t))}{\Delta t} = \lim_{\Delta t\to 0} \frac{e(x(t+\Delta t), y(t+\Delta t))}{\sqrt{(\Delta x)^2 + (\Delta y)^2}} \left(\frac{\sqrt{(\Delta x)^2 + (\Delta y)^2}}{\Delta t} \right)$$

$$= \underbrace{\left(\lim_{\Delta t\to 0} \frac{e(x(t+\Delta t), y(t+\Delta t))}{\sqrt{(\Delta x)^2 + (\Delta y)^2}} \right)}_{\text{Zero}} \lim_{\Delta t\to 0} \left(\sqrt{\left(\frac{\Delta x}{\Delta t}\right)^2 + \left(\frac{\Delta y}{\Delta t}\right)^2} \right) = 0$$

The first limit is zero because a differentiable function is locally linear (Section 15.4). The second limit is equal to $\sqrt{x'(t)^2 + y'(t)^2}$, so the product is zero. ■

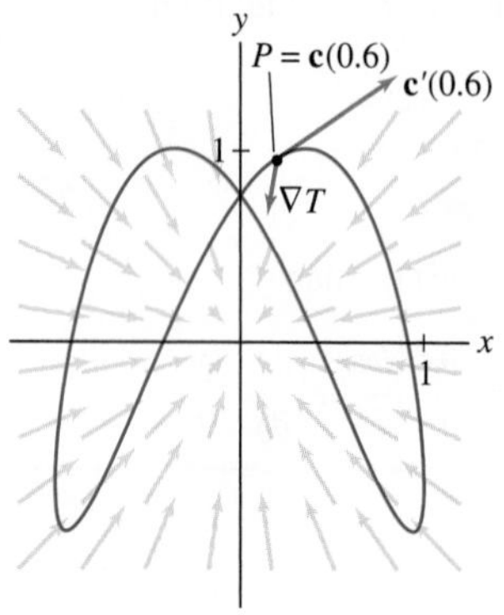

FIGURE 4 Gradient vectors ∇T and the path $\mathbf{c}(t) = (\cos(t-2), \sin 2t)$.

EXAMPLE 4 The temperature at location (x, y) is $T(x, y) = 20 + 10e^{-0.3(x^2+y^2)}$°C. A bug carries a tiny thermometer along the path

$$\mathbf{c}(t) = (\cos(t-2), \sin 2t)$$

(t in seconds) as in Figure 4. How fast is the temperature changing at $t = 0.6$ s?

Solution At $t = 0.6$ s, the bug is at location

$$\mathbf{c}(0.6) = (\cos(-1.4), \sin 0.6) \approx (0.170, 0.932)$$

By the Chain Rule for Paths, the rate of change of temperature is the dot product

$$\left.\frac{dT}{dt}\right|_{t=0.6} = \nabla T_{\mathbf{c}(0.6)} \cdot \mathbf{c}'(0.6)$$

We compute the vectors

$$\nabla T = \left\langle -6xe^{-0.3(x^2+y^2)}, -6ye^{-0.3(x^2+y^2)} \right\rangle$$
$$\mathbf{c}'(t) = \langle -\sin(t-2), 2\cos 2t \rangle$$

and evaluate at $\mathbf{c}(0.6) = (0.170, 0.932)$ using a calculator:

$$\nabla T_{\mathbf{c}(0.6)} \approx \langle -0.779, -4.272 \rangle$$
$$\mathbf{c}'(0.6) \approx \langle 0.985, 0.725 \rangle$$

Therefore, the rate of change is

$$\left.\frac{dT}{dt}\right|_{t=0.6} \nabla T_{\mathbf{c}(0.6)} \cdot \mathbf{c}'(t) \approx \langle -0.779, -4.272 \rangle \cdot \langle 0.985, 0.725 \rangle \approx -3.87\text{°C/s}$$ ■

In the next example, we apply the Chain Rule for Paths to a function of three variables. In general, if $f(x_1, \dots, x_n)$ is a differentiable function of n variables and $\mathbf{c}(t) = (x_1(t), \dots, x_n(t))$ is a differentiable path, then

$$\frac{d}{dt} f(\mathbf{c}(t)) = \nabla f \cdot \mathbf{c}'(t) = \frac{\partial f}{\partial x_1}\frac{dx_1}{dt} + \frac{\partial f}{\partial x_2}\frac{dx_2}{dt} + \cdots + \frac{\partial f}{\partial x_n}\frac{dx_n}{dt}$$

EXAMPLE 5 Calculate $\left.\frac{d}{dt} f(\mathbf{c}(t))\right|_{t=\pi/2}$, where

$$f(x, y, z) = xy + z^2 \qquad \text{and} \qquad \mathbf{c}(t) = (\cos t, \sin t, t)$$

Solution We have $\mathbf{c}\left(\frac{\pi}{2}\right) = \left(\cos\frac{\pi}{2}, \sin\frac{\pi}{2}, \frac{\pi}{2}\right) = \left(0, 1, \frac{\pi}{2}\right)$. Compute the gradient:

$$\nabla f = \left\langle \frac{\partial f}{\partial x}, \frac{\partial f}{\partial y}, \frac{\partial f}{\partial z} \right\rangle = \langle y, x, 2z \rangle, \qquad \nabla f_{\mathbf{c}(\pi/2)} = \nabla f\left(0, 1, \frac{\pi}{2}\right) = \langle 1, 0, \pi \rangle$$

Then compute the tangent vector:

$$\mathbf{c}'(t) = \langle -\sin t, \cos t, 1 \rangle, \qquad \mathbf{c}'\left(\frac{\pi}{2}\right) = \left\langle -\sin\frac{\pi}{2}, \cos\frac{\pi}{2}, 1 \right\rangle = \langle -1, 0, 1 \rangle$$

By the Chain Rule,

$$\left.\frac{d}{dt} f(\mathbf{c}(t))\right|_{t=\pi/2} = \nabla f_{\mathbf{c}(\pi/2)} \cdot \mathbf{c}'\left(\frac{\pi}{2}\right) = \langle 1, 0, \pi \rangle \cdot \langle -1, 0, 1 \rangle = \pi - 1$$ ■

Directional Derivatives

We come now to one of the most important applications of the Chain Rule for Paths. Consider a line through a point $P = (a, b)$ in the direction of a unit vector $\mathbf{u} = \langle h, k\rangle$ (see Figure 5):

$$\mathbf{c}(t) = (a + th, b + tk)$$

The derivative of $f(\mathbf{c}(t))$ at $t = 0$ is called the **directional derivative of f with respect to u at P**, and is denoted $D_{\mathbf{u}}f(P)$ or $D_{\mathbf{u}}f(a, b)$:

$$D_{\mathbf{u}}f(a, b) = \left.\frac{d}{dt}f(\mathbf{c}(t))\right|_{t=0} = \lim_{t\to 0}\frac{f(a + th, b + tk) - f(a, b)}{t}$$

Directional derivatives of functions of three or more variables are defined in a similar way.

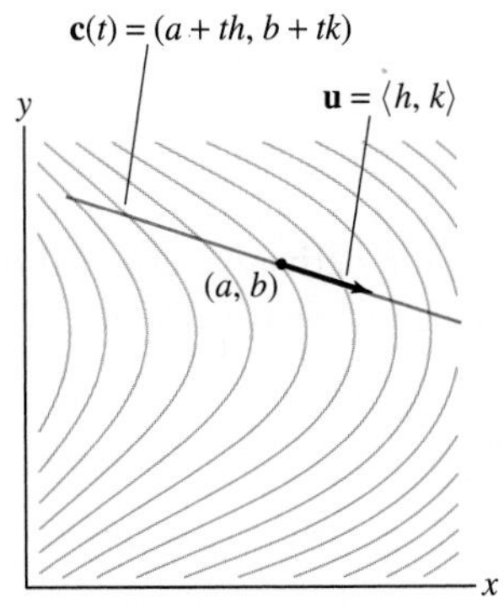

Contour map of $f(x, y)$

FIGURE 5 The directional derivative $D_{\mathbf{u}}f(a, b)$ is the rate of change of f along the linear path through P with direction vector $\mathbf{u}$.

DEFINITION Directional Derivative The directional derivative in the direction of a unit vector $\mathbf{u} = \langle h, k\rangle$ is the limit (assuming it exists)

$$D_{\mathbf{u}}f(P) = D_{\mathbf{u}}f(a, b) = \lim_{t\to 0}\frac{f(a + th, b + tk) - f(a, b)}{t}$$

Note that the partial derivatives are the directional derivatives with respect to the standard unit vectors $\mathbf{i} = \langle 1, 0\rangle$ and $\mathbf{j} = \langle 0, 1\rangle$. For example,

$$D_{\mathbf{i}}f(a, b) = \lim_{t\to 0}\frac{f(a + t(1), b + t(0)) - f(a, b)}{t} = \lim_{t\to 0}\frac{f(a + t, b) - f(a, b)}{t}$$
$$= f_x(a, b)$$

Thus we have

$$f_x(a, b) = D_{\mathbf{i}}f(a, b), \qquad f_y(a, b) = D_{\mathbf{j}}f(a, b)$$

CONCEPTUAL INSIGHT The directional derivative $D_{\mathbf{u}}f(P)$ is the rate of change of f per *unit change* in the horizontal direction of $\mathbf{u}$ at P (Figure 6). This is the slope of the tangent line at Q to the trace curve obtained when we intersect the graph with the vertical plane through P in the direction $\mathbf{u}$.

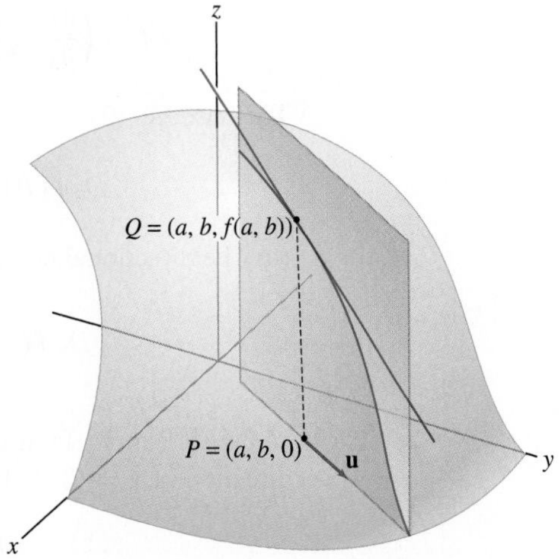

FIGURE 6 $D_{\mathbf{u}}f(a, b)$ is the slope of the tangent line to the trace curve through Q in the vertical plane through P in the direction $\mathbf{u}$.

To evaluate directional derivatives, it is convenient to define $D_{\mathbf{v}} f(a, b)$ even when $\mathbf{v} = \langle h, k \rangle$ is not a unit vector:

$$D_{\mathbf{v}} f(a, b) = \left. \frac{d}{dt} f(\mathbf{c}(t)) \right|_{t=0} = \lim_{t \to 0} \frac{f(a + th, b + tk) - f(a, b)}{t}$$

We call $D_{\mathbf{v}} f$ the **derivative with respect to v**.

If we set $\mathbf{c}(t) = (a + th, b + tk)$, then $D_{\mathbf{v}} f(a, b)$ is the derivative at $t = 0$ of the composite function $f(\mathbf{c}(t))$, where $\mathbf{c}(t) = (a + th, b + tk)$, and we can evaluate it using the Chain Rule for Paths. We have $\mathbf{c}'(t) = \langle h, k \rangle = \mathbf{v}$, so

$$D_{\mathbf{v}} f(a, b) = \nabla f_{(a,b)} \cdot \mathbf{c}'(0) = \nabla f_{(a,b)} \cdot \mathbf{v}$$

This yields the basic formula:

$$D_{\mathbf{v}} f(a, b) = \nabla f_{(a,b)} \cdot \mathbf{v} \qquad \boxed{2}$$

Similarly, in three variables, $D_{\mathbf{v}} f(a, b, c) = \nabla f_{(a,b,c)} \cdot \mathbf{v}$.

For any scalar λ, $D_{\lambda \mathbf{v}} f(P) = \nabla f_P \cdot (\lambda \mathbf{v}) = \lambda \nabla f_P \cdot \mathbf{v}$. Therefore,

$$D_{\lambda \mathbf{v}} f(P) = \lambda D_{\mathbf{v}} f(P) \qquad \boxed{3}$$

If $\mathbf{v} \neq \mathbf{0}$, then $\mathbf{u} = \frac{1}{\|\mathbf{v}\|}\mathbf{v}$ is a unit vector in the direction of $\mathbf{v}$. Applying Eq. (3) with $\lambda = 1/\|\mathbf{u}\|$ gives us a formula for the directional derivative $D_{\mathbf{u}} f(P)$ in terms of $D_{\mathbf{v}} f(P)$.

THEOREM 3 Computing the Directional Derivative If $\mathbf{v} \neq \mathbf{0}$, then $\mathbf{u} = \mathbf{v}/\|\mathbf{v}\|$ is the unit vector in the direction of $\mathbf{v}$, and the directional derivative is given by

$$D_{\mathbf{u}} f(P) = \frac{1}{\|\mathbf{v}\|} \nabla f_P \cdot \mathbf{v} \qquad \boxed{4}$$

EXAMPLE 6 Let $f(x, y) = xe^y$, $P = (2, -1)$, and $\mathbf{v} = \langle 2, 3 \rangle$.

(a) Calculate $D_{\mathbf{v}} f(P)$.

(b) Then calculate the directional derivative in the direction of $\mathbf{v}$.

Solution (a) First compute the gradient at $P = (2, -1)$:

$$\nabla f = \left\langle \frac{\partial f}{\partial x}, \frac{\partial f}{\partial y} \right\rangle = \langle e^y, xe^y \rangle \quad \Rightarrow \quad \nabla f_P = \nabla f_{(2,-1)} = \left\langle e^{-1}, 2e^{-1} \right\rangle$$

Then use Eq. (2):

$$D_{\mathbf{v}} f(P) = \nabla f_P \cdot \mathbf{v} = \left\langle e^{-1}, 2e^{-1} \right\rangle \cdot \langle 2, 3 \rangle = 8e^{-1} \approx 2.94$$

(b) The directional derivative is $D_{\mathbf{u}} f(P)$, where $\mathbf{u} = \mathbf{v}/\|\mathbf{v}\|$. By Eq. 4,

$$D_{\mathbf{u}} f(P) = \frac{1}{\|\mathbf{v}\|} D_{\mathbf{v}} f(P) = \frac{8e^{-1}}{\sqrt{2^2 + 3^2}} = \frac{8e^{-1}}{\sqrt{13}} \approx 0.82 \qquad \blacksquare$$

EXAMPLE 7 Find the rate of change of pressure at the point $Q = (1, 2, 1)$ in the direction of $\mathbf{v} = \langle 0, 1, 1 \rangle$, assuming that the pressure (in millibars) is given by

$$f(x, y, z) = 1000 + 0.01(yz^2 + x^2 z - xy^2) \qquad (x, y, z \text{ in kilometers})$$

Solution First compute the gradient at $Q = (1, 2, 1)$:

$$\nabla f = 0.01\left\langle 2xz - y^2, z^2 - 2xy, 2yz + x^2\right\rangle$$

$$\nabla f_Q = \nabla f_{(1,2,1)} = \langle -0.02, -0.03, 0.05\rangle$$

Then use Eq. (2) to compute the derivative with respect to $\mathbf{v}$:

$$D_{\mathbf{v}} f(Q) = \nabla f_Q \cdot \mathbf{v} = \langle -0.02, -0.03, 0.05\rangle \cdot \langle 0, 1, 1\rangle = 0.01(-3+5) = 0.02$$

The rate of change per kilometer is the directional derivative. The unit vector in the direction of $\mathbf{v}$ is $\mathbf{u} = \mathbf{v}/\|\mathbf{v}\|$. Since $\|\mathbf{v}\| = \sqrt{2}$, Eq. (4) yields

$$D_{\mathbf{u}} f(Q) = \frac{1}{\|\mathbf{v}\|} D_{\mathbf{v}} f(Q) = \frac{0.02}{\sqrt{2}} \approx 0.014 \text{ mb/km}$$ ■

Properties of the Gradient

REMINDER For any vectors $\mathbf{u}$ *and* $\mathbf{v}$,

$$\mathbf{v}\cdot\mathbf{u} = \|\mathbf{v}\|\|\mathbf{u}\|\cos\theta$$

where θ *is the angle between* $\mathbf{v}$ *and* $\mathbf{u}$. *If* $\mathbf{u}$ *is a unit vector, then*

$$\mathbf{v}\cdot\mathbf{u} = \|\mathbf{v}\|\cos\theta$$

We are now in a position to draw some interesting and important conclusions about the gradient. First, suppose that $\nabla f_P \neq \mathbf{0}$ and let $\mathbf{u}$ be a unit vector (Figure 7). By the properties of the dot product,

$$D_{\mathbf{u}} f(P) = \nabla f_P \cdot \mathbf{u} = \|\nabla f_P\|\cos\theta \qquad \boxed{5}$$

where θ is the angle between ∇f_P and $\mathbf{u}$. In other words, *the rate of change in a given direction varies with the cosine of the angle* θ *between the gradient and the direction.*

Because the cosine takes values between -1 and 1, we have

$$-\|\nabla f_P\| \le D_{\mathbf{u}} f(P) \le \|\nabla f_P\|$$

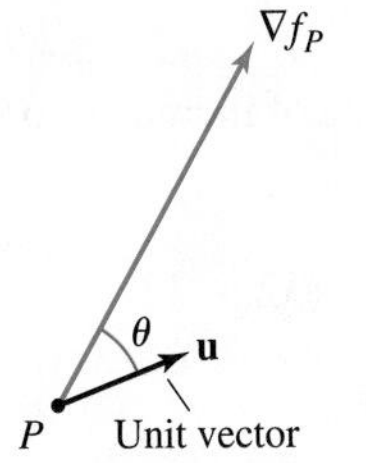

∇f_P
θ
$\mathbf{u}$
P
Unit vector

FIGURE 7 $D_{\mathbf{u}} f(P) = \|\nabla f_P\|\cos\theta$.

Since $\cos 0 = 1$, the maximum value of $D_{\mathbf{u}} f(P)$ occurs for $\theta = 0$—that is, when $\mathbf{u}$ points in the direction of ∇f_P. In other words the *gradient vector points in the direction of the maximum rate of increase, and this maximum rate is* $\|\nabla f_P\|$. Similarly, f decreases most rapidly in the opposite direction, $-\nabla f_P$, because $\cos\theta = -1$ for $\theta = \pi$. The rate of maximum decrease is $-\|\nabla f_P\|$. The directional derivative is zero in directions orthogonal to the gradient because $\cos\frac{\pi}{2} = 0$.

In the earlier scenario where the biker Chloe rides along a path (Figure 8), the temperature T changes at a rate that depends on the cosine of the angle θ between ∇T and the direction of motion.

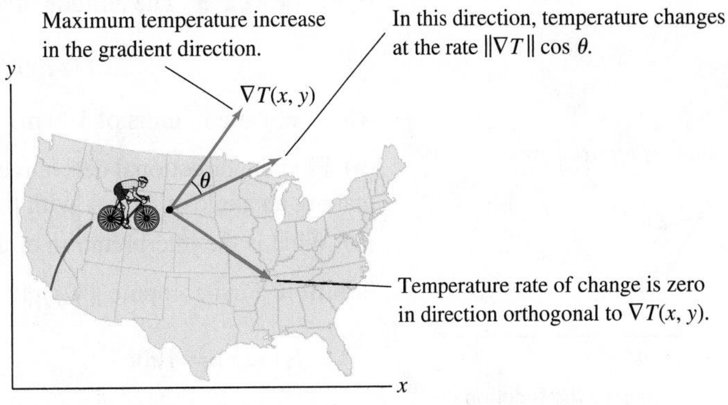

Maximum temperature increase in the gradient direction.
In this direction, temperature changes at the rate $\|\nabla T\|\cos\theta$.
y
$\nabla T(x, y)$
θ
Temperature rate of change is zero in direction orthogonal to $\nabla T(x, y)$.
x

FIGURE 8

REMINDER

- *The words "normal" and "orthogonal" both mean "perpendicular."*
- *We say that a vector is normal to a curve at a point P if it is normal to the tangent line to the curve at P.*

Another key property is that gradient vectors are normal to level curves (Figure 9). To prove this, suppose that P lies on the level curve $f(x, y) = k$. We parametrize this level curve by a path $\mathbf{c}(t)$ such that $\mathbf{c}(0) = P$ and $\mathbf{c}'(0) \neq \mathbf{0}$ (this is possible whenever $\nabla f_P \neq \mathbf{0}$). Then $f(\mathbf{c}(t)) = k$ for all t, so by the Chain Rule,

$$\nabla f_P \cdot \mathbf{c}'(0) = \frac{d}{dt} f(\mathbf{c}(t))\bigg|_{t=0} = \frac{d}{dt}k = 0$$

This proves that ∇f_P is orthogonal to $\mathbf{c}'(0)$, and since $\mathbf{c}'(0)$ is tangent to the level curve, we conclude that ∇f_P is normal to the level curve (Figure 9). For functions of three variables, a similar argument shows that ∇f_P is normal to the level surface $f(x, y, z) = k$ through P.

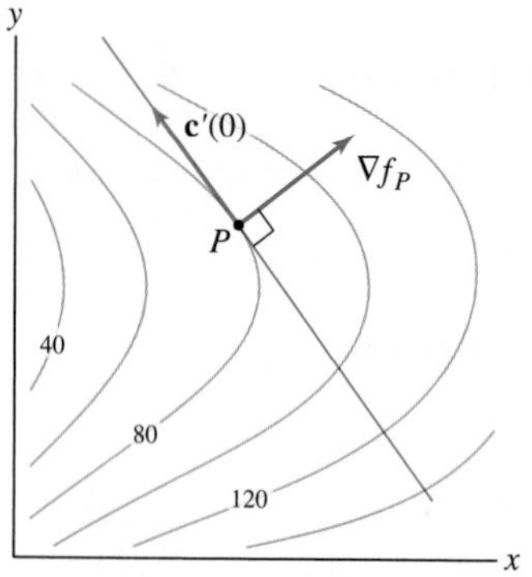

FIGURE 9 Contour map of $f(x, y)$. The gradient at P is orthogonal to the level curve through P.

THEOREM 4 Interpretation of the Gradient Assume that $\nabla f_P \neq \mathbf{0}$. Let $\mathbf{u}$ be a unit vector making an angle θ with ∇f_P. Then

$$D_{\mathbf{u}} f(P) = \|\nabla f_P\| \cos\theta \qquad (6)$$

- ∇f_P points in the direction of maximum rate of increase of f at P.
- $-\nabla f_P$ points in the direction of maximum rate of decrease at P.
- ∇f_P is normal to the level curve (or surface) of f at P.

GRAPHICAL INSIGHT At each point P, there is a unique direction in which $f(x, y)$ increases most rapidly (per unit distance). Theorem 4 tells us that this chosen direction is perpendicular to the level curves and that it is specified by the gradient vector (Figure 10). For most functions, however, the direction of maximum rate of increase varies from point to point.

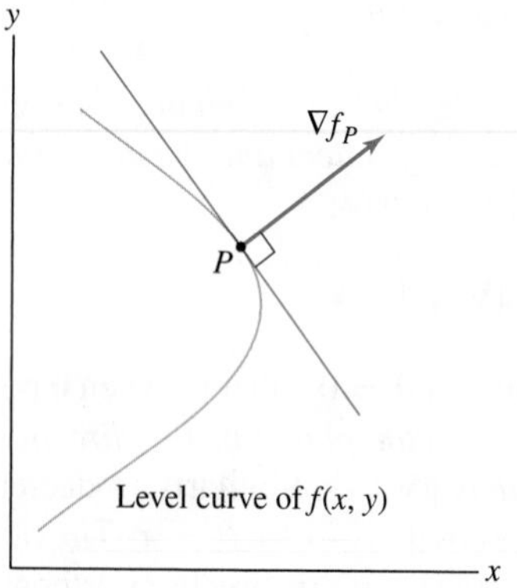

FIGURE 10 The gradient points in the direction of maximum increase.

■ **EXAMPLE 8** Let $f(x, y) = x^4 y^{-2}$ and $P = (2, 1)$. Find the unit vector that points in the direction of maximum rate of increase at P.

Solution The gradient points in the direction of maximum rate of increase, so we evaluate the gradient at P:

$$\nabla f = \left\langle 4x^3 y^{-2}, -2x^4 y^{-3} \right\rangle, \qquad \nabla f_{(2,1)} = \langle 32, -32\rangle$$

The unit vector in this direction is

$$\mathbf{u} = \frac{\langle 32, -32\rangle}{\|\langle 32, -32\rangle\|} = \frac{\langle 32, -32\rangle}{32\sqrt{2}} = \left\langle \frac{\sqrt{2}}{2}, -\frac{\sqrt{2}}{2} \right\rangle$$ ■

■ **EXAMPLE 9** The altitude of a mountain at (x, y) is

$$f(x, y) = 2500 + 100(x + y^2)e^{-0.3y^2}$$

where x, y are in units of 100 m.

(a) Find the directional derivative of f at $P = (-1, -1)$ in the direction of unit vector $\mathbf{u}$ making an angle of $\theta = \frac{\pi}{4}$ with the gradient (Figure 11).

(b) What is the interpretation of this derivative?

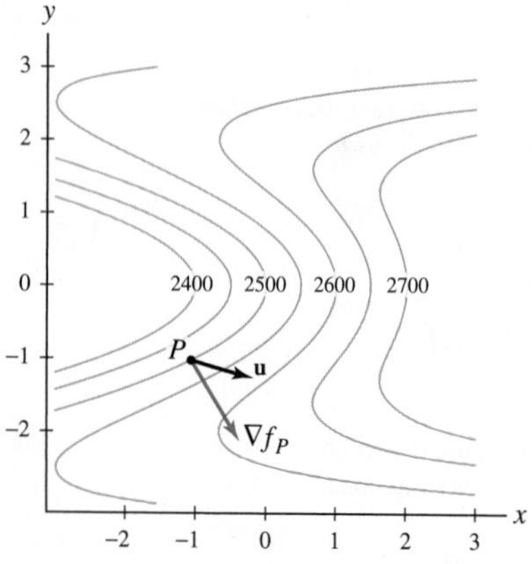

FIGURE 11 Contour map of the function $f(x, y)$ in Example 9.

Solution First compute $\|\nabla f_P\|$:

$$f_x(x, y) = 100e^{-0.3y^2}, \qquad f_y(x, y) = 100y(2 - 0.6x - 0.6y^2)e^{-0.3y^2}$$

$$f_x(-1, -1) = 100e^{-0.3} \approx 74, \qquad f_y(-1, -1) = -200e^{-0.3} \approx -148$$

Hence, $\nabla f_P \approx \langle 74, -148\rangle$ and

$$\|\nabla f_P\| \approx \sqrt{74^2 + (-148)^2} \approx 165.5$$

Apply Eq. (6) with $\theta = \pi/4$:

$$D_{\mathbf{u}} f(P) = \|\nabla f_P\| \cos\theta \approx 165.5 \left(\frac{\sqrt{2}}{2}\right) \approx 117$$

Recall that x and y are measured in units of 100 meters. Therefore, the interpretation is: If you stand on the mountain at the point lying above $(-1, -1)$ and begin climbing so that your horizontal displacement is in the direction of **u**, then your altitude increases at a rate of 117 meters per 100 meters of horizontal displacement, or 1.17 meters per meter of horizontal displacement. ■

The symbol ψ (pronounced "p-sigh" or "p-see") is the lowercase Greek letter psi.

CONCEPTUAL INSIGHT The directional derivative is related to the **angle of inclination** ψ in Figure 12. Think of the graph of $z = f(x, y)$ as a mountain lying over the xy-plane. Let Q be the point on the mountain lying above a point $P = (a, b)$ in the xy-plane. If you start moving up the mountain so that your horizontal displacement is in the direction of **u**, then you will actually be moving up the mountain at an angle of inclination ψ defined by

$$\tan\psi = D_{\mathbf{u}} f(P) \qquad \boxed{7}$$

The steepest direction up the mountain is the direction for which the horizontal displacement is in the direction of ∇f_P.

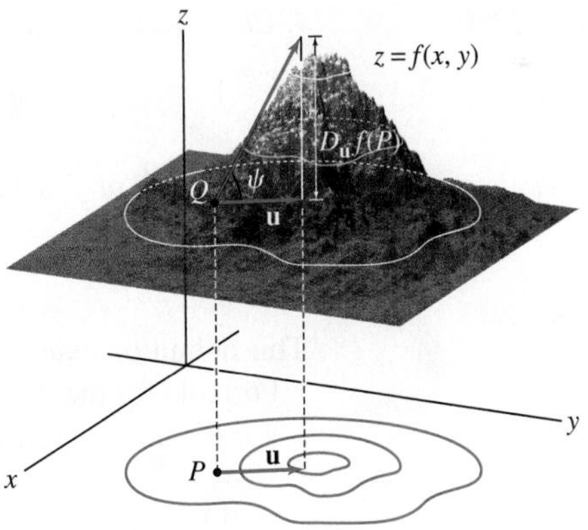

FIGURE 12

■ **EXAMPLE 10 Angle of Inclination** You are standing on the side of a mountain in the shape $z = f(x, y)$, at a point $Q = (a, b, f(a, b))$, where $\nabla f_{(a,b)} = \langle 0.4, 0.02\rangle$. Find the angle of inclination in a direction making an angle of $\theta = \frac{\pi}{3}$ with the gradient.

Solution The gradient has length $\|\nabla f_{(a,b)}\| = \sqrt{(0.4)^2 + (0.02)^2} \approx 0.4$. If **u** is a unit vector making an angle of $\theta = \frac{\pi}{3}$ with $\nabla f_{(a,b)}$, then

$$D_{\mathbf{u}} f(a, b) = \|\nabla f_{(a,b)}\| \cos\frac{\pi}{3} \approx (0.4)(0.5) = 0.2$$

The angle of inclination at Q in the direction of **u** satisfies $\tan\psi = 0.2$. It follows that $\psi \approx \tan^{-1} 0.2 \approx 0.197$ rad or approximately 11.3°. ■

Another use of the gradient is in finding normal vectors on a surface with equation $F(x, y, z) = k$, where k is a constant. Let $P = (a, b, c)$ and assume that $\nabla F_P \neq \mathbf{0}$. Then ∇F_P is normal to the level surface $F(x, y, z) = k$ by Theorem 4. The tangent plane at P has equation

$$\nabla F_P \cdot \langle x - a, y - b, z - c \rangle = 0$$

Expanding the dot product, we obtain

$$F_x(a, b, c)(x - a) + F_y(a, b, c)(y - b) + F_z(a, b, c)(z - c) = 0$$

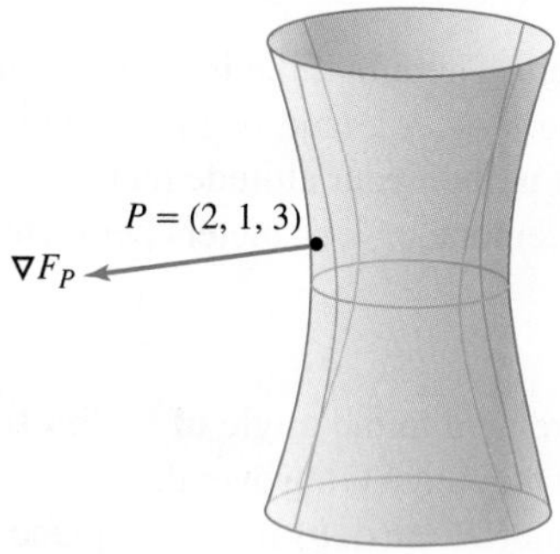

FIGURE 13 The gradient vector ∇F_P is normal to the surface at P.

■ **EXAMPLE 11 Normal Vector and Tangent Plane** Find an equation of the tangent plane to the surface $4x^2 + 9y^2 - z^2 = 16$ at $P = (2, 1, 3)$.

Solution Let $F(x, y, z) = 4x^2 + 9y^2 - z^2$. Then

$$\nabla F = \langle 8x, 18y, -2z \rangle, \qquad \nabla F_P = \nabla F_{(2,1,3)} = \langle 16, 18, -6 \rangle$$

The vector $\langle 16, 18, -6 \rangle$ is normal to the surface $F(x, y, z) = 16$ (Figure 13), so the tangent plane at P has equation

$$16(x - 2) + 18(y - 1) - 6(z - 3) = 0 \qquad \text{or} \qquad 16x + 18y - 6z = 32$$ ■

15.5 SUMMARY

- The *gradient* of a function f is the vector of partial derivatives:

$$\nabla f = \left\langle \frac{\partial f}{\partial x}, \frac{\partial f}{\partial y} \right\rangle \qquad \text{or} \qquad \nabla f = \left\langle \frac{\partial f}{\partial x}, \frac{\partial f}{\partial y}, \frac{\partial f}{\partial z} \right\rangle$$

- *Chain Rule for Paths:*

$$\frac{d}{dt} f(\mathbf{c}(t)) = \nabla f_{\mathbf{c}(t)} \cdot \mathbf{c}'(t)$$

- *Derivative of f with respect to $\mathbf{v} = \langle h, k \rangle$:*

$$D_{\mathbf{v}} f(a, b) = \lim_{t \to 0} \frac{f(a + th, b + tk) - f(a, b)}{t}$$

This definition extends to three or more variables.

- Formula for the derivative with respect to $\mathbf{v}$: $D_{\mathbf{v}} f(a, b) = \nabla f_{(a,b)} \cdot \mathbf{v}$.
- For $\mathbf{u}$ a unit vector, $D_{\mathbf{u}} f$ is called the *directional derivative*.
  - If $\mathbf{u} = \dfrac{\mathbf{v}}{\|\mathbf{v}\|}$, then $D_{\mathbf{u}} f(a, b) = \dfrac{1}{\|\mathbf{v}\|} D_{\mathbf{v}} f(a, b)$.
  - $D_{\mathbf{u}} f(a, b) = \|\nabla f_{(a,b)}\| \cos\theta$, where θ is the angle between $\nabla f_{(a,b)}$ and $\mathbf{u}$.
- Basic geometric properties of the gradient (assume $\nabla f_P \neq \mathbf{0}$):
  - ∇f_P points in the direction of maximum rate of increase. The maximum rate of increase is $\|\nabla f_P\|$.
  - $-\nabla f_P$ points in the direction of maximum rate of decrease. The maximum rate of decrease is $-\|\nabla f_P\|$.
  - ∇f_P is orthogonal to the level curve (or surface) through P.
- Equation of the tangent plane to the level surface $F(x, y, z) = k$ at $P = (a, b, c)$:

$$\nabla F_P \cdot \langle x - a, y - b, z - c \rangle = 0$$

$$F_x(a, b, c)(x - a) + F_y(a, b, c)(y - b) + F_z(a, b, c)(z - c) = 0$$

15.5 EXERCISES

Preliminary Questions

1. Which of the following is a possible value of the gradient ∇f of a function $f(x, y)$ of two variables?

(a) 5 **(b)** $\langle 3, 4 \rangle$ **(c)** $\langle 3, 4, 5 \rangle$

2. True or false? A differentiable function increases at the rate $\|\nabla f_P\|$ in the direction of ∇f_P.

3. Describe the two main geometric properties of the gradient ∇f.

4. You are standing at a point where the temperature gradient vector is pointing in the northeast (NE) direction. In which direction(s) should you walk to avoid a change in temperature?

(a) NE **(b)** NW **(c)** SE **(d)** SW

5. What is the rate of change of $f(x, y)$ at $(0, 0)$ in the direction making an angle of 45° with the x-axis if $\nabla f(0, 0) = \langle 2, 4 \rangle$?

Exercises

1. Let $f(x, y) = xy^2$ and $\mathbf{c}(t) = \left(\frac{1}{2}t^2, t^3\right)$.

(a) Calculate ∇f and $\mathbf{c}'(t)$.

(b) Use the Chain Rule for Paths to evaluate $\frac{d}{dt} f(\mathbf{c}(t))$ at $t = 1$ and $t = -1$.

2. Let $f(x, y) = e^{xy}$ and $\mathbf{c}(t) = (t^3, 1 + t)$.

(a) Calculate ∇f and $\mathbf{c}'(t)$.

(b) Use the Chain Rule for Paths to calculate $\frac{d}{dt} f(\mathbf{c}(t))$.

(c) Write out the composite $f(\mathbf{c}(t))$ as a function of t and differentiate. Check that the result agrees with part (b).

3. Figure 14 shows the level curves of a function $f(x, y)$ and a path $\mathbf{c}(t)$, traversed in the direction indicated. State whether the derivative $\frac{d}{dt} f(\mathbf{c}(t))$ is positive, negative, or zero at points A–D.

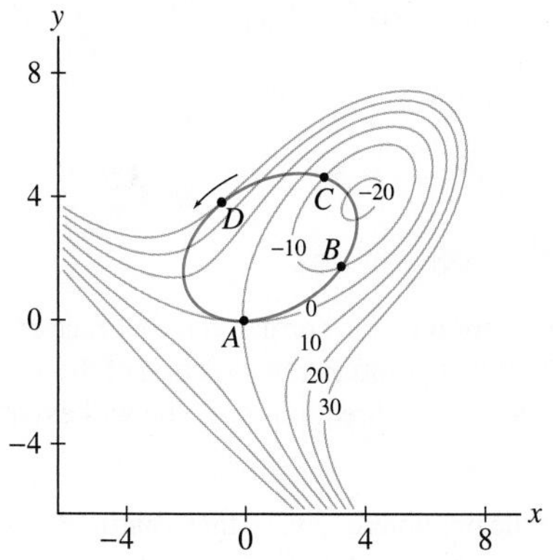

FIGURE 14

4. Let $f(x, y) = x^2 + y^2$ and $\mathbf{c}(t) = (\cos t, \sin t)$.

(a) Find $\frac{d}{dt} f(\mathbf{c}(t))$ without making any calculations. Explain.

(b) Verify your answer to (a) using the Chain Rule.

In Exercises 5–8, calculate the gradient.

5. $f(x, y) = \cos(x^2 + y)$

6. $g(x, y) = \dfrac{x}{x^2 + y^2}$

7. $h(x, y, z) = xyz^{-3}$

8. $r(x, y, z, w) = xze^{yw}$

In Exercises 9–20, use the Chain Rule to calculate $\frac{d}{dt} f(\mathbf{c}(t))$.

9. $f(x, y) = 3x - 7y, \quad \mathbf{c}(t) = (\cos t, \sin t), \quad t = 0$

10. $f(x, y) = 3x - 7y, \quad \mathbf{c}(t) = (t^2, t^3), \quad t = 2$

11. $f(x, y) = x^2 - 3xy, \quad \mathbf{c}(t) = (\cos t, \sin t), \quad t = 0$

12. $f(x, y) = x^2 - 3xy, \quad \mathbf{c}(t) = (\cos t, \sin t), \quad t = \frac{\pi}{2}$

13. $f(x, y) = \sin(xy), \quad \mathbf{c}(t) = (e^{2t}, e^{3t}), \quad t = 0$

14. $f(x, y) = \cos(y - x), \quad \mathbf{c}(t) = (e^t, e^{2t}), \quad t = \ln 3$

15. $f(x, y) = x - xy, \quad \mathbf{c}(t) = (t^2, t^2 - 4t), \quad t = 4$

16. $f(x, y) = xe^y, \quad \mathbf{c}(t) = (t^2, t^2 - 4t), \quad t = 0$

17. $f(x, y) = \ln x + \ln y, \quad \mathbf{c}(t) = (\cos t, t^2), \quad t = \frac{\pi}{4}$

18. $g(x, y, z) = xye^z, \quad \mathbf{c}(t) = (t^2, t^3, t - 1), \quad t = 1$

19. $g(x, y, z) = xyz^{-1}, \quad \mathbf{c}(t) = (e^t, t, t^2), \quad t = 1$

20. $g(x, y, z, w) = x + 2y + 3z + 5w, \quad \mathbf{c}(t) = (t^2, t^3, t, t-2), \quad t = 1$

In Exercises 21–30, calculate the directional derivative in the direction of $\mathbf{v}$ at the given point. Remember to normalize the direction vector or use Eq. (4).

21. $f(x, y) = x^2 + y^3, \quad \mathbf{v} = \langle 4, 3 \rangle, \quad P = (1, 2)$

22. $f(x, y) = x^2y^3, \quad \mathbf{v} = \mathbf{i} + \mathbf{j}, \quad P = (-2, 1)$

23. $f(x, y) = x^2y^3, \quad \mathbf{v} = \mathbf{i} + \mathbf{j}, \quad P = \left(\frac{1}{6}, 3\right)$

24. $f(x, y) = \sin(x - y), \quad \mathbf{v} = \langle 1, 1 \rangle, \quad P = \left(\frac{\pi}{2}, \frac{\pi}{6}\right)$

25. $f(x, y) = \tan^{-1}(xy), \quad \mathbf{v} = \langle 1, 1 \rangle, \quad P = (3, 4)$

26. $f(x, y) = e^{xy - y^2}, \quad \mathbf{v} = \langle 12, -5 \rangle, \quad P = (2, 2)$

27. $f(x, y) = \ln(x^2 + y^2), \quad \mathbf{v} = 3\mathbf{i} - 2\mathbf{j}, \quad P = (1, 0)$

28. $g(x, y, z) = z^2 - xy^2, \quad \mathbf{v} = \langle -1, 2, 2 \rangle, \quad P = (2, 1, 3)$

29. $g(x, y, z) = xe^{-yz}, \quad \mathbf{v} = \langle 1, 1, 1 \rangle, \quad P = (1, 2, 0)$

30. $g(x, y, z) = x\ln(y + z), \quad \mathbf{v} = 2\mathbf{i} - \mathbf{j} + \mathbf{k}, \quad P = (2, e, e)$

31. Find the directional derivative of $f(x, y) = x^2 + 4y^2$ at $P = (3, 2)$ in the direction pointing to the origin.

32. Find the directional derivative of $f(x, y, z) = xy + z^3$ at $P = (3, -2, -1)$ in the direction pointing to the origin.

33. A bug located at $(3, 9, 4)$ begins walking in a straight line toward $(5, 7, 3)$. At what rate is the bug's temperature changing if the temperature is $T(x, y, z) = xe^{y-z}$? Units are in meters and degrees Celsius.

34. The temperature at location (x, y) is $T(x, y) = 20 + 0.1(x^2 - xy)$ (degrees Celsius). Beginning at $(200, 0)$ at time $t = 0$ (seconds), a bug travels along a circle of radius 200 cm centered at the origin, at a speed of 3 cm/s. How fast is the temperature changing at time $t = \pi/3$?

35. Suppose that $\nabla f_P = \langle 2, -4, 4 \rangle$. Is f increasing or decreasing at P in the direction $\mathbf{v} = \langle 2, 1, 3 \rangle$?

36. Let $f(x, y) = xe^{x^2 - y}$ and $P = (1, 1)$.

(a) Calculate $\|\nabla f_P\|$.

(b) Find the rate of change of f in the direction ∇f_P.

(c) Find the rate of change of f in the direction of a vector making an angle of 45° with ∇f_P.

37. Let $f(x, y, z) = \sin(xy + z)$ and $P = (0, -1, \pi)$. Calculate $D_{\mathbf{u}} f(P)$, where $\mathbf{u}$ is a unit vector making an angle $\theta = 30°$ with ∇f_P.

38. Let $T(x, y)$ be the temperature at location (x, y). Assume that $\nabla T = \langle y - 4, x + 2y \rangle$. Let $\mathbf{c}(t) = (t^2, t)$ be a path in the plane. Find the values of t such that

$$\frac{d}{dt} T(\mathbf{c}(t)) = 0$$

39. Find a vector normal to the surface $x^2 + y^2 - z^2 = 6$ at $P = (3, 1, 2)$.

40. Find a vector normal to the surface $3z^3 + x^2y - y^2x = 1$ at $P = (1, -1, 1)$.

41. Find the two points on the ellipsoid

$$\frac{x^2}{4} + \frac{y^2}{9} + z^2 = 1$$

where the tangent plane is normal to $\mathbf{v} = \langle 1, 1, -2 \rangle$.

In Exercises 42–45, find an equation of the tangent plane to the surface at the given point.

42. $x^2 + 3y^2 + 4z^2 = 20, \quad P = (2, 2, 1)$

43. $xz + 2x^2y + y^2z^3 = 11, \quad P = (2, 1, 1)$

44. $x^2 + z^2e^{y-x} = 13, \quad P = \left(2, 3, \frac{3}{\sqrt{e}}\right)$

45. $\ln[1 + 4x^2 + 9y^4] - 0.1z^2 = 0, \quad P = (3, 1, 6.1876)$

46. Verify what is clear from Figure 15: Every tangent plane to the cone $x^2 + y^2 - z^2 = 0$ passes through the origin.

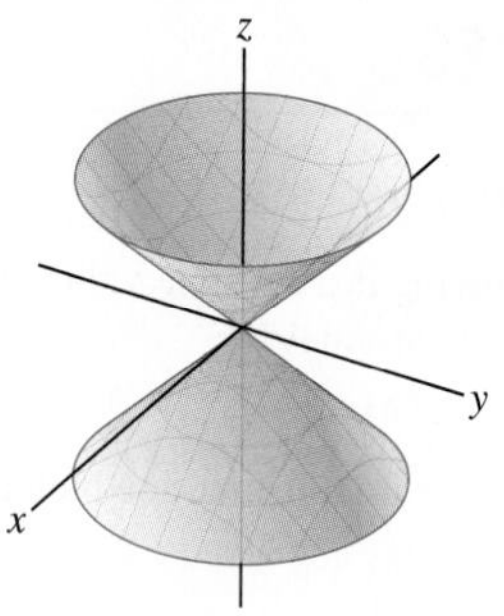

FIGURE 15 Graph of $x^2 + y^2 - z^2 = 0$.

47. CAS Use a computer algebra system to produce a contour plot of $f(x, y) = x^2 - 3xy + y - y^2$ together with its gradient vector field on the domain $[-4, 4] \times [-4, 4]$.

48. Find a function $f(x, y, z)$ such that ∇f is the constant vector $\langle 1, 3, 1 \rangle$.

49. Find a function $f(x, y, z)$ such that $\nabla f = \langle 2x, 1, 2 \rangle$.

50. Find a function $f(x, y, z)$ such that $\nabla f = \langle x, y^2, z^3 \rangle$.

51. Find a function $f(x, y, z)$ such that $\nabla f = \langle z, 2y, x \rangle$.

52. Find a function $f(x, y)$ such that $\nabla f = \langle y, x \rangle$.

53. Show that there does not exist a function $f(x, y)$ such that $\nabla f = \langle y^2, x \rangle$. *Hint:* Use Clairaut's Theorem $f_{xy} = f_{yx}$.

54. Let $\Delta f = f(a + h, b + k) - f(a, b)$ be the change in f at $P = (a, b)$. Set $\Delta \mathbf{v} = \langle h, k \rangle$. Show that the linear approximation can be written

$$\Delta f \approx \nabla f_P \cdot \Delta \mathbf{v} \qquad \boxed{8}$$

55. Use Eq. (8) to estimate

$$\Delta f = f(3.53, 8.98) - f(3.5, 9)$$

assuming that $\nabla f_{(3.5,9)} = \langle 2, -1 \rangle$.

56. Find a unit vector $\mathbf{n}$ that is normal to the surface $z^2 - 2x^4 - y^4 = 16$ at $P = (2, 2, 8)$ that points in the direction of the xy-plane (in other words, if you travel in the direction of $\mathbf{n}$, you will eventually cross the xy-plane).

57. Suppose, in the previous exercise, that a particle located at the point $P = (2, 2, 8)$ travels toward the xy-plane in the direction normal to the surface.

(a) Through which point Q on the xy-plane will the particle pass?

(b) Suppose the axes are calibrated in centimeters. Determine the path $\mathbf{c}(t)$ of the particle if it travels at a constant speed of 8 cm/s. How long will it take the particle to reach Q?

58. Let $f(x, y) = \tan^{-1} \dfrac{x}{y}$ and $\mathbf{u} = \left\langle \dfrac{\sqrt{2}}{2}, \dfrac{\sqrt{2}}{2} \right\rangle$.

(a) Calculate the gradient of f.

(b) Calculate $D_{\mathbf{u}} f(1, 1)$ and $D_{\mathbf{u}} f(\sqrt{3}, 1)$.

(c) Show that the lines $y = mx$ for $m \neq 0$ are level curves for f.

(d) Verify that ∇f_P is orthogonal to the level curve through P for $P = (x, y) \neq (0, 0)$.

59. Suppose that the intersection of two surfaces $F(x, y, z) = 0$ and $G(x, y, z) = 0$ is a curve $\mathcal{C}$, and let P be a point on $\mathcal{C}$. Explain why the vector $\mathbf{v} = \nabla F_P \times \nabla G_P$ is a direction vector for the tangent line to $\mathcal{C}$ at P.

60. Let $\mathcal{C}$ be the curve of intersection of the spheres $x^2 + y^2 + z^2 = 3$ and $(x-2)^2 + (y-2)^2 + z^2 = 3$. Use the result of Exercise 59 to find parametric equations of the tangent line to $\mathcal{C}$ at $P = (1, 1, 1)$.

61. Let $\mathcal{C}$ be the curve obtained by intersecting the two surfaces $x^3 + 2xy + yz = 7$ and $3x^2 - yz = 1$. Find the parametric equations of the tangent line to $\mathcal{C}$ at $P = (1, 2, 1)$.

62. Verify the linearity relations for gradients:

(a) $\nabla(f + g) = \nabla f + \nabla g$

(b) $\nabla(cf) = c\nabla f$

63. Prove the Chain Rule for Gradients (Theorem 1).

64. Prove the Product Rule for Gradients (Theorem 1).

Further Insights and Challenges

65. Let $\mathbf{u}$ be a unit vector. Show that the directional derivative $D_{\mathbf{u}}f$ is equal to the component of ∇f along $\mathbf{u}$.

66. Let $f(x, y) = (xy)^{1/3}$.

(a) Use the limit definition to show that $f_x(0, 0) = f_y(0, 0) = 0$.

(b) Use the limit definition to show that the directional derivative $D_{\mathbf{u}}f(0, 0)$ does not exist for any unit vector $\mathbf{u}$ other than $\mathbf{i}$ and $\mathbf{j}$.

(c) Is f differentiable at $(0, 0)$?

67. Use the definition of differentiability to show that if $f(x, y)$ is differentiable at $(0, 0)$ and

$$f(0, 0) = f_x(0, 0) = f_y(0, 0) = 0$$

then

$$\lim_{(x,y)\to(0,0)} \frac{f(x, y)}{\sqrt{x^2 + y^2}} = 0 \qquad \boxed{9}$$

68. This exercise shows that there exists a function that is not differentiable at $(0, 0)$ even though all directional derivatives at $(0, 0)$ exist. Define $f(x, y) = x^2y/(x^2 + y^2)$ for $(x, y) \neq 0$ and $f(0, 0) = 0$.

(a) Use the limit definition to show that $D_{\mathbf{v}}f(0, 0)$ exists for all vectors $\mathbf{v}$. Show that $f_x(0, 0) = f_y(0, 0) = 0$.

(b) Prove that f is *not* differentiable at $(0, 0)$ by showing that Eq. (9) does not hold.

69. Prove that if $f(x, y)$ is differentiable and $\nabla f_{(x,y)} = \mathbf{0}$ for all (x, y), then f is constant.

70. Prove the following Quotient Rule, where f, g are differentiable:

$$\nabla\left(\frac{f}{g}\right) = \frac{g\nabla f - f\nabla g}{g^2}.$$

In Exercises 71–73, a path $\mathbf{c}(t) = (x(t), y(t))$ *follows the gradient of a function* $f(x, y)$ *if the tangent vector* $\mathbf{c}'(t)$ *points in the direction of* ∇f *for all* t. *In other words,* $\mathbf{c}'(t) = k(t)\nabla f_{\mathbf{c}(t)}$ *for some positive function* $k(t)$. *Note that in this case,* $\mathbf{c}(t)$ *crosses each level curve of* $f(x, y)$ *at a right angle.*

71. Show that if the path $\mathbf{c}(t) = (x(t), y(t))$ follows the gradient of $f(x, y)$, then

$$\frac{y'(t)}{x'(t)} = \frac{f_y}{f_x}$$

72. Find a path of the form $\mathbf{c}(t) = (t, g(t))$ passing through $(1, 2)$ that follows the gradient of $f(x, y) = 2x^2 + 8y^2$ (Figure 16). *Hint:* Use Separation of Variables.

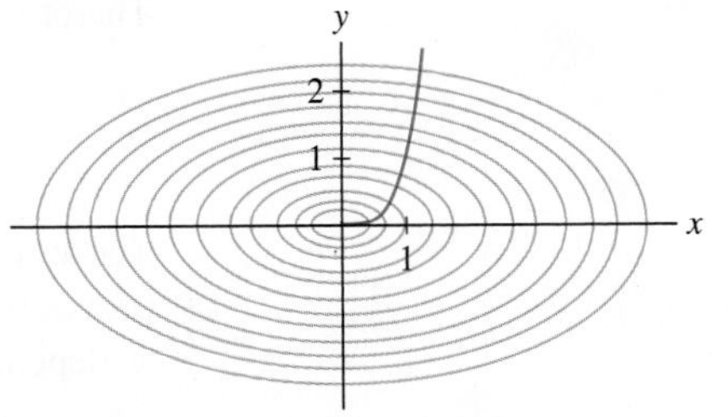

FIGURE 16 The path $\mathbf{c}(t)$ is orthogonal to the level curves of $f(x, y) = 2x^2 + 8y^2$.

73. CAS Find the curve $y = g(x)$ passing through $(0, 1)$ that crosses each level curve of $f(x, y) = y \sin x$ at a right angle. If you have a computer algebra system, graph $y = g(x)$ together with the level curves of f.

15.6 The Chain Rule

The Chain Rule for Paths that we derived in the previous section can be extended to general composite functions. Suppose, for example, that x, y, z are differentiable functions of s and t—say $x = x(s, t)$, $y = y(s, t)$, and $z = z(s, t)$. The composite

$$f(x(s, t), y(s, t), z(s, t)) \qquad \boxed{1}$$

is then a function of s and t. We refer to s and t as the **independent variables**.

EXAMPLE 1 Find the composite function where $f(x, y, z) = xy + z$ and $x = s^2$, $y = st$, $z = t^2$.

Solution The composite function is

$$f(x(s,t), y(s,t), z(s,t)) = xy + z = (s^2)(st) + t^2 = s^3t + t^2$$ ■

The Chain Rule expresses the derivatives of f with respect to the independent variables. For example, the partial derivatives of $f(x(s,t), y(s,t), z(s,t))$ are

$$\frac{\partial f}{\partial s} = \frac{\partial f}{\partial x}\frac{\partial x}{\partial s} + \frac{\partial f}{\partial y}\frac{\partial y}{\partial s} + \frac{\partial f}{\partial z}\frac{\partial z}{\partial s} \qquad \boxed{2}$$

$$\frac{\partial f}{\partial t} = \frac{\partial f}{\partial x}\frac{\partial x}{\partial t} + \frac{\partial f}{\partial y}\frac{\partial y}{\partial t} + \frac{\partial f}{\partial z}\frac{\partial z}{\partial t} \qquad \boxed{3}$$

To prove these formulas, we observe that $\partial f/\partial s$, when evaluated at a point (s_0, t_0), is equal to the derivative with respect to the path

$$\mathbf{c}(s) = (x(s, t_0), y(s, t_0), z(s, t_0))$$

In other words, we fix $t = t_0$ and take the derivative with respect to s:

$$\frac{\partial f}{\partial s}(s_0, t_0) = \frac{d}{ds} f(\mathbf{c}(s))\bigg|_{s=s_0}$$

The tangent vector is

$$\mathbf{c}'(s) = \left\langle \frac{\partial x}{\partial s}(s, t_0), \frac{\partial y}{\partial s}(s, t_0), \frac{\partial z}{\partial s}(s, t_0) \right\rangle$$

Therefore, by the Chain Rule for Paths,

$$\frac{\partial f}{\partial s}\bigg|_{(s_0,t_0)} = \frac{d}{ds} f(\mathbf{c}(s))\bigg|_{s=s_0} = \nabla f \cdot \mathbf{c}'(s_0) = \frac{\partial f}{\partial x}\frac{\partial x}{\partial s} + \frac{\partial f}{\partial y}\frac{\partial y}{\partial s} + \frac{\partial f}{\partial z}\frac{\partial z}{\partial s}$$

The derivatives on the right are evaluated at (s_0, t_0). This proves Eq. (2). A similar argument proves Eq. (3), as well as the general case of a function $f(x_1, \dots, x_n)$, where the variables x_i depend on independent variables $t_1, \dots, t_m$.

THEOREM 1 General Version of Chain Rule Let $f(x_1, \dots, x_n)$ be a differentiable function of n variables. Suppose that each of the variables $x_1, \dots, x_n$ is a differentiable function of m independent variables $t_1, \dots, t_m$. Then, for $k = 1, \dots, m$,

$$\frac{\partial f}{\partial t_k} = \frac{\partial f}{\partial x_1}\frac{\partial x_1}{\partial t_k} + \frac{\partial f}{\partial x_2}\frac{\partial x_2}{\partial t_k} + \cdots + \frac{\partial f}{\partial x_n}\frac{\partial x_n}{\partial t_k} \qquad \boxed{4}$$

As an aid to remembering the Chain Rule, we will refer to

$$\frac{\partial f}{\partial x_1}, \quad \dots, \quad \frac{\partial f}{\partial x_n}$$

The term "primary derivative" is not standard. We use it in this section only, to clarify the structure of the Chain Rule.

as the **primary derivatives**. They are the components of the gradient ∇f. By Eq. (4), the derivative of f with respect to the independent variable t_k is equal to a sum of n terms:

$$j\text{th term: } \quad \frac{\partial f}{\partial x_j}\frac{\partial x_j}{\partial t_k} \quad \text{for } j = 1, 2, \dots, n$$

Note that we can write Eq. (4) as a dot product:

$$\frac{\partial f}{\partial t_k} = \left\langle \frac{\partial f}{\partial x_1}, \frac{\partial f}{\partial x_2}, \dots, \frac{\partial f}{\partial x_n} \right\rangle \cdot \left\langle \frac{\partial x_1}{\partial t_k}, \frac{\partial x_2}{\partial t_k}, \dots, \frac{\partial x_n}{\partial t_k} \right\rangle \qquad \boxed{5}$$

EXAMPLE 2 Using the Chain Rule Let $f(x, y, z) = xy + z$. Calculate $\partial f/\partial s$, where

$$x = s^2, \quad y = st, \quad z = t^2$$

Solution

Step 1. **Compute the primary derivatives.**

$$\frac{\partial f}{\partial x} = y, \qquad \frac{\partial f}{\partial y} = x, \qquad \frac{\partial f}{\partial z} = 1$$

Step 2. **Apply the Chain Rule.**

$$\begin{aligned} \frac{\partial f}{\partial s} = \frac{\partial f}{\partial x}\frac{\partial x}{\partial s} + \frac{\partial f}{\partial y}\frac{\partial y}{\partial s} + \frac{\partial f}{\partial z}\frac{\partial z}{\partial s} &= y\frac{\partial}{\partial s}(s^2) + x\frac{\partial}{\partial s}(st) + \frac{\partial}{\partial s}(t^2) \\ &= (y)(2s) + (x)(t) + 0 \\ &= 2sy + xt \end{aligned}$$

This expresses the derivative in terms of both sets of variables. If desired, we can substitute $x = s^2$ and $y = st$ to write the derivative in terms of s and t:

$$\frac{\partial f}{\partial s} = 2ys + xt = 2(st)s + (s^2)t = 3s^2 t$$

To check this result, recall that in Example 1, we computed the composite function:

$$f(x(s,t), y(s,t), z(s,t)) = f(s^2, st, t^2) = s^3 t + t^2$$

From this we see directly that $\partial f/\partial s = 3s^2 t$, confirming our result. ■

EXAMPLE 3 Evaluating the Derivative Let $f(x, y) = e^{xy}$. Evaluate $\partial f/\partial t$ at $(s, t, u) = (2, 3, -1)$, where $x = st$, $y = s - ut^2$.

Solution We can use either Eq. (4) or Eq. (5). We'll use the dot product form in Eq. (5). We have

$$\nabla f = \left\langle \frac{\partial f}{\partial x}, \frac{\partial f}{\partial y} \right\rangle = \left\langle ye^{xy}, xe^{xy} \right\rangle, \qquad \left\langle \frac{\partial x}{\partial t}, \frac{\partial y}{\partial t} \right\rangle = \langle s, -2ut \rangle$$

and the Chain Rule gives us

$$\begin{aligned} \frac{\partial f}{\partial t} = \nabla f \cdot \left\langle \frac{\partial x}{\partial t}, \frac{\partial y}{\partial t} \right\rangle &= \left\langle ye^{xy}, xe^{xy} \right\rangle \cdot \langle s, -2ut \rangle \\ &= ye^{xy}(s) + xe^{xy}(-2ut) \\ &= (ys - 2xut)e^{xy} \end{aligned}$$

To finish the problem, we do not have to rewrite $\partial f/\partial t$ in terms of s, t, u. For $(s, t, u) = (2, 3, -1)$, we have

$$x = st = 2(3) = 6, \qquad y = s - ut^2 = 2 - (-1)(3^2) = 11$$

With $(s, t, u) = (2, 3, -1)$ and $(x, y) = (6, 11)$, we have

$$\left.\frac{\partial f}{\partial t}\right|_{(2,3,-1)} = (ys - 2xut)e^{xy}\Big|_{(2,3,-1)} = \Big((11)(2) - 2(6)(-1)(3)\Big)e^{6(11)} = 58e^{66} \quad ■$$

EXAMPLE 4 Polar Coordinates Let $f(x, y)$ be a function of two variables, and let (r, θ) be polar coordinates.

(a) Express $\partial f/\partial\theta$ in terms of $\partial f/\partial x$ and $\partial f/\partial y$.

(b) Evaluate $\partial f/\partial\theta$ at $(x, y) = (1, 1)$ for $f(x, y) = x^2 y$.

Solution

(a) Since $x = r\cos\theta$ and $y = r\sin\theta$,

$$\frac{\partial x}{\partial\theta} = -r\sin\theta, \qquad \frac{\partial y}{\partial\theta} = r\cos\theta$$

By the Chain Rule,

$$\frac{\partial f}{\partial\theta} = \frac{\partial f}{\partial x}\frac{\partial x}{\partial\theta} + \frac{\partial f}{\partial y}\frac{\partial y}{\partial\theta} = -r\sin\theta\frac{\partial f}{\partial x} + r\cos\theta\frac{\partial f}{\partial y}$$

*If you have studied quantum mechanics, you may recognize the right-hand side of Eq. (6) as the **angular momentum** operator (with respect to the z-axis).*

Since $x = r\cos\theta$ and $y = r\sin\theta$, we can write $\partial f/\partial\theta$ in terms of x and y alone:

$$\frac{\partial f}{\partial\theta} = x\frac{\partial f}{\partial y} - y\frac{\partial f}{\partial x} \qquad \boxed{6}$$

(b) Apply Eq. (6) to $f(x, y) = x^2 y$:

$$\frac{\partial f}{\partial\theta} = x\frac{\partial}{\partial y}(x^2 y) - y\frac{\partial}{\partial x}(x^2 y) = x^3 - 2xy^2$$

$$\left.\frac{\partial f}{\partial\theta}\right|_{(x,y)=(1,1)} = 1^3 - 2(1)(1^2) = -1$$

Implicit Differentiation

In single-variable calculus, we used implicit differentiation to compute dy/dx when y is defined implicitly as a function of x through an equation $f(x, y) = 0$. This method also works for functions of several variables. Suppose that z is defined implicitly by an equation

$$F(x, y, z) = 0$$

Thus $z = z(x, y)$ is a function of x and y. We may not be able to solve explicitly for $z(x, y)$, but we can treat $F(x, y, z)$ as a composite function with x and y as independent variables, and use the Chain Rule to differentiate with respect to x:

$$\frac{\partial F}{\partial x}\frac{\partial x}{\partial x} + \frac{\partial F}{\partial y}\frac{\partial y}{\partial x} + \frac{\partial F}{\partial z}\frac{\partial z}{\partial x} = 0$$

We have $\partial x/\partial x = 1$, and also $\partial y/\partial x = 0$ since y does not depend on x. Thus

$$\frac{\partial F}{\partial x} + \frac{\partial F}{\partial z}\frac{\partial z}{\partial x} = F_x + F_z\frac{\partial z}{\partial x} = 0$$

If $F_z \neq 0$, we may solve for $\partial z/\partial x$ (we compute $\partial z/\partial y$ similarly):

$$\boxed{\frac{\partial z}{\partial x} = -\frac{F_x}{F_z}, \qquad \frac{\partial z}{\partial y} = -\frac{F_y}{F_z}} \qquad \boxed{7}$$

■ **EXAMPLE 5** Calculate $\partial z/\partial x$ and $\partial z/\partial y$ at $P = (1, 1, 1)$, where

$$F(x, y, z) = x^2 + y^2 - 2z^2 + 12x - 8z - 4 = 0$$

What is the graphical interpretation of these partial derivatives?

Solution We have

$$F_x = 2x + 12, \qquad F_y = 2y, \qquad F_z = -4z - 8$$

and hence,

$$\frac{\partial z}{\partial x} = -\frac{F_x}{F_z} = \frac{2x+12}{4z+8}, \qquad \frac{\partial z}{\partial y} = -\frac{F_y}{F_z} = \frac{2y}{4z+8}$$

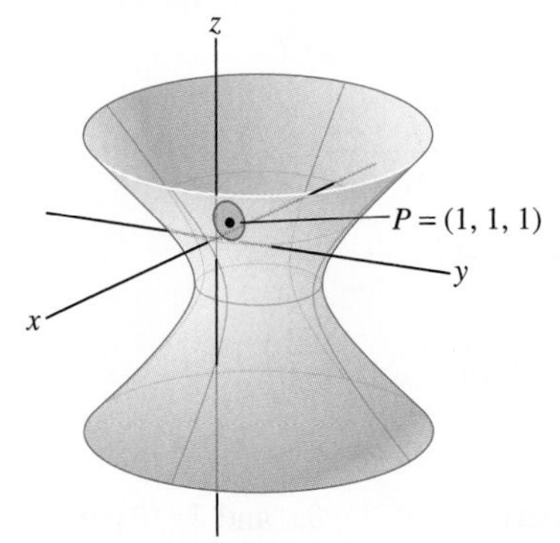

FIGURE 1 The surface $x^2 + y^2 - 2z^2 + 12x - 8z - 4 = 0$. A small patch of the surface around P can be represented as the graph of a function of x and y.

The derivatives at $P = (1, 1, 1)$ are

$$\left.\frac{\partial z}{\partial x}\right|_{(1,1,1)} = \frac{2(1)+12}{4(1)+8} = \frac{14}{12} = \frac{7}{6}, \qquad \left.\frac{\partial z}{\partial y}\right|_{(1,1,1)} = \frac{2(1)}{4(1)+8} = \frac{2}{12} = \frac{1}{6}$$

Figure 1 shows the surface $F(x, y, z) = 0$. The surface as a whole is not the graph of a function because it fails the Vertical Line Test. However, a small patch near P may be represented as a graph of a function $z = f(x, y)$, and the partial derivatives $\partial z/\partial x$ and $\partial z/\partial y$ are equal to f_x and f_y. Implicit differentiation has enabled us to compute these partial derivatives without finding $f(x, y)$ explicitly. ■

Assumptions Matter Implicit differentiation is based on the assumption that we can solve the equation $F(x, y, z) = 0$ for z in the form $z = f(x, y)$. Otherwise, the partial derivatives $\partial z/\partial x$ and $\partial z/\partial y$ would have no meaning. The Implicit Function Theorem of advanced calculus guarantees that this can be done (at least near a point P) if F has continuous partial derivatives and $F_z(P) \neq 0$. Why is this condition necessary? Recall that the gradient vector $\nabla F_P = \langle F_x(P), F_y(P), F_z(P)\rangle$ is normal to the surface at P, so $F_z(P) = 0$ means that the tangent plane at P is vertical. To see what can go wrong, consider the cylinder (shown in Figure 2):

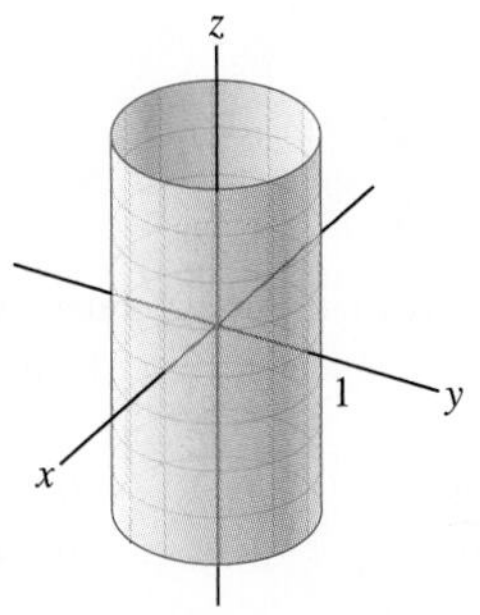

FIGURE 2 Graph of the cylinder $x^2 + y^2 - 1 = 0$.

$$F(x, y, z) = x^2 + y^2 - 1 = 0$$

In this extreme case, $F_z = 0$. The z-coordinate on the cylinder does not depend on x or y, so it is impossible to represent the cylinder as a graph $z = f(x, y)$ and the derivatives $\partial z/\partial x$ and $\partial z/\partial y$ do not exist.

15.6 SUMMARY

- If $f(x, y, z)$ is a function of x, y, z, and if x, y, z depend on two other variables, say s and t, then

$$f(x, y, z) = f(x(s, t), y(s, t), z(s, t))$$

is a composite function of s and t. We refer to s and t as the *independent variables*.
- The *Chain Rule* expresses the partial derivatives with respect to the independent variables s and t in terms of the *primary derivatives:*

$$\frac{\partial f}{\partial x}, \quad \frac{\partial f}{\partial y}, \quad \frac{\partial f}{\partial z}$$

Namely,

$$\frac{\partial f}{\partial s} = \frac{\partial f}{\partial x}\frac{\partial x}{\partial s} + \frac{\partial f}{\partial y}\frac{\partial y}{\partial s} + \frac{\partial f}{\partial z}\frac{\partial z}{\partial s}, \qquad \frac{\partial f}{\partial t} = \frac{\partial f}{\partial x}\frac{\partial x}{\partial t} + \frac{\partial f}{\partial y}\frac{\partial y}{\partial t} + \frac{\partial f}{\partial z}\frac{\partial z}{\partial t}$$

- In general, if $f(x_1, \dots, x_n)$ is a function of n variables and if $x_1, \dots, x_n$ depend on the independent variables $t_1, \dots, t_m$, then

$$\frac{\partial f}{\partial t_k} = \frac{\partial f}{\partial x_1}\frac{\partial x_1}{\partial t_k} + \frac{\partial f}{\partial x_2}\frac{\partial x_2}{\partial t_k} + \cdots + \frac{\partial f}{\partial x_n}\frac{\partial x_n}{\partial t_k}$$

- The Chain Rule can be expressed as a dot product:

$$\frac{\partial f}{\partial t_k} = \underbrace{\left\langle \frac{\partial f}{\partial x_1}, \frac{\partial f}{\partial x_2}, \dots, \frac{\partial f}{\partial x_n} \right\rangle}_{\nabla f} \cdot \left\langle \frac{\partial x_1}{\partial t_k}, \frac{\partial x_2}{\partial t_k}, \dots, \frac{\partial x_n}{\partial t_k} \right\rangle$$

- Implicit differentiation is used to find the partial derivatives $\partial z/\partial x$ and $\partial z/\partial y$ when z is defined implicitly by an equation $F(x, y, z) = 0$:

$$\frac{\partial z}{\partial x} = -\frac{F_x}{F_z}, \qquad \frac{\partial z}{\partial y} = -\frac{F_y}{F_z}$$

15.6 EXERCISES

Preliminary Questions

1. Let $f(x, y) = xy$, where $x = uv$ and $y = u + v$.

(a) What are the primary derivatives of f?

(b) What are the independent variables?

In Questions 2 and 3, suppose that $f(u, v) = ue^v$, where $u = rs$ and $v = r + s$.

2. The composite function $f(u, v)$ is equal to:

(a) rse^{r+s} (b) re^s (c) rse^{rs}

3. What is the value of $f(u, v)$ at $(r, s) = (1, 1)$?

4. According to the Chain Rule, $\partial f/\partial r$ is equal to (choose the correct answer):

(a) $\frac{\partial f}{\partial x}\frac{\partial x}{\partial r} + \frac{\partial f}{\partial x}\frac{\partial x}{\partial s}$

(b) $\frac{\partial f}{\partial x}\frac{\partial x}{\partial r} + \frac{\partial f}{\partial y}\frac{\partial y}{\partial r}$

(c) $\frac{\partial f}{\partial r}\frac{\partial r}{\partial x} + \frac{\partial f}{\partial s}\frac{\partial s}{\partial x}$

5. Suppose that x, y, z are functions of the independent variables u, v, w. Which of the following terms appear in the Chain Rule expression for $\partial f/\partial w$?

(a) $\frac{\partial f}{\partial v}\frac{\partial x}{\partial v}$ (b) $\frac{\partial f}{\partial w}\frac{\partial w}{\partial x}$ (c) $\frac{\partial f}{\partial z}\frac{\partial z}{\partial w}$

6. With notation as in the previous question, does $\partial x/\partial v$ appear in the Chain Rule expression for $\partial f/\partial u$?

Exercises

1. Let $f(x, y, z) = x^2y^3 + z^4$ and $x = s^2$, $y = st^2$, and $z = s^2t$.

(a) Calculate the primary derivatives $\frac{\partial f}{\partial x}, \frac{\partial f}{\partial y}, \frac{\partial f}{\partial z}$.

(b) Calculate $\frac{\partial x}{\partial s}, \frac{\partial y}{\partial s}, \frac{\partial z}{\partial s}$.

(c) Compute $\frac{\partial f}{\partial s}$ using the Chain Rule:

$$\frac{\partial f}{\partial s} = \frac{\partial f}{\partial x}\frac{\partial x}{\partial s} + \frac{\partial f}{\partial y}\frac{\partial y}{\partial s} + \frac{\partial f}{\partial z}\frac{\partial z}{\partial s}$$

Express the answer in terms of the independent variables s, t.

2. Let $f(x, y) = x\cos(y)$ and $x = u^2 + v^2$ and $y = u - v$.

(a) Calculate the primary derivatives $\frac{\partial f}{\partial x}, \frac{\partial f}{\partial y}$.

(b) Use the Chain Rule to calculate $\partial f/\partial v$. Leave the answer in terms of both the dependent and the independent variables.

(c) Determine (x, y) for $(u, v) = (2, 1)$ and evaluate $\partial f/\partial v$ at $(u, v) = (2, 1)$.

In Exercises 3–10, use the Chain Rule to calculate the partial derivatives. Express the answer in terms of the independent variables.

3. $\frac{\partial f}{\partial s}, \frac{\partial f}{\partial r}$; $f(x, y, z) = xy + z^2$, $x = s^2$, $y = 2rs$, $z = r^2$

4. $\frac{\partial f}{\partial r}, \frac{\partial f}{\partial t}$; $f(x, y, z) = xy + z^2$, $x = r + s - 2t$, $y = 3rt$, $z = s^2$

5. $\frac{\partial g}{\partial u}, \frac{\partial g}{\partial v}$; $g(x, y) = \cos(x - y)$, $x = 3u - 5v$, $y = -7u + 15v$

6. $\frac{\partial R}{\partial u}, \frac{\partial R}{\partial v}$; $R(x, y) = (3x + 4y)^5$, $x = u^2$, $y = uv$

7. $\dfrac{\partial F}{\partial y}$; $F(u, v) = e^{u+v}$, $u = x^2$, $v = xy$

8. $\dfrac{\partial f}{\partial u}$; $f(x, y) = x^2 + y^2$, $x = e^{u+v}$, $y = u + v$

9. $\dfrac{\partial h}{\partial t_2}$; $h(x, y) = \dfrac{x}{y}$, $x = t_1 t_2$, $y = t_1^2 t_2$

10. $\dfrac{\partial f}{\partial \theta}$; $f(x, y, z) = xy - z^2$, $x = r\cos\theta$, $y = \cos^2\theta$, $z = r$

In Exercises 11–16, use the Chain Rule to evaluate the partial derivative at the point specified.

11. $\partial f/\partial u$ and $\partial f/\partial v$ at $(u, v) = (-1, -1)$, where $f(x, y, z) = x^3 + yz^2$, $x = u^2 + v$, $y = u + v^2$, $z = uv$.

12. $\partial f/\partial s$ at $(r, s) = (1, 0)$, where $f(x, y) = \ln(xy)$, $x = 3r + 2s$, $y = 5r + 3s$.

13. $\partial g/\partial \theta$ at $(r, \theta) = \left(2\sqrt{2}, \frac{\pi}{4}\right)$, where $g(x, y) = 1/(x + y^2)$, $x = r\sin\theta$, $y = r\cos\theta$.

14. $\partial g/\partial s$ at $s = 4$, where $g(x, y) = x^2 - y^2$, $x = s^2 + 1$, $y = 1 - 2s$.

15. $\partial g/\partial u$ at $(u, v) = (0, 1)$, where $g(x, y) = x^2 - y^2$, $x = e^u \cos v$, $y = e^u \sin v$.

16. $\dfrac{\partial h}{\partial q}$ at $(q, r) = (3, 2)$, where $h(u, v) = ue^v$, $u = q^3$, $v = qr^2$.

17. Jessica and Matthew are running toward the point P along the straight paths that make a fixed angle of θ (Figure 3). Suppose that Matthew runs with velocity v_a m/s and Jessica with velocity v_b m/s. Let $f(x, y)$ be the distance from Matthew to Jessica when Matthew is x meters from P and Jessica is y meters from P.

(a) Show that $f(x, y) = \sqrt{x^2 + y^2 - 2xy\cos\theta}$.

(b) Assume that $\theta = \pi/3$. Use the Chain Rule to determine the rate at which the distance between Matthew and Jessica is changing when $x = 30$, $y = 20$, $v_a = 4$ m/s, and $v_b = 3$ m/s.

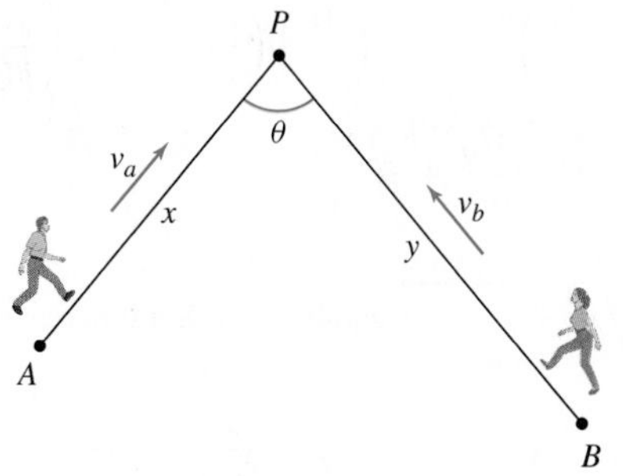

FIGURE 3

18. The Law of Cosines states that $c^2 = a^2 + b^2 - 2ab\cos\theta$, where a, b, c are the sides of a triangle and θ is the angle opposite the side of length c.

(a) Compute $\partial\theta/\partial a$, $\partial\theta/\partial b$, and $\partial\theta/\partial c$ using implicit differentiation.

(b) Suppose that $a = 10$, $b = 16$, $c = 22$. Estimate the change in θ if a and b are increased by 1 and c is increased by 2.

19. Let $u = u(x, y)$, and let (r, θ) be polar coordinates. Verify the relation

$$\|\nabla u\|^2 = u_r^2 + \frac{1}{r^2}u_\theta^2 \qquad \boxed{8}$$

Hint: Compute the right-hand side by expressing u_θ and u_r in terms of u_x and u_y.

20. Let $u(r, \theta) = r^2\cos^2\theta$. Use Eq. (8) to compute $\|\nabla u\|^2$. Then compute $\|\nabla u\|^2$ directly by observing that $u(x, y) = x^2$, and compare.

21. Let $x = s + t$ and $y = s - t$. Show that for any differentiable function $f(x, y)$,

$$\left(\frac{\partial f}{\partial x}\right)^2 - \left(\frac{\partial f}{\partial y}\right)^2 = \frac{\partial f}{\partial s}\frac{\partial f}{\partial t}$$

22. Express the derivatives

$$\frac{\partial f}{\partial \rho}, \frac{\partial f}{\partial \theta}, \frac{\partial f}{\partial \phi} \quad \text{in terms of} \quad \frac{\partial f}{\partial x}, \frac{\partial f}{\partial y}, \frac{\partial f}{\partial z}$$

where (ρ, θ, ϕ) are spherical coordinates.

23. Suppose that z is defined implicitly as a function of x and y by the equation $F(x, y, z) = xz^2 + y^2z + xy - 1 = 0$.

(a) Calculate F_x, F_y, F_z.

(b) Use Eq. (7) to calculate $\dfrac{\partial z}{\partial x}$ and $\dfrac{\partial z}{\partial y}$.

24. Calculate $\partial z/\partial x$ and $\partial z/\partial y$ at the points $(3, 2, 1)$ and $(3, 2, -1)$, where z is defined implicitly by the equation $z^4 + z^2x^2 - y - 8 = 0$.

In Exercises 25–30, calculate the partial derivative using implicit differentiation.

25. $\dfrac{\partial z}{\partial x}$, $\quad x^2y + y^2z + xz^2 = 10$

26. $\dfrac{\partial w}{\partial z}$, $\quad x^2w + w^3 + wz^2 + 3yz = 0$

27. $\dfrac{\partial z}{\partial y}$, $\quad e^{xy} + \sin(xz) + y = 0$

28. $\dfrac{\partial r}{\partial t}$ and $\dfrac{\partial t}{\partial r}$, $\quad r^2 = te^{s/r}$

29. $\dfrac{\partial w}{\partial y}$, $\quad \dfrac{1}{w^2 + x^2} + \dfrac{1}{w^2 + y^2} = 1$ at $(x, y, w) = (1, 1, 1)$

30. $\partial U/\partial T$ and $\partial T/\partial U$, $\quad (TU - V)^2 \ln(W - UV) = 1$ at $(T, U, V, W) = (1, 1, 2, 4)$

31. Let $\mathbf{r} = \langle x, y, z\rangle$ and $e_{\mathbf{r}} = \mathbf{r}/\|\mathbf{r}\|$. Show that if a function $f(x, y, z) = F(r)$ depends only on the distance from the origin $r = \|\mathbf{r}\| = \sqrt{x^2 + y^2 + z^2}$, then

$$\nabla f = F'(r)e_{\mathbf{r}} \qquad \boxed{9}$$

32. Let $f(x, y, z) = e^{-x^2-y^2-z^2} = e^{-r^2}$, with r as in Exercise 31. Compute ∇f directly and using Eq. (9).

33. Use Eq. (9) to compute $\nabla\left(\frac{1}{r}\right)$.

34. Use Eq. (9) to compute $\nabla(\ln r)$.

35. Figure 4 shows the graph of the equation

$$F(x, y, z) = x^2 + y^2 - z^2 - 12x - 8z - 4 = 0$$

(a) Use the quadratic formula to solve for z as a function of x and y. This gives two formulas, depending on the choice of sign.

(b) Which formula defines the portion of the surface satisfying $z \geq -4$? Which formula defines the portion satisfying $z \leq -4$?

(c) Calculate $\partial z/\partial x$ using the formula $z = f(x, y)$ (for both choices of sign) and again via implicit differentiation. Verify that the two answers agree.

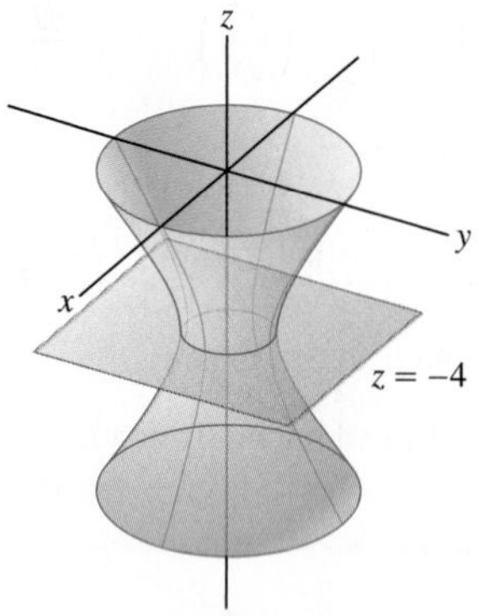

FIGURE 4 Graph of $x^2 + y^2 - z^2 - 12x - 8z - 4 = 0$.

36. For all $x > 0$, there is a unique value $y = r(x)$ that solves the equation $y^3 + 4xy = 16$.

(a) Show that $dy/dx = -4y/(3y^2 + 4x)$.

(b) Let $g(x) = f(x, r(x))$, where $f(x, y)$ is a function satisfying

$$f_x(1, 2) = 8, \quad f_y(1, 2) = 10$$

Use the Chain Rule to calculate $g'(1)$. Note that $r(1) = 2$ because $(x, y) = (1, 2)$ satisfies $y^3 + 4xy = 16$.

37. The pressure P, volume V, and temperature T of a van der Waals gas with n molecules (n constant) are related by the equation

$$\left(P + \frac{an^2}{V^2}\right)(V - nb) = nRT$$

where a, b, and R are constant. Calculate $\partial P/\partial T$ and $\partial V/\partial P$.

38. When x, y, and z are related by an equation $F(x, y, z) = 0$, we sometimes write $(\partial z/\partial x)_y$ in place of $\partial z/\partial x$ to indicate that in the differentiation, z is treated as a function of x with y held constant (and similarly for the other variables).

(a) Use Eq. (7) to prove the **cyclic relation**

$$\left(\frac{\partial z}{\partial x}\right)_y \left(\frac{\partial x}{\partial y}\right)_z \left(\frac{\partial y}{\partial z}\right)_x = -1 \qquad \boxed{10}$$

(b) Verify Eq. (10) for $F(x, y, z) = x + y + z = 0$.

(c) Verify the cyclic relation for the variables P, V, T in the ideal gas law $PV - nRT = 0$ (n and R are constants).

39. Show that if $f(x)$ is differentiable and $c \neq 0$ is a constant, then $u(x, t) = f(x - ct)$ satisfies the so-called **advection equation**

$$\frac{\partial u}{\partial t} + c\frac{\partial u}{\partial x} = 0$$

Further Insights and Challenges

In Exercises 40–43, a function $f(x, y, z)$ is called ***homogeneous of degree n*** *if $f(\lambda x, \lambda y, \lambda z) = \lambda^n f(x, y, z)$ for all $\lambda \in \mathbf{R}$.*

40. Show that the following functions are homogeneous and determine their degree.

(a) $f(x, y, z) = x^2 y + xyz$

(b) $f(x, y, z) = 3x + 2y - 8z$

(c) $f(x, y, z) = \ln\left(\frac{xy}{z^2}\right)$

(d) $f(x, y, z) = z^4$

41. Prove that if $f(x, y, z)$ is homogeneous of degree n, then $f_x(x, y, z)$ is homogeneous of degree $n - 1$. *Hint:* Either use the limit definition or apply the Chain Rule to $f(\lambda x, \lambda y, \lambda z)$.

42. Prove that if $f(x, y, z)$ is homogeneous of degree n, then

$$x\frac{\partial f}{\partial x} + y\frac{\partial f}{\partial y} + z\frac{\partial f}{\partial z} = nf \qquad \boxed{11}$$

Hint: Let $F(t) = f(tx, ty, tz)$ and calculate $F'(1)$ using the Chain Rule.

43. Verify Eq. (11) for the functions in Exercise 40.

44. Suppose that $x = g(t, s)$, $y = h(t, s)$. Show that f_{tt} is equal to

$$f_{xx}\left(\frac{\partial x}{\partial t}\right)^2 + 2f_{xy}\left(\frac{\partial x}{\partial t}\right)\left(\frac{\partial y}{\partial t}\right) + f_{yy}\left(\frac{\partial y}{\partial t}\right)^2$$
$$+ f_x\frac{\partial^2 x}{\partial t^2} + f_y\frac{\partial^2 y}{\partial t^2} \qquad \boxed{12}$$

45. Let $r = \sqrt{x_1^2 + \cdots + x_n^2}$ and let $g(r)$ be a function of r. Prove the formulas

$$\frac{\partial g}{\partial x_i} = \frac{x_i}{r}g_r, \qquad \frac{\partial^2 g}{\partial x_i^2} = \frac{x_i^2}{r^2}g_{rr} + \frac{r^2 - x_i^2}{r^3}g_r$$

46. Prove that if $g(r)$ is a function of r as in Exercise 45, then

$$\frac{\partial^2 g}{\partial x_1^2} + \cdots + \frac{\partial^2 g}{\partial x_n^2} = g_{rr} + \frac{n-1}{r}g_r$$

In Exercises 47–51, the **Laplace operator** *is defined by* $\Delta f = f_{xx} + f_{yy}$. *A function* $f(x, y)$ *satisfying the Laplace equation* $\Delta f = 0$ *is called* **harmonic**. *A function* $f(x, y)$ *is called* **radial** *if* $f(x, y) = g(r)$, *where* $r = \sqrt{x^2 + y^2}$.

47. Use Eq. (12) to prove that in polar coordinates (r, θ),

$$\Delta f = f_{rr} + \frac{1}{r^2} f_{\theta\theta} + \frac{1}{r} f_r \qquad \boxed{13}$$

48. Use Eq. (13) to show that $f(x, y) = \ln r$ is harmonic.

49. Verify that $f(x, y) = x$ and $f(x, y) = y$ are harmonic using both the rectangular and polar expressions for Δf.

50. Verify that $f(x, y) = \tan^{-1} \frac{y}{x}$ is harmonic using both the rectangular and polar expressions for Δf.

51. Use the Product Rule to show that

$$f_{rr} + \frac{1}{r} f_r = r^{-1} \frac{\partial}{\partial r}\left(r \frac{\partial f}{\partial r}\right)$$

Use this formula to show that if f is a radial harmonic function, then $r f_r = C$ for some constant C. Conclude that $f(x, y) = C \ln r + b$ for some constant b.

15.7 Optimization in Several Variables

Recall that optimization is the process of finding the extreme values of a function. This amounts to finding the highest and lowest points on the graph over a given domain. As we saw in the one-variable case, it is important to distinguish between *local* and *global* extreme values. A local extreme value is a value $f(a, b)$ that is a maximum or minimum in some small open disk around (a, b) (Figure 1).

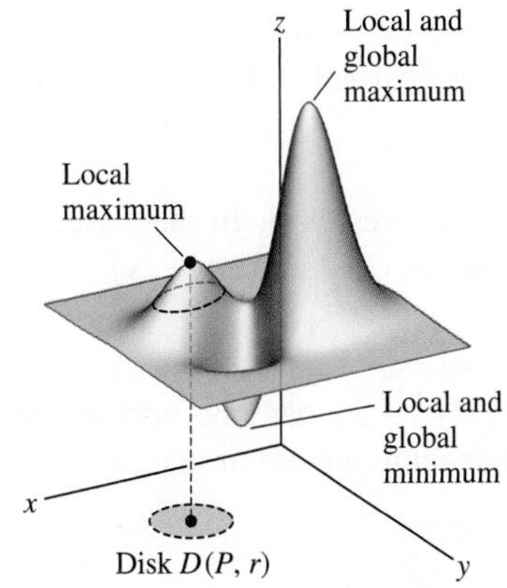

FIGURE 1 $f(x, y)$ has a local maximum at P.

DEFINITION Local Extreme Values A function $f(x, y)$ has a **local extremum** at $P = (a, b)$ if there exists an open disk $D(P, r)$ such that:

- **Local maximum:** $f(x, y) \le f(a, b)$ for all $(x, y) \in D(P, r)$
- **Local minimum:** $f(x, y) \ge f(a, b)$ for all $(x, y) \in D(P, r)$

Fermat's Theorem states that if $f(a)$ is a local extreme value, then a is a critical point and thus the tangent line (if it exists) is horizontal at $x = a$. We can expect a similar result for functions of two variables, but in this case, it is the *tangent plane* that must be horizontal (Figure 2). The tangent plane to $z = f(x, y)$ at $P = (a, b)$ has equation

$$z = f(a, b) + f_x(a, b)(x - a) + f_y(a, b)(y - b)$$

← REMINDER The term "extremum" (the plural is "extrema") means a minimum or maximum value.

Thus, the tangent plane is horizontal if $f_x(a, b) = f_y(a, b) = 0$—that is, if the equation reduces to $z = f(a, b)$. This leads to the following definition of a critical point, where we take into account the possibility that one or both partial derivatives do not exist.

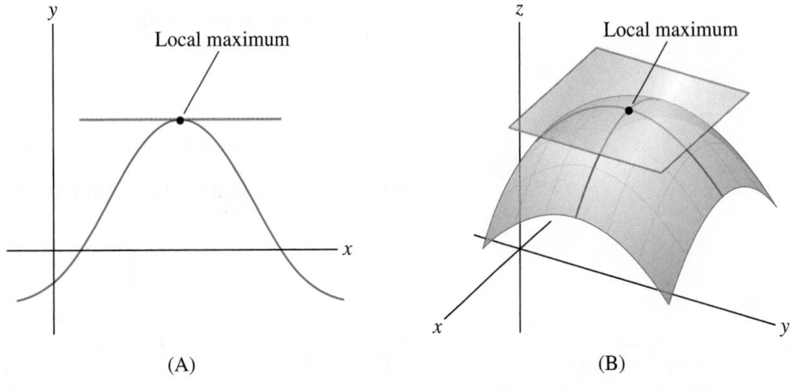

FIGURE 2 The tangent line or plane is horizontal at a local extremum.

- *More generally, $(a_1, \ldots, a_n)$ is a critical point of $f(x_1, \ldots, x_n)$ if each partial derivative satisfies*

$$f_{x_j}(a_1, \ldots, a_n) = 0$$

or does not exist.
- *Theorem 1 holds in any number of variables: Local extrema occur at critical points.*

DEFINITION Critical Point A point $P = (a, b)$ in the domain of $f(x, y)$ is called a **critical point** if:

- $f_x(a, b) = 0$ or $f_x(a, b)$ does not exist, and
- $f_y(a, b) = 0$ or $f_y(a, b)$ does not exist.

As in the single-variable case, we have

THEOREM 1 Fermat's Theorem If $f(x, y)$ has a local minimum or maximum at $P = (a, b)$, then (a, b) is a critical point of $f(x, y)$.

Proof If $f(x, y)$ has a local minimum at $P = (a, b)$, then $f(x, y) \geq f(a, b)$ for all (x, y) near (a, b). In particular, there exists $r > 0$ such that $f(x, b) \geq f(a, b)$ if $|x - a| < r$. In other words, $g(x) = f(x, b)$ has a local minimum at $x = a$. By Fermat's Theorem for functions of one variable, either $g'(a) = 0$ or $g'(a)$ does not exist. Since $g'(a) = f_x(a, b)$, we conclude that either $f_x(a, b) = 0$ or $f_x(a, b)$ does not exist. Similarly, $f_y(a, b) = 0$ or $f_y(a, b)$ does not exist. Therefore, $P = (a, b)$ is a critical point. The case of a local maximum is similar. ■

Usually, we deal with functions whose partial derivatives exist. In this case, finding the critical points amounts to solving the simultaneous equations $f_x(x, y) = 0$ and $f_y(x, y) = 0$.

■ **EXAMPLE 1** Show that $f(x, y) = 11x^2 - 2xy + 2y^2 + 3y$ has one critical point. Use Figure 3 to determine whether it corresponds to a local minimum or maximum.

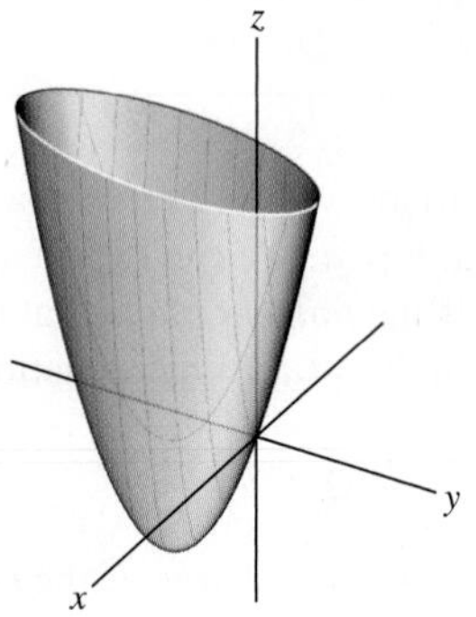

FIGURE 3 Graph of $f(x, y) = 11x^2 - 2xy + 2y^2 + 3y$.

Solution Set the partial derivatives equal to zero and solve:

$$f_x(x, y) = 22x - 2y = 0$$

$$f_y(x, y) = -2x + 4y + 3 = 0$$

By the first equation, $y = 11x$. Substituting $y = 11x$ in the second equation gives

$$-2x + 4y + 3 = -2x + 4(11x) + 3 = 42x + 3 = 0$$

Thus $x = -\frac{1}{14}$ and $y = -\frac{11}{14}$. There is just one critical point, $P = \left(-\frac{1}{14}, -\frac{11}{14}\right)$. Figure 3 shows that $f(x, y)$ has a local minimum at P. ■

It is not always possible to find the solutions exactly, but we can use a computer to find numerical approximations.

■ **EXAMPLE 2** CAS **Numerical Example** Use a computer algebra system to approximate the critical points of

$$f(x, y) = \frac{x - y}{2x^2 + 8y^2 + 3}$$

Are they local minima or maxima? Refer to Figure 4.

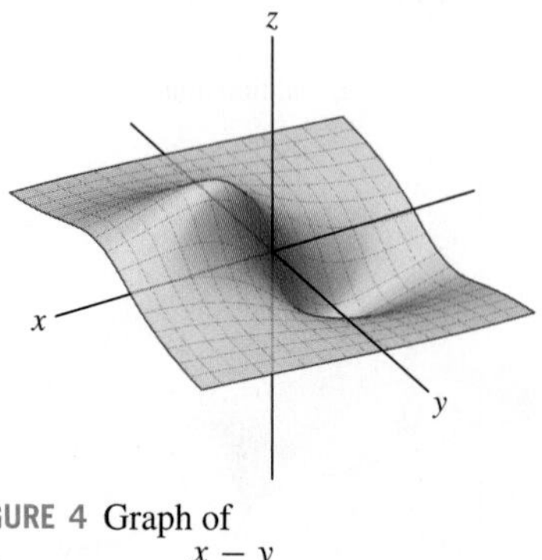

FIGURE 4 Graph of $f(x, y) = \dfrac{x - y}{2x^2 + 8y^2 + 3}$.

Solution We use a CAS to compute the partial derivatives and solve

$$f_x(x, y) = \frac{-2x^2 + 8y^2 + 4xy + 3}{(2x^2 + 8y^2 + 3)^2} = 0$$

$$f_y(x, y) = \frac{-2x^2 + 8y^2 - 16xy - 3}{(2x^2 + 8y^2 + 3)^2} = 0$$

To solve these equations, set the numerators equal to zero. Figure 4 suggests that $f(x, y)$ has a local max with $x > 0$ and a local min with $x < 0$. The following Mathematica command searches for a solution near $(1, 0)$:

```
FindRoot[{-2x^2+8y^2+4xy+3 == 0, -2x^2+8y^2-16xy-3 == 0},
 {{x,1},{y,0}}]
```

The result is

```
{x -> 1.095, y -> -0.274}
```

Thus, $(1.095, -0.274)$ is an approximate critical point where, by Figure 4, f takes on a local maximum. A second search near $(-1, 0)$ yields $(-1.095, 0.274)$, which approximates the critical point where $f(x, y)$ takes on a local minimum. ■

We know that in one variable, a function $f(x)$ may have a point of inflection rather than a local extremum at a critical point. A similar phenomenon occurs in several variables. Each of the functions in Figure 5 has a critical point at $(0, 0)$. However, the function in Figure 5(C) has a saddle point, which is neither a local minimum nor a local maximum. If you stand at the saddle point and begin walking, some directions take you uphill and other directions take you downhill.

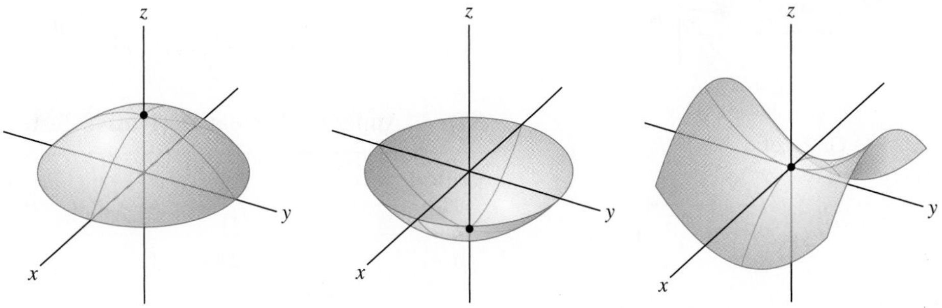

(A) Local maximum (B) Local minimum (C) Saddle

FIGURE 5

As in the one-variable case, there is a Second Derivative Test for determining the type of a critical point (a, b) of a function $f(x, y)$ in two variables. This test relies on the sign of the **discriminant** $D = D(a, b)$, defined as follows:

The discriminant is also referred to as the "Hessian determinant."

$$D = D(a, b) = f_{xx}(a, b) f_{yy}(a, b) - f_{xy}^2(a, b)$$

If $D > 0$, then $f_{xx}(a, b)$ and $f_{yy}(a, b)$ must have the same sign, so the sign of $f_{yy}(a, b)$ also determines whether $f(a, b)$ is a local minimum or a local maximum.

THEOREM 2 Second Derivative Test Let $P = (a, b)$ be a critical point of $f(x, y)$. Assume that f_{xx}, f_{yy}, f_{xy} are continuous near P. Then:

(i) If $D > 0$ and $f_{xx}(a, b) > 0$, then $f(a, b)$ is a local minimum.
(ii) If $D > 0$ and $f_{xx}(a, b) < 0$, then $f(a, b)$ is a local maximum.
(iii) If $D < 0$, then f has a saddle point at (a, b).
(iv) If $D = 0$, the test is inconclusive.

A proof of this theorem is discussed at the end of this section.

■ EXAMPLE 3 Applying the Second Derivative Test Find the critical points of

$$f(x, y) = (x^2 + y^2)e^{-x}$$

and analyze them using the Second Derivative Test.

Solution

***Step 1.* Find the critical points.**

Set the partial derivatives equal to zero and solve:

$$f_x(x, y) = -(x^2 + y^2)e^{-x} + 2xe^{-x} = (2x - x^2 - y^2)e^{-x} = 0$$

$$f_y(x, y) = 2ye^{-x} = 0 \quad \Rightarrow \quad y = 0$$

Substituting $y = 0$ in the first equation then gives

$$(2x - x^2 - y^2)e^{-x} = (2x - x^2)e^{-x} = 0 \quad \Rightarrow \quad x = 0, 2$$

The critical points are (0, 0) and (2, 0) [Figure 6].

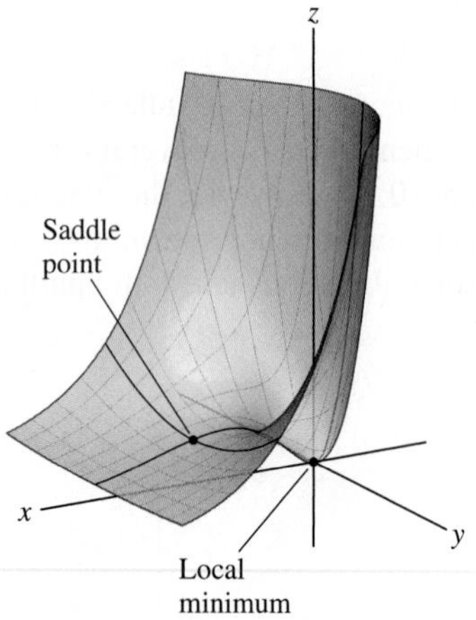

FIGURE 6 Graph of $f(x, y) = (x^2 + y^2)e^{-x}$.

***Step 2.* Compute the second-order partials.**

$$f_{xx}(x, y) = \frac{\partial}{\partial x}\left((2x - x^2 - y^2)e^{-x}\right) = (2 - 4x + x^2 + y^2)e^{-x}$$

$$f_{yy}(x, y) = \frac{\partial}{\partial y}(2ye^{-x}) = 2e^{-x}$$

$$f_{xy}(x, y) = f_{yx}(x, y) = \frac{\partial}{\partial x}(2ye^{-x}) = -2ye^{-x}$$

***Step 3.* Apply the Second Derivative Test.**

Critical Point	f_{xx}	f_{yy}	f_{xy}	Discriminant $D = f_{xx}f_{yy} - f_{xy}^2$	Type
(0, 0)	2	2	0	$2(2) - 0^2 = 4$	Local minimum since $D > 0$ and $f_{xx} > 0$
(2, 0)	$-2e^{-2}$	$2e^{-2}$	0	$-2e^{-2}(2e^{-2}) - 0^2 = -4e^{-4}$	Saddle since $D < 0$

■

GRAPHICAL INSIGHT We can also read off the type of critical point from the contour map. Notice that the level curves in Figure 7 encircle the local minimum at P, with f increasing in all directions emanating from P. By contrast, f has a saddle point at Q: The neighborhood near Q is divided into four regions in which $f(x, y)$ alternately increases and decreases.

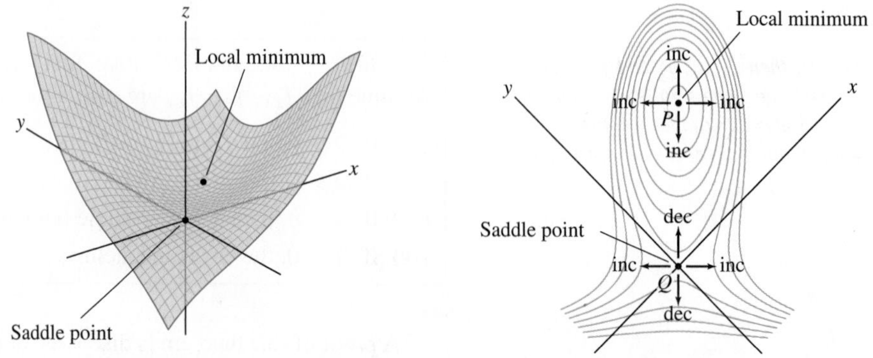

FIGURE 7 $f(x, y) = x^3 + y^3 - 12xy$.

■ **EXAMPLE 4** Analyze the critical points of $f(x, y) = x^3 + y^3 - 12xy$.

Solution Again, we set the partial derivatives equal to zero and solve:

$$f_x(x, y) = 3x^2 - 12y = 0 \quad \Rightarrow \quad y = \frac{1}{4}x^2$$

$$f_y(x, y) = 3y^2 - 12x = 0$$

Substituting $y = \frac{1}{4}x^2$ in the second equation yields

$$3y^2 - 12x = 3\left(\frac{1}{4}x^2\right)^2 - 12x = \frac{3}{16}x(x^3 - 64) = 0 \quad \Rightarrow \quad x = 0, 4$$

Since $y = \frac{1}{4}x^2$, the critical points are (0, 0) and (4, 4).

We have

$$f_{xx}(x, y) = 6x, \qquad f_{yy}(x, y) = 6y, \qquad f_{xy}(x, y) = -12$$

The Second Derivative Test confirms what we see in Figure 7: f has a local min at (4, 4) and a saddle at (0, 0).

Critical Point	f_{xx}	f_{yy}	f_{xy}	Discriminant $D = f_{xx}f_{yy} - f_{xy}^2$	Type
(0, 0)	0	0	−12	$0(0) - 12^2 = -144$	Saddle since $D < 0$
(4, 4)	24	24	−12	$24(24) - 12^2 = 432$	Local minimum since $D > 0$ and $f_{xx} > 0$

■

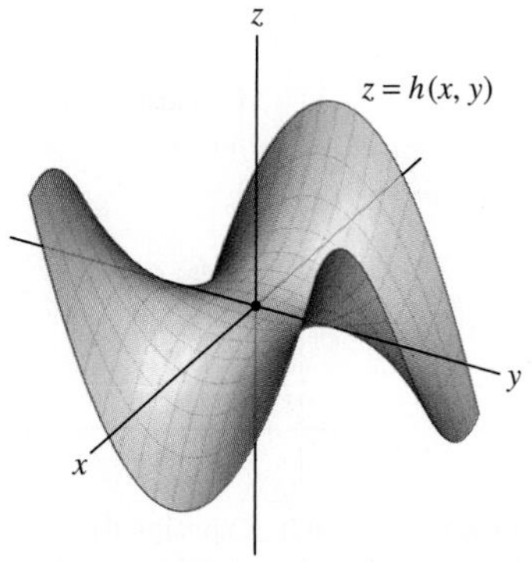

FIGURE 8 Graph of a "monkey saddle" with equation $h(x, y) = 3xy^2 - x^3$.

GRAPHICAL INSIGHT A graph can take on a variety of different shapes at a saddle point. The graph of $h(x, y)$ in Figure 8 is called a "monkey saddle" (because a monkey can sit on this saddle with room for his tail in the back).

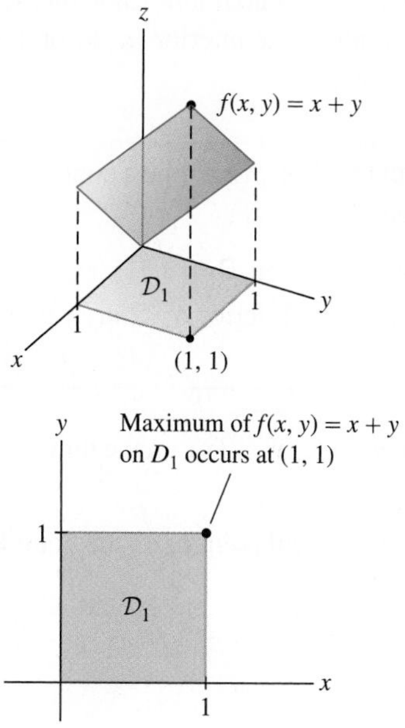

FIGURE 9

Global Extrema

Often we are interested in finding the minimum or maximum value of a function f on a given domain $\mathcal{D}$. These are called **global** or **absolute extreme values**. However, global extrema do not always exist. The function $f(x, y) = x + y$ has a maximum value on the unit square $\mathcal{D}_1$ in Figure 9 (the max is $f(1, 1) = 2$), but it has no maximum value on the entire plane $\mathbf{R}^2$.

To state conditions that guarantee the existence of global extrema, we need a few definitions. First, we say that a domain $\mathcal{D}$ is **bounded** if there is a number $M > 0$ such that $\mathcal{D}$ is contained in a disk of radius M centered at the origin. In other words, no point of $\mathcal{D}$ is more than a distance M from the origin [Figures 11(A) and 11(B)]. Next, a point P is called:

- An **interior point** of $\mathcal{D}$ if $\mathcal{D}$ contains some open disk $D(P, r)$ centered at P.
- A **boundary point** of $\mathcal{D}$ if every disk centered at P contains points in $\mathcal{D}$ and points not in $\mathcal{D}$.

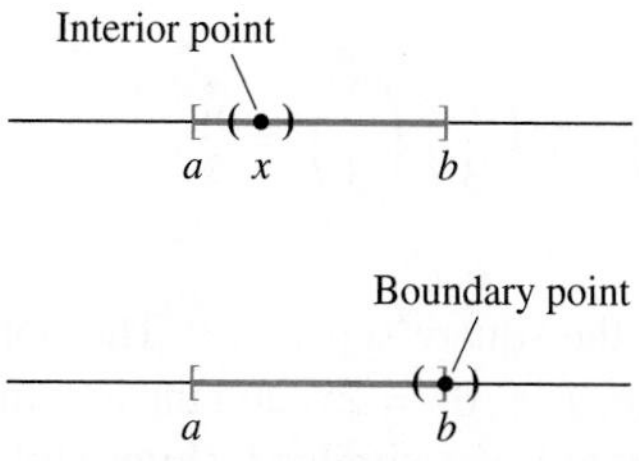

FIGURE 10 Interior and boundary points of an interval $[a, b]$.

CONCEPTUAL INSIGHT To understand the concept of interior and boundary points, think of the familiar case of an interval $I = [a, b]$ in the real line $\mathbf{R}$ (Figure 10). Every point x in the open interval (a, b) is an *interior point* of I (because there exists a small open interval around x entirely contained in I). The two endpoints a and b are *boundary points* (because every open interval containing a or b also contains points not in I).

The **interior** of $\mathcal{D}$ is the set of all interior points, and the **boundary** of $\mathcal{D}$ is the set of all boundary points. In Figure 11(C), the boundary is the curve surrounding the domain. The interior consists of all points in the domain not lying on the boundary curve.

A domain $\mathcal{D}$ is called **closed** if $\mathcal{D}$ contains all its boundary points (like a closed interval in $\mathbf{R}$). A domain $\mathcal{D}$ is called **open** if every point of $\mathcal{D}$ is an interior point (like an open interval in $\mathbf{R}$). The domain in Figure 11(A) is closed because the domain includes its boundary curve. In Figure 11(C), some boundary points are included and some are excluded, so the domain is neither open nor closed.

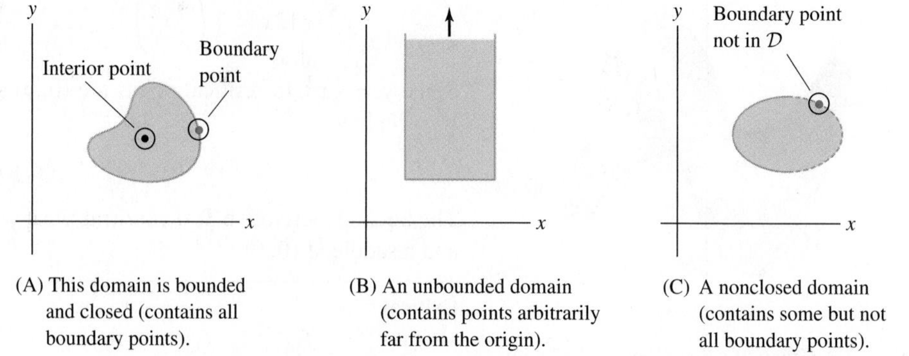

y
Interior point
Boundary point
x
(A) This domain is bounded and closed (contains all boundary points).
y
x
(B) An unbounded domain (contains points arbitrarily far from the origin).
y
Boundary point not in $\mathcal{D}$
x
(C) A nonclosed domain (contains some but not all boundary points).

FIGURE 11 Domains in $\mathbf{R}^2$.

In Section 4.2, we stated two basic results. First, a continuous function $f(x)$ on a *closed, bounded interval* $[a, b]$ takes on both a minimum and a maximum value on $[a, b]$. Second, these extreme values occur either at critical points in the interior (a, b) or at the endpoints. Analogous results are valid in several variables.

> **THEOREM 3 Existence and Location of Global Extrema** Let $f(x, y)$ be a continuous function on a closed, bounded domain $\mathcal{D}$ in $\mathbf{R}^2$. Then:
>
> **(i)** $f(x, y)$ takes on both a minimum and a maximum value on $\mathcal{D}$.
>
> **(ii)** The extreme values occur either at critical points in the interior of $\mathcal{D}$ or at points on the boundary of $\mathcal{D}$.

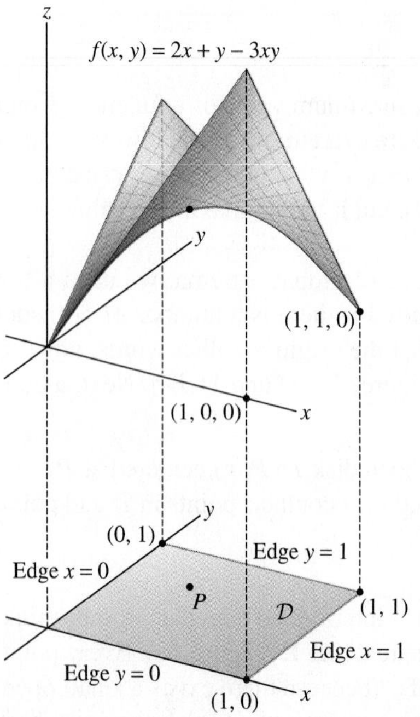

FIGURE 12

■ **EXAMPLE 5** Find the maximum value of $f(x, y) = 2x + y - 3xy$ on the unit square $\mathcal{D} = \{(x, y) : 0 \le x, y \le 1\}$.

Solution By Theorem 3, the maximum occurs either at a critical point or on the boundary of the square (Figure 12).

***Step 1.* Examine the critical points.**
Set the partial derivatives equal to zero and solve:

$$f_x(x, y) = 2 - 3y = 0 \quad \Rightarrow \quad y = \frac{2}{3}, \qquad f_y(x, y) = 1 - 3x = 0 \quad \Rightarrow \quad x = \frac{1}{3}$$

There is a unique critical point $P = \left(\frac{1}{3}, \frac{2}{3}\right)$ and

$$f(P) = f\left(\frac{1}{3}, \frac{2}{3}\right) = 2\left(\frac{1}{3}\right) + \left(\frac{2}{3}\right) - 3\left(\frac{1}{3}\right)\left(\frac{2}{3}\right) = \frac{2}{3}$$

***Step 2.* Check the boundary.**
We do this by checking each of the four edges of the square separately. The bottom edge is described by $y = 0$, $0 \le x \le 1$. On this edge, $f(x, 0) = 2x$, and the maximum value occurs at $x = 1$, where $f(1, 0) = 2$. Proceeding in a similar fashion with the other edges, we obtain

Edge	Restriction of $f(x, y)$ to Edge	Maximum of $f(x, y)$ on Edge
Lower: $y = 0, 0 \le x \le 1$	$f(x, 0) = 2x$	$f(1, 0) = 2$
Upper: $y = 1, 0 \le x \le 1$	$f(x, 1) = 1 - x$	$f(0, 1) = 1$
Left: $x = 0, 0 \le y \le 1$	$f(0, y) = y$	$f(0, 1) = 1$
Right: $x = 1, 0 \le y \le 1$	$f(1, y) = 2 - 2y$	$f(1, 0) = 2$

Step 3. **Compare.**

The maximum of f on the boundary is $f(1, 0) = 2$. This is larger than the value $f(P) = \frac{2}{3}$ at the critical point, so the maximum of f on the unit square is 2. ■

■ **EXAMPLE 6 Box of Maximum Volume** Find the maximum volume of a box inscribed in the tetrahedron bounded by the coordinate planes and the plane $\frac{1}{3}x + y + z = 1$.

Solution

Step 1. **Find a function to be maximized.**

Let $P = (x, y, z)$ be the corner of the box lying on the front face of the tetrahedron (Figure 13). Then the box has sides of lengths x, y, z and volume $V = xyz$. Using $\frac{1}{3}x + y + z = 1$, or $z = 1 - \frac{1}{3}x - y$, we express V in terms of x and y:

$$V(x, y) = xyz = xy\left(1 - \frac{1}{3}x - y\right) = xy - \frac{1}{3}x^2y - xy^2$$

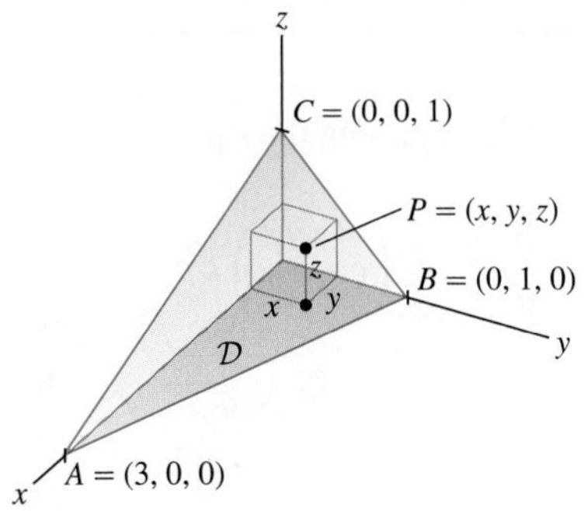

FIGURE 13 The shaded triangle is the domain of $V(x, y)$.

Our problem is to maximize V, but which domain $\mathcal{D}$ should we choose? We let $\mathcal{D}$ be the shaded triangle $\triangle OAB$ in the xy-plane in Figure 13. Then the corner point $P = (x, y, z)$ of each possible box lies above a point (x, y) in $\mathcal{D}$. Because $\mathcal{D}$ is closed and bounded, the maximum occurs at a critical point inside $\mathcal{D}$ or on the boundary of $\mathcal{D}$.

Step 2. **Examine the critical points.**

First, set the partial derivatives equal to zero and solve:

$$\frac{\partial V}{\partial x} = y - \frac{2}{3}xy - y^2 = y\left(1 - \frac{2}{3}x - y\right) = 0$$

$$\frac{\partial V}{\partial y} = x - \frac{1}{3}x^2 - 2xy = x\left(1 - \frac{1}{3}x - 2y\right) = 0$$

If $x = 0$ or $y = 0$, then (x, y) lies on the boundary of $\mathcal{D}$, so assume that x and y are both nonzero. Then the first equation gives us

$$1 - \frac{2}{3}x - y = 0 \quad \Rightarrow \quad y = 1 - \frac{2}{3}x$$

The second equation yields

$$1 - \frac{1}{3}x - 2y = 1 - \frac{1}{3}x - 2\left(1 - \frac{2}{3}x\right) = 0 \quad \Rightarrow \quad x - 1 = 0 \quad \Rightarrow \quad x = 1$$

For $x = 1$, we have $y = 1 - \frac{2}{3}x = \frac{1}{3}$. Therefore, $\left(1, \frac{1}{3}\right)$ is a critical point, and

$$V\left(1, \frac{1}{3}\right) = (1)\frac{1}{3} - \frac{1}{3}(1)^2\frac{1}{3} - (1)\left(\frac{1}{3}\right)^2 = \frac{1}{9}$$

Step 3. **Check the boundary.**

We have $V(x, y) = 0$ for all points on the boundary of $\mathcal{D}$ (because the three edges of the boundary are defined by $x = 0$, $y = 0$, and $1 - \frac{1}{3}x - y = 0$). Clearly, then, the maximum occurs at the critical point, and the maximum volume is $\frac{1}{9}$. ■

Proof of the Second Derivative Test The proof is based on "completing the square" for quadratic forms. A **quadratic form** is a function

$$Q(h,k) = ah^2 + 2bhk + ck^2$$

where a, b, c are constants (not all zero). The discriminant of Q is the quantity

$$D = ac - b^2$$

Some quadratic forms take on only positive values for $(h,k) \neq (0,0)$, and others take on both positive and negative values. According to the next theorem, the sign of the discriminant determines which of these two possibilities occurs.

To illustrate Theorem 4, consider

$$Q(h,k) = h^2 + 2hk + 2k^2$$

It has a positive discriminant

$$D = (1)(2) - 1 = 1$$

We can see directly that $Q(h,k)$ takes on only positive values for $(h,k) \neq (0,0)$ by writing $Q(h,k)$ as

$$Q(h,k) = (h+k)^2 + k^2$$

THEOREM 4 With $Q(h,k)$ and D as above:

(i) If $D > 0$ and $a > 0$, then $Q(h,k) > 0$ for $(h,k) \neq (0,0)$.
(ii) If $D > 0$ and $a < 0$, then $Q(h,k) < 0$ for $(h,k) \neq (0,0)$.
(iii) If $D < 0$, then $Q(h,k)$ takes on both positive and negative values.

Proof Assume first that $a \neq 0$ and rewrite $Q(h,k)$ by "completing the square":

$$Q(h,k) = ah^2 + 2bhk + ck^2 = a\left(h + \frac{b}{a}k\right)^2 + \left(c - \frac{b^2}{a}\right)k^2$$

$$= a\left(h + \frac{b}{a}k\right)^2 + \frac{D}{a}k^2 \qquad \boxed{1}$$

If $D > 0$ and $a > 0$, then $D/a > 0$ and both terms in Eq. (1) are nonnegative. Furthermore, if $Q(h,k) = 0$, then each term in Eq. (1) must equal zero. Thus $k = 0$ and $h + \frac{b}{a}k = 0$, and then, necessarily, $h = 0$. This shows that $Q(h,k) > 0$ if $(h,k) \neq 0$, and (i) is proved. Part (ii) follows similarly. To prove (iii), note that if $a \neq 0$ and $D < 0$, then the coefficients of the squared terms in Eq. (1) have opposite signs and $Q(h,k)$ takes on both positive and negative values. Finally, if $a = 0$ and $D < 0$, then $Q(h,k) = 2bhk + ck^2$ with $b \neq 0$. In this case, $Q(h,k)$ again takes on both positive and negative values. ■

Now assume that $f(x,y)$ has a critical point at $P = (a,b)$. We shall analyze f by considering the restriction of $f(x,y)$ to the line (Figure 14) through $P = (a,b)$ in the direction of a unit vector $\langle h,k\rangle$:

$$F(t) = f(a+th, b+tk)$$

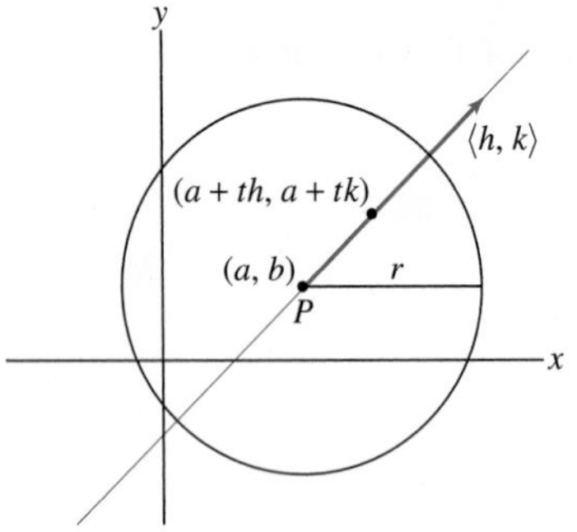

FIGURE 14 Line through P in the direction of $\langle h,k\rangle$.

Then $F(0) = f(a,b)$. By the Chain Rule,

$$F'(t) = f_x(a+th, b+tk)h + f_y(a+th, b+tk)k$$

Because P is a critical point, we have $f_x(a,b) = f_y(a,b) = 0$, and therefore,

$$F'(0) = f_x(a,b)h + f_y(a,b)k = 0$$

Thus $t = 0$ is a critical point of $F(t)$.

Now apply the Chain Rule again:

$$F''(t) = \frac{d}{dt}\Big(f_x(a+th, b+tk)h + f_y(a+th, b+tk)k\Big)$$
$$= \Big(f_{xx}(a+th, b+tk)h^2 + f_{xy}(a+th, b+tk)hk\Big)$$
$$+ \Big(f_{yx}(a+th, b+tk)kh + f_{yy}(a+th, b+tk)k^2\Big)$$
$$= f_{xx}(a+th, b+tk)h^2 + 2f_{xy}(a+th, b+tk)hk + f_{yy}(a+th, b+tk)k^2 \qquad \boxed{2}$$

We see that $F''(t)$ is the value at (h, k) of a quadratic form whose discriminant is equal to $D(a+th, b+tk)$. Here, we set

$$D(r, s) = f_{xx}(r, s)f_{yy}(r, s) - f_{xy}(r, s)^2$$

Note that the discriminant of $f(x, y)$ at the critical point $P = (a, b)$ is $D = D(a, b)$.

Case 1: $D(a, b) > 0$ and $f_{xx}(a, b) > 0$. We must prove that $f(a, b)$ is a local minimum. Consider a small disk of radius r around P (Figure 14). Because the second derivatives are continuous near P, we can choose $r > 0$ so that for every unit vector $\langle h, k\rangle$,

$$D(a+th, b+tk) > 0 \qquad \text{for } |t| < r$$
$$f_{xx}(a+th, b+tk) > 0 \qquad \text{for } |t| < r$$

Then $F''(t)$ is positive for $|t| < r$ by Theorem 4(i). This tells us that $F(t)$ is concave up, and hence $F(0) < F(t)$ if $0 < |t| < |r|$ (see Exercise 64 in Section 4.4). Because $F(0) = f(a, b)$, we may conclude that $f(a, b)$ is the minimum value of f along each segment of radius r through (a, b). Therefore, $f(a, b)$ is a local minimum value of f as claimed. The case that $D(a, b) > 0$ and $f_{xx}(a, b) < 0$ is similar.

Case 2: $D(a, b) < 0$. For $t = 0$, Eq. (2) yields

$$F''(0) = f_{xx}(a, b)h^2 + 2f_{xy}(a, b)hk + f_{yy}(a, b)k^2$$

Since $D(a, b) < 0$, this quadratic form takes on both positive and negative values by Theorem 4(iii). Choose $\langle h, k\rangle$ for which $F''(0) > 0$. By the Second Derivative Test in one variable, $F(0)$ is a local minimum of $F(t)$, and hence, there is a value $r > 0$ such that $F(0) < F(t)$ for all $0 < |t| < r$. But we can also choose $\langle h, k\rangle$ so that $F''(0) < 0$, in which case $F(0) > F(t)$ for $0 < |t| < r$ for some $r > 0$. Because $F(0) = f(a, b)$, we conclude that $f(a, b)$ is a local min in some directions and a local max in other directions. Therefore, f has a saddle point at $P = (a, b)$.

15.7 SUMMARY

- We say that $P = (a, b)$ is a *critical point* of $f(x, y)$ if
  - $f_x(a, b) = 0$ or $f_x(a, b)$ does not exist, and
  - $f_y(a, b) = 0$ or $f_y(a, b)$ does not exist.

In n-variables, $P = (a_1, \dots, a_n)$ is a critical point of $f(x_1, \dots, x_n)$ if each partial derivative $f_{x_j}(a_1, \dots, a_n)$ either is zero or does not exist.

- The local minimum or maximum values of f occur at critical points.
- The *discriminant* of $f(x, y)$ at $P = (a, b)$ is the quantity

$$D(a, b) = f_{xx}(a, b)f_{yy}(a, b) - f_{xy}^2(a, b)$$

- *Second Derivative Test:* If $P = (a, b)$ is a critical point of $f(x, y)$, then

$$D(a, b) > 0, \quad f_{xx}(a, b) > 0 \quad \Rightarrow \quad f(a, b) \text{ is a local minimum}$$

$$D(a, b) > 0, \quad f_{xx}(a, b) < 0 \quad \Rightarrow \quad f(a, b) \text{ is a local maximum}$$

$$D(a, b) < 0 \quad \Rightarrow \quad \text{saddle point}$$

$$D(a, b) = 0 \quad \Rightarrow \quad \text{test inconclusive}$$

- A point P is an *interior* point of a domain $\mathcal{D}$ if $\mathcal{D}$ contains some open disk $D(P, r)$ centered at P. A point P is a *boundary point* of $\mathcal{D}$ if every open disk $D(P, r)$ contains points in $\mathcal{D}$ and points not in $\mathcal{D}$. The *interior* of $\mathcal{D}$ is the set of all interior points, and the *boundary* is the set of all boundary points. A domain is *closed* if it contains all of its boundary points and *open* if it is equal to its interior.
- *Existence and location of global extrema:* If f is continuous and $\mathcal{D}$ is closed and bounded, then
  - f takes on both a minimum and a maximum value on $\mathcal{D}$.
  - The extreme values occur either at critical points in the interior of $\mathcal{D}$ or at points on the boundary of $\mathcal{D}$.

To determine the extreme values, first find the critical points in the interior of $\mathcal{D}$. Then compare the values of f at the critical points with the minimum and maximum values of f on the boundary.

15.7 EXERCISES

Preliminary Questions

1. The functions $f(x, y) = x^2 + y^2$ and $g(x, y) = x^2 - y^2$ both have a critical point at $(0, 0)$. How is the behavior of the two functions at the critical point different?

2. Identify the points indicated in the contour maps as local minima, local maxima, saddle points, or neither (Figure 15).

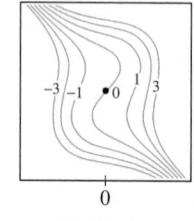

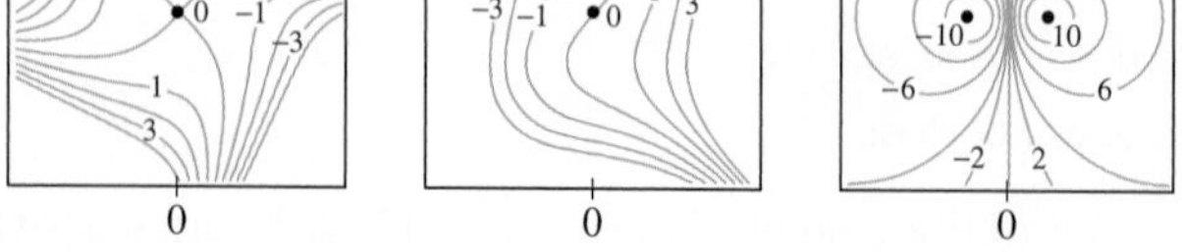

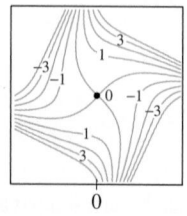

FIGURE 15

3. Let $f(x, y)$ be a continuous function on a domain $\mathcal{D}$ in $\mathbf{R}^2$. Determine which of the following statements are true:

(a) If $\mathcal{D}$ is closed and bounded, then f takes on a maximum value on $\mathcal{D}$.

(b) If $\mathcal{D}$ is neither closed nor bounded, then f does not take on a maximum value of $\mathcal{D}$.

(c) $f(x, y)$ need not have a maximum value on the domain $\mathcal{D}$ defined by $0 \le x \le 1, 0 \le y \le 1$.

(d) A continuous function takes on neither a minimum nor a maximum value on the open quadrant

$$\{(x, y) : x > 0, y > 0\}$$

Exercises

1. Let $P = (a, b)$ be a critical point of $f(x, y) = x^2 + y^4 - 4xy$.

(a) First use $f_x(x, y) = 0$ to show that $a = 2b$. Then use $f_y(x, y) = 0$ to show that $P = (0, 0)$, $(2\sqrt{2}, \sqrt{2})$, or $(-2\sqrt{2}, -\sqrt{2})$.

(b) Referring to Figure 16, determine the local minima and saddle points of $f(x, y)$ and find the absolute minimum value of $f(x, y)$.

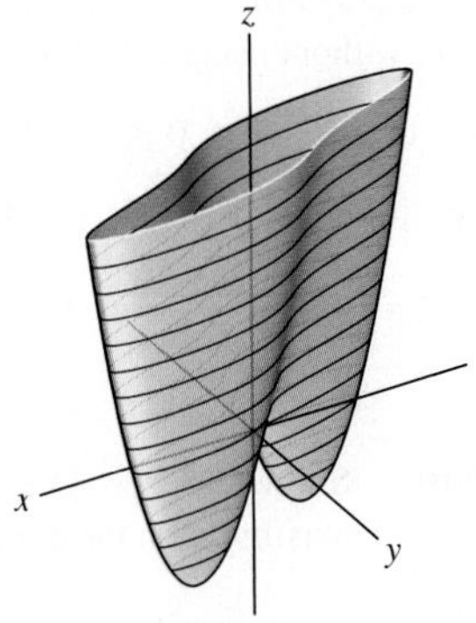

FIGURE 16

2. Find the critical points of the functions

$$f(x, y) = x^2 + 2y^2 - 4y + 6x, \qquad g(x, y) = x^2 - 12xy + y$$

Use the Second Derivative Test to determine the local minimum, local maximum, and saddle points. Match $f(x, y)$ and $g(x, y)$ with their graphs in Figure 17.

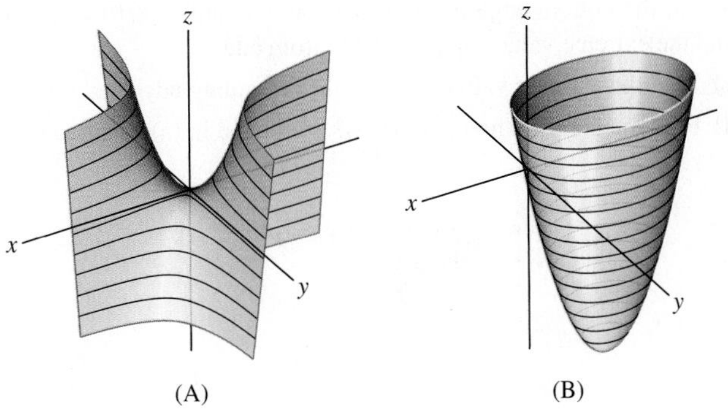

FIGURE 17

3. Find the critical points of

$$f(x, y) = 8y^4 + x^2 + xy - 3y^2 - y^3$$

Use the contour map in Figure 18 to determine their nature (local minimum, local maximum, or saddle point).

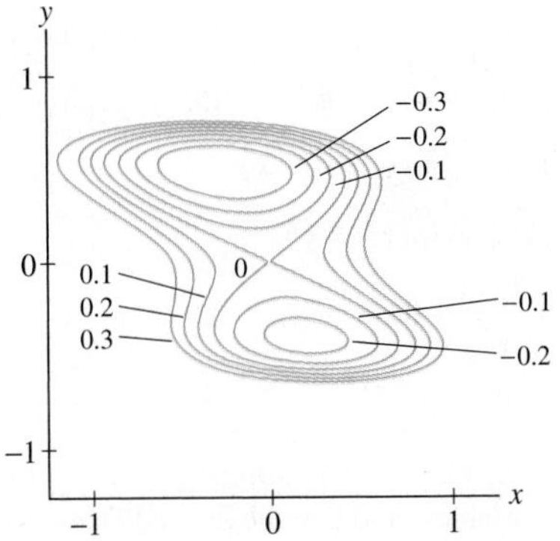

FIGURE 18 Contour map of $f(x, y) = 8y^4 + x^2 + xy - 3y^2 - y^3$.

4. Use the contour map in Figure 19 to determine whether the critical points A, B, C, D are local minima, local maxima, or saddle points.

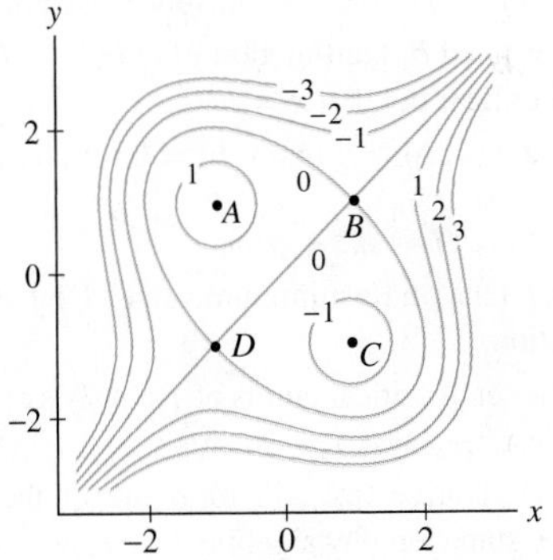

FIGURE 19

5. Let $f(x, y) = y^2x - yx^2 + xy$.

(a) Show that the critical points (x, y) satisfy the equations

$$y(y - 2x + 1) = 0, \qquad x(2y - x + 1) = 0$$

(b) Show that f has three critical points.

(c) Use the second derivative to determine the nature of the critical points.

6. Show that $f(x, y) = \sqrt{x^2 + y^2}$ has one critical point P and that f is nondifferentiable at P. Does f take on a minimum, maximum, or saddle point at P?

In Exercises 7–23, find the critical points of the function. Then use the Second Derivative Test to determine whether they are local minima, local maxima, or saddle points (or state that the test fails).

7. $f(x, y) = x^2 + y^2 - xy + x$

8. $f(x, y) = x^3 - xy + y^3$

9. $f(x, y) = x^3 + 2xy - 2y^2 - 10x$

10. $f(x, y) = x^3y + 12x^2 - 8y$

11. $f(x, y) = 4x - 3x^3 - 2xy^2$

12. $f(x, y) = x^3 + y^4 - 6x - 2y^2$

13. $f(x, y) = x^4 + y^4 - 4xy$

14. $f(x, y) = e^{x^2 - y^2 + 4y}$

15. $f(x, y) = xye^{-x^2 - y^2}$

16. $f(x, y) = e^x - xe^y$

17. $f(x, y) = \sin(x + y) - \cos x$

18. $f(x, y) = x \ln(x + y)$

19. $f(x, y) = \ln x + 2 \ln y - x - 4y$

20. $f(x, y) = (x + y)\ln(x^2 + y^2)$

21. $f(x, y) = x - y^2 - \ln(x + y)$

22. $f(x, y) = (x - y)e^{x^2 - y^2}$

23. $f(x, y) = (x + 3y)e^{y - x^2}$

24. Show that $f(x, y) = x^2$ has infinitely many critical points (as a function of two variables) and that the Second Derivative Test fails for all of them. What is the minimum value of f? Does $f(x, y)$ have any local maxima?

25. Prove that the function $f(x, y) = \frac{1}{3}x^3 + \frac{2}{3}y^{3/2} - xy$ satisfies $f(x, y) \geq 0$ for $x \geq 0$ and $y \geq 0$.

(a) First, verify that the set of critical points of f is the parabola $y = x^2$ and that the Second Derivative Test fails for these points.

(b) Show that for fixed b, the function $g(x) = f(x, b)$ is concave up for $x > 0$ with a critical point at $x = b^{1/2}$.

(c) Conclude that $f(a, b) \geq f(b^{1/2}, b) = 0$ for all $a, b \geq 0$.

26. Let $f(x, y) = (x^2 + y^2)e^{-x^2 - y^2}$.

(a) Where does f take on its minimum value? Do not use calculus to answer this question.

(b) Verify that the set of critical points of f consists of the origin $(0, 0)$ and the unit circle $x^2 + y^2 = 1$.

(c) The Second Derivative Test fails for points on the unit circle (this can be checked by some lengthy algebra). Prove, however, that f takes on its maximum value on the unit circle by analyzing the function $g(t) = te^{-t}$ for $t > 0$.

27. CAS Use a computer algebra system to find a numerical approximation to the critical point of

$$f(x, y) = (1 - x + x^2)e^{y^2} + (1 - y + y^2)e^{x^2}$$

Apply the Second Derivative Test to confirm that it corresponds to a local minimum as in Figure 20.

28. Which of the following domains are closed and which are bounded?

(a) $\{(x, y) \in \mathbf{R}^2 : x^2 + y^2 \leq 1\}$

(b) $\{(x, y) \in \mathbf{R}^2 : x^2 + y^2 < 1\}$

(c) $\{(x, y) \in \mathbf{R}^2 : x \geq 0\}$

(d) $\{(x, y) \in \mathbf{R}^2 : x > 0, y > 0\}$

(e) $\{(x, y) \in \mathbf{R}^2 : 1 \leq x \leq 4, 5 \leq y \leq 10\}$

(f) $\{(x, y) \in \mathbf{R}^2 : x > 0, x^2 + y^2 \leq 10\}$

In Exercises 29–32, determine the global extreme values of the function on the given set without using calculus.

29. $f(x, y) = x + y, \quad 0 \leq x \leq 1, \quad 0 \leq y \leq 1$

30. $f(x, y) = 2x - y, \quad 0 \leq x \leq 1, \quad 0 \leq y \leq 3$

31. $f(x, y) = (x^2 + y^2 + 1)^{-1}, \quad 0 \leq x \leq 3, \quad 0 \leq y \leq 5$

32. $f(x, y) = e^{-x^2 - y^2}, \quad x^2 + y^2 \leq 1$

33. Assumptions Matter Show that $f(x, y) = xy$ does not have a global minimum or a global maximum on the domain

$$\mathcal{D} = \{(x, y) : 0 < x < 1, 0 < y < 1\}$$

Explain why this does not contradict Theorem 3.

34. Find a continuous function that does not have a global maximum on the domain $\mathcal{D} = \{(x, y) : x + y \geq 0, x + y \leq 1\}$. Explain why this does not contradict Theorem 3.

35. Find the maximum of

$$f(x, y) = x + y - x^2 - y^2 - xy$$

on the square, $0 \leq x \leq 2, 0 \leq y \leq 2$ (Figure 21).

(a) First, locate the critical point of f in the square, and evaluate f at this point.

(b) On the bottom edge of the square, $y = 0$ and $f(x, 0) = x - x^2$. Find the extreme values of f on the bottom edge.

(c) Find the extreme values of f on the remaining edges.

(d) Find the largest among the values computed in (a), (b), and (c).

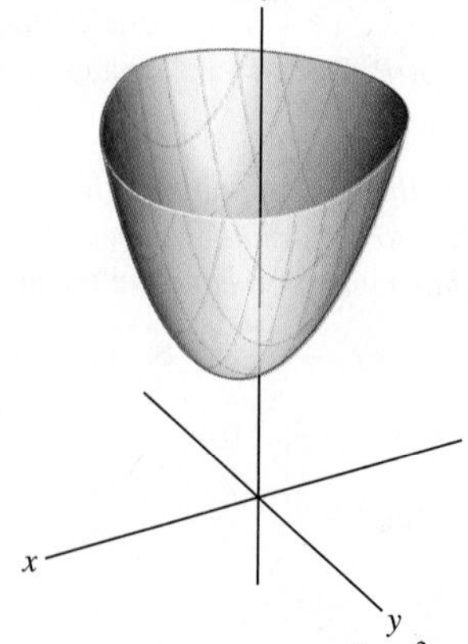

FIGURE 20 Plot of $f(x, y) = (1 - x + x^2)e^{y^2} + (1 - y + y^2)e^{x^2}$.

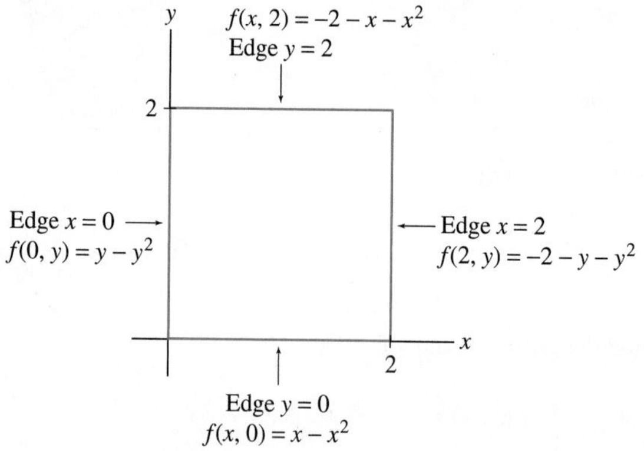

FIGURE 21 The function $f(x, y) = x + y - x^2 - y^2 - xy$ on the boundary segments of the square $0 \leq x \leq 2, 0 \leq y \leq 2$.

36. Find the maximum of $f(x, y) = y^2 + xy - x^2$ on the square $0 \le x \le 2, 0 \le y \le 2$.

In Exercises 37–43, determine the global extreme values of the function on the given domain.

37. $f(x, y) = x^3 - 2y, \quad 0 \le x \le 1, \quad 0 \le y \le 1$

38. $f(x, y) = 5x - 3y, \quad y \ge x - 2, \quad y \ge -x - 2, \quad y \le 3$

39. $f(x, y) = x^2 + 2y^2, \quad 0 \le x \le 1, \quad 0 \le y \le 1$

40. $f(x, y) = x^3 + x^2y + 2y^2, \quad x, y \ge 0, \quad x + y \le 1$

41. $f(x, y) = x^3 + y^3 - 3xy, \quad 0 \le x \le 1, \quad 0 \le y \le 1$

42. $f(x, y) = x^2 + y^2 - 2x - 4y, \quad x \ge 0, \quad 0 \le y \le 3, \quad y \ge x$

43. $f(x, y) = (4y^2 - x^2)e^{-x^2-y^2}, \quad x^2 + y^2 \le 2$

44. Find the maximum volume of a box inscribed in the tetrahedron bounded by the coordinate planes and the plane

$$x + \frac{1}{2}y + \frac{1}{3}z = 1$$

45. Find the maximum volume of the largest box of the type shown in Figure 22, with one corner at the origin and the opposite corner at a point $P = (x, y, z)$ on the paraboloid

$$z = 1 - \frac{x^2}{4} - \frac{y^2}{9} \quad \text{with } x, y, z \ge 0$$

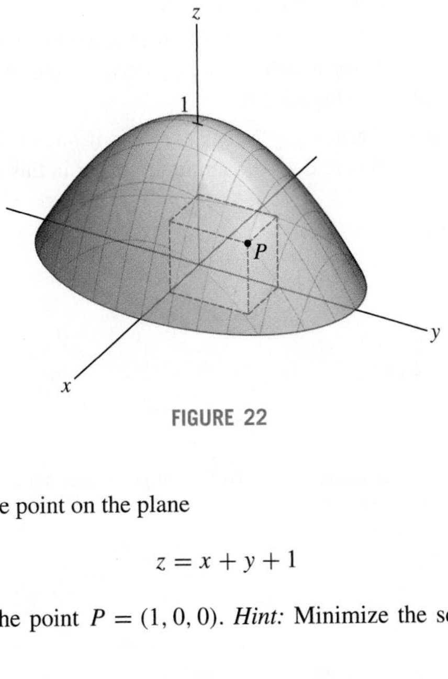

FIGURE 22

46. Find the point on the plane

$$z = x + y + 1$$

closest to the point $P = (1, 0, 0)$. *Hint:* Minimize the square of the distance.

47. Show that the sum of the squares of the distances from a point $P = (c, d)$ to n fixed points $(a_1, b_1), \dots, (a_n, b_n)$ is minimized when c is the average of the x-coordinates a_i and d is the average of the y-coordinates b_i.

48. Show that the rectangular box (including the top and bottom) with fixed volume $V = 27$ m^3 and smallest possible surface area is a cube (Figure 23).

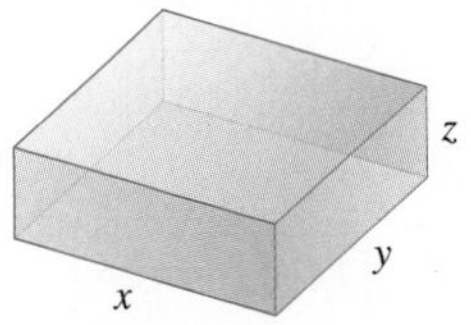

FIGURE 23 Rectangular box with sides x, y, z.

49. Consider a rectangular box B that has a bottom and sides but no top and has minimal surface area among all boxes with fixed volume V.

(a) Do you think B is a cube as in the solution to Exercise 48? If not, how would its shape differ from a cube?

(b) Find the dimensions of B and compare with your response to (a).

50. Given n data points $(x_1, y_1), \dots, (x_n, y_n)$, the **linear least-squares fit** is the linear function

$$f(x) = mx + b$$

that minimizes the sum of the squares (Figure 24):

$$E(m, b) = \sum_{j=1}^{n} (y_j - f(x_j))^2$$

Show that the minimum value of E occurs for m and b satisfying the two equations

$$m\left(\sum_{j=1}^{n} x_j\right) + bn = \sum_{j=1}^{n} y_j$$

$$m\sum_{j=1}^{n} x_j^2 + b\sum_{j=1}^{n} x_j = \sum_{j=1}^{n} x_j y_j$$

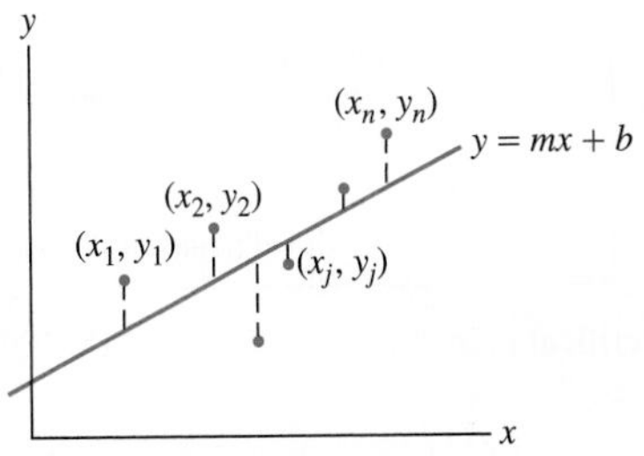

FIGURE 24 The linear least-squares fit minimizes the sum of the squares of the vertical distances from the data points to the line.

51. The power (in microwatts) of a laser is measured as a function of current (in milliamps). Find the linear least-squares fit (Exercise 50) for the data points.

Current (mA)	1.0	1.1	1.2	1.3	1.4	1.5
Laser power (μW)	0.52	0.56	0.82	0.78	1.23	1.50

52. Let $A = (a, b)$ be a fixed point in the plane, and let $f_A(P)$ be the distance from A to the point $P = (x, y)$. For $P \neq A$, let $\mathbf{e}_{AP}$ be the unit vector pointing from A to P (Figure 25):

$$\mathbf{e}_{AP} = \frac{\overrightarrow{AP}}{\|\overrightarrow{AP}\|}$$

Show that

$$\nabla f_A(P) = \mathbf{e}_{AP}$$

Note that we can derive this result without calculation: Because $\nabla f_A(P)$ points in the direction of maximal increase, it must point directly away from A at P, and because the distance $f_A(x, y)$ increases at a rate of one as you move away from A along the line through A and P, $\nabla f_A(P)$ must be a unit vector.

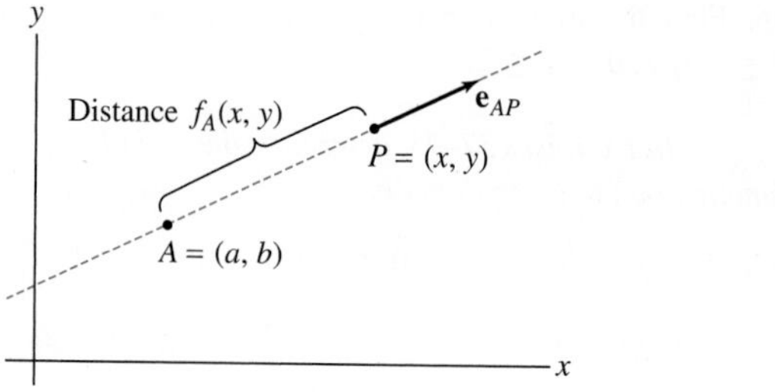

FIGURE 25 The distance from A to P increases most rapidly in the direction $\mathbf{e}_{AP}$.

Further Insights and Challenges

53. In this exercise, we prove that for all $x, y \geq 0$:

$$\frac{1}{\alpha}x^\alpha + \frac{1}{\beta}x^\beta \geq xy$$

where $\alpha \geq 1$ and $\beta \geq 1$ are numbers such that $\alpha^{-1} + \beta^{-1} = 1$. To do this, we prove that the function

$$f(x, y) = \alpha^{-1}x^\alpha + \beta^{-1}y^\beta - xy$$

satisfies $f(x, y) \geq 0$ for all $x, y \geq 0$.

(a) Show that the set of critical points of $f(x, y)$ is the curve $y = x^{\alpha-1}$ (Figure 26). Note that this curve can also be described as $x = y^{\beta-1}$. What is the value of $f(x, y)$ at points on this curve?

(b) Verify that the Second Derivative Test fails. Show, however, that for fixed $b > 0$, the function $g(x) = f(x, b)$ is concave up with a critical point at $x = b^{\beta-1}$.

(c) Conclude that for all $x > 0$, $f(x, b) \geq f(b^{\beta-1}, b) = 0$.

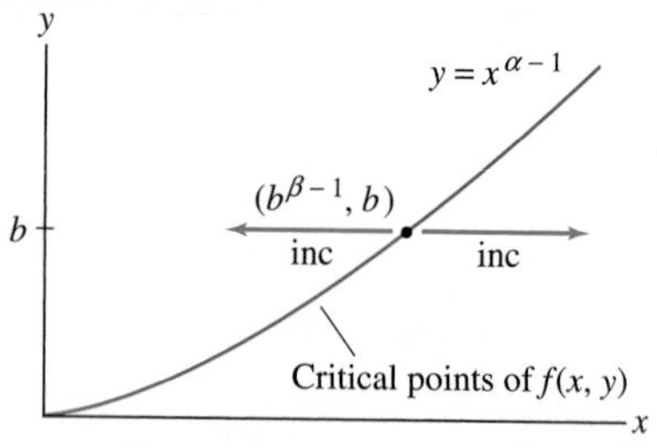

FIGURE 26 The critical points of $f(x, y) = \alpha^{-1}x^\alpha + \beta^{-1}y^\beta - xy$ form a curve $y = x^{\alpha-1}$.

54. The following problem was posed by Pierre de Fermat: Given three points $A = (a_1, a_2)$, $B = (b_1, b_2)$, and $C = (c_1, c_2)$ in the plane, find the point $P = (x, y)$ that minimizes the sum of the distances

$$f(x, y) = AP + BP + CP$$

Let $\mathbf{e}, \mathbf{f}, \mathbf{g}$ be the unit vectors pointing from P to the points A, B, C as in Figure 27.

(a) Use Exercise 52 to show that the condition $\nabla f(P) = 0$ is equivalent to

$$\mathbf{e} + \mathbf{f} + \mathbf{g} = 0 \qquad \boxed{3}$$

(b) Show that $f(x, y)$ is differentiable except at points A, B, C. Conclude that the minimum of $f(x, y)$ occurs either at a point P satisfying Eq. (3) or at one of the points A, B, or C.

(c) Prove that Eq. (3) holds if and only if P is the **Fermat point**, defined as the point P for which the angles between the segments $\overline{AP}$, $\overline{BP}$, $\overline{CP}$ are all 120° (Figure 27).

(d) Show that the Fermat point does not exist if one of the angles in $\triangle ABC$ is $> 120°$. Where does the minimum occur in this case?

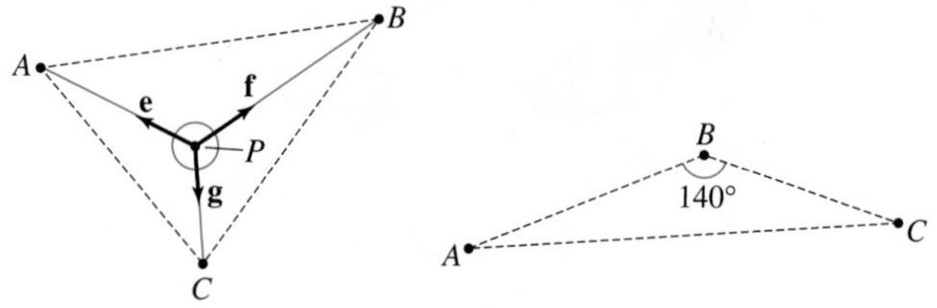

(A) P is the Fermat point (the angles between **e**, **f**, and **g** are all 120°).

(B) Fermat point does not exist.

FIGURE 27

15.8 Lagrange Multipliers: Optimizing with a Constraint

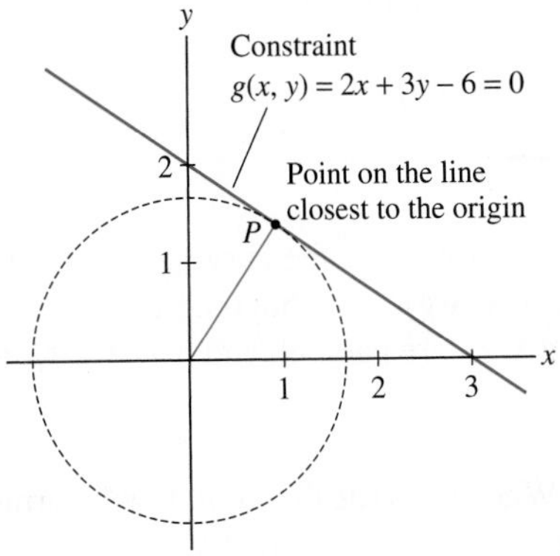

FIGURE 1 Finding the minimum of

$$f(x, y) = \sqrt{x^2 + y^2}$$

on the line $2x + 3y = 6$.

Some optimization problems involve finding the extreme values of a function $f(x, y)$ subject to a constraint $g(x, y) = 0$. Suppose that we want to find the point on the line $2x + 3y = 6$ closest to the origin (Figure 1). The distance from (x, y) to the origin is $f(x, y) = \sqrt{x^2 + y^2}$, so our problem is

$$\text{Minimize } f(x, y) = \sqrt{x^2 + y^2} \quad \text{subject to} \quad g(x, y) = 2x + 3y - 6 = 0$$

We are not seeking the minimum value of $f(x, y)$ (which is 0), but rather the minimum among all points (x, y) that lie on the line.

The method of **Lagrange multipliers** is a general procedure for solving optimization problems with a constraint. Here is a description of the main idea.

GRAPHICAL INSIGHT Imagine standing at point Q in Figure 2(A). We want to increase the value of f while remaining on the constraint curve. The gradient vector ∇f_Q points in the direction of *maximum* increase, but we cannot move in the gradient direction because that would take us off the constraint curve. However, the gradient points to the right, and so we can still increase f somewhat by moving to the right along the constraint curve.

We keep moving to the right until we arrive at the point P, where ∇f_P is orthogonal to the constraint curve [Figure 2(B)]. Once at P, we cannot increase f further by moving either to the right or to the left along the constraint curve. Thus $f(P)$ is a local maximum subject to the constraint.

Now, the vector ∇g_P is also orthogonal to the constraint curve, so ∇f_P and ∇g_P must point in the same or opposite directions. In other words, $\nabla f_P = \lambda \nabla g_P$ for some scalar λ (called a **Lagrange multiplier**). Graphically, this means that a local max subject to the constraint occurs at points P where the level curves of f and g are tangent.

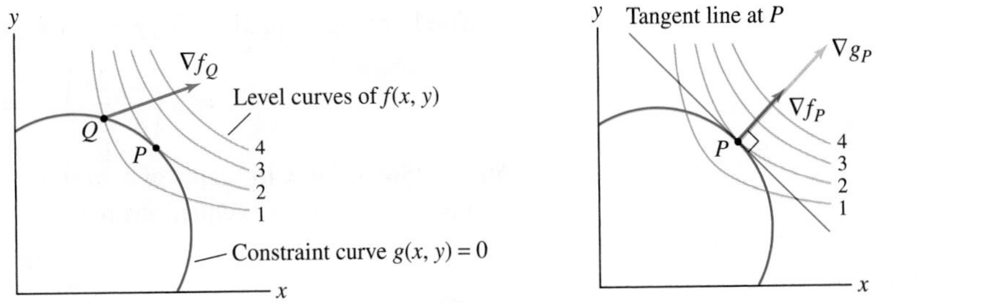

(A) f increases as we move to the right along the constraint curve.

(B) The local maximum of f on the constraint curve occurs where ∇f_P and ∇g_P are parallel.

FIGURE 2

THEOREM 1 Lagrange Multipliers Assume that $f(x, y)$ and $g(x, y)$ are differentiable functions. If $f(x, y)$ has a local minimum or a local maximum on the constraint curve $g(x, y) = 0$ at $P = (a, b)$, and if $\nabla g_P \neq \mathbf{0}$, then there is a scalar λ such that

$$\nabla f_P = \lambda \nabla g_P \qquad \boxed{1}$$

In Theorem 1, the assumption $\nabla g_P \neq \mathbf{0}$ guarantees (by the Implicit Function Theorem of advanced calculus) that we can parametrize the curve $g(x, y) = 0$ near P by a path $\mathbf{c}$ such that $\mathbf{c}(0) = P$ and $\mathbf{c}'(0) \neq \mathbf{0}$.

Proof Let $\mathbf{c}(t)$ be a parametrization of the constraint curve $g(x, y) = 0$ near P, chosen so that $\mathbf{c}(0) = P$ and $\mathbf{c}'(0) \neq \mathbf{0}$. Then $f(\mathbf{c}(0)) = f(P)$, and by assumption, $f(\mathbf{c}(t))$ has a local min or max at $t = 0$. Thus, $t = 0$ is a critical point of $f(\mathbf{c}(t))$ and

$$\underbrace{\left.\frac{d}{dt} f(\mathbf{c}(t))\right|_{t=0} = \nabla f_P \cdot \mathbf{c}'(0)}_{\text{Chain Rule}} = 0$$

This shows that ∇f_P is orthogonal to the tangent vector $\mathbf{c}'(0)$ to the curve $g(x, y) = 0$. The gradient ∇g_P is also orthogonal to $\mathbf{c}'(0)$ (because ∇g_P is orthogonal to the level curve $g(x, y) = 0$ at P). We conclude that ∇f_P and ∇g_P are parallel, and hence ∇f_P is a multiple of ∇g_P as claimed. ■

*← **REMINDER** Eq. (1) states that if a local min or max of $f(x, y)$ subject to a constraint $g(x, y) = 0$ occurs at $P = (a, b)$, then*

$$\nabla f_P = \lambda \nabla g_P$$

provided that $\nabla g_P \neq \mathbf{0}$.

We refer to Eq. (1) as the **Lagrange condition**. When we write this condition in terms of components, we obtain the **Lagrange equations**:

$$f_x(a, b) = \lambda g_x(a, b)$$
$$f_y(a, b) = \lambda g_y(a, b)$$

A point $P = (a, b)$ satisfying these equations is called a **critical point** for the optimization problem with constraint and $f(a, b)$ is called a **critical value**.

■ **EXAMPLE 1** Find the extreme values of $f(x, y) = 2x + 5y$ on the ellipse

$$\left(\frac{x}{4}\right)^2 + \left(\frac{y}{3}\right)^2 = 1$$

Solution

***Step 1.* Write out the Lagrange equations.**

The constraint curve is $g(x, y) = 0$, where $g(x, y) = (x/4)^2 + (y/3)^2 - 1$. We have

$$\nabla f = \langle 2, 5 \rangle, \qquad \nabla g = \left\langle \frac{x}{8}, \frac{2y}{9} \right\rangle$$

The Lagrange equations $\nabla f_P = \lambda \nabla g_P$ are:

$$\langle 2, 5 \rangle = \lambda \left\langle \frac{x}{8}, \frac{2y}{9} \right\rangle \quad \Rightarrow \quad 2 = \frac{\lambda x}{8}, \qquad 5 = \frac{\lambda(2y)}{9} \qquad \boxed{2}$$

***Step 2.* Solve for λ in terms of x and y.**

Eq. (2) gives us two equations for λ:

$$\lambda = \frac{16}{x}, \qquad \lambda = \frac{45}{2y} \qquad \boxed{3}$$

To justify dividing by x and y, note that x and y must be nonzero, because $x = 0$ or $y = 0$ would violate Eq. (2).

***Step 3.* Solve for x and y using the constraint.**

The two expressions for λ must be equal, so we obtain $\frac{16}{x} = \frac{45}{2y}$ or $y = \frac{45}{32}x$. Now substitute this in the constraint equation and solve for x:

$$\left(\frac{x}{4}\right)^2 + \left(\frac{\frac{45}{32}x}{3}\right)^2 = 1$$

$$x^2\left(\frac{1}{16} + \frac{225}{1024}\right) = x^2\left(\frac{289}{1024}\right) = 1$$

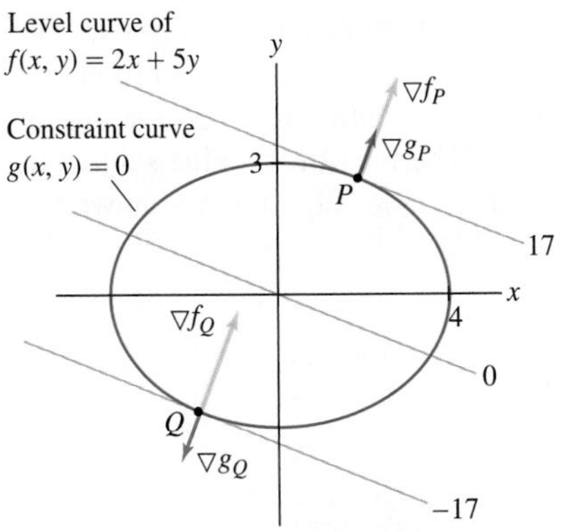

FIGURE 3 The min and max occur where a level curve of f is tangent to the constraint curve

$$g(x, y) = \left(\frac{x}{4}\right)^2 + \left(\frac{y}{3}\right)^2 - 1 = 0$$

Thus $x = \pm\sqrt{\frac{1024}{289}} = \pm\frac{32}{17}$, and since $y = \frac{45x}{32}$, the critical points are $P = \left(\frac{32}{17}, \frac{45}{17}\right)$ and $Q = \left(-\frac{32}{17}, -\frac{45}{17}\right)$.

Step 4. **Calculate the critical values.**

$$f(P) = f\left(\frac{32}{17}, \frac{45}{17}\right) = 2\left(\frac{32}{17}\right) + 5\left(\frac{45}{17}\right) = 17$$

and $f(Q) = -17$. We conclude that the maximum of $f(x, y)$ on the ellipse is 17 and the minimum is -17 (Figure 3). ■

Assumptions Matter According to Theorem 3 in Section 15.7, a continuous function on a closed, bounded domain takes on extreme values. This tells us that if the constraint curve is *bounded* (as in the previous example, where the constraint curve is an ellipse), then every continuous function $f(x, y)$ takes on both a minimum and a maximum value subject to the constraint. Be aware, however, that extreme values need not exist if the constraint curve is not bounded. For example, the constraint $x - y = 0$ is an unbounded line. The function $f(x, y) = x$ has neither a minimum nor a maximum subject to $x - y = 0$ because $P = (a, a)$ satisfies the constraint, yet $f(a, a) = a$ can be arbitrarily large or small.

FIGURE 4 Economist Paul Douglas, working with mathematician Charles Cobb, arrived at the production functions $P(x, y) = Cx^a y^b$ by fitting data gathered on the relationships between labor, capital, and output in an industrial economy. Douglas was a professor at the University of Chicago and also served as U.S. senator from Illinois from 1949 to 1967.

■ **EXAMPLE 2 Cobb–Douglas Production Function** By investing x units of labor and y units of capital, a low-end watch manufacturer can produce $P(x, y) = 50x^{0.4}y^{0.6}$ watches. (See Figure 4.) Find the maximum number of watches that can be produced on a budget of \$20,000 if labor costs \$100 per unit and capital costs \$200 per unit.

Solution The total cost of x units of labor and y units of capital is $100x + 200y$. Our task is to maximize the function $P(x, y) = 50x^{0.4}y^{0.6}$ subject to the following budget constraint (Figure 5):

$$g(x, y) = 100x + 200y - 20{,}000 = 0 \qquad \boxed{4}$$

Step 1. **Write out the Lagrange equations.**

$$P_x(x, y) = \lambda g_x(x, y): \quad 20x^{-0.6}y^{0.6} = 100\lambda$$

$$P_y(x, y) = \lambda g_y(x, y): \quad 30x^{0.4}y^{-0.4} = 200\lambda$$

Step 2. **Solve for λ in terms of x and y.**
These equations yield two expressions for λ that must be equal:

$$\lambda = \frac{1}{5}\left(\frac{y}{x}\right)^{0.6} = \frac{3}{20}\left(\frac{y}{x}\right)^{-0.4} \qquad \boxed{5}$$

Step 3. **Solve for x and y using the constraint.**
Multiply Eq. (5) by $5(y/x)^{0.4}$ to obtain $y/x = 15/20$, or $y = \frac{3}{4}x$. Then substitute in Eq. (4):

$$100x + 200y = 100x + 200\left(\frac{3}{4}x\right) = 20{,}000 \quad \Rightarrow \quad 250x = 20{,}000$$

We obtain $x = \frac{20{,}000}{250} = 80$ and $y = \frac{3}{4}x = 60$. The critical point is $A = (80, 60)$.

Step 4. **Calculate the critical values.**
Since $P(x, y)$ is increasing as a function of x and y, ∇P points to the northeast, and it is clear that $P(x, y)$ takes on a maximum value at A (Figure 5). The maximum is $P(80, 60) = 50(80)^{0.4}(60)^{0.6} = 3365.87$, or roughly 3365 watches, with a cost per watch of $\frac{20{,}000}{3365}$ or about \$5.94. ■

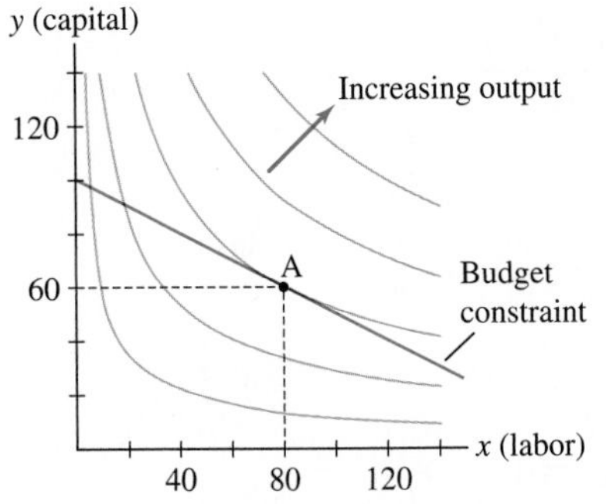

FIGURE 5 Contour plot of the Cobb–Douglas production function $P(x, y) = 50x^{0.4}y^{0.6}$. The level curves of a production function are called *isoquants*.

GRAPHICAL INSIGHT In an ordinary optimization problem without constraint, the global maximum value is the height of the highest point on the surface $z = f(x, y)$ (point Q in Figure 6). When a constraint is given, we restrict our attention to the curve on the surface lying above the constraint curve $g(x, y) = 0$. The maximum value subject to the constraint is the height of the highest point on this curve. Figure 6(B) shows the optimization problem solved in Example 1.

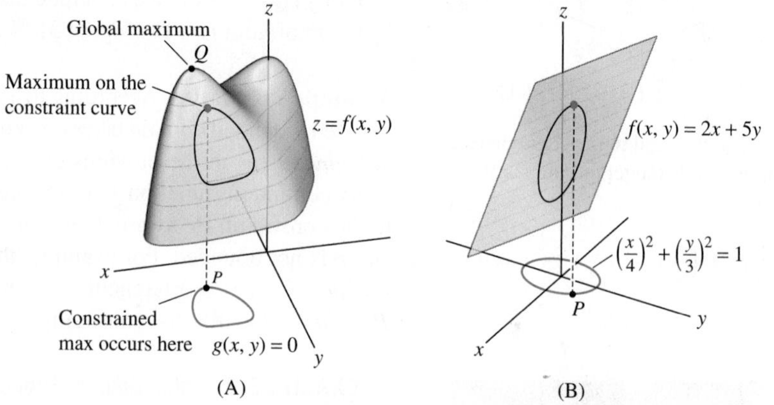

FIGURE 6

The method of Lagrange multipliers is valid in any number of variables. In the next example, we consider a problem in three variables.

EXAMPLE 3 Lagrange Multipliers in Three Variables Find the point on the plane $\frac{x}{2} + \frac{y}{4} + \frac{z}{4} = 1$ closest to the origin in $\mathbf{R}^3$.

Solution Our task is to minimize the distance $d = \sqrt{x^2 + y^2 + z^2}$ subject to the constraint $\frac{x}{2} + \frac{y}{4} + \frac{z}{4} = 1$. But finding the minimum distance d is the same as finding the minimum square of the distance d^2, so our problem can be stated:

$$\text{Minimize } f(x, y, z) = x^2 + y^2 + z^2 \quad \text{subject to} \quad g(x, y, z) = \frac{x}{2} + \frac{y}{4} + \frac{z}{4} - 1 = 0$$

The Lagrange condition is

$$\underbrace{\langle 2x, 2y, 2z \rangle}_{\nabla f} = \lambda \underbrace{\left\langle \frac{1}{2}, \frac{1}{4}, \frac{1}{4} \right\rangle}_{\nabla g}$$

This yields

$$\lambda = 4x = 8y = 8z \quad \Rightarrow \quad z = y = \frac{x}{2}$$

Substituting in the constraint equation, we obtain

$$\frac{x}{2} + \frac{y}{4} + \frac{z}{4} = \frac{2z}{2} + \frac{z}{4} + \frac{z}{4} = \frac{3z}{2} = 1 \quad \Rightarrow \quad z = \frac{2}{3}$$

Thus, $x = 2z = \frac{4}{3}$ and $y = z = \frac{2}{3}$. This critical point must correspond to the minimum of f (because f has no maximum on the constraint plane). Hence, the point on the plane closest to the origin is $P = \left(\frac{4}{3}, \frac{2}{3}, \frac{2}{3}\right)$ (Figure 7). ■

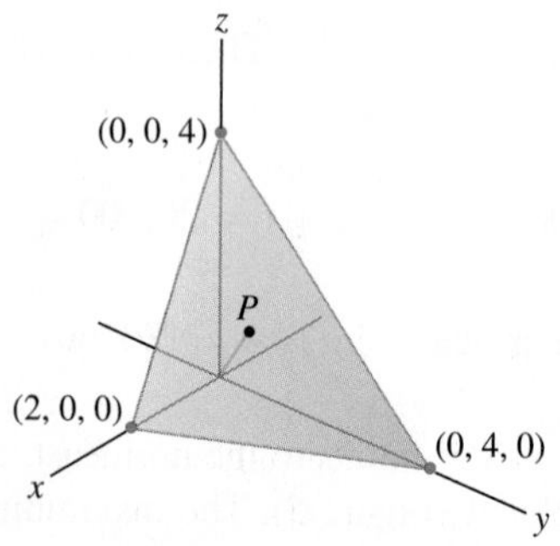

FIGURE 7 Point P closest to the origin on the plane.

The method of Lagrange multipliers can be used when there is more than one constraint equation, but we must add another multiplier for each additional constraint. For example, if the problem is to minimize $f(x, y, z)$ subject to constraints $g(x, y, z) = 0$ and $h(x, y, z) = 0$, then the Lagrange condition is

$$\nabla f = \lambda \nabla g + \mu \nabla h$$

The intersection of a sphere with a plane through its center is called a ***great circle****.*

■ **EXAMPLE 4 Lagrange Multipliers with Multiple Constraints** The intersection of the plane $x + \frac{1}{2}y + \frac{1}{3}z = 0$ with the unit sphere $x^2 + y^2 + z^2 = 1$ is a great circle (Figure 8). Find the point on this great circle with the largest x coordinate.

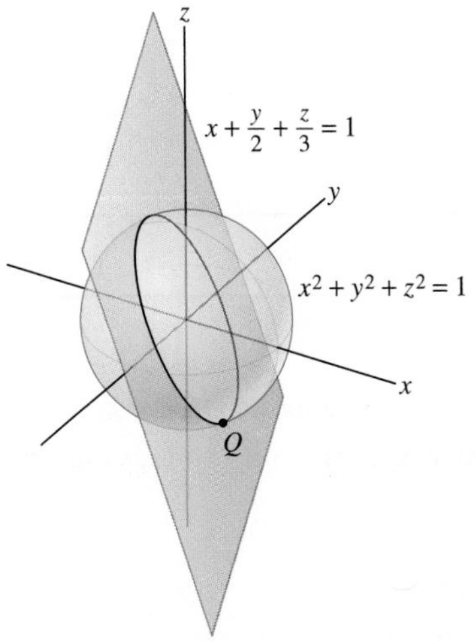

FIGURE 8 The plane intersects the sphere in a great circle. Q is the point on this great circle with the largest x-coordinate.

Solution Our task is to maximize the function $f(x, y, z) = x$ subject to the two constraint equations

$$g(x, y, z) = x + \frac{1}{2}y + \frac{1}{3}z = 0, \qquad h(x, y, z) = x^2 + y^2 + z^2 - 1 = 0$$

The Lagrange condition is

$$\nabla f = \lambda \nabla g + \mu \nabla h$$
$$\langle 1, 0, 0 \rangle = \lambda \left\langle 1, \frac{1}{2}, \frac{1}{3} \right\rangle + \mu \langle 2x, 2y, 2z \rangle$$

Note that μ cannot be zero. The Lagrange condition would become $\langle 1, 0, 0 \rangle = \lambda \langle 1, \frac{1}{2}, \frac{1}{3} \rangle$, and this equation is not satisfed for any value of λ. Now, the Lagrange condition gives us three equations:

$$\lambda + 2\mu x = 1, \qquad \frac{1}{2}\lambda + 2\mu y = 0, \qquad \frac{1}{3}\lambda + 2\mu z = 0$$

The last two equations yield $\lambda = -4\mu y$ and $\lambda = -6\mu z$. Because $\mu \neq 0$,

$$-4\mu y = -6\mu z \quad \Rightarrow \quad \boxed{y = \frac{3}{2}z}$$

Now use this relation in the first constraint equation:

$$x + \frac{1}{2}y + \frac{1}{3}z = x + \frac{1}{2}\left(\frac{3}{2}z\right) + \frac{1}{3}z = 0 \quad \Rightarrow \quad \boxed{x = -\frac{13}{12}z}$$

Finally, we can substitute in the second constraint equation:

$$x^2 + y^2 + z^2 - 1 = \left(-\frac{13}{12}z\right)^2 + \left(\frac{3}{2}z\right)^2 + z^2 - 1 = 0$$

to obtain $\frac{637}{144}z^2 = 1$ or $z = \pm\frac{12}{7\sqrt{13}}$. Since $x = -\frac{13}{12}z$ and $y = \frac{3}{2}z$, the critical points are

$$P = \left(-\frac{\sqrt{13}}{7}, \frac{18}{7\sqrt{13}}, \frac{12}{7\sqrt{13}}\right), \qquad Q = \left(\frac{\sqrt{13}}{7}, -\frac{18}{7\sqrt{13}}, -\frac{12}{7\sqrt{13}}\right)$$

The critical point with the largest x-coordinate (the maximum value of $f(x, y, z)$) is Q with x-coordinate $\frac{\sqrt{13}}{7} \approx 0.515$. ■

15.8 SUMMARY

- *Method of Lagrange multipliers*: The local extreme values of $f(x, y)$ subject to a constraint $g(x, y) = 0$ occur at points P (called critical points) satisfying the Lagrange condition $\nabla f_P = \lambda \nabla g_P$. This condition is equivalent to the *Lagrange equations*

$$f_x(x, y) = \lambda g_x(x, y), \qquad f_y(x, y) = \lambda g_y(x, y)$$

- If the constraint curve $g(x, y) = 0$ is bounded [e.g., if $g(x, y) = 0$ is a circle or ellipse], then global minimum and maximum values of f subject to the constraint exist.
- Lagrange condition for a function of three variables $f(x, y, z)$ subject to two constraints $g(x, y, z) = 0$ and $h(x, y, z) = 0$:

$$\nabla f = \lambda \nabla g + \mu \nabla h$$

15.8 EXERCISES

Preliminary Questions

1. Suppose that the maximum of $f(x, y)$ subject to the constraint $g(x, y) = 0$ occurs at a point $P = (a, b)$ such that $\nabla f_P \neq 0$. Which of the following statements is true?

(a) ∇f_P is tangent to $g(x, y) = 0$ at P.

(b) ∇f_P is orthogonal to $g(x, y) = 0$ at P.

2. Figure 9 shows a constraint $g(x, y) = 0$ and the level curves of a function f. In each case, determine whether f has a local minimum, a local maximum, or neither at the labeled point.

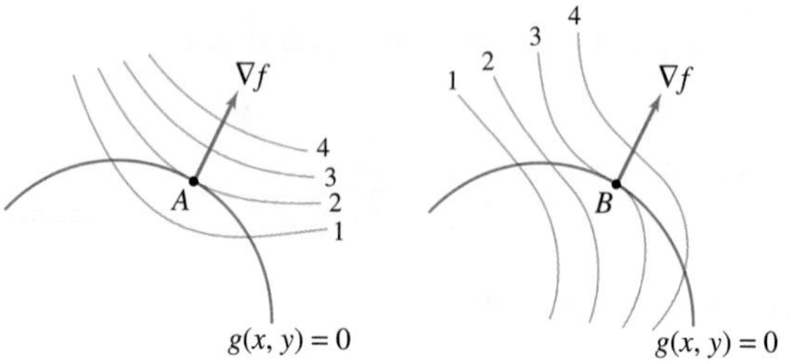

FIGURE 9

3. On the contour map in Figure 10:

(a) Identify the points where $\nabla f = \lambda \nabla g$ for some scalar λ.

(b) Identify the minimum and maximum values of $f(x, y)$ subject to $g(x, y) = 0$.

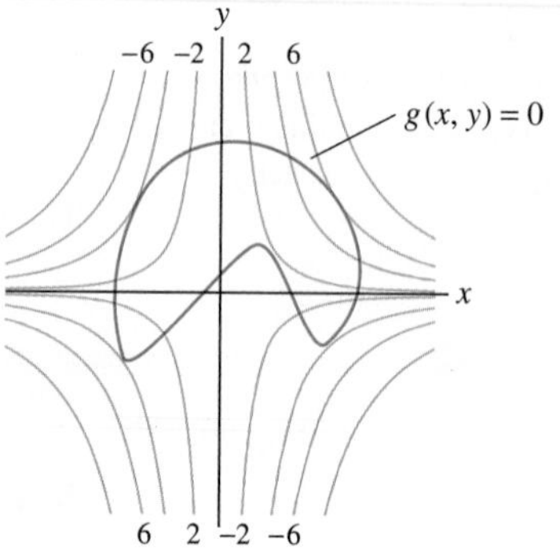

Contour plot of $f(x, y)$
(contour interval 2)

FIGURE 10 Contour map of $f(x, y)$; contour interval 2.

Exercises

In this exercise set, use the method of Lagrange multipliers unless otherwise stated.

1. Find the extreme values of the function $f(x, y) = 2x + 4y$ subject to the constraint $g(x, y) = x^2 + y^2 - 5 = 0$.

(a) Show that the Lagrange equation $\nabla f = \lambda \nabla g$ gives $\lambda x = 1$ and $\lambda y = 2$.

(b) Show that these equations imply $\lambda \neq 0$ and $y = 2x$.

(c) Use the constraint equation to determine the possible critical points (x, y).

(d) Evaluate $f(x, y)$ at the critical points and determine the minimum and maximum values.

2. Find the extreme values of $f(x, y) = x^2 + 2y^2$ subject to the constraint $g(x, y) = 4x - 6y = 25$.

(a) Show that the Lagrange equations yield $2x = 4\lambda$, $4y = -6\lambda$.

(b) Show that if $x = 0$ or $y = 0$, then the Lagrange equations give $x = y = 0$. Since $(0, 0)$ does not satisfy the constraint, you may assume that x and y are nonzero.

(c) Use the Lagrange equations to show that $y = -\frac{3}{4}x$.

(d) Substitute in the constraint equation to show that there is a unique critical point P.

(e) Does P correspond to a minimum or maximum value of f? Refer to Figure 11 to justify your answer. *Hint:* Do the values of $f(x, y)$ increase or decrease as (x, y) moves away from P along the line $g(x, y) = 0$?

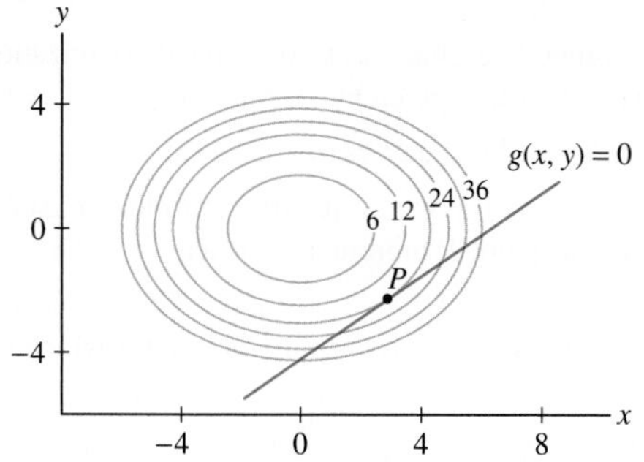

FIGURE 11 Level curves of $f(x, y) = x^2 + 2y^2$ and graph of the constraint $g(x, y) = 4x - 6y - 25 = 0$.

3. Apply the method of Lagrange multipliers to the function $f(x, y) = (x^2 + 1)y$ subject to the constraint $x^2 + y^2 = 5$. *Hint:* First show that $y \neq 0$; then treat the cases $x = 0$ and $x \neq 0$ separately.

In Exercises 4–13, find the minimum and maximum values of the function subject to the given constraint.

4. $f(x, y) = 2x + 3y, \quad x^2 + y^2 = 4$

5. $f(x, y) = x^2 + y^2, \quad 2x + 3y = 6$

6. $f(x, y) = 4x^2 + 9y^2, \quad xy = 4$

7. $f(x, y) = xy, \quad 4x^2 + 9y^2 = 32$

8. $f(x, y) = x^2y + x + y, \quad xy = 4$

9. $f(x, y) = x^2 + y^2, \qquad x^4 + y^4 = 1$

10. $f(x, y) = x^2y^4, \quad x^2 + 2y^2 = 6$

11. $f(x, y, z) = 3x + 2y + 4z, \quad x^2 + 2y^2 + 6z^2 = 1$

12. $f(x, y, z) = x^2 - y - z, \quad x^2 - y^2 + z = 0$

13. $f(x, y, z) = xy + 3xz + 2yz, \quad 5x + 9y + z = 10$

14. Let

$$f(x, y) = x^3 + xy + y^3, \qquad g(x, y) = x^3 - xy + y^3$$

(a) Show that there is a unique point $P = (a, b)$ on $g(x, y) = 1$ where $\nabla f_P = \lambda \nabla g_P$ for some scalar λ.

(b) Refer to Figure 12 to determine whether $f(P)$ is a local minimum or a local maximum of f subject to the constraint.

(c) Does Figure 12 suggest that $f(P)$ is a global extremum subject to the constraint?

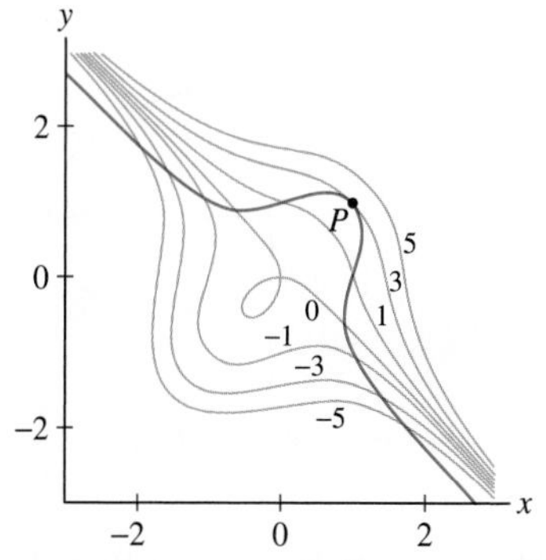

FIGURE 12 Contour map of $f(x, y) = x^3 + xy + y^3$ and graph of the constraint $g(x, y) = x^3 - xy + y^3 = 1$.

15. Find the point (a, b) on the graph of $y = e^x$ where the value ab is as small as possible.

16. Find the rectangular box of maximum volume if the sum of the lengths of the edges is 300 cm.

17. The surface area of a right-circular cone of radius r and height h is $S = \pi r\sqrt{r^2 + h^2}$, and its volume is $V = \frac{1}{3}\pi r^2 h$.

(a) Determine the ratio h/r for the cone with given surface area S and maximum volume V.

(b) What is the ratio h/r for a cone with given volume V and minimum surface area S?

(c) Does a cone with given volume V and maximum surface area exist?

18. In Example 1, we found the maximum of $f(x, y) = 2x + 5y$ on the ellipse $(x/4)^2 + (y/3)^2 = 1$. Solve this problem again without using Lagrange multipliers. First, show that the ellipse is parametrized by $x = 4\cos t$, $y = 3\sin t$. Then find the maximum value of $f(4\cos t, 3\sin t)$ using single-variable calculus. Is one method easier than the other?

19. Find the point on the ellipse

$$x^2 + 6y^2 + 3xy = 40$$

with largest x-coordinate (Figure 13).

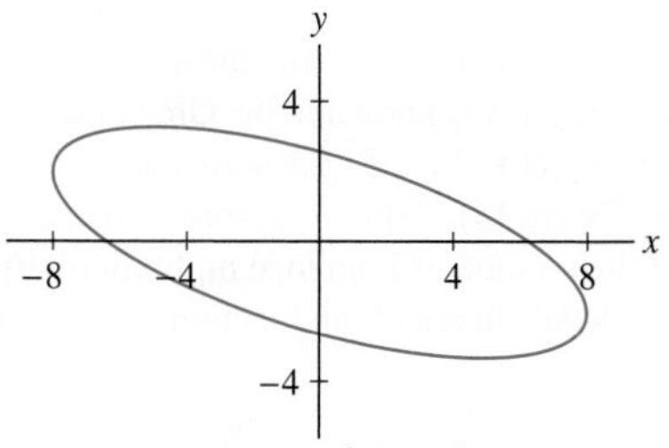

FIGURE 13 Graph of $x^2 + 6y^2 + 3xy = 40$

20. Find the maximum area of a rectangle inscribed in the ellipse (Figure 14):

$$\frac{x^2}{a^2} + \frac{y^2}{b^2} = 1$$

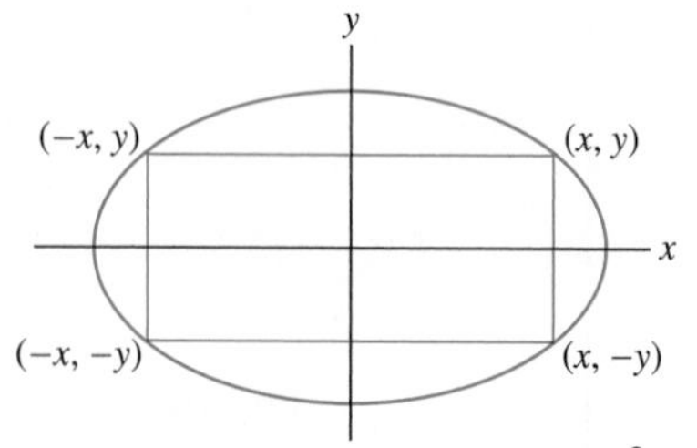

FIGURE 14 Rectangle inscribed in the ellipse $\frac{x^2}{a^2} + \frac{y^2}{b^2} = 1$.

21. Find the point (x_0, y_0) on the line $4x + 9y = 12$ that is closest to the origin.

22. Show that the point (x_0, y_0) closest to the origin on the line $ax + by = c$ has coordinates

$$x_0 = \frac{ac}{a^2 + b^2}, \qquad y_0 = \frac{bc}{a^2 + b^2}$$

23. Find the maximum value of $f(x, y) = x^a y^b$ for $x \geq 0$, $y \geq 0$ on the line $x + y = 1$, where $a, b > 0$ are constants.

24. Show that the maximum value of $f(x, y) = x^2 y^3$ on the unit circle is $\frac{6}{25}\sqrt{\frac{3}{5}}$.

25. Find the maximum value of $f(x, y) = x^a y^b$ for $x \geq 0$, $y \geq 0$ on the unit circle, where $a, b > 0$ are constants.

26. Find the maximum value of $f(x, y, z) = x^a y^b z^c$ for $x, y, z \geq 0$ on the unit sphere, where $a, b, c > 0$ are constants.

27. Show that the minimum distance from the origin to a point on the plane $ax + by + cz = d$ is

$$\frac{|d|}{\sqrt{a^2 + b^2 + d^2}}$$

28. Antonio has \$5.00 to spend on a lunch consisting of hamburgers (\$1.50 each) and French fries (\$1.00 per order). Antonio's satisfaction from eating x_1 hamburgers and x_2 orders of French fries is measured by a function $U(x_1, x_2) = \sqrt{x_1 x_2}$. How much of each type of food should he purchase to maximize his satisfaction? (Assume that fractional amounts of each food can be purchased.)

29. Let Q be the point on an ellipse closest to a given point P outside the ellipse. It was known to the Greek mathematician Apollonius (third century BCE) that $\overline{PQ}$ is perpendicular to the tangent to the ellipse at Q (Figure 15). Explain in words why this conclusion is a consequence of the method of Lagrange multipliers. *Hint:* The circles centered at P are level curves of the function to be minimized.

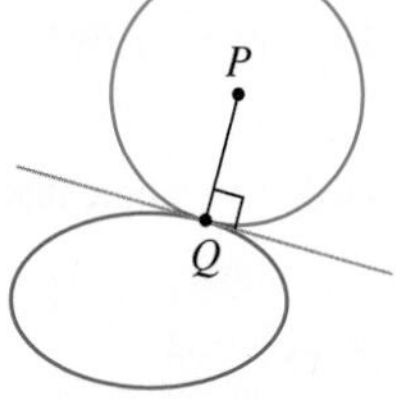

FIGURE 15

30. In a contest, a runner starting at A must touch a point P along a river and then run to B in the shortest time possible (Figure 16). The runner should choose the point P that minimizes the total length of the path.

(a) Define a function

$$f(x, y) = AP + PB, \quad \text{where } P = (x, y)$$

Rephrase the runner's problem as a constrained optimization problem, assuming that the river is given by an equation $g(x, y) = 0$.

(b) Explain why the level curves of $f(x, y)$ are ellipses.

(c) Use Lagrange multipliers to justify the following statement: The ellipse through the point P minimizing the length of the path is tangent to the river.

(d) Identify the point on the river in Figure 16 for which the length is minimal.

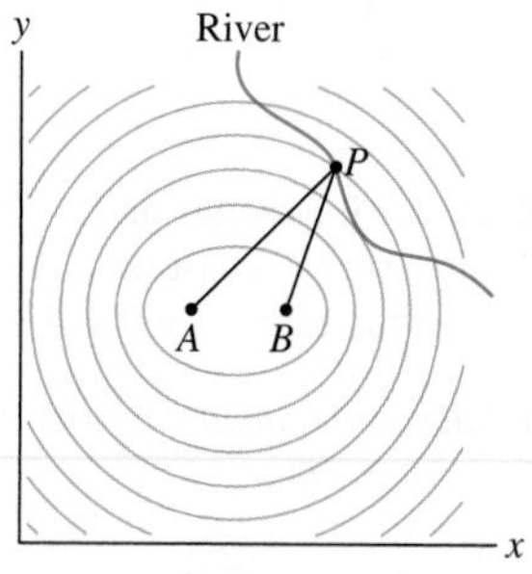

FIGURE 16

In Exercises 31 and 32, let V be the volume of a can of radius r and height h, and let S be its surface area (including the top and bottom).

31. Find r and h that minimize S subject to the constraint $V = 54\pi$.

32. Show that for both of the following two problems, $P = (r, h)$ is a Lagrange critical point if $h = 2r$:

- Minimize surface area S for fixed volume V.
- Maximize volume V for fixed surface area S.

Then use the contour plots in Figure 17 to explain why S has a minimum for fixed V but no maximum and, similarly, V has a maximum for fixed S but no minimum.

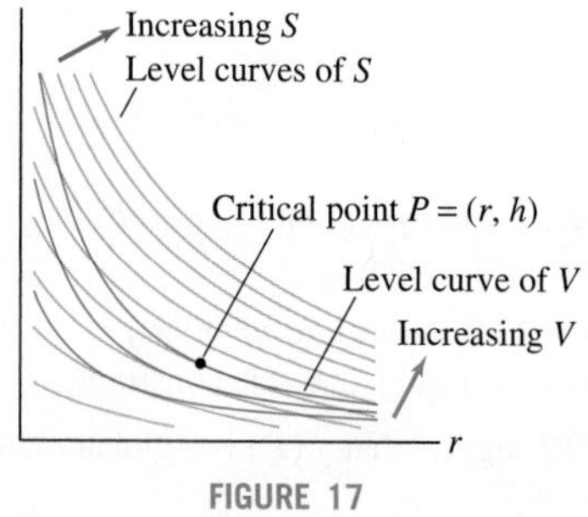

FIGURE 17

33. A plane with equation $\frac{x}{a} + \frac{y}{b} + \frac{z}{c} = 1$ $(a, b, c > 0)$ together with the positive coordinate planes forms a tetrahedron of volume $V = \frac{1}{6}abc$ (Figure 18). Find the minimum value of V among all planes passing through the point $P = (1, 1, 1)$.

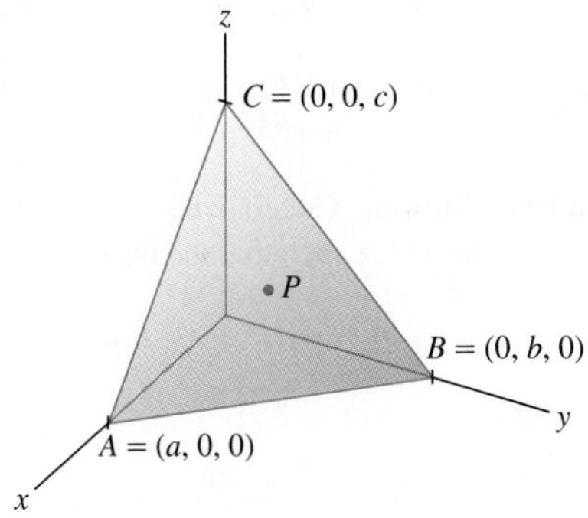

FIGURE 18

34. With the same set-up as in the previous problem, find the plane that minimizes V if the plane is constrained to pass through a point $P = (\alpha, \beta, \gamma)$ with $\alpha, \beta, \gamma > 0$.

35. Show that the Lagrange equations for $f(x, y) = x + y$ subject to the constraint $g(x, y) = x + 2y = 0$ have no solution. What can you conclude about the minimum and maximum values of f subject to $g = 0$? Show this directly.

36. Show that the Lagrange equations for $f(x, y) = 2x + y$ subject to the constraint $g(x, y) = x^2 - y^2 = 1$ have a solution but that f has no min or max on the constraint curve. Does this contradict Theorem 1?

37. Let L be the minimum length of a ladder that can reach over a fence of height h to a wall located a distance b behind the wall.

(a) Use Lagrange multipliers to show that $L = (h^{2/3} + b^{2/3})^{3/2}$ (Figure 19). *Hint:* Show that the problem amounts to minimizing $f(x, y) = (x + b)^2 + (y + h)^2$ subject to $y/b = h/x$ or $xy = bh$.

(b) Show that the value of L is also equal to the radius of the circle with center $(-b, -h)$ that is tangent to the graph of $xy = bh$.

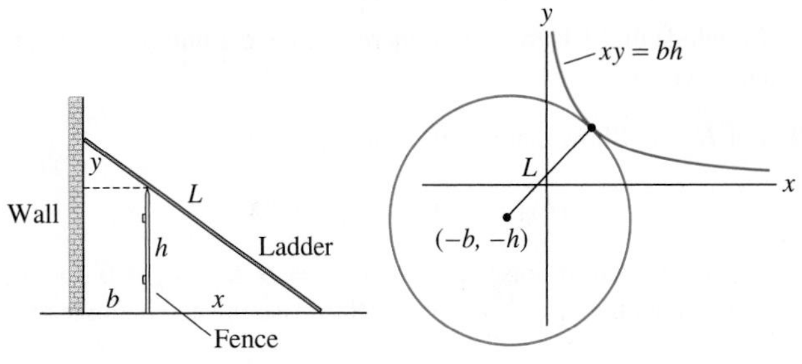

FIGURE 19

38. Find the maximum value of $f(x, y, z) = xy + xz + yz - xyz$ subject to the constraint $x + y + z = 1$, for $x \geq 0$, $y \geq 0$, $z \geq 0$.

39. Find the point lying on the intersection of the plane $x + \frac{1}{2}y + \frac{1}{4}z = 0$ and the sphere $x^2 + y^2 + z^2 = 9$ with the largest z-coordinate.

40. Find the maximum of $f(x, y, z) = x + y + z$ subject to the two constraints $x^2 + y^2 + z^2 = 9$ and $\frac{1}{4}x^2 + \frac{1}{4}y^2 + 4z^2 = 9$.

41. The cylinder $x^2 + y^2 = 1$ intersects the plane $x + z = 1$ in an ellipse. Find the point on that ellipse that is farthest from the origin.

42. Find the minimum and maximum of $f(x, y, z) = y + 2z$ subject to two constraints, $2x + z = 4$ and $x^2 + y^2 = 1$.

43. Find the minimum value of $f(x, y, z) = x^2 + y^2 + z^2$ subject to two constraints, $x + 2y + z = 3$ and $x - y = 4$.

Further Insights and Challenges

44. Suppose that both $f(x, y)$ and the constraint function $g(x, y)$ are linear. Use contour maps to explain why $f(x, y)$ does not have a maximum subject to $g(x, y) = 0$ unless $g = af + b$ for some constants a, b.

45. Assumptions Matter Consider the problem of minimizing $f(x, y) = x$ subject to $g(x, y) = (x - 1)^3 - y^2 = 0$.

(a) Show, without using calculus, that the minimum occurs at $P = (1, 0)$.

(b) Show that the Lagrange condition $\nabla f_P = \lambda \nabla g_P$ is not satisfied for any value of λ.

(c) Does this contradict Theorem 1?

46. Marginal Utility Goods 1 and 2 are available at dollar prices of p_1 per unit of good 1 and p_2 per unit of good 2. A utility function $U(x_1, x_2)$ is a function representing the **utility** or benefit of consuming x_j units of good j. The **marginal utility** of the jth good is $\partial U/\partial x_j$, the rate of increase in utility per unit increase in the jth good. Prove the following law of economics: Given a budget of L dollars, utility is maximized at the consumption level (a, b) where the ratio of marginal utility is equal to the ratio of prices:

$$\frac{\text{Marginal utility of good 1}}{\text{Marginal utility of good 2}} = \frac{U_{x_1}(a, b)}{U_{x_2}(a, b)} = \frac{p_1}{p_2}$$

47. Consider the utility function $U(x_1, x_2) = x_1 x_2$ with budget constraint $p_1 x_1 + p_2 x_2 = c$.

(a) Show that the maximum of $U(x_1, x_2)$ subject to the budget constraint is equal to $c^2/(4p_1 p_2)$.

(b) Calculate the value of the Lagrange multiplier λ occurring in (a).

(c) Prove the following interpretation: λ is the rate of increase in utility per unit increase in total budget c.

48. This exercise shows that the multiplier λ may be interpreted as a rate of change in general. Assume that the maximum of $f(x, y)$ subject to $g(x, y) = c$ occurs at a point P. Then P depends on the value of c, so we may write $P = (x(c), y(c))$ and we have $g(x(c), y(c)) = c$.

(a) Show that

$$\nabla g(x(c), y(c)) \cdot \langle x'(c), y'(c) \rangle = 1$$

Hint: Differentiate the equation $g(x(c), y(c)) = c$ with respect to c using the Chain Rule.

(b) Use the Chain Rule and the Lagrange condition $\nabla f_P = \lambda \nabla g_P$ to show that

$$\frac{d}{dc} f(x(c), y(c)) = \lambda$$

(c) Conclude that λ is the rate of increase in f per unit increase in the "budget level" c.

49. Let $B > 0$. Show that the maximum of

$$f(x_1, \ldots, x_n) = x_1 x_2 \cdots x_n$$

subject to the constraints $x_1 + \cdots + x_n = B$ and $x_j \geq 0$ for $j = 1, \ldots, n$ occurs for $x_1 = \cdots = x_n = B/n$. Use this to conclude that

$$(a_1 a_2 \cdots a_n)^{1/n} \leq \frac{a_1 + \cdots + a_n}{n}$$

for all positive numbers $a_1, \ldots, a_n$.

50. Let $B > 0$. Show that the maximum of $f(x_1, \ldots, x_n) = x_1 + \cdots + x_n$ subject to $x_1^2 + \cdots + x_n^2 = B^2$ is $\sqrt{n}B$. Conclude that

$$|a_1| + \cdots + |a_n| \leq \sqrt{n}(a_1^2 + \cdots + a_n^2)^{1/2}$$

for all numbers $a_1, \ldots, a_n$.

51. Given constants E, E_1, E_2, E_3, consider the maximum of

$$S(x_1, x_2, x_3) = x_1 \ln x_1 + x_2 \ln x_2 + x_3 \ln x_3$$

subject to two constraints:

$$x_1 + x_2 + x_3 = N, \qquad E_1 x_1 + E_2 x_2 + E_3 x_3 = E$$

Show that there is a constant μ such that $x_i = A^{-1} e^{\mu E_i}$ for $i = 1, 2, 3$, where $A = N^{-1}(e^{\mu E_1} + e^{\mu E_2} + e^{\mu E_3})$.

52. Boltzmann Distribution Generalize Exercise 51 to n variables: Show that there is a constant μ such that the maximum of

$$S = x_1 \ln x_1 + \cdots + x_n \ln x_n$$

subject to the constraints

$$x_1 + \cdots + x_n = N, \qquad E_1 x_1 + \cdots + E_n x_n = E$$

occurs for $x_i = A^{-1} e^{\mu E_i}$, where

$$A = N^{-1}(e^{\mu E_1} + \cdots + e^{\mu E_n})$$

This result lies at the heart of statistical mechanics. It is used to determine the distribution of velocities of gas molecules at temperature T; x_i is the number of molecules with kinetic energy E_i; $\mu = -(kT)^{-1}$, where k is Boltzmann's constant. The quantity S is called the **entropy**.

CHAPTER REVIEW EXERCISES

1. Given $f(x, y) = \dfrac{\sqrt{x^2 - y^2}}{x + 3}$:

(a) Sketch the domain of f.

(b) Calculate $f(3, 1)$ and $f(-5, -3)$.

(c) Find a point satisfying $f(x, y) = 1$.

2. Find the domain and range of:

(a) $f(x, y, z) = \sqrt{x - y} + \sqrt{y - z}$

(b) $f(x, y) = \ln(4x^2 - y)$

3. Sketch the graph $f(x, y) = x^2 - y + 1$ and describe its vertical and horizontal traces.

4. CAS Use a graphing utility to draw the graph of the function $\cos(x^2 + y^2)e^{1-xy}$ in the domains $[-1, 1] \times [-1, 1]$, $[-2, 2] \times [-2, 2]$, and $[-3, 3] \times [-3, 3]$, and explain its behavior.

5. Match the functions (a)–(d) with their graphs in Figure 5.

(a) $f(x, y) = x^2 + y$

(b) $f(x, y) = x^2 + 4y^2$

(c) $f(x, y) = \sin(4xy)e^{-x^2-y^2}$

(d) $f(x, y) = \sin(4x)e^{-x^2-y^2}$

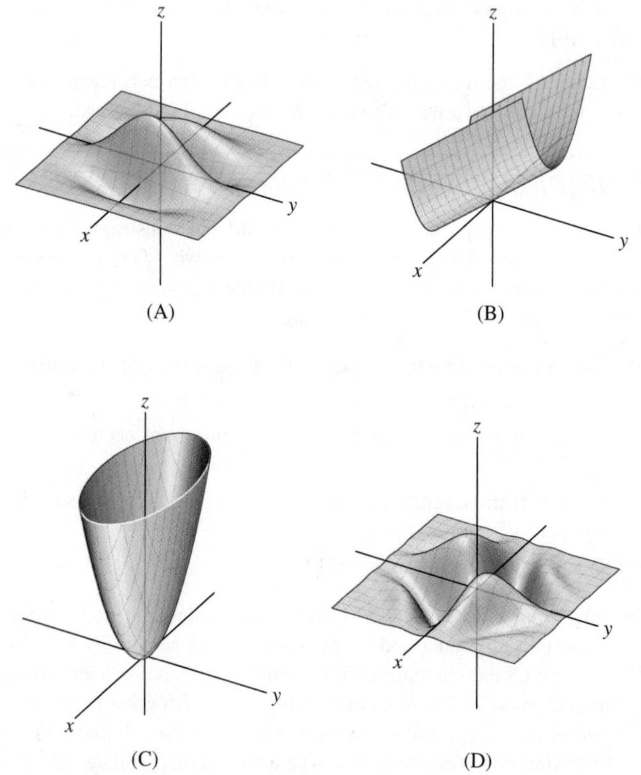

FIGURE 1

6. Referring to the contour map in Figure 2:

(a) Estimate the average rate of change of elevation from A to B and from A to D.

(b) Estimate the directional derivative at A in the direction of $\mathbf{v}$.

(c) What are the signs of f_x and f_y at D?

(d) At which of the labeled points are both f_x and f_y negative?

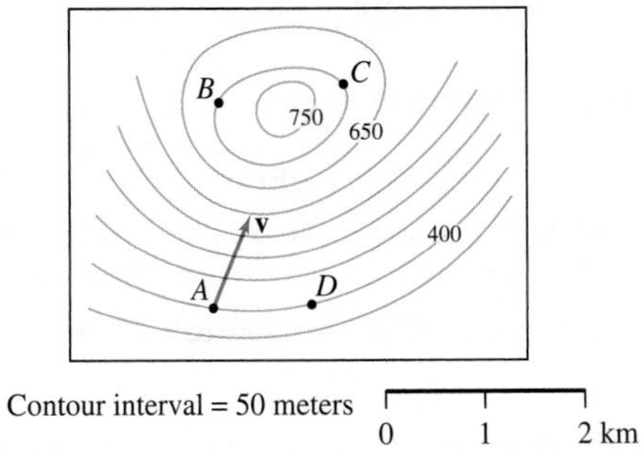

FIGURE 2

7. Describe the level curves of:

(a) $f(x, y) = e^{4x-y}$ **(b)** $f(x, y) = \ln(4x - y)$

(c) $f(x, y) = 3x^2 - 4y^2$ **(d)** $f(x, y) = x + y^2$

8. Match each function (a)–(c) with its contour graph (i)–(iii) in Figure 3:

(a) $f(x, y) = xy$

(b) $f(x, y) = e^{xy}$

(c) $f(x, y) = \sin(xy)$

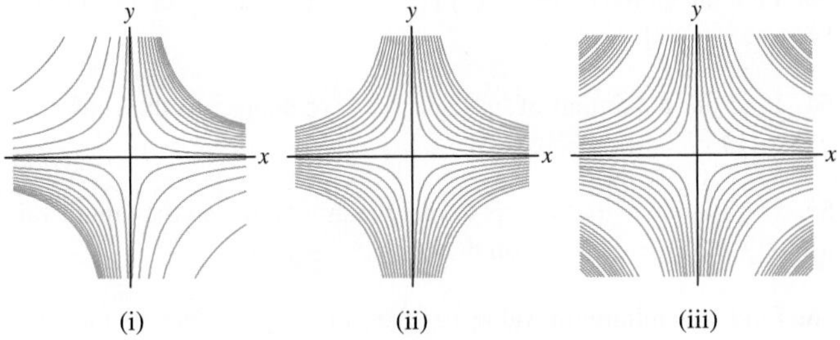

FIGURE 3

In Exercises 9–14, evaluate the limit or state that it does not exist.

9. $\lim_{(x,y)\to(1,-3)} (xy + y^2)$ **10.** $\lim_{(x,y)\to(1,-3)} \ln(3x + y)$

11. $\lim_{(x,y)\to(0,0)} \dfrac{xy + xy^2}{x^2 + y^2}$ **12.** $\lim_{(x,y)\to(0,0)} \dfrac{x^3y^2 + x^2y^3}{x^4 + y^4}$

13. $\lim_{(x,y)\to(1,-3)} (2x + y)e^{-x+y}$

14. $\lim_{(x,y)\to(0,2)} \dfrac{(e^x - 1)(e^y - 1)}{x}$

15. Let

$$f(x, y) = \begin{cases} \dfrac{(xy)^p}{x^4 + y^4} & (x, y) \neq (0, 0) \\ 0 & (x, y) = (0, 0) \end{cases}$$

Use polar coordinates to show that $f(x, y)$ is continuous at all (x, y) if $p > 2$ but is discontinuous at $(0, 0)$ if $p \le 2$.

16. Calculate $f_x(1, 3)$ and $f_y(1, 3)$ for $f(x, y) = \sqrt{7x + y^2}$.

In Exercises 17–20, compute f_x and f_y.

17. $f(x, y) = 2x + y^2$ **18.** $f(x, y) = 4xy^3$

19. $f(x, y) = \sin(xy)e^{-x-y}$ **20.** $f(x, y) = \ln(x^2 + xy^2)$

21. Calculate f_{xxyz} for $f(x, y, z) = y\sin(x + z)$.

22. Fix $c > 0$. Show that for any constants α, β, the function $u(t, x) = \sin(\alpha ct + \beta)\sin(\alpha x)$ satisfies the wave equation

$$\frac{\partial^2 u}{\partial t^2} = c^2\frac{\partial^2 u}{\partial x^2}$$

23. Find an equation of the tangent plane to the graph of $f(x, y) = xy^2 - xy + 3x^3y$ at $P = (1, 3)$.

24. Suppose that $f(4, 4) = 3$ and $f_x(4, 4) = f_y(4, 4) = -1$. Use the linear approximation to estimate $f(4.1, 4)$ and $f(3.88, 4.03)$.

25. Use a linear approximation of $f(x, y, z) = \sqrt{x^2 + y^2 + z}$ to estimate $\sqrt{7.1^2 + 4.9^2 + 69.5}$. Compare with a calculator value.

26. The plane $z = 2x - y - 1$ is tangent to the graph of $z = f(x, y)$ at $P = (5, 3)$.

(a) Determine $f(5, 3)$, $f_x(5, 3)$, and $f_y(5, 3)$.

(b) Approximate $f(5.2, 2.9)$.

27. Figure 4 shows the contour map of a function $f(x, y)$ together with a path $\mathbf{c}(t)$ in the counterclockwise direction. The points $\mathbf{c}(1)$, $\mathbf{c}(2)$, and $\mathbf{c}(3)$ are indicated on the path. Let $g(t) = f(\mathbf{c}(t))$. Which of statements (i)–(iv) are true? Explain.

(i) $g'(1) > 0$.

(ii) $g(t)$ has a local minimum for some $1 \le t \le 2$.

(iii) $g'(2) = 0$.

(iv) $g'(3) = 0$.

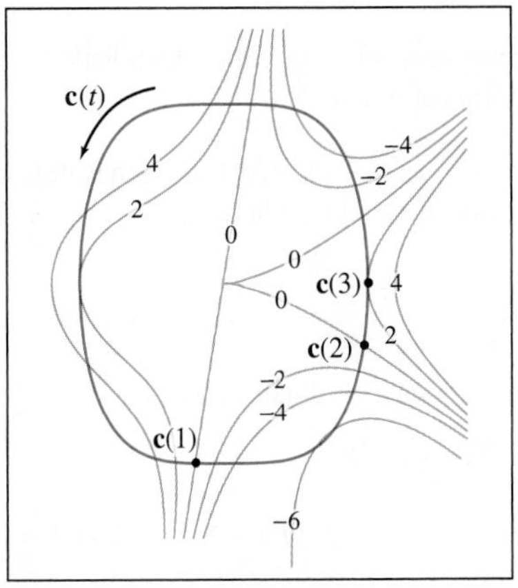

FIGURE 4

28. Jason earns $S(h, c) = 20h\left(1 + \frac{c}{100}\right)^{1.5}$ dollars per month at a used car lot, where h is the number of hours worked and c is the number of cars sold. He has already worked 160 hours and sold 69 cars. Right now Jason wants to go home but wonders how much more he might earn if he stays another 10 minutes with a customer who is considering buying a car. Use the linear approximation to estimate how much extra money Jason will earn if he sells his 70th car during these 10 minutes.

In Exercises 29–32, compute $\frac{d}{dt} f(\mathbf{c}(t))$ *at the given value of* t.

29. $f(x, y) = x + e^y, \quad \mathbf{c}(t) = (3t - 1, t^2)$ at $t = 2$

30. $f(x, y, z) = xz - y^2, \quad \mathbf{c}(t) = (t, t^3, 1 - t)$ at $t = -2$

31. $f(x, y) = xe^{3y} - ye^{3x}, \quad \mathbf{c}(t) = (e^t, \ln t)$ at $t = 1$

32. $f(x, y) = \tan^{-1}\frac{y}{x}, \quad \mathbf{c}(t) = (\cos t, \sin t), t = \frac{\pi}{3}$

In Exercises 33–36, compute the directional derivative at P *in the direction of* $\mathbf{v}$.

33. $f(x, y) = x^3y^4, \quad P = (3, -1), \quad \mathbf{v} = 2\mathbf{i} + \mathbf{j}$

34. $f(x, y, z) = zx - xy^2, \quad P = (1, 1, 1), \quad \mathbf{v} = \langle 2, -1, 2\rangle$

35. $f(x, y) = e^{x^2+y^2}, \quad P = \left(\frac{\sqrt{2}}{2}, \frac{\sqrt{2}}{2}\right), \quad \mathbf{v} = \langle 3, -4\rangle$

36. $f(x, y, z) = \sin(xy + z), \quad P = (0, 0, 0), \quad \mathbf{v} = \mathbf{j} + \mathbf{k}$

37. Find the unit vector $\mathbf{e}$ at $P = (0, 0, 1)$ pointing in the direction along which $f(x, y, z) = xz + e^{-x^2+y}$ increases most rapidly.

38. Find an equation of the tangent plane at $P = (0, 3, -1)$ to the surface with equation

$$ze^x + e^{z+1} = xy + y - 3$$

39. Let $n \neq 0$ be an integer and r an arbitrary constant. Show that the tangent plane to the surface $x^n + y^n + z^n = r$ at $P = (a, b, c)$ has equation

$$a^{n-1}x + b^{n-1}y + c^{n-1}z = r$$

40. Let $f(x, y) = (x - y)e^x$. Use the Chain Rule to calculate $\partial f/\partial u$ and $\partial f/\partial v$ (in terms of u and v), where $x = u - v$ and $y = u + v$.

41. Let $f(x, y, z) = x^2y + y^2z$. Use the Chain Rule to calculate $\partial f/\partial s$ and $\partial f/\partial t$ (in terms of s and t), where

$$x = s + t, \quad y = st, \quad z = 2s - t$$

42. Let P have spherical coordinates $(\rho, \theta, \phi) = \left(2, \frac{\pi}{4}, \frac{\pi}{4}\right)$. Calculate $\left.\frac{\partial f}{\partial \phi}\right|_P$ assuming that

$$f_x(P) = 4, \quad f_y(P) = -3, \quad f_z(P) = 8$$

Recall that $x = \rho\cos\theta\sin\phi, \ y = \rho\sin\theta\sin\phi, \ z = \rho\cos\phi$.

43. Let $g(u, v) = f(u^3 - v^3, v^3 - u^3)$. Prove that

$$v^2\frac{\partial g}{\partial u} - u^2\frac{\partial g}{\partial v} = 0$$

44. Let $f(x, y) = g(u)$, where $u = x^2 + y^2$ and $g(u)$ is differentiable. Prove that

$$\left(\frac{\partial f}{\partial x}\right)^2 + \left(\frac{\partial f}{\partial y}\right)^2 = 4u\left(\frac{dg}{du}\right)^2$$

45. Calculate $\partial z/\partial x$, where $xe^z + ze^y = x + y$.

46. Let $f(x, y) = x^4 - 2x^2 + y^2 - 6y$.

(a) Find the critical points of f and use the Second Derivative Test to determine whether they are a local minima or a local maxima.

(b) Find the minimum value of f without calculus by completing the square.

In Exercises 47–50, find the critical points of the function and analyze them using the Second Derivative Test.

47. $f(x, y) = x^4 - 4xy + 2y^2$

48. $f(x, y) = x^3 + 2y^3 - xy$

49. $f(x, y) = e^{x+y} - xe^{2y}$

50. $f(x, y) = \sin(x + y) - \frac{1}{2}(x + y^2)$

51. Prove that $f(x, y) = (x + 2y)e^{xy}$ has no critical points.

52. Find the global extrema of $f(x, y) = x^3 - xy - y^2 + y$ on the square $[0, 1] \times [0, 1]$.

53. Find the global extrema of $f(x, y) = 2xy - x - y$ on the domain $\{y \leq 4, y \geq x^2\}$.

54. Find the maximum of $f(x, y, z) = xyz$ subject to the constraint $g(x, y, z) = 2x + y + 4z = 1$.

55. Use Lagrange multipliers to find the minimum and maximum values of $f(x, y) = 3x - 2y$ on the circle $x^2 + y^2 = 4$.

56. Find the minimum value of $f(x, y) = xy$ subject to the constraint $5x - y = 4$ in two ways: using Lagrange multipliers and setting $y = 5x - 4$ in $f(x, y)$.

57. Find the minimum and maximum values of $f(x, y) = x^2y$ on the ellipse $4x^2 + 9y^2 = 36$.

58. Find the point in the first quadrant on the curve $y = x + x^{-1}$ closest to the origin.

59. Find the extreme values of $f(x, y, z) = x + 2y + 3z$ subject to the two constraints $x + y + z = 1$ and $x^2 + y^2 + z^2 = 1$.

60. Find the minimum and maximum values of $f(x, y, z) = x - z$ on the intersection of the cylinders $x^2 + y^2 = 1$ and $x^2 + z^2 = 1$ (Figure 5).

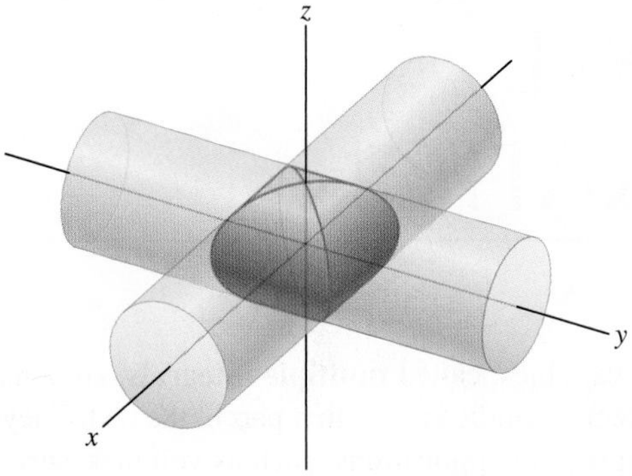

FIGURE 5

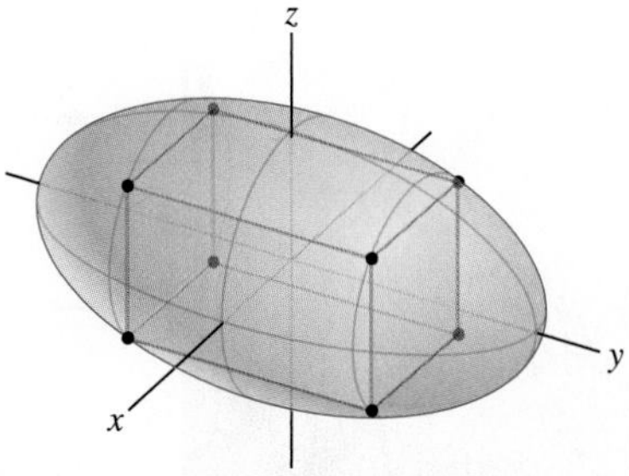

FIGURE 6

61. Use Lagrange multipliers to find the dimensions of a cylindrical can with a bottom but no top, of fixed volume V with minimum surface area.

62. Find the dimensions of the box of maximum volume with its sides parallel to the coordinate planes that can be inscribed in the ellipsoid (Figure 6)

$$\left(\frac{x}{a}\right)^2 + \left(\frac{y}{b}\right)^2 + \left(\frac{z}{c}\right)^2 = 1$$

63. Given n nonzero numbers $\sigma_1, \dots, \sigma_n$, show that the minimum value of

$$f(x_1, \dots, x_n) = x_1^2\sigma_1^2 + \cdots + x_n^2\sigma_n^2$$

subject to $x_1 + \cdots + x_n = 1$ is c, where $c = \left(\sum_{j=1}^{n} \sigma_j^{-2}\right)^{-1}$.

16 MULTIPLE INTEGRATION

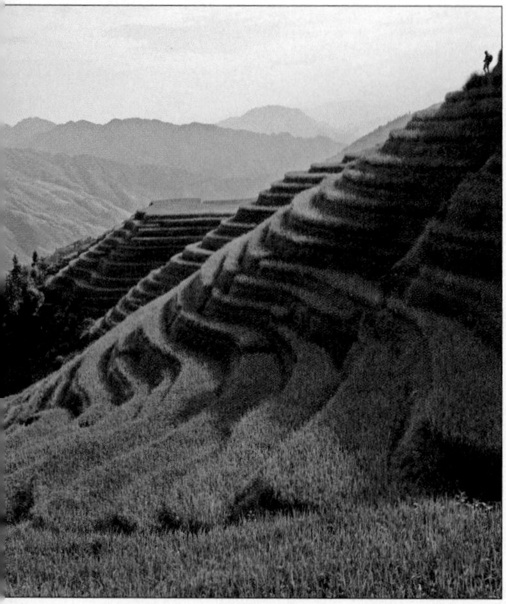

These rice terraces illustrate how volume under a graph is computed using iterated integration.

Integrals of functions of several variables, called **multiple integrals**, are a natural extension of the single-variable integrals studied in the first part of the text. They are used to compute many quantities that appear in applications, such as volumes, surface areas, centers of mass, probabilities, and average values.

16.1 Integration in Two Variables

The integral of a function of two variables $f(x, y)$, called a **double integral**, is denoted

$$\iint_{\mathcal{D}} f(x, y)\, dA$$

It represents the *signed volume* of the solid region between the graph of $f(x, y)$ and a domain $\mathcal{D}$ in the xy-plane (Figure 1), where the volume is positive for regions above the xy-plane and negative for regions below.

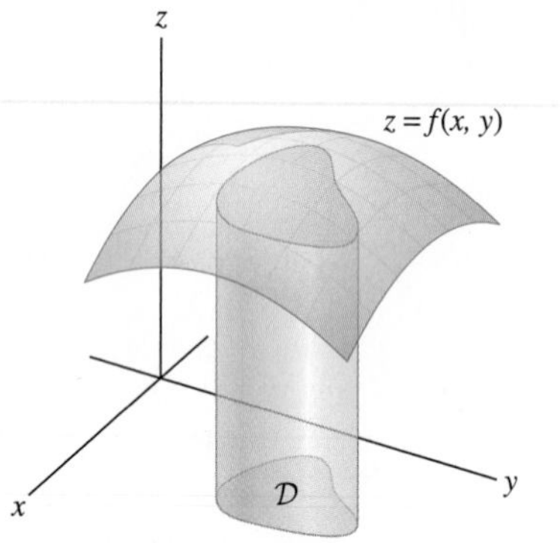

FIGURE 1 The double integral computes the volume of the solid region between the graph of $f(x, y)$ and the xy-plane over a domain $\mathcal{D}$.

There are many similarities between double integrals and the single integrals:

- Double integrals are defined as limits of sums.
- Double integrals are evaluated using the Fundamental Theorem of Calculus (but we have to use it twice—see the discussion of iterated integrals below).

An important difference, however, is that the domain of integration plays a more prominent role in the multivariable case. In one variable, the domain of integration is simply an interval $[a, b]$. In two variables, the domain $\mathcal{D}$ is a plane region whose boundary may be curved (Figure 1).

In this section, we focus on the simplest case where the domain is a rectangle, leaving more general domains for Section 16.2. Let

$$\mathcal{R} = [a, b] \times [c, d]$$

denote the rectangle in the plane (Figure 2) consisting of all points (x, y) such that

$$\mathcal{R}: \quad a \le x \le b, \qquad c \le y \le d$$

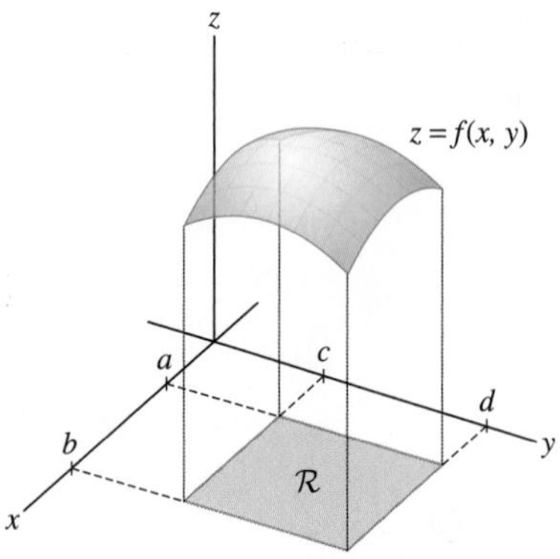

FIGURE 2

Like integrals in one variable, double integrals are defined through a three-step process: subdivision, summation, and passage to the limit. Figure 3 illustrates how the rectangle $\mathcal{R}$ is subdivided:

1. Subdivide $[a, b]$ and $[c, d]$ by choosing partitions:

$$a = x_0 < x_1 < \cdots < x_N = b, \qquad c = y_0 < y_1 < \cdots < y_M = d$$

where N and M are positive integers.

2. Create an $N \times M$ grid of subrectangles $\mathcal{R}_{ij}$.
3. Choose a sample point P_{ij} in each $\mathcal{R}_{ij}$.

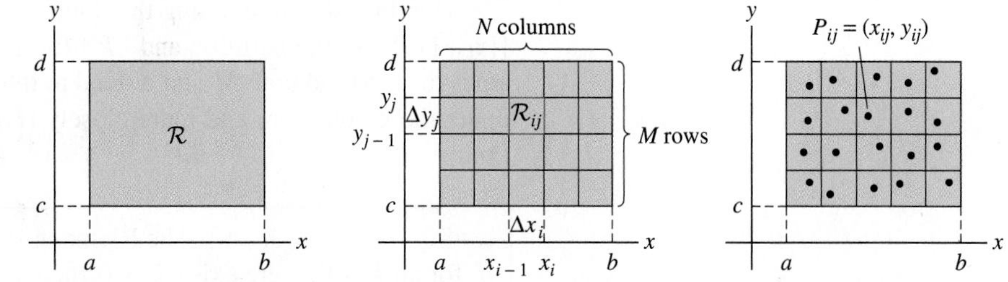

(A) Rectangle $\mathcal{R} = [a, b] \times [c, d]$ (B) Create $N \times M$ grid (C) Sample point P_{ij}

FIGURE 3

Note that $\mathcal{R}_{ij} = [x_{i-1}, x_i] \times [y_{j-1}, y_j]$, so $\mathcal{R}_{ij}$ has area

$$\Delta A_{ij} = \Delta x_i \, \Delta y_j$$

where $\Delta x_i = x_i - x_{i-1}$ and $\Delta y_j = y_j - y_{j-1}$.

Next, we form the Riemann sum with the function values $f(P_{ij})$:

$$S_{N,M} = \sum_{i=1}^{N} \sum_{j=1}^{M} f(P_{ij}) \, \Delta A_{ij} = \sum_{i=1}^{N} \sum_{j=1}^{M} f(P_{ij}) \, \Delta x_i \, \Delta y_j$$

Keep in mind that a Riemann sum depends on the choice of partition and sample points. It would be more proper to write

$$S_{N,M}(\{P_{ij}\}, \{x_i\}, \{y_j\})$$

but we write $S_{N,M}$ to keep the notation simple.

The double summation runs over all i and j in the ranges $1 \le i \le N$ and $1 \le j \le M$, a total of NM terms.

The geometric interpretation of $S_{N,M}$ is shown in Figure 4. Each individual term $f(P_{ij}) \, \Delta A_{ij}$ of the sum is equal to the signed volume of the narrow box of height $f(P_{ij})$ above $\mathcal{R}_{ij}$:

$$f(P_{ij}) \, \Delta A_{ij} = f(P_{ij}) \, \Delta x_i \, \Delta y_j = \underbrace{\text{height} \times \text{area}}_{\text{Signed volume of box}}$$

When $f(P_{ij})$ is negative, the box lies below the xy-plane and has negative signed volume. The sum $S_{N,M}$ of the signed volumes of these narrow boxes approximates volume in the same way that Riemann sums in one variable approximate area by rectangles [Figure 4(A)].

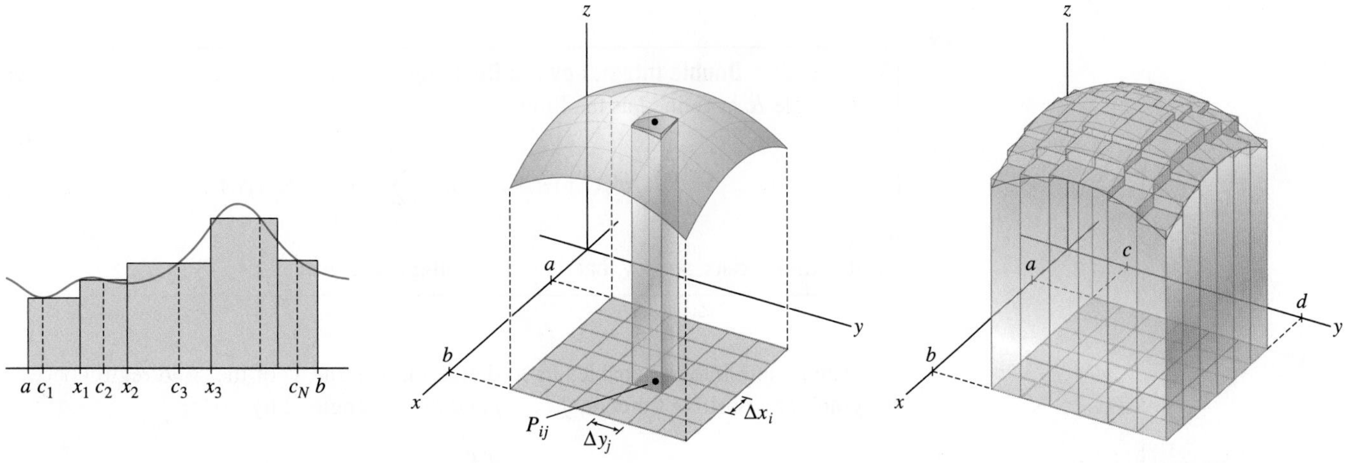

(A) In one variable, a Riemann sum approximates the area under the curve by a sum of areas of rectangles.

(B) The volume of the box is $f(P_{ij})\Delta A_{ij}$, where $\Delta A_{ij} = \Delta x_i \Delta y_j$.

(C) The Riemann sum $S_{N,M}$ is the sum of the volumes of the boxes.

FIGURE 4

The final step in defining the double integral is passing to the limit. We write $\mathcal{P} = \{\{x_i\}, \{y_j\}\}$ for the partition and $\|\mathcal{P}\|$ for the maximum of the widths Δx_i, Δy_j. As $\|\mathcal{P}\|$ tends to zero (and both M and N tend to infinity), the boxes approximate the solid region under the graph more and more closely (Figure 5). Here is the precise definition of the limit:

Limit of Riemann Sums The Riemann sum $S_{N,M}$ approaches a limit L as $\|\mathcal{P}\| \to 0$ if, for all $\epsilon > 0$, there exists $\delta > 0$ such that

$$|L - S_{N,M}| < \epsilon$$

for all partitions satisfying $\|\mathcal{P}\| < \delta$ and all choices of sample points.

In this case, we write

$$\lim_{\|\mathcal{P}\|\to 0} S_{N,M} = \lim_{\|\mathcal{P}\|\to 0} \sum_{i=1}^{N}\sum_{j=1}^{M} f(P_{ij})\,\Delta A_{ij} = L$$

This limit L, if it exists, is the double integral $\iint_{\mathcal{R}} f(x, y)\,dA$.

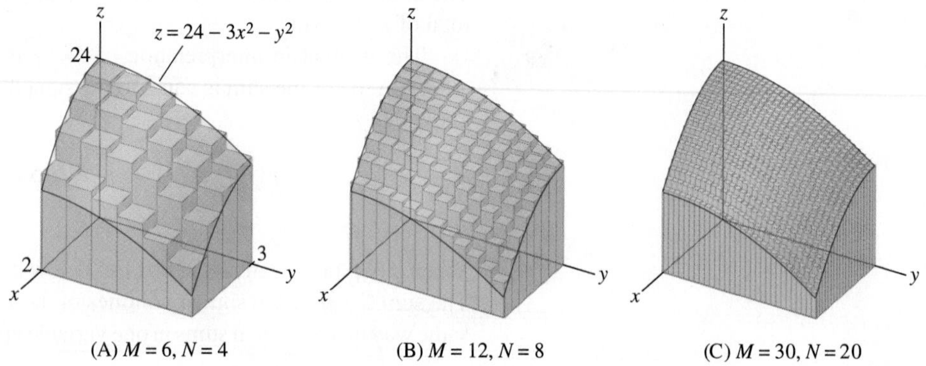

FIGURE 5 Midpoint approximations to the volume under $z = 24 - 3x^2 - y^2$.

DEFINITION Double Integral over a Rectangle The double integral of $f(x, y)$ over a rectangle $\mathcal{R}$ is defined as the limit

$$\iint_{\mathcal{R}} f(x, y)\,dA = \lim_{\|\mathcal{P}\|\to 0} \sum_{i=1}^{N}\sum_{j=1}^{M} f(P_{ij})\Delta A_{ij}$$

If this limit exists, we say that $f(x, y)$ is **integrable** over $\mathcal{R}$.

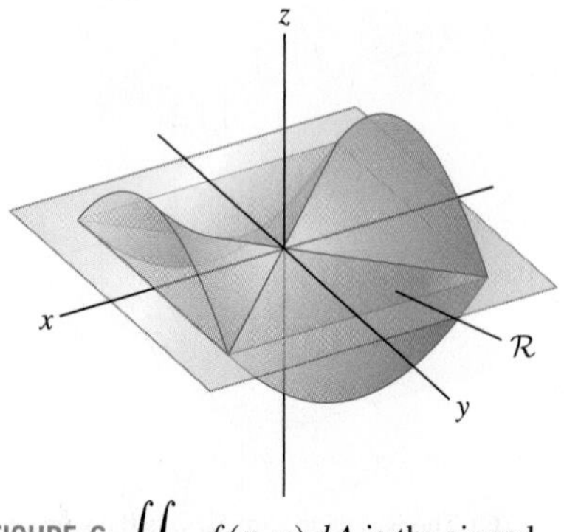

FIGURE 6 $\iint_{\mathcal{R}} f(x, y)\,dA$ is the signed volume of the region between the graph of $z = f(x, y)$ and the rectangle $\mathcal{R}$.

The double integral enables us to define the volume V of the solid region between the graph of a positive function $f(x, y)$ and the rectangle $\mathcal{R}$ by

$$V = \iint_{\mathcal{R}} f(x, y)\,dA$$

If $f(x, y)$ takes on both positive and negative values, the double integral defines the signed volume (Figure 6).

In computations, we often assume that the partition $\mathcal{P}$ is **regular**, meaning that the intervals $[a, b]$ and $[c, d]$ are both divided into subintervals of equal length. In other words, the partition is regular if $\Delta x_i = \Delta x$ and $\Delta y_j = \Delta y$, where

$$\Delta x = \frac{b-a}{N}, \qquad \Delta y = \frac{d-c}{M}$$

For a regular partition, $\|\mathcal{P}\|$ tends to zero as N and M tend to ∞.

EXAMPLE 1 Estimating a Double Integral Let $\mathcal{R} = [1, 2.5] \times [1, 2]$. Calculate $S_{3,2}$ for the integral (Figure 7)

$$\iint_{\mathcal{R}} xy\, dA$$

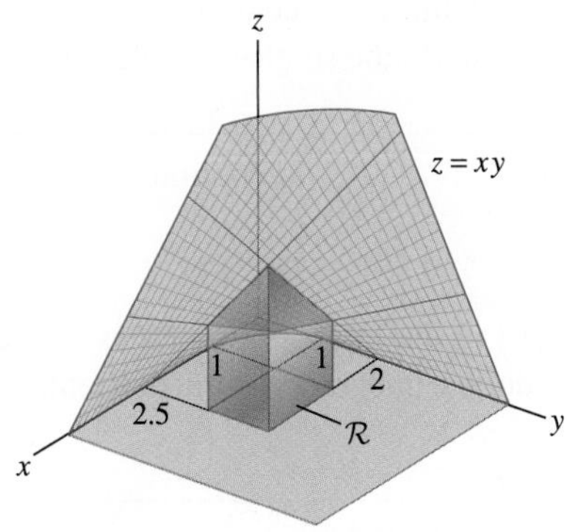

FIGURE 7 Graph of $z = xy$.

using the following two choices of sample points:

(a) Lower-left vertex **(b)** Midpoint of rectangle

Solution Since we use the regular partition to compute $S_{3,2}$, each subrectangle (in this case they are squares) has sides of length

$$\Delta x = \frac{2.5-1}{3} = \frac{1}{2}, \qquad \Delta y = \frac{2-1}{2} = \frac{1}{2}$$

and area $\Delta A = \Delta x\, \Delta y = \frac{1}{4}$. The corresponding Riemann sum is

$$S_{3,2} = \sum_{i=1}^{3}\sum_{j=1}^{2} f(P_{ij})\, \Delta A = \frac{1}{4}\sum_{i=1}^{3}\sum_{j=1}^{2} f(P_{ij})$$

where $f(x, y) = xy$.

(a) If we use the lower-left vertices shown in Figure 8(A), the Riemann sum is

$$S_{3,2} = \tfrac{1}{4}\Big(f(1,1) + f\big(1,\tfrac{3}{2}\big) + f\big(\tfrac{3}{2},1\big) + f\big(\tfrac{3}{2},\tfrac{3}{2}\big) + f(2,1) + f\big(2,\tfrac{3}{2}\big)\Big)$$
$$= \tfrac{1}{4}\big(1 + \tfrac{3}{2} + \tfrac{3}{2} + \tfrac{9}{4} + 2 + 3\big) = \tfrac{1}{4}\big(\tfrac{45}{4}\big) = 2.8125$$

(b) Using the midpoints of the rectangles shown in Figure 8(B), we obtain

$$S_{3,2} = \tfrac{1}{4}\Big(f\big(\tfrac{5}{4},\tfrac{5}{4}\big) + f\big(\tfrac{5}{4},\tfrac{7}{4}\big) + f\big(\tfrac{7}{4},\tfrac{5}{4}\big) + f\big(\tfrac{7}{4},\tfrac{7}{4}\big) + f\big(\tfrac{9}{4},\tfrac{5}{4}\big) + f\big(\tfrac{9}{4},\tfrac{7}{4}\big)\Big)$$
$$= \tfrac{1}{4}\big(\tfrac{25}{16} + \tfrac{35}{16} + \tfrac{35}{16} + \tfrac{49}{16} + \tfrac{45}{16} + \tfrac{63}{16}\big) = \tfrac{1}{4}\big(\tfrac{252}{16}\big) = 3.9375$$ ■

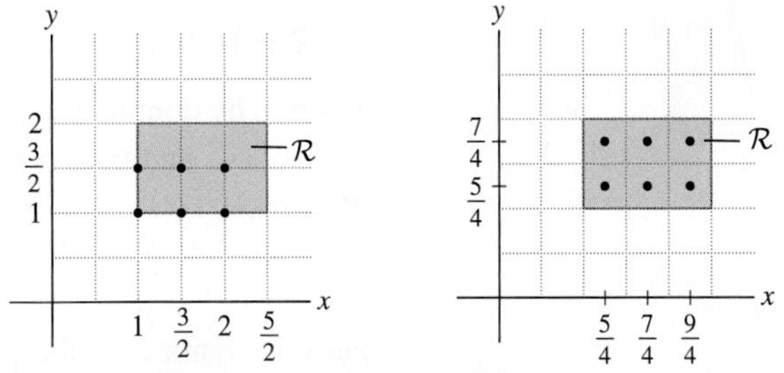

(A) Sample points are the lower-left vertices.

(B) Sample points are midpoints.

FIGURE 8

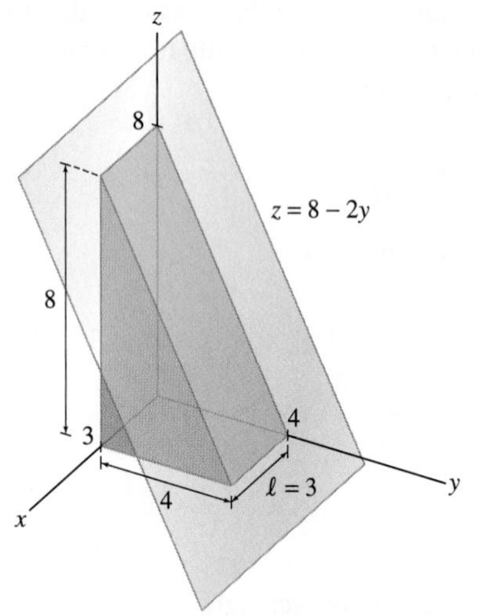

FIGURE 9 Solid wedge under the graph of $z = 8 - 2y$.

EXAMPLE 2 Evaluate $\iint_{\mathcal{R}} (8 - 2y)\, dA$, where $\mathcal{R} = [0, 3] \times [0, 4]$.

Solution Figure 9 shows the graph of $z = 8 - 2y$. The double integral is equal to the volume V of the solid wedge underneath the graph. The triangular face of the wedge has area $A = \frac{1}{2}(8)4 = 16$. The volume of the wedge is equal to the area A times the length $\ell = 3$; that is, $V = \ell A = 3(16) = 48$. Therefore,

$$\iint_{\mathcal{R}} (8 - 2y)\, dA = 48$$

The next theorem assures us that continuous functions are integrable. Since we have not yet defined continuity at boundary points of a domain, for the purposes of the next theorem, we define continuity on $\mathcal{R}$ to mean that f is defined and continuous on some open set containing $\mathcal{R}$. We omit the proof, which is similar to the single-variable case.

THEOREM 1 Continuous Functions Are Integrable If $f(x, y)$ is continuous on a rectangle $\mathcal{R}$, then $f(x, y)$ is integrable over $\mathcal{R}$.

As in the single-variable case, we often make use of the linearity properties of the double integral. They follow from the definition of the double integral as a limit of Riemann sums.

THEOREM 2 Linearity of the Double Integral Assume that $f(x, y)$ and $g(x, y)$ are integrable over a rectangle $\mathcal{R}$. Then:

(i) $\iint_{\mathcal{R}} \big(f(x, y) + g(x, y)\big)\, dA = \iint_{\mathcal{R}} f(x, y)\, dA + \iint_{\mathcal{R}} g(x, y)\, dA$

(ii) For any constant C, $\iint_{\mathcal{R}} Cf(x, y)\, dA = C \iint_{\mathcal{R}} f(x, y)\, dA$

If $f(x, y) = C$ is a constant function, then

$$\iint_{\mathcal{R}} C\, dA = C \cdot \text{Area}(\mathcal{R})$$

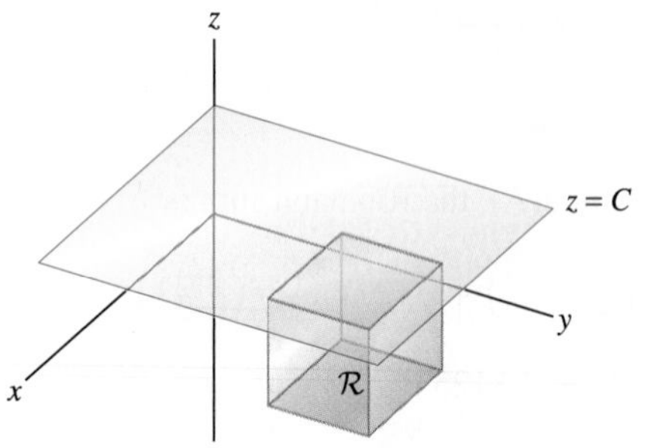

FIGURE 10 The double integral of $f(x, y) = C$ over a rectangle $\mathcal{R}$ is $C \cdot \text{Area}(\mathcal{R})$.

The double integral is the signed volume of the box of base $\mathcal{R}$ and height C (Figure 10). If $C < 0$, then the rectangle lies below the xy-plane, and the integral is equal to the signed volume, which is negative.

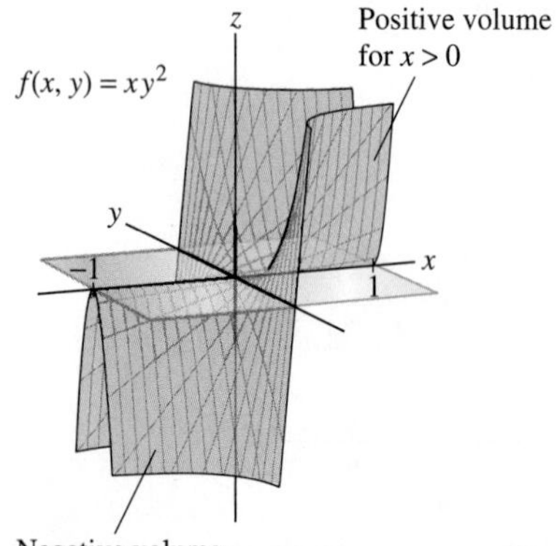

FIGURE 11

EXAMPLE 3 Arguing by Symmetry Use symmetry to show that $\iint_{\mathcal{R}} xy^2\, dA = 0$, where $\mathcal{R} = [-1, 1] \times [-1, 1]$.

Solution The double integral is the signed volume of the region between the graph of $f(x, y) = xy^2$ and the xy-plane (Figure 11). However, $f(x, y)$ takes opposite values at (x, y) and $(-x, y)$:

$$f(-x, y) = -xy^2 = -f(x, y)$$

Because of symmetry, the (negative) signed volume of the region below the xy-plane where $-1 \le x \le 0$ cancels with the (positive) signed volume of the region above the xy-plane where $0 \le x \le 1$. The net result is $\iint_{\mathcal{R}} xy^2\, dA = 0$.

Iterated Integrals

We often omit the parentheses in the notation for an iterated integral:

$$\int_a^b \int_c^d f(x, y)\, dy\, dx$$

The order of the variables in $dy\,dx$ tells us to integrate first with respect to y between the limits $y = c$ and $y = d$.

Our main tool for evaluating double integrals is the Fundamental Theorem of Calculus (FTC), as in the single-variable case. To use the FTC, we express the double integral as an **iterated integral**, which is an expression of the form

$$\int_a^b \left(\int_c^d f(x, y)\, dy \right) dx$$

Iterated integrals are evaluated in two steps. Step One: Hold x constant and evaluate the inner integral with respect to y. This gives us a function of x alone:

$$S(x) = \int_c^d f(x, y)\, dy$$

Step Two: Integrate the resulting function $S(x)$ with respect to x.

■ **EXAMPLE 4** Evaluate $\displaystyle\int_2^4 \left(\int_1^9 ye^x\, dy \right) dx$.

Solution First evaluate the inner integral, treating x as a constant:

$$S(x) = \int_1^9 ye^x\, dy = e^x \int_1^9 y\, dy = e^x \left(\frac{1}{2} y^2 \right) \bigg|_{y=1}^{9} = e^x \left(\frac{81 - 1}{2} \right) = 40e^x$$

Then integrate $S(x)$ with respect to x:

$$\int_2^4 \left(\int_1^9 ye^x\, dy \right) dx = \int_2^4 40e^x\, dx = 40e^x \bigg|_2^4 = 40(e^4 - e^2)$$ ■

In an iterated integral where dx precedes dy, integrate first with respect to x:

$$\int_c^d \int_a^b f(x, y)\, dx\, dy = \int_{y=c}^{d} \left(\int_{x=a}^{b} f(x, y)\, dx \right) dy$$

Sometimes for clarity, as on the right-hand side here, we include the variables in the limits of integration.

■ **EXAMPLE 5** Evaluate $\displaystyle\int_{y=0}^{4} \int_{x=0}^{3} \frac{dx\, dy}{\sqrt{3x + 4y}}$.

Solution We evaluate the inner integral first, treating y as a constant. Since we are integrating with respect to x, we need an antiderivative of $1/\sqrt{3x + 4y}$ as a function of x. We can use $\frac{2}{3}\sqrt{3x + 4y}$ because

$$\frac{\partial}{\partial x} \left(\frac{2}{3} \sqrt{3x + 4y} \right) = \frac{1}{\sqrt{3x + 4y}}$$

Thus we have

$$\int_{x=0}^{3} \frac{dx}{\sqrt{3x + 4y}} = \frac{2}{3} \sqrt{3x + 4y} \bigg|_{x=0}^{3} = \frac{2}{3} \left(\sqrt{4y + 9} - \sqrt{4y} \right)$$

$$\int_{y=0}^{4} \int_{x=0}^{3} \frac{dx\, dy}{\sqrt{3x + 4y}} = \frac{2}{3} \int_{y=0}^{4} \left(\sqrt{4y + 9} - \sqrt{4y} \right) dy$$

Therefore, we have:

$$\int_{y=0}^{4}\int_{x=0}^{3}\frac{dx\,dy}{\sqrt{3x+4y}} = \frac{2}{3}\left(\frac{1}{6}(4y+9)^{3/2} - \frac{1}{6}(4y)^{3/2}\right)\Bigg|_{y=0}^{4}$$
$$= \frac{1}{9}\left(25^{3/2} - 16^{3/2} - 9^{3/2}\right) = \frac{34}{9}$$ ■

■ **EXAMPLE 6 Reversing the Order of Integration** Verify that

$$\int_{y=0}^{4}\int_{x=0}^{3}\frac{dx\,dy}{\sqrt{3x+4y}} = \int_{x=0}^{3}\int_{y=0}^{4}\frac{dy\,dx}{\sqrt{3x+4y}}$$

Solution We evaluated the iterated integral on the left in the previous example. We compute the integral on the right and verify that the result is also $\frac{34}{9}$:

$$\int_{y=0}^{4}\frac{dy}{\sqrt{3x+4y}} = \frac{1}{2}\sqrt{3x+4y}\Bigg|_{y=0}^{4} = \frac{1}{2}(\sqrt{3x+16} - \sqrt{3x})$$

$$\int_{x=0}^{3}\int_{y=0}^{4}\frac{dy\,dx}{\sqrt{3x+4y}} = \frac{1}{2}\int_{0}^{3}(\sqrt{3x+16} - \sqrt{3x})\,dy$$
$$= \frac{1}{2}\left(\frac{2}{9}(3x+16)^{3/2} - \frac{2}{9}(3x)^{3/2}\right)\Bigg|_{x=0}^{3}$$
$$= \frac{1}{9}\left(25^{3/2} - 9^{3/2} - 16^{3/2}\right) = \frac{34}{9}$$ ■

The previous example illustrates a general fact: The value of an iterated integral does not depend on the order in which the integration is performed. This is part of Fubini's Theorem. Even more important, Fubini's Theorem states that a double integral over a rectangle can be evaluated as an iterated integral.

CAUTION When you reverse the order of integration in an iterated integral, remember to interchange the limits of integration (the inner limits become the outer limits).

THEOREM 3 Fubini's Theorem The double integral of a continuous function $f(x, y)$ over a rectangle $\mathcal{R} = [a, b] \times [c, d]$ is equal to the iterated integral (in either order):

$$\iint_{\mathcal{R}} f(x, y)\,dA = \int_{x=a}^{b}\int_{y=c}^{d} f(x, y)\,dy\,dx = \int_{y=c}^{d}\int_{x=a}^{b} f(x, y)\,dx\,dy$$

Proof We sketch the proof. We can compute the double integral as a limit of Riemann sums that use a regular partition of $\mathcal{R}$ and sample points $P_{ij} = (x_i, y_j)$, where $\{x_i\}$ are sample points for a regular partition on $[a, b]$, and $\{y_j\}$ are sample points for a regular partition of $[c, d]$:

$$\iint_{\mathcal{R}} f(x, y)\,dA = \lim_{N,M\to\infty}\sum_{i=1}^{N}\sum_{j=1}^{M} f(x_i, y_j)\Delta y\Delta x$$

Here $\Delta x = (b-a)/N$ and $\Delta y = (d-c)/M$. Fubini's Theorem stems from the elementary fact that we can add up the values in the sum in any order. So if we list the values $f(P_{ij})$ in an $N \times M$ array as shown in the margin, we can add up the columns first and then add up the column sums. This yields

3	$f(P_{13})$	$f(P_{23})$	$f(P_{33})$
2	$f(P_{12})$	$f(P_{22})$	$f(P_{32})$
1	$f(P_{11})$	$f(P_{21})$	$f(P_{31})$
j / i	1	2	3

$$\iint_{\mathcal{R}} f(x, y)\,dA = \lim_{N,M\to\infty}\sum_{i=1}^{N}\underbrace{\left(\sum_{j=1}^{M} f(x_i, y_j)\Delta y\right)}_{\text{First sum the columns; then add up the column sums.}}\Delta x$$

For fixed i, $f(x_i, y)$ is a continuous function of y and the inner sum on the right is a Riemann sum that approaches the single integral $\int_c^d f(x_i, y)\,dy$. In other words, setting $S(x) = \int_c^d f(x, y)\,dy$, we have

$$\lim_{M\to\infty} \sum_{j=1}^{M} f(x_i, y_j) = \int_c^d f(x_i, y)\,dy = S(x_i)$$

To complete the proof, we take two facts for granted. First, that $S(x)$ is a continuous function for $a \le x \le b$. Second, that the limit as $N, M \to \infty$ may be computed by taking the limit first with respect to M and then with respect to N. Granting this,

$$\iint_{\mathcal{R}} f(x, y)\,dA = \lim_{N\to\infty} \sum_{i=1}^{N} \left(\lim_{M\to\infty} \sum_{j=1}^{M} f(x_i, y_j)\Delta y \right) \Delta x = \lim_{N\to\infty} \sum_{i=1}^{N} S(x_i)\Delta x$$

$$= \int_a^b S(x)\,dx = \int_a^b \left(\int_c^d f(x, y)\,dy \right) dx$$

Note that the sums on the right in the first line are Riemann sums for $S(x)$ that converge to the integral of $S(x)$ in the second line. This proves Fubini's Theorem for the order $dy\,dx$. A similar argument applies to the order $dx\,dy$. ■

GRAPHICAL INSIGHT When we write a double integral as an iterated integral in the order $dy\,dx$, then for each fixed value $x = x_0$, the inner integral is the area of the cross section of S in the vertical plane $x = x_0$ perpendicular to the x-axis (Figure 12(A)):

$$S(x_0) = \int_c^d f(x_0, y)\,dy = \begin{array}{l}\text{area of cross section in vertical plane}\\ x = x_0 \text{ perpendicular to the } x\text{-axis}\end{array}$$

What Fubini's Theorem says is that the volume V of S can be calculated as the integral of cross-sectional area $S(x)$:

$$V = \int_a^b \int_c^d f(x, y)\,dy\,dx = \int_a^b S(x)\,dx = \text{integral of cross-sectional area}$$

Similarly, the iterated integral in the order $dx\,dy$ calculates V as the integral of cross sections perpendicular to the y-axis (Figure 12(B)).

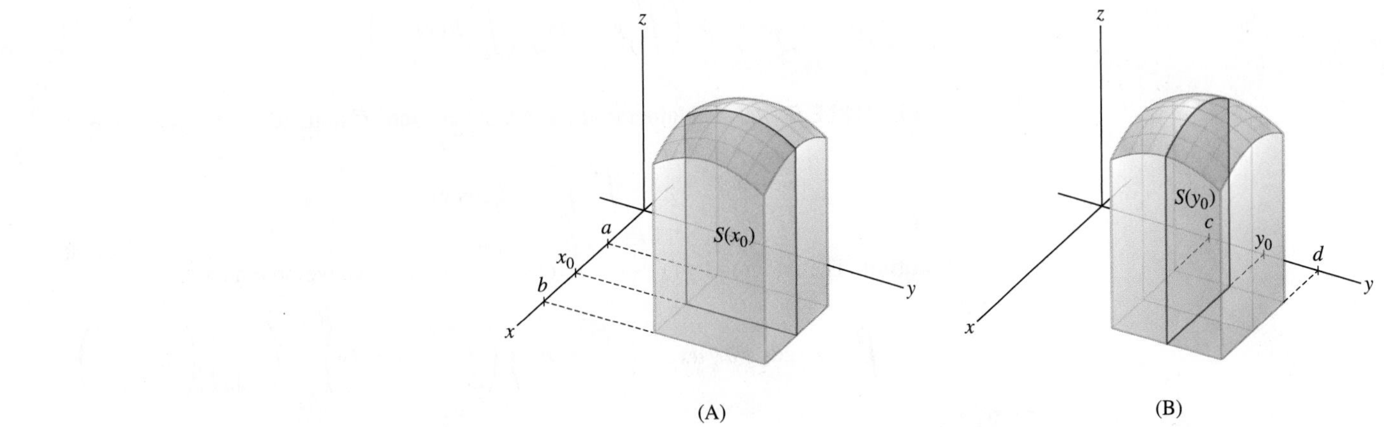

FIGURE 12

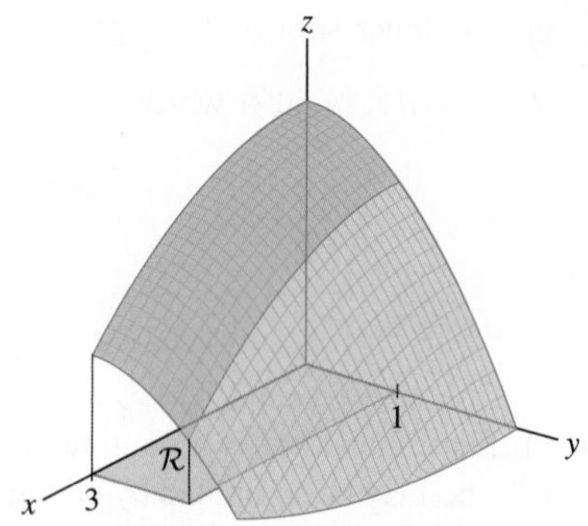

FIGURE 13 Graph of $f(x, y) = 16 - x^2 - 3y^2$ over $\mathcal{R} = [0, 3] \times [0, 1]$.

EXAMPLE 7 Find the volume V between the graph of $f(x, y) = 16 - x^2 - 3y^2$ and the rectangle $\mathcal{R} = [0, 3] \times [0, 1]$ (Figure 13).

Solution The volume V is equal to the double integral of $f(x, y)$, which we write as an iterated integral:

$$V = \iint_{\mathcal{R}} (16 - x^2 - 3y^2)\, dA = \int_{x=0}^{3} \int_{y=0}^{1} (16 - x^2 - 3y^2)\, dy\, dx$$

We evaluate the inner integral first and then compute V:

$$\int_{y=0}^{1} (16 - x^2 - 3y^2)\, dy = (16y - x^2y - y^3)\Big|_{y=0}^{1} = 15 - x^2$$

$$V = \int_{x=0}^{3} (15 - x^2)\, dx = \left(15x - \frac{1}{3}x^3\right)\Big|_0^3 = 36$$ ■

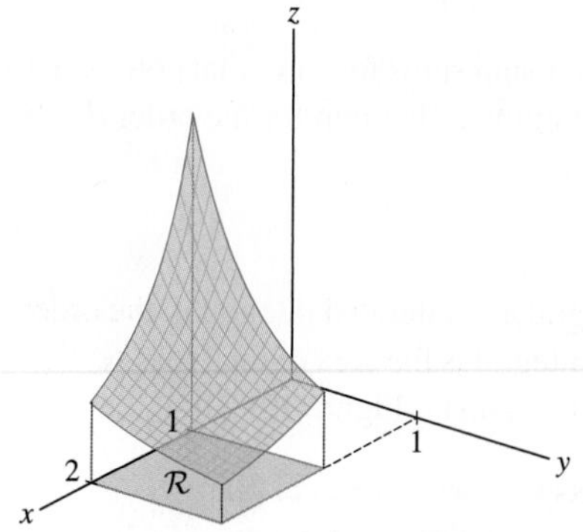

FIGURE 14 Graph of $z = (x + y)^{-2}$ over $\mathcal{R} = [1, 2] \times [0, 1]$.

EXAMPLE 8 Calculate $\displaystyle\iint_{\mathcal{R}} \frac{dA}{(x + y)^2}$, where $\mathcal{R} = [1, 2] \times [0, 1]$ (Figure 14).

Solution

$$\iint_{\mathcal{R}} \frac{dA}{(x + y)^2} = \int_{x=1}^{2} \left(\int_{y=0}^{1} \frac{dy}{(x + y)^2}\right) dx = \int_1^2 \left(-\frac{1}{x + y}\Big|_{y=0}^{1}\right) dx$$

$$= \int_1^2 \left(-\frac{1}{x + 1} + \frac{1}{x}\right) dx = \left(\ln x - \ln(x + 1)\right)\Big|_1^2$$

$$= (\ln 2 - \ln 3) - (\ln 1 - \ln 2) = 2\ln 2 - \ln 3 = \ln \frac{4}{3}$$ ■

When the function is a product $f(x, y) = g(x)h(y)$, the double integral over a rectangle is simply the product of the single integrals. We verify this by writing the double integral as an iterated integral. If $\mathcal{R} = [a, b] \times [c, d]$,

$$\iint_{\mathcal{R}} g(x)h(y)\, dA = \int_a^b \left(\int_c^d g(x)h(y)\, dy\right) dx = \int_a^b g(x) \left(\int_c^d h(y)\, dy\right) dx$$

$$= \left(\int_a^b g(x)\, dx\right)\left(\int_c^d h(y)\, dy\right) \qquad \boxed{1}$$

EXAMPLE 9 Iterated Integral of a Product Function Calculate

$$\int_0^2 \int_0^{\pi/2} e^x \cos y\, dy\, dx$$

Solution The integrand $f(x, y) = e^x \cos y$ is a product, so we obtain

$$\int_0^2 \int_0^{\pi/2} e^x \cos y\, dy\, dx = \left(\int_0^2 e^x\, dx\right)\left(\int_0^{\pi/2} \cos y\, dy\right) = \left(e^x\Big|_0^2\right)\left(\sin y\Big|_0^{\pi/2}\right)$$

$$= (e^2 - 1)(1) = e^2 - 1$$ ■

16.1 SUMMARY

- A *Riemann sum* for $f(x, y)$ on a rectangle $\mathcal{R} = [a, b] \times [c, d]$ is a sum of the form

$$S_{N,M} = \sum_{i=1}^{N} \sum_{j=1}^{M} f(P_{ij})\, \Delta x_i\, \Delta y_j$$

corresponding to partitions of $[a, b]$ and $[c, d]$, and choice of sample points P_{ij} in the subrectangle $\mathcal{R}_{ij}$.
- The double integral of $f(x, y)$ over $\mathcal{R}$ is defined as the limit (if it exists):

$$\iint_{\mathcal{R}} f(x, y)\, dA = \lim_{M,N \to \infty} \sum_{i=1}^{N} \sum_{j=1}^{M} f(P_{ij})\, \Delta x_i\, \Delta y_j$$

We say that $f(x, y)$ is *integrable* over $\mathcal{R}$ if this limit exists.
- A continuous function on a rectangle $\mathcal{R}$ is integrable.
- The double integral is equal to the *signed volume* of the region between the graph of $z = f(x, y)$ and the rectangle $\mathcal{R}$. The signed volume of a region is positive if it lies above the xy-plane and negative if it lies below the xy-plane.
- If $f(x, y) = C$ is a constant function, then

$$\iint_{\mathcal{R}} C\, dA = C \cdot \text{Area}(\mathcal{R})$$

- Fubini's Theorem: The double integral of a continuous function $f(x, y)$ over a rectangle $\mathcal{R} = [a, b] \times [c, d]$ can be evaluated as an iterated integral (in either order):

$$\iint_{\mathcal{R}} f(x, y)\, dA = \int_{x=a}^{b} \int_{y=c}^{d} f(x, y)\, dy\, dx = \int_{y=c}^{d} \int_{x=a}^{b} f(x, y)\, dx\, dy$$

16.1 EXERCISES

Preliminary Questions

1. If $S_{8,4}$ is a Riemann sum for a double integral over $\mathcal{R} = [1, 5] \times [2, 10]$ using a regular partition, what is the area of each subrectangle? How many subrectangles are there?

2. Estimate the double integral of a continuous function f over the small rectangle $\mathcal{R} = [0.9, 1.1] \times [1.9, 2.1]$ if $f(1, 2) = 4$.

3. What is the integral of the constant function $f(x, y) = 5$ over the rectangle $[-2, 3] \times [2, 4]$?

4. What is the interpretation of $\iint_{\mathcal{R}} f(x, y)\, dA$ if $f(x, y)$ takes on both positive and negative values on $\mathcal{R}$?

5. Which of (a) or (b) is equal to $\int_1^2 \int_4^5 f(x, y)\, dy\, dx$?

(a) $\int_1^2 \int_4^5 f(x, y)\, dx\, dy$ **(b)** $\int_4^5 \int_1^2 f(x, y)\, dx\, dy$

6. For which of the following functions is the double integral over the rectangle in Figure 15 equal to zero? Explain your reasoning.

(a) $f(x, y) = x^2 y$ **(b)** $f(x, y) = xy^2$

(c) $f(x, y) = \sin x$ **(d)** $f(x, y) = e^x$

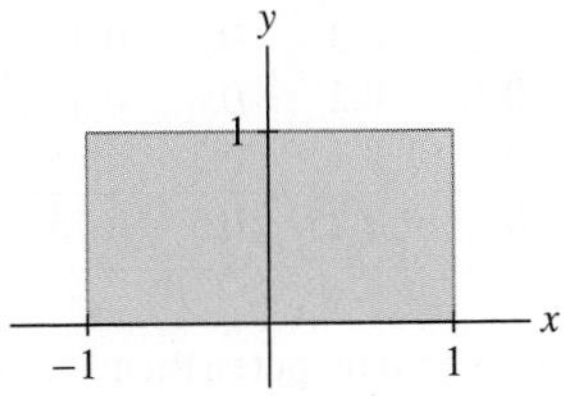

FIGURE 15

Exercises

1. Compute the Riemann sum $S_{4,3}$ to estimate the double integral of $f(x, y) = xy$ over $\mathcal{R} = [1, 3] \times [1, 2.5]$. Use the regular partition and upper-right vertices of the subrectangles as sample points.

2. Compute the Riemann sum with $N = M = 2$ to estimate the integral of $\sqrt{x+y}$ over $\mathcal{R} = [0, 1] \times [0, 1]$. Use the regular partition and midpoints of the subrectangles as sample points.

In Exercises 3–6, compute the Riemann sums for the double integral $\iint_{\mathcal{R}} f(x, y)\, dA$, *where* $\mathcal{R} = [1, 4] \times [1, 3]$, *for the grid and two choices of sample points shown in Figure 16.*

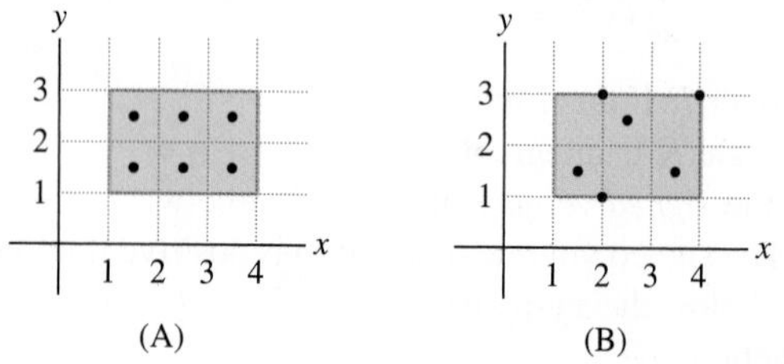

FIGURE 16

3. $f(x, y) = 2x + y$

4. $f(x, y) = 7$

5. $f(x, y) = 4x$

6. $f(x, y) = x - 2y$

7. Let $\mathcal{R} = [0, 1] \times [0, 1]$. Estimate $\iint_{\mathcal{R}} (x + y)\, dA$ by computing two different Riemann sums, each with at least six rectangles.

8. Evaluate $\iint_{\mathcal{R}} 4\, dA$, where $\mathcal{R} = [2, 5] \times [4, 7]$.

9. Evaluate $\iint_{\mathcal{R}} (15 - 3x)\, dA$, where $\mathcal{R} = [0, 5] \times [0, 3]$, and sketch the corresponding solid region (see Example 2).

10. Evaluate $\iint_{\mathcal{R}} (-5)\, dA$, where $\mathcal{R} = [2, 5] \times [4, 7]$.

11. The following table gives the approximate height at quarter-meter intervals of a mound of gravel. Estimate the volume of the mound by computing the average of the two Riemann sums $S_{4,3}$ with lower-left and upper-right vertices of the subrectangles as sample points.

0.75	0.1	0.2	0.2	0.15	0.1
0.5	0.2	0.3	0.5	0.4	0.2
0.25	0.15	0.2	0.4	0.3	0.2
0	0.1	0.15	0.2	0.15	0.1
y \ x	0	0.25	0.5	0.75	1

12. Use the following table to compute a Riemann sum $S_{3,3}$ for $f(x, y)$ on the square $\mathcal{R} = [0, 1.5] \times [0.5, 2]$. Use the regular partition and sample points of your choosing.

Values of $f(x, y)$

2	2.6	2.17	1.86	1.62	1.44
1.5	2.2	1.83	1.57	1.37	1.22
1	1.8	1.5	1.29	1.12	1
0.5	1.4	1.17	1	0.87	0.78
0	1	0.83	0.71	0.62	0.56
y \ x	0	0.5	1	1.5	2

13. CAS Let $S_{N,N}$ be the Riemann sum for $\int_0^1 \int_0^1 e^{x^3 - y^3}\, dy\, dx$ using the regular partition and the lower left-hand vertex of each subrectangle as sample points. Use a computer algebra system to calculate $S_{N,N}$ for $N = 25, 50, 100$.

14. CAS Let $S_{N,M}$ be the Riemann sum for

$$\int_0^4 \int_0^2 \ln(1 + x^2 + y^2)\, dy\, dx$$

using the regular partition and the upper right-hand vertex of each subrectangle as sample points. Use a computer algebra system to calculate $S_{2N,N}$ for $N = 25, 50, 100$.

In Exercises 15–18, use symmetry to evaluate the double integral.

15. $\iint_{\mathcal{R}} x^3\, dA, \quad \mathcal{R} = [-4, 4] \times [0, 5]$

16. $\iint_{\mathcal{R}} 1\, dA, \quad \mathcal{R} = [2, 4] \times [-7, 7]$

17. $\iint_{\mathcal{R}} \sin x\, dA, \quad \mathcal{R} = [0, 2\pi] \times [0, 2\pi]$

18. $\iint_{\mathcal{R}} (2 + x^2 y)\, dA, \quad \mathcal{R} = [0, 1] \times [-1, 1]$

In Exercises 19–36, evaluate the iterated integral.

19. $\int_1^3 \int_0^2 x^3 y\, dy\, dx$

20. $\int_0^2 \int_1^3 x^3 y\, dx\, dy$

21. $\int_4^9 \int_{-3}^8 1\, dx\, dy$

22. $\int_{-4}^{-1} \int_4^8 (-5)\, dx\, dy$

23. $\int_{-1}^1 \int_0^\pi x^2 \sin y\, dy\, dx$

24. $\int_{-1}^1 \int_0^\pi x^2 \sin y\, dx\, dy$

25. $\int_2^6 \int_1^4 x^2\, dx\, dy$

26. $\int_2^6 \int_1^4 y^2\, dx\, dy$

27. $\int_0^1 \int_0^2 (x + 4y^3)\, dx\, dy$

28. $\int_0^2 \int_0^2 (x^2 - y^2)\, dy\, dx$

29. $\int_0^4 \int_0^9 \sqrt{x + 4y}\, dx\, dy$

30. $\int_0^{\pi/4} \int_{\pi/4}^{\pi/2} \cos(2x + y)\, dy\, dx$

31. $\int_1^2 \int_0^4 \frac{dy\, dx}{x + y}$

32. $\int_1^2 \int_2^4 e^{3x - y}\, dy\, dx$

33. $\int_0^4 \int_0^5 \frac{dy\, dx}{\sqrt{x + y}}$

34. $\displaystyle\int_0^8 \int_1^2 \frac{x\,dx\,dy}{\sqrt{x^2+y}}$

35. $\displaystyle\int_1^2 \int_1^3 \frac{\ln(xy)\,dy\,dx}{y}$

36. $\displaystyle\int_0^1 \int_2^3 \frac{1}{(x+4y)^3}\,dx\,dy$

In Exercises 37–42, use Eq. (1) to evaluate the integral.

37. $\displaystyle\iint_{\mathcal{R}} \frac{x}{y}\,dA, \quad \mathcal{R} = [-2, 4] \times [1, 3]$

38. $\displaystyle\iint_{\mathcal{R}} x^2 y\,dA, \quad \mathcal{R} = [-1, 1] \times [0, 2]$

39. $\displaystyle\iint_{\mathcal{R}} \cos x \sin 2y\,dA, \quad \mathcal{R} = \left[0, \tfrac{\pi}{2}\right] \times \left[0, \tfrac{\pi}{2}\right]$

40. $\displaystyle\iint_{\mathcal{R}} \frac{y}{x+1}\,dA, \quad \mathcal{R} = [0, 2] \times [0, 4]$

41. $\displaystyle\iint_{\mathcal{R}} e^x \sin y\,dA, \quad \mathcal{R} = [0, 2] \times \left[0, \tfrac{\pi}{4}\right]$

42. $\displaystyle\iint_{\mathcal{R}} e^{3x+4y}\,dA, \quad \mathcal{R} = [0, 1] \times [1, 2]$

43. Let $f(x, y) = mxy^2$, where m is a constant. Find a value of m such that $\displaystyle\iint_{\mathcal{R}} f(x, y)\,dA = 1$, where $\mathcal{R} = [0, 1] \times [0, 2]$.

44. Evaluate $I = \displaystyle\int_1^3 \int_0^1 ye^{xy}\,dy\,dx$. You will need Integration by Parts and the formula

$$\int e^x(x^{-1} - x^{-2})\,dx = x^{-1}e^x + C$$

Then evaluate I again using Fubini's Theorem to change the order of integration (that is, integrate first with respect to x). Which method is easier?

45. Evaluate $\displaystyle\int_0^1 \int_0^1 \frac{y}{1+xy}\,dy\,dx$. *Hint:* Change the order of integration.

46. Calculate a Riemann sum $S_{3,3}$ on the square $\mathcal{R} = [0, 3] \times [0, 3]$ for the function $f(x, y)$ whose contour plot is shown in Figure 17. Choose sample points and use the plot to find the values of $f(x, y)$ at these points.

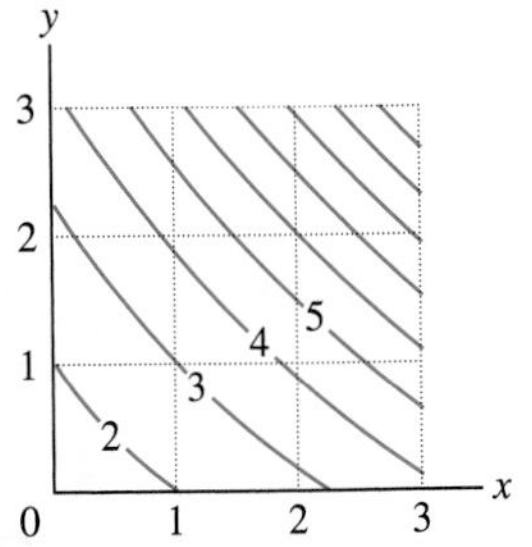

FIGURE 17 Contour plot of $f(x, y)$.

47. Using Fubini's Theorem, argue that the solid in Figure 18 has volume AL, where A is the area of the front face of the solid.

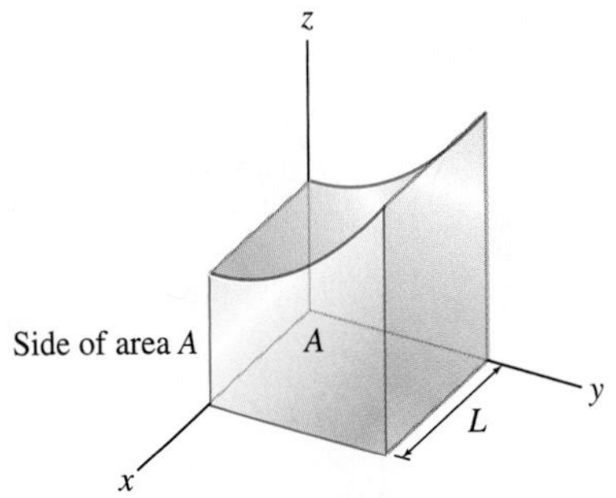

FIGURE 18

Further Insights and Challenges

48. Prove the following extension of the Fundamental Theorem of Calculus to two variables: If $\dfrac{\partial^2 F}{\partial x\,\partial y} = f(x, y)$, then

$$\iint_{\mathcal{R}} f(x, y)\,dA = F(b, d) - F(a, d) - F(b, c) + F(a, c)$$

where $\mathcal{R} = [a, b] \times [c, d]$.

49. Let $F(x, y) = x^{-1}e^{xy}$. Show that $\dfrac{\partial^2 F}{\partial x\,\partial y} = ye^{xy}$ and use the result of Exercise 48 to evaluate $\displaystyle\iint_{\mathcal{R}} ye^{xy}\,dA$ for the rectangle $\mathcal{R} = [1, 3] \times [0, 1]$.

50. Find a function $F(x, y)$ satisfying $\dfrac{\partial^2 F}{\partial x\,\partial y} = 6x^2 y$ and use the result of Exercise 48 to evaluate $\displaystyle\iint_{\mathcal{R}} 6x^2 y\,dA$ for the rectangle $\mathcal{R} = [0, 1] \times [0, 4]$.

51. In this exercise, we use double integration to evaluate the following improper integral for $a > 0$ a positive constant:

$$I(a) = \int_0^\infty \frac{e^{-x} - e^{-ax}}{x}\,dx$$

(a) Use L'Hôpital's Rule to show that $f(x) = \dfrac{e^{-x} - e^{-ax}}{x}$, though not defined at $x = 0$, can be made continuous by assigning the value $f(0) = a - 1$.

(b) Prove that $|f(x)| \le e^{-x} + e^{-ax}$ for $x > 1$ (use the triangle inequality), and apply the Comparison Theorem to show that $I(a)$ converges.

(c) Show that $I(a) = \displaystyle\int_0^\infty \int_1^a e^{-xy}\,dy\,dx$.

(d) Prove, by interchanging the order of integration, that

$$I(a) = \ln a - \lim_{T\to\infty} \int_1^a \frac{e^{-Ty}}{y}\, dy \qquad \boxed{2}$$

(e) Use the Comparison Theorem to show that the limit in Eq. (2) is zero. *Hint:* If $a \geq 1$, show that $e^{-Ty}/y \leq e^{-T}$ for $y \geq 1$, and if $a < 1$, show that $e^{-Ty}/y \leq e^{-aT}/a$ for $a \leq y \leq 1$. Conclude that $I(a) = \ln a$ (Figure 19).

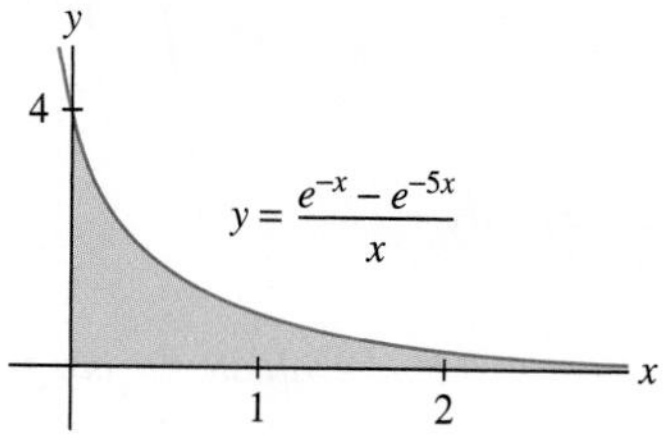

FIGURE 19 The shaded region has area ln 5.

16.2 Double Integrals over More General Regions

In the previous section, we restricted our attention to rectangular domains. Now we shall treat the more general case of domains $\mathcal{D}$ whose boundaries are simple closed curves (a curve is *simple* if it does not intersect itself). We assume that the boundary of $\mathcal{D}$ is smooth as in Figure 1(A) or consists of finitely many smooth curves, joined together with possible corners, as in Figure 1(B). A boundary curve of this type is called **piecewise smooth**. We also assume that $\mathcal{D}$ is a closed domain; that is, $\mathcal{D}$ contains its boundary.

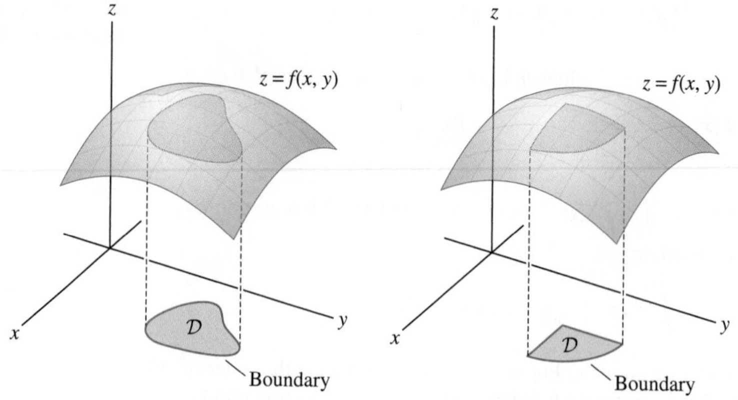

(A) $\mathcal{D}$ has a smooth boundary.

(B) $\mathcal{D}$ has a piecewise smooth boundary, consisting of three smooth curves joined at the corners.

FIGURE 1

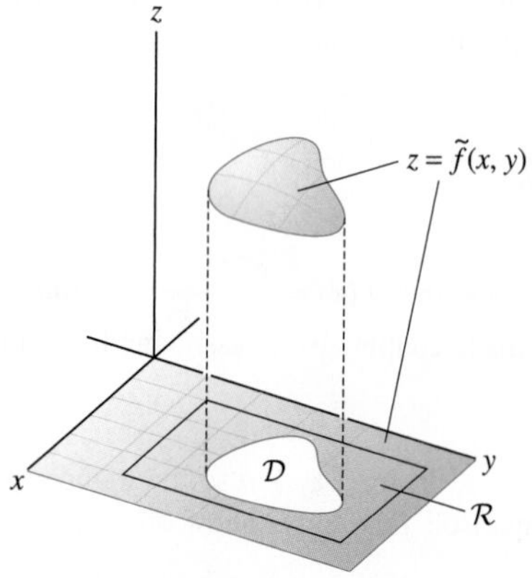

FIGURE 2 The function $\tilde{f}$ is zero outside of $\mathcal{D}$.

Fortunately, we do not need to start from the beginning to define the double integral over a domain $\mathcal{D}$ of this type. Given a function $f(x, y)$ on $\mathcal{D}$, we choose a rectangle $\mathcal{R} = [a, b] \times [c, d]$ containing $\mathcal{D}$ and define a new function $\tilde{f}(x, y)$ that agrees with $f(x, y)$ on $\mathcal{D}$ and is zero outside of $\mathcal{D}$ (Figure 2):

$$\tilde{f}(x, y) = \begin{cases} f(x, y) & \text{if } (x, y) \in \mathcal{D} \\ 0 & \text{if } (x, y) \notin \mathcal{D} \end{cases}$$

The double integral of f over $\mathcal{D}$ is defined as the integral of $\tilde{f}$ over $\mathcal{R}$:

$$\iint_{\mathcal{D}} f(x, y)\, dA = \iint_{\mathcal{R}} \tilde{f}(x, y)\, dA \qquad \boxed{1}$$

We say that f is **integrable** over $\mathcal{D}$ if the integral of $\tilde{f}$ over $\mathcal{R}$ exists. The value of the integral does not depend on the particular choice of $\mathcal{R}$ because $\tilde{f}$ is zero outside of $\mathcal{D}$.

This definition seems reasonable because the integral of $\tilde{f}$ only "picks up" the values of f on $\mathcal{D}$. However, $\tilde{f}$ is likely to be discontinuous because its values jump suddenly to zero beyond the boundary. Despite this possible discontinuity, the next theorem guarantees that the integral of $\tilde{f}$ over $\mathcal{R}$ exists if our original function f is continuous.

THEOREM 1 If $f(x, y)$ is continuous on a closed domain $\mathcal{D}$ whose boundary is a closed, simple, piecewise smooth curve, then $\iint_{\mathcal{D}} f(x, y)\,dA$ exists.

In Theorem 1, we define continuity on $\mathcal{D}$ to mean that f is defined and continuous on some open set containing $\mathcal{D}$.

As in the previous section, the double integral defines the signed volume between the graph of $f(x, y)$ and the xy-plane, where regions below the xy-plane are assigned negative volume.

We can approximate the double integral by Riemann sums for the function $\tilde{f}$ on a rectangle $\mathcal{R}$ containing $\mathcal{D}$. Because $\tilde{f}(P) = 0$ for points P in $\mathcal{R}$ that do not belong to $\mathcal{D}$, any such Riemann sum reduces to a sum over those sample points that lie in $\mathcal{D}$:

$$\iint_{\mathcal{D}} f(x, y)\,dA \approx \sum_{i=1}^{N}\sum_{j=1}^{M} \tilde{f}(P_{ij})\,\Delta x_i\,\Delta y_j = \underbrace{\sum f(P_{ij})\,\Delta x_i\,\Delta y_j}_{\text{Sum only over points } P_{ij} \text{ that lie in } \mathcal{D}} \qquad \boxed{2}$$

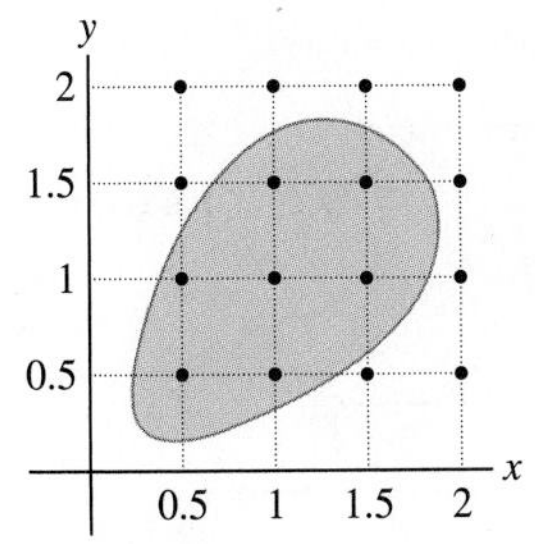

FIGURE 3 Domain $\mathcal{D}$

■ **EXAMPLE 1** Compute $S_{4,4}$ for the integral $\iint_{\mathcal{D}} (x + y)\,dA$, where $\mathcal{D}$ is the shaded domain in Figure 3. Use the upper right-hand corners of the squares as sample points.

Solution Let $f(x, y) = x + y$. The subrectangles in Figure 3 have sides of length $\Delta x = \Delta y = \frac{1}{2}$ and area $\Delta A = \frac{1}{4}$. Only 7 of the 16 sample points lie in $\mathcal{D}$, so

$$S_{4,4} = \sum_{i=1}^{4}\sum_{j=1}^{4} \tilde{f}(P_{ij})\,\Delta x\,\Delta y = \frac{1}{4}\big(f(0.5, 0.5) + f(1, 0.5) + f(0.5, 1) + f(1, 1)$$

$$+ f(1.5, 1) + f(1, 1.5) + f(1.5, 1.5)\big)$$

$$= \frac{1}{4}\big(1 + 1.5 + 1.5 + 2 + 2.5 + 2.5 + 3\big) = \frac{7}{2}$$ ■

The linearity properties of the double integral carry over to general domains: If $f(x, y)$ and $g(x, y)$ are integrable and C is a constant, then

$$\iint_{\mathcal{D}} (f(x, y) + g(x, y))\,dA = \iint_{\mathcal{D}} f(x, y)\,dA + \iint_{\mathcal{D}} g(x, y)\,dA$$

$$\iint_{\mathcal{D}} Cf(x, y)\,dA = C\iint_{\mathcal{D}} f(x, y)\,dA$$

Although we usually think of double integrals as representing volumes, it is worth noting that we can express the *area* of a domain $\mathcal{D}$ in the plane as the double integral of the constant function $f(x, y) = 1$:

$$\boxed{\text{Area}(\mathcal{D}) = \iint_{\mathcal{D}} 1\,dA} \qquad \boxed{3}$$

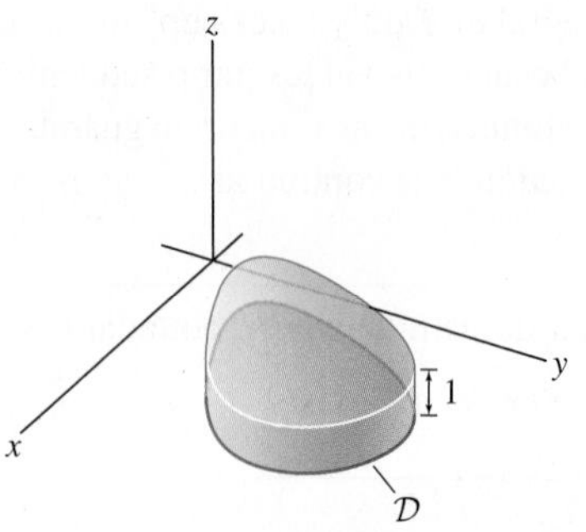

FIGURE 4 The volume of the cylinder of height 1 with $\mathcal{D}$ as base is equal to the area of $\mathcal{D}$.

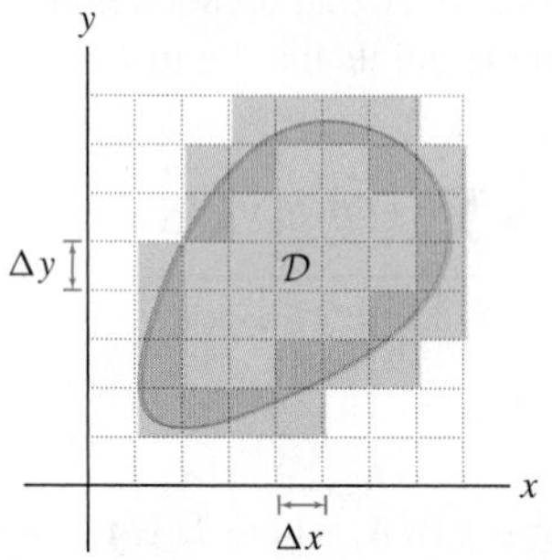

FIGURE 5 The area of $\mathcal{D}$ is approximated by the sum of the areas of the rectangles contained in $\mathcal{D}$.

When you write a double integral over a vertically simple region as an iterated integral, the inner integral is an integral over the dashed segment shown in Figure 6(A). For a horizontally simple region, the inner integral is an integral over the dashed segment shown in Figure 6(B).

Indeed, as we see in Figure 4, the the area of $\mathcal{D}$ is equal to the volume of the "cylinder" of height 1 with $\mathcal{D}$ as base. More generally, for any constant C,

$$\iint_{\mathcal{D}} C\, dA = C\, \text{Area}(\mathcal{D}) \qquad \boxed{4}$$

CONCEPTUAL INSIGHT Eq. (3) tells us that we can approximate the area of a domain $\mathcal{D}$ by a Riemann sum for $\iint_{\mathcal{D}} 1\, dA$. In this case, $f(x, y) = 1$, and we obtain a Riemann sum by adding up the areas $\Delta x_i\, \Delta y_j$ of those rectangles in a grid that are contained in $\mathcal{D}$ or that intersects the boundary of $\mathcal{D}$ (Figure 5). The finer the grid, the better the approximation. The exact area is the limit as the sides of the rectangles tend to zero.

Regions between Two Graphs

When $\mathcal{D}$ is a region between two graphs in the xy-plane, we can evaluate double integrals over $\mathcal{D}$ as iterated integrals. We say that $\mathcal{D}$ is **vertically simple** if it is the region between the graphs of two continuous functions $y = g_1(x)$ and $y = g_2(x)$ (Figure 6):

$$\mathcal{D} = \{(x, y) : a \le x \le b, \quad g_1(x) \le y \le g_2(x)\}$$

Similarly, $\mathcal{D}$ is **horizontally simple** if

$$\mathcal{D} = \{(x, y) : c \le y \le d, \quad g_1(y) \le x \le g_2(y)\}$$

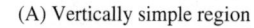

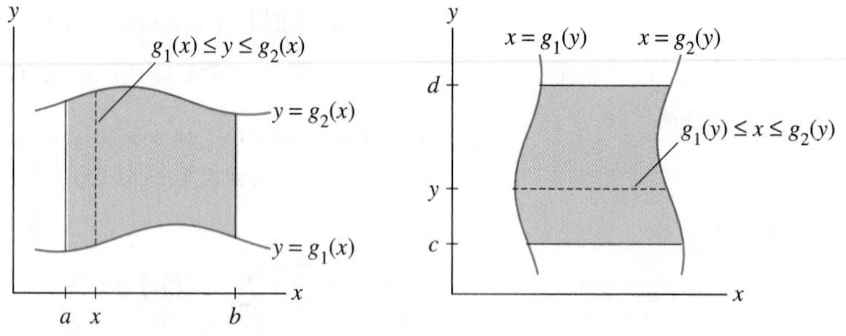

FIGURE 6

THEOREM 2 If $\mathcal{D}$ is vertically simple with description

$$a \le x \le b, \qquad g_1(x) \le y \le g_2(x)$$

then

$$\iint_{\mathcal{D}} f(x, y)\, dA = \int_a^b \int_{g_1(x)}^{g_2(x)} f(x, y)\, dy\, dx$$

If $\mathcal{D}$ is a horizontally simple region with description

$$c \le y \le d, \qquad g_1(y) \le x \le g_2(y)$$

then

$$\iint_{\mathcal{D}} f(x, y)\, dA = \int_c^d \int_{g_1(y)}^{g_2(y)} f(x, y)\, dx\, dy$$

Proof We sketch the proof, assuming that $\mathcal{D}$ is vertically simple (the horizontally simple case is similar). Choose a rectangle $\mathcal{R} = [a, b] \times [c, d]$ containing $\mathcal{D}$. Then

Although $\tilde{f}$ need not be continuous, the use of Fubini's Theorem in Eq. (5) can be justified. In particular, the integral $\int_c^d \tilde{f}(x, y)\, dy$ exists and is a continuous function of x.

$$\iint_{\mathcal{D}} f(x, y)\, dA = \int_a^b \int_c^d \tilde{f}(x, y)\, dy\, dx \qquad \boxed{5}$$

By definition, $\tilde{f}(x, y)$ is zero outside $\mathcal{D}$, so for fixed x, $\tilde{f}(x, y)$ is zero unless y satisfies $g_1(x) \le y \le g_2(x)$. Therefore,

$$\int_c^d \tilde{f}(x, y)\, dy = \int_{g_1(x)}^{g_2(x)} f(x, y)\, dy$$

Substituting in Eq. (5), we obtain the desired equality:

$$\iint_{\mathcal{D}} f(x, y)\, dA = \int_a^b \int_{g_1(x)}^{g_2(x)} f(x, y)\, dy\, dx \qquad ■$$

Integration over a simple region is similar to integration over a rectangle with one difference: The limits of the inner integral may be functions instead of constants.

■ **EXAMPLE 2** Evaluate $\iint_{\mathcal{D}} x^2 y\, dA$, where $\mathcal{D}$ is the region in Figure 7.

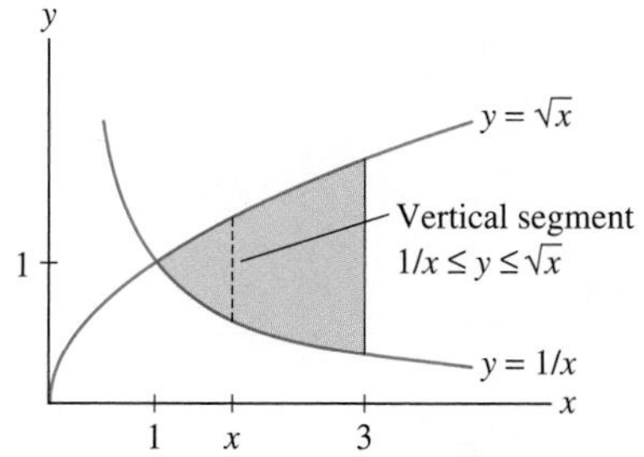

FIGURE 7 Domain between $y = \sqrt{x}$ and $y = 1/x$.

Solution

***Step 1.* Describe $\mathcal{D}$ as a vertically simple region.**

$$\underbrace{1 \le x \le 3}_{\text{Limits of outer integral}}, \qquad \underbrace{\frac{1}{x} \le y \le \sqrt{x}}_{\text{Limits of inner integral}}$$

In this case, $g_1(x) = 1/x$ and $g_2(x) = \sqrt{x}$.

***Step 2.* Set up the iterated integral.**

$$\iint_{\mathcal{D}} x^2 y\, dA = \int_1^3 \int_{y=1/x}^{\sqrt{x}} x^2 y\, dy\, dx$$

Notice that the inner integral is an integral over a vertical segment between the graphs of $y = 1/x$ and $y = \sqrt{x}$.

***Step 3.* Compute the iterated integral.**

As usual, we evaluate the inner integral by treating x as a constant, but now the upper and lower limits depend on x:

$$\int_{y=1/x}^{\sqrt{x}} x^2 y\, dy = \frac{1}{2} x^2 y^2 \bigg|_{y=1/x}^{\sqrt{x}} = \frac{1}{2} x^2 (\sqrt{x})^2 - \frac{1}{2} x^2 \left(\frac{1}{x}\right)^2 = \frac{1}{2} x^3 - \frac{1}{2}$$

We complete the calculation by integrating with respect to x:

$$\iint_{\mathcal{D}} x^2 y\, dA = \int_1^3 \left(\frac{1}{2} x^3 - \frac{1}{2}\right) dx = \left(\frac{1}{8} x^4 - \frac{1}{2} x\right)\bigg|_1^3$$

$$= \frac{69}{8} - \left(-\frac{3}{8}\right) = 9 \qquad ■$$

■ **EXAMPLE 3 Horizontally Simple Description Better** Find the volume V of the region between the plane $z = 2x + 3y$ and the triangle $\mathcal{D}$ in Figure 8.

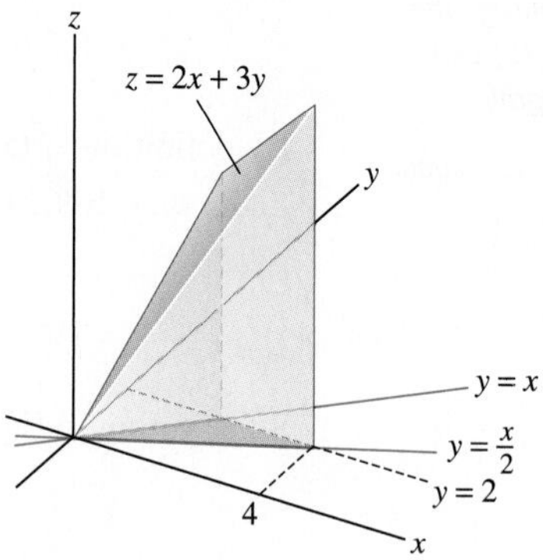

FIGURE 8

Solution The triangle $\mathcal{D}$ is bounded by the lines $y = x/2$, $y = x$, and $y = 2$. We see in Figure 9 that $\mathcal{D}$ is vertically simple, but the upper curve is not given by a single formula: The formula switches from $y = x$ to $y = 2$. Therefore, it is more convenient to describe $\mathcal{D}$ as a horizontally simple region (Figure 9):

$$\mathcal{D}: 0 \le y \le 2, \quad y \le x \le 2y$$

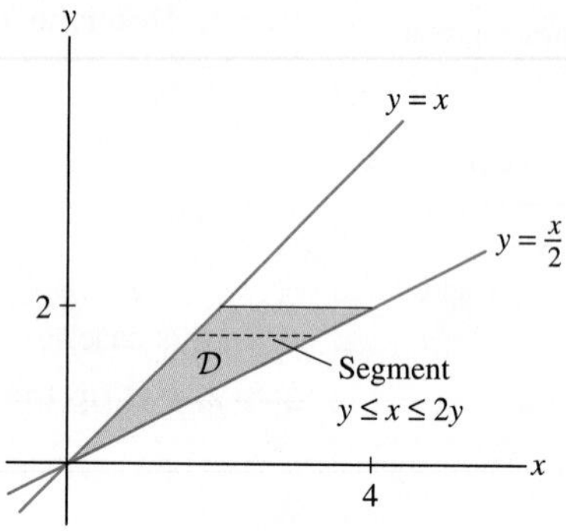

FIGURE 9

The volume is equal to the double integral of $f(x, y) = 2x + 3y$ over $\mathcal{D}$,

$$V = \iint_{\mathcal{D}} f(x, y)\, dA = \int_0^2 \int_{x=y}^{2y} (2x + 3y)\, dx\, dy$$

$$= \int_0^2 \left(x^2 + 3yx\right)\Big|_{x=y}^{2y} dy = \int_0^2 \left((4y^2 + 6y^2) - (y^2 + 3y^2)\right) dy$$

$$= \int_0^2 6y^2\, dy = 2y^3\Big|_0^2 = 16$$

■

The next example shows that in some cases, one iterated integral is easier to evaluate than the other.

■ **EXAMPLE 4 Choosing the Best Iterated Integral** Evaluate $\displaystyle\iint_{\mathcal{D}} e^{y^2}\, dA$ for $\mathcal{D}$ in Figure 10.

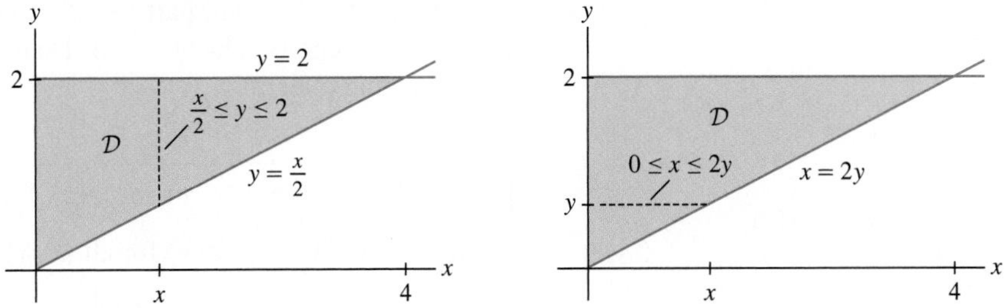

(A) $\mathcal{D}$ as a vertically simple domain: $0 \le x \le 4,\ x/2 \le y \le 2$

(B) $\mathcal{D}$ as a horizontally simple domain: $0 \le y \le 2,\ 0 \le x \le 2y$

FIGURE 10 The region $\mathcal{D}$ is horizontally and vertically simple.

Solution First, let's try describing $\mathcal{D}$ as a vertically simple domain. Referring to Figure 10(A), we have

$$\mathcal{D}: 0 \le x \le 4, \quad \frac{1}{2}x \le y \le 2 \quad \Rightarrow \quad \iint_{\mathcal{D}} e^{y^2}\, dA = \int_{x=0}^{4} \int_{y=x/2}^{2} e^{y^2}\, dy\, dx$$

The inner integral cannot be evaluated because we have no explicit antiderivative for e^{y^2}. Therefore, we try describing $\mathcal{D}$ as horizontally simple [Figure 10(B)]:

$$\mathcal{D}: 0 \le y \le 2, \quad 0 \le x \le 2y$$

This leads to an iterated integral that can be evaluated:

$$\int_0^2 \int_{x=0}^{2y} e^{y^2}\, dx\, dy = \int_0^2 \left(x e^{y^2} \Big|_{x=0}^{2y} \right) dy = \int_0^2 2y e^{y^2}\, dy$$

$$= e^{y^2} \Big|_0^2 = e^4 - 1$$

■

■ **EXAMPLE 5** **Changing the Order of Integration** Sketch the domain of integration $\mathcal{D}$ corresponding to

$$\int_1^9 \int_{\sqrt{y}}^{3} x e^y\, dx\, dy$$

Then change the order of integration and evaluate.

Solution The limits of integration give us inequalities that describe the domain $\mathcal{D}$ (as a horizontally simple region since dx precedes dy):

$$1 \le y \le 9, \qquad \sqrt{y} \le x \le 3$$

We sketch the region in Figure 11. Now observe that $\mathcal{D}$ is also vertically simple:

$$1 \le x \le 3, \qquad 1 \le y \le x^2$$

so we can rewrite our integral and evaluate:

$$\int_1^9 \int_{x=\sqrt{y}}^{3} x e^y\, dx\, dy = \int_1^3 \int_{y=1}^{x^2} x e^y\, dy\, dx = \int_1^3 \left(\int_{y=1}^{x^2} x e^y\, dy \right) dx$$

$$= \int_1^3 \left(x e^y \Big|_{y=1}^{x^2} \right) dx = \int_1^3 (x e^{x^2} - ex)\, dx = \frac{1}{2}(e^{x^2} - ex^2) \Big|_1^3$$

$$= \frac{1}{2}(e^9 - 9e) - 0 = \frac{1}{2}(e^9 - 9e)$$

■

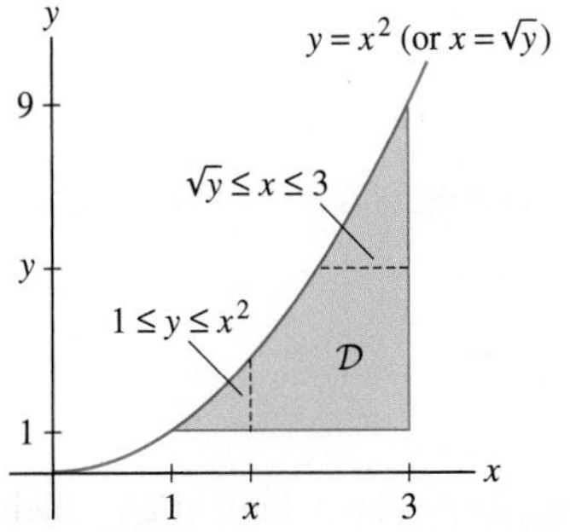

FIGURE 11 Describing $\mathcal{D}$ as a horizontally or vertically simple region.

In the next theorem, part (a) is a formal statement of the fact that larger functions have larger integrals, a fact that we also noted in the single-variable case. Part (b) is useful for estimating integrals.

THEOREM 3 Let $f(x, y)$ and $g(x, y)$ be integrable functions on $\mathcal{D}$.

(a) If $f(x, y) \le g(x, y)$ for all $(x, y) \in \mathcal{D}$, then

$$\iint_{\mathcal{D}} f(x, y)\, dA \le \iint_{\mathcal{D}} g(x, y)\, dA \tag{6}$$

(b) If $m \le f(x, y) \le M$ for all $(x, y) \in \mathcal{D}$, then

$$m \operatorname{Area}(\mathcal{D}) \le \iint_{\mathcal{D}} f(x, y)\, dA \le M \operatorname{Area}(\mathcal{D}) \tag{7}$$

Proof If $f(x, y) \le g(x, y)$, then every Riemann sum for $f(x, y)$ is less than or equal to the corresponding Riemann sum for g:

$$\sum f(P_{ij})\, \Delta x_i\, \Delta y_j \le \sum g(P_{ij})\, \Delta x_i\, \Delta y_j$$

We obtain (6) by taking the limit. Now suppose that $f(x, y) \le M$ and apply (6) with $g(x, y) = M$:

$$\iint_{\mathcal{D}} f(x, y)\, dA \le \iint_{\mathcal{D}} M\, dA = M \operatorname{Area}(\mathcal{D})$$

This proves half of (7). The other half follows similarly. ■

■ **EXAMPLE 6** Estimate $\displaystyle\iint_{\mathcal{D}} \frac{dA}{\sqrt{x^2 + (y-2)^2}}$ where $\mathcal{D}$ is the disk of radius 1 centered at the origin.

Solution The quantity $\sqrt{x^2 + (y-2)^2}$ is the distance d from (x, y) to $(0, 2)$, and we see from Figure 12 that $1 \le d \le 3$. Taking reciprocals, we have

$$\frac{1}{3} \le \frac{1}{\sqrt{x^2 + (y-2)^2}} \le 1$$

We apply (7) with $m = \frac{1}{3}$ and $M = 1$, using the fact that $\operatorname{Area}(\mathcal{D}) = \pi$, to obtain

$$\frac{\pi}{3} \le \iint_{\mathcal{D}} \frac{dA}{\sqrt{x^2 + (y-2)^2}} \le \pi$$ ■

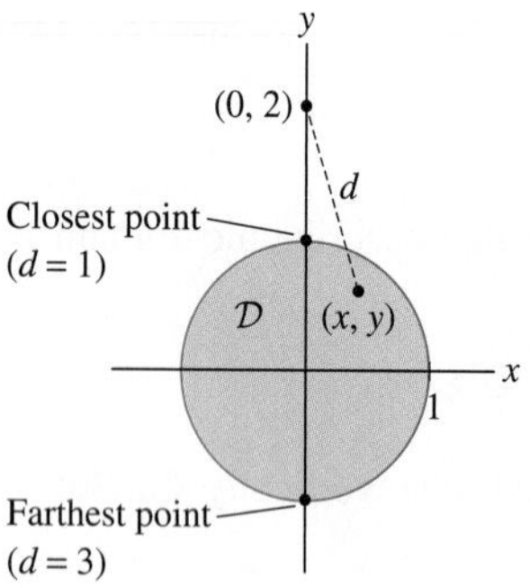

FIGURE 12 The distance d from (x, y) to $(0, 2)$ varies from 1 to 3 for (x, y) in the unit disk.

The **average value** (or **mean value**) of a function $f(x, y)$ on a domain $\mathcal{D}$, which we denote by $\overline{f}$, is the quantity

$$\overline{f} = \frac{1}{\operatorname{Area}(\mathcal{D})} \iint_{\mathcal{D}} f(x, y)\, dA = \frac{\iint_{\mathcal{D}} f(x, y)\, dA}{\iint_{\mathcal{D}} 1\, dA} \tag{8}$$

← REMINDER Equation (8) is similar to the definition of an average value in one variable:

$$\overline{f} = \frac{1}{b-a} \int_a^b f(x)\, dx = \frac{\int_a^b f(x)\, dx}{\int_a^b 1\, dx}$$

Equivalently, $\overline{f}$ is the value satisfying the relation

$$\iint_{\mathcal{D}} f(x, y)\, dA = \overline{f} \cdot \text{Area}(\mathcal{D})$$

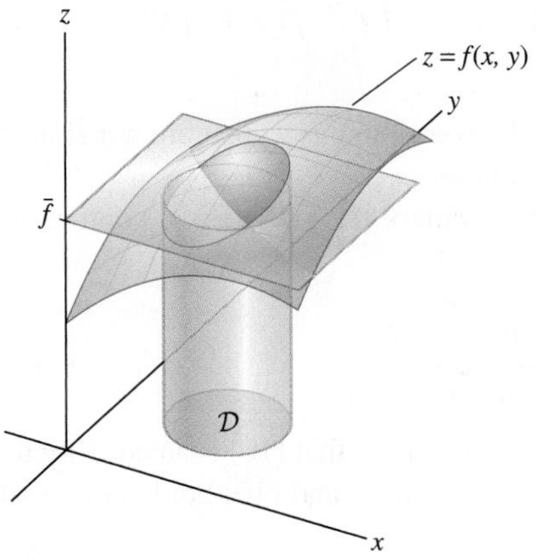

FIGURE 13

GRAPHICAL INSIGHT The solid region under the graph has the same (signed) volume as the cylinder with base $\mathcal{D}$ of height $\overline{f}$ (Figure 13).

■ **EXAMPLE 7** An architect needs to know the average height $\overline{H}$ of the ceiling of a pagoda whose base $\mathcal{D}$ is the square $[-4, 4] \times [-4, 4]$ and roof is the graph of

$$H(x, y) = 32 - x^2 - y^2$$

where distances are in feet (Figure 14). Calculate $\overline{H}$.

Solution First, we compute the integral of $H(x, y)$ over $\mathcal{D}$:

$$\iint_{\mathcal{D}} (32 - x^2 - y^2)\, dA = \int_{-4}^{4} \int_{-4}^{4} (32 - x^2 - y^2)\, dy\, dx$$

$$= \int_{-4}^{4} \left(\left(32y - x^2 y - \frac{1}{3} y^3 \right) \Big|_{-4}^{4} \right) dx = \int_{-4}^{4} \left(\frac{640}{3} - 8x^2 \right) dx$$

$$= \left(\frac{640}{3} x - \frac{8}{3} x^3 \right) \Big|_{-4}^{4} = \frac{4096}{3}$$

The area of $\mathcal{D}$ is $8 \times 8 = 64$, so the average height of the pagoda's ceiling is

$$\overline{H} = \frac{1}{\text{Area}(\mathcal{D})} \iint_{\mathcal{D}} H(x, y)\, dA = \frac{1}{64} \left(\frac{4096}{3} \right) = \frac{64}{3} \approx 21.3 \text{ ft} \qquad ■$$

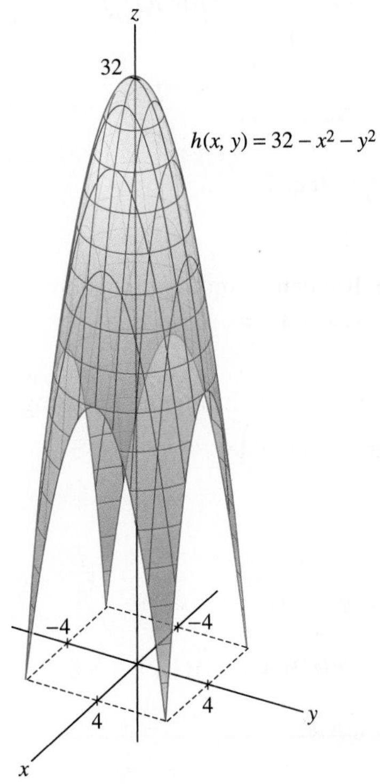

FIGURE 14 Pagoda with ceiling $H(x, y) = 32 - x^2 - y^2$.

The Mean Value Theorem states that a continuous function on a domain $\mathcal{D}$ must take on its average value at some point P in $\mathcal{D}$, provided that $\mathcal{D}$ is closed, bounded, and also **connected** (see Exercise 63 for a proof). By definition, $\mathcal{D}$ is connected if any two points in $\mathcal{D}$ can be joined by a curve in $\mathcal{D}$ (Figure 15).

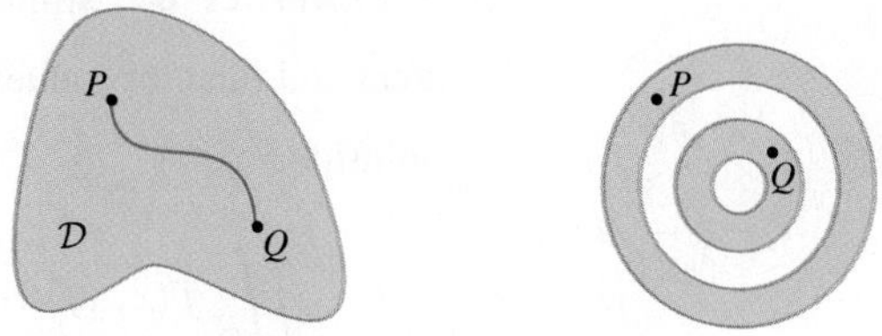

(A) Connected domain: Any two points can be joined by a curve lying entirely in $\mathcal{D}$.

(B) Nonconnected domain.

FIGURE 15

THEOREM 4 Mean Value Theorem for Double Integrals If $f(x, y)$ is continuous and $\mathcal{D}$ is closed, bounded, and connected, then there exists a point $P \in \mathcal{D}$ such that

$$\iint_{\mathcal{D}} f(x, y)\, dA = f(P)\, \text{Area}(\mathcal{D}) \qquad \boxed{9}$$

Equivalently, $f(P) = \overline{f}$, where $\overline{f}$ is the average value of f on $\mathcal{D}$.

Decomposing the Domain into Smaller Domains

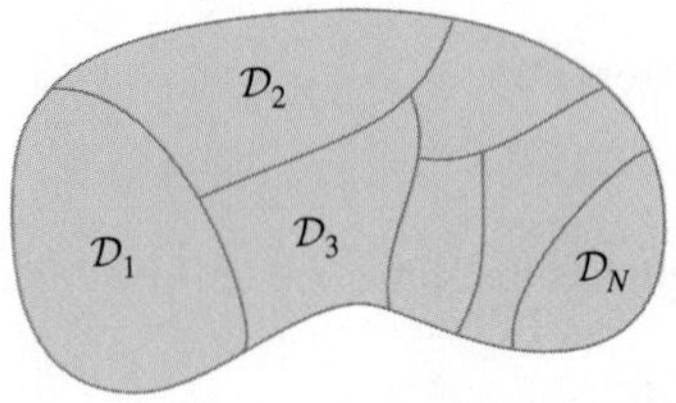

FIGURE 16 The region $\mathcal{D}$ is a union of smaller domains.

Double integrals are additive with respect to the domain: If $\mathcal{D}$ is the union of domains $\mathcal{D}_1, \mathcal{D}_2, \dots, \mathcal{D}_N$ that do not overlap except possibly on boundary curves (Figure 16), then

$$\iint_{\mathcal{D}} f(x, y)\, dA = \iint_{\mathcal{D}_1} f(x, y)\, dA + \cdots + \iint_{\mathcal{D}_N} f(x, y)\, dA$$

Additivity may be used to evaluate double integrals over domains $\mathcal{D}$ that are not simple but can be decomposed into finitely many simple domains.

We close this section with a simple but useful remark. If $f(x, y)$ is a continuous function on a *small* domain $\mathcal{D}$, then

In general, the approximation (10) is useful only if $\mathcal{D}$ is small in both width and length, that is, if $\mathcal{D}$ is contained in a circle of small radius. If $\mathcal{D}$ has small area but is very long and thin, then f may be far from constant on $\mathcal{D}$.

$$\iint_{\mathcal{D}} f(x, y)\, dA \approx \underbrace{f(P)\,\text{Area}(\mathcal{D})}_{\text{Function value} \times \text{area}} \qquad \boxed{10}$$

where P is any sample point in $\mathcal{D}$. In fact, we can choose P so that (10) is an equality by Theorem 4. But if $\mathcal{D}$ is small, then f is nearly constant on $\mathcal{D}$, and (10) holds as a good approximation for all $P \in \mathcal{D}$.

If the domain $\mathcal{D}$ is not small, we may partition it into N smaller subdomains $\mathcal{D}_1, \dots, \mathcal{D}_N$ and choose sample points P_j in $\mathcal{D}_j$. By additivity,

$$\iint_{\mathcal{D}} f(x, y)\, dA = \sum_{j=1}^{N} \iint_{\mathcal{D}_j} f(x, y)\, dA \approx \sum_{j=1}^{N} f(P_j)\,\text{Area}(\mathcal{D}_j)$$

and thus we have the approximation

$$\iint_{\mathcal{D}} f(x, y)\, dA \approx \sum_{j=1}^{N} f(P_j)\,\text{Area}(\mathcal{D}_j) \qquad \boxed{11}$$

We can think of Eq. (11) as a generalization of the Riemann sum approximation. In a Riemann sum, $\mathcal{D}$ is partitioned by rectangles $\mathcal{R}_{ij}$ of area $\Delta A_{ij} = \Delta x_i\, \Delta y_j$.

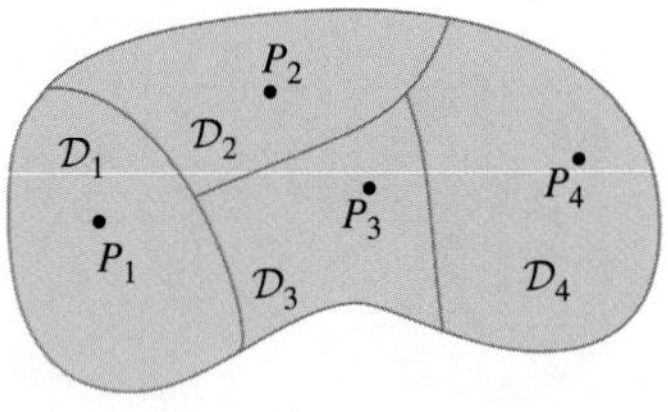

FIGURE 17

j	1	2	3	4
Area$(\mathcal{D}_j)$	1	1	0.9	1.2
$f(P_j)$	1.8	2.2	2.1	2.4

■ **EXAMPLE 8** Estimate $\iint_{\mathcal{D}} f(x, y)\, dA$ for the domain $\mathcal{D}$ in Figure 17, using the areas and function values given there and the accompanying table.

Solution

$$\iint_{\mathcal{D}} f(x, y)\, dA \approx \sum_{j=1}^{4} f(P_j)\,\text{Area}(\mathcal{D}_j)$$

$$= (1.8)(1) + (2.2)(1) + (2.1)(0.9) + (2.4)(1.2) \approx 8.8$$ ■

16.2 SUMMARY

- We assume that $\mathcal{D}$ is a closed, bounded domain whose boundary is a simple closed curve that either is smooth or has a finite number of corners. The double integral is defined by

$$\iint_{\mathcal{D}} f(x, y)\, dA = \iint_{\mathcal{R}} \tilde{f}(x, y)\, dA$$

where $\mathcal{R}$ is a rectangle containing $\mathcal{D}$ and $\tilde{f}(x, y) = f(x, y)$ if $(x, y) \in \mathcal{D}$, and $\tilde{f}(x, y) = 0$ otherwise. The value of the integral does not depend on the choice of $\mathcal{R}$.

- The double integral defines the signed volume between the graph of $f(x, y)$ and the xy-plane, where regions below the xy-plane are assigned negative volume.
- For any constant C, $\iint_{\mathcal{D}} C\,dA = C \cdot \text{Area}(\mathcal{D})$.
- If $\mathcal{D}$ is vertically or horizontally simple, $\iint_{\mathcal{D}} f(x, y)\,dA$ can be evaluated as an iterated integral:

Vertically simple domain $a \le x \le b, \quad g_1(x) \le y \le g_2(x)$	$\int_a^b \int_{g_1(x)}^{g_2(x)} f(x, y)\,dy\,dx$
Horizontally simple domain $c \le y \le d, \quad g_1(y) \le x \le g_2(y)$	$\int_c^d \int_{g_1(y)}^{g_2(y)} f(x, y)\,dx\,dy$

- If $f(x, y) \le g(x, y)$ on $\mathcal{D}$, then $\iint_{\mathcal{D}} f(x, y)\,dA \le \iint_{\mathcal{D}} g(x, y)\,dA$.
- If m is the minimum value and M the maximum value of f on $\mathcal{D}$, then

$$m\,\text{Area}(\mathcal{D}) \le \iint_{\mathcal{D}} f(x, y)\,dA \le \iint_{\mathcal{D}} M\,dA = M\,\text{Area}(\mathcal{D})$$

- The *average value* of f on $\mathcal{D}$ is

$$\overline{f} = \frac{1}{\text{Area}(\mathcal{D})} \iint_{\mathcal{D}} f(x, y)\,dA = \frac{\iint_{\mathcal{D}} f(x, y)\,dA}{\iint_{\mathcal{D}} 1\,dA}$$

- *Mean Value Theorem for Integrals*: If $f(x, y)$ is continuous and $\mathcal{D}$ is closed, bounded, and connected, then there exists a point $P \in \mathcal{D}$ such that

$$\iint_{\mathcal{D}} f(x, y)\,dA = f(P)\,\text{Area}(\mathcal{D})$$

Equivalently, $f(P) = \overline{f}$, where $\overline{f}$ is the average value of f on $\mathcal{D}$.
- *Additivity with respect to the domain:* If $\mathcal{D}$ is a union of nonoverlapping (except possibly on their boundaries) domains $\mathcal{D}_1, \dots, \mathcal{D}_N$, then

$$\iint_{\mathcal{D}} f(x, y)\,dA = \sum_{j=1}^{N} \iint_{\mathcal{D}_j} f(x, y)\,dA$$

- If the domains $\mathcal{D}_1, \dots, \mathcal{D}_N$ are small and P_j is a sample point in $\mathcal{D}_j$, then

$$\iint_{\mathcal{D}} f(x, y)\,dA \approx \sum_{j=1}^{N} f(P_j)\text{Area}(\mathcal{D}_j)$$

16.2 EXERCISES

Preliminary Questions

1. .Which of the following expressions do not make sense?

(a) $\int_0^1 \int_1^x f(x, y)\,dy\,dx$

(b) $\int_0^1 \int_1^y f(x, y)\,dy\,dx$

(c) $\int_0^1 \int_x^y f(x, y)\,dy\,dx$

(d) $\int_0^1 \int_x^1 f(x, y)\,dy\,dx$

2. Draw a domain in the plane that is neither vertically nor horizontally simple.

3. Which of the four regions in Figure 18 is the domain of integration for $\int_{-\sqrt{2}/2}^{0} \int_{-x}^{\sqrt{1-x^2}} f(x, y)\,dy\,dx$?

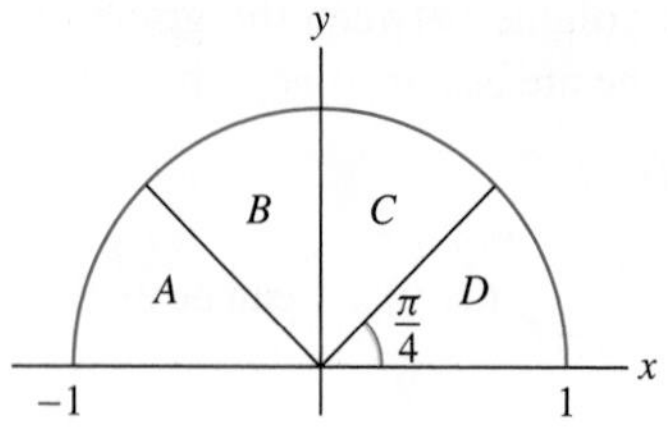

FIGURE 18

4. Let $\mathcal{D}$ be the unit disk. If the maximum value of $f(x, y)$ on $\mathcal{D}$ is 4, then the largest possible value of $\iint_{\mathcal{D}} f(x, y)\, dA$ is (choose the correct answer):

(a) 4 **(b)** 4π **(c)** $\frac{4}{\pi}$

Exercises

1. Calculate the Riemann sum for $f(x, y) = x - y$ and the shaded domain $\mathcal{D}$ in Figure 19 with two choices of sample points, • and ∘. Which do you think is a better approximation to the integral of f over $\mathcal{D}$? Why?

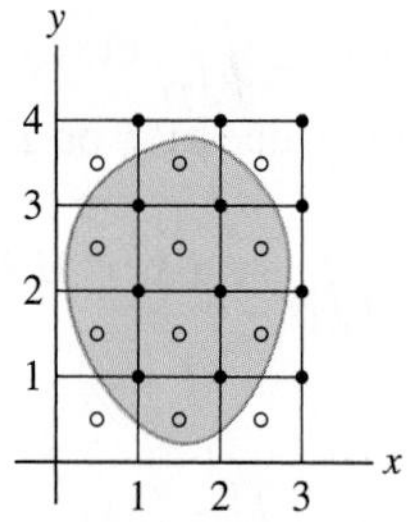

FIGURE 19

2. Approximate values of $f(x, y)$ at sample points on a grid are given in Figure 20. Estimate $\iint_{\mathcal{D}} f(x, y)\, dx\, dy$ for the shaded domain by computing the Riemann sum with the given sample points.

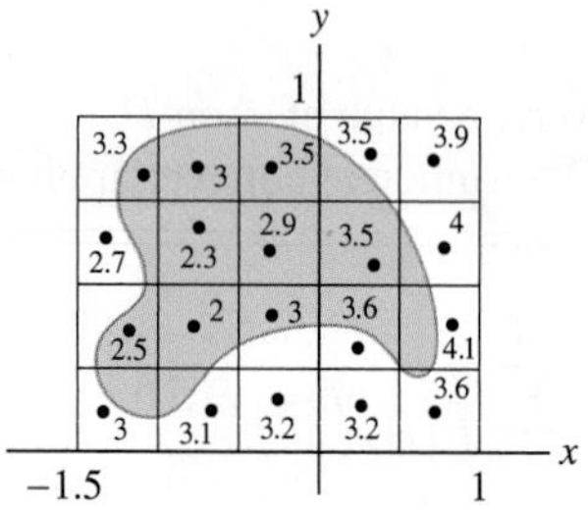

FIGURE 20

3. Express the domain $\mathcal{D}$ in Figure 21 as both a vertically simple region and a horizontally simple region, and evaluate the integral of $f(x, y) = xy$ over $\mathcal{D}$ as an iterated integral in two ways.

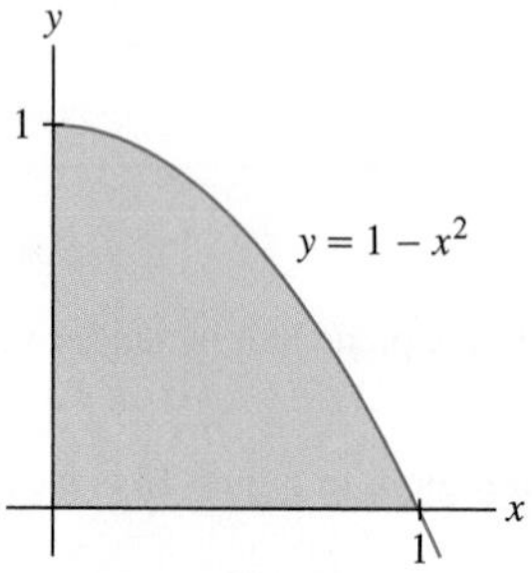

FIGURE 21

4. Sketch the domain

$$\mathcal{D} : 0 \le x \le 1, \quad x^2 \le y \le 4 - x^2$$

and evaluate $\iint_{\mathcal{D}} y\, dA$ as an iterated integral.

In Exercises 5–7, compute the double integral of $f(x, y) = x^2 y$ over the given shaded domain in Figure 22.

5. (A) **6.** (B) **7.** (C)

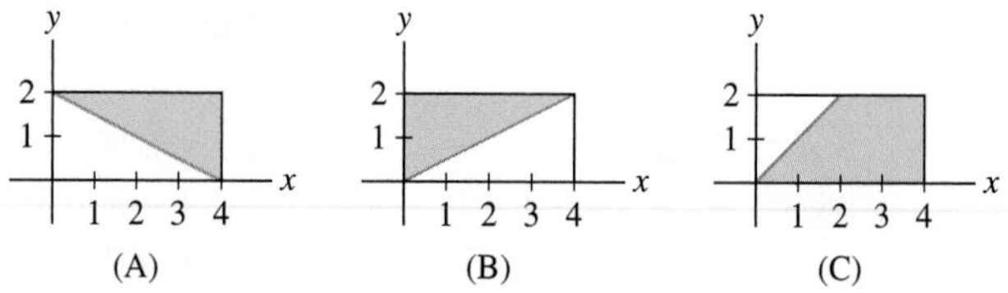

FIGURE 22

8. Sketch the domain $\mathcal{D}$ defined by $x + y \le 12$, $x \ge 4$, $y \ge 4$ and compute $\iint_{\mathcal{D}} e^{x+y}\, dA$.

9. Integrate $f(x, y) = x$ over the region bounded by $y = x^2$ and $y = x + 2$.

10. Sketch the region $\mathcal{D}$ between $y = x^2$ and $y = x(1 - x)$. Express $\mathcal{D}$ as a simple region and calculate the integral of $f(x, y) = 2y$ over $\mathcal{D}$.

11. Evaluate $\iint_{\mathcal{D}} \frac{y}{x}\, dA$, where $\mathcal{D}$ is the shaded part of the semicircle of radius 2 in Figure 23.

12. Calculate the double integral of $f(x, y) = y^2$ over the rhombus $\mathcal{R}$ in Figure 24.

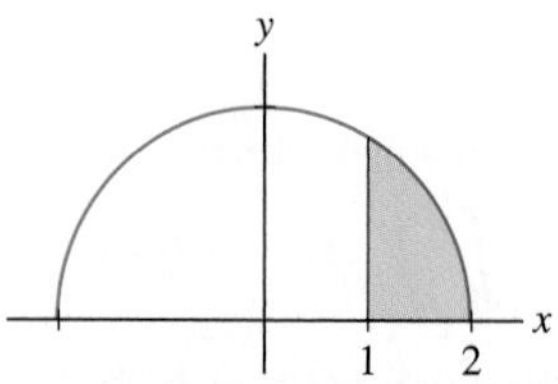

FIGURE 23 $y = \sqrt{4 - x^2}$

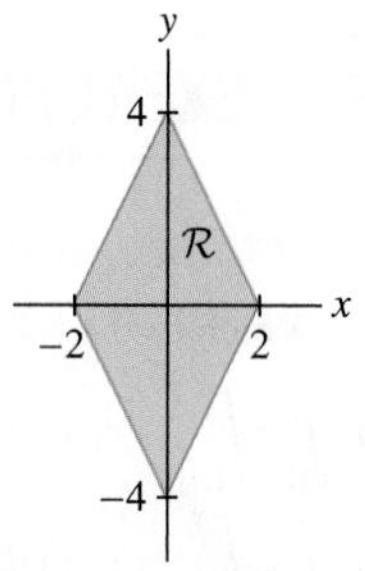

FIGURE 24 $|x| + \frac{1}{2}|y| \le 1$

13. Calculate the double integral of $f(x, y) = x + y$ over the domain $\mathcal{D} = \{(x, y) : x^2 + y^2 \le 4,\ y \ge 0\}$.

14. Integrate $f(x, y) = (x + y + 1)^{-2}$ over the triangle with vertices $(0, 0)$, $(4, 0)$, and $(0, 8)$.

15. Calculate the integral of $f(x, y) = x$ over the region $\mathcal{D}$ bounded above by $y = x(2 - x)$ and below by $x = y(2 - y)$. *Hint:* Apply the quadratic formula to the lower boundary curve to solve for y as a function of x.

16. Integrate $f(x, y) = x$ over the region bounded by $y = x$, $y = 4x - x^2$, and $y = 0$ in two ways: as a vertically simple region and as a horizontally simple region.

In Exercises 17–24, compute the double integral of $f(x, y)$ over the domain $\mathcal{D}$ indicated.

17. $f(x, y) = x^2 y;\quad 1 \le x \le 3,\quad x \le y \le 2x + 1$

18. $f(x, y) = 1;\quad 0 \le x \le 1,\quad 1 \le y \le e^x$

19. $f(x, y) = x;\quad 0 \le x \le 1,\quad 1 \le y \le e^{x^2}$

20. $f(x, y) = \cos(2x + y);\quad \frac{1}{2} \le x \le \frac{\pi}{2},\quad 1 \le y \le 2x$

21. $f(x, y) = 2xy$; bounded by $x = y$, $x = y^2$

22. $f(x, y) = \sin x$; bounded by $x = 0$, $x = 1$, $y = \cos x$

23. $f(x, y) = e^{x+y}$; bounded by $y = x - 1$, $y = 12 - x$ for $2 \le y \le 4$

24. $f(x, y) = (x + y)^{-1}$; bounded by $y = x$, $y = 1$, $y = e$, $x = 0$

In Exercises 25–28, sketch the domain of integration and express as an iterated integral in the opposite order.

25. $\displaystyle\int_0^4 \int_x^4 f(x, y)\, dy\, dx$

26. $\displaystyle\int_4^9 \int_{\sqrt{y}}^3 f(x, y)\, dx\, dy$

27. $\displaystyle\int_4^9 \int_2^{\sqrt{y}} f(x, y)\, dx\, dy$

28. $\displaystyle\int_0^1 \int_{e^x}^e f(x, y)\, dy\, dx$

29. Sketch the domain $\mathcal{D}$ corresponding to

$$\int_0^4 \int_{\sqrt{y}}^2 \sqrt{4x^2 + 5y}\, dx\, dy$$

Then change the order of integration and evaluate.

30. Change the order of integration and evaluate

$$\int_0^1 \int_0^{\pi/2} x \cos(xy)\, dx\, dy$$

Explain the simplification achieved by changing the order.

31. Compute the integral of $f(x, y) = (\ln y)^{-1}$ over the domain $\mathcal{D}$ bounded by $y = e^x$ and $y = e^{\sqrt{x}}$. *Hint:* Choose the order of integration that enables you to evaluate the integral.

32. Evaluate by changing the order of integration:

$$\int_0^9 \int_0^{\sqrt{y}} \frac{x\, dx\, dy}{(3x^2 + y)^{1/2}}$$

In Exercises 33–36, sketch the domain of integration. Then change the order of integration and evaluate. Explain the simplification achieved by changing the order.

33. $\displaystyle\int_0^1 \int_y^1 \frac{\sin x}{x}\, dx\, dy$

34. $\displaystyle\int_0^4 \int_{\sqrt{y}}^2 \sqrt{x^3 + 1}\, dx\, dy$

35. $\displaystyle\int_0^1 \int_{y=x}^1 x e^{y^3}\, dy\, dx$

36. $\displaystyle\int_0^1 \int_{y=x^{2/3}}^1 x e^{y^4}\, dy\, dx$

37. Sketch the domain $\mathcal{D}$ where $0 \le x \le 2$, $0 \le y \le 2$, and x *or* y is greater than 1. Then compute $\displaystyle\iint_{\mathcal{D}} e^{x+y}\, dA$.

38. Calculate $\displaystyle\iint_{\mathcal{D}} e^x\, dA$, where $\mathcal{D}$ is bounded by the lines $y = x + 1$, $y = x$, $x = 0$, and $x = 1$.

In Exercises 39–42, calculate the double integral of $f(x, y)$ over the triangle indicated in Figure 25.

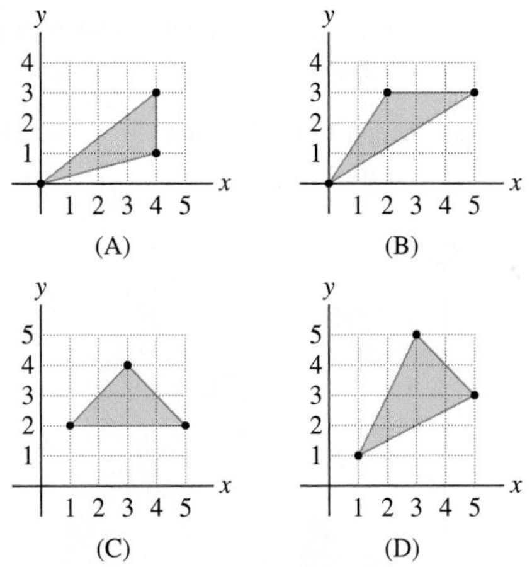

FIGURE 25

39. $f(x, y) = e^{x^2}$, (A)

40. $f(x, y) = 1 - 2x$, (B)

41. $f(x, y) = \dfrac{x}{y^2}$, (C)

42. $f(x, y) = x + 1$, (D)

43. Calculate the double integral of $f(x, y) = \dfrac{\sin y}{y}$ over the region $\mathcal{D}$ in Figure 26.

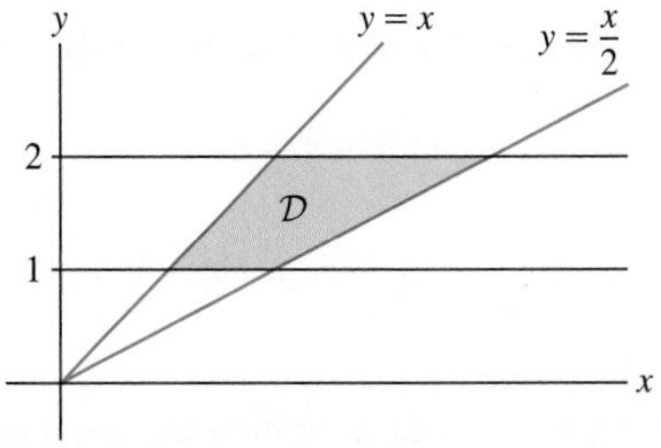

FIGURE 26

44. Evaluate $\iint_{\mathcal{D}} x\,dA$ for $\mathcal{D}$ in Figure 27.

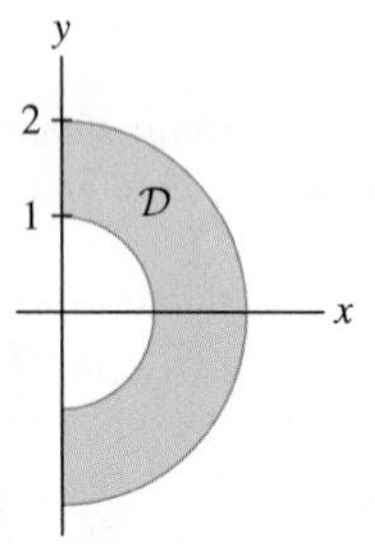

FIGURE 27

45. Find the volume of the region bounded by $z = 40 - 10y$, $z = 0$, $y = 0$, $y = 4 - x^2$.

46. Find the volume of the region enclosed by $z = 1 - y^2$ and $z = y^2 - 1$ for $0 \le x \le 2$.

47. Calculate the average value of $f(x, y) = e^{x+y}$ on the square $[0, 1] \times [0, 1]$.

48. Calculate the average height above the x-axis of a point in the region $0 \le x \le 1$, $0 \le y \le x^2$.

49. Find the average height of the "ceiling" in Figure 28 defined by $z = y^2 \sin x$ for $0 \le x \le \pi$, $0 \le y \le 1$.

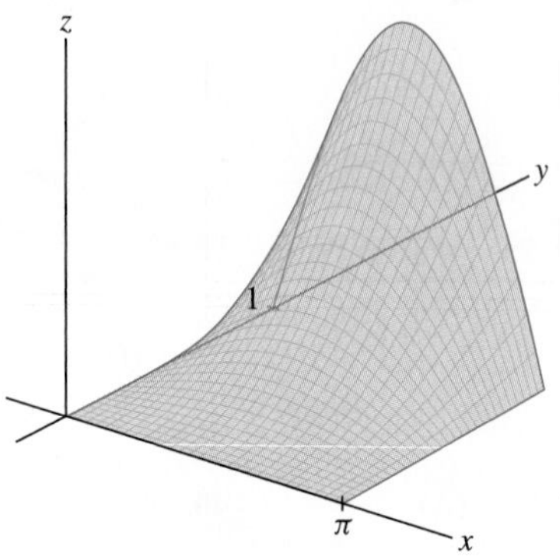

FIGURE 28

50. Calculate the average value of the x-coordinate of a point on the semicircle $x^2 + y^2 \le R^2$, $x \ge 0$. What is the average value of the y-coordinate?

51. What is the average value of the linear function

$$f(x, y) = mx + ny + p$$

on the ellipse $\left(\frac{x}{a}\right)^2 + \left(\frac{y}{b}\right)^2 \le 1$? Argue by symmetry rather than calculation.

52. Find the average square distance from the origin to a point in the domain $\mathcal{D}$ in Figure 29.

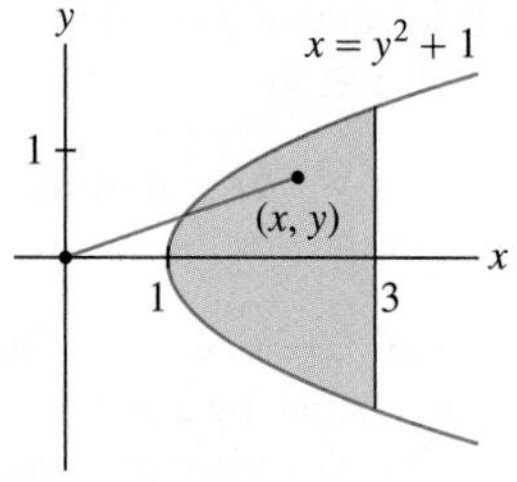

FIGURE 29

53. Let $\mathcal{D}$ be the rectangle $0 \le x \le 2$, $-\frac{1}{8} \le y \le \frac{1}{8}$, and let $f(x, y) = \sqrt{x^3 + 1}$. Prove that

$$\iint_{\mathcal{D}} f(x, y)\,dA \le \frac{3}{2}$$

54. (a) Use the inequality $0 \le \sin x \le x$ for $x \ge 0$ to show that

$$\int_0^1 \int_0^1 \sin(xy)\,dx\,dy \le \frac{1}{4}$$

(b) Use a computer algebra system to evaluate the double integral to three decimal places.

55. Prove the inequality $\iint_{\mathcal{D}} \frac{dA}{4 + x^2 + y^2} \le \pi$, where $\mathcal{D}$ is the disk $x^2 + y^2 \le 4$.

56. Let $\mathcal{D}$ be the domain bounded by $y = x^2 + 1$ and $y = 2$. Prove the inequality

$$\frac{4}{3} \le \iint_{\mathcal{D}} (x^2 + y^2)\,dA \le \frac{20}{3}$$

57. Let $\overline{f}$ be the average of $f(x, y) = xy^2$ on $\mathcal{D} = [0, 1] \times [0, 4]$. Find a point $P \in \mathcal{D}$ such that $f(P) = \overline{f}$ (the existence of such a point is guaranteed by the Mean Value Theorem for Double Integrals).

58. Verify the Mean Value Theorem for Double Integrals for $f(x, y) = e^{x-y}$ on the triangle bounded by $y = 0$, $x = 1$, and $y = x$.

In Exercises 59 and 60, use (11) to estimate the double integral.

59. The following table lists the areas of the subdomains $\mathcal{D}_j$ of the domain $\mathcal{D}$ in Figure 30 and the values of a function $f(x, y)$ at sample points $P_j \in \mathcal{D}_j$. Estimate $\iint_{\mathcal{D}} f(x, y)\,dA$.

j	1	2	3	4	5	6
Area($\mathcal{D}_j$)	1.2	1.1	1.4	0.6	1.2	0.8
$f(P_j)$	9	9.1	9.3	9.1	8.9	8.8

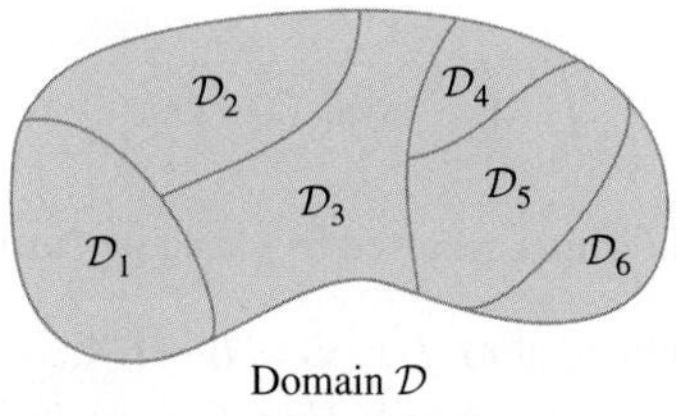

FIGURE 30

60. The domain $\mathcal{D}$ between the circles of radii 5 and 5.2 in the first quadrant in Figure 31 is divided into six subdomains of angular width $\Delta\theta = \frac{\pi}{12}$, and the values of a function $f(x, y)$ at sample points are given. Compute the area of the subdomains and estimate $\iint_{\mathcal{D}} f(x, y)\, dA$.

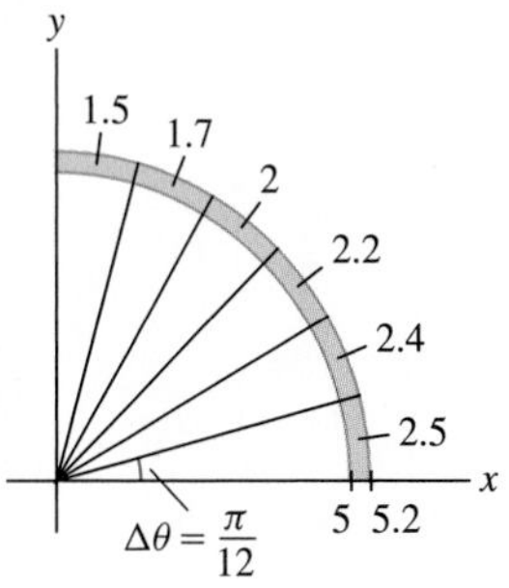

FIGURE 31

61. According to Eq. (3), the area of a domain $\mathcal{D}$ is equal to $\iint_{\mathcal{D}} 1\, dA$. Prove that if $\mathcal{D}$ is the region between two curves $y = g_1(x)$ and $y = g_2(x)$ with $g_2(x) \le g_1(x)$ for $a \le x \le b$, then

$$\iint_{\mathcal{D}} 1\, dA = \int_a^b (g_1(x) - g_2(x))\, dx$$

Further Insights and Challenges

62. Let $\mathcal{D}$ be a closed connected domain and let $P, Q \in \mathcal{D}$. The Intermediate Value Theorem (IVT) states that if f is continuous on $\mathcal{D}$, then $f(x, y)$ takes on every value between $f(P)$ and $f(Q)$ at some point in $\mathcal{D}$.

(a) Show, by constructing a counterexample, that the IVT is false if $\mathcal{D}$ is not connected.

(b) Prove the IVT as follows: Let $\mathbf{c}(t)$ be a path such that $\mathbf{c}(0) = P$ and $\mathbf{c}(1) = Q$ (such a path exists because $\mathcal{D}$ is connected). Apply the IVT in one variable to the composite function $f(\mathbf{c}(t))$.

63. Use the fact that a continuous function on a closed domain $\mathcal{D}$ attains both a minimum value m and a maximum value M, together with Theorem 3, to prove that the average value $\overline{f}$ lies between m and M. Then use the IVT in Exercise 62 to prove the Mean Value Theorem for Double Integrals.

64. Let $f(y)$ be a function of y alone and set $G(t) = \int_0^t \int_0^x f(y)\, dy\, dx$.

(a) Use the Fundamental Theorem of Calculus to prove that $G''(t) = f(t)$.

(b) Show, by changing the order in the double integral, that $G(t) = \int_0^t (t - y) f(y)\, dy$. This shows that the "second antiderivative" of $f(y)$ can be expressed as a single integral.

16.3 Triple Integrals

Triple integrals of functions $f(x, y, z)$ of three variables are a fairly straightforward generalization of double integrals. Instead of a rectangle in the plane, our domain is a box (Figure 1)

$$\mathcal{B} = [a, b] \times [c, d] \times [p, q]$$

consisting of all points (x, y, z) in $\mathbf{R}^3$ such that

$$a \le x \le b, \qquad c \le y \le d, \qquad p \le z \le q$$

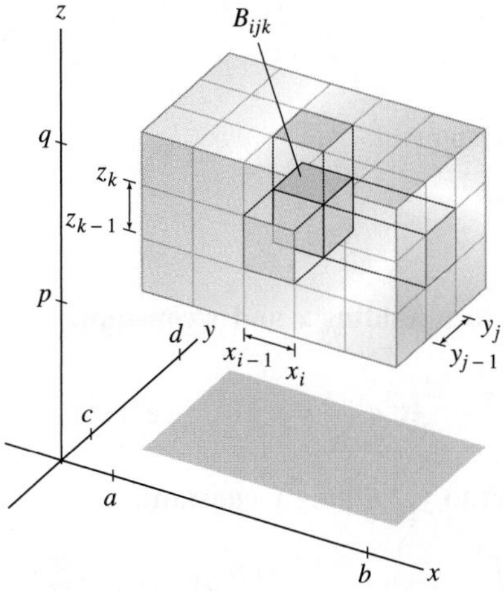

FIGURE 1 The box $\mathcal{B} = [a, b] \times [c, d] \times [p, q]$ decomposed into smaller boxes.

To integrate over this box, we subdivide the box (as usual) into "sub"-boxes

$$\mathcal{B}_{ijk} = [x_{i-1}, x_i] \times [y_{j-1}, y_j] \times [z_{k-1}, z_k]$$

by choosing partitions of the three intervals

$$a = x_0 < x_1 < \cdots < x_N = b$$

$$c = y_0 < y_1 < \cdots < y_M = d$$

$$p = z_0 < z_1 < \cdots < z_L = q$$

Here N, M, and L are positive integers. The volume of $\mathcal{B}_{ijk}$ is $\Delta V_{ijk} = \Delta x_i\, \Delta y_j\, \Delta z_k$ where

$$\Delta x_i = x_i - x_{i-1}, \qquad \Delta y_j = y_j - y_{j-1}, \qquad \Delta z_k = z_k - z_{k-1}$$

Then, we choose a sample point P_{ijk} in each box $\mathcal{B}_{ijk}$ and form the Riemann sum

$$S_{N,M,L} = \sum_{i=1}^{N} \sum_{j=1}^{M} \sum_{k=1}^{L} f(P_{ijk})\, \Delta V_{ijk}$$

As in the previous section, we write $\mathcal{P} = \{\{x_i\}, \{y_j\}, \{z_k\}\}$ for the partition and let $\|\mathcal{P}\|$ be the maximum of the widths Δx_i, Δy_j, Δz_k. If the sums $S_{N,M,L}$ approach a limit as $\|\mathcal{P}\| \to 0$ for arbitrary choices of sample points, we say that f is **integrable** over $\mathcal{B}$. The limit value is denoted

$$\iiint_{\mathcal{B}} f(x, y, z)\, dV = \lim_{\|\mathcal{P}\| \to 0} S_{N,M,L}$$

Triple integrals have many of the same properties as double and single integrals. The linear properties are satisfied, and continuous functions are integrable over a box $\mathcal{B}$. Furthermore, triple integrals can be evaluated as iterated integrals.

The notation dA, used in the previous section, suggests area and occurs in double integrals over domains in the plane. Similarly, dV suggests volume and is used in the notation for triple integrals.

THEOREM 1 Fubini's Theorem for Triple Integrals The triple integral of a continuous function $f(x, y, z)$ over a box $\mathcal{B} = [a, b] \times [c, d] \times [p, q]$ is equal to the iterated integral:

$$\iiint_{\mathcal{B}} f(x, y, z)\, dV = \int_{x=a}^{b} \int_{y=c}^{d} \int_{z=p}^{q} f(x, y, z)\, dz\, dy\, dx$$

Furthermore, the iterated integral may be evaluated in any order.

As noted in the theorem, we are free to evaluate the iterated integral in any order (there are six different orders). For instance,

$$\int_{x=a}^{b} \int_{y=c}^{d} \int_{z=p}^{q} f(x, y, z)\, dz\, dy\, dx = \int_{z=p}^{q} \int_{y=c}^{d} \int_{x=a}^{b} f(x, y, z)\, dx\, dy\, dz$$

EXAMPLE 1 Integration over a Box Calculate the integral $\iiint_{\mathcal{B}} x^2 e^{y+3z}\, dV$, where $\mathcal{B} = [1, 4] \times [0, 3] \times [2, 6]$.

Solution We write this triple integral as an iterated integral:

$$\iiint_{\mathcal{B}} x^2 e^{y+3z}\, dV = \int_1^4 \int_0^3 \int_2^6 x^2 e^{y+3z}\, dz\, dy\, dx$$

***Step 1.* Evaluate the inner integral with respect to *z*, holding *x* and *y* constant.**

$$\int_{z=2}^{6} x^2 e^{y+3z}\, dz = \frac{1}{3} x^2 e^{y+3z} \Big|_2^6 = \frac{1}{3} x^2 e^{y+18} - \frac{1}{3} x^2 e^{y+6} = \frac{1}{3}(e^{18} - e^6) x^2 e^y$$

***Step 2.* Evaluate the middle integral with respect to *y*, holding *x* constant.**

$$\int_{y=0}^{3} \frac{1}{3}(e^{18} - e^6) x^2 e^y\, dy = \frac{1}{3}(e^{18} - e^6) x^2 \int_{y=0}^{3} e^y\, dy = \frac{1}{3}(e^{18} - e^6)(e^3 - 1) x^2$$

***Step 3.* Evaluate the outer integral with respect to *x*.**

$$\iiint_{\mathcal{B}} (x^2 e^{y+3z})\, dV = \frac{1}{3}(e^{18} - e^6)(e^3 - 1) \int_{x=1}^{4} x^2\, dx = 7(e^{18} - e^6)(e^3 - 1) \quad \blacksquare$$

Note that in the previous example, the integrand factors as a product of three functions $f(x, y, z) = g(x)h(y)k(z)$—namely,

$$f(x, y, z) = x^2 e^{y+3z} = x^2 e^y e^{3z}$$

Because of this, the triple integral can be evaluated simply as the product of three single integrals:

$$\iiint_{\mathcal{B}} x^2 e^y e^{3z}\, dV = \left(\int_1^4 x^2\, dx\right)\left(\int_0^3 e^y\, dy\right)\left(\int_2^6 e^{3z}\, dz\right)$$

$$= (21)(e^3 - 1)\left(\frac{e^{18} - e^6}{3}\right) = 7(e^{18} - e^6)(e^3 - 1)$$

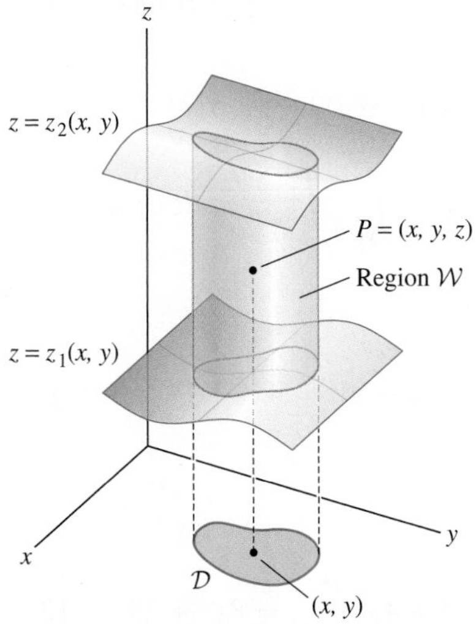

FIGURE 2 The point $P = (x, y, z)$ in the simple region $\mathcal{W}$ if $(x, y) \in \mathcal{D}$ and $z_1(x, y) \le z \le z_2(x, y)$.

Next, instead of a box, we integrate over a solid region $\mathcal{W}$ that is *simple* as in Figure 2. In other words, $\mathcal{W}$ is the region between two surfaces $z = z_1(x, y)$ and $z = z_2(x, y)$ over a domain $\mathcal{D}$ in the xy-plane. In this case,

$$\mathcal{W} = \{(x, y, z) : (x, y) \in \mathcal{D} \quad \text{and} \quad z_1(x, y) \le z \le z_2(x, y)\} \qquad \boxed{1}$$

The domain $\mathcal{D}$ is the **projection** of $\mathcal{W}$ onto the xy-plane.

As a formal matter, as in the case of double integrals, we define the triple integral of $f(x, y, z)$ over $\mathcal{W}$ by

$$\iiint_{\mathcal{W}} f(x, y, z)\, dV = \iiint_{\mathcal{B}} \tilde{f}(x, y, z)\, dV$$

where $\mathcal{B}$ is a box containing $\mathcal{W}$, and $\tilde{f}$ is the function that is equal to f on $\mathcal{W}$ and equal to zero outside of $\mathcal{W}$. The triple integral exists, assuming that $z_1(x, y)$, $z_2(x, y)$, and the integrand f are continuous. In practice, however, we evaluate triple integrals as iterated integrals. This is justified by the following theorem, whose proof is similar to that of Theorem 2 in Section 16.2.

THEOREM 2 The triple integral of a continuous function f over the region

$$\mathcal{W} : (x, y) \in \mathcal{D}, \quad z_1(x, y) \le z \le z_2(x, y)$$

is equal to the iterated integral

$$\iiint_{\mathcal{W}} f(x, y, z)\, dV = \iint_{\mathcal{D}} \left(\int_{z=z_1(x,y)}^{z_2(x,y)} f(x, y, z)\, dz\right) dA$$

More generally, integrals of functions of n variables (for any n) arise naturally in many different contexts. For example, the average distance between two points in a ball is expressed as a six-fold integral because we integrate over all possible coordinates of the two points. Each point has three coordinates for a total of six variables.

One thing missing from our discussion so far is a geometric interpretation of triple integrals. A double integral represents the signed volume of the three-dimensional region between a graph $z = f(x, y)$ and the xy-plane. The graph of a function $f(x, y, z)$ of three variables lives in *four-dimensional space*, and thus a triple integral represents a four-dimensional volume. This volume is hard or impossible to visualize. On the other hand, triple integrals represent many other types of quantities. Some examples are total mass, average value, probabilities, and centers of mass (see Section 16.5).

Furthermore, the volume V of a region $\mathcal{W}$ is defined as the triple integral of the constant function $f(x, y, z) = 1$:

$$V = \iiint_{\mathcal{W}} 1\, dV$$

In particular, if $\mathcal{W}$ is a simple region between $z = z_1(x, y)$ and $z = z_2(x, y)$, then

$$\iiint_{\mathcal{W}} 1\,dV = \iint_{\mathcal{D}} \left(\int_{z=z_1(x,y)}^{z_2(x,y)} 1\,dz \right) dA = \iint_{\mathcal{D}} (z_2(x, y) - z_1(x, y))\,dA$$

Thus, the triple integral is equal to the double integral defining the volume of the region between the two surfaces.

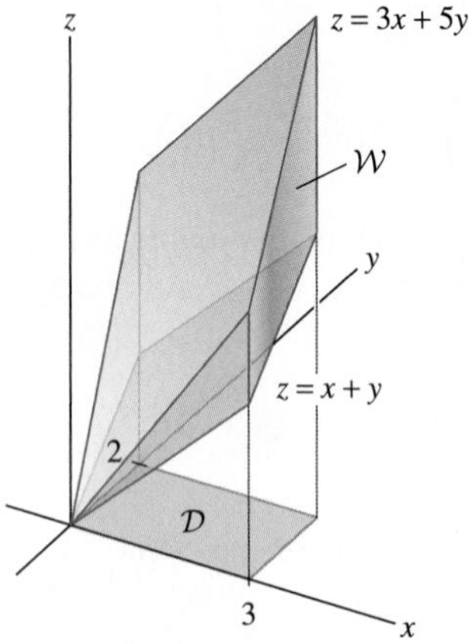

FIGURE 3 Region $\mathcal{W}$ between the planes $z = x + y$ and $z = 3x + 5y$ lying over $\mathcal{D} = [0, 3] \times [0, 2]$.

■ **EXAMPLE 2 Solid Region with a Rectangular Base** Evaluate $\iiint_{\mathcal{W}} z\,dV$, where $\mathcal{W}$ is the region between the planes $z = x + y$ and $z = 3x + 5y$ lying over the rectangle $\mathcal{D} = [0, 3] \times [0, 2]$ (Figure 3).

Solution Apply Theorem 2 with $z_1(x, y) = x + y$ and $z_2(x, y) = 3x + 5y$:

$$\iiint_{\mathcal{W}} z\,dV = \iint_{\mathcal{D}} \left(\int_{z=x+y}^{3x+5y} z\,dz \right) dA = \int_{x=0}^{3} \int_{y=0}^{2} \int_{z=x+y}^{3x+5y} z\,dz\,dy\,dx$$

***Step 1.* Evaluate the inner integral with respect to *z*.**

$$\int_{z=x+y}^{3x+5y} z\,dz = \frac{1}{2}z^2 \bigg|_{z=x+y}^{3x+5y} = \frac{1}{2}(3x + 5y)^2 - \frac{1}{2}(x + y)^2 = 4x^2 + 14xy + 12y^2 \qquad \boxed{2}$$

***Step 2.* Evaluate the result with respect to *y*.**

$$\int_{y=0}^{2} (4x^2 + 14xy + 12y^2)\,dy = (4x^2 y + 7xy^2 + 4y^3) \bigg|_{y=0}^{2} = 8x^2 + 28x + 32$$

***Step 3.* Evaluate the result with respect to *x*.**

$$\iiint_{\mathcal{W}} z\,dV = \int_{x=0}^{3} (8x^2 + 28x + 32)\,dx = \left(\frac{8}{3}x^3 + 14x^2 + 32x \right) \bigg|_0^3$$
$$= 72 + 126 + 96 = 294$$ ■

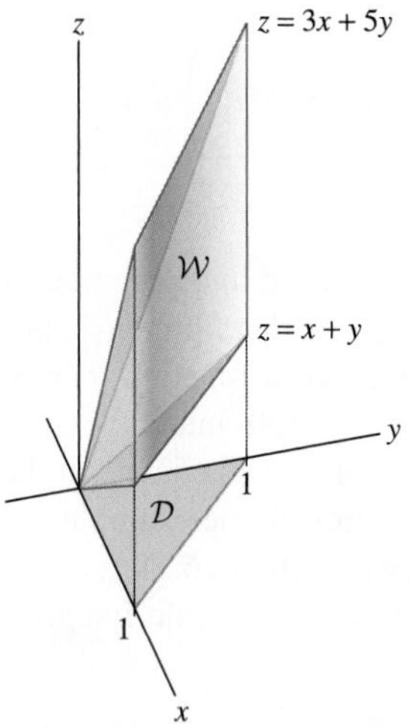

FIGURE 4 Region $\mathcal{W}$ between the planes $z = x + y$ and $z = 3x + 5y$ lying over the triangle $\mathcal{D}$.

■ **EXAMPLE 3 Solid Region with a Triangular Base** Evaluate $\iiint_{\mathcal{W}} z\,dV$, where $\mathcal{W}$ is the region in Figure 4.

Solution This is similar to the previous example, but now $\mathcal{W}$ lies over the triangle $\mathcal{D}$ in the xy-plane defined by

$$0 \le x \le 1, \qquad 0 \le y \le 1 - x$$

Thus, the triple integral is equal to the iterated integral:

$$\iiint_{\mathcal{W}} z\,dV = \iint_{\mathcal{D}} \left(\int_{z=x+y}^{3x+5y} z\,dz \right) dA = \underbrace{\int_{x=0}^{1} \int_{y=0}^{1-x}}_{\text{Integral over triangle}} \int_{z=x+y}^{3x+5y} z\,dz\,dy\,dx$$

We computed the inner integral in the previous example [see Eq. (2)]:

$$\int_{z=x+y}^{3x+5y} z\,dz = \frac{1}{2}z^2 \bigg|_{x+y}^{3x+5y} = 4x^2 + 14xy + 12y^2$$

Next, we integrate with respect to y:

$$\int_{y=0}^{1-x} (4x^2 + 14xy + 12y^2)\,dy = (4x^2y + 7xy^2 + 4y^3)\Big|_{y=0}^{1-x}$$
$$= 4x^2(1-x) + 7x(1-x)^2 + 4(1-x)^3$$
$$= 4 - 5x + 2x^2 - x^3$$

And finally,

$$\iiint_{\mathcal{W}} z\,dV = \int_{x=0}^{1} (4 - 5x + 2x^2 - x^3)\,dx$$
$$= 4 - \frac{5}{2} + \frac{2}{3} - \frac{1}{4} = \frac{23}{12}$$

■

■ **EXAMPLE 4 Region between Intersecting Surfaces** Integrate $f(x, y, z) = x$ over the region $\mathcal{W}$ bounded above by $z = 4 - x^2 - y^2$ and below by $z = x^2 + 3y^2$ in the octant $x \geq 0$, $y \geq 0$, $z \geq 0$.

Solution The region $\mathcal{W}$ is simple, so

$$\iiint_{\mathcal{W}} x\,dV = \iint_{\mathcal{D}} \int_{z=x^2+3y^2}^{4-x^2-y^2} x\,dz\,dA$$

where $\mathcal{D}$ is the projection of $\mathcal{W}$ onto the xy-plane. To evaluate the integral over $\mathcal{D}$, we must find the equation of the curved part of the boundary of $\mathcal{D}$.

***Step 1.* Find the boundary of $\mathcal{D}$.**
The upper and lower surfaces intersect where they have the same height:

$$z = x^2 + 3y^2 = 4 - x^2 - y^2 \qquad \text{or} \qquad x^2 + 2y^2 = 2$$

Therefore, as we see in Figure 5, $\mathcal{W}$ projects onto the domain $\mathcal{D}$ consisting of the quarter of the ellipse $x^2 + 2y^2 = 2$ in the first quadrant. This ellipse hits the axes at $(\sqrt{2}, 0)$ and $(0, 1)$.

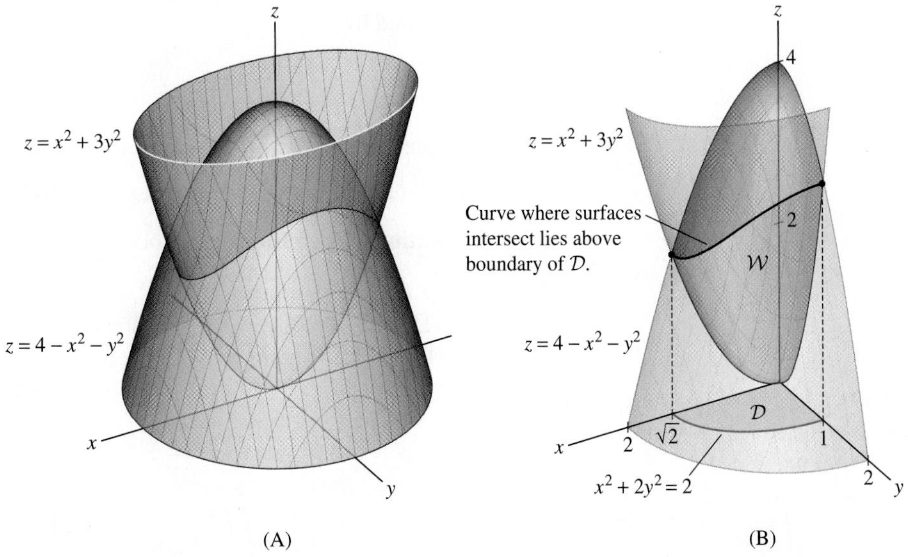

FIGURE 5 Region $x^2 + 3y^2 \leq z \leq 4 - x^2 - y^2$.

Step 2. **Express $\mathcal{D}$ as a simple domain.**
We can integrate in either the order $dy\,dx$ or $dx\,dy$. If we choose $dx\,dy$, then y varies from 0 to 1 and the domain is described by

$$\mathcal{D}: 0 \le y \le 1, \quad 0 \le x \le \sqrt{2-2y^2}$$

Step 3. **Write the triple integral as an iterated integral.**

$$\iiint_{\mathcal{W}} x\,dV = \int_{y=0}^{1} \int_{x=0}^{\sqrt{2-2y^2}} \int_{z=x^2+3y^2}^{4-x^2-y^2} x\,dz\,dx\,dy$$

Step 4. **Evaluate.**
Here are the results of evaluating the integrals in order:

Inner integral: $\displaystyle\int_{z=x^2+y^2}^{4-x^2-y^2} x\,dz = xz\Big|_{z=x^2+3y^2}^{4-x^2-y^2} = 4x - 2x^3 - 4y^2x$

Middle integral:

$$\int_{x=0}^{\sqrt{2-2y^2}} (4x - 2x^3 - 4y^2x)\,dx = \left(2x^2 - \frac{1}{2}x^4 - 2x^2y^2\right)\Bigg|_{x=0}^{\sqrt{2-2y^2}}$$
$$= 2 - 4y^2 + 2y^4$$

Triple integral: $\displaystyle\iiint_{\mathcal{W}} x\,dV = \int_0^1 (2 - 4y^2 + 2y^4)\,dy = 2 - \frac{4}{3} + \frac{2}{5} = \frac{16}{15}$ ■

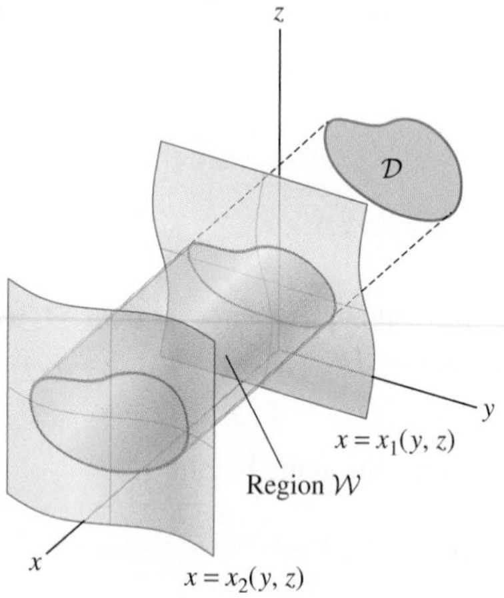

FIGURE 6 $\mathcal{D}$ is the projection of $\mathcal{W}$ onto the yz-plane.

So far, we have evaluated triple integrals by projecting the region $\mathcal{W}$ onto a domain $\mathcal{D}$ in the xy-plane. We can integrate equally well by projecting onto domains in the xz- or yz-plane. For example, if $\mathcal{W}$ is the simple region between the graphs of $x = x_1(y, z)$ and $x = x_2(y, z)$ lying over a domain $\mathcal{D}$ in the yz-plane (Figure 6), then

$$\iiint_{\mathcal{W}} f(x, y, z)\,dV = \iint_{\mathcal{D}} \left(\int_{x=x_1(y,z)}^{x_2(y,z)} f(x, y, z)\,dx \right) dA$$

■ **EXAMPLE 5 Writing a Triple Integral in Three Ways** The region $\mathcal{W}$ in Figure 7 is bounded by

$$z = 4 - y^2, \qquad y = 2x, \qquad z = 0, \qquad x = 0$$

Express $\displaystyle\iiint_{\mathcal{W}} xyz\,dV$ as an iterated integral in three ways, by projecting onto each of the three coordinate planes (but do not evaluate).

Solution We consider each coordinate plane separately.

Step 1. **The xy-plane.**
The upper face $z = 4 - y^2$ intersects the first quadrant of the xy-plane ($z = 0$) in the line $y = 2$ [Figure 7(A)]. Therefore, the projection of $\mathcal{W}$ onto the xy-plane is a triangle $\mathcal{D}$ defined by $0 \le x \le 1$, $2x \le y \le 2$, and

$$\mathcal{W}: 0 \le x \le 1, \quad 2x \le y \le 2, \quad 0 \le z \le 4 - y^2$$

$$\iiint_{\mathcal{W}} xyz\,dV = \int_{x=0}^{1} \int_{y=2x}^{2} \int_{z=0}^{4-y^2} xyz\,dz\,dy\,dx$$

You can check that all three ways of writing the triple integral in Example 5 yield the same answer:

$$\iiint_{\mathcal{W}} xyz\,dV = \frac{2}{3}$$

3

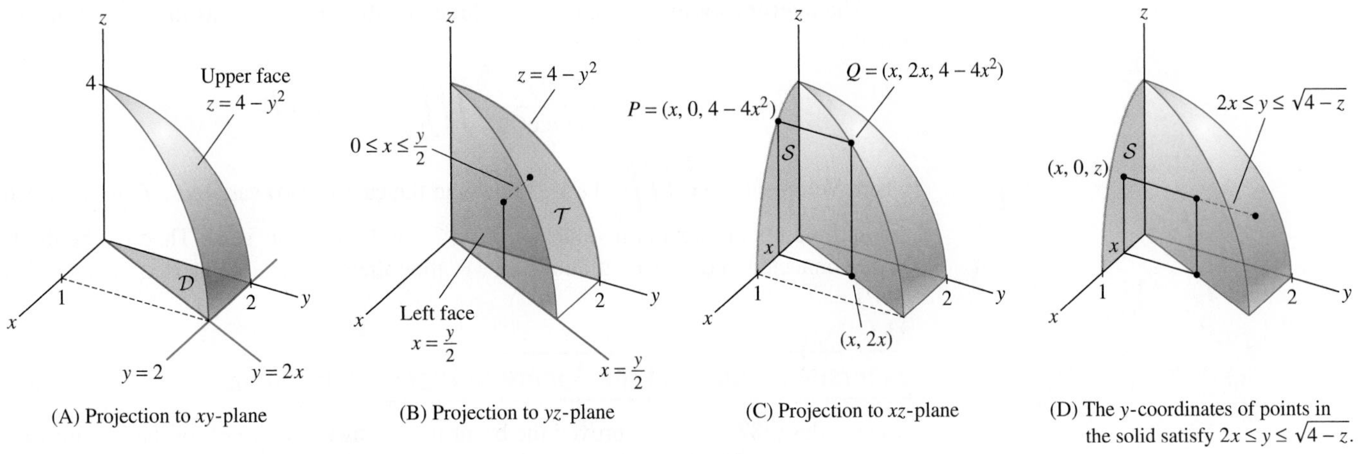

FIGURE 7

Step 2. **The yz-plane.**

The projection of $\mathcal{W}$ onto the yz-plane is the domain $\mathcal{T}$ [Figure 7(B)]:

$$\mathcal{T}: 0 \le y \le 2, \quad 0 \le z \le 4 - y^2$$

The region $\mathcal{W}$ consists of all points lying between $\mathcal{T}$ and the "left face" $x = \frac{1}{2}y$. In other words, the x-coordinate must satisfy $0 \le x \le \frac{1}{2}y$. Thus,

$$\mathcal{W}: 0 \le y \le 2, \quad 0 \le z \le 4 - y^2, \qquad 0 \le x \le \frac{1}{2}y$$

$$\iiint_{\mathcal{W}} xyz\, dV = \int_{y=0}^{2} \int_{z=0}^{4-y^2} \int_{x=0}^{y/2} xyz\, dx\, dz\, dy$$

Step 3. **The xz-plane.**

The challenge in this case is to determine the projection of $\mathcal{W}$ onto the xz-plane, that is, the region S in Figure 7(C). We need to find the equation of the boundary curve of S. A point P on this curve is the projection of a point $Q = (x, y, z)$ on the boundary of the left face. Since Q lies on both the plane $y = 2x$ and the surface $z = 4 - y^2$, $Q = (x, 2x, 4 - 4x^2)$. The projection of Q is $P = (x, 0, 4 - 4x^2)$. We see that the projection of $\mathcal{W}$ onto the xz-plane is the domain

$$S: 0 \le x \le 1, \quad 0 \le z \le 4 - 4x^2$$

This gives us limits for x and z variables, so the triple integral can be written

$$\iiint_{\mathcal{W}} xyz\, dV = \int_{x=0}^{1} \int_{z=0}^{4-4x^2} \int_{y=??}^{??} xyz\, dy\, dz\, dx$$

What are the limits for y? The equation of the upper face $z = 4 - y^2$ can be written $y = \sqrt{4 - z}$. Referring to Figure 7(D), we see that $\mathcal{W}$ is bounded by the left face $y = 2x$ and the upper face $y = \sqrt{4 - z}$. In other words, the y-coordinate of a point in $\mathcal{W}$ satisfies

$$2x \le y \le \sqrt{4 - z}$$

Now we can write the triple integral as the following iterated integral:

$$\iiint_{\mathcal{W}} xyz\, dV = \int_{x=0}^{1} \int_{z=0}^{4-4x^2} \int_{y=2x}^{\sqrt{4-z}} xyz\, dy\, dz\, dx \qquad \blacksquare$$

The **average value** of a function of three variables is defined as in the case of two variables:

$$\overline{f} = \frac{1}{\text{Volume}(\mathcal{W})} \iiint_{\mathcal{W}} f(x, y, z)\, dV \qquad \boxed{4}$$

where $\text{Volume}(\mathcal{W}) = \iiint_{\mathcal{W}} 1\, dV$. And, as in the case of two variables, $\overline{f}$ lies between the minimum and maximum values of f on $\mathcal{D}$, and the Mean Value Theorem holds: If $\mathcal{W}$ is connected and f is continuous on $\mathcal{W}$, then there exists a point $P \in \mathcal{W}$ such that $f(P) = \overline{f}$.

Excursion: Volume of the Sphere in Higher Dimensions

Archimedes (287–212 BCE) proved the beautiful formula $V = \frac{4}{3}\pi r^3$ for the volume of a sphere nearly 2000 years before calculus was invented, by means of a brilliant geometric argument showing that the volume of a sphere is equal to two-thirds the volume of the circumscribed cylinder. According to Plutarch (ca. 45–120 CE), Archimedes valued this achievement so highly that he requested that a sphere with circumscribed cylinder be engraved on his tomb.

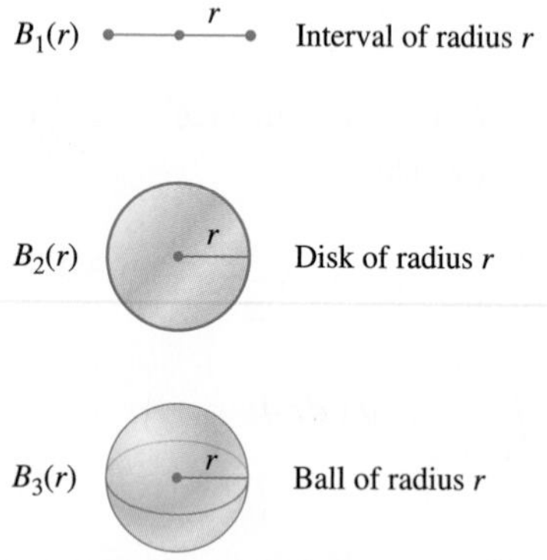

FIGURE 8 Balls of radius r in dimensions $n = 1, 2, 3$.

We can use integration to generalize Archimedes' formula to n dimensions. The ball of radius r in $\mathbf{R}^n$, denoted $B_n(r)$, is the set of points $(x_1, \dots, x_n)$ in $\mathbf{R}^n$ such that

$$x_1^2 + x_2^2 + \cdots + x_n^2 \le r^2$$

The balls $B_n(r)$ in dimensions 1, 2, and 3 are the interval, disk, and ball shown in Figure 8. In dimensions $n \ge 4$, the ball $B_n(r)$ is difficult, if not impossible, to visualize, but we can compute its volume. Denote this volume by $V_n(r)$. For $n = 1$, the "volume" $V_1(r)$ is the length of the interval $B_1(r)$, and for $n = 2$, $V_2(r)$ is the area of the disk $B_2(r)$. We know that

$$V_1(r) = 2r, \qquad V_2(r) = \pi r^2, \qquad V_3(r) = \frac{4}{3}\pi r^3$$

For $n \ge 4$, $V_n(r)$ is sometimes called the **hypervolume**.

The key idea is to determine $V_n(r)$ from the formula for $V_{n-1}(r)$ by integrating cross-sectional volume. Consider the case $n = 3$, where the horizontal slice at height $z = c$ is a two-dimensional ball (a disk) of radius $\sqrt{r^2 - c^2}$ (Figure 9). The volume $V_3(r)$ is equal to the integral of these horizontal slices:

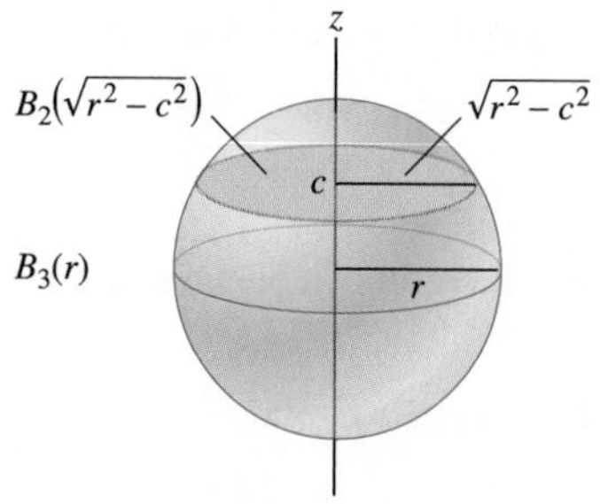

FIGURE 9 The volume $V_3(r)$ is the integral of cross-sectional area $V_2(\sqrt{r^2 - c^2})$.

$$V_3(r) = \int_{z=-r}^{r} V_2\left(\sqrt{r^2 - z^2}\right) dz = \int_{z=-r}^{r} \pi(r^2 - z^2)\, dz = \frac{4}{3}\pi r^3$$

By induction, we can show that for all $n \ge 1$, there is a constant A_n (equal to the volume of the n-dimensional unit ball) such that

$$\boxed{V_n(r) = A_n r^n} \qquad \boxed{5}$$

The slice of $B_n(r)$ at height $x_n = c$ has equation

$$x_1^2 + x_2^2 + \cdots + x_{n-1}^2 + c^2 = r^2$$

This slice is the ball $B_{n-1}(\sqrt{r^2 - c^2})$ of radius $\sqrt{r^2 - c^2}$, and $V_n(r)$ is obtained by integrating the volume of these slices:

$$V_n(r) = \int_{x_n=-r}^{r} V_{n-1}\left(\sqrt{r^2 - x_n^2}\right) dx_n = A_{n-1} \int_{x_n=-r}^{r} \left(\sqrt{r^2 - x_n^2}\right)^{n-1} dx_n$$

Using the substitution $x_n = r\sin\theta$ and $dx_n = r\cos\theta\, d\theta$, we have

$$V_n(r) = A_{n-1}r^n \int_{-\pi/2}^{\pi/2} \cos^n\theta\, d\theta = A_{n-1}C_n r^n$$

where $C_n = \int_{\theta=-\pi/2}^{\pi/2} \cos^n\theta\, d\theta$. This proves Eq. (5) with

$$\boxed{A_n = A_{n-1}C_n} \qquad (6)$$

In Exercise 39, you are asked to use Integration by Parts to verify the relation

$$C_n = \left(\frac{n-1}{n}\right) C_{n-2} \qquad (7)$$

It is easy to check directly that $C_0 = \pi$ and $C_1 = 2$. By Eq. (7), $C_2 = \frac{1}{2}C_0 = \frac{\pi}{2}$, $C_3 = \frac{2}{3}(2) = \frac{4}{3}$, and so on. Here are the first few values of C_n:

n	0	1	2	3	4	5	6	7
C_n	π	2	$\frac{\pi}{2}$	$\frac{4}{3}$	$\frac{3\pi}{8}$	$\frac{16}{15}$	$\frac{5\pi}{16}$	$\frac{32}{35}$

TABLE 1

n	A_n
1	2
2	$\pi \approx 3.14$
3	$\frac{4}{3}\pi \approx 4.19$
4	$\frac{\pi^2}{2} \approx 4.93$
5	$\mathbf{\frac{8\pi^2}{15} \approx 5.26}$
6	$\frac{\pi^3}{6} \approx 5.17$
7	$\frac{16\pi^3}{105} \approx 4.72$

We also have $A_1 = 2$ and $A_2 = \pi$, so we can use the values of C_n together with Eq. (6) to obtain the values of A_n in Table 1. We see, for example, that the ball of radius r in four dimensions has volume $V_4(r) = \frac{1}{2}\pi^2 r^4$. The general formula depends on whether n is even or odd. Using induction and formulas (6) and (7), we can prove that

$$\boxed{A_{2m} = \frac{\pi^m}{m!}, \qquad A_{2m+1} = \frac{2^{m+1}\pi^m}{1\cdot 3\cdot 5\cdot\cdots\cdot(2m+1)}}$$

This sequence of numbers A_n has a curious property. Setting $r = 1$ in Eq. (5), we see that A_n is the volume of the unit ball in n dimensions. From Table 1, it appears that the volumes increase up to dimension 5 and then begin to decrease. In Exercise 40, you are asked to verify that the five-dimensional unit ball has the largest volume. Furthermore, the volumes A_n tend to 0 as $n \to \infty$.

16.3 SUMMARY

- The triple integral over a box $\mathcal{B} = [a,b]\times[c,d]\times[p,q]$ is equal to the iterated integral

$$\iiint_{\mathcal{B}} f(x,y,z)\, dV = \int_{x=a}^{b}\int_{y=c}^{d}\int_{z=p}^{q} f(x,y,z)\, dz\, dy\, dx$$

The iterated integral may be written in any one of six possible orders—for example,

$$\int_{z=p}^{q}\int_{y=c}^{d}\int_{x=a}^{b} f(x,y,z)\, dx\, dy\, dz$$

- A *simple region* $\mathcal{W}$ in $\mathbf{R}^3$ is a region consisting of the points (x,y,z) between two surfaces $z = z_1(x,y)$ and $z = z_2(x,y)$, where $z_1(x,y) \le z_2(x,y)$, lying over a domain $\mathcal{D}$ in the xy-plane. In other words, $\mathcal{W}$ is defined by

$$(x,y) \in \mathcal{D}, \qquad z_1(x,y) \le z \le z_2(x,y)$$

The triple integral over $\mathcal{W}$ is equal to an iterated integral:

$$\iiint_{\mathcal{W}} f(x,y,z)\,dV = \iint_{\mathcal{D}} \left(\int_{z=z_1(x,y)}^{z_2(x,y)} f(x,y,z)\,dz \right) dA$$

- The *average value* of $f(x,y,z)$ on a region $\mathcal{W}$ of volume V is the quantity

$$\overline{f} = \frac{1}{V} \iiint_{\mathcal{W}} f(x,y,z)\,dV, \qquad V = \iiint_{\mathcal{W}} 1\,dV$$

16.3 EXERCISES

Preliminary Questions

1. Which of (a)–(c) is not equal to $\int_0^1 \int_3^4 \int_6^7 f(x,y,z)\,dz\,dy\,dx$?

(a) $\int_6^7 \int_0^1 \int_3^4 f(x,y,z)\,dy\,dx\,dz$

(b) $\int_3^4 \int_0^1 \int_6^7 f(x,y,z)\,dz\,dx\,dy$

(c) $\int_0^1 \int_3^4 \int_6^7 f(x,y,z)\,dx\,dz\,dy$

2. Which of the following is not a meaningful triple integral?

(a) $\int_0^1 \int_0^x \int_{x+y}^{2x+y} e^{x+y+z}\,dz\,dy\,dx$

(b) $\int_0^1 \int_0^z \int_{x+y}^{2x+y} e^{x+y+z}\,dz\,dy\,dx$

3. Describe the projection of the region of integration $\mathcal{W}$ onto the xy-plane:

(a) $\int_0^1 \int_0^x \int_0^{x^2+y^2} f(x,y,z)\,dz\,dy\,dx$

(b) $\int_0^1 \int_0^{\sqrt{1-x^2}} \int_2^4 f(x,y,z)\,dz\,dy\,dx$

Exercises

In Exercises 1–8, evaluate $\iiint_{\mathcal{B}} f(x,y,z)\,dV$ *for the specified function f and box* $\mathcal{B}$.

1. $f(x,y,z) = z^4;\quad 2 \le x \le 8,\quad 0 \le y \le 5,\quad 0 \le z \le 1$

2. $f(x,y,z) = xz^2;\quad [-2,3] \times [1,3] \times [1,4]$

3. $f(x,y,z) = xe^{y-2z};\quad 0 \le x \le 2,\quad 0 \le y \le 1,\quad 0 \le z \le 1$

4. $f(x,y,z) = \dfrac{x}{(y+z)^2};\quad [0,2] \times [2,4] \times [-1,1]$

5. $f(x,y,z) = (x-y)(y-z);\quad [0,1] \times [0,3] \times [0,3]$

6. $f(x,y,z) = \dfrac{z}{x};\quad 1 \le x \le 3,\quad 0 \le y \le 2,\quad 0 \le z \le 4$

7. $f(x,y,z) = (x+z)^3;\quad [0,a] \times [0,b] \times [0,c]$

8. $f(x,y,z) = (x+y-z)^2;\quad [0,a] \times [0,b] \times [0,c]$

In Exercises 9–14, evaluate $\iiint_{\mathcal{W}} f(x,y,z)\,dV$ *for the function f and region* $\mathcal{W}$ *specified.*

9. $f(x,y,z) = x+y;\quad \mathcal{W}: y \le z \le x,\quad 0 \le y \le x,\quad 0 \le x \le 1$

10. $f(x,y,z) = e^{x+y+z};\quad \mathcal{W}: 0 \le z \le 1,\quad 0 \le y \le x,\quad 0 \le x \le 1$

11. $f(x,y,z) = xyz;\quad \mathcal{W}: 0 \le z \le 1,\quad 0 \le y \le \sqrt{1-x^2},\quad 0 \le x \le 1$

12. $f(x,y,z) = x;\quad \mathcal{W}: x^2 + y^2 \le z \le 4$

13. $f(x,y,z) = e^z;\quad \mathcal{W}: x+y+z \le 1,\quad x \ge 0,\quad y \ge 0,\quad z \ge 0$

14. $f(x,y,z) = z;\quad \mathcal{W}: x^2 \le y \le 2,\quad 0 \le x \le 1,\quad x - y \le z \le x + y$

15. Calculate the integral of $f(x,y,z) = z$ over the region $\mathcal{W}$ in Figure 10 below the hemisphere of radius 3 and lying over the triangle $\mathcal{D}$ in the xy-plane bounded by $x = 1$, $y = 0$, and $x = y$.

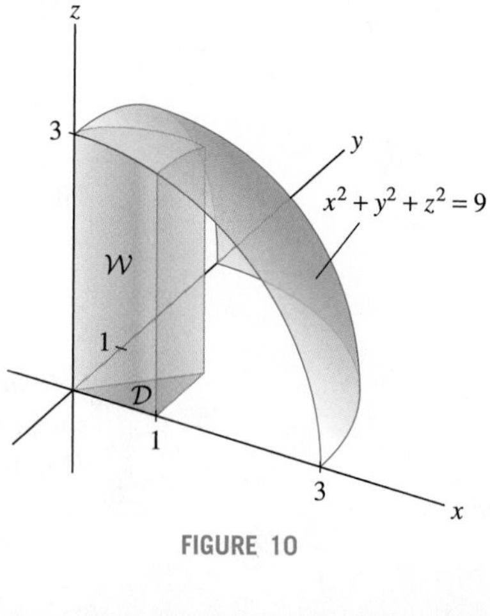

FIGURE 10

16. Calculate the integral of $f(x, y, z) = e^z$ over the tetrahedron $\mathcal{W}$ in Figure 11.

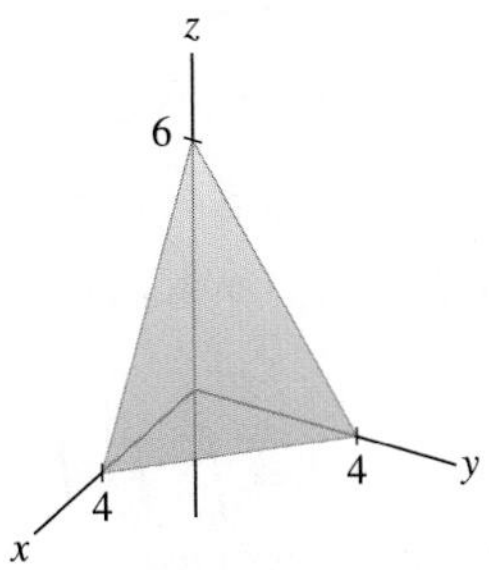

FIGURE 11

17. Integrate $f(x, y, z) = x$ over the region in the first octant ($x \geq 0, y \geq 0, z \geq 0$) above $z = y^2$ and below $z = 8 - 2x^2 - y^2$.

18. Compute the integral of $f(x, y, z) = y^2$ over the region within the cylinder $x^2 + y^2 = 4$ where $0 \leq z \leq y$.

19. Find the triple integral of the function z over the ramp in Figure 12. Here, z is the height above the ground.

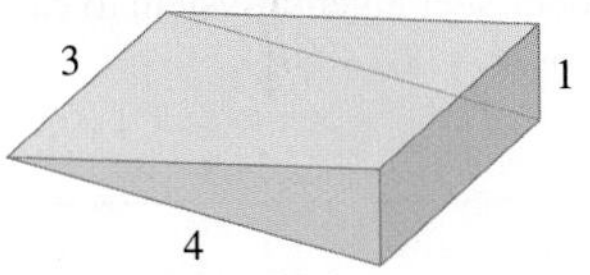

FIGURE 12

20. Find the volume of the solid in $\mathbf{R}^3$ bounded by $y = x^2$, $x = y^2$, $z = x + y + 5$, and $z = 0$.

21. Find the volume of the solid in the octant $x \geq 0, y \geq 0, z \geq 0$ bounded by $x + y + z = 1$ and $x + y + 2z = 1$.

22. Calculate $\iiint_{\mathcal{W}} y\, dV$, where $\mathcal{W}$ is the region above $z = x^2 + y^2$ and below $z = 5$, and bounded by $y = 0$ and $y = 1$.

23. Evaluate $\iiint_{\mathcal{W}} xz\, dV$, where $\mathcal{W}$ is the domain bounded by the elliptic cylinder $\frac{x^2}{4} + \frac{y^2}{9} = 1$ and the sphere $x^2 + y^2 + z^2 = 16$ in the first octant $x \geq 0, y \geq 0, z \geq 0$ (Figure 13).

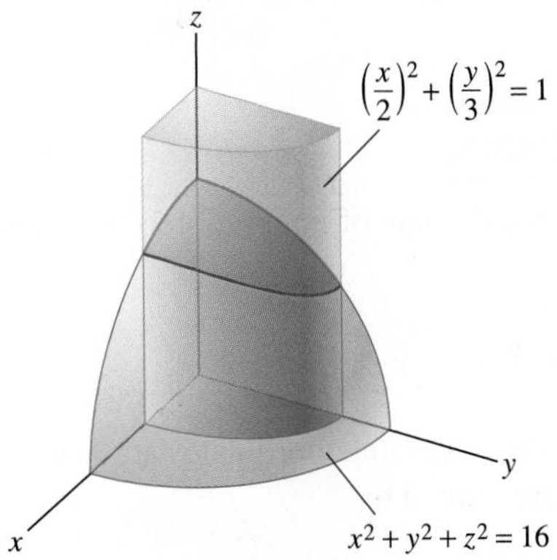

FIGURE 13

24. Describe the domain of integration and evaluate:

$$\int_0^3 \int_0^{\sqrt{9-x^2}} \int_0^{\sqrt{9-x^2-y^2}} xy\, dz\, dy\, dx$$

25. Describe the domain of integration of the following integral:

$$\int_{-2}^{2} \int_{-\sqrt{4-z^2}}^{\sqrt{4-z^2}} \int_{1}^{\sqrt{5-x^2-z^2}} f(x, y, z)\, dy\, dx\, dz$$

26. Let $\mathcal{W}$ be the region below the paraboloid

$$x^2 + y^2 = z - 2$$

that lies above the part of the plane $x + y + z = 1$ in the first octant ($x \geq 0, y \geq 0, z \geq 0$). Express

$$\iiint_{\mathcal{W}} f(x, y, z)\, dV$$

as an iterated integral (for an arbitrary function f).

27. In Example 5, we expressed a triple integral as an iterated integral in the three orders

$$dz\, dy\, dx, \quad dx\, dz\, dy, \quad \text{and} \quad dy\, dz\, dx$$

Write this integral in the three other orders:

$$dz\, dx\, dy, \quad dx\, dy\, dz, \quad \text{and} \quad dy\, dx\, dz$$

28. Let $\mathcal{W}$ be the region bounded by

$$y + z = 2, \quad 2x = y, \quad x = 0, \quad \text{and } z = 0$$

(Figure 14). Express and evaluate the triple integral of $f(x, y, z) = z$ by projecting $\mathcal{W}$ onto the:

(a) xy-plane **(b)** yz-plane **(c)** xz-plane

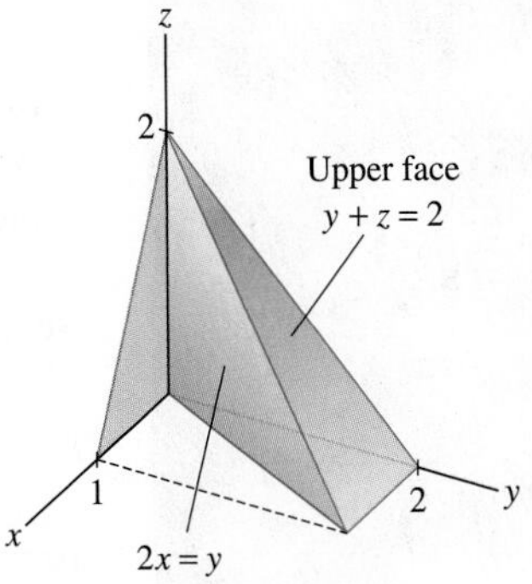

FIGURE 14

29. Let

$$\mathcal{W} = \left\{(x, y, z) : \sqrt{x^2 + y^2} \leq z \leq 1\right\}$$

(see Figure 15). Express $\iiint_{\mathcal{W}} f(x, y, z)\, dV$ as an iterated integral in the order $dz\, dy\, dx$ (for an arbitrary function f).

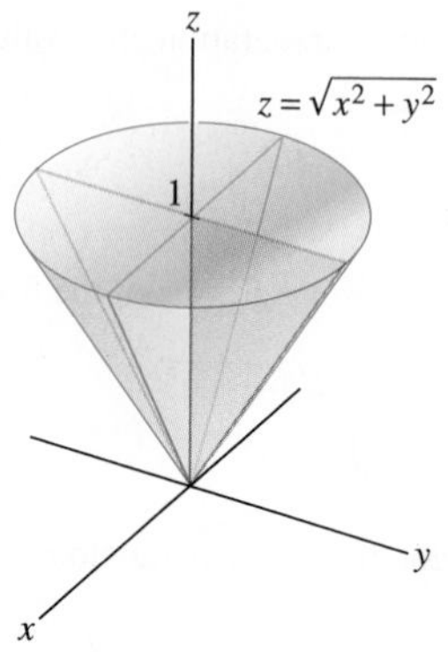

FIGURE 15

30. Repeat Exercise 29 for the order $dx\,dy\,dz$.

31. Let $\mathcal{W}$ be the region bounded by $z = 1 - y^2$, $y = x^2$, and the planes $z = 0$, $y = 1$. Calculate the volume of $\mathcal{W}$ as a triple integral in the order $dz\,dy\,dx$.

32. Calculate the volume of the region $\mathcal{W}$ in Exercise 31 as a triple integral in the following orders:

(a) $dx\,dz\,dy$ (b) $dy\,dz\,dx$

In Exercises 33–36, compute the average value of $f(x, y, z)$ over the region $\mathcal{W}$.

33. $f(x, y, z) = xy\sin(\pi z)$; $\mathcal{W} = [0, 1] \times [0, 1] \times [0, 1]$

34. $f(x, y, z) = xyz$; $\mathcal{W} : 0 \le z \le y \le x \le 1$

35. $f(x, y, z) = e^y$; $\mathcal{W} : 0 \le y \le 1 - x^2$, $0 \le z \le x$

36. $f(x, y, z) = x^2 + y^2 + z^2$; $\mathcal{W}$ bounded by the planes $2y + z = 1$, $x = 0$, $x = 1$, $z = 0$, and $y = 0$.

In Exercises 37 and 38, let $I = \int_0^1 \int_0^1 \int_0^1 f(x, y, z)\,dV$ and let $S_{N,N,N}$ be the Riemann sum approximation

$$S_{N,N,N} = \frac{1}{N^3} \sum_{i=1}^{N} \sum_{j=1}^{N} \sum_{k=1}^{N} f\left(\frac{i}{N}, \frac{j}{N}, \frac{k}{N}\right)$$

37. CAS Calculate $S_{N,N,N}$ for $f(x, y, z) = e^{x^2 - y - z}$ for $N = 10$, 20, 30. Then evaluate I and find an N such that $S_{N,N,N}$ approximates I to two decimal places.

38. CAS Calculate $S_{N,N,N}$ for $f(x, y, z) = \sin(xyz)$ for $N = 10$, 20, 30. Then use a computer algebra system to calculate I numerically and estimate the error $|I - S_{N,N,N}|$.

Further Insights and Challenges

39. Use Integration by Parts to verify Eq. (7).

40. Compute the volume A_n of the unit ball in $\mathbf{R}^n$ for $n = 8, 9, 10$. Show that $C_n \le 1$ for $n \ge 6$ and use this to prove that of all unit balls, the five-dimensional ball has the largest volume. Can you explain why A_n tends to 0 as $n \to \infty$?

16.4 Integration in Polar, Cylindrical, and Spherical Coordinates

In single-variable calculus, a well-chosen substitution (also called a change of variables) often transforms a complicated integral into a simpler one. Change of variables is also useful in multivariable calculus, but the emphasis is different. In the multivariable case, we are usually interested in simplifying not just the integrand, but also the domain of integration.

This section treats three of the most useful changes of variables, in which an integral is expressed in polar, cylindrical, or spherical coordinates. The general Change of Variables Formula is discussed in Section 16.6.

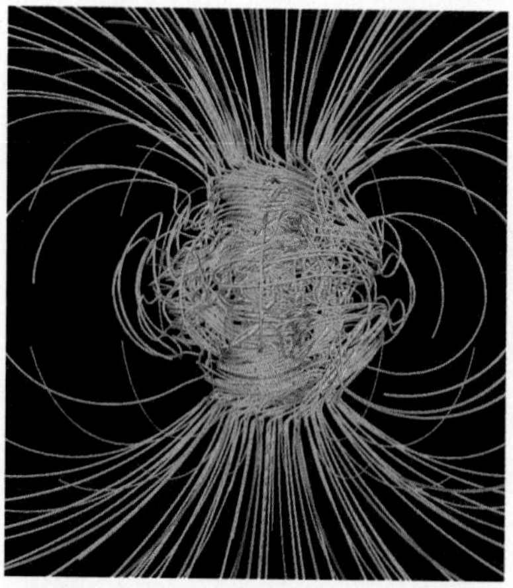

FIGURE 1 Spherical coordinates are used in mathematical models of the earth's magnetic field. This computer simulation, based on the Glatzmaier–Roberts model, shows the magnetic lines of force, representing inward and outward directed field lines in blue and yellow, respectively.

Double Integrals in Polar Coordinates

Polar coordinates are convenient when the domain of integration is an angular sector or a **polar rectangle** (Figure 2):

$$\mathcal{R} : \theta_1 \le \theta \le \theta_2, \quad r_1 \le r \le r_2 \qquad \boxed{1}$$

We assume throughout that $r_1 \ge 0$ and that all radial coordinates are nonnegative. Recall that rectangular and polar coordinates are related by

$$x = r\cos\theta, \qquad y = r\sin\theta$$

Eq. (2) expresses the integral of $f(x, y)$ over the polar rectangle in Figure 2 as the integral of a new function $rf(r\cos\theta, r\sin\theta)$ over the ordinary rectangle $[\theta_1, \theta_2] \times [r_1, r_2]$. In this sense, the change of variables "simplifies" the domain of integration.

Thus, we write a function $f(x, y)$ in polar coordinates as $f(r\cos\theta, r\sin\theta)$. The Change of Variables Formula for a polar rectangle $\mathcal{R}$ is:

$$\iint_{\mathcal{R}} f(x, y)\, dA = \int_{\theta_1}^{\theta_2} \int_{r_1}^{r_2} f(r\cos\theta, r\sin\theta)\, r\, dr\, d\theta \qquad \boxed{2}$$

Notice the extra factor r in the integrand on the right.

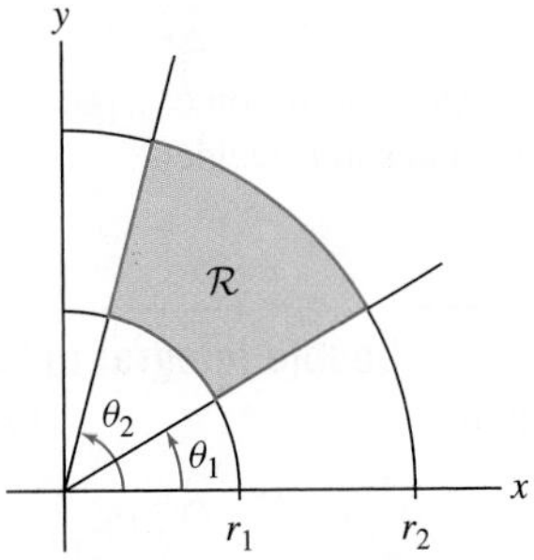

FIGURE 2 Polar rectangle.

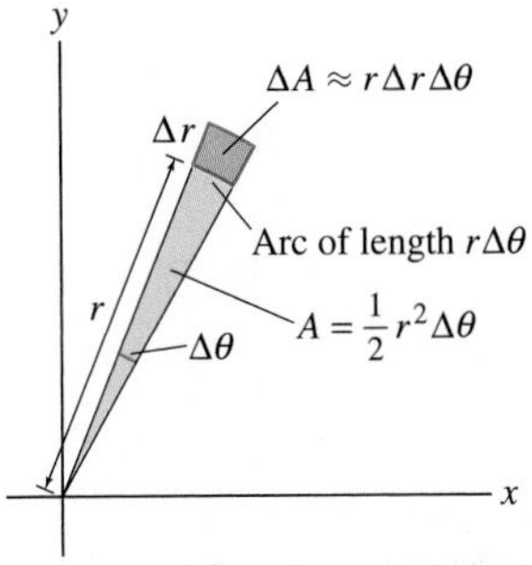

FIGURE 3 Small polar rectangle.

To derive Eq. (2), the key step is to estimate the area ΔA of the small polar rectangle shown in Figure 3. If Δr and $\Delta\theta$ are small, then this polar rectangle is very nearly an ordinary rectangle of sides Δr and $r\Delta\theta$, and therefore $\Delta A \approx r\,\Delta r\,\Delta\theta$. In fact, ΔA is the difference of areas of two sectors:

REMINDER The length of the arc subtended by an angle θ is θ, and the area of a sector is $\frac{1}{2}r^2\theta$.

$$\Delta A = \frac{1}{2}(r + \Delta r)^2\, \Delta\theta - \frac{1}{2}r^2\, \Delta\theta = r(\Delta r\, \Delta\theta) + \frac{1}{2}(\Delta r)^2\Delta\theta \approx r\, \Delta r\, \Delta\theta$$

The error in our approximation is the term $\frac{1}{2}(\Delta r)^2\Delta\theta$, which has smaller order of magnitude than $\Delta r\, \Delta\theta$ when Δr and $\Delta\theta$ are both small.

Now, decompose $\mathcal{R}$ into an $N \times M$ grid of small polar subrectangles $\mathcal{R}_{ij}$ as in Figure 4, and choose a sample point P_{ij} in $\mathcal{R}_{ij}$. If $\mathcal{R}_{ij}$ is small and $f(x, y)$ is continuous, then

REMINDER In Eq. (3). we use the approximation (10) in Section 16.2: If f is continuous and $\mathcal{D}$ is a small domain,

$$\iint_{\mathcal{D}} f(x, y)\, dA \approx f(P)\,\text{Area}(\mathcal{D})$$

where P is any sample point in $\mathcal{D}$.

$$\iint_{\mathcal{R}_{ij}} f(x, y)\, dx\, dy \approx f(P_{ij})\,\text{Area}(\mathcal{R}_{ij}) \approx f(P_{ij})\, r_{ij}\, \Delta r\, \Delta\theta \qquad \boxed{3}$$

Note that each polar rectangle $\mathcal{R}_{ij}$ has angular width $\Delta\theta = (\theta_2 - \theta_1)/N$ and radial width $\Delta r = (r_2 - r_1)/M$. The integral over $\mathcal{R}$ is the sum:

$$\iint_{\mathcal{R}} f(x, y)\, dx\, dy = \sum_{i=1}^{N}\sum_{j=1}^{M} \iint_{\mathcal{R}_{ij}} f(x, y)\, dx\, dy$$

$$\approx \sum_{i=1}^{N}\sum_{j=1}^{M} f(P_{ij})\,\text{Area}(\mathcal{R}_{ij})$$

$$\approx \sum_{i=1}^{N}\sum_{j=1}^{M} f(r_{ij}\cos\theta_{ij}, r_{ij}\sin\theta_{ij})\, r_{ij}\, \Delta r\, \Delta\theta$$

This is a Riemann sum for the double integral of $rf(r\cos\theta, r\sin\theta)$ over the region $r_1 \le r \le r_2$, $\theta_1 \le \theta \le \theta_2$, and we can prove that it approaches the double integral as $N, M \to \infty$. A similar derivation is valid for domains (Figure 5) that can be described as the region between two polar curves $r = r_1(\theta)$ and $r = r_2(\theta)$. This gives us Theorem 1.

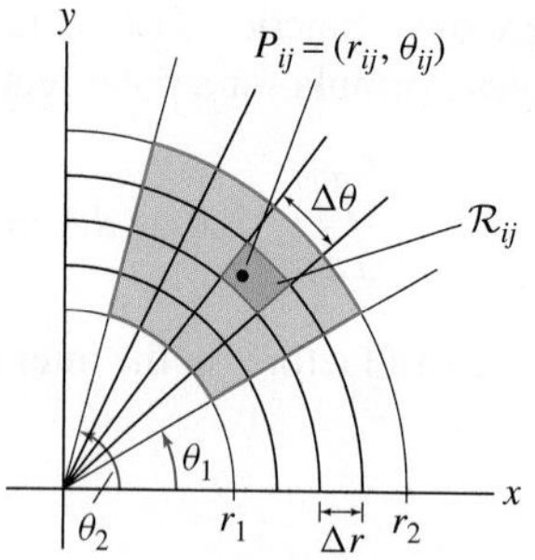

FIGURE 4 Decomposition of a polar rectangle into subrectangles.

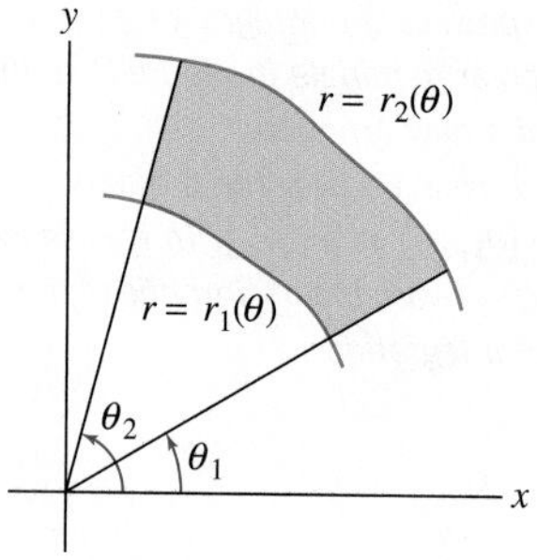

FIGURE 5 General polar region.

> **THEOREM 1 Double Integral in Polar Coordinates** For a continuous function f on the domain
>
> $$\mathcal{D} : \theta_1 \le \theta \le \theta_2, \quad r_1(\theta) \le r \le r_2(\theta)$$
>
> $$\iint_{\mathcal{D}} f(x, y)\, dA = \int_{\theta_1}^{\theta_2} \int_{r=r_1(\theta)}^{r_2(\theta)} f(r\cos\theta, r\sin\theta)\, r\, dr\, d\theta \qquad \boxed{4}$$

Eq. (4) is summarized in the symbolic expression for the "area element" dA in polar coordinates:

$$dA = r\, dr\, d\theta$$

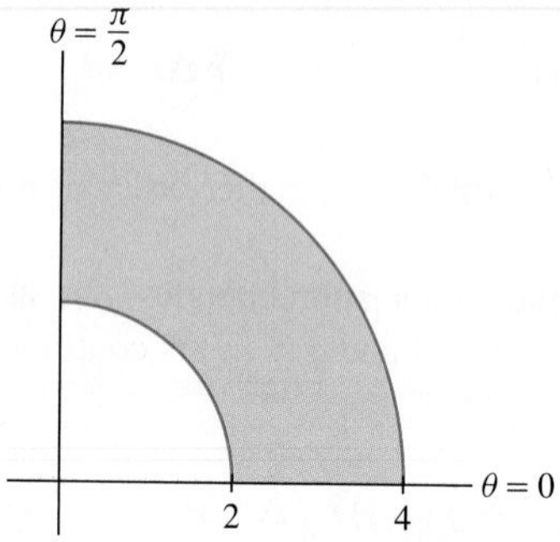

FIGURE 6 Quarter annulus $0 \le \theta \le \frac{\pi}{2}$, $2 \le r \le 4$.

EXAMPLE 1 Compute $\iint_{\mathcal{D}} (x + y)\, dA$, where $\mathcal{D}$ is the quarter annulus in Figure 6.

Solution

Step 1. **Describe $\mathcal{D}$ and f in polar coordinates.**

The quarter annulus $\mathcal{D}$ is defined by the inequalities (Figure 6)

$$\mathcal{D} : 0 \le \theta \le \frac{\pi}{2}, \qquad 2 \le r \le 4$$

In polar coordinates,

$$f(x, y) = x + y = r\cos\theta + r\sin\theta = r(\cos\theta + \sin\theta)$$

Step 2. **Change variables and evaluate.**

To write the integral in polar coordinates, we replace dA by $r\, dr\, d\theta$:

$$\iint_{\mathcal{D}} (x + y)\, dA = \int_0^{\pi/2} \int_{r=2}^{4} r(\cos\theta + \sin\theta)\, r\, dr\, d\theta$$

The inner integral is

$$\int_{r=2}^{4} (\cos\theta + \sin\theta) r^2\, dr = (\cos\theta + \sin\theta)\left(\frac{4^3}{3} - \frac{2^3}{3}\right) = \frac{56}{3}(\cos\theta + \sin\theta)$$

and

$$\iint_{\mathcal{D}} (x + y)\, dA = \frac{56}{3} \int_0^{\pi/2} (\cos\theta + \sin\theta)\, d\theta = \frac{56}{3}(\sin\theta - \cos\theta)\Big|_0^{\pi/2} = \frac{112}{3} \quad ■$$

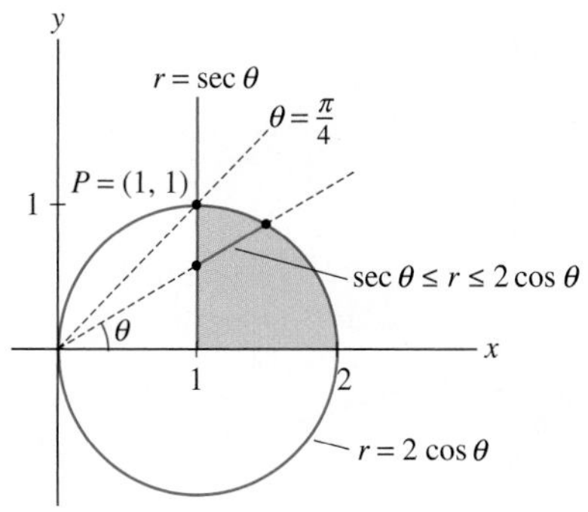

FIGURE 7

■ **EXAMPLE 2** Calculate $\iint_{\mathcal{D}} (x^2 + y^2)^{-2}\, dA$ for the shaded domain $\mathcal{D}$ in Figure 7.

Solution

***Step 1.* Describe $\mathcal{D}$ and f in polar coordinates.**

The quarter circle lies in the angular sector $0 \le \theta \le \frac{\pi}{4}$ because the line through $P = (1, 1)$ makes an angle of $\frac{\pi}{4}$ with the x-axis (Figure 7).

To determine the limits on r, recall from Section 11.3 (Examples 5 and 7) that:

- The vertical line $x = 1$ has polar equation $r \cos\theta = 1$ or $r = \sec\theta$.
- The circle of radius 1 and center $(1, 0)$ has polar equation $r = 2\cos\theta$.

Therefore, a ray of angle θ intersects $\mathcal{D}$ in the segment where r ranges from $\sec\theta$ to $2\cos\theta$. In other words, our domain has polar description

$$\mathcal{D} : 0 \le \theta \le \frac{\pi}{4}, \quad \sec\theta \le r \le 2\cos\theta$$

The function in polar coordinates is

$$f(x, y) = (x^2 + y^2)^{-2} = (r^2)^{-2} = r^{-4}$$

***Step 2.* Change variables and evaluate.**

← REMINDER

$$\int \cos^2\theta\, d\theta = \frac{1}{2}\left(\theta + \frac{1}{2}\sin 2\theta\right) + C$$

$$\int \sec^2\theta\, d\theta = \tan\theta + C$$

$$\iint_{\mathcal{D}} (x^2 + y^2)^{-2}\, dA = \int_0^{\pi/4} \int_{r=\sec\theta}^{2\cos\theta} r^{-4}\, r\, dr\, d\theta = \int_0^{\pi/4} \int_{r=\sec\theta}^{2\cos\theta} r^{-3}\, dr\, d\theta$$

The inner integral is

$$\int_{r=\sec\theta}^{2\cos\theta} r^{-3}\, dr = -\frac{1}{2} r^{-2}\Big|_{r=\sec\theta}^{2\cos\theta} = -\frac{1}{8}\sec^2\theta + \frac{1}{2}\cos^2\theta$$

Therefore,

$$\begin{aligned}\iint_{\mathcal{D}} (x^2 + y^2)^{-2}\, dA &= \int_0^{\pi/4} \left(\frac{1}{2}\cos^2\theta - \frac{1}{8}\sec^2\theta\right) d\theta \\ &= \left(\frac{1}{4}\left(\theta + \frac{1}{2}\sin 2\theta\right) - \frac{1}{8}\tan\theta\right)\Big|_0^{\pi/4} \\ &= \frac{1}{4}\left(\frac{\pi}{4} + \frac{1}{2}\sin\frac{\pi}{2}\right) - \frac{1}{8}\tan\frac{\pi}{4} = \frac{\pi}{16}\end{aligned}$$ ■

Triple Integrals in Cylindrical Coordinates

Cylindrical coordinates, introduced in Section 13.7, are useful when the domain has **axial symmetry**—that is, symmetry with respect to an axis. In cylindrical coordinates (r, θ, z), the axis of symmetry is the z-axis. Recall the relations (Figure 8)

$$x = r\cos\theta, \qquad y = r\sin\theta, \qquad z = z$$

To set up a triple integral in cylindrical coordinates, we assume that the domain of integration $\mathcal{W}$ can be described as the region between two surfaces (Figure 9)

$$z_1(r, \theta) \le z \le z_2(r, \theta)$$

lying over a domain $\mathcal{D}$ in the xy-plane with polar description

$$\mathcal{D} : \theta_1 \le \theta \le \theta_2, \quad r_1(\theta) \le r \le r_2(\theta)$$

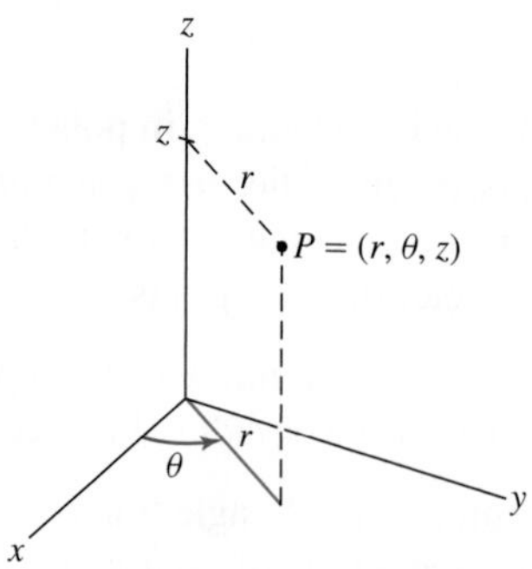

FIGURE 8 Cylindrical coordinates.

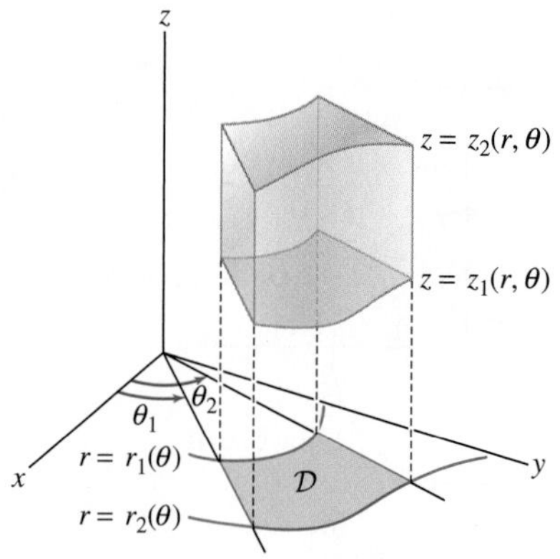

FIGURE 9 Region described in cylindrical coordinates.

A triple integral over $\mathcal{W}$ can be written as an iterated integral (Theorem 2 of Section 16.3):

$$\iiint_{\mathcal{W}} f(x, y, z)\, dV = \iint_{\mathcal{D}} \left(\int_{z=z_1(r,\theta)}^{z_2(r,\theta)} f(x, y, z)\, dz \right) dA$$

By expressing the integral over $\mathcal{D}$ in polar coordinates, we obtain the following Change of Variables Formula.

Eq. (5) is summarized in the symbolic expression for the "volume element" dV in cylindrical coordinates:

$$dV = r\, dz\, dr\, d\theta$$

THEOREM 2 Triple Integrals in Cylindrical Coordinates For a continuous function f on the region

$$\theta_1 \le \theta \le \theta_2, \qquad r_1(\theta) \le r \le r_2(\theta), \qquad z_1(r, \theta) \le z \le z_2(r, \theta),$$

the triple integral $\iiint_{\mathcal{W}} f(x, y, z)\, dV$ is equal to

$$\int_{\theta_1}^{\theta_2} \int_{r=r_1(\theta)}^{r_2(\theta)} \int_{z=z_1(r,\theta)}^{z_2(r,\theta)} f(r\cos\theta, r\sin\theta, z)\, r\, dz\, dr\, d\theta \qquad \boxed{5}$$

EXAMPLE 3 Integrate $f(x, y, z) = z\sqrt{x^2 + y^2}$ over the cylinder $x^2 + y^2 \le 4$ for $1 \le z \le 5$ (Figure 10).

Solution The domain of integration $\mathcal{W}$ lies above the disk of radius 2, so in cylindrical coordinates,

$$\mathcal{W}: 0 \le \theta \le 2\pi, \quad 0 \le r \le 2, \quad 1 \le z \le 5$$

We write the function in cylindrical coordinates:

$$f(x, y, z) = z\sqrt{x^2 + y^2} = zr$$

and integrate with respect to $dV = r\, dz\, dr\, d\theta$. The function f is a product zr, so the resulting triple integral is a product of single integrals:

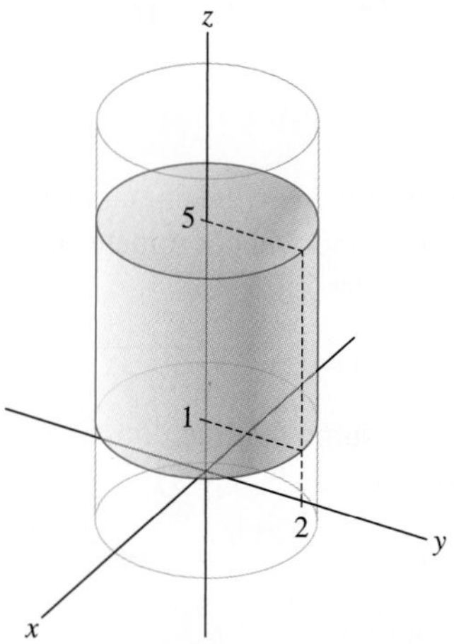

FIGURE 10 The cylinder $x^2 + y^2 \le 4$.

$$\iiint_{\mathcal{W}} z\sqrt{x^2+y^2}\,dV = \int_0^{2\pi}\int_{r=0}^{2}\int_{z=1}^{5} (zr)r\,dz\,dr\,d\theta$$
$$= \left(\int_0^{2\pi} d\theta\right)\left(\int_{r=0}^{2} r^2\,dr\right)\left(\int_{z=1}^{5} z\,dz\right)$$
$$= (2\pi)\left(\frac{2^3}{3}\right)\left(\frac{5^2-1^2}{2}\right) = 64\pi \qquad \blacksquare$$

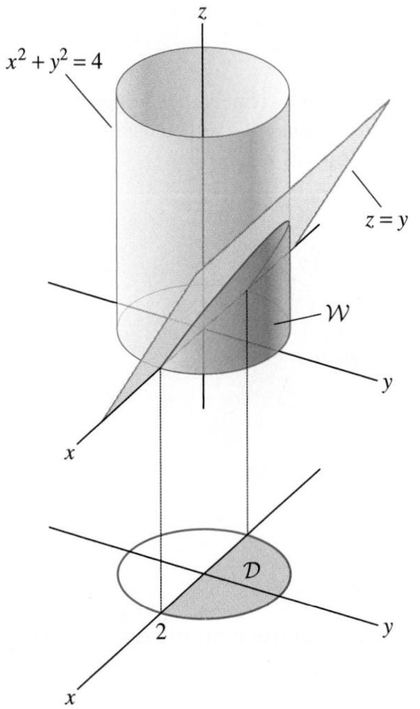

FIGURE 11

EXAMPLE 4 Compute the integral of $f(x, y, z) = z$ over the region $\mathcal{W}$ within the cylinder $x^2 + y^2 \le 4$ where $0 \le z \le y$.

Solution

***Step 1.* Express $\mathcal{W}$ in cylindrical coordinates.**
The condition $0 \le z \le y$ tells us that $y \ge 0$, so $\mathcal{W}$ projects onto the semicircle $\mathcal{D}$ in the xy-plane of radius 2 where $y \ge 0$ shown in Figure 11. In polar coordinates,

$$\mathcal{D} : 0 \le \theta \le \pi, \quad 0 \le r \le 2$$

The z-coordinate in $\mathcal{W}$ varies from $z = 0$ to $z = y$, and in polar coordinates $y = r\sin\theta$, so the region has the description

$$\mathcal{W} : 0 \le \theta \le \pi, \quad 0 \le r \le 2, \quad 0 \le z \le r\sin\theta$$

***Step 2.* Change variables and evaluate.**

$$\iiint_{\mathcal{W}} f(x, y, z)\,dV = \int_0^{\pi}\int_{r=0}^{2}\int_{z=0}^{r\sin\theta} zr\,dz\,dr\,d\theta$$
$$= \int_0^{\pi}\int_{r=0}^{2} \frac{1}{2}(r\sin\theta)^2 r\,dr\,d\theta$$
$$= \frac{1}{2}\left(\int_0^{\pi} \sin^2\theta\,d\theta\right)\left(\int_0^{2} r^3\,dr\right)$$
$$= \frac{1}{2}\left(\frac{\pi}{2}\right)\frac{2^4}{4} = \pi \qquad \blacksquare$$

← REMINDER

$$\int \sin^2\theta\,d\theta = \frac{1}{2}\left(\theta - \frac{1}{2}\sin 2\theta\right) + C$$
$$\int_0^{\pi} \sin^2\theta\,d\theta = \frac{\pi}{2}$$

Triple Integrals in Spherical Coordinates

We noted that the Change of Variables Formula in cylindrical coordinates is summarized by the symbolic equation $dV = r\,dr\,d\theta\,dz$. In spherical coordinates (introduced in Section 13.7), the analog is the formula

$$\boxed{dV = \rho^2 \sin\phi\,d\rho\,d\phi\,d\theta}$$

Recall (Figure 12) that

$$x = \rho\cos\theta\sin\phi, \qquad y = \rho\sin\theta\sin\phi, \qquad z = \rho\cos\phi$$

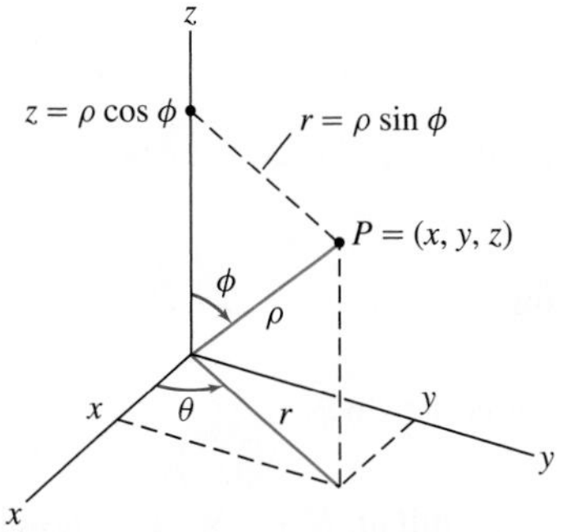

FIGURE 12 Spherical coordinates.

The key step in deriving this formula is estimating the volume of a small **spherical wedge** $\mathcal{W}$, defined by the inequalities

$$\mathcal{W} : \theta_1 \le \theta \le \theta_2, \quad \phi_1 \le \phi \le \phi_2, \quad \rho_1 \le \rho \le \rho_2$$

6

Referring to Figure 13, we see that when the increments

$$\Delta\theta = \theta_2 - \theta_1, \qquad \Delta\phi = \phi_2 - \phi_1, \qquad \Delta\rho = \rho_2 - \rho_1$$

are small, the spherical wedge is nearly a box with sides $\Delta\rho$, $\rho_1\Delta\phi$, and $\rho_1 \sin\phi_1 \Delta\theta$ and volume

$$\text{Volume}(\mathcal{W}) \approx \rho_1^2 \sin\phi_1\, \Delta\rho\, \Delta\phi\, \Delta\theta \qquad \boxed{7}$$

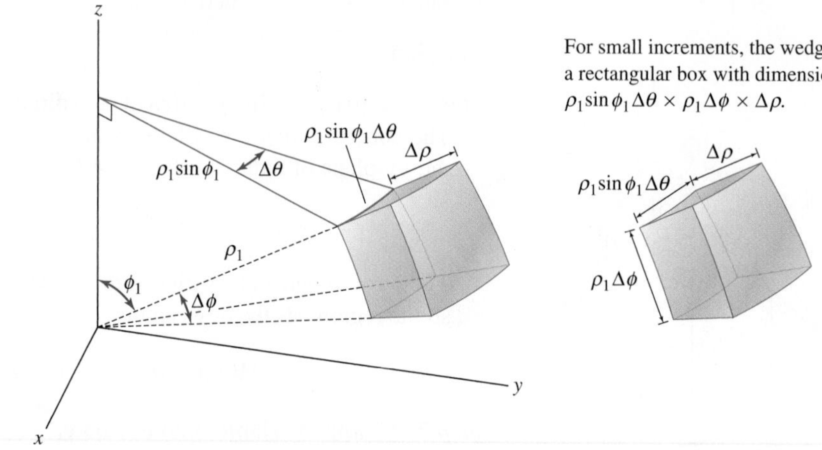

FIGURE 13 Spherical wedge.

Following the usual steps, we decompose $\mathcal{W}$ into N^3 spherical subwedges $\mathcal{W}_i$ (Figure 14) with increments

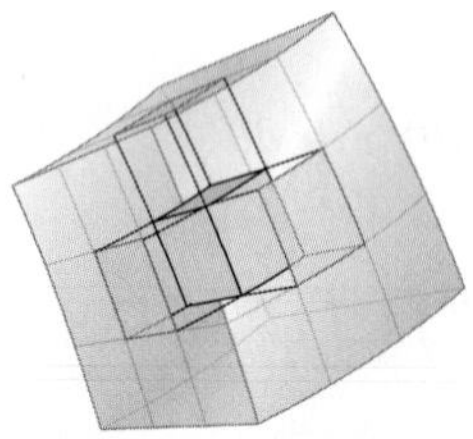

FIGURE 14 Decomposition of a spherical wedge into subwedges.

$$\Delta\theta = \frac{\theta_2 - \theta_1}{N}, \qquad \Delta\phi = \frac{\phi_2 - \phi_1}{N}, \qquad \Delta\rho = \frac{\rho_2 - \rho_1}{N}$$

and choose a sample point $P_i = (\rho_i, \theta_i, \phi_i)$ in each $\mathcal{W}_i$. Assuming f is continuous, the following approximation holds for large N (small $\mathcal{W}_i$):

$$\iiint_{\mathcal{W}_i} f(x, y, z)\, dV \approx f(P_i)\text{Volume}(\mathcal{W}_i)$$

$$\approx f(P_i)\rho_i^2 \sin\phi_i\, \Delta\rho\, \Delta\theta\, \Delta\phi$$

Taking the sum over i, we obtain

$$\iiint_{\mathcal{W}} f(x, y, z)\, dV \approx \sum_i f(P_i)\rho_i^2 \sin\phi_i\, \Delta\rho\, \Delta\theta\, \Delta\phi \qquad \boxed{8}$$

The sum on the right is a Riemann sum for the function

$$f(\rho\cos\theta\sin\phi,\ \rho\sin\theta\sin\phi,\ \rho\cos\phi)\, \rho^2 \sin\phi$$

on the domain $\mathcal{W}$. Eq. (9) below follows by passing to the limit an $N \to \infty$ (and showing that the error in Eq. (8) tends to zero). This argument applies more generally to regions defined by an inequality $\rho_1(\theta, \phi) \le \rho \le \rho_2(\theta, \phi)$.

THEOREM 3 Triple Integrals in Spherical Coordinates For a region $\mathcal{W}$ defined by

$$\theta_1 \le \theta \le \theta_2, \qquad \phi_1 \le \phi \le \phi_2, \qquad \rho_1(\theta,\phi) \le \rho \le \rho_2(\theta,\phi)$$

the triple integral $\iiint_{\mathcal{W}} f(x,y,z)\,dV$ is equal to

$$\int_{\theta_1}^{\theta_2}\int_{\phi=\phi_1}^{\phi_2}\int_{\rho=\rho_1(\theta,\phi)}^{\rho_2(\theta,\phi)} f(\rho\cos\theta\sin\phi,\ \rho\sin\theta\sin\phi,\ \rho\cos\phi)\,\rho^2\sin\phi\,d\rho\,d\phi\,d\theta \tag{9}$$

Eq. (9) is summarized in the symbolic expression for the "volume element" dV in spherical coordinates:

$$dV = \rho^2 \sin\phi\,d\rho\,d\phi\,d\theta$$

■ **EXAMPLE 5** Compute the integral of $f(x,y,z) = x^2 + y^2$ over the sphere S of radius 4 centered at the origin (Figure 15).

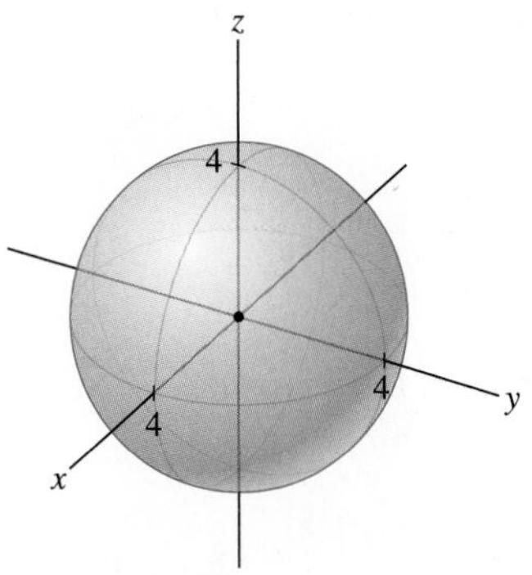

FIGURE 15 Sphere of radius 4.

Solution First, write $f(x,y,z)$ in spherical coordinates:

$$f(x,y,z) = x^2 + y^2 = (\rho\cos\theta\sin\phi)^2 + (\rho\sin\theta\sin\phi)^2$$
$$= \rho^2\sin^2\phi(\cos^2\theta + \sin^2\theta) = \rho^2\sin^2\phi$$

Since we are integrating over the entire sphere S of radius 4, ρ varies from 0 to 4, θ from 0 to 2π, and ϕ from 0 to π. In the following computation, we integrate first with respect to θ:

$$\iiint_S (x^2+y^2)\,dV = \int_0^{2\pi}\int_{\phi=0}^{\pi}\int_{\rho=0}^{4} (\rho^2\sin^2\phi)\,\rho^2\sin\phi\,d\rho\,d\phi\,d\theta$$
$$= 2\pi\int_{\phi=0}^{\pi}\int_{\rho=0}^{4} \rho^4\sin^3\phi\,d\rho\,d\phi = 2\pi\int_0^{\pi}\left(\frac{\rho^5}{5}\bigg|_0^4\right)\sin^3\phi\,d\phi$$
$$= \frac{2048\pi}{5}\int_0^{\pi}\sin^3\phi\,d\phi$$
$$= \frac{2048\pi}{5}\left(\frac{1}{3}\cos^3\phi - \cos\phi\right)\bigg|_0^{\pi} = \frac{8192\pi}{15}$$ ■

← *REMINDER*

$$\int \sin^3\phi\,d\phi = \frac{1}{3}\cos^3\phi - \cos\phi + C$$

[write $\sin^3\phi = \sin\phi(1-\cos^2\phi)$]

■ **EXAMPLE 6** Integrate $f(x,y,z) = z$ over the ice cream cone–shaped region $\mathcal{W}$ in Figure 16, lying above the cone and below the sphere.

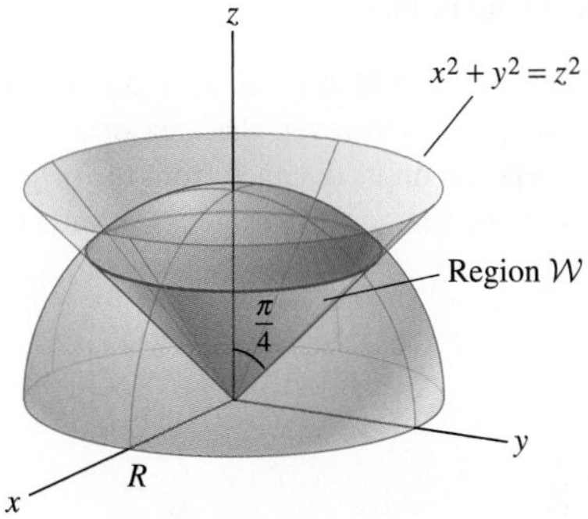

FIGURE 16 Ice cream cone defined by $0 \le \rho \le R$, $0 \le \phi \le \pi/4$.

Solution The cone has equation $x^2 + y^2 = z^2$, which in spherical coordinates is

$$(\rho\cos\theta\sin\phi)^2 + (\rho\sin\theta\sin\phi)^2 = (\rho\cos\phi)^2$$
$$\rho^2\sin^2\phi(\cos^2\theta + \sin^2\theta) = \rho^2\cos^2\phi$$
$$\sin^2\phi = \cos^2\phi$$
$$\sin\phi = \pm\cos\phi \quad\Rightarrow\quad \phi = \frac{\pi}{4}, \frac{3\pi}{4}$$

The upper branch of the cone has the simple equation $\phi = \frac{\pi}{4}$. On the other hand, the sphere has equation $\rho = R$, so the ice cream cone has the description

$$\mathcal{W}: 0 \le \theta \le 2\pi, \quad 0 \le \phi \le \frac{\pi}{4}, \quad 0 \le \rho \le R$$

We have

$$\iiint_{\mathcal{W}} z\,dV = \int_0^{2\pi}\int_{\phi=0}^{\pi/4}\int_{\rho=0}^{R} (\rho\cos\phi)\rho^2\sin\phi\,d\rho\,d\phi\,d\theta$$

$$= 2\pi\int_{\phi=0}^{\pi/4}\int_{\rho=0}^{R} \rho^3\cos\phi\sin\phi\,d\rho\,d\phi = \frac{\pi R^4}{2}\int_0^{\pi/4}\sin\phi\cos\phi\,d\phi = \frac{\pi R^4}{8}$$

■

16.4 SUMMARY

In symbolic form:

$$dA = r\,dr\,d\theta$$

$$dV = r\,dz\,dr\,d\theta$$

$$dV = \rho^2\sin\phi\,d\rho\,d\phi\,d\theta$$

- Double integral in *polar coordinates:*

$$\iint_{\mathcal{D}} f(x,y)\,dA = \int_{\theta_1}^{\theta_2}\int_{r=r_1(\theta)}^{r_2(\theta)} f(r\cos\theta, r\sin\theta)\,r\,dr\,d\theta$$

- Triple integral $\iiint_{\mathcal{R}} f(x,y,z)\,dV$

  – *In cylindrical coordinates:*

$$\int_{\theta_1}^{\theta_2}\int_{r=r_1(\theta)}^{r_2(\theta)}\int_{z=z_1(r,\theta)}^{z_2(r,\theta)} f(r\cos\theta, r\sin\theta, z)\,r\,dz\,dr\,d\theta$$

  – *In spherical coordinates:*

$$\int_{\theta_1}^{\theta_2}\int_{\phi=\phi_1}^{\phi_2}\int_{\rho=\rho_1(\theta,\phi)}^{\rho_2(\theta,\phi)} f(\rho\cos\theta\sin\phi, \rho\sin\theta\sin\phi, \rho\cos\phi)\,\rho^2\sin\phi\,d\rho\,d\phi\,d\theta$$

16.4 EXERCISES

Preliminary Questions

1. Which of the following represent the integral of $f(x,y) = x^2 + y^2$ over the unit circle?

(a) $\int_0^1\int_0^{2\pi} r^2\,dr\,d\theta$ **(b)** $\int_0^{2\pi}\int_0^1 r^2\,dr\,d\theta$

(c) $\int_0^1\int_0^{2\pi} r^3\,dr\,d\theta$ **(d)** $\int_0^{2\pi}\int_0^1 r^3\,dr\,d\theta$

2. What are the limits of integration in $\iiint f(r,\theta,z)r\,dr\,d\theta\,dz$ if the integration extends over the following regions?

(a) $x^2 + y^2 \le 4, \quad -1 \le z \le 2$

(b) Lower hemisphere of the sphere of radius 2, center at origin

3. What are the limits of integration in

$$\iiint f(\rho,\phi,\theta)\rho^2\sin\phi\,d\rho\,d\phi\,d\theta$$

if the integration extends over the following spherical regions centered at the origin?

(a) Sphere of radius 4

(b) Region between the spheres of radii 4 and 5

(c) Lower hemisphere of the sphere of radius 2

4. An ordinary rectangle of sides Δx and Δy has area $\Delta x\,\Delta y$, no matter where it is located in the plane. However, the area of a polar rectangle of sides Δr and $\Delta\theta$ depends on its distance from the origin. How is this difference reflected in the Change of Variables Formula for polar coordinates?

Exercises

In Exercises 1–6, sketch the region $\mathcal{D}$ indicated and integrate $f(x,y)$ over $\mathcal{D}$ using polar coordinates.

1. $f(x,y) = \sqrt{x^2+y^2}, \quad x^2 + y^2 \le 2$

2. $f(x,y) = x^2 + y^2; \quad 1 \le x^2 + y^2 \le 4$

3. $f(x,y) = xy; \quad x \ge 0, \quad y \ge 0, \quad x^2 + y^2 \le 4$

4. $f(x,y) = y(x^2+y^2)^3; \quad y \ge 0, \quad x^2 + y^2 \le 1$

5. $f(x, y) = y(x^2 + y^2)^{-1}; \quad y \geq \frac{1}{2}, \quad x^2 + y^2 \leq 1$

6. $f(x, y) = e^{x^2+y^2}; \quad x^2 + y^2 \leq R$

In Exercises 7–14, sketch the region of integration and evaluate by changing to polar coordinates.

7. $\displaystyle\int_{-2}^{2} \int_{0}^{\sqrt{4-x^2}} (x^2 + y^2)\, dy\, dx$

8. $\displaystyle\int_{0}^{3} \int_{0}^{\sqrt{9-y^2}} \sqrt{x^2 + y^2}\, dx\, dy$

9. $\displaystyle\int_{0}^{1/2} \int_{\sqrt{3}x}^{\sqrt{1-x^2}} x\, dy\, dx$

10. $\displaystyle\int_{0}^{4} \int_{0}^{\sqrt{16-x^2}} \tan^{-1} \frac{y}{x}\, dy\, dx$

11. $\displaystyle\int_{0}^{5} \int_{0}^{y} x\, dx\, dy$

12. $\displaystyle\int_{0}^{2} \int_{x}^{\sqrt{3}x} y\, dy\, dx$

13. $\displaystyle\int_{-1}^{2} \int_{0}^{\sqrt{4-x^2}} (x^2 + y^2)\, dy\, dx$

14. $\displaystyle\int_{1}^{2} \int_{0}^{\sqrt{2x-x^2}} \frac{1}{\sqrt{x^2 + y^2}}\, dy\, dx$

In Exercises 15–20, calculate the integral over the given region by changing to polar coordinates.

15. $f(x, y) = (x^2 + y^2)^{-2}; \quad x^2 + y^2 \leq 2, \quad x \geq 1$

16. $f(x, y) = x; \quad 2 \leq x^2 + y^2 \leq 4$

17. $f(x, y) = |xy|; \quad x^2 + y^2 \leq 1$

18. $f(x, y) = (x^2 + y^2)^{-3/2}; \quad x^2 + y^2 \leq 1, \quad x + y \geq 1$

19. $f(x, y) = x - y; \quad x^2 + y^2 \leq 1, \quad x + y \geq 1$

20. $f(x, y) = y; \quad x^2 + y^2 \leq 1, \quad (x - 1)^2 + y^2 \leq 1$

21. Find the volume of the wedge-shaped region (Figure 17) contained in the cylinder $x^2 + y^2 = 9$, bounded above by the plane $z = x$ and below by the xy-plane.

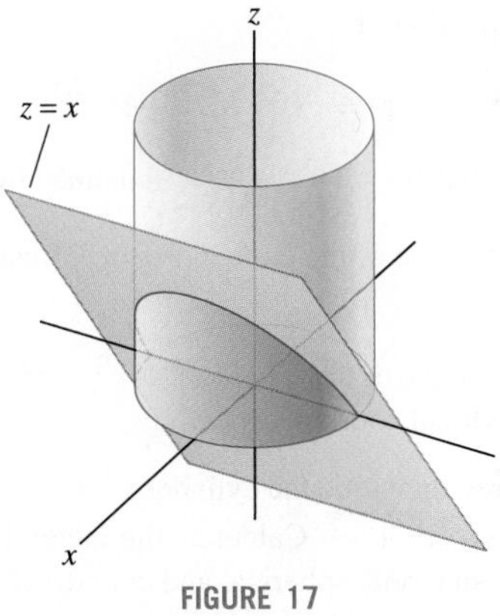

FIGURE 17

22. Let $\mathcal{W}$ be the region above the sphere $x^2 + y^2 + z^2 = 6$ and below the paraboloid $z = 4 - x^2 - y^2$.

(a) Show that the projection of $\mathcal{W}$ on the xy-plane is the disk $x^2 + y^2 \leq 2$ (Figure 18).

(b) Compute the volume of $\mathcal{W}$ using polar coordinates.

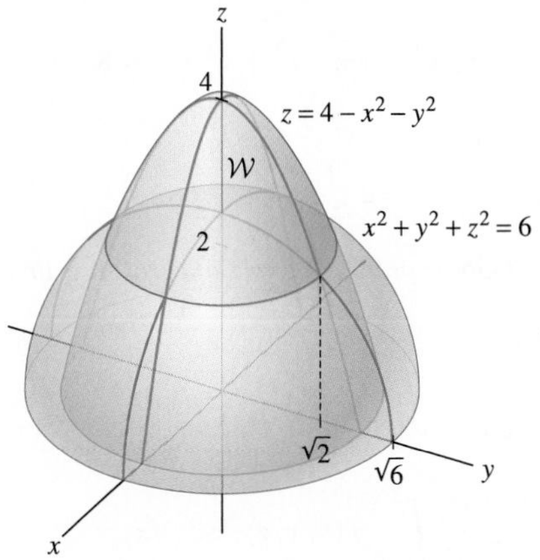

FIGURE 18

23. Evaluate $\displaystyle\iint_{\mathcal{D}} \sqrt{x^2 + y^2}\, dA$, where $\mathcal{D}$ is the domain in Figure 19. *Hint:* Find the equation of the inner circle in polar coordinates and treat the right and left parts of the region separately.

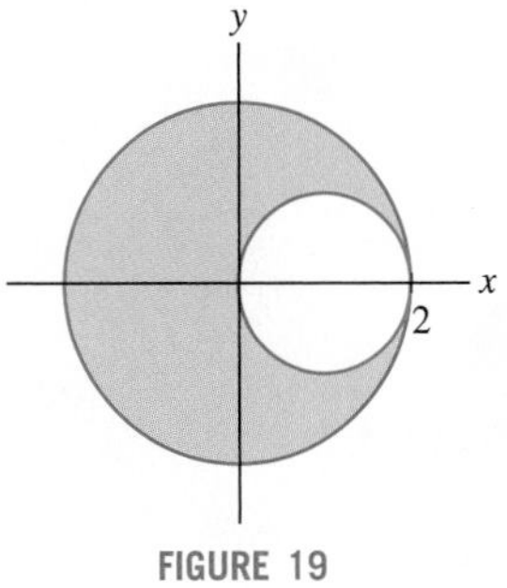

FIGURE 19

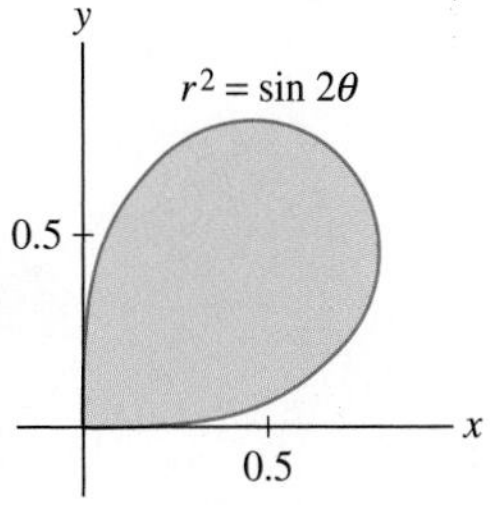

FIGURE 20

24. Evaluate $\displaystyle\iint_{\mathcal{D}} x\sqrt{x^2 + y^2}\, dA$, where $\mathcal{D}$ is the shaded region enclosed by the lemniscate curve $r^2 = \sin 2\theta$ in Figure 20.

25. Let $\mathcal{W}$ be the region between the paraboloids $z = x^2 + y^2$ and $z = 8 - x^2 - y^2$.

(a) Describe $\mathcal{W}$ in cylindrical coordinates.

(b) Use cylindrical coordinates to compute the volume of $\mathcal{W}$.

26. Use cylindrical coordinates to calculate the integral of the function $f(x, y, z) = z$ over the region above the disk $x^2 + y^2 = 1$ in the xy-plane and below the surface $z = 4 + x^2 + y^2$.

In Exercises 27–32, use cylindrical coordinates to calculate $\iiint_{\mathcal{W}} f(x, y, z)\, dV$ for the given function and region.

27. $f(x, y, z) = x^2 + y^2; \quad x^2 + y^2 \le 9, \quad 0 \le z \le 5$

28. $f(x, y, z) = xz; \quad x^2 + y^2 \le 1, \quad x \ge 0, \quad 0 \le z \le 2$

29. $f(x, y, z) = y; \quad x^2 + y^2 \le 1, \quad x \ge 0, \quad y \ge 0, \quad 0 \le z \le 2$

30. $f(x, y, z) = z\sqrt{x^2 + y^2}; \quad x^2 + y^2 \le z \le 8 - (x^2 + y^2)$

31. $f(x, y, z) = z; \quad x^2 + y^2 \le z \le 9$

32. $f(x, y, z) = z; \quad 0 \le z \le x^2 + y^2 \le 9$

In Exercises 33–36, express the triple integral in cylindrical coordinates.

33. $\displaystyle\int_{-1}^{1} \int_{y=-\sqrt{1-x^2}}^{y=\sqrt{1-x^2}} \int_{z=0}^{4} f(x, y, z)\, dz\, dy\, dx$

34. $\displaystyle\int_{0}^{1} \int_{y=-\sqrt{1-x^2}}^{y=\sqrt{1-x^2}} \int_{z=0}^{4} f(x, y, z)\, dz\, dy\, dx$

35. $\displaystyle\int_{-1}^{1} \int_{y=0}^{y=\sqrt{1-x^2}} \int_{z=0}^{x^2+y^2} f(x, y, z)\, dz\, dy\, dx$

36. $\displaystyle\int_{0}^{2} \int_{y=0}^{y=\sqrt{2x-x^2}} \int_{z=0}^{\sqrt{x^2+y^2}} f(x, y, z)\, dz\, dy\, dx$

37. Find the equation of the right-circular cone in Figure 21 in cylindrical coordinates and compute its volume.

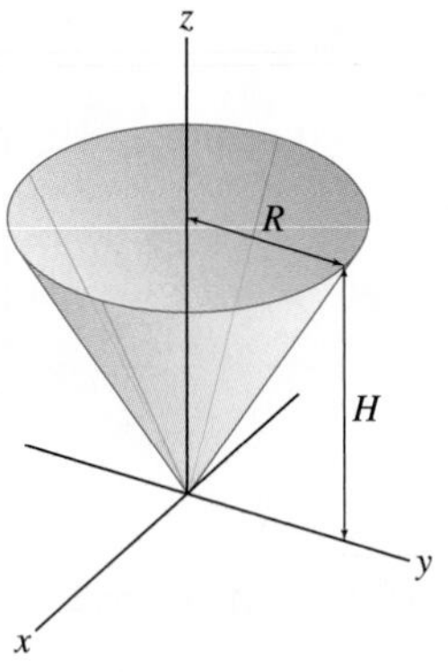

FIGURE 21

38. Use cylindrical coordinates to integrate $f(x, y, z) = z$ over the intersection of the hemisphere $x^2 + y^2 + z^2 = 4$, $z \ge 0$, and the cylinder $x^2 + y^2 = 1$.

39. Use cylindrical coordinates to calculate the volume of the solid obtained by removing a central cylinder of radius b from a sphere of radius a where $b < a$.

40. Find the volume of the region in Figure 22.

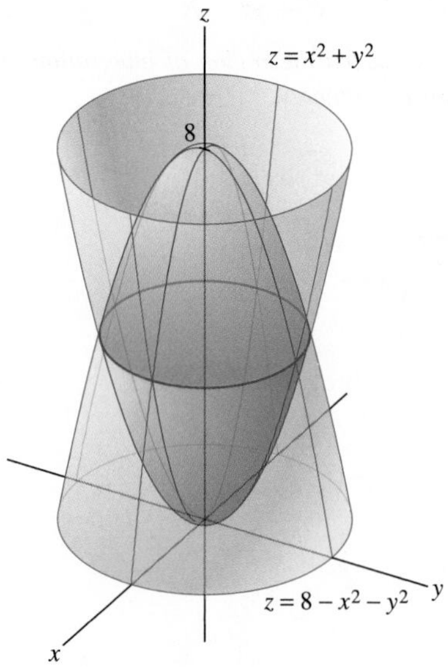

FIGURE 22

In Exercises 41–46, use spherical coordinates to calculate the triple integral of $f(x, y, z)$ over the given region.

41. $f(x, y, z) = y; \quad x^2 + y^2 + z^2 \le 1, \quad x, y, z \le 0$

42. $f(x, y, z) = \rho^{-3}; \quad 2 \le x^2 + y^2 + z^2 \le 4$

43. $f(x, y, z) = x^2 + y^2; \quad \rho \le 1$

44. $f(x, y, z) = 1; \quad x^2 + y^2 + z^2 \le 4z, \quad z \ge \sqrt{x^2 + y^2}$

45. $f(x, y, z) = \sqrt{x^2 + y^2 + z^2}; \quad x^2 + y^2 + z^2 \le 2z$

46. $f(x, y, z) = \rho; \quad x^2 + y^2 + z^2 \le 4, \quad z \le 1, \quad x \ge 0$

47. Use spherical coordinates to evaluate the triple integral of $f(x, y, z) = z$ over the region

$$0 \le \theta \le \frac{\pi}{3}, \qquad 0 \le \phi \le \frac{\pi}{2}, \qquad 1 \le \rho \le 2$$

48. Find the volume of the region lying above the cone $\phi = \phi_0$ and below the sphere $\rho = R$.

49. Calculate the integral of

$$f(x, y, z) = z(x^2 + y^2 + z^2)^{-3/2}$$

over the part of the ball $x^2 + y^2 + z^2 \le 16$ defined by $z \ge 2$.

50. Calculate the volume of the cone in Figure 21 using spherical coordinates.

51. Calculate the volume of the sphere $x^2 + y^2 + z^2 = a^2$, using both spherical and cylindrical coordinates.

52. Let $\mathcal{W}$ be the region within the cylinder $x^2 + y^2 = 2$ between $z = 0$ and the cone $z = \sqrt{x^2 + y^2}$. Calculate the integral of $f(x, y, z) = x^2 + y^2$ over $\mathcal{W}$, using both spherical and cylindrical coordinates.

53. Bell-Shaped Curve One of the key results in calculus is the computation of the area under the bell-shaped curve (Figure 23):

$$I = \int_{-\infty}^{\infty} e^{-x^2}\,dx$$

This integral appears throughout engineering, physics, and statistics, and although e^{-x^2} does not have an elementary antiderivative, we can compute I using multiple integration.

(a) Show that $I^2 = J$, where J is the improper double integral

$$J = \int_{-\infty}^{\infty}\int_{-\infty}^{\infty} e^{-x^2-y^2}\,dx\,dy$$

Hint: Use Fubini's Theorem and $e^{-x^2-y^2} = e^{-x^2}e^{-y^2}$.

(b) Evaluate J in polar coordinates.

(c) Prove that $I = \sqrt{\pi}$.

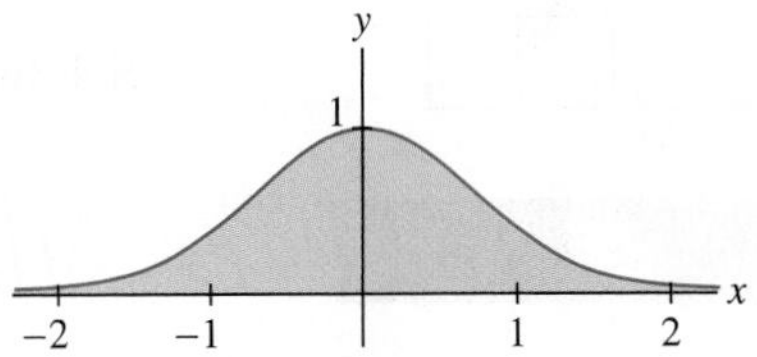

FIGURE 23 The bell-shaped curve $y = e^{-x^2}$.

Further Insights and Challenges

54. An Improper Multiple Integral Show that a triple integral of $(x^2+y^2+z^2+1)^{-2}$ over all of $\mathbf{R}^3$ is equal to π^2. This is an improper integral, so integrate first over $\rho \le R$ and let $R \to \infty$.

55. Prove the formula

$$\iint_{\mathcal{D}} \ln r\,dA = -\frac{\pi}{2}$$

where $r = \sqrt{x^2+y^2}$ and $\mathcal{D}$ is the unit disk $x^2+y^2 \le 1$. This is an improper integral since $\ln r$ is not defined at $(0,0)$, so integrate first over the annulus $a \le r \le 1$ where $0 < a < 1$, and let $a \to 0$.

56. Recall that the improper integral $\int_0^1 x^{-a}\,dx$ converges if and only if $a < 1$. For which values of a does $\iint_{\mathcal{D}} r^{-a}\,dA$ converge, where $r = \sqrt{x^2+y^2}$ and $\mathcal{D}$ is the unit disk $x^2+y^2 \le 1$?

16.5 Applications of Multiple Integrals

This section discusses some applications of multiple integrals. First, we consider quantities (such as mass, charge, and population) that are distributed with a given density ρ in $\mathbf{R}^2$ or $\mathbf{R}^3$. In single-variable calculus, we saw that the "total amount" is defined as the integral of density. Similarly, the total amount of a quantity distributed in $\mathbf{R}^2$ or $\mathbf{R}^3$ is defined as the double or triple integral:

$$\text{Total amount} = \iint_{\mathcal{D}} \rho(x,y)\,dA \quad \text{or} \quad \iiint_{\mathcal{W}} \rho(x,y,z)\,dV \qquad \boxed{1}$$

The density function ρ has units of "amount per unit area" (or per unit volume).

The intuition behind Eq. (1) is similar to that of the single variable case. Suppose, for example, that $\rho(x,y)$ is population density (Figure 1). When density is constant, the total population is simply density times area:

$$\text{Population} = \text{density (people/km}^2) \times \text{area (km}^2)$$

To treat variable density in the case, say, of a rectangle $\mathcal{R}$, we divide $\mathcal{R}$ into smaller rectangles $\mathcal{R}_{ij}$ of area $\Delta x\,\Delta y$ on which ρ is nearly constant (assuming that ρ is continuous on $\mathcal{R}$). The population in $\mathcal{R}_{ij}$ is approximately $\rho(P_{ij})\,\Delta x\,\Delta y$ for any sample point P_{ij} in $\mathcal{R}_{ij}$, and the sum of these approximations is a Riemann sum that converges to the double integral:

$$\int_{\mathcal{R}} \rho(x,y)\,dA \approx \sum_i \sum_j \rho(P_{ij})\Delta x\,\Delta y$$

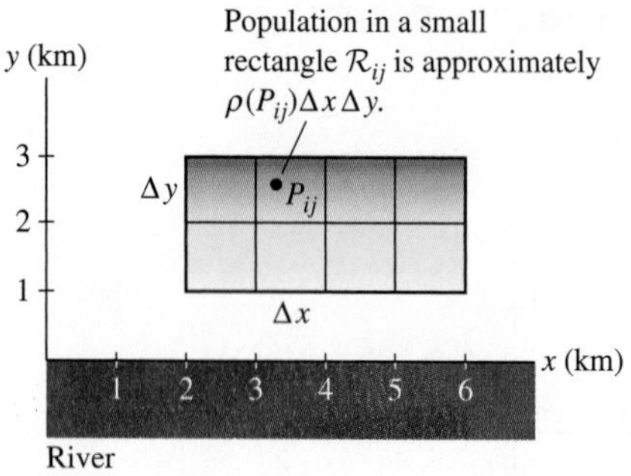

FIGURE 1

■ **EXAMPLE 1 Population Density** The population in a rural area near a river has density

$$\rho(x, y) = 40xe^{0.1y} \text{ people per km}^2$$

How many people live in the region $\mathcal{R}$: $2 \le x \le 6$, $1 \le y \le 3$ (Figure 1)?

Solution The total population is the integral of population density:

$$\iint_{\mathcal{R}} 40xe^{0.1y}\, dA = \int_2^6 \int_1^3 40xe^{0.1y}\, dx\, dy$$

$$= \left(\int_2^6 40x\, dx\right)\left(\int_1^3 e^{0.1y}\, dy\right)$$

$$= \left(20x^2\Big|_{x=2}^{6}\right)\left(10e^{0.1y}\Big|_{y=1}^{3}\right) \approx (640)(2.447) \approx 1566 \text{ people} \quad ■$$

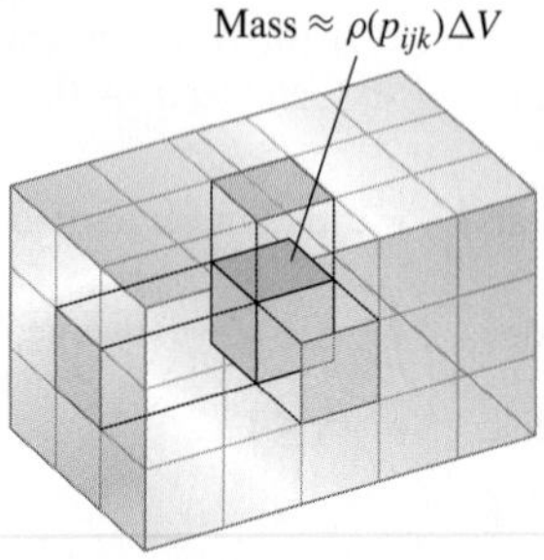

FIGURE 2 The mass of a small box is approximately $\rho(P_{ijk})\,\Delta V$.

In the next example, we compute the mass of an object as the integral of mass density. In three dimensions, we justify this computation by dividing $\mathcal{W}$ into boxes $\mathcal{B}_{ijk}$ of volume ΔV that are so small that the mass density is nearly constant on $\mathcal{B}_{ijk}$ (Figure 2). The mass of $\mathcal{B}_{ijk}$ is approximately $\rho(P_{ijk})\,\Delta V$, where P_{ijk} is any sample point in $\mathcal{B}_{ijk}$, and the sum of these approximations is a Riemann sum that converges to the triple integral:

$$\iiint_{\mathcal{W}} \rho(x, y, z)\, dV \approx \sum_i \sum_j \sum_k \underbrace{\rho(P_{ijk})\Delta V}_{\text{Approximate mass of } \mathcal{B}_{ijk}}$$

When ρ is constant, we say that the solid has a **uniform** mass density. In this case, the triple integral has the value ρV and the mass is simply $M = \rho V$.

■ **EXAMPLE 2** Let $a > 0$. Find the mass of the "solid bowl" $\mathcal{W}$ consisting of points inside the paraboloid $z = a(x^2 + y^2)$ for $0 \le z \le H$ (Figure 3). Assume a mass density of $\rho(x, y, z) = z$.

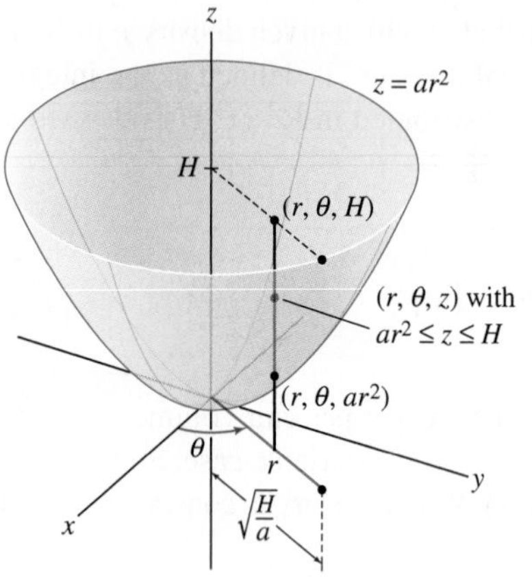

FIGURE 3 The paraboloid $z = a(x^2 + y^2)$.

Solution Because the bowl is symmetric with respect to the z-axis, we use cylindrical coordinates (r, θ, z). Recall that $r^2 = x^2 + y^2$, so the polar equation of the paraboloid is $z = ar^2$. A point (r, θ, z) lies *above* the paraboloid if $z \ge ar^2$, so it lies *in* the bowl if $ar^2 \le z \le H$. In other words, the bowl is described by

$$0 \le \theta \le 2\pi, \qquad 0 \le r \le \sqrt{\frac{H}{a}}, \qquad ar^2 \le z \le H$$

The mass of the bowl is the integral of mass density:

$$M = \iiint_{\mathcal{W}} \rho(x, y, z)\, dV = \int_{\theta=0}^{2\pi} \int_{r=0}^{\sqrt{H/a}} \int_{z=ar^2}^{H} zr\, dz\, dr\, d\theta$$

$$= 2\pi \int_{r=0}^{\sqrt{H/a}} \left(\frac{1}{2}H^2 - \frac{1}{2}a^2r^4\right) r\,dr$$

$$= 2\pi \left(\frac{H^2r^2}{4} - \frac{a^2r^6}{12}\right)\Bigg|_{r=0}^{\sqrt{H/a}}$$

$$= 2\pi \left(\frac{H^3}{4a} - \frac{H^3}{12a}\right) = \frac{\pi H^3}{3a} \quad ■$$

Next, we compute centers of mass. In Section 9.3, we computed centers of mass of laminas (thin plates in the plane), but we had to assume that the mass density is constant. Multiple integration enables us to treat variable mass density. We define the moments of a lamina $\mathcal{D}$ with respect to the coordinate axes:

$$M_y = \iint_{\mathcal{D}} x\rho(x, y)\, dA, \qquad M_x = \iint_{\mathcal{D}} y\rho(x, y)\, dA$$

The **center of mass** (COM) is the point $P_{CM} = (x_{CM}, y_{CM})$ where

$$x_{CM} = \frac{M_y}{M}, \qquad y_{CM} = \frac{M_x}{M} \qquad \boxed{2}$$

You can think of the coordinates x_{CM} and y_{CM} as **weighted averages**—they are the averages of x and y in which the factor ρ assigns a larger coefficient to points with larger mass density.

If $\mathcal{D}$ has uniform mass density (ρ constant), then the factors of ρ in the numerator and denominator in Eq. (2) cancel, and the center of mass coincides with the **centroid**, defined as the point whose coordinates are the averages of the coordinates over the domain:

$$\overline{x} = \frac{1}{A}\iint_{\mathcal{D}} x\, dA, \qquad \overline{y} = \frac{1}{A}\iint_{\mathcal{D}} y\, dA$$

Here $A = \iint_{\mathcal{D}} 1\, dA$ is the area of $\mathcal{D}$.

- *In* $\mathbf{R}^2$, *we write* M_y *for the integral of* $x\rho(x, y)$ *because* x *is the* distance *to the* y*-axis.*
- *In* $\mathbf{R}^3$, *we write* M_{yz} *for the integral of* $x\rho(x, y, z)$ *because in* $\mathbf{R}^3$, x *is the* distance *to the* yz*-plane.*

In $\mathbf{R}^3$, the moments of a solid region $\mathcal{W}$ are defined not with respect to the axes as in $\mathbf{R}^2$, but with respect to the coordinate planes:

$$M_{yz} = \iiint_{\mathcal{W}} x\rho(x, y, z)\, dV$$

$$M_{xz} = \iiint_{\mathcal{W}} y\rho(x, y, z)\, dV$$

$$M_{xy} = \iiint_{\mathcal{W}} z\rho(x, y, z)\, dV$$

The center of mass is the point $P_{CM} = (x_{CM}, y_{CM}, z_{CM})$ with coordinates

$$x_{CM} = \frac{M_{yz}}{M}, \qquad y_{CM} = \frac{M_{xz}}{M}, \qquad z_{CM} = \frac{M_{xy}}{M}$$

The centroid of $\mathcal{W}$ is the point $P = (\overline{x}, \overline{y}, \overline{z})$, which, as before, coincides with the center of mass when ρ is constant:

$$\overline{x} = \frac{1}{V}\iiint_{\mathcal{W}} x\, dV, \qquad \overline{y} = \frac{1}{V}\iiint_{\mathcal{W}} y\, dV, \qquad \overline{z} = \frac{1}{V}\iiint_{\mathcal{W}} z\, dV$$

where $V = \iiint_{\mathcal{W}} 1\, dV$ is the volume of $\mathcal{W}$.

Symmetry can often be used to simplify COM calculations. We say that a region $\mathcal{W}$ in $\mathbf{R}^3$ is symmetric with respect to the xy-plane if $(x, y, -z)$ lies in $\mathcal{W}$ whenever (x, y, z) lies in $\mathcal{W}$. The density ρ is symmetric with respect to the xy-plane if

$$\rho(x, y, -z) = \rho(x, y, z)$$

In other words, the mass density is the same at points located symmetrically with respect to the xy-plane. If both $\mathcal{W}$ and ρ have this symmetry, then $M_{xy} = 0$ and the COM lies on the xy-plane—that is, $z_{CM} = 0$. Similar remarks apply to the other coordinate axes and to domains in the plane.

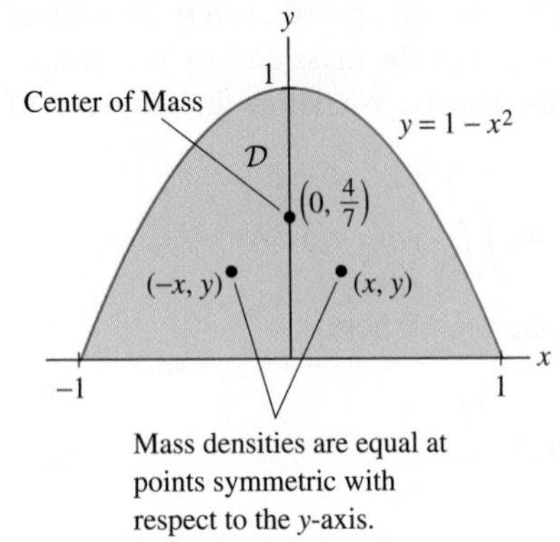

FIGURE 4

EXAMPLE 3 Center of Mass Find the center of mass of the domain $\mathcal{D}$ bounded by $y = 1 - x^2$ and the x-axis, assuming a mass density of $\rho(x, y) = y$ (Figure 4).

Solution The domain $\mathcal{D}$ is symmetric with respect to the y-axis, and so too is the mass density because $\rho(x, y) = \rho(-x, y) = y$. Therefore, $x_{CM} = 0$. We need only compute y_{CM}:

$$M_x = \iint_{\mathcal{D}} y\rho(x, y)\,dA = \int_{x=-1}^{1} \int_{y=0}^{1-x^2} y^2\,dy\,dx = \int_{x=-1}^{1} \left(\frac{1}{3}y^3 \Big|_{y=0}^{1-x^2} \right) dx$$

$$= \frac{1}{3} \int_{x=-1}^{1} \left(1 - 3x^2 + 3x^4 - x^6\right) dx = \frac{1}{3}\left(2 - 2 + \frac{6}{5} - \frac{2}{7}\right) = \frac{32}{105}$$

$$M = \iint_{\mathcal{D}} \rho(x, y)\,dA = \int_{x=-1}^{1} \int_{y=0}^{1-x^2} y\,dy\,dx = \int_{x=-1}^{1} \left(\frac{1}{2}y^2 \Big|_{y=0}^{1-x^2} \right) dx$$

$$= \frac{1}{2} \int_{x=-1}^{1} \left(1 - 2x^2 + x^4\right) dx = \frac{1}{2}\left(2 - \frac{4}{3} + \frac{2}{5}\right) = \frac{8}{15}$$

Therefore, $y_{CM} = \dfrac{M_x}{M} = \dfrac{32}{105}\left(\dfrac{8}{15}\right)^{-1} = \dfrac{4}{7}$. ■

EXAMPLE 4 Find the center of mass of the solid bowl $\mathcal{W}$ in Example 2 consisting of points inside the paraboloid $z = a(x^2 + y^2)$ for $0 \le z \le H$, assuming a mass density of $\rho(x, y, z) = z$.

Solution The domain is shown in Figure 3 above.

Step 1. **Use symmetry.**

The bowl $\mathcal{W}$ and the mass density are both symmetric with respect to the z-axis, so we can expect the COM to lie on the z-axis. In fact, the density satisfies both $\rho(-x, y, z) = \rho(x, y, z)$ and $\rho(x, -y, z) = \rho(x, y, z)$, and thus we have $M_{xz} = M_{yz} = 0$. It remains to compute the moment M_{xy}.

Step 2. **Compute the moment.**

In Example 2, we described the bowl in cylindrical coordinates as

$$0 \le \theta \le 2\pi, \qquad 0 \le r \le \sqrt{\frac{H}{a}}, \qquad ar^2 \le z \le H$$

and we computed the bowl's mass as $M = \dfrac{\pi H^3}{3a}$. The moment is

$$M_{xy} = \iiint_{\mathcal{W}} z\rho(x, y, z)\,dV = \iiint_{\mathcal{W}} z^2\,dV = \int_{\theta=0}^{2\pi} \int_{r=0}^{\sqrt{H/a}} \int_{z=ar^2}^{H} z^2 r\,dz\,dr\,d\theta$$

$$= 2\pi \int_{r=0}^{\sqrt{H/a}} \left(\frac{1}{3}H^3 - \frac{1}{3}a^3r^6\right) r\,dr$$

$$= 2\pi \left(\frac{1}{6}H^3r^2 - \frac{1}{24}a^3r^8\right) \Big|_{r=0}^{\sqrt{H/a}}$$

$$= 2\pi \left(\frac{H^4}{6a} - \frac{a^3H^4}{24a^4}\right) = \frac{\pi H^4}{4a}$$

The z-coordinate of the center of mass is

$$z_{CM} = \frac{M_{xy}}{V} = \frac{\pi H^4/(4a)}{\pi H^3/(3a)} = \frac{3}{4}H$$

and the center of mass itself is $\left(0, 0, \frac{3}{4}H\right)$. ■

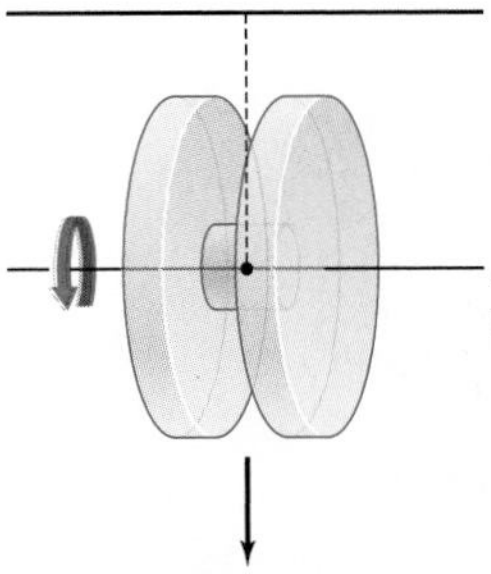

FIGURE 5 A spinning yo-yo has rotational kinetic energy $\frac{1}{2}I\omega^2$, where I is the moment of inertia and ω is the angular velocity. See Exercise 47.

Moments of inertia are used to analyze rotation about an axis. For example, the spinning yo-yo in Figure 5 rotates about its center as it falls downward, and according to physics, it has a rotational kinetic energy equal to

$$\text{Rotational KE} = \frac{1}{2}I\omega^2$$

Here, ω is the angular velocity (in radians per second) about this axis and I is the **moment of inertia** with respect to the axis of rotation. The quantity I is a rotational analog of the mass m, which appears in the expression $\frac{1}{2}mv^2$ for linear kinetic energy.

By definition, the moment of inertia with respect to an axis L is the integral of "distance squared from the axis," weighted by mass density. We confine our attention to the coordinate axes. Thus, for a lamina in the plane $\mathbf{R}^2$, we define the moments of inertia

$$I_x = \iint_{\mathcal{D}} y^2\rho(x, y)\, dA$$
$$I_y = \iint_{\mathcal{D}} x^2\rho(x, y)\, dA \qquad \boxed{3}$$
$$I_0 = \iint_{\mathcal{D}} (x^2 + y^2)\rho(x, y)\, dA$$

The quantity I_0 is called the **polar moment of inertia.** It is the moment of inertia relative to the z-axis, because $x^2 + y^2$ is the square of the distance from a point in the xy-plane to the z-axis. Notice that $I_0 = I_x + I_y$.

For a solid object occupying the region $\mathcal{W}$ in $\mathbf{R}^3$,

$$I_x = \iiint_{\mathcal{W}} (y^2 + z^2)\rho(x, y, z)\, dV$$
$$I_y = \iiint_{\mathcal{W}} (x^2 + z^2)\rho(x, y, z)\, dV$$
$$I_z = \iiint_{\mathcal{W}} (x^2 + y^2)\rho(x, y, z)\, dV$$

Moments of inertia have units of mass times length-squared.

EXAMPLE 5 A lamina $\mathcal{D}$ of uniform mass density and total mass M kg occupies the region between $y = 1 - x^2$ and the x-axis (in meters). Calculate the rotational KE if $\mathcal{D}$ rotates with angular velocity $\omega = 4$ rad/s about:

(a) the x-axis **(b)** the z-axis

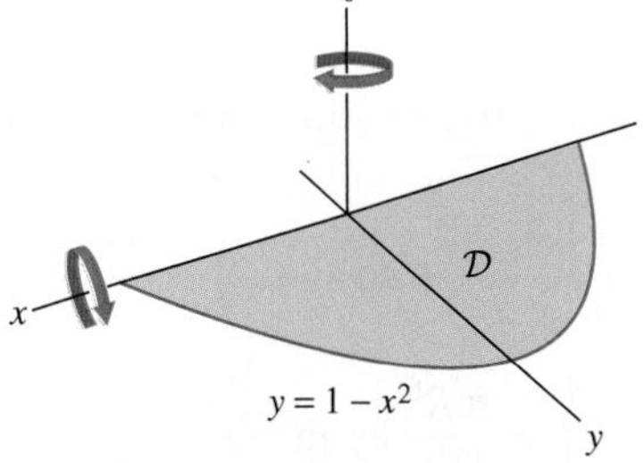

FIGURE 6 Rotating about the z-axis, the plate remains in the xy-plane. About the x-axis, it rotates out of the xy-plane.

Solution The lamina is shown in Figure 6. To find the rotational kinetic energy about the x- and z-axes, we need to compute I_x and I_0, respectively.

Step 1. **Find the mass density.**

The mass density is uniform (that is, ρ is constant), but this does not mean that $\rho = 1$. In fact, the area of $\mathcal{D}$ is $\int_{-1}^{1} (1 - x^2)\, dx = \frac{4}{3}$, so the mass density (mass per unit area) is

$$\rho = \frac{\text{mass}}{\text{area}} = \frac{M}{\frac{4}{3}} = \frac{3M}{4} \text{ kg/m}^2$$

***Step 2.* Calculate the moments.**

$$I_x = \int_{-1}^{1}\int_{y=0}^{1-x^2} y^2\rho\,dy\,dx = \int_{-1}^{1}\frac{1}{3}(1-x^2)^3\left(\frac{3M}{4}\right)dx$$

$$= \frac{M}{4}\int_{-1}^{1}(1-3x^2+3x^4-x^6)\,dx = \frac{8M}{35}\ \text{kg-m}^2$$

To calculate I_0, we use the relation $I_0 = I_x + I_y$. We have

$$I_y = \int_{-1}^{1}\int_{y=0}^{1-x^2} x^2\rho\,dy\,dx = \left(\frac{3M}{4}\right)\int_{-1}^{1}x^2(1-x^2)\,dx = \frac{M}{5}$$

CAUTION *The relation*

$$I_0 = I_x + I_y$$

is valid for a lamina in the xy-plane. However, there is no relation of this type for solid objects in $\mathbf{R}^3$.

and thus

$$I_0 = I_x + I_y = \frac{8M}{35} + \frac{M}{5} = \frac{3M}{7} \qquad \boxed{4}$$

***Step 3.* Calculate kinetic energy.**
Assuming an angular velocity of $\omega = 4$ rad/s,

$$\text{Rotational KE about } x\text{-axis} = \frac{1}{2}I_x\omega^2 = \frac{1}{2}\left(\frac{8M}{35}\right)4^2 \approx 1.8M\ \text{J}$$

$$\text{Rotational KE about } z\text{-axis} = \frac{1}{2}I_0\omega^2 = \frac{1}{2}\left(\frac{3M}{7}\right)4^2 \approx 3.4M\ \text{J}$$

The unit of energy is the joule (J), equal to 1 kg-m^2/s^2. ■

A point mass m located a distance r from an axis has moment of inertia $I = mr^2$ with respect to that axis. Given an extended object of total mass M (not necessarily a point mass) whose moment of inertia with respect to the axis is I, we define the **radius of gyration** by $r_g = (I/M)^{1/2}$. With this definition, the moment of inertia would not change if all of the mass of the object were concentrated at a point located a distance r_g from the axis.

■ **EXAMPLE 6 Radius of Gyration of a Hemisphere** Find the radius of gyration about the z-axis of the solid hemisphere $\mathcal{W}$ defined by $x^2+y^2+z^2 = R^2, 0 \le z \le 1$, assuming a mass density of $\rho(x,y,z) = z$ kg/m^3.

Solution To compute the radius of gyration about the z-axis, we must compute I_z and the total mass M. We use spherical coordinates:

$$x^2+y^2 = (\rho\cos\theta\sin\phi)^2 + (\rho\sin\theta\sin\phi)^2 = \rho^2\sin^2\phi, \qquad z = \rho\cos\phi$$

$$I_z = \iiint_{\mathcal{W}}(x^2+y^2)z\,dV = \int_{\theta=0}^{2\pi}\int_{\phi=0}^{\pi/2}\int_{\rho=0}^{R}(\rho^2\sin^2\phi)(\rho\cos\phi)\rho^2\sin\phi\,d\rho\,d\phi\,d\theta$$

$$= 2\pi\left(\int_0^R \rho^5\,d\rho\right)\left(\int_{\phi=0}^{\pi/2}\sin^3\phi\cos\phi\,d\phi\right)$$

$$= 2\pi\left(\frac{R^6}{6}\right)\left(\left.\frac{\sin^4\phi}{4}\right|_0^{\pi/2}\right) = \frac{\pi R^6}{12}\ \text{kg-m}^2$$

$$M = \iiint_{\mathcal{W}} z\,dV = \int_{\theta=0}^{2\pi}\int_{\phi=0}^{\pi/2}\int_{\rho=0}^{R}(\rho\cos\phi)\rho^2\sin\phi\,d\rho\,d\phi\,d\theta$$

$$= \left(\int_{\rho=0}^{R} \rho^3 \, d\rho \right) \left(\int_{\phi=0}^{\pi/2} \cos\phi \sin\phi \, d\phi \right) \left(\int_{\theta=0}^{2\pi} d\theta \right) = \frac{\pi R^4}{4} \text{ kg}$$

The radius of gyration is $r_g = (I_z/M)^{1/2} = (R^2/3)^{1/2} = R/\sqrt{3}$ m.

Probability Theory

In Section 8.7, we discussed how probabilities can be represented as areas under curves (Figure 7). Recall that a *random variable* X is defined as the outcome of an experiment or measurement whose value is not known in advance. The probability that the value of X lies between a and b is denoted $P(a \le X \le b)$. Furthermore, X is a *continuous random variable* if there is a continuous function $p(x)$, called the *probability density function*, such that (Figure 7),

$$P(a \le X \le b) = \int_a^b p(x)\, dx$$

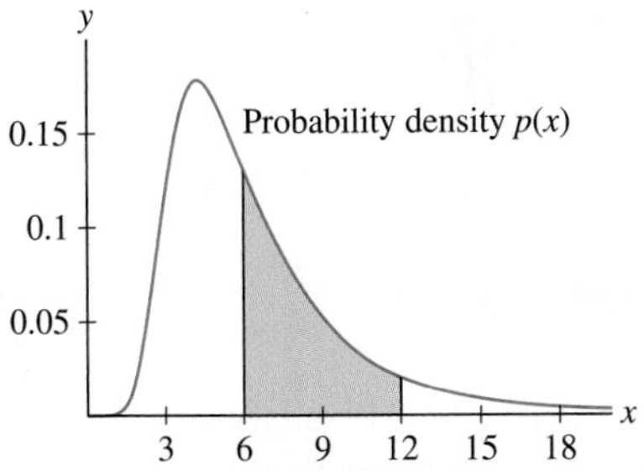

FIGURE 7 The shaded area is the probability that X lies between 6 and 12.

Double integration enters the picture when we compute "joint probabilities" of two random variables X and Y. We let

$$P(a \le X \le b;\ c \le Y \le d)$$

denote the probability that X and Y satisfy

$$a \le X \le b, \qquad c \le Y \le d$$

REMINDER Conditions on a probability density function:

- $p(x) \ge 0$
- $p(x)$ satisfies $\int_{-\infty}^{\infty} p(x) = 1$

For example, if X is the height (in centimeters) and Y is the weight (in kilograms) in a certain population, then

$$P(160 \le X \le 170;\ 52 \le Y \le 63)$$

is the probability that a person chosen at random has height between 160 and 170 cm and weight between 52 and 63 kg.

We say that X and Y are jointly continuous if there is a continuous function $p(x, y)$, called the **joint probability density function** (or simply the joint density), such that for all intervals $[a, b]$ and $[c, d]$ (Figure 8),

$$P(a \le X \le b; c \le Y \le d) = \int_{x=a}^{b} \int_{y=c}^{d} p(x, y)\, dy\, dx$$

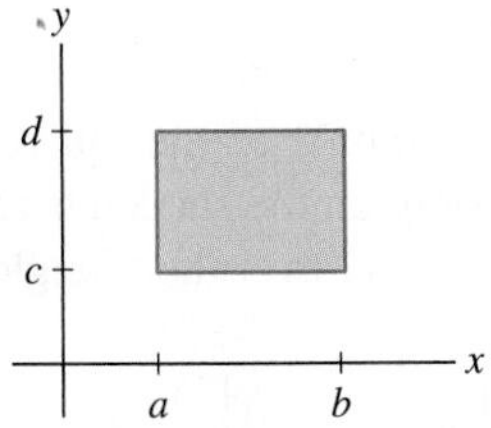

FIGURE 8 The probability $P(a \le X \le b; c \le Y \le d)$ is equal to the integral of $p(x, y)$ over the rectangle.

In the margin, we recall two conditions that a probability density function must satisfy. Joint density functions must satisfy similar conditions: First, $p(x, y) \ge 0$ for all x and y (because probabilities cannot be negative), and second,

$$\int_{-\infty}^{\infty} \int_{-\infty}^{\infty} p(x, y)\, dy\, dx = 1 \qquad \boxed{5}$$

This is often called the **normalization condition**. It holds because it is certain (the probability is 1) that X and Y take on some value between $-\infty$ and ∞.

EXAMPLE 7 Without proper maintenance, the time to failure (in months) of two sensors in an aircraft are random variables X and Y with joint density

$$p(x, y) = \begin{cases} \dfrac{1}{864} e^{-x/24 - y/36} & \text{for } x \ge 0,\ y \ge 0 \\ 0 & \text{otherwise} \end{cases}$$

What is the probability that neither sensor functions after two years?

Solution The problem asks for the probability $P(0 \le X \le 24;\ 0 \le Y \le 24)$:

$$\begin{aligned}\int_{x=0}^{24}\int_{y=0}^{24} p(x,y)\,dy\,dx &= \frac{1}{864}\int_{x=0}^{24}\int_{y=0}^{24} e^{-x/24-y/36}\,dy\,dx\\ &= \frac{1}{864}\left(\int_{x=0}^{24} e^{-x/24}\,dx\right)\left(\int_{y=0}^{24} e^{-y/36}\,dy\right)\\ &= \frac{1}{864}\left(-24e^{-x/24}\Big|_0^{24}\right)\left(-36e^{-y/36}\Big|_0^{24}\right)\\ &= \left(1-e^{-1}\right)\left(1-e^{-24/36}\right) \approx 0.31\end{aligned}$$

There is a 31% chance that neither sensor will function after two years. ■

More generally, we can compute the probability that X and Y satisfy conditions of various types. For example, $P(X+Y \le M)$ denotes the probability that the sum $X+Y$ is at most M. This probability is equal to the integral

$$P(X+Y \le M) = \iint_{\mathcal{D}} p(x,y)\,dy\,dx$$

where $\mathcal{D} = \{(x,y) : x+y \le M\}$.

■ **EXAMPLE 8** Calculate the probability that $X+Y \le 3$, where X and Y have joint probability density

$$p(x,y) = \begin{cases} \dfrac{1}{81}(2xy+2x+y) & \text{for } 0 \le x \le 3,\ 0 \le y \le 3\\ 0 & \text{otherwise}\end{cases}$$

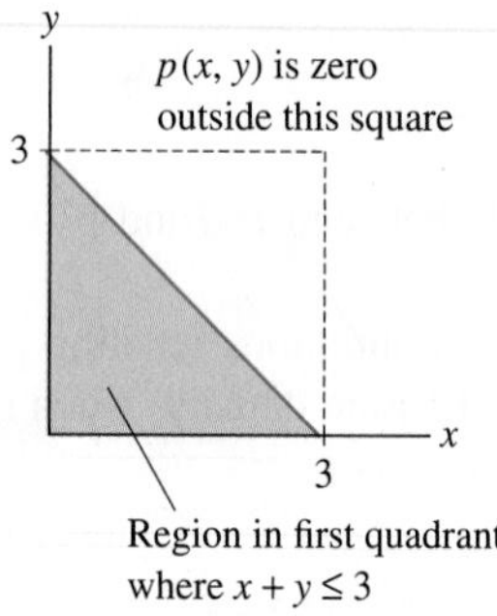

FIGURE 9

Solution The probability density function $p(x,y)$ is nonzero only on the square in Figure 9. Within that square, the inequality $x+y \le 3$ holds only on the shaded triangle, so the probability that $X+Y \le 3$ is equal to the integral of $p(x,y)$ over the triangle:

$$\begin{aligned}\int_{x=0}^{3}\int_{y=0}^{3-x} p(x,y)\,dy\,dx &= \frac{1}{81}\int_{x=0}^{3}\left(xy^2+\frac{1}{2}y^2+2xy\right)\Big|_{y=0}^{3-x}\,dx\\ &= \frac{1}{81}\int_{x=0}^{3}\left(x^3-\frac{15}{2}x^2+12x+\frac{9}{2}\right)dx\\ &= \frac{1}{81}\left(\frac{1}{4}3^4-\frac{5}{2}3^3+6(3^2)+\frac{9}{2}(3)\right) = \frac{1}{4}\end{aligned}$$

■

16.5 SUMMARY

- If the mass density is constant, then the center of mass coincides with the *centroid*, whose coordinates $\overline{x}$, $\overline{y}$ (and $\overline{z}$ in three dimensions) are the average values of x, y, and z over the domain. For a domain in $\mathbf{R}^2$,

$$\overline{x} = \frac{1}{A}\iint_{\mathcal{D}} x\,dA, \qquad \overline{y} = \frac{1}{A}\iint_{\mathcal{D}} y\,dA, \qquad A = \iint_{\mathcal{D}} 1\,dA$$

	In $\mathbf{R}^2$	In $\mathbf{R}^3$
Total mass	$M = \iint_{\mathcal{D}} \rho(x, y)\, dA$	$M = \iiint_{\mathcal{W}} \rho(x, y, z)\, dV$
Moments	$M_x = \iint_{\mathcal{D}} y\rho(x, y)\, dA$ $M_y = \iint_{\mathcal{D}} x\rho(x, y)\, dA$	$M_{yz} = \iiint_{\mathcal{W}} x\rho(x, y, z)\, dV$ $M_{xz} = \iiint_{\mathcal{W}} y\rho(x, y, z)\, dV$ $M_{xy} = \iiint_{\mathcal{W}} z\rho(x, y, z)\, dV$
Center of Mass	$x_{CM} = \frac{M_y}{M}, \quad y_{CM} = \frac{M_x}{M}$	$x_{CM} = \frac{M_{yz}}{M}, \quad y_{CM} = \frac{M_{xz}}{M}, \quad z_{CM} = \frac{M_{xy}}{M}$
Moments of Inertia	$I_x = \iint_{\mathcal{D}} y^2\rho(x, y)\, dA$ $I_y = \iint_{\mathcal{D}} x^2\rho(x, y)\, dA$ $I_0 = \iint_{\mathcal{D}} (x^2 + y^2)\rho(x, y)\, dA$ $(I_0 = I_x + I_y)$	$I_x = \iiint_{\mathcal{W}} (y^2 + z^2)\rho(x, y, z)\, dV$ $I_y = \iiint_{\mathcal{W}} (x^2 + z^2)\rho(x, y, z)\, dV$ $I_z = \iiint_{\mathcal{W}} (x^2 + y^2)\rho(x, y, z)\, dV$

- Radius of gyration: $r_g = (I/M)^{1/2}$
- Random variables X and Y have joint probability density function $p(x, y)$ if

$$P(a \le X \le b;\, c \le Y \le d) = \int_{x=a}^{b} \int_{y=c}^{d} p(x, y)\, dy\, dx$$

- A joint probability density function must satisfy $p(x, y) \ge 0$ and

$$\int_{x=-\infty}^{\infty} \int_{y=-\infty}^{\infty} p(x, y)\, dy\, dx = 1$$

16.5 EXERCISES

Preliminary Questions

1. What is the mass density $\rho(x, y, z)$ of a solid of volume 5 m^3 with uniform mass density and total mass 25 kg?

2. A domain $\mathcal{D}$ in $\mathbf{R}^2$ with uniform mass density is symmetric with respect to the y-axis. Which of the following are true?

(a) $x_{CM} = 0$ **(b)** $y_{CM} = 0$ **(c)** $I_x = 0$ **(d)** $I_y = 0$

3. If $p(x, y)$ is the joint probability density function of random variables X and Y, what does the double integral of $p(x, y)$ over $[0, 1] \times [0, 1]$ represent? What does the integral of $p(x, y)$ over the triangle bounded by $x = 0$, $y = 0$, and $x + y = 1$ represent?

Exercises

1. Find the total mass of the square $0 \le x \le 1$, $0 \le y \le 1$ assuming a mass density of

$$\rho(x, y) = x^2 + y^2$$

2. Calculate the total mass of a plate bounded by $y = 0$ and $y = x^{-1}$ for $1 \le x \le 4$ (in meters) assuming a mass density of $\rho(x, y) = y/x$ kg/m^2.

3. Find the total charge in the region under the graph of $y = 4e^{-x^2/2}$ for $0 \le x \le 10$ (in centimeters) assuming a charge density of $\rho(x, y) = 10^{-6}xy$ coulombs per square centimeter.

4. Find the total population within a 4-kilometer radius of the city center (located at the origin) assuming a population density of $\rho(x, y) = 2000(x^2 + y^2)^{-0.2}$ people per square kilometer.

5. Find the total population within the sector $2|x| \le y \le 8$ assuming a population density of $\rho(x, y) = 100e^{-0.1y}$ people per square kilometer.

6. Find the total mass of the solid region $\mathcal{W}$ defined by $x \ge 0$, $y \ge 0$, $x^2 + y^2 \le 4$, and $x \le z \le 32 - x$ (in centimeters) assuming a mass density of $\rho(x, y, z) = 6y$ g/cm^3.

7. Calculate the total charge of the solid ball $x^2 + y^2 + z^2 \le 5$ (in centimeters) assuming a charge density (in coulombs per cubic centimeter) of

$$\rho(x, y, z) = (3 \cdot 10^{-8})(x^2 + y^2 + z^2)^{1/2}$$

8. Compute the total mass of the plate in Figure 10 assuming a mass density of $f(x, y) = x^2/(x^2 + y^2)$ g/cm^2.

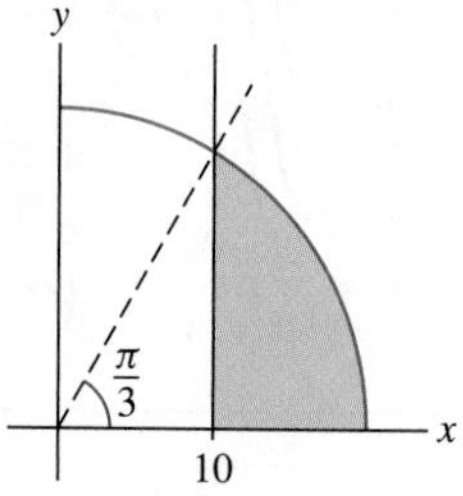

FIGURE 10

9. Assume that the density of the atmosphere as a function of altitude h (in km) above sea level is $\rho(h) = ae^{-bh}$ kg/km^3, where $a = 1.225 \times 10^9$ and $b = 0.13$. Calculate the total mass of the atmosphere contained in the cone-shaped region $\sqrt{x^2 + y^2} \le h \le 3$.

10. Calculate the total charge on a plate $\mathcal{D}$ in the shape of the ellipse with the polar equation

$$r^2 = \left(\frac{1}{6}\sin^2\theta + \frac{1}{9}\cos^2\theta\right)^{-1}$$

with the disk $x^2 + y^2 \le 1$ removed (Figure 11) assuming a charge density of $\rho(r, \theta) = 3r^{-4}$ C/cm^2.

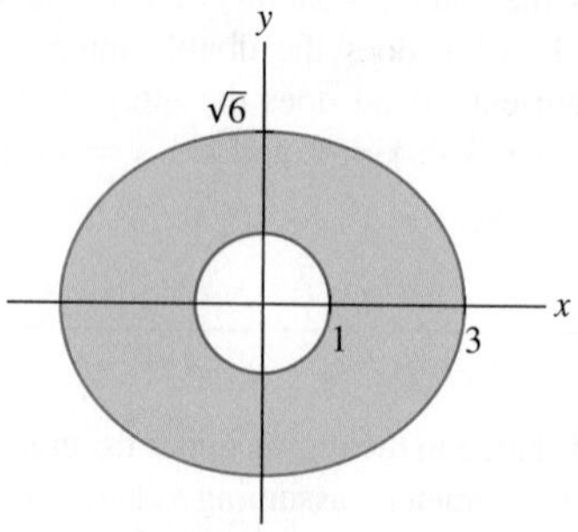

FIGURE 11

In Exercises 11–14, find the centroid of the given region.

11. Region bounded by $y = 1 - x^2$ and $y = 0$

12. Region bounded by $y^2 = x + 4$ and $x = 4$

13. Quarter circle $x^2 + y^2 \le R^2$, $x \ge 0$, $y \ge 0$

14. Infinite lamina bounded by the x- and y-axes and the graph of $y = e^{-x}$

15. CAS Use a computer algebra system to compute numerically the centroid of the shaded region in Figure 12 bounded by $r^2 = \cos 2\theta$ for $x \ge 0$.

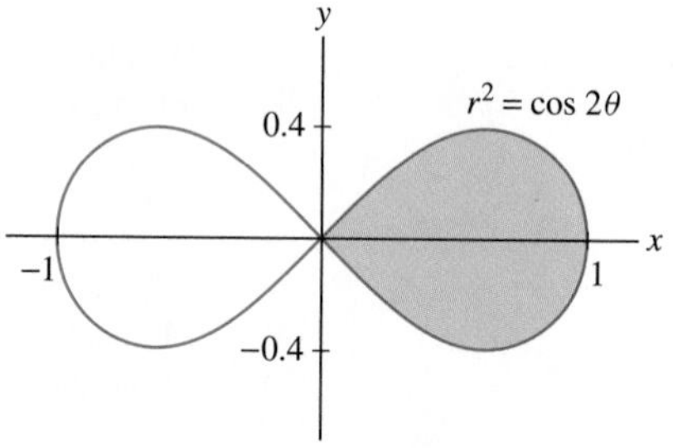

FIGURE 12

16. Show that the centroid of the sector in Figure 13 has y-coordinate

$$\overline{y} = \left(\frac{2R}{3}\right)\left(\frac{\sin\theta}{\theta}\right)$$

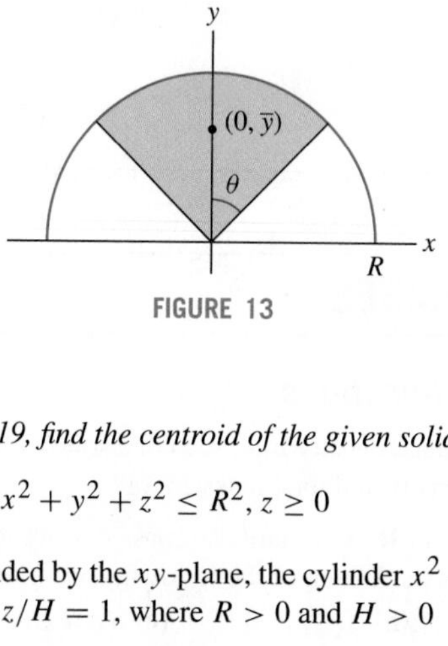

FIGURE 13

In Exercises 17–19, find the centroid of the given solid region.

17. Hemisphere $x^2 + y^2 + z^2 \le R^2$, $z \ge 0$

18. Region bounded by the xy-plane, the cylinder $x^2 + y^2 = R^2$, and the plane $x/R + z/H = 1$, where $R > 0$ and $H > 0$

19. The "ice cream cone" region $\mathcal{W}$ bounded, in spherical coordinates, by the cone $\phi = \pi/3$ and the sphere $\rho = 2$

20. Show that the z-coordinate of the centroid of the tetrahedron bounded by the coordinate planes and the plane

$$\frac{x}{a} + \frac{y}{b} + \frac{z}{c} = 1$$

in Figure 14 is $\overline{z} = c/4$. Conclude by symmetry that the centroid is $(a/4, b/4, c/4)$.

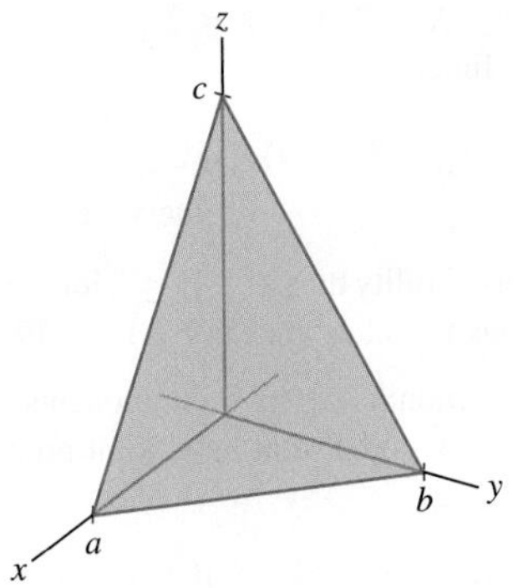

FIGURE 14

21. Find the centroid of the region $\mathcal{W}$ lying above the unit sphere $x^2 + y^2 + z^2 = 6$ and below the paraboloid $z = 4 - x^2 - y^2$ (Figure 15).

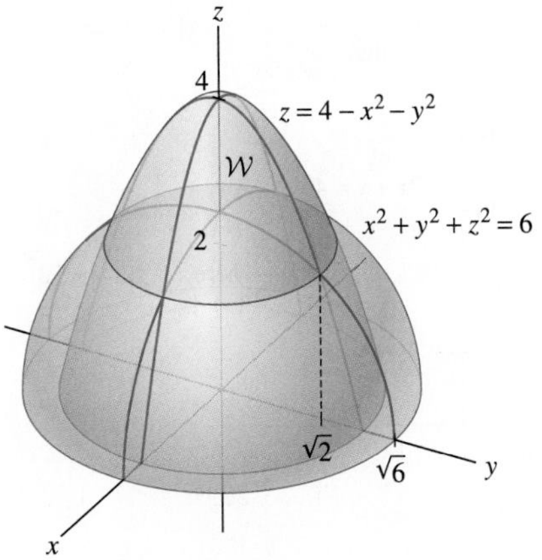

FIGURE 15

22. Let $R > 0$ and $H > 0$, and let $\mathcal{W}$ be the upper half of the ellipsoid $x^2 + y^2 + (Rz/H)^2 = R^2$ where $z \geq 0$ (Figure 16). Find the centroid of $\mathcal{W}$ and show that it depends on the height H but not on the radius R.

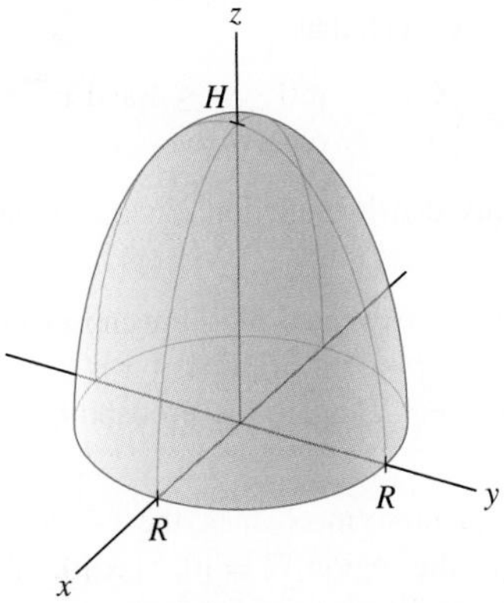

FIGURE 16 Upper half of ellipsoid $x^2 + y^2 + (Rz/H)^2 = R^2$, $z \geq 0$.

In Exercises 23–26, find the center of mass of the region with the given mass density ρ.

23. Region bounded by $y = 4 - x$, $x = 0$, $y = 0$; $\rho(x, y) = x$

24. Region bounded by $y^2 = x + 4$ and $x = 0$; $\rho(x, y) = y$

25. Region $|x| + |y| \leq 1$; $\rho(x, y) = x^2$

26. Semicircle $x^2 + y^2 \leq R^2$, $y \geq 0$; $\rho(x, y) = y$

27. Find the z-coordinate of the center of mass of the first octant of the unit sphere with mass density $\rho(x, y, z) = y$ (Figure 17).

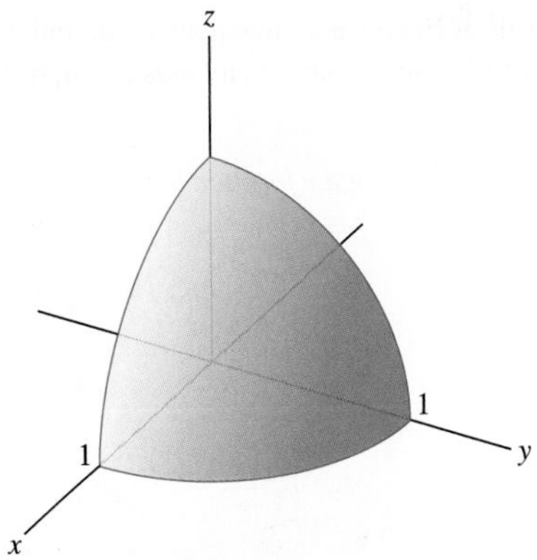

FIGURE 17

28. Find the center of mass of a cylinder of radius 2 and height 4 and mass density e^{-z}, where z is the height above the base.

29. Let $\mathcal{R}$ be the rectangle $[-a, a] \times [b, -b]$ with uniform density and total mass M. Calculate:

(a) The mass density ρ of $\mathcal{R}$

(b) I_x and I_0

(c) The radius of gyration about the x-axis

30. Calculate I_x and I_0 for the rectangle in Exercise 29 assuming a mass density of $\rho(x, y) = x$.

31. Calculate I_0 and I_x for the disk $\mathcal{D}$ defined by $x^2 + y^2 \leq 16$ (in meters), with total mass 1000 kg and uniform mass density. *Hint:* Calculate I_0 first and observe that $I_0 = 2I_x$. Express your answer in the correct units.

32. Calculate I_x and I_y for the half-disk $x^2 + y^2 \leq R^2$, $x \geq 0$ (in meters), of total mass M kg and uniform mass density.

In Exercises 33–36, let $\mathcal{D}$ be the triangular domain bounded by the coordinate axes and the line $y = 3 - x$, with mass density $\rho(x, y) = y$. Compute the given quantities.

33. Total mass

34. Center of Mass

35. I_x

36. I_0

In Exercises 37–40, let $\mathcal{D}$ be the domain between the line $y = bx/a$ and the parabola $y = bx^2/a^2$ where $a, b > 0$. Assume the mass density is $\rho(x, y) = xy$. Compute the given quantities.

37. Centroid

38. Center of Mass

39. I_x

40. I_0

41. Calculate the moment of inertia I_x of the disk $\mathcal{D}$ defined by $x^2 + y^2 \leq R^2$ (in meters) with total mass M kg. How much kinetic energy (in joules) is required to rotate the disk about the x-axis with angular velocity 10 rad/s?

42. Calculate the moment of inertia I_z of the box $\mathcal{W} = [-a, a] \times [-a, a] \times [0, H]$ assuming that $\mathcal{W}$ has total mass M.

43. Show that the moment of inertia of a sphere of radius R of total mass M with uniform mass density about any axis passing through the center of the sphere is $\frac{2}{5}MR^2$. Note that the mass density of the sphere is $\rho = M/(\frac{4}{3}\pi R^3)$.

44. Use the result of Exercise 43 to calculate the radius of gyration of a uniform sphere of radius R about any axis through the center of the sphere.

In Exercises 45 and 46, prove the formula for the right circular cylinder in Figure 18.

45. $I_z = \frac{1}{2}MR^2$

46. $I_x = \frac{1}{4}MR^2 + \frac{1}{12}MH^2$

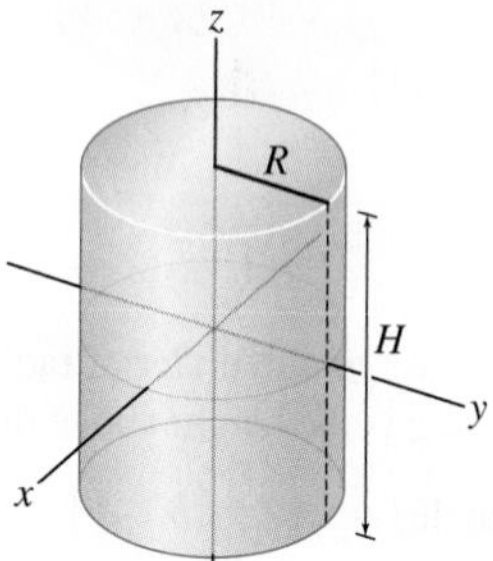

FIGURE 18

47. The yo-yo in Figure 19 is made up of two disks of radius $r = 3$ cm and an axle of radius $b = 1$ cm. Each disk has mass $M_1 = 20$ g, and the axle has mass $M_2 = 5$ g.

(a) Use the result of Exercise 45 to calculate the moment of inertia I of the yo-yo with respect to the axis of symmetry. Note that I is the sum of the moments of the three components of the yo-yo.

(b) The yo-yo is released and falls to the end of a 100-cm string, where it spins with angular velocity ω. The total mass of the yo-yo is $m = 45$ g, so the potential energy lost is PE $= mgh = (45)(980)100$ g-cm^2/s^2. Find ω under the assumption that one-third of this potential energy is converted into rotational kinetic energy.

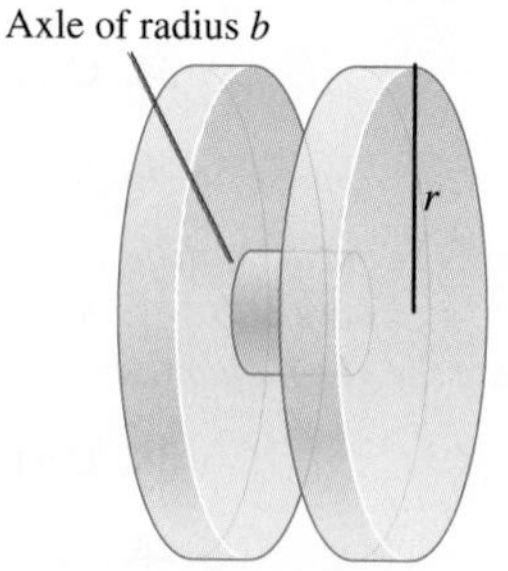

FIGURE 19

48. Calculate I_z for the solid region $\mathcal{W}$ inside the hyperboloid $x^2 + y^2 = z^2 + 1$ between $z = 0$ and $z = 1$.

49. Calculate $P(0 \le X \le 2;\ 1 \le Y \le 2)$, where X and Y have joint probability density function

$$p(x, y) = \begin{cases} \frac{1}{72}(2xy + 2x + y) & \text{if } 0 \le x \le 4 \text{ and } 0 \le y \le 2 \\ 0 & \text{otherwise} \end{cases}$$

50. Calculate the probability that $X + Y \le 2$ for random variables with joint probability density function as in Exercise 49.

51. The lifetime (in months) of two components in a certain device are random variables X and Y that have joint probability distribution function

$$p(x, y) = \begin{cases} \frac{1}{9216}(48 - 2x - y) & \text{if } x \ge 0,\ y \ge 0,\ 2x + y \le 48 \\ 0 & \text{otherwise} \end{cases}$$

Calculate the probability that both components function for at least 12 months without failing. Note that $p(x, y)$ is nonzero only within the triangle bounded by the coordinate axes and the line $2x + y = 48$ shown in Figure 20.

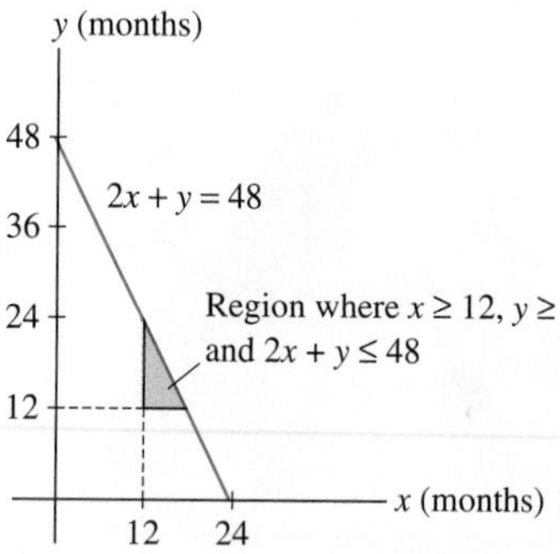

FIGURE 20

52. Find a constant C such that

$$p(x, y) = \begin{cases} Cxy & \text{if } 0 \le x \text{ and } 0 \le y \le 1 - x \\ 0 & \text{otherwise} \end{cases}$$

is a joint probability density function. Then calculate

(a) $P(X \le \frac{1}{2};\ Y \le \frac{1}{4})$ **(b)** $P(X \ge Y)$

53. Find a constant C such that

$$p(x, y) = \begin{cases} Cy & \text{if } 0 \le x \le 1 \text{ and } x^2 \le y \le x \\ 0 & \text{otherwise} \end{cases}$$

is a joint probability density function. Then calculate the probability that $Y \ge X^{3/2}$.

54. Numbers X and Y between 0 and 1 are chosen randomly. The joint probability density is $p(x, y) = 1$ if $0 \le x \le 1$ and $0 \le y \le 1$, and $p(x, y) = 0$ otherwise. Calculate the probability P that the product XY is at least $\frac{1}{2}$.

55. According to quantum mechanics, the x- and y-coordinates of a particle confined to the region $\mathcal{R} = [0, 1] \times [0, 1]$ are random variables with joint probability density function

$$p(x, y) = \begin{cases} C\sin^2(2\pi\ell x)\sin^2(2\pi n y) & \text{if } (x, y) \in \mathcal{R} \\ 0 & \text{otherwise} \end{cases}$$

The integers ℓ and n determine the energy of the particle, and C is a constant.

(a) Find the constant C.

(b) Calculate the probability that a particle with $\ell = 2$, $n = 3$ lies in the region $\left[0, \frac{1}{4}\right] \times \left[0, \frac{1}{8}\right]$.

56. The wave function for the 1s state of an electron in the hydrogen atom is

$$\psi_{1s}(\rho) = \frac{1}{\sqrt{\pi a_0^3}} e^{-\rho/a_0}$$

where a_0 is the Bohr radius. The probability of finding the electron in a region $\mathcal{W}$ of $\mathbf{R}^3$ is equal to

$$\iiint_{\mathcal{W}} p(x, y, z)\, dV$$

where, in spherical coordinates,

$$p(\rho) = |\psi_{1s}(\rho)|^2$$

Use integration in spherical coordinates to show that the probability of finding the electron at a distance greater than the Bohr radius is equal to $5/e^2 \approx 0.677$. The Bohr radius is $a_0 = 5.3 \times 10^{-11}$ m, but this value is not needed.

57. According to Coulomb's Law, the force between two electric charges of magnitude q_1 and q_2 separated by a distance r is kq_1q_2/r^2 (k is a negative constant). Let F be the net force on a charged particle P of charge Q coulombs located d centimeters above the center of a circular disk of radius R with a uniform charge distribution of density ρ C/m^2 (Figure 21). By symmetry, F acts in the vertical direction.

(a) Let $\mathcal{R}$ be a small polar rectangle of size $\Delta r \times \Delta\theta$ located at distance r. Show that $\mathcal{R}$ exerts a force on P whose vertical component is

$$\left(\frac{k\rho Q d}{(r^2 + d^2)^{3/2}}\right) r\, \Delta r\, \Delta\theta$$

(b) Explain why F is equal to the following double integral, and evaluate:

$$F = k\rho Q d \int_0^{2\pi} \int_0^R \frac{r\, dr\, d\theta}{(r^2 + d^2)^{3/2}}$$

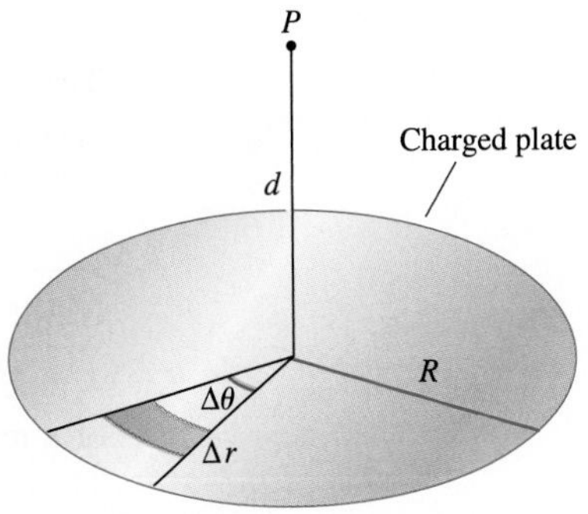

FIGURE 21

58. Let $\mathcal{D}$ be the annular region

$$-\frac{\pi}{2} \le \theta \le \frac{\pi}{2}, \quad a \le r \le b$$

where $b > a > 0$. Assume that $\mathcal{D}$ has a uniform charge distribution of ρ C/m^2. Let F be the net force on a charged particle of charge Q coulombs located at the origin (by symmetry, F acts along the x-axis).

(a) Argue as in Exercise 57 to show that

$$F = k\rho Q \int_{\theta=-\pi/2}^{\pi/2} \int_{r=a}^{b} \left(\frac{\cos\theta}{r^2}\right) r\, dr\, d\theta$$

(b) Compute F.

Further Insights and Challenges

59. Let $\mathcal{D}$ be the domain in Figure 22. Assume that $\mathcal{D}$ is symmetric with respect to the y-axis; that is, both $g_1(x)$ and $g_2(x)$ are even functions.

(a) Prove that the centroid lies on the y-axis—that is, that $\overline{x} = 0$.

(b) Show that if the mass density satisfies $\rho(-x, y) = -\rho(x, y)$, then $M_y = 0$ and $x_{CM} = 0$.

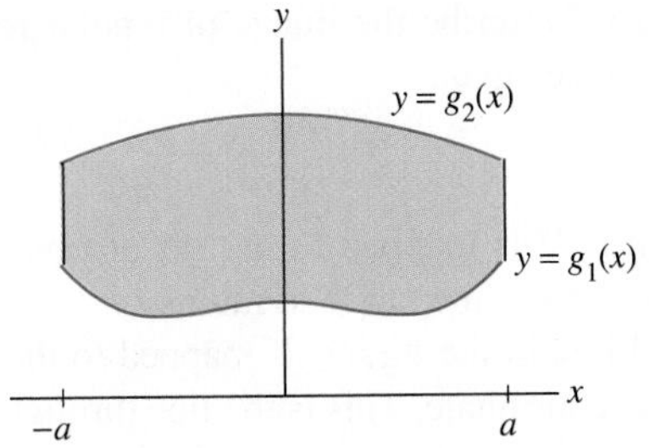

FIGURE 22

60. Pappus's Theorem Let A be the area of the region $\mathcal{D}$ between two graphs $y = g_1(x)$ and $y = g_2(x)$ over the interval $[a, b]$, where $g_2(x) \ge g_1(x) \ge 0$. Prove Pappus's Theorem: The volume of the solid obtained by revolving $\mathcal{D}$ about the x-axis is $V = 2\pi A\overline{y}$, where $\overline{y}$ is the y-coordinate of the centroid of $\mathcal{D}$ (the average of the y-coordinate). *Hint:* Show that

$$A\overline{y} = \int_{x=a}^{b} \int_{y=g_1(x)}^{g_2(x)} y\, dy\, dx$$

61. Use Pappus's Theorem in Exercise 60 to show that the torus obtained by revolving a circle of radius b centered at $(0, a)$ about the x-axis (where $b < a$) has volume $V = 2\pi^2 ab^2$.

62. Use Pappus's Theorem to compute $\overline{y}$ for the upper half of the disk $x^2 + y^2 \le a^2$, $y \ge 0$. *Hint:* The disk revolved about the x-axis is a sphere.

63. Parallel-Axis Theorem Let $\mathcal{W}$ be a region in $\mathbf{R}^3$ with center of mass at the origin. Let I_z be the moment of inertia of $\mathcal{W}$ about the z-axis, and let I_h be the moment of inertia about the vertical axis through a point $P = (a, b, 0)$, where $h = \sqrt{a^2 + b^2}$. By definition,

$$I_h = \iiint_{\mathcal{W}} ((x - a)^2 + (y - b)^2)\rho(x, y, z)\, dV$$

Prove the Parallel-Axis Theorem: $I_h = I_z + Mh^2$.

64. Let $\mathcal{W}$ be a cylinder of radius 10 cm and height 20 cm, with total mass $M = 500$ g. Use the Parallel-Axis Theorem (Exercise 63) and the result of Exercise 45 to calculate the moment of inertia of $\mathcal{W}$ about an axis that is parallel to and at a distance of 30 cm from the cylinder's axis of symmetry.

16.6 Change of Variables

The formulas for integration in polar, cylindrical, and spherical coordinates are important special cases of the general Change of Variables Formula for multiple integrals. In this section we discuss the general formula.

Maps from $\mathbf{R}^2$ to $\mathbf{R}^2$

A function $G : X \to Y$ from a set X (the domain) to another set Y is often called a **map** or a **mapping**. For $x \in X$, the element $G(x)$ belongs to Y and is called the **image** of x. The set of all images $G(x)$ is called the image or **range** of G. We denote the image by $G(X)$.

In this section, we consider maps $G : \mathcal{D} \to \mathbf{R}^2$ defined on a domain $\mathcal{D}$ in $\mathbf{R}^2$ (Figure 1). To prevent confusion, we'll often use u, v as our domain variables and x, y for the range. Thus, we will write $G(u, v) = (x(u, v), y(u, v))$, where the components x and y are functions of u and v:

$$x = x(u, v), \qquad y = y(u, v)$$

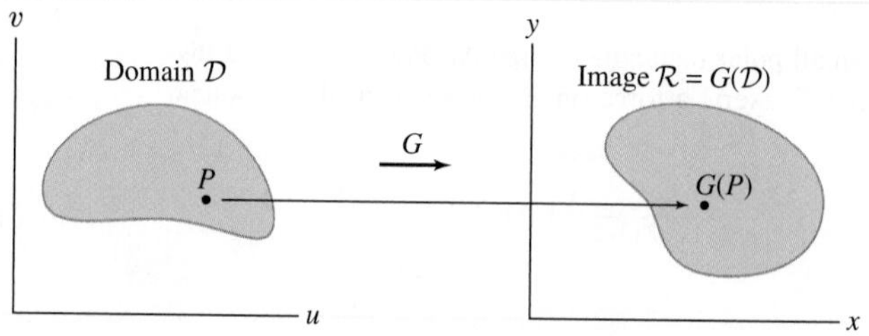

FIGURE 1 G maps $\mathcal{D}$ to $\mathcal{R}$.

One map we are familiar with is the map defining polar coordinates. For this map, we use variables r, θ instead of u, v. The **polar coordinates map** $G : \mathbf{R}^2 \to \mathbf{R}^2$ is defined by

$$G(r, \theta) = (r\cos\theta, r\sin\theta)$$

■ **EXAMPLE 1 Polar Coordinates Map** Describe the image of a polar rectangle $\mathcal{R} = [r_1, r_2] \times [\theta_1, \theta_2]$ under the polar coordinates map.

Solution Referring to Figure 2, we see that:

- A vertical line $r = r_1$ (shown in red) is mapped to the set of points with radial coordinate r_1 and arbitrary angle. This is the circle of radius r_1.
- A horizontal line $\theta = \theta_1$ (dashed line in the figure) is mapped to the set of points with polar angle θ and arbitrary r-coordinate. This is the line through the origin of angle θ_1.

The image of $\mathcal{R} = [r_1, r_2] \times [\theta_1, \theta_2]$ under the polar coordinates map $G(r, \theta) = (r\cos\theta, r\sin\theta)$ is the polar rectangle in the xy-plane defined by $r_1 \le r \le r_2, \theta_1 \le \theta \le \theta_2$. ■

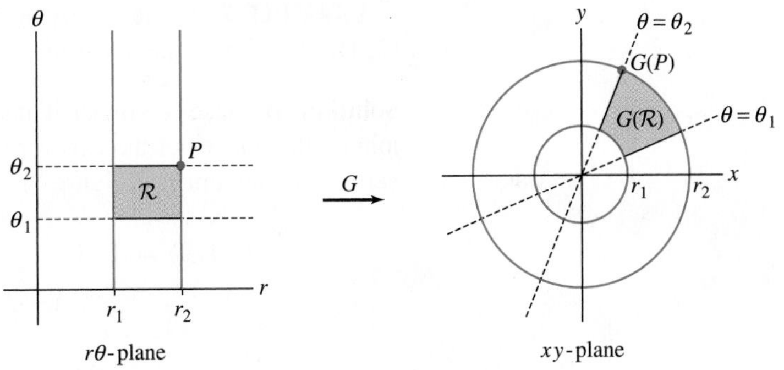

FIGURE 2 The polar coordinates map $G(r, \theta) = (r\cos\theta, r\sin\theta)$.

General mappings can be quite complicated, so it is useful to study the simplest case—linear maps—in detail. A map $G(u, v)$ is **linear** if it has the form

$$G(u, v) = (Au + Cv, Bu + Dv) \quad (A, B, C, D \text{ constants})$$

We can get a clear picture of this linear map by thinking of G as a map from vectors in the uv-plane to vectors in the xy-plane. Then G has the following linearity properties (see Exercise 46):

$$G(u_1 + u_2, v_1 + v_2) = G(u_1, v_1) + G(u_2, v_2) \tag{1}$$

$$G(cu, cv) = cG(u, v) \quad (c \text{ any constant}) \tag{2}$$

A consequence of these properties is that G maps the parallelogram spanned by any two vectors **a** and **b** in the uv-plane to the parallelogram spanned by the images $G(\mathbf{a})$ and $G(\mathbf{b})$, as shown in Figure 3.

More generally, *G maps the segment joining any two points P and Q to the segment joining $G(P)$ and $G(Q)$* (see Exercise 47). The grid generated by basis vectors $\mathbf{i} = \langle 1, 0\rangle$ and $\mathbf{j} = \langle 0, 1\rangle$ is mapped to the grid generated by the image vectors (Figure 3)

$$\mathbf{r} = G(1, 0) = \langle A, B\rangle$$

$$\mathbf{s} = G(0, 1) = \langle C, D\rangle$$

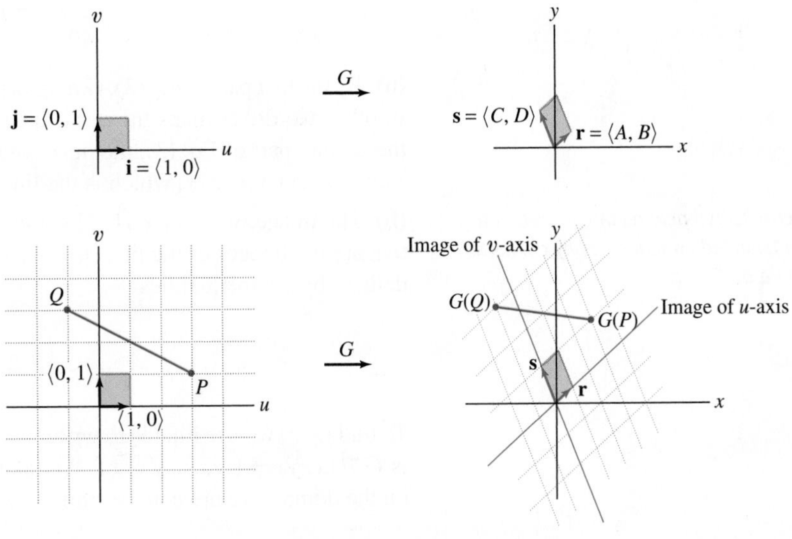

FIGURE 3 A linear mapping G maps a parallelogram to a parallelogram.

EXAMPLE 2 **Image of a Triangle** Find the image of the triangle $\mathcal{T}$ with vertices (1, 2), (2, 1), (3, 4) under the linear map $G(u, v) = (2u - v, u + v)$.

Solution Because G is linear, it maps the segment joining two vertices of $\mathcal{T}$ to the segment joining the images of the two vertices. Therefore, the image of $\mathcal{T}$ is the triangle whose vertices are the images (Figure 4)

$$G(1, 2) = (0, 3), \qquad G(2, 1) = (3, 3), \qquad G(3, 4) = (2, 7)$$ ■

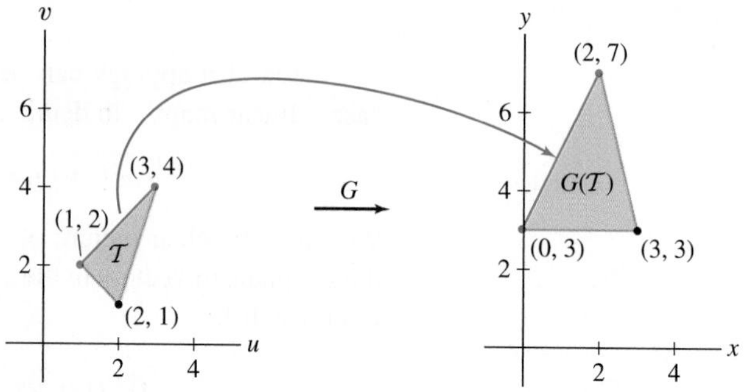

FIGURE 4 The map $G(u, v) = (2u - v, u + v)$.

To understand a nonlinear map, it is usually helpful to determine the images of horizontal and vertical lines, as we did for the polar coordinate mapping.

EXAMPLE 3 Let $G(u, v) = (uv^{-1}, uv)$ for $u > 0$, $v > 0$. Determine the images of:

(a) The lines $u = c$ and $v = c$ **(b)** $[1, 2] \times [1, 2]$

Find the inverse map G^{-1}.

Solution In this map, we have $x = uv^{-1}$ and $y = uv$. Thus

$$xy = u^2, \qquad \frac{y}{x} = v^2 \qquad \boxed{3}$$

(a) By the first part of Eq. (3), G maps a point (c, v) to a point in the xy-plane with $xy = c^2$. In other words, G maps the vertical line $u = c$ to the hyperbola $xy = c^2$. Similarly, by the second part of Eq. (3), the horizontal line $v = c$ is mapped to the set of points where $x/y = c^2$, or $y = c^2x$, which is the line through the origin of slope c^2. See Figure 5.

The term "curvilinear rectangle" refers to a region bounded on four sides by curves as in Figure 5.

(b) The image of $[1, 2] \times [1, 2]$ is the *curvilinear* rectangle bounded by the four curves that are the images of the lines $u = 1$, $u = 2$, and $v = 1$, $v = 2$. By Eq. (3), this region is defined by the inequalities

$$1 \le xy \le 4, \qquad 1 \le \frac{y}{x} \le 4$$

To find G^{-1}, we use Eq. (3) to write $u = \sqrt{xy}$ and $v = \sqrt{y/x}$. Therefore, the inverse map is $G^{-1}(x, y) = \left(\sqrt{xy}, \sqrt{y/x}\right)$. We take positive square roots because $u > 0$ and $v > 0$ on the domain we are considering. ■

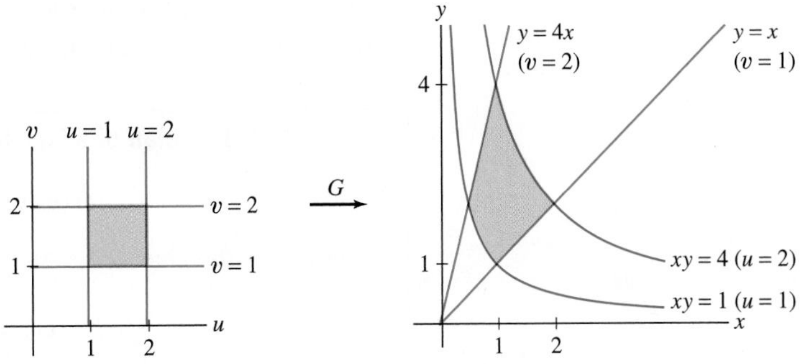

FIGURE 5 The mapping $G(u, v) = (uv^{-1}, uv)$.

How Area Changes under a Mapping: The Jacobian Determinant

The **Jacobian determinant** (or simply the Jacobian) of a map

$$G(u, v) = (x(u, v), y(u, v))$$

is the determinant

$$\text{Jac}(G) = \begin{vmatrix} \dfrac{\partial x}{\partial u} & \dfrac{\partial x}{\partial v} \\ \dfrac{\partial y}{\partial u} & \dfrac{\partial y}{\partial v} \end{vmatrix} = \frac{\partial x}{\partial u}\frac{\partial y}{\partial v} - \frac{\partial x}{\partial v}\frac{\partial y}{\partial u}$$

← *REMINDER The definition of a* 2×2 *determinant is*

$$\begin{vmatrix} a & b \\ c & d \end{vmatrix} = ad - bc \qquad \boxed{4}$$

The Jacobian Jac(G) is also denoted $\dfrac{\partial(x, y)}{\partial(u, v)}$. Note that Jac$(G)$ is a function of u and v.

■ **EXAMPLE 4** Evaluate the Jacobian of $G(u, v) = (u^3 + v, uv)$ at $(u, v) = (2, 1)$.

Solution We have $x = u^3 + v$ and $y = uv$, so

$$\text{Jac}(G) = \frac{\partial(x, y)}{\partial(u, v)} = \begin{vmatrix} \dfrac{\partial x}{\partial u} & \dfrac{\partial x}{\partial v} \\ \dfrac{\partial y}{\partial u} & \dfrac{\partial y}{\partial v} \end{vmatrix}$$

$$= \begin{vmatrix} 3u^2 & 1 \\ v & u \end{vmatrix} = 3u^3 - v$$

The value of the Jacobian at (2, 1) is $\text{Jac}(G)(2, 1) = 3(2)^3 - 1 = 23$. ■

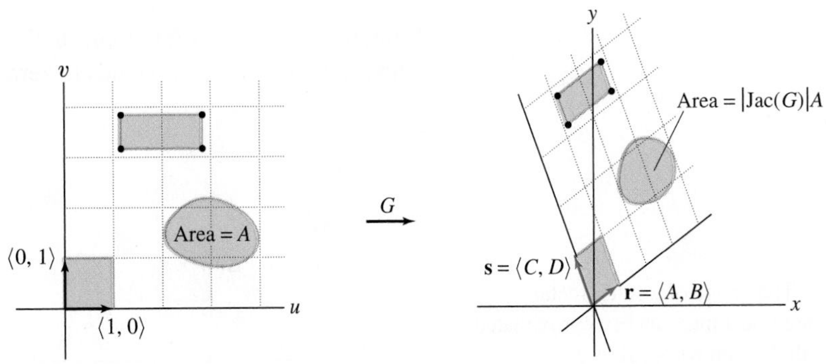

FIGURE 6 A linear map G expands (or shrinks) area by the factor $|\text{Jac}(G)|$.

The Jacobian tells us how area changes under a map G. We can see this most directly in the case of a linear map $G(u, v) = (Au + Cv, Bu + Dv)$.

THEOREM 1 Jacobian of a Linear Map The Jacobian of a linear map

$$G(u, v) = (Au + Cv, Bu + Dv)$$

is *constant* with value

$$\text{Jac}(G) = \begin{vmatrix} A & C \\ B & D \end{vmatrix} = AD - BC \qquad (5)$$

Under G, the area of a region $\mathcal{D}$ is multiplied by the factor $|\text{Jac}(G)|$; that is,

$$\text{Area}(G(\mathcal{D})) = |\text{Jac}(G)|\text{Area}(\mathcal{D}) \qquad (6)$$

Proof Eq. (5) is verified by direct calculation: Because

$$x = Au + Cv \quad \text{and} \quad y = Bu + Dv$$

the partial derivatives in the Jacobian are the constants A, B, C, D.

We sketch a proof of Eq. (6). It certainly holds for the unit rectangle $\mathcal{D} = [1, 0] \times [0, 1]$ because $G(\mathcal{D})$ is the parallelogram spanned by the vectors $\langle A, B\rangle$ and $\langle C, D\rangle$ (Figure 6) and this parallelogram has area

$$|\text{Jac}(G)| = |AD - BC|$$

by Theorem 3 in Section 13.4. Similarly, we can check directly that Eq. (6) holds for arbitrary parallelograms (see Exercise 48). To verify Eq. (6) for a general domain $\mathcal{D}$, we use the fact that $\mathcal{D}$ can be approximated as closely as desired by a union of rectangles in a fine grid of lines parallel to the u- and v-axes. ■

We cannot expect Eq. (6) to hold for a nonlinear map. In fact, it would not make sense as stated because the value $\text{Jac}(G)(P)$ may vary from point to point. However, it is *approximately true* if the domain $\mathcal{D}$ is small and P is a sample point in $\mathcal{D}$:

$$\text{Area}(G(\mathcal{D})) \approx |\text{Jac}(G)(P)|\text{Area}(\mathcal{D}) \qquad (7)$$

This result may be stated more precisely as the limit relation:

$$|\text{Jac}(G)(P)| = \lim_{|\mathcal{D}|\to 0} \frac{\text{Area}(G(\mathcal{D}))}{\text{Area}(\mathcal{D})} \qquad (8)$$

Here, we write $|\mathcal{D}| \to 0$ to indicate the limit as the diameter of $\mathcal{D}$ (the maximum distance between two points in $\mathcal{D}$) tends to zero.

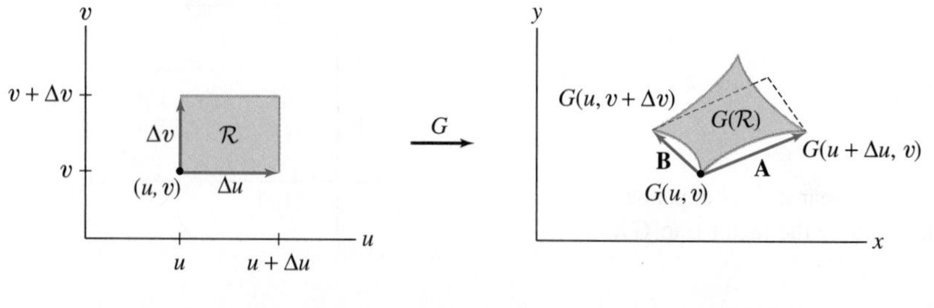

FIGURE 7 The image of a small rectangle under a nonlinear map can be approximated by a parallelogram whose sides are determined by the linear approximation.

CONCEPTUAL INSIGHT Although a rigorous proof of Eq. (8) is too technical to include here, we can understand Eq. (7) as an application of linear approximation. Consider a rectangle $\mathcal{R}$ with vertex at $P = (u, v)$ and sides of lengths Δu and Δv, assumed to be small as in Figure 7. The image $G(\mathcal{R})$ is not a parallelogram, but it is approximated well by the parallelogram spanned by the vectors **A** and **B** in the figure:

$$\begin{aligned}\mathbf{A} &= G(u + \Delta u, v) - G(u, v)\\ &= (x(u + \Delta u, v) - x(u, v), y(u + \Delta u, v) - y(u, v))\\ \mathbf{B} &= G(u, v + \Delta v) - G(u, v)\\ &= (x(u, v + \Delta v) - x(u, v), y(u, v + \Delta v) - y(u, v))\end{aligned}$$

The linear approximation applied to the components of G yields

$$\mathbf{A} \approx \left\langle \frac{\partial x}{\partial u}\Delta u, \frac{\partial y}{\partial u}\Delta u \right\rangle \tag{9}$$

$$\mathbf{B} \approx \left\langle \frac{\partial x}{\partial v}\Delta v, \frac{\partial y}{\partial v}\Delta v \right\rangle \tag{10}$$

REMINDER Equations (9) and (10) use the linear approximations

$$x(u + \Delta u, v) - x(u, v) \approx \frac{\partial x}{\partial u}\Delta u$$

$$y(u + \Delta u, v) - y(u, v) \approx \frac{\partial y}{\partial u}\Delta u$$

and

$$x(u, v + \Delta v) - x(u, v) \approx \frac{\partial x}{\partial v}\Delta v$$

$$y(u, v + \Delta v) - y(u, v) \approx \frac{\partial y}{\partial v}\Delta v$$

This yields the desired approximation:

$$\begin{aligned}\text{Area}(G(\mathcal{R})) &\approx \left|\det\begin{pmatrix}\mathbf{A}\\ \mathbf{B}\end{pmatrix}\right| = \left|\det\begin{pmatrix}\frac{\partial x}{\partial u}\Delta u & \frac{\partial y}{\partial u}\Delta u\\ \frac{\partial x}{\partial v}\Delta v & \frac{\partial y}{\partial v}\Delta v\end{pmatrix}\right|\\ &= \left|\frac{\partial x}{\partial u}\frac{\partial y}{\partial v} - \frac{\partial y}{\partial u}\frac{\partial x}{\partial v}\right| \Delta u\,\Delta v\\ &= |\text{Jac}(G)(P)|\text{Area}(\mathcal{R})\end{aligned}$$

since the area of $\mathcal{R}$ is $\Delta u\,\Delta v$.

The Change of Variables Formula

Recall the formula for integration in polar coordinates:

$$\iint_{\mathcal{D}} f(x, y)\,dx\,dy = \int_{\theta_1}^{\theta_2}\int_{r_1}^{r_2} f(r\cos\theta, r\sin\theta)\,r\,dr\,d\theta \tag{11}$$

Here, $\mathcal{D}$ is the polar rectangle consisting of points $(x, y) = (r\cos\theta, r\sin\theta)$ in the xy-plane (see Figure 2 above). The domain of integration on the right is the rectangle $\mathcal{R} = [\theta_1, \theta_2] \times [r_1, r_2]$ in the $r\theta$-plane. Thus, $\mathcal{D}$ is the image of the domain on the right under the polar coordinates map.

The general Change of Variables Formula has a similar form. Given a map

$$G: \underset{\text{in } uv\text{-plane}}{\mathcal{D}_0} \to \underset{\text{in } xy\text{-plane}}{\mathcal{D}}$$

REMINDER G is called "one-to-one" if $G(P) = G(Q)$ only for $P = Q$.

from a domain in the uv-plane to a domain in the xy-plane (Figure 8), our formula expresses an integral over $\mathcal{D}$ as an integral over $\mathcal{D}_0$. The Jacobian plays the role of the factor r on the right-hand side of Eq. (11).

A few technical assumptions are necessary. First, we assume that G is one-to-one, at least on the interior of $\mathcal{D}_0$, because we want G to cover the target domain $\mathcal{D}$ just once. We also assume that G is a C^1 **map**, by which we mean that the component functions x and y have continuous partial derivatives. Under these conditions, if $f(x, y)$ is continuous, we have the following result.

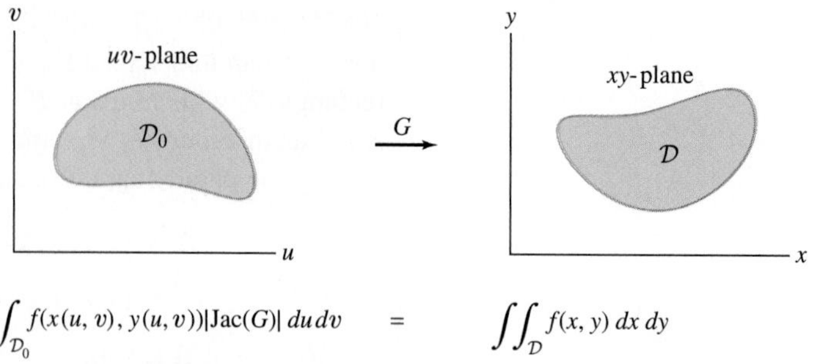

FIGURE 8 The Change of Variables Formula expresses a double integral over $\mathcal{D}$ as a double integral over $\mathcal{D}_0$.

Eq. (12) is summarized by the symbolic equality

$$dx\,dy = \left|\frac{\partial(x,y)}{\partial(u,v)}\right| du\,dv$$

Recall that $\dfrac{\partial(x,y)}{\partial(u,v)}$ *denotes the Jacobian* $\text{Jac}(G)$.

REMINDER *If $\mathcal{D}$ is a domain of small diameter, $P \in \mathcal{D}$ is a sample point, and $f(x, y)$ is continuous, then (see Section 16.2)*

$$\iint_{\mathcal{D}} f(x,y)\,dx\,dy \approx f(P)\text{Area}(\mathcal{D})$$

THEOREM 2 Change of Variables Formula Let $G : \mathcal{D}_0 \to \mathcal{D}$ be a C^1 mapping that is one-to-one on the interior of $\mathcal{D}_0$. If $f(x, y)$ is continuous, then

$$\iint_{\mathcal{D}} f(x,y)\,dx\,dy = \iint_{\mathcal{D}_0} f(x(u,v), y(u,v)) \left|\frac{\partial(x,y)}{\partial(u,v)}\right| du\,dv \qquad 12$$

Proof We sketch the proof. Observe first that Eq. (12) is *approximately* true if the domains $\mathcal{D}_0$ and $\mathcal{D}$ are small. Let $P = G(P_0)$ where P_0 is any sample point in $\mathcal{D}_0$. Since $f(x, y)$ is continuous, the approximation recalled in the margin together with Eq. (7) yield

$$\begin{aligned}\iint_{\mathcal{D}} f(x,y)\,dx\,dy &\approx f(P)\text{Area}(\mathcal{D})\\ &\approx f(G(P_0))\,|\text{Jac}(G)(P_0)|\,\text{Area}(\mathcal{D}_0)\\ &\approx \iint_{\mathcal{D}_0} f(G(u,v))\,|\text{Jac}(G)(u,v)|\,du\,dv\end{aligned}$$

If $\mathcal{D}$ is not small, divide it into small subdomains $D_j = G(\mathcal{D}_{0j})$ (Figure 9 shows a rectangle divided into smaller rectangles), apply the approximation to each subdomain, and sum:

$$\begin{aligned}\iint_{\mathcal{D}} f(x,y)\,dx\,dy &= \sum_j \iint_{\mathcal{D}_j} f(x,y)\,dx\,dy\\ &\approx \sum_j \iint_{\mathcal{D}_{0j}} f(G(u,v)))\,|\text{Jac}(G)(u,v)|\,du\,dv\\ &= \iint_{\mathcal{D}_0} f(G(u,v))\,|\text{Jac}(G)(u,v)|\,du\,dv\end{aligned}$$

Careful estimates show that the error tends to zero as the maximum of the diameters of the subdomains $\mathcal{D}_j$ tends to zero. This yields the Change of Variables Formula. ■

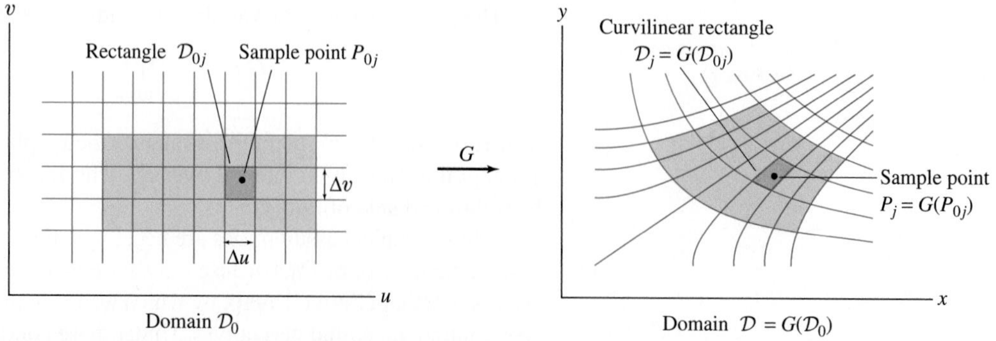

FIGURE 9 G maps a rectangular grid on $\mathcal{D}_0$ to a curvilinear grid on $\mathcal{D}$.

■ **EXAMPLE 5 Polar Coordinates Revisited** Use the Change of Variables Formula to derive the formula for integration in polar coordinates.

Solution The Jacobian of the polar coordinate map $G(r,\theta) = (r\cos\theta, r\sin\theta)$ is

$$\operatorname{Jac}(G) = \begin{vmatrix} \dfrac{\partial x}{\partial r} & \dfrac{\partial x}{\partial \theta} \\ \dfrac{\partial y}{\partial r} & \dfrac{\partial y}{\partial \theta} \end{vmatrix} = \begin{vmatrix} \cos\theta & -r\sin\theta \\ \sin\theta & r\cos\theta \end{vmatrix} = r(\cos^2\theta + \sin^2\theta) = r$$

Let $\mathcal{D} = G(\mathcal{R})$ be the image under the polar coordinates map G of the rectangle $\mathcal{R}$ defined by $r_0 \le r \le r_1, \theta_0 \le \theta \le \theta_1$ (see Figure 2). Then Eq. (12) yields the expected formula for polar coordinates:

$$\iint_{\mathcal{D}} f(x,y)\,dx\,dy = \int_{\theta_0}^{\theta_1}\int_{r_0}^{r_1} f(r\cos\theta, r\sin\theta)\,r\,dr\,d\theta \qquad \boxed{13}$$

■

Assumptions Matter In the Change of Variables Formula, we assume that G is one-to-one on the interior but not necessarily on the boundary of the domain. Thus, we can apply Eq. (12) to the polar coordinates map G on the rectangle $\mathcal{D}_0 = [0,1]\times[0,2\pi]$. In this case, G is one-to-one on the interior but not on the boundary of $\mathcal{D}_0$ since $G(0,\theta) = (0,0)$ for all θ and $G(r,0) = G(r,2\pi)$ for all r. On the other hand, Eq. (12) cannot be applied to G on the rectangle $[0,1]\times[0,4\pi]$ because it is not one-to-one on the interior.

■ **EXAMPLE 6** Use the Change of Variables Formula to calculate $\displaystyle\iint_{\mathcal{P}} e^{4x-y}\,dx\,dy$, where $\mathcal{P}$ is the parallelogram spanned by the vectors $\langle 4,1\rangle$, $\langle 3,3\rangle$ in Figure 10.

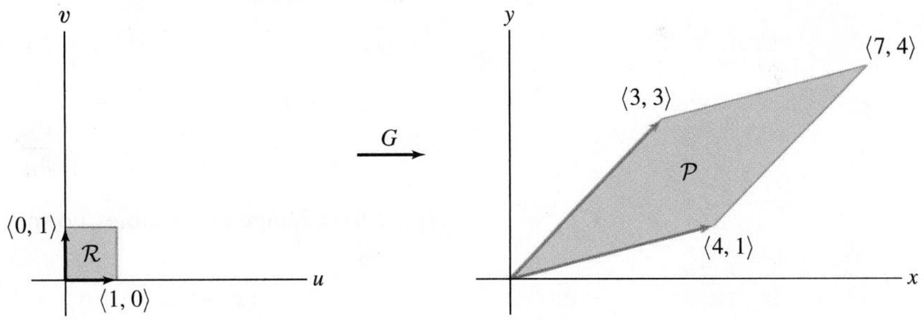

FIGURE 10 The map $G(u,v) = (4u+3v, u+3v)$.

Recall that the linear map

$$G(u,v) = (Au + Cv, Bu + Dv)$$

satisfies

$$G(1,0) = (A,B), \quad G(0,1) = (C,D)$$

Solution

***Step 1.* Define a map.**

We can convert our double integral to an integral over the unit square $\mathcal{R} = [0,1]\times[0,1]$ if we can find a map that sends $\mathcal{R}$ to $\mathcal{P}$. The following linear map does the job:

$$G(u,v) = (4u+3v, u+3v)$$

Indeed, $G(1,0) = (4,1)$ and $G(0,1) = (3,3)$, so it maps $\mathcal{R}$ to $\mathcal{P}$ because linear maps map parallelograms to parallelograms.

***Step 2.* Compute the Jacobian.**

$$\operatorname{Jac}(G) = \begin{vmatrix} \dfrac{\partial x}{\partial u} & \dfrac{\partial x}{\partial v} \\ \dfrac{\partial y}{\partial u} & \dfrac{\partial y}{\partial v} \end{vmatrix} = \begin{vmatrix} 4 & 3 \\ 1 & 3 \end{vmatrix} = 9$$

Step 3. **Express $f(x, y)$ in terms of the new variables.**
Since $x = 4u + 3v$ and $y = u + 3v$, we have

$$e^{4x-y} = e^{4(4u+3v)-(u+3v)} = e^{15u+9v}$$

Step 4. **Apply the Change of Variables Formula.**
The Change of Variables Formula tells us that $dx\,dy = 9\,du\,dv$:

$$\iint_{\mathcal{P}} e^{4x-y}\,dx\,dy = \iint_{\mathcal{R}} e^{15u+9v}\,|\text{Jac}(G)|\,du\,dv = \int_0^1\int_0^1 e^{15u+9v}\,(9\,du\,dv)$$

$$= 9\left(\int_0^1 e^{15u}\,du\right)\left(\int_0^1 e^{9v}\,dv\right) = \frac{1}{15}(e^{15}-1)(e^9-1) \qquad ■$$

■ **EXAMPLE 7** Use the Change of Variables Formula to compute

$$\iint_{\mathcal{D}} (x^2 + y^2)\,dx\,dy$$

where $\mathcal{D}$ is the domain $1 \le xy \le 4$, $1 \le y/x \le 4$ (Figure 11).

Solution In Example 3, we studied the map $G(u, v) = (uv^{-1}, uv)$, which can be written

$$x = uv^{-1}, \qquad y = uv$$

We showed (Figure 11) that G maps the rectangle $\mathcal{D}_0 = [1, 2] \times [1, 2]$ to our domain $\mathcal{D}$. Indeed, because $xy = u^2$ and $xy^{-1} = v^2$, the two conditions $1 \le xy \le 4$ and $1 \le y/x \le 4$ that define $\mathcal{D}$ become $1 \le u \le 2$ and $1 \le v \le 2$.

The Jacobian is

$$\text{Jac}(G) = \frac{\partial(x, y)}{\partial(u, v)} = \begin{vmatrix} \dfrac{\partial x}{\partial u} & \dfrac{\partial x}{\partial v} \\[2ex] \dfrac{\partial y}{\partial u} & \dfrac{\partial y}{\partial v} \end{vmatrix} = \begin{vmatrix} v^{-1} & -uv^{-2} \\ v & u \end{vmatrix} = \frac{2u}{v}$$

To apply the Change of Variables Formula, we write $f(x, y)$ in terms of u and v:

$$f(x, y) = x^2 + y^2 = \left(\frac{u}{v}\right)^2 + (uv)^2 = u^2(v^{-2} + v^2)$$

By the Change of Variables Formula,

$$\iint_{\mathcal{D}} (x^2 + y^2)\,dx\,dy = \iint_{\mathcal{D}_0} u^2(v^{-2} + v^2)\left|\frac{2u}{v}\right|\,du\,dv$$

$$= 2\int_{v=1}^{2}\int_{u=1}^{2} u^3(v^{-3} + v)\,du\,dv$$

$$= 2\left(\int_{u=1}^{2} u^3\,du\right)\left(\int_{v=1}^{2} (v^{-3} + v)\,dv\right)$$

$$= 2\left(\frac{1}{4}u^4\bigg|_1^2\right)\left(-\frac{1}{2}v^{-2} + \frac{1}{2}v^2\bigg|_1^2\right) = \frac{225}{16} \qquad ■$$

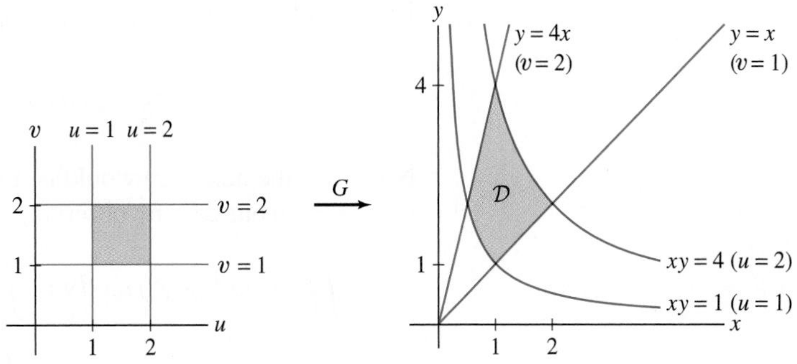

FIGURE 11

Keep in mind that the Change of Variables Formula turns an xy-integral into a uv-integral, but the map G goes from the uv-domain to the xy-domain. Sometimes, it is easier to find a map F going in the *wrong direction*, from the xy-domain to the uv-domain. The desired map G is then the inverse $G = F^{-1}$. The next example shows that in some cases, we can evaluate the integral without solving for G. The key fact is that the Jacobian of F is the reciprocal of Jac(G) (see Exercises 49–51):

Eq. (14) can be written in the suggestive form

$$\frac{\partial(x, y)}{\partial(u, v)} = \left(\frac{\partial(u, v)}{\partial(x, y)}\right)^{-1}$$

$$\text{Jac}(G) = \text{Jac}(F)^{-1} \qquad \text{where } F = G^{-1} \tag{14}$$

■ **EXAMPLE 8 Using the Inverse Map** Integrate $f(x, y) = xy(x^2 + y^2)$ over

$$\mathcal{D}: -3 \le x^2 - y^2 \le 3, \quad 1 \le xy \le 4$$

Solution There is a simple map F that goes in the *wrong* direction. Let $u = x^2 - y^2$ and $v = xy$. Then our domain is defined by the inequalities $-3 \le u \le 3$ and $1 \le v \le 4$, and we can define a map from $\mathcal{D}$ to the rectangle $\mathcal{R} = [-3, 3] \times [1, 4]$ in the uv-plane (Figure 12):

$$F : \mathcal{D} \to \mathcal{R}$$

$$(x, y) \to (x^2 - y^2, xy)$$

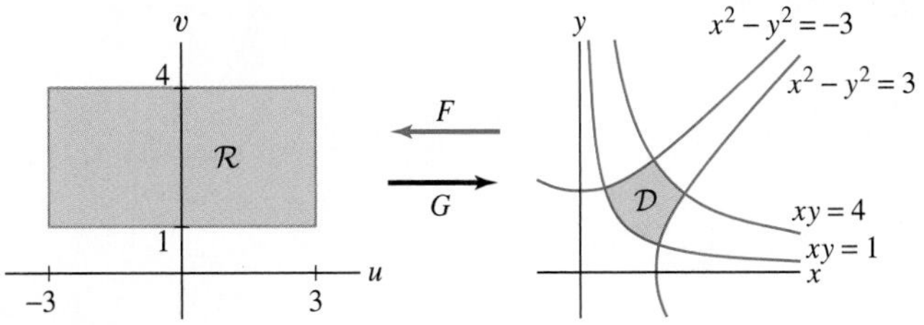

FIGURE 12 The map F goes in the "wrong" direction.

To convert the integral over $\mathcal{D}$ into an integral over the rectangle $\mathcal{R}$, we have to apply the Change of Variables Formula to the inverse mapping:

$$G = F^{-1} : \mathcal{R} \to \mathcal{D}$$

We will see that it is not necessary to find G explicitly. Since $u = x^2 - y^2$ and $v = xy$, the Jacobian of F is

$$\text{Jac}(F) = \begin{vmatrix} \dfrac{\partial u}{\partial x} & \dfrac{\partial u}{\partial y} \\ \dfrac{\partial v}{\partial x} & \dfrac{\partial v}{\partial y} \end{vmatrix} = \begin{vmatrix} 2x & -2y \\ y & x \end{vmatrix} = 2(x^2 + y^2)$$

By Eq. (14),

$$\text{Jac}(G) = \text{Jac}(F)^{-1} = \frac{1}{2(x^2+y^2)}$$

Normally, the next step would be to express $f(x, y)$ in terms of u and v. We can avoid doing this in our case by observing that the Jacobian cancels with one factor of $f(x, y)$:

$$\begin{aligned}\iint_{\mathcal{D}} xy(x^2+y^2)\,dx\,dy &= \iint_{\mathcal{R}} f(x(u,v), y(u,v))\,|\text{Jac}(G)|\,du\,dv \\ &= \iint_{\mathcal{R}} xy(x^2+y^2)\frac{1}{2(x^2+y^2)}\,du\,dv \\ &= \frac{1}{2}\iint_{\mathcal{R}} xy\,du\,dv \\ &= \frac{1}{2}\iint_{\mathcal{R}} v\,du\,dv \qquad \text{(because } v = xy\text{)} \\ &= \frac{1}{2}\int_{-3}^{3}\int_{1}^{4} v\,dv\,du = \frac{1}{2}(6)\left(\frac{1}{2}4^2 - \frac{1}{2}1^2\right) = \frac{45}{2}\end{aligned}$$

■

Change of Variables in Three Variables

The Change of Variables Formula has the same form in three (or more) variables as in two variables. Let

$$G : \mathcal{W}_0 \to \mathcal{W}$$

be a mapping from a three-dimensional region $\mathcal{W}_0$ in (u, v, w)-space to a region $\mathcal{W}$ in (x, y, z)-space, say,

$$x = x(u, v, w), \qquad y = y(u, v, w), \qquad z = z(u, v, w)$$

REMINDER 3×3*-determinants are defined in Eq. (2) of Section 13.4.*

The Jacobian Jac(G) is the 3×3 determinant:

$$\text{Jac}(G) = \frac{\partial(x, y, z)}{\partial(u, v, w)} = \begin{vmatrix} \frac{\partial x}{\partial u} & \frac{\partial x}{\partial v} & \frac{\partial x}{\partial w} \\ \frac{\partial y}{\partial u} & \frac{\partial y}{\partial v} & \frac{\partial y}{\partial w} \\ \frac{\partial z}{\partial u} & \frac{\partial z}{\partial v} & \frac{\partial z}{\partial w} \end{vmatrix} \qquad \boxed{15}$$

The Change of Variables Formula states

$$dx\,dy\,dz = \left|\frac{\partial(x, y, z)}{\partial(u, v, w)}\right| du\,dv\,dw$$

More precisely, if G is C^1 and one-to-one on the interior of $\mathcal{W}_0$, and if f is continuous, then

$$\iiint_{\mathcal{W}} f(x, y, z)\,dx\,dy\,dz$$
$$= \iiint_{\mathcal{W}_0} f(x(u, v, w), y(u, v, w), z(u, v, w)) \left|\frac{\partial(x, y, z)}{\partial(u, v, w)}\right| du\,dv\,dw \qquad \boxed{16}$$

In Exercises 42 and 43, you are asked to use the general Change of Variables Formula to derive the formulas for integration in cylindrical and spherical coordinates developed in Section 16.4.

16.6 SUMMARY

- Let $G(u, v) = (x(u, v), y(u, v))$ be a mapping. We write $x = x(u, v)$ or $x = x(u, v)$ and, similarly, $y = y(u, v)$ or $y = y(u, v)$. The Jacobian of G is the determinant

$$\text{Jac}(G) = \frac{\partial(x, y)}{\partial(u, v)} = \begin{vmatrix} \frac{\partial x}{\partial u} & \frac{\partial x}{\partial v} \\ \frac{\partial y}{\partial u} & \frac{\partial y}{\partial v} \end{vmatrix}$$

- $\text{Jac}(G) = \text{Jac}(F)^{-1}$ where $F = G^{-1}$.
- Change of Variables Formula: If $G : \mathcal{D}_0 \to \mathcal{D}$ is C^1 and one-to-one on the interior of $\mathcal{D}_0$, and if f is continuous, then

$$\iint_{\mathcal{D}} f(x, y)\, dx\, dy = \iint_{\mathcal{D}_0} f(x(u, v), y(u, v)) \left| \frac{\partial(x, y)}{\partial(u, v)} \right| du\, dv$$

- The Change of Variables Formula is written symbolically in two and three variables as

$$dx\, dy = \left| \frac{\partial(x, y)}{\partial(u, v)} \right| du\, dv, \qquad dx\, dy\, dz = \left| \frac{\partial(x, y, z)}{\partial(u, v, w)} \right| du\, dv\, dw$$

16.6 EXERCISES

Preliminary Questions

1. Which of the following maps is linear?

(a) (uv, v) **(b)** $(u + v, u)$ **(c)** $(3, e^u)$

2. Suppose that G is a linear map such that $G(2, 0) = (4, 0)$ and $G(0, 3) = (-3, 9)$. Find the images of:

(a) $G(1, 0)$ **(b)** $G(1, 1)$ **(c)** $G(2, 1)$

3. What is the area of $G(\mathcal{R})$ if $\mathcal{R}$ is a rectangle of area 9 and G is a mapping whose Jacobian has constant value 4?

4. Estimate the area of $G(\mathcal{R})$, where $\mathcal{R} = [1, 1.2] \times [3, 3.1]$ and G is a mapping such that $\text{Jac}(G)(1, 3) = 3$.

Exercises

1. Determine the image under $G(u, v) = (2u, u + v)$ of the following sets:

(a) The u- and v-axes

(b) The rectangle $\mathcal{R} = [0, 5] \times [0, 7]$

(c) The line segment joining $(1, 2)$ and $(5, 3)$

(d) The triangle with vertices $(0, 1)$, $(1, 0)$, and $(1, 1)$

2. Describe [in the form $y = f(x)$] the images of the lines $u = c$ and $v = c$ under the mapping $G(u, v) = (u/v, u^2 - v^2)$.

3. Let $G(u, v) = (u^2, v)$. Is G one-to-one? If not, determine a domain on which G is one-to-one. Find the image under G of:

(a) The u- and v-axes

(b) The rectangle $\mathcal{R} = [-1, 1] \times [-1, 1]$

(c) The line segment joining $(0, 0)$ and $(1, 1)$

(d) The triangle with vertices $(0, 0)$, $(0, 1)$, and $(1, 1)$

4. Let $G(u, v) = (e^u, e^{u+v})$.

(a) Is G one-to-one? What is the image of G?

(b) Describe the images of the vertical lines $u = c$ and the horizontal lines $v = c$.

In Exercises 5–12, let $G(u, v) = (2u + v, 5u + 3v)$ be a map from the uv-plane to the xy-plane.

5. Show that the image of the horizontal line $v = c$ is the line $y = \frac{5}{2}x + \frac{1}{2}c$. What is the image (in slope-intercept form) of the vertical line $u = c$?

6. Describe the image of the line through the points $(u, v) = (1, 1)$ and $(u, v) = (1, -1)$ under G in slope-intercept form.

7. Describe the image of the line $v = 4u$ under G in slope-intercept form.

8. Show that G maps the line $v = mu$ to the line of slope $(5 + 3m)/(2 + m)$ through the origin in the xy-plane.

9. Show that the inverse of G is

$$G^{-1}(x, y) = (3x - y, -5x + 2y)$$

Hint: Show that $G(G^{-1}(x, y)) = (x, y)$ and $G^{-1}(G(u, v)) = (u, v)$.

10. Use the inverse in Exercise 9 to find:

(a) A point in the uv-plane mapping to $(2, 1)$

(b) A segment in the uv-plane mapping to the segment joining $(-2, 1)$ and $(3, 4)$

11. Calculate $\text{Jac}(G) = \dfrac{\partial(x, y)}{\partial(u, v)}$.

12. Calculate $\text{Jac}(G^{-1}) = \dfrac{\partial(u, v)}{\partial(x, y)}$.

In Exercises 13–18, compute the Jacobian (at the point, if indicated).

13. $G(u, v) = (3u + 4v, u - 2v)$

14. $G(r, s) = (rs, r + s)$

15. $G(r, t) = (r \sin t, r - \cos t), \quad (r, t) = (1, \pi)$

16. $G(u, v) = (v \ln u, u^2 v^{-1}), \quad (u, v) = (1, 2)$

17. $G(r, \theta) = (r \cos\theta, r \sin\theta), \quad (r, \theta) = \left(4, \frac{\pi}{6}\right)$

18. $G(u, v) = (ue^v, e^u)$

19. Find a linear mapping G that maps $[0, 1] \times [0, 1]$ to the parallelogram in the xy-plane spanned by the vectors $\langle 2, 3\rangle$ and $\langle 4, 1\rangle$.

20. Find a linear mapping G that maps $[0, 1] \times [0, 1]$ to the parallelogram in the xy-plane spanned by the vectors $\langle -2, 5\rangle$ and $\langle 1, 7\rangle$.

21. Let $\mathcal{D}$ be the parallelogram in Figure 13. Apply the Change of Variables Formula to the map $G(u, v) = (5u + 3v, u + 4v)$ to evaluate $\iint_{\mathcal{D}} xy\, dx\, dy$ as an integral over $\mathcal{D}_0 = [0, 1] \times [0, 1]$.

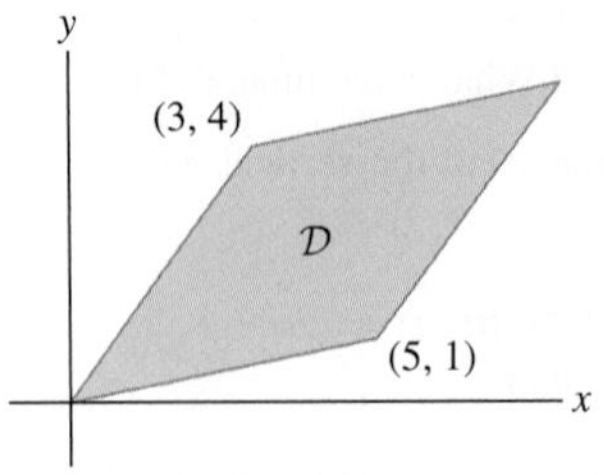

FIGURE 13

22. Let $G(u, v) = (u - uv, uv)$.

(a) Show that the image of the horizontal line $v = c$ is $y = \dfrac{c}{1 - c}x$ if $c \neq 1$, and is the y-axis if $c = 1$.

(b) Determine the images of vertical lines in the uv-plane.

(c) Compute the Jacobian of G.

(d) Observe that by the formula for the area of a triangle, the region $\mathcal{D}$ in Figure 14 has area $\frac{1}{2}(b^2 - a^2)$. Compute this area again, using the Change of Variables Formula applied to G.

(e) Calculate $\iint_{\mathcal{D}} xy\, dx\, dy$.

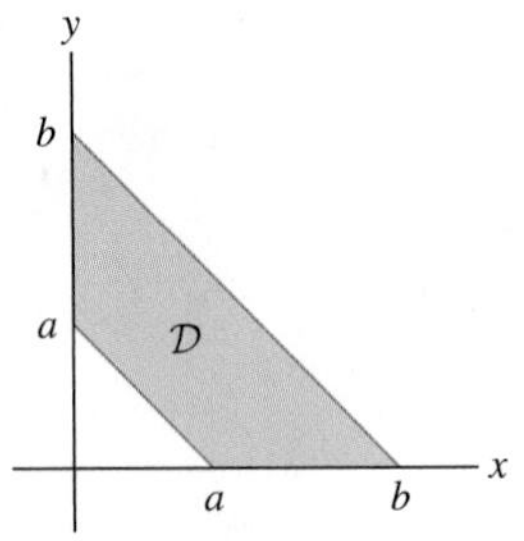

FIGURE 14

23. Let $G(u, v) = (3u + v, u - 2v)$. Use the Jacobian to determine the area of $G(\mathcal{R})$ for:

(a) $\mathcal{R} = [0, 3] \times [0, 5]$ **(b)** $\mathcal{R} = [2, 5] \times [1, 7]$

24. Find a linear map T that maps $[0, 1] \times [0, 1]$ to the parallelogram $\mathcal{P}$ in the xy-plane with vertices $(0, 0)$, $(2, 2)$, $(1, 4)$, $(3, 6)$. Then calculate the double integral of e^{2x-y} over $\mathcal{P}$ via change of variables.

25. With G as in Example 3, use the Change of Variables Formula to compute the area of the image of $[1, 4] \times [1, 4]$.

In Exercises 26–28, let $\mathcal{R}_0 = [0, 1] \times [0, 1]$ be the unit square. The translate of a map $G_0(u, v) = (\phi(u, v), \psi(u, v))$ is a map

$$G(u, v) = (a + \phi(u, v), b + \psi(u, v))$$

where a, b are constants. Observe that the map G_0 in Figure 15 maps $\mathcal{R}_0$ to the parallelogram $\mathcal{P}_0$ and that the translate

$$G_1(u, v) = (2 + 4u + 2v, 1 + u + 3v)$$

maps $\mathcal{R}_0$ to $\mathcal{P}_1$.

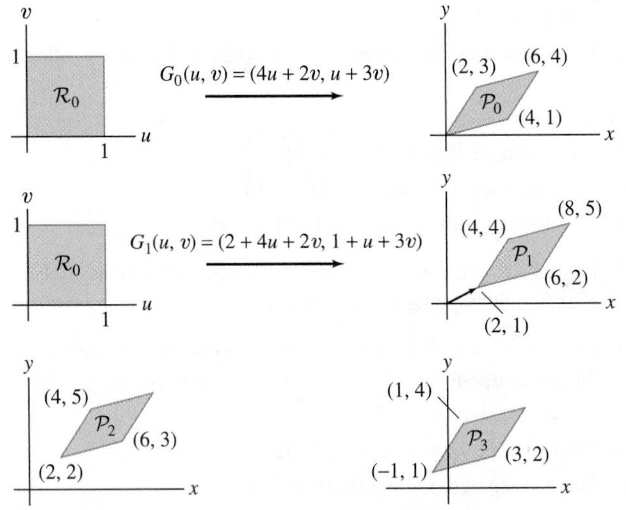

FIGURE 15

26. Find translates G_2 and G_3 of the mapping G_0 in Figure 15 that map the unit square $\mathcal{R}_0$ to the parallelograms $\mathcal{P}_2$ and $\mathcal{P}_3$.

27. Sketch the parallelogram $\mathcal{P}$ with vertices $(1, 1), (2, 4), (3, 6), (4, 9)$ and find the translate of a linear mapping that maps $\mathcal{R}_0$ to $\mathcal{P}$.

28. Find the translate of a linear mapping that maps $\mathcal{R}_0$ to the parallelogram spanned by the vectors $\langle 3, 9\rangle$ and $\langle -4, 6\rangle$ based at $(4, 2)$.

29. Let $\mathcal{D} = G(\mathcal{R})$, where $G(u, v) = (u^2, u + v)$ and $\mathcal{R} = [1, 2] \times [0, 6]$. Calculate $\iint_{\mathcal{D}} y\, dx\, dy$. *Note:* It is not necessary to describe $\mathcal{D}$.

30. Let $\mathcal{D}$ be the image of $\mathcal{R} = [1, 4] \times [1, 4]$ under the map $G(u, v) = (u^2/v, v^2/u)$.

(a) Compute $\text{Jac}(G)$.

(b) Sketch $\mathcal{D}$.

(c) Use the Change of Variables Formula to compute $\text{Area}(\mathcal{D})$ and $\iint_{\mathcal{D}} f(x, y)\, dx\, dy$, where $f(x, y) = x + y$.

31. Compute $\iint_{\mathcal{D}} (x + 3y)\, dx\, dy$, where $\mathcal{D}$ is the shaded region in Figure 16. *Hint:* Use the map $G(u, v) = (u - 2v, v)$.

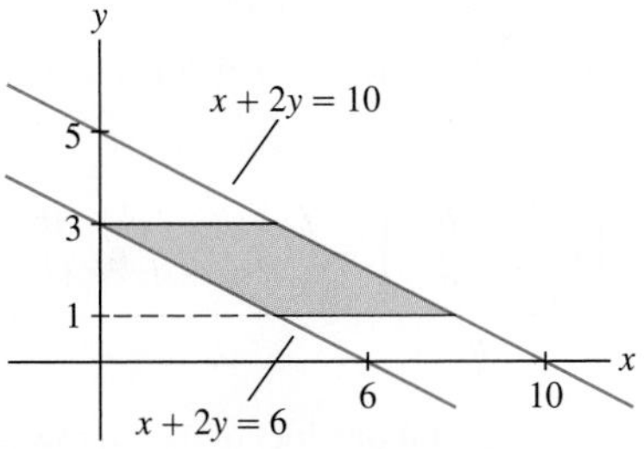

FIGURE 16

32. Use the map $G(u, v) = \left(\dfrac{u}{v + 1}, \dfrac{uv}{v + 1}\right)$ to compute

$$\iint_{\mathcal{D}} (x + y)\, dx\, dy$$

where $\mathcal{D}$ is the shaded region in Figure 17.

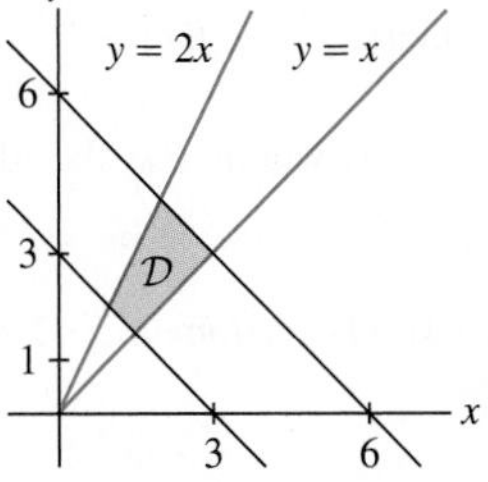

FIGURE 17

33. Show that $T(u, v) = (u^2 - v^2, 2uv)$ maps the triangle $\mathcal{D}_0 = \{(u, v) : 0 \le v \le u \le 1\}$ to the domain $\mathcal{D}$ bounded by $x = 0$, $y = 0$, and $y^2 = 4 - 4x$. Use T to evaluate

$$\iint_{\mathcal{D}} \sqrt{x^2 + y^2}\, dx\, dy$$

34. Find a mapping G that maps the disk $u^2 + v^2 \le 1$ onto the interior of the ellipse $\left(\frac{x}{a}\right)^2 + \left(\frac{y}{b}\right)^2 \le 1$. Then use the Change of Variables Formula to prove that the area of the ellipse is πab.

35. Calculate $\iint_{\mathcal{D}} e^{9x^2+4y^2}\, dx\, dy$, where $\mathcal{D}$ is the interior of the ellipse $\left(\frac{x}{2}\right)^2 + \left(\frac{y}{3}\right)^2 \le 1$.

36. Compute the area of the region enclosed by the ellipse $x^2 + 2xy + 2y^2 - 4y = 8$ as an integral in the variables $u = x + y$, $v = y - 2$.

37. Sketch the domain $\mathcal{D}$ bounded by $y = x^2$, $y = \frac{1}{2}x^2$, and $y = x$. Use a change of variables with the map $x = uv$, $y = u^2$ to calculate

$$\iint_{\mathcal{D}} y^{-1}\, dx\, dy$$

This is an improper integral since $f(x, y) = y^{-1}$ is undefined at $(0, 0)$, but it becomes proper after changing variables.

38. Find an appropriate change of variables to evaluate

$$\iint_{\mathcal{R}} (x + y)^2 e^{x^2 - y^2}\, dx\, dy$$

where $\mathcal{R}$ is the square with vertices $(1, 0), (0, 1), (-1, 0), (0, -1)$.

39. Let G be the inverse of the map $F(x, y) = (xy, x^2y)$ from the xy-plane to the uv-plane. Let $\mathcal{D}$ be the domain in Figure 18. Show, by applying the Change of Variables Formula to the inverse $G = F^{-1}$, that

$$\iint_{\mathcal{D}} e^{xy}\, dx\, dy = \int_{10}^{20} \int_{20}^{40} e^u v^{-1}\, dv\, du$$

and evaluate this result. *Hint:* See Example 8.

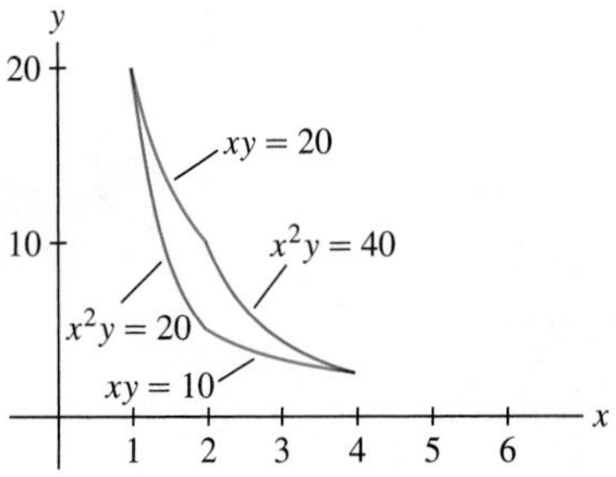

FIGURE 18

40. Sketch the domain

$$\mathcal{D} = \{(x, y) : 1 \le x + y \le 4,\ -4 \le y - 2x \le 1\}$$

(a) Let F be the map $u = x + y$, $v = y - 2x$ from the xy-plane to the uv-plane, and let G be its inverse. Use Eq. (14) to compute $\text{Jac}(G)$.

(b) Compute $\iint_{\mathcal{D}} e^{x+y}\, dx\, dy$ using the Change of Variables Formula with the map G. *Hint:* It is not necessary to solve for G explicitly.

41. Let $I = \iint_{\mathcal{D}} (x^2 - y^2)\,dx\,dy$, where

$$\mathcal{D} = \{(x, y) : 2 \le xy \le 4, 0 \le x - y \le 3, x \ge 0, y \ge 0\}$$

(a) Show that the mapping $u = xy$, $v = x - y$ maps $\mathcal{D}$ to the rectangle $\mathcal{R} = [2, 4] \times [0, 3]$.
(b) Compute $\partial(x, y)/\partial(u, v)$ by first computing $\partial(u, v)/\partial(x, y)$.
(c) Use the Change of Variables Formula to show that I is equal to the integral of $f(u, v) = v$ over $\mathcal{R}$ and evaluate.

42. Derive formula (5) in Section 16.4 for integration in cylindrical coordinates from the general Change of Variables Formula.

43. Derive formula (9) in Section 16.4 for integration in spherical coordinates from the general Change of Variables Formula.

44. Use the Change of Variables Formula in three variables to prove that the volume of the ellipsoid $\left(\frac{x}{a}\right)^2 + \left(\frac{y}{b}\right)^2 + \left(\frac{z}{c}\right)^2 = 1$ is equal to $abc \times$ the volume of the unit sphere.

Further Insights and Challenges

45. Use the map

$$x = \frac{\sin u}{\cos v}, \qquad y = \frac{\sin v}{\cos u}$$

to evaluate the integral

$$\int_0^1 \int_0^1 \frac{dx\,dy}{1 - x^2y^2}$$

This is an improper integral since the integrand is infinite if $x = \pm 1$, $y = \pm 1$, but applying the Change of Variables Formula shows that the result is finite.

46. Verify properties (1) and (2) for linear functions and show that any map satisfying these two properties is linear.

47. Let P and Q be points in $\mathbf{R}^2$. Show that a linear map $G(u, v) = (Au + Cv, Bu + Dv)$ maps the segment joining P and Q to the segment joining $G(P)$ to $G(Q)$. *Hint:* The segment joining P and Q has parametrization

$$(1 - t)\overrightarrow{OP} + t\overrightarrow{OQ} \quad \text{for} \quad 0 \le t \le 1$$

48. Let G be a linear map. Prove Eq. (6) in the following steps.
(a) For any set $\mathcal{D}$ in the uv-plane and any vector $\mathbf{u}$, let $\mathcal{D} + \mathbf{u}$ be the set obtained by translating all points in $\mathcal{D}$ by $\mathbf{u}$. By linearity, G maps $\mathcal{D} + \mathbf{u}$ to the translate $G(\mathcal{D}) + G(\mathbf{u})$ [Figure 19(C)]. Therefore, if Eq. (6) holds for $\mathcal{D}$, it also holds for $\mathcal{D} + \mathbf{u}$.
(b) In the text, we verified Eq. (6) for the unit rectangle. Use linearity to show that Eq. (6) also holds for all rectangles with vertex at the origin and sides parallel to the axes. Then argue that it also holds for each triangular half of such a rectangle, as in Figure 19(A).
(c) Figure 19(B) shows that the area of a parallelogram is a difference of the areas of rectangles and triangles covered by steps (a) and (b). Use this to prove Eq. (6) for arbitrary parallelograms.

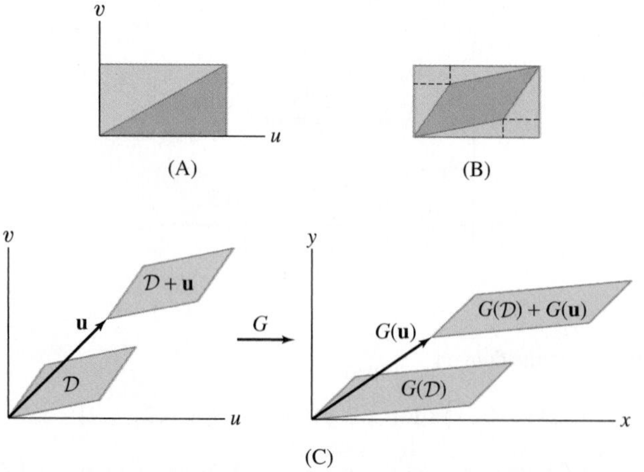

FIGURE 19

49. The product of 2×2 matrices A and B is the matrix AB defined by

$$\underbrace{\begin{pmatrix} a & b \\ c & d \end{pmatrix}}_{A} \underbrace{\begin{pmatrix} a' & b' \\ c' & d' \end{pmatrix}}_{B} = \underbrace{\begin{pmatrix} aa' + bc' & ab' + bd' \\ ca' + dc' & cb' + dd' \end{pmatrix}}_{AB}$$

The (i, j)-entry of A is the **dot product** of the ith row of A and the jth column of B. Prove that $\det(AB) = \det(A)\det(B)$.

50. Let $G_1 : \mathcal{D}_1 \to \mathcal{D}_2$ and $G_2 : \mathcal{D}_2 \to \mathcal{D}_3$ be C^1 maps, and let $G_2 \circ G_1 : \mathcal{D}_1 \to \mathcal{D}_3$ be the composite map. Use the Multivariable Chain Rule and Exercise 49 to show that

$$\text{Jac}(G_2 \circ G_1) = \text{Jac}(G_2)\text{Jac}(G_1)$$

51. Use Exercise 50 to prove that

$$\text{Jac}(G^{-1}) = \text{Jac}(G)^{-1}$$

Hint: Verify that $\text{Jac}(I) = 1$, where I is the identity map $I(u, v) = (u, v)$.

52. Let $(\overline{x}, \overline{y})$ be the centroid of a domain $\mathcal{D}$. For $\lambda > 0$, let $\lambda\mathcal{D}$ be the **dilate** of $\mathcal{D}$, defined by

$$\lambda\mathcal{D} = \{(\lambda x, \lambda y) : (x, y) \in \mathcal{D}\}$$

Use the Change of Variables Formula to prove that the centroid of $\lambda\mathcal{D}$ is $(\lambda\overline{x}, \lambda\overline{y})$.

CHAPTER REVIEW EXERCISES

1. Calculate the Riemann sum $S_{2,3}$ for $\int_1^4 \int_2^6 x^2 y\,dx\,dy$ using two choices of sample points:

(a) Lower-left vertex

(b) Midpoint of rectangle

Then calculate the exact value of the double integral.

2. Let $S_{N,N}$ be the Riemann sum for $\int_0^1 \int_0^1 \cos(xy)\,dx\,dy$ using midpoints as sample points.

(a) Calculate $S_{4,4}$.

(b) CAS Use a computer algebra system to calculate $S_{N,N}$ for $N = 10, 50, 100$.

3. Let $\mathcal{D}$ be the shaded domain in Figure 1.

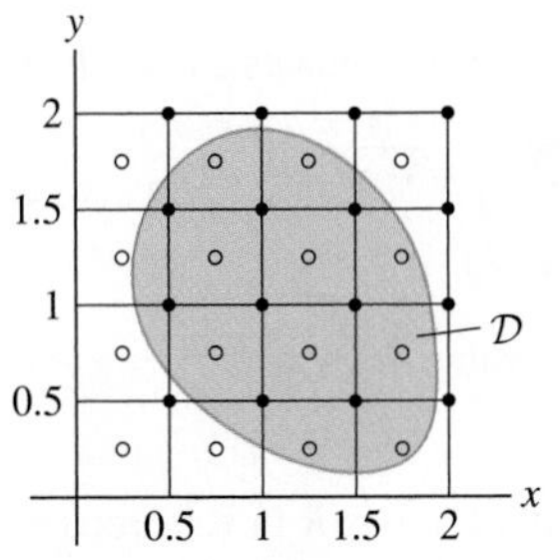

FIGURE 1

Estimate $\iint_{\mathcal{D}} xy\,dA$ by the Riemann sum whose sample points are the midpoints of the squares in the grid.

4. Explain the following:

(a) $\int_{-1}^1 \int_{-1}^1 \sin(xy)\,dx\,dy = 0$

(b) $\int_{-1}^1 \int_{-1}^1 \cos(xy)\,dx\,dy > 0$

In Exercises 5–8, evaluate the iterated integral.

5. $\int_0^2 \int_3^5 y(x-y)\,dx\,dy$

6. $\int_{1/2}^0 \int_0^{\pi/6} e^{2y} \sin 3x\,dx\,dy$

7. $\int_0^{\pi/3} \int_0^{\pi/6} \sin(x+y)\,dx\,dy$

8. $\int_1^2 \int_1^2 \frac{y\,dx\,dy}{x+y^2}$

In Exercises 9–14, sketch the domain $\mathcal{D}$ and calculate $\iint_{\mathcal{D}} f(x,y)\,dA$.

9. $\mathcal{D} = \{0 \le x \le 4,\ 0 \le y \le x\}$, $f(x,y) = \cos y$

10. $\mathcal{D} = \{0 \le x \le 2,\ 0 \le y \le 2x - x^2\}$, $f(x,y) = \sqrt{x}y$

11. $\mathcal{D} = \{0 \le x \le 1,\ 1 - x \le y \le 2 - x\}$, $f(x,y) = e^{x+2y}$

12. $\mathcal{D} = \{1 \le x \le 2,\ 0 \le y \le 1/x\}$, $f(x,y) = \cos(xy)$

13. $\mathcal{D} = \{0 \le y \le 1,\ 0.5y^2 \le x \le y^2\}$, $f(x,y) = ye^{1+x}$

14. $\mathcal{D} = \{1 \le y \le e,\ y \le x \le 2y\}$, $f(x,y) = \ln(x+y)$

15. Express $\int_{-3}^3 \int_0^{9-x^2} f(x,y)\,dy\,dx$ as an iterated integral in the order $dx\,dy$.

16. Let $\mathcal{W}$ be the region bounded by the planes $y = z$, $2y + z = 3$, and $z = 0$ for $0 \le x \le 4$.

(a) Express the triple integral $\iiint_{\mathcal{W}} f(x,y,z)\,dV$ as an iterated integral in the order $dy\,dx\,dz$ (project $\mathcal{W}$ onto the xz-plane).

(b) Evaluate the triple integral for $f(x,y,z) = 1$.

(c) Compute the volume of $\mathcal{W}$ using geometry and check that the result coincides with the answer to (b).

17. Let $\mathcal{D}$ be the domain between $y = x$ and $y = \sqrt{x}$. Calculate $\iint_{\mathcal{D}} xy\,dA$ as an iterated integral in the order $dx\,dy$ and $dy\,dx$.

18. Find the double integral of $f(x,y) = x^3 y$ over the region between the curves $y = x^2$ and $y = x(1-x)$.

19. Change the order of integration and evaluate $\int_0^9 \int_0^{\sqrt{y}} \frac{x\,dx\,dy}{(x^2+y)^{1/2}}$.

20. Verify directly that

$$\int_2^3 \int_0^2 \frac{dy\,dx}{1+x-y} = \int_0^2 \int_2^3 \frac{dx\,dy}{1+x-y}$$

21. Prove the formula

$$\int_0^1 \int_0^y f(x)\,dx\,dy = \int_0^1 (1-x) f(x)\,dx$$

Then use it to calculate $\int_0^1 \int_0^y \frac{\sin x}{1-x}\,dx\,dy$.

22. Rewrite $\int_0^1 \int_{-\sqrt{1-y^2}}^{\sqrt{1-y^2}} \frac{y\,dx\,dy}{(1+x^2+y^2)^2}$ by interchanging the order of integration, and evaluate.

23. Use cylindrical coordinates to compute the volume of the region defined by $4 - x^2 - y^2 \le z \le 10 - 4x^2 - 4y^2$.

24. Evaluate $\iint_{\mathcal{D}} x\,dA$, where $\mathcal{D}$ is the shaded domain in Figure 2.

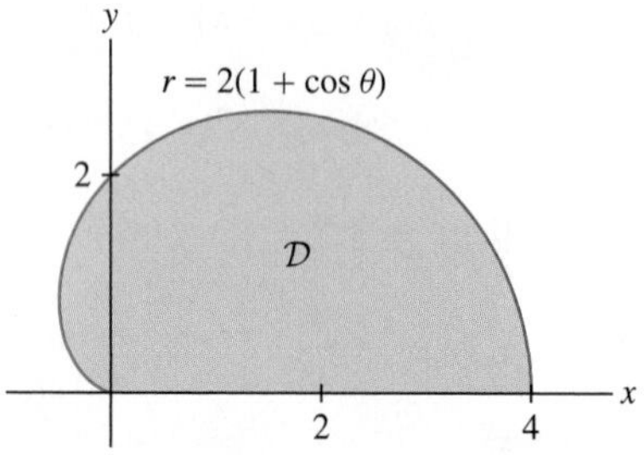

FIGURE 2

25. Find the volume of the region between the graph of the function $f(x, y) = 1 - (x^2 + y^2)$ and the xy-plane.

26. Evaluate $\int_0^3 \int_1^4 \int_2^4 (x^3 + y^2 + z)\,dx\,dy\,dz$.

27. Calculate $\iiint_{\mathcal{B}} (xy + z)\,dV$, where

$$\mathcal{B} = \{0 \le x \le 2,\ 0 \le y \le 1,\ 1 \le z \le 3\}$$

as an iterated integral in two different ways.

28. Calculate $\iiint_{\mathcal{W}} xyz\,dV$, where

$$\mathcal{W} = \{0 \le x \le 1,\ x \le y \le 1,\ x \le z \le x + y\}$$

29. Evaluate $I = \int_{-1}^{1} \int_0^{\sqrt{1-x^2}} \int_0^1 (x + y + z)\,dz\,dy\,dx$.

30. Describe a region whose volume is equal to:

(a) $\int_0^{2\pi} \int_0^{\pi/2} \int_4^9 \rho^2 \sin\phi\,d\rho\,d\phi\,d\theta$

(b) $\int_{-2}^{1} \int_{\pi/3}^{\pi/4} \int_0^2 r\,dr\,d\theta\,dz$

(c) $\int_0^{2\pi} \int_0^3 \int_{-\sqrt{9-r^2}}^{0} r\,dz\,dr\,d\theta$

31. Find the volume of the solid contained in the cylinder $x^2 + y^2 = 1$ below the curve $z = (x + y)^2$ and above the curve $z = -(x - y)^2$.

32. Use polar coordinates to evaluate $\iint_{\mathcal{D}} x\,dA$, where $\mathcal{D}$ is the shaded region between the two circles of radius 1 in Figure 3.

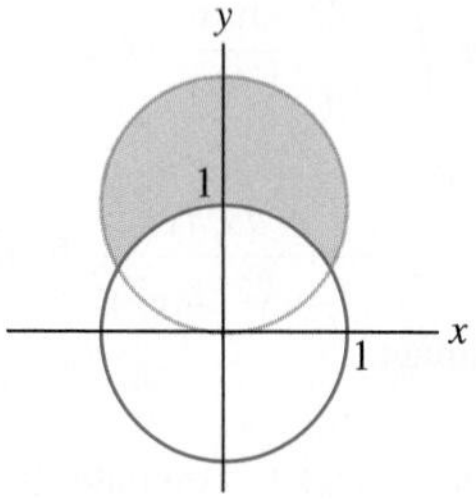

FIGURE 3

33. Use polar coordinates to calculate $\iint_{\mathcal{D}} \sqrt{x^2 + y^2}\,dA$, where $\mathcal{D}$ is the region in the first quadrant bounded by the spiral $r = \theta$, the circle $r = 1$, and the x-axis.

34. Calculate $\iint_{\mathcal{D}} \sin(x^2 + y^2)\,dA$, where

$$\mathcal{D} = \left\{\frac{\pi}{2} \le x^2 + y^2 \le \pi\right\}$$

35. Express in cylindrical coordinates and evaluate:

$$\int_0^1 \int_0^{\sqrt{1-x^2}} \int_0^{\sqrt{x^2+y^2}} z\,dz\,dy\,dx$$

36. Use spherical coordinates to calculate the triple integral of $f(x, y, z) = x^2 + y^2 + z^2$ over the region

$$1 \le x^2 + y^2 + z^2 \le 4$$

37. Convert to spherical coordinates and evaluate:

$$\int_{-2}^{2} \int_{-\sqrt{4-x^2}}^{\sqrt{4-x^2}} \int_0^{\sqrt{4-x^2-y^2}} e^{-(x^2+y^2+z^2)^{3/2}}\,dz\,dy\,dx$$

38. Find the average value of $f(x, y, z) = xy^2z^3$ on the box $[0, 1] \times [0, 2] \times [0, 3]$.

39. Let $\mathcal{W}$ be the ball of radius R in $\mathbf{R}^3$ centered at the origin, and let $P = (0, 0, R)$ be the North Pole. Let $d_P(x, y, z)$ be the distance from P to (x, y, z). Show that the average value of d_P over the sphere $\mathcal{W}$ is equal to $\overline{d} = 6R/5$. *Hint:* Show that

$$\overline{d} = \frac{1}{\frac{4}{3}\pi R^3} \int_{\theta=0}^{2\pi} \int_{\rho=0}^{R} \int_{\phi=0}^{\pi} \rho^2 \sin\phi \sqrt{R^2 + \rho^2 - 2\rho R \cos\phi}\,d\phi\,d\rho\,d\theta$$

and evaluate.

40. CAS Express the average value of $f(x, y) = e^{xy}$ over the ellipse $\frac{x^2}{2} + y^2 = 1$ as an iterated integral, and evaluate numerically using a computer algebra system.

41. Use cylindrical coordinates to find the mass of the solid bounded by $z = 8 - x^2 - y^2$ and $z = x^2 + y^2$, assuming a mass density of $f(x, y, z) = (x^2 + y^2)^{1/2}$.

42. Let $\mathcal{W}$ be the portion of the half-cylinder $x^2 + y^2 \le 4$, $x \ge 0$ such that $0 \le z \le 3y$. Use cylindrical coordinates to compute the mass of $\mathcal{W}$ if the mass density is $\rho(x, y, z) = z^2$.

43. Use cylindrical coordinates to find the mass of a cylinder of radius 4 and height 10 if the mass density at a point is equal to the square of the distance from the cylinder's central axis.

44. Find the centroid of the region $\mathcal{W}$ bounded, in spherical coordinates, by $\phi = \phi_0$ and the sphere $\rho = R$.

45. Find the centroid of the solid bounded by the xy-plane, the cylinder $x^2 + y^2 = R^2$, and the plane $x/R + z/H = 1$.

46. Using cylindrical coordinates, prove that the centroid of a right circular cone of height h and radius R is located at height $\frac{h}{4}$ on the central axis.

47. Find the centroid of solid (A) in Figure 4 defined by $x^2 + y^2 \le R^2$, $0 \le z \le H$, and $\frac{\pi}{6} \le \theta \le 2\pi$, where θ is the polar angle of (x, y).

48. Calculate the coordinate y_{CM} of the centroid of solid (B) in Figure 4 defined by $x^2 + y^2 \le 1$ and $0 \le z \le \frac{1}{2}y + \frac{3}{2}$.

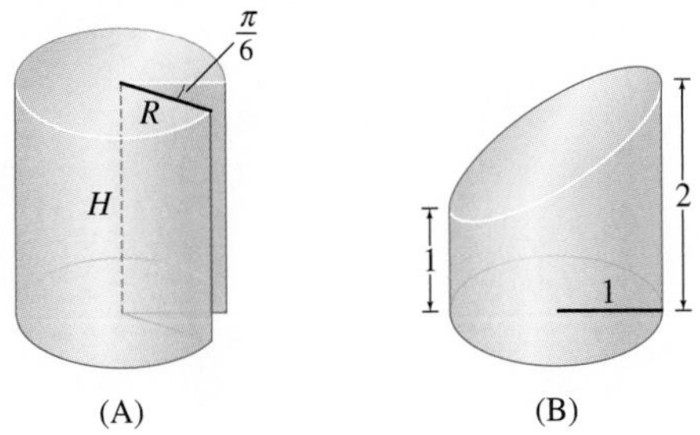

FIGURE 4

49. Find the center of mass of the cylinder $x^2 + y^2 = 1$ for $0 \le z \le 1$, assuming a mass density of $\rho(x, y, z) = z$.

50. Find the center of mass of the sector of central angle $2\theta_0$ (symmetric with respect to the y-axis) in Figure 5, assuming that the mass density is $\rho(x, y) = x^2$.

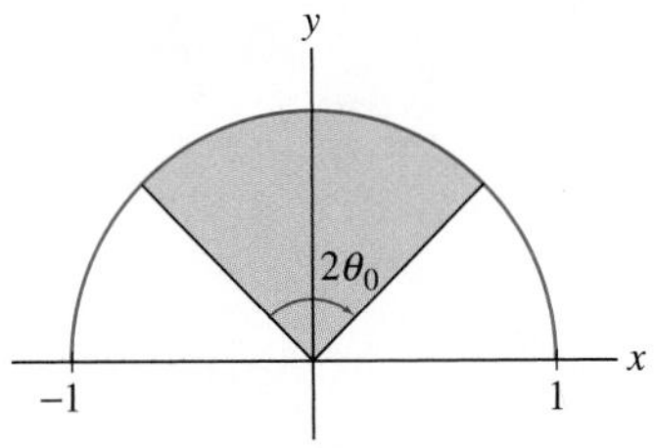

FIGURE 5

51. Find the center of mass of the first octant of the ball $x^2 + y^2 + z^2 = 1$, assuming a mass density of $\rho(x, y, z) = x$.

52. Find a constant C such that

$$p(x, y) = \begin{cases} C(4x - y + 3) & \text{if } 0 \le x \le 2 \text{ and } 0 \le y \le 3 \\ 0 & \text{otherwise} \end{cases}$$

is a probability distribution and calculate $P(X \le 1; Y \le 2)$.

53. Calculate $P(3X + 2Y \ge 6)$ for the probability density in Exercise 52.

54. The lifetimes X and Y (in years) of two machine components have joint probability density

$$p(x, y) = \begin{cases} \frac{6}{125}(5 - x - y) & \text{if } 0 \le x \le 5 - y \text{ and } 0 \le y \le 5 \\ 0 & \text{otherwise} \end{cases}$$

What is the probability that both components are still functioning after 2 years?

55. An insurance company issues two kinds of policies A and B. Let X be the time until the next claim of type A is filed, and let Y be the time (in days) until the next claim of type B is filed. The random variables have joint probability density

$$p(x, y) = 12e^{-4x-3y}$$

Find the probability that $X \le Y$.

56. Compute the Jacobian of the map

$$G(r, s) = \left(e^r \cosh(s), e^r \sinh(s)\right)$$

57. Find a linear mapping $G(u, v)$ that maps the unit square to the parallelogram in the xy-plane spanned by the vectors $\langle 3, -1 \rangle$ and $\langle 1, 4 \rangle$. Then, use the Jacobian to find the area of the image of the rectangle $\mathcal{R} = [0, 4] \times [0, 3]$ under G.

58. Use the map

$$G(u, v) = \left(\frac{u + v}{2}, \frac{u - v}{2}\right)$$

to compute $\iint_{\mathcal{R}} \left((x - y)\sin(x + y)\right)^2 \, dx\, dy$, where $\mathcal{R}$ is the square with vertices $(\pi, 0)$, $(2\pi, \pi)$, $(\pi, 2\pi)$, and $(0, \pi)$.

59. Let $\mathcal{D}$ be the shaded region in Figure 6, and let F be the map

$$u = y + x^2, \qquad v = y - x^3$$

(a) Show that F maps $\mathcal{D}$ to a rectangle $\mathcal{R}$ in the uv-plane.

(b) Apply Eq. (7) in Section 16.6 with $P = (1, 7)$ to estimate Area($\mathcal{D}$).

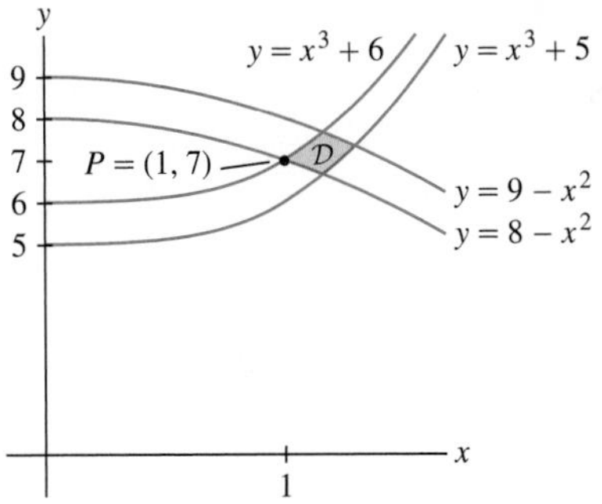

FIGURE 6

60. Calculate the integral of $f(x, y) = e^{3x-2y}$ over the parallelogram in Figure 7.

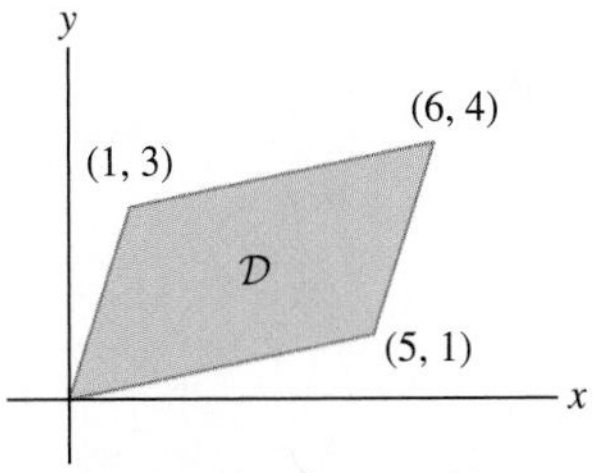

FIGURE 7

61. Sketch the region $\mathcal{D}$ bounded by the curves $y = 2/x$, $y = 1/(2x)$, $y = 2x$, $y = x/2$ in the first quadrant. Let F be the map $u = xy$, $v = y/x$ from the xy-plane to the uv-plane.

(a) Find the image of $\mathcal{D}$ under F.

(b) Let $G = F^{-1}$. Show that $|\text{Jac}(G)| = \dfrac{1}{2|v|}$.

(c) Apply the Change of Variables Formula to prove the formula

$$\iint_{\mathcal{D}} f\left(\frac{y}{x}\right) dx\,dy = \frac{3}{4}\int_{1/2}^{2} \frac{f(v)\,dv}{v}$$

(d) Apply (c) to evaluate $\displaystyle\iint_{\mathcal{D}} \frac{ye^{y/x}}{x}\,dx\,dy$.

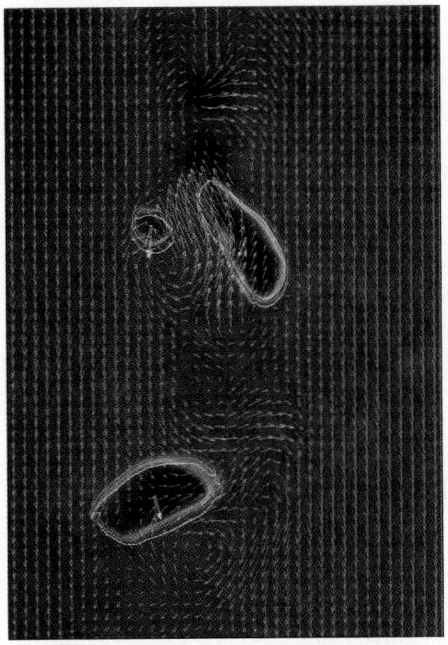

This fluid velocity vector field, from a study of turbulent flow, was produced using PIV (particle image velocimetry) in which the motion of tracer particles is captured by a high-speed digital camera.

17 LINE AND SURFACE INTEGRALS

In the previous chapter, we generalized integration from one variable to several variables. In this chapter, we generalize still further to include integration over curves and surfaces, and we will integrate not just functions but also vector fields. Integrals of vector fields are used in the study of phenomena such as electromagnetism, fluid dynamics, and heat transfer. To lay the groundwork, the chapter begins with a discussion of vector fields.

17.1 Vector Fields

How can we describe a physical object such as the wind, that consists of a large number of molecules moving in a region of space? What we need is a new mathematical object called a **vector field**. A vector field **F** assigns to each point P a vector $\mathbf{F}(P)$ that represents the velocity (speed and direction) of the wind at that point (Figure 1). Another velocity field is shown in Figure 2. However, vector fields describe many other types of quantities, such as forces, and electric and magnetic fields.

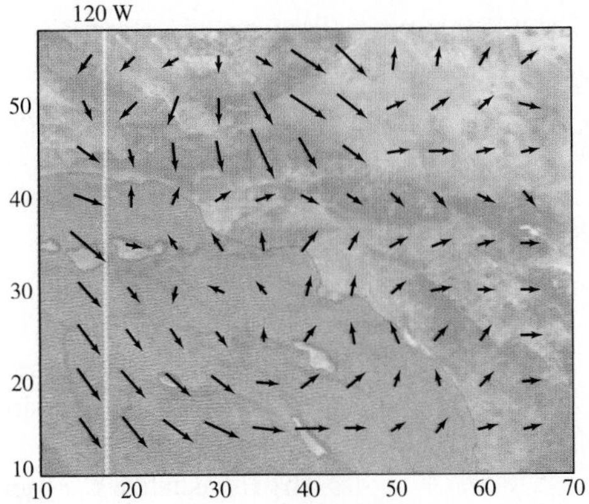

FIGURE 1 Velocity vector field of wind velocity off the coast at Los Angeles.

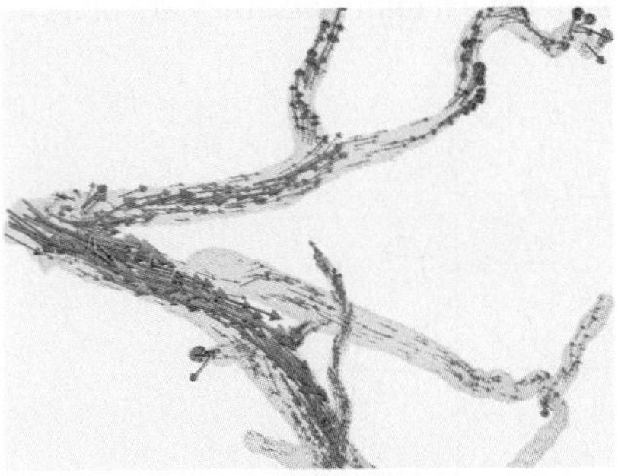

FIGURE 2 Blood flow in an artery represented by a vector field.

Mathematically, a vector field in $\mathbf{R}^3$ is represented by a vector whose components are functions:

$$\mathbf{F}(x, y, z) = \langle F_1(x, y, z), F_2(x, y, z), F_3(x, y, z)\rangle$$

To each point $P = (a, b, c)$ is associated the vector $\mathbf{F}(a, b, c)$, which we also denote by $\mathbf{F}(P)$. Alternatively,

$$\mathbf{F} = F_1\mathbf{i} + F_2\mathbf{j} + F_3\mathbf{k}$$

When drawing a vector field, we draw $\mathbf{F}(P)$ as a vector based at P (rather than the origin). The **domain** of **F** is the set of points P for which $\mathbf{F}(P)$ is defined. Vector fields in the plane are written in a similar fashion:

$$\mathbf{F}(x, y) = \langle F_1(x, y), F_2(x, y)\rangle = F_1\mathbf{i} + F_2\mathbf{j}$$

Throughout this chapter, we assume that the component functions F_j are smooth—that is, they have partial derivatives of all orders on their domains.

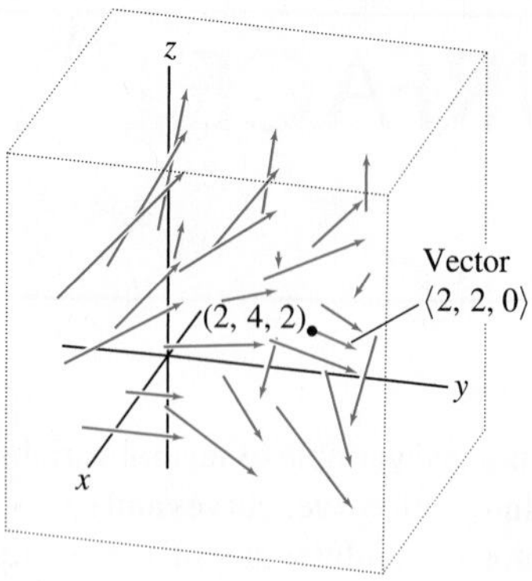

FIGURE 3

EXAMPLE 1 Which vector is attached to the point $P = (2, 4, 2)$ by the vector field $\mathbf{F} = \langle y - z, x, z - \sqrt{y}\rangle$?

Solution The vector attached to P is

$$\mathbf{F}(2, 4, 2) = \left\langle 4 - 2, 2, 2 - \sqrt{4}\right\rangle = \langle 2, 2, 0\rangle$$

This is the red vector in Figure 3.

Although it is not practical to sketch complicated vector fields in three dimensions by hand, computer algebra systems can produce useful visual representations (Figure 4). The vector field in Figure 4(B) is an example of a **constant vector field**. It assigns the same vector $\langle 1, -1, 3\rangle$ to every point in $\mathbf{R}^3$.

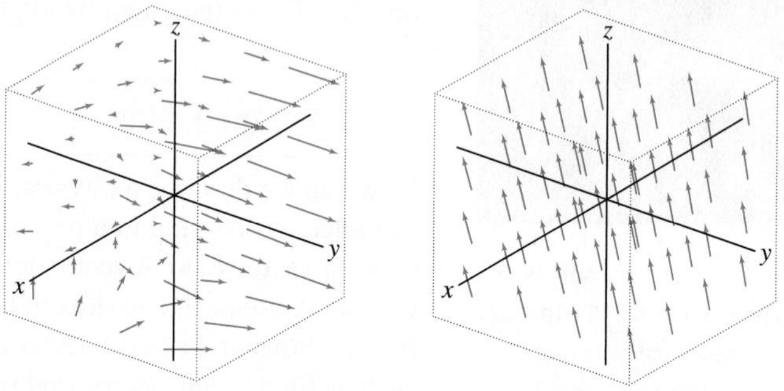

(A) $\mathbf{F} = \langle x \sin z, y^2, x/(z^2+1)\rangle$ (B) Constant vector field $\mathbf{F} = \langle 1, -1, 3\rangle$

FIGURE 4

In the next example, we analyze two vector fields in the plane "qualitatively."

EXAMPLE 2 Describe the following vector fields:

(a) $\mathbf{G} = \mathbf{i} + x\mathbf{j}$ **(b)** $\mathbf{F} = \langle -y, x\rangle$

Solution (a) The vector field $\mathbf{G} = \mathbf{i} + x\mathbf{j}$ assigns the vector $\langle 1, a\rangle$ to the point (a, b). In particular, it assigns the same vector to all points with the same x-coordinate [Figure 5(A)]. Notice that $\langle 1, a\rangle$ has slope a and length $\sqrt{1 + a^2}$. We may describe $\mathbf{G}$ as follows: $\mathbf{G}$ assigns a vector of slope a and length $\sqrt{1 + a^2}$ to all points with $x = a$.

(b) To visualize $\mathbf{F}$, observe that $\mathbf{F}(a, b) = \langle -b, a\rangle$ has length $r = \sqrt{a^2 + b^2}$. It is perpendicular to the radial vector $\langle a, b\rangle$ and points counterclockwise. Thus $\mathbf{F}$ has the following description: The vectors along the circle of radius r all have length r and they are tangent to the circle, pointing counterclockwise [Figure 5(B)]. ■

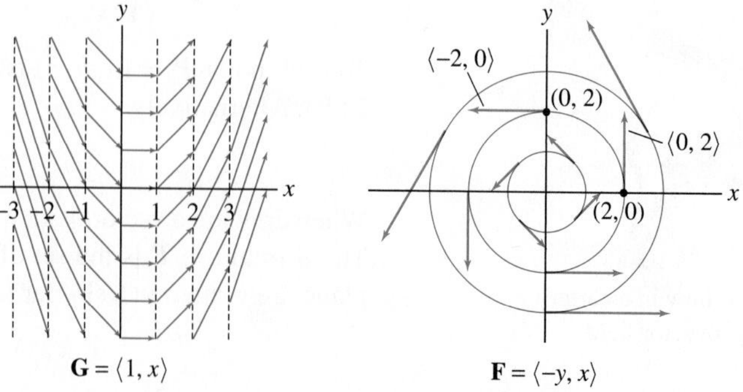

$\mathbf{G} = \langle 1, x\rangle$ $\mathbf{F} = \langle -y, x\rangle$

(A) (B)

FIGURE 5

The English physicist and Nobel laureate Paul Dirac (1902–1984) introduced a generalization of vectors called "spinors" to unify the special theory of relativity with quantum mechanics. This led to the discovery of the positron, an elementary particle used today in PET-scan imaging.

A **unit vector field** is a vector field $\mathbf{F}$ such that $\|\mathbf{F}(P)\| = 1$ for all points P. A vector field $\mathbf{F}$ is called a **radial vector field** if $\mathbf{F}(P)$ depends only on the distance r from P to the origin. Here we use the notation $r = (x^2 + y^2)^{1/2}$ for $n = 2$ and $r = (x^2 + y^2 + z^2)^{1/2}$ for $n = 3$. Two important examples are the unit radial vector fields in two and three dimensions (Figures 6):

$$\mathbf{e}_r = \left\langle \frac{x}{r}, \frac{y}{r} \right\rangle = \left\langle \frac{x}{\sqrt{x^2+y^2}}, \frac{y}{\sqrt{x^2+y^2}} \right\rangle \tag{1}$$

$$\mathbf{e}_r = \left\langle \frac{x}{r}, \frac{y}{r}, \frac{z}{r} \right\rangle = \left\langle \frac{x}{\sqrt{x^2+y^2+z^2}}, \frac{y}{\sqrt{x^2+y^2+z^2}}, \frac{z}{\sqrt{x^2+y^2+z^2}} \right\rangle \tag{2}$$

Observe that $\mathbf{e}_r(P)$ is a unit vector pointing away from the origin at P. Note, however, that $\mathbf{e}_r$ is not defined at the origin where $r = 0$.

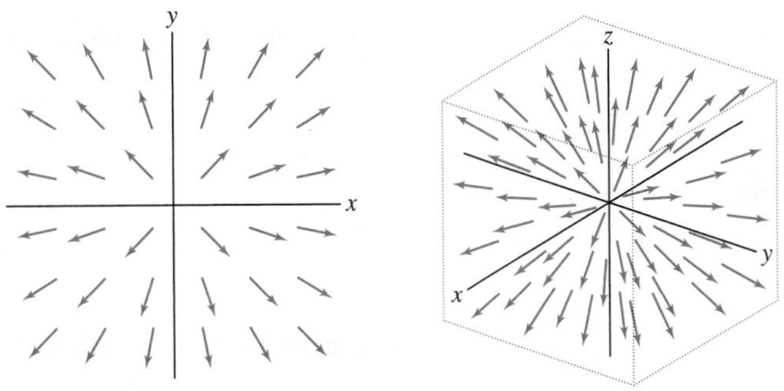

(A) Unit radial vector field in the plane $\mathbf{e}_r = \langle x/r, y/r\rangle$

(B) Unit radial vector field in 3-space $\mathbf{e}_r = \langle x/r, y/r, z/r\rangle$

FIGURE 6

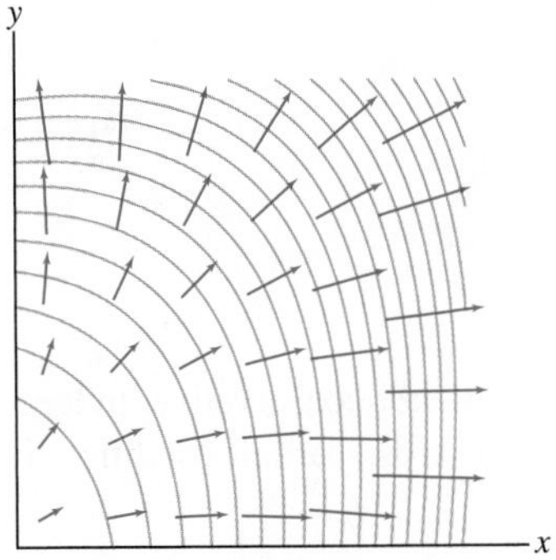

FIGURE 7 A conservative vector field is orthogonal to the level curves of the potential function.

Conservative Vector Fields

We already encountered one type of vector field in Chapter 16—namely, the gradient vector field of a differentiable function $V(x, y, z)$:

$$\mathbf{F} = \nabla V = \left\langle \frac{\partial V}{\partial x}, \frac{\partial V}{\partial y}, \frac{\partial V}{\partial z} \right\rangle$$

A vector field of this type is called a **conservative vector field**, and the function V is called a **potential function** (or scalar potential function) for $\mathbf{F}$.

The same terms apply in two variables and, more generally, in n variables. Recall that the gradient vectors are orthogonal to the level curves, and thus in a conservative vector field, the vector at every point P is orthogonal to the level curve through P (Figure 7).

- *The term "conservative" comes from physics and the law of conservation of energy (see Section 17.3).*
- *Any letter can be used to denote a potential function. We use V, which suggests "volt," the unit of electric potential. Some textbooks use $\phi(x, y, z)$ or $U(x, y, z)$, or simply $f(x, y, z)$.*
- *Theorem 1 is valid for a vector field in the plane $\mathbf{F} = \langle F_1, F_2\rangle$. If $\mathbf{F} = \nabla V$, then $\frac{\partial F_1}{\partial y} = \frac{\partial F_2}{\partial x}$.*

■ **EXAMPLE 3** Verify that $V(x, y, z) = xy + yz^2$ is a potential function for the vector field $\mathbf{F} = \langle y, x + z^2, 2yz\rangle$.

Solution We compute the gradient of V:

$$\frac{\partial V}{\partial x} = y, \qquad \frac{\partial V}{\partial y} = x + z^2, \qquad \frac{\partial V}{\partial z} = 2yz$$

Thus, $\nabla V = \langle y, x + z^2, 2yz\rangle = \mathbf{F}$ as claimed. ■

Conservative vector fields have a special property: They satisfy the cross-partial condition.

THEOREM 1 Cross-Partial Property of a Conservative Vector Field If the vector field $\mathbf{F} = \langle F_1, F_2, F_3 \rangle$ is conservative, then

$$\frac{\partial F_1}{\partial y} = \frac{\partial F_2}{\partial x}, \qquad \frac{\partial F_2}{\partial z} = \frac{\partial F_3}{\partial y}, \qquad \frac{\partial F_3}{\partial x} = \frac{\partial F_1}{\partial z}$$

Proof If $\mathbf{F} = \nabla V$, then

$$F_1 = \frac{\partial V}{\partial x}, \qquad F_2 = \frac{\partial V}{\partial y}, \qquad F_3 = \frac{\partial V}{\partial z}$$

Now compare the "cross"-partial derivatives:

$$\frac{\partial F_1}{\partial y} = \frac{\partial}{\partial y}\left(\frac{\partial V}{\partial x}\right) = \frac{\partial^2 V}{\partial y \partial x}$$

$$\frac{\partial F_2}{\partial x} = \frac{\partial}{\partial x}\left(\frac{\partial V}{\partial y}\right) = \frac{\partial^2 V}{\partial x \partial y}$$

Clairaut's Theorem (Section 15.3) tells us that $\dfrac{\partial^2 V}{\partial y\,\partial x} = \dfrac{\partial^2 V}{\partial x\,\partial y}$, and thus

$$\frac{\partial F_1}{\partial y} = \frac{\partial F_2}{\partial x}$$

Similarly, $\dfrac{\partial F_2}{\partial z} = \dfrac{\partial F_3}{\partial y}$ and $\dfrac{\partial F_3}{\partial x} = \dfrac{\partial F_1}{\partial z}$. ■

From Theorem 1, we can see that most vector fields are *not* conservative. Indeed, an arbitrary triple of functions $\langle F_1, F_2, F_3 \rangle$ does not satisfy the cross-partials condition. Here is an example.

■ **EXAMPLE 4** Show that $\mathbf{F} = \langle y, 0, 0 \rangle$ is not conservative.

Solution We have

$$\frac{\partial F_1}{\partial y} = \frac{\partial}{\partial y} y = 1, \qquad \frac{\partial F_2}{\partial x} = \frac{\partial}{\partial x} 0 = 0$$

Thus $\dfrac{\partial F_1}{\partial y} \neq \dfrac{\partial F_2}{\partial x}$. By Theorem 1, $\mathbf{F}$ is not conservative, even though the other cross-partials agree:

$$\frac{\partial F_3}{\partial x} = \frac{\partial F_1}{\partial z} = 0 \qquad \text{and} \qquad \frac{\partial F_2}{\partial z} = \frac{\partial F_3}{\partial y} = 0$$ ■

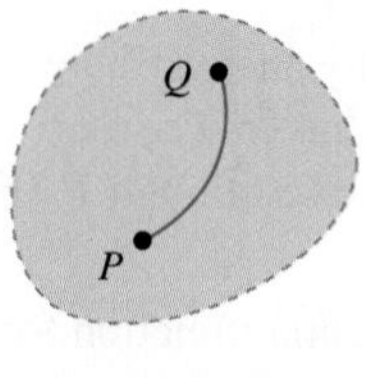

Connected domain

FIGURE 8 In a connected open domain $\mathcal{D}$, any two points in $\mathcal{D}$ can be joined by a path entirely contained in $\mathcal{D}$.

Potential functions, like antiderivatives in one variable, are unique to within an additive constant. To state this precisely, we must assume that the domain $\mathcal{D}$ of the vector field is open and connected (Figure 8). "Connected" means that any two points can be joined by a path within the domain.

THEOREM 2 Uniqueness of Potential Functions If **F** is conservative on an open connected domain, then any two potential functions of **F** differ by a constant.

Proof If both V_1 and V_2 are potential functions of **F**, then

$$\nabla(V_1 - V_2) = \nabla V_1 - \nabla V_2 = \mathbf{F} - \mathbf{F} = \mathbf{0}$$

However, a function whose gradient is zero on an open connected domain is a constant function (this generalizes the fact from single-variable calculus that a function on an interval with zero derivative is a constant function—see Exercise 36). Thus $V_1 - V_2 = C$ for some constant C, and hence $V_1 = V_2 + C$. ■

The next two examples consider two important radial vector fields.

The result of Example 5 is valid in $\mathbf{R}^2$*: The function*

$$V(x, y) = \sqrt{x^2 + y^2} = r$$

is a potential function for $\mathbf{e}_r$.

■ **EXAMPLE 5 Unit Radial Vector Fields** Show that

$$V(x, y, z) = r = \sqrt{x^2 + y^2 + z^2}$$

is a potential function for $\mathbf{e}_r$. That is, $\mathbf{e}_r = \nabla r$.

Solution We have

$$\frac{\partial r}{\partial x} = \frac{\partial}{\partial x}\sqrt{x^2 + y^2 + z^2} = \frac{x}{\sqrt{x^2 + y^2 + z^2}} = \frac{x}{r}$$

Similarly, $\frac{\partial r}{\partial y} = \frac{y}{r}$ and $\frac{\partial r}{\partial z} = \frac{z}{r}$. Therefore, $\nabla r = \left\langle \frac{x}{r}, \frac{y}{r}, \frac{z}{r} \right\rangle = \mathbf{e}_r$. ■

REMINDER

$$\mathbf{e}_r = \left\langle \frac{x}{r}, \frac{y}{r}, \frac{z}{r} \right\rangle$$

where

$$r = (x^2 + y^2 + z^2)^{1/2}$$

In $\mathbf{R}^2$,

$$\mathbf{e}_r = \left\langle \frac{x}{r}, \frac{y}{r} \right\rangle$$

where $r = (x^2 + y^2)^{1/2}$.

The gravitational force exerted by a point mass m is described by an inverse-square force field (Figure 9). A point mass located at the origin exerts a gravitational force **F** on a unit mass located at (x, y, z) equal to

$$\mathbf{F} = -\frac{Gm}{r^2}\mathbf{e}_r = -Gm\left\langle \frac{x}{r^3}, \frac{y}{r^3}, \frac{z}{r^3} \right\rangle$$

where G is the universal gravitation constant. The minus sign indicates that the force is attractive (it pulls in the direction of the origin). The electrostatic force field due to a charged particle is also an inverse-square vector field. The next example shows that these vector fields are conservative.

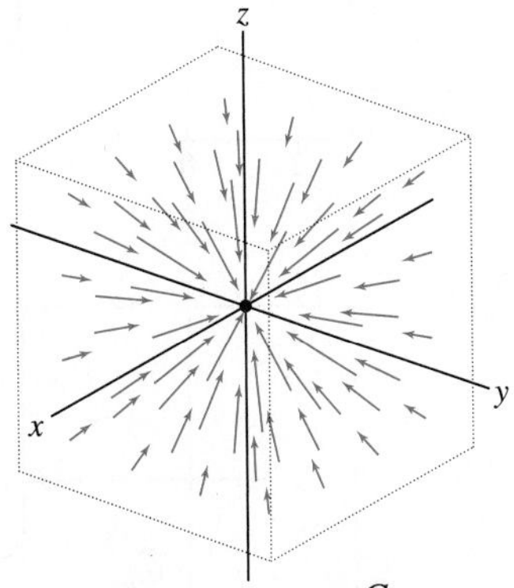

FIGURE 9 The vector field $-\frac{Gm\mathbf{e}_r}{r^2}$ represents the force of gravitational attraction due to a point mass located at the origin.

■ **EXAMPLE 6 Inverse-Square Vector Field** Show that

$$\frac{\mathbf{e}_r}{r^2} = \nabla\left(\frac{-1}{r}\right)$$

Solution Use the Chain Rule for gradients (Theorem 1 in Section 15.5) and Example 5:

$$\nabla(-r^{-1}) = r^{-2}\nabla r = r^{-2}\mathbf{e}_r$$ ■

17.1 SUMMARY

- A *vector field* assigns a vector to each point in a domain. A vector field in $\mathbf{R}^3$ is represented by a triple of functions

$$\mathbf{F} = \langle F_1, F_2, F_3 \rangle$$

A vector field in $\mathbf{R}^2$ is represented by a pair of functions $\mathbf{F} = \langle F_1, F_2 \rangle$. We always assume that the components F_j are smooth functions on their domains.

- If $\mathbf{F} = \nabla V$, then V is called a *potential function* for $\mathbf{F}$.
- $\mathbf{F}$ is called *conservative* if it has a potential function.
- Any two potential functions for a conservative vector field differ by a constant (on an open, connected domain).
- A conservative vector field $\mathbf{F}$ satisfies the cross-partial condition:

$$\frac{\partial F_1}{\partial y} = \frac{\partial F_2}{\partial x}, \qquad \frac{\partial F_2}{\partial z} = \frac{\partial F_3}{\partial y}, \qquad \frac{\partial F_3}{\partial x} = \frac{\partial F_1}{\partial z}$$

- We define

$$r = \sqrt{x^2 + y^2} \quad (\text{in } \mathbf{R}^2), \qquad r = \sqrt{x^2 + y^2 + z^2} \quad (\text{in } \mathbf{R}^3)$$

- The radial unit vector field and the inverse-square vector field are conservative:

$$\mathbf{e}_r = \left\langle \frac{x}{r}, \frac{y}{r}, \frac{z}{r} \right\rangle = \nabla r, \qquad \frac{\mathbf{e}_r}{r^2} = \left\langle \frac{x}{r^3}, \frac{y}{r^3}, \frac{z}{r^3} \right\rangle = \nabla(-r^{-1})$$

17.1 EXERCISES

Preliminary Questions

1. Which of the following is a unit vector field in the plane?

(a) $\mathbf{F} = \langle y, x \rangle$

(b) $\mathbf{F} = \left\langle \dfrac{y}{\sqrt{x^2 + y^2}}, \dfrac{x}{\sqrt{x^2 + y^2}} \right\rangle$

(c) $\mathbf{F} = \left\langle \dfrac{y}{x^2 + y^2}, \dfrac{x}{x^2 + y^2} \right\rangle$

2. Sketch an example of a nonconstant vector field in the plane in which each vector is parallel to $\langle 1, 1 \rangle$.

3. Show that the vector field $\mathbf{F} = \langle -z, 0, x \rangle$ is orthogonal to the position vector $\overrightarrow{OP}$ at each point P. Give an example of another vector field with this property.

4. Give an example of a potential function for $\langle yz, xz, xy \rangle$ other than $f(x, y, z) = xyz$.

Exercises

1. Compute and sketch the vector assigned to the points $P = (1, 2)$ and $Q = (-1, -1)$ by the vector field $\mathbf{F} = \langle x^2, x \rangle$.

2. Compute and sketch the vector assigned to the points $P = (1, 2)$ and $Q = (-1, -1)$ by the vector field $\mathbf{F} = \langle -y, x \rangle$.

3. Compute and sketch the vector assigned to the points $P = (0, 1, 1)$ and $Q = (2, 1, 0)$ by the vector field $\mathbf{F} = \langle xy, z^2, x \rangle$.

4. Compute the vector assigned to the points $P = (1, 1, 0)$ and $Q = (2, 1, 2)$ by the vector fields $\mathbf{e}_r$, $\dfrac{\mathbf{e}_r}{r}$, and $\dfrac{\mathbf{e}_r}{r^2}$.

In Exercises 5–12, sketch the following planar vector fields by drawing the vectors attached to points with integer coordinates in the rectangle $-3 \le x \le 3$, $-3 \le y \le 3$. Instead of drawing the vectors with their true lengths, scale them if necessary to avoid overlap.

5. $\mathbf{F} = \langle 1, 0 \rangle$

6. $\mathbf{F} = \langle 1, 1 \rangle$

7. $\mathbf{F} = x\mathbf{i}$

8. $\mathbf{F} = y\mathbf{i}$

9. $\mathbf{F} = \langle 0, x \rangle$

10. $\mathbf{F} = x^2\mathbf{i} + y\mathbf{j}$

11. $\mathbf{F} = \left\langle \dfrac{x}{x^2 + y^2}, \dfrac{y}{x^2 + y^2} \right\rangle$

12. $\mathbf{F} = \left\langle \dfrac{-y}{\sqrt{x^2 + y^2}}, \dfrac{x}{\sqrt{x^2 + y^2}} \right\rangle$

In Exercises 13–16, match each of the following planar vector fields with the corresponding plot in Figure 10.

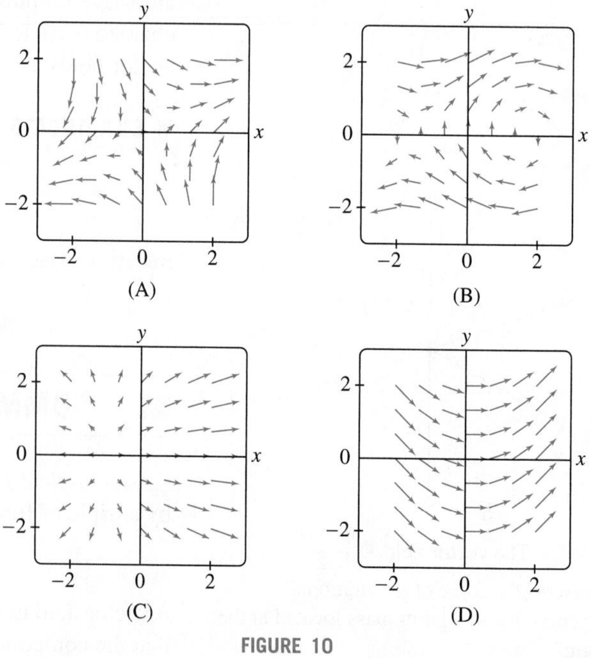

FIGURE 10

13. $\mathbf{F} = \langle 2, x \rangle$

14. $\mathbf{F} = \langle 2x + 2, y \rangle$

15. $\mathbf{F} = \langle y, \cos x \rangle$

16. $\mathbf{F} = \langle x + y, x - y \rangle$

In Exercises 17–20, match each three-dimensional vector field with the corresponding plot in Figure 11.

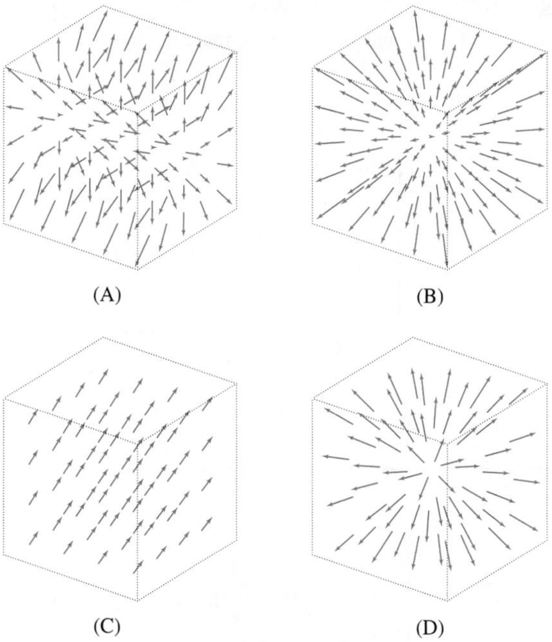

FIGURE 11

17. $\mathbf{F} = \langle 1, 1, 1 \rangle$

18. $\mathbf{F} = \langle x, 0, z \rangle$

19. $\mathbf{F} = \langle x, y, z \rangle$

20. $\mathbf{F} = \mathbf{e}_r$

21. Find (by inspection) a potential function for $\mathbf{F} = \langle x, 0 \rangle$ and prove that $\mathbf{G} = \langle y, 0 \rangle$ is not conservative.

22. Prove that $\mathbf{F} = \langle yz, xz, y \rangle$ is not conservative.

In Exercises 23–26, find a potential function for the vector field **F** *by inspection.*

23. $\mathbf{F} = \langle x, y \rangle$

24. $\mathbf{F} = \langle ye^{xy}, xe^{xy} \rangle$

25. $\mathbf{F} = \langle yz^2, xz^2, 2xyz \rangle$

26. $\mathbf{F} = \langle 2xze^{x^2}, 0, e^{x^2} \rangle$

27. Find potential functions for $\mathbf{F} = \dfrac{\mathbf{e}_r}{r^3}$ and $\mathbf{G} = \dfrac{\mathbf{e}_r}{r^4}$ in $\mathbf{R}^3$. *Hint:* See Example 6.

28. Show that $\mathbf{F} = \langle 3, 1, 2 \rangle$ is conservative. Then prove more generally that any constant vector field $\mathbf{F} = \langle a, b, c \rangle$ is conservative.

29. Let $\varphi = \ln r$, where $r = \sqrt{x^2 + y^2}$. Express $\nabla\varphi$ in terms of the unit radial vector $\mathbf{e}_r$ in $\mathbf{R}^2$.

30. For $P = (a, b)$, we define the unit radial vector field based at P:

$$\mathbf{e}_P = \frac{\langle x - a, y - b \rangle}{\sqrt{(x-a)^2 + (y-b)^2}}$$

(a) Verify that $\mathbf{e}_P$ is a unit vector field.

(b) Calculate $\mathbf{e}_P(1, 1)$ for $P = (3, 2)$.

(c) Find a potential function for $\mathbf{e}_P$.

31. Which of (A) or (B) in Figure 12 is the contour plot of a potential function for the vector field **F**? Recall that the gradient vectors are perpendicular to the level curves.

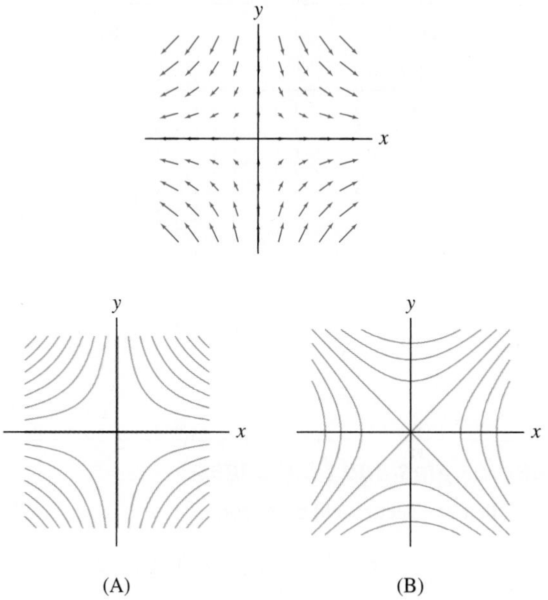

FIGURE 12

32. Which of (A) or (B) in Figure 13 is the contour plot of a potential function for the vector field **F**?

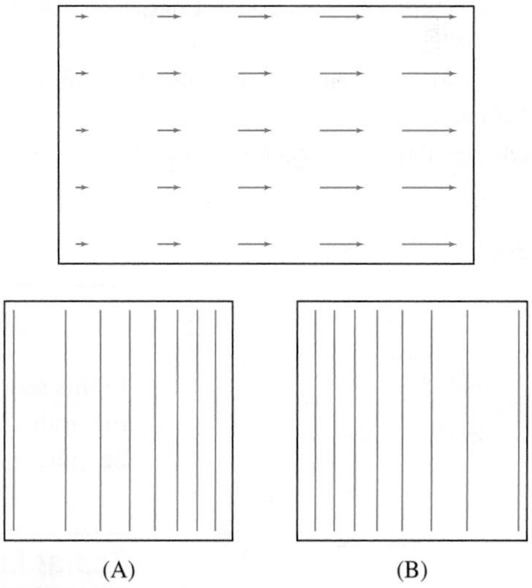

FIGURE 13

33. Match each of these descriptions with a vector field in Figure 14:

(a) The gravitational field created by two planets of equal mass located at P and Q.

(b) The electrostatic field created by two equal and opposite charges located at P and Q (representing the force on a negative test charge; opposite charges attract and like charges repel).

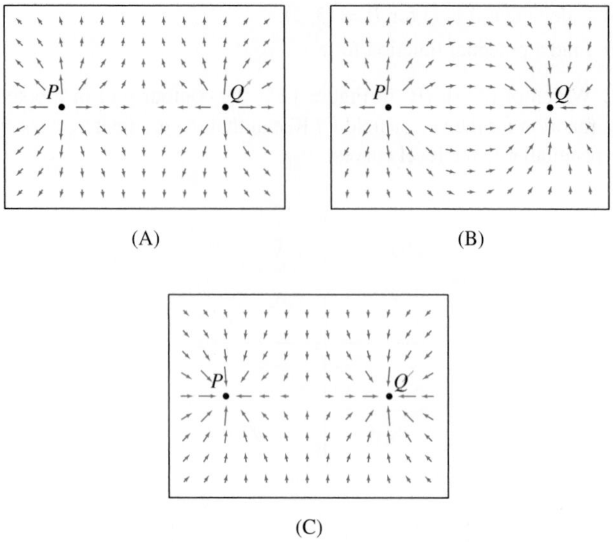

FIGURE 14

34. In this exercise, we show that the vector field **F** in Figure 15 is not conservative. Explain the following statements:

(a) If a potential function V for **F** exists, then the level curves of V must be vertical lines.

(b) If a potential function V for **F** exists, then the level curves of V must grow farther apart as y increases.

(c) Explain why (a) and (b) are incompatible, and hence V cannot exist.

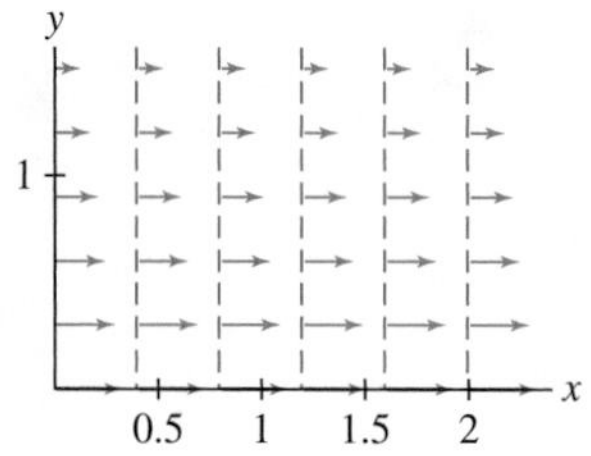

FIGURE 15

Further Insights and Challenges

35. Show that any vector field of the form

$$\mathbf{F} = \langle f(x), g(y), h(z) \rangle$$

has a potential function. Assume that f, g, and h are continuous.

36. Let $\mathcal{D}$ be a disk in $\mathbf{R}^2$. This exercise shows that if

$$\nabla V(x, y) = \mathbf{0}$$

for all (x, y) in $\mathcal{D}$, then V is constant. Consider points $P = (a, b)$, $Q = (c, d)$ and $R = (c, b)$ as in Figure 16.

(a) Use single-variable calculus to show that V is constant along the segments $\overline{PR}$ and $\overline{RQ}$.

(b) Conclude that $V(P) = V(Q)$ for any two points $P, Q \in \mathcal{D}$.

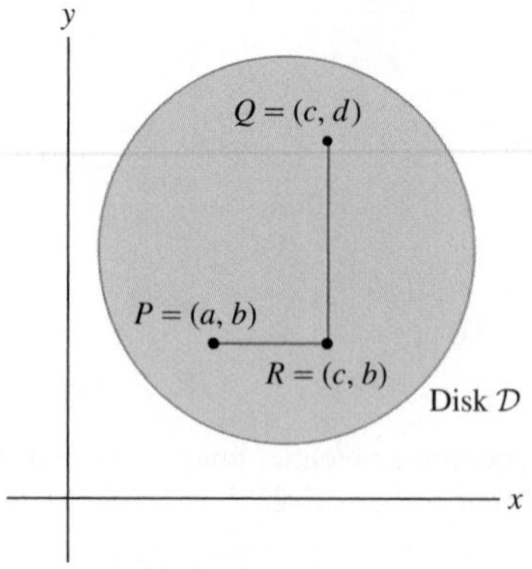

FIGURE 16

17.2 Line Integrals

In this section we introduce two types of integrals over curves: integrals of functions and integrals of vector fields. These are traditionally called **line integrals**, although it would be more appropriate to call them "curve" or "path" integrals.

Scalar Line Integrals

We begin by defining the **scalar line integral** $\int_{\mathcal{C}} f(x, y, z)\, ds$ of a function f over a curve $\mathcal{C}$. We will see how integrals of this type represent total mass and charge, and how they can be used to find electric potentials.

Like all integrals, this line integral is defined through a process of subdivision, summation, and passage to the limit. We divide $\mathcal{C}$ into N consecutive arcs $\mathcal{C}_1, \dots, \mathcal{C}_N$, choose a sample point P_i in each arc $\mathcal{C}_i$, and form the Riemann sum (Figure 1)

$$\sum_{i=1}^{N} f(P_i)\,\text{length}(\mathcal{C}_i) = \sum_{i=1}^{N} f(P_i)\,\Delta s_i$$

where Δs_i is the length of $\mathcal{C}_i$.

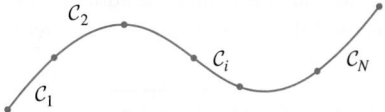

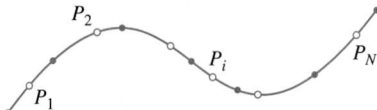

$\mathcal{C}_1$
$\mathcal{C}_2$
$\mathcal{C}_i$
$\mathcal{C}_N$
Partition of $\mathcal{C}$ into N small arcs
P_1
P_2
P_i
P_N
Choice of sample points P_i in each arc

FIGURE 1 The curve $\mathcal{C}$ is divided into N small arcs.

The line integral of f over $\mathcal{C}$ is the limit (if it exists) of these Riemann sums as the maximum of the lengths Δs_i approaches zero:

In Eq. (1), we write $\{\Delta s_i\} \to 0$ to indicate that the limit is taken over all Riemann sums as the maximum of the lengths Δs_i tends to zero.

$$\int_{\mathcal{C}} f(x, y, z)\,ds = \lim_{\{\Delta s_i\}\to 0} \sum_{i=1}^{N} f(P_i)\,\Delta s_i \qquad \boxed{1}$$

This definition also applies to functions $f(x, y)$ of two variables.

The scalar line integral of the function $f(x, y, z) = 1$ is simply the length of $\mathcal{C}$. In this case, all the Riemann sums have the same value:

$$\sum_{i=1}^{N} 1\,\Delta s_i = \sum_{i=1}^{N} \text{length}(\mathcal{C}_i) = \text{length}(\mathcal{C})$$

and thus

$$\int_{\mathcal{C}} 1\,ds = \text{length}(\mathcal{C})$$

In practice, line integrals are computed using parametrizations. Suppose that $\mathcal{C}$ has a parametrization $\mathbf{c}(t)$ for $a \le t \le b$ with continuous derivative $\mathbf{c}'(t)$. Recall that the derivative is the tangent vector

$$\mathbf{c}'(t) = \langle x'(t), y'(t), z'(t)\rangle$$

We divide $\mathcal{C}$ into N consecutive arcs $\mathcal{C}_1, \dots, \mathcal{C}_N$ corresponding to a partition of the interval $[a, b]$:

$$a = t_0 < t_1 < \cdots < t_{N-1} < t_N = b$$

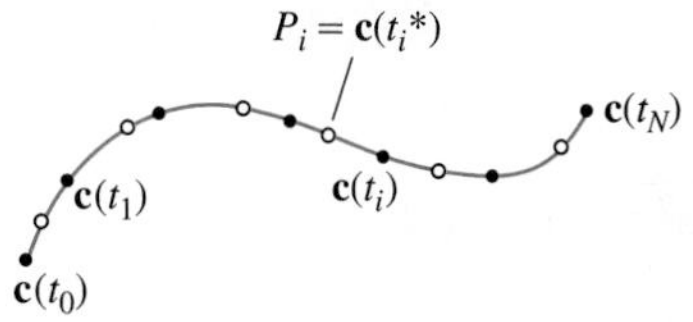

FIGURE 2 Partition of parametrized curve $\mathbf{c}(t)$.

so that $\mathcal{C}_i$ is parametrized by $\mathbf{c}(t)$ for $t_{i-1} \le t \le t_i$ (Figure 2), and choose sample points $P_i = \mathbf{c}(t_i^*)$ with t_i^* in $[t_{i-1}, t_i]$. According to the arc length formula (Section 14.3),

$$\text{Length}(\mathcal{C}_i) = \Delta s_i = \int_{t_{i-1}}^{t_i} \|\mathbf{c}'(t)\|\,dt$$

Because $\mathbf{c}'(t)$ is continuous, the function $\|\mathbf{c}'(t)\|$ is nearly constant on $[t_{i-1}, t_i]$ if the length $\Delta t_i = t_i - t_{i-1}$ is small, and thus $\int_{t_{i-1}}^{t_i} \|\mathbf{c}'(t)\|\,dt \approx \|\mathbf{c}'(t_i^*)\|\Delta t_i$. This gives us the approximation

← REMINDER Arc length formula: The length s of a path $\mathbf{c}(t)$ for $a \le t \le b$ is

$$s = \int_a^b \|\mathbf{c}'(t)\|\,dt$$

$$\sum_{i=1}^{N} f(P_i)\,\Delta s_i \approx \sum_{i=1}^{N} f(\mathbf{c}(t_i^*))\|\mathbf{c}'(t_i^*)\|\,\Delta t_i \qquad \boxed{2}$$

The sum on the right is a Riemann sum that converges to the integral

$$\int_a^b f(\mathbf{c}(t))\|\mathbf{c}'(t)\|\,dt \qquad \boxed{3}$$

as the maximum of the lengths Δt_i tends to zero. By estimating the errors in this approximation, we can show that the sums on the left-hand side of (2) also approach (3). This gives us the following formula for the scalar line integral.

THEOREM 1 Computing a Scalar Line Integral Let $\mathbf{c}(t)$ be a parametrization of a curve $\mathcal{C}$ for $a \le t \le b$. If $f(x, y, z)$ and $\mathbf{c}'(t)$ are continuous, then

$$\int_{\mathcal{C}} f(x, y, z)\,ds = \int_a^b f(\mathbf{c}(t))\|\mathbf{c}'(t)\|\,dt \qquad \boxed{4}$$

The symbol ds is intended to suggest arc length s and is often referred to as the **line element** or **arc length differential**. In terms of a parametrization, we have the symbolic equation

$$ds = \|\mathbf{c}'(t)\|\,dt$$

where

$$\|\mathbf{c}'(t)\| = \sqrt{x'(t)^2 + y'(t)^2 + z'(t)^2}$$

Eq. (4) tells us that to evaluate a scalar line integral, we replace the integrand $f(x, y, z)\,ds$ with $f(\mathbf{c}(t))\,\|\mathbf{c}'(t)\|\,dt$.

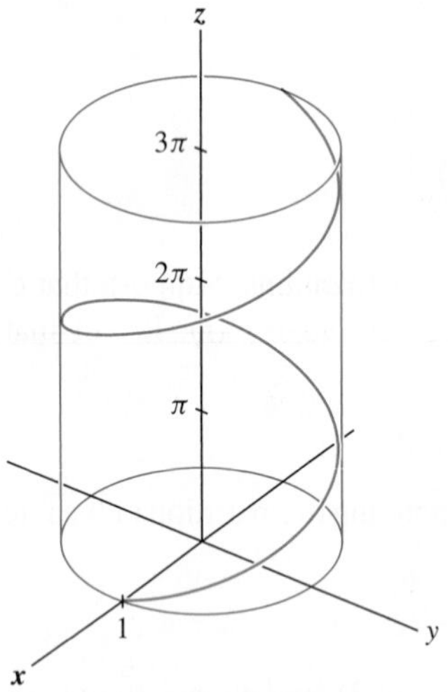

FIGURE 3 The helix $\mathbf{c}(t) = (\cos t, \sin t, t)$.

EXAMPLE 1 Integrating along a Helix Calculate

$$\int_{\mathcal{C}} (x + y + z)\,ds$$

where $\mathcal{C}$ is the helix $\mathbf{c}(t) = (\cos t, \sin t, t)$ for $0 \le t \le \pi$ (Figure 3).

Solution

***Step 1.* Compute ds.**

$$\mathbf{c}'(t) = \langle -\sin t, \cos t, 1\rangle$$

$$\|\mathbf{c}'(t)\| = \sqrt{(-\sin t)^2 + \cos^2 t + 1} = \sqrt{2}$$

$$ds = \|\mathbf{c}'(t)\|dt = \sqrt{2}\,dt$$

***Step 2.* Write out the integrand and evaluate.**

We have $f(x, y, z) = x + y + z$, and so

$$f(\mathbf{c}(t)) = f(\cos t, \sin t, t) = \cos t + \sin t + t$$

$$f(x, y, z)\,ds = f(\mathbf{c}(t))\,\|\mathbf{c}'(t)\|\,dt = (\cos t + \sin t + t)\sqrt{2}\,dt$$

By Eq. (4),

$$\int_C f(x,y,z)\,ds = \int_0^{\pi} f(\mathbf{c}(t))\,\|\mathbf{c}'(t)\|\,dt = \int_0^{\pi} (\cos t + \sin t + t)\sqrt{2}\,dt$$

$$= \sqrt{2}\left(\sin t - \cos t + \frac{1}{2}t^2\right)\Big|_0^{\pi}$$

$$= \sqrt{2}\left(0 + 1 + \frac{1}{2}\pi^2\right) - \sqrt{2}\,(0 - 1 + 0) = 2\sqrt{2} + \frac{\sqrt{2}}{2}\pi^2$$ ■

■ **EXAMPLE 2** **Arc Length** Calculate $\int_C 1\,ds$ for the helix in the previous example: $\mathbf{c}(t) = (\cos t, \sin t, t)$ for $0 \le t \le \pi$. What does this integral represent?

Solution In the previous example, we showed that $ds = \sqrt{2}\,dt$ and thus

$$\int_C 1\,ds = \int_0^{\pi} \sqrt{2}\,dt = \pi\sqrt{2}$$

This is the length of the helix for $0 \le t \le \pi$. ■

Applications of the Scalar Line Integral

In Section 16.5 we discussed the general principle that "the integral of a density is the total quantity." This applies to scalar line integrals. For example, we can view the curve $\mathcal{C}$ as a wire with continuous **mass density** $\rho(x,y,z)$, given in units of mass per unit length. The total mass is defined as the integral of mass density:

$$\text{Total mass of } \mathcal{C} = \int_{\mathcal{C}} \rho(x,y,z)\,ds \qquad (5)$$

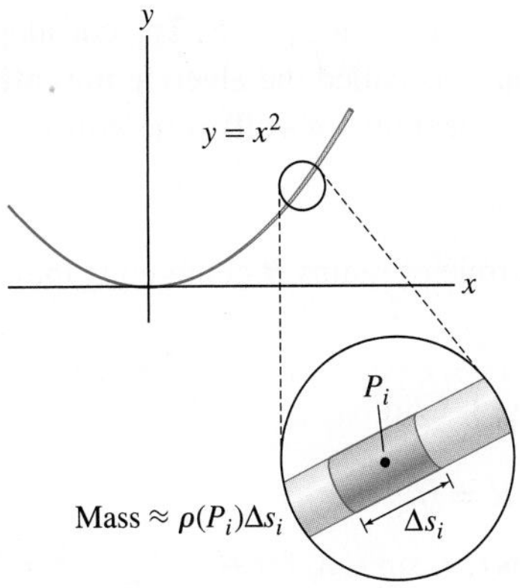

FIGURE 4

A similar formula for total charge is valid if $\rho(x,y,z)$ is the charge density along the curve. As in Section 16.5, we justify this interpretation by dividing $\mathcal{C}$ into N arcs $\mathcal{C}_i$ of length Δs_i with N large. The mass density is nearly constant on $\mathcal{C}_i$, and therefore, the mass of $\mathcal{C}_i$ is approximately $\rho(P_i)\,\Delta s_i$, where P_i is any sample point on $\mathcal{C}_i$ (Figure 4). The total mass is the sum

$$\text{Total mass of } \mathcal{C} = \sum_{i=1}^{N} \text{mass of } \mathcal{C}_i \approx \sum_{i=1}^{N} \rho(P_i)\,\Delta s_i$$

As the maximum of the lengths Δs_i tends to zero, the sums on the right approach the line integral in Eq. (5).

■ **EXAMPLE 3** **Scalar Line Integral as Total Mass** Find the total mass of a wire in the shape of the parabola $y = x^2$ for $1 \le x \le 4$ (in centimeters) with mass density given by $\rho(x,y) = y/x$ g/cm.

Solution The arc of the parabola is parametrized by $\mathbf{c}(t) = (t, t^2)$ for $1 \le t \le 4$.

***Step 1.* Compute ds.**

$$\mathbf{c}'(t) = \langle 1, 2t \rangle$$

$$ds = \|\mathbf{c}'(t)\|\,dt = \sqrt{1 + 4t^2}\,dt$$

Step 2. **Write out the integrand and evaluate.**
We have $\rho(\mathbf{c}(t)) = \rho(t, t^2) = t^2/t = t$, and thus

$$\rho(x, y)\, ds = \rho(\mathbf{c}(t))\sqrt{1 + 4t^2}\, dt = t\sqrt{1 + 4t^2}\, dt$$

We evaluate the line integral of mass density using the substitution $u = 1 + 4t^2$, $du = 8t\, dt$:

$$\int_C \rho(x, y)\, ds = \int_1^4 \rho(\mathbf{c}(t))\|\mathbf{c}'(t)\|\, dt = \int_1^4 t\sqrt{1 + 4t^2}\, dt$$

$$= \frac{1}{8}\int_5^{65} \sqrt{u}\, du = \frac{1}{12}u^{3/2}\Big|_5^{65}$$

$$= \frac{1}{12}(65^{3/2} - 5^{3/2}) \approx 42.74$$

Note that after the substitution, the limits of integration become $u(1) = 5$ and $u(4) = 65$. The total mass of the wire is approximately 42.7 g. ■

Scalar line integrals are also used to compute electric potentials. When an electric charge is distributed continuously along a curve $\mathcal{C}$, with charge density $\rho(x, y, z)$, the charge distribution sets up an electrostatic field $\mathbf{E}$ that is a conservative vector field. Coulomb's Law tells us that $\mathbf{E} = -\nabla V$ where

By definition, $\mathbf{E}$ is the vector field with the property that the electrostatic force on a point charge q placed at location $P = (x, y, z)$ is the vector $q\mathbf{E}(x, y, z)$.

$$V(P) = k\int_C \frac{\rho(x, y, z)\, ds}{r_P} \qquad (6)$$

In this integral, $r_P = r_P(x, y, z)$ denotes the distance from (x, y, z) to P. The constant k has the value $k = 8.99 \times 10^9$ N-m²/C². The function V is called the **electric potential**. It is defined for all points P that do not lie on $\mathcal{C}$ and has units of volts (one volt is one N-m/C).

The constant k is usually written as $\frac{1}{4\pi\epsilon_0}$ where ϵ_0 is the vacuum permittivity.

■ **EXAMPLE 4 Electric Potential** A charged semicircle of radius R centered at the origin in the xy-plane (Figure 5) has charge density

$$\rho(x, y, 0) = 10^{-8}\left(2 - \frac{x}{R}\right) \text{ C/m}$$

Find the electric potential at a point $P = (0, 0, a)$ if $R = 0.1$ m.

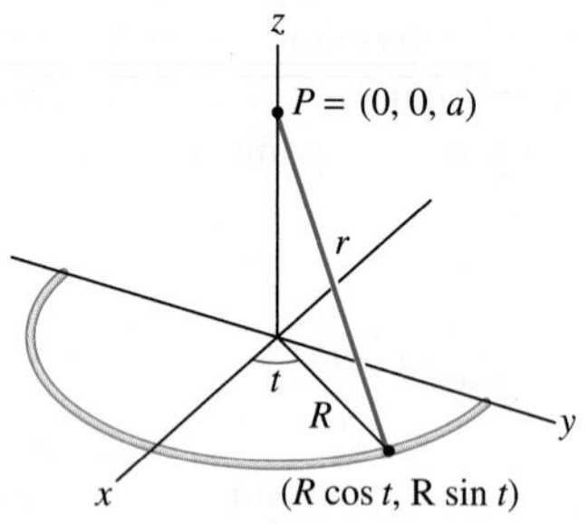

FIGURE 5

Solution Parametrize the semicircle by $\mathbf{c}(t) = (R\cos t, R\sin t, 0)$ for $-\pi/2 \le t \le \pi/2$:

$$\|\mathbf{c}'(t)\| = \|\langle -R\sin t, R\cos t, 0\rangle\| = \sqrt{R^2\sin^2 t + R^2\cos^2 t + 0} = R$$

$$ds = \|\mathbf{c}'(t)\|\, dt = R\, dt$$

$$\rho(\mathbf{c}(t)) = \rho(R\cos t, R\sin t, 0) = 10^{-8}\left(2 - \frac{R\cos t}{R}\right) = 10^{-8}(2 - \cos t)$$

In our case, the distance r_P from P to a point $(x, y, 0)$ on the semicircle has the constant value $r_P = \sqrt{R^2 + a^2}$ (Figure 5). Thus,

$$V(P) = k\int_C \frac{\rho(x, y, z)\, ds}{r_P} = k\int_C \frac{10^{-8}(2 - \cos t)\, R dt}{\sqrt{R^2 + a^2}}$$

$$= \frac{10^{-8}kR}{\sqrt{R^2 + a^2}}\int_{-\pi/2}^{\pi/2}(2 - \cos t)\, dt = \frac{10^{-8}kR}{\sqrt{R^2 + a^2}}(2\pi - 2)$$

With $R = 0.1$ m and $k = 8.99 \times 10^9$, we then obtain $10^{-8}kR(2\pi - 2) \approx 38.5$ and $V(P) \approx \dfrac{38.5}{\sqrt{0.01 + a^2}}$ volts. ■

Vector Line Integrals

When you carry a backpack up a mountain, you do work against the Earth's gravitational field. The work, or energy expended, is one example of a quantity represented by a vector line integral.

An important difference between vector and scalar line integrals is that vector line integrals depend on a *direction* along the curve. This is reasonable if you think of the vector line integral as work, because the work performed going down the mountain is the negative of the work performed going up.

A specified direction along a path curve C is called an **orientation** (Figure 6). We refer to this direction as the **positive** direction along C, the opposite direction is the **negative** direction, and C itself is called an **oriented curve**. In Figure 6(A), if we reversed the orientation, the positive direction would become the direction from Q to P.

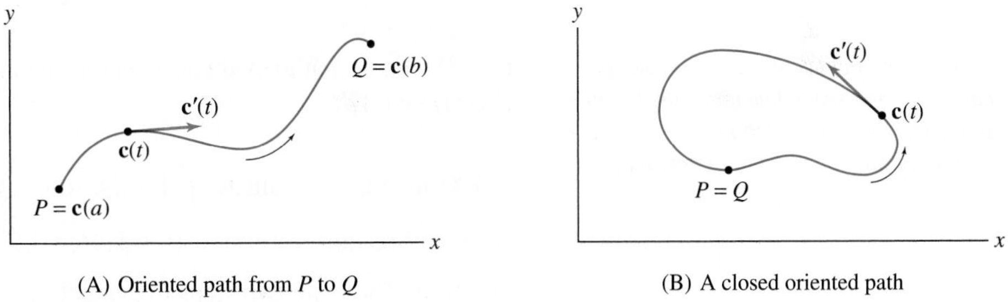

FIGURE 6

The line integral of a vector field $\mathbf{F}$ over a curve C is defined as the scalar line integral of the tangential component of $\mathbf{F}$. More precisely, let $\mathbf{T} = \mathbf{T}(P)$ denote the unit tangent vector at a point P on C pointing in the positive direction. The **tangential component** of $\mathbf{F}$ at P is the dot product (Figure 7)

The unit tangent vector $\mathbf{T}$ *varies from point to point along the curve. When it is necessary to stress this dependence, we write* $\mathbf{T}(P)$.

$$\mathbf{F}(P) \cdot \mathbf{T}(P) = \|\mathbf{F}(P)\| \, \|\mathbf{T}(P)\| \cos\theta = \|\mathbf{F}(P)\| \cos\theta$$

where θ is the angle between $\mathbf{F}(P)$ and $\mathbf{T}(P)$. The vector line integral of $\mathbf{F}$ is the scalar line integral of the scalar function $\mathbf{F} \cdot \mathbf{T}$. We make the standing assumption that C is piecewise smooth (it consists of finitely many smooth curves joined together with possible corners).

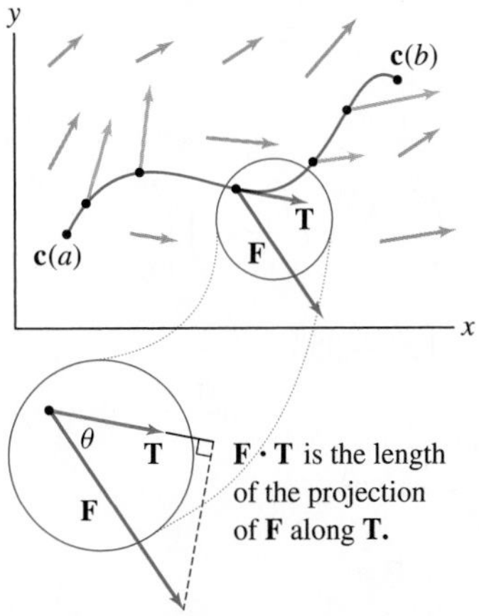

FIGURE 7 The line integral is the integral of the tangential component of $\mathbf{F}$ along C.

DEFINITION Vector Line Integral The line integral of a vector field $\mathbf{F}$ along an oriented curve C is the integral of the tangential component of $\mathbf{F}$:

$$\int_C \mathbf{F} \cdot d\mathbf{s} = \int_C (\mathbf{F} \cdot \mathbf{T})\, ds \qquad \boxed{7}$$

We use parametrizations to evaluate vector line integrals, but there is one important difference with the scalar case: The parametrization $\mathbf{c}(t)$ must be *positively oriented*; that is, $\mathbf{c}(t)$ must trace C in the positive direction. We assume also that $\mathbf{c}(t)$ is regular; that is, $\mathbf{c}'(t) \neq \mathbf{0}$ for $a \le t \le b$. Then $\mathbf{c}'(t)$ is a nonzero tangent vector pointing in the positive direction, and

$$\mathbf{T} = \frac{\mathbf{c}'(t)}{\|\mathbf{c}'(t)\|}$$

In terms of the arc length differential $ds = \|\mathbf{c}'(t)\|\,dt$, we have

$$(\mathbf{F}\cdot\mathbf{T})\,ds = \left(\mathbf{F}(\mathbf{c}(t))\cdot\frac{\mathbf{c}'(t)}{\|\mathbf{c}'(t)\|}\right)\|\mathbf{c}'(t)\|\,dt = \mathbf{F}(\mathbf{c}(t))\cdot\mathbf{c}'(t)\,dt$$

Therefore, the integral on the right-hand side of Eq. (7) is equal to the right-hand side of Eq. (8) in the next theorem.

THEOREM 2 Computing a Vector Line Integral If $\mathbf{c}(t)$ is a regular parametrization of an oriented curve C for $a \le t \le b$, then

$$\int_C \mathbf{F}\cdot d\mathbf{s} = \int_a^b \mathbf{F}(\mathbf{c}(t))\cdot\mathbf{c}'(t)\,dt \qquad \boxed{8}$$

It is useful to think of $d\mathbf{s}$ as a "vector line element" or "vector differential" that is related to the parametrization by the symbolic equation

$$d\mathbf{s} = \mathbf{c}'(t)\,dt$$

Vector line integrals are usually easier to calculate than scalar line integrals, because the length $\|\mathbf{c}'(t)\|$, which involves a square root, does not appear in the integrand.

Eq. (8) tells us that to evaluate a vector line integral, we replace the integrand $\mathbf{F}\cdot d\mathbf{s}$ with $\mathbf{F}(\mathbf{c}(t))\cdot\mathbf{c}'(t)\,dt$.

■ **EXAMPLE 5** Evaluate $\int_C \mathbf{F}\cdot d\mathbf{s}$, where $\mathbf{F} = \langle z, y^2, x\rangle$ and C is parametrized (in the positive direction) by $\mathbf{c}(t) = (t+1, e^t, t^2)$ for $0 \le t \le 2$.

Solution There are two steps in evaluating a line integral.

Step 1. **Calculate the integrand.**

$$\mathbf{c}(t) = (t+1, e^t, t^2)$$

$$\mathbf{F}(\mathbf{c}(t)) = \langle z, y^2, x\rangle = \langle t^2, e^{2t}, t+1\rangle$$

$$\mathbf{c}'(t) = \langle 1, e^t, 2t\rangle$$

The integrand (as a differential) is the dot product:

$$\mathbf{F}(\mathbf{c}(t))\cdot\mathbf{c}'(t)dt = \langle t^2, e^{2t}, t+1\rangle\cdot\langle 1, e^t, 2t\rangle\,dt = (e^{3t} + 3t^2 + 2t)dt$$

Step 2. **Evaluate the line integral.**

$$\int_C \mathbf{F}\cdot d\mathbf{s} = \int_0^2 \mathbf{F}(\mathbf{c}(t))\cdot\mathbf{c}'(t)\,dt$$

$$= \int_0^2 (e^{3t} + 3t^2 + 2t)\,dt = \left(\frac{1}{3}e^{3t} + t^3 + t^2\right)\Big|_0^2$$

$$= \left(\frac{1}{3}e^6 + 8 + 4\right) - \frac{1}{3} = \frac{1}{3}\left(e^6 + 35\right) \qquad ■$$

Another standard notation for the line integral $\int_C \mathbf{F}\cdot d\mathbf{s}$ is

$$\int_C F_1\,dx + F_2\,dy + F_3\,dz$$

In this notation, we write $d\mathbf{s}$ as a vector differential

$$d\mathbf{s} = \langle dx, dy, dz \rangle$$

so that

$$\mathbf{F} \cdot d\mathbf{s} = \langle F_1, F_2, F_3 \rangle \cdot \langle dx, dy, dz \rangle = F_1\,dx + F_2\,dy + F_3\,dz$$

In terms of a parametrization $\mathbf{c}(t) = (x(t), y(t), z(t))$,

$$d\mathbf{s} = \left\langle \frac{dx}{dt}, \frac{dy}{dt}, \frac{dz}{dt} \right\rangle dt$$

$$\mathbf{F} \cdot d\mathbf{s} = \left(F_1(\mathbf{c}(t))\frac{dx}{dt} + F_2(\mathbf{c}(t))\frac{dy}{dt} + F_3(\mathbf{c}(t))\frac{dz}{dt} \right) dt$$

So we have the following formula:

$$\int_C F_1\,dx + F_2\,dy + F_3\,dz = \int_a^b \left(F_1(\mathbf{c}(t))\frac{dx}{dt} + F_2(\mathbf{c}(t))\frac{dy}{dt} + F_3(\mathbf{c}(t))\frac{dz}{dt} \right) dt$$

GRAPHICAL INSIGHT The magnitude of a vector line integral (or even whether it is positive or negative) depends on the angles between **F** and **T** along the path. Consider the line integral of $\mathbf{F} = \langle 2y, -3 \rangle$ around the ellipse in Figure 8.

- In Figure 8(A), the angles θ between **F** and **T** appear to be mostly obtuse along the top part of the ellipse. Consequently, $\mathbf{F} \cdot \mathbf{T} \le 0$ and the line integral is negative.
- In Figure 8(B), the angles θ appear to be mostly acute along the bottom part of the ellipse. Consequently, $\mathbf{F} \cdot \mathbf{T} \ge 0$ and the line integral is positive.
- We can guess that the line integral around the entire ellipse is negative because $\|\mathbf{F}\|$ is larger in the top half, so the negative contribution of $\mathbf{F} \cdot \mathbf{T}$ from the top half appears to dominate the positive contribution of the bottom half. We verify this in Example 6.

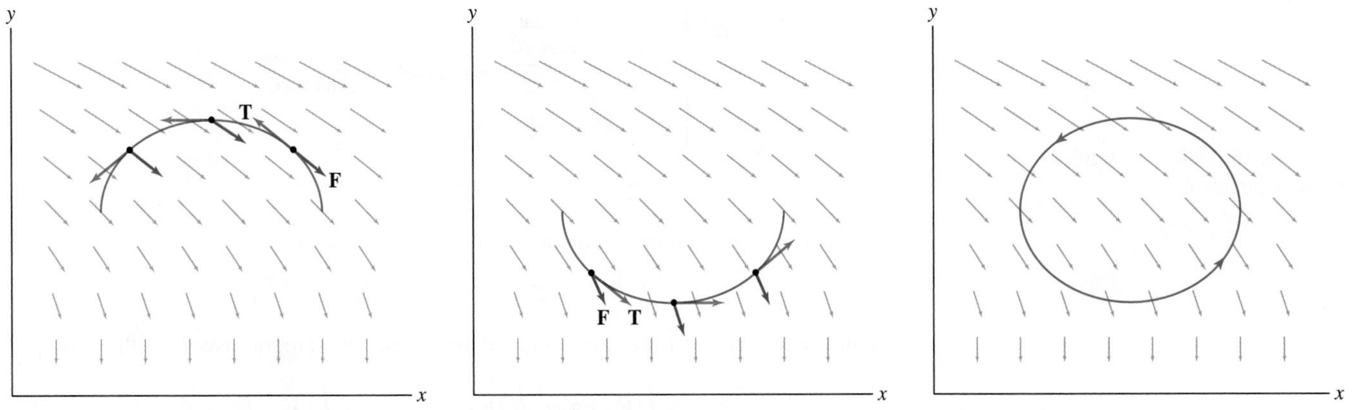

(A) Most of the dot products **F** • **T** are negative because the angles between the vectors are obtuse. Therefore: the line integral is negative

(B) Most of the dot products **F** • **T** are positive because the angles between the vectors are acute. Therefore: the line integral is positive

(C) Total line integral is negative.

FIGURE 8 The vector field $\mathbf{F} = \langle 2y, -3 \rangle$.

■ **EXAMPLE 6** The ellipse $\mathcal{C}$ in Figure 8(C) with counterclockwise orientation is parametrized by $\mathbf{c}(\theta) = (5 + 4\cos\theta, 3 + 2\sin\theta)$ for $0 \le \theta < 2\pi$. Calculate

$$\int_{\mathcal{C}} 2y\,dx - 3\,dy$$

Solution We have $x(\theta) = 5 + 4\cos\theta$ and $y(\theta) = 3 + 2\sin\theta$, and

$$\frac{dx}{d\theta} = -4\sin\theta, \qquad \frac{dy}{d\theta} = 2\cos\theta$$

The integrand of the line integral is

$$2y\,dx - 3\,dy = \left(2y\frac{dx}{d\theta} - 3\frac{dy}{d\theta}\right)d\theta$$

$$= \bigl(2(3 + 2\sin\theta)(-4\sin\theta) - 3(2\cos\theta)\bigr)\,d\theta$$

$$= -\bigl(24\sin\theta + 16\sin^2\theta + 6\cos\theta\bigr)\,d\theta$$

Since the integrals of $\cos\theta$ and $\sin\theta$ over $[0, 2\pi]$ are zero,

$$\int_{\mathcal{C}} 2y\,dx - 3\,dy = -\int_0^{2\pi} \bigl(24\sin\theta + 16\sin^2\theta + 6\cos\theta\bigr)\,d\theta$$

$$= -16\int_0^{2\pi} \sin^2\theta\,d\theta = -16\pi \qquad ■$$

In Example 6, keep in mind that

$$\int_{\mathcal{C}} 2y\,dx - 3\,dy$$

is another notation for the line integral of $\mathbf{F} = \langle 2y, -3 \rangle$ *over $\mathcal{C}$. Formally,*

$$\mathbf{F} \cdot d\mathbf{s} = \langle 2y, -3 \rangle \cdot \langle dx, dy \rangle$$

$$= 2y\,dx - 3\,dy$$

◀·· **REMINDER**

- $\displaystyle\int \sin^2\theta\,d\theta = \frac{1}{2}\theta - \frac{1}{4}\sin 2\theta$
- $\displaystyle\int_0^{2\pi} \sin^2\theta\,d\theta = \pi$

We now state some basic properties of vector line integrals. First, given an oriented curve $\mathcal{C}$, we write $-\mathcal{C}$ to denote the curve $\mathcal{C}$ with the opposite orientation (Figure 9). The unit tangent vector changes sign from $\mathbf{T}$ to $-\mathbf{T}$ when we change orientation, so the tangential component of $\mathbf{F}$ and the line integral also change sign:

$$\int_{-\mathcal{C}} \mathbf{F} \cdot d\mathbf{s} = -\int_{\mathcal{C}} \mathbf{F} \cdot d\mathbf{s}$$

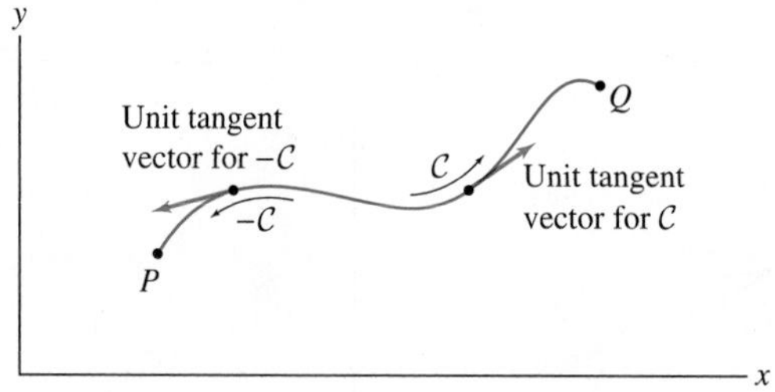

FIGURE 9 The path from P to Q has two possible orientations.

Next, if we are given n oriented curves $\mathcal{C}_1, \dots, \mathcal{C}_n$, we write

$$\mathcal{C} = \mathcal{C}_1 + \cdots + \mathcal{C}_n$$

to indicate the union of the paths, and we define the line integral over $\mathcal{C}$ as the sum

$$\int_{\mathcal{C}} \mathbf{F} \cdot d\mathbf{s} = \int_{\mathcal{C}_1} \mathbf{F} \cdot d\mathbf{s} + \cdots + \int_{\mathcal{C}_n} \mathbf{F} \cdot d\mathbf{s}$$

We use this formula to define the line integral when $\mathcal{C}$ is **piecewise smooth**, meaning that $\mathcal{C}$ is a union of smooth curves $\mathcal{C}_1, \dots, \mathcal{C}_n$. For example, the triangle in Figure 10 is piecewise smooth but not smooth. The next theorem summarizes the main properties of vector line integrals.

THEOREM 3 Properties of Vector Line Integrals Let $\mathcal{C}$ be a smooth oriented curve, and let $\mathbf{F}$ and $\mathbf{G}$ be vector fields.

(i) Linearity: $\displaystyle\int_{\mathcal{C}} (\mathbf{F}+\mathbf{G}) \cdot d\mathbf{s} = \int_{\mathcal{C}} \mathbf{F} \cdot d\mathbf{s} + \int_{\mathcal{C}} \mathbf{G} \cdot d\mathbf{s}$

$$\int_{\mathcal{C}} k\mathbf{F} \cdot d\mathbf{s} = k \int_{\mathcal{C}} \mathbf{F} \cdot d\mathbf{s} \quad (k \text{ a constant})$$

(ii) Reversing orientation: $\displaystyle\int_{-\mathcal{C}} \mathbf{F} \cdot d\mathbf{s} = -\int_{\mathcal{C}} \mathbf{F} \cdot d\mathbf{s}$

(iii) Additivity: If $\mathcal{C}$ is a union of n smooth curves $\mathcal{C}_1 + \cdots + \mathcal{C}_n$, then

$$\int_{\mathcal{C}} \mathbf{F} \cdot d\mathbf{s} = \int_{\mathcal{C}_1} \mathbf{F} \cdot d\mathbf{s} + \cdots + \int_{\mathcal{C}_n} \mathbf{F} \cdot d\mathbf{s}$$

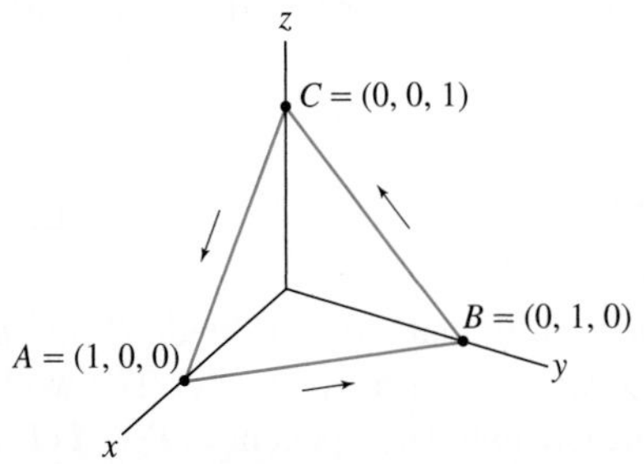

FIGURE 10 The triangle is piecewise smooth: It is the union of its three edges, each of which is smooth.

EXAMPLE 7 Compute $\int_{\mathcal{C}} \mathbf{F} \cdot d\mathbf{s}$, where $\mathbf{F} = \langle e^z, e^y, x+y \rangle$ and $\mathcal{C}$ is the triangle joining $(1,0,0)$, $(0,1,0)$, and $(0,0,1)$ oriented in the counterclockwise direction when viewed from above (Figure 10).

Solution The line integral is the sum of the line integrals over the edges of the triangle:

$$\int_{\mathcal{C}} \mathbf{F} \cdot d\mathbf{s} = \int_{\overline{AB}} \mathbf{F} \cdot d\mathbf{s} + \int_{\overline{BC}} \mathbf{F} \cdot d\mathbf{s} + \int_{\overline{CA}} \mathbf{F} \cdot d\mathbf{s}$$

Segment $\overline{AB}$ is parametrized by $\mathbf{c}(t) = (1-t, t, 0)$ for $0 \le t \le 1$. We have

$$\mathbf{F}(\mathbf{c}(t)) \cdot \mathbf{c}'(t) = \mathbf{F}(1-t, t, 0) \cdot \langle -1, 1, 0 \rangle = \langle e^0, e^t, 1 \rangle \cdot \langle -1, 1, 0 \rangle = -1 + e^t$$

$$\int_{\overline{AB}} \mathbf{F} \cdot d\mathbf{s} = \int_0^1 (e^t - 1)\, dt = (e^t - t)\Big|_0^1 = (e-1) - 1 = e - 2$$

Similarly, $\overline{BC}$ is parametrized by $\mathbf{c}(t) = (0, 1-t, t)$ for $0 \le t \le 1$, and

$$\mathbf{F}(\mathbf{c}(t)) \cdot \mathbf{c}'(t) = \langle e^t, e^{1-t}, 1-t \rangle \cdot \langle 0, -1, 1 \rangle = -e^{1-t} + 1 - t$$

$$\int_{\overline{BC}} \mathbf{F} \cdot d\mathbf{s} = \int_0^1 (-e^{1-t} + 1 - t)\, dt = \left(e^{1-t} + t - \frac{1}{2}t^2 \right)\Bigg|_0^1 = \frac{3}{2} - e$$

Finally, $\overline{CA}$ is parametrized by $\mathbf{c}(t) = (t, 0, 1-t)$ for $0 \le t \le 1$, and

$$\mathbf{F}(\mathbf{c}(t)) \cdot \mathbf{c}'(t) = \langle e^{1-t}, 1, t \rangle \cdot \langle 1, 0, -1 \rangle = e^{1-t} - t$$

$$\int_{\overline{CA}} \mathbf{F} \cdot d\mathbf{s} = \int_0^1 (e^{1-t} - t)\, dt = \left(-e^{1-t} - \frac{1}{2}t^2 \right)\Bigg|_0^1 = -\frac{3}{2} + e$$

The total line integral is the sum

$$\int_{\mathcal{C}} \mathbf{F} \cdot d\mathbf{s} = (e-2) + \left(\frac{3}{2} - e\right) + \left(-\frac{3}{2} + e\right) = e - 2$$ ■

Applications of the Vector Line Integral

Recall that in physics, "work" refers to the energy expended when a force is applied to an object as it moves along a path. By definition, the work W performed along the straight segment from P to Q by applying a constant force $\mathbf{F}$ at an angle θ [Figure 11(A)] is

$$W = (\text{tangential component of } \mathbf{F}) \times \text{distance} = (\|\mathbf{F}\| \cos\theta) \times PQ$$

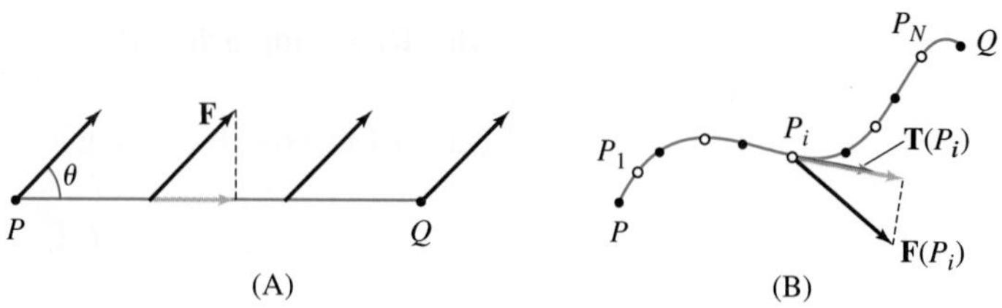

FIGURE 11

When the force acts on the object along a curved path $\mathcal{C}$, it makes sense to define the work W performed as the line integral [Figure 11(B)]:

$$W = \int_{\mathcal{C}} \mathbf{F} \cdot d\mathbf{s} \qquad \boxed{9}$$

This is the work "performed by the field $\mathbf{F}$." The idea is that we can divide $\mathcal{C}$ into a large number of short consecutive arcs $\mathcal{C}_1, \ldots, \mathcal{C}_N$, where $\mathcal{C}_i$ has length Δs_i. The work W_i performed along $\mathcal{C}_i$ is approximately equal to the tangential component $\mathbf{F}(P_i) \cdot \mathbf{T}(P_i)$ times the length Δs_i, where P_i is a sample point in $\mathcal{C}_i$. Thus we have

$$W = \sum_{i=1}^{N} W_i \approx \sum_{i=1}^{N} (\mathbf{F}(P_i) \cdot \mathbf{T}(P_i))\Delta s_i$$

The right-hand side approaches $\int_{\mathcal{C}} \mathbf{F} \cdot d\mathbf{s}$ as the lengths Δs_i tend to zero.

REMINDER Work has units of energy. The SI unit of force is the newton, and the unit of energy is the joule, defined as 1 newton-meter. The British unit is the foot-pound.

Often, we are interested in calculating the work required to move an object along a path in the presence of a force field $\mathbf{F}$ (such as an electrical or gravitational field). In this case, $\mathbf{F}$ acts on the object and we must work *against* the force field to move the object. The work required is the negative of the line integral in Eq. (9):

$$\text{Work performed against } \mathbf{F} = -\int_{\mathcal{C}} \mathbf{F} \cdot d\mathbf{s}$$

■ **EXAMPLE 8 Calculating Work** Calculate the work performed moving a particle from $P = (0, 0, 0)$ to $Q = (4, 8, 1)$ along the path

$$\mathbf{c}(t) = (t^2, t^3, t) \text{ (in meters)} \qquad \text{for } 1 \le t \le 2$$

in the presence of a force field $\mathbf{F} = \langle x^2, -z, -yz^{-1} \rangle$ in newtons.

Solution We have

$$\mathbf{F}(\mathbf{c}(t)) = \mathbf{F}(t^2, t^3, t) = \left\langle t^4, -t, -t^2 \right\rangle$$

$$\mathbf{c}'(t) = \left\langle 2t, 3t^2, 1 \right\rangle$$

$$\mathbf{F}(\mathbf{c}(t)) \cdot \mathbf{c}'(t) = \langle t^4, -t, -t^2 \rangle \cdot \langle 2t, 3t^2, 1 \rangle = 2t^5 - 3t^3 - t^2$$

The work performed against the force field in joules is

$$W = -\int_C \mathbf{F} \cdot d\mathbf{s} = -\int_1^2 (2t^5 - 3t^3 - t^2)\,dt = \frac{89}{12}$$

Line integrals are also used to compute the "**flux across a plane curve**," defined as the integral of the *normal component* of a vector field, rather than the tangential component (Figure 12). Suppose that a plane curve $\mathcal{C}$ is parametrized by $\mathbf{c}(t)$ for $a \le t \le b$, and let

$$\mathbf{n} = \mathbf{n}(t) = \left\langle y'(t), -x'(t) \right\rangle, \qquad \mathbf{e}_n(t) = \frac{\mathbf{n}(t)}{\|\mathbf{n}(t)\|}$$

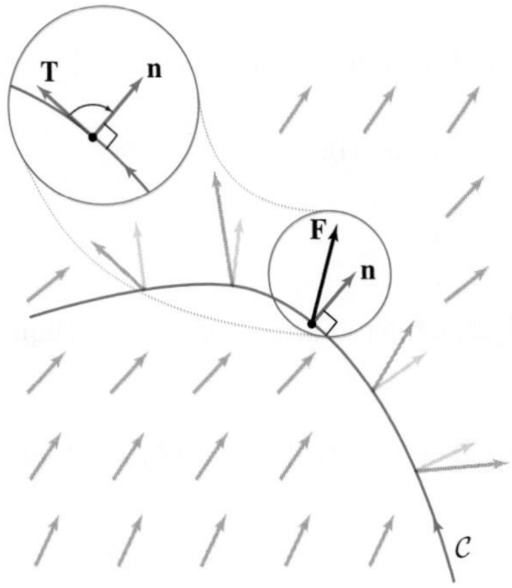

FIGURE 12

These vectors are normal to $\mathcal{C}$ and point to the right as you follow the curve in the direction of $\mathbf{c}$. The flux across $\mathcal{C}$ is the integral of the normal component $\mathbf{F} \cdot \mathbf{e}_n$, obtained by integrating $\mathbf{F}(\mathbf{c}(t)) \cdot \mathbf{n}(t)$:

$$\text{Flux across } \mathcal{C} = \int_{\mathcal{C}} (\mathbf{F} \cdot \mathbf{e}_n)\,ds = \int_a^b \mathbf{F}(\mathbf{c}(t)) \cdot \mathbf{n}(t)\,dt \qquad \boxed{10}$$

If $\mathbf{F}$ is the velocity field of a fluid (modeled as a two-dimensional fluid), then the flux is the quantity of water flowing across the curve per unit time.

EXAMPLE 9 Flux across a Curve Calculate the flux of the velocity vector field $\mathbf{v} = \left\langle 3 + 2y - y^2/3, 0 \right\rangle$ (in centimeters per second) across the quarter ellipse $\mathbf{c}(t) = \langle 3\cos t, 6\sin t\rangle$ for $0 \le t \le \frac{\pi}{2}$ (Figure 13).

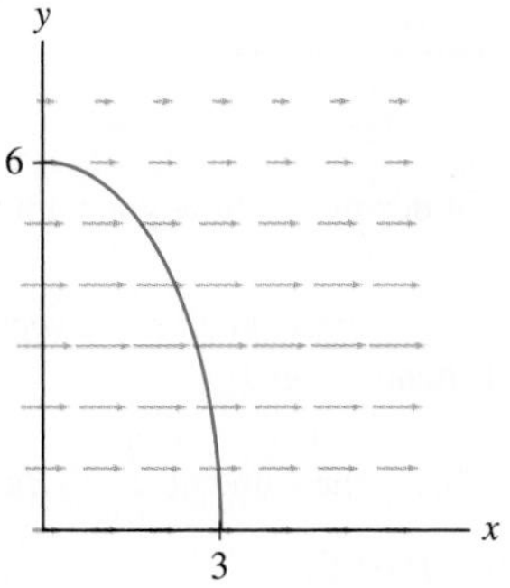

FIGURE 13

Solution The vector field along the path is

$$\mathbf{v}(\mathbf{c}(t)) = \left\langle 3 + 2(6\sin t) - (6\sin t)^2/3, 0 \right\rangle = \left\langle 3 + 12\sin t - 12\sin^2 t, 0 \right\rangle$$

The tangent vector is $\mathbf{c}'(t) = \langle -3\sin t, 6\cos t\rangle$, and thus $\mathbf{n}(t) = \langle 6\cos t, 3\sin t, \rangle$. We integrate the dot product

$$\begin{aligned} \mathbf{v}(\mathbf{c}(t)) \cdot \mathbf{n}(t) &= \left\langle 3 + 12\sin t - 12\sin^2 t, 0 \right\rangle \cdot \langle 6\cos t, 3\sin t, \rangle \\ &= (3 + 12\sin t - 12\sin^2 t)(6\cos t) \\ &= 18\cos t + 72\sin t\cos t - 72\sin^2 t\cos t \end{aligned}$$

to obtain the flux:

$$\begin{aligned} \int_a^b \mathbf{v}(\mathbf{c}(t)) \cdot \mathbf{n}(t)\,dt &= \int_0^{\pi/2} (18\cos t + 72\sin t\cos t - 72\sin^2 t\cos t)\,dt \\ &= 18 + 36 - 24 = 30 \text{ cm}^2/\text{s} \end{aligned}$$

17.2 SUMMARY

- Line integral over a curve with parametrization $\mathbf{c}(t)$ for $a \le t \le b$:

Scalar line integral: $\displaystyle\int_{\mathcal{C}} f(x, y, z)\,ds = \int_a^b f(\mathbf{c}(t))\,\|\mathbf{c}'(t)\|\,dt$

Vector line integral: $\displaystyle\int_{\mathcal{C}} \mathbf{F} \cdot d\mathbf{s} = \int_{\mathcal{C}} (\mathbf{F} \cdot \mathbf{T})\,ds = \int_a^b \mathbf{F}(\mathbf{c}(t)) \cdot \mathbf{c}'(t)\,dt$

- Arc length differential: $ds = \|\mathbf{c}'(t)\|\, dt$. To evaluate a scalar line integral, replace $f(x, y, z)\, ds$ with $f(\mathbf{c}(t))\, \|\mathbf{c}'(t)\|\, dt$.
- Vector differential: $d\mathbf{s} = \mathbf{c}'(t)\, dt$. To evaluate a vector line integral, replace $\mathbf{F} \cdot d\mathbf{s}$ with $F(\mathbf{c}(t)) \cdot \mathbf{c}'(t)\, dt$.
- An *oriented curve* C is a curve in which one of the two possible directions along C (called the *positive direction*) is chosen.
- The vector line integral depends on the orientation of the curve C. The parametrization $\mathbf{c}(t)$ must be regular, and it must trace C in the positive direction.
- We write $-C$ for the curve C with the opposite orientation. Then

$$\int_{-C} \mathbf{F} \cdot d\mathbf{s} = -\int_C \mathbf{F} \cdot d\mathbf{s}$$

- If $\rho(x, y, z)$ is the mass or charge density along C, then the total mass or charge is equal to the scalar line integral $\int_C \rho(x, y, z)\, ds$.
- The vector line integral is used to compute the work W exerted on an object along a curve C:

$$W = \int_C \mathbf{F} \cdot d\mathbf{s}$$

The work performed *against* $\mathbf{F}$ is the quantity $-\int_C \mathbf{F} \cdot d\mathbf{s}$.

17.2 EXERCISES

Preliminary Questions

1. What is the line integral of the constant function $f(x, y, z) = 10$ over a curve C of length 5?

2. Which of the following have a zero line integral over the vertical segment from $(0, 0)$ to $(0, 1)$?

(a) $f(x, y) = x$ **(b)** $f(x, y) = y$
(c) $\mathbf{F} = \langle x, 0\rangle$ **(d)** $\mathbf{F} = \langle y, 0\rangle$
(e) $\mathbf{F} = \langle 0, x\rangle$ **(f)** $\mathbf{F} = \langle 0, y\rangle$

3. State whether each statement is true or false. If the statement is false, give the correct statement.

(a) The scalar line integral does not depend on how you parametrize the curve.

(b) If you reverse the orientation of the curve, neither the vector line integral nor the scalar line integral changes sign.

4. Suppose that C has length 5. What is the value of $\int_C \mathbf{F} \cdot d\mathbf{s}$ if:

(a) $\mathbf{F}(P)$ is normal to C at all points P on C?

(b) $\mathbf{F}(P)$ is a unit vector pointing in the negative direction along the curve?

Exercises

1. Let $f(x, y, z) = x + yz$, and let C be the line segment from $P = (0, 0, 0)$ to $(6, 2, 2)$.

(a) Calculate $f(\mathbf{c}(t))$ and $ds = \|\mathbf{c}'(t)\|\, dt$ for the parametrization $\mathbf{c}(t) = (6t, 2t, 2t)$ for $0 \le t \le 1$.

(b) Evaluate $\int_C f(x, y, z)\, ds$.

2. Repeat Exercise 1 with the parametrization $\mathbf{c}(t) = (3t^2, t^2, t^2)$ for $0 \le t \le \sqrt{2}$.

3. Let $\mathbf{F} = \langle y^2, x^2\rangle$, and let C be the curve $y = x^{-1}$ for $1 \le x \le 2$, oriented from left to right.

(a) Calculate $\mathbf{F}(\mathbf{c}(t))$ and $d\mathbf{s} = \mathbf{c}'(t)\, dt$ for the parametrization of C given by $\mathbf{c}(t) = (t, t^{-1})$.

(b) Calculate the dot product $\mathbf{F}(\mathbf{c}(t)) \cdot \mathbf{c}'(t)\, dt$ and evaluate $\int_C \mathbf{F} \cdot d\mathbf{s}$.

4. Let $\mathbf{F} = \langle z^2, x, y\rangle$ and let C be the path $\mathbf{c}(t) = \langle 3 + 5t^2, 3 - t^2, t\rangle$ for $0 \le t \le 2$.

(a) Calculate $\mathbf{F}(\mathbf{c}(t))$ and $d\mathbf{s} = \mathbf{c}'(t)\, dt$.

(b) Calculate the dot product $\mathbf{F}(\mathbf{c}(t)) \cdot \mathbf{c}'(t)\, dt$ and evaluate $\int_C \mathbf{F} \cdot d\mathbf{s}$.

In Exercises 5–8, compute the integral of the scalar function or vector field over $\mathbf{c}(t) = (\cos t, \sin t, t)$ for $0 \le t \le \pi$.

5. $f(x, y, z) = x^2 + y^2 + z^2$

6. $f(x, y, z) = xy + z$

7. $\mathbf{F} = \langle x, y, z\rangle$

8. $\mathbf{F} = \langle xy, 2, z^3\rangle$

In Exercises 9–16, compute $\int_C f\,ds$ for the curve specified.

9. $f(x, y) = \sqrt{1 + 9xy}$, $y = x^3$ for $0 \le x \le 1$

10. $f(x, y) = \dfrac{y^3}{x^7}$, $y = \frac{1}{4}x^4$ for $1 \le x \le 2$

11. $f(x, y, z) = z^2$, $\mathbf{c}(t) = (2t, 3t, 4t)$ for $0 \le t \le 2$

12. $f(x, y, z) = 3x - 2y + z$, $\mathbf{c}(t) = (2 + t, 2 - t, 2t)$ for $-2 \le t \le 1$

13. $f(x, y, z) = xe^{z^2}$, piecewise linear path from $(0, 0, 1)$ to $(0, 2, 0)$ to $(1, 1, 1)$

14. $f(x, y, z) = x^2 z$, $\mathbf{c}(t) = (e^t, \sqrt{2}t, e^{-t})$ for $0 \le t \le 1$

15. $f(x, y, z) = 2x^2 + 8z$, $\mathbf{c}(t) = (e^t, t^2, t)$, $0 \le t \le 1$

16. $f(x, y, z) = 6xz - 2y^2$, $\mathbf{c}(t) = \left(t, \dfrac{t^2}{\sqrt{2}}, \dfrac{t^3}{3}\right)$, $0 \le t \le 2$

17. Calculate $\int_C 1\,ds$, where the curve C is parametrized by $\mathbf{c}(t) = (4t, -3t, 12t)$ for $2 \le t \le 5$. What does this integral represent?

18. Calculate $\int_C 1\,ds$, where the curve C is parametrized by $\mathbf{c}(t) = (e^t, \sqrt{2}t, e^{-t})$ for $0 \le t \le 2$.

In Exercises 19–26, compute $\int_C \mathbf{F} \cdot d\mathbf{s}$ for the oriented curve specified.

19. $\mathbf{F} = \langle x^2, xy \rangle$, line segment from $(0, 0)$ to $(2, 2)$

20. $\mathbf{F} = \langle 4, y \rangle$, quarter circle $x^2 + y^2 = 1$ with $x \le 0$, $y \le 0$, oriented counterclockwise

21. $\mathbf{F} = \langle x^2, xy \rangle$, part of circle $x^2 + y^2 = 9$ with $x \le 0$, $y \ge 0$, oriented clockwise

22. $\mathbf{F} = \langle e^{y-x}, e^{2x} \rangle$, piecewise linear path from $(1, 1)$ to $(2, 2)$ to $(0, 2)$

23. $\mathbf{F} = \langle 3zy^{-1}, 4x, -y \rangle$, $\mathbf{c}(t) = (e^t, e^t, t)$ for $-1 \le t \le 1$

24. $\mathbf{F} = \left\langle \dfrac{-y}{(x^2 + y^2)^2}, \dfrac{x}{(x^2 + y^2)^2} \right\rangle$, circle of radius R with center at the origin oriented counterclockwise

25. $\mathbf{F} = \left\langle \dfrac{1}{y^3 + 1}, \dfrac{1}{z + 1}, 1 \right\rangle$, $\mathbf{c}(t) = (t^3, 2, t^2)$ for $0 \le t \le 1$

26. $\mathbf{F} = \langle z^3, yz, x \rangle$, quarter of the circle of radius 2 in the yz-plane with center at the origin where $y \ge 0$ and $z \ge 0$, oriented clockwise when viewed from the positive x-axis

In Exercises 27–32, evaluate the line integral.

27. $\int_C y\,dx - x\,dy$, parabola $y = x^2$ for $0 \le x \le 2$

28. $\int_C y\,dx + z\,dy + x\,dz$, $\mathbf{c}(t) = (2 + t^{-1}, t^3, t^2)$ for $0 \le t \le 1$

29. $\int_C (x - y)\,dx + (y - z)\,dy + z\,dz$, line segment from $(0, 0, 0)$ to $(1, 4, 4)$

30. $\int_C z\,dx + x^2\,dy + y\,dz$, $\mathbf{c}(t) = (\cos t, \tan t, t)$ for $0 \le t \le \frac{\pi}{4}$

31. $\int_C \dfrac{-y\,dx + x\,dy}{x^2 + y^2}$, segment from $(1, 0)$ to $(0, 1)$.

32. $\int_C y^2\,dx + z^2\,dy + (1 - x^2)\,dz$, quarter of the circle of radius 1 in the xz-plane with center at the origin in the quadrant $x \ge 0$, $z \le 0$, oriented counterclockwise when viewed from the positive y-axis.

33. CAS Let $f(x, y, z) = x^{-1}yz$, and let C be the curve parametrized by $\mathbf{c}(t) = (\ln t, t, t^2)$ for $2 \le t \le 4$. Use a computer algebra system to calculate $\int_C f(x, y, z)\,ds$ to four decimal places.

34. CAS Use a CAS to calculate $\int_C \langle e^{x-y}, e^{x+y} \rangle \cdot d\mathbf{s}$ to four decimal places, where C is the curve $y = \sin x$ for $0 \le x \le \pi$, oriented from left to right.

In Exercises 35 and 36, calculate the line integral of $\mathbf{F} = \langle e^z, e^{x-y}, e^y \rangle$ over the given path.

35. The blue path from P to Q in Figure 14

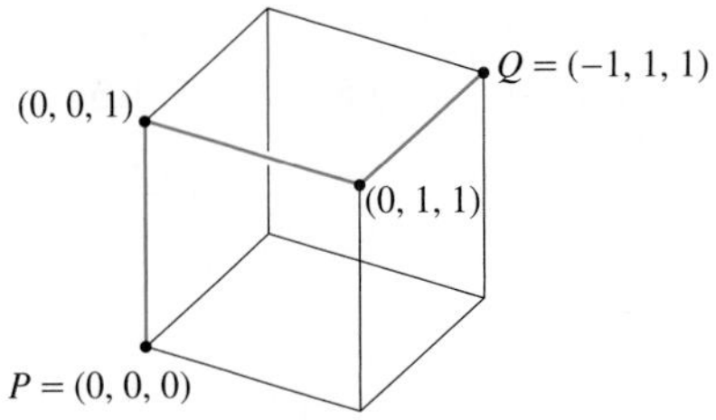

FIGURE 14

36. The closed path $ABCA$ in Figure 15

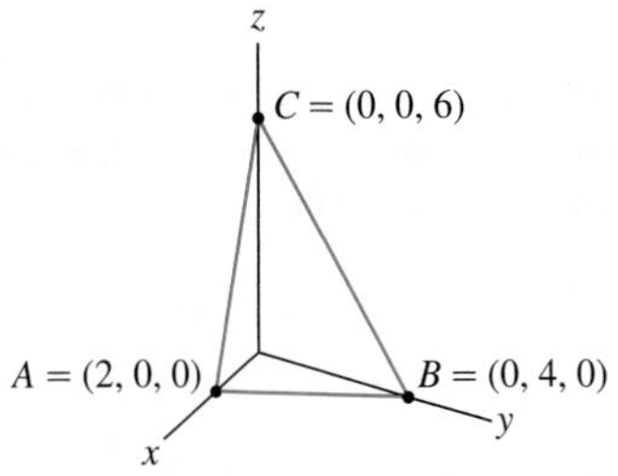

FIGURE 15

In Exercises 37 and 38, C is the path from P to Q in Figure 16 that traces C_1, C_2, and C_3 in the orientation indicated, and **F** *is a vector field such that*

$$\int_C \mathbf{F}\cdot d\mathbf{s} = 5, \qquad \int_{C_1} \mathbf{F}\cdot d\mathbf{s} = 8, \qquad \int_{C_3} \mathbf{F}\cdot d\mathbf{s} = 8$$

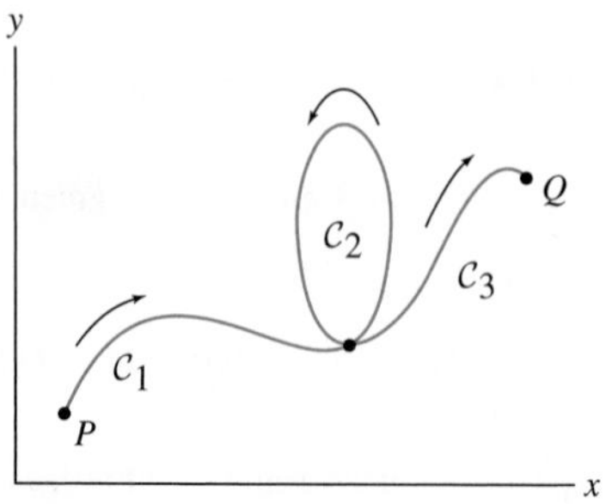

FIGURE 16

37. Determine:

(a) $\displaystyle\int_{-C_3} \mathbf{F}\cdot d\mathbf{s}$ **(b)** $\displaystyle\int_{C_2} \mathbf{F}\cdot d\mathbf{s}$ **(c)** $\displaystyle\int_{-C_1-C_3} \mathbf{F}\cdot d\mathbf{s}$

38. Find the value of $\displaystyle\int_{C'} \mathbf{F}\cdot d\mathbf{s}$, where C' is the path that traverses the loop C_2 four times in the clockwise direction.

39. The values of a function $f(x, y, z)$ and vector field $\mathbf{F}(x, y, z)$ are given at six sample points along the path ABC in Figure 17. Estimate the line integrals of f and $\mathbf{F}$ along ABC.

Point	$f(x, y, z)$	$\mathbf{F}(x, y, z)$
$(1, \frac{1}{6}, 0)$	3	$\langle 1, 0, 2\rangle$
$(1, \frac{1}{2}, 0)$	3.3	$\langle 1, 1, 3\rangle$
$(1, \frac{5}{6}, 0)$	3.6	$\langle 2, 1, 5\rangle$
$(1, 1, \frac{1}{6})$	4.2	$\langle 3, 2, 4\rangle$
$(1, 1, \frac{1}{2})$	4.5	$\langle 3, 3, 3\rangle$
$(1, 1, \frac{5}{6})$	4.2	$\langle 5, 3, 3\rangle$

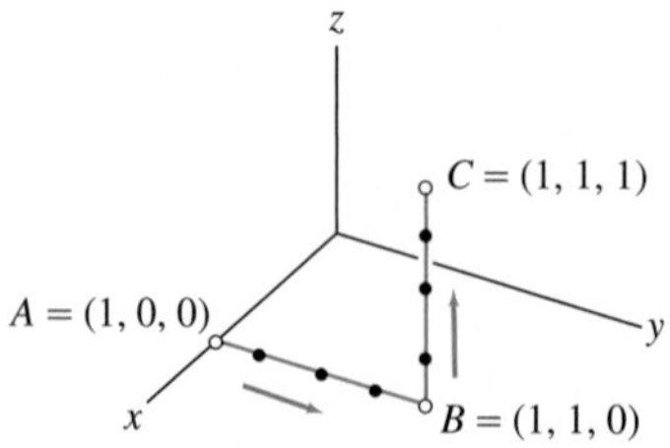

FIGURE 17

40. Estimate the line integrals of $f(x, y)$ and $\mathbf{F}(x, y)$ along the quarter circle (oriented counterclockwise) in Figure 18 using the values at the three sample points along each path.

Point	$f(x, y)$	$\mathbf{F}(x, y)$
A	1	$\langle 1, 2\rangle$
B	-2	$\langle 1, 3\rangle$
C	4	$\langle -2, 4\rangle$

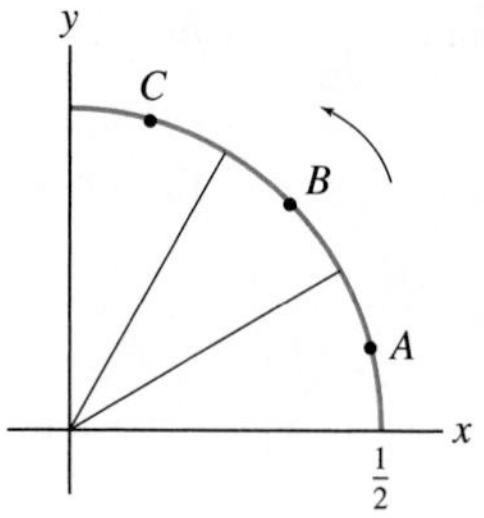

FIGURE 18

41. Determine whether the line integrals of the vector fields around the circle (oriented counterclockwise) in Figure 19 are positive, negative, or zero.

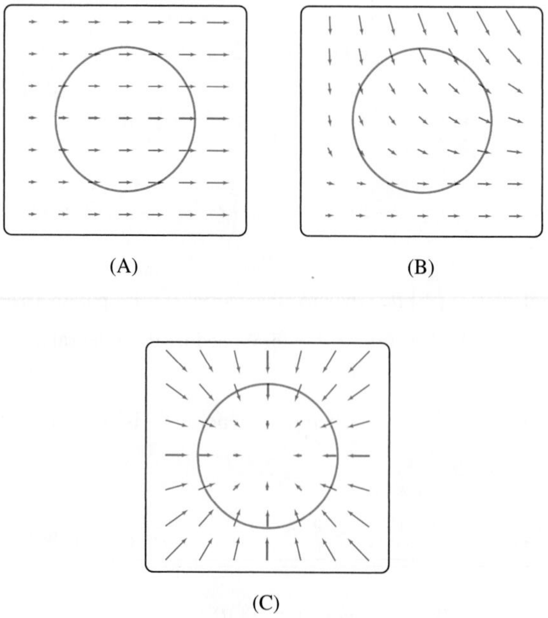

FIGURE 19

42. Determine whether the line integrals of the vector fields along the oriented curves in Figure 20 are positive or negative.

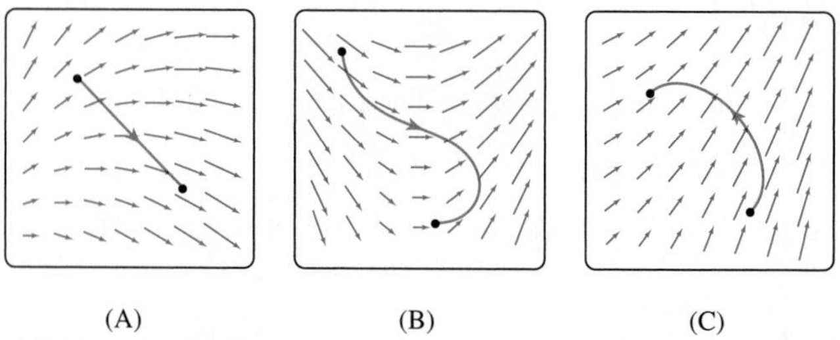

FIGURE 20

43. Calculate the total mass of a circular piece of wire of radius 4 cm centered at the origin whose mass density is $\rho(x, y) = x^2$ g/cm.

44. Calculate the total mass of a metal tube in the helical shape $\mathbf{c}(t) = (\cos t, \sin t, t^2)$ (distance in centimeters) for $0 \le t \le 2\pi$ if the mass density is $\rho(x, y, z) = \sqrt{z}$ g/cm.

45. Find the total charge on the curve $y = x^{4/3}$ for $1 \le x \le 8$ (in cm) assuming a charge density of $\rho(x, y) = x/y$ (in units of 10^{-6} C/cm).

46. Find the total charge on the curve $\mathbf{c}(t) = (\sin t, \cos t, \sin^2 t)$ in centimeters for $0 \le t \le \frac{\pi}{8}$ assuming a charge density of $\rho(x, y, z) = xy(y^2 - z)$ (in units of 10^{-6} C/cm).

In Exercises 47–50, use Eq. (6) to compute the electric potential $V(P)$ at the point P for the given charge density (in units of 10^{-6} C).

47. Calculate $V(P)$ at $P = (0, 0, 12)$ if the electric charge is distributed along the quarter circle of radius 4 centered at the origin with charge density $\rho(x, y, z) = xy$.

48. Calculate $V(P)$ at the origin $P = (0, 0)$ if the negative charge is distributed along $y = x^2$ for $1 \le x \le 2$ with charge density $\rho(x, y) = -y\sqrt{x^2 + 1}$.

49. Calculate $V(P)$ at $P = (2, 0, 2)$ if the negative charge is distributed along the y-axis for $1 \le y \le 3$ with charge density $\rho(x, y, z) = -y$.

50. Calculate $V(P)$ at the origin $P = (0, 0)$ if the electric charge is distributed along $y = x^{-1}$ for $\frac{1}{2} \le x \le 2$ with charge density $\rho(x, y) = x^3 y$.

51. Calculate the work done by a field $\mathbf{F} = \langle x + y, x - y \rangle$ when an object moves from $(0, 0)$ to $(1, 1)$ along each of the paths $y = x^2$ and $x = y^2$.

52. Calculate the work done by the force field $\mathbf{F} = \langle x, y, z \rangle$ along the path $(\cos t, \sin t, t)$ for $0 \le t \le 3\pi$.

53. Figure 21 shows a force field $\mathbf{F}$.

(a) Over which of the two paths, ADC or ABC, does $\mathbf{F}$ perform less work?

(b) If you have to work against $\mathbf{F}$ to move an object from C to A, which of the paths, CBA or CDA, requires less work?

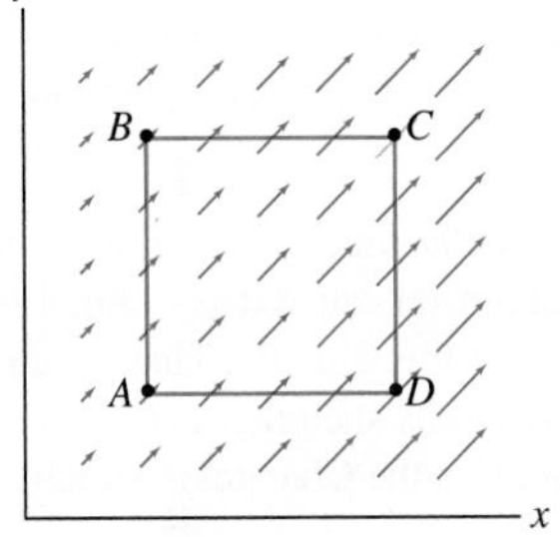

FIGURE 21

54. Verify that the work performed along the segment $\overline{PQ}$ by the constant vector field $\mathbf{F} = \langle 2, -1, 4 \rangle$ is equal to $\mathbf{F} \cdot \overrightarrow{PQ}$ in these cases:

(a) $P = (0, 0, 0)$, $Q = (4, 3, 5)$

(b) $P = (3, 2, 3)$, $Q = (4, 8, 12)$

55. Show that work performed by a constant force field $\mathbf{F}$ over any path C from P to Q is equal to $\mathbf{F} \cdot \overrightarrow{PQ}$.

56. Note that a curve C in polar form $r = f(\theta)$ is parametrized by $\mathbf{c}(\theta) = (f(\theta)\cos\theta, f(\theta)\sin\theta))$ because the x- and y-coordinates are given by $x = r\cos\theta$ and $y = r\sin\theta$.

(a) Show that $\|\mathbf{c}'(\theta)\| = \sqrt{f(\theta)^2 + f'(\theta)^2}$.

(b) Evaluate $\int_C (x - y)^2\, ds$, where C is the semicircle in Figure 22 with polar equation $r = 2\cos\theta$, $0 \le \theta \le \frac{\pi}{2}$.

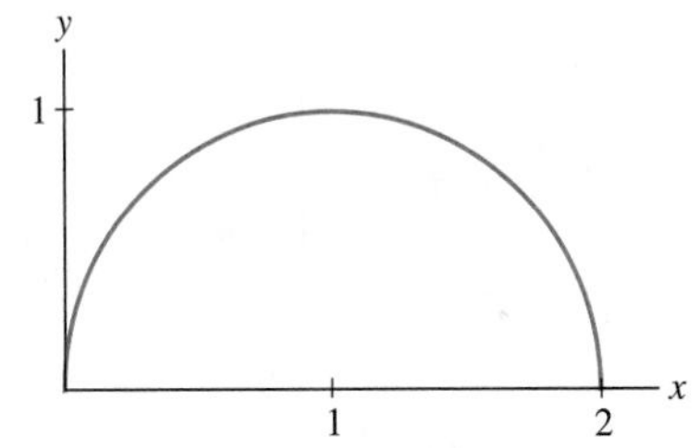

FIGURE 22 Semicircle $r = 2\cos\theta$.

57. Charge is distributed along the spiral with polar equation $r = \theta$ for $0 \le \theta \le 2\pi$. The charge density is $\rho(r, \theta) = r$ (assume distance is in centimeters and charge in units of 10^{-6} C/cm). Use the result of Exercise 56(a) to compute the total charge.

In Exercises 58–61, let $\mathbf{F}$ *be the* **vortex field** *(so-called because it swirls around the origin as in Figure 23):*

$$\mathbf{F} = \left\langle \frac{-y}{x^2 + y^2}, \frac{x}{x^2 + y^2} \right\rangle$$

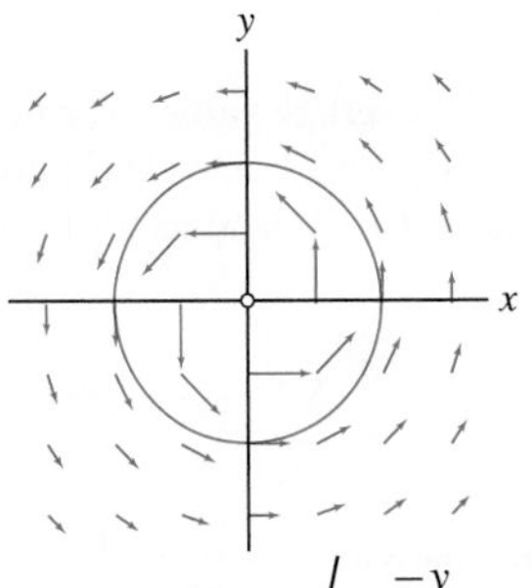

FIGURE 23 Vortex field $\mathbf{F} = \left\langle \frac{-y}{x^2 + y^2}, \frac{x}{x^2 + y^2} \right\rangle$.

58. Calculate $I = \int_C \mathbf{F} \cdot d\mathbf{s}$, where C is the circle of radius 2 centered at the origin. Verify that I changes sign when C is oriented in the clockwise direction.

59. Show that the value of $\int_{C_R} \mathbf{F} \cdot d\mathbf{s}$, where C_R is the circle of radius R centered at the origin and oriented counterclockwise, does not depend on R.

60. Let $a > 0$, $b < c$. Show that the integral of $\mathbf{F}$ along the segment [Figure 24(A)] from $P = (a, b)$ to $Q = (a, c)$ is equal to the angle $\angle POQ$.

61. Let C be a curve in polar form $r = f(\theta)$ for $\theta_1 \le \theta \le \theta_2$ [Figure 24(B)], parametrized by $\mathbf{c}(\theta) = (f(\theta)\cos\theta, f(\theta)\sin\theta)$ as in Exercise 56.

(a) Show that the vortex field in polar coordinates is written $\mathbf{F} = r^{-1}\langle -\sin\theta, \cos\theta\rangle$.

(b) Show that $\mathbf{F} \cdot \mathbf{c}'(\theta)\, d\theta = d\theta$.

(c) Show that $\int_C \mathbf{F} \cdot d\mathbf{s} = \theta_2 - \theta_1$.

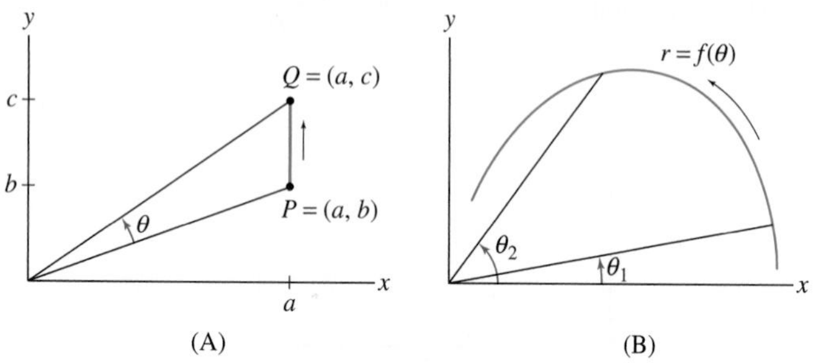

FIGURE 24

In Exercises 62–65, use Eq. (10) to calculate the flux of the vector field across the curve specified.

62. $\mathbf{F} = \langle -y, x\rangle$; upper half of the unit circle, oriented clockwise

63. $\mathbf{F} = \langle x^2, y^2\rangle$; segment from (3, 0) to (0, 3), oriented upward

64. $\mathbf{v} = \left\langle \dfrac{x+1}{(x+1)^2 + y^2}, \dfrac{y}{(x+1)^2 + y^2}\right\rangle$; segment $1 \le y \le 4$ along the y-axis, oriented upward

65. $\mathbf{v} = \langle e^y, 2x - 1\rangle$; parabola $y = x^2$ for $0 \le x \le 1$, oriented left to right

66. Let $I = \int_C f(x, y, z)\, ds$. Assume that $f(x, y, z) \ge m$ for some number m and all points (x, y, z) on C. Which of the following conclusions is correct? Explain.

(a) $I \ge m$

(b) $I \ge mL$, where L is the length of C

Further Insights and Challenges

67. Let $\mathbf{F} = \langle x, 0\rangle$. Prove that if C is any path from (a, b) to (c, d), then

$$\int_C \mathbf{F} \cdot d\mathbf{s} = \frac{1}{2}(c^2 - a^2)$$

68. Let $\mathbf{F} = \langle y, x\rangle$. Prove that if C is any path from (a, b) to (c, d), then

$$\int_C \mathbf{F} \cdot d\mathbf{s} = cd - ab$$

69. We wish to define the **average value** $\mathrm{Av}(f)$ of a continuous function f along a curve C of length L. Divide C into N consecutive arcs $C_1, \dots, C_N$, each of length L/N, and let P_i be a sample point in C_i (Figure 25). The sum

$$\frac{1}{N}\sum_{i=1} f(P_i)$$

may be considered an approximation to $\mathrm{Av}(f)$, so we define

$$\mathrm{Av}(f) = \lim_{N\to\infty} \frac{1}{N}\sum_{i=1} f(P_i)$$

Prove that

$$\mathrm{Av}(f) = \frac{1}{L}\int_C f(x, y, z)\, ds \qquad \boxed{11}$$

Hint: Show that $\dfrac{L}{N}\sum_{i=1} f(P_i)$ is a Riemann sum approximation to the line integral of f along C.

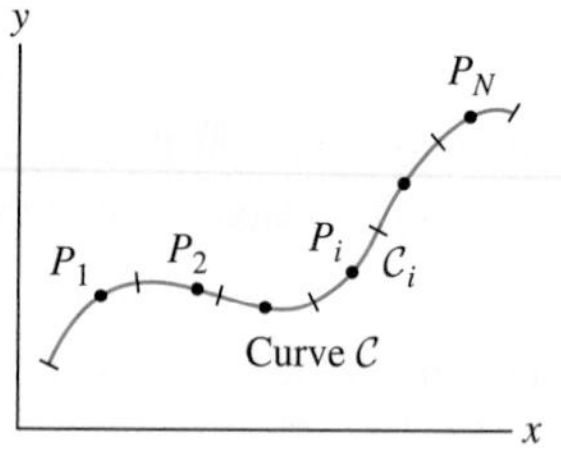

FIGURE 25

70. Use Eq. (11) to calculate the average value of $f(x, y) = x - y$ along the segment from $P = (2, 1)$ to $Q = (5, 5)$.

71. Use Eq. (11) to calculate the average value of $f(x, y) = x$ along the curve $y = x^2$ for $0 \le x \le 1$.

72. The temperature (in degrees centigrade) at a point P on a circular wire of radius 2 cm centered at the origin is equal to the square of the distance from P to $P_0 = (2, 0)$. Compute the average temperature along the wire.

73. The value of a scalar line integral does not depend on the choice of parametrization (because it is defined without reference to a parametrization). Prove this directly. That is, suppose that $\mathbf{c}_1(t)$ and $\mathbf{c}(t)$ are two parametrizations such that $\mathbf{c}_1(t) = \mathbf{c}(\varphi(t))$, where $\varphi(t)$ is an increasing function. Use the Change of Variables Formula to verify that

$$\int_c^d f(\mathbf{c}_1(t))\|\mathbf{c}_1'(t)\|\, dt = \int_a^b f(\mathbf{c}(t))\|\mathbf{c}'(t)\|\, dt$$

where $a = \varphi(c)$ and $b = \varphi(d)$.

17.3 Conservative Vector Fields

REMINDER

- *A vector field* $\mathbf{F}$ *is conservative if* $\mathbf{F} = \nabla V$ *for some function* $V(x, y, z)$.
- V *is called a potential function.*

In this section we study conservative vector fields in greater depth. For convenience, when a particular parametrization $\mathbf{c}(t)$ of an oriented curve C is specified, we will denote the line integral $\int_C \mathbf{F} \cdot d\mathbf{s}$ by

$$\int_{\mathbf{c}} \mathbf{F} \cdot d\mathbf{s}$$

When the curve C is *closed*, we often refer to the line integral as the **circulation** of $\mathbf{F}$ around C (Figure 1) and denote it with the symbol $\oint$:

$$\oint_C \mathbf{F} \cdot d\mathbf{s}$$

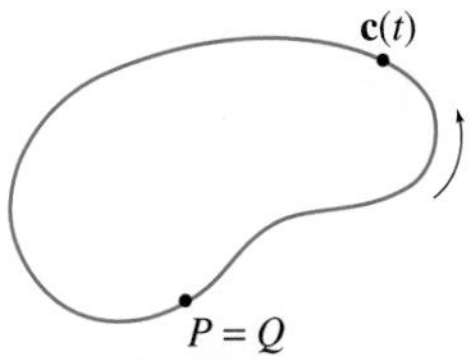

FIGURE 1 The circulation around a closed path is denoted $\oint \mathbf{F} \cdot d\mathbf{s}$.

Our first result establishes the fundamental **path independence** of conservative vector fields, which means that the line integral of $\mathbf{F}$ along a path from P to Q depends only on the endpoints P and Q, not on the particular path followed (Figure 2).

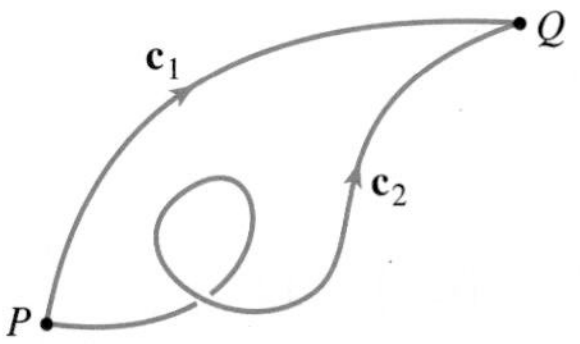

FIGURE 2 Path independence: If $\mathbf{F}$ is conservative, then the line integrals over $\mathbf{c}_1$ and $\mathbf{c}_2$ are equal.

THEOREM 1 Fundamental Theorem for Conservative Vector Fields Assume that $\mathbf{F} = \nabla V$ on a domain $\mathcal{D}$.

1. If $\mathbf{c}$ is a path from P to Q in $\mathcal{D}$, then

$$\int_{\mathbf{c}} \mathbf{F} \cdot d\mathbf{s} = V(Q) - V(P) \qquad \boxed{1}$$

In particular, $\mathbf{F}$ is path-independent.

2. The circulation around a closed path $\mathbf{c}$ (that is, $P = Q$) is zero:

$$\oint_{\mathbf{c}} \mathbf{F} \cdot d\mathbf{s} = 0$$

Proof Let $\mathbf{c}(t)$ be a path in $\mathcal{D}$ for $a \le t \le b$ with $\mathbf{c}(a) = P$ and $\mathbf{c}(b) = Q$. Then

$$\int_{\mathbf{c}} \mathbf{F} \cdot d\mathbf{s} = \int_{\mathbf{c}} \nabla V \cdot d\mathbf{s} = \int_a^b \nabla V(\mathbf{c}(t)) \cdot \mathbf{c}'(t)\, dt$$

However, by the Chain Rule for Paths (Theorem 2 in Section 15.5),

$$\frac{d}{dt} V(\mathbf{c}(t)) = \nabla V(\mathbf{c}(t)) \cdot \mathbf{c}'(t)$$

Thus we can apply the Fundamental Theorem of Calculus:

$$\int_{\mathbf{c}} \mathbf{F} \cdot d\mathbf{s} = \int_a^b \frac{d}{dt} V(\mathbf{c}(t))\, dt = V(\mathbf{c}(t))\Big|_a^b = V(\mathbf{c}(b)) - V(\mathbf{c}(a)) = V(Q) - V(P)$$

This proves Eq. (1). It also proves path independence, because the quantity $V(Q) - V(P)$ depends on the endpoints but not on the path $\mathbf{c}$. If $\mathbf{c}$ is a closed path, then $P = Q$ and $V(Q) - V(P) = 0$. ■

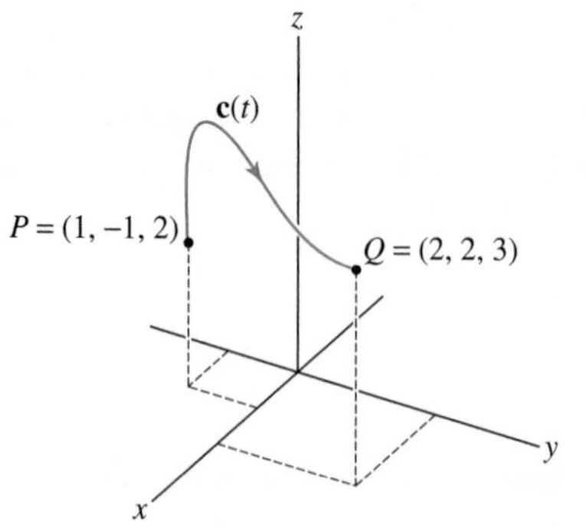

FIGURE 3 An arbitrary path from $(1, -1, 2)$ to $(2, 2, 3)$.

EXAMPLE 1 Let $\mathbf{F} = \langle 2xy + z, x^2, x \rangle$.

(a) Verify that $V(x, y, z) = x^2y + xz$ is a potential function.

(b) Evaluate $\int_{\mathbf{c}} \mathbf{F} \cdot d\mathbf{s}$, where $\mathbf{c}$ is a path from $P = (1, -1, 2)$ to $Q = (2, 2, 3)$.

Solution (a) The partial derivatives of $V(x, y, z) = x^2y + xz$ are the components of $\mathbf{F}$:

$$\frac{\partial V}{\partial x} = 2xy + z, \qquad \frac{\partial V}{\partial y} = x^2, \qquad \frac{\partial V}{\partial z} = x$$

Therefore, $\nabla V = \langle 2xy + z, x^2, x \rangle = \mathbf{F}$.
(b) By Theorem 1, the line integral over any path $\mathbf{c}(t)$ from $P = (1, -1, 2)$ to $Q = (2, 2, 3)$ [Figure 3] has the value

$$\int_{\mathbf{c}} \mathbf{F} \cdot d\mathbf{s} = V(Q) - V(P)$$

$$= V(2, 2, 3) - V(1, -1, 2)$$

$$= \left(2^2(2) + 2(3)\right) - \left(1^2(-1) + 1(2)\right) = 13$$ ■

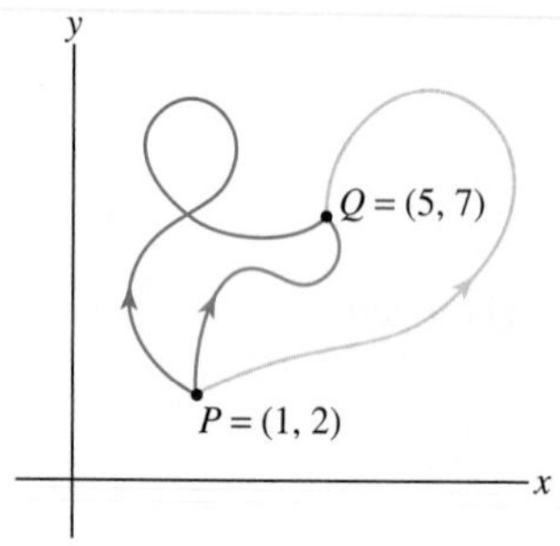

FIGURE 4 Paths from $(1, 2)$ to $(5, 7)$.

EXAMPLE 2 Find a potential function for $\mathbf{F} = \langle 2x + y, x \rangle$ and use it to evaluate $\int_{\mathbf{c}} \mathbf{F} \cdot d\mathbf{s}$, where $\mathbf{c}$ is any path (Figure 4) from $(1, 2)$ to $(5, 7)$.

Solution We will develop a general method for finding potential functions. At this point we can see by inspection that $V(x, y) = x^2 + xy$ satisfies $\nabla V = \mathbf{F}$:

$$\frac{\partial V}{\partial x} = \frac{\partial}{\partial x}(x^2 + xy) = 2x + y,$$

$$\frac{\partial V}{\partial y} = \frac{\partial}{\partial y}(x^2 + xy) = x$$

Therefore, for any path $\mathbf{c}$ from $(1, 2)$ to $(5, 7)$,

$$\int_{\mathbf{c}} \mathbf{F} \cdot d\mathbf{s} = V(5, 7) - V(1, 2)$$

$$= (5^2 + 5(7)) - (1^2 + 1(2)) = 57$$ ■

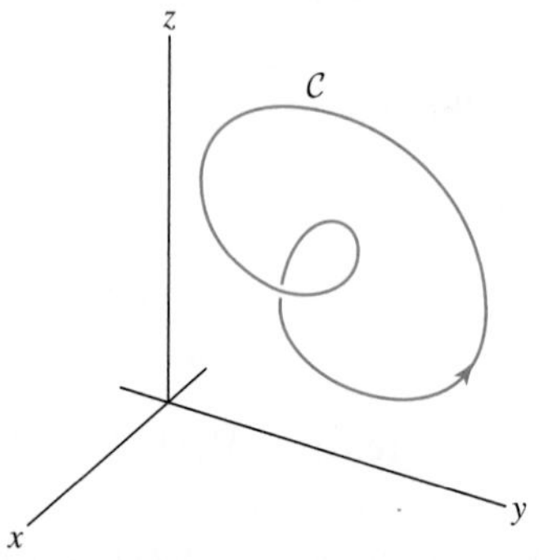

FIGURE 5 The line integral of a conservative vector field around a closed curve is zero.

EXAMPLE 3 Integral around a Closed Path Let $V(x, y, z) = xy\sin(yz)$. Evaluate $\oint_C \nabla V \cdot d\mathbf{s}$, where C is the closed curve in Figure 5.

Solution By Theorem 1, the integral of a gradient vector around any closed path is zero. In other words, $\oint_C \nabla V \cdot d\mathbf{s} = 0$. ■

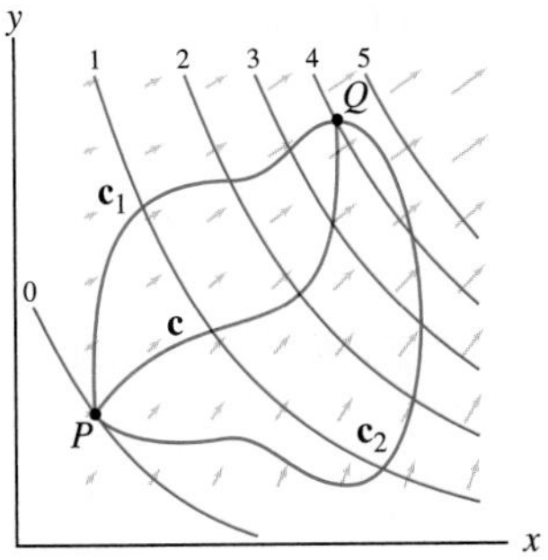

FIGURE 6 Vector field $\mathbf{F} = \nabla V$ with the contour lines of V.

CONCEPTUAL INSIGHT A good way to think about path independence is in terms of the contour map of the potential function. Consider a vector field $\mathbf{F} = \nabla V$ in the plane (Figure 6). The level curves of V are called **equipotential curves**, and the value $V(P)$ is called the potential at P.

When we integrate $\mathbf{F}$ along a path $\mathbf{c}(t)$ from P to Q, the integrand is

$$\mathbf{F}(\mathbf{c}(t)) \cdot \mathbf{c}'(t) = \nabla V(\mathbf{c}(t)) \cdot \mathbf{c}'(t)$$

Now recall that by the Chain Rule for paths,

$$\nabla V(\mathbf{c}(t)) \cdot \mathbf{c}'(t)) = \frac{d}{dt} V(\mathbf{c}(t))$$

In other words, the integrand is the rate at which the potential changes along the path, and thus the integral itself is the net change in potential:

$$\int \mathbf{F} \cdot d\mathbf{s} = \underbrace{V(Q) - V(P)}_{\text{Net change in potential}}$$

So informally speaking, what the line integral does is count the net number of equipotential curves crossed as you move along any path P to Q. By "net number" we mean that crossings in the opposite direction are counted with a minus sign. This net number is independent of the particular path.

We can also interpret the line integral in terms of the graph of the potential function $z = V(x, y)$. The line integral computes the change in height as we move up the surface (Figure 7). Again, this change in height does not depend on the path from P to Q. Of course, these interpretations apply only to conservative vector fields—otherwise, there is no potential function.

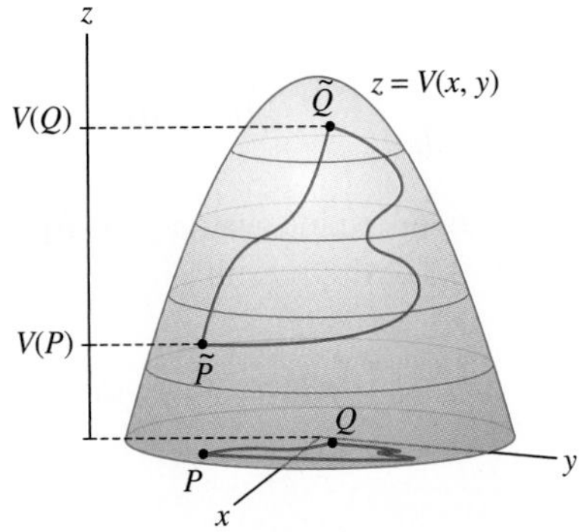

FIGURE 7 The potential surface $z = V(x, y)$.

You might wonder whether there exist any path-independent vector fields other than the conservative ones. The answer is no. By the next theorem, a path-independent vector field is necessarily conservative.

THEOREM 2 A vector field $\mathbf{F}$ on an open connected domain $\mathcal{D}$ is path-independent if and only if it is conservative.

Proof We have already shown that conservative vector fields are path-independent. So we assume that $\mathbf{F}$ is path-independent and prove that $\mathbf{F}$ has a potential function.

To simplify the notation, we treat the case of a planar vector field $\mathbf{F} = \langle F_1, F_2 \rangle$. The proof for vector fields in $\mathbf{R}^3$ is similar. Choose a point P_0 in $\mathcal{D}$, and for any point $P = (x, y) \in \mathcal{D}$, define

$$V(P) = V(x, y) = \int_{\mathbf{c}} \mathbf{F} \cdot d\mathbf{s}$$

where $\mathbf{c}$ is any path in $\mathcal{D}$ from P_0 to P (Figure 8). Note that this definition of $V(P)$ is meaningful only because we are assuming that the line integral does not depend on the path $\mathbf{c}$.

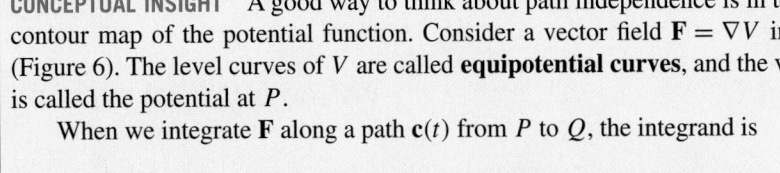

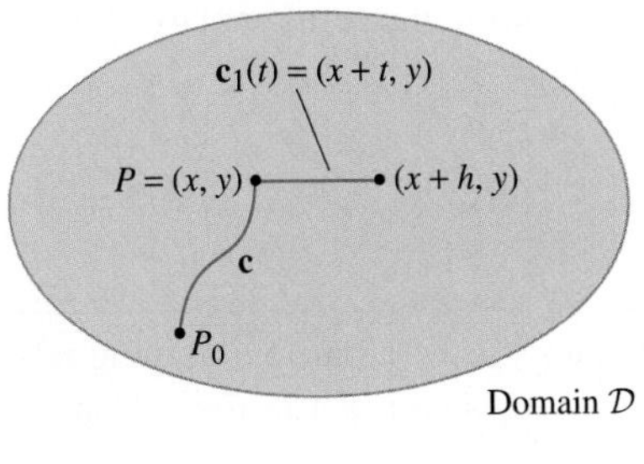

FIGURE 8

We will prove that $\mathbf{F} = \nabla V$, which involves showing that $\frac{\partial V}{\partial x} = F_1$ and $\frac{\partial V}{\partial y} = F_2$. We will only verify the first equation, as the second can be checked in a similar manner. Let $\mathbf{c}_1$ be the horizontal path $\mathbf{c}_1(t) = (x + t, y)$ for $0 \le t \le h$. For $|h|$ small enough, $\mathbf{c}_1$

lies inside $\mathcal{D}$. Let $\mathbf{c} + \mathbf{c}_1$ denote the path $\mathbf{c}$ followed by $\mathbf{c}_1$. It begins at P_0 and ends at $(x + h, y)$, so

$$V(x+h, y) - V(x, y) = \int_{\mathbf{c}+\mathbf{c}_1} \mathbf{F} \cdot d\mathbf{s} - \int_{\mathbf{c}} \mathbf{F} \cdot d\mathbf{s}$$

$$\left(\int_{\mathbf{c}} \mathbf{F} \cdot d\mathbf{s} + \int_{\mathbf{c}_1} \mathbf{F} \cdot d\mathbf{s}\right) - \int_{\mathbf{c}} \mathbf{F} \cdot d\mathbf{s} = \int_{\mathbf{c}_1} \mathbf{F} \cdot d\mathbf{s}$$

The path $\mathbf{c}_1$ has tangent vector $\mathbf{c}_1'(t) = \langle 1, 0 \rangle$, so

$$\mathbf{F}(\mathbf{c}_1(t)) \cdot \mathbf{c}_1'(t) = \langle F_1(x+t, y), F_2(x+t, y) \rangle \cdot \langle 1, 0 \rangle = F_1(x+t, y)$$

$$V(x+h, y) - V(x, y) = \int_{\mathbf{c}_1} \mathbf{F} \cdot d\mathbf{s} = \int_0^h F_1(x+t, y)\, dt$$

Using the substitution $u = x + t$, we have

$$\frac{V(x+h, y) - V(x, y)}{h} = \frac{1}{h} \int_0^h F_1(x+t, y)\, dt = \frac{1}{h} \int_x^{x+h} F_1(u, y)\, du$$

The integral on the right is the average value of $F_1(u, y)$ over the interval $[x, x + h]$. It converges to the value $F_1(x, y)$ as $h \to 0$, and this yields the desired result:

$$\frac{\partial V}{\partial x} = \lim_{h \to 0} \frac{V(x+h, y) - V(x, y)}{h} = \lim_{h \to 0} \frac{1}{h} \int_x^{x+h} F_1(u, y)\, du = F_1(x, y) \qquad ■$$

Conservative Fields in Physics

The Conservation of Energy principle says that the sum $KE + PE$ of kinetic and potential energy remains constant in an isolated system. For example, a falling object picks up kinetic energy as it falls to earth, but this gain in kinetic energy is offset by a loss in gravitational potential energy (g times the change in height), such that the sum $KE + PE$ remains unchanged.

We show now that conservation of energy is valid for the motion of a particle of mass m under a force field $\mathbf{F}$ if $\mathbf{F}$ has a potential function. This explains why the term "conservative" is used to describe vector fields that have a potential function.

We follow the convention in physics of writing the potential function with a minus sign:

$$\mathbf{F} = -\nabla V$$

When the particle is located at $P = (x, y, z)$, it is said to have **potential energy** $V(P)$. Suppose that the particle moves along a path $\mathbf{c}(t)$. The particle's velocity is $\mathbf{v} = \mathbf{c}'(t)$, and its kinetic energy is $KE = \frac{1}{2} m \|\mathbf{v}\|^2 = \frac{1}{2} m \mathbf{v} \cdot \mathbf{v}$. By definition, the **total energy** at time t is the sum

In a conservative force field, the work W against $\mathbf{F}$ required to move the particle from P to Q is equal to the change in potential energy:

$$W = -\int_{\mathbf{c}} \mathbf{F} \cdot d\mathbf{s} = V(Q) - V(P)$$

$$E = KE + PE = \frac{1}{2} m \mathbf{v} \cdot \mathbf{v} + V(\mathbf{c}(t))$$

THEOREM 3 Conservation of Energy The total energy E of a particle moving under the influence of a conservative force field $\mathbf{F} = -\nabla V$ is constant in time. That is, $\dfrac{dE}{dt} = 0$.

Potential functions first appeared in 1774 in the writings of Jean-Louis Lagrange (1736–1813). One of the greatest mathematicians of his time, Lagrange made fundamental contributions to physics, analysis, algebra, and number theory. He was born in Turin, Italy, to a family of French origin but spent most of his career first in Berlin and then in Paris. After the French Revolution, Lagrange was required to teach courses in elementary mathematics, but apparently he spoke above the heads of his audience. A contemporary wrote, "whatever this great man says deserves the highest degree of consideration, but he is too abstract for youth."

Proof Let $\mathbf{a} = \mathbf{v}'(t)$ be the particle's acceleration and let m be its mass. According to Newton's Second Law of Motion, $\mathbf{F}(\mathbf{c}(t)) = m\mathbf{a}(t)$, and thus

$$\begin{aligned}\frac{dE}{dt} &= \frac{d}{dt}\left(\frac{1}{2}m\mathbf{v}\cdot\mathbf{v} + V(\mathbf{c}(t))\right)\\ &= m\mathbf{v}\cdot\mathbf{a} + \nabla V(\mathbf{c}(t))\cdot\mathbf{c}'(t) && \text{(Product and Chain Rules)}\\ &= \mathbf{v}\cdot m\mathbf{a} - \mathbf{F}\cdot\mathbf{v} && (\text{since } \mathbf{F} = -\nabla V \text{ and } \mathbf{c}'(t) = \mathbf{v})\\ &= \mathbf{v}\cdot(m\mathbf{a} - \mathbf{F}) = 0 && (\text{since } \mathbf{F} = m\mathbf{a})\end{aligned}$$

■

In Example 6 of Section 17.1, we verified that inverse-square vector fields are conservative:

$$\mathbf{F} = k\frac{\mathbf{e}_r}{r^2} = -\nabla V \quad \text{with} \quad V = \frac{k}{r}$$

Basic examples of inverse-square vector fields are the gravitational and electrostatic forces due to a point mass or charge. By convention, these fields have units of force *per unit mass or unit charge*. Thus, if $\mathbf{F}$ is a gravitational field, the force on a particle of mass m is $m\mathbf{F}$ and its potential energy is mV, where $\mathbf{F} = -\nabla V$.

■ **EXAMPLE 4 Work against Gravity** Compute the work W against the earth's gravitational field required to move a satellite of mass $m = 600$ kg along any path from an orbit of altitude 2000 km to an orbit of altitude 4000 km.

Example 6 of Section 17.1 showed that

$$\frac{\mathbf{e}_r}{r^2} = -\nabla\left(\frac{1}{r}\right)$$

The constant k is equal to GM_e where $G \approx 6.67\cdot 10^{-11}$ m^3 kg^{-1} s^{-2} and the mass of the earth is $M_e \approx 5.98\cdot 10^{24}$ kg:

$$k = GM_e \approx 4\cdot 10^{14}\ m^3\, s^{-2}$$

Solution The earth's gravitational field is the inverse-square field

$$\mathbf{F} = -k\frac{\mathbf{e}_r}{r^2} = -\nabla V, \qquad V = -\frac{k}{r}$$

where r is the distance from the center of the earth and $k = 4\cdot 10^{14}$ (see marginal note). The radius of the earth is approximately $6.4\cdot 10^6$ meters, so the satellite must be moved from $r = 8.4\cdot 10^6$ meters to $r = 10.4\cdot 10^6$ meters. The force on the satellite is $m\mathbf{F} = 600\mathbf{F}$, and the work W required to move the satellite along a path $\mathbf{c}$ is

$$\begin{aligned}W &= -\int_{\mathbf{c}} m\mathbf{F}\cdot d\mathbf{s} = 600\int_{\mathbf{c}} \nabla V\cdot d\mathbf{s}\\ &= -\frac{600k}{r}\bigg|_{8.4\cdot 10^6}^{10.4\times 10^6}\\ &\approx -\frac{2.4\cdot 10^{17}}{10.4\cdot 10^6} + \frac{2.4\cdot 10^{17}}{8.4\cdot 10^6} \approx 5.5\cdot 10^9 \text{ joules}\end{aligned}$$

■

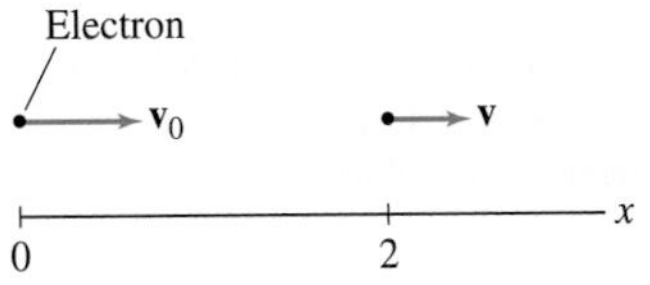

FIGURE 9 An electron moving in an electric field.

■ **EXAMPLE 5** An electron is traveling in the positive x-direction with speed $v_0 = 10^7$ m/s. When it passes $x = 0$, a horizontal electric field $\mathbf{E} = 100x\mathbf{i}$ (in newtons per coulomb) is turned on. Find the electron's velocity after it has traveled 2 meters. Assume that $q_e/m_e = -1.76\cdot 10^{11}$ C/kg, where q_e and m_e are the mass and charge of the electron, respectively.

Solution We have $\mathbf{E} = -\nabla V$ where $V(x, y, z) = -50x^2$, so the electric field is conservative. Since V depends only on x, we write $V(x)$ for $V(x, y, z)$. By the Law of Conservation of Energy, the electron's total energy E is constant and therefore E has the same value when the electron is at $x = 0$ and at $x = 2$:

$$E = \frac{1}{2}m_e v_0^2 + q_e V(0) = \frac{1}{2}m_e v^2 + q_e V(2)$$

Since $V(0) = 0$, we obtain

$$\frac{1}{2}m_e v_0^2 = \frac{1}{2}m_e v^2 + q_e V(2) \quad \Rightarrow \quad v = \sqrt{v_0^2 - 2(q_e/m_e)V(2)}$$

Using the numerical value of q_e/m_e, we obtain

$$v \approx \sqrt{10^{14} - 2(-1.76 \cdot 10^{11})(-50(2)^2)} \approx \sqrt{2.96 \cdot 10^{13}} \approx 5.4 \cdot 10^6 \text{ m/s}$$

Note that the velocity has decreased. This is because **E** exerts a force in the negative x-direction on a negative charge. ■

Finding Potential Functions

We do not yet have an effective way of telling whether a given vector field is conservative. By Theorem 1 in Section 17.1, every conservative vector field satisfies the cross-partials condition:

$$\frac{\partial F_1}{\partial y} = \frac{\partial F_2}{\partial x}, \qquad \frac{\partial F_2}{\partial z} = \frac{\partial F_3}{\partial y}, \qquad \frac{\partial F_3}{\partial x} = \frac{\partial F_1}{\partial z} \qquad \boxed{2}$$

But does this condition guarantee that **F** is conservative? The answer is a qualified yes; the cross-partials condition does guarantee that **F** is conservative, but only on domains $\mathcal{D}$ with a property called simple-connectedness.

Roughly speaking, a domain $\mathcal{D}$ in the plane is **simply-connected** if it does not have any "holes" (Figure 10). More precisely, $\mathcal{D}$ is simply-connected if every loop in $\mathcal{D}$ can be drawn down, or "contracted," to a point *while staying within* $\mathcal{D}$ as in Figure 11(A). Examples of simply-connected regions in $\mathbf{R}^2$ are disks, rectangles, and the entire plane $\mathbf{R}^2$. By contrast, the disk with a point removed in Figure 11(B) is not simply-connected: The loop cannot be drawn down to a point without passing through the point that was removed. In $\mathbf{R}^3$, the interiors of balls and boxes are simply-connected, as is the entire space $\mathbf{R}^3$.

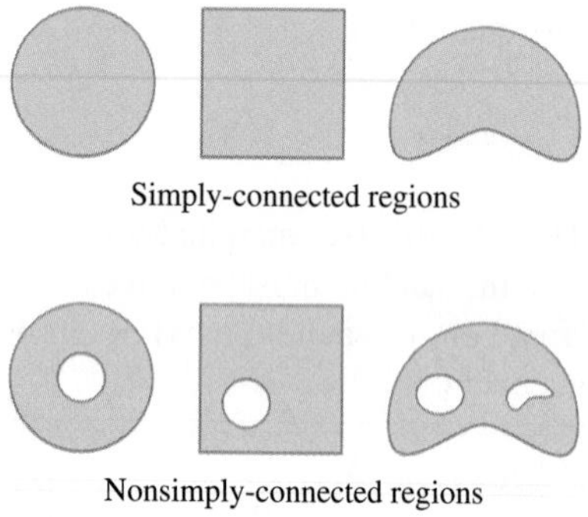

FIGURE 10 Simple connectedness means "no holes."

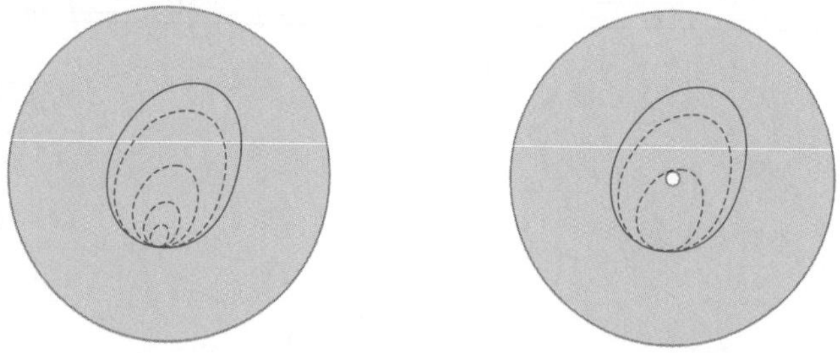

(A) Simply-connected region: Any loop can be drawn down to a point within the region.

(B) Nonsimply-connected region: a loop around the missing hole cannot be drawn tight without passing through the hole.

FIGURE 11

THEOREM 4 Existence of a Potential Function Let **F** be a vector field on a simply-connected domain $\mathcal{D}$. If **F** satisfies the cross-partials condition (2), then **F** is conservative.

Rather than prove Theorem 4, we illustrate a practical procedure for finding a potential function when the cross-partials condition is satisfied. The proof itself involves Stokes' Theorem and is somewhat technical because of the role played by the simply-connected property of the domain.

EXAMPLE 6 Finding a Potential Function Show that

$$\mathbf{F} = \langle 2xy + y^3, x^2 + 3xy^2 + 2y \rangle$$

is conservative and find a potential function.

Solution First we observe that the cross-partial derivatives are equal:

$$\frac{\partial F_1}{\partial y} = \frac{\partial}{\partial y}(2xy + y^3) \qquad = 2x + 3y^2$$

$$\frac{\partial F_2}{\partial x} = \frac{\partial}{\partial x}(x^2 + 3xy^2 + 2y) = 2x + 3y^2$$

Furthermore, $\mathbf{F}$ is defined on all of $\mathbf{R}^2$, which is a simply-connected domain. Therefore, a potential function exists by Theorem 4.

Now, the potential function V satisfies

$$\frac{\partial V}{\partial x} = F_1(x, y) = 2xy + y^3$$

This tells us that V is an antiderivative of $F_1(x, y)$, regarded as a function of x alone:

$$\begin{aligned} V(x, y) &= \int F_1(x, y)\, dx \\ &= \int \left(2xy + y^3\right) dx \\ &= x^2 y + xy^3 + g(y) \end{aligned}$$

Note that to obtain the general antiderivative of $F_1(x, y)$ with respect to x, we must add on an arbitrary function $g(y)$ depending on y alone, rather than the usual constant of integration. Similarly, we have

$$\begin{aligned} V(x, y) &= \int F_2(x, y)\, dy \\ &= \int \left(x^2 + 3xy^2 + 2y\right) dy \\ &= x^2 y + xy^3 + y^2 + h(x) \end{aligned}$$

The two expressions for $V(x, y)$ must be equal:

$$x^2 y + xy^3 + g(y) = x^2 y + xy^3 + y^2 + h(x)$$

This tells us that $g(y) = y^2$ and $h(x) = 0$, up to the addition of an arbitrary numerical constant C. Thus we obtain the general potential function

$$V(x, y) = x^2 y + xy^3 + y^2 + C$$

■

The same method works for vector fields in three-space.

EXAMPLE 7 Find a potential function for

$$\mathbf{F} = \left\langle 2xyz^{-1}, z + x^2z^{-1}, y - x^2yz^{-2} \right\rangle$$

Solution If a potential function V exists, then it satisfies

$$V(x, y, z) = \int 2xyz^{-1}\, dx \qquad = x^2yz^{-1} + f(y, z)$$

$$V(x, y, z) = \int (z + x^2z^{-1})\, dy = zy + x^2z^{-1}y + g(x, z)$$

$$V(x, y, z) = \int (y - x^2yz^{-2})\, dz = yz + x^2yz^{-1} + h(x, y)$$

These three ways of writing $V(x, y, z)$ must be equal:

$$x^2yz^{-1} + f(y, z) = zy + x^2z^{-1}y + g(x, z) = yz + x^2yz^{-1} + h(x, y)$$

In Example 7, $\mathbf{F}$ is only defined for $z \neq 0$, so the domain has two halves: $z > 0$ and $z < 0$. We are free to choose different constants C on the two halves, if desired.

These equalities hold if $f(y, z) = yz$, $g(x, z) = 0$, and $h(x, y) = 0$. Thus $\mathbf{F}$ is conservative and, for any constant C, a potential function is

$$V(x, y, z) = x^2yz^{-1} + yz + C$$

■

Assumptions Matter We cannot expect the method for finding a potential function to work if $\mathbf{F}$ does not satisfy the cross-partials condition (because in this case, no potential function exists). What goes wrong? Consider $\mathbf{F} = \langle y, 0 \rangle$. If we attempted to find a potential function, we would calculate

$$V(x, y) = \int y\, dx = xy + g(y)$$

$$V(x, y) = \int 0\, dy = 0 + h(x)$$

However, there is no choice of $g(y)$ and $h(x)$ for which $xy + g(y) = h(x)$. If there were, we could differentiate this equation twice, once with respect to x and once with respect to y. This would yield $1 = 0$, which is a contradiction. The method fails in this case because $\mathbf{F}$ does not satisfy the cross-partials condition and thus is not conservative.

The Vortex Field Why does Theorem 4 assume that the domain is simply-connected? This is an interesting question that we can answer by examining the vortex field (Figure 12):

$$\mathbf{F} = \left\langle \frac{-y}{x^2 + y^2}, \frac{x}{x^2 + y^2} \right\rangle$$

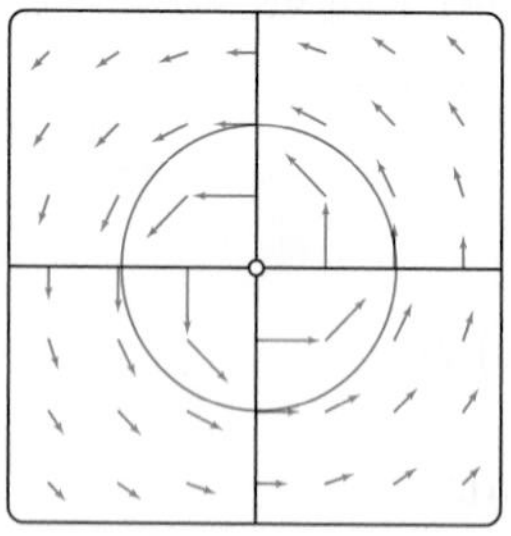

FIGURE 12 The vortex field.

EXAMPLE 8 Show that the vortex field satisfies the cross-partials condition but is not conservative. Does this contradict Theorem 4?

Solution We check the cross-partials condition directly:

$$\frac{\partial}{\partial x}\left(\frac{x}{x^2 + y^2}\right) = \frac{(x^2 + y^2) - x(\partial/\partial x)(x^2 + y^2)}{(x^2 + y^2)^2} = \frac{y^2 - x^2}{(x^2 + y^2)^2}$$

$$\frac{\partial}{\partial y}\left(\frac{-y}{(x^2 + y^2)}\right) = \frac{-(x^2 + y^2) + y(\partial/\partial y)(x^2 + y^2)}{(x^2 + y^2)^2} = \frac{y^2 - x^2}{(x^2 + y^2)^2}$$

Now consider the line integral of $\mathbf{F}$ around the unit circle $\mathcal{C}$ parametrized by $\mathbf{c}(t) = (\cos t, \sin t)$:

$$F(\mathbf{c}(t)) \cdot \mathbf{c}'(t) = \langle -\sin t, \cos t \rangle \cdot \langle -\sin t, \cos t \rangle = \sin^2 t + \cos^2 t = 1$$

$$\oint_{\mathbf{c}} \mathbf{F} \cdot d\mathbf{s} = \int_0^{2\pi} \mathbf{F}(\mathbf{c}(t)) \cdot \mathbf{c}'(t)\, dt = \int_0^{2\pi} dt = 2\pi \neq 0 \qquad \boxed{3}$$

If **F** were conservative, its circulation around every closed curve would be zero by Theorem 1. Thus **F** cannot be conservative, even though it satisfies the cross-partials condition.

This result does not contradict Theorem 4 because the domain of **F** does not satisfy the simply-connected condition of the theorem. Because **F** is not defined at $(x, y) = (0, 0)$, its domain is $\mathcal{D} = \{(x, y) \neq (0, 0)\}$, and this domain is not simply-connected (Figure 13). ■

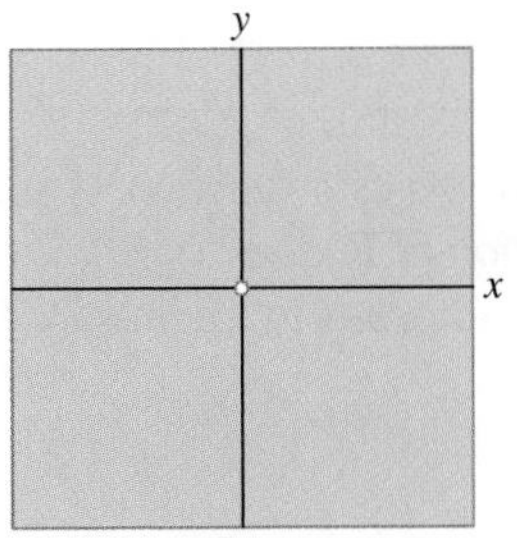

FIGURE 13 The domain $\mathcal{D}$ of the vortex **F** is the plane with the origin removed. This domain is not simply-connected.

CONCEPTUAL INSIGHT Although the vortex field **F** is not conservative on its domain, it is conservative on any smaller, simply-connected domain, such as the upper half-plane $\{(x, y) : y > 0\}$. In fact, we can show (see the marginal note) that $\mathbf{F} = \nabla V$, where

$$V(x, y) = \theta = \tan^{-1} \frac{y}{x}$$

By definition, then, the potential $V(x, y)$ at any point (x, y) is the angle θ of the point in polar coordinates (Figure 14). The line integral of **F** along a path **c** is equal to the change in potential θ along the path [Figures 15(A) and (B)]:

$$\int_{\mathbf{c}} \mathbf{F} \cdot d\mathbf{s} = \theta_2 - \theta_1 = \text{the change in angle } \theta \text{ along } \mathbf{c}$$

Now we can see what is preventing **F** from being conservative on all of its domain. The angle θ is defined only up to integer multiples of 2π. The angle along a path that goes all the way around the origin does not return to its original value but rather increases by 2π. This explains why the line integral around the unit circle (Eq. 3) is 2π rather than 0. And it shows that $V(x, y) = \theta$ cannot be defined as a continuous function on the entire domain $\mathcal{D}$. However, θ is continuous on any domain that does not enclose the origin, and on such domains we have $\mathbf{F} = \nabla\theta$.

In general, if a closed path **c** winds around the origin n times (where n is negative if the curve winds in the clockwise direction), then [Figures 15(C) and (D)]:

$$\oint_{\mathbf{c}} \mathbf{F} \cdot d\mathbf{s} = 2\pi n$$

The number n is called the *winding number* of the path. It plays an important role in the mathematical field of topology.

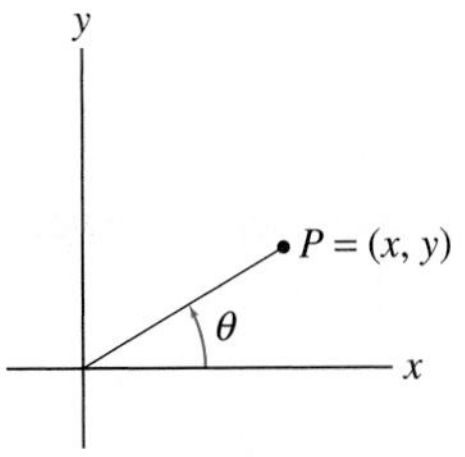

FIGURE 14 The potential function $V(x, y)$ takes the value θ at (x, y).

Using the Chain Rule and the formula

$$\frac{d}{dt} \tan^{-1} t = \frac{1}{1 + t^2}$$

we can check that $\mathbf{F} = \nabla V$

$$\frac{\partial \theta}{\partial x} = \frac{\partial}{\partial x} \tan^{-1} \frac{y}{x} = \frac{-y/x^2}{1 + (y/x)^2} = \frac{-y}{x^2 + y^2}$$

$$\frac{\partial \theta}{\partial y} = \frac{\partial}{\partial y} \tan^{-1} \frac{y}{x} = \frac{1/x}{1 + (y/x)^2} = \frac{x}{x^2 + y^2}$$

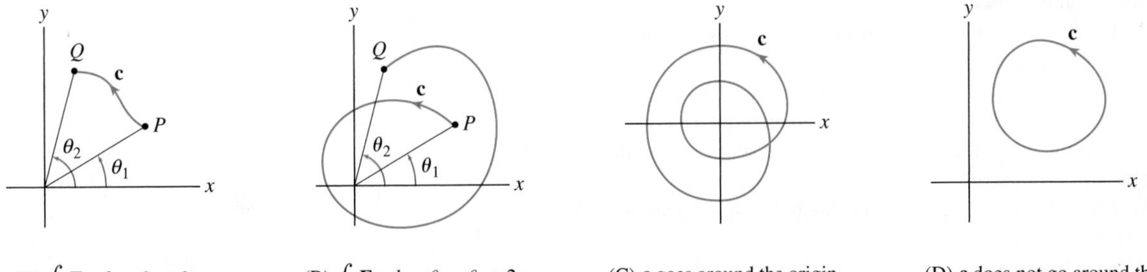

(A) $\int_{\mathbf{c}} \mathbf{F} \cdot d\mathbf{s} = \theta_2 - \theta_1$

(B) $\int_{\mathbf{c}} \mathbf{F} \cdot d\mathbf{s} = \theta_2 - \theta_1 + 2\pi$

(C) **c** goes around the origin twice, so $\int_{\mathbf{c}} \mathbf{F} \cdot d\mathbf{s} = 4\pi$.

(D) **c** does not go around the origin, so $\int_{\mathbf{c}} \mathbf{F} \cdot d\mathbf{s} = 0$.

FIGURE 15 The line integral of the vortex field $\mathbf{F} = \nabla\theta$ is equal to the change in θ along the path.

17.3 SUMMARY

- A vector field $\mathbf{F}$ on a domain $\mathcal{D}$ is conservative if there exists a function V such that $\nabla V = \mathbf{F}$ on $\mathcal{D}$. The function V is called a *potential function* of $\mathbf{F}$.
- A vector field $\mathbf{F}$ on a domain $\mathcal{D}$ is called *path-independent* if for any two points $P, Q \in \mathcal{D}$, we have

$$\int_{\mathbf{c}_1} \mathbf{F} \cdot d\mathbf{s} = \int_{\mathbf{c}_2} \mathbf{F} \cdot d\mathbf{s}$$

for any two paths $\mathbf{c}_1$ and $\mathbf{c}_2$ in $\mathcal{D}$ from P to Q.
- The *Fundamental Theorem for Conservative Vector Fields*: If $\mathbf{F} = \nabla V$, then

$$\int_{\mathbf{c}} \mathbf{F} \cdot d\mathbf{s} = V(Q) - V(P)$$

for any path $\mathbf{c}$ from P to Q in the domain of $\mathbf{F}$. This shows that conservative vector fields are path-independent. In particular, if $\mathbf{c}$ is a *closed path* ($P = Q$), then

$$\oint_{\mathbf{c}} \mathbf{F} \cdot d\mathbf{s} = 0$$

- The converse is also true: On an open, connected domain, a path-independent vector field is conservative.
- Conservative vector fields satisfy the cross-partial condition

$$\frac{\partial F_1}{\partial y} = \frac{\partial F_2}{\partial x}, \qquad \frac{\partial F_2}{\partial z} = \frac{\partial F_3}{\partial y}, \qquad \frac{\partial F_3}{\partial x} = \frac{\partial F_1}{\partial z} \qquad \boxed{4}$$

- Equality of the cross-partials guarantees that $\mathbf{F}$ is conservative if the domain $\mathcal{D}$ is simply connected—that is, if any loop in $\mathcal{D}$ can be drawn down to a point within $\mathcal{D}$.

17.3 EXERCISES

Preliminary Questions

1. The following statement is false. *If* $\mathbf{F}$ *is a gradient vector field, then the line integral of* $\mathbf{F}$ *along every curve is zero.* Which single word must be added to make it true?

2. Which of the following statements are true for all vector fields, and which are true only for conservative vector fields?

(a) The line integral along a path from P to Q does not depend on which path is chosen.

(b) The line integral over an oriented curve $\mathcal{C}$ does not depend on how $\mathcal{C}$ is parametrized.

(c) The line integral around a closed curve is zero.

(d) The line integral changes sign if the orientation is reversed.

(e) The line integral is equal to the difference of a potential function at the two endpoints.

(f) The line integral is equal to the integral of the tangential component along the curve.

(g) The cross-partials of the components are equal.

3. Let $\mathbf{F}$ be a vector field on an open, connected domain $\mathcal{D}$. Which of the following statements are always true, and which are true under additional hypotheses on $\mathcal{D}$?

(a) If $\mathbf{F}$ has a potential function, then $\mathbf{F}$ is conservative.

(b) If $\mathbf{F}$ is conservative, then the cross-partials of $\mathbf{F}$ are equal.

(c) If the cross-partials of $\mathbf{F}$ are equal, then $\mathbf{F}$ is conservative.

4. Let $\mathcal{C}$, $\mathcal{D}$, and $\mathcal{E}$ be the oriented curves in Figure 16 and let $\mathbf{F} = \nabla V$ be a gradient vector field such that $\int_{\mathcal{C}} \mathbf{F} \cdot d\mathbf{s} = 4$. What are the values of the following integrals?

(a) $\int_{\mathcal{D}} \mathbf{F} \cdot d\mathbf{s}$ **(b)** $\int_{\mathcal{E}} \mathbf{F} \cdot d\mathbf{s}$

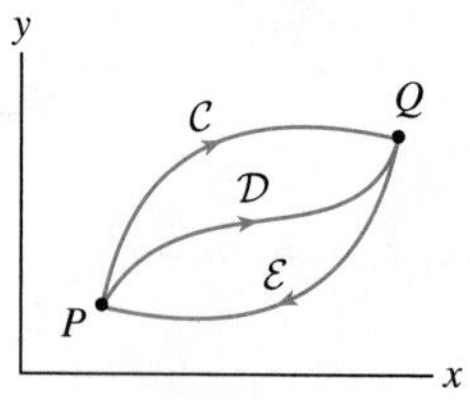

FIGURE 16

Exercises

1. Let $V(x, y, z) = xy\sin(yz)$ and $\mathbf{F} = \nabla V$. Evaluate $\int_{\mathbf{c}} \mathbf{F} \cdot d\mathbf{s}$, where **c** is any path from $(0, 0, 0)$ to $(1, 1, \pi)$.

2. Let $\mathbf{F} = \langle x^{-1}z, y^{-1}z, \log(xy) \rangle$.

(a) Verify that $\mathbf{F} = \nabla V$, where $V(x, y, z) = z\ln(xy)$.

(b) Evaluate $\int_{\mathbf{c}} \mathbf{F} \cdot d\mathbf{s}$, where $\mathbf{c}(t) = \langle e^t, e^{2t}, t^2 \rangle$ for $1 \le t \le 3$.

(c) Evaluate $\int_{\mathbf{c}} \mathbf{F} \cdot d\mathbf{s}$ for any path **c** from $P = (\frac{1}{2}, 4, 2)$ to $Q = (2, 2, 3)$ contained in the region $x > 0$, $y > 0$.

(d) Why is it necessary to specify that the path lie in the region where x and y are positive?

In Exercises 3–6, verify that $\mathbf{F} = \nabla V$ *and evaluate the line integral of* **F** *over the given path.*

3. $\mathbf{F} = \langle 3, 6y \rangle$, $V(x, y, z) = 3x + 3y^2$; $\mathbf{c}(t) = (t, 2t^{-1})$ for $1 \le t \le 4$

4. $\mathbf{F} = \langle \cos y, -x\sin y \rangle$, $V(x, y) = x\cos y$; upper half of the unit circle centered at the origin, oriented counterclockwise

5. $\mathbf{F} = ye^z\mathbf{i} + xe^z\mathbf{j} + xye^z\mathbf{k}$, $V(x, y, z) = xye^z$; $\mathbf{c}(t) = (t^2, t^3, t - 1)$ for $1 \le t \le 2$

6. $\mathbf{F} = \dfrac{z}{x}\mathbf{i} + \mathbf{j} + \ln x\mathbf{k}$, $V(x, y, z) = y + z\ln x$; circle $(x - 4)^2 + y^2 = 1$ in the clockwise direction

In Exercises 7–16, find a potential function for **F** *or determine that* **F** *is not conservative.*

7. $\mathbf{F} = \langle z, 1, x \rangle$

8. $\mathbf{F} = x\mathbf{j} + y\mathbf{k}$

9. $\mathbf{F} = y^2\mathbf{i} + (2xy + e^z)\mathbf{j} + ye^z\mathbf{k}$

10. $\mathbf{F} = \langle y, x, z^3 \rangle$

11. $\mathbf{F} = \langle \cos(xz), \sin(yz), xy\sin z \rangle$

12. $\mathbf{F} = \langle \cos z, 2y, -x\sin z \rangle$

13. $\mathbf{F} = \langle z\sec^2 x, z, y + \tan x \rangle$

14. $\mathbf{F} = \langle e^x(z + 1), -\cos y, e^x \rangle$

15. $\mathbf{F} = \langle 2xy + 5, x^2 - 4z, -4y \rangle$

16. $\mathbf{F} = \langle yze^{xy}, xze^{xy} - z, e^{xy} - y \rangle$

17. Evaluate

$$\int_{\mathbf{c}} 2xyz\,dx + x^2z\,dy + x^2y\,dz$$

over the path $\mathbf{c}(t) = (t^2, \sin(\pi t/4), e^{t^2-2t})$ for $0 \le t \le 2$.

18. Evaluate

$$\oint_{C} \sin x\,dx + z\cos y\,dy + \sin y\,dz$$

where C is the ellipse $4x^2 + 9y^2 = 36$, oriented clockwise.

19. A vector field **F** and contour lines of a potential function for **F** are shown in Figure 17. Calculate the common value of $\int_{C} \mathbf{F} \cdot d\mathbf{s}$ for the curves shown in Figure 17 oriented in the direction from P to Q.

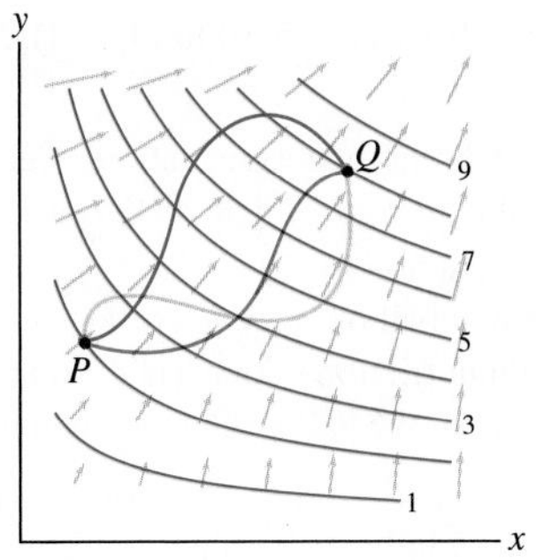

FIGURE 17

20. Give a reason why the vector field **F** in Figure 18 is not conservative.

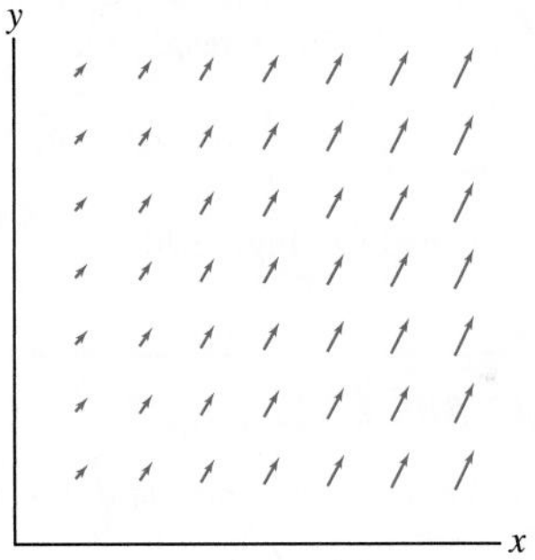

FIGURE 18

21. Calculate the work expended when a particle is moved from O to Q along segments $\overline{OP}$ and $\overline{PQ}$ in Figure 19 in the presence of the force field $\mathbf{F} = \langle x^2, y^2 \rangle$. How much work is expended moving in a complete circuit around the square?

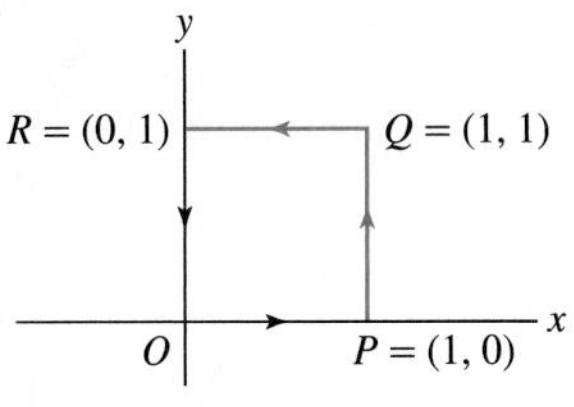

FIGURE 19

22. Let $\mathbf{F} = \left\langle \dfrac{1}{x}, \dfrac{-1}{y} \right\rangle$. Calculate the work against F required to move an object from $(1, 1)$ to $(3, 4)$ along any path in the first quadrant.

23. Compute the work W against the earth's gravitational field required to move a satellite of mass $m = 1000$ kg along any path from an orbit of altitude 4000 km to an orbit of altitude 6000 km.

24. An electric dipole with dipole moment $p = 4 \times 10^{-5}$ C-m sets up an electric field (in newtons per coulomb)

$$\mathbf{F}(x, y, z) = \frac{kp}{r^5}\left\langle 3xz, 3yz, 2z^2 - x^2 - y^2\right\rangle$$

where $r = (x^2 + y^2 + z^2)^{1/2}$ with distance in meters and $k = 8.99 \times 10^9$ N-m²/C². Calculate the work against $\mathbf{F}$ required to move a particle of charge $q = 0.01$ C from $(1, -5, 0)$ to $(3, 4, 4)$. *Note:* The force on q is $q\mathbf{F}$ newtons.

25. On the surface of the earth, the gravitational field (with z as vertical coordinate measured in meters) is $\mathbf{F} = \langle 0, 0, -g\rangle$.

(a) Find a potential function for $\mathbf{F}$.

(b) Beginning at rest, a ball of mass $m = 2$ kg moves under the influence of gravity (without friction) along a path from $P = (3, 2, 400)$ to $Q = (-21, 40, 50)$. Find the ball's velocity when it reaches Q.

26. An electron at rest at $P = (1, 1, 1)$ moves along a path ending at $Q = (5, 3, 7)$ under the influence of the electric field (in newtons per coulomb)

$$\mathbf{F}(x, y, z) = 400(x^2 + z^2)^{-1}\langle x, 0, z\rangle$$

(a) Find a potential function for $\mathbf{F}$.

(b) What is the electron's speed at point Q? Use Conservation of Energy and the value $q_e/m_e = -1.76 \times 10^{11}$ C/kg, where q_e and m_e are the charge and mass on the electron, respectively.

27. Let $\mathbf{F} = \left\langle \frac{-y}{x^2+y^2}, \frac{x}{x^2+y^2}\right\rangle$ be the vortex field. Determine $\int_{\mathbf{c}} \mathbf{F} \cdot d\mathbf{s}$ for each of the paths in Figure 20.

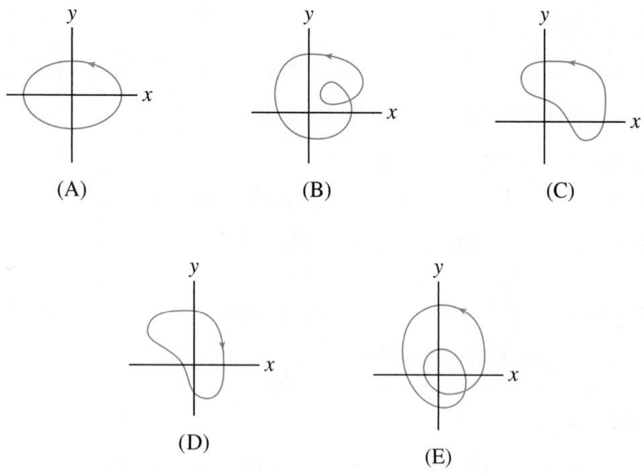

FIGURE 20

28. The vector field $\mathbf{F} = \left\langle \frac{x}{x^2+y^2}, \frac{y}{x^2+y^2}\right\rangle$ is defined on the domain $\mathcal{D} = \{(x, y) \neq (0, 0)\}$.

(a) Is $\mathcal{D}$ simply-connected?

(b) Show that $\mathbf{F}$ satisfies the cross-partial condition. Does this guarantee that $\mathbf{F}$ is conservative?

(c) Show that $\mathbf{F}$ is conservative on $\mathcal{D}$ by finding a potential function.

(d) Do these results contradict Theorem 4?

Further Insights and Challenges

29. Suppose that $\mathbf{F}$ is defined on $\mathbf{R}^3$ and that $\oint_{\mathbf{c}} \mathbf{F} \cdot d\mathbf{s} = 0$ for all closed paths $\mathbf{c}$ in $\mathbf{R}^3$. Prove:

(a) $\mathbf{F}$ is path-independent; that is, for any two paths $\mathbf{c}_1$ and $\mathbf{c}_2$ in $\mathcal{D}$ with the same initial and terminal points,

$$\int_{\mathbf{c}_1} \mathbf{F} \cdot d\mathbf{s} = \int_{\mathbf{c}_2} \mathbf{F} \cdot d\mathbf{s}$$

(b) $\mathbf{F}$ is conservative.

17.4 Parametrized Surfaces and Surface Integrals

The basic idea of an integral appears in several guises. So far, we have defined single, double, and triple integrals and, in the previous section, line integrals over curves. Now we consider one last type on integral: integrals over surfaces. We treat scalar surface integrals in this section and vector surface integrals in the following section.

Just as parametrized curves are a key ingredient in the discussion of line integrals, surface integrals require the notion of a **parametrized surface**—that is, a surface $\mathcal{S}$ whose points are described in the form

$$G(u, v) = (x(u, v), y(u, v), z(u, v))$$

The variables u, v (called parameters) vary in a region $\mathcal{D}$ called the **parameter domain**. Two parameters u and v are needed to parametrize a surface because the surface is two-dimensional.

Figure 1 shows a plot of the surface $\mathcal{S}$ with the parametrization

$$G(u, v) = (u + v, u^3 - v, v^3 - u)$$

This surface consists of all points (x, y, z) in $\mathbf{R}^3$ such that

$$x = u + v, \qquad y = u^3 - v, \qquad z = v^3 - u$$

for (u, v) in $\mathcal{D} = \mathbf{R}^2$.

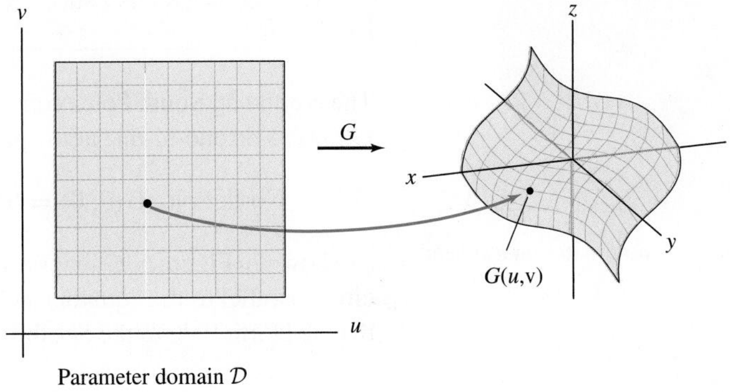

FIGURE 1 The parametric surface $G(u, v) = (u + v, u^3 - v, v^3 - u)$.

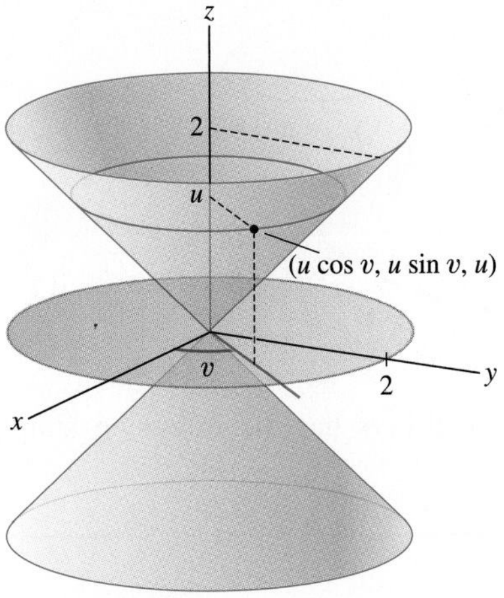

FIGURE 2 The cone $x^2 + y^2 = z^2$.

■ **EXAMPLE 1 Parametrization of a Cone** Find a parametrization of the portion $\mathcal{S}$ of the cone with equation $x^2 + y^2 = z^2$ lying above or below the disk $x^2 + y^2 \le 4$. Specify the domain $\mathcal{D}$ of the parametrization.

Solution This surface $x^2 + y^2 = z^2$ is a cone whose horizonal cross section at height $z = u$ is the circle $x^2 + y^2 = u^2$ of radius u (Figure 2). So a point on the cone at height u has coordinates $(u \cos v, u \sin v, u)$ for some angle v. Thus, the cone has the parametrization

$$G(u, v) = (u \cos v, u \sin v, u)$$

Since we are interested in the portion of the cone where $x^2 + y^2 = u^2 \le 4$, the height variable u satisfies $-2 \le u \le 2$. The angular variable v varies in the interval $[0, 2\pi)$, and therefore, the parameter domain is $\mathcal{D} = [-2, 2] \times [0, 2\pi)$. ■

If necessary, review cylindrical and spherical coordinates in Section 13.7. They are used often in surface calculations.

Three standard parametrizations arise often in computations. First, the **cylinder of radius** R with equation $x^2 + y^2 = R^2$ is conveniently parametrized in cylindrical coordinates (Figure 3). Points on the cylinder have cylindrical coordinates (R, θ, z), so we use θ and z as parameters (with fixed R).

Parametrization of a Cylinder:

$$G(\theta, z) = (R \cos\theta, R \sin\theta, z), \qquad 0 \le \theta < 2\pi, \qquad -\infty < z < \infty$$

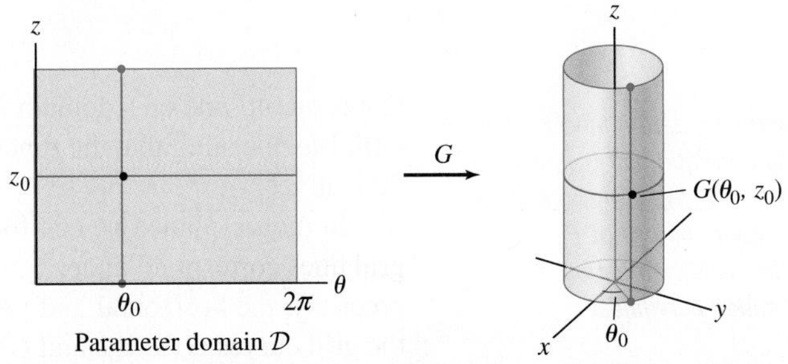

FIGURE 3 The parametrization of a cylinder by cylindrical coordinates amounts to wrapping the rectangle around the cylinder.

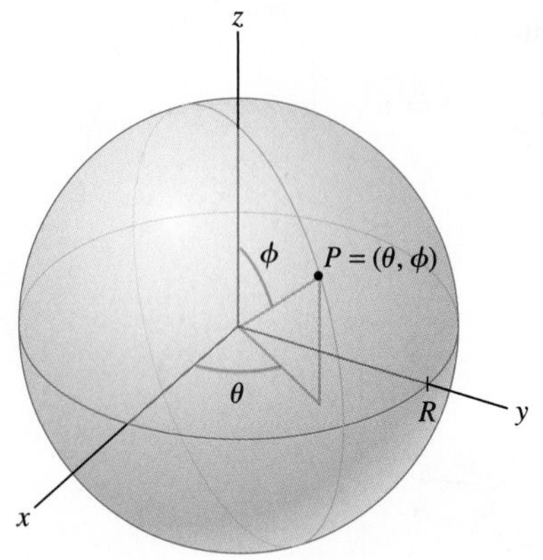

FIGURE 4 Spherical coordinates on a sphere of radius R.

The sphere of radius R with center at the origin is parametrized conveniently using spherical coordinates (ρ, θ, ϕ) with $\rho = R$ (Figure 4).

Parametrization of a Sphere:

$$G(\theta, \phi) = (R\cos\theta\sin\phi, R\sin\theta\sin\phi, R\cos\phi), \quad 0 \le \theta < 2\pi, \quad 0 \le \phi \le \pi$$

The North and South Poles correspond to $\phi = 0$ and $\phi = \pi$ with any value of θ (the map G fails to be one-to-one at the poles):

$$\text{North Pole: } G(\theta, 0) = (0, 0, R), \qquad \text{South Pole: } G(\theta, \pi) = (0, 0, -R)$$

As shown in Figure 5, G maps each horizontal segment $\phi = c\ (0 < c < \pi)$ to a latitude (a circle parallel to the equator) and each vertical segment $\theta = c$ to a longitudinal arc from the the North Pole to the South Pole.

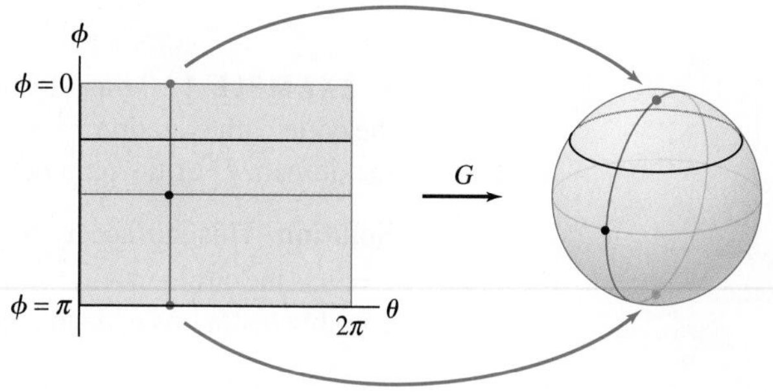

FIGURE 5 The parametrization by spherical coordinates amounts to wrapping the rectangle around the sphere. The top and bottom edges of the rectangle are collapsed to the North and South Poles.

Finally, the **graph of a function** $z = f(x, y)$ always has the following simple parametrization (Figure 6).

Parametrization of a Graph:

$$G(x, y) = (x, y, f(x, y))$$

In this case, the parameters are x and y.

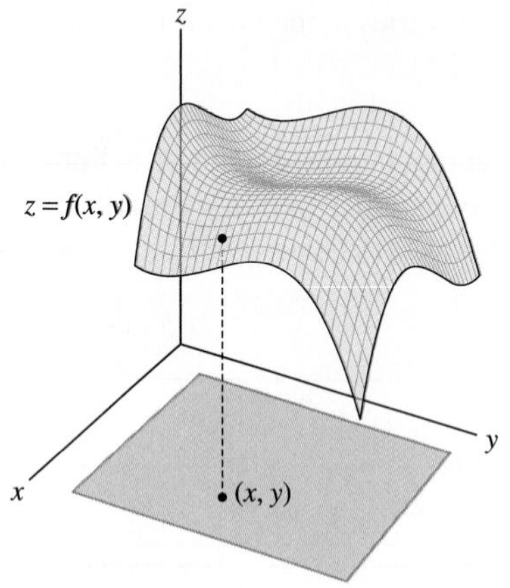

FIGURE 6

Grid Curves, Normal Vectors, and the Tangent Plane

Suppose that a surface $\mathcal{S}$ has a parametrization

$$G(u, v) = (x(u, v), y(u, v), z(u, v))$$

that is one-to-one on a domain $\mathcal{D}$. We shall always assume that G is **continuously differentiable**, meaning that the functions $x(u, v)$, $y(u, v)$, and $z(u, v)$ have continuous partial derivatives.

In essence, a parametrization labels each point P on $\mathcal{S}$ by a unique pair (u_0, v_0) in the parameter domain. We can think of (u_0, v_0) as the "coordinates" of P determined by the parametrization. They are sometimes called ***curvilinear coordinates****.*

In the uv-plane, we can form a grid of lines parallel to the coordinates axes. These grid lines correspond under G to a system of **grid curves** on the surface (Figure 7). More precisely, the horizontal and vertical lines through (u_0, v_0) in the domain correspond to the grid curves $G(u, v_0)$ and $G(u_0, v)$ that intersect at the point $P = G(u_0, v_0)$.

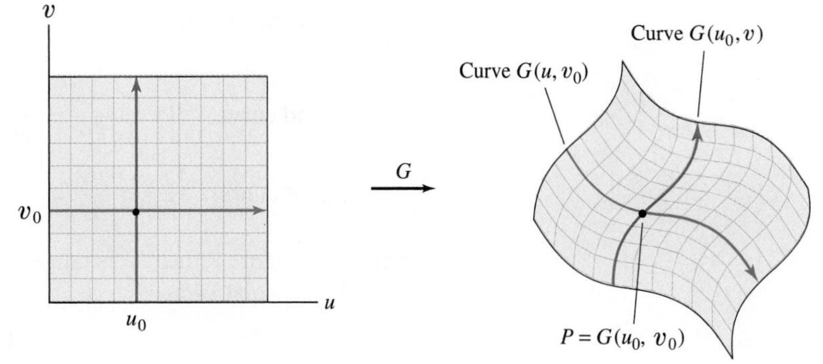

FIGURE 7 Grid curves.

Now consider the tangent vectors to these grid curves (Figure 8):

$$\text{For } G(u, v_0): \quad \mathbf{T}_u(P) = \frac{\partial G}{\partial u}(u_0, v_0) = \left\langle \frac{\partial x}{\partial u}(u_0, v_0), \frac{\partial y}{\partial u}(u_0, v_0), \frac{\partial z}{\partial u}(u_0, v_0) \right\rangle$$

$$\text{For } G(u_0, v): \quad \mathbf{T}_v(P) = \frac{\partial G}{\partial v}(u_0, v_0) = \left\langle \frac{\partial x}{\partial v}(u_0, v_0), \frac{\partial y}{\partial v}(u_0, v_0), \frac{\partial z}{\partial v}(u_0, v_0) \right\rangle$$

The parametrization G is called **regular** at P if the following cross product is nonzero:

$$\mathbf{n}(P) = \mathbf{n}(u_0, v_0) = \mathbf{T}_u(P) \times \mathbf{T}_v(P)$$

In this case, $\mathbf{T}_u$ and $\mathbf{T}_v$ span the tangent plane to $\mathcal{S}$ at P and $\mathbf{n}(P)$ is a **normal vector** to the tangent plane. We call $\mathbf{n}(P)$ a normal to the surface $\mathcal{S}$.

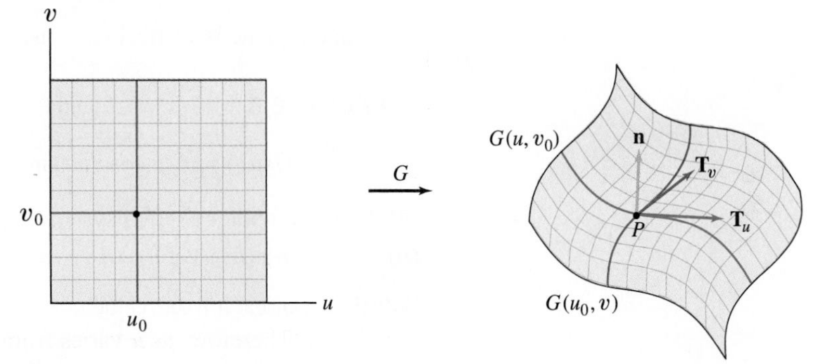

FIGURE 8 The vectors $\mathbf{T}_u$ and $\mathbf{T}_v$ are tangent to the grid curves through $P = G(u_0, v_0)$.

At each point on a surface, the normal vector points in one of two opposite directions. If we change the parametrization, the length of $\mathbf{n}$ *may change and its direction may be reversed.*

We often write $\mathbf{n}$ instead of $\mathbf{n}(P)$ or $\mathbf{n}(u, v)$, but it is understood that the vector $\mathbf{n}$ varies from point to point on the surface. Similarly, we often denote the tangent vectors by $\mathbf{T}_u$ and $\mathbf{T}_v$. Note that $\mathbf{T}_u$, $\mathbf{T}_v$, and $\mathbf{n}$ need not be unit vectors (thus the notation here differs from that in Sections 14.4, 14.5, and 17.2, where $\mathbf{T}$ and $\mathbf{n}$ denote unit vectors).

EXAMPLE 2 Consider the parametrization $G(\theta, z) = (2\cos\theta, 2\sin\theta, z)$ of the cylinder $x^2 + y^2 = 4$:

(a) Describe the grid curves.

(b) Compute $\mathbf{T}_\theta$, $\mathbf{T}_z$, and $\mathbf{n}(\theta, z)$.

(c) Find an equation of the tangent plane at $P = G(\frac{\pi}{4}, 5)$.

Solution

(a) The grid curves on the cylinder through $P = (\theta_0, z_0)$ are (Figure 9)

θ-grid curve: $G(\theta, z_0) = (2\cos\theta, 2\sin\theta, z_0)$ (circle of radius 2 at height $z = z_0$)

z-grid curve: $G(\theta_0, z) = (2\cos\theta_0, 2\sin\theta_0, z)$ (vertical line through P with $\theta = \theta_0$)

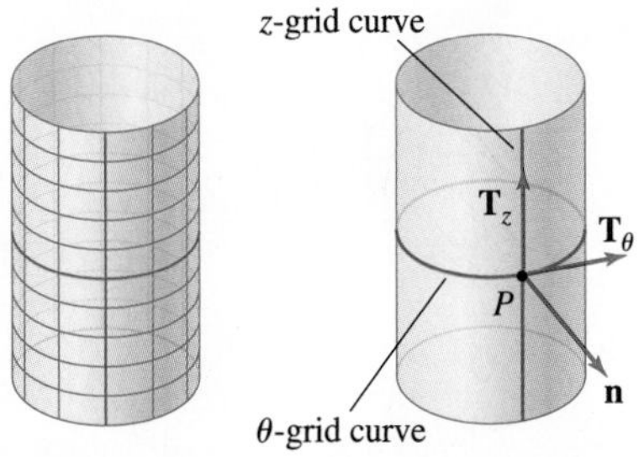

FIGURE 9 Grid curves on the cylinder.

(b) The partial derivatives of $G(\theta, z) = (2\cos\theta, 2\sin\theta, z)$ give us the tangent vectors at P:

$$\theta\text{-grid curve:}\quad \mathbf{T}_\theta = \frac{\partial G}{\partial \theta} = \frac{\partial}{\partial \theta}(2\cos\theta, 2\sin\theta, z) = \langle -2\sin\theta, 2\cos\theta, 0\rangle$$

$$z\text{-grid curve:}\quad \mathbf{T}_z = \frac{\partial G}{\partial z} = \frac{\partial}{\partial z}(2\cos\theta, 2\sin\theta, z) = \langle 0, 0, 1\rangle$$

Observe in Figure 9 that $\mathbf{T}_\theta$ is tangent to the θ-grid curve and $\mathbf{T}_z$ is tangent to the z-grid curve. The normal vector is

$$\mathbf{n}(\theta, z) = \mathbf{T}_\theta \times \mathbf{T}_z = \begin{vmatrix} \mathbf{i} & \mathbf{j} & \mathbf{k} \\ -2\sin\theta & 2\cos\theta & 0 \\ 0 & 0 & 1 \end{vmatrix} = 2\cos\theta\mathbf{i} + 2\sin\theta\mathbf{j}$$

The coefficient of $\mathbf{k}$ is zero, so $\mathbf{n}$ points directly out of the cylinder.

← *REMINDER An equation of the plane through $P = (x_0, y_0, z_0)$ with normal vector $\mathbf{n}$ is*

$$\langle x - x_0, y - y_0, z - z_0\rangle \cdot \mathbf{n} = 0$$

(c) For $\theta = \frac{\pi}{4}$, $z = 5$,

$$P = G\left(\frac{\pi}{4}, 5\right) = \langle\sqrt{2}, \sqrt{2}, 5\rangle, \qquad \mathbf{n} = \mathbf{n}\left(\frac{\pi}{4}, 5\right) = \langle\sqrt{2}, \sqrt{2}, 0\rangle$$

The tangent plane through P has normal vector $\mathbf{n}$ and thus has equation

$$\langle x - \sqrt{2}, y - \sqrt{2}, z - 5\rangle \cdot \langle\sqrt{2}, \sqrt{2}, 0\rangle = 0$$

This can be written

$$\sqrt{2}(x - \sqrt{2}) + \sqrt{2}(y - \sqrt{2}) = 0 \qquad \text{or} \qquad x + y = 2\sqrt{2}$$

The tangent plane is vertical (because z does not appear in the equation). ■

EXAMPLE 3 *CAS* **Helicoid Surface** Describe the surface $\mathcal{S}$ with parametrization

$$G(u, v) = (u\cos v, u\sin v, v), \qquad -1 \le u \le 1, \quad 0 \le v < 2\pi$$

(a) Use a computer algebra system to plot $\mathcal{S}$.

(b) Compute $\mathbf{n}(u, v)$ at $u = \frac{1}{2}$, $v = \frac{\pi}{2}$.

Solution For each fixed value $u = a$, the curve $G(a, v) = (a\cos v, a\sin v, v)$ is a helix of radius a. Therefore, as u varies from -1 to 1, $G(u, v)$ describes a family of helices of radius u. The resulting surface is a "helical ramp."

(a) Here is a typical command for a computer algebra system that generates the helicoid surface shown on the right-hand side of Figure 10.

```
ParametricPlot3D[{u*Cos[v],u*Sin[v],v},{u,-1,1},{v,0,2Pi}]
```

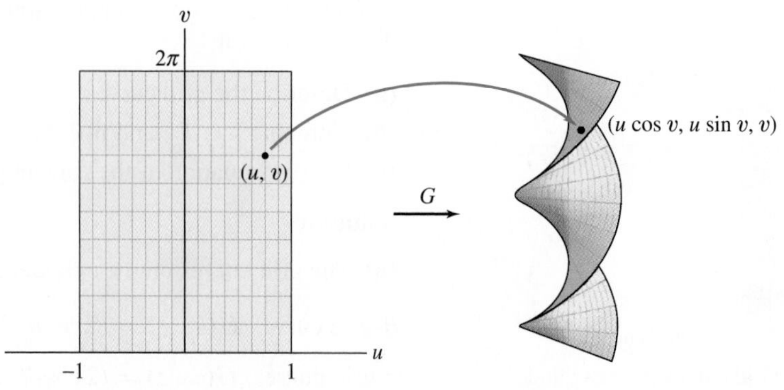

FIGURE 10 Helicoid.

(b) The tangent and normal vectors are

$$\mathbf{T}_u = \frac{\partial G}{\partial u} = \langle \cos v, \sin v, 0\rangle$$

$$\mathbf{T}_v = \frac{\partial G}{\partial v} = \langle -u\sin v, u\cos v, 1\rangle$$

$$\mathbf{n}(u, v) = \mathbf{T}_u \times \mathbf{T}_v = \begin{vmatrix} \mathbf{i} & \mathbf{j} & \mathbf{k} \\ \cos v & \sin v & 0 \\ -u\sin v & u\cos v & 1 \end{vmatrix} = (\sin v)\mathbf{i} - (\cos v)\mathbf{j} + u\mathbf{k}$$

At $u = \frac{1}{2}$, $v = \frac{\pi}{2}$, we have $\mathbf{n} = \mathbf{i} + \frac{1}{2}\mathbf{k}$. ■

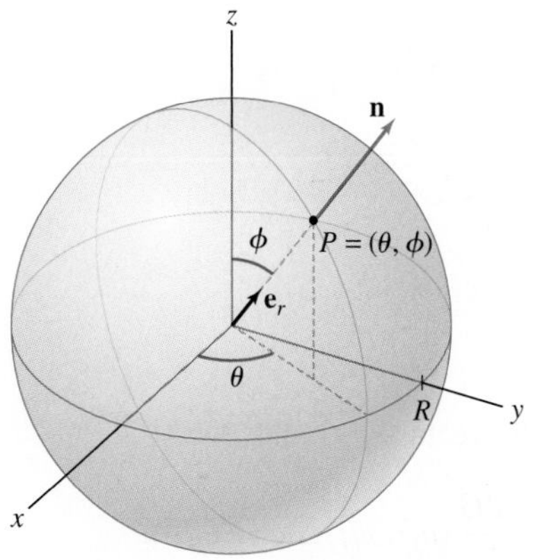

FIGURE 11 The normal vector **n** points in the radial direction $\mathbf{e}_r$.

For future reference, we compute the outward-pointing normal vector in the standard parametrization of the sphere of radius R centered at the origin (Figure 11):

$$G(\theta, \phi) = (R\cos\theta\sin\phi, R\sin\theta\sin\phi, R\cos\phi)$$

Note first that since the distance from $G(\theta, \phi)$ to the origin is R, the *unit* radial vector at $G(\theta, \phi)$ is obtained by dividing by R:

$$\mathbf{e}_r = \langle \cos\theta\sin\phi, \sin\theta\sin\phi, \cos\phi\rangle$$

Furthermore,

$$\mathbf{T}_\theta = \langle -R\sin\theta\sin\phi, R\cos\theta\sin\phi, 0\rangle$$

$$\mathbf{T}_\phi = \langle R\cos\theta\cos\phi, R\sin\theta\cos\phi, -R\sin\phi\rangle$$

$$\mathbf{n} = \mathbf{T}_\theta \times \mathbf{T}_\phi = \begin{vmatrix} \mathbf{i} & \mathbf{j} & \mathbf{k} \\ -R\sin\theta\sin\phi & R\cos\theta\sin\phi & 0 \\ R\cos\theta\cos\phi & R\sin\theta\cos\phi & -R\sin\phi \end{vmatrix}$$

$$= -R^2\cos\theta\sin^2\phi\,\mathbf{i} - R^2\sin\theta\sin^2\phi\,\mathbf{j} - R^2\cos\phi\sin\phi\,\mathbf{k}$$

$$= -R^2\sin\phi\,\langle \cos\theta\sin\phi, \sin\theta\sin\phi, \cos\phi\rangle \qquad \boxed{1}$$

$$= -(R^2\sin\phi)\,\mathbf{e}_r$$

This is an inward-pointing normal vector. However, in most computations it is standard to use the outward-pointing normal vector:

$$\boxed{\mathbf{n} = \mathbf{T}_\phi \times \mathbf{T}_\theta = (R^2\sin\phi)\,\mathbf{e}_r, \qquad \|\mathbf{n}\| = R^2\sin\phi} \qquad \boxed{2}$$

Surface Area

The length $\|\mathbf{n}\|$ of the normal vector in a parametrization has an important interpretation in terms of area. Assume, for simplicity, that $\mathcal{D}$ is a rectangle (the argument also applies to more general domains). Divide $\mathcal{D}$ into a grid of small rectangles $\mathcal{R}_{ij}$ of size $\Delta u \times \Delta v$, as in Figure 12, and compare the area of $\mathcal{R}_{ij}$ with the area of its image under G. This image is a "curved" parallelogram $\mathcal{S}_{ij} = G(\mathcal{R}_{ij})$.

First, we note that if Δu and Δv in Figure 12 are small, then the curved parallelogram $\mathcal{S}_{ij}$ has approximately the same area as the "genuine" parallelogram with sides $\overrightarrow{PQ}$ and $\overrightarrow{PS}$. Recall that the area of the parallelogram spanned by two vectors is the length of their cross product, so

$$\text{Area}(\mathcal{S}_{ij}) \approx \|\overrightarrow{PQ} \times \overrightarrow{PS}\|$$

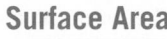

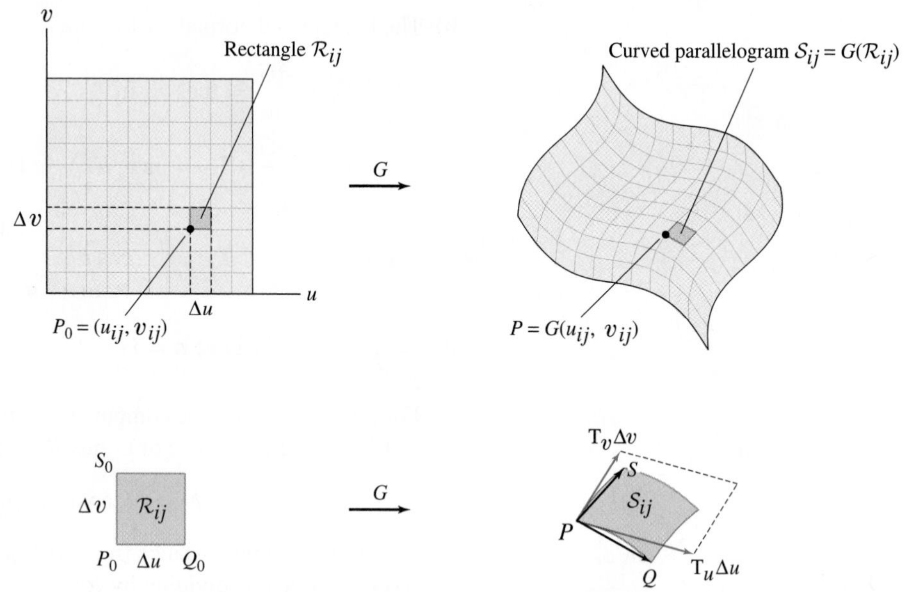

FIGURE 12

◀·· *REMINDER* *By Theorem 3 in Section 13.4, the area of the parallelogram spanned by vectors* $\mathbf{v}$ *and* $\mathbf{w}$ *in* $\mathbf{R}^3$ *is equal to* $\|\mathbf{v} \times \mathbf{w}\|$.

Next, we use the linear approximation to estimate the vectors $\overrightarrow{PQ}$ and $\overrightarrow{PS}$:

$$\overrightarrow{PQ} = G(u_{ij} + \Delta u, v_{ij}) - G(u_{ij}, v_{ij}) \approx \frac{\partial G}{\partial u}(u_{ij}, v_{ij})\Delta u = \mathbf{T}_u \Delta u$$

$$\overrightarrow{PS} = G(u_{ij}, v_{ij} + \Delta v) - G(u_{ij}, v_{ij}) \approx \frac{\partial G}{\partial v}(u_{ij}, v_{ij})\Delta v = \mathbf{T}_v \Delta v$$

Thus we have

The approximation (3) is valid for any small region $\mathcal{R}$ *in the* uv*-plane:*

$$\text{Area}(\mathcal{S}) \approx \|\mathbf{n}(u_0, v_0))\| \text{Area}(\mathcal{R})$$

where $\mathcal{S} = G(\mathcal{R})$ *and* (u_0, v_0) *is any sample point in* $\mathcal{R}$. *Here, "small" means contained in a small disk. We do not allow* $\mathcal{R}$ *to be very thin and wide.*

$$\text{Area}(\mathcal{S}_{ij})) \approx \|\mathbf{T}_u \Delta u \times \mathbf{T}_v \Delta v\| = \|\mathbf{T}_u \times \mathbf{T}_v\| \, \Delta u \, \Delta v$$

Since $\mathbf{n}(u_{ij}, v_{ij}) = \mathbf{T}_u \times \mathbf{T}_v$ and $\text{Area}(\mathcal{R}_{ij}) = \Delta u \Delta v$, we obtain

$$\text{Area}(\mathcal{S}_{ij}) \approx \|\mathbf{n}(u_{ij}, v_{ij})\| \text{Area}(\mathcal{R}_{ij}) \qquad \boxed{3}$$

Our conclusion: $\|\mathbf{n}\|$ *is a distortion factor that measures how the area of a small rectangle* $\mathcal{R}_{ij}$ *is altered under the map* G.

To compute the surface area of $\mathcal{S}$, we assume that G is one-to-one, except possibly on the boundary of $\mathcal{D}$. We also assume that G is regular, except possibly on the boundary of $\mathcal{D}$. Recall that "regular" means that $\mathbf{n}(u, v)$ is nonzero.

Note: We require only that G *be one-to-one on the interior of* $\mathcal{D}$. *Many common parametrizations (such as the parametrizations by cylindrical and spherical coordinates) fail to be one-to-one on the boundary of their domains.*

The entire surface $\mathcal{S}$ is the union of the small patches $\mathcal{S}_{ij}$, so we can apply the approximation on each patch to obtain

$$\text{Area}(\mathcal{S}) = \sum_{i,j} \text{Area}(\mathcal{S}_{ij}) \approx \sum_{i,j} \|\mathbf{n}(u_{ij}, v_{ij})\| \Delta u \Delta v \qquad \boxed{4}$$

The sum on the right is a Riemann sum for the double integral of $\|\mathbf{n}(u, v)\|$ over the parameter domain $\mathcal{D}$. As Δu and Δv tend to zero, these Riemann sums converge to a double integral, which we take as the definition of surface area:

$$\text{Area}(\mathcal{S}) = \iint_{\mathcal{D}} \|\mathbf{n}(u, v)\| \, du \, dv$$

Surface Integral

Now we can define the surface integral of a function $f(x, y, z)$:

$$\iint_{\mathcal{S}} f(x, y, z)\, dS$$

This is similar to the definition of the line integral of a function over a curve. Choose a sample point $P_{ij} = G(u_{ij}, v_{ij})$ in each small patch $\mathcal{S}_{ij}$ and form the sum:

$$\sum_{i,j} f(P_{ij})\text{Area}(\mathcal{S}_{ij}) \qquad \boxed{5}$$

The limit of these sums as Δu and Δv tend to zero (if it exists) is the **surface integral**:

$$\iint_{\mathcal{S}} f(x, y, z)\, dS = \lim_{\Delta u, \Delta v \to 0} \sum_{i,j} f(P_{ij})\text{Area}(\mathcal{S}_{ij})$$

To evaluate the surface integral, we use Eq. (3) to write

$$\sum_{i,j} f(P_{ij})\text{Area}(\mathcal{S}_{ij}) \approx \sum_{i,j} f(G(u_{ij}, v_{ij}))\|\mathbf{n}(u_{ij}, v_{ij})\|\, \Delta u\, \Delta v \qquad \boxed{6}$$

On the right we have a Riemann sum for the double integral of

$$f(G(u, v))\|\mathbf{n}(u, v)\|$$

over the parameter domain $\mathcal{D}$. Under the assumption that G is continuously differentiable, we can show these the sums in Eq. (6) approach the same limit. This yields the next theorem.

It is interesting to note that Eq. (7) includes the Change of Variables Formula for double integrals (Theorem 1 in Section 16.6) as a special case. If the surface $\mathcal{S}$ is a domain in the xy-plane [in other words, $z(u, v) = 0$], then the integral over $\mathcal{S}$ reduces to the double integral of the function $f(x, y, 0)$. We may view $G(u, v)$ as a mapping from the uv-plane to the xy-plane, and we find that $\|\mathbf{n}(u, v)\|$ is the Jacobian of this mapping.

THEOREM 1 Surface Integrals and Surface Area Let $G(u, v)$ be a parametrization of a surface $\mathcal{S}$ with parameter domain $\mathcal{D}$. Assume that G is continuously differentiable, one-to-one, and regular (except possibly at the boundary of $\mathcal{D}$). Then

$$\iint_{\mathcal{S}} f(x, y, z)\, dS = \iint_{\mathcal{D}} f(G(u, v))\|\mathbf{n}(u, v)\|\, du\, dv \qquad \boxed{7}$$

For $f(x, y, z) = 1$, we obtain the surface area of $\mathcal{S}$:

$$\text{Area}(\mathcal{S}) = \iint_{\mathcal{D}} \|\mathbf{n}(u, v)\|\, du\, dv$$

Equation (7) is summarized by the symbolic expression for the "surface element":

$$dS = \|\mathbf{n}(u, v)\|\, du\, dv$$

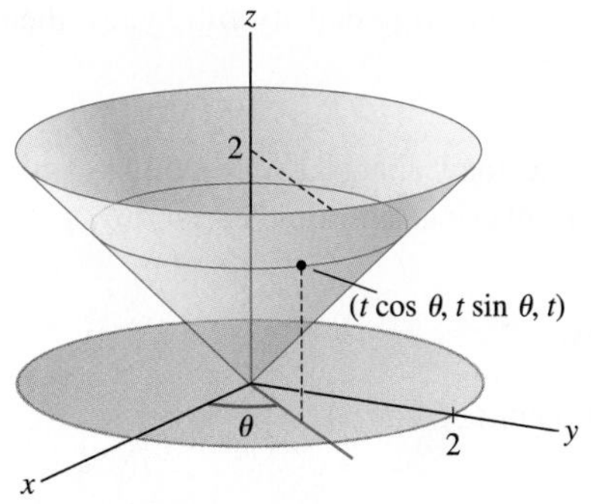

FIGURE 13 Portion $\mathcal{S}$ of the cone $x^2 + y^2 = z^2$ lying over the disk $x^2 + y^2 \le 4$.

■ **EXAMPLE 4** Calculate the surface area of the portion $\mathcal{S}$ of the cone $x^2 + y^2 = z^2$ lying above the disk $x^2 + y^2 \le 4$ (Figure 13). Then calculate $\iint_{\mathcal{S}} x^2 z\, dS$.

Solution A parametrization of the cone was found in Example 1. Using the variables θ and t, this parametrization is

$$G(\theta, t) = (t\cos\theta, t\sin\theta, t), \qquad 0 \le t \le 2, \quad 0 \le \theta < 2\pi$$

REMINDER In this example,

$$G(\theta, t) = (t\cos\theta, t\sin\theta, t)$$

Step 1. **Compute the tangent and normal vectors.**

$$\mathbf{T}_\theta = \frac{\partial G}{\partial \theta} = \langle -t\sin\theta, t\cos\theta, 0\rangle, \qquad \mathbf{T}_t = \frac{\partial G}{\partial t} = \langle \cos\theta, \sin\theta, 1\rangle$$

$$\mathbf{n} = \mathbf{T}_\theta \times \mathbf{T}_t = \begin{vmatrix} \mathbf{i} & \mathbf{j} & \mathbf{k} \\ -t\sin\theta & t\cos\theta & 0 \\ \cos\theta & \sin\theta & 1 \end{vmatrix} = t\cos\theta\mathbf{i} + t\sin\theta\mathbf{j} - t\mathbf{k}$$

The normal vector has length

$$\|\mathbf{n}\| = \sqrt{t^2\cos^2\theta + t^2\sin^2\theta + (-t)^2} = \sqrt{2t^2} = \sqrt{2}\,|t|$$

Thus, $dS = \sqrt{2}|t|\,d\theta\,dt$. Since $t \ge 0$ on our domain, we drop the absolute value.

Step 2. **Calculate the surface area.**

$$\text{Area}(\mathcal{S}) = \iint_{\mathcal{D}} \|\mathbf{n}\|\,du\,dv = \int_0^2 \int_0^{2\pi} \sqrt{2}\,t\,d\theta\,dt = \sqrt{2}\pi t^2\Big|_0^2 = 4\sqrt{2}\pi$$

Step 3. **Calculate the surface integral.**

We express $f(x, y, z) = x^2 z$ in terms of the parameters t and θ and evaluate:

$$f(G(\theta, t)) = f(t\cos\theta, t\sin\theta, t) = (t\cos\theta)^2 t = t^3\cos^2\theta$$

$$\iint_S f(x, y, z)\,dS = \int_{t=0}^{2} \int_{\theta=0}^{2\pi} f(G(\theta, t))\,\|\mathbf{n}(\theta, t)\|\,d\theta\,dt$$

$$= \int_{t=0}^{2} \int_{\theta=0}^{2\pi} (t^3\cos^2\theta)(\sqrt{2}t)\,d\theta\,dt$$

$$= \sqrt{2}\left(\int_0^2 t^4\,dt\right)\left(\int_0^{2\pi} \cos^2\theta\,d\theta\right)$$

$$= \sqrt{2}\left(\frac{32}{5}\right)(\pi) = \frac{32\sqrt{2}\pi}{5}$$

■

REMINDER

$$\int_0^{2\pi} \cos^2\theta\,d\theta = \int_0^{2\pi} \frac{1 + \cos 2\theta}{2}\,d\theta = \pi$$

In previous discussions of multiple and line integrals, we applied the principle that the integral of a density is the total quantity. This applies to surface integrals as well. For example, a surface with mass density $\rho(x, y, z)$ [in units of mass per area] is the surface integral of the mass density:

$$\text{Mass of } \mathcal{S} = \iint_{\mathcal{S}} \rho(x, y, z)\,dS$$

Similarly, if an electric charge is distributed over $\mathcal{S}$ with charge density $\rho(x, y, z)$, then the surface integral of $\rho(x, y, z)$ is the total charge on $\mathcal{S}$.

■ **EXAMPLE 5** **Total Charge on a Surface** Find the total charge (in coulombs) on a sphere S of radius 5 cm whose charge density in spherical coordinates is $\rho(\theta, \phi) = 0.003\cos^2\phi$ C/cm^2.

Solution We parametrize S in spherical coordinates:

$$G(\theta, \phi) = (5\cos\theta\sin\phi, 5\sin\theta\sin\phi, 5\cos\phi)$$

By Eq. (2), $\|\mathbf{n}\| = 5^2\sin\phi$ and

$$\text{Total charge} = \iint_S \rho(\theta, \phi)\,dS = \int_{\theta=0}^{2\pi} \int_{\phi=0}^{\pi} \rho(\theta, \phi)\|\mathbf{n}\|\,d\phi\,d\theta$$

$$= \int_{\theta=0}^{2\pi} \int_{\phi=0}^{\pi} (0.003 \cos^2 \phi)(25 \sin \phi)\, d\phi\, d\theta$$

$$= (0.075)(2\pi) \int_{\phi=0}^{\pi} \cos^2 \phi \sin \phi\, d\phi$$

$$= 0.15\pi \left(-\frac{\cos^3 \phi}{3} \right) \Big|_0^{\pi} = 0.15\pi \left(\frac{2}{3} \right) \approx 0.1\pi \text{ C} \qquad ■$$

When a graph $z = g(x, y)$ is parametrized by $G(x, y) = (x, y, g(x, y))$, the tangent and normal vectors are

$$\mathbf{T}_x = (1, 0, g_x), \qquad \mathbf{T}_y = (0, 1, g_y)$$

$$\mathbf{n} = \mathbf{T}_x \times \mathbf{T}_y = \begin{vmatrix} \mathbf{i} & \mathbf{j} & \mathbf{k} \\ 1 & 0 & g_x \\ 0 & 1 & g_y \end{vmatrix} = -g_x\mathbf{i} - g_y\mathbf{j} + \mathbf{k}, \qquad \|\mathbf{n}\| = \sqrt{1 + g_x^2 + g_y^2} \qquad \boxed{8}$$

The surface integral over the portion of a graph lying over a domain $\mathcal{D}$ in the xy-plane is

$$\text{Surface integral over a graph} = \iint_{\mathcal{D}} f(x, y, g(x, y))\sqrt{1 + g_x^2 + g_y^2}\, dx\, dy \qquad \boxed{9}$$

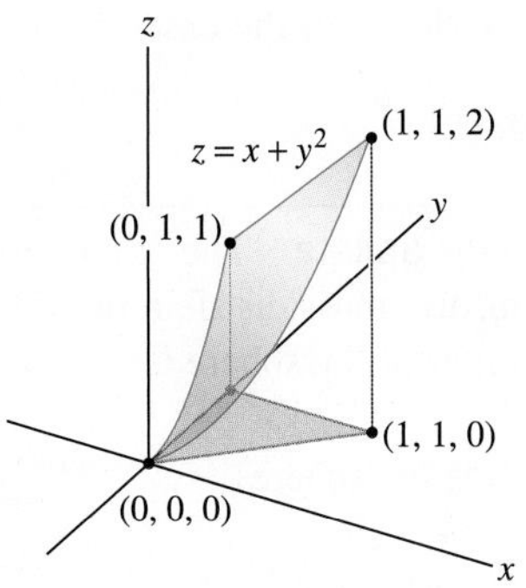

FIGURE 14 The surface $z = x + y^2$ over $0 \le x \le y \le 1$.

■ **EXAMPLE 6** Calculate $\iint_{\mathcal{S}} (z - x)\, dS$, where $\mathcal{S}$ is the portion of the graph of $z = x + y^2$ where $0 \le x \le y$, $0 \le y \le 1$ (Figure 14).

Solution Let $z = g(x, y) = x + y^2$. Then $g_x = 1$ and $g_y = 2y$, and

$$dS = \sqrt{1 + g_x^2 + g_y^2}\, dx\, dy = \sqrt{1 + 1 + 4y^2}\, dx\, dy = \sqrt{2 + 4y^2}\, dx\, dy$$

On the surface $\mathcal{S}$, we have $z = x + y^2$, and thus

$$f(x, y, z) = z - x = (x + y^2) - x = y^2$$

By Eq. (9),

$$\iint_{\mathcal{S}} f(x, y, z)\, dS = \int_{y=0}^{1} \int_{x=0}^{y} y^2\sqrt{2 + 4y^2}\, dx\, dy$$

$$= \int_{y=0}^{1} \left(y^2\sqrt{2 + 4y^2} \right) x \Big|_{x=0}^{y} dy = \int_0^1 y^3\sqrt{2 + 4y^2}\, dy$$

Now use the substitution $u = 2 + 4y^2$, $du = 8y\, dy$. Then $y^2 = \frac{1}{4}(u - 2)$, and

$$\int_0^1 y^3\sqrt{2 + 4y^2}\, dy = \frac{1}{8}\int_2^6 \frac{1}{4}(u - 2)\sqrt{u}\, du = \frac{1}{32}\int_2^6 (u^{3/2} - 2u^{1/2})\, du$$

$$= \frac{1}{32}\left(\frac{2}{5}u^{5/2} - \frac{4}{3}u^{3/2} \right) \Big|_2^6 = \frac{1}{30}(6\sqrt{6} + \sqrt{2}) \approx 0.54 \qquad ■$$

Excursion

The French mathematician Pierre Simon, Marquis de Laplace (1749–1827) showed that the gravitational potential satisfies the Laplace equation $\Delta V = 0$, *where* Δ *is the* Laplace operator

$$\Delta V = \frac{\partial^2 V}{\partial x^2} + \frac{\partial^2 V}{\partial y^2} + \frac{\partial^2 V}{\partial z^2}$$

This equation plays an important role in more advanced branches of math and physics.

In physics it is an important fact that the gravitational field $\mathbf{F}$ corresponding to any arrangement of masses is conservative; that is, $\mathbf{F} = -\nabla V$ (recall that the minus sign is a convention of physics). The field at a point P due to a mass m located at point Q is $\mathbf{F} = -\frac{Gm}{r^2}\mathbf{e}_r$, where $\mathbf{e}_r$ is the unit vector pointing from Q to P and r is the distance from P to Q, which we denote by $|P - Q|$. As we saw in Example 4 of Section 17.3,

$$V(P) = -\frac{Gm}{r} = -\frac{Gm}{|P - Q|}$$

If, instead of a single mass, we have N point masses $m_1, \dots, m_N$ located at $Q_1, \dots, Q_N$, then the gravitational potential is the sum

$$V(P) = -G\sum_{i=1}^{N} \frac{m_i}{|P - Q_i|} \tag{10}$$

If mass is distributed continuously over a thin surface $\mathcal{S}$ with mass density function $\rho(x, y, z)$, we replace the sum by the surface integral

$$V(P) = -G\iint_{\mathcal{S}} \frac{\rho(x, y, z)\,dS}{|P - Q|} = -G\iint_{\mathcal{S}} \frac{\rho(x, y, z)\,dS}{\sqrt{(x-a)^2 + (y-b)^2 + (z-c)^2}} \tag{11}$$

where $P = (a, b, c)$. However, this surface integral cannot be evaluated explicitly unless the surface and mass distribution are sufficiently symmetric, as in the case of a hollow sphere of uniform mass density (Figure 15).

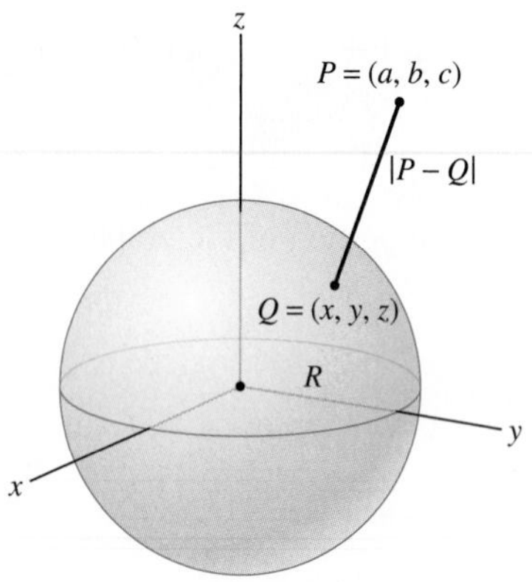

FIGURE 15

THEOREM 2 Gravitational Potential of a Uniform Hollow Sphere The gravitational potential V due to a hollow sphere of radius R with uniform mass distribution of total mass m at a point P located at a distance r from the center of the sphere is

$$V(P) = \begin{cases} \dfrac{-Gm}{r} & \text{if } r > R \quad (P \text{ outside the sphere}) \\[2mm] \dfrac{-Gm}{R} & \text{if } r < R \quad (P \text{ inside the sphere}) \end{cases} \tag{12}$$

We leave this calculation as an exercise (Exercise 48), because we will derive it again with much less effort using Gauss's Law in Section 18.3.

In his magnum opus, *Principia Mathematica*, Isaac Newton proved that a sphere of uniform mass density (whether hollow or solid) attracts a particle outside the sphere as if the entire mass were concentrated at the center. In other words, a uniform sphere behaves like a point mass as far as gravity is concerned. Furthermore, if the sphere is hollow, then the sphere exerts no gravitational force on a particle inside it. Newton's result follows from Eq. (12). Outside the sphere, V has the same formula as the potential due to a point mass. Inside the sphere, the potential is *constant* with value $-Gm/R$. But constant potential means zero force because the force is the (negative) gradient of the potential. This discussion applies equally well to the electrostatic force. In particular, a uniformly charged sphere behaves like a point charge (when viewed from outside the sphere).

17.4 SUMMARY

- A *parametrized surface* is a surface $\mathcal{S}$ whose points are described in the form

$$G(u, v) = (x(u, v), y(u, v), z(u, v))$$

where the *parameters* u and v vary in a domain $\mathcal{D}$ in the uv-plane.
- Tangent and normal vectors:

$$\mathbf{T}_u = \frac{\partial G}{\partial u} = \left\langle \frac{\partial x}{\partial u}, \frac{\partial y}{\partial u}, \frac{\partial z}{\partial u} \right\rangle, \qquad \mathbf{T}_v = \frac{\partial G}{\partial v} = \left\langle \frac{\partial x}{\partial v}, \frac{\partial y}{\partial v}, \frac{\partial z}{\partial v} \right\rangle$$

$$\mathbf{n} = \mathbf{n}(u, v) = \mathbf{T}_u \times \mathbf{T}_v$$

The parametrization is *regular* at (u, v) if $\mathbf{n}(u, v) \neq \mathbf{0}$.
- The quantity $\|\mathbf{n}\|$ is an "area distortion factor." If $\mathcal{D}$ is a small region in the uv-plane and $\mathcal{S} = G(\mathcal{D})$, then

$$\text{Area}(\mathcal{S}) \approx \|\mathbf{n}(u_0, v_0)\| \text{Area}(\mathcal{D})$$

where (u_0, v_0) is any sample point in $\mathcal{D}$.
- Formulas:

$$\text{Area}(\mathcal{S}) = \iint_{\mathcal{D}} \|\mathbf{n}(u, v)\| \, du \, dv$$

$$\iint_{\mathcal{S}} f(x, y, z) \, dS = \iint_{\mathcal{D}} f(G(u, v)) \, \|\mathbf{n}(u, v)\| \, du \, dv$$

- Some standard parametrizations:
  - Cylinder of radius R (z-axis as central axis):

$$G(\theta, z) = (R\cos\theta, R\sin\theta, z)$$

Outward normal: $\mathbf{n} = \mathbf{T}_\theta \times \mathbf{T}_z = R\langle \cos\theta, \sin\theta, 0 \rangle$

$$dS = \|\mathbf{n}\| \, d\theta \, dz = R \, d\theta \, dz$$

  - Sphere of radius R, centered at the origin:

$$G(\theta, \phi) = (R\cos\theta\sin\phi, R\sin\theta\sin\phi, R\cos\phi)$$

Unit radial vector: $\mathbf{e}_r = \langle \cos\theta\sin\phi, \sin\theta\sin\phi, \cos\phi \rangle$

Outward normal: $\mathbf{n} = \mathbf{T}_\phi \times \mathbf{T}_\theta = (R^2 \sin\phi)\, \mathbf{e}_r$

$$dS = \|\mathbf{n}\| \, d\phi \, d\theta = R^2 \sin\phi \, d\phi \, d\theta$$

  - Graph of $z = g(x, y)$:

$$G(x, y) = (x, y, g(x, y))$$

$$\mathbf{n} = \mathbf{T}_x \times \mathbf{T}_y = \langle -g_x, -g_y, 1 \rangle$$

$$dS = \|\mathbf{n}\| \, dx \, dy = \sqrt{1 + g_x^2 + g_y^2} \, dx \, dy$$

17.4 EXERCISES

Preliminary Questions

1. What is the surface integral of the function $f(x, y, z) = 10$ over a surface of total area 5?

2. What interpretation can we give to the length $\|\mathbf{n}\|$ of the normal vector for a parametrization $G(u, v)$?

3. A parametrization maps a rectangle of size 0.01×0.02 in the uv-plane onto a small patch $\mathcal{S}$ of a surface. Estimate Area$(\mathcal{S})$ if $\mathbf{T}_u \times \mathbf{T}_v = \langle 1, 2, 2\rangle$ at a sample point in the rectangle.

4. A small surface $\mathcal{S}$ is divided into three small pieces, each of area 0.2. Estimate $\iint_{\mathcal{S}} f(x, y, z)\, dS$ if $f(x, y, z)$ takes the values 0.9, 1, and 1.1 at sample points in these three pieces.

5. A surface $\mathcal{S}$ has a parametrization whose domain is the square $0 \le u, v \le 2$ such that $\|\mathbf{n}(u, v)\| = 5$ for all (u, v). What is Area$(\mathcal{S})$?

6. What is the outward-pointing unit normal to the sphere of radius 3 centered at the origin at $P = (2, 2, 1)$?

Exercises

1. Match each parametrization with the corresponding surface in Figure 16.

(a) $(u, \cos v, \sin v)$

(b) $(u, u + v, v)$

(c) (u, v^3, v)

(d) $(\cos u \sin v, 3\cos u \sin v, \cos v)$

(e) $(u, u(2 + \cos v), u(2 + \sin v))$

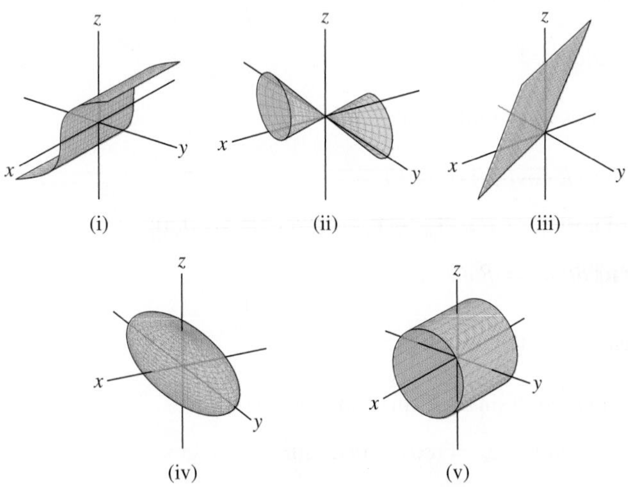

FIGURE 16

2. Show that $G(r, \theta) = (r\cos\theta, r\sin\theta, 1 - r^2)$ parametrizes the paraboloid $z = 1 - x^2 - y^2$. Describe the grid curves of this parametrization.

3. Show that $G(u, v) = (2u + 1, u - v, 3u + v)$ parametrizes the plane $2x - y - z = 2$. Then:

(a) Calculate $\mathbf{T}_u$, $\mathbf{T}_v$, and $\mathbf{n}(u, v)$.

(b) Find the area of $\mathcal{S} = G(\mathcal{D})$, where $\mathcal{D} = \{(u, v) : 0 \le u \le 2, 0 \le v \le 1\}$.

(c) Express $f(x, y, z) = yz$ in terms of u and v, and evaluate $\iint_{\mathcal{S}} f(x, y, z)\, dS$.

4. Let $\mathcal{S} = G(\mathcal{D})$, where $\mathcal{D} = \{(u, v) : u^2 + v^2 \le 1, u \ge 0, v \ge 0\}$ and G is as defined in Exercise 3.

(a) Calculate the surface area of $\mathcal{S}$.

(b) Evaluate $\iint_{\mathcal{S}} (x - y)\, dS$. *Hint:* Use polar coordinates.

5. Let $G(x, y) = (x, y, xy)$.

(a) Calculate $\mathbf{T}_x$, $\mathbf{T}_y$, and $\mathbf{n}(x, y)$.

(b) Let S be the part of the surface with parameter domain $\mathcal{D} = \{(x, y) : x^2 + y^2 \le 1, x \ge 0, y \ge 0\}$. Verify the following formula and evaluate using polar coordinates:

$$\iint_{S} 1\, dS = \iint_{\mathcal{D}} \sqrt{1 + x^2 + y^2}\, dx\, dy$$

(c) Verify the following formula and evaluate:

$$\iint_{S} z\, dS = \int_0^{\pi/2}\int_0^1 (\sin\theta\cos\theta) r^3 \sqrt{1 + r^2}\, dr\, d\theta$$

6. A surface $\mathcal{S}$ has a parametrization $G(u, v)$ whose domain $\mathcal{D}$ is the square in Figure 17. Suppose that G has the following normal vectors:

$$\mathbf{n}(A) = \langle 2, 1, 0\rangle, \quad \mathbf{n}(B) = \langle 1, 3, 0\rangle$$
$$\mathbf{n}(C) = \langle 3, 0, 1\rangle, \quad \mathbf{n}(D) = \langle 2, 0, 1\rangle$$

Estimate $\iint_{\mathcal{S}} f(x, y, z)\, dS$, where f is a function such that $f(G(u, v)) = u + v$.

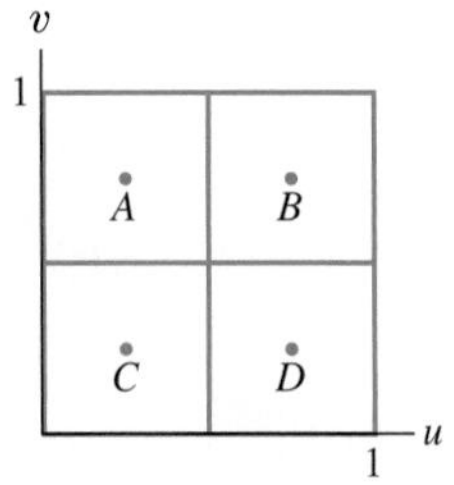

FIGURE 17

In Exercises 7–10, calculate $\mathbf{T}_u$, $\mathbf{T}_v$, and $\mathbf{n}(u, v)$ for the parametrized surface at the given point. Then find the equation of the tangent plane to the surface at that point.

7. $G(u, v) = (2u + v, u - 4v, 3u)$; $\quad u = 1, \quad v = 4$

8. $G(u, v) = (u^2 - v^2, u + v, u - v)$; $\quad u = 2, \quad v = 3$

9. $G(\theta, \phi) = (\cos\theta \sin\phi, \sin\theta \sin\phi, \cos\phi)$; $\quad \theta = \frac{\pi}{2}, \quad \phi = \frac{\pi}{4}$

10. $G(r, \theta) = (r\cos\theta, r\sin\theta, 1 - r^2)$; $\quad r = \frac{1}{2}, \quad \theta = \frac{\pi}{4}$

11. Use the normal vector computed in Exercise 8 to estimate the area of the small patch of the surface $G(u, v) = (u^2 - v^2, u + v, u - v)$ defined by

$$2 \le u \le 2.1, \qquad 3 \le v \le 3.2$$

12. Sketch the small patch of the sphere whose spherical coordinates satisfy

$$\frac{\pi}{2} - 0.15 \le \theta \le \frac{\pi}{2} + 0.15, \qquad \frac{\pi}{4} - 0.1 \le \phi \le \frac{\pi}{4} + 0.1$$

Use the normal vector computed in Exercise 9 to estimate its area.

In Exercises 13–26, calculate $\iint_S f(x, y, z)\, dS$ for the given surface and function.

13. $G(u, v) = (u\cos v, u\sin v, u)$, $\quad 0 \le u \le 1, \quad 0 \le v \le 1$;
$f(x, y, z) = z(x^2 + y^2)$

14. $G(r, \theta) = (r\cos\theta, r\sin\theta, \theta)$, $\quad 0 \le r \le 1, \quad 0 \le \theta \le 2\pi$;
$f(x, y, z) = \sqrt{x^2 + y^2}$

15. $y = 9 - z^2$, $\quad 0 \le x \le 3, 0 \le z \le 3$; $\quad f(x, y, z) = z$

16. $y = 9 - z^2$, $\quad 0 \le x \le z \le 3$; $\quad f(x, y, z) = 1$

17. $x^2 + y^2 + z^2 = 1$, $\; x, y, z \ge 0$; $\quad f(x, y, z) = x^2$.

18. $z = 4 - x^2 - y^2$, $\quad 0 \le z \le 3$; $\quad f(x, y, z) = x^2/(4 - z)$

19. $x^2 + y^2 = 4$, $\quad 0 \le z \le 4$; $\quad f(x, y, z) = e^{-z}$

20. $G(u, v) = (u, v^3, u + v)$, $\quad 0 \le u \le 1, 0 \le v \le 1$;
$f(x, y, z) = y$

21. Part of the plane $x + y + z = 1$, where $x, y, z \ge 0$;
$f(x, y, z) = z$

22. Part of the plane $x + y + z = 0$ contained in the cylinder $x^2 + y^2 = 1$; $\quad f(x, y, z) = z^2$

23. $x^2 + y^2 + z^2 = 4, 1 \le z \le 2$;
$f(x, y, z) = z^2(x^2 + y^2 + z^2)^{-1}$

24. $x^2 + y^2 + z^2 = 4, 0 \le y \le 1$; $\quad f(x, y, z) = y$

25. Part of the surface $z = x^3$, where $0 \le x \le 1$, $0 \le y \le 1$;
$f(x, y, z) = z$

26. Part of the unit sphere centered at the origin, where $x \ge 0$ and $|y| \le x$; $\quad f(x, y, z) = x$

27. A surface $\mathcal{S}$ has a parametrization $G(u, v)$ with domain $0 \le u \le 2, 0 \le v \le 4$ such that the following partial derivatives are constant:

$$\frac{\partial G}{\partial u} = \langle 2, 0, 1 \rangle, \qquad \frac{\partial G}{\partial v} = \langle 4, 0, 3 \rangle$$

What is the surface area of $\mathcal{S}$?

28. Let S be the sphere of radius R centered at the origin. Explain using symmetry:

$$\iint_S x^2\, dS = \iint_S y^2\, dS = \iint_S z^2\, dS$$

Then show that $\iint_S x^2\, dS = \frac{4}{3}\pi R^4$ by adding the integrals.

29. Calculate $\iint_S (xy + e^z)\, dS$, where $\mathcal{S}$ is the triangle in Figure 18 with vertices $(0, 0, 3)$, $(1, 0, 2)$, and $(0, 4, 1)$.

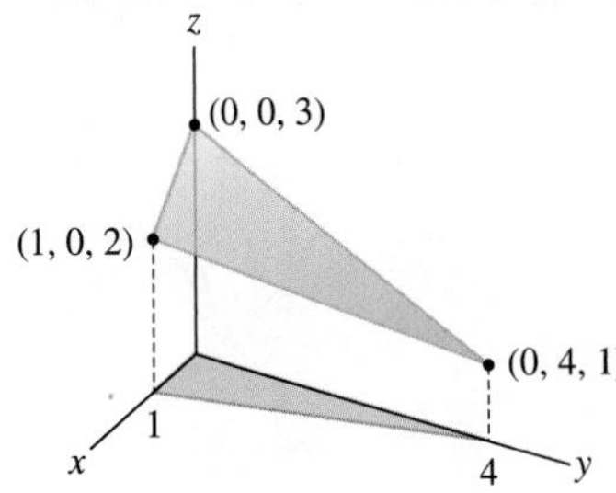

FIGURE 18

30. Use spherical coordinates to compute the surface area of a sphere of radius R.

31. Use cylindrical coordinates to compute the surface area of a sphere of radius R.

32. CAS Let $\mathcal{S}$ be the surface with parametrization

$$G(u, v) = \big((3 + \sin v)\cos u, (3 + \sin v)\sin u, v\big)$$

for $0 \le u \le 2\pi, 0 \le v \le 2\pi$. Using a computer algebra system:

(a) Plot $\mathcal{S}$ from several different viewpoints. Is $\mathcal{S}$ best described as a "vase that holds water" or a "bottomless vase"?

(b) Calculate the normal vector $\mathbf{n}(u, v)$.

(c) Calculate the surface area of $\mathcal{S}$ to four decimal places.

33. CAS Let $\mathcal{S}$ be the surface $z = \ln(5 - x^2 - y^2)$ for $0 \le x \le 1$, $0 \le y \le 1$. Using a computer algebra system:

(a) Calculate the surface area of $\mathcal{S}$ to four decimal places.

(b) Calculate $\iint_S x^2 y^3\, dS$ to four decimal places.

34. Find the area of the portion of the plane $2x + 3y + 4z = 28$ lying above the rectangle $1 \le x \le 3, 2 \le y \le 5$ in the xy-plane.

35. What is the area of the portion of the plane $2x + 3y + 4z = 28$ lying above the domain $\mathcal{D}$ in the xy-plane in Figure 19 if Area($\mathcal{D}$) = 5?

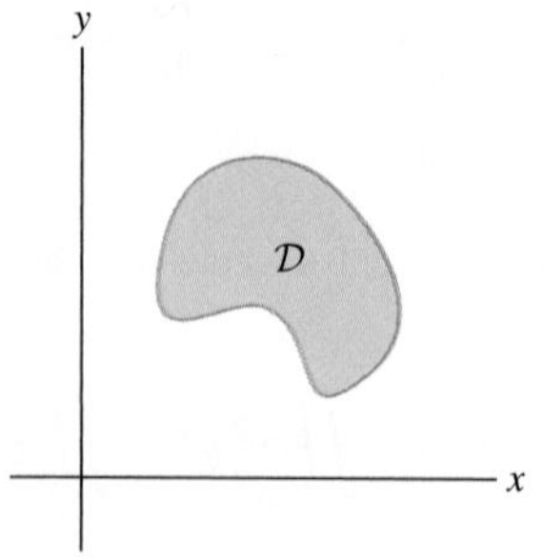

FIGURE 19

36. Find the surface area of the part of the cone $x^2 + y^2 = z^2$ between the planes $z = 2$ and $z = 5$.

37. Find the surface area of the portion S of the cone $z^2 = x^2 + y^2$, where $z \geq 0$, contained within the cylinder $y^2 + z^2 \leq 1$.

38. Calculate the integral of ze^{2x+y} over the surface of the box in Figure 20.

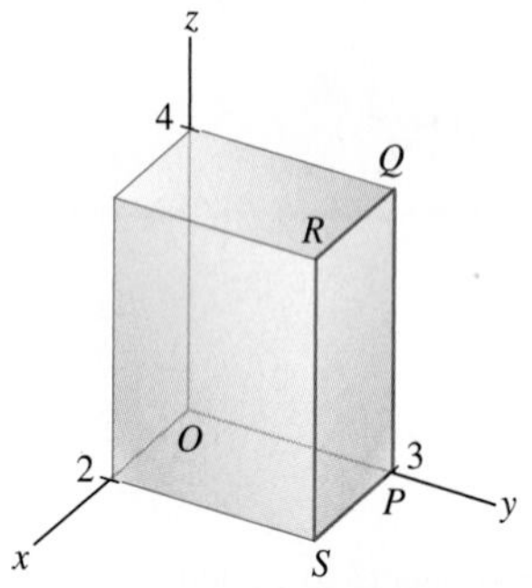

FIGURE 20

39. Calculate $\iint_G x^2 z\, dS$, where G is the cylinder (including the top and bottom) $x^2 + y^2 = 4$, $0 \leq z \leq 3$.

40. Let S be the portion of the sphere $x^2 + y^2 + z^2 = 9$, where $1 \leq x^2 + y^2 \leq 4$ and $z \geq 0$ (Figure 21). Find a parametrization of S in polar coordinates and use it to compute:

(a) The area of S **(b)** $\iint_S z^{-1}\, dS$

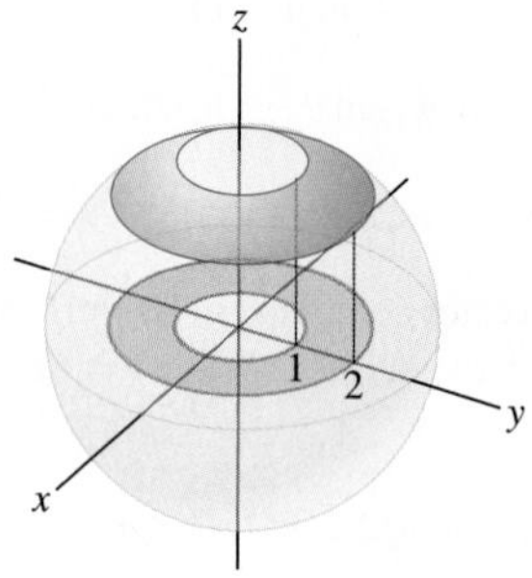

FIGURE 21

41. Prove a famous result of Archimedes: The surface area of the portion of the sphere of radius R between two horizontal planes $z = a$ and $z = b$ is equal to the surface area of the corresponding portion of the circumscribed cylinder (Figure 22).

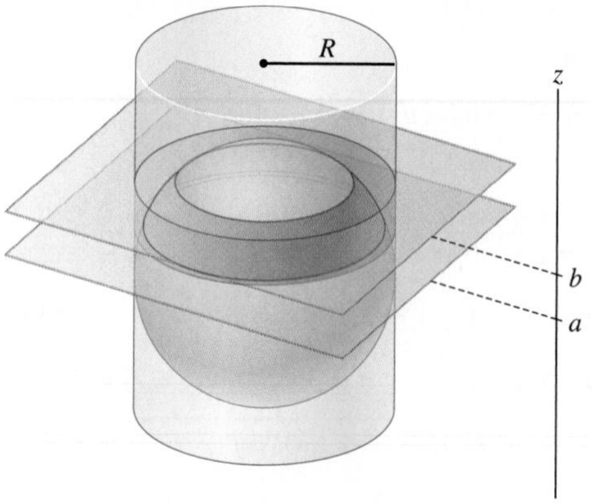

FIGURE 22

Further Insights and Challenges

42. Surfaces of Revolution Let $\mathcal{S}$ be the surface formed by rotating the region under the graph $z = g(y)$ in the yz-plane for $c \leq y \leq d$ about the z-axis, where $c \geq 0$ (Figure 23).

(a) Show that the circle generated by rotating a point $(0, a, b)$ about the z-axis is parametrized by

$$(a\cos\theta, a\sin\theta, b), \quad 0 \leq \theta \leq 2\pi$$

(b) Show that $\mathcal{S}$ is parametrized by

$$G(y, \theta) = (y\cos\theta, y\sin\theta, g(y)) \qquad \boxed{13}$$

for $c \leq y \leq d$, $0 \leq \theta \leq 2\pi$.

(c) Use Eq. (13) to prove the formula

$$\text{Area}(\mathcal{S}) = 2\pi \int_c^d y\sqrt{1 + g'(y)^2}\, dy \qquad \boxed{14}$$

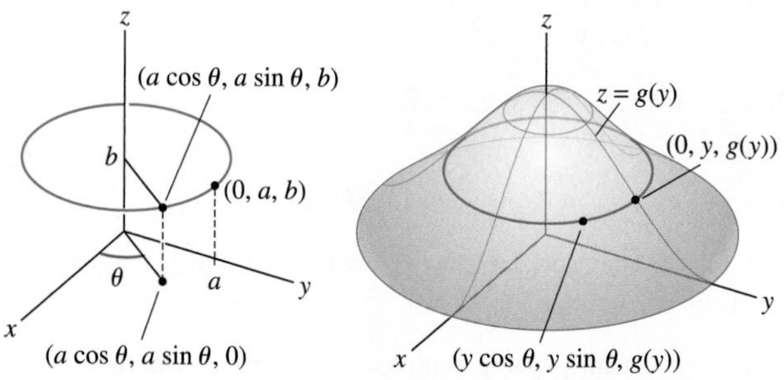

FIGURE 23

43. Use Eq. (14) to compute the surface area of $z = 4 - y^2$ for $0 \le y \le 2$ rotated about the z-axis.

44. Describe the upper half of the cone $x^2 + y^2 = z^2$ for $0 \le z \le d$ as a surface of revolution (Figure 2) and use Eq. (14) to compute its surface area.

45. Area of a Torus Let T be the torus obtained by rotating the circle in the yz-plane of radius a centered at $(0, b, 0)$ about the z-axis (Figure 24). We assume that $b > a > 0$.

(a) Use Eq. (14) to show that

$$\text{Area(T)} = 4\pi \int_{b-a}^{b+a} \frac{ay}{\sqrt{a^2 - (b-y)^2}}\,dy$$

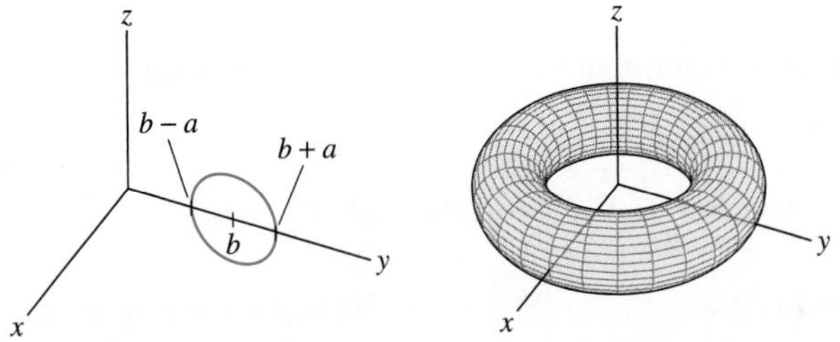

FIGURE 24 The torus obtained by rotating a circle of radius a.

(b) Show that $\text{Area(T)} = 4\pi^2 ab$.

46. Pappus's Theorem (also called **Guldin's Rule**) states that the area of a surface of revolution $\mathcal{S}$ is equal to the length L of the generating curve times the distance traversed by the center of mass. Use Eq. (14) to prove Pappus's Theorem. If $\mathcal{C}$ is the graph $z = g(y)$ for $c \le y \le d$, then the center of mass is defined as the point $(\overline{y}, \overline{z})$ with

$$\overline{y} = \frac{1}{L}\int_{\mathcal{C}} y\,ds, \qquad \overline{z} = \frac{1}{L}\int_{\mathcal{C}} z\,ds$$

47. Compute the surface area of the torus in Exercise 45 using Pappus's Theorem.

48. **Potential Due to a Uniform Sphere** Let $\mathcal{S}$ be a hollow sphere of radius R with center at the origin with a uniform mass distribution of total mass m [since $\mathcal{S}$ has surface area $4\pi R^2$, the mass density is $\rho = m/(4\pi R^2)$]. The gravitational potential $V(P)$ due to $\mathcal{S}$ at a point $P = (a, b, c)$ is equal to

$$-G \iint_{\mathcal{S}} \frac{\rho\,dS}{\sqrt{(x-a)^2 + (y-b)^2 + (z-c)^2}}$$

(a) Use symmetry to conclude that the potential depends only on the distance r from P to the center of the sphere. Therefore, it suffices to compute $V(P)$ for a point $P = (0, 0, r)$ on the z-axis (with $r \ne R$).

(b) Use spherical coordinates to show that $V(0, 0, r)$ is equal to

$$\frac{-Gm}{4\pi} \int_0^{\pi} \int_0^{2\pi} \frac{\sin\phi\,d\theta\,d\phi}{\sqrt{R^2 + r^2 - 2Rr\cos\phi}}$$

(c) Use the substitution $u = R^2 + r^2 - 2Rr\cos\phi$ to show that

$$V(0, 0, r) = \frac{-mG}{2Rr}\big(|R + r| - |R - r|\big)$$

(d) Verify Eq. (12) for V.

49. Calculate the gravitational potential V for a hemisphere of radius R with uniform mass distribution.

50. The surface of a cylinder of radius R and length L has a uniform mass distribution ρ (the top and bottom of the cylinder are excluded). Use Eq. (11) to find the gravitational potential at a point P located along the axis of the cylinder.

51. Let S be the part of the graph $z = g(x, y)$ lying over a domain $\mathcal{D}$ in the xy-plane. Let $\phi = \phi(x, y)$ be the angle between the normal to S and the vertical. Prove the formula

$$\text{Area}(S) = \iint_{\mathcal{D}} \frac{dA}{|\cos\phi|}$$

17.5 Surface Integrals of Vector Fields

*The word **flux** is derived from the Latin word **fluere**, which means "to flow."*

The last integrals we will consider are surface integrals of vector fields. These integrals represent flux or rates of flow through a surface. One example is the flux of molecules across a cell membrane (number of molecules per unit time).

Because flux through a surface goes from one side of the surface to the other, we need to specify a *positive direction* of flow. This is done by means of an **orientation**, which is a choice of unit normal vector $\mathbf{e_n}(P)$ at each point P of $\mathcal{S}$, chosen in a continuously varying manner (Figure 1). There are two normal directions at each point, so the orientation serves to specify one of the two "sides" of the surface in a consistent manner. The unit vectors $-\mathbf{e_n}(P)$ define the *opposite orientation*. For example, if $\mathbf{e_n}$ are outward-pointing unit normal vectors on a sphere, then a flow from the inside of the sphere to the outside is a positive flux.

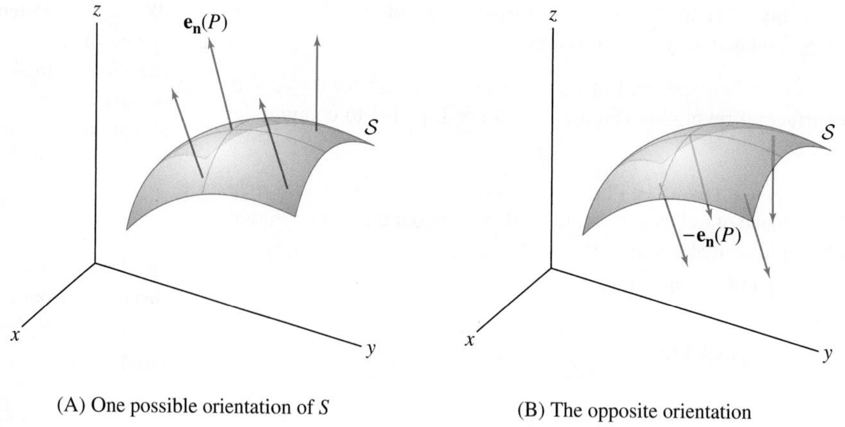
z
$\mathbf{e_n}(P)$
$\mathcal{S}$
x
y
(A) One possible orientation of S
z
$\mathcal{S}$
$-\mathbf{e_n}(P)$
x
y
(B) The opposite orientation

FIGURE 1 The surface $\mathcal{S}$ has two possible orientations.

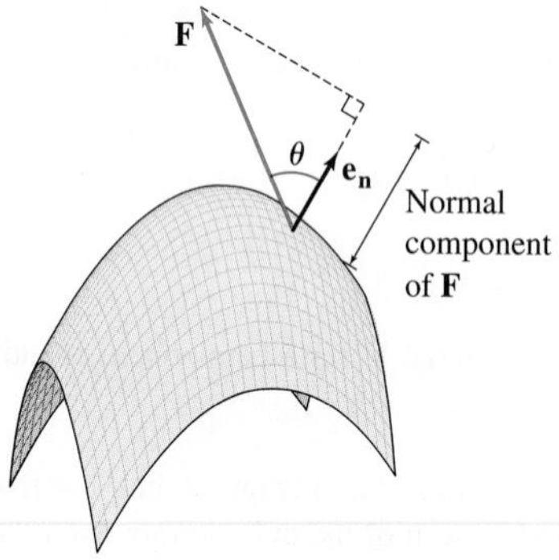

FIGURE 2 The normal component of a vector to a surface.

The **normal component** of a vector field $\mathbf{F}$ at a point P on an oriented surface $\mathcal{S}$ is the dot product

$$\text{Normal component at } P = \mathbf{F}(P)\cdot \mathbf{e_n}(P) = \|\mathbf{F}(P)\|\cos\theta$$

where θ is the angle between $\mathbf{F}(P)$ and $\mathbf{e_n}(P)$ (Figure 2). Often, we write $\mathbf{e_n}$ instead of $\mathbf{e_n}(P)$, but it is understood that $\mathbf{e_n}$ varies from point to point on the surface. The vector surface integral, denoted $\iint_{\mathcal{S}} \mathbf{F}\cdot d\mathbf{S}$, is defined as the integral of the normal component:

$$\text{Vector surface integral:}\quad \iint_{\mathcal{S}} \mathbf{F}\cdot d\mathbf{S} = \iint_{\mathcal{S}} (\mathbf{F}\cdot \mathbf{e_n})\, dS$$

This quantity is also called the **flux** of $\mathbf{F}$ across or through $\mathcal{S}$.

An oriented parametrization $G(u, v)$ is a regular parametrization (meaning that $\mathbf{n}(u, v)$ is nonzero for all u, v) whose unit normal vector defines the orientation:

$$\mathbf{e_n} = \mathbf{e_n}(u, v) = \frac{\mathbf{n}(u, v)}{\|\mathbf{n}(u, v)\|}$$

REMINDER Formula for a scalar surface integral in terms of an oriented parametrization:

$$\iint_{\mathcal{S}} f(x, y, z)\, dS = \iint f(G(u, v))\|\mathbf{n}(u, v)\|\, du\, dv \qquad \boxed{1}$$

Applying Eq. (1) in the margin to $\mathbf{F}\cdot\mathbf{e}_n$, we obtain

$$\begin{aligned}\iint_{\mathcal{S}} \mathbf{F}\cdot d\mathbf{S} &= \iint_{\mathcal{D}} (\mathbf{F}\cdot\mathbf{e_n})\|\mathbf{n}(u, v)\|\, du\, dv \\ &= \iint_{\mathcal{D}} \mathbf{F}(G(u, v))\cdot\left(\frac{\mathbf{n}(u, v)}{\|\mathbf{n}(u, v)\|}\right)\|\mathbf{n}(u, v)\|\, du\, dv \\ &= \iint_{\mathcal{D}} \mathbf{F}(G(u, v))\cdot \mathbf{n}(u, v)\, du\, dv \qquad \boxed{2}\end{aligned}$$

This formula remains valid even if $\mathbf{n}(u, v)$ is zero at points on the boundary of the parameter domain $\mathcal{D}$. If we reverse the orientation of $\mathcal{S}$ in a vector surface integral, $\mathbf{n}(u, v)$ is replaced by $-\mathbf{n}(u, v)$ and the integral changes sign.

We can think of $d\mathbf{S}$ as a "vector surface element" that is related to a parametrization by the symbolic equation

$$d\mathbf{S} = \mathbf{n}(u, v)\, du\, dv$$

THEOREM 1 Vector Surface Integral Let $G(u, v)$ be an oriented parametrization of an oriented surface $\mathcal{S}$ with parameter domain $\mathcal{D}$. Assume that G is one-to-one and regular, except possibly at points on the boundary of $\mathcal{D}$. Then

$$\iint_{\mathcal{S}} \mathbf{F} \cdot d\mathbf{S} = \iint_{\mathcal{D}} \mathbf{F}(G(u, v)) \cdot \mathbf{n}(u, v)\, du\, dv \tag{3}$$

If the orientation of $\mathcal{S}$ is reversed, the surface integral changes sign.

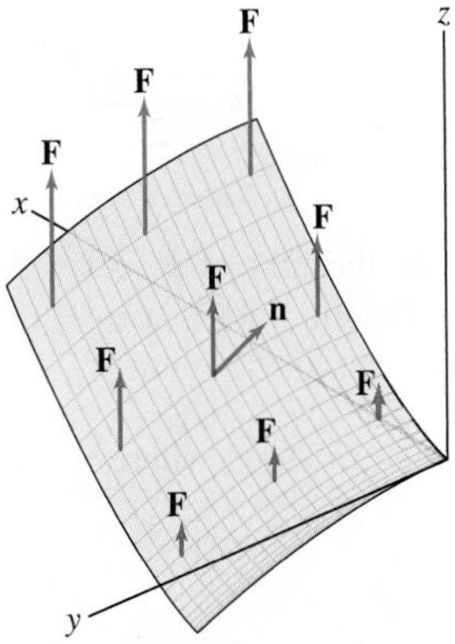

FIGURE 3 The surface $G(u, v) = (u^2, v, u^3 - v^2)$ with an upward-pointing normal. The vector field $\mathbf{F} = \langle 0, 0, x\rangle$ points in the vertical direction.

■ **EXAMPLE 1** Calculate $\iint_{\mathcal{S}} \mathbf{F} \cdot d\mathbf{S}$, where $\mathbf{F} = \langle 0, 0, x\rangle$ and $\mathcal{S}$ is the surface with parametrization $G(u, v) = (u^2, v, u^3 - v^2)$ for $0 \le u \le 1,\ 0 \le v \le 1$ and oriented by upward-pointing normal vectors.

Solution

***Step 1.* Compute the tangent and normal vectors.**

$$\mathbf{T}_u = \langle 2u, 0, 3u^2\rangle, \qquad T_v = \langle 0, 1, -2v\rangle$$

$$\mathbf{n}(u, v) = \mathbf{T}_u \times \mathbf{T}_v = \begin{vmatrix} \mathbf{i} & \mathbf{j} & \mathbf{k} \\ 2u & 0 & 3u^2 \\ 0 & 1 & -2v \end{vmatrix}$$

$$= -3u^2\mathbf{i} + 4uv\mathbf{j} + 2u\mathbf{k} = \langle -3u^2, 4uv, 2u\rangle$$

The z-component of $\mathbf{n}$ is positive on the domain $0 \le u \le 1$, so $\mathbf{n}$ is the upward-pointing normal (Figure 3).

***Step 2.* Evaluate F · n.**

Write $\mathbf{F}$ in terms of the parameters u and v. Since $x = u^2$,

$$\mathbf{F}(G(u, v)) = \langle 0, 0, x\rangle = \langle 0, 0, u^2\rangle$$

and

$$\mathbf{F}(G(u, v)) \cdot \mathbf{n}(u, v) = \langle 0, 0, u^2\rangle \cdot \langle -3u^2, 4uv, 2u\rangle = 2u^3$$

***Step 3.* Evaluate the surface integral.**

The parameter domain is $0 \le u \le 1, 0 \le v \le 1$, so

$$\iint_{\mathcal{S}} \mathbf{F} \cdot d\mathbf{S} = \int_{u=0}^{1}\int_{v=0}^{1} \mathbf{F}(G(u, v)) \cdot \mathbf{n}(u, v)\, dv\, du$$

$$= \int_{u=0}^{1}\int_{v=0}^{1} 2u^3\, dv\, du = \int_{u=0}^{1} 2u^3\, du = \frac{1}{2}$$ ■

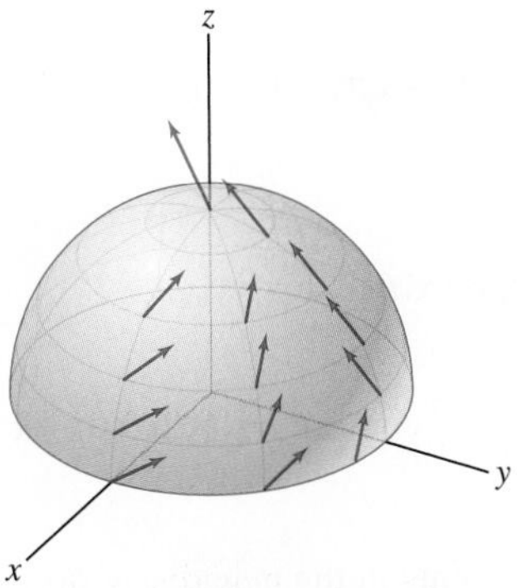

FIGURE 4 The vector field $\mathbf{F} = \langle z, x, 1\rangle$.

■ **EXAMPLE 2 Integral over a Hemisphere** Calculate the flux of $\mathbf{F} = \langle z, x, 1\rangle$ across the upper hemisphere $\mathcal{S}$ of the sphere $x^2 + y^2 + z^2 = 1$, oriented with outward-pointing normal vectors (Figure 4).

Solution Parametrize the hemisphere by spherical coordinates:

$$G(\theta, \phi) = (\cos\theta \sin\phi, \sin\theta \sin\phi, \cos\phi), \qquad 0 \le \phi \le \frac{\pi}{2}, \quad 0 \le \theta < 2\pi$$

***Step 1.* Compute the normal vector.**

According to Eq. (2) in Section 17.4, the outward-pointing normal vector is

$$\mathbf{n} = \mathbf{T}_\phi \times \mathbf{T}_\theta = \sin\phi\langle \cos\theta \sin\phi, \sin\theta \sin\phi, \cos\phi\rangle$$

Step 2. **Evaluate F · n.**

$$\mathbf{F}(G(\theta,\phi)) = \langle z, x, 1\rangle = \langle \cos\phi, \cos\theta\sin\phi, 1\rangle$$

$$\mathbf{F}(G(\theta,\phi))\cdot\mathbf{n}(\theta,\phi) = \langle \cos\phi, \cos\theta\sin\phi, 1\rangle\cdot\langle\cos\theta\sin^2\phi, \sin\theta\sin^2\phi, \cos\phi\sin\phi\rangle$$
$$= \cos\theta\sin^2\phi\cos\phi + \cos\theta\sin\theta\sin^3\phi + \cos\phi\sin\phi$$

Step 3. **Evaluate the surface integral.**

$$\iint_S \mathbf{F}\cdot d\mathbf{S} = \int_{\phi=0}^{\pi/2}\int_{\theta=0}^{2\pi} \mathbf{F}(G(\theta,\phi))\cdot\mathbf{n}(\theta,\phi)\,d\theta\,d\phi$$
$$= \int_{\phi=0}^{\pi/2}\int_{\theta=0}^{2\pi} (\underbrace{\cos\theta\sin^2\phi\cos\phi + \cos\theta\sin\theta\sin^3\phi}_{\text{Integral over }\theta\text{ is zero}} + \cos\phi\sin\phi)\,d\theta\,d\phi$$

The integrals of $\cos\theta$ and $\cos\theta\sin\theta$ over $[0, 2\pi]$ are both zero, so we are left with

$$\int_{\phi=0}^{\pi/2}\int_{\theta=0}^{2\pi} \cos\phi\sin\phi\,d\theta\,d\phi = 2\pi\int_{\phi=0}^{\pi/2}\cos\phi\sin\phi\,d\phi = -2\pi\left.\frac{\cos^2\phi}{2}\right|_0^{\pi/2} = \pi \quad ■$$

■ **EXAMPLE 3 Surface Integral over a Graph** Calculate the flux of $\mathbf{F} = x^2\mathbf{j}$ through the surface $\mathcal{S}$ defined by $y = 1 + x^2 + z^2$ for $1 \le y \le 5$. Orient $\mathcal{S}$ with normal pointing in the negative y-direction.

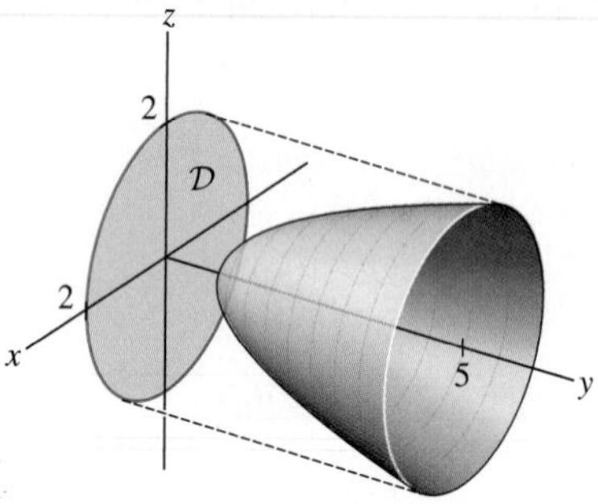

FIGURE 5

Solution This surface is the graph of the function $y = 1 + x^2 + z^2$, where x and z are the independent variables (Figure 5).

Step 1. **Find a parametrization.**
It is convenient to use x and z because y is given explicitly as a function of x and z. Thus we define

$$G(x, z) = (x, 1 + x^2 + z^2, z)$$

What is the parameter domain? The condition $1 \le y \le 5$ is equivalent to $1 \le 1 + x^2 + z^2 \le 5$ or $0 \le x^2 + z^2 \le 4$. Therefore, the parameter domain is the disk of radius 2 in the xz-plane—that is, $\mathcal{D} = \{(x, z) : x^2 + z^2 \le 4\}$.

Because the parameter domain is a disk, it makes sense to use the polar variables r and θ in the xz-plane. In other words, we write $x = r\cos\theta$, $z = r\sin\theta$. Then

$$y = 1 + x^2 + z^2 = 1 + r^2$$

$$G(r, \theta) = (r\cos\theta, 1 + r^2, r\sin\theta), \quad 0 \le \theta \le 2\pi, \quad 0 \le r \le 2$$

Step 2. **Compute the tangent and normal vectors.**

$$\mathbf{T}_r = \langle\cos\theta, 2r, \sin\theta\rangle, \qquad \mathbf{T}_\theta = \langle -r\sin\theta, 0, r\cos\theta\rangle$$

$$\mathbf{n} = \mathbf{T}_r\times\mathbf{T}_\theta = \begin{vmatrix} \mathbf{i} & \mathbf{j} & \mathbf{k} \\ \cos\theta & 2r & \sin\theta \\ -r\sin\theta & 0 & r\cos\theta \end{vmatrix} = 2r^2\cos\theta\mathbf{i} - r\mathbf{j} + 2r^2\sin\theta\mathbf{k}$$

The coefficient of $\mathbf{j}$ is $-r$. Because it is negative, $\mathbf{n}$ points in the negative y-direction, as required.

Step 3. **Evaluate F · n.**

CAUTION In Step 3, we integrate $\mathbf{F}\cdot\mathbf{n}$ *with respect to* $dr\,d\theta$*, and* not $r\,dr\,d\theta$*. The factor of* r *in* $r\,dr\,d\theta$ *is a Jacobian factor that we add only when changing variables in a double integral. In surface integrals, the Jacobian factor is incorporated into the magnitude of* $\mathbf{n}$ *(recall that* $\|\mathbf{n}\|$ *is the area "distortion factor").*

$$\mathbf{F}(G(r,\theta)) = x^2\mathbf{j} = r^2\cos^2\theta\,\mathbf{j} = \langle 0, r^2\cos^2\theta, 0\rangle$$

$$\mathbf{F}(G(r,\theta))\cdot\mathbf{n} = \langle 0, r^2\cos^2\theta, 0\rangle \cdot \langle 2r^2\cos\theta, -r, 2r^2\sin\theta\rangle = -r^3\cos^2\theta$$

$$\iint_{\mathcal{S}} \mathbf{F}\cdot d\mathbf{S} = \iint_{\mathcal{D}} \mathbf{F}(G(r,\theta))\cdot\mathbf{n}\,dr\,d\theta = \int_0^{2\pi}\int_0^2 (-r^3\cos^2\theta)\,dr\,d\theta$$

$$= -\left(\int_0^{2\pi}\cos^2\theta\,d\theta\right)\left(\int_0^2 r^3\,dr\right)$$

$$= -(\pi)\left(\frac{2^4}{4}\right) = -4\pi$$

■

CONCEPTUAL INSIGHT Since a vector surface integral depends on the orientation of the surface, this integral is defined only for surfaces that have two sides. However, some surfaces, such as the Möbius strip (discovered in 1858 independently by August Möbius and Johann Listing), cannot be oriented because they are one-sided. You can construct a Möbius strip M with a rectangular strip of paper: Join the two ends of the strip together with a 180° twist. Unlike an ordinary two-sided strip, the Möbius strip M has only one side, and it is impossible to specify an outward direction in a consistent manner (Figure 6). If you choose a unit normal vector at a point P and carry that unit vector continuously around M, when you return to P, the vector will point in the opposite direction. Therefore, we cannot integrate a vector field over a Möbius strip, and it is not meaningful to speak of the "flux" across M. On the other hand, it is possible to integrate a scalar function. For example, the integral of mass density would equal the total mass of the Möbius strip.

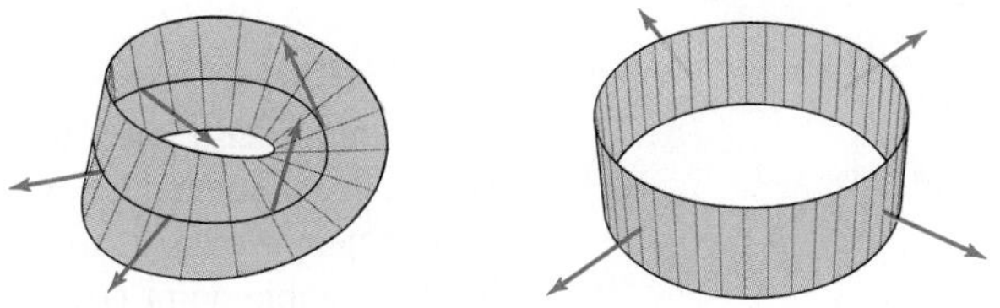

FIGURE 6 It is not possible to choose a continuously varying unit normal vector on a Möbius strip.

Fluid Flux

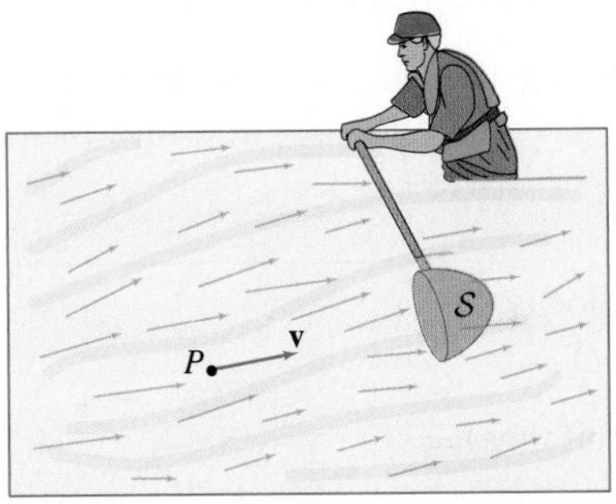

FIGURE 7 Velocity field of a fluid flow.

Imagine dipping a net into a stream of flowing water (Figure 7). The **flow rate** is the volume of water that flows through the net per unit time.

To compute the flow rate, let $\mathbf{v}$ be the velocity vector field. At each point P, $\mathbf{v}(P)$ is velocity vector of the fluid particle located at the point P. We claim that *the flow rate through a surface $\mathcal{S}$ is equal to the surface integral of* $\mathbf{v}$ *over* $\mathcal{S}$.

To explain why, suppose first that $\mathcal{S}$ is a rectangle of area A and that $\mathbf{v}$ is a constant vector field with value $\mathbf{v}_0$ perpendicular to the rectangle. The particles travel at speed $\|\mathbf{v}_0\|$, say in meters per second, so a given particle flows through $\mathcal{S}$ within a one-second time interval if its distance to $\mathcal{S}$ is at most $\|\mathbf{v}_0\|$ meters—in other words, if its velocity vector passes through $\mathcal{S}$ (see Figure 8). Thus the block of fluid passing through $\mathcal{S}$ in a one-second interval is a box of volume $\|\mathbf{v}_0\|A$ (Figure 9), and

$$\text{Flow rate} = (\text{velocity})(\text{area}) = \|\mathbf{v}_0\|A$$

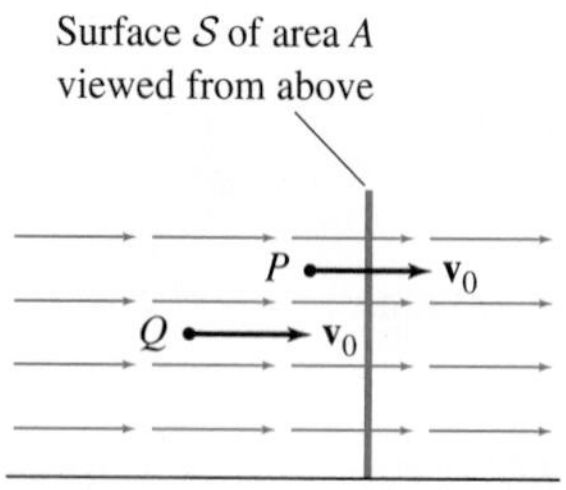

FIGURE 8 The particle P flows through $\mathcal{S}$ within a one-second interval, but Q does not.

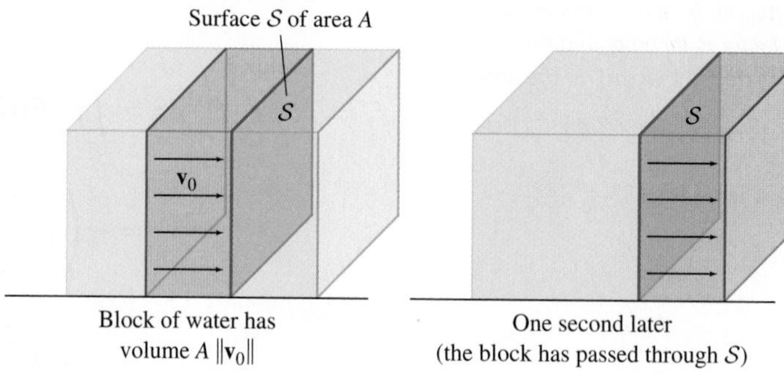

FIGURE 9

If the fluid flows at an angle θ relative to $\mathcal{S}$, then the block of water is a parallelepiped (rather than a box) of volume $A\|\mathbf{v}_0\|\cos\theta$ (Figure 10). If $\mathbf{n}$ is a vector normal to $\mathcal{S}$ of length equal to the area A, then we can write the flow rate as a dot product:

$$\text{Flow rate} = A\|\mathbf{v}_0\|\cos\theta = \mathbf{v}_0 \cdot \mathbf{n}$$

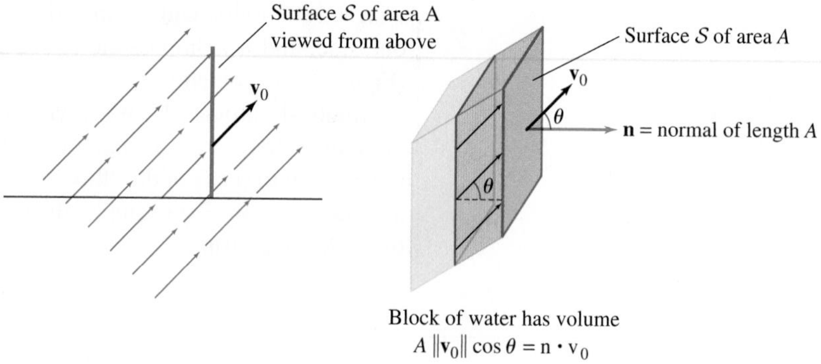

FIGURE 10 Water flowing at constant velocity $\mathbf{v}_0$, making an angle θ with a rectangular surface.

In the general case, the velocity field $\mathbf{v}$ is not constant, and the surface $\mathcal{S}$ may be curved. To compute the flow rate, we choose a parametrization $G(u, v)$ and we consider a small rectangle of size $\Delta u \times \Delta v$ that is mapped by G to a small patch $\mathcal{S}_0$ of $\mathcal{S}$ (Figure 11). For any sample point $G(u_0, v_0)$ in $\mathcal{S}_0$, the vector $\mathbf{n}(u_0, v_0)\,\Delta u\,\Delta v$ is a normal vector of length approximately equal to the area of $\mathcal{S}_0$ [Eq. (3) in Section 17.4]. This patch is nearly rectangular, so we have the approximation

$$\text{Flow rate through } \mathcal{S}_0 \approx \mathbf{v}(u_0, v_0) \cdot \mathbf{n}(u_0, v_0)\,\Delta u\,\Delta v$$

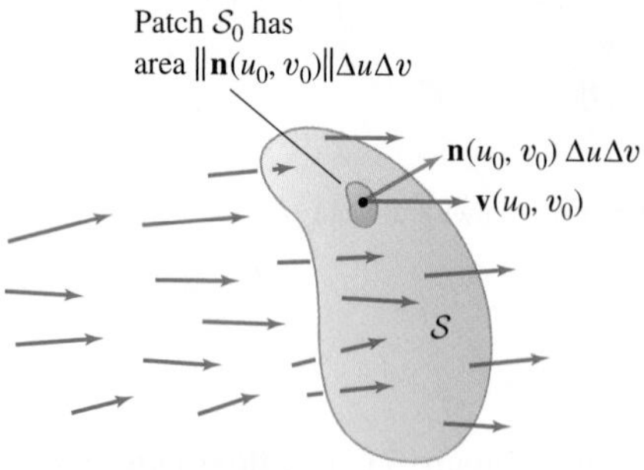

FIGURE 11 The flow rate across the small patch $\mathcal{S}_0$ is approximately $\mathbf{v}(u_0, v_0) \cdot \mathbf{n}(u_0, v_0)\,\Delta u\,\Delta v$.

The total flow per second is the sum of the flows through the small patches. As usual, the limit of the sums as Δu and Δv tend to zero is the integral of $\mathbf{v}(u, v) \cdot \mathbf{n}(u, v)$, which is the surface integral of $\mathbf{v}$ over $\mathcal{S}$.

Flow Rate through a Surface For a fluid with velocity vector field $\mathbf{v}$,

$$\text{Flow rate across the } \mathcal{S} \text{ (volume per unit time)} = \iint_{\mathcal{S}} \mathbf{v} \cdot d\mathbf{S} \qquad \boxed{4}$$

■ **EXAMPLE 4** Let $\mathbf{v} = \langle x^2 + y^2, 0, z^2 \rangle$ be the velocity field (in centimeters per second) of a fluid in $\mathbf{R}^3$. Compute the flow rate through the upper hemisphere $\mathcal{S}$ of the unit sphere centered at the origin.

Solution We use spherical coordinates:

$$x = \cos\theta \sin\phi, \qquad y = \sin\theta \sin\phi, \qquad z = \cos\phi$$

The upper hemisphere corresponds to the ranges $0 \le \phi \le \frac{\pi}{2}$ and $0 \le \theta \le 2\pi$. By Eq. (2) in Section 17.4, the upward-pointing normal is

$$\mathbf{n} = \mathbf{T}_\phi \times \mathbf{T}_\theta = \sin\phi \langle \cos\theta \sin\phi, \sin\theta \sin\phi, \cos\phi \rangle$$

We have $x^2 + y^2 = \sin^2\phi$, so

$$\mathbf{v} = \langle x^2 + y^2, 0, z^2 \rangle = \langle \sin^2\phi, 0, \cos^2\phi \rangle$$

$$\mathbf{v} \cdot \mathbf{n} = \sin\phi \langle \sin^2\phi, 0, \cos^2\phi \rangle \cdot \langle \cos\theta \sin\phi, \sin\theta \sin\phi, \cos\phi \rangle$$

$$= \sin^4\phi \cos\theta + \sin\phi \cos^3\phi$$

$$\iint_{\mathcal{S}} \mathbf{v} \cdot d\mathbf{S} = \int_{\phi=0}^{\pi/2} \int_{\theta=0}^{2\pi} (\sin^4\phi \cos\theta + \sin\phi \cos^3\phi)\, d\theta\, d\phi$$

The integral of $\sin^4\phi \cos\theta$ with respect to θ is zero, so we are left with

$$\int_{\phi=0}^{\pi/2} \int_{\theta=0}^{2\pi} \sin\phi \cos^3\phi \, d\theta\, d\phi = 2\pi \int_{\phi=0}^{\pi/2} \cos^3\phi \sin\phi \, d\phi$$

$$= 2\pi \left(-\frac{\cos^4\phi}{4} \right) \Bigg|_{\phi=0}^{\pi/2} = \frac{\pi}{2} \text{ cm}^3/\text{s}$$

Since $\mathbf{n}$ is an upward-pointing normal, this is the rate at which fluid flows across the hemisphere from below to above. ■

Electric and Magnetic Fields

The laws of electricity and magnetism are expressed in terms of two vector fields, the electric field $\mathbf{E}$ and the magnetic field $\mathbf{B}$, whose properties are summarized in Maxwell's four equations. One of these equations is **Faraday's Law of Induction**, which can be formulated either as a partial differential equation or in the following "integral form":

$$\int_{\mathcal{C}} \mathbf{E} \cdot d\mathbf{s} = -\frac{d}{dt} \iint_{\mathcal{S}} \mathbf{B} \cdot d\mathbf{S} \qquad \boxed{5}$$

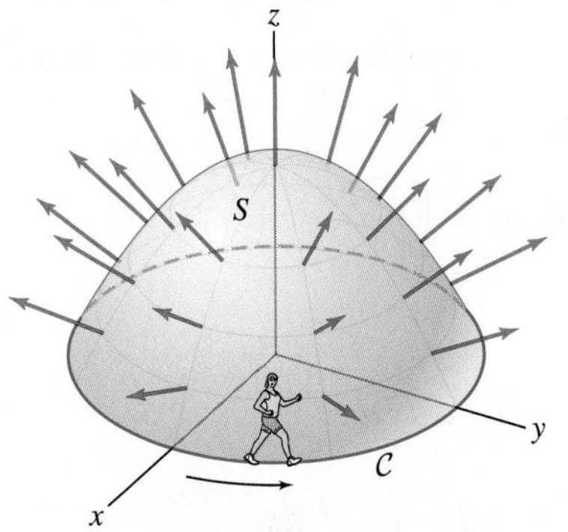

FIGURE 12 The positive direction along the boundary curve $\mathcal{C}$ is defined so that if a pedestrian walks in the positive direction with the surface to her left, then her head points in the outward (normal) direction.

In this equation, $\mathcal{S}$ is an oriented surface with boundary curve $\mathcal{C}$, oriented as indicated in Figure 12. The line integral of $\mathbf{E}$ is equal to the voltage drop around the boundary curve (the work performed by $\mathbf{E}$ moving a positive unit charge around $\mathcal{C}$).

To illustrate Faraday's Law, consider an electric current of i amperes flowing through a straight wire. According to the Biot-Savart Law, this current produces a magnetic field $\mathbf{B}$ of magnitude $B(r) = \dfrac{\mu_0 |i|}{2\pi r}$ T, where r is the distance (in meters) from the wire and $\mu_0 = 4\pi \cdot 10^{-7}$ T-m/A. At each point P, $\mathbf{B}$ is tangent to the circle through P perpendicular to the wire as in Figure 13(A), with direction determined by the right-hand rule: If the thumb of your right hand points in the direction of the current, then your fingers curl in the direction of $\mathbf{B}$.

*The **tesla** (T) is the SI unit of magnetic field strength. A one-coulomb point charge passing through a magnetic field of 1 T at 1 m/s experiences a force of 1 newton.*

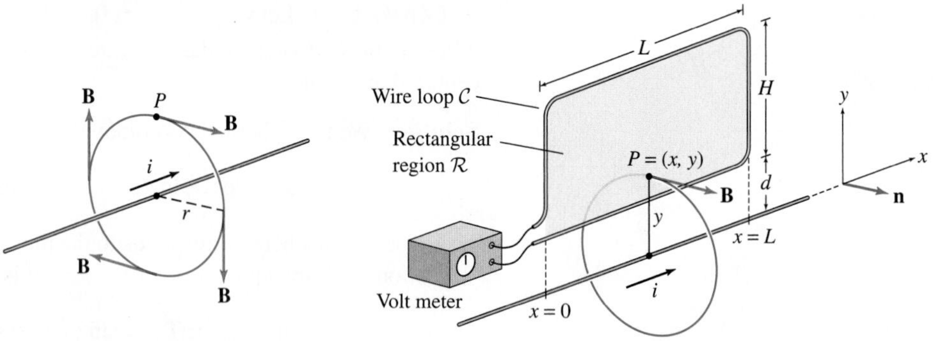

(A) Magnetic field B due to a current in a wire

(B) The magnetic field B points in the direction **n** normal to $\mathcal{R}$

FIGURE 13

The electric field **E** *is conservative when the charges are stationary or, more generally, when the magnetic field* **B** *is constant. When* **B** *varies in time, the integral on the right in Eq. (5) is nonzero for some surface, and hence the circulation of* **E** *around the boundary curve* $\mathcal{C}$ *is also nonzero. This shows that* **E** *is not conservative when* **B** *varies in time.*

Magnetic flux as a function of time is often denoted by the Greek letter Φ:

$$\Phi(t) = \iint_S \mathbf{B} \cdot d\mathbf{S}$$

■ **EXAMPLE 5** A varying current of magnitude (t in seconds)

$$i = 28\cos(400t) \text{ amperes}$$

flows through a straight wire [Figure 13(B)]. A rectangular wire loop $\mathcal{C}$ of length $L = 1.2$ m and width $H = 0.7$ m is located a distance $d = 0.1$ m from the wire as in the figure. The loop encloses a rectangular surface $\mathcal{R}$, which is oriented by normal vectors pointing out of the page.

(a) Calculate the flux $\Phi(t)$ of **B** through $\mathcal{R}$.

(b) Use Faraday's Law to determine the voltage drop (in volts) around the loop $\mathcal{C}$.

Solution We choose coordinates (x, y) on rectangle $\mathcal{R}$ as in Figure 13, so that y is the distance from the wire and $\mathcal{R}$ is the region

$$0 \le x \le L, \qquad d \le y \le H + d$$

Our parametrization of $\mathcal{R}$ is simply $G(x, y) = (x, y)$, for which the normal vector **n** is the unit vector perpendicular to $\mathcal{R}$, pointing out of the page. The magnetic field **B** at $P = (x, y)$ has magnitude $\dfrac{\mu_0 |i|}{2\pi y}$. It points out of the page in the direction of **n** when i is positive and into the page when i is negative. Thus,

$$\mathbf{B} = \frac{\mu_0 i}{2\pi y}\mathbf{n} \qquad \text{and} \qquad \mathbf{B} \cdot \mathbf{n} = \frac{\mu_0 i}{2\pi y}$$

(a) The flux $\Phi(t)$ of **B** through $\mathcal{R}$ at time t is

$$\begin{aligned}
\Phi(t) &= \iint_{\mathcal{R}} \mathbf{B} \cdot d\mathbf{S} = \int_{x=0}^{L} \int_{y=d}^{H+d} \mathbf{B} \cdot \mathbf{n}\, dy\, dx \\
&= \int_{x=0}^{L} \int_{y=d}^{H+d} \frac{\mu_0 i}{2\pi y}\, dy\, dx = \frac{\mu_0 L i}{2\pi} \int_{y=d}^{H+d} \frac{dy}{y} \\
&= \frac{\mu_0 L}{2\pi}\left(\ln \frac{H+d}{d}\right) i \\
&= \frac{\mu_0 (1.2)}{2\pi}\left(\ln \frac{0.8}{0.1}\right) 28\cos(400t)
\end{aligned}$$

With $\mu_0 = 4\pi \cdot 10^{-7}$, we obtain

$$\Phi(t) \approx 1.4 \times 10^{-5} \cos(400t) \text{ T-m}^2$$

(b) By Faraday's Law [Eq. (5)], the voltage drop around the rectangular loop $\mathcal{C}$, oriented in the counterclockwise direction, is

$$\int_{\mathcal{C}} \mathbf{E} \cdot d\mathbf{s} = -\frac{d\Phi}{dt} \approx -(1.4 \times 10^{-5})\,(400)\sin(400t) = -0.0056\sin(400t)\ \text{V} \quad ■$$

17.5 SUMMARY

- A surface $\mathcal{S}$ is *oriented* if a continuously varying unit normal vector $\mathbf{e_n}(P)$ is specified at each point on $\mathcal{S}$. This distinguishes an "outward" direction on the surface.
- The integral of a vector field $\mathbf{F}$ over an oriented surface $\mathcal{S}$ is defined as the integral of the normal component $\mathbf{F} \cdot \mathbf{e_n}$ over $\mathcal{S}$.
- Vector surface integrals are computed using the formula

$$\iint_{\mathcal{S}} \mathbf{F} \cdot d\mathbf{S} = \iint_{\mathcal{D}} \mathbf{F}(G(u, v)) \cdot \mathbf{n}(u, v)\, du\, dv$$

Here, $G(u, v)$ is a parametrization of $\mathcal{S}$ such that $\mathbf{n}(u, v) = \mathbf{T}_u \times \mathbf{T}_v$ points in the direction of the unit normal vector specified by the orientation.
- The surface integral of a vector field $\mathbf{F}$ over $\mathcal{S}$ is also called the *flux* of $\mathbf{F}$ through G. If $\mathbf{F}$ is the velocity field of a fluid, then $\iint_{\mathcal{S}} \mathbf{F} \cdot d\mathbf{S}$ is the rate at which fluid flows through $\mathcal{S}$ per unit time.

17.5 EXERCISES

Preliminary Questions

1. Let $\mathbf{F}$ be a vector field and $G(u, v)$ a parametrization of a surface $\mathcal{S}$, and set $\mathbf{n} = \mathbf{T}_u \times \mathbf{T}_v$. Which of the following is the normal component of $\mathbf{F}$?

(a) $\mathbf{F} \cdot \mathbf{n}$ **(b)** $\mathbf{F} \cdot \mathbf{e_n}$

2. The vector surface integral $\iint_{\mathcal{S}} \mathbf{F} \cdot d\mathbf{S}$ is equal to the scalar surface integral of the function (choose the correct answer):

(a) $\|\mathbf{F}\|$

(b) $\mathbf{F} \cdot \mathbf{n}$, where $\mathbf{n}$ is a normal vector

(c) $\mathbf{F} \cdot \mathbf{e_n}$, where $\mathbf{e_n}$ is the unit normal vector

3. $\iint_{\mathcal{S}} \mathbf{F} \cdot d\mathbf{S}$ is zero if (choose the correct answer):

(a) $\mathbf{F}$ is tangent to $\mathcal{S}$ at every point.

(b) $\mathbf{F}$ is perpendicular to $\mathcal{S}$ at every point.

4. If $\mathbf{F}(P) = \mathbf{e_n}(P)$ at each point on $\mathcal{S}$, then $\iint_{\mathcal{S}} \mathbf{F} \cdot d\mathbf{S}$ is equal to (choose the correct answer):

(a) Zero **(b)** Area($\mathcal{S}$) **(c)** Neither

5. Let $\mathcal{S}$ be the disk $x^2 + y^2 \le 1$ in the xy-plane oriented with normal in the positive z-direction. Determine $\iint_{\mathcal{S}} \mathbf{F} \cdot d\mathbf{S}$ for each of the following vector constant fields:

(a) $\mathbf{F} = \langle 1, 0, 0 \rangle$ **(b)** $\mathbf{F} = \langle 0, 0, 1 \rangle$ **(c)** $\mathbf{F} = \langle 1, 1, 1 \rangle$

6. Estimate $\iint_{\mathcal{S}} \mathbf{F} \cdot d\mathbf{S}$, where $\mathcal{S}$ is a tiny oriented surface of area 0.05 and the value of $\mathbf{F}$ at a sample point in $\mathcal{S}$ is a vector of length 2 making an angle $\frac{\pi}{4}$ with the normal to the surface.

7. A small surface $\mathcal{S}$ is divided into three pieces of area 0.2. Estimate $\iint_{\mathcal{S}} \mathbf{F} \cdot d\mathbf{S}$ if $\mathbf{F}$ is a unit vector field making angles of 85°, 90°, and 95° with the normal at sample points in these three pieces.

Exercises

1. Let $\mathbf{F} = \langle z, 0, y \rangle$ and let $\mathcal{S}$ be the oriented surface parametrized by $G(u, v) = (u^2 - v, u, v^2)$ for $0 \le u \le 2$, $-1 \le v \le 4$. Calculate:

(a) $\mathbf{n}$ and $\mathbf{F} \cdot \mathbf{n}$ as functions of u and v

(b) The normal component of $\mathbf{F}$ to the surface at $P = (3, 2, 1) = G(2, 1)$

(c) $\iint_{\mathcal{S}} \mathbf{F} \cdot d\mathbf{S}$

2. Let $\mathbf{F} = \langle y, -x, x^2 + y^2 \rangle$ and let $\mathcal{S}$ be the portion of the paraboloid $z = x^2 + y^2$ where $x^2 + y^2 \le 3$.

(a) Show that if $\mathcal{S}$ is parametrized in polar variables $x = r\cos\theta$, $y = r\sin\theta$, then $\mathbf{F} \cdot \mathbf{n} = r^3$.

(b) Show that $\iint_S \mathbf{F} \cdot d\mathbf{S} = \int_0^{2\pi} \int_0^3 r^3\, dr\, d\theta$ and evaluate.

3. Let $\mathcal{S}$ be the unit square in the xy-plane shown in Figure 14, oriented with the normal pointing in the positive z-direction. Estimate

$$\iint_S \mathbf{F} \cdot d\mathbf{S}$$

where $\mathbf{F}$ is a vector field whose values at the labeled points are

$$\mathbf{F}(A) = \langle 2, 6, 4\rangle, \qquad \mathbf{F}(B) = \langle 1, 1, 7\rangle$$

$$\mathbf{F}(C) = \langle 3, 3, -3\rangle, \qquad \mathbf{F}(D) = \langle 0, 1, 8\rangle$$

4. Suppose that $\mathcal{S}$ is a surface in $\mathbf{R}^3$ with a parametrization G whose domain $\mathcal{D}$ is the square in Figure 14. The values of a function f, a vector field $\mathbf{F}$, and the normal vector $\mathbf{n} = \mathbf{T}_u \times \mathbf{T}_v$ at $G(P)$ are given for the four sample points in $\mathcal{D}$ in the following table. Estimate the surface integrals of f and $\mathbf{F}$ over $\mathcal{S}$.

Point P in $\mathcal{D}$	f	$\mathbf{F}$	$\mathbf{n}$
A	3	$\langle 2, 6, 4\rangle$	$\langle 1, 1, 1\rangle$
B	1	$\langle 1, 1, 7\rangle$	$\langle 1, 1, 0\rangle$
C	2	$\langle 3, 3, -3\rangle$	$\langle 1, 0, -1\rangle$
D	5	$\langle 0, 1, 8\rangle$	$\langle 2, 1, 0\rangle$

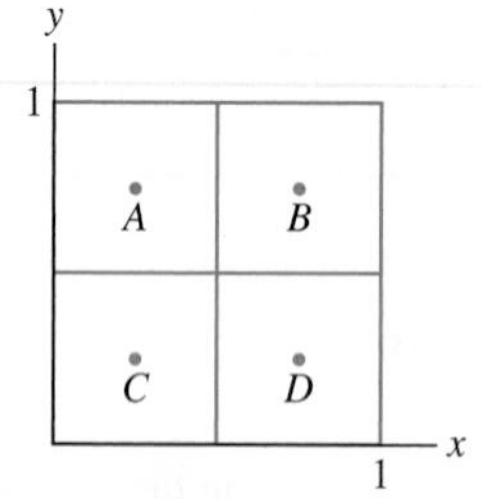

FIGURE 14

In Exercises 5–17, compute $\iint_S \mathbf{F} \cdot d\mathbf{S}$ *for the given oriented surface.*

5. $\mathbf{F} = \langle y, z, x\rangle$, plane $3x - 4y + z = 1$, $0 \le x \le 1, 0 \le y \le 1$, upward-pointing normal

6. $\mathbf{F} = \langle e^z, z, x\rangle$, $G(r, s) = (rs, r + s, r)$, $0 \le r \le 1, 0 \le s \le 1$, oriented by $\mathbf{T}_r \times \mathbf{T}_s$

7. $\mathbf{F} = \langle 0, 3, x\rangle$, part of sphere $x^2 + y^2 + z^2 = 9$, where $x \ge 0,\ y \ge 0,\ z \ge 0$ outward-pointing normal

8. $\mathbf{F} = \langle x, y, z\rangle$, part of sphere $x^2 + y^2 + z^2 = 1$, where $\frac{1}{2} \le z \le \frac{\sqrt{3}}{2}$, inward-pointing normal

9. $\mathbf{F} = \langle z, z, x\rangle$, $z = 9 - x^2 - y^2$, $x \ge 0,\ y \ge 0,\ z \ge 0$ upward-pointing normal

10. $\mathbf{F} = \langle \sin y, \sin z, yz\rangle$, rectangle $0 \le y \le 2,\ 0 \le z \le 3$ in the (y, z)-plane, normal pointing in negative x-direction

11. $\mathbf{F} = y^2\mathbf{i} + 2\mathbf{j} - x\mathbf{k}$, portion of the plane $x + y + z = 1$ in the octant $x, y, z \ge 0$, upward-pointing normal

12. $\mathbf{F} = \langle x, y, e^z\rangle$, cylinder $x^2 + y^2 = 4$, $1 \le z \le 5$, outward-pointing normal

13. $\mathbf{F} = \langle xz, yz, z^{-1}\rangle$, disk of radius 3 at height 4 parallel to the xy-plane, upward-pointing normal

14. $\mathbf{F} = \langle xy, y, 0\rangle$, cone $z^2 = x^2 + y^2$, $x^2 + y^2 \le 4$, $z \ge 0$, downward-pointing normal

15. $\mathbf{F} = \langle 0, 0, e^{y+z}\rangle$, boundary of unit cube $0 \le x \le 1, 0 \le y \le 1, 0 \le z \le 1$, outward-pointing normal

16. $\mathbf{F} = \langle 0, 0, z^2\rangle$, $G(u, v) = (u\cos v, u \sin v, v)$, $0 \le u \le 1$, $0 \le v \le 2\pi$, upward-pointing normal

17. $\mathbf{F} = \langle y, z, 0\rangle$, $G(u, v) = (u^3 - v, u + v, v^2)$, $0 \le u \le 2$, $0 \le v \le 3$, downward-pointing normal

18. Let $\mathcal{S}$ be the oriented half-cylinder in Figure 15. In (a)–(f), determine whether $\iint_S \mathbf{F} \cdot d\mathbf{S}$ is positive, negative, or zero. Explain your reasoning.

(a) $\mathbf{F} = \mathbf{i}$ **(b)** $\mathbf{F} = \mathbf{j}$ **(c)** $\mathbf{F} = \mathbf{k}$

(d) $\mathbf{F} = y\mathbf{i}$ **(e)** $\mathbf{F} = -y\mathbf{j}$ **(f)** $\mathbf{F} = x\mathbf{j}$

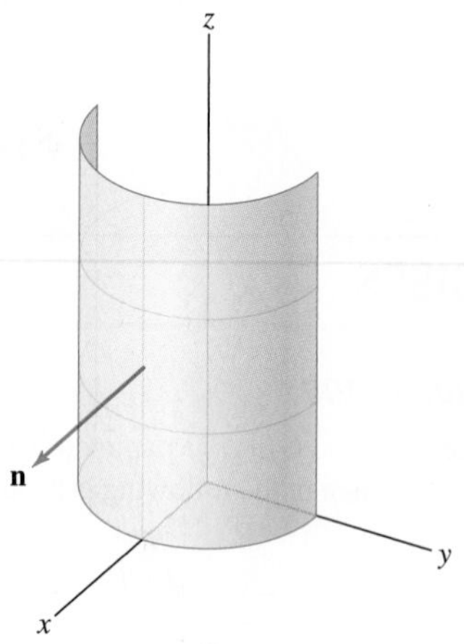

FIGURE 15

19. Let $\mathbf{e_r} = \langle x/r, y/r, z/r\rangle$ be the unit radial vector, where $r = \sqrt{x^2 + y^2 + z^2}$. Calculate the integral of $\mathbf{F} = e^{-r}\mathbf{e_r}$ over:

(a) The upper hemisphere of $x^2 + y^2 + z^2 = 9$, outward-pointing normal.

(b) The octant $x \ge 0, y \ge 0, z \ge 0$ of the unit sphere centered at the origin.

20. Show that the flux of $\mathbf{F} = \dfrac{\mathbf{e}_r}{r^2}$ through a sphere centered at the origin does not depend on the radius of the sphere.

21. The electric field due to a point charge located at the origin in $\mathbf{R}^3$ is $\mathbf{E} = k\dfrac{\mathbf{e}_r}{r^2}$, where $r = \sqrt{x^2 + y^2 + z^2}$ and k is a constant. Calculate the flux of $\mathbf{E}$ through the disk D of radius 2 parallel to the xy-plane with center $(0, 0, 3)$.

22. Let $\mathcal{S}$ be the ellipsoid $\left(\frac{x}{4}\right)^2 + \left(\frac{y}{3}\right)^2 + \left(\frac{z}{2}\right)^2 = 1$. Calculate the flux of $\mathbf{F} = z\mathbf{i}$ over the portion of $\mathcal{S}$ where $x, y, z \le 0$ with upward-pointing normal. *Hint:* Parametrize $\mathcal{S}$ using a modified form of spherical coordinates (θ, ϕ).

23. Let $\mathbf{v} = z\mathbf{k}$ be the velocity field (in meters per second) of a fluid in $\mathbf{R}^3$. Calculate the flow rate (in cubic meters per second) through the upper hemisphere ($z \geq 0$) of the sphere $x^2 + y^2 + z^2 = 1$.

24. Calculate the flow rate of a fluid with velocity field $\mathbf{v} = \langle x, y, x^2 y \rangle$ (in m/s) through the portion of the ellipse $\left(\frac{x}{2}\right)^2 + \left(\frac{y}{3}\right)^2 = 1$ in the xy-plane, where $x, y \geq 0$, oriented with the normal in the positive z-direction.

In Exercises 25–26, let $\mathcal{T}$ be the triangular region with vertices $(1, 0, 0)$, $(0, 1, 0)$, *and* $(0, 0, 1)$ *oriented with upward-pointing normal vector (Figure 16). Assume distances are in meters.*

25. A fluid flows with constant velocity field $\mathbf{v} = 2\mathbf{k}$ (m/s). Calculate:

(a) The flow rate through $\mathcal{T}$

(b) The flow rate through the projection of $\mathcal{T}$ onto the xy-plane [the triangle with vertices $(0, 0, 0)$, $(1, 0, 0)$, and $(0, 1, 0)$]

26. Calculate the flow rate through $\mathcal{T}$ if $\mathbf{v} = -\mathbf{j}$ m/s.

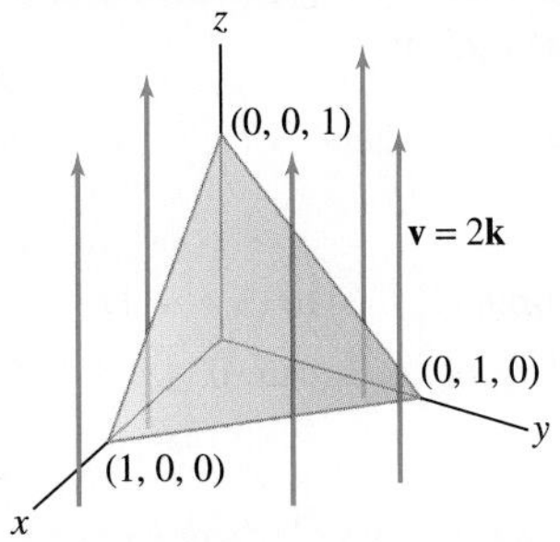

FIGURE 16

27. Prove that if $\mathcal{S}$ is the part of a graph $z = g(x, y)$ lying over a domain $\mathcal{D}$ in the xy-plane, then

$$\iint_{\mathcal{S}} \mathbf{F} \cdot d\mathbf{S} = \iint_{\mathcal{D}} \left(-F_1 \frac{\partial g}{\partial x} - F_2 \frac{\partial g}{\partial y} + F_3 \right) dx\, dy$$

In Exercises 28–29, a varying current $i(t)$ flows through a long straight wire in the xy-plane as in Example 5. The current produces a magnetic field $\mathbf{B}$ *whose magnitude at a distance r from the wire is* $B = \dfrac{\mu_0 i}{2\pi r}$ *T, where* $\mu_0 = 4\pi \cdot 10^{-7}$ T-m/A. *Furthermore,* $\mathbf{B}$ *points into the page at points P in the xy-plane.*

28. Assume that $i(t) = t(12 - t)$ A (t in seconds). Calculate the flux $\Phi(t)$, at time t, of $\mathbf{B}$ through a rectangle of dimensions $L \times H = 3 \times 2$ m whose top and bottom edges are parallel to the wire and whose bottom edge is located $d = 0.5$ m above the wire, similar to Figure 13(B). Then use Faraday's Law to determine the voltage drop around the rectangular loop (the boundary of the rectangle) at time t.

29. Assume that $i = 10e^{-0.1t}$ A (t in seconds). Calculate the flux $\Phi(t)$, at time t, of $\mathbf{B}$ through the isosceles triangle of base 12 cm and height 6 cm whose bottom edge is 3 cm from the wire, as in Figure 17. Assume the triangle is oriented with normal vector pointing out of the page. Use Faraday's Law to determine the voltage drop around the triangular loop (the boundary of the triangle) at time t.

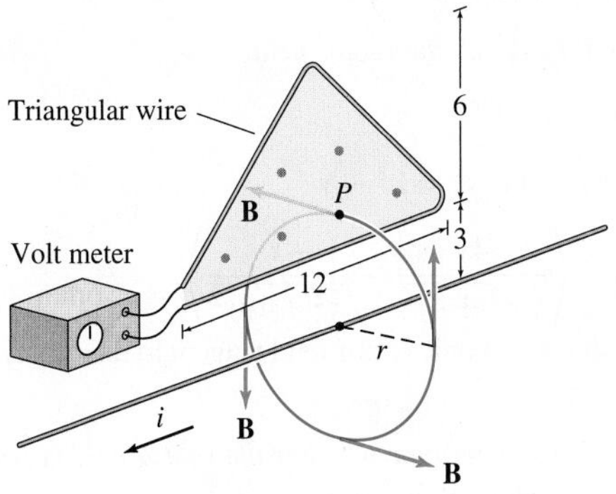

FIGURE 17

Further Insights and Challenges

30. A point mass m is located at the origin. Let Q be the flux of the gravitational field $\mathbf{F} = -Gm\dfrac{\mathbf{e}_r}{r^2}$ through the cylinder $x^2 + y^2 = R^2$ for $a \leq z \leq b$, including the top and bottom (Figure 18). Show that $Q = -4\pi Gm$ if $a < 0 < b$ (m lies inside the cylinder) and $Q = 0$ if $0 < a < b$ (m lies outside the cylinder).

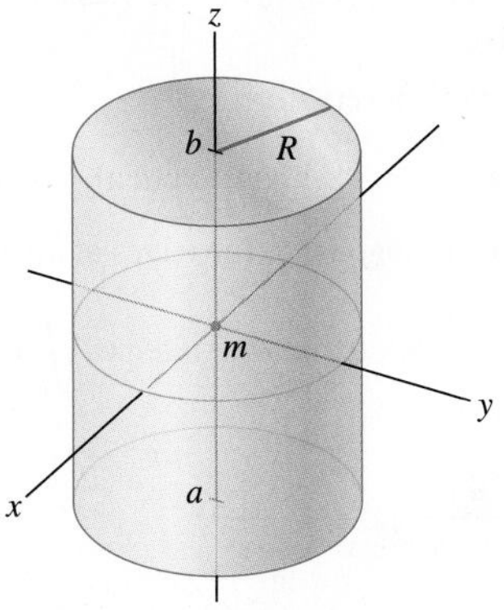

FIGURE 18

In Exercises 31 and 32, let $\mathcal{S}$ be the surface with parametrization

$$G(u, v) = \left(\left(1 + v \cos \frac{u}{2}\right) \cos u,\ \left(1 + v \cos \frac{u}{2}\right) \sin u,\ v \sin \frac{u}{2} \right)$$

for $0 \leq u \leq 2\pi$, $-\frac{1}{2} \leq v \leq \frac{1}{2}$.

31. CAS Use a computer algebra system.

(a) Plot $\mathcal{S}$ and confirm visually that $\mathcal{S}$ is a Möbius strip.

(b) The intersection of $\mathcal{S}$ with the xy-plane is the unit circle $G(u, 0) = (\cos u, \sin u, 0)$. Verify that the normal vector along this circle is

$$\mathbf{n}(u, 0) = \left\langle \cos u \sin \frac{u}{2}, \sin u \sin \frac{u}{2}, -\cos \frac{u}{2} \right\rangle$$

(c) As u varies from 0 to 2π, the point $G(u, 0)$ moves once around the unit circle, beginning and ending at $G(0, 0) = G(2\pi, 0) = (1, 0, 0)$. Verify that $\mathbf{n}(u, 0)$ is a unit vector that varies continuously but that $\mathbf{n}(2\pi, 0) = -\mathbf{n}(0, 0)$. This shows that $\mathcal{S}$ is not orientable—that is, it is not possible to choose a nonzero normal vector at each point on $\mathcal{S}$

in a continuously varying manner (if it were possible, the unit normal vector would return to itself rather than to its negative when carried around the circle).

32. CAS We cannot integrate vector fields over $\mathcal{S}$ because $\mathcal{S}$ is not orientable, but it is possible to integrate functions over $\mathcal{S}$. Using a computer algebra system:

(a) Verify that

$$\|\mathbf{n}(u, v)\|^2 = 1 + \frac{3}{4}v^2 + 2v\cos\frac{u}{2} + \frac{1}{2}v^2\cos u$$

(b) Compute the surface area of $\mathcal{S}$ to four decimal places.

(c) Compute $\iint_{\mathcal{S}} (x^2 + y^2 + z^2)\, dS$ to four decimal places.

CHAPTER REVIEW EXERCISES

1. Compute the vector assigned to the point $P = (-3, 5)$ by the vector field:

(a) $\mathbf{F} = \langle xy, y - x\rangle$

(b) $\mathbf{F} = \langle 4, 8\rangle$

(c) $\mathbf{F} = \langle 3^{x+y}, \log_2(x + y)\rangle$

2. Find a vector field $\mathbf{F}$ in the plane such that $\|\mathbf{F}(x, y)\| = 1$ and $\mathbf{F}(x, y)$ is orthogonal to $\mathbf{G}(x, y) = \langle x, y\rangle$ for all x, y.

In Exercises 3–6, sketch the vector field.

3. $\mathbf{F}(x, y) = \langle y, 1\rangle$

4. $\mathbf{F}(x, y) = \langle 4, 1\rangle$

5. ∇V, where $V(x, y) = x^2 - y$

6. $\mathbf{F}(x, y) = \left\langle \frac{4y}{\sqrt{x^2 + 4y^2}}, \frac{-x}{\sqrt{x^2 + 16y^2}} \right\rangle$

Hint: Show that $\mathbf{F}$ is a unit vector field tangent to the family of ellipses $x^2 + 4y^2 = c^2$.

In Exercises 7–15, determine whether the vector field is conservative, and if so, find a potential function.

7. $\mathbf{F}(x, y) = \langle x^2y, y^2x\rangle$

8. $\mathbf{F}(x, y) = \langle 4x^3y^5, 5x^4y^4\rangle$

9. $\mathbf{F}(x, y, z) = \langle \sin x, e^y, z\rangle$

10. $\mathbf{F}(x, y, z) = \langle 2, 4, e^z\rangle$

11. $\mathbf{F}(x, y, z) = \langle xyz, \frac{1}{2}x^2z, 2z^2y\rangle$

12. $\mathbf{F}(x, y) = \langle y^4x^3, x^4y^3\rangle$

13. $\mathbf{F}(x, y, z) = \left\langle \frac{y}{1 + x^2}, \tan^{-1} x, 2z \right\rangle$

14. $\mathbf{F}(x, y, z) = \left\langle \frac{2xy}{x^2 + z}, \ln(x^2 + z), \frac{y}{x^2 + z} \right\rangle$

15. $\mathbf{F}(x, y, z) = \langle xe^{2x}, ye^{2z}, ze^{2y}\rangle$

16. Find a conservative vector field of the form $\mathbf{F} = \langle g(y), h(x)\rangle$ such that $\mathbf{F}(0, 0) = \langle 1, 1\rangle$, where $g(y)$ and $h(x)$ are differentiable functions. Determine all such vector fields.

In Exercises 17–20, compute the line integral $\int_C f(x, y)\, ds$ for the given function and path or curve.

17. $f(x, y) = xy$, the path $\mathbf{c}(t) = (t, 2t - 1)$ for $0 \le t \le 1$

18. $f(x, y) = x - y$, the unit semicircle $x^2 + y^2 = 1$, $y \ge 0$

19. $f(x, y, z) = e^x - \frac{y}{2\sqrt{2}z}$, the path $\mathbf{c}(t) = \left(\ln t, \sqrt{2}t, \frac{1}{2}t^2\right)$ for $1 \le t \le 2$

20. $f(x, y, z) = x + 2y + z$, the helix $\mathbf{c}(t) = (\cos t, \sin t, t)$ for $-1 \le t \le 3$

21. Find the total mass of an L-shaped rod consisting of the segments $(2t, 2)$ and $(2, 2 - 2t)$ for $0 \le t \le 1$ (length in centimeters) with mass density $\rho(x, y) = x^2y$ g/cm.

22. Calculate $\mathbf{F} = \nabla V$, where $V(x, y, z) = xye^z$, and compute $\int_C \mathbf{F} \cdot d\mathbf{s}$, where:

(a) C is any curve from $(1, 1, 0)$ to $(3, e, -1)$.

(b) C is the boundary of the square $0 \le x \le 1$, $0 \le y \le 1$ oriented counterclockwise.

23. Calculate $\int_{C_1} y\, dx + x^2y\, dy$, where C_1 is the oriented curve in Figure 1(A).

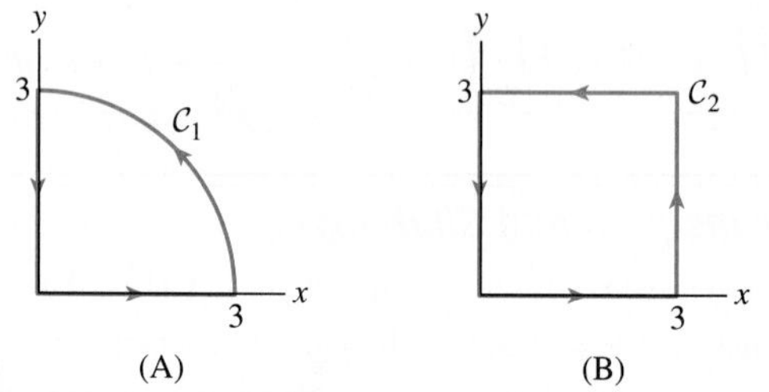

FIGURE 1

24. Let $\mathbf{F}(x, y) = \langle 9y - y^3, e^{\sqrt{y}}(x^2 - 3x)\rangle$ and let C_2 be the oriented curve in Figure 1(B).

(a) Show that $\mathbf{F}$ is not conservative.

(b) Show that $\int_{C_2} \mathbf{F} \cdot d\mathbf{s} = 0$ without explicitly computing the integral. *Hint:* Show that $\mathbf{F}$ is orthogonal to the edges along the square.

In Exercises 25–28, compute the line integral $\int_{\mathbf{c}} \mathbf{F} \cdot d\mathbf{s}$ for the given vector field and path.

25. $\mathbf{F}(x, y) = \left\langle \frac{2y}{x^2 + 4y^2}, \frac{x}{x^2 + 4y^2} \right\rangle$,

the path $\mathbf{c}(t) = \left(\cos t, \frac{1}{2}\sin t\right)$ for $0 \le t \le 2\pi$

26. $\mathbf{F}(x, y) = \langle 2xy, x^2 + y^2 \rangle$, the part of the unit circle in the first quadrant oriented counterclockwise.

27. $\mathbf{F}(x, y) = \langle x^2 y, y^2 z, z^2 x \rangle$, the path $\mathbf{c}(t) = \left(e^{-t}, e^{-2t}, e^{-3t}\right)$ for $0 \le t < \infty$

28. $\mathbf{F} = \nabla V$, where $V(x, y, z) = 4x^2 \ln(1 + y^4 + z^2)$, the path $\mathbf{c}(t) = \left(t^3, \ln(1 + t^2), e^t\right)$ for $0 \le t \le 1$

29. Consider the line integrals $\int_{\mathbf{c}} \mathbf{F} \cdot d\mathbf{s}$ for the vector fields $\mathbf{F}$ and paths $\mathbf{c}$ in Figure 2. Which two of the line integrals appear to have a value of zero? Which of the other two appears to have a negative value?

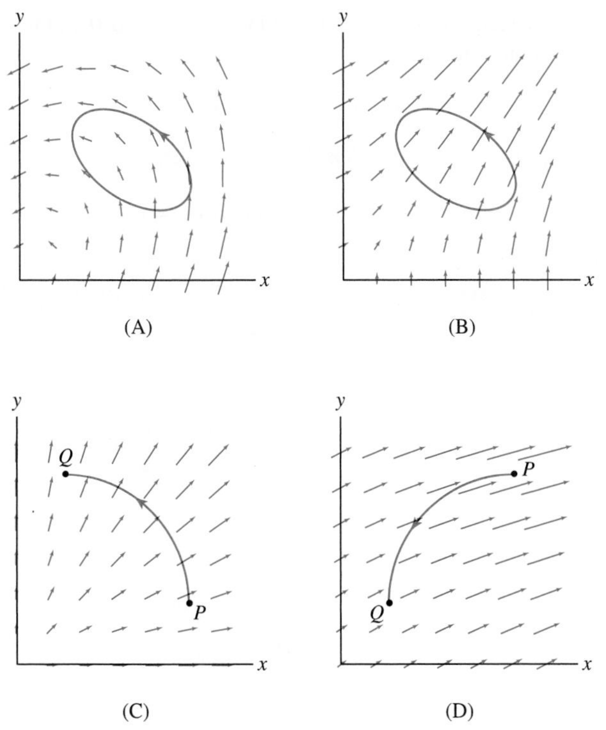

FIGURE 2

30. Calculate the work required to move an object from $P = (1, 1, 1)$ to $Q = (3, -4, -2)$ against the force field $\mathbf{F}(x, y, z) = -12r^{-4}\langle x, y, z \rangle$ (distance in meters, force in newtons), where $r = \sqrt{x^2 + y^2 + z^2}$. *Hint:* Find a potential function for $\mathbf{F}$.

31. Find constants a, b, c such that

$$G(u, v) = (u + av, bu + v, 2u - c)$$

parametrizes the plane $3x - 4y + z = 5$. Calculate $\mathbf{T}_u$, $\mathbf{T}_v$, and $\mathbf{n}(u, v)$.

32. Calculate the integral of $f(x, y, z) = e^z$ over the portion of the plane $x + 2y + 2z = 3$, where $x, y, z \ge 0$.

33. Let $\mathcal{S}$ be the surface parametrized by

$$G(u, v) = \left(2u \sin \frac{v}{2}, 2u \cos \frac{v}{2}, 3v\right)$$

for $0 \le u \le 1$ and $0 \le v \le 2\pi$.

(a) Calculate the tangent vectors $\mathbf{T}_u$ and $\mathbf{T}_v$ and the normal vector $\mathbf{n}(u, v)$ at $P = G(1, \frac{\pi}{3})$.
(b) Find the equation of the tangent plane at P.
(c) Compute the surface area of $\mathcal{S}$.

34. CAS Plot the surface with parametrization

$$G(u, v) = (u + 4v, 2u - v, 5uv)$$

for $-1 \le v \le 1$, $-1 \le u \le 1$. Express the surface area as a double integral and use a computer algebra system to compute the area numerically.

35. CAS Express the surface area of the surface $z = 10 - x^2 - y^2$ for $-1 \le x \le 1, -3 \le y \le 3$ as a double integral. Evaluate the integral numerically using a CAS.

36. Evaluate $\iint_{\mathcal{S}} x^2 y \, dS$, where $\mathcal{S}$ is the surface $z = \sqrt{3}x + y^2$, $-1 \le x \le 1, 0 \le y \le 1$.

37. Calculate $\iint_{\mathcal{S}} \left(x^2 + y^2\right) e^{-z} \, dS$, where $\mathcal{S}$ is the cylinder with equation $x^2 + y^2 = 9$ for $0 \le z \le 10$.

38. Let $\mathcal{S}$ be the upper hemisphere $x^2 + y^2 + z^2 = 1, z \ge 0$. For each of the functions (a)–(d), determine whether $\iint_{\mathcal{S}} f \, dS$ is positive, zero, or negative (without evaluating the integral). Explain your reasoning.

(a) $f(x, y, z) = y^3$ **(b)** $f(x, y, z) = z^3$
(c) $f(x, y, z) = xyz$ **(d)** $f(x, y, z) = z^2 - 2$

39. Let $\mathcal{S}$ be a small patch of surface with a parametrization $G(u, v)$ for $0 \le u \le 0.1, 0 \le v \le 0.1$ such that the normal vector $\mathbf{n}(u, v)$ for $(u, v) = (0, 0)$ is $\mathbf{n} = \langle 2, -2, 4 \rangle$. Use Eq. (3) in Section 17.4 to estimate the surface area of $\mathcal{S}$.

40. The upper half of the sphere $x^2 + y^2 + z^2 = 9$ has parametrization $G(r, \theta) = (r \cos\theta, r \sin\theta, \sqrt{9 - r^2})$ in cylindrical coordinates (Figure 3).

(a) Calculate the normal vector $\mathbf{n} = \mathbf{T}_r \times \mathbf{T}_\theta$ at the point $G\left(2, \frac{\pi}{3}\right)$.
(b) Use Eq. (3) in Section 17.4 to estimate the surface area of $G(\mathcal{R})$, where $\mathcal{R}$ is the small domain defined by

$$2 \le r \le 2.1, \qquad \frac{\pi}{3} \le \theta \le \frac{\pi}{3} + 0.05$$

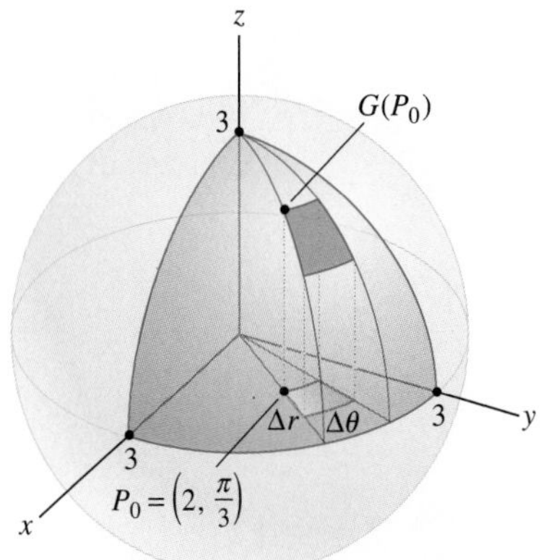

FIGURE 3

In Exercises 41–46, compute $\iint_S \mathbf{F} \cdot d\mathbf{S}$ *for the given oriented surface or parametrized surface.*

41. $\mathbf{F}(x, y, z) = \langle y, x, e^{xz} \rangle$, $x^2 + y^2 = 9, x \geq 0, y \geq 0$, $-3 \leq z \leq 3$, outward-pointing normal

42. $\mathbf{F}(x, y, z) = \langle -y, z, -x \rangle$, $G(u, v) = (u + 3v, v - 2u, 2v + 5)$, $0 \leq u \leq 1, 0 \leq v \leq 1$, upward-pointing normal

43. $\mathbf{F}(x, y, z) = \langle 0, 0, x^2 + y^2 \rangle$, $x^2 + y^2 + z^2 = 4$, $z \geq 0$, outward-pointing normal

44. $\mathbf{F}(x, y, z) = \langle z, 0, z^2 \rangle$, $G(u, v) = (v \cosh u, v \sinh u, v)$, $0 \leq u \leq 1, 0 \leq v \leq 1$, upward-pointing normal

45. $\mathbf{F}(x, y, z) = \langle 0, 0, xze^{xy} \rangle$, $z = xy$, $0 \leq x \leq 1, 0 \leq y \leq 1$, upward-pointing normal

46. $\mathbf{F}(x, y, z) = \langle 0, 0, z \rangle$, $3x^2 + 2y^2 + z^2 = 1$, $z \geq 0$, upward-pointing normal

47. Calculate the total charge on the cylinder

$$x^2 + y^2 = R^2, \qquad 0 \leq z \leq H$$

if the charge density in cylindrical coordinates is $\rho(\theta, z) = Kz^2 \cos^2 \theta$, where K is a constant.

48. Find the flow rate of a fluid with velocity field $\mathbf{v} = \langle 2x, y, xy \rangle$ m/s across the part of the cylinder $x^2 + y^2 = 9$ where $x \geq 0$, $y \geq 0$, and $0 \leq z \leq 4$ (distance in meters).

49. With $\mathbf{v}$ as in Exercise 48, calculate the flow rate across the part of the elliptic cylinder $\frac{x^2}{4} + y^2 = 1$ where $x \geq 0$, $y \geq 0$, and $0 \leq z \leq 4$.

50. Calculate the flux of the vector field $\mathbf{E}(x, y, z) = \langle 0, 0, x \rangle$ through the part of the ellipsoid

$$4x^2 + 9y^2 + z^2 = 36$$

where $z \geq 3$, $x \geq 0$, $y \geq 0$. *Hint:* Use the parametrization

$$G(r, \theta) = \left(3r \cos \theta, 2r \sin \theta, 6\sqrt{1 - r^2}\right)$$

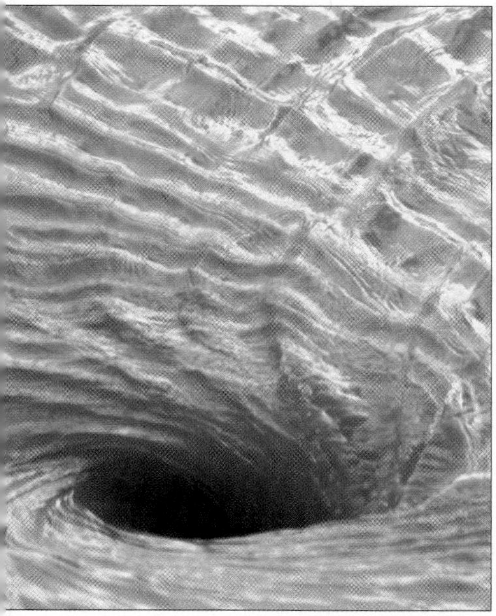

Fluid flows, such as this water vortex, are analyzed using the fundamental theorems of vector analysis.

18 FUNDAMENTAL THEOREMS OF VECTOR ANALYSIS

In this final chapter, we study three generalizations of the Fundamental Theorem of Calculus, known as Green's Theorem, Stokes' Theorem, and the Divergence Theorem. This is a culmination of our efforts to extend the ideas of single-variable calculus to the multivariable setting. However, vector analysis is not so much an endpoint as a gateway to a host of applications, not only in the traditional domains of physics and engineering, but also in biological, earth, and environmental sciences, where an understanding of fluid and aerodynamics, and continuous matter is required.

18.1 Green's Theorem

In Section 17.3, we showed that the circulation of a conservative vector field **F** around every closed path is zero. For vector fields in the plane, Green's Theorem tells us what happens when **F** is not conservative.

To formulate Green's Theorem, we need some notation. Consider a domain $\mathcal{D}$ whose boundary $\mathcal{C}$ is a **simple closed curve**—that is, a closed curve that does not intersect itself (Figure 1). We follow standard usage and denote the boundary curve $\mathcal{C}$ by $\partial\mathcal{D}$. The counterclockwise orientation of $\partial\mathcal{D}$ is called the **boundary orientation**. When you traverse the boundary in this direction, the domain lies to your left (Figure 1).

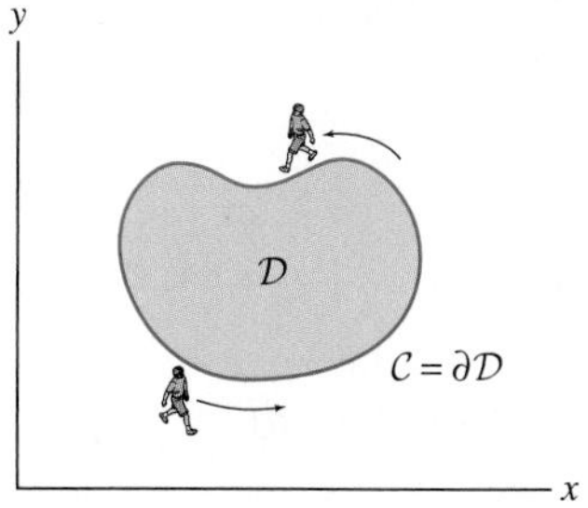

FIGURE 1 The boundary of $\mathcal{D}$ is a simple closed curve $\mathcal{C}$ that is denoted $\partial\mathcal{D}$. The boundary is oriented in the counterclockwise direction.

Recall that we have two notations for the line integral of $\mathbf{F} = \langle F_1, F_2 \rangle$:

$$\int_{\mathcal{C}} \mathbf{F} \cdot d\mathbf{s} \qquad \text{and} \qquad \int_{\mathcal{C}} F_1\,dx + F_2\,dy$$

If $\mathcal{C}$ is parametrized by $\mathbf{c}(t) = (x(t), y(t))$ for $a \le t \le b$, then

$$dx = x'(t)\,dt, \qquad dy = y'(t)\,dt$$

$$\int_{\mathcal{C}} F_1\,dx + F_2\,dy = \int_a^b \big(F_1(x(t), y(t))x'(t) + F_2(x(t), y(t))y'(t)\big)\,dt \qquad \boxed{1}$$

Throughout this chapter, we assume that the components of all vector fields have continuous partial derivatives, and also that $\mathcal{C}$ is smooth ($\mathcal{C}$ has a parametrization with derivatives of all orders) or piecewise smooth (a finite union of smooth curves joined together at corners).

REMINDER The line integral of a vector field over a closed curve is called the "circulation" and is often denoted with the symbol $\oint$.

THEOREM 1 Green's Theorem Let $\mathcal{D}$ be a domain whose boundary $\partial\mathcal{D}$ is a simple closed curve, oriented counterclockwise. Then

$$\oint_{\partial\mathcal{D}} F_1\,dx + F_2\,dy = \iint_{\mathcal{D}} \left(\frac{\partial F_2}{\partial x} - \frac{\partial F_1}{\partial y}\right) dA \qquad \boxed{2}$$

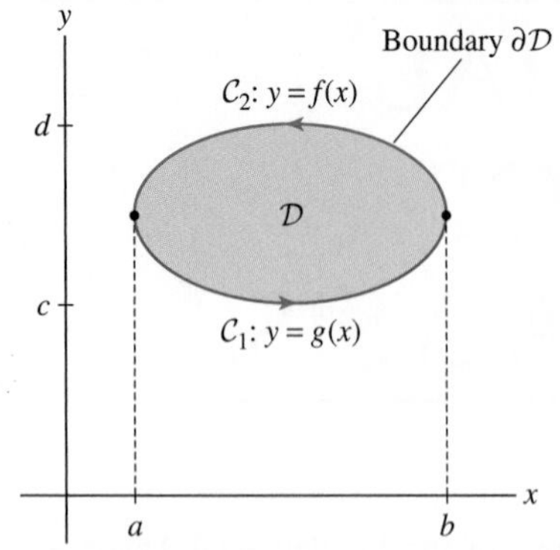

FIGURE 2 The boundary curve $\partial\mathcal{D}$ is the union of the graphs of $y = g(x)$ and $y = f(x)$ oriented counterclockwise.

Proof Because a complete proof is quite technical, we shall make the simplifying assumption that the boundary of $\mathcal{D}$ can be described as the union of two graphs $y = g(x)$ and $y = f(x)$ with $g(x) \le f(x)$ as in Figure 2 and also as the union of two graphs $x = g_1(y)$ and $x = f_1(y)$, with $g_1(y) \le f_1(y)$ as in Figure 3.

Green's Theorem splits up into two equations, one for F_1 and one for F_2:

$$\oint_{\partial\mathcal{D}} F_1\,dx = -\iint_{\mathcal{D}} \frac{\partial F_1}{\partial y}\,dA \qquad \boxed{3}$$

$$\oint_{\partial\mathcal{D}} F_2\,dy = \iint_{\mathcal{D}} \frac{\partial F_2}{\partial x}\,dA \qquad \boxed{4}$$

In other words, Green's Theorem is obtained by adding these two equations. To prove Eq. (3), we write

$$\oint_{\partial\mathcal{D}} F_1\,dx = \oint_{\mathcal{C}_1} F_1\,dx + \oint_{\mathcal{C}_2} F_1\,dx$$

where $\mathcal{C}_1$ is the graph of $y = g(x)$ and $\mathcal{C}_2$ is the graph of $y = f(x)$, oriented as in Figure 2. To compute these line integrals, we parameterize the graphs from left to right using t as parameter:

$$\text{Graph of } y = g(x): \quad \mathbf{c}_1(t) = (t, g(t)), \quad a \le t \le b$$

$$\text{Graph of } y = f(x): \quad \mathbf{c}_2(t) = (t, f(t)), \quad a \le t \le b$$

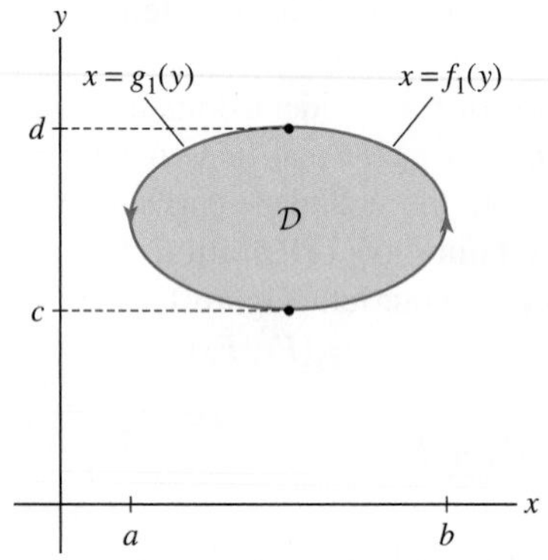

FIGURE 3 The boundary curve $\partial\mathcal{D}$ is also the union of the graphs of $x = g_1(x)$ and $y = f(x)$ oriented counterclockwise.

Since $\mathcal{C}_2$ is oriented from right to left, the line integral over $\partial\mathcal{D}$ is the difference

$$\oint_{\partial\mathcal{D}} F_1\,dx = \int_{\mathbf{c}_1} F_1\,dx - \int_{\mathbf{c}_2} F_1\,dx$$

In both parametrizations, $x = t$, so $dx = dt$, and by Eq. (1)

$$\oint_{\partial\mathcal{D}} F_1\,dx = \int_{t=a}^{b} F_1(t, g(t))\,dt - \int_{t=a}^{b} F_1(t, f(t))\,dt \qquad \boxed{5}$$

Now, the key step is to apply the Fundamental Theorem of Calculus to $\dfrac{\partial F_1}{\partial y}(t, y)$ as a function of y with t held constant:

$$F_1(t, f(t)) - F_1(t, g(t)) = \int_{y=g(t)}^{f(t)} \frac{\partial F_1}{\partial y}(t, y)\,dy$$

Substituting the integral on the right in Eq. (5), we obtain Eq. (3):

$$\oint_{\partial\mathcal{D}} F_1\,dx = -\int_{t=a}^{b}\int_{y=g(t)}^{f(t)} \frac{\partial F_1}{\partial y}(t, y)\,dy\,dt = -\iint_{\mathcal{D}} \frac{\partial F_1}{\partial y}\,dA$$

Eq. (4) is proved in a similar fashion, by expressing $\partial\mathcal{D}$ as the union of the graphs of $x = f_1(y)$ and $x = g_1(y)$ (Figure 3). ■

Recall that if $\mathbf{F} = \nabla V$, then the cross-partial condition is satisfied:

$$\frac{\partial F_2}{\partial x} - \frac{\partial F_1}{\partial y} = 0$$

In this case, Green's Theorem merely confirms what we already know: The line integral of a conservative vector field around any closed curve is zero.

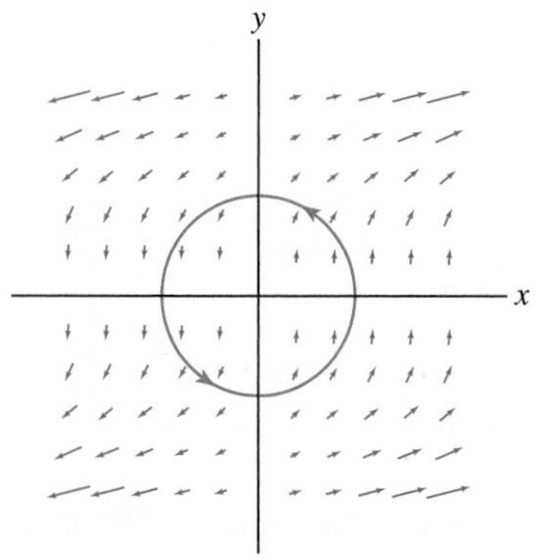

FIGURE 4 The vector field $\mathbf{F} = \langle xy^2, x \rangle$.

Green's Theorem states:

$$\oint_{\partial\mathcal{D}} F_1\,dx + F_2\,dy = \iint_{\mathcal{D}} \left(\frac{\partial F_2}{\partial x} - \frac{\partial F_1}{\partial y}\right) dA$$

REMINDER To integrate $\cos^2\theta$, use the identity $\cos^2\theta = \frac{1}{2}(1+\cos 2\theta)$.

EXAMPLE 1 Verifying Green's Theorem Verify Green's Theorem for the line integral along the unit circle $\mathcal{C}$, oriented counterclockwise (Figure 4):

$$\oint_{\mathcal{C}} xy^2\,dx + x\,dy$$

Solution

***Step 1.* Evaluate the line integral directly.**

We use the standard parametrization of the unit circle:

$$x = \cos\theta, \qquad y = \sin\theta$$
$$dx = -\sin\theta\,d\theta, \qquad dy = \cos\theta\,d\theta$$

The integrand in the line integral is

$$xy^2\,dx + x\,dy = \cos\theta\sin^2\theta(-\sin\theta\,d\theta) + \cos\theta(\cos\theta\,d\theta)$$
$$= \left(-\cos\theta\sin^3\theta + \cos^2\theta\right)d\theta$$

and

$$\oint_{\mathcal{C}} xy^2\,dx + x\,dy = \int_0^{2\pi}\left(-\cos\theta\sin^3\theta + \cos^2\theta\right)d\theta$$
$$= -\frac{\sin^4\theta}{4}\bigg|_0^{2\pi} + \frac{1}{2}\left(\theta + \frac{1}{2}\sin 2\theta\right)\bigg|_0^{2\pi}$$
$$= 0 + \frac{1}{2}(2\pi + 0) = \boxed{\pi}$$

***Step 2.* Evaluate the line integral using Green's Theorem.**

In this example, $F_1 = xy^2$ and $F_2 = x$, so

$$\frac{\partial F_2}{\partial x} - \frac{\partial F_1}{\partial y} = \frac{\partial}{\partial x}x - \frac{\partial}{\partial y}xy^2 = 1 - 2xy$$

According to Green's Theorem,

$$\oint_{\mathcal{C}} xy^2\,dx + x\,dy = \iint_{\mathcal{D}}\left(\frac{\partial F_2}{\partial x} - \frac{\partial F_1}{\partial y}\right)dA = \iint_{\mathcal{D}}(1 - 2xy)\,dA$$

where $\mathcal{D}$ is the disk $x^2 + y^2 \le 1$ enclosed by $\mathcal{C}$. The integral of $2xy$ over $\mathcal{D}$ is zero by symmetry—the contributions for positive and negative x cancel. We can check this directly:

$$\iint_{\mathcal{D}}(-2xy)\,dA = -2\int_{x=-1}^{1}\int_{y=-\sqrt{1-x^2}}^{\sqrt{1-x^2}} xy\,dy\,dx = -\int_{x=-1}^{1} xy^2\bigg|_{y=-\sqrt{1-x^2}}^{\sqrt{1-x^2}} dx = 0$$

Therefore,

$$\iint_{\mathcal{D}}\left(\frac{\partial F_2}{\partial x} - \frac{\partial F_1}{\partial y}\right)dA = \iint_{\mathcal{D}} 1\,dA = \text{Area}(\mathcal{D}) = \boxed{\pi}$$

This agrees with the value in Step 1, so Green's Theorem is verified in this case. ■

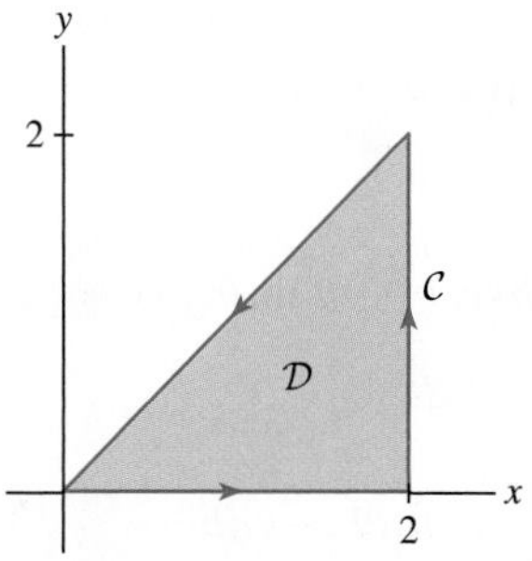

FIGURE 5 The region $\mathcal{D}$ is described by $0 \le x \le 2,\ 0 \le y \le x$.

EXAMPLE 2 Computing a Line Integral Using Green's Theorem Compute the circulation of $\mathbf{F} = \left\langle \sin x, x^2y^3 \right\rangle$ around the triangular path $\mathcal{C}$ in Figure 5.

Solution To compute the line integral directly, we would have to parametrize all three sides of the triangle. Instead, we apply Green's Theorem to the domain $\mathcal{D}$ enclosed by the triangle. This domain is described by $0 \le x \le 2,\ 0 \le y \le x$.

Applying Green's Theorem, we obtain

$$\frac{\partial F_2}{\partial x} - \frac{\partial F_1}{\partial y} = \frac{\partial}{\partial x}x^2y^3 - \frac{\partial}{\partial y}\sin x = 2xy^3$$

$$\oint_{\mathcal{C}} \sin x\,dx + x^2y^3\,dy = \iint_{\mathcal{D}} 2xy^3\,dA = \int_0^2 \int_{y=0}^{x} 2xy^3\,dy\,dx$$

$$= \int_0^2 \left(\frac{1}{2}xy^4\Big|_0^x\right) dx = \frac{1}{2}\int_0^2 x^5\,dx = \frac{1}{12}x^6\Big|_0^2 = \frac{16}{3} \quad ■$$

Green's Theorem applied to $\mathbf{F} = \langle -y, x\rangle$ leads to a formula for the area of the domain $\mathcal{D}$ enclosed by a simple closed curve $\mathcal{C}$ (Figure 6). We have

$$\frac{\partial F_2}{\partial x} - \frac{\partial F_1}{\partial y} = \frac{\partial}{\partial x}x - \frac{\partial}{\partial y}(-y) = 2$$

By Green's Theorem, $\oint_{\mathcal{C}} -y\,dx + x\,dy = \iint_{\mathcal{D}} 2\,dx\,dy = 2\,\text{Area}(\mathcal{D})$. We obtain

$$\text{Area enclosed by } \mathcal{C} = \frac{1}{2}\oint_{\mathcal{C}} x\,dy - y\,dx \qquad \boxed{6}$$

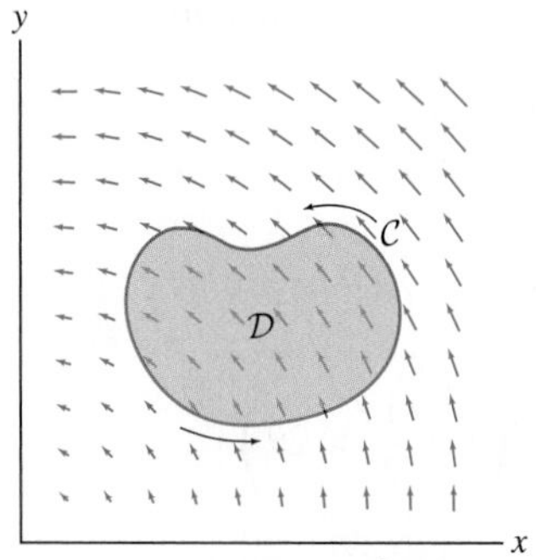

y
x
C
D

FIGURE 6 The line integral $\oint_{\mathcal{C}} x\,dy - y\,dx$ is equal to twice the area enclosed by $\mathcal{C}$.

This remarkable formula tells us how to compute an enclosed area by making measurements only along the boundary. It is the mathematical basis of the **planimeter**, a device that computes the area of an irregular shape when you trace the boundary with a pointer at the end of a movable arm (Figure 7).

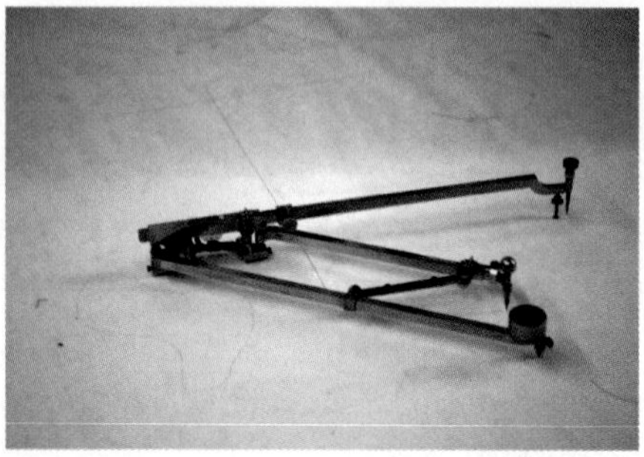

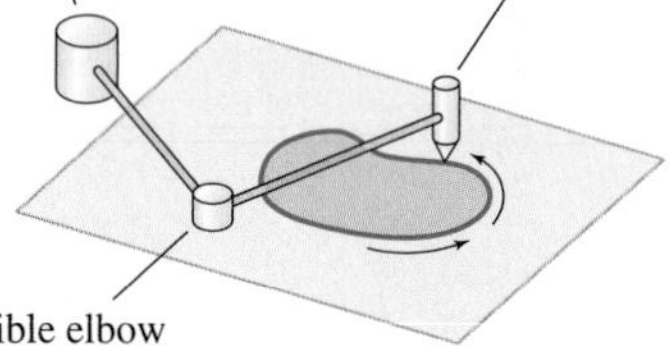

This end of the planimeter is fixed in place.
This end of the planimeter traces the shape.
Flexible elbow

FIGURE 7 A planimeter is a mechanical device used for measuring the areas of irregular shapes.

"Fortunately (for me), I was the only one in the local organization who had even heard of Green's Theorem,...although I was not able to make constructive contributions, I could listen, nod my head and exclaim in admiration at the right places."
John M. Crawford, geophysicist and director of research at Conoco Oil, 1951–1971, writing about his first job interview in 1943, when a scientist visiting the company began speaking about applications of mathematics to oil exploration.

■ **EXAMPLE 3 Computing Area via Green's Theorem** Compute the area of the ellipse $\left(\frac{x}{a}\right)^2 + \left(\frac{y}{b}\right)^2 = 1$ using a line integral.

Solution We parametrize the boundary of the ellipse by

$$x = a\cos\theta, \qquad y = b\sin\theta, \qquad 0 \le \theta < 2\pi$$

and use Eq. (6):

$$x\,dy - y\,dx = (a\cos\theta)(b\cos\theta\,d\theta) - (b\sin\theta)(-a\sin\theta\,d\theta)$$

$$= ab(\cos^2\theta + \sin^2\theta)\,d\theta = ab\,d\theta$$

$$\text{Enclosed area} = \frac{1}{2}\oint_{\mathcal{C}} x\,dy - y\,dx = \frac{1}{2}\int_0^{2\pi} ab\,d\theta = \pi ab$$

This is the standard formula for the area of an ellipse. ■

CONCEPTUAL INSIGHT What is the meaning of the integrand in Green's Theorem? For convenience, we denote this integrand by $\text{curl}_z(\mathbf{F})$:

$$\text{curl}_z(\mathbf{F}) = \frac{\partial F_2}{\partial x} - \frac{\partial F_1}{\partial y}$$

In Section 18.3, we will see that $\text{curl}_z(\mathbf{F})$ is the z-component of a vector field $\text{curl}(\mathbf{F})$ called the "curl" of $\mathbf{F}$. Now apply Green's Theorem to a small domain $\mathcal{D}$ with simple closed boundary curve and let P be a point in $\mathcal{D}$. Because $\text{curl}_z(\mathbf{F})$ is a continuous function, its value does not change much on D if $\mathcal{C}$ is sufficiently small, so to a first approximation, we can replace $\text{curl}_z(\mathbf{F})$ by the constant value $\text{curl}_z(\mathbf{F})(P)$ (Figure 8). Green's Theorem yields the following approximation for the circulation:

$$\begin{aligned} \oint_{\mathcal{C}} \mathbf{F} \cdot d\mathbf{s} &= \iint_{\mathcal{D}} \text{curl}_z(\mathbf{F})\, dA \\ &\approx \text{curl}_z(\mathbf{F})(P) \iint_{\mathcal{D}} dA \\ &\approx \text{curl}_z(\mathbf{F})(P) \cdot \text{Area}(\mathcal{D}) \end{aligned} \qquad \boxed{7}$$

In other words, *the circulation around a small, simple closed curve $\mathcal{C}$ is, to a first-order approximation, equal to the curl times the enclosed area.* Thus, we can think of $\text{curl}_z(\mathbf{F})(P)$ as the **circulation per unit of enclosed area**.

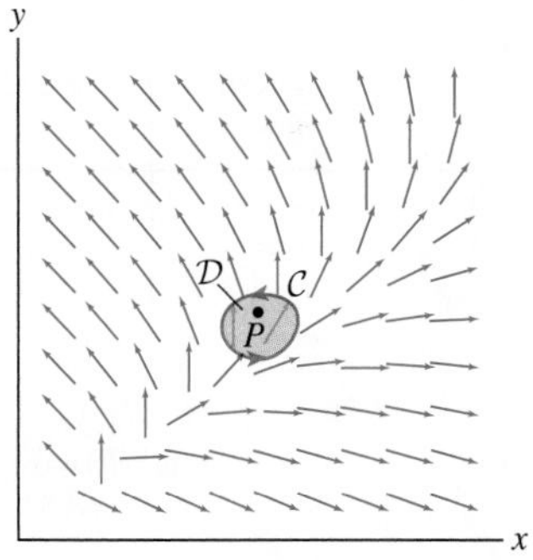

FIGURE 8 The circulation of $\mathbf{F}$ around $\mathcal{C}$ is approximately $\text{curl}_z(\mathbf{F})(P) \cdot \text{Area}(\mathcal{D})$.

GRAPHICAL INSIGHT If we think of $\mathbf{F}$ as the velocity field of a fluid, then we can measure the curl by placing a small paddle wheel in the stream at a point P and observing how fast it rotates (Figure 9). Because the fluid pushes each paddle to move with a velocity equal to the tangential component of $\mathbf{F}$, we can assume that the wheel itself rotates with a velocity v_a equal to the *average tangential component* of $\mathbf{F}$. If the paddle is a circle C of radius r (and hence length $2\pi r$), then the average tangential component of velocity is

$$v_a = \frac{1}{2\pi r} \oint_{\mathcal{C}_r} \mathbf{F} \cdot d\mathbf{s}$$

On the other hand, the paddle encloses an area of πr^2, and for small r, we can apply the approximation (7):

$$v_a \approx \frac{1}{2\pi r}(\pi r^2)\text{curl}_z(\mathbf{F})(P) = \left(\frac{1}{2} r\right) \text{curl}_z(\mathbf{F})(P)$$

Now if an object moves along a circle of radius r with speed v_a, then its angular velocity (in radians per unit time) is $v_a/r \approx \frac{1}{2}\text{curl}_z(\mathbf{F})(P)$. Therefore, *the angular velocity of the paddle wheel is approximately one-half the curl.*

Angular Velocity *An arc of ℓ meters on a circle of radius r meters has radian measure ℓ/r. Therefore, an object moving along the circle with a speed of v meters per second travels v/r radians per second. In other words, the object has angular velocity v/r.*

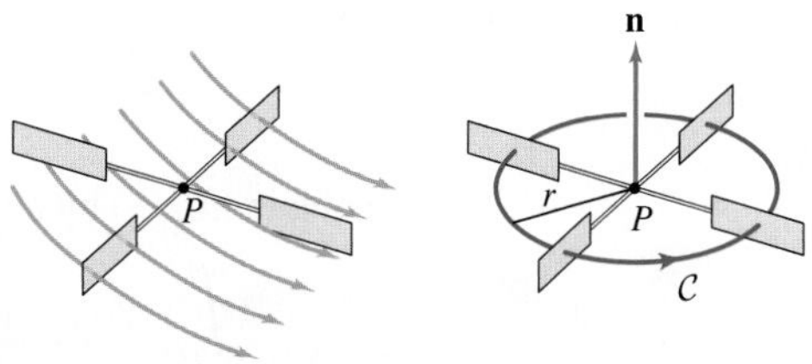

FIGURE 9 The curl is approximately equal to one-half the angular velocity of a small paddle wheel placed at P.

Figure 10 shows vector fields with constant curl. Field (A) describes a fluid rotating counterclockwise around the origin, and field (B) describes a fluid that spirals into the origin. However, a nonzero curl does not mean that the fluid itself is necessarily rotating. It means only that a small paddle wheel would rotate if placed in the fluid. Field (C) is an example of **shear flow** (also known as a Couette flow). It has nonzero curl but does not rotate about any point. Compare with the fields in Figures (D) and (E), which have zero curl.

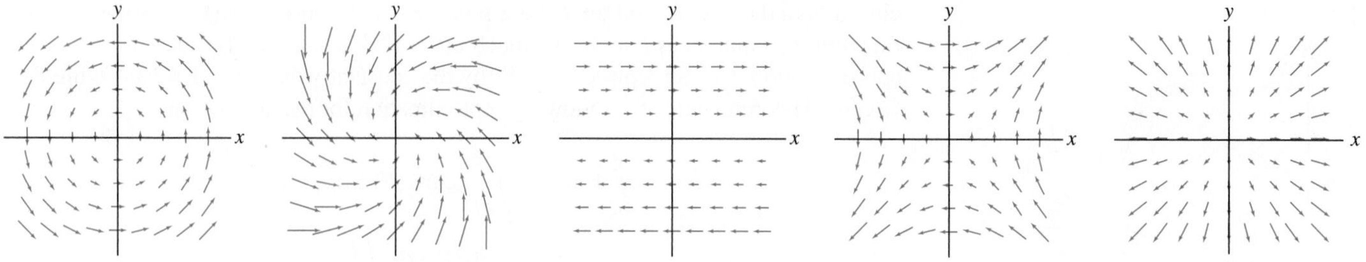

(A) $\mathbf{F} = \langle -y, x \rangle$ $\text{curl}_z(\mathbf{F}) = 2$

(B) $\mathbf{F} = \langle -x - y, x - y \rangle$ $\text{curl}_z(\mathbf{F}) = 2$

(C) $\mathbf{F} = \langle y, 0 \rangle$ $\text{curl}_z(\mathbf{F}) = -1$

(D) $\mathbf{F} = \langle y, x \rangle$ $\text{curl}_z(\mathbf{F}) = 0$

(E) $\mathbf{F} = \langle x, y \rangle$ $\text{curl}_z(\mathbf{F}) = 0$

FIGURE 10

Additivity of Circulation

Circulation around a closed curve has an important additivity property: If we decompose a domain $\mathcal{D}$ into two (or more) non-overlapping domains $\mathcal{D}_1$ and $\mathcal{D}_2$ that intersect only on part of their boundaries as in Figure 11, then

$$\oint_{\partial \mathcal{D}} \mathbf{F} \cdot d\mathbf{s} = \oint_{\partial \mathcal{D}_1} \mathbf{F} \cdot d\mathbf{s} + \oint_{\partial \mathcal{D}_2} \mathbf{F} \cdot d\mathbf{s} \qquad \boxed{8}$$

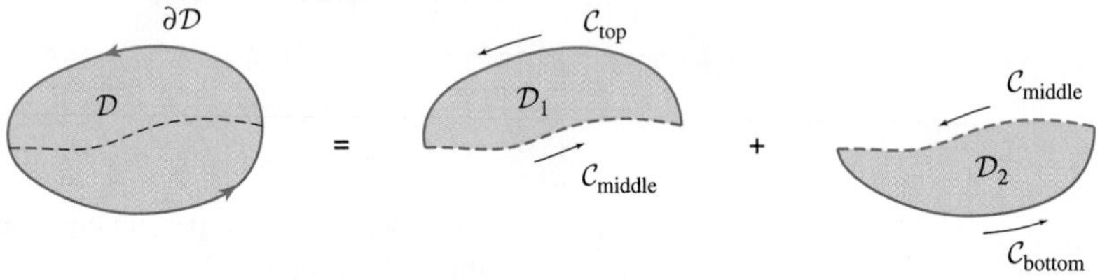

FIGURE 11 The domain $\mathcal{D}$ is the union of $\mathcal{D}_1$ and $\mathcal{D}_2$.

To verify this equation, note first that

$$\oint_{\partial \mathcal{D}} \mathbf{F} \cdot d\mathbf{s} = \int_{\mathcal{C}_{\text{top}}} \mathbf{F} \cdot d\mathbf{s} + \int_{\mathcal{C}_{\text{bot}}} \mathbf{F} \cdot d\mathbf{s}$$

with $\mathcal{C}_{\text{top}}$ and $\mathcal{C}_{\text{bot}}$ as in Figure 11. Then observe that the dashed segment $\mathcal{C}_{\text{middle}}$ occurs in both $\partial \mathcal{D}_1$ and $\partial \mathcal{D}_2$ but with opposite orientations. If $\mathcal{C}_{\text{middle}}$ is oriented right to left, then

$$\oint_{\partial \mathcal{D}_1} \mathbf{F} \cdot d\mathbf{s} = \int_{\mathcal{C}_{\text{top}}} \mathbf{F} \cdot d\mathbf{s} - \int_{\mathcal{C}_{\text{middle}}} \mathbf{F} \cdot d\mathbf{s}$$

$$\oint_{\partial \mathcal{D}_2} \mathbf{F} \cdot d\mathbf{s} = \int_{\mathcal{C}_{\text{bot}}} \mathbf{F} \cdot d\mathbf{s} + \int_{\mathcal{C}_{\text{middle}}} \mathbf{F} \cdot d\mathbf{s}$$

We obtain Eq. (8) by adding these two equations:

$$\oint_{\partial \mathcal{D}_1} \mathbf{F} \cdot d\mathbf{s} + \oint_{\partial \mathcal{D}_2} \mathbf{F} \cdot d\mathbf{s} = \int_{\mathcal{C}_{\text{top}}} \mathbf{F} \cdot d\mathbf{s} + \int_{\mathcal{C}_{\text{bot}}} \mathbf{F} \cdot d\mathbf{s} = \oint_{\partial \mathcal{D}} \mathbf{F} \cdot d\mathbf{s}$$

More General Form of Green's Theorem

Consider a domain $\mathcal{D}$ whose boundary consists of more than one simple closed curve as in Figure 12. As before, $\partial\mathcal{D}$ denotes the boundary of $\mathcal{D}$ with its boundary orientation. In other words, *the region lies to the left as the curve is traversed in the direction specified by the orientation*. For the domains in Figure 12,

$$\partial\mathcal{D}_1 = \mathcal{C}_1 + \mathcal{C}_2, \qquad \partial\mathcal{D}_2 = \mathcal{C}_3 + \mathcal{C}_4 - \mathcal{C}_5$$

The curve $\mathcal{C}_5$ occurs with a minus sign because it is oriented counterclockwise, but the boundary orientation requires a clockwise orientation.

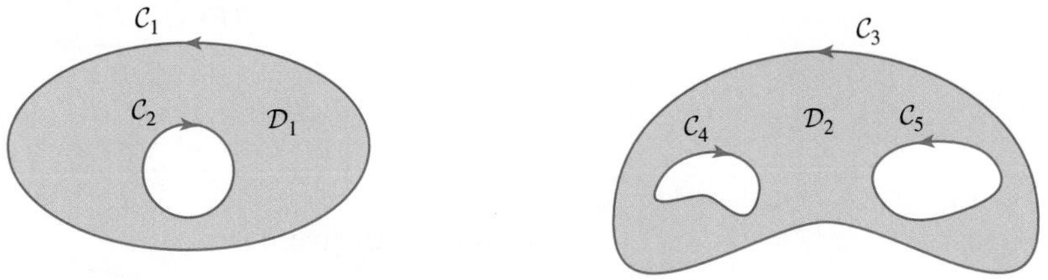

(A) Oriented boundary of $\mathcal{D}_1$ is $\mathcal{C}_1 + \mathcal{C}_2$ (B) Oriented boundary of $\mathcal{D}_2$ is $\mathcal{C}_3 + \mathcal{C}_4 - \mathcal{C}_5$

FIGURE 12

Green's Theorem remains valid for more general domains of this type:

$$\oint_{\partial\mathcal{D}} \mathbf{F}\cdot d\mathbf{s} = \iint_{\mathcal{D}} \left(\frac{\partial F_2}{\partial x} - \frac{\partial F_1}{\partial y}\right) dA \qquad \boxed{9}$$

This equality is proved by decomposing $\mathcal{D}$ into smaller domains each of which is bounded by a simple closed curve. To illustrate, consider the region $\mathcal{D}$ in Figure 13. We decompose $\mathcal{D}$ into domains $\mathcal{D}_1$ and $\mathcal{D}_2$. Then

$$\partial\mathcal{D} = \partial\mathcal{D}_1 + \partial\mathcal{D}_2$$

because the edges common to $\partial\mathcal{D}_1$ and $\partial\mathcal{D}_2$ occur with opposite orientation and therefore cancel. The previous version of Green's Theorem applies to both $\mathcal{D}_1$ and $\mathcal{D}_2$, and thus

$$\begin{aligned}\oint_{\partial\mathcal{D}} \mathbf{F}\cdot d\mathbf{s} &= \int_{\partial\mathcal{D}_1} \mathbf{F}\cdot d\mathbf{s} + \int_{\partial\mathcal{D}_2} \mathbf{F}\cdot d\mathbf{s}\\ &= \iint_{\mathcal{D}_1} \left(\frac{\partial F_2}{\partial x} - \frac{\partial F_1}{\partial y}\right) dA + \iint_{\mathcal{D}_2} \left(\frac{\partial F_2}{\partial x} - \frac{\partial F_1}{\partial y}\right) dA\\ &= \iint_{\mathcal{D}} \left(\frac{\partial F_2}{\partial x} - \frac{\partial F_1}{\partial y}\right) dA\end{aligned}$$

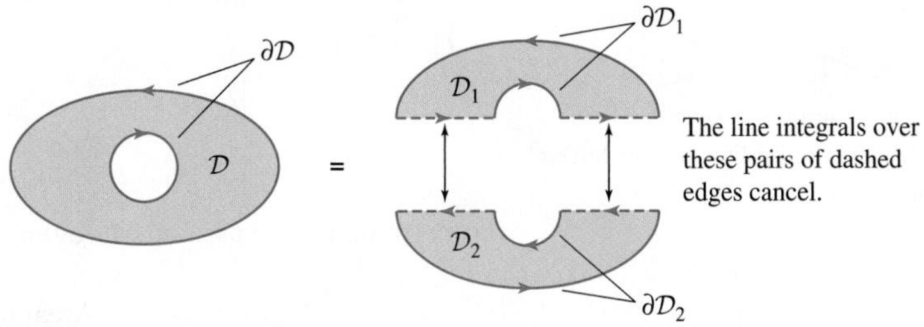

FIGURE 13 The boundary of $\partial\mathcal{D}$ is the sum $\partial\mathcal{D}_1 + \partial\mathcal{D}_2$ because the straight edges cancel.

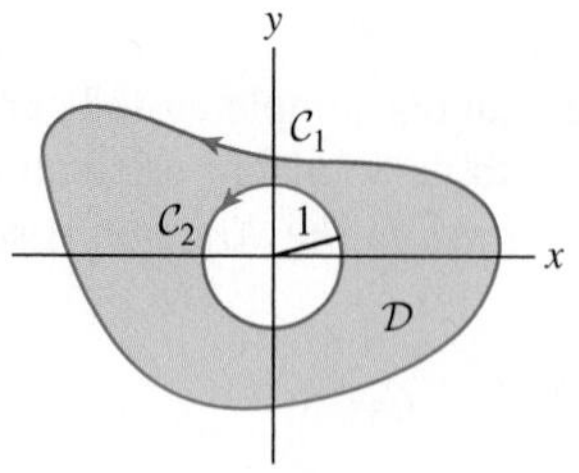

FIGURE 14 $\mathcal{D}$ has area 8, and C_2 is a circle of radius 1.

■ **EXAMPLE 4** Calculate $\oint_{C_1} \mathbf{F} \cdot d\mathbf{s}$, where $\mathbf{F} = \langle x - y, x + y^3 \rangle$ and C_1 is the outer boundary curve oriented counterclockwise. Assume that the domain $\mathcal{D}$ in Figure 14 has area 8.

Solution We cannot compute the line integral over C_1 directly because the curve C_1 is not specified. However, $\partial\mathcal{D} = C_1 - C_2$, so Green's Theorem yields

$$\oint_{C_1} \mathbf{F} \cdot d\mathbf{s} - \oint_{C_2} \mathbf{F} \cdot d\mathbf{s} = \iint_{\mathcal{D}} \left(\frac{\partial F_2}{\partial x} - \frac{\partial F_1}{\partial y} \right) dA \qquad \boxed{10}$$

We have

$$\frac{\partial F_2}{\partial x} - \frac{\partial F_1}{\partial y} = \frac{\partial}{\partial x}(x + y^3) - \frac{\partial}{\partial y}(x - y) = 1 - (-1) = 2$$

$$\iint_{\mathcal{D}} \left(\frac{\partial F_2}{\partial x} - \frac{\partial F_1}{\partial y} \right) dA = \iint_{\mathcal{D}} 2\, dA = 2\,\text{Area}(\mathcal{D}) = 2(8) = 16$$

Thus Eq. (10) gives us

$$\oint_{C_1} \mathbf{F} \cdot d\mathbf{s} - \oint_{C_2} \mathbf{F} \cdot d\mathbf{s} = 16 \qquad \boxed{11}$$

To compute the second integral, parametrize the unit circle C_2 by $\mathbf{c}(t) = (\cos\theta, \sin\theta)$. Then

$$\begin{aligned}
\mathbf{F} \cdot \mathbf{c}'(t) &= \langle \cos\theta - \sin\theta, \cos\theta + \sin^3\theta \rangle \cdot \langle -\sin\theta, \cos\theta \rangle \\
&= -\sin\theta\cos\theta + \sin^2\theta + \cos^2\theta + \sin^3\theta\cos\theta \\
&= 1 - \sin\theta\cos\theta + \sin^3\theta\cos\theta
\end{aligned}$$

The integrals of $\sin\theta\cos\theta$ and $\sin^3\theta\cos\theta$ over $[0, 2\pi]$ are both zero, so

$$\oint_{C_2} \mathbf{F} \cdot d\mathbf{s} = \int_0^{2\pi} (1 - \sin\theta\cos\theta + \sin^3\theta\cos\theta)\, d\theta = \int_0^{2\pi} d\theta = 2\pi$$

Eq. (11) yields $\oint_{C_1} \mathbf{F} \cdot d\mathbf{s} = 16 + 2\pi$. ■

18.1 SUMMARY

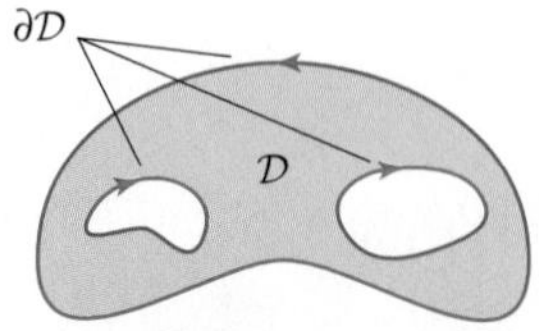

FIGURE 15 The boundary orientation is chosen so that the region lies to your left as you walk along the curve.

- We have two notations for the line integral of a vector field:

$$\int_C \mathbf{F} \cdot d\mathbf{s} \qquad \text{and} \qquad \int_C F_1\, dx + F_2\, dy$$

- $\partial\mathcal{D}$ denotes the boundary of $\mathcal{D}$ with its boundary orientation (Figure 15).
- Green's Theorem:

$$\oint_{\partial\mathcal{D}} F_1\, dx + F_2\, dy = \iint_{\mathcal{D}} \left(\frac{\partial F_2}{\partial x} - \frac{\partial F_1}{\partial y} \right) dA$$

- Formula for the area of the region $\mathcal{D}$ enclosed by C:

$$\text{Area}(\mathcal{D}) = \frac{1}{2} \oint_C x\, dy - y\, dx$$

• The quantity

$$\text{curl}_z(\mathbf{F}) = \frac{\partial F_2}{\partial x} - \frac{\partial F_1}{\partial y}$$

is interpreted as *circulation per unit area*. If $\mathcal{D}$ is a small domain with boundary $\mathcal{C}$, then for any $P \in \mathcal{D}$,

$$\oint_{\mathcal{C}} F_1\,dx + F_2\,dy \approx \text{curl}_z(\mathbf{F})(P) \cdot \text{Area}(\mathcal{D})$$

18.1 EXERCISES

Preliminary Questions

1. Which vector field **F** is being integrated in the line integral $\oint x^2\,dy - e^y\,dx$?

2. Draw a domain in the shape of an ellipse and indicate with an arrow the boundary orientation of the boundary curve. Do the same for the annulus (the region between two concentric circles).

3. The circulation of a conservative vector field around a closed curve is zero. Is this fact consistent with Green's Theorem? Explain.

4. Indicate which of the following vector fields possess the following property: For every simple closed curve $\mathcal{C}$, $\int_{\mathcal{C}} \mathbf{F} \cdot d\mathbf{s}$ is equal to the area enclosed by $\mathcal{C}$.

(a) $\mathbf{F} = \langle -y, 0\rangle$

(b) $\mathbf{F} = \langle x, y\rangle$

(c) $\mathbf{F} = \langle \sin(x^2), x + e^{y^2}\rangle$

Exercises

1. Verify Green's Theorem for the line integral $\oint_{\mathcal{C}} xy\,dx + y\,dy$, where $\mathcal{C}$ is the unit circle, oriented counterclockwise.

2. Let $I = \oint_{\mathcal{C}} \mathbf{F} \cdot d\mathbf{s}$, where $\mathbf{F} = \langle y + \sin x^2, x^2 + e^{y^2}\rangle$ and $\mathcal{C}$ is the circle of radius 4 centered at the origin.

(a) Which is easier, evaluating I directly or using Green's Theorem?

(b) Evaluate I using the easier method.

In Exercises 3–10, use Green's Theorem to evaluate the line integral. Orient the curve counterclockwise unless otherwise indicated.

3. $\oint_{\mathcal{C}} y^2\,dx + x^2\,dy$, where $\mathcal{C}$ is the boundary of the unit square $0 \le x \le 1, 0 \le y \le 1$

4. $\oint_{\mathcal{C}} e^{2x+y}\,dx + e^{-y}\,dy$, where $\mathcal{C}$ is the triangle with vertices $(0, 0)$, $(1, 0)$, and $(1, 1)$

5. $\oint_{\mathcal{C}} x^2 y\,dx$, where $\mathcal{C}$ is the unit circle centered at the origin

6. $\oint_{\mathcal{C}} \mathbf{F} \cdot d\mathbf{s}$, where $\mathbf{F} = \langle x + y, x^2 - y\rangle$ and $\mathcal{C}$ is the boundary of the region enclosed by $y = x^2$ and $y = \sqrt{x}$ for $0 \le x \le 1$

7. $\oint_{\mathcal{C}} \mathbf{F} \cdot d\mathbf{s}$, where $\mathbf{F} = \langle x^2, x^2\rangle$ and $\mathcal{C}$ consists of the arcs $y = x^2$ and $y = x$ for $0 \le x \le 1$

8. $\oint_{\mathcal{C}} (\ln x + y)\,dx - x^2\,dy$, where $\mathcal{C}$ is the rectangle with vertices $(1, 1)$, $(3, 1)$, $(1, 4)$, and $(3, 4)$

9. The line integral of $\mathbf{F} = \langle e^{x+y}, e^{x-y}\rangle$ along the curve (oriented clockwise) consisting of the line segments by joining the points $(0, 0)$, $(2, 2)$, $(4, 2)$, $(2, 0)$, and back to $(0, 0)$ (note the orientation).

10. $\int_{\mathcal{C}} xy\,dx + (x^2 + x)\,dy$, where $\mathcal{C}$ is the path in Figure 16

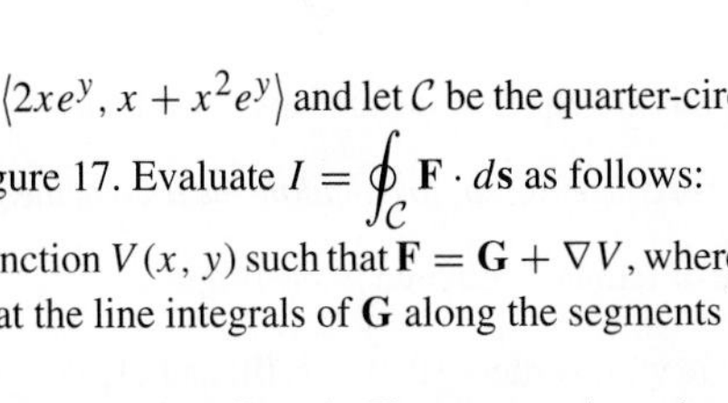

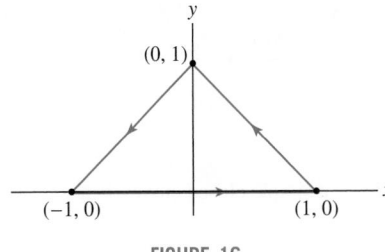

FIGURE 16

11. Let $\mathbf{F} = \langle 2xe^y, x + x^2 e^y\rangle$ and let $\mathcal{C}$ be the quarter-circle path from A to B in Figure 17. Evaluate $I = \oint_{\mathcal{C}} \mathbf{F} \cdot d\mathbf{s}$ as follows:

(a) Find a function $V(x, y)$ such that $\mathbf{F} = \mathbf{G} + \nabla V$, where $\mathbf{G} = \langle 0, x\rangle$.

(b) Show that the line integrals of $\mathbf{G}$ along the segments $\overline{OA}$ and $\overline{OB}$ are zero.

(c) Evaluate I. *Hint:* Use Green's Theorem to show that

$$I = V(B) - V(A) + 4\pi$$

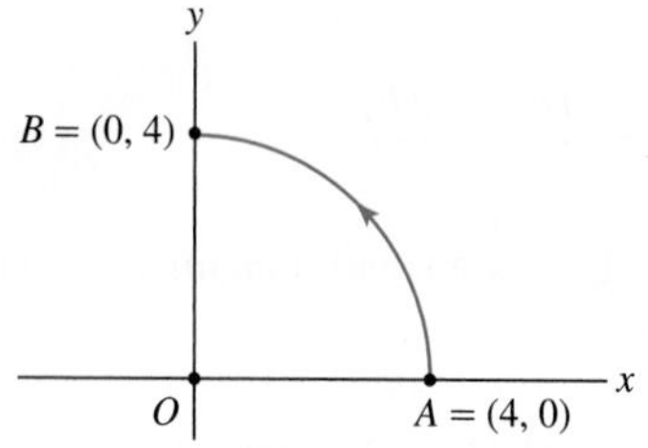

FIGURE 17

12. Compute the line integral of $\mathbf{F} = \langle x^3, 4x \rangle$ along the path from A to B in Figure 18. To save work, use Green's Theorem to relate this line integral to the line integral along the vertical path from B to A.

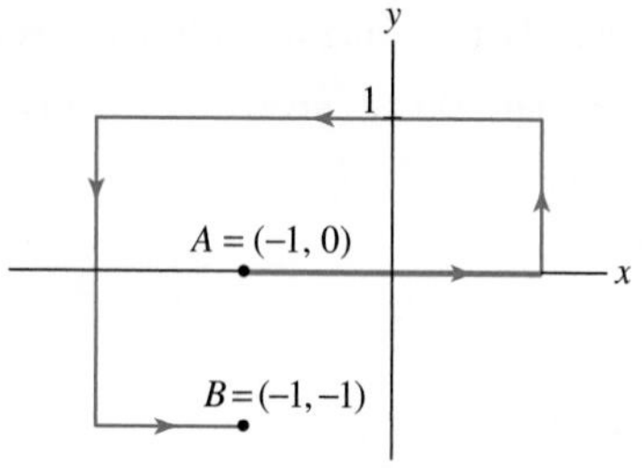

FIGURE 18

13. Evaluate $I = \int_C (\sin x + y)\, dx + (3x + y)\, dy$ for the nonclosed path $ABCD$ in Figure 19. Use the method of Exercise 12.

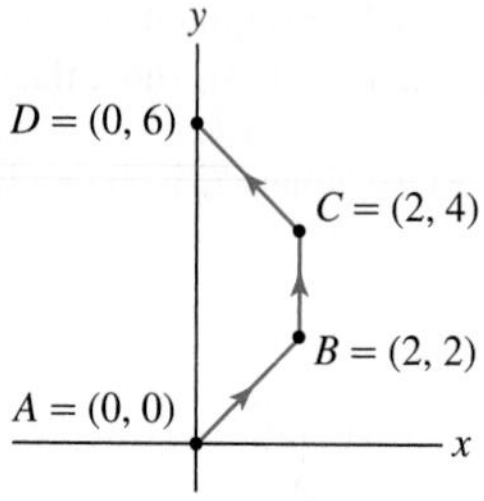

FIGURE 19

14. Show that if $\mathcal{C}$ is a simple closed curve, then

$$\oint_{\mathcal{C}} -y\, dx = \oint_{\mathcal{C}} x\, dy$$

and both integrals are equal to the area enclosed by $\mathcal{C}$.

In Exercises 15–18, use Eq. (6) to calculate the area of the given region.

15. The circle of radius 3 centered at the origin

16. The triangle with vertices $(0, 0)$, $(1, 0)$, and $(1, 1)$

17. The region between the x-axis and the cycloid parametrized by $\mathbf{c}(t) = (t - \sin t, 1 - \cos t)$ for $0 \le t \le 2\pi$ (Figure 20)

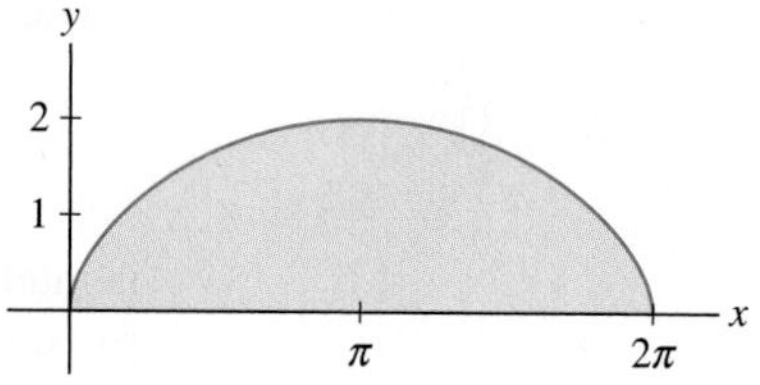

FIGURE 20 Cycloid.

18. The region between the graph of $y = x^2$ and the x-axis for $0 \le x \le 2$

19. Let $x^3 + y^3 = 3xy$ be the **folium of Descartes** (Figure 21).

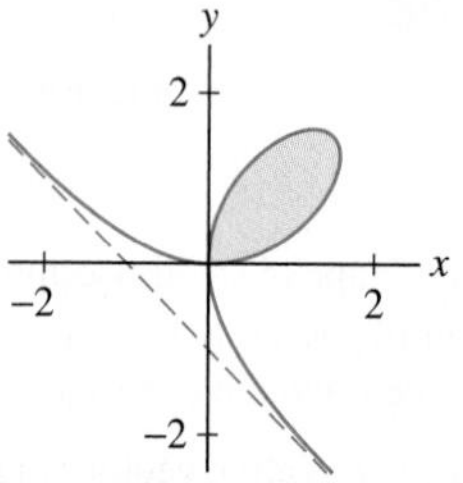

FIGURE 21 Folium of Descartes.

(a) Show that the folium has a parametrization in terms of $t = y/x$ given by

$$x = \frac{3t}{1 + t^3}, \qquad y = \frac{3t^2}{1 + t^3} \qquad (-\infty < t < \infty) \quad (t \neq -1)$$

(b) Show that

$$x\, dy - y\, dx = \frac{9t^2}{(1 + t^3)^2}\, dt$$

Hint: By the Quotient Rule,

$$x^2\, d\left(\frac{y}{x}\right) = x\, dy - y\, dx$$

(c) Find the area of the loop of the folium.

20. Find a parametrization of the lemniscate $(x^2 + y^2)^2 = xy$ (see Figure 22) by using $t = y/x$ as a parameter (see Exercise 19). Then use Eq. (6) to find the area of one loop of the lemniscate.

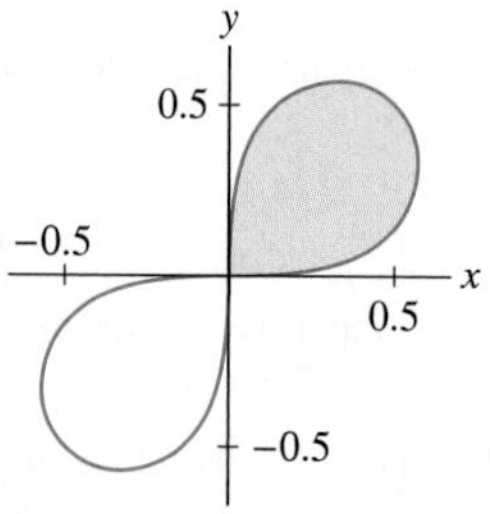

FIGURE 22 Lemniscate.

21. The Centroid via Boundary Measurements The centroid (see Section 16.5) of a domain $\mathcal{D}$ enclosed by a simple closed curve $\mathcal{C}$ is the point with coordinates $(\overline{x}, \overline{y}) = (M_y/M, M_x/M)$, where M is the area of $\mathcal{D}$ and the moments are defined by

$$M_x = \iint_{\mathcal{D}} y\, dA, \qquad M_y = \iint_{\mathcal{D}} x\, dA$$

Show that $M_x = \oint_{\mathcal{C}} xy\, dy$. Find a similar expression for M_y.

22. Use the result of Exercise 21 to compute the moments of the semicircle $x^2 + y^2 = R^2$, $y \geq 0$ as line integrals. Verify that the centroid is $(0, 4R/(3\pi))$.

23. Let $\mathcal{C}_R$ be the circle of radius R centered at the origin. Use the general form of Green's Theorem to determine $\oint_{\mathcal{C}_2} \mathbf{F} \cdot d\mathbf{s}$, where $\mathbf{F}$ is a vector field such that $\oint_{\mathcal{C}_1} \mathbf{F} \cdot d\mathbf{s} = 9$ and $\dfrac{\partial F_2}{\partial x} - \dfrac{\partial F_1}{\partial y} = x^2 + y^2$ for (x, y) in the annulus $1 \leq x^2 + y^2 \leq 4$.

24. Referring to Figure 23, suppose that $\oint_{\mathcal{C}_2} \mathbf{F} \cdot d\mathbf{s} = 12$. Use Green's Theorem to determine $\int_{\mathcal{C}_1} \mathbf{F} \cdot d\mathbf{s}$, assuming that $\dfrac{\partial F_2}{\partial x} - \dfrac{\partial F_1}{\partial x} = -3$ in $\mathcal{D}$.

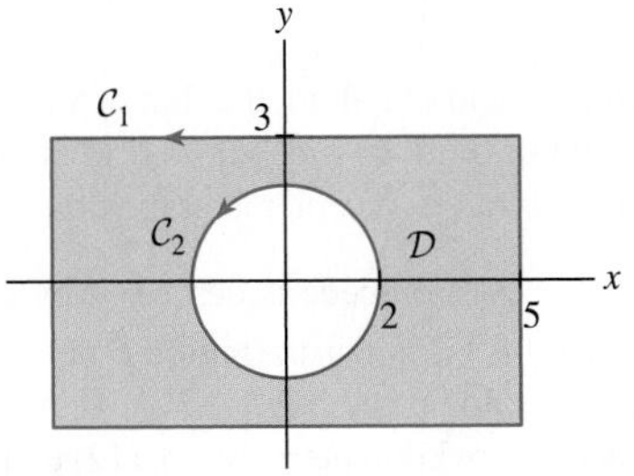

FIGURE 23

25. Referring to Figure 24, suppose that

$$\oint_{\mathcal{C}_2} \mathbf{F} \cdot d\mathbf{s} = 3\pi, \qquad \oint_{\mathcal{C}_3} \mathbf{F} \cdot d\mathbf{s} = 4\pi$$

Use Green's Theorem to determine the circulation of $\mathbf{F}$ around $\mathcal{C}_1$, assuming that $\dfrac{\partial F_2}{\partial x} - \dfrac{\partial F_1}{\partial x} = 9$ on the shaded region.

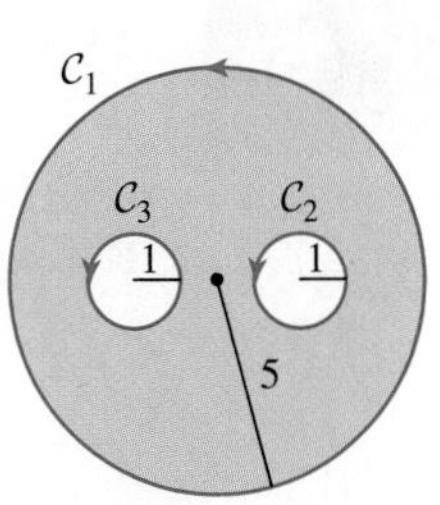

FIGURE 24

26. Let $\mathbf{F}$ be the vortex vector field

$$\mathbf{F} = \left\langle \frac{-y}{x^2 + y^2}, \frac{x}{x^2 + y^2} \right\rangle$$

In Section 16.3 we verified that $\int_{\mathcal{C}_R} \mathbf{F} \cdot d\mathbf{s} = 2\pi$, where $\mathcal{C}_R$ is the circle of radius R centered at the origin. Prove that $\oint_{\mathcal{C}} \mathbf{F} \cdot d\mathbf{s} = 2\pi$ for any simple closed curve $\mathcal{C}$ whose interior contains the origin (Figure 25). *Hint:* Apply the general form of Green's Theorem to the domain between $\mathcal{C}$ and $\mathcal{C}_R$, where R is so small that $\mathcal{C}_R$ is contained in $\mathcal{C}$.

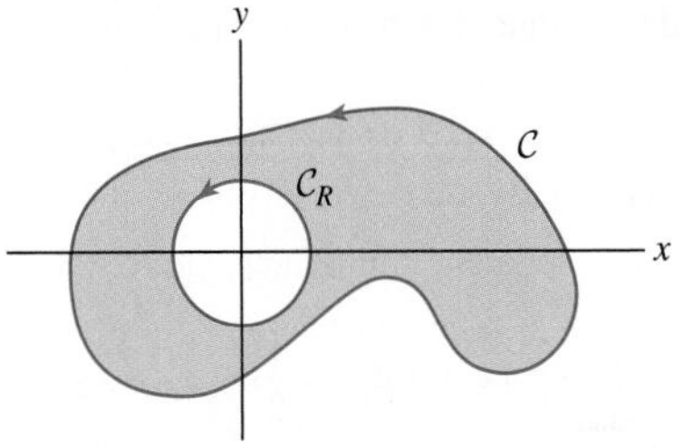

FIGURE 25

In Exercises 27–30, refer to the Conceptual Insight that discusses the curl, defined by

$$\text{curl}_z(\mathbf{F}) = \frac{\partial F_2}{\partial x} - \frac{\partial F_1}{\partial y}$$

27. For the vector fields (A)–(D) in Figure 26, state whether the curl_z at the origin appears to be positive, negative, or zero.

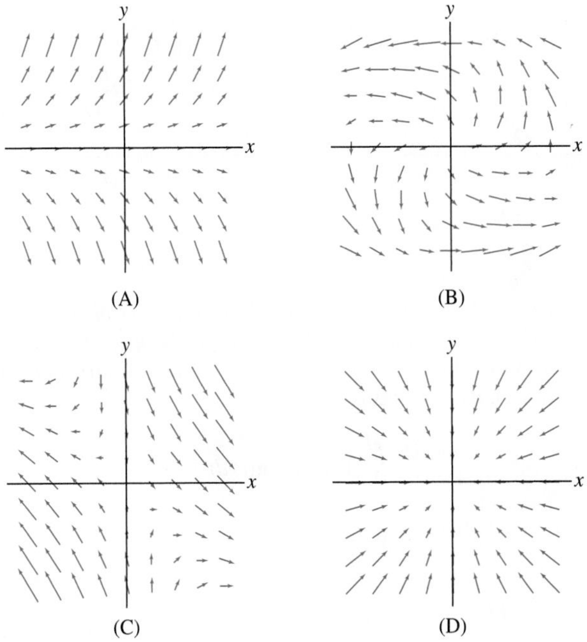

FIGURE 26

28. Estimate the circulation of a vector field **F** around a circle of radius $R = 0.1$, assuming that $\text{curl}_z(\mathbf{F})$ takes the value 4 at the center of the circle.

29. Estimate $\oint_C \mathbf{F} \cdot d\mathbf{s}$, where $\mathbf{F} = \langle x + 0.1y^2, y - 0.1x^2 \rangle$ and C encloses a small region of area 0.25 containing the point $P = (1, 1)$.

30. Let **F** be the velocity field. Estimate the circulation of **F** around a circle of radius $R = 0.05$ with center P, assuming that $\text{curl}_z(\mathbf{F})(P) = -3$. In which direction would a small paddle placed at P rotate? How fast would it rotate (in radians per second) if **F** is expressed in meters per second?

31. Let C_R be the circle of radius R centered at the origin. Use Green's Theorem to find the value of R that maximizes $\oint_{C_R} y^3\, dx + x\, dy$.

32. Area of a Polygon Green's Theorem leads to a convenient formula for the area of a polygon.

(a) Let C be the line segment joining (x_1, y_1) to (x_2, y_2). Show that

$$\frac{1}{2}\int_C -y\, dx + x\, dy = \frac{1}{2}(x_1 y_2 - x_2 y_1)$$

(b) Prove that the area of the polygon with vertices (x_1, y_1), (x_2, y_2), $\dots$, (x_n, y_n) is equal [where we set $(x_{n+1}, y_{n+1}) = (x_1, y_1)$] to

$$\frac{1}{2}\sum_{i=1}^{n}(x_i y_{i+1} - x_{i+1} y_i)$$

33. Use the result of Exercise 32 to compute the areas of the polygons in Figure 27. Check your result for the area of the triangle in (A) using geometry.

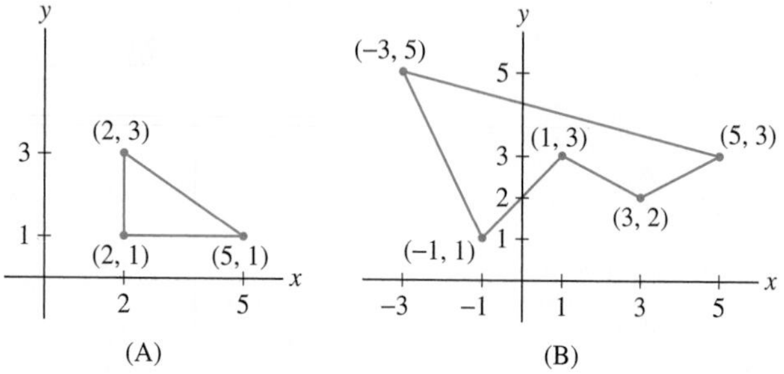

FIGURE 27

Exercises 34–39: In Section 17.2, we defined the flux of **F** *across a curve C (Figure 28) as the integral of the normal component of* **F** *along C, and we showed that if $\mathbf{c}(t) = (x(t), y(t))$ is a parametrization of C for $a \le t \le b$, then the flux is equal to*

$$\int_a^b \mathbf{F}(\mathbf{c}(t)) \cdot \mathbf{n}(t)\, dt$$

where $\mathbf{n}(t) = \langle y'(t), -x'(t)\rangle$.

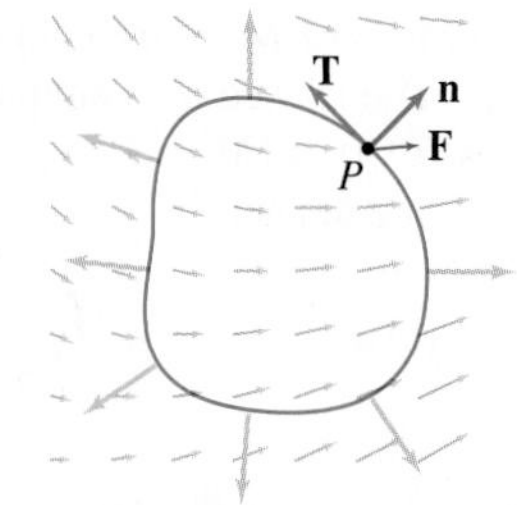

FIGURE 28 The flux of **F** is the integral of the normal component $\mathbf{F} \cdot \mathbf{n}$ around the curve.

34. Show that the flux of $\mathbf{F} = \langle P, Q\rangle$ across C is equal to $\oint_C P\, dy - Q\, dx$.

35. Define $\text{div}(\mathbf{F}) = \dfrac{\partial P}{\partial x} + \dfrac{\partial Q}{\partial y}$. Use Green's Theorem to prove that for any simple closed curve C,

$$\text{Flux across } C = \iint_{\mathcal{D}} \text{div}(\mathbf{F})\, dA \qquad \boxed{12}$$

where $\mathcal{D}$ is the region enclosed by C. This is a two-dimensional version of the **Divergence Theorem** discussed in Section 18.3.

36. Use Eq. (12) to compute the flux of $\mathbf{F} = \langle 2x + y^3, 3y - x^4\rangle$ across the unit circle.

37. Use Eq. (12) to compute the flux of $\mathbf{F} = \langle \cos y, \sin y\rangle$ across the square $0 \le x \le 2$, $0 \le y \le \frac{\pi}{2}$.

38. If **v** is the velocity field of a fluid, the flux of **v** across C is equal to the flow rate (amount of fluid flowing across C in m^2/s). Find the flow rate across the circle of radius 2 centered at the origin if $\text{div}(\mathbf{v}) = x^2$.

39. A buffalo (Figure 29) stampede is described by a velocity vector field $\mathbf{F} = \langle xy - y^3, x^2 + y\rangle$ km/h in the region $\mathcal{D}$ defined by $2 \le x \le 3$, $2 \le y \le 3$ in units of kilometers (Figure 30). Assuming a density is $\rho = 500$ buffalo per square kilometer, use Eq. (12) to determine the net number of buffalo leaving or entering $\mathcal{D}$ per minute (equal to ρ times the flux of **F** across the boundary of $\mathcal{D}$).

FIGURE 29 Buffalo stampede.

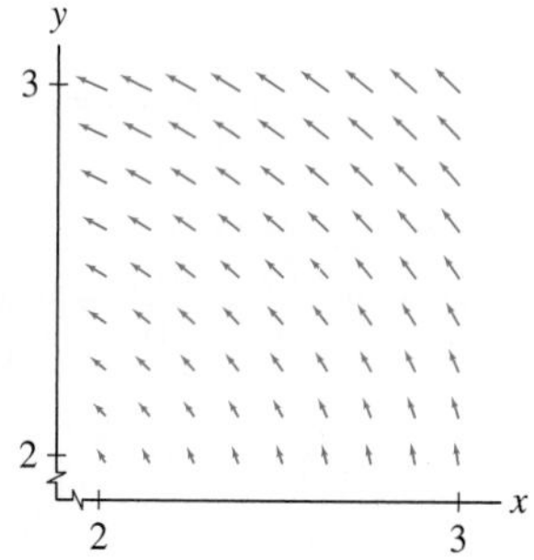

FIGURE 30 The vector field $\mathbf{F} = \langle xy - y^3, x^2 + y\rangle$.

Further Insights and Challenges

In Exercises 40–43, the **Laplace operator** Δ *is defined by*

$$\Delta\varphi = \frac{\partial^2\varphi}{\partial x^2} + \frac{\partial^2\varphi}{\partial y^2} \qquad \boxed{13}$$

For any vector field $\mathbf{F} = \langle F_1, F_2\rangle$*, define the conjugate vector field* $\mathbf{F}^* = \langle -F_2, F_1\rangle$.

40. Show that if $\mathbf{F} = \nabla\varphi$, then $\text{curl}_z(\mathbf{F}^*) = \Delta\varphi$.

41. Let $\mathbf{n}$ be the outward-pointing unit normal vector to a simple closed curve $\mathcal{C}$. The **normal derivative** of a function φ, denoted $\frac{\partial\varphi}{\partial\mathbf{n}}$, is the directional derivative $D_{\mathbf{n}}(\varphi) = \nabla\varphi\cdot\mathbf{n}$. Prove that

$$\oint_{\mathcal{C}} \frac{\partial\varphi}{\partial\mathbf{n}}\,ds = \iint_{\mathcal{D}} \Delta\varphi\,dA$$

where $\mathcal{D}$ is the domain enclosed by a simple closed curve $\mathcal{C}$. *Hint:* Let $\mathbf{F} = \nabla\varphi$. Show that $\frac{\partial\varphi}{\partial\mathbf{n}} = \mathbf{F}^*\cdot\mathbf{T}$ where $\mathbf{T}$ is the unit tangent vector, and apply Green's Theorem.

42. Let $P = (a, b)$ and let $\mathcal{C}_r$ be the circle of radius r centered at P. The average value of a continuous function φ on $\mathcal{C}_r$ is defined as the integral

$$I_\varphi(r) = \frac{1}{2\pi}\int_0^{2\pi} \varphi(a + r\cos\theta, b + r\sin\theta)\,d\theta$$

(a) Show that

$$\frac{\partial\varphi}{\partial\mathbf{n}}(a + r\cos\theta, b + r\sin\theta) = \frac{\partial\varphi}{\partial r}(a + r\cos\theta, b + r\sin\theta)$$

(b) Use differentiation under the integral sign to prove that

$$\frac{d}{dr}I_\varphi(r) = \frac{1}{2\pi r}\int_{\mathcal{C}_r} \frac{\partial\varphi}{\partial\mathbf{n}}\,ds$$

(c) Use Exercise 41 to conclude that

$$\frac{d}{dr}I_\varphi(r) = \frac{1}{2\pi r}\iint_{\mathcal{D}(r)} \Delta\varphi\,dA$$

where $\mathcal{D}(r)$ is the interior of $\mathcal{C}_r$.

43. Prove that $m(r) \le I_\varphi(r) \le M(r)$, where $m(r)$ and $M(r)$ are the minimum and maximum values of φ on $\mathcal{C}_r$. Then use the continuity of φ to prove that $\lim_{r\to 0} I_\varphi(r) = \varphi(P)$.

In Exercises 44 and 45, let $\mathcal{D}$ *be the region bounded by a simple closed curve* $\mathcal{C}$*. A function* $\varphi(x, y)$ *on* $\mathcal{D}$ *(whose second-order partial derivatives exist and are continuous) is called* **harmonic** *if* $\Delta\varphi = 0$*, where* $\Delta\varphi$ *is the Laplace operator defined in Eq. (13).*

44. Use the results of Exercises 42 and 43 to prove the **mean-value property** of harmonic functions: If φ is harmonic, then $I_\varphi(r) = \varphi(P)$ for all r.

45. Show that $f(x, y) = x^2 - y^2$ is harmonic. Verify the mean-value property for $f(x, y)$ directly [expand $f(a + r\cos\theta, b + r\sin\theta)$ as a function of θ and compute $I_\varphi(r)$]. Show that $x^2 + y^2$ is not harmonic and does not satisfy the mean-value property.

18.2 Stokes' Theorem

Stokes' Theorem is an extension of Green's Theorem to three dimensions in which circulation is related to a surface integral in $\mathbf{R}^3$ (rather than to a double integral in the plane). In order to state it, we introduce some definitions and terminology.

Figure 1 shows three surfaces with different types of boundaries. The boundary of a surface is denoted $\partial\mathcal{S}$. Observe that the boundary in (A) is a single, simple closed curve and the boundary in (B) consists of three closed curves. The surface in (C) is called a **closed surface** because its boundary is empty. In this case, we write $\partial\mathcal{S} = \emptyset$.

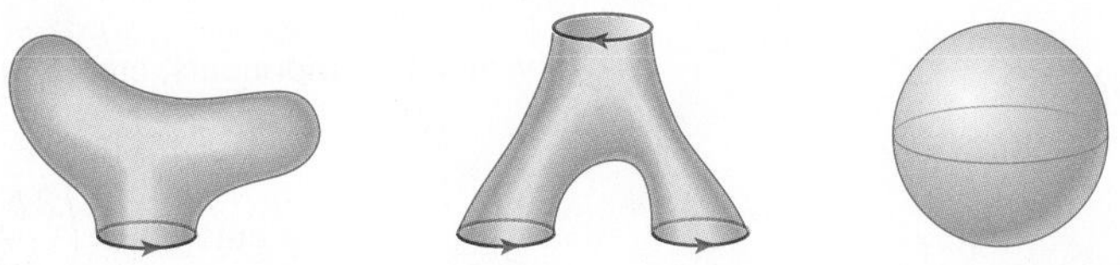

(A) Boundary consists of a single closed curve

(B) Boundary consists of three closed curves

(C) Closed surface (the boundary is empty)

FIGURE 1 Surfaces and their boundaries.

Recall from Section 17.5 that an orientation is a continuously varying choice of unit normal vector at each point of a surface $\mathcal{S}$. When $\mathcal{S}$ is oriented, we can specify an orientation of $\partial\mathcal{S}$, called the **boundary orientation**. Imagine that you are a unit normal

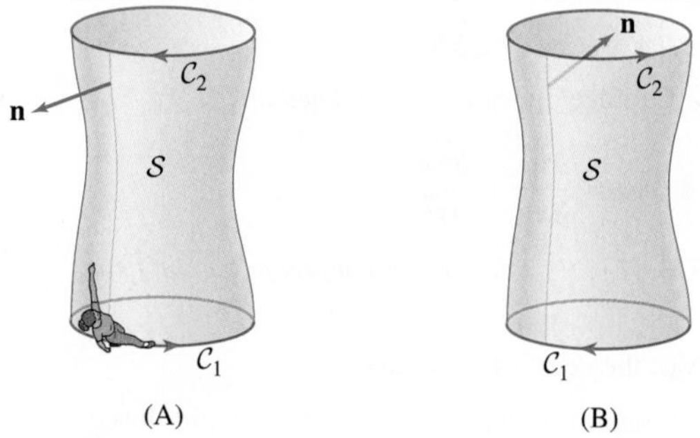

FIGURE 2 The orientation of the boundary $\partial\mathcal{S}$ for each of the two possible orientations of the surface $\mathcal{S}$.

vector walking along the boundary curve. The boundary orientation is the direction for which the surface is on your left as you walk. For example, the boundary of the surface in Figure 2 consists of two curves, $\mathcal{C}_1$ and $\mathcal{C}_2$. In (A), the normal vector points to the outside. The woman (representing the normal vector) is walking along $\mathcal{C}_1$ and has the surface to her left, so she is walking in the positive direction. The curve $\mathcal{C}_2$ is oriented in the opposite direction because she would have to walk along $\mathcal{C}_2$ in that direction to keep the surface on her left. The boundary orientations in (B) are reversed because the opposite normal has been selected to orient the surface.

All that's left is to define curl. The **curl** of the vector field $\mathbf{F} = \langle F_1, F_2, F_3 \rangle$ is the vector field defined by the symbolic determinant

$$\operatorname{curl}(\mathbf{F}) = \begin{vmatrix} \mathbf{i} & \mathbf{j} & \mathbf{k} \\ \dfrac{\partial}{\partial x} & \dfrac{\partial}{\partial y} & \dfrac{\partial}{\partial z} \\ F_1 & F_2 & F_3 \end{vmatrix}$$
$$= \left(\frac{\partial F_3}{\partial y} - \frac{\partial F_2}{\partial z}\right)\mathbf{i} - \left(\frac{\partial F_3}{\partial x} - \frac{\partial F_1}{\partial z}\right)\mathbf{j} + \left(\frac{\partial F_2}{\partial x} - \frac{\partial F_1}{\partial y}\right)\mathbf{k}$$

In more compact form, the curl is the symbolic cross product

$$\boxed{\operatorname{curl}(\mathbf{F}) = \nabla \times \mathbf{F}}$$

where ∇ is the del "operator" (also called "nabla"):

$$\nabla = \left\langle \frac{\partial}{\partial x}, \frac{\partial}{\partial y}, \frac{\partial}{\partial z} \right\rangle$$

In terms of components, curl(**F**) is the vector field

$$\boxed{\operatorname{curl}(\mathbf{F}) = \left\langle \frac{\partial F_3}{\partial y} - \frac{\partial F_2}{\partial z}, \frac{\partial F_1}{\partial z} - \frac{\partial F_3}{\partial x}, \frac{\partial F_2}{\partial x} - \frac{\partial F_1}{\partial y} \right\rangle}$$

It is straightforward to check that curl obeys the **linearity** rules:

$$\operatorname{curl}(\mathbf{F} + \mathbf{G}) = \operatorname{curl}(\mathbf{F}) + \operatorname{curl}(\mathbf{G})$$
$$\operatorname{curl}(c\mathbf{F}) = c\operatorname{curl}(\mathbf{F}) \qquad (c \text{ any constant})$$

EXAMPLE 1 Calculating the Curl Calculate the curl of $\mathbf{F} = \langle xy, e^x, y+z \rangle$.

Solution We compute the curl as a symbolic determinant:

$$\operatorname{curl}(\mathbf{F}) = \begin{vmatrix} \mathbf{i} & \mathbf{j} & \mathbf{k} \\ \dfrac{\partial}{\partial x} & \dfrac{\partial}{\partial y} & \dfrac{\partial}{\partial z} \\ xy & e^x & y+z \end{vmatrix}$$

$$= \left(\frac{\partial}{\partial y}(y+z) - \frac{\partial}{\partial z}e^x\right)\mathbf{i} - \left(\frac{\partial}{\partial x}(y+z) - \frac{\partial}{\partial z}xy\right)\mathbf{j} + \left(\frac{\partial}{\partial x}e^x - \frac{\partial}{\partial y}xy\right)\mathbf{k}$$

$$= \mathbf{i} + (e^x - x)\mathbf{k}$$ ■

EXAMPLE 2 Conservative Vector Fields Have Zero Curl Verify:

$$\text{If } \mathbf{F} = \nabla V, \text{ then } \operatorname{curl}(\mathbf{F}) = \mathbf{0}. \quad \text{That is,} \quad \operatorname{curl}(\nabla V) = \mathbf{0}. \qquad (1)$$

Solution The curl of a vector field is zero if

$$\frac{\partial F_3}{\partial y} - \frac{\partial F_2}{\partial z} = 0, \qquad \frac{\partial F_1}{\partial z} - \frac{\partial F_3}{\partial x} = 0, \qquad \frac{\partial F_2}{\partial x} - \frac{\partial F_1}{\partial y} = 0$$

But these equations are equivalent to the cross-partials condition that is satisfied by every conservative vector field $\mathbf{F} = \nabla V$. ■

In the next theorem, we assume that $\mathcal{S}$ is an oriented surface with parametrization $G : \mathcal{D} \to \mathcal{S}$, where $\mathcal{D}$ is a domain in the plane bounded by smooth, simple closed curves, and G is one-to-one and regular, except possibly on the boundary of $\mathcal{D}$. More generally, $\mathcal{S}$ may be a finite union of surfaces of this type. The surfaces in applications we consider, such as spheres, cubes, and graphs of functions, satisfy these conditions.

THEOREM 1 Stokes' Theorem For surfaces $\mathcal{S}$ as described above,

$$\oint_{\partial \mathcal{S}} \mathbf{F} \cdot d\mathbf{s} = \iint_{\mathcal{S}} \operatorname{curl}(\mathbf{F}) \cdot d\mathbf{S} \qquad (2)$$

The integral on the left is defined relative to the boundary orientation of $\partial \mathcal{S}$. If $\mathcal{S}$ is closed (that is, $\partial \mathcal{S}$ is empty), then the surface integral on the right is zero.

The curl measures the extent to which $\mathbf{F}$ *fails to be conservative. If* $\mathbf{F}$ *is conservative, then* $\operatorname{curl}(\mathbf{F}) = \mathbf{0}$ *and Stokes' Theorem merely confirms what we already know: The circulation of a conservative vector field around a closed path is zero.*

Proof Each side of Eq. (2) is equal to a sum over the components of $\mathbf{F}$:

$$\oint_C \mathbf{F} \cdot d\mathbf{s} = \oint_C F_1\,dx + F_2\,dy + F_3\,dz$$

$$\iint_{\mathcal{S}} \operatorname{curl}(\mathbf{F}) \cdot d\mathbf{S} = \iint_{\mathcal{S}} \operatorname{curl}(F_1\mathbf{i}) \cdot d\mathbf{S} + \iint_{\mathcal{S}} \operatorname{curl}(F_2\mathbf{j}) \cdot d\mathbf{S} + \iint_{\mathcal{S}} \operatorname{curl}(F_3\mathbf{k}) \cdot d\mathbf{S}$$

The proof consists of showing that the F_1-, F_2-, and F_3-terms are separately equal.

Because a complete proof is quite technical, we will prove it under the simplifying assumption that $\mathcal{S}$ is the graph of a function $z = f(x, y)$ lying over a domain $\mathcal{D}$ in the xy-plane. Furthermore, we will carry the details only for the F_1-terms. The calculation for F_2-components is similar, and we leave as an exercise the equality of the F_3-terms (Exercise 31). Thus, we shall prove that

$$\oint_{\mathcal{C}} F_1\,dx = \iint_{\mathcal{S}} \operatorname{curl}(F_1(x, y, z)\mathbf{i}) \cdot d\mathbf{S} \qquad \boxed{3}$$

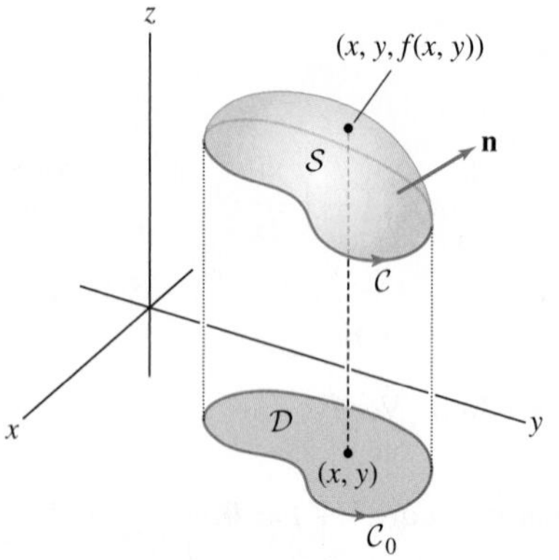

FIGURE 3

Orient $\mathcal{S}$ with upward-pointing normal as in Figure 3 and let $\mathcal{C} = \partial\mathcal{S}$ be the boundary curve. Let $\mathcal{C}_0$ be the boundary of $\mathcal{D}$ in the xy-plane, and let $\mathbf{c}_0(t) = (x(t), y(t))$ (for $a \le t \le b$) be a counterclockwise parametrization of $\mathcal{C}_0$ as in Figure 3. The boundary curve $\mathcal{C}$ projects onto $\mathcal{C}_0$, so $\mathcal{C}$ has parametrization

$$\mathbf{c}(t) = \big(x(t), y(t), f(x(t), y(t))\big)$$

and thus

$$\oint_{\mathcal{C}} F_1(x, y, z)\,dx = \int_a^b F_1\big(x(t), y(t), f(x(t), y(t))\big)\frac{dx}{dt}\,dt$$

The integral on the right is precisely the integral we obtain by integrating $F_1\big(x, y, f(x, y)\big)\,dx$ over the curve $\mathcal{C}_0$ in the plane $\mathbf{R}^2$. In other words,

$$\oint_{\mathcal{C}} F_1(x, y, z)\,dx = \int_{\mathcal{C}_0} F_1\big(x, y, f(x, y)\big)\,dx$$

By Green's Theorem applied to the integral on the right,

$$\oint_{\mathcal{C}} F_1(x, y, z)\,dx = -\iint_{\mathcal{D}} \frac{\partial}{\partial y} F_1(x, y, f(x, y))\,dA$$

By the Chain Rule,

$$\frac{\partial}{\partial y} F_1\big(x, y, f(x, y)\big) = F_{1y}\big(x, y, f(x, y)\big) + F_{1z}\big(x, y, f(x, y)\big) f_y(x, y)$$

so finally we obtain

$$\oint_{\mathcal{C}} F_1\,dx = -\iint_{\mathcal{D}} \Big(F_{1y}\big(x, y, f(x, y)\big) + F_{1z}\big(x, y, f(x, y)\big) f_y(x, y)\Big)\,dA \qquad \boxed{4}$$

REMINDER *Calculating a surface integral:*

$$\iint_{\mathcal{S}} \mathbf{F} \cdot d\mathbf{S} = \iint_{\mathcal{D}} \mathbf{F}(u, v) \cdot \mathbf{n}(u, v)\,du\,dv$$

If $\mathcal{S}$ is a graph $z = f(x, y)$, parametrized by $G(x, y) = (x, y, f(x, y))$, then

$$\mathbf{n}(x, y) = \langle -f_x(x, y), -f_y(x, y), 1\rangle$$

To finish the proof, we compute the surface integral of $\operatorname{curl}(\mathbf{F}_1\mathbf{i})$ using the parametrization $G(x, y) = (x, y, f(x, y))$ of $\mathcal{S}$:

$$\mathbf{n} = \big\langle -f_x(x, y), -f_y(x, y), 1\big\rangle \qquad \text{(upward-pointing normal)}$$

$$\begin{aligned}\operatorname{curl}(F_1\mathbf{i}) \cdot \mathbf{n} &= \big\langle 0, F_{1z}, -F_{1y}\big\rangle \cdot \big\langle -f_x(x, y), -f_y(x, y), 1\big\rangle \\ &= -F_{1z}\big(x, y, f(x, y)\big) f_y(x, y) - F_{1y}\big(x, y, f(x, y)\big)\end{aligned}$$

$$\iint_{\mathcal{S}} \operatorname{curl}(F_1\mathbf{i}) \cdot d\mathbf{S} = -\iint_{\mathcal{D}} \Big(F_{1z}(x, y, z) f_y(x, y) + F_{1y}\big(x, y, f(x, y)\big)\Big)\,dA \qquad \boxed{5}$$

The right-hand sides of Eq. (4) and Eq. (5) are equal. This proves Eq. (3). ■

■ **EXAMPLE 3** **Verifying Stokes' Theorem** Verify Stokes' Theorem for

$$\mathbf{F} = \langle -y, 2x, x + z\rangle$$

and the upper hemisphere with outward-pointing normal vectors (Figure 4):

$$\mathcal{S} = \{(x, y, z) : x^2 + y^2 + z^2 = 1, z \ge 0\}$$

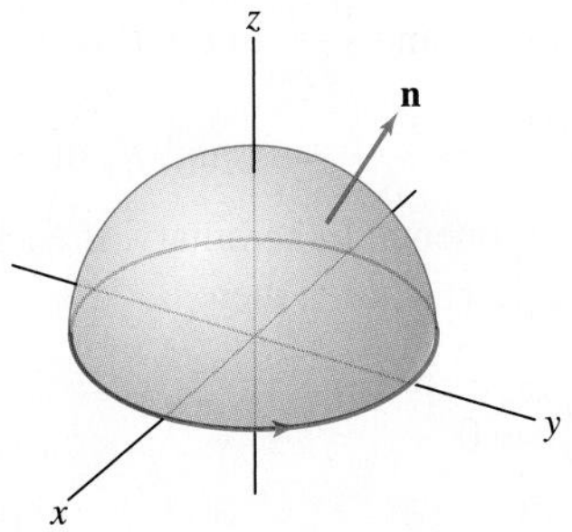

FIGURE 4 Upper hemisphere with oriented boundary.

Solution We will show that both the line integral and the surface integral in Stokes' Theorem are equal to 3π.

***Step 1.* Compute the line integral around the boundary curve.**
The boundary of $\mathcal{S}$ is the unit circle oriented in the counterclockwise direction with parametrization $\mathbf{c}(t) = (\cos t, \sin t, 0)$. Thus,

$$\mathbf{c}'(t) = \langle -\sin t, \cos t, 0\rangle$$

$$\mathbf{F}(\mathbf{c}(t)) = \langle -\sin t, 2\cos t, \cos t\rangle$$

$$\mathbf{F}(\mathbf{c}(t)) \cdot \mathbf{c}'(t) = \langle -\sin t, 2\cos t, \cos t\rangle \cdot \langle -\sin t, \cos t, 0\rangle$$

$$= \sin^2 t + 2\cos^2 t = 1 + \cos^2 t$$

$$\oint_{\partial \mathcal{S}} \mathbf{F} \cdot d\mathbf{s} = \int_0^{2\pi} (1 + \cos^2 t)\, dt = 2\pi + \pi = \boxed{3\pi} \qquad \boxed{6}$$

REMINDER In Eq. (6), we use

$$\int_0^{2\pi} \cos^2 t\, dt = \int_0^{2\pi} \frac{1 + \cos 2t}{2}\, dt = \pi$$

***Step 2.* Compute the curl.**

$$\mathrm{curl}(\mathbf{F}) = \begin{vmatrix} \mathbf{i} & \mathbf{j} & \mathbf{k} \\ \dfrac{\partial}{\partial x} & \dfrac{\partial}{\partial y} & \dfrac{\partial}{\partial z} \\ -y & 2x & x + z \end{vmatrix}$$

$$= \left(\frac{\partial}{\partial y}(x + z) - \frac{\partial}{\partial z}2x\right)\mathbf{i} - \left(\frac{\partial}{\partial x}(x + z) - \frac{\partial}{\partial z}(-y)\right)\mathbf{j}$$

$$+ \left(\frac{\partial}{\partial x}2x - \frac{\partial}{\partial y}(-y)\right)\mathbf{k}$$

$$= \langle 0, -1, 3\rangle$$

***Step 3.* Compute the surface integral of the curl.**
We parametrize the hemisphere using spherical coordinates:

$$G(\theta, \phi) = (\cos\theta \sin\phi, \sin\theta \sin\phi, \cos\phi)$$

REMINDER Stokes' Theorem states

$$\oint_{\partial \mathcal{S}} \mathbf{F} \cdot d\mathbf{s} = \iint_{\mathcal{S}} \mathrm{curl}(\mathbf{F}) \cdot d\mathbf{S}$$

By Eq. (2) of Section 17.4, the outward-pointing normal vector is

$$\mathbf{n} = \sin\phi\, \langle \cos\theta \sin\phi, \sin\theta \sin\phi, \cos\phi\rangle$$

Therefore,

$$\mathrm{curl}(\mathbf{F}) \cdot \mathbf{n} = \sin\phi\, \langle 0, -1, 3\rangle \cdot \langle \cos\theta \sin\phi, \sin\theta \sin\phi, \cos\phi\rangle$$

$$= -\sin\theta \sin^2\phi + 3\cos\phi \sin\phi$$

The upper hemisphere $\mathcal{S}$ corresponds to $0 \le \phi \le \frac{\pi}{2}$, so

$$\iint_{\mathcal{S}} \mathrm{curl}(\mathbf{F}) \cdot d\mathbf{S} = \int_{\phi=0}^{\pi/2} \int_{\theta=0}^{2\pi} (-\sin\theta \sin^2\phi + 3\cos\phi \sin\phi)\, d\theta\, d\phi$$

$$= 0 + 2\pi \int_{\phi=0}^{\pi/2} 3\cos\phi \sin\phi\, d\phi = 2\pi \left(\frac{3}{2}\sin^2\phi\right)\bigg|_{\phi=0}^{\pi/2}$$

$$= \boxed{3\pi}$$

■

Notice that curl(**F**) contains the partial derivatives $\frac{\partial F_1}{\partial y}$ and $\frac{\partial F_1}{\partial z}$ but not the partial $\frac{\partial F_1}{\partial x}$. So if $F_1 = F_1(x)$ is a function of x alone, then $\frac{\partial F_1}{\partial y} = \frac{\partial F_1}{\partial z} = 0$, and F_1 does not contribute to the curl. The same holds for the other components. In summary, if each of F_1, F_2, and F_3 depends only on its corresponding variable x, y, or z, then

$$\text{curl}\left(\langle F_1(x), F_2(y), F_3(z)\rangle\right) = 0$$

■ **EXAMPLE 4** Use Stokes' Theorem to show that $\oint_C \mathbf{F} \cdot d\mathbf{s} = 0$, where

$$\mathbf{F} = \left\langle \sin(x^2), e^{y^2} + x^2, z^4 + 2x^2 \right\rangle$$

and C is the boundary of the triangle in Figure 5 with the indicated orientation.

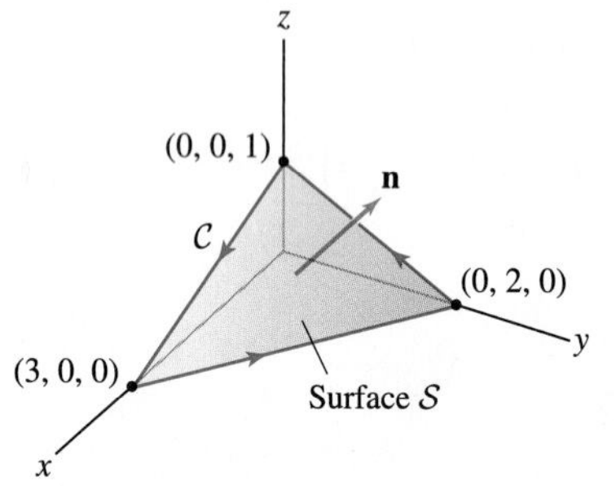

FIGURE 5

Solution We apply Stokes' Theorem

$$\oint_C \mathbf{F} \cdot d\mathbf{s} = \iint_{\mathcal{S}} \text{curl}(\mathbf{F}) \cdot d\mathbf{S}$$

and show that the integral on the right is zero.

By the preceding remark, the first component $\sin(x^2)$ does not contribute to the curl since it depends only on x. Similarly, e^{y^2} and z^4 drop out of the curl, and we have

$$\text{curl}\left(\left\langle \sin x^2, e^{y^2} + x^2, z^4 + 2x^2 \right\rangle\right) = \overbrace{\text{curl}\left(\left\langle \sin x^2, e^{y^2}, z^4 \right\rangle\right)}^{\text{Automatically zero}} + \text{curl}\left(\left\langle 0, x^2, 2x^2 \right\rangle\right)$$

$$= \left\langle 0, -\frac{\partial}{\partial x}2x^2, \frac{\partial}{\partial x}x^2 \right\rangle = \langle 0, -4x, 2x\rangle$$

Now, it turns out (by the author's design) that we can show the surface integral is zero without actually computing it. Referring to Figure 5, we see that C is the boundary of the triangular surface $\mathcal{S}$ contained in the plane

$$\frac{x}{3} + \frac{y}{2} + z = 1$$

Therefore, $\mathbf{u} = \left\langle \frac{1}{3}, \frac{1}{2}, 1 \right\rangle$ is a normal vector to this plane. But **u** and curl(**F**) are orthogonal:

$$\text{curl}(\mathbf{F}) \cdot \mathbf{u} = \langle 0, -4x, 2x\rangle \cdot \left\langle \frac{1}{3}, \frac{1}{2}, 1 \right\rangle = -2x + 2x = 0$$

In other words, the normal component of curl(**F**) along $\mathcal{S}$ is zero. Since the surface integral of a vector field is equal to the surface integral of the normal component, we conclude that $\iint_{\mathcal{S}} \text{curl}(\mathbf{F}) \cdot d\mathbf{S} = 0$. ■

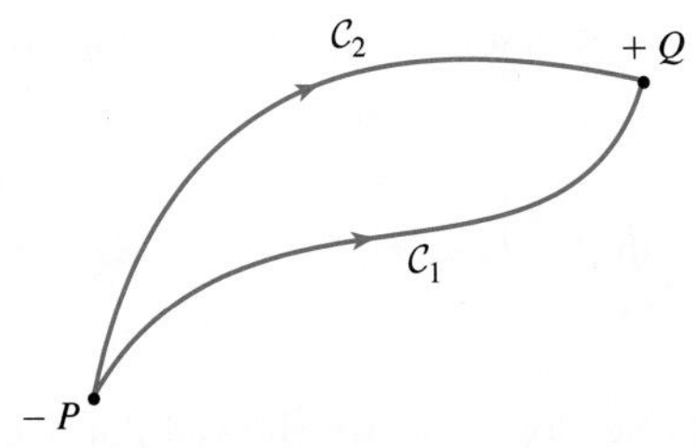

FIGURE 6 Two paths with the same boundary $Q - P$.

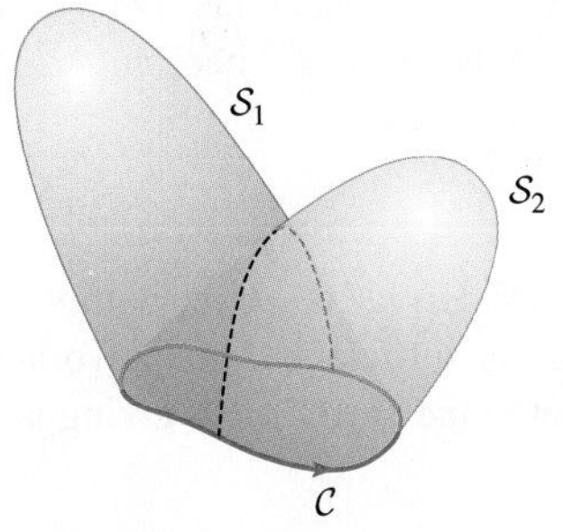

FIGURE 7 Surfaces $\mathcal{S}_1$ and $\mathcal{S}_2$ have the same oriented boundary.

CONCEPTUAL INSIGHT Recall that if **F** is conservative—that is, $\mathbf{F} = \nabla V$—then for any two paths $\mathcal{C}_1$ and $\mathcal{C}_2$ from P to Q (Figure 6),

$$\int_{\mathcal{C}_1} \mathbf{F} \cdot d\mathbf{s} = \int_{\mathcal{C}_2} \mathbf{F} \cdot d\mathbf{s} = V(Q) - V(P)$$

In other words, the line integral is path independent. In particular, $\oint_{\mathcal{C}} \mathbf{F} \cdot d\mathbf{s}$ is zero if $\mathcal{C}$ is closed ($P = Q$).

Analogous facts are true for surface integrals when $\mathbf{F} = \text{curl}(\mathbf{A})$. The vector field **A** is called a **vector potential** for **F**. Stokes' Theorem tells us that for any two surfaces $\mathcal{S}_1$ and $\mathcal{S}_2$ with the same oriented boundary $\mathcal{C}$ (Figure 7),

$$\iint_{\mathcal{S}_1} \mathbf{F} \cdot d\mathbf{S} = \iint_{\mathcal{S}_2} \mathbf{F} \cdot d\mathbf{S} = \oint_{\mathcal{C}} \mathbf{A} \cdot d\mathbf{s}$$

In other words, *the surface integral of a vector field with vector potential* **A** *is surface independent*, just as a vector field with a potential function V is path independent.

If the surface is closed, then the boundary curve is empty and the surface integral is zero:

$$\iint_{\mathcal{S}} \mathbf{F} \cdot d\mathbf{S} = 0 \quad \text{if} \quad \mathbf{F} = \text{curl}(\mathbf{A}) \text{ and } \mathcal{S} \text{ is closed}$$

Vector potentials are not unique: If $\mathbf{F} = \text{curl}(\mathbf{A})$*, then* $\mathbf{F} = \text{curl}(\mathbf{A} + \mathbf{B})$ *for any vector field* **B** *such that* $\text{curl}(\mathbf{B}) = \mathbf{0}$.

THEOREM 2 Surface Independence for Curl Vector Fields
If $\mathbf{F} = \text{curl}(\mathbf{A})$, then the flux of **F** through a surface $\mathcal{S}$ depends only on the oriented boundary $\partial\mathcal{S}$ and not on the surface itself:

$$\iint_{\mathcal{S}} \mathbf{F} \cdot d\mathbf{S} = \oint_{\partial\mathcal{S}} \mathbf{A} \cdot d\mathbf{s} \qquad \boxed{7}$$

In particular, if $\mathcal{S}$ is *closed* (that is, $\partial\mathcal{S}$ is empty), then $\iint_{\mathcal{S}} \mathbf{F} \cdot d\mathbf{S} = 0$.

REMINDER By the ***flux*** *of a vector field through a surface, we mean the surface integral of the vector field.*

EXAMPLE 5 Let $\mathbf{F} = \text{curl}(\mathbf{A})$, where $\mathbf{A} = \langle y + z, \sin(xy), e^{xyz} \rangle$. Find the flux of **F** through the surfaces $\mathcal{S}_1$ and $\mathcal{S}_2$ in Figure 8 whose common boundary $\mathcal{C}$ is the unit circle in the xz-plane.

Solution With $\mathcal{C}$ oriented in the direction of the arrow, $\mathcal{S}_1$ lies to the left, and by Eq. (7),

$$\iint_{\mathcal{S}_1} \mathbf{F} \cdot d\mathbf{S} = \oint_{\mathcal{C}} \mathbf{A} \cdot d\mathbf{s}$$

We shall compute the line integral on the right. The parametrization $\mathbf{c}(t) = (\cos t, 0, \sin t)$ traces $\mathcal{C}$ in the direction of the arrow because it begins at $\mathbf{c}(0) = (1, 0, 0)$ and moves in the direction of $\mathbf{c}\left(\frac{\pi}{2}\right) = (0, 0, 1)$. We have

$$\mathbf{A}(\mathbf{c}(t)) = \langle 0 + \sin t, \sin(0), e^0 \rangle = \langle \sin t, 0, 1 \rangle$$

$$\mathbf{A}(\mathbf{c}(t)) \cdot \mathbf{c}'(t) = \langle \sin t, 0, 1 \rangle \cdot \langle -\sin t, 0, \cos t \rangle = -\sin^2 t + \cos t$$

$$\oint_{\mathcal{C}} \mathbf{A} \cdot d\mathbf{s} = \int_0^{2\pi} (-\sin^2 t + \cos t)\, dt = -\pi$$

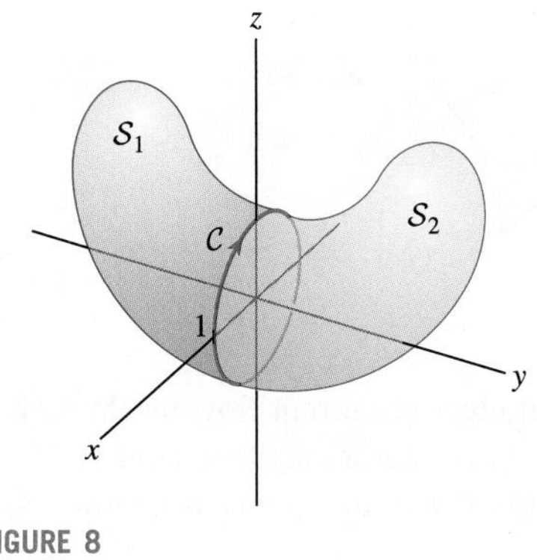

FIGURE 8

We conclude that $\iint_{\mathcal{S}_1} \mathbf{F} \cdot d\mathbf{S} = -\pi$. On the other hand, $\mathcal{S}_2$ lies on the right as you traverse $\mathcal{C}$. Therefore $\mathcal{S}_2$ has oriented boundary $-\mathcal{C}$, and

$$\iint_{\mathcal{S}_2} \mathbf{F} \cdot d\mathbf{S} = \oint_{-\mathcal{C}} \mathbf{A} \cdot d\mathbf{s} = -\oint_{\mathcal{C}} \mathbf{A} \cdot d\mathbf{s} = \pi$$ ■

CONCEPTUAL INSIGHT Interpretation of the Curl In Section 18.1, we showed that the quantity $\frac{\partial F_2}{\partial x} - \frac{\partial F_1}{\partial y}$ in Green's Theorem is the "circulation per unit of enclosed area." A similar interpretation is valid in $\mathbf{R}^3$.

Consider a plane through a point P with unit normal vector $\mathbf{e_n}$ and let $\mathcal{D}$ be a small domain containing P with boundary curve $\mathcal{C}$ (Figure 9). By Stokes' Theorem,

$$\oint_{\mathcal{C}} \mathbf{F} \cdot d\mathbf{s} \approx \iint_{\mathcal{D}} (\mathrm{curl}(\mathbf{F}) \cdot \mathbf{e_n})\, dS \qquad \boxed{8}$$

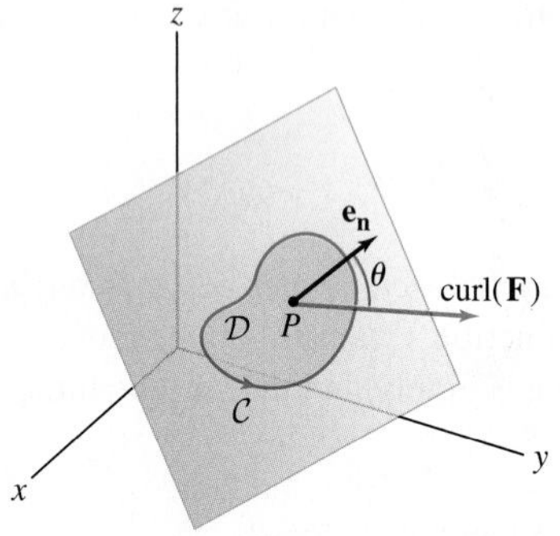

FIGURE 9 The curve $\mathcal{C}$ around P lies in the plane through P with normal vector $\mathbf{e_n}$.

The vector field curl(**F**) is continuous (its components are derivatives of the components of **F**), so its value does not change much on $\mathcal{D}$ if $\mathcal{D}$ is sufficiently small. To a first approximation, we can replace curl(**F**) by the constant value curl(**F**)(P), giving us the approximation

$$\iint_{\mathcal{D}} (\mathrm{curl}(\mathbf{F}) \cdot \mathbf{e_n})\, dS \approx \iint_{\mathcal{D}} (\mathrm{curl}(\mathbf{F})(P) \cdot \mathbf{e_n})\, dS$$
$$\approx (\mathrm{curl}(\mathbf{F})(P) \cdot \mathbf{e_n})\, \mathrm{Area}(\mathcal{D}) \qquad \boxed{9}$$

Furthermore, $\mathrm{curl}(\mathbf{F})(P) \cdot \mathbf{e_n} = \|\mathrm{curl}(\mathbf{F})(P)\| \cos\theta$, where θ is the angle between curl(**F**) and $\mathbf{e_n}$. Together, Eq. (8) and Eq. (9) give us

$$\oint_{\mathcal{C}} \mathbf{F} \cdot d\mathbf{s} \approx \|\mathrm{curl}(\mathbf{F})(P)\| (\cos\theta)\, \mathrm{Area}(\mathcal{D}) \qquad \boxed{10}$$

This is a remarkable result. It tells us that curl(**F**) encodes the the circulation per unit of enclosed area in every plane through P in a simple way—namely, as the dot product $\mathrm{curl}(\mathbf{F})(P) \cdot \mathbf{e_n}$. In particular, the circulation rate varies (to a first-order approximation) as the cosine of the angle θ between curl(**F**)(P) and $\mathbf{e_n}$.

We can also argue (as in Section 18.1 for vector fields in the plane) that if **F** is the velocity field of a fluid, then a small paddle wheel with normal $\mathbf{e_n}$ will rotate with an angular velocity of approximately $\frac{1}{2}\mathrm{curl}(\mathbf{F})(P) \cdot \mathbf{e_n}$ (see Figure 10).

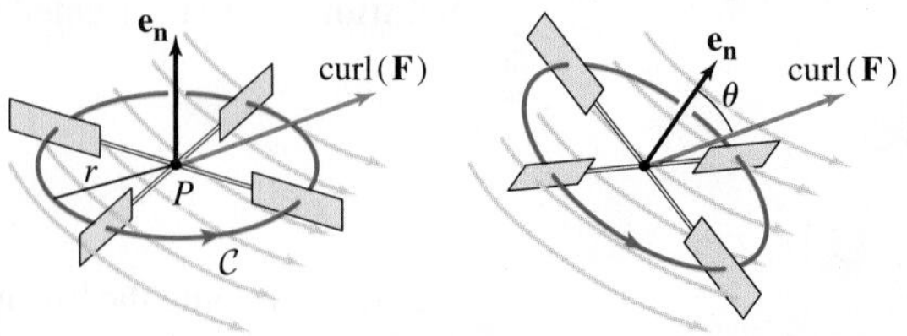

FIGURE 10 The paddle wheel can be oriented in different ways, as specified by the normal vector $\mathbf{e_n}$.

■ **EXAMPLE 6 Vector Potential for a Solenoid** An electric current flowing through a solenoid (a tightly wound spiral of wire; see Figure 11) creates a magnetic field **B**. If we assume that the solenoid is infinitely long, with radius R and the z-axis as central axis, then

$$\mathbf{B} = \begin{cases} \mathbf{0} & \text{if } r > R \\ B\mathbf{k} & \text{if } r < R \end{cases}$$

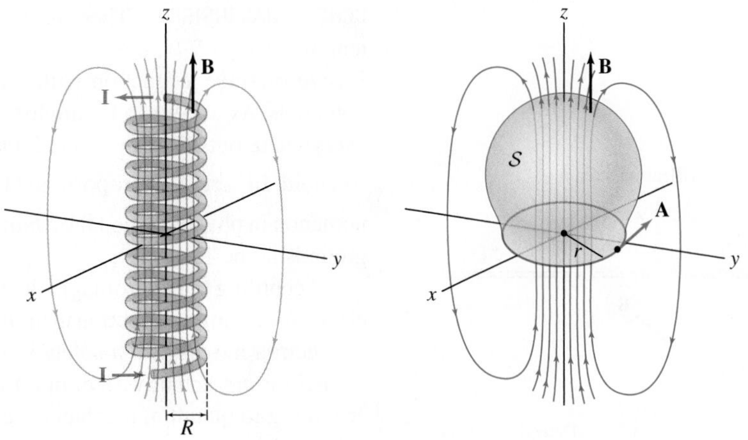

FIGURE 11 The magnetic field of a long solenoid is nearly uniform inside and weak outside. In practice, we treat the solenoid as "infinitely long" if it is very long in comparison with its radius.

where $r = (x^2 + y^2)^{1/2}$ and B is a constant that depends on the current strength and the spacing of the turns of wire.

(a) Show that a vector potential for **B** is

$$\mathbf{A} = \begin{cases} \frac{1}{2} R^2 B \left\langle -\frac{y}{r^2}, \frac{x}{r^2}, 0 \right\rangle & \text{if } r > R \\ \frac{1}{2} B \langle -y, x, 0 \rangle & \text{if } r < R \end{cases}$$

(b) Calculate the flux of **B** through the surface $\mathcal{S}$ (with upward-pointing normal) in Figure 11 whose boundary is a circle of radius r where $r > R$.

The vector potential **A** *is continuous but not differentiable on the cylinder $r = R$, that is, on the solenoid itself (Figure 12). The magnetic field* $\mathbf{B} = \text{curl}(\mathbf{A})$ *has a jump discontinuity where $r = R$. We take for granted the fact that Stokes' Theorem remains valid in this setting.*

Solution

(a) For any functions f and g,

$$\text{curl}(\langle f, g, 0\rangle) = \langle -g_z, f_z, g_x - f_y \rangle$$

Applying this to **A** for $r < R$, we obtain

$$\text{curl}(\mathbf{A}) = \frac{1}{2} B \left\langle 0, 0, \frac{\partial}{\partial x} x - \frac{\partial}{\partial y}(-y) \right\rangle = \langle 0, 0, B \rangle = B\mathbf{k} = \mathbf{B}$$

We leave it as an exercise [Exercise 29] to show that $\text{curl}(\mathbf{A}) = \mathbf{B} = \mathbf{0}$ for $r > R$.

(b) The boundary circle of $\mathcal{S}$ with counterclockwise parametrization $\mathbf{c}(t) = (r \cos t, r \sin t, 0)$, so

$$\mathbf{c}'(t) = \langle -r \sin t, r \cos t, 0 \rangle$$

$$\mathbf{A}(\mathbf{c}(t)) = \frac{1}{2} R^2 B r^{-1} \langle -\sin t, \cos t, 0 \rangle$$

$$\mathbf{A}(\mathbf{c}(t)) \cdot \mathbf{c}'(t) = \frac{1}{2} R^2 B \left((-\sin t)^2 + \cos^2 t \right) = \frac{1}{2} R^2 B$$

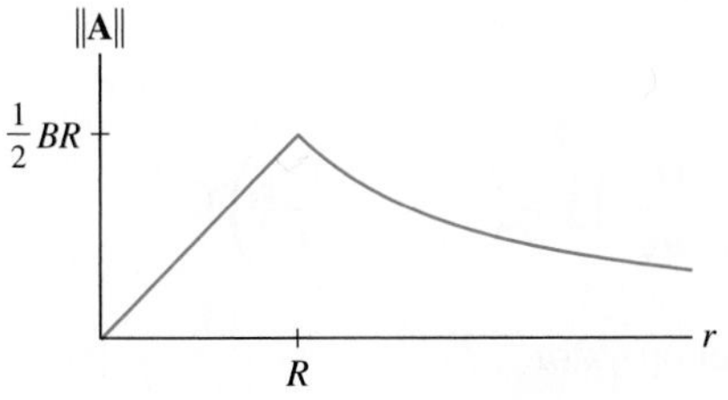

FIGURE 12 The magnitude $\|\mathbf{A}\|$ of the vector potential as a function of distance r to the z-axis.

By Stokes' Theorem, the flux of **B** through $\mathcal{S}$ is equal to

$$\iint_{\mathcal{S}} \mathbf{B} \cdot d\mathbf{S} = \oint_{\partial \mathcal{S}} \mathbf{A} \cdot d\mathbf{s} = \int_0^{2\pi} \mathbf{A}(\mathbf{c}(t)) \cdot \mathbf{c}'(t)\, dt = \frac{1}{2} R^2 B \int_0^{2\pi} dt = \pi R^2 B \quad \blacksquare$$

CONCEPTUAL INSIGHT There is an interesting difference between scalar and vector potentials. If $\mathbf{F} = \nabla V$, then the scalar potential V is constant in regions where the field $\mathbf{F}$ is zero (since a function with zero gradient is constant). This is not true for vector potentials. As we saw in Example 6, the magnetic field $\mathbf{B}$ produced by a solenoid is zero everywhere outside the solenoid, but the vector potential $\mathbf{A}$ is not constant outside the solenoid. In fact, $\mathbf{A}$ is proportional to $\left\langle -\frac{y}{r^2}, \frac{x}{r^2}, 0\right\rangle$. This is related to an intriguing phenomenon in physics called the *Aharonov-Bohm (AB) effect*, first proposed on theoretical grounds in the 1940s.

According to electromagnetic theory, a magnetic field $\mathbf{B}$ exerts a force on a moving electron, causing a deflection in the electron's path. We do not expect any deflection when an electron moves past a solenoid because $\mathbf{B}$ is zero outside the solenoid (in practice, the field is not actually zero, but it is very small—we ignore this difficulty). However, according to quantum mechanics, electrons have both particle and wave properties. In a double-slit experiment, a stream of electrons passing through two small slits creates a wavelike interference pattern on a detection screen (Figure 13). The AB effect predicts that if we place a small solenoid between the slits as in the figure (the solenoid is so small that the electrons never pass through it), then the interference pattern will shift slightly. It is as if the electrons are "aware" of the magnetic field inside the solenoid, even though they never encounter the field directly.

The AB effect was hotly debated until it was confirmed definitively in 1985, in experiments carried out by a team of Japanese physicists led by Akira Tonomura. The AB effect appeared to contradict "classical" electromagnetic theory, according to which the trajectory of an electron is determined by $\mathbf{B}$ alone. There is no such contradiction in quantum mechanics, because the behavior of the electrons is governed not by $\mathbf{B}$ but by a "wave function" derived from the nonconstant vector potential $\mathbf{A}$.

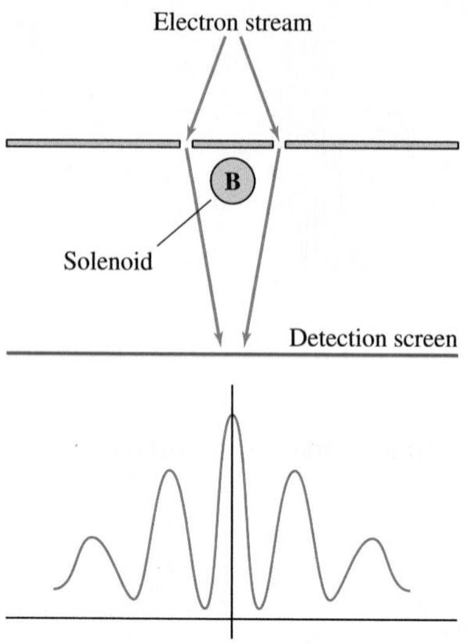

FIGURE 13 A stream of electrons passing through a double slit produces an interference pattern on the detection screen. The pattern shifts slightly when an electric current flows through the solenoid.

18.2 SUMMARY

- The *boundary* of a surface $\mathcal{S}$ is denoted $\partial\mathcal{S}$. We say that $\mathcal{S}$ is *closed* if $\partial\mathcal{S}$ is empty.
- Suppose that $\mathcal{S}$ is oriented (a continuously varying unit normal is specified at each point of $\mathcal{S}$). The *boundary orientation* of $\partial\mathcal{S}$ is defined as follows: If you walk along the boundary in the positive direction with your head pointing in the normal direction, then the surface is on your left.
-

$$\text{curl}(\mathbf{F}) = \begin{vmatrix} \mathbf{i} & \mathbf{j} & \mathbf{k} \\ \frac{\partial}{\partial x} & \frac{\partial}{\partial y} & \frac{\partial}{\partial z} \\ F_1 & F_2 & F_3 \end{vmatrix}$$

$$= \left(\frac{\partial F_3}{\partial y} - \frac{\partial F_2}{\partial z}\right)\mathbf{i} - \left(\frac{\partial F_3}{\partial x} - \frac{\partial F_1}{\partial z}\right)\mathbf{j} + \left(\frac{\partial F_2}{\partial x} - \frac{\partial F_1}{\partial y}\right)\mathbf{k}$$

Symbolically, $\text{curl}(\mathbf{F}) = \nabla \times \mathbf{F}$ where ∇ is the del operator

$$\nabla = \left\langle \frac{\partial}{\partial x}, \frac{\partial}{\partial y}, \frac{\partial}{\partial z}\right\rangle$$

- Stokes' Theorem relates the circulation around the boundary to the surface integral of the curl:

$$\oint_{\partial\mathcal{S}} \mathbf{F}\cdot d\mathbf{s} = \iint_{\mathcal{S}} \operatorname{curl}(\mathbf{F})\cdot d\mathbf{S}$$

- If $\mathbf{F} = \nabla V$, then $\operatorname{curl}(\mathbf{F}) = \mathbf{0}$.
- Surface Independence: If $\mathbf{F} = \operatorname{curl}(\mathbf{A})$, then the flux of $\mathbf{F}$ through a surface $\mathcal{S}$ depends only on the oriented boundary $\partial\mathcal{S}$ and not on the surface itself:

$$\iint_{\mathcal{S}} \mathbf{F}\cdot d\mathbf{S} = \oint_{\partial\mathcal{S}} \mathbf{A}\cdot d\mathbf{s}$$

In particular, if $\mathcal{S}$ is *closed* (that is, $\partial\mathcal{S}$ is empty) and $\mathbf{F} = \operatorname{curl}(\mathbf{A})$, then $\iint_{\mathcal{S}} \mathbf{F}\cdot d\mathbf{S} = 0$.

- The curl is interpreted as a vector that encodes circulation per unit area: If P is any point and $\mathbf{e_n}$ is a unit normal vector, then

$$\int_{\mathcal{C}} \mathbf{F}\cdot d\mathbf{s} \approx (\operatorname{curl}(\mathbf{F})(P)\cdot \mathbf{e_n})\,\text{Area}(\mathcal{D})$$

where $\mathcal{C}$ is a small, simple closed curve around P in the plane through P with normal vector $\mathbf{e_n}$, and $\mathcal{D}$ is the enclosed region.

18.2 EXERCISES

Preliminary Questions

1. Indicate with an arrow the boundary orientation of the boundary curves of the surfaces in Figure 14, oriented by the outward-pointing normal vectors.

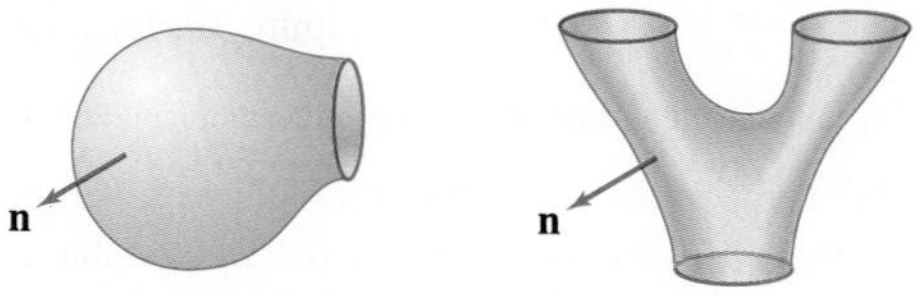

(A) (B)

FIGURE 14

2. Let $\mathbf{F} = \operatorname{curl}(\mathbf{A})$. Which of the following are related by Stokes' Theorem?

(a) The circulation of $\mathbf{A}$ and flux of $\mathbf{F}$.

(b) The circulation of $\mathbf{F}$ and flux of $\mathbf{A}$.

3. What is the definition of a vector potential?

4. Which of the following statements is correct?

(a) The flux of $\operatorname{curl}(\mathbf{A})$ through every oriented surface is zero.

(b) The flux of $\operatorname{curl}(\mathbf{A})$ through every closed, oriented surface is zero.

5. Which condition on $\mathbf{F}$ guarantees that the flux through $\mathcal{S}_1$ is equal to the flux through $\mathcal{S}_2$ for any two oriented surfaces $\mathcal{S}_1$ and $\mathcal{S}_2$ with the same oriented boundary?

Exercises

In Exercises 1–4, calculate $\operatorname{curl}(\mathbf{F})$.

1. $\mathbf{F} = \langle z - y^2, x + z^3, y + x^2\rangle$

2. $\mathbf{F} = \left\langle \dfrac{y}{x}, \dfrac{y}{z}, \dfrac{z}{x}\right\rangle$

3. $\mathbf{F} = \langle e^y, \sin x, \cos x\rangle$

4. $\mathbf{F} = \left\langle \dfrac{x}{x^2+y^2}, \dfrac{y}{x^2+y^2}, 0\right\rangle$

In Exercises 5–8, verify Stokes' Theorem for the given vector field and surface, oriented with an upward-pointing normal.

5. $\mathbf{F} = \langle 2xy, x, y + z\rangle$, the surface $z = 1 - x^2 - y^2$ for $x^2 + y^2 \le 1$

6. $\mathbf{F} = \langle yz, 0, x\rangle$, the portion of the plane $\dfrac{x}{2} + \dfrac{y}{3} + z = 1$ where $x, y, z \ge 0$

7. $\mathbf{F} = \langle e^{y-z}, 0, 0\rangle$, the square with vertices $(1, 0, 1)$, $(1, 1, 1)$, $(0, 1, 1)$, and $(0, 0, 1)$

8. $\mathbf{F} = \langle y, x, x^2 + y^2\rangle$, the upper hemisphere $x^2 + y^2 + z^2 = 1$, $z \ge 0$

In Exercises 9 and 10, calculate $\operatorname{curl}(\mathbf{F})$ *and then use Stokes' Theorem to compute the flux of* $\operatorname{curl}(\mathbf{F})$ *through the given surface as a line integral.*

9. $\mathbf{F} = \left\langle e^{z^2} - y, e^{z^3} + x, \cos(xz)\right\rangle$, the upper hemisphere $x^2 + y^2 + z^2 = 1$, $z \ge 0$ with outward-pointing normal

10. $\mathbf{F} = \left\langle x + y, z^2 - 4, x\sqrt{y^2 + 1}\right\rangle$, surface of the wedge-shaped box in Figure 15 (bottom included, top excluded) with outward-pointing normal

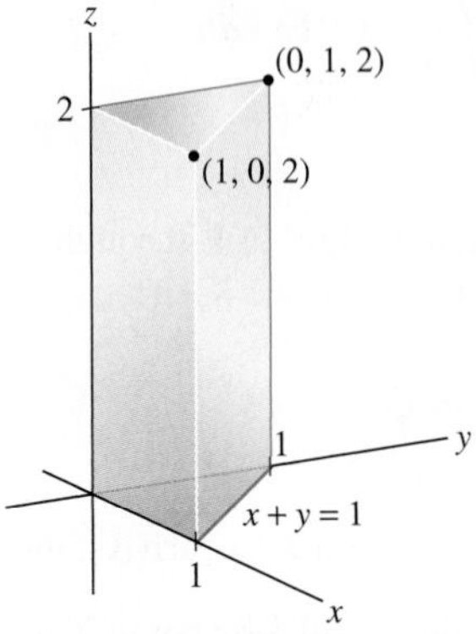

FIGURE 15

11. Let $\mathcal{S}$ be the surface of the cylinder (not including the top and bottom) of radius 2 for $1 \le z \le 6$, oriented with outward-pointing normal (Figure 16).

(a) Indicate with an arrow the orientation of $\partial\mathcal{S}$ (the top and bottom circles).

(b) Verify Stokes' Theorem for $\mathcal{S}$ and $\mathbf{F} = \langle yz^2, 0, 0\rangle$.

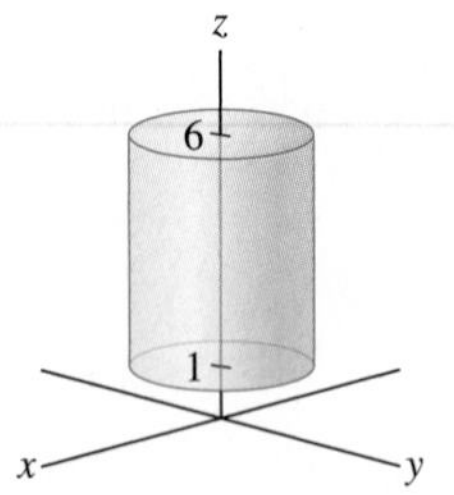

FIGURE 16

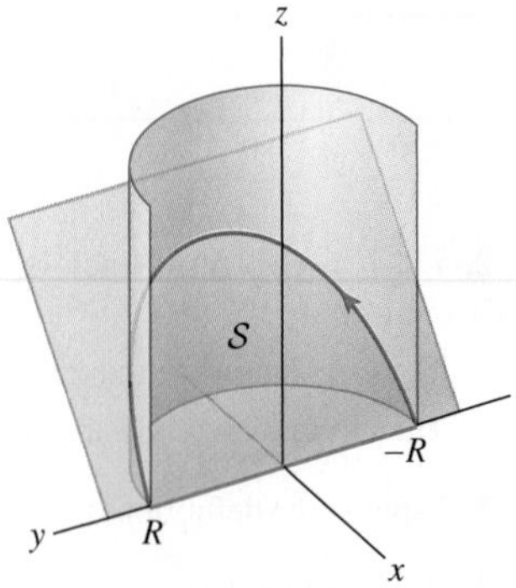

FIGURE 17

12. Let $\mathcal{S}$ be the portion of the plane $z = x$ contained in the half-cylinder of radius R depicted in Figure 17. Use Stokes' Theorem to calculate the circulation of $\mathbf{F} = \langle z, x, y + 2z\rangle$ around the boundary of $\mathcal{S}$ (a half-ellipse) in the counterclockwise direction when viewed from above. *Hint:* Show that $\text{curl}(\mathbf{F})$ is orthogonal to the normal vector to the plane.

13. Let I be the flux of $\mathbf{F} = \langle e^y, 2xe^{x^2}, z^2\rangle$ through the upper hemisphere $\mathcal{S}$ of the unit sphere.

(a) Let $\mathbf{G} = \langle e^y, 2xe^{x^2}, 0\rangle$. Find a vector field $\mathbf{A}$ such that $\text{curl}(\mathbf{A}) = \mathbf{G}$.

(b) Use Stokes' Theorem to show that the flux of $\mathbf{G}$ through $\mathcal{S}$ is zero. *Hint:* Calculate the circulation of $\mathbf{A}$ around $\partial\mathcal{S}$.

(c) Calculate I. *Hint:* Use (b) to show that I is equal to the flux of $\langle 0, 0, z^2\rangle$ through $\mathcal{S}$.

14. Let $\mathbf{F} = \langle 0, -z, 1\rangle$. Let $\mathcal{S}$ be the spherical cap $x^2 + y^2 + z^2 \le 1$, where $z \ge \frac{1}{2}$. Evaluate $\iint_{\mathcal{S}} \mathbf{F} \cdot d\mathbf{S}$ directly as a surface integral. Then verify that $\mathbf{F} = \text{curl}(\mathbf{A})$, where $\mathbf{A} = \langle 0, x, xz\rangle$ and evaluate the surface integral again using Stokes' Theorem.

15. Let $\mathbf{A}$ be the vector potential and $\mathbf{B}$ the magnetic field of the infinite solenoid of radius R in Example 6. Use Stokes' Theorem to compute:

(a) The flux of $\mathbf{B}$ through a circle in the xy-plane of radius $r < R$

(b) The circulation of $\mathbf{A}$ around the boundary $\mathcal{C}$ of a surface lying outside the solenoid

16. The magnetic field $\mathbf{B}$ due to a small current loop (which we place at the origin) is called a **magnetic dipole** (Figure 18). Let $\rho = (x^2 + y^2 + z^2)^{1/2}$. For ρ large, $\mathbf{B} = \text{curl}(\mathbf{A})$, where

$$\mathbf{A} = \left\langle -\frac{y}{\rho^3}, \frac{x}{\rho^3}, 0\right\rangle$$

(a) Let $\mathcal{C}$ be a horizontal circle of radius R with center $(0, 0, c)$, where c is large. Show that $\mathbf{A}$ is tangent to $\mathcal{C}$.

(b) Use Stokes' Theorem to calculate the flux of $\mathbf{B}$ through $\mathcal{C}$.

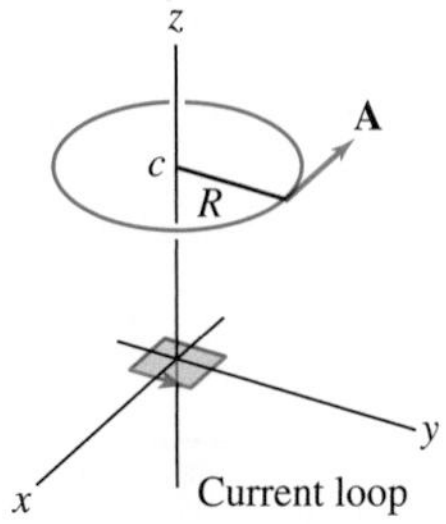

FIGURE 18

17. A uniform magnetic field $\mathbf{B}$ has constant strength b in the z-direction [that is, $\mathbf{B} = \langle 0, 0, b\rangle$].

(a) Verify that $\mathbf{A} = \frac{1}{2}\mathbf{B} \times \mathbf{r}$ is a vector potential for $\mathbf{B}$, where $\mathbf{r} = \langle x, y, 0\rangle$.

(b) Calculate the flux of $\mathbf{B}$ through the rectangle with vertices A, B, C, and D in Figure 19.

18. Let $\mathbf{F} = \langle -x^2 y, x, 0\rangle$. Referring to Figure 19, let $\mathcal{C}$ be the closed path $ABCD$. Use Stokes' Theorem to evaluate $\int_{\mathcal{C}} \mathbf{F} \cdot d\mathbf{s}$ in two ways. First, regard $\mathcal{C}$ as the boundary of the rectangle with vertices A, B, C, and D. Then treat $\mathcal{C}$ as the boundary of the wedge-shaped box with open top.

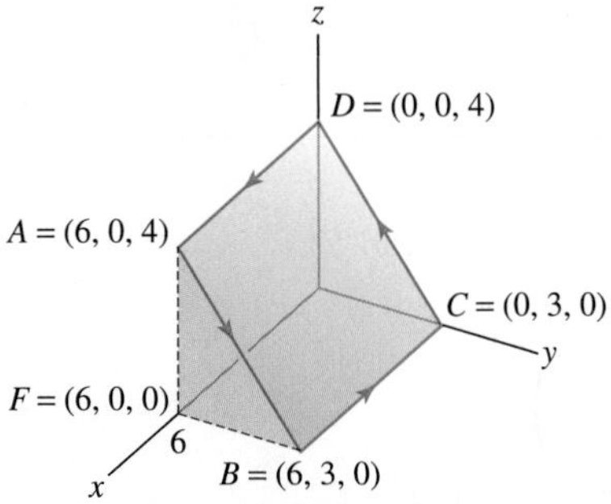

FIGURE 19

19. Let $\mathbf{F} = \langle y^2, 2z + x, 2y^2\rangle$. Use Stokes' Theorem to find a plane with equation $ax + by + cz = 0$ (where a, b, c are not all zero) such that $\oint_{\mathcal{C}} \mathbf{F} \cdot d\mathbf{s} = 0$ for every closed $\mathcal{C}$ lying in the plane. *Hint:* Choose a, b, c so that $\text{curl}(\mathbf{F})$ lies in the plane.

20. Let $\mathbf{F} = \langle -z^2, 2zx, 4y - x^2\rangle$ and let $\mathcal{C}$ be a simple closed curve in the plane $x + y + z = 4$ that encloses a region of area 16 (Figure 20).

Calculate $\oint_C \mathbf{F} \cdot d\mathbf{s}$, where $\mathcal{C}$ is oriented in the counterclockwise direction (when viewed from above the plane).

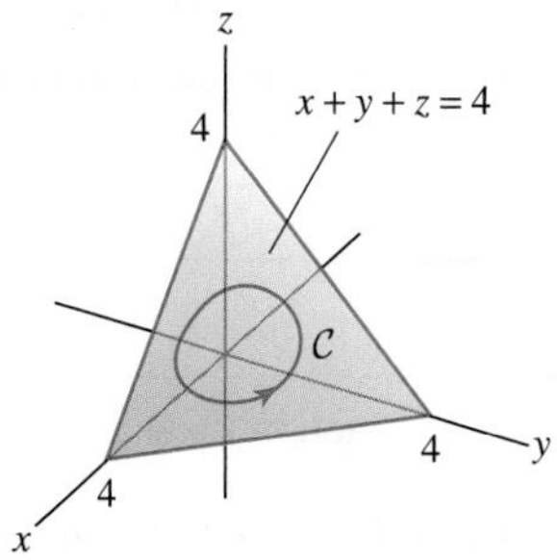

FIGURE 20

21. Let $\mathbf{F} = \langle y^2, x^2, z^2 \rangle$. Show that

$$\int_{\mathcal{C}_1} \mathbf{F} \cdot d\mathbf{s} = \int_{\mathcal{C}_2} \mathbf{F} \cdot d\mathbf{s}$$

for any two closed curves lying on a cylinder whose central axis is the z-axis (Figure 21).

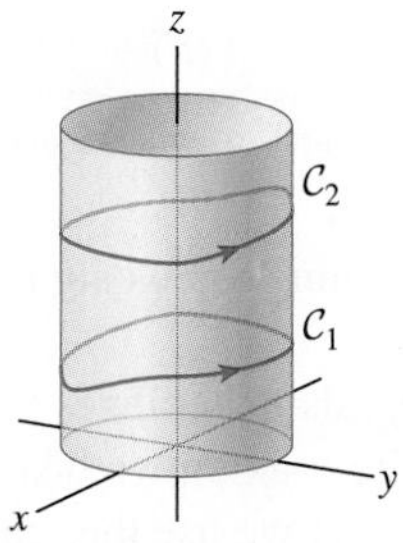

FIGURE 21

22. The curl of a vector field $\mathbf{F}$ at the origin is $\mathbf{v}_0 = \langle 3, 1, 4 \rangle$. Estimate the circulation around the small parallelogram spanned by the vectors $\mathbf{A} = \langle 0, \frac{1}{2}, \frac{1}{2} \rangle$ and $\mathbf{B} = \langle 0, 0, \frac{1}{3} \rangle$.

23. You know two things about a vector field $\mathbf{F}$:

(i) $\mathbf{F}$ has a vector potential $\mathbf{A}$ (but $\mathbf{A}$ is unknown).

(ii) The circulation of $\mathbf{A}$ around the unit circle (oriented counterclockwise) is 25.

Determine the flux of $\mathbf{F}$ through the surface $\mathcal{S}$ in Figure 22, oriented with upward pointing normal.

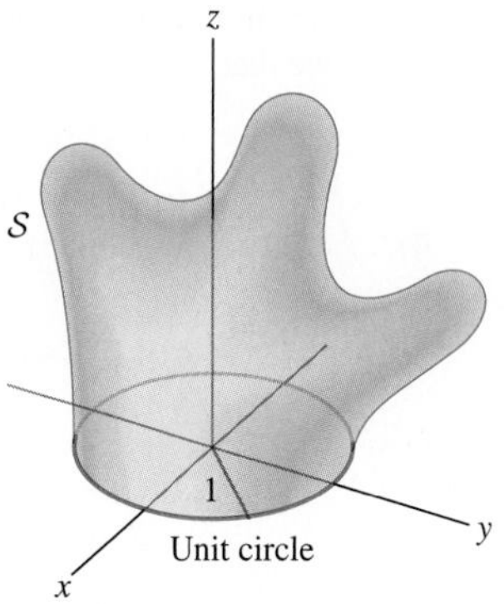

FIGURE 22 Surface $\mathcal{S}$ whose boundary is the unit circle.

24. Suppose that $\mathbf{F}$ has a vector potential and that $\mathbf{F}(x, y, 0) = \mathbf{k}$. Find the flux of $\mathbf{F}$ through the surface $\mathcal{S}$ in Figure 22, oriented with upward pointing normal.

25. Prove that $\text{curl}(f\mathbf{a}) = \nabla f \times \mathbf{a}$, where f is a differentiable function and $\mathbf{a}$ is a constant vector.

26. Show that $\text{curl}(\mathbf{F}) = \mathbf{0}$ if $\mathbf{F}$ is **radial**, meaning that $\mathbf{F} = f(\rho)\langle x, y, z \rangle$ for some function $f(\rho)$, where $\rho = \sqrt{x^2 + y^2 + z^2}$. *Hint:* It is enough to show that one component of curl($\mathbf{F}$) is zero, because it will then follow for the other two components by symmetry.

27. Prove the following Product Rule:

$$\text{curl}(f\mathbf{F}) = f\,\text{curl}(\mathbf{F}) + \nabla f \times \mathbf{F}$$

28. Assume that f and g have continuous partial derivatives of order 2. Prove that

$$\oint_{\partial \mathcal{S}} f \nabla(g) \cdot d\mathbf{s} = \iint_{\mathcal{S}} \nabla(f) \times \nabla(g) \cdot d\mathbf{S}$$

29. Verify that $\mathbf{B} = \text{curl}(\mathbf{A})$ for $r > R$ in the setting of Example 6.

30. Explain carefully why Green's Theorem is a special case of Stokes' Theorem.

Further Insights and Challenges

31. In this exercise, we use the notation of the proof of Theorem 1 and prove

$$\oint_{\mathcal{C}} F_3(x, y, z)\mathbf{k} \cdot d\mathbf{s} = \iint_{\mathcal{S}} \text{curl}(F_3(x, y, z)\mathbf{k}) \cdot d\mathbf{S} \qquad \boxed{11}$$

In particular, $\mathcal{S}$ is the graph of $z = f(x, y)$ over a domain $\mathcal{D}$, and $\mathcal{C}$ is the boundary of $\mathcal{S}$ with parametrization $(x(t), y(t), f(x(t), y(t)))$.

(a) Use the Chain Rule to show that

$$F_3(x, y, z)\mathbf{k} \cdot d\mathbf{s} = F_3(x(t), y(t), f(x(t), y(t))) \Big(f_x(x(t), y(t))x'(t) + f_y(x(t), y(t))y'(t)\Big)\,dt$$

and verify that

$$\oint_{\mathcal{C}} F_3(x, y, z)\mathbf{k} \cdot d\mathbf{s} = \oint_{\mathcal{C}_0} \big\langle F_3(x, y, z) f_x(x, y), F_3(x, y, z) f_y(x, y) \big\rangle \cdot d\mathbf{s}$$

where $\mathcal{C}_0$ has parametrization $(x(t), y(t))$.

(b) Apply Green's Theorem to the line integral over $\mathcal{C}_0$ and show that the result is equal to the right-hand side of Eq. (11).

32. Let $\mathbf{F}$ be a continuously differentiable vector field in $\mathbf{R}^3$, Q a point, and $\mathcal{S}$ a plane containing Q with unit normal vector $\mathbf{e}$. Let $\mathcal{C}_r$ be a circle of radius r centered at Q in $\mathcal{S}$, and let $\mathcal{S}_r$ be the disk enclosed by $\mathcal{C}_r$. Assume $\mathcal{S}_r$ is oriented with unit normal vector $\mathbf{e}$.

(a) Let $m(r)$ and $M(r)$ be the minimum and maximum values of $\mathrm{curl}(\mathbf{F}(P)) \cdot \mathbf{e}$ for $P \in \mathcal{S}_r$. Prove that

$$m(r) \le \frac{1}{\pi r^2} \iint_{\mathcal{S}_r} \mathrm{curl}(\mathbf{F}) \cdot d\mathbf{S} \le M(r)$$

(b) Prove that

$$\mathrm{curl}(\mathbf{F}(Q)) \cdot \mathbf{e} = \lim_{r \to 0} \frac{1}{\pi r^2} \int_{\mathcal{C}_r} \mathbf{F} \cdot d\mathbf{s}$$

This proves that $\mathrm{curl}(\mathbf{F}(Q)) \cdot \mathbf{e}$ is the circulation per unit area in the plane $\mathcal{S}$.

18.3 Divergence Theorem

We have studied several "Fundamental Theorems." Each of these is a relation of the type:

$$\text{Integral of a derivative on an oriented domain} = \text{Integral over the } \textit{oriented boundary} \text{ of the domain}$$

Here are the examples we have seen so far:

- In single-variable calculus, the Fundamental Theorem of Calculus (FTC) relates the integral of $f'(x)$ over an interval $[a, b]$ to the "integral" of $f(x)$ over the boundary of $[a, b]$ consisting of two points a and b:

$$\underbrace{\int_a^b f'(x)\, dx}_{\text{Integral of derivative over } [a,b]} = \underbrace{f(b) - f(a)}_{\text{"Integral" over the boundary of } [a,b]}$$

  The boundary of $[a, b]$ is oriented by assigning a plus sign to b and a minus sign to a.

- The Fundamental Theorem for Line Integrals generalizes the FTC: Instead of an interval $[a, b]$ (a path from a to b along the x-axis), we take any path from points P to Q in $\mathbf{R}^3$ (Figure 1), and instead of $f'(x)$ we use the gradient:

$$\underbrace{\int_{\mathcal{C}} \nabla V \cdot d\mathbf{s}}_{\text{Integral of derivative over a curve}} = \underbrace{V(Q) - V(P)}_{\text{"Integral" over the boundary } \partial\mathcal{C} = Q - P}$$

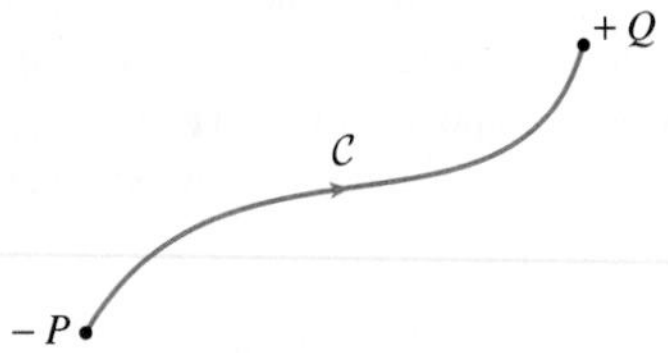

FIGURE 1 The oriented boundary of $\mathcal{C}$ is $\partial\mathcal{C} = Q - P$.

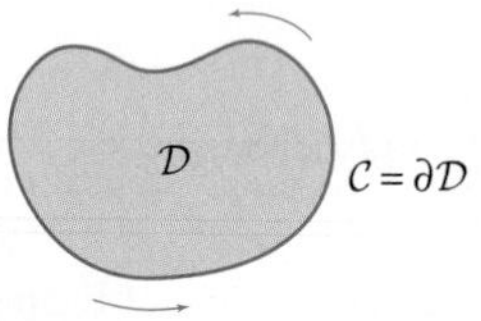

FIGURE 2 Domain $\mathcal{D}$ in $\mathbf{R}^2$ with boundary curve $\mathcal{C} = \partial\mathcal{D}$.

- Green's Theorem is a two-dimensional version of the FTC that relates the integral of a derivative over a domain $\mathcal{D}$ in the plane to an integral over its boundary curve $\mathcal{C} = \partial\mathcal{D}$ (Figure 2):

$$\underbrace{\iint_{\mathcal{D}} \left(\frac{\partial F_2}{\partial y} - \frac{\partial F_1}{\partial x} \right) dA}_{\text{Integral of derivative over domain}} = \underbrace{\int_{\mathcal{C}} \mathbf{F} \cdot d\mathbf{s}}_{\text{Integral over boundary curve}}$$

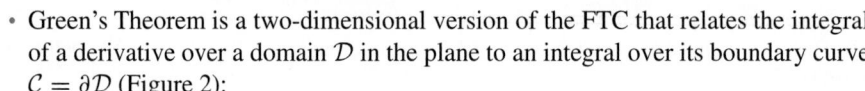

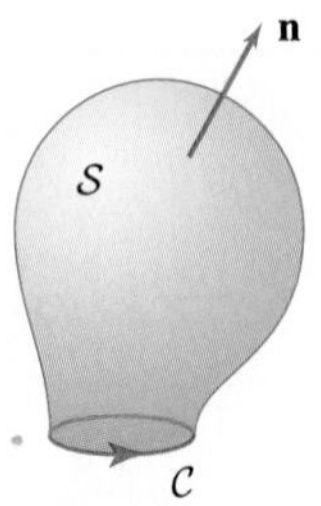

FIGURE 3 The oriented boundary of $\mathcal{S}$ is $\mathcal{C} = \partial\mathcal{S}$.

- Stokes' Theorem extends Green's Theorem: Instead of a domain in the plane (a flat surface), we allow any surface in $\mathbf{R}^3$ (Figure 3). The appropriate derivative is the curl:

$$\underbrace{\iint_{\mathcal{S}} \mathrm{curl}(\mathbf{F}) \cdot d\mathbf{S}}_{\text{Integral of derivative over surface}} = \underbrace{\int_{\mathcal{C}} \mathbf{F} \cdot d\mathbf{s}}_{\text{Integral over boundary curve}}$$

Our last theorem—the Divergence Theorem—follows this pattern:

$$\underbrace{\iiint_{\mathcal{W}} \mathrm{div}(\mathbf{F})\, d\mathcal{W}}_{\text{Integral of derivative over 3-D region}} = \underbrace{\iint_{\mathcal{S}} \mathbf{F} \cdot d\mathbf{S}}_{\text{Integral over boundary surface}}$$

Here, $\mathcal{S}$ is a *closed* surface that encloses a 3-D region $\mathcal{W}$. In other words, $\mathcal{S}$ is the boundary of $\mathcal{W}$: $\mathcal{S} = \partial\mathcal{W}$. Recall that a closed surface is a surface that "holds air." Figure 4 shows two examples of regions and boundary surfaces that we will consider.

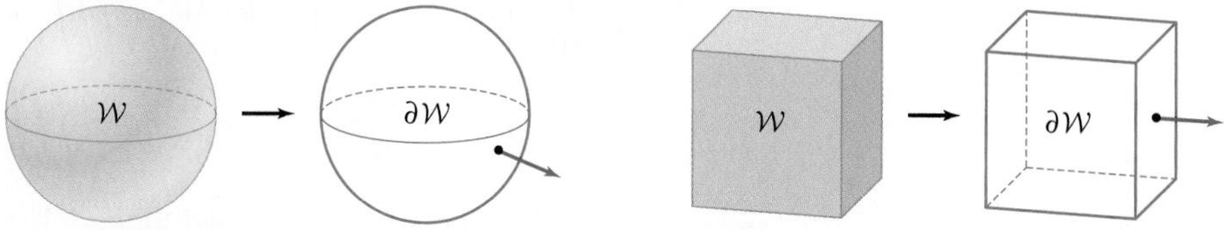

FIGURE 4

More advanced treatments of vector calculus use the theory of "differential forms" to formulate a general version of Stokes' Theorem that is valid in all dimensions and includes each of our main theorems (Green's, Stokes', Divergence) as a special case.

The derivative appearing in the Divergence Theorem is the **divergence** of a vector field $\mathbf{F} = \langle F_1, F_2, F_3\rangle$, defined by

$$\operatorname{div}(\mathbf{F}) = \frac{\partial F_1}{\partial x} + \frac{\partial F_2}{\partial y} + \frac{\partial F_3}{\partial z} \qquad \boxed{1}$$

We often write the divergence as a symbolic dot product:

$$\nabla \cdot \mathbf{F} = \left\langle \frac{\partial}{\partial x}, \frac{\partial}{\partial y}, \frac{\partial}{\partial z} \right\rangle \cdot \langle F_1, F_2, F_3\rangle = \frac{\partial F_1}{\partial x} + \frac{\partial F_2}{\partial y} + \frac{\partial F_3}{\partial z}$$

Note that, unlike the gradient and curl, the divergence is a scalar function. Like the gradient and curl, the divergence obeys the **linearity** rules:

$$\operatorname{div}(\mathbf{F} + \mathbf{G}) = \operatorname{div}(\mathbf{F}) + \operatorname{div}(\mathbf{G})$$
$$\operatorname{div}(c\mathbf{F}) = c\,\operatorname{div}(\mathbf{F}) \qquad (c \text{ any constant})$$

■ **EXAMPLE 1** Evaluate the divergence of $\mathbf{F} = \left\langle e^{xy}, xy, z^4 \right\rangle$ at $P = (1, 0, 2)$.

Solution

$$\operatorname{div}(\mathbf{F}) = \frac{\partial}{\partial x} e^{xy} + \frac{\partial}{\partial y} xy + \frac{\partial}{\partial z} z^4 = ye^{xy} + x + 4z^3$$

$$\operatorname{div}(\mathbf{F})(P) = \operatorname{div}(\mathbf{F})(1, 0, 2) = 0 \cdot e^0 + 1 + 4 \cdot 2^3 = 33$$ ■

THEOREM 1 Divergence Theorem Let $\mathcal{S}$ be a closed surface that encloses a region $\mathcal{W}$ in $\mathbf{R}^3$. Assume that $\mathcal{S}$ is piecewise smooth and is oriented by normal vectors pointing to the outside of $\mathcal{W}$. Let $\mathbf{F}$ be a vector field whose domain contains $\mathcal{W}$. Then

$$\iint_{\mathcal{S}} \mathbf{F} \cdot d\mathbf{S} = \iiint_{\mathcal{W}} \operatorname{div}(\mathbf{F})\, dV \qquad \boxed{2}$$

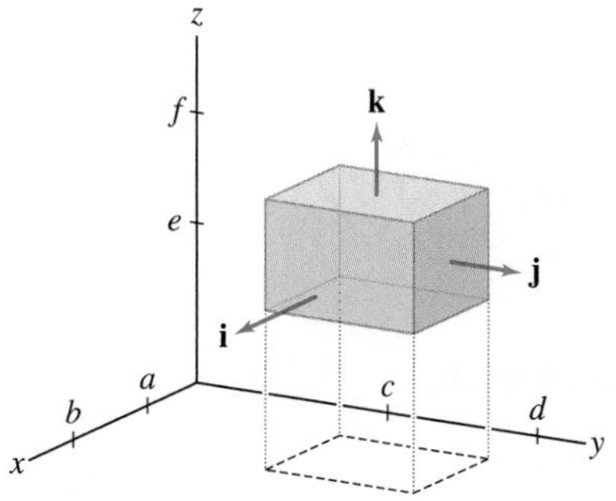

FIGURE 5 A box $\mathcal{W} = [a, b] \times [c, d] \times [e, f]$.

Proof We prove the Divergence Theorem in the special case that $\mathcal{W}$ is a box $[a, b] \times [c, d] \times [e, f]$ as in Figure 5. The proof can be modified to treat more general regions such as the interiors of spheres and cylinders.

REMINDER The Divergence Theorem states

$$\iint_S \mathbf{F}\cdot d\mathbf{S} = \iiint_{\mathcal{W}} \operatorname{div}(\mathbf{F})\,dV$$

We write each side of Eq. (2) as a sum over components:

$$\iint_{\partial\mathcal{W}} (F_1\mathbf{i}+F_2\mathbf{j}+F_3\mathbf{k})\cdot d\mathbf{S} = \iint_{\partial\mathcal{W}} F_1\mathbf{i}\cdot d\mathbf{S} + \iint_{\partial\mathcal{W}} F_2\mathbf{j}\cdot d\mathbf{S} + \iint_{\partial\mathcal{W}} F_3\mathbf{k}\cdot d\mathbf{S}$$

$$\iiint_{\mathcal{W}} \operatorname{div}(F_1\mathbf{i}+F_2\mathbf{j}+F_3\mathbf{k})\,dV = \iiint_{\mathcal{W}} \operatorname{div}(F_1\mathbf{i})\,dV + \iiint_{\mathcal{W}} \operatorname{div}(F_2\mathbf{j})\,dV + \iiint_{\mathcal{W}} \operatorname{div}(F_3\mathbf{k})\,dV$$

As in the proofs of Green's and Stokes' Theorems, we show that corresponding terms are equal. It will suffice to carry out the argument for the **i**-component (the other two components are similar). Thus we assume that $\mathbf{F} = F_1\mathbf{i}$.

The surface integral over boundary $\mathcal{S}$ of the box is the sum of the integrals over the six faces. However, $\mathbf{F} = F_1\mathbf{i}$ is orthogonal to the normal vectors to the top and bottom as well as the two side faces because $\mathbf{F}\cdot\mathbf{j} = \mathbf{F}\cdot\mathbf{k} = 0$. Therefore, the surface integrals over these faces are zero. Nonzero contributions come only from the front and back faces, which we denote $\mathcal{S}_f$ and $\mathcal{S}_b$ (Figure 6):

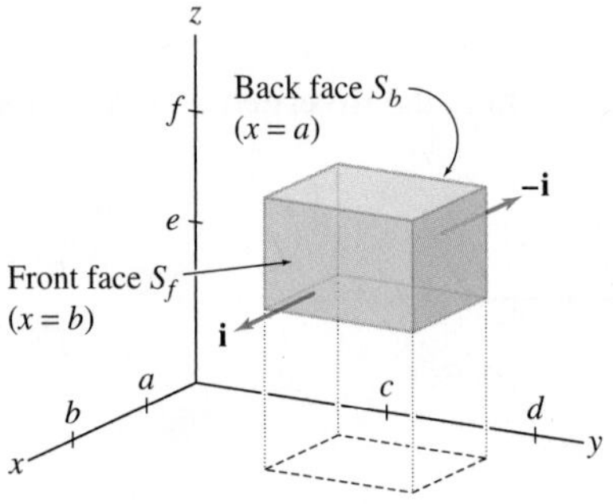

FIGURE 6

$$\iint_{\mathcal{S}} \mathbf{F}\cdot d\mathbf{S} = \iint_{\mathcal{S}_f} \mathbf{F}\cdot d\mathbf{S} + \iint_{\mathcal{S}_b} \mathbf{F}\cdot d\mathbf{S}$$

To evaluate these integrals, we parametrize $\mathcal{S}_f$ and $\mathcal{S}_b$ by

$$G_f(y,z) = (b,y,z), \qquad c \le y \le d,\ e \le z \le f$$

$$G_b(y,z) = (a,y,z), \qquad c \le y \le d,\ e \le z \le f$$

The normal vectors for these parametrizations are

$$\frac{\partial G_f}{\partial y} \times \frac{\partial G_f}{\partial z} = \mathbf{j}\times\mathbf{k} = \mathbf{i}$$

$$\frac{\partial G_b}{\partial y} \times \frac{\partial G_b}{\partial z} = \mathbf{j}\times\mathbf{k} = \mathbf{i}$$

However, the outward-pointing normal for $\mathcal{S}_b$ is $-\mathbf{i}$, so a minus sign is needed in the surface integral over $\mathcal{S}_b$ using the parametrization G_b:

$$\iint_{\mathcal{S}_f} \mathbf{F}\cdot d\mathbf{S} + \iint_{\mathcal{S}_b} \mathbf{F}\cdot d\mathbf{S} = \int_e^f \int_c^d F_1(b,y,z)\,dy\,dz - \int_e^f \int_c^d F_1(a,y,z)\,dy\,dz$$

$$= \int_e^f \int_c^d \Big(F_1(b,y,z) - F_1(a,y,z)\Big)\,dy\,dz$$

By the FTC in one variable,

$$F_1(b,y,z) - F_1(a,y,z) = \int_a^b \frac{\partial F_1}{\partial x}(x,y,z)\,dx$$

Since $\operatorname{div}(\mathbf{F}) = \operatorname{div}(F_1\mathbf{i}) = \dfrac{\partial F_1}{\partial x}$, we obtain the desired result:

$$\iint_{\mathcal{S}} \mathbf{F}\cdot d\mathbf{S} = \int_e^f \int_c^d \int_a^b \frac{\partial F_1}{\partial x}(x,y,z)\,dx\,dy\,dz = \iiint_{\mathcal{W}} \operatorname{div}(\mathbf{F})\,dV \qquad \blacksquare$$

The names attached to mathematical theorems often conceal a more complex historical development. What we call Green's Theorem was stated by Augustin Cauchy in 1846 but it was never stated by George Green himself (he published a result that implies Green's Theorem in 1828). Stokes' Theorem first appeared as a problem on a competitive exam written by George Stokes at Cambridge University, but William Thomson (Lord Kelvin) had previously stated the theorem in a letter to Stokes. Gauss published special cases of the Divergence Theorem in 1813 and later in 1833 and 1839, while the general theorem was stated and proved by the Russian mathematician Michael Ostrogradsky in 1826. For this reason, the Divergence Theorem is also referred to as "Gauss's Theorem" or the "Gauss-Ostrogradsky Theorem."

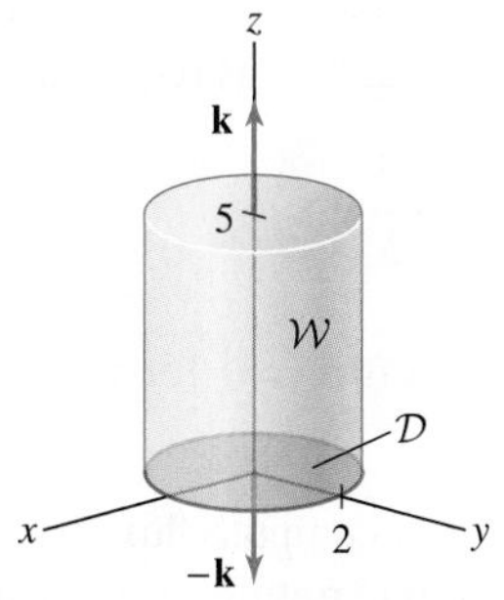

FIGURE 7 Cylinder of radius 2 and height 5.

■ **EXAMPLE 2 Verifying the Divergence Theorem** Verify Theorem 1 for $\mathbf{F} = \langle y, yz, z^2 \rangle$ and the cylinder in Figure 7.

Solution We must verify that the flux $\iint_{\mathcal{S}} \mathbf{F} \cdot d\mathbf{S}$, where $\mathcal{S}$ is the boundary of the cylinder, is equal to the integral of $\text{div}(\mathcal{W})$ over the cylinder. We compute the flux through $\mathcal{S}$ first: It is the sum of three surface integrals over the side, the top, and the bottom.

***Step 1.* Integrate over the side of the cylinder.**
We use the standard parametrization of the cylinder:

$$G(\theta, z) = (2\cos\theta, 2\sin\theta, z), \qquad 0 \le \theta < 2\pi, \quad 0 \le z \le 5$$

The normal vector is

$$\mathbf{n} = \mathbf{T}_\theta \times \mathbf{T}_z = \langle -2\sin\theta, 2\cos\theta, 0 \rangle \times \langle 0, 0, 1 \rangle = \langle 2\cos\theta, 2\sin\theta, 0 \rangle$$

and $\mathbf{F}(G(\theta, z)) = \langle y, yz, z^2 \rangle = \langle 2\sin\theta, 2z\sin\theta, z^2 \rangle$. Thus

$$\mathbf{F} \cdot d\mathbf{S} = \langle 2\sin\theta, 2z\sin\theta, z^2 \rangle \cdot \langle 2\cos\theta, 2\sin\theta, 0 \rangle \, d\theta \, dz$$

$$= 4\cos\theta\sin\theta + 4z\sin^2\theta \, d\theta \, dz$$

$$\iint_{\text{side}} \mathbf{F} \cdot d\mathbf{S} = \int_0^5 \int_0^{2\pi} (4\cos\theta\sin\theta + 4z\sin^2\theta) \, d\theta \, dz$$

$$= 0 + 4\pi \int_0^5 z \, dz = 4\pi \left(\frac{25}{2} \right) = 50\pi \qquad \boxed{3}$$

← REMINDER In Eq. (3), we use

$$\int_0^{2\pi} \cos\theta \sin\theta \, d\theta = 0$$

$$\int_0^{2\pi} \sin^2\theta \, d\theta = \pi$$

***Step 2.* Integrate over the top and bottom of the cylinder.**
The top of the cylinder is at height $z = 5$, so we can parametrize the top by $G(x, y) = (x, y, 5)$ for (x, y) in the disk $\mathcal{D}$ of radius 2:

$$\mathcal{D} = \{(x, y) : x^2 + y^2 \le 4\}$$

Then

$$\mathbf{n} = \mathbf{T}_x \times \mathbf{T}_y = \langle 1, 0, 0 \rangle \times \langle 0, 1, 0 \rangle = \langle 0, 0, 1 \rangle$$

and since $\mathbf{F}(G(x, y)) = \mathbf{F}(x, y, 5) = \langle y, 5y, 5^2 \rangle$, we have

$$\mathbf{F}(G(x, y)) \cdot \mathbf{n} = \langle y, 5y, 5^2 \rangle \cdot \langle 0, 0, 1 \rangle = 25$$

$$\iint_{\text{top}} \mathbf{F} \cdot d\mathbf{S} = \iint_{\mathcal{D}} 25 \, dA = 25 \, \text{Area}(\mathcal{D}) = 25(4\pi) = 100\pi$$

Along the bottom disk of the cylinder, we have $z = 0$ and $\mathbf{F}(x, y, 0) = \langle y, 0, 0 \rangle$. Thus $\mathbf{F}$ is orthogonal to the vector $-\mathbf{k}$ normal to the bottom disk, and the integral along the bottom is zero.

***Step 3.* Find the total flux.**

$$\iint_{\mathcal{S}} \mathbf{F} \cdot d\mathbf{S} = \text{sides} + \text{top} + \text{bottom} = 50\pi + 100\pi + 0 = \boxed{150\pi}$$

***Step 4.* Compare with the integral of divergence.**

$$\text{div}(\mathbf{F}) = \text{div}(\langle y, yz, z^2 \rangle) = \frac{\partial}{\partial x} y + \frac{\partial}{\partial y}(yz) + \frac{\partial}{\partial z} z^2 = 0 + z + 2z = 3z$$

The cylinder $\mathcal{W}$ consists of all points (x, y, z) for $0 \le z \le 5$ and (x, y) in the disk $\mathcal{D}$. We see that the integral of the divergence is equal to the total flux as required:

$$\iiint_{\mathcal{W}} \operatorname{div}(\mathbf{F})\, dV = \iint_{\mathcal{D}} \int_{z=0}^{5} 3z\, dV = \iint_{\mathcal{D}} \frac{75}{2}\, dA$$

$$= \left(\frac{75}{2}\right)(\text{Area}(\mathcal{D})) = \left(\frac{75}{2}\right)(4\pi) = \boxed{150\pi} \qquad \blacksquare$$

In many applications, the Divergence Theorem is used to compute flux. In the next example, we reduce a flux computation (that would involve integrating over six sides of a box) to a more simple triple integral.

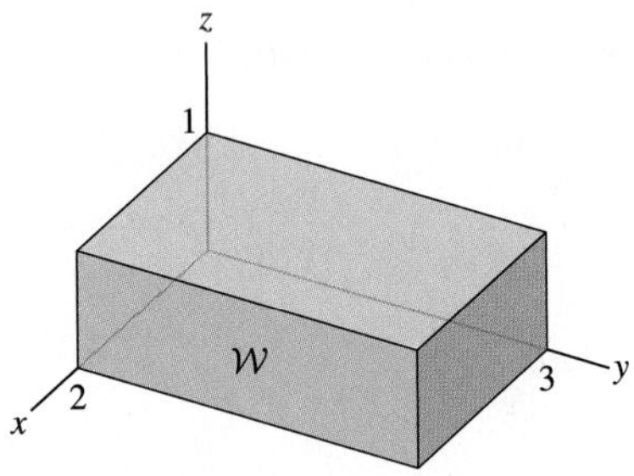

FIGURE 8

■ **EXAMPLE 3 Using the Divergence Theorem** Use the Divergence Theorem to evaluate $\iint_{\mathcal{S}} \langle x^2, z^4, e^z \rangle \cdot d\mathbf{S}$, where $\mathcal{S}$ is the boundary of the box $\mathcal{W}$ in Figure 8.

Solution First, compute the divergence:

$$\operatorname{div}(\langle x^2, z^4, e^z \rangle) = \frac{\partial}{\partial x} x^2 + \frac{\partial}{\partial y} z^4 + \frac{\partial}{\partial z} e^z = 2x + e^z$$

Then apply the Divergence Theorem and use Fubini's Theorem:

$$\iint_{\mathcal{S}} \langle x^2, z^4, e^z \rangle \cdot d\mathbf{S} = \iiint_{\mathcal{W}} (2x + e^z)\, dV = \int_0^2 \int_0^3 \int_0^1 (2x + e^z)\, dz\, dy\, dx$$

$$= 3\int_0^2 2x\, dx + 6\int_0^1 e^z\, dz = 12 + 6(e - 1) = 6e + 6 \qquad \blacksquare$$

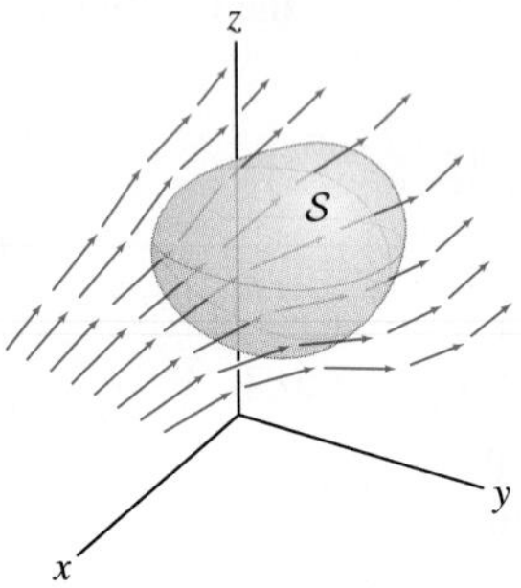

FIGURE 9

■ **EXAMPLE 4 A Vector Field with Zero Divergence** Compute the flux of

$$\mathbf{F} = \left\langle z^2 + xy^2, \cos(x + z), e^{-y} - zy^2 \right\rangle$$

through the boundary of the surface $\mathcal{S}$ in Figure 9.

Solution Although $\mathbf{F}$ is rather complicated, its divergence is zero:

$$\operatorname{div}(\mathbf{F}) = \frac{\partial}{\partial x}(z^2 + xy^2) + \frac{\partial}{\partial y}\cos(x + z) + \frac{\partial}{\partial z}(e^{-y} - zy^2) = y^2 - y^2 = 0$$

The Divergence Theorem shows that the flux is zero. Letting $\mathcal{W}$ be the region enclosed by $\mathcal{S}$, we have

$$\iint_{\mathcal{S}} \mathbf{F} \cdot d\mathbf{S} = \iiint_{\mathcal{W}} \operatorname{div}(\mathbf{F})\, dV = \iiint_{\mathcal{W}} 0\, dV = 0 \qquad \blacksquare$$

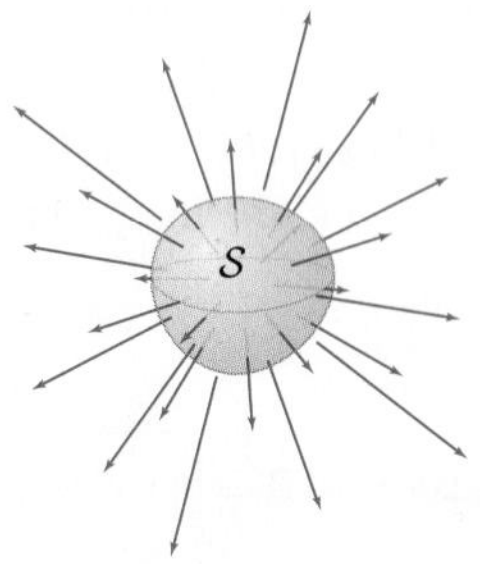

FIGURE 10 For a velocity field, the flux through a surface is the flow rate (in volume per time) of fluid across the surface.

GRAPHICAL INSIGHT Interpretation of Divergence Let's assume again that $\mathbf{F}$ is the velocity field of a fluid (Figure 10). Then the flux of $\mathbf{F}$ through a surface $\mathcal{S}$ is the flow rate (volume of fluid passing through $\mathcal{S}$ per unit time). If $\mathcal{S}$ encloses the region $\mathcal{W}$, then by the Divergence Theorem,

$$\text{Flow rate across } \mathcal{S} = \iiint_{\mathcal{W}} \operatorname{div}(\mathbf{F})\, dV \qquad \boxed{4}$$

Now assume that $\mathcal{S}$ is a small surface containing a point P. Because $\operatorname{div}(\mathbf{F})$ is continuous (it is a sum of derivatives of the components of $\mathbf{F}$), its value does not

Do the units match up in Eq. (5)? The flow rate has units of volume per unit time. On the other hand, the divergence is a sum of derivatives of velocity with respect to distance. Therefore, the divergence has units of "distance per unit time per distance," or unit time^{-1}, and the right-hand side of Eq. (5) also has units of volume per unit time.

change much on $\mathcal{W}$ if $\mathcal{S}$ is sufficiently small and to a first approximation, we can replace $\operatorname{div}(\mathbf{F})$ by the constant value $\operatorname{div}(\mathbf{F})(P)$. This gives us the approximation

$$\text{Flow rate across } \mathcal{S} = \iiint_{\mathcal{W}} \operatorname{div}(\mathbf{F})\, dV \approx \operatorname{div}(\mathbf{F})(P) \cdot \operatorname{Vol}(\mathcal{W}) \qquad \boxed{5}$$

In other words, the *flow rate through a small closed surface containing P is approximately equal to the divergence at P times the enclosed volume*, and thus $\operatorname{div}(\mathbf{F})(P)$ has an interpretation as "flow rate (or flux) per unit volume":

- If $\operatorname{div}(\mathbf{F})(P) > 0$, there is a net outflow of fluid across any small closed surface enclosing P, or, in other words, a net "creation" of fluid near P.
- If $\operatorname{div}(\mathbf{F})(P) < 0$, there is a net inflow of fluid across any small closed surface enclosing P, or, in other words, a net "destruction" of fluid near P.

Because of this, $\operatorname{div}(\mathbf{F})$ is sometimes called the *source density* of the field.

- If $\operatorname{div}(\mathbf{F})(P) = 0$, then to a first-order approximation, the net flow across any small closed surface enclosing P is equal to zero.

A vector field such that $\operatorname{div}(\mathbf{F}) = 0$ everywhere is called *incompressible.*

To visualize these cases, consider the two-dimensional situation, where we define

$$\operatorname{div}(\langle F_1, F_2 \rangle) = \frac{\partial F_1}{\partial x} + \frac{\partial F_2}{\partial y}$$

In Figure 11, field (A) has positive divergence. There is a positive net flow of fluid across every circle per unit time. Similarly, field (B) has negative divergence. By contrast, field (C) is *incompressible*. The fluid flowing into every circle is balanced by the fluid flowing out.

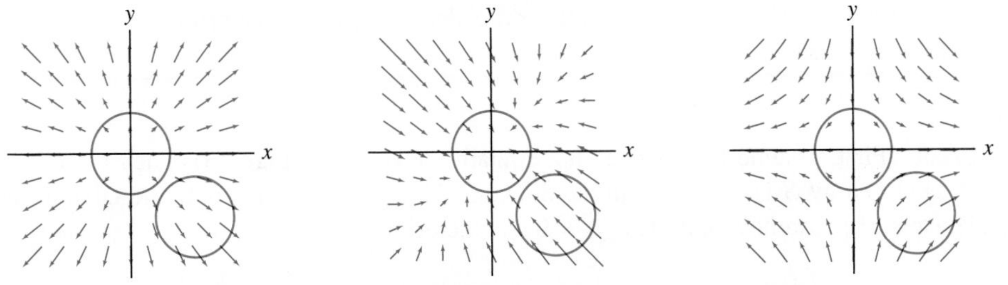

(A) The field $\mathbf{F} = \langle x, y \rangle$ with $\operatorname{div}(\mathbf{F}) = 2$. There is a net outflow through every circle.

(B) The field $\mathbf{F} = \langle y - 2x, x - 2y \rangle$ with $\operatorname{div}(\mathbf{F}) = -4$. There is a net inflow into every circle.

(C) The field $\mathbf{F} = \langle x, -y \rangle$ with $\operatorname{div}(\mathbf{F}) = 0$. The flux through every circle is zero.

FIGURE 11

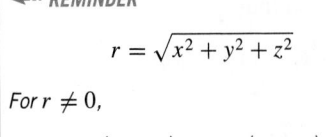

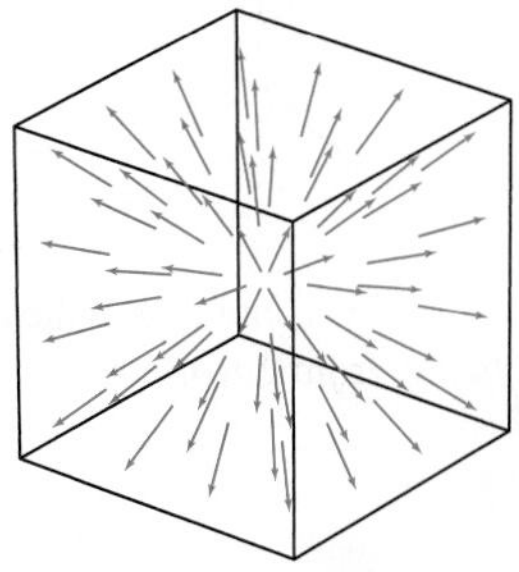

FIGURE 12 Unit radial vector field $\mathbf{e}_r$.

Applications to Electrostatics

The Divergence Theorem is a powerful tool for computing the flux of electrostatic fields. This is due to the special properties of the inverse-square vector field (Figure 12). In this section, we denote the inverse-square vector field by $\mathbf{F}_{\text{i-sq}}$:

$$\boxed{\mathbf{F}_{\text{i-sq}} = \frac{\mathbf{e}_r}{r^2}}$$

← ***REMINDER***

$$r = \sqrt{x^2 + y^2 + z^2}$$

For $r \neq 0$,

$$\mathbf{e}_r = \frac{\langle x, y, z \rangle}{r} = \frac{\langle x, y, z \rangle}{\sqrt{x^2 + y^2 + z^2}}$$

Recall that $\mathbf{F}_{\text{i-sq}}$ is defined for $r \neq 0$. The next example verifies the key property that $\operatorname{div}(\mathbf{F}_{\text{i-sq}}) = 0$.

■ **EXAMPLE 5 The Inverse-Square Vector Field** Verify that $\mathbf{F}_{\text{i-sq}} = \dfrac{\mathbf{e}_r}{r^2}$ has zero divergence:

$$\operatorname{div}\left(\frac{\mathbf{e}_r}{r^2}\right) = 0$$

Solution Write the field as

$$\mathbf{F}_{\text{i-sq}} = \langle F_1, F_2, F_3 \rangle = \frac{1}{r^2}\left\langle \frac{x}{r}, \frac{y}{r}, \frac{z}{r} \right\rangle = \langle xr^{-3}, yr^{-3}, zr^{-3} \rangle$$

We have

$$\frac{\partial r}{\partial x} = \frac{\partial}{\partial x}(x^2 + y^2 + z^2)^{1/2} = \frac{1}{2}(x^2 + y^2 + z^2)^{-1/2}(2x) = \frac{x}{r}$$

$$\frac{\partial F_1}{\partial x} = \frac{\partial}{\partial x} xr^{-3} = r^{-3} - 3xr^{-4}\frac{\partial r}{\partial x} = r^{-3} - (3xr^{-4})\frac{x}{r} = \frac{r^2 - 3x^2}{r^5}$$

The derivatives $\dfrac{\partial F_2}{\partial y}$ and $\dfrac{\partial F_3}{\partial z}$ are similar, so

$$\operatorname{div}(\mathbf{F}_{\text{i-sq}}) = \frac{r^2 - 3x^2}{r^5} + \frac{r^2 - 3y^2}{r^5} + \frac{r^2 - 3z^2}{r^5} = \frac{3r^2 - 3(x^2 + y^2 + z^2)}{r^5} = 0 \quad ■$$

The next theorem shows that the flux of $\mathbf{F}_{\text{i-sq}}$ through a closed surface $\mathcal{S}$ depends only on whether $\mathcal{S}$ contains the origin.

THEOREM 2 Flux of the Inverse-Square Field The flux of $\mathbf{F}_{\text{i-sq}} = \dfrac{\mathbf{e}_r}{r^2}$ through closed surfaces has the following remarkable description:

$$\iint_{\mathcal{S}} \left(\frac{\mathbf{e}_r}{r^2}\right) \cdot d\mathbf{S} = \begin{cases} 4\pi & \text{if } \mathcal{S} \text{ encloses the origin} \\ 0 & \text{if } \mathcal{S} \text{ does not enclose the origin} \end{cases}$$

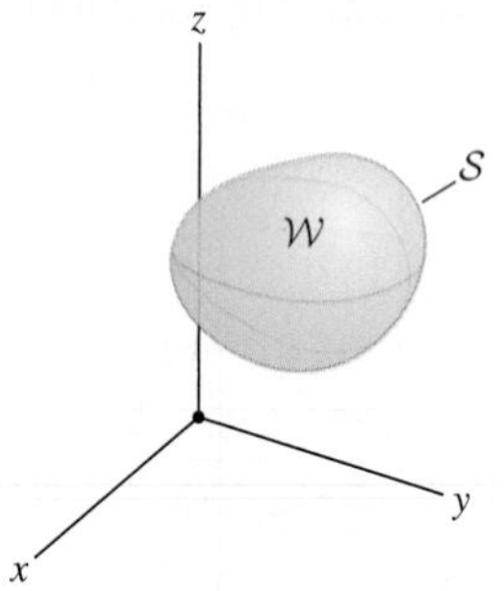

FIGURE 13 $\mathcal{S}$ is contained in the domain of $\mathbf{F}_{\text{i-sq}}$ (away from the origin).

Proof First, assume that $\mathcal{S}$ does not contain the origin (Figure 13). Then the region $\mathcal{W}$ enclosed by $\mathcal{S}$ is contained in the domain of $\mathbf{F}_{\text{i-sq}}$ and we can apply the Divergence Theorem. By Example 5, $\operatorname{div}(\mathbf{F}_{\text{i-sq}}) = 0$ and therefore

$$\iint_{\mathcal{S}} \left(\frac{\mathbf{e}_r}{r^2}\right) \cdot d\mathbf{S} = \iiint_{\mathcal{W}} \operatorname{div}(\mathbf{F}_{\text{i-sq}})\, dV = \iiint_{\mathcal{W}} 0\, dV = 0$$

Next, let $\mathcal{S}_R$ be the sphere of radius R centered at the origin (Figure 14). We cannot use the Divergence Theorem because $\mathcal{S}_R$ contains a point (the origin) where $\mathbf{F}_{\text{i-sq}}$ is not defined. However, we can compute the flux of $\mathbf{F}_{\text{i-sq}}$ through $\mathcal{S}_R$ using spherical coordinates. Recall from Section 17.4 [Eq. (5)] that the outward-pointing normal vector in spherical coordinates is

$$\mathbf{n} = \mathbf{T}_\phi \times \mathbf{T}_\theta = (R^2 \sin\phi)\mathbf{e}_r$$

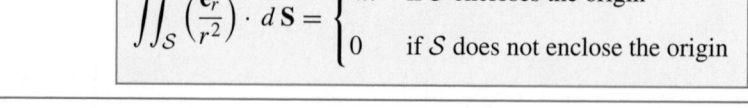

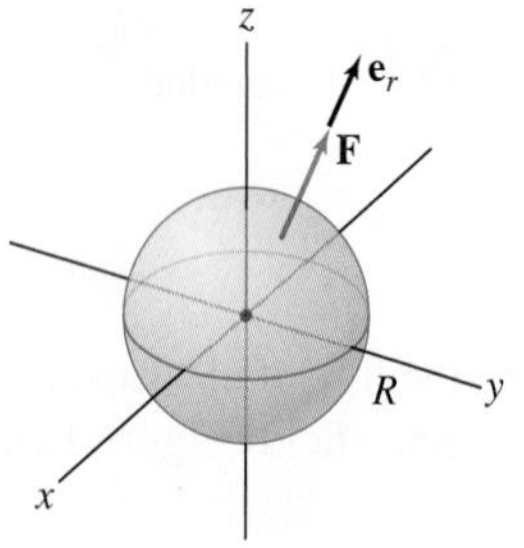

FIGURE 14

The inverse-square field on $\mathcal{S}_R$ is simply $\mathbf{F}_{\text{i-sq}} = R^{-2}\mathbf{e}_r$, and thus

$$\mathbf{F}_{\text{i-sq}} \cdot \mathbf{n} = (R^{-2}\mathbf{e}_r) \cdot (R^2 \sin\phi\, \mathbf{e}_r) = \sin\phi\,(\mathbf{e}_r \cdot \mathbf{e}_r) = \sin\phi$$

$$\iint_{\mathcal{S}_R} \mathbf{F}_{\text{i-sq}} \cdot d\mathbf{S} = \int_0^{2\pi} \int_0^{\pi} \mathbf{F}_{\text{i-sq}} \cdot \mathbf{n}\, d\phi\, d\theta$$

$$= \int_0^{2\pi} \int_0^{\pi} \sin\phi \, d\phi \, d\theta$$

$$= 2\pi \int_0^{\pi} \sin\phi \, d\phi = 4\pi$$

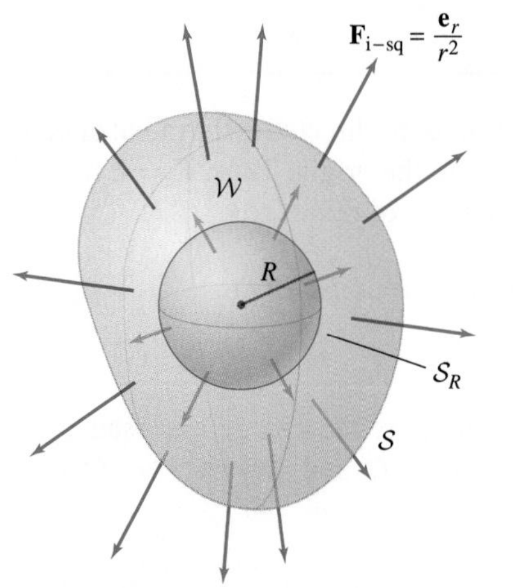

FIGURE 15 $\mathcal{W}$ is the region between $\mathcal{S}$ and the sphere $\mathcal{S}_R$.

To extend this result to *any* surface $\mathcal{S}$ containing the origin, choose a sphere $\mathcal{S}_R$ whose radius $R > 0$ is so small that $\mathcal{S}_R$ is contained inside $\mathcal{S}$. Let $\mathcal{W}$ be the region *between* $\mathcal{S}_R$ and $\mathcal{S}$ (Figure 15). The oriented boundary of $\mathcal{W}$ is the difference

$$\partial\mathcal{W} = \mathcal{S} - \mathcal{S}_R$$

This means that $\mathcal{S}$ is oriented by outward-pointing normals and $\mathcal{S}_R$ by inward-pointing normals. By the Divergence Theorem,

$$\iint_{\partial\mathcal{W}} \mathbf{F}_{\text{i-sq}} \cdot d\mathbf{S} = \iint_{\mathcal{S}} \mathbf{F}_{\text{i-sq}} \cdot d\mathbf{S} - \iint_{\mathcal{S}_R} \mathbf{F}_{\text{i-sq}} \cdot d\mathbf{S}$$

$$= \iiint_{\mathcal{W}} \operatorname{div}(\mathbf{F}_{\text{i-sq}}) \, dV \qquad \text{(Divergence Theorem)}$$

$$= \iiint_{\mathcal{W}} 0 \, dV = 0 \qquad (\text{Because } \operatorname{div}(\mathbf{F}_{\text{i-sq}}) = 0)$$

This proves that the fluxes through $\mathcal{S}$ and $\mathcal{S}_R$ are equal, and hence both equal 4π.

Notice that we just applied the Divergence Theorem to a region $\mathcal{W}$ that lies *between two surfaces, one contained in the other*. This is a more general form of the theorem than the one we stated formally in Theorem 1 above. The marginal comment explains why this is justified. ■

To verify that the Divergence Theorem remains valid for regions between two surfaces, such as the region $\mathcal{W}$ in Figure 15, we cut $\mathcal{W}$ down the middle. Each half is a region enclosed by a surface, so the the Divergence Theorem as we have stated it applies. By adding the results for the two halves, we obtain the Divergence Theorem for $\mathcal{W}$. This uses the fact that the fluxes through the common face of the two halves cancel.

This result applies directly to the electric field $\mathbf{E}$ of a point charge, which is a multiple of the inverse-square vector field. For a charge of q coulombs at the origin,

$$\mathbf{E} = \left(\frac{q}{4\pi\epsilon_0}\right)\frac{\mathbf{e}_r}{r^2}$$

where $\epsilon_0 = 8.85 \times 10^{-12}\ \text{C}^2/\text{N-m}^2$ is the permittivity constant. Therefore,

$$\text{Flux of } \mathbf{E} \text{ through } \mathcal{S} = \begin{cases} \dfrac{q}{\epsilon_0} & \text{if } q \text{ is inside } \mathcal{S} \\ 0 & \text{if } q \text{ is outside } \mathcal{S} \end{cases}$$

Now, instead of placing just one point charge at the origin, we may distribute a finite number N of point charges q_i at different points in space. The resulting electric field $\mathbf{E}$ is the sum of the fields $\mathbf{E}_i$ due to the individual charges, and

$$\iint_{\mathcal{S}} \mathbf{E} \cdot d\mathbf{S} = \iint_{\mathcal{S}} \mathbf{E}_1 \cdot d\mathbf{S} + \cdots + \iint_{\mathcal{S}} \mathbf{E}_N \cdot d\mathbf{S}$$

Each integral on the right is either 0 or q_i/ϵ_0, according to whether or not $\mathcal{S}$ contains q_i, so we conclude that

$$\iint_{\mathcal{S}} \mathbf{E} \cdot d\mathbf{S} = \frac{\text{total charge enclosed by } \mathcal{S}}{\epsilon_0} \qquad \boxed{6}$$

This fundamental relation is called **Gauss's Law**. A limiting argument shows that Eq. (6) remains valid for the electric field due to a *continuous* distribution of charge.

The next theorem, describing the electric field due to a uniformly charged sphere, is a classic application of Gauss's Law.

THEOREM 3 Uniformly Charged Sphere The electric field due to a uniformly charged hollow sphere $\mathcal{S}_R$ of radius R, centered at the origin and of total charge Q, is

$$\mathbf{E} = \begin{cases} \dfrac{Q}{4\pi\epsilon_0 r^2}\mathbf{e}_r & \text{if } r > R \\ \mathbf{0} & \text{if } r < R \end{cases} \qquad \boxed{7}$$

where $\epsilon_0 = 8.85 \times 10^{-12}\ \mathrm{C^2/N\text{-}m^2}$.

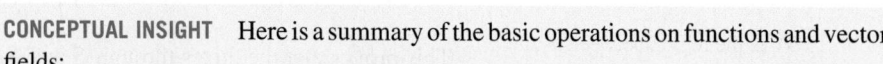

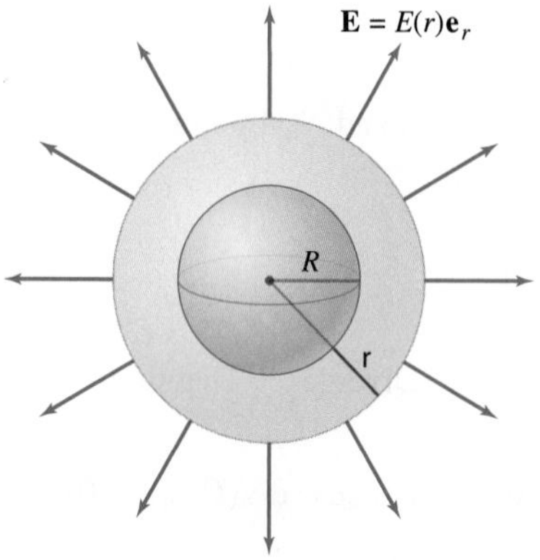

FIGURE 16 The electric field due to a uniformly charged sphere.

Proof By symmetry (Figure 16), the electric field $\mathbf{E}$ must be directed in the radial direction $\mathbf{e}_r$ with magnitude depending only on the distance r to the origin. Thus, $\mathbf{E} = E(r)\mathbf{e}_r$ for some function $E(r)$. The flux of $\mathbf{E}$ through the sphere $\mathcal{S}_r$ of radius r is

$$\iint_{\mathcal{S}_r} \mathbf{E} \cdot d\mathbf{S} = E(r) \underbrace{\iint_{\mathcal{S}_r} \mathbf{e}_r \cdot d\mathbf{S}}_{\text{Surface area of sphere}} = 4\pi r^2 E(r)$$

By Gauss's Law, this flux is equal to C/ϵ_0, where C is the charge enclosed by $\mathcal{S}_r$. If $r < R$, then $C = 0$ and $\mathbf{E} = \mathbf{0}$. If $r > R$, then $C = Q$ and $4\pi r^2 E(r) = Q/\epsilon_0$, or $E(r) = Q/(\epsilon_0 4\pi r^2)$. This proves Eq. (7).

We proved Theorem 3 in the analogous case of a gravitational field (also a radial inverse-square field) by a laborious calculation in Exercise 48 of Section 17.4. Here, we have derived it from Gauss's Law and a simple appeal to symmetry.

CONCEPTUAL INSIGHT Here is a summary of the basic operations on functions and vector fields:

$$\underset{\text{function}}{f} \xrightarrow{\nabla} \underset{\text{vector field}}{\mathbf{F}} \xrightarrow{\text{curl}} \underset{\text{vector field}}{\mathbf{G}} \xrightarrow{\text{div}} \underset{\text{function}}{g}$$

One basic fact is that the result of two consecutive operations in this diagram is zero:

$$\operatorname{curl}(\nabla(f)) = \mathbf{0}, \qquad \operatorname{div}(\operatorname{curl}(\mathbf{F})) = 0$$

We verified the first identity in Example 1 of Section 18.2. The second identity is left as an exercise (Exercise 6).

An interesting question is whether every vector field satisfying $\operatorname{curl}(\mathbf{F}) = \mathbf{0}$ is necessarily conservative—that is, $\mathbf{F} = \nabla V$ for some function V. The answer is yes, but only if the domain $\mathcal{D}$ is simply connected (every path can be drawn down to a point in $\mathcal{D}$). We saw in Section 17.3 that the vortex vector satisfies $\operatorname{curl}(\mathbf{F}) = \mathbf{0}$ and yet cannot be conservative because its circulation around the unit circle is nonzero (the circulation of a conservative vector field is always zero). However, the domain of the vortex vector field is $\mathbf{R}^2$ with the origin removed, and this domain is not simply-connected.

The situation for vector potentials is similar. Can every vector field $\mathbf{G}$ satisfying $\operatorname{div}(\mathbf{G}) = \mathbf{0}$ be written in the form $\mathbf{G} = \operatorname{curl}(\mathbf{A})$ for some vector potential $\mathbf{A}$? Again, the answer is yes—provided that the domain is a region $\mathcal{W}$ in $\mathbf{R}^3$ that has "no holes," like a ball, cube, or all of $\mathbf{R}^3$. The inverse-square field $\mathbf{F}_{\text{i-sq}} = \mathbf{e}_r/r^2$ plays the role of vortex field in this setting: Although $\operatorname{div}(\mathbf{F}_{\text{i-sq}}) = 0$, $\mathbf{F}_{\text{i-sq}}$ cannot have a vector potential because its flux through the unit sphere is nonzero as shown in Theorem 2 (the flux over a closed surface of a vector field with a vector potential is always zero by Theorem 2 of Section 18.2). In this case, the domain of $\mathbf{e}_r/r^2$ is $\mathbf{R}^3$ with the origin removed, which "has a hole."

These properties of the vortex and inverse-square vector fields are significant because they relate line and surface integrals to "topological" properties of the domain, such as whether the domain is simply-connected or has holes. They are a first hint of the important and fascinating connections between vector analysis and the area of mathematics called topology.

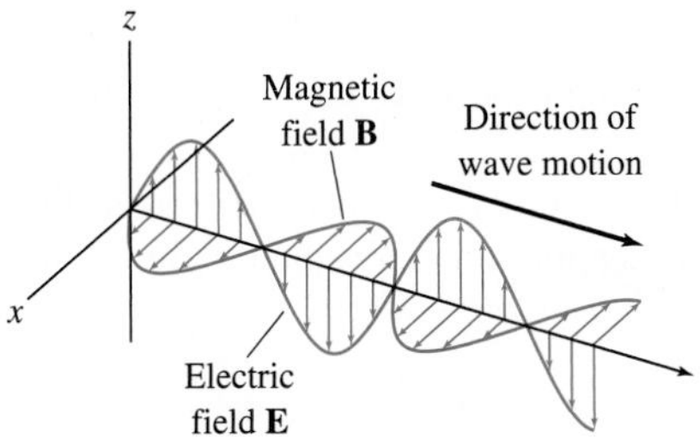

FIGURE 17 The **E** and **B** fields of an electromagnetic wave along an axis of motion.

This is not just mathematical elegance... but beauty.
It is so simple and yet it describes something so complex.

Francis Collins (1950–), leading geneticist and former director of the Human Genome Project, speaking of the Maxwell Equations.

HISTORICAL PERSPECTIVE

James Clerk Maxwell
(1831–1879)

Vector analysis was developed in the nineteenth century, in large part, to express the laws of electricity and magnetism. Electromagnetism was studied intensively in the period 1750–1890, culminating in the famous Maxwell Equations, which provide a unified understanding in terms of two vector fields: the electric field **E** and the magnetic field **B**. In a region of empty space (where there are no charged particles), the Maxwell Equations are

$$\operatorname{div}(\mathbf{E}) = 0, \qquad \operatorname{div}(\mathbf{B}) = 0$$

$$\operatorname{curl}(\mathbf{E}) = -\frac{\partial \mathbf{B}}{\partial t}, \qquad \operatorname{curl}(\mathbf{B}) = \mu_0\epsilon_0\frac{\partial \mathbf{E}}{\partial t}$$

where μ_0 and ϵ_0 are experimentally determined constants. In SI units,

$$\mu_0 = 4\pi \times 10^{-7} \text{ henries/m}$$

$$\epsilon_0 \approx 8.85 \times 10^{-12} \text{ farads/m}$$

These equations led Maxwell to make two predictions of fundamental importance: (1) that electromagnetic waves exist (this was confirmed by H. Hertz in 1887), and (2) that light is an electromagnetic wave.

How do the Maxwell Equations suggest that electromagnetic waves exist? And why did Maxwell conclude that light is an electromagnetic wave? It was known to mathematicians in the eighteenth century that waves traveling with velocity c may be described by functions $\varphi(x, y, z, t)$ that satisfy the *wave equation*

$$\Delta\varphi = \frac{1}{c^2}\frac{\partial^2\varphi}{\partial t^2} \qquad \boxed{8}$$

where Δ is the Laplace operator (or "Laplacian")

$$\Delta\varphi = \frac{\partial^2\varphi}{\partial x^2} + \frac{\partial^2\varphi}{\partial y^2} + \frac{\partial^2\varphi}{\partial z^2}$$

We will show that the components of **E** satisfy this wave equation. Take the curl of both sides of Maxwell's third equation:

$$\operatorname{curl}(\operatorname{curl}(\mathbf{E})) = \operatorname{curl}\left(-\frac{\partial \mathbf{B}}{\partial t}\right) = -\frac{\partial}{\partial t}\operatorname{curl}(\mathbf{B})$$

Then apply Maxwell's fourth equation to obtain

$$\operatorname{curl}(\operatorname{curl}(\mathbf{E})) = -\frac{\partial}{\partial t}\left(\mu_0\epsilon_0\frac{\partial \mathbf{E}}{\partial t}\right)$$

$$= -\mu_0\epsilon_0\frac{\partial^2 \mathbf{E}}{\partial t^2} \qquad \boxed{9}$$

Finally, let us define the Laplacian of a vector field

$$\mathbf{F} = \langle F_1, F_2, F_3\rangle$$

by applying the Laplacian Δ to each component, $\Delta\mathbf{F} = \langle \Delta F_1, \Delta F_2, \Delta F_3\rangle$. Then the following identity holds (see Exercise 36):

$$\operatorname{curl}(\operatorname{curl}(\mathbf{F})) = \nabla(\operatorname{div}(\mathbf{F})) - \Delta\mathbf{F}$$

Applying this identity to **E**, we obtain $\operatorname{curl}(\operatorname{curl}(\mathbf{E})) = -\Delta\mathbf{E}$ because $\operatorname{div}(\mathbf{E}) = 0$ by Maxwell's first equation. Thus, Eq. (9) yields

$$\boxed{\Delta\mathbf{E} = \mu_0\epsilon_0\frac{\partial^2 \mathbf{E}}{\partial t^2}}$$

In other words, each component of the electric field satisfies the wave equation (8), with $c = (\mu_0\epsilon_0)^{-1/2}$. This tells us that the **E**-field (and similarly the **B**-field) can propagate through space like a wave, giving rise to electromagnetic radiation (Figure 17).

Maxwell computed the velocity c of an electromagnetic wave:

$$c = (\mu_0\epsilon_0)^{-1/2} \approx 3 \times 10^8 \text{ m/s}$$

and observed that the value is suspiciously close to the velocity of light (first measured by Olaf Römer in 1676). This had to be more than a coincidence, as Maxwell wrote in 1862: "We can scarcely avoid the conclusion that light consists in the transverse undulations of the same medium which is the cause of electric and magnetic phenomena." Needless to say, the wireless technologies that drive our modern society rely on the unseen electromagnetic radiation whose existence Maxwell first predicted on mathematical grounds.

18.3 SUMMARY

- Divergence of $\mathbf{F} = \langle F_1, F_2, F_3 \rangle$:

$$\operatorname{div}(\mathbf{F}) = \nabla \cdot \mathbf{F} = \frac{\partial F_1}{\partial x} + \frac{\partial F_2}{\partial y} + \frac{\partial F_3}{\partial z}$$

- The Divergence Theorem: If $\mathcal{W}$ is a region in $\mathbf{R}^3$ whose boundary $\partial\mathcal{W}$ is a surface, oriented by normal vectors pointing outside $\mathcal{W}$, then

$$\iint_{\partial\mathcal{W}} \mathbf{F} \cdot d\mathbf{S} = \iiint_{\mathcal{W}} \operatorname{div}(\mathbf{F})\, dV$$

- Corollary: If $\operatorname{div}(\mathbf{F}) = 0$, then $\mathbf{F}$ has zero flux through the boundary $\partial\mathcal{W}$ of any $\mathcal{W}$ contained in the domain of $\mathbf{F}$.
- The divergence $\operatorname{div}(\mathbf{F})$ is interpreted as "flux per unit volume," which means that the flux through a small closed surface containing a point P is approximately equal to $\operatorname{div}(\mathbf{F})(P)$ times the enclosed volume.
- Basic operations on functions and vector fields:

$$\underset{\text{function}}{f} \xrightarrow{\nabla} \underset{\text{vector field}}{\mathbf{F}} \xrightarrow{\text{curl}} \underset{\text{vector field}}{\mathbf{G}} \xrightarrow{\text{div}} \underset{\text{function}}{g}$$

- The result of two consecutive operations is zero:

$$\operatorname{curl}(\nabla(f)) = \mathbf{0}, \qquad \operatorname{div}(\operatorname{curl}(\mathbf{F})) = 0$$

- The inverse-square field $\mathbf{F} = \mathbf{e}_r/r^2$, defined for $r \neq 0$, satisfies $\operatorname{div}(\mathbf{F}) = 0$. The flux of $\mathbf{F}$ through a closed surface $\mathcal{S}$ is 4π if $\mathcal{S}$ contains the origin and is zero otherwise.

18.3 EXERCISES

Preliminary Questions

1. What is the flux of $\mathbf{F} = \langle 1, 0, 0 \rangle$ through a closed surface?

2. Justify the following statement: The flux of $\mathbf{F} = \langle x^3, y^3, z^3 \rangle$ through every closed surface is positive.

3. Which of the following expressions are meaningful (where $\mathbf{F}$ is a vector field and f is a function)? Of those that are meaningful, which are automatically zero?

(a) $\operatorname{div}(\nabla f)$ **(b)** $\operatorname{curl}(\nabla f)$ **(c)** $\nabla\operatorname{curl}(f)$
(d) $\operatorname{div}(\operatorname{curl}(\mathbf{F}))$ **(e)** $\operatorname{curl}(\operatorname{div}(\mathbf{F}))$ **(f)** $\nabla(\operatorname{div}(\mathbf{F}))$

4. Which of the following statements is correct (where $\mathbf{F}$ is a continuously differentiable vector field defined everywhere)?

(a) The flux of $\operatorname{curl}(\mathbf{F})$ through all surfaces is zero.
(b) If $\mathbf{F} = \nabla\varphi$, then the flux of $\mathbf{F}$ through all surfaces is zero.
(c) The flux of $\operatorname{curl}(\mathbf{F})$ through all closed surfaces is zero.

5. How does the Divergence Theorem imply that the flux of $\mathbf{F} = \langle x^2, y - e^z, y - 2zx \rangle$ through a closed surface is equal to the enclosed volume?

Exercises

In Exercises 1–4, compute the divergence of the vector field.

1. $\mathbf{F} = \langle xy, yz, y^2 - x^3 \rangle$ **2.** $x\mathbf{i} + y\mathbf{j} + z\mathbf{k}$

3. $\mathbf{F} = \langle x - 2zx^2, z - xy, z^2x^2 \rangle$ **4.** $\sin(x + z)\mathbf{i} - ye^{xz}\mathbf{k}$

5. Find a constant c for which the velocity field

$$\mathbf{v} = (cx - y)\mathbf{i} + (y - z)\mathbf{j} + (3x + 4cz)\mathbf{k}$$

of a fluid is incompressible [meaning that $\operatorname{div}(\mathbf{v}) = 0$].

6. Verify that for any vector field $\mathbf{F} = \langle F_1, F_2, F_3 \rangle$,

$$\operatorname{div}(\operatorname{curl}(\mathbf{F})) = 0$$

In Exercises 7–10, verify the Divergence Theorem for the vector field and region.

7. $\mathbf{F} = \langle z, x, y \rangle$, the box $[0, 4] \times [0, 2] \times [0, 3]$

8. $\mathbf{F} = \langle y, x, z \rangle$, the region $x^2 + y^2 + z^2 \le 4$

9. $\mathbf{F} = \langle 2x, 3z, 3y \rangle$, the region $x^2 + y^2 \le 1, 0 \le z \le 2$

10. $\mathbf{F} = \langle x, 0, 0 \rangle$, the region $x^2 + y^2 \le z \le 4$

In Exercises 11–18, use the Divergence Theorem to evaluate the flux $\iint_{\mathcal{S}} \mathbf{F} \cdot d\mathbf{S}$.

11. $\mathbf{F} = \left\langle 0, 0, z^3/3 \right\rangle$, $\mathcal{S}$ is the sphere $x^2 + y^2 + z^2 = 1$.

12. $\mathbf{F} = \langle y, z, x \rangle$, $\mathcal{S}$ is the sphere $x^2 + y^2 + z^2 = 1$.

13. $\mathbf{F} = \left\langle x^3, 0, z^3 \right\rangle$, $\mathcal{S}$ is the octant of the sphere $x^2 + y^2 + z^2 = 4$, in the first octant $x \ge 0, y \ge 0, z \ge 0$.

14. $\mathbf{F} = \left\langle e^{x+y}, e^{x+z}, e^{x+y} \right\rangle$, $\mathcal{S}$ is the boundary of the unit cube $0 \le x \le 1, 0 \le y \le 1, 0 \le z \le 1$.

15. $\mathbf{F} = \left\langle x, y^2, z + y \right\rangle$, $\mathcal{S}$ is the boundary of the region contained in the cylinder $x^2 + y^2 = 4$ between the planes $z = x$ and $z = 8$.

16. $\mathbf{F} = \left\langle x^2 - z^2, e^{z^2} - \cos x, y^3 \right\rangle$, $\mathcal{S}$ is the boundary of the region bounded by $x + 2y + 4z = 12$ and the coordinate planes in the first octant.

17. $\mathbf{F} = \langle x + y, z, z - x \rangle$, $\mathcal{S}$ is the boundary of the region between the paraboloid $z = 9 - x^2 - y^2$ and the xy-plane.

18. $\mathbf{F} = \left\langle e^{z^2}, 2y + \sin(x^2 z), 4z + \sqrt{x^2 + 9y^2} \right\rangle$, $\mathcal{S}$ is the region $x^2 + y^2 \le z \le 8 - x^2 - y^2$.

19. Calculate the flux of the vector field $\mathbf{F} = 2xy\mathbf{i} - y^2\mathbf{j} + \mathbf{k}$ through the surface $\mathcal{S}$ in Figure 18. *Hint:* Apply the Divergence Theorem to the closed surface consisting of $\mathcal{S}$ and the unit disk.

20. Let $\mathcal{S}_1$ be the closed surface consisting of $\mathcal{S}$ in Figure 18 together with the unit disk. Find the volume enclosed by $\mathcal{S}_1$, assuming that

$$\iint_{\mathcal{S}_1} \langle x, 2y, 3z \rangle \cdot d\mathbf{S} = 72$$

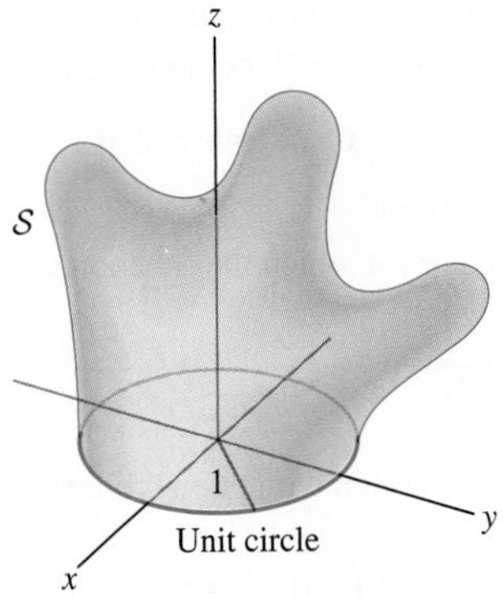

FIGURE 18 Surface $\mathcal{S}$ whose boundary is the unit circle.

21. Let $\mathcal{S}$ be the half-cylinder $x^2 + y^2 = 1, x \ge 0, 0 \le z \le 1$. Assume that $\mathbf{F}$ is a horizontal vector field (the z-component is zero) such that $\mathbf{F}(0, y, z) = zy^2\mathbf{i}$. Let $\mathcal{W}$ be the solid region enclosed by $\mathcal{S}$, and assume that

$$\iiint_{\mathcal{W}} \text{div}(\mathbf{F})\, dV = 4$$

Find the flux of $\mathbf{F}$ through the curved side of $\mathcal{S}$.

22. Volume as a Surface Integral Let $\mathbf{F} = \langle x, y, z \rangle$. Prove that if $\mathcal{W}$ is a region $\mathbf{R}^3$ with a smooth boundary $\mathcal{S}$, then

$$\text{Volume}(\mathcal{W}) = \frac{1}{3} \iint_{\mathcal{S}} \mathbf{F} \cdot d\mathbf{S} \qquad \boxed{10}$$

23. Use Eq. (10) to calculate the volume of the unit ball as a surface integral over the unit sphere.

24. Verify that Eq. (10) applied to the box $[0, a] \times [0, b] \times [0, c]$ yields the volume $V = abc$.

25. Let $\mathcal{W}$ be the region in Figure 19 bounded by the cylinder $x^2 + y^2 = 4$, the plane $z = x + 1$, and the xy-plane. Use the Divergence Theorem to compute the flux of $\mathbf{F} = \left\langle z, x, y + z^2 \right\rangle$ through the boundary of $\mathcal{W}$.

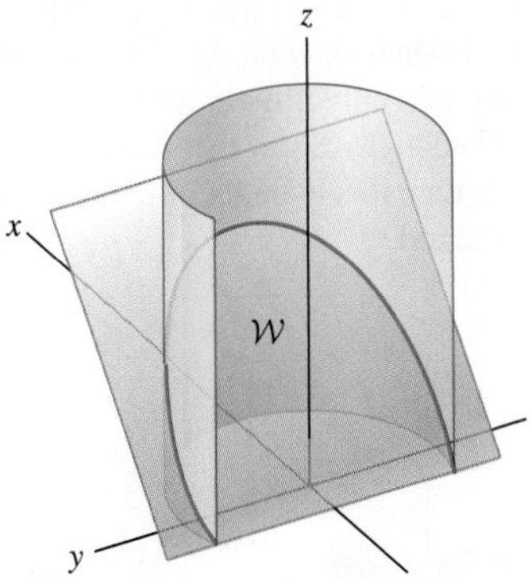

FIGURE 19

26. Let $I = \iint_{\mathcal{S}} \mathbf{F} \cdot d\mathbf{S}$, where

$$\mathbf{F} = \left\langle \frac{2yz}{r^2}, -\frac{xz}{r^2}, -\frac{xy}{r^2} \right\rangle$$

($r = \sqrt{x^2 + y^2 + z^2}$) and $\mathcal{S}$ is the boundary of a region $\mathcal{W}$.

(a) Check that $\mathbf{F}$ is divergence-free.

(b) Show that $I = 0$ if $\mathcal{S}$ is a sphere centered at the origin. Explain, however, why the Divergence Theorem cannot be used to prove this.

27. The velocity field of a fluid $\mathbf{v}$ (in meters per second) has divergence $\text{div}(\mathbf{v})(P) = 3$ at the point $P = (2, 2, 2)$. Estimate the flow rate out of the sphere of radius 0.5 centered at P.

28. A hose feeds into a small screen box of volume 10 cm^3 that is suspended in a swimming pool. Water flows across the surface of the box at a rate of 12 cm^3/s. Estimate $\text{div}(\mathbf{v})(P)$, where $\mathbf{v}$ is the velocity field of the water in the pool and P is the center of the box. What are the units of $\text{div}(\mathbf{v})(P)$?

29. The electric field due to a unit electric dipole oriented in the $\mathbf{k}$-direction is $\mathbf{E} = \nabla(z/r^3)$, where $r = (x^2 + y^2 + z^2)^{1/2}$ (Figure 20). Let $\mathbf{e}_r = r^{-1} \langle x, y, z \rangle$.

(a) Show that $\mathbf{E} = r^{-3}\mathbf{k} - 3zr^{-4}\mathbf{e}_r$.

(b) Calculate the flux of $\mathbf{E}$ through a sphere centered at the origin.

(c) Calculate div(**E**).

(d) Can we use the Divergence Theorem to compute the flux of **E** through a sphere centered at the origin?

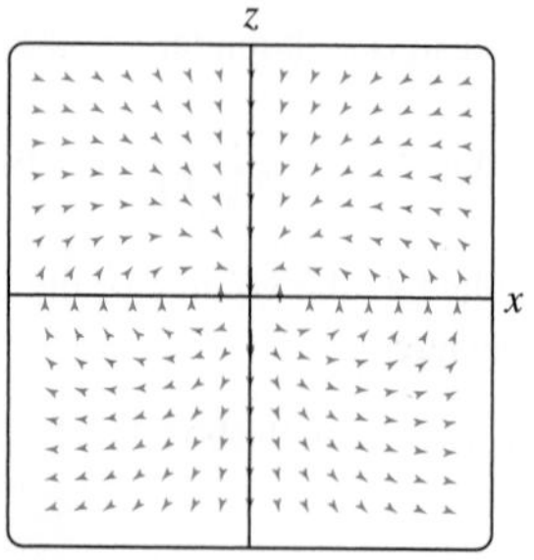

FIGURE 20 The dipole vector field restricted to the xz-plane.

30. Let **E** be the electric field due to a long, uniformly charged rod of radius R with charge density δ per unit length (Figure 21). By symmetry, we may assume that **E** is everywhere perpendicular to the rod and its magnitude $E(d)$ depends only on the distance d to the rod (strictly speaking, this would hold only if the rod were infinite, but it is nearly true if the rod is long enough). Show that $E(d) = \delta/2\pi\epsilon_0 d$ for $d > R$. *Hint:* Apply Gauss's Law to a cylinder of radius R and of unit length with its axis along the rod.

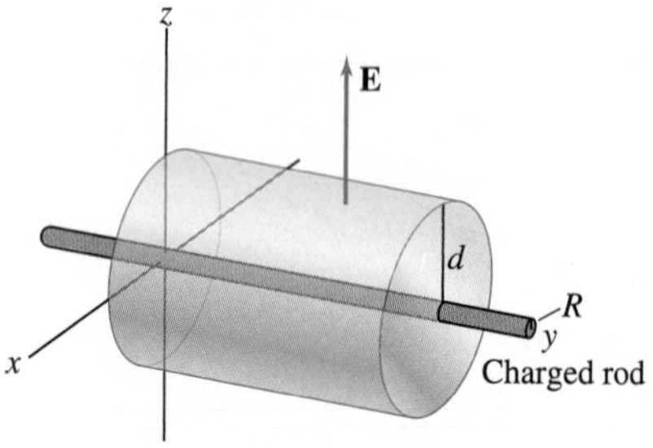

FIGURE 21

31. Let $\mathcal{W}$ be the region between the sphere of radius 4 and the cube of side 1, both centered at the origin. What is the flux through the boundary $\mathcal{S} = \partial\mathcal{W}$ of a vector field **F** whose divergence has the constant value $\text{div}(\mathbf{F}) = -4$?

32. Let $\mathcal{W}$ be the region between the sphere of radius 3 and the sphere of radius 2, both centered at the origin. Use the Divergence Theorem to calculate the flux of $\mathbf{F} = x\mathbf{i}$ through the boundary $\mathcal{S} = \partial\mathcal{W}$.

33. Find and prove a Product Rule expressing $\text{div}(f\mathbf{F})$ in terms of $\text{div}(\mathbf{F})$ and ∇f.

34. Prove the identity

$$\text{div}(\mathbf{F} \times \mathbf{G}) = \text{curl}(\mathbf{F}) \cdot \mathbf{G} - \mathbf{F} \cdot \text{curl}(\mathbf{G})$$

Then prove that the cross product of two irrotational vector fields is incompressible [**F** is called **irrotational** if $\text{curl}(\mathbf{F}) = 0$ and **incompressible** if $\text{div}(\mathbf{F}) = 0$].

35. Prove that $\text{div}(\nabla f \times \nabla g) = 0$.

In Exercises 36–38, Δ denotes the Laplace operator defined by

$$\Delta\varphi = \frac{\partial^2\varphi}{\partial x^2} + \frac{\partial^2\varphi}{\partial y^2} + \frac{\partial^2\varphi}{\partial z^2}$$

36. Prove the identity

$$\text{curl}(\text{curl}(\mathbf{F})) = \nabla(\text{div}(\mathbf{F})) - \Delta\mathbf{F}$$

where $\Delta\mathbf{F}$ denotes $\langle \Delta F_1, \Delta F_2, \Delta F_3 \rangle$.

37. A function φ satisfying $\Delta\varphi = 0$ is called **harmonic**.

(a) Show that $\Delta\varphi = \text{div}(\nabla\varphi)$ for any function φ.

(b) Show that φ is harmonic if and only if $\text{div}(\nabla\varphi) = 0$.

(c) Show that if **F** is the gradient of a harmonic function, then $\text{curl}(F) = 0$ and $\text{div}(F) = 0$.

(d) Show that $\mathbf{F} = \left\langle xz, -yz, \frac{1}{2}(x^2 - y^2) \right\rangle$ is the gradient of a harmonic function. What is the flux of **F** through a closed surface?

38. Let $\mathbf{F} = r^n \mathbf{e}_r$, where n is any number, $r = (x^2 + y^2 + z^2)^{1/2}$, and $\mathbf{e}_r = r^{-1}\langle x, y, z \rangle$ is the unit radial vector.

(a) Calculate div(**F**).

(b) Calculate the flux of **F** through the surface of a sphere of radius R centered at the origin. For which values of n is this flux independent of R?

(c) Prove that $\nabla(r^n) = n\, r^{n-1}\mathbf{e}_r$.

(d) Use (c) to show that **F** is conservative for $n \neq -1$. Then show that $\mathbf{F} = r^{-1}\mathbf{e}_r$ is also conservative by computing the gradient of $\ln r$.

(e) What is the value of $\int_C \mathbf{F} \cdot d\mathbf{s}$, where C is a closed curve that does not pass through the origin?

(f) Find the values of n for which the function $\varphi = r^n$ is harmonic.

Further Insights and Challenges

39. Let $\mathcal{S}$ be the boundary surface of a region $\mathcal{W}$ in $\mathbf{R}^3$ and let $D_{\mathbf{e_n}}\varphi$ denote the directional derivative of φ, where $\mathbf{e_n}$ is the outward unit normal vector. Let Δ be the Laplace operator defined earlier.

(a) Use the Divergence Theorem to prove that

$$\iint_{\mathcal{S}} D_{\mathbf{e_n}}\varphi\, dS = \iiint_{\mathcal{W}} \Delta\varphi\, dV$$

(b) Show that if φ is a harmonic function (defined in Exercise 37), then

$$\iint_{\mathcal{S}} D_{\mathbf{e_n}}\varphi\, dS = 0$$

40. Assume that φ is harmonic. Show that $\text{div}(\varphi\nabla\varphi) = \|\nabla\varphi\|^2$ and conclude that

$$\iint_{\mathcal{S}} \varphi D_{\mathbf{e_n}}\varphi\, dS = \iiint_{\mathcal{W}} \|\nabla\varphi\|^2\, dV$$

41. Let $\mathbf{F} = \langle P, Q, R \rangle$ be a vector field defined on $\mathbf{R}^3$ such that $\text{div}(\mathbf{F}) = 0$. Use the following steps to show that $\mathbf{F}$ has a vector potential.

(a) Let $\mathbf{A} = \langle f, 0, g \rangle$. Show that

$$\text{curl}(\mathbf{A}) = \left\langle \frac{\partial g}{\partial y}, \frac{\partial f}{\partial z} - \frac{\partial g}{\partial x}, -\frac{\partial f}{\partial y} \right\rangle$$

(b) Fix any value y_0 and show that if we define

$$f(x, y, z) = -\int_{y_0}^{y} R(x, t, z)\, dt + \alpha(x, z)$$

$$g(x, y, z) = \int_{y_0}^{y} P(x, t, z)\, dt + \beta(x, z)$$

where α and β are any functions of x and z, then $\partial g/\partial y = P$ and $-\partial f/\partial y = R$.

(c) It remains for us to show that α and β can be chosen so $Q = \partial f/\partial z - \partial g/\partial x$. Verify that the following choice works (for any choice of z_0):

$$\alpha(x, z) = \int_{z_0}^{z} Q(x, y_0, t)\, dt, \qquad \beta(x, z) = 0$$

Hint: You will need to use the relation $\text{div}(\mathbf{F}) = 0$.

42. Show that

$$\mathbf{F} = \langle 2y - 1, 3z^2, 2xy \rangle$$

has a vector potential and find one.

43. Show that

$$\mathbf{F} = \langle 2ye^z - xy, y, yz - z \rangle$$

has a vector potential and find one.

44. In the text, we observed that although the inverse-square radial vector field $\mathbf{F} = \frac{\mathbf{e}_r}{r^2}$ satisfies $\text{div}(\mathbf{F}) = 0$, $\mathbf{F}$ cannot have a vector potential on its domain $\{(x, y, z) \neq (0, 0, 0)\}$ because the flux of $\mathbf{F}$ through a sphere containing the origin is nonzero.

(a) Show that the method of Exercise 41 produces a vector potential $\mathbf{A}$ such that $\mathbf{F} = \text{curl}(\mathbf{A})$ on the restricted domain $\mathcal{D}$ consisting of $\mathbf{R}^3$ with the y-axis removed.

(b) Show that $\mathbf{F}$ also has a vector potential on the domains obtained by removing either the x-axis or the z-axis from $\mathbf{R}^3$.

(c) Does the existence of a vector potential on these restricted domains contradict the fact that the flux of $\mathbf{F}$ through a sphere containing the origin is nonzero?

CHAPTER REVIEW EXERCISES

1. Let $\mathbf{F}(x, y) = \langle x + y^2, x^2 - y \rangle$ and let $\mathcal{C}$ be the unit circle, oriented counterclockwise. Evaluate $\oint_{\mathcal{C}} \mathbf{F} \cdot d\mathbf{s}$ directly as a line integral and using Green's Theorem.

2. Let $\partial\mathcal{R}$ be the boundary of the rectangle in Figure 1 and let $\partial\mathcal{R}_1$ and $\partial\mathcal{R}_2$ be the boundaries of the two triangles, all oriented counterclockwise.

(a) Determine $\oint_{\partial\mathcal{R}_1} \mathbf{F} \cdot d\mathbf{s}$ if $\oint_{\partial\mathcal{R}} \mathbf{F} \cdot d\mathbf{s} = 4$ and $\oint_{\partial\mathcal{R}_2} \mathbf{F} \cdot d\mathbf{s} = -2$.

(b) What is the value of $\oint_{\partial\mathcal{R}} \mathbf{F}\, d\mathbf{s}$ if $\partial\mathcal{R}$ is oriented clockwise?

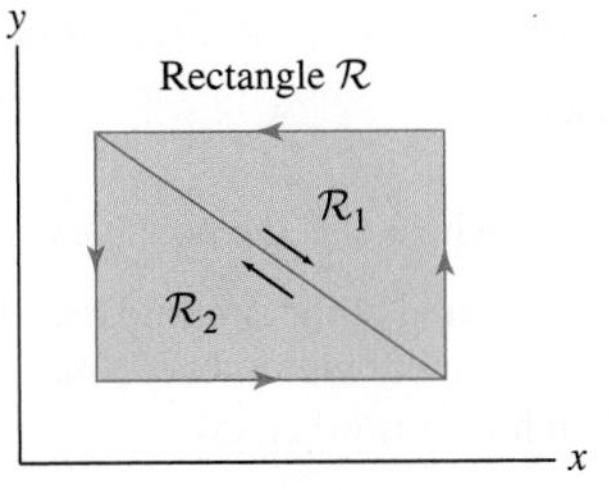

FIGURE 1

In Exercises 3–6, use Green's Theorem to evaluate the line integral around the given closed curve.

3. $\oint_{\mathcal{C}} xy^3\, dx + x^3 y\, dy$, where $\mathcal{C}$ is the rectangle $-1 \le x \le 2$, $-2 \le y \le 3$, oriented counterclockwise.

4. $\oint_{\mathcal{C}} (3x + 5y - \cos y)\, dx + x \sin y\, dy$, where $\mathcal{C}$ is any closed curve enclosing a region with area 4, oriented counterclockwise.

5. $\oint_{\mathcal{C}} y^2\, dx - x^2\, dy$, where $\mathcal{C}$ consists of the arcs $y = x^2$ and $y = \sqrt{x}$, $0 \le x \le 1$, oriented clockwise.

6. $\oint_{\mathcal{C}} ye^x\, dx + xe^y\, dy$, where $\mathcal{C}$ is the triangle with vertices $(-1, 0)$, $(0, 4)$, and $(0, 1)$, oriented counterclockwise.

7. Let $\mathbf{c}(t) = \left(t^2(1 - t), t(t - 1)^2\right)$.

(a) GU Plot the path $\mathbf{c}(t)$ for $0 \le t \le 1$.

(b) Calculate the area A of the region enclosed by $\mathbf{c}(t)$ for $0 \le t \le 1$ using the formula $A = \frac{1}{2} \oint_{\mathcal{C}} (x\, dy - y\, dx)$.

8. In (a)–(d), state whether the equation is an identity (valid for all $\mathbf{F}$ or V). If it is not, provide an example in which the equation does not hold.

(a) $\text{curl}(\nabla V) = 0$

(b) $\text{div}(\nabla V) = 0$

(c) $\text{div}(\text{curl}(\mathbf{F})) = 0$

(d) $\nabla(\text{div}(\mathbf{F})) = 0$

In Exercises 9–12, calculate the curl and divergence of the vector field.

9. $\mathbf{F} = y\mathbf{i} - z\mathbf{k}$

10. $\mathbf{F} = \langle e^{x+y}, e^{y+z}, xyz \rangle$

11. $\mathbf{F} = \nabla(e^{-x^2-y^2-z^2})$

12. $\mathbf{e}_r = r^{-1}\langle x, y, z \rangle$ $\left(r = \sqrt{x^2 + y^2 + z^2}\right)$

13. Recall that if F_1, F_2, and F_3 are differentiable functions of one variable, then

$$\text{curl}\,(\langle F_1(x), F_2(y), F_3(z)\rangle) = \mathbf{0}$$

Use this to calculate the curl of

$$\mathbf{F} = \left\langle x^2 + y^2, \ln y + z^2, z^3 \sin(z^2)e^{z^3} \right\rangle$$

14. Give an example of a nonzero vector field $\mathbf{F}$ such that $\text{curl}(\mathbf{F}) = \mathbf{0}$ and $\text{div}(\mathbf{F}) = 0$.

15. Verify the identities of Exercises 6 and 34 in Section 18.3 for the vector fields $\mathbf{F} = \langle xz, ye^x, yz\rangle$ and $\mathbf{G} = \langle z^2, xy^3, x^2y\rangle$.

16. Suppose that $\mathcal{S}_1$ and $\mathcal{S}_2$ are surfaces with the same oriented boundary curve $\mathcal{C}$. Which of the following conditions guarantees that the flux of $\mathbf{F}$ through $\mathcal{S}_1$ is equal to the flux of $\mathbf{F}$ through $\mathcal{S}_2$?

(i) $\mathbf{F} = \nabla V$ for some function V

(ii) $\mathbf{F} = \text{curl}(\mathbf{G})$ for some vector field $\mathbf{G}$

17. Prove that if $\mathbf{F}$ is a gradient vector field, then the flux of $\text{curl}(\mathbf{F})$ through a smooth surface $\mathcal{S}$ (whether closed or not) is equal to zero.

18. Verify Stokes' Theorem for $\mathbf{F} = \langle y, z - x, 0\rangle$ and the surface $z = 4 - x^2 - y^2$, $z \ge 0$, oriented by outward-pointing normals.

19. Let $\mathbf{F} = \langle z^2, x + z, y^2\rangle$ and let $\mathcal{S}$ be the upper half of the ellipsoid

$$\frac{x^2}{4} + y^2 + z^2 = 1$$

oriented by outward-pointing normals. Use Stokes' Theorem to compute $\iint_{\mathcal{S}} \text{curl}(\mathbf{F}) \cdot d\mathbf{S}$.

20. Use Stokes' Theorem to evaluate $\oint_{\mathcal{C}} \langle y, z, x\rangle \cdot d\mathbf{s}$, where $\mathcal{C}$ is the curve in Figure 2.

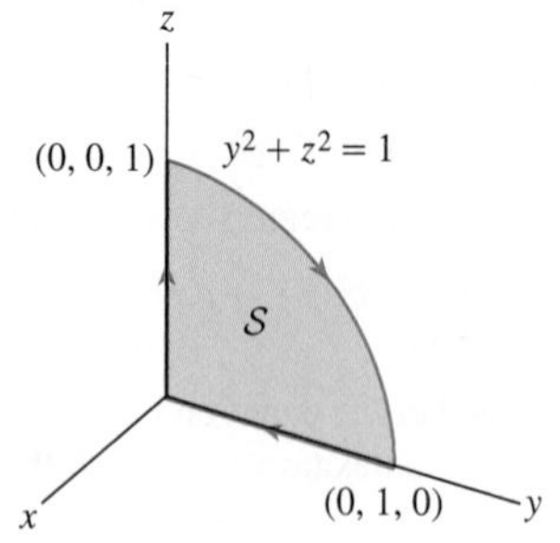

FIGURE 2

21. Let $\mathcal{S}$ be the side of the cylinder $x^2 + y^2 = 4$, $0 \le z \le 2$ (not including the top and bottom of the cylinder). Use Stokes' Theorem to compute the flux of $\mathbf{F} = \langle 0, y, -z\rangle$ through $\mathcal{S}$ (with outward pointing normal) by finding a vector potential $\mathbf{A}$ such that $\text{curl}(\mathbf{A}) = \mathbf{F}$.

22. Verify the Divergence Theorem for $\mathbf{F} = \langle 0, 0, z\rangle$ and the region $x^2 + y^2 + z^2 = 1$.

In Exercises 23–26, use the Divergence Theorem to calculate $\iint_{\mathcal{S}} \mathbf{F} \cdot d\mathbf{S}$ *for the given vector field and surface.*

23. $\mathbf{F} = \langle xy, yz, x^2z + z^2\rangle$, $\mathcal{S}$ is the boundary of the box $[0, 1] \times [2, 4] \times [1, 5]$.

24. $\mathbf{F} = \langle xy, yz, x^2z + z^2\rangle$, $\mathcal{S}$ is the boundary of the unit sphere.

25. $\mathbf{F} = \left\langle xyz + xy, \frac{1}{2}y^2(1 - z) + e^x, e^{x^2+y^2}\right\rangle$, $\mathcal{S}$ is the boundary of the solid bounded by the cylinder $x^2 + y^2 = 16$ and the planes $z = 0$ and $z = y - 4$.

26. $\mathbf{F} = \left\langle \sin(yz), \sqrt{x^2 + z^4}, x\cos(x - y)\right\rangle$, $\mathcal{S}$ is any smooth closed surface that is the boundary of a region in $\mathbf{R}^3$.

27. Find the volume of a region $\mathcal{W}$ if

$$\iint_{\partial\mathcal{W}} \left\langle x + xy + z, x + 3y - \frac{1}{2}y^2, 4z\right\rangle \cdot d\mathbf{S} = 16$$

28. Show that the circulation of $\mathbf{F} = \langle x^2, y^2, z(x^2 + y^2)\rangle$ around any curve $\mathcal{C}$ on the surface of the cone $z^2 = x^2 + y^2$ is equal to zero (Figure 3).

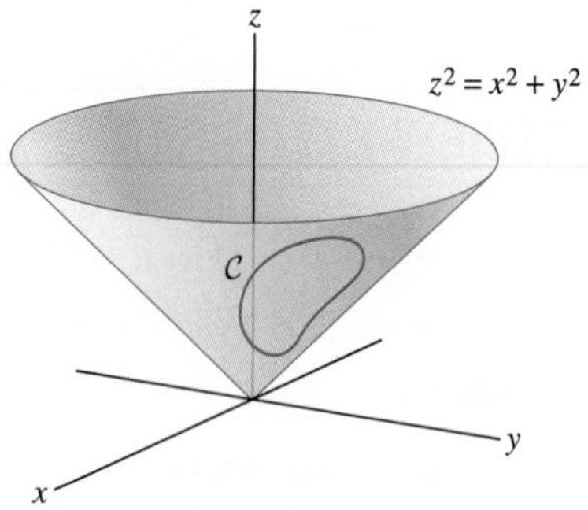

FIGURE 3

In Exercises 29–32, let $\mathbf{F}$ *be a vector field whose curl and divergence at the origin are*

$$\text{curl}(\mathbf{F})(0, 0, 0) = \langle 2, -1, 4\rangle, \qquad \text{div}(\mathbf{F})(0, 0, 0) = -2$$

29. Estimate $\oint_{\mathcal{C}} \mathbf{F} \cdot d\mathbf{s}$, where $\mathcal{C}$ is the circle of radius 0.03 in the xy-plane centered at the origin.

30. Estimate $\oint_{\mathcal{C}} \mathbf{F} \cdot d\mathbf{s}$, where $\mathcal{C}$ is the boundary of the square of side 0.03 in the yz-plane centered at the origin. Does the estimate depend on how the square is oriented within the yz-plane? Might the actual circulation depend on how it is oriented?

31. Suppose that $\mathbf{v}$ is the velocity field of a fluid and imagine placing a small paddle wheel at the origin. Find the equation of the plane in which the paddle wheel should be placed to make it rotate as quickly as possible.

32. Estimate the flux of $\mathbf{F}$ through the box of side 0.5 in Figure 4. Does the result depend on how the box is oriented relative to the coordinate axes?

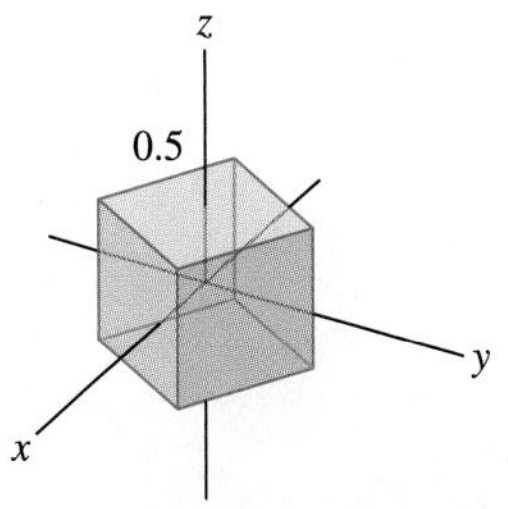

FIGURE 4

33. The velocity vector field of a fluid (in meters per second) is

$$\mathbf{F} = \langle x^2 + y^2, 0, z^2 \rangle$$

Let $\mathcal{W}$ be the region between the hemisphere

$$\mathcal{S} = \{(x, y, z) : x^2 + y^2 + z^2 = 1, \quad x, y, z \geq 0\}$$

and the disk $\mathcal{D} = \{(x, y, 0) : x^2 + y^2 \leq 1\}$ in the xy-plane. Recall that the flow rate of a fluid across a surface is equal to the flux of $\mathbf{F}$ through the surface.

(a) Show that the flow rate across $\mathcal{D}$ is zero.

(b) Use the Divergence Theorem to show that the flow rate across $\mathcal{S}$, oriented with outward-pointing normal, is equal to $\iiint_{\mathcal{W}} \operatorname{div}(\mathbf{F})\, dV$. Then compute this triple integral.

34. The velocity field of a fluid (in meters per second) is

$$\mathbf{F} = (3y - 4)\mathbf{i} + e^{-y(z+1)}\mathbf{j} + (x^2 + y^2)\mathbf{k}$$

(a) Estimate the flow rate (in cubic meters per second) through a small surface $\mathcal{S}$ around the origin if $\mathcal{S}$ encloses a region of volume $0.01\ \text{m}^3$.

(b) Estimate the circulation of $\mathbf{F}$ about a circle in the xy-plane of radius $r = 0.1$ m centered at the origin (oriented counterclockwise when viewed from above).

(c) Estimate the circulation of $\mathbf{F}$ about a circle in the yz-plane of radius $r = 0.1$ m centered at the origin (oriented counterclockwise when viewed from the positive x-axis).

35. Let $V(x, y) = x + \dfrac{x}{x^2 + y^2}$. The vector field $\mathbf{F} = \nabla V$ (Figure 5) provides a model in the plane of the velocity field of an incompressible, irrotational fluid flowing past a cylindrical obstacle (in this case, the obstacle is the unit circle $x^2 + y^2 = 1$).

(a) Verify that $\mathbf{F}$ is irrotational [by definition, $\mathbf{F}$ is irrotational if $\operatorname{curl}(\mathbf{F}) = \mathbf{0}$].

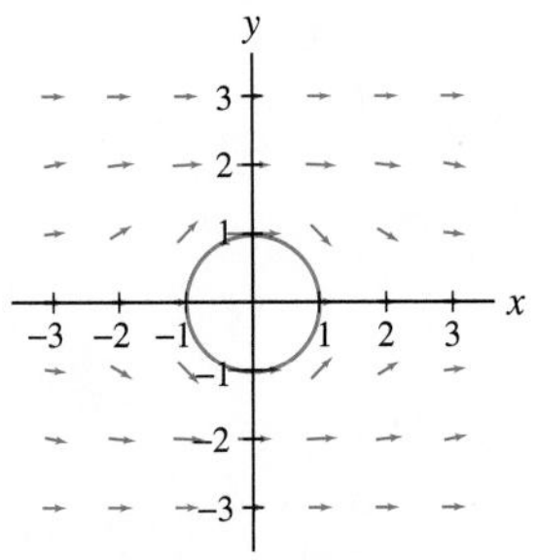

FIGURE 5 The vector field ∇V for $V(x, y) = x + \dfrac{x}{x^2 + y^2}$.

(b) Verify that $\mathbf{F}$ is tangent to the unit circle at each point along the unit circle except $(1, 0)$ and $(-1, 0)$ (where $\mathbf{F} = \mathbf{0}$).

(c) What is the circulation of $\mathbf{F}$ around the unit circle?

(d) Calculate the line integral of $\mathbf{F}$ along the upper and lower halves of the unit circle separately.

36. Figure 6 shows the vector field $\mathbf{F} = \nabla V$, where

$$V(x, y) = \ln\left(x^2 + (y - 1)^2\right) + \ln\left(x^2 + (y + 1)^2\right)$$

which is the velocity field for the flow of a fluid with sources of equal strength at $(0, \pm 1)$ (note that V is undefined at these two points). Show that $\mathbf{F}$ is both irrotational and incompressible—that is, $\operatorname{curl}_z(\mathbf{F}) = 0$ and $\operatorname{div}(\mathbf{F}) = 0$ [in computing $\operatorname{div}(\mathbf{F})$, treat $\mathbf{F}$ as a vector field in $\mathbf{R}^3$ with a zero z-component]. Is it necessary to compute $\operatorname{curl}_z(\mathbf{F})$ to conclude that it is zero?

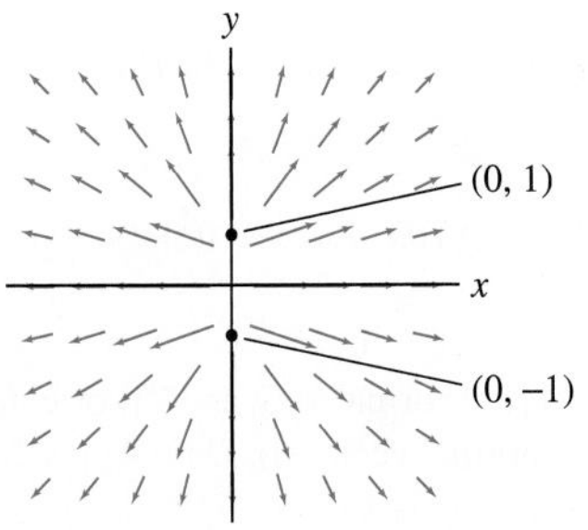

FIGURE 6 The vector field ∇V for $V(x, y) = \ln(x^2 + (y - 1)^2) + \ln(x^2 + (y + 1)^2)$.

37. In Section 18.1, we showed that if $\mathcal{C}$ is a simple closed curve, oriented counterclockwise, then the line integral is

$$\text{Area enclosed by } \mathcal{C} = \frac{1}{2}\oint_{\mathcal{C}} x\, dy - y\, dx \qquad \boxed{1}$$

Suppose that $\mathcal{C}$ is a path from P to Q that is not closed but has the property that every line through the origin intersects $\mathcal{C}$ in at most one point, as in Figure 7. Let $\mathcal{R}$ be the region enclosed by $\mathcal{C}$ and the two radial segments joining P and Q to the origin. Show that the line integral in Eq. (1) is equal to the area of $\mathcal{R}$. *Hint:* Show that the line integral of $\mathbf{F} = \langle -y, x \rangle$ along the two radial segments is zero and apply Green's Theorem.

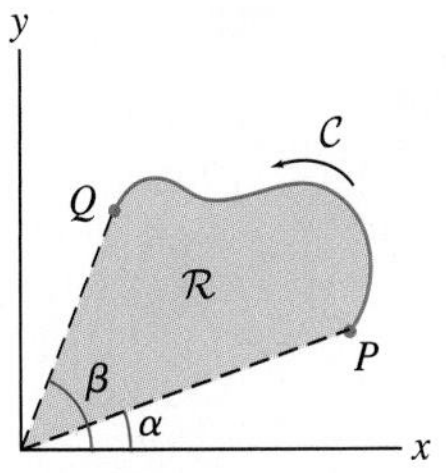

FIGURE 7

38. Suppose that the curve $\mathcal{C}$ in Figure 7 has the polar equation $r = f(\theta)$.

(a) Show that $\mathbf{c}(\theta) = (f(\theta)\cos\theta, f(\theta)\sin\theta)$ is a counterclockwise parametrization of $\mathcal{C}$.

(b) In Section 12.4, we showed that the area of the region $\mathcal{R}$ is given by the formula

$$\text{Area of } \mathcal{R} = \frac{1}{2}\int_{\alpha}^{\beta} f(\theta)^2\, d\theta$$

Use the result of Exercise 37 to give a new proof of this formula. *Hint:* Evaluate the line integral in Eq. (1) using $\mathbf{c}(\theta)$.

39. Prove the following generalization of Eq. (1). Let $\mathcal{C}$ be a simple closed curve in the plane (Figure 8)

$$\mathcal{S}: \quad ax + by + cz + d = 0$$

Then the area of the region R enclosed by $\mathcal{C}$ is equal to

$$\frac{1}{2\|\mathbf{n}\|}\oint_{\mathcal{C}} (bz - cy)\, dx + (cx - az)\, dy + (ay - bx)\, dz$$

where $\mathbf{n} = \langle a, b, c\rangle$ is the normal to $\mathcal{S}$, and $\mathcal{C}$ is oriented as the boundary of $\mathcal{R}$ (relative to the normal vector $\mathbf{n}$). *Hint:* Apply Stokes' Theorem to $\mathbf{F} = \langle bz - cy, cx - az, ay - bx\rangle$.

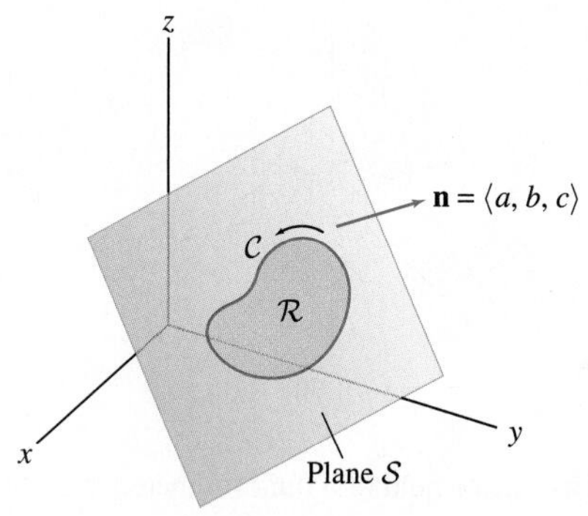

FIGURE 8

40. Use the result of Exercise 39 to calculate the area of the triangle with vertices (1, 0, 0), (0, 1, 0), and (0, 0, 1) as a line integral. Verify your result using geometry.

41. Show that $G(\theta, \phi) = (a\cos\theta\sin\phi, b\sin\theta\sin\phi, c\cos\phi)$ is a parametrization of the ellipsoid

$$\left(\frac{x}{a}\right)^2 + \left(\frac{y}{b}\right)^2 + \left(\frac{z}{c}\right)^2 = 1$$

Then calculate the volume of the ellipsoid as the surface integral of $\mathbf{F} = \frac{1}{3}\langle x, y, z\rangle$ (this surface integral is equal to the volume by the Divergence Theorem).

A THE LANGUAGE OF MATHEMATICS

One of the challenges in learning calculus is growing accustomed to its precise language and terminology, especially in the statements of theorems. In this section, we analyze a few details of logic that are helpful, and indeed essential, in understanding and applying theorems properly.

Many theorems in mathematics involve an **implication**. If A and B are statements, then the implication $A \implies B$ is the assertion that A implies B:

$$A \implies B: \qquad \textit{If } A \textit{ is true, then } B \textit{ is true.}$$

Statement A is called the **hypothesis** (or premise) and statement B the **conclusion** of the implication. Here is an example: *If m and n are even integers, then $m + n$ is an even integer*. This statement may be divided into a hypothesis and conclusion:

$$\underbrace{m \text{ and } n \text{ are even integers}}_{A} \implies \underbrace{m + n \text{ is an even integer}}_{B}$$

In everyday speech, implications are often used in a less precise way. An example is: *If you work hard, then you will succeed.* Furthermore, some statements that do not initially have the form $A \implies B$ may be restated as implications. For example, the statement "Cats are mammals" can be rephrased as follows:

$$\text{Let } X \text{ be an animal.} \quad \underbrace{X \text{ is a cat}}_{A} \implies \underbrace{X \text{ is a mammal}}_{B}$$

When we say that an implication $A \implies B$ is true, we do not claim that A or B is necessarily true. Rather, we are making the conditional statement that *if* A happens to be true, *then* B is also true. In the above, if X does not happen to be a cat, the implication tells us nothing.

The **negation** of a statement A is the assertion that A is false and is denoted $\neg A$.

Statement A	Negation $\neg A$
X lives in California.	X does not live in California.
$\triangle ABC$ is a right triangle.	$\triangle ABC$ is not a right triangle.

The negation of the negation is the original statement: $\neg(\neg A) = A$. To say that X does *not not live in California* is the same as saying that X *lives in California.*

EXAMPLE 1 State the negation of each statement.

(a) The door is open and the dog is barking.

(b) The door is open or the dog is barking (or both).

Solution

(a) The first statement is true if two conditions are satisfied (door open and dog barking), and it is false if at least one of these conditions is not satisfied. So the negation is

Either the door is not open *OR* the dog is not barking *(or both)*.

(b) The second statement is true if at least one of the conditions (door open or dog barking) is satisfied, and it is false if neither condition is satisfied. So the negation is

The door is not open *AND* the dog is not barking. ■

Contrapositive and Converse

Two important operations are the formation of the contrapositive and the formation of the converse of a statement. The **contrapositive** of $A \implies B$ is the statement "If B is false, then A is false":

Keep in mind that when we form the contrapositive, we reverse the order of A and B. The contrapositive of $A \implies B$ is NOT $\neg A \implies \neg B$.

The contrapositive of $A \implies B$ is $\neg B \implies \neg A$.

Here are some examples:

Statement	Contrapositive
If X is a cat, then X is a mammal.	If X is not a mammal, then X is not a cat.
If you work hard, then you will succeed.	If you did not succeed, then you did not work hard.
If m and n are both even, then $m + n$ is even.	If $m + n$ is not even, then m and n are not both even.

A key observation is this:

The contrapositive and the original implication are equivalent.

The fact that $A \implies B$ is equivalent to its contrapositive $\neg B \implies \neg A$ is a general rule of logic that does not depend on what A and B happen to mean. This rule belongs to the subject of "formal logic," which deals with logical relations between statements without concern for the actual content of these statements.

In other words, if an implication is true, then its contrapositive is automatically true, and vice versa. In essence, an implication and its contrapositive are two ways of saying the same thing. For example, the contrapositive "If X is not a mammal, then X is not a cat" is a roundabout way of saying that cats are mammals.

The **converse** of $A \implies B$ is the *reverse* implication $B \implies A$:

Implication: $A \implies B$	Converse $B \implies A$
If A is true, then B is true.	If B is true, then A is true.

The converse plays a very different role than the contrapositive because *the converse is NOT equivalent to the original implication.* The converse may be true or false, even if the original implication is true. Here are some examples:

True Statement	Converse	Converse True or False?
If X is a cat, then X is a mammal.	If X is a mammal, then X is a cat.	False
If m is even, then m^2 is even.	If m^2 is even, then m is even.	True

■ **EXAMPLE 2 An Example Where the Converse Is False** Show that the converse of "If m and n are even, then $m+n$ is even" is false.

A counterexample is an example that satisfies the hypothesis but not the conclusion of a statement. If a single counterexample exists, then the statement is false. However, we cannot prove that a statement is true merely by giving an example.

Solution The converse is "If $m+n$ is even, then m and n are even." To show that the converse is false, we display a counterexample. Take $m=1$ and $n=3$ (or any other pair of odd numbers). The sum is even (since $1+3=4$) but neither 1 nor 3 is even. Therefore, the converse is false. ■

■ **EXAMPLE 3 An Example Where the Converse Is True** State the contrapositive and converse of the Pythagorean Theorem. Are either or both of these true?

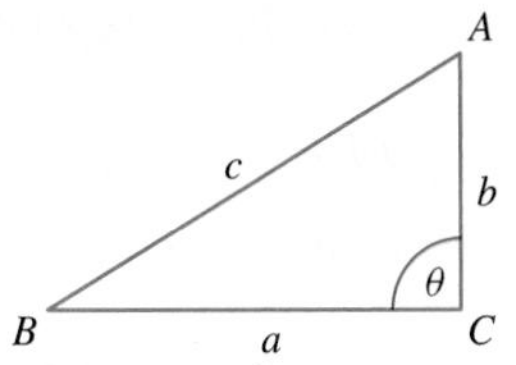

FIGURE 1

Solution Consider a triangle with sides a, b, and c, and let θ be the angle opposite the side of length c as in Figure 1. The Pythagorean Theorem states that if $\theta = 90°$, then $a^2+b^2=c^2$. Here are the contrapositive and converse:

Pythagorean Theorem	$\theta = 90° \Longrightarrow a^2+b^2=c^2$	True
Contrapositive	$a^2+b^2 \neq c^2 \Longrightarrow \theta \neq 90°$	Automatically true
Converse	$a^2+b^2=c^2 \Longrightarrow \theta = 90°$	True (but not automatic)

The contrapositive is automatically true because it is just another way of stating the original theorem. The converse is not automatically true since there could conceivably exist a nonright triangle that satisfies $a^2+b^2=c^2$. However, the converse of the Pythagorean Theorem is, in fact, true. This follows from the Law of Cosines (see Exercise 38). ■

When both a statement $A \Longrightarrow B$ and its converse $B \Longrightarrow A$ are true, we write $A \Longleftrightarrow B$. In this case, A and B are **equivalent**. We often express this with the phrase

$A \Longleftrightarrow B$ $\quad$ A is true *if and only if* B is true.

For example,

$a^2+b^2=c^2$	if and only if	$\theta = 90°$
It is morning	if and only if	the sun is rising.

We mention the following variations of terminology involving implications that you may come across:

Statement	Is Another Way of Saying
A is true if B is true.	$B \Longrightarrow A$
A is true only if B is true.	$A \Longrightarrow B$ (A cannot be true unless B is also true.)
For A to be true, it is necessary that B be true.	$A \Longrightarrow B$ (A cannot be true unless B is also true.)
For A to be true, it is sufficient that B be true.	$B \Longrightarrow A$
For A to be true, it is necessary and sufficient that B be true.	$B \Longleftrightarrow A$

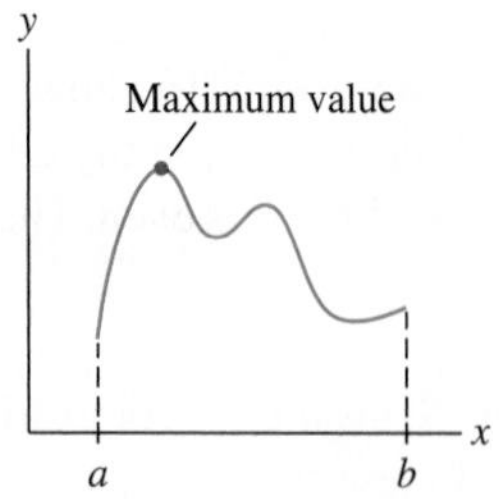

FIGURE 2 A continuous function on a closed interval $I = [a, b]$ has a maximum value.

Analyzing a Theorem

To see how these rules of logic arise in calculus, consider the following result from Section 4.2:

> **THEOREM 1 Existence of a Maximum on a Closed Interval** If $f(x)$ is a continuous function on a closed (bounded) interval $I = [a, b]$, then $f(x)$ takes on a maximum value on I (Figure 2).

To analyze this theorem, let's write out the hypotheses and conclusion separately:

Hypotheses A: $f(x)$ is continuous and I is closed.

Conclusion B: $f(x)$ takes on a maximum value on I.

A first question to ask is: "Are the hypotheses necessary?" Is the conclusion still true if we drop one or both assumptions? To show that both hypotheses are necessary, we provide counterexamples:

- **The continuity of $f(x)$ is a necessary hypothesis.** Figure 3(A) shows the graph of a function on a closed interval $[a, b]$ that is not continuous. This function has no maximum value on $[a, b]$, which shows that the conclusion may fail if the continuity hypothesis is not satisfied.
- **The hypothesis that I is closed is necessary.** Figure 3(B) shows the graph of a continuous function on an *open interval* (a, b). This function has no maximum value, which shows that the conclusion may fail if the interval is not closed.

We see that both hypotheses in Theorem 1 are necessary. In stating this, we do not claim that the conclusion *always* fails when one or both of the hypotheses are not satisfied. We claim only that the conclusion *may* fail when the hypotheses are not satisfied. Next, let's analyze the contrapositive and converse:

- **Contrapositive $\neg B \Longrightarrow \neg A$ (automatically true):** If $f(x)$ does not have a maximum value on I, then either $f(x)$ is not continuous or I is not closed (or both).
- **Converse $B \Longrightarrow A$ (in this case, false):** If $f(x)$ has a maximum value on I, then $f(x)$ is continuous and I is closed. We prove this statement false with a counterexample [Figure 3(C)].

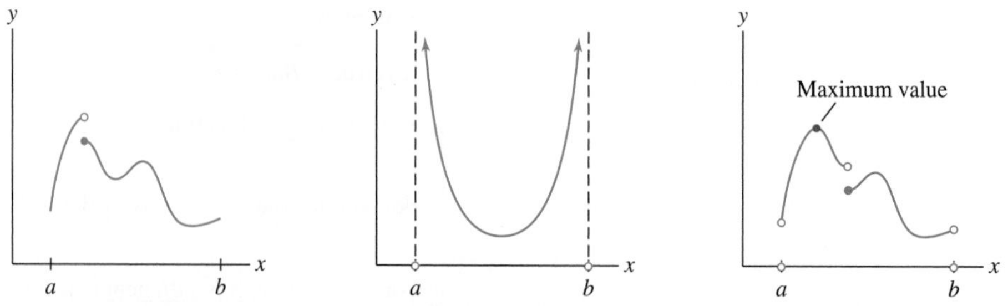

(A) The interval is closed but the function is not continuous. The function has no maximum value.

(B) The function is continuous but the interval is open. The function has no maximum value.

(C) This function is not continuous and the interval is not closed, but the function does have a maximum value.

FIGURE 3

The technique of proof by contradiction *is also known by its Latin name* reductio ad absurdum *or "reduction to the absurd." The ancient Greek mathematicians used proof by contradiction as early as the fifth century* BC*, and Euclid (325–265* BC*) employed it in his classic treatise on geometry entitled* The Elements. *A famous example is the proof that $\sqrt{2}$ is irrational in Example 4. The philosopher Plato (427–347* BC*) wrote: "He is unworthy of the name of man who is ignorant of the fact that the diagonal of a square is incommensurable with its side."*

As we know, the contrapositive is merely a way of restating the theorem, so it is automatically true. The converse is not automatically true, and in fact, in this case it is false. The function in Figure 3(C) provides a counterexample to the converse: $f(x)$ has a maximum value on $I = (a, b)$, but $f(x)$ is not continuous and I is not closed.

Mathematicians have devised various general strategies and methods for proving theorems. The method of proof by induction is discussed in Appendix C. Another important method is **proof by contradiction**, also called **indirect proof**. Suppose our goal is to prove statement A. In a proof by contradiction, we start by assuming that A is false, and then show that this leads to a contradiction. Therefore, A must be true (to avoid the contradiction).

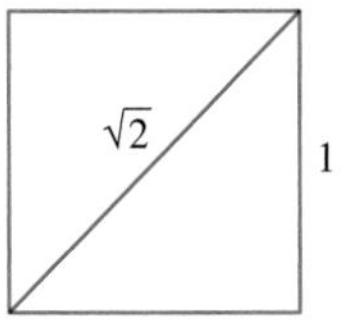

FIGURE 4 The diagonal of the unit square has length $\sqrt{2}$.

■ **EXAMPLE 4 Proof by Contradiction** The number $\sqrt{2}$ is irrational (Figure 4).

Solution Assume that the theorem is false, namely that $\sqrt{2} = p/q$, where p and q are whole numbers. We may assume that p/q is in lowest terms, and therefore, at most one of p and q is even. Note that if the square m^2 of a whole number is even, then m itself must be even.

The relation $\sqrt{2} = p/q$ implies that $2 = p^2/q^2$ or $p^2 = 2q^2$. This shows that p must be even. But if p is even, then $p = 2m$ for some whole number m, and $p^2 = 4m^2$. Because $p^2 = 2q^2$, we obtain $4m^2 = 2q^2$, or $q^2 = 2m^2$. This shows that q is also even. But we chose p and q so that at most one of them is even. This contradiction shows that our original assumption, that $\sqrt{2} = p/q$, must be false. Therefore, $\sqrt{2}$ is irrational. ■

One of the most famous problems in mathematics is known as "Fermat's Last Theorem." It states that the equation

$$x^n + y^n = z^n$$

has no solutions in positive integers if $n \geq 3$. In a marginal note written around 1630, Fermat claimed to have a proof, and over the centuries, that assertion was verified for many values of the exponent n. However, only in 1994 did the British-American mathematician Andrew Wiles, working at Princeton University, find a complete proof.

CONCEPTUAL INSIGHT The hallmark of mathematics is precision and rigor. A theorem is established, not through observation or experimentation, but by a proof that consists of a chain of reasoning with no gaps.

This approach to mathematics comes down to us from the ancient Greek mathematicians, especially Euclid, and it remains the standard in contemporary research. In recent decades, the computer has become a powerful tool for mathematical experimentation and data analysis. Researchers may use experimental data to discover potential new mathematical facts, but the title "theorem" is not bestowed until someone writes down a proof.

This insistence on theorems and proofs distinguishes mathematics from the other sciences. In the natural sciences, facts are established through experiment and are subject to change or modification as more knowledge is acquired. In mathematics, theories are also developed and expanded, but previous results are not invalidated. The Pythagorean Theorem was discovered in antiquity and is a cornerstone of plane geometry. In the nineteenth century, mathematicians began to study more general types of geometry (of the type that eventually led to Einstein's four-dimensional space-time geometry in the Theory of Relativity). The Pythagorean Theorem does not hold in these more general geometries, but its status in plane geometry is unchanged.

A. SUMMARY

- The implication $A \Longrightarrow B$ is the assertion "If A is true, then B is true."
- The *contrapositive* of $A \Longrightarrow B$ is the implication $\neg B \Longrightarrow \neg A$, which says "If B is false, then A is false." An implication and its contrapositive are equivalent (one is true if and only if the other is true).
- The *converse* of $A \Longrightarrow B$ is $B \Longrightarrow A$. An implication and its converse are not necessarily equivalent. One may be true and the other false.
- A and B are *equivalent* if $A \Longrightarrow B$ and $B \Longrightarrow A$ are both true.

- In a proof by contradiction (in which the goal is to prove statement A), we start by assuming that A is false and show that this assumption leads to a contradiction.

A. EXERCISES

Preliminary Questions

1. Which is the contrapositive of $A \implies B$?

(a) $B \implies A$ **(b)** $\neg B \implies A$

(c) $\neg B \implies \neg A$ **(d)** $\neg A \implies \neg B$

2. Which of the choices in Question 1 is the converse of $A \implies B$?

3. Suppose that $A \implies B$ is true. Which is then automatically true, the converse or the contrapositive?

4. Restate as an implication: "A triangle is a polygon."

Exercises

1. Which is the negation of the statement "The car and the shirt are both blue"?

(a) Neither the car nor the shirt is blue.

(b) The car is not blue and/or the shirt is not blue.

2. Which is the contrapositive of the implication "If the car has gas, then it will run"?

(a) If the car has no gas, then it will not run.

(b) If the car will not run, then it has no gas.

In Exercises 3–8, state the negation.

3. The time is 4 o'clock.

4. $\triangle ABC$ is an isosceles triangle.

5. m and n are odd integers.

6. Either m is odd or n is odd.

7. x is a real number and y is an integer.

8. $f(x)$ is a linear function.

In Exercises 9–14, state the contrapositive and converse.

9. If m and n are odd integers, then mn is odd.

10. If today is Tuesday, then we are in Belgium.

11. If today is Tuesday, then we are not in Belgium.

12. If $x > 4$, then $x^2 > 16$.

13. If m^2 is divisible by 3, then m is divisible by 3.

14. If $x^2 = 2$, then x is irrational.

In Exercise 15–18, give a counterexample to show that the converse of the statement is false.

15. If m is odd, then $2m + 1$ is also odd.

16. If $\triangle ABC$ is equilateral, then it is an isosceles triangle.

17. If m is divisible by 9 and 4, then m is divisible by 12.

18. If m is odd, then $m^3 - m$ is divisible by 3.

In Exercise 19–22, determine whether the converse of the statement is false.

19. If $x > 4$ and $y > 4$, then $x + y > 8$.

20. If $x > 4$, then $x^2 > 16$.

21. If $|x| > 4$, then $x^2 > 16$.

22. If m and n are even, then mn is even.

In Exercises 23 and 24, state the contrapositive and converse (it is not necessary to know what these statements mean).

23. If $f(x)$ and $g(x)$ are differentiable, then $f(x)g(x)$ is differentiable.

24. If the force field is radial and decreases as the inverse square of the distance, then all closed orbits are ellipses.

In Exercises 25–28, the ***inverse*** *of $A \implies B$ is the implication $\neg A \implies \neg B$.*

25. Which of the following is the inverse of the implication "If she jumped in the lake, then she got wet"?

(a) If she did not get wet, then she did not jump in the lake.

(b) If she did not jump in the lake, then she did not get wet.

Is the inverse true?

26. State the inverses of these implications:

(a) If X is a mouse, then X is a rodent.

(b) If you sleep late, you will miss class.

(c) If a star revolves around the sun, then it's a planet.

27. Explain why the inverse is equivalent to the converse.

28. State the inverse of the Pythagorean Theorem. Is it true?

29. Theorem 1 in Section 2.4 states the following: "If $f(x)$ and $g(x)$ are continuous functions, then $f(x) + g(x)$ is continuous." Does it follow logically that if $f(x)$ and $g(x)$ are not continuous, then $f(x) + g(x)$ is not continuous?

30. Write out a proof by contradiction for this fact: There is no smallest positive rational number. Base your proof on the fact that if $r > 0$, then $0 < r/2 < r$.

31. Use proof by contradiction to prove that if $x + y > 2$, then $x > 1$ or $y > 1$ (or both).

In Exercises 32–35, use proof by contradiction to show that the number is irrational.

32. $\sqrt{\frac{1}{2}}$ **33.** $\sqrt{3}$ **34.** $\sqrt[3]{2}$ **35.** $\sqrt[4]{11}$

36. An isosceles triangle is a triangle with two equal sides. The following theorem holds: If Δ is a triangle with two equal angles, then Δ is an isosceles triangle.

(a) What is the hypothesis?

(b) Show by providing a counterexample that the hypothesis is necessary.

(c) What is the contrapositive?

(d) What is the converse? Is it true?

37. Consider the following theorem: Let $f(x)$ be a quadratic polynomial with a positive leading coefficient. Then $f(x)$ has a minimum value.

(a) What are the hypotheses?

(b) What is the contrapositive?

(c) What is the converse? Is it true?

Further Insights and Challenges

38. Let a, b, and c be the sides of a triangle and let θ be the angle opposite c. Use the Law of Cosines (Theorem 1 in Section 1.4) to prove the converse of the Pythagorean Theorem.

39. Carry out the details of the following proof by contradiction that $\sqrt{2}$ is irrational (This proof is due to R. Palais). If $\sqrt{2}$ is rational, then $n\sqrt{2}$ is a whole number for some whole number n. Let n be the smallest such whole number and let $m = n\sqrt{2} - n$.

(a) Prove that $m < n$.

(b) Prove that $m\sqrt{2}$ is a whole number.

Explain why (a) and (b) imply that $\sqrt{2}$ is irrational.

40. Generalize the argument of Exercise 39 to prove that $\sqrt{A}$ is irrational if A is a whole number but not a perfect square. *Hint:* Choose n as before and let $m = n\sqrt{A} - n[\sqrt{A}]$, where $[x]$ is the greatest integer function.

41. Generalize further and show that for any whole number r, the rth root $\sqrt[r]{A}$ is irrational unless A is an rth power. *Hint:* Let $x = \sqrt[r]{A}$. Show that if x is rational, then we may choose a smallest whole number n such that nx^j is a whole number for $j = 1, \dots, r-1$. Then consider $m = nx - n[x]$ as before.

42. Given a finite list of prime numbers $p_1, \dots, p_N$, let $M = p_1 \cdot p_2 \cdots p_N + 1$. Show that M is not divisible by any of the primes $p_1, \dots, p_N$. Use this and the fact that every number has a prime factorization to prove that there exist infinitely many prime numbers. This argument was advanced by Euclid in *The Elements*.

B PROPERTIES OF REAL NUMBERS

"The ingenious method of expressing every possible number using a set of ten symbols (each symbol having a place value and an absolute value) emerged in India. The idea seems so simple nowadays that its significance and profound importance is no longer appreciated. Its simplicity lies in the way it facilitated calculation and placed arithmetic foremost amongst useful inventions. The importance of this invention is more readily appreciated when one considers that it was beyond the two greatest men of Antiquity, Archimedes and Apollonius."

—Pierre-Simon Laplace,
one of the great French mathematicians
of the eighteenth century

In this appendix, we discuss the basic properties of real numbers. First, let us recall that a real number is a number that may be represented by a finite or infinite decimal (also called a decimal expansion). The set of all real numbers is denoted **R** and is often visualized as the "number line" (Figure 1).

Thus, a real number a is represented as

$$a = \pm n.a_1a_2a_3a_4\ldots,$$

where n is any whole number and each digit a_j is a whole number between 0 and 9. For example, $10\pi = 31.41592\ldots$. Recall that a is rational if its expansion is finite or repeating, and is irrational if its expansion is nonrepeating. Furthermore, the decimal expansion is unique apart from the following exception: Every finite expansion is equal to an expansion in which the digit 9 repeats. For example, $0.5 = 0.4999\cdots = 0.4\bar{9}$.

We shall take for granted that the operations of addition and multiplication are defined on **R**—that is, on the set of all decimals. Roughly speaking, addition and multiplication of infinite decimals are defined in terms of finite decimals. For $d \geq 1$, define the dth truncation of $a = n.a_1a_2a_3a_4\ldots$ to be the finite decimal $a(d) = a.a_1a_2\ldots a_d$ obtained by truncating at the dth place. To form the sum $a + b$, assume that both a and b are infinite (possibly ending with repeated nines). This eliminates any possible ambiguity in the expansion. Then the nth digit of $a + b$ is equal to the nth digit of $a(d) + b(d)$ for d sufficiently large (from a certain point onward, the nth digit of $a(d) + b(d)$ no longer changes, and this value is the nth digit of $a + b$). Multiplication is defined similarly. Furthermore, the Commutative, Associative, and Distributive Laws hold (Table 1).

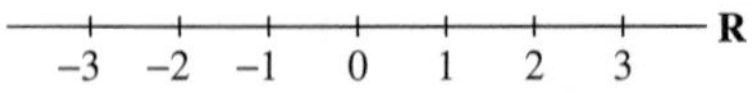

FIGURE 1 The real number line.

TABLE 1 Algebraic Laws

Commutative Laws:	$a + b = b + a, \quad ab = ba$
Associative Laws:	$(a + b) + c = a + (b + c), \quad (ab)c = a(bc)$
Distributive Law:	$a(b + c) = ab + ac$

Every real number x has an additive inverse $-x$ such that $x + (-x) = 0$, and every nonzero real number x has a multiplicative inverse x^{-1} such that $x(x^{-1}) = 1$. We do not regard subtraction and division as separate algebraic operations because they are defined in terms of inverses. By definition, the difference $x - y$ is equal to $x + (-y)$, and the quotient x/y is equal to $x(y^{-1})$ for $y \neq 0$.

In addition to the algebraic operations, there is an **order relation** on **R**: For any two real numbers a and b, precisely one of the following is true:

$$\text{Either} \quad a = b, \quad \text{or} \quad a < b, \quad \text{or} \quad a > b$$

To distinguish between the conditions $a \leq b$ and $a < b$, we often refer to $a < b$ as a **strict inequality**. Similar conventions hold for $>$ and $\geq$. The rules given in Table 2 allow us to manipulate inequalities. The last order property says that an inequality reverses direction when multiplied by a negative number c. For example,

$$-2 < 5 \quad \text{but} \quad (-3)(-2) > (-3)5$$

TABLE 2 Order Properties

If $a < b$ and $b < c$,	then $a < c$.
If $a < b$ and $c < d$,	then $a + c < b + d$.
If $a < b$ and $c > 0$,	then $ac < bc$.
If $a < b$ and $c < 0$,	then $ac > bc$.

The algebraic and order properties of real numbers are certainly familiar. We now discuss the less familiar **Least Upper Bound (LUB) Property** of the real numbers. This property is one way of expressing the so-called **completeness** of the real numbers. There are other ways of formulating completeness (such as the so-called nested interval property discussed in any book on analysis) that are equivalent to the LUB Property and serve the same purpose. Completeness is used in calculus to construct rigorous proofs of basic theorems about continuous functions, such as the Intermediate Value Theorem, (IVT) or the existence of extreme values on a closed interval. The underlying idea is that the real number line "has no holes." We elaborate on this idea below. First, we introduce the necessary definitions.

Suppose that S is a nonempty set of real numbers. A number M is called an **upper bound** for S if

$$x \le M \qquad \text{for all } x \in S$$

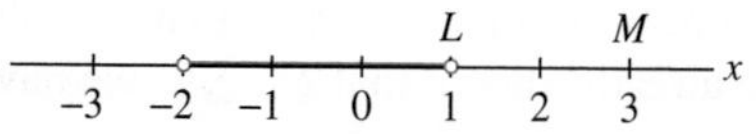

FIGURE 2 $M = 3$ is an upper bound for the set $S = (-2, 1)$. The LUB is $L = 1$.

If S has an upper bound, we say that S is **bounded above**. A **least upper bound** L is an upper bound for S such that every other upper bound M satisfies $M \ge L$. For example (Figure 2),

- $M = 3$ is an upper bound for the open interval $S = (-2, 1)$.
- $L = 1$ is the LUB for $S = (-2, 1)$.

We now state the LUB Property of the real numbers.

THEOREM 1 Existence of a Least Upper Bound Let S be a nonempty set of real numbers that is bounded above. Then S has an LUB.

In a similar fashion, we say that a number B is a **lower bound** for S if $x \ge B$ for all $x \in S$. We say that S is **bounded below** if S has a lower bound. A **greatest lower bound** (GLB) is a lower bound M such that every other lower bound B satisfies $B \le M$. The set of real numbers also has the GLB Property: If S is a nonempty set of real numbers that is bounded below, then S has a GLB. This may be deduced immediately from Theorem 1. For any nonempty set of real numbers S, let $-S$ be the set of numbers of the form $-x$ for $x \in S$. Then $-S$ has an upper bound if S has a lower bound. Consequently, $-S$ has an LUB L by Theorem 1, and $-L$ is a GLB for S.

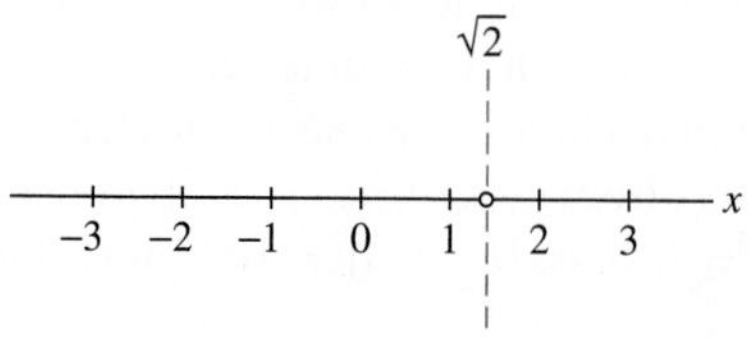

FIGURE 3 The rational numbers have a "hole" at the location $\sqrt{2}$.

CONCEPTUAL INSIGHT Theorem 1 may appear quite reasonable, but perhaps it is not clear why it is useful. We suggested above that the LUB Property expresses the idea that **R** is "complete" or "has no holes." To illustrate this idea, let's compare **R** to the set of rational numbers, denoted **Q**. Intuitively, **Q** is not complete because the irrational numbers are missing. For example, **Q** has a "hole" where the irrational number $\sqrt{2}$ should be located (Figure 3). This hole divides **Q** into two halves that are not connected to each other (the half to the left and the half to the right of $\sqrt{2}$). Furthermore, the half on the left is bounded above but no rational number is an LUB, and the half on the right is bounded below but no rational number is a GLB. The LUB and GLB are both equal to the irrational number $\sqrt{2}$, which exists in only **R** but not **Q**. So unlike **R**, the rational numbers **Q** do not have the LUB property.

■ **EXAMPLE 1** Show that 2 has a square root by applying the LUB Property to the set

$$S = \{x : x^2 < 2\}$$

Solution First, we note that S is bounded with the upper bound $M = 2$. Indeed, if $x > 2$, then x satisfies $x^2 > 4$, and hence x does not belong to S. By the LUB Property, S has a least upper bound. Call it L. We claim that $L = \sqrt{2}$, or, equivalently, that $L^2 = 2$. We prove this by showing that $L^2 \geq 2$ and $L^2 \leq 2$.

If $L^2 < 2$, let $b = L + h$, where $h > 0$. Then

$$b^2 = L^2 + 2Lh + h^2 = L^2 + h(2L + h) \qquad \boxed{1}$$

We can make the quantity $h(2L + h)$ as small as desired by choosing $h > 0$ small enough. In particular, we may choose a positive h so that $h(2L + h) < 2 - L^2$. For this choice, $b^2 < L^2 + (2 - L^2) = 2$ by Eq. (1). Therefore, $b \in S$. But $b > L$ since $h > 0$, and thus L is not an upper bound for S, in contradiction to our hypothesis on L. We conclude that $L^2 \geq 2$.

If $L^2 > 2$, let $b = L - h$, where $h > 0$. Then

$$b^2 = L^2 - 2Lh + h^2 = L^2 - h(2L - h)$$

Now choose h positive but small enough so that $0 < h(2L - h) < L^2 - 2$. Then $b^2 > L^2 - (L^2 - 2) = 2$. But $b < L$, so b is a smaller lower bound for S. Indeed, if $x \geq b$, then $x^2 \geq b^2 > 2$, and x does not belong to S. This contradicts our hypothesis that L is the LUB. We conclude that $L^2 \leq 2$, and since we have already shown that $L^2 \geq 2$, we have $L^2 = 2$ as claimed. ■

We now prove three important theorems, the third of which is used in the proof of the LUB Property below.

> **THEOREM 2 Bolzano–Weierstrass Theorem** Let S be a bounded, infinite set of real numbers. Then there exists a sequence of distinct elements $\{a_n\}$ in S such that the limit $L = \lim_{n\to\infty} a_n$ exists.

Proof For simplicity of notation, we assume that S is contained in the unit interval [0, 1] (a similar proof works in general). If $k_1, k_2, \ldots, k_n$ is a sequence of n digits (that is, each k_j is a whole number and $0 \leq k_j \leq 9$), let

$$S(k_1, k_2, \ldots, k_n)$$

be the set of $x \in S$ whose decimal expansion begins $0.k_1k_2 \ldots k_n$. The set S is the union of the subsets $S(0), S(1), \ldots, S(9)$, and since S is infinite, at least one of these subsets must be infinite. Therefore, we may choose k_1 so that $S(k_1)$ is infinite. In a similar fashion, at least one of the set $S(k_1, 0), S(k_2, 1), \ldots, S(k_1, 9)$ must be infinite, so we may choose k_2 so that $S(k_1, k_2)$ is infinite. Continuing in this way, we obtain an infinite sequence $\{k_n\}$ such that $S(k_1, k_2, \ldots, k_n)$ is infinite for all n. We may choose a sequence of elements $a_n \in S(k_1, k_2, \ldots, k_n)$ with the property that a_n differs from $a_1, \ldots, a_{n-1}$ for all n. Let L be the infinite decimal $0.k_1k_2k_3\ldots$. Then $\lim_{n\to\infty} a_n = L$ since $|L - a_n| < 10^{-n}$ for all n. ■

We use the Bolzano–Weierstrass Theorem to prove two important results about sequences $\{a_n\}$. Recall that an upper bound for $\{a_n\}$ is a number M such that $a_j \leq M$ for all j. If an upper bound exists, $\{a_n\}$ is said to be bounded from above. Lower bounds are defined similarly and $\{a_n\}$ is said to be bounded from below if a lower bound exists.

A sequence is bounded if it is bounded from above and below. A **subsequence** of $\{a_n\}$ is a sequence of elements $a_{n_1}, a_{n_2}, a_{n_3}, \dots$, where $n_1 < n_2 < n_3 < \cdots$.

Now consider a bounded sequence $\{a_n\}$. If infinitely many of the a_n are distinct, the Bolzano–Weierstrass Theorem implies that there exists a subsequence $\{a_{n_1}, a_{n_2}, \dots\}$ such that $\lim_{n\to\infty} a_{n_k}$ exists. Otherwise, infinitely many of the a_n must coincide, and these terms form a convergent subsequence. This proves the next result.

| Section 11.1

THEOREM 3 Every bounded sequence has a convergent subsequence.

THEOREM 4 Bounded Monotonic Sequences Converge

- If $\{a_n\}$ is increasing and $a_n \le M$ for all n, then $\{a_n\}$ converges and $\lim_{n\to\infty} a_n \le M$.
- If $\{a_n\}$ is decreasing and $a_n \ge M$ for all n, then $\{a_n\}$ converges and $\lim_{n\to\infty} a_n \ge M$.

Proof Suppose that $\{a_n\}$ is increasing and bounded above by M. Then $\{a_n\}$ is automatically bounded below by $m = a_1$ since $a_1 \le a_2 \le a_3 \cdots$. Hence, $\{a_n\}$ is bounded, and by Theorem 3, we may choose a convergent subsequence $a_{n_1}, a_{n_2}, \dots$. Let

$$L = \lim_{k\to\infty} a_{n_k}$$

Observe that $a_n \le L$ for all n. For if not, then $a_n > L$ for some n and then $a_{n_k} \ge a_n > L$ for all k such that $n_k \ge n$. But this contradicts that $a_{n_k} \to L$. Now, by definition, for any $\epsilon > 0$, there exists $N_\epsilon > 0$ such that

$$|a_{n_k} - L| < \epsilon \qquad \text{if } n_k > N_\epsilon$$

Choose m such that $n_m > N_\epsilon$. If $n \ge n_m$, then $a_{n_m} \le a_n \le L$, and therefore,

$$|a_n - L| \le |a_{n_m} - L| < \epsilon \qquad \text{for all } n \ge n_m$$

This proves that $\lim_{n\to\infty} a_n = L$ as desired. It remains to prove that $L \le M$. If $L > M$, let $\epsilon = (L - M)/2$ and choose N so that

$$|a_n - L| < \epsilon \qquad \text{if } k > N$$

Then $a_n > L - \epsilon = M + \epsilon$. This contradicts our assumption that M is an upper bound for $\{a_n\}$. Therefore, $L \le M$ as claimed. ∎

Proof of Theorem 1 We now use Theorem 4 to prove the LUB Property (Theorem 1). As above, if x is a real number, let $x(d)$ be the truncation of x of length d. For example,

$$\text{If } x = 1.41569, \text{ then } x(3) = 1.415$$

We say that x is a *decimal of length d* if $x = x(d)$. Any two distinct decimals of length d differ by at least 10^{-d}. It follows that for any two real numbers $A < B$, there are at most finitely many decimals of length d between A and B.

Now let S be a nonempty set of real numbers with an upper bound M. We shall prove that S has an LUB. Let $S(d)$ be the set of truncations of length d:

$$S(d) = \{x(d) : x \in S\}$$

We claim that $S(d)$ has a maximum element. To verify this, choose any $a \in S$. If $x \in S$ and $x(d) > a(d)$, then

$$a(d) \le x(d) \le M$$

Thus, by the remark of the previous paragraph, there are at most finitely many values of $x(d)$ in $S(d)$ larger than $a(d)$. The largest of these is the maximum element in $S(d)$.

For $d = 1, 2, \dots,$ choose an element x_d such that $x_d(d)$ is the maximum element in $S(d)$. By construction, $\{x_d(d)\}$ is an increasing sequence (since the largest dth truncation cannot get smaller as d increases). Furthermore, $x_d(d) \le M$ for all d. We now apply Theorem 4 to conclude that $\{x_d(d)\}$ converges to a limit L. We claim that L is the LUB of S. Observe first that L is an upper bound for S. Indeed, if $x \in S$, then $x(d) \le L$ for all d and thus $x \le L$. To show that L is the LUB, suppose that M is an upper bound such that $M < L$. Then $x_d \le M$ for all d and hence $x_d(d) \le M$ for all d. But then

$$L = \lim_{d\to\infty} x_d(d) \le M$$

This is a contradiction since $M < L$. Therefore, L is the LUB of S. ■

As mentioned above, the LUB Property is used in calculus to establish certain basic theorems about continuous functions. As an example, we prove the IVT. Another example is the theorem on the existence of extrema on a closed interval (see Appendix D).

> **THEOREM 5 Intermediate Value Theorem** If $f(x)$ is continuous on a closed interval $[a, b]$ and $f(a) \ne f(b)$, then for every value M between $f(a)$ and $f(b)$, there exists at least one value $c \in (a, b)$ such that $f(c) = M$.

Proof Assume first that $M = 0$. Replacing $f(x)$ by $-f(x)$ if necessary, we may assume that $f(a) < 0$ and $f(b) > 0$. Now let

$$S = \{x \in [a, b] : f(x) < 0\}$$

Then $a \in S$ since $f(a) < 0$ and thus S is nonempty. Clearly, b is an upper bound for S. Therefore, by the LUB Property, S has an LUB L. We claim that $f(L) = 0$. If not, set $r = f(L)$. Assume first that $r > 0$.

Since $f(x)$ is continuous, there exists a number $\delta > 0$ such that

$$|f(x) - f(L)| = |f(x) - r| < \frac{1}{2}r \qquad \text{if} \qquad |x - L| < \delta$$

Equivalently,

$$\frac{1}{2}r < f(x) < \frac{3}{2}r \qquad \text{if} \qquad |x - L| < \delta$$

The number $\frac{1}{2}r$ is positive, so we conclude that

$$f(x) > 0 \qquad \text{if} \qquad L - \delta < x < L + \delta$$

By definition of L, $f(x) \ge 0$ for all $x \in [a, b]$ such that $x > L$, and thus $f(x) \ge 0$ for all $x \in [a, b]$ such that $x > L - \delta$. Thus, $L - \delta$ is an upper bound for S. This is a contradiction since L is the LUB of S, and it follows that $r = f(L)$ cannot satisfy $r > 0$. Similarly, r cannot satisfy $r < 0$. We conclude that $f(L) = 0$ as claimed.

Now, if M is nonzero, let $g(x) = f(x) - M$. Then 0 lies between $g(a)$ and $g(b)$, and by what we have proved, there exists $c \in (a, b)$ such that $g(c) = 0$. But then $f(c) = g(c) + M = M$, as desired. ■

C INDUCTION AND THE BINOMIAL THEOREM

The Principle of Induction is a method of proof that is widely used to prove that a given statement $P(n)$ is valid for all natural numbers $n = 1, 2, 3, \ldots$. Here are two statements of this kind:

- $P(n)$: The sum of the first n odd numbers is equal to n^2.
- $P(n)$: $\dfrac{d}{dx}x^n = nx^{n-1}$.

The first statement claims that for all natural numbers n,

$$\underbrace{1 + 3 + \cdots + (2n - 1)}_{\text{Sum of first } n \text{ odd numbers}} = n^2 \qquad \boxed{1}$$

We can check directly that $P(n)$ is true for the first few values of n:

$$P(1) \text{ is the equality:} \qquad 1 = 1^2 \quad \text{(true)}$$

$$P(2) \text{ is the equality:} \qquad 1 + 3 = 2^2 \quad \text{(true)}$$

$$P(3) \text{ is the equality:} \qquad 1 + 3 + 5 = 3^2 \quad \text{(true)}$$

The Principle of Induction may be used to establish $P(n)$ for all n.

The Principle of Induction applies if $P(n)$ is an assertion defined for $n \geq n_0$, where n_0 is a fixed integer. Assume that

*(i) **Initial step:** $P(n_0)$ is true.*

*(ii) **Induction step:** If $P(n)$ is true for $n = k$, then $P(n)$ is also true for $n = k + 1$.*

Then $P(n)$ is true for all $n \geq n_0$.

THEOREM 1 Principle of Induction Let $P(n)$ be an assertion that depends on a natural number n. Assume that

(i) Initial step: $P(1)$ is true.

(ii) Induction step: If $P(n)$ is true for $n = k$, then $P(n)$ is also true for $n = k + 1$.

Then $P(n)$ is true for all natural numbers $n = 1, 2, 3, \ldots$.

EXAMPLE 1 Prove that $1 + 3 + \cdots + (2n - 1) = n^2$ for all natural numbers n.

Solution As above, we let $P(n)$ denote the equality

$$P(n): \qquad 1 + 3 + \cdots + (2n - 1) = n^2$$

***Step 1.* Initial step: Show that $P(1)$ is true.**

We checked this above. $P(1)$ is the equality $1 = 1^2$.

***Step 2.* Induction step: Show that if $P(n)$ is true for $n = k$, then $P(n)$ is also true for $n = k + 1$.**

Assume that $P(k)$ is true. Then

$$1 + 3 + \cdots + (2k - 1) = k^2$$

Add $2k + 1$ to both sides:

$$\big[1 + 3 + \cdots + (2k - 1)\big] + (2k + 1) = k^2 + 2k + 1 = (k + 1)^2$$

$$1 + 3 + \cdots + (2k + 1) = (k + 1)^2$$

This is precisely the statement $P(k+1)$. Thus, $P(k+1)$ is true whenever $P(k)$ is true. By the Principle of Induction, $P(k)$ is true for all k. ■

The intuition behind the Principle of Induction is the following. If $P(n)$ were not true for all n, then there would exist a smallest natural number k such that $P(k)$ is false. Furthermore, $k > 1$ since $P(1)$ is true. Thus $P(k-1)$ is true [otherwise, $P(k)$ would not be the smallest "counterexample"]. On the other hand, if $P(k-1)$ is true, then $P(k)$ is also true by the induction step. This is a contradiction. So $P(k)$ must be true for all k.

■ **EXAMPLE 2** Use Induction and the Product Rule to prove that for all whole numbers n,

$$\frac{d}{dx}x^n = nx^{n-1}$$

Solution Let $P(n)$ be the formula $\frac{d}{dx}x^n = nx^{n-1}$.

***Step 1.* Initial step: Show that $P(1)$ is true.**
We use the limit definition to verify $P(1)$:

$$\frac{d}{dx}x = \lim_{h\to 0}\frac{(x+h)-x}{h} = \lim_{h\to 0}\frac{h}{h} = \lim_{h\to 0} 1 = 1$$

***Step 2.* Induction step: Show that if $P(n)$ is true for $n = k$, then $P(n)$ is also true for $n = k+1$.**
To carry out the induction step, assume that $\frac{d}{dx}x^k = kx^{k-1}$, where $k \geq 1$. Then, by the Product Rule,

$$\frac{d}{dx}x^{k+1} = \frac{d}{dx}(x\cdot x^k) = x\frac{d}{dx}x^k + x^k\frac{d}{dx}x = x(kx^{k-1}) + x^k$$
$$= kx^k + x^k = (k+1)x^k$$

This shows that $P(k+1)$ is true.

By the Principle of Induction, $P(n)$ is true for all $n \geq 1$. ■

In Pascal's Triangle, *the nth row displays the coefficients in the expansion of $(a+b)^n$:*

n	
0	1
1	1 1
2	1 2 1
3	1 3 3 1
4	1 4 6 4 1
5	1 5 10 10 5 1
6	1 6 15 20 15 6 1

The triangle is constructed as follows: Each entry is the sum of the two entries above it in the previous line. For example, the entry 15 in line $n = 6$ is the sum $10 + 5$ of the entries above it in line $n = 5$. The recursion relation guarantees that the entries in the triangle are the binomial coefficients.

As another application of induction, we prove the Binomial Theorem, which describes the expansion of the binomial $(a+b)^n$. The first few expansions are familiar:

$$(a+b)^1 = a + b$$
$$(a+b)^2 = a^2 + 2ab + b^2$$
$$(a+b)^3 = a^3 + 3a^2b + 3ab^2 + b^3$$

In general, we have an expansion

$$(a+b)^n = a^n + \binom{n}{1}a^{n-1}b + \binom{n}{2}a^{n-2}b^2 + \binom{n}{3}a^{n-3}b^3 + \cdots + \binom{n}{n-1}ab^{n-1} + b^n \qquad \boxed{2}$$

where the coefficient of $x^{n-k}x^k$, denoted $\binom{n}{k}$, is called the **binomial coefficient**. Note that the first term in Eq. (2) corresponds to $k = 0$ and the last term to $k = n$; thus,

$\binom{n}{0} = \binom{n}{n} = 1$. In summation notation,

$$(a+b)^n = \sum_{k=0}^{n} \binom{n}{k} a^k b^{n-k}$$

Pascal's Triangle (described in the marginal note on page A14) can be used to compute binomial coefficients if n and k are not too large. The Binomial Theorem provides the following general formula:

$$\binom{n}{k} = \frac{n!}{k!\,(n-k)!} = \frac{n(n-1)(n-2)\cdots(n-k+1)}{k(k-1)(k-2)\cdots 2\cdot 1} \qquad \boxed{3}$$

Before proving this formula, we prove a recursion relation for binomial coefficients. Note, however, that Eq. (3) is certainly correct for $k=0$ and $k=n$ (recall that by convention, $0! = 1$):

$$\binom{n}{0} = \frac{n!}{(n-0)!\,0!} = \frac{n!}{n!} = 1, \qquad \binom{n}{n} = \frac{n!}{(n-n)!\,n!} = \frac{n!}{n!} = 1$$

THEOREM 2 Recursion Relation for Binomial Coefficients

$$\binom{n}{k} = \binom{n-1}{k} + \binom{n-1}{k-1} \qquad \text{for } 1 \le k \le n-1$$

Proof We write $(a+b)^n$ as $(a+b)(a+b)^{n-1}$ and expand in terms of binomial coefficients:

$$(a+b)^n = (a+b)(a+b)^{n-1}$$

$$\sum_{k=0}^{n} \binom{n}{k} a^{n-k} b^k = (a+b) \sum_{k=0}^{n-1} \binom{n-1}{k} a^{n-1-k} b^k$$

$$= a \sum_{k=0}^{n-1} \binom{n-1}{k} a^{n-1-k} b^k + b \sum_{k=0}^{n-1} \binom{n-1}{k} a^{n-1-k} b^k$$

$$= \sum_{k=0}^{n-1} \binom{n-1}{k} a^{n-k} b^k + \sum_{k=0}^{n-1} \binom{n-1}{k} a^{n-(k+1)} b^{k+1}$$

Replacing k by $k-1$ in the second sum, we obtain

$$\sum_{k=0}^{n} \binom{n}{k} a^{n-k} b^k = \sum_{k=0}^{n-1} \binom{n-1}{k} a^{n-k} b^k + \sum_{k=1}^{n} \binom{n-1}{k-1} a^{n-k} b^k$$

On the right-hand side, the first term in the first sum is a^n and the last term in the second sum is b^n. Thus, we have

$$\sum_{k=0}^{n} \binom{n}{k} a^{n-k} b^k = a^n + \left(\sum_{k=1}^{n-1} \left(\binom{n-1}{k} + \binom{n-1}{k-1} \right) a^{n-k} b^k \right) + b^n$$

The recursion relation follows because the coefficients of $a^{n-k}b^k$ on the two sides of the equation must be equal. ■

We now use induction to prove Eq. (3). Let $P(n)$ be the claim

$$\binom{n}{k} = \frac{n!}{k!\,(n-k)!} \qquad \text{for } 0 \le k \le n$$

We have $\binom{1}{0} = \binom{1}{1} = 1$ since $(a+b)^1 = a+b$, so $P(1)$ is true. Furthermore, $\binom{n}{n} = \binom{n}{0} = 1$ as observed above, since a^n and b^n have coefficient 1 in the expansion of $(a+b)^n$. For the inductive step, assume that $P(n)$ is true. By the recursion relation, for $1 \le k \le n$, we have

$$\binom{n+1}{k} = \binom{n}{k} + \binom{n}{k-1} = \frac{n!}{k!\,(n-k)!} + \frac{n!}{(k-1)!\,(n-k+1)!}$$

$$= n!\left(\frac{n+1-k}{k!\,(n+1-k)!} + \frac{k}{k!\,(n+1-k)!}\right) = n!\left(\frac{n+1}{k!\,(n+1-k)!}\right)$$

$$= \frac{(n+1)!}{k!\,(n+1-k)!}$$

Thus, $P(n+1)$ is also true and the Binomial Theorem follows by induction.

■ **EXAMPLE 3** Use the Binomial Theorem to expand $(x+y)^5$ and $(x+2)^3$.

Solution The fifth row in Pascal's Triangle yields

$$(x+y)^5 = x^5 + 5x^4y + 10x^3y^2 + 10x^2y^3 + 5xy^4 + y^5$$

The third row in Pascal's Triangle yields

$$(x+2)^3 = x^3 + 3x^2(2) + 3x(2)^2 + 2^3 = x^3 + 6x^2 + 12x + 8$$ ■

C. EXERCISES

In Exercises 1–4, use the Principle of Induction to prove the formula for all natural numbers n.

1. $1 + 2 + 3 + \cdots + n = \dfrac{n(n+1)}{2}$

2. $1^3 + 2^3 + 3^3 + \cdots + n^3 = \dfrac{n^2(n+1)^2}{4}$

3. $\dfrac{1}{1\cdot 2} + \dfrac{1}{2\cdot 3} + \cdots + \dfrac{1}{n(n+1)} = \dfrac{n}{n+1}$

4. $1 + x + x^2 + \cdots + x^n = \dfrac{1-x^{n+1}}{1-x}$ for any $x \ne 1$

5. Let $P(n)$ be the statement $2^n > n$.

(a) Show that $P(1)$ is true.

(b) Observe that if $2^n > n$, then $2^n + 2^n > 2n$. Use this to show that if $P(n)$ is true for $n = k$, then $P(n)$ is true for $n = k+1$. Conclude that $P(n)$ is true for all n.

6. Use induction to prove that $n! > 2^n$ for $n \ge 4$.

Let $\{F_n\}$ be the Fibonacci sequence, defined by the recursion formula

$$F_n = F_{n-1} + F_{n-2}, \qquad F_1 = F_2 = 1$$

The first few terms are 1, 1, 2, 3, 5, 8, 13, ... In Exercises 7–10, use induction to prove the identity.

7. $F_1 + F_2 + \cdots + F_n = F_{n+2} - 1$

8. $F_1^2 + F_2^2 + \cdots + F_n^2 = F_{n+1}F_n$

9. $F_n = \dfrac{R_+^n - R_-^n}{\sqrt{5}}$, where $R_\pm = \dfrac{1 \pm \sqrt{5}}{2}$

10. $F_{n+1}F_{n-1} = F_n^2 + (-1)^n$. *Hint:* For the induction step, show that

$$F_{n+2}F_n = F_{n+1}F_n + F_n^2$$
$$F_{n+1}^2 = F_{n+1}F_n + F_{n+1}F_{n-1}$$

11. Use induction to prove that $f(n) = 8^n - 1$ is divisible by 7 for all natural numbers n. *Hint:* For the induction step, show that

$$8^{k+1} - 1 = 7 \cdot 8^k + (8^k - 1)$$

12. Use induction to prove that $n^3 - n$ is divisible by 3 for all natural numbers n.

13. Use induction to prove that $5^{2n} - 4^n$ is divisible by 7 for all natural numbers n.

14. Use Pascal's Triangle to write out the expansions of $(a+b)^6$ and $(a-b)^4$.

15. Expand $(x + x^{-1})^4$.

16. What is the coefficient of x^9 in $(x^3 + x)^5$?

17. Let $S(n) = \sum_{k=0}^{n} \binom{n}{k}$.

(a) Use Pascal's Triangle to compute $S(n)$ for $n = 1, 2, 3, 4$.

(b) Prove that $S(n) = 2^n$ for all $n \geq 1$. *Hint:* Expand $(a+b)^n$ and evaluate at $a = b = 1$.

18. Let $T(n) = \sum_{k=0}^{n} (-1)^k \binom{n}{k}$.

(a) Use Pascal's Triangle to compute $T(n)$ for $n = 1, 2, 3, 4$.

(b) Prove that $T(n) = 0$ for all $n \geq 1$. *Hint:* Expand $(a+b)^n$ and evaluate at $a = 1, b = -1$.

D ADDITIONAL PROOFS

In this appendix, we provide proofs of several theorems that were stated or used in the text.

| Section 2.3

THEOREM 1 Basic Limit Laws Assume that $\lim_{x\to c} f(x)$ and $\lim_{x\to c} g(x)$ exist. Then:

(i) $\lim_{x\to c} \big(f(x)+g(x)\big) = \lim_{x\to c} f(x) + \lim_{x\to c} g(x)$

(ii) For any number k, $\lim_{x\to c} kf(x) = k \lim_{x\to c} f(x)$

(iii) $\lim_{x\to c} f(x)g(x) = \left(\lim_{x\to c} f(x)\right)\left(\lim_{x\to c} g(x)\right)$

(iv) If $\lim_{x\to c} g(x) \neq 0$, then

$$\lim_{x\to c} \frac{f(x)}{g(x)} = \frac{\lim_{x\to c} f(x)}{\lim_{x\to c} g(x)}$$

Proof Let $L = \lim_{x\to c} f(x)$ and $M = \lim_{x\to c} g(x)$. The Sum Law (i) was proved in Section 2.9. Observe that (ii) is a special case of (iii), where $g(x) = k$ is a constant function. Thus, it will suffice to prove the Product Law (iii). We write

$$f(x)g(x) - LM = f(x)(g(x) - M) + M(f(x) - L)$$

and apply the Triangle Inequality to obtain

$$|f(x)g(x) - LM| \le |f(x)(g(x) - M)| + |M(f(x) - L)| \qquad \boxed{1}$$

By the limit definition, we may choose $\delta > 0$ so that

$$|f(x) - L| < 1 \qquad \text{if } 0 < |x - c| < \delta$$

If follows that $|f(x)| < |L| + 1$ for $0 < |x - c| < \delta$. Now choose any number $\epsilon > 0$. Applying the limit definition again, we see that by choosing a smaller δ if necessary, we may also ensure that if $0 < |x - c| < \delta$, then

$$|f(x) - L| \le \frac{\epsilon}{2(|M| + 1)} \qquad \text{and} \qquad |g(x) - M| \le \frac{\epsilon}{2(|L| + 1)}$$

Using Eq. (1), we see that if $0 < |x - c| < \delta$, then

$$\begin{aligned} |f(x)g(x) - LM| &\le |f(x)|\,|g(x) - M| + |M|\,|f(x) - L| \\ &\le (|L| + 1)\frac{\epsilon}{2(|L| + 1)} + |M|\frac{\epsilon}{2(|M| + 1)} \\ &\le \frac{\epsilon}{2} + \frac{\epsilon}{2} = \epsilon \end{aligned}$$

Since ϵ is arbitrary, this proves that $\lim_{x\to c} f(x)g(x) = LM$. To prove the Quotient Law (iv), it suffices to verify that if $M \neq 0$, then

$$\lim_{x\to c} \frac{1}{g(x)} = \frac{1}{M} \qquad \boxed{2}$$

For if Eq. (2) holds, then we may apply the Product Law to $f(x)$ and $g(x)^{-1}$ to obtain the Quotient Law:

$$\lim_{x\to c} \frac{f(x)}{g(x)} = \lim_{x\to c} f(x)\frac{1}{g(x)} = \left(\lim_{x\to c} f(x)\right)\left(\lim_{x\to c} \frac{1}{g(x)}\right)$$
$$= L\left(\frac{1}{M}\right) = \frac{L}{M}$$

We now verify Eq. (2). Since $g(x)$ approaches M and $M \neq 0$, we may choose $\delta > 0$ so that $|g(x)| \geq |M|/2$ if $0 < |x - c| < \delta$. Now choose any number $\epsilon > 0$. By choosing a smaller δ if necessary, we may also ensure that

$$|M - g(x)| < \epsilon|M|\left(\frac{|M|}{2}\right) \qquad \text{for } 0 < |x - c| < \delta$$

Then

$$\left|\frac{1}{g(x)} - \frac{1}{M}\right| = \left|\frac{M - g(x)}{Mg(x)}\right| \leq \left|\frac{M - g(x)}{M(M/2)}\right| \leq \frac{\epsilon|M|(|M|/2)}{|M|(|M|/2)} = \epsilon$$

Since ϵ is arbitrary, the limit in Eq. (2) is proved. ∎

The following result was used in the text.

THEOREM 2 Limits Preserve Inequalities Let (a, b) be an open interval and let $c \in (a, b)$. Suppose that $f(x)$ and $g(x)$ are defined on (a, b), except possibly at c. Assume that

$$f(x) \leq g(x) \qquad \text{for } x \in (a, b), \quad x \neq c$$

and that the limits $\lim_{x\to c} f(x)$ and $\lim_{x\to c} g(x)$ exist. Then

$$\lim_{x\to c} f(x) \leq \lim_{x\to c} g(x)$$

Proof Let $L = \lim_{x\to c} f(x)$ and $M = \lim_{x\to c} g(x)$. To show that $L \leq M$, we use proof by contradiction. If $L > M$, let $\epsilon = \frac{1}{2}(L - M)$. By the formal definition of limits, we may choose $\delta > 0$ so that the following two conditions are satisfied:

$$|M - g(x)| < \epsilon \qquad \text{if } |x - c| < \delta$$
$$|L - f(x)| < \epsilon \qquad \text{if } |x - c| < \delta$$

But then

$$f(x) > L - \epsilon = M + \epsilon > g(x)$$

This is a contradiction since $f(x) \leq g(x)$. We conclude that $L \leq M$. ∎

THEOREM 3 Limit of a Composite Function Assume that the following limits exist:

$$L = \lim_{x \to c} g(x) \qquad \text{and} \qquad M = \lim_{x \to L} f(x)$$

Then $\lim_{x \to c} f(g(x)) = M$.

Proof Let $\epsilon > 0$ be given. By the limit definition, there exists $\delta_1 > 0$ such that

$$|f(x) - M| < \epsilon \qquad \text{if } 0 < |x - L| < \delta_1 \tag{3}$$

Similarly, there exists $\delta > 0$ such that

$$|g(x) - L| < \delta_1 \qquad \text{if } 0 < |x - c| < \delta \tag{4}$$

We replace x by $g(x)$ in Eq. (3) and apply Eq. (4) to obtain

$$|f(g(x)) - M| < \epsilon \qquad \text{if } 0 < |x - c| < \delta$$

Since ϵ is arbitrary, this proves that $\lim_{x \to c} f(g(x)) = M$. ■

| Section 2.4

THEOREM 4 Continuity of Composite Functions Let $F(x) = f(g(x))$ be a composite function. If g is continuous at $x = c$ and f is continuous at $x = g(c)$, then $F(x)$ is continuous at $x = c$.

Proof By definition of continuity,

$$\lim_{x \to c} g(x) = g(c) \qquad \text{and} \qquad \lim_{x \to g(c)} f(x) = f(g(c))$$

Therefore, we may apply Theorem 3 to obtain

$$\lim_{x \to c} f(g(x)) = f(g(c))$$

This proves that $f(g(x))$ is continuous at $x = c$. ■

| Section 2.9

THEOREM 5 Squeeze Theorem Assume that for $x \neq c$ (in some open interval containing c),

$$l(x) \le f(x) \le u(x) \qquad \text{and} \qquad \lim_{x \to c} l(x) = \lim_{x \to c} u(x) = L$$

Then $\lim_{x \to c} f(x)$ exists and

$$\lim_{x \to c} f(x) = L$$

Proof Let $\epsilon > 0$ be given. We may choose $\delta > 0$ such that

$$|l(x) - L| < \epsilon \qquad \text{and} \qquad |u(x) - L| < \epsilon \qquad \text{if } 0 < |x - c| < \delta$$

In principle, a different δ may be required to obtain the two inequalities for $l(x)$ and $u(x)$, but we may choose the smaller of the two deltas. Thus, if $0 < |x - c| < \delta$, we have

$$L - \epsilon < l(x) < L + \epsilon$$

and

$$L - \epsilon < u(x) < L + \epsilon$$

Since $f(x)$ lies between $l(x)$ and $u(x)$, it follows that

$$L - \epsilon < l(x) \le f(x) \le u(x) < L + \epsilon$$

and therefore $|f(x) - L| < \epsilon$ if $0 < |x - c| < \delta$. Since ϵ is arbitrary, this proves that $\lim_{x \to c} f(x) = L$ as desired. ■

| Section 7.2

THEOREM 6 Derivative of the Inverse Assume that $f(x)$ is differentiable and one-to-one on an open interval (r, s) with inverse $g(x)$. If b belongs to the domain of $g(x)$ and $f'(g(b)) \neq 0$, then $g'(b)$ exists and

$$g'(b) = \frac{1}{f'(g(b))}$$

Proof The function $f(x)$ is one-to-one and continuous (since it is differentiable). It follows that $f(x)$ is monotonic increasing or decreasing on (r, s). For if not, then $f(x)$ would have a local minimum or maximum at some point $x = x_0$. But then $f(x)$ would not be one-to-one in a small interval around x_0 by the IVT.

Suppose that $f(x)$ is increasing (the decreasing case is similar). We shall prove that $g(x)$ is continuous at $x = b$. Let $a = g(b)$, so that $f(a) = b$. Fix a small number $\epsilon > 0$. Since $f(x)$ is an increasing function, it maps the open interval $(a - \epsilon, a + \epsilon)$ to the open interval $(f(a - \epsilon), f(a + \epsilon))$ containing $f(a) = b$. We may choose a number $\delta > 0$ so that $(b - \delta, b + \delta)$ is contained in $(f(a - \epsilon), f(a + \epsilon))$. Then $g(x)$ maps $(b - \delta, b + \delta)$ back into $(a - \epsilon, a + \epsilon)$. It follows that

$$|g(y) - g(b)| < \epsilon \qquad \text{if } 0 < |y - b| < \delta$$

This proves that g is continuous at $x = b$.

To complete the proof, we must show that the following limit exists and is equal to $1/f'(g(b))$:

$$g'(a) = \lim_{y \to b} \frac{g(y) - g(b)}{y - b}$$

By the inverse relationship, if $y = f(x)$, then $g(y) = x$, and since $g(y)$ is continuous, x approaches a as y approaches b. Thus, since $f(x)$ is differentiable and $f'(a) \neq 0$,

$$\lim_{y \to b} \frac{g(y) - g(b)}{y - b} = \lim_{x \to a} \frac{x - a}{f(x) - f(a)} = \frac{1}{f'(a)} = \frac{1}{f'(g(b))}$$ ■

| Section 4.2

THEOREM 7 Existence of Extrema on a Closed Interval If $f(x)$ is a continuous function on a closed (bounded) interval $I = [a, b]$, then $f(x)$ takes on a minimum and a maximum value on I.

Proof We prove that $f(x)$ takes on a maximum value in two steps (the case of a minimum is similar).

***Step 1.* Prove that $f(x)$ is bounded from above.**

We use proof by contradiction. If $f(x)$ is not bounded from above, then there exist points $a_n \in [a, b]$ such that $f(a_n) \ge n$ for $n = 1, 2, \ldots$. By Theorem 3 in Appendix B, we may choose a subsequence of elements $a_{n_1}, a_{n_2}, \ldots$ that converges to a limit in $[a, b]$—say, $\lim_{k \to \infty} a_{n_k} = L$. Since $f(x)$ is continuous, there exists $\delta > 0$ such that

$$|f(x) - f(L)| < 1 \qquad \text{if} \quad x \in [a, b] \quad \text{and} \quad |x - L| < \delta$$

Therefore,

$$f(x) < f(L) + 1 \quad \text{if} \quad x \in [a, b] \quad \text{and} \quad x \in (L - \delta, L + \delta) \qquad \boxed{5}$$

For k sufficiently large, a_{n_k} lies in $(L - \delta, L + \delta)$ because $\lim_{k\to\infty} a_{n_k} = L$. By Eq. (5), $f(a_{n_k})$ is bounded by $f(L) + 1$. However, $f(a_{n_k}) = n_k$ tends to infinity as $k \to \infty$. This is a contradiction. Hence, our assumption that $f(x)$ is not bounded from above is false.

***Step 2.* Prove that $f(x)$ takes on a maximum value.**
The range of $f(x)$ on $I = [a, b]$ is the set

$$S = \{f(x) : x \in [a, b]\}$$

By the previous step, S is bounded from above and therefore has a least upper bound M by the LUB Property. Thus $f(x) \le M$ for all $x \in [a, b]$. To complete the proof, we show that $f(c) = M$ for some $c \in [a, b]$. This will show that $f(x)$ attains the maximum value M on $[a, b]$.

By definition, $M - 1/n$ is not an upper bound for $n \ge 1$, and therefore, we may choose a point b_n in $[a, b]$ such that

$$M - \frac{1}{n} \le f(b_n) \le M$$

Again by Theorem 3 in Appendix B, there exists a subsequence of elements $\{b_{n_1}, b_{n_2}, \dots\}$ in $\{b_1, b_2, \dots\}$ that converges to a limit—say,

$$\lim_{k\to\infty} b_{n_k} = c$$

Let $\epsilon > 0$. Since $f(x)$ is continuous, we may choose k so large that the following two conditions are satisfied: $|f(c) - f(b_{n_k})| < \epsilon/2$ and $n_k > 2/\epsilon$. Then

$$|f(c) - M| \le |f(c) - f(b_{n_k})| + |f(b_{n_k}) - M| \le \frac{\epsilon}{2} + \frac{1}{n_k} \le \frac{\epsilon}{2} + \frac{\epsilon}{2} = \epsilon$$

Thus, $|f(c) - M|$ is smaller than ϵ for all positive numbers ϵ. But this is not possible unless $|f(c) - M| = 0$. Thus $f(c) = M$ as desired. ■

Section 5.2

THEOREM 8 Continuous Functions Are Integrable If $f(x)$ is continuous on $[a, b]$, then $f(x)$ is integrable over $[a, b]$.

Proof We shall make the simplifying assumption that $f(x)$ is differentiable and that its derivative $f'(x)$ is bounded. In other words, we assume that $|f'(x)| \le K$ for some constant K. This assumption is used to show that $f(x)$ cannot vary too much in a small interval. More precisely, let us prove that if $[a_0, b_0]$ is any closed interval contained in $[a, b]$ and if m and M are the minimum and maximum values of $f(x)$ on $[a_0, b_0]$, then

$$|M - m| \le K|b_0 - a_0| \qquad \boxed{6}$$

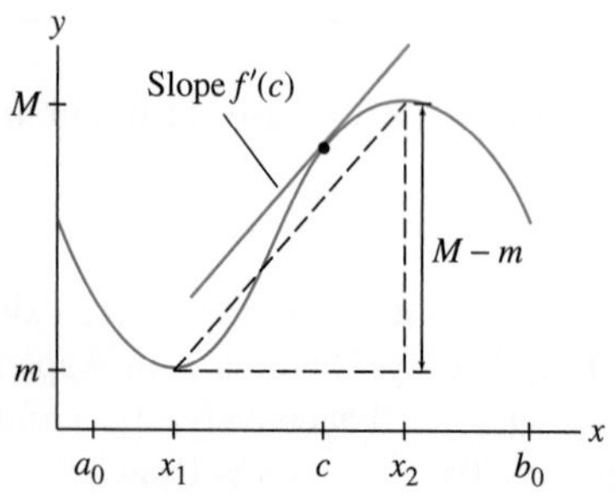

FIGURE 1 Since $M - m = f'(c)(x_2 - x_1)$, we conclude that $M - m \le K(b_0 - a_0)$.

Figure 1 illustrates the idea behind this inequality. Suppose that $f(x_1) = m$ and $f(x_2) = M$, where x_1 and x_2 lie in $[a_0, b_0]$. If $x_1 \ne x_2$, then by the Mean Value Theorem (MVT), there is a point c between x_1 and x_2 such that

$$\frac{M - m}{x_2 - x_1} = \frac{f(x_2) - f(x_1)}{x_2 - x_1} = f'(c)$$

Since x_1, x_2 lie in $[a_0, b_0]$, we have $|x_2 - x_1| \le |b_0 - a_0|$, and thus,

$$|M - m| = |f'(c)|\,|x_2 - x_1| \le K|b_0 - a_0|$$

This proves Eq. (6).

We divide the rest of the proof into two steps. Consider a partition P:

$$P: \qquad x_0 = a < x_1 < \quad \cdots \quad < x_{N-1} < x_N = b$$

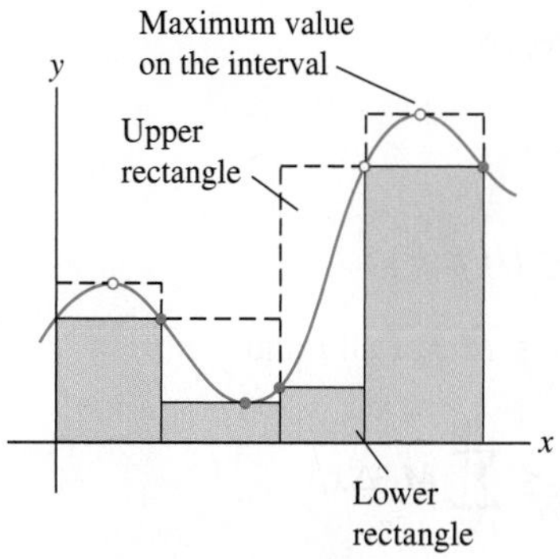

FIGURE 2 Lower and upper rectangles for a partition of length $N = 4$.

Let m_i be the minimum value of $f(x)$ on $[x_{i-1}, x_i]$ and M_i the maximum on $[x_{i-1}, x_i]$. We define the *lower* and *upper* Riemann sums

$$L(f, P) = \sum_{i=1}^{N} m_i \Delta x_i, \qquad U(f, P) = \sum_{i=1}^{N} M_i \Delta x_i$$

These are the particular Riemann sums in which the intermediate point in $[x_{i-1}, x_i]$ is the point where $f(x)$ takes on its minimum or maximum on $[x_{i-1}, x_i]$. Figure 2 illustrates the case $N = 4$.

Step 1. **Prove that the lower and upper sums approach a limit.**

We observe that

$$L(f, P_1) \le U(f, P_2) \quad \text{for any two partitions } P_1 \text{ and } P_2 \qquad \boxed{7}$$

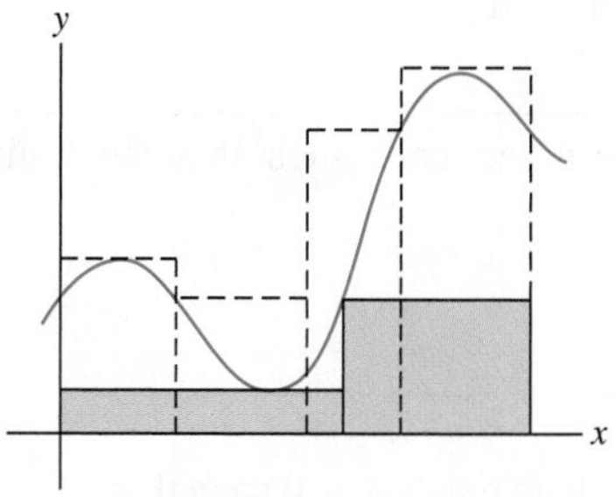

FIGURE 3 The lower rectangles always lie below the upper rectangles, even when the partitions are different.

Indeed, if a subinterval I_1 of P_1 overlaps with a subinterval I_2 of P_2, then the minimum of f on I_1 is less than or equal to the maximum of f on I_2 (Figure 3). In particular, the lower sums are bounded above by $U(f, P)$ for all partitions P. Let L be the least upper bound of the lower sums. Then for all partitions P,

$$L(f, P) \le L \le U(f, P) \qquad \boxed{8}$$

According to Eq. (6), $|M_i - m_i| \le K\Delta x_i$ for all i. Since $\|P\|$ is the largest of the widths Δx_i, we see that $|M_i - m_i| \le K\|P\|$ and

$$|U(f, P) - L(f, P)| \le \sum_{i=1}^{N} |M_i - m_i|\,\Delta x_i$$

$$\le K\|P\| \sum_{i=1}^{N} \Delta x_i = K\|P\|\,|b - a| \qquad \boxed{9}$$

Let $c = K\,|b - a|$. Using Eq. (8) and Eq. (9), we obtain

$$|L - U(f, P)| \le |U(f, P) - L(f, P)| \le c\|P\|$$

We conclude that $\lim\limits_{||P|| \to 0} |L - U(f, P)| = 0$. Similarly,

$$|L - L(f, P)| \le c\|P\|$$

and

$$\lim_{||P|| \to 0} |L - L(f, P)| = 0$$

Thus, we have

$$\lim_{\|P\|\to 0} U(f, P) = \lim_{\|P\|\to 0} L(f, P) = L$$

Step 2. **Prove that $\int_a^b f(x)\,dx$ exists and has value L.**

Recall that for any choice C of intermediate points $c_i \in [x_{i-1}, x_i]$, we define the Riemann sum

$$R(f, P, C) = \sum_{i=1}^{N} f(c_i)\Delta x_i$$

We have

$$L(f, P) \le R(f, P, C) \le U(f, P)$$

Indeed, since $c_i \in [x_{i-1}, x_i]$, we have $m_i \le f(c_i) \le M_i$ for all i and

$$\sum_{i=1}^{N} m_i\,\Delta x_i \le \sum_{i=1}^{N} f(c_i)\,\Delta x_i \le \sum_{i=1}^{N} M_i\,\Delta x_i$$

It follows that

$$|L - R(f, P, C)| \le |U(f, P) - L(f, P)| \le c\|P\|$$

This shows that $R(f, P, C)$ converges to L as $\|P\| \to 0$. ■

| Section 11.1

THEOREM 9 If $f(x)$ is continuous and $\{a_n\}$ is a sequence such that the limit $\lim_{n\to\infty} a_n = L$ exists, then

$$\lim_{n\to\infty} f(a_n) = f(L)$$

Proof Choose any $\epsilon > 0$. Since $f(x)$ is continuous, there exists $\delta > 0$ such that

$$|f(x) - f(L)| < \epsilon \qquad \text{if } 0 < |x - L| < \delta$$

Since $\lim_{n\to\infty} a_n = L$, there exists $N > 0$ such that $|a_n - L| < \delta$ for $n > N$. Thus,

$$|f(a_n) - f(L)| < \epsilon \qquad \text{for } n > N$$

It follows that $\lim_{n\to\infty} f(a_n) = f(L)$. ■

| Section 15.3

THEOREM 10 Clairaut's Theorem If f_{xy} and f_{yx} are both continuous functions on a disk D, then $f_{xy}(a, b) = f_{yx}(a, b)$ for all $(a, b) \in D$.

Proof We prove that both $f_{xy}(a, b)$ and $f_{yx}(a, b)$ are equal to the limit

$$L = \lim_{h\to 0} \frac{f(a+h, b+h) - f(a+h, b) - f(a, b+h) + f(a, b)}{h^2}$$

Let $F(x) = f(x, b+h) - f(x, b)$. The numerator in the limit is equal to

$$F(a+h) - F(a)$$

and $F'(x) = f_x(x, b+h) - f_x(x, b)$. By the MVT, there exists a_1 between a and $a+h$ such that

$$F(a+h) - F(a) = hF'(a_1) = h(f_x(a_1, b+h) - f_x(a_1, b))$$

By the MVT applied to f_x, there exists b_1 between b and $b+h$ such that

$$f_x(a_1, b+h) - f_x(a_1, b) = hf_{xy}(a_1, b_1)$$

Thus,

$$F(a+h) - F(a) = h^2 f_{xy}(a_1, b_1)$$

and

$$L = \lim_{h\to 0} \frac{h^2 f_{xy}(a_1, b_1)}{h^2} = \lim_{h\to 0} f_{xy}(a_1, b_1) = f_{xy}(a, b)$$

The last equality follows from the continuity of f_{xy} since (a_1, b_1) approaches (a, b) as $h \to 0$. To prove that $L = f_{yx}(a, b)$, repeat the argument using the function $F(y) = f(a+h, y) - f(a, y)$, with the roles of x and y reversed. ■

| Section 15.4

THEOREM 11 Criterion for Differentiability If $f_x(x, y)$ and $f_y(x, y)$ exist and are continuous on an open disk D, then $f(x, y)$ is differentiable on D.

Proof Let $(a, b) \in D$ and set

$$L(x, y) = f(a, b) + f_x(a, b)(x - a) + f_y(a, b)(y - b)$$

It is convenient to switch to the variables h and k, where $x = a + h$ and $y = b + k$. Set

$$\Delta f = f(a+h, b+k) - f(a, b)$$

Then

$$L(x, y) = f(a, b) + f_x(a, b)h + f_y(a, b)k$$

and we may define the function

$$e(h, k) = f(x, y) - L(x, y) = \Delta f - (f_x(a, b)h + f_y(a, b)k)$$

To prove that $f(x, y)$ is differentiable, we must show that

$$\lim_{(h,k)\to(0,0)} \frac{e(h, k)}{\sqrt{h^2 + k^2}} = 0$$

To do this, we write Δf as a sum of two terms:

$$\Delta f = (f(a+h, b+k) - f(a, b+k)) + (f(a, b+k) - f(a, b))$$

and apply the MVT to each term separately. We find that there exist a_1 between a and $a+h$, and b_1 between b and $b+k$, such that

$$f(a+h, b+k) - f(a, b+k) = hf_x(a_1, b+k)$$
$$f(a, b+k) - f(a, b) = kf_y(a, b_1)$$

Therefore,

$$e(h,k) = h(f_x(a_1, b+k) - f_x(a,b)) + k(f_y(a,b_1) - f_y(a,b))$$

and for $(h,k) \neq (0,0)$,

$$\begin{aligned}
\left|\frac{e(h,k)}{\sqrt{h^2+k^2}}\right| &= \left|\frac{h(f_x(a_1,b+k) - f_x(a,b)) + k(f_y(a,b_1) - f_y(a,b))}{\sqrt{h^2+k^2}}\right| \\
&\le \left|\frac{h(f_x(a_1,b+k) - f_x(a,b))}{\sqrt{h^2+k^2}}\right| + \left|\frac{k(f_y(a,b_1) - f_y(a,b))}{\sqrt{h^2+k^2}}\right| \\
&= |f_x(a_1,b+k) - f_x(a,b)| + |f_y(a,b_1) - f_y(a,b)|
\end{aligned}$$

In the second line, we use the Triangle Inequality (see Eq. (1) in Section 1.1), and we may pass to the third line because $|h/\sqrt{h^2+k^2}|$ and $|k/\sqrt{h^2+k^2}|$ are both less than 1. Both terms in the last line tend to zero as $(h,k) \to (0,0)$ because f_x and f_y are assumed to be continuous. This completes the proof that $f(x,y)$ is differentiable. ■

ANSWERS TO ODD-NUMBERED EXERCISES

Chapter 1

Section 1.1 Preliminary Questions

1. $a = -3$ and $b = 1$

2. The numbers $a \geq 0$ satisfy $|a| = a$ and $|-a| = a$. The numbers $a \leq 0$ satisfy $|a| = -a$.

3. $a = -3$ and $b = 1$ **4.** $(9, -4)$

5. **(a)** First quadrant. **(b)** Second quadrant.

(c) Fourth quadrant. **(d)** Third quadrant.

6. 3 **7.** **(b)** **8.** Symmetry with respect to the origin

Section 1.1 Exercises

1. $r = \frac{12337}{1250}$ **3.** $|x| \leq 2$ **5.** $|x - 2| < 2$ **7.** $|x - 3| \leq 2$
9. $-8 < x < 8$ **11.** $-3 < x < 2$ **13.** $(-4, 4)$ **15.** $(2, 6)$
17. $\left[-\frac{7}{4}, \frac{9}{4}\right]$ **19.** $(-\infty, 2) \cup (6, \infty)$
21. $(-\infty, -\sqrt{3}) \cup (\sqrt{3}, \infty)$

23. **(a)** **(i)** **(b)** **(iii)** **(c)** **(v)** **(d)** **(vi)** **(e)** **(ii)** **(f)** **(iv)**

25. $-3 < x < 1$

29. $|a + b - 13| = |(a - 5) + (b - 8)| \leq |a - 5| + |b - 8| < \frac{1}{2} + \frac{1}{2} = 1$

31. **(a)** 11 **(b)** 1

33. $r_1 = \frac{3}{11}$ and $r_2 = \frac{4}{15}$

35. Let $a = 1$ and $b = .\bar{9}$ (see the discussion before Example 1). The decimal expansions of a and b do not agree, but $|1 - .\bar{9}| < 10^{-k}$ for all k.

37. **(a)** $(x - 2)^2 + (y - 4)^2 = 9$

(b) $(x - 2)^2 + (y - 4)^2 = 26$

39. $D = \{r, s, t, u\}$; $R = \{A, B, E\}$

41. D : all reals; R : all reals

43. D : all reals; R : all reals

45. D : all reals; $R : \{y : y \geq 0\}$

47. $D : \{x : x \neq 0\}$; $R : \{y : y > 0\}$

49. On the interval $(-1, \infty)$

51. On the interval $(0, \infty)$

53. Zeros: ± 2; Increasing: $x > 0$; Decreasing: $x < 0$; Symmetry: $f(-x) = f(x)$, so y-axis symmetry.

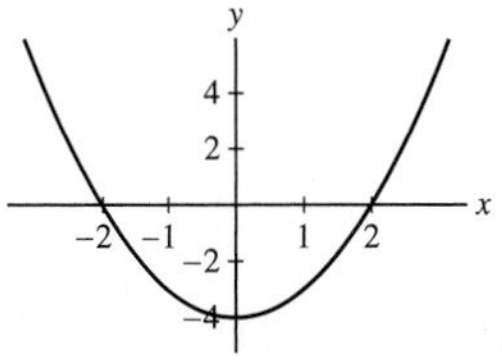

55. Zeros: $0, \pm 2$; Symmetry: $f(-x) = -f(x)$, so origin symmetry.

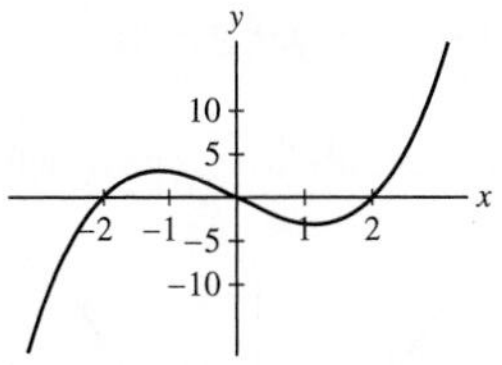

57. This is an x-axis reflection of x^3 translated up 2 units. There is one zero at $x = \sqrt[3]{2}$.

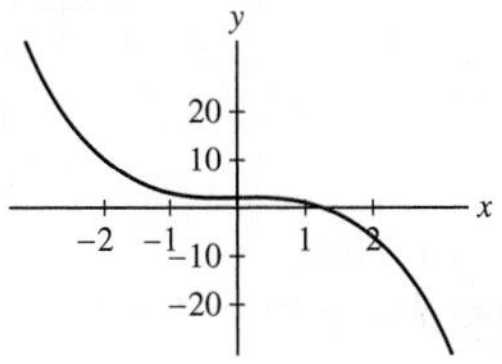

59. (B)

61. **(a)** Odd **(b)** Odd **(c)** Neither odd nor even **(d)** Even

63. It is decreasing everywhere it is defined, i.e. for $x \neq 4$.

65. $D : [0, 4]$; $R : [0, 4]$

67.

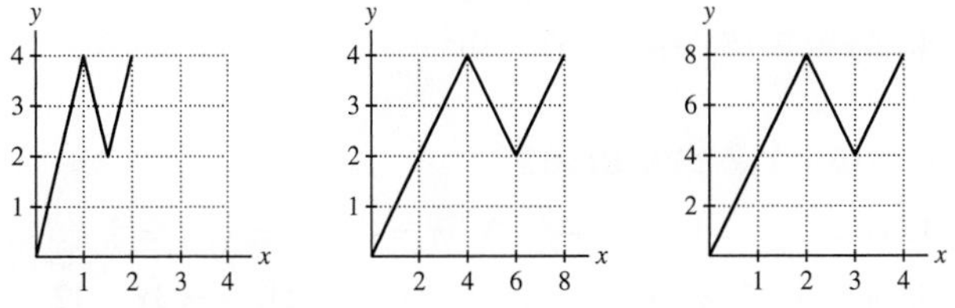

$f(2x)$ $f(x/2)$ $2f(x)$

69.

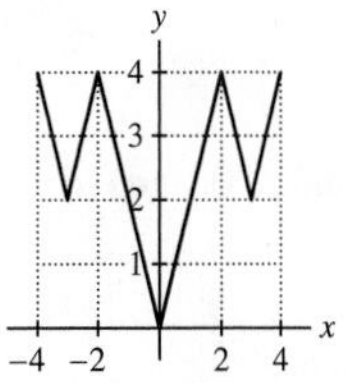

71. **(a)** $D : [4, 8]$, $R : [5, 9]$. **(b)** $D : [1, 5]$, $R : [2, 6]$.
(c) $D : \left[\frac{4}{3}, \frac{8}{3}\right]$, $R : [2, 6]$. **(d)** $D : [4, 8]$, $R : [6, 18]$.

73. **(a)** $h(x) = \sin(2x - 10)$ **(b)** $h(x) = \sin(2x - 5)$

75.

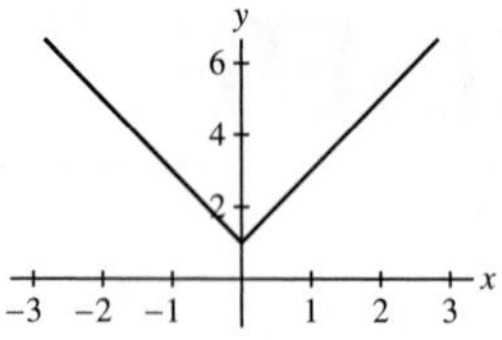

$f(2x)$

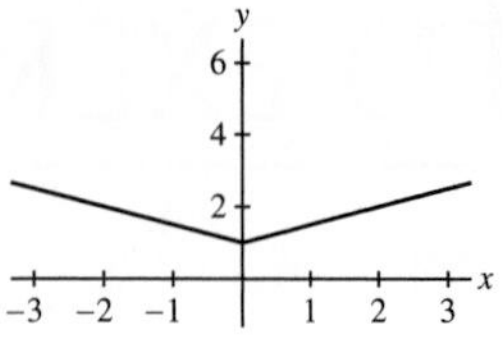

$f(x/2)$

77.

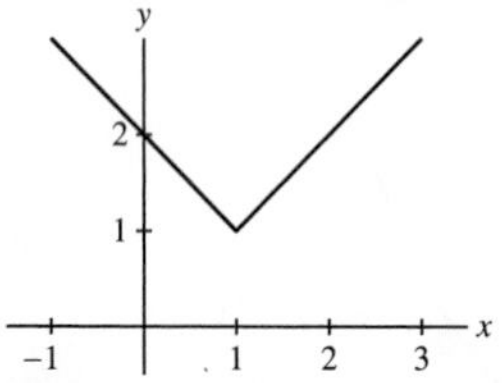

D : all reals; $R : \{y \mid y \geq 1\}$; $f(x) = |x - 1| + 1$

79. Even:
$(f + g)(-x) = f(-x) + g(-x) \stackrel{\text{even}}{=} f(x) + g(x) = (f + g)(x)$
Odd: $(f + g)(-x) = f(-x) + g(-x) \stackrel{\text{odd}}{=} -f(x) + -g(x) = -(f + g)(x)$

85. **(a)** There are many possibilities, one of which is

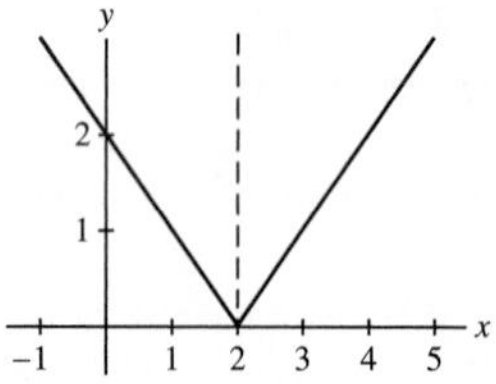

$y = |x - 2|$

(b) Let $g(x) = f(x + a)$. Then
$g(-x) = f(-x + a) = f(a - x) = f(a + x) = g(x)$

Section 1.2 Preliminary Questions

1. -4 **2.** No.

3. Parallel to the y-axis when $b = 0$; parallel to the x-axis when $a = 0$

4. $\Delta y = 9$ **5.** -4 **6.** $(x - 0)^2 + 1$

Section 1.2 Exercises

1. $m = 3$; $y = 12$; $x = -4$ **3.** $m = -\frac{4}{9}$; $y = \frac{1}{3}$; $x = \frac{3}{4}$

5. $m = 3$ **7.** $m = -\frac{3}{4}$ **9.** $y = 3x + 8$ **11.** $y = 3x - 12$

13. $y = -2$ **15.** $y = 3x - 2$ **17.** $y = \frac{5}{3}x - \frac{1}{3}$ **19.** $y = 4$

21. $y = -2x + 9$ **23.** $3x + 4y = 12$

25. **(a)** $c = -\frac{1}{4}$ **(b)** $c = -2$

(c) No value for c that will make this slope equal to 0 **(d)** $c = 0$

27. **(a)** 40.0248 cm **(b)** 64.9597 in
(c) $L = 65(1 + \alpha(T - 100))$

29. $b = 4$

31. No, because the slopes between consecutive data points are not equal.

33. **(a)** 1 or $-\frac{1}{4}$ **(b)** $1 \pm \sqrt{2}$

35. Minimum value is 0 **37.** Minimum value is -7
39. Maximum value is $\frac{137}{16}$ **41.** Maximum value is $\frac{1}{3}$

43.

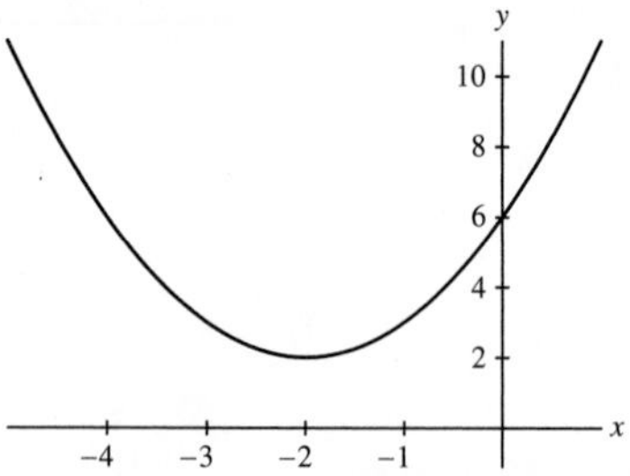

45. A double root occurs when $c = \pm 2$. There are no real roots when $-2 < c < 2$.

47. For all $x \geq 0$, $0 \leq \left(x^{1/2} - x^{-1/2}\right)^2 = x - 2 + \frac{1}{x}$.

51. $4 + 2\sqrt{2}$ and $4 - 2\sqrt{2}$

55. For x^2, $\frac{\Delta y}{\Delta x} = \frac{x_2^2 - x_1^2}{x_2 - x_1} = x_2 + x_1$.

59.
$(x - \alpha)(x - \beta) = x^2 - \alpha x - \beta x + \alpha\beta = x^2 + (-\alpha - \beta)x + \alpha\beta$

Section 1.3 Preliminary Questions

1. One example is $\frac{3x^2 - 2}{7x^3 + x - 1}$

2. $|x|$ is not a polynomial; $|x^2 + 1|$ is a polynomial

3. The domain of $f(g(x))$ is the empty set.

4. Decreasing

5. One possibility is $f(x) = e^x - \sin x$

Section 1.3 Exercises

1. $x \geq 0$ **3.** All reals **5.** $t \neq -2$ **7.** $u \neq \pm 2$ **9.** $x \neq 0, 1$
11. $y > 0$ **13.** Polynomial **15.** Algebraic **17.** Transcendental
19. Rational **21.** Transcendental **23.** Rational **25.** Yes

27. $f(g(x)) = \sqrt{x + 1}$; D: $x \geq -1$, $g(f(x)) = \sqrt{x} + 1$; D: $x \geq 0$

29. $f(g(x)) = 2^{x^2}$; D: $\mathbf{R}$, $g(f(x)) = (2^x)^2 = 2^{2x}$; D: $\mathbf{R}$

31. $f(g(x)) = \cos(x^3 + x^2)$; D: $\mathbf{R}$, $g(f(\theta)) = \cos^3\theta + \cos^2\theta$; D: $\mathbf{R}$

33. $f(g(t)) = \frac{1}{\sqrt{-t^2}}$; D: Not valid for any t,
$g(f(t)) = -\left(\frac{1}{\sqrt{t}}\right)^2 = -\frac{1}{t}$; D: $t > 0$

35.
$P(t + 10) = 30 \cdot 2^{0.1(t+10)} = 30 \cdot 2^{0.1t+1} = 2(30 \cdot 2^{0.1t}) = 2P(t)$;
$g\left(t + \frac{1}{k}\right) = a2^{k(t+1/k)} = a2^{kt+1} = 2a2^{kt} = 2g(t)$

37. $f(x) = x^2$:
$\delta f(x) = f(x + 1) - f(x) = (x + 1)^2 - x^2 = 2x + 1$
$f(x) = x$: $\delta f(x) = x + 1 - x = 1$
$f(x) = x^3$: $\delta f(x) = (x + 1)^3 - x^3 = 3x^2 + 3x + 1$

39.

$$\delta(f+g) = (f(x+1)+g(x+1)) - (f(x)-g(x))$$
$$= (f(x+1)-f(x)) + (g(x+1)-g(x)) = \delta f(x) + \delta g(x)$$
$$\delta(cf) = cf(x+1) - cf(x) = c(f(x+1)-f(x)) = c\delta f(x).$$

Section 1.4 Preliminary Questions

1. Two rotations that differ by a whole number of full revolutions will have the same ending radius.

2. $\frac{9\pi}{4}$ and $\frac{41\pi}{4}$ **3.** $-\frac{5\pi}{3}$ **4.** **(a)**

5. Let O denote the center of the unit circle, and let P be a point on the unit circle such that the radius $\overline{OP}$ makes an angle θ with the positive x-axis. Then, $\sin\theta$ is the y-coordinate of the point P.

6. Let O denote the center of the unit circle, and let P be a point on the unit circle such that the radius $\overline{OP}$ makes an angle θ with the positive x-axis. The angle $\theta + 2\pi$ is obtained from the angle θ by making one full revolution around the circle. The angle $\theta + 2\pi$ will therefore have the radius $\overline{OP}$ as its terminal side.

Section 1.4 Exercises

1. $5\pi/4$

3. **(a)** $\frac{180^\circ}{\pi} \approx 57.3^\circ$ **(b)** 60° **(c)** $\frac{75^\circ}{\pi} \approx 23.87^\circ$ **(d)** -135°

5. $s = r\theta = 3.6$; $s = r\phi = 8$

7.

θ	$(\cos\theta, \sin\theta)$	θ	$(\cos\theta, \sin\theta)$
$\frac{\pi}{2}$	$(0, 1)$	$\frac{5\pi}{4}$	$\left(\frac{-\sqrt{2}}{2}, \frac{-\sqrt{2}}{2}\right)$
$\frac{2\pi}{3}$	$\left(\frac{-1}{2}, \frac{\sqrt{3}}{2}\right)$	$\frac{4\pi}{3}$	$\left(\frac{-1}{2}, \frac{-\sqrt{3}}{2}\right)$
$\frac{3\pi}{4}$	$\left(\frac{-\sqrt{2}}{2}, \frac{\sqrt{2}}{2}\right)$	$\frac{3\pi}{2}$	$(0, -1)$
$\frac{5\pi}{6}$	$\left(\frac{-\sqrt{3}}{2}, \frac{1}{2}\right)$	$\frac{5\pi}{3}$	$\left(\frac{1}{2}, \frac{-\sqrt{3}}{2}\right)$
π	$(-1, 0)$	$\frac{7\pi}{4}$	$\left(\frac{\sqrt{2}}{2}, \frac{-\sqrt{2}}{2}\right)$
$\frac{7\pi}{6}$	$\left(\frac{-\sqrt{3}}{2}, \frac{-1}{2}\right)$	$\frac{11\pi}{6}$	$\left(\frac{\sqrt{3}}{2}, \frac{-1}{2}\right)$

9. $\theta = \frac{\pi}{3}, \frac{5\pi}{3}$ **11.** $\theta = \frac{3\pi}{4}, \frac{7\pi}{4}$ **13.** $x = \frac{\pi}{3}, \frac{2\pi}{3}$

15.

θ	$\frac{\pi}{6}$	$\frac{\pi}{4}$	$\frac{\pi}{3}$	$\frac{\pi}{2}$	$\frac{2\pi}{3}$	$\frac{3\pi}{4}$	$\frac{5\pi}{6}$
$\tan\theta$	$\frac{1}{\sqrt{3}}$	1	$\sqrt{3}$	und	$-\sqrt{3}$	-1	$-\frac{1}{\sqrt{3}}$
$\sec\theta$	$\frac{2}{\sqrt{3}}$	$\sqrt{2}$	2	und	-2	$-\sqrt{2}$	$-\frac{2}{\sqrt{3}}$

17. $\cos\theta = \frac{1}{\sec\theta} = \frac{1}{\sqrt{1+\tan^2\theta}} = \frac{1}{\sqrt{1+c^2}}$

19. $\sin\theta = \frac{12}{13}$ and $\tan\theta = \frac{12}{5}$

21. $\sin\theta = \frac{2\sqrt{53}}{53}$, $\sec\theta = \frac{\sqrt{53}}{7}$ and $\cot\theta = \frac{7}{2}$

23. $23/25$

25. $\cos\theta = -\frac{\sqrt{21}}{5}$ and $\tan\theta = -\frac{2\sqrt{21}}{21}$

27. $\cos\theta = -\frac{4}{5}$

29. Let's start with the four points in Figure 23(A).

- The point in the first quadrant:

$$\sin\theta = 0.918,\ \cos\theta = 0.3965,\ \text{and}\ \tan\theta = \frac{0.918}{0.3965} = 2.3153.$$

- The point in the second quadrant:

$$\sin\theta = 0.3965,\ \cos\theta = -0.918,\ \text{and}$$
$$\tan\theta = \frac{0.3965}{-0.918} = -0.4319.$$

- The point in the third quadrant:

$$\sin\theta = -0.918,\ \cos\theta = -0.3965,\ \text{and}$$
$$\tan\theta = \frac{-0.918}{-0.3965} = 2.3153.$$

- The point in the fourth quadrant:

$$\sin\theta = -0.3965,\ \cos\theta = 0.918,\ \text{and}$$
$$\tan\theta = \frac{-0.3965}{0.918} = -0.4319.$$

Now consider the four points in Figure 23(B).

- The point in the first quadrant:

$$\sin\theta = 0.918,\ \cos\theta = 0.3965,\ \text{and}$$
$$\tan\theta = \frac{0.918}{0.3965} = 2.3153.$$

- The point in the second quadrant:

$$\sin\theta = 0.918,\ \cos\theta = -0.3965,\ \text{and}$$
$$\tan\theta = \frac{0.918}{0.3965} = -2.3153.$$

- The point in the third quadrant:

$$\sin\theta = -0.918,\ \cos\theta = -0.3965,\ \text{and}$$
$$\tan\theta = \frac{-0.918}{-0.3965} = 2.3153.$$

- The point in the fourth quadrant:

$$\sin\theta = -0.918,\ \cos\theta = 0.3965,\ \text{and}$$
$$\tan\theta = \frac{-0.918}{0.3965} = -2.3153.$$

31. $\cos\psi = 0.3$, $\sin\psi = \sqrt{0.91}$, $\cot\psi = \frac{0.3}{\sqrt{0.91}}$ and $\csc\psi = \frac{1}{\sqrt{0.91}}$

33. $\cos\left(\frac{\pi}{3} + \frac{\pi}{4}\right) = \frac{\sqrt{2}-\sqrt{6}}{4}$

35.

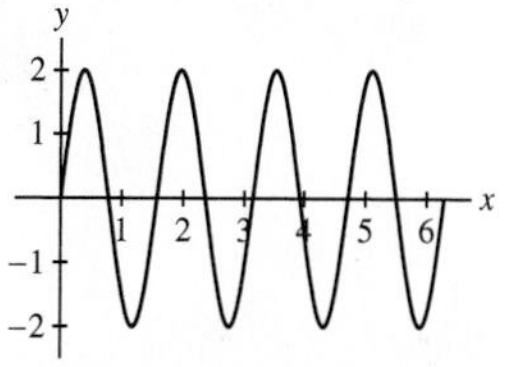

37.

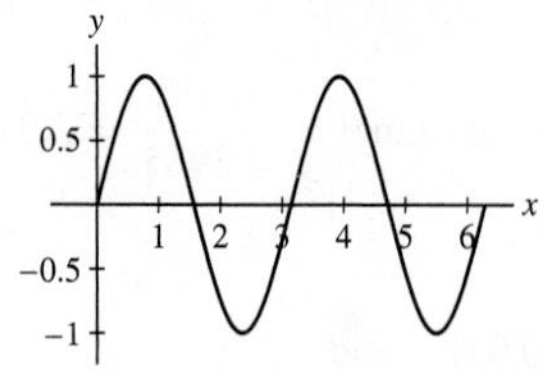

39. If $|c| > 1$, no points of intersection; if $|c| = 1$, one point of intersection; if $|c| < 1$, two points of intersection.

41. $\theta = 0, \frac{2\pi}{5}, \frac{4\pi}{5}, \pi, \frac{6\pi}{5}, \frac{8\pi}{5}$

43. $\theta = \frac{\pi}{6}, \frac{\pi}{2}, \frac{5\pi}{6}, \frac{7\pi}{6}, \frac{3\pi}{2}, \frac{11\pi}{6}$

45. Starting from the double angle formula for cosine, $\cos^2\theta = \frac{1}{2}(1 + \cos 2\theta)$, solve for $\cos 2\theta$.

47. Substitute $x = \theta/2$ into the double angle formula for sine, $\sin^2 x = \frac{1}{2}(1 - \cos 2x)$, then take the square root of both sides.

49. $\cos(\theta + \pi) = \cos\theta\cos\pi - \sin\theta\sin\pi = \cos\theta(-1) = -\cos\theta$

51. $\tan(\pi - \theta) = \frac{\sin(\pi-\theta)}{\cos(\pi-\theta)} = \frac{-\sin(-\theta)}{-\cos(-\theta)} = \frac{\sin\theta}{-\cos\theta} = -\tan\theta.$

53. $\frac{\sin 2x}{1+\cos 2x} = \frac{2\sin x\cos x}{1+2\cos^2 x-1} = \frac{2\sin x\cos x}{2\cos^2 x} = \frac{\sin x}{\cos x} = \tan x$

57. 16.928

Section 1.5 Preliminary Questions

1. No

2. (a) The screen will display nothing.

(b) The screen will display the portion of the parabola between the points $(0, 3)$ and $(1, 4)$.

3. No

4. Experiment with the viewing window to zoom in on the lowest point on the graph of the function. The y-coordinate of the lowest point on the graph is the minimum value of the function.

Section 1.5 Exercises

1.

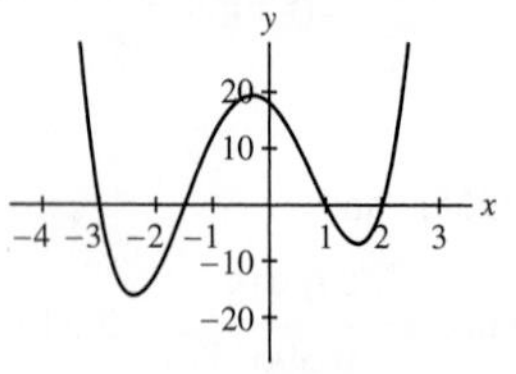

$x = -3$, $x = -1.5$, $x = 1$, and $x = 2$

3. Two positive solutions **5.** There are no solutions

7. Nothing. An appropriate viewing window: [50, 150] by [1000, 2000]

9.

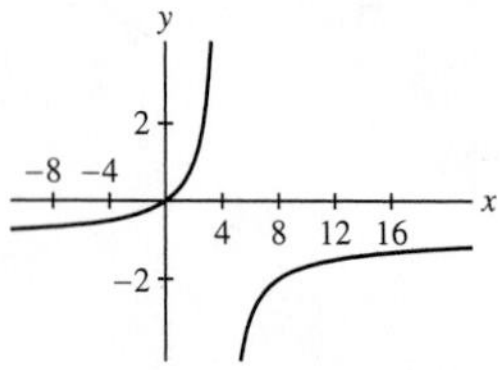

11.

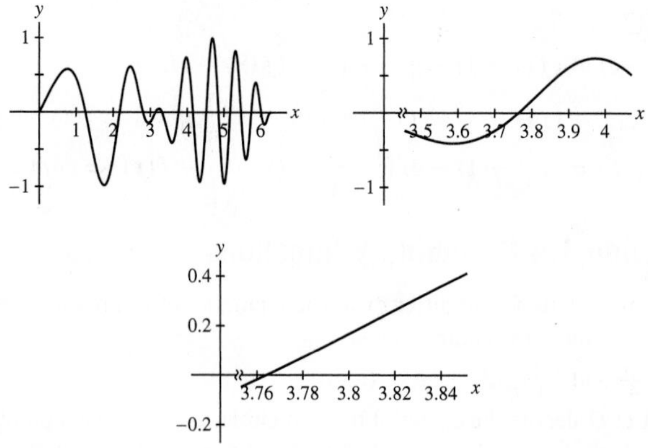

13. The table and graphs below suggest that as n gets large, $n^{1/n}$ approaches 1.

n	$n^{1/n}$
10	1.258925412
10^2	1.047128548
10^3	1.006931669
10^4	1.000921458
10^5	1.000115136
10^6	1.000013816

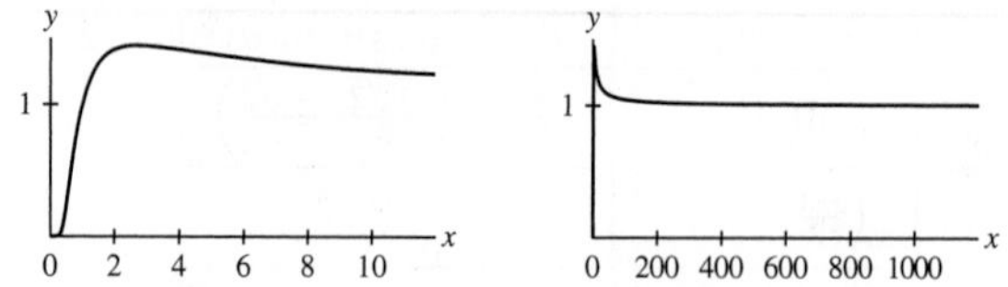

15. The table and graphs below suggest that as n gets large, $f(n)$ tends toward ∞.

n	$\left(1 + \frac{1}{n}\right)^{n^2}$
10	13780.61234
10^2	$1.635828711 \times 10^{43}$
10^3	$1.195306603 \times 10^{434}$
10^4	$5.341783312 \times 10^{4342}$
10^5	$1.702333054 \times 10^{43429}$
10^6	$1.839738749 \times 10^{434294}$

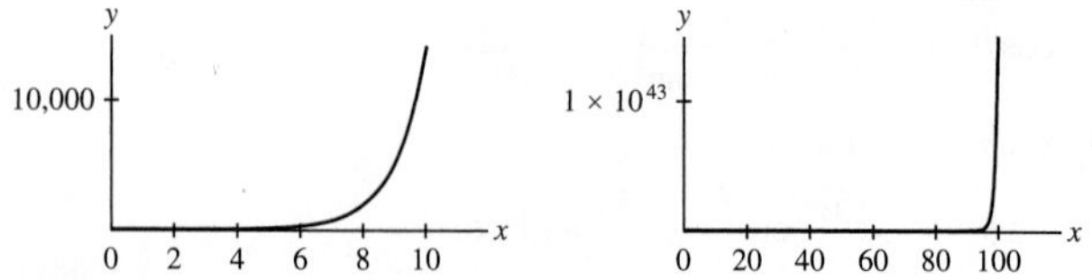

17. The table and graphs below suggest that as x gets large, $f(x)$ approaches 1.

x	$\left(x \tan \frac{1}{x}\right)^x$
10	1.033975759
10^2	1.003338973
10^3	1.000333389
10^4	1.000033334
10^5	1.000003333
10^6	1.000000333

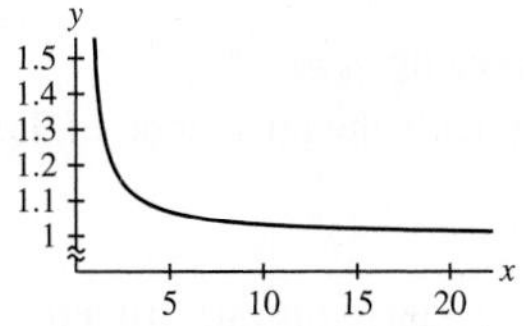
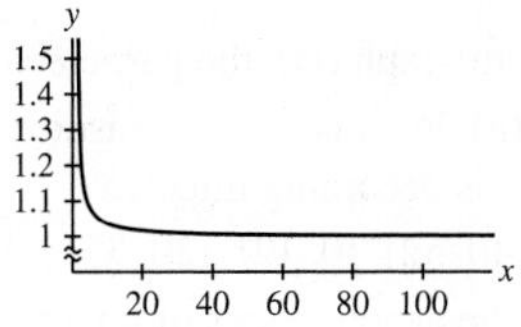

19.

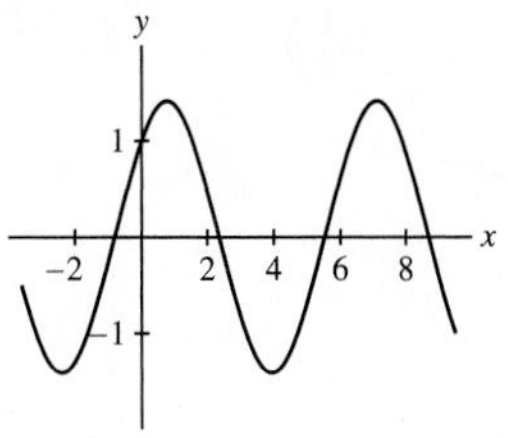

$(A, B) = (1, 1)$

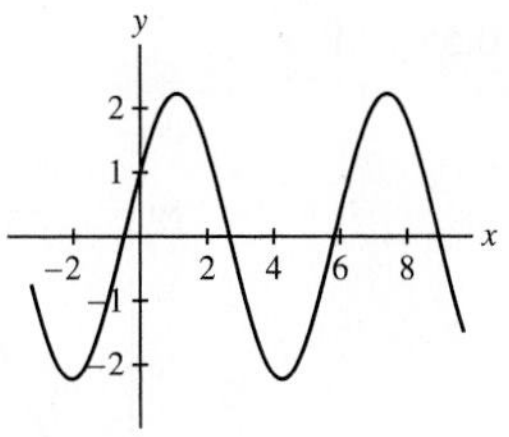

$(A, B) = (1, 2)$

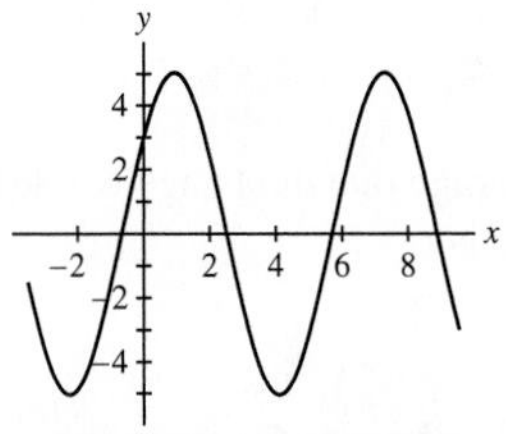

$(A, B) = (3, 4)$

21. $x \in (-2, 0) \cup (3, \infty)$

23.

$$f_3(x) = \frac{1}{2}\left(\frac{1}{2}(x+1) + \frac{x}{\frac{1}{2}(x+1)}\right) = \frac{x^2+6x+1}{4(x+1)}$$

$$f_4(x) = \frac{1}{2}\left(\frac{x^2+6x+1}{4(x+1)} + \frac{x}{\frac{x^2+6x+1}{4(x+1)}}\right) = \frac{x^4+28x^3+70x^2+28x+1}{8(1+x)(1+6x+x^2)}$$

and

$$f_5(x) = \frac{1+120x+1820x^2+8008x^3+12870x^4+8008x^5+1820x^6+120x^7+x^8}{16(1+x)(1+6x+x^2)(1+28x+70x^2+28x^3+x^4)}.$$

It appears as if the f_n are asymptotic to $\sqrt{x}$.

Chapter 1 Review

1. $\{x : |x-7| < 3\}$ **3.** $[-5, -1] \cup [3, 7]$

5. $(x, 0)$ with $x \geq 0$; $(0, y)$ with $y < 0$

7.

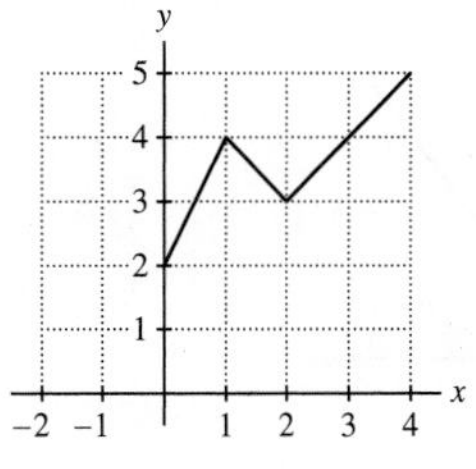

$f(x) + 2$

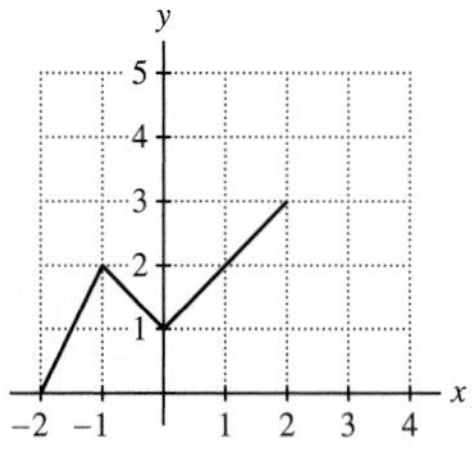

$f(x + 2)$

9.

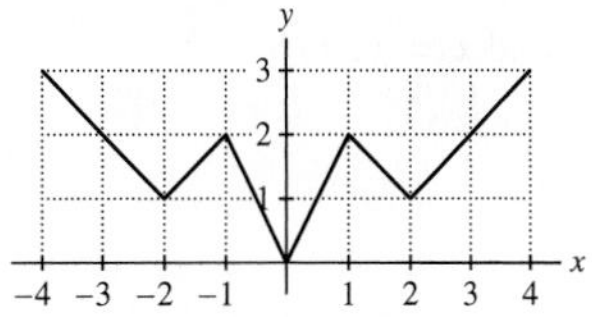

11. $D : \{x : x \geq -1\}$; $R : \{y : y \geq 0\}$

13. $D : \{x : x \neq 3\}$; $R : \{y : y \neq 0\}$

15. **(a)** Decreasing **(b)** Neither **(c)** Neither **(d)** Increasing

17. $2x - 3y = -14$ **19.** $6x - y = 53$

21. $x + y = 5$ **23.** Yes

25. Roots: $x = -2$, $x = 0$ and $x = 2$; decreasing: $x < -1.4$ and $0 < x < 1.4$

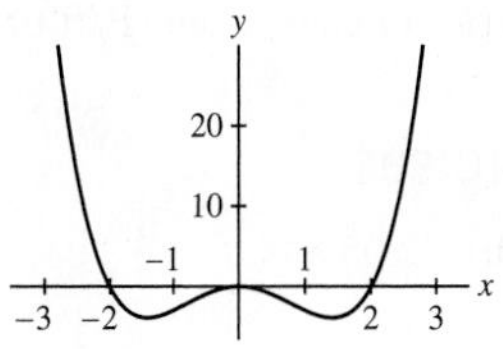

27. $f(x) = 10x^2 + 2x + 5$; minimum value is $\frac{49}{10}$

29.

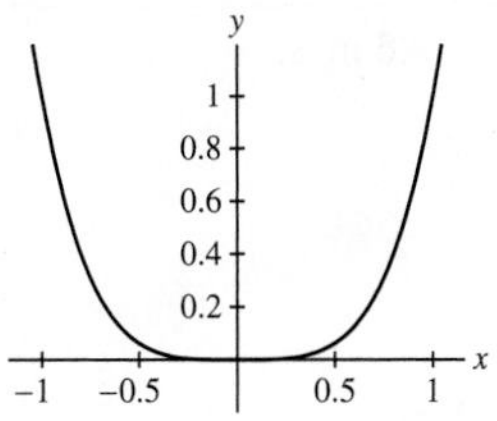

31.

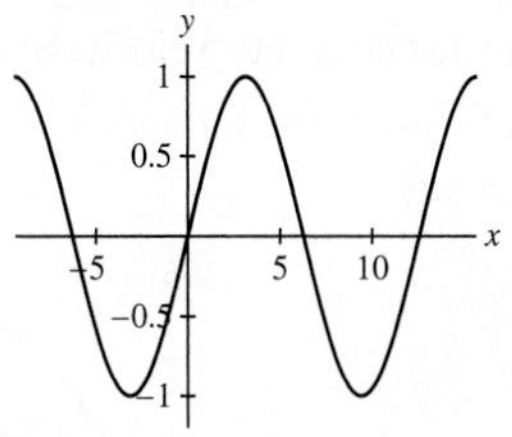

33.

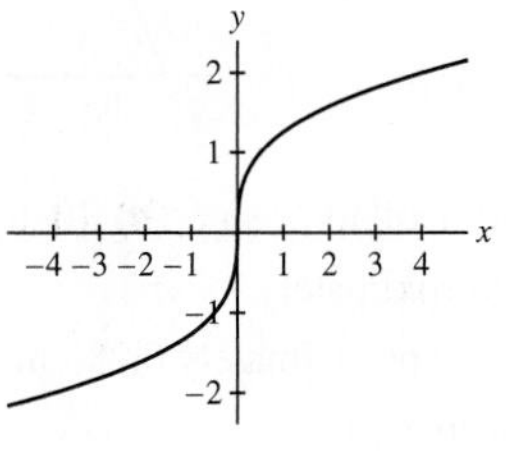

35. Let $g(x) = f\left(\frac{1}{3}x\right)$. Then $g(x - 3b) = f\left(\frac{1}{3}(x - 3b)\right) = f\left(\frac{1}{3}x - b\right)$. The graph of $y = \left|\frac{1}{3}x - 4\right|$:

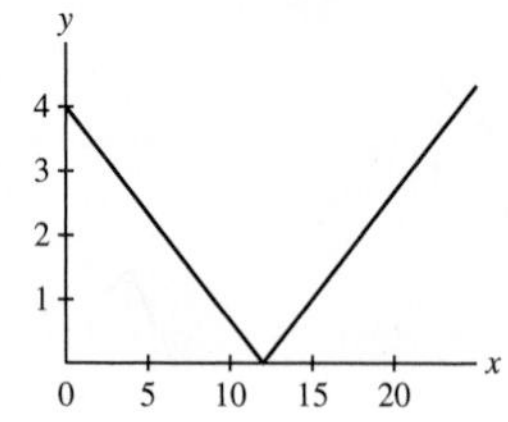

37. $f(t) = t^4$ and $g(t) = 12t + 9$ **39.** 4π

41. **(a)** $a = b = \pi/2$ **(b)** $a = \pi$

43. $x = \pi/2, x = 7\pi/6, x = 3\pi/2$ and $x = 11\pi/6$

45. There are no solutions

Chapter 2

Section 2.1 Preliminary Questions

1. The graph of position as a function of time

2. No. Instantaneous velocity is defined as the limit of average velocity as time elapsed shrinks to zero.

3. The slope of the line tangent to the graph of position as a function of time at $t = t_0$

4. The slope of the secant line over the interval $[x_0, x_1]$ approaches the slope of the tangent line at $x = x_0$.

5. The graph of atmospheric temperature as a function of altitude. Possible units for this rate of change are °F/ft or °C/m.

Section 2.1 Exercises

1. **(a)** 11.025 m **(b)** 22.05 m/s

(c)

time interval	[2, 2.01]	[2, 2.005]	[2, 2.001]	[2, 2.00001]
average velocity	19.649	19.6245	19.6049	19.600049

The instantaneous velocity at $t = 2$ is 19.6 m/s.

3. 0.57735 m/(s · K)

5. 0.3 m/s

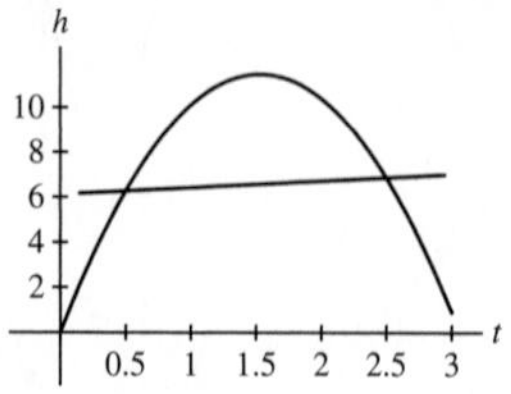

7. **(a)** Dollars/year **(b)** [0, 0.5]: 7.8461; [0, 1]: 8

(c) Approximately $8/yr

9. **(a)** Approximately 0.283 million Internet users per year.

(b) Decreases

(c) Approximately 0.225 million Internet users per year.

(d) Greater than

11. 12 **13.** −0.06 **15.** 1.105 **17.** 0.864

19. **(a)** [0, 0.1]: −144.721 cm/s; [3, 3.5]: 0 cm/s **(b)** 0 cm/s

21. **(a)** Seconds per meter; measures the sensitivity of the period of the pendulum to a change in the length of the pendulum.

(b) B: average rate of change in T from $L = 1$ m to $L = 3$ m; A: instantaneous rate of change of T at $L = 3$ m.

(c) 0.4330 s/m.

23. Sales decline more slowly as time increases.

25.
- In graph (A), the particle is (c) slowing down.
- In graph (B), the particle is (b) speeding up and then slowing down.
- In graph (C), the particle is (d) slowing down and then speeding up.
- In graph (D), the particle is (a) speeding up.

27. **(a)** Percent /day; measures how quickly the population of flax plants is becoming infected.

(b) [40, 52], [0, 12], [20, 32]

(c) The average rates of infection over the intervals [30, 40], [40, 50], [30, 50] are .9, .5, .7 %/d, respectively.

(d) 0.55%/d

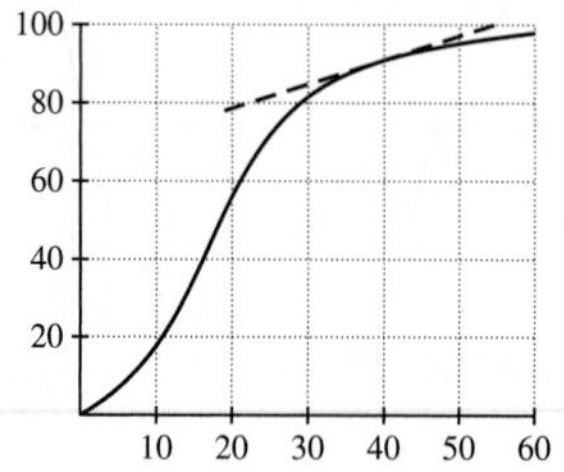

31. (B)

33. Interval $[1, t]$: average rate of change is $t + 1$; interval $[2, t]$: average rate of change is $t + 2$

35. $x^2 + 2x + 4$

Section 2.2 Preliminary Questions

1. 1 **2.** π **3.** 20 **4.** Yes

5. $\lim_{x\to 1-} f(x) = \infty$ and $\lim_{x\to 1+} f(x) = 3$

6. No **7.** Yes

Section 2.2 Exercises

1.

x	0.998	0.999	0.9995	0.99999
$f(x)$	1.498501	1.499250	1.499625	1.499993

x	1.00001	1.0005	1.001	1.002
$f(x)$	1.500008	1.500375	1.500750	1.501500

The limit as $x \to 1$ is $\frac{3}{2}$.

3.

y	1.998	1.999	1.9999
$f(y)$	0.59984	0.59992	0.599992

y	2.0001	2.001	2.02
$f(y)$	0.600008	0.60008	0.601594

The limit as $y \to 2$ is $\frac{3}{5}$.

5. 1.5 **7.** 21 **9.** $|3x - 12| = 3|x - 4|$

11. $|(5x + 2) - 17| = |5x - 15| = 5|x - 3|$

13. Suppose $|x| < 1$, so that
$|x^2 - 0| = |x + 0||x - 0| = |x||x| < |x|$

15. If $|x| < 1$, $|4x + 2|$ can be no bigger than 6, so
$|4x^2 + 2x + 5 - 5| = |4x^2 + 2x| = |x||4x + 2| < 6|x|$

17. $\frac{1}{2}$ **19.** $\frac{5}{3}$ **21.** 2 **23.** 0

25. As $x \to 4-$, $f(x) \to -\infty$; similarly, as $x \to 4+$, $f(x) \to \infty$

27. $-\infty$ **29.** 0 **31.** 1

33. 2 **35.** $\frac{1}{2}$

37.

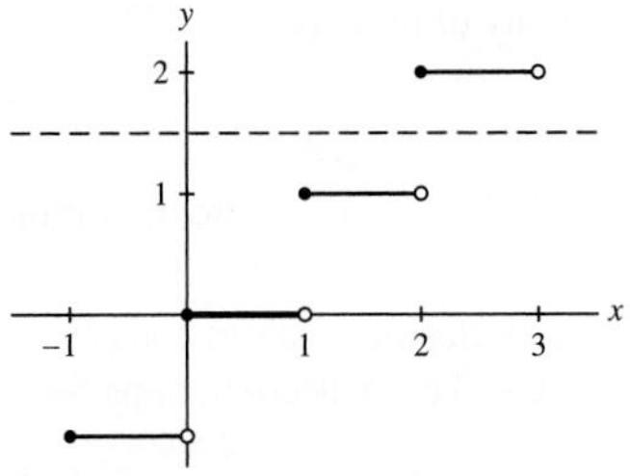

(a) $c - 1$ **(b)** c

39. $\lim_{x\to 0-} f(x) = -1$, $\lim_{x\to 0+} f(x) = 1$

41. $\lim_{x\to 0-} f(x) = \infty$, $\lim_{x\to 0+} f(x) = \dfrac{1}{6}$

43. $\lim_{x\to -2-} \dfrac{4x^2 + 7}{x^3 + 8} = -\infty$, $\lim_{x\to -2+} \dfrac{4x^2 + 7}{x^3 + 8} = \infty$

45. $\lim_{x\to 1\pm} \dfrac{x^5 + x - 2}{x^2 + x - 2} = 2$

47.
- $\lim_{x\to 2-} f(x) = \infty$ and $\lim_{x\to 2+} f(x) = \infty$.
- $\lim_{x\to 4-} f(x) = -\infty$ and $\lim_{x\to 4+} f(x) = 10$.

The vertical asymptotes are the vertical lines $x = 2$ and $x = 4$.

49.

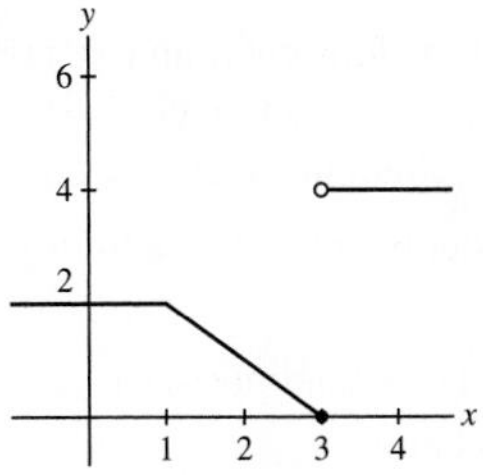

51.

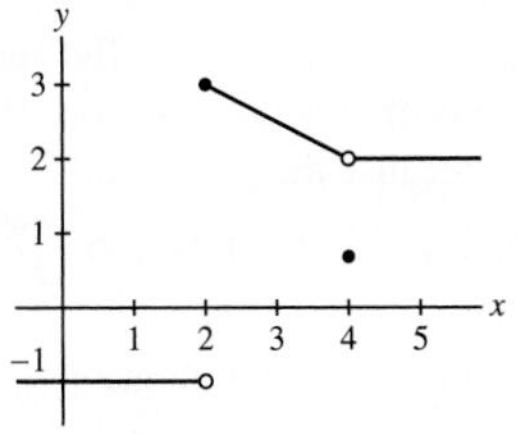

53.
- $\lim_{x\to 1-} f(x) = \lim_{x\to 1+} f(x) = 3$
- $\lim_{x\to 3-} f(x) = -\infty$
- $\lim_{x\to 3+} f(x) = 4$
- $\lim_{x\to 5-} f(x) = 2$
- $\lim_{x\to 5+} f(x) = -3$
- $\lim_{x\to 6-} f(x) = \lim_{x\to 6+} f(x) = \infty$

55. $\frac{5}{2}$

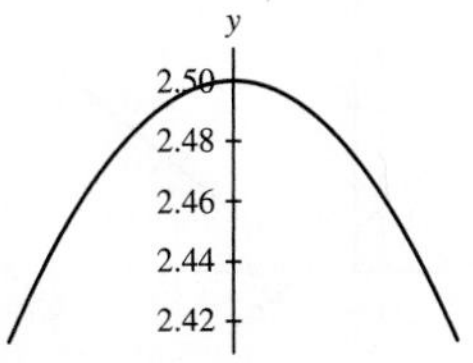

57. 0.693 (The exact answer is ln 2.)

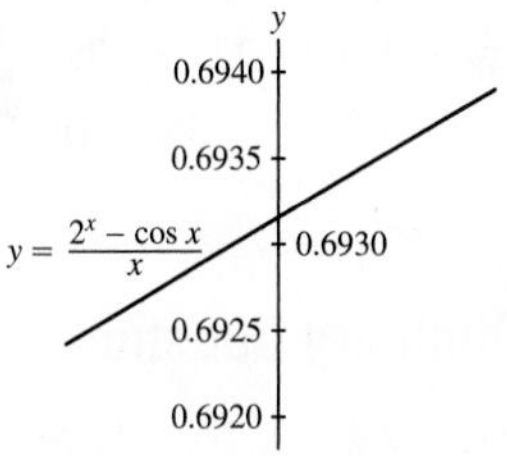

59. -12

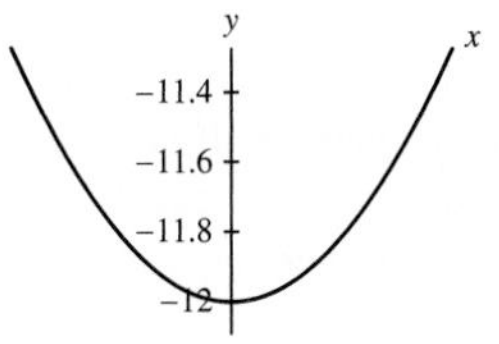

61. For n even

63. **(a)** No **(b)** $f\left(\frac{1}{2n}\right) = 1$ for all integers n.

(c) At $x = 1, \frac{1}{3}, \frac{1}{5}, \ldots$, the value of $f(x)$ is always -1.

65. $\lim_{\theta\to 0} \dfrac{\sin n\theta}{\theta} = n$

67. $\frac{1}{2}, 2, \frac{3}{2}, \frac{2}{3}$; $\lim_{x\to 1} \dfrac{x^n - 1}{x^m - 1} = \dfrac{n}{m}$

69. **(a)**

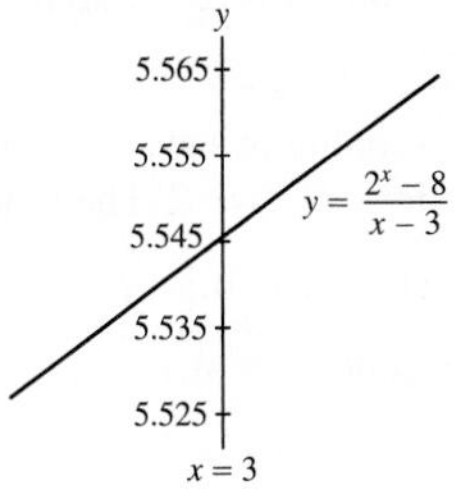

(b) $L = 5.545$.

Section 2.3 Preliminary Questions

1. Suppose $\lim_{x\to c} f(x)$ and $\lim_{x\to c} g(x)$ both exist. The Sum Law states that

$$\lim_{x\to c} (f(x) + g(x)) = \lim_{x\to c} f(x) + \lim_{x\to c} g(x).$$

Provided $\lim_{x\to c} g(x) \neq 0$, the Quotient Law states that

$$\lim_{x\to c} \frac{f(x)}{g(x)} = \frac{\lim_{x\to c} f(x)}{\lim_{x\to c} g(x)}.$$

2. **(b)** **3.** **(a)**

Section 2.3 Exercises

1. 9 **3.** $\frac{1}{16}$ **5.** $\frac{1}{2}$ **7.** 4.6 **9.** 1 **11.** 9 **13.** $-\frac{2}{5}$ **15.** 10
17. $\frac{1}{5}$ **19.** $\frac{1}{5}$ **21.** $\frac{2}{5}$ **23.** 64 **27.** 3 **29.** $\frac{1}{16}$ **31.** No
33. $f(x) = 1/x$ and $g(x) = -1/x$ **35.** Write $g(t) = \frac{tg(t)}{t}$
37. **(b)**

Section 2.4 Preliminary Questions

1. Continuity **2.** $f(3) = \frac{1}{2}$ **3.** No **4.** No; Yes

5. **(a)** False. The correct statement is "$f(x)$ is continuous at $x = a$ if the left- and right-hand limits of $f(x)$ as $x \to a$ exist and equal $f(a)$."

(b) True.

(c) False. The correct statement is "If the left- and right-hand limits of $f(x)$ as $x \to a$ are equal but not equal to $f(a)$, then f has a removable discontinuity at $x = a$."

(d) True.

(e) False. The correct statement is "If $f(x)$ and $g(x)$ are continuous at $x = a$ and $g(a) \neq 0$, then $f(x)/g(x)$ is continuous at $x = a$."

Section 2.4 Exercises

1.
- The function f is discontinuous at $x = 1$; it is right-continuous there.
- The function f is discontinuous at $x = 3$; it is neither left-continuous nor right-continuous there.
- The function f is discontinuous at $x = 5$; it is left-continuous there.

None of these discontinuities is removable.

3. $x = 3$; redefine $g(3) = 4$

5. The function f is discontinuous at $x = 0$, at which $\lim_{x\to 0-} f(x) = \infty$ and $\lim_{x\to 0+} f(x) = 2$. The function f is also discontinuous at $x = 2$, at which $\lim_{x\to 2-} f(x) = 6$ and $\lim_{x\to 2+} f(x) = 6$. The discontinuity at $x = 2$ is removable. Assigning $f(2) = 6$ makes f continuous at $x = 2$.

7. x and $\sin x$ are continuous, so is $x + \sin x$ by Continuity Law (i)

9. Since x and $\sin x$ are continuous, so are $3x$ and $4 \sin x$ by Continuity Law (ii). Thus $3x + 4\sin x$ is continuous by Continuity Law (i).

11. Since x is continuous, so is x^2 by Continuity Law (iii). Recall that constant functions, such as 1, are continuous. Thus $x^2 + 1$ is continuous by Continuity Law (i). Finally, $\dfrac{1}{x^2+1}$ is continuous by Continuity Law (iv) because $x^2 + 1$ is never 0.

13. The function $f(x)$ is a composite of two continuous functions: $\cos x$ and x^2, so $f(x)$ is continuous by Theorem 5.

15. 2^x and $\cos 3x$ are continuous, so $2^x \cos 3x$ is continuous by Continuity Law (iii).

17. Discontinuous at $x = 0$, at which there is an infinite discontinuity. The function is neither left- nor right-continuous at $x = 0$.

19. Discontinuous at $x = 1$, at which there is an infinite discontinuity. The function is neither left- nor right-continuous at $x = 1$.

21. Discontinuous at even integers, at which there are jump discontinuities. Function is right-continuous at the even integers but not left-continuous.

23. Discontinuous at $x = \frac{1}{2}$, at which there is an infinite discontinuity. The function is neither left- nor right-continuous at $x = \frac{1}{2}$.

25. Continuous for all x

27. Jump discontinuity at $x = 2$. Function is left-continuous at $x = 2$ but not right-continuous.

29. Discontinuous whenever $t = \frac{(2n+1)\pi}{4}$, where n is an integer. At every such value of t there is an infinite discontinuity. The function is neither left- nor right-continuous at any of these points of discontinuity.

31. Continuous everywhere

33. Discontinuous at $x = 0$, at which there is an infinite discontinuity. The function is neither left- nor right-continuous at $x = 0$.

35. The domain is all real numbers. Both $\sin x$ and $\cos x$ are continuous on this domain, so $2\sin x + 3\cos x$ is continuous by Continuity Laws (i) and (ii).

37. Domain is $x \geq 0$. Since $\sqrt{x}$ and $\sin x$ are continuous, so is $\sqrt{x} \sin x$ by Continuity Law (iii).

39. Domain is all real numbers. Both $x^{2/3}$ and 2^x are continuous on this domain, so $x^{2/3}2^x$ is continuous by Continuity Law (iii).

41. Domain is $x \neq 0$. Because the function $x^{4/3}$ is continuous and not equal to zero for $x \neq 0$, $x^{-4/3}$ is continuous for $x \neq 0$ by Continuity Law (iv).

43. Domain is all $x \neq \pm(2n - 1)\pi/2$ where n is a positive integer. Because $\tan x$ is continuous on this domain, it follows from Continuity Law (iii) that $\tan^2 x$ is also continuous on this domain.

45. Domain of $(x^4 + 1)^{3/2}$ is all real numbers. Because $x^{3/2}$ and the polynomial $x^4 + 1$ are both continuous, so is the composite function $(x^4 + 1)^{3/2}$.

47. Domain is all $x \neq \pm 1$. Because the functions $\cos x$ and x^2 are continuous on this domain, so is the composite function $\cos(x^2)$. Finally, because the polynomial $x^2 - 1$ is continuous and not equal to zero for $x \neq \pm 1$, the function $\frac{\cos(x^2)}{x^2-1}$ is continuous by Continuity Law (iv).

49. $f(x)$ is right-continuous at $x = 1$; $f(x)$ is continuous at $x = 2$

51. The function f is continuous everywhere.

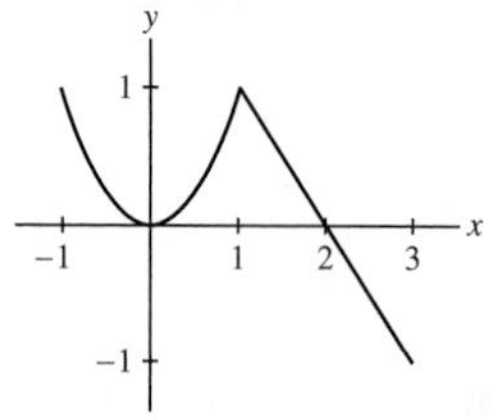

53. The function f is neither left- nor right-continuous at $x = 2$.

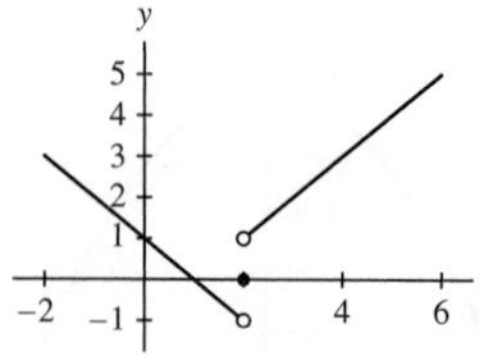

55. $\lim_{x\to 4} \frac{x^2-16}{x-4} = \lim_{x\to 4} (x+4) = 8 \neq 10 = f(4)$

57. $c = \frac{5}{3}$ **59.** $a = 2$ and $b = 1$

61. **(a)** No **(b)** $g(1) = 0$

63.

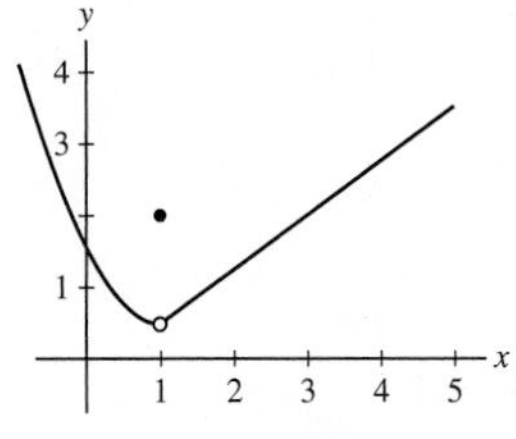

65.

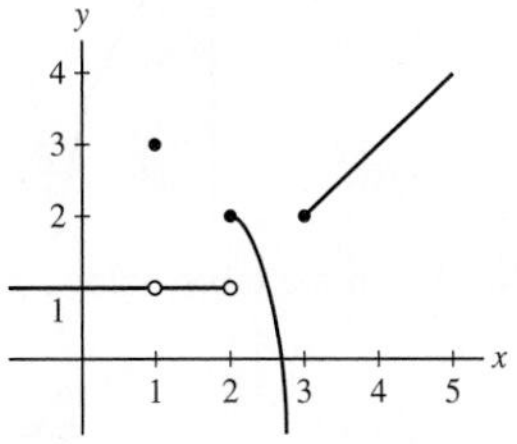

67. -6 **69.** $\frac{1}{3}$ **71.** -1 **73.** $\frac{1}{32}$ **75.** 27 **77.** 1000 **79.** $\frac{1}{2}$

81. No. Take $f(x) = -x^{-1}$ and $g(x) = x^{-1}$

83. $f(x) = |g(x)|$ is a composition of the continuous functions $g(x)$ and $|x|$

85. No.

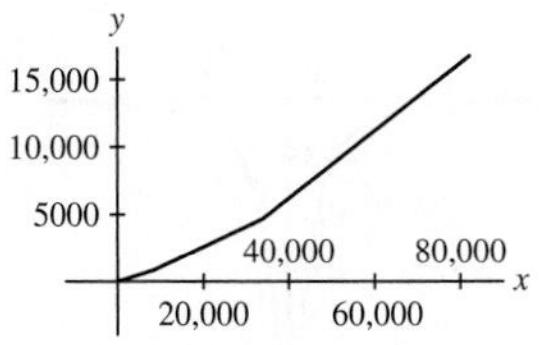

87. $f(x) = 3$ and $g(x) = [x]$

Section 2.5 Preliminary Questions

1. $\frac{x^2-1}{\sqrt{x+3}-2}$

2. **(a)** $f(x) = \frac{x^2-1}{x-1}$ **(b)** $f(x) = \frac{x^2-1}{x-1}$ **(c)** $f(x) = \frac{1}{x}$

3. The "simplify and plug-in" strategy is based on simplifying a function which is indeterminate to a continuous function. Once the simplification has been made, the limit of the remaining continuous function is obtained by evaluation.

Section 2.5 Exercises

1. $\lim_{x\to 6} \frac{x^2-36}{x-6} = \lim_{x\to 6} \frac{(x-6)(x+6)}{x-6} = \lim_{x\to 6} (x+6) = 12$

3. 0 **5.** $\frac{1}{14}$ **7.** -1 **9.** $\frac{11}{10}$ **11.** 2 **13.** 1 **15.** 2 **17.** $\frac{1}{8}$

19. $\frac{7}{17}$

21. Limit does not exist.

- As $h \to 0+$, $\dfrac{\sqrt{h+2}-2}{h} \to -\infty$.
- As $h \to 0-$, $\dfrac{\sqrt{h+2}-2}{h} \to \infty$.

23. 2 **25.** $\frac{1}{4}$ **27.** 1 **29.** 9 **31.** $\frac{\sqrt{2}}{2}$ **33.** $\frac{1}{2}$

35. $\lim_{x\to 4} f(x) \approx 2.00$; to two decimal places, this matches the value of 2 obtained in Exercise 23.

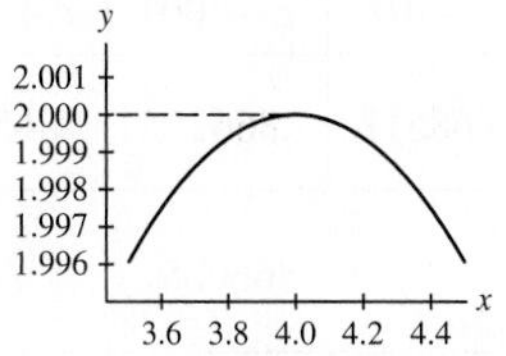

37. 12 **39.** -1 **41.** $\frac{4}{3}$ **43.** $\frac{1}{4}$ **45.** $2a$ **47.** $-4+5a$ **49.** $4a$

51. $\frac{1}{2\sqrt{a}}$ **53.** $3a^2$ **55.** $c = -1$ and $c = 6$ **57.** $c = 3$ **59.** $+$

Section 2.6 Preliminary Questions

1. $\lim_{x\to 0} f(x) = 0$; No

2. Assume that for $x \neq c$ (in some open interval containing c),

$$l(x) \le f(x) \le u(x)$$

and that $\lim_{x\to c} l(x) = \lim_{x\to c} u(x) = L$. Then $\lim_{x\to c} f(x)$ exists and

$$\lim_{x\to c} f(x) = L.$$

3. **(a)**

Section 2.6 Exercises

1. For all $x \neq 1$ on the open interval $(0, 2)$ containing $x = 1$, $\ell(x) \le f(x) \le u(x)$. Moreover,

$$\lim_{x\to 1} \ell(x) = \lim_{x\to 1} u(x) = 2.$$

Therefore, by the Squeeze Theorem,

$$\lim_{x\to 1} f(x) = 2.$$

3. $\lim_{x\to 7} f(x) = 6$

5. **(a)** *not* sufficient information **(b)** $\lim_{x\to 1} f(x) = 1$ **(c)** $\lim_{x\to 1} f(x) = 3$

7. $\lim_{x\to 0} x^2 \cos\frac{1}{x} = 0$ **9.** $\lim_{x\to 1} (x-1)\sin\frac{\pi}{x-1} = 0$

11. $\lim_{t\to 0} (2^t - 1)\cos\frac{1}{t} = 0$

13. $\lim_{t\to 2} (t^2 - 4)\cos\frac{1}{t-2} = 0$

15. $\lim_{\theta\to \frac{\pi}{2}} \cos\theta \cos(\tan\theta) = 0$

17. 1 **19.** 3 **21.** 1 **23.** 0 **25.** $\frac{2\sqrt{2}}{\pi}$ **27.** **(b)** $L = 14$ **29.** 9

31. $\frac{1}{5}$ **33.** $\frac{7}{3}$ **35.** $\frac{1}{25}$ **37.** 6 **39.** $-\frac{3}{4}$ **41.** $\frac{1}{2}$ **43.** $\frac{6}{5}$ **45.** 0

47. 0 **49.** -1 **53.** $-\frac{9}{2}$

55. $\lim_{t\to 0+} \frac{\sqrt{1-\cos t}}{t} = \frac{\sqrt{2}}{2}$; $\lim_{t\to 0-} \frac{\sqrt{1-\cos t}}{t} = -\frac{\sqrt{2}}{2}$

59. **(a)**

x	$c - .01$	$c - .001$	$c + .001$	$c + .01$
$\frac{\sin x - \sin c}{x - c}$	.999983	.99999983	.99999983	.999983

Here $c = 0$ and $\cos c = 1$.

x	$c-.01$	$c-.001$	$c+.001$	$c+.01$
$\frac{\sin x-\sin c}{x-c}$	.868511	.866275	.865775	.863511

Here $c=\frac{\pi}{6}$ and $\cos c=\frac{\sqrt{3}}{2}\approx .866025$.

x	$c-.01$	$c-.001$	$c+.001$	$c+.01$
$\frac{\sin x-\sin c}{x-c}$	.504322	.500433	.499567	.495662

Here $c=\frac{\pi}{3}$ and $\cos c=\frac{1}{2}$.

x	$c-.01$	$c-.001$	$c+.001$	$c+.01$
$\frac{\sin x-\sin c}{x-c}$	.710631	.707460	.706753	.703559

Here $c=\frac{\pi}{4}$ and $\cos c=\frac{\sqrt{2}}{2}\approx 0.707107$.

x	$c-.01$	$c-.001$	$c+.001$	$c+.01$
$\frac{\sin x-\sin c}{x-c}$	.005000	.000500	−.000500	−.005000

Here $c=\frac{\pi}{2}$ and $\cos c=0$.

(b) $\lim_{x\to c}\frac{\sin x-\sin c}{x-c}=\cos c$.

(c)

x	$c-.01$	$c-.001$	$c+.001$	$c+.01$
$\frac{\sin x-\sin c}{x-c}$	−.411593	−.415692	−.416601	−.420686

Here $c=2$ and $\cos c=\cos 2\approx -.416147$.

x	$c-.01$	$c-.001$	$c+.001$	$c+.01$
$\frac{\sin x-\sin c}{x-c}$	.863511	.865775	.866275	.868511

Here $c=-\frac{\pi}{6}$ and $\cos c=\frac{\sqrt{3}}{2}\approx .866025$.

Section 2.7 Preliminary Questions

1. **(a)** Correct **(b)** Not correct **(c)** Not correct **(d)** Correct

2. **(a)** $\lim_{x\to\infty}x^3=\infty$ **(b)** $\lim_{x\to-\infty}x^3=-\infty$
(c) $\lim_{x\to-\infty}x^4=\infty$

3.

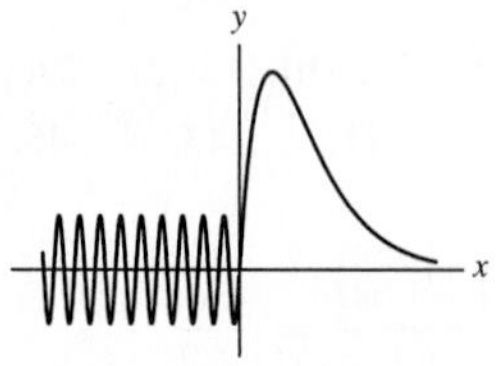

4. Negative **5.** Negative

6. As $x\to\infty$, $\frac{1}{x}\to 0$, so

$$\lim_{x\to\infty}\sin\frac{1}{x}=\sin 0=0.$$

On the other hand, $\frac{1}{x}\to\pm\infty$ as $x\to 0$, and as $\frac{1}{x}\to\pm\infty$, $\sin\frac{1}{x}$ oscillates infinitely often.

Section 2.7 Exercises

1. $y=1$ and $y=2$

3.

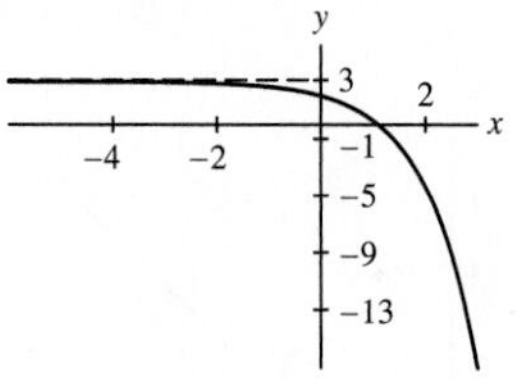

5. **(a)** From the table below, it appears that

$$\lim_{x\to\pm\infty}\frac{x^3}{x^3+x}=1.$$

x	±50	±100	±500	±1000
$f(x)$	.999600	.999900	.999996	.999999

(b) From the graph below, it also appears that

$$\lim_{x\to\pm\infty}\frac{x^3}{x^3+x}=1.$$

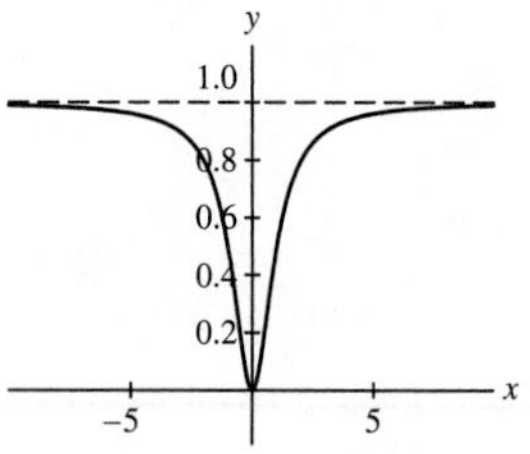

(c) The horizontal asymptote of $f(x)$ is $y=1$.

7. 1 **9.** 0 **11.** $\frac{7}{4}$ **13.** $-\infty$ **15.** ∞ **17.** $y=\frac{1}{4}$ **19.** $y=\frac{2}{3}$ and $y=-\frac{2}{3}$ **21.** $y=0$ **23.** 0 **25.** 2 **27.** $\frac{1}{16}$ **29.** 0

31. 1; the graph of $y=5^{-1/t^2}$ has a horizontal asymptote at $y=1$.

33. **(a)** $\lim_{s\to\infty}R(s)=\lim_{s\to\infty}\frac{As}{K+s}=\lim_{s\to\infty}\frac{A}{1+\frac{K}{s}}=A$.

(b) $R(K)=\frac{AK}{K+K}=\frac{AK}{2K}=\frac{A}{2}$ half of the limiting value.

(c) 3.75 mM

35. 0 **37.** ∞ **39.** $-\sqrt{3}$

43. $\lim_{x\to\infty}\frac{3x^2-x}{2x^2+5}=\lim_{t\to 0+}\frac{3-t}{2+5t^2}=\frac{3}{2}$

45.
- $b=0.2$:

x	5	10	50	100
$f(x)$	1.000064	1.000000	1.000000	1.000000

It appears that $G(0.2)=1$.

- $b=0.8$:

x	5	10	50	100
$f(x)$	1.058324	1.010251	1.000000	1.000000

It appears that $G(0.8)=1$.

- $b=2$:

x	5	10	50	100
$f(x)$	2.012347	2.000195	2.000000	2.000000

It appears that $G(2)=2$.

- $b = 3$:

x	5	10	50	100
$f(x)$	3.002465	3.000005	3.000000	3.000000

It appears that $G(3) = 3$.

- $b = 5$:

x	5	10	50	100
$f(x)$	5.000320	5.000000	5.000000	5.000000

It appears that $G(5) = 5$.

Based on these observations we conjecture that $G(b) = 1$ if $0 \le b \le 1$ and $G(b) = b$ for $b > 1$. The graph of $y = G(b)$ is shown below; the graph does appear to be continuous.

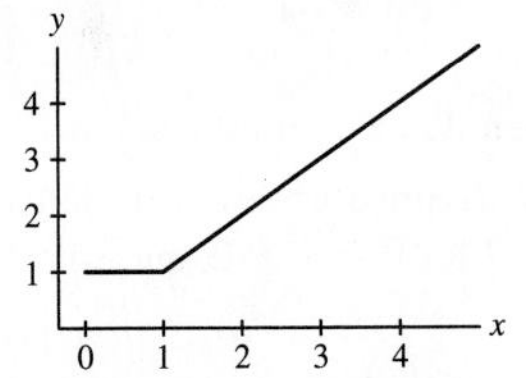

Section 2.8 Preliminary Questions

1. Observe that $f(x) = x^2$ is continuous on [0, 1] with $f(0) = 0$ and $f(1) = 1$. Because $f(0) < 0.5 < f(1)$, the Intermediate Value Theorem guarantees there is a $c \in [0, 1]$ such that $f(c) = 0.5$.

2. We must assume that temperature is a continuous function of time.

3. If f is continuous on $[a, b]$, then the horizontal line $y = k$ for every k between $f(a)$ and $f(b)$ intersects the graph of $y = f(x)$ at least once.

4.

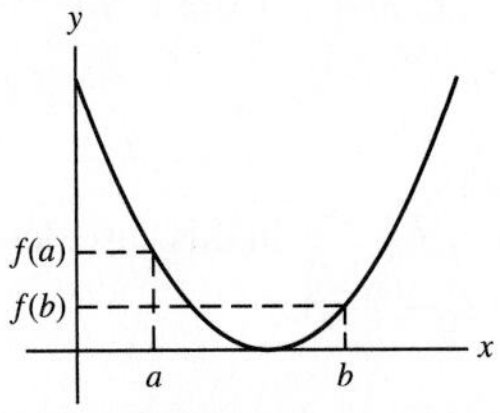

5. **(a)** Sometimes true. **(b)** Always true. **(c)** Never true. **(d)** Sometimes true.

Section 2.8 Exercises

1. Observe that $f(1) = 2$ and $f(2) = 10$. Since f is a polynomial, it is continuous everywhere; in particular on [1, 2]. Therefore, by the IVT there is a $c \in [1, 2]$ such that $f(c) = 9$.

3. $g(0) = 0$ and $g\left(\frac{\pi}{4}\right) = \frac{\pi^2}{16}$. $g(t)$ is continuous for all t between 0 and $\frac{\pi}{4}$, and $0 < \frac{1}{2} < \frac{\pi^2}{16}$; therefore, by the IVT, there is a $c \in \left[0, \frac{\pi}{4}\right]$ such that $g(c) = \frac{1}{2}$.

5. Let $f(x) = x - \cos x$. Observe that f is continuous with $f(0) = -1$ and $f(1) = 1 - \cos 1 \approx .46$. Therefore, by the IVT there is a $c \in [0, 1]$ such that $f(c) = c - \cos c = 0$.

7. Let $f(x) = \sqrt{x} + \sqrt{x+2} - 3$. Note that f is continuous on $\left[\frac{1}{4}, 2\right]$ with $f\left(\frac{1}{4}\right) = -1$ and $f(2) = \sqrt{2} - 1 \approx .41$. Therefore, by the IVT there is a $c \in \left[\frac{1}{4}, 2\right]$ such that $f(c) = \sqrt{c} + \sqrt{c+2} - 3 = 0$.

9. Let $f(x) = x^2$. Observe that f is continuous with $f(1) = 1$ and $f(2) = 4$. Therefore, by the IVT there is a $c \in [1, 2]$ such that $f(c) = c^2 = 2$.

11. For each positive integer k, let $f(x) = x^k - \cos x$. Observe that f is continuous on $\left[0, \frac{\pi}{2}\right]$ with $f(0) = -1$ and $f\left(\frac{\pi}{2}\right) = \left(\frac{\pi}{2}\right)^k > 0$. Therefore, by the IVT there is a $c \in \left[0, \frac{\pi}{2}\right]$ such that $f(c) = c^k - \cos(c) = 0$.

13. Let $f(x) = 2^x + 3^x - 4^x$. Observe that f is continuous on [0, 2] with $f(0) = 1 > 0$ and $f(2) = -3 < 0$. Therefore, by the IVT, there is a $c \in (0, 2)$ such that $f(c) = 2^c + 3^c - 4^c = 0$.

15. Let $f(x) = 2^x + \frac{1}{x} + 4$. Note that f is continuous on $[-\infty, 0)$. $f(-1) = 2^{-1} + \frac{1}{-1} + 4 = \frac{7}{2} > 0$, and $f\left(-\frac{1}{10}\right) = \frac{1}{2^{1/10}} - 10 + 4 < 0$. Therefore, by the IVT, there is a $c \in \left(-1, -\frac{1}{10}\right)$ such that $f(c) = 2^c + \frac{1}{c} + 4 = 0$, so that $2^c + \frac{1}{c} = -4$.

17. **(a)** $f(1) = 1$, $f(1.5) = 2^{1.5} - (1.5)^3 < 3 - 3.375 < 0$. Hence, $f(x) = 0$ for some x between 1 and 1.5.

(b) $f(1.25) \approx 0.4253 > 0$ and $f(1.5) < 0$. Hence, $f(x) = 0$ for some x between 1.25 and 1.5.

(c) $f(1.375) \approx -0.0059$. Hence, $f(x) = 0$ for some x between 1.25 and 1.375.

19. [0, .25]

21.

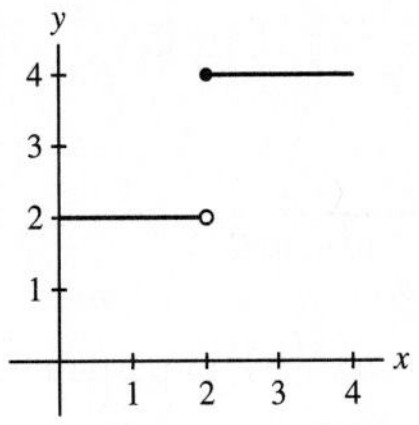

23.

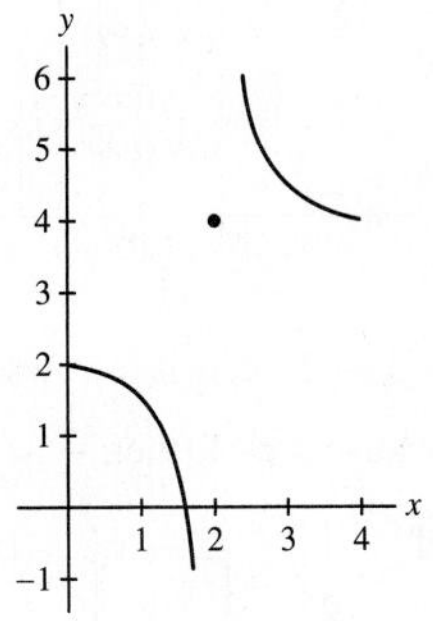

25. No; no

Section 2.9 Preliminary Questions

1. **(c)**

2. **(b)** and **(d)** are true

Section 2.9 Exercises

1. $L = 4$, $\epsilon = .8$, and $\delta = .1$

3. **(a)** $|f(x) - 35| = |8x + 3 - 35| = |8x - 32| = |8(x - 4)| = 8\,|x - 4|$

(b) Let $\epsilon > 0$. Let $\delta = \epsilon/8$ and suppose $|x - 4| < \delta$. By part **(a)**, $|f(x) - 35| = 8|x - 4| < 8\delta$. Substituting $\delta = \epsilon/8$, we see $|f(x) - 35| < 8\epsilon/8 = \epsilon$.

5. **(a)** If $0 < |x - 2| < \delta = .01$, then $|x| < 3$ and $\left|x^2 - 4\right| = |x - 2||x + 2| \le |x - 2|\,(|x| + 2) < 5|x - 2| < .05$.

(b) If $0 < |x-2| < \delta = .0002$, then $|x| < 2.0002$ and

$$\left|x^2 - 4\right| = |x-2||x+2| \leq |x-2|\,(|x|+2) < 4.0002|x-2|$$

$$< .00080004 < .0009$$

(c) $\delta = 10^{-5}$

7. $\delta = 6 \times 10^{-4}$

9. $\delta = 0.25$

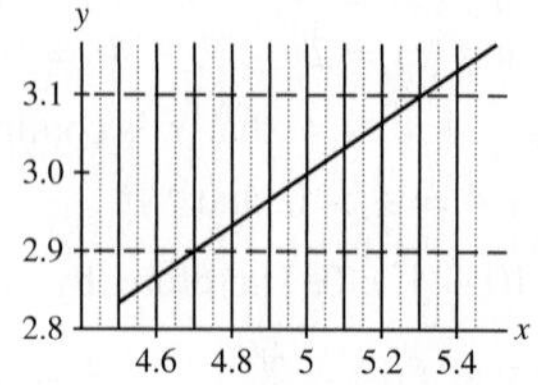

11. $\delta = 0.05$

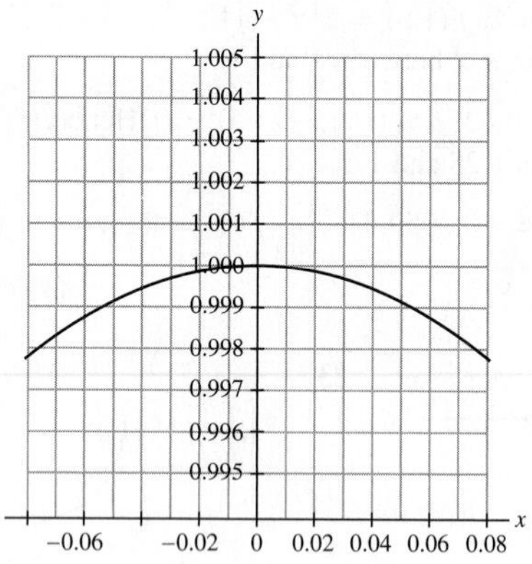

13. (a) Since $|x-2| < 1$, it follows that $1 < x < 3$, in particular that $x > 1$. Because $x > 1$, then $\frac{1}{x} < 1$ and

$$\left|\frac{1}{x} - \frac{1}{2}\right| = \left|\frac{2-x}{2x}\right| = \frac{|x-2|}{2x} < \frac{1}{2}|x-2|.$$

(b) Let $\delta = \min\{1, 2\epsilon\}$ and suppose that $|x-2| < \delta$. Then by part (a) we have

$$\left|\frac{1}{x} - \frac{1}{2}\right| < \frac{1}{2}|x-2| < \frac{1}{2}\delta < \frac{1}{2}\cdot 2\epsilon = \epsilon.$$

(c) Choose $\delta = .02$.

(d) Let $\epsilon > 0$ be given. Then whenever $0 < |x-2| < \delta = \min\{1, 2\epsilon\}$, we have

$$\left|\frac{1}{x} - \frac{1}{2}\right| < \frac{1}{2}\delta \leq \epsilon.$$

15.

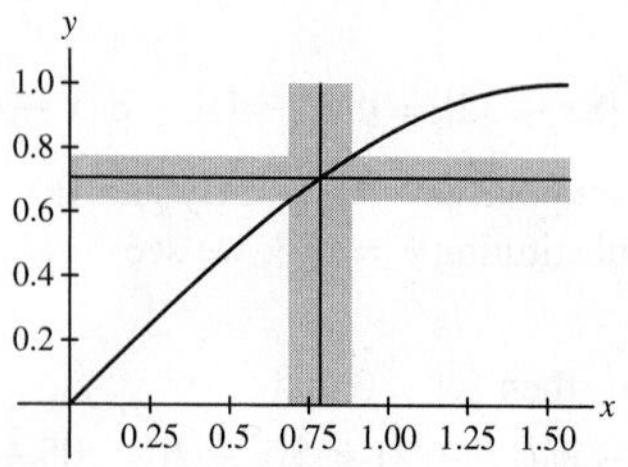

17. Given $\epsilon > 0$, we let

$$\delta = \min\left\{|c|, \frac{\epsilon}{3|c|}\right\}.$$

Then, for $|x-c| < \delta$, we have

$$|x^2 - c^2| = |x-c|\,|x+c| < 3|c|\delta < 3|c|\frac{\epsilon}{3|c|} = \epsilon.$$

19. Let $\epsilon > 0$ be given. Let $\delta = \min(1, 3\epsilon)$. If $|x-4| < \delta$,

$$|\sqrt{x} - 2| = |x-4|\left|\frac{1}{\sqrt{x}+2}\right| < |x-4|\frac{1}{3} < \delta\frac{1}{3} < 3\epsilon\frac{1}{3} = \epsilon.$$

21. Let $\epsilon > 0$ be given. Let $\delta = \min\left(1, \frac{\epsilon}{7}\right)$, and assume $|x-1| < \delta$. Since $\delta < 1$, $0 < x < 2$. Since $x^2 + x + 1$ increases as x increases for $x > 0$, $x^2 + x + 1 < 7$ for $0 < x < 2$, and so

$$\left|x^3 - 1\right| = |x-1|\left|x^2 + x + 1\right| < 7|x-1| < 7\frac{\epsilon}{7} = \epsilon.$$

23. Let $\epsilon > 0$ be given. Let $\delta = \min\left(1, \frac{4}{5}\epsilon\right)$, and suppose $|x-2| < \delta$. Since $\delta < 1$, $|x-2| < 1$, so $1 < x < 3$. This means that $4x^2 > 4$ and $|2+x| < 5$, so that $\frac{2+x}{4x^2} < \frac{5}{4}$. We get:

$$\left|x^{-2} - \frac{1}{4}\right| = |2-x|\left|\frac{2+x}{4x^2}\right| < \frac{5}{4}|x-2| < \frac{5}{4}\cdot\frac{4}{5}\epsilon = \epsilon.$$

25. Let L be any real number. Let $\delta > 0$ be any small positive number. Let $x = \frac{\delta}{2}$, which satisfies $|x| < \delta$, and $f(x) = 1$. We consider two cases:

- $\left(|f(x) - L| \geq \frac{1}{2}\right)$: we are done.
- $\left(|f(x) - L| < \frac{1}{2}\right)$: This means $\frac{1}{2} < L < \frac{3}{2}$. In this case, let $x = -\frac{\delta}{2}$. $f(x) = -1$, and so $\frac{3}{2} < L - f(x)$.

In either case, there exists an x such that $|x| < \frac{\delta}{2}$, but $|f(x) - L| \geq \frac{1}{2}$.

27. Let $\epsilon > 0$ and let $\delta = \min(1, \frac{\epsilon}{2})$. Then, whenever $|x-1| < \delta$, it follows that $0 < x < 2$. If $1 < x < 2$, then $\min(x, x^2) = x$ and

$$|f(x) - 1| = |x-1| < \delta < \frac{\epsilon}{2} < \epsilon.$$

On the other hand, if $0 < x < 1$, then $\min(x, x^2) = x^2$, $|x+1| < 2$ and

$$|f(x) - 1| = |x^2 - 1| = |x-1|\,|x+1| < 2\delta < \epsilon.$$

Thus, whenever $|x-1| < \delta$, $|f(x) - 1| < \epsilon$.

31. Suppose that $\lim_{x\to c} f(x) = L$. Let $\epsilon > 0$ be given. Since $\lim_{x\to c} f(x) = L$, we know there is a $\delta > 0$ such that $|x-c| < \delta$ forces $|f(x) - L| < \epsilon/|a|$. Suppose $|x-c| < \delta$. Then $|af(x) - aL| = |a||f(x) - aL| < |a|(\epsilon/|a|) = \epsilon$.

Chapter 2 Review

1. average velocity approximately 0.954 m/s; instantaneous velocity approximately 0.894 m/s.

3. $\frac{200}{9}$ **5.** 1.50 **7.** 1.69 **9.** 2.00

11. 5 **13.** $-\frac{1}{2}$ **15.** $\frac{1}{6}$ **17.** 2

19. Does not exist;

$$\lim_{t\to 9-} \frac{t-6}{\sqrt{t}-3} = -\infty \quad \text{and} \quad \lim_{t\to 9+} \frac{t-6}{\sqrt{t}-3} = \infty$$

21. ∞

23. Does not exist;

$$\lim_{x\to 1-} \frac{x^3-2x}{x-1} = \infty \quad \text{and} \quad \lim_{x\to 1+} \frac{x^3-2x}{x-1} = -\infty$$

25. 2 **27.** $\frac{2}{3}$ **29.** $-\frac{1}{2}$ **31.** $3b^2$ **33.** $\frac{1}{9}$ **35.** ∞

37. Does not exist;

$$\lim_{\theta\to \frac{\pi}{2}-} \theta \sec\theta = \infty \quad \text{and} \quad \lim_{\theta\to \frac{\pi}{2}+} \theta \sec\theta = -\infty$$

39. Does not exist;

$$\lim_{\theta\to 0-} \frac{\cos\theta - 2}{\theta} = \infty \quad \text{and} \quad \lim_{\theta\to 0+} \frac{\cos\theta - 2}{\theta} = -\infty$$

41. ∞ **43.** ∞

45. Does not exist;

$$\lim_{x\to \frac{\pi}{2}-} \tan x = \infty \quad \text{and} \quad \lim_{x\to \frac{\pi}{2}+} \tan x = -\infty$$

47. 0 **49.** 0

51. According to the graph of $f(x)$,

$$\lim_{x\to 0-} f(x) = \lim_{x\to 0+} f(x) = 1$$

$$\lim_{x\to 2-} f(x) = \lim_{x\to 2+} f(x) = \infty$$

$$\lim_{x\to 4-} f(x) = -\infty$$

$$\lim_{x\to 4+} f(x) = \infty.$$

The function is both left- and right-continuous at $x = 0$ and neither left- nor right-continuous at $x = 2$ and $x = 4$.

53. At $x = 0$, the function has an infinite discontinuity but is left-continuous.

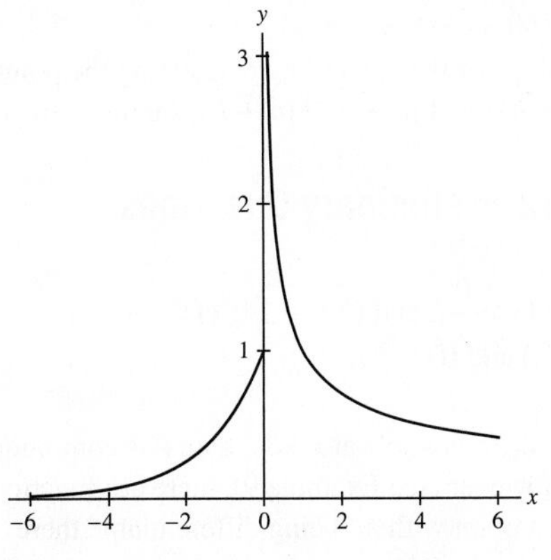

55. $g(x)$ has a jump discontinuity at $x = -1$; $g(x)$ is left-continuous at $x = -1$.

57. $b = 7$; $h(x)$ has a jump discontinuity at $x = -2$

59. Does not have any horizontal asymptotes

61. $y = 2$ **63.** $y = 1$

65.

$$B = B \cdot 1 = B \cdot L =$$

$$\lim_{x\to a} g(x) \cdot \lim_{x\to a} \frac{f(x)}{g(x)} = \lim_{x\to a} g(x)\frac{f(x)}{g(x)} = \lim_{x\to a} f(x) = A.$$

67. $f(x) = \frac{1}{(x-a)^3}$ and $g(x) = \frac{1}{(x-a)^5}$

71. Let $f(x) = x^2 - \cos x$. Now, $f(x)$ is continuous over the interval $[0, \frac{\pi}{2}]$, $f(0) = -1 < 0$ and $f(\frac{\pi}{2}) = \frac{\pi^2}{4} > 0$. Therefore, by the Intermediate Value Theorem, there exists a $c \in (0, \frac{\pi}{2})$ such that $f(c) = 0$; consequently, the curves $y = x^2$ and $y = \cos x$ intersect.

73. Let $f(x) = 2^{-x^2} - x$. Observe that f is continuous on $[0, 1]$ with $f(0) = 2^0 - 0 = 1 > 0$ and $f(1) = 2^{-1} - 1 < 0$. Therefore, the IVT guarantees there exists a $c \in (0, 1)$ such that $f(c) = 2^{-c^2} - c = 0$.

75. $g(x) = [x]$; On the interval

$$x \in \left[\frac{a}{2 + 2\pi a}, \frac{a}{2}\right] \subset [-a, a],$$

$\frac{1}{x}$ runs from $\frac{2}{a}$ to $\frac{2}{a} + 2\pi$, so the sine function covers one full period and clearly takes on every value from $-\sin a$ through $\sin a$.

77. $\delta = 0.55$;

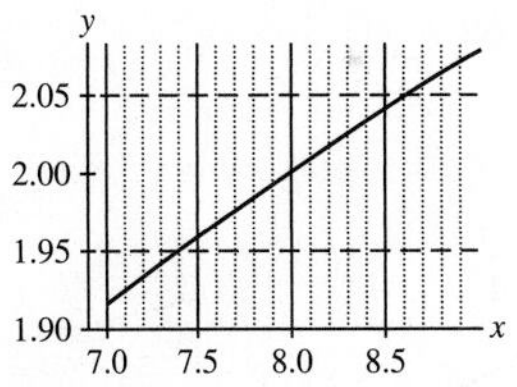

79. Let $\epsilon > 0$ and take $\delta = \epsilon/8$. Then, whenever $|x - (-1)| = |x + 1| < \delta$,

$$|f(x) - (-4)| = |4 + 8x + 4| = 8|x + 1| < 8\delta = \epsilon.$$

Chapter 3

Section 3.1 Preliminary Questions

1. B and D

2. $\frac{f(x) - f(a)}{x - a}$ and $\frac{f(a+h) - f(a)}{h}$

3. $a = 3$ and $h = 2$

4. Derivative of the function $f(x) = \tan x$ at $x = \frac{\pi}{4}$

5. **(a)** The difference in height between the points $(0.9, \sin 0.9)$ and $(1.3, \sin 1.3)$.

(b) The slope of the secant line between the points $(0.9, \sin 0.9)$ and $(1.3, \sin 1.3)$.

(c) The slope of the tangent line to the graph at $x = 0.9$.

Section 3.1 Exercises

1. $f'(3) = 30$ **3.** $f'(0) = 9$ **5.** $f'(-1) = -2$

7. Slope of the secant line = 1; the secant line through $(2, f(2))$ and $(2.5, f(2.5))$ has a larger slope than the tangent line at $x = 2$.

9. $f'(1) \approx 0$; $f'(2) \approx 0.8$

11. $f'(1) = f'(2) = 0$; $f'(4) = \frac{1}{2}$; $f'(7) = 0$

13. $f'(5.5)$ **15.** $f'(x) = 7$ **17.** $g'(t) = -3$ **19.** $y = 2x - 1$

21. The tangent line at any point is the line itself

23. $f(-2+h) = \dfrac{1}{-2+h}$; $-\dfrac{1}{3}$

25. $f'(5) = -\dfrac{1}{10\sqrt{5}}$

27. $f'(3) = 22$; $y = 22x - 18$

29. $f'(3) = -11$; $y = -11t + 18$

31. $f'(0) = 1$; $y = x$

33. $f'(8) = -\dfrac{1}{64}$; $y = -\dfrac{1}{64}x + \dfrac{1}{4}$

35. $f'(-2) = -1$; $y = -x - 1$

37. $f'(1) = \dfrac{1}{2\sqrt{5}}$; $y = \dfrac{1}{2\sqrt{5}}x + \dfrac{9}{2\sqrt{5}}$

39. $f'(4) = -\dfrac{1}{16}$; $y = -\dfrac{1}{16}x + \dfrac{3}{4}$

41. $f'(3) = \dfrac{3}{\sqrt{10}}$; $y = \dfrac{3}{\sqrt{10}}t + \dfrac{1}{\sqrt{10}}$

43. $f'(0) = 0$; $y = 1$

45. $W'(4) \approx 0.9$ kg/year; slope of the tangent is zero at $t = 10$ and at $t = 11.6$; slope of the tangent line is negative for $10 < t < 11.6$.

47. **(a)** $f'(0) \approx -0.68$

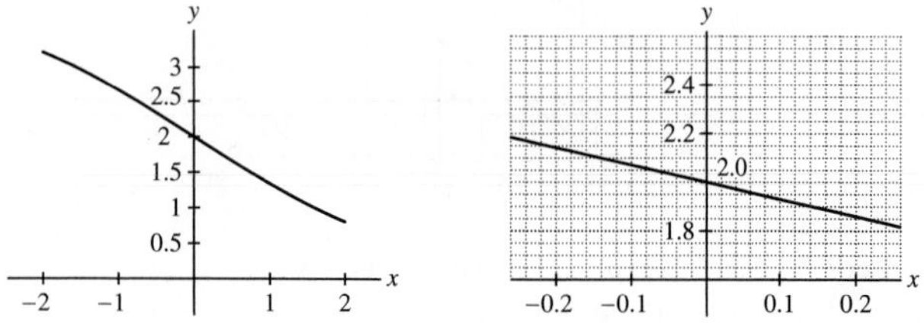

(b) $y = -0.68x + 2$

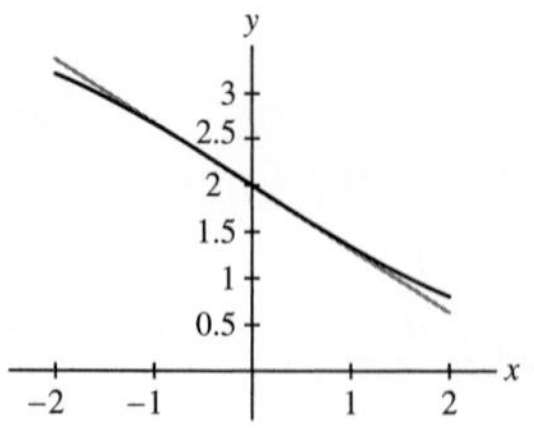

49. For $1 < x < 2.5$ and for $x > 3.5$

51. $f(x) = x^3$ and $a = 5$

53. $f(x) = \sin x$ and $a = \frac{\pi}{6}$

55. $f(x) = 5^x$ and $a = 2$

57. $f'\left(\frac{\pi}{4}\right) \approx 0.7071$

59.
- On curve (A), $f'(1)$ is larger than $\dfrac{f(1+h) - f(1)}{h}$; the curve is bending downwards, so that the secant line to the right is at a lower angle than the tangent line.
- On curve (B), $f'(1)$ is smaller than $\dfrac{f(1+h) - f(1)}{h}$; the curve is bending upwards, so that the secant line to the right is at a steeper angle than the tangent line.

61. **(b)** $f'(4) \approx 20.0000$
(c) $y = 20x - 48$

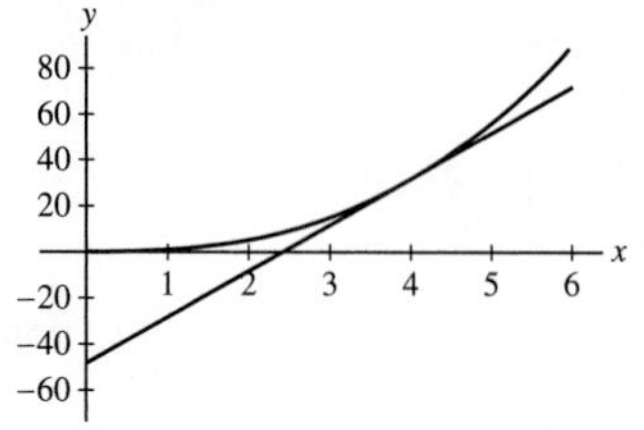

63. $c \approx 0.37$.

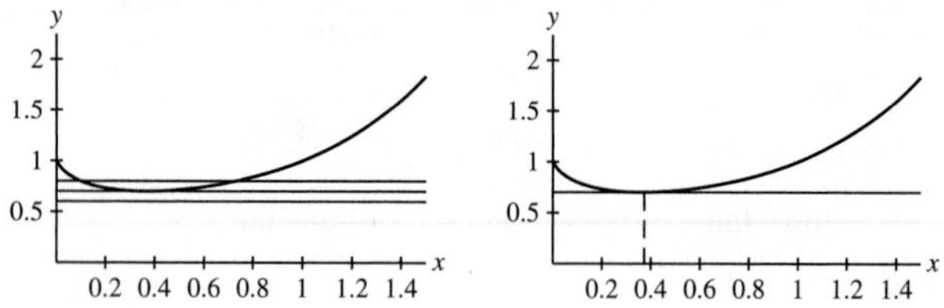

65.

$$P'(303) \approx \frac{P(313) - P(293)}{20} = \frac{0.0808 - 0.0278}{20} = 0.00265 \text{ atm/K};$$

$$P'(313) \approx \frac{P(323) - P(303)}{20} = \frac{0.1311 - 0.0482}{20} = 0.004145 \text{ atm/K};$$

$$P'(323) \approx \frac{P(333) - P(313)}{20} = \frac{0.2067 - 0.0808}{20} = 0.006295 \text{ atm/K};$$

$$P'(333) \approx \frac{P(343) - P(323)}{20} = \frac{0.3173 - 0.1311}{20} = 0.00931 \text{ atm/K};$$

$$P'(343) \approx \frac{P(353) - P(333)}{20} = \frac{0.4754 - 0.2067}{20} = 0.013435 \text{ atm/K}$$

67. -0.39375 kph·km/car

69. $i(3) = 0.06$ amperes

71. $v'(4) \approx 160$; $C \approx 0.2$ farads

73. It is the slope of the secant line connecting the points $(a - h, f(a - h))$ and $(a + h, f(a + h))$ on the graph of f.

Section 3.2 Preliminary Questions

1. 8

2. $(f - g)'(1) = -2$ and $(3f + 2g)'(1) = 19$

3. **(a)**, **(b)**, **(c)** and **(f)**

4. **(b)**

5. If $f(x)$ is differentiable at $x = c$, then it is continuous at $x = c$ by Theorem 4. However (see Example 9), there are functions that are continuous at a point without being differentiable there.

Section 3.2 Exercises

1. $f'(x) = 3$ **3.** $f'(x) = 3x^2$ **5.** $f'(x) = 1 - \dfrac{1}{2\sqrt{x}}$

7. $\left.\dfrac{d}{dx}x^4\right|_{x=-2} = 4(-2)^3 = -32$

9. $\left.\dfrac{d}{dt}t^{2/3}\right|_{t=8} = \dfrac{2}{3}(8)^{-1/3} = \dfrac{1}{3}$

11. $0.35x^{-0.65}$ **13.** $\sqrt{17}t^{\sqrt{17}-1}$

15. $f'(x) = 4x^3$; $y = 32x - 48$

17. $f'(x) = 5 - 16x^{-1/2}$; $y = -3x - 32$

19. $f'(x) = -\dfrac{1}{2x^{3/2}}$; $y - \frac{1}{3} = -\frac{1}{54}(x - 9)$; $y = \frac{1}{2} - \frac{1}{54}x$

21. $f'(x) = 6x^2 - 6x$ **23.** $f'(x) = \dfrac{20}{3}x^{2/3} + 6x^{-3}$

25. $g'(z) = -\dfrac{5}{2}z^{-19/14} - 5z^{-6}$

27. $f'(s) = \dfrac{1}{4}s^{-3/4} + \dfrac{1}{3}s^{-2/3}$

29. $g'(x) = 0$ **31.** $h'(t) = 2t^{\sqrt{2}-1}$ **33.** $P'(s) = 32s - 24$

35. $g'(x) = -6x^{-5/2}$ **37.** 1 **39.** -60 **41.** $\frac{47}{16}$

43.
- The graph in (A) matches the derivative in (III).
- The graph in (B) matches the derivative in (I).
- The graph in (C) matches the derivative in (II).
- The graph in (D) matches the derivative in (III).

(A) and (D) have the same derivative because the graph in (D) is just a vertical translation of the graph in (A).

45. Label the graph in (A) as $f(x)$, the graph in (B) as $h(x)$, and the graph in (C) as $g(x)$

47. (B) might be the graph of the derivative of $f(x)$

49. **(a)** $\dfrac{d}{dt}ct^3 = 3ct^2$.

(b) $\dfrac{d}{dz}(5z + 4cz^2) = 5 + 8cz$.

(c) $\dfrac{d}{dy}(9c^2y^3 - 24c) = 27c^2y^2$.

51. $x = \frac{1}{2}$

53. $a = 2$ and $b = -3$

55.
- $f'(x) = 3x^2 - 3 \geq -3$ since $3x^2$ is nonnegative.
- The two parallel tangent lines with slope 2 are shown with the graph of $f(x)$ here.

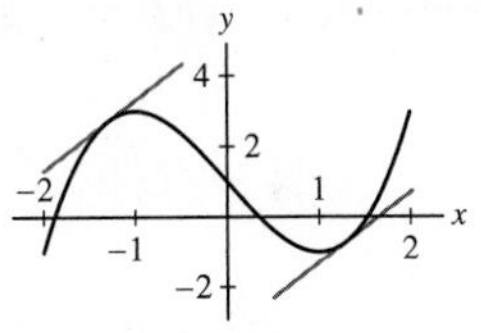

57. $f'(x) = \dfrac{3}{2}x^{1/2}$

59. $\displaystyle\lim_{h\to 0}\frac{f(2+h) - f(2)}{h} = \lim_{h\to 0}\frac{|h^2 + 2h|}{h} = \lim_{h\to 0}\frac{(h+2)|h|}{h}.$
This limit does not exist, since the $h + 2$ term becomes 2 as $h \to 0$, but the remaining term, $|h|/h$, is 1 for $h > 0$ and -1 for $h < 0$. Thus the two one-sided limits are not equal.

61. Decreasing; $y = -0.63216(m - 33) + 83.445$;
$y = -0.25606(m - 68) + 69.647$

63.

$P'(303) \approx \dfrac{P(313) - P(293)}{20} = \dfrac{0.0808 - 0.0278}{20} = 0.00265$ atm/K;

$P'(313) \approx \dfrac{P(323) - P(303)}{20} = \dfrac{0.1311 - 0.0482}{20} = 0.004145$ atm/K;

$P'(323) \approx \dfrac{P(333) - P(313)}{20} = \dfrac{0.2067 - 0.0808}{20} = 0.006295$ atm/K;

$P'(333) \approx \dfrac{P(343) - P(323)}{20} = \dfrac{0.3173 - 0.1311}{20} = 0.00931$ atm/K;

$P'(343) \approx \dfrac{P(353) - P(333)}{20} = \dfrac{0.4754 - 0.2067}{20} = 0.013435$ atm/K

$\dfrac{T^2}{P}\dfrac{dP}{dT}$ is roughly constant, suggesting that the Clausius–Clapeyron law is valid, and that $k \approx 5000$

67.

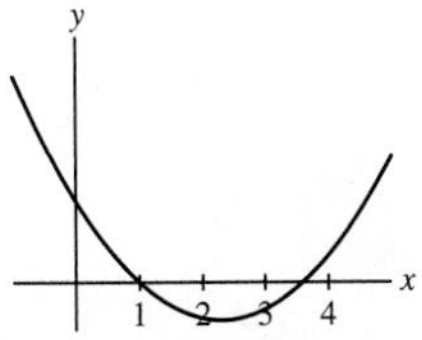

69. For $x < 0$, $f(x) = -x^2$, and $f'(x) = -2x$. For $x > 0$, $f(x) = x^2$, and $f'(x) = 2x$. Thus, $f'(0) = 0$.

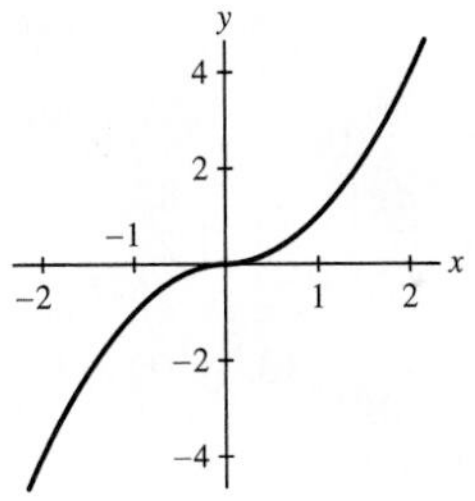

71. $c = 1$ **73.** $c = 0$ **75.** $c = \pm 1$

77. It appears that f is not differentiable at $a = 0$. Moreover, the tangent line does not exist at this point.

79. It appears that f is not differentiable at $a = 3$. Moreover, the tangent line appears to be vertical.

81. It appears that f is not differentiable at $a = 0$. Moreover, the tangent line does not exist at this point.

83. The graph of $f'(x)$ is shown in the figure below at the left and it is clear that $f'(x) > 0$ for all $x > 0$. The positivity of $f'(x)$ tells us that the graph of $f(x)$ is increasing for $x > 0$.

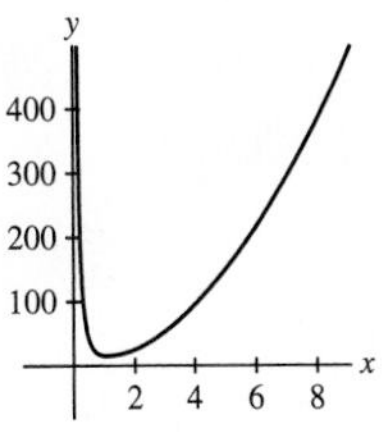

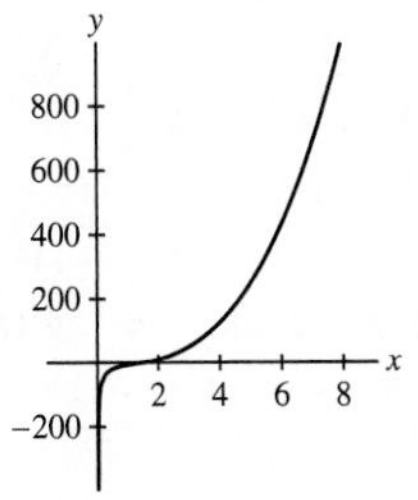

85. $\dfrac{10}{7}$

87. The normal line intersects the x-axis at the point T with coordinates $(x + f(x)f'(x), 0)$. The point R has coordinates $(a, 0)$, so the subnormal is $|x + f(x)f'(x) - x| = |f(x)f'(x)|$.

89. The tangent line to f at $x = a$ is $y = 2ax - a^2$. The x-intercept of this line is $\frac{a}{2}$ so the subtangent is $a - a/2 = a/2$.

91. The subtangent is $\frac{1}{n}a$. **93.** $r \le \frac{1}{2}$

Section 3.3 Preliminary Questions

1. (a) False. The notation fg denotes the function whose value at x is $f(x)g(x)$.

(b) True.

(c) False. The derivative of a product fg is $f'(x)g(x) + f(x)g'(x)$.

(d) False. $\left.\frac{d}{dx}(fg)\right|_{x=4} = f(4)g'(4) + g(4)f'(4)$.

(e) True.

2. -1 **3.** 5

Section 3.3 Exercises

1. $f'(x) = 10x^4 + 3x^2$

3. $f'(x) = \sqrt{x}(-3x^2) + \frac{1}{2\sqrt{x}}(1 - x^3) = \frac{1 - 7x^3}{2\sqrt{x}}$

5. $\frac{dh}{ds} = -\frac{7}{2}s^{-3/2} + \frac{3}{2}s^{-5/2} + 14$; $\left.\frac{dh}{ds}\right|_{s=4} = \frac{871}{64}$

7. $f'(x) = \frac{-2}{(x-2)^2}$

9. $\frac{dg}{dt} = -\frac{4t}{(t^2-1)^2}$; $\left.\frac{dg}{dt}\right|_{t=-2} = \frac{8}{9}$

11. $g'(x) = -\frac{3\sqrt{x}}{2(1 + x^{3/2})^2}$ **13.** $f'(t) = 6t^2 + 2t - 4$

15. $h'(t) = 1$ for $t \ne 1$

17. $f'(x) = 6x^5 + 4x^3 + 18x^2 + 5$

19. $\frac{dy}{dx} = -\frac{1}{(x+10)^2}$; $\left.\frac{dy}{dx}\right|_{x=3} = -\frac{1}{169}$

21. $f'(x) = 1$

23. $\frac{dy}{dx} = \frac{2x^5 - 20x^3 + 8x}{(x^2-5)^2}$; $\left.\frac{dy}{dx}\right|_{x=2} = -80$

25. $\frac{dz}{dx} = -\frac{3x^2}{(x^3+1)^2}$; $\left.\frac{dz}{dx}\right|_{x=1} = -\frac{3}{4}$

27. $h'(t) = \frac{-2t^3 - t^2 + 1}{(t^3 + t^2 + t + 1)^2}$

29. $f'(t) = 0$

31. $f'(x) = 3x^2 - 6x - 13$

33. $f'(x) = \frac{\sqrt{x}(5x^3 + 7x^2 + x + 3)}{2(x+1)^2}$

35. For $z \ne -2$ and $z \ne 1$, $g'(z) = 2z - 1$

37. $f'(t) = \frac{-xt^2 + 8t - x^2}{(t^2 - x)^2}$

39. $(fg)'(4) = -20$ and $(f/g)'(4) = 0$

41. $G'(4) = -10$ **43.** $F'(0) = -7$

45.

$$\begin{aligned}(x^3)' &= (x \cdot x \cdot x)' = x' \cdot (x \cdot x) + x \cdot (x \cdot x)' \\ &= 1 \cdot (x \cdot x) + x \cdot (x' \cdot x + x \cdot x') \\ &= x \cdot x + x \cdot (1 \cdot x + x \cdot 1) = 3x \cdot x = 3x^2\end{aligned}$$

47. From the plot of $f(x)$ shown below, we see that $f(x)$ is decreasing on its domain $\{x : x \ne \pm 1\}$. Consequently, $f'(x)$ must be negative. Using the quotient rule, we find

$$f'(x) = \frac{(x^2-1)(1) - x(2x)}{(x^2-1)^2} = -\frac{x^2+1}{(x^2-1)^2},$$

which is negative for all $x \ne \pm 1$.

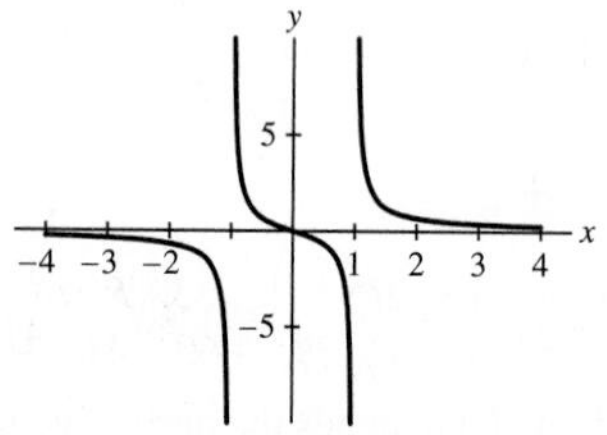

49. $a = -2, 4$

51. (a) Given $R(t) = N(t)S(t)$, it follows that

$$\frac{dR}{dt} = N(t)S'(t) + S(t)N'(t).$$

(b) $\left.\frac{dR}{dt}\right|_{t=0} = 1,250,000$

(c) The term $5S(0)$ is larger than the term $10,000N(0)$. Thus, if only one leg of the campaign can be implemented, it should be part A: increase the number of stores by 5 per month.

53.
- At $x = -1$, the tangent line is $y = \frac{1}{2}x + 1$
- At $x = 1$, the tangent line is $y = -\frac{1}{2}x + 1$

55. Let $g = f^2 = ff$. Then $g' = \left(f^2\right)' = (ff)' = ff' + ff' = 2ff'$.

57. Let $p = fgh$. Then $p' = (fgh)' = f\left(gh' + hg'\right) + ghf' = f'gh + fg'h + fgh'$.

61.

$$\begin{aligned}\frac{d}{dx}(xf(x)) &= \lim_{h\to 0}\frac{(x+h)f(x+h) - f(x)}{h} \\ &= \lim_{h\to 0}\left(x\frac{f(x+h)-f(x)}{h} + f(x+h)\right) \\ &= x\lim_{h\to 0}\frac{f(x+h)-f(x)}{h} + \lim_{h\to 0} f(x+h) \\ &= xf'(x) + f(x).\end{aligned}$$

65. (a) Is a multiple root **(b)** Not a multiple root

Section 3.4 Preliminary Questions

1. **(a)** atmospheres/meter. **(b)** moles/(liter·hour).

2. 90 mph

3. $f(26) \approx 43.75$

4. **(a)** $P'(2009)$ measures the rate of change of the population of Freedonia in the year 2009.

(b) $P(2010) \approx 5.2$ million.

Section 3.4 Exercises

1. 10 square units per unit increase

3.

c	ROC of $f(x)$ with respect to x at $x = c$.
1	$f'(1) = \frac{1}{3}$
8	$f'(8) = \frac{1}{12}$
27	$f'(27) = \frac{1}{27}$

5. $d' = 2$ **7.** $dV/dr = 3\pi r^2$

9. **(a)** 100 km/hour **(b)** 100 km/hour **(c)** 0 km/hour **(d)** -50 km/hour

11. **(a)** (i) **(b)** (ii) **(c)** (iii)

13. $\dfrac{dT}{dt} \approx -1.5625°\text{C/hour}$

15. -8×10^{-6} 1/s

17. $\left.\dfrac{dT}{dh}\right|_{h=30} \approx 2.94°\text{C/km}$; $\left.\dfrac{dT}{dh}\right|_{h=70} \approx -3.33°\text{C/km}$; $\dfrac{dT}{dh} = 0$ over the interval $[13, 23]$, and near the points $h = 50$ and $h = 90$.

19. $v'_{\text{esc}}(r) = -1.41 \times 10^7 r^{-3/2}$

21. $t = \frac{5}{2}$ s

23. The particle passes through the origin when $t = 0$ seconds and when $t = 3\sqrt{2} \approx 4.24$ seconds. The particle is instantaneously motionless when $t = 0$ seconds and when $t = 3$ seconds.

25. Maximum velocity: 200 m/s; maximum height: 2040.82 m

27. Initial velocity: $v_0 = 19.6$ m/s; maximum height: 19.6 m

31. **(a)** $\dfrac{dV}{dv} = -1$ **(b)** -4

35. Rate of change of BSA with respect to mass: $\dfrac{\sqrt{5}}{20\sqrt{m}}$; $m = 70$ kg, rate of change is $\approx 0.0133631 \frac{\text{m}^2}{\text{kg}}$; $m = 80$ kg, rate of change is $\frac{1}{80} \frac{\text{m}^2}{\text{kg}}$; BSA increases more rapidly at lower body mass.

37. 2

39. $\sqrt{2} - \sqrt{1} \approx \frac{1}{2}$; the actual value, to six decimal places, is 0.414214. $\sqrt{101} - \sqrt{100} \approx .05$; the actual value, to six decimal places, is 0.0498756.

41.
- $F(65) = 282.75$ ft
- Increasing speed from 65 to 66 therefore increases stopping distance by approximately 7.6 ft.
- The actual increase in stopping distance when speed increases from 65 mph to 66 mph is $F(66) - F(65) = 290.4 - 282.75 = 7.65$ feet, which differs by less than one percent from the estimate found using the derivative.

43. The cost of producing 2000 bagels is \$796. The cost of the 2001st bagel is approximately \$0.244, which is indistinguishable from the estimated cost.

45. An increase in oil prices of a dollar leads to a decrease in demand of 0.5625 barrels a year, and a decrease of a dollar leads to an *increase* in demand of 0.5625 barrels a year.

47. $\dfrac{dB}{dI} = \dfrac{2k}{3I^{1/3}}$; $\dfrac{dH}{dW} = \dfrac{3k}{2}W^{1/2}$

(a) As I increases, $\frac{dB}{dI}$ shrinks, so that the rate of change of perceived intensity decreases as the actual intensity increases.

(b) As W increases, $\frac{dH}{dW}$ increases as well, so that the rate of change of perceived weight increases as weight increases.

49. **(a)** The average income among households in the bottom rth part is

$$\frac{F(r)T}{rN} = \frac{F(r)}{r} \cdot \frac{T}{N} = \frac{F(r)}{r}A.$$

(b) The average income of households belonging to an interval $[r, r + \Delta r]$ is equal to

$$\frac{F(r + \Delta r)T - F(r)T}{\Delta r N} = \frac{F(r + \Delta r) - F(r)}{\Delta r} \cdot \frac{T}{N} = \frac{F(r + \Delta r) - F(r)}{\Delta r}A$$

(c) Take the result from part (b) and let $\Delta r \to 0$. Because

$$\lim_{\Delta r \to 0} \frac{F(r + \Delta r) - F(r)}{\Delta r} = F'(r),$$

we find that a household in the 100rth percentile has income $F'(r)A$.

(d) The point P in Figure 14(B) has an r-coordinate of 0.6, while the point Q has an r-coordinate of roughly 0.75. Thus, on curve L_1, 40% of households have $F'(r) > 1$ and therefore have above-average income. On curve L_2, roughly 25% of households have above-average income.

53. By definition, the slope of the line through $(0, 0)$ and $(x, C(x))$ is

$$\frac{C(x) - 0}{x - 0} = \frac{C(x)}{x} = C_{\text{avg}}(x).$$

- At point A, average cost is greater than marginal cost.
- At point B, average cost is greater than marginal cost.
- At point C, average cost and marginal cost are nearly the same.
- At point D, average cost is less than marginal cost.

Section 3.5 Preliminary Questions

1. The first derivative of stock prices must be positive, while the second derivative must be negative.

2. True

3. All quadratic polynomials

4. e^x

Section 3.5 Exercises

1. $y'' = 28$ and $y''' = 0$

3. $y'' = 12x^2 - 50$ and $y''' = 24x$

5. $y'' = 8\pi r$ and $y''' = 8\pi$

7. $y'' = -\frac{16}{5}t^{-6/5} + \frac{4}{3}t^{-4/3}$ and $y''' = \frac{96}{25}t^{-11/15} - \frac{16}{9}t^{-7/3}$

9. $y'' = -8z^{-3}$ and $y''' = 24z^{-4}$

11. $y'' = 12\theta + 14$ and $y''' = 12$

13. $y'' = -8x^{-3}$ and $y''' = 24x^{-4}$

15. $y'' = \frac{1}{4}(3s^{-5/2} - s^{-3/2})$ and $y''' = \frac{1}{8}(3s^{-5/2} - 15s^{-7/2})$

17. $f^{(4)}(1) = 24$ **19.** $\left.\frac{d^2y}{dt^2}\right|_{t=1} = 54$

21. $\left.\frac{d^4x}{dt^4}\right|_{t=16} = \frac{3465}{134217728}$ **23.** $f'''(-3) = -\frac{62}{9}$

25. $h''(1) = \frac{1}{8}$

27. $y^{(0)}(0) = d$, $y^{(1)}(0) = c$, $y^{(2)}(0) = 2b$, $y^{(3)}(0) = 6a$, $y^{(4)}(0) = 24$, and $y^{(5)}(0) = 0$

29. $\frac{d^6}{dx^6}x^{-1} = 720x^{-7}$

31. $f^{(n)}(x) = (-1)^n(n+1)!x^{-(n+2)}$

33. $f^{(n)}(x) = (-1)^n \frac{(2n-1)\times(2n-3)\times\ldots\times 1}{2^n} x^{-(2n+1)/2}$

35. $f^{(n)}(x) = (-1)^n \frac{n!(x+n+1)}{x^{n+2}}$

37. (a) $a(5) = -120 \text{ m/min}^2$

(b) The acceleration of the helicopter for $0 \le t \le 6$ is shown in the figure below. As the acceleration of the helicopter is negative, the velocity of the helicopter must be decreasing. Because the velocity is positive for $0 \le t \le 6$, the helicopter is slowing down.

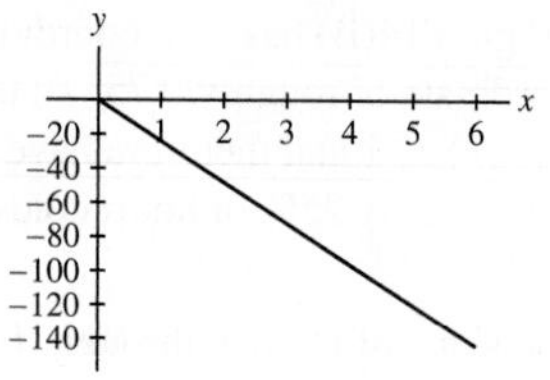

39. (a) f'' (b) f' (c) f

41. Roughly from time 10 to time 20 and from time 30 to time 40

43. $n = -1, 4$

45. (a) $v(t) = -5.12$ m/s **(b)** $v(t) = -7.25$ m/s

47. A possible plot of the drill bit's vertical velocity follows:

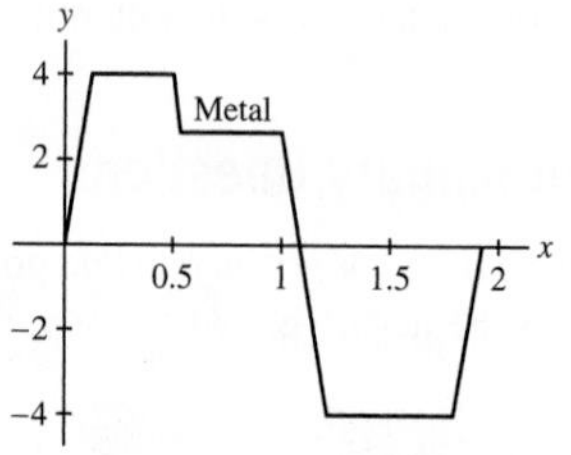

A graph of the acceleration is extracted from this graph:

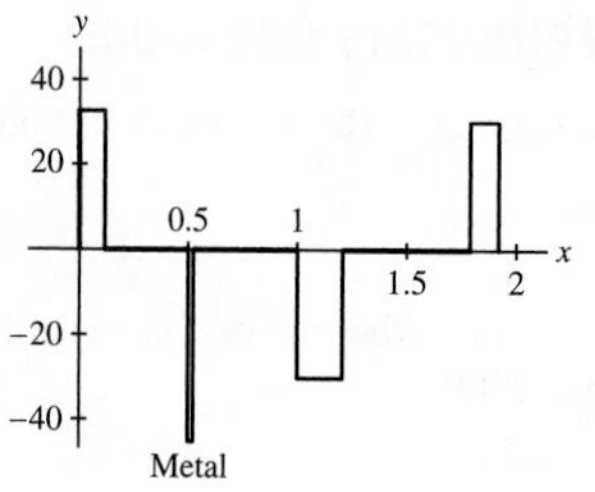

49. (a) Traffic speed must be reduced when the road gets more crowded so we expect $\frac{dS}{dQ}$ to be negative.

(b) The decrease in speed due to a one-unit increase in density is approximately $\frac{dS}{dQ}$ (a negative number). Since $\frac{d^2S}{dQ^2} = 5764Q^{-3} > 0$ is positive, this tells us that $\frac{dS}{dQ}$ gets larger as Q increases.

(c) dS/dQ is plotted below. The fact that this graph is increasing shows that $d^2S/dQ^2 > 0$.

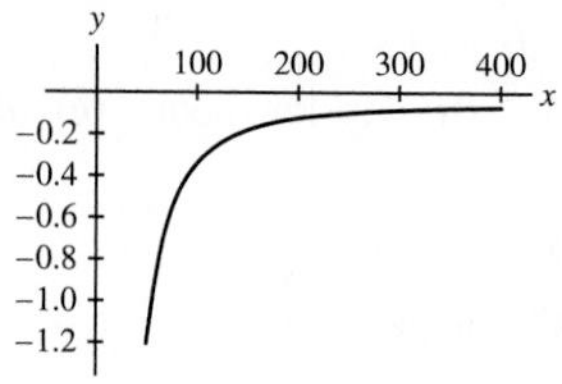

51.

$$f'(x) = -\frac{3}{(x-1)^2} = (-1)^1\frac{3\cdot 1}{(x-1)^{1+1}};$$
$$f''(x) = \frac{6}{(x-1)^3} = (-1)^2\frac{3\cdot 2\cdot 1}{(x-1)^{2+1}};$$
$$f'''(x) = -\frac{18}{(x-1)^4} = (-1)^3\frac{3\cdot 3!}{(x-1)^{3+1}}; \text{ and}$$
$$f^{(4)}(x) = \frac{72}{(x-1)^5} = (-1)^4\frac{3\cdot 4!}{(x-1)^{4+1}}.$$

From the pattern observed above, we conjecture

$$f^{(k)}(x) = (-1)^k\frac{3\cdot k!}{(x-1)^{k+1}}.$$

53. 99!

55. $(fg)''' = f'''g + 3f''g' + 3f'g'' + fg'''$;

$$(fg)^{(n)} = \sum_{k=0}^{n}\binom{n}{k}f^{(n-k)}g^{(k)}$$

57.

$$f'(x) = x^2e^x + 2xe^x = (x^2+2x)e^x;$$
$$f''(x) = (x^2+2x)e^x + (2x+2)e^x = (x^2+4x+2)e^x;$$
$$f'''(x) = (x^2+4x+2)e^x + (2x+4)e^x = (x^2+6x+6)e^x;$$
$$f^{(4)}(x) = (x^2+6x+6)e^x + (2x+6)e^x = (x^2+8x+12)e^x.$$

From this information, we conjecture that the general formula is

$$f^{(n)}(x) = (x^2 + 2nx + n(n-1))e^x.$$

Section 3.6 Preliminary Questions

1. (a) $\frac{d}{dx}(\sin x + \cos x) = -\sin x + \cos x$

(b) $\frac{d}{dx}\sec x = \sec x \tan x$

(c) $\frac{d}{dx}\cot x = -\csc^2 x$

2. (a) This function can be differentiated using the Product Rule.

(b) We have not yet discussed how to differentiate a function like this.

(c) This function can be differentiated using the Product Rule.

3. 0

4. The difference quotient for the function $\sin x$ involves the expression $\sin(x+h)$. The addition formula for the sine function is used to expand this expression as $\sin(x+h) = \sin x \cos h + \sin h \cos x$.

Section 3.6 Exercises

1. $y = \frac{\sqrt{2}}{2}x + \frac{\sqrt{2}}{2}\left(1 - \frac{\pi}{4}\right)$ **3.** $y = 2x + 1 - \frac{\pi}{2}$

5. $f'(x) = -\sin^2 x + \cos^2 x$ **7.** $f'(x) = 2\sin x \cos x$

9. $H'(t) = 2\sin t \sec^2 t \tan t + \sec t$

11. $f'(\theta) = \left(\tan^2\theta + \sec^2\theta\right)\sec\theta$

13. $f'(x) = (2x^4 - 4x^{-1})\sec x \tan x + \sec x(8x^3 + 4x^{-2})$

15. $y' = \dfrac{\theta\sec\theta\tan\theta - \sec\theta}{\theta^2}$ **17.** $R'(y) = \dfrac{4\cos y - 3}{\sin^2 y}$

19. $f'(x) = \dfrac{2\sec^2 x}{(1-\tan x)^2}$ **21.** $f'(x) = \dfrac{-2\cos x}{(\sin x - 1)^2}$

23. $R'(\theta) = \dfrac{-4\sin\theta}{(4+\cos\theta)^2}$

25. $y = 1$ **27.** $y = x + 3$

29. $y = (1-\sqrt{3})\left(x - \frac{\pi}{3}\right) + 1 + \sqrt{3}$

31. $y = -\frac{7}{6}t + \frac{3\sqrt{3}+7\pi}{18}$ **33.** $y = 6\pi x - \frac{9}{2}\pi^2$

35. $\cot x = \frac{\cos x}{\sin x}$; use the quotient rule

37. $\csc x = \frac{1}{\sin x}$; use the quotient rule

39. $f''(\theta) = -\theta\sin\theta + 2\cos\theta$

41.

$$y'' = 2\sec^2 x \tan x$$
$$y''' = 2\sec^4 x + 4\sec^2 x \tan^2 x.$$

43.
- Then $f'(x) = -\sin x$, $f''(x) = -\cos x$, $f'''(x) = \sin x$, $f^{(4)}(x) = \cos x$, and $f^{(5)}(x) = -\sin x$.
- Accordingly, the successive derivatives of f cycle among

$$\{-\sin x, -\cos x, \sin x, \cos x\}$$

in that order. Since 8 is a multiple of 4, we have $f^{(8)}(x) = \cos x$.
- Since 36 is a multiple of 4, we have $f^{(36)}(x) = \cos x$. Therefore, $f^{(37)}(x) = -\sin x$.

45. $x = \frac{\pi}{4}, \frac{3\pi}{4}, \frac{5\pi}{4}, \frac{7\pi}{4}$

47. (a)

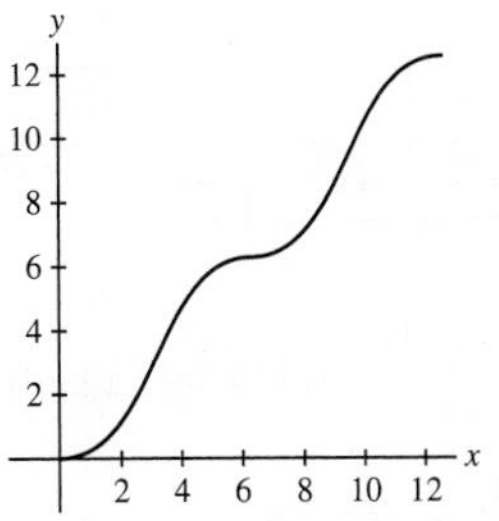

(b) Since $g'(t) = 1 - \cos t \geq 0$ for all t, the slope of the tangent line to g is always nonnegative.

(c) $t = 0, 2\pi, 4\pi$

49. $f'(x) = \sec^2 x = \frac{1}{\cos^2 x}$. Note that $f'(x) = \frac{1}{\cos^2 x}$ has numerator 1; the equation $f'(x) = 0$ therefore has no solution. The least slope for a tangent line to $\tan x$ is 1. Here is a graph of f'.

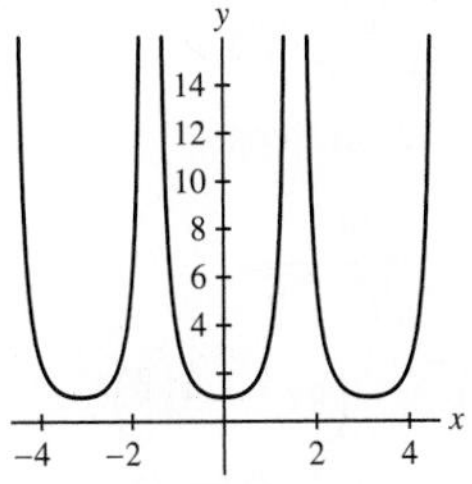

51. $\frac{dR}{d\theta} = (v_0^2/9.8)(-\sin^2\theta + \cos^2\theta)$; if $\theta = 7\pi/24$, increasing the angle will *decrease* the range.

53.

$$f'(x) = \lim_{h\to 0}\frac{\cos(x+h) - \cos x}{h} = \lim_{h\to 0}\frac{\cos x\cos h - \sin x \sin h - \cos x}{h}$$
$$= \lim_{h\to 0}\left((-\sin x)\frac{\sin h}{h} + (\cos x)\frac{\cos h - 1}{h}\right)$$
$$= (-\sin x)\cdot 1 + (\cos x)\cdot 0 = -\sin x.$$

Section 3.7 Preliminary Questions

1. (a) The outer function is $\sqrt{x}$, and the inner function is $4x + 9x^2$.

(b) The outer function is $\tan x$, and the inner function is $x^2 + 1$.

(c) The outer function is x^5, and the inner function is $\sec x$.

(d) The outer function is x^4, and the inner function is $1 + x^{12}$.

2. The function $\frac{x}{x+1}$ can be differentiated using the Quotient Rule, and the functions $\sqrt{x}\cdot\sec x$ and xe^x can be differentiated using the Product Rule. The functions $\tan(7x^2+2)$, $\sqrt{x\cos x}$ and $e^{\sin x}$ require the Chain Rule

3. (b)

4. We do not have enough information to compute $F'(4)$. We are missing the value of $f'(1)$.

Section 3.7 Exercises

1.

$f(g(x))$	$f'(u)$	$f'(g(x))$	$g'(x)$	$(f \circ g)'$
$(x^4+1)^{3/2}$	$\frac{3}{2}u^{1/2}$	$\frac{3}{2}(x^4+1)^{1/2}$	$4x^3$	$6x^3(x^4+1)^{1/2}$

3.

$f(g(x))$	$f'(u)$	$f'(g(x))$	$g'(x)$	$(f \circ g)'$
$\tan(x^4)$	$\sec^2 u$	$\sec^2(x^4)$	$4x^3$	$4x^3 \sec^2(x^4)$

5. $4(x+\sin x)^3(1+\cos x)$

7. **(a)** $2x\sin(9-x^2)$ **(b)** $\dfrac{\sin(x^{-1})}{x^2}$ **(c)** $-\sec^2 x \sin(\tan x)$

9. 12 **11.** $12x^3(x^4+5)^2$ **13.** $\dfrac{7}{2\sqrt{7x-3}}$

15. $-2(x^2+9x)^{-3}(2x+9)$ **17.** $-4\cos^3\theta\sin\theta$

19. $9(2\cos\theta+5\sin\theta)^8(5\cos\theta-2\sin\theta)$

21. $(x+1)(x^2+2x+9)^{-1/2}\cos\sqrt{x^2+2x+9}$.

23. $2\cos(2x+1)$ **25.** $\sec^2 x - \csc^2 x$

27.

$$\frac{d}{dx}f(g(x)) = -\sin(x^2+1)(2x) = -2x\sin(x^2+1)$$

$$\frac{d}{dx}g(f(x)) = -2\sin x\cos x$$

29. $2x\cos\left(x^2\right)$ **31.** $\dfrac{t}{\sqrt{t^2+9}}$

33. $\dfrac{2}{3}\left(x^4-x^3-1\right)^{-1/3}\left(4x^3-3x^2\right)$

35. $\dfrac{8(1+x)^3}{(1-x)^5}$ **37.** $-\dfrac{\sec(1/x)\tan(1/x)}{x^2}$

39. $(1-\sin\theta)\sec^2(\theta+\cos\theta)$

41. $4\theta\cot(2\theta^2-9)\csc(2\theta^2-9)$

43. $(2x+4)\sec^2(x^2+4x)$

45. $3x\sin(1-3x)+\cos(1-3x)$

47. $2(4t+9)^{-1/2}$ **49.** $4(\sin x-3x^2)(x^3+\cos x)^{-5}$

51. $\dfrac{\cos 2x}{\sqrt{2\sin 2x}}$ **53.** $\dfrac{x\cos(x^2)-3\sin 6x}{\sqrt{\cos 6x+\sin(x^2)}}$

55. $3\left(x^2\sec^2(x^3)+\sec^2 x\tan^2 x\right)$ **57.** $\dfrac{-1}{\sqrt{z+1}\,(z-1)^{3/2}}$

59. $\dfrac{\sin(-1)-\sin(1+x)}{(1+\cos x)^2}$ **61.** $-35x^4\cot^6\left(x^5\right)\csc^2\left(x^5\right)$

63. $-180x^3\cot^4\left(x^4+1\right)\csc^2\left(x^4+1\right)\left(1+\cot^5\left(x^4+1\right)\right)^8$

65. $-36x^2(1-\csc^2(1-x^3))^5\csc^2(1-x^3)\cot(1-x^3)$

67. $\dfrac{1-x^2}{2x^2}\left(x+\dfrac{1}{x}\right)^{-3/2}$

69. $\dfrac{1}{8\sqrt{x}\sqrt{1+\sqrt{x}}\sqrt{1+\sqrt{1+\sqrt{x}}}}$

71. $-\dfrac{k}{3}(kx+b)^{-4/3}$ **73.** $2\cos\left(x^2\right)-4x^2\sin\left(x^2\right)$

75. $-336(9-x)^5$

77. $\left.\dfrac{dv}{dP}\right|_{P=1.5} = \dfrac{290\sqrt{3}}{3}\,\dfrac{\text{m}}{\text{s}\cdot\text{atmospheres}}$

79. **(a)** When $r=3$, $\dfrac{dV}{dt} = 1.6\pi(3)^2 \approx 45.24$ cm/s.

(b) When $t=3$, we have $r=1.2$. Hence $\dfrac{dV}{dt} = 1.6\pi(1.2)^2 \approx 7.24$ cm/s.

81. $W'(10) \approx 0.3600$ kg/yr **83.** **(a)** $\dfrac{\pi}{360}$ **(b)** $1+\dfrac{\pi}{90}$

85. $5\sqrt{3}$ **87.** 12 **89.** $\dfrac{1}{16}$

91. $\left.\dfrac{dP}{dt}\right|_{t=3} = -0.727\ \dfrac{\text{dollars}}{\text{year}}$

93. $\dfrac{dP}{dh} = -4.03366\times10^{-16}(288.14-0.000649\,h)^{4.256}$; for each additional meter of altitude, $\Delta P \approx -1.15\times10^{-2}$ Pa

95. 0.0973 kelvins/yr

97. $f'(g(x))g''(x)+f''(g(x))\left(g'(x)\right)^2$

99. Let $u=h(x)$, $v=g(u)$, and $w=f(v)$. Then

$$\frac{dw}{dx} = \frac{df}{dv}\frac{dv}{dx} = \frac{df}{dv}\frac{dv}{du}\frac{du}{dx} = f'(g(h(x))g'(h(x))h'(x)$$

103. For $n=1$, we find

$$\frac{d}{dx}\sin x = \cos x = \sin\left(x+\frac{\pi}{2}\right),$$

as required. Now, suppose that for some positive integer k,

$$\frac{d^k}{dx^k}\sin x = \sin\left(x+\frac{k\pi}{2}\right).$$

Then

$$\frac{d^{k+1}}{dx^{k+1}}\sin x = \frac{d}{dx}\sin\left(x+\frac{k\pi}{2}\right)$$

$$= \cos\left(x+\frac{k\pi}{2}\right) = \sin\left(x+\frac{(k+1)\pi}{2}\right).$$

Section 3.8 Preliminary Questions

1. The chain rule

2. **(a)** This is correct **(b)** This is correct

(c) This is incorrect. Because the differentiation is with respect to the variable x, the chain rule is needed to obtain

$$\frac{d}{dx}\sin(y^2) = 2y\cos(y^2)\frac{dy}{dx}.$$

3. There are two mistakes in Jason's answer. First, Jason should have applied the product rule to the second term to obtain

$$\frac{d}{dx}(2xy) = 2x\frac{dy}{dx}+2y.$$

Second, he should have applied the general power rule to the third term to obtain

$$\frac{d}{dx}y^3 = 3y^2\frac{dy}{dx}.$$

4. **(b)**

Section 3.8 Exercises

1. $(2, 1), \dfrac{dy}{dx} = -\dfrac{2}{3}$

3. $\dfrac{d}{dx}\left(x^2y^3\right) = 3x^2y^2y' + 2xy^3$

5. $\dfrac{d}{dx}\left(\left(x^2+y^2\right)^{3/2}\right) = 3\left(x + yy'\right)\sqrt{x^2+y^2}$

7. $\dfrac{d}{dx}\dfrac{y}{y+1} = \dfrac{y'}{(y+1)^2}$ **9.** $y' = -\dfrac{2x}{9y^2}$

11. $y' = \dfrac{1-2xy-6x^2y}{x^2+2x^3-1}$ **13.** $R' = -\dfrac{3R}{5x}$

15. $y' = \dfrac{y(y^2-x^2)}{x(y^2-x^2-2xy^2)}$ **17.** $y' = \dfrac{9}{4}x^{1/2}y^{5/3}$

19. $y' = \dfrac{(2x+1)y^2}{y^2-1}$ **21.** $y' = \dfrac{1-\cos(x+y)}{\cos(x+y)+\sin y}$

23. $y' = -\dfrac{\tan^2(x+y)-\tan^2(x)}{\tan^2(x+y)-\tan^2(y)}$

25. $y' = \dfrac{y+3\sin(3x-y)-1}{\sin(3x-y)-x}$ **29.** $y' = \dfrac{1}{4}$ **31.** $y = -\dfrac{1}{2}x + 2$

33. $y = -2x + 2$ **35.** $y = -\dfrac{12}{5}x + \dfrac{32}{5}$ **37.** $y = 2x + \pi$

39. The tangent is horizontal at the points $(-1, \sqrt{3})$ and $(-1, -\sqrt{3})$

41. The tangent line is horizontal at

$$\left(\frac{2\sqrt{78}}{13}, -\frac{4\sqrt{78}}{13}\right) \quad\text{and}\quad \left(-\frac{2\sqrt{78}}{13}, \frac{4\sqrt{78}}{13}\right).$$

43.
- When $y = 2^{1/4}$, we have

$$y' = \frac{-2^{1/4}-1}{4\left(2^{3/4}\right)} = -\frac{\sqrt{2}+\sqrt[4]{2}}{8} \approx -0.3254.$$

When $y = -2^{1/4}$, we have

$$y' = \frac{2^{1/4}-1}{-4\left(2^{3/4}\right)} = -\frac{\sqrt{2}-\sqrt[4]{2}}{8} \approx -0.02813.$$

- At the point $(1, 1)$, the tangent line is $y = \dfrac{1}{5}x + \dfrac{4}{5}$.

45. $(2^{1/3}, 2^{2/3})$ **47.** $x = \dfrac{1}{2}, 1 \pm \sqrt{2}$

49.
- At $(1, 2)$, $y' = \dfrac{1}{3}$
- At $(1, -2)$, $y' = -\dfrac{1}{3}$
- At $(1, \frac{1}{2})$, $y' = \dfrac{11}{12}$
- At $(1, -\frac{1}{2})$, $y' = -\dfrac{11}{12}$

51. $\dfrac{dx}{dy} = \dfrac{y(2y^2-1)}{x}$; The tangent line is vertical at:

$$(1,0), (-1,0), \left(\frac{\sqrt{3}}{2}, \frac{\sqrt{2}}{2}\right),$$
$$\left(-\frac{\sqrt{3}}{2}, \frac{\sqrt{2}}{2}\right), \left(\frac{\sqrt{3}}{2}, -\frac{\sqrt{2}}{2}\right), \left(-\frac{\sqrt{3}}{2}, -\frac{\sqrt{2}}{2}\right).$$

53. $\dfrac{dx}{dy} = \dfrac{2y}{3x^2-4}$; it follows that $\frac{dx}{dy} = 0$ when $y = 0$, so the tangent line to this curve is vertical at the points where the curve intersects the x-axis.

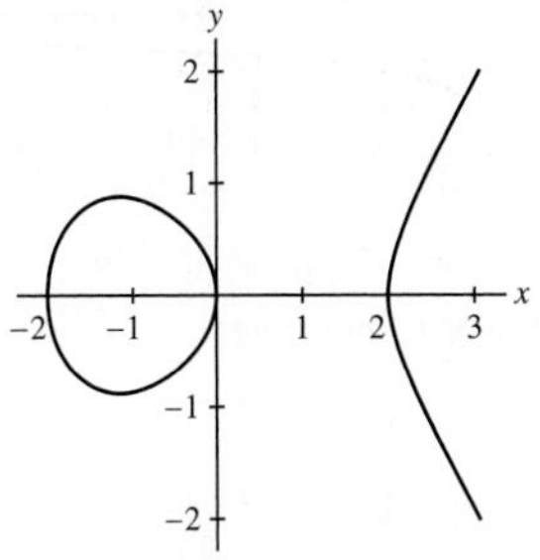

55. **(b)**: $y'' = \dfrac{y^3-2x^2}{y^5}$ **57.** $y'' = \dfrac{10}{27}$

59. $x\dfrac{dy}{dt} + y\dfrac{dx}{dt} = 0$, and $\dfrac{dy}{dt} = -\dfrac{y}{x}\dfrac{dx}{dt}$

61. **(a)** $\dfrac{dy}{dt} = \dfrac{x^2}{y^2}\dfrac{dx}{dt}$ **(b)** $\dfrac{dy}{dt} = -\dfrac{x+y}{2y^3+x}\dfrac{dx}{dt}$

63. Let $C1$ be the curve described by $x^2 - y^2 = c$, and let $C2$ be the curve described by $xy = d$. Suppose that $P = (x_0, y_0)$ lies on the intersection of the two curves $x^2 - y^2 = c$ and $xy = d$. Since $x^2 - y^2 = c$, $y' = \frac{x}{y}$. The slope to the tangent line to $C1$ is $\frac{x_0}{y_0}$. On the curve $C2$, since $xy = d$, $y' = -\frac{y}{x}$. Therefore the slope to the tangent line to $C2$ is $-\frac{y_0}{x_0}$. The two slopes are negative reciprocals of one another, hence the tangents to the two curves are perpendicular.

65.
- Upper branch:

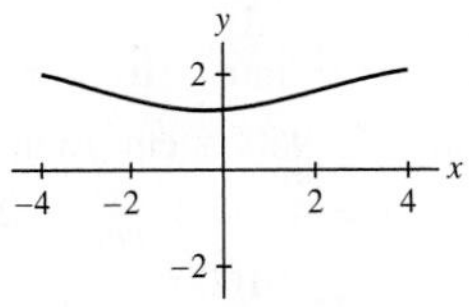

- Lower part of lower left curve:

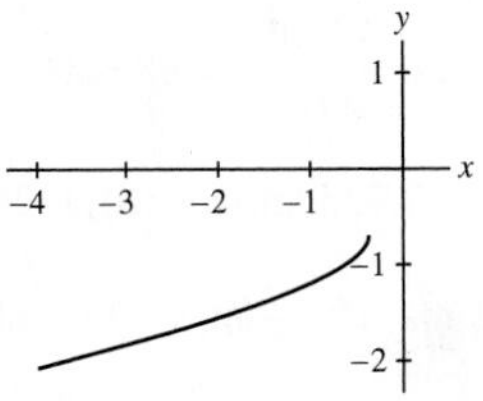

- Upper part of lower left curve:

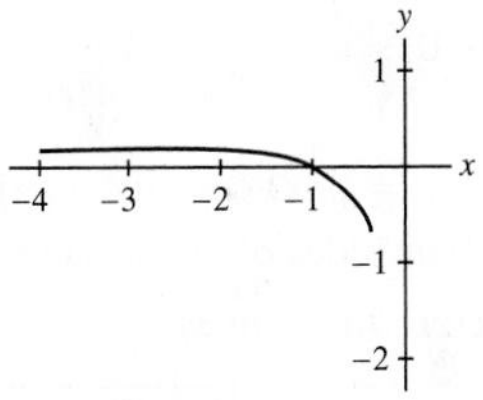

- Upper part of lower right curve:

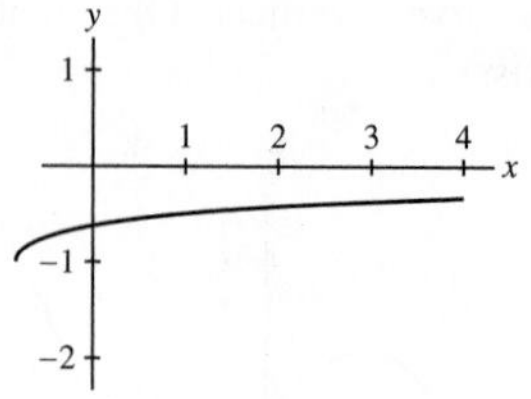

- Lower part of lower right curve:

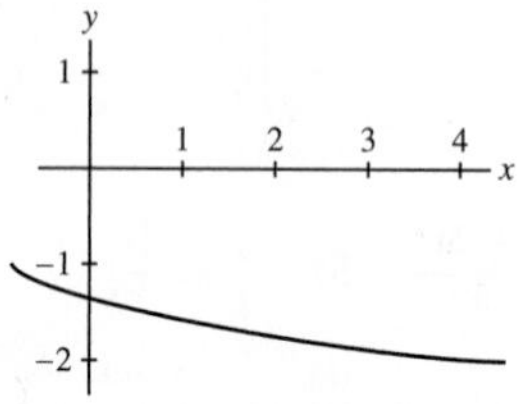

Thus,

$$L^2 = (x - r\cos\theta)^2 + r^2\sin^2\theta.$$

(b) From (a) we have

$$0 = 2(x - r\cos\theta)\left(\frac{dx}{dt} + r\sin\theta\frac{d\theta}{dt}\right) + 2r^2\sin\theta\cos\theta\frac{d\theta}{dt}.$$

(c) $-80\pi \approx -251.33$ cm/min

Section 3.9 Preliminary Questions

1. Let s and V denote the length of the side and the corresponding volume of a cube, respectively. Determine $\frac{dV}{dt}$ if $\frac{ds}{dt} = 0.5$ cm/s.

2. $\frac{dV}{dt} = 4\pi r^2\frac{dr}{dt}$

3. Determine $\frac{dh}{dt}$ if $\frac{dV}{dt} = 2$ cm^3/min

4. Determine $\frac{dV}{dt}$ if $\frac{dh}{dt} = 1$ cm/min

Section 3.9 Exercises

1. 0.039 ft/min

3. **(a)** $100\pi \approx 314.16$ m^2/min **(b)** $24\pi \approx 75.40$ m^2/min

5. 27000π cm^3/min **7.** 9600π cm^2/min

9. -0.632 m/s **11.** $x = 4.737$ m; $\frac{dx}{dt} \approx 0.405$ m/s

13. $\frac{9}{8\pi} \approx 0.36$ m/min **15.** $\frac{1000\pi}{3} \approx 1047.20$ cm^3/s

17. 0.675 meters per second

19. **(a)** 799.91 km/h **(b)** 0 km/h

21. 1.22 km/min **23.** $\frac{1200}{241} \approx 4.98$ rad/hr

25. **(a)** $\frac{100\sqrt{13}}{13} \approx 27.735$ km/h **(b)** 112.962 km/h

27. $\sqrt{16.2} \approx 4.025$ m **29.** $\frac{5}{3}$ m/s **31.** -1.92kPa/min

33. $-\frac{1}{8}$ rad/s

35. **(b)**: when $x = 1$, $L'(t) = 0$; when $x = 2$, $L'(t) = \frac{16}{3}$

37. $-4\sqrt{5} \approx -8.94$ ft/s

39. -0.79 m/min

41. Let the equation $y = f(x)$ describe the shape of the roller coaster track. Taking $\frac{d}{dt}$ of both sides of this equation yields $\frac{dy}{dt} = f'(x)\frac{dx}{dt}$.

43. **(a)** The distance formula gives

$$L = \sqrt{(x - r\cos\theta)^2 + (-r\sin\theta)^2}.$$

Chapter 3 Review

1. 3; the slope of the secant line through the points (2, 7) and (0, 1) on the graph of $f(x)$

3. $\frac{8}{3}$; the value of the difference quotient should be larger than the value of the derivative

5. $f'(1) = 1$; $y = x - 1$

7. $f'(4) = -\frac{1}{16}$; $y = -\frac{1}{16}x + \frac{1}{2}$

9. $-2x$ **11.** $\frac{1}{(2-x)^2}$ **13.** $f'(1)$ where $f(x) = \sqrt{x}$

15. $f'(\pi)$ where $f(t) = \sin t\cos t$ **17.** $f(4) = -2$; $f'(4) = 3$

19. (C) is the graph of $f'(x)$

21. **(a)** 8.05 cm/year **(b)** Larger over the first half **(c)** $h'(3) \approx 7.8$ cm/year; $h'(8) \approx 6.0$ cm/year

23. $A'(t)$ measures the rate of change in automobile production in the United States; $A'(1971) \approx 0.25$ million automobiles/year; $A'(1974)$ would be negative

25. $15x^4 - 14x$ **27.** $-7.3t^{-8.3}$ **29.** $\frac{1 - 2x - x^2}{(x^2+1)^2}$

31. $6(4x^3 - 9)(x^4 - 9x)^5$ **33.** $27x(2 + 9x^2)^{1/2}$

35. $\frac{2-z}{2(1-z)^{3/2}}$ **37.** $2x - \frac{3}{2}x^{-5/2}$

39.
$$\frac{1}{2}\left(x + \sqrt{x + \sqrt{x}}\right)^{-1/2}\left(1 + \frac{1}{2}\left(x + \sqrt{x}\right)^{-1/2}\left(1 + \frac{1}{2}x^{-1/2}\right)\right)$$

41. $-3t^{-4}\sec^2(t^{-3})$

43. $-6\sin^2 x\cos^2 x + 2\cos^4 x$

45. $\frac{1 + \sec t - t\sec t\tan t}{(1+\sec t)^2}$ **47.** $\frac{8\csc^2\theta}{(1+\cot\theta)^2}$

49. $-\frac{\sec^2(\sqrt{1+\csc\theta})\csc\theta\cot\theta}{2(\sqrt{1+\csc\theta})}$

51. -27 **53.** $-\frac{57}{16}$ **55.** -18

57. $(1, 1)$ and $(-3^{1/3}, 3^{2/3})$ **59.** $a = \frac{1}{6}$

61. $1 \pm \frac{\sqrt{6}}{3}$ **63.** $k = -2$ (at $x = 1$)

65. $72x - 10$ **67.** $-(2x+3)^{-3/2}$

69. $8x^2\sec^2(x^2)\tan(x^2) + 2\sec^2(x^2)$ **71.** $\frac{dy}{dx} = \frac{x^2}{y^2}$

73. $\frac{dy}{dx} = \frac{y^2 + 4x}{1 - 2xy}$ **75.** $\frac{dy}{dx} = \frac{\cos(x+y)}{1 - \cos(x+y)}$

77. For the plot on the left, the red, green and blue curves, respectively, are the graphs of f, f' and f''. For the plot on the right, the green, red and blue curves, respectively, are the graphs of f, f' and f''.

79. $\frac{dR}{dp} = p\frac{dq}{dp} + q = q\frac{p}{q}\frac{dq}{dp} + q = q(E + 1)$

81. $E(150) = -3$; number of passengers increases 3% when the ticket price is lowered 1%

83. $\frac{-11\pi}{360} \approx -0.407$ cm/min

85. $\frac{640}{(336)^2} \approx 0.00567$ cm/s **87.** 0.284 m/s

Chapter 4

Section 4.1 Preliminary Questions

1. True **2.** $g(1.2) - g(1) \approx 0.8$ **3.** $f(2.1) \approx 1.3$

4. The Linear Approximation tells us that up to a small error, the change in output Δf is directly proportional to the change in input Δx when Δx is small.

Section 4.1 Exercises

1. $\Delta f \approx 0.12$ **3.** $\Delta f \approx -0.00222$ **5.** $\Delta f \approx 0.003333$

7. $\Delta f \approx 0.0074074$

9. $\Delta f \approx 0.049390$; error is 0.000610; percentage error is 1.24%

11. $\Delta f \approx -0.0245283$; error is 0.0054717; percentage error is 22.31%

13. $\Delta y \approx -0.007$ **15.** $\Delta y \approx -0.026667$

17. $\Delta f \approx 0.1$; error is 0.000980486

19. $\Delta f \approx -0.0005$; error is 3.71902×10^{-6}

21. $\Delta f \approx 0.083333$; error is 3.25×10^{-3}

23. $\Delta f \approx .023$; error is 2.03×10^{-6} **25.** $f(4.03) \approx 2.01$

27. $\sqrt{2.1} - \sqrt{2}$ is larger than $\sqrt{9.1} - \sqrt{9}$

29. $R(9) = 25110$ euros; if p is raised by 0.5 euros, then $\Delta R \approx 585$ euros; on the other hand, if p is lowered by 0.5 euros, then $\Delta R \approx -585$ euros.

31. $\Delta L \approx -0.00171$ cm

33. $P \approx 5.5 + 0.5 \cdot (-.87) = 5.065$ kilopascals

35. (a) $\Delta W \approx W'(R)\Delta x = -\frac{2wR^2}{R^3}h = -\frac{2wh}{R} \approx -0.0005wh$

(b) $\Delta W \approx -0.7$ pounds

37. (a) $\Delta h \approx 0.71$ cm **(b)** $\Delta h \approx 1.02$ cm.

(c) There is a bigger effect at higher velocities.

39. (a) If $\theta = 34°$ (i.e., $t = \frac{17}{90}\pi$), then

$$\Delta s \approx s'(t)\Delta t = \frac{625}{16}\cos\left(\frac{17}{45}\pi\right)\Delta t$$
$$= \frac{625}{16}\cos\left(\frac{17}{45}\pi\right)\Delta\theta \cdot \frac{\pi}{180} \approx 0.255\Delta\theta.$$

(b) If $\Delta\theta = 2°$, this gives $\Delta s \approx 0.51$ ft, in which case the shot would not have been successful, having been off half a foot.

41. $\Delta V \approx 4\pi(25)^2(0.5) \approx 3927$ cm^3; $\Delta S \approx 8\pi(25)(0.5) \approx 314.2$ cm^2

43. $P = 6$ atmospheres; $\Delta P \approx \pm 0.45$ atmospheres

45. $L(x) = 4x - 3$ **47.** $L(\theta) = \theta - \frac{\pi}{4} + \frac{1}{2}$

49. $L(x) = -\frac{1}{2}x + 1$ **51.** $L(x) = 1$ **53.** $L(x) = -\frac{4}{\pi^2}x + \frac{4}{\pi}$

55. $f(2) = 8$

57. $\sqrt{16.2} \approx L(16.2) = 4.025$. Graphs of $f(x)$ and $L(x)$ are shown below. Because the graph of $L(x)$ lies above the graph of $f(x)$, we expect that the estimate from the Linear Approximation is too large.

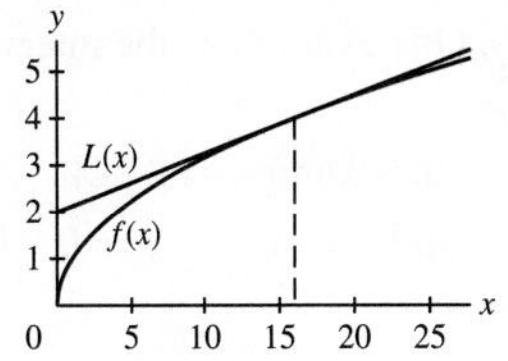

59. $\frac{1}{\sqrt{17}} \approx L(17) \approx 0.24219$; the percentage error is 0.14%

61. $\frac{1}{(10.03)^2} \approx L(10.03) = 0.00994$; the percentage error is 0.0027%

63. $(64.1)^{1/3} \approx L(64.1) \approx 4.002083$; the percentage error is 0.000019%

65. $\tan(0.04) \approx L(0.04) = 0.04$; the percentage error is .053%

67. $\frac{3.1/2}{\sin(3.1/2)} \approx L(3.1/2) = 1.55$; the percentage error is 0.0216%

69. Let $f(x) = \sqrt{x}$. Then $f(9) = 3$, $f'(x) = \frac{1}{2}x^{-1/2}$ and $f'(9) = \frac{1}{6}$. Therefore, by the Linear Approximation,

$$f(9 + h) - f(9) = \sqrt{9 + h} - 3 \approx \frac{1}{6}h.$$

Moreover, $f''(x) = -\frac{1}{4}x^{-3/2}$, so $|f''(x)| = \frac{1}{4}x^{-3/2}$. Because this is a decreasing function, it follows that for $x \geq 9$,

$$K = \max|f''(x)| \leq |f''(9)| = \frac{1}{108} < 0.01.$$

From the following table, we see that for $h = 10^{-n}$, $1 \leq n \leq 4$, $E \leq \frac{1}{2}Kh^2$.

h	$E = \lvert\sqrt{9+h} - 3 - \frac{1}{6}h\rvert$	$\frac{1}{2}Kh^2$
10^{-1}	4.604×10^{-5}	5.00×10^{-5}
10^{-2}	4.627×10^{-7}	5.00×10^{-7}
10^{-3}	4.629×10^{-9}	5.00×10^{-9}
10^{-4}	4.627×10^{-11}	5.00×10^{-11}

71. $\left.\frac{dy}{dx}\right|_{(2,1)} = -\frac{1}{3}$; $y \approx L(2.1) = 0.967$

73. $L(x) = -\frac{14}{25}x + \frac{36}{25}$; $y \approx L(-1.1) = 2.056$

75. Let $f(x) = x^2$. Then

$$\Delta f = f(5 + h) - f(5) = (5 + h)^2 - 5^2 = h^2 + 10h$$

and

$$E = |\Delta f - f'(5)h| = |h^2 + 10h - 10h| = h^2 = \frac{1}{2}(2)h^2 = \frac{1}{2}Kh^2.$$

Section 4.2 Preliminary Questions

1. A critical point is a value of the independent variable x in the domain of a function f at which either $f'(x) = 0$ or $f'(x)$ does not exist.

2. (b) **3.** (b)

4. Fermat's Theorem claims: If $f(c)$ is a local extreme value, then either $f'(c) = 0$ or $f'(c)$ does not exist.

Section 4.2 Exercises

1. (a) 3 (b) 6 (c) Local maximum of 5 at $x = 5$

(d) Answers may vary. One example is the interval [4, 8]. Another is [2, 6].

(e) Answers may vary. One example is [0, 2].

3. $x = 1$ **5.** $x = -3$ and $x = 6$ **7.** $x = 0$ **9.** $x = \pm 1$

11. $t = 3$ and $t = -1$

13. $x = 0, x = \pm\sqrt{2/3}, x = \pm 1$

15. $\theta = \dfrac{n\pi}{2}$

17. (a) $c = 2$ (b) $f(0) = f(4) = 1$

(c) Maximum value: 1; minimum value: −3.

(d) Maximum value: 1; minimum value: −2.

19. $x = \dfrac{\pi}{4}$; Maximum value: $\sqrt{2}$; minimum value: 1

21. Maximum value: 1

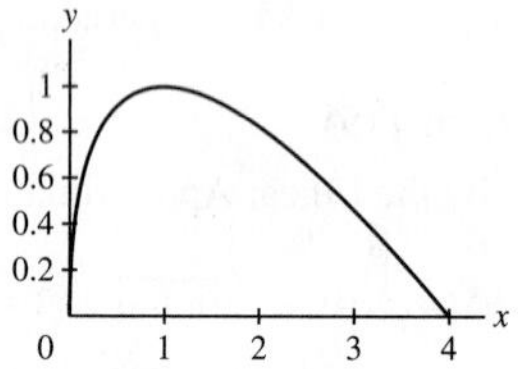

23. Local maximum at $(x, y) \approx (.860334, 0.561096)$; local minimum at $(x, y) \approx (3.425618, -3.288371)$.

25. Minimum: $f(-1) = 3$, maximum: $f(2) = 21$

27. Minimum: $f(0) = 0$, maximum: $f(3) = 9$

29. Minimum: $f(4) = -24$, maximum: $f(6) = 8$

31. Minimum: $f(1) = 5$, maximum: $f(2) = 28$

33. Minimum: $f(2) = -128$, maximum: $f(-2) = 128$

35. Minimum: $f(6) = 18.5$, maximum: $f(5) = 26$

37. Minimum: $f(1) = -1$, maximum: $f(0) = f(3) = 0$

39. Minimum: $f(0) = 2\sqrt{6} \approx 4.9$, maximum: $f(2) = 4\sqrt{2} \approx 5.66$

41. Minimum: $f\left(\dfrac{\sqrt{3}}{2}\right) \approx -0.589980$, maximum: $f(4) \approx 0.472136$

43. Minimum: $f(0) = f\left(\dfrac{\pi}{2}\right) = 0$, maximum: $f\left(\dfrac{\pi}{4}\right) = \dfrac{1}{2}$

45. Minimum: $f(0) = -1$, maximum: $f\left(\dfrac{\pi}{4}\right) = \sqrt{2}\left(\dfrac{\pi}{4} - 1\right) \approx -0.303493$

47. Minimum: $g\left(\dfrac{\pi}{3}\right) = \dfrac{\pi}{3} - \sqrt{3} \approx -0.685$, maximum: $g\left(\dfrac{5}{3}\pi\right) = \dfrac{5}{3}\pi + \sqrt{3} \approx 6.968$

49. Minimum: $f\left(\dfrac{\pi}{4}\right) = 1 - \dfrac{\pi}{2} \approx -0.570796$, maximum: $f(0) = 0$

51. (d) $\frac{\pi}{6}, \frac{\pi}{2}, \frac{5\pi}{6}, \frac{7\pi}{6}, \frac{3\pi}{2}$, and $\frac{11\pi}{6}$; the maximum value is $f\left(\frac{\pi}{6}\right) = f\left(\frac{7\pi}{6}\right) = \frac{3\sqrt{3}}{2}$ and the minimum value is $f\left(\frac{5\pi}{6}\right) = f\left(\frac{11\pi}{6}\right) = -\frac{3\sqrt{3}}{2}$

(e) We can see that there are six flat points on the graph between 0 and 2π, as predicted. There are 4 local extrema, and two points at $\left(\frac{\pi}{2}, 0\right)$ and $\left(\frac{3\pi}{2}, 0\right)$ where the graph has neither a local maximum nor a local minimum.

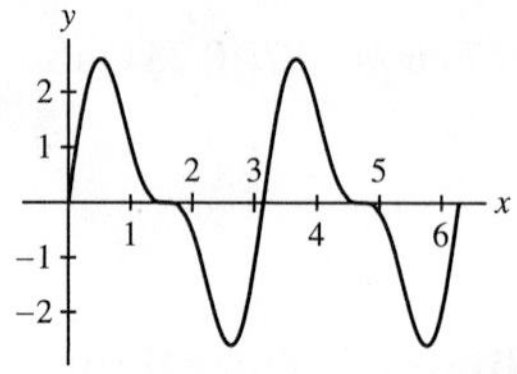

53. Critical point: $x = 2$; minimum value: $f(2) = 0$, maximum: $f(0) = f(4) = 2$

55. Critical point: $x = 2$; minimum value: $f(2) = 0$, maximum: $f(4) = 20$

57. $c = 1$ **59.** $c = \dfrac{15}{4}$

61. $f(0) < 0$ and $f(2) > 0$ so there is at least one root by the Intermediate Value Theorem; there cannot be another root because $f'(x) \geq 4$ for all x.

63. There cannot be a root $c > 0$ because $f'(x) > 4$ for all $x > 0$.

67. (a) $F = \dfrac{1}{2}\left(1 - \dfrac{v_2^2}{v_1^2}\right)\left(1 + \dfrac{v_2}{v_1}\right)$

(b) $F(r)$ achieves its maximum value when $r = 1/3$

(c) If v_2 were 0, then no air would be passing through the turbine, which is not realistic.

71.
- The maximum value of f on [0, 1] is
$$f\left(\left(\frac{a}{b}\right)^{1/(b-a)}\right) = \left(\frac{a}{b}\right)^{a/(b-a)} - \left(\frac{a}{b}\right)^{b/(b-a)}.$$
- $\dfrac{1}{4}$

73. Critical points: $x = 1$, $x = 4$ and $x = \frac{5}{2}$; maximum value: $f(1) = f(4) = \frac{5}{4}$, minimum value: $f(-5) = \frac{17}{70}$

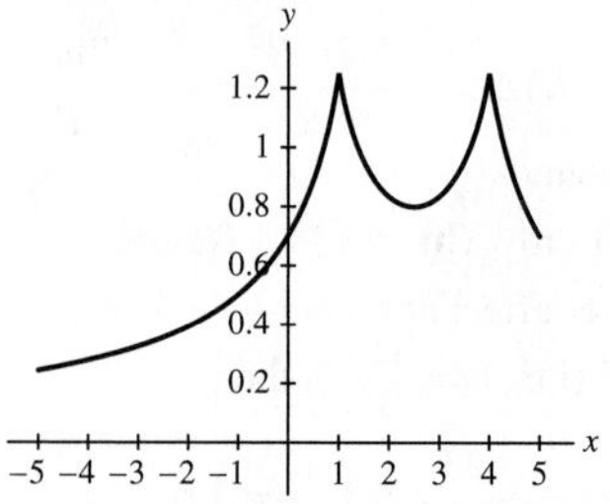

75. (a) There are four points at which the derivative is zero:
$$(-1, -\sqrt{2}), (-1, \sqrt{2}), (1, -\sqrt{2}), (1, \sqrt{2}).$$
There are also critical points where the derivative does not exist:
$$(0, 0), (\pm\sqrt[4]{27}, 0).$$

(b) The curve $27x^2 = (x^2 + y^2)^3$ and its horizontal tangents are plotted below.

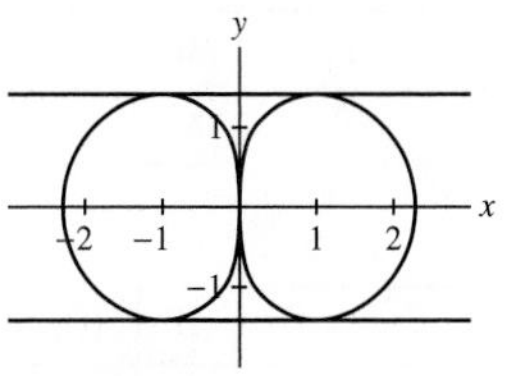

77.

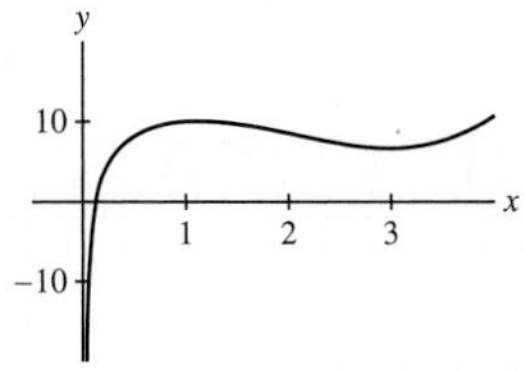

79.

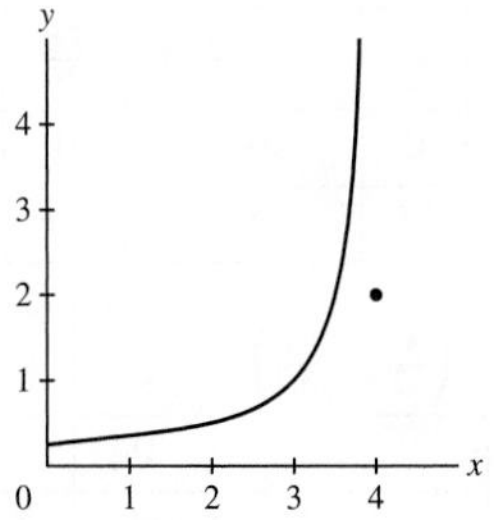

81. If $f(x) = a\sin x + b\cos x$, then $f'(x) = a\cos x - b\sin x$, so that $f'(x) = 0$ implies $a\cos x - b\sin x = 0$. This implies $\tan x = \frac{a}{b}$. Then,

$$\sin x = \frac{\pm a}{\sqrt{a^2 + b^2}} \quad \text{and} \quad \cos x = \frac{\pm b}{\sqrt{a^2 + b^2}}.$$

Therefore

$$\begin{aligned} f(x) = a\sin x + b\cos x &= a\frac{\pm a}{\sqrt{a^2 + b^2}} + b\frac{\pm b}{\sqrt{a^2 + b^2}} \\ &= \pm\frac{a^2 + b^2}{\sqrt{a^2 + b^2}} = \pm\sqrt{a^2 + b^2}. \end{aligned}$$

83. Let $f(x) = x^2 + rx + s$ and suppose that $f(x)$ takes on both positive and negative values. This will guarantee that f has two real roots. By the quadratic formula, the roots of f are

$$x = \frac{-r \pm \sqrt{r^2 - 4s}}{2}.$$

Observe that the midpoint between these roots is

$$\frac{1}{2}\left(\frac{-r + \sqrt{r^2 - 4s}}{2} + \frac{-r - \sqrt{r^2 - 4s}}{2}\right) = -\frac{r}{2}.$$

Next, $f'(x) = 2x + r = 0$ when $x = -\frac{r}{2}$ and, because the graph of $f(x)$ is an upward opening parabola, it follows that $f(-\frac{r}{2})$ is a minimum.

85. $b > \frac{1}{4}a^2$

87.
- Let $f(x)$ be a continuous function with $f(a)$ and $f(b)$ local minima on the interval $[a, b]$. By Theorem 1, $f(x)$ must take on both a minimum and a maximum on $[a, b]$. Since local minima occur at $f(a)$ and $f(b)$, the maximum must occur at some other point in the interval, call it c, where $f(c)$ is a local maximum.
- The function graphed here is discontinuous at $x = 0$.

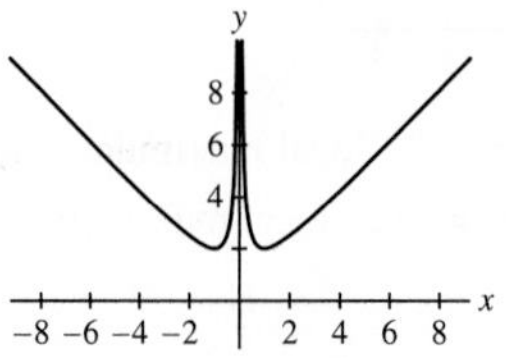

Section 4.3 Preliminary Questions

1. $m = 3$ **2.** **(c)**

3. Yes. The figure below displays a function that takes on only negative values but has a positive derivative.

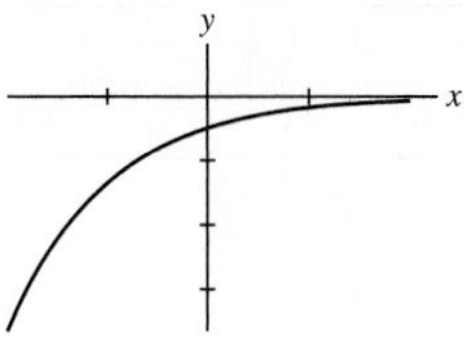

4. **(a)** $f(c)$ must be a local maximum. **(b)** No.

Section 4.3 Exercises

1. $c = 4$ **3.** $c = \frac{7\pi}{4}$ **5.** $c = \pm\sqrt{7}$ **7.** $c = 0$

9. The slope of the secant line between $x = 0$ and $x = 1$ is

$$\frac{f(1) - f(0)}{1 - 0} = \frac{2 - 0}{1} = 2.$$

It appears that the x-coordinate of the point of tangency is approximately 0.62.

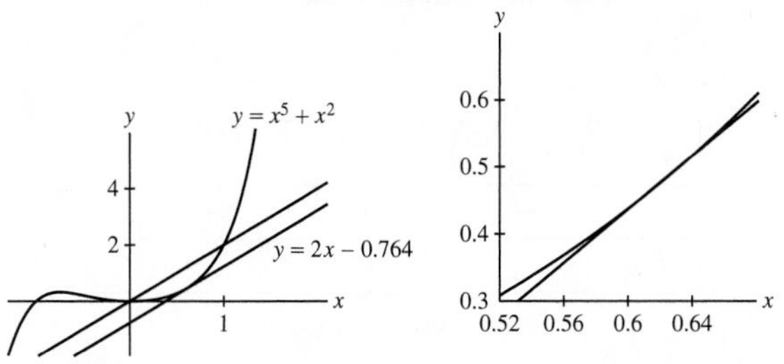

11. The derivative is positive on the intervals (0, 1) and (3, 5) and negative on the intervals (1, 3) and (5, 6).

13. $f(2)$ is a local maximum; $f(4)$ is a local minimum

15.

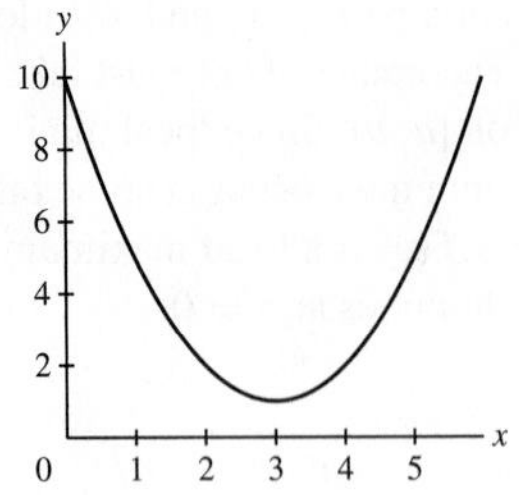

17.

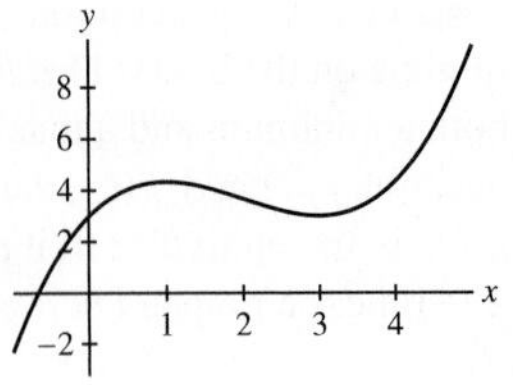

19. critical point: $x = 3$ - local maximum

21. critical point: $x = -2$ - local maximum; critical point: $x = 0$ - local minimum

23. $c = \frac{7}{2}$

x	$\left(-\infty, \frac{7}{2}\right)$	$7/2$	$\left(\frac{7}{2}, \infty\right)$
f'	+	0	−
f	↗	M	↘

25. $c = 0, 8$

x	$(-\infty, 0)$	0	$(0, 8)$	8	$(8, \infty)$
f'	+	0	−	0	+
f	↗	M	↘	m	↗

27. $c = -2, -1, 1$

x	$(-\infty, -2)$	-2	$(-2, -1)$	-1	$(-1, 1)$	1	$(1, \infty)$
f'	−	0	+	0	−	0	+
f	↘	m	↗	M	↘	m	↗

29. $c = -2, -1$

x	$(-\infty, -2)$	-2	$(-2, -1)$	-1	$(-1, \infty)$
f'	+	0	−	0	+
f	↗	M	↘	m	↗

31. $c = 0$

x	$(-\infty, 0)$	0	$(0, \infty)$
f'	+	0	+
f	↗	$\neq$	↗

33. $c = \left(\frac{3}{2}\right)^{2/5}$

x	$\left(0, \left(\frac{3}{2}\right)^{2/5}\right)$	$\frac{3}{2}^{2/5}$	$\left(\left(\frac{3}{2}\right)^{2/5}, \infty\right)$
f'	−	0	+
f	↘	m	↗

35. $c = 1$

x	$(0, 1)$	1	$(1, \infty)$
f'	−	0	+
f	↘	m	↗

37. $c = 0$

x	$(-\infty, 0)$	0	$(0, \infty)$
f'	+	0	−
f	↗	M	↘

39. $c = 0$

x	$(-\infty, 0)$	0	$(0, \infty)$
f'	+	0	+
f	↗	$\neg$	↗

41. $c = \frac{\pi}{2}$ and $c = \pi$

x	$\left(0, \frac{\pi}{2}\right)$	$\frac{\pi}{2}$	$\left(\frac{\pi}{2}, \pi\right)$	π	$(\pi, 2\pi)$
f'	+	0	−	0	+
f	↗	M	↘	m	↗

43. $c = \frac{\pi}{2}, \frac{7\pi}{6}, \frac{3\pi}{2}$, and $\frac{11\pi}{6}$

x	$\left(0, \frac{\pi}{2}\right)$	$\frac{\pi}{2}$	$\left(\frac{\pi}{2}, \frac{7\pi}{6}\right)$	$\frac{7\pi}{6}$	$\left(\frac{7\pi}{6}, \frac{3\pi}{2}\right)$
f'	+	0	−	0	+
f	↗	M	↘	m	↗

x	$\frac{3\pi}{2}$	$\left(\frac{3\pi}{2}, \frac{11\pi}{6}\right)$	$\frac{11\pi}{6}$	$\left(\frac{11\pi}{6}, 2\pi\right)$
f'	0	−	0	+
f	M	↘	m	↗

45. $\left(\frac{1}{e}\right)^{1/e} \approx 0.692201$ **55.** $f'(x) > 0$ for all x

49. The graph of $h(x)$ is shown below at the left. Because $h(x)$ is negative for $x < -1$ and for $0 < x < 1$, it follows that $f(x)$ is decreasing for $x < -1$ and for $0 < x < 1$. Similarly, $f(x)$ is increasing for $-1 < x < 0$ and for $x > 1$ because $h(x)$ is positive on these intervals. Moreover, $f(x)$ has local minima at $x = -1$ and $x = 1$ and a local maximum at $x = 0$. A plausible graph for $f(x)$ is shown below at the right.

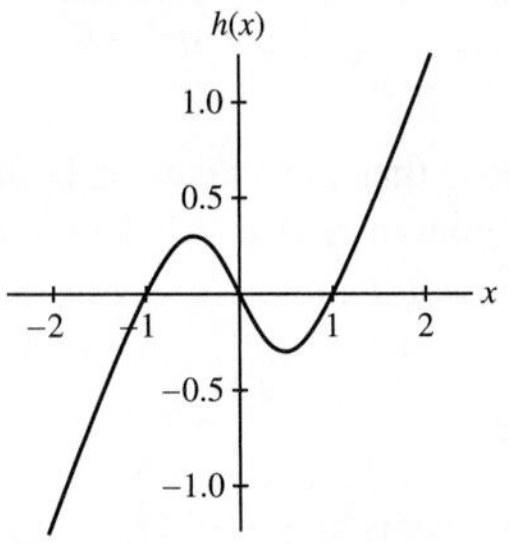

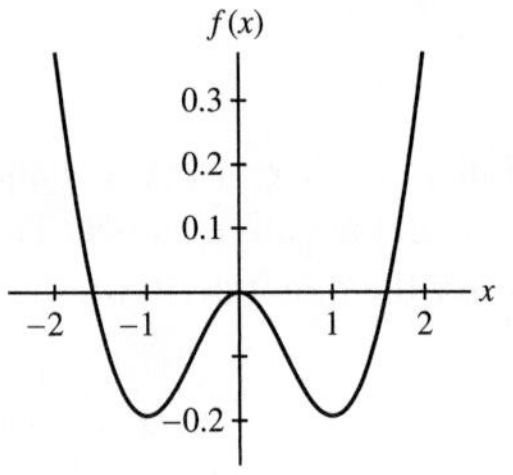

51. $f'(x) < 0$ as long as $x < 500$; so, $800^2 + 200^2 = f(200) > f(400) = 600^2 + 400^2$.

53. *every* point $c \in (a, b)$

61. **(a)** Let $g(x) = \cos x$ and $f(x) = 1 - \frac{1}{2}x^2$. Then $f(0) = g(0) = 1$ and $g'(x) = -\sin x \geq -x = f'(x)$ for $x \geq 0$ by Exercise 67. Now apply Exercise 67 to conclude that $\cos x \geq 1 - \frac{1}{2}x^2$ for $x \geq 0$.

(b) Let $g(x) = \sin x$ and $f(x) = x - \frac{1}{6}x^3$. Then $f(0) = g(0) = 0$ and $g'(x) = \cos x \ge 1 - \frac{1}{2}x^2 = f'(x)$ for $x \ge 0$ by part (a). Now apply Exercise 67 to conclude that $\sin x \ge x - \frac{1}{6}x^3$ for $x \ge 0$.

(c) Let $g(x) = 1 - \frac{1}{2}x^2 + \frac{1}{24}x^4$ and $f(x) = \cos x$. Then $f(0) = g(0) = 1$ and $g'(x) = -x + \frac{1}{6}x^3 \ge -\sin x = f'(x)$ for $x \ge 0$ by part (b). Now apply Exercise 67 to conclude that $\cos x \le 1 - \frac{1}{2}x^2 + \frac{1}{24}x^4$ for $x \ge 0$.

(d) The next inequality in the series is $\sin x \le x - \frac{1}{6}x^3 + \frac{1}{120}x^5$, valid for $x \ge 0$.

63. Let $f''(x) = 0$ for all x. Then $f'(x) =$ constant for all x. Since $f'(0) = m$, we conclude that $f'(x) = m$ for all x. Now let $g(x) = f(x) - mx$. Then $g'(x) = f'(x) - m = m - m = 0$ which implies that $g(x) =$ constant for all x and consequently $f(x) - mx =$ constant for all x. Rearranging the statement, $f(x) = mx +$ constant. Since $f(0) = b$, we conclude that $f(x) = mx + b$ for all x.

65. (a) Let $g(x) = f(x)^2 + f'(x)^2$. Then

$$g'(x) = 2f(x)f'(x) + 2f'(x)f''(x)$$
$$= 2f(x)f'(x) + 2f'(x)(-f(x))$$
$$= 0,$$

Because $g'(0) = 0$ for all x, $g(x) = f(x)^2 + f'(x)^2$ must be a constant function. To determine the value of C, we can substitute any number for x. In particular, for this problem, we want to substitute $x = 0$ and find $C = f(0)^2 + f'(0)^2$. Hence,

$$f(x)^2 + f'(x)^2 = f(0)^2 + f'(0)^2.$$

(b) Let $f(x) = \sin x$. Then $f'(x) = \cos x$ and $f''(x) = -\sin x$, so $f''(x) = -f(x)$. Next, let $f(x) = \cos x$. Then $f'(x) = -\sin x$, $f''(x) = -\cos x$, and we again have $f''(x) = -f(x)$. Finally, if we take $f(x) = \sin x$, the result from part (a) guarantees that

$$\sin^2 x + \cos^2 x = \sin^2 0 + \cos^2 0 = 0 + 1 = 1.$$

Section 4.4 Preliminary Questions

1. (a) increasing **2.** $f(c)$ is a local maximum

3. False **4.** False

Section 4.4 Exercises

1. (a) In C, we have $f''(x) < 0$ for all x.

(b) In A, $f''(x)$ goes from + to −.

(c) In B, we have $f''(x) > 0$ for all x.

(d) In D, $f''(x)$ goes from − to +.

3. concave up everywhere; no points of inflection

5. concave up for $x < -\sqrt{3}$ and for $0 < x < \sqrt{3}$; concave down for $-\sqrt{3} < x < 0$ and for $x > \sqrt{3}$; point of inflection at $x = 0$ and at $x = \pm\sqrt{3}$

7. concave up for $0 < \theta < \pi$; concave down for $\pi < \theta < 2\pi$; point of inflection at $\theta = \pi$

9. concave down for $0 < x < 9$; concave up for $x > 9$; point of inflection at $x = 9$

11. concave up on $(0, 1)$; concave down on $(-\infty, 0) \cup (1, \infty)$; point of inflection at both $x = 0$ and $x = 1$

13. concave up for $|x| > 1$; concave down for $|x| < 1$; point of inflection at both $x = -1$ and $x = 1$

15. The point of inflection in Figure 15 appears to occur at $t = 40$ days. The growth rate at the point of inflection is approximately 5.5 cm/day. Because the logistic curve changes from concave up to concave down at $t = 40$, the growth rate at this point is the maximum growth rate for the sunflower plant. Sketches of the first and second derivative of $h(t)$ are shown below at the left and at the right, respectively.

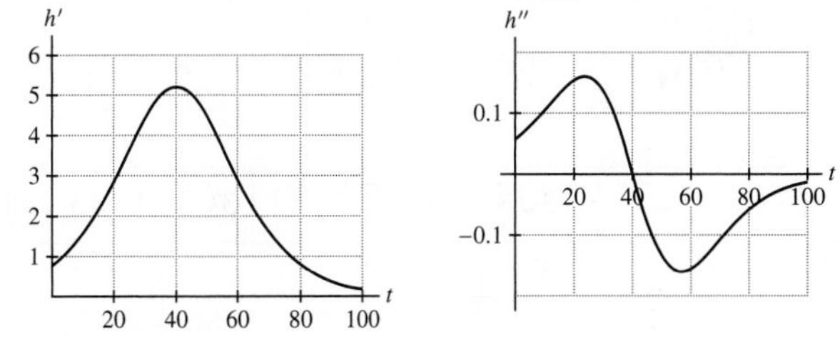

17. $f(x)$ has an inflection point at $x = b$ and another at $x = e$; $f(x)$ is concave down for $b < x < e$.

19. (a) f is increasing on $(0, 0.4)$.

(b) f is decreasing on $(0.4, 1) \cup (1, 1.2)$.

(c) f is concave up on $(0, 0.17) \cup (0.64, 1)$.

(d) f is concave down on $(0.17, 0.64) \cup (1, 1.2)$.

21. critical points are $x = 3$ and $x = 5$; $f(3) = 54$ is a local maximum, and $f(5) = 50$ is a local minimum

23. critical points are $x = 0$ and $x = 1$; $f(0) = 0$ is a local minimum, Second derivative test is inconclusive at $x = 1$

25. critical points are $x = -4$ and $x = 2$; $f(-4) = -16$ is a local maximum and $f(2) = -4$ is a local minimum

27. critical points are $x = 0$ and $x = \frac{2}{9}$; $f\left(\frac{2}{9}\right)$ is a local minimum; $f''(x)$ is undefined at $x = 0$, so the Second Derivative Test cannot be applied there

29. critical point in $(0, \infty)$ is $x = 2$; $f(2) = 32$ is a local minimum ($f''(2) = 24$).

31. critical points are $x = 0$, $x = \frac{\pi}{3}$ and $x = \pi$; $f(0)$ is a local minimum, $f\left(\frac{\pi}{3}\right)$ is a local maximum and $f(\pi)$ is a local minimum

33. critical point is $x = 0$; local minimum ($f''(0) = 2$).

35.

x	$(-\infty, \frac{1}{3})$	$\frac{1}{3}$	$(\frac{1}{3}, 1)$	1	$(1, \infty)$
f'	+	0	−	0	+
f	↗	M	↘	m	↗

x	$(-\infty, \frac{2}{3})$	$\frac{2}{3}$	$(\frac{2}{3}, \infty)$
f''	−	0	+
f	⌢	I	⌣

37.

t	$(-\infty, 0)$	0	$(0, \frac{2}{3})$	$\frac{2}{3}$	$(\frac{2}{3}, \infty)$
f'	$-$	0	$+$	0	$-$
f	↘	m	↗	M	↘

t	$(-\infty, \frac{1}{3})$	$\frac{1}{3}$	$(\frac{1}{3}, \infty)$
f''	$+$	0	$-$
f	⌣	I	⌢

39. $f''(x) > 0$ for all $x \geq 0$, which means there are no inflection points

x	0	$(0, (2)^{2/3})$	$(2)^{2/3}$	$((2)^{2/3}, \infty)$
f'	U	$-$	0	$+$
f	M	↘	m	↗

41.

x	$(-\infty, -3\sqrt{3})$	$-3\sqrt{3}$	$(-3\sqrt{3}, 3\sqrt{3})$	$3\sqrt{3}$	$(3\sqrt{3}, \infty)$
f'	$-$	0	$+$	0	$-$
f	↘	m	↗	M	↘

x	$(-\infty, -9)$	-9	$(-9, 0)$	0	$(0, 9)$	9	$(9, \infty)$
f''	$-$	0	$+$	0	$-$	0	$+$
f	⌢	I	⌣	I	⌢	I	⌣

43.

θ	$(0, \pi)$	π	$(\pi, 2\pi)$
f'	$+$	0	$+$
f	↗	¬	↗

θ	0	$(0, \pi)$	π	$(\pi, 2\pi)$	2π
f''	0	$-$	0	$+$	0
f	¬	⌢	I	⌣	¬

45.

x	$(-\frac{\pi}{2}, \frac{\pi}{2})$
f'	$+$
f	↗

x	$(-\frac{\pi}{2}, 0)$	0	$(0, \frac{\pi}{2})$
f''	$-$	0	$+$
f	⌢	I	⌣

47.

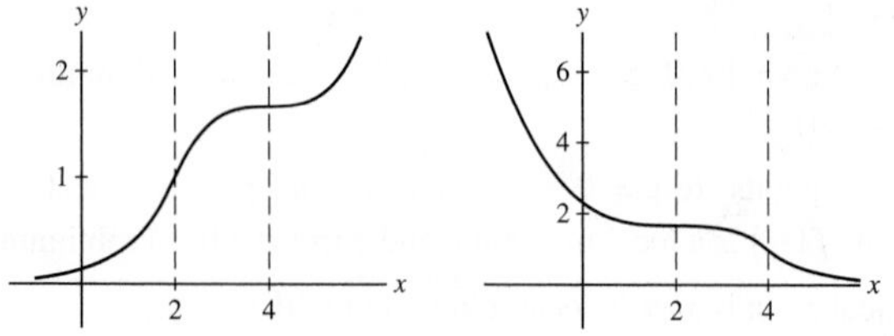

49.

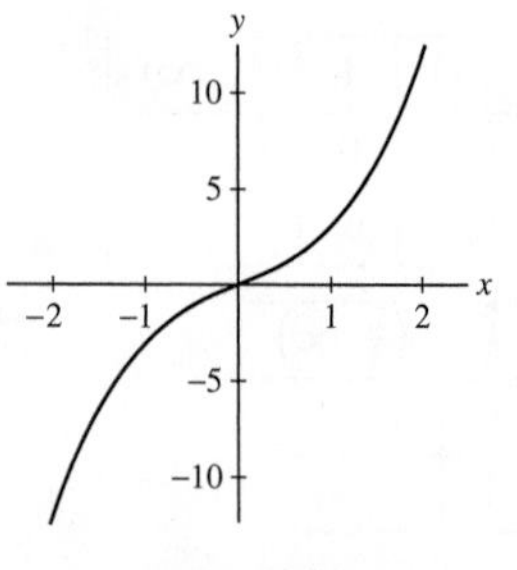

51. **(a)** Near the beginning of the epidemic, the graph of R is concave up. Near the epidemic's end, R is concave down.

(b) "Epidemic subsiding: number of new cases declining."

53. The point of inflection should occur when the water level is equal to the radius of the sphere. A possible graph of $V(t)$ is shown below.

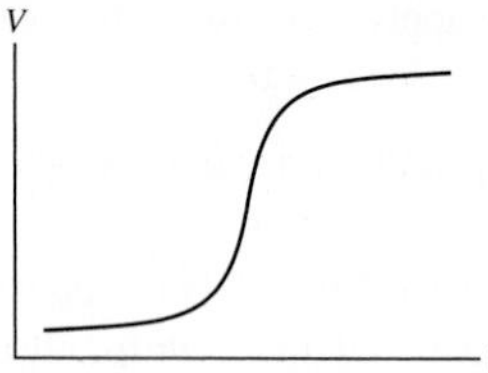

55. **(a)** From the definition of the derivative, we have

$$f''(c) = \lim_{h \to 0} \frac{f'(c+h) - f'(c)}{h} = \lim_{h \to 0} \frac{f'(c+h)}{h}.$$

(b) We are given that $f''(c) > 0$. By part (a), it follows that

$$\lim_{h \to 0} \frac{f'(c+h)}{h} > 0;$$

in other words, for sufficiently small h,

$$\frac{f'(c+h)}{h} > 0.$$

Now, if h is sufficiently small but negative, then $f'(c+h)$ must also be negative (so that the ratio $f'(c+h)/h$ will be positive) and $c + h < c$. On the other hand, if h is sufficiently small but positive, then $f'(c+h)$ must also be positive and $c + h > c$. Thus, there exists an open interval (a, b) containing c such that $f'(x) < 0$ for $a < x < c$ and $f'(c) > 0$ for $c < x < b$. Finally, because $f'(x)$ changes from negative to positive at $x = c$, $f(c)$ must be a local minimum.

57. **(b)** $f(x)$ has a point of inflection at $x = 0$ and at $x = \pm 1$. The figure below shows the graph of $y = f(x)$ and its tangent lines at each of the points of inflection. It is clear that each tangent line crosses the graph of $f(x)$ at the inflection point.

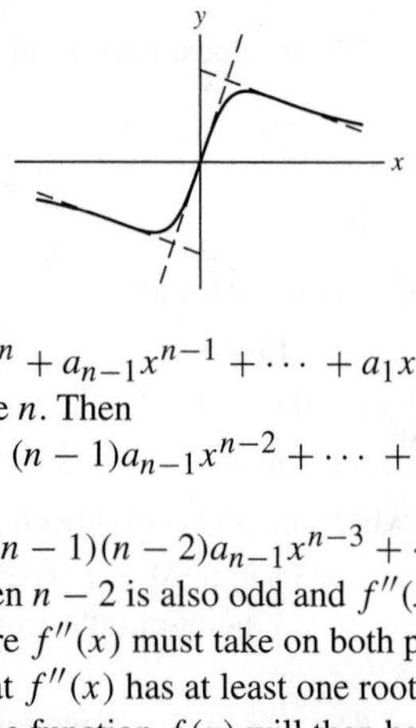

59. Let $f(x) = a_n x^n + a_{n-1} x^{n-1} + \cdots + a_1 x + a_0$ be a polynomial of degree n. Then $f'(x) = n a_n x^{n-1} + (n-1) a_{n-1} x^{n-2} + \cdots + 2a_2 x + a_1$ and $f''(x) = n(n-1) a_n x^{n-2} + (n-1)(n-2) a_{n-1} x^{n-3} + \cdots + 6a_3 x + 2a_2$. If $n \geq 3$ and is odd, then $n - 2$ is also odd and $f''(x)$ is a polynomial of odd degree. Therefore $f''(x)$ must take on both positive and negative values. It follows that $f''(x)$ has at least one root c such that $f''(x)$ changes sign at c. The function $f(x)$ will then have a point of inflection at $x = c$. On the other hand, the functions $f(x) = x^2$, x^4 and x^8 are polynomials of even degree that do not have any points of inflection.

Section 4.5 Preliminary Questions

1. An arc with the sign combination ++ (increasing, concave up) is shown below at the left. An arc with the sign combination −+ (decreasing, concave up) is shown below at the right.

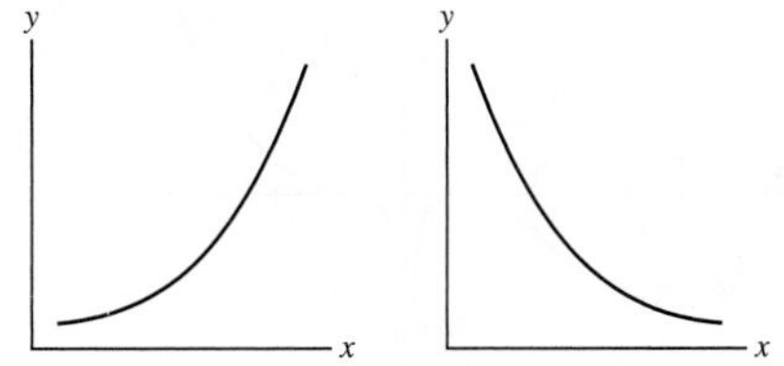

2. (c)

3. $x = 4$ is not in the domain of f

Section 4.5 Exercises

1.
- In A, f is decreasing and concave up, so $f' < 0$ and $f'' > 0$.
- In B, f is increasing and concave up, so $f' > 0$ and $f'' > 0$.
- In C, f is increasing and concave down, so $f' > 0$ and $f'' < 0$.
- In D, f is decreasing and concave down, so $f' < 0$ and $f'' < 0$.
- In E, f is decreasing and concave up, so $f' < 0$ and $f'' > 0$.
- In F, f is increasing and concave up, so $f' > 0$ and $f'' > 0$.
- In G, f is increasing and concave down, so $f' > 0$ and $f'' < 0$.

3. This function changes from concave up to concave down at $x = -1$ and from increasing to decreasing at $x = 0$.

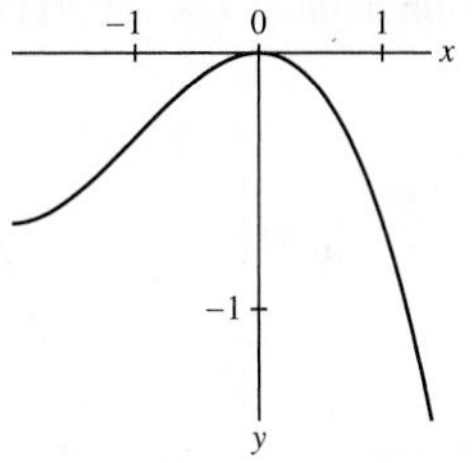

5. The function is decreasing everywhere and changes from concave up to concave down at $x = -1$ and from concave down to concave up at $x = -\frac{1}{2}$.

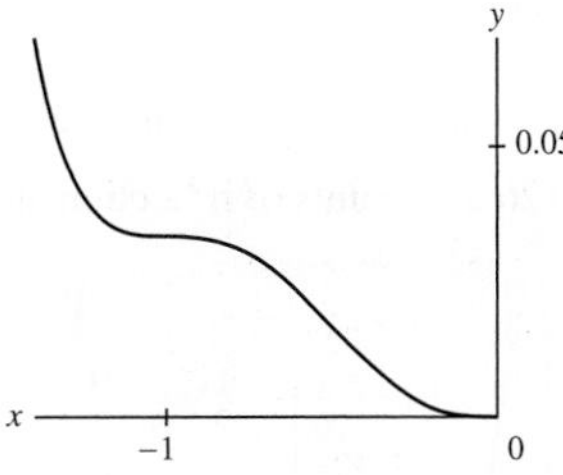

7.

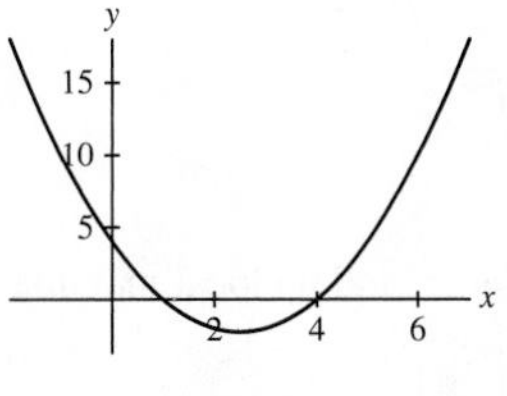

9.

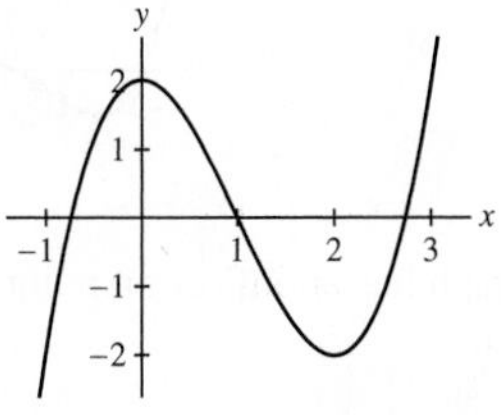

11.

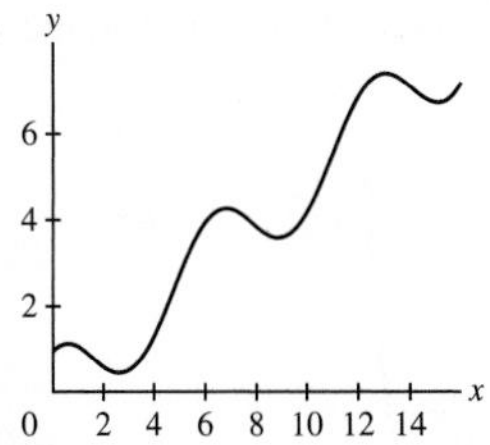

13. Local maximum at $x = -16$, a local minimum at $x = 0$, and an inflection point at $x = -8$.

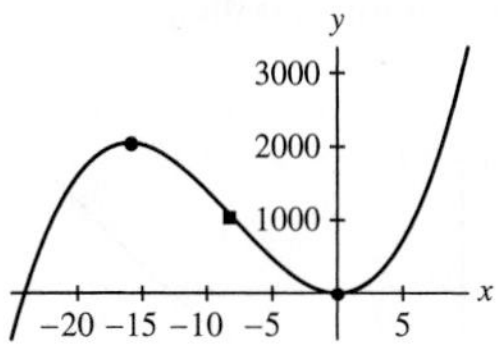

15. $f(0)$ is a local minimum, $f\left(\frac{1}{6}\right)$ is a local maximum, and there is a point of inflection at $x = \frac{1}{12}$.

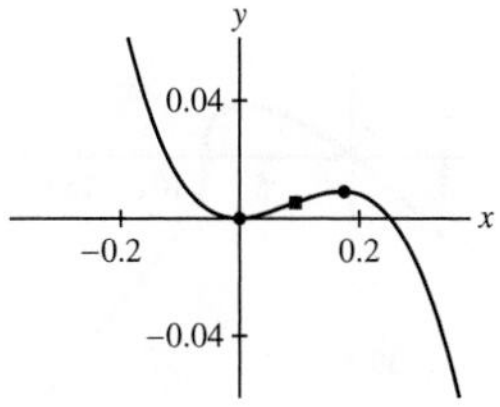

17. f has local minima at $x = \pm\sqrt{6}$, a local maximum at $x = 0$, and inflection points at $x = \pm\sqrt{2}$.

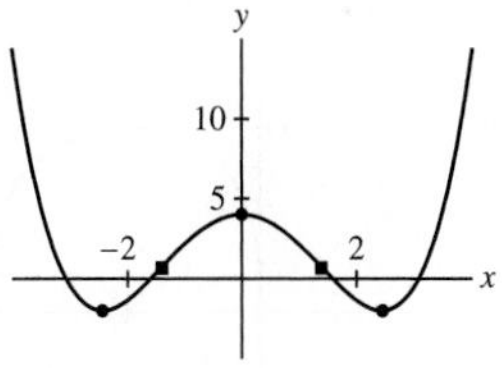

19. Graph has no critical points and is always increasing, inflection point at (0, 0).

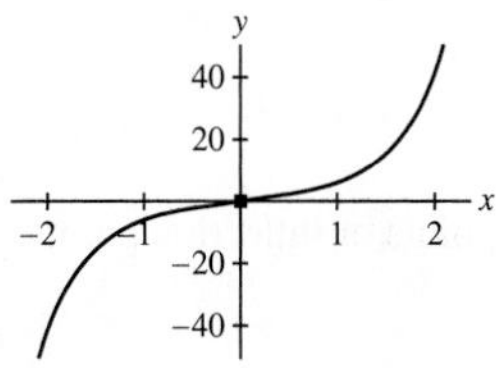

21. $f\left(\frac{1-\sqrt{33}}{8}\right)$ and $f(2)$ are local minima, and $f\left(\frac{1+\sqrt{33}}{8}\right)$ is a local maximum; points of inflection both at $x = 0$ and $x = \frac{3}{2}$.

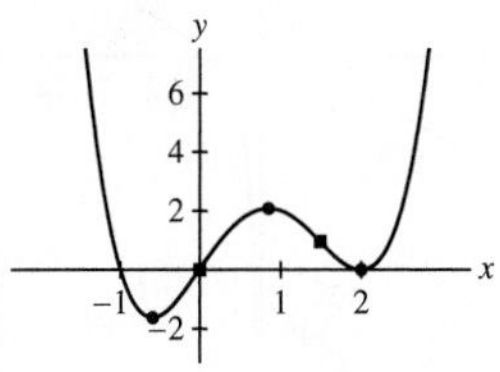

23. $f(0)$ is a local maximum, $f(12)$ is a local minimum, and there is a point of inflection at $x = 10$.

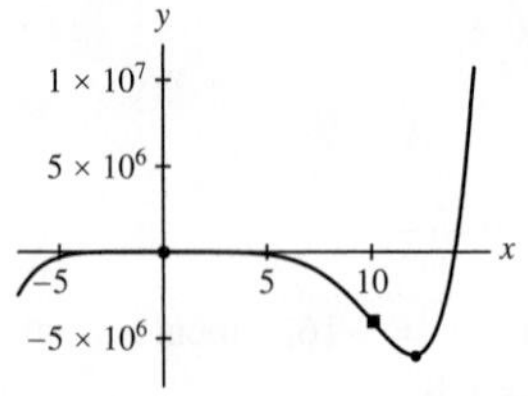

25. $f(4)$ is a local minimum, and the graph is always concave up.

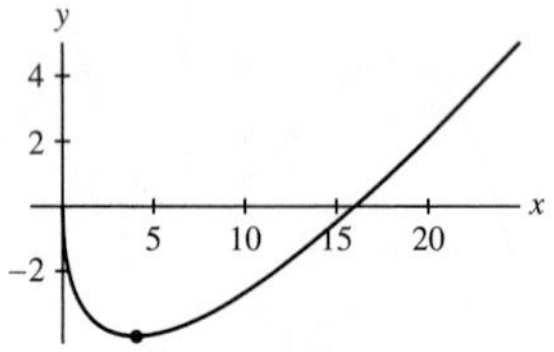

27. f has a local maximum at $x = 6$ and inflection points at $x = 8$ and $x = 12$.

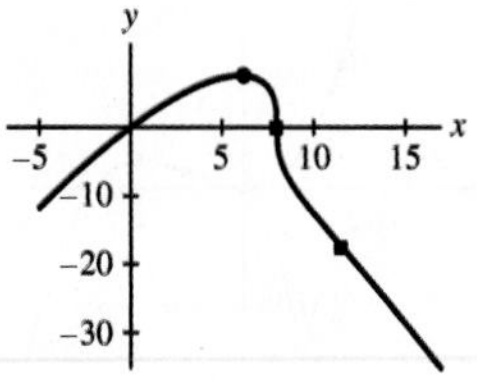

29. f has a local maximum at $x = 1$ and no inflection points. There are also critical points at $x = 0$ and $x = 2$ where the derivative does not exist.

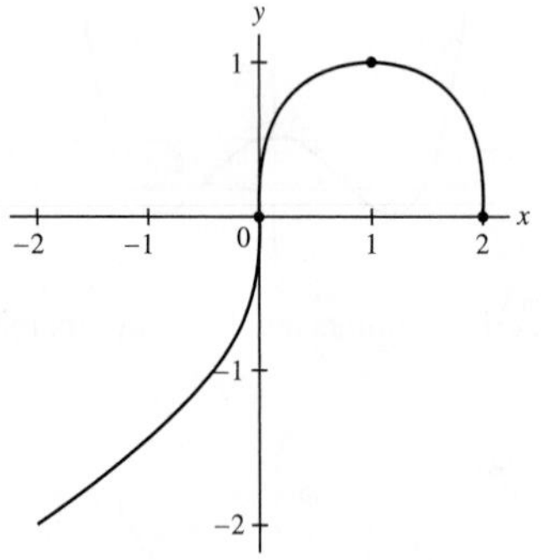

31. f has no critical points or inflection points.

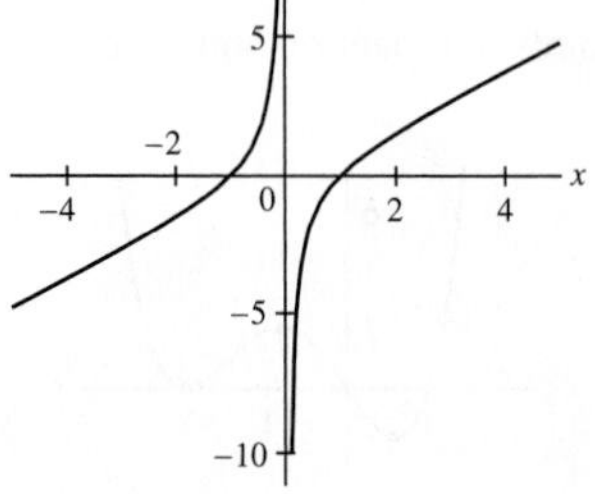

33. f has a local maximum at $x = -2$ and an inflection point at $\sqrt[5]{48}$.

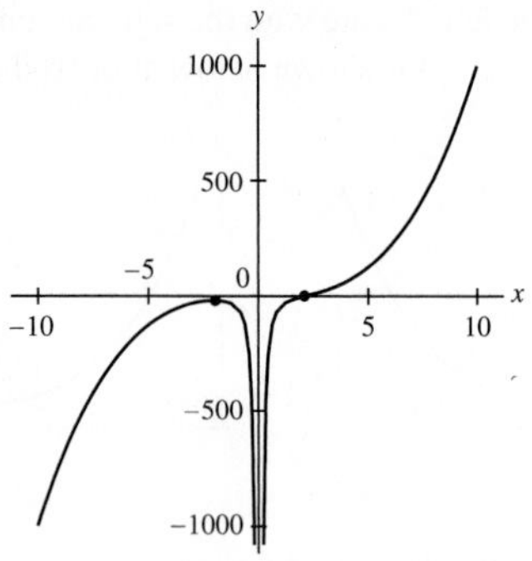

35. Graph has an inflection point at $x = \frac{3}{5}$, a local maximum at $x = 1$ (at which the graph has a cusp), and a local minimum at $x = \frac{9}{5}$.

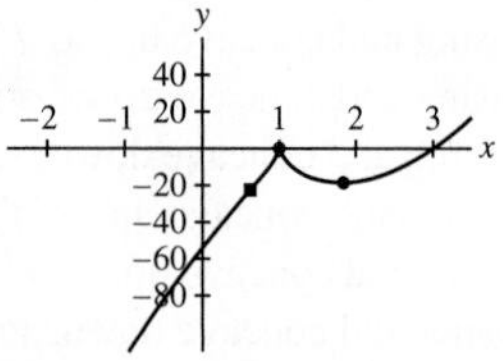

37. f has a local maximum at $x \approx -0.86776$ and a local minimum at $x = 0$. It has an inflection point at $x \approx -0.41119$.

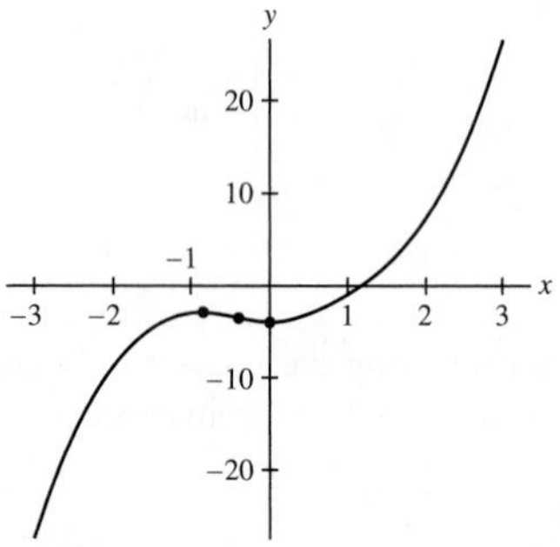

39. f has local minima at $x = -1.473$ and $x = 1.347$, a local maximum at $x = 0.126$ and points of inflection at $x = \pm\sqrt{\frac{2}{3}}$.

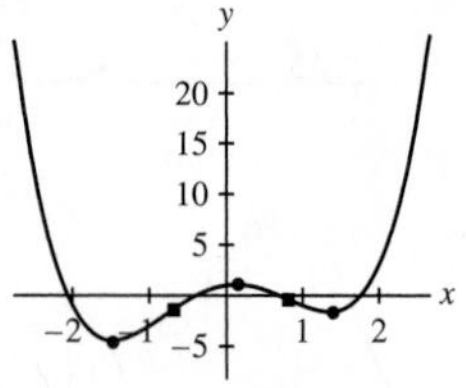

41. Graph has an inflection point at $x = \pi$, and no local maxima or minima.

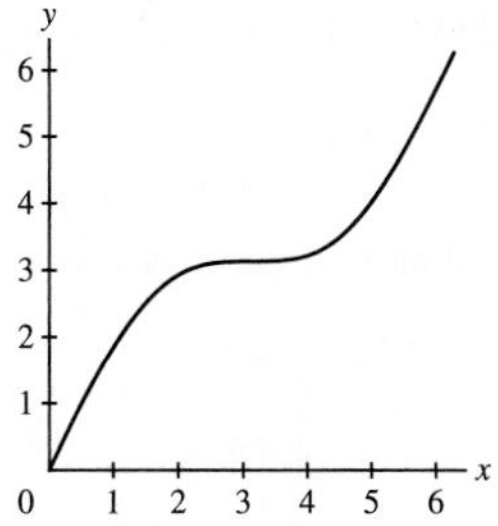

43. Local maximum at $x = \frac{\pi}{2}$, a local minimum at $x = \frac{3\pi}{2}$, and inflection points at $x = \frac{\pi}{6}$ and $x = \frac{5\pi}{6}$.

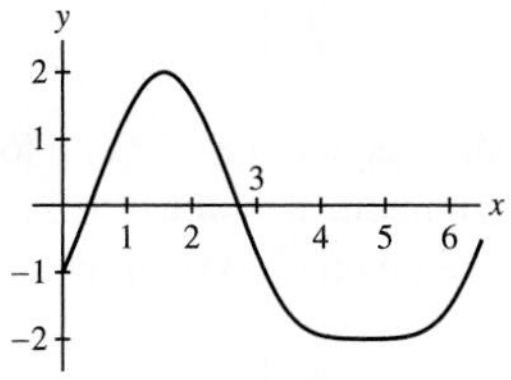

45. Local maximum at $x = \frac{\pi}{6}$ and a point of inflection at $x = \frac{2\pi}{3}$.

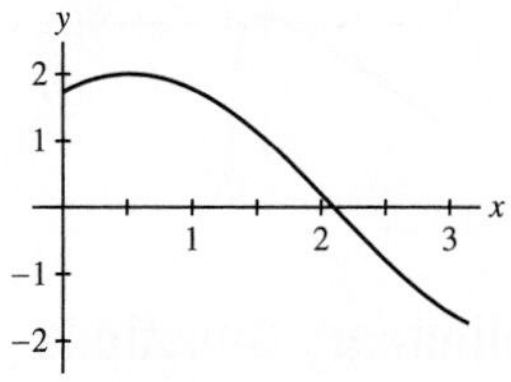

47. In both cases, there is a point where f is not differentiable at the transition from increasing to decreasing or decreasing to increasing.

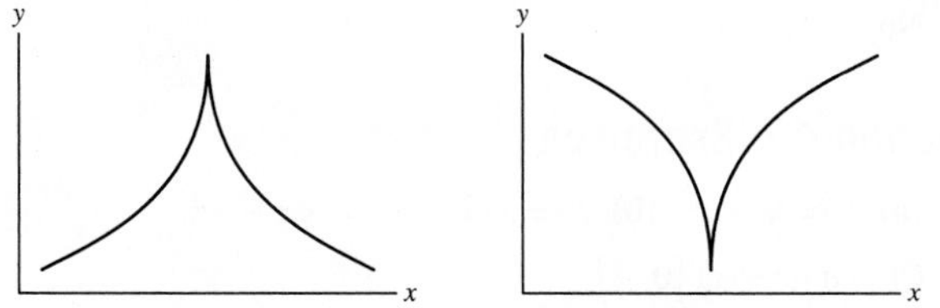

49. Graph (B) cannot be the graph of a polynomial.

51. (B) is the graph of $f(x) = \dfrac{3x^2}{x^2 - 1}$; (A) is the graph of $f(x) = \dfrac{3x}{x^2 - 1}$.

53. f is decreasing for all $x \neq \frac{1}{3}$, concave up for $x > \frac{1}{3}$, concave down for $x < \frac{1}{3}$, has a horizontal asymptote at $y = 0$ and a vertical asymptote at $x = \frac{1}{3}$.

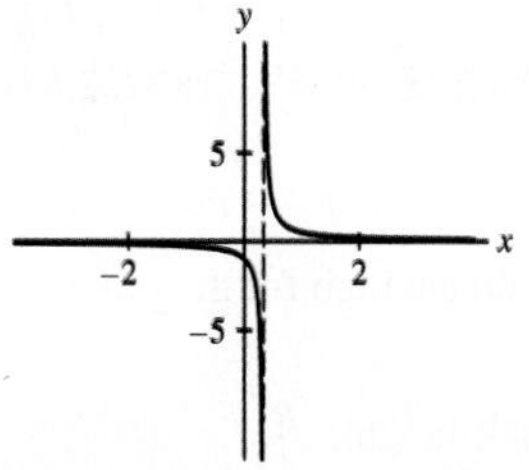

55. f is decreasing for all $x \neq 2$, concave up for $x > 2$, concave down for $x < 2$, has a horizontal asymptote at $y = 1$ and a vertical asymptote at $x = 2$.

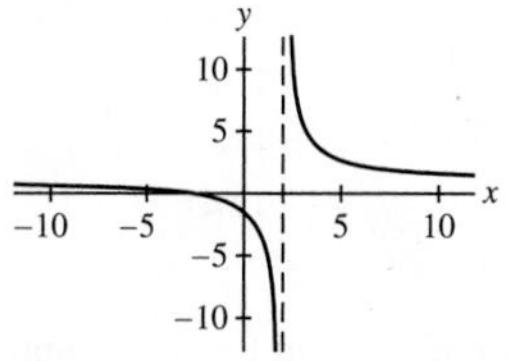

57. f is decreasing for all $x \neq 0, 1$, concave up for $0 < x < \frac{1}{2}$ and $x > 1$, concave down for $x < 0$ and $\frac{1}{2} < x < 1$, has a horizontal asymptote at $y = 0$ and vertical asymptotes at $x = 0$ and $x = 1$.

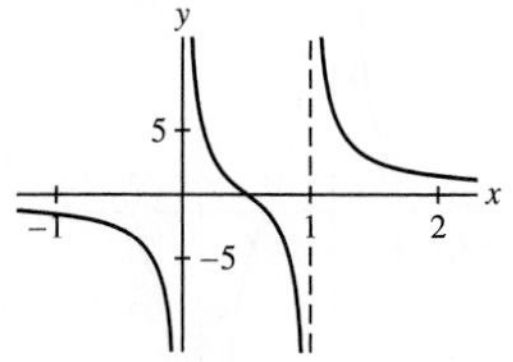

59. f is increasing for $x < 0$ and $0 < x < 1$ and decreasing for $1 < x < 2$ and $x > 2$; f is concave up for $x < 0$ and $x > 2$ and concave down for $0 < x < 2$; f has a horizontal asymptote at $y = 0$ and vertical asymptotes at $x = 0$ and $x = 2$.

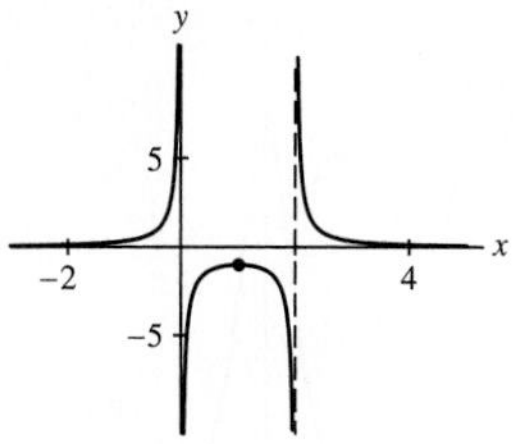

61. f is increasing for $x < 2$ and for $2 < x < 3$, is decreasing for $3 < x < 4$ and for $x > 4$, and has a local maximum at $x = 3$; f is concave up for $x < 2$ and for $x > 4$ and is concave down for $2 < x < 4$; f has a horizontal asymptote at $y = 0$ and vertical asymptotes at $x = 2$ and $x = 4$.

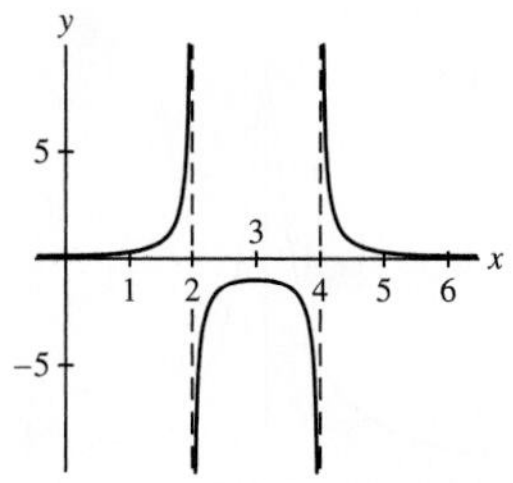

63. f is increasing for $|x| > 2$ and decreasing for $-2 < x < 0$ and for $0 < x < 2$; f is concave down for $-2\sqrt{2} < x < 0$ and for $x > 2\sqrt{2}$ and concave up for $x < -2\sqrt{2}$ and for $0 < x < 2\sqrt{2}$; f has a horizontal asymptote at $y = 1$ and a vertical asymptote at $x = 0$.

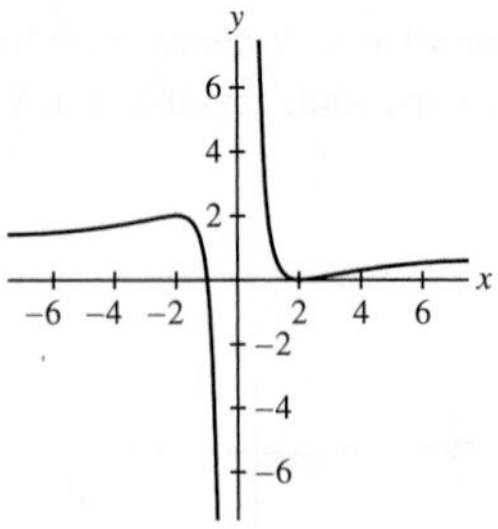

65. f is increasing for $x < 0$ and for $x > 2$ and decreasing for $0 < x < 2$; f is concave up for $x < 0$ and for $0 < x < 1$, is concave down for $1 < x < 2$ and for $x > 2$, and has a point of inflection at $x = 1$; f has a horizontal asymptote at $y = 0$ and vertical asymptotes at $x = 0$ and $x = 2$.

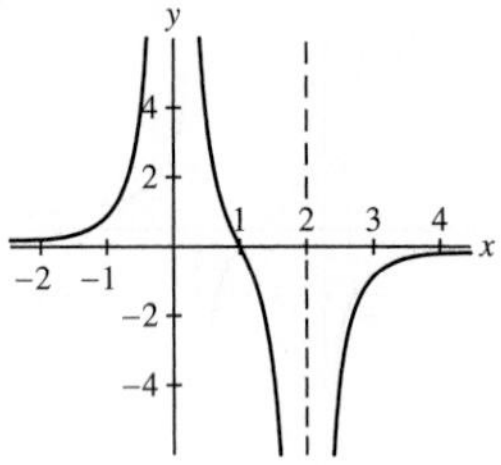

67. f is increasing for $x < 0$, decreasing for $x > 0$ and has a local maximum at $x = 0$; f is concave up for $|x| > 1/\sqrt{5}$, is concave down for $|x| < 1/\sqrt{5}$, and has points of inflection at $x = \pm 1/\sqrt{5}$; f has a horizontal asymptote at $y = 0$ and no vertical asymptotes.

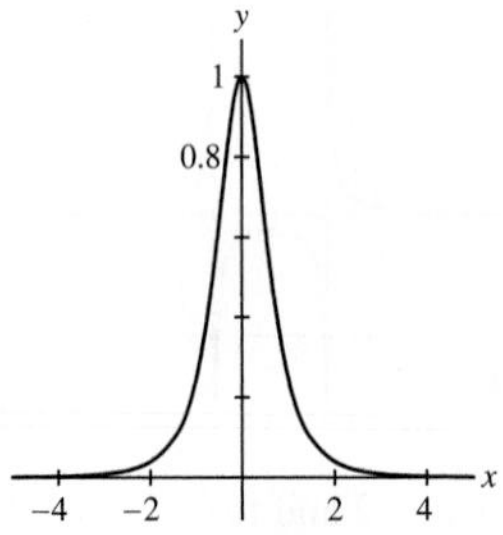

69. f is increasing for $x < 0$ and decreasing for $x > 0$; f is concave down for $|x| < \frac{\sqrt{2}}{2}$ and concave up for $|x| > \frac{\sqrt{2}}{2}$; f has a horizontal asymptote at $y = 0$ and no vertical asymptotes.

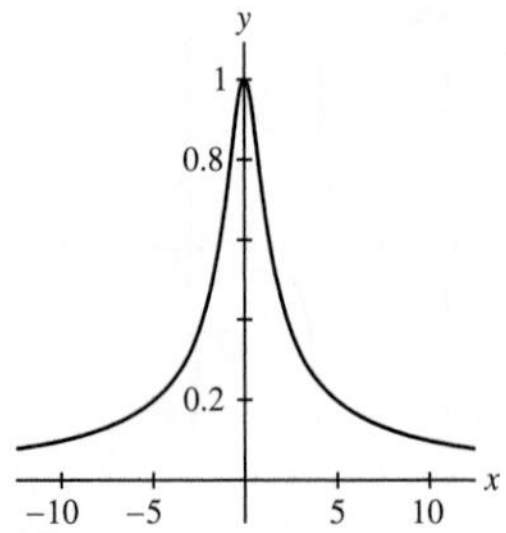

73. f is increasing for $x < -2$ and for $x > 0$, is decreasing for $-2 < x < -1$ and for $-1 < x < 0$, has a local minimum at $x = 0$, has a local maximum at $x = -2$, is concave down on $(-\infty, -1)$ and concave up on $(-1, \infty)$; f has a vertical asymptote at $x = -1$; by polynomial division, $f(x) = x - 1 + \frac{1}{x+1}$ and

$$\lim_{x \to \pm\infty} \left(x - 1 + \frac{1}{x+1} - (x - 1) \right) = 0,$$

which implies that the slant asymptote is $y = x - 1$.

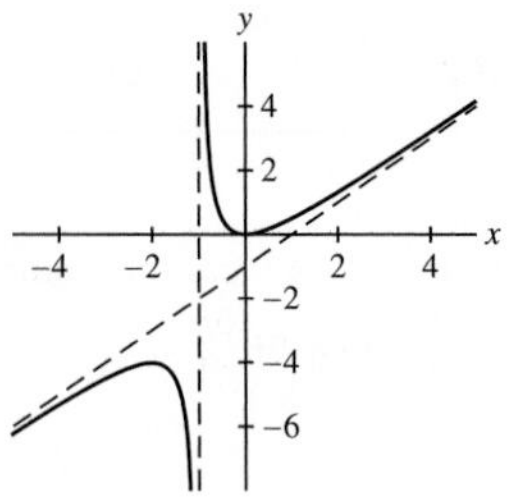

75. $y = x + 2$ is the slant asymptote of $f(x)$; local minimum at $x = 2 + \sqrt{3}$, a local maximum at $x = 2 - \sqrt{3}$ and f is concave down on $(-\infty, 2)$ and concave up on $(2, \infty)$; vertical asymptote at $x = 2$.

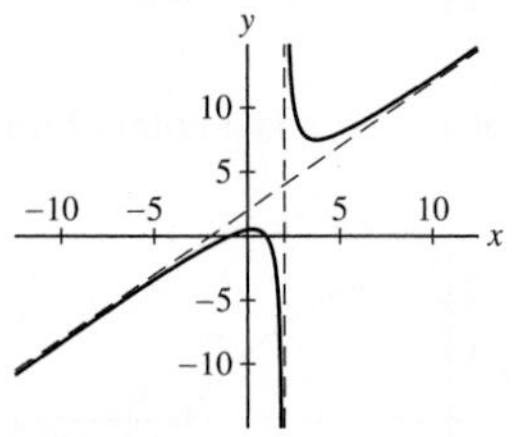

Section 4.6 Preliminary Questions

1. $b + h + \sqrt{b^2 + h^2} = 10$

2. If the function tends to infinity at the endpoints of the interval, then the function must take on a minimum value at a critical point.

3. No

Section 4.6 Exercises

1. **(a)** $y = \frac{3}{2} - x$ **(b)** $A = x(\frac{3}{2} - x) = \frac{3}{2}x - x^2$

(c) Closed interval $\left[0, \frac{3}{2}\right]$

(d) The maximum area $0.5625\ \text{m}^2$ is achieved with $x = y = \frac{3}{4}$ m.

3. Allot approximately 5.28 m of the wire to the circle.

5. The middle of the wire

7. The corral of maximum area has dimensions

$$x = \frac{300}{1 + \pi/4}\ \text{m} \quad \text{and} \quad y = \frac{150}{1 + \pi/4}\ \text{m},$$

where x is the width of the corral and therefore the diameter of the semicircle and y is the height of the rectangular section

9. Square of side length $4\sqrt{2}$ **11.** $\left(\frac{1}{2}, \frac{1}{2}\right)$

13. $x = 2^{1/4} \approx 1.1892$; $(x, y) \approx (1.1892, 2.8710)$. **15.** $\theta = \frac{\pi}{2}$

17. $\frac{3\sqrt{3}}{4} r^2$

19. 60 cm wide by 100 cm high for the full poster (48 cm by 80 cm for the printed matter)

21. Radius: $\sqrt{\frac{2}{3}}R$; half-height: $\frac{R}{\sqrt{3}}$

23. $x = 10\sqrt{5} \approx 22.36$ m and $y = 20\sqrt{5} \approx 44.72$ m where x is the length of the brick wall and y is the length of an adjacent side

25. 1.0718 **27.** $LH + \frac{1}{2}(L^2 + H^2)$ **29.** $y = -3x + 24$

33. $s = 3\sqrt[3]{4}$ m and $h = 2\sqrt[3]{4}$ m, where s is the length of the side of the square bottom of the box and h is the height of the box

35. **(a)** Each compartment has length of 600 m and width of 400 m.

(b) 240000 square meters.

37. $N \approx 58.14$ pounds and $P \approx 77.33$ pounds

39. \$990

41. 1.2 million euros in equipment and 600000 euros in labor

43. Brandon swims diagonally to a point located 20.2 m downstream and then runs the rest of the way.

45. $h = 3$; dimensions are $9 \times 18 \times 3$

47. $A = B = 30$ cm **49.** $x = \sqrt{bh + h^2}$

51. There are N shipments per year, so the time interval between shipments is $T = 1/N$ years. Hence, the total storage costs per year are sQ/N. The yearly delivery costs are dN and the total costs is $C(N) = dN + sQ/N$. Solving,

$$C'(N) = d - \frac{sQ}{N^2} = 0$$

for N yields $N = \sqrt{sQ/d}$. $N = 9$.

53. **(a)** If $b < \sqrt{3}a$, then $d = a - b/\sqrt{3} > 0$ and the minimum occurs at this value of d. On the other hand, if $b \geq \sqrt{3}a$, then the minimum occurs at the endpoint $d = 0$.

(b) Plots of $S(d)$ for $b = 0.5$, $b = \sqrt{3}$ and $b = 3$ are shown below. For $b = 0.5$, the results of (a) indicate the minimum should occur for $d = 1 - 0.5/\sqrt{3} \approx 0.711$, and this is confirmed in the plot. For both $b = \sqrt{3}$ and $b = 3$, the results of (a) indicate that the minimum should occur at $d = 0$, and both of these conclusions are confirmed in the plots.

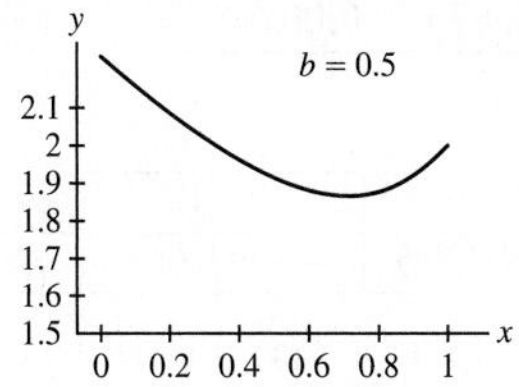

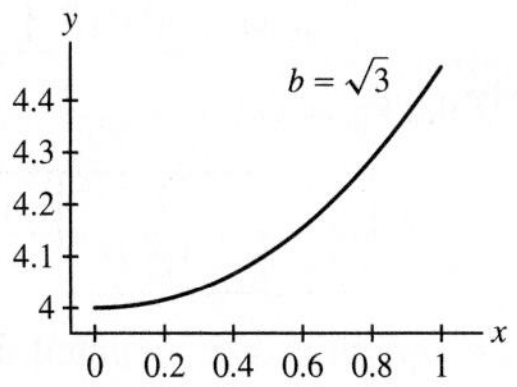

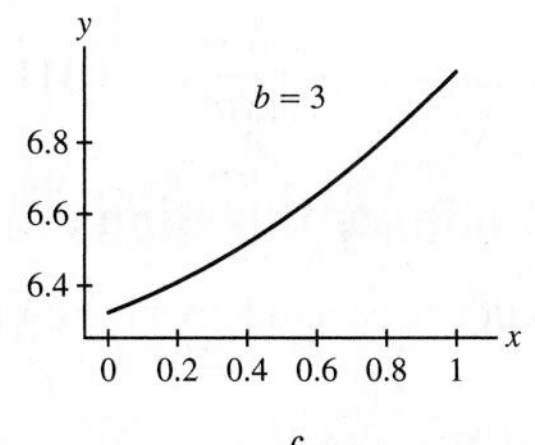

55. minimum value of $F(\theta)$ is $\dfrac{fmg}{\sqrt{1+f^2}}$.

57. $s \approx 30.07$ **59.** $15\sqrt{5}$ **61.** $\ell = (b^{2/3} + h^{2/3})^{3/2}$ ft

63. **(a)** $\alpha = 0$ corresponds to shooting the ball directly at the basket while $\alpha = \pi/2$ corresponds to shooting the ball directly upward. In neither case is it possible for the ball to go into the basket. If the angle α is extremely close to 0, the ball is shot almost directly at the basket; on the other hand, if the angle α is extremely close to $\pi/2$, the ball is launched almost vertically. In either one of these cases, the ball has to travel at an enormous speed.

(b) The minimum clearly occurs where $\theta = \pi/3$.

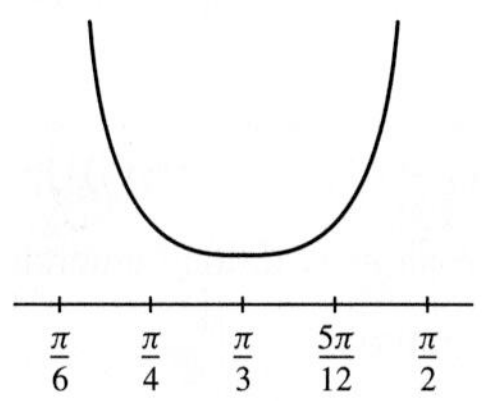

(c) $v^2 = \dfrac{16d}{F(\theta)}$; hence v^2 is smallest whenever $F(\theta)$ is greatest.

(d) A critical point of $F(\theta)$ occurs where $\cos(\alpha - 2\theta) = 0$, so that $\alpha - 2\theta = -\frac{\pi}{2}$ (negative because $2\theta > \theta > \alpha$), and this gives us $\theta = \alpha/2 + \pi/4$. The minimum value $F(\theta_0)$ takes place at $\theta_0 = \alpha/2 + \pi/4$.

(e) Plug in $\theta_0 = \alpha/2 + \pi/4$. From Figure 34 we see that

$$\cos\alpha = \frac{d}{\sqrt{d^2+h^2}} \quad \text{and} \quad \sin\alpha = \frac{h}{\sqrt{d^2+h^2}}.$$

(f) This shows that the minimum velocity required to launch the ball to the basket drops as shooter height increases. This shows one of the ways height is an advantage in free throws; a taller shooter need not shoot the ball as hard to reach the basket.

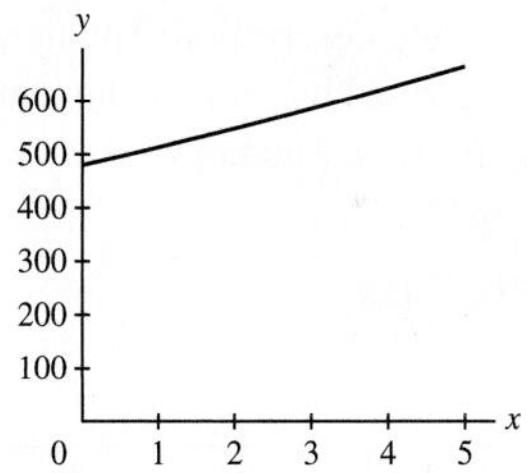

65. **(a)** From the figure, we see that

$$\theta(x) = \tan^{-1}\frac{c - f(x)}{x} - \tan^{-1}\frac{b - f(x)}{x}.$$

Then

$$\begin{aligned}\theta'(x) &= \frac{b - (f(x) - xf'(x))}{x^2 + (b - f(x))^2} - \frac{c - (f(x) - xf'(x))}{x^2 + (c - f(x))^2}\\ &= (b-c)\frac{x^2 - bc + (b+c)(f(x) - xf'(x)) - (f(x))^2 + 2xf(x)f'(x)}{(x^2 + (b - f(x))^2)(x^2 + (c - f(x))^2)}\\ &= (b-c)\frac{(x^2 + (xf'(x))^2 - (bc - (b+c)(f(x) - xf'(x)) + (f(x) - xf'(x))^2)}{(x^2 + (b - f(x))^2)(x^2 + (c - f(x))^2)}\\ &= (b-c)\frac{(x^2 + (xf'(x))^2 - (b - (f(x) - xf'(x)))(c - (f(x) - xf'(x)))}{(x^2 + (b - f(x))^2)(x^2 + (c - f(x))^2)}.\end{aligned}$$

(b) The point Q is the y-intercept of the line tangent to the graph of $f(x)$ at point P. The equation of this tangent line is

$$Y - f(x) = f'(x)(X - x).$$

The y-coordinate of Q is then $f(x) - xf'(x)$.

(c) From the figure, we see that

$$BQ = b - (f(x) - xf'(x)),$$
$$CQ = c - (f(x) - xf'(x))$$

and

$$PQ = \sqrt{x^2 + (f(x) - (f(x) - xf'(x)))^2} = \sqrt{x^2 + (xf'(x))^2}.$$

Comparing these expressions with the numerator of $d\theta/dx$, it follows that $\dfrac{d\theta}{dx} = 0$ is equivalent to

$$PQ^2 = BQ \cdot CQ.$$

(d) The equation $PQ^2 = BQ \cdot CQ$ is equivalent to

$$\frac{PQ}{BQ} = \frac{CQ}{PQ}.$$

In other words, the sides CQ and PQ from the triangle ΔQCP are proportional in length to the sides PQ and BQ from the triangle ΔQPB. As $\angle PQB = \angle CQP$, it follows that triangles ΔQCP and ΔQPB are similar.

Section 4.7 Preliminary Questions

1. One

2. Every term in the Newton's Method sequence will remain x_0.

3. Newton's Method will fail.

4. Yes, that is a reasonable description. The iteration formula for Newton's Method was derived by solving the equation of the tangent line to $y = f(x)$ at x_0 for its x-intercept.

Section 4.7 Exercises

1.

n	1	2	3
x_n	2.5	2.45	2.44948980

3.

n	1	2	3
x_n	2.16666667	2.15450362	2.15443469

5.

n	1	2	3
x_n	0.28540361	0.24288009	0.24267469

7. We take $x_0 = -1.4$, based on the figure, and then calculate

n	1	2	3
x_n	−1.330964467	−1.328272820	−1.328268856

9. $x \approx 0.5739277833$.

11. $\sqrt{11} \approx 3.317$; a calculator yields 3.31662479.

13. $2^{7/3} \approx 5.040$; a calculator yields 5.0396842.

15. 2.093064358 **17.** −2.225 **19.** 1.202

21. $x = 4.49341$, which is approximately 1.4303π

23. (2.7984, −0.941684)

25. **(a)** $P \approx \$156.69$

(b) $b \approx 1.02121$; the interest rate is around 25.45%

27. **(a)** The sector SAB is the slice OAB with the triangle OPS removed. OAB is a central sector with arc θ and radius $\overline{OA} = a$, and therefore has area $\frac{a^2\theta}{2}$. OPS is a triangle with height $a\sin\theta$ and base length $\overline{OS} = ea$. Hence, the area of the sector is

$$\frac{a^2}{2}\theta - \frac{1}{2}ea^2\sin\theta = \frac{a^2}{2}(\theta - e\sin\theta).$$

(b) Since Kepler's second law indicates that the area of the sector is proportional to the time t since the planet passed point A, we get

$$\pi a^2(t/T) = a^2/2\,(\theta - e\sin\theta)$$
$$2\pi\frac{t}{T} = \theta - e\sin\theta.$$

(c) From the point of view of the Sun, Mercury has traversed an angle of approximately 1.76696 radians $= 101.24°$. Mercury has therefore traveled more than one fourth of the way around (from the point of view of central angle) during this time.

29. The sequence of iterates diverges spectacularly, since $x_n = (-2)^n x_0$.

31. **(a)** Let $f(x) = \frac{1}{x} - c$. Then

$$x - \frac{f(x)}{f'(x)} = x - \frac{\frac{1}{x} - c}{-x^{-2}} = 2x - cx^2.$$

(b) For $c = 10.3$, we have $f(x) = \frac{1}{x} - 10.3$ and thus $x_{n+1} = 2x_n - 10.3x_n^2$.

- Take $x_0 = 0.1$.

n	1	2	3
x_n	0.097	0.0970873	0.09708738

- Take $x_0 = 0.5$.

n	1	2	3
x_n	−1.575	−28.7004375	−8541.66654

(c) The graph is disconnected. If $x_0 = .5$, $(x_1, f(x_1))$ is on the other portion of the graph, which will never converge to any point under Newton's Method.

33. $\theta \approx 1.2757$; hence, $h = L\dfrac{1 - \cos\theta}{2\sin\theta} \approx 1.11181$

Section 4.8 Preliminary Questions

1. Any constant function is an antiderivative for the function $f(x) = 0$.

2. No difference **3.** No

4. **(a)** False. Even if $f(x) = g(x)$, the antiderivatives F and G may differ by an additive constant.

(b) True. This follows from the fact that the derivative of any constant is 0.

(c) False. If the functions f and g are different, then the antiderivatives F and G differ by a linear function: $F(x) - G(x) = ax + b$ for some constants a and b.

5. No

Section 4.8 Exercises

1. $6x^3 + C$ **3.** $\frac{2}{5}x^5 - 8x^3 + 12\ln|x| + C$

5. $2\sin x + 9\cos x + C$ **7.** $-\frac{1}{2}\cos 2x + 4\sin 3x + C$

9. **(a)** **(ii)** **(b)** **(iii)** **(c)** **(i)** **(d)** **(iv)**

11. $4x - 9x^2 + C$ **13.** $\frac{11}{5}t^{5/11} + C$

15. $3t^6 - 2t^5 - 14t^2 + C$

17. $5z^{1/5} - \frac{3}{5}z^{5/3} + \frac{4}{9}z^{9/4} + C$

19. $\frac{3}{2}x^{2/3} + C$ **21.** $-\frac{18}{t^2} + C$

23. $\frac{2}{5}t^{5/2} + \frac{1}{2}t^2 + \frac{2}{3}t^{3/2} + t + C$

25. $\frac{1}{2}x^2 + 3x + 4x^{-1} + C$

27. $12\sec x + C$ **29.** $-\csc t + C$

31. $-\frac{1}{3}\tan(7 - 3x) + C$ **33.** $\frac{25}{3}\tan(3z + 1) + C$

35. $3\sin 4x - 3\cos 3x + C$

37. $\frac{5}{8}\ln(\sec^2(4\theta + 3)) + C$

39. $\frac{1}{3}\sin(3\theta) - 2\tan\left(\frac{\theta}{4}\right) + C$

41. Graph (B) does not have the same local extrema as indicated by $f(x)$ and therefore is *not* an antiderivative of $f(x)$.

43. $\frac{d}{dx}\left(\frac{1}{7}(x + 13)^7 + C\right) = (x + 13)^6$

45. $\frac{d}{dx}\left(\frac{1}{12}(4x + 13)^3 + C\right) = \frac{1}{4}(4x + 13)^2(4) = (4x + 13)^2$

47. $y = \frac{1}{4}x^4 + 4$ **49.** $y = t^2 + 3t^3 - 2$ **51.** $y = \frac{2}{3}t^{3/2} + \frac{1}{3}$

53. $y = \frac{1}{12}(3x + 2)^4 - \frac{1}{3}$ **55.** $y = 1 - \cos x$

57. $y = 3 + \frac{1}{5}\sin 5x$ **59.** $y = 2\sec(3\theta) - 4 - 2\sqrt{2}$

61. $y = -2\sin\left(\frac{1}{2}\theta\right) + 6$

63. $f'(x) = 6x^2 + 1$; $f(x) = 2x^3 + x + 2$

65. $f'(x) = \frac{1}{4}x^4 - x^2 + x + 1$; $f(x) = \frac{1}{20}x^5 - \frac{1}{3}x^3 + \frac{1}{2}x^2 + x$

67. $f'(t) = -2t^{-1/2} + 2$; $f(t) = -4t^{1/2} + 2t + 4$

69. $f'(t) = \frac{1}{2}t^2 - \sin t + 2$; $f(t) = \frac{1}{6}t^3 + \cos t + 2t - 3$

71. The differential equation satisfied by $s(t)$ is

$$\frac{ds}{dt} = v(t) = 6t^2 - t,$$

and the associated initial condition is $s(1) = 0$; $s(t) = 2t^3 - \frac{1}{2}t^2 - \frac{3}{2}$.

73. The differential equation satisfied by $s(t)$ is

$$\frac{ds}{dt} = v(t) = \sin(\pi t/2),$$

and the associated initial condition is $s(0) = 0$; $s(t) = \frac{2}{\pi}(1 - \cos(\pi t/2))$

75. 6.25 seconds; 78.125 meters

77. 300 m/s **81.** $c_1 = -1$ and $c_2 = 1$

83. **(a)** By the Chain Rule, we have

$$\frac{d}{dx}\left(\frac{1}{2}F(2x)\right) = \frac{1}{2}F'(2x) \cdot 2 = F'(2x) = f(2x).$$

Thus $\frac{1}{2}F(2x)$ is an antiderivative of $f(2x)$.

(b) $\frac{1}{k}F(kx) + C$

Chapter 4 Review

1. $8.1^{1/3} - 2 \approx 0.00833333$; error is 3.445×10^{-5}

3. $625^{1/4} - 624^{1/4} \approx 0.002$; error is 1.201×10^{-6}

5. $\frac{1}{1.02} \approx 0.98$; error is 3.922×10^{-4}

7. $L(x) = 5 + \frac{1}{10}(x - 25)$ **9.** $L(r) = 36\pi(r - 2)$

11. $L(\theta) = 3\theta - \pi$ **13.** $\Delta s \approx 0.632$

15. **(a)** An increase of \$1500 in revenue.

(b) A small increase in price would result in a decrease in revenue.

17. 9%

19. $\frac{f(8) - f(1)}{7} = -\frac{1}{14} = f'\left(\left(\frac{14}{3}\right)^{3/4}\right).$

21. $\frac{f(5) - f(2)}{3} = \frac{9}{10} = f'(\sqrt{10})$ since $f'(x) = 1 - \frac{1}{x^2}$.

23. Let $x > 0$. Because f is continuous on $[0, x]$ and differentiable on $(0, x)$, the Mean Value Theorem guarantees there exists a $c \in (0, x)$ such that

$$f'(c) = \frac{f(x) - f(0)}{x - 0} \quad \text{or} \quad f(x) = f(0) + xf'(c).$$

Now, we are given that $f(0) = 4$ and that $f'(x) \leq 2$ for $x > 0$. Therefore, for all $x \geq 0$,

$$f(x) \leq 4 + x(2) = 2x + 4.$$

25. $x = \frac{2}{3}$ and $x = 2$ are critical points; $f\left(\frac{2}{3}\right)$ is a local maximum while $f(2)$ is a local minimum.

27. $x = 0$, $x = -2$ and $x = -\frac{4}{5}$ are critical points; $f(-2)$ is neither a local maximum nor a local minimum, $f\left(-\frac{4}{5}\right)$ is a local maximum and $f(0)$ is a local minimum.

29. $\theta = \frac{3\pi}{4} + n\pi$ is a critical point for all integers n; $g\left(\frac{3\pi}{4} + n\pi\right)$ is neither a local maximum nor a local minimum for any integer n.

31. Maximum value is 21; minimum value is -11.

33. Minimum value is -1; maximum value is $\frac{5}{4}$.

35. Minimum value is -1; maximum value is 3.

37. Minimum value is $2 - 2^{3/2}$ at $x = 2$; maximum value is $\frac{4}{27}$ at $x = \frac{4}{9}$.

39. Minimum value is 2; maximum value is 17.

41. $x = \frac{4}{3}$ **43.** $x = \pm\frac{2}{\sqrt{3}}$ **45.** $x = 0$ and $x = \pm\sqrt{3}$

47. No horizontal asymptotes; no vertical asymptotes

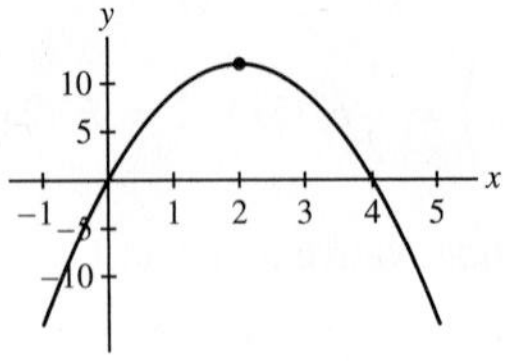

49. No horizontal asymptotes; no vertical asymptotes

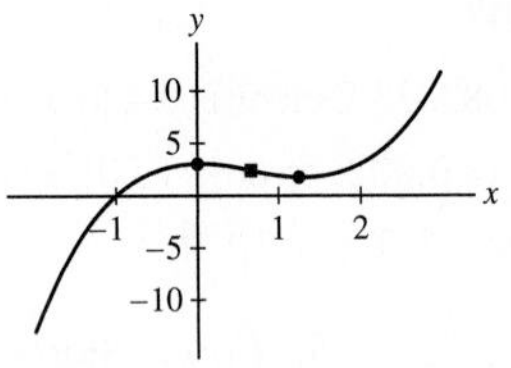

51. $y = 0$ is a horizontal asymptote; $x = -1$ is a vertical asymptote

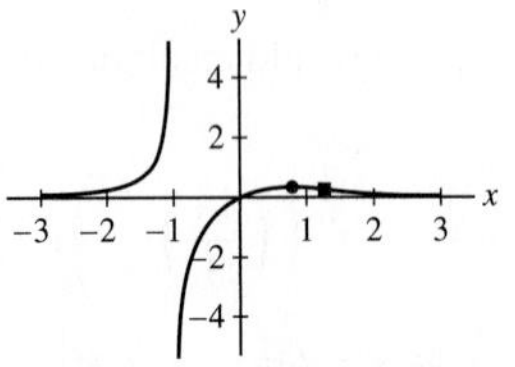

53. horizontal asymptote of $y = 0$; no vertical asymptotes

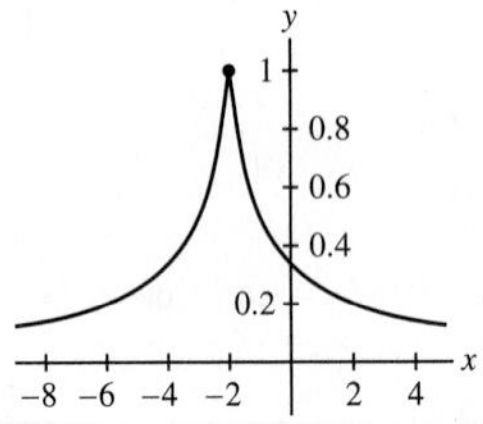

55.

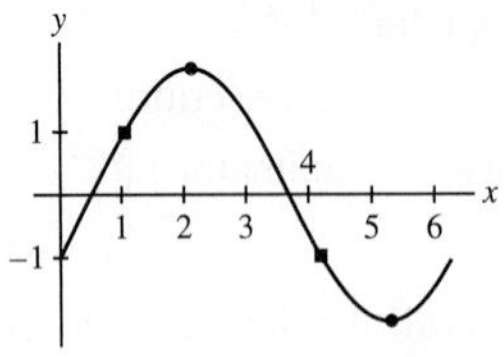

57.

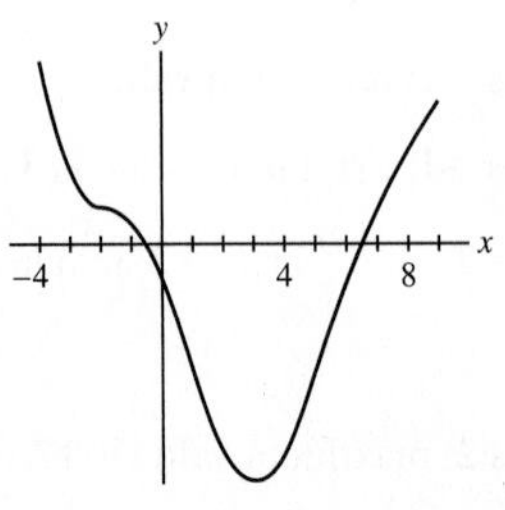

59. $b = \sqrt[3]{12}$ meters and $h = \frac{1}{3}\sqrt[3]{12}$ meters

61. $t = 8$ days

63. $\dfrac{16}{9}\pi$ **69.** $\sqrt[3]{25} = 2.9240$

71. $x^4 - \dfrac{2}{3}x^3 + C$ **73.** $-\cos(\theta - 8) + C$

75. $-2t^{-2} + 4t^{-3} + C$

77. $\tan x + C$ **79.** $\dfrac{1}{5}(y + 2)^5 + C$ **81.** $\sin\theta - \dfrac{1}{2}\theta + C$

83. $-4x^{-2} + C$ **85.** $y(x) = x^4 + 3$ **87.** $y(x) = 2x^{1/2} - 1$

89. $y(t) = t - \frac{\pi}{3}(\cos(3t) - 1)$ **91.** $f(t) = \dfrac{1}{2}t^2 - \dfrac{1}{3}t^3 - t + 2$

Chapter 5

Section 5.1 Preliminary Questions

1. The right endpoints of the subintervals are then $\frac{5}{2}, 3, \frac{7}{2}, 4, \frac{9}{2}, 5$, while the left endpoints are $2, \frac{5}{2}, 3, \frac{7}{2}, 4, \frac{9}{2}$.

2. **(a)** $\dfrac{9}{2}$ **(b)** $\dfrac{3}{2}$ and 2

3. **(a)** *Are* the same **(b)** *Not* the same

(c) *Are* the same **(d)** *Are* the same

4. The first term in the sum $\sum_{j=0}^{100} j$ is equal to zero, so it may be dropped; on the other hand, the first term in $\sum_{j=0}^{100} 1$ is not zero.

5. On [3, 7], the function $f(x) = x^{-2}$ is a decreasing function.

Section 5.1 Exercises

1. Over the interval [0, 3]: 0.96 km; over the interval [1, 2.5]: 0.5 km

3. 28.5 cm; The figure below is a graph of the rainfall as a function of time. The area of the shaded region represents the total rainfall.

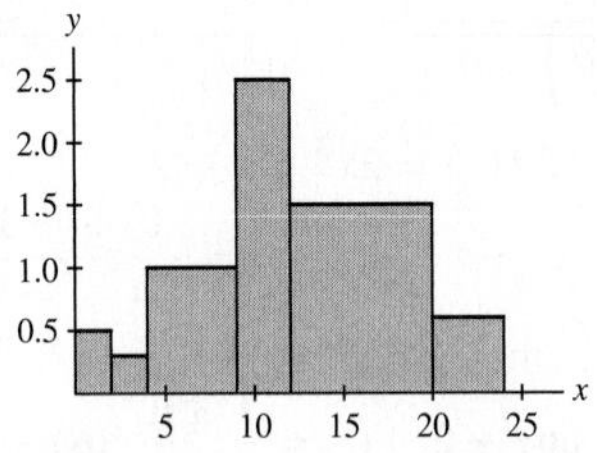

5. $L_5 = 46$; $R_5 = 44$

7. **(a)** $L_6 = 16.5$; $R_6 = 19.5$

(b) Via geometry (see figure below), the exact area is $A = 18$. Thus, L_6 underestimates the true area ($L_6 - A = -1.5$), while R_6 overestimates the true area ($R_6 - A = +1.5$).

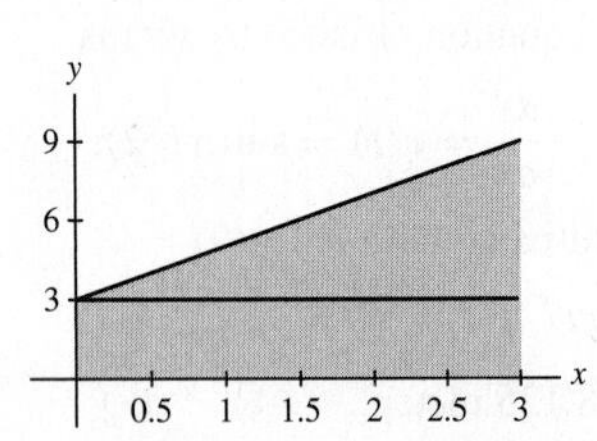

9. $R_3 = 32$; $L_3 = 20$; the area under the graph is larger than L_3 but smaller than R_3

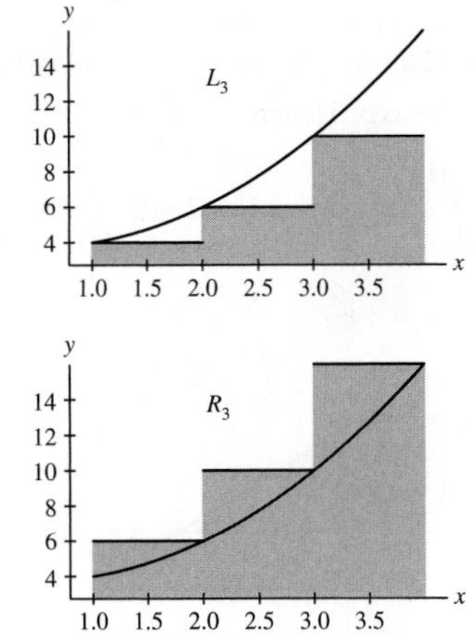

11. $R_3 = 2.5$; $M_3 = 2.875$; $L_6 = 3.4375$ **13.** $R_3 = \frac{16}{3}$

15. $M_6 = 87$ **17.** $L_6 = 12.125$ **19.** $L_4 \approx 0.410236$ **21.** $\sum_{k=4}^{8} k^7$

23. $\sum_{k=2}^{5}(2^k + 2)$ **25.** $\sum_{i=1}^{n} \frac{i}{(i+1)(i+2)}$

27. **(a)** 45 **(b)** 24 **(c)** 99

29. **(a)** -1 **(b)** 13 **(c)** 12

31. 15050 **33.** 352800 **35.** 1093350 **37.** 41650

39. -123165 **41.** $\frac{1}{2}$ **43.** $\frac{1}{3}$

45. 18; the region under the graph is a triangle with base 2 and height 18

47. 12; the region under the curve is a trapezoid with base width 4 and heights 2 and 4

49. 2; the region under the curve over $[0, 2]$ is a triangle with base and height 2

51. $\lim_{N\to\infty} R_N = 16$

53. $R_N = \frac{1}{3} + \frac{1}{2N} + \frac{1}{6N^2}; \frac{1}{3}$

55. $R_N = 222 + \frac{189}{N} + \frac{27}{N^2}; 222$

57. $R_N = 2 + \frac{6}{N} + \frac{8}{N^2}; 2$

59. $R_N = (b-a)(2a+1) + (b-a)^2 + \frac{(b-a)^2}{N}$; $(b^2 + b) - (a^2 + a)$

61. The area between the graph of $f(x) = x^4$ and the x-axis over the interval $[0, 1]$

63. The area between the graph of $y = x^4$ and the x-axis over the interval $[-2, 3]$

65. $\lim_{N\to\infty} R_N = \lim_{N\to\infty} \frac{\pi}{N} \sum_{k=1}^{N} \sin\left(\frac{k\pi}{N}\right)$

67. $\lim_{N\to\infty} L_N = \lim_{N\to\infty} \frac{4}{N} \sum_{j=0}^{N-1} \sqrt{15 + \frac{8j}{N}}$

69. $\lim_{N\to\infty} M_N = \lim_{N\to\infty} \frac{1}{2N} \sum_{j=1}^{N} \tan\left(\frac{1}{2} + \frac{1}{2N}\left(j - \frac{1}{2}\right)\right)$

71. Represents the area between the graph of $y = f(x) = \sqrt{1 - x^2}$ and the x-axis over the interval $[0, 1]$. This is the portion of the circular disk $x^2 + y^2 \le 1$ that lies in the first quadrant. Accordingly, its area is $\frac{\pi}{4}$.

73. Of the three approximations, R_N is the least accurate, then L_N and finally M_N is the most accurate.

75. The area A under the curve is somewhere between $L_4 \approx 0.518$ and $R_4 \approx 0.768$.

77. $f(x)$ is increasing over the interval $[0, \pi/2]$, so $0.79 \approx L_4 \le A \le R_4 \approx 1.18$.

79. $L_{100} = 0.793988$; $R_{100} = 0.80399$; $L_{200} = 0.797074$; $R_{200} = 0.802075$; thus, $A = 0.80$ to two decimal places.

81.

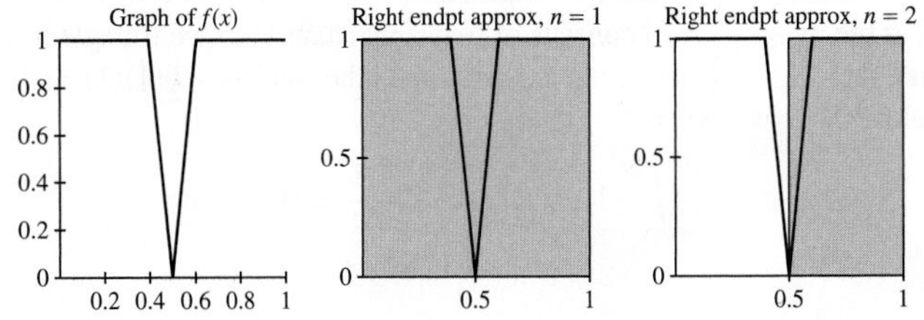

83. When f' is large, the graph of f is steeper and hence there is more gap between f and L_N or R_N.

87. $N > 30000$

Section 5.2 Preliminary Questions

1. 2

2. **(a)** False. $\int_a^b f(x)\,dx$ is the *signed* area between the graph and the x-axis.

(b) True. **(c)** True.

3. Because $\cos(\pi - x) = -\cos x$, the "negative" area between the graph of $y = \cos x$ and the x-axis over $\left[\frac{\pi}{2}, \pi\right]$ exactly cancels the "positive" area between the graph and the x-axis over $\left[0, \frac{\pi}{2}\right]$.

4. $\int_{-1}^{-5} 8\,dx$

Section 5.2 Exercises

1. The region bounded by the graph of $y = 2x$ and the x-axis over the interval $[-3, 3]$ consists of two right triangles. One has area $\frac{1}{2}(3)(6) = 9$ below the axis, and the other has area $\frac{1}{2}(3)(6) = 9$ above the axis. Hence,

$$\int_{-3}^{3} 2x\,dx = 9 - 9 = 0.$$

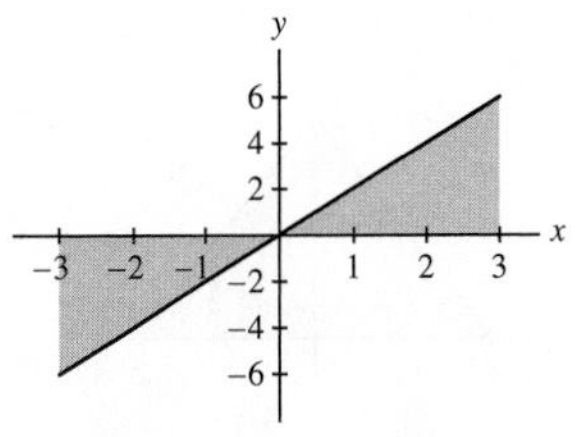

3. The region bounded by the graph of $y = 3x + 4$ and the x-axis over the interval $[-2, 1]$ consists of two right triangles. One has area $\frac{1}{2}\left(\frac{2}{3}\right)(2) = \frac{2}{3}$ below the axis, and the other has area $\frac{1}{2}\left(\frac{7}{3}\right)(7) = \frac{49}{6}$ above the axis. Hence,

$$\int_{-2}^{1} (3x + 4)\, dx = \frac{49}{6} - \frac{2}{3} = \frac{15}{2}.$$

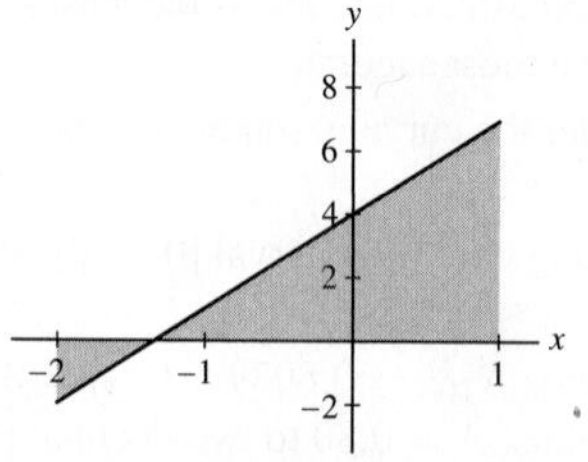

5. The region bounded by the graph of $y = 7 - x$ and the x-axis over the interval $[6, 8]$ consists of two right triangles. One triangle has area $\frac{1}{2}(1)(1) = \frac{1}{2}$ above the axis, and the other has area $\frac{1}{2}(1)(1) = \frac{1}{2}$ below the axis. Hence,

$$\int_{6}^{8} (7 - x)\, dx = \frac{1}{2} - \frac{1}{2} = 0.$$

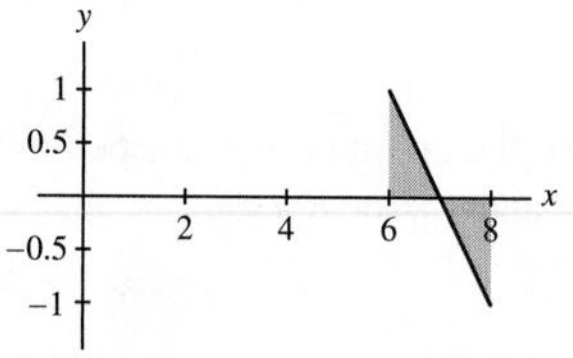

7. The region bounded by the graph of $y = \sqrt{25 - x^2}$ and the x-axis over the interval $[0, 5]$ is one-quarter of a circle of radius 5. Hence,

$$\int_{0}^{5} \sqrt{25 - x^2}\, dx = \frac{1}{4}\pi(5)^2 = \frac{25\pi}{4}.$$

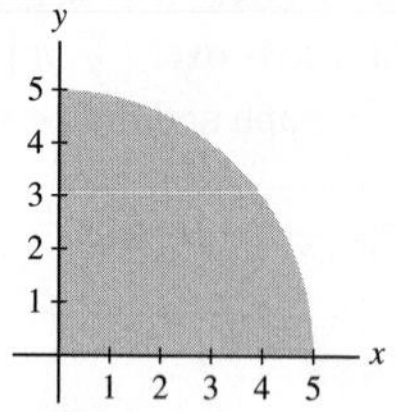

9. The region bounded by the graph of $y = 2 - |x|$ and the x-axis over the interval $[-2, 2]$ is a triangle above the axis with base 4 and height 2. Consequently,

$$\int_{-2}^{2} (2 - |x|)\, dx = \frac{1}{2}(2)(4) = 4.$$

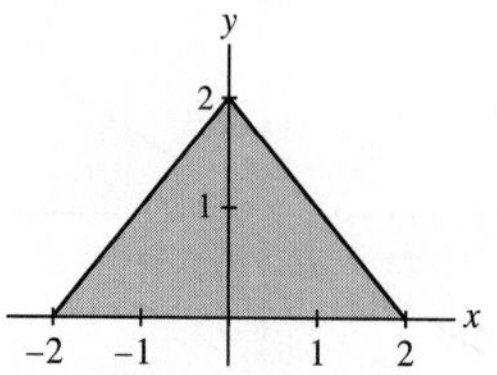

11. **(a)** $\lim_{N\to\infty} R_N = \lim_{N\to\infty} \left(30 - \frac{50}{N}\right) = 30$

(b) The region bounded by the graph of $y = 8 - x$ and the x-axis over the interval $[0, 10]$ consists of two right triangles. One triangle has area $\frac{1}{2}(8)(8) = 32$ above the axis, and the other has area $\frac{1}{2}(2)(2) = 2$ below the axis. Hence,

$$\int_{0}^{10} (8 - x)\, dx = 32 - 2 = 30.$$

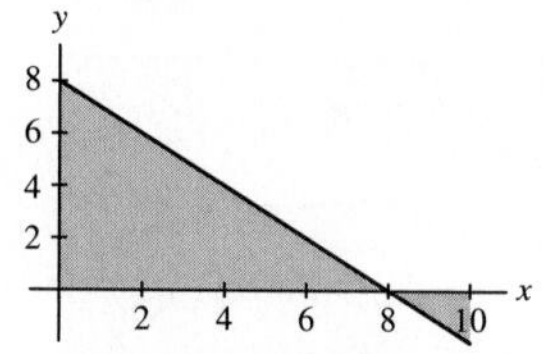

13. **(a)** $-\frac{\pi}{2}$ **(b)** $\frac{3\pi}{2}$

15. $\int_{0}^{3} g(t)\, dt = \frac{3}{2}$; $\int_{3}^{5} g(t)\, dt = 0$

17. The partition P is defined by

$$x_0 = 0 \quad < \quad x_1 = 1 \quad < \quad x_2 = 2.5 \quad < \quad x_3 = 3.2 \quad < \quad x_4 = 5$$

The set of sample points is given by $C = \{c_1 = 0.5, c_2 = 2, c_3 = 3, c_4 = 4.5\}$. Finally, the value of the Riemann sum is

$$34.25(1 - 0) + 20(2.5 - 1) + 8(3.2 - 2.5) + 15(5 - 3.2) = 96.85.$$

19. $R(f, P, C) = 1.59$; Here is a sketch of the graph of f and the rectangles.

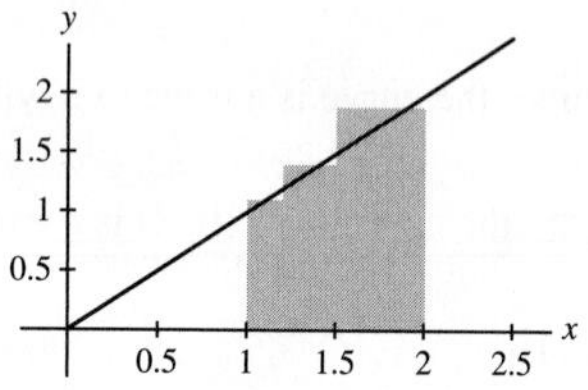

21. $R(f, P, C) = 44.625$; Here is a sketch of the graph of f and the rectangles.

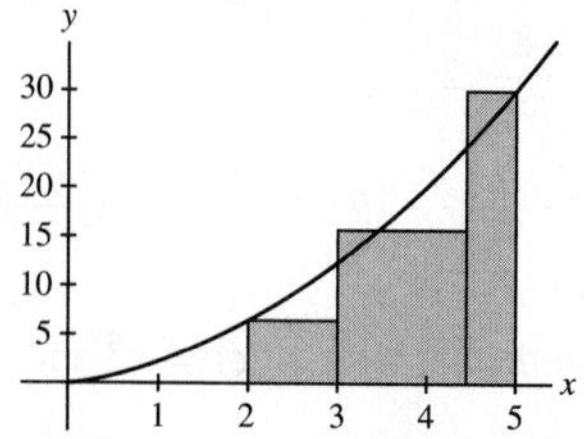

23.

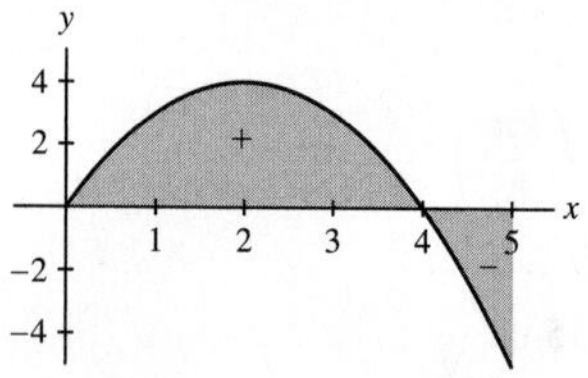

25.

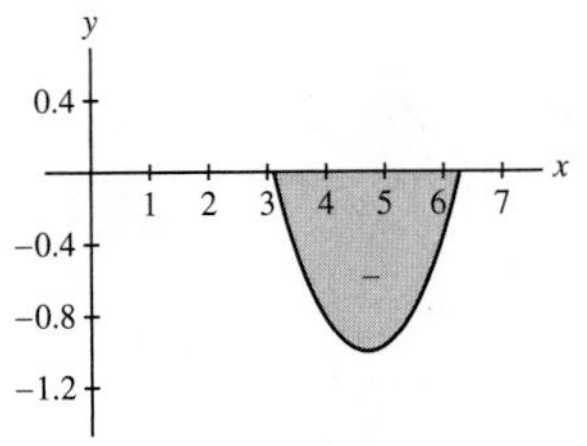

27.

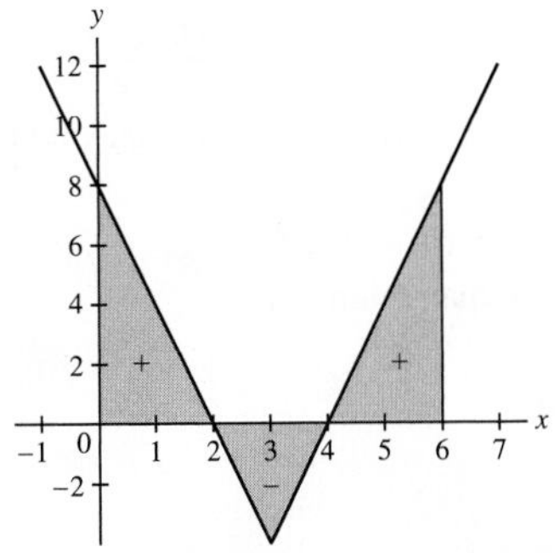

29. The integrand is always positive. The integral must therefore be positive, since the signed area has only a positive part.

31. The area below the axis is greater than the area above the axis. Thus, the definite integral is negative.

33. 36 **35.** 243 **37.** $-\frac{2}{3}$ **39.** $\frac{196}{3}$ **41.** $\frac{1}{3}a^3 - \frac{1}{2}a^2 + \frac{5}{6}$

43. 17 **45.** -12 **47.** No. **49.** $\frac{81}{4}$ **51.** $-\frac{63}{4}$ **53.** 7 **55.** 8

57. -7 **59.** $\int_0^7 f(x)\,dx$ **61.** $\int_5^9 f(x)\,dx$ **63.** $\frac{4}{5}$ **65.** $-\frac{35}{2}$

67. When $f(x)$ takes on both positive and negative values on $[a, b]$, $\int_a^b f(x)\,dx$ represents the signed area between $f(x)$ and the x-axis, whereas $\int_a^b |f(x)|\,dx$ represents the total (unsigned) area between $f(x)$ and the x-axis. Any negatively signed areas that were part of $\int_a^b f(x)\,dx$ are regarded as positive areas in $\int_a^b |f(x)|\,dx$.

69. $[-1, \sqrt{2}]$ or $[-\sqrt{2}, 1]$ **71.** 9 **73.** $\frac{1}{2}$

75. On the interval $[0, 1]$, $x^5 \le x^4$; On the other hand, $x^4 \le x^5$ for $x \in [1, 2]$.

77. $\sin x$ is increasing on $[0.2, 0.3]$. Accordingly, for $0.2 \le x \le 0.3$, we have

$$m = 0.198 \le 0.19867 \approx \sin 0.2 \le \sin x \le \sin 0.3$$
$$\approx 0.29552 \le 0.296 = M$$

Therefore, by the Comparison Theorem, we have

$$0.0198 = m(0.3 - 0.2) = \int_{0.2}^{0.3} m\,dx \le \int_{0.2}^{0.3} \sin x\,dx \le \int_{0.2}^{0.3} M\,dx$$
$$= M(0.3 - 0.2) = 0.0296.$$

79. $f(x)$ is decreasing and non-negative on the interval $[\pi/4, \pi/2]$. Therefore $0 \le f(x) \le f(\pi/4) = \frac{2\sqrt{2}}{\pi}$ for all x in $[\pi/4, \pi/2]$.

81. The assertion $f'(x) \le g'(x)$ is false. Consider $a = 0$, $b = 1$, $f(x) = x$, $g(x) = 2$. $f(x) \le g(x)$ for all x in the interval $[0, 1]$, but $f'(x) = 1$ while $g'(x) = 0$ for all x.

83. If f is an odd function, then $f(-x) = -f(x)$ for all x. Accordingly, for every positively signed area in the right half-plane where f is above the x-axis, there is a corresponding negatively signed area in the left half-plane where f is below the x-axis. Similarly, for every negatively signed area in the right half-plane where f is below the x-axis, there is a corresponding positively signed area in the left half-plane where f is above the x-axis.

Section 5.3 Preliminary Questions

1. (a) 4

(b) The signed area between $y = f(x)$ and the x-axis.

2. 3

3. (a) False. The FTC I is valid for continuous functions.

(b) False. The FTC I works for any antiderivative of the integrand.

(c) False. If you cannot find an antiderivative of the integrand, you cannot use the FTC I to evaluate the definite integral, but the definite integral may still exist.

4. 0

Section 5.3 Exercises

1. $A = \frac{1}{3}$

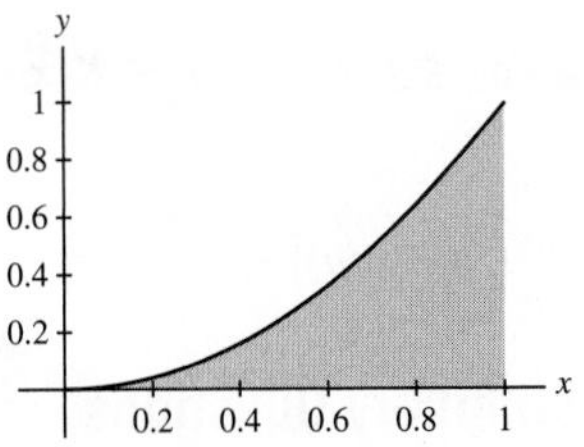

3. $A = \frac{1}{2}$

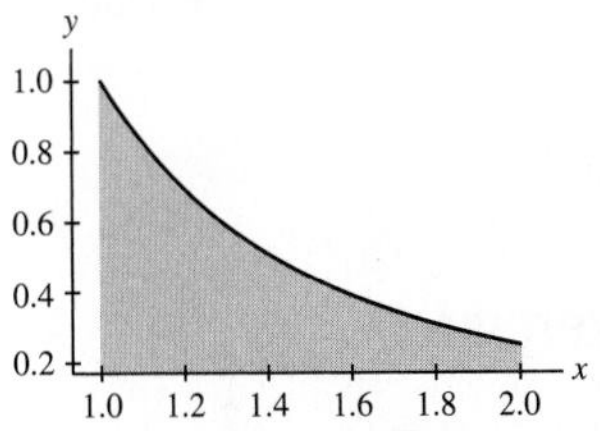

5. $\frac{27}{2}$ **7.** -8 **9.** -1 **11.** 128 **13.** $\frac{27}{2}$ **15.** $\frac{16}{3}$ **17.** $\frac{31}{40}$

19. $\frac{2}{3}$ **21.** 12 **23.** $\frac{11}{6}$ **25.** $60\sqrt{3} - \frac{8}{3}$ **27.** $\sqrt{2}$ **29.** $\frac{3}{2}$

31. $\frac{4}{3\sqrt{3}}$ **33.** $\frac{1}{5}(\sqrt{2} - 1)$ **35.** $\frac{5}{2}$ **37.** $\frac{97}{4}$ **39.** 2

41. $\frac{1}{4}\left(b^4 - 1\right)$ **43.** $\frac{1}{6}(b^6 - 1)$ **45.** $\frac{707}{12}$

47. Graphically speaking, for an odd function, the positively signed area from $x = 0$ to $x = 1$ cancels the negatively signed area from $x = -1$ to $x = 0$.

49. 24

51. $\int_0^1 x^n\,dx$ represents the area between the positive curve $f(x) = x^n$ and the x-axis over the interval $[0, 1]$. This area gets smaller as n gets larger, as is readily evident in the following graph, which shows curves for several values of n.

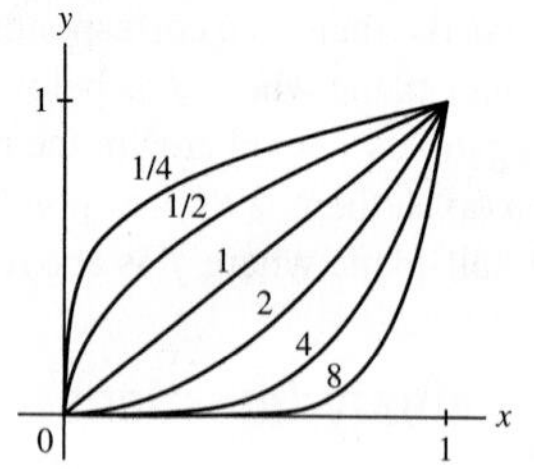

57. Let $a > b$ be real numbers, and let $f(x)$ be such that $|f'(x)| \leq K$ for $x \in [a, b]$. By FTC,

$$\int_a^x f'(t)\,dt = f(x) - f(a).$$

Since $f'(x) \geq -K$ for all $x \in [a, b]$, we get:

$$f(x) - f(a) = \int_a^x f'(t)\,dt \geq -K(x - a).$$

Since $f'(x) \leq K$ for all $x \in [a, b]$, we get:

$$f(x) - f(a) = \int_a^x f'(t)\,dt \leq K(x - a).$$

Combining these two inequalities yields

$$-K(x - a) \leq f(x) - f(a) \leq K(x - a),$$

so that, by definition,

$$|f(x) - f(a)| \leq K|x - a|.$$

Section 5.4 Preliminary Questions

1. **(a)** No **(b)** Yes

2. **(c)**

3. Yes. All continuous functions have an antiderivative, namely $\int_a^x f(t)\,dt$.

4. **(b)**, **(e)**, and **(f)**

Section 5.4 Exercises

1. $A(x) = \int_{-2}^x (2t + 4)\,dt = (x + 2)^2$.

3. $G(1) = 0$; $G'(1) = -1$ and $G'(2) = 2$; $G(x) = \frac{1}{3}x^3 - 2x + \frac{5}{3}$

5. $G(1) = 0$; $G'(0) = 0$ and $G'(\frac{\pi}{4}) = 1$

7. $\frac{1}{5}x^5 - \frac{32}{5}$ **9.** $1 - \cos x$ **11.** $\frac{1}{2} - \frac{1}{\sqrt{x}}$ **13.** $\frac{1}{2}x^4 - \frac{1}{2}$

15. $\frac{1}{4}(x^6 - 81x^2)$ **17.** $F(x) = \int_5^x \sqrt{t^3 + 1}\,dt$

19. $F(x) = \int_0^x \sec t\,dt$ **21.** $x^5 - 9x^3$ **23.** $\sec(5t - 9)$

25. **(a)** $A(2) = 4$; $A(3) = 6.5$; $A'(2) = 2$ and $A'(3) = 3$.
(b)

$$A(x) = \begin{cases} 2x, & 0 \leq x < 2 \\ \frac{1}{2}x^2 + 2, & 2 \leq x \leq 4 \end{cases}$$

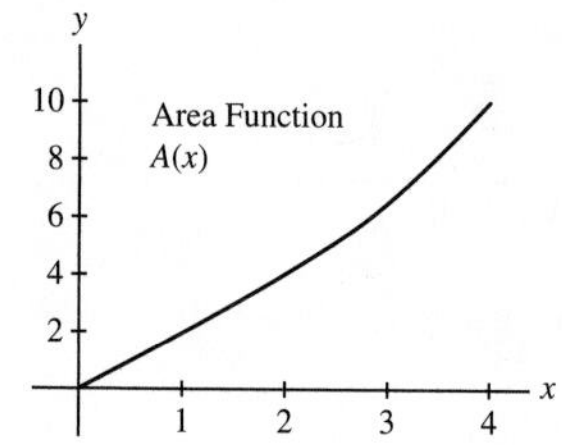

29. $\frac{2x^3}{x^2 + 1}$ **31.** $-\cos^4 s \sin s$ **33.** $2x\tan(x^2) - \frac{\tan(\sqrt{x})}{2\sqrt{x}}$

35. The minimum value of $A(x)$ is $A(1.5) = -1.25$; the maximum value of $A(x)$ is $A(4.5) = 1.25$.

37. $A(x) = (x - 2) - 1$ and $B(x) = (x - 2)$

39. **(a)** $A(x)$ does not have a local maximum at P.
(b) $A(x)$ has a local minimum at R.
(c) $A(x)$ has a local maximum at S.
(d) True.

41. $g(x) = 2x + 1$; $c = 2$ or $c = -3$

43. **(a)** If $x = c$ is an inflection point of $A(x)$, then $A''(c) = f'(c) = 0$.
(b) If $A(x)$ is concave up, then $A''(x) > 0$. Since $A(x)$ is the area function associated with $f(x)$, $A'(x) = f(x)$ by FTC II, so $A''(x) = f'(x)$. Therefore $f'(x) > 0$, so $f(x)$ is increasing.
(c) If $A(x)$ is concave down, then $A''(x) < 0$. Since $A(x)$ is the area function associated with $f(x)$, $A'(x) = f(x)$ by FTC II, so $A''(x) = f'(x)$. Therefore, $f'(x) < 0$ and so $f(x)$ is decreasing.

45. **(a)** $A(x)$ is increasing on the intervals $(0, 4)$ and $(8, 12)$ and is decreasing on the intervals $(4, 8)$ and $(12, \infty)$.
(b) Local minimum: $x = 8$; local maximum: $x = 4$ and $x = 12$.
(c) $A(x)$ has inflection points at $x = 2$, $x = 6$, and $x = 10$.
(d) $A(x)$ is concave up on the intervals $(0, 2)$ and $(6, 10)$ and is concave down on the intervals $(2, 6)$ and $(10, \infty)$.

47. The graph of one such function is:

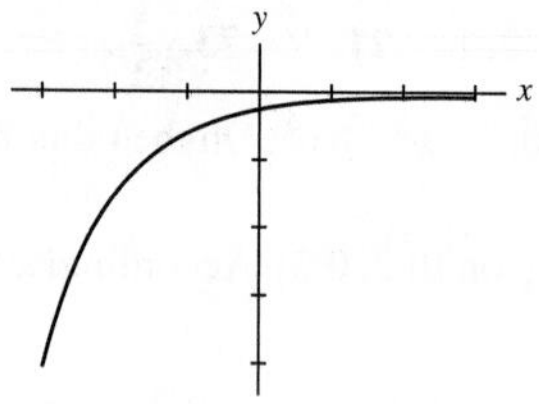

49. Smallest positive critical point: $x = (\pi/2)^{2/3}$ corresponds to a local maximum; smallest positive inflection point: $x = \pi^{2/3}$, $F(x)$ changes from concave down to concave up.

51. **(a)** Then by the FTC, Part II, $A'(x) = f(x)$ and thus $A(x)$ and $F(x)$ are both antiderivatives of $f(x)$. Hence $F(x) = A(x) + C$ for some constant C.
(b)

$$F(b) - F(a) = (A(b) + C) - (A(a) + C) = A(b) - A(a)$$
$$= \int_a^b f(t)\,dt - \int_a^a f(t)\,dt$$
$$= \int_a^b f(t)\,dt - 0 = \int_a^b f(t)\,dt$$

which proves the FTC, Part I.

53. Write

$$\int_{u(x)}^{v(x)} f(x)\,dx = \int_{u(x)}^{0} f(x)\,dx + \int_{0}^{v(x)} f(x)\,dx$$
$$= \int_{0}^{v(x)} f(x)\,dx - \int_{0}^{u(x)} f(x)\,dx.$$

Then, by the Chain Rule and the FTC,

$$\frac{d}{dx}\int_{u(x)}^{v(x)} f(x)\,dx = \frac{d}{dx}\int_{0}^{v(x)} f(x)\,dx - \frac{d}{dx}\int_{0}^{u(x)} f(x)\,dx$$
$$= f(v(x))v'(x) - f(u(x))u'(x).$$

Section 5.5 Preliminary Questions

1. The total drop in temperature of the metal object in the first T minutes after being submerged in the cold water.

2. 560 km

3. Quantities **(a)** and **(c)** would naturally be represented as derivatives; quantities **(b)** and **(d)** would naturally be represented as integrals.

Section 5.5 Exercises

1. 15250 gallons **3.** 3,660,000 **5.** 33 meters **7.** 3.675 meters

9. Displacement: 10 meters; distance: 26 meters

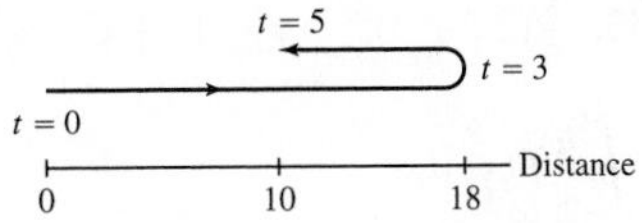

11. Displacement: 0 meters; distance: 1 meter

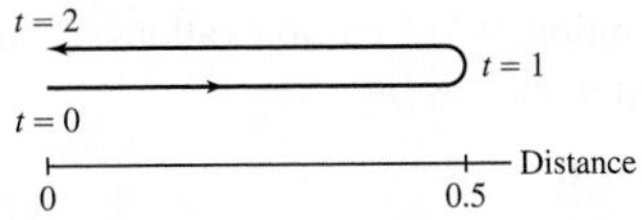

13. 39 m/s **15.** 9200 cars

17. Total cost: \$650; average cost of first 10: \$37.50; average cost of last 10: \$27.50

19. 112.5 feet

21. The area under the graph in Figure 5 represents the total power consumption over one day in California; 3.627×10^{11} joules

23. (a) 2.916×10^{10}

(b) Approximately 240526 asteroids of diameter 50 km

25. $\int_0^{365} R(t)\,dt \approx 605.05$ billion cubic feet

27. $100 \le t \le 150$: 404.968 families; $350 \le t \le 400$: 245.812 families

29. The particle's velocity is $v(t) = s'(t) = t^{-2}$, an antiderivative for which is $F(t) = -t^{-1}$. Hence, the particle's position at time t is

$$s(t) = \int_1^t s'(u)\,du = F(u)\Big|_1^t = F(t) - F(1) = 1 - \frac{1}{t} < 1$$

for all $t \ge 1$. Thus, the particle will never pass $x = 1$, which implies it will never pass $x = 2$ either.

Section 5.6 Preliminary Questions

1. (a) and **(b)**

2. (a) $u(x) = x^2 + 9$ **(b)** $u(x) = x^3$ **(c)** $u(x) = \cos x$

3. (c)

Section 5.6 Exercises

1. $du = (3x^2 - 2x)\,dx$

3. $du = -2x\sin(x^2)\,dx$

5. $du = 4\sin^3\theta\cos\theta\,d\theta$

7. $\int (x-7)^3\,dx = \int u^3\,du = \frac{1}{4}u^4 + C = \frac{1}{4}(x-7)^4 + C$

9.
$\int t\sqrt{t^2+1}\,dt = \frac{1}{2}\int u^{1/2}\,du = \frac{1}{3}u^{3/2} + C = \frac{1}{3}(t^2+1)^{3/2} + C$

11. $\int \frac{t^3}{(4-2t^4)^{11}}\,dt = -\frac{1}{8}\int u^{-11}\,du = \frac{1}{80}u^{-10} + C = \frac{1}{80}(4-2t^4)^{-10} + C$

13.

$$\int x(x+1)^9\,dx = \int (u-1)u^9\,du = \int (u^{10} - u^9)\,du$$
$$= \frac{1}{11}u^{11} - \frac{1}{10}u^{10} + C = \frac{1}{11}(x+1)^{11} - \frac{1}{10}(x+1)^{10} + C.$$

15.

$$\int x^2\sqrt{x+1}\,dx = \int (u-1)^2u^{1/2}\,du = \int (u^{5/2} - 2u^{3/2} + u^{1/2})\,du$$
$$= \frac{2}{7}u^{7/2} - \frac{4}{5}u^{5/2} + \frac{2}{3}u^{3/2} + C$$
$$= \frac{2}{7}(x+1)^{7/2} - \frac{4}{5}(x+1)^{5/2} + \frac{2}{3}(x+1)^{3/2} + C.$$

17. $\int \sin^2\theta\cos\theta\,d\theta = \int u^2\,du = \frac{1}{3}u^3 + C = \frac{1}{3}\sin^3\theta + C$

19.
$\int x\sec^2(x^2)\,dx = \frac{1}{2}\int \sec^2 u\,du = \frac{1}{2}\tan u + C = \frac{1}{2}\tan(x^2) + C$

21. $u = x^4$; $\frac{1}{4}\sin(x^4) + C$ **23.** $u = x^{3/2}$; $\frac{2}{3}\sin(x^{3/2}) + C$

25. $\frac{1}{40}(4x+5)^{10} + C$ **27.** $2\sqrt{t+12} + C$

29. $-\frac{1}{4(x^2+2x)^2} + C$ **31.** $\sqrt{x^2+9} + C$ **33.** $\frac{1}{3}(x^3+x)^3 + C$

35. $\frac{1}{36}(3x+8)^{12} + C$ **37.** $\frac{2}{9}(x^3+1)^{3/2} + C$

39. $-\frac{1}{2}(x+5)^{-2} + C$ **41.** $\frac{1}{39}(z^3+1)^{13} + C$

43. $\frac{4}{9}(x+1)^{9/4} + \frac{4}{5}(x+1)^{5/4} + C$ **45.** $\frac{1}{3}\cos(8-3\theta) + C$

47. $2\sin\sqrt{t} + C$ **49.** $\frac{2}{3}\sqrt{\sin x + 1}(\sin(x) - 2) + C$

51. $3\tan^4 x - 2\tan^3 x + C$ **53.** $\frac{1}{4}\tan(4x+9) + C$

55. $2\tan(\sqrt{x}) + C$ **57.** $-\frac{1}{6}(\cos 4x + 1)^{3/2} + C$

59. $\frac{1}{2}(\sec\theta - 1)^2 + C$ **61.** $-\frac{2}{1+\sqrt{x}} + \frac{1}{(1+\sqrt{x})^2} + C$

63. With $u = \sin x$, $\frac{1}{2}\sin^2 x + C_1$; with $u = \cos x$, $-\frac{1}{2}\cos^2 x + C_2$; the two results differ by a constant.

65. $u = \pi$ and $u = 4\pi$ **67.** 136 **69.** $\frac{3}{16}$ **71.** $\frac{98}{3}$ **73.** $\frac{243}{4}$

75. $2\sqrt{2}$ **77.** $\frac{1}{4}$ **79.** $\frac{20}{3}\sqrt{5} - \frac{32}{5}\sqrt{3}$

81. $\frac{1}{4}f(x)^4 + C$

83. Let $u = \sin\theta$. Then $u(\pi/6) = 1/2$ and $u(0) = 0$, as required. Furthermore, $du = \cos\theta\, d\theta$, so that

$$d\theta = \frac{du}{\cos\theta}.$$

If $\sin\theta = u$, then $u^2 + \cos^2\theta = 1$, so that $\cos\theta = \sqrt{1 - u^2}$. Therefore $d\theta = du/\sqrt{1 - u^2}$. This gives

$$\int_0^{\pi/6} f(\sin\theta)\, d\theta = \int_0^{1/2} f(u)\frac{1}{\sqrt{1 - u^2}}\, du.$$

85. $I = \pi/4$

Chapter 5 Review

1. $L_4 = \frac{23}{4}$; $M_4 = 7$

3. In general, R_N is larger than $\int_a^b f(x)\,dx$ on any interval $[a, b]$ over which $f(x)$ is increasing. Given the graph of $f(x)$, we may take $[a, b] = [0, 2]$. In order for L_4 to be larger than $\int_a^b f(x)\,dx$, $f(x)$ must be decreasing over the interval $[a, b]$. We may therefore take $[a, b] = [2, 3]$.

5. $R_6 = \frac{625}{8}$

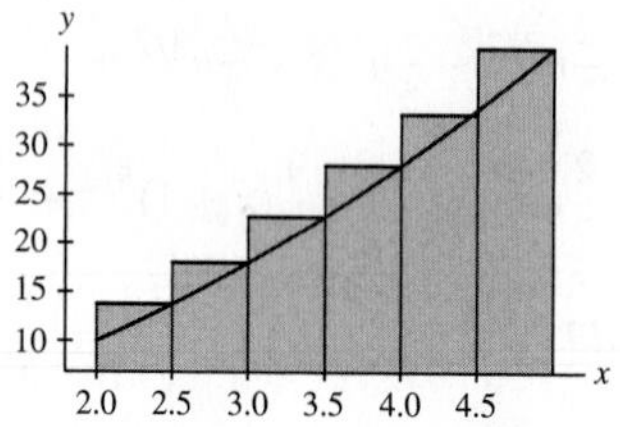

$M_6 = \frac{1127}{16}$

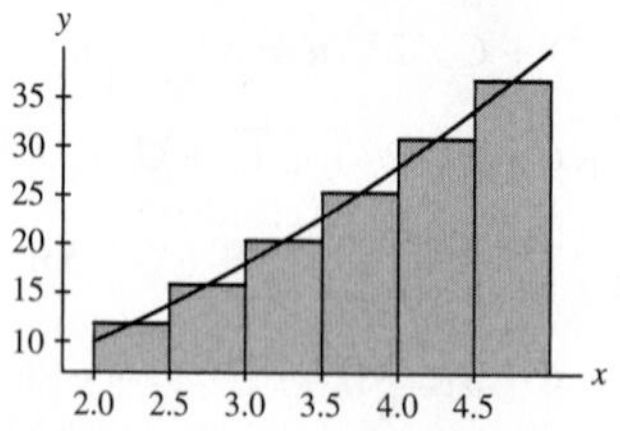

$L_6 = \frac{505}{8}$ The rectangles corresponding to this approximation are shown below.

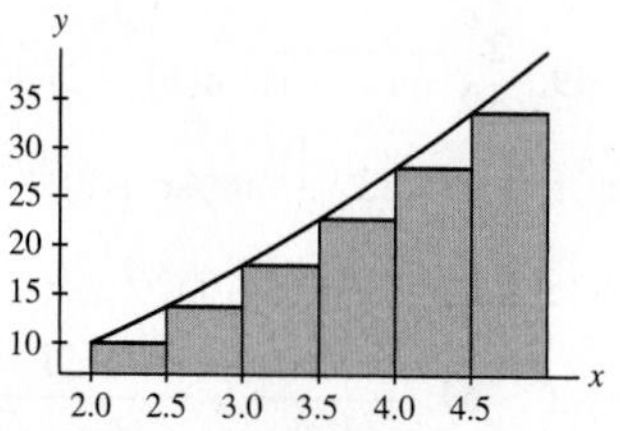

7. $R_N = \frac{141}{2} + \frac{45}{N} + \frac{9}{2N^2}$; $\frac{141}{2}$

9. $R_5 \approx 0.733732$; $M_5 \approx 0.786231$; $L_5 \approx 0.833732$

11. The area represented by the shaded rectangles is R_5; $R_5 = 90$; $L_5 = 90$

13. $\lim_{N\to\infty} \frac{\pi}{6N} \sum_{j=1}^{N} \sin\left(\frac{\pi}{3} + \frac{\pi j}{6N}\right) = \int_{\pi/3}^{\pi/2} \sin x\, dx = \frac{1}{2}$

15. $\lim_{N\to\infty} \frac{5}{N} \sum_{j=1}^{N} \sqrt{4 + 5j/N} = \int_4^9 \sqrt{x}\, dx = \frac{38}{3}$

17. $\frac{1}{120}$ **19.** $\frac{1}{5}\left(1 - \frac{9\sqrt{3}}{32}\right)$

21. $4x^5 - \frac{9}{4}x^4 - x^2 + C$ **23.** $\frac{4}{5}x^5 - 3x^4 + 3x^3 + C$

25. $\frac{1}{4}x^4 + x^3 + C$ **27.** $\frac{46}{3}$ **29.** 3

31. $\frac{1}{150}(10t - 7)^{15} + C$ **33.** $-\frac{1}{24}(3x^4 + 9x^2)^{-4} + C$

35. 506 **37.** $-\frac{3\sqrt{3}}{2\pi}$

39. $\frac{1}{27}\tan(9t^3 + 1) + C$ **41.** $\frac{1}{2}\cot(9 - 2\theta) + C$

43. $3 - \frac{3\sqrt[3]{4}}{2}$ **45.** $\frac{1}{5}(2y + 3)^{3/2}(y - 1) + C$

47. $\tan(1)$ **49.** $\int_{-2}^{6} f(x)\, dx$

51. Local minimum at $x = 0$, no local maxima, inflection points at $x = \pm 1$

53. Daily consumption: 9.312 million gallons; From 6 PM to midnight: 1.68 million gallons

55. $208,245 **57.** 0

59. The function $f(x) = 2^x$ is increasing, so $1 \le x \le 2$ implies that $2 = 2^1 \le 2^x \le 2^2 = 4$. Consequently,

$$2 = \int_1^2 2\, dx \le \int_1^2 2^x\, dx \le \int_1^2 4\, dx = 4.$$

On the other hand, the function $f(x) = 3^{-x}$ is decreasing, so $1 \le x \le 2$ implies that

$$\frac{1}{9} = 3^{-2} \le 3^{-x} \le 3^{-1} = \frac{1}{3}.$$

It then follows that

$$\frac{1}{9} = \int_1^2 \frac{1}{9}\, dx \le \int_1^2 3^{-x}\, dx \le \int_1^2 \frac{1}{3}\, dx = \frac{1}{3}.$$

61. $\frac{4}{3} \le \int_0^1 f(x)\, dx \le \frac{5}{3}$ **63.** $-\frac{1}{1 + \pi}$

65. $\sin^3 x \cos x$ **67.** -2

69. Consider the figure below, which displays a portion of the graph of a linear function.

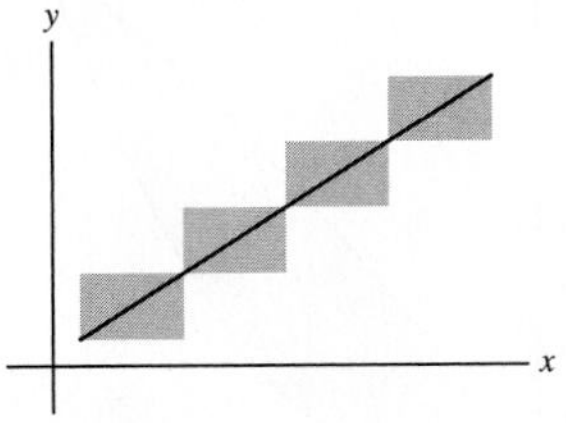

The shaded rectangles represent the differences between the right-endpoint approximation R_N and the left-endpoint approximation L_N. Because the graph of $y = f(x)$ is a line, the lower portion of each shaded rectangle is exactly the same size as the upper portion. Therefore, if we average L_N and R_N, the error in the two approximations will exactly cancel, leaving

$$\frac{1}{2}(R_N + L_N) = \int_a^b f(x)\,dx.$$

Chapter 6

Section 6.1 Preliminary Questions

1. Area of the region between the graphs of $y = f(x)$ and $y = g(x)$, bounded on the left by the vertical line $x = a$ and on the right by the vertical line $x = b$.

2. Yes

3. $\int_0^3 (f(x) - g(x))\,dx - \int_3^5 (g(x) - f(x))\,dx$

4. Negative

Section 6.1 Exercises

1. 102 **3.** $\frac{32}{3}$

5. $\sqrt{2} - 1$

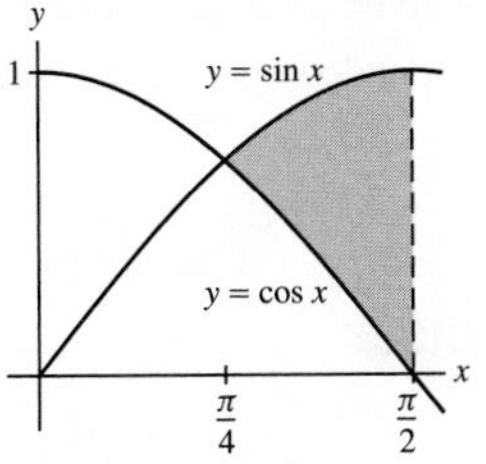

7. $\frac{343}{3}$

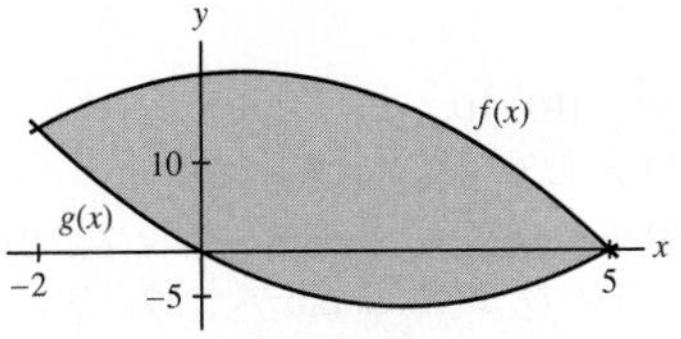

9. The curves intersect at $(\pm 1, 0)$:

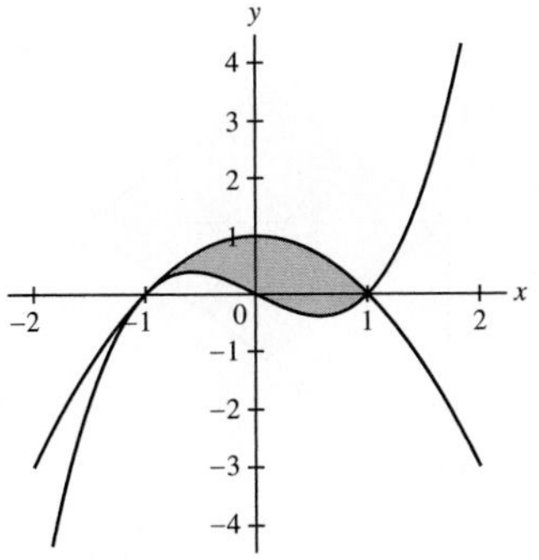

$$\int_{-1}^{1} (1 - x^2) - x(x^2 - 1)\,dx = \frac{4}{3}$$

11. $\pi - 2$

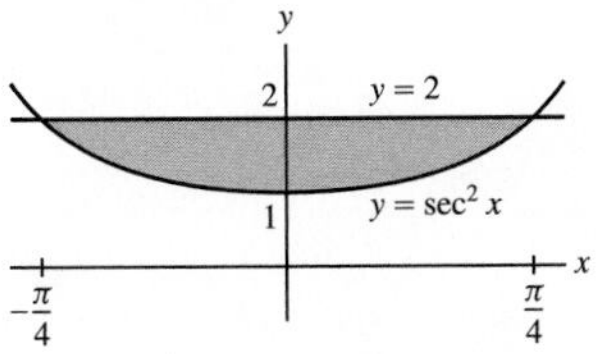

13. $\frac{160}{3}$ **15.** $\frac{12\sqrt{3}-12+(\sqrt{3}-2)\pi}{24}$ **17.** $2 - \frac{\pi}{2}$ **19.** $\frac{1,331}{6}$

21. 256 **23.** $\frac{32}{3}$ **25.** $\frac{64}{3}$

27. $\frac{64}{3}$

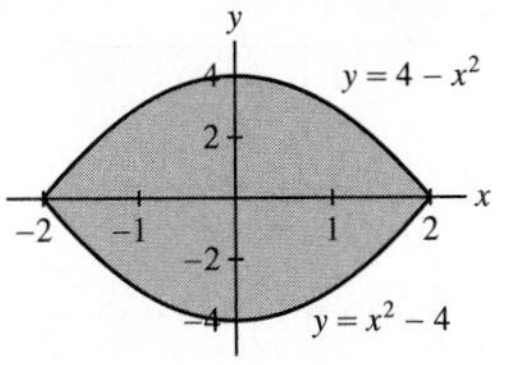

29. 2

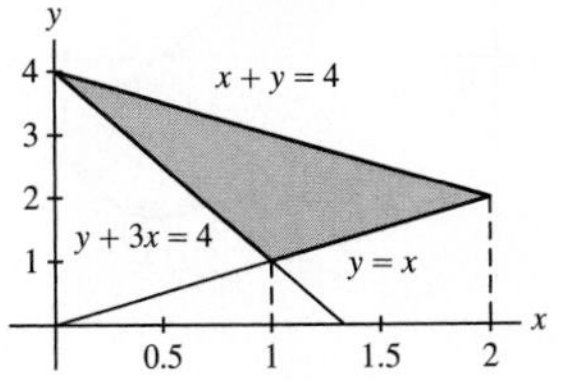

31. $\frac{128}{3}$

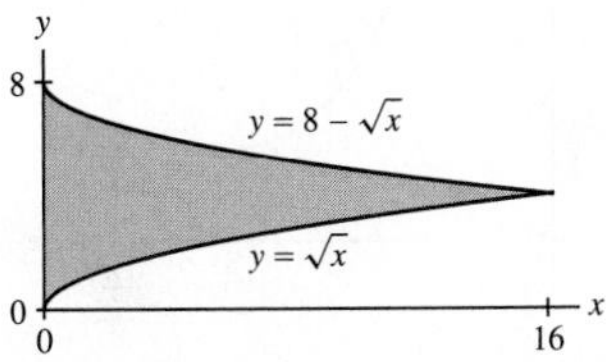

33. $\frac{1}{2}$

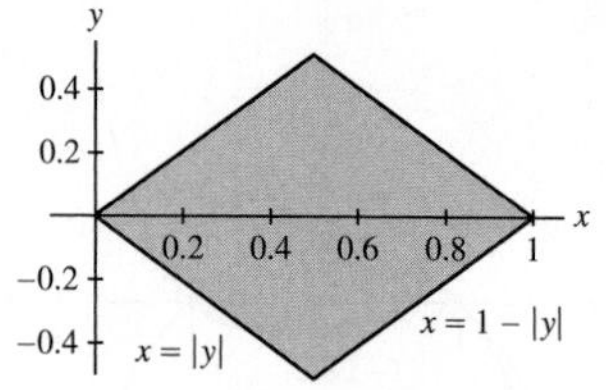

35. $\frac{1{,}225}{8}$

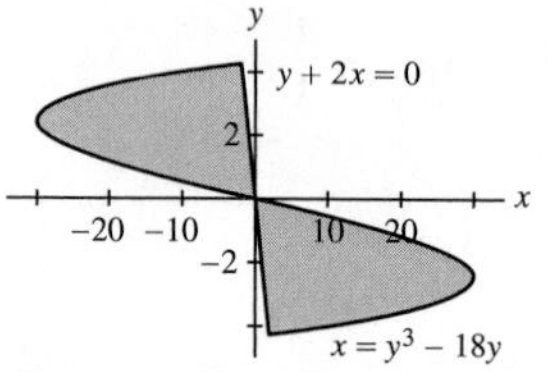

37. $\frac{32}{3}$

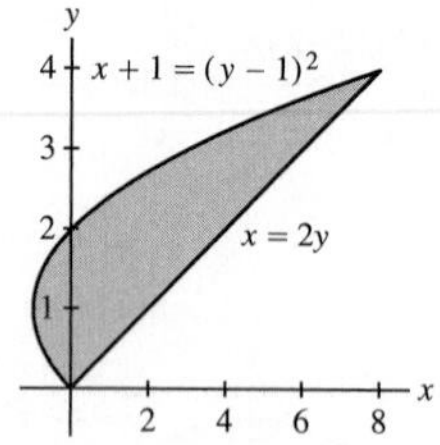

39. $\frac{3\sqrt{3}}{4}$

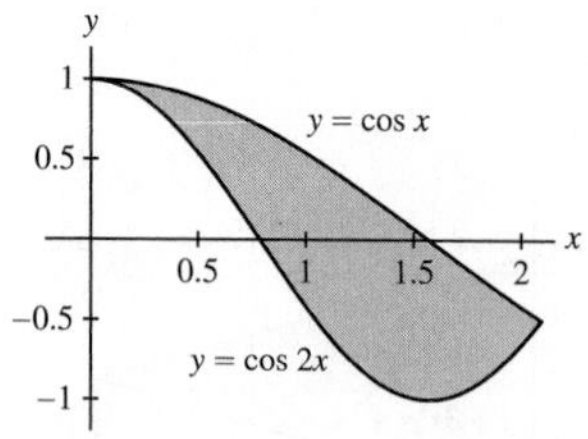

41. $\frac{2-\sqrt{2}}{2}$

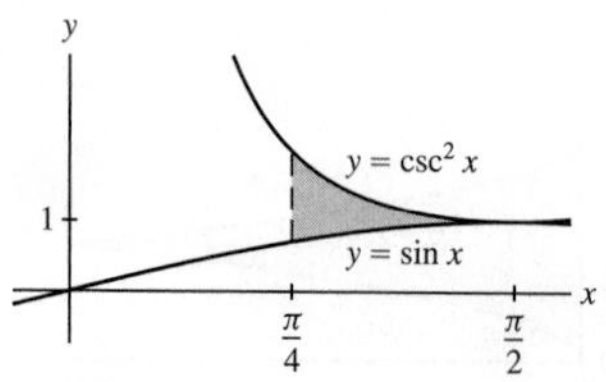

43. $\frac{1}{2}(\cos 1 - 1) \approx 0.2298$

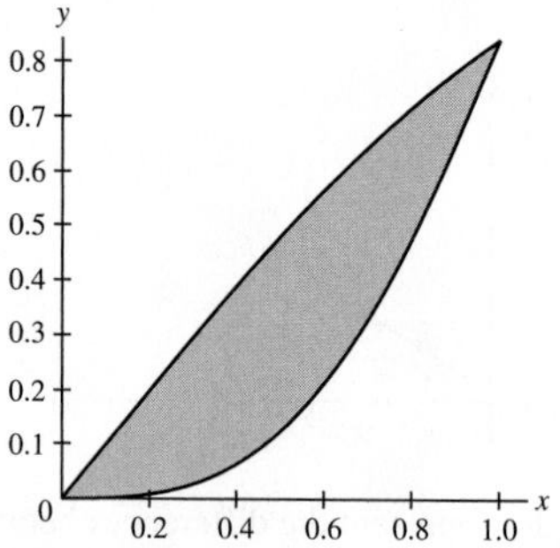

45. ≈ 0.7567130951

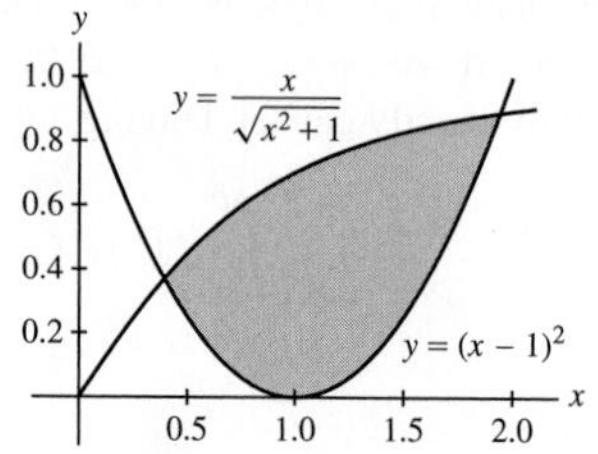

47. (a) (ii) **(b)** No
(c) At 10 seconds, athlete 1; at 25 seconds, athlete 2.

49. $\frac{8}{3}c^{3/2}$; $c = \frac{9^{1/3}}{4} \approx 0.520021$.

51. $\int_{-\sqrt{(-1+\sqrt{5})/2}}^{\sqrt{(-1+\sqrt{5})/2}} \left[(1+x^2)^{-1} - x^2\right] dx$

53. 0.8009772242 **55.** 214.75 in^2

57. (b) $\frac{1}{3}$ **(c)** 0 **(d)** 1

59. $m = 1 - \left(\frac{1}{2}\right)^{1/3} \approx 0.206299$

Section 6.2 Preliminary Questions

1. 3 **2.** 15

3. Flow rate is the volume of fluid that passes through a cross-sectional area at a given point per unit time.

4. The fluid velocity depended only on the radial distance from the center of the tube.

5. 15

Section 6.2 Exercises

1. (a) $\frac{4}{25}(20 - y)^2$
(b) $\frac{1{,}280}{3}$

3. $\frac{\pi R^2 h}{3}$ **5.** $\pi\left(Rh^2 - \frac{h^3}{3}\right)$ **7.** $\frac{1}{6}abc$ **9.** $\frac{8}{3}$ **11.** 36 **13.** 18
15. $\frac{\pi}{3}$ **17.** 96π

21. (a) $2\sqrt{r^2 - y^2}$ **(b)** $4(r^2 - y^2)$ **(c)** $\frac{16}{3}r^3$

23. 160π **25.** 5 kg **27.** 0.36 g

29. $P \approx 4{,}423.59$ thousand **31.** $L_{10} = 442.24$, $R_{10} = 484.71$

33. $P \approx 61$ deer **35.** $Q = 128\pi\ \text{cm}^3/\text{s}$ **37.** $Q = \frac{8\pi}{3}\ \text{cm}^3/\text{s}$

39. 16 **41.** $\frac{3}{\pi}$ **43.** $\frac{1}{10}$ **45.** -4 **47.** $\frac{1}{n+1}$

49. Over [0,24], the average temperature is 20; over [2,6] the average temperature is $20 + \frac{15}{2\pi} \approx 22.387325$.

51. $\frac{17}{2}$ m/s

53. Average acceleration $= -80$ m/s^2; average speed $= 20\sqrt{5} + 104$ m/s ≈ 148.7213596 m/s

55. $\frac{3}{5^{1/4}} \approx 2.006221$

57. Mean Value Theorem for Integrals; $c = \frac{A}{\sqrt[3]{4}}$

59. Over [0, 1], $f(x)$; over [1, 2], $g(x)$.

61. Many solutions exist. One could be:

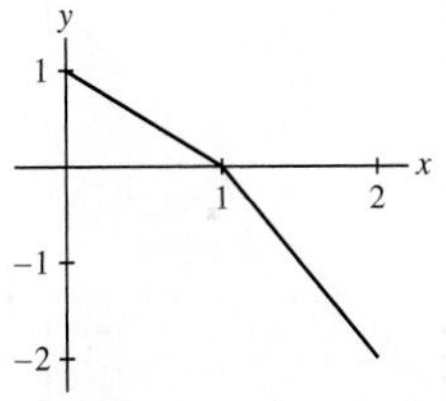

63. $v_0/2$

Section 6.3 Preliminary Questions

1. (a), (c) **2.** True

3. False, the cross sections will be washers.

4. (b)

Section 6.3 Exercises

1. (a)

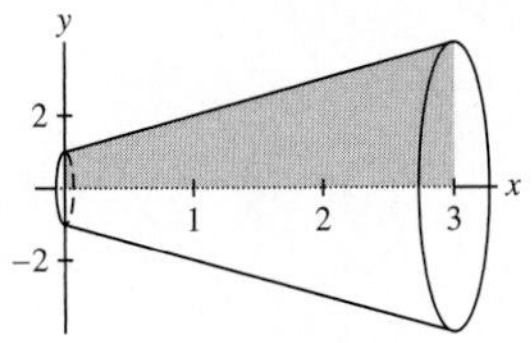

(b) Disk with radius $x + 1$

(c) $V = 21\pi$

3. (a)

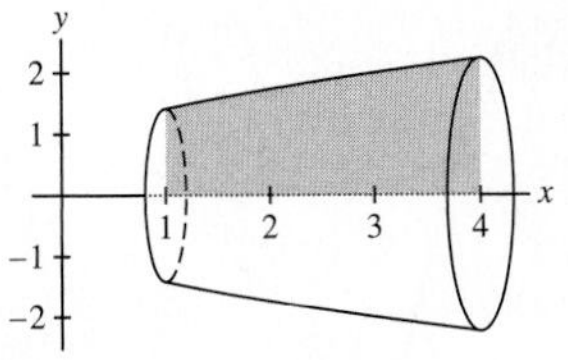

(b) Disk with radius $\sqrt{x+1}$

(c) $V = \frac{21\pi}{2}$

5. $V = \frac{81\pi}{10}$ **7.** $V = \frac{24{,}573\pi}{13}$ **9.** $V = \pi$

11. $V = 2\pi$ **13.** (iv)

15. (a)

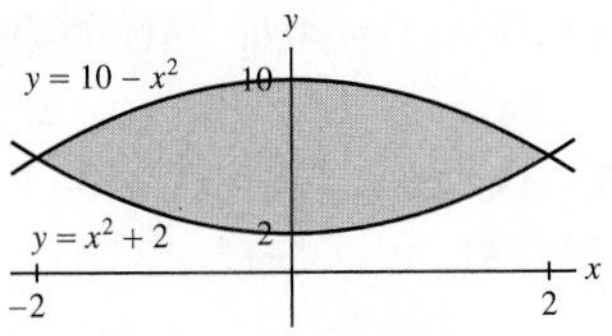

(b) A washer with outer radius $R = 10 - x^2$ and inner radius $r = x^2 + 2$.

(c) $V = 256\pi$

17. (a)

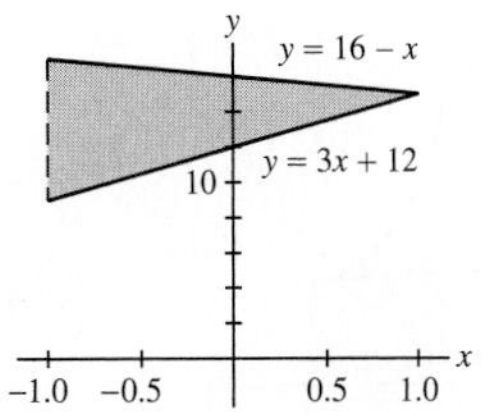

(b) A washer with outer radius $R = 16 - x$ and inner radius $r = 3x + 12$.

(c) $V = \frac{656\pi}{3}$

19. (a)

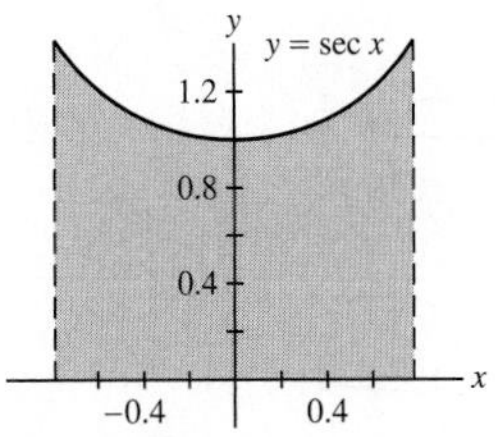

(b) A circular disk with radius $R = \sec x$.

(c) $V = 2\pi$

21. $V = \frac{15\pi}{2}$ **23.** $V = \frac{3\pi}{10}$ **25.** $V = 32\pi$ **27.** $V = \frac{704\pi}{15}$

29. $V = \frac{128\pi}{5}$ **31.** $V = 40\pi$ **33.** $V = \frac{376\pi}{15}$ **35.** $V = \frac{824\pi}{15}$

37. $V = \frac{32\pi}{3}$ **39.** $V = \frac{1{,}872\pi}{5}$ **41.** $V = \frac{1{,}400\pi}{3}$

43. $V = \pi\left(\frac{7\pi}{9} - \sqrt{3}\right)$ **45.** $V = \frac{96\pi}{5}$ **47.** $V = \frac{32\pi}{35}$

49. $V = \frac{1184\pi}{15}$ **51.** $V = \frac{9}{8}\pi$ **53.** $V \approx 43{,}000$ cm^3

55. $V = \frac{1}{3}\pi r^2 h$

57. $V = \frac{32\pi}{105}$

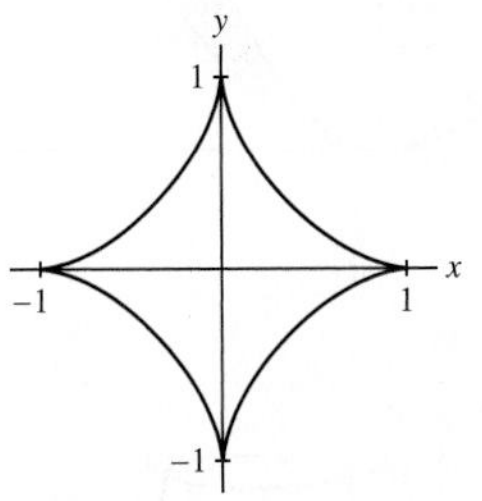

59. $V = 4\pi\sqrt{3}$ **61.** $V = \frac{4}{3}\pi a^2 b$

Section 6.4 Preliminary Questions

1. (a) Radius h and height r. **(b)** Radius r and height h.

2. (a) With respect to x. **(b)** With respect to y.

Section 6.4 Exercises

1. $V = \frac{2}{5}\pi$

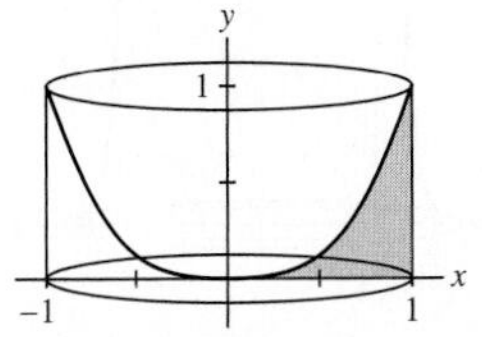

3. $V = 4\pi$

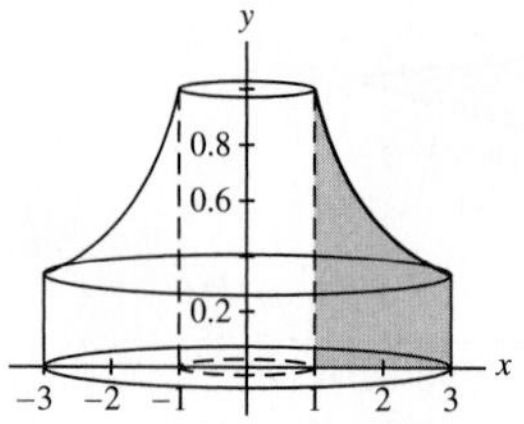

5. $V = 18\pi\left(2\sqrt{2} - 1\right)$

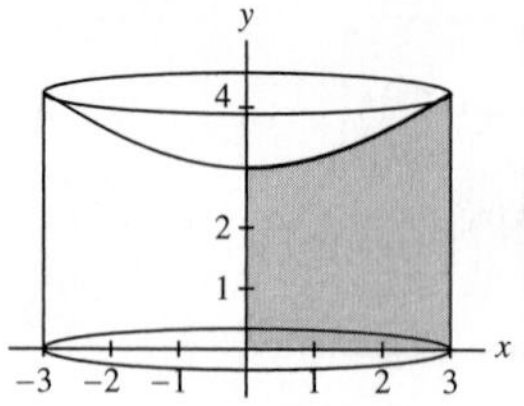

7. $V = \frac{32\pi}{3}$ **9.** $V = 16\pi$ **11.** $V = \frac{32\pi}{5}$

13. The points of intersection are $x = 0$, $y = 0$ and $x \approx \pm 1.376769504, 0.9477471335$, and $V \approx 1.321975576$.

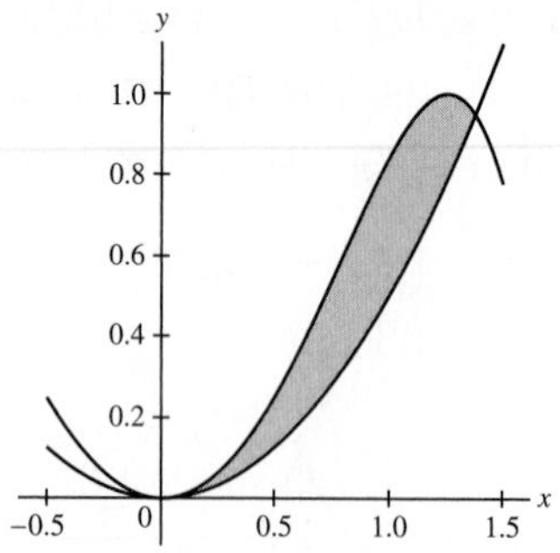

15. $V = \frac{3\pi}{5}$

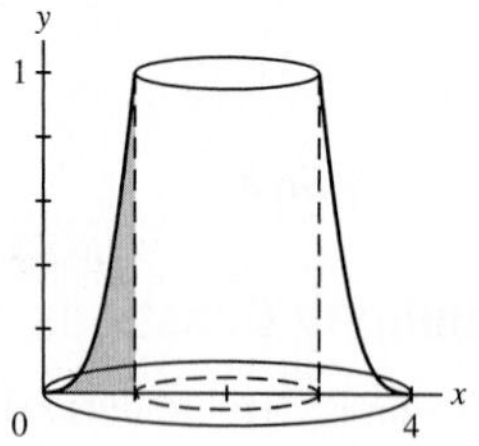

17. $V = \frac{280\pi}{81}$

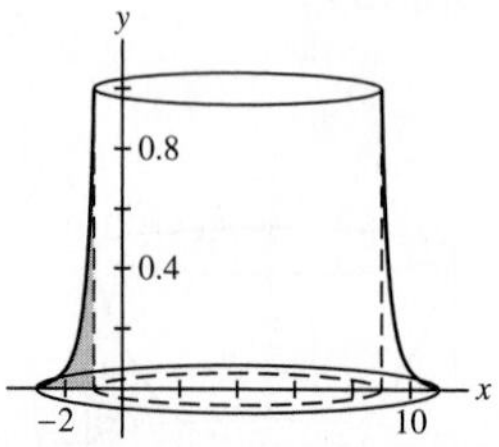

19. $V = \frac{1}{3}\pi a^3 + \pi a^2$

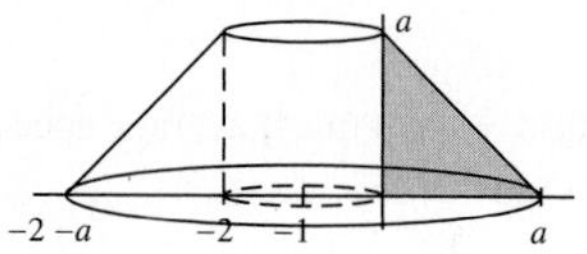

21. $V = \frac{\pi}{3}$

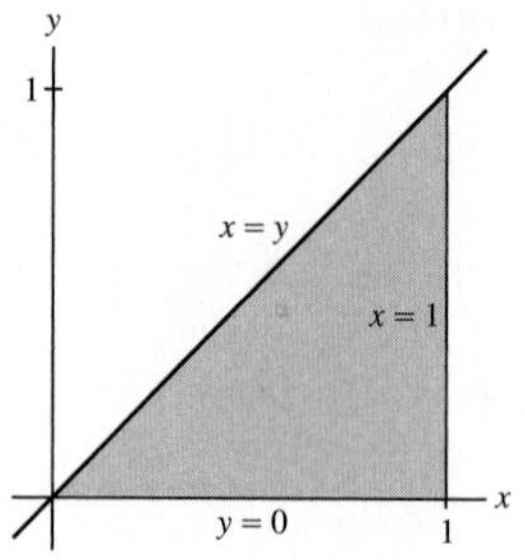

23. $V = \frac{128\pi}{3}$

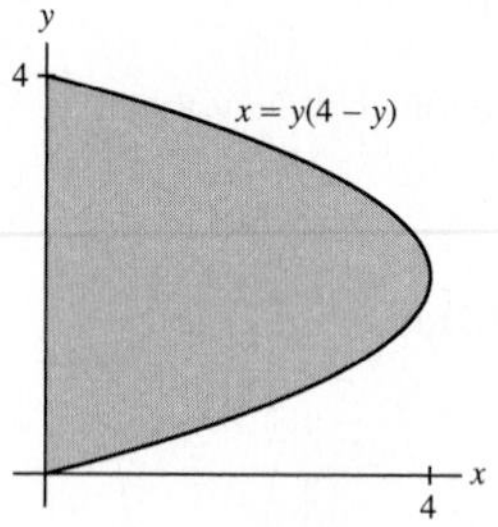

25. $V = 8\pi$

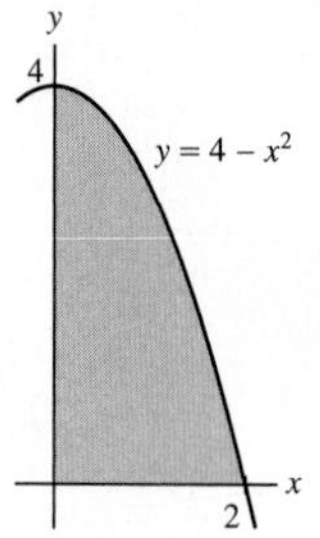

27. **(a)** $V = \frac{576\pi}{7}$ **(b)** $V = \frac{96\pi}{5}$

29. **(a)** $\overline{AB}$ generates a disk with radius $R = h(y)$; $\overline{CB}$ generates a shell with radius x and height $f(x)$.

(b) Shell, $V = 2\pi \int_0^2 x f(x)\,dx$; Disk, $V = \pi \int_0^{1.3} (h(y))^2\,dy$.

31. $V = \frac{602\pi}{5}$ **33.** $V = 8\pi$ **35.** $V = \frac{40\pi}{3}$ **37.** $V = \frac{1{,}024\pi}{15}$
39. $V = 16\pi$ **41.** $V = \frac{32\pi}{3}$ **43.** $V = \frac{776\pi}{15}$ **45.** $V = \frac{625\pi}{6}$
47. $V = \frac{121\pi}{525}$ **49.** $V = \frac{563\pi}{30}$ **51.** $V = \frac{4}{3}\pi r^3$
53. $V = 2\pi^2 ab^2$

55. **(b)** $V \approx 4\pi\left(\frac{R}{N}\right)\sum_{k=1}^{N}\left(\frac{kR}{N}\right)^2$ **(c)** $V = \frac{4}{3}\pi R^3$

Section 6.5 Preliminary Questions

1. Because the required force is not constant through the stretching process.

2. The force involved in lifting the tank is the weight of the tank, which is constant.

3. $\frac{1}{2}kx^2$

Section 6.5 Exercises

1. $W = 627.2$ J **3.** $W = 5.76$ J **5.** $W = 8$ J **7.** $W = 11.25$ J

9. $W = 3.800$ J **11.** $W = 105,840$ J

13. $W = \frac{56,448\pi}{5}$ J $\approx 3.547 \times 10^4$ J **15.** $W \approx 1.842 \times 10^{12}$ J

17. $W = 3.92 \times 10^{-6}$ J **19.** $W \approx 1.18 \times 10^8$ J

21. $W = 9800\pi \ell r^3$ J **23.** $W = 2.94 \times 10^6$ J

25. $W \approx 1.222 \times 10^6$ J **27.** $W = 3920$ J **29.** $W = 529.2$ J

31. $W = 1,470$ J **33.** $W = 374.85$ J

37. $W \approx 5.16 \times 10^9$ J **41.** $\sqrt{2GM_e\left(\frac{1}{R_e} - \frac{1}{r+R_e}\right)}$ m/s

43. $v_{\text{esc}} = \sqrt{\frac{2GM_e}{R_e}}$ m/s

Chapter 6 Review

1. $\frac{32}{3}$ **3.** $\frac{1}{2}$ **5.** 24 **7.** $\frac{1}{2}$ **9.** $3\sqrt{2} - 1$ **11.** $\frac{4}{\pi}\left(1 - 2\tan\left(\frac{\pi}{8}\right)\right)$

13. Intersection points $x \approx \pm 0.8241323123$; Area ≈ 1.094753609

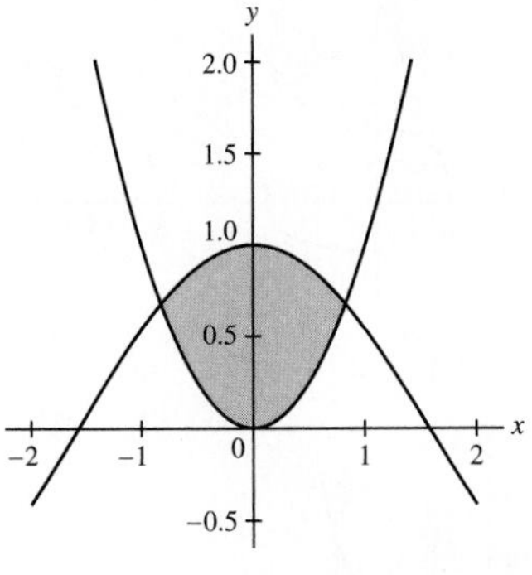

15. $V = 4\pi$ **17.** 2.7552 kg **19.** $\frac{9}{4}$ **21.** $\frac{1}{72}(625 \cdot 5^{3/5} - 5)$

23. $\frac{3\pi}{4}$ **25.** 27 **27.** $\frac{2\pi m^5}{15}$ **29.** $V = \frac{162\pi}{5}$ **31.** $V = 64\pi$

33. $V = 8\pi$ **35.** $V = \frac{56\pi}{15}$ **37.** $V = \frac{128\pi}{15}$ **39.** $V = \pi$

41. $V = 2\pi\left(c + \frac{c^3}{3}\right)$ **43.** $V = c\pi$

45. **(a)** $\int_0^1 \left(\sqrt{1-(x-1)^2} - (1 - \sqrt{1-x^2})\right) dx$

(b) $\pi \int_0^1 \left[(1-(x-1)^2) - (1-\sqrt{1-x^2})^2\right] dx$

47. $W = 1.08$ J **49.** 0.75 ft

51. $W = 117600\pi$ J $\approx 3.695 \times 10^5$ J **53.** $W = 98,000$ J

Chapter 7

Section 7.1 Preliminary Questions

1. Equations (a), (b), and (d) are correct, but equation (c) is not: $3^2 \cdot 2^3 = 9 \cdot 8 = 72$.

2. The domain of $\ln x$ is $(0, \infty)$; its range is $(-\infty, \infty)$. $\ln x < 0$ when $0 < x < 1$.

3. (a), (c), and (f). The others do not have a variable base and a constant exponent.

4. $0 < b < 1$. **5.** For all $b > 0$ except $b = 1$.

6. $(0, 1)$. **7.** (c).

Section 7.1 Exercises

1. (a) 1 (b) 29 (c) 1 (d) 81 (e) 16 (f) 0 **3.** $x = 4$

5. $x = -\frac{1}{2}$ **7.** $x = -\frac{1}{3}$

9. $k = 9$ **11.** ∞ **13.** ∞

15. $y = 4x + 4$ **17.** $y = e(x+1) + e$

19. $14e^{2x} + 12e^{4x}$ **21.** $\pi e^{\pi x}$

23. $-4e^{-4x+9}$ **25.** $e^{x^2}\frac{2x^2-1}{x^2}$

27. $4(1+e^x)^3 e^x$ **29.** $(2x+2)e^{x^2+2x-3}$

31. $\cos x \cdot e^{\sin x}$ **33.** $e^\theta \cos(e^\theta)$ **35.** $-\frac{3e^{-3t}}{(1-e^{-3t})^2}$

37. $e^x \frac{3x-2}{(3x+1)^2}$ **39.** $\frac{2e^x - 2e^x x - e^{x+1} - 1}{(2e^x - 1)^2}$

41. $16e^{4x-3}$ **43.** $2e^t \cos t$

45. $((1-2t)^2 - 2)e^{t-t^2}$

47. Local minimum at $x = 0$

49. Local minimum at $x = 1$

51. Critical point at $t = 1$, neither maximum nor minimum

53. Critical point, local maximum, at $x = 1$; point of inflection at $x = 2$.

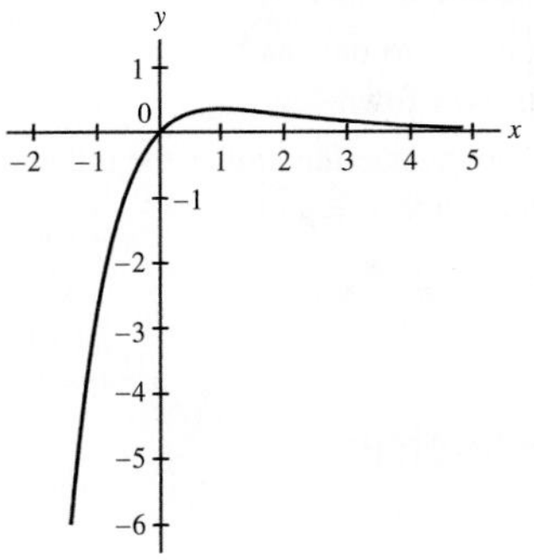

55. Critical point, local maximum, at $x = -\frac{\pi}{4}$; inflection point at $x = 0$.

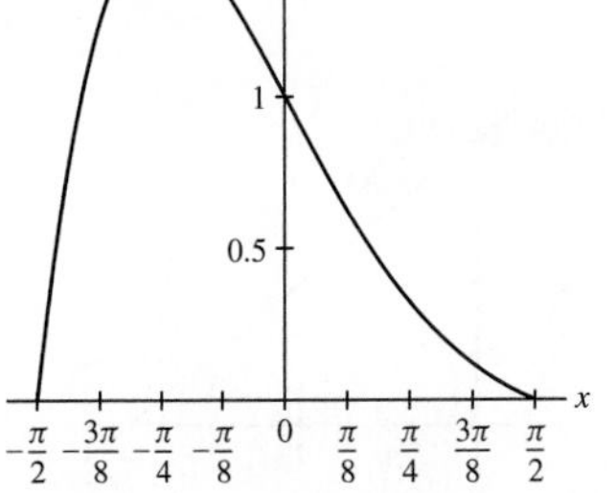

57. Critical point, local minimum, at $x = 0$; no inflection points.

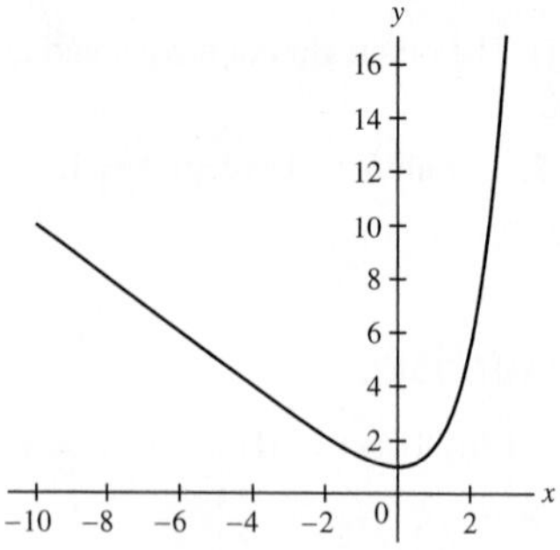

59. $a = 1$ **61.** $y = x$ **63.** $y = x + 1$; $e^{-0.1} \approx 0.99$

65. (a) 1.29 cm/yr (b) 2.33 yr (c) 32 cm

67. $kb^{1/m-1}(b - a)$ **69.** $e^x + 2x + C$

71. $\frac{1}{3}(1 - e^{-3})$ **73.** $\frac{1}{6}(e - e^{-17})$

75. $\frac{1}{4}e^{4x} + x + C$ **77.** $1 - e^{-1/2}$

79. $\frac{2}{3}(e^t + 1)^{3/2} + C$ **81.** $e^x - \frac{1}{3}e^{3x} + C$ **83.** $2\sqrt{e^x + 1} + C$

85. $2e^{\sqrt{x}} + C$ **87.** $\frac{1}{2} + \frac{1}{2}e^2 - e$ **89.** $e^2 + 1$

91. (b) $e^{-1/16} - e^{-25/64} \approx 0.263$

97. $f'(x) = e^x + xe^x$; $f''(x) = 2e^x + xe^x$; $f'''(x) = 3e^x + xe^x$; $f^{(n)}(x) = ne^x + xe^x$

Section 7.2 Preliminary Questions

1. **(a), (b), (f)** **2.** No

3. Many different teenagers will have the same last name, so this function will not be one-to-one.

4. This function is one-to-one, and $f^{-1}(6{:}27) =$ Hamilton Township.

5. The graph of the inverse function is the reflection of the graph of $y = f(x)$ through the line $y = x$.

6. 2

7. 1/3

Section 7.2 Exercises

1. $f^{-1}(x) = \frac{x+4}{7}$ **3.** $[-\pi/2, \pi/2]$

5.
- $f(g(x)) = \left((x-3)^{1/3}\right)^3 + 3 = x - 3 + 3 = x.$
- $g(f(x)) = \left(x^3 + 3 - 3\right)^{1/3} = \left(x^3\right)^{1/3} = x.$

7. $v^{-1}(R) = \frac{2GM}{R^2}$

9. $f^{-1}(x) = 4 - x.$

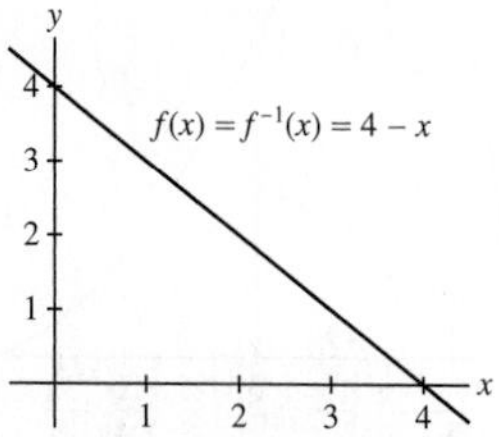

11. $f^{-1}(x) = \frac{1}{7x} + \frac{3}{7}$

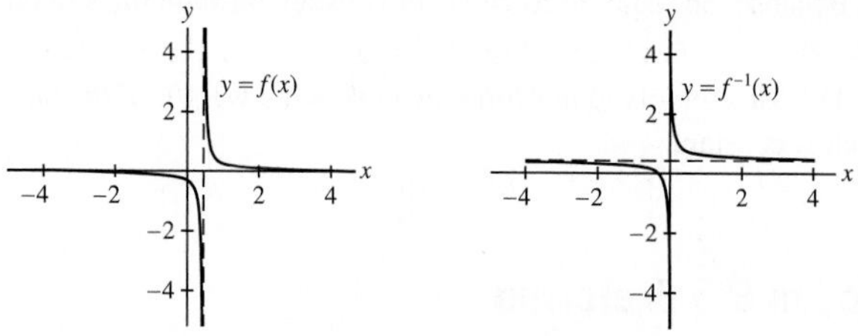

13. Domain $\{x : x \geq 0\}$: $f^{-1}(x) = \frac{\sqrt{1-x^2}}{x}$; domain $\{x : x \leq 0\}$: $f^{-1}(x) = -\frac{\sqrt{1-x^2}}{x}$

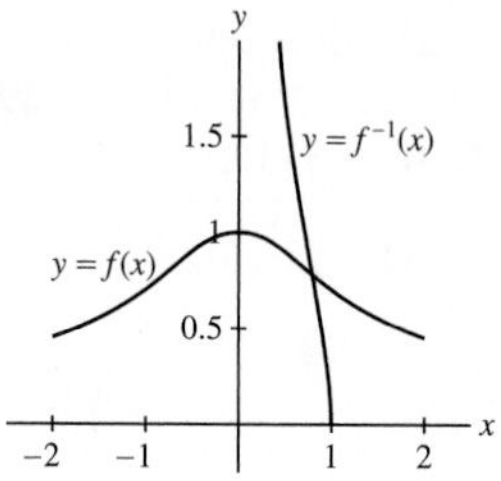

15. $f^{-1}(x) = (x^2 - 9)^{1/3}$

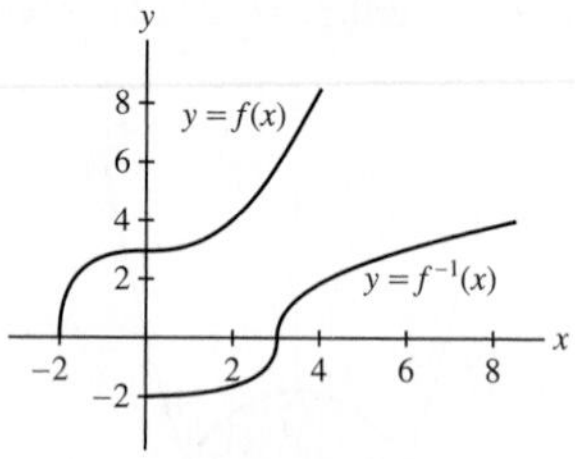

17. Figures (B) and (C)

19. **(a)**

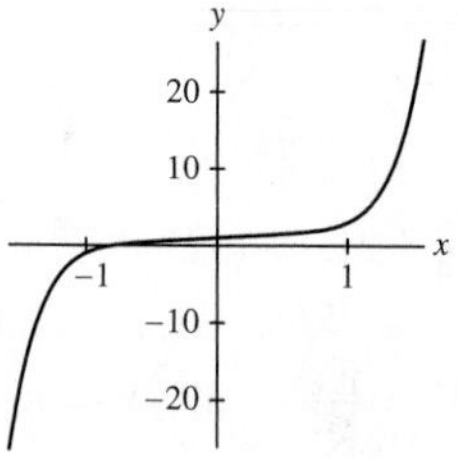

(b) $(-\infty, \infty)$. **(c)** $f^{-1}(3) = 1$.

21. Domain $x \leq 1$: $f^{-1}(x) = 1 - \sqrt{x+1}$; domain $x \geq 1$: $f^{-1}(x) = 1 + \sqrt{x+1}$

23. $g(x) = f^{-1}(x) = \sqrt{x^2 - 9}$; $g'(x) = \dfrac{x}{\sqrt{x^2 - 9}}$

25. $g'(x) = \dfrac{1}{7}$ **27.** $g'(x) = -\dfrac{1}{5}x^{-6/5}$

29. $g'(x) = \dfrac{1}{(1-x)^2}$ **31.** $g(7) = 1$; $g'(7) = \dfrac{1}{5}$ **33.** $g(1) = 0$; $g'(1) = 1$ **35.** $g(4) = 2$; $g'(4) = \dfrac{4}{5}$ **37.** $g(1/4) = 3$; $g'(1/4) = -16$

41. (a) T (b) T (c) F (d) F (e) T (f) F (g) T

Section 7.3 Preliminary Questions

1. $\log_{b^2}(b^4) = 2$ **2.** For $0 < x < 1$ **3.** $\ln(-3)$ is not defined

4. This phrase is a verbal description of the general property of logarithms that states $\log(ab) = \log a + \log b$.

5. D: $x > 0$; R: real numbers **6.** $\ln(-x)$. **7.** $\ln 4$ **8.** $\frac{1}{10}$

Section 7.3 Exercises

1. 3 **3.** 0 **5.** $\frac{5}{3}$ **7.** $\frac{1}{3}$ **9.** $\frac{5}{6}$ **11.** 1 **13.** 7 **15.** 29

17. **(a)** $\ln 1600$ **(b)** $\ln(9x^{7/2})$

19. $t = \frac{1}{5}\ln\left(\frac{100}{7}\right)$ **21.** $x = -1$ or $x = 3$

23. $x = e$ **25.** Let $a = e^2$ and $b = e^3$

27. June 2012 (about 11.55 years)

29. $\frac{d}{dx} x\ln x = \ln x + 1$ **31.** $\frac{d}{dx}(\ln x)^2 = \frac{2}{x}\ln x$

33. $\frac{d}{dx}\ln(9x^2 - 8) = \frac{18x}{9x^2 - 8}$

35. $\frac{d}{dt}\ln(\sin t + 1) = \frac{\cos t}{\sin t + 1}$

37. $\frac{d}{dx}\frac{\ln x}{x} = \frac{1 - \ln x}{x^2}$ **39.** $\frac{d}{dx}\ln(\ln x) = \frac{1}{x\ln x}$

41. $\frac{d}{dx}(\ln(\ln x))^3 = \frac{3(\ln(\ln x))^2}{x\ln x}$

43. $\frac{d}{dx}\ln((x+1)(2x+9)) = \frac{4x + 11}{(x+1)(2x+9)}$

45. $\frac{d}{dx}11^x = \ln 11 \cdot 11^x$

47. $\frac{d}{dx}\frac{2^x - 3^{-x}}{x} = \frac{x(2^x\ln 2 + 3^{-x}\ln 3) - (2^x - 3^{-x})}{x^2}$

49. $f'(x) = \frac{1}{x}\cdot\frac{1}{\ln 2}$ **51.** $\frac{d}{dt}\log_3(\sin t) = \frac{\cot t}{\ln 3}$

53. $y = 36\ln 6(x - 2) + 36$

55. $y = 3^{20}\ln 3(t - 2) + 3^{18}$

57. $y = 5^{-1}$ **59.** $y = -1(t - 1) + \ln 4$

61. $y = \frac{12}{25\ln 5}(z - 3) + 2$ **63.** $y = \frac{8}{\ln 2}\left(w - \frac{1}{8}\right) - 3$

65. $y' = 2x + 14$ **67.** $y' = 3x^2 - 12x - 79$

69. $y' = \frac{x(x^2+1)}{\sqrt{x+1}}\left(\frac{1}{x} + \frac{2x}{x^2+1} - \frac{1}{2(x+1)}\right)$

71.
$$y' = \frac{1}{2}\sqrt{\frac{x(x+2)}{(2x+1)(3x+2)}}\cdot\left(\frac{1}{x} + \frac{1}{x+2} - \frac{2}{2x+1} - \frac{3}{3x+2}\right)$$

73. $\frac{d}{dx}x^{3x} = x^{3x}(3 + 3\ln x)$

75. $\frac{d}{dx}x^{e^x} = x^{e^x}\left(\frac{e^x}{x} + e^x\ln x\right)$

77. $\frac{d}{dx}x^{3^x} = x^{3^x}\left(\frac{3^x}{x} + (\ln x)(\ln 3)3^x\right)$

79. $g(e)$ is a local maximum.

81. $g(e^{1/3})$ is a local maximum.

83. There is a local maximum at $x = e^{1/2} \approx 1.65$, and an inflection point at $x = e^{5/6} \approx 2.301$.

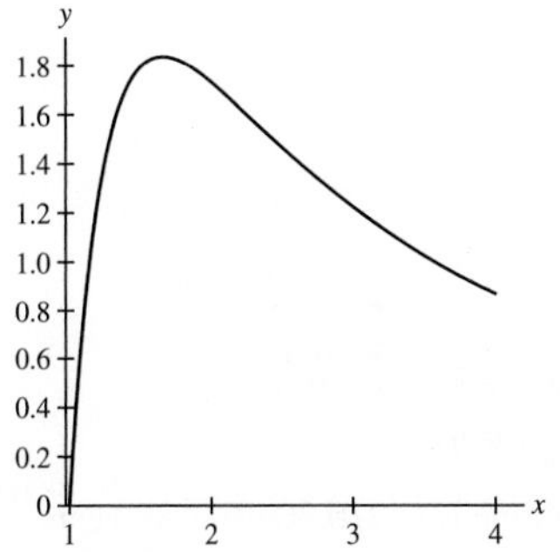

85. $7\ln x + C$ **87.** $\frac{1}{2}\ln(2x+4) + C$

89. $\frac{1}{2}\ln(t^2+4) + C$ **91.** $\frac{1}{2}\ln(9 - 2x + 3x^2) + C$

93. $\ln|\sin x| + C$

95. $\frac{1}{2}\ln^2 x + C$ **97.** $\frac{1}{3}\ln^3 x + C$ **99.** $\frac{1}{4}\ln(\ln(8x - 2)) + C$

101. $\frac{1}{2}\ln^2(\sin x) + C$ **103.** $\frac{1}{2\ln 3}3^{x^2} + C$

105. $-\frac{1}{3\ln 2}\cdot\left(\frac{1}{2}\right)^{3x+2}$ **107.** $\ln 3$

109. $\ln 2 - \frac{1}{3}\ln 5$ **111.** $\ln 2$

113. $f(e^{-1}) = (e^{-1})^{e^{-1}} = \frac{1}{e^{-e}}$

115. 1.22 cents per year

117. **(a)** $\frac{dP}{dT} = -\frac{1}{T\ln 10}$ **(b)** $\Delta P \approx -0.054$

119. Show the equivalent $\log_a x = \log_a b\log_b x$. But $a^{\log_a b\log_b x} = (a^{\log_a b})^{\log_b x} = b^{\log_b x} = x = a^{\log_a x}$, so this is true.

Section 7.4 Preliminary Questions

1. Doubling time is inversely proportional to the growth constant. Consequently, the quantity with $k = 3.4$ doubles more rapidly.

2. It takes longer for the population to increase from one cell to two cells.

3. $\frac{dS}{dn} = -\ln 2S(n)$

4. They would be too ancient, as we would have overestimated the actual amount of C^{14} and thus overestimated the time it took to decay.

Section 7.4 Exercises

1. **(a)** 2000 bacteria initially **(b)** $t = \frac{1}{1.3}\ln 5 \approx 1.24$ hours

3. $f(t) = 5e^{t\ln 7}$

5. $N'(t) = \frac{\ln 2}{3}N(t)$; 1048576 molecules after one hour

7. $y(t) = Ce^{-5t}$ for some constant C; $y(t) = 3.4e^{-5t}$

9. $y(t) = 1000e^{3(t-2)}$ **11.** 5.33 years

13. $k \approx 0.023$ hours^{-1}; $P_0 \approx 332$

15. Double: 11.55 years; triple: 18.31 years; seven-fold: 32.43 years

17. One-half: 1.98 days; one-third: 3.14 days; one-tenth: 6.58 days

19. Set I **21.** **(a)** 26.39 years **(b)** 1969

23. 7600 years **25.** 2.34×10^{-13} to 2.98×10^{-13}

27. 2.55 hours

29. **(a)** Yes, the graph looks like an exponential graph especially towards the latter years; $k \approx 0.369$ years^{-1}.

(b)

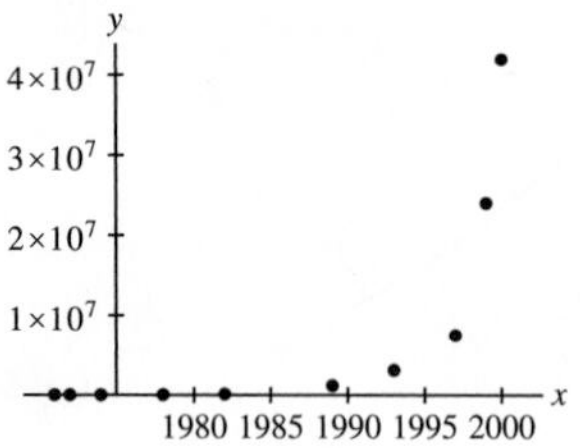

(c) $N(t) = 2250e^{0.369t}$

(d) The doubling time is $\ln 2/0.369 \approx 1.88$ years.

(e) $\approx 2.53 \times 10^{10}$ transistors

(f) No, you can't make a microchip smaller than an atom.

31. With $t_0 = 10$, the doubling time is then 14; with $t_0 = 20$, the doubling time is 24.

33. $T = -\frac{1}{k} \ln\left(1 - \frac{d}{L}\right)$

35. $P(t) = 204e^{ae^{0.15t}}$ with $a \approx -0.02$; 136 rats after 20 months

37. For m-fold growth, $P(t) = mP_0$ for some t. Solving $mP_0 = P_0e^{kt}$ for t, we find $t = \frac{\ln m}{k}$.

Section 7.5 Preliminary Questions

1. 12% compounded quarterly is an annual rate of 12.55%; 11% compounded continuously is 11.63%, so 12% compounded quarterly.

2. 1.0942; 1.0931 **3.** **(b)**

4. If the interest rate goes up, the present value of \$1 a year from now will decrease.

5. He will be sad, because at 7%, the amount someone would need to invest today for \$1000 in a year is less than it would be at 6%. Thus today's value of that gift is smaller.

Section 7.5 Exercises

1. **(a)** $P(10) = \$4870.38$ **(b)** $P(10) = \$4902.71$ **(c)** $P(10) = \$4919.21$

3. **(a)** 1.0508 **(b)** 1.0513

5. \$12,752.56

7. In 3 years:

(a) $PV = \$4176.35$

(b) $PV = \$3594.62$

In 5 years:

(a) $PV = \$3704.09$

(b) $PV = \$2884.75$

9. 9.16%

11. **(a)** The present value of the reduced labor costs is

$$7000(e^{-0.08} + e^{-0.16} + e^{-0.24} + e^{-0.32} + e^{-0.4}) = \$27{,}708.50.$$

This is more than the \$25,000 cost of the computer system, so the computer system should be purchased.

(b) The present value of the savings is

$$\$27{,}708.50 - \$25{,}000 = \$2708.50.$$

13. \$39,346.93 **15.** \$41,906.75 **19.** $R = \$1200$

21. \$71,460.53

23. e^6 **25.** e^6

27. Start by expressing

$$\ln\left(1 + \frac{x}{n}\right) = \int_1^{1+x/n} \frac{dt}{t}.$$

Following the proof in the text, we note that

$$\frac{x}{n+x} \le \ln\left(1 + \frac{x}{n}\right) \le \frac{x}{n}$$

provided $x > 0$, while

$$\frac{x}{n} \le \ln\left(1 + \frac{x}{n}\right) \le \frac{x}{n+x}$$

when $x < 0$. Multiplying both sets of inequalities by n and passing to the limit as $n \to \infty$, the squeeze theorem guarantees that

$$\lim_{n\to\infty} \left(\ln\left(1 + \frac{x}{n}\right)\right)^n = x.$$

Finally,

$$\lim_{n\to\infty} \left(1 + \frac{x}{n}\right)^n = e^x.$$

29. **(a)** 9.38%

(b) In general,

$$P_0(1 + r/M)^{Mt} = P_0(1 + r_e)^t,$$

so $(1 + r/M)^{Mt} = (1 + r_e)^t$ or $r_e = (1 + r/M)^M - 1$. If interest is compounded continuously, then $P_0e^{rt} = P_0(1 + r_e)^t$ so $e^{rt} = (1 + r_e)^t$ or $r_e = e^r - 1$.

(c) 11.63%

(d) 18.26%

Section 7.6 Preliminary Questions

1. $y(t) = 5 - ce^{4t}$ for any positive constant c

2. No **3.** True

4. The difference in temperature between a cooling object and the ambient temperature is decreasing. Hence the rate of cooling, which is proportional to this difference, is also decreasing in magnitude.

Section 7.6 Exercises

1. General solution: $y(t) = 10 + ce^{2t}$; solution satisfying $y(0) = 25$: $y(t) = 10 + 15e^{2t}$; solution satisfying $y(0) = 5$: $y(t) = 10 - 5e^{2t}$

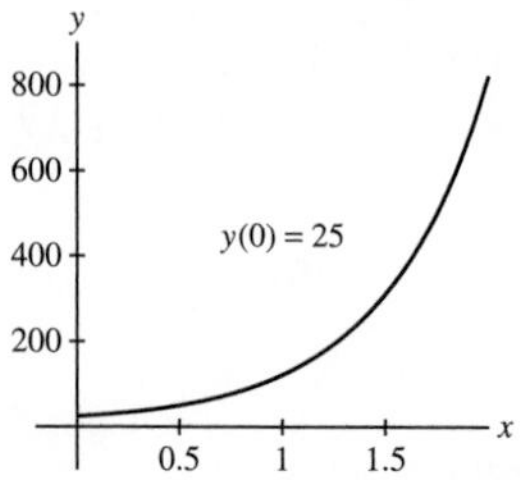

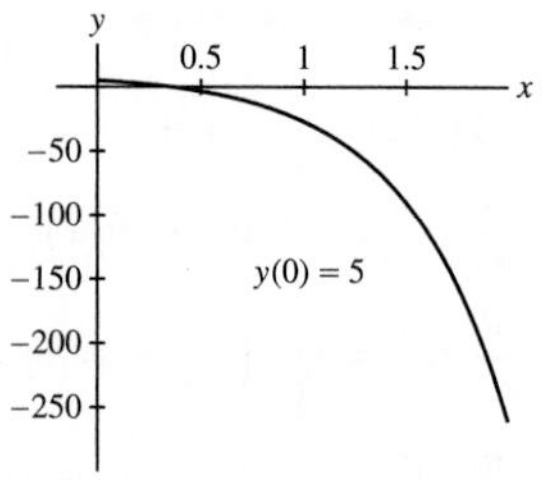

3. $y = -6 + 11e^{4x}$

5. **(a)** $y' = -0.02(y - 10)$ **(b)** $y = 10 + 90e^{-\frac{1}{50}t}$

(c) $100 \ln 3$ s ≈ 109.8 s

7. $\approx$ 5:50 AM **9.** ≈ 0.77 min $= 46.6$ s

11. $500 \ln \frac{3}{2}$ s ≈ 203 s $= 3$ min 23 s

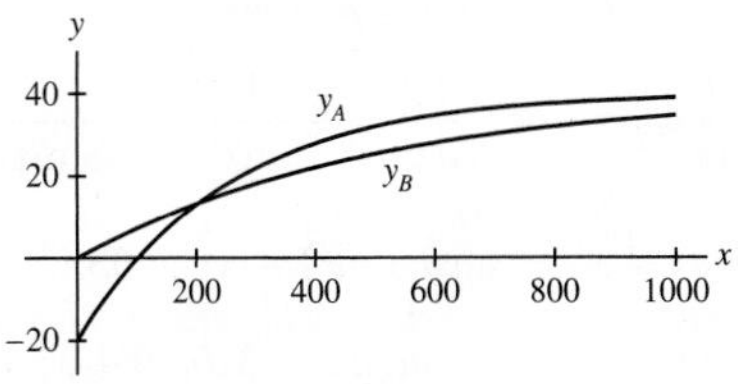

13. -58.8 m/s **15.** -11.8 m/s

17. **(a)** \$17, 563.94 **(b)** 13.86 yr

19. \$120, 000 **21.** 8%

23. **(b)** $t = \frac{1}{0.09} \ln\left(\frac{13{,}333.33}{3{,}333.33}\right) \approx 15.4$ yr **(c)** No

25. **(a)** $N'(t) = k(1 - N(t)) = -k(N(t) - 1)$
(b) $N(t) = 1 - e^{-kt}$ **(c)** $\approx 64.63\%$

29. **(a)** $v(t) = \dfrac{-g}{k} + \left(v_0 + \dfrac{g}{k}\right)e^{-kt}$

Section 7.7 Preliminary Questions

1. Not of the form $\frac{0}{0}$ or $\frac{\infty}{\infty}$

2. No

Section 7.7 Exercises

1. L'Hôpital's Rule does not apply.

3. L'Hôpital's Rule does not apply.

5. L'Hôpital's Rule does not apply.

7. L'Hôpital's Rule does not apply.

9. 0

11. Quotient is of the form $\frac{\infty}{\infty}$; $-\dfrac{9}{2}$

13. Quotient is of the form $\frac{\infty}{\infty}$; 0

15. Quotient is of the form $\frac{\infty}{\infty}$; 0

17. $\dfrac{5}{6}$ **19.** $-\dfrac{3}{5}$ **21.** $-\dfrac{7}{3}$ **23.** $\dfrac{9}{7}$ **25.** $\dfrac{2}{7}$ **27.** 1 **29.** 2

31. -1 **33.** $\dfrac{1}{2}$ **35.** 0 **37.** $-\dfrac{2}{\pi}$ **39.** 1 **41.** Does not exist

43. 0 **45.** $\ln a$ **47.** e **49.** $e^{-3/2}$

51.

$$\lim_{x\to\pi/2} \frac{\cos mx}{\cos nx} = \begin{cases} (-1)^{(m-n)/2}, & m, n \text{ even} \\ \text{does not exist}, & m \text{ even}, n \text{ odd} \\ 0 & m \text{ odd}, n \text{ even} \\ (-1)^{(m-n)/2}\frac{m}{n}, & m, n \text{ odd} \end{cases}$$

53.

$$\lim_{x\to 0} \ln\left((1+x)^{1/x}\right) = \lim_{x\to 0} \frac{1}{x}\ln(1+x) = \lim_{x\to 0} \frac{\ln(1+x)}{x} = 1,$$

so $\lim_{x\to 0}(1+x)^{1/x} = e^1 = e$; $x = 0.0005$

55. **(a)** $\lim_{x\to 0+} f(x) = 0$; $\lim_{x\to\infty} f(x) = e^0 = 1$.
(b) f is increasing for $0 < x < e$, is decreasing for $x > e$ and has a maximum at $x = e$. The maximum value is $f(e) = e^{1/e} \approx 1.444668$.

57. Neither

59. $\displaystyle\lim_{x\to\infty} \frac{\ln x}{x^a} = \lim_{x\to\infty} \frac{x^{-1}}{ax^{a-1}} = \lim_{x\to\infty} \frac{1}{a}x^{-a} = 0$

63. **(a)** $1 \le 2 + \sin x \le 3$, so

$$\frac{x}{x^2+1} \le \frac{x(2+\sin x)}{x^2+1} \le \frac{3x}{x^2+1};$$

it follows by the Squeeze Theorem that

$$\lim_{x\to\infty} \frac{x(2+\sin x)}{x^2+1} = 0.$$

(b) $\lim_{x\to\infty} f(x) = \lim_{x\to\infty} x(2+\sin x) \ge \lim_{x\to\infty} x = \infty$ and $\lim_{x\to\infty} g(x) = \lim_{x\to\infty}(x^2+1) = \infty$, but

$$\lim_{x\to\infty} \frac{f'(x)}{g'(x)} = \lim_{x\to\infty} \frac{x(\cos x) + (2+\sin x)}{2x}$$

does not exist since $\cos x$ oscillates. This does not violate L'Hôpital's Rule since the theorem clearly states

$$\lim_{x\to\infty} \frac{f(x)}{g(x)} = \lim_{x\to\infty} \frac{f'(x)}{g'(x)}$$

"provided the limit on the right exists."

65. **(a)** Using Exercise 64, we see that $G(b) = e^{H(b)}$. Thus, $G(b) = 1$ if $0 \le b \le 1$ and $G(b) = b$ if $b > 1$.

(b)

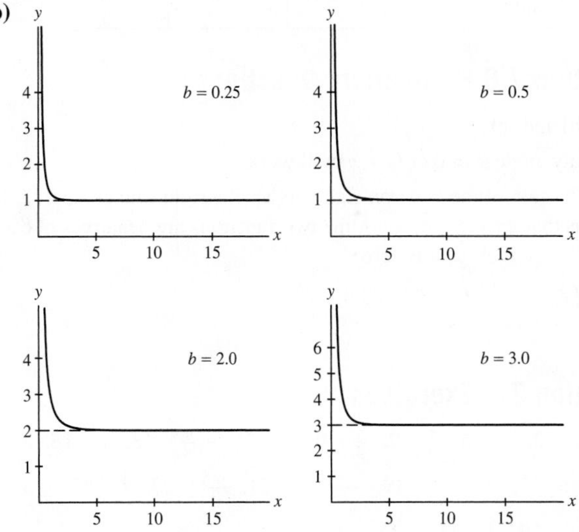

67. $\lim_{x\to 0} \frac{f(x)}{x^k} = \lim_{x\to 0} \frac{1}{x^k e^{1/x^2}}$. Let $t = 1/x$. As $x \to 0$, $t \to \infty$. Thus,

$$\lim_{x\to 0} \frac{1}{x^k e^{1/x^2}} = \lim_{t\to\infty} \frac{t^k}{e^{t^2}} = 0$$

by Exercise 66.

69. For $x \ne 0$, $f'(x) = e^{-1/x^2}\left(\frac{2}{x^3}\right)$. Here $P(x) = 2$ and $r = 3$. Assume $f^{(k)}(x) = \frac{P(x)e^{-1/x^2}}{x^r}$. Then

$$f^{(k+1)}(x) = e^{-1/x^2}\left(\frac{x^3P'(x) + (2 - rx^2)P(x)}{x^{r+3}}\right)$$

which is of the form desired.

Moreover, from Exercise 67, $f'(0) = 0$. Suppose $f^{(k)}(0) = 0$. Then

$$f^{(k+1)}(0) = \lim_{x\to 0} \frac{f^{(k)}(x) - f^{(k)}(0)}{x - 0} = \lim_{x\to 0} \frac{P(x)e^{-1/x^2}}{x^{r+1}}$$
$$= P(0) \lim_{x\to 0} \frac{f(x)}{x^{r+1}} = 0.$$

75. $\lim_{x\to 0} \frac{\sin x}{x} = \lim_{x\to 0} \frac{\cos x}{1} = 1$. To use L'Hôpital's Rule to evaluate $\lim_{x\to 0} \frac{\sin x}{x}$, we must know that the derivative of $\sin x$ is $\cos x$, but to determine the derivative of $\sin x$, we must be able to evaluate $\lim_{x\to 0} \frac{\sin x}{x}$.

77. **(a)** $e^{-1/6} \approx 0.846481724$

x	1	0.1	0.01
$\left(\frac{\sin x}{x}\right)^{1/x^2}$	0.841471	0.846435	0.846481

(b) $1/3$

x	± 1	± 0.1	± 0.01
$\frac{1}{\sin^2 x} - \frac{1}{x^2}$	0.412283	0.334001	0.333340

Section 7.8 Preliminary Questions

1. **(b)** and **(c)**
2. Any angle $\theta < 0$ or $\theta > \pi$ will work.
3. The sum of the two non-right angles in a right triangle is $\pi/2$. It tells us that the derivatives of the two functions are negatives of each other (since their sum is zero).
4. $\sqrt{3}$
5. $x = \pm 4u$

Section 7.8 Exercises

1. 0 **3.** $\frac{\pi}{4}$ **5.** $\frac{\pi}{3}$ **7.** $\frac{\pi}{3}$ **9.** $\frac{\pi}{2}$ **11.** $-\frac{\pi}{4}$ **13.** π **15.** No inverse **17.** $\frac{\sqrt{1-x^2}}{x}$ **19.** $\frac{1}{\sqrt{x^2-1}}$ **21.** $\frac{\sqrt{5}}{3}$ **23.** $\frac{4}{3}$ **25.** $\sqrt{3}$

27. $\frac{1}{20}$ **29.** $\frac{5}{4}$ **31.** $\frac{1}{4\sqrt{15}}$

33. $\frac{d}{dx}\sin^{-1}(7x) = \frac{7}{\sqrt{1-(7x)^2}}$

35. $\frac{d}{dx}\cos^{-1}(x^2) = \frac{-2x}{\sqrt{1-x^4}}$

37. $\frac{d}{dx} x\tan^{-1} x = x\left(\frac{1}{1+x^2}\right) + \tan^{-1} x$

39. $\frac{d}{dx}\sin^{-1}(e^x) = \frac{e^x}{\sqrt{1-e^{2x}}}$

41. $\frac{d}{dt}\left(\sqrt{1-t^2} + \sin^{-1} t\right) = \frac{1-t}{\sqrt{1-t^2}}$

43. $\frac{d}{dx}\left((\tan^{-1} x)^3\right) = \frac{3(\tan^{-1} x)^2}{x^2+1}$

45. $\frac{d}{dt}(\cos^{-1} t^{-1} - \sec^{-1} t) = 0$

47. $\frac{d}{dt}\cos^{-1}(\ln x) = \frac{-1}{x\sqrt{1-\ln^2 x}}$

49. Let $\theta = \cos^{-1} x$. Then $\cos\theta = x$ and

$$-\sin\theta \frac{d\theta}{dx} = 1 \quad \text{or} \quad \frac{d\theta}{dx} = -\frac{1}{\sin\theta} = -\frac{1}{\sin(\cos^{-1} x)}.$$

Moreover, $\sin(\cos^{-1} x) = \sin\theta = \sqrt{1-x^2}$. **53.** 7 **55.** $\frac{\pi}{6}$

57. Let $u = x/3$. Then, $x = 3u$, $dx = 3\,du$, $9 + x^2 = 9(1+u^2)$, and

$$\int \frac{dx}{9+x^2} = \int \frac{3\,du}{9(1+u^2)} = \frac{1}{3}\int \frac{du}{1+u^2}$$
$$= \frac{1}{3}\tan^{-1} u + C = \frac{1}{3}\tan^{-1}\frac{x}{3} + C.$$

59. $\frac{\pi}{3\sqrt{3}}$ **61.** $\frac{1}{4}\sin^{-1}(4t) + C$ **63.** $\frac{1}{\sqrt{3}}\sin^{-1}\sqrt{\frac{3}{5}}t + C$

65. $\frac{1}{\sqrt{3}}\sec^{-1}(2x) + C = -\frac{1}{\sqrt{3}}\csc^{-1}(2x) + C$

67. $\frac{1}{2}\sec^{-1} x^2 + C = -\frac{1}{2}\csc^{-1} x^2 + C$

69. $-\frac{1}{2}\ln^2(\cos^{-1} x) + C$ **71.** $2\ln 2 - \ln 3 = \ln\frac{4}{3}$

73. $\frac{1}{2}e^{y^2} + C$ **75.** $\frac{1}{4}\sqrt{4x^2+9} + C$ **77.** $-\frac{7^{-x}}{\ln 7} + C$

79. $\frac{1}{8}\tan^8\theta + C$ **81.** $-\sqrt{7-t^2} + C$

83. $\frac{3}{2}\ln(x^2+4) + \tan^{-1}(x/2) + C$ **85.** $\frac{1}{4}\sin^{-1}(4x) + C$

87. $-e^{-x} - 2x^2 + C$ **89.** $e^x - \frac{e^{3x}}{3} + C$

91. $-\sqrt{4-x^2} + 5\sin^{-1}(x/2) + C$ **93.** $\sin(e^x) + C$

95. $\frac{1}{4}\sin^{-1}\left(\frac{4x}{3}\right) + C$ **97.** $\frac{e^{7x}}{7} + \frac{3e^{5x}}{5} + e^{3x} + e^x + C$

99. $\frac{1}{3}\ln|x^3+2| + C$ **101.** $\ln|\sin x| + C$

103. $\frac{1}{8}(4\ln x + 5)^2 + C$ **105.** $\frac{3^{x^2}}{2\ln 3} + C$ **107.** $\frac{(\ln(\sin x))^2}{2} + C$

109. $\frac{2}{7}(t-3)^{7/2} + \frac{12}{5}(t-3)^{5/2} + 6(t-3)^{3/2} + C$

111. The definite integral $\int_0^x \sqrt{1-t^2}\,dt$ represents the area of the region under the upper half of the unit circle from 0 to x. The region consists of a sector of the circle and a right triangle. The sector has a central angle of $\frac{\pi}{2} - \theta$, where $\cos\theta = x$, and the right triangle has a base of length x and a height of $\sqrt{1-x^2}$.

113. Show that $\frac{d}{dt}\left(\sqrt{1-t^2} + t\sin^{-1} t\right) = \sin^{-1} t$.

Section 7.9 Preliminary Questions

1. $\cosh x$ and $\operatorname{sech} x$ **2.** $\sinh x$ and $\tanh x$
3. Parity, identities and derivative formulas
4. $\cosh x$, $\sinh x$

Section 7.9 Exercises

1.

x	-3	0	5
$\sinh x = \dfrac{e^x - e^{-x}}{2}$	-10.0179	0	74.203
$\cosh x = \dfrac{e^x + e^{-x}}{2}$	10.0677	1	74.210

3. $\sinh x$ is increasing everywhere and strictly increasing for $x \neq 0$); $\cosh x$ is increasing for $x \geq 0$ and strictly increasing for $x > 0$.

7.

$$\begin{aligned}\cosh x \cosh y + \sinh x \sinh y \\ &= \frac{e^x + e^{-x}}{2} \cdot \frac{e^y + e^{-y}}{2} + \frac{e^x - e^{-x}}{2} \cdot \frac{e^y - e^{-y}}{2} \\ &= \frac{2e^{x+y} + 2e^{-(x+y)}}{4} = \frac{e^{x+y} + e^{-(x+y)}}{2} = \cosh(x+y)\end{aligned}$$

9. $\dfrac{d}{dx}\sinh(9x) = 9\cosh(9x)$

11. $\dfrac{d}{dt}\cosh^2(9-3t) = -6\cosh(9-3t)\sinh(9-3t)$

13. $\dfrac{d}{dx}\sqrt{\cosh x + 1} = \dfrac{1}{2}(\cosh x + 1)^{-1/2}\sinh x$

15. $\dfrac{d}{dt}\dfrac{\coth t}{1 + \tanh t} = -\dfrac{(\operatorname{csch} t^2 + 2\operatorname{csch} t \operatorname{sech} t)}{(1 + \tanh t)^2}$

17. $\dfrac{d}{dx}\sinh(\ln x) = \dfrac{\cosh(\ln x)}{x}$

19. $\dfrac{d}{dx}\tanh(e^x) = e^x \operatorname{sech}^2(e^x)$

21. $\dfrac{d}{dx}\operatorname{sech}(\sqrt{x}) = -\dfrac{1}{2}x^{-1/2}\operatorname{sech}\sqrt{x}\tanh\sqrt{x}$

23. $\dfrac{d}{dx}\operatorname{sech} x \coth x = -\operatorname{csch} x \coth x$

25. $\dfrac{d}{dx}\cosh^{-1}(3x) = \dfrac{3}{\sqrt{9x^2 - 1}}$

27. $\dfrac{d}{dx}(\sinh^{-1}(x^2))^3 = 3(\sinh^{-1}(x^2))^2 \dfrac{2x}{\sqrt{x^4+1}}$

29. $\dfrac{d}{dx}e^{\cosh^{-1} x} = e^{\cosh^{-1} x}\left(\dfrac{1}{\sqrt{x^2-1}}\right)$

31. $\dfrac{d}{dt}\tanh^{-1}(\ln t) = \dfrac{1}{t(1-(\ln t)^2)}$

35. $\frac{1}{3}\sinh 3x + C$ **37.** $\frac{1}{2}\cosh(x^2+1) + C$

39. $-\frac{1}{2}\tanh(1-2x) + C$ **41.** $\frac{1}{2}\tanh^2 x + C$

43. $\ln\cosh x + C$ **45.** $\frac{1}{2}x + \frac{1}{4}e^{-2x} + C$

47.

$$\frac{d}{dx}\tanh x = \frac{d}{dx}\frac{\sinh x}{\cosh x} = \frac{\cosh^2 x - \sinh^2 x}{\cosh^2 x} = \frac{1}{\sinh^2 x} = \operatorname{sech}^2 x$$

49. Note

$$\begin{aligned}\frac{d}{dt}\cosh(\sinh^{-1} t) &= \sinh(\sinh^{-1} t)\frac{1}{\sqrt{t^2+1}} \\ &= \frac{t}{\sqrt{t^2+1}} = \frac{d}{dt}\sqrt{t^2+1}\end{aligned}$$

so $\cosh(\sinh^{-1} t)$ and $\sqrt{t^2+1}$ differ by a constant; substituting $t = 0$, we find that the constant is 0. Hence, $\cosh(\sinh^{-1} t) = \sqrt{t^2+1}$.

53. $\cosh^{-1} 4 - \cosh^{-1} 2$ **55.** $\sinh^{-1}\frac{x}{3} + C$

57. $\tanh^{-1}\frac{1}{2} - \tanh^{-1}\frac{1}{3}$ **59.** $\frac{1}{4}\ln\frac{95}{63}$

61. Let $x = \sinh^{-1} t$. Then $\sinh x = t$. It follows from the identity $\cosh^2 x - \sinh^2 x = 1$ that $\cosh x = \sqrt{1 + \sinh^2 x} = \sqrt{1+t^2}$. Recall that $\cosh x > 0$ for all x. Now,

$$\sinh x + \cosh x = \frac{e^x - e^{-x}}{2} + \frac{e^x + e^{-x}}{2} = e^x$$

so $x = \ln(\sinh x + \cosh x)$. Finally, $\sinh^{-1} t = \ln(t + \sqrt{1+t^2})$.

63. Let $A = \tanh^{-1} t$. Then

$$t = \tanh A = \frac{\sinh A}{\cosh A} = \frac{e^A - e^{-A}}{e^A + e^{-A}}$$

Solving for A yields

$$A = \tanh^{-1} t = \frac{1}{2}\ln\frac{1+t}{1-t}$$

65. **(a)** By Galileo's law, $w = 500 + 10 = 510$ m/s. Using Einstein's law, $w = c \cdot \tanh(1.7 \times 10^{-6}) \approx 510$ m/s.
(b) By Galileo's law, $u + v = 10^7 + 10^6 = 1.1 \times 10^7$ m/s. By Einstein's law, $w \approx c \cdot \tanh(0.036679) \approx 1.09988 \times 10^7$ m/s.

69. **(d)**

$$\lim_{t\to\infty} v(t) = \lim_{t\to\infty}\sqrt{\frac{mg}{k}}\tanh\left(\sqrt{\frac{kg}{m}}t\right) = \sqrt{\frac{mg}{k}} \cdot 1 = \sqrt{\frac{mg}{k}}$$

(e) $k = \frac{150(78{,}545.5)}{100^2} = 1{,}178.18$ lbs/mi

71.

$$\begin{aligned}s &= y(M) - y(0) \\ &= a\cosh\left(\frac{M}{a}\right) + C - (a\cosh 0 + C) = a\cosh\left(\frac{M}{a}\right) - a\end{aligned}$$

(a) $\frac{ds}{da} = \cosh\left(\frac{M}{a}\right) - \frac{M}{a}\sinh\left(\frac{M}{a}\right) - 1$

(b) $\frac{da}{dL} = \left(2\sinh\left(\frac{M}{a}\right) - \frac{2M}{a}\cosh\left(\frac{M}{a}\right)\right)^{-1}$

(c) By the Chain Rule, $\frac{ds}{dL} = \frac{ds}{da} \cdot \frac{da}{dL}$. The formula for $\frac{ds}{dL}$ follows upon substituting the results from parts (a) and (b).

Chapter 7 Review

1. **(a)** No match **(b)** No match **(c) (i):** $(2^a)^b = 2^{ab}$ **(d) (iii):** $2^{a-b}3^{b-a} = 2^{a-b}\left(\frac{1}{3}\right)^{a-b} = \left(\frac{2}{3}\right)^{a-b}$

3. **(b):** $(\ln 2)2^x$

5. $f^{-1}(x) = \frac{x-2}{x-1}$; domain $\{x : x \neq 1\}$; range: $\{y : y \neq 1\}$

7. $g(g(x)) = \frac{\frac{x}{x-1}}{\frac{x}{x-1} - 1} = \frac{x}{x-(x-1)} = x$

9. **(a)** (iii) **(b)** (iv) **(c)** (ii) **(d)** (i) **11.** $\frac{1}{2}e$

13. $-36e^{-4x}$ **15.** $-\frac{e^{-x}(x+1)}{x^2}$ **17.** $\frac{2}{s}\ln s$

19. $(4-2t)e^{4t-t^2}$ **21.** $\cot\theta$

23. $\frac{e^x-4}{e^x-4x}$ **25.** $\left(1+\frac{1}{x}\right)e^{x+\ln x}$ **27.** $-2^{1-y}\ln 2$

29. $-2\ln 7\cdot 7^{-2x}$ **31.** $\frac{1}{\sqrt{s^4-s^2}}$

33. $-\frac{1}{|x|\csc^{-1}x\sqrt{x^2-1}}$ **35.** $1+\ln s$

37. $(\sin^{2t}t)(2\ln\sin t+2t\cot t)$ **39.** $2t\cosh(t^2)$ **41.** $\frac{e^x}{1-e^{2x}}$

43. $y=-\frac{1}{2}x+6$

47. Global minimum at $x=\ln(2)$; no maximum.

49. Local minimum at $x=e^{-1}$; no points of inflection ; $\lim_{x\to 0+} x\ln x=0$; $\lim_{x\to\infty} x\ln x=\infty$.

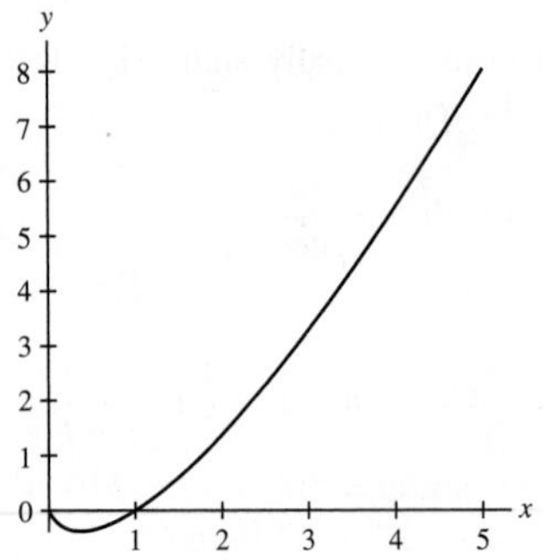

51. Local minimum at $x=1$; local maximum at $x=e^{-2}$; point of inflection at $x=\frac{1}{e}$; $\lim_{x\to 0+} x(\log x)^2=0$; $\lim_{x\to\infty} x(\log x)^2=\infty$.

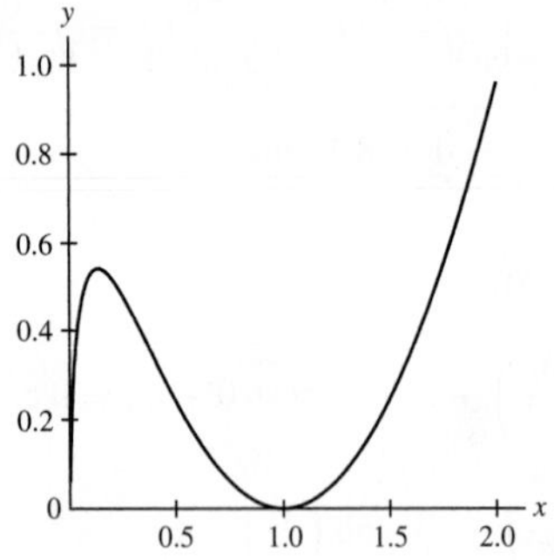

53. $y'=\frac{(x+1)^3}{(4x-2)^2}\left(\frac{3}{x+1}-\frac{8}{4x-2}\right)$

55. $y'=4e^{(x-1)^2}e^{(x-3)^2}(x-2)$

57. $y'=\frac{e^{3x}(x-2)^2}{(x+1)^2}\left(3+\frac{2}{x-2}-\frac{2}{x+1}\right)$

59. **(a)** $f'(u)=\dfrac{be^{b(a-u)}}{(1+e^{b(a-u)})^2}>0$

(b) $u=a+\dfrac{1}{b}\ln 2$

61. $\frac{1}{3}(\ln x)^3+C$ **63.** $-\sin^{-1}(e^{-x})+C$

65. $\tan^{-1}(\ln t)+C$ **67.** $-\frac{1}{2}e^{9-2x}+C$

69. $\frac{1}{2}\cos(e^{-2x})+C$ **71.** $\frac{1}{4}(e^9-e)$ **73.** $\frac{1}{2}$

75. $\sin^{-1}\left(\frac{2}{3}\right)-\sin^{-1}\left(\frac{1}{3}\right)$ **77.** $\frac{1}{2}\sinh 2$ **79.** $\frac{1}{2}\ln 2$

81. $\frac{1}{2}\sin^{-1}x^2+C$ **83.** $\frac{1}{2(e^{-x}+2)^2}$ **85.** $\frac{1}{2}\ln 2$

87. $\frac{1}{2}(\sin^{-1}x)^2+C$ **89.** $\frac{1}{4}\sinh^4 x+C$ **91.** $\frac{\sqrt{2}}{2}\tan^{-1}(4\sqrt{2})$

93. **(a)** $y'(t)=-\frac{\ln 2}{24.5}y(t)$ **(b)** $y(365)\approx 0.0655$ g

95. 3,938.5 yr

97. $\frac{dk}{dT}\approx 12.27\ \text{hr}^{-1}\text{-K}^{-1}$; approximate change in k when T is raised from 500 to 510: $122.7\ \text{hr}^{-1}$.

99. Solution satisfying $y(0)=3$: $y(t)=4-e^{-2t}$; solution satisfying $y(0)=4$: $y(t)=4$

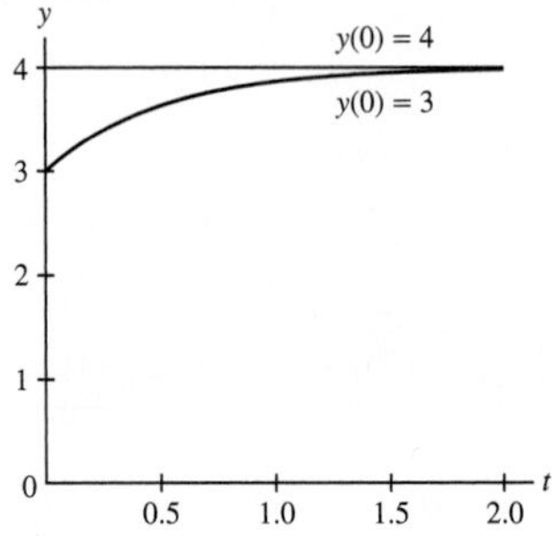

101. **(a)** 12

(b) ∞, if $y(0)>12$; 12, if $y(0)=12$; $-\infty$, if $y(0)<12$

(c) -3

103. $P'(t)=.05P(t)-2{,}000$ **105.** $20\ln\left(\frac{20}{19}\right)\approx 1.025$ yr

107. Yes; the present value of those savings is \$1,134,704. Largest interest rate at which it is a good investment is about 10.66%.

109. 4 **111.** 0 **113.** 3 **115.** $\ln 2$ **117.** $\frac{1}{6}$ **119.** 2

121. As $x\to\infty$, both $2x-\sin x$ and $3x+\cos 2x$ tend toward infinity, so L'Hôpital's Rule applies to $\lim_{x\to\infty}\frac{2x-\sin x}{3x+\cos 2x}$; however, the resulting limit, $\lim_{x\to\infty}\frac{2-\cos x}{3-2\sin 2x}$, does not exist due to the oscillation of $\sin x$ and $\cos x$. $\lim_{x\to\infty}\frac{2x-\sin x}{3x+2\cos x}=\frac{2}{3}$.

123. e^4

127. Let $gd(y)=\tan^{-1}(\sinh y)$. Then

$$\frac{d}{dy}gd(y)=\frac{1}{1+\sinh^2 y}\cosh y=\frac{1}{\cosh y}=\operatorname{sech} y$$

129. Let $x=gd(y)=\tan^{-1}(\sinh y)$. Solving for y yields $y=\sinh^{-1}(\tan x)$. Therefore,

$$gd^{-1}(y)=\sinh^{-1}(\tan y)$$

Chapter 8

Section 8.1 Preliminary Questions

1. The Integration by Parts formula is derived from the Product Rule.

3. Transforming $v'=x$ into $v=\frac{1}{2}x^2$ increases the power of x and makes the new integral harder than the original.

Section 8.1 Exercises

1. $-x\cos x + \sin x + C$ **3.** $e^x(2x+7)+C$

5. $\frac{x^4}{16}(4\ln x - 1) + C$ **7.** $-e^{-x}(4x+1)+C$

9. $\frac{1}{25}(5x-1)e^{5x+2}+C$ **11.** $\frac{1}{2}x\sin 2x + \frac{1}{4}\cos 2x + C$

13. $-x^2\cos x + 2x\sin x + 2\cos x + C$

15. $-\frac{1}{2}e^{-x}(\sin x + \cos x) + C$

17. $-\frac{1}{26}e^{-5x}(\cos(x) + 5\sin(x)) + C$

19. $\frac{1}{4}x^2(2\ln x - 1) + C$ **21.** $\frac{x^3}{3}\left(\ln x - \frac{1}{3}\right) + C$

23. $x\left[(\ln x)^2 - 2\ln x + 2\right] + C$

25. $x\tan x - \ln|\sec x| + C$

27. $x\cos^{-1}x - \sqrt{1-x^2} + C$

29. $x\sec^{-1}x - \ln|x + \sqrt{x^2-1}| + C$

31. $\dfrac{3^x(\sin x + \ln 3\cos x)}{1 + (\ln 3)^2} + C$

33. $(x^2+2)\sinh x - 2x\cosh x + C$

35. $x\tanh^{-1}4x + \frac{1}{8}\ln|1-16x^2| + C$

37. $2e^{\sqrt{x}}(\sqrt{x}-1)+C$

39. $\frac{1}{4}x\sin 4x + \frac{1}{16}\cos 4x + C$

41. $\frac{2}{3}(x+1)^{3/2} - 2(x+1)^{1/2} + C$

43. $\sin x\ln(\sin x) - \sin x + C$

45. $2xe^{\sqrt{x}} - 4\sqrt{x}e^{\sqrt{x}} + 4e^{\sqrt{x}} + C$

47. $\frac{1}{4}(\ln x)^2[2\ln(\ln x) - 1] + C$

49. $\frac{1}{16}(11e^{12}+1)$

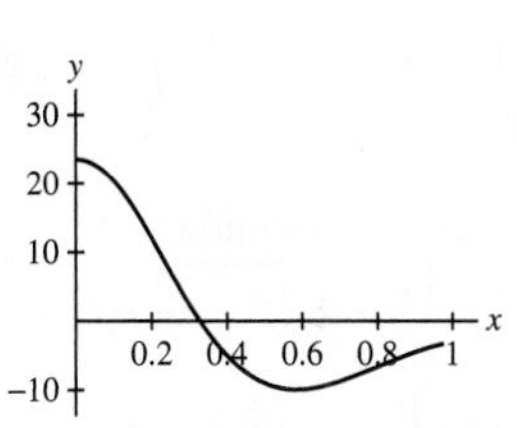

51. $2\ln 2 - \frac{3}{4}$ **53.** $\frac{e^{\pi}+1}{2}$

55. $e^x(x^4 - 4x^3 + 12x^2 - 24x + 24) + C$.

57. $\int x^n e^{-x}\,dx = -x^n e^{-x} + n\int x^{n-1}e^{-x}\,dx$

59. Use Integration by Parts, with $u = \ln x$ and $v' = \sqrt{x}$.

61. Use substitution, followed by algebraic manipulation, with $u = 4 - x^2$ and $du = -2x\,dx$.

63. Use substitution with $u = x^2 + 4x + 3$, $\frac{du}{2} = x + 2\,dx$.

65. Use Integration by Parts, with $u = x$ and $v' = \sin(3x+4)$.

67. $x(\sin^{-1}x)^2 + 2\sqrt{1-x^2}\sin^{-1}x - 2x + C$

69. $\frac{1}{4}x^4\sin(x^4) + \frac{1}{4}\cos(x^4) + C$

71. $2\pi(e^2+1)$ **73.** \$42, 995

75. For $k=2$: $x(\ln x)^2 - 2x\ln x + 2x + C$; for $k=3$: $x(\ln x)^3 - 3x(\ln x)^2 + 6x\ln x - 6x + C$.

77. Use Integration by Parts with $u = x$ and $v' = b^x$.

79. **(b)** $V(x) = \frac{1}{2}x^2 + \frac{1}{2}$ is simpler, and yields $\frac{1}{2}(x^2\tan^{-1}x - x + \tan^{-1}x) + C$.

81. An example of a function satisfying these properties for some λ is $f(x) = \sin\pi x$.

83. **(a)** $I_n = \frac{1}{2}x^{n-1}\sin(x^2) - \frac{n-1}{2}J_{n-2}$;
(c) $\frac{1}{2}x^2\sin(x^2) + \frac{1}{2}\cos(x^2) + C$

Section 8.2 Preliminary Questions

1. Rewrite $\sin^5 x = \sin x\sin^4 x = \sin x(1-\cos^2 x)^2$ and then substitute $u = \cos x$.

3. No, a reduction formula is not needed because the sine function is raised to an odd power.

5. The second integral requires the use of reduction formulas, and therefore more work.

Section 8.2 Exercises

1. $\sin x - \frac{1}{3}\sin^3 x + C$

3. $-\frac{1}{3}\cos^3\theta + \frac{1}{5}\cos^5\theta + C$

5. $-\frac{1}{4}\cos^4 t + \frac{1}{6}\cos^6 t + C$ **7.** 2

9. $\frac{1}{4}\cos^3 y\sin y + \frac{3}{8}\cos y\sin y + \frac{3}{8}y + C$

11. $\frac{1}{6}\sin^5 x\cos x - \frac{1}{24}\sin^3 x\cos x - \frac{1}{16}\sin x\cos x + \frac{1}{16}x + C$

13. $\frac{1}{5}\sin^4 x\cos x - \frac{1}{15}\sin^2 x\cos x - \frac{2}{15}\cos x + C$

15. $\frac{1}{3}\sec^3 x - \sec x + C$

17. $\frac{1}{5}\tan x\sec^4 x - \frac{1}{15}\tan(x)\sec^2 x - \frac{2}{15}\tan x + C$

19. $-\frac{1}{2}\cot^2 x + \ln|\csc x| + C$ **21.** $-\frac{1}{6}\cot^6 x + C$

23. $-\frac{1}{6}\cos^6 x + C$

25. $\frac{1}{12}(\cos^3 x\sin x + \frac{3}{2}(x + \sin x\cos x)) + C$

27. $\frac{1}{5\pi}\sin^5(\pi\theta) - \frac{1}{7\pi}\sin^7(\pi\theta) + C$

29. $-\frac{1}{12}\sin^3(3x)\cos(3x) - \frac{1}{8}\sin(3x)\cos(3x) + \frac{9}{8}x + C$

31. $\frac{1}{2}\cot(3-2x) + C$ **33.** $\frac{1}{2}\tan^2 x + C$

35. $\frac{1}{8}\sec^8 x - \frac{1}{3}\sec^6 x + \frac{1}{4}\sec^4 x + C$

37. $\frac{1}{9}\tan^9 x + \frac{1}{7}\tan^7 x + C$

39. $-\frac{1}{9}\csc^9 x + \frac{2}{7}\csc^7 x - \frac{1}{5}\csc^5 x + C$

41. $\frac{1}{4}\sin^2 2x + C$

43. $\frac{1}{6}\cos^2(t^2)\sin(t^2) + \frac{1}{3}\sin(t^2) + C$

45. $\frac{1}{2}\cos(\sin t)\sin(\sin t) + \frac{1}{2}\sin t + C$

47. π **49.** $\frac{8}{15}$ **51.** $\ln\left(\sqrt{2}+1\right)$ **53.** $\ln 2$

55. $\frac{8}{3}$ **57.** $-\frac{6}{7}$ **59.** $\frac{8}{15}$

61. First, observe $\sin 4x = 2\sin 2x\cos 2x = 2\sin 2x(1 - 2\sin^2 x) = 2\sin 2x - 4\sin 2x\sin^2 x = 2\sin 2x - 8\sin^3 x\cos x$. Then $\frac{1}{32}(12x - 8\sin 2x + \sin 4x) + C = \frac{3}{8}x - \frac{3}{16}\sin 2x - \frac{1}{4}\sin^3 x\cos x + C = \frac{3}{8}x - \frac{3}{8}\sin x\cos x - \frac{1}{4}\sin^3 x\cos x + C$.

63. $\frac{\pi^2}{2}$ **65.** $\frac{1}{8}x - \frac{1}{16}\sin 2x\cos 2x + C$

67. $\frac{1}{16}x - \frac{1}{48}\sin 2x - \frac{1}{32}\sin 2x\cos 2x + \frac{1}{48}\cos^2 2x\sin 2x + C$

69. Use the identity $\tan^2 x = \sec^2 x - 1$ and the substitution $u = \tan x$, $du = \sec^2 x\,dx$.

71. **(a)** $I_0 = \int_0^{\pi/2}\sin^0 x\,dx = \frac{\pi}{2}$; $I_1 = \int_0^{\pi/2}\sin x\,dx = 1$

(b) $\frac{m-1}{m}\int_0^{\pi/2}\sin^{m-2}x\,dx$
(c) $I_2 = \frac{\pi}{4}$; $I_3 = \frac{2}{3}$; $I_4 = \frac{3\pi}{16}$; $I_5 = \frac{8}{15}$
73. $\cos(x) - \cos(x)\ln(\sin(x)) + \ln|\csc(x) - \cot(x)| + C$
77. Use Integration by Parts with $u = \sec^{m-2}x$ and $v' = \sec^2 x$.

Section 8.3 Preliminary Questions

1. (a) $x = 3\sin\theta$ **(b)** $x = 4\sec\theta$ **(c)** $x = 4\tan\theta$ **(d)** $x = \sqrt{5}\sec\theta$
3. $2x\sqrt{1-x^2}$

Section 8.3 Exercises

1. (a) $\theta + C$ **(b)** $\sin^{-1}\left(\frac{x}{3}\right) + C$
3. (a) $\int \frac{dx}{\sqrt{4x^2+9}} = \frac{1}{2}\int \sec\theta\,d\theta$
(b) $\frac{1}{2}\ln|\sec\theta + \tan\theta| + C$
(c) $\ln|\sqrt{4x^2+9} + 2x| + C$
5. $\frac{8}{\sqrt{5}}\arccos(\frac{\sqrt{16-5x^2}}{4}) + \frac{x\sqrt{16-5x^2}}{2} + C$
7. $\frac{1}{3}\sec^{-1}\left(\frac{x}{3}\right) + C$ **9.** $\frac{-x}{4\sqrt{x^2-4}} + C$
11. $\sqrt{x^2-4} + C$
13. (a) $-\sqrt{1-x^2}$ **(b)** $\frac{1}{8}(\arcsin x - x\sqrt{1-x^2}(1-2x^2))$
(c) $-\frac{1}{3}(1-x^2)^{\frac{3}{2}} + \frac{1}{5}(1-x^2)^{\frac{5}{2}}$
(d) $\sqrt{1-x^2}\left(-\frac{x^3}{4} - \frac{3x}{8}\right) + \frac{3}{8}\arcsin(x)$
15. $\frac{9}{2}\sin^{-1}\left(\frac{x}{3}\right) - \frac{1}{2}x\sqrt{9-x^2} + C$
17. $\frac{1}{4}\ln\left|\frac{\sqrt{x^2+16}-4}{x}\right| + C$ **19.** $\ln\left|x + \sqrt{x^2-9}\right| + C$
21. $-\frac{\sqrt{5-y^2}}{5y} + C$ **23.** $\frac{1}{5}\ln\sqrt{25x^2+25x} + C$
25. $\frac{1}{16}\sec^{-1}\left(\frac{z}{2}\right) + \frac{\sqrt{z^2-4}}{8z^2} + C$
27. $\frac{1}{12}x\sqrt{6x^2-49} + \frac{1}{2}\ln x + \sqrt{x^2-1} + C$
29. $\frac{1}{54}\tan^{-1}\left(\frac{t}{3}\right) + \frac{t}{18(t^2+9)} + C$
31. $\frac{x}{\sqrt{x^2-1}} + \ln x + \sqrt{x^2-1} + C$
33. Use the substitution $x = \sqrt{a}\,u$.
35. (a) $x^2 - 4x + 8 = x^2 - 4x + 4 + 4 = (x-2)^2 + 4$
(b) $\ln\left|\sqrt{u^2+4} + u\right| + C$
(c) $\ln\left|\sqrt{(x-2)^2+4} + x - 2\right| + C$
37. $\ln\left|\sqrt{x^2+4x+13} + x + 2\right| + C$
39. $\frac{1}{\sqrt{6}}\ln 12x + 1 + 2\sqrt{6}\sqrt{x+6x^2} + C$
41. $\frac{1}{2}(x-2)\sqrt{x^2-4x+3} + \frac{7}{2}\ln\left|x - 2 + \sqrt{x^2-4x+3}\right| + C$
43. Begin by multiplying by -1, then completing the square, and then follow up with u-substitution ($u = (x+3)$; $du = dx$) and then trigonometric substitution.
45. Use one of the following trigonometric methods: rewrite $\sin^3 x = (1-\cos^2 x)\sin x$ and let $u = \cos x$, or rewrite $\cos^3 x = (1-\sin^2 x)\cos x$ and let $u = \sin x$.
47. Use trigonometric substitution, with $x = 3\sin\theta$ or substitution with $x = 3u$ and $dx = 3\,du$.
49. The techniques learned thus far are insufficient to solve this integral.
51. The techniques we have covered thus far are not sufficient to treat this integral. This integral requires a technique known as partial fractions.
53. $x\sec^{-1}x - \ln\left|x + \sqrt{x^2-1}\right| + C$
55. $x(\ln(x^2+1) - 2) + 2\tan^{-1}x + C$
57. $\frac{\pi}{4}$ **59.** $4\pi\left[\sqrt{3} - \ln\left|2+\sqrt{3}\right|\right]$
61. $\frac{1}{2}\ln|x-1| - \frac{1}{2}\ln|x+1| + C$
63. (a) $1.789 \times 10^6\ \frac{V}{m}$ **(b)** $3.526 \times 10^6\ \frac{V}{m}$

Section 8.4 Preliminary Questions

1. (a) $x = \sinh t$ **(b)** $x = 3\sinh t$ **(c)** $3x = \sinh t$
3. $\frac{1}{2}\ln\left|\frac{1+x}{1-x}\right|$

Section 8.4 Exercises

1. $\frac{1}{3}\sinh 3x + C$. **3.** $\frac{1}{2}\cosh(x^2+1) + C$
5. $-\frac{1}{2}\tanh(1-2x) + C$ **7.** $\frac{\tanh^2 x}{2} + C$ **9.** $\ln\cosh x + C$
11. $\ln|\sinh x| + C$ **13.** $\frac{1}{16}\sinh(8x-18) - \frac{1}{2}x + C$
15. $\frac{1}{32}\sinh 4x - \frac{1}{8}x + C$ **17.** $\cosh^{-1}x + C$
19. $\frac{1}{5}\sinh^{-1}\left(\frac{5x}{4}\right) + C$ **21.** $\frac{1}{2}x\sqrt{x^2-1} - \frac{1}{2}\cosh^{-1}x + C$
23. $2\tanh^{-1}\left(\frac{1}{2}\right)$ **25.** $\sinh^{-1}1$
27. $\frac{1}{4}\left(\operatorname{csch}^{-1}\left(-\frac{1}{4}\right) - \operatorname{csch}^{-1}\left(-\frac{3}{4}\right)\right)$
29. $\cosh^{-1}x - \frac{\sqrt{x^2-1}}{x} + C$
31. Let $x = \sinh t$ for the first formula and $x = \cosh t$ for the second.
33. $\frac{1}{2}x\sqrt{x^2+16} + 8\ln\left|\frac{x}{4} + \sqrt{\left(\frac{x}{4}\right)^2 + 1}\right| + C$
35. Using Integration by Parts with $u = \cosh^{n-1}x$ and $v' = \cosh x$ to begin proof.
37. $-\frac{1}{2}\left(\tanh^{-1}x\right)^2 + C$
39. $x\tanh^{-1}x + \frac{1}{2}\ln|1-x^2| + C$
41. $u = \sqrt{\frac{\cosh x - 1}{\cosh x + 1}}$. From this it follows that $\cosh x = \frac{1+u^2}{1-u^2}$, $\sinh x = \frac{2u}{1-u^2}$ and $dx = \frac{2du}{1-u^2}$.
43. $\int du = u + C = \tanh\frac{x}{2} + C$
45. Let $gd(y) = \tan^{-1}(\sinh y)$. Then

$$\frac{d}{dy}gd(y) = \frac{1}{1+\sinh^2 y}\cosh y = \frac{1}{\cosh y} = \operatorname{sech} y,$$

where we have used the identity $1 + \sinh^2 y = \cosh^2 y$.
47. Let $x = gd(y) = \tan^{-1}(\sinh y)$. Solving for y yields $y = \sinh^{-1}(\tan x)$. Therefore, $gd^{-1}(y) = \sinh^{-1}(\tan y)$.
49. Let $x = it$. Then $\cosh^2 x = (\cosh(it))^2 = \cos^2 t$ and $\sinh^2 x = (\sinh(it))^2 = i^2\sin^2 t = -\sin^2 t$. Thus, $1 = \cosh^2(it) - \sinh^2(it) = \cos^2 t - (-\sin^2 t) = \cos^2 t + \sin^2 t$, as desired.

Section 8.5 Preliminary Questions

1. No, $f(x)$ cannot be a rational function because the integral of a rational function cannot contain a term with a non-integer exponent such as $\sqrt{x+1}$

3. (a) Square is already completed; irreducible.
(b) Square is already completed; factors as $(x-\sqrt{5})(x+\sqrt{5})$.
(c) $x^2+4x+6=(x+2)^2+2$; irreducible.
(d) $x^2+4x+2=(x+2)^2-2$; factors as $(x+2-\sqrt{2})(x+2+\sqrt{2})$.

Section 8.5 Exercises

1. (a) $\dfrac{x^2+4x+12}{(x+2)(x^2+4)}=\dfrac{1}{x+2}+\dfrac{4}{x^2+4}$.
(b) $\dfrac{2x^2+8x+24}{(x+2)^2(x^2+4)}=\dfrac{1}{x+2}+\dfrac{2}{(x+2)^2}+\dfrac{-x+2}{x^2+4}$.
(c) $\dfrac{x^2-4x+8}{(x-1)^2(x-2)^2}=\dfrac{-8}{x-2}+\dfrac{4}{(x-2)^2}+\dfrac{8}{x-1}+\dfrac{5}{(x-1)^2}$.
(d) $\dfrac{x^4-4x+8}{(x+2)(x^2+4)}=x-2+\dfrac{4}{x+2}-\dfrac{4x-4}{x^2+4}$.

3. -2 **5.** $\frac{1}{9}(3x+4\ln(3x-4))+C$

7. $\frac{x^3}{3}+\ln(x+2)+C$ **9.** $-\frac{1}{2}\ln|x-2|+\frac{1}{2}\ln|x-4|+C$

11. $\ln|x|-\ln|2x+1|+C$

13. $x-3\arctan\frac{x}{3}+C$

15. $2\ln|x+3|-\ln|x+5|-\frac{2}{3}\ln|3x-2|+C$

17. $3\ln|x-1|-2\ln|x+1|-\frac{5}{x+1}+C$

19. $2\ln|x-1|-\frac{1}{x-1}-2\ln|x-2|-\frac{1}{x-2}+C$

21. $\ln(x)-\ln(x+2)+\frac{2}{x+2}+\frac{2}{(x+2)^2}+C$

23. $\frac{1}{2\sqrt{6}}\ln\left|\sqrt{2}x-\sqrt{3}\right|-\frac{1}{2\sqrt{6}}\ln\left|\sqrt{2}x+\sqrt{3}\right|+C$

25. $\frac{5}{2x+5}-\frac{5}{4(2x+5)^2}+\frac{1}{2}\ln(2x+5)+C$

27. $-\ln|x|+\ln|x-1|+\frac{1}{x-1}-\frac{1}{2(x-1)^2}+C$

29. $x+\ln|x|-3\ln|x+1|+C$

31. $2\ln|x-1|+\frac{1}{2}\ln|x^2+1|-3\tan^{-1}x+C$

33. $\frac{1}{25}\ln|x|-\frac{1}{50}\ln|x^2+25|+C$

35. $6x-14\ln x+3+2\ln x-1+C$

37. $-\frac{1}{5}\ln|x-1|-\frac{1}{x-1}+\frac{1}{10}\ln|x^2+9|-\frac{4}{15}\tan^{-1}\left(\frac{x}{3}\right)+C$

39. $\frac{1}{64}\ln|x|-\frac{1}{128}\ln|x^2+8|+\frac{1}{16(x^2+8)}+C$

41. $\frac{1}{6}\ln|x+2|-\frac{1}{12}\ln|x^2+4x+10|+C$

43. $\ln|x|-\frac{1}{2}\ln|x^2+2x+5|-5\dfrac{5}{2(x^2+2x+5)}-3\tan^{-1}\left(\frac{x+1}{2}\right)+C$

45. $\frac{1}{2}\arctan(x^2)+C$ **47.** $\ln(e^x-1)-x+C$

49. $2\sqrt{x}+\ln|\sqrt{x}-1|-\ln|\sqrt{x}+1|+C$

51. $\ln\left|\frac{x}{\sqrt{x^2-1}}-\frac{1}{\sqrt{x^2-1}}\right|+C=\ln\left|\frac{x-1}{\sqrt{x^2-1}}\right|+C$

53. $-\frac{1}{4}\left(\frac{\sqrt{4-x^2}}{x}\right)+C=-\frac{\sqrt{4-x^2}}{4x}+C$

55. $\frac{1}{2}x+\frac{1}{8}\sin 4x\cos 4x+C$

57. $\frac{1}{54}\tan^{-1}\left(\frac{x}{3}\right)+\frac{x}{18(x^2+9)}+C$

59. $\frac{1}{5}\sec^5 x-\frac{2}{3}\sec^3 x+\sec x+C$

61. $x\ln(x^2+1)+(x+1)\ln(x+1)+(x-1)\ln(x-1)-4x-2\arctan x+C$

63. $\ln\left|x+\sqrt{x^2-1}\right|-\frac{x}{\sqrt{x^2-1}}+C$

65. $\frac{2}{3}\tan^{-1}(x^{3/2})+C$

67. If $\theta=2\tan^{-1}t$, then $d\theta=2\,dt/(1+t^2)$. We also have that $\cos(\frac{\theta}{2})=1/\sqrt{1+t^2}$ and $\sin(\frac{\theta}{2})=t/\sqrt{1+t^2}$. To find $\cos\theta$, we use the double angle identity $\cos\theta=1-2\sin^2(\frac{\theta}{2})$. This gives us $\cos\theta=\frac{1-t^2}{1+t^2}$. To find $\sin\theta$, we use the double angle identity $\sin\theta=2\sin(\frac{\theta}{2})\cos(\frac{\theta}{2})$. This gives us $\sin\theta=\frac{2t}{1+t^2}$. It follows then that $\displaystyle\int\frac{d\theta}{\cos\theta+\frac{3}{4}\sin\theta}=$
$-\dfrac{4}{5}\ln\left|2-\tan\left(\dfrac{\theta}{2}\right)\right|+\dfrac{4}{5}\ln\left|1+2\tan\left(\dfrac{\theta}{2}\right)\right|+C.$

69. Partial fraction decomposition shows $\frac{1}{(x-a)(x-b)}=\frac{\frac{1}{a-b}}{x-a}+\frac{\frac{1}{b-a}}{x-b}$. This can be used to show $\int\frac{dx}{(x-a)(x-b)}=\frac{1}{a-b}\ln\left|\frac{x-a}{x-b}\right|+C$.

71. $\frac{2}{x-6}+\frac{1}{x+2}$

Section 8.6 Preliminary Questions

1. (a) The integral is converges.
(b) The integral is diverges.
(c) The integral is diverges.
(d) The integral is converges.

3. Any value of b satisfying $|b|\ge 2$ will make this an improper integral.

5. Knowing that an integral is smaller than a divergent integral does not allow us to draw any conclusions using the comparison test.

Section 8.6 Exercises

1. (a) Improper. The function $x^{-1/3}$ is infinite at 0.
(b) Improper. Infinite interval of integration.
(c) Improper. Infinite interval of integration.
(d) Proper. The function e^{-x} is continuous on the finite interval [0, 1].
(e) Improper. The function $\sec x$ is infinite at $\frac{\pi}{2}$.
(f) Improper. Infinite interval of integration.
(g) Proper. The function $\sin x$ is continuous on the finite interval [0, 1].
(h) Proper. The function $1/\sqrt{3-x^2}$ is continuous on the finite interval [0, 1].
(i) Improper. Infinite interval of integration.
(j) Improper. The function $\ln x$ is infinite at 0.

3. $\int_1^\infty x^{-2/3}\,dx=\lim_{R\to\infty}\int_1^R x^{-2/3}\,dx=\lim_{R\to\infty}3\left(R^{1/3}-1\right)=\infty$

5. The integral does not converge.

7. The integral converges; $I = 10{,}000e^{0.0004}$.

9. The integral does not converge.

11. The integral converges; $I = 4$.

13. The integral converges; $I = \frac{1}{8}$.

15. The integral converges; $I = 2$.

17. The integral converges; $I = 1.25$.

19. The integral converges; $I = \frac{1}{3e^{12}}$.

21. The integral converges; $I = \frac{1}{3}$.

23. The integral converges; $I = 2\sqrt{2}$.

25. The integral does not converge.

27. The integral converges; $I = \frac{1}{2}$.

29. The integral converges; $I = \frac{1}{2}$.

31. The integral converges; $I = \frac{\pi}{2}$.

33. The integral does not converge.

35. The integral does not converge.

37. The integral converges; $I = -1$.

39. The integral does not converge.

41. (a) Partial fractions yields $\frac{dx}{(x-2)(x-3)} = \frac{dx}{x-3} - \frac{dx}{x-2}$. This yields $\int_4^R \frac{dx}{(x-2)(x-3)} = \ln\left|\frac{R-3}{R-2}\right| - \ln\frac{1}{2}$

(b) $I = \lim_{R\to\infty}\left(\ln\left|\frac{R-3}{R-2}\right| - \ln\frac{1}{2}\right) = \ln 1 - \ln\frac{1}{2} = \ln 2$

43. The integral does not converge.

45. The integral does not converge.

47. The integral converges; $I = 0$.

49. $\displaystyle\int_{-1}^{1}\frac{dx}{x^{1/3}} = \int_{-1}^{0}\frac{dx}{x^{1/3}} + \int_{0}^{1}\frac{dx}{x^{1/3}} = 0$

51. The integral converges for $a < 0$.

53. $\int_{-\infty}^{\infty}\frac{dx}{1+x^2} = \pi$.

55. $\frac{1}{x^3+4} \le \frac{1}{x^3}$. Therefore, by the comparison test, the integral converges.

57. For $x \ge 1$, $x^2 \ge x$, so $-x^2 \le -x$ and $e^{-x^2} \le e^{-x}$. Now $\int_1^\infty e^{-x}\,dx$ converges, so $\int_1^\infty e^{-x^2}\,dx$ converges by the comparison test. We conclude that our integral converges by writing it as a sum: $\int_0^\infty e^{-x^2}\,dx = \int_0^1 e^{-x^2}\,dx + \int_1^\infty e^{-x^2}\,dx$.

59. Let $f(x) = \dfrac{1-\sin x}{x^2}$. Since $f(x) \le \dfrac{2}{x^2}$ and $\displaystyle\int_1^\infty 2x^{-2}\,dx = 2$, it follows that $\displaystyle\int_1^\infty \frac{1-\sin x}{x^2}\,dx$ converges by the comparison test.

61. The integral converges.

63. The integral does not converge.

65. The integral converges.

67. The integral does not converge.

69. The integral converges.

71. The integral converges.

73. The integral does note converge.

75. $\displaystyle\int_0^1 \frac{dx}{x^{1/2}(x+1)}$ and $\displaystyle\int_1^\infty \frac{dx}{x^{1/2}(x+1)}$ both converge, therefore J converges.

77. $\frac{250}{0.07}$ **79.** \$2,000,000

81. (a) π **(b)** $\int_1^\infty \frac{1}{x}\sqrt{1+\frac{1}{x^4}}\,dx$ diverges.

83. $W = \lim_{T\to\infty} CV^2\left(\frac{1}{2} - e^{-T/RC} + \frac{1}{2}e^{-2T/RC}\right) = CV^2\left(\frac{1}{2} - 0 + 0\right) = \frac{1}{2}CV^2$

85. The integrand is infinite at the upper limit of integration, $x = \sqrt{2E/k}$, so the integral is improper.
$T = \lim_{R\to\sqrt{2E/k}} T(R) = 4\sqrt{\frac{m}{k}}\sin^{-1}(1) = 2\pi\sqrt{\frac{m}{k}}$.

87. $Lf(s) = \dfrac{-1}{s^2+\alpha^2}\lim_{t\to\infty} e^{-st}(s\sin(\alpha t) + \alpha\cos(\alpha t)) - \alpha$.

89. $\frac{s}{s^2+\alpha^2}$ **91.** $J_n = \frac{n}{\alpha}J_{n-1} = \frac{n}{\alpha}\cdot\frac{(n-1)!}{\alpha^n} = \frac{n!}{\alpha^{n+1}}$

93. $E = \frac{8\pi h}{c^3}\int_0^\infty \frac{\nu^3}{e^{\alpha\nu}-1}\,d\nu$. Because $\alpha > 0$ and $8\pi h/c^3$ is a constant, we know E is finite by Exercise 92.

95. Because $t > \ln t$ for $t > 2$, $F(x) = \displaystyle\int_2^x \frac{dt}{\ln t} > \int_2^x \frac{dt}{t} > \ln x$. Thus, $F(x) \to \infty$ as $x \to \infty$. Moreover, $\lim_{x\to\infty} G(x) = \lim_{x\to\infty}\frac{1}{1/x} = \lim_{x\to\infty} x = \infty$. Thus, $\lim_{x\to\infty}\frac{F(x)}{G(x)}$ is of the form ∞/∞, and L'Hôpital's Rule applies. Finally,
$$L = \lim_{x\to\infty}\frac{F(x)}{G(x)} = \lim_{x\to\infty}\frac{\frac{1}{\ln x}}{\frac{\ln x - 1}{(\ln x)^2}} = \lim_{x\to\infty}\frac{\ln x}{\ln x - 1} = 1.$$

97. The integral is absolutely convergent. Use the comparison test with $\frac{1}{x^2}$.

Section 8.7 Preliminary Questions

1. No, $p(x) \ge 0$ fails. **3.** $p(x) = 4e^{-4x}$

Section 8.7 Exercises

1. $C = 2$; $P(0 \le X \le 1) = \frac{3}{4}$

3. $C = \frac{1}{\pi}$; $P(-\frac{1}{2} \le X \le \frac{1}{2}) = \frac{1}{3}$

5. $C = \frac{2}{\pi}$; $P(-\frac{1}{2} \le X \le 1) = \frac{2}{3} + \frac{\sqrt{3}}{4\pi}$

7. $\int_1^\infty 3x^{-4} = 1$; $\mu = \frac{3}{2}$

9. Integration confirms $\int_0^\infty \frac{1}{50}e^{-t/50} = 1$

11. $e^{-\frac{3}{2}} \approx 0.2231$ **13.** $\frac{1}{2}\left(2 - 10e^{-2}\right) \approx 0.32$

15. $F(-\frac{2}{3}) - F(-\frac{13}{6}) \approx 0.2374$

17. (a) ≈ 0.8849 **(b)** ≈ 0.6554

19. $1 - F(z)$ and $F(-z)$ are the same area on opposite tails of the distribution function. Simple algebra with the standard normal cumulative distribution function shows $P(\mu - r\sigma \le X \le \mu + r\sigma) = 2F(r) - 1$

21. ≈ 0.0062 **23.** $\mu = 5/3$; $\sigma = \sqrt{10/3}$ **25.** $\mu = 3$; $\sigma = 3$

27. (a) $f(t)$ is the fraction of initial atoms present at time t. Therefore, the fraction of atoms that decay is going to be the rate of change of the total number of atoms. Over a small interval, this is simply $-f'(t)\Delta t$.

(b) The fraction of atoms that decay over an arbitrarily small interval is equivalent to the probability that an individual atom will decay over that same interval. Thus, the probability density function becomes $-f'(t)$. (c) $\int_0^\infty -tf'(t)\,dt = \frac{1}{k}$

Section 8.8 Preliminary Questions

1. $T_1 = 6$; $T_2 = 7$

3. The Trapezoidal Rule integrates linear functions exactly, so the error will be zero.

5. The two graphical interpretations of the Midpoint Rule are the sum of the areas of the midpoint rectangles and the sum of the areas of the tangential trapezoids.

Section 8.8 Exercises

1. $T_4 = 2.75$; $M_4 = 2.625$

3. $T_6 = 64.6875$; $M_6 \approx 63.2813$

5. $T_6 \approx 1.4054$; $M_6 \approx 1.3769$

7. $T_6 = 1.1703$; $M_6 = 1.2063$

9. $T_4 \approx 0.3846$; $M_5 \approx 0.3871$

11. $T_5 = 1.4807$; $M_5 = 1.4537$

13. $S_4 \approx 5.2522$ **15.** $S_6 \approx 1.1090$ **17.** $S_4 \approx 0.7469$
19. $S_8 \approx 2.5450$ **21.** $S_10 \approx 0.3466$ **23.** ≈ 2.4674
25. ≈ 1.8769 **27.** ≈ 608.611

29. **(a)** Assuming the speed of the tsunami is a continuous function, at x miles from the shore, the speed is $\sqrt{15f(x)}$. Covering an infinitesimally small distance, dx, the time T required for the tsunami to cover that distance becomes $\frac{dx}{\sqrt{15f(x)}}$. It follows from this that $T = \int_0^M \frac{dx}{\sqrt{15f(x)}}$.

(b) ≈ 3.347 hours.

31. **(a)** Since x^3 is concave up on $[0, 2]$, T_6 is too large.

(b) We have $f'(x) = 3x^2$ and $f''(x) = 6x$. Since $|f''(x)| = |6x|$ is *increasing* on $[0, 2]$, its maximum value occurs at $x = 2$ and we may take $K_2 = |f''(2)| = 12$. Thus, $\text{Error}(T_6) \le \frac{2}{9}$.

(c) $\text{Error}(T_6) \approx 0.1111 < \frac{2}{9}$

33. T_10 will overestimate the integral. $\text{Error}(T_{10}) \le 0.045$.

35. M_10 will overestimate the integral. $\text{Error}(M_{10}) \le 0.0113$

37. $N \ge 10^3$; Error $\approx 3.333 \times 10^{-7}$

39. $N \ge 750$; Error $\approx 2.805 \times 10^{-7}$

41. $\text{Error}(T_{10}) \le 0.0225$; $\text{Error}(M_{10}) \le 0.01125$

43. $S_8 \approx 4.0467$; $N \ge 23$

45. $\text{Error}(S_{40}) \le 1.017 \times 10^{-4}$.

47. $N \ge 305$ **49.** $N \ge 186$

51. **(a)** The maximum value of $|f^{(4)}(x)|$ on the interval $[0, 1]$ is 24.

(b) $N \ge 20$; $S_{20} \approx 0.785398$; $|0.785398 - \frac{\pi}{4}| \approx 1.55 \times 10^{-10}$.

53. **(a)** Notice $|f''(x)| = |2\cos(x^2) - 4x^2\sin(x^2)|$; proof follows.

(b) When $K_2 = 2$, $\text{Error}(M_N) \le \frac{1}{4N^2}$.

(c) $N \ge 16$

55. $\text{Error}(T_4) \approx 0.1039$; $\text{Error}(T_8) \approx 0.0258$; $\text{Error}(T_{16}) \approx 0.0064$; $\text{Error}(T_{32}) \approx 0.0016$; $\text{Error}(T_{64}) \approx 0.0004$. Thes are about twice as large as the error in M_N.

57. $S_2 = \frac{1}{4}$. This is the exact value of the integral.

59. $T_N = \dfrac{r(b^2 - a^2)}{2} + s(b - a) = \displaystyle\int_a^b f(x)\,dx$

61. **(a)** This result follows because the even-numbered interior endpoints overlap:

$$\sum_{i=0}^{(N-2)/2} S_2^{2j} = \frac{b-a}{6}\left[(y_0 + 4y_1 + y_2) + (y_2 + 4y_3 + y_4) + \cdots\right]$$

$$= \frac{b-a}{6}\left[y_0 + 4y_1 + 2y_2 + 4y_3 + 2y_4 + \cdots + 4y_{N-1} + y_N\right] = S_N.$$

(b) If $f(x)$ is a quadratic polynomial, then by part (a) we have

$$S_N = S_2^0 + S_2^2 + \cdots + S_2^{N-2} = \int_a^b f(x)\,dx.$$

63. Let $f(x) = ax^3 + bx^2 + cx + d$, with $a \ne 0$, be any cubic polynomial. Then, $f^{(4)}(x) = 0$, so we can take $K_4 = 0$. This yields $\text{Error}(S_N) \le \frac{0}{180N^4} = 0$. In other words, S_N is exact for all cubic polynomials for all N.

Chapter 8 Review

1. **(a)** (v) **(b)** (iv) **(c)** (iii) **(d)** (i) **(e)** (ii)

3. $\frac{\sin^9\theta}{9} - \frac{\sin^{11}\theta}{11} + C.$

5.
$\frac{\tan\theta\sec^5\theta}{6} - \frac{7\tan\theta\sec^3\theta}{24} + \frac{\tan\theta\sec\theta}{16} + \frac{1}{16}\ln|\sec\theta + \tan\theta| + C$

7. $-\frac{1}{\sqrt{x^2-1}} - \sec^{-1}x + C$ **9.** $2\tan^{-1}\sqrt{x} + C$

11. $-\frac{\tan^{-1}x}{x} + \ln|x| - \frac{1}{2}\ln\left(1 + x^2\right) + C.$

13. $\frac{5}{32}e^4 - \frac{1}{32} \approx 8.50$ **15.** $\frac{\cos^{12}6\theta}{72} - \frac{\cos^{10}6\theta}{60} + C$

17. $5\ln|x - 1| + \ln|x + 1| + C$

19. $\frac{\tan^3\theta}{3} + \tan\theta + C$ **21.** ≈ 1.0794

23. $-\frac{\cos^5\theta}{5} + \frac{2\cos^3\theta}{3} - \cos\theta + C$ **25.** $-\frac{1}{4}$

27. $\frac{2}{3}(\tan x)^{3/2} + C$

29. $\frac{\sin^6\theta}{6} - \frac{\sin^8\theta}{8} + C$ **31.** $-\frac{1}{3}u^3 + C = -\frac{1}{3}\cot^3 x + C$

33. ≈ 0.4202 **35.** $\frac{1}{49}\ln\left|\frac{t+4}{t-3}\right| - \frac{1}{7}\cdot\frac{1}{t-3} + C$

37. $\frac{1}{2}\sec^{-1}\frac{x}{2} + C$

39. $$\int \frac{dx}{x^{3/2} + ax^{1/2}} = \begin{cases} \frac{2}{\sqrt{a}}\tan^{-1}\sqrt{\frac{x}{a}} + C & a > 0 \\ \frac{1}{\sqrt{-a}}\ln\left|\frac{\sqrt{x}-\sqrt{-a}}{\sqrt{x}+\sqrt{-a}}\right| + C & a < 0 \\ -\frac{2}{\sqrt{x}} + C & a = 0 \end{cases}$$

41. $\ln|x + 2| + \frac{5}{x+2} - \frac{3}{(x+2)^2} + C$

43. $-\ln|x - 2| - 2\frac{1}{x-2} + \frac{1}{2}\ln\left(x^2 + 4\right) + C$

45. $\frac{1}{3}\tan^{-1}\left(\frac{x+4}{3}\right) + C$ **47.** $\ln|x + 2| + \frac{5}{x+2} - \frac{3}{(x+2)^2} + C$

49. $-\frac{\left(x^2+4\right)^{3/2}}{48x^3} + \frac{\sqrt{x^2+4}}{16x} + C$ **51.** $-\frac{1}{9}e^{4-3x}(3x + 4) + C$

53. $\frac{1}{2}x^2\sin x^2 + \frac{1}{2}\cos x^2 + C$

55. $\frac{x^2}{2}\tanh^{-1}x + \frac{x}{2} - \frac{1}{4}\ln\left|\frac{1+x}{1-x}\right| + C$

57. $x\ln\left(x^2 + 9\right) - 2x + 6\tan^{-1}\left(\frac{x}{3}\right) + C$

59. $\frac{1}{2}\sinh 2$ **61.** $t + \frac{1}{4}\coth(1 - 4t) + C$ **63.** $\frac{\pi}{3}$

65. $\tan^{-1}(\tanh x) + C$

67. (a) $I_n = \int \frac{x^n}{x^2+1}\,dx = \int \frac{x^{n-2}(x^2+1-1)}{x^2+1}\,dx = \int x^{n-2}\,dx - \int \frac{x^{n-2}}{x^2+1}\,dx = \frac{x^{n-1}}{n-1} - I_{n-2}$

(b) $I_0 = \tan^{-1}x + C$; $I_1 = \frac{1}{2}\ln\left(x^2+1\right) + C$;
$I_2 = x - \tan^{-1}x + C$; $I_3 = \frac{x^2}{2} - \frac{1}{2}\ln\left(x^2+1\right) + C$;
$I_4 = \frac{x^3}{3} - x + \tan^{-1}x + C$; $I_5 = \frac{x^4}{4} - \frac{x^2}{2} + \frac{1}{2}\ln\left(x^2+1\right) + C$

(c) Prove by induction; show it works for $n = 1$, then assume it works for $n = k$ and use that to show it works for $n = k+1$.

69. $\frac{3}{4}$ **71.** $C = 2$; $p(0 \le X \le 1) = 1 - \frac{2}{e}$

73. (a) 0.1587 **(b)** 0.49997

75. Integral converges; $I = \frac{1}{2}$.

77. Integral converges; $I = 3\sqrt[3]{4}$.

79. Integral converges; $I = \frac{\pi}{2}$.

81. The integral does not converge.

83. The integral does not converge.

85. The integral converges.

87. The integral converges.

89. The integral converges. **91.** π **95.** $\frac{2}{(s-\alpha)^3}$

97. (a) T_N is smaller and M_N is larger than the integral.
(b) M_N is smaller and T_N is larger than the integral.
(c) M_N is smaller and T_N is larger than the integral.
(d) T_N is smaller and M_N is larger than the integral.

99. $M_5 \approx 0.7481$ **101.** $M_4 \approx 0.7450$ **103.** $S_6 \approx 0.7469$
105. $V \approx T_9 \approx 20$ hectare-ft $= 871{,}200$ ft^3 **107.** Error $\le \frac{3}{128}$.
109. $N \ge 29$

Chapter 9

Section 9.1 Preliminary Questions

1. $\int_0^{\pi} \sqrt{1+\sin^2 x}\,dx$

2. The graph of $y = f(x) + C$ is a vertical translation of the graph of $y = f(x)$; hence, the two graphs should have the same arc length. We can explicitly establish this as follows:

$$\text{Length of } y = f(x)+C = \int_a^b \sqrt{1+\left[\frac{d}{dx}(f(x)+C)\right]^2}\,dx = \int_a^b \sqrt{1+[f'(x)]^2}\,dx = \text{length of } y = f(x).$$

3. Since $\sqrt{1+f'(x)^2} \ge 1$ for any function f, we have

$$\text{Length of graph of } f(x) \text{ over } [1,4] = \int_1^4 \sqrt{1+f'(x)^2}\,dx \ge \int_1^4 1\,dx = 3$$

Section 9.1 Exercises

1. $L = \int_2^6 \sqrt{1+16x^6}\,dx$ **3.** $\frac{13}{12}$ **5.** $3\sqrt{10}$
7. $\frac{1}{27}(22\sqrt{22} - 13\sqrt{13})$ **9.** $e^2 + \frac{\ln 2}{2} + \frac{1}{4}$
11. $\int_1^2 \sqrt{1+x^6}\,dx \approx 3.957736$
13. $\int_1^2 \sqrt{1+\frac{1}{x^4}}\,dx \approx 1.132123$
15. 6
19. $a = \sinh^{-1}(5) = \ln(5+\sqrt{26})$
23. Let s denote the arc length. Then $s = \frac{a}{2}\sqrt{1+4a^2} + \frac{1}{4}\ln\left|\sqrt{1+4a^2}+2a\right|$. Thus, when $a = 1$, $s = \frac{1}{2}\sqrt{5} + \frac{1}{4}\ln(\sqrt{5}+2) \approx 1.478943$.
25. $\sqrt{1+e^{2a}} + \frac{1}{2}\ln\frac{\sqrt{1+e^{2a}}-1}{\sqrt{1+e^{2a}}+1} - \sqrt{2} + \frac{1}{2}\ln\frac{1+\sqrt{2}}{\sqrt{2}-1}$
27. $\ln(1+\sqrt{2})$ **31.** 1.552248 **33.** $16\pi\sqrt{2}$
35. $\frac{\pi}{27}(145^{3/2} - 1)$ **37.** $\frac{384\pi}{5}$ **39.** $\frac{\pi}{16}(e^4 - 9)$
41. $2\pi\int_1^3 x^{-1}\sqrt{1+x^{-4}}\,dx \approx 7.60306$
43. $2\pi\int_0^2 e^{-x^2/2}\sqrt{1+x^2e^{-x^2}}\,dx \approx 8.222696$
45. $2\pi\ln 2 + \frac{15\pi}{8}$ **47.** $4\pi^2 br$
49. $2\pi b^2 + \frac{2\pi b a^2}{\sqrt{b^2-a^2}}\ln\left|\frac{\sqrt{b^2-a^2}}{a} + \frac{b}{a}\right|$

Section 9.2 Preliminary Questions

1. Pressure is defined as force per unit area.

2. The factor of proportionality is the weight density of the fluid, $w = \rho g$.

3. Fluid force acts in the direction perpendicular to the side of the submerged object.

4. Pressure depends only on depth and does not change horizontally at a given depth.

5. When a plate is submerged vertically, the pressure is not constant along the plate, so the fluid force is not equal to the pressure times the area.

Section 9.2 Exercises

1. (a) Top: $F = 176{,}500$ N; bottom: $F = 705{,}600$ N
(b) $F \approx \sum_{j=1}^{N} \rho g 3 y_j \,\Delta y$ **(c)** $F = \int_2^8 \rho g 3y\,dy$
(d) $F = 882{,}000$ N

3. (a) The width of the triangle varies linearly from 0 at a depth of $y = 3$ m to 1 at a depth of $y = 5$ m. Thus, $f(y) = \frac{1}{2}(y-3)$.
(b) The area of the strip at depth y is $\frac{1}{2}(y-3)\Delta y$, and the pressure at depth y is $\rho g y$, where $\rho = 10^3$ kg/m^3 and $g = 9.8$. Thus, the fluid force acting on the strip at depth y is approximately equal to $\rho g \frac{1}{2} y(y-3)\Delta y$.
(c) $F \approx \sum_{j=1}^{N} \rho g \frac{1}{2} y_j(y_j - 3)\,\Delta y \to \int_3^5 \rho g \frac{1}{2} y(y-3)\,dy$
(d) $F = \frac{127{,}400}{3}$ N

5. (b) $F = \frac{19{,}600}{3} r^3$ N

7. $F = \frac{19{,}600}{3} r^3 + 4{,}900\pi m r^2$ N

9. $F \approx 321{,}250{,}000$ lb

11. $F = \frac{815360}{3}$ N **13.** $F \approx 5593.804$ N **15.** $F \approx 5652.4$ N
17. $F = 940,800$ N
19. $F = 4{,}532{,}500{,}000 \sec\left(\frac{7\pi}{36}\right) \approx 5.53316 \times 10^9$ N
21. $F = (15b + 30a)h^2$ lb
23. Front and back: $F = \frac{62.5\sqrt{3}}{9}H^3$; slanted sides: $F = \frac{62.5\sqrt{3}}{3}\ell H^2$.

Section 9.3 Preliminary Questions

1. $M_x = M_y = 0$ **2.** $M_x = 21$ **3.** $M_x = 5$; $M_y = 10$

4. Because a rectangle is symmetric with respect to both the vertical line and the horizontal line through the center of the rectangle, the Symmetry Principle guarantees that the centroid of the rectangle must lie along both these lines. The only point in common to both lines of symmetry is the center of the rectangle, so the centroid of the rectangle must be the center of the rectangle.

Section 9.3 Exercises

1. **(a)** $M_x = 4m$; $M_y = 9m$; center of mass: $\left(\frac{9}{4}, 1\right)$
(b) $\left(\frac{46}{17}, \frac{14}{17}\right)$

5. A sketch of the lamina is shown here.

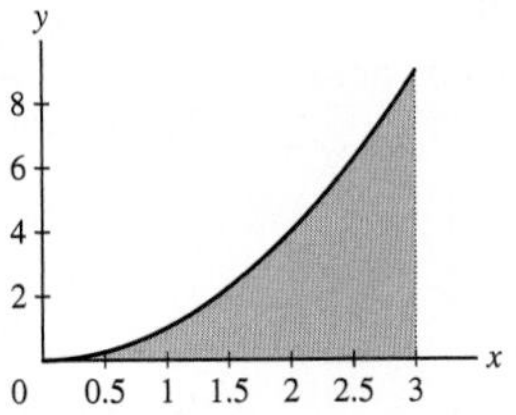

(a) $M_x = \frac{729}{10}$; $M_y = \frac{243}{4}$
(b) Area $= 9$ cm^2; center of mass: $\left(\frac{9}{4}, \frac{27}{10}\right)$

7. $M_x = \frac{64\rho}{7}$; $M_y = \frac{32\rho}{5}$; center of mass : $\left(\frac{8}{5}, \frac{16}{7}\right)$

9. **(a)** $M_x = 24$
(b) $M = 12$, so $y_{cm} = 2$; center of mass: $(0, 2)$

11. $\left(\frac{93}{35}, \frac{45}{56}\right)$ **13.** $\left(\frac{9}{8}, \frac{18}{5}\right)$

15. $\left(\frac{1-5e^{-4}}{1-e^{-4}}, \frac{1-e^{-8}}{4(1-e^{-4})}\right)$ **17.** $\left(\frac{\pi}{2}, \frac{\pi}{8}\right)$

19. A sketch of the region is shown here.

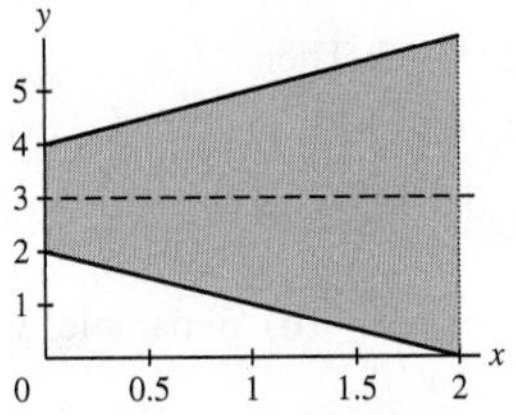

The region is clearly symmetric about the line $y = 3$, so we expect the centroid of the region to lie along this line. We find $M_x = 24$, $M_y = \frac{28}{3}$, centroid: $\left(\frac{7}{6}, 3\right)$.

21. $\left(\frac{9}{20}, \frac{9}{20}\right)$ **23.** $\left(\frac{1}{2(e-2)}, \frac{e^2-3}{4(e-2)}\right)$

25. $\left(\frac{\pi\sqrt{2}-4}{4(\sqrt{2}-1)}, \frac{1}{4(\sqrt{2}-1)}\right)$

27. A sketch of the region is shown here. Centroid: $\left(0, \frac{2}{7}\right)$

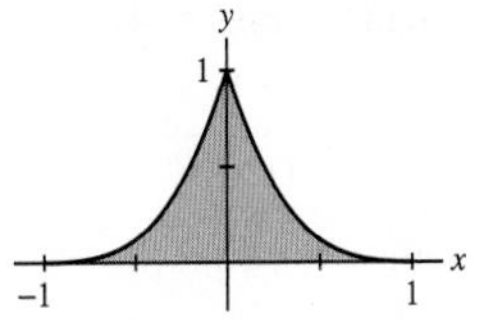

29. $\left(0, \frac{4b}{3\pi}\right)$ **31.** $\left(\frac{4}{3\pi}, \frac{4}{3\pi}\right)$

33. $\left(0, \frac{\frac{2}{3}(r^2-h^2)^{3/2}}{r^2\sin^{-1}\sqrt{1-h^2/r^2} - h\sqrt{r^2-h^2}}\right)$; with $r = 1$ and $h = \frac{1}{2}$: $\left(0, \frac{3\sqrt{3}}{4\pi - 3\sqrt{3}}\right) \approx (0, 0.71)$

35. $\left(0, \frac{49}{24}\right)$ **37.** $\left(-\frac{4}{9\pi}, \frac{4}{9\pi}\right)$

39. For the square on the left: $(4, 4)$; for the square on the right: $\left(4, \frac{25}{7}\right)$.

Section 9.4 Preliminary Questions

1. $T_3(x) = 9 + 8(x-3) + 2(x-3)^2 + 2(x-3)^3$

2. The polynomial graphed on the right is a Maclaurin polynomial.

3. A Maclaurin polynomial gives the value of $f(0)$ exactly.

4. The correct statement is **(b)**: $|T_3(2) - f(2)| \le \frac{2}{3}$

Section 9.4 Exercises

1. $T_2(x) = x$; $T_3(x) = x - \frac{x^3}{6}$

3. $T_2(x) = \frac{1}{3} - \frac{1}{9}(x-2) + \frac{1}{27}(x-2)^2$;
$T_3(x) = \frac{1}{3} - \frac{1}{9}(x-2) + \frac{1}{27}(x-2)^2 - \frac{1}{81}(x-2)^3$

5. $T_2(x) = 75 + 106(x-3) + 54(x-3)^2$;
$T_3(x) = 75 + 106(x-3) + 54(x-3)^2 + 12(x-3)^3$

7. $T_2(x) = x$; $T_3(x) = x + \frac{x^3}{3}$

9. $T_2(x) = 2 - 3x + \frac{5x^2}{2}$; $T_3(x) = 2 - 3x + \frac{5x^2}{2} - \frac{3x^3}{2}$

11. $T_2(x) = \frac{1}{e} + \frac{1}{e}(x-1) - \frac{1}{2e}(x-1)^2$;
$T_3(x) = \frac{1}{e} + \frac{1}{e}(x-1) - \frac{1}{2e}(x-1)^2 - \frac{1}{6e}(x-1)^3$

13. $T_2(x) = (x-1) - \frac{3(x-1)^2}{2}$;
$T_3(x) = (x-1) - \frac{3(x-1)^2}{2} + \frac{11(x-1)^3}{6}$

15. Let $f(x) = e^x$. Then, for all n,

$$f^{(n)}(x) = e^x \quad \text{and} \quad f^{(n)}(0) = 1.$$

It follows that

$$T_n(x) = 1 + \frac{x}{1!} + \frac{x^2}{2!} + \cdots + \frac{x^n}{n!}.$$

19. $T_n(x) = 1 + x + x^2 + x^3 + \cdots + x^n$

21. $T_n(x) = e + e(x-1) + \frac{e(x-1)^2}{2!} + \cdots + \frac{e(x-1)^n}{n!}$

23.
$T_n(x) = \frac{1}{\sqrt{2}} - \frac{1}{\sqrt{2}}\left(x - \frac{\pi}{4}\right) - \frac{1}{2\sqrt{2}}\left(x - \frac{\pi}{4}\right)^2 + \frac{1}{6\sqrt{2}}\left(x - \frac{\pi}{4}\right)^3 \cdots$
In general, the coefficient of $(x - \pi/4)^n$ is

$$\pm\frac{1}{(\sqrt{2})n!}$$

with the pattern of signs $+, -, -, +, +, -, -, \ldots$.

25. $T_2(x) = 1 + x + \frac{x^2}{2}$; $|T_2(-0.5) - f(-0.5)| \approx 0.018469$

27. $T_2(x) = 1 - \frac{2}{3}(x-1) + \frac{5}{9}(x-1)^2$; $|f(1.2) - T_2(1.2)| \approx 0.00334008$

29. $T_3(x) = 1 + \frac{1}{2}(x-1) - \frac{1}{8}(x-1)^2 + \frac{1}{16}(x-1)^3$

31. $\frac{e^{1.1}|1.1|^4}{4!}$

33. $T_5(x) = 1 - \frac{x^2}{2} + \frac{x^4}{24}$; maximum error $= \frac{(0.25)^6}{6!}$

35. $T_3(x) = \frac{1}{2} - \frac{1}{16}(x-4) + \frac{3}{256}(x-4)^2 - \frac{5}{2048}(x-4)^3$; maximum error $= \frac{35(0.3)^4}{65{,}536}$

37. $T_3(x) = x - \frac{x^3}{3}$; $T_3\left(\frac{1}{2}\right) = \frac{11}{24}$. With $K = 5$,

$$\left|T_3\left(\tfrac{1}{2}\right) - \tan^{-1}\tfrac{1}{2}\right| \le \frac{5\left(\frac{1}{2}\right)^4}{4!} = \frac{5}{384}.$$

39. $T_3(x) = \cos(0.25) - \sin(0.25)(x-0.5) - \frac{\cos(0.25)+2\sin(0.25)}{2}(x-0.5)^2 + \frac{\sin(0.25)-6\cos(0.25)}{6}(x-0.5)^3$;
$|T_3(0.6) - f(0.6)| \le \frac{K(0.0001)}{24}$; $K = 10$ is acceptable.

41. $n = 4$

43. $n = 6$

47. $n = 4$

51. $T_{4n}(x) = 1 - \frac{x^4}{2} + \frac{x^8}{4!} + \cdots + (-1)^n\frac{x^{4n}}{(2n)!}$

53. At $a = 0$,

$$T_1(x) = -4 - x$$
$$T_2(x) = -4 - x + 2x^2$$
$$T_3(x) = -4 - x + 2x^2 + 3x^3 = f(x)$$
$$T_4(x) = T_3(x)$$
$$T_5(x) = T_3(x)$$

At $a = 1$,

$$T_1(x) = 12(x-1)$$
$$T_2(x) = 12(x-1) + 11(x-1)^2$$
$$T_3(x) = 12(x-1) + 11(x-1)^2 + 3(x-1)^3 = -4 - x + 2x^2 + 3x^3 = f(x)$$
$$T_4(x) = T_3(x)$$
$$T_5(x) = T_3(x)$$

55. $T_2(t) = 60 + 24t - \frac{3}{2}t^2$; truck's distance from intersection after 4 s is ≈ 132 m

57. **(a)** $T_3(x) = -\frac{k}{R^3}x + \frac{3k}{2R^5}x^3$

65. $T_4(x) = 1 - x^2 + \frac{1}{2}x^4$; the error is approximately $|0.461458 - 0.461281| = 0.000177$

67. **(b)** $\int_0^{1/2} T_4(x)\,dx = \frac{1841}{3840}$; error bound:

$$\left|\int_0^{1/2} \cos x\,dx - \int_0^{1/2} T_4(x)\,dx\right| < \frac{\left(\frac{1}{2}\right)^7}{6!}$$

69. **(a)** $T_6(x) = x^2 - \frac{1}{6}x^6$

Chapter 9 Review

1. $\frac{779}{240}$ **3.** $4\sqrt{17}$ **7.** $24\pi\sqrt{2}$ **9.** $\frac{67\pi}{36}$

11. $12\pi + 4\pi^2$ **13.** 176,400 N

15. Fluid force on a triangular face: $183{,}750\sqrt{3} + 306{,}250$ N; fluid force on a slanted rectangular edge: $122{,}500\sqrt{3} + 294{,}000$ N

17. $M_x = 20480$; $M_y = 25600$; center of mass: $\left(2, \frac{8}{5}\right)$

19. $\left(0, \frac{2}{\pi}\right)$

21. $T_3(x) = 1 + 3(x-1) + 3(x-2)^2 + (x-1)^3$

23. $T_4(x) = (x-1) + \frac{1}{2}(x-1)^2 - \frac{1}{6}(x-1)^3 + \frac{1}{12}(x-1)^4$

25. $T_4(x) = x - x^3$

27. $T_n(x) = 1 + 3x + \frac{1}{2!}(3x)^2 + \frac{1}{3!}(3x)^3 + \cdots + \frac{1}{n!}(3x)^n$

29. $T_3(1.1) = 0.832981496$; $\left|T_3(1.1) - \tan^{-1}1.1\right| = 2.301 \times 10^{-7}$

31. $n = 11$ is sufficient.

33. The nth Maclaurin polynomial for $g(x) = \frac{1}{1+x}$ is $T_n(x) = 1 - x + x^2 - x^3 + \cdots + (-x)^n$.

Chapter 10

Section 10.1 Preliminary Questions

1. **(a)** First order **(b)** First order **(c)** Order 3 **(d)** Order 2

2. Yes **3.** Example: $y' = y^2$ **4.** Example: $y' = y^2$

5. Example: $y' + y = x$

Section 10.1 Exercises

1. **(a)** First order **(b)** Not first order **(c)** First order **(d)** First order **(e)** Not first order **(f)** First order

3. Let $y = 4x^2$. Then $y' = 8x$ and $y' - 8x = 8x - 8x = 0$.

5. Let $y = 25e^{-2x^2}$. Then $y' = -100xe^{-2x^2}$ and $y' + 4xy = -100xe^{-2x^2} + 4x(25e^{-2x^2}) = 0$

7. Let $y = 4x^4 - 12x^2 + 3$. Then

$$y'' - 2xy' + 8y = (48x^2 - 24) - 2x(16x^3 - 24x) + 8(4x^4 - 12x^2 + 3)$$
$$= 48x^2 - 24 - 32x^4 + 48x^2 + 32x^4 - 96x^2 + 24 = 0$$

9. **(a)** Separable: $y' = \frac{9}{x}y^2$ **(b)** Separable: $y' = \frac{\sin x}{\sqrt{4 - x^2}}e^{3y}$

(c) Not separable **(d)** Separable: $y' = (1)(9 - y^2)$

11. $C = 4$

13. $y = \left(2x^2 + C\right)^{-1}$, where C is an arbitrary constant.

15. $y = \ln\left(4t^5 + C\right)$, where C is an arbitrary constant.

17. $y = Ce^{-(5/2)x} + \frac{4}{5}$, where C is an arbitrary constant.

19. $y = Ce^{-\sqrt{1-x^2}}$, where C is an arbitrary constant.

21. $y = \pm\sqrt{x^2 + C}$, where C is an arbitrary constant.

23. $x = \tan\left(\frac{1}{2}t^2 + t + C\right)$, where C is an arbitrary constant.

25. $y = \sin^{-1}\left(\frac{1}{2}x^2 + C\right)$, where C is an arbitrary constant.

27. $y = C \sec t$, where C is an arbitrary constant.

29. $y = 75e^{-2x}$ **31.** $y = -\sqrt{\ln\left(x^2 + e^4\right)}$

33. $y = 2 + 2e^{x(x-2)/2}$ **35.** $y = \tan\left(x^2/2\right)$ **37.** $y = e^{1-e^{-t}}$

39. $y = \frac{et}{e^{1/t}} - 1$ **41.** $y = \sin^{-1}\left(\frac{1}{2}e^x\right)$ **43.** $a = -3, 4$

45. $t = \pm\sqrt{\pi + 4}$

47. (a) ≈ 1145 s or 19.1 min (b) ≈ 3910 s or 65.2 min

49. $y = 8 - (8 + 0.0002215t)^{2/3}$; $t_e \approx 66000$ s or 18 hr, 20 min

53. **(a)** $q(t) = CV\left(1 - e^{-t/RC}\right)$

(b)
$$\lim_{t\to\infty} q(t) = \lim_{t\to\infty} CV\left(1 - e^{-t/RC}\right) = \lim_{t\to\infty} CV(1-0) = CV$$

(c) $q(RC) = CV\left(1 - e^{-1}\right) \approx (0.63)\,CV$

55. $V = (kt/3 + C)^3$, V increases roughly with the cube of time.

57. $g(x) = Ce^{(3/2)x}$, where C is an arbitrary constant; $g(x) = \frac{C}{x-1}$, where C is an arbitrary constant.

59. $y = Cx^3$ and $y = \pm\sqrt{A - \frac{x^2}{3}}$

61. **(b)** $v(t) = -9.8t + 100(\ln(50) - \ln(50 - 4.75t))$;
$v(10) = -98 + 100(\ln(50) - \ln(2.5)) \approx 201.573$ m/s

67. **(c)** $C = \frac{7\pi}{60B}R^{5/2}$

Section 10.2 Preliminary Questions

1. 7 **2.** $y = \pm\sqrt{1+t}$ **3.** (b) **4.** 20

Section 10.2 Exercises

1.

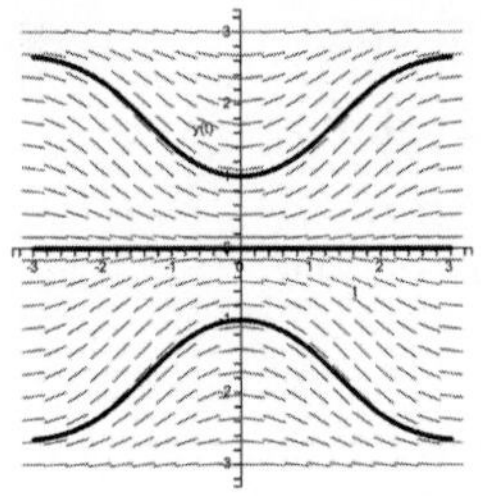

3.

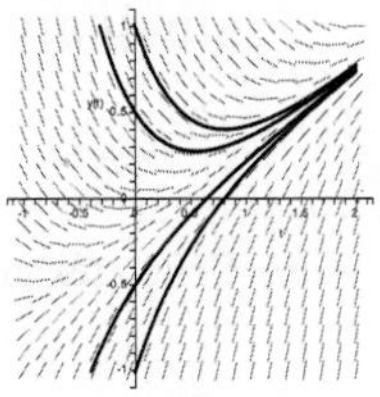

5. **(a)**

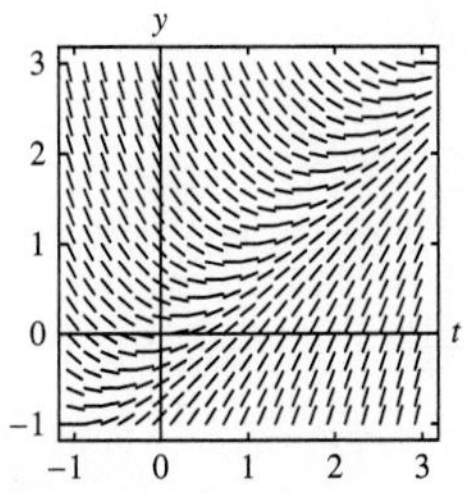

7. For $y' = t$, y' only depends on t. The isoclines of any slope c will be the vertical lines $t = c$.

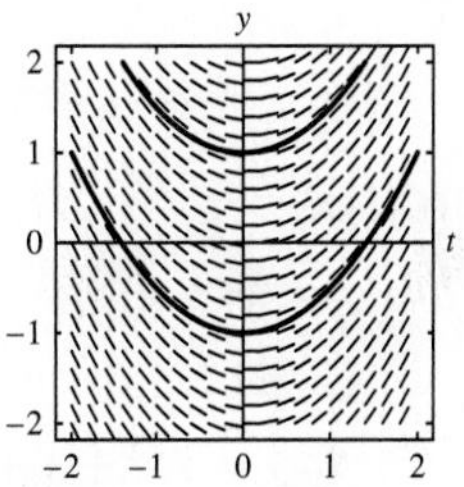

9. (i) C (ii) B (iii) F (iv) D (v) A (vi) E

11. **(a)**

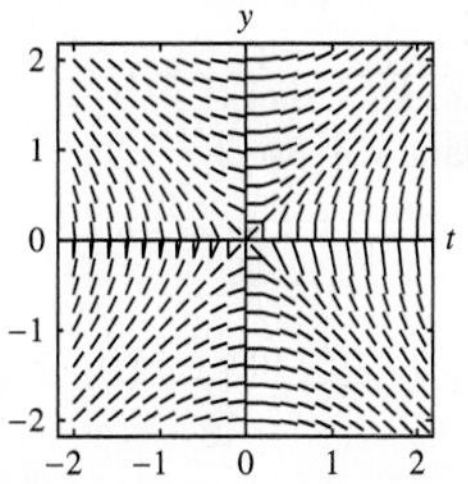

13. **(a)** $y_1 = 3.1$ **(b)** $y_2 = 3.231$
(c) $y_3 = 3.3919$, $y_4 = 3.58171$, $y_5 = 3.799539$, $y_6 = 4.0445851$
(d) $y(2.2) \approx 3.231$, $y(2.5) \approx 3.799539$

15. $y(0.5) \approx 1.7210$ **17.** $y(3.3) \approx 3.3364$

19. $y(2) \approx 2.8838$ **23.** $y(0.5) \approx 1.794894$

25. $y(0.25) \approx 1.094871$

Section 10.3 Preliminary Questions

1. **(a)** No **(b)** Yes **(c)** No **(d)** Yes

2. No **3.** Yes

Section 10.3 Exercises

1. $y = \dfrac{5}{1 - e^{-3t}/C}$ and $y = \dfrac{5}{1 + (3/2)e^{-3t}}$

3. $\lim_{t\to\infty} y(t) = 2$

5. **(a)** $P(t) = \dfrac{2000}{1 + 3e^{-0.6t}}$ **(b)** $t = \frac{1}{0.6}\ln 3 \approx 1.83$ yrs

7. $k = \ln\frac{81}{31} \approx 0.96$ yrs^{-1}; $t = \frac{\ln 9}{2\ln 9 - \ln 31} \approx 2.29$ yrs

9. After $t = 8$ hours, or at 4:00 PM

11. **(a)** $y_1(t) = \frac{10}{10 - 9e^{-t}}$ and $y_2(t) = \frac{1}{1 - 2e^{-t}}$
(b) $t = \ln\frac{9}{8}$ **(c)** $t = \ln 2$

13. **(a)** $A(t) = 16(1 - \frac{5}{3}e^{t/40})^2/(1 + \frac{5}{3}e^{t/40})^2$
(b) $A(10) \approx 2.1$
(c)

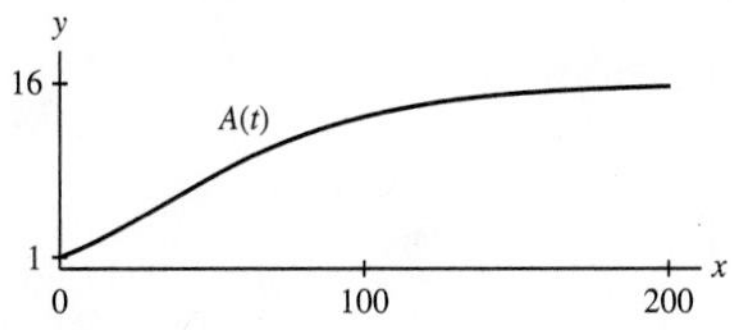

15. ≈ 943 million

17. **(d)** $t = -\frac{1}{k}(\ln y_0 - \ln(A - y_0))$

Section 10.4 Preliminary Questions

1. **(a)** Yes **(b)** No **(c)** Yes **(d)** No

2. (b)

Section 10.4 Exercises

1. **(c)** $y = \frac{x^4}{5} + \frac{C}{x}$ **(d)** $y = \frac{x^4}{5} - \frac{1}{5x}$

5. $y = \frac{1}{2}x + \frac{C}{x}$

7. $y = -\frac{1}{4}x^{-1} + Cx^{1/3}$ **9.** $y = \frac{1}{5}x^2 + \frac{1}{3} + Cx^{-3}$

11. $y = -x \ln x + Cx$ **13.** $y = \frac{1}{2}e^x + Ce^{-x}$

15. $y = x \cos x + C \cos x$ **17.** $y = x^x + Cx^x e^{-x}$

19. $y = \frac{1}{5}e^{2x} - \frac{6}{5}e^{-3x}$ **21.** $y = \frac{\ln|x|}{x+1} - \frac{1}{x(x+1)} + \frac{5}{x+1}$

23. $y = -\cos x + \sin x$ **25.** $y = \tanh x + 3\operatorname{sech} x$

27. For $m \neq -n$: $y = \frac{1}{m+n}e^{mx} + Ce^{-nx}$; for $m = -n$: $y = (x + C)e^{-nx}$

29. **(a)** $y' = 4000 - \frac{40y}{500+40t}$; $y = 1000\frac{4t^2+100t+125}{2t+25}$
(b) 40 g/L

31. 50 g/L

33. **(a)** $\frac{dV}{dt} = \frac{20}{1+t} - 5$ and $V(t) = 20\ln(1 + t) - 5t + 100$
(b) The maximum value is $V(3) = 20\ln 4 - 15 + 100 \approx 112.726$
(c)

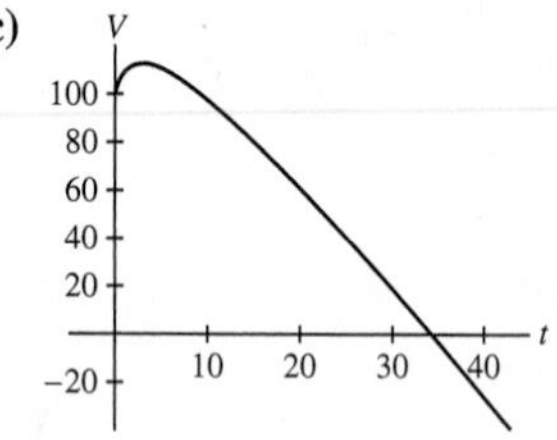

35. $I(t) = \frac{1}{10}\left(1 - e^{-20t}\right)$

37. **(a)** $I(t) = \frac{V}{R} - \frac{V}{R}e^{-(R/L)t}$ **(c)** Approximately 0.0184 s

39. **(b)** $c_1(t) = 10e^{-t/6}$

Chapter 10 Review

1. **(a)** No, first order **(b)** Yes, first order **(c)** No, order 3 **(d)** Yes, second order

3. $y = \pm\left(\frac{4}{3}t^3 + C\right)^{1/4}$, where C is an arbitrary constant

5. $y = Cx - 1$, where C is an arbitrary constant

7. $y = \frac{1}{2}\left(x + \frac{1}{2}\sin 2x\right) + \frac{\pi}{4}$ **9.** $y = \frac{2}{2-x^2}$

11.

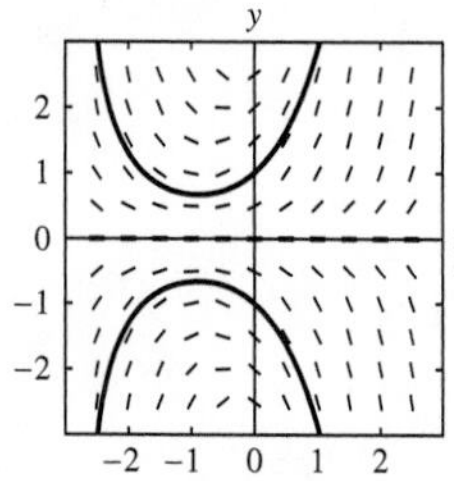

13. $y(t) = \tan t$

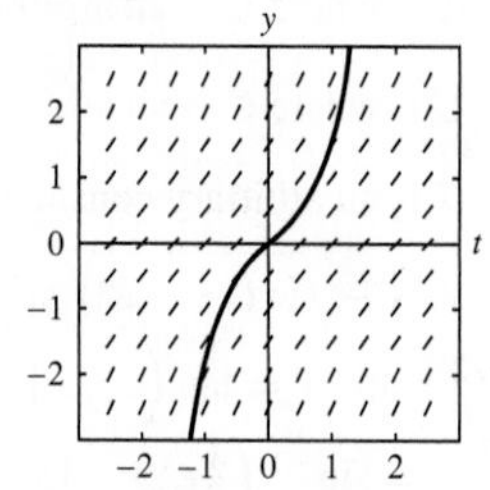

15. $y(0.1) \approx 1.1$; $y(0.2) \approx 1.209890$; $y(0.3) \approx 1.329919$

17. $y = x^2 + 2x$ **19.** $y = \frac{1}{2} + e^{-x} - \frac{11}{2}e^{-2x}$

21. $y = \frac{1}{2}\sin 2x - 2\cos x$ **23.** $y = 1 - \sqrt{t^2 + 15}$

25. $w = \tan\left(k \ln x + \frac{\pi}{4}\right)$

27. $y = -\cos x + \frac{\sin x}{x} + \frac{C}{x}$, where C is an arbitrary constant

29. Solution satisfying $y(0) = 3$: $y(t) = 4 - e^{-2t}$; solution satisfying $y(0) = 4$: $y(t) = 4$

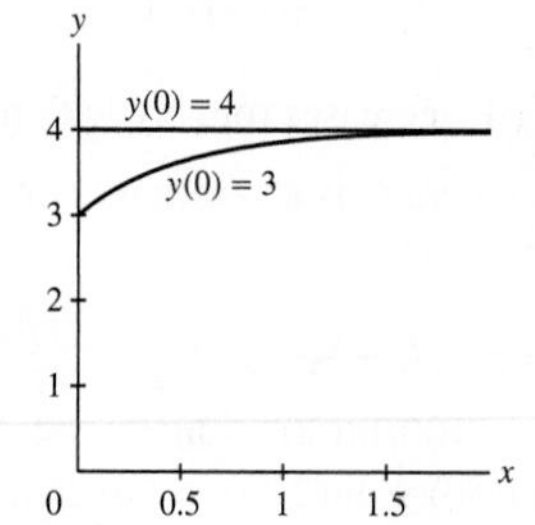

31. $y = 1$

35. $\frac{-1.77\sqrt{y}}{240y+64800}$; $t = 9198$ s about 2.56 hours.

37. 2 **39.** $t = 5\ln 441 \approx 30.45$ days

41. **(a)** $\frac{dc_1}{dt} = -\frac{2}{5}c_1$ **(b)** $c_1(t) = 8e^{(-2/5)t}$ g/L

Chapter 11

Section 11.1 Preliminary Questions

1. $a_4 = 12$ **2.** (c) **3.** $\lim_{n\to\infty} a_n = \sqrt{2}$ **4.** (b)

5. **(a)** False. Counterexample: $a_n = \cos \pi n$
(b) True **(c)** False. Counterexample: $a_n = (-1)^n$

Section 11.1 Exercises

1. **(a)** (iv) **(b)** (i) **(c)** (iii) **(d)** (ii)

3. $c_1 = 3, c_2 = \frac{9}{2}, c_3 = \frac{9}{2}, c_4 = \frac{27}{8}$

5. $a_1 = 2, a_2 = 5, a_3 = 47, a_4 = 4415$

7. $b_1 = 4, b_2 = 6, b_3 = 4, b_4 = 6$

9. $c_1 = 1, c_2 = \frac{3}{2}, c_3 = \frac{11}{6}, c_4 = \frac{25}{12}$

11. $b_1 = 2, b_2 = 3, b_3 = 8, b_4 = 19$

13. **(a)** $a_n = \frac{(-1)^{n+1}}{n^3}$ **(b)** $a_n = \frac{n+1}{n+5}$

15. $\lim_{n\to\infty} 12 = 12$ **17.** $\lim_{n\to\infty} \frac{5n-1}{12n+9} = \frac{5}{12}$

19. $\lim_{n\to\infty} \left(-2^{-n}\right) = 0$ **21.** The sequence diverges.

23. $\lim_{n\to\infty} \frac{n}{\sqrt{n^2+1}} = 1$ **25.** $\lim_{n\to\infty} \ln\left(\frac{12n+2}{-9+4n}\right) = \ln 3$

27. $\lim_{n\to\infty} \sqrt{4+\frac{1}{n}} = 2$ **29.** $\lim_{n\to\infty} \cos^{-1}\left(\frac{n^3}{2n^3+1}\right) = \frac{\pi}{3}$

31. **(a)** $M = 999$ **(b)** $M = 99999$

35. $\lim_{n\to\infty} \left(10 + \left(-\frac{1}{9}\right)^n\right) = 10$ **37.** The sequence diverges.

39. $\lim_{n\to\infty} 2^{1/n} = 1$ **41.** $\lim_{n\to\infty} \frac{9^n}{n!} = 0$

43. $\lim_{n\to\infty} \frac{3n^2+n+2}{2n^2-3} = \frac{3}{2}$ **45.** $\lim_{n\to\infty} \frac{\cos n}{n} = 0$

47. The sequence diverges. **49.** $\lim_{n\to\infty} \left(2+\frac{4}{n^2}\right)^{1/3} = 2^{1/3}$

51. $\lim_{n\to\infty} \ln\left(\frac{2n+1}{3n+4}\right) = \ln\frac{2}{3}$ **53.** The sequence diverges.

55. $\lim_{n\to\infty} \frac{e^n+(-3)^n}{5^n} = 0$ **57.** $\lim_{n\to\infty} n \sin\frac{\pi}{n} = \pi$

59. $\lim_{n\to\infty} \frac{3-4^n}{2+7\cdot 4^n} = -\frac{1}{7}$ **61.** $\lim_{n\to\infty} \left(1+\frac{1}{n}\right)^n = e$

63. $\lim_{n\to\infty} \frac{(\ln n)^2}{n} = 0$ **65.** $\lim_{n\to\infty} n\left(\sqrt{n^2+1} - n\right) = \frac{1}{2}$

67. $\lim_{n\to\infty} \frac{1}{\sqrt{n^4+n^8}} = 0$ **69.** $\lim_{n\to\infty} (2^n + 3^n)^{1/n} = 3$ **71.** (b)

73. Any number greater than or equal to 3 is an upper bound.

75. Example: $a_n = (-1)^n$ **79.** Example: $f(x) = \sin \pi x$

87. (e) $AGM\left(1, \sqrt{2}\right) \approx 1.198$

Section 11.2 Preliminary Questions

1. The sum of an infinite series is defined as the limit of the sequence of partial sums. If the limit of this sequence does not exist, the series is said to diverge.

2. $S = \frac{1}{2}$

3. The result is negative, so the result is not valid: a series with all positive terms cannot have a negative sum. The formula is not valid because a geometric series with $|r| \geq 1$ diverges.

4. No **5.** No **6.** $N = 13$

7. No, S_N is increasing and converges to 1, so $S_N \leq 1$ for all N.

8. Example: $\sum_{n=1}^{\infty} \frac{1}{n^{9/10}}$

Section 11.2 Exercises

1. **(a)** $a_n = \frac{1}{3^n}$ **(b)** $a_n = \left(\frac{5}{2}\right)^{n-1}$

(c) $a_n = (-1)^{n+1}\frac{n^n}{n!}$ **(d)** $a_n = \frac{1+\frac{(-1)^{n+1}+1}{2}}{n^2+1}$

3. $S_2 = \frac{5}{4}$, $S_4 = \frac{205}{144}$, $S_6 = \frac{5369}{3600}$

5. $S_2 = \frac{2}{3}$, $S_4 = \frac{4}{5}$, $S_6 = \frac{6}{7}$

7. $S_6 = 1.24992$

9. $S_{10} = 0.03535167962$, $S_{100} = 0.03539810274$, $S_{500} = 0.03539816290$, $S_{1000} = 0.03539816334$. Yes.

11. $S_3 = \frac{3}{10}$, $S_4 = \frac{1}{3}$, $S_5 = \frac{5}{14}$, $\sum_{n=1}^{\infty}\left(\frac{1}{n+1} - \frac{1}{n+2}\right) = \frac{1}{2}$

13. $S_3 = \frac{3}{7}$, $S_4 = \frac{4}{9}$, $S_5 = \frac{5}{11}$, $\sum_{n=1}^{\infty} \frac{1}{4n^2-1} = \frac{1}{2}$

15. $S = \frac{1}{2}$ **17.** $\lim_{n\to\infty} \frac{n}{10n+12} = \frac{1}{10} \neq 0$

19. $\lim_{n\to\infty} (-1)^n \left(\frac{n-1}{n}\right)$ does not exist.

21. $\lim_{n\to\infty} a_n = \cos\frac{1}{n+1} = 1 \neq 0$

23. $S = \frac{8}{7}$ **25.** The series diverges. **27.** $S = \frac{59049}{3328}$

29. $S = \frac{1}{e-1}$ **31.** $S = \frac{35}{3}$ **33.** $S = 4$ **35.** $S = \frac{7}{15}$

37. (b) and (c)

41. **(a)** Counterexample: $\sum_{n=1}^{\infty} \left(\frac{1}{2}\right)^n = 1$.

(b) Counterexample: If $a_n = 1$, then $S_N = N$.

(c) Counterexample: $\sum_{n=1}^{\infty} \frac{1}{n}$ diverges.

(d) Counterexample: $\sum_{n=1}^{\infty} \cos 2\pi n \neq 1$.

43. The total area is $\frac{1}{4}$.

45. The total length of the path is $2 + \sqrt{2}$.

Section 11.3 Preliminary Questions

1. (b)

2. A function $f(x)$ such that $a_n = f(n)$ must be positive, decreasing, and continuous for $x \geq 1$.

3. Convergence of p-series or integral test

4. Comparison Test

5. No; $\sum_{n=1}^{\infty} \frac{1}{n}$ diverges, but since $\frac{e^{-n}}{n} < \frac{1}{n}$ for $n \geq 1$, the Comparison Test tells us nothing about the convergence of $\sum_{n=1}^{\infty} \frac{e^{-n}}{n}$.

Section 11.3 Exercises

1. $\int_1^{\infty} \frac{dx}{x^4}\, dx$ converges, so the series converges.

3. $\int_1^{\infty} x^{-1/3}\, dx = \infty$, so the series diverges.

5. $\int_{25}^{\infty} \frac{x^2}{(x^3+9)^{5/2}}\, dx$ converges, so the series converges.

7. $\int_1^{\infty} \frac{dx}{x^2+1}$ converges, so the series converges.

9. $\int_1^{\infty} \frac{dx}{x(x+1)}$ converges, so the series converges.

11. $\int_2^{\infty} \frac{1}{x(\ln x)^2}\, dx$ converges, so the series converges.

13. $\int_1^{\infty} \frac{dx}{2^{\ln x}} = \infty$, so the series diverges.

15. $\frac{1}{n^3+8n} \leq \frac{1}{n^3}$, so the series converges.

19. $\frac{1}{n2^n} \leq \left(\frac{1}{2}\right)^n$, so the series converges.

21. $\frac{1}{n^{1/3}+2^n} \leq \left(\frac{1}{2}\right)^n$, so the series converges.

23. $\frac{4}{m!+4^m} \leq 4\left(\frac{1}{4}\right)^m$, so the series converges.

25. $0 \leq \frac{\sin^2 k}{k^2} \leq \frac{1}{k^2}$, so the series converges.

27. $\frac{2}{3^n+3^{-n}} \leq 2\left(\frac{1}{3}\right)^n$, so the series converges.

29. $\frac{1}{(n+1)!} \le \frac{1}{n^2}$, so the series converges.

31. $\frac{\ln n}{n^3} \le \frac{1}{n^2}$ for $n \ge 1$, so the series converges.

33. $\frac{(\ln n)^{100}}{n^{1.1}} \le \frac{1}{n^{1.09}}$ for n sufficiently large, so the series converges.

35. $\frac{n}{3^n} \le \left(\frac{2}{3}\right)^n$ for $n \ge 1$, so the series converges.

39. The series converges. **41.** The series diverges.

43. The series converges. **45.** The series diverges.

47. The series converges. **49.** The series converges.

51. The series diverges. **53.** The series converges.

55. The series diverges. **57.** The series converges.

59. The series diverges. **61.** The series diverges.

63. The series diverges. **65.** The series converges.

67. The series diverges. **69.** The series diverges.

71. The series converges. **73.** The series converges.

75. The series diverges. **77.** The series converges.

79. The series converges for $a > 1$ and diverges for $a \le 1$.

87. $\sum_{n=1}^{\infty} n^{-5} \approx 1.0369540120$.

91. $\sum_{n=1}^{1000} \frac{1}{n^2} = 1.6439345667$ and $1 + \sum_{n=1}^{100} \frac{1}{n^2(n+1)} =$ 1.6448848903. The second sum is a better approximation to $\frac{\pi^2}{6} \approx 1.6449340668$.

Section 11.4 Preliminary Questions

1. Example: $\sum \frac{(-1)^n}{\sqrt[3]{n}}$ **2.** (b) **3.** No.

4. $|S - S_{100}| \le 10^{-3}$, and S is larger than S_{100}.

Section 11.4 Exercises

3. Converges conditionally

5. Converges absolutely

7. Converges conditionally

9. Converges conditionally

11. (a)

n	S_n	n	S_n
1	1	6	0.899782407
2	0.875	7	0.902697859
3	0.912037037	8	0.900744734
4	0.896412037	9	0.902116476
5	0.904412037	10	0.901116476

13. $S_5 = 0.947$ **15.** $S_{44} = 0.06567457397$

17. Converges (by geometric series)

19. Converges (by Comparison Test)

21. Converges (by Limit Comparison Test)

23. Diverges (by Limit Comparison Test)

25. Converges (by geometric series and linearity)

27. Converges absolutely (by Integral Test)

29. Converges conditionally (by Leibniz Test)

31. Converges (by Integral Test)

33. Converges conditionally

Section 11.5 Preliminary Questions

1. $\rho = \lim_{n\to\infty} \left|\frac{a_{n+1}}{a_n}\right|$

2. The Ratio Test is conclusive for $\sum_{n=1}^{\infty} \frac{1}{2^n}$ and inconclusive for $\sum_{n=1}^{\infty} \frac{1}{n}$.

3. No.

Section 11.5 Exercises

1. Converges absolutely **3.** Converges absolutely

5. The ratio test is inconclusive. **7.** Diverges

9. Converges absolutely **11.** Converges absolutely

13. Diverges **15.** The ratio test is inconclusive.

17. Converges absolutely **19.** Converges absolutely

21. $\rho = \frac{1}{3} < 1$ **23.** $\rho = 2|x|$

25. $\rho = |r|$ **29.** Converges absolutely

31. The ratio test is inconclusive, so the series may converge or diverge.

33. Converges absolutely **35.** The ratio test is inconclusive.

37. Converges absolutely **39.** Converges absolutely

41. Converges absolutely

43. Converges (by geometric series and linearity)

45. Converges (by the Ratio Test)

47. Converges (by the Limit Comparison Test)

49. Diverges (by p-series) **51.** Converges (by geometric series)

53. Converges (by Limit Comparison Test)

55. Diverges (by Divergence Test)

Section 11.6 Preliminary Questions

1. Yes. The series must converge for both $x = 4$ and $x = -3$.

2. (a), (c) **3.** $R = 4$

4. $F'(x) = \sum_{n=1}^{\infty} n^2 x^{n-1}$; $R = 1$

Section 11.6 Exercises

1. $R = 2$. It does not converge at the endpoints.

3. $R = 3$ for all three series.

9. $(-1, 1)$ **11.** $[-\sqrt{2}, \sqrt{2}]$ **13.** $[-1, 1]$ **15.** $(-\infty, \infty)$

17. $\left[-\frac{1}{4}, \frac{1}{4}\right)$ **19.** $(-1, 1]$ **21.** $(-1, 1)$ **23.** $[-1, 1)$ **25.** $(2, 4)$

27. $(6, 8)$ **29.** $\left[-\frac{7}{2}, -\frac{5}{2}\right)$ **31.** $(-\infty, \infty)$ **33.** $\left(2 - \frac{1}{e}, 2 + \frac{1}{e}\right)$

35. $\sum_{n=0}^{\infty} 3^n x^n$ on the interval $\left(-\frac{1}{3}, \frac{1}{3}\right)$

37. $\sum_{n=0}^{\infty} \frac{x^n}{3^{n+1}}$ on the interval $(-3, 3)$

39. $\sum_{n=0}^{\infty}(-1)^n x^{2n}$ on the interval $(-1, 1)$

43. $\sum_{n=0}^{\infty}(-1)^{n+1}(x-5)^n$ on the interval $(4, 6)$

47. (c) $S_4 = \frac{69}{640}$ and $|S - S_4| \approx 0.000386 < a_5 = \frac{1}{1920}$

49. $R = 1$ **51.** $\sum_{n=1}^{\infty} \frac{n}{2^n} = 2$ **53.** $F(x) = \frac{1-x-x^2}{1-x^3}$

55. $-1 \le x \le 1$ **57.** $P(x) = \sum_{n=0}^{\infty} (-1)^n \frac{x^n}{n!}$

59. N must be at least 5; $S_5 = 0.3680555556$

61. $P(x) = 1 - \frac{1}{2}x^2 - \sum_{n=2}^{\infty} \frac{1\cdot 3\cdot 5\cdots(2n-3)}{(2n)!}x^{2n}$; $R = \infty$

Section 11.7 Preliminary Questions

1. $f(0) = 3$ and $f'''(0) = 30$

2. $f(-2) = 0$ and $f^{(4)}(-2) = 48$

3. Substitute x^2 for x in the Maclaurin series for $\sin x$.

4. $f(x) = 4 + \sum_{n=1}^{\infty} \frac{(x-3)^{n+1}}{n(n+1)}$ **5.** (c)

Section 11.7 Exercises

1. $f(x) = 2 + 3x + 2x^2 + 2x^3 + \cdots$

3. $\frac{1}{1-2x} = \sum_{n=0}^{\infty} 2^n x^n$ on the interval $\left(-\frac{1}{2}, \frac{1}{2}\right)$

5. $\cos 3x = \sum_{n=0}^{\infty} (-1)^n \frac{9^n x^{2n}}{(2n)!}$ on the interval $(-\infty, \infty)$

7. $\sin(x^2) = \sum_{n=0}^{\infty}(-1)^n \frac{x^{4n+2}}{(2n+1)!}$ on the interval $(-\infty, \infty)$

9. $\ln(1-x^2) = -\sum_{n=1}^{\infty} \frac{x^{2n}}{n}$ on the interval $(-1, 1)$

11. $\tan^{-1}(x^2) = \sum_{n=0}^{\infty} (-1)^n \frac{x^{4n+2}}{2n+1}$ on the interval $[-1, 1]$

13. $e^{x-2} = \sum_{n=0}^{\infty} \frac{x^n}{e^2 n!}$ on the interval $(-\infty, \infty)$

15. $\ln(1-5x) = -\sum_{n=1}^{\infty} \frac{5^n x^n}{n}$ on the interval $\left[-\frac{1}{5}, \frac{1}{5}\right)$

17. $\sinh x = \sum_{k=0}^{\infty} \frac{x^{2k+1}}{(2k+1)!}$ on the interval $(-\infty, \infty)$

19. $e^x \sin x = x + x^2 + \frac{x^3}{3} - \frac{x^5}{30} + \cdots$

21. $\frac{\sin x}{1-x} = x + x^2 + \frac{5x^3}{6} + \frac{5x^4}{6} + \cdots$

23. $(1+x)^{1/4} = 1 + \frac{1}{4}x - \frac{3}{32}x^2 + \frac{7}{128}x^3 + \cdots$

25. $e^x \tan^{-1} x = x + x^2 + \frac{1}{6}x^3 - \frac{1}{6}x^4 + \cdots$

27. $e^{\sin x} = 1 + x + \frac{1}{2}x^2 - \frac{1}{8}x^4 + \cdots$

29. $\frac{1}{x} = \sum_{n=0}^{\infty}(-1)^n(x-1)^n$ on the interval $(0, 2)$

31. $\frac{1}{1-x} = \sum_{n=0}^{\infty} (-1)^{n+1} \frac{(x-5)^n}{4^{n+1}}$ on the interval $(1, 9)$

33. $21 + 35(x-2) + 24(x-2)^2 + 8(x-2)^3 + (x-2)^4$ on the interval $(-\infty, \infty)$

35. $\frac{1}{x^2} = \sum_{n=0}^{\infty}(-1)^n(n+1)\frac{(x-4)^n}{4^{n+2}}$ on the interval $(0, 8)$

37. $\frac{1}{1-x^2} = \sum_{n=0}^{\infty} \frac{(-1)^{n+1}(2^{n+1}-1)}{2^{2n+3}}(x-3)^n$ on the interval $(1, 5)$

39. $\cos^2 x = \frac{1}{2} + \frac{1}{2}\sum_{n=0}^{\infty} (-1)^n \frac{(4)^n x^{2n}}{(2n)!}$

45. $S_4 = 0.1822666667$

47. (a) 4 (b) $S_4 = 0.7474867725$

49. $\int_0^1 \cos(x^2)\,dx = \sum_{n=0}^{\infty} \frac{(-1)^n}{(2n)!(4n+1)}$; $S_3 = 0.9045227920$

51. $\int_0^1 e^{-x^3}\,dx = \sum_{n=0}^{\infty} \frac{(-1)^n}{n!(3n+1)}$; $S_5 = 0.8074461996$

53. $\int_0^x \frac{1-\cos(t)}{t}\,dt = \sum_{n=1}^{\infty}(-1)^{n+1}\frac{x^{2n}}{(2n)!2n}$

55. $\int_0^x \ln(1+t^2)\,dt = \sum_{n=1}^{\infty}(-1)^{n-1}\frac{x^{2n+1}}{n(2n+1)}$

57. $\frac{1}{1+2x}$ **63.** e^{x^3} **65.** $1 - 5x + \sin 5x$

67. $\frac{1}{(1-2x)(1-x)} = \sum_{n=0}^{\infty}\left(2^{n+1}-1\right)x^n$

69. $I(t) = \frac{V}{R}\sum_{n=1}^{\infty} \frac{(-1)^{n+1}}{n!}\left(\frac{Rt}{L}\right)^n$

71. $f(x) = \sum_{n=0}^{\infty} \frac{(-1)^n x^{6n}}{(2n)!}$ and $f^{(6)}(0) = -360$.

73. $e^{20x} = 1 + x^{20} + \frac{x^{40}}{2} + \cdots$ **75.** No.

81. $\lim_{x\to 0} \frac{\sin x - x + \frac{x^3}{6}}{x^5} = \frac{1}{120}$

83. $\lim_{x\to 0}\left(\frac{\sin(x^2)}{x^4} - \frac{\cos x}{x^2}\right) = \frac{1}{2}$

85. $S = \frac{\pi}{4} - \frac{1}{2}\ln 2$ **89.** $L \approx 28.369$

Chapter 11 Review

1. (a) $a_1^2 = 4, a_2^2 = \frac{1}{4}, a_3^2 = 0$

(b) $b_1 = \frac{1}{24}, b_2 = \frac{1}{60}, b_3 = \frac{1}{240}$

(c) $a_1 b_1 = -\frac{1}{12}, a_2 b_2 = -\frac{1}{120}, a_3 b_3 = 0$

(d) $2a_2 - 3a_1 = 5, 2a_3 - 3a_2 = \frac{3}{2}, 2a_4 - 3a_3 = \frac{1}{12}$

3. $\lim_{n\to\infty}(5a_n - 2a_n^2) = 2$ **5.** $\lim_{n\to\infty} e^{a_n} = e^2$

7. $\lim_{n\to\infty}(-1)^n a_n$ does not exist.

9. $\lim_{n\to\infty}\left(\sqrt{n+5} - \sqrt{n+2}\right) = 0$ **11.** $\lim_{n\to\infty} 2^{1/n^2} = 1$

13. The sequence diverges.

15. $\lim_{n\to\infty} \tan^{-1}\left(\frac{n+2}{n+5}\right) = \frac{\pi}{4}$

17. $\lim_{n\to\infty}\left(\sqrt{n^2+n} - \sqrt{n^2+1}\right) = \frac{1}{2}$

19. $\lim_{m\to\infty}\left(1 + \frac{1}{m}\right)^{3m} = e^3$ **21.** $\lim_{n\to\infty}(n\ln(n+1) - \ln n) = 1$

25. $\lim_{n\to\infty}\frac{a_{n+1}}{a_n} = 3$ **27.** $S_4 = -\frac{11}{60}$, $S_7 = \frac{41}{630}$

29. $\sum_{n=2}^{\infty}\left(\frac{2}{3}\right)^n = \frac{4}{3}$ **31.** $\sum_{n=-1}^{\infty}\frac{2^{n+3}}{3^n} = 36$

33. Example: $a_n = \left(\frac{1}{2}\right)^n + 1$, $b_n = -1$

35. $S = \frac{47}{180}$ **37.** The series diverges.

39. $\int_1^{\infty}\frac{1}{(x+2)(\ln(x+2))^3}\,dx = \frac{1}{2(\ln(3))^2}$, so the series converges.

41. $\frac{1}{(n+1)^2} < \frac{1}{n^2}$, so the series converges.

43. $\sum_{n=0}^{\infty}\frac{1}{n^{1.5}}$ converges, so the series converges.

45. $\frac{n}{\sqrt{n^5+2}} < \frac{1}{n^{3/2}}$, so the series converges.

47. $\sum_{n=0}^{\infty}\left(\frac{10}{11}\right)^n$ converges, so the series converges.

49. Converges

53. **(b)** $0.3971162690 \le S \le 0.3971172688$, so the maximum size of the error is 10^{-6}.

55. Converges absolutely **57.** Diverges

59. **(a)** 500 **(b)** $K \approx \sum_{n=0}^{499}\frac{(-1)^k}{(2k+1)^2} = 0.9159650942$

61. **(a)** Converges **(b)** Converges **(c)** Diverges **(d)** Converges

63. Converges **65.** Converges **67.** Diverges

69. Diverges **71.** Converges **73.** Converges

75. Converges (by geometric series)

77. Converges (by geometric series)

79. Converges (by the Leibniz Test)

81. Converges (by the Leibniz Test)

83. Converges (by the Comparison Test)

85. Converges using partial sums (the series is telescoping)

87. Diverges (by the Comparison Test)

89. Converges (by the Comparison Test)

91. Converges (by the Comparison Test)

93. Converges on the interval $(-\infty, \infty)$

95. Converges on the interval $[2, 4]$

97. Converges at $x = 0$

99. $\frac{2}{4-3x} = \frac{1}{2}\sum_{n=0}^{\infty}\left(\frac{3}{4}\right)^n x^n$. The series converges on the interval $\left(\frac{-4}{3}, \frac{4}{3}\right)$

101. **(c)**

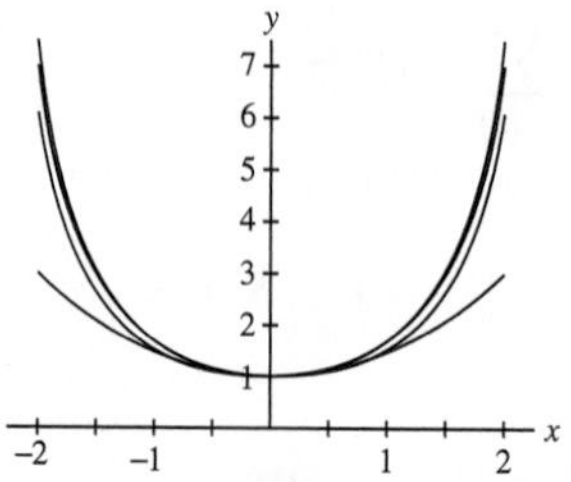

103. $e^{4x} = \sum_{n=0}^{\infty}\frac{4^n}{n!}x^n$

105. $x^4 = 16 + 32(x-2) + 24(x-2)^2 + 8(x-2)^3 + (x-2)^4$

107. $\sin x = \sum_{n=0}^{\infty}\frac{(-1)^{n+1}(x-\pi)^{2n+1}}{(2n+1)!}$

109. $\frac{1}{1-2x} = \sum_{n=0}^{\infty}\frac{2^n}{5^{n+1}}(x+2)^n$ **111.** $\ln\frac{x}{2} = \sum_{n=1}^{\infty}\frac{(-1)^{n+1}(x-2)^n}{n2^n}$

113. $(x^2 - x)e^{x^2} = \sum_{n=0}^{\infty}\left(\frac{x^{2n+2} - x^{2n+1}}{n!}\right)$ so $f^{(3)}(0) = -6$

115. $\frac{1}{1+\tan x} = -x + x^2 - \frac{4}{3}x^3 + \frac{2}{3}x^4 + \cdots$ so $f^{(3)}(0) = -8$

117. $\frac{\pi}{2} - \frac{\pi^3}{2^3 3!} + \frac{\pi^5}{2^5 5!} - \frac{\pi^7}{2^7 7!} + \cdots = \sin\frac{\pi}{2} = 1$

Chapter 12

Section 12.1 Preliminary Questions

1. A circle of radius 3 centered at the origin.

2. The center is at (4, 5) **3.** Maximum height: 4

4. Yes; no **5.** (a) ↔ (iii), (b) ↔ (ii), (c) ↔ (i)

Section 12.1 Exercises

1. $(t = 0)(1, 9)$; $(t = 2)(9, -3)$; $(t = 4)(65, -39)$

5. **(a)**

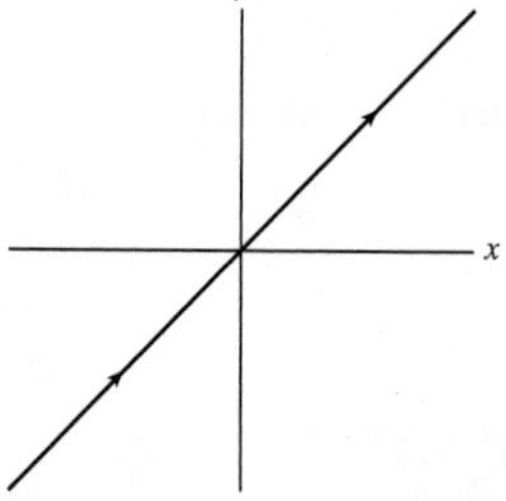

(b)

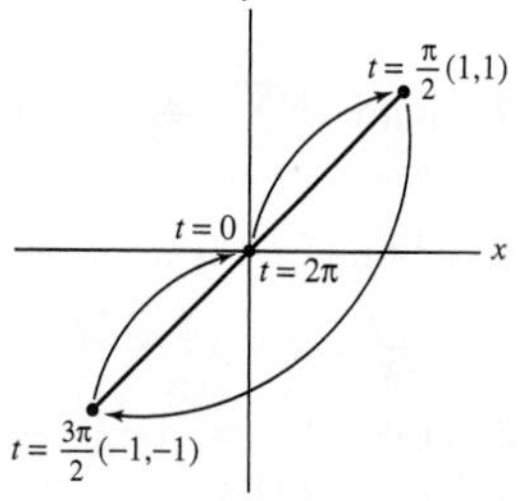

(c)

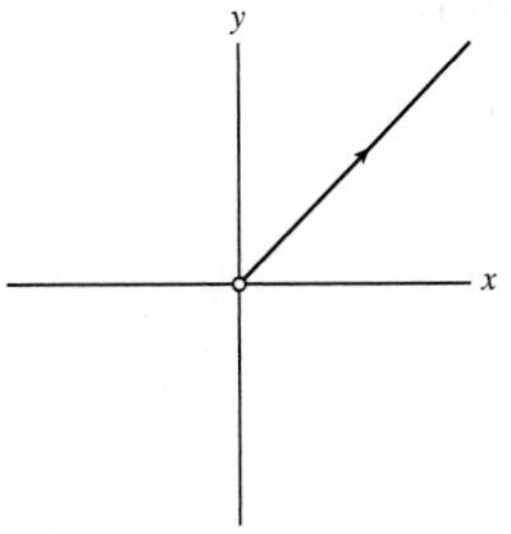

(d)

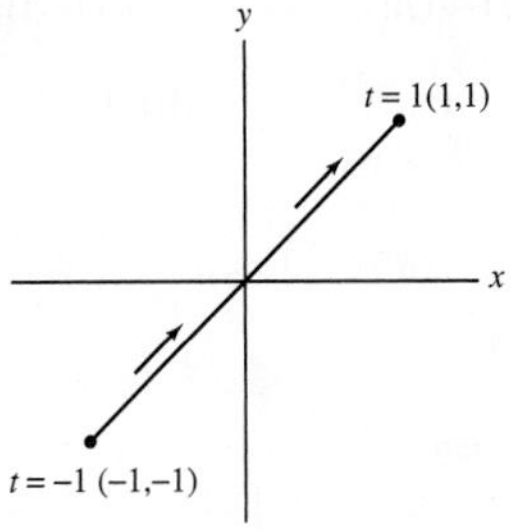

7. $y = 4x - 12$ **9.** $y = \tan^{-1}\left(x^3 + e^x\right)$

11. $y = \frac{6}{x^2}$ (where $x > 0$) **13.** $y = 2 - e^x$

15.

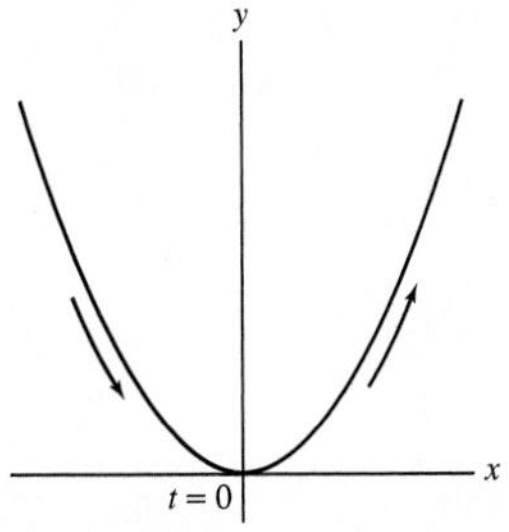

17.

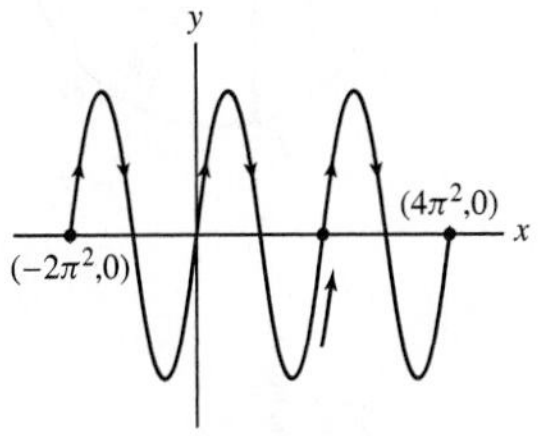

19. (a) $\leftrightarrow$ (iv), (b) $\leftrightarrow$ (ii), (c) $\leftrightarrow$ (iii), (d) $\leftrightarrow$ (i)

21. $\pi \le t \le 2\pi$ **23.** $c(t) = (t, 9 - 4t)$ **25.** $c(t) = \left(\frac{5+t^2}{4}, t\right)$

27. $c(t) = (-9 + 7\cos t, 4 + 7\sin t)$ **29.** $c(t) = (-4 + t, 9 + 8t)$

31. $c(t) = (3 - 8t, 1 + 3t)$ **33.** $c(t) = (1 + t, 1 + 2t)$ $(0 \le t \le 1)$

35. $c(t) = (3 + 4\cos t, 9 + 4\sin t)$ **37.** $c(t) = \left(-4 + t, -8 + t^2\right)$

39. $c(t) = (2 + t, 2 + 3t)$ **41.** $c(t) = \left(3 + t, (3 + t)^2\right)$

43. $y = \sqrt{x^2 - 1}$ $(1 \le x < \infty)$ **45.** Plot III.

47.

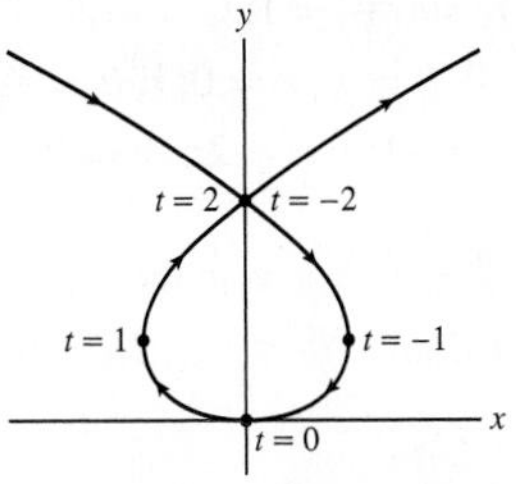

49. $\left.\frac{dy}{dx}\right|_{t=-4} = -\frac{1}{6}$ **51.** $\left.\frac{dy}{dx}\right|_{s=-1} = -\frac{3}{4}$

53. $y = -\frac{9}{2}x + \frac{11}{2}$; $\frac{dy}{dx} = -\frac{9}{2}$

55. $y = x^2 + x^{-1}$; $\frac{dy}{dx} = 2x - \frac{1}{x^2}$

57. $(0, 0)$, $(96, 180)$

59.

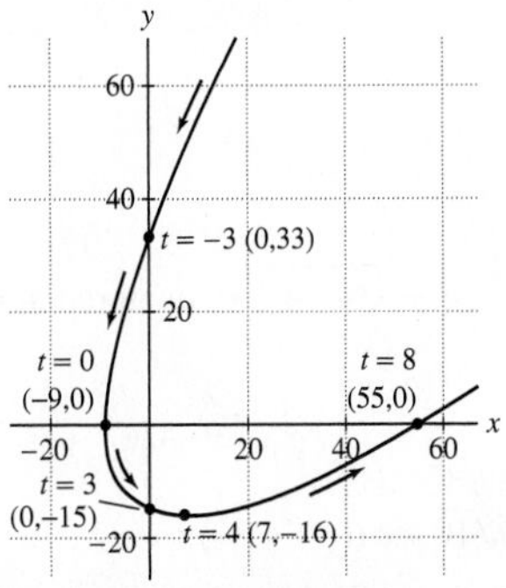

The graph is in: quadrant (i) for $t < -3$ or $t > 8$, quadrant (ii) for $-3 < t < 0$, quadrant (iii) for $0 < t < 3$, quadrant (iv) for $3 < t < 8$.

61. $(55, 0)$

63. The coordinates of P, $(R\cos\theta, r\sin\theta)$, describe an ellipse for $0 \le \theta \le 2\pi$.

67. $c(t) = (3 - 9t + 24t^2 - 16t^3, 2 + 6t^2 - 4t^3)$, $0 \le t \le 1$

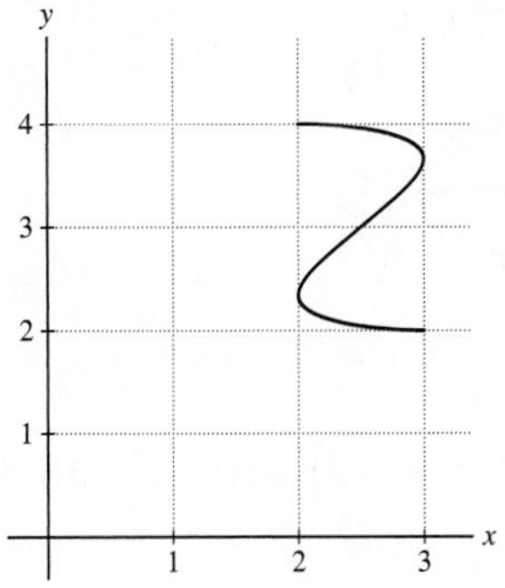

71. $y = -\sqrt{3}x + \frac{\sqrt{3}}{2}$

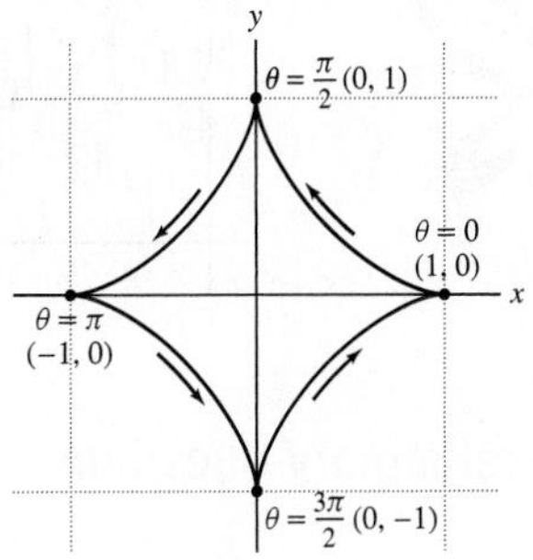

73. $((2k - 1)\pi, 2)$, $k = 0, \pm 1, \pm 2, \ldots$

83. $\left.\frac{d^2y}{dx^2}\right|_{t=2} = -\frac{21}{512}$ **85.** $\left.\frac{d^2y}{dx^2}\right|_{t=-3} = 0$ **87.** Concave up: $t > 0$

Section 12.2 Preliminary Questions

1. $S = \int_a^b \sqrt{x'(t)^2 + y'(t)^2}\, dt$ **2.** The speed at time t

3. Displacement: 5; no **4.** $L = 180$ cm

Section 12.2 Exercises

1. $S = 10$ **3.** $S = 16\sqrt{13}$ **5.** $S = \frac{1}{2}(65^{3/2} - 5^{3/2}) \approx 256.43$

7. $S = 3\pi$ **9.** $S = -8\left(\frac{\sqrt{2}}{2} - 1\right) \approx 2.34$

13. $S = \ln(\cosh(A))$ **15.** $\left.\frac{ds}{dt}\right|_{t=2} = 4\sqrt{10} \approx 12.65$ m/s

17. $\left.\frac{ds}{dt}\right|_{t=9} = \sqrt{41} \approx 6.4$ m/s **19.** $\left(\frac{ds}{dt}\right)_{\min} \approx \sqrt{4.89} \approx 2.21$

21. $\frac{ds}{dt} = 8$

23.

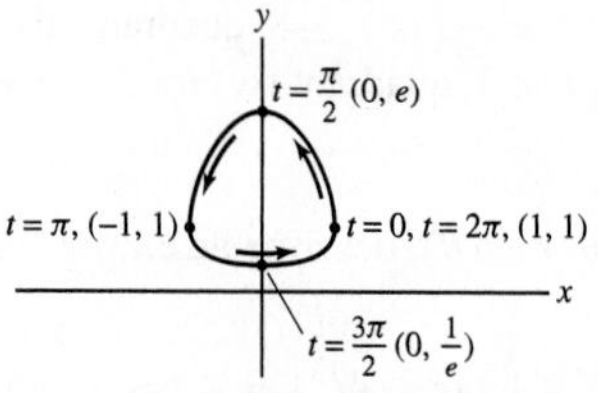

$M_{10} = 6.903734$, $M_{20} = 6.915035$, $M_{30} = 6.914949$, $M_{50} = 6.914951$

25.

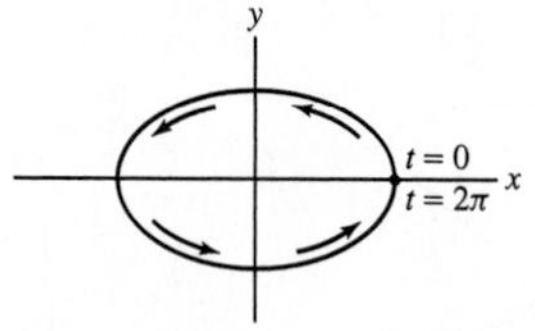

$M_{10} = 25.528309$, $M_{20} = 25.526999$, $M_{30} = 25.526999$, $M_{50} = 25.526999$

27. $S = 2\pi^2 R$ **29.** $S = m\sqrt{1+m^2}\pi A^2$ **31.** $S = \frac{64\pi}{3}$

33. (a)

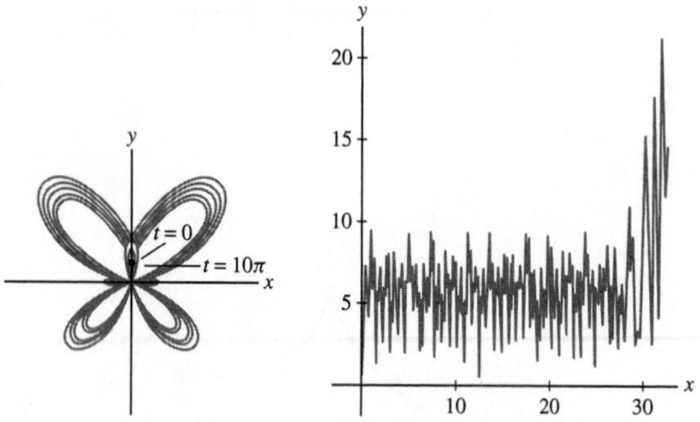

(b) $L \approx 212.09$

Section 12.3 Preliminary Questions

1. (b)
2. Positive: $(r,\theta) = \left(1, \frac{\pi}{2}\right)$; Negative: $(r,\theta) = \left(-1, \frac{3\pi}{2}\right)$
3. **(a)** Equation of the circle of radius 2 centered at the origin.
(b) Equation of the circle of radius $\sqrt{2}$ centered at the origin.
(c) Equation of the vertical line through the point (2, 0).
4. (a)

Section 12.3 Exercises

1. (A): $\left(3\sqrt{2}, \frac{3\pi}{4}\right)$; (B): $(3, \pi)$; (C): $\left(\sqrt{5}, \pi + 0.46\right) \approx \left(\sqrt{5}, 3.60\right)$; (D): $\left(\sqrt{2}, \frac{5\pi}{4}\right)$; (E): $\left(\sqrt{2}, \frac{\pi}{4}\right)$; (F): $\left(4, \frac{\pi}{6}\right)$; (G): $\left(4, \frac{11\pi}{6}\right)$

3. (a) $(1, 0)$ **(b)** $\left(\sqrt{12}, \frac{\pi}{6}\right)$ **(c)** $\left(\sqrt{8}, \frac{3\pi}{4}\right)$ **(d)** $\left(2, \frac{2\pi}{3}\right)$

5. (a) $\left(\frac{3\sqrt{3}}{2}, \frac{3}{2}\right)$ **(b)** $\left(-\frac{6}{\sqrt{2}}, \frac{6}{\sqrt{2}}\right)$ **(c)** $(0, 0)$ **(d)** $(0, -5)$

7. (A): $0 \le r \le 3, \pi \le \theta \le 2\pi$, (B): $0 \le r \le 3, \frac{\pi}{4} \le \theta \le \frac{\pi}{2}$, (C): $3 \le r \le 5, \frac{3\pi}{4} \le \theta \le \pi$

9. $m = \tan\frac{3\pi}{5} \approx -3.1$ **11.** $x^2 + y^2 = 7^2$

13. $x^2 + (y-1)^2 = 1$ **15.** $y = x - 1$ **17.** $r = \sqrt{5}$

19. $r = \tan\theta\sec\theta$

21. (a)↔(iii), (b)↔(iv), (c)↔(i), (d)↔(ii)

23. (a) $(r, 2\pi - \theta)$ **(b)** $(r, \theta + \pi)$ **(c)** $(r, \pi - \theta)$ **(d)** $\left(r, \frac{\pi}{2} - \theta\right)$

25. $r\cos\left(\theta - \frac{\pi}{3}\right) = d$

27.

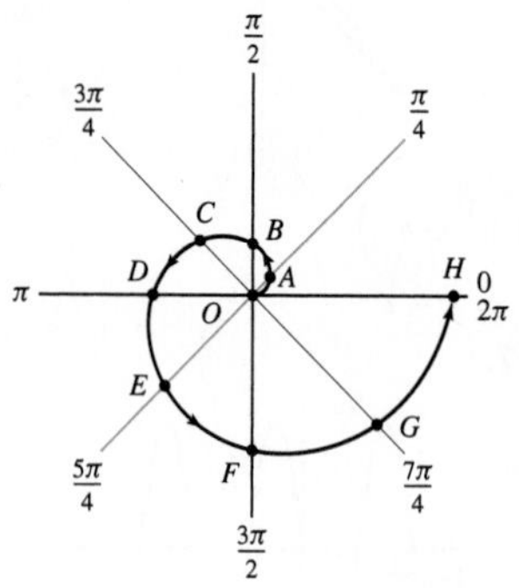

29.

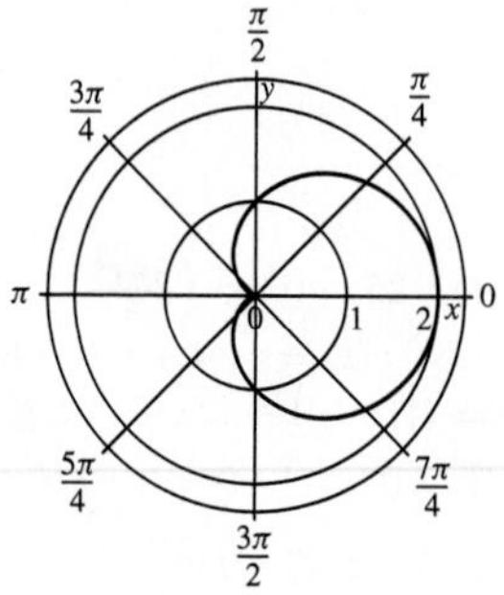

31. (a) A, $\theta = 0, r = 0$; B, $\theta = \frac{\pi}{4}, r = \sin\frac{2\pi}{4} = 1$; C, $\theta = \frac{\pi}{2}$, $r = 0$; D, $\theta = \frac{3\pi}{4}, r = \sin\frac{2\cdot 3\pi}{4} = -1$; E, $\theta = \pi, r = 0$; F, $\theta = \frac{5\pi}{4}$, $r = 1$; G, $\theta = \frac{3\pi}{2}, r = 0$; H, $\theta = \frac{7\pi}{4}, r = -1$; I, $\theta = 2\pi, r = 0$

(b) $0 \le \theta \le \frac{\pi}{2}$ is in the first quadrant. $\frac{\pi}{2} \le \theta \le \pi$ is in the fourth quadrant. $\pi \le \theta \le \frac{3\pi}{2}$ is in the third quadrant. $\frac{3\pi}{2} \le \theta \le 2\pi$ is in the second quadrant.

33.

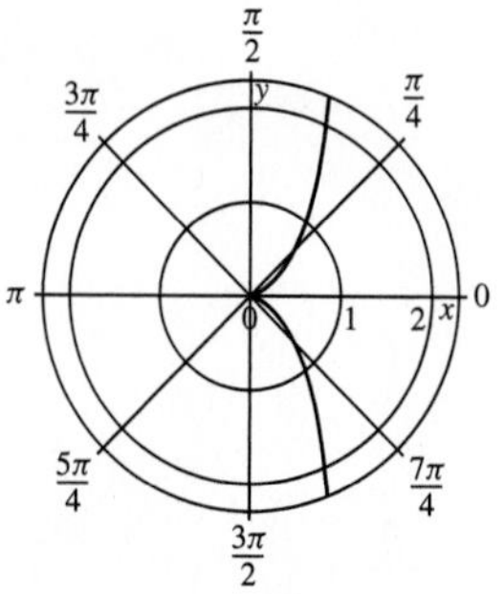

35. $\left(x - \frac{a}{2}\right)^2 + \left(y - \frac{b}{2}\right)^2 = \frac{a^2+b^2}{4}$, $r = \sqrt{a^2+b^2}$, centered at the point $\left(\frac{a}{2}, \frac{b}{2}\right)$

37. $r^2 = \sec 2\theta$ **39.** $(x^2 + y^2) = x^3 - 3y^2x$

41. $r = 2\sec\left(\theta - \frac{\pi}{9}\right)$ **43.** $r = 2\sqrt{10}\sec(\theta - 4.39)$

47. $r^2 = 2a^2 \cos 2\theta$

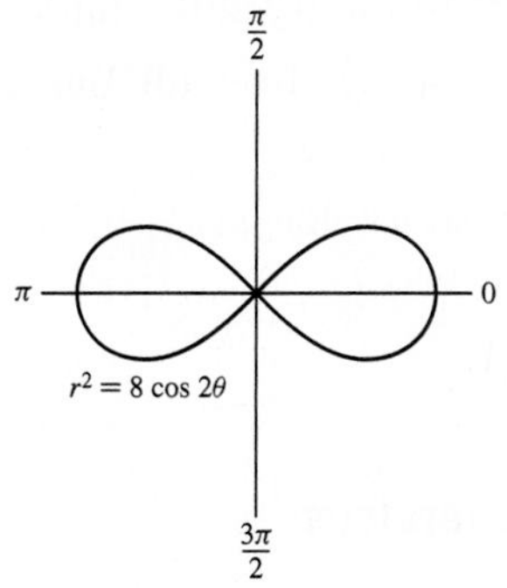

51. $\theta = \frac{\pi}{2}, m = -\frac{2}{\pi}; \theta = \pi, m = \pi$

53. $\left(\frac{\sqrt{2}}{2}, \frac{\pi}{6}\right), \left(\frac{\sqrt{2}}{2}, \frac{5\pi}{6}\right), \left(\frac{\sqrt{2}}{2}, \frac{7\pi}{6}\right), \left(\frac{\sqrt{2}}{2}, \frac{11\pi}{6}\right)$

55. A: $m = 1$, B: $m = -1$, C: $m = 1$

Section 12.4 Preliminary Questions

1. (b) **2.** Yes **3.** (c)

Section 12.4 Exercises

1. $A = \frac{1}{2}\int_{\pi/2}^{\pi} r^2\, d\theta = \frac{25\pi}{4}$

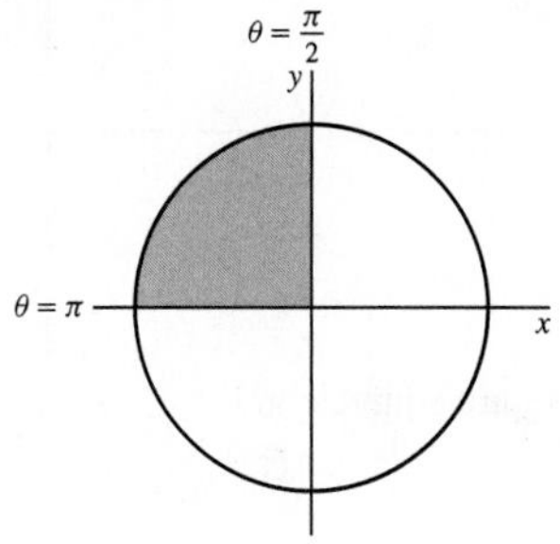

3. $A = \frac{1}{2}\int_0^{\pi} r^2\, d\theta = 4\pi$ **5.** $A = 16$

7. $A = \frac{3\pi}{2}$ **9.** $A = \frac{\pi}{8} \approx 0.39$

11. $A = \frac{\pi^3}{48}$

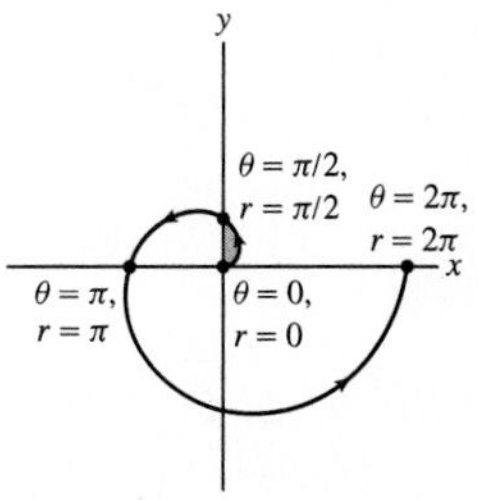

13. $A = \frac{\sqrt{15}}{2} + 7\cos^{-1}\left(\frac{1}{4}\right) \approx 11.163$

15. $A = \pi - \frac{3\sqrt{3}}{2} \approx 0.54$ **17.** $A = \frac{\pi}{8} - \frac{1}{4} \approx 0.14$ **19.** $A = 4\pi$

21. $A = \frac{9\pi}{2} - 4\sqrt{2}$ **23.** $A = 4\pi$

25. $L = \frac{1}{3}\left(\left(\pi^2 + 4\right)^{3/2} - 8\right) \approx 14.55$

27. $L = \sqrt{2}\left(e^{2\pi} - 1\right) \approx 755.9$ **29.** $L = 8$

31. $L = \int_0^{2\pi} \sqrt{5 - 4\cos\theta}\,(2 - \cos\theta)^{-2}\, d\theta$ **33.** $L \approx 6.682$

35. $L \approx 79.564$

Section 12.5 Preliminary Questions

1. **(a)** Hyperbola **(b)** Parabola **(c)** Ellipse
(d) Not a conic section

2. Hyperbolas **3.** The points $(0, c)$ and $(0, -c)$

4. $\pm\frac{b}{a}$ are the slopes of the two asymptotes of the hyperbola.

Section 12.5 Exercises Questions

1. $F_1 = \left(-\sqrt{65}, 0\right)$, $F_2 = \left(\sqrt{65}, 0\right)$. The vertices are $(9, 0)$, $(-9, 0)$, $(0, 4)$ and $(0, -4)$.

3. $F_1 = \left(\sqrt{97}, 0\right)$, $F_2 = \left(\sqrt{97}, 0\right)$. The vertices are $(4, 0)$ and $(-4, 0)$.

5. $F_1 = \left(\sqrt{65} + 3, -1\right)$, $F_2 = \left(-\sqrt{65} + 3, -1\right)$. The vertices are $(10, -1)$ and $(-4, -1)$.

7. $\frac{x^2}{6^2} + \frac{y^2}{3^2} = 1$ **9.** $\frac{(x-14)^2}{6^2} + \frac{(y+4)^2}{3^2} = 1$

11. $\frac{x^2}{5^2} + \frac{y^2}{7^2} = 1$ **13.** $\frac{x^2}{(40/3)^2} + \frac{y^2}{(50/3)^2} = 1$

15. $\left(\frac{x}{3}\right)^2 - \left(\frac{y}{4}\right)^2 = 1$ **17.** $\frac{x^2}{2^2} + \frac{y^2}{\left(2\sqrt{3}\right)^2} = 1$

19. $\left(\frac{x-2}{5}\right)^2 - \left(\frac{y}{10\sqrt{2}}\right)^2 = 1$ **21.** $y = 3x^2$

23. $y = \frac{1}{20}x^2$ **25.** $y = \frac{1}{16}x^2$ **27.** $x = \frac{1}{8}y^2$

29. Vertices: $(\pm 4, 0)$, $(0, \pm 2)$. Foci: $\left(\pm\sqrt{12}, 0\right)$. Centered at the origin.

31. Vertices: $(7, -5)$, $(-1, -5)$. Foci: $\left(\sqrt{65} + 3, -5\right)$, $\left(-\sqrt{65} + 3, -5\right)$. Center: $(3, -5)$. Asymptotes: $y = \frac{4}{7}x + \frac{47}{7}$ and $y = -\frac{4}{7}x + \frac{23}{7}$.

33. Vertices: $(5, 5)$, $(-7, 5)$. Foci: $\left(\sqrt{84} - 1, 5\right)$, $\left(-\sqrt{84} - 1, 5\right)$. Center: $(-1, 5)$. Asymptotes: $y = \frac{\sqrt{48}}{6}(x + 1) + 5 \approx 1.15x + 6.15$ and $y = -\frac{\sqrt{48}}{6}(x + 1) + 5 \approx -1.15x + 3.85$.

35. Vertex: $(0, 0)$. Focus: $\left(0, \frac{1}{16}\right)$.

37. Vertices: $\left(1 \pm \frac{5}{2}, \frac{1}{5}\right)$, $\left(1, \frac{1}{5} \pm 1\right)$. Foci: $\left(-\frac{\sqrt{21}}{2} + 1, \frac{1}{5}\right)$, $\left(\frac{\sqrt{21}}{2} + 1, \frac{1}{5}\right)$. Centered at $\left(1, \frac{1}{5}\right)$.

39. $D = -87$; ellipse **41.** $D = 40$; hyperbola

47. Focus: $(0, c)$. Directrix: $y = -c$. **49.** $A = \frac{8}{3}c^2$

51. $r = \frac{3}{2+\cos\theta}$ **53.** $r = \frac{4}{1+\cos\theta}$

55. Hyperbola, $e = 4$, directrix $x = 2$

57. Ellipse, $e = \frac{3}{4}$, directrix $x = \frac{8}{3}$ **59.** $r = \frac{12}{5-6\cos\theta}$

61. $\left(\frac{x+3}{5}\right)^2 + \left(\frac{y}{16/5}\right)^2 = 1$

63. 4.5 billion miles

Chapter 12 Review

1. (a), (c)

3. $c(t) = (1 + 2\cos t, 1 + 2\sin t)$. The intersection points with the y-axis are $\left(0, 1 \pm \sqrt{3}\right)$. The intersection points with the x-axis are $\left(1 \pm \sqrt{3}, 0\right)$.

5. $c(\theta) = (\cos(\theta + \pi), \sin(\theta + \pi))$ **7.** $c(t) = (1 + 2t, 3 + 4t)$

9. $y = -\frac{x}{4} + \frac{37}{4}$ **11.** $y = \frac{8}{(3-x)^2} + \frac{3-x}{2}$

13. $\left.\frac{dy}{dx}\right|_{t=3} = \frac{3}{14}$ **15.** $\left.\frac{dy}{dx}\right|_{t=0} = \frac{\cos 20}{e^{20}}$

17. $(0, 1)$, $(\pi, 2)$, $(0.13, 0.40)$, and $(1.41, 1.60)$

19. $x(t) = -2t^3 + 4t^2 - 1$, $y(t) = 2t^3 - 8t^2 + 6t - 1$

21. $\frac{ds}{dt} = \sqrt{3 + 2(\cos t - \sin t)}$; maximal speed: $\sqrt{3 + 2\sqrt{2}}$

23. $s = \sqrt{2}$

25.

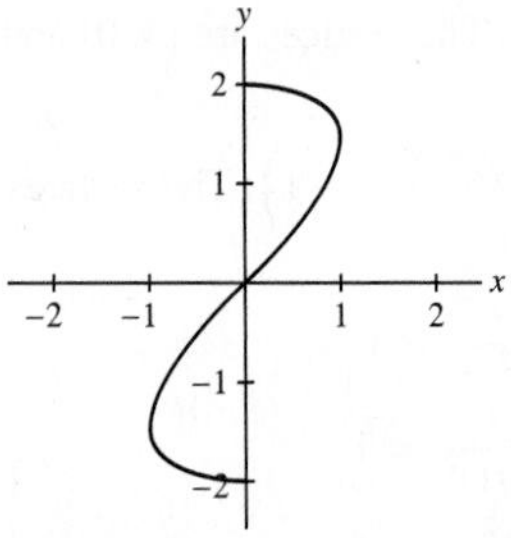

$s = 2\int_0^{\pi} \sqrt{\cos^2 2t + \sin^2 t}\, dt \approx 6.0972$

27. $\left(1, \frac{\pi}{6}\right)$ and $\left(3, \frac{5\pi}{4}\right)$ have rectangular coordinates $\left(\frac{\sqrt{3}}{2}, \frac{1}{2}\right)$ and $\left(-\frac{3\sqrt{2}}{2}, -\frac{3\sqrt{2}}{2}\right)$.

29. $\sqrt{x^2 + y^2} = \frac{2x}{x-y}$ **31.** $r = 3 + 2\sin\theta$

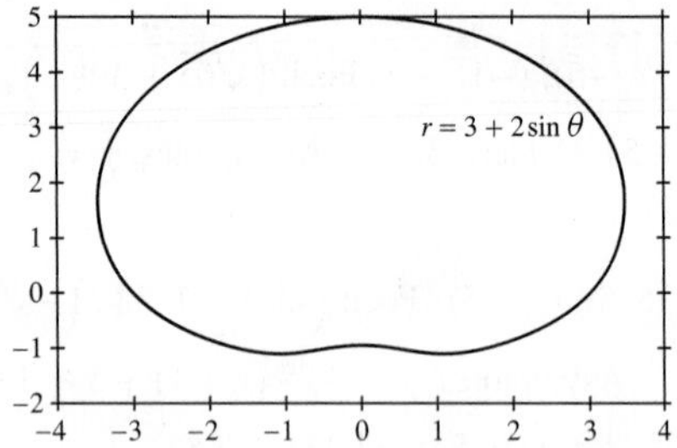

33. $A = \frac{\pi}{16}$ **35.** $e - \frac{1}{e}$

Note: One needs to double the integral from $-\frac{\pi}{2}$ *to* $\frac{\pi}{2}$ *in order to account for both sides of the graph.*

37. $A = \frac{3\pi a^2}{2}$

39. Outer: $L \approx 36.121$, inner: $L \approx 7.5087$, difference: 28.6123

41. Ellipse. Vertices: $(\pm 3, 0)$, $(0, \pm 2)$. Foci: $(\pm\sqrt{5}, 0)$.

43. Ellipse. Vertices: $\left(\pm\frac{2}{\sqrt{5}}, 0\right)$, $\left(0, \pm\frac{4}{\sqrt{5}}\right)$. Foci: $\left(0, \pm\sqrt{\frac{12}{5}}\right)$.

45. $\left(\frac{x}{8}\right)^2 + \left(\frac{y}{\sqrt{61}}\right)^2 = 1$ **47.** $\left(\frac{x}{8}\right)^2 - \left(\frac{y}{6}\right)^2 = 1$ **49.** $x = \frac{1}{32}y^2$

51. $y = \sqrt{3}x + \left(\sqrt{3} - 5\right)$ and $y = -\sqrt{3}x + \left(-\sqrt{3} - 5\right)$

CHAPTER 13

Section 13.1 Preliminary Questions

1. **(a)** True **(b)** False **(c)** True **(d)** True

2. $\|-3\mathbf{a}\| = 15$

3. The components are not changed.

4. $(0, 0)$

5. **(a)** True **(b)** False

Section 13.1 Exercises

1. $\mathbf{v}_1 = \langle 2, 0\rangle$, $\|\mathbf{v}_1\| = 2$ $\mathbf{v}_2 = \langle 2, 0\rangle$, $\|\mathbf{v}_2\| = 2$

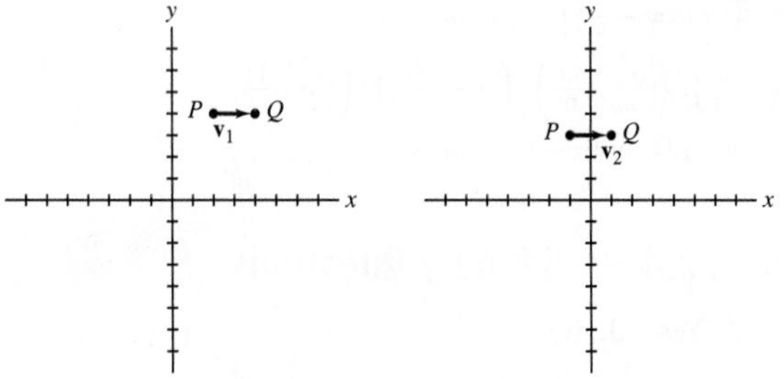

$\mathbf{v}_3 = \langle 3, 1\rangle$, $\|\mathbf{v}_3\| = \sqrt{10}$ $\mathbf{v}_4 = \langle 2, 2\rangle$, $\|\mathbf{v}_4\| = 2\sqrt{2}$

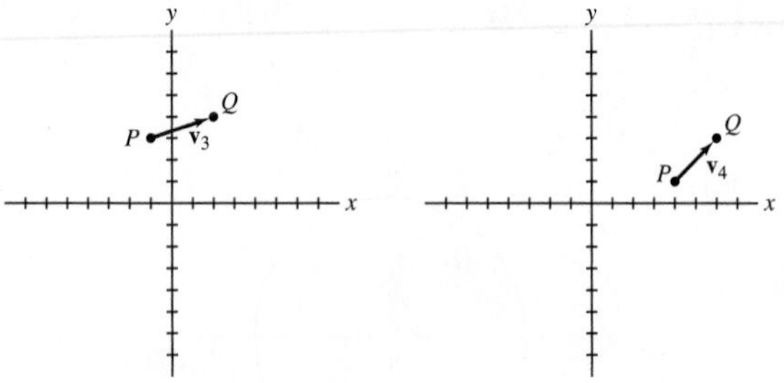

Vectors $\mathbf{v}_1$ and $\mathbf{v}_2$ are equivalent.

3. $(3, 5)$

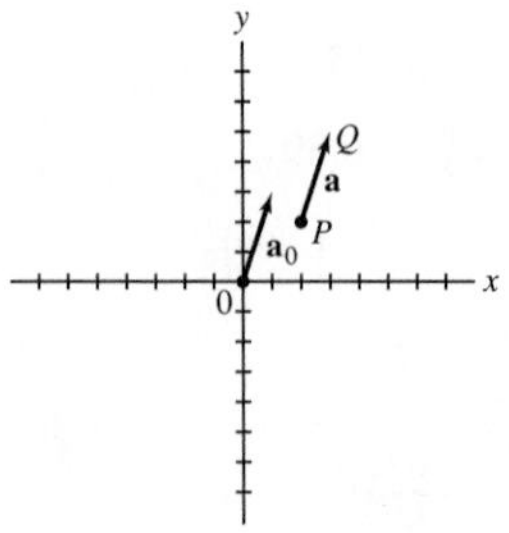

5. $\overrightarrow{PQ} = \langle -1, 5\rangle$ **7.** $\overrightarrow{PQ} = \langle -2, -9\rangle$ **9.** $\langle 5, 5\rangle$

11. $\langle 30, 10\rangle$ **13.** $\left\langle \frac{5}{2}, 5\right\rangle$

15. Vector (B)

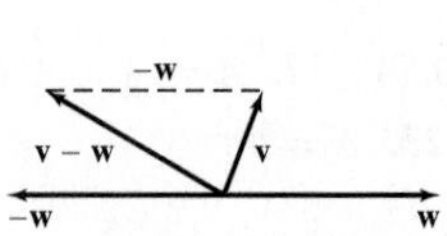

17. $2\mathbf{v} = \langle 4,\ 6\rangle$ $\quad -\mathbf{w} = \langle -4,\ -1\rangle$

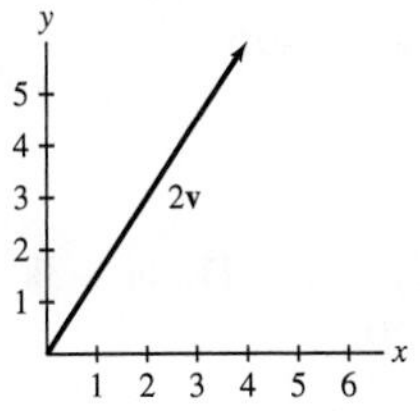

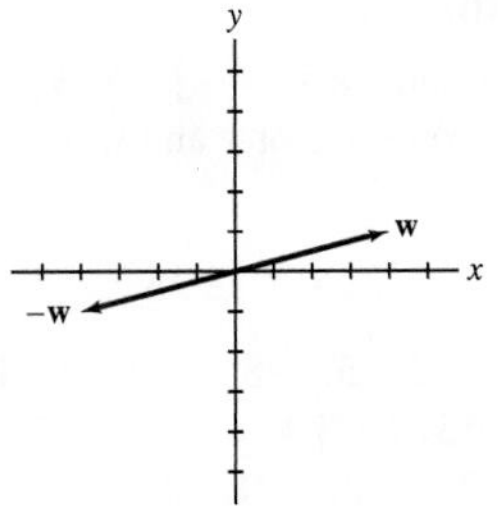

$2\mathbf{v} - \mathbf{w} = \langle 0,\ 5\rangle$ $\quad \mathbf{v} + \mathbf{w} = \langle 6,\ 4\rangle$

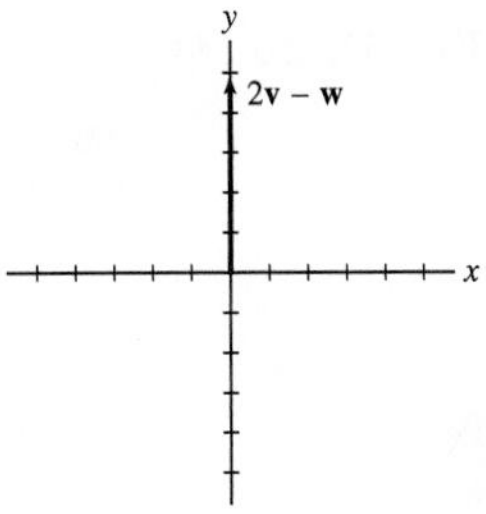

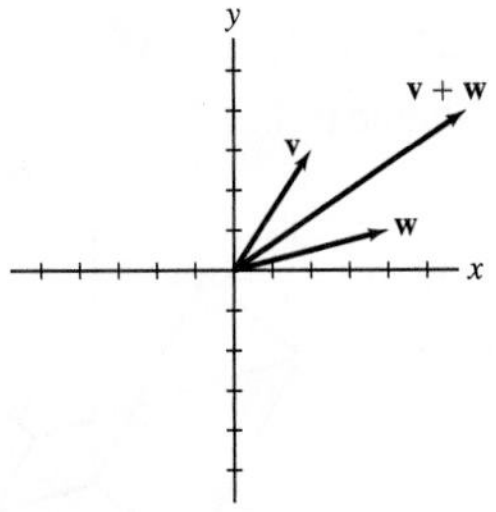

19. $3\mathbf{v} + \mathbf{w} = \langle -2,\ 10\rangle$, $2\mathbf{v} - 2\mathbf{w} = \langle 4,\ -4\rangle$

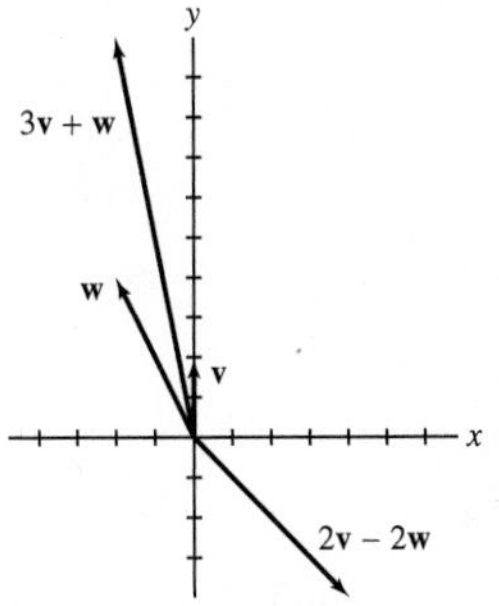

21.

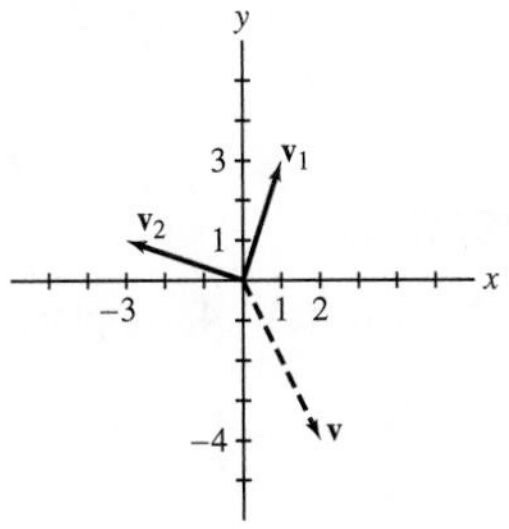

23. (b) and (c)

25. $\overrightarrow{AB} = \langle 2,\ 6\rangle$ and $\overrightarrow{PQ} = \langle 2,\ 6\rangle$; equivalent

27. $\overrightarrow{AB} = \langle 3, -2\rangle$ and $\overrightarrow{PQ} = \langle 3, -2\rangle$; equivalent

29. $\overrightarrow{AB} = \langle 2,\ 3\rangle$ and $\overrightarrow{PQ} = \langle 6,\ 9\rangle$; parallel and point in the same direction

31. $\overrightarrow{AB} = \langle -8,\ 1\rangle$ and $\overrightarrow{PQ} = \langle 8, -1\rangle$; parallel and point in opposite directions

33. $\left\|\overrightarrow{OR}\right\| = \sqrt{53}$ **35.** $P = (0,\ 0)$ **37.** $\mathbf{e_v} = \frac{1}{5}\langle 3,\ 4\rangle$

39. $4\mathbf{e_u} = \left\langle -2\sqrt{2},\ -2\sqrt{2}\right\rangle$

41. $\mathbf{e} = \left\langle \cos\frac{4\pi}{7},\ \sin\frac{4\pi}{7}\right\rangle = \langle -0.22,\ 0.97\rangle$

43. $\lambda = \pm\frac{1}{\sqrt{13}}$ **45.** $P = (4,\ 6)$

47. (a) → (ii), (b) → (iv), (c) → (iii), (d) → (i) **49.** $9\mathbf{i} + 7\mathbf{j}$

51. $-5\mathbf{i} - \mathbf{j}$

53.

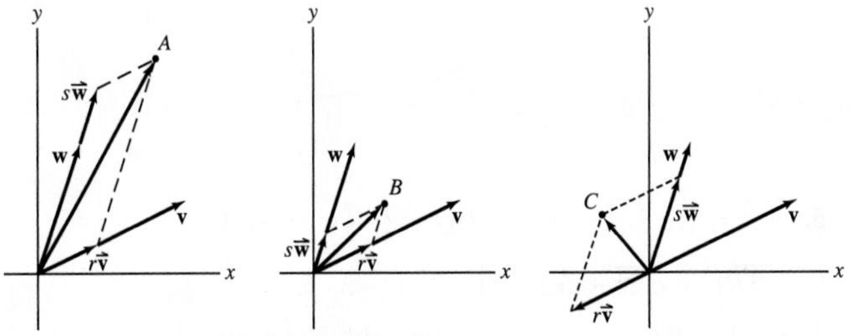

55. $\mathbf{u} = 2\mathbf{v} - \mathbf{w}$

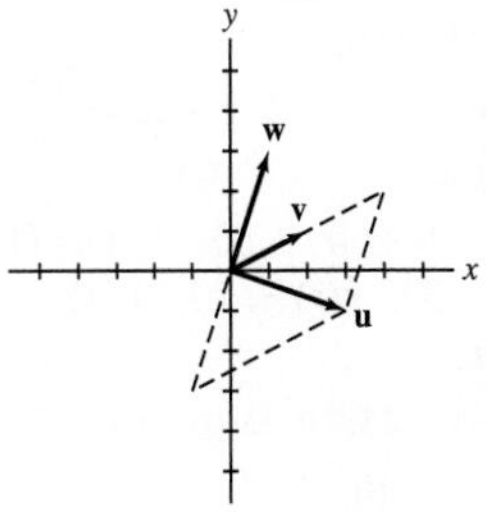

57. The force on cable 1 is ≈ 45 lb, and force on cable 2 is ≈ 21 lb.

59. 230 km/hr **61.** $\mathbf{r} = \langle 6.45,\ 0.38\rangle$

Section 13.2 Preliminary Questions

1. (4, 3, 2) **2.** ⟨3, 2, 1⟩ **3.** (a) **4.** (c)

5. Infinitely many direction vectors **6.** True

Section 13.2 Exercises

1. $\|\mathbf{v}\| = \sqrt{14}$

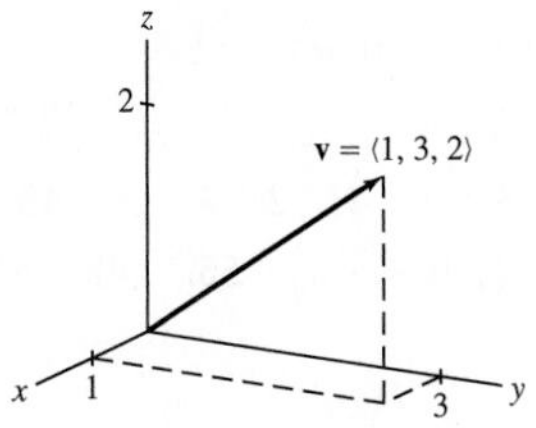

3. The head of $\mathbf{v} = \overrightarrow{PQ}$ is $Q = (1,\ 2,\ 1)$.

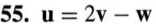

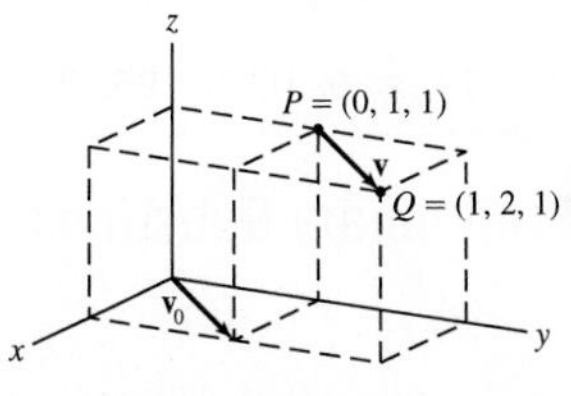

$\mathbf{v}_0 = \overrightarrow{OS}$, where $S = (1, 1, 0)$

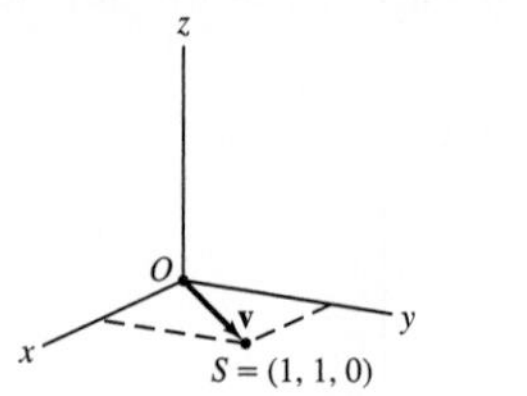

5. $\overrightarrow{PQ} = \langle 1, 1, -1 \rangle$ **7.** $\overrightarrow{PQ} = \left\langle -\frac{9}{2}, -\frac{3}{2}, 1 \right\rangle$

9. $\left\| \overrightarrow{OR} \right\| = \sqrt{26} \approx 5.1$ **11.** $P = (-2, 6, 0)$

13. **(a)** Parallel and same direction **(b)** Not parallel
(c) Parallel and opposite directions **(d)** Not parallel

15. Not equivalent **17.** Not equivalent **19.** $\langle -8, -18, -2 \rangle$

21. $\langle -2, -2, 3 \rangle$ **23.** $\langle 16, -1, 9 \rangle$

25. $\mathbf{e_w} = \left\langle \frac{4}{\sqrt{21}}, \frac{-2}{\sqrt{21}}, \frac{-1}{\sqrt{21}} \right\rangle$ **27.** $-\mathbf{e_v} = \left\langle \frac{2}{3}, -\frac{2}{3}, -\frac{1}{3} \right\rangle$

29. $\mathbf{r}(t) = \langle 1 + 2t, 2 + t, -8 + 3t \rangle$

31. $\mathbf{r}(t) = \langle 4 + 7t, 0, 8 + 4t \rangle$ **33.** $\mathbf{r}(t) = \langle 1 + 2t, 1 - 6t, 1 + t \rangle$

35. $\mathbf{r}(t) = \langle 4t, t, t \rangle$ **37.** $\mathbf{r}(t) = \langle 0, 0, t \rangle$

39. $\mathbf{r}(t) = \langle -t, -2t, 4 - 2t \rangle$

41. (c) **43.** $(3, 4, 3)$ **45.** $R = (6, 13, 15)$

49. $\mathbf{r}_1(t) = \langle 5, 5, 2 \rangle + t\langle 0, -2, 1 \rangle$;
$\mathbf{r}_2(t) = \langle 5, 5, 2 \rangle + t\langle 0, -20, 10 \rangle$

53. $(3, 4, 7)$ **55.** $\mathbf{v} = \left\langle 0, \frac{1}{2}, -\frac{1}{2} \right\rangle$ **59.** $\frac{x-1}{-3} = \frac{y-1}{3} = \frac{z-2}{-2}$

61. $\mathbf{r}(t) = \langle 5, -3, 10 \rangle + t\langle 9, 7, 1 \rangle$

Section 13.3 Preliminary Questions

1. Scalar **2.** Obtuse **3.** Distributive Law

4. **(a)** $\mathbf{v}$ **(b)** $\mathbf{v}$

5. (b); (c) **6.** (c)

Section 13.3 Exercises

1. 15 **3.** 41 **5.** 5 **7.** 0 **9.** 1 **11.** 0 **13.** Obtuse

15. Orthogonal **17.** Acute **19.** 0 **21.** $\frac{1}{\sqrt{10}}$ **23.** $\pi/4$

25. ≈ 0.615 **27.** $2\pi/3$

29. **(a)** $b = -\frac{1}{2}$ **(b)** $b = 0$ or $b = \frac{1}{2}$

31. $\mathbf{v}_1 = \langle 0, 1, 0 \rangle$, $\mathbf{v}_2 = \langle 3, 2, 2 \rangle$ **33.** $-\frac{3}{2}$ **35.** $\|\mathbf{v}\|^2$

37. $\|\mathbf{v}\|^2 - \|\mathbf{w}\|^2$ **39.** 8 **41.** 2 **43.** π **45.** (b) 7 **49.** 51.91°

51. $\left\langle \frac{7}{2}, \frac{7}{2} \right\rangle$ **53.** $\left\langle -\frac{4}{5}, 0, -\frac{2}{5} \right\rangle$ **55.** $-4\mathbf{k}$ **57.** $a\mathbf{i}$ **59.** $2\sqrt{2}$

61. $\sqrt{17}$

63. $\mathbf{a} = \left\langle \frac{1}{2}, \frac{1}{2} \right\rangle + \left\langle \frac{1}{2}, -\frac{1}{2} \right\rangle$

65. $\mathbf{a} = \left\langle 0, -\frac{1}{2}, -\frac{1}{2} \right\rangle + \left\langle 4, -\frac{1}{2}, \frac{1}{2} \right\rangle$

67. $\left\langle \frac{x-y}{2}, \frac{y-x}{2} \right\rangle + \left\langle \frac{x+y}{2}, \frac{y+x}{2} \right\rangle$

71. $\approx 35°$ **73.** $\overrightarrow{AD}$ **77.** ≈ 68.07 N **95.** $2x + 2y - 2z = 1$

Section 13.4 Preliminary Questions

1. $\begin{vmatrix} -5 & -1 \\ 4 & 0 \end{vmatrix}$

2. $\|\mathbf{e} \times \mathbf{f}\| = \frac{1}{2}$ **3.** $\mathbf{u} \times \mathbf{v} = \langle -2, -2, -1 \rangle$

4. **(a)** 0 **(b)** 0

5. $\mathbf{i} \times \mathbf{j} = \mathbf{k}$ and $\mathbf{i} \times \mathbf{k} = -\mathbf{j}$ **6.** $\mathbf{v} \times \mathbf{w} = \mathbf{0}$ if either $\mathbf{v}$ or $\mathbf{w}$ (or both) is the zero vector or $\mathbf{v}$ and $\mathbf{w}$ are parallel vectors.

Section 13.4 Exercises

1. -5 **3.** -15 **5.** -8 **7.** 0 **9.** $\mathbf{i} + 2\mathbf{j} - 5\mathbf{k}$ **11.** $6\mathbf{i} - 8\mathbf{k}$

13. $-\mathbf{j} + \mathbf{i}$ **15.** $\mathbf{i} + \mathbf{j} + \mathbf{k}$ **17.** $\langle -1, -1, 0 \rangle$

19. $\langle -2, -2, -2 \rangle$ **21.** $\langle 4, 4, 0 \rangle$

23. $\mathbf{v} \times \mathbf{i} = \langle 0, c, -b \rangle$; $\mathbf{v} \times \mathbf{j} = \langle -c, 0, a \rangle$;
$\mathbf{v} \times \mathbf{k} = \langle b, -a, 0 \rangle$

25. $-\mathbf{u}$ **27.** $\langle 0, 3, 3 \rangle$ **31.** $\mathbf{e}'$ **33.** $\mathbf{F}_1$ **37.** $2\sqrt{138}$

39. The volume is 4.

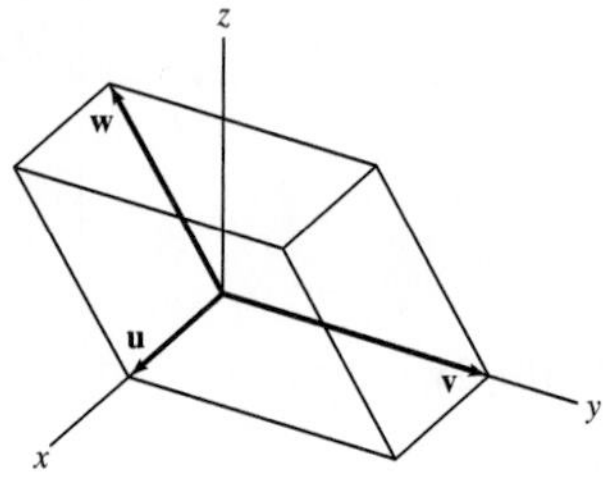

41. $\sqrt{35} \approx 5.92$

43.

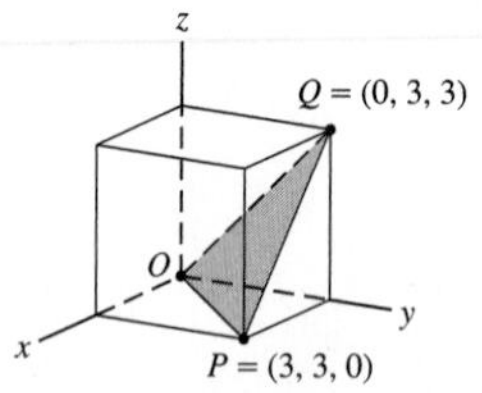

The area of the triangle is $\frac{9\sqrt{3}}{2} \approx 7.8$.

55. $\mathbf{X} = \langle a, a, a + 1 \rangle$

59. $\tau = 250 \sin 125° \, \mathbf{k} \approx 204.79 \, \mathbf{k}$

Section 13.5 Preliminary Questions

1. $3x + 4y - z = 0$ **2.** (c): $z = 1$ **3.** Plane (c) **4.** xz-plane

5. (c): $x + y = 0$ **6.** Statement (a)

Section 13.5 Exercises

1. $\langle 1, 3, 2 \rangle \cdot \langle x, y, z \rangle = 3$
$x + 3y + 2z = 3$
$(x - 4) + 3(y + 1) + 2(z - 1) = 0$

3. $\langle -1, 2, 1 \rangle \cdot \langle x, y, z \rangle = 3$
$-x + 2y + z = 3$
$-(x - 4) + 2(y - 1) + (z - 5) = 0$

5. $\langle 1, 0, 0 \rangle \cdot \langle x, y, z \rangle = 3$
$x = 3$
$(x - 3) + 0(y - 1) + 0(z + 9) = 0$

7. $\langle 0, 0, 1 \rangle \cdot \langle x, y, z \rangle = 2$
$z = 2$
$0(x - 6) + 0(y - 7) + 1(z - 2) = 0$

9. $x = 0$ **11.** Statements (b) and (d) **13.** $\langle 9, -4, -11\rangle$
15. $\langle 3, -8, 11\rangle$ **17.** $6x + 9y + 4z = 19$ **19.** $x + 2y - z = 1$
21. $4x - 9y + z = 0$ **23.** $x = 4$ **25.** $x + z = 3$
27. $13x + y - 5z = 27$ **29.** Yes, the planes are parallel.
31. $10x + 15y + 6z = 30$ **33.** $(1, 5, 8)$ **35.** $(-2, 3, 12)$
37. $-9y + 4z = 5$ **39.** $x = -\frac{2}{3}$ **41.** $x = -4$

43. The two planes have no common points.

45. $y - 4z = 0$
$x + y - 4z = 0$

47. $(3\lambda)x + by + (2\lambda)z = 5\lambda,\ \lambda \neq 0$ **49.** $\theta = \pi/2$
51. $\theta = 1.143$ rad or $\theta = 65.49°$ **53.** $\theta \approx 55.0°$
55. $x + y + z = 1$ **57.** $x - y - z = f$
59. $x = \frac{9}{5} + 2t,\ y = -\frac{6}{5} - 3t,\ z = 2 + 5t$ **61.** $\pm 24\langle 1, 2, -2\rangle$
67. $\left(\frac{2}{3}, -\frac{1}{3}, \frac{2}{3}\right)$ **69.** $\frac{6}{\sqrt{30}} \approx 1.095$ **71.** $|a|$

Section 13.6 Preliminary Questions

1. True, mostly, except at $x = \pm a,\ y = \pm b,$ or $z = \pm c$.

2. False **3.** Hyperbolic paraboloid

4. No **5.** Ellipsoid

6. All vertical lines passing through a parabola c in the xy-plane.

Section 13.6 Exercises

1. Ellipsoid **3.** Ellipsoid

5. Hyperboloid of one sheet **7.** Elliptic paraboloid

9. Hyperbolic paraboloid **11.** Hyperbolic paraboloid

13. Ellipsoid, the trace is a circle on the xz-plane

15. Ellipsoid, the trace is an ellipse on the xy-plane

17. Hyperboloid of one sheet, the trace is a hyperbola.

19. Parabolic cylinder, the trace is the parabola $y = 3x^2$

21. (a) ↔ Figure b; (b) ↔Figure c; (c) ↔ Figure a

23. $y = \left(\frac{x}{2}\right)^2 + \left(\frac{z}{4}\right)^2$

25.

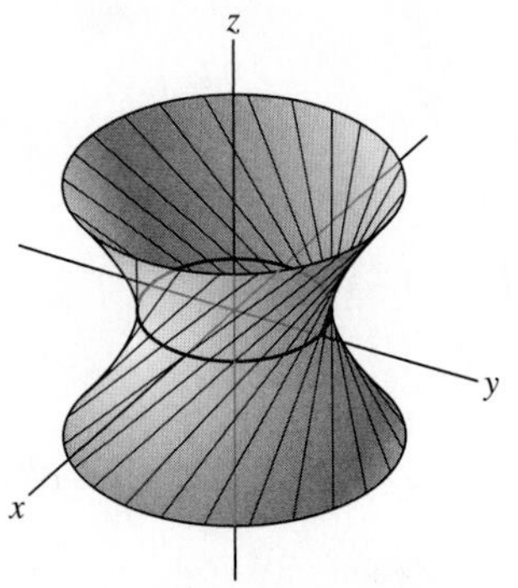

Graph of $x^2 + y^2 - z^2 = 1$

27.

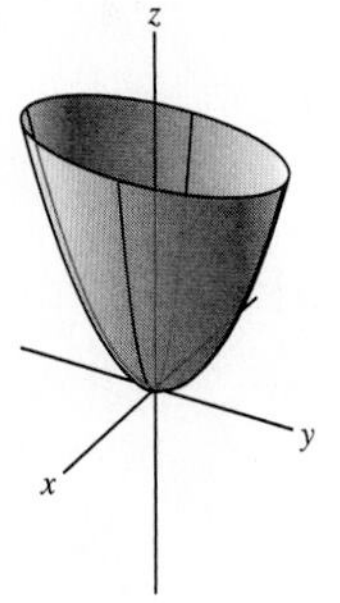

29.

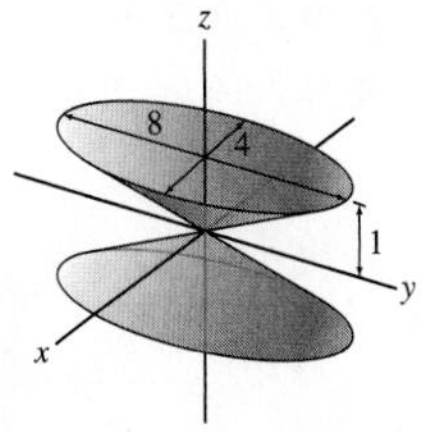

31. $\left(\frac{x}{2}\right)^2 + \left(\frac{y}{4}\right)^2 + \left(\frac{z}{6}\right)^2 = 1$ **33.** $\left(\frac{x}{4}\right)^2 + \left(\frac{y}{6}\right)^2 - \left(\frac{z}{3\sqrt{3}}\right)^2 = 1$

35. One or two vertical lines, or an empty set

37. The upper part of an elliptic cone

Section 13.7 Preliminary Questions

1. Cylinder of radius R whose axis is the z-axis, sphere of radius R centered at the origin.

2. (b) **3.** (a) **4.** $\phi = 0,\ \pi$ **5.** $\phi = \frac{\pi}{2}$, the xy-plane

Section 13.7 Exercises

1. $(-4, 0, 4)$ **3.** $\left(0, 0, \frac{1}{2}\right)$ **5.** $\left(\sqrt{2}, \frac{7\pi}{4}, 1\right)$ **7.** $\left(2, \frac{\pi}{3}, 7\right)$

9. $\left(5, \frac{\pi}{4}, 2\right)$ **11.** $r^2 \le 1$

13. $r^2 + z^2 \le 4,\ \theta = \frac{\pi}{2}$ or $\theta = \frac{3\pi}{2}$

15. $r^2 \le 9,\ \frac{5\pi}{4} \le \theta \le 2\pi$ and $0 \le \theta \le \frac{\pi}{4}$

17.

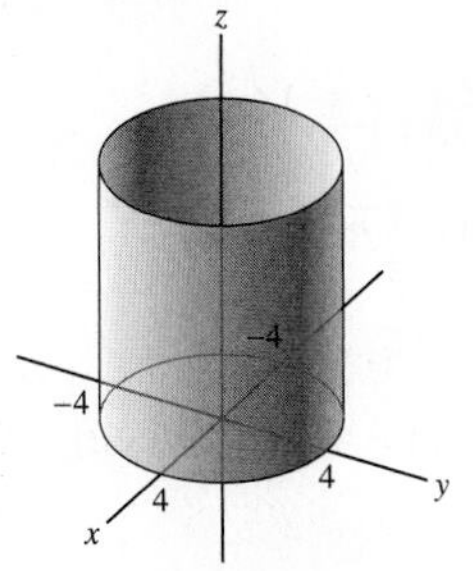

19.

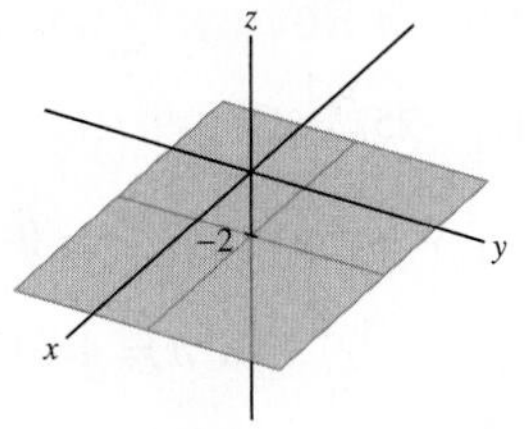

21.

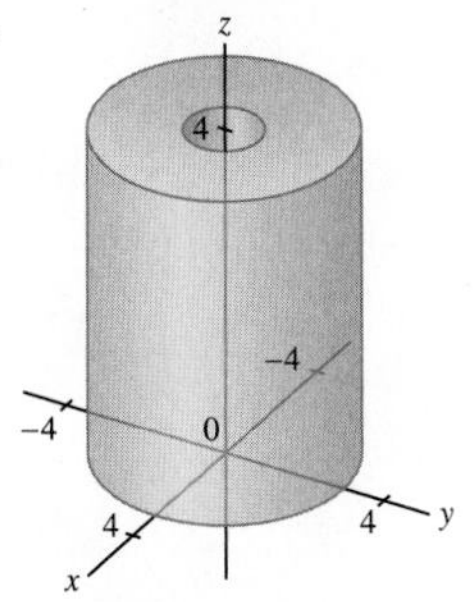

23.

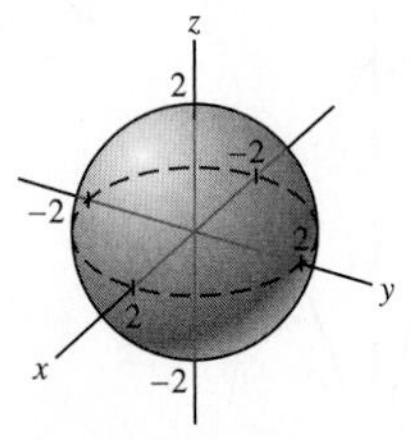

25. $r = \frac{z}{\cos\theta + \sin\theta}$ **27.** $r = \frac{z\tan\theta}{\cos\theta}$ **29.** $r = 2$ **31.** $(3, 0, 0)$

33. $(0, 0, 3)$ **35.** $\left(\frac{3\sqrt{3}}{2}, \frac{3}{2}, -3\sqrt{3}\right)$ **37.** $\left(2, 0, \frac{\pi}{3}\right)$

39. $\left(\sqrt{3}, \frac{\pi}{4}, 0.955\right)$ **41.** $\left(2, \frac{\pi}{3}, \frac{\pi}{6}\right)$ **43.** $\left(2\sqrt{2}, 0, \frac{\pi}{4}\right)$

45. $\left(2\sqrt{2}, 0, 2\sqrt{2}\right)$ **47.** $0 \le \rho \le 1$

49. $\rho = 1,\ 0 \le \theta \le \frac{\pi}{2},\ 0 \le \phi \le \frac{\pi}{2}$

51. $\left\{(\rho,\ \theta,\ \phi) : 0 \le \rho \le 2,\ \theta = \frac{\pi}{2} \text{ or } \theta = \frac{3\pi}{2}\right\}$

53.

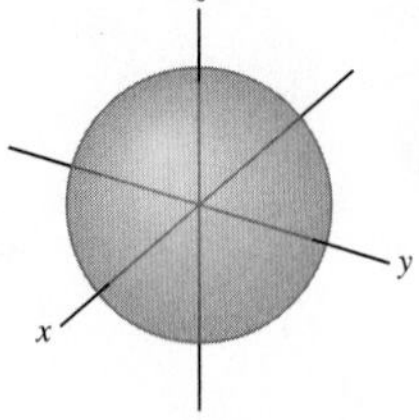

55.

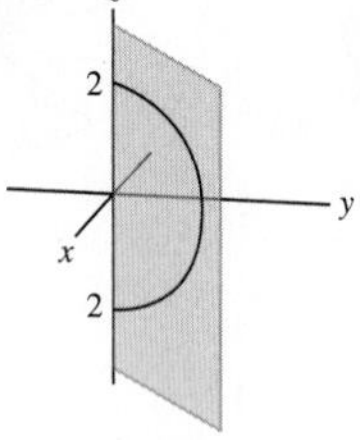

57.

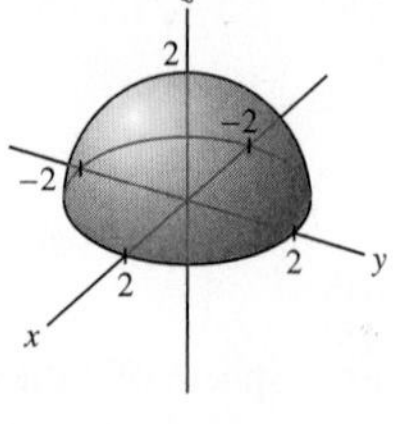

59.

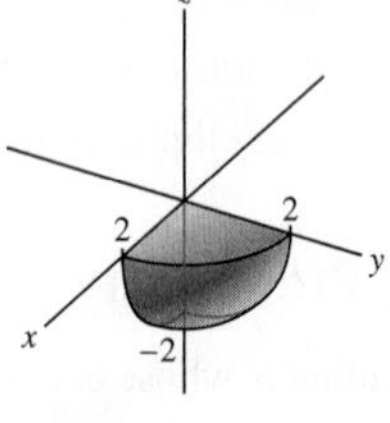

61. $\rho = \frac{2}{\cos\phi}$ **63.** $\rho = \frac{\cos\theta\tan\phi}{\cos\phi}$ **65.** $\rho = \frac{2}{\sin\phi\sqrt{\cos 2\theta}}$ **67.** (b)

69. Helsinki:(25.0°, 29.9°), Sao Paulo: (313.48°, 113.52°)

71. Sydney: (−4618.8, 2560.3, − 3562.1), Bogota: (1723.7, − 6111.7, 503.1)

73. $z = \pm r\sqrt{\cos 2\theta}$

77. $r = \sqrt{z^2 + 1}$ and $\rho = \sqrt{-\frac{1}{\cos 2\phi}}$; no points; $\frac{\pi}{4} < \phi < \frac{3\pi}{4}$

Chapter 13 Review

1. $\langle 21,\ -25\rangle$ and $\langle -19,\ 31\rangle$ **3.** $\left\langle \frac{-2}{\sqrt{29}},\ \frac{5}{\sqrt{29}}\right\rangle$

5. $\mathbf{i} = \frac{2}{11}\mathbf{v} + \frac{5}{11}\mathbf{w}$ **7.** $\overrightarrow{PQ} = \langle -4,\ 1\rangle$; $\left\|\overrightarrow{PQ}\right\| = \sqrt{17}$

9. $\left\langle \frac{3}{\sqrt{2}},\ -\frac{3}{\sqrt{2}}\right\rangle$ **11.** $\beta = \frac{3}{2}$ **13.** $\mathbf{u} = \left\langle \frac{1}{3},\ -\frac{11}{6},\ \frac{7}{6}\right\rangle$

15. $\mathbf{r}_1(t) = \langle 1 + 3t,\ 4 + t,\ 5 + 6t\rangle$; $\mathbf{r}_2(t) = \langle 1 + 3t,\ t,\ 6t\rangle$

17. $a = -2,\ b = 2$

19.

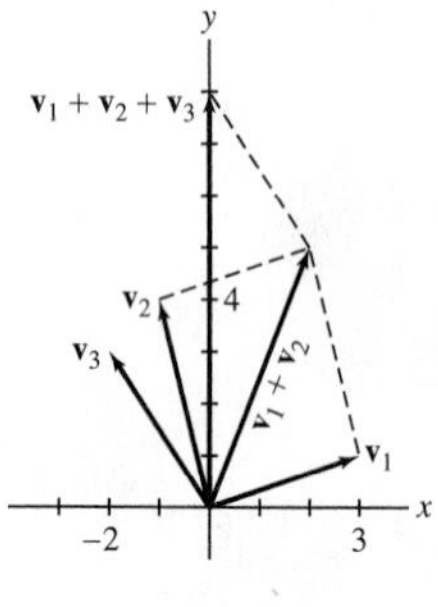

21. $\mathbf{v}\cdot\mathbf{w} = -9$ **23.** $\mathbf{v}\times\mathbf{w} = \langle 10,\ -8,\ -7\rangle$

25. $V = 48$ **29.** $\frac{5}{3}$

31. $\|\mathbf{F}_1\| = \frac{2\|\mathbf{F}_2\|}{\sqrt{3}}$; $\|\mathbf{F}_1\| = 980$ N

33. $\mathbf{v}\times\mathbf{w} = \langle -6,\ 7,\ -2\rangle$

35. −47 **37.** $5\sqrt{2}$ **41.** $\|\mathbf{e} - 4\mathbf{f}\| = \sqrt{13}$

47. $(x - 0) + 4(y - 1) - 3(z + 1) = 0$

49. $17x - 21y - 13z = -28$ **51.** $3x - 2y = 4$ **53.** Ellipsoid

55. Elliptic paraboloid **57.** Elliptic cone

59. **(a)** Empty set **(b)** Hyperboloid of one sheet

(c) Hyperboloid of two sheets

61. $(r,\ \theta,\ z) = \left(5,\ \tan^{-1}\frac{4}{3},\ -1\right)$, $(\rho,\ \theta,\ \phi) = \left(\sqrt{26},\ \tan^{-1}\frac{4}{3},\ \cos^{-1}\left(\frac{-1}{\sqrt{26}}\right)\right)$

63. $(r,\ \theta,\ z) = \left(\frac{3\sqrt{3}}{2},\ \frac{\pi}{6},\ \frac{3}{2}\right)$

65. $z = 2x$

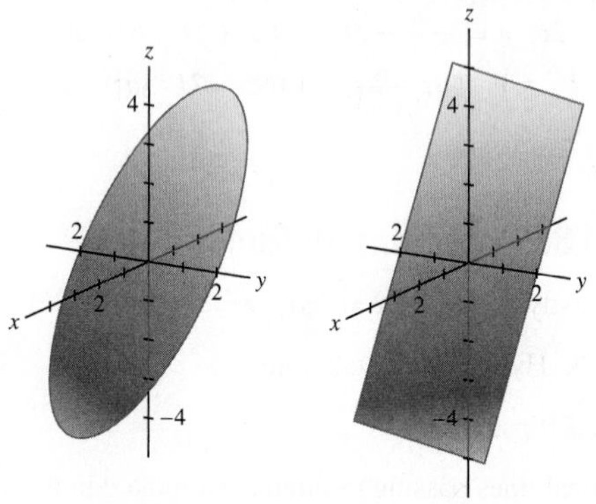

69. $A < -1$: Hyperboloid of one sheet
$A = -1$: Cylinder with the z-axis as its central axis
$A > -1$: Ellipsoid
$A = 0$: Sphere

CHAPTER 14

Section 14.1 Preliminary Questions

1. (c) **2.** The curve $z = e^x$

3. The projection onto the xz-plane

4. The point (−2, 2, 3)

5. As t increases from 0 to 2π, a point on $\sin t\mathbf{i} + \cos t\mathbf{j}$ moves clockwise and a point on $\cos t\mathbf{i} + \sin t\mathbf{j}$ moves counterclockwise.

6. (a), (c), and (d)

Section 14.1 Exercises

1. $D = \{t \in \mathbf{R},\ t \ne 0,\ t \ne -1\}$

3. $\mathbf{r}(2) = \left\langle 0,\ 4,\ \frac{1}{5}\right\rangle$; $\mathbf{r}(-1) = \left\langle -1,\ 1,\ \frac{1}{2}\right\rangle$

5. $\mathbf{r}(t) = (3 + 3t)\mathbf{i} - 5\mathbf{j} + (7 + t)\mathbf{k}$

7. A ↔ ii, B ↔ i, C ↔ iii

9. (a) = (v), (b) = (i), (c) = (ii), (d) = (vi), (e) = (iv), (f) = (iii)

11. C ↔ i, A ↔ ii, B ↔ iii

13. Radius 9, center (0, 0, 0), xy-plane

15. Radius 1, center (0, 0, 4), xz-plane

17. (b)

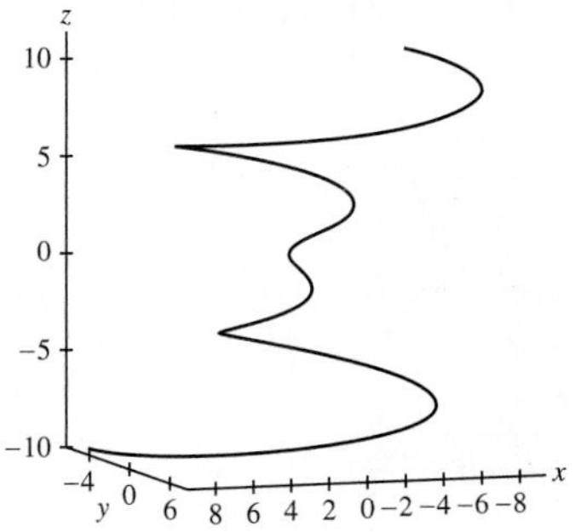

19. $(0,\ 1,\ 0), (0,\ -1,\ 0), \left(\frac{1}{\sqrt{2}},\ \frac{1}{\sqrt{2}},\ 0\right), \left(\frac{1}{\sqrt{2}},\ -\frac{1}{\sqrt{2}},\ 0\right),$
$\left(-\frac{1}{\sqrt{2}},\ -\frac{1}{\sqrt{2}},\ 0\right), \left(-\frac{1}{\sqrt{2}},\ \frac{1}{\sqrt{2}},\ 0\right)$

21. $\mathbf{r}(t) = \left\langle 2t^2 - 7,\ t,\ \pm\sqrt{9 - t^2}\right\rangle$, for $-3 \le t \le 3$

23. (a) $\mathbf{r}(t) = \left\langle \pm t\sqrt{1 - t^2},\ t^2,\ t\right\rangle$ for $-1 \le t \le 1$

(b) The projection is a circle in the xy-plane with radius $\frac{1}{2}$ and centered at the xy-point $\left(0,\ \frac{1}{2}\right)$.

25. $\mathbf{r}(t) = \langle \cos t,\ \pm \sin t,\ \sin t\rangle$; the projection of the curve onto the xy-plane is traced by $\langle \cos t,\ \pm \sin t,\ 0\rangle$, which is the unit circle in this plane; the projection of the curve onto the xz-plane is traced by $\langle \cos t,\ 0,\ \sin t\rangle$, which is the unit circle in this plane; the projection of the curve onto the yz-plane is traced by $\langle 0,\ \pm \sin t,\ \sin t\rangle$, which is the two segments $z = y$ and $z = -y$for $-1 \le y \le 1$.

27. $\mathbf{r}(t) = \left\langle \cos t,\ \sin t,\ 4\cos t^2\right\rangle, 0 \le t \le 2\pi$

29. Collide at the point $(12,\ 4,\ 2)$ and intersect at the points $(4,\ 0,\ -6)$ and $(12,\ 4,\ 2)$

31. $\mathbf{r}(t) = \langle 3,\ 2,\ t\rangle,\ -\infty < t < \infty$

33. $\mathbf{r}(t) = \langle t,\ 3t,\ 15t\rangle,\quad -\infty < t < \infty$

35. $\mathbf{r}(t) = \langle 1,\ 2 + 2\cos t,\ 5 + 2\sin t\rangle,\quad 0 \le t \le 2\pi$

37. $\mathbf{r}(t) = \left\langle \frac{\sqrt{3}}{2}\cos t,\ \frac{1}{2},\ \frac{\sqrt{3}}{2}\sin t\right\rangle,\quad 0 \le t \le 2\pi$

39. $\mathbf{r}(t) = \langle 3 + 2\cos t,\ 1,\ 5 + 3\sin t\rangle,\quad 0 \le t \le 2\pi$

41.

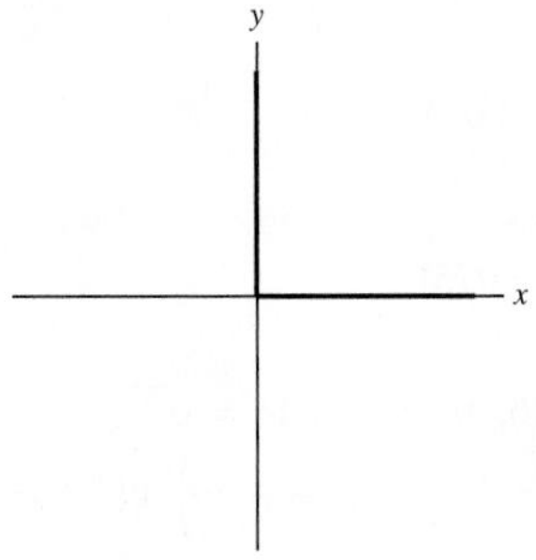

$\mathbf{r}(t) = \langle |t| + t, |t| - t\rangle$

Section 14.2 Preliminary

1. $\frac{d}{dt}(f(t)\mathbf{r}(t)) = f(t)\mathbf{r}'(t) + f'(t)\mathbf{r}(t)$

$\frac{d}{dt}(\mathbf{r}_1(t)\cdot\mathbf{r}_2(t)) = \mathbf{r}_1(t)\cdot\mathbf{r}'_2(t) + \mathbf{r}'_1(t)\cdot\mathbf{r}_2(t)$

$\frac{d}{dt}(\mathbf{r}_1(t)\times\mathbf{r}_2(t)) = \mathbf{r}_1(t)\times\mathbf{r}'_2(t) + \mathbf{r}'_1(t)\times\mathbf{r}_2(t)$

2. True **3.** False **4.** True **5.** False **6.** False

7. (a) Vector **(b)** Scalar **(c)** Vector

Section 14.2 Exercises

1. $\lim_{t\to 3}\left\langle t^2,\ 4t,\ \frac{1}{t}\right\rangle = \left\langle 9,\ 12,\ \frac{1}{3}\right\rangle$

3. $\lim_{t\to 0}(e^{2t}\mathbf{i} + \ln(t+1)\mathbf{j} + 4\mathbf{k}) = \mathbf{i} + 4\mathbf{k}$

5. $\lim_{h\to 0}\frac{\mathbf{r}(t+h)-\mathbf{r}(t)}{h} = \left\langle -\frac{1}{t^2},\ \cos t,\ 0\right\rangle$

7. $\frac{d\mathbf{r}}{dt} = \left\langle 1,\ 2t,\ 3t^2\right\rangle$

9. $\frac{d\mathbf{r}}{ds} = \left\langle 3e^{3s},\ -e^{-s},\ 4s^3\right\rangle$

11. $\mathbf{c}'(t) = -t^{-2}\mathbf{i} - 2e^{2t}\mathbf{k}$

13. $\mathbf{r}'(t) = \left\langle 1,\ 2t,\ 3t^2\right\rangle, \mathbf{r}''(t) = \langle 0,\ 2,\ 6t\rangle$

15.

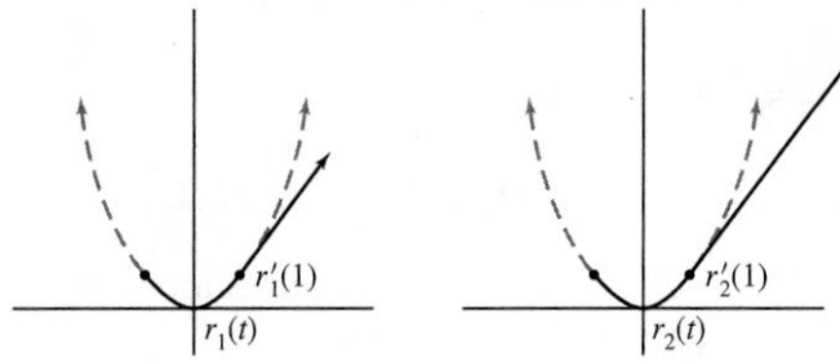

17. $\frac{d}{dt}(\mathbf{r}_1(t)\cdot\mathbf{r}_2(t)) =$
$2t^3e^{2t} + 3t^2e^{3t} + 2te^{3t} + 3t^2e^{2t} + te^t + e^t$

19. $\frac{d}{dt}(\mathbf{r}_1(t)\times\mathbf{r}_2(t)) =$
$\left\langle \begin{array}{l} 3t^2e^t - 2te^{2t} - e^{2t} + t^3e^t,\ e^{3t} + 3te^{3t} - t^2e^t - 2te^t, \\ 2te^{2t} + 2t^2e^{2t} - 3t^2e^{3t} - 3t^3e^{3t} \end{array}\right\rangle$

21. $\frac{d}{dt}(\mathbf{r}_1(t)\cdot\mathbf{r}_2(t)) = 2t + 2e^t + 2te^t$

23. $\frac{d}{dt}\mathbf{r}(g(t)) = \left\langle 2e^{2t},\ -e^t\right\rangle$

25. $\frac{d}{dt}\mathbf{r}(g(t)) = \left\langle 4e^{4t+9},\ 8e^{8t+18},\ 0\right\rangle$

27. $\frac{d}{dt}(\mathbf{r}(t)\cdot\mathbf{a}(t))|_{t=2} = 13$

29. $\ell(t) = \langle 4 - 4t,\ 16 - 32t\rangle$

31. $\ell(t) = \langle -3 - 4t,\ 10 + 5t,\ 16 + 24t\rangle$

33. $\ell(t) = \left\langle 2 - t,\ 0,\ -\frac{1}{3} + \frac{1}{2}t\right\rangle$

35. $\frac{d}{dt}(\mathbf{r}\times\mathbf{r}') = \left\langle (t^2 - 2)e^t,\ -te^t,\ 2t\right\rangle$

39. $\left\langle \frac{212}{3},\ 124\right\rangle$ **41.** $\langle 0,\ 0\rangle$

43. $\left\langle 1,\ 2,\ -\frac{\sin 3}{3}\right\rangle$ **45.** $(\ln 4)\mathbf{i} + \frac{56}{3}\mathbf{j} - \frac{496}{5}\mathbf{k}$

47. $\mathbf{r}(t) = \left\langle -t^2 + t + 3,\ 2t^2 + 1\right\rangle$

49. $\mathbf{r}(t) = \left(\frac{1}{3}t^3\right)\mathbf{i} + \left(\frac{5t^2}{2}\right)\mathbf{j} + t\mathbf{k} + \mathbf{c}$; with initial conditions,
$\mathbf{r}(t) = \frac{1}{3}t^3\mathbf{i} + \left(\frac{5t^2}{2} + 1\right)\mathbf{j} + (t+2)\mathbf{k}$

51. $\mathbf{r}(t) = (8t^2)\mathbf{k} + \mathbf{c}_1 t + \mathbf{c}_2$; with initial conditions,
$\mathbf{r}(t) = \mathbf{i} + t\mathbf{j} + (8t^2)\mathbf{k}$

53. $\mathbf{r}(t) = \left\langle 0,\ t^2,\ 0\right\rangle + \mathbf{c}_1 t + \mathbf{c}_2$; with initial conditions,
$\mathbf{r}(t) = \left\langle 1,\ t^2 - 6t + 10,\ t - 3\right\rangle$

55. $\mathbf{r}(3) = \left\langle \frac{45}{4},\ 5\right\rangle$

57. Only at time $t = 3$ can the pilot hit a target located at the origin.

59. $\mathbf{r}(t) = (t-1)\mathbf{v} + \mathbf{w}$ **61.** $\mathbf{r}(t) = e^{2t}\mathbf{c}$

Section 14.3 Preliminary Questions

1. $2\mathbf{r}' = \langle 50,\ -70,\ 20\rangle$, $-\mathbf{r}' = \langle -25,\ 35,\ -10\rangle$

2. Statement (b) is true.

3. **(a)** $L'(2) = 4$

(b) $L(t)$ is the distance along the path traveled, which is usually different from the distance from the origin.

4. 6

Section 14.3 Exercises

1. $L = 3\sqrt{61}$ **3.** $L = 15 + \ln 4$

5. $L = \pi\sqrt{4\pi^2 + 10} + 5\ln \frac{2\pi+\sqrt{4\pi^2+10}}{\sqrt{10}} \approx 29.3$

7. $s(t) = \frac{1}{27}\left((20 + 9t^2)^{3/2} - 20^{3/2}\right)$

9. $v(4) \approx 4.58$ **11.** $v\left(\frac{\pi}{2}\right) = 5$ **13.** $\mathbf{r}' = \left\langle \frac{20}{\sqrt{17}},\ \frac{-5}{\sqrt{17}}\right\rangle$

15. **(c)** $L_1 \approx 132.0, L_2 \approx 125.7$; the first spring uses more wire.

17. **(a)** $t = \pi$

19. **(a)** $s(t) = \sqrt{29}t$ **(b)** $t = \phi(s) = \frac{s}{\sqrt{29}}$

21. $\left\langle 1 + \frac{3s}{\sqrt{50}},\ 2 + \frac{4s}{\sqrt{50}},\ 3 + \frac{5s}{\sqrt{50}}\right\rangle$

23. $\mathbf{r}_1(s) = \langle 2 + 4\cos(2s),\ ,10,\ -3 + 4\sin(2s)\rangle$

25. $\mathbf{r}_1(s) = \left\langle \frac{1}{9}(27s+8)^{2/3} - \frac{4}{9},\ \pm\frac{1}{27}\left((27s+8)^{2/3} - 4\right)^{3/2}\right\rangle$

27. $\left\langle \frac{s}{\sqrt{1+m^2}},\ \frac{sm}{\sqrt{1+m^2}}\right\rangle$

29. **(a)** $\sqrt{17}e^t$ **(b)** $\frac{s}{\sqrt{17}}\left\langle \cos\left(4\ln\frac{s}{\sqrt{17}}\right),\ \sin\left(4\ln\frac{s}{\sqrt{17}}\right)\right\rangle$

31. $L = \int_{-\infty}^{\infty}\|\mathbf{r}'(t)\| = 2\int_{-\infty}^{\infty}\frac{dt}{1+t^2} = 2\pi$

Section 14.4 Preliminary Questions

1. $\left\langle \frac{2}{3},\ \frac{1}{3},\ -\frac{2}{3}\right\rangle$ **2.** $\frac{1}{4}$

3. The curvature of a circle of radius 2 **4.** Zero curvature

5. $\kappa = \sqrt{14}$ **6.** 4 **7.** $\frac{1}{9}$

Section 14.4 Exercises

1. $\|\mathbf{r}'(t)\| = \sqrt{64t^2 + 81}$, $\mathbf{T}(t) = \frac{1}{\sqrt{64t^2+81}}\langle 8t,\ 9\rangle$,

$\mathbf{T}(1) = \left\langle \frac{8}{\sqrt{145}},\ \frac{9}{\sqrt{145}}\right\rangle$

3. $\|\mathbf{r}'(t)\| = \sqrt{122}$, $\mathbf{T}(t) = \left\langle \frac{4}{\sqrt{122}},\ -\frac{5}{\sqrt{122}},\ \frac{9}{\sqrt{122}}\right\rangle$, $\mathbf{T}(1) = \mathbf{T}(t)$

5. $\|\mathbf{r}'(t)\| = \sqrt{\pi^2 + 1}$,

$\mathbf{T}(t) = \frac{1}{\sqrt{\pi^2+1}}\langle -\pi\sin\pi t,\ \pi\cos\pi t,\ 1\rangle$,

$\mathbf{T}(1) = \left\langle 0,\ -\frac{\pi}{\sqrt{\pi^2+1}},\ \frac{1}{\sqrt{\pi^2+1}}\right\rangle$

7. $\kappa(t) = \frac{e^t}{(1+e^{2t})^{3/2}}$ **9.** $\kappa(t) = 0$ **11.** $\kappa = \frac{2\sqrt{74}}{27}$

13. $\kappa = \frac{\sqrt{\pi^2+5}}{(\pi^2+1)^{3/2}} \approx 0.108$ **15.** $\kappa(3) \approx 0.0025$

17. $\kappa(2) \approx 0.0015$ **19.** $\kappa\left(\frac{\pi}{3}\right) \approx 4.54$, $\kappa\left(\frac{\pi}{2}\right) = 0.2$

23. $\alpha = \pm\sqrt{2}$ **29.** $\kappa(2) \approx 0.012$ **31.** $\kappa(\pi) \approx 1.11$

35. $\kappa(t) = t^2$

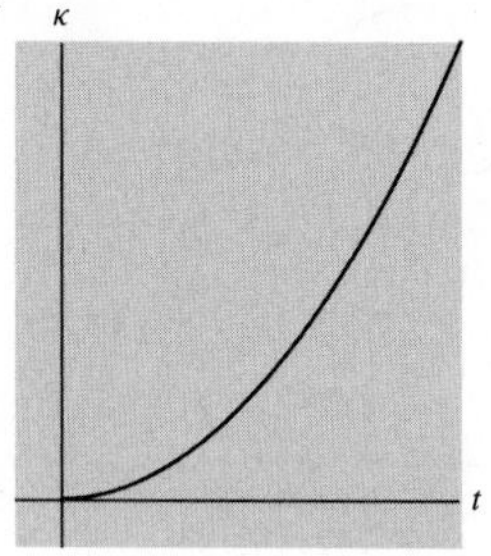

37. $\mathbf{N}(t) = \langle 0,\ -\sin 2t,\ -\cos 2t\rangle$

39. $\mathbf{T}'\left(\frac{\pi}{4}\right) = \left\langle -\frac{\sqrt{2}}{3\sqrt{3}},\ -\frac{2}{3\sqrt{3}}\right\rangle$, $\mathbf{T}'\left(\frac{3\pi}{4}\right) = \left\langle \frac{\sqrt{2}}{3\sqrt{3}},\ \frac{2}{3\sqrt{3}}\right\rangle$

41. $\mathbf{N}\left(\pi^{1/3}\right) = \left\langle \frac{1}{2},\ -\frac{\sqrt{3}}{2}\right\rangle$ **43.** $\mathbf{N}(1) = \frac{1}{\sqrt{13}}\langle -3,\ 2\rangle$

45. $\mathbf{N}(1) = \frac{1}{\sqrt{2}}\langle 0,\ 1,\ -1\rangle$ **47.** $\mathbf{N}(0) = \frac{1}{6}\left\langle -\sqrt{6},\ 2\sqrt{6},\ -\sqrt{6}\right\rangle$

51. $\langle \cos t,\ \sin t\rangle$, that is, the unit circle itself.

53. $\mathbf{c}(t) = \left\langle -4,\ -\frac{7}{2}\right\rangle + \frac{5^{3/2}}{2}\langle \cos t,\ \sin t\rangle$

55. $\mathbf{c}(t) = \langle \pi,\ -2\rangle + 4\langle \cos t,\ \sin t\rangle$

57. $\mathbf{c}(t) = \left\langle -1 - 2\cos t,\ \frac{2\sin t}{\sqrt{2}},\ \frac{2\sin t}{\sqrt{2}}\right\rangle$

65. $\kappa(\theta) = 1$ **67.** $\kappa(\theta) = \frac{1}{\sqrt{2}}e^{-\theta}$

Section 14.5 Preliminary Questions

1. No, since the particle may change its direction. **2.** $\mathbf{a}(t)$

3. Statement (a), their velocity vectors point in the same direction.

4. The velocity vector always points in the direction of motion. Since the vector $\mathbf{N}(t)$ is orthogonal to the direction of motion, the vectors $\mathbf{a}(t)$ and $\mathbf{v}(t)$ are orthogonal.

5. Description (b), parallel **6.** $\|\mathbf{a}(t)\| = 8\ \text{cm/s}^2$ **7.** $a_{\mathbf{N}}$

Section 14.5 Exercises

1. $h = -0.2$: $\langle -0.085,\ 1.91,\ 2.635\rangle$
$h = -0.1$: $\langle -0.19,\ 2.07,\ 2.97\rangle$
$h = 0.1$: $\langle -0.41,\ 2.37,\ 4.08\rangle$
$h = 0.2$: $\langle -0.525,\ 2.505,\ 5.075\rangle$
$\mathbf{v}(1) \approx \langle -0.3,\ 2.2,\ 3.5\rangle$, $v(1) \approx 4.1$

3. $\mathbf{v}(1) = \langle 3,\ -1,\ 8\rangle$, $\mathbf{a}(1) = \langle 6,\ 0,\ 8\rangle$, $v(1) = \sqrt{74}$

5. $\mathbf{v}\left(\frac{\pi}{3}\right) = \left\langle \frac{1}{2},\ -\frac{\sqrt{3}}{2},\ 0\right\rangle$, $\mathbf{a}\left(\frac{\pi}{3}\right) = \left\langle -\frac{\sqrt{3}}{2},\ -\frac{1}{2},\ 9\right\rangle$, $v\left(\frac{\pi}{3}\right) = 1$

7. $\mathbf{a}(t) = -2\left\langle \cos\frac{t}{2},\ \sin\frac{t}{2}\right\rangle$; $\mathbf{a}\left(\frac{\pi}{4}\right) \approx \langle -1.85,\ -.077\rangle$

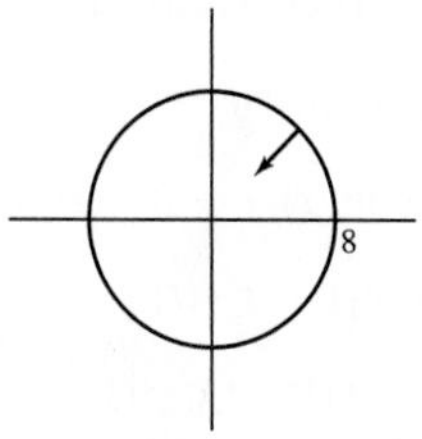

$R(t) = 8\left\langle \cos\frac{t}{2},\ \sin\frac{t}{2}\right\rangle$

9.

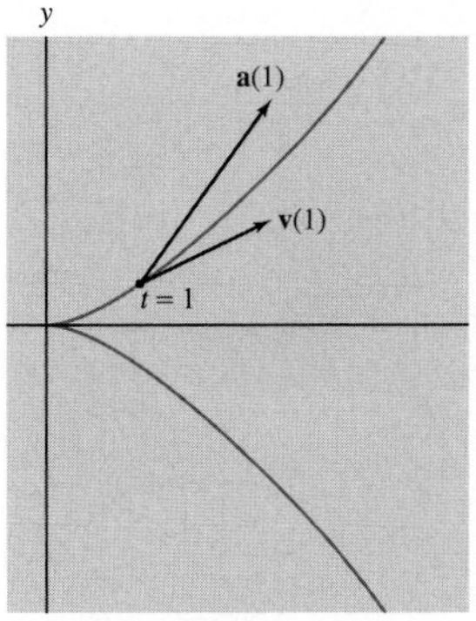

$\mathbf{r}(t) = \langle t^2, t^3\rangle$

11. $\mathbf{v}(t) = \left\langle \frac{3t^2+2}{6},\ 4t-2\right\rangle$ **13.** $\mathbf{v}(t) = \mathbf{i} + t\mathbf{k}$

15. $\mathbf{v}(t) = \left\langle \frac{t^2}{2}+3,\ 4t-2\right\rangle,\ \mathbf{r}(t) = \left\langle \frac{t^3}{6}+3t,\ 2t^2-2t\right\rangle$

17. $\mathbf{v}(t) = \mathbf{i} + \frac{t^2}{2}\mathbf{k},\ \ \mathbf{r}(t) = t\mathbf{i} + \mathbf{j} + \frac{t^3}{6}\mathbf{k}$

19. $v_0 = \sqrt{5292} \approx 72.746$ m/s **23.** $H = 355$ m

25. $\mathbf{r}(10) = \langle 45, \ -20\rangle$

27. **(a)** At its original position **(b)** No

29. The speed is decreasing.

31. $a_{\mathbf{T}} = 0,\ \ a_{\mathbf{N}} = 1$ **33.** $a_{\mathbf{T}} = \frac{7}{\sqrt{6}},\ \ a_{\mathbf{N}} = \sqrt{\frac{53}{6}}$

35. $\mathbf{a}(-1) = -\frac{2}{\sqrt{10}}\mathbf{T} + \frac{6}{\sqrt{10}}\mathbf{N}$ with $\mathbf{T} = \frac{1}{\sqrt{10}}\langle 1, \ -3\rangle$ and $\mathbf{N} = \frac{1}{\sqrt{10}}\langle -3, \ -1\rangle$

37. $a_{\mathbf{T}}(4) = 4,\ \ a_{\mathbf{N}}(4) = 1$, so $\mathbf{a} = 4\mathbf{T} + \mathbf{N}$, with $\mathbf{T} = \left\langle \frac{1}{9},\ \frac{4}{9},\ \frac{8}{9}\right\rangle$ and $\mathbf{N} = \left\langle -\frac{4}{9},\ -\frac{7}{9},\ \frac{4}{9}\right\rangle$

39. $\mathbf{a}(0) = \sqrt{3}\mathbf{T} + \sqrt{2}\mathbf{N}$, with $\mathbf{T} = \frac{1}{\sqrt{3}}\langle 1,\ 1,\ 1\rangle$ and $\mathbf{N} = \frac{1}{\sqrt{2}}\langle -1,\ 0,\ 1\rangle$

41. $\mathbf{a}\left(\frac{\pi}{2}\right) = -\frac{\pi}{2\sqrt{3}}\mathbf{T} + \frac{\pi}{\sqrt{6}}\mathbf{N}$, with $\mathbf{T} = \frac{1}{\sqrt{3}}\langle 1,\ -1,\ 1\rangle$ and $\mathbf{N} = \frac{1}{\sqrt{6}}\langle 1,\ -1,\ -2\rangle$

43. $a_{\mathbf{T}} = 0,\ \ a_{\mathbf{N}} = 0.25\ \text{cm/s}^2$

45. The tangential acceleration is $\frac{50}{\sqrt{2}} \approx 35.36\ \text{m/min}^2$, $v = \sqrt{35.36(30)} \approx 32.56$ m/min

47. $\|\mathbf{a}\| = 1.157 \times 10^5\ \text{km/h}^2$

49. $\mathbf{a} = \left\langle -\frac{1}{6},\ -1,\ \frac{1}{6}\right\rangle$

51. (A) slowing down, (B) speeding up, (C) slowing down 57. After 139.91 s the car will begin to skid.

59. $R \approx 105$ m

Section 14.6 Preliminary Questions

1. $\frac{dA}{dt} = \frac{1}{2}\|\mathbf{J}\|$

3. The period is increased eightfold.

Section 14.6 Exercises

1. The data supports Kepler's prediction; $T \approx \sqrt{a^3 \cdot 3 \cdot 10^{-4}} \approx 11.9$ years **3.** $M \approx 1.897 \times 10^{27}$ kg

5. $M \approx 2.6225 \times 10^{41}$ kg **11.** $\{(x,\ y,\ z)\ :\ 2x - y = 0\}$

Chapter 14 Review

1. (a) $-1 < t < 0$ or $0 < t \le 1$ (b) $0 < t \le 2$

3. $\mathbf{r}(t) = \left\langle t^2,\ t,\ \sqrt[3]{3-t^4}\right\rangle,\ \ -\infty < t < \infty$

5. $\mathbf{r}'(t) = \left\langle -1,\ -2t^{-3},\ \frac{1}{t}\right\rangle$ **7.** $\mathbf{r}'(0) = \langle 2,\ 0,\ 6\rangle$

9. $\frac{d}{dt}e^t\left\langle 1,\ t,\ t^2\right\rangle = e^t\left\langle 1,\ 1+t,\ 2t+t^2\right\rangle$

11. $\frac{d}{dt}\left(6\mathbf{r}_1(t) - 4\mathbf{r}_2(t)\right)|_{t=3} = \langle 0,\ -8,\ -10\rangle$

13. $\frac{d}{dt}\left(\mathbf{r}_1(t)\cdot\mathbf{r}_2(t)\right)|_{t=3} = 2$

15. $\int_0^3\left\langle 4t+3,\ t^2,\ -4t^3\right\rangle\,dt = \langle 27,\ 9,\ -81\rangle$

17. $\left(3,\ 3,\ \frac{16}{3}\right)$ **19.** $\mathbf{r}(t) = \left\langle 2t^2 - \frac{8}{3}t^3 + t,\ t^4 - \frac{1}{6}t^3 + 1\right\rangle$

21. $L = 2\sqrt{13}$ **23.** $\left\langle 5\cos\frac{2\pi s}{5\sqrt{1+4\pi^2}},\ 5\sin\frac{2\pi s}{5\sqrt{1+4\pi^2}},\ \frac{s}{\sqrt{1+4\pi^2}}\right\rangle$

25. $v_0 \approx 67.279$ m/s **27.** $(0,\ -1,\ -2)$

29. $\mathbf{T}(\pi) = \left\langle \frac{-1}{\sqrt{2}},\ \frac{1}{\sqrt{2}},\ 0\right\rangle$ **31.** $\kappa(1) = \frac{1}{2^{3/2}}$

33. $\mathbf{a} = \frac{1}{\sqrt{2}}\mathbf{T} + 4\mathbf{N}$, where $\mathbf{T} = \langle -1,\ 0\rangle$ and $\mathbf{N} = \langle 0,\ -1\rangle$

35. $\kappa = \frac{13}{16}$ **37.** $\mathbf{c}(t) = \left\langle -\frac{9}{2},\ 36\right\rangle + \frac{17^{3/2}}{2}\langle \cos t,\ \sin t\rangle$

Chapter 15

Section 15.1 Preliminary Questions

1. Same shape, but located in parallel planes

2. The parabola $z = x^2$ in the xz-plane **3.** Not possible

4. The vertical lines $x = c$ with distance of 1 unit between adjacent lines

5. In the contour map of $g(x,\ y) = 2x$, the distance between two adjacent vertical lines is $\frac{1}{2}$.

Section 15.1 Exercises

1. $f(2,\ 2) = 18,\ f(-1,\ 4) = -5$

3. $h(3,\ 8,\ 2) = 6; h(3,\ -2,\ -6) = -\frac{1}{6}$

5. The domain is the entire xy-plane.

7.

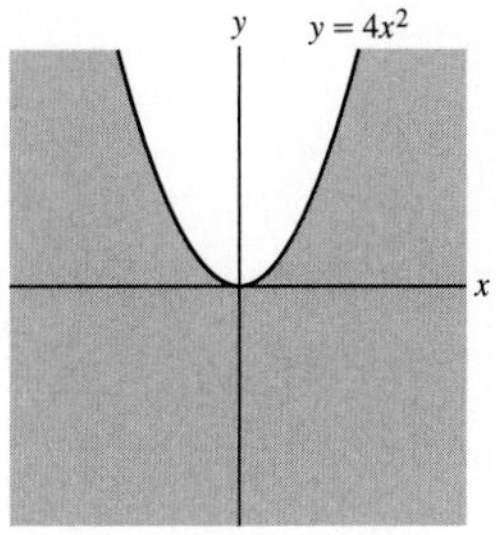

9. $\mathcal{D} = \left\{(y, z) : z \neq -y^2\right\}$

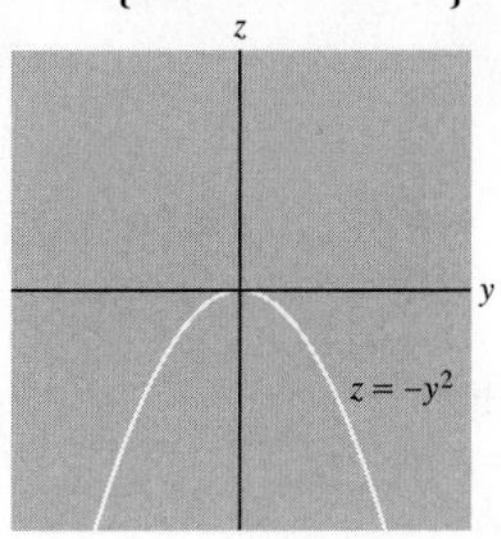

$z + y^2 \neq 0$

11.

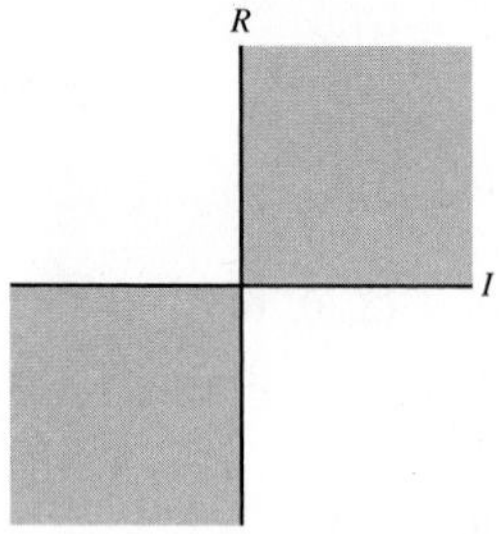

$IR \geq 0$

13. Domain: entire (x, y, z)−space; range: entire real line

15. Domain: $\{(r, s, t) : |rst| \leq 4\}$; range: $\{w : 0 \leq w \leq 4\}$

17. $f \leftrightarrow$ (B), $g \leftrightarrow$ (A)

19. **(a)** D **(b)** C **(c)** E **(d)** B **(e)** A **(f)** F

21.

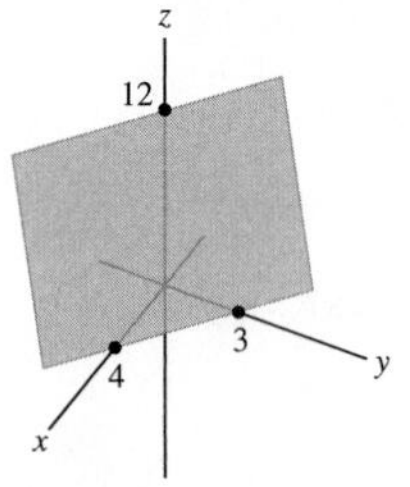

Horizontal trace: $3x + 4y = 12 - c$ in the plane $z = c$

Vertical trace: $z = (12 - 3a) - 4y$ and $z = -3x + (12 - 4a)$ in the planes $x = a$, and $y = a$, respectively

23.

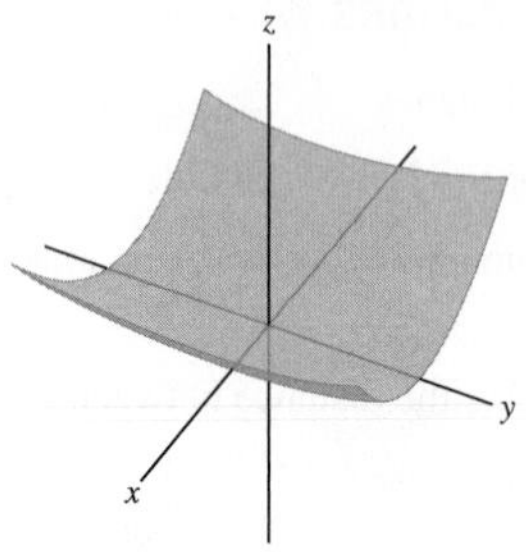

The horizontal traces are ellipses for $c > 0$.

The vertical trace in the plane $x = a$ is the parabola $z = a^2 + 4y^2$.

The vertical trace in the plane $y = a$ is the parabola $z = x^2 + 4a^2$.

25.

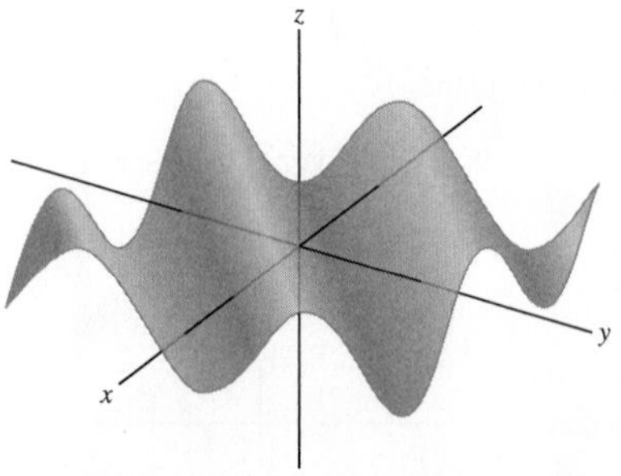

The horizontal traces in the plane $z = c$, $|c| \leq 1$, are the lines $x - y = \sin^{-1} c + 2k\pi$ and $x - y = \pi - \sin^{-1} c + 2k\pi$, for integer k

The vertical trace in the plane $x = a$ is $z = \sin(a - y)$.

The vertical trace in the plane $y = a$ is $z = \sin(x - a)$.

27. $m = 1 : m = 2 :$

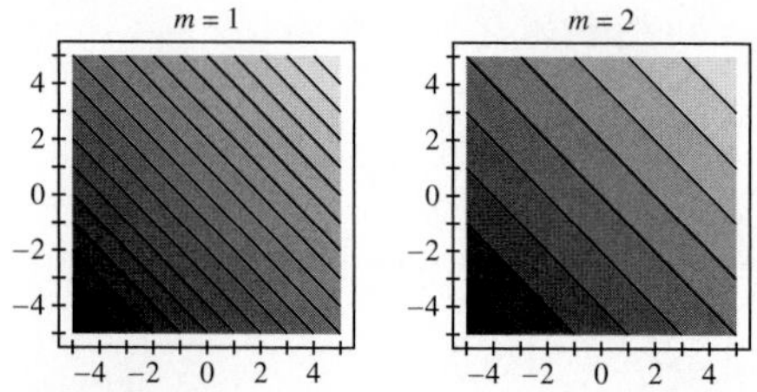

29.

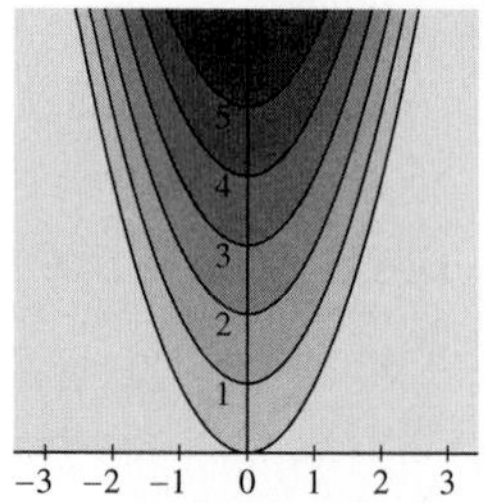

31.

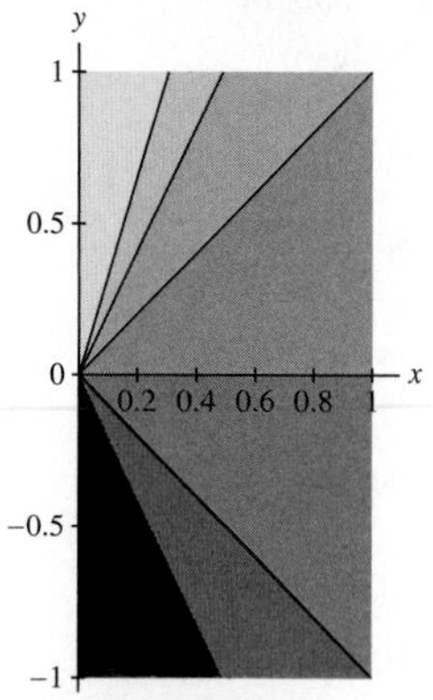

33.

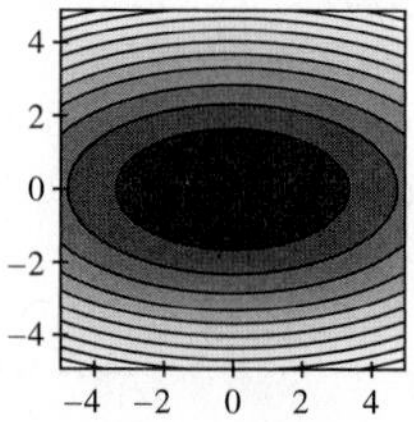

35.

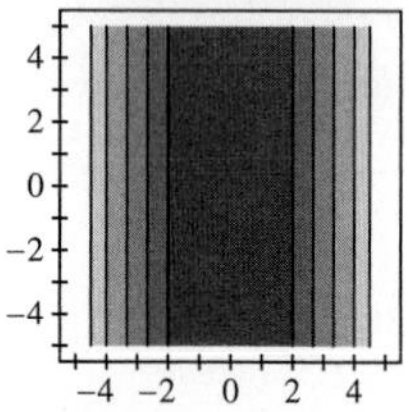

37. $m = 6 :\ f(x, y) = 2x + 6y + 6$

$m = 3 :\ f(x, y) = x + 3y + 3$

39. **(a)** Only at (A) **(b)** Only at (C) **(c)** West

41. Average ROC from B to $C = 0.000625\ \text{kg/m}^3 \cdot \text{ppt}$

43. At point A

45. Average ROC from A to $B \approx 0.0737$, average ROC from A to $C \approx 0.0457$

47.

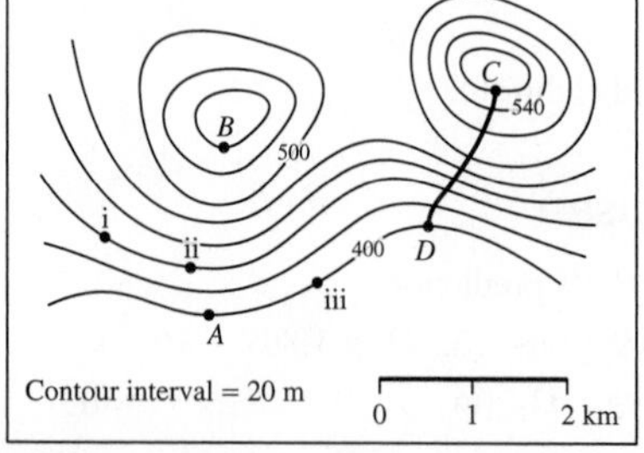

49. $f(r, \theta) = \cos\theta$; the level curves are $\theta = \pm\cos^{-1}(c)$for $|c| < 1, \quad c \neq 0$;
the $y-$axis for $c = 0$;
the positive $x-$axis for $c = 1$;
the negative $x-$axis for $c = -1$.

Section 15.2 Preliminary Questions

1. $D^*(p, r)$ consists of all points in $D(p, r)$ other than p itself.
2. $f(2, 3) = 27$
3. All three statements are true
4. $\lim_{(x, y)\to(0, 0)} f(x, y)$ does not exist.

Section 15.2 Exercises

1. $\lim_{(x, y)\to(1, 2)} (x^2 + y) = 3$
3. $\lim_{(x, y)\to(2, -1)} (xy - 3x^2y^3) = 10$
5. $\lim_{(x, y)\to(\frac{\pi}{4}, 0)} \tan x \cos y = 1$
7. $\lim_{(x, y)\to(1, 1)} \frac{e^{x^2} - e^{-y^2}}{x+y} = \frac{1}{2}(e - e^{-1})$
9. $\lim_{(x, y)\to(2, 5)} (g(x, y) - 2f(x, y)) = 1$
11. $\lim_{(x, y)\to(2, 5)} e^{f(x, y)^2 - g(x, y)} = e^2$
13. No; the limit along the x-axis and the limit along the y-axis are different.
17. $\lim_{(x, y)\to(4, 0)} (x^2 - 16)\cos\left(\frac{1}{(x-4)^2 + y^2}\right) = 0$
19. $\lim_{(z, w)\to(-2, 1)} \frac{z^4\cos(\pi w)}{e^{z+w}} = -16e$
21. $\lim_{(x, y)\to(4, 2)} \frac{y-2}{\sqrt{x^2-4}} = 0$
23. $\lim_{(x, y)\to(3, 4)} \frac{1}{\sqrt{x^2+y^2}} = \frac{1}{5}$
25. $\lim_{(x, y)\to(1, -3)} e^{x-y}\ln(x - y) = e^4\ln(4)$
27. $\lim_{(x, y)\to(-3, -2)} (x^2y^3 + 4xy) = -48$
29. $\lim_{(x, y)\to(0, 0)} \tan(x^2 + y^2)\tan^{-1}\left(\frac{1}{x^2+y^2}\right) = 0$
31. $\lim_{(x, y)\to(0, 0)} \frac{x^2+y^2}{\sqrt{x^2+y^2+1}-1} = 2$
35. $\lim_{(x, y)\to Q} g(x, y) = 4$
37. Yes
41. **(b)** $f(10^{-1}, 10^{-2}) = \frac{1}{2}$, $f(10^{-5}, 10^{-10}) = \frac{1}{2}$, $f(10^{-20}, 10^{-40}) = \frac{1}{2}$

Section 15.3 Preliminary Questions

1. $\frac{\partial}{\partial x}(x^2y^2) = 2xy^2$
2. In this case, the Constant Multiple Rule can be used. In the second part, since y appears in both the numerator and the denominator, the Quotient Rule is preferred.
3. (a), (c) **4.** $f_x = 0$ **5.** (a), (d)

Section 15.3 Exercises

3. $\frac{\partial}{\partial y}\frac{y}{z+y} = \frac{x}{(x+y)^2}$ **5.** $f_z(2, 3, 1) = 6$
7. $m = 10$ **9.** $f_x(A) \approx 8$, $f_y(A) \approx -16.7$ **11.** NW
13. $\frac{\partial}{\partial x}(x^2 + y^2) = 2x$, $\frac{\partial}{\partial y}(x^2 + y^2) = 2y$
15. $\frac{\partial}{\partial x}(x^4y + xy^{-2}) = 4x^3y + y^{-2}$, $\frac{\partial}{\partial y}(x^4y + xy^{-2}) = x^4 - 2xy^{-3}$
17. $\frac{\partial}{\partial x}\left(\frac{x}{y}\right) = \frac{1}{y}$, $\frac{\partial}{\partial y}\left(\frac{x}{y}\right) = \frac{-x}{y^2}$
19. $\frac{\partial}{\partial x}\left(\sqrt{9 - x^2 - y^2}\right) = \frac{-x}{\sqrt{9-x^2-y^2}}$, $\frac{\partial}{\partial y}\left(\sqrt{9 - x^2 - y^2}\right) = \frac{-y}{\sqrt{9-x^2-y^2}}$
21. $\frac{\partial}{\partial x}(\sin x \sin y) = \sin y \cos x$, $\frac{\partial}{\partial y}(\sin x \sin y) = \sin x \cos y$
23. $\frac{\partial}{\partial x}\left(\tan\frac{x}{y}\right) = \frac{1}{y\cos^2\left(\frac{x}{y}\right)}$, $\frac{\partial}{\partial y}\left(\tan\frac{x}{y}\right) = \frac{-x}{y^2\cos^2\left(\frac{x}{y}\right)}$
25. $\frac{\partial}{\partial x}\ln(x^2 + y^2) = \frac{2x}{x^2+y^2}$, $\frac{\partial}{\partial y}\ln(x^2 + y^2) = \frac{2y}{x^2+y^2}$
27. $\frac{\partial}{\partial r}e^{r+s} = e^{r+s}$, $\frac{\partial}{\partial s}e^{r+s} = e^{r+s}$
29. $\frac{\partial}{\partial x}e^{xy} = ye^{xy}$, $\frac{\partial}{\partial y}e^{xy} = xe^{xy}$
31. $\frac{\partial z}{\partial y} = -2xe^{-x^2-y^2}$, $\frac{\partial z}{\partial y} = -2ye^{-x^2-y^2}$
33. $\frac{\partial U}{\partial t} = -e^{-rt}$, $\frac{\partial U}{\partial r} = \frac{-e^{-rt}(rt+1)}{r^2}$
35. $\frac{\partial}{\partial x}\sinh(x^2y) = 2xy\cosh(x^2y)$, $\frac{\partial}{\partial y}\sinh(x^2y) = x^2\cosh(x^2y)$
37. $\frac{\partial w}{\partial x} = y^2z^3$, $\frac{\partial w}{\partial y} = 2xz^3y$, $\frac{\partial w}{\partial z} = 3xy^2z^2$
39. $\frac{\partial Q}{\partial L} = \frac{M - Lt}{M^2}e^{-Lt/M}$, $\frac{\partial Q}{\partial M} = \frac{L(Lt - M)}{M^3}e^{-Lt/M}$, $\frac{\partial Q}{\partial t} = -\frac{L^2}{M^2}e^{-Lt/M}$
41. $f_x(1, 2) = -164$ **43.** $g_u(1, 2) = \ln 3 + \frac{1}{3}$
45. $N = 2865.058$, $\Delta N \approx -217.74$
47. **(a)** $I(95, 50) \approx 73.1913$ **(b)** $\frac{\partial I}{\partial T}$; 1.66
49. A 1-cm increase in r
51. $\frac{\partial W}{\partial E} = -\frac{1}{kT}e^{-E/kT}$, $\frac{\partial W}{\partial T} = \frac{E}{kT^2}e^{-E/kT}$
55. (a), (b) **57.** $\frac{\partial^2 f}{\partial x^2} = 6y$, $\frac{\partial^2 f}{\partial y^2} = -72xy^2$
59. $h_{vv} = \frac{32u}{(u + 4v)^3}$ **61.** $f_{yy}(2, 3) = -\frac{4}{9}$
63. $f_{xyxzy} = 0$ **65.** $f_{uuv} = 2v\sin(u + v^2)$
67. $F_{rst} = 0$ **69.** $F_{uu\theta} = \cosh(uv + \theta^2) \cdot 2\theta v^2$
71. $g_{xyz} = \frac{3xyz}{(x^2+y^2+z^2)^{5/2}}$ **73.** $f(x, y) = x^2y$
77. $B = A^2$

Section 15.4 Preliminary Questions

1. $L(x, y) = f(a, b) + f_x(a, b)(x - a) + f_y(a, b)(y - b)$
2. $f(x, y) - L(x, y) = \epsilon(x, y)\sqrt{(x - a)^2 + (y - b)^2}$
3. (b) **4.** $f(2, 3, 1) \approx 8.7$ **5.** $\Delta f \approx -0.1$
6. Criterion for Differentiability

Section 15.4 Exercises

1. $z = -34 - 20x + 16y$ **3.** $z = 5x + 10y - 14$

5. $z = 8x - 2y - 13$ **7.** $z = 4r - 5s + 2$

9. $z = \left(\frac{4}{5} + \frac{12}{25}\ln 2\right) - \frac{12}{25}x + \frac{12}{25}y$ **11.** $\left(-\frac{1}{4}, \frac{1}{8}, \frac{1}{8}\right)$

13. **(a)** $f(x, y) = -16 + 4x + 12y$

(b) $f(2.01, 1.02) \approx 4.28$; $f(1.97, 1.01) \approx 4$

15. $\Delta f \approx 3.56$ **17.** $f(0.01, -0.02) \approx 0.98$

19. $L(x, y, z) = -8.66025 + 0.721688x + 0.721688y + 3.4641z$

21. 5.07 **23.** 8.44 **25.** 4.998 **27.** 3.945

29. $z = 3x - 3y + 13$ **31.** $\Delta I \approx 0.5644$

33. **(b)** $\Delta H \approx 0.022\text{m}$

35. **(b)** 6% **(c)** 1% error in r

37. **(a)** \$7.10 **(b)** \$28.85, \$57.69 **(c)** −\$74.24

39. Maximum error in V is about 8.948 m.

Section 15.5 Preliminary Questions

1. (b) $\langle 3, 4\rangle$

2. False

3. ∇f points in the direction of maximum rate of increase of f and is normal to the level curve of f.

4. **(b)** NW and **(c)** SE

5. $3\sqrt{2}$

Section 15.5 Exercises

1. (a) $\nabla f = \langle y^2, 2xy\rangle, \mathbf{c}'(t) = \langle t, 3t^2\rangle$

(b) $\frac{d}{dt}(f(\mathbf{c}(t)))\Big|_{t=1} = 4$; $\frac{d}{dt}(f(\mathbf{c}(t)))\Big|_{t=-1} = -4$

3. A: zero, B: negative, C: positive, D: zero

5. $\nabla f = -\sin(x^2 + y)\langle 2x, 1\rangle$

7. $\nabla h = \langle yz^{-3}, xz^{-3}, -3xyz^{-4}\rangle$

9. $\frac{d}{dt}(f(\mathbf{c}(t)))\Big|_{t=0} = -7$ **11.** $\frac{d}{dt}(f(\mathbf{c}(t)))\Big|_{t=0} = -3$

13. $\frac{d}{dt}(f(\mathbf{c}(t)))\Big|_{t=0} = 5\cos 1 \approx 2.702$

15. $\frac{d}{dt}(f(\mathbf{c}(t)))\Big|_{t=4} = -56$

17. $\frac{d}{dt}(f(\mathbf{c}(t)))\Big|_{t=\pi/4} = -1 + \frac{\pi}{8} \approx 1.546$

19. $\frac{d}{dt}(g(\mathbf{c}(t)))\Big|_{t=1} = 0$

21. $D_{\mathbf{u}}f(1, 2) = 8.8$ **23.** $D_{\mathbf{u}}f\left(\frac{1}{6}, 3\right) = \frac{39}{4\sqrt{2}}$

25. $D_{\mathbf{u}}f(3, 4) = \frac{7\sqrt{2}}{290}$ **27.** $D_{\mathbf{u}}f(1, 0) = \frac{6}{\sqrt{13}}$

29. $D_{\mathbf{u}}f(1, 2, 0) = -\frac{1}{\sqrt{3}}$ **31.** $D_{\mathbf{u}}f(3, 2) = \frac{-50}{\sqrt{13}}$

33. $D_{\mathbf{u}}f(P) = -\frac{e^5}{3} \approx -49.47$

35. f is increasing at P in the direction of v.

37. $D_{\mathbf{u}}f(P) = \frac{\sqrt{6}}{2}$ **39.** $\langle 6, 2, -4\rangle$

41. $\left(\frac{4}{\sqrt{17}}, \frac{9}{\sqrt{17}}, -\frac{2}{\sqrt{17}}\right)$ and $\left(-\frac{4}{\sqrt{17}}, -\frac{9}{\sqrt{17}}, \frac{2}{\sqrt{17}}\right)$

43. $9x + 10y + 5z = 33$

45. $0.5217x + 0.7826y - 1.2375z = -5.309$

47.

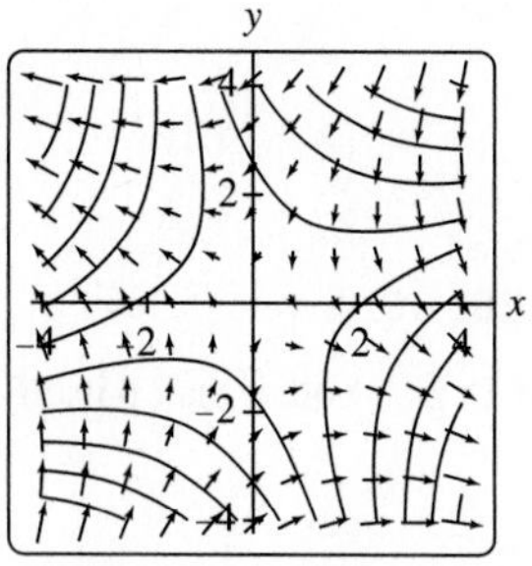

49. $f(x, y, z) = x^2 + y + 2z$

51. $f(x, y, z) = xz + y^2$ **55.** $\Delta f \approx 0.08$

57. **(a)** $\langle 34, 18, 0\rangle$

(b) $\left\langle 2 + \frac{32}{\sqrt{21}}t, 2 + \frac{16}{\sqrt{21}}t, 8 - \frac{8}{\sqrt{21}}t\right\rangle$; ≈ 4.58 s

61. $x = 1 - 4t, \quad y = 2 + 26t, \quad z = 1 - 25t$

73. $y = \sqrt{1 - \ln(\cos^2 x)}$

Section 15.6 Preliminary Questions

1. **(a)** $\frac{\partial f}{\partial x}$ and $\frac{\partial f}{\partial y}$ **(b)** u and v

2. (a) **3.** $f(u, v)|_{(r,s)=(1,1)} = e^2$ **4.** (b) **5.** (c) **6.** No

Section 15.6 Exercises

1. **(a)** $\frac{\partial f}{\partial x} = 2xy^3, \ \frac{\partial f}{\partial y} = 3x^2y^2, \ \frac{\partial f}{\partial z} = 4z^3$

(b) $\frac{\partial x}{\partial s} = 2s, \ \frac{\partial y}{\partial s} = 2t^2, \ \frac{\partial z}{\partial s} = 2st$

(c) $\frac{\partial f}{\partial s} = 7s^6t^6 + 8s^7t^4$

3. $\frac{\partial f}{\partial s} = 6rs^2, \ \frac{\partial f}{\partial r} = 2s^3 + 4r^3$

5. $\frac{\partial g}{\partial u} = -10\sin(10u - 20v), \ \frac{\partial g}{\partial v} = 20\sin(10u - 20v)$

7. $\frac{\partial F}{\partial y} = xe^{x^2+xy}$ **9.** $\frac{\partial h}{\partial t_2} = 0$

11. $\frac{\partial f}{\partial u}\Big|_{(u,v)=(-1,-1)} = 1, \ \frac{\partial f}{\partial v}\Big|_{(u,v)=(-1,-1)} = -2$

13. $\frac{\partial g}{\partial \theta}\Big|_{(r,\theta)=\left(2\sqrt{2}, \pi/4\right)} = \frac{1}{6}$ **15.** $\frac{\partial f}{\partial v}\Big|_{(u,v)=(0,1)} = 2\cos 2$

17. **(b)** $\frac{\partial f}{\partial t} = \frac{19}{2\sqrt{7}}$

23. **(a)** $F_x = z^2 + y, \ F_y = 2yz + x, \ F_z = 2xz + y^2$

(b) $\frac{\partial z}{\partial x} = -\frac{z^2 + y}{2xz + y^2}, \ \frac{\partial z}{\partial y} = -\frac{2yz + x}{2xz + y^2}$

25. $\frac{\partial z}{\partial x} = -\frac{2xy + z^2}{2xz + y^2}$ **27.** $\frac{\partial z}{\partial y} = -\frac{xe^{xy} + 1}{x\cos(xz)}$

29. $\frac{\partial w}{\partial y} = \frac{-y(w^2 + x^2)^2}{w\left((w^2 + y^2)^2 + (w^2 + x^2)^2\right)}$; at (1, 1, 1), $\frac{\partial w}{\partial y} = -\frac{1}{2}$

33. $\nabla\left(\frac{1}{r}\right) = -\frac{1}{r^3}\mathbf{r}$ **35.** (c) $\frac{\partial z}{\partial x} = \frac{x - 6}{z + 4}$

37. $\frac{\partial P}{\partial T} = -\frac{nR}{V - nb}, \ \frac{\partial V}{\partial P} = \frac{nbV^3 - V^4}{PV^3 + 2an^3b - an^2V}$

Section 15.7 Preliminary Questions

1. f has a local (and global) min at $(0,\ 0)$; g has a saddle point at $(0,\ 0)$.

2.

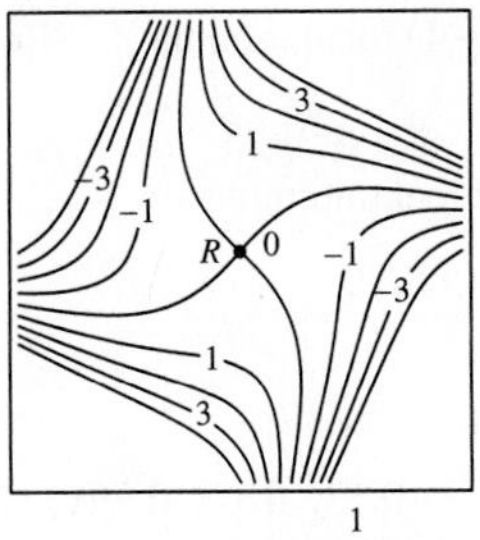

Point R is a saddle point.

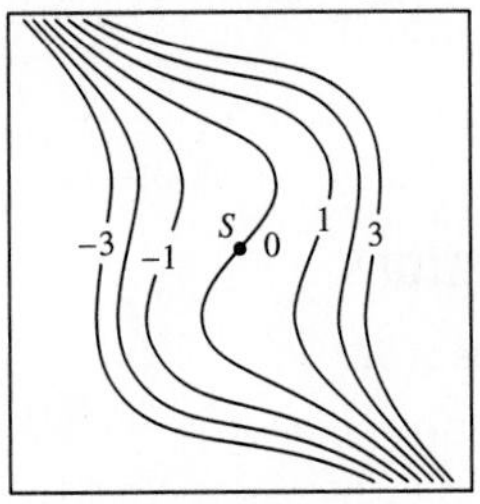

Point S is neither a local extremum nor a saddle point.

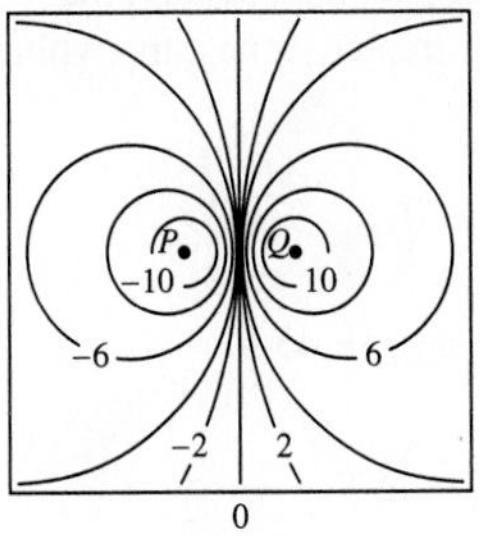

Point P is a local minimum and point Q is a local maximum.

3. Statement (a)

Section 15.7 Exercises

1. **(b)** $P_1 = (0, 0)$ is a saddle point, $P_2 = \left(2\sqrt{2}, \sqrt{2}\right)$ and $P_3 = \left(-2\sqrt{2}, -\sqrt{2}\right)$ are local minima; absolute minimum value of f is -4.

3. $(0,\ 0)$ saddle point, $\left(\frac{13}{64}, -\frac{13}{32}\right)$ and $\left(-\frac{1}{4}, \frac{1}{2}\right)$ local minima

5. **(c)** $(0,\ 0)$, $(1,\ 0)$, and $(0,\ -1)$ saddle points, $\left(\frac{1}{3}, -\frac{1}{3}\right)$ local minimum.

7. $\left(-\frac{2}{3}, -\frac{1}{3}\right)$ local minimum

9. $(-2, -1)$ local maximum, $\left(\frac{5}{3}, \frac{5}{6}\right)$ saddle point

11. $\left(0, \pm\sqrt{2}\right)$ saddle points, $\left(\frac{2}{3}, 0\right)$ local maximum, $\left(-\frac{2}{3},\ 0\right)$ local minimum

13. $(0, 0)$ saddle point, $(1, 1)$ and $(-1, -1)$ local minima

15. $(0, 0)$ saddle point, $\left(\frac{1}{\sqrt{2}}, \frac{1}{\sqrt{2}}\right)$ and $\left(-\frac{1}{\sqrt{2}},\ -\frac{1}{\sqrt{2}}\right)$ local maximum, $\left(\frac{1}{\sqrt{2}}, -\frac{1}{\sqrt{2}}\right)$ and $\left(-\frac{1}{\sqrt{2}}, \frac{1}{\sqrt{2}}\right)$ local minimum

17. Critical points are $\left(j\pi,\ k\pi + \frac{\pi}{2}\right)$, for
j,k even: saddle points
j,k odd: local maxima
j even, k odd: local minima
j odd, k even: saddle points

19. $\left(1, \frac{1}{2}\right)$ local maximum **21.** $\left(\frac{3}{2}, -\frac{1}{2}\right)$ saddle point

23. $\left(-\frac{1}{6},\ -\frac{17}{18}\right)$ local minimum

27. $x = y = 0.27788$ local minimum

29. Global maximum 2, global minimum 0

31. Global maximum 1, global minimum $\frac{1}{35}$

35. Maximum value $\frac{1}{3}$

37. Global minimum $f(0,\ 1) = -2$, global maximum $f(1,\ 0) = 1$

39. Global maximum 3, global minimum 0

41. Global minimum $f(1,\ 1) = -1$, global maximum $f(1,\ 0) = f(0,\ 1) = 1$

43. Global minimum $f(1,\ 0) = f(-1,\ 0) = -0.368$, global maximum $f(0,\ -1) = f(0,\ 1) = 1.472$

45. Maximum volume $\frac{3}{4}$

49. **(a)** No. In the box B with minimal surface area, z is smaller than $\sqrt[3]{V}$, which is the side of a cube with volume V.
(b) Width: $x = (2V)^{1/3}$; length: $y = (2V)^{1/3}$;
height: $z = \left(\frac{V}{4}\right)^{1/3}$

51. $f(x) = 1.9629x - 1.5519$

Section 15.8 Preliminary Questions

1. Statement (b)

2. f had a local maximum 2, under the constraint, at A; $f(B)$ is neither a local minimum nor a local maximum of f.

3. **(a)**

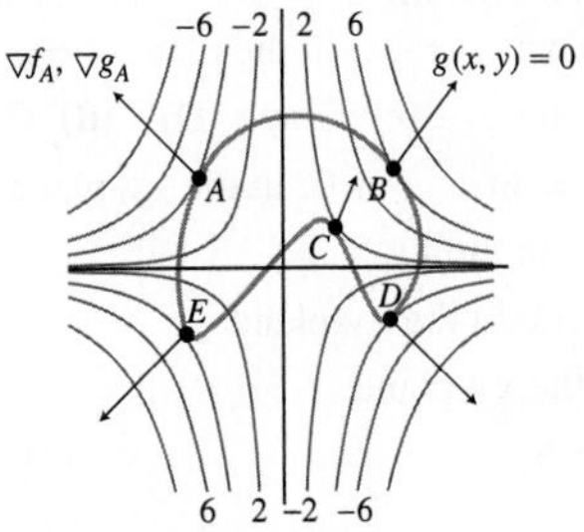

Contour plot of $f(x, y)$
(contour interval 2)

(b) Global minimum -4, global maximum 6

Section 15.8 Exercises

1. **(c)** Critical points $(-1, -2)$ and $(1, 2)$
(d) Maximum 10, minimum -10

3. Maximum $4\sqrt{2}$, minimum $-4\sqrt{2}$

5. Minimum $\frac{36}{13}$, no maximum value

7. Maximum $\frac{8}{3}$, minimum $-\frac{8}{3}$

9. Maximum $\sqrt{2}$, minimum 1

11. Maximum 3.7, minimum -3.7

13. No maximum and minimum values

15. $(-1,\ e^{-1})$

17. (a) $h = \sqrt{\frac{2}{\sqrt{3}\pi}} \approx 0.6$, $r = \sqrt{\frac{1}{\sqrt{3}\pi}} \approx 0.43$ **(b)** $\frac{h}{r} = \sqrt{2}$

(c) There is no cone of volume 1 and maximal surface area.

19. $(8, \ -2)$ **21.** $\left(\frac{48}{97}, \frac{108}{97}\right)$ **23.** $\frac{a^a b^b}{(a+b)^{a+b}}$ **25.** $\sqrt{\frac{a^a b^b}{(a+b)^{a+b}}}$

31. $r = 3, \ h = 6$ **33.** $x + y + z = 3$

39. $\left(\frac{-6}{\sqrt{105}}, \frac{-3}{\sqrt{105}}, \frac{30}{\sqrt{105}}\right)$ **41.** $(-1, \ 0, \ 2)$

43. Minimum $\frac{138}{11} \approx 12.545$, no maximum value

47. (b) $\lambda = \frac{c}{2p_1 p_2}$

Chapter 15 Review

1. (a)

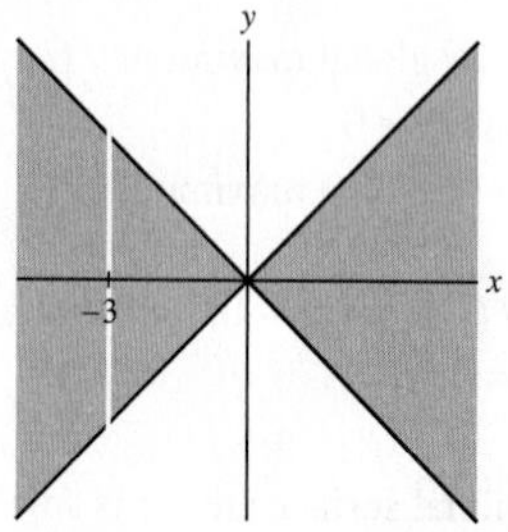

(b) $f(3, \ 1) = \frac{\sqrt{2}}{3}$, $f(-5, \ -3) = -2$ **(c)** $\left(-\frac{5}{3}, 1\right)$

3.

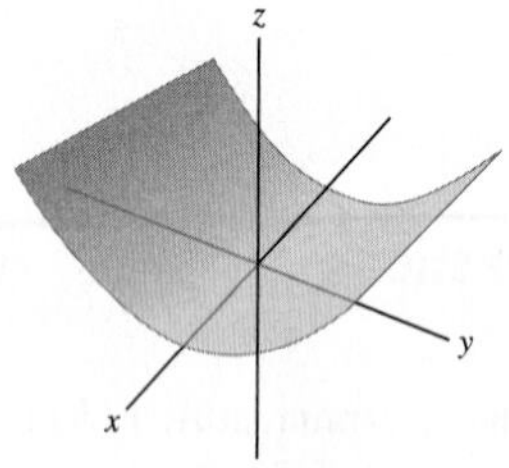

Vertical and horizontal traces: the line $z = (c^2 + 1) - y$ in the plane $x = c$, the parabola $z = x^2 - c + 1$ in the plane $y = c$.

5. (a) Graph (B) **(b)** Graph (C) **(c)** Graph (D) **(d)** Graph (A)

7. (a) Parallel lines $4x - y = \ln c, \ c > 0$, in the xy-plane

(b) Parallel lines $4x - y = e^c$ in the xy-plane

(c) Hyperbolas $3x^2 - 4y^2 = c$ in the xy-plane

(d) Parabolas $x = c - y^2$ in the xy-plane

9. $\lim\limits_{(x,y)\to(1,-3)} (xy + y^2) = 6$

11. The limit does not exist.

13. $\lim\limits_{(x,y)\to(1,-3)} (2x + y)e^{-x+y} = -e^{-4}$

17. $f_x = 2, \ f_y = 2y$

19. $f_x = e^{-x-y}(y\cos(xy) - \sin(xy))$
$f_y = e^{-x-y}(x\cos(yx) - \sin(yx))$

21. $f_{xxyz} = -\cos(x + z)$ **23.** $z = 33x + 8y - 42$

25. Estimate, 12.146; calculator value to three places, 11.996.

27. Statements (ii) and (iv) are true.

29. $\frac{d}{dt}\left(f(\mathbf{c}(t))\right)\Big|_{t=2} = 3 + 4e^4 \approx 221.4$

31. $\frac{d}{dt}\left(f(\mathbf{c}(t))\right)\Big|_{t=1} = 4e - e^{3e} \approx -3469.3$

33. $D_{\mathbf{u}}f(3, \ -1) = -\frac{54}{\sqrt{5}}$

35. $D_{\mathbf{u}}f(P) = -\frac{\sqrt{2}e}{5}$ **37.** $\left\langle\frac{1}{\sqrt{2}}, \frac{1}{\sqrt{2}}, 0\right\rangle$

41. $\frac{\partial f}{\partial s} = 3s^2 t + 4st^2 + t^3 - 2st^3 + 6s^2 t^2$
$\frac{\partial f}{\partial t} = 4s^2 t + 3st^2 + s^3 + 4s^3 t - 3s^2 t^2$

45. $\frac{\partial z}{\partial x} = -\frac{e^z - 1}{xe^z + e^y}$

47. $(0, 0)$ saddle point, $(1, 1)$ and $(-1, -1)$ local minima

49. $\left(\frac{1}{2}, \frac{1}{2}\right)$ saddle point

53. Global maximum $f(2, \ 4) = 10$, global minimum $f(-2, \ 4) = -18$

55. Maximum $\frac{26}{\sqrt{13}}$, minimum $-\frac{26}{\sqrt{13}}$

57. Maximum $\frac{12}{\sqrt{3}}$, minimum $-\frac{12}{\sqrt{3}}$

59. $f(0.8, \ 0.52, \ -0.32) = 0.88$ and $f(-0.13, \ 0.15, \ 0.99) = 3.14$

61. $r = \left(\frac{V}{2\pi}\right)^{1/3}$, $h = 2\left(\frac{V}{2\pi}\right)^{1/3}$

Chapter 16

Section 16.1 Preliminary Questions

1. $\Delta A = 1$, the number of subrectangles is 32.

2. $\iint_R f\, dA \approx S_{1,1} = 0.16$

3. $\iint_R 5\, dA = 50$

4. The signed volume between the graph $z = f(x, \ y)$ and the xy-plane. The region below the xy-plane is treated as negative volume.

5. (b) **6. (b), (c)**

Section 16.1 Exercises

1. $S_{4,3} = 13.5$ **3. (A)** $S_{3,2} = 42$, **(B)** $S_{3,2} = 43.5$

5. (A) $S_{3,2} = 60$, **(B)** $S_{3,2} = 62$

7. Two possible solutions are $S_{3,2} = \frac{77}{72}$ and $S_{3,2} = \frac{79}{72}$.

9. $\frac{225}{2}$

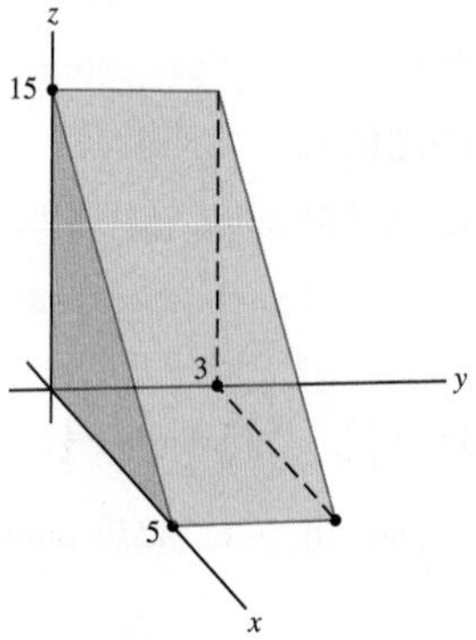

11. 0.19375 **13.** 1.0731, 1.0783, 1.0809 **15.** 0 **17.** 0 **19.** 40

21. 55 **23.** $\frac{4}{3}$ **25.** 84 **27.** 4 **29.** $\frac{1858}{15}$

31. $6\ln 6 - 2\ln 2 - 5\ln 5 \approx 1.317$ **33.** $\frac{4}{3}\left(19 - 5\sqrt{5}\right) \approx 10.426$

35. $\frac{1}{2}(\ln 3)(-2 + \ln 48) \approx 1.028$ **37.** $6\ln 3 \approx 6.592$

39. 1 **41.** $\left(e^2 - 1\right)\left(1 - \frac{\sqrt{2}}{2}\right) \approx 1.871$

43. $m = \frac{3}{4}$ **45.** $2\ln 2 - 1 \approx 0.386$

49. $\frac{e^3}{3} - \frac{1}{3} - e + 1 \approx 4.644$

Section 16.2 Preliminary Questions

1. (b), (c)

2.

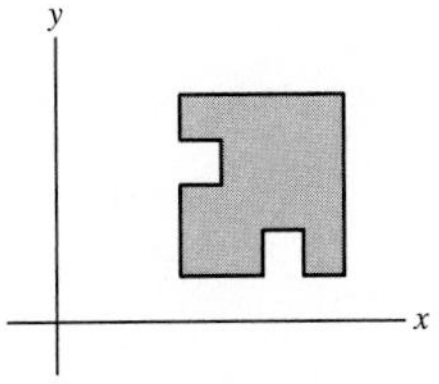

3.

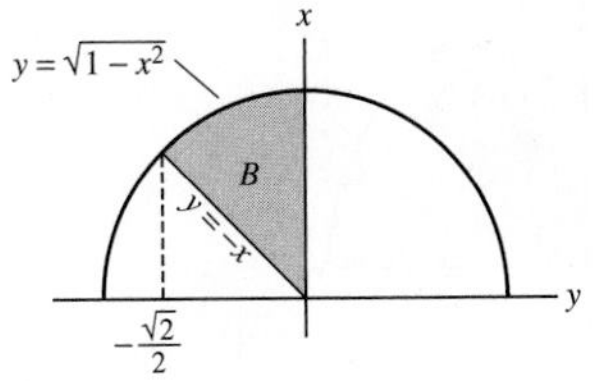

4. (b)

Section 16.2 Exercises

1. (a) Sample points •, $S_{3,4} = -3$
(b) Sample points ∘, $S_{3,4} = -4$

3. As a vertically simple region: $0 \le x \le 1,\ 0 \le y \le 1 - x^2$; as a horizontally simple region: $0 \le y \le 1,\ 0 \le x \le \sqrt{1-y}$

5. $\frac{192}{5} = 38.4$ **7.** $\frac{608}{15} \approx 40.53$ **9.** $2\frac{1}{4}$ **11.** $-\frac{3}{4} + \ln 4$

13. $\frac{16}{3} \approx 5.33$ **15.** $\frac{11}{60}$ **17.** $\frac{1754}{15} \approx 116.93$ **19.** $\frac{e-2}{2} \approx 0.359$

21. $\frac{1}{12}$ **23.** $2e^{12} - \frac{1}{2}e^9 + \frac{1}{2}e^5 \approx 321,532.2$

25.

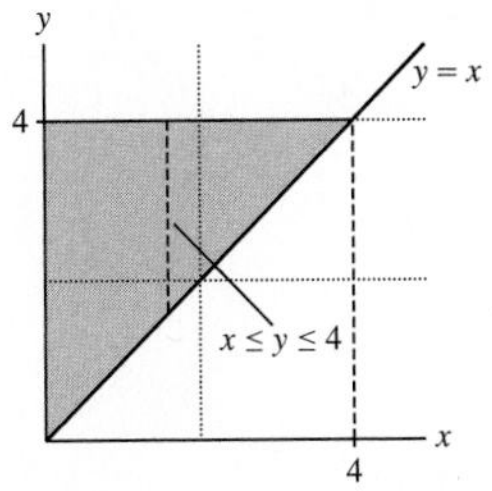

$$\int_0^4 \int_x^4 f(x, y)\, dy\, dx = \int_0^4 \int_0^y f(x, y)\, dx\, dy$$

27.

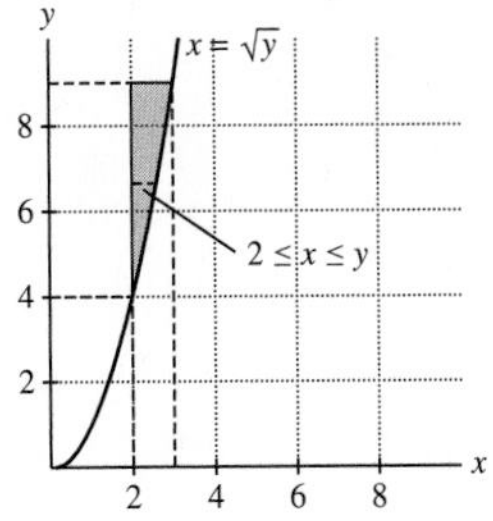

$$\int_4^9 \int_2^{\sqrt{y}} f(x, y)\, dx\, dy = \int_2^3 \int_{x^2}^9 f(x, y)\, dy\, dx$$

29.

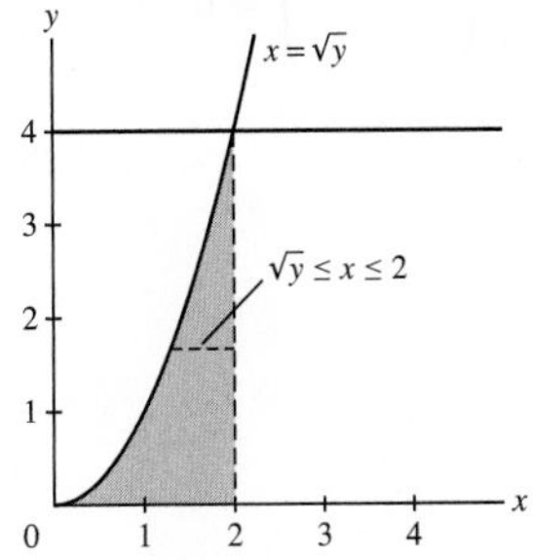

$$\int_0^2 \int_0^{x^2} \sqrt{4x^2 + 5y}\, dy\, dx = \frac{152}{15}$$

31. $\int_1^e \int_{\ln^2 y}^{\ln y} (\ln y)^{-1}\, dx\, dy = e - 2 \approx 0.718$

33.

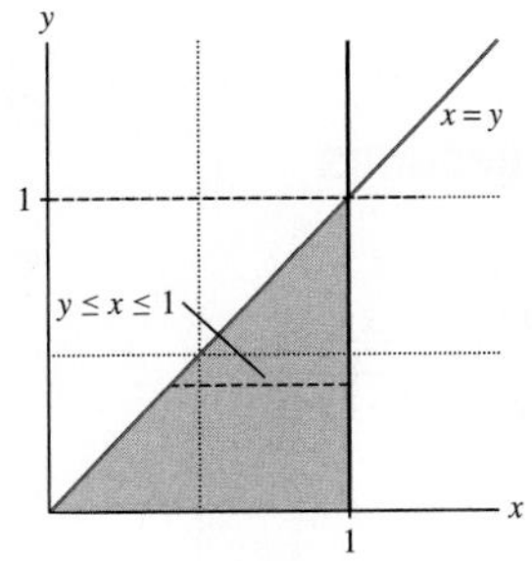

$$\int_0^1 \int_0^x \frac{\sin x}{x}\, dy\, dx = 1 - \cos 1 \approx 0.460$$

35.

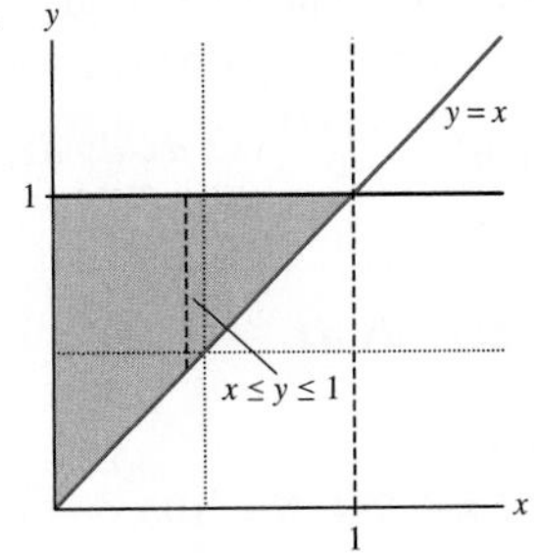

$$\int_0^1 \int_0^y x e^{y^3}\, dx\, dy = \frac{e-1}{6} \approx 0.286$$

37.

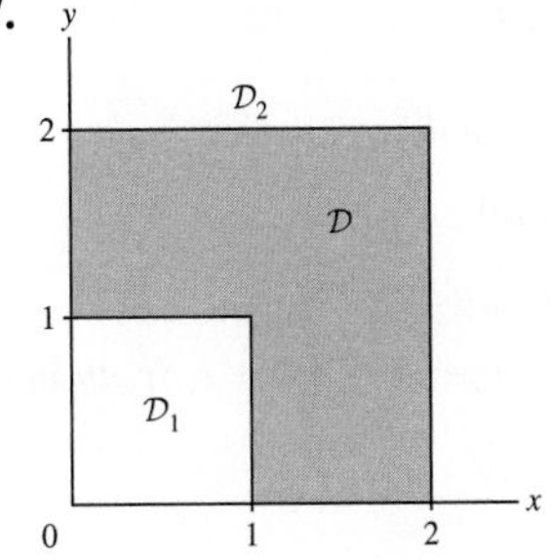

$$\iint_D e^{x+y}\, dA = e^4 - 3e^2 + 2e \approx 37.878$$

39. $\int_0^4 \int_{x/4}^{3x/4} e^{x^2}\, dy\, dx = \frac{1}{4}\left(e^{16} - 1\right)$

41. $\int_2^4 \int_{y-1}^{7-y} \frac{x}{y^2}\, dx\, dy = 6 - 6\ln 2 \approx 1.841$

43. $\iint_D \frac{\sin y}{y}\, dA = \cos 1 - \cos 2 \approx 0.956$

45. $\int_{-2}^2 \int_0^{4-x^2} (40 - 10y)\, dy\, dx = 256$

47. $\int_0^1 \int_0^1 e^{x+y}\, dx\, dy = e^2 - 2e + 1 \approx 2.952$

49. $\frac{1}{\pi} \int_0^1 \int_0^\pi y^2 \sin x\, dx\, dy = \frac{2}{3\pi}$ **51.** $\bar{f} = p$

57. One possible solution is $P = \left(\frac{2}{3},\ 2\right)$

59. $\iint_D f(x,\ y)\,dA \approx 57.01$

Section 16.3 Preliminary Questions

1. (c) **2.** (b)

3. **(a)** $D = \{(x,\ y) : 0 \le x \le 1,\ 0 \le y \le x\}$

(b) $D = \left\{(x,\ y) : 0 \le x \le 1,\ 0 \le y \le \sqrt{1-x^2}\right\}$

Section 16.3 Exercises

1. 6 **3.** $(e-1)(1-e^{-2})$ **5.** $-\frac{27}{4} = -6.75$

7. $\frac{b}{20}\left[(a+c)^5 - a^5 - c^5\right]$ **9.** $\frac{1}{6}$ **11.** $\frac{1}{16}$ **13.** $e - \frac{5}{2}$

15. $2\frac{1}{12}$ **17.** $\frac{128}{15}$ **19.** 2 **21.** $\frac{1}{12}$ **23.** $\frac{126}{5}$

25. The region bounded by the plane $y = 1$ and the paraboloid $y = 5 - x^2 - z^2$ lying over the disk $x^2 + z^2 \le 4$ in the xz-plane.

27. $\int_0^2 \int_0^{y/2} \int_0^{4-y^2} xyz\,dz\,dx\,dy$, $\int_0^4 \int_0^{\sqrt{4-z}} \int_0^{y/2} xyz\,dx\,dy\,dz$, and $\int_0^4 \int_0^{\sqrt{1-(z/4)}} \int_{2x}^{\sqrt{4-z}} xyz\,dy\,dx\,dz$

29. $\int_{-1}^{1} \int_{-\sqrt{1-x^2}}^{\sqrt{1-x^2}} \int_{\sqrt{x^2+y^2}}^{1} f(x,\ y,\ z)\,dz\,dy\,dx$

31. $\frac{16}{21}$ **33.** $\frac{1}{2\pi}$ **35.** $2e - 4 \approx 1.437$

37. $S_{N,N,N} \approx 0.561,\ 0.572,\ 0.576;\ I \approx 0.584;\ N = 100$

Section 16.4 Preliminary Questions

1. (d)

2. **(a)** $\int_{-1}^{2} \int_0^{2\pi} \int_0^{2} f(P)\,r\,dr\,d\theta\,dz$

(b) $\int_{-2}^{0} \int_0^{2\pi} \int_0^{\sqrt{4-z^2}} r\,dr\,d\theta\,dz$

3. **(a)** $\int_0^{2\pi} \int_0^{\pi} \int_0^{4} f(P)\,\rho^2 \sin\phi\,d\rho\,d\phi\,d\theta$

(b) $\int_0^{2\pi} \int_0^{\pi} \int_4^{5} f(P)\,\rho^2 \sin\phi\,d\rho\,d\phi\,d\theta$

(c) $\int_0^{2\pi} \int_{\pi/2}^{\pi} \int_0^{2} f(P)\,\rho^2 \sin\phi\,d\rho\,d\phi\,d\theta$

4. $\Delta A \approx r(\Delta r \Delta\theta)$, and the factor r appears in $dA = r\,dr\,d\theta$ in the Change of Variables formula.

Section 16.4 Exercises Questions

1.

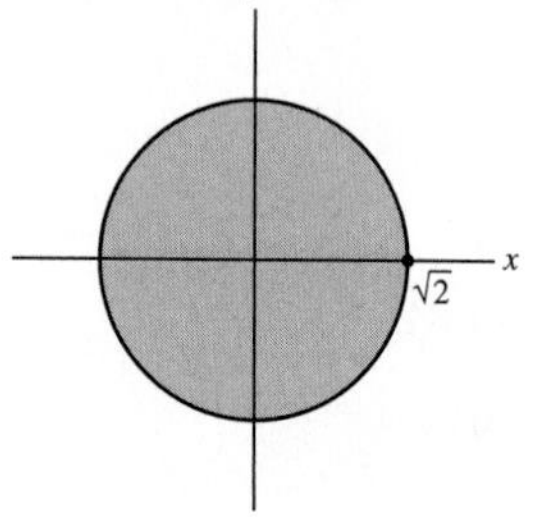

$$\iint_D \sqrt{x^2+y^2}\,dA = \frac{4\sqrt{2}\pi}{3}$$

3.

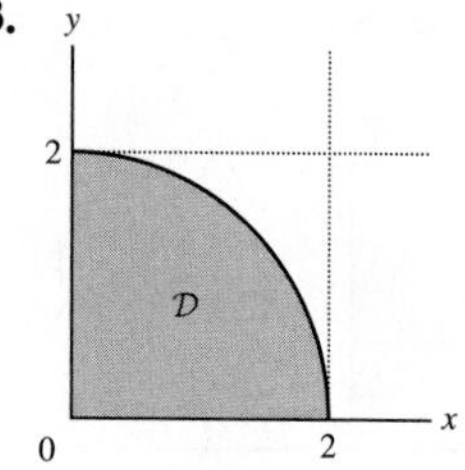

$$\iint_D xy\,dA = 2$$

5.

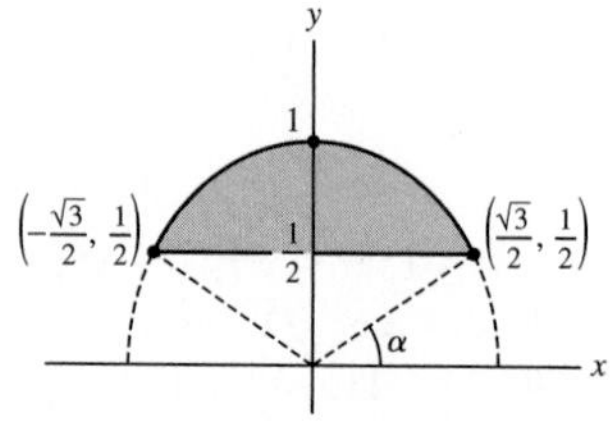

$$\iint_D y\left(x^2+y^2\right)^{-1} dA = \sqrt{3} - \frac{\pi}{3} \approx 0.685$$

7.

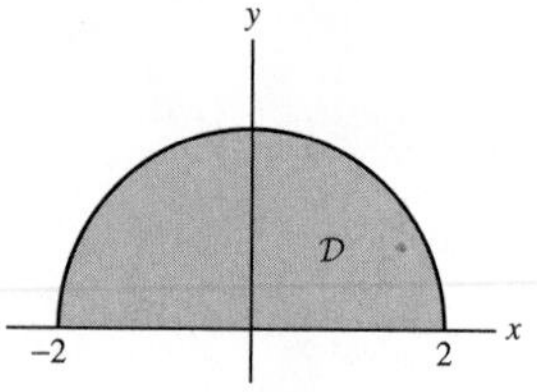

$$\int_{-2}^{2} \int_0^{\sqrt{4-x^2}} (x^2+y^2)\,dy\,dx = 4\pi$$

9.

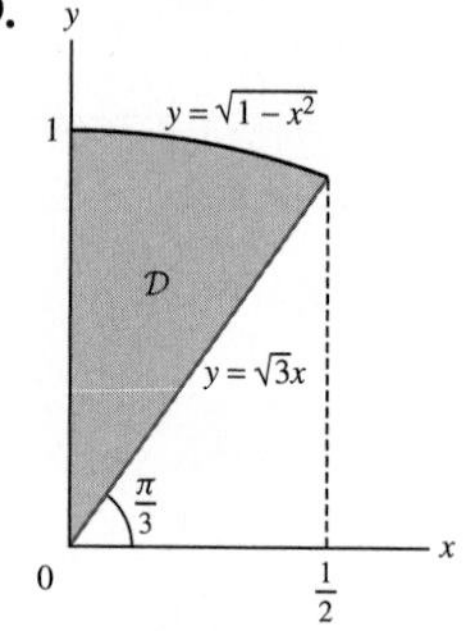

$$\int_0^{1/2} \int_{\sqrt{3}x}^{\sqrt{1-x^2}} x\,dy\,dx = \frac{1}{3}\left(1 - \frac{\sqrt{3}}{2}\right) \approx 0.045$$

11.

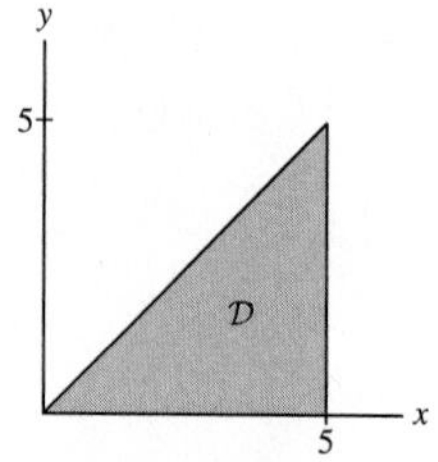

$$\int_0^5 \int_0^y x\,dx\,dy = \frac{125}{6}$$

13.

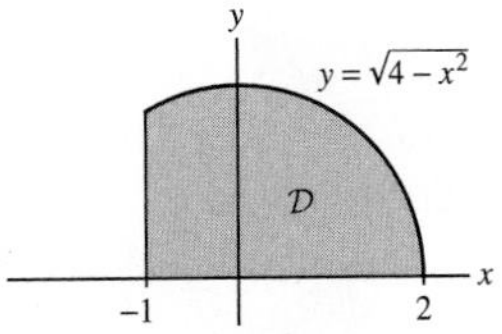

$$\int_{-1}^{2}\int_{0}^{\sqrt{4-x^2}}(x^2+y^2)\,dy\,dx = \frac{\sqrt{3}}{2}+\frac{8\pi}{3}\approx 9.244$$

15. $\frac{1}{4}$ **17.** $\frac{1}{2}$ **19.** 0 **21.** 18 **23.** $\frac{48\pi-32}{9}\approx 13.2$

25. **(a)** $W: 0\le\theta\le 2\pi,\ 0\le r\le 2,\ r^2\le z\le 8-r^2$

(b) 16π

27. $\frac{405\pi}{2}\approx 636.17$ **29.** $\frac{2}{3}$ **31.** 243π

33. $\int_0^{2\pi}\int_0^1\int_0^4 f(r\cos\theta,\ r\sin\theta,\ z)\,r\,dz\,dr\,d\theta$

35. $\int_0^{\pi}\int_0^1\int_0^{r^2} f(r\cos\theta,\ r\sin\theta,\ z)\,r\,dz\,dr\,d\theta$

37. $z=\frac{H}{R}r$; $V=\frac{\pi R^2 H}{3}$ **39.** $\frac{4}{3}\pi\left(a^2-b^2\right)^{3/2}$ **41.** $-\frac{\pi}{16}$

43. $\frac{8\pi}{15}$ **45.** $\frac{8\pi}{5}$ **47.** $\frac{5\pi}{8}$ **49.** π **51.** $\frac{4\pi a^3}{3}$

Section 16.5 Preliminary Questions

1. 5 kg/m^3 **2.** (a)

3. The probability that $0\le X\le 1$ and $0\le Y\le 1$; the probability that $0\le X+Y\le 1$

Section 16.5 Exercises

1. $\frac{2}{3}$

3. $4\left(1-e^{-100}\right)\times 10^{-6}\ \text{C}\approx 4\times 10^{-6}\ \text{C}$

5. $10{,}000-18{,}000e^{-4/5}\approx 1912$

7. $25\pi\left(3\times 10^{-8}\ \text{C}\right)\approx 2.356\times 10^{-6}\ \text{C}$

9. $\approx 2.593\times 10^{10}$ kg **11.** $\left(0,\ \frac{2}{5}\right)$ **13.** $\left(\frac{4R}{3\pi},\ \frac{4R}{3\pi}\right)$

15. $(0.555,\ 0)$ **17.** $\left(0,\ 0,\ \frac{3R}{8}\right)$ **19.** $\left(0,\ 0,\ \frac{9}{8}\right)$

21. $\left(0,\ 0,\ \frac{13}{2\left(17-6\sqrt{6}\right)}\right)$ **23.** $(1,\ 2)$ **25.** $(0,\ 0)$ **27.** $\frac{16}{15\pi}$

29. **(a)** $\frac{M}{4ab}$ **(b)** $I_x=\frac{Mb^2}{3}$; $I_0=\frac{M(a^2+b^2)}{3}$ **(c)** $\frac{b}{\sqrt{3}}$

31. $I_0=8000\ \text{kg}\cdot\text{m}^2$; $I_x=4000\ \text{kg}\cdot\text{m}^2$

33. $\frac{9}{2}$ **35.** $\frac{243}{20}$ **37.** $\left(\frac{24a}{35},\ \frac{3b}{5}\right)$ **39.** $\frac{a^2b^4}{60}$

41. $I_x=\frac{MR^2}{4}$; kinetic energy required is $\frac{25MR^2}{2}$ J

47. **(a)** $I=182.5\ \text{g}\cdot\text{cm}^2$ **(b)** $\omega\approx 126.92$ rad/s

49. $\frac{13}{72}$ **51.** $\frac{1}{64}$

53. $C=15$; probability is $\frac{5}{8}$.

55. **(a)** $C=4$ **(b)** $\frac{1}{48\pi}+\frac{1}{32}\approx 0.038$

Section 16.6 Preliminary Questions

1. (b)

2. **(a)** $G(1,\ 0)=(2,\ 0)$ **(b)** $G(1,\ 1)=(1,\ 3)$

(c) $G(2,\ 1)=(3,\ 3)$

3. Area $(G(R))=36$ **4.** Area $(G(R))=0.06$

Section 16.6 Exercises

1. **(a)** Image of the u-axis is the line $y=\frac{1}{2}x$; image of the v-axis is the y-axis.

(b) The parallelogram with vertices $(0,\ 0)$, $(10,\ 5)$ $(10,\ 2)$, $(0,\ 7)$.

(c) The segment joining the points $(2,\ 3)$ and $(10,\ 8)$.

(d) The triangle with vertices $(0,\ 1)$, $(2,\ 1)$, and $(2,\ 2)$.

3. G is not one-to-one; G is one-to-one on the domain $\{(u,\ v): u\ge 0\}$, and G is one-to-one on the domain $\{(u,\ v): u\le 0\}$.

(a) The positive x-axis including the origin and the y-axis, respectively.

(b) The rectangle $[0,\ 1]\times[-1,\ 1]$.

(c) The curve $y=\sqrt{x}$ for $0\le x\le 1$.

(d)

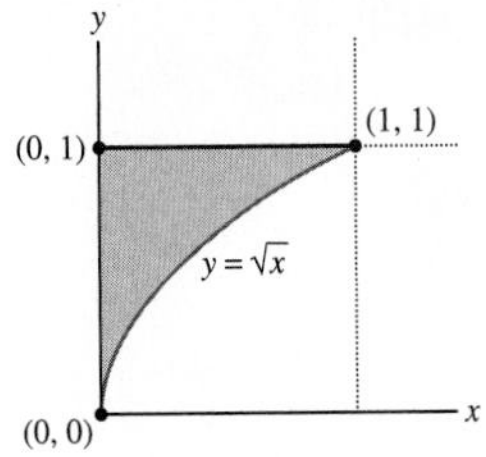

5. $y=3x-c$ **7.** $y=\frac{17}{6}x$ **11.** $\text{Jac}(G)=1$

13. $\text{Jac}(G)=-10$ **15.** $\text{Jac}(G)=1$ **17.** $\text{Jac}(G)=4$

19. $G(u,\ v)=(4u+2v,\ u+3v)$

21. $\frac{2329}{12}\approx 194.08$

23. **(a)** Area $(G(R))=105$ **(b)** Area $(G(R))=126$

25. $\text{Jac}(G)=\frac{2u}{v}$; for $R=[1,\ 4]\times[1,\ 4]$, area $(G(R))=15\ln 4$

27.

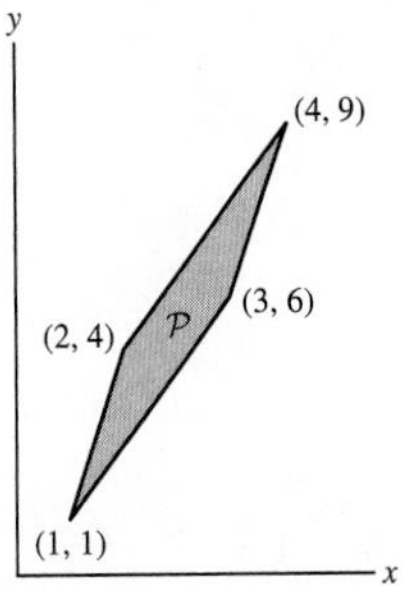

$G(u,\ v)=(1+2u+v,\ 1+5u+3v)$

29. 82 **31.** 80 **33.** $\frac{56}{45}$ **35.** $\frac{\pi(e^{36}-1)}{6}$

37.

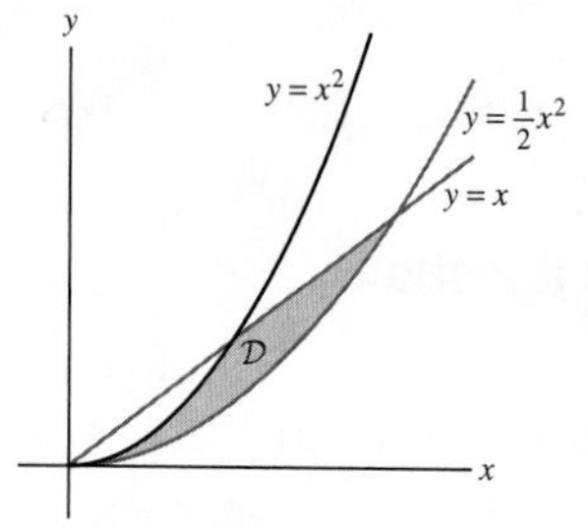

$$\iint_D y^{-1}\,dx\,dy = 1$$

39. $\iint_D e^{xy}\,dA = (e^{20} - e^{10})\ln 2$

41. **(b)** $-\frac{1}{x+y}$ **(c)** $I = 9$ **45.** $\frac{\pi^2}{8}$

Chapter 16 Review

1. **(a)** $S_{2,3} = 240$ **(b)** $S_{2,3} = 510$ **(c)** 520

3. $S_{4,4} = 2.9375$ **5.** $\frac{32}{3}$ **7.** $\frac{\sqrt{3}-1}{2}$

9.

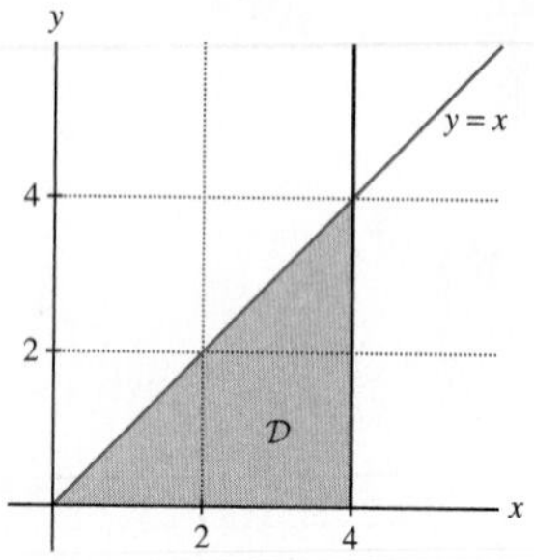

$$\iint_D \cos y\,dA = 1 - \cos 4$$

11.

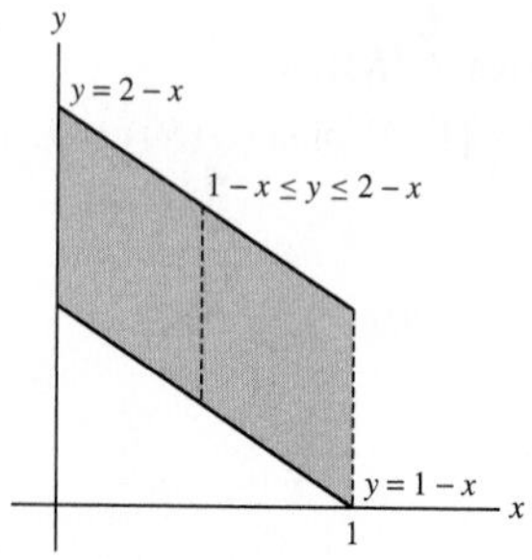

$$\iint_D e^{x+2y}\,dA = \tfrac{1}{2}e(e+1)(e-1)^2$$

13.

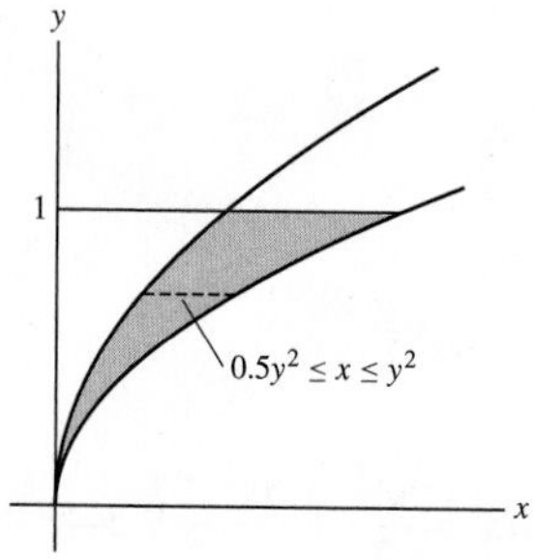

$$\iint_D ye^{1+x}\,dA = 0.5(e^2 - 2e^{1.5} + e)$$

15. $\int_0^9 \int_{-\sqrt{9-y}}^{\sqrt{9-y}} f(x, y)\,dx\,dy$ **17.** $\frac{1}{24}$ **19.** $18(\sqrt{2}-1)$

21. $1 - \cos 1$ **23.** 6π **25.** $\pi/2$ **27.** 10 **29.** $\frac{\pi}{4} + \frac{2}{3}$ **31.** π

33. $\frac{1}{4}$ **35.** $\int_0^{\pi/2}\int_0^1\int_0^r zr\,dz\,dr\,d\theta = \pi/16$ **37.** $\frac{2\pi(-1+e^8)}{3e^8}$

41. $\frac{256\pi}{15} \approx 53.62$ **43.** 1280π **45.** $\left(-\frac{1}{4}R,\ 0,\ \frac{5}{8}H\right)$

47. $\left(-\frac{2}{11\pi}R,\ -\frac{2}{11\pi}R(2-\sqrt{3}),\ \frac{1}{2}H\right)$. **49.** $\left(0,\ 0,\ \frac{2}{3}\right)$

51. $\left(\frac{8}{15},\ \frac{16}{15\pi},\ \frac{16}{15\pi}\right)$ **53.** $\frac{19}{33}$ **55.** $\frac{4}{7}$

57. $G(u, v) = (3u + v,\ -u + 4v)$; Area $(G(R)) = 156$

59. Area$(D) \approx \frac{1}{5}$

61. **(a)**

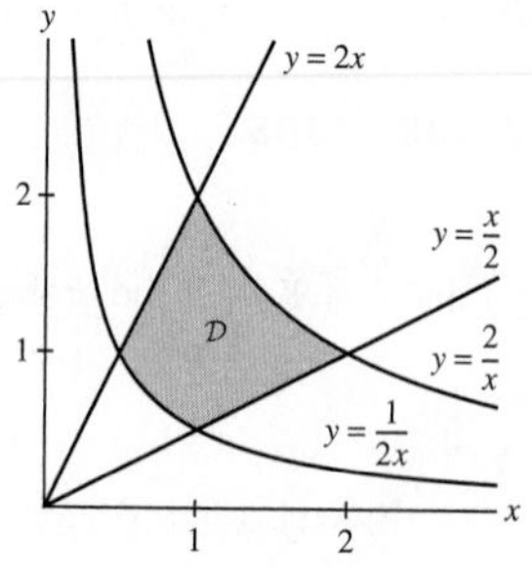

(d) $\frac{3}{4}(e^2 - \sqrt{e})$

Chapter 17

Section 17.1 Preliminary Questions

1. (b)

2.

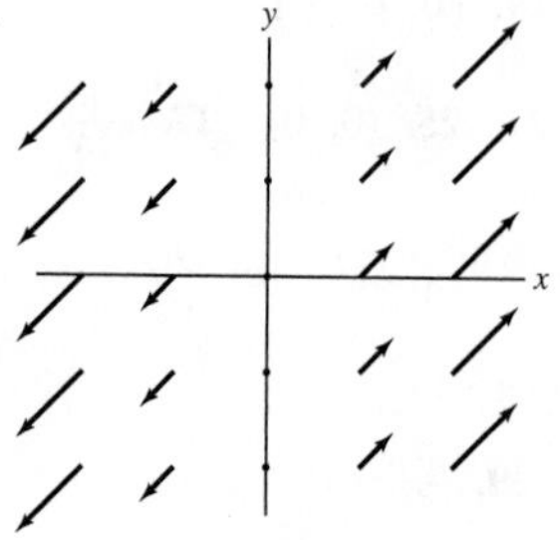

3. $\mathbf{F} = \langle 0,\ -z,\ y\rangle$

4. $f_1(x, y, z) = xyz + 1$

Section 17.1 Exercises

1. $\mathbf{F}(1, 2) = \langle 1, 1 \rangle$, $\mathbf{F}(-1, -1) = \langle 1, -1 \rangle$

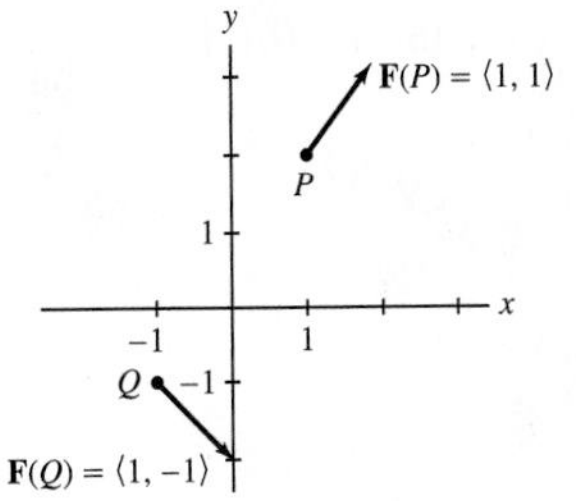

3. $\mathbf{F}(P) = \langle 0, 1, 0 \rangle$, $\mathbf{F}(Q) = \langle 2, 0, 2 \rangle$

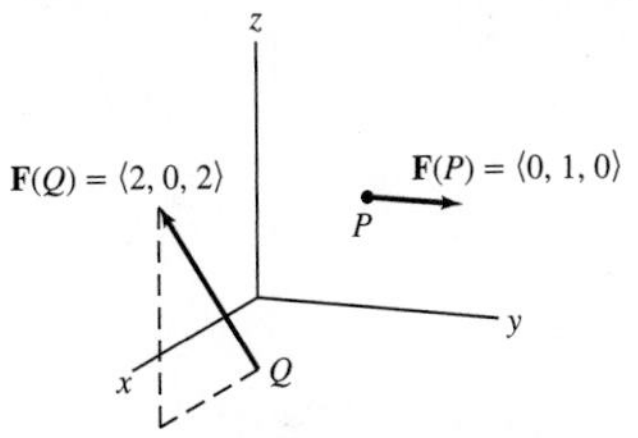

5. $\mathbf{F} = \langle 1, 0 \rangle$

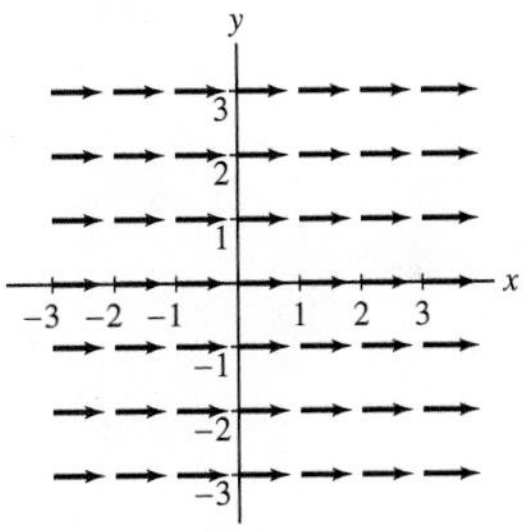

7. $\mathbf{F} = x\mathbf{i}$

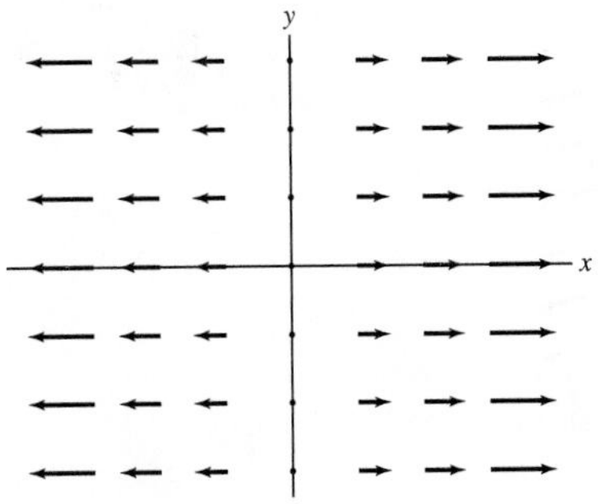

9. $\mathbf{F}(x, y) = \langle 0, x \rangle$

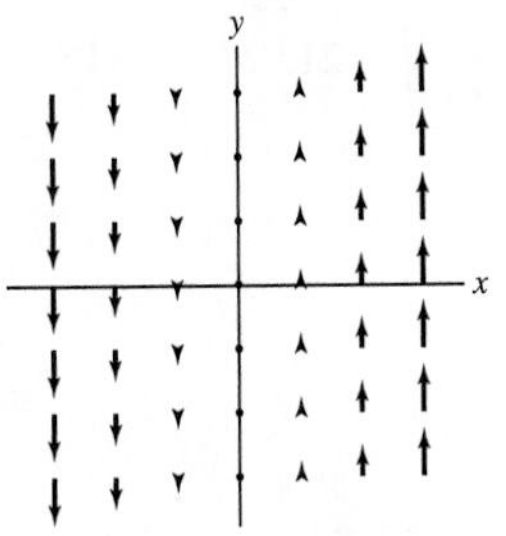

11. $\mathbf{F} = \left\langle \frac{x}{x^2 + y^2}, \frac{y}{x^2 + y^2} \right\rangle$

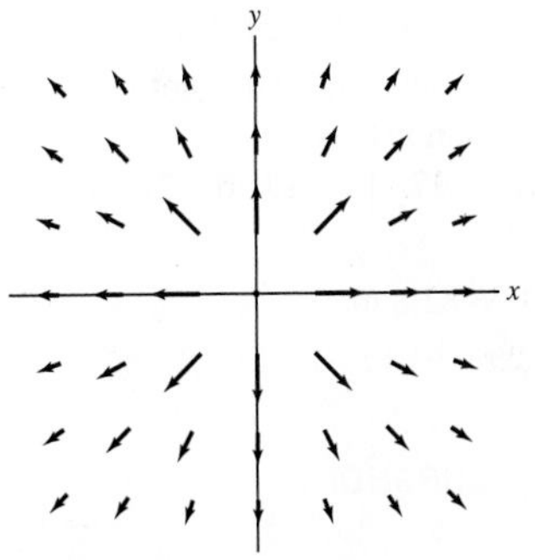

13. Plot (D) **15.** Plot (B) **17.** Plot (C) **19.** Plot (B)
21. $f(x, y) = \frac{1}{2}x^2$ **23.** $f(x, y) = \frac{1}{2}x^2 + \frac{1}{2}y^2$
25. $f(x, y, z) = xyz^2$ **27.** $f_1(r) = -\frac{1}{2r^2}$; $f_2(r) = -\frac{1}{3r^3}$
29. $\nabla\phi = \frac{\mathbf{e}_r}{r}$ **31.** Plot (B)
33. **(a)** Plot (C) **(b)** Plot (B)

Section 17.2 Preliminary Questions

1. 50 **2.** (a), (c), (d), (e)
3. **(a)** True
(b) False. Reversing the orientation of the curve changes the sign of the vector line integral.
4. **(a)** 0 **(b)** −5

Section 17.2 Exercises

1. **(a)** $f(\mathbf{c}(t)) = 6t + 4t^2$, $ds = 2\sqrt{11}\,dt$ **(b)** $\frac{26\sqrt{11}}{3}$
3. **(a)** $\mathbf{F}(\mathbf{c}(t)) = \left\langle t^{-2}, t^2 \right\rangle$, $d\mathbf{s} = \left\langle 1, -t^{-2} \right\rangle dt$ **(b)** $-\frac{1}{2}$
5. $\sqrt{2}\left(\pi + \frac{\pi^3}{3}\right)$ **7.** $\pi^2/2$ **9.** 2.8 **11.** $\frac{128\sqrt{29}}{3} \approx 229.8$
13. $\frac{\sqrt{3}}{2}(e - 1) \approx 1.488$ **15.** $\frac{2}{3}\left((e^2 + 5)^{3/2} - 2^{3/2}\right)$
17. 39; the distance between $(8, -6, 24)$ and $(20, -15, 60)$
19. $\frac{16}{3}$ **21.** 0 **23.** $2(e^2 - e^{-2}) - (e - e^{-1}) \approx 12.157$ **25.** $\frac{10}{9}$
27. $-\frac{8}{3}$ **29.** $\frac{13}{2}$ **31.** $\frac{\pi}{2}$ **33.** 339.5587 **35.** $2 - e - \frac{1}{e}$
37. **(a)** −8 **(b)** −11 **(c)** −16
39. ≈ 7.6; $\approx 4\frac{2}{3}$ **41.** (A) Zero, (B) Negative, (C) Zero **43.** 64π g
45. $\approx 10.4 \times 10^{-6}$ C **47.** ≈ 22743.10 volts **49.** ≈ -10097 volts
51. 1
53. **(a)** *ABC* **(b)** *CBA*
57. $\frac{1}{3}\left((4\pi^2 + 1)^{3/2} - 1\right) \approx 85.5 \times 10^{-6}$ C
63. 18 **65.** $e - 1$ **71.** 0.574

Section 17.3 Preliminary Questions

1. Closed
2. **(a)** Conservative vector fields **(b)** All vector fields
(c) Conservative vector fields **(d)** All vector fields
(e) Conservative vector fields **(f)** All vector fields
(g) Conservative vector fields and *some* other vector fields
3. **(a)** Always true **(b)** Always true
(c) True under additional hypotheses on D
4. **(a)** 4 **(b)** −4

Section 17.3 Exercises

1. 0 **3.** $-\frac{9}{4}$ **5.** $32e - 1$ **7.** $V(x, y, z) = zx + y$
9. $V(x, y, z) = y^2x + e^z y$ **11.** The vector field is not conservative. **13.** $V(x, y, z) = z \tan x + zy$
15. $V(x, y, z) = x^2y + 5x - 4zy$ **17.** 16 **19.** 6 **21.** $-\frac{2}{3}$; 0
23. 6.2×10^9 J
25. (a) $V(x, y, z) = -gz$ **(b)** ≈ 82.8 m/s
27. (A) 2π, (B) 2π, (C) 0, (D) -2π, (E) 4π

Section 17.4 Preliminary Questions

1. 50
2. A distortion factor that indicates how much the area of R_{ij} is altered under the map G.
3. Area$(S) \approx 0.0006$ **4.** $\iint_S f(x, y, z)\, dS \approx 0.6$
5. Area$(S) = 20$ **6.** $\left\langle \frac{2}{3}, \frac{2}{3}, \frac{1}{3} \right\rangle$

Section 17.4 Exercises

1. (a) v **(b)** iii **(c)** i **(d)** iv **(e)** ii
3. (a) $\mathbf{T}_u = \langle 2, 1, 3\rangle$, $\mathbf{T}_v = \langle 0, -1, 1\rangle$, $\mathbf{n}(u, v) = \langle 4, -2, -2\rangle$
(b) Area$(S) = 4\sqrt{6}$ **(c)** $\iint_S f(x, y, z)\, dS = \frac{32\sqrt{6}}{3}$
5. (a) $\mathbf{T}_x = \langle 1, 0, y\rangle$, $\mathbf{T}_y = \langle 0, 1, x\rangle$, $\mathbf{n}(x, y) = \langle -y, -x, 1\rangle$
(b) $\frac{(2\sqrt{2} - 1)\pi}{6}$ **(c)** $\frac{\sqrt{2} + 1}{15}$
7. $\mathbf{T}_u = \langle 2, 1, 3\rangle$, $\mathbf{T}_v = \langle 1, -4, 0\rangle$, $\mathbf{n}(u, v) = 3\langle 4, 1, -3\rangle$, $4x + y - 3z = 0$
9. $\mathbf{T}_\theta = \langle -\sin\theta \sin\phi, \cos\theta \sin\phi, 0\rangle$,
$\mathbf{T}_\phi = \langle \cos\theta\cos\phi, \sin\theta\cos\phi, -\sin\phi\rangle$,
$\mathbf{n}(u, v) = -(\cos\theta \sin^2\phi)\mathbf{i} - (\sin\theta \sin^2\phi)\mathbf{j} - (\sin\phi\cos\phi)\mathbf{k}$,
$y + z = \sqrt{2}$
11. Area$(S) \approx 0.2078$ **13.** $\frac{\sqrt{2}}{5}$ **15.** $\frac{37\sqrt{37} - 1}{4} \approx 56.02$ **17.** $\frac{\pi}{6}$
19. $4\pi(1 - e^{-4})$ **21.** $\frac{\sqrt{3}}{6}$ **23.** $\frac{7\pi}{3}$ **25.** $\frac{5\sqrt{10}}{27} - \frac{1}{54}$
27. Area$(S) = 16$ **29.** $3e^3 - 6e^2 + 3e + 1 \approx 25.08$
31. Area$(S) = 4\pi R^2$
33. (a) Area$(S) \approx 1.0780$ **(b)** ≈ 0.09814
35. Area$(S) = \frac{5\sqrt{29}}{4} \approx 6.73$ **37.** Area$(S) = \pi$ **39.** 48π
43. Area$(S) = \frac{\pi}{6}\left(17\sqrt{17} - 1\right) \approx 36.18$ **47.** $4\pi^2 ab$
49. $V(r) = -\frac{Gm}{2Rr}\left(\sqrt{R^2 + r^2} - |R - r|\right)$

Section 17.5 Preliminary Questions

1. (b) **2.** (c) **3.** (a) **4.** (b)
5. (a) 0 **(b)** π **(c)** π
6. $\approx 0.05\sqrt{2} \approx 0.0707$ **7.** 0

Section 17.5 Exercises

1. (a) $\mathbf{n} = \langle 2v, -4uv, 1\rangle$, $\mathbf{F}\cdot\mathbf{n} = 2v^3 + u$
(b) $\frac{4}{\sqrt{69}}$ **(c)** 265
3. 4 **5.** -4 **7.** $\frac{27}{12}(3\pi + 4)$ **9.** $\frac{693}{5}$ **11.** $\frac{11}{12}$ **13.** $\frac{9\pi}{4}$
15. $(e - 1)^2$ **17.** 270 **19. (a)** $18\pi e^{-3}$ **(b)** $\frac{\pi}{2}e^{-1}$
21. $\left(2 - \frac{6}{\sqrt{13}}\right)\pi k$ **23.** $\frac{2\pi}{3}$ m^3/s **25. (a)** 1 **(b)** 1
29. $\Phi(t) = -1.56 \times 10^{-5} e^{-0.1t}$ T–m^2;
voltage drop $= -1.56 \times 10^{-6} e^{-0.1t}$ V
31. (a)

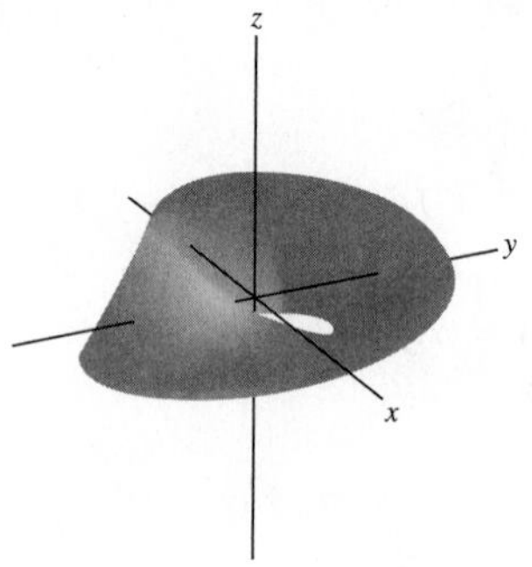

Chapter 17 Review

1. (a) $\langle -15, 8\rangle$ **(b)** $\langle 4, 8\rangle$ **(c)** $\langle 9, 1\rangle$
3.

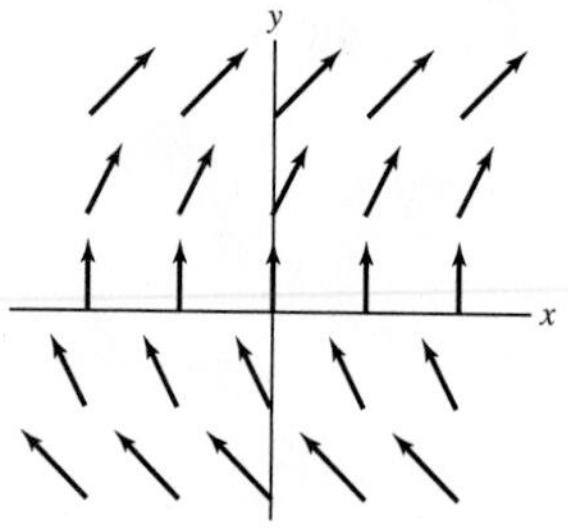

$\mathbf{F} = \langle y, 1\rangle$

5. $\mathbf{F}(x, y) = \langle 2x, -1\rangle$

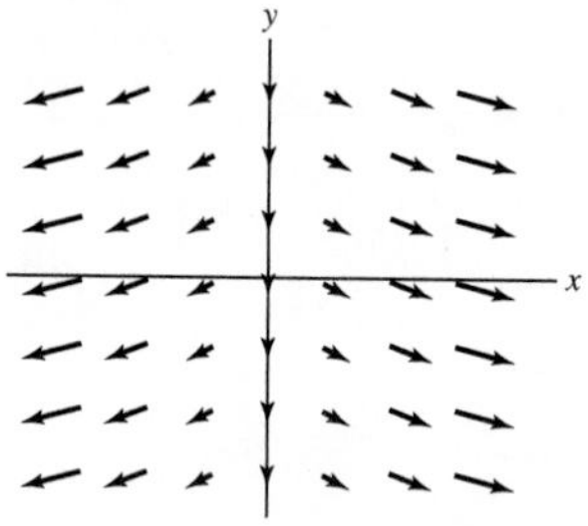

$\nabla V = \langle 2x, -1\rangle$

7. $\mathbf{F}$ is not conservative. **9.** $-\cos x + e^y + \frac{z^2}{2}$
11. $\mathbf{F}$ is not conservative. **13.** $y\tan^{-1}(x) + z^2$
15. $\mathbf{F}$ is not conservative. **17.** $\frac{\sqrt{5}}{6}$ **19.** $\frac{11}{6}$ **21.** $M = 13\frac{1}{3}$
23. $\frac{81 - 9\pi}{4}$ **25.** $-\frac{\pi}{2}$ **27.** $-\frac{13}{18}$
29. (B) and (C) Zero, (D) Negative
31. $a = \frac{4}{3}$, $b = \frac{5}{4}$, $c = -5$,
$\mathbf{T}_u = \left\langle 1, \frac{5}{4}, 2\right\rangle$,
$\mathbf{T}_v = \left\langle \frac{4}{3}, 1, 0\right\rangle$,
$\mathbf{n} = \left\langle -2, \frac{8}{3}, -\frac{2}{3}\right\rangle$

33. **(a)** $\mathbf{T}_u\left(1, \frac{\pi}{3}\right) = \left\langle 1, \sqrt{3}, 0\right\rangle$,

$\mathbf{T}_v\left(1, \frac{\pi}{3}\right) = \left\langle \frac{\sqrt{3}}{2}, -\frac{1}{2}, 3\right\rangle$,

$\mathbf{n}\left(1, \frac{\pi}{3}\right) = \left\langle 3\sqrt{3}, -3, -2\right\rangle$

(b) $3\sqrt{3}x - 3y - 2z + 2\pi = 0$ **(c)** Area$(S) \approx 38.4$

35. Area$(S) \approx 41.8525$ **37.** $54\pi(e^{-10} + 1) \approx 54\pi$

39. Area$(S) = 0.02\sqrt{6} \approx 0.049$ **41.** 54 **43.** 8π **45.** $3 - e$

47. $\frac{\pi}{3}KH^3R$ **49.** 6π

Chapter 18

Section 18.1 Preliminary Questions

1. $\mathbf{F} = \left\langle -e^y, x^2\right\rangle$

2.

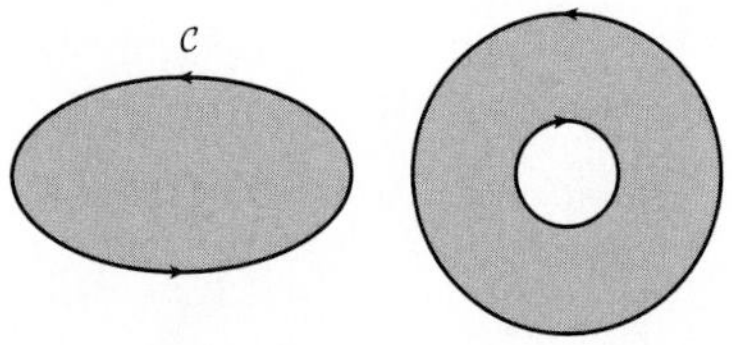

3. Yes **4.** (a), (c)

Section 18.1 Exercises

3. 0 **5.** $-\frac{\pi}{4}$ **7.** $\frac{1}{6}$ **9.** $\frac{(e^2-1)(e^4-5)}{2}$

11. (a) $V(x, y) = x^2e^y$ **13.** $I = 34$ **15.** $A = 9\pi$ **17.** $A = 3\pi$

19. (c) $A = \frac{3}{2}$ **23.** $9 + \frac{15\pi}{2}$ **25.** 214π

27. (A) Zero (B) Positive (C) Negative (D) Zero

29. -0.10 **31.** $R = \sqrt{\frac{2}{3}}$ **33.** Triangle (A), 3; Polygon (B), 12

37. 2 **39.** 0.021 buffalos per second

Section 18.2 Preliminary Questions

1.

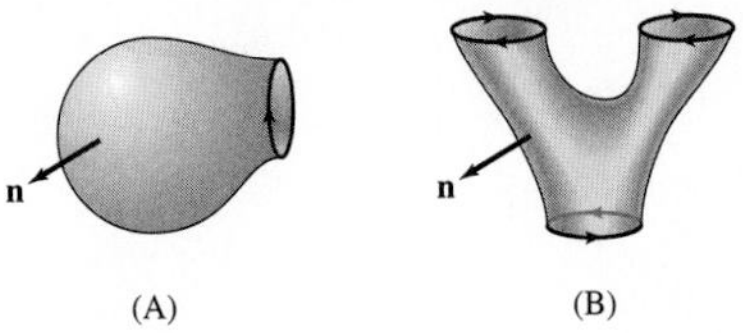

(A) (B)

2. (a)

3. A vector field **A** such that $\mathbf{F} = \text{curl}(\mathbf{A})$is a vector potential for **F**.

4. (b)

5. If the two oriented surfaces S_1 and S_2 have the same oriented boundary curve, C.

Section 18.2 Exercises

1. $\left\langle 1 - 3z^2, 1 - 2x, 1 + 2y\right\rangle$ **3.** $\left\langle 0, \sin x, \cos x - e^y\right\rangle$

5. $\iint_C \mathbf{F} \cdot d\mathbf{s} = \iint_S \text{curl}(\mathbf{F}) \cdot d\mathbf{S} = \pi$

7. $\iint_C \mathbf{F} \cdot d\mathbf{s} = \iint_S \text{curl}(\mathbf{F}) \cdot d\mathbf{S} = e^{-1} - 1$

9. $\left\langle -3z^2e^{z^3}, 2ze^{z^2} + z\sin(xz), 2\right\rangle$; 2π

11. **(a)**

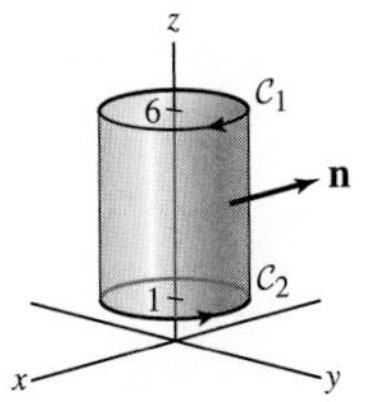

(b) 140π

13. **(a)** $\mathbf{A} = \left\langle 0, 0, e^y - e^{x^2}\right\rangle$ **(c)** $\iint_S \mathbf{F} \cdot d\mathbf{S} = \frac{\pi}{2}$

15. **(a)** $\iint_S \mathbf{B} \cdot d\mathbf{S} = r^2B\pi$ **(b)** $\int_{\partial S} \mathbf{A} \cdot d\mathbf{s} = 0$

17. $\iint_S \mathbf{B} \cdot d\mathbf{S} = b\pi$ **19.** $c = 2a$ and b is arbitrary.

23. $\iint_S \mathbf{F} \cdot d\mathbf{S} = 25$

Section 18.3 Preliminary Questions

1. $\iint_S \mathbf{F} \cdot d\mathbf{S} = 0$

2. Since the integrand is positive for all $(x, y, z) \neq (0, 0, 0)$, the triple integral, hence also the flux, is positive.

3. (a), (b), (d), (f) are meaningful; (b) and (d) are automatically zero.

4. (c) **5.** $\text{div}(\mathbf{F}) = 1$ and flux $= \int \text{div}(\mathbf{F})\, dV =$ volume

Section 18.3 Exercises

1. $y + z$ **3.** $1 - 4zx - x + 2zx^2$ **5.** $c = -\frac{1}{5}$

7. $\iint_S \mathbf{F} \cdot d\mathbf{S} = \iiint_R \text{div}(\mathbf{F})\, dV = \iiint_R 0\, dV = 0$

9. $\iint_S \mathbf{F} \cdot d\mathbf{S} = \iiint_R \text{div}(\mathbf{F})\, dV = 4\pi$

11. $\frac{4\pi}{15}$ **13.** $\frac{32\pi}{5}$ **15.** 64π **17.** 81π **19.** 0 **21.** $\frac{13}{3}$ **23.** $\frac{4\pi}{3}$

25. $\frac{16\pi}{3} + \frac{9\sqrt{3}}{2} \approx 24.549$ **27.** $\approx 1.57\ \text{m}^3/\text{s}$

29. **(b)** 0 **(c)** 0

(d) Since **E** is not defined at the origin, which is inside the ball W, we cannot use the Divergence Theorem.

31. $(-4) \cdot \left[\frac{256\pi}{3} - 1\right] \approx -1068.33$

33. $\text{div}(f\mathbf{F}) = f\,\text{div}(\mathbf{F}) + \mathbf{F} \cdot \nabla f$

Chapter 18 Review

1. 0 **3.** -30 **5.** $\frac{3}{5}$

7. **(a)**

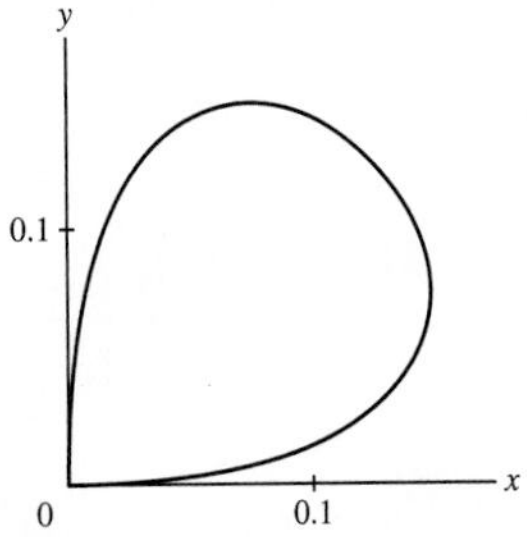

(b) $A = \frac{1}{60}$

9. $\text{curl}(\mathbf{F}) = -\mathbf{k}$, $\text{div}(\mathbf{F}) = -1$

11. $\text{curl}(\mathbf{F}) = \mathbf{0}$, $\text{div}(\mathbf{F}) = 2e^{-x^2-y^2-z^2}\left(2(x^2 + y^2 + z^2) - 3\right)$

13. $\text{curl}(\mathbf{F}) = \langle -2z, 0, -2y\rangle$ **19.** 2π

21. $\mathbf{A} = \langle yz, 0, 0\rangle$ and the flux is 8π. **23.** $\frac{296}{3}$ **25.** -128π

27. Volume$(W) = 2$ **29.** $4 \cdot 0.0009\pi \approx 0.0113$

31. $2x - y + 4z = 0$

33. **(b)** $\frac{\pi}{2}$

35. **(c)** 0 **(d)** $\int_{C_1} \mathbf{F} \cdot d\mathbf{s} = -4$, $\int_{C_2} \mathbf{F} \cdot d\mathbf{s} = 4$

REFERENCES

The online source MacTutor History of Mathematics Archive **www-history.mcs.st-and.ac.uk** has been a valuable source of historical information.

Section 1.1

(EX 77) Adapted from *Calculus Problems for a New Century*, Robert Fraga, ed., Mathematical Association of America, Washington, DC, 1993, p. 9.

Section 1.2

(EX 25) Adapted from *Calculus Problems for a New Century*, Robert Fraga, ed., Mathematical Association of America, Washington, DC, 1993, p. 9.

Section 1.5

(EXMP 4) Adapted from B. Waits and F. Demana, "The Calculator and Computer Pre-Calculus Project, " in *The Impact of Calculators on Mathematics Instruction*, University of Houston, 1994.

(EX 12) Adapted from B. Waits and F. Demana, "The Calculator and Computer Pre-Calculus Project, " in *The Impact of Calculators on Mathematics Instruction*, University of Houston, 1994.

Section 2.2

(EX 61) Adapted from *Calculus Problems for a New Century*, Robert Fraga, ed., Mathematical Association of America, Washington, DC, 1993, Note 28.

Section 2.3

(EX 38) Adapted from *Calculus Problems for a New Century*, Robert Fraga, ed., Mathematical Association of America, Washington, DC, 1993, Note 28.

Chapter 2 Review

(EX 68) Adapted from *Calculus Problems for a New Century*, Robert Fraga, ed., Mathematical Association of America, Washington, DC, 1993, Note 28.

Section 3.1

(EX 75) Problem suggested by Dennis DeTurck, University of Pennsylvania.

Section 3.2

(EX 92) Problem suggested by Chris Bishop, SUNY Stony Brook.

(EX 93) Problem suggested by Chris Bishop, SUNY Stony Brook.

Section 3.4

(PQ 2) Adapted from *Calculus Problems for a New Century*, Robert Fraga, ed., Mathematical Association of America, Washington, DC, 1993, p. 25.

(EX 48) Karl J. Niklas and Brian J. Enquist, "Invariant Scaling Relationships for Interspecific Plant Biomass Production Rates and Body Size," *Proc. Natl. Acad. Sci.* 98, no. 5:2922-2927 (February 27, 2001)

Section 3.5

(EX 47) Adapted from a contribution by Jo Hoffacker, University of Georgia.

(EX 48-49) Adapted from a contribution by Thomas M. Smith, University of Illinois at Chicago, and Cindy S. Smith, Plainfield High School.

(EX 45) Adapted from Walter Meyer, *Falling Raindrops,* in *Applications of Calculus*, P. Straffin, ed., Mathematical Association of America, Washington, DC, 1993

(EX 52, 56) Problems suggested by Chris Bishop, SUNY Stony Brook.

Section 3.9

(EX 32) Adapted from *Calculus Problems for a New Century*, Robert Fraga, ed., Mathematical Association of America, Washington, DC, 1993.

(EX 34) Problem suggested by Kay Dundas.

(EX 38, 44) Adapted from *Calculus Problems for a New Century*, Robert Fraga, ed., Mathematical Association of America, Washington, DC, 1993.

Chapter 3 Review

(EX 64, 83) Problems suggested by Chris Bishop, SUNY Stony Brook.

Section 4.2

(MN p. 184) Adapted from "Stories about Maxima and Minima," V. M. Tikhomirov, AMS, (1990).

(MN p. 189) From Pierre Fermat, On Maxima and Minima and on Tangents, translated by D.J. Struik (ed.), A Source Book in Mathematics, 1200-1800, Princeton University Press, Princeton, NJ, 1986.

Section 4.5

(EX 26-27) Adapted from *Calculus Problems for a New Century*, Robert Fraga, ed., Mathematical Association of America, Washington, DC, 1993.

(EX 40) Problem suggested by John Haverhals, Bradley University. *Source:* Illinois Agrinews.

(EX 42) Adapted from *Calculus Problems for a New Century*, Robert Fraga, ed., Mathematical Association of America, Washington, DC, 1993.

(EX 66-68) Adapted from B. Noble, Applications of Undergraduate Mathematics in Engineering, Macmillan, New York, 1967.

(EX 70) Adapted from Roger Johnson, "A Problem in Maxima and Minima," *American Mathematical Monthly*, 35:187-188 (1928).

Section 4.6

(EX 67) Adapted from Robert J. Bumcrot, "Some Subtleties in L' Hôpital's Rule," in *A Century of Calculus*, Part II, Mathematical Association of America, Washington, DC, 1992.

(EX 28) Adapted from "Calculus for a Real and Complex World" by Frank Wattenberg, PWS Publishing, Boston, 1995.

Section 4.7

(EX 20) Adapted from *Calculus Problems for a New Century*, Robert Fraga, ed., Mathematical Association of America, Washington, DC, 1993, p. 52.

(EX 32-33) Adapted from E. Packel and S. Wagon, *Animating Calculus*, Springer-Verlag, New York, 1997, p. 79.

Chapter 4 Review

(EX 68) Adapted from *Calculus Problems for a New Century*, Robert Fraga, ed., Mathematical Association of America, Washington, DC, 1993.

Section 5.1

(EX 3) Problem suggested by John Polhill, Bloomsburg University.

Section 5.2

(FI&C 84) Problem suggested by Chris Bishop, SUNY Stony Brook.

Section 5.4

(EX 40-41) Adapted from *Calculus Problems for a New Century*, Robert Fraga, ed., Mathematical Association of America, Washington, DC, 1993, p. 102.

(EX 42) Problem suggested by Dennis DeTurck, University of Pennsylvania.

Section 5.5

(EX 25-26) M. Newman and G. Eble, "Decline in Extinction Rates and Scale Invariance in the Fossil Record." *Paleobiology* 25:434-439 (1999).

(EX 28) From H. Flanders, R. Korfhage, and J. Price, *Calculus,* Academic Press, New York, 1970.

Section 5.6

(EX 62) Adapted from *Calculus Problems for a New Century*, Robert Fraga, ed., Mathematical Association of America, Washington, DC, 1993, p. 121.

Section 6.1

(EX 48) Adapted from Tom Farmer and Fred Gass, "Miami University: An Alternative Calculus" in *Priming the Calculus Pump*, Thomas Tucker, ed., Mathematical Association of America, Washington, DC, 1990, Note 17.

(EX 61) Adapted from *Calculus Problems for a New Century*, Robert Fraga, ed., Mathematical Association of America, Washington, DC, 1993.

Section 6.3

(EX 60, 62) Adapted from G. Alexanderson and L. Klosinski, "Some Surprising Volumes of Revolution, " *Two-Year College Mathematics Journal* 6, 3:13-15 (1975).

Section 7.7

(EX 48, 79) Adapted from *Calculus Problems for a New Century*, Robert Fraga, ed., Mathematical Association of America, Washington, DC, 1993.

Section 8.1

(EX 56-58, 59, 60-62, 65) Problems suggested by Brian Bradie, Christopher Newport University.

(EX 70) Adapted from *Calculus Problems for a New Century*, Robert Fraga, ed., Mathematical Association of America, Washington, DC, 1993.

(FI&C 79) Adapted from J. L. Borman, "A Remark on Integration by Parts," *American Mathematical Monthly* 51:32-33 (1944).

Section 8.3

(EX 43-47, 51) Problems suggested by Brian Bradie, Christopher Newport University.

(EX 62) Adapted from *Calculus Problems for a New Century*, Robert Fraga, ed., Mathematical Association of America, Washington, DC, 1993, p. 118.

Section 8.6

(EX 81) Problem suggested by Chris Bishop, SUNY Stony Brook.

Section 8.8

See R. Courant and F. John, Introduction to Calculus and Analysis, Vol. 1, Springer-Verlag, New York, 1989.

Section 9.1

(FI&C 52) Adapted from G. Klambauer, *Aspects of Calculus*, Springer-Verlag, New York, 1986, Ch 6.

Section 10.1

(EX 55) Adapted from E. Batschelet, *Introduction to Mathematics for Life Scientists*, Springer-Verlag, New York, 1979.

(EX 57) Adapted from *Calculus Problems for a New Century*, Robert Fraga, ed., Mathematical Association of America, Washington, DC, 1993.

(EX 58, 63) Adapted from M. Tenenbaum and H. Pollard, *Ordinary Differential Equations*, Dover, New York, 1985.

Section 11.1

(EX 68) Adapted from G. Klambauer, *Aspects of Calculus*, Springer-Verlag, New York, 1986, p. 393.

(EX 92) Adapted from Apostol and Mnatsakanian, "New Insights into Cycloidal Areas," *American Math Monthly*, August-September 2009.

Section 11.2

(EX 42) Adapted from *Calculus Problems for a New Century*, Robert Fraga, ed., Mathematical Association of America, Washington, DC, 1993, p. 137.

(EX 43) Adapted from *Calculus Problems for a New Century*, Robert Fraga, ed., Mathematical Association of America, Washington, DC, 1993, p. 138.

(FI&C 51) Adapted from George Andrews, "The Geometric Series in Calculus," *American Mathematical Monthly* 105, 1:36-40 (1998)

(FI&C 54) Adapted from Larry E. Knop, "Cantor's Disappearing Table," *The College Mathematics Journal* 16, 5:398-399 (1985).

Section 11.4

(EX 33) Adapted from *Calculus Problems for a New Century*, Robert Fraga, ed., Mathematical Association of America, Washington, DC, 1993, p. 145.

Section 12.2

(FI&C 35) Adapted from Richard Courant and Fritz John, *Differential and Integral Calculus*, Wiley-Interscience, New York, 1965.

Section 12.3

(EX 56) Adapted from *Calculus Problems for a New Century*, Robert Fraga, ed., Mathematical Association of America, Washington, DC, 1993.

Section 13.4

(EX 61) Adapted from Ethan Berkove and Rich Marchand, "The Long Arm of Calculus," *The College Mathematics Journal* 29, 5:376-386 (November 1998).

Section 14.3

(EX 15) Adapted from *Calculus Problems for a New Century*, Robert Fraga, ed., Mathematical Association of America, Washington, DC, 1993.

Section 14.4

(EX 60) Damien Gatinel, Thanh Hoang-Xuan, and Dimitri T. Azar, "Determination of Corneal Asphericity After Myopia Surgery with the Excimer Laser: A Mathematical Model," *Investigative Opthalmology and Visual Science* 42: 1736-1742 (2001).

Section 14.5

(EX 55, 58) Adapted from notes to the course "Dynamics and Vibrations" at Brown University **http://www.engin.brown.edu/courses/en4/**.

Section 15.8

(EX 42) Adapted from C. Henry Edwards, "Ladders, Moats, and Lagrange Multipliers," *Mathematica Journal* 4, Issue 1 (Winter 1994).

Section 16.3

(FIGURE 9 COMPUTATION) The computation is based on Jeffrey Nunemacher, "The Largest Unit Ball in Any Euclidean Space," in *A Century of Calculus*, Part II, Mathematical Association of America, Washington DC, 1992.

Section 16.6

(CONCEPTUAL INSIGHT) See R. Courant and F. John, Introduction to Calculus and Analysis, Springer-Verlag, New York, 1989, p. 534.

Section 17.2

(FIGURE 9) Inspired by Tevian Dray and Corinne A. Manogue, "The Murder Mystery Method for Determining Whether a Vector Field Is Conservative," *The College Mathematics Journal*, May 2003.

Section 17.3

(EX 19) Adapted from *Calculus Problems for a New Century*, Robert Fraga, ed., Mathematical Association of America, Washington, DC, 1993.

Appendix D

(PROOF OF THEOREM 7) A proof without this simplifying assumption can be found in R. Courant and F. John, *Introduction to Calculus and Analysis*, Vol. 1, Springer-Verlag, New York, 1989.

PHOTO CREDITS

PAGE 1, CO1
Roger Whiteway/ iStockphoto.com

PAGE 34
left Eric Dietz/ http://www.wrongway.org/java/ mandel.php

PAGE 34
right Eric Dietz/ http://www.wrongway.org/java/ mandel.php

PAGE 40, CO2
Scott Camazine/Photo Researchers

PAGE 41
right NASA/JPL-Caltech/STScI

PAGE 41
left Wayne Boucher

PAGE 81
NASA

PAGE 86
right University Archives, University of Pittsburgh

PAGE 86
left Rockefeller Archive Center

PAGE 101, CO3
United States Department of Defense

PAGE 111
The Granger Collection, New York

PAGE 127
Brian Jackson/ iStockphoto.com

PAGE 134
right Bettmann/Corbis

PAGE 134
left Franca Principe/ IMSS - Florence

PAGE 150
Bettmann/Corbis

PAGE 155
top A & J Visage / Alamy

PAGE 155
bottom Doug James/ Dreamstime.com

PAGE 175, CO4
John Derbyshire

PAGE 177
Space Age Control, Inc.

PAGE 189
Bettmann/Corbis

PAGE 189
Stapleton Collection/Corbis

PAGE 193
Daniel Grill/ iStockphoto.com

PAGE 244, CO5
Aerial Archives / Alamy

PAGE 253
Bettmann/Corbis

PAGE 254
Daryl Balfour/Gallo Images/ Alamy

PAGE 259
The Granger Collection

PAGE 296, CO6
Mason Morfit/ Peter Arnold Images / Photolibrary

PAGE 310
Corbis/ Alamy

PAGE 334
Elvele Images / Alamy

PAGE 339, CO7
Pierre Vauthey/Corbis SYGMA

PAGE 339
AP Photo/Paul Sakuma

PAGE 355
Romeo Gacad/AFP/ Getty Images

PAGE 365
Dr. Gary Gaugler / Photo Researchers, Inc

PAGE 366
courtesy Michael Zingale

PAGE 367
Copyright © 2010 The Regents of the University of California. All rights reserved. Use with permission.

PAGE 368
Bettmann/Corbis

PAGE 368
Pierre Vauthey/Corbis SYGMA

PAGE 369
G.D. Carr

PAGE 378
Mr_Jamsey/ iStockphoto

PAGE 378
Hector Mandel/ iStockphoto

PAGE 399
Corbis

PAGE 410
left Library of Congress

PAGE 410
right Library of Congress

PAGE 413, CO8
Courtesy, Steven N. Ward, University of California Santa Cruz http://es.ucsc.edu/ ward

PAGE 478, CO9
NASA Ames Research Center / Photo Researchers

PAGE 485
Jeff Rotman

PAGE 487
Lester Lefkowitz/ Corbis

PAGE 490
REUTERS/STR /Landov

PAGE 491
Mark Dadswell/Getty Images

PAGE 496
The Granger Collection, New York

PAGE 498
Roger Viollet Collection/Getty Images

PAGE 502
The Granger Collection

PAGE 503
Alexandr Ozerov/iStockphoto

PAGE 513, CO10
Elizabeth Kreutz/Corbis

PAGE 543, CO11
NASA and J.J. Hester (Arizona State University)

PAGE 561
Mechanics Magazine London, 1824

PAGE 613, CO12
Mircea Bezergheanu/ iStockphoto

PAGE 620
Renault Communication /ALL RIGHTS RESERVED

PAGE 654
Stockbyte

PAGE 655
Architect of the Capitol

PAGE 663, CO13
Ryan Pyle

PAGE 694
LBNL / Science Source / Photo Researchers

PAGE 729, CO14
Spectral-Design/ iStockphoto

PAGE 752
bottom Alfred Pasieka / Photo Researchers, Inc.

PAGE 752
top Robin Smith/ Stone/ Getty Images

PAGE 762
NASA

PAGE 766
Bernd Kappelmeyer/ zefa/Corbis

PAGE 775
NASA, ESA, and the Hubble Heritage Team (STScI/AURA)-ESA/Hubble Collaboration

PAGE 780, CO15
top Tom Dempsey / Photoseek.com

PAGE 780
bottom Krista Longnecker/Woods Hole Oceanographic Institution

PAGE 786
Galen Rowell/Corbis

PAGE 786
USGS/http://www.topozone.com

PAGE 805
SSPL / The Image Works

PAGE 807
left Hulton Archive/Getty Images

PAGE 807
right Wikipedia

PAGE 855
Bettmann/Corbis

PAGE 866, CO16
scottygo / Alamy

PAGE 902
Gary A Glatzmaier (University of California, Santa Cruz) and Paul H Roberts (University of California, Los Angeles)

PAGE 945, CO17
Honkanen M, Jung J, Kuo C-J, Peles Y, Amitay M 2010, Two-phase PIV/PTV measurement of bubbly flow across pin fins in a micro-channel. In Proc. of 7th Int. Symp. on Multiphase Flow, Tampa, US, 30.5.-4.6.2010.

PAGE 945
Michelle Borkin (Harvard University)

PAGE 947
Bettmann/Corbis

PAGE 1009, CO18
Alamy

PAGE 1012
planimeter courtesy of John D. Eggers UCSD/ photo by Adriene Hughes/ UC San Diego Media Lab

PAGE 1020
Paul Horsted/Stock Connection Distribution / Alamy

PAGE 1043
SSPL / The Image Works

INDEX

ELEMENTARY FUNCTIONS

Power Functions $f(x) = x^a$

$f(x) = x^n$, n a positive integer

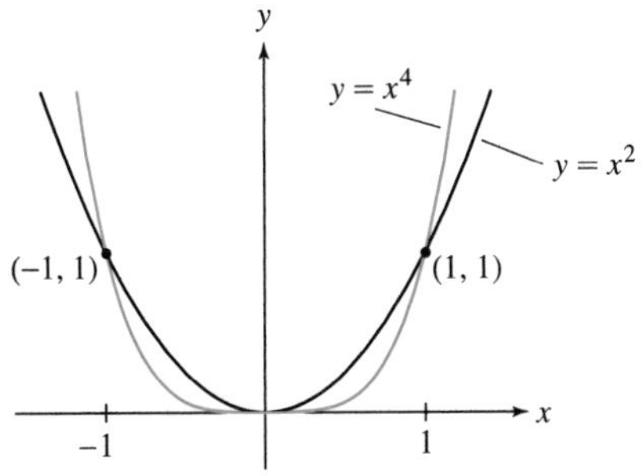

n even

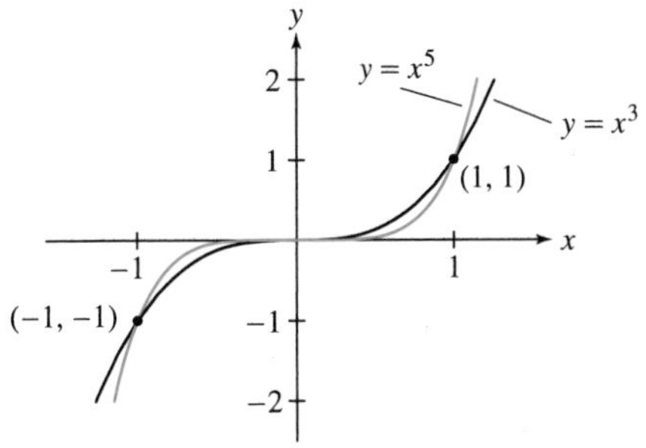

n odd

Asymptotic behavior of a polynomial function of even degree and positive leading coefficient

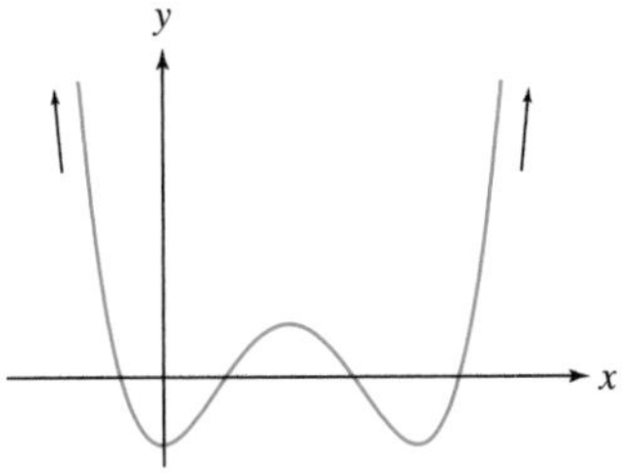

n even

Asymptotic behavior of a polynomial function of odd degree and positive leading coefficient

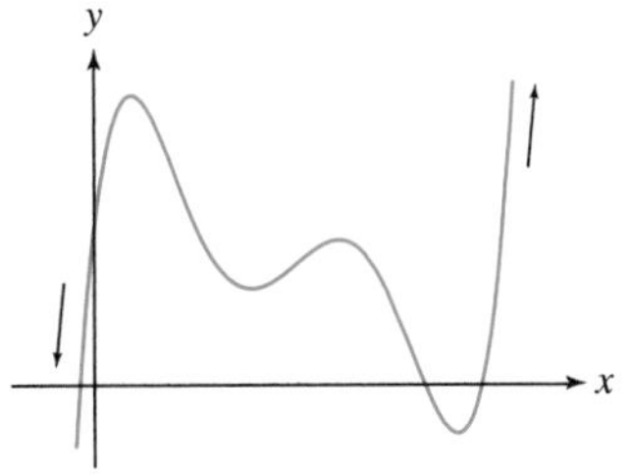

n odd

$f(x) = x^{-n} = \dfrac{1}{x^n}$

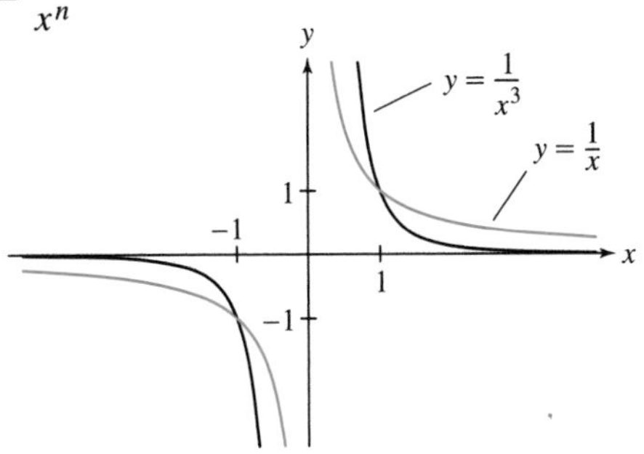

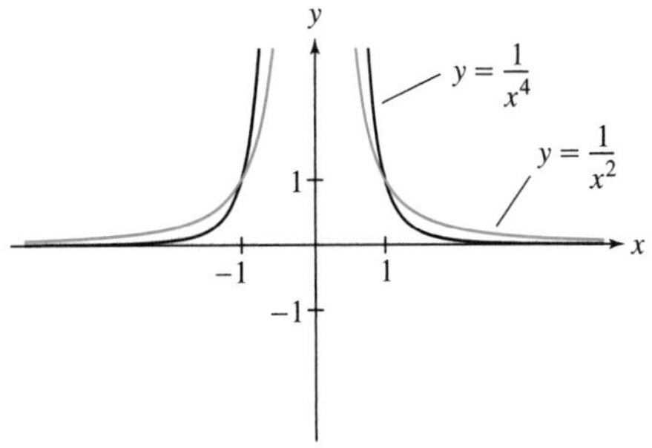

Inverse Trigonometric Functions

$\arcsin x = \sin^{-1} x = \theta$

$\Leftrightarrow \quad \sin\theta = x, \quad -\dfrac{\pi}{2} \le \theta \le \dfrac{\pi}{2}$

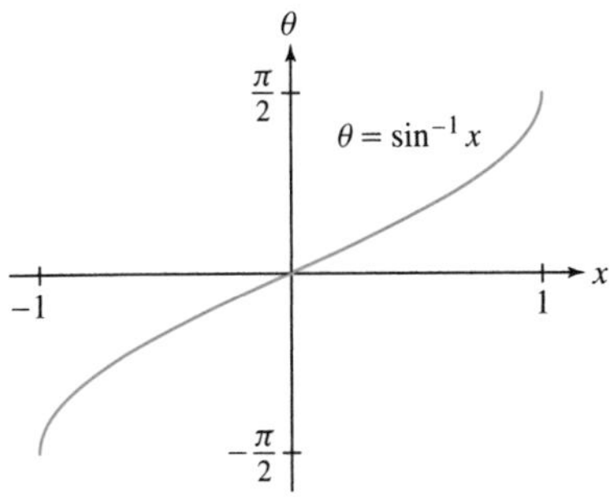

$\arccos x = \cos^{-1} x = \theta$

$\Leftrightarrow \quad \cos\theta = x, \quad 0 \le \theta \le \pi$

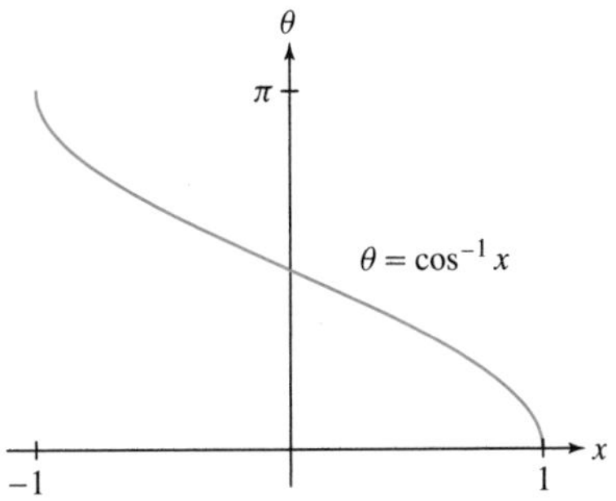

$\arctan x = \tan^{-1} x = \theta$

$\Leftrightarrow \quad \tan\theta = x, \quad -\dfrac{\pi}{2} < \theta < \dfrac{\pi}{2}$

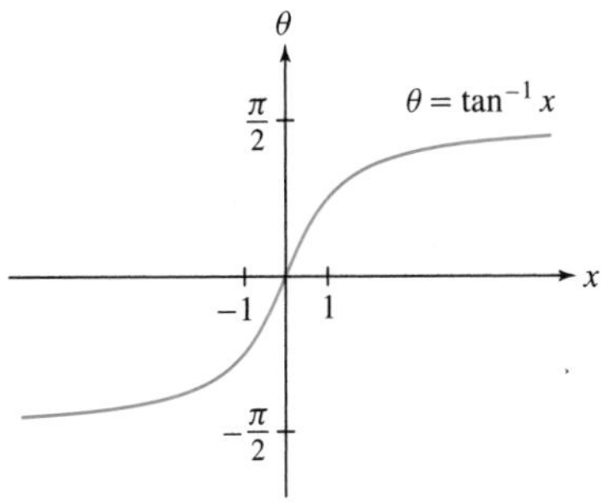

Exponential and Logarithmic Functions

$$\log_a x = y \iff a^y = x$$

$$\log_a(a^x) = x \qquad a^{\log_a x} = x$$

$$\log_a 1 = 0 \qquad \log_a a = 1$$

$$\ln x = y \iff e^y = x$$

$$\ln(e^x) = x \qquad e^{\ln x} = x$$

$$\ln 1 = 0 \qquad \ln e = 1$$

$$\log_a(xy) = \log_a x + \log_a y$$

$$\log_a\left(\frac{x}{y}\right) = \log_a x - \log_a y$$

$$\log_a(x^r) = r\log_a x$$

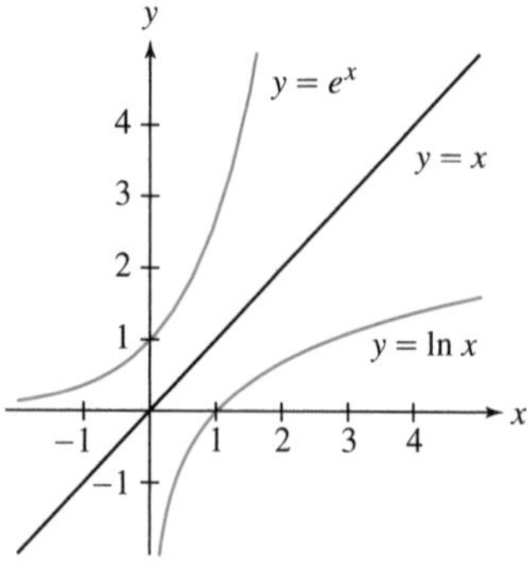

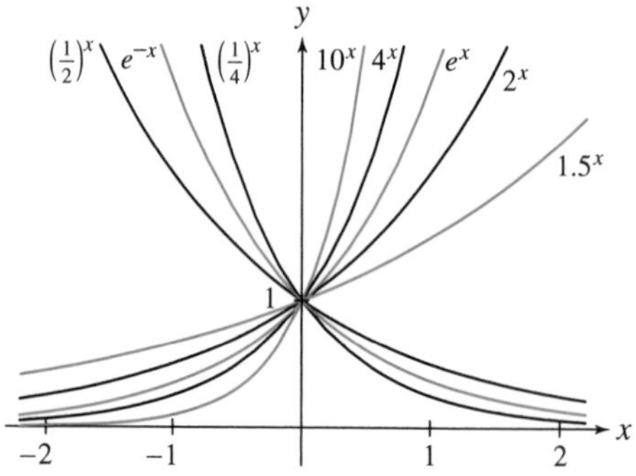

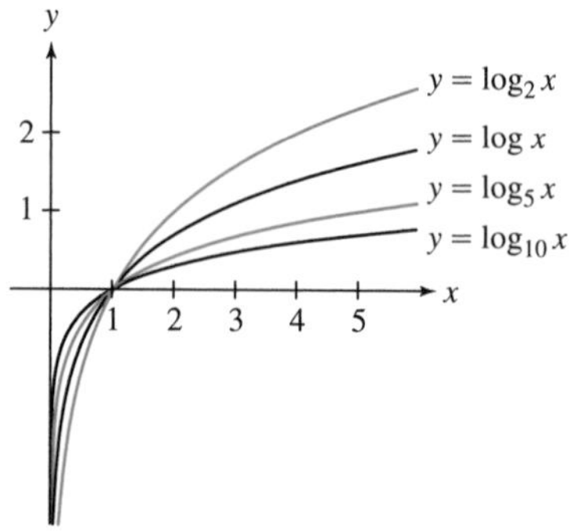

$$\lim_{x\to\infty} a^x = \infty, \quad a > 1$$

$$\lim_{x\to\infty} a^x = 0, \quad 0 < a < 1$$

$$\lim_{x\to-\infty} a^x = 0, \quad a > 1$$

$$\lim_{x\to-\infty} a^x = \infty, \quad 0 < a < 1$$

$$\lim_{x\to 0^+} \log_a x = -\infty$$

$$\lim_{x\to\infty} \log_a x = \infty$$

Hyperbolic Functions

$$\sinh x = \frac{e^x - e^{-x}}{2} \qquad \operatorname{csch} x = \frac{1}{\sinh x}$$

$$\cosh x = \frac{e^x + e^{-x}}{2} \qquad \operatorname{sech} x = \frac{1}{\cosh x}$$

$$\tanh x = \frac{\sinh x}{\cosh x} \qquad \coth x = \frac{\cosh x}{\sinh x}$$

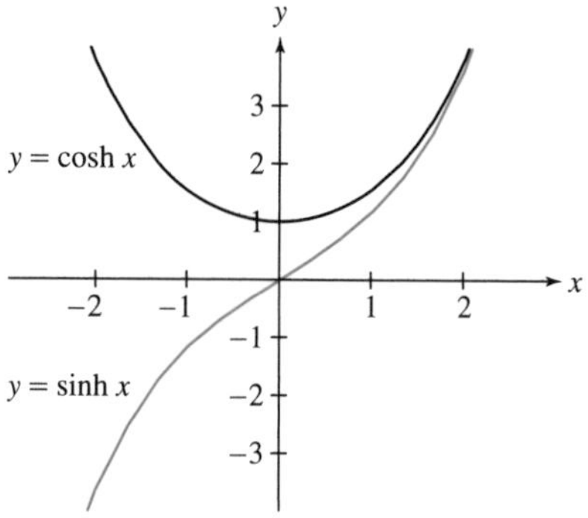

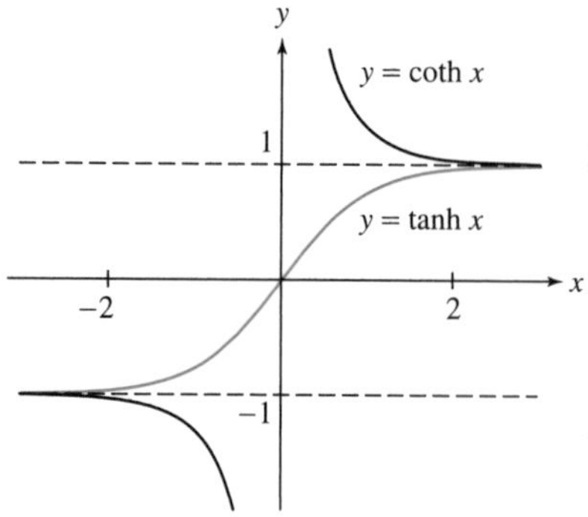

$$\sinh(x + y) = \sinh x\cosh y + \cosh x\sinh y$$

$$\cosh(x + y) = \cosh x\cosh y + \sinh x\sinh y$$

$$\sinh 2x = 2\sinh x\cosh x$$

$$\cosh 2x = \cosh^2 x + \sinh^2 x$$

Inverse Hyperbolic Functions

$$y = \sinh^{-1} x \iff \sinh y = x$$

$$y = \cosh^{-1} x \iff \cosh y = x \text{ and } y \geq 0$$

$$y = \tanh^{-1} x \iff \tanh y = x$$

$$\sinh^{-1} x = \ln\left(x + \sqrt{x^2 + 1}\right)$$

$$\cosh^{-1} x = \ln\left(x + \sqrt{x^2 - 1}\right) \quad x > 1$$

$$\tanh^{-1} x = \frac{1}{2}\ln\left(\frac{1 + x}{1 - x}\right) \quad -1 < x < 1$$

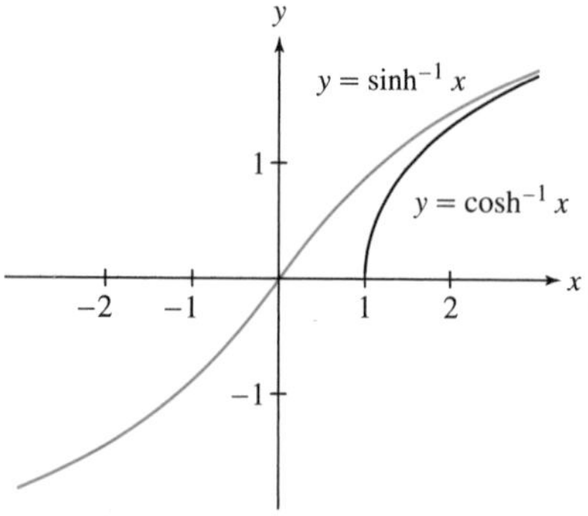

DIFFERENTIATION

Differentiation Rules

1. $\frac{d}{dx}(c) = 0$

2. $\frac{d}{dx}x = 1$

3. $\frac{d}{dx}(x^n) = nx^{n-1}$ (Power Rule)

4. $\frac{d}{dx}[cf(x)] = cf'(x)$

5. $\frac{d}{dx}[f(x) + g(x)] = f'(x) + g'(x)$

6. $\frac{d}{dx}[f(x)g(x)] = f(x)g'(x) + g(x)f'(x)$ (Product Rule)

7. $\frac{d}{dx}\left[\frac{f(x)}{g(x)}\right] = \frac{g(x)f'(x) - f(x)g'(x)}{[g(x)]^2}$ (Quotient Rule)

8. $\frac{d}{dx}f(g(x)) = f'(g(x))g'(x)$ (Chain Rule)

9. $\frac{d}{dx}f(x)^n = nf(x)^{n-1}f'(x)$ (General Power Rule)

10. $\frac{d}{dx}f(kx + b) = kf'(kx + b)$

11. $g'(x) = \frac{1}{f'(g(x))}$ where $g(x)$ is the inverse $f^{-1}(x)$

12. $\frac{d}{dx}\ln f(x) = \frac{f'(x)}{f(x)}$

Trigonometric Functions

13. $\frac{d}{dx}\sin x = \cos x$

14. $\frac{d}{dx}\cos x = -\sin x$

15. $\frac{d}{dx}\tan x = \sec^2 x$

16. $\frac{d}{dx}\csc x = -\csc x \cot x$

17. $\frac{d}{dx}\sec x = \sec x \tan x$

18. $\frac{d}{dx}\cot x = -\csc^2 x$

Inverse Trigonometric Functions

19. $\frac{d}{dx}(\sin^{-1} x) = \frac{1}{\sqrt{1 - x^2}}$

20. $\frac{d}{dx}(\cos^{-1} x) = -\frac{1}{\sqrt{1 - x^2}}$

21. $\frac{d}{dx}(\tan^{-1} x) = \frac{1}{1 + x^2}$

22. $\frac{d}{dx}(\csc^{-1} x) = -\frac{1}{|x|\sqrt{x^2 - 1}}$

23. $\frac{d}{dx}(\sec^{-1} x) = \frac{1}{|x|\sqrt{x^2 - 1}}$

24. $\frac{d}{dx}(\cot^{-1} x) = -\frac{1}{1 + x^2}$

Exponential and Logarithmic Functions

25. $\frac{d}{dx}(e^x) = e^x$

26. $\frac{d}{dx}(a^x) = (\ln a)a^x$

27. $\frac{d}{dx}\ln|x| = \frac{1}{x}$

28. $\frac{d}{dx}(\log_a x) = \frac{1}{(\ln a)x}$

Hyperbolic Functions

29. $\frac{d}{dx}(\sinh x) = \cosh x$

30. $\frac{d}{dx}(\cosh x) = \sinh x$

31. $\frac{d}{dx}(\tanh x) = \text{sech}^2 x$

32. $\frac{d}{dx}(\text{csch}\, x) = -\text{csch}\, x \coth x$

33. $\frac{d}{dx}(\text{sech}\, x) = -\text{sech}\, x \tanh x$

34. $\frac{d}{dx}(\coth x) = -\text{csch}^2 x$

Inverse Hyperbolic Functions

35. $\frac{d}{dx}(\sinh^{-1} x) = \frac{1}{\sqrt{1 + x^2}}$

36. $\frac{d}{dx}(\cosh^{-1} x) = \frac{1}{\sqrt{x^2 - 1}}$

37. $\frac{d}{dx}(\tanh^{-1} x) = \frac{1}{1 - x^2}$

38. $\frac{d}{dx}(\text{csch}^{-1} x) = -\frac{1}{|x|\sqrt{x^2 + 1}}$

39. $\frac{d}{dx}(\text{sech}^{-1} x) = -\frac{1}{x\sqrt{1 - x^2}}$

40. $\frac{d}{dx}(\coth^{-1} x) = \frac{1}{1 - x^2}$

INTEGRATION

Substitution

If an integrand has the form $f(u(x))u'(x)$, then rewrite the entire integral in terms of u and its differential $du = u'(x)\,dx$:

$$\int f(u(x))u'(x)\,dx = \int f(u)\,du$$

Integration by Parts Formula

$$\int u(x)v'(x)\,dx = u(x)v(x) - \int u'(x)v(x)\,dx$$

TABLE OF INTEGRALS

Basic Forms

1. $\displaystyle\int u^n\,du = \frac{u^{n+1}}{n+1} + C, \quad n \neq -1$

2. $\displaystyle\int \frac{du}{u} = \ln|u| + C$

3. $\displaystyle\int e^u\,du = e^u + C$

4. $\displaystyle\int a^u\,du = \frac{a^u}{\ln a} + C$

5. $\displaystyle\int \sin u\,du = -\cos u + C$

6. $\displaystyle\int \cos u\,du = \sin u + C$

7. $\displaystyle\int \sec^2 u\,du = \tan u + C$

8. $\displaystyle\int \csc^2 u\,du = -\cot u + C$

9. $\displaystyle\int \sec u \tan u\,du = \sec u + C$

10. $\displaystyle\int \csc u \cot u\,du = -\csc u + C$

11. $\displaystyle\int \tan u\,du = \ln|\sec u| + C$

12. $\displaystyle\int \cot u\,du = \ln|\sin u| + C$

13. $\displaystyle\int \sec u\,du = \ln|\sec u + \tan u| + C$

14. $\displaystyle\int \csc u\,du = \ln|\csc u - \cot u| + C$

15. $\displaystyle\int \frac{du}{\sqrt{a^2 - u^2}} = \sin^{-1}\frac{u}{a} + C$

16. $\displaystyle\int \frac{du}{a^2 + u^2} = \frac{1}{a}\tan^{-1}\frac{u}{a} + C$

Exponential and Logarithmic Forms

17. $\displaystyle\int ue^{au}\,du = \frac{1}{a^2}(au - 1)e^{au} + C$

18. $\displaystyle\int u^n e^{au}\,du = \frac{1}{a}u^n e^{au} - \frac{n}{a}\int u^{n-1}e^{au}\,du$

19. $\displaystyle\int e^{au}\sin bu\,du = \frac{e^{au}}{a^2 + b^2}(a\sin bu - b\cos bu) + C$

20. $\displaystyle\int e^{au}\cos bu\,du = \frac{e^{au}}{a^2 + b^2}(a\cos bu + b\sin bu) + C$

21. $\displaystyle\int \ln u\,du = u\ln u - u + C$

22. $\displaystyle\int u^n \ln u\,du = \frac{u^{n+1}}{(n+1)^2}[(n+1)\ln u - 1] + C$

23. $\displaystyle\int \frac{1}{u\ln u}\,du = \ln|\ln u| + C$

Hyperbolic Forms

24. $\displaystyle\int \sinh u\,du = \cosh u + C$

25. $\displaystyle\int \cosh u\,du = \sinh u + C$

26. $\displaystyle\int \tanh u\,du = \ln\cosh u + C$

27. $\displaystyle\int \coth u\,du = \ln|\sinh u| + C$

28. $\displaystyle\int \operatorname{sech} u\,du = \tan^{-1}|\sinh u| + C$

29. $\displaystyle\int \operatorname{csch} u\,du = \ln\left|\tanh\frac{1}{2}u\right| + C$

30. $\displaystyle\int \operatorname{sech}^2 u\,du = \tanh u + C$

31. $\displaystyle\int \operatorname{csch}^2 u\,du = -\coth u + C$

32. $\displaystyle\int \operatorname{sech} u \tanh u\,du = -\operatorname{sech} u + C$

33. $\displaystyle\int \operatorname{csch} u \coth u\,du = -\operatorname{csch} u + C$

Trigonometric Forms

34. $\displaystyle\int \sin^2 u\,du = \frac{1}{2}u - \frac{1}{4}\sin 2u + C$

35. $\displaystyle\int \cos^2 u\,du = \frac{1}{2}u + \frac{1}{4}\sin 2u + C$

36. $\displaystyle\int \tan^2 u\,du = \tan u - u + C$

37. $\displaystyle\int \cot^2 u\,du = -\cot u - u + C$

38. $\displaystyle\int \sin^3 u\,du = -\frac{1}{3}(2 + \sin^2 u)\cos u + C$

39. $\displaystyle\int \cos^3 u\,du = \frac{1}{3}(2 + \cos^2 u)\sin u + C$

40. $\displaystyle\int \tan^3 u\,du = \frac{1}{2}\tan^2 u + \ln|\cos u| + C$

41. $\displaystyle\int \cot^3 u\,du = -\frac{1}{2}\cot^2 u - \ln|\sin u| + C$

42. $\displaystyle\int \sec^3 u\,du = \frac{1}{2}\sec u \tan u + \frac{1}{2}\ln|\sec u + \tan u| + C$

43. $\int \csc^3 u\,du = -\frac{1}{n}\csc u \cot u + \frac{1}{n}\ln|\csc u - \cot u| + C$

44. $\int \sin^n u\,du = -\frac{1}{n}\sin^{n-1} u \cos u + \frac{n-1}{n}\int \sin^{n-2} u\,du$

45. $\int \cos^n u\,du = \frac{1}{n}\cos^{n-1} u \sin u + \frac{n-1}{n}\int \cos^{n-2} u\,du$

46. $\int \tan^n u\,du = \frac{1}{n-1}\tan^{n-1} u - \int \tan^{n-2} u\,du$

47. $\int \cot^n u\,du = \frac{-1}{n-1}\cot^{n-1} u - \int \cot^{n-2} u\,du$

48. $\int \sec^n u\,du = \frac{1}{n-1}\tan u \sec^{n-2} u + \frac{n-2}{n-1}\int \sec^{n-2} u\,du$

49. $\int \csc^n u\,du = \frac{-1}{n-1}\cot u \csc^{n-2} u + \frac{n-2}{n-1}\int \csc^{n-2} u\,du$

50. $\int \sin au \sin bu\,du = \frac{\sin(a-b)u}{2(a-b)} - \frac{\sin(a+b)u}{2(a+b)} + C$

51. $\int \cos au \cos bu\,du = \frac{\sin(a-b)u}{2(a-b)} + \frac{\sin(a+b)u}{2(a+b)} + C$

52. $\int \sin au \cos bu\,du = -\frac{\cos(a-b)u}{2(a-b)} - \frac{\cos(a+b)u}{2(a+b)} + C$

53. $\int u \sin u\,du = \sin u - u\cos u + C$

54. $\int u \cos u\,du = \cos u + u \sin u + C$

55. $\int u^n \sin u\,du = -u^n \cos u + n\int u^{n-1}\cos u\,du$

56. $\int u^n \cos u\,du = u^n \sin u - n\int u^{n-1}\sin u\,du$

57. $\int \sin^n u \cos^m u\,du$

$$= -\frac{\sin^{n-1} u \cos^{m+1} u}{n+m} + \frac{n-1}{n+m}\int \sin^{n-2} u \cos^m u\,du$$

$$= \frac{\sin^{n+1} u \cos^{m-1} u}{n+m} + \frac{m-1}{n+m}\int \sin^n u \cos^{m-2} u\,du$$

Inverse Trigonometric Forms

58. $\int \sin^{-1} u\,du = u\sin^{-1} u + \sqrt{1-u^2} + C$

59. $\int \cos^{-1} u\,du = u\cos^{-1} u - \sqrt{1-u^2} + C$

60. $\int \tan^{-1} u\,du = u\tan^{-1} u - \frac{1}{2}\ln(1+u^2) + C$

61. $\int u\sin^{-1} u\,du = \frac{2u^2-1}{4}\sin^{-1} u + \frac{u\sqrt{1-u^2}}{4} + C$

62. $\int u\cos^{-1} u\,du = \frac{2u^2-1}{4}\cos^{-1} u - \frac{u\sqrt{1-u^2}}{4} + C$

63. $\int u\tan^{-1} u\,du = \frac{u^2+1}{2}\tan^{-1} u - \frac{u}{2} + C$

64. $\int u^n \sin^{-1} u\,du = \frac{1}{n+1}\left[u^{n+1}\sin^{-1} u - \int \frac{u^{n+1}\,du}{\sqrt{1-u^2}}\right], \quad n \neq -1$

65. $\int u^n \cos^{-1} u\,du = \frac{1}{n+1}\left[u^{n+1}\cos^{-1} u + \int \frac{u^{n+1}\,du}{\sqrt{1-u^2}}\right], \quad n \neq -1$

66. $\int u^n \tan^{-1} u\,du = \frac{1}{n+1}\left[u^{n+1}\tan^{-1} u - \int \frac{u^{n+1}\,du}{1+u^2}\right], \quad n \neq -1$

Forms Involving $\sqrt{a^2 - u^2}$, $a > 0$

67. $\int \sqrt{a^2-u^2}\,du = \frac{u}{2}\sqrt{a^2-u^2} + \frac{a^2}{2}\sin^{-1}\frac{u}{a} + C$

68. $\int u^2\sqrt{a^2-u^2}\,du = \frac{u}{8}(2u^2-a^2)\sqrt{a^2-u^2} + \frac{a^4}{8}\sin^{-1}\frac{u}{a} + C$

69. $\int \frac{\sqrt{a^2-u^2}}{u}\,du = \sqrt{a^2-u^2} - a\ln\left|\frac{a+\sqrt{a^2-u^2}}{u}\right| + C$

70. $\int \frac{\sqrt{a^2-u^2}}{u^2}\,du = -\frac{1}{u}\sqrt{a^2-u^2} - \sin^{-1}\frac{u}{a} + C$

71. $\int \frac{u^2\,du}{\sqrt{a^2-u^2}} = -\frac{u}{2}\sqrt{a^2-u^2} + \frac{a^2}{2}\sin^{-1}\frac{u}{a} + C$

72. $\int \frac{du}{u\sqrt{a^2-u^2}} = -\frac{1}{a}\ln\left|\frac{a+\sqrt{a^2-u^2}}{u}\right| + C$

73. $\int \frac{du}{u^2\sqrt{a^2-u^2}} = -\frac{1}{a^2u}\sqrt{a^2-u^2} + C$

74. $\int (a^2-u^2)^{3/2}\,du = -\frac{u}{8}(2u^2-5a^2)\sqrt{a^2-u^2} + \frac{3a^4}{8}\sin^{-1}\frac{u}{a} + C$

75. $\int \frac{du}{(a^2-u^2)^{3/2}} = \frac{u}{a^2\sqrt{a^2-u^2}} + C$

Forms Involving $\sqrt{u^2 - a^2}$, $a > 0$

76. $\int \sqrt{u^2-a^2}\,du = \frac{u}{2}\sqrt{u^2-a^2} - \frac{a^2}{2}\ln\left|u+\sqrt{u^2-a^2}\right| + C$

77. $\int u^2\sqrt{u^2-a^2}\,du$

$$= \frac{u}{8}(2u^2-a^2)\sqrt{u^2-a^2} - \frac{a^4}{8}\ln\left|u+\sqrt{u^2-a^2}\right| + C$$

78. $\int \frac{\sqrt{u^2-a^2}}{u}\,du = \sqrt{u^2-a^2} - a\cos^{-1}\frac{a}{|u|} + C$

79. $\int \frac{\sqrt{u^2-a^2}}{u}\,du = -\frac{\sqrt{u^2-a^2}}{u} + \ln\left|u+\sqrt{u^2-a^2}\right| + C$

80. $\int \frac{du}{\sqrt{u^2-a^2}} = \ln\left|u+\sqrt{u^2-a^2}\right| + C$

81. $\int \frac{u^2\,du}{\sqrt{u^2-a^2}} = \frac{u}{2}\sqrt{u^2-a^2} + \frac{a^2}{2}\ln\left|u+\sqrt{u^2-a^2}\right| + C$

82. $\int \frac{du}{u^2\sqrt{u^2-a^2}} = \frac{\sqrt{u^2-a^2}}{a^2u} + C$

83. $\int \frac{du}{(u^2-a^2)^{3/2}} = -\frac{u}{a^2\sqrt{u^2-a^2}} + C$

Forms Involving $\sqrt{a^2 + u^2}$, $a > 0$

84. $\int \sqrt{a^2+u^2}\,du = \frac{u}{2}\sqrt{a^2+u^2} + \frac{a^2}{2}\ln\left(u+\sqrt{a^2+u^2}\right) + C$

85. $\int u^2\sqrt{a^2+u^2}\,du$

$$= \frac{u}{8}(a^2+2u^2)\sqrt{a^2+u^2} - \frac{a^4}{8}\ln\left(u+\sqrt{a^2+u^2}\right) + C$$

86. $\int \frac{\sqrt{a^2+u^2}}{u}\,du = \sqrt{a^2+u^2} - a\ln\left|\frac{a+\sqrt{a^2+u^2}}{u}\right| + C$

87. $\int \frac{\sqrt{a^2+u^2}}{u^2}\,du = -\frac{\sqrt{a^2+u^2}}{u} + \ln\left(u+\sqrt{a^2+u^2}\right) + C$

88. $\displaystyle\int \frac{du}{\sqrt{a^2+u^2}} = \ln\left(u+\sqrt{a^2+u^2}\right)+C$

89. $\displaystyle\int \frac{u^2\,du}{\sqrt{a^2+u^2}} = \frac{u}{2}\sqrt{a^2+u^2} - \frac{a^2}{2}\ln\left(u+\sqrt{a^2+u^2}\right)+C$

90. $\displaystyle\int \frac{du}{u\sqrt{a^2+u^2}} = -\frac{1}{a}\ln\left|\frac{\sqrt{a^2+u^2}+a}{u}\right|+C$

91. $\displaystyle\int \frac{du}{u^2\sqrt{a^2+u^2}} = -\frac{\sqrt{a^2+u^2}}{a^2u}+C$

92. $\displaystyle\int \frac{du}{(a^2+u^2)^{3/2}} = \frac{u}{a^2\sqrt{a^2+u^2}}+C$

Forms Involving $a + bu$

93. $\displaystyle\int \frac{u\,du}{a+bu} = \frac{1}{b^2}\left(a+bu-a\ln|a+bu|\right)+C$

94. $\displaystyle\int \frac{u^2\,du}{a+bu} = \frac{1}{2b^3}\left[(a+bu)^2-4a(a+bu)+2a^2\ln|a+bu|\right]+C$

95. $\displaystyle\int \frac{du}{u(a+bu)} = \frac{1}{a}\ln\left|\frac{u}{a+bu}\right|+C$

96. $\displaystyle\int \frac{du}{u^2(a+bu)} = -\frac{1}{au}+\frac{b}{a^2}\ln\left|\frac{a+bu}{u}\right|+C$

97. $\displaystyle\int \frac{u\,du}{(a+bu)^2} = \frac{a}{b^2(a+bu)}+\frac{1}{b^2}\ln|a+bu|+C$

98. $\displaystyle\int \frac{du}{u(a+bu)^2} = \frac{1}{a(a+bu)}-\frac{1}{a^2}\ln\left|\frac{a+bu}{u}\right|+C$

99. $\displaystyle\int \frac{u^2\,du}{(a+bu)^2} = \frac{1}{b^3}\left(a+bu-\frac{a^2}{a+bu}-2a\ln|a+bu|\right)+C$

100. $\displaystyle\int u\sqrt{a+bu}\,du = \frac{2}{15b^2}(3bu-2a)(a+bu)^{3/2}+C$

101. $\displaystyle\int u^n\sqrt{a+bu}\,du$

$$= \frac{2}{b(2n+3)}\left[u^n(a+bu)^{3/2}-na\int u^{n-1}\sqrt{a+bu}\,du\right]$$

102. $\displaystyle\int \frac{u\,du}{\sqrt{a+bu}} = \frac{2}{3b^2}(bu-2a)\sqrt{a+bu}+C$

103. $\displaystyle\int \frac{u^n\,du}{\sqrt{a+bu}} = \frac{2u^n\sqrt{a+bu}}{b(2n+1)}-\frac{2na}{b(2n+1)}\int\frac{u^{n-1}\,du}{\sqrt{a+bu}}$

104. $\displaystyle\int \frac{du}{u\sqrt{a+bu}} = \frac{1}{\sqrt{a}}\ln\left|\frac{\sqrt{a+bu}-\sqrt{a}}{\sqrt{a+bu}+\sqrt{a}}\right|+C, \quad \text{if } a>0$

$$= \frac{2}{\sqrt{-a}}\tan^{-1}\sqrt{\frac{a+bu}{-a}}+C, \quad \text{if } a<0$$

105. $\displaystyle\int \frac{du}{u^n\sqrt{a+bu}} = -\frac{\sqrt{a+bu}}{a(n-1)u^{n-1}}-\frac{b(2n-3)}{2a(n-1)}\int\frac{du}{u^{n-1}\sqrt{a+bu}}$

106. $\displaystyle\int \frac{\sqrt{a+bu}}{u}\,du = 2\sqrt{a+bu}+a\int\frac{du}{u\sqrt{a+bu}}$

107. $\displaystyle\int \frac{\sqrt{a+bu}}{u^2}\,du = -\frac{\sqrt{a+bu}}{u}+\frac{b}{2}\int\frac{du}{u\sqrt{a+bu}}$

Forms Involving $\sqrt{2au-u^2}$, $a > 0$

108. $\displaystyle\int \sqrt{2au-u^2}\,du = \frac{u-a}{2}\sqrt{2au-u^2}+\frac{a^2}{2}\cos^{-1}\left(\frac{a-u}{a}\right)+C$

109. $\displaystyle\int u\sqrt{2au-u^2}\,du$

$$= \frac{2u^2-au-3a^2}{6}\sqrt{2au-u^2}+\frac{a^3}{2}\cos^{-1}\left(\frac{a-u}{a}\right)+C$$

110. $\displaystyle\int \frac{du}{\sqrt{2au-u^2}} = \cos^{-1}\left(\frac{a-u}{a}\right)+C$

111. $\displaystyle\int \frac{du}{u\sqrt{2au-u^2}} = -\frac{\sqrt{2au-u^2}}{au}+C$

ESSENTIAL THEOREMS

Intermediate Value Theorem

If $f(x)$ is continuous on a closed interval $[a, b]$ and $f(a) \neq f(b)$, then for every value M between $f(a)$ and $f(b)$, there exists at least one value $c \in (a, b)$ such that $f(c) = M$.

Mean Value Theorem

If $f(x)$ is continuous on a closed interval $[a, b]$ and differentiable on (a, b), then there exists at least one value $c \in (a, b)$ such that

$$f'(c) = \frac{f(b)-f(a)}{b-a}$$

Extreme Values on a Closed Interval

If $f(x)$ is continuous on a closed interval $[a, b]$, then $f(x)$ attains both a minimum and a maximum value on $[a, b]$. Furthermore, if $c \in [a, b]$ and $f(c)$ is an extreme value (min or max), then c is either a critical point of $f(x)$ in (a, b) or one of the endpoints a or b.

The Fundamental Theorem of Calculus, Part I

Assume that $f(x)$ is continuous on $[a, b]$ and let $F(x)$ be an antiderivative of $f(x)$ on $[a, b]$. Then

$$\int_a^b f(x)\,dx = F(b) - F(a)$$

Fundamental Theorem of Calculus, Part II

Assume that $f(x)$ is a continuous function on $[a, b]$. Then the area function $A(x) = \int_a^x f(t)\,dt$ is an antiderivative of $f(x)$, that is,

$$A'(x) = f(x) \quad \text{or equivalently} \quad \frac{d}{dx}\int_a^x f(t)\,dt = f(x)$$

Furthermore, $A(x)$ satisfies the initial condition $A(a) = 0$.